Immunobiology of HLA

Bo Dupont
Editor

Immunobiology of HLA

Volume I

Histocompatibility Testing 1987

With 362 Illustrations

Springer-Verlag
New York Berlin Heidelberg
London Paris Tokyo Hong Kong

Bo Dupont
Member, Sloan-Kettering Institute for Cancer Research;
Professor of Immunology
Cornell University Graduate School of Medical Science
New York, New York 10021, USA

Library of Congress Cataloging-in-Publication Data
Immunobiology of HLA / Bo Dupont, editor.
p. cm.
Proceedings of the Tenth International Histocompatibility Workshop and Conference, held in Princeton, N.J. and New York City, Nov. 12–23, 1987.
Dedicated to the memory of Ruggero Ceppellini.
Contents: v. 1. Histocompatibility testing 1987 — v. 2. Immunogenetics and histocompatibility.
Includes bibliographies and indexes.
1. Histocompatibility testing—Congresses. 2. HLA histocompatibility antigens—Congresses. I. Dupont, Bo. II. Ceppellini, Ruggero. III. International Histocompatibility Workshop and Conference (10th : 1987 : Princeton, N.J. and New York, N.Y.)
[DNLM: 1. Histocompatibility Testing—congresses. 2. HLA Antigens—immunology—congresses. WO 680 I3237 1987]
QR187.H57I45 1989
616.07'9—dc19
DNLM/DLC
88-39761

Typeset by Publishers Service, Bozeman, Montana.
Printed and bound by Arcata Graphics/Halliday, West Hanover, Massachusetts.
Printed in the United States of America.

9 8 7 6 5 4 3 2 1

ISBN-13: 978-1-4612-8154-2 e-ISBN-13: 978-1-4612-3552-1
DOI: 10.1007/978-1-4612-3552-1

Dedication

If this is my day of harvest, in what fields
have I sowed the seed, and in what unremembered seasons?
If this indeed be the hour in which I lift up my lantern,
it is not my flame that shall burn therein.

Kahlil Gibran

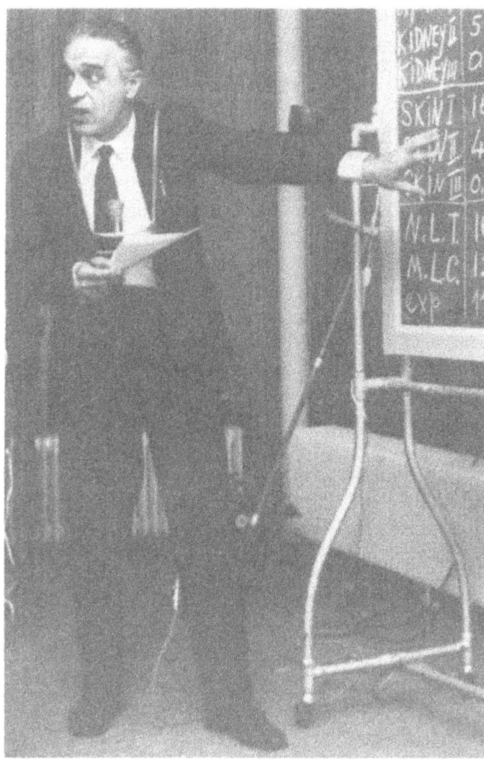

Courtesy of C. Zmijewski.

Ruggero Ceppellini died on May 5, 1988, after many months of struggle against lung cancer. His death leaves everyone, who knew him as a person or through his writings, with sadness and with an awareness of great loss.

It was natural for Ceppellini, an innovator and problem solver in genetics, to synthesize new concepts from observations in different fields of science that appeared to have little relationship to each other. Above all he scored by applying his quantitative sense as a biometrician to problems posed to him by colleagues who did not think in terms of segregation analysis or quantitative variation. Ceppellini was to immunogenetics what Mozart was to the keyboard, revelling in its virtuosity, ever fresh and new in the variety of concepts. He threw away in casual conversation ideas that others would have hoarded as prize-winning treasures. Ceppellini had good insight into his own relationship with others. He knew he was headstrong and abrasive; he referred to himself—and to the rest of our little clique in the field—as a Baron. He loved center stage and could occupy that position for hours, sometimes mainly to retaliate against mediocrity, but always to stimulate and to enlighten. A man who could not lecture because his broken arm was in a plaster cast, a racing driver in a rented car, a genius, a friend—Ruggero.

We were with you from the beginning,
when youth affected our resolve
with meaningful moments of living
and toasted to joy, our main goal!

You are amongst us, even today
after the enemy of life
destroyed the body
but not the spirit and mind . . .

You once said that living is learning
to ask questions and obtain answers:
the interminable sequence
of science and experimenting
and that, we are bound to grow old,
to seed, struggle, enjoy and harvest.
To leave an imprint and fold
following a road with burning lanterns . . .
Keep the lights bright and do not cease
to focus your designs and objectives
to conquer genetic basis of disease
and embellish our incentives.

"My life stopped too soon
when I still filled front stage,
wanting to be here with you
and to live to old age.

Why visit me at this place
of my source, where we come to rest?
There is nothing but empty space
without bright figures of life's zest.

There is no reason to mourn.
Rejoice our moment, this moment
when your hearts and voices
stop all of life's torments."

We wish your return from your dreams
at dawn near the sun
with vigorous songs
to overcome exhaustion.

Raise your cup and sip the wine
many times mixed with storms,
elegance and love, to combine
with labour to discover the unknown.

Oh traveler of the clouds,
from distant constellations,

dispeller of unfulfilled wishes
and ephemeral illusions.

Your arrival was heralded
by music and lightning
from igniting moons and stars
flowers and tunes of spring.

We remember your profile,
the voice and its force
its thunder and style
and the unyielding course.

But, why must we be deprived
of your luminous guide
with relentless passion,
a force to give us pride?

Our hearts trembling
and our eyes blurred by tears.
Life is nothing but continuing
to learn to face our fears . . .

Farewell our friend!
Your spirit remains unchanged
by memories and feelings unbend,
of energy and unparalleled mind . . .

Edmund Yunis and Bernard Amos

International Histocompatibility Workshop Council

Chairman
B. Dupont, New York, New York, USA

Presidents
E.D. Albert, Munich, Federal Republic of Germany
W. Mayr, Vienna, Austria

Past Presidents
D.B. Amos, Durham, North Carolina, USA
J.J. Van Rood, Leiden, the Netherlands
R. Ceppellini, Turin, Italy
J. Dausset, Paris, France
F. Kissmeyer-Nielsen, Aarhus, Denmark
W.F. Bodmer, London, United Kingdom
P.I. Terasaki, Los Angeles, California, USA

Councillors
M. Aizawa, Sapporo, Japan
F. Bach, Minneapolis, Minnesota, USA
H. Bashir, Sydney, Australia
J.R. Batchelor, London, United Kingdom
J.G. Bodmer, London, United Kingdom
C.B. Carpenter, Boston, Massachusetts, USA
L. Degos, Paris, France
C. Engelfriet, Amsterdam, the Netherlands
H. Festenstein, London, United Kingdom
P. Ivanyi, Amsterdam, the Netherlands
Z. Layrisse, Caracas, Venezuela
P.J. Morris, Oxford, United Kingdom
R. Payne, Stanford, California USA
A. Piazza, Torino, Italy
P. Rubenstein, New York, New York, USA
A. Svejgaard, Copenhagen, Denmark
E.J. Thorsby, Oslo, Norway
R. Tosi, Rome, Italy
K. Tsuji, Isehara, Japan
R. Walford, Los Angeles, California, USA
E.J. Yunis, Boston, Massachusetts, USA

Organizing Committee

Chairman
B. Dupont, New York, New York, USA

Data Analysis
J.M. Lalouel, Salt Lake City, Utah, USA

Serology
E.L. Milford, Boston, Massachusetts, USA
E.J. Yunis, Boston, Massachusetts, USA

Biochemistry I
S.Y. Yang, New York, New York, USA

Biochemistry II
R.W. Knowles, New York, New York, USA

T-Cell I
J.A. Hansen, Seattle, Washington, USA

T-Cell II
N. Flomenberg, New York, New York, USA

Southern Blot
D. Cohen, Paris, France

cDNA
J.S. Lee, New York, New York, USA

Preface to Volume I

The Tenth International Histocompatibility Workshop and Conference was held in Princeton, New Jersey and New York City, New York, in November 1987.

The International Histocompatibility Workshops date back to 1964 when D. Bernard Amos organized the first workshop on the HLA system for a small group of investigators who met at Duke University, North Carolina. The following nine workshops were held in 1965, 1967, 1970, 1972, 1975, 1977, 1980, 1984, and 1987. These workshops have become a central and unifying forum for research and development in the field of human histocompatibility testing and immunogenetics.

We have seen a rapid and sustained growth in active participation and involvement in this unique example of international collaborative research. Investigators from around the world have, for more than two decades, developed an open community, and lasting friendships have evolved. A sense of common goals, cooperative spirit, and mutual goodwill permeated not only the actual workshop and conference, but was also evident during the preceding three to four years of workshop planning and workshop studies.

It has been evident since the early days of the first histocompatibility workshops that most of the scientific developments in the field of histocompatibility and immunogenetics occur outside the framework of the workshops. Many individual investigators have, during the years, made the critical scientific contributions from which the histocompatibility community has benefitted. It has been, and continues to be, the responsibility of the International Histocompatibility Workshop Council to select the programmatic theme of each of the workshops. The past history of the workshops has clearly documented the symbiotic relationship between scientific progress in basic immunogenetics and the timely selection of subject matters for these large collaborative research programs.

The workshop concept has from time to time been challenged and the wisdom of continuing these very expensive and time-consuming projects has been questioned. Such concern is clearly legitimate and the question is also seriously considered prior to the organization of a new histocompatibility workshop.

The decision to initiate the Tenth International Histocompatibility Workshop was made in Munich and Vienna in May 1984. I offered at that time to undertake the organization of the next workshop with the general programmatic subject, "The Molecular Genetic Basis for HLA Polymorphism." The intent was to focus the attention of the histocompatibility testing community on modern techniques and approaches for biochemical, molecular genetic, and cellular analysis of HLA genes and gene products. One of the major goals of this workshop has been to develop a well-defined common source of HLA genes and gene products to be used as reference reagents. The panel of B-lymphoblastoid cell lines, now called the "Reference Panel of B-lymphoblastoid Cell Lines for Factors of the HLA System," consists of 107 cell lines, most of which are homozygous for HLA determinants. The B-cell lines and the DNA obtained from them constituted the central core around which most of the workshop studies evolved. These cell lines are now available to interested investigators and can be obtained from cell repositories in different parts of the world.

The present publication, *Immunobiology of HLA*, is divided into two volumes. Volume I, entitled *Histocompatibility Testing 1987,* contains the reports from the Tenth International Histocompatibility Workshop and describes the experimental designs of the workshop studies as well as the results obtained in these collaborative projects. It is our hope that this volume will serve as a practical guide in histocompatibility testing laboratories. The Joint Reports from the individual workshop components are presented in considerable detail, and the accompanying figures, tables, and pictures of gels from the biochemical studies, as well as Southern Blots, constitute an "Atlas for Factors of the HLA System."

Volume II, entitled, *Immunogenetics and Histocompatibility,* contains 253 invited papers presented at the International Conference in New York City. We felt, however, that this volume would best serve as an up-to-date authoritative publication on molecular biology of the HLA system if the brief communications were supplemented with a series of review papers covering the major subjects in this rapidly evolving scientific field.

Volume II therefore consists of two parts: Part 1 contains six overview papers, and Part 2 contains the 253 brief communications. We hope this volume will serve as a useful handbook on the "state of the art" of basic research on HLA.

The publication of *Immunobiology of HLA* signals the completion of four years of a project which has engaged a large number of participants and involved 362 laboratories worldwide. The workshop meeting in Princeton had 412 active participants and more than 1,400 attended the Conference in New York City. The workshop studies required very substantial investments in manpower and laboratory resources by the participating laboratories. The involvement in this workshop has, for many participants, meant sacrifices of personal research objectives for the benefit of the global interest of our scientific community. The completion of this project has—as has always been the case—required that large and scientifically very productive laboratories be willing to participate and facilitate the technology transfer into laboratories with less experience. An endeavor of this magnitude and complexity necessitates this level of scientific cooperation. But this is precisely the uniqueness of the HLA community and the reason for the professional pride and friendship that is characteristic for our community.

The credit for this tradition should be given to the founding members of the International Histocompatibility Workshops. I have felt it a great privilege to have been responsible for the present workshop project. As a member of a younger generation who has been welcomed and encouraged to contribute to this tradition, I felt throughout the last four years a strong commitment to implement the project in this spirit.

We were all saddened when one of the original members of the HLA community, Ruggero Ceppellini, became seriously ill and recently died. We have decided to dedicate this publication to his memory. Ruggero was intensely interested in the scientific and clinical importance of the HLA system right up to his death. He frequently visited me throughout 1986–1987 and was keenly interested in how the workshop proceeded. He will be deeply missed.

New York, New York, USA

Bo Dupont
Chairman
Tenth International Histocompatibility
Workshop and Conference

Acknowledgments

The organization, implementation, and completion of the Tenth International Histocompatibility Workshop and Conference has depended upon the help, support, and goodwill of numerous investigators and staff members in the participating laboratories. The procurement of reagents, antibodies, antisera, cell lines, etc., and their shipment, sorting, and redistribution required considerable dedication, meticulous care, and many long hours of work, not only by the staff in the organizers' laboratories, but also by staff in the laboratories of the participants. We wish to express our thanks and appreciation for this essential effort.

It is obvious that a major effort has been required by the staff in the laboratories of the organizing committee members. We wish to express our thanks to all of our associates who, in one way or another, have made this project possible to complete. There are, however, some who should receive special thanks: Lisa Ratner (Boston), Elizabeth May (Paris), Lorna Kennedy (London), Bernadetta Kienzle, John Klein, and Halina Korsun (New York).

The central data management, data collection, and statistical analysis posed an unprecedented challenge in this workshop due to the complexity of the experimental design and the introduction of a large number of experimental procedures. The dedication and commitment of the staff members of the Howard Hughes Medical Institute and the Department of Human Genetics, University of Utah, Salt Lake City, must be gratefully acknowledged. We wish to thank in particular Mark Ferguson, Robert Wheeler, Nancy King, and Rosemarie Plätke for their continuous efforts on these projects.

The central administration, organization, and management of the workshop and conference was the task of Mary Ann Barletta, with the assistance of Sally Krell. The decentralization of the workshop organization with component chairpersons located in Boston, Paris, Seattle, Salt Lake City, and New York imposed considerable additional demands on their time, efforts, and skills. It was, in retrospect, not reasonable to anticipate that an organization of this magnitude could be managed by only two individuals. The events during the last four years have, however, clearly shown that they were capable of getting the work done. We wish, in particular, to acknowledge the indispensable assistance of Mary Ann Barletta, who devoted herself to this endeavor over a four-year period. Without her inexhaustable effort and skill, the task of completing the Tenth International Workshop and Conference would have been formidable.

The major effort in editing these two volumes and coordinating the layout, proofreading, and revisions has been done by Sally Krell. Her careful attention to all the many details involved has been essential for completion of this publication. We have also been fortunate in developing this publication with Springer-Verlag New York Inc. The friendly help, advice, and understanding that we have received from the staff has been essential for the timely completion of this project.

The Organizing Committee

The Tenth International Histocompatibility Workshop and Conference expresses its gratitude for the generous support provided by its major sponsors and contributors:

National Institute of Allergy and Infectious Diseases (NIAID)
National Heart, Lung and Blood Institute (NHLBI)
National Cancer Institute (NCI)
National Institute of Diabetes and Digestive and Kidney Diseases (NIDDKD)
Howard Hughes Medical Institute
Memorial Sloan-Kettering Cancer Center
Fresenius AG (sponsor)
Pel-Freez Clinical Systems (sponsor)
Becton Dickenson Immunocytometry Systems (contributor)
Sandoz Pharmaceuticals (contributor)
Ortho Pharmaceutical Corporation (contributor)
One Lambda (contributor)

The Conference also acknowledges the generous contributions provided by the following:

Accurate Chemical & Scientific Corporation
Bio-Rad Laboratories
Biotest Diagnostics Corporation
Cryomed
E.I. Dupont De Nemour & Company
Digital Equipment Corporation
Dynal, Inc.
E. Leitz, Inc.
Elsevier Science Publishing Company
Genetic Systems
Gen Trak, Inc.
International Biotechnologies, Inc.
International Business Machine Corporation
Inotech Biosystems
Integrated Diagnostics
S. Karger Publishers, Inc.
Eastman Kodax Company
One Lambda, Inc.
Organon Teknika-Cappel
Robbins Scientific Corporation
Saxon Micro
Sepratech Corporation
Skatron, Inc.
Springer-Verlag New York Inc.
Xerox Corporation

Contents of Volume I

Nomenclature for Factors of the HLA System

Joint Reports: Data Collection and Central Analysis

Joint Reports: Serology

Joint Reports: HLA Class I Polymorphism Detected by One-Dimensional Isoelectric Focusing

Joint Reports: T-Cell Recognition of HLA Class II Molecules

Joint Reports: Restriction Fragment Length Polymorphism (RFLP) with HLA Region Probes (Southern Blot)

RFLP Standardization Reports

RFLP Summary Reports

Contents of Volume II

Section II Class I Histocompatibility Antigens: Genes, Structure, and Allotypes

Section III Class I Histocompatibility Antigens: Transfection and Gene Expression

Section IV Class I Histocompatibility Antigens: Regulation

Section V Class I Histocompatibility Antigens: New Class I, TL/Qa-Like, and Others

Section VI Class I Histocompatibility Antigens: HLA-DRB1

Section XXI Cell Surface Antigens Other than HLA

Section XXII Histocompatibility Antigens in Nonhuman Species

Section XXIII New Techniques

Overview

Overview of Experimental Design for the Tenth International Histocompatibility Workshop

Bo Dupont

It was essential in planning the experimental design and the organizational structure for the Tenth International Histocompatibility Workshop that the Workshop effort had a well-defined scientific and practical purpose. From the outset in 1984, we could anticipate that there would be large scale participation in the collaborative research projects and that investments in man-power and laboratory resources would be considerable.

We chose as the programmatic theme "Molecular Genetic Basis for HLA Polymorphism." This umbrella description of the Workshop reflected our intent to focus the projects on an in-depth analysis of HLA genes and their gene products. The molecular genetic exploration of the HLA class I and class II genes had, during the years 1980–1984, rapidly become a very active field of investigation. The first HLA class I cDNA clones were obtained in 1980 (1) and 1981 (2) and cDNA for HLA DRα (3) and DRβ (4,5) were isolated in 1982. A series of annual HLA/H2 cloning workshops were initiated in 1982 by Walter Bodmer and colleagues in the form of small meetings, which rapidly grew in size. By 1984, representative genes for all of the classical HLA loci had been identified and the class III region genes for C2 (6), BF (7), C4 (8) and 21-hydroxylase (9) had been obtained (reviewed in 10–12).

The Ninth International Histocompatibility Workshop held in Munich, May 6–11, 1984, followed the third HLA/H2 Cloning Workshop held in Strasbourg the preceding week. The initial plans for the Tenth Workshop were presented to the Council of the International Histocompatibility Workshops during the meetings in Munich and were in part inspired by the results presented at the HLA/H2 cloning meeting.

The international network of investigators studying HLA polymorphism had evolved during 20 years within the framework of the International Histocompatibility Workshop. The major emphasis within the Workshops during these years, by necessity, had to be placed on the serological definition of the HLA specificities and was from 1975 supplemented with collaborative projects on HLA polymorphism detected by in vitro assays for alloreactivity. The introduction of anti-HLA monoclonal antibodies as typing reagents started in 1980 and preliminary collaborative studies on characterization of HLA class II gene products by 2D gel analysis was introduced as a satellite project organized by Michael Crumpton and Julia Bodmer in the 1984 workshop.

It became evident that the introduction of molecular genetic techniques, biochemical characterization of HLA gene products, and utilization of T-lymphocyte clones for characterization of HLA genes and their gene products would provide the definitive definition of HLA alleles.

The experimental design for the Tenth International Histocompatibility Workshop had, as its most important objective, to provide a mechanism by which investigators with longstanding interest in characterization of HLA polymorphism would continue to play an active leadership role in the elucidation of the HLA system also at the molecular genetic level.

The plans for the Workshop took advantage of the fact that many of the investigators who had contributed substantially to the recent progress in the molecular biology of HLA also had longstanding active participation in the HLA workshops. Furthermore, many other investigators who had not previously participated in such collaborative projects indicated an interest in such studies and had already established collaboration with individual HLA laboratories.

The experimental design for the Workshop had five specific aims:

1. Development of standardized assays for molecular, genetic, biochemical and cellular characterization of HLA genes and gene products.
2. Development of a Reference Panel of B-lymphoblastoid cell lines for Factors of the HLA System.
3. Characterization of the Reference Cell Line Panel for Factors of the HLA System by the standardized assays.
4. Cross-correlations between HLA specificities defined by conventional serological and cellular typing procedures with HLA specificities defined by the new standardized assays.
5. Transfer of new technology from laboratories with prior experience to other interested investigators.

Experimental Procedures

HLA Serology. Serological characterization and definition of HLA specificities continued in this Workshop to provide the common and unifying data base with which to compare the results obtained in all other Workshop studies.

The Serology Component was divided into serological core exercises, supplemented with additional in-depth analysis with sets of supplemental antibodies. The serological core exercises involved a common "core serum set" which was distributed to a group of core serology laboratories. Based on the individual participating laboratory's specific interest in selected antigens, the studies were organized within 34 Antigen Groups.

Each Antigen Group contained a set of serologically related antigens. The core serology studies were supplemented with additional antisera representing the HLA antigens in the particular Antigen Group. These "Antigen Society" sera were only tested by the laboratories within the Antigen Group/Antigen Society. The serological HLA antigen reports presented in this volume have been prepared by the participants in each Antigen Group/Antigen Society.

The Serology component also included serological studies with anti-HLA monoclonal antibodies. These studies included binding assays utilizing an ELISA assay specifically designed for this Workshop.

Each participating HLA serology laboratory was responsible for selection of panel cells to be tested with the core serum set and with the Antigen Society sera. The participants were encouraged to provide representative panel cells for testing in the Biochemistry class I component of the Workshop.

The cells included in the Reference Panel of B-lymphoblastoid cell lines were also tested in several laboratories with the core serum set. A separate report on the results of these repeat testings are provided in this volume.

One-Dimensional Isoelectric Focusing (1D-IEF) Gel Electrophoresis for Determination of HLA Class I Alleles. This workshop component constituted the Biochemistry Component for HLA class I Antigens. A simple and standardized 1D-IEF technique was designed for this Workshop component. HLA class I antigens were obtained by sequential immunoprecipitation with the murine anti-HLA monoclonal antibodies 4E and W6/32. 1D-IEF gel electrophoresis was performed for the desialated HLA class I heavy chains and the band position for the individual HLA antigens was determined.

The studies were performed on 918 target cells provided by the different Antigen Groups. Specific emphasis was placed on testing the representative cells from different ethnic groups and representation of cells with rare HLA antigens or "splits" of common antigens. In addition, all cells in the Reference Panel of B-lymphoblastoid cell lines were tested.

The collaborative studies in this Workshop component focused on the characterization of HLA-A and HLA-B alleles. A total of 34 HLA-A alleles and 52 HLA-B alleles were defined by the 1D-IEF technique.

Two-Dimensional (2D) Gel Analysis of HLA Class II α and β Chains. A standardized method for 2-D gel electrophoresis was developed for the analysis of structural polymorphism of HLA class II antigens. The standardized Workshop technique required the purchase of common equipment and reagents by each laboratory. The monoclonal antibodies and additional reagents for immunoprecipitation were prepared centrally and distributed with radioactive standard proteins to each laboratory. The α chains were analyzed on a relatively acidic gradient while a more basic gradient was used for the β chains.

A total of 71 of the B-lymphoblastoid cell lines from the Reference Panel were studied in this Workshop component and every cell line was analyzed independently in two laboratories.

The studies were completed for the α and β chains of DR and DQ and for the DPα chain.

Analysis of HLA Class I Epitopes by T Lymphocyte Clones. Investigators with access to T-lymphocyte clones recognizing HLA class I antigens were encouraged to undertake an analysis of the T-cell reactivity pattern on a panel of target cells obtained from the relevant Antigen Group/Antigen Society.

These studies were, due to the small number of available T-cell clones, limited to the analysis of HLA-B27, HLA-B7 and HLA-Bw62.

A standard protocol for testing of available proliferative and cytolytic T-cell clones for reactivity against the Reference Panel of B-cell lines was designed. The protocol was similar to the experimental design for T-cell recognition of HLA class II molecules with the exception that monoclonal antibody studies were not performed. The same method for data reduction, whereby a specific proliferative or cytotoxic activity was converted into simple numerical scores ("1," negative; "6" or "8," strong positive) was used.

Functional Polymorphisms of HLA Class II Gene Products Detected by T-Lymphocyte Clones. T-cell defined (TCD) HLA class II polymorphism was studied using proliferative or cytolytic T-lymphocyte clones. The specificity was defined by the reactivity pattern of each T-cell clone on a panel of 75 of the Reference B-lymphoblastoid cell lines. Secondly, the specificity was mapped to the appropriate HLA class II family (DR, DQ, DP) through monoclonal antibody blocking experiments.

The experimental protocol included in addition to the 75 Reference Panel B-cell lines, eight blind duplicate cell lines. Duplicate testing was required in all cases to insure reproducibility of results. Local positive and negative control cells were required in each experiment as a means of normalizing the data and providing an independent quality control assessment. A panel of selected monoclonal antibodies was distributed for blocking studies. These antibodies were overlapping with the antibodies used in the 2D gel analysis of HLA class II α and β chain patterns.

Seventy-one of the 75 Reference Panel B-cell lines were identical with the cell lines used in the 2D gel study and 69 of these were also studied in the Southern Blot component.

Procedures for data reduction and statistical analysis were designed and a value of "8," "6," "4," "2," or "1" was assigned to each clone/target cell reaction.

The results indicated that it was possible to develop objective means for combining data obtained in multiple experiments in multiple laboratories.

A total of 134 T-lymphocyte clones were centrally analyzed. Forty-six new T-cell defined HLA specificities were detected in this Workshop component.

Southern Blot Analysis of Restriction Fragment Length Polymorphism (RFLP) with HLA-DNA Probes. The Southern Blot component of the Workshop had two objectives: to introduce Southern Blot technology into a large number of HLA laboratories and second, to define a standard set of RFLPs for the HLA complex.

Standardization of the Southern Blot protocol was undertaken in pre-Workshop studies for the purpose of designing robust experimental assay conditions allowing for comparison of data between laboratories.

Each digested DNA was systematically subjected to two different migration conditions: L for optimal resolution of large fragments and S for optimal resolution of small fragments.

The collaborative studies were divided into two components: the characterization of RFLPs for the Reference Panel of B-lymphoblastoid cell lines and other studies performed on local cell donor material such as family studies and disease-association studies.

All studies utilized the standardized Workshop assay. Thirteen HLA-DNA probes were prepared centrally and distributed to the participants together with 12 common restriction enzymes. The DNA probes selected were: Class I, C4, 210H, C2, BF, DRA, DQA, DNA, DPA, DRB, DQB, DOB, and DPB. Thus, a total of 156 enzyme/probe combinations (called systems) were studied independently in 2 to 5 laboratories.

DNA from 70 of the Reference Panel cells and two blind duplicate cells was prepared and coded at CEPH, Paris. This material was distributed to the participants together with the DNA probes, restriction enzymes, Klenow-enzyme, and DNA markers. A comb with 30 wells was prepared centrally and used by all participants for loading the gels. A standard nylon membrane obtained from a single manufacturer and pretested prior to distribution was used in all laboratories. Defined conditions for voltage, amperage, time, and agarose concentration was established. The migration included a visible marker (pAT153) and was stopped when the marker had migrated the predetermined distance.

Data was entered locally on floppy disks through an IBM-PC compatible software program developed by the central data analysis group (Salt Lake City). All diskettes and autoradiograms were collected in 12 standardization centers, each center responsible for 13 systems. A computer-assisted standardization procedure was developed for the definitive assignment of standard molecular weight to each fragment.

RFLPs were defined for the HLA class I, class III, and class II region genes. A comprehensive data base for HLA-RFLP has been compiled and DNA fragments corresponding to many of the serologically defined HLA specificities have been identified.

Nucleotide Sequence Determination of HLA DR and DQ Subregion Alleles. A Workshop component on nucleotide sequence analysis of cDNA clones for HLA-DR and DQ genes from several of the cells in the Reference Panel of B-lymphoblastoid cell lines was planned.

A series of cDNA clones were prepared centrally and distributed to interested investigators. A report on "First Domain Sequence Diversity of DR and DQ Subregion Alleles" is included in this volume. This article summarizes the deduced amino acid sequences for all presently published HLA-DR and DQ alleles and also contains the data for 16 of the Reference Panel Cells.

Central Data Processing and Analysis. The experimental design for this Workshop required a novel approach for data collection, processing, and analysis. The many new technologies introduced into the collaborative Workshop format presented the individual investigators, as well as the central data analysis group, with an unprecedented challenge.

The number of participating laboratories was substantially increased relative to previous workshops and the total central data base was very large. This required that data collection and verification and some components of data analysis be decentralized. The personal computer (IBM-PC or compatible) was chosen as the standard for participating laboratories. For each component of the Workshop, specialized software was designed centrally in the Pascal computer language and distributed to the participants. The software allowed entry, review, and update of the data locally in each laboratory.

Analytical programs were developed centrally and provided to laboratories involved in analysis.

The computer programs were operational on small personal computers as well as on the VAX 8650 superminicomputer used for central analysis.

Computer programs were implemented for data entry for (1) Serology, (2) Biochemistry Class I, (3) T-cell clone studies for Class I and Class II and for (4) Southern Blot experiments. The data entry and analysis for the 2D gel patterns in the Biochemistry Class II component was undertaken by Dr. Knowles and was not handled centrally. Nucleotide sequence data collection was organized into GenBank format.

The final central data analysis and cross-correlation of all data obtained in the Workshop for characterization of the Reference Panel of B-lymphoblastoid cell lines was performed. In this analysis, the data from each workshop component was reduced into simple numerical scores: "1," "2" (negative), "4" (positive), and "6," "8" (strong positive), similar to the conventional scoring system in HLA serology. This allowed for the application of computer programs and standard statistical methods that the experimenters traditionally have used to examine their data (13).

Workshop Organization

The responsibility for the organization of each of the Workshop components was delegated to the individual component chairperson(s) (Table 1). The organizing committee consisted of these component chairpersons and the Workshop chairman. The overall experimental design was developed jointly by the organizing committee and then presented to the Workshop community at a planning conference in Oakland, California, June 27–30, 1985. Ninety-five active Workshop participants, representing all regions who have traditionally participated in HLA workshops, were presented with the preliminary experimental design for the Workshop. These presentations also included preliminary descriptions of progress made in the standardization of new assays. The plans for implementing data collection and the decen-

Table 1. Tenth International Histocompatibility Workshop Organizing Committee. Bo Dupont, Chairman

Component	Chairperson	Affiliation
Central Data Analysis	J.M. Lalouel	Howard Hughes Medical Institute Department of Human Genetics University of Utah School of Medicine Salt Lake City, Utah, USA
Serology	E.J. Yunis	Dana Farber Cancer Institute Department of Pathology, Harvard Medical School American Red Cross Boston Massachusetts, USA
	E.L. Milford	Brigham and Women's Hospital Department of Medicine, Harvard Medical School Boston, Massachusetts, USA
Biochemistry Class I	S.Y. Yang	Sloan-Kettering Institute for Cancer Research Immunology Program, Memorial Hospital, MSKCC Cornell Univ. Graduate School of Med. Sciences New York, New York, USA
Biochemistry Class II	R.W. Knowles	Sloan-Kettering Institute for Cancer Research Immunology Program Cornell Univ. Graduate School of Med. Sciences New York, New York, USA
T Cell Class I	J.A. Hansen	Fred Hutchinson Cancer Research Center Department of Medicine, University of Washington Seattle, Washington, USA
T Cell Class II	N. Flomenberg	Sloan-Kettering Institute for Cancer Research Immunology Program, Memorial Hospital, MSKCC Cornell Univ. Graduate School of Med. Sciences New York, New York, USA
Southern Blot	D. Cohen	Centre d'Etude du Polymorphism Humain (CEPH) University of Paris, France
cDNA HLA Class II	J.S. Lee	Sloan-Kettering Institute for Cancer Research Immunology Program Cornell Univ. Graduate School of Med. Sciences New York, New York, USA

tralization of parts of the Workshop analysis using personal computers were also presented.

The final experimental plans were developed following the Oakland meeting and reflected the concensus opinions of the community.

Workshop Participation

A total of 362 laboratory groups participated actively in the individual Workshop components and submitted data for the central analysis.

The distribution of laboratory groups within each of the projects is shown in Table 2.

Participation in any of the component projects involved a considerable effort by the individual laboratories. It was evident that each investigator chose to participate in only the component of specific interest.

It was therefore essential for the successful completion of the total Workshop project that the experimental design centered around the analysis of the Reference Panel of B-lymphoblastoid cell lines. This allowed for a comprehensive cross-analysis of data obtained in all workshop components in spite of the large differences in number of participants in each individual component.

It should, however, be noted that the Reference Panel of B-lymphoblastoid cell lines consisted only of 107 cells and not all of these were studied in all components. (For details see "Description of the Reference Panel of B-lymphoblastoid Cell Lines for Factors of the HLA System", this volume.)

In light of these limitations, it was very important that each individual Workshop component project contained a series of self-contained objectives. It should be evident from reading the summary reports for each Workshop project, as well as the individual laboratory reports, that this goal was achieved.

Data Reporting and Presentation of Workshop Results

This volume of the Proceedings from the Tenth International Histocompatibility Workshop and Conference is solely dedicated to the presentation of the Workshop studies, the results and conclusions.

The publication is organized into three main sections: (1) Reference Panel of B-lymphoblastoid cell lines for factors of the HLA system; (2) Nomenclature for factors of the HLA system; and (3) Joint Reports from the individual Workshop components.

Table 2. Summary of Tenth International Histocompatibility Workshop

Project	Participating Laboratories		Reagents
Serology	157	> 2600 antisera and monoclonal antibodies	
		Core sera: A, B, C = 549	
		DR, DQ, DP = 244	
		C'MCA A, B, C = 82	
		DR, DQ, DP = 74	
ELISA	49	107	Reference Panel B-cell Lines
Biochem I	14	2	Anti-Class I monoclonal antibodies
		58	Serology Laboratories contributed 918 test cells
		107	Reference Panel B-cell Lines
Biochem II	27	41	D-region monoclonal antibodies
		71	Reference Panel B-cell Lines
T-Cell I	4		Local target cells
		90	Reference Panel B-cell lines
		13	anti-HLA Class I T-cell clones
T-Cell II	23	75	Reference Panel B-cell Lines plus 8 blind duplicate lines
		47	Monoclonal antibodies plus
		3	blind duplicate antibodies
		134	anti-HLA Class II T-cell clones
Southern Blot	80	70	Reference Panel B-cell lines plus
		2	blind duplicate
			Local family material and others
		13	HLA DNA Probes
		12	restriction enzymes
cDNA Sequencing	8	5	cDNA Libraries for DR and DQ genes.

Section 1 describes in an abbreviated form the Reference Panel of B-cell lines and the characterization of each of the cell lines for factors of the HLA system. The assignment of HLA specificities to the panel cells includes not only the HLA determinants presently recognized by the WHO Nomenclature Committee, but also contains the assignment of provisional HLA specificities identified by the new assay systems used in this Workshop. Application of molecular genetic techniques and biochemical assays for characterization of HLA genes and gene products requires access to relevant cell sources. The chapter describing the Reference Panel should provide interested investigators with important information describing the starting material from which further characterization of HLA alleles can be made.

The second section of the Proceedings contains the chapter on Nomenclature for Factors of the HLA system. This chapter consists of brief concise reports prepared by each of the Workshop component chairpersons summarizing the new HLA specificities identified during the 10th Workshop. These reports specify which HLA specifities are recognized by the WHO Nomenclature Committee and list in addition the provisional (10th Workshop) HLA specificities. The final section of this chapter consists of "Nomenclature for Factors of the HLA System, 1987" (Report by the WHO Nomenclature Committee on Leukocyte Antigens).

The third section of the Proceedings contains all the Joint Reports and Individual Laboratory Reports describing the studies performed within the 10th Workshop. Each of the Joint Reports contains a Summary Report that reflects the concensus opinion of the workshop participants in the component project. The last paper in this volume contains a display of cross-correlations of Workshop data on the reference panel cells.

Summary of Workshop Project

A total of 362 laboratories participated in 8 collaborative research projects on the characterization of HLA genes and their gene products. New standardized assay systems were designed for these studies. This included a standardized ELISA assay for studies of monoclonal antibody binding to target cells; 1D-IEF gel electrophoresis for determination of HLA-A and -B alleles; 2D gel analysis of HLA class II α and β chains; standardization of experimental conditions and statistical methods for analysis of HLA class I and class II epitopes detected by T-cell clones and a standardized assay for characterization of HLA region RFLPs by Southern Blot analysis. IBM and compatible personal computers were introduced into the Workshop format for data entry, review of data, and decentralized analysis.

A reference panel of 107 B-lymphoblastoid cell lines was developed and characterized for factors of the HLA system (14). This included the assignment of HLA specificities recognized by the WHO Nomenclature Committee as well as the provisional (10th Workshop defined) HLA specificities.

A total of 7 new HLA class I and 10 new class II serological specificities were defined. An additional 11

HLA-A specificities, 24 HLA-B specificities and 3 HLA-C specificities were identified as provisional new serological specificities. Evidence for four new HLA-DR specificities, one DQw2 split, two splits of DRw5, and one DRw53 variant was obtained in the serological Workshop studies (15).

Thirty-four HLA-A 1D-IEF alleles and fifty-two HLA-B 1D-IEF alleles were identified in the Biochemistry Class I studies. An additional three provisional HLA-A variants and five provisional HLA-B variants were detected (16).

A total of 19 DRβ chain patterns were found associated with the DRB1 locus with five additional patterns associated with a second expressed DRB gene. For the DQ molecules, 10 β chain patterns were found along with 9 DQα chain patterns. Two DP α chain patterns were observed (17).

Seven new T-cell defined (TCD) HLA-Dw specificities were recognized following this workshop. The Dw20–Dw23 were identified as major stimulating determinants in primary MLC. The Dw24–Dw26 specificities represent TCD specificities associated with HLA-DRw52 subtypes. An additional 18 TCD specificities were defined and found to be HLA-DR associated and 16 additional TCD specificities were identified for the DQ antigens. Five new TCD supertypic DR specificities were described (18).

The Southern Blot technique was used to describe the RFLP patterns on 70 of the cells from the Reference Panel of B-cell lines. The analysis demonstrated a number of unique RFLPs which correlated with the serologically defined HLA class II alleles. The pattern analysis of the RFLP has resulted in the identification of RFLP clusters which can be applied for DNA typing of HLA.

Acknowledgments. I wish to express sincere gratitude to my colleagues on the Organizing Committee for their tireless and dedicated efforts in completing the Workshop projects. We all realized from the outset in 1984 that our plans were very ambitious, but none of us expected the task to become as demanding, time consuming, and draining, both physically and mentally, as it turned out to be.

It was essential for the completion of these projects that each component of the Workshop was chaired by individuals who themselves were experts in the respective subject matter. Their scientific and technical knowledge and insights permeated all stages of planning and implementation of the work. Their willingness to participate in this endeavor is a tribute to their sincere dedication to the idea of collective pursuit of knowledge.

I consider it a great privilege to have had the opportunity to undertake this project with colleagues who could maintain their professional dedication and personal friendships intact during these 4 years.

References

1. Ploegh HL, Orr HT, Strominger JL. Molecular cloning of a human histocompatibility antigen cDNA fragment. Proc Natl Acad Sci USA 1980;77:6081.
2. Sood AK, Pereira D, Weissman SM. Isolation and partial nucleotide sequence of a cDNA clone for human histocompatibility antigen HLA-B7 by use of an oligodeoxynucleotide primer. Proc Natl Acad Sci USA 1981;78:616.
3. Lee JS, Trowsdale J, Bodmer WF. cDNA clones coding for the heavy chain of human HLA-DR antigen. Proc Natl Acad Sci USA 1982;79:545.
4. Long EO, Wake CT, Gorski J, Mach B. Complete sequence of an HLA-DRβ chain deduced from a cDNA clone and identification of multiple non-allelic DRβ chain genes. EMBO J 1982;2:389.
5. Wiman K, Larhammer D, Claesson L, Gustafsson K, Schenning L, Bill P, Böhme J, Denaro M, Dobberstein B, Hammerling U, Kvist S, Servenius B, Sundelin J, Peterson P, Rask L. Isolation and identification of a cDNA clone corresponding to an HLA-DR antigen β chain. Proc Natl Acad Sci USA 1982; 79:1703.
6. Bentley DR, Porter RR. Isolation of cDNA clones for human complement component C2. Proc Natl Acad Sci USA 1984;81:1212.
7. Campbell RD, Porter RR. Molecular cloning and characterization of the gene coding for human complement protein Factor B. Proc Natl Acad Sci USA 1983;80:4464.
8. Carroll MC, Porter RR. Cloning of a human complement component C4 gene. Proc Natl Acad Sci USA 1983;80: 264.
9. White PC, New MI, Dupont B. HLA-linked congenital adrenal hyperplasia results from a defective gene encoding a cytochrome P-450. Proc Natl Acad Sci USA 1984; 81:7505.
10. Molecular genetics of Class I and II MHL antigens. Moller G (ed) Immunol Rev 1985;84:29.
11. Molecular genetics of Class I and II MHL antigens. Moller G (ed) Immunol Rev 1985;85:5.
12. Molecular genetics of Class III MHL antigens. Moller G (ed) Immunol Rev 1985;87:7.
13. Lalouel JM, Ferguson M, Wheeler R, King N. Overview of data processing and analysis. In: Dupont B (ed). Immunobiology of HLA, Volume I: Histocompatibility Testing 1987. New York: Springer-Verlag Inc., 1989.
14. Yang SY, Milford EL, Hammerling U, Dupont B. Description of the reference panel of B-lymphoblastoid cell lines for factors of the HLA system. In: Dupont B (ed). Immunobiology of HLA, Volume I: Histocompatibility Testing 1987. New York: Springer-Verlag Inc., 1989.
15. Milford EL, Park MS, Lalouel JM, Kennedy L, Yunis EJ. Serologic exercises of the Tenth Workshop. In: Dupont B (ed). Immunobiology of HLA, Volume I: Histocompatibility Testing 1987. New York: Springer-Verlag Inc., 1989.
16. Yang SY. Population analysis of Class I HLA antigens by one-dimensional isoelectric focusing gel electrophoresis. In: Dupont B (ed). Immunobiology of HLA, Volume I: Histocompatibility Testing 1987. New York: Springer-Verlag Inc., 1989.
17. Knowles RW. Structural polymorphism of the HLA class II α and β chains: Summary of the Tenth Workshop 2-D gel analysis. In: Dupont B (ed). Immunobiology of HLA, Volume I: Histocompatibility Testing 1987. New York: Springer-Verlag Inc., 1989.
18. Flomenberg N. Functional polymorphisms of HLA Class II gene products detected by T-lymphocyte clones: Summary of the Tenth International Histocompatibility Workshop Cellular Studies. In: Dupont B (ed). Immunobiology of HLA, Volume I: Histocompatibility Testing 1987. New York: Springer-Verlag Inc., 1989.

Reference Panel of B-Lymphoblastoid Cell Lines for Factors of the HLA System

Description of the Reference Panel of B-Lymphoblastoid Cell Lines for Factors of the HLA System: The B-Cell Line Panel Designed for the Tenth International Histocompatibility Workshop

Soo Young Yang,[1] Edgar Milford,[2] Ulrich Hämmerling,[3] and Bo Dupont[4]

The major objective for the collaborative studies within the Tenth International Histocompatibility Workshop was to compare biochemical and molecular genetic techniques for characterization of polymorphic HLA determinants with HLA specificities defined by conventional serological and cellular techniques. It was, for this purpose, attractive to develop an experimental design that would focus on the in-depth characterization of a limited number of target cells common for all Workshop components and shared by the participating laboratories.

Epstein Barr virus (EBV)-transformed B-lymphoblastoid cell lines have for many years provided an effective source for HLA antigens and HLA genes, and many individual investigators have prepared cell lines for this purpose. The HLA Workshop community was at the outset of the Tenth Workshop in a unique position for obtaining a highly selective panel of cell donors from which EBV-transformed cell lines could be prepared if representative B-cell lines were not already available.

The development, design, and composition of the "Reference Panel of B-Lymphoblastoid Cell Lines for Factors of the HLA System" was made possible due to the unprecedented cooperative efforts of many colleagues in all parts of the world. It is our hope that this cell line panel will be supplemented in the future with new cell lines to make it as representative as possible for the genes and gene products of the HLA region.

The cell line panel as it exists at the completion of the Tenth International Histocompatibility Workshop consists of 107 cell lines. The present papers describe the development of the cell panel, the selection criteria that were applied, and the origin of the cell line donors or cell line sources. It should be realized that a panel of only 107 cell lines cannot be fully representative for all known HLA alleles. The composition of the present cell panel reflects the major emphasis in this Workshop on the characterization of the HLA-DR/DQ region of the HLA complex. It should also be noted that the ethnic representation on the cell panel is biased toward Caucasian haplotypes. This deficiency in non-caucasoid cell lines was due to the limited number of such cell lines offered by the contributing laboratories.

The present cell panel is also limited with regard to representation of HLA class I alleles and HLA-DP specificities. These limitations are due to the logistic problems of preparing large batches of cells for distribution to more than 100 laboratories, the feasibility of undertaking in-depth studies of more than 100 individual cells, and the limited number of HLA-DP typed donor cells offered to the Workshop.

A major emphasis was placed on four selection criteria: 1) HLA homozygosity, 2) consanguineous origin, 3) common HLA haplotypes in genetic linkage disequilibrium, and 4) representation of cells from donors homozygous for autosomal recessive diseases in genetic linkage with HLA.

This paper contains a summary table (Table 4) of the 107 cell lines in the reference cell line panel. This table provides the baseline information for each cell line, as well as an updated listing of HLA specificities defined on each cell. The experimental data from which the serologically defined HLA specificities are derived are provided in the accompanying paper. The following brief communications provide listings of the cellularly defined HLA-Dw and HLA-DPw specificities identified on individual cell lines, as well as additional HLA specificities defined in the Workshop studies of HLA class I 1D-IEF antigens (Biochemistry Class I Workshop component), HLA class II two-dimensional gel studies (Biochemistry Class II Workshop component), and the HLA class II analysis by T-lymphocyte clones (T-cell Recognition of Class II Workshop component).

Selection of Panel Cells

To obtain the most varied composition of the Reference Cell Line Panel, inquiries were mailed from New York to all laboratories that had participated in the 9th International Histocompatibility Workshop 1984. The responses to this inquiry were reviewed and requests for submission of already established, EBV-transformed B-cell lines (B-LCLs) and/or peripheral blood lymphocytes (PBL) were forwarded to the offering laboratories.

Approximately 200 different cells were offered to the Workshop. One hundred and fifty of these were requested for submission to the organizing laboratory at Sloan-Kettering Institute (SKI). A total of 120 different cells were received as B-LCLs and/or as PBL.

It was essential that the cell lines selected for the final Reference B-Cell Line Panel were healthy and demonstrated good growth patterns. Cells received as EBV-transformed cell lines were tested for viability, exponential growth patterns, and mycoplasma status. Only mycoplasma-free cell lines were accepted.

In selecting cell lines for the reference panel, primary emphasis was placed on HLA homozygous, consanguineous cell line donors. All submitted consanguineous cell lines were included in the panel if the cells were

---

Participating Laboratories: US7SYY,[1] NCYCAR,[2] US7UHA,[3] US7DUP[4]

Table 1. B-Cell Line Panel – Submitting Laboratories

Region	Lab	Number of Lines	Workshop Cell ID#
ANZ	BSH	6	9060,9076,9077,9083,9085,9098
ANZ	DAW	6	9008,9011,9018,9041,9086,9090
BEN	REE	2	9025,9102
BEN	ROO	8	9020,9032,9039,9062,9065,9070,9075,9097
EAE	KAS	2	9003,9009
FRA	FAU	4	9027,9055,9082,9088
FRA	BET	3	9023,9034,9042
FRA	MYE	4	9010,9013,9045,9056
FRA	OHA	4	9019,9084,9087,9094
GER	GOL	5	9007,9057,9080,9093,9095
GER	MUC	1	9089
IT1	CEP	6	9006,9015,9017,9029,9061,9063
IT2	FER	7	9033,9038,9040,9043,9044,9068,9092
JAP	TSU	3	9001,9053,9074
JAP	KSH	4	9024,9073,9103,9107
LAT	LAY	3	9064,9071,9072
SAF	DUT	4	9021,9022,9051,9058
SAF	HAM	1	9106
SCA	KIS	2	9004,9031
SCA	SVE	3	9047,9050,9069
SCA	TSB	3	9054,9067,9081
UKI	BOD	5	9016,9046,9099,9100,9101
UKI	FES	7	9005,9030,9048,9091,9096,9104,9105
NCY	BIA	1	9052
NCY	YUN	1	9046
US7	SYY	4	9002,9014,9026,9079
US7	HAN	8	9012,9028,9036,9037,9059,9066,9078,9097
US8	RUB	1	9035

mycoplasma free. PBL from consanguineous donors were transformed with EBV in cases where the submitted cell line was of poor quality or mycoplasma-infected or where only PBL were available.

The panel was then supplemented with non-consanguineous, HLA homozygous cells. HLA heterozygous cells were also included in the final cell panel to provide representation of antigens that would otherwise not be represented.

The final B-Cell Line Panel for the Workshop consists of 107 cell lines. Table 1 demonstrates the origin of the individual cells given by the submitting laboratory. A total of 28 laboratories have cell lines included in the reference panel. The panel, however, represents cells from several additional laboratories since some of the submitting laboratories had obtained the cells from other collaborating laboratories.

Table 2. B-Cell Line Panel – Preparation of Cells for Distribution

Region	Lab	Number of Cell Lines
UKI	BOD	5
BEN	ROO	7
UKI	FES	3
US7	HAN	8
US7	NFL	8*
US7	SYY	84

*Control cell lines for cellular studies.

Preparation of Cell Panel for Distribution

Each of the 107 cell lines was prepared in quantities of at least 220 vials containing 10^7 cells each. These cells were prepared in six laboratories (Table 2) and all cells were then collected at SKI for distribution. Of the 107 cell lines included in the final panel, 61 cell lines were prepared from PBL during 1985–1986 for inclusion in the Workshop. A total of 14 laboratories provided the EBV-transformed cell lines as shown in Table 3.

Pre-Workshop Data Collection and Characterization of Panel Cells

Inquiries were mailed to each investigator who provided cells that were included in the Reference Panel of B-Cell Lines. The investigator was asked to complete a form describing previously collected information for each of their cell line donors. These forms were submitted to SKI during 1987, and the information was included in the description of the reference panel. Furthermore, it

Table 3. B-Cell Line Panel–EBV Transformation of Cells

Region	Lab	Number of Cell Lines
GER	GOL	4
GER	MUC	1
ANZ	BSH	2
ANZ	DAW	5
BEN	REE	2
BEN	ROO	8
UKI	BOD	5
UKI	FES	6
FRA	OHA	2*
JAP	KSH	1
US7	NFL	4
US7	HAN	8
US7	UHA	38*
US7	SYY	21*

*EBV transformation performed in 1985–1986 for inclusion in Workshop Panel.

was requested that family studies be performed using the 10th Workshop Core Serum Set for each of the cell line donors and that the HLA-linked genetic markers (e.g., C4, C2, BF, GLO) be tested in each family.

The data presented in Table 4A contains all the baseline information that was collectable. Fifty-three of the 107 cell lines originated from cell donors who were of consanguineous origin. Two of these cell lines were derived from a pair of HLA identical siblings (WSNO 9064 and 9099). The cell panel also contains three cells that are duplicated on the panel WSNO 9071 = 9100, 9072 = 9101, and 9048 = 9096. These three pairs of identical cells were derived from six independent EBV transformations.

HLA Antigen Assignments for Panel Cells

The HLA antigen assignments for each of the panel cells are provided in Table 4B. The data shown in this table represent a consensus typing report, which contains the results derived from all of the Workshop components. The data are limited to the HLA specificities and provisional specificities recognized by the WHO Nomenclature Committee for Factors of the HLA System.

The HLA-Dw listing contains a few cases of assignment of new specificities defined by the laboratory submitting the cell or cell line (i.e., DKT2, Dw19v, SYD3, and FS), as well as three instances where a "new" specificity has been identified in cellular testings. These data are included in Table 4B since they demonstrate the existence of HLA-Dw determinants that are different from the presently recognized Dw specificities Dw1 to Dw26. The DB1, DB2, and DB6 specificities were recognized in the Eighth International Histocompatibility Workshop, but HLA-Dw numbers have not yet been assigned.

The following cell lines are heterozygous for HLA-A antigens: 9018, 9021, 9024, 9045, 9073, 9078, 9094, 9097, and 9103; 9002, 9024, and 9074 are heterozygous for HLA-C antigens, and 9024, 9028, 9044, 9073, 9074 and 9103 are heterozygous for HLA-B antigens. It should be noted that the serological typing of the cell line 9028 is Bw60 or Bw60+Bw61 and 9074 by serology is Bw61. The assignment of both of these antigens to those two cell lines is based on the one-dimensional isoelectric focusing (1D-IEF) analysis.

Heterozygosity for HLA-DR antigens was detected on four cell lines: 9007, 9045, 9056, and 9077. Heterozygosity for HLA-DQ antigens was determined on three cell lines: 9007, 9095, and possibly 9106.

One additional cell line, 9074, was found to be heterozygous in the DQ beta region by two dimensional (2-D) gel analysis. This heterozygosity could not be detected serologically.

A total of 16 cell lines were detected as heterozygous in the HLA-DP region (Table 5). It is, however, possible that several additional cell lines are DP heterozygous, but the typing reagents for HLA-DP are not yet sufficiently well developed to allow for a definitive HLA-DP assignment on all cell lines. The Southern blot analysis of all the panel cells may allow for a more complete description of the HLA-DP region on the individual cell lines (see: "RFLP-Sequence Interrelations at the DPA and DPB Loci" by Simons and Erlich, this volume).

Four panel cells were identified that were heterozygous for HLA class III region allotypes: 9027, 9080, 9095, and 9107. Table 4 demonstrates, however, that allotyping of the panel cell for the HLA class III region determinants is incomplete. It is, therefore, possible that more cells will be found to be heterozygous in the class III region when such typings become available.

A listing of all panel cells for which heterozygosity in the HLA region had been identified is shown in Table 5.

Workshop Studies with the Reference Panel of B-Cell Lines

All of the 107 Reference B-Cell Lines were tested with the 10th Workshop Core Serum Set. A report containing the complete serological data set, including all repeat testings and submitted for Central Analysis, is presented in the accompanying paper entitled "Serological Characterization of the Reference Panel of B-Lymphoblastoid Cell Lines for Factors of the HLA System." The data presented in Table 4B represent a "consensus" typing for each of the cell lines. This chapter of the proceedings describing the Reference Panel of B-Cell Lines also contains brief summary reports, with additional antigen assignments to the cell lines tested in the different Workshop components. An in-depth description of these Workshop components is presented in the respective chapters of this volume.

1. HLA Polymorphism Detected by RFLP (Southern Blot Studies). Seventy of the cell lines were selected by Daniel Cohen (of Paris) to constitute the Core Cell

Table 4A. Characterization of Reference Panel Cells for Factors of the HLA System

WSNO	Name	Reg-Lab	Ethnic	Sex	Consang	Comments	BF	C2	C4A	C4B
9001	SA	JAP-TSU	O(120)							
9002	MZ070782	US7-SYY	C(565)	M	N	21-Hydroxylase Deficiency	S	C	A2	B1,B2
9003	KAS116	EAE-KAS	C(595)	F	N					
9004	JESTHOM	SCA-KIS	C(575)	M	Y		S			
9005	HOM2	UKI-FES	C(510)	F	Y					
9006	WT100BIS	IT1-CEP	C(560)		Y		F	C	A2,A3	BQ0
9007	DEM	GER-GOL	C(550)				S		A6	B1
9008	D0208915	ANZ-DAW	C(520)	F	N	C2 Deficiency	S	CQ0	A4	B2
9009	KAS011	EAE-KAS	C(595)	F	N		S			
9010	AMAL	FRA-MYE	C(599)	M	Y					
9011	E4181324	ANZ-DAW	C(520)	M	N		S		A3	B1
9012	WJR076	US7-HAN	C(515)	M	N		S	C	A6	B1
9013	SCHU	FRA-MYE	C(545)	M	N	Idiopathic Hemochromatosis				
9014	MGAR	US7-SYY	C(517)	F			S			
9015	WT24	IT1-CEP	C(560)	M	Y					
9016	RML	UKI-BOD	AI(220)	M	Y					
9017	WT8	IT1-CEP	C(560)	F	Y					
9018	L0081785	ANZ-DAW	C(520)	M	N	IDDM	F1	C	A3	BQ0
9019	DUCAF	FRA-OHA	C(545)	M	N		F1	C	A3	BQ0
9020	QBL	BEN-ROO	C(535)	M	Y		F1	C	A3	BQ0
9021	RSH	SAF-DUT	N(10)	M	N		F	C	A1	BQ0
9022	COX	SAF-DUT	C(580)	M	Y		S	C	AQ0	B1
9023	VAVY	FRA-BET	C(545)	F	Y		S	C	AQ0	B1
9024	KT17	JAP-KSH	O(120)	F	N					
9025	DEU	BEN-REE	C(535)		N					
9026	YAR	US7-SYY	C(565)	M			S	C	A2	B1
9027	PF97387	FRA-FAU	C(545)	M			S	C	A3	B1/BQ0
9028	PE117	US7-HAN	AI(210)	M	N					
9029	WT51	IT1-CEP	C(560)	M	Y					
9030	JHAF	UKI-FES	C(540)	M	Y					
9031	BOLETH	SCA-KIS	C(575)	M	Y		S	C	A3	B3
9032	BSM	BEN-ROO	C(535)	F	Y		S	C	A3	B3
9033	BM14	IT2-FER	C(560)	M	Y					
9034	SAVC	FRA-BET	C(545)	M	Y		S		A3	B1
9035	JBUSH	US8-RUB	C(515)	F	Y		S			
9036	SP0010	US7-HAN	C(560)	F						
9037	SWEIG007	US7-HAN	C(515)	M	N					
9038	BM16	IT2-FER	C(560)	F	Y					
9039	JVM	BEN-ROO	C(535)	M	Y					
9040	BM15	IT2-FER	C(560)	M	Y					
9041	J0528239	ANZ-DAW	C(560)	M			S		A3	B1
9042	TISI	FRA-BET	C(545)	F	Y		S		A3	B1
9043	BM21	IT2-FER	C(560)	M	Y					
9044	BRIP	IT2-FER	C(560)		Y					
9045	TUBO	FRA-MYE	C(545)	F	N		S			
9046	BH	NCY-YUN	C(515)	F	N					
9047	PLH	SCA-SVE	C(575)		Y	21-Hydroxylase Deficiency	F	C	A91	BQ0
9048	LBUF	UKI-FES	C(540)	M	Y	= WSNO 9096	S		A3	B1
9049	IBW9	UKI-BOD	C(560)	M	Y					
9050	MOU	SCA-SVE	C(575)	F	Y				AQ0	B1

Ethnic Codes: N = Negro, O = Oriental, AI = American Indian, SEA = Southeast Asian, c = Caucasian. For ethnic subgroups, enclosed in ( )'s, see Table 2 in "Population Analysis of Class I HLA Antigens by 1D-IEF Gel Electrophoresis."
Consang: Y = known consanguinity; N = known not to be of consanguineous origin; blank = not known
"/" = heterozygosity; "," = gene duplication

Table 4A. *Continued*

WSNO	Name	Reg-Lab	Ethnic	Sex	Consang	Comments	BF	C2	C4A	C4B
9051	PITOUT	SAF-DUT	C(580)	M	Y		F	C	A3	B1
9052	DBB	NCY-BIA	C(516)	M	Y		S	C	A6	B1
9053	HOR	JAP-TSU	O(120)		N					
9054	EK	SCA-TSB	C(575)	F	N					
9055	HO301	FRA-FAU	C(545)		N					
9056	KOSE	FRA-MYE	C(550)	F	Y		S			
9057	TEM	GER-GOL	C(599)	M	N					
9058	OMW	SAF-DUT	N(10)	M	N		F	C	A3	B1
9059	SLE005	US7-HAN	C(515)	F	N					
9060	CB6B	ANZ-BSH	C(520)	F	Y		F		A3	B1
9061	31227ABO	IT1-CEP	C(560)	M	Y					
9062	WDV	BEN-ROO	C(535)	F	Y			C	A2	B1
9063	WT47	IT1-CEP	C(560)		Y					
9064	AMALA	LAT-LAY	AI(220)	F	Y	Ident. sib. of WSNO 9099				
9065	HHKB	BEN-ROO	C(535)	F	Y		S	C	A3	B1
9066	TAB089	US7-HAN	O(120)	M	N		S	C	A4	B2
9067	BTB	SCA-TSB	C(575)	M	N	Ankylosing Spondylitis				
9068	BM9	IT2-FER	C(560)	F	Y					
9069	MADURA	SCA-SVE	C(575)	F	Y	21-Hydroxylase Deficiency				
9070	LUY	BEN-ROO	C(535)	F	N					
9071	OLGA	LAT-LAY	AI(220)	F	Y	= WSNO 9100				
9072	SPACH	LAT-LAY	AI(220)	M	Y	= WSNO 9101				
9073	KT12	JAP-KSH	O(120)	M	N					
9074	HID	JAP-TSU	O(120)							
9075	DKB	BEN-ROO	C(535)	F	Y		S	C	A4	B2
9076	T7526	ANZ-BSH	O(110)							
9077	T7527	ANZ-BSH	O(110)							
9078	PMG075	US7-HAN	C(515)	F	N		S	C		
9079	LWAGS	US7-SYY	C(565)	M		21-Hydroxylase Deficiency	S			
9080	EHM	GER-GOL	C(550)	M	N		F/S		A2/A3	B1
9081	EA	SCA-TSB	C(575)	M	N					
9082	HO104	FRA-FAU	C(545)				S	C	A3	B1
9083	LD2B	ANZ-BSH	C(520)	M	N	Idiopathic Hemochromotosis	S	C	A3	B1
9084	CALOGERO	FRA-OHA	C(560)	M	Y	21-Hydroxylase Deficiency				
9085	EJ32B	ANZ-BSH	C(520)	F	N	Brother has IDDM	F1	C	A3	BQ0
9086	LO541265	ANZ-DAW	C(520)	M	N	Mother has Rheumatoid Arthritis	S	C	AQ0	B1
9087	STEINLIN	FRA-OHA	C(545)	F	N		S		AQ0	B1
9088	PFO4015	FRA-FAU	C(545)	F	N		S	C	AQ0	B1
9089	BOB	GER-MUC	C(550)	F						
9090	AWELLS	ANZ-DAW	C(520)	F	N		S	C	A3	BQ0
9091	MLF	UKI-FES	C(540)	F	Y	Rheumatoid Arthritis				
9092	BM92	IT2-FER	C(560)	M	Y					
9093	BER	GER-GOL	C(550)	M	N					
9094	CF996	FRA-OHA	C(545)	M	N		S		A3	B1
9095	WIN	GER-GOL	C(550)	M	N		S		A5/A6	B1
9096	LBF	UKI-FES	C(540)	M	Y	= WSNO 9048				
9097	EMJ	US7-HAN	C(515)	F	N					
9098	MT14B	ANZ-BSH	C(520)	M	Y		S		A3	B1
9099	LZL	UKI-BOD	AI(220)	F	Y	Ident. sib. of WSNO 9064				
9100	OLL	UKI-BOD	AI(220)	F	Y	= WSNO 9071				
9101	SPL	UKI-BOD	AI(220)	M	Y	= WSNO 9072				
9102	ARBO	BEN-REE	C(535)		Y					
9103	KT14	JAP-KSH	O(120)	M	N					
9104	DHIF	UKI-FES	C(540)	F	Y		S		A3	B1
9105	FPAF	UKI-FES	C(565)	M	Y		S	C	A3	B1
9106	MANIKA	SAF-HAM	C(580)	F	Y		S1	C	A2	B1/B18
9107	LKT3	JAP-KSH	O(120)	M						

Table 4B. HLA Antigen Assignments to Reference Panel Cells

WSNO	A	C	B w4/w6	B	DR	DRw 52/53	DQ	Dw	TCD DRw52	DP	GLO	WSNO
9001	24	7	6	7	1		1	1		4		9001
9002	24	2/8	6	14	1		5	20		4	1/2	9002
9003	24	-	4	51	1		5	1		-		9003
9004	2	1	4	27	1		5	1		4		9004
9005	3	1	4	27	1		5	1		-		9005
9006	11	4	6	35	1		5	1		1		9006
9007	2	6	4	57	16/4	53	5/8	-		3/4	1/2	9007
9008	25	-	6	18	15		6	2		2/4		9008
9009	1	6	4	37	2		1	21		3/4		9009
9010	28	4	4	53	15		6	2		-		9010
9011	1	-	4	52	15		6	12		2/4		9011
9012	2	7	4	57	16		6	21		2	1/2	9012
9013	3	7	6	7	15		6	2		-		9013
9014	26	7	6	8	15		6	2		4	1/2	9014
9015	2	2	4	27	16		5	21		3		9015
9016	2	-	4	51	16		7	22		-		9016
9017	3	7	6	7	15		1	2		2/4		9017
9018	3/24	5	6	18	3	52	2	3	25	3		9018
9019	30	5	6	18	3	52	2	3	25	-		9019
9020	26	?5	6	18	3	52	2	3	25	2		9020
9021	68/30	2	6	42	3	52	4	New	24	1	1	9021
9022	1	7	6	8	3	52	2	3	24	3	2	9022
9023	1	7	6	8	3	52	2	3	24	1	1	9023
9024	2/11	9/4	6	62/35	4	53	8	DKT2		5		9024
9025	31	4	6	35	4	53	7	4		4	1/2	9025
9026	26	-	4	38	4	53	8	10		4		9026
9027	29	-	4	44	4	53	7	4		4		9027
9028	24	10	6	60/61	4	53	8	14		4		9028
9029	23	8	6	65	4	53	8	4		2		9029
9030	31	8	4	51	4	53	7	13		3		9030
9031	2	10	6	62(75)	4	53	8	4		4		9031
9032	2	9	6	62(75)	4	53	8	4		2		9032
9033	3	7	6	7	4	53	8	4		4		9033
9034	3	7	6	7	4	53	8	4		-	1	9034
9035	32	-	4	38	11	52	7	5	25	4		9035
9036	2	5	4	44	11	52	5	DB2	25	2		9036
9037	29	2	6	61	11	52	7	5	25	-		9037
9038	2	7	6	18	12	52	7	DB6	25	2		9038
9039	2	5	6	18	11	52	7	New	25	2		9039
9040	1	7	4	49	11	52	7	5	25	3		9040
9041	1	4	6	35	11	52	7	5	25	-		9041
9042	24	4	6	35	11	52	7	New	25	-	2	9042
9043	1	-	6	41	11	52	7	5	25	-		9043
9044	24	-	4	51/63	11	52	7	5	25	2		9044
9045	2/3	7	4	51	11/12	52	7	5	25	2/4		9045
9046	2	6	4	13	7	53	2	DB1		4		9046
9047	3	6	4	47	7	53	2	DB1		-		9047
9048	30	6	4	13	7	53	2	DB1		-		9048
9049	33	8	6	65	7	53	2	17		1		9049
9050	29	-	4	44	7	53	2	17		2		9050

Ethnic Codes: N = Negro; O = Oriental; AI = American Indian, SEA = Southeast Asian, C = Caucasian. For ethnic subgroups, enclosed in ( )'s, see Table 2 in "Population Analysis of Class I HLA Antigens by 1D-IEF Gel Electrophoresis."
"/" = heterozygosity; "," = gene duplication
Note: For clarity, "w" has been deleted from all HLA specificities listed (e.g. Aw68 = A68)
9031, 9032, 9064, 9099: Bw62 possible Bw75

Table 4B. *Continued*

WSNO	A	C	B w4/w6	B	DR	DRw 52/53	DQ	Dw	TCD DRw52	DP	GLO	WSNO
9051	29	-	4	44	7	53	2	17		4	2	9051
9052	2	6	4	57	7	53	9	11		4	1	9052
9053	33	-	4	44	13	52	6	19	26	4		9053
9054	2	5	4	44	14	52	5	9	25	-		9054
9055	3	8	6	14	6	52	6	19	26	5		9055
9056	2	-	6	35	13/14	52	5	9/19	25	2		9056
9057	26	-	4	38	14	52	1	9	25	4		9057
9058	2	-	6	45	13	52	6	18	24	1	2	9058
9059	2	10	6	60	13	52	6	19	26	3		9059
9060	1	9	6	62	13	52	5	18	25	-		9060
9061	2	7	6	18	14	52	5	9	25	4		9061
9062	2	-	4	38	13	52	6	18	24	4		9062
9063	32	5	4	44	13	52	6	19	26	2		9063
9064	2	9	6	62(75)	14	52	7	16	24	-		9064
9065	3	7	6	7	13	52	6	18	24	4		9065
9066	2	11	6	46	8	52	6	8.3		-	2	9066
9067	2	1	4	27	8	52	4	8.1		4		9067
9068	2	4	6	35	8	52	4	8.1		2		9068
9069	2	10	6	60	8	52	4	8.1		4		9069
9070	2	-	4	51	8	52	7	8.3		1/4		9070
9071	31	11	6	62	8	52	4	8.2		3		9071
9072	31	11	6	62	8	52	4	8.2		4		9072
9073	24/31	4	4/6	35/52	9	53	9	23		5		9073
9074	2	10/8	6	61/60	9	53	9	23		2/5		9074
9075	24	10	6	60	9	53	9	23		4		9075
9076	2	11	6	46	9	53	3	-		-		9076
9077	2	11	6	46	9/12	52/53	3	-		5		9077
9078	3/33	8	6	65	1		5	20		3/4	1/2	9078
9079	33	8	6	14	1		5	20		3/4	1/2	9079
9080	3	4	6	35	1		5	1		4	1/2	9080
9081	3	7	6	7	15		6	2		4		9081
9082	3	7	6	7	15		6	-		4		9082
9083	3	7	6	7	15		6	2		4		9083
9084	2	2	6	61	16		5	-		4		9084
9085	30	5	6	18	3	52	2	SYD3	25	2		9085
9086	1	7	6	8	3	52	2	-	24	1/4		9086
9087	1	-	6	8	3	52	2	3	24	3/4	1	9087
9088	1	7	6	8	3	52	2	-	24	1/4		9088
9089	24	-	4	51	11	52	7	-	25	4		9089
9090	2	5	4	44	4	53	7	4		3/4		9090
9091	2	9	6	62	4	53	7	4		-		9091
9092	25	1	4	51	4	53	8	14		-		9092
9093	2	6	4	13	7	53	2	7		2/4		9093
9094	2/3	8	6	14	7	53	2	7		4	1/2	9094
9095	1	6	4	57	7	53	2/9	-		4	1/2	9095
9096	30	6	4	13	7	53	2	DB1		-		9096
9097	2/3	10	6	60	13	52	6	19	26	3/4		9097
9098	31	10	6	60	4	53	8	14		-		9098
9099	2	9	6	62(75)	14	52	7	16	24	-		9099
9100	31	11	6	62	8	52	4	8.2		3		9100
9101	31	11	6	62	8	52	4	8.2		4		9101
9102	3	6	4	57	9	53	2	-	24/25	1		9102
9103	24/26	8	4/6	51/61	9	53	9	23		2		9103
9104	31			38	11	52	7	5		NT		9104
9105	1	4	6	35	11	52	6	FS		2		9105
9106	3	6	6	50	7	53	2/?3	-		NT		9106
9107	24	1	6	54	4	53	4	15		5		9107

Table 5. HLA Heterozygosity on Panel Cells

WSNO	HLA-A	HLA-C	HLA-B	Class III	HLA-Dw	HLA-DR	HLA-DQ	HLA-DP	WSNO
9002		+							9002
9007						+	+	+	9007
9008								+	9008
9009								+	9009
9011								+	9011
9017								+	9017
9018	+								9018
9021	+								9021
9024	+	+	+						9024
9027				+					9027
9028			+*						9028
9037					+?				9037
9044			+						9044
9045	+				+?	+		+	9045
9056					+	+			9056
9070								+	9070
9073	+		+						9073
9074		+	+*				+*	+	9074
9077						+			9077
9078	+							+	9078
9079								+	9079
9080				+					9080
9083					+?				9083
9086								+	9086
9087								+	9087
9088								+	9088
9090								+	9090
9093								+	9093
9094	+								9094
9095				+			+		9095
9097	+							+	9097
9102					+				9102
9103	+		+						9103
9106							+		9106
9107				+					9107

\+ denotes HLA heterozygosity of this locus
* denotes HLA heterozygosity detected only in biochemical studies
+? denotes possible heterozygosity for HLA-Dw determinants

Lines to be used for the analysis of RFLP obtained with 12 restriction enzymes and 13 HLA probes (see Southern Blot component, this volume). Two vials of EBV-transformed cell lines from the complete B-Cell Line Reference Panel were provided to CEPH (Paris) during the spring of 1986. More than 10^9 cells from each cell line were prepared and the DNA extracted for distribution to the participating laboratories. Two of the cell lines (WSNO 9032 and WSNO 9056) were included as duplicate cells in this Workshop component.

2. T-Cell Recognition of HLA Class II Molecules (T-Cell-Defined Epitope Studies). A total of 75 of the reference B-cell lines and eight blinded duplicates were selected as the core cells for this Workshop component. Sixty-nine of the 75 cell lines in this Workshop component were identical to the core cell lines used in the Southern Blot component. All 71 cell lines used in the Biochemistry Class II component were also included in the T-cell Class II component. The cell lines were used as stimulating antigens or as target cells for cytotoxic effector cells.

3. Structural Polymorphism of HLA Class II Alpha and Beta Chains (Biochemistry Class II Studies). Seventy-one reference B-cell lines, which were also used in the T-cell class II studies, were selected for the characterization of HLA Class II antigens in this Workshop component.

4. Structural Polymorphism of HLA Class I Antigens Directed by 1D-IEF (Biochemistry Class I Studies). All 107 cell lines in the B-cell line panel served as reference cells for the HLA class I antigens in this Workshop component.

5. HLA Class II Polymorphism Detected by Nucleotide Sequence Analysis (cDNA Sequencing Studies). cDNA libraries for HLA class II cDNA clones were prepared for nucleotide sequence analysis of DR beta, DQ beta, and DQ alpha. The cell lines used for the studies within the Workshop were #9021 (RSH); #9023 (VAVY); #9046 (BH); #9047 (PLH); and #9052 (DBB). The paper entitled "First Domain Sequence Diversity of DR and DQ subregion alleles" by Gregersen et al. (this volume) provides a listing of the nucleotide sequences that presently are published or in press. These listings also contain sequence data for 16 of the reference panel cell lines: 9021, 9023, 9029, 9037, 9039, 9046, 9047, 9052, 9064, 9065, 9066, 9069, 9070, 9075, 9101, and 9107.

Summary

A panel of the 107 EBV-transformed B-lymphoblastoid cell lines has been prepared for the purpose of serving as a common set of target cells for the different component projects within the Tenth International Histocompatibility Workshop. These B-cell lines have been extensively tested with serological and cellular assays for HLA typing. In addition, the cell lines have been included in biochemical, cellular, and molecular genetic studies within the Workshop. A detailed report on the results obtained in these studies is presented in Histocompatibility Testing 1987. These cell lines, therefore, represent a Reference Panel of B-Lymphoblastoid Cell Lines for Factors of the HLA System.

Acknowledgment. We wish to acknowledge the help and support provided by all our colleagues who have assisted us in collecting cell lines and lymphocytes from their cell donors. The exceptional quality of the cell panel depended completely on their efforts.

Many colleagues have assisted us in preparing the cells. We wish in particular to acknowledge the contributions made by the staff in the laboratories of Drs. W.F. Bodmer, J. Bodmer, H. Festenstein, N. Flomenberg, J.A. Hansen, J. van Rood, and A. Termijtelen. Our colleagues at CEPH must also be acknowledged for preparing the large quantities of cells necessary for the Southern blot studies.

This work was supported at SKI by grants from the Department of Health and Human Services, NIH-NIAID-RO1-AI22960, and NIH-NCI-CA22507.

Serologic Characterization of the Reference Panel of B-Lymphoblastoid Cell Lines for Factors of the HLA System

E.L. Milford,[1] L.J. Kennedy,[2] S.Y. Yang,[3] B. Dupont,[4] J.-M. Lalouel,[5] and E.J. Yunis[6]

Serologic characterization of the B-lymphoblastoid cell lines (LCLs) described in the preceding paper was one of the major goals for the 10th International Histocompatibility Workshop. The individual Workshop components on structural polymorphism of HLA class I and class II antigens, the T-cell recognition of HLA class II molecules, and the analysis of restriction fragment length polymorphism (RFLP) were designed with the intent of correlating the data obtained in these component studies with the classic HLA alleles defined by conventional serologic and cellular techniques. The present paper describes the results obtained in the serologic studies of the Reference Cell Lines.

An average of three laboratories tested each line with the core serology set by cytotoxicity. The original (submitted) phenotype, the results of the core serology, and the definitions provided by the antigen societies in their reports were used to assign HLA class I and class II phenotypes to the cells. Most of the assignments could be soundly supported by the serologic data. Others, in particular some of the assignments for new splits, rested on a small number of relevant sera or equivocal reaction patterns. As the Workshop cell lines constitute a permanent resource, we have included a large amount of the primary serologic reaction data on these lines as a reference to help the reader to determine the degree of certainty which should be placed on a given assignment.

Format of Tables 1 to 21

Tables 1 through 21 display matrices of cytotoxic scores for the Workshop LCLs sorted by phenotype of the LCL. Each row represents a lymphoblastoid cell line tested in a specific laboratory (lab code omitted). Heavy

---

Participating Laboratories: NCYCAR,[1] UKIBOD,[2] US7SYY,[3] US7DUP,[4] US7LAL,[5] NCYYUN[6]

```
                |HMMDDDKECMCBTTTAZMGDKMGGWFMDDDKTBCBODMTTTGJMJMCKAEMWGMSMCGG
                |AIIDPPANRBNREEOIHLSDWRSAAABUUDUEREEHUUSSESEILTEAINIESRTYESE
                |MTDCTTPGOCGARRMZASCCCVCNRUCTTCERNPRATCUURCRDETPPZGTGCIERPCL
                |
                |             2 2  2   2222   2 2     22222      2   22  222
                |            10 0  0   0000   010     00000      0   00  000
                |   1 1111   00645505441100545141555165666567677667566676444
                |65748235194363952647873059190477945017452881576012843007124

9009 A1         |2 8888688888 0 0  0   0000 62080   8 00000      0   00  000
9011 A1         |8688 6 88888                     008 888           0
9011 A1         | 88888888888 88 4488       88 868888 8888        0    8
9011 A1         |868 888 8888 0 06 0   0000 880802  8 00000      0   00  000
9022 A1         | 8880808888888  06            88  48     8         6
9022 A1         |888888888888 8 0  8           84  48   004
9022 A1         |668886288888 0 0  0   0000   0 0   8 00000      0   00  000
9023 A1         | 888 828888828   8            88  886    8         0
9023 A1         | 48688668888 840 68          88866686   08         60   68
9023 A1         |6888888882860  0 20   0000   080  68 00000      0  200  000
9040 A1         |  8888 88888 8    8          688 8 8 88888
9040 A1         |888 8886888888 0  8          6 6 8 8 888 8      8  808
9040 A1         |22888888888880 0  0   0000   080 4 8 00000      0   00  000
9041 A1         |488808 88888      8           2   08               0
9041 A1         |  88888 888888               68    8        4
9041 A1         |88888888888880802 06   00002   080 8 8 000006 42  0   00  000
9043 A1         |8888884888888880868       2  6884 48    4   4    2    4
9043 A1         |64888888888808 0 2864  22  2  88   8   22         2
9043 A1         |8888888888 880 02 0   0000   080 2 8 00000      0   00  000
9060 A1         |4 8848 886 8   0  8                6     4
9060 A1         |88888888888820 0 206  00002880808888 00000     20   00  000
9086 A1         |8 8808 88888 0 0000   0000  0040 008 00000      0   00  000
9086 A1         |  888888888848    86          88   8     8
9086 A1         |6888888888886840268          688   8     8         6
9086 A1         |660866 8880  0 0  0   0000   080   8 00000      0   00  000
9087 A1         | 4886848888868 0 68        0 888  68     8         4     4
9087 A1         |680888668808 0 0  0   0000   080   8 00000      0   00  000
9088 A1         |88880868888868   8 8        0888  88     6         8
9088 A1         | 8888868888848 0 68          888  68     8              46
9088 A1         | 60888888808 0 0  0   0000   080   8 00000      0  200  000
9095 A1         |888688 888888840648 0        888 688688888  4    468     6
9095 A1         |88888868888880 0  0   0000   080 228 00000      0 8600  000
9005 A3         |            62      888888888882 2   88888      2       888
9005 A3         | 2        2  820    888888888888 2   88888        2     888
9005 A3         |            80 0  0 880000888080 6   00000      0   00  000
9013 A3         |          8 2060  0 880000888080444  00000      0   00  000
9013 A3         |          6 6  0  0 888808888888 68     0           2   808
9013 A3         | 0        6 40    0 888888888888        0               808
9017 A3         |    0     8  88 00  88888888 888800                  8  888
9017 A3         |  24   8  8 4880  6 888888888880688      6       2  8   888
9017 A3         |          82 020 20 880000888080844  00000      0   00  000
9018 A3   A24   |    0           0 8 88888828848 8   866828  8888888084  888
9018 A3   A24   |  88    688  8    8 888888888888   8888888 68888888888  088
9018 A3   A24   | 6 6    686  8604 8 88888888888864468888884 88868 8888  888
9018 A3   A24   |  84     8   0 0  0 880000882080    800000  8888088800  000
9033 A3         | 8  0   2 888880    888888888888 86       2          8  888
9033 A3         | 2 2   4  888080  0 880000888080 46  000006     0   00  000
9034 A3         |        0 888880    888888888888 86                     888
9034 A3         |          888080  0 880000888080  2  000002     0   00  000
9045 A2         |         62 40 0  0 880000886080  42 00000      0   00  000
9045 A2         | 06       6    0    888888888888     00000      0   00  808
9045 A2         | 0       88  0 0  0 888688868 8      8880               888
9047 A3         | 8             0    888888888 86     888                888
9047 A3         | 88          8 0    88888888888  68  888           8    888
9047 A3         |88      6   88 0    88888888888  88  888                888
9047 A3         | 80       0  0 0  0 880000888080     00000      04  00  000
9055 A3         |2        64  860    888888888888 26 2               4   888
9055 A3         |         4  6020  0 880000888080 2   00000      0   00  000
9065 A3         |  28   6 68 60 0220 880000888080886  00000      0   00  000
9065 A3         |  28   6 68 60 0220 880000888080886  00000      0   00  000
9078 A3   AW33  |            4  0    88888868448                         888
9078 A3   AW33  |             84 00   888888888888 00                  0  888
9078 A3   AW33  |             0 0  0 880000888080 2   00000      0   00  000
9080 A3         |             82     888888888888 6    00             8  888
9080 A3         |          2 68002   688888888888 6                      888
9080 A3         |             0 0  0 880000888080     00000      0   00  000
9081 A3         |          8 6860    88 8888888886 824      2        4   888
9081 A3         |       4  8 4040  0 8800008880 0444 200000      0   00  000
9082 A3         |       2  8  860    888888888888626      2              888
9082 A3         |        2 8 4060  0 880000888080 44  00000      0 6 00  000
```

Table 1. Lymphoblastoid Cell Line Reactions with Informative Sera. Rows represent LCL tested in an individual lab. LCL number and phenotype shown on left. Columns represent individual sera whose submitting lab code and workshop number appears at top. Contents of table and subsequent tables are standard microcytoxicity scores, however, scores of 1 are shown as blanks.

Table 1. *Continued*

			HMMDDDKECMCBTTTAZMGDKMGGWFMDDDKTBCBODMTTTGJMJMCKAEMWGMSMCGG
			AIIDPPANRBNREEOIHLSDWRSAAABUUDUEREEHUUSSESEILTEAINIESRTYESE
			MTDCTTPGOCGARRMZASCCCVCNRUCTTCERNPRATCUURCRDETPPZGTGCIERPCL
			2 2 2 2222 2 2 22222 2 22 222
			10 0 0 0000 010 00000 0 00 000
			1 1111 00645505441100545141555165666567677667566676444
			65748235194363952647873059190477945017452881576012843007124
9083	A3		0 868888 86 8 0 888
9083	A3		2 8 6860 88888888888648 22 6 888
9083	A3		8 2060 0 880000888080888 00000 0 00 000
9094	A3	A2	48 6 6 8 0 888888888888 0666 462 868
9094	A3	A2	0 0 0 880000 88060 2 2000002 0 00 000
9097	A3	A2	0 86888868628 888
9097	A3	A2	2 82 880 8 8880888888886844688 4 4 6 888
9097	A3	A2	22 2 6 020 0 880000888080 242000002 0 00 000
9102	A3		642 6 8 488868888888 88 488888 8 6 888
9102	A3		6 4 68 0 888888888888 88888 824 888
9102	A3		8 888868888888 88888 6 888
9106	A3		888888888 8 8 888
9106	A3		2 4 0 0 0 880000888080 8 000004 0 400 000
9006	A11		88 8886 88888 882 8 68
9006	A11		68 0648 88888 822 88
9006	A11		2 4 80 08806 40000 88080 864400000 0 00 2000
9024	A11	A2	888880868 888 8 88 8 888
9024	A11	A2	888880848 888 8686 888
9024	A11	A2	42 8880808808 200002 080 888600000 2 0 00 000
9029	A23		68688 8 8 088888888 8 88 88 888
9029	A23		28060 0 0000 020 8000008822240 880048000
9001	A24		22 0 0 0 0000 88060 480000028888 0 8800 000
9002	A24		0 8 0 8 88 8888888 88808 8888
9002	A24		4860 8 4 8 4888888 888888888 6
9002	A24		66 0 0 0 0000 880 06 66800000828888088800 000
9003	A24		6 0 0 0 00006880208 868000008 888068800 000
9028	A24		668860 8 68 488888886888 8488884
9028	A24		888 888 68 8864 8888888888888888888 0
9028	A24		4 2 46080 0 0000 0206 800000648886068800 000
9042	A24		84 8 0 8 8 2888888 8888888888
9042	A24		8 4 88 6860 8 8 6888888468 88888888
9042	A24		84 0 0 0 0000 0 08 2800000 888088800 000
9044	A24		88 80 0 0 00000 0 0448680000068888088800 2 000
9044	A24		88 80 0 0 00000 0 0448680000068888088800 2 000
9073	A24	A31	6 8 8 4 888888 8888888888 888
9073	A24	A31	4 48 2 8 844 2 888888228888888888 888
9075	A24		848 8 8 8 8888888 8888888888
9075	A24		642 20 0 0 0000 0602 288000008888606 8800 000
9089	A24		6 88 6880 8 6 8288888862888868888882 8
9089	A24		88 40 0 0 00002 0 0 868000006 888088800 000
9103	A24	A26	8 8 68 88 866888888888 08888888888 8
9103	A24	A26	8 8 68 88 866888888888 08888888888 8
9103	A24	A26	88 0 88 6 8 888888 888888888
9103	A24	A26	6 68 0 88 288 268888888 86468 8888 88
9103	A24	A26	2 2 0 0 08 0000 0808 8800000 88828088800 000
9107	A24		6 86 6844 8 8 888888 6888888888 8
9107	A24		62688080 0 0000 0602 880000086888608 8800 000
9107	A24		62688080 0 0000 0602 880000086888608 8800 000

horizontal lines separate cells with a different identity. In the left margin, we have listed the cell line identification codes and phenotypes of the cell. Each column represents scores with the serum whose Workshop identification code appears at the top margin.

Sera were selected based on the optimal reagents for the definition of each antigen, including all new WHO-designated antigens and all new splits that could be defined with the core set: Whenever possible, the order of the sera is identical to that used in the serograms displayed in the individual antigen society reports (see this volume). A serum analysis was done centrally with the LCL only and the core set; however, this analysis in no instance identified a group of sera to use for the definition of an antigen with the lymphoblastoid cell lines that are different from those that are appropriate for lymphocytes.

Results

A Locus. Cell line 9018, submitted as a homozygous A3, has clear cut A3 plus A24 patterns of reactivity

Table 2. Lymphoblastoid Cell Line Reactions with Informative Sera

Cell line	Antigen	Antigen	Reactions
			CMMBGMMTGVTDRJDBMTHBMGGBFDDDKM
			RRREOYBESIEUILURRAAEIVROAUUURL
			SVATLECRCVRPCETSVTMTTYUUUTTTES
			22222
			111111100001 11111 1 111111
			122222255561900111891999000133
			901234568526907245383345589001
9008	A25		888888888884 4 8 88888
9008	A25		888888 888886 8 6 88888888
9008	A25		8 888 888884 6 888686
9008	A25		28 88 00000 888 8
9092	A25		884868 88888 8 88 888888
9092	A25		2 8 8 88888 2 288 66
9092	A25		44286800000024 4402 88 288
9014	A26		8 6 28 4 8862 68888888 8
9014	A26		668 888886888888886
9014	A26		8 00000828864 88888888 4
9020	A26		8 88888 888888
9020	A26		4 4 886662 2688686
9020	A26		2 4 00000686862 4888082
9026	A26		8 2 2864 6888888 4
9026	A26		2 2 8266686224688886
9057	A26		000008 88 88 88882
9103	A26	A24	8888 88 0888888
9103	A26	A24	8888 88 0888888
9103	A26	A24	88 8 88886 8888888
9103	A26	A24	0 8888 624 888886
9103	A26	A24	0000088888 6888888

Table 3. Lymphoblastoid Cell Line Reactions with Informative Sera

Cell line	Antigen	Antigen	Reactions
			MGMKBMBCNKSWFDBRFBM
			LLURABRRWUTAEUAOERY
			SALETCTOOEENRTTORUR
			2
			1151111111101111111
			3363444334545565505
			1250123457168905427
9027	A29		6888 8 828 8
9027	A29		8888 88 8 0 84 4
9037	A29		8888 8 6 8
9037	A29		8888 8 8 8 8 8
9037	A29		8888 8 8 8
9037	A29		8888 8 8 0
9050	A29		8888 8 0 8 66
9050	A29		8888 8 0 8 66
9051	A29		8888 84 8 6 48 8
9051	A29		8888 84 8 6 48 8
9051	A29		8888 8 8 0 84
9019	A30		0848888 8 6
9019	A30		8 288 0 0
9021	A30	AW68	66 88
9021	A30	AW68	8888888
9021	A30	AW68	8 868888 64
9021	A30	AW68	4 8 8688886 2
9021	A30	AW68	8888888
9021	A30	Aw68	8 88 88880 0 2
9048	A30		8 8888888 82 8
9048	A30		688888880 4 2
9085	A30		866888 4
9085	A30		8 0 88 8
9085	A30		8 888682 6
9085	A30		64868868640 6
9096	A30		686888880 2 2
9096	A30		686888880 2 2
9025	A31		888 888 8 4
9025	A31		48684 888 8 8
9025	A31		828 880 0
9030	A31		868 888 88 68
9030	A31		8882 880 88 8
9071	A31		8686 888 8 8
9071	A31		8 8242 888 4
9071	A31		2 888 880 4 4
9072	A31		8886 886 86
9072	A31		68868 8882
9072	A31		228888 880 08 2
9073	A24	A31	8 8 888 88 8
9073	A24	A31	828 888 8 4
9098	A31		888 888 8
9098	A31		664 886
9098	A31		8 8 0888 8
9098	A31		866 860 8 2
9100	A31		4 826 888 4
9100	A31		82 8 0 828 2
9101	A31		42866 886
9101	A31		42866 886
9101	A31		2 8862 880 42 4
9035	A32		68 8 888
9035	A32		28 806880 0
9063	A32		0888
9049	AW33		2 64 2820 48888
9049	AW33		2 64 2820 48888
9053	AW33		4244 840 8688
9079	AW33		8 8 888 88 88
9079	AW33		8 6 888 88888
9079	AW33		8 8 888 88 88
9079	AW33		2 8 8888
9079	AW33		6 8 2888 88888
9078	AW33	A3	2 8 4 8888
9078	AW33	A3	0 0 808 00808
9078	AW33	A3	8 2 860 60688
9078	AW33	A3	2 8 4 8888
9078	AW33	A3	0 0 808 00808
9078	AW33	A3	8 2 860 60688

Table 4. Lymphoblastoid Cell Line Reactions with Informative Sera

```
                   |MBWBPGGMGLLFMMKTDMGGGABFDCGBDBG
                   |IOEOEOEAEAAAUURAUUOAABSEURAAAOS
                   |TDGDRLLRLWWUCCEKPCLNNSHRTONTWDC
                   |
                   |      222222222 222          22
                   |      000000000 000          00
                   |3322122222223341333333343433133
                   |5792600156793409246328609210879

9004 A2            |8888 8888828888888888 6  688888
9004 A2            |8888888888888882888 0    688888
9004 A2            |888888000000000200088   8 88800
9007 A2            |6888888888888888 888 0   488888
9007 A2            |6488888888888882488 8     84888
9012 A2            |888888888888888688888   2288888
9012 A2            |62648800000000 000 8   2 88800
9015 A2            |6888880000000008000 8     88800
9016 A2            |8888 088 8088888 8868  8  88888
9016 A2            | 4888888888888 888 866  888888
9016 A2            |868888000000000200088 8 2888800
9024 A2      A11   |  888888888888 888 8      88888
9024 A2      A11   |648888888888888 888 0     88888
9024 A2      A11   |688888000000000600028   2 88800
9031 A2            |888888888888888 888 0   6 88888
9031 A2            |868888000000000400068 2 2288800
9032 A2            |8888888888888888888888 80 888888
9032 A2            |0 888 888888888 8888086  688888
9032 A2            |888888000000000800088 8 8888800
9036 A2            |888888888888888488868 6  688888
9036 A2            |888888888888888888880 008688888
9036 A2            |888888000000000600088 6 86 8800
9038 A2            |888888888888888 88880   8888888
9038 A2            |888888000000000800088 2 6888800
9039 A2            |8 888888888888 88880    8888888
9039 A2            |8888888888888886888 0 8 8888888
9039 A2            |6 848800000000 00028    4688800
9045 A2      A3    |84868800000000 00088    2688800
9045 A2      A3    |888888800888880888888     88888
9045 A2      A3    |6 8 888008888 0   660    888886
9046 A2            |4888868888 88888088 6 8   888 8
9046 A2            |888888000000000800088    688800
9052 A2            |8888888888888888888 8   6688888
9052 A2            |888888000000000800088   8888800
9054 A2            |888888000000000200088 8 6 88800
9056 A2            |88880888888888888888020 8888888
9056 A2            |64 88800000000 00028    8 88800
9056 A2            |888888800888880888888  24 88888
9056 A2            |8808888  88888   8888     0088
9058 A2            |68888888888088 868868    6888 0
9058 A2            |888888888888888888888 2 2 88888
9058 A2            |888888000000000600088   6 88800
9059 A2            |88888 888888888888888     88888
9059 A2            |8688880000000002000 8  22488800
```

```
                   |MBWBPGGMGLLFMMKTDMGGGABFDCGBDBG
                   |IOEOEOEAEAAAUURAUUOAABSEURAAAOS
                   |TDGDRLLRLWWUCCEKPCLNNSHRTONTWDC
                   |
                   |      222222222 222          22
                   |      000000000 000          00
                   |3322122222223341333333343433133
                   |5792600156793409246328609210879

9061 A2            |668888000000000200028   2488800
9061 A2            |668888000000000200028   2488800
9062 A2            |888888000000000800088  28488800
9064 A2            |888888000000000800088  28488800
9064 A2            |88888 88880888888888 8   888888
9064 A2            |888888888888888 888 0 2 8888888
9064 A2            |868888000000000400088   8888800
9066 A2            |8888888888888888888888 2 688888
9066 A2            |888888000000000800048   6288800
9067 A2            |8 8888888808880 88828     88888
9067 A2            |668888888888888 888 0    888888
9067 A2            |688688000000000200066     88800
9068 A2            |8288868888888882888 0    688888
9068 A2            |6468880000000002000 8    688800
9069 A2            |8688888888 88886886 8    888888
9069 A2            |688888888888888 888 024  888888
9069 A2            |88888 000000006000       888800
9070 A2            |888888000000000800088 888888800
9074 A2            |888888888888888 888 0     88888
9074 A2            |4888688888 88880888 8    688888
9074 A2            |82888800000000 00088    2 86800
9076 A2            |648888888888888 88800    688888
9076 A2            |668888000000000200068 2 2488800
9077 A2            |668888888888888 888 0    888888
9077 A2            |8 888888888888 88828    2488888
9084 A2            |8688888888 8888 88868    888888
9084 A2            |  4888888888888 888 8     88888
9084 A2            |6488880000000004000262  2288800
9090 A2            |688888888888888 88880   8888888
9090 A2            |888888000000000800088     88800
9091 A2            |688888000000000400048     88800
9093 A2            |448888888888888 888 0     88888
9093 A2            |88888800000000 00088    2 88800
9094 A2      A3    |888888868886686888880    688888
9094 A2      A3    |8888880000000008000 8     88400
9097 A2      A3    |88888888888882868 2888     88888
9097 A2      A3    |688888888888888688880 2 4888888
9097 A2      A3    |6488880000000002000 8  26288800
9099 A2            |888888888808888688880     88888
9099 A2            |  8888888888888 888 8 6 6888888
9099 A2            |826888000000000200048 2 8888800
9010 A28           |6 4   888848888 888 68868648888
9010 A28           |2 82  000000000 000 82868888800
9010 A28           |      000000000 000  88888  200
9010 A28           |8     800888880  8  6 8 48 68
9021 AW68    A30   |      88888888   8    8 28 8
9021 AW68    A30   |      888888888 888  68888
9021 AW68    A30   |      886868888 888  88688888
9021 AW68    A30   |      888888888 888 8 8488882 8
9021 AW68    A30   |      888888888 888  68888
9021 AW68    A30   |      000000000 000 8888888800
```

Table 5. Lymphoblastoid Cell Line Reactions with Informative Sera

```
                   |FJGMJDEFJTPNAPWMMBKK
                   |EERBOANEUETWMNOBIRAA
                   |RRUCYWGSJRROYWLCTAPP
                   |
                   |
                   |22221111111111211111
                   |77716687778889199989
                   |10218750793212518485

9003 B51           |   808888888   8  88
9016 B51           |   48888 8        8
9016 B51           |4 28688888  24 4  8
9016 B51           |8  8088888888  6 68
9030 B51           |4  888888888   8  88
9030 B51           |8  8888888428     888
9044 B51           |   8888888888  8 484
9044 B51           |   8888888888  8 484
9045 B51           |6  8888888 46     242
9045 B51           |   8888888       888
9045 B51           |   8888 66 6
9070 B51           |8  88888888 8   8888
9089 B51           |6248888888668 22  84
9089 B51           |8  8888888866     88
9092 B51      BW62 |   8888888 88      8
9092 B51      BW62 |   8688888
9092 B51      BW62 |   808888822     466
9103 B51      BW61 |   08888808    4
9103 B51      BW61 |   08888808    4
9103 B51      BW61 |   8888888       488
9103 B51      BW61 |   88888884       64
9103 B51      BW61 |   88888882 6    884
9011 BW52          |   82880  6
9011 BW52          | 0 8888 28886888
9011 BW52          |   8488   8444 2
9008 B18           |88
9008 B18           |88    68
9008 B18           |882   42
9008 B18           |886       2 8
9018 B18           |828      0
9018 B18           |8         6    6
9018 B18           |8  2 4446464 4 4
9018 B18           |86
9019 B18           |886
9019 B18           |84
9020 B18           |8        0
9020 B18           |8 4          2
9020 B18           |862    2
9038 B18           |888  4 0          4
9038 B18           |8884 6 6    4
9039 B18           |88     4 0     2
9039 B18           |8868   6    6     4
9039 B18           |84     2       2
9061 B18           |8642
9061 B18           |8642
9085 B18           |886
9085 B18           |84     000000
9085 B18           |846  2    4 4
9085 B18           |888  2 22   6   2  4
```

```
                   |FJGMJDEFJTPNAPWMMBKK
                   |EERBOANEUETWMNOBIRAA
                   |RRUCYWGSJRROYWLCTAPP
                   |
                   |
                   |22221111111111211111
                   |77716687778889199989
                   |10218750793212518485

9006 B35           |   8         8888888
9006 B35           |   6   8      6   2
9006 B35           |   6   4     6824886
9024 B35      BW62 |       8 0   88 8888
9024 B35      BW62 |       4     64   8
9024 B35      BW62 |4      6     8648688
9025 B35           |4      6 0     8 48
9025 B35           |6 2    6     4846886
9025 B35           |             46 228
9041 B35           |             28 8888
9041 B35           |8           488 6
9041 B35           |82 6 228    68868888
9042 B35           |            6884888
9042 B35           |646    8  8  868  84
9042 B35           |            6884888
9056 B35           |   2   0     8888888
9056 B35           |              4   6
9056 B35           |             8888888
9056 B35           |             882866
9068 B35           |   6   6     64  682
9068 B35           |             66  28
9080 B35           |   4     0      2888
9080 B35           |   8   6     6444288
9080 B35           |   2
9010 BW53          |    466      62   6
9010 BW53          |  2  88      4 8 688
9010 BW53          |     2           68
9010 BW53          |     46          64
```

Table 6. Lymphoblastoid Cell Line Reactions with Informative Sera

```
          |MJMJMMBMCMDBBDWBMGMMJGYASDH
          |YLRLYBERRYDREDORIRIRLEUBTAA
          |REVERCRVORCSRCLSTUTVELNSEWM
          |
          |
          |444442222222252444444414444
          |343333332320341444111332322
          |608972439080105871798238047

9040 B49  |  28  6 8 8      88     6 4
9040 B49  |86688484868   86 6664   8 6
9040 B49  |8 688 2 8      8 82 6     4
9106 BW50 |88888888688888 8
9106 BW50 |42684888848888 82
9027 B44  |        88        888    88
9027 B44  |                  888 8  88
9036 B44  |      6           888 6  88
9036 B44  |8     8           888 8 688
9036 B44  |   44 8           888   488
9050 B44  |                  888 8  88
9050 B44  |                  888 8  88
9051 B44  |  2 6             888 8 288
9051 B44  |  2 6             888 8 288
9051 B44  |                  888 8  88
9053 B44  |                  888 4  88
9054 B44  |                  888    88
9063 B44  |        6      22  88    88
9090 B44  |  8 8   88        888 8  88
9090 B44  |                  888    88
9058 B45  |   6   6 8     8  8888668
9058 B45  |6        8  8  8  8888888
9058 B45  |         8  8  8  88888 8
```

Table 7. Lymphoblastoid Cell Line Reactions with Informative Sera

		PMBAMTMCBGBHDJMTMG
		EYESIORREASAULREOS
		RRTTTKVOTNHMTEVRRC
		222
		344444433424444000
		900110099060000888
		279355389231043679
9002	B14	8888 84 8
9002	B14	888 8 4 6 4
9002	B14	88886 88 000
9055	B14	888882288 26 2 8
9055	B14	88888 886 000
9079	B14	8888 8 80
9079	B14	88884 28 0 84
9079	B14	888 8 2 20 2
9079	B14	88888 80
9094	B14	8888844 66 4 4 8
9094	B14	88888 2 000
9029	BW65	888 6 4 8
9029	BW65	88864 2 000
9049	BW65	88828 2882 2 2000
9078	BW65	888 8 4 4
9014	B8	6686 288888622888
9014	B8	8 6 888888888888
9014	B8	8 82 8868882 000
9022	B8	8 2 88888 888
9022	B8	6 488888884008
9022	B8	2 888888 000
9023	B8	6 2 888888 888
9023	B8	6 6 62 886886 888
9023	B8	8 6 8888866 000
9086	B8	8888888 000
9086	B8	0888888 0888
9086	B8	6 8 8 288888842888
9086	B8	4 88 886 000
9087	B8	4 80888 888
9087	B8	28 888888888000
9088	B8	888 8888888 888
9088	B8	4 8 2 888884 888
9088	B8	26 888886 000

Table 8. Lymphoblastoid Cell Line Reactions with Informative Sera

			MPTMOEMDAGHWZPBMBDTTTMKTBTJGGLMSKBDPWMMFJT
			INEYHNYCBEAAHNRREDEEEYAERHLAEAUFRRPNOBIEUE
			NWREAGRHSLMLAWAVRCRRRRPRSPENLWERENTWLCTSJR
			2 2 22
			22222222442221222222212241441451145121111 1
			34155435231059033222266640990900090919977 7
			82620806822852534823566785852307692251807 9
9031	BW62		8888 88 888 888 8 8 88 888888 8
9031	BW62		88882882 686 848 2 6 0260 80088 26
9032	BW62		88888888 888 888 8 88 6 88888 8 88 0
9032	BW62		8888 868 66 884 8 88 8 6
9032	BW62		88888888 8882 888 8 8 0880 8008824 86
9060	BW62		8868488 868 4 88888
9060	BW62		8888 882 868888 8 8 0 80 800886 8
9064	BW62		8888888 6 888 8 88888 8
9064	BW62		8888 886 88 68868 8888 62 888886 8
9064	BW62		88882862 864 8888 6 86 0 60280088 6 22
9071	BW62		88888888 868 8 88888 8 6
9071	BW62		8868 842 68886 6 8 26 888886
9071	BW62		8888208 448888 6 8 0 60 8008842
9072	BW62		8888 88 6 888 8 88888 8 6
9072	BW62		8868 88 4 48868 8 8 8 888884 6
9072	BW62		8888 88 2 888888 8 880280 800888 82
9091	BW62		88882884 6 8848 4 2 0 20 0088 2 26
9099	BW62		8888888 8 888 8 88888 8
9099	BW62		8668 866 86 68868 26488 4888886 08 6
9099	BW62		8868 862 862 8888 6 66 0 40 80088 2 4
9100	BW62		8 8 86 66884 6 6 888886 46
9100	BW62		88884882 468888 8 8 0280 8008864 8
9101	BW62		8868404 6826 4 6 4 88888 2
9101	BW62		8868404 6826 4 6 4 88888 2
9101	BW62		8888 88 268888 82 8 0680 800886
9006	B35		8 88 88888 6 888888888
9006	B35		6 88 6886 2 86888 6 8
9006	B35		8 884688082 62 0 20 8008868244
9025	B35		8 8 6 888 8888 8 6 688888 8 6 0
9025	B35		4 082 88648888 08 666 48888848466
9025	B35		8 88248848 2 0 0 8008846 2
9041	B35		2 88 28868 8888828 8
9041	B35		6 8 88 88888 8 6 8888888 6
9041	B35		8 88 88888888 8 6 066048008888688
9042	B35		8 88 88888 8 2 68888888888848
9042	B35		886 86 88 88888 88 8 68888888888868 8
9042	B35		8 88 88888 8 2 08808800888848
9056	B35		246 8 08488888 8 86 2 8 8888888880
9056	B35		6 8 88 8888 8 88 0 20 80088 4
9056	B35		8 88 88888 8 2 880888888
9056	B35		88 88888 22 8 888828
9068	B35		8 644 88668888 6488 8 6 8888864 6
9068	B35		6 88268888 2 48 0 0 8008866
9073	BW52	B35	8 88 88888888 888888888 4480 0
9073	BW52	B35	22 88 488886688 8888888684426
9080	B35		8 88 8888 88888 2 0
9080	B35		4 6 88268868 4 6 4 8888864446
9080	B35		88 6888 02 0 80088
9003	B51		8 8 88882 088088000 8 888
9016	B51		86 6868 88888 8 8
9016	B51		84 68 28 888482 2488884 88882 464 4 888
9016	B51		2 8 888 8 2 88888 08808 00 8 6 888
9030	B51		8 888 8 808888 4 88888 8 888
9030	B51		2 6 888 86 48888 2 08808800 8 888
9044	B51		8866 8 888 8 8 888888 08808 00 6 8 888
9045	B51		686 6 6 64887 08806200 888
9045	B51		88 6 8888 08808 00 888
9045	B51		88 8 8 4488 888 66
9070	B51		8 4 888 8 8 888888 08808800 8 8888
9089	B51		428 0 84 886 8 828888 4 88886 28 22 888
9089	B51		28 8 888 8 6 8888 8 0880840026 888
9092	B51		8 84 882 88888 8 888888 8 888
9092	B51		6 4 22 68 4 88 888686 6 888
9092	B51		8 24 886 8 8888 08808800 2 888
9103	B51	BW61	88 8888 888888868 4 880
9103	B51	BW61	88 8888 888888868 4 880
9103	B51	BW61	4 886 8888 888888888 888
9103	B51	BW61	2 88 4 88 888880888 888
9103	B51	BW61	4 888 8888 0880880088 888

Table 9. Lymphoblastoid Cell Line Reactions with Informative Sera

```
                 |BDNDEMRSKBBTEFBYGTMTMDGNDBJ
                 |SCACNUIFRREENARAOOROIDREERU
                 |HHIHGECRENRRGUSGRKVKTCUUCSJ
                 |
                 |       22
                 |2222255114525555255555555255
                 |6556600009010111512222223623
                 |3781201769468234904567900082

9066 BW46  CW11  |88 88888   888888888888888
9066 BW46  CW11  |888888 00 4 888888888888888
9076 BW46  CW11  |04 8 8 6   48864  8   6666
9076 BW46  CW11  |848848 00   8886  864 4668
9077 BW46  CW11  |8  888688 8888868 864 8 88
9077 BW46  CW11  |866888 88 6 8886448   4668
```

Table 10. Lymphoblastoid Cell Line Reactions with Informative Sera

```
           |GCRMDKFTBGJGGLMSKBD
           |ORIRAATEEOLAEAUFRRP
           |ESCVWPWRRLENLWERENT
           |
           |            2  22
           |4444424244441451145
           |7777867688990900090
           |1246668703852307692

9007 BW57  |468888888 8888
9012 BW57  |88888888888888
9012 BW57  |68888828848808 00
9052 BW57  |888888888 88888
9052 BW57  |88888888888808200
9095 BW57  |888888888888884  6
9095 BW57  |8888888886  04800
9102 BW57  |888848888 8888   28
9102 BW57  |68888888868888    6
9102 BW57  |888888888 8888
```

Table 11. Lymphoblastoid Cell Line Reactions with Informative Sera

```
          |BFMPZJMDADGBDBCTJNDTJGGLMMSKBD
          |SEYEHETUBUOEURRKLAUOLAEABUFRRP
          |HRERARTQSQLRPSSBEITKENLWCERENT
          |
          |                      2   22
          |444444444444444443334414451145
          |899955555666666764249909900090
          |812089532012756094358523607692

9009 B37  |8888  8    8    4   88048 00
9026 B38  | 8 286888 86  846   888888   8
9026 B38  | 2 288886 8842848 2 888668  2
9035 B38  | 6  88888 8888888 8 888888   8
9035 B38  | 4  8888228888866   88088600 6
9057 B38  |    8888648882828   88088800 8
9062 B38  |66  8888648886868   88088800 8
9062 B38  |66  8888648886868   88088800 8
9001 B7   |                    6808680086
9013 B7   |    248    2    8 8   0  80088
9013 B7   |     4 24                88884
9013 B7   |   68   6       6     6  8 888
9017 B7   |2   8 8   000   8     8  88888
9017 B7   |6   6462   4  6 8 6   86 88888
9017 B7   |2   648    2    8 4   0  80088
9033 B7   |    88    88  8   8   8  88888
9033 B7   |    68    24  2   8   02680088
9034 B7   |    88     4  4   8      88 88
9034 B7   |    220       2   6   0  80088
9065 B7   |64  88        8 4 8   0  80088
9065 B7   |64  88        8 4 8   0  80088
9081 B7   |4   6 6    2    4 6   64 88888
9081 B7   |2   6660   2  2 6  0  04680068
9082 B7   |    8 6    2      4   8  88888
9082 B7   |2   488    2    6 8   0  80088
9083 B7   |                         88888
9083 B7   |2   6 8 4  6    4 4   8  88888
9083 B7   |2   288    2    8   2 0  80088
```

Table 12. Lymphoblastoid Cell Line Reactions with Informative Sera

```
           |JTJERAMAJKMJDEKDCCCJDGAOMBCDDDDMKME
           |UKLNIMRMUSILPNAUERRRDLMHBERPPDUIAIN
           |JBEGCYVYJHTETGPTPOOMCAYACROTTCTTPTG
           |
           |
           |33233323333333332333233232233333333
           |00911030103238829228901903244224803
           |45827633082511537606921614923427373

9107 BW54  |88888868888     4   684 8
9107 BW54  |68664848684  2  2    22 2  62   0
9107 BW54  |68664848684  2  2    22 2  62   0
9021 BW42  |     4  88 888 8 8       8 86882884
9021 BW42  |   4 86 8848888888884 68 2 08888888
9021 BW42  |     88 88688888 888 8 4 4 88888886
9021 BW42  |     84 88 886882888   2   6688 682
9021 BW42  |   4 86 8848888888884 68 2 08888888
9021 BW42  |2    8668888888888888 888 2 84888888
9001 B7    |     224 8 8884 8    6 2  2888 8888
9013 B7    |   2  2268688486   2 2    288888888
9013 B7    |        44 888             08888888
9013 B7    |6  66  668 888       6     08888888
9017 B7    |    0   68 8882         0 488888888
9017 B7    |       6686888666    6    888888888
9017 B7    |8  2 248888888648    828  688888888
9033 B7    |      48884888886  6      888888888
9033 B7    |8  64668888888888  2 6 6  888888088
9034 B7    |       8 88888888         8888 8888
9034 B7    |      2 88288866   2 2    688868088
9065 B7    |2  222688 88888888 8 8 2 2888888088
9065 B7    |2  222688 88888888 8 8 2 2888888088
9081 B7    |   6  24686888 6   4      688888888
9081 B7    |   2  228868882 2  2 2    488888088
9082 B7    | 4 2    682688 4            88668888
9082 B7    |2  2  4468488888     6 2  688888888
9083 B7    |         6 888             88888888
9083 B7    | 0   0  486888 4   6      688888888
9083 B7    |       48868888 8  2 4    888888888
```

Table 13. Lymphoblastoid Cell Line Reactions with Informative Sera

		KFRMJMBJBCTCHPRRDMRMEEJGGLMMSKBD
		WAIILLRLSRARFROIPYIYNNLAEABUFRRP
		CUCDESSEHOKSPROCTRCRGGENLWCERENT
		2 22
		22323222223333333333334414451145
		77372888986577344334389909900090
		68555023288350913680318523607692
9001	B7	8 688 8 882888 8886808680086
9013	B7	828882 284688822888 0 80088
9013	B7	82688 88866888 88884
9013	B7	86888 888 888 6 8 888
9017	B7	8 888 0 8 888 8888 8 88888
9017	B7	848888 2 82 88868888 86 88888
9017	B7	868882 844 8866888 0 80088
9033	B7	8 688 8 80888888888 8 88888
9033	B7	88888 8 88688888888 02680088
9034	B7	84888 8 88888888888 88 88
9034	B7	84888 8 84288866888 0 80088
9065	B7	888884 4 886880888888 0 80088
9065	B7	888884 4 886880888888 0 80088
9081	B7	866888 866888 888 64 88888
9081	B7	848882 2 86488824888 04680068
9082	B7	8 6866 8 288848888 8 88888
9082	B7	888882 2 82488844888 0 80088
9083	B7	88888 0 888 888 88888
9083	B7	828886 826888 888 8 88888
9083	B7	888888 2 848888 28882 0 80088
9004	B27	88886 86 8886 888888888
9004	B27	2 88888 86 888 8 88888862 2
9004	B27	88888 8628886 88888088800
9005	B27	2 88888 28886 8888888886
9005	B27	22 88 88 6 6888 8 8888886
9005	B27	88888 6488882 88888068400 4
9015	B27	88888 86 888 88888088 00
9067	B27	88888 8 888 8 888848
9067	B27	4 88488 8 2888628088886886 24
9067	B27	88888 688 62688028 00
9047	BW47	4 86 88 888 8 8
9047	BW47	8 882888 88888 4
9047	BW47	68888888 88888 4 8
9047	BW47	8 88 888 8 8 88088 00
9021	BW42	6 68 8 48 6 848 88888
9021	BW42	8 488 8 8 888 88888
9021	BW42	8 88 8 8 8 868 88888
9021	BW42	6 28 8 68 6 826 8 88888
9021	BW42	8 488 8 8 888 88888
9021	BW42	8 8848 668 4 888 0 280088

Table 14. Lymphoblastoid Cell Line Reactions with Informative Sera

			ACPHBFPDTFGMGTTRCTGDMODAKKJEMDDEMJCKG
			SRRFSAEUEAOUSESIRSOPLHASAALNYPDNYRRAO
			TSRPHURTRULCCRUCOURTSAWTPPEGRTCGRMOPE
			222222 2
			3333333200010033333333333333330333333
			8577999798905865756755568868449988289
			4305165320414171156387632301036076654
9046	B13		86868 8868888 26886 8 4 642 8 8
9046	B13		8888888000000 288 8 8 8 6 406 868
9048	B13		8868888868888 8886 88 644 84 868
9048	B13		8888888000000 26866 8 842 06 828
9093	B13		2888868888888 4 6 8 8 6 48 8
9093	B13		8888888000000 26 4 8 24 6 06 8 8
9096	B13		2888888000000 666882846644 06 848
9096	B13		2888888000000 666882846644 06 848
9028	BW60		8888 8 8 88 88 88888888 888 68
9028	BW60		8888 8 8 88 888888888888888 8 88
9028	BW60		8868 6000000 486626866888 640
9059	BW60		8888 8 8888888888888 88 8 8
9059	BW60		8888 80000002286488888868242042 8 2
9069	BW60		888 4 4 88 888 6888 6 8 84
9069	BW60		8888 4 82 6 888868888886688 66 2
9069	BW60		8888 60000000888 88888 8 20
9075	BW60		8888 8 8 888 880888888888888 848
9075	BW60		8888 8000000 88 8888888886880 8 6
9097	BW60		8888 8888888888886868
9097	BW60		8888 60 8 88886888888888662668
9097	BW60		8888 00000064886888888844 04 28
9098	BW60		8888 8888 8 8 88 8
9098	BW60		8868 2 668886844806 8 0 8
9098	BW60		8868 8 2 88266866888 88 6
9098	BW60		8888 60000002 86 888488682480 2 2
9037	BW61		8888 4 8888882 88 88 6 8 88
9037	BW61		8888 8 0 888 8 8 888 888 8 888
9037	BW61		888 86 6 8 888868 6688 86 88
9037	BW61		8888 8000000 888 8 8 8688 80 882
9074	BW61		8888 8 8 8888888888888888 8 8
9074	BW61		888 6 668 8888 8888 888 6 866
9074	BW61		8888 2000000 86 82666888 6602 8 4
9084	BW61		88 00 4 88 6 688 8
9084	BW61		88 8 8 4 8 684 68 4 8
9084	BW61		8888 000000026466 446288 20 422
9103	BW61	B51	8888 6 8 888 408 88 88 48 8
9103	BW61	B51	8888 6 8 888 408 88 88 48 8
9103	BW61	B51	08888 8 88 848888 488 46844 8 8
9103	BW61	B51	8868 8 8 886 2 666 86 88 68 2
9103	BW61	B51	8888 4000000 68868 88 602 8 6
9043	BW41		8886 448 6 6 6 8 6 88888686
9043	BW41		8666 28 0008 6 0 2 42 28888888
9043	BW41		8686 80000000 8 2 6088688
9001	B7		6828 2000000 4 4 88 8880 44
9013	B7		2864 6000000 6 8 288 8880 2 48
9013	B7		80 88 888 4
9013	B7		0 88 8888
9017	B7		28 8 6 00 88 8888 26
9017	B7		68 2 8 4 6 4 2 88 8886 68
9017	B7		6844 6000000 868 2 8868880 68
9033	B7		8880 08 88 0 88 8888 6 80
9033	B7		6868 6000000 8628 2 28028880 2 88
9034	B7		8888 8 88 88 8888 88
9034	B7		824 6000000 8 80 88800 2 6
9065	B7		8886 28000000 82626 80488806 8888
9065	B7		8886 28000000 82626 80488806 8888
9081	B7		6866 8 6 4 8828888 4 6
9081	B7		4846 2 8000000 8 4 8028880 2 26
9082	B7		482 8 6 2 88 8888 6
9082	B7		6842 6000000 6 2 88 8880 88
9083	B7		6 88 8886
9083	B7		6862 84 2 6 4)828888 6 6
9083	B7		6884 8000000 6 6 38 8880 2 88

Table 15. Lymphoblastoid Cell Line Reactions with Informative Sera

			LGJTTGMWBMKSHMJSMTMDMJEGBFD	
			AALSSEUARURFAYETYOIPRLNSAFU	
			WNEUULCNNEERMRRERMDTIEGCTRT	
			22222 22 2 2	
			44400100451141 50 0111	
			999660599000266766706676655	
			358452789067768079120523089	
9029	A23		2888888888882 8888888 8	BW6
9029	A23		2888888888882 8888888 8	BW6
9029	A23		4880000068002 84868802 0	BW6
9001	A24		886000008800 2 8608 06	BW6
9002	A24		48888888 8880 888 88	BW6
9002	A24		88 8 8880 6 8888844	BW6
9002	A24		8880000088002 8 2808808	BW6
9028	A24		888888888888 84 6688848	BW6
9028	A24		888888888888 88 8888888	BW6
9042	A24		888888888888 88888	BW6
9042	A24		888888888888 84 6688888	BW6
9075	A24		888888888888 888888	BW6
9107	A24		8688848 8888 4688888	BW6
9006	A11		88888 88 8	BW6
9006	A11		88866 8 6	BW6
9024	A11	A2	8 8888 8 88	BW6
9024	A11	A2	6 8888 4 8 8	BW6

Table 16. Lymphoblastoid Cell Line Reactions with Informative Sera

```
                |EFBYGTMTMDGNDBJMYCKBPDDTDKLCMMGFPB
                |NARAOOROIDREERUYURALRDDSDAAEYBRTUR
                |GUSGRKVKTCUUCSJRNSPTRCCUCPYPRCUWRN
                |
                |
                |5555255555555255555555555555555555
                |0111512222236235555555555611212126 2
                |8234904567900823104567891870529133

9004 CW1        |888688    6  8       4     846  8
9004 CW1        |8886 2    4  6  2 0    6   8 62 4
9004 CW1        |888682    2  8  2    6 8   8 8  2
9005 CW1        |888808  4 82 2     2       88   2
9005 CW1        |8864 2          2  2       2 62 2
9005 CW1        |888866                     846  4
9067 CW1        |888                    4
9067 CW1        |888866 2  4  6       4 602 6 0   2
9067 CW1        |8886 6                     2
9092 CW1        |8888668             8288     6  8
9092 CW1        |88644 2      2         4     2
9092 CW1        |8888842   2          2 80       2
9107 CW1        |888888    64              4
9107 CW1        |888864                   24
9107 CW1        |888864                   24
9071 CW11       |88888888 88888
9071 CW11       |88868686468888       8 66
9071 CW11       |88828 88888888    42 6 80     0
9100 CW11       |888 6286 48888  442    66
9100 CW11       |88848288888888  228  2 88
9072 CW11       |888888 6688686          8
9072 CW11       |888884    8288   6    080
9072 CW11       |888888    4 88  4886 8 88
9101 CW11       |888886    4 86   6   6 66
9101 CW11       |888886    4 86   6   6 66
9101 CW11       |888684      88    6  8 88    2
9028 CW3   CW10 |      888888688         226
9028 CW3   CW10 |      888888888         8 8  6
9028 CW3   CW10 |      8 448848          4 2
9031 CW3   CW10 |    4 888888886 8   6  88    6
9031 CW3   CW10 |    8 886888882 2      22
9032 CW3   CW9  |    8 888888888     8  88    8
9032 CW3   CW9  |      888888884
9032 CW3   CW9  |    8 888888888 2 6644 88    8
9059 CW3   CW10 |      888888 88        88
9059 CW3   CW10 | 2    8 4488 8     2 2 66
9060 CW3   CW9  |    4 884688888     0     0
9060 CW3   CW9  |    8 888888888         8    8
9064 CW3   CW9  |    8 888888888
9064 CW3   CW9  | 4    886888888      4 864   4   4
9064 CW3   CW9  |    6 88888888802  2 2286    2
9069 CW3   CW10 |      88868868                   2
9069 CW3   CW10 |2  6 4888888684        486   6 442
9069 CW3   CW10 | 8  8 88268888         68
9075 CW3   CW10 |      888888888   4       8
9075 CW3   CW10 |      88668868          8 4
9091 CW3   CW9  |    6 884688882        22    22
9097 CW3   CW10 |      88888888          6
9097 CW3   CW10 | 4    888888884    6 6 80 66
9097 CW3   CW10 | 4    8 668868         26       2
9098 CW3   CW10 |      888888 88        88
9098 CW3   CW10 |      86648868
9098 CW3   CW10 |    0 86668868   6 2 6 26  2
9098 CW3   CW10 |      84448848   22  2 64
9099 CW3   CW9  |    8 888888888
9099 CW3   CW9  |  4 46886888888     48 6  2  6 4
9099 CW3   CW9  |    6 886888888    2 4 66
9018 CW5        |               888              4
9018 CW5        |             8 688      8 8     6
9018 CW5        |       4     8 8880    6  84    6
9018 CW5        |             8 888       4      8
9019 CW5        |             8 8886     8       8
9019 CW5        |             8 888
9020 CW5        |             8  88
9020 CW5        | 22          8  88 2
9020 CW5        |  2          8 686 4   2        4
9036 CW5        |             8 888     8        8
9036 CW5        |             8 888 2   8        8
9036 CW5        |             8 688 2 2 8        8
9039 CW5        |             8 888   8 80    8  8
9039 CW5        | 42 0     4  8 888   8 8     8  8
9039 CW5        |             8 888     62    4  2
9054 CW5        |             8 888   6 88
9063 CW5        |             8 888   6 8
9085 CW5        |             2 8864
9085 CW5        |             8 8884     0       8
9085 CW5        |         4   8 888              8
9085 CW5        | 2 2      22 8 88862    6 2 6   84
9090 CW5        |               888     8  6  8  8
9090 CW5        |               888     82
```

```
                |EFBYGTMTMDGNDBJMYCKBPDDTDKLCMMGFPB
                |NARAOOROIDREERUYURALRDDSDAAEYBRTUR
                |GUSGRKVKTCUUCSJRNSPTRCCUCPYPRCUWRN
                |
                |
                |5555255555555255555555555555555555
                |0111512222236235555555555611212126 2
                |8234904567900823104567891870529133

9029 CW8        |               8  8
9029 CW8        |               8  8
9029 CW8        |      2        0  8
9049 CW8        |4 2   42     8 8 88   22022 2
9049 CW8        |4 2   42     8 8 88   22022 2
9055 CW8        | 8   4       8 84 6 2 2        2
9055 CW8        |        2    8 82 8
9074 CW8   CW10 |      88888888    8    88    8
9074 CW8   CW10 |      88468866    8    46
9074 CW8   CW10 |      884488 8    66 6 84
9078 CW8        |               8  8
9078 CW8        |         0   008
9078 CW8        |               8  8     0
9079 CW8        |      00       8  8  4 4      00
9079 CW8        |4     4     4  8  8 0  6
9079 CW8        |               8  8     0
9079 CW8        |0 00 0         8  8    2000
9094 CW8        |8 2   66     6 82682   66
9094 CW8        |               8  8    6
9103 CW8        |   6  8      8    8 2    68
9103 CW8        |   6  8      8    8 2    68
9103 CW8        |      6   86 8    8       8 6   46
9103 CW8        |             4    4 2     6
9103 CW8        |      2      8 2  8       8
9001 CW7        |                      222 6 2
9012 CW7        |            8      8 688
9012 CW7        |      2     6    6 668686 2  2  4
9013 CW7        |        2       2  842886 2 2
9013 CW7        |        4  2       688886  20
9013 CW7        |0         6        6 8686   8
9014 CW7        |                   888888
9014 CW7        |                   888888
9014 CW7        |                   888888
9017 CW7        |0 00 00000   00     88888000 0000
9017 CW7        |  44 4448     62   888880 6826 2
9017 CW7        |                   886888 222
9022 CW7        |             8  8  088888
9022 CW7        |   0               888888      0
9022 CW7        |             8  2  28 424       6
9023 CW7        |             8  220 88888
9023 CW7        | 6  4 62     8   4 886886
9023 CW7        |    2        8     88266
9033 CW7        |    0   8          888888    4
9033 CW7        |  4     2        2 888888  64   2
9034 CW7        |        8          888888
9034 CW7        |        2          882886
9038 CW7        |  2          8 26  888888 2  6
9038 CW7        |            48 242 8888882   2
9040 CW7        |             6     888888
9040 CW7        |    6   6    8      8448
9040 CW7        |             8     888884    2
9045 CW7        |             8 0  246446        2
9045 CW7        |0                  8  68
9045 CW7        |0            8
9061 CW7        |             8     464286       2
9061 CW7        |             8     464286       2
9065 CW7        |      2 6    8  8  888888 2826  8
9065 CW7        |      2 6    8  8  888888 2826  8
9081 CW7        |       2            8 88624  2
9081 CW7        |        6          884888 24
9082 CW7        |        6          882686 4
9082 CW7        |                   884888 2     2
9086 CW7        |0 00 00000   00     8 880000 000
9086 CW7        |             8     68  88
9086 CW7        |             8  6  684886
9086 CW7        |             8      8 266
9015 CW2        |                     2 6 888888688
9021 CW2        |             8   2              42
9021 CW2        |  2          4     4 2 6 8888 8 88
9021 CW2        |  4          6           8886 4 88
9021 CW2        |             8  4  4     888 68 88
9021 CW2        |  2          4     4 2 6 8888 8 88
9021 CW2        |             8    22 2 6 8884 6 88
9037 CW2        |        6              2 888888888
9037 CW2        |        4    8     4   8 888888888
9037 CW2        |             8     8     888888 88
9037 CW2        |             4      22  2888888488
9084 CW2        |                       6 888888 84
9084 CW2        |                  0  6 6 888288 86
9084 CW2        |                       62666426242
```

Table 17A. Lymphoblastoid Cell Line Reactions with Informative Sera

WS #			Tenth Workshop core reagents
			:11111111111111111111113111111111111111111131199311139111111111111111111111111111153331113
			:000000000000000000000000000000000110001011900111001000000000000000000000000001122230002220
			:0011000232222223343342333344454555499459130195632326167777888896656655611229900086241014
			:63018597056349123790565488560461714936463390836410189075567917672389086018915743582967
			<--DR1-><--DR2--><--DR3--><---DR4---><--DR5--><DRW6><DR7><DR8><-DRW9-><10><-DRW52->DRW53
9001	DR1	DW1	:8 248 8 0 0 2000 08 0 08 6 0006 000 20
9002	DR1	DW20	:8 888 8 0 2 8 6 8 4 8
9002	DR1	DW20	:8 888 8 0 0 0 2 0 0 8 6 8 4 000 0
9003	DR1	DW1	:8 88 8 8 0 2 0 2 0 0 64 2 8 000 0
9004	DR1	DW1	:8 886868 6 8 8 8 8 6 6 6 86 8
9004	DR1	DW1	:8 48 46 4 4 6 4 8 4 8 4 8 8 4 62
9004	DR1	DW1	:8 248 62 02 2 20 60 00 0 62 0008 000 60
9005	DR1	DW1	:6 8 4 2
9005	DR1	DW1	:8644684 2 6 2 62 2 4 4 4 2 244
9005	DR1	DW1	:0 86 2 4 2 2 4 244 4 6 4 6 2 6
9005	DR1	DW1	:8424844 0 0 4 0 8600 04 0 0 2 0 082 000 260
9006	DR1	DW1	:8 80 8 0 8 0 0 6 6 8 0
9006	DR1	DW1	:8 8 2 2 4 6 2 22 4 2
9006	DR1	DW1	:82228 8 0 22 2 0 0 00 04 0 2 4 06 000 0
9078	DR1	DW20	:8888864 8 4 484 4 2
9078	DR1	DW20	:8 68 8 4 8 0 8 8 8 8
9078	DR1	DW20	:8424626 2 02 2 0 022220 2202 2 8 2 2 4 000 2 0
9079	DR1	DW20	:8 6800 0 0 00 0000 000 0000 60 0 80 0 0 0 0 0 8
9079	DR1	DW20	:86 48 6 2 6 6 6 8 644 6 6 8 4 6 2 8 6
9079	DR1	DW20	:822 848 4 4
9079	DR1	DW20	:8 2482 6 6 20 2
9080	DR1	DW1	:8 88 8 8 8 6
9080	DR1	DW1	:8 468 6 82 2 2 2 2 2 22 6 4 6426 8 6
9080	DR1	DW1	:8 8 0 0 2 0 0 4 0 6 000 0
9008	DRW15	DW2	: 888888888 0 0 4
9008	DRW15	DW2	: 88 88 888888 4 0 6 8 8
9008	DRW15	DW2	: 24 88 888886 4 6 0 6 8
9008	DRW15	DW2	: 8 882846888 0 2 20 620 0 6 2 0 8 000 0
9010	DRW15	DW2	: 6 886868886 4 2 28 4
9010	DRW15	DW2	: 28 88 868886 0 6 2 2 0 220 0 2 8 8 000 0
9010	DRW15	DW2	: 888888868 0
9010	DRW15	DW2	: 8868 8864 68 8
9011	DRW15	DW12	: 0 88882 888 0 0 2 0
9011	DRW15	DW12	: 0 88 86 888 6 8 0 8 8 0 6
9011	DRW15	DW12	: 2 88 8 888 0 0 0 20 6 0 2 000 0
9013	DRW15	DW2	: 8 886886888 0 6 04 260 24200 202 2 4 622440 08 2 266 000 60
9013	DRW15	DW2	: 868848848 4 48 0 8
9013	DRW15	DW2	: 6 886806688 6 6 6 8 8 6
9014	DRW15	DW2	: 66 88 866886 2 4 6 64 4 8 4 8 8 2 2
9014	DRW15	DW2	:2 8 888888888 0 4 2 40 6 0 06 8 8 6 000 0
9017	DRW15	DW2	:80 8 88208 886 84 8 004 0 80 2 64 0 0 022 0 8 8 6 0408 0 8 0 8
9017	DRW15	DW2	:4 46 2888866886 284 6 24 4 66 2 64 22 4 6 8 86 4 6 480 444 8 64
9017	DRW15	DW2	: 8 886886888 0 6 2 60 24 000 02 2 8 4420608 2 2 000 60
9081	DRW15		:2 28 4884886886 84 8 4 6 4 466 6 2 86 62 6 8 42 2 6 84 8
9081	DRW15		:2 4 288 828888 02 8 80 2 00 20 0 8 2 206 000 20
9083	DRW15	DW2	: 0 884846886 0 0 80
9083	DRW15	DW2	:4 6 884468886 6 6 4 2 2222 2 8 4 6 6 6 4
9083	DRW15	DW2	:2 8 886868888 0 4 2 0 80 00 0 2 0 08 000 2 0
9009	DR2	DW21	: 24 88644 648 0 0 02 4 0 04 2 0 6 22 42 000 0
9012	DRW16	DW21	: 66646682 6 4 84 6 8
9012	DRW16	DW21	:2 0 886862886 02 8 8 802 22000 0 020 0008 42 2 000 20
9015	DR2	DW21	: 8 88884 882 02 02 2 808 4000 04 2 0 0 6 880 08 8 000 20
9084	DRW16		: 6 88684 84 6 4 2
9084	DRW16		:8 6 2806666864 4 6 02 0 66424 4 2 6 66 4 6 4
9084	DRW16		:4 4 884642842 0 20 0 0 2 2 2 6 000 0
9007	DRW16	DR4	:4 6 88686 8 6 4 8 8888886888 6 6 4 4 8 8 4 6 888
9007	DRW16	DR4	:2 4 8888668 0 2 828888808888 02 00 0 2 8 0006 2 000 880
9016	DRW16	DW22	: 88688 86 8 8 6 66
9016	DRW16	DW22	: 88 8 48 8 28
9016	DRW16	DW22	: 6 80 6 86 6 6 6 6 66 66 62 28 6 6 4 66 88
9016	DRW16	DW22	:2 48 88888486 0 2 04 6 2 0 820 02 2 8 242 48 84288 000 280
9082	DR2		: 22 68 668882 4 26 2 2 2 6 44 4 8 4 6 8 4 242 24
9082	DR2		: 4 884866888 0 8 2 0 2 00 4600 02 0 2 8 22204062 2 000 60

Table 17B. Lymphoblastoid Cell Line Reactions with Informative Sera

WS #			Tenth Workshop core reagents
			:1111111111111111111111131111111111111111111311993111391111111111111111111111111111153331113
			:0000000000000000000000000000000000110001011900111001000000000000000000000000001122230002220
			:0011000232222233433423333444545554994591301956323261677778888966566556112299000862410 14
			:6301859705634912379056548856046171493646339083641018907556791767238908601891574358296 7
			<--DR1-><--DR2--><--DR3--><---DR4---><--DR5--><DRW6><DR7><DR8><-DRW9-><10><-DRW52->DRW53
9018	DR3	DW3	: 8888886 8 8 0 0 0 88888888 2 8
9018	DR3	DW3	: 2 8 808 46 8 2 4 488888888 8
9018	DR3	DW3	: 2 0 4 86868 8 68 2 488686888 6 8
9018	DR3	DW3	: 2 88880 02 2 0 0 882888000 0
9019	DR3	DW3	: 80888 860 8 8 88888888 8
9019	DR3	DW3	: 8 820 0 2 0 08 4 888888000 0
9020	DR3	DW3	: 2 4 4 6468686 2 2 2 2 6 4 4 6 26 4 28 2 688888686 4 6
9020	DR3	DW3	: 88888 8 0 4 88888888 8
9020	DR3	DW3	: 86848 4 4 4 4 4 88868888 8
9020	DR3	DW3	: 448 0 0 0 2 00 0 0 0 0 0 882868000 0
9021	DRW18	RML	: 4 66 2 64628808 84 2 26 6 0 480860066 0 000 000 8 000 046 68 4868000 6 0
9021	DRW18	RML	: 848 8 8 2 682868888
9021	DRW18	RML	: 88888668 4 4 866 6 8 888888888 6
9021	DRW18	RML	: 4 8 8 4 62 6 4 8 88 248888 246
9021	DRW18	RML	: 4 86668 4 6 88886688 4
9021	DRW18	RML	: 8 8 0222 0 0 2 0 6 0 8 0 888888000 60
9022	DR3	DW3	: 48 682 0 6 8 0 8 0 886888888 8
9022	DR3	DW3	: 88888 666 2 06 0 042 8 8 28888480802 20
9022	DR3	DW3	: 84820 2 04 0 0 8 886868000 0
9023	DR3	DW3	: 8 888 0 0 8 0 888888888 8
9023	DR3	DW3	: 4 66 4 86868 6 6 66 68 6 8 2 6 6 48888868888 6
9023	DR3	DW3	: 2 4 2 888802646 6 2 02 2 0 066 2 86 26688888800068 0
9085	DR3	SYD3	: 0 888886646 0 0 2 4 0 00 6888888826 8
9085	DR3	SYD3	: 8 8 8 6 88888888
9085	DR3	SYD3	: 6 8 8 4 8 48 62 6 0 88286888 2 8
9085	DR3	SYD3	: 88860 224 6 0 00 022 06 88888800026 0
9086	DR3		:0 84888 4 8 0 8888888880 4
9086	DR3		: 88888 64 8 4 8 4 2 888888888 8
9086	DR3		: 868882228 4 26 2646 2 2 44 82 8 8888868888 2 8
9086	DR3		: 868802 44 0 220 02 84 888 82 2 888000 0
9087	DR3	DW3	: 24 2 68868 2 6 4 86 6888888888 8
9087	DR3	DW3	: 4 666608888 0 088 04 86 8888880008 0
9088	DR3		: 0000 20 0 8888088462004 26200 00 0 02 0000 00 2 80 0002 08880000000000
9088	DR3		: 0 88888888 0 8 84 4888880088 8
9088	DR3		: 86888 2 44 8 2 4888868888 8
9088	DR3		: 888802624 0 04 0 86 8888880004 0
9007	DRW16	DR4	:4 6 88686 8 6 4 8 8888886888 6 6 4 4 8 8 4 6 888
9007	DRW16	DR4	:2 4 8888668 0 2 828888808888 02 00 0 2 8 0006 2 000 880
9025	DR4	DW4	: 8 4 60 8888808 0 8 4 6 6 8 4888
9025	DR4	DW4	: 4 4 666868888868 2 64 24 8 84 2 22 66 4 888
9025	DR4	DW4	: 0 488428848868 0 220 0 2 2 8 4 0002880
9027	DR4	DW4	: 888888888888 28 22 8 22 8888
9027	DR4	DW4	: 0 888888888888 0 280 022 8 22 0008880
9029	DR4	DW4	: 4 4 8 8888888888 8 8 8 6 48 4 8888
9029	DR4	DW4	: 4 4 8 8 86888 8 8 8 4 888
9029	DR4	DW4	: 0 646668886868 0 2 0 0 4 8 8 2 2 6 0002880
9031	DR4	DW4	: 6 6 8888 8888888 6 8 88 88 888
9031	DR4	DW4	:2 2 26 0 6688688088882 0 460 0 6 2 6 8 20206 6 26 0006880
9032	DR4	DW4	:0 2 8866 6 2 0 6 488688 8466886 2 6 24 8 8 6 2 26 8 4486
9032	DR4	DW4	: 4 0 8 888888888888 8 0 8 888 8 8 8888
9032	DR4	DW4	: 6 4 6 4 6 888888888888 6 06 6 64 464 86 84 66844 06884 6886
9032	DR4	DW4	: 6 2 8 0 888888808888 20 4 0 0 6 8 8 20608 8 28 0008880
9033	DR4	DW4	: 8 4 8 8 8888888888 6 8 82 848 8 6 8 888
9033	DR4	DW4	: 0 8688 0808888 0 000 0 8 0 0 06 8 20202 24 0006880
9034	DR4	DW4	: 6 88 88888 8 8 8 8 8 6 6 888
9034	DR4	DW4	: 2 0 24 808848 0 00 0 0 8 8 28 2 0002880
9090	DR4	DW4	: 8 8 8 8888888888 8 44 8 8 88 48 8 888
9090	DR4	DW4	: 2 8 2 0 888888888888 0 4 0 0 8 8 2 2 4 000 880
9091	DR4	DW4	: 2 00 088868808888 0 00 0 2 8 020 6 242 000 880
9026	DR4	DW10	: 6 6 4 8 286 8888 8 6 6 88 4888 8 88 8 8 888
9026	DR4	DW10	: 22 26 0 62628808888 02 6200 006 2 0220 2840 608 48266240006880
9024	DR4	KT2	: 8 80 8888888 8 8 6 8888
9024	DR4	KT2	: 4 2 88 6888 42 2 2 8 2 4 6 2 6 888
9024	DR4	KT2	: 4 4 0 0 82 8808888 00 2200 0 0 0 0 2 8 4 4086 8 68 40006880
9030	DR4	DW13	: 2 4 4 2228888868 8 8 8 2 86 888
9030	DR4	DW13	: 8 8 88 8888888 8 6 8 888 8 8 8888
9030	DR4	DW13	: 4 6 0 888288808888 08 2 00 0 8 0 0 84268 0002880
9028	DR4	DW14	: 6 66 8888 88 8 8 8 888
9028	DR4	DW14	: 8 688888868888 8 888
9028	DR4	DW14	:6 8 6 0 868888888888 0 20 0 4 8 2 6 8 0008880
9092	DR4	DW14	: 8 8 2 8 8 0 0 88288 8888888 2 8 8 8 88 2 888
9092	DR4	DW14	: 264 262 246 8646628 2 2 42 8 2 2 6 842 268 22244 22 888
9092	DR4	DW14	: 86 0 6828808848 0 22008 0 8 8 20 08 82 2 000 880
9098	DR4	DW14	: 6 6 8 888888888888 6 8 68 8 8 888
9098	DR4	DW14	: 6 0 8 668888888888 4 6040 4 0 6886
9098	DR4	DW14	: 4 6 8466846 44 4 8 4 888
9098	DR4	DW14	: 4 0 64 8666868 0 0 0 2 8 2 2 2 000 880
9107	DR4	DW15	: 8 6 6 8 0 888888888888 0 0 0 8 6888 66 80008880
9107	DR4	DW15	: 4 4 0 8288 88 8848 0 004 0 2 0 8 0 44 000 820

Table 17C. Lymphoblastoid Cell Line Reactions with Informative Sera

WS #			Tenth Workshop core reagents
			:1111111111111111111113111111111111111111113119931113911111111111111111111111111111111153331113
			:000000000000000000000000000000000011000101190011100100000000000000000000000001122230002220
			:00110002322222334334233334445455499459130195632326167777888896656655611229900086241014
			:63018597056349123790565488560461714936463390836410189075567917672389086018915743582967
			<--DR1-><--DR2--><--DR3--><---DR4---><--DR5--><DRW6><DR7><DR8><-DRW9-><10><-DRW52->DRW53
9035	DRW11	DW5	: 6 8 8 88860 8 88888888 88
9035	DRW11	DW5	: 2 2 0 02 66084 06 0 6 886688000 260
9036	DRW11	DB2	:4 4 88 46 6 6 888884 8 6 6 8 8 64 88488888 6
9036	DRW11	DB2	: 8 4 6 88888 2 82 8 6 6 88888888 4 6
9036	DRW11	DB2	: 8 8 8 88880 8 8 8 8 88888888 8
9036	DRW11	DB2	: 02 2 880844 082 068 6 6 886888000 0
9037	DRW11	DW5	: 8 6 88886 864 8 4 88888888 484
9037	DRW11	DW5	: 8 8 8 88888 8 6 88888888 88
9037	DRW11	DW5	:6 66 24 8 6 2 4648 8 88886 468868 8 8888 64 6 688888888 88
9037	DRW11	DW5	:2 2 2 8 0 2 826 880888208 026 8 2 8888888000 280
9039	DRW11		: 2 4 6 4 6 644 66 226 4888 82 288 4 6 2 6 4 222448688866688
9039	DRW11		: 8 68882 86 8 2 2 88888888 88
9039	DRW11		: 8 8 8 0 888884 88 8 8 8468 0 4888888888 888
9039	DRW11		: 6 2 8 28 8 08 4 2 6 0 8 2 8 2 888888888 886
9039	DRW11		: 0 02 002 00 0 4 0 0 0 0 8 6888000 280
9040	DRW11	DW5	: 2 8 2688 4 8 888888884 82
9040	DRW11	DW5	: 4 8 8 688 8 4 4 888888888 88
9040	DRW11	DW5	: 6688668 8 688 6 6 4 8888668686 8 6 6668 8 6 8 6888688888 66
9040	DRW11	DW5	: 8 6 2 0 0 02 8808664000 0 8 0 0 0 0 688888 000 480
9041	DRW11	DW5	: 0 0 8 0 88880 2 0 0 0 88888888 206
9041	DRW11	DW5	: 8 8 06888 8 8 44 88888888 688
9041	DRW11	DW5	: 2 0 2 2 40824 680886 00 0 8 0 2 6 2 8888888000 480
9042	DRW11		: 8 8 8 2 88888 8 8 8 8262 288888888 888
9042	DRW11		: 8 44 2 8 8 46 86424 8 868 2 644 62 2 6888888886 688
9042	DRW11		: 2 42 8 0 4404 2 88086860846028 4 22246 6 0 682888888000 880
9043	DRW11	DW5	: 8 4 68862 68 88488888 86
9043	DRW11	DW5	: 46 66 6 6 4 02 84 828 6 8608828086 068 6 4846 86 46 2 28888888000 880
9043	DRW11	DW5	: 4 2 0 8 2 2 88888 886 8 2 04 2 2288888888 482
9043	DRW11	DW5	: 4 0 2 8 6 880888208 028 6 8 22 888888000 260
9044	DRW11	DW5	: 8 2 88888 8 08 88888888 684
9044	DRW11	DW5	: 6 8 2 6 0 02 880826 06 028 4 2 888888000 880
9089	DRW11		: 4 6 22 8 6 84 6 888 668 46 8 44886888886 688
9089	DRW11		: 44 4 0 82 880886 08 028 8 6 8 4888888000 880
9105	DRW11	FS	: 2 6 8 8888 66 8 6 4888888882 8
9045	DRW11	DRW12	: 8 6888 8888 2 888888888 688
9045	DRW11	DRW12	: 0 0 880868804 0 8 4 2886688000 260
9045	DRW11	DRW12	: 8 4 64646 6 6 88888848
9045	DRW11	DRW12	: 0 8 848 0 4 8 888888888 8 8 88888888
9038	DRW12	DB6	: 8 8 8 8 6 8888 8 88 44 4 4888888 88 888
9038	DRW12	BD6	: 262 2 0 06 2 2 0 8806220220 6 2 020 888888000 680
9077	DRW9	DRW12	:2 2 8 0 44 6466 02 880 06 6 86088 486 626668480006880
9077	DRW9	DRW12	: 2 0 22 02 860 60 2 842884282 44 6680002880
9058	DRW13	DW18	: 6 6 668 4 6 6688 86 6 664 6 2 88686 8884 4
9058	DRW13	DW18	: 8 68268 8 888888886 6 8
9058	DRW13	DW18	: 2 02 2 0 0 22 00 0 62 000 886888000 0
9060	DRW13	DW18	: 4 4 0 86 0 6 06680 8 4 40 688888000 6 0
9060	DRW13	DW18	: 2 2 6 0 0 2 0 2 2006 0 0 888888000 0
9062	DRW13	DW18	: 2 66 28 88 88 8 4 6002666 6 8888 6 4 4 4 0 6 4800848888 6
9062	DRW13	DW18	: 4 8 6 2 66 8 22 886888888
9062	DRW13	DW18	: 2 6 828806 8 0 6 20 444008404 6 84 4 0008 888888000 0
9065	DRW13	DW18	: 8 62 68 48 6 4 24 684 2 2 2482 82 6 4 4 26888886688 468
9065	DRW13	DW18	: 22 2 608406 6 6 60 286002608 0 0 88 88 808 888888000 80
9053	DRW13	DW19	: 4 0 0 4608620 2 886868000 0
9059	DRW13	DW19	: 6 6 8 6 88 8 6 8 8 888888888 68
9059	DRW13	DW19	: 2 2 6 4 0 62 08222086202 2 2 2 2 888888000 0
9063	DRW13	DW19	: 6 0 0 84 086 0 2 888888000 0
9097	DRW13	DW19	: 4 6 8 8 86888 4 2 4 888888888 8
9097	DRW13	DW19	: 8 2 8 8 282 8 66 4 882888888 8
9097	DRW13	DW19	: 2 6 4 4 4 628 4 2 8 4 6466 82 64 06 6 6 2 8 6686868888 8
9097	DRW13	DW19	: 2 4 2 222 8 04 262 20824808640 2 2 42 2 4 22 888888000 0
9056	DRW13,14	DW9,19	: 2 2 02 8 4 2 2 88640 6 8 8 2 4 6 888888886
9056	DRW13,14	DW9,19	:0 0 0 0 0 0 00020 6 20 2020 0 0 8 288 000 20
9056	DRW13,14	DW9,19	: 8 6 888 8 0 88888808
9056	DRW13,14	DW9,19	: 0 88 8 646 88888888
9054	DRW14	DW9	: 0 0 2 0 068 2 8 888888000 0
9057	DRW14	DW9	: 0 0 0 00 0 8 2 0 0 0 888888000 0
9061	DRW14	DW9	: 2 8 02 2 4 422 88688888 2
9061	DRW14	DW9	: 0 0 0 0 2 484866000 0
9064	DRW14	DW16	: 6 8 4 66 6 8 66 8 6 6 6 668688 6 68
9064	DRW14	DW16	:4 6 6 62 8 6 8 6 88 6 66 66 2 8886088888 888
9064	DRW14	DW16	: 2 22 2 4 42460 2 2 4246 0 4202 402 2 2 4 2 2 2 4 886888000 680
9099	DRW14	DW16	: 66 4 66 8 6 8 6 6 68886666 666
9099	DRW14	DW16	: 4 2 8 8 8 888888 684
9099	DRW14	DW16	: 4 4 6 6 8 84 4 64 6 688828888 666
9099	DRW14	DW16	: 2 2 2 420 8 2 20 4 02 02 22 0 882668000 480
9055	DRW6	DWN	:0 4 2 0 8 2 62 8 82 488 668888 8
9055	DRW6	DWN	: 6 0 0 0 0 2 000 0 0 0 8 0 0 8 688 000 20

Table 17D. Lymphoblastoid Cell Line Reactions with Informative Sera

WS #			Tenth Workshop core reagents
			:11111111111111111111113111111111111111111113119931113911111111111111111111111111153331113
			:000000000000000000000000000000000110001011900111001000000000000000000000001122230002220
			:00110002322222334334233334445455499459130195632326167777888896656655611229900086241014
			:63018597056349123790565488560461714936463390836410189075567917672389086018915743582967
			<--DR1-><--DR2--><--DR3--><---DR4---><--DR5--><DRW6><DR7><DR8><-DRW9-><10><-DRW52->DRW53
9046	DR7	DB1	: 4 4 4 4 8 88 88 6 6 66 6 4 48 864
9046	DR7	DB1	: 0 828 0 8 0 8608 86868 4 888 628 6 0008880
9047	DR7	DB1	: 8 0 8 6 8 8 88288 4 888 8 6 48 888888
9047	DR7	DB1	: 8 6 6 88888 88 000 888888
9047	DR7	DB1	: 6 0 6 88 0 4 0 0 8 88 888 6 22 0008880
9048	DR7	DB1	: 6 8 6 06 8 466 4 66 6 8 888 886 6 66 6 888888
9048	DR7	DB1	: 2 2 0 2 6 82 04 2 0 202 88888 2864 26 22 2 440006880
9049	DR7	DW17	: 8 2 2688 88 4 888888
9049	DR7	DW17	: 2 6 0 2222 2 0 4 4 0 002 0 886882 2 2886 0002 22 40008880
9050	DR7	DW17	: 8 6 8 88888 88 4 888888
9050	DR7	DW17	: 2 0 8 8 2 8 0 0 0 88888 888 8 2 0008880
9051	DR7	DW17	: 2 2 8 24 8 2 8 888 2 884 882 26 8888888
9051	DR7	DW17	: 88888 6 88 4 48828
9051	DR7	DW17	: 2 0 8 20 2 0 0 88888 888 682 0008880
9052	DR7	DW11	: 8 8 6 8 08 88068 8 88 68 8 8 888888
9052	DR7	DW11	: 2 2 0 2 8 20 68000 0 86888 2 0 88 0 06 88224 0008680
9093	DR7	DW7	: 4 2 8 622 2 4 8 668 46 88 2 6 2 882888
9093	DR7	DW7	: 4 0 424 0 2 0 220 88688 26 884 6 6 6 0008880
9094	DR7	DW7	: 8 4 8 68 4 88 4 864886
9094	DR7	DW7	: 0 2 2 8 0 0 0 88886 882 8 0008480
9095	DR7	DW7	: 4 4 8 6 4 6 44 2 6 8 688 86 88446 4 2 8 68 888888
9095	DR7	DW7	: 2 22 0 2 0 000 0 82688 88 000 282 4 0008880
9096	DR7	DB1	: 2 8 6 8 8 4 8 886882 886 4 8 2888888
9096	DR7	DB1	:2 0 4 42 06 0 8 0 888882 4 846 2 2 6 220006880
9106	DR7		: 8 4 88888 88 48888
9106	DR7		: 2 0 4 8 0 20 02 88488 884 2 2 2 40008880
9066	DRW8	8.3	: 8 6 6 648 8 88888 68 886
9066	DRW8	8.3	: 8 6 0 0 4 0 88002 0 4 88488 0 08 882888000 0
9067	DRW8	8.1	: 6 6 4 8 4 4 42 4 4 6 888 8 4 4 8 488 8 686666 4 88
9067	DRW8	8.1	: 0 0 0 88000 0 82080 0 0 88 226000 80
9068	DRW8	8.1	: 4 6 2 22 0 888 2 2 82486 44 6 884464 2
9068	DRW8	8.1	: 2 0 0 880 6 022 84888 2 6 88 268000 0
9069	DRW8	8.3	: 6 468 4 8 486 882446
9069	DRW8	8.3	: 4 8 886 64 86688 886
9069	DRW8	8.3	: 8 6 4 4 84 8 46 04 2 2 2 4 846 04688048 062 8 886 6 8 8886868000 260
9069	DRW8	8.3	: 6 2 0 8 08 880688088 2 88888 26 2 8 882888000 60
9070	DRW8	8.3	: 42 228 4 24 628 4 8 2 4 86 4 6248 6 4 662 4 26666 624 2 64 88668 86
9070	DRW8	8.3	: 2 2 2 2 2 02 2 2 6 48860866 8 2 88888 0 06 888888 886
9071	DRW8	8.2	: 6 66686 866644 6
9071	DRW8	8.2	: 2 6 2 6 2 6 688 6 8 888 2 4 68 666 26
9071	DRW8	8.2	: 6 6 2 0 0 4 08 8800820 86888 00604 8026880002 0
9072	DRW8	8.2	: 6 8 88688 8688
9072	DRW8	8.2	: 6 8 2 446 6 264 2 684666
9072	DRW8	8.2	: 2 8 2 0 22 2 02 8802880 82888 8 88 888000 0
9100	DRW8	8.2	: 4 2 6 6 8 86 86 884
9100	DRW8	8.2	: 4 6 0 4 8 2 6 8 2 66 446 6
9100	DRW8	8.2	: 2 0 08 88006 0 84488 0 02 88 688000 0
9101	DRW8	8.2	: 2 44 6 88464 88 488 224
9101	DRW8	8.2	: 4 4 8 442 2 42 442
9101	DRW8	8.2	: 0 6 0 240 0 0 2 8 080 000 86 664000 0
9073	DRW9	DW23	: 8 48 6 88 86 88 8 888
9073	DRW9	DW23	: 2226222222 22 2 2222 224 2 2648 22 2 2 2262 2 2 884842488 42 2428 84888
9074	DRW9	DW23	: 0 8 68 8 8 88 8 688 8 888
9074	DRW9	DW23	: 6 66 8 8 866888486 8 888
9074	DRW9	DW23	: 8 4 0 4 468 0 0 02 2 8848886882 8 2 0008880
9075	DRW9	DW23	: 2 2 8 42 4 426 686 6 46 4 68 8 888 42 26 2 8866
9075	DRW9	DW23	: 8 0 88 8 888888888 88888
9075	DRW9	DW23	: 28 0 8 686 0 0 0 2 888888688 2 0008880
9076	DRW9	DW23	: 2 6 4 44 8 2 88 866 68 48 8 888
9076	DRW9	DW23	:2 6 6 4 0 6 868 0 4 0 02 6 2 8888888884228 4 0008880
9077	DRW9	DRW12	:2 2 8 0 44 6466 02 880 06 6 86088 486 626668480006880
9077	DRW9	DRW12	: 2 0 22 02 860 60 2 842884282 44 6680002880
9102	DRW9		: 2 2 6 466 6 8 886888 882 862 6 284888
9102	DRW9		: 6 2 4 646 444 6 488 866 86 8 4 4 848 8
9102	DRW9		: 8 8 2 8 848 2 4 6 888888888 8 288888
9103	DRW9	DW23	: 8 688 4 4 6 888888888 8 88888
9103	DRW9	DW23	: 6 86 8 8 8 888
9103	DRW9	DW23	: 0 8 88 8 88 8 8 888
9103	DRW9	DW23	: 6 8 0 8 6 8 0 64 0 888888886 8 86888
9103	DRW9	DW23	: 6 4 4 6 4 2 6 86206 6062 4 8 888
9103	DRW9	DW23	: 8 6 0 88 8888 0 4 0 0 8888884882 8 000 880

Table 18. Lymphoblastoid Cell Line Reactions with Informative Sera

WS #			Tenth Workshop core reagents.
			:31111133399333311111111111111333311113111333331 1
			:011112000000111111111111111110111111110111111100
			:956341898559000777766767781896111187915001221089
			:232617806561567042397586946836125877044679018132
			<--DQW1-----><---DQW2--><-------DQW3--------->DQW4
9018	DQW2		: 288888 88846 8 8 8 8
9018	DQW2		: 88888888 84 6 8 0 0
9018	DQW2		: 2 66 88884 02 4 8 6 6
9018	DQW2		:0 0 0000 000088882 260 0000 0 00000
9019	DQW2		: 888888808 4 8 8
9019	DQW2		:0 0000 0000 682 60 2 0000 80 00000
9020	DQW2		:4 24 46 4468866448 6 662 4 264 2 2 6
9020	DQW2		: 88888888 8 8 8 0 4 8
9020	DQW2		: 88884884 6 8 8
9020	DQW2		:0 0000 0000846 2 8 20000 2 0 00000
9022	DQW2		: 8888 88 8 6 8 8
9022	DQW2		: 2 00 88886 88 24 0
9022	DQW2		:0 0000 000084 82 2 2 0000 0 00000
9023	DQW2		: 8884 8 40 0 0
9023	DQW2		: 86 88886686646 68 8 6 26 8 6
9023	DQW2		:0 0000880000888882 8 8 2 00004 4 0 26000004
9046	DQW2		: 0 4 8668 686 66 4 0 0
9046	DQW2		:0 82 00008800008888888 8 60000 2 08 00000
9047	DQW2		: 8 8888 888888 8 0 4 8
9047	DQW2		: 88 88886888868 8 8 0 0 6
9047	DQW2		:0 2 00006800008688888 8 60000 0 00000 2
9048	DQW2		: 6 6 66648886680888 8 066 8 0 68
9048	DQW2		:02 00002200006688886 6 0000 04 00000 2
9049	DQW2		: 8888 888888 8 08
9049	DQW2		:0 0 0000 000082888888 8 0000224 0 200000
9050	DQW2		: 2 8888 888888 8 08
9050	DQW2		:0 00008600008688888 8 0000 0 00000
9051	DQW2		: 8 48 8888 8888886 8 08 6
9051	DQW2		: 6 88686488888
9051	DQW2		:0 2 0000 000088888886 6 20000 0 00000
9085	DQW2		:0 0 688888486644 662 4 0248 6
9085	DQW2		: 88880880 4 8 8
9085	DQW2		: 8888 68 8 8 8
9085	DQW2		:0 0000 000088888 8 6 0000 0 00000
9086	DQW2		: 88888 8462 6 8 8
9086	DQW2		: 8088888888 2 6 8 8 8
9086	DQW2		: 66 888868884 6628 8 6 6 0
9086	DQW2		:0 0000 00008868 4 8 4 0000 0 00000
9087	DQW2		: 22 8888828 6 8 6 8 4 4
9087	DQW2		:0 0000 000088888888 286 0000 0888000008
9088	DQW2		:06 000000 000000086006 000000 000000000004
9088	DQW2		: 8888 88 8 4 408 6 0 2
9088	DQW2		: 44 88886884 4 26 6 4 2 2
9088	DQW2		:0 0000 000088884 8 6 2 0000 0 000002
9093	DQW2		: 4 8888 2888484 8 2 8
9093	DQW2		:0 00002 000084888888 8 0000 02 00000 2
9094	DQW2		: 4 8888 666848 8 8
9094	DQW2		:0 0 00002200008488888 8 0000 0 00000
9095	DQW2	DQW3	: 084888888 8888 88886888888868888 48 88 8 8
9095	DQW2	DQW3	:0 0 00008800008 60888888886860000 26 04 600000
9096	DQW2		: 2 2 88884888888 8 08 2 8 8
9096	DQW2		:0 00004200008688868 28 0000 2 06 00000 2
9102	DQW2		: 0 8 82 088888886 6 46 0 00 2
9102	DQW2		: 6 44 6888 664 46 8 2 6 6
9102	DQW2		: 8 0 82 08888882880 82 2 0 8 00
9106	DQW2		: 66 8 888 8888888 886866 8 8
9106	DQW2		:0 00008800008 8888888888 0000 6 04 600000

Table 19. Lymphoblastoid Cell Line Reactions with Informative Sera

WS	#	Tenth Workshop core reagents.
		:311111333993333111111111111111333311113111333311
		:011112000000111111111111111110111111101111111100
		:956341898559000777766767781896111187915001221089
		:232617806561567042397586946836125877044679018132
		<--DQW1-----><---DQW2--><-------DQW3--------->DQW4
9021	DQW4	:044 0000880000484 866 2 44 8000066 20 62000008
9021	DQW4	: 8888 8 86 888
9021	DQW4	:8 6888888 66 6 8 4 06 0 68
9021	DQW4	:6 8888888 8 6 4 4 8 4 824
9021	DQW4	: 48 448 88 8 686
9021	DQW4	:0 2 000088000086 2 2 0000 0 0000082
9067	DQW4	:84 8688888 2 6 6 62 8 2 6 2 62 888
9067	DQW4	:0 2800000880000 0 2 0000 0 020000088
9068	DQW4	: 4 6886888 2 6 84 6 888
9068	DQW4	:022 0000880000 0 4 00002 0 0000088
9069	DQW4	: 6888888 2 2 60 888
9069	DQW4	:8 48686668 6 0 0 0 888
9069	DQW4	:06 40000880000 8 648 8244000064680 660000088
9069	DQW4	:088 80000880000242 4 8 200008 0 820000088
9071	DQW4	: 66866 6 4 866
9071	DQW4	:62 8886888 08 2 8 2 6 888
9071	DQW4	:0244 0000880000 2 0 2 000022 06 0000088
9072	DQW4	: 888868 888
9072	DQW4	:4 2 8888868 6 6 868
9072	DQW4	:0 6 80000880000 0 8 008 2 0000 0 0000088
9100	DQW4	: 2888888 8 2 8 888
9100	DQW4	: 2 4888688 6 2 6 888
9100	DQW4	:0 20 0000880000 0 4 0000 02 0000088
9101	DQW4	:2 2 682 48 22 4 448
9101	DQW4	: 46446 8 2 2 2 648
9101	DQW4	:0 0 0000640000 8 0000 0 0000088
9107	DQW4	:0 0000660000 4866848 46 6000068 60 0000086
9107	DQW4	:0 0000860000 000088 0 000008
9107	DQW4	:0 0000860000 000088 0 000008

Table 20. Lymphoblastoid Cell Line Reactions with Informative Sera

WS #	Tenth Workshop core reagents.
	:311111333993333111111111111111333311113111333311
	:011112000000111111111111111110111111110111111100
	:956341898559000777766767781896111187915001221089
	:232617806561567042397586946836125877044679018132
	<--DQW1-----><---DQW2--><-------DQW3--------->DQW4
9008 DQW6	:888 88888882 4 0 4 0
9008 DQW6	:888 888888880 6 0 4 004
9008 DQW6	:888 28888888 4
9008 DQW6	:088060000880000 0 8 20000 6 0 00000
9010 DQW6	:886 88888688 6 00 0
9010 DQW6	:088 60000880000 6 868 0000 6 0 00000
9010 DQW6	: 88 48888 4 8 4
9010 DQW6	:888 8888686 0 6
9011 DQW6	:888 8888888 0 4 4 2 02
9011 DQW6	:888 88888880 8 0 4 008
9011 DQW6	:088 0000880000 0 0000 8 0 00000
9012 DQW6	:664 44846 8 4
9012 DQW6	:088020000880000 0 2 6 0000 2 0 00000
9013 DQW6	:088060000880000 2 2 4 6 660000 8 0 00000
9013 DQW6	:488 8888888 0 4
9013 DQW6	:888 48888888 64 0 8
9014 DQW6	:888 68888888 4 6 2 4 86 846 2 6
9014 DQW6	:088 80000880000 4 8800002 0 00000
9014 DQW6	:888 68888888 4 6 2 4 86 846 2 6
9014 DQW6	:088 80000880000 4 8800002 0 00000
9053 DQW6	:088 600008800064 0000 0 00000
9055 DQW6	:088000000680000000004200 2000000 00000000000044
9055 DQW6	:888 8888888 4 2 2 4 8
9055 DQW6	:088 600008800008 2 0000 0 00000 2
9058 DQW6	:886 68888888 866664466 2446 604 66 4 06666 6
9058 DQW6	:888 68888888 62
9058 DQW6	:0880600008800008 2 0000 0 00000
9059 DQW6	:888 8888 888 8 88
9059 DQW6	:088 400008800008 2 2 0000 02 200000
9062 DQW6	:860488888848 88 64 000 2 2 2 282
9062 DQW6	:888 8888888 82 2
9062 DQW6	:08808000088000086 24 0000 6 0 00000
9063 DQW6	:088 400008800008 0000 0 00000
9065 DQW6	:8862888682 8 26 2 2 2 8 4
9065 DQW6	:08826000088000086 0 2 6 0000 6 0 0 00000
9066 DQW6	:888 8868888 6 8
9066 DQW6	:088060000880000 4 0000 0 0000028
9081 DQW6	:888 28888888 6 2 6244 8 4 86
9081 DQW6	:088 40000880000 2 0000 8 0 00000
9082 DQW6	:888 48888888 6 8 24 64
9082 DQW6	:088040000880000 2 0000 6 0 00000
9083 DQW6	:068 60000460000 0 000 0 000000
9083 DQW6	:888 48888888 4 2 8 8
9083 DQW6	:088080000880000 0 2 0000 8 0 00000
9097 DQW6	:888 68888888 2 8
9097 DQW6	:888 88888888 8 4 0 8 8
9097 DQW6	:888 8888888 646 22 2 80 2
9097 DQW6	:088 6000088000082 2 000024 06 00000 4
9105 DQW6	:888 88888888 0 4 88 4 28
9001 DQW1	:088040000880000 2 0000 0 0 00000
9003 DQW1	:0888800008400002 2 8 6 0000 0 00000
9009 DQW1	:0880400008800002 4 0000 02 00000
9057 DQW1	:0880 0000880000 0000 0 00000

Table 21. Lymphoblastoid Cell Line Reactions with Informative Sera

			MDTTMMGAJHTMMMGDFCJJAFHF RAEEBBAIUAEBRBSAAXUUIRJE VWRRCCNZJNRCVCCWUHJJZNHR 1111113333111131113333133 1111110111111101111111100 7781896111187915001221038 6946836125877044679018858
9016	DQW3	DQW7	6666868888666686666688688
9016	DQW3	DQW7	8886888888 86 8868888888
9016	DQW3	DQW7	88868 88886 8 88 8888888
9016	DQW3	DQW7	888 8800008 8608 80000600
9025	DQW3	DQW7	88888 8888888 8860888888
9025	DQW3	DQW7	8886868888868688688888888
9025	DQW3	DQW7	8886860000668408 80000 00
9027	DQW3	DQW7	0888888888888888888808888
9027	DQW3	DQW7	0888880000888808880000800
9030	DQW3	DQW7	88848 8808686888 8888888
9030	DQW3	DQW7	8888888888888848888808888
9030	DQW3	DQW7	8888880000888808 80000 00
9035	DQW3	DQW7	8888848888888 888888 888
9035	DQW3	DQW7	68866600006 6 06680000 00
9037	DQW3	DQW7	86888688888 868886888888
9037	DQW3	DQW7	88888888 888 888 8 88
9037	DQW3	DQW7	88888880 4868848 6068 86
9037	DQW3	DQW7	8888880000868808880000800
9038	DQW3	DQW7	8888868888888 8888888888
9038	DQW3	DQW7	888888000086 8 0868000000 0
9039	DQW3	DQW7	464888 8486448886866 88
9039	DQW3	DQW7	88888888888 8 8888888888
9039	DQW3	DQW7	8888888888888 8888808888
9039	DQW3	DQW7	88888688888 8 8888888888
9039	DQW3	DQW7	88 8600004 6 08 0000600
9040	DQW3	DQW7	8884848888 4 8888888888
9040	DQW3	DQW7	888 8 88888 8 8888888888
9040	DQW3	DQW7	8888866088868468860888886
9040	DQW3	DQW7	8888800008 8408 0000000
9041	DQW3	DQW7	888 868888 08088888888 8
9041	DQW3	DQW7	88868 8888868 888880 888
9041	DQW3	DQW7	888886000088 8 0884000080 0
9042	DQW3	DQW7	6840000000000000000000000
9042	DQW3	DQW7	8888888888888 8888888888
9042	DQW3	DQW7	8886888088868688688888888
9042	DQW3	DQW7	8888880000848 08880000 00
9043	DQW3	DQW7	8886848888 4 8848888888
9043	DQW3	DQW7	8888880000888608880000800
9043	DQW3	DQW7	8886888888888888888888488
9043	DQW3	DQW7	8888880000888608880000800
9044	DQW3	DQW7	88888 88888 8 8888888888
9043	DQW3	DQW7	8888880000888608880000800
9043	DQW3	DQW7	8886888888888888888888488
9043	DQW3	DQW7	8888880000888608880000800
9044	DQW3	DQW7	88888 88888 8 8888888888
9044	DQW3	DQW7	8888800008 8 08880000 00
9045	DQW3	DQW7	6886888888 6 88 8888888
9045	DQW3	DQW7	8884600006 4 08880000( )0
9045	DQW3	DQW7	664 4 446444684 48
9045	DQW3	DQW7	8808888888 8 8888888888
9064	DQW3	DQW7	668866 6668866686868 8686
9064	DQW3	DQW7	8886868888860488 8888888
9064	DQW3	DQW7	6888860000848408 80000600
9070	DQW3	DQW7	846 848888 88 8888886
9070	DQW3	DQW7	8888888488868888888888866
9089	DQW3	DQW7	8888848888668688888888888
9089	DQW3	DQW7	888888000088860888000 0 00
9090	DQW3	DQW7	8888888088888888848800888
9090	DQW3	DQW7	8888880000868808880000 00
9091	DQW3	DQW7	8886880000888808 80000000
9099	DQW3	DQW7	688666686666668 68666 86
9099	DQW3	DQW7	888 8 8808 44 88 8888888
9099	DQW3	DQW7	88888 888866848848888888
9099	DQW3	DQW7	888484000064 8 08 80000 00

			MDTTMMGAJHTMMMGDFCJJAFHF RAEEBBAIUAEBRBSAAXUUIRJE VWRRCCNZJNRCVCCWUHJJZNHR 1111113333111131113333133 1111110111111101111111100 7781896111187915001221038 6946836125877044679018858
9103	DQW3	DQW3	8888888808 848 6868
9007	DQW5	DQW8	6888888888846486 6 8868
9007	DQW5	DQW8	8888800008 6406 0000 00
9024	DQW3	DQW8	8888868888888 88 8 0 8888
9024	DQW3	DQW8	8886868888 68 88 8 8888
9024	DQW3	DQW8	888880000888408 0000800
9026	DQW3	DQW8	8888888888 88888 6 88888
9026	DQW3	DQW8	8888880000688808 0000800
9028	DQW3	DQW8	8888688886 8 8 88 8
9028	DQW3	DQW8	8888888888888 8 8868
9028	DQW3	DQW8	8886860000888606 80000800
9029	DQW3	DQW8	888 888888688688 8868
9029	DQW3	DQW8	8888868888688 88 0 8888
9029	DQW3	DQW8	68888600008 460 40000000
9031	DQW3	DQW8	8888888888888 88 8 0 8888
9031	DQW3	DQW8	8888880000888808 80000 00
9032	DQW3	DQW8	88 66 888 8866888 8488 8
9032	DQW3	DQW8	888 888888888 88 000 8848
9032	DQW3	DQW8	8888868088888688080 88 8
9032	DQW3	DQW8	8888880000888808 80000 00
9033	DQW3	DQW8	8888888888888 88 8 008888
9033	DQW3	DQW8	088880000888808 80000800
9034	DQW3	DQW8	8888888888888 88 0 8888
9034	DQW3	DQW8	888 880000668 06 0000 00
9092	DQW3	DQW8	8888888888888 88 8 8888
9092	DQW3	DQW8	8886848888 48 88 8 48888
9092	DQW3	DQW8	8888860000888406 80000000
9098	DQW3	DQW8	88888888888888886 8 8
9098	DQW3	DQW8	888 68800006860 800000600
9098	DQW3	DQW8	88888 8888 6 88 6 8868
9098	DQW3	DQW8	88868600004 6 04 40000800
9052	DQW3	DQW9	8888888888 8 88 0 0 8888
9052	DQW3	DQW9	8888880000 8 08080000000
9073	DQW3	DQW9	8888 8888 8 8 0 88 8
9073	DQW3	DQW9	8886888888 46 86 6 6848
9074	DQW3	DQW9	8888888088688 8 8 008868
9074	DQW3	DQW9	8888688088 86 84 60 88 8
9074	DQW3	DQW9	8888880000 48 08 80000800
9075	DQW3	DQW9	88686 4888 6 6 4 8 66 8
9075	DQW3	DQW9	8888888888 8 80 8 8808
9075	DQW3	DQW9	8888880000 8 0 80000800
9095	DQW2	DQW9	8888868888 48 88 8 8888
9095	DQW2	DQW9	8886860000 6 04 60000 00
9103	DQW3	DQW9	888 8 8888 86 8 8888
9103	DQW3	DQW9	888 888888 88 88 8 8888
9103	DQW3	DQW9	8888888088448 0 0 88 0
9103	DQW3	DQW9	8886868888 86 8 68888
9103	DQW3	DQW9	88888800008 6608 80000800
9076	DQW3		8886868888 68488 8 68888
9076	DQW3		8888880000 8 06 80000800
9077	DQW3		66888600006 6408 60000800
9077	DQW3		6886640000 6 06 80000800

(Table 1). Three of the A1 cells (9011, 9040, 9095) and four of the A3 cells (9005, 9108, 9047, 9102) reacted strongly with the five monoclonals 2057, 2058, 2062, 2064, and 2065. This could be accounted for on the basis of cross-reactivity of these monoclonals with BW4. The A24 cell lines could be nicely distinguished from the one A23 line (9029) on the basis of negative reactions with sera 67 and 70 and positive reactions with sera 65. The two A25- and five A26-positive lines were well defined by available sera and monoclonal antibodies 2056, 2058, 2052, 2062, and 2116, which reacted with A25 but not 26, and the A25-specific sera 119 through 124 (Table 2). The Workshop core sera gave good definition of the A19 cross-reactive group (Table 3), as well as the A2 and A28 group (Table 4).

B Locus. Splits of B5 were clearly defined on the LCL using the core set. These included nine B51-positive cells, one BW52 cell, and a BW53 cell line (Table 5). There were no LCLs submitted that bore the BW70, W71, W72 antigens. Table 8 again illustrates the cross-reactivity of some BW4 sera with a consanguineous homozygous BW6-positive cell (LCL9042) that has the A24 antigen. This confirms the shared epitope found on these molecules. There were no BW63, BW77, BW76, BW15.3 cell lines or lines representing the BW4-associated B15 antigens. There was a suggestion that LCLs 9031, 9032, 9064, and 9099 could represent the BW75 split. There were three LCLs that were positive for BW46 (9066, 9076, and 9077). While 9066 had a reaction pattern characteristic of CW11, 9076 and 9077 had shorter patterns, which have been described as CW11 variants. Table 9 illustrates this for the cells 9076 and 9077 in the context of BW46, while reactions with the same CW1 and CW3 sera are shown in Table 16 for the BW62-positive cells 9072 and 9101.

There were five cell lines positive for BW57 (Table 10), and none expressing the BW58 split of B17. Table 11 shows the reaction patterns of the one B37- and four B38- positive cells, including reactions with the BW4 and B7 sera. There were no B39 or BW67 cells. There was only one BW54 cell, which reacted well with the cluster of sera that define this specificity. BW55 and W56 cells were absent from the panel. Table 12 shows the relevant reaction patterns with the BW54 cell 9107, with the BW42 cell 9021, and a set of B7 controls. A number of B7 and B27 LCLs were tested on the core set and had well defined reaction patterns. Neither the variant of B7 nor of B27 described by the Antigen Society could be distinguished (Table 13). Sera 356 and 358 were the only ones useful in splitting B40 into BW60 and BW61 on the core cell lines (Table 14). Six monoclonal antibodies with very good *r* values reacted well with the B13-positive cells. There were no BW48- and only one BW41-positive LCL. Of the six cells which type as BW60, all were also CW10. Table 15 again illustrates that there is a shared epitope between BW4 and A9. The BW4 monoclonal antibodies 2057, 2064, 2065, and 2102 and alloantisera 493, 495, 498 react with A9-positive, BW6-positive LCLs but not the A11, BW6 controls. These antibodies also react with A32-positive cells (data not shown).

C Locus. CW1 could clearly be distinguished from CW3 with a number of sera (Table 16). The short CW3 (now called CW10) was identified by negative reactions with only one serum (532). There were seven LCLs that clearly bear the CW1/3 hybrid antigen. Of interest, two of these CW11 LCLs were associated with BW62 rather than BW46. One was WSH9071 (duplicated as 9100). The other was WSH9072 (duplicated as 9101), both from native South Americans. WSH9071 and 9101 had shorter reactivity patterns with the CW3 sera, suggestive of a CW11 variant on this cell. CW5 LCLs were well defined by at least three core sera (Table 16). The definition of CW8, in contrast, rested on only two core sera (553 and 554). One of the LCLs (CWH9074) clearly is CW10, but its additional CW8 assignment could not be confirmed with certainty by serology. CW2 and CW7 serologic definitions were excellent. Four of the cell lines that were previously not phenotypes for the C locus had good patterns for CW8 (9030), CW7 (9083, 9088), and CW6 (9106).

Class II Serology

DR Serology. Tables 17A through 20 display the serologic reactions of the LCLs with informative class II core sera. Reactions of anti-DR1 sera with DR1-positive LCL were mediocre, but were adequate to define this specificity (Table 17A). DR2 sera reacted well with the LCLs, and the assignment of DRW15 and 16 rested solely on the basis of weak reactions with two core sera (1031 and 1032). No cells bearing the DRW10 antigen were found. Table 17B shows the well defined clusters for DR3 and DR4, associated respectively with their positive reactions with anti-DRW52 and DRW53 sera. The other associations with DRW52 and 53 are displayed in Tables 17A through 17D and were as expected (DR3, DR5, DRW6, DRW8, DR5, DRW6, and DR3 with DRW52, and DR4, 7 and 9 with DRW53). DRW13 was defined by a cluster of DRW6 sera in the presence of DRW52, while DRW14 failed to react with these sera (Table 17C). Although the new specificity DRW18 could not be assigned with certainty (Table 17B), there was a suggestion of a short DR3 pattern in WSH9021 and 9085. DR7, DRW8, and DRW9 were quite well defined by cytotoxicity on the core cell lines, and the "8.1" and "8.3" variants of DRW8 could be distinguished from the "8.2" on the basis of reactions with some of the DR5 sera (Table 17D).

DQ Serology. The DQW2 antigen was well defined. Monoclonal antibodies did not reveal heterogeneity in the DQW2-positive LCLs by cytotoxicity (Table 18). The majority of the DQw2 cells were either DR3 or DR7. Some short patterns of DQW2 allosera can be appreciated, but they are more related to the DR3 or DR7 profile of the cells and may be related to differential anti-DR activity in these sera. The new specificity DQW4, previously DQ"WA," is well represented on the LCL panel as shown in Table 19. Particularly informative were the reagents

3101, 1083, and 1092. The distinction of the new split of DQW1 called DQW6 from DQW5 rested solely on the absence of reactivity with serum 1136 and positive reactions with the other DQW1 sera (Table 20). The newly designated splits of DQW3, which have been named DQW7, DQW8, and DQW9, were assigned to the LCL on the basis of reaction patterns with three clusters of sera as detailed in the Antigen Society #28 report. The first cluster including six alloantibodies and four monoclonals react broadly with DQW1 (Table 21, first 10 columns). The second cluster of five reagents ending with monoclonal antibody 3114 are negative only on DQW9 cells, whereas the third group of antibodies, ending with monoclonal 3118 are positive only on DQW7 cells. The last three antibodies displayed are broad DQW3 plus DQW1 reagents.

Summary

This paper describes the primary serologic reaction data on the Reference Panel of B-lymphoblastoid Cell Lines for Factors of the HLA System. An average of three laboratories tested each cell line with the core serology set by cytotoxicity. The tables provide a detailed listing of serologic scores. Data interpretation and representation of a "consensus" typing for each cell line is presented in Table 4B in the preceding article.

HLA-Dw and HLA-DP Typing of the Reference Panel of B-Lymphoblastoid Cell Lines

E. Mickelson,[1] N. Reinsmoen,[2] F.M. Robbins,[3] R. Hartzman,[3] N. Ødum,[4] A. Svejgaard,[4] C. Farrell,[2] M. Honeyman,[5] N. Flomenberg,[6] T. Sell,[7] and D. Eckels[7]

Cellular typing for HLA-Dw and HLA-DPw within the Tenth International Histocompatibility Workshop was incorporated into the Workshop component on T-Cell Recognition of HLA Class II Molecules. Separate Workshop components on Dw and DPw typing for identification of new specificities using homozygous typing cells (HTC) or primed lymphocyte typing (PLT) were not designed as part of the Tenth Workshop. It was, however, essential that accurate information was obtained for HLA-Dw and HLA-DPw specificities for the Reference Panel of B-Lymphoblastoid Cell Lines. A working group consisting of the authors of this paper was established for this purpose. Whenever possible, any residual PBL from the cell line donors was collected and distributed to several interested laboratories for cellular typing studies. In the case of DP, typing was also performed using the B-cell lines themselves. DP typing was performed by four groups: Sell and Eckels, Hartzman and Robbins, Ødum and Svejgaard, and Farrell and Honeyman. The results of the DP typing studies are shown in Table 1. Consensus assignments of HLA-Dw and HLA-DPw specificities are incorporated into Table 4B in the paper entitled "Description of the Reference Panel of B-Lymphoblastoid Cell Lines for Factors of the HLA System" by Yang et al. (this volume).

Dw assignments were made jointly by Reinsmoen and Mickelson. Assignments were made based on data from several sources: 1) assignments in the Tenth Workshop reference laboratories in Minneapolis (US2BAC) and Seattle (US7HAN), 2) assignments made in previous International or Asia-Oceania Histocompatibility Workshops, 3) analyses with cytolytic and proliferative T-cell clones, the patterns of reactivity of which were well established versus other well-typed cell lines, 4) assignments made by local laboratories, and 5) assignments made based on the Workshop analyses. The assignments, together with their source are listed in Table 2.

Participating Laboratories: US7HAN,[1] US2BAC,[2] US1HJH,[3] SCASVE,[4] ANZBSH,[5] US7NFL,[6] US2MBC[7]

Table 1. Composite Summary of Workshop DP Typing

WS ID#	Cell Name	DP[a]	DP[b]	DP[c]	DP[d]	Consensus
9001	SA	4B?	?	4^s		4
9002	MZ070782	4	?	4^s		4
9003	KAS116	?	?	CDP-HEI		-
9004	JESTHOM	4	4#			4
9005	HOM2	4	?			-
9006	WT7100BIS	1	4			1
9007	DEM	3,4	3,4	4		3,4
9008	D02089125	2,4	?			2,4
9009	KAS011	3,4	?			3,4
9010	AMAI	?	?	4^s		-
9011	E4181324	2,4	4,2			2,4
9012	WJR076	2	2?	2,JET		2
9013	SCHU	4B?	?	4^s		-
9014	MGAR	4	4	4		4
9015	WT24	1,3,2B?	3?			3
9016	RML	4B?	?			-
9017	WT8	2,4	4?			2,4
9018	L0081785	2	3	3		3
9019	DUCAF	2B?	?	2,JET	1,4	-
9020	QBL	2wk	2#	2		2
9021	RSH	1	1#			1
9022	COX	3	3#			3
9023	VAVY	1	1			1
9024	KT17	5	?	5		5
9025	DEU	4	4			4
9026	YAR	4	4	4		4
9027	PF97387	4	4			4
9028	PE117	4	4	4^s,PET	6	4
9029	WT51	2	2?			2
9030	JHAF	3	?	1?,3	1	3
9031	BOLETH	4	4?			4
9032	BSM	2	NT	2		2
9033	BM14	4	4			4
9034	SAVC	7	?	JET,CDP-HEI		-
9035	JBUSH	4	4		2,4	4
9036	SP0010	2	?		1,2,?4	2
9037	SWEIG007	4B?	4?#		6	-
9038	BM16	2	2#			2
9039	JVM	2	?		2 weak	2
9040	BM15	3	3?#			3
9041	J0528239	4B?	7#			-
9042	TISI	4B?	?		4	-
9043	BM21	2B?	?#			-
9044	BRIP	2,4B?	2#			2
9045	TUBO	2,4	?	$2,4^s$	?	2,4
9046	BH	2wk,4	4		3	4
9047	PLH	4B?	?		1	-
9048	LBUF	?	6	6?,JET,PET	?3,?cp63	-
9049	IBW9	1	1		1	1
9050	MOU	2	2,6?	2,JET	2	2
9051	PITOUT	4	4#			4
9052	DBB	4	4		4	4
9053	HOR	4	4		4	4
9054	EK	4B?,+?	?		1	-
9055	H0301	5	5#			5
9056	KOSE	2	2		1,2	2
9057	TEM	4,+?	4		1,4,?2	4
9058	OMW	1	1#			1
9059	SLE005	3	3,1?	1?,3	1,2	3
9060	CB6B	?	?#		?2,?cp63	-

Table 1. *Continued*

WS ID#	Cell Name	DP[a]	DP[b]	DP[c]	DP[d]	Consensus
9061	31227ABO	4	4		4	4
9062	WDV	2wk,4	4,2?		1,4	4
9063	WT47	2	2,6?			2
9064	AMALA	?	?#			-
9065	HHKB	4	4		4	4
9066	TAB089	7	?	?	2,?5	-
9067	BTB	4	4			4
9068	BM9	2B?	2#			2
9069	MADURA	4	4#		4	4
9070	LUY	1,4	1,4			1,4
9071	OLGA	3,4?	3,1?4?#			3
9072	SPACH	4B?	4?			4
9073	KT12	5,4B?	5			5
9074	HID	2,5	5,2?			2,5
9075	DKB	4	4	4^s,PET	2,4	4
9076	T7526	?	?			-
9077	T7527	5	?			5
9078	PMG075	3,4	?			3,4
9079	LWAGS	3,4	3,4			3,4
9080	EHM	4	4			4
9081	EA	4	4			4
9082	H0104	4	4			4
9083	LD2B	4	4?		4	4
9084	CALOGERO	4	NT			4
9085	EJ32B	2	2#		2	2
9086	10541265	1,4	1,4#			1,4
9087	STEINLIN	3,4	3,4			3,4
9088	PF04015	1,4	1,4#			1,4
9089	BOB	4B?	NT			4
9090	AWELLS	3,4	4?			3,4
9091	MLF	4B?	?	4^s?		-
9092	BM92	4B?	?			-
9093	BER	2,4	2,4	2		2,4
9094	CF996	4	4			4
9095	WIN	4	4	4		4
9096	LBF	7	?			-
9097	EMJ	3,4	3,4		3	3,4
9098	MT14B	4B?	?		4	-
9099	LZL	4B?	?		4	-
9100	OLL	3,4B?	3,1?			3
9101	SPL	4B?	?			4
9102	ARBO	1	1,4?			1
9103	KT14	2	2		2,cp63	2
9104	DHIF	NT	NT			NT
9105	FPAF	NT	NT			2[e]
9106	MANIKA	NT	NT			NT
9107	LKT3	NT	NT			5[e]

[a]Sell, T.W. and Eckels, D.D. (US2MBC).
[b]Hartzman, R.J. and Robbins, F.M. (USIHJH).
[c]Ødum, N., Morling, N. and Svejgaard, A. (SCASVE).
[d]Farrell, C. and Honeyman, M. (ANZBSH).
[e]Data obtained from laboratory submitting the cell line.
? = Unidentified or uncertain HLA specificity
= PBL and BLCL were typed and results were consistent
– = No consensus
Note: A DPw4 "split" CDP-4^s, a "new" DP specificity, "CDP-HEI" different from DPw1-w6 and SB7 were locally defined. A "new" supertypic specificity "JET" related to DPw2,DPw6 and DP blank and a yet undefined Class II-related determinant "PET" were also defined in laboratory SCASVE.

Table 2. Summary of Workshop Dw Typing

WS ID#	Cell Name	HLA-DR	HLA-Dw	Comments or Prev. WS #
9001	SA	1	Dw1	Oriental variant
9002	MZ070782	1	Dw20	
9003	KAS116	1	Dw1	
9004	JESTHOM	1	Dw1	
9005	HOM2	1	Dw1	6WS
9006	WT7100BIS	1	Dw1	
9007	DEM	2/4	Dw-	Cell is serologically heterozygous
9008	D02089125	2	Dw2	
9009	KAS011	2	Dw21	tb24, FJO, MN2, WJR-like specificity
9010	AMAI	2	Dw2	
9011	E4181324	2	Dw12	
9012	WJR076	3	Dw21	
9013	SCHU	2	Dw2	
9014	MGAR	2	Dw2	
9015	WT24	2	Dw21	
9016	RML	2	Dw22	7,8,9WS
9017	WT8	2	Dw2	
9018	L0081785	3	Dw3	(B-18var)
9019	DUCAF	3	Dw3	(B-18var)
9020	QBL	3	Dw3	(B-18var) 8WS
9021	RSH	DRwl8	DRw18;Dw"new"	African Black
9022	COX	3	Dw3	6,9WS
9023	VAVY	3	Dw3	
9024	KT17	4	DKT2	3AOHWS
9025	DEU	4	Dw4	
9026	YAR	4	Dw10	
9027	PF97387	4	Dw4	
9028	PE117	4	Dw14	
9029	WT51	4	Dw4	
9030	JHAF	4	Dw13	6,8,9WS
9031	BOLETH	4	Dw4	
9032	BSM	4	Dw4	6WS
9033	BM14	4	Dw4	7WS
9034	SAVC	4	Dw4	
9035	JBUSH	5	Dw5	
9036	SP0010	5	DB2	
9037	SWEIG007	5	Dw5	8WS
9038	BM16	5	DB6	6,9WS
9039	JVM	5	Dw"new"	9WS
9040	BM15	5	Dw5	6WS
9041	J0528239	5	Dw5	
9042	TISI	5	Dw"new"	
9043	BM21	5	Dw5	7WS
9044	BRIP	5	Dw5	9WS
9045	TUBO	5	Dw5	
9046	BH	7	DB1	6,8,9WS
9047	PLH	7	DB1	
9048	LBUF	7	DB1	6,7,8,9WS
9049	IBW9	7	Dw17	
9050	MOU	7	Dw17	6,7,8WS
9051	PITOUT	7	Dw17	7,9WS
9052	DBB	7	Dw11	
9053	HOR	w6	Dw19	
9054	EK/OH	w6	Dw9	6,7,8,9WS

Table 2. *Continued*

WS ID#	Cell Name	HLA-DR	HLA-Dw	Comments or Prev. WS #
9055	H0301	w6	Dw19	B14-associated variant Dw19
9056	KOSE	w6	Dw9+Dw19	Cell may be heterozygous
9057	TEM	w6	Dw9	
9058	OMW	w6	Dw18	
9059	SLE005	w6	Dw19	8,9WS
9060	CB6B	w6	Dw18	
9061	31227ABO	w6	Dw9	
9062	WDV	w6	Dw18	9WS
9063	WT47	w6	Dw19	
9064	AMALA	w6	Dw16	9WS
9065	HHKB	w6	Dw18	9WS
9066	TAB089	w8	Dw8 ("8.3")	9WS; Oriental
9067	BTB	w8	Dw8 ("8.1")	7,8,9WS
9068	BM9	w8	Dw8 ("8.1")	8WS
9069	MADURA	w8	Dw8 ("8.1")	7,8,9WS
9070	LUY	w8	Dw8 ("8.3")	9WS
9071	OLGA	w8	Dw8 ("8.2")	9WS; Amer. Indian
9072	SPACH	w8	Dw8 ("8.2")	Amer. Indian
9073	KT12	w9	Dw23	3AOHWS
9074	HID	w9	Dw23	3AOHWS
9075	DKB	w9	Dw23	
9076	T7526	w9	Dw ?	
9077	T7527	w9	Dw ?	
9078	PMG075	1	Dw20	
9079	LWAGS	1	Dw20	
9080	EHM	1	Dw1	
9081	EA	2	Dw2	
9082	H0104	?	Dw ?	
9083	LD2B	2	Dw2	
9084	CALOGERO	2	Dw ?	
9085	EJ32B	3	SYD3	
9086	I0541265	3	Dw ?	
9087	STEINLIN	3	Dw3	
9088	PF04015	3	Dw ?	
9089	BOB	5	Dw ?	
9090	AWELLS	4	Dw4	
9091	MLF	4	Dw4	
9092	BM92	4	Dw14	
9093	BER	7	Dw7	
9094	CF996	7	Dw7	
9095	WIN	7	Dw ?	
9096	LBF	7	DB1	(=9048)
9097	EMJ	w6	Dw19	
9098	MT14B	4	Dw14	
9099	LZL	w14	Dw16	
9100	OLL	w8	Dw8 ("8.2")	(=9071)
9101	SPL	w8	Dw8 ("8.2")	(=9072)
9102	ARBO	w9	Dw ?	
9103	KT14	w9	Dw23	
9104	DHIF	w11	Dw5	
9105	FPAF	w11	FS	
9106	MANIKA	7	Dw ?	
9107	LKT3	4	Dw15	

Assignment of HLA-A and HLA-B Antigens for the Reference Panel of B-Lymphoblastoid Cell Lines Determined by One-Dimensional Isoelectric Focusing (1D-IEF) Gel Electrophoresis

Soo Young Yang[1]

The one-dimensional isoelectric focusing (1D-IEF) patterns for HLA-A and HLA-B antigens for the cells in the Reference Panel of B-lymphoblastoid cell lines provide a permanent record for individual cell lines that can be used as reference cells in future studies. Table 1 shows the phenotyping of the cell line panel assigned by 1D-IEF typing. Designations of the IEF assignments are explained in detail in the Workshop summary report in the HLA Class I Biochemistry component in this volume. Briefly, decimal numbers were used to designate subtypes or variants for each antigen, starting from the top of the gel (cathode) to the bottom (anode) of the gel, keeping the original serologic specificities. For example, B17.2, Bw57.2, and Bw58.2 have the same IEF band and designate the second band from the top of the gel among the bands that belong to the B17 antigen group.

The Reference Cell Line Panel represents, in most instances, the commonly occurring subtypes for serologically defined HLA-A and HLA-B alleles. Many IEF variants detected in the Class I Biochemistry cell panel and described in "Population Analysis of Class I HLA Antigens by 1D-IEF Gel Electrophoresis" (this volume) were underrepresented. Although the IEF band patterns were consistent with serologic typings in the majority of the cell lines, 1D-IEF has detected extra HLA class I bands on three cell lines: WSNO 9028 and WSNO 9074 expressed both the Bw60 band and the Bw61 band, and WSNO 9044 expressed the B51 band plus the B15.1 band (Bw63). It is interesting to note that WSNO 9044 (BRIP) is of consanguineous origin. The 9089 (BOB) was submitted as A2, Bw62, DR4, but the EBV-transformed cells were typed serologically as A24, B51, DR5, and IEF typing was consistent with these serologic data.

Rare variants were carried by five cell lines: WSNO 9103 with A26*, WSNO 9045 with A2.3, WSNO 9016 and WSNO 9064 with A2.4, WSNO 9030 with B51*, and WSNO 9102 with Bw57.3. The cells with A26* (WSNO 9103) showed extra reactions with A25-specific monoclonal antibodies 2056, 2058, 2062 in addition to all the A26-specific sera. The cell with Bw57.3 (WSNO 9102) had the same serologic reaction patterns for Bw57 as with Bw57.2.

An unusual HLA haplotype A30.1, B13.2 was observed in the cell line LBUF (WSNO 9048 = WSNO 9096). The Caucasian IEF subtype B13.2 is primarily associated with A30.3, and not with A30.1.

Participating Laboratory: US7SYY[1]

Table 1. IEF Assignment of HLA-A and HLA-B Antigens on Core Cell Lines*

ID	Cell Name	Ethnic	HLA-A		HLA-B	
9001	SA	O(120)	A24.1		B7.2	
9002	MZO70782	C(565)	A24.1		B14	
9003	KAS116	C(595)	A24.1		B51	
9004	JESTHOM	C(575)	A2.2		B27.5	
9005	HOM2	C(510)	A3.2		B27.5	
9006	WT100BIS	C(560)	A11.2		B35.3	
9007	DEM	C(550)	A2.2		Bw57.2	
9008	DO208915	C(520)	A25		B18.1	
9009	KASO11	C(595)	A1		B37	
9010	AMA1	C(599)	A28.1		Bw53	
9011	E4181324	C(520)	A1		Bw52	
9012	WJRO76	C(515)	A2.2		Bw57.2	
9013	SCHU	C(545)	A3.2		B7.2	
9014	MGAR	C(517)	A26.1		B8	
9015	WT24	C(560)	A2.2		B27.5	
9016	RML	AI(220)	A2.4		B51	
9017	WT8	C(560)	A3.2		B7.2	
9018	LO081785	C(520)	A3.2	A24.1	B18.1	
9019	DUCAF	C(545)	A30.1		B18.1	
9020	QBL	C(535)	A26.1		B18.1	
9021	RSH	N(10)	A68.1	A30.3	Bw42	
9022	COX	C(580)	A1		B8	
9023	VAVY	C(545)	A1		B8	
9024	KT17	O(120)	A2.2	A11.2	B35.3	Bw62.3
9025	DEU	C(535)	A31		B35.3	
9026	YAR	C(565)	A26.1		B38.1	
9027	PF97387	C(545)	A29.2		B44.1	
9028	PE117	AI(210)	A24.1		Bw60	Bw61
9029	WT51	C(560)	A23		Bw65	
9030	JHAF	C(540)	A31		Bw51*	
9031	BOLETH	C(575)	A2.2		Bw62.3	
9032	BSM	C(535)	A2.2		Bw62.3	
9033	BM14	C(560)	A3.2		B7.2	
9034	SAVC	C(545)	A3.2		B7.2	
9035	JBUSH	C(515)	A32		B38.1	
9036	SPOO10	C(560)	A2.2		B44.2	
9037	SWEIGOO7	C(515)	A29.2		Bw61	
9038	BM16	C(560)	A2.2		B18.1	
9039	JVM	C(535)	A2.2		B18.1	
9040	BM15	C(560)	A1		Bw49.2	
9041	JO528239	C(560)	A1		B35.2	
9042	TISI	C(545)	A24.1		B35.2	
9043	BM21	C(560)	A1		B41.1	
9044	BRIP	C(560)	A24.1		B51	B15.1
9045	TUBO	C(545)	A2.3	A3.2	B51	
9046	BH	C(515)	A2.2		B13.2	
9047	PLH	C(575)	A3.2		Bw47	
9048	LBUF	C(540)	A30.3		B13.2	
9049	IBW9	C(560)	Aw33.1		Bw65	
9050	MOU	C(575)	A29.2		B44.1	
9051	PITOUT	C(580)	A29.2		B44.1	
9052	DBB	C(516)	A2.2		Bw57.2	
9053	HOR	O(120)	Aw33.2		B44.1	
9054	EK	C(575)	A2.2		B44.2	
9055	HO301	C(545)	A3.2		B14	
9056	KOSE	C(550)	A2.2		B35.3	
9057	TEM	C(599)	A26.2		B38.1	

Table 1. *Continued*

ID	Cell Name	Ethnic	HLA-A		HLA-B		ID	Cell Name	Ethnic	HLA-A		HLA-B	
9058	OMW	N(10)	A2.2		B45		9083	LD2B	C(520)	A3.2		B7.2	
9059	SLE005	C(515)	A2.2		Bw60		9084	CALOGERO	C(560)	A2.2		Bw61	
9060	CB6B	C(520)	A1		Bw62.3		9085	EJ32B	C(520)	A30.1		B18.1	
9061	31227ABO	C(560)	A2.2		B18.1		9086	LO541265	C(520)	A1		B8	
9062	WDV	C(535)	A2.2		B38.1		9087	STEINLIN	C(545)	A1		B8	
9063	WT47	C(560)	A32		B44.2		9088	PF04015	C(545)	A1		B8	
9064	AMALA	AI(220)	A2.4		Bw62.3		9089	BOB	C(550)	A24.1		B51	
9065	HHKB	C(535)	A3.2		B7.2		9090	AWELLS	C(520)	A2.2		B44.2	
9066	TAB089	O(120)	A2.2		Bw46		9091	MLF	C(540)	A2.2		Bw62.3	
9067	BTB	C(575)	A2.2		B27.5		9092	BM92	C(560)	A25		B51	
9068	BM9	C(560)	A2.2		B35.3		9093	BER	C(550)	A2.2		B13.2	
9069	MADURA	C(575)	A2.2		Bw60		9094	CF996	C(545)	A2.2	A3.2	B14	
9070	LUY	C(535)	A2.2		B51		9095	WIN	C(550)	A1		B57.2	
9071	OLGA	AI(220)	A31		Bw62.3		9096	LBF	C(540)	A30.3		B13.2	
9072	SPACH	AI(220)	A31		Bw62.3		9097	EMJ	C(515)	A2.2	A3.2	Bw60	
9073	KT12	O(120)	A24.1	A31	B51	B35.3	9098	MT14B	C(520)	A31		Bw60	
9074	HID	O(120)	A2.2		Bw60	Bw61	9099	LZL	AI(220)	A2.4		Bw62.3	
9075	DKB	C(535)	A24.1		Bw60		9100	OLL	AI(220)	A31		Bw62.3	
9076	T7526	C(110)	A2.2		Bw46		9101	SPL	AI(220)	A31		Bw62.3	
9077	T7527	C(110)	A2.2		Bw46		9102	ARBO	C(535)	A3.2		Bw57.3	
9078	PMG075	C(515)	A3.2	Aw33.1	Bw65		9103	KT14	O(120)	A24.1	A26*	B51	Bw61
9079	LWAGS	C(565)	Aw33.1		B14		9104	DHIF	C(540)	A31		B38	
9080	EHM	C(550)	A3.2		B35.3		9105	FPAF	C(565)	A1		B35.2	
9081	EA	C(575)	A3.2		B7.2		9106	MANIKA	C(580)	A3.2		Bw50.2	
9082	HO104	C(545)	A3.2		B7.2		9107	LKT3	O(120)	A24.1		Bw54.1	

#Designation of class I antigens with more than one band was made by sequential decimal numbers starting from the top of the gel (cathode), after the serologic assignment.

*Rare variant.

Ethnic Code: N = Negro, AI = American Indian and C = Caucasian. For ethnic subgroups, enclosed in ( )'s, see Table 2 in "Population Analysis of Class I HLA Antigens by 1D-IEF Gel Electrophoresis," this volume.

Assignment of HLA-Class II α and β Chain 2-D Gel Patterns for the Workshop Reference Panel of B-Lymphoblastoid Cell Lines

Robert W. Knowles[1]

A comprehensive analysis of structural polymorphism in the HLA class II antigens has been performed using a standardized two-dimensional (2-D) gel electrophoresis system. Details of the analysis are presented in these proceedings (Volume I). Polymorphic patterns have been defined for the α and β chains of the DR and DQ molecules isolated from 71 of the Workshop core B-cell lines representing 34 functionally distinct HLA class II haplotypes (Table 1). For the DR molecules, a total of 19 DRβ chain patterns were found associated with the DRB1 locus, with five additional patterns associated with a second expressed DRB gene. For the DQ molecules, 10 β chain patterns were found, along with 9 DQα chain patterns. For the DP molecules, two DPα chain patterns were found, but the DPβ chains were not resolved. This corresponds well to the extent of polymorphism found in the first domain sequences of cloned DR and DQ genes from these haplotypes. While some of the known sequence variation does not result in distinct 2-D gel patterns, additional polymorphism has been found in this study, where first domain sequences are identical. Although there is extensive linkage disequilibrium between these genes, it is also clear from this study that recombination events within the class II region have played a significant role in the generation of functionally distinct haplotypes.

Participating Laboratory: US7KNO[1]

Immunobiology of HLA
Volume I

Table 1. Complete Listing of HLA Class II 2-D Gel Patterns for the Tenth Workshop Core Cell Lines

		DR				DQ			DP	
		DRB1		DRB-(B3/B4)			DQB1	DQA1	DPA1	
CELL #	NAME	Serol. DR[a]	2-D Gels DRβ[b]	Serol. w52/53	2-D Gels DRβ[b]	Serol. DQ[a]	2-D Gels DQβ[b]	2-D Gels DQα[b]	T cell DP[c]	2-D Gels DPα[b]
9001	SA	1	I-1.1		-	w5*	- (1.1*)[d]	1.1	4	1.1+* [g]
9002	MZO7O782	1	I-1.1		-	w5	- (1.1*)	-	4	(-)+*
9003	KAS116	1	I-1.1		-	w5	- (1.1*)	1.1	-	(-)+*
9004	JESTHOM	1	I-1.1		-	w5	- (1.1*)	-	4	1.1+*
9005	HOM2	1	I-1.1		-	w5	- (1.1*)	1.1	-	1.1+*
9006	WT1OOBIS	1	I-1.1		-	w5	- (1.1*)	1.1	1	1.1+*
9008	DO2O8915	w15	a-2.2		b-2.1	w6	1.2	1.2	2+4	1.1
9009	KASO11	w16	a-2.16		b-2.1	w5*	- (1.21*)[d]	1.2	3+4	1.1
9010	AMAI	w15	a-2.2		(b-*)[e]	w6	1.2	1.2	-	1.1
9011	E4181324	w15	a-2.12		b-2.1	w6	1.12	1.18	2+4	1.1
9012	WJRO76	w16	a-2.16		b-2.1	w5*	- (1.21*)	1.2	2	1.1
9013	SCHU	w15	a-2.2		b-2.1	w6	1.2	1.2	-	1.1
9014	MGAR	w15	a-2.2		b-2.1	w6	1.2	1.2	4	1.1
9015	WT24	w16	a-2.16		b-2.1	w5	- (1.21*)	-	3	-
9016	RML	w16	a-2.16		b-2.1	w7	3.1	2	-	1.1
9017	WT8	w15	a-2.2		b-2.1	w6*	1.2	1.2	2+4	-
9018	LOO81785	w17*	I-3	w52	III-1	w2	2	-	3	1.1
9019	DUCAF	w17*	I-3	w52	III-1	w2	2	2	-	1.1
9020	QBL	w17*	I-3	w52	III-1	w2	-	-	2	1.1
9021	RSH	w18*	I-3	w52	III-3	w4	Wa	3.8	1	1.1+1.2
9022	COX	w17*	I-3	w52	III-3	w2	-	2	3	1.1
9023	VAVY	w17*	I-3	w52	III-3	w2	2	2	1	1.2
9024	KT17	4	I-4.2	w53	IV-1	w8	3.2	3.2	5	1.2
9025	DEU	4	I-4.1	w53	IV-1	w7	3.1	3.1	4	-
9026	YAR	4	I-4.3	w53	IV-1	w8	3.2	3.2	4	1.1
9027	PF97387	4	I-4.1	w53	IV-1	w7	3.1	3.1	4	1.1
9028	PE117	4	I-4.1	w53	IV-1	w8	3.2	3.2	4	1.1
9029	WT51	4	I-4.1	w53	IV-1	w8	3.2	3.2	2	1.1
9030	JHAF	4	I-4.2	w53	IV-1	w7	3.1	3.1	3	1.1
9031	BOLETH	4	I-4.1	w53	IV-1	w8	3.2	3.2	4	-
9032	BSM	4	I-4.1	w53	IV-1	w8	3.2	3.2	2	1.1
9033	BM14	4	I-4.1	w53	IV-1	w8	3.2	3.2	4	1.1
9034	SAVC	4	I-4.1	w53	IV-1	w8	3.2	3.1	-	1.2
9035	JBUSH	w11	I-11.1	w52	III-1	w7	3.1	2	4	1.1
9036	SPOO1O	w11	I-11.1	w52	III-1	w5	1.12* [f]	1.2	2	1.1
9037	SWEIGOO7	w11	I-11.1	w52	III-1	w7	3.1	2	-	1.1
9038	BM16	w12	I-12	w52	III-1	w7	3.1	2	2	1.1
9039	JVM	w11	I-11.2	w52	III-1	w7	-	-	2	-
9040	BM15	w11	I-11.2	w52	III-1	w7	3.1	-	3	-
9042	TISI	w11	I-11.2	w52	III-1	w7	3.1	2	-	1.1
9043	BM21	w11	I-11.1	w52	III-1	w7	3.1	2	-	1.2
9044	BRIP	w11	I-11.1	w52	III-1	w7	3.1	2	2	1.1
9045	TUBO	w11+12	I-11.1+12	w52	III-1	w7	3.1	-	2+4	1.1
9046	BH	7	I-7	w53	IV-1	w2	2	-	4	-
9047	PLH	7	I-7	w53	IV-1	w2	2	3.7	-	1.1
9048	LBUF	7	I-7	w53	IV-1	w2	2	3.7	-	1.2
9049	IBW9	7	I-7	w53	IV-1	w2	2	-	1	-
9050	MOU	7	I-7	w53	IV-1	w2	2	3.7	2	1.1
9051	PITOUT	7	I-7	w53	IV-1	w2	2	3.7	4	1.1
9052	DBB	7	I-7	w53	-	w9	3.3	3.7	4	1.1
9053	HOR	w13	I-13	w52	-	w6	-	1.2	4	1.1
9054	EK	w14	I-14.1	w52	III-1	w5	- (1.9*)[d]	1.14	-	1.1
9055	HO3O1	w13	I-13	w52	III-2	w6	1.19	1.2	5	-
9056	KOSE	w13+14	I-(13+14.1)	w52	III-1+2	w5+6*	-	1.2+1.14	2+N	1.1
9057	TEM	w14	I-14.1	w52	III-1	w5*	- (1.9*)	1.14	4	-
9058	OMW	w13	I-13	w52	III-3	w6	1.18	1.18	1	1.2
9059	SLEOO5	w13	I-13	w52	III-2	w6	1.19	1.2	3	1.1
9060	CB6B	w13	I-13	w52	III-1	w6*	1.18	1.18	-	1.2
9061	31227ABO	w14	I-14.1	w52	III-1	w5	- (1.9*)	1.14	4	1.1
9062	WDV	w13	I-13	w52	III-3	w6	1.18	1.18	4	-
9063	WT47	w13	I-13	w52	III-2	w6	1.19	1.2	2	-
9064	AMALA	w14	I-14.2	w52	III-3	w7	3.1	2	-	1.1
9065	HHKB	w13	I-13	w52	III-3	w6	1.18	1.18	4	1.1
9066	TABO89	w8	I-8.1	w52	-	w6	1.12	1.18	-	1.2
9068	BM9	w8	I-8.1	w52	-	w4	Wa	3.8	2	1.1
9069	MADURA	w8	I-8.1	w52	-	w4	Wa	3.8	4	-
9070	LUY	w8	I-8.1	w52	-	w7	3.1	3.8	1+4	1.1+1.2
9071	OLGA	w8	I-8.2	w52	-	w4	Wa	3.8	3	1.1
9073	KT12	9	I-9	w53	IV-1	w9	3.3	3.1	5	1.1+1.2
9074	HID	9	I-9	w53	IV-1	w9+w1*	3.3+1.12*	3.1	2+5	1.1+1.2
9075	DKB	9	I-9	w53	IV-1	w9	3.3	3.1	4	1.1

continued on next page

Table 1. *Continued*

[a]DR and DQ specificities (using the new WHO nomenclature) are based on the results of Tenth Workshop serologic typing on these cell lines, except where indicated (*).
[b]The nomenclature for the 2-D gel patterns is based in the results described later in these Proceedings (Volume I). The designations "DRβ-a" and "DRβ-b" are used since the patterns in DR2 cells have not been assigned to DRB1 or a second, DRB- locus. In some cell lines, 2-D patterns for the second DRβ chain, the DQβ chain, or the DQα chain were not determined, indicated as "-".
[c]DP typing (DP1-5) is based on the Workshop analysis of these cell lines, where DP-blank is indicated as: "-", a new DP type is indicated as "N", but DP6 and DP7 were not assigned.
[d]The DQβ-(1.1*), DQβ-(1.9*), and DQβ-(1.21*) patterns were not characterized by Workshop 2-D gel analysis.
[e]DRβ-(b-*) was not determined in this study, but appears to be a distinct pattern from DRβ-b-2.1.
[f]The DQβ-1.12* appears similar to the DQβ-1.12, but may be distinct.
[g]The DR1 cell lines all express DQα-1.1, which comigrates with the DPα-1.2 pattern. These cell lines could only be analyzed for the presence or absence of the DPα-1.1 pattern, as indicated, "1.1+*" or "(−)+*, where * indicates the presence of DQα-1.1, with or without DPα-1.2.

Assignment of T-Cell-Defined (TCD) HLA Class II Specificities for the Reference Panel of B-Lymphoblastoid Cell Lines

Neal Flomenberg[1]

T-cell-defined (TCD) polymorphic determinants on HLA class II molecules have been assigned to the Reference Panel of B-lymphoblastoid cell lines. The HLA-Dw and HLA-DPw specificities recognized by the WHO Nomenclature Committee are listed in Table 4B of the report "Description of the Reference Panel of B-Lymphoblastoid Cell Lines for Factors of the HLA System" (this volume). Additional details are provided in the report "HLA-Dw and HLA-DP Typing of the Reference Panel of B-Lymphoblastoid Cell Lines" (this volume).

The Workshop component on "T-cell Recognition of HLA Class II Molecules" defined 18 additional TCD specificities associated with HLA-DR molecules, 16 additional TCD specificities associated with HLA-DQ molecules, and 5 new DR-associated supertypic TCD specificities.

These new TCD specificities have been assigned to the individual cells in the Reference Panel of B-lymphoblastoid cell lines and are presented in Table 1.

A detailed description of these TCD specificities and the experimental data supporting the assignments are provided in the Workshop section entitled "T-Cell Recognition of HLA Class II Molecules" and summarized in the report "Functional Polymorphisms of HLA Class II Gene Products Detected by T-Lymphocyte Clones" (this volume).

Participating Laboratory: US7NFL[1]

Immunobiology of HLA
Volume I

Table 1. Cellular Typing of Workshop B-LCL Panel by T-Lymphocyte Clones

					TCD Workshop Assignments					
Cell Name	Cell #	HLA-DR	HLA-Dw	Dw-DRw52	Dw Related	DR Related		DQ Related		
SA	9001	1	w1					DQw5(Dw1+Dw20)+		
MZ070782	9002	1	w20					DQw5(Dw1+Dw20)+		
KAS116	9003	1	w1					DQw5(Dw1+Dw20)+		
JESTHOM	9004	1	w1					DQw5(Dw1+Dw20)+		
HOM2	9005	1	w1					DQw5(Dw1+Dw20)+		
WT100BIS	9006	1	w1					DQw5(Dw1+Dw20)+		
DEM	9007	2				DRw2.2+				
DO208915	9008	2	w2			DRw15.1-		DQw6(Dw2)+	DQw6(Dw2+Dw12)+	
KAS011	9009	2	w21			DRw16.1+	DRw16.2+			
AMAI	9010	2	w2			DRw15.1-	DRw15.2+	DQw6(Dw2)+	DQw6(Dw2+Dw12)-	
E4181324	9011	2	w12			DRw15.1+			DQw6(Dw2+Dw12)+	
WJR076	9012	2	w21			DRw16.1-				
SCHU	9013	2	w2			DRw15.1+		DQw6(Dw2)+	DQw6(Dw2+Dw12)+	
MGAR	9014	2	w2			DRw15.1-		DQw6(Dw2)+	DQw6(Dw2+Dw12)+	
WT24	9015	2	w21			DRw16.1+	DRw16.2+			
RML	9016	2	w22			DRw16.1+	DRw16.2-	DQw3.1+		
WT8	9017	2	w2			DRw15.1+		DQw6(Dw2)+	DQw6(Dw2+Dw12)+	
L0081785	9018	3	w3	w25						
DUCAF	9019	3	w3	w25						
QBL	9020	3	w3	w25						
RSH	9021	3	NEW	w24		DRw18.0+		DQw4.1+	DQw4.2-	DQw4.3+
COX	9022	3	w3	w24						
VAVY	9023	3	w3	w24						
KT17	9024	4	DKT2					DQw3.1+	DQw8.1+	
DEU	9025	4	w4					DQw7.1+		
YAR	9026	4	w10					DQw3.1+	DQw8.1+	
PF97387	9027	4	w4					DQw7.1+		
PE117	9028	4	w14					DQw3.1+		
WT51	9029	4	w4					DQw3.1+	DQw8.1+	
JHAF	9030	4	w13					DQw7.1+		
BOLETH	9031	4	w4					DQw3.1+	DQw8.1+	
BSM	9032	4	w4					DQw3.1+		
BM14	9033	4	w4					DQw3.1+	DQw8.1+	
SAVC	9034	4	w4					DQw3.1+	DQw8.1- (DQw8.0+)	
JBUSH	9035	5(w11)	w5	w25		DRw11.3+		DQw3.1+		
SPO010	9036	5(w11)	DB2	w25		DRw11.3+		DQw3.1+		
SWEIG007	9037	5(w11)	w5	w25		DRw11.3+		DQw3.1+		
BM16	9038	5(w12)	DB6	w25				DQw3.1+		
JVM	9039	5(w11)	NEW	w25		DRw11.2+		DQw3.1+		
BM15	9040	5(w11)	w5	w25		DRw11.1+		DQw3.1+		
J0528239	9041	5(w11)	w5	w25		DRw11.4+		DQw3.1+		
TISI	9042	5(w11)	NEW	w25		DRw11.1+		DQw3.1+		
BM21	9043	5(w11)	w5	w25	Dw25.1+	DRw11.3+		DQw3.1+		
BRIP	9044	5(w11)	w5	w25		DRw11.3+		DQw3.1+		
TUBO	9045	5(w11+w12)	w5	w25		DRw11.4+		DQw3.1+		
BH	9046	7	DB1					DQw2(DR7)+	DQA(DR7)+	
PLH	9047	7	DB1					DQw2(DR7)+	DQA(DR7)+	
LBUF	9048	7	DB1					DQw2(DR7)+	DQA(DR7)+	
IBW9	9049	7	w17					DQw2(DR7)+	DQA(DR7)+	
MOU	9050	7	w17					DQw2(DR7)+	DQA(DR7)+	
PITOUT	9051	7	w17					DQw2(DR7)+	DQA(DR7)+	
DBB	9052	7	w11		Dw11.1+	DRw53.1+		DQw3.1+	DQA(DR7)+	

Table 1. *Continued*

Cell Name	Cell #	HLA-DR	HLA-Dw	Dw-DRw52	TCD Workshop Assignments				
					Dw Related	DR Related	DQ Related		
HOR	9053	w6(w13)	w19	w26			DQw6(Dw19)+		
EK	9054	w9(w14)	w9	w25					
HO301	9055	w6(w13)	w19V	w26			DQw6(Dw19)-		
KOSE	9056	w6	w9/w19	w25/w26					
TEM	9057	w6(w14)	w9	w25					
OMW	9058	w6(w13)	w18	w24			DQw6(Dw18)+		
SLE005	9059	w6(w13)	w19	w26			DQw6(Dw19)+		
CB6B	9060	w6(w13)	w18	w25			DQw6(Dw18)+		
31227ABO	9061	w6(w14)	w9	w25					
WDV	9062	w6(w13)	w18	w24			DQw6(Dw18)+		
WT47	9063	w6(w13)	w19	w26			DQw6(Dw19)+		
AMALA	9064	w6(w14)	w16	w24			DQw3.1+		
HHKB	9065	w6(w13)	w18	w24			DQw6(Dw18)+		
TAB089	9066	w8	w8.3			DRw8.3+			
BTB	9067	w8	w8.1			DRw8.1+	DQw4.1+	DQw4.2+	DQw4.3+
BM9	9068	w8	w8.1			DRw8.1a+	DQw4.1+	DQw4.2+	DQw4.3+
MADURA	9069	w8	w8.1			DRw8.1+	DQw4.1+	DQw4.2+	DQw4.3+
LUY	9070	w8	w8.3			DRw8.3+	DQw3.1+		
OLGA	9071	w8	w8.2			DRw8.2+	DQw4.1+	DQw4.2+	
SPACH	9072	w8	w8.2			DRw8.2+	DQw4.1+	DQw4.2+	
KT12	9073	w9	w23				DQw3.1+		
HID	9074	w9	w23				DQw3.1+		
DKB	9075	w9	w23				DQw3.1+		
T7526	9076	w9					DQw3.1+		
T7527	9077	w9					DQw3.1+		
PMG075	9078	1	w20				DQw5(Dw1+Dw20)+		
LWAGS	9079	1	w20				DQw5(Dw1+Dw20)+		
EHM	9080	1	w1				DQw5(Dw1+Dw20)+		
EA	9081	2	w2			DRw15.1-			
HO104	9082	?							
LD2B	9083	2	w2			?DRw15/w16 heterozygote			
CALOGERO	9084	2				DR2.1+			
EJ32B	9085	3	SYD3	w25					
LO541265	9086	3		w24					
STEINLIN	9087	3	w3	w24					
PF04015	9088	3		w24					
BOB	9089	5(w11)		w25					
AWELLS	9090	4	w4				DQw3.1- (DQw7.1?)		
MLF	9091	4	w4				DQw3.1- (DQw7.1?)		
BM92	9092	4	w14				DQw3.1+		
BER	9093	7	w7				DQw2(DR7)+	DQA(DR7)+	
CF996	9094	7	w7					DQA(DR7)+	
WIN	9095	7	w11				DQw3.1+	DQA(DR7)+	
LBF	9096	7	DB1				DQw2(DR7)+	DQA(DR7)+	
EMJ	9097	w6(w13)	w19	w26			DQw6(Dw19)+		
MT14B	9098	4	w14				DQw3.1+		
LZL	9099	w6(w14)	w16	w24			DQw3.1+		
OLL	9100	w8	8.2			DRw8.2+	DQw4.1+	DQw4.2+	
SPL	9101	w8	8.2			DRw8.2+	DQw4.1+	DQw4.2+	
ARBO	9102	w9					DQw3.1-		
KT14	9103	w9	w23				DQw3.1+		

Nomenclature for Factors of the HLA System

New HLA Specificities and Provisional HLA Determinants Defined by the Tenth International Histocompatibility Workshop

Bo Dupont[1]

This section describes the new HLA alleles that were identified at the completion of the Tenth International Histocompatibility Workshop. The report "Nomenclature for Factors of the HLA System, 1987," presented elsewhere in this volume, represents the consensus opinion of the Nomenclature Committee on Leukocyte Antigens.

The central analysis data generated within the individual Workshop components resulted, however, in the identification of several additional provisional HLA specificities for HLA class I and class II antigens.

A summary of all new serologically recognized HLA antigens and provisional new "splits" of antigens identified in the Antigen Society reports is provided in the report "New HLA Antigens Identified in the Serological Studies in the Tenth International Histocompatibility Workshop, 1987" (1).

The results from the Workshop studies utilizing biochemical techniques (i.e., 1D-IEF gel electrophoresis and 2D gel analysis) have been used by the Nomenclature Committee to support the description of new HLA specificities. It was recognized that the biochemical techniques provide definitions of gene products that may be closely related to the amino acid sequence. It is therefore a reasonable conclusion not to introduce a new, official, but intermediate nomenclature for the presently known biochemical variants, but to await the definitive sequence data for the definition of each HLA allele. It is, however, advantageous for the purpose of communication between investigators to provide an interim provisional nomenclature for biochemically detectable polymorphic HLA gene products. The reports entitled "Nomenclature for HLA-A and HLA-B Alleles Detected by 1D-IEF Gel Electrophoresis" (2) and "Nomenclature for HLA Class II Alpha and Beta Chains Detected by 2D Gel Electrophoresis" (3) provide a provisional nomenclature for biochemically detectable HLA gene products.

The Workshop component "T Cell Recognition of HLA Class II Molecules" provided the experimental data for the definition of new T-cell-defined (TCD) specificities Dw20 to Dw26. The final central analysis of this Workshop component identified a large number of additional new TCD specificities. A summary of all TCD specificities identified in the Tenth International Histocompatibility Workshop is provided in the report "Nomenclature for T-Cell-Defined HLA Class II Determinants" (4).

Nomenclature for restriction fragment length polymorphism (RFLP) with HLA gene probes presents a particularly difficult problem. It is well recognized that RFLP is frequently due to restriction sites located outside of the exons encoding allelic gene products. RFLP analysis with HLA gene probes is, however, very effective in defining HLA haplotypes. This was clearly demonstrated in the Workshop studies of the 70 Core Cell Lines. Thirty-four different DR, DQ, Dw haplotypes could be clearly recognized also by the Southern blot technique. It is therefore evident that RFLP typing for HLA will now be included as one of the "tissue-typing techniques." For the practical purpose of providing guidelines for HLA-RFLP typing, we have designed a system for communication of HLA-RFLP data. This is described in the report "Nomenclature for HLA-RFLP Typing" (5). It should be recognized that this nomenclature for HLA-RFLP is fundamentally different from all other designations for factors of the HLA system. The individual RFLPs do not provide a determination of an allelic gene or gene product, but may have very significant correlation with (and in some instances demonstrate complete concordance) with the existence of a particular allelic gene product.

References

1. Milford EL, Yunis EJ. New HLA antigens identified in the serological studies of the Tenth International Histocompatibility Workshop, 1987. In: Dupont B (ed). Immunobiology of HLA, Volume I: Histocompatibility Testing 1987. New York: Springer-Verlag Inc., 1989.
2. Yang SY. Nomenclature for HLA-A and HLA-B alleles detected by 1D-IEF gel electrophoresis. In: Dupont B (ed). Immunobiology of HLA: Volume I: Histocompatibility Testing 1987. New York: Springer-Verlag, 1989.
3. Knowles RW. Nomenclature for 2D gel patterns of the HLA class II alpha and beta chains defined in the 10th Workshop. In: Dupont B (ed). Immunobiology of HLA: Volume I: Histocompatibility Testing 1987. New York: Springer-Verlag, 1989.
4. Flomenberg N. Nomenclature for T-cell defined HLA class II determinants. In: Dupont B (ed). Immunobiology of HLA: Volume I: Histocompatibility Testing 1987. New York: Springer-Verlag, 1989.
5. Cohen D, Simons MJ, Lalouel JM, Dupont B. Nomenclature for HLA-RFLP typing. In: Dupont B (ed). Immunobiology of HLA: Volume I: Histocompatibility Testing 1987. New York: Springer-Verlag, 1989.

Participating Laboratory: US7DUP[1]

New HLA Antigens Identified in the Serologic Studies of the Tenth International Histocompatibility Workshop, 1987

Edgar L. Milford[1] and Edmond J. Yunis[2]

The serologic studies within the Tenth International Histocompatibility Workshop included more than 2600 serologic reagents from 157 laboratories. The serologic exercises were decentralized into "Antigen Groups/Antigen Societies" with specialized interest in analyzing a related set of antibodies. A summary of the experimental design and the results obtained is presented in "Serologic Exercises of the Tenth Workshop" (1).

The details from the individual "Antigen Groups/Antigen Societies" are provided in the 35 Antigen Society reports (this volume).

On the basis of both historical analyses and preliminary analysis of the Tenth Workshop serology data at Princeton, New Jersey, the WHO Nomenclature Committee approved seven new class I provisional "W" designations: AW74 (TH); BW75, BW76; BW77 (B15); CW9, CW10 (CW3); and CW11 (Cx46, CW1+3). Ten new class II provisional specificities were also approved as follows: DRW15 and DRW16 (DR2); DRW17 and DRW18 (DR3); DQW4 (DQ"Wa"); DQW5 and DQw6 (DQw1); DQW7, DQW8 and DQW9 (DQW3) (Table 1).

An in-depth analysis of the serologic data was undertaken during the spring of 1988, and the results are incorporated into the Antigen Society Reports. The reports provide evidence for an additional 11 HLA-A specificities, 24 HLA-B specificities, and 3 HLA-C specificities (Table 2).

Evidence for four new HLA-DR specificities, one DQw2 split, two splits of DRw5, and one variant of DRw53 is also presented (Table 3).

It should be noted that the strength of the experimental evidence for these additional, provisional serologic specificities varies substantially. The present summary of the serologic studies within this Workshop contains a complete listing of new provisional serologic specificities since they may serve the purpose of guiding investigators toward interesting antigens that need further attention.

Table 1. Nomenclature Committee WHO Approved Designations and Specificities

Antigen	Comments
Aw74	Also called Th
Bw75	Bw62.1,Bw6 associated B15 split,SH7, ST1
Bw76	Bw6 associated B15 split,B15S,B15.4
Bw77	Bw4 associated B15 split,B15T,B15.2
Cw9	Long form of Cw3, Cw3.1, associated with Bw55
Cw10	Short form of Cw3, Cw3.2, associated with Bw60
Cw11	Hybrid short Cw1 + short Cw3, associated with Bw46(Bw62), also called Cx46, Cw1+3, Cw1×3, C-Bangkok, CSH1. There are several variants of Cw11
DRw15	Long version of DR2
DRw16	Short version of DR2
DRw17	Long version of DR3
DRw18	Short version of DR3 assoc. with Bw42
DQw4	Also DQ"Wa" associated with DRw8, Dw15
DQw5	Also DQ"1.1" associated with DR1,DRw10,DRw14,DRw16
DQw6	Also DQ"1.2" associated with DRw15, DRw13
DQw7	Also DQ"3.1" is positive with TA10 sera, negative with 11B3 sera, associated with Drw11,DRw12,DR9
DQw8	Also DQ"3.2" is positive with 11B3 and negative with TA10 sera, associated with DR4
DQw9	Also DQ"3.3" is positive with 11B3 and negative with TA10 sera but associated with DR7, DR9

Reference

1. Milford EL, Park MS, Lalouel J-M, Kennedy L, and Yunis EJ. Serologic exercises of the Tenth Workshop. In: Dupont B (ed). Immunobiology of HLA, Volume I, Histocompatibility Testing 1987. New York: Springer-Verlag, 1989.

Participating Laboratories: NCYCAR,[1] NCYYUN[2]

Table 2. Other Class I Antigens

Antigen	Comments	Society Report
A3 black	based on differences in strengths of reactions (4037,4045)	S1
A23	short A9 (0065)	S1
A24 long	in Orientals based on one serum (4092)	S1
A26.1	in Japanese based on one serum (9159)	S2
Aw34 short	in S. African blacks, Australian Aborigines (93,94,95,105)	S2
A30.1	(136,187) B18 associated	S3
A30.3	(151) B13 associated	S3
A2 short	in Orientals (4237,4219,4210,4224)	S4
A2 short.a	in Orientals (19,2032,2034,2036)	S4
A2 short.b	in Orientals (32,33)	S4
A2 short.c	in Orientals (32,33,19,2032,2034,2036)	S4
BN21	Bw6 associated, Bw50 short in Native and Mexican American	S2
B44 short	? assoc. with Cw5 (4379,4380,4370,4400)	S6
Bw72.1	short Bw71 (198,4478,205,9201,9217)	S8
Bw72.2	short Bw71 (200,4484,9195,9206)	S8
B15.3	short Bw75,Bw6 associated (212,208,255,192,205)	S9
B39 short	(4562,467) ?SH6 (ST-16)	S11
Bw56.1	short Bw56 in Caucasians (4584,4617,4614,4593)	S12
Bw56.2	long Bw56 (Japanese)	S12
Bpot	short B7 (340,333,381)	S13
B7.1	B7 split (4636,4637), CTL defined	S13
B7.2	B7 split (4655,4669,4685), CTL defined	S13
B27.5,B27.2	B27 splits (292,284,4661), CTL defined	S13
B27.4,B27.6	B27 splits (9333), CTL defined	S13
Bw60a	long variant of Bw60	S14
Bw60b	short variant of Bw60 (382,383)	S14
Bw61a	long variant of Bw61	S14
Bw61b	short variant of Bw61 (382,366,373)	S14
Bw41.2	short Bw41 in Brazil, Hungary (326,385,386,9658)	S14
Bw48 short	short Bw48 (343,4725,4729)	S14
FU	very short Bw48 in Japanese (381,360,382,4743)	S14
B13b	Chinese short B13 (2101,2054,2081)	S14
B13c	Thai short B13 (2080,2094)	S14
BRI	B40 like antigen (232,357,353,370,375)	S14
BST40	B40 like antigen (273,287,288,292,326,351,357,375,394)	S14
Cw11 var 2	Japanese short Cw3 (508,512,513,525,526)	S16
Cw11 var 3	Caucasian short Cw11 (510,524,525)	S16
Cw9 short	Intermediate between Cw9 and Cw10 (9064,532)	S16

Numbers listed in brackets indicate key sera or monoclonal antibodies. The numbers refer to the workshop serum number listed in "Serologic Exercises of the Tenth Workshop" (this volume).

Table 3. Other Class II Antigens

Antigen	Comments	Society Report
DR1 short	short DR1 (5011,5018,5015)	S21
DR2.3	Oriental (1029,5071)	S22
DR2.4	S. African blacks (LUM) (1026,1023,1024)	S22
DR4.1	Dw4,Dw13,Dw14,Dw15 associated (3009,3008,5119,3009,1046,1045)	S23
DQw2 split	Subset of DR7 or DR3 positive DQw2 cells monoclonal antibodies (3106,3107,3105)	S26
DRw52R3	Not DRw8 or DRw12 (3063 monoclonal)	S29
DRw52a+c	LB-Q4,DR7,DRw9,DRw12 (3048 monoclonal)	S29
DRw53 var	Blacks (DR4.1 association) (1212)	S30

See footnote in Table 2.

Nomenclature for HLA-A and HLA-B Alleles Detected by One-Dimensional Isoelectric Focusing (1D-IEF) Gel Electrophoresis

Soo Young Yang[1]

A nomenclature for one-dimensional isoelectric focusing (1D-IEF) alleles identified for the HLA-A and HLA-B antigens has been developed by the participants in the HLA Class I Biochemistry component of the Tenth International Histocompatibility Workshop.

The gel position of the individual HLA Class I heavy chains depends on the pI of the denatured protein, which in turn reflects the net charge of the amino acid composition.

The definitive HLA allele designation for each class I antigen awaits the completion of the amino acid sequence, as described in the "WHO Nomenclature Report for Factors of the HLA System" (this volume). The provisional nomenclature for HLA-A and HLA-B 1D-IEF alleles is designed for the purpose of assisting investigators in the communication of their results and to serve as a common source of reference.

Sequential decimal numbers from the top of the gel (cathode) were adopted from the serologic specificity numbers for antigens with more than one band. For example, three bands that correspond to the A11 antigen are designated as A11.1, A11.2, and A11.3, starting from the basic (top) to the acidic (bottom) part of the gel. In the case of a serologic split with no correlation with IEF bands, supertypic serologic specificities were used. For example, a total of three IEF bands corresponding to the B17 antigen were observed, but no correlation could be made between the Bw57 or Bw58 antigens and the three IEF bands. These three B17 bands were named B17.1, B17.2, and B17.3, starting from the top of the gels.

A total of 34 HLA-A 1D-IEF alleles were recognized in the Tenth Workshop (Table 1). Three additional HLA-A alleles are indicated with an asterisk (*), i.e., Al*, A24*, and A26*. The asterisks indicate IEF variants seen only on one or two cells among 918 tested in the Workshop. It should be noted that the allele designations depend on correct serologic antigen assignment to the

Table 1. HLA-A Alleles Detected by 1D-IEF

Serologic Specificity		IEF Allele Designation
A1		A1, A1*
A2		A2.1, **A2.2**, A2.3, A2.4, A2.5
A28		**A28.1, A28.2**
A3		A3.1, **A3.2**
A9:	A23	A23
	A24	**A24.1**, A24.2, A24.3, A24*
A10:	A25	A25
	A26	**A26.1**, A26.2, A26*
	Aw34	Aw34.1, **Aw34.2**
	Aw66	Aw66
A11		A11.1, **A11.2**, A11.3
Aw19:	A29	A29.1, **A29.2**
	A30	**A30.1**, A30.2, **A30.3**
	A31	A31
	A32	A32
	Aw33	**Aw33.1, Aw33.2**
Aw36		Aw36
Aw43		Aw43

*Rare variants.
IEF alleles shown in **bold type** represent common subtypes.

Table 2. HLA-B Alleles Detected by 1D-IEF

Serologic Specificity		IEF Allele Designation
B5:	B51	B51, B51*
	Bw52	Bw52
	Bw53	Bw53
B7		B7.1, **B7.2**
B12:	B44	**B44.1, B44.2**
	B45	B45
B13		**B13.1, B13.2**
B14:	Bw64	B14
	Bw65	B14
B15:	Bw63	B15.1
	Bw62	**B15.2, B15.3,** B15.4
B16:	B38	B16.1
	B39	**B16.2,** B16.3
B17:		B17.1, B17.2, B17.3
	Bw57	**B17.2**
	Bw58	**B17.2**
B18		**B18.1,** B18.2
B21:		B21.1, B21.3
	B49	**B21.2**
	Bw50	**B21.2**
Bw22:	Bw54	Bw22.1
	Bw55	Bw22.3
	Bw56	Bw22.2
B27		B27.1, B27.2, B27.3, B27.4, **B27.5,** B27.6, B27*
B35		B35.1, B35.2, **B35.3,** B35*
B37		B37
B40:	Bw60	Bw60
	Bw61	Bw61
Bw41		**Bw41.1,** Bw41.2
Bw42		Bw42, Bw42*
Bw46		Bw46, Bw46*
Bw47		Bw47
Bw48		Bw48
Bw59		Bw59
Bw67		Bw67
Bw70:	Bw71	**Bw70.2**
	Bw72	**Bw70.1, Bw70.3**

IEF alleles shown in **bold type** represent common subtypes.

Participating Laboratory: US7SYY[1]

Immunobiology of HLA
Volume I

Table 3. Two-Locus HLA Class I IEF Haplotypes

HLA-A	HLA-B	HLA-C
A2	B44.2	
A28	B44.2	
A2.1	Bw50	
A11.1	Bw46	
A11.3	Bw46	
A29.1	B7.1	
A29.2	B44.1	
A30.1	B18.1	
A30.3	B13.2	
Aw33.1	B14	
Aw33.2	B44.1	
Aw33.2	B17.2	
	B15.2	Cw8
	B15.3	Cw3
	B27.2	Cw2

test cell. IEF variants with an asterisk indicate this uncertainty. The IEF alleles shown in bold type in Tables 1 and 2 represent common subtypes. The A11.2 is, for example, the common HLA-A11 IEF allele. When two IEF alleles for the same serologic specificity are shown in bold, they represent two IEF alleles that are both common (e.g., Aw33.1 and Aw33.2).

Fifty-two (52) HLA-B IEF alleles were identified in the Workshop studies (Table 2). Five additional IEF alleles are indicated by an asterisk (i.e., B51*, B27*, B35*, Bw46*, and Bw42*).

The distribution of IEF alleles within different ethnic groups varies considerably. For example, B13.1 is the predominant Oriental B13 IEF allele, while B13.2 is the corresponding Caucasian B13 IEF allele. Both B13.1 and B13.2 are, however, common IEF alleles (as indicated by bold type in Table 2).

A detailed description of the ethnic distribution of HLA-A and HLA-B IEF alleles is given in "Population Analysis of Class I HLA Antigens by One-Dimensional Isoelectric Focusing Gel Electrophoresis" (1).

Genetic linkage disequilibrium for HLA-A, HLA-B, and HLA-C alleles is also observed for many of the IEF alleles (e.g., A2, B44.2 and A29.2, B44.1). A summary of some of the most common class I haplotypes demonstrated in the Tenth Workshop is shown in Table 3.

The relative band position for each of the HLA-A and HLA-B 1D-IEF alleles is shown in Figures 1 and 2. (For clarity, "w" is not listed in the figures for antigens with provisional WHO designations.) Further details and a complete description of the Workshop studies are presented in "HLA Class I Polymorphism Detected by 1D-IEF" (this volume).

Reference

1. Yang SY. Population analysis of class I HLA antigens by one-dimensional isoelectric focusing gel electrophoresis. In: Dupont B (ed). Immunobiology of HLA, Volume 1: Histocompatibility Testing 1987. New York: Springer-Verlag, 1989.

	A 1	A 2	A 3	A 11	A 23	A 24	A 25	A 26	A 28	A 29	A 30	A 31	A 32	A W 33	A W 34	A W 36	A W 43	A W 66
1#		A2.1																
2										A29.1								
3		**A2.2**																
4									**A28.1**									
5		A2.3																
6		A2.4																
7		A2.5*																
8																		
9										**A29.2**								
10								A26*						**A33.1**				
11									**A28.2**									
12																		
13														**A33.2**				
14																		
15																		
16																		
17	A1						A25	**A26.1**										
18																		
19																		
20																		
21				A11.1														
22																		
23													A32					
24												A31						
25																		
26																		
27															A34.1	A36	A43	
28						**A24.1**												A66
29								A26.2										
30					A23	A24.2					**A30.1**							
31																		
32																		
33						A24.3												
34			A3.1	**A11.2**											**A34.2**			
35																		
36				A11.3							A30.2							
37			**A3.2**															
38											**A30.3**							
39																		
40																		
41																		
42																		

| | A 1 | A 2 | A 3 | A 11 | A 23 | A 24 | A 25 | A 26 | A 28 | A 29 | A 30 | A 31 | A 32 | A W 33 | A W 34 | A W 36 | A W 43 | A W 66 |

Figure 1. IEF band patterns for HLA-A locus products.

	B7	B8	B13	B14	B15	B16	B17	B18	B21	BW22	B27	B35	B37	BW41	BW42	B44	B45	BW46	BW47	BW48	B51	BW52	BW53	BW60	BW61	BW67	BW70
1#																											
2																											
3																											
4																											
5																											
6																											
7																											
8								**B18.1**																			
9																											
10																											
11																											
12																								B60			
13																											
14																		B46									
15												B35.1															
16												B35.2															
17																											
18								B18.2																	B61		
19																											
20														**B41.1**													
21									B21.1			B35*															
22					**B15.1**																						
23																											
24																											
25												**B35.3**											B53				
26						B16.1	B17.1																				
27									**B21.2**								B45										
28					B15.2							B35*									B51						
29			**B13.1**		**B15.3**		**B17.2**							B41.2													
30						**B16.2**				B22.1 (54)																	B70.1
31																										B67	
32																						B52					
33																**B44.1**											
34	B7.1	**B13.2**							B21.3										B47								
35					B15.4	_	B17.3			B22.2 (56)			B37							B48							B70.2
36						B16.3					B27.1					**B44.2**											
37	**B7.2**																										B70.3
38				B14						B22.3	B27.2				B42												
39										(55)	B27.3																
40											B27.4																
41		B8									**B27.5**																
42											B27.6																
	B7	B8	B13	B14	B15	B16	B17	B18	B21	BW22	B27	B35	B37	BW41	BW42	B44	B45	BW46	BW47	BW48	B51	BW52	BW53	BW60	BW61	BW67	BW70

Figure 2. IEF band patterns for HLA-B locus products.

Nomenclature for Two-Dimensional (2-D) Gel Patterns of the HLA Class II α and β Chains Defined in the Tenth Workshop

Robert W. Knowles[1]

The provisional Tenth Workshop two-dimensional (2-D) gel nomenclature indicates the conclusion of the participants as to whether the spots found in 2-D patterns in different cell lines have identical or distinct pI and molecular weight. Details of the analysis are presented later in these Workshop Proceedings (Volume I). This nomenclature should be considered provisional, since analysis of the 2-D patterns in this study alone does not formally demonstrate that a given pattern represents the product of a specific class II gene. The alleles for each class II locus will eventually be given official WHO nomenclature based on confirmed cDNA sequences. Therefore, the Tenth Workshop nomenclature for the 2-D gel patterns will remain provisional until the corresponding alleles are given official WHO designations. A complete listing of the 2-D gel patterns assigned to specific class II DR and DQ loci is shown in Table 1 with the DNA sequence of the most closely associated allele.

The observed association between specific DRβ chain patterns and the DR and Dw typing of the Workshop cell lines is shown in Table 2, and the 19 distinct DQ α-β dimers observed between the DQα and DQβ alleles found in the Workshop cell lines are listed in Table 3. Serologic and cellular specificities associated with these 2-D gel patterns are from the W.H.O. nomenclature based on the Tenth Workshop studies of the Reference B Cell Line Panel.

The 34 distinct haplotypes defined by the combination of DR and DQ α and β chain patterns are shown in Table 4. Although the DPβ chain 2-D patterns were not defined in this study, the association between two distinct DPα chain patterns and cellular DP typing is shown in Table 5. There was no correlation between these DPα chain patterns and the haplotypes defined by the DR and DQ chain patterns.

Participating Laboratory: US7KNO[1]

Immunobiology of HLA
Volume I

Table 1. Complete Listing of Allelic 2-D Gel Patterns and Their Associated First Domain Amino Acid Sequences

DRβI 2-D Patterns[1]	**DRB1 Sequences**[2]
DRβI-1.1	DR1 βI
DRβI-1.1	DR1 NASC
DRβ-a-2.2[3]	DR2 Dw2βI
DRβ-a-2.12[3]	DR2 Dw12βI
DRβ-a-2.16[3]	DR2 AZHβI
DRβI-3	DR3 βI
DRβI-3	DR3 DQWaβI
DRβI-4.1	DR4 Dw14βI
DRβI-4.1	DR4 Dw4βI
DRβI-4.2	DR4 Dw13βI
DRβI-4.2	DR4 KT2βI
DRβI-4.3	DR4 Dw10βI
DRβI-(4.4*)[4]	DR4 Dw15βI[4]
DRβI-7	DR7 βI
DRβI-8.1	DRw8 βI
DRβI-8.2	DRw8 SPLβI
DRβI-8.1	DRw8 TABβI
DRβI-9	DRw9 βI
DRβI-11.1	DR5 Dw5βI
DRβI-11.2	DR5 JVMβI
DRβI-12	-[6]
DRβI-13	DRw6aβI
DRβI-14.1	DRw6bβI
DRβI-14.2	DRw6 AMALAβI
DRβ-*[4]	DR1 CETUS[4]
DRβ-*[4]	DRw10βI[4]
DRβ-[7] **2-D Patterns**	**DRB-**[7] **Sequences**
DRβ-b-2.1[3]	Dw2 βIII
DRβ-b-2.1[3]	Dw12 βIII
DRβ-b-2.1[3]	AZH βIII
DRβIII-1	DRw52bβIII
DRβIII-2	DRw52cβIII
DRβIII-3	DRw52aβIII
DRβIV-1	DR4 DRw53βIV

DQβ 2-D Patterns	**DQB1 Sequences**
DQβ-(1.1*)[5]	DR1 DQw1.1
DQβ-1.2	DR2 DQw1.2
DQβ-(1.9*)[5]	DRw6 DQw1.9
DQβ-1.12	DR2 DQw1.12
DQβ-1.18	DRw6 DQw1.18
DQβ-1.19	DRw6 DQw1.19
DQβ-(1.21*)[5]	DR2 DQw1.AZH
DQβ-2	DR3 DQw2
DQβ-3.1	DR4 DQw3.1
DQβ-3.2	DR4 DQw3.2
DQβ-3.3	DRw9 DQw3.3
DQβ-Wa	DRw8 DQ WA
DQβ-1.12*	-[6]
DQα 2-D Patterns	**DQA1 Sequences**
DQα-1.1	DR1 DQw1.1*[8]
DQα-1.2	DR2 DQw1.2
DQα-1.14	DRw6 DQw1.9*[8]
DQα-1.18	DRw6 DQw1.18
DQα-2	DR5 DQw3.1
DQα-3.1	DRw9 DQw3*[9]
DQα-3.2	DR4 DQw3.2*[9]
DQα-3.7	DR7 DQw2 and w3
DQα-3.8	DRw8 ARC (WA)
DQα-3.8	DRw8 LUY (DQw3)

[1]Associations of 2-D patterns with specific first domain sequences is based on the criteria discussed later in the Workshop Proceedings (Knowles, volume I).
[2]Official WHO nomenclature has not been established for these alleles. Nomenclature is from first domain sequences (Gregersen et al, volume I).
[3]The designations "DRβ-a" and "DRβ-b" are used since these patterns have not been assigned to DRB1 or a second DRB- locus.
[4]No DR4,Dw15, DR1 CETUS, or DRw10 cell lines were included in the Workshop panel.
[5]DQβ-1.1, DQβ-1.21, DQβ-1.9, and DQβ-Wa15 were not characterized by Workshop 2-D gel analysis.
[6](-) No known sequence is associated with this pattern.
[7]DRβ- is used to designate the second expressed DRβ chain, encoded by the second expressed DRB-gene, which can be the unnamed DRB gene in DR2 cells, the DRB3 or the DRB4 gene.
[8]DR1 DQw1.1 and DRw6 DQw1.9 have identical DQA1 first domain DNA sequences.
[9]DR4 DQw3.2 and DRw9 DQw3 have identical DQA1 first domain DNA sequences.

Table 2. DRβ Chain 2-D Gel Patterns and Their Associated DR and Dw Specificities Observed in Workshop Cell Lines

DRβ 2-D Patterns	DR Typing[1]	Dw Typing[1]
DRβI-1.1	DR1	Dw1, Dw20
DRβ-a-2.2[2]	DRw15(2)	Dw2
DRβ-a-2.12[2]	DRw15(2)	Dw12
DRβ-a-2.16[2]	DRw16(2)	Dw21, Dw22
DRβI-3	DRw17(3), DRw18(3)	Dw3, Dw"new"
DRβI-4.1	DR4	Dw4, Dw14
DRβI-4.2	DR4	Dw13, DwKT2
DRβI-4.3	DR4	Dw10
DRβI-(4.4*)[3]	DR4	Dw15
DRβI-7	DR7	Dw17, Dw11, DB1
DRβI-8.1	DRw8	Dw8.1, Dw8.3
DRβI-8.2	DRw8	Dw8.2
DRβI-9	DR9	Dw23
DRβI-11.1	DRw11(5)	Dw5, DB2
DRβI-11.2	DRw11(5)	Dw5, Dw"new"
DRβI-12	DRw12(5)	DB6
DRβI-13	DRw13	Dw18, Dw19
DRβI-14.1	DRw14	Dw9
DRβI-14.2	DRw14	Dw16
DRβ-b-2.1[2]	DRw15(2), DRw16(2)	Dw2, Dw12, Dw21,Dw22
DRβIII-1	DRw52	Dw25
DRβIII-2	DRw52	Dw26
DRβIII-3	DRw52	Dw24
DRβIV-1	DRw53	-

[1]Associations of the 2-D patterns with DR and Dw specificities is based on the results of the 10th Workshop serological and cellular typing of these cell lines.

[2]The designations "DRβ-a" and DRβ-b" are used since these patterns have not been assigned to DRB1 or a second, DRB- locus.

[3]No DR4,Dw15 cell lines were included in the Workshop.

Table 3. DQ α-β Dimers and Their Association with DQ and Dw Specificities Observed in the Workshop Cell Lines

DQ α-β Dimers		Associated Serological DQ and Cellular Dw Specificities[1]	
DQα-chain	DQβ-chain	DQ Typing	Dw Typing
DQβ-(1.1*)[2]	DQα-1.1	DQw5 (w1)	Dw1, Dw20
DQβ-(1.9*)[2]	DQα-1.14	DQw5 (w1)	Dw9
DQβ-(1.21*)[2]	DQα-1.2	DQw5 (w1)	Dw21
DQβ-1.2	DQα-1.2	DQw6 (w1)	Dw2
DQβ-1.12	DQα-1.18	DQw6 (w1)	Dw12
DQβ-1.12*	DQα-1.2	DQw1	DB2
DQβ-1.18	DQα-1.18	DQw6 (w1)	Dw18
DQβ-1.19	DQα-1.2	DQw6 (w1)	Dw19
DQβ-2	DQα-2	DQw2	Dw3
DQβ-2	DQα-3.7	DQw2	Dw17, DB1
DQβ-3.1	DQα-3.1	DQw7 (w3)	Dw4, Dw13
DQβ-3.1	DQα-2	DQw7 (w3)	Dw5, DB6, Dw"new5" Dw16, Dw22
DQβ-3.1	DQα-3.8	DQw7 (w3)	Dw8.3(L)
DQβ-3.2	DQα-3.2	DQw8 (w3)	Dw4, Dw14, DwKT2
DQβ-3.2	DQα-3.1	DQw8 (w3)	Dw4
DQβ-3.3	DQα-3.1	DQw9 (w3)	Dw23
DQβ-3.3	DQα-3.7	DQw9 (w3)	Dw11
DQβ-Wa	DQα-3.8	DQw4	Dw8.1, Dw8.1, Dw"new3"
DQβ-(Wa15*)[3]	DQα-(3.1*)	DQw4	Dw15

[1]Associations of 2-D patterns with DQ and Dw specificities is based on the results of the 10th Workshop serological and cellular typing on these cell lines.

[2]DQβ-1.1, DQβ-1.21, DQβ-1.9, and DQβ-Wa15 allele designations are based on first domain DNA sequences, but were not characterized by 2-D gels.

[3]No Dw15 cell lines were included in the Workshop panel.

Table 4. HLA Class II Haplotypes Defined by the 10th Workshop 2-D Gel Patterns[a]

Serol. DQ	2-D DQβ	2-D DQα	2-D DRβ	2-D DRβ	Serol. DR	Serol. DRw52/53	T cell Dw	T cell DRw52
DQw5*	- (1.1*)	1.1	I-1.1	-	DR1		Dw1	
DQw5	- (1.1*)	1.1	I-1.1	-	DR1		Dw20	
DQw6*	1.2	1.2	a-2.2	b-2.1	DRw15		Dw2	
DQw6	1.12	1.18	a-2.12	b-2.1	DRw15		Dw12	
DQw5*	- (1.21*)	1.2	a-2.16	b-2.1	DRw16		Dw21	
DQw7	3.1	2	a-2.16	b-2.1	DRw16		Dw22	
DQw6	1.19	1.2	I-13	III-2	DRw13	DRw52	Dw19	Dw26
DQw6	1.18	1.18	I-13	III-3	DRw13	DRw52	Dw18	Dw24
DQw6*	1.18	1.18	I-13	III-1	DRw13	DRw52	Dw18	Dw25
DQw7	3.1	2	I-14.2	III-3	DRw14	DRw52	Dw16	Dw24
DQw5	- (1.9*)	1.14	I-14.1	III-1	DRw14	DRw52	Dw9	Dw25
DQw2	2	2	I-3	III-1	DRw17*	DRw52	Dw3	Dw25
DQw2	2	2	I-3	III-3	DRw17*	DRw52	Dw3	Dw24
DQw4	Wa	3.8	I-3	III-3	DRw18*	DRw52	Dw"new"	Dw24
DQw5	1.12*	1.2	I-11.1	III-1	DRw11	DRw52	DB2	Dw25
DQw7	3.1	2	I-11.1	III-1	DRw11	DRw52	Dw5	Dw25
DQw7	3.1	2	I-11.2	III-1	DRw11	DRw52	Dw5	Dw25
DQw7	3.1	2	I-11.2	III-1	DRw11	DRw52	Dw"new"	Dw25
DQw7	3.1	2	I-12	III-1	DRw12	DRw52	DB6	Dw25
DQw2	2	3.7	I-7	IV-1	DR7	DRw53	Dw17	
DQw2	2	3.7	I-7	IV-1	DR7	DRw53	DB1	
DQw9	3.3	3.7	I-7	IV-null	DR7	DRw53	Dw11	
DQw9	3.3	3.1	I-9	IV-1	DR9	DRw53	Dw23	
DQw8	3.2	3.2	I-4.1	IV-1	DR4	DRw53	Dw14	
DQw8	3.2	3.2	I-4.1	IV-1	DR4	DRw53	Dw4	
DQw8	3.2	3.1	I-4.1	IV-1	DR4	DRw53	Dw4	
DQw7	3.1	3.1	I-4.1	IV-1	DR4	DRw53	Dw4	
DQw7	3.1	3.1	I-4.2	IV-1	DR4	DRw53	Dw13	
DQw8	3.2	3.2	I-4.2	IV-1	DR4	DRw53	KT2	
DQw8	3.2	3.2	I-4.3	IV-1	DR4	DRw53	Dw10	
DQw7	3.1	3.8	I-8.1	-	DRw8	DRw52	Dw8.3(L)	
DQw4	Wa	3.8	I-8.1	-	DRw8	DRw52	Dw8.1	
DQw4	Wa	3.8	I-8.2	-	DRw8	DRw52	Dw8.2	
DQw6	1.12	1.18	I-8.1	-	DRw8	DRw52	Dw8.3(T)	

[a]Details regarding the nomenclature for the 2-D gel patterns defined in this study are discussed later in these Workshop Proceedings, volume I. The WHO nomenclature for DR, DQ, and Dw specificities is based on the results of 10th Workshop serological and cellular typing of these cell lines, except where indicated (*).

Table 5. Association Between the Expression of the DPα Chain and DP Cellular Typing on 10th Workshop Core Cell Lines

DPα 2-D gel patterns[1]	DP Cellular Typing[2]
DPα-1.1	DP2, DP3, DP4
DPα-1.2	DP1, DP5

[1]Workshop reference panel B cell lines were typed for either the DPα-1.1 or DPα-1.2 2-D gel patterns. DRI cell lines could not be typed because they all contained the DQα-1.1 chain which comigrates with DPα-1.2. Details regarding the nomenclature for the 2-D gel patterns are discussed later in these Workshop Proceedings (Knowles, volume I).
[2]Associations of the DPα 2-D patterns with DP specificities is based on the results of 10th Workshop cellular typing on these cell lines.

Nomenclature for T-Cell-Defined HLA Class II Determinants

Neal Flomenberg[1]

The Workshop component on T-Cell Recognition of HLA Class II Molecules utilized 125 T-lymphocyte clones as typing reagents. The analysis of the reactivity pattern on the cells from the Reference Panel of B-Cell Lines demonstrated clear clusters that segregate with each of the serologically defined DR and DQ specificities. Within these large groups, there were a large number of subclusters. Indeed, it is probably the exception rather than the rule when several clones exhibit exactly the same pattern of reaction across the entire cell panel. Moreover, it is often the subtle differences that the clones exhibit which prove to be most informative.

Blocking studies were performed with monoclonal antibodies recognizing DR, DQ, and DP products. Of the 89 T-cell clones for which monoclonal antibody blocking could be interpreted, 77% recognized DR products, 18% recognized DQ products, while only 3% recognized DP products.

As a result of these investigations, the Workshop serologic studies, and cDNA sequence analysis, new HLA class II specificities were recognized by the World Health Organization Nomenclature Committee. When serologic reagents were capable of recognizing the products, a DR (or DQ) designation was given, irrespective of the independent recognition of the product by cellular reagents. When a serologic equivalent was not available, a Dw designation was assigned. Thus, Dw designations continue to be used for cellularly defined epitopes, although prior to the Tenth Workshop, they have reflected cellular epitopes stimulating strong responses in primary mixed lymphocyte cultures (see "Nomenclature for Factors of the HLA System," this volume).

A summary of the cellularly defined HLA-Dw specificities and their common HLA class II haplotypes is provided in Tables 1 and 2. The DR1 specificity is associated with Dw1 and Dw20. The Dw20 specificity is in linkage disequilibrium with B14 in the Ashkenaic population. The DR2 specificity is associated with Dw2, Dw12, Dw21 and Dw22. Dw21 and Dw22 haplotypes express the DRw16 (DR2) serologic specificity, while Dw21 and Dw22 express DRw15 (DR2). The serologic specificity DR3 is associated with three haplotypes: B8;DRw17(DR3);Dw3, B18;DRw17(DR3);Dw3-variant, and Bw42;DRw18(DR3);Dw-new.

Two new Dw specificities were identified by individual laboratories: a Dw-new specificity associated with DRw10, DQw5 and another Dw-new specificity associated with the DR specificity "BON" and DQw1.

The Dw24, Dw25, and Dw26 specificities identify cellularly defined subtypes of DRw52. Their haplotype associations are summarized in Table 2. These cellularly defined specificities were identified by T-cell clones. These specificities can also be distinguished by protein chemistry and molecular genetic techniques.

Evidence for other class II splits or subtypes was obtained from the Workshop studies. The following nomenclature is unofficial and is intended solely for the purpose of facilitating future communication. The designations are preceded by "TCD" (T-lymphocyte clone defined or T-cell defined) to differentiate this nomenclature from official W.H.O. specificities. The designations incorporate the most closely related DR, DQ, or Dw specificity. Specificities that correlate precisely with serologic equivalents are, in general, not redefined, except in those cases where they are differentiated from closely related but distinct specificities. The decimal addition ".0" is used in these rare instances to indicate complete agreement with the serologic counterpart. Other decimal extensions (".1", ".2", ".3", etc.) indicate a cellularly defined subtype. In the case of DQ products, some of the subtypes identified exhibit concordance with particular DR or Dw specificities, in which case the DR or Dw type is included in parenthesis. In one instance, a DQ specificity exhibited concordance with the presence of a particular DQ alpha chain, irrespective of the serologic DQ type. In this case, the designation TCD-DQA was utilized to reflect the DQ alpha specificity. These specificities are described in detail below and are summarized in Tables 3 and 4.

Additional cellularly defined supertypic specificities were identified in this Workshop. In defining nomenclature for supertypic specificities, we have followed the WHO example of DRw52 and DRw53, using higher numbers for the supertypic determinants. Since the Nomenclature Committee may choose to add a DRw54 specificity at some future time, we have begun the numbering system in the 80s to avoid future confusion. These new supertypic specificities are listed in Table 5.

TCD-DR81 and TCD-DR81.1. This designation is used to represent the cross-reactivities detected between DR1, DR4, and DRw6. Occasionally, DRw16 is included as well. The prototypic example DR1 + DRw14 + Dw16 probably reflects the large stretch of common amino acid sequence between residues 38 through 86 of these DR beta-I molecules. Because of its strong structural basis, this example is given the specific designation TCD-DR81.1, while the entire group (containing several different patterns of DR1, DR4, DRw6 + DRw16 cross reactivity) is referred to as TCD-DR81.

TCD-DR82. This refers to the cross-reactive group formed between DR2 (particularly DRw16), DR5, and

Participating Laboratory: US7NFL[1]

Immunobiology of HLA
Volume I

Table 1. Summary of HLA Class II Haplotypes

HLA-DR		HLA-Dw	HLA-DRw52/53/Dw	HLA-DQ	HLA-B	Ethnic	Comments
HLA-DR1		Dw1		DQw5 (DQw1)	B35		
		Dw20		DQw5 (DQw1)	B14	Jewish	
						Black	
HLA-DR2*	DRw15	Dw2		DRw6 (DQw1)	B7		
	DRw15	Dw12		DQw6 (DQw1)	Bw52	Japanese	
	DRw16	Dw21		DQw5 (DQw1)			MN2, FJO, AZH
	DRw16	Dw22		DQw7 (DQw3)		Warao Indian	tb24, LD-5a
HLA-DR3	DRw17	Dw3	DRw52a, Dw24	DQw2	B8		
	DRw17	Dw3 variant	DRw52b, Dw25	DQw2	B18		
	DRw18	Dw new	DRw52a, Dw24	DQw4	Bw42	Black	
HLA-DR4		Dw4	DRw53	DQw7 (DQw3)	B44(12)		Dw4.1
				DQw8 (DQw3)	Bw62(15)		Dw4.2
		Dw10	DRw53	DQw8 (DQw3)	B38	Jewish	
		Dw13	DRw53	DQw7 (DQw3)			
				DQw8 (DQw3)			
		Dw14	DRw53	DQw8 (DQw3)	Bw60(w40)		
		Dw15	DRw53	DQw4		Japanese	
HLA-DR5*	DRw11	Dw5	DRw52b, Dw25	DQw7 (DQw3)			
	DRw11	DB2	DRw52b, Dw25	DQw1			
	DRw12	DB6	DRw52b, Dw25	DQw7 (DQw3)			
HLA-DRw6	DRw13	Dw18	DRw52a, Dw24	DQw6 (DQw1)			
	DRw13	Dw18	DRw52b, Dw25	DQw1			
	DRw13	Dw19	DRw52c, Dw26	DQw6 (DQw1)			
	DRw14	Dw9	DRw52b, Dw25	DQw5 (DQw1)			
	DRw14	Dw16	DRw52a, Dw24	DQw7 (DQw3)	Bw62	American Indian	
HLA-DR7		Dw7	DRw53	DQw2			
		Dw17	DRw53	DQw2	B44		
		Dw11	blank	DQw9 (DQw3)	Bw57		
		Dw11	DRw53	DQw9 (DQw3)	Bw57		
		DB1	DRw53	DQw2	B13		
HLA-DRw8		Dw8.1		DQw4		Caucasian	
		Dw8.2		DQw4		American Indian	
		Dw8.3		DQw1		Oriental	
		Dw8.3		DQw7 (DQw3)		Dutch	LUY
HLA-DR9		Dw23	DRw53	DQw9 (DQw3)		Oriental	
HLA-DRw10		Dw new		DQw5 (DQw1)		Jewish	SHY
HLA-DR Bon		Dw new		DQw1		French	27-CH3

DRw8. These cross-reactivities may be based on common sequences between residues 26-31 and 67-74 of these DR beta-I molecules.

TCD-DR83. This represents the DR1 + DRw8 cross-reactive group.

TCD-DR84. TCD-DR84 is a supertypic specificity including DR1 + Dw25 + DW16.

Further details describing the cellularly defined HLA class II determinants are provided in "T Cell Recognition of HLA Class II Molecules" (this volume) and summarized in the section "Functional Polymorphisms of HLA Class II Gene Products Detected by T-lymphocyte Clones."

Table 2. T-Cell-Defined HLA-Dw Subtypes of HLA-DRw52

HLA-Dw	HLA-DRw52	Associations		
HLA-Dw24	DRw52a	DR3	Dw3	(B8 associated)
		DR3	Dw new	(Bw42 associated)
		DRw13	Dw18	
		DRw14	Dw16	
HLA-Dw25	DRw52b	DR3	Dw3 variant	(B18 associated)
		DRw11	Dw5	
		DRw11	DB2	
		DRw12	DB6	
		DRw13	Dw18	
		DRw14	Dw9	
HLA-Dw26	DRw52c	DRw13	Dw19	

Table 3. HLA-DR-Associated Specificities Recognized in the Tenth International Histocompatibility Workshop Cellular Studies

Cluster Designation	Specificity and/or Prototypic Cell Line
TCD-DRw15.1	Dw12 + subset of Dw2
TCD-DRw15.2	DRw15;Dw2 variant - DRβ also exhibits more acidic pattern - Cell line 9010 (AMAI)
TCD-DRw16.1	DRw16 subset - Associated with an additional DRβ product in cells of either Dw21 or Dw22 type
TCD-DRw16.2	DRw16 subset - Associated with an additional DRβ product in cells of Dw21 type only
TCD-DR2.1	DR2 variant - Cell line 9084 (CALOGERO)
TCD-DR2.2	DR2 variant - Cell line 9007 (DEM)
TCD-DRw18.0	Cellular equivalent of DRw18 Cell line 9021 (RSH)
TCD-DRw11.1	DR5(w11) variant - Cell line 9042 (TISI)
TCD-DRw11.2	DR5(w11) variant - Cell line 9039 (JVM)
TCD-DRw11.3	DB2 + Dw5 subset
TCD-DRw11.4	DR5 subset (TCD-DRw11.1-, TCD-DRw11.2-, TCD-DRw11.3-)
TCD-Dw11.1	Dw11 cells lacking DRB4 product
TCD-Dw11.2	Dw11 cells expressing DRB4 product
TCD-DRw8.1, TCD-DRw8.2, and TCD-DRw8.3	DR specificities associated with 3 subtypes of Dw8. Additional heterogeneity in the TCD-DRw8.1 group.
TCD-Dw25.1	Dw25 (DRw52b) variant - Cell line 9043 (BM21)
TCD-DRw53.1	DRw53-like epitope expressed on the DRB1 gene product of DR7 cells

Table 4. HLA-DQ-Associated Specificities Recognized in the Tenth International Histocompatibility Workshop Cellular Studies

Cluster Designation	Specificity and/or Prototypic Cell Line
TCD-DQw5	Similar to serologically defined DQw5 Associated with DQβ1.3 RFLP Common sequence (residues 58-78 of DQβ)
TCD-DQw5(Dw1+Dw20)	DQw5 subset limited to DR1 cells
TCD-DQw6(Dw2)	DQw6 subset limited to Dw2 cells
TCD-DQw6(Dw2+Dw12)	DQw6 subset limited to Dw2 and Dw12 cells
TCD-DQw6(Dw18)	DQw6 subset limited to Dw18 cells
TCD-DQw6(Dw19)	DQw6 subset limited to Dw19 cells Associated with DQβ1.6 RFLP
TCD-DQw6(Dw2+Dw18)	DQw6 subset limited to Dw2 and Dw18 cells Associated with DQβ1.5 RFLP Common sequence (residues 31-78 of DQβ)
TCD-DQw2(DR7)	DQw2 subset limited to DR7;DQw2 cells
TCD-DQA(DR7)	DQ specificity common to all DR7 cells irrespective of DQw2 or DQw3 typing
TCD-DQw7.1	DQw3 subset limited to cells of DR4;DQw7 type
TCD-DQw3.1	DQw3 subset including all cells but DR4;DQw7 homozygotes
TCD-DQw8.0	Equivalent to serologically defined DQw8 Dependent upon presence of DQβ-3.2 chain only
TCD-DQw8.1	DQw8 subset - expressed only in presence of both DQβ-3.2 and DQα-3.2 chains
TCD-DQw4.1	DQw4 subset common to DRw8 and DRw18 (lacking in Dw15)
TCD-DQw4.2	DQw4 subset limited to DRw8
TCD-DQw4.3	DQw4 subset limited to Dw8.1 and DRw18

Table 5. Supertypic Specificities Recognized in the Tenth International Histocompatibility Workshop Cellular Studies

Cluster Designation	Specificity and/or Prototypic Cell Line
TCD-DR81	Supertypic specificity associated with DR1, DR4, DRw6, ± DRw16
TCD-DR81.1	Supertypic specificity associated with DR1 + Dw14 + Dw16 Common sequence (38-86 of DRβ)
TCD-DR82	Supertypic specificity associated with DR2 (particularly DRw16), DR5, and DRw8
TCD-DR83	Supertypic specificity associated with DR1 and DRw8
TCD-DR84	DR1 + Dw25 + Dw16

Nomenclature for HLA-RFLP

Daniel Cohen,[1] Malcolm J. Simons,[2] Jean-Marc Lalouel,[3] and Bo Dupont[4]

The Southern Blot component of the Tenth Workshop focused on the study of 70 cell lines. Thirty-four different DR, DQ, Dw haplotypes emerged from serologic and cellular typings. The RFLP analysis demonstrated that these haplotypes could be clearly defined using the Southern blot technique.

HLA typing by RFLP is likely to be incorporated as a standard method for histocompatibility testing and has already been in use in many laboratories for some time.

The aim of this paper is to provide a nomenclature for HLA-RFLP typing for the purpose of communication between investigators. Although RFLP analysis by Southern blotting is a direct approach to gene detection, it is at best an indirect method of identifying gene sequence variation. Restriction sites do occur within exons, but the large majority of RFLP sites occur in non-exon regions. Thus, the relationship between exon sequence variation and RFLP is a function of linkage disequilibrium between intron and flanking region fragments bearing polymorphic restriction sites and exon segments with allelic variability.

Unlike serology, there are no "blank" alleles in Southern blot analysis. All fragments generated by a particular endonuclease can be expected to be present in the corresponding Southern blot lane. Furthermore, intensity of detectable fragments provides information on gene dosage.

RFLP-HLA typing is not based only on specific fragments (i.e., those correlating with a single DR or DQ specificity), but rather on combinations of fragments that constitute patterns specific for single haplotypes.

Recommendations

Any RFLP fragment will be reported by its system name and its Workshop molecular weight (the probe preceding the enzyme; the molecular weight in Kb, to two decimal points, and without the letters "Kb"): e.g., DRB/Taq I 7.00. It is convenient to refer to each pattern of fragments by its probable immunologically defined corresponding type, preceded by D- (for *D*NA). For example, the pattern that associates the five fragments, DRB/Taq I 9.79, 7.00, 4.12 and DQB/Taq I 4.62 and 0.87, is specific for the haplotype DRw17, DQw2, Dw3 (24). Therefore, this pattern will be referred to as D-DRw17, D-DQw2, D-Dw3 (24). The pattern which associates the fragments, DRB/Taq I 5.80, 4.40, 2.08, DQB/Taq I 5.34, 1.98, DQB/BamH I 5.76, 3.03 and DRB/Pst I 3.73, will be referred to as D-DR1, D-DQw5, D-Dw20.

A large number of HLA specificity-associated RFLP fragments were identified in the Workshop studies of the Reference Panel Cells. A detailed description of these results is presented in the section G (Southern Blot) of the Workshop Proceedings and summarized in the paper

Participating Laboratories: FRACOH,[1] US5SIM,[2] US7LAL,[3] US7DUP[4]

by Simons et al. (1). Application of RFLP for HLA class II typing is also presented (2).

Three summary tables are included in this nomenclature report and describe the "unique" HLA specificity-associated fragments identified in the Tenth Workshop analysis of the core cell lines. A detailed description of these data is provided in reference 1.

Table 1 lists the "unique" fragments associated with HLA class I specificities when genomic DNA is probed with the HLA class I probe.

Table 2 lists the "unique" HLA-DR fragments, and Table 3 lists the unique HLA-DQ fragments.

These tables serve only as a guide for data interpretation. It is clearly demonstrated in reference 3 that HLA-RFLP typing using a combination of unique fragments combined with a pattern analysis including other fragments can be designed by which the number of systems (probe × enzymes) can be reduced to three to four systems.

Summary

Reporting of HLA-RFLP data should consist of probe, restriction enzyme, and molecular weight in Kb, to two

Table 1. HLA-A, -B, -C Unique Fragments

HLA Specificity A	BAM	BGL	ECI	ECV	HID	KPN	MSP	PST	PVU	SST	TAQ
A1					(476)						
A2			[1179]	[1355]							
A3			287						(1482)		62
			1083								96
A23,A24		(602)			(498)	(562)				(592)	
A23,<A24						1270					
<A24				664							
A25,A26						512		430			
A11	825	1123	1299			1570					
A30			1381				262		1181	797	
A31	536				(707)			143	549		
A<31									(1619)		
HLA Specificity B											
B8	2356			(786)							
Bw41					1491						
Bw41,Bw42	250							(79)			
<B44	778	948								528	
Bw47								172			
<B51										278	
<Bw57		(747)									
Bw60					993						
Bw61,<Bw60					2644						
HLA Specificity C											
Cw4				1096							
Cw7								101			
<Cw7	1343										
C blank							427				

The fragments listed in Tables 1 to 3 were defined in the Tenth Workshop analysis of Reference Panel Cells. A few cells show fragment patterns intermediate between those of established specificities. Rather than identify the presence or absence in these atypical cells in conjunction with unique fragments, thereby complicating the table, these fragments are indicated using brackets: ( ) indicates one additional cell reaction; [ ] indicates one non-reactive cell.

Legend: HLA Class I fragments obtained with the pan-class I workshop DNA probe.

HLA Class II fragments obtained with DRB (RB); DQB (QB) and DQA (QA) DNA probes.

Table 2. HLA-DR/Dw Unique Fragments

HLA Specificity D	BAM	BGL	ECI	ECV	HIC	HID	KPN	MSP	PST	PVU	SST	TAQ
DR1, Dw1		(RB696)	RB1006	RB183		RB302	QB1775	(RB493)		RB657	RB1159	(RB580)
DR1, Dw20				RB630		RB343						
						RB610						
DR1, Dw1									RB142			
DR1, Dw20									RB373			
DR2			RB628			RB332				RB447	(RB1319)	
<DR2		RB1455										
DR15, Dw2								RB381			(RB615)	(RB142)
DR15, Dw12												(RB212)
DRw15, Dw2												(RB123)
DRw15, Dw12												
DRw16, Dw21												
DRw16, Dw21											RB1030	RB169
DRw16, Dw22												
DRw16, Dw22		(QA382)										
DRw17, Dw3				(QB1514)		QB1142	QA2520					
DRw17, Dw3/24											RB2396	
DRw17, Dw3/25						RB255						
<DRw11, Dw5									(RB538)			
DRw11, DB2												
DRw12						RB183				RB592		
DRw13, Dw18						(QA768)					RB1287	
											QA1553	
DRw13, Dw19	(QA2393)			RB1096		RB270	[QB1175]		RB632	RB1483	(RB1375)	
									QB643			
DRw14, Dw16							(QA1836)					
							(QB1544)					
DRw8												RB866

Table 2. *Continued*

HLA Specificity D	BAM	BGL	ECI	ECV	HIC	HID	KPN	MSP	PST	PVU	SST	TAQ
DR7								QA605	(RB215)	RB475		
DR7, DB1 DR7, Dw17			(RB1533)	(RB2047) (QB2001)		(QB1280)	RB943 (QB920)		RB490	(QB478)	QB540	
<DR7, DB1 DR7, Dw17												RB260 RB723
DR7, Dw11				(QB1643)		(RB966)						
<DR7, DB1 DR7, Dw11												QA2355
DRw9, Dw23						RB172			(QA912)			
DR4	RB526	RB1027		RB2298 (QB1452)	[RB1077]	(RB158)		RB368 (QA215)	RB205 RB355 RB411	RB565		RB522
DR4, DQw7					(RB411)							
DR4, DQw8									[QB396] QB684			
DRw52				RB1824		RB296						
DRw52* [DRw8]	[RB1133]	RB916	RB946	RB752 [RB1534]							[RB1125]	
DRw52, Dw24							RB272 RB769		RB107			
DRw52, Dw25				[RB2004]								[RB1148]
DRw52, Dw26	QA2393			RB1096		RB270	[QB1175]		RB632 [QB643]	RB1483	(RB1375)	
DRw53	RB428 RB463 RB659 QA763	RB564	RB373 RB398 RB410 RB1255 (RB1198) QA634	RB1648		RB144 RB313 RB717 QA483	QA1176	RB115 RB176	QA232	RB388 RB517 RB616 RB1350 [RB460]	RB519 RB1095 RB2112	RB284

* Signifies DR3, 5, 6, excluding DRw8

Table 3. HLA-DQ Unique Fragments

HLA Specificity D	BAM	BGL	ECI	ECV	HIC	HID	KPN	MSP	PST	PVU	SST	TAQ
DQw1			(QB383)			QB644	(QB515)					
<DQw1 #				QB673	(QB850)						(QB228)	QB534
DQw5	QB576			QB26 QB27			[QA1042]		QB136			QA245
DQw6		(QB310)										
<6.Dw2/18/19*			QB226 QB922	QB524								
<6.Dw2/18*	(QB289)								QB1068	QB217		(QB287) (QB261)
<6.Dw12/8.3*	QB872			QB1182	QB185 QB1536					QB202		QB450
DQw2	QB383									[QB63] (QB166)		[QB158] (QB87)
DQw2.DR3 +				(QB1514)		QB1142						
DQw2.DR7 +				(QB2001)	[QB766]	QB1280	QB920			[QB478]	[QB540]	(QB669)
DQw7	[QB355]					QB341	[QB120] [QB533]			QB372 [QB34]	[QB222] [QB286]	[QB100]
DQw8									[QB396] QB684			

#A subgroup of DQw1, designated DQw1.1, includes four fragments that are absent in cells 9011 and 9066. These two cells comprise the DQw6.3 subgroup (see below).

*There are unique RFLP fragments for three DQw6 subtypes. For convenience, they have been distinguished as 6.1, 6.2 and 6.3. No relation to any HLA nomenclature is implied.

+There are unique RFLP fragments for two DQw2 subtypes corresponding to the DR3 and the DR7 haplotypes.

decimal points (e.g., DRB/Taq I 7.00). If RFLP typing is applied as the assay system for designation of the corresponding serologically or cellularly defined HLA specificity, this should be indicated by D- (for *D*NA typing). For example, HLA-B8 identified by CL1/BamH I 23.56 should be named D-B8.

References

1. Simons MJ, Wheeler R, Cohen D, Lalouel J-M, Dupont B. Restriction fragment length polymorphism of HLA genes: summary of the Tenth International Workshop Southern Blot Analysis. In: Dupont B (ed). Immunobiology of HLA, Volume I: Histocompatibility Testing 1987. New York: Springer-Verlag, 1989.
2. Martell R, Marcadet A, Cohen D. HLA class II typing by the workshop-defined RFLP specificities in unrelated individuals. In: Dupont B (ed). Immunobiology of HLA, Volume I: Histocompatibility Testing 1987. New York: Springer-Verlag, 1989.

Nomenclature for Factors of the HLA System, 1987

W.F. Bodmer,[1] E. Albert,[2] J.G. Bodmer,[1] B. Dupont,[3] B. Mach,[4] W.R. Mayr,[5] T. Sasazuki,[6] G.M.Th. Schreuder,[7] A. Svejgaard,[8] and P.I. Terasaki[9]

The HLA Nomenclature Committee met after the Tenth International Histocompatibility Testing Workshop to consider revisions and additions to the nomenclature of specificities identified by serologic and cellular typing, and the naming of genes based on molecular techniques following the principles established in previous reports (1–7). Only nomenclature for HLA "class I" and "class II" products was considered. Special attention was given to the naming of genes and alleles in the HLA-D region.

Molecular data have now identified many of the genes and their arrangement in the HLA region. It was, therefore, decided that a number of these genes, including especially those coding for the new products known to be expressed in the HLA-D region, could be given official designations following the general principles described in a previous report (7). All genes in the HLA-D region are: 1) prefixed by the letter D; 2) followed by a letter P etc. for the subregion (defined by the position and similarity of genes within a sub-region); 3) followed by letters A or B for alpha or beta chains (or related sequences in the case of pseudogenes), and 4) a number when there is more than one alpha or beta chain gene (or pseudogene) in a subregion. Thus, for example, the expressed alpha chain gene in the DQ region, which is coding for part of the DQ product, is called HLA-DQA1, while the DP beta pseudogene in the DP region is called HLA-DPB2 and the three or four DR β chain loci found in most haplotypes will be called HLA-DRB1, B2, B3, and B4 (8,9). Two new subregions have been called DO (as originally defined by a single beta chain gene, called DOβ) and DN (defined by a single alpha chain gene formerly known as DZα or DOα). New genes in the class I region will simply be given single letter designations in sequence, omitting D, and prefixed by HLA. Thus, the gene defined by "clone 6.2" based on a 6.2 kilobase Hind III fragment detected by an HLA class I probe (clone pHLA-6.2) is called HLA-E (10). A complete listing of new gene names with their previous designations and molecular characterization is given in Table 1. The order of the named genes in the HLA-D region is given in Table 2 (11,12).

Nucleotide sequence data are revealing a wide variety of alleles with, in many cases, several alleles corresponding to a single serologic or cellular specificity. Some HLA-D region specificities, moreover, such as

Participating Laboratories: UKIBOD,[1] GERALB,[2] US7DUP,[3] University of Geneva, Geneva, Switzerland,[4] EAEMYR,[5] JAPSAS,[6] BENROO,[7] SCASVE,[8] US5TER[9]

Table 1. New Names for Genes in the HLA Region

Name	Previous Equivalents	Molecular Characteristics
HLA-E	E, "6.2"	Associated with class I 6.2 kB Hind III fragment (10)
HLA-DRA	DRα	DR α chain
HLA-DRB1	DRβI, DR1B	DR β1 chain determining specificities DR1, DR3, DR4, DR5 etc.
HLA-DRB2	DRβII, DR2B	Pseudogene with DRβ-like sequences
HLA-DRB3	DRβIII, DR3B	DR β3 chain determining DRw52 and Dw24, Dw25, Dw26 specificities (8)
HLA-DRB4	DRβIV, DR4B	DR β4 chain determining DRw53 (9)
HLA-DQA1	DQα1, DQ1A	DQ α chain as expressed
HLA-DQB1	DQβ1, DQ1B	DQ β chain as expressed
HLA-DQA2	DXα, DQ2A	DQ α chain related sequence, not known to be expressed
HLA-DQB2	DXβ, DQ2B	DQ β chain related sequence, not known to be expressed
HLA-DOB	DOβ	DO β chain
HLA-DNA	DZα,DOα	DN α chain
HLA-DPA1	DPα1, DP1A	DP α chain as expressed
HLA-DPB1	DPβ1, DP1B	DP β chain as expressed
HLA-DPA2	DPα2, DP2A	DP α chain related pseudogene
HLA-DPB2	DPβ2, DP2B	DP β chain related pseudogene

Immunobiology of HLA
Volume I

Table 2. Sequence of Named Genes in the HLA-D Region

	DP	DN	DO	DQ	DR
(centromere)	–+—+—+—+–	–+–	–+–	–+—+—+—+–	–+—+—+——+—+–
	B2 A2 B1 A1	A	B	B2 A2 B1 A1	B1 B2 B3 or B4 A

DQw1, correspond to clearly associated pairs of DQ α and β chain variants. The ultimate definition of an allele must be the nucleotide sequence of a gene, possibly including controlling regions and introns. Since in practice it is mainly the expressed product that is relevant for biological function, it was decided that, to start with, at least, an allele should be defined by the confirmed amino acid sequence of the protein product of a given gene. Each allele so defined will be given a unique number to be placed after an asterisk following the gene name, according to the convention adopted by the Human Gene Mapping Nomenclature Committee (13). Alleles with confirmed amino acid or nucleotide sequences will in general be numbered using four digits, according to the following principles. The first two digits describe the most closely associated serologic specificity and the other two complete the allele number. This makes it possible to retain as far as possible the relationship between alleles and serologic specificities. Thus, the formal designation of the allele corresponding to the B27-related specificity identified by one-dimensional isoelectric focusing, T-cell clones and a unique amino acid sequence is HLA-B*2701 (Table 3). However, following previous practice for the naming of HLA gene symbols, the alternative designation *B2701* using italics for the combination of gene and allele symbols without an asterisk, is also acceptable for alleles of the HLA class I region genes. The asterisk will, however, be necessary in the designation of HLA-D region alleles, for example HLA-DRB4*0101, to separate clearly the number identifying a gene from that identifying an allele. It was decided that a systematic hierarchical nomenclature for the HLA-D region genes took precedence over the difficulty that might arise in naming alleles without the use of the asterisk. This, moreover, allowed the HLA nomenclature to parallel as closely as possible, that being used for other human genes (13). Tables 3 and 4 give the new names for some HLA-A, B, and D region alleles.

It may become impossible in the future to retain the relationship between alleles and specificities, as the complexity of this relation increases. If, in the future, alleles with different nucleotide sequences but the same amino acid sequence are to be defined, then those could be given related numbers using extra digits, for example *B27011*. Haplotypes are designated as described in previous reports, with the option of using specificity designations as a shorthand for complete allele names.

Consideration was also given to the naming of specificities or products based on restriction fragment length polymorphism (RFLP) or biochemical techniques such as one- and two-dimensional isoelectric focusing electrophoresis. It was decided that while RFLP "clusters" could be very helpful in defining phenotypes and genotypes with respect both to known or possibly new specificities and alleles, they did not yet contribute a basis for a formal designation. This may, in the future, come from, for example, extended restriction maps for one or more related haplotyptes that correspond to well defined RFLP clusters. In the meantime, we recommend the adoption of the informal naming of clusters with the prefix 10W adopted by the Tenth Workshop (14).

Table 3. New Designations of HLA-A and B Alleles

New HLA Alleles	HLA Specificity	10W IEF[a] Variants	Previous Equivalents (ref.)	References for Sequence Data
A*0201	A2	A2.1	A2.2 (16)	(17)
A*0202	A2	A2.2	A2.1/A2.4 (16)	(17,18)
A*0203	A2	A2.3	A2.3? (16)	(17)
A*0204	A2	A2.4	A2.3? (16)	(17)
B*0701	B7	B7.1		(19)
B*0702	B7	B7.2		(20)
B*2701	B27	B27.1	B27f (21)	(24)
B*2702	B27	B27.2	B27e (21), 27K(22), B27.2 (23)	(25)
B*2703	B27	B27.3	B27d (21), 27J (22)	(26,27)
B*2704	B27	B27.4	B27b (21), 27C (22), B27.3 (23)	(28)
B*2705	B27	B27.5	B27a (21), 27W (22), B27.1 (23)	(25,29)
B*2706	B27	B27.6	B27D (22)	(30)

[a]One-dimensional isoelectric focusing electrophoresis performed in the Tenth Workshop.

Table 4. New Designations of HLA-D Region Alleles

New HLA Alleles	HLA-DR Specificities	HLA-D Associated (T-Cell-Defined) Specificities	References
DRB1*0401	DR4	Dw4	(31)
DRB1*0402	DR4	Dw10	(31)
DRB1*0403	DR4	Dw13	(32)
DRB1*0404	DR4	Dw14	(31, 32)
DRB1*0405	DR4	Dw15	(31)
DRB3*0101	DRw52a	Dw24	(8)
DRB3*0201	DRw52b	Dw25	(8)
DRB3*0301	DRw52c	Dw26	(8)
DRB4*0101	DRw53	Dw4, Dw10, Dw13, Dw14, Dw15, Dw17, Dw23	(33,34)

Biochemical techniques have been used, whenever possible, to support the designation of new specificities. These techniques clearly give definitions of products that may be more closely related to the amino acid sequence than either serologic or cellular typing. The latter are, however, well established, and serology in particular is still by far the most widely used technique for HLA typing. It was decided, therefore, that while it was justified to continue the naming of serologic and cellular specificities as before, biochemically defined variants should not be given separate designations. Many amino acid sequences are, or will be soon, available for the definition of alleles, and these will subsume the presently known biochemical variants, making it unnecessary and perhaps confusing to introduce a new intermediate nomenclature. Formal procedures are being established for the criteria for the acceptance of an amino acid sequence to be named as an allele, and the assignment of an allele number.

The criteria for acceptance of a confirmed sequence include depositing in publicly accessible repositories or data bases:

1. the clone(s) from which the sequence was derived;
2. the cell source(s) from which the clone(s) was(were) derived; and
3. the nucleotide sequence or, if not available, the amino acid sequence.

Provisional designations, indicated by a "w" before the allele number, may be given to partial sequences; for example, of the majority of an N-terminal domain of a DR beta chain.

The use of numbers or names for alleles, genes, or specificities which preempt formal designations such as DRB1*0102, DM or HLA-F before considerations by the Nomenclature Committee is strongly discouraged.

HLA-A, B, and C Specificities

As before, the HLA-A and B specificities are numbered jointly with nonoverlapping numbers. New provisional designations for HLA-A, B and C specificities are listed in Table 5. The only new HLA-A specificity is Aw74, formerly known as Th, a split of Aw19 found mainly in Black populations and clearly defined by sera in the Aw19 Antigen Society.

Three new HLA-B locus specificities were identified: HLA-Bw75, Bw76, and Bw77. These are splits of B15. Bw75, formerly Bw62.1, SH7, or TS1, is a short version of Bw62 in both Caucasoids and Orientals, showing cross-reactivity with B35 sera. Bw76, formerly 15S, is mainly found in the Thai population characterized by cross-reactivity with B45 sera. Both Bw75 and Bw76 are Bw6-associated. HLA-Bw77, formerly B15T, is a Bw4-associated split of B15 and is distinguished from Bw63 by reactivity to Bw53 defining sera. Bw77 is

Table 5. New Provisional Designations of HLA-A, -B and -C Specificities

New	Previous Equivalents	Associated With
Aw74	Th	-
Bw75	Bw62.1 (15), SH7, ST1	Bw6
Bw76	15S, 15.4	Bw6
Bw77	15T, 15.2	Bw4
Cw9	Cw3.1	Bw55
Cw10	Cw3.2	Bw60
Cw11	Cx46, Cw1+3, Cw1x3, C-Bangkok, CSH1	Bw46

Table 6. New Provisional HLA-DR and DQ Specificities

New[a]	Previous Designations	Associated With	Biochemical Variant
DRw15	DR2 long		
DRw16	DR2 short, FT31		
DRw17	DR3 long		
DRw18	DR3 short	Bw42	
DQw4	DQ "Wa"	DRw8, Dw15	DQβ "Wa"
DQw5	DQ "1.1"	DR1, DRw10, DRw14, DRw16	DQβ "1.1"
DQw6	DQ "1.2"	DRw15, DRw13	DQβ "1.2", "1.12", "1.18"
DQw7	DQ"3.1", TA10+, IIB3-	DRw11, DRw12, DR4	DQβ "3.1"
DQw8	DQ"3.2", TA10-, IIB3+	DR4	DQβ "3.2"
DQw9	DQ"3.3", TA10-, IIB3+	DR7, DR9	DQβ "3.3"

[a]T-cell clones identify: DQw5(DR1) and DQw6(DRw13,DRw15); subsets of DRw13(Dw18), DRw13(Dw19), and DRw14(Dw9); and the DQw3 variants DQw7(DR4,"3.1") and DQw8(DR4,"3.2").

found in both Thai and other southeast Asian populations.

Three new HLA-C locus specificities were identified: HLA-Cw9, Cw10, and Cw11. HLA-Cw9 and Cw10 are splits of Cw3. HLA-Cw9, formerly known as Cw3.1, is in linkage disequilibrium with Bw55, and Cw10, formerly known as Cw3.2, is in linkage disequilibrium with Bw60. Bw62 is found in association with both Cw9 and Cw10. HLA-Cw11 is the antigen associated with Bw46, formerly known as Cx46 Thai, Cw1x3, Cw1+3, or C Bangkok. Sequence data have shown that this antigen combines parts of the Cw1 and Cw3 sequences to form a new specificity. Parts of this sequence are also found in the associated Bw46 sequence (15).

HLA-D Region Specificities

No new provisional designations were assigned to the HLA-DP locus because few antibodies were available for this identification and no PLT testing was carried out in the Workshop. New provisional designations were assigned to four HLA-DR specificities, six HLA-DQ specificities (Table 6), and seven HLA-Dw specificities (Table 7).

HLA-DR Specificities

Of the HLA-DR specificities, one antigen HLA-DRw9, was upgraded to HLA-DR9, dropping the "w" designation according to previous criteria. The new provisional HLA-DR specificities are HLA-DRw15, DRw16, DRw17, and DRw18. HLA-DRw15 and DRw16 are splits of HLA-DR3. HLA-DRw15, formerly known as DR2 long or DR2.1, is associated with the newly designated DQ antigen DQw6 (DQ1.2). DRw16, formerly known as DR2 short, is associated with DQw5 (DQ1.1). It is distinguished from DRw15 by lack of reactivity to long DR2 sera. HLA-DRw17 and DRw18 are splits of DR3. DRw17 includes most of the originally defined DR3 specificity usually associated with DQw2. DRw18 is a short DR3 specificity found in Black populations, and associated with Bw42 and DQw4 ("Wa"). This specificity is also defined by T-cell clones, but is not assigned a Dw specificity since serologic reagents exist.

HLA-DQ Specificities

Six new provisional DQ designations were assigned. It was noticeable that most of these specificities were identifiable by monoclonal antibodies that made the definition of antigens very clear. DQw4, formerly Wa, appears to be allelic to DQw1, 2, and 3, and in positive linkage disequilibrium with DRw8, DR4(Dw15), and DRw18. It is identified clearly both by specific monoclonal antibodies and antibodies also reacting with DQw1 and/or DQw8. The DQw4 β chain shows a characteristic electrophoretic pattern. DQw5 and DQw6 are splits of DQw1 identified by monoclonal antibodies

Table 7. New Provisional HLA-D (T-Cell-Defined) Specificities

New	Previous Designations	Associated With
Dw20	LD14, LVA	DR1, DQw5, B14
Dw21	LD2s, MN2, FJO, AZH	DRw16, DQw5
Dw22	DB9, LD5a	DRw16, DQw7
Dw23	DB5	DR9, DQw9
Dw24	52a, LB-Q4, ßIII-3	B8,DR3; DRw13,Dw18; DRw14,Dw16
Dw25	52b, LB-Q1, ßIII-1, BO-1	B18,DR3; DRw11,Dw5; DRw12,DB6; DRw13,Dw18; DRw14,Dw9
Dw26	52c, ßIII-2	DRw13,Dw19

Table 8. HLA-D (T-Cell-Defined) and DR Relationships

HLA-D Specificities	Associated DR Specificities
Dw1, Dw20	DR1
Dw2, Dw12	DRw15 (2)
Dw21, Dw22	DRw16 (2)
Dw3	DR3
Dw4, Dw10, Dw13, Dw14, Dw15	DR4
Dw5	DRw11 (5)
Dw6, Dw18, Dw19	DRw13 (w6)
Dw9, Dw16	DRw14 (w6)
Dw7, Dw11, Dw17	DR7
Dw8	DRw8
Dw23	DR9
Dw24, Dw25, Dw26	DRw52

and antisera and are characterized in part by different electrophoretic DQβ chain patterns. DQw5, formerly DQ"1.1," is associated with DR1, DRw16 (DR2 short), DRw10, and DRw14. A characteristic electrophoretic pattern of the DQβ chain is seen in DR1, DQw5 cells. DQw6, formerly DQ"1.2," the other split of DQw1, is in positive linkage disequilibrium with DRw15 (DR2 long) and DRw13. A characteristic electrophoretic pattern is seen for the DQβ chain associated with the three haplotypes DRw15, Dw2; DRw15, Dw12; and DRw13, Dw18.

Three splits of DQw3 have been identified: DQw7, DQw8, and DQw9. DQw7, formerly TA10, "3.1," is identified by specific monoclonal antibodies being TA10-positive and negative with the monoclonal antibody IIB3. It is associated with DRw11, DRw12, and a subset of DR4. DQw7 shows a characteristic DQβ pattern "3.1" in 2D gel electrophoresis. DQw8, formerly "3.2," is another split of DQw3, being TA10-negative and positive to IIB3. It is in linkage disequilibrium with DR4. The DQβ chain has a characteristic 2D electrophoretic pattern "3.2." DQw9 is the third split of DQw3, characterized also as being positive to IIB3 and negative to TA10. DQw9 is in linkage disequilibrium with DR9 and those DR7 (Dw11), which are not DQw2 associated. DQw9 has a characteristic electrophoretic DQβ pattern, "3.3."

HLA-D (T-Cell-Defined) Specificities

Cellular typing reagents are capable of identifying HLA class II specificities which may not be detectable using alloantisera or for which appropriate serologic reagents have not yet been identified. HLA-Dw specificities have traditionally been defined with HLA-D homozygous typing cells (HTCs) in primary in vitro mixed lymphocyte cultures (MLCs). HLA-DP and other class II specificities (e.g., LB-Q1 and LB-Q4) have been defined using secondary MLCs. In recent years, it has become clear that cloned populations of T lymphocytes can be generated which recognize allospecificities similar or identical to those detected by primary or secondary MLC typing. Furthermore, T-cell clones may be identified whose pattern of reactivity corresponds to the presence of a unique amino acid sequence in the class II molecule. It was, therefore, decided to extend the definition of HLA-Dw specificities to include T-cell-defined determinants using either bulk culture or clonal cellular reagents including either cytotoxic (CTLc) or proliferative (PTLc) T-cell clones. New HLA-Dw specificities were defined when more than one independently derived T-cell reagent identified the particular HLA class II specificity. New HLA-Dw specificities were assigned when equivalent serologic reagents were not available (e.g., Dw24, Dw25, Dw26). Conversely new HLA-Dw specificities were not assigned when equivalent serologic reagents were identified (e.g., DRw18). Monoclonal antibody blocking of T-lymphocyte clones was used to discriminate between HLA-DP specificities and the other class II T-cell-defined determinants (Dw).

New provisional designations for HLA-Dw specificities are listed in Table 7. Dw20 (LD14) Dw21 (previously LD2sh, "AZH", "FJO", "MN2"), Dw22 (previously DB9 or LD5a; found in Amerindians), and Dw23 (previously DB5; found in Caucasoids, but more commonly in Orientals) are T-cell-defined specificities associated with the DRB1 gene. Dw20 is associated with the B14, DR1 haplotype. Dw21 and Dw22 are associated with DRw16 (DR2 short), and Dw23 is associated with DR9.

New provisional designations for HLA-Dw specificities were given to the alleles of the DRB3 (DRw52) gene. HLA-Dw24 (previously 52a, LB-Q4, βIII-3), Dw25 (previously 52b, LB-Q1, βIII-1, BO-1), and Dw26 (previously 52c, βIII-2) were clearly identified by two or more PTLc or CTLc. Dw24 is associated with the haplotypes B8,DR3,Dw3; DRw13,Dw18; and DRw14,Dw16. Dw25 is associated with B18,DR3,Dw3; DRw11,DW5; DRw12,DB6; DRw13,Dw18; and DRw14,Dw9. Dw26 is associated with DRw13,Dw19. The DRw13,Dw18 haplotype was thus subdivided according to the DRw52 variants: DRw13,Dw18,Dw24 and DRw13,Dw18,Dw25.

The relationship between T-cell defined Dw specificities and DR alleles is depicted in Table 8. A complete listing of the recognized HLA specificities is given in Table 9. Further background to the basis for the decisions reported here can be found in Histocompatibility Testing 1987 (14).

Acknowledgments

The committee gratefully acknowledges the financial support for its meeting in New York, November 21-23, 1987, given by the Biotest Company and the contributions of Dr. Daniel Cohen, Dr. Neal Flomenberg, Dr. John A. Hansen, Dr. Robert W. Knowles, Dr. Jean-Marc Lalouel, Dr. Edgar L. Milford, Dr. Thomas Shows, Dr. Soo Young Yang, and Dr. Edmond J. Yunis, who acted as co-opted members of the committee that produced this report.

Table 9. Complete Listing of Recognized HLA Specificities

A	B	C	D	DR	DQ	DP
A1	B5	Cw1	Dw1	DR1	DQw1	DPw1
A2	B7	Cw2	Dw2	DR2	DQw2	DPw2
A3	B8	Cw3	Dw3	DR3	DQw3	DPw3
A9	B12	Cw4	Dw4	DR4	DQw4	DPw4
A10	B13	Cw5	Dw5	DR5	DQw5(w1)	DPw5
A11	B14	Cw6	Dw6	DRw6	DQw6(w1)	DPw6
Aw19	B15	Cw7	Dw7	DR7	DQw7(w3)	
A23(9)	B16	Cw8	Dw8	DRw8	DQw8(w3)	
A24(9)	B17	Cw9(w3)	Dw9	DR9	DQw9(w3)	
A25(10)	B18	Cw10(w3)	Dw10	DRw10		
A26(10)	B21	Cw11	Dw11(w7)	DRw11(5)		
A28	Bw22		Dw12	DRw12(5)		
A29(w19)	B27		Dw13	DRw13(w6)		
A30(w19)	B35		Dw14	DRw14(w6)		
A31(w19)	B37		Dw15	DRw15(2)		
A32(w19)	B38(16)		Dw16	DRw16(2)		
Aw33(w19)	B39(16)		Dw17(w7)	DRw17(3)		
Aw34(10)	B40		Dw18(w6)	DRw18(3)		
Aw36	Bw41		Dw19(w6)			
Aw43	Bw42		Dw20	DRw52		
Aw66(10)	B44(12)		Dw21			
Aw68(28)	B45(12)		Dw22	DRw53		
Aw69(28)	Bw46		Dw23			
Aw74(w19)	Bw47		Dw24			
	Bw48		Dw25			
	B49(21)		Dw26			
	Bw50(21)					
	B51(5)					
	Bw52(5)					
	Bw53					
	Bw54(w22)					
	Bw55(w22)					
	Bw56(w22)					
	Bw57(17)					
	Bw58(17)					
	Bw59					
	Bw60(40)					
	Bw61(40)					
	Bw62(15)					
	Bw63(15)					
	Bw64(14)					
	Bw65(14)					
	Bw67					
	Bw70					
	Bw71(w70)					
	Bw72(w70)					
	Bw73					
	Bw75(15)					
	Bw76(15)					
	Bw77(15)					
	Bw4					
	Bw6					

The listings of broad specificities in parenthesis after a narrow specificity, e.g. HLA-A23(9) is optional. The following is a listing of these specificities which arose as clear-cut splits of other specificities.

Original Broad Specificities	Splits
A9	A23,A24
A10	A25,A26,Aw34,Aw66
Aw19	A29,A30,A31,A32,Aw33,Aw74
A28	Aw68,Aw69
B5	B51,Bw52
B12	B44,B45
B14	Bw64,Bw65
B15	Bw62,Bw63,Bw75,Bw76,Bw77
B16	B38,B39
B17	Bw57,Bw58
B21	B49,Bw50
Bw22	Bw54,Bw55,Bw56
B40	Bw60,Bw61
Bw70	Bw71,Bw72
Cw3	Cw9,Cw10
DR2	DRw15,DRw16
DR3	DRw17,DRw18
DR5	DRw11,DRw12
DRw6	DRw13,DRw14
DQw1	DQw5,DQw6
DQw3	DQw7,DQw8,DQw9
Dw6	Dw18,Dw19
Dw7	Dw11,Dw17

continued on next page

Table 9. *Continued*

The following specificities are generally agreed inclusions of HLA-B specificities Bw4 and Bw6.

Bw4:	B5,B13,B17,B27,B37;B38(16),B44(12),Bw47,B49(21),B51(5),Bw52(5),Bw53,Bw57(17),Bw58(17),Bw59,Bw63(15), Bw77(15)
Bw6:	B7,B8,B14,B18,Bw22,B35,B39(16),B40,Bw41,Bw42,B45(12),Bw46,Bw48,Bw50(21),Bw54(w22),Bw55(w22), Bw56(w22),Bw60(40),Bw61(40),Bw62(15),Bw64(14),Bw65(14),Bw67,Bw70,Bw71(w70),Bw72(w70),Bw73, Bw75(15),Bw76(15)

The following specificities are generally agreed to be associated with DRw52 and DRw53:

DRw52:	DR3,DR5,DRw6,DRw8,DRw11(5),DRw12(5),DRw13(w6),DRw14(w6),DRw17(3),DRw18(3)
DRw53:	DR4,DR7,DR9

References

1. Bulletin of the World Health Organization, 39: 483 (1968).
2. Terasaki PI, ed., Histocompatibility Testing, 1970, Copenhagen, Munksgaard, 1970, p 49.
3. Bulletin of the World Health Organization, 47: 659 (1972).
4. Bulletin of the World Health Organization, 52: 261 (1975).
5. Bulletin of the World Health Organization, 56: 461 (1978).
6. Terasaki PI, ed., Histocompatibility Testing, 1980. Los Angeles, UCLA Tissue Typing Laboratory, 1980, pp 18–20.
7. Albert ED, et al. (eds). Histocompatibility Testing 1984. Heidelberg: Springer-Verlag, 1984.
8. Rollini P, Mach B, Gorski J. Linkage map of three HLA-DR β-chain genes: evidence for a recent duplication event. Proc Natl Acad Sci USA 1985; 82:7197–7201.
9. Gorski J, Rollini P, Mach B. Structural comparison of the genes of two HLA-DR supertypic groups: the loci encoding DRw52 and DRw53 are not truly allelic. Immunogenetics 1987; 25:397–402.
10. Koller BH, Geraghty D, Orr HT, Shimizu Y, DeMars R. Organization of the human class I major histocompatibility complex genes. Immunological Research 1987; 6:1–10.
11. Dunham I, Sargent CA, Trowsdale J, Campbell RD. Molecular mapping of the human major histocompatibility complex by pulsed-field gel electrophoresis. Proc Natl Acad Sci USA 1987; 84:7237–7241.
12. Carroll MC, Katzman P, Alicot EM, et al. Linkage map of the human major histocompatibility complex including the tumor necrosis factor genes. Proc Natl Acad Sci USA 1987; 84:8535–8539.
13. Shows TB, Alper CA, Bootsma D, et al. International system for human gene nomenclature (1979). Cytogenet Cell Genet 1979; 25:96–116.
14. Dupont B (ed). Immunobiology of HLA, Volume I: Histocompatibility Testing 1987. New York: Springer-Verlag, 1989.
15. Parham P, et al. In: Dupont B (ed). Immunobiology of HLA, Volume I: Histocompatibility Testing 1987. New York: Springer-Verlag, 1989.
16. Brenner MB, McLean J, Yang SY, van der Poel JJ, Pious D, Strominger JL. Clonal T lymphocyte recognition of the fine structure of the HLA-A2 molecule. J Immunol 1985; 135:384–390.
17. Mattson DH, Handy DE, Bradley DA, Coligan JE, Cowan EP, Biddison WE. DNA sequences of the genes that encode the CTL-defined HLA-A2 variants M7 and DK1. Immunogenetics 1987; 26:190–192.
18. Ezquerra A, Domenech N, van der Poel JJ, Strominger JL, Vega MA, Lopez de Castro JA. Molecular analysis of an HLA-A2 functional variant, CLA, defined by cytotoxic T lymphocytes. J Immunol 1986; 137:1642–1649.
19. Taketani S, Krangel MS, Spits H, de Vries J, Strominger JL. Structural analysis of an HLA-B7 antigen variant detected by cytotoxic T lymphocytes. J Immunol 1984;133:816–821.
20. Biro PA, Pan J, Sood AK, Kole R, Reddy VB, Weissman SM. The human major histocompatibility complex. Cold Springs Harbor Symposium on Quantitative Biology 1983; 47:1079–1086.
21. Choo SY, Antonelli P, Nisperos B, Nepom GT, Hansen JA. Six variants of HLA-B27 identified by isoelectric focusing. Immunogenetics 1986; 23:24–29.
22. Breur-Vriesendorp BS, Neefjes JC, Huis B, van Seventer GA, Ploegh HL, Ivanyi P. Identification of new B27 subtypes (B27C and B27D) prevalent in Oriental populations. Hum Immunol 1986; 16:163–168. (B27J: personal communication).
23. Toubert A, Gromard E, Grumet FC, Amor B, Muller J-Y, Levy J-P. Identification of several functional subgroups of HLA-B27 by restriction of the activity of antiviral T killer lymphocyte. Immunogenetics 1984; 20:513–525.
24. Rojo S, Aparicio P, Choo SY, Hansen JA, Lopez de Castro JA. Structural analysis of an HLA-B27 population variant, B27f. Multiple patterns of amino acid changes within a single polypeptide segment generate polymorphism in HLAB27. J Immunol 1987; 139:831–836.
25. Seemann GHA, Rein RS, Brown CS, Ploegh HL. Gene conversion-like mechanisms may generate polymorphism in human class I genes. EMBO J 1986; 5:547–552.
26. Rojo S, Aparicio P, Hansen JA, Choo SY, Lopez de Castro JA. Structural analysis of an HLA-B27 functional variant, B27d, detected in American Blacks. J Immunol 1987; 139:3396–3401.
27. Choo SY, St. John T, Orr HT, Hansen JA. Molecular analysis of the variant alloantigen HLA-B27d identifies a unique single amino acid substitution. Hum Immunol 1988;21: 209–219.
28. Vega M, Wallace L, Rojo S, Bragado R, Aparicio P, Lopez de Castro JA. Delineation of functional sites in HLA-B27 antigens. Molecular analysis of HLA-B27 variant Wewak I defined by cytotoxic T lymphocytes. J Immunol 1985; 135:3323–3332.
29. Weiss EH, Kuon W, Dorner C, Lang M, Riethmuller G. Organization, sequence, and expression of the HLA-B27 gene: A molecular approach to analyze HLA and disease associations. Immunobiology 1985; 170:367–380.

30. Vega MA, Bragado R, Ivanyi P, Pelaez JL, Lopez de Castro JA. Molecular analysis of a functional subtype of HLA-B27. A possible evolutionary pathway for HLA-B27 polymorphism. J Immunol 1986; 137:3557–3565.
31. Gregersen PK, Shen M, Song Q, et al. Molecular diversity of HLA-DR4 haplotypes. Proc Natl Acad Sci USA 1986; 83:2642–2646.
32. Cairns JS, Curtsinger JM, Dahl CA, Freeman S, Alter BJ, Bach F. Sequence polymorphism of HLA DRβ1 chain alleles relating to T-cell-recognized determinants. Nature 1985; 317:166–168.
33. Gregersen PK, Moriuchi T, Karr RW, et al. Polymorphism of HLA-DR β chains in DR4, -7, and -9 haplotypes: implications for the mechanisms of allelic variation. Proc Natl Acad Sci USA 1986; 83:9149–9153.
34. Young JAT, Wilkinson D, Bodmer WF, Trowsdale J. Sequence and evolution of HLA-DR7- and -DRw53-associated β-chain genes. Proc Natl Acad Sci USA 1987;84:4929–4933.

Joint Reports: Data Collection and Central Analysis

Tenth International Histocompatibility Workshop: Overview of Data Processing and Analysis

Jean-Marc Lalouel,[1] Mark Ferguson,[1] Robert Wheeler,[1] and Nancy King[1]

International Histocompatibility Workshops generate a vast amount of data at multiple locations. These data must be assembled into computer files, corrected or updated, merged into common files, and submitted to analysis.

As in past workshops, the Tenth Workshop included a series of serologic experiments, where a battery of Workshop sera were used to characterize HLA antigens in a large number of cells. Moreover, detailed investigations of subsets of antigens were carried out by Antigen Societies, each including several laboratories. Each of 33 such societies generated data on distinct target cells with society-specific sets of sera.

The Tenth Workshop also had as its focus the thorough characterization of a panel of lymphoblastoid cell lines. The experiments performed on these cells included the following components: 1) serologic analysis; 2) analysis of class I products by isoelectric focusing (Biochemistry I); 3) analysis of class II products by two-dimensional gel electrophoresis (Biochemistry II); 4) functional assays with T-cell clones; 5) Southern blot analysis with a battery of probes and restriction enzymes; and 6) sequencing of selected class II cDNAs.

The multiplicity of the participating laboratories and experimental components was a singular challenge. For data collection, verification, and analysis to be at all possible with the necessarily limited team available centrally for this project, a new conceptual approach was required. The overall strategy for data handling and analysis called for a greater role of the participating laboratories in data entry, quality control, and analysis.

To implement this distributed, rather than central, responsibility in data processing, the following protocol was implemented. The personal computer (IBM PC or compatible) was chosen as a de facto standard for participating laboratories. For each component of the Workshop, specialized software designed and implemented centrally in the Pascal computer language was distributed to all participating laboratories. This software allowed entry, review, and update of the data in the laboratory. It was designed to follow as much as possible the natural layout of each experiment, and numerous checks were provided to control the quality of the data. Each laboratory mailed computer files on floppy disks to the central laboratory. All data pertaining to a common component were assembled into a single set of files. Whenever appropriate, data were subdivided into subcomponents that were sent back to designated leaders; thus, data for a large number of Antigen Groups and Antigen Societies were assembled and mailed back to various laboratories. Lastly, analytical programs were developed centrally and mailed to all laboratories involved in analysis. In designing these analytical tools, the guiding principle was to make available for operation on personal computers the very same programs developed for central analysis. This required that all computer programs implemented be operational on small personal computers, as well as our VAX 8650 superminicomputer.

Each component generated a set of data on the Reference Workshop Cell Lines. A unique data reduction converted all types of data to a common format. A particular program was developed to combine data across components. Once this was performed, the same analytical tools used for analysis within a component were used to analyze such combined data.

Computer programs were implemented for entry of the following types of data: 1) serology, 2) biochemistry class I, 3) T-cell clones, and 4) Southern blot experiments. Because the analysis of two-dimensional (2D) gels for biochemistry class II required scanning and spot analysis with specialized computer equipment, this component was not handled centrally. Rather, the data we received from Dr. Robert Knowles were already reduced to a format compatible for analysis across components. With regard to the sequencing data, we only provided assistance to a group of investigators led by Dr. Janet Lee for assembly into GenBank format, and provided remote access to our Intelligenetics software. A brief description of the methods implemented for processing each type of data is provided in relevant sections of this volume. In this overview, we will only summarize the overall architecture of the analytical tools developed. Further descriptions of these programs of analysis are provided in separate sections.

After examination of past Workshop reports, we chose to implement simple proven techniques of data manipulation. The emphasis was more on the examination and display of critical features of the data than on rigorous statistical tests of hypothesis. Indeed, the multitude of the items scored, often on a limited number of subjects, defies traditional significance tests. We have essentially attempted to implement into computer programs the methods that experimenters have traditionally used to examine their data. Implementation of these steps into computer programs should alleviate the laborious task of examining data structures without losing the immediate significance conveyed by the raw data.

The analytical package, called SERAN, consists of a suite of computer programs implemented in Pascal, access to each function being provided by a series of command files driven through a general menu. Figure 1 reproduces the menu as it appears on the screen. These programs operate on three input files named by default

Participating Laboratory: US7LAL[1]

Immunobiology of HLA
Volume I

```
TENTH INTERNATIONAL HISTOCOMPATIBILITY WORKSHOP

SEROLOGY DATA ANALYSIS SYSTEM

(C) 1987 Howard Hughes Medical Institute for the
Tenth International Histocompatibility Workshop

1 - SELECT          7 - TYPING          13 - CLUS
2 - EXTRACT         8 - TYPE FAMILY     14 - SXS
3 - TAIL            9 - VIEWTYPE        15 - CXS
4 - QUAL           10 - BLDCELL         16 - CELL
5 - VIEWQUAL       11 - PATTERN         17 - CORECELL
6 - PREPTYPE       12 - CORR

Enter Selection (hit return to exit): █
```

Figure 2. Selection menu of EXTRACT for analysis across components.

cellfile.dat, serafile.dat, and scorefile.dat. Scorefile.dat is a two-dimensional array of experimental reactions obtained on individual target cells (rows) for a battery of items (columns); the latter can be sera, DNA fragments, T-cell clones, etc. Cellfile.dat defined the row entries, while serafile.dat defines the column entries. The analytical functions provided by the procedures fall into three classes: assembly of a selected set of data (1 and 2); serum analysis and cell typing (3 to 11); display of data and correlation structures (12 to 17). Because the data from each component are transformed to a structure equivalent to that of serologic reactions, the analytical tools can be described as applied to serologic data. All the procedures are interactive, providing flexibility of use through a variety of parameter settings and selection criteria.

A subset of the data included in these three files can be selected for analysis through the procedure SELECT (1). This procedure allows the user to select class I or class II data, specific cells, and/or specific sera according to serologic specificities or identification numbers, laboratory codes, ethnic origin, cell types, or locus. The procedure EXTRACT (2) is used to combine data across components, as illustrated in Figure 2. Thus, data on the core cell lines can be assembled to include the results of the serology, Southern blot, biochemistry class II, and T-cell clones experiments, allowing the user to correlate experimental data across various components.

Procedures (3) to (11) are used for serologic analysis. They are described in greater details in a separate section. Note that such analysis can also be applied to nonserologic data (e.g., the results of the Southern blot experiments can be used, yielding correlations between DNA fragments and serologic antigens or cell typing based on the presence/absence of such fragments). TAIL (3) performs sequential tail analysis in a manner similar to the methods used in previous Workshops. For each sera, it yields the list of antigens in the recognition set of a serum, together with the corresponding statistics of association. QUAL (4) performs quality analysis on the serum, and VIEWQUAL (5) is used for interactive examination and search of the serum quality tables. PREPTYPE (6) initializes a working file used by the cell typing procedures. The command TYPING (7) controls the procedure CELLTYPE, which allows to type cells with various user-selected controls. The procedure VIEWTYPE (9) is used to review cell typing, with multiple criteria to restrict the examination to subset of the cells; it can be used to reassign antigens to cells. Various cycles of typing can be applied, using alternately (7) and (9). Once completed, the antigens assigned to the cells can be revised with the procedure BLDCELL (10), which constructs a new file of cell definition, and a new cycle of tail analysis and cell typing (3 to 10) can be applied. At any stage of the typing process, the procedure PATTERN (11) can be used to tabulate patterns of cell reactions for selected serum contrasts, as was done in past workshops.

The procedures (12) to (15) are used to examine the correlation structure of cells and sera on the basis of the experimental reactions generated in the laboratory.

```
Cell and Component Extraction Utility V2.0

Select the components that you want extracted

Serology                              [No] ? y
Use selection criteria                [No] ? n
Elisa                                 [No] ? y
Use selection criteria                [No] ? n
Biochem                               [No] ? y
Use selection criteria                [No] ? n
Southern Core                         [No] ? y
Use selection criteria                [No] ? n
Southern Family                       [No] ? y
Use selection criteria                [No] ? n
T-Cell Clone Proliferation            [No] ? y
Use selection criteria                [No] ? n
T-Cell Clone Cytotoxicity             [No] ? y
Use selection criteria                [No] ? n
```

Figure 1. Menu offered by the SERAN system of analysis developed for the Tenth Workshop.

CELL (16) is used for direct review of the cells, as defined in the file cellfile.dat. Correlations between sera are constructed by the procedure CORR (12). The resulting correlation matrix can be submitted to cluster analysis (CLUS, 13). The output of this procedure, after traversal of the tree diagram generated by cluster analysis, is a file used by subsequent display procedures (14 and 15) to present sera in a sequence reflecting the similarity of their patterns of cell reactions. SXS (14) is used to display the matrix of correlations between all pairs of sera after permutations to display clustering. CXS (15) displays individuals cell reactions on the sera, the latter being ordered to display clusters. On personal computers, the cells are sorted according to a list of antigens selected by the user. On larger computers, clustering can be applied to the cells themselves. SXS and CXS operate as specialized screen editors which can be used by the user to inspect the data, to move or to delete blocks of data to highlight important features of the correlational structures, and to produce printouts of part or all of the data.

The burden of processing and analyzing the vast amount of data generated by the Workshop did not allow us to implement cell typing in families. The procedures outlined in this report met significant needs of the Workshop. They will be maintained and made available to investigators on request. We acknowledge gladly the contribution of all members of the Workshop Organizing Committee, particularly Bo Dupont, Edgar Milford, Neal Flomenberg, Soo Young Yang, and Daniel Cohen, as well as the tremendous help generously extended by Malcolm Simons in the central analysis.

Display of Correlation Structures

Nancy King,[1] Robert Wheeler,[1] Mark Ferguson,[1] and Jean-Marc Lalouel[1]

When a panel of individuals is tested for a battery of items, questions arise regarding the mutual dependencies between individuals based on their scores on these items, as well as between items as a function of individual scores. Are there identifiable correlation structures that describe the relationship between individuals, or between items?

The necessary experimental data consist of a two-dimensional array of experimental scores, whether discrete or continuous traits, where the element at row i and column j is the score of individual i on item j. Individuals (rows) and items (columns) are further specified by appropriate identifiers. The data can be described as individual points in a multidimensional space generated by the items, or conversely they can be perceived as item points in a multidimensional space generated by the individuals. Similarities between individuals, or similarities between items, can be measured by various statistics. In past Workshops, the preferred statistics has been the correlation in a two by two contingency table. Thus, an array of correlation coefficients between all pairs of individuals, or conversely between all pairs of items, can be readily derived from the experimental data.

So many correlations are thus generated that conventional significance tests are impractical. Rather, descriptive statistics are in order. What is desired is a descriptive method that yields a summary of the most significant associations. While a variety of algebraic methods can provide a description of multidimensional structures by projection in spaces of low dimensionality, cluster analysis is the preferred approach in serologic analyses. The purpose of cluster analysis is to approximate a multidimensional array of correlation by defining subsets of the data that account for most of the observed correlations. The easiest algorithms which produce such clustering are hierarchical: rather than a flat partition of the data into discrete subsets, the elements subject to clustering are collected into greater and greater subsets. For the Tenth Workshop, clustering was generated by the average pair group method, which proceeds as follows: 1) from a correlation matrix between n elements, select the pair of elements which generate the highest correlation coefficient, say j and k; then 2) reduce the correlation array by replacing all correlations of j and k with any other elements by their average. The matrix has been reduced to n-1 elements, j and k being pooled. The process is repeated until all elements are clustered into a single set.

The outcome of cluster analysis can be represented in several ways. The tree diagram, which reflects each step of the clustering process, can be traced. A given clustering level can be selected to display the subsets into which the elements fall at this level of association. We followed another alternative. The clustering process is used to generate an ordering of the elements correlated so as to best reveal clustering. This ordering is obtained by traversing the tree generated by cluster analysis, as illustrated in a simple example (Fig. 1). As is clear from that example, the relative orientation of elements in contiguous subsets of the tree is arbitrary, as each branch can be rotated without altering the structure of the clusters. Thus, in Figure 1, the sequence chosen to represent the data could be 1,4,2,3,5 as well as 4,1,2,3,5, and so on.

Participating Laboratory: US7LAL[1]

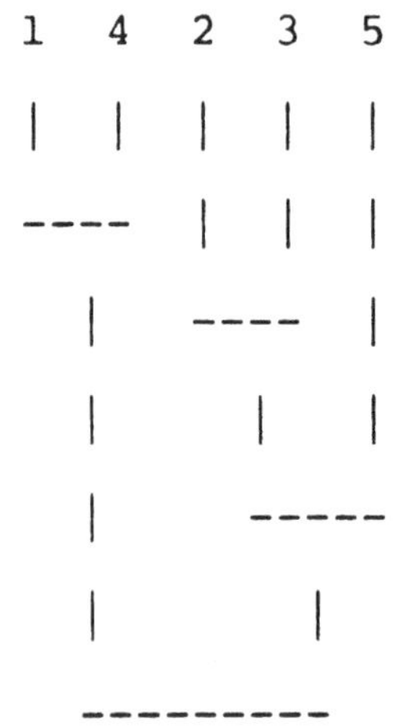

Figure 1. Tree representation of a cluster analysis.

The sequence so generated can be used to rearrange the original matrix of correlations and display it.

In SERAN, cluster analysis can be applied to a matrix of correlation between items. The input consists of the three default files defined in the Overview (scorefile, defining the array of experimental scores, cellfile, defining the individual row entries, and serafile, defining the column entries). Correlations between items are computed (menu command 12). Cluster analysis and tree traversal are performed by two programs, CLUS and TRAV, initiated by a single command (menu command 13).

This generates a new sequence for the display of the column items. Several procedures were implemented to display the experimental data according to such a sequence. Two procedures were included in the SERAN package distributed to Workshop participants. They are operated as special purpose full-screen editors. The rearranged matrix of correlations between items (sera, DNA fragments, T-cell clones, etc.) can be displayed with the program SXS (command 14), as shown in Figure 2.

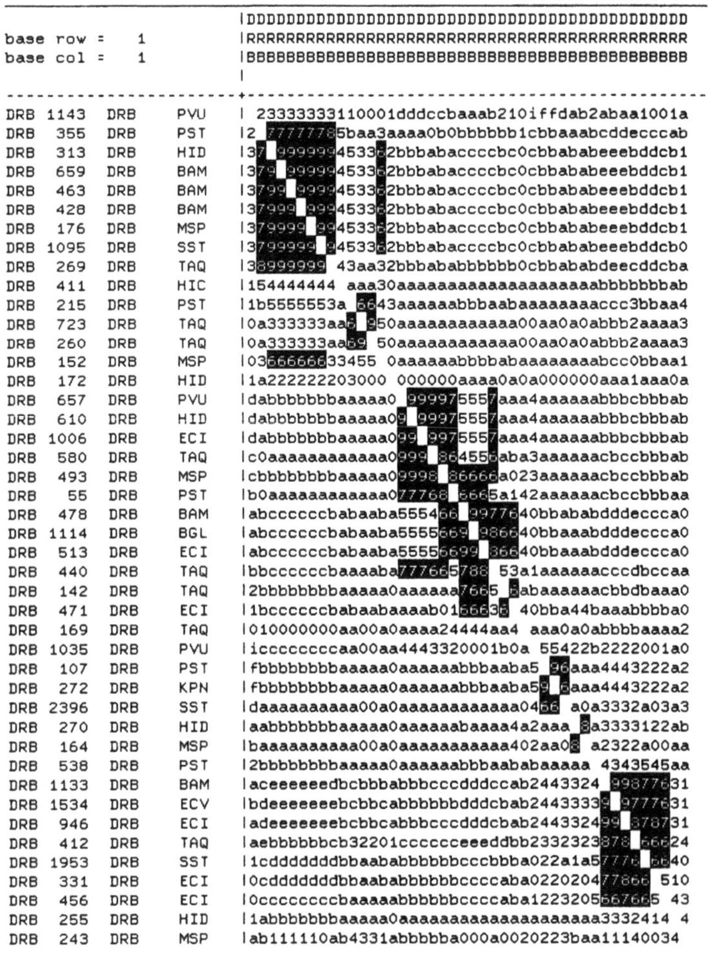

Figure 2. Correlation matrix between DR beta DNA fragments after permutations are applied to reveal clustering (procedure SXS).

Another useful display consists of a presentation of the individual scores of the individuals on the items after the latter have been so resorted (Fig. 3). This function is implemented in the program CXS (command 15). In both SXS and CXS, the video monitor functions as a window on a large two-dimensional array. This window can be moved to display any part of the array. Selected screens or all the screens can be printed by keyboard commands. The whole array of sorted experimental data can also be sent to the printer. In the display of experimental data generated by CXS, the column items, such as sera, are ordered according to the outcome of clustering. Because the number of row entries is usually very large (individual cells, etc.), the version of SERAN distributed to Workshop participants displayed row entries sorted by serologic antigens rather than formal clustering. The order of the antigens is defined by the user. Other functions are available in the procedures SXS and CXS, as illustrated by the printout of the on-line help menu of CXS in Figure 4, including permutations or deletions of rows and/or columns and sorting of the row items according to specific reaction patterns in selected columns.

In the central analysis of the Workshop cell lines, the use of a larger computer made clustering analysis feasible on column entries for up to 1000 items. The limited number of row entries (cells) made cluster analysis also feasible on the row entries. Consequently, the displays generated for analysis of the core cell lines used clustering on both columns and rows, clustering items as well as cells according to their similarities.

To facilitate the work of antigen group leaders, another display technique was used. Large array of experimental data on sera and cells with antigens included in a given antigen group could be printed at high density after cluster analysis of the sera and sorting of the cells by antigen. Thus, up to 180 sera and 240 cells

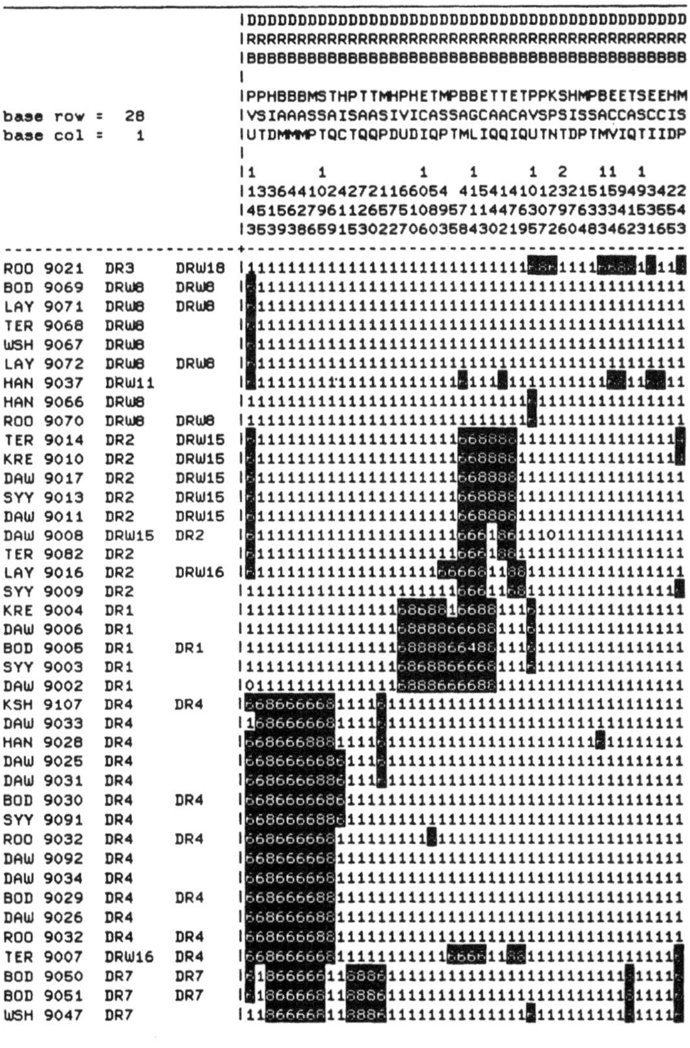

Figure 3. Distribution of DR beta DNA fragments among core cells after permutations are applied to both cells and fragments to reveal clustering (procedure CXS).

```
        arrow keys : move cursor one position
        prev screen, next screen,
        ctrl-L, ctrl-R : shift window ( up, down, left, right)
        ctrl-J : cursor jump - prompts for magnitude and direction of jump
        ctrl-A : absolute jump - prompts for new base row & base column

        ctrl-F : prompts for and finds first occurrence of an id or specificity
        ctrl-N : finds next occurrence of id/specificity

        ctrl-B : select current row/column as a range bound
        ctrl-U : unselect - clears range bounds
        I (upper case) : insert range after current row
        ctrl-K : kill range - removes rows from display
        ctrl-G : range scan - scans rows/columns for matching pattern or
                 high score within range
        ctrl-M : continues scanning
        ctrl-P : pattern sort - sorts rows by similarity to segment of
                 current row as delimited by vertical range bounds

        ctrl-Z : switch to display of cell labels
        ctrl-D : write data to external file for printing
        ctrl-W : write current order of cells & sera to permutation files
        ctrl-E : exit
   Press carriage return to continue>
```

Figure 4. On-line menu of the display procedure CXS (left and above).

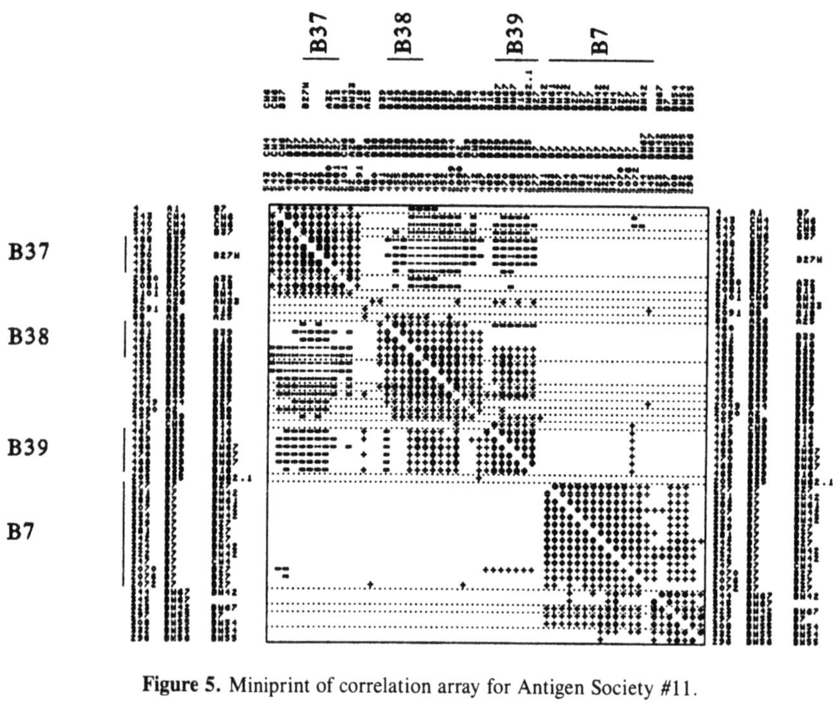

Figure 5. Miniprint of correlation array for Antigen Society #11.

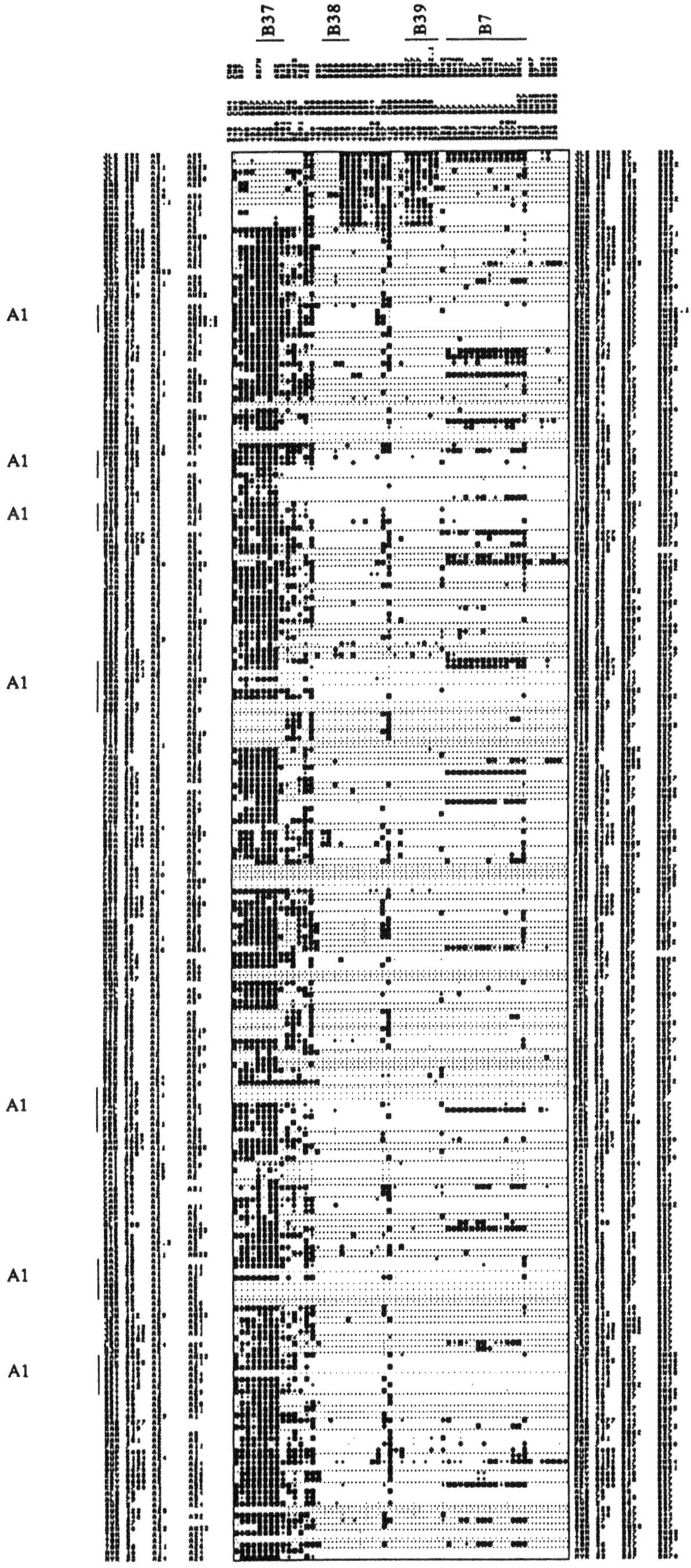

Figure 6. Sample miniprint of data array for Antigen Society #11.

could be displayed on a single transparency, or 43,200 data items, as well as row and column labels identifying cells and sera. This was accomplished by addressing each dot of a laser printer with a resolution of 300 dots per inch; each character, digit, and symbol summarizing degree of correlation was individually programmed to occupy a 9×9 array of dots. This process was used to generate such miniprints of all antigen groups at the Workshop in Princeton. The transparencies could be examined on overhead projectors, or directly with the help of a magnifying glass.

This series of procedures implemented traditional methods of inspection and display of correlation arrays usually requiring tedious manipulations of the data. Various applications can be found throughout these proceedings.

Joint Reports: Serology

Serologic Exercises of the Tenth Workshop

E.L. Milford,[1] M.S. Park,[2] J.-M. Lalouel,[3] L.J. Kennedy,[4] and E.J. Yunis[5]

The serologic exercises of the Tenth International Histocompatibility Workshop were conducted with the following goals:

1. To produce a central or "core" serology set of excellent reagents for the purpose of phenotyping the families and homozygous cell lines used in the Workshop.
2. To provide serologic analysis on a large set of class I and class II monoclonal antibodies.
3. To study the utility of a noncytotoxic serology technique (ELISA assay) for the definition of HLA specificities with mouse monoclonal antibodies, and to compare the results with cytotoxicity.
4. To permit intensive study of a large number of sera, in addition to the core set, which exhibit reactivity with a related set of antigens by establishing "Antigen Societies" or groups of laboratories with a particular interest in that set of antigens.
5. To develop, in collaboration with the Howard Hughes Medical Institute, Salt Lake City, Utah, permanent software resources that would permit the collection, merging, sharing, and flexible analysis of serologic and phenotypic data in individual laboratories and groups of labs.

Organization and Data Collection

Laboratories were canvassed in writing to determine the quantity and quality of sera and monoclonal antibodies that they wished to offer. Fifty ml quantities of sera were required for submission to the core set and 6 ml for the Antigen Societies. Seven laboratories prescreened all submitted reagents to determine if they were of sufficient merit to be included in the core set. The prescreening laboratories were DUP, MRV, BOD, TER, HAN, GSC, and CAR. In addition, laboratories BOD, TER, and GSC screened the monoclonal antibodies to determine whether dilutions other than those submitted were optimal for use in the Workshop, and to determine which subset of monoclonals should be used in the ELISA exercises.

Prescreening results were submitted to Boston using a uniform format and a commercial database entry program. The prescreening data was submitted to UCLA Tissue Typing Laboratory for analysis by Dr. Min Sik Park using his established empiric serum analysis programs, and to Salt Lake City, where it was analyzed by Dr. Jean-Marc Lalouel with the preliminary versions of the newly developed SERAN utilities. Results of the empiric analyses from both groups were identical. Sera were included in the core and Antigen Society sets if they exhibited strong reaction patterns with a subset of cells, even if there was no apparent correlation with established HLA antigens. This was done so that we would not bias against the detection of new specificities.

Sera and monoclonal antibodies were submitted to the central storage facilities in Rye, New York, prelabeled and aliquoted into small plastic tubes. Based on the prescreening and available volumes of sera, a "core" set (Table 1) and 33 "Antigen Society" sets were established. This involved the collation of 572,000 individual tubes into over 1250 separate core or Antigen Society serum sets. Ultimately, over 2600 serologic reagents from 157 laboratories were used in the Tenth International Histocompatibility Workshop exercises (Table 2).

Based on preliminary questionnaires, groups of laboratories were assigned to specific Antigen Societies, which specialized in a related group of antigens (Table 3). Wherever possible, councillor laboratories took the leadership of societies in which they had historical expertise. Because of the geographic dispersion of the Antigen Societies, several laboratories were given responsibility for leadership of each group.

Work Scope

Between two and five leaders were selected for each antigen group. These leaders were responsible for ensuring that they tested the sera of the Tenth Workshop core set on at least 50 previously phenotyped individuals selected to represent each of the HLA antigens in the antigen group. The leaders were also instructed to test the appropriate Antigén Society serum sets with the same cells, so that a reasonable selection of cells would be available for cross-correlating Antigen Society and core sera. Each Antigen Society was, as a group, to also test a larger group (200 to 1000 individuals) with the Antigen Society plus core sets or the Antigen Society sets alone. Antigen Society leaders who were participating in exercises involving the core homozygous lymphoblastoid cell lines were also encouraged to test these lines with the core set and Antigen Society sets. Each laboratory that submitted a lymphoblastoid cell line was also to type the donors with the core set.

Data Entry

The format of trays for the core serology set was determined centrally and was fixed. Provision was made for using either 60 well or 72 well trays. Leaders of the Antigen Societies submitted fixed tray formats, which were agreed upon by all members of the respective Society. This permitted the Antigen Societies to put additional sera onto their tray sets after the closing date for receipt of sera centrally.

[1]NCYCAR, [2]US5TER, [3]US7LAL, [4]UKIBOD, [5]NCYYUN

Immunobiology of HLA
Volume I

Table 1. International Histocompatibility Workshop Core Sera

1	DUP	POS,
2	DUP	NEG,
3	BRA	A1,
4	CNG	A1,
5	MIT	A1,
6	HAM	A1,
7	MID	A1,
8	DPT	A1,AW36,
9	MBC	A1,AW36,
10	OHA	A1,AW36,
11	CRO	A1,AW36,
12	DPT	A1,AW36,
13	KAP	AW36,A1,
14	DDC	A1,A11,AW36,
15	ENG	AW36,A1,
16	PER	A2,
18	DAW	A2,
19	TAK	A2,
20	GOL	A2,
21	BOD	A2,
22	BOD	A2,
23	MRA	A2,A28,
24	GOL	A2,A28,
25	LAW	A2,A28,
26	MBC	A2,A28,
27	MBC	A2,A28,
28	BET	A2,A28,
29	WEG	A2,AW69,AW68,
30	BAT	A2,AW69,
31	GAN	A2,AW69,
32	GAN	A2,AW69,
33	GAN	A2,AW69,
34	BOD	A2,AW69,B8,
35	MIT	A2,AW69,
36	BSH	AW68,
37	BOD	A2,
38	ABS	A28,
39	DUT	A28,
40	FER	A28,
41	CRO	A28,
42	CRO	A28,
43	WOL	A28,AW33,
44	DUT	A28,AW34,
45	CRO	A28,AW34,AW33,
46	DPT	A28,AW33,AW34,AW66,DR2,
47	MRV	A3,
48	KWC	A3,
49	DUT	A3,
50	DUT	A3,
51	MBC	A3,
52	ZHA	A11,
53	MRA	A11,
54	CEP	A11,
55	BER	A11,
56	MLS	A11,
57	DDC	A11,A10,AW34,
58	MIT	A9,
59	BRN	A9,
60	KAP	A9,
61	DUT	A9,
62	OHA	A9,
63	BET	A23,A24,
64	WEG	A24,A23,
65	JLE	A24,A9.3,
66	BET	A23,A24,A9.3,
67	MYR	A23,
68	JER	A23,
69	TOM	A23,
70	STE	A23,
71	MID	A23,
72	ENG	A24,
73	DUP	POS,
74	DUP	NEG,
75	CON	A24,
76	CEP	A24,
77	MTT	A24,
78	JUJ	A24,
79	CEP	A10,
80	BRN	A10,AW33,
82	KAP	A26,A25,AW33,TH,A32,
83	HAM	A10,AW33,AW34,
84	DUP	UNK,
85	HAM	A10,AW33,AW34,
86	MYR	A10,AW34,
87	MLS	A10,AW34,
88	KRE	A10,AW34,
89	HAM	A10,AW34,
90	TOM	A25,A26,AW34,
91	MRV	A25,A26,AW34,
92	MBC	A25,A26,AW34,
93	GVY	A25,A26,AW34,
94	GRU	A25,A26,AW34,
95	BOU	A25,A26,AW66,
96	BER	A25,A26,AW66,AW34,
97	CRO	A25,A26,AW34,AW66,
98	BET	AW34,
99	RIC	A26,AW34,AW66,
100	JLE	A26,AW66,
101	DPT	A10,AW34,A9,AW66,DQW1,
102	BRU	AW66,AW34,AW33,A28,
103	BRU	AW66,AW34,AW33,A28,
104	WEG	A26,A25,AW34,AW66,AW43,
105	FAU	A26,A25,AW34,AW43,
106	TER	AW34,AW43,A26,A11,
107	DUT	A26,AW43,
108	DUT	A26,AW43,
109	DUT	A26,AW43,
110	DUT	AW43,
111	NAI	A26,
112	BRS	A26,
113	MIT	A26,
114	MRV	A26,
115	TAT	A26,
118	MRI	A26,
119	CRS	A25,
120	MRV	A25,
121	MRA	A25,
122	BET	A25,
123	GOL	A25,
124	MYE	A25,
125	MBC	A25,
126	TAT	A25,
127	PER	A29,
128	PNW	A29,
129	FER	A29,
130	KRE	A29,
131	MLS	A29,
132	GLA	A29,
133	YUN	A29,B45,
134	CRO	A30,
135	NWO	A30,
136	TAI	A30,
137	CRS	A30,A31,
138	ABS	A30,A31,
139	PRR	A30,A31,
140	MYE	A30,A31,
141	BAT	A31,A30,
142	MBC	A30,A31,
143	BRT	A30,A31,
144	BRS	A30,A31,AW33,
145	DUP	POS,
146	DUP	NEG,
147	KUE	A3,A11,A29,A30,A31,AW33,
148	BIG	A31,
149	JLE	A31,
150	FER	A31,
151	STE	A31,
152	BRN	A31,AW33,
153	OHA	AW33,
154	FER	AW33,
155	ROO	AW33,
156	TOM	AW33,AW34,
157	MYR	AW33,B14,
158	FER	A32,
159	DUT	A32,
160	BAT	A32,
161	DDC	A32,TH,
162	BRS	A32,B27,
163	LAY	AN,
164	DAW	B5,
165	MYR	B5,
166	MYR	B5,
167	DAW	B5,
168	JOY	B5,
169	KAS	B51,
170	FES	B51,
171	BIG	B51,
172	CRS	B51,
173	NAI	B51,
174	FAU	B51,
175	SHA	B51,
176	HAM	B51,
177	JUJ	B51,
178	TER	B51,TE76,
179	TER	B51,TE76,
180	DUT	B51,
181	AMY	BW52,
182	NWO	BW52,
183	PTR	BW52,
184	NWO	BW52,
185	ENG	B51,BW52,BW53,
186	OHA	B51,BW52,BW53,
187	BRU	B5,B51,BW52,BW53,
188	KAP	BW53,BW52,B35,B51,
189	KAP	BW53,BW52,B51,B35,
190	GOR	B51,BW53,
191	MBC	B35,BW53,
192	PNW	BW53,B35,
193	BRU	BW53,B35,
194	BRA	B35,B51,BW53,
195	KAP	B51,BW53,B35,
196	BET	BW53,B35,
198	MIT	BW71,B35,BW53,B15,
199	LAW	B7,BW22,BW71,
200	BRS	B35,BW62,BW71,BW72,
201	DPT	B5,B35,BW53,BW72,
202	ROO	BW72,BW62,B5,B35,BW53,B18,
203	TSU	B35,
204	CON	CW4,
205	BRA	B35,
206	GAN	B35,
207	BRN	B35,
208	WAL	B35,
209	MIT	B35,
210	MBC	B51,BW52,8W58,
211	MBC	B51,BW52,BW59,8W58,
212	HAM	B51,B35,
213	MBC	B35,B51,BW52,BW53,BW70,8W58,BW62
214	WOL	B35,B51,BW52,BW53,BW70,TE79,CW4,
215	WOL	B35,BW62,BW50,TE79,BW70,B17,
216	TER	BW62,BW63,BW57,TE79,
217	DUP	POS,
218	DUP	NEG,
219	TER	BW62,BW63,BW57,TE79,
220	TER	BW57,BW46,B15,TE77,TE79,
221	TER	BW62,BW63,TE77,TE79,

able 1. *Continued*

22	TER	BW53,B49,BW78,B51,BW52,
23	TER	BW53,B49,BW78,B51,BW52,
24	TER	B49,B51,BW52,TE78,
25	TER	B49,BW53,BW78,
26	TER	B49,B5,TE78,
28	DDC	B35,BW62,BW70,
29	CRO	BW62,BW70,BW50,BW56,B35,
30	MYR	B15,BW50,BW70,
31	BER	B15,B17,B35,BW70,
32	MBC	BW50,BW41,BW62,BW70,
33	MRV	BW62,B15.3,BW50,BW70,BW56,B35,
34	BER	B35,BW70,B15.3,BW62,BW56,
35	MRV	B35,B15.3,BW53,B51,
36	BRN	BW62,B15.3,SH.7,
37	GOE	B15,
38	MIN	B15,
39	NWO	B15,
40	IVS	B15,
41	BRN	B15,
42	PNW	B15,
43	TAT	B15,
44	MRA	B15,
45	PRR	BW62,BW63,
46	ENG	BW62,
47	MYR	BW62,
48	ENG	BW62,
49	CRO	BW62,
50	OHA	BW62,
52	MYE	BW62,
53	MBC	BW62,
54	ZHA	B15,SH.7,
55	ZHA	B15,SH.7,
56	DCH	B15S,B15,
57	DCH	BW46,
58	NAI	BW46,
59	GOR	BW46,CW1,
60	DEC	BW46,B15,B17,CW3.1,
61	DCH	B15,BW46,B13,
62	ENG	BW62,BW57,BW63,BW46,
63	BSH	B8,BW46,
64	WEG	BW62,BW63,BW57,BW46,
65	HJH	B17,BW63,
66	KAP	B17,BW63,
67	TER	BW63,BW57,BW58,
68	ENG	B51,BW52,BW53,B49,BW63,
69	DUT	B18,
70	JER	B18,
71	FER	B18,
72	GRU	B18,
73	DUT	B18,
74	DAW	B7,
75	MID	B7,
76	KWC	B7,
77	CEP	B7,
78	FAU	B7,
79	DAW	B27,
80	MLS	B27,
81	KWC	B27,
82	BRS	B27,
83	JLE	B27,
84	ENG	B27,
85	YUN	B27,B13,BW47,B7,
86	FAU	BW47,B27,
87	MRV	B27,BW47,B13,
88	CRO	B27,BW47,
89	DUP	POS,
90	DUP	NEG,
91	GAN	B27,BW47,
92	BSH	B27,BW47,
93	PER	B27,BW47,CW2,
94	CEP	BW22,
95	MIN	BW22,
296	OHA	BW22,
297	CEP	BW22,
298	JLE	BW22,
299	DDC	BW22,BW42,
300	DUP	UNK,
301	MBC	BW55,BW54,BW56,
302	GLA	BW55,BW56,BW54,
303	AMY	BW54,
304	JUJ	BW54,
305	TKB	BW54,
306	AMY	BW54,BW55,
307	MIT	BW54,BW55,
308	KSH	BW54,BW56,
309	GOL	BW55,
310	JUJ	BW55,
311	AMY	BW55,
312	ENG	BW55,
313	MID	BW55,
315	TAK	BW55,BW42,
316	FOT	BW55,BW46,
317	RIC	BW56,BW55,
318	JUJ	BW56,
319	ROO	BW56,
320	CRO	BW42,
321	HJH	BW42,
322	DUT	BW42,
323	DUT	BW42,B7,
324	DDC	B7,BW42,
325	JLE	B7,BW42,
326	CRO	BW42,BW41,CW2,
329	TUL	BW41,BW42,
330	MYE	BW42,BW73,B7,
331	DPT	B7,BW42,BW73
332	MIT	BW73,B7,BW42,
333	ENG	BW73,B7,BW42,
334	DDC	B7,BW73,
335	RIC	B7,BW73,
336	MYR	BW73,
337	KIS	BW73,
338	RIC	BW73,
339	ROO	7C,BW73,
340	MYR	B7,B27,BW73,
341	RIC	B7,B27,BW60,BW73,
342	DPT	B7,B27,BW22,BW42,BW73
343	DPT	B7,B27,BW22,B40,BW42,BW73
344	NAI	BW67,
345	TOK	BW67,
346	GOR	BW67,
347	MIT	BW67,B7,BW42,CW4,
348	HAM	B40,
349	HAM	B40,
350	MID	B40,
351	RIC	B40,
352	ABS	B40,
353	CRS	BW60,BW61,
354	GEL	BW60,BW61,
355	TSU	BW60,BW61,
356	DAW	BW60,
357	OHA	BW60,
358	MLS	BW60,
359	GAZ	BW60,
360	JLE	BW60,
361	DUP	POS,
362	DUP	NEG,
363	AST	BW60,
364	MEH	BW61,
365	RAD	BW61,
366	GOR	BW61,
367	TSU	BW61,B13,
368	TAK	BW60,BW61,B13,
369	GAN	B40,B13,BW47,
370	PRR	B13,B40,BW47,
371	CRO	B40,BW47,BW73,
372	FAU	BW61,BW60,B13,BW47,BW48,
373	DPT	B40,B13,BW47,BW48,
374	DPT	B40,B13,BW47,BW48,
375	HFP	B40,B13,BW47,BW48,BW41,
376	ENG	BW61,BW60,BW48,
377	TSU	BW60,BW48,
378	JLE	BW48,
379	DPT	BW48,B51,BW52,BW63,
380	MYE	BW48,B7,A3,
381	ENG	BW48,B7,B27,BW42,
382	KAP	B7,B27,BW48,
383	KAP	B7,B40,BW42,BW48,
384	AST	B40,BW41,
385	KAP	BW41,BW42,
386	JRM	BW41,BW42,
387	MYR	BW41,
388	RAD	BW41,
389	RAD	BW41,
390	ENG	BW41,
391	BSH	B13,
392	PER	B14,
393	BRU	B13,
394	GOE	B13,
395	PER	B13,
396	FAU	B13,
397	SHA	B13,
398	CRO	B8,
399	BET	B8,
400	DUT	B8,
401	HAM	B8,
402	GAN	B8,
403	MRV	B8,BW59,
404	JLE	B8,BW59,
405	TOK	BW59,
406	JLE	BW59,
407	MYR	B14,
408	MLS	B14,
409	BET	B14,
411	GOL	B14,
412	MBC	B14,B39,
413	AST	BW64,
414	KAP	BW64,
415	MIT	BW65
416	DUQ	BW65,
417	MIT	B12,
418	JLE	B12,
419	MRV	B12,
420	HAM	B12,
421	CEP	B12,
422	KWC	B12,B44,
423	UAB	B44,
424	DAW	B44,
425	MRV	B44,
426	MIT	B44,
427	HAM	B44,
428	ABS	B45,
429	TOM	B45,
430	STE	B45,
431	DRK	B45,
432	GEL	B45,
433	DUP	POS,
434	DUP	NEG,
435	MRV	B21,
436	MYR	B21,
437	MYR	B21,
438	MRV	B21,
439	JLE	B21,
440	JLE	B21,
441	GRU	B49,
442	DDC	B49,
443	BER	B49,

Table 1. *Continued*

444	TRP	B49,
445	FER	B49,
446	MYE	B49,BW52,
447	MIT	BW50,
448	BRS	BW50,
450	GOL	BW50,
451	VER	BW50,
452	ABS	B16,
453	DUQ	B16,
454	DUQ	B16,
455	MTT	B16,
456	YAG	B16,
457	OHA	B16,
458	ZHA	B16,ST.16,
459	JER	B39,B38,
460	DUQ	B38,
461	GOL	B38,
462	BER	B38,
463	POL	B38,
464	GAB	B38,
465	BRS	B39,
466	CRS	B39,
467	DUP	B39,
468	CEP	B39,
469	JLE	B39,
470	TKB	B39,
471	GOE	B17,
472	CRS	B17,
473	MYR	B17,
474	RIC	B17,
475	BRN	B17,
476	MRV	B17,
477	BET	BW57,BW58,
478	FTW	BW57,BW58,BW63,BW62,
479	STA	BW58,BW57,
480	BER	BW57,
481	SIN	BW57,
482	SND	BW57,
483	GOL	BW57,
484	DAW	BW57,
485	DAW	BW57,
486	DAW	BW57,
487	CON	BW58,
488	BSH	B37,
489	KWC	B37,
490	PER	B37,
491	FER	B37,
492	MYE	B37,
493	LAW	BW4,
494	GOE	BW4,
495	GAN	BW4,
496	MBC	BW4,
497	DPT	BW4,
498	JLE	BW4,
499	BRN	BW6,
500	MUE	BW6,
501	RIC	BW6,
502	DPT	BW6,
504	BER	BW6,
505	DUP	POS,
506	DUP	NEG,
507	FAU	BW6,
508	ENG	CW1,
509	ROO	BW6,AN,
510	TOK	CW1,
511	LAY	CW1,
512	FAU	CW1,
513	BRS	CW1,
514	YAG	CW1,
515	MYR	CW2,
516	ENG	CW2,
517	LAY	CW2,
518	KAP	CW2,
519	GRU	CW2,
520	CEP	CW2,
521	FTW	CW2,
522	MBC	CW2,
523	BRN	CW2,
524	MRV	CW3,
525	TOK	CW3,
526	MIT	CW3,
527	DDC	CW3,
528	BRS	CW3,
529	GRU	CW3,
530	NEU	CW3,
531	TSU	CW3,CN,
532	JUJ	CW3.1,
533	ENG	CW4,
534	FES	CW4,
535	MYR	CW4,
536	CRO	CW4,
537	BRS	CW4,
538	BOU	CW4,
539	ZAR	CW4,
540	DDC	CW4,CW6,
541	MRV	CW4,
542	MTT	CW4,CW6,
543	JLE	CW4,CW6,
544	TER	CW4,CW6,TEC10,
545	TER	CW4,CW6,TEC10,
546	KAP	CW6,
547	MUE	CW6,
548	HAN	CW6,
549	MBC	CW6,
550	CRS	CW5,
551	YUN	CW5,
552	MBC	CW5,CW8,
553	MYR	CW5,CW8,
554	KAP	CW8,
555	BLT	CW7,
556	PRR	CW7,
557	DDC	CW7,
558	DDC	CW7,
559	TSU	CW7,
560	POL	CW7,
561	DDC	CW7,
562	PUR	CW4,
563	PUR	CW2,
564	MUL	B51,A25,
565	MUL	A29,
1001	DUP	POS,
1002	DUP	NEG,
1003	BSH	DR1,
1004	MOO	DR1,
1005	MRV	DR1,
1006	GOL	DR1,
1007	GOL	DR1,
1008	MTT	DR1,
1009	FER	DR1,
1010	BET	DR1,
1011	MBC	DR1,
1012	MBC	DR1,DRW10,
1013	WEG	DR1,DRW10,
1014	KIS	DRW10,DR1,
1015	FAU	DRW10,
1016	MUE	DRW10,
1017	VIV	DRW10,
1018	HJH	DRW10,
1019	PER	DRW10,
1020	WOL	DRW10,
1021	FAU	DRW10,DRN,
1022	DAW	DR2,
1023	MRV	DR2,
1024	MRV	DR2,
1025	BRN	DR2,
1026	FAU	DR2,
1027	FAU	DR2,
1028	NAI	DR2,
1029	YUN	DR2,
1030	BAC	DR2,
1031	MRV	DR2S,
1032	BRN	DR2,
1033	NAI	DR2,DQN,DPN,
1034	MRV	DR3,
1035	MRV	DR3,
1036	FES	DR3,
1037	HAM	DR3,
1038	RUB	DR3,
1039	BRN	DR3,
1040	HAM	DR3,
1041	TAT	DR3,
1042	THP	DR3,
1043	DRK	DR3,
1044	MRV	DR4,
1045	MRV	DR4,
1046	GOE	DR4,
1047	FES	DR4,
1048	MRV	DR4,
1049	AST	DR4,
1050	RUB	DR4,
1051	BRN	DR4,
1052	DAW	DR4,
1053	DAW	DR4,
1054	DAW	DR4,
1055	BAC	DR1,DR4,
1056	JUJ	DR4.1,
1057	BSH	DR4,DRW9,
1058	MRV	DRW9,
1059	TER	DRW9,
1060	FAU	DRW9,
1061	NAI	DRW9,
1062	HJH	DRW9,
1063	NAI	DRW9,
1064	PRR	DRW9,
1065	AMY	DRW9,DRW10
1066	FAU	DR7,DRW9,
1067	FAU	DR7,DRW9,
1068	BSH	DR7,
1069	HAM	DR7,
1070	DAW	DR7,
1071	MRA	DR7,
1072	AST	DR7,
1073	DUP	POS,
1074	DUP	NEG,
1075	GAN	DR7,
1076	GOL	DR7,
1077	NEU	DR7,
1078	DRK	DR7,
1079	BET	DR7,
1080	BRN	DR7,
1081	CEP	DR7,
1082	TER	DRW8,
1083	THP	DRW8,
1084	DUP	UNK,
1085	STA	DRW8,
1086	SEK	DRW8,
1087	HAN	DRW8,
1088	GOR	DRW8,
1089	JLE	DRW8,
1090	FER	DRW8,
1091	MOO	DRW8,
1092	DRK	DRW8,
1093	FAU	DR1,DRW8,
1094	OHA	DRW8,DR5,DR2
1095	FAU	DR5,
1096	SWE	DR5,

Table 1. *Continued*

1097	MRA	DR5,
1098	HAM	DR5,
1099	AST	DR5,
1100	MTT	DR5,
1102	GOL	DR5,
1103	WAL	DR5,
1104	CEP	DR5,
1105	FAU	DR4,DRW11,
1106	FAU	DR5,TA10,
1107	CXH	DRW11,DRW12,DR4,TA10,
1108	HJH	DR5,DRW8,DQW3,
1109	WTZ	DRW11,TA10,
1110	DUQ	DR5,
1111	BER	DRW11,
1112	BET	DR4,DRW11,
1113	BRN	DRW11,
1114	BET	DRW11,
1115	TER	DRW12,DRW11,
1116	TER	DRW12,DRN,
1117	TER	DPW1,
1118	TER	DRW12,
1119	RIC	DR2,DRW11,
1120	KAP	DR5,DRW6,
1121	KAP	DR5,DRW6,
1122	DUQ	DR5,DRW6,
1123	BER	DR5,DRW6,
1124	TER	DRW6,
1125	TER	DRW6,
126	HAN	DRW6,
127	DAW	DRW6,
128	GOR	DRW6,
129	BER	DRW6,
130	DUQ	DRW6,
131	BRN	DRW6,
132	BSH	DRW13,
133	ROO	DRW13,
134	ROO	DRW13,DRW12,
135	BER	DRW13,DRW14,
136	TKB	DRW14,
137	BRN	DR2,DRW6,
138	DUQ	DR2,DRW6,
139	TER	DR2,DRW6,
140	TER	DR2,DRW6,
141	BRN	DR2,DRW6,
143	PRR	DR2,DRW6,DRW10,DR1,DQW1,
144	YUN	DR3,DR6,
145	DUP	POS,
146	DUP	NEG,
147	MYE	DR3,DRW6,
149	TER	DRW6,
150	BAC	DR3,DRW6,
151	JEA	DR3,DRW6,
52	JOY	DR3,DRW6,DRW8,
153	DAW	DQW1,
54	BSH	DQW1,
55	BER	DQW1,
57	HAM	DQW1,
58	FER	DQW1,
59	STA	DQW1,
60	MRV	DQW1,
61	MOR	DQW1,
62	MOR	DQW1,
63	MBC	DQW1,
64	JUJ	DQW1,
65	LAW	DR1,DR2,DRW6,DQW1,
66	AST	DR1,DR2,DRW6,
67	MRV	DQW2,
68	BRN	DQW2,
69	MTT	DQW2,
70	BRA	DQW2,
71	DUQ	DQW2,
72	BER	DQW2,
1173	FAU	DQW2,
1174	MTT	DQW2,
1175	MBC	DQW2,
1176	MRV	DQW3,
1177	MRV	DQW3,
1178	THP	DQW3,
1179	DAW	DQW3,
1180	MTT	DQW3,
1181	MBC	DQW3,
1182	TER	DQW3,
1183	TER	DQW3,
1184	TER	DQW3,
1185	BAC	DQW3,
1186	FER	DQW3,
1187	MBC	DQW3,
1188	MBC	DQW3,
1189	MBC	DQW3,
1190	MBC	DQW3,
1191	MBC	DQW3,
1192	BRN	DQW3,
1193	MBC	DQW3,
1194	DAW	DR4,DR5,
1195	DAW	DR4,DR5,
1196	DAW	DR4,DR5,
1197	DAW	DR4,DR5,
1198	MRV	DRW52,
1199	FES	DRW52,
1200	MTT	DRW52,
1201	MRV	DRW52,
1202	GOE	DRW52,
1203	STA	DRW52,
1204	TAK	DRW52,
1205	SIN	DRW52,
1206	DUP	DRW52,
1207	MYE	DRW52,
1208	DRK	DR3,DR5,DRW6,DRW52,
1209	FAU	DRW53,
1210	HAM	DRW53,
1211	HAM	DRW53,
1212	MRV	DRW53,
1213	FER	DRW53,
1214	POL	DRW53,
1215	GOL	DRW53,
1216	JLE	DRW53,
1217	FAL	DQW1,
1218	DUP	DRW53,
1219	DUP	DR7,DR5,
1220	DUP	B13,
1250	WAN	DQN,
2001	DUP	POS,
2002	DUP	NEG,
2003	TER	A1,
2004	GSC	A1,A11,A23,A24,A25,A26,AW36,
2005	WAR	A3,
2006	TER	A3,
2007	GOL	A3,
2009	FAU	A3,
2010	GAN	A3,
2011	GAN	A3,
2012	RUB	A3,
2013	GSC	A3,
2014	DDC	A3,A11,
2015	WAN	A11,AW19,B44,B45,
2016	GRU	A11,A24,
2017	TER	A11,
2018	GEL	A11,
2019	WAN	A2,
2020	GEL	A2,
2021	MAR	A2,A28,
2022	GRU	A2,A28,
2023	WAN	A2,A28,
2024	FRN	A2,A28,
2025	GEL	A2,A28,
2026	LAW	A2,A28,
2027	LAW	A2,A28,
2028	LAW	A2,A28,
2029	FAU	A2,A28,
2030	DDC	A2,A28,
2031	GSC	A2,A28,
2032	DUP	A2,A28,
2033	MUC	A2,A28,
2034	MUC	A2,AW68,
2035	MUC	A2,AW69,
2036	GOL	A2,AW69,
2037	BOD	A2,AW69,
2039	GSC	A2,AW69,
2040	KRE	AW69,
2041	CEP	A2,A3,A28,A29,A30,A31,AW33,
2042	GSC	A2,A3,A28,A29,A30,A31,AW33,
2043	ROO	A3,A28,A11,
2044	GEL	A3,A28,AW19,
2045	AIZ	AW19,A10,A11,A28,
2046	WAN	A2,A31,
2047	FAU	A29,
2048	DDC	A29,
2049	FAU	A29,B12,
2050	TER	A30,A31,
2051	WAN	A31,A29,
2052	TER	A32,
2053	CNG	A32,B15,B40,BW41,B17,
2054	GSC	A32,B27,
2055	VIV	A32,A25,
2056	TER	A25,A32,
2057	MUC	A32,A23,A24,BW4,
2058	GSC	A23,A24,A25,A32,
2059	WAN	A28,A32,A25,
2060	MRI	A9,
2061	AIZ	A9,
2062	TER	A9,A10,
2063	GSC	A24,
2064	TSU	A24,BW4,
2065	TSU	A24,BW4,
2066	TER	A10,
2067	MRI	A10,AW33,
2068	BOD	B7,
2069	FER	B7,
2070	FRN	B27,B7,
2071	GRU	B27,
2072	FAU	B27,
2073	DUP	POS,
2074	DUP	NEG,
2075	GSC	B27,
2076	FRN	B27,
2077	GRU	B27,BW47,
2078	TER	B27,BW47,
2079	DDC	B40,BW47,
2080	FAU	B27,B37,B13,
2081	TER	B37,B27,B13,
2082	WAN	A2,B7,B27,
2083	GSC	B7,B27,B17,BW22,BW42,BW63,
2084	GSC	B7,B27,BW54,BW55,BW56,BW57,BW58,
2085	WAN	B7,B8,BW60,BW61,
2086	TER	B8,
2087	MOR	B8,
2088	GAN	B8,
2089	GSC	B8,
2090	CEP	A2,B8,B16,
2091	FAU	B14,B8,B18,B5,
2092	TER	B13,
2093	MUC	B13,
2094	GOL	B13,
2096	DDC	B40,B12,
2097	WAN	B44,B45,B51,BW52,
2098	WAN	B44,B45,B51,B13,BW57,

Table 1. *Continued*

2099	CNG	B15,B17,
2100	MIT	BW4,
2101	MUC	BW4,
2102	GEL	BW4,
2103	GEL	BW6,
2104	GSC	BW6,
2105	THP	BW6,
2106	KRE	BW6,
2107	SFR	BW6,
2108	BRN	CW1,CW3,
2109	TSU	CL1,
2110	TSU	CL1,
2111	AIZ	CL1,
2112	HAN	CL1,
2113	ENG	CL1,
2114	GEL	UNK,
2115	TSU	UNK,
2116	DUP	A25,
3001	DUP	POS,
3002	DUP	NEG,
3003	SFR	DR1,DR2,DR4,DR5,DRW6,
3004	MOR	DR1,DR2,DR4,DR5,DRW6,DRW8,
3005	WAN	DR1,DR3,DR4,DRW8,
3006	WAN	DR1,DR3,DR4,DRW8,
3007	WAN	DR1,DR3,DR4,
3008	BRN	DR1,DR4,
3009	CEP	DR1,DR4,DRW6,
3010	DDC	DR1,DR7,
3011	AIZ	DR2,DR1,
3012	MOR	DR2,
3013	CEL	DR2,
3014	CEP	DR2,DR5,
3015	AIZ	DR2,DRW8,DR5,DRW9,
3016	WAN	DR2,DR3,DR4,DR5,DRW13,
3017	WAN	DR3,DR2,DRW13,
3018	WAN	DR3,DRW13,DR1,
3019	WAN	DR3,DRW13,
3020	WAN	DR3,DRW13,
3021	WAN	DRW13,
3022	WAN	DR3,DRW14,
3023	DDC	DR3,DRW6,
3024	SFR	DRW6,
3025	ROO	DR3,DR5,DRW6,
3026	MOR	DR5,DRW6,DR3,
3027	MOR	DR3,DR5,DRW6,DRW8,
3028	MOR	DR3,DR5,DRW6,DRW8,
3029	SFR	DR3,DR5,DRW8,DRW12,
3031	FES	DR3,
3032	GOL	DR3,
3033	TER	DR3,
3034	DDC	DR5,
3035	FER	DR5,
3036	GSC	DR5,
3037	SFR	DR5,
3038	TER	DPN,
3039	WAN	DRW13,DRW8,DR5,
3040	AIZ	DRW8,DRW12,
3041	DDC	DRW8,
3042	DDC	DRW8,
3043	MAR	DR4,
3044	AIZ	DR4,DRW11,
3045	GSC	DRW9,DR4,DR3,DR7,
3046	WAN	DR3,DR4,DR7,
3047	MAR	DR4,DRW9,DR7,DRW53,
3048	JUJ	DR7,DRW9,DRW12,
3049	BOD	DR7,
3050	SFR	DR7,
3051	GSC	DR7,DRW10,
3052	TER	DRW9,
3053	FRN	DRW9,
3054	FAL	DRN,
3055	GSC	DR1,DR2,DR3,DR4,DR5,DRW6, DRW8,DRW9,
3056	AIZ	DRN,
3057	AIZ	DRN,
3058	SFR	DRN,
3059	ROO	DRN,
3060	ROO	DRN,
3061	WAN	DRW52,
3062	WAN	DRW52,
3063	BOD	DRW52,
3064	BOD	DPW2,DPW4,
3065	DDC	DRW52,
3066	GAN	DRW53,
3067	RJW	DRW53,
3068	TER	DRN,
3071	WAN	DRN,
3072	CNG	DRN,
3073	DUP	POS,
3074	DUP	NEG,
3075	ENG	DRN,
3076	ENG	DRN,
3077	STA	DRN,
3078	WAN	DRN,
3079	TSB	DRN,
3080	TKB	DRN,
3081	TKB	DRN,
3082	TKB	DRN,
3083	SND	DRN,
3084	SND	DRN,
3085	WAN	DQW1,DR1,DR2,DRW6,DR4,DRW
3086	WAN	DQW1,DR1,DR2,DRW6,DR4,DRW
3087	WAN	DQW1,DR2,DRW13,
3088	ROO	DR2,DR4,DRW13,DR7,DRW9,
3089	AIZ	DQW1,TE23,
3090	ROO	DQW1,DR4,
3091	FRN	DR4,DQW1,
3092	BOD	DQW1,
3093	MAR	DQW1,
3094	BOD	DQW1,
3095	BOD	DWQ1,
3096	GSC	DQW1,
3097	AIZ	DQW1,
3098	THP	DQW1,
3099	GSC	DQW1,DQN,
3100	BOD	DQW1,DQN,
3101	AIZ	DQN,
3102	GSC	DQW2,DQW3,DQN,
3103	GSC	DQW3,DQN,
3104	WAN	DQW2,
3105	WAN	DQW2,
3106	FER	DQW2,
3107	GAN	DQW2,
3108	STA	DQW3,
3109	TER	DPW1,DPW4,
3110	MRI	DQW3,
3111	AIZ	DQW3,
3112	JUJ	DQW3,
3113	GSC	DQW3,
3114	GSC	DQW3,
3115	HAN	DQW3,
3116	SFR	DQW3,
3117	SFR	DQW3,
3118	FRN	DQW3,
3119	JUJ	DQW3,TA10,
3120	JUJ	DQW3,TA10,
3121	AIZ	TA10,DQN,
3122	JUJ	DQN,
3123	SFR	DQN,
3124	RJW	DQN,
3125	CNG	DQN,

Core sera are listed by workshop serum number, submitting laboratory identification code, and submitted antigen specificity.

Table 2. Tenth Workshop Sera Submitted

REG	LAB	1ALL	2ALL	1MON	2MON	SOCI	TOT	REG	LAB	1ALL	2ALL	1MON	2MON	SOCI	TOT
US5	ABS	5	0	0	0	25	30	GER	KUE	1	0	0	0	0	1
JAP	AIZ	0	0	3	10	0	13	US5	KWC	5	0	0	0	9	14
JAP	AMY	4	1	0	0	3	8	UK1	LAW	2	1	3	0	5	11
UK1	AST	3	4	0	0	12	19	LAT	LAY	2	0	0	0	7	9
US2	BAC	0	4	0	0	15	19	FRA	LEP	0	0	0	0	2	2
UK1	BAT	3	0	0	0	2	5	UK1	MAC	0	0	0	0	6	6
BEN	BER	8	6	0	0	6	20	NCY	MAR	0	0	1	3	0	4
FRA	BET	7	4	0	0	22	33	US2	MBC	19	10	0	0	61	90
FRA	BIG	2	0	0	0	10	12	ANZ	MEH	0	0	0	0	5	5
CHI	BLT	1	0	0	0	2	3	UK1	MID	5	0	0	0	17	22
UK1	BOD	2	0	2	7	19	30	EAE	MIN	2	0	0	0	1	3
BEN	BOU	2	0	0	0	10	12	US5	MIT	14	0	1	0	11	26
GER	BRA	3	1	0	0	11	15	US5	MLS	6	0	0	0	0	6
US6	BRN	9	11	1	1	26	48	US2	MOO	0	2	0	0	2	4
UK1	BRS	10	0	0	0	10	20	UK1	MOR	0	2	1	5	6	14
EAE	BRT	1	0	0	0	7	8	IT1	MRA	4	2	0	0	2	8
BEN	BRU	5	0	0	0	22	27	JAP	MRI	1	0	2	1	4	8
ANZ	BSH	5	5	0	0	16	26	NCY	MRV	15	17	0	0	108	140
NCY	CAR	0	0	0	0	1	1	IT1	MTT	2	6	0	0	15	23
IT1	CEP	9	1	2	2	20	34	GER	MUC	0	0	5	0	0	5
ANZ	CHI	0	0	0	0	4	4	GER	MUE	2	1	0	0	1	4
IT1	CNG	1	0	0	0	6	7	FRA	MUL	2	0	0	0	4	6
IT1	CON	3	0	0	0	1	4	FRA	MYE	7	2	0	0	25	34
US5	CRO	13	0	0	0	9	22	EAE	MYR	17	0	0	0	1	18
ANZ	CRS	7	0	0	0	11	18	JAP	NAI	4	4	0	0	7	15
UK1	CXH	0	1	0	0	5	6	JAP	NAK	0	0	0	0	2	2
ANZ	DAW	10	12	0	0	16	38	GER	NEU	1	1	0	0	3	5
ANZ	DCH	3	0	0	0	12	15	IT1	NUS	0	0	0	0	3	3
FRA	DDC	13	2	5	6	8	34	US1	NWO	3	0	0	0	8	11
US8	DEC	1	0	0	0	1	2	FRA	OHA	8	1	0	0	24	33
BEN	DPT	12	0	0	0	22	34	US3	PER	6	1	0	0	5	12
UK1	DRK	1	4	0	0	11	16	US5	PNW	3	0	0	0	17	20
US7	DUP	20	8	12	6	217	263	US7	POL	0	1	0	0	13	14
US1	DUQ	4	5	0	0	24	33	FRA	PRR	4	2	0	0	16	22
SAF	DUT	16	0	0	0	12	28	EAE	PTR	1	0	0	0	5	6
UK1	DYR	0	0	0	0	5	5	IT1	PUR	2	0	0	0	4	6
BEN	ENG	14	0	0	2	9	25	US4	RAD	0	0	0	0	4	4
NCY	FAL	0	0	0	2	0	2	BEN	REE	0	0	0	0	5	5
FRA	FAU	8	13	7	0	39	67	EAE	RIC	8	1	0	0	9	18
IT2	FER	8	5	1	2	21	37	US8	RJW	0	0	0	1	0	1
UK1	FES	2	3	0	1	3	9	BEN	ROO	5	2	0	0	25	32
ANZ	FON	0	0	0	0	2	2	US8	RUB	0	2	1	0	0	3
US8	FOT	1	0	0	0	1	2	JAP	SEK	0	1	0	0	2	3
NCY	FRN	0	0	3	3	0	6	NCY	SFR	0	0	1	9	0	10
US5	FTW	2	0	0	0	5	7	EAE	SHA	2	0	0	0	7	9
IT1	GAB	1	0	0	0	2	3	US5	SIN	1	1	0	0	8	10
IT2	GAN	8	1	3	2	9	23	US4	SND	1	0	0	0	10	11
US6	GAT	0	0	0	0	1	1	US4	STA	1	3	0	2	9	15
GER	GAZ	1	0	0	0	10	11	US7	STE	3	0	0	0	18	21
UK1	GEL	2	0	7	0	9	18	GER	SUD	0	0	0	0	1	1
UK1	GLA	2	0	0	0	14	16	SCA	SVE	0	0	0	0	16	16
NCY	GOE	4	2	0	0	16	22	US5	SWE	0	1	0	0	4	5
GER	GOL	8	5	3	1	14	31	ANZ	TAI	1	0	0	0	10	11
LAT	GOR	4	2	0	0	12	18	JAP	TAK	3	1	0	0	6	10
FRA	GRA	0	0	0	0	1	1	UK1	TAT	3	1	0	0	7	11
US3	GRU	5	0	4	0	3	12	US5	TER	15	14	12	5	24	70
US7	GSC	0	0	12	10	0	22	US1	THP	0	3	1	1	2	7
US3	GVY	1	0	0	0	4	5	JAP	TKB	2	1	0	3	0	6
SAF	HAM	11	7	0	0	42	60	JAP	TOK	4	0	0	0	0	4
US7	HAN	1	2	1	1	9	14	NCY	TOM	4	0	0	0	6	10
US2	HFP	1	0	0	0	13	14	US6	TRP	1	0	0	0	4	5
US1	HJH	0	3	0	0	15	18	SCA	TSB	0	0	0	1	1	2
GER	HOP	0	0	0	0	4	4	JAP	TSU	6	0	5	0	9	20
US1	IND	0	0	0	0	16	16	US1	TUL	1	0	0	0	13	14
EAE	IVS	1	0	0	0	11	12	US5	UAB	0	0	0	0	8	8
UK1	JAM	0	0	0	0	3	3	BEN	VER	1	0	0	0	11	12
FRA	JEA	0	1	0	0	1	2	FRA	VIV	0	1	1	0	0	2
SCA	JER	3	0	0	0	1	4	US5	WAL	1	1	0	0	0	2
US8	JLE	15	2	0	0	25	42	GER	WAN	0	0	9	20	0	29
LAT	JOB	0	0	0	0	1	1	US1	WAR	0	0	1	0	0	1
UK1	JOY	1	1	0	0	8	10	EAE	WEG	4	1	0	0	6	11
LAT	JRM	1	0	0	0	0	1	US2	WOL	3	1	0	0	7	11
JAP	JUJ	5	1	0	5	5	16	EAE	WTZ	0	1	0	0	0	1
NCY	KAP	14	2	0	0	41	57	NCY	YAG	2	0	0	0	4	6
EAE	KAS	1	0	0	0	5	6	CHI	YEG	0	0	0	0	12	12
GER	KEY	0	0	0	0	1	1	NCY	YUN	3	2	0	0	16	21
SCA	KIS	1	1	0	0	2	4	EAE	ZAR	1	0	0	0	7	8
FRA	KRE	2	0	2	0	5	9	CHI	ZHA	4	0	0	0	2	6
JAP	KSH	1	0	0	0	2	3	***	TOTAL	530	211	112	112	1660	2625

Number of sera submitted to the workshop are listed by submitting laboratory. (1)-class I; (2)-class II; (ALL)-allosera; (MON)-monoclonals; (SOCI)-antigen society; (TOT)-total.

Table 3. List of Antigen Societies

Society No.	Antigens Included in the Society
1	A1, A3, A11, AW36, A9, A23, A24
2	A10, A25, A26, AW34, AW66, AW43
3	A29, A30, A31, A32, AW33, AW74
4	A2, AW68, AW69, A28
5	B5, B51, B52, BW53, B18, B35
6	B12, B44, B45, B21, B49, BW50
7	B14, BW64, BW65, B8, BW59
8	BW70, BW71, BW72
9	B15, BW62, BW63, BW46, BW75, BW76, BW77
10	B17, BW57, BW58
11	B16, B38, B39, B37, BW67
12	B22, BW54, BW55, BW56, BW42
13	B7, B27, BW47, BW73
14	B40, BW60, BW61, B13, BW41, BW48
15	BW4, BW6
16	CW1, CW3, CW9, CW10, CW11
17	CW5, CW8
18	CW4, CW6
19	CW7
20	CW2
21	DR1, DRW10
22	DR2, DRW15, DRW16
23	DR4, DRW9
24	DRW8, DR5, DRW11, DRW12
25	DRW6, DRW13, DRW14
26	DR3, DRW17, DRW18, DR7, DQW2
27	DQW1, DQW5, DQW6
28	DQW3, DQW7, DQW8, DQW9
29	DRW52
30	DRW53
31	Qa and TL-like antigens
32	DP Serology
33	Interlocus and public
34	DQW4
35	ELISA exercises

Antigen groups or societies and the HLA antigens that are included are listed by society number.

Cell phenotypes and cytotoxic score data were entered onto 5.25 inch IBM compatible floppy disks, using Tenth International Histocompatibility Workshop Data Entry Software from Howard Hughes Medical Institute. Disks were sent directly to Salt Lake City, where the data files were sorted and merged.

Selection of Data for Analysis

Quality of data submitted was checked by assessing the consistency of reactions between five hidden duplicate sera provided by Min Sik Park of US5/TER, and by noting the pattern of mean cytotoxic scores across all cells tested in a given laboratory with hidden serial dilutions of a strong multispecific serum. Laboratories fell into two broad categories. Those laboratories with mean correlation coefficients between hidden duplicates which were 0.7 or higher tended to have appropriate patterns of decreasing mean scores with increased dilution of the hidden control serum. The laboratories with poor correlation coefficients also tended to have erratic patterns of mean scores upon dilution of the control serum (Table 4). Cells coming from laboratories where there was poor (mean correlation between hidden duplicates < 0.7) reproducibility of reactions were flagged. Provision was made to be able to run analyses with or without the exclusion of the flagged cells.

The primary purpose of the five class II hidden controls was to have a means by which suitable cells could be selected for serum analysis so that the core sera set could be accurately characterized. Therefore, use of these controls was not just an exercise to test the proficiency of the Workshop participants. In this Workshop, there were no hidden duplicates for class I; however, in the class II sera set, one alloantibody and four different monoclonal antibodies adjusted to their working dilutions were duplicated, and the correlation coefficients of the duplicates were checked. Serial dilution of a multispecific serum (1:1,1:2,1:4,1:8) was used as another control to check B-cell purity and complement toxicity. The mean values given in Table 3 are actually the mean correlations within duplicates and not just the average of the five different r values. For the serial dilutions, the value in the matrix represents the mean reaction strength for each dilution over all cells tested with that serum. If the score for the undiluted multispecific serum is less than 2.0, then the purity of B-cell preparations may be questionable in that laboratory, and these cells might be inappropriate for analysis. If the score for the multispecific serum at the 1:8 dilution is more than 2.0, then this may indicate that the complement may be toxic to the B cells. From the data we tabulated, it appeared that 29 of 126 (23%) of the laboratories displayed a complement

Table 4. Summary of Reproducibility of Class II Testing

Classification	Correlations Between Duplicates							Mean Scores			
	N	Dupl	Dup2	Dup3	Dup4	Dup5	Mean	Und.	1:2	1:4	1:8
A	1857	0.79	0.91	0.94	0.82	0.85	0.89	3.20	1.90	1.39	1.24
B	2645	0.64	0.74	0.87	0.70	0.77	0.79	3.21	1.98	1.39	1.24
C	1639	0.46	0.65	0.83	0.56	0.66	0.71	3.15	1.89	1.41	1.27
D	2083	0.36	0.50	0.71	0.56	0.55	0.59	2.84	1.95	1.45	1.30
F	2143	0.30	0.42	0.50	0.46	0.52	0.46	3.67	3.04	2.67	2.64
Total	10367	0.47	0.60	0.73	0.59	0.64	0.65	3.36	2.27	1.77	1.65

Correlation coefficients between hidden duplicate wells of five broad class-II antibodies appear in left panel. Mean reaction scores for serial dilutions of a hidden class-II antibody appear on right for five categories of labs (see text).

toxicity problem in their class II screenings. If there is no purity or toxicity problem, then the reactions should be decreasing at higher dilutions. Based on the controls, laboratories were classified into categories ranging from A to F. The laboratories in the "A" category obtained a mean r value over 0.85 and the appropriate decreasing score in the dilution test. Laboratories classified in the "B" category obtained mean r values of 0.75 to 0.85 and a decreasing score in the dilution test. Laboratories in the "C" category obtained mean r values of 0.65 to 0.75 and a decreasing score in the dilution test. "D" category laboratories had mean r values ranging from 0.5 to 0.65 and decreasing dilution scores, while laboratories with mean r values less than 0.5 were placed in the "F" category.

Data Analysis: The SERAN Utilities

Serum analysis was done using the set of analytic utilities, called SERAN, developed by Dr. Jean-Marc Lalouel and his colleagues at the Howard Hughes Medical Institute, University of Utah Medical Center, Salt Lake City. The use of these utilities is fully described in the *Seran Manual* and the *Seran Update Manual*. The programs, which can be used on an IBM PC or PC compatible microcomputer, consist of separate modules accessed through a menu. Briefly, these include the following functions:

1. SELECT: This module permits the user to define the subset of cells and sera that will be used in the other modules. The user can select specific cells by laboratory and cell identification code, by laboratory only, by racial code, or by phenotype.
2. TAIL: This is an iterative antibody analysis module that empirically defines the set of antigens which explain the reaction patterns of each serum on the cells with which it was tested. The user is able to set parameters for inclusion, minimum chi-square, and number of antigen-positive cells with a positive reaction score for the program to consider a specific antigen or to continue with the iterative process.
3. QUAL: The quality calculator uses the empirically defined antibody specificities and measures the strength of the serum for each of the specificities by tabulating the fraction of cells with a given reaction score with that antigen. A separate VIEWQUAL utility is used to view the output tables.
4. TYPING: This permits the automated phenotyping of cells. When used with a complete set of sera, it is possible to do a complete phenotyping. Even when using a partial set of sera, this program provides a set of likelihoods for each antigen recognized by the serum set, a statistic which is helpful in identifying variant cells.
5. PATTERN: This is one of the most powerful of the utilities in SERAN. It lists and provides valuable statistics about all cells that have arbitrary permutations of phenotypic and serologic reaction characteristics supplied by the user. The user specifies which antigens and antibodies will be the variables of interest. The program will then give information about all possible permutations of these variables, or the user can restrict the search to any specific subset of permutations.
6. CORR and CLUS: This can perform correlations between all sera selected based on the reaction scores. Sera are then clustered on the basis of similarity of their reaction patterns with the cells tested.
7. SxS and CxS: These can display the correlation coefficients among sera or the reaction scores of cells with sera, respectively. The CxS program allows the user to display cells by specific antigen phenotype, to sort the display by patterns of reaction scores, and to print the display or save it as a textfile. The order of sera in the display can be taken from the clustered order determined by running the CORR and CLUS utilities, or the order can be taken from a user-defined file. Some practical examples of the use of the CORR, CLUS, SxS, and CxS programs are given below.

Decentralized Data Analysis

The assignment of serum specificities and detection of new antigens were the primary tasks of the Antigen Societies. In addition to the preliminary SxS cluster information, CxS displays, and first cycle of empiric analysis provided to the Workshop participants in Princeton, each Antigen Society leader was sent a compressed set of program and data disks containing 1) all data on cells tested with both the core and Antigen Society (the "DAT" files), and 2) all data on the subset of non-Caucasian cells tested on the core set only (the "ETH" file). With this information, the Antigen Societies were able to define the specificity of the sera, detect new antigens and splits, and prepare comprehensive reports on their subset of both the core and Antigen Society sets (see individual Antigen Society reports).

Two cycles of central data analysis were performed on the core serum set. The analysis was done for all cells, and separately for Caucasians, Blacks, and Orientals. The first cycle was done prior to the Tenth International Histocompatibility Conference in Princeton. A cycle of retyping was done, and the phenotype of cells which had discrepancy of 7 log units or more in stated versus assigned phenotype was visually inspected and the phenotype changed when appropriate. There was not a significant difference in the serum statistics in the second run of empiric analysis.

Use of the SERAN Programs with Antigen Society/Group Sets

The SxS Display. Figure 1 illustrates the results of running the CORR, CLUS, and SxS display programs with the "DAT" dataset from Antigen Society 12 (B22, BW54, BW55, BW56, BW42). The rows and columns represent the specific core or antigen society sera indicated in the row and column headers, while the elements

```
                                  |        444 4 4    4 444    44444442224    44   44 444 4       44   444444444 444      99     2
                                  |2333333666363633362666333666666600063335523553555353332355325555555556255533330033230333432222222
                                  |7238042001313142817000844010011077702248990783888080019189197988899919999928800829870005123233 0
                                  |5511734012354372317794220246870506230340792912456631808221178278905366894465756613493456793840 5
-------------------------------------------------------------------------------------------------------------------------------------
MID  275 B7           | 776666666666666554555545545543443433 3     --      ---  --       --                           3-----   -
JLE  325 B7      BW42 |7 777667666777666466455555665444 333 4     --- -   --c --c-      --     -          -   3 3      3-----
DPT  331 B7      BW42 |77 776676767666565664555555554443443 3     ---     ---  ---      --                 4 3   3       -----   -
ENG  381 BW48    B7   |677 8767776665656655456576555454344 3 4     -      ---  --       -                           3-----   -      CLUSTER
MIT  307 BW54    BW55 |6778 76777676675755535546565444 4 33334     ---     --c ----     --     -          -   3        3-----
DPT  343 B7      B27  |66677 5666666664566555656554534334 33       --- --  ---  --       --                              ----   -
DDC  324 B7      BW42 |666665 78877767764555648677776655555 4633             -  --     4   443 3                        ----
SEK 4600 B7      BW55 |6777767 8877666676655645677766663444 35                ---  --     3   33                3        3-----  --      1
KAP 4601 B7      BW42 |66677688 977666665555646677776655554353                  --        3   443   3          3        3----
KAP 4612 B7      BW42 |667776889 886666756556566887766655543 53       -     ---  ---     3   34                 3     3 3-----
ENG  333 BW73    B7   |6666667778 87665656656466666654544543 3       -     ---  --      3                                -----   -
TUL 4615 B7      BW42 |67767677788 776565664656577665654544 4       -     ---  ---     3   3           -    3   3       -----
DDC  334 B7      BW73 |676666766677 866546646455565544434 34 3     --- -   ---  ----    --     -       -                ----    ---
MBC 4613 B7      BW42 |6765666666678 555567564546655444343 4       --- --  ---  -----    -     -                        --------- --
MIT  347 BW67    B7   |66667676666665 66444455566665544 34434 3     -     ---  --      3     3                   -      ---
DUT  322 BW42         |565554766655656 5344 43445654543333 4454    ---     ---  ----    4     3          -   3 33       3---  --
KAP  383 B7      B40  |5666756767665565 654454555544444 3 3 4      --      --- ----     --                             44-----   ---
MYE 4611 BW42    BW73 |4456564655554543 6 54444455434444 3                 -------                                     344-----   --
CEP  277 B7           |56655656566666 4455 7664445544443 3 3          -- --  -------     --   --                        3   -----
YEG 4607 B7           |5665555555666744447 653434444333        - -   -------------   - --   --    -          -          -----    -
IND 4609 B7      BW22 |544435555554454 4466 545454553554343    3333              3   33 3   3    333    3                        -
KAP 4604 B7      BW55 |5555556666666654546 55 45567555553344 3 3              ---  --      3          3                    3----
KAP  382 B7      B27  |4556564445454453444344 554444333  33                   -     -                      -      -     ----   --  -
-------------------------------------------------------------------------------------------------------------------------------------
DPT  342 B7      B27  |5555456566665554544455 5 676775656655 3 33 33333 3    3      3433454454                          ----    ----
MYR  340 B7      B27  |5557666666655464554345 56 6666656455434        33 3         4   453333                          3----    -
VER 4602 BW55    B7   |4556557778675665555456476 887777555635333 3333434    3    33533554344                            ----           2
ENG 4614 B7      BW42 |5655657778676666545447466 8 7776654464533         3          35   444 33                          ----        -
MRV 4606 B7      BW22 |5555477776655654344554768 7 877765563533  3333344    4     3543554444 3                        - ----      --
YEG 4608 B7      BW22 |455545767756555444445547677 8 776555635333  333344    3    33434565445                           ----     ---
POL 4617 B7      BW42 |3444436666454455444333535677 77 654346453343343 5444   33    446335654564      3 3               ---
MYE 4610 BW42    BW55 |44454466665644444443553657677 6 7666534 44334444553   34334434335543443                                  -
IND 4605 B7      BW22 |444443565645444344335535676765 7 5545 3 33 3333444    33 333 433544343            -    -
-------------------------------------------------------------------------------------------------------------------------------------
FRN 2070 B27     B7   |3 33 353554433 3      43 645565465 864 3 3   3334445544554453 443543343           - -          - 333   ------
FRN 2076 B27          |4344345455554433333 33 6554553658 63 3 3    333333     4    3 33 54   33     -             -          ----       3
FAU 2072 B27          |3344335455443343       4435554554646 6 4 3 4333333333    343      43 443333      3           -              -
-------------------------------------------------------------------------------------------------------------------------------------
IND 4603 B7      BW22 |33333354443444443 3 343546666655434            333 33    43      4534333333   3      -          - ---- - - --
CRO  320 BW42         |     3 4333      34     -      3343343          643 3       33     -  ---    4   344    33      3353554     -----
DUT  323 BW42    B7   |34344 6555343 454      3 345555543333 6 5333       3333    -   -     4   5643443      3 4 33      -----         4
NAI  344 BW67         |       3 33      4     -       33333          45 3        3 3               3    43                    -
-------------------------------------------------------------------------------------------------------------------------------------
FTW 4580 BW22         |       3           3        33 3 3333343334 333 444534443   3343343335345446433433
DYR 4597 BW55         |                            3   3 3   3443   3   3 4 444445443 334333444 34444543
DDC  299 BW22    BW42 |                           -3            33      3 33 44 565544433434333434334343333 3 3-          33
GLA  302 BW55    BW56 | -- --         -- -        -3   3 3 3 3333 3     445 66556554645554554454454 54 3434 --        -   3333   -
PNW 4579 BW22         |--- --         -- -- --      3 3 334433333      5466 76665545465545546555566 5 3333- -         -       3-  --
TRP 4581 BW22         |------     -------- --       3 3 333433333      34567 566565656555554575566 55  3 3----      --        3-  --  -
MIT  332 BW73    B7   |                      -      333 335434333 33445565 666555555545444455544 53                            33     --      5
FTW 4584 BW22    BW42 | -    -       --     --      334 33444433 33 4545666 76645565556555465656 63                            3 3
ENG 4585 BW55    BW42 |      -        -     --       3 44454433333344466667 76556655564445766655                              33   - --
MBC 4586 BW55    BW42 |                      -      3343444545333 3 344555667 65565565544446554 44         --                344   ---- -
AMY  306 BW54    BW55 |------ - ----------- -             3 5         335565666 877667654344434433         ---      ---- 6653 ------
TAK 4583 BW54    BW55 |------ ------------ --             4            344554558 76466643   33 33          ------- -- 6763 ---- -
MBC  301 BW55    BW54 |-c--c-------------- -             4     -- 334656555577 66765443343 3333   3   --c------- 45533---- -
KSH  308 BW54    BW56 |                  ---          3335 34      3334455566766 566534433434333          ---      --- 554   ------
-------------------------------------------------------------------------------------------------------------------------------------
JUJ  310 BW55         | -  -           - -----3   3 3 433435443      44456656656465 77654464555554 43 3---        ---- 33 4
JLE  298 BW22         |-------- ----------- -           3 4 3 -- 333555555566767 76443533 333  4333c----------45433-   - -
ENG  312 BW55         |-c------ ------------ --          334      -  3335555556766677 643 54333444 33 3---------  45534-    -          6
HFP 4582 BW22         | -- -     - --- --     3          435      -  4334454555665566 6 5  434334433   3----- -     443
-------------------------------------------------------------------------------------------------------------------------------------
ROO 4591 BW55         |            --       3    3  344333        3445555665544354 45 33445445553                              -
AMY  311 BW55         |                    -    3 333343     4    343555454443444 43 3 64534545433333
CEP  297 BW22         |       433333   34      33 44555464443454433444444544 3 3443   36 4455645433433       3
BOU 4578 BW22         |------          --   - --    3 3 433334333    5 346545444 336554444 6456564 544 - -          ------4     -
SIN 4592 BW55         |--- --           -    - --    3 3 343333   4    33355744544343434345 46 456664         -           ------ ------
MBC 4587 BW55    BW42 |       4343 3           3   44545555555 54335 4444555676433453345 3544 765563                             -         7
MRV 4588 BW55    BW42 |       4344      33          55545665444434645434555565 3  35 33445557 76663                        ----
YAG 4589 BW55    BW42 |       3 3          --      4344455443 3344344455656654334533345666 67 666                        - ----  -  - -
BLT 4590 BW55         | -  -        --      --      433 444333 33 3 4434664564433353445445656 6 66                           ----  - ---
BAT 4595 BW55         |       3               3    5343445444333 4 65356546543 335344555665666 643333                       ----
MAC 4593 BW55         |         3             3    434345643333334 44345556543 334 435444466666 5    3                      -
-------------------------------------------------------------------------------------------------------------------------------------
POL 4616 B7      BW42 |                        33        43        33 333     33            3333   33   45      3
OHA  296 BW22         |                       -        3            3   33            443   335        3    544---          ----   3 3   -
MRV 4598 BW56         |                                       3     4 3433         3 333   344        3   5 64---                             8
MRV 4599 BW56         | -   -      --                          -     3   33            3    334        3   46 4              ---
TSU 4594 BW55         |                     -     3            3    3 3433         3333 33           333444       -
-------------------------------------------------------------------------------------------------------------------------------------
GOE  394 B13          | 4                            3          33   - --     ------c--     -           --   45334433   -----   -
CRO  326 BW42    BW41 |                       -         --  -3         - -    ---------       -          --   4 6434435   ----
KAP  385 BW41    BW42 | 33 3   333 3    3                3          54   ---     --c-----     -          --   56 4445433 -----
MYR  387 BW41         |                                 -   3           -      --   ---                        344 5443    --   -----
DUP 9005 BW41         | 3               3   -                   53              --   ---     3                -3345 64       ----         9
DUP 9006 BW41         |  3          3    3                       53              --   --                        44446 433     ----
JRM  386 BW41    BW42 |                                -   4             ---   ---                          445444 3      -
HJH  321 BW42         |                   -     -                             - -----                       3343 33 33 -----
PER  293 B27     BW47 |            3        33                              -      -------                       353   3 3    ----
AST  384 B40     BW41 |                    44         -    ----        ---     ------        --  -    -       3     3  4     --  -
DDC 2079 B40     BW47 |33 33  333        344    3  3                                 --         --           -         4      -  -
-------------------------------------------------------------------------------------------------------------------------------------
AMY  303 BW54         |----- ------ ------- ---------  3  ----              36645 444       -- ----- - - ----    --     873 ----
JUJ  304 BW54         |------------------- ---------  3  ---     3   3 3467553554      -- ----    - - ---------   8 84 ----
TKB  305 BW54         |------------------- ---------  3  ---   33   333456543453       -- ----      - --- -- --   78 4 - -            10
YAG  456 B16          |-------------   ---- --------       ---   33        333   33       -- ----      --- -- ---  344   ----
-------------------------------------------------------------------------------------------------------------------------------------
RIC  317 BW56    BW55 |------ - ----- -----                --    333 3      3 434      4          3    - ---- - --        44333
CRO  229 BW62    BW70 |             - -                   -  -         --   ------ --      -   --           -          ----4 77664
MRV  233 BW62    B15.:|             -       -             -                  -----          -          3      -          -- -47 7663    11
DDC  228 B35     BW62 |-  -- -  -   -   -   --   -  ---    - - -- -       ---- ------    -    --   -       -    -       ------377 763
BER  234 B35     BW70 |        -     -    --     -     -     --          --- ------ -        -   --           -          -- -3667 54
MYR  230 B15     BW50 |             --   -  -     -    --     -- -              -  -            -    -                        36665 4
BRA  205 B35          |             --         --  --     ----        -     ----- --       --   -    -                          43344
```

Figure 1. Antigen Society 12 serum x serum (SxS) correlation matrix of "SXS" SERAN utility output. The elements are the r values between sera listed in rows and column heads (see text).

of the matrix represent correlation coefficients ranging from −1 to 1. For clarity of viewing, all negative r values have been changed to a "−" sign, and positive correlation coefficients less than 3 were changed to blanks. The order of the sera in the rows is identical to the order in the columns, but the row headers display additional information (i.e., the lab of origin and the submitted specificity of the serum). A great deal of information can be obtained from such a display. First, it is evident which sera exhibit high mutual correlations with each other and should be considered together in further analyses. Second, the stated (or empirically analyzed) serum specificities for the sera in a given cluster can provide an immediate suggestion of which antigens the cluster of sera are detecting in common, while discrepancies bear further investigation. An interesting example is serum 394 in cluster 9 of Figure 1. While submitted as a B13, this serum actually has good BW42 and BW41 activity as well. The small cluster of sera in the lower right corner of Figure 1, labeled cluster 11, appears to be negatively correlated with the BW22 sera in the rest of the table, and represents BW62, BW75, plus some B35 and BW70. Last, looking in the vertical columns above one diagonal cluster, one can also observe clusters of negative correlation with specific other groups. The BW54 cluster 10, for example, is negatively correlated with the other clusters which relate to the BW22 group, except for clusters 5 and 6, which also contain anti-BW54 activity. These groups of negative correlations point to an allelic relation between the antigens that the clusters represent, or alternatively, a negative linkage disequilibrium.

The CxS Display. Figure 2 is a display of the cytotoxic scores of selected cells from Antigen Society 12 with the same sera which were displayed in Figure 1. Figure 2 was generated with the CxS program and stripped of reaction scores of "1" with a text editor. The cells are sorted by phenotype and the submitted B locus antigen assignments are displayed in the left margin along with the laboratory code and cell identification number. Well defined patterns of reactivity are easily seen, which confirm the specificity of the sera. Because the sorted cells also contain a variety of other specificities on the other haplotype, it is easy to determine whether a given serum has additional specificity, and if so what it is likely to be.

The ELISA Exercises

One of the goals of the Workshop was to determine whether measurement of the binding of monoclonal mouse anti-class I and anti-class II antibodies by ELISA technique would yield qualitatively or quantitatively different results from the cytotoxic technique where additional requirements for the fixation of complement are present. A small quality control exercise was run by laboratories YUN, CAR, DUP, and GSC to assess the feasibility of using the Genetic Systems microELISA reader for this exercise. It was found that sufficient reproducibility could be achieved to proceed with a comparison using the Genetic Systems hardware and accompanying software for data collection on PC floppy disks. The report by Karen Nelson et al. in this volume demonstrates clearly that a good degree of concordance between cytotoxicity and ELISA techniques was found, and that some particularly good correlations can be achieved with some of the anti-class II monoclonals.

Results

The Tenth International Histocompatibility Workshop serologic exercises were decentralized into "Antigen Groups/Societies" with specialized interest in analyzing a related set of antibodies (Table 3). On the basis of both historical analyses and preliminary analysis of the Tenth Workshop serology data at Princeton, the World Health Organization (WHO) approved seven new class I provisional "W" designations: AW74 (TH); BW75, BW76, BW77 (B15); CW9, CW10 (CW3); and CW11 (Cx46, CW1+3). Ten new class II provisional specificities were also approved as follows: DRW15 and DRW16 (DR2); DRW17 and DRW18 (DR3); DQW4 (DQ"Wa"); DQW5 and DQW6 (DQW1); DQW7, DQW8 and DQW9 (DQW3). Numerous other splits and antigen variants which were suggested by further analysis of the data after the Princeton meeting are reported below.

A Locus. The A26.1 variant of A26 was defined by positive reactions with the good A26 sera as well as serum 9159 of the Antigen Society set. A previously described short AW34 found in Black South Africans could be defined by positive reactions with the AW34 sera 89, 98, and 113 and negative reactions with 108, 93, 94, 95, and 105. AW66 could not be clearly defined in this Workshop.

There was serologic evidence for two A30 splits, based on reciprocal patterns of reactions with sera 136 and 4187 on the one hand, and serum 151 on the other. These apparent splits were respectively associated with B18 and B13 haplotypes. AW74, the new WHO Nomenclature Committee designation for TH could be defined by positive reactions with 4192, 4188, 4189, 4195, and 4198, but negative reactions with 158, 159, 160, and 9013.

There were different short versions of A2 appreciated on selected Caucasian and Oriental cells. The three Caucasian short A2 varieties were characterized by negative reactions with (4229, 4231, and 4239), (19 and 2034), or both groups of sera. The Oriental A2 short patterns were negative with (19, 2032, 2034, and 2036), (32 and 33), or both.

B Locus. The existence of two serologically detectable splits of B35 and of B51 was suggested, and two findings of BW6-associated short B51 antigens were noted in Black and Moroccan Berber families.

A number of good operationally monospecific sera in this Workshop were able to define B49 (441, 4391, 4395, and 4396). Some of the other B49 sera had additional BW50, BN21, B5, or B44 activity. The short BW50 antigen called BN21, associated with BW6 and found in Native Americans, could readily be detected in this

```
                       ************************       *****    **************    ***********      ***********    *******
       SERUM CLUSTER: <------1-------------><---2---><-3><-<4><-----5-----><-6-><----7-----><8><---9----><-10-><-11->
                      |       444 4 4    4 444    44444442224    44   44 444 4      44  444444444 444      99     2
       SERUM NUMBER:  |2333333666363633362666333666666600063335523553555353332355325555555562555333003323033343222222
                      |7238042001313142817000844010011077702248990783888080019189197988899919999928800829870005123233O
                      |55117340123543723177942202468705062303407929124566318082211782789053668944657566134934567938405
        CELLS          ************************       *****    **************    ***********      ***********    *******
TSU  211 B7      BW54 |886888888888888868888888888888882888688            68828   86    4       8          842
TSU  308 B7      BW54 |88888888888888880888888888888880880888244884086808008808 88808 2804  8  4280    0 022800004 0  8
KSH 6005 B7      BW54 |886888888888888868888888888888886888 88      4  44668888 648                     6 6686
TSU  203 B7      BW54 |888888888888888888886888888888886888888       2 6888 84 286   8                  888
TSU  101 B27     BW54 |8 8 88848888 88 888 8888882886880888  68868888888888888888888  08 4 8  6 6        8 888888 6  88
TSU  404 BW54    BW60 |   6    2       8       4    88088    482864888888888888888   42          6     2  0888808 0  8
TSU  405 BW54    BW60 |  8            8  8 8 8     88088     6688 8888888888888 6            4   8      88888 88   88
TSU  406 BW54    BW60 |  888 6      8  8  8 8 8     88086     860888888088080888   6    26            88088 8 0  88
TSU  408 BW54    BW46 |            4                8 04    46   4 688888088888                         88808
TSU  409 BW54    BW61 |               68            88048          8688888088888                        8888848    4
TSU  410 BW54    BW52 |                            88 886     868888888888888888844  8    4    6          88888    68
TSU  411 BW54    BW52 |                            68 888     6688688888888888882   6    6    4 6    6   888888  6 46
TSU  413 BW54    B51  | 2             6  6          8 888    686888888888888888888   8    6  2       4   888888     8
TSU  414 BW54    B44  |                            686888     868888888888888888   8 8       6 6        888888    86
TSU  415 BW54    BW48 |  888     4 8 8  8  8 8 88    88886     6688848888 8888888   8              4  8888888 4  88
TSU  416 BW54    BW61 |  6             8  8    8   888888     868888888888888888   8     6            8888888 8  8
NEU   22 B38     BW55 |                      484  88848862   4 42888888 6884888  88888268                  26
NEU   24 B39     BW55 |  2    4             2482 88 868888   4  244288 8488648862282886628 2              4   8 2  4
NEU   25 B44     BW55 |      22      6      24 4 884488882   42 28 68882428828882 8 886448            2     8
NEU   27 B13     BW55 | 44   8 4           2 2886 862088882   2  448888868484 888  88686888      8          4   8
TSU  402 BW55    BW61 |      6868     4488  88 8888888880 88886888888888888888888888888888888888448    2  8  88    8 8  8
TSU  403 BW55    BW60 |      8888     8688  886888888888068088488888888888888888888888888886 8          88   08 8  8
TSU  407 BW55         |   6  8     0    8  686888888888088868688888888888088088888 888888888648 8            08808 0  8
TSU  103 B35     BW55 |   4  6   2   6      848888888880488  688 8888888888888888888 8888864868              2    888888
TSU  106 BW55    BW61 |2      66244      88  8448886880880 86 00680888888888888888888888888844246 00    0 488    8 6  8
MRV  767 B18     BW55 |                     02 248 886888868   22   888888828886888 84882  280             8
MRV  791 B18     BW55 |      6              0248886888888848   28 28888888888888888888888888      2             4
MRV  810 B18     BW55 |           2         0  828486 888866   422 8826886 4662888 268862668
MRV  837 B8      BW55 |        4            02 8884888888888    2 288 8888 8888888488888888880            8    6   2
MRV  854 B44     BW55 |                     0 6868288 888888   424488 888868888888 888888888     4              2   2
NEU    1 B44     BW56 |           8           6  8  44884     42  4          8868   8        8             8 844
NEU    3 BW56    BW56 |     4  2            4  4  8  468846    6            2 88 8   8 2      62 2           8 86 2
NEU    5 B44     BW56 |                           8  048868    2    6        8888   8        8  8        6   8 68
NEU    9 BW57    BW56 |  2      4          8222 6  688882    8   2 2 44    68884   8 42     82 4     2    2  84822
NEU   10 BW56    B14  |                    8      6  088884     2  2          88 8   8        4 28              888  2
NEU   11 BW62    BW56 |  2   44 2           4 2688 86 888882    46 464888822 884806  86884848               8    8 84
TSU  412 BW56    B51  |           8          888 888 8888 80   88 88  868  888888  88      6  8888 6 6      8   688  8
NEU   53 BW62    BW56 |        2            2   82 8  84888       4      2     8466   8  2     6  8      2  8   846642
REE   27 B8      BW41 | 44 4   42          2646    2 206    4 62 22     2        22                     488862886686     262 46
REE   28 BW41    B49  |  2 2               2          02  24 6                2   4                    68888 684486    2 24 66
REE   29 BW41    B14  |  2 2   2           2 2         204 226 2  2      4        224  2      22   26686 244486     22  26
REE   32 B5      BW41 |  2 6   6           22           0222 2 2   4              622                4686 248688    6 42 64
REE   34 B44     BW41 |        22 4      2  2       2204      4   2        2       22      2   2   48868 66648
REE  262 BW41    BW70 |                   4 2         202      2                   22                26664 24428      2462 42
REE   60 BW58    BW41 |                     2         20   2  2  24       42       22     2   222    6686 244 82          6
REE   61 BW41    BW41 |2 2 2    2    2    4646          042 22 62   4       2       44              2   6886 666686      24  2
MRV  806 B51     BW41 |                     0           0   2                                      0   8888888  88
HAM   13 B13     BW41 |0                6     6         000 6                           8            8888688 880          64
REE   64 BW42    BW70 |46448286882     6646   2 2482688066888 86 22   2 2642     2    4 22 662426      688   6648442     2466 64
REE   65 BW42    B44  |68468 8888462 866 2 4 2682888088886 88 462     6642 2     26 62 8866664      248   2  642            6
MRV  817 BW50    BW42 |48448 888824 24882   02    88888888888884 2      68 2   22   22 822804228      8886868 2 8        888882
MRV  879 BW42    8W58 | 8268 888882  688    0     888888888888888            46     6    4 8  822 260      888 2 8   8       8 8
MRV  908 B8      BW42 |682086888868 2888     064888888888888888  6      686    86    8 8  8862880      8884688 408
DUT  507 BW42    B45  |8888888888 866 4868 48888888888888888888882846628888 2 6    888668888488      88888848682      4  82
DUT  511 BW42    BW71 |888888888866648888    88888888888088886 8     68464  6     48  86  466      8888668 888        888884
DUT  524 B8      BW42 |8868868888 6  8884    846888888888888888 264     46664   6   6 8  88   8       84  8 688
DUT  526 BW42    B45  |8888868888868 8882    2668888888888888888444    488886   64 4 248  8842 8     6888668 668          6  8
DUT  529 BW42         |8888888888868888888 88 8888888888888888868446888886   66   4868 888886844  44688648 868
REE  184 B35     BW67 |     00062604020002 2660006688086000   44 6    0642      6400   64 64660  4    00   0 0    02266 20
REE  240 BW67    BW62 | 226000668462200066 620008688086000   6646  2  6462     486  4 66 6662  2  2      0 0     4888264
REE  241 BW67    B49  |24260008686846000662860008688088000 288684 6 486642  22688 66 8866666  6          0 02  4  26262
MRV  906 B35     BW67 |   8  8888 2 8886    0 6 688888888888 66       866   8   6  8  80 2 20          0  8     888888
                       ************************       *****    **************    ***********      ***********    *******

       ANTIGEN                       PATTERNS OF ANTIBODY REACTIVITY WITH ANTIGENS WITHIN GROUP
    B7 OR B27:  |+++++++++++++++++++++++++++++++?????-------------------------------------------------------------|
         BW54:  |------------------------------+++++----+++++++++++++++++---------------------------------++++-------|
         BW55:  |-----------------------++++++++++++---+++++++++++++++++++++++++++++-------------------------------|
         BW56:  |------------------------------+++++-------------------++++------------------------------------------|
         BW41:  |----------------------------------------------------------------------------+++++++++++----------|
         BW42:  |+++++++++++++++++++++++++++++++++++++++(++++++++++++)---+++++++++++-----(+++++++++)-----------|
```

Figure 2. Antigen Society 12 cell x serum (CxS) matrix of primary reaction scores of cells with the sera listed in column heads. The panel below indicates the conclusions that can be drawn about the specificity of each serum displayed.

Workshop. BN21 cells are positive for the broad (B49+BW50+BN21) and (BW50+BN21) sera, but negative with the (B49+BW50) sera 4408, 4419, 4427, and 4426. Good evidence for a short B44 was found in the Antigen Society sera, possibly in linkage disequilibrium with CW5.

Although they found no new splits of B14, Antigen Society 7 noted a correlation of BW64 and BW65 with the congenital and late onset form of 21-hydroxylase deficiency, respectively.

Two possible splits of BW72 were appreciated based on negative reactions with 198, 4478, 205, 9201, 4483, and 9217 only (BW72.1), or negative reactions with 200, 4484, 9195, and 9206 (BW72.2).

Six splits of B15 were recognized in the Tenth International Histocompatibility Workshop, four BW6-associated and two BW4-associated. The newly WHO-designated BW75 antigen (BW6) has been previously called BW62.1, TS1, B15 ShortThai, B15Kemp, SH7, and BW62S. Te79 may also be equivalent to BW75. Another B15 split, previously B15 (Siamese), has been named BW76 by the WHO Nomenclature Committee. Like BW75, this BW6-associated antigen is selectively found in the Southeast Asian population. A variant of the BW4-associated BW63 antigen was defined on analysis of Negroid populations (BW63.1). The WHO Nomenclature Committee also found sufficient evidence to assign the BW4-associated B15Thai antigen the designation BW77.

Two sera (467 and 4562) are negative on a set of B39 cells and may define a new split within the B16 group, possibly the previously described SH6 or ST-16 antigens found in Orientals and Mexicans, respectively.

The BW56 antigen was split into a long BW56.2 (Japanese) and short BW56.1 (Caucasian) variant on the basis of reactions with sera (4584, 4593, 4617, and 4614). Of interest was the observation that a number of BW56 core sera had cross-reactivity outside of the BW22 complex.

A new antigen, called B-POT by Reekers, could be defined by eight of the Tenth Workshop sera. Several sera were able to define the splits of B7 and B27 recognized by cytotoxic lymphocytes.

There were variants of both BW60 and BW61, designated "a" and "b," respectively. As described in the Ninth International Histocompatibility Workshop, a short BW41 was seen in Hungarian gypsies, Brazilians, and some other individuals. A short BW48, corresponding to the Japanese "FU," was also described in this Workshop. Two B13 variants in addition to the common long B13 were identified by negative reactions with five monoclonal antibodies (2080, 2094, 2101, 2054, and 2081).

Analysis of BW4 and BW6 sera yielded results similar to previous Workshops. A number of BW6 sera were monospecific, including two monoclonals (2106, 2107). Conclusive evidence was provided for an epitope shared between BW4, A32, and A9 as several monoclonals (2057, 2064, 2065, 2102) as well as allosera (493, 498, 4758) shared similar cross-reactive patterns of cytotoxicity on cells bearing these antigens.

C Locus. New WHO assignments were given to C locus antigens, including two splits of CW3 that were designated CW9 (the long variant) and CW10 (defined by negative reactions with 9064 and 532). Another short variant of CW3 was also indicated, which was only negative with the serum 532. The new WHO assignment CW11 was given to the CW1/3 variant that is associated with BW46 and shares CW1 and partial CW3 reactivity.

DR Locus. Evidence for a DR1 short variant was found. This short variant constitutes 16 to 17% of DR1-positive cells that fail to react with the serum cluster (5011, 5018, 5015).

There was sufficient proof to give separate WHO designations to DR2 long (DRW15) and DR2 short (DRW16). DRW16 can be defined by negative reactions with nine workshop reagents, including the core sera 1031 and 1032. DR2.3, which was found in Orientals, and DR2.4 or LUM, which was found in South African Blacks, were based on positivity with most DRW15 sera and short reactions with some of the DRW16 sera.

There was a good monospecific DRW13 serum (1134) and DRW14 serum (9060) on the core set, but both were somewhat weak in their reactivity.

DQ Locus. Two serum clusters made it possible to define the new splits of DQW1, which were designated DQW5 and DQW6 by the WHO. DQW5, defined by sera 9358 and 1136, is associated with DR1, DRW10, DRW14, and DRW16. DQW6, the other split of DQW1, is associated with DRW13 and DRW15 and can be defined by sera 9354, 5343, and 5337.

Three newly WHO-named subtypes of DQW3 were approved. Each can be defined serologically. There also appear to be DNA restriction fragment and two-dimensional gel DQ-beta chain biochemical variants that correspond to each of these splits. The new splits are called DQW7, DQW8, and DQW9. A set of allosera and four monoclonals gave good definition of DQW7 (formerly DQW3-TA10). Two sera (5360, 5361) may define a further subtype of DQW7 on both the DR4 and DR5 haplotypes. The DQ "Wa" antigen, associated with DRW8 and DRW15 haplotypes, was given the new WHO designation of DQW4.

DRW52 and 53. DRW52 represents serologic epitopes that can be found on the DRB1 or DRB3 gene product, depending on the haplotype. By direct binding assay, it appeared that the oligonucleotide-defined variants DRW52a and DRW52b could be detected by the serum clusters (5372, 5373, 5382, 5388) and (5377, 5385, 5390), respectively. By cytotoxicity, the allosera did not give clear cut correlation with the oligonucleotide variants. Of interest, the monoclonal 3063 correlated highly by cytotoxicity with the DRW52a (LB-Q4) variant associated with the B8-DR3 haplotype.

Among the DRW53 sera, 1212 failed to react with 62% of Black DRW53-positive panel members. These negative reactions were selectively found on DR4-positive rather than DR7-positive haplotypes. Many of the DRW53 sera appeared to exhibit a bias for either the DR4 or the DR7 haplotype, raising the question of possible differential reactivity with the DRB1 gene product of the two haplotypes.

T-gamma/TCA. A miniworkshop was conducted in Leiden, Netherlands to characterize a variety of sera without classic anti HLA reactivity, and to compare methods for screening these antibodies. The exercise included 1) seven "anti-TCA" sera, which define a diallelic system on T-gamma lymphocytes; 2) 57 platelet absorbed anti-PHA blast sera; and 3) seven anti-T cell subset sera. PHA blasts or T-gamma targets were used with the appropriate sera. It was clear that the TCA system is different from the specificities defined by the anti-PHA blast sera, and that the T-cell subset submitted by Ferrara can recognize variants within the TCA system. The report of the larger workshop on T-cell activation antigens is also reported on in this volume by Gazit et al.

Data Processing and Serum Analysis

Mark Ferguson,[1] Edgar Milford,[2] Robert Wheeler,[1] Bo Dupont,[3] and Jean-Marc Lalouel[1]

Data Processing

The serologic component of the workshop included a set of core exercises and numerous antigen societies. In both instances, each participating laboratory determined the serologic reactions of a panel of cells for a battery of sera. Specialized software was made available to each contributor, allowing him or her to enter, verify, and update their data on a personal computer.

The serologic data were stored into three files: 1) scorefile.dat, the array of serologic reactions of the cells (rows) for the sera (columns); 2) cellfile.dat, a file of descriptors identifying each cell tested; 3) serafile.dat, a file defining each serum.

Individual functions are controlled through a menu (Fig. 1). The attributes recorded for each cell (Fig. 2) included individual identification, family relationship, ethnic origin, cell type, and 14 antigens presumed to be present at 7 MHC loci. The serologic reactions were entered through an interactive program that paralleled the design of the serologic experiment: for two types of trays (72 or 60 wells), the data were entered according to the actual layout of the sera on the trays, individual reactions being entered in a serpentine fashion. A typical screen is shown in Figure 3. Serum reactions were scored according to a conventional semiquantitative scale (1 = strong negative; 2 = weak negative; 0 = not done or inconclusive; 4, 6, and 8 = three degrees of strength of a positive reaction). Both cell attributes and serologic data were entered through a specialized screen editor, testing each individual entry against a list or a range of allowable values. The files defining the sera for the core exercises as well as the antigen societies were created centrally because the experiments were carried out blindly. The definition of each serum included identification number and serum name, antigens presumably detected, and the monoclonal or polyclonal origin of the serum; this information was collected in the file serafile.dat, which eventually also included relevant output from tail analysis (Fig. 4 and Table 1).

Participating Laboratories: US7LAL,[1] NCYCAR,[2] US7DUP[3]

```
HLA Serology Data Base    (C) 1987 Howard Hughes Medical Institute for the
                Tenth International Histocompatibility Workshop

                          STANDARD CORE SEROLOGY

Region Code : US5                Number of wells per tray (60 or 72): 72
Lab Code    : TER                Disk Number                        : 1

              1 - Cell Definition

              2 - Core Serology Score (Tray Reading) Entry

              3 - Print Cell Information

              4 - Print Tray Reading Information

           Enter Selection (hit return to exit): █
```

Figure 1. Menu of the program for serologic data entry.

Immunobiology of HLA
Volume I

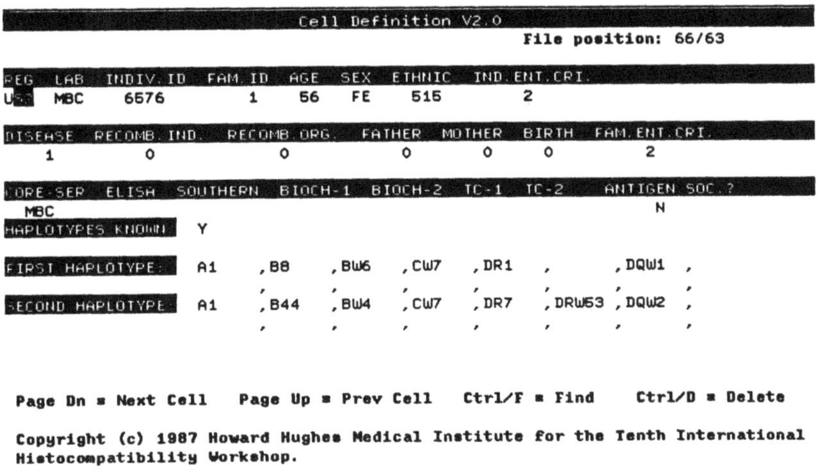

Figure 2. Screen of the procedure for cell definition.

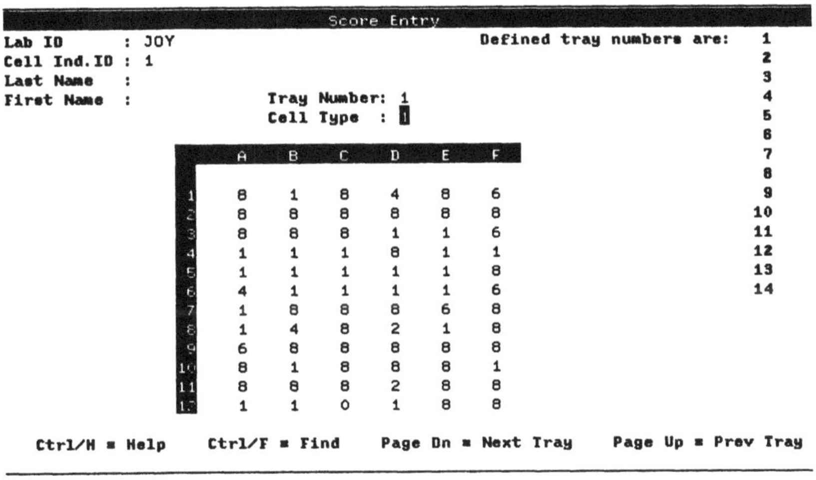

Figure 3. Screen of the procedure for capture of serologic reactions.

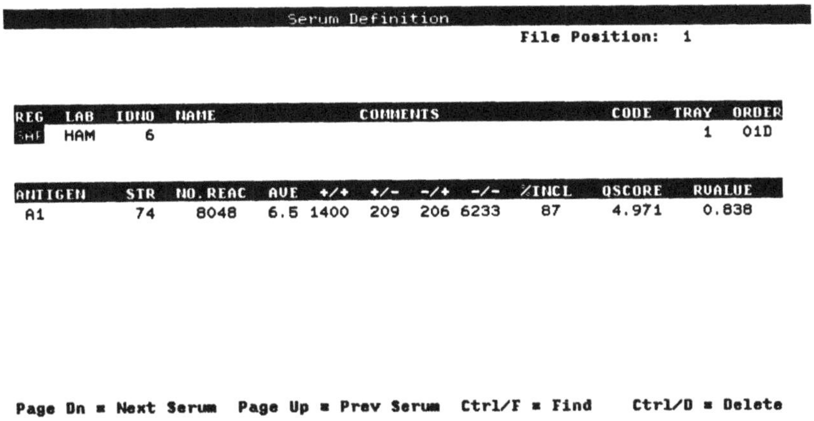

Figure 4. Screen displaying the definition of a serum.

Table 1. A Section of the Tail Analysis Output

	65 US8 JLE HAM												
	A24	93	98	7.6	97	8021	1668	52	98	6203	7.723	0.945	7160.246
+	A9	85	99	5.7	69	6301	25	11	73	6192	4.147	0.416	1091.594
*	A9.3	87	99	7.0	86	6265	6	1	67	6191	6.391	0.265	439.971
	OR	93	99	7.6	96	8021	1693	63	73	6192	8.028	0.950	7241.839
	RES	41	A2 = 10%		CW7 = 6%								
	72 BEN ENG CLB9												
	A24	94	97	7.6	97	8044	1681	52	166	6145	7.997	0.922	6840.307
+	A23	23	98	2.0	21	6311	76	284	90	5861	0.830	0.285	511.206
+	A9	91	99	5.1	57	5951	20	15	70	5846	3.982	0.352	738.657
	OR	91	99	6.7	84	8044	1777	351	70	5846	5.685	0.863	5992.095
	RES	40	A2 = 10%		CW7 = 8%		CW3 = 6%						
	77 IT1 MTT F1095												
	A24	92	95	7.4	94	8038	1645	97	286	6010	5.964	0.867	6036.305
+	A23	45	98	3.5	45	6296	160	195	126	5815	1.910	0.476	1424.470
+	A9	82	98	6.1	77	5941	27	8	99	5807	4.564	0.401	954.539
	OR	88	98	6.8	86	8038	1832	300	99	5807	5.628	0.870	6088.449
	RES	39	CW7 = 8%		A2 = 7%								

These files specified the serologic data contributed by each laboratory. They were assembled centrally into master files for each component, whether core exercises or antigen societies. The antigen society data so assembled were in turn made available to society leaders, together with subsets of the core exercises for cells and sera belonging to an antigen group. The core exercises were analyzed centrally. Because of the large amount of data generated for this component, efficient computing required substantial computing power and memory.

Tail Analysis

Any given serum may react with several antigens. In addition to one or several primary antigens for which the serum exhibits strong reactivity, additional antigens may elicit weaker reactions. The recognition set of a serum is determined in a stepwise fashion by analysis of the serologic reactions through simple two-way contingency tables. The method implemented for the Tenth Workshop follows the approach used in the past two workshops (1,2), with specific features outlined below.

Serum analysis can be applied repeatedly in an iterative fashion. Given the presumed antigen specificities assigned to the cells, analysis of the serologic reactions leads to the definition of a recognition set for each serum. Each cell can then be typed on the basis of its serologic reactions on the sera, taking into account the presently defined recognition sets of the sera. As a result, the antigens assigned to the cells may be redefined. Depending on the extent of retyping performed, a new cycle of serum analysis and cell typing can be applied.

Various selection criteria can be applied to define the subsets of cells eligible for analysis, including ethnic origin and cell type. Only cells from independent subjects are retained for analysis; thus, cells from children in defined families were not included. In addition, flags can be set during cell typing to eliminate particular cells from a subsequent round of serum analysis. The analysis can be restricted to class 1 or class 2 loci, to subsets of loci, or to subsets of antigens. The set of eligible antigens is defined by the user in a file, specfile.dat. This list includes both broad and narrow specificities.

Each serum is analyzed individually. For a given serum, the analysis proceeds through the following steps:

1. *Screen of eligible antigens.* The number of cells with positive and negative reactions is tabulated for each eligible antigen. In general, this set is very large and includes very many antigens that are not part of the recognition set of the serum.
2. *Selection of candidate antigens.* From the tabulations generated for each eligible antigen, a subset of candidate antigens is selected by applying the following two selection criteria: a minimum inclusion ratio, X_1 (taken as 25% in our central analyses), and a minimum number of positive cell reactions, X_2 (taken as 10).
3. *Selection of leading antigen.* For each candidate antigen, two by two contingency tables, chi-squares, and correlations are computed. The antigen with a positive correlation and the greatest chi-square is selected. If it is greater than a minimum value, X_3 (taken as 25), the corresponding antigen is included in the recognition set of the serum, and the following statistics are computed and included in the output file (as shown in the example of Table 1): strength, statistical specificity, average score, inclusion (proportion of positive cells presenting the antigen), number of reactions, true positives (+ +), "missings" (+ −), "extras" (− +), true negatives (− −), q-score, correlation, and chi-square. These statistics are computed as defined elsewhere (1,2).
4. *Data reduction.* The cells presenting the antigen selected at this step are masked.
5. *Cycling test.* If the number of cells positive for any remaining candidate antigen exceeds a certain value, X_4 (chosen as 10), steps (3) to (5) are repeated until the minimum values X_3 and X_4 are no longer exceeded.
6. *Analysis of the pooled recognition set.* Once the iterative cycle is completed, the same statistics are computed for the union of all antigens included on the recognition set ("OR" in Table 1).

Serum Quality Review V1.0

wsno : 67 Position : 476 Masked : FALSE

	1	2	4	6	8	Strength	Inclusion	Count
A23	0.0881	0.0085	0.0540	0.1534	0.6960	78%	90%	352
* A9	0.7879	0.0303	0.0000	0.0303	0.1515			33
* A24	0.9810	0.0074	0.0054	0.0020	0.0041			1477
* A9.3	1.0000	0.0000	0.0000	0.0000	0.0000			7
Other	0.9766	0.0109	0.0068	0.0032	0.0025			5302
							Total :	7171

View next sera (Y/N/RETURN)?

Figure 5. Example of a serum quality table.

7. *Extra reactions.* The remaining extra reactions are examined, computing their average strength and identifying the most common antigens in this remainder.
8. *Stated antigens unidentified in the analysis.* For each antigen reputed to be in the recognition set of the serum, as initially stated, which was not detected in the present analysis, the above statistics are computed and included in the output file. For each antigen included in this output, a symbol indicates the reason for its entry: "+" flags an antigen found significant, which was not included in the list of stated antigens for this serum; "*" identifies an antigen in the list of stated antigens, which was not found significant in the analysis; a blank identifies an antigen both stated and significant.

For each serum, the list of significant antigens and their corresponding contingency tables are stored in the file defining the sera, serafile.dat, for further analyses.

Quality Analysis

For each serum, cell typing will require estimates of the probabilities of each reaction grade (1, 2, 4, 6, or 8) for each antigen in the recognition set. A description of quality analysis is given by Albert and coworkers (2, pp 24–25). A typical output is presented in Figure 5. In addition to the proportion of cells scored for each reaction grade, strength, inclusion, and number of cells with the antigen are also reported.

Some flexibility in handling cross-reactions and serologic splits was introduced by allowing the user to define splits and superspecificities. The user defines splits and superspecificities in a file (splitfile.dat). The relations so specified are taken into account to generate recognition sets and quality tables to be used in cell typing. For example, assume that A23 and A24 have been defined as splits of A9. If A23 was included in the recognition set of a serum by tail analysis, probabilities of reaction grades will be computed not only for A23, but also for A24 and A9, even if they were not significant enough to be detected in tail analysis (Fig. 5).

The tables so generated can be reviewed interactively through several search criteria. Thus, the procedural VIEWQUAL allows the user to review all sera detecting a particular antigen, or all sera detecting that antigen which did not identify another particular antigen (Fig. 6).

Serum Quality Review V1.0

You can search through the sera based on any of the following criteria:

1. by workshop number
2. by searching for a certain specificity
3. by searching for sera with one specificity excluding another
4. all sera in the file

Enter type of search

Figure 6. Options available for review of serum quality tables.

References

1. Mickey MR, Terasaki PI. The serological data of the 8th Workshop and summary analyses. In: Terasaki PI (ed) Histocompatibility Testing 1980. Los Angeles: UCLA Tissue Typing Laboratory, 1980:21–136.
2. Albert ED, Barbalho T, Baur MP, et al. The serological analysis of the data of the Ninth Histocompatibility Workshop. In: Albert ED, Baur MP, Mayr WR (eds). Histocompatibility Testing 1984. Berlin: Springer-Verlag, 1984:72–74.

Pattern Analysis

Robert Wheeler,[1] Edgar Milford,[2] and Jean-Marc Lalouel[1]

The assignment of antigens to cells is not a straightforward process for several reasons: 1) a defined antigen may encompass several unknown immunologic epitopes; 2) most sera are multispecific, recognizing multiple unknown epitopes or multiple defined antigens; 3) most cells are heterozygous, presenting two distinct antigens at a given locus; and 4) experimental error may occur in determining serologic reactions.

This one to many correspondence implies that the characterization of an antigen often requires the consideration of patterns of serologic reactions for a selected set of sera. Moreover, discrimination between two related antigens may be feasible only by identifying contrasts of serologic reactions on a subset of sera. The examination of patterns of serologic reactions across selected sera has been the favored approach to define and resolve antigens in previous workshops. Particularly critical is the definition of subsets of sera that resolve closely related antigens.

We have implemented a computer program, PATTERN, selected through the general menu of SERAN, to tabulate patterns of reactions across selected sera and extract cells that present particular patterns of serologic reactions. This function complements tail analysis and cell typing functions, as well as the displays procedures implemented in SERAN, SXS, and CXS, as described elsewhere in this volume. It operates on the same general input files used by other serologic procedures. The particular functions implemented in PATTERN are displayed in Figure 1.

Participating Laboratories: US7LAL,[1], NCYCAR[2]

To illustrate these functions, we shall use the antigen group consisting of the antigens A1, Aw36, A3, and A11. Tail analysis, quality analysis, and inspection of data with the display procedures SXS and CXS have suggested that a subset of 10 sera provides adequate discrimination of these antigens. PATTERN allows the user to generate patterns of reaction not only for selected sera, but also for selected antigens. As will be seen, this feature provides flexibility in the identification of reaction patterns and their association to particular antigens. Upon selection of function (1) in the menu of Figure 1, tabulations of antigens and serological reactions for selected items are performed internally. Thereafter, other functions can be selected for various contrasts of the antigens and sera selected through function (1); for each antigen or serum, the user can specify require positivity with the symbol "+," negativity with "−," or either one by simple carriage return. Thus, selecting function (4) and striking the carriage return for all selected items lead to the patterns presented in Figure 2. The first pattern corresponds to cells having none of the items selected. The second pattern, with positive reactions for sera 48, 2005, and 2009, is also positive for antigen A3; therefore, it is the most common serum pattern found in cells presenting the A3 antigen. Likewise, the third pattern is most common in A1 cells; the fourth one is most common in A11 cells.

Any such pattern can be confirmed in one of two ways. One can use function (4), selecting the presence of A3 and absence of other antigens, with no particular selection on the sera, leading to the display in Figure 3. Or one may select function (3), selecting no particular antigen, but rather a specific pattern of serum reactions to

```
Cell Reaction Patterns with Selected Serum  V2.2

1. Compute new pattern reactions for cells
2. Create listing file of cells with specified patterns
3. Show information about a specific pattern
4. Show most common reaction patterns
5. Exit


Enter selection    :
```

Figure 1. General menu of PATTERN.

Immunobiology of HLA
Volume I

```
              Cell Reaction Patterns with Selected Serum  V2.2

              AAAA3678142255  Total Cells                  :  8108
              1W31     580045  Total Matching Selection :  8108
               3 1         00
               6           59
                               Count       Percent      Percent of
                                                        All Cells
Selection                       8108       100.0%       100.0%

              --------------    3563        43.9%        43.9%
              --+------+++--     902        11.1%        11.1%
              +---+++++-----     886        10.9%        10.9%
              ---+--------++     666         8.2%         8.2%
              ----+---------     155         1.9%         1.9%
              +-+-++++++++--     142         1.8%         1.8%
              +---+-+++-----      97         1.2%         1.2%
              +--++++++---++      79         1.0%         1.0%
              -----+--------      74         0.9%         0.9%
              --++-----+++++      66         0.8%         0.8%
              ------+-------      66         0.8%         0.8%
              ------------+-      56         0.7%         0.7%
              +---+++-+-----      45         0.6%         0.6%
                    Press RETURN to continue
```

Figure 2. All patterns of antigens and serum reactions for the antigen group A1, Aw36, A3, A11.

examine the distribution of antigens in cells presenting this pattern of serologic reactions, as shown in Figure 4 for pattern 2 of Figure 2. It reveals that the most common antigen associated to this pattern is indeed A3 – over 51% of the antigens present at locus A in cells having this pattern are A3.

Inspection of Figure 2 reveals that none of the leading patterns include the antigen Aw36. Selection of function (4) with required positivity for Aw36 shows that there are only 57 cells with the Aw36 antigen. To identify pattern(s) associated with the Aw36 antigen, function (4) can be selected, requiring the presence of Aw36 but the absence of all other antigens belonging to this antigen group, leading to the display in Figure 5. The leading pattern in this display is a typical A1 pattern. The second pattern has been identified as a typical Aw36 by the leaders of this antigen group at this Workshop. While resolution of A1 and Aw36 would be based on the patterns + + + + + and − − + + + on sera 3, 6, 7, 8, and 15 for the core exercise data, it is clear that the available sera do not provide accurate discrimination, as attested by the other patterns displayed in Figure 5. Lacking a serum specifically identifying Aw36, the latter antigen is "included" in the A1 recognition sets of all A1 sera but two. The cells with stated antigens (Aw36+, A1−) having a typical A1 reaction pattern can be listed with the use of function (2), as shown in Table 1. The stated antigens used here are the assignments originally given by

```
              Cell Reaction Patterns with Selected Serum  V2.2

              AAAA3678142255  Total Cells                  :  8108
              1W31     580045  Total Matching Selection :  1139
               3 1         00
               6           59
                               Count       Percent      Percent of
                                                        All Cells
Selection     --+-              1139       100.0%        14.0%

              --+------+++--     902        79.2%        11.1%
              --+-+----+++--      38         3.3%         0.5%
              --+-------++--      25         2.2%         0.3%
              --+---+--+++--      23         2.0%         0.3%
              --+------++---      19         1.7%         0.2%
              --+------+++-+      18         1.6%         0.2%
              --+------+----      15         1.3%         0.2%
              --+--+---+++--      15         1.3%         0.2%
              --+------+-+--      14         1.2%         0.2%
              --+-----++++--       9         0.8%         0.1%
              --+------++++-       9         0.8%         0.1%
              --+------+++++       6         0.5%         0.1%
              --+----+-+++--       4         0.4%         0.0%
                    Press RETURN to continue
```

Figure 3. Patterns of serum reactions for A3 cells.

```
                 Cell Reaction Patterns with Selected Serum   V2.2

Reaction Pattern:
 A1      AW36   A3      A11          3       6       7       8       15+     48+   2005
+  2009      54      55

Total Cells          :  8108
Cells with Pattern :    1374     16.9%

Locus    Total          Most Common Specificity Associated with Pattern

A        2604   A3      51.7%   A2      13.1%   A1       7.9%   A24      5.4%
B        2570   B7      18.1%   B35     10.7%   B44      9.2%   B8       5.7%
BW       1969   BW6     60.6%   BW4     39.4%
C        1853   CW7     27.3%   CW4     16.3%   CW6     13.4%   CW3     12.8%
DR       2157   DR2     17.5%   DR1     13.0%   DR4     12.9%   DR7     12.0%
DRW      1272   DRW52   59.0%   DRW53   39.3%   DR-      1.7%
DQ       1784   DQW1    46.6%   DQW3    26.8%   DQW2    21.6%   DQW3TP   2.0%
DP        247   DPW3    37.2%   DPW4    27.5%   DPW1    12.1%   DPW2    11.7%
DW        139   DW2     20.9%   DW1     10.1%   DW7     10.1%   DW3      7.9%


                          Press RETURN to continue
```

Figure 4. Distribution of antigens among cells presenting a typical A3 pattern of serum reactions.

```
                 Cell Reaction Patterns with Selected Serum   V2.2

            AAAA3678142255   Total Cells                :  8108
            1W31     580045   Total Matching Selection :    52
             3 1       00
             6         59
                             Count       Percent       Percent of
                                                       All Cells
Selection   -+--                52        100.0%          0.6%

            -+--+++++-----       8         15.4%          0.1%
            -+----+++-----       8         15.4%          0.1%
            -+--+-+++-----       6         11.5%          0.1%
            -+----+-+-----       6         11.5%          0.1%
            -+---++++-----       3          5.8%          0.0%
            -+-----++-----       3          5.8%          0.0%
            -+----+++-+---       2          3.8%          0.0%
            -+----+++---+-       2          3.8%          0.0%
            -+------+-----       2          3.8%          0.0%
            -+--+---------       1          1.9%          0.0%
            -+--+++-+-----       1          1.9%          0.0%
            -+--+++++-----       1          1.9%          0.0%
            -+---+++++----+      1          1.9%          0.0%
                     Press RETURN to continue
```

Figure 5. Patterns of serum reactions in cells presenting Aw36, but not A1, A3, or A11.

```
Menu                         VT220 Terminal                                   KB
                         Cell Typing Review   V2.1

    Locus : A    Region :       Lab :       Cell ID : 1030  Ethnic Group :  20

    Cell Type :  0   Quality : 0.9190    Father  :    0  Mother :     0  pattern8

    Computed Specs   Likelihood   Pos/Total        Computed Specificities
    A2       AW36     63.1393      42/  43   A2     B35     BW4     CW3
                                             AW36   BW53    BW6     CW4
    Probable Specs
                                                   Stated Specificities
 1. A2       A1       71.0469 *    48/  50   A2     B35     BW4     CW3
 2. A2       AW36     63.1393      42/  43   AW36   BW53    BW6     CW4
 3. A2.1     A1       62.5403      44/  48
 4. A2.1     AW36     54.6327      38/  41    Counts    Round typed        :     1
 5. A2       AW19     51.5774      46/  91   0 :     4  Typed last round :  TRUE
 6. A2       TH       51.4441      41/  88   1 :   369  Type next round  : FALSE
 7. A2       A2       51.0664      34/  35   2 :     8  Soft deleted     : FALSE
 8. A2       AW69     50.9973      42/  47   4 :    13
 9. A2       A28      50.1514      40/  46   6 :    22  Status :
10. A2       AW68     49.2857      42/  47   8 :   119  likely mistype

     Option (T/C/D/E/RETURN) ?
```

Figure 6. Example of a cell with declared presence of Aw36 and absence of A1, but presenting a typical A1 pattern: display of cell typing screen.

Table 1. Listing of Aw36+, A1- Cells that Present a Typical A1 Pattern of Serological Reactions

-A1		+AW36	-A3	-A11	+		3 +	6 +	7 +		8 +	15 -	48 -	2005 -	2009 -		54 -	55			
Lab	Cell	Ethnic	Type	A	A	B	B	BW	BW	C	C	DR	DR	DRW	DRW	DQ	DQ	DP	DP	DW	DW
	282	0	3, 5	A31	AW36	B8	B39		BW6			DR3	DRW6	DRW52		DQW1	DQW2				
	4676	20	4, 6	AW69	AW36	BW53	BW58		BW4	CW4		DR3	DRW13	DRW52		DQW1	DQW2				
	2	20	4, 6	A2	AW36	BW53	BW58		BW4	CW4		DRW10	DRW11	DRW52		DQW1		DPW2			
	10	20	3, 5	A28	AW36	BW53	B14	BW6	BW4	CW4	CW8	DRW10	DR5	DRW52		DQW1					
	11	20	3, 5	AW34	AW36	BW58	B35	BW6	BW4	CW4	CW7	DR1	DR2			DQW1					
	15	20	1, 5	A28	AW36	BW52	B27		BW4	CW2		DRW6	DR2	DRW52		DQW1					
	6210	20	0, 0	A23	AW36	B7	BW53	BW6	BW4	CW4	CW7	DR7	DRW14	DRW52	DRW53	DQW1	DQW2				
	1030	20	0, 0	A2	AW36	B35	BW53	BW6	BW4	CW4	CW3										
-A1		+AW36	-A3	-A11	+		3 +	6 +	7 +		8 +	15 -	48 -	2005 -	2009 +		54 -	55			
Lab	Cell	Ethnic	Type	A	A	B	B	BW	BW	C	C	DR	DR	DRW	DRW	DQ	DQ	DP	DP	DW	DW
TER	1089	20	0, 0	AW34	AW36	BW53	B45	BW4	BW6	CW4	CW6										
-A1		+AW36	-A3	+A11	+		3 +	6 +	7 +		8 +	15 -	48 -	2005 -	2009 +		54 +	55			
Lab	Cell	Ethnic	Type	A	A	B	B	BW	BW	C	C	DR	DR	DRW	DRW	DQ	DQ	DP	DP	DW	DW
KYL	61	20	1, 1	A11	AW36	B35	BW53	BW4	BW6	CW2	CW4	DR1				DQW1					

the contributing laboratories (the latter codes being masked in the present displays). The cells listed in Table 1 may therefore have been mistyped. This is confirmed by inspecting the typing data corresponding to the first cycle of tail analysis and cell typing: indeed, these cells were identified by the typing program as possibly mistyped, as shown in Figure 6 for the cell 1030 listed in Table 1. This cell exhibited positive reactions on 34 of 35 sera recognizing A2, 8 of 8 sera recognizing Aw36, and 14 of 15 sera recognizing A1; positive reactions were noted for both sera (3 and 6), which typically give negative reactions for Aw36.

These examples illustrate how the PATTERN procedure could be used by antigen group and antigen society leaders in preparing their report. In conjunction with other functions provided by SERAN, it permitted extensive examination of serologic reactions for selected cells and sera.

Cell Typing

Robert Wheeler,[1] Edgar Milford,[2] Lorna Kennedy,[3] and Jean-Marc Lalouel[1]

Design Overview

Given serologic reactions of cells on a battery of sera with known antigen recognition sets, cell typing can be attempted with the help of a computer. A number of factors limits the efficiency of cell typing by computer, however. The battery of available sera may not include some antigens in their recognition sets, or may not provide a proper contrast of sera to distinguish two closely related antigens. Serologic reactions are subject to error; were a single serum available for a given antigen, the assignment of this antigen to a cell could be erroneous. In practice, computer typing can at most provide some assistance in cell typing, with definitive assignments requiring the careful consideration of patterns of reactions for selected sets of sera.

A series of computer programs has been implemented for cell typing. This series was designed to allow the user to interact in the process. Each program is selected from the main menu of SERAN and offers a particular function. The first step (PREPTYPE) involves the initialization of the file typefile.dat, which will hold all information accumulated through rounds of typing. Various cycles of typing can then be performed on subsets of loci, cells, or antigens with the program CELLTYPE. All results are accumulated in the file typefile.dat without alteration to the original assignment of antigens to cells included in the file cellfile.dat. At any time in this process, the current status of cell typing can be reviewed with the program VIEWTYPE. During this review, the user can individually select new antigen assignments. Once retyping is deemed complete, the program BLDCELL generates a new file of cell definitions, cellfile.dat, which will update cells in all instances flagged by the user. If necessary, a new cycle of serum analysis and cell typing can be performed.

Participating Laboratories: US7LAL,[1] NCYCAR[2], UKIBOD[3]

These programs operate on a series of data files. The serologic reactions are contained in the file scorefile.dat. Probability calculations require the tabulated probabilities of each reaction grade for each antigen recognized by each serum; they are generated by quality analysis (QUAL) and included in the file qualfile.dat. The file typespec.dat contains the list of all antigens that can be present in the cells tested. For each antigen, this file also includes a rough estimate of its frequency; these frequencies are used as weights in the probability calculations of the typing program. All known splits or superspecificities are defined in a file created by the user, splitfile.dat, as described in the section on serum analysis. Further flexibility is afforded by allowing the user to restrict retyping to a subsets of antigens defined in the file subfile.dat. Thus, if no tested serum recognizes a given antigen, that antigen will not be subjected to retyping. This feature allowed antigen society leaders to partially retype cells on the basis of a select set of sera with a narrow range of recognition. More flexibility is provided in the typing procedures to be described now.

Typing the Cells

Typing may be restricted to subsets of cells or antigenic specificities. Thus, selection criteria defined with the procedure SELECT can be applied; typing can be restricted to a subsets of antigens, cells, or loci.

At a given locus, cell typing consists of assigning a probability to each possible genotype defined by a set of antigens. The elementary probabilities used in these calculations are the prior probabilities of each antigen, defined by the user in the file typespec.dat, and the probabilities of a given reaction grade for a given serum and a given antigen, obtained from the file qualfile.dat, constructed by quality analysis (QUAL).

Serum analysis is performed serum by serum, using all selected cells, neglecting the presumed antibody specificities of the sera, but taking antigenic specificities of

Immunobiology of HLA
Volume I

```
                         Cell Typing Review V1.1

Lab to restrict searches to :
Locus to display : A

You can search through the cells based on any of the following criteria:

    1.  by lab number and cell id
    2.  by stated specificity being a given type
    3.  by computed specificity being a given type
    4.  by most probable specificity being a give type
    5.  all cells with likely mistyped specificities
    6.  all cells with likely split specificities
    7.  all cells with a non detectable specificity
    8.  all cells with non stated specificities
    9.  all cells whose stated and computed specificities differ
    10. cells that were typed last round
    11. cells currently selected for typing next round
    12. all cells in the file

    Enter type of search : █
```

Figure 1. Options available for review of cell typing.

the cells as stated. By contrast, cell typing is performed cell by cell, using all selected sera, neglecting the presumed antigenic specificities of the cells, but using the antibody specificities assigned to the sera by the previous serum analysis. For each cell, the serological reactions are used to compute the probability of each genotype. These calculations were implemented following the logic introduced by Mickey and Terasaki (1) for the Eighth Workshop; some refinements introduced by Deppe et al. (2) were also included. For a given cell, the probability of a particular genotype, G, given the set of all serologic reactions, X, can be expressed as:

$$P(G \mid X) = P(X \mid G)\, P(G) / P(X).$$

$P(X)$, an unknown constant for all genotypes, can be ignored. When prior genotype probabilities $P(G)$ are ignored, the term $P(X \mid G)$ is the likelihood of genotype G given the observed serological reactions, X. $P(G)$ is computed from the allelic frequencies provided by the user in the file typespec.dat. For a given serum, i, the probability of the observed reaction grade, $x(i)$, is provided by the corresponding serum quality table generated by quality analysis. If the genotype under consideration, G, includes one antigen which belongs to the recognition set of serum I, $P(x(i) \mid G)$ is taken from the row including this antigen and the column for which the reaction grade is equal to $x(i)$; if genotype G includes two antigens in the recognition set of serum i, $P(x(i) \mid G)$ is taken from the antigen with the highest inclusion. If the antigens defined by genotype G do not belong to the recognition set of serum i, $P(x(i) \mid G)$ is taken from the last row of the quality table, labeled "other," which tabulates the frequency of each reaction grade for all cells which do not include an antigen belonging to the recognition set of the serum.

$P(X \mid G)$ is obtained by calculating the product of all individual terms $P(x(i) \mid G)$ over all sera. A computing scheme introduced by Mickey and Terasaki (1) greatly simplifies this calculation. In their notations, $P(x(i) \mid G)$ is denoted $p(i)$ if G is recognized by serum i; it is denoted $q(i)$ otherwise. After division by the products of all $q(i)$ over all sera, which is constant for all genotypes, the term $P(X \mid G)$ reduces to the products of the ratios $p(i)/q(i)$ for only those sera which recognize G. Further simplification is introduced by using logarithms to transform multiplications to additions.

Calculations are performed for all possible genotypes, and the decimal logarithms of the probability (or the likelihood) of the 10 leading genotypes are stored in the ancillary file typefile.dat. These genotypes are contrasted to the genotype presently assumed for the cell. This comparison yields one of several possible results: 1) the most likely genotype is indeed the presumed genotype; 2) the most likely genotype involves a split of a broad specificity assumed for the cell; 3) the most likely genotype indeed includes a broad specificity presumed present in the cell, but the next likely genotype which does not include that broad specificity includes one of its splits; 4) the most likely genotype does not include any of the presumed antigens; 5) the presumed genotype includes an antigen which cannot be recognized with the available sera; and 6) no serologic data are available. These situations are flagged by the procedure as correctly typed (1), suggested split (2,3), mistyped (4), or untypable (5,6).

Review of Cell Typing

The results of cell typing can be examined through an interactive procedure, VIEWTYPE. Multiple options are available to restrict the review to subsets of cells, antigens, or other criteria as shown by the screen in Figure 1. Thus, one may only review the cells which appear to be mistyped, or only the cells for which a split is suggested. This review can be further restricted to these cells for which the most likely genotype exceeds the likelihood of its closest competitor by a minimum

```
                         Cell Typing Review V1.1

  Locus : B    Region :        Lab :        Cell ID :        Ethnic Group : 515

  Cell Type :  0   Quality : 0.9444    Father  :    0  Mother :     0

  Computed Specs  Likelihood   Pos/Total         Computed Specificities
  B45      B13    -14.8322     10/  38   AW34   B45     BW4     CW6
                                         A30    B13     BW6     CW8
  Probable Specs
                                                Stated Specificities
 1. B44     BW58    38.5738 *   25/  34   AW34   B45     BW4     CW6
 2. B44     BW57    35.8891     27/  38   A30    B13     BW6     CW8
 3. B44     B17     35.7718     26/  35
 4. B12     BW58    35.1567     24/  33    Counts    Round typed        :     1
 5. B12     BW57    32.5148     26/  37   0 :   26   Typed last round  :  TRUE
 6. B12     B17     31.8984     24/  33   1 :  375   Type next round   : FALSE
 7. B45     BW58    25.4760     23/  34   2 :    5   Family data       : FALSE
 8. B45     BW57    23.3263     25/  38   4 :   14   Soft deleted      :  TRUE
 9. B45     B17     22.6121     23/  34   6 :    7
10. B44     B44     21.7328     13/  18   8 :   67   Status :
                                                     likely mistype
   Option (T/C/D/E/RETURN) ?
```

Figure 2. Typical displays in cell typing review.

amount; thus if the latter parameter is set equal to 5, only these cells for which the odds favoring the most likely genotype against any other exceeds 100,000:1 will be examined. During this review, the user has the opportunity to flag out cells that should not be used in the next round of serum analysis or that should be retyped. In the latter case, a new presumed specificity is selected by the user. Typical outputs are presented in Figure 2. For each genotype, the display includes both the logarithm of the likelihood and the proportion of the sera that recognize this genotype for which the cell exhibited a positive reaction.

Various cycles of typing can be performed. Once retyping is completed, the procedure BLDCELL reconstructs a file of cell definition, where the new presumed genotypes of the cells are updated in all instances flagged by the user. A new round of serum analysis may then be deemed necessary.

References

1. Mickey MR, Terasaki PI. The serological data of the 8th Workshop and summary analyses. In: Terasaki PI (ed). Histocompatibility Testing 1980. Los Angeles: UCLA Tissue Typing laboratory, 1980:21–136.
2. Deppe H, Albert ED, Baur MP. Cell typing by computer. In: Albert ED, Baur MP, Mayr WR (eds). Histocompatibility Testing 1984. Berlin: Springer-Verlag, 1984:60–63.

Antigen Society #1 Report (A1, Aw36, A3, A11, A9, A23, A24)

Kilian W.K. Cheung,[1] Linda S. King,[1] Mary Jane N.M.N. Coffing,[1] Leonard I. Boral,[1] John W. Oldfather,[2] Sanking Chen,[3] Rosemarie Strothman,[3] Shyuan Hsia,[4] Ulf Johnson,[5] Berit Clarkholm,[5] Bengt Low,[5] Edward K. Yang,[6] Helena Mervart,[6] Judith Masniuk,[7] David Rush,[7] Pascale Perrier,[8] John Marwick,[9] Gen-yao Ye,[10] Derek Middleton,[11] Robert A. Good,[12] Jill K. Benson,[12] C. Bernard Carpenter,[13] Rick Belcher,[14] John Lin,[15] and Michael Riley[16]

History

HLA A1. HLA A1 was designated in 1967 as Da-11 by Dausset et al. (6) and in 1964 as LA-1 by Payne et al. (14). It is common in Caucasians, less frequent in blacks, and very rare in Orientals. It received its present designation at the Sixth International Histocompatibility Workshop and was well defined in the Eighth and Ninth

Reporting Laboratories: US5KWC,[1] USIKYL[2]
Participating Laboratories: US5FTW,[3] US4SND,[4] SCALOW,[5] NCYMRV,[6] US5TIL,[7] FRAPRR,[8] UKICXH,[9] CHIYEG,[10] UKIMID,[11] NCYRAG,[12] NCYCAR,[13] US6CBC,[14] US5LIN,[15] USICHR[16]

Immunobiology of HLA
Volume I:

Table 1. Serology Antigen Society #1 Society Cell Panel Makeup

Sera submitted:	98	
Cell panel used:	Total	889 (100%)
	Cau	613 (69%)
	Black	178 (20%)
	Oriental	49 (5.5%)
	Others	49 (5.5%)

Workshops with no evidence of split specificity (8,11). HLA Aw36 was originally demonstrated in African blacks (19) and closely associated with A1.

The most common haplotypes in Caucasians observed were: two-loci association, e.g., A1-B8, A1-Bw57, A1-Cw6, A1-Cw7 (8), A1-DR3, A1-DR7, A1-DR2, A1-DR5 (11); three-loci association, e.g., A1-B8-DR3, A1-Cw6-DR7, A1-Bw57-DR7, A1-Cw1-B35, A1-Bw41-DRw8, and A1-Bw61-DRw8 (11).

HLA A3. HLA A3 was first assigned in 1968 in the "WHO Report on Nomenclature" (13) and did not show polymorphism at the International Histocompatibility Workshops 1980 (10) and 1984 (12). The gene frequency is highest in Caucasians and lowest in Orientals (10). Linkage disequilibrium has been demonstrated in Caucasians with alleles B7, B35, B14, Bw47, Cw4, DR1, and DR2.

HLA A11. HLA A11 was originally defined by complement-fixation, and was considered as part of the LA system (2). It was included in the "HU-1 system" by Dausset et al. (6). It was well defined in 1980 (15) and 1984 (12), and is common in Caucasians.

HLA A9. HLA A9 was demonstrated to have split specificities A23 and A24 by Dausset in 1971 (7). It was confirmed in the 1972 (3), as well in the 1980 (18) Workshops.

HLA A23. HLA A23 has been well defined by high correlation coefficients (1,3,16). It is most common in blacks and very rare in Orientals and Native Americans. In Caucasians, it has positive linkage disequilibrium with B44 and DR7 (1). In blacks, it has lower positive linkage disequilibrium with B7-DR1 (1).

Table 3. Core and Antigen Society #1 Cell Panel Composition

Caucasian	64%
Black	29%
Oriental	4%
Others	3%

HLA A24. HLA A24 was first defined by Dausset in 1971 (7) and was confirmed in 1972 as the split of A9 (3). It has been well defined through the later International Histocompatibility Workshops (17,18). A24 has the highest gene frequency in Orientals and the lowest in blacks. In Caucasians, it has positive linkage disequilibrium as A24-Bw61-DR2 (18). It has negative linkage disequilibrium in Orientals as A24-B7-DR2 (18).

HLA A9.3. HLA A9.3 was originally described as the third subset of A9 in Caucasians and African blacks in 1975 (4). However, it has not been actively considered as valid in the last three Workshops. It was briefly described as a variant of A9 that is distinct from A23 and A24 in an Arab family (9). At the Third Asia-Oceania Histocompatibility Workshop, a variant of A9 that is not A23 or A24, found in Thais, Chinese, Caucasoid Jews, and Caucasoid Indians, was confirmed with family studies and was named A9V2(A9.3?) (5). The A9V2(A9.3?) cells also demonstrated a short pattern with some of the A9 sera (5). It is not clear that the later two observations described the same variant mentioned in 1975.

Serology

Ninety-eight sera were submitted to the Antigen Serology Society #1 and tested with 889 cells (69% Caucasian, 20% black, 5.5% Oriental, and 5.5% others) (Table 1). A total of 171 sera were included as both Antigen Society #1 sera and Core Serological sera. The breakdown of the total sera specificities is depicted in Table 2. In general, the antibody specificities were well covered by either monospecific or multispecific sera with the exception of anti-Aw36 and A9. The cell panels used to examine all sera consisted of 64% Caucasian, 29% black, 4% Oriental, and 3% others (Table 3).

Table 2. Antigen Society #1 and Core Sera Specificities

	Total number of sera submitted: 171		
	Antibody Specificity		
Antigen	Monosp	Polysp	Total
A1	16	20	36
Aw36	1	13	14
A3	29	5	34
A11	21	10	31
A9	2	45	47
A23	13	37	50
A24	3	49	52

Table 4. Antigen Society #1 Sera Reactivity Pattern

	Society Sera									
Antigen	4005	4007	4010	4012	4014	4025	4033	4038	4039	4040
A1	+	+	+	+	+	−	−	−	−	−
Aw36	−	+/−	−	−	−	−	−	−	−	−
A3	−	−	−	−	−	−	−	+	+	+
A11	−	−	−	−	−	+	+	−	−	−

Table 5. Antigen Society #1 Core Sera Reactivity Pattern

	Antigen			
Sera	A1	Aw36	A3	A11
0003	+	−	−	−
0006	+	−	−	−
0007	+	+/−	−	−
0008	+	+/−	−	−
0015	+	+/−	−	−
0048	−	−	+	−
2005	−	−	+	−
2009	−	−	+	−
0054	−	−	−	+
0055	−	−	−	+

Based on the Tail Analysis, Correlation, Cluster, CxS and SxS patterns, the following ten society sera were chosen to define A1, Aw36, A3, and A11: 4005, 4007, 4010, 4012, 4014, 4025, 4033, 4038, 4039, and 4040 (Table 4). Ten core sera: 0003, 0006, 0007, 0008, 0015, 0048, 2005, 2009, 0054, and 0055 (Table 5) were also chosen from the analysis. It appears that all antigen specificities can be precisely recognized with the exception of Aw36—to which no monospecific sera can be defined. The unusual reaction patterns of society sera (4003, 4008, 4006, 4056, 5453) and core sera (0014, 0009, 0010, 0011, 0012, 2010, 2013, 0049, 0056) (Table 6) warrant further investigation.

Although no further split in the A3 antigen specificity can be found after the analysis, two sera (4037 and 4045) have been identified to show striking differences in the strength of reactivity between Caucasian and black populations (Table 7). Serum 4037 appears to be the best for defining A3 in the black population.

Eleven society sera (4061, 4063, 4068, 4072, 4075, 4083, 4084, 4086, 4087, 4094, and 4095) (Table 8) and eight core sera (0058, 0064, 0067, 0068, 0070, 0069, 0072, and 0065) (Table 9) were selected to define the A9 group. These sera have A9 reactivity with the exception of (4083, 4084, 4086, 4087) of the society sera and (0067, 0068, 0069, 0070) of the core sera. Society serum 4087 and core sera 0067, 0068, and 0070 define A23 specificity. Society sera (4094, 4095) and core serum 0065 define A24 specificity. Society sera (4092, 4099) and core sera (0065, 0077) need additional review (Table 10).

Serum 0065 demonstrates slight A9.3 activity. Serum 4092, detecting A24 specificity, reacts well (r=0.845) with the Oriental population while cells from Caucasian and black populations exhibit a short reactivity pattern. Society serum 4068 and core sera (0058, 0060, 0061, 0064), exhibit a short reactivity pattern to A9 as compared to the rest of the other anti-A9 sera, and may indicate the presence of a new variant of the A9 group (Table 11).

Table 8. Antigen Society #1 Sera with A9 Group Reactivity Pattern

	Antigen		
Sera	A9	A23	A24
4061	+	+	+
4063	+	+	+
4068	+	+	+
4072	+	+	+
4075	+	+	+
4083	−	+	+/−
4084	−	+	+/−
4086	−	I	+/−
4087	−	+	−
4094	+	−	+
4095	+	−	+

Table 6. Serology Society #1 Sera Need Further Clarification

Antigen	Core Sera	Society Sera
A1	0014	4003, 4008
Aw36	0009, 0010, 0011, 0012	4006, 4056, 5453
A3	2010, 2013, 0049	
A11	0056	

Table 7. Society #1 Anti-A3 Sera Reactivity Pattern in Caucasian and Black Populations

Cell Source	# of Cells Tested	Sera			
		4037		4045	
		Scores (%)			
		8+	<4 (neg)	8+	<4 (neg)
Cau	145	35	30	49	23
Blk	14	86	14	64	7

Table 9. Antigen Society #1 Core Sera A9 Group Reactivity Pattern

	Antigen		
Sera	A9	A23	A24
0058	+	+	+
0064	+	+	+
0067	−	+	−
0068	−	+	−
0070	−	+	−
0069	−	+	+/−
0072	+/−	+/−	+
0065	+/−	−	+

Table 10. Serology Society #1 Sera Need Further Clarification

Antigen	Core Sera	Society Sera
A9	0065	
A23	0077	
A24		4092, 4099

Table 11. Serology Society Sera with a Short A9 Reactivity Pattern

Core Sera	0058, 0060, 0061, 0064
Society Sera	4068

Conclusions

HLA A1 is well defined in the core and antigen serology society studies.

HLA Aw36 cannot be precisely recognized because of lack of monospecific sera and well-studied cell panels.

HLA A3 further splits cannot be found; however, there is an indication of a possible variant in the black population.

HLA A9 is defined with splits of A23 and A24; the presence of a variant with a short reactivity pattern cannot be ruled out.

A23 can be precisely recognized in the core and serology society studies.

A24 can easily be identified as the split of A9. One serum expresses a long reactivity pattern in the Oriental as compared to the Caucasian and black populations and may indicate the presence of a variant.

A9.3 remains elusive. Only one serum shows slight activity of this specificity. Perhaps further studies can be considered to look into its possible relationship in the unusual reactivity pattern of the A9 group.

References

1. Arnaiz-Villena A. Aw23. In: Terasaki PI (ed): Histocompatibility Testing 1980. Tissue Typing Lab, Los Angeles, 1980, p 302.
2. Bodmer W, Bodmer J, Adler S, Payne R, Bialek J. Genetics of '4' and 'LA' human leukocyte groups. An NY Acad Sci 1966; 129:473.
3. Bodmer JG. HL-A9. In: Dausset J, Colombani J (eds): Histocompatibility Testing 1972. Munksgaard, Copenhagen, 1972, p 637.
4. Bodmer JG. The ABC of HLA. In: Kissmeyer-Nielsen F (ed): Histocompatibility Testing 1975. Munksgaard, Copenhagen, 1975, p 21.
5. Chandanayingyong D, Koktathong K, Wattirangkul S, Udee S, Rungroung E, Bejrachandra S. Is this A9 variant A9.3?. Tenth International Histocompatibility Workshop Newsletter 1987;1:2-3.
6. Dausset J, Ivanyi P, Colombani J, Feingold N, Legrand L. The Hu-1 system. In: Curtoni ES, et al (eds): Histocompatibility Testing 1967. Munksgaard, Copenhagen, 1967, p 190.
7. Dausset J. The genetics of transplantation antigens. Transplant Proc 1971;3:8-14.
8. Festenstein H, Carpenter CB, Alonso A, Williams E. HLA-A1. In: Albert ED, Baur MP, Mayr WR (eds): Histocompatibility Testing 1984. Munksgaard, Copenhagen, 1984, p 118.
9. Gebuhrer L, Betuel H, Labonne MP, et al. A third subset of HLA-A9. Ninth International Histocompatibility Workshop Newsletter 1983;3:9-13.
10. Howson W, Stiller CR, StC. Sinclair N. A3. In: Terasaki P (ed): Histocompatibility Testing 1980. Munksgaard, Copenhagen, 1980, p 298.
11. Ivaskova E. A1. In: Terasaki P (ed): Histocompatibility Testing 1980. Munskgaard, Copenhagen, 1980, p 292.
12. Kissmeyer-Nielsen F, Walford RL. HLA-A3. In: Albert ED, Baur MP, Mayr WR (eds): Histocompatibility Testing 1984. Munksgaard, Copenhagen, 1984, p 119.
13. Nomenclature Committee. 1968. WHO Bulletin.
14. Payne R, Tripp M, Bodmer W, Bodmer J. A new leucocyte isoantigen system in man. Cold Spring Harbor Sym Quant Biol 1964;29:285-295.
15. Rauffoux C, Streiff F. A11. In: Albert ED, Baur MP, Mayr WR (eds): Histocompatibility Testing 1984. Munksgaard, Copenhagen, 1984, p 301.
16. Singal DP, Souza M, Betuel H, Gebuhrer L. HLA-A24. In: Albert ED, Baur MP, Mayr WR (eds): Histocompatibility Testing 1984. Munksgaard, Copenhagen, 1984, p 121.
17. Singal DP, Souza M, Betuel H, Gebuhrer L. A9.3. In: Albert ED, Baur MP, Mayr WR (eds): Histocompatibility Testing 1984. Munksgaard, Copenhagen, 1984, p 122.
18. Tait BD. AW24. In: Terasaki P (ed): Histocompatibility Testing 1980. Munksgaard, Copenhagen, 1980, p 307.
19. Wolf E, Festenstein H, Pritchard J, Watson B, Sachs J, Traub N. Further HL-A heterogeneity in Zambian and Caucasoid populations. In: Kissmeyer-Nielsen F (ed): Histocompatibility Testing 1975. Munksgaard, Copenhagen, 1975, p 179.

Antigen Society #2 Report (A25, A26, AW34, AW43 and AW66)

V. Joysey,[1] E.D. du Toit,[2] R.W. Martell,[2] D.G. Taljaard,[2] B. Crichton,[3] T. Doran,[4] Y. Morishima,[5] P. Wheeler,[6] V. Lazda,[7] S.M. Tait,[1] and A. Cambon-Thomsen[8]

The HLA antigens A25, A26, AW34, AW43, and AW66 form a strongly cross-reacting group as most sera to these antigens react with two or more of these specificities. Monospecific sera against some of these specificities are rare and therefore, most antigens in this group need to be defined by patterns of positive and negative reactions (Tables 1 and 2).

The R values quoted were defined on the post-Workshop corrected data with flagged cells excluded,

Reporting Laboratories: UKIJOY[1] SAFDUT[2]
Participating Laboratories: UKIGLA,[3] ANZBSH,[4] JAPMRI,[5] US5CRO,[6] US1JON,[7] FRAOHA[8]

Table 1. HLA A10. Key Sera–Core Set

	A25							A26			AW34				AW34s				AW43				
	119	120	121	122	123	124	125	100	107	112	83	98	113	108	93	94	95	105	109	110	130	131	88°
A25	+	+	+	+	+	+	+	–	w	–	–	w	–	+	+	+	+	+	–	–	–	–	+
A26	–	–	–	–	–	–	–	+	+	+	+	w	+	+	+	+	+	+	–/w	–	–	–	+
AW34	–	–	–	–	–	–	–	+/–	–	+/–	+	+	+	–	+	+	+	+	–	–	–	–	+
AW34 short	–	–	–	–	–	–	–	–	–	–	+	+	+	–	–	–	–	–	–	–	–	–	+
AW43	–	–	–	–	–	–	–	–	+	+	–	–	w	+	+	+	+	+	+	+	+	+	–
AW66	–	–	–	–	–	–	–	+/–	–	+/–	–	–	+	+	+	+	+	+	–	–	–	–	+
AW33	–	–	–	–	–	–	–	–	–	–	+	–	w	–	–	–	–	–	–	–	–	–	–
A29	–	–	–	–	–	–	–	–	–	–	–	–	–	–	–	–	–	–	–	–	+	+	–
AW19	–	–	–	–	–	–	–	–	–	–	–	–	–	–	–	–	–	–	–	–	–	–	–

Key: ° = negative with index antigen; + = positive; w = weak; – = negative

Table 2. HLA A10. Key Sera–Antigen Society Set

	A25			A26		AW34		AW43	
	4115	4120	4131	9162	4137	9132	4153	9172	9174
A25	+	+	+	−	−	w	−	−	−
A26	w/−	−	−	+	+	+	w	−	+
AW34	−	w/−	−	−	w	+	+	−	−
AW34 short	−	−	−	−	w	+	+	−	−
AW43	−	−	−	−	−	−	−	+	+
AW66	−	−	−	w/−	−	−	−	−	−
AW33	−	−	−	−	−	−	+	−	−

and were computed with the following parameters:

X1 (percentage inclusion) = 25
X2 (minimum number of posit vely reacting cell's) = 5
X3 (minimum chi square for iterations) = 50
X4 (minimum positively reacting cells for iterations) = 5

HLA-A25

HLA-A25 was clearly defined by both Core and Antigen Society serum sets by a number of excellent operationally monospecific sera. Those with the highest R values are shown in Table 3a. The R and Q values indicate that the best sera were 124 (Core) and 4120 (Antigen Society). There were five monoclonal antibodies (MAbs) reacting with A25; of these, four also reacted against other specificities (Table 4). MAb 2116 was the most specific but the least sensitive and in performance did not rival the best of the A25 sera.

HLA-A26

There were no operationally monospecific sera in the Core or Antigen Society sets. Sera with R values greater than 0.6 are listed in Table 5.

There was some evidence for a split of A26. In the Japanese population, an antigen identical with A26.1 reported in the Ninth Japanese Workshop of 1984, appeared definable by a positive reaction with serum 9159 in addition to the other A26 sera.

AW34

Antibody to the original "antigen Malay" (AW33+AW34) recurred in sera of "primary" specificity A10, AW19, and A28. Antigen AW34 cross reacts with A26 and AW33, and A28 (1). Many anti-AW34 sera contain anti-AW33 and discrimination between these antigens can be difficult. There were no sera in either the Core or the Antigen Society sets with high R values. In the present Workshop, anti-AW34 was present in 29 sera of the Core set and 33 sera of the Antigen Society set, but all also contain other specificities. Those with the highest R values are listed in Table 6. Patterns of reactivity with

Table 3. Excellent Anti-A25 Sera

Core	r value	Q score	Tail	Res (of A10 CREG)	
119	0.861	8.234	−	A26	¥
120	0.817	6.416	−	AW68	¥
121	0.838	7.874	−	A26	¥
122	0.927	8.878	−	AW34	¥
123	0.872	9.084	−	A26	¥
124*	0.986	11.211	−	−	
125	0.920	10.477	−	A29	¥
Antigen Society					
4115	0.901	7.779	−	A26,A32	¥
*4120	0.934	9.356	−	AW34	¥
4131	0.924	8.637	−	−	¥
9138	0.904	6.504	−	A26,AW34	¥
9141	0.940	8.550	−	−	
9143	0.955	8.740	−	−	

Key: ¥ = extra reactions not of the A10 CREG.

Core and Antigen Society sera are shown in Tables 1 and 2. Serum 108 was useful for the definition of AW34 in that it reacted with the other antigens of the A10 group but was negative with AW34 cells. MAb 2066 had an R value of .607 but also reacted with A25, A26 and AW43.

A split of AW34 (AW34 short) in South African blacks described by Martell et al. (2) was demonstrable with Core sera but was not consistently shown with the Antigen Society set. The short variant of AW34 could be defined by negative reactions with Core sera 93, 94, 95, and 105 and positive reactions with the other AW34 sera. This reaction pattern was seen in 5 of 23 unrelated individuals and was shown to segregate in one family. This variant was also seen in Australian aborigines.

HLA-AW43

HLA-AW43 was clearly defined in South African blacks and Cape coloreds by a characteristic pattern of positive and negative reactions using both Core and Antigen Society sera. AW43 reacts with both A10 and A29 sera as described by Botha et al. (3) when they postulated an "HLA-A10-W29 allele" in the African population. In the present Workshop, AW43 reacted with 28 Core sera, 39 Antigen Society sera, and two MAbs. Antigen AW43 was assigned on the basis of strong positive reactions with Core sera 130 and 131, which are anti-A29 sera, and also reactions with 109 and 110. Serum 110 is

Table 4. MAbs Reacting With A25

MAb #	r value	Q score	Tail	Res (of A10 CREG)	
2056	0.667	3.644	A32	−	¥
2058	0.487	3.288	A24,A23,A32	−	¥
2055	0.525	4.917	A32	−	¥
2062	0.497	3.190	A24,A9,A10	−	¥
2116	0.765	4.363	−	AW33	¥

Key: ¥ = extra reactions not of the A10 CREG.

Table 5. Selected A26 Sera

Core	r value	Q score	Tail	Res (of A10 CREG)	
99	0.639	4.088	AW34,AW66	A25,AW43	¥
100	0.637	2.845	AW66	AW34	¥
107	0.628	3.252	AW43,A25	AW68,AW34	¥
112	0.695	5.243	AW43,AW66	AW34	¥
114	0.622	2.897	AW43	AW34	
115	0.652	2.778	–	AW68	¥
Antigen Society					
4110	0.675	4.501	A25	AW34	¥
4112	0.630	3.287	A25,AW43,A10,B27	AW34,A11	¥
4136	0.630	2.702	–	A11	¥
4137	0.799	4.882	–	AW34	¥
4147	0.721	4.238	AW66,AW43	AW34	¥
9148	0.608	2.301	A25	AW34,AW43	¥
9162	0.751	3.832	–	AW66	¥
9159 (Japanese A26.1)	0.226	0.786	–	AW34	¥

Key: ¥ = extra reactions not of the A10 CREG.

Table 6. Selected AW34 Sera

Core	r value	Q score	Tail	Res (of A10 CREG)	
79	0.468	2.558	A26,A25,AW43,A10	–	¥
83	0.507	2.309	AW33,A26,A10	A29,A30	¥
87	0.514	2.447	A26,A25,AW43,AW66,A10	–	¥
88	0.575	3.476	A26,A25,A10,AW66	A30	¥
91	0.556	2.439	A26,AW43,A25,AW66	AW68	¥
93	0.703	3.349	A26,A25,AW43,AW66	–	¥
94	0.683	2.874	A26,A25,AW43,AW66	–	¥
95	0.659	2.765	A26,A25,AW43,AW66	A29	¥
98	0.544	3.106	A26,A25	AW33	¥
103	0.506	1.646	AW33,AW66,A28	A26	¥
105	0.721	3.198	A26,A25,AW43,AW66	A10	¥
113	0.514	3.408	A26	–	¥
155	0.507	2.202	AW33	A26,A25	¥
MAB Moab 2066	0.607	2.087	A26,AW43,A25,A10	A28	¥
Antigen Society					
4101	0.549	1.688	A26,A25,AW43,A10	–	¥
4106	0.526	1.870	A26,A25,AW43,A10	AW66	¥
4139	0.705	3.302	A26,AW43	–	¥
4151	0.542	2.216	A26,AW43,A25,A10,AW66	AW68	¥
4153	0.550	2.788	AW68,AW33,A28	A26	¥
4155	0.597	2.162	A26,AW43,AW66,A11	–	¥
4158	0.581	1.865	A26,A25,AW43,AW66	–	¥
4159	0.535	2.149	A26,AW66	–	¥
9132	0.557	2.736	A26,A10	A25	¥
9149	0.785	3.796	A26,A25,AW66	A10	¥

Key: ¥ = extra reactions not of the A10 CREG.

Table 7. Selected AW43 Sera

Core	r value	Q score	Tail		Res (of A10 CREG)	
90	0.731	4.909	A26,AW34,A25,AW66		–	¥
92	0.747	6.411	A26,AW34,A25		–	¥
96	0.782	5.777	A26,AW34,A25,AW66		–	¥
97	0.782	6.831	A26,AW34,A25,AW66		A29	¥
104	0.756	6.670	A26,AW34,A25,AW66		AW33	¥
108	0.751	5.435	A26,A25,AW66	Th	AW34,A30	¥
109	0.684	3.888	A26		–	¥
110	0.794	5.233	–		–	
130	0.490	4.241	A29	Th	A11	¥
131	0.741	5.168	A29		–	¥
MAb						
2014	0.720	3.569	A3,A11		AW34	¥
2066	0.523	4.335	A26,A25,AW34,A10		A28	¥
Antigen Society						
4107	0.700	5.066	A26,AW34,A25,AW66,A10		AW68	¥
9172	0.703	6.479	–		–	¥
9173	0.870	5.564	–		–	¥
9174	0.660	5.635	A26,A10		A11	¥

Key: ¥ = extra reactions not of the A10 CREG.

monospecific for AW43 but reacts weakly and produces false negative reactions. Serum 88 is useful in the definition of AW43 in that it reacts strongly with all the A10 cross-reacting group of antigens A25, A26, AW34, and AW66 but does not react with AW43.

The reactions of selected sera of the Core and Antigen Society set defining AW43 are shown in Table 7.

HLA-AW66

HLA-AW66 could not be clearly identified in the present Workshop due to a lack of well-defined cells as well as specific sera. In serum by serum cluster analysis no clustering was observed. The following sera were said to contain anti-AW66: 112, 108, 99, 113, 93, 94, 95, and 105 in the Core set and 9132, 4153, and 9168 in the Antigen Society set. Apart from serum 9132, all these sera contain anti-A26.

References

1. Joysey VC, Roger JH, Abbas A, Bland C, Bolton M, Chan, et al: Study of a Malay population. In: Dausset J, Colombani J (eds), Histocompatibility Testing 1972. Munksgaard, Copenhagen, 1972, pp 251-260.
2. Martell RW, Taljaard DG, Du Toit ED. A split of AW34 in South African Negroes. Tenth International Histocompatibility Workshop Newsletter, 1987;1(5):21-22.
3. Botha MC, Du Toit ED, Jenkins T, Van Leeuwen A, D'Amaro J, Meera Khan P, et al. The HL-A System in Bushman (San) and Hottentot (Khôikoi) populations of South West Africa. In: Dausset J, Colombani J (eds), Histocompatibility Testing 1972. Munksgaard, Copenhagen, 1972, pp 421-434.

Antigen Society #3 Report (AW19 and A32/Aw74)

C.B. Carpenter,[1] P. Dyer,[2] H. Festenstein,[3] J. Moraes,[4] J. Gagne,[1] D.J. McCloskey,[3] J. Marwick,[5] D. Tate,[2] T. Barbalho,[6] M. Beach,[1] M. Lima,[7] P. Mackintosh,[8] M. Moraes,[4] and L. Ratner[1]

History of the AW19 Complex

The AW19 complex was first described in the Fourth International Histocompatibility Workshop (1). In the Fifth Workshop, the four components of this complex were sufficiently defined to be given a "w" designation: Aw29 (Aw19.1), Aw30 (Aw19.3), Aw31 (Aw19.4), and Aw32 (Aw19.5) with the residuals still being referred to as Aw19.6 (2).

A19.1 : A29 well defined by the Sixth I.H.W. (3)
A19.3 : A30 well defined by the Sixth I.H.W. (3)
A19.4 : A31 well defined by the Sixth I.H.W. (3)
A19.5 : A32 well defined by the Seventh I.H.W. (4)
A19.6 : Aw33 well defined by the Sixth I.H.W. (5,6)
Aw74 was first well characterized as Th in the Eighth I.H.W. (7)

The data in this antigen society were found to have numerous errors of duplicate and blank entries. A considerable amount of time and effort was spent in removing erroneous data and retyping, by hand, all the cells using CxS printouts. This considerably reduced the database for analysis:

Antigen Society Data (AS3) 1756 to 1127 cells
Core Data (ethnic) 2203 cells to 2065 cells
AS3 and Core Data 1006 cells to 600 cells

The number of cells from the various ethnic groups was insufficient for a detailed analysis of the AW19 antigens in these populations.

HLA-A29. This antigen has been well-defined since 1975 (3). In the Tenth I.H.W. there were six good monospecific reagents for this specificity:

Core Sera : 131, 132
AS3 Sera : 4161, 4167, 4168, 4169

The additional Core sera (130 and 565), despite extra inconsistent reactions, were useful in defining this specificity. There was no serological evidence of any splits of this antigen.

	XW 131	XW 132	XW 565	XW 130	XW 4161	XW 4167	XW 4168	XW 4169
(A29)	+	+	+	+	+	+	+	+
(Aw43)	+	−	−	+	NT	NT	NT	NT

HLA-A30. This specificity was clearly defined by two monospecific Core sea (134 and 135) when used on conjunction with the duospecific A30/A31 sera, which had predominantly A30 activity:

Core sera : 141, 142, 143
AS3 sera : 4178, 4179, 4184, 4191

There appears to be serological evidence of two A30 splits, defined by sera 136, 4187, and 151, which may correlate with the Tenth I.H.W. biochemically designated A30.1 (associated with B18) and A30.3 (associated with B13). Serum 4187 (9w133) was tested previously in the Ninth I.H.W. where it suggested a split of A30 in linkage disequilibrium with B13 by negative association (8). In this Workshop, one lab (UKI/DYR) has shown this pattern segregating in families. Serum 4187 was submitted to this Workshop as anti-A30,Cw5. Of the 17 cells that showed the pattern of positive with sera 136 and 4187 but negative with 151, only six were Cw5 positive. However, 4187 did react with approximately 50% of all Cw5 cells so it cannot be used to define this split except in conjunction with 136. A third pattern, in which all three sera are positive, may indicate the third rarer biochemically-defined split A30.2. However, these three patterns only constitute approximately 50% of the A30 positive cells that may be due to technical inconsistencies, or the three sera are not truly specific for the splits they define.

	Excellent A30 Sera									?Splits			
	XW 141	XW 142	XW 143	XW 134	XW 135	XW 4178	XW 4179	XW 4184	XW 4191	XW 136	XW 4187	XW 151	
(A30)	+	+	+	+	+	+	+	+	+	+	+	−	B18 associated
(A30)	+	+	+	+	+	+	+	+	+	−	−	+	B13 associated
(A30)	+	+	+	+	+	+	+	+	+	+	+	+	
(A31)	+	+	+	−	−	+	+	+	+	−	−	+	

Additional specificities: #143 occasionally Aw33; #4178 A29,Aw33; #4179 occasionally Aw33; #4191 occasionally Cw5.

Reporting Laboratories: NCYCAR,[1] UKIDYR,[2] UKIFES,[3] LATJRM[4]
Participating Laboratories: UKICXH,[5] LATTBA,[6] LATEPM,[7] UKIMAC[8]

HLA-A31. This specificity can be clearly defined by the monospecific AS3 serum 4180 together with the good A30/A31 sera, which had predominantly A31 activity.

Core sera : 142, 143, 147, 151
AS3 sera : 4178, 4191, 4192

Sera 2046 (monoclonal) and 4162 showed strong inconsistent reactions with A31 cells. These sera however did not complement each other and did not provide firm evidence of A31 splits.

	Excellent A31 Sera								?Splits	
	XW142	XW143	XW4178	XW4191	XW151	XW147	XW4180	XW4192	XW4162	XW2046
(A31)	+	+	+	+	+	+	+	+	+	−
(A31)	+	+	+	+	+	+	+	+	−	+
(A31)	+	+	+	+	+	+	+	+	+	+
(A31)	+	+	+	+	+	+	+	+	−	−
(A30)	+	+	+	+	(+)	+	−	−	(+)	(+)

Additional specificities: #147 A29,Aw33; #4192 occasionally A29; #4162 A29,Bw60.

HLA-A32 and Aw74. The specificity A32 was defined by the monospecific Core serum 158 together with other good but longer AW19 sera:

Core sera : 159, 160
AS3 sera : 4192, 4188, 9013

At the November 1987 meeting of the Tenth I.H.W. Nomenclature Committee, the specificity formerly known as Th was given the new designation Aw74. Aw74 can be clearly defined on the basis of consistent positive reactions with sera 4192 and 4188 and negative reactions with the A32 sera 158, 159, 160, and 9013. The additional sera 4189, 4195, and 4198 were useful in the definition but were not 100% consistent. Although Aw74 has been primarily described as a Negroid antigen, it has been found in other racial groups. Due to the relatively high frequency of AW19 specificities in the Negroid population, there was a large number of cells expressing both Aw74 and another Aw19 specificity, which, when excluded, left few Aw74 cells for analysis on both the Core and AS3 sera (n=19). One lab (UKI/FES) gathered data showing segregation of Aw74 in families.

	Excellent Sera							
	XW158	XW159	XW160	XW9013	XW4192	XW4188	XW4189	XW4195
(A32)	+	+	+	+	+	+	+	+
(TH)	−	−	−	−	+	+	(+)	(+)
(A25)	+	+	+	+	−	−	−	−

Additional specificities: #4188 occasionally A29,A30,A31; #4189 occasionally A29,A30,A31,Aw33,B38; #4195 occasionally A29,A30,A31,Aw33,B38.

HLA-AW33. No monospecific AW33 sera were found; however, AW33 can be defined by its reaction with the sera:

Core sera : 155, 154, 102, 157
AS3 sera : 4188, 4192, 4197, 9010, 9011

In the Seventh and Ninth I.H.W. there were two proposed splits of AW33 (4,9,10); however, there was no serological evidence for these splits using the Tenth I.H.W. reagents.

	XW4192	XW4188	XW155	XW102	XW151	XW154	XW4197	XW9017	XW9010
(AW33)	+	+	+	+	+	+	+	+	+
(AW34)	+	+	+	+	+	(+)	(+)	−	+
(A25)	+	+	−	−	−	−	−	+	+
(A26)	+	+	−	−	−	−	−	+	+
(A28)	+	+	−	(+)	(+)	(+)	+	(+)	+

Additional specificities: #154 occasionally A29,A31,A32; #4197 occasionally A30,A31; #9011 occasionally A29,A31.

Guide to tables: + = strong positive reaction; − = clear negative reactions; (+) = occasional positive reaction, sporadic; NT = no cells of that specificity tested against that serum.

References

1. Jeannet M, Vassalli P. Studies on new HLA antigens of the first segregant series. Vox Sang 1971; 20:317.
2. Bodmer JG, Rocques P, Bodmer WF, et al. Joint report of the Fifth International Histocompatibility Workshop. In: Dausset J, Colombani J (eds): Histocompatibility Testing 1972. Munksgaard, Copenhagen, 1972, p 621.
3. Bodmer JG, Curtoni ES, van Leeuwen A, et al. The ABC of HLA. A serological report of the Sixth Histocompatibility Testing Workshop. In: Kissmeyer-Nielsen F (ed): Histocompatibility Testing 1975. Munksgaard, Copenhagen, 1975, p 21.
4. Dick M. HLA-A, B and C serology and antigen reports. In: Bodmer WF (eds): Histocompatibility Testing 1977. Munksgaard, Copenhagen, 1977, p 157.
5. Ferrara GB, Tosi RM, Azzolina G, Carminati C. A new antigen at the LA locus positively associated with the W14 (Maki) specificity of the four locus. Tissue Antigens 1973; 3:130.
6. Joysey BC, Roger JH, Abbas A, Bland C, et al. Study of Malay population. In: Dausset J, Colombani J (eds): Histocompatibility Testing 1972. Munksgaard, Copenhagen, 1972, p 251.
7. Wolf E, Pritchard J, Watson B, Festenstein H. Characterization of a new Negroid A locus specificity Th. In: Terasaki PI (ed): Histocompatibility Testing 1980. Munksgaard, Copenhagen, 1980, p 744.
8. Raffoux C, Savi M. Antigen report: HLA-A30/HLA-A31. In: Albert ED, Baur MP, Mayr WR, et al (eds): Histocompatibility Testing 1984. Springer-Verlag, Berlin, 1984, p 127.

9. France OT. Regional reports I and II. In: Bodmer WF, et al (eds): Histocompatibility Testing 1977. Munksgaard, Copenhagen, 1977, p 441.

10. Mazzilli MC, et al. Antigen report: HLA-AW33. In: Albert ED, et al (eds): Histocompatibility Testing 1984. Springer-Verlag, Berlin, 1984, p 129.

Antigen Society #4 Report (A2, AW68, AW69 and A28)

J.M. Kreisler,[1] M. Morellini,[2] M.E. Moreno,[1] W.F. Wakeling,[3] S. Marsh,[4] L.J. Kennedy,[4] P. Hall,[5] S.C. Knight,[5] B. Tait,[6] C. Vermylen,[7] Y.S. Kao,[8] P. Saunders,[9] K.W.K. Cheung,[10] C. Giraudo,[11] N. Prigoshin,[12] S. Trabace,[2] A. Salerno,[13] L. Salvaneschi,[14] A. Amoroso,[15] and H. Mervart[16]

Two sets of cells and sera were used for antigen definition in this group. The Core Set contained approximately 4379 A2 cells, 337 AW68 cells, 54 AW69 cells, and 235 A28 cells. These were tested with #23 alloantisera and 17 monoclonal antibodies submitted with similar group specificities. Serological studies with the Society Set were done on a total of 484 cells (413 Caucasian). A2 cells: 243 (196 Caucasian), AW68 cells: 64 (61 Caucasian), AW69 cells: 8 (8 Caucasian), A28 cells: 44 (38 Caucasian). These were tested with 76 alloantisera of similar submitted specificities; there were no monoclonal antibodies in this set.

A2 and A28 were easily defined as in the Ninth International Histocompatibility Workshop (1,2). The best sera to define A2 and A28 in the Core Set and Society Set are presented in Tables 1 and 2.

The correlation coefficients among the alloantisera in both serum sets were good, whereas the correlation coefficients among the monoclonal reagents in the Core Set were poor. This poor correlation among the monoclonals may be explained by the significant (r > 0.3) tail reactions that were computed for several sera in this group. They were recognizing similar (perhaps public) epitopes within the antigen group A3, A10, A11, A19, Table 3. There were no monospecific sera AW68 or AW69.

Serum reactions from Core and Society datasets together indicate short A2 antigens on selected Caucasian cells, Table 4.

The Core dataset also indicated patterns of short A2 on Oriental cells, Table 5. The Society Oriental cell file was

Reporting Laboratories: FRAKRE,[1] IT2GAN,[2] UKILAW[3]
Participating Laboratories: UKIBOD,[4] UKIKNT,[5] ANZTAI,[6] BENVER,[7] US1NWO,[8] US4SND,[9] US5KWC,[10] LATGIR,[11] LATPRI,[12] IT1SAL,[13] IT1MRA,[14] IT1CEP,[15] NCYMRV[16]

Table 1. Selected Reagents from Core Set

A2	#	Incl	Corr	S×S
	35	98	95	99999
	37	95	95	9999
	29	98	94	999
	22	96	94	99
	16	95	94	9
	20	97	92	
A28	38	79	78	8877
	36	90	74	878
	40	86	71	87
	39	81	66	7
	42	93	63	

Table 2. Selected Alloantisera from Society Set

A2	#	Incl	Corr	S×S
	4211	97	93	99999989999998989999
	4218	97	91	9999989999988989888
	4215	97	91	999988989988988998
	4209	97	91	99989989889999998
	4201	92	87	9889989888988888
	4199	99	84	989988888888888
	4216	92	84	98888888888888
	4223	90	80	8888888878878
	4203	90	84	999898989888
	4208	93	88	99899989888
	4205	90	84	9889988888
	4236	98	89	999988889
	4240	92	81	98888998
	4230	90	82	9989888
	4206	95	82	998988
	4214	92	87	99888
	4220	90	81	9888
	4222	94	85	888
	4210	95	83	98
	4224	91	81	8
	4228	86	84	
A28	4244	84	53	7
	4243	84	53	

Table 3. Core Set Monoclonals

2030	A2 mono
2020, 2021, 2022, 2023, 2025, 2026, 2027, 2029, 2032, 2033, 2034, 2040	A2 + A28
2037, 2039	A2,AW69
2041, 2042	A2, A3, A28, A30, A31, AW33, A29
2044	A3, A28
2046	A2, A31
2067	A26, A25, A28
2090	A2, B39, B8
2036	A2, A28, A26, AW34, AW43

limited to 38 A2 positive cells; within this group a possible short A2 was indicated, sera 4237, 4219, 4210, 4224 negative, Table 6. The serum patterns for A2 on Oriental cells do not correlate with those found for A2 on Caucasian cells. No variation was shown in the Core dataset between black and Caucasian cells for the antigens reported by this Society.

AW69 was well-defined on Caucasian cells by the Core serum set, Table 7. Similarly, definition of AW69 on Caucasian cells was possible with the Society alloantisera, Table 8.

Table 4. A2 Pattern in Caucasians

		4220	4211	4218	4215	2032	4217	19	2034	4229	4231	4239
A2 long		+	+	+	+	+	+	+	+	+	+	±
A2 short	a	+	+	+	+	+	±	+	+	−	−	−
	b	+	+	+	+	±	+	−	−	+	+	±
	c	+	+	+	+	±	±	−	−	−	−	−

Table 5. A2 Pattern in Orientals–Core Set

		35	37	29	32	33	19	2032	2034	2036
A2 long		+	+	+	+	+	+	+	+	+
A2 short	a	+	+	+	+	+	−	−	−	−
	b	+	+	+	−	−	+	+	+	+
	c	+	+	+	−	−	−	−	−	−

Table 6. A2 Pattern in Orientals–Society Set

	4237	4219	4210	4224	4211	4208	4218
A2 long	+	+	+	+	+	+	+
A2 short*	−	−	−	−	+	+	+

**A2 short pattern only found in cells from TSU Lab.*

Table 7. AW69 Definition–Core Set

	A2+(AW69)					A2 mono				A28 mono		
	31	30	18	2037	2039	35	37	29	22	38	36	40
A2	+	+	+	+	+	+	+	+	+	−	−	−
AW69	+	+	+	+	+	−	−	−	−	+	+	+
AW68	−	−	−	−	−	−	−	−	−	+	+	+

Table 8. AW69 Definition in Caucasions–Society Set

	A2+(AW69)							A2 mono				A28 mono	
	4209	4210	4206	4219	4224	4216	4230	4208	4211	4203	4236	4244	4243
A2	+	+	+	+	+	+	+	+	+	+	+	−	−
AW69	+	+	+	+	+	+	±	−	−	−	−	+	+
AW68	−	−	−	−	−	−	−	−	−	−	−	+	+

Acknowledgments. We would like to thank our collaborators in the many laboratories worldwide for their support and hard work, which enabled the preparation of this report. E.H. Jones

References

1. Curtoni ES, Lawler SD, Kennedy LJ, Marshall WH, Wakeling W. Antigen report: HLA-A2. In: Histocompatibility Testing 1984. Springer-Verlag, 1984, p 119.
2. Curtoni ES, Lawler SD, Kennedy LJ, Marshall WH, Wakeling W. Antigen report: HLA-A28. In: Histocompatibility Testing 1984. Springer-Verlag, 1984, p 125.

Antigen Society #5 Report (B5, B51, Bw52, B18, B35, Bw53)

Clara Gorodezky,[1] Pablo Rubinstein,[2] Bruce Barger,[3] Carol Carpenter,[2] James Wolf,[4] Rhonda Andersen,[4] Josefina Flores,[1] Frances E. Ward,[5] Bento Arce-Gomez,[6] Marc de Bruyere,[7] Benjamin Bradley,[8] Sanking Cheng,[9] Donald E. Cross,[10] Maria DiLonardo,[11] Philippe Dyer,[12] Leonardo Fainboim,[13] Luis F. Garcia,[14] Maria Gerbase de Lima,[15] Emilio Haas,[16] Eugene Heise,[17] Ram Kakaiya,[4] Andrija Kastelan,[18] Pauline Mackintosh,[19] Helena Mervart,[20] Norma Prigoshin,[21] Ma Carmen Samaniego,[22] Argimiro Suarez,[23] Marlis Schroeder,[24] Paul Terasaki,[25] Luis Teran,[26] and Kimiyoshi Tsuji[27]

The Antigen Society dealt with subtypic specificities of the antigen cluster "4c" (1). B5 was recognized in the 1967 Workshop; B18, B35 and Bw53 (formerly HR), were identified in 1972 (2,3). The B5 "splits," B51 and Bw52, were recognized in the 1975 Workshop (4) but did not achieve official status until the 1977 Workshop (5), together with Bw53. An additional specificity, Bw70 (formerly Bu-Sv), and its two "splits" Bw71 and Bw72, all of them Bw6-associated, were included in the B5 CREG and are now officially established.

Most difficult within this group has been the serological definition of Bw53, first described by Engelfriet et al. in 1972 (3), and officially recognized during the Seventh Workshop although it could only be defined by extra reactions in some B5 and B35 sera. No monospecific sera were available during the Eighth and Ninth Workshops, but Bw53 could be differentiated from B51, Bw52, B35, and Bw70 by exclusion, with the Workshop anti-sera (6,7).

Reporting Laboratories: LATGOR,[1] US8RUB,[2] US2WOL,[3] US1WAR[5]

Participating Laboratories: LATBAG,[6] BENBRU,[7] UKIBRS,[8] US5FTW,[9] US5CRO,[10] LATDIL,[11] UKIDYR,[12] LATFAI,[13] LATGAR,[14] LATGER,[15] LATHAA,[16] US1HEI,[17] EAEKAS,[18] UKIMAC,[19] NCYMRV,[20] LATPRI,[21] LATSAM,[22] LATSUA,[23] US5MLA,[24] US5TER,[25] LATTRN,[26] JAPTSU[27]

Table 1. Antigen Society #5: Best Conventional Reagents in the Core-Serum Set

Antigen	Serum sets: Core			
	Serum #	R value	Q score	"extras"
B51	170	0.832	5.515	B5
	177	0.810	4.999	B5
	179	0.788	4.834	B5,C−
Bw52	183	0.650	4.636	B5
	182	0.593	4.206	B5
	181	0.434	2.895	B5
B5(51+52)	211	51(0.686) w52(0.632)	5.914	
	168	51(0.743) w52(0.500)	5.498	
	167	51(0.673) w52(0.567)	5.585	
	185	51(0.683) w52(0.587)	7.438	Bw53,Bw63
B35	192	0.753	4.093	Bw62
	215	0.692	3.947	Bw62,Bw62,1,Bw71
	191	0.731	3.679	Bw53
	198	0.644	3.688	Bw62,Bw71
Bw53	194	0.659	4.655	
	188	0.652	6.573	
	195	0.586	4.399	
Bw53	All contained strong Anti-B35 or Anti-B5			
B18	271	0.895	7.947	
	270	0.896	7.581	
	272	0.894	7.342	

Immunobiology of HLA
Volume I

Table 2. Antigen Society #5: Best Conventional Reagents in the Antigen Society Serum Set

Antigen	Serum sets: Antigen Society			
	Serum #	R value	Q score	"extras"
B51	4291	0.721	3.539	TE10,8w58
	4292	0.685	3.337	B5
	4294	0.671	3.014	TE10
	4290	0.688	2.854	B5
Bw52	4271	0.442	3.729	TE10
B5(51+52)	4275	51(0.637) w52(0.552)	2.330	8w58
	4276	51(0.639) w52(0.583)	1.733	8w58
	4281	51(0.626) w52(0.623)		
B35	4310	0.624	2.060	
	4307	0.557	1.595	
	4306	0.540	1.545	
Bw53	4313	0.570	3.433	
	4329	0.539	3.965	
	4323	0.391	2.687	
Bw53	All contained strong Anti-B35 or Anti-B5			
B18	4299	0.789	6.222	
	4301	0.795	4.580	
	4298	0.787	4.583	
	4300	0.767	4.929	
	4303	0.730	4.291	
	4302	0.716	3.616	

Conventional Serology

Antigens B5, B51, Bw52, B18, B35, and Bw53 were analyzed taking into account the cell reactions recorded with all anti-sera and monoclonals included in the Core serology specifically recognizing these specificities (64 anti-sera and 3 monoclonals), as well as the sera selected for the Antigen Society set (74 sera). A total of 502 cells belonging to different ethnic origins were tested with both sets of reagents: Caucasoids, Orientals, blacks, and Latin American Mestizo cells, while 7702 cells were tested with core sera only.

Some of the core sera also detected the related antigens Bw70 and its subdivisions Bw71 and Bw72: (10W) 198, 200, 201, 202, 213, 214, 215, 228, 229, 231, and 234. A number of other variants and splits proposed have been monitored for this report.

The best Core sera are shown in Table 1, together with some of their characteristics. A very clear identification of B18 and B51 was achieved with the Core reagents. Weaker but still reliable sera defined Bw52, and several anti-sera contained anti-B5 antibodies, which included B51 and Bw52 activities. It was not possible, however, to define the latter in the presence of B51 and difficult when B35 was present. As in previous Workshops, no monospecific sera were found for Bw53; all Bw53 sera contained other specificities, specially anti-B35 or anti-B5, but a fair definition of this antigen was usually possible.

Table 2 shows the best sera in the Antigen Society set. The results are consistent with those described for the Core set. The B18 and B51 anti-sera were very good monospecific reagents, particularly those defining B18. Reactivity by races of the Core set can be seen in Table 3. Interestingly, the sera that identified an antigen in one ethnic group, in general, behaved in a similar way in the others. There were exceptions, however, which provide leads for further analysis at the biochemical and/or molecular level. For example, 10W205 reacted well with B35 in Orientals but poorly in blacks. Sera 10W225, 222, and 223 defined Bw53 in blacks very well but were weak in Orientals, and anti-Bw4 was selected as their "primary" specificity in that population.

Reactivity differences among races was more pronounced with the Antigen Society (Table 4). B51 definition was excellent with sera 4295, 4294, 4292, 4291, 4290, and 4288 in Orientals, fair in Caucasians and very poor in blacks. Bw52 was detected by serum 4271 in all populations (Table 2), and by 4287 and 4279 in blacks only, while 4271 reacted best in Orientals and Caucasians. The only B5 serum that reacted fairly well with the two subtypes in the three ethnic groups was 4281, while 4275 and 4276 were good in Caucasoids only. B35 was best defined with the four sera in Table 4, best in Orientals, fairly in Caucasoids, and poorly in blacks; as already stated, no monospecific reagents were available for Bw53. While containing other antibodies, two anti-

Table 3. Reactivity by Races of Antigen Society #5 Core Serum Set

		Race											
		Caucasians				Orientals				Blacks			
Antigen	10W Serum	Spec	R value	Q score	Extras	Spec	R value	Q score	Extras	Spec	R value	Q score	Extras
B51	170		0.83	5.5	B5		0.85	4.41			0.90	6.80	
	174		0.78	4.41	B5		0.72	3.22			0.90	5.46	
	177		0.81	4.99	B5,Te76		0.85	4.87			0.86	5.19	
	179		0.79	4.83	B5,Te76		0.85	4.87			0.82	4.96	Te76
	178		0.77	4.52	B5,Te76		0.85	4.53	Te76		0.68	4.05	Te76
	173		0.76	4.51	B5		0.72	3.83			0.68	4.33	B5
Bw52	183		0.65	4.64			0.83	4.88			0.82	7.24	
	182		0.59	4.21			0.69	3.38			0.73	5.90	
	181		0.43	2.90			0.69	2.14			0.53	2.38	
B5	211	B51	0.71	3.86	Bw58,Bw59	B51	0.52	2.13	Bw58,Bw59	B51	0.66	4.69	Bw58,Bw59
		Bw52	0.79	5.79	B5	Bw52	0.41	1.89		Bw52	0.79	5.85	
	164	B51	0.65	3.16		B5	−0.02	8.63		B51	0.63	4.17	B5
		Bw52	0.66	4.58						Bw52	0.69	3.96	
	185	B51	0.69	4.32	Bw53	B51	0.40	7.81	(Bw4,1st spec)	B51	0.54	3.61	Bw53
		Bw52	0.71	5.70		Bw52	0.60	7.81		Bw52	0.68	5.16	
B35	233		0.59	2.33	B(w)62,71,72 15.3,50,70,56		0.79	4.83	B(w)62,15.3,50 70,56		0.79	4.83	B(w)62,15.3,50,70,56
	228		0.50	1.59	B(w)62,72,71,70		0.77	3.66	B(w)62,70		0.77	3.66	B(w)70
	234		0.60	2.59	B(w)70,15.3,62		0.79	4.16	B(w)62,70,15.3,56		0.77	4.13	B(w)62,15.3,70,56
	191		0.58	5.01	(Cw4,1st spec)Bw53								
Bw53	225		0.48	3.40	B51,w52,49(1st spec)		0.18	14.57	Bw4,1st spec,		0.66	2.54	w52,49,w78,51
	222		0.65	4.75	B51,w52,49(1st spec)		0.20	17.66	Bw4,1st spec, B49,w78,51,w52		0.64	1.79	w52,49,w78,51
	223		0.61	4.20	51,52,49(1st spec)		0.12	14.34	w4,1st spec, 49,278,51,w52		0.57	0.38	w52,49,78,51
B18	271		0.90	7.6			0.85	4.87			0.90	6.05	
	270		0.90	7.6			0.84	4.72			0.90	5.84	
	272		0.89	7.3			0.85	4.87			0.85	4.65	
	273		0.59	4.5			0.73	4.53			0.85	4.63	
	269		0.58	2.5			0.69	4.03			0.80	3.71	

Table 4. Reactivity of Races of Antigen Society #5 Antigen Society Serum Set

		Race								
		Caucasian			Orientals			Blacks		
Antigens	Serum:10W	R value	Q score	Extras	R value	Q score	Extras	R value	Q score	Extras
B51	4295	0.60	2.25		0.87	4.71		0.36	1.88	
	4294	0.72	3.21		0.75	3.46		0.34	2.28	
	4292	0.75	3.88		0.75	3.33		0.22	1.14	
	4291	0.81	4.17		0.75	4.36		0.32	2.23	
	4290	0.74	3.56		0.73	4.65		0.48	1.88	
	4288	0.73	3.18		0.75	4.71		0.11	11.77	B5 only
Bw52	4287	0.45	2.60	B51,49	0	9.30	DR3	0.83	5.73	B5,DR3
	4279	0.60	5.26	B51(1st)	0.39	1.96	B51	0.68	3.97	B51
	4271	0.76	3.75	B51(1st)	0.73	3.0	Bw4,B5	0.63	4.27	B5
	4344	0.69	5.07	B51(1st)	B5 0.40	7.81	Bw4,1st spec	0.53	3.51	B5,Bw66
	4284	0.44	2.18	B51	0.70	3.55	B51	0.40	2.08	B51
B5	4281	41(0.66) 52(0.69)	3.59 5.61		51(0.40) w52(0.40)		Bw4,1st spec	52(0.51) 51(0.39)	3.07 1.93	A28
B35	4307	0.69	2.83	Cw4(1st)	0.80	4.74		0.35	1.17	
	4306	0.69	3.01	Cw4	0.69	4.55		0.29	1.05	
	4310	0.68	2.74		0.65	3.74		0.39	1.45	Aw33
	4308	0.41	2.87	Cw4(1st)	0.65	3.51		0.18	0.51	
Bw53	4323	0.31	4.76	Cw4,51,35(1st)	0.143	1.65	B35,51,1st spec	0.64	3.69	Cw4,1st spec B35,B51
	4274			B51	B5−0.149	10.8		0.59	2.88	w52,51,5
	4327	0.35	4.60	Cw4,B51,B35(1st)	−0.24	1.99	B35,1st spec	0.59	5.47	Cw4,B35,1st spec
	4316			B18	B5,35(−0.15,0.09)		B5,35,18	0.42	1.41	5,35,18
	4329	0.50	2.96	B35,B51(1st)	−0.29	2.47	B35,1st spec	0.38	2.16	Cw4,1st spec 35 2nd 51
	4313				B5,35(−0.15,0.09)		51	Cw4,Bw52,5,35		
B18	4302	0.79	3.13		0.70	4.37		0.93	6.48	
	4300	0.79	5.75		0.70	4.66		0.89	7.90	
	4301	0.89	6.01		0.86	4.66		0.85	4.41	
	4303	0.60	3.19		0.85	4.68		0.83	5.13	
	4298	0.82	6.40		0.86	4.37		0.84	7.34	
	4299	0.75	5.90		0.73	4.36		0.72	4.39	

Table 5. Antigen Society #5: Monoclonal Antibodies

MAb #	Specificities detected	
	a) In ELISA	b) Cytotoxicity tests
2091	B14, B18, B51, (r=0.651)	B14,B8,B18,B5, B18 (r=0.317 Q=2.27) (r values between −0.19 and 0.21)
2097	B44,B51 (r=0.598)	B44,B45,B51,Bw52 B52(r=0.254 Q=1.038) (r values between 0.08 and 0.59)
2101	B51 (r=0.606)	Bw4 (r=0.177)

Bw53 sera were cells fairly good: 4323 for blacks and 4327 for Caucasoids, but their reactions with Oriental cells were very weak. The recognition of B18 was excellent in all races.

Monoclonal Antibodies

A small number of monoclonal antibodies detected specificities in this group. They are listed in Table 5, indicating the range of reactivity in the ELISA and cytotoxicity techniques. Differences in reactivity and specificity were noticed in these tests, and it is noteworthy that the reactions were "cleaner" and the R values higher in ELISA.

Isoelectric Focusing Characterization and Possible Splits

Two B51 variants were identified (B51.0 and B51.1); three B35 variants (B35.0, B35.1, B35.2); and two B18 variants (B18.1 and B18.2) by IEF. It was thus suggestive that some patterns of reactivity may have emerged with selected sera in the Antigen Society serum set, hinting at the possible existence of serologically detectable "splits." These patterns showed up for B35 and B51 as shown in Table 6.

Two B5 variants were suggested independently with the 10W sera. One by Lee et al. (8). They studied 25 members of a black family, and 11 individuals reacted with the Core sera 165, 191, and 215 and were associated to Bw6. They called this variant B5CV. These three sera define (in blacks) antigens Bw53,B5; Bw53,B35; and B35, 62, 70, Te79, 17 respectively. The other variant named SNA was suggested by Andrien et al. (9) also with 10W Core reagents. They found it in two families of Moroccan Berber origin. B51 sera (10W) 178, 179, 175 were negative, while the rest of the B5 (and/or) B5 CREG were positive, except for serum 225. The antigen is also Bw6-associated. The isoelectric point of the novel antigen is the same as that of Bw52 and different from other B5 CREG molecules.

In spite of these findings, the central analysis of the Core serology did not provide evidence supporting new "splits." Monoclonals or sera from exotic populations may succeed in serologically detecting the biochemical variants confirmed in this Workshop. The search for these antibodies is therefore important and should be encouraged.

References

1. Bodmer WF, Payne R. Theoretical considerations of leukocyte grouping using multispecific anti-sera. In: Bodmer W, Cleton FJ, Ermisse JG (eds): Histocompatibility Testing 1966. Munksgaard, Copenhagen, 1966, p 1941.
2. Albert ED, Mickey MR, Terasaki PI. Serology and the genetics of Te58 (w18) and other specificities included in the 4c complex. Tissue Antigens 1972; 1:47.
3. Engelfriet CP, Veenhoven von Riesz E, Kort-Bakker M, van der Berg-Loanen PM. Some studies with anti-4c, anti-R, anti-HL-A5, anti-w5,w18 and description of a new antigen in the FOUR segregant series, called HR. In: Dausset J, Colombani J (eds): Histocompatibility Testing 1973. Munksgaard, Copenhagen, 1972, p 475.
4. Bodmer J. The ABC of HLA. In: Kissmeyer-Nielsen F (ed): Histocompatibility Testing 1975. Munksgaard, Copenhagen, 1975, p 42.
5. Nomenclature for factors of the HLA system 1977. In: Bodmer WF (ed): Histocompatibility Testing 1977. Munksgaard, Copenhagen, 1978, p 14.
6. Hammond MG Bw53. Joint Report. In: Terasaki PI (ed): Histocompatibility Testing 1980. UCLA Tissue Typing Laboratory. Los Angeles, California, 1980, p 432.
7. Taylor C, Ting A, Hammond MG, de Wall LP, de Lange GG, Engelfriet P. Antigen Report: HLA-Bw53. In: Albert ED, Baur MP, Mayr WR (eds): Histocompatibility Testing 1984. Springer-Verlag, Berlin, 1984, p. 161.
8. Lee TD, Andersen R, Dutson GM. A new cross-reactive variant of B5 identified in extended black Panamian family. (In press).
9. Andrien M, Reekers P, Dupont E. Analysis of SNA: A possible "new" split of HLA B5 antigen Bw6-associated. (In press.)

Table 6. Cell Reaction Pattern With B51 and B35 Sera of the Antigen Society Set

Sera	4294	4269	4288	4290	4291	4292	Sera	4320	4307	4329	4327	4320	4310
	+	+	+	+	+	+		+	+	+	+	+	+
B51	+	+	+	−	−	−	B35	+	−	+	+	+	+
	−	+	+	+	+	+		+	−	+	+	+	−

Antigen Society #6 Report (B21, B49, Bw50, BN21, B12, B44, B45)

R.C. Williams,[1] S.N. Chen,[2] D.K. Gill,[3] J.T. Lane,[1] J.E. McAuley,[1] R. Strothman,[2] and K.K. Mittal[3]

Antigen Society and Group 6 was the result of a collaboration of fifteen laboratories. They typed persons for the B12 and B21 antigens using a combination of antigen society and core trays. A total of 546 cells and 171 anti-sera was included on the final dataset that was incorporated into the present analysis, 443 of these cells having a reaction profile greater than 0.75, all having been tested against both the antigen society and core anti-sera. Antigen B12 and its splits were represented by 186 of the "non-flagged" cells while B21 and its subtypes were present on 135 cells. Twenty ethnic groups were represented in Antigen Society and Group 6. Only the 443 cells with the high reaction profiles were used for the analysis whereas anti-sera were chosen for their high serum-by-serum correlations and because they gave clear patterns in the cell-by-serum matrices. A complete listing of the sera which were submitted to Antigen Society and Group 6 as containing B12 and B21 specificities is contained in the Appendix.

B21 Antigens

Introduction. HLA-B21 was recognized to contain two components in the Sixth International Workshop (1). The Bw4-related split of B21 (Bw21.1 or SL-ET) was designated Bw49, and the Bw6-related split (Bw21.2 or ET*) was named Bw50 in the Seventh Workshop (2). In the Ninth Workshop "w" status was dropped for B49 (3). A third split of B21 was suggested by Mittal et al. who found a Caucasian donor who typed as B21 but who lacked B49 and Bw50 (4). R.C. Williams et al. were unable to split the B21 antigen in either the Navajo or the Hopi while Ensroth et al. and Ward et al have reported a similar finding in the Pima Indians of the Gila River Indian Community (5,6,7). At the general session of the Ninth Workshop in Munich, one of us (RCW) presented data indicating that a third subtype of B21 existed that is associated with Bw6 and which is found primarily in Native and Mexican Americans.

B49

Antigen B49 is clearly defined by positive reactions with three sets of anti-sera: those with B49, Bw50, and BN21 specificities, those with B49 and Bw50 specificities, and the operationally monospecific B49 anti-sera. It is also defined by negative reactions with two sets of anti-sera, one containing Bw50 alone, and the second having both Bw50 and BN21 specificities (Figures 1 and 2). These anti-sera were selected from 49 of the 63 Core and Antigen Society 6 anti-sera that were submitted as containing B49 specificity. They correlated with an R value of greater than 0.5 with B49 specificity when examined against 486 cells tested with both Core and Antigen Society 6 anti-sera.

The operationally monospecific anti-sera were 441, 444, 445, 4391, 4395, 4396, 4397, 4394, 4393, and 4399. Anti-sera 441, 4391, 4395, and 4396 form a tight cluster for B49 and clearly define this antigen (Figure 1). Of the remaining anti-sera, nine contained additional Bw50 reactivity (4408, 4419, 4427, 4426, 4413, 4414, 4415, 4421, and 4422), sixteen contained additional Bw50 and BN21 reactivity (435, 436, 437, 438, 439, 440, 450, 4407, 4403, 4409, 4410, 4412, 4416, 4417, 4418, and 4423), and seven contained additional B5-related reactivity (222, 223, 224, 225, 4404, 4405, and 9030). Four anti-sera, 4369, 4374, 4382, and 4390 reacted with both B44 and B49 positive cells, possibly due to sharing of a common epitope by both antigens. None of the monoclonal antibodies showed B21 or B49 specificity. Selected cells and anti-sera shown in the cell by serum matrix in Figure 2 illustrate that B49 is clearly defined. The pattern for B49 is found in Table 1. It has been reported in previous Workshops that B49 is associated with the Bw4 antigen. In this Workshop, B49 was found widely distributed in all ethnic groups.

Bw50

Antigen Bw50 is defined by four sets of anti-sera. Eight sera recognize the Bw50 and BN21 antigens: 232, 234, 233, 228, 229, 230, 231, and 200. The highest correlations are found between serum 233 and the four sera 234, 228, 229, and 230. A second cluster of sera, which recognizes only the Bw50 antigen, is represented by 540, 215, 448, and 447. As can be seen in Figure 1, however, the correlations between the sera are low, the highest value being between the serum 447 and sera 215 and 448. Figure 1 also presents the correlation coefficients between the Core and Antigen Society sera that are "long" B21, that is, they react with antigens B49, Bw50, and BN21. They are sera 436, 440, 439, 438, 437, 4407, 4412, 4418, 4416, 4409, 4410, 4417, 4403, 451, and 4423. Core sera 436, 440, and 439 have the highest coefficients while Antigen Society reagents 4407, 4412, and 4418 form a second cluster with high correlations. Finally, four Antigen Society 6 anti-sera recognize only the B49 and Bw50 antigens, 4408, 4419, 4427, and 4426.

These four clusters of anti-sera provide a very good definition of Bw50 in the Tenth Workshop, which is evidenced by the clear pattern in the cell-by-serum matrix

---

Reporting Laboratories: US5ABS,[1] US5FTW,[2] US5MIT[3]
Participating Laboratories: US5KWC, SAFDUT, US8FOT, GERGAZ, FRAKRE, EAELEV, FRAMYE, NCYMRV, UKIMID, GERMUC, BENVER

```
         |    4444444 44                4444 4444
         |444444444444444422222222252444444443334
         |343331021111500333232034144012249990      Specificity
         |608972930867173243908010587897611654
---------------------------------------------------------------------------
MYR  436 | 8788775876657644333423115467615443 3
JLE  440 |8 787765776647544333422024467514443 4
MRV  438 |77 77664766556533233322125455424432 3
JLE  439 |887 7785777657644233422014467525443 4
MYR  437 |8777 674766666644333413114356534332 3
CRO 4412 |77676 75787758533232322013378615655 5
KIS 4409 |766877 5777657644223312113366624443 4
ABS 4423 |5545455 555535422111211012244423333 3     B49, Bw50, BN21
PRR 4410 |87777775 77747634323423114377724453 4
SVE 4418 |776768757 7748533222322013378615555 5
DOU 4416 |6667677577 658633222322013366524545 5
GOR 4417 |66566765776 46534333312012366524443 4
VER  451 |545565534454 5522121211023334433323 3
FES 4407 |7767687578865 544333422013367624555 5
MYE 4403 |65566564656555 23222331321425563443 33
MBC  232 |4434434233332 42 7776655 24442320baaaa
BER  234 |4434434243342437 87776626332320a00a0
MRV  233 |33223221322313278 8885525331 22abaaba
CRO  229 |333333212223232778 775525341320aaaa0
MYR  230 |3333323132231336787 74624431230aaaa0     Bw50, BN21
DDC  228 |44344332433324367877 562533242a00000
BRS  200 |222212112221121565545 53524121abaaba
BER  231 |3222322132221235655665 34432230a00a0
DDC  540 |10101010100000222222233 3330220bbbba
WOL  215 |122111111111211465545543 351211aaaa0
BRS  448 |5454433243323344333432433 5445310000     Bw50
MIT  447 |44443332333333243343343355 343200a00
TOM 4408 |666565764776636522111212014 376256555
STE 4419 |77576864786647533232422222447 6245445    B49, Bw50
BRA 4427 |654556647655466222232132153 66 444444
HAM 4426 |1122312221223230 0a00aa00132224 11112
GRU  441 |544545434544344babaa0baba105441 8776
BRS 4391 |444436434554354a0aaa0a0ba0065418 987
GRU 4396 |443435435544253a0aaa0a0ba0a544179 77     B49
CRO 4395 |332325333553353aabaa0baba005441787 9
MTT 4404 |343435434554353a0a000a0a0005542 6779
```

Figure 1. A serum-by-serum correlation matrix for the 36 antisera that were employed by Antigen Society and Group 6 to define the antigens contained in B21. The correlations were performed against 443 cells that were typed against both the Core and the Antigen Society sera and that had, in addition, a reaction profile of greater than 0.75 ("non-flagged"). Five clusters of antisera are defined in the figure. They also appear in the cell-by-serum matrix in Figure 2.

of Figure 2. As has been reported in previous Workshops, Bw50 is associated with the Bw6 antigen. A search of the ETHNIC dataset revealed six cells that corresponded to the Bw50 pattern in the Core anti-sera, in addition to those that were typed in Antigen Society, Group 6. They are NIK 454 (Race: Other Southeast Asian), STE 52 (American black), STE 55 (American black), DAK 16 (Other Mongoloid), DAK 26 (Chinese), and TER 1032 (American black). The pattern for Bw50 is presented in Table 1.

BN21

The new antigen, BN21, is a Bw6-associated "short" Bw50, which has been found in Native and Mexican Americans. It was the subject of intensive investigation by one of us (RCW) during the Ninth International Workshop during which family studies in the Yaqui Indian Community were performed (8). It segregates as a codominant allele at the HLA-B locus and is clearly defined by the anti-sera in the present Workshop. Two sets of reagents, previously discussed in the Bw50 section, recognize the BN21 antigen. Sera 436, 440, 439, 438, 437, 4407, 4412, 4418, 4416, 4409, 4410, 4417, 4403, 451, and 4423 are part of the broad, B21 anti-sera that recognize all three antigens, B49, Bw50, and BN21 while anti-sera 232, 234, 233, 228, 229, 230, 231, and 200 share specificities Bw50 and BN21. An additional three sets of anti-sera type BN21 by subtraction. Sera 540, 215, 448, and 447 recognize only the Bw50 epitope in the B21 complex while 4408, 4419, 4427 and 4426, recognize both B49 and Bw50 but not BN21. Sera 441,

Abbreviations: The correlation coefficients in Figures 1 and 3 are abbreviated as follows: 0 (0.0 to 0.05), 1 (0.06 to 0.15), 2 (0.16 to 0.25), 3 (0.26 to 0.35), 4 (0.36 to 0.45), 5 (0.46 to 0.55), 6 (0.56 to 0.65), 7 (0.66 to 0.75), 8 (0.76 to 0.85), and 9 (0.86 to 1.00). The lower case letters a, b, c, . . . , j correspond to the respective negative intervals.

```
                              MJMJMCKAPSBGVFMMBMCMDBBDWBMTSBHGBGCM
                              YLRLYRIBRVOOEEYBERRYDREDORIOTRARRRRT
                              REVEROSSREURRSECRVORCSRCLSTMEAMUSUOT

                                 4444444 44              4444 4444
                              444444444444444222222225244444443334
                              343331021111500333232034144012249990
                              608972930867173243908010587897611654
------------------------------------------------------------------------------------
FTW    23 B18     B49         888888888686888111111111114888888888
FTW    26 B44     B49         868888888888888141111141161888888688
FTW    30 B44     B49         861188888484888111111111111888888816
FTW    35 B44     B49         888888888688888141111141161888088888
FTW    37 B49     BW55        888888888888888241166141181888888888
FTW    40 B8      B49         868818888686888111111114111868188888
FTW    41 B39     B49         888888080888888111166141111188888888
FTW    48 B7      B49         662686888484848111141111111818188886
FTW    49 B44     B49         888888888686868111114111111888866868
FTW    50 B17     B49         888888888888888111116188164088888888      B49
FTW    64 B49     B51         888888888888888111111121161088888888
FTW     1 B49     BW60        888888888888888111111111181888888888
FTW     2 B7      B49         888888888886888121144111142888688888
FTW     3 B49     BW58        868888888888888111116118162888688888
FTW     4 B49     BW52        888888888888888111118468164888888888
FTW     6 B37     B49         868888888886888146111128141888888888
FTW     7 B18     B49         888888888888888111161111111888688888
FTW    31 B44     BW50        888888888888888888888888888888811111
FTW    34 B14     BW50        888886888686888888888888486888811114
FTW    38 BW63    BW50        888888888888888888888888888688811111
FTW    44 B7      BW50        888888888886868888888888184088111111
FTW    45 B8      BW50        824888888886888888888888884888811111
FTW    55 B44     BW50        868888888888888888888888888888811111
FTW    56 B38     BW50        118886888684488168888888611888811116
FTW    57 B18     BW50        808884888886888888888888884888811111
FTW    45 B8      BW50        824888888886888888888888884888811111      Bw50
FTW    55 B44     BW50        868888888888888888888888888888811111
FTW    56 B38     BW50        118886888684488168888888611888811116
FTW    57 B18     BW50        808884888886888888888888884888811111
FTW    58 B8      BW50        868888888688888888888888880888811111
FTW    67 B7      BW50        886881818141848888888888181114811111
FTW    10 B44     BW50        888888888888888888888888888888811111
FTW    13 BW50    BW62        888888888888888888888888880888811211
FTW    17 BW50    BW56        888888888888881888888888888888816411
ABS  2904 BN21    BW48        888884888888888888888888111111111111
ABS  2840 B5      BN21        888884888688181888888888111111118414
ABS  2838 B5      BN21        888886888888188888818888111811168112
ABS  2839 B35     BN21        888881888688881688888888881111111111
ABS  2868 B35     BN21        864661888266886688818886881111111111
ABS  2864 B5      BN21        888681888180881108188821111111011101
ABS  2866 B5      BN21        888881888188848888848888141111111111      BN21
ABS  2867 B5      BN21        888081888188044168188800111111111111
ABS  2865 B5      BN21        888886888486088208488841111111111111
ABS  2946 BN21                888888888888888888888888111118411111
ABS  5422 B27     BN21        888888888688888888888888111111211111
ABS  2847 BN21    BW61        888888888488818688881 8111111111111
ABS  2944 B39     BN21        888888888688888888888886111111111111
ABS  2874 BN21    BW48        888888880181810888148461211111111111
```

Figure 2. A cell-by-serum matrix for selected cells and antisera from Antigen Society and Group 6 that define the three antigens in B21: B49, Bw50, and BN21. To construct the matrix, all of the cells from the laboratory of the American Red Cross, Fort Wayne Region, were chosen for B49 and Bw50. All of the cells that typed for BN21 were included. Three distinct patterns are evident.

4391, 4396, 4395, and 4404 are strong B49 reagents that do not have BN21 activity. The cell-by-serum matrix in Figure 2 illustrates the discrete pattern for the BN21 antigen that is given by these anti-sera.

Fourteen cells were identified in the Tenth Workshop that had the BN21 antigen (Figure 2). Cells ABS 2946, ABS 5422, ABS 2847, ABS 2944, and ABS 2874 are Pima Indians from the Gila River Indian Community in Arizona. In addition, members of two families from the Yaqui Indian Community in Guadalupe, Arizona, who were also included in the Ninth International Workshop, are found in the data. Cell ABS 2868 comes from the father of four HLA identical siblings, ABS 2867, ABS 2866, ABS 2865, and ABS 2864, all of whom inherited the BN21 antigen from him. Three siblings from a second family who share the BN21 specificity are ABS 2840, ABS 2838, and ABS 2839. Cell ABS 2904 is from a from a Mexican American. A search of the ETHNIC

Table 1. Pattern Analysis for Antigens B49, Bw50, and BN21

	436	440	438	439	437	4412	4409	4423	4410	4418	4416	4417	451	4407	4403	232	234	233	229	230	228	200	231	540	215	448	447	4408	4419	4427	4426	441	4391	4396	4395	4404
B49	+	+	+	+	+	+	+	+	+	+	+	+	+	+	+	−	−	−	−	−	−	−	−	−	−	−	−	+	+	+	+	+	+	+	+	+
Bw50	+	+	+	+	+	+	+	+	+	+	+	+	+	+	+	+	+	+	+	+	+	+	+	+	+	+	+	+	+	+	+	−	−	−	−	−
BN21	+	+	+	+	+	+	+	+	+	+	+	+	+	+	+	+	+	+	+	+	+	+	+	−	−	−	−	−	−	−	−	−	−	−	−	−
Additional Specificities								A2							52	1	35 62	62 70 35		70	35 62 70	35 71	35	Cw4 Cw6	35 62											52

dataset revealed one cell, TRP 12, also a Native American, that had the BN21 antigen. It is from the laboratory of Dr. Gary Troup in Albuquerque, New Mexico. The pattern for BN21 is presented in Table 1.

B12 Antigens

Introduction. The history of B44 and B45 is associated each with the other and with the broad supertypic antigen B12, which was initially recognized in 1970 (9). The common subdivisions of B12 were recognized at that time by one alloantiserum, TT*, from Aarhus (9). Those cells that did not react with TT* were subsequently shown to be associated with Bw4 and were officially named B44 in 1975 (1). Those cells that did react with TT* were shown to be associated with Bw6 and were officially named Bw45 in 1975 (1). The Workshop designation for both antigens was dropped after the Ninth Workshop (10,11). Both antigens are considered easy to identify and a large number of reagents were submitted to this Workshop with specificity for B12, B44, or B45. Serological heterogeneity of B44 has been suggested in each of the last three Workshops. The "long" pattern was considered more common. In the Eighth Workshop a "short" pattern was thought to correlate with a subpopulation of blacks (12). In the Ninth Workshop there was a suggestion of correlation between "short" B44 serological patterns and subdivisions of B44 based upon cell mediated lympholysis (CML). The evidence, however, was not considered strong enough to be conclusive (10). The evidence for serological heterogeneity of B45 has never been strong.

B12

The best alloantisera in this Workshop identifying B12 were three core anti-sera (417, 419, 418) and six Antigen Society alloantisera (4352, 4365, 4355, 4353, 4356, 4348). This group of anti-sera were excellent B12 reagents identifying B44 and B45 equally well. These six anti-sera correlated well with each other (Figure 3) and exhibited high coefficients of correlation with the antigens giving very few false positive or negative reactions (R values, 0.949 to 0.888, see Appendix). These anti-sera almost always gave strong reactions when they were cytotoxic, and were employed in the analysis of HLA-B44 and HLA-B45.

B44

Six anti-sera specific for B44 were chosen for their clear-cut ability to distinguish B44 from B45. Of these 6 alloantisera two were from the Core set and four from the Antigen Society sera. They are 424, 427, 4367,

```
         |  444444  44444444444  4
         |444333333444333333333441344
         |111565554226776867877033823      Specificity
         |798255368477650369003023780
---------------------------------------------------------
MIT  417 | 999999997677766655451443333
MRV  419 |9 99999997677766655451443443
JLE  418 |99 8999987677766655451443333     B44"long", B44"short",
MRV 4352 |998 99999767777775555144333              B45
FES 4365 |9999 9999768887775555144333
MID 4355 |99999 999768887775555144333
MRV 4353 |999999 99768887775555144333
DPT 4356 |9999999 9767776675555144332
MIT 4348 |99899999 667776675555144333
DAW  424 |777777776 9988878665620la0a
HAM  427 |6666666669 88776766552a0b0a
DUQ 4367 |77778887798 99879776720la0a
PNW 4376 |777788877889 9889766720la0a
BET 4375 |7777888778799 878666620la00     B44"long", B44"short"
HAM 4360 |66677776687888 888766202000
BAC 4383 |666777766767878 77666312010
BRN 4366 |6667777778799887 7766202a0a
ZAR 4379 |55555555566776877 665302a0a
DUP 4380 |555555555667667676 66302a0a
POL 4370 |44455555555566666666 6301010     B44"long"
ABS 4373 |5555555556577666656 62ala0a
CEP 4400 |1111111112222222323332 11000
GEL  432 |4444444440a000010000a1 5675
YUN  133 |44444444410111222221115 444
BRN 4387 |333333333abaaa00aaa0a064 55     B45
ABS  428 |3433333330000001000100745 5
STE  430 |333333323aaaa000aaa0a05455
```

Figure 3. A serum-by-serum corrrelation matrix for the 27 antisera that were employed by Antigen Society and Group 6 to define the antigens contained in B12. The correlations were performed against 443 cells that were typed against both the Core and Antigen Society sera and that had, in addition, a reaction profile of greater than 0.75 ("non-flagged"). Four clusters of sera are defined. They also are shown in the cell-by-serum matrices of Figures 4 and 5.

				MMJMFMMDMDHDPBHBBZDPAC	
				IRLREIRPIAAUNEAARAUOBE	
				TVEVSDVTTWMQWTMCNRPLSP	
				444444 44444444444	
				4443333333443333333334	
				1115655542267768678770	
				7982553684776503690030	
FTW	21	B35	B44	8888888888888888886668	
FTW	22	B44	BW60	8888888888888888811618	
FTW	25	B44	BW58	8888888888888688818888	
FTW	26	B44	B49	8888888888886888888818	
FTW	29	B44	BW52	8888888888888888818888	
MOO	889	B44	BW50	8888888888886888688418	
MIT	4542	B35	D44	1868888888808888888414	
MID	8	B44	BW61	8888886888888816814661	
MIT	4545	B37	B44	8488888188088188864182	
MID	4	B44	BW41	8888888888888888816841	
MIT	4551	B8	B44	8888488888288868888188	B44 "long"
MIT	4552	B44	BW72	8884884888088868864468	
MIT	4554	B27	B44	8888884488086488818864	
MIT	4561	B44	BW50	8848888888088868688688	
KWC	56	B44	B49	8888888888888288646144	
KRE	42	B27	B44	8888888888888886814662	
KRE	82	B44	B18	8888888888888888888884	
KRE	81	B44	B27	8888888888888888888881	
ABS	2829	B44	BW50	8888888888888866866166	
ABS	5304	B44	BW50	8688888888888888802861	
MRV	857	B44	B18	0008888880088888841421	
KWC	15	B8	B44	8486866488288888614141	
MIT	4563	B8	B44	8888888888088816812224	
KWC	28	B44		8884484818288448211141	
FTW	65	B44	BW61	8888888888868886841214	
KRE	95	B44	B35	8888886888888844642220	
MIT	4548	B44	B51	8888884886084644421141	
FOT	4	B44	BW60	8088644848848401411411	
MYE	143	B44	B49	8688888888888824814144	
MYE	22	B44	BW62	8888486818688611211441	
LEV	71	B44	BN	8888886448888681842241	
MRV	798	B44	BW60	8888688888888421441211	
KRE	91	B44	B14	8888888888868882811141	
HAM	108	B44	BW41	8008888880888868841111	
HAM	132	B39	B44	8888888888888888811111	
HAM	25	B44	BW58	8888888888686888841111	
MRV	841	B44	B39	8888888888888886812221	
MRV	801	B44	B39	8888888888888826611121	B44 "short"
MYE	34	B7	B44	8888884888886216411141	
MYE	44	B44	B51	8488888888888118111121	
MYE	89	B44		8488886888848418110121	
MYE	96	B44	BW47	8888888888808624111141	
ABS	1588	B44	B38	8888884688811811611114	
MRV	826	B44	BW61	8888888888888822811111	
MRV	859	B44	BW73	8688888888668811611111	
MID	7	B44	BW41	8888881868818811611111	
KRE	67	B44	BW70	8888888888888841411111	
MRV	870	B44	BW50	0008888880088882821221	
MRV	787	B44	B35	8888888868868864211111	
MID	7	B44	BW41	8888881868818811611111	
KRE	67	B44	BW70	8888888888888841411111	
MRV	870	B44	BW50	0008888880088882821221	
MRV	787	B44	B35	8888888868868864211111	
MYE	131	B18	B44	8488684688188614100111	

Figure 4. A cell-by-serum matrix for selected cells and antisera from Antigen Society and Group 6 that define the antigens B44 "long" and B44 "short." Cells for the B44 "long" pattern were chosen at random by laboratory code. All cells that expressed the B44 "short" pattern were included in the matrix. Two distinct patterns are evident.

Table 2. Pattern Analysis for Antigens B44 and B45

	417	419	418	4352	4365	4355	4353	4356	4348	424	427	4367	4376	4375	4360	4383	4366	4379	4380	4370	4373	4400	432	133	4387	428	430
B44	+	+	+	+	+	+	+	+	+	+	+	+	+	+	+	+	+	+	+	+	+	+	–	–	–	–	–
B45	+	+	+	+	+	+	+	+	+	–	–	–	–	–	–	–	–	–	–	–	–	–	+	+	+	+	+
B44 'short'	+	+	+	+	+	+	+	+	+	+	+	+	+	+	+	+	+	–	–	–	–	–	–	–	–	–	–
Additional															2			1	2					29			
Specificities																			13								

				MMJMFMMDMGYBAS
				IRLREIRPIEURBT
				TVEVSDVTTLNNSE
				444444 4
				44433333341344
				11156555433823
				79825536823780
FTW	33	B8	B45	88888888888686
FTW	46	B45	B35	88888888888886
FTW	60	B45	BW63	88888888888888
FTW	12	B7	B45	88888888888888
MOO	890	B7	B45	88888888888888
MOO	714	B7	B45	88888888888888
ABS	3046	B45	BW42	88888888888888
MYE	92	B45	BW50	88888888888846
DUT	513	B45		88888888888888
GAZ	27	B35	B45	88888888888888
ABS	3071	B7	B45	88888888888888
DUT	526	BW42	B45	88888888888688
ABS	2496	B8	B45	88888888888688
MIT	4568	B45	BW62	88888888888888
KRE	71	B13	B45	88888888888688
KRE	85	BW65	B45	88888868888488
ABS	3044	B7	B45	88888888888888
FTW	8	B35	B45	88888888888841
MYE	147	B14	B45	88888888888661
ABS	2973	B45	B18	88888888888661
HAM	139	B45	BW70	88888888888416
HAM	38	BW42	B45	88888888888611
HAM	72	B7	B45	88888888888811
MYE	26	B7	B45	88888888886411
FTW	15	B45	BW57	88888888888821
FTW	42	B13	B45	88888888888611
HAM	15	BW42	B45	88888888888181
FTW	54	B7	B45	88888888888181
MYE	97	B45	BW58	88888868864116
MYE	95	B13	B45	88888888884114
HAM	44	B8	B45	88888888888111
FTW	39	B45	B45	88888888888111
HAM	10	B45	BW58	88888888888111

B45

Figure 5. A cell-by-serum matrix for selected cells and antisera from Antigen Society and Group 6 that define the antigen B45. See text for details.

4376, 4375, and 4360. These six alloantisera correlated well with each other (Figure 3) and give strong cytotoxic reactions with B44 positive cells.

One hundred thiry-one cells with unambiguous phenotypes and clear-cut patterns of reactivity with the B12 alloantisera and B44 alloantisera were identified and used to investigate possible heterogeneity within B44. The possibility of "short" serological variants of B44 appears quite strong. Several anti-sera of the Antigen Society set were thought to give negative reactions with short variants of B44. Seven alloantisera were informative for this study (Figures 3 and 4). These anti-sera (4383, 4366, 4379, 4380, 4370, 4373, and 4400) exhibited correlation coefficients for B44 cells ranging from 0.821 to 0.482 (see Appendix). These alloantisera were distinguished from poor anti-sera by their ability to yield either strong (8+) cytotoxic reactions or negative reactions and their lack of many intermediate reactions.

Of the 131 B44 cells chosen, 33 cells exhibited various "short" patterns with these seven alloantisera (Figure 4). The remainder of the cells gave consistently strong reactions with these seven antisera. We will refer to these cells as having the "long" pattern of B44. As in previous Workshops, we found that the "long" pattern was the most common. Figure 4 illustrates the "long" and "short" serological variants of B44 in a cell by serum matrix. Although only 20 of the 98 cells exhibiting the "long" pattern are actually shown, all 33 of the cells exhibiting the "short" variations are depicted. Twenty-nine out of 33 of the "short" B44 cells studied in this Workshop were of Caucasian origin. Three were African and one was from Southeast Asia. Sixteen of these 33 cells were found to have Cw5 suggesting that the "short" B44 gene may be in greater linkage disequilibrium with this antigen than with the "long" B44.

Unfortunately there were no cytotoxic anti-sera submitted that reacted positively with the short B44-positive cells and negatively with the long B44-positive cells. The patterns for B44 "long" and B44 "short" are presented in Table 2.

B45

Fewer good cytotoxic reagents were submitted for B45 than for B12 or B44. The best cytotoxic alloantisera (432, 133, 4387, 428, 430) were moderately well correlated with each other (Figure 3) yielding correlation coefficients for B45 ranging from 0.792 to 0.585. All five anti-sera exhibited some residual reactivity for B44 (14 to 17%) (see Appendix). Intermediate grades of cytotoxicity were more likely to be seen with these anti-sera. A total of 33 cells (typed with both Core and Antigen Socity sera) were chosen for study of B45 heterogeneity (Figure 5). All cells were strongly reactive with the nine B12 anti-sera. Only 17 out of the 33 cells were consistently and strongly reactive with all five B45 alloantisera (Figure 5). However, all 33 cells were reactive with the Core sera 432 and 133. Serum 133 is a duospecific serum recognizing both A29 and B45. It appears that 16 out of 33 B45 cells were variably reactive with the anti-sera 4387, 428, and 130 (Figure 5). Eleven of these cells were missed by at least two of these three alloantisera. No distinction could be made regarding ethnic restrictions between the long and short patterns of B45. Black and Caucasian cells exhibited both long and short patterns. Since the anti-sera were showing some intermediate grades of reactivity, the long and short patterns of B45 might be an artifact of weakening anti-sera. The pattern for B45 is presented in Table 2.

Acknowledgments. The authors wish to thank the following investigators for their contribution to Antigen Society and Group 6: W-K Cheung, E.D. du Toit, M. Fotino, E. Gazit, J.M. Kreisler, S. Leverenz, S. Mayer, H. Mervart, D. Middleton, S.B. Moore, C. Mueller, and C. Vermylen. We also wish to thank W.C. Knowler and D.J. Pettitt of the Diabetes and Arthritis Epidemiology Section, National Institute of Diabetes and Digestive and Kidney Diseases, for their help with the samples from the Gila River Indian Community. Finally, we would like to express our appreciations to the Yaqui Indian Community of Guadalupe, Arizona for their continuing assistance.

References

1. Bodmer J. The A B C of HLA: A serological report of the Sixth Histocompatibility Testing Workshop. In: Kissinger-Nielsen F (ed): Histocompatibility Testing 1975. Munksgaard, Copenhagen, Denmark, 1975, p 54, 55.
2. Dick HM. HLA-A, B and C serology and antigen reports. In: Bodmer WF, Batchelor JR, Bodmer JG, Festenstein H, Morris PJ (eds): Histocompatibility Testing 1977. Munksgaard, Copenhagen, Denmark, 1977, p 175–177.
3. Ness D, Grumet FC, Graugaard G, Kissmeyer-Nielsen F. Antigen report: HLA-B49. In: Albert ED, Baur MP, Mayr WR (eds): Histocompatibility Testing 1984. Springer-Verlag, Berlin, 1984, p 157.
4. Mittal K, Thompson J, Mason J, Kerman R, Poulik M, Bull R. B12, B21 Group. In: Duquesnoy RJ, Schacter B (eds): The Third American Histocompatibility Workshop. The American Association for Clinical Histocompatibility Testing, New York, 1980, p 23–28.
5. Williams RC, Morse HG, Bonnell MD, Rate RG, Kuberski TT. The HLA loci of the Hopi and Navajo. Am J Phys Anthropol 1981;56:291–296.
6. Ensroth AF, Mann DL, Johnson AH, Knowler WC, Pettitt DJ, Bennett PH. HLA and B-lymphocyte alloantigens in Gila River Indians. Tissue Antigens 1983;21:198–207.
7. Ward FE, Kostyu DD, Stewart AD, Amos DB. HLA antigens in the Pima Indians. In: Aizawa M (ed): Proceedings of the 3rd Asia-Oceania Histocompatibility Workshop and Conference. Hokkaido University Press, Sapporo, Japan, 1986, p 531–534.
8. Unpublished data.
9. Svejgaard A, Kissmeyer-Nielsen F, Thorsby E. HLA typing of platelets. In: Terasaki PI (ed), Histocompatibility Testing 1970. Munksgaard, Copenhage, Denmark, 1970, p 153.
10. Mittal KK. Antigen report: HLA-B44. In: Albert ED, Baur MP, Mayr WR (eds): Histocompatibility Testing 1984. Springer-Verlag, Berlin, 1984, p 151.
11. Moller E. Antigen report: HLA-B45. In: Albert ED, Baur MP, Mayr WR (eds): Histocompatibility Testing 1984. Springer-Verlag, Berlin, 1984, p 152.
12. Mittal K. Joint report–Bw44. In: Histocompatibility Testing 1980. UCLA Tissue Typing Laboratory, Los Angeles, 1980, p 398.

Appendix

The Appendix lists the results of the TAIL analysis for all anti-sera that were part of Antigen Society and Group 6, and which expressed specificity for either B12 or B21 antigens. It was performed using the SERAN software furnished for the Tenth International Workshop. Parameters chosen for the TAIL analysis were the following: minimum inclusion ratio, 75%; minimum number of positively reacting cells, 7; minimum Chi-squared value, 100; and minimum number of positively reacting cells necessary for repeated iterations on a serum, 7. Abbreviations in the Appendix are as follows: %8, strength of the serum; %Inc., percent inclusion; + +, true positives; + –, false negatives; – +, false positives; – –, true negatives; r, the correlation coefficient; OR, the logical union of all specificities identified in the analysis; and RES, residual reactions. Only cells with a reaction profile greater than 0.75 were used in the analysis.

	%B	%Inc.	++	+-	-+	--	r
421 ITI CEP TORP180							
B12	80	83	148	30	31	221	0.705
RES	57	A2 = 15% Cw6 = 7%					
540 FRA DDC PIERSON							
Cw6	95	96	106	4	90	198	0.579
Cw4	98	96	48	2	42	196	0.634
OR	97	96	154	6	42	196	0.767
RES	82	Bw50 = 9% A2 = 9% A3 = 6% B45 = 6%					
552 US2 MBC BC.SH.GOST							
B44	89	95	121	6	50	220	0.719
Cw5	69	71	5	2	45	218	0.223
Cw8	84	100	11	0	39	220	0.424
RES	63	A1 = 100%					
222 US5 TER TER.B5349A							
B49	91	95	52	3	32	306	0.714
Bw53	69	100	5	0	27	306	0.371
Bw78	33	0	0	0	32	306	0.074
B51	83	91	10	1	22	305	0.503
Bw52	89	100	8	0	24	306	0.472
RES	73	A28 = 100%					
223 US5 TER TER.B5349B							
B49	79	85	47	8	32	301	0.652
Bw53	82	80	4	1	28	300	0.295
Bw78	33	0	0	0	32	301	0.074
B51	91	91	10	1	22	300	0.502
Bw52	89	100	8	0	24	301	0.472
RES	84	A28 = 100%					
224 US5 TER TER.B7849B							
B49	72	73	41	15	28	301	0.591
B51	93	92	12	1	57	315	0.358
Bw52	91	100	10	0	59	316	0.344
TE78	33	0	0	0	69	316	0.042
RES	79	B49 = 12% A2 = 12% Cw7 = 7%					
225 US5 TER TER.B5349C							
B49	89	95	53	3	41	290	0.670
Bw53	54	100	5	0	36	290	0.323
Bw78	33	0	0	0	41	290	0.062
RES	60	A28 = 100%					
425 NCY MRV 11146.4							
B44	79		115	11	9	254	0.877
RES	5	B49 = 15% A2 = 12% B18 = 6% Cw4 = 6% A26 = 6%					
424 ANZ DAW HOULISTON R8008							
B44	97	97	121	4	11	258	0.909
RES	28	B49 = 9% A2 = 6% A26 = 6% Cw4 = 6% A23 = 6%					
427 SAF HAM N137							
B44	85	90	104	12	16	236	0.821
RES	31	B49 = 10% A2 = 10% B27 = 6%					
430 US7 STE OF.SHAW							
B45	65	57	24	18	9	339	0.602
RES	43	B44 = 14% A2 = 9% Cw6 = 9% A1 = 7% Cw2 = 7%					
428 US5 ABS AZ.13							
B45	79	64	27	15	12	343	0.624
RES	26	B44 = 17% A2 = 12% Bw50 = 7%					
432 UK1 GEL MUS22289							
B45	93	100	42	0	20	335	0.792
RES	40	B44 = 14% Bw50 = 9% A2 = 8% Cw4 = 7% Cw6 = 6%					
426 US5 MIT C5324.1							
B44	95	100	125	0	30	240	0.842
RES	57	B45 = 18% A2 = 10% Cw6 = 9% A29 = 6%					
417 US5 MIT B5749.1							
B12	98	99	163	2	7	224	0.949
RES	88	A2 = 12% Bw50 = 9% B18 = 6% Cw6 = 6% Bw71 = 6%					
418 US8 JLE YEAG							
B12	96	100	164	0	14	211	0.925
RES	16	A1 = 12% Bw50 = 9% Cw6 = 7% A26 = 7% Cw3 = 6%					
419 NCY MRV 3034.7							
B12	88	98	159	4	4	223	0.953
RES	27	A30 = 10% Cw3 = 10% A24 = 10%					
441 US3 GRU SIMPKINS.1							
B49	79	91	50	5	12	329	0.824
RES	26	A2 = 15% B44 = 7% A30 = 7%					
444 US6 TRP MS119							
B49	38	61	34	22	6	335	0.676
RES	20	A2 = 14% Cw3 = 7% Bw50 = 7% A29 = 7% Cw7 = 7%					
445 IT2 FER FE99							
B49	68	64	36	20	3	336	0.734
RES	33	A1 = 7% Bw52 = 7% Bw50 = 7% Cw6 = 7% A2 = 7%					

442 FRA DDC GRAUPNER
B49 81 73 41 15 29 312 0.587
RES 41 Bw50 = 15% A2 = 14% Cw6 = 7%

448 UK1 BRS 11235.LCS
Bw50 90 80 47 12 43 288 0.563
RES 46 B45 = 12% A2 = 10% B49 = 7% Cw6 = 6%

447 US5 MIT C4503.1
Bw50 58 59 34 24 10 321 0.620
RES 13 B49 = 8% A2 = 8% Cw4 = 6% A26 = 6% Cw2 = 6%

232 US2 MBC BC.MA.SCHL
Bw50 80 92 54 5 57 281 0.588
Bw41 64 57 4 3 53 278 0.162
Bw62 71 53 9 8 48 273 0.225
Bw70 26 40 12 18 45 263 0.196
RES 63 A1 = 100%

435 NCY MRV 5736.10
B21 81 82 93 20 11 270 0.799
RES 44 A2 = 13% B45 = 7% A24 = 7% BN21 = 7%

436 EAE MYR SCHERZER
B21 84 93 107 8 29 249 0.786
RES 61 A2 = 14% B45 = 9% A24 = 8% BN21 = 8% Cw3 = 6%

440 US8 JLE BENG
B21 73 85 104 19 6 259 0.845
RES 60 B35 = 10% B44 = 10% A26 = 10% Cw4 = 7% Cw7 = 7%

450 GER GOL 15984II
Bw50 79 88 51 7 49 289 0.594
RES 64 B49 = 14% A2 = 12% Cw7 = 8%

437 EAE MYR KELLNER
B21 78 100 3 0 173 223 0.088
RES 64 A2 = 11% Bw50 = 7% B44 = 6% Cw6 = 6% B49 = 6%

451 BEN VER A1674
Bw50 89 78 46 13 131 203 0.276
RES 71 A2 = 9% B44 = 8% B49 = 6% Cw7 = 6%

229 US5 CRO DOSS
Bw62 70 100 22 0 121 255 0.318
Bw70 45 80 24 6 119 249 0.260
Bw50 74 88 52 7 91 248 0.451
Bw56 71 100 2 0 141 255 0.086
B35 71 73 19 7 124 248 0.203
RES 69 A2 = 10% Bw50 = 8% Cw6 = 7%

230 EAE MYR KOPKE
B15 33 0 0 0 181 216 0.004
Bw50 83 97 57 2 124 214 0.424
Bw70 80 93 28 2 153 214 0.270
RES 72 A2 = 10% Bw50 = 7% Cw6 = 7%

233 NCY MRV 74.3
Bw62 73 100 21 0 134 240 0.290
B15.3 60 100 1 0 154 240 0.052
Bw50 91 90 52 6 103 234 0.425
Bw70 73 93 28 2 127 238 0.314
Bw56 71 100 2 0 153 240 0.079
B35 81 96 25 1 130 239 0.305
RES 81 A2 = 10% Bw50 = 8% Cw6 = 7%

234 BEN BER MS25
B35 84 92 24 2 111 260 0.322
Bw70 45 50 15 15 120 247 0.097
B15.3 60 100 1 0 134 262 0.061
Bw62 54 73 16 6 119 256 0.196
Bw56 71 100 2 0 133 262 0.090
RES 68 A2 = 10% Bw50 = 8% Cw6 = 6%

231 BEN BER MS36
B15 33 0 0 0 209 188 -0.003
B17 71 100 2 0 207 188 0.054
B35 80 92 23 2 186 186 0.200
Bw70 69 60 18 12 191 176 0.041
RES 81 A2 = 10% Cw6 = 8% Bw50 = 6%

228 FRA DDC BERTIN
B35 88 88 23 3 143 229 0.247
Bw62 86 91 20 2 146 230 0.237
Bw70 80 97 29 1 137 231 0.314
RES 86 A2 = 11% Bw50 = 8% Cw6 = 7%

215 US2 WOL NW 10,696
B35 89 96 25 1 48 324 0.526
Bw62 41 32 7 15 41 309 0.148
Bw50 64 45 26 32 22 292 0.408
TE79 33 0 0 0 48 324 0.057
Bw70 29 24 7 22 41 302 0.104
B17 33 0 0 2 48 322 0.010
RES 56 A2 = 100%

200 UK1 BRS 10555.DUN
B35 96 85 22 4 113 260 0.280
Bw62 48 55 12 10 123 254 0.106
Bw71 84 100 11 0 124 264 0.230
Bw72 71 67 2 1 133 263 0.061
RES 72 A2 = 10% Cw6 = 8% Bw50 = 7%

133 NCY YUN REED, S.
B45 88 98 41 1 55 300 0.585
A29 77 87 34 5 21 295 0.689
OR 84 93 75 6 21 295 0.803
RES 42 B44 = 13% A26 = 9% Cw6 = 6% A2 = 6%

422 US5 KWC AS070
B12 91 96 159 7 13 218 0.893
RES 38 A2 = 7% B49 = 7% Bw50 = 7% Cw6 = 7% A1 = 7%

438 NCY MRV 2722.4
B21 60 33 1 2 123 269 0.013
RES 69 A2 = 12% Bw50 = 8% B49 = 7% Cw6 = 6%

439 US8 JLE SHAF
B21 88 97 111 4 31 252 0.805
RES 54 A2 = 9% A24 = 9% B44 = 7% BN21 = 7% Cw3 = 6%

443 BEN BER MS10
B49 48 60 33 22 7 335 0.659
RES 29 B44 = 11% Cw6 = 11% Aw33 = 6% Cw4 = 6% A2 = 6%

2049 FRA FAU H287
A29 81 100 36 0 129 121 0.319
B12 33 0 0 0 165 121 -0.009
RES 78 B44 = 10% A2 = 9% Cw6 = 6%

2097 GER WAN KT4
B44 60 38 41 67 21 167 0.315
B45 52 47 15 17 47 217 0.223
B51 43 15 2 11 60 223 -0.018
Bw52 33 30 3 7 59 227 0.051
RES 59 B44 = 13% A2 = 10% Cw6 = 6% Cw7 = 6%

2098 GER WAN S4
B44 58 45 49 60 69 119 0.081
B45 27 12 4 28 114 151 -0.188
B51 44 85 11 2 107 177 0.192
B13 41 75 12 4 106 175 0.169
Bw57 54 71 5 2 113 177 0.098
RES 59 A2 = 10% B44 = 8% B49 = 7% Cw6 = 6%

4348 US5 MIT B5837.1
B12 94 96 177 7 17 231 0.884
RES 62 Cw3 = 11% A2 = 6% A24 = 6% Bw60 = 6% Cw2 = 6%

4349 US7 POL 026
B12 81 85 151 27 10 234 0.817
RES 65 A2 = 12% Cw3 = 10% A1 = 6% Bw60 = 6% Cw6 = 6%

4350 US7 POL 185
B44 85 91 129 13 38 248 0.745
RES 72 A1 = 100%

4351 US1 HJH 3228B
B12 33 0 0 0 149 283 0.016
RES 77 B44 = 14% A2 = 9% B45 = 6%

4352 NCY MRV 3500.1
B12 93 97 178 6 16 233 0.894
RES 43 Cw3 = 13% Bw60 = 8% Cw6 = 6% A2 = 6% A23 =6%

4353 NCY MRV 7836.3
B12 84 96 176 8 10 239 0.911
RES 39 Cw3 = 12% A24 = 8% Bw60 = 8% Cw6 = 6% A23 =6%

4354 US5 ABS AZ.353
B12 84 91 167 17 16 232 0.840
RES 54 Cw6 = 9% Cw3 = 9% A2 = 8% B13 = 8%

4355 UK1 MID BEL60A
B12 96 97 179 5 11 238 0.921
RES 84 Cw3 = 13% Bw60 = 9% A24 = 7% B7 = 7%

4356 BEN DPT MILOUTA.29
B12 96 95 175 9 15 234 0.883
RES 58 Cw3 = 11% A2 = 7% A24 = 7% Bw60 = 7%

4357 NCY KAP CWR.PLA.1517
B12 33 0 0 0 206 89 -0.025
RES 85 A2 = 10% B44 = 10%

4358 NCY KAP CWR.PLA.1299
B12 33 0 0 0 135 281 0.018
RES 70 B44 = 15% A2 = 9%

4359 USI RTUL TUL.P207
B12 90 97 178 6 30 219 0.834
RES 46 A2 = 9% Cw3 = 9% B49 = 8% Cw6 = 6%

4360 SAF HAM N312
B44 73 77 108 33 5 281 0.794
B12 33 0 0 0 5 281 0.198
DR7 33 0 0 0 5 281 0.198
RES 23 A2 = 100%

4361 FRA OHA DIOT
B44 86 94 134 9 32 258 0.796
B45 33 37 15 26 17 232 0.331
RES 46 A1 = 100%

4362 FRA BET E933
B12 91 97 179 5 39 208 0.802
RES 63 A23 = 12% Cw3 = 6% Cw6 = 6%

4363 ANZ CRS PERRY
B44 92 94 134 8 87 200 0.600
B45 98 95 39 2 48 198 0.570
RES 52 A3 = 100%

4364 ANZ TAI MARINAL
B12 82 87 160 23 29 219 0.751
RES 41 A2 = 10% Cw3 = 8% B49 = 7% Cw7 = 6%

4365 UK1 FES LECHLEY
B12 92 97 178 6 12 237 0.911
RES 70 Cw3 = 13% Bw60 = 8% A1 = 7% A23 = 7% A24 = 7%

4366 US6 BRN CC487.1
B44 77 83 119 24 9 281 0.821
RES 52 Cw3 = 12% A24 = 10% A1 = 7% Bw60 = 7% A23 = 7%

4367 US1 DUQ FARMERIE
B44 89 93 131 10 9 281 0.895
RES 62 Cw3 = 11% A24 = 9% A2 = 7% Cw6 = 7% Bw60 = 7%

4368 JAP NAI FD3952
B12 88 93 155 12 22 198 0.819
RES 45 A2 = 8% Cw3 = 7% B49 = 6% A23 = 6% Cw6 = 6%

4369 GER GOL 13491
B44 91 93 130 10 49 239 0.718
RES 59 B49 = 12% A2 = 11% Cw7 = 7%

4370 US7 POL 146
B44 51 63 90 53 30 258 0.549
RES 8 A2 = 9% B49 = 6% Cw7 = 6% B7 = 6%

4371 NCY MRV 20002.1
B44 86 94 134 9 35 255 0.783
RES 58 B45 = 10% A2 = 10% B7 = 6%

4373 US5 ABS AZ.361
B44 37 57 81 61 10 278 0.613
RES 13 Cw3 = 8% A23 = 8% B7 = 8% Cw7 = 6% A24 = 6%

4374 UK1 GEL CRO20511
B44 93 94 135 8 55 235 0.711
RES 53 B49 = 10% A2 = 10% Cw7 = 6% A23 = 6%

4375 FRA BET E1017
B44 84 91 130 13 16 274 0.845
RES 27 Cw3 = 9% A2 = 7% A24 = 7% B7 = 6% Bw60 = 6%

4376 US5 PNW PULL
B44 85 92 131 12 9 281 0.885
RES 43 Cw3 = 12% A24 = 10% B7 = 10% Bw60 = 7% A23 = 7%

4377 JAP TAK KS0222
B44 61 69 98 45 27 262 0.611
RES 40 B45 = 18% A2 = 10% A30 = 6%

4378 BEN VER L301
B44 68 68 96 46 8 282 0.709
RES 47 Cw3 = 12% A24 = 8% A30 = 8% Bw60 = 8% A2 = 8%

4379 EAE ZAR M01869
B44 53 56 77 61 8 280 0.617
RES 26 A1 = 11% B49 = 8% A23 = 8% Cw6 = 8% Bw52 = 6%

4380 US7 DUP NJ7503
B44 51 53 74 66 2 285 0.636
RES 43 A2 = 20% Bw41 = 20% A29 = 10% Bw62 = 10% Cw3=10%

4381 US4 SND S15L61
B44 80 86 122 20 17 273 0.801
RES 46 A2 = 11% B13 = 11% Cw6 = 10% A29 = 7%

4382 US2 HPP SIMPSON
B44 75 85 121 22 34 256 3 0.712
RES 42 B49 = 13% A2 = 9% A23 = 8% Cw7 = 6% Cw3 = 6%

4383 US2 BAC SOUNDERS
B44 78 78 112 31 22 268 0.716
RES 32 B49 = 9% A2 = 9% A23 = 8% B45 = 8%

4384 NCY MRV 19400.0
B44 78 83 118 24 7 282 0.831
RES 29 Cw3 = 11% A2 = 9% A23 = 9% A29 = 6% B45 = 6%

4385 UK1 BRS 7297.LCS
B44 89 92 130 12 36 253 0.760
RES 54 A2 = 12% B49 = 9% B13 = 8% Cw6 = 8%

4386 SAF HAM N1143
B44 78 68 75 36 36 229 0.537
DR7 33 0 0 0 111 265 0.023
RES 69 B44 = 13% A2 = 9% Cw6 = 6%

4387 US6 BRN CC425.2
B45 51 66 29 15 10 369 0.662
RES 65 A2 = 14% B44 = 14% Cw6 = 9%

4388 US1 TUL TUL.P193
B45 84 91 41 4 102 283 0.417
RES 74 B44 = 13% A2 = 10% B45 = 6%

4389 FRA OHA CHRISTODOULOU
B45 14 20 9 36 25 360 0.159
Bw50 65 12 7 51 27 345 0.067
RES 61 A2 = 9% B44 = 7% Cw6 = 6% B45 = 6%

4390 SAF HAM N1219
B44 72 79 113 30 66 224 0.535
B49 81 80 36 9 30 215 0.580
RES 56 A3 = 100%

4391 UK1 BRS 11262.MBT
B49 81 91 50 5 21 357 0.762
RES 60 Cw4 = 7% A2 = 7% A1 = 7%

4392 US5 FTW 38BUTS
B49 49 67 36 18 25 351 0.567
RES 47 B44 = 7% A2 = 6% Cw6 = 6% A1 = 6% Cw4 = 6%

4393 US3 GUY 39207M
B49 39 65 36 19 8 356 0.691
RES 37 A1 = 11% B44 = 8% Cw6 = 8%

4394 US5 ABS AZ.358
B49 65 72 39 15 13 365 0.693
RES 52 Cw4 = 8% A1 = 8% Cw6 = 8% B44 = 6% Cw7 = 6%

4395 US5 CRO POOL1.VINCENT
B49 78 87 48 7 33 344 0.666
RES 86 A2 = 8% Cw3 = 8% A1 = 7% Bw60 = 6% Cw6 = 6%

4396 US3 GRU SIMPKINS
B49 87 83 44 9 16 359 0.741
RES 54 A2 = 6% A1 = 6% Cw7 = 6%

4397 US8 JLE MORR
B49 48 64 35 20 4 374 0.721
RES 45 Cw7 = 11%

4398 US7 HAN SEA1245
B49 48 57 31 23 16 361 0.561
RES 31 A2 = 7% B44 = 7% Cw6 = 7% Cw4 = 6% Cw7 = 6%

4399 US1 IND STU.152
B49 45 73 40 15 20 358 0.645
RES 30 A2 = 11% B44 = 6%

4400 IT1 CEP TORP1129
B49 78 87 48 7 79 296 0.482
RES 48 B44 = 13% A2 = 10%

4401 BEN VER L25.3
B49 68 85 46 8 50 328 0.569
RES 51 A23 = 9% B44 = 7% Cw6 = 7% A2 = 7%

4402 FRA OHA SAUVAGEON
B49 86 90 47 5 47 310 0.607
RES 48 A2 = 8% B44 = 7% Cw6 = 7%

4403 FRA MYE GEOFFORY.COLET
B49 93 100 55 0 168 208 0.365
Bw52 81 90 9 1 214 207 0.112
RES 71 A2 = 10% B44 = 7% Cw6 = 6%

4404 IT1 MTT F10121
B49 81 85 45 8 31 347 0.656
Bw52 64 50 4 4 27 343 0.233
RES 65 A24 = 100%

4405 USB JLE H350
B49 66 69 37 17 22 350 0.599
Bw52 68 80 8 2 51 365 0.296
RES 62 B49 = 14% A2 = 13% Cw7 = 9%

4406 US7 STE OF4951
B49 59 74 40 14 12 363 0.714
DRN 33 0 0 0 52 377 0.055
RES 53 B49 = 16% A2 = 12% Cw7 = 9% A1 = 6%

4407 UK1 FES FERENDES
B21 78 86 95 16 29 287 0.734
RES 87 Cw3 = 11% A24 = 8% A2 = 8% Bw60 = 6% BN21 = 6%

4408 NCY TOM MIALLA
B21 70 79 82 22 13 301 0.766
RES 59 A2 = 8% B44 = 8% B45 = 8% A30 = 6%

4409 SCA KIS 1115/78
B49 92 96 50 2 110 263 0.447
Bw50 95 95 54 3 106 262 0.460
RES 78 A2 = 10% B49 = 9% Cw7 = 7% B44 = 6%

4410 FRA PRR MES1472
B21 96 95 105 5 42 272 0.752
RES 61 B45 = 8% A2 = 8% A24 = 7%

4411 US2 WOL NW72083
B44 97 86 116 19 158 123 0.291
B45 97 90 37 4 237 138 0.166
B49 98 94 49 3 225 139 0.222
Bw50 98 95 54 3 220 139 0.238
A9 33 0 0 0 274 142 -0.016
RES 95 A2 = 10% B44 = 9%

4412 US5 CRO 6.20.78GUMLIA
B49 70 91 49 5 65 308 0.547
B21 43 67 2 1 63 307 0.122
RES 58 A2 = 100%

4413 UK1 MID BEL97A
B21 20 33 1 2 80 343 0.045
RES 53 A2 = 11% Bw50 = 8% Cw6 = 8% B49 = 7%

4414 IT1 MTT F101298
B21 33 0 0 3 62 302 -0.013
RES 51 A2 = 11% B49 = 9% Cw6 = 8%

4415 EAE IVS 5043
B21 20 33 1 2 93 328 0.036
RES 54 A2 = 12% B49 = 8% Cw6 = 7% Bw50 = 6% Cw7 = 6%

4416 BEN BOU 64916 11.9.82
B21 71 67 2 1 163 257 0.047
RES 78 A2 = 11% B49 = 7% Bw50 = 6% Cw6 = 6% Cw7 = 6%

4417 LAT GOR BM269
B21 71 67 2 1 123 301 0.071
RES 55 A2 = 13% B49 = 7% Bw50 = 7% Cw6 = 6% B44 = 6%

4418 SCA SVE DO329.83
B49 54 85 46 8 60 312 0.528
B21 60 33 1 2 59 310 0.060
RES 65 A2 = 100%

4419 US7 STE OF.21A
B21 68 82 94 21 13 306 0.791
RES 59 A1 = 10% A31 = 8% Cw6 = 6% Cw3 = 6% B35 = 6%

4420 US1 DUQ RADELET
B21 33 0 0 2 53 372 0.010
RES 52 A2 = 14% B49 = 12% Bw50 = 6% Cw6 = 6%

4421 USB JLE STEI
B21 60 33 1 2 107 323 0.028
RES 49 A2 = 12% Bw50 = 8% B49 = 8% Cw6 = 7% Cw7 = 6%

4422 US7 DUP NJ8024
B21 72 84 95 18 33 285 0.706
RES 62 A2 = 10% Cw6 = 7% Cw3 = 6% Cw7 = 6%

4423 US5 ABS AZ.151
A2 86 93 167 13 60 189 0.676
B21 43 67 2 1 58 188 0.133
RES 78 A3 = 100%

4424 NCY KAP CWR.PLA.1464
B44 91 96 137 5 131 159 0.493
RES 63 A3 = 100%

4425 US6 TRP ABQ84.468
Bw50 35 58 34 25 38 336 0.436
RES 57 A2 = 11% B49 = 9% Cw7 = 8% Cw6 = 6%

4426 SAF HAM N1161
Bw50 72 75 41 14 205 163 0.126
B12 33 0 0 0 246 177 -0.008
RES 76 B44 = 11% A2 = 10%

4427 GER BRA 1291
Bw50 93 95 53 3 112 232 0.434
B49 84 89 48 6 117 229 0.379
RES 84 A2 = 11% Bw50 = 9% Cw6 = 9% B44 = 6%

5455 IT1 PUR CNTS.68
B12 33 0 0 0 220 183 -0.005
RES 74 B44 = 10% A2 = 10%

9014 US7 DUP SCOTT
B12 33 0 0 0 128 142 0.003
RES 82 B44 = 14% A2 = 11% Cw5 = 6%

9015 US7 DUP BENJAMIN
B12 33 0 0 0 148 112 -0.009
RES 89 B44 = 13% A2 = 10%

9016 US7 DUP BLISH
B12 33 0 0 0 75 171 0.027
RES 62 B44 = 15% A2 = 10%

9017 US7 DUP KESTER
B12 33 0 0 0 94 158 0.016
B21 60 33 1 2 93 156 0.000
RES 63 B44 = 13% A2 = 12%

9018 US7 DUP GOLDBERG
B44 84 82 84 18 32 125 0.604
RES 70 B45 = 12% A2 = 12% Cw6 = 8% A29 = 6%

9019 US7 DUP MOORE
B44 72 64 67 37 39 119 0.393
RES 58 A2 = 15% B49 = 14% Cw7 = 6% Cw6 = 6%

9020 US7 DUP COSTA
B21 60 33 1 2 122 145 -0.020
RES 70 A2 = 12% B49 = 7% Bw50 = 6% B44 = 6%

9021 NCY MRV MRV2837
B44 84 96 108 5 41 125 0.692
B12 33 0 0 0 41 125 0.045
RES 74 A1 = 100%

9022 NCY MRV MRV7197
B12 33 0 0 0 111 164 0.012
RES 63 B44 = 14% A2 = 11% Cw5 = 6%

9023 NCY MRV MRV7395
B12 33 0 0 0 144 133 -0.002
RES 75 B44 = 14% A2 = 10%

9024 NCY MRV MRV18489
B12 33 0 0 0 149 127 -0.005
RES 90 B44 = 14% A2 = 10%

9025 NCY MRV MRV19469
B12 33 0 0 0 140 137 -0.001
RES 79 B44 = 14% A2 = 10%

9026 NCY MRV MRV23300
B12 33 0 0 0 118 154 0.008
RES 74 B44 = 14% A2 = 10%

9027 NCY MRV MRV21784
B45 58 47 15 17 10 236 0.475
RES 65 B44 = 16% A2 = 12% A26 = 10% Cw4 = 6% Bw50 = 6%

9028 NCY MRV MRV10776
B21 43 67 2 1 120 154 0.045
RES 70 A2 = 11% Bw50 = 8% B49 = 7% Cw6 = 7%

9029 NCY MRV MRV22898
B21 20 33 1 2 109 165 -0.006
RES 67 A2 = 12% B49 = 8% Bw50 = 8% Cw6 = 6%

9030 NCY MRV MRV11087
B49 75 90 46 5 24 202 0.703
Bw52 68 100 8 0 16 202 0.542
Bw53 71 67 2 1 22 201 0.220
Bw63 71 67 2 1 22 201 0.220
RES 66 A1 = 100%

Antigen Society #7 Report (B8, w59, B14, w64, w65)

G. Mueller-Eckhardt,[1] M.L. Schroeder,[2] C. Darke,[3] R. Khan,[4] R. Wank,[5] G. Dejour,[6] R. Fauchet,[6] F. Christiansen,[7] E. Du Toit,[8] H. Mervart,[9] F. Schunter,[10] A. Stratton,[11] and G. Sudmeyer[12]

History

HLA B8 and HLA B14 have been well established specificities since 1965 (1) and 1970 (2), respectively. Bw59, restricted to the Oriental population, was defined in 1980 (3), and the B14 subtypes Bw64 and Bw65 were accepted as specificities in 1984 (4). Associations were observed between Bw64, Aw32, and DR7 on one hand, and between Bw65, Aw33, and DR1 on the other (5–7). Specific complotypes were found to be in strong linkage disequilibrium with Bw64 or Bw65, respectively (8,9).

Serology

Cells of 540 individuals were tested within Antigen Society 7; these included 273 Caucasoids, 223 blacks, 3 Orientals, and 26 others. Fifty-five patients with 21 hydroxylase (21-OH) deficiency, provided by Professor Bouée, INSERM U73, Paris, were included in the study. The majority of the cells were tested by both Core and Antigen Society sera. Complement C4* and Bf allotypes were determined in a subset of the individuals. In addition, 2338 cells of the ETHNIC ARC file tested exclusively with the Core set were available for analysis of antigen group specificities.

Bw59. No Bw59 positive cells or monospecific antisera were available for study in the Antigen Society. Core sera with Bw59 specificity are listed in Table 1.

B8, B14. Many good to excellent B8 and broad B14 sera were included in both the Antigen Society and Core sets (Tables 2 and 3). Five monoclonal antibodies were included in the Core set (Table 4), three of these (2086, 2087, 2089) were very good typing reagents.

Reporting Laboratories: GERMUE,[1] US5MLS,[2] UKIDRK,[3] US7DUP,[4] GERWAN,[5] FRAFAU[6]
Participating Laboratories: ANZDAW,[7] SAFDUT,[8] NCYMRV,[9] GERSCN,[10] UKIAST,[11] GERSUD[12]

Table 1. Ten WS Core Sera With Bw59 Specificity

WS No.	Specificity	Ave score	Q-score	R value
405	Bw59	5.0	3.093	0.436
9063	Bw59	7.6	8.055	0.811
	B8	7.6	6.523	0.734
403	Bw59	6.0	4.644	0.632
	B8	6.5	4.288	0.627

Bw64,Bw65. The serological definition of the B14 subtypes in the Tenth International Histocompatibility Workshop (IHW) was provided by almost the same set of anti-sera (Table 5) as in the Ninth IHW. It appeared from the Tenth IHW that 10 WS 413 (Bw64), which was the original B14.1 serum (10) and has been included in three IHWs, is still the most reliable reagent to subdivide B14. The specificity of Bw65 sera appears to be very dependent on the level of test sensitivity and on cell quality. Reaction patterns within the Core and Antigen Society sets are shown in Tables 6 and 7, respectively.

Eighty-four percent of the B14 positive cells could easily be assigned to the Bw64 or Bw65 specificity. The majority of cells reacting on Bw64+/Bw65+ were either Bw64/Bw65 heterozygous or homozygous for either sub-type. B14 positive, Bw64−, Bw65− typing cells were considered to be weakly reacting cells of Bw64 or Bw65 specificity.

Table 2. Ten WS Sera With B8 Specificity Listed in Order of Quality Scores

	WS No.	Ave score	Q-score	R value
Core sera	398	7.8	9.728	0.946
	399	7.9	8.241	0.938
	402	7.2	8.238	0.923
	263	7.2	7.736	0.935
	9063	7.0	7.068	0.886
	401	7.1	6.399	0.895
	400	6.2	6.324	0.886
	404	6.2	5.492	0.862
	403	6.4	3.693	0.761
Antigen Society sera	4438	7.8	9.441	0.960
	4444	7.7	8.989	0.952
	4434	7.6	8.872	0.945
	4441	7.5	8.779	0.945
	4443	7.6	8.650	0.907
	4440	7.5	8.409	0.930
	4429	7.7	8.200	0.904
	4431	7.8	8.177	0.935
	4432	7.9	8.020	0.906
	4435	7.3	7.634	0.913
	4439	7.8	7.489	0.918
	4437	7.5	7.489	0.908
	4428	7.5	7.473	0.929
	4446	7.6	7.456	0.887
	4442	7.8	7.091	0.878
	4430	7.8	7.002	0.890
	5457	7.6	6.951	0.901
	4447	7.4	6.925	0.881
	4433	5.9	5.434	0.837
	4445	6.8	5.134	0.826

Table 3. Ten WS Sera With B14 Specificity Listed in Order of Quality Scores

	WS No.	Ave score	Q-score	R value
Core sera	392	8.0	7.855	0.755
	411	8.0	7.573	0.755
	408	8.0	6.812	0.627
	412	5.6	5.308	0.730
	407	6.3	5.255	0.582
	409	6.3	4.469	0.509
Antigen Society sera	4473	7.7	8.066	0.884
	4459	7.9	8.044	0.911
	4466	7.6	7.808	0.853
	4471	7.9	6.803	0.726
	4465	7.5	6.619	0.904
	4461	7.1	6.221	0.884
	4467	7.7	5.922	0.781
	4452	7.5	5.867	0.823
	4468	7.8	5.756	0.711
	4470	7.5	5.617	0.696
	4463	6.9	5.528	0.821
	4456	6.9	5.381	0.843
	5456	5.4	5.034	0.787
	4469	5.7	4.649	0.775
	4457	5.7	4.465	0.721
	4455	5.2	4.179	0.759
	4475	7.5	3.351	0.538
	4460	4.7	3.298	0.509
	4476	5.4	3.228	0.534
	4472	6.1	2.748	0.451

Linkage Disequilibrium With Complement Allotypes

Analysis of complement data for associations with Bw64 and/or Bw65 was possible in 58 individuals (Table 8). Earlier reported significant associations of C4*A2 and C4*B2 with Bw65 (8) were confirmed by the Tenth IHW material. However, the correlation of C4*A3 with Bw64 (8) appeared to be less striking, and an increased frequency of BfF in Bw65 individuals (7) did not reach the level of significance.

B14 and 21-OH Deficiency

Among the 55 B14 positive patients studied, four were Bw64 and 51 Bw65. Three individuals in each group were homozygous for the respective B14 split.

Table 4. B8/B14 Specific Monoclonal Antibodies in the Ten WS Core Set

WS No.	Specificity	Ave score	Q-score	R value
2086	B8	7.2	7.106	0.897
2087	B8	6.2	4.258	0.819
2089	B8	7.8	7.881	0.953
2090	A2	7.0	1.632	0.541
	B8	4.5	0.537	0.275
2091	B14	7.5	0.287	0.092
	B8	7.4	0.547	0.232
	B18	7.8	0.664	0.149

Table 5. Ten WS B14 Splitting Sera

WS No.	Specificity	Ave score	Q-score	R value	Remarks[1]
413[2]	Bw64	6.9	3.210	0.516	C
9034[2]	Bw64	7.4	5.230	0.759	AS
9032	Bw65	6.9	4.587	0.819	AS
9031	Bw65	6.6	3.864	0.741	AS
9033	Bw65	6.6	3.859	0.766	AS
4477	Bw65	7.4	4.567	0.711	AS

[1]C = core serum; AS = Antigen Society serum
[2]413 = 9034

When patients with the congenital form (n=17) and those with the late onset form (n=38) of 21-OH deficiency were considered separately (Table 9), it appeared that the Bw64 phenotype was present only in individuals with the congenital form, while all patients with the late onset form were Bw65 positive (P = 0.007). Thus, significant associations of Bw64 with the congenital form on one hand and of Bw65 with the late onset form of the disease on the other became obvious.

B14 Cell Exchange

A cell exchange was organized by Antigen Society 7 to confirm clear definition of the B14 subtypes by Tenth IHW sera and to evaluate interlaboratory reproducibility. A total of 12 B14 cell samples were sent out by four laboratories (DRK, MLS, MUE, SUD) and tested by eight participants (AST, DAW, DRK, FAU, MLS, MUE, SUD, WAN). In all cases where subtypes were assigned, no discrepancies in the assignment of Bw64/Bw65 were observed among the participating laboratories.

Conclusion

HLA B8 and B14 were well presented in the serological part of the Tenth IHW, while too few Bw59 sera as well as cells could be investigated.

As in the Ninth IHW, Bw64 and Bw65 can be clearly discriminated by the Antigen Society serum set but also by Core sera. There is no evidence for further splitting of the B14 antigen.

A striking differential correlation was found for the B14 subtypes with the congenital (Bw64) and the late onset form (Bw65) of 21 hydroxylase deficiency, respectively.

Table 6. Reactions Patterns and Split Definition of B14 Cells (n=98) by Core Sera

10 WS No. Specificity	329 14	407 14	409 14	413 w64	415 w65	n	%	Split design
	+	+	+	+	–	24	24.5	Bw64
	+	+	+	–	+	58	59.2	Bw65
	+	+	+	+	+	9	9.2	B14[1]
	+	+	+	–	–	7	7.1	B14[1]

[1]See text

Table 7. Reaction Patterns and Split Definition of B14 Cells (n=83) by Antigen Society Sera

10 WS No. Specificity	4473 14	4459 14	9034 64	9032 65	9031 65	9033 65	4477 65	n	%	Split design
	+	+	+	−	−	−	−	16	19.3	Bw64
	+	+	+	−	−	−	+	6	7.2	Bw64
	+	+	+	−	−	+	+	3	3.6	Bw64
	+	+	−	+	+	+	+	36	43.4	Bw65
	+	+	−	−	+	+	+	3	3.6	Bw65
	+	+	−	+	−	+	+	5	6.0	Bw65
	+	+	−	+	+	−	+	4	4.8	Bw65
	+	+	−	+	+	+	−	3	3.6	Bw65
	+	+	+	+	+	+	+	7	8.4	B14[1]

[1]See text

Table 8. Phenotype Frequencies (%) of Complement. C4*A, C4*B, and Bf Allotypes in Bw64 (n=20) vs. Bw65 (n=38) Positive Individuals

		Bw64	Bw65	P-value[1]
C4*A	1	0.0	0.0	n.s.
	2	5.6	63.2	0.00003
	3	94.4	73.7	0.0652
	4	11.1	15.8	n.s.
	5	0.0	0.0	n.s.
	6	11.1	0.0	n.s.
	Q0	27.8	13.2	n.s.
C4*B	1	100.0	92.1	n.s.
	2	27.8	63.2	0.0138
	3	0.0	2.6	n.s.
	Q0	5.6	2.6	n.s.
Bf	S	100.0	97.4	n.s.
	F	25.0	36.8	n.s.
	F[1]	5.0	0.0	n.s.

[1]Fisher's Exact Test

Table 9. Bw64/Bw65 Distribution in 55 B14 Positive Patients With 21-OH Deficiency

	Congenital form (n=17)	Late onset form (n=38)	Controls
Bw64	4	0	20%
Bw65	13	38	80%

P = 0.007 (Fisher's Exact Test)

References

1. Dausset J, Ivanyi P, Ivanyi D. Tissue alloantigens in humans: Identification of a complex system (Hu-1). In: Balner H, Cleton FJ, Eernisse JG (eds): Histocompatibility Testing 1965. Munksgaard, Copenhagen, 1965; p 51.
2. Albert ED, Mickey MR, McNicholas AC, Terasaki PI. Seven new HL-A specificities and their distribution in three races. In: Terasaki PI (ed): Histocompatibility Testing 1970. Munksgaard, Copenhagen, 1970; p 221.
3. Aizawa M, Itakura K. Joint Report Bw59. In: Terasaki PI (ed): Histocompatibility Testing 1980. UCLA Tissue Typing Laboratory, Los Angeles, 1980; p 450.
4. Mueller-Eckhardt G, Darke C, Fauchet R, Dewar PJ. HLA-B14: Definite splitting into Bw64 (14.1) and Bw65 (14.2). In: Albert ED, Baur MP, Mayr WR (eds): Histocompatibility Testing 1984. Springer-Verlag, Berlin, Heidelberg, New York, Tokyo, 1984; p 141.
5. Mueller-Eckhardt G, Mayr WR, Prel L, Mueller-Eckhardt C. New evidence for the existence of HLA-B14 subtypes. Linkage with different DR antigens. Tissue Antigens 1983;21:58.
6. Dewar PJ. HLA-B14.1 in linkage disequilibrium with HLA-DR7. Tissue Antigens 1983;22:310.
7. Darke C. Linkage disequilibrium between the HLA-B14 subdivisions and HLA-A, DR and properdin factor B alleles. Tissue Antigens 1984;23:68.
8. O'Neill GJ, Mueller-Eckhardt G, Prel L, Schendel DJ. C4 allotyping distinguishes HLA-B14.1 and B14.2 subgroups. Immunogenetics 1984;19:335.
9. Darke C, O'Neill GJ. Association of HLA-Bw65 with two major complotypes. Tissue Antigens 1987;30:11.
10. Dewar PJ, Coates EW, Murray S. Subdivision of the antigen B14. Tissue Antigens 1977;10:345.

Antigen Society #8 Report (Bw70, Bw71, Bw72)

Godfrey J. Laundy,[1] Marilyn S. Pollack,[2] Martin G. Guttridge,[1] and Peter T. Klouda[1]

History

The Bw6-associated antigens Bw71 (Bu) and Bw72 (SV) were originally identified at low frequencies in English and Dutch Caucasoid populations (1,2). That Bu and SV were two closely related but discrete cross-reacting antigens was established in a two-center collaborative study (3) and in an investigation of East African blacks (4). The associations of A28 and Cw3 with Bu and of A23 and Cw2 with SV were confirmed in a further study of an East African black population (5).

At the Eighth Histocompatibility Workshop, Bu and SV were included in the broad 8w59 specificity and were shown to be two independently segregating antigens (6).

Data from the extensive multiracial studies of the Ninth Histocompatibility Workshop (7) demonstrated a clear serological definition of Bu and SV. The broad Bu-SV antigen was subsequently awarded provisional Workshop status as Bw70; the associated splits were called Bw71 (=Bu) and Bw72 (=SV).

The Ninth Workshop data highlighted the considerable variation of antigen frequencies in different populations. The frequency of the Bw70 antigen varied from 29% in African blacks and 11% in American blacks to 2% in the Japanese and 0.9% in West European Caucasoids. Bw72 was twice as frequent as Bw71 in Caucasians and Negroes but only half as frequent as Bw71 in Orientals.

The serological distinction between Bw71 and Bw72 is problematical. No operationally monospecific reagent has been described and definition of the antigens has depended on reagents that show extensive cross-reaction with antigens of the B5,B35,B18,B15,B21 cross-reactive group (CREG) and, on occasion, with B17, Bw22, and B12. These sera have often shown reaction frequencies in excess of 40% and the definition of the Bw70 antigen, particularly in the presence of B35 or Bw62, has proved difficult.

Data from previous international investigations (6,7) have suggested antigen heterogeneity for both Bw71 and Bw72. The results of the Tenth Workshop also indicate that Bw72 is heterogenous. Because of the complexity of the reagents used, some of the previously described Bw71 and Bw72 reaction patterns may have been due to inter-laboratory variation in technique sensitivity; a reagent identifying Bw71 in one laboratory may define the broad Bw70 in a laboratory with greater technique sensitivity. The Tenth Workshop data emphasize the difficulty associated with the serological definition of antigens defined solely by multispecific cross-reacting sera. They further emphasize that international serological pattern recognition of such antigens is dependent on the reliability of the typing sera and the differences in the technique sensitivities between the laboratories. In consequence, the distinction between Bw71 and Bw72 and the definition of further antigen splits are likely to remain problematical.

Tenth Workshop Serology

Inter-laboratory variation in technique sensitivity, measured as a function of the average reaction frequency of Antigen Society 8 sera against Bw70, Bw71, and Bw72 cells, was considerable.

In five laboratories with complete datasets for a minimum of ten Bw70 cells, the average reaction frequency of 39 Antigen Society 8 sera varied from 57% to 86% (Table 1). This had a measurable influence on antigen definition and was not due to the ethnic composition of the Bw70 cells submitted. Data from the laboratory with the "lowest" sensitivity showed clear discrimination between Bw71 and Bw72 (laboratory 1). Furthermore, these data, when combined with Core data and data from Antigen Societies 5 and 9, defined the existence of two possible variants of Bw72.

The data from laboratory 3 of "medium" sensitivity could reasonably discriminate between Bw71 and Bw72, but data from laboratory 5 ("high" sensitivity) showed no serological distinction between Bw71 and Bw72. This caused a confusing situation where the reaction pattern for Bw72 positive cells submitted by laboratory 5 was highly similar to that of the Bw71 positive cells submitted by laboratory 1.

Allowing for the inter-laboratory variation, it remained evident that the broad Bw70 antigen could be identified by five Core sera (230,233,229,231,228) and five Antigen Society sera (4481,9202,4479,9216,9200) (Table 2). All the sera in this group reacted with Bw62 and Bw35 (Table 3) and several were also positive with antigens of the 4C and SL groups (B5,B18,B49,B17). In the absence of family studies the Bw70 antigen was impossible to define in the presence of Bw62 or B35.

Table 1. Variation in Technique Sensitivity in Five Laboratories

Laboratory	1	2	3	4	5
Bw70 cells tested	23	16	15	14	17
Reaction frequency (Rf)	57%	67%	73%	82%	86%

Rf: Average reaction frequency of 39 Bw70 AS#8 sera tested against Bw70 (Bw71+Bw72) positive cells.

Reporting Laboratories: UKIBRS,[1] US7POL[2]
Participating Laboratories: BENDPT, BENREE, FRABIG, NCYMRV, SAFHAM, US1HFP, US1JON, US1TUL, US2MBC

Immunobiology of HLA
Volume I

Table 2. The Serological Definition of Bw71 and Two Variants of Bw72

	230			
	233			
Core Sera	229			
	231	198		
	228	205	200	
Bw71	+	+	+	+
Bw72.1	+	−	+	−
Bw72.2	+	−	−	+
	4481	4478*	4484§	4483
Antigen	9202	9201	9195	
Society	4479	9217	9206	
Sera	9216			
	9200			

*Two sera from Antigen Society 9 (4512 and 9240) also defined Bw71

§Serum 4514 from Antigen Society 9 also defined Bw71 + Bw72.2

A further group of seven sera, including two from Antigen Society 9, were positive with Bw71 cells and generally negative with Bw72 cells (198,205,4478, 9201,9217,AS#9 sera 4512 and 9240). Serum 198 was shown in the Ninth Workshop (9w179) to react with Bw71 and not Bw72 (7). This serum has apparent identical specificity (Bw71 + B35) to the only other previously reported duospecific serum, 8W1040 (6), and with the confirmatory Tenth Workshop serology, this serum could be regarded as an important international standard. Antigen Society 8 serum 4478 has a similar specificity to serum 198 but also reacts weakly with some Bw62 positive cells (Table 3). Sera that identify Bw71 and are negative with Bw72 tend to include fewer antibody specificities than reagents which identify the broad Bw70 antigen.

A possible split of Bw72 (Bw72.1) is identified with a group of five sera, which includes the Core serum 200, AS#8 sera 4484,9195,9206, and AS#9 serum 4514 (Table 2). The sera in this group reacted with Bw72.1, B35, Bw53, and Bw62 positive cells (Table 3). No additional antibody specificities were identified in Core serum 200. The three AS#8 sera were also positive with B51 and Bw52 positive cells and had further variable antibody activity. The cells defined as Bw72.2 were negative with this group of sera but were positive with the broad Bw70 sera and with a single serum from AS#8, 4483. This serum also reacted with B35, B51, Bw52, and Bw58 positive cells and was weakly positive with Bw62 and B18 cells.

Biochemical evidence (8) supports the serological dissection of Bw70 into Bw71 and Bw72 with a single IEF band being almost unique to the Bw71 molecule. However, the heterogeneity of Bw72 revealed by biochemistry does not correlate strongly with the serological splits Bw72.1 and Bw72.2.

Previously reported antigen frequencies (7) in different ethnic groups cannot be confirmed as most of the cells typed were selected for this study. There is evidence, however, of a further haplotype in blacks—Aw74,Bw72,Cw2,DR3 and a probable association of Cw7 with Bw70 in Caucasoids.

References

1. Laundy GJ, Entwistle CC. Bu: A new antigen at the HLA-B locus. Tissue Antigens 1977;10:180.

Table 3. Specificity Inclusions in Core and Antigen Society 8 Sera Defining Bw71 and Bw72

	Core and Antigen Society 8 Sera																			
	233	229	228	4481	231	230	9202	4479	9216	9200	198	4478	205	9201	9217	200	4484	9195	9206	4483
Bw7.1	+	+	+	+	+	+	+	+	+	+	+	+	+	+	+	+	+	+	+	+
Bw72.1	+	+	+	+	+	+	+	+	+	+	−	−	−	−	−	+	+	+	+	−
Bw72.2	+	+	+	+	+	+	+	+	+	+	−	−	−	−	−	−	−	−	−	+
B35	+	+	+	+	+	+	+	+	+	+	+	+	+	+	+	+	+	+	+	+
Bw62	+	+	+	+	+	+	+	+	+	+	−	±	±	±	±	+	+	+	+	±
Bw63	−	−	−	−	−	+	+	+	−	+	−	−	−	−	+	−	−	−	−	−
Bw53	−	−	−	−	−	*	+	−	+	+	−	−	−	+	+	+	+	+	+	−
B18	−	−	−	−	−	−	−	*	+	+	−	−	−	−	−	−	+	*	−	*
B51	−	−	−	−	−	−	+	−	+	+	−	−	+	+	+	−	+	+	+	+
Bw52	−	−	−	−	*	*	+	−	+	+	−	−	−	+	+	−	+	+	+	+
B49§	−	−	−	−	−	−	+	−	−	+	−	−	−	−	+	−	+	−	−	−
Bw57	−	−	−	+	−	+	+	+	+	−	−	−	−	−	−	−	+	−	+	−
Bw58	−	−	−	−	−	+	−	+	+	+	−	−	*	−	+	−	−	−	−	+
Bw55	−	−	−	−	*	*	−	±	−	−	−	−	−	−	±	−	−	−	−	−

§Insufficient data for Bw50
*Weak activity
±"Short" specificity

2. Schreuder I, Bos A, van de Berg-Loonen E. SV=PR: A new B-locus antigen discovered. In: Bodmer WF, Batchelor JR, Bodmer JG, Festenstein H, Morris PJ (eds): Histocompatibility Testing 1977. Munksgaard, Copenhagen, 1977, p 408.
3. Schreuder I, Nieman HG, Laundy GJ, Entwistle CC. Bu and SV: Two closely related B-locus specificities. Tissue Antigens 1980;16:169.
4. Hall PJ, Levin AG, Entwistle CC, Knight SC, Wasunna A, Brubacker G. B15 heterogeneity in East African blacks. Tissue Antigens 1980;16:326.
5. Hall PJ, Levin AG, Entwistle CC, Knight SC, Wasunna A, Kung'n A, Brubacker G. HLA antigens in African black patients with Burkitt's lymphoma or nasopharyngeal carcinoma and in controls. A pilot study. Hum Immunol 1982;5:91.
6. Entwistle CC, Laundy GJ. 8w58: Joint report. In: Terasaki PI (ed): Histocompatibility Testing 1980. UCLA Tissue Typing Laboratory, Los Angeles, 1980, p 475.
7. Laundy GJ, Raffoux C, Schreuder GMT, Klouda PT. Antigen Report: HLA-Bw70, HLA-Bw71 and HLA-Bw72. In: Albert ED, Baur MP, Mayr WR (eds): Histocompatibility Testing 1984. Springer-Verlag, Berlin, 1984, p 173.
8. Guttridge MG, Laundy GJ, Klouda PT. 1988. Biochemical variants of the Bw70 antigen (Bw71, Bw72). In: Dupont B (ed): Immunobiology of HLA, Volume I: Histocompatibility Testing 1987. New York: Springer-Verlag, 1989.

Antigen Society #9 Report (Bw46 and the Subgroups of B15)

E.D. Albert,[1] D. Chandanayingyong,[2] J.S. Thompson,[3] T. Zhao,[4] M.G. Hammond,[5] S. Naito,[6] and G.M. Sierp[1]

History

The antigen HLA-B15, first described as LND (Thorsby et al. 1970 (29)) or TE15 (Albert et al. 1970 (1)), was first recognized to be heterogeneous at the Fourth International Workshop (Thorsby et al. 1970b (30)). In the following Workshops, a growing number of reports documented the heterogeneity of B15 (Richiardi et al. 1974 (23), Joysey et al. 1975 (18), Dick et al. 1978 (15), Singal et al. 1980 (27), Danilovs and Pollock 1980 (13), Saueracker et al. 1981 (24), Alonso et al. 1983 (2), Zhao and Shiraki 1986 (33)). It was becoming clear that most of the variants of B15 are found in Southeast Asian populations. During the Ninth International Histocompatibility Workshop, the complexity of B15 was discussed in nine different Newsletter contributions (3,6,9,11,12, 14,17,25,32) and summarized by Chandanayingyong et al. (10) and Cambon-Thomsen et al. (5). From this summary it appears that next to the classical Bw62 antigen there exists a Bw6-associated short Bw62 antigen, which has been observed by several different authors (6,9,11 17,25,34) mostly in Asian populations, and named Bw62.1, TS1, B15short Thai, B15 Kemp, SH7, Bw62S. This antigen is characterized by cross-reactions with B35 sera. There was evidence for an even shorter split of B15, which is also Bw6-associated and which was named B15.3 or B15SL1. Two further antigens described B15SAU and B15G (4) may or may not be equivalent with B15.3. In the Thai population, there exists one further split of B15, named B15S, with a characteristic cross-reaction with anti-B45 sera (9).

Reporting Laboratories: GERALB,[1] ANZDCH,[2] US1THP,[3] CHIZHA,[4] SAFHAM,[5] JAPNAI[6]
Participating Laboratories: ANZCRS, ANZTAT, BENBER, SAFDUT, UKIAST, UKIBRS, UKIFES, US7POL, NCYMRV

Among the Bw4-associated splits of B15, there is, in addition to the classical Bw63, an antigen found in Negroid populations and named 8w66 (31), which has a shorter reaction pattern than Bw63. One further split of B15 with a clearly shorter reaction pattern than Bw63 occurs in the Thai population and is named B15T (9).

B15 Serology X. International Histocompatibility Workshop

Using Core sera and Antigen Society sera, seven different subtypes of B15 could be identified. The reaction patterns and the key antisera for the definition of the various subtypes are given in Tables 1 and 2. It is quite likely that several more subtypes of B15 do exist; however, in the absence of family segregation data it was difficult to assess slight differences in reactivity.

Antigen Bw62. This antigen is by far the most common subtype of B15. It is characterized by association with Bw6 and it can be very well defined using a large number of antisera, of which only the best examples were chosen for the reaction patterns given in Tables 1 and 2. There are many antisera reacting with all B15 subtypes, but only two antisera (nos. 250 and 252) distinguish Bw62 and Bw76 from the rest of the B15 group.

Antigen Bw75 (Equivalents: Bw62.1, TS1, B15short Thai, B15 Kemp, SH7, Bw62S). This antigen was already clearly defined in the Ninth International Histocompatibility Workshop (10) as a short, Bw6-associated variant of Bw62, which is characterized by a cross-reactivity with B35 (i.e., in the presence of Bw75 and the absence of B35 and Bw53 some of the anti-B35, Bw53 antisera react positively). Among the cells, whose reaction pattern corresponds closely to the Bw75 pattern given in Table 1, we have found the following codings:

Table 1. Reaction Pattern of B15 Antisera from the Core Serum Set

Specificities	All B15			Bw62		Bw62 Bw75 Bw76			Bw76 B45		B35 Bw53 Bw75					B35 Bw75 Bw70			B51 Bw53 B49 Bw77				Bw63 Bw57 Bw58					
WS-Sera No.	238	242	216	252	250	248	230	256	428	432	212	208	255	192	205	233	234	228	222	223	225	166	266	267	448	2105	Bw4	Bw6
Bw62	+	+	+	+	+	+	+	−	−	−	−	−	−	−	−	+	+	+	−	−	−	−	−	−	−	−	−	+
Bw75	+	+	+	−	−	+	+	−	−	−	+	+	+	+	+	+	+	+	−	−	−	−	−	−	−	+	−	+
Bw76	+	+	+	+	+	+	+	+	+	+	−	−	−	−	−	−	−	−	−	−	−	−	−	−	−	+	−	+
B15.3	+	+	+	−	−	−	−	−	−	−	−	−	−	−	−	+	+	+	−	−	−	−	−	−	−	−	−	+
Bw77	+	+	+	−	−	−	−	−	−	−	+	+	+	−	−	−	−	−	+	+	+	+	−	−	−	+	+	−
Bw63	+	+	+	−	−	−	−	−	−	−	−	−	−	−	−	−	−	−	−	−	−	−	+	+	+	−	+	−
Bw63.1	+	+	+	−	−	−	−	−	−	−	−	−	−	−	−	−	−	−	+	+	+	+	+	+	+	−	+	−

Table 2. Reaction Patterns of B15 on Tenth Workshop Antigen Society No. 9 Sera Set

Specificities of Antisera:

Antigen	w77 w53 w63 9275	w77 w63 15.3 9303	w77 w53 9289	w77 w53 15.3 w75 B35 9288	w77 w53 9293	w62 w63 w57 9310	w46 A31 B15 4508	w75 w77 9314	w76 15.3 w75 4507	all B15 9306	w62 w76 4493	w62 w76 9313	w62 w76 4487	w71 B35 w75 9241	w76 w75 9235	w76 9245	w76 B44 9274	Bw4	Bw6
Bw62	−	−	−	−	−	+	+	+	+	+	−	+	w	−	−	−	−	−	+
Bw75	−	−	−	+	−	+	+	+	+	+	−	−	−	+	+	−	−	−	+
Bw76	−	−	−	+	−	+	+	+	+	+	+	+	+	+	+	+	+	−	+
Bw63	+	+	−	−	+	+	+	+	+	+	−	−	−	−	−	−	+	+	−
Bw77	+	+	+	+	+	+	+	+	+	+	−	−	−	−	−	−	−	+	−

Bw62, Bw62.1, B15Sw6, B15, TE79, B15K. As all four cells coded TE79 and all 13 cells coded B15K show a reaction pattern that is very similar to that of Bw75, it is possible that one should add TE79 and B15K to the list of equivalents for Bw75. For Te79 this is also borne out by the fact that the three antisera (nos. 215, 216, and 221) submitted as anti-Te79 react with all cells positive for Bw75. For the specificity called B15SL1, there is different coding in different laboratories. The cells from Dr. Zhao in Shanghai, which have been called B15SL1, correspond exactly to the Bw75 reaction pattern, while one cell from Dr. Festenstein's laboratory has a different reaction pattern. Also in the Ninth International Histocompatibility Workshop, B15SL1 showed a reaction pattern clearly different from that of Bw75 (then Bw62.1) (10). Bw75 is found mostly in Chinese, Thais, and other Southeast Asian populations and only very occasionally in cells coded as Caucasian. The relatively frequent occurrence of this antigen in the Cape colored population of South Africa probably reflects the contribution of Southeast Asian genes to the gene pool in this population.

Antigen Bw76 (Equivalent: B15S (Siamese)). This Bw6-associated split of B15 was first described during the Ninth International Histocompatibility Workshop (9). It is characterized by a strong cross-reactivity with anti-B45 sera. This antigen seems to be restricted to the Southeast Asian populations. The characteristic reaction pattern for Bw76 is given in Tables 1 and 2.

Antigen B15.3. This designation characterizes a Bw6-associated antigen with a short Bw62 reaction pattern. It has been noted that in this workshop there was an inconsistent use of this designation, as cells belonging to clearly distinct subtypes of B15 were called B15.3. Among the cells coded as B15.3, there is a group characterized by a short Bw62 reaction pattern (see Tables 1 and 2), with reactivity with long B35/Bw62/Bw70 antisera. It must be stressed however, that this specificity is ill-defined and there is considerable indication for the existence of several more Bw6-associated short variants of B15.

Antigen Bw63. This antigen is characterized by its association with Bw4 and by a short reaction pattern (see Tables 1 and 2). Reactivity with Bw63 cells is frequently found in anti-Bw57 and anti-Bw58 sera. There appears to be a variant of Bw63, tentatively called Bw63.1, found in Negroid populations, which reacts, in addition to the Bw63 typical pattern, with antisera directed against B51, Bw53, and B49. This antigen may be identical to the specificity 8w66 (20,21). Unfortunately, there were too few cells coded for this specificity in order to determine identity or non-identity with what has been called Bw63.1.

Antigen Bw77 (Equivalent B15T). This antigen was first described in the Ninth International Histocompatibility Workshop in the Thai population (9). It is associated with Bw4 and characterized by a short B15 reaction pattern and cross-reactivity with antisera containing anti-Bw53 activity (see Tables 1 and 2).

Table 3. Reaction Pattern of Bw46 on the Tenth Workshop Core Sera Set

Specificities	Bw46								Bw6					
WS Sera No	263	257	258	261	262	9246	9250	9251	500	507	501	499	504	216
Bw46	+	+	+	+	+	+	+	+	+	+	+	−	−	−

Antigen Bw46. This antigen does not belong to the B15 complex even though in almost all long B15 antisera, anti-Bw46 activity can be detected. The antigen occurs almost exclusively in Chinese, Japanese, Thais, or other Southeast Asians. Bw46 is strongly associated with Cw11 and DRw8 among Japanese and DR9 among Chinese. At the Ninth International Histocompatibility Workshop, there was a suspicion of a split of Bw46 (7) that could not be substantiated in the Third Asia-Oceania Histocompatibility Workshop (8).

In this Workshop, Bw46 was very well defined with three narrow antisera (263, 257, and 258) and two broad sera (261 and 262) as well as with three broad sera from the Antigen Society set (nos. 9246, 9250, and 9251). With anti-serum 258, there was some suspicion of a mixup, as apparently this serum was a perfect anti-Bw46 in some laboratories and completely negative in other laboratories (Kennedy et al., Newsletter No. 1 (19)). In cells that possess in addition to Bw46 a Bw4-associated B-locus antigen, it has been observed that there is a short reaction pattern for Bw6 as given in Table 3. This may be in accordance with the finding from DNA sequencing of the Bw47 gene that the Bw46 gene carries in the position covering aminoacid 79-83, which is responsible for the Bw6/Bw4 variability—a sequence coming from a Cw3 gene (Parham et al., this volume (22)).

Computer Cell Typing

Using the computer cell-typing procedure developed for the Ninth International Histocompatibility Workshop (26), HLA-A,B,C computer cell-typing was performed on the basis of local assignments. For the antigen Bw75, cells that were coded as B15SL1 or SH7 and corresponded in their reaction pattern to the reaction pattern of Bw75 were recoded as Bw75. As is shown in Table 4 for the antigens discussed in this report, there is a very high percentage of cells for which the local assignment and the computer assignment is identical. It can be seen that the two least well defined and probably heterogeneous antigen groups, Bw63 and B15.3, show the lowest R value between local and computer assignment.

Table 4. Computer-Cell Typing and Correlation with Local Assignment

Antigen	No. Cells Tested	Lab + Prog. +	Lab + Prog. −	Lab − Prog. +	R value
Bw46	738	110	10	8	0.91
Bw62	738	169	17	11	0.90
Bw63	738	16	5	14	0.63
Bw75	562	55	2	6	0.93
Bw76	562	18	0	0	1.00
Bw77	717	10	1	1	0.91
B15.3	562	9	2	2	0.82

Listing of Sera Typing Information of All Sera Relevant to the Antigens of the Antigen Society and of all Antigen Society Sera

Sera typing was performed using the procedure and the format developed for the Ninth International Histocompatibility Workshop. The definition of most of the B15 subgroups is dependent on the reactions of many broad antisera and therefore it is important to investigate the inclusion of narrow specificities into the broad ones. For a restricted analysis, this is only possible if the narrow specificities are analyzed first. The sera typing of the relevant Core sera is given in Table 5 and those of the Antigen Society sera in Table 6.

T-Cell Defined Bw62 Variants

Information obtained from Dr. Beatty, Seattle, indicates that a cytolytic T-cell clone (HAN 4) and a proliferative T-cell clone LAY-1 recognize all core cell lines expressing Bw62. In addition, the T-cell clone HAN 4, when tested with Bw62 variants, appears to be recognizing subtypes, which are positive for the Workshop sera 250 and 252 (Bw62 and Bw76).

Correlation Between Biochemical Subtypes and Serology

Information from Dr. Choo, Seattle, shows that the biochemical variant B15.1 corresponds to Bw75, B15.2 corresponds to Bw76, and B15.3 corresponds to Bw62.

Table 5. Seratyping of Core Serology Sera Defining B15 and Its Subgroups

SERUM	ANTIGEN	NO. REAC	AVE	++	MISS +-	EXTR -+	--	PERCENTAGE STR	MISS	EXTR	INCL	R	CHI	QSCORE	QNORM	ANTIGEN
10W165	TRAY: 6	POS: 6			LOCAL SPECIFICITY: B5											
10W165	B51	530	7.9	39	0	73	418	76	0	65	100	0.54	157.1	7.73	0.85	B51
10W165	BW52	491	8.0	15	0	58	418	65	0	79	100	0.42	88.6	6.52	0.84	BW52
10W165	BW53	476	7.7	6	0	52	418	56	0	89	100	0.30	43.8	5.30	0.81	BW53
10W165	BW77	470	7.7	7	1	45	417	53	12	86	88	0.32	48.3	4.48	0.65	BW77
10W165	BW63	462	7.2	5	4	40	413	48	44	88	56	0.22	21.9	3.03	0.43	BW63
10W165	B15.3	453	6.0	4	3	36	410	47	42	90	58	0.21	20.6	2.75	0.41	B15.3
10W165	B18	446	6.7	17	14	19	396	50	45	52	55	0.47	98.2	2.47	0.29	B18
10W165	B35	415	6.4	11	30	8	366	42	73	42	27	0.35	51.6	0.74	0.08	B35
10W166	TRAY: 4	POS: 35			LOCAL SPECIFICITY: B5											
10W166	B51	524	8.0	40	0	59	425	84	0	59	100	0.60	185.9	7.35	0.81	B51
10W166	BW52	484	7.7	15	0	44	425	74	0	74	100	0.48	111.5	5.95	0.77	BW52
10W166	BW77	469	8.0	8	1	36	424	70	11	81	89	0.38	68.2	4.37	0.62	BW77
10W166	BW63	460	8.0	4	5	32	419	63	55	88	45	0.19	17.1	3.36	0.48	BW63
10W166	BW53	451	6.0	4	2	28	417	59	33	87	67	0.27	32.7	2.93	0.45	BW53
10W166	BW57	445	7.1	7	16	21	401	60	69	75	31	0.23	24.0	0.85	0.11	BW57
10W166	BW58	422	7.2	10	34	11	367	57	77	52	23	0.28	32.7	0.61	0.07	BW58
10W166	BW4	378	6.4	11	141	0	226	45	92	0	8	0.21	16.8	1.43	0.13	BW4
10W185	TRAY: 6	POS: 5			LOCAL SPECIFICITY: B51,BW52,BW53											
10W185	B51	529	8.0	40	0	33	456	83	0	45	100	0.71	270.3	9.15	0.98	B51
10W185	BW52	489	7.7	15	0	18	456	63	0	54	100	0.66	213.8	7.40	0.93	BW52
10W185	BW53	474	7.0	4	1	14	455	44	20	77	80	0.41	80.3	5.59	0.86	BW53
10W185	BW77	469	6.7	6	1	8	454	35	14	57	86	0.60	167.9	4.82	0.69	BW77
10W185	BW4	462	5.6	5	229	3	225	12	97	37	3	0.03	0.5	0.83	0.08	BW4
10W189	TRAY: 6	POS: 9			LOCAL SPECIFICITY: BW53,BW52,B51,B35											
10W189	BW77	532	7.5	11	1	112	408	52	8	91	92	0.25	32.5	5.18	0.72	BW77
10W189	BW75	520	6.8	42	18	70	390	49	30	62	70	0.43	94.3	3.44	0.37	BW75
10W189	B15.3	460	6.8	5	4	65	386	38	44	92	56	0.16	11.6	2.44	0.36	B15.3
10W189	B35	451	6.6	28	14	37	372	36	33	56	67	0.48	102.5	3.06	0.35	B35
10W189	B51	409	6.7	17	16	20	356	32	48	54	52	0.44	78.7	2.65	0.32	B51
10W192	TRAY: 6	POS: 11			LOCAL SPECIFICITY: BW53,B35											
10W192	B35	530	7.6	39	8	57	426	66	17	59	83	0.53	146.3	5.61	0.61	B35
10W192	BW75	483	7.0	42	13	15	413	57	23	26	77	0.72	248.6	4.36	0.47	BW75
10W198	TRAY: 6	POS: 23			LOCAL SPECIFICITY: BW71,B35,BW53,B15											
10W198	B35	531	7.6	47	0	125	359	63	0	72	100	0.45	107.6	8.15	0.92	B35
10W198	BW75	484	7.3	56	0	69	359	56	0	55	100	0.61	181.9	8.11	0.89	BW75
10W198	BW62	428	6.6	44	62	25	297	47	58	36	42	0.40	67.2	1.70	0.17	BW62
10W198	B51	322	6.7	6	24	19	273	44	80	76	20	0.15	6.9	1.22	0.15	B51
10W198	BW70	292	6.3	13	33	6	240	36	71	31	29	0.38	42.5	0.95	0.11	BW70
10W198	BW6	246	5.2	5	150	1	90	16	96	16	4	0.07	1.1	0.91	0.09	BW6

Table 5. *Continued*

10W200	TRAY:	6	POS: 22		LOCAL	SPECIFICITY: B35,BW62,BW71,BW72										
10W200	BW75	532	7.9	60	0	166	306	69	0	73	100	0.41	91.6	6.96	0.78	BW75
10W200	B35	472	8.0	39	4	127	302	60	9	76	91	0.37	64.0	4.71	0.55	B35
10W200	BW62	429	6.6	60	47	67	255	48	43	52	57	0.33	47.9	2.23	0.23	BW62
10W200	BW58	322	6.5	23	20	44	235	47	46	65	54	0.32	32.2	1.74	0.21	BW58
10W200	BW70	279	6.7	22	18	22	217	50	45	50	55	0.44	54.1	1.58	0.19	BW70
10W200	BW57	239	6.4	5	11	17	206	54	68	77	32	0.20	10.0	1.09	0.15	BW57
10W200	B51	223	6.9	7	21	10	185	52	75	58	25	0.25	13.7	0.45	0.06	B51
10W200	BW6	195	6.8	10	138	0	47	50	93	0	7	0.13	3.3	0.51	0.10	BW6
10W202	TRAY:	6	POS: 24		LOCAL	SPECIFICITY: BW72,BW62,B5,B35										
10W202	BW75	521	7.9	60	0	254	207	80	0	80	100	0.29	44.7	6.93	0.78	BW75
10W202	B35	461	7.9	43	0	211	207	77	0	83	100	0.29	38.6	6.58	0.78	B35
10W202	BW76	418	8.0	14	0	197	207	73	0	93	100	0.18	14.2	5.34	0.77	BW76
10W202	BW52	404	8.0	13	0	184	207	72	0	93	100	0.19	14.1	5.25	0.77	BW52
10W202	B51	391	8.0	34	1	150	206	70	2	81	98	0.31	38.7	5.69	0.70	B51
10W202	B18	356	7.5	24	1	126	205	63	4	84	96	0.30	32.0	4.80	0.62	B18
10W202	B15.3	331	8.0	6	1	120	204	60	14	95	86	0.14	6.9	3.42	0.57	B15.3
10W202	BW77	324	8.0	5	1	115	203	58	16	95	84	0.13	5.6	3.23	0.55	BW77
10W202	BW62	318	6.9	70	12	45	191	56	14	39	86	0.60	115.9	3.88	0.42	BW62
10W202	BW70	236	7.1	28	15	17	176	53	34	37	66	0.55	72.2	3.34	0.40	BW70
10W202	B37	193	7.0	4	5	13	171	35	55	76	45	0.28	14.9	1.41	0.22	B37
10W205	TRAY:	6	POS: 10		LOCAL	SPECIFICITY: B35										
10W205	BW75	531	7.6	60	0	124	347	63	0	67	100	0.49	127.6	6.59	0.77	BW75
10W205	B35	471	7.7	42	1	82	346	54	2	66	98	0.51	124.2	5.52	0.68	B35
10W205	BW41	428	6.0	4	3	78	343	40	42	95	58	0.12	6.6	1.93	0.33	BW41
10W205	BW70	421	6.3	23	23	55	320	39	50	70	50	0.28	33.9	1.59	0.20	BW70
10W205	BW42	375	6.7	6	10	49	310	38	62	89	38	0.14	7.0	1.01	0.15	BW42
10W205	BW76	359	6.5	4	10	45	300	36	71	91	29	0.09	2.8	0.89	0.13	BW76
10W205	B39	345	6.8	5	15	40	285	37	75	88	25	0.09	2.7	0.57	0.08	B39
10W205	BW6	325	5.9	32	189	8	96	35	85	20	15	0.10	3.0	0.81	0.08	BW6
10W206	TRAY:	6	POS: 7		LOCAL	SPECIFICITY: B35										
10W206	BW53	532	6.0	5	2	57	468	58	28	91	72	0.22	24.6	4.41	0.80	BW53
10W206	B35	525	7.2	36	10	21	458	61	21	36	79	0.67	236.7	4.34	0.47	B35
10W206	BW75	479	7.0	12	44	9	414	52	78	42	22	0.30	43.9	0.73	0.08	BW75
10W208	TRAY:	6	POS: 21		LOCAL	SPECIFICITY: B35										
10W208	B35	532	8.0	47	0	118	367	80	0	71	100	0.46	114.7	8.16	0.91	B35
10W208	BW53	485	8.0	6	0	112	367	72	0	94	100	0.20	18.9	5.59	0.89	BW53
10W208	BW75	479	7.3	55	1	57	366	71	1	50	99	0.64	198.2	6.96	0.75	BW75
10W208	BW77	423	8.0	10	1	47	365	73	9	82	91	0.37	58.1	5.23	0.74	BW77
10W208	B51	412	7.9	31	3	16	362	68	8	34	92	0.75	233.3	5.73	0.67	B51
10W212	TRAY:	6	POS: 16		LOCAL	SPECIFICITY: B51,B35										
10W212	BW75	517	7.1	58	1	80	378	79	1	57	99	0.58	174.5	6.78	0.73	BW75
10W212	BW77	458	8.0	10	1	70	377	88	9	87	91	0.30	42.2	5.00	0.71	BW77
10W212	B35	447	7.9	38	3	32	374	87	7	45	93	0.67	202.8	5.75	0.65	B35
10W212	B51	406	7.9	21	12	11	362	78	36	34	64	0.62	153.8	3.13	0.37	B51

10W213	TRAY:	6	POS: 8		LOCAL	SPECIFICITY:	B35,B51,BW52,BW53									
10W213	B35	531	7.4	46	1	115	369	68	2	71	98	0.46	111.4	7.59	0.84	B35
10W213	BW77	484	8.0	10	1	105	368	65	9	91	91	0.24	28.0	5.08	0.72	BW77
10W213	BW53	473	6.4	5	1	100	367	61	16	95	84	0.17	13.2	4.34	0.69	BW53
10W213	BW52	467	6.6	10	3	90	364	63	23	90	77	0.23	24.5	4.00	0.55	BW52
10W213	B51	454	7.0	24	11	66	353	65	31	73	69	0.35	56.7	3.79	0.45	B51
10W213	BW75	419	7.7	33	21	33	332	65	38	50	62	0.48	96.1	2.80	0.31	BW75
10W213	BW62	365	6.3	31	60	2	272	39	65	6	35	0.50	92.3	1.55	0.16	BW62
10W214	TRAY:	8	POS: 29		LOCAL	SPECIFICITY:	B35,B51,BW52,BW53									
10W214	BW75	533	7.6	60	0	143	330	81	0	70	100	0.45	109.9	5.59	0.60	BW75
10W214	B35	473	8.0	42	1	101	329	80	2	70	98	0.46	102.0	4.67	0.53	B35
10W214	BW77	430	8.0	10	2	91	327	73	16	90	84	0.24	24.6	3.22	0.45	BW77
10W214	B51	418	7.3	25	9	66	318	70	26	72	74	0.37	58.2	2.77	0.33	B51
10W214	CW4	384	7.6	43	4	23	314	66	8	34	92	0.74	207.7	2.79	0.28	CW4
10W214	BW52	337	5.0	4	6	19	308	30	60	82	40	0.23	17.8	1.66	0.23	BW52
10W214	BW62	327	6.0	12	60	7	248	31	83	36	17	0.25	19.9	0.59	0.06	BW62
10W214	BW6	255	6.7	6	172	1	76	42	96	14	4	0.06	0.9	0.72	0.14	BW6
10W215	TRAY:	6	POS: 12		LOCAL	SPECIFICITY:	B35,BW62,BW50,TE79									
10W215	B35	529	7.5	45	2	99	383	66	4	68	96	0.48	122.3	5.73	0.64	B35
10W215	BW75	482	7.4	53	3	46	380	61	5	46	95	0.67	213.2	5.62	0.61	BW75
10W215	BW70	426	5.8	13	33	33	347	36	71	71	29	0.20	16.3	0.66	0.07	BW70
10W215	BW62	380	5.8	26	79	7	268	36	75	21	25	0.35	47.3	0.66	0.07	BW62
10W215	BW6	275	6.0	5	160	2	108	42	96	28	4	0.04	0.4	0.82	0.08	BW6
10W216	TRAY:	6	POS: 51		LOCAL	SPECIFICITY:	BW62,BW63,BW57,TE79									
10W216	BW75	530	8.0	60	0	185	285	91	0	75	100	0.39	78.7	9.06	0.98	BW75
10W216	BW62	470	7.9	115	0	70	285	89	0	37	100	0.71	234.6	9.78	0.98	BW62
10W216	BW76	355	8.0	13	0	57	285	77	0	81	100	0.39	54.9	7.02	0.97	BW76
10W216	BW77	342	8.0	11	0	46	285	71	0	80	100	0.41	56.8	6.78	0.97	BW77
10W216	BW63	331	7.3	9	1	37	284	65	10	80	90	0.39	49.9	6.11	0.89	BW63
10W216	B15.3	321	8.0	5	2	32	282	62	28	86	72	0.28	25.2	4.69	0.73	B15.3
10W216	BW57	314	6.7	9	8	23	274	56	47	71	53	0.34	35.9	2.44	0.32	BW57
10W216	BW46	297	7.2	20	19	3	255	56	48	13	52	0.63	119.1	2.41	0.28	BW46
10W220	TRAY:	6	POS: 55		LOCAL	SPECIFICITY:	BW57,BW46,B15,TE77									
10W220	BW75	523	8.0	59	0	236	228	87	0	80	100	0.31	51.4	8.10	0.92	BW75
10W220	B15.3	464	8.0	9	0	227	228	84	0	96	100	0.14	8.9	5.80	0.92	B15.3
10W220	BW77	455	8.0	12	0	215	228	83	0	94	100	0.16	12.4	6.14	0.92	BW77
10W220	BW76	443	8.0	14	0	201	228	82	0	93	100	0.19	15.3	6.33	0.91	BW76
10W220	BW62	429	8.0	104	1	97	227	81	0	48	100	0.60	152.1	7.95	0.84	BW62
10W220	BW57	324	7.8	18	1	79	226	63	5	81	95	0.35	40.4	5.61	0.77	BW57
10W220	BW63	305	8.0	7	1	72	225	58	12	91	88	0.23	16.2	4.51	0.73	BW63
10W220	BW46	297	7.1	34	5	38	220	54	12	52	88	0.57	96.8	4.71	0.57	BW46
10W220	BW70	258	6.0	28	18	10	202	39	39	26	61	0.61	94.9	2.98	0.36	BW70
10W220	B35	212	6.7	6	24	4	178	50	80	40	20	0.29	18.2	0.70	0.09	B35

Table 5. *Continued*

10W221	TRAY:	6 POS: 52			LOCAL SPECIFICITY: BW62,BW63,TE77,TE79											
10W221	BW75	529	7.9	60	0	192	277	85	0	76	100	0.37	74.4	7.91	0.89	BW75
10W221	BW76	469	8.0	14	0	178	277	82	0	92	100	0.21	20.8	6.19	0.89	BW76
10W221	BW77	455	8.0	12	0	166	277	80	0	93	100	0.21	19.2	5.99	0.89	BW77
10W221	BW62	443	7.9	110	3	56	274	79	2	33	98	0.72	232.1	8.46	0.88	BW62
10W221	B15.3	330	7.7	6	1	50	273	50	14	89	86	0.27	24.0	5.07	0.84	B15.3
10W221	BW63	323	6.5	8	2	42	271	46	20	84	80	0.32	32.8	3.46	0.53	BW63
10W221	BW57	313	6.9	7	10	35	261	45	58	83	42	0.20	11.9	1.47	0.20	BW57
10W221	B35	296	6.9	9	23	26	238	42	71	74	29	0.18	9.1	1.19	0.15	B35
10W221	BW70	264	5.8	13	32	13	206	34	71	50	29	0.29	22.2	1.17	0.14	BW70
10W221	BW46	219	6.9	9	29	4	177	46	76	30	24	0.34	25.9	0.63	0.08	BW46
10W222	TRAY:	4 POS: 31			LOCAL SPECIFICITY: BW53,B49,BW78,B51											
10W222	BW52	531	8.0	15	0	72	444	87	0	82	100	0.39	78.8	6.02	0.76	BW52
10W222	BW63	516	6.6	7	2	65	442	84	22	90	78	0.25	31.1	4.70	0.75	BW63
10W222	BW53	507	7.7	7	0	58	442	87	0	89	100	0.31	48.3	5.15	0.74	BW53
10W222	B51	500	7.9	37	2	21	440	87	5	36	95	0.76	286.0	5.70	0.62	B51
10W222	BW77	461	8.0	7	2	14	438	71	22	66	78	0.50	113.2	3.61	0.50	BW77
10W222	BW4	452	6.8	13	213	1	225	57	94	7	6	0.15	10.6	1.10	0.10	BW4
10W223	TRAY:	4 POS: 32			LOCAL SPECIFICITY: BW53,B49,BW78,B51											
10W223	BW52	526	7.9	15	0	73	438	87	0	82	100	0.38	76.9	5.27	0.66	BW52
10W223	BW53	511	7.3	6	1	67	437	86	14	91	86	0.24	29.6	4.44	0.64	BW53
10W223	B51	504	8.0	38	1	29	436	86	2	43	98	0.72	259.6	5.63	0.61	B51
10W223	BW77	465	7.4	7	2	22	434	68	22	75	78	0.42	80.3	3.66	0.50	BW77
10W223	BW4	456	7.2	19	211	3	223	63	91	13	9	0.16	11.9	1.13	0.10	BW4
10W225	TRAY:	4 POS: 34			LOCAL SPECIFICITY: B49,BW53,BW78											
10W225	BW52	524	8.0	15	0	86	423	80	0	85	100	0.35	64.7	6.01	0.77	BW52
10W225	BW53	509	8.0	7	0	79	423	76	0	91	100	0.26	34.9	5.10	0.75	BW53
10W225	BW63	502	7.0	8	1	71	422	74	11	89	89	0.27	37.0	5.11	0.72	BW63
10W225	B51	493	7.9	38	1	33	421	76	2	46	98	0.69	236.9	6.32	0.69	B51
10W225	BW77	454	8.0	8	1	25	420	51	11	75	89	0.45	90.8	4.29	0.60	BW77
10W225	B15.3	445	7.5	4	3	21	417	36	42	84	58	0.28	35.6	2.49	0.36	B15.3
10W225	B13	438	5.2	5	28	16	389	28	84	76	16	0.14	8.4	0.64	0.07	B13
10W225	BW6	405	5.7	14	335	2	54	37	95	12	5	0.01	0.0	1.03	0.17	BW6
10W226	TRAY:	6 POS: 56			LOCAL SPECIFICITY: B49,B5,TE78											
10W226	BW77	531	8.0	12	0	223	296	90	0	94	100	0.17	15.5	6.98	0.98	BW77
10W226	BW62	519	7.9	116	0	107	296	89	0	47	100	0.62	198.3	9.76	0.97	BW62
10W226	BW75	403	7.9	58	0	49	296	83	0	45	100	0.68	187.4	8.92	0.97	BW75
10W226	BW63	345	7.6	10	0	39	296	69	0	79	100	0.42	62.2	6.34	0.92	BW63
10W226	B15.3	335	7.3	6	1	33	295	66	14	84	86	0.34	38.1	4.57	0.71	B15.3
10W226	BW57	328	7.1	9	8	24	287	66	47	72	53	0.33	36.4	2.60	0.36	BW57
10W226	BW46	311	7.2	21	21	3	266	66	50	12	50	0.63	121.9	2.33	0.26	BW46
10W228	TRAY:	5 POS: 46			LOCAL SPECIFICITY: B35,BW62,BW70											
10W228	BW75	531	7.8	60	0	209	262	85	0	77	100	0.35	65.9	6.75	0.73	BW75
10W228	B35	471	7.8	43	0	166	262	84	0	79	100	0.35	59.3	6.42	0.73	B35
10W228	BW70	428	7.2	44	1	122	261	83	2	73	98	0.41	73.7	5.48	0.62	BW70
10W228	BW62	383	7.8	102	4	20	257	88	3	16	97	0.85	279.8	5.63	0.57	BW62
10W228	BW56	277	7.7	6	2	14	255	65	25	70	75	0.45	56.5	2.90	0.44	BW56
10W228	BW6	269	7.1	9	151	5	104	57	94	35	6	0.02	0.1	1.52	0.28	BW6
10W228	BW4	109	6.4	5	101	0	3	40	95	0	5	0.04	0.1	0.93	0.17	BW4

10W229	TRAY:	5	POS: 41		LOCAL	SPECIFICITY:	BW62,BW70,BW50,BW56									
10W229	BW75	534	7.9	60	0	211	263	81	0	77	100	0.35	65.6	7.96	0.87	BW75
10W229	BW62	474	7.8	114	1	97	262	78	0	45	100	0.62	183.4	7.90	0.79	BW62
10W229	BW77	359	7.6	11	1	86	261	62	8	88	92	0.27	26.3	5.02	0.71	BW77
10W229	BW56	347	6.6	7	2	79	259	60	22	91	78	0.20	13.9	3.86	0.58	BW56
10W229	B35	338	7.1	29	4	50	255	60	12	63	88	0.50	85.0	4.87	0.58	B35
10W229	BW70	305	7.0	41	5	9	250	60	10	18	90	0.83	209.1	5.07	0.57	BW70
10W229	B15.3	259	7.0	4	2	5	248	66	33	55	67	0.53	73.1	3.33	0.54	B15.3
10W229	BW6	253	6.8	5	149	0	99	60	96	0	4	0.11	3.3	1.42	0.14	BW6
10W230	TRAY:	5	POS: 42		LOCAL	SPECIFICITY:	B15,BW50,BW70									
10W230	BW76	527	7.9	18	0	262	247	83	0	93	100	0.18	16.4	5.46	0.73	BW76
10W230	BW75	509	7.8	54	2	208	245	82	3	79	97	0.32	50.9	5.43	0.61	BW75
10W230	BW70	453	7.8	45	2	163	243	80	4	78	96	0.34	52.4	5.18	0.59	BW70
10W230	B35	406	7.5	34	4	129	239	77	10	79	90	0.32	42.4	4.68	0.57	B35
10W230	BW62	368	7.7	99	6	30	233	76	5	23	95	0.78	226.4	5.07	0.52	BW62
10W230	B45	263	7.0	10	2	20	231	46	16	66	84	0.49	64.4	3.21	0.47	B45
10W230	BW41	251	6.0	4	3	16	228	40	42	80	58	0.31	23.7	2.09	0.34	BW41
10W230	BW6	244	6.7	15	125	1	103	43	89	6	11	0.19	9.3	1.26	0.24	BW6
10W231	TRAY:	5	POS: 45		LOCAL	SPECIFICITY:	B15,B17,B35,BW70									
10W231	BW75	533	8.0	60	0	268	205	87	0	81	100	0.28	42.3	7.81	0.91	BW75
10W231	BW76	473	8.0	14	0	254	205	85	0	94	100	0.15	11.0	6.05	0.90	BW76
10W231	BW77	459	8.0	13	0	241	205	84	0	94	100	0.15	10.8	5.96	0.90	BW77
10W231	B15.3	446	7.6	9	0	232	205	83	0	96	100	0.13	7.8	5.39	0.88	B15.3
10W231	BW62	437	8.0	110	1	122	204	84	0	52	100	0.54	126.5	7.80	0.83	BW62
10W231	B35	326	7.7	32	2	90	202	70	5	73	95	0.40	52.1	6.04	0.76	B35
10W231	BW63	292	8.0	7	2	83	200	65	22	92	78	0.18	9.6	4.33	0.71	BW63
10W231	B45	283	7.1	13	2	70	198	62	13	84	87	0.30	25.1	4.14	0.61	B45
10W231	BW57	268	7.7	13	3	57	195	61	18	81	82	0.32	26.8	3.89	0.56	BW57
10W231	BW70	252	7.0	31	11	26	184	56	26	45	74	0.55	75.5	3.28	0.40	BW70
10W231	BW58	210	7.1	15	17	11	167	53	53	42	47	0.44	41.4	1.64	0.21	BW58
10W231	BW60	178	6.3	6	21	5	146	36	77	45	23	0.28	14.1	0.71	0.09	BW60
10W233	TRAY:	5	POS: 43		LOCAL	SPECIFICITY:	BW62,B15.3,BW50,BW70									
10W233	BW75	531	8.0	60	0	209	262	92	0	77	100	0.35	65.9	6.97	0.76	BW75
10W233	BW70	471	7.6	46	1	163	261	90	2	77	98	0.36	60.5	5.67	0.64	BW70
10W233	B35	424	8.0	40	2	123	259	93	4	75	96	0.39	63.5	5.30	0.61	B35
10W233	BW62	382	7.9	103	4	20	255	91	3	16	97	0.86	279.4	5.76	0.59	BW62
10W233	BW56	275	8.0	7	1	13	254	70	12	65	88	0.53	78.6	3.72	0.57	BW56
10W233	BW6	267	6.5	12	147	1	107	53	92	7	8	0.15	6.1	2.36	0.22	BW6
10W234	TRAY:	5	POS: 44		LOCAL	SPECIFICITY:	B35,BW70,B15.3,BW62									
10W234	BW75	532	7.9	60	0	172	300	72	0	74	100	0.41	87.4	7.92	0.86	BW75
10W234	B35	472	7.8	40	3	132	297	65	6	76	94	0.37	65.4	5.73	0.65	B35
10W234	BW56	429	7.0	6	3	126	294	57	33	95	67	0.11	5.6	3.62	0.63	BW56
10W234	BW62	420	6.9	89	16	37	278	57	15	29	85	0.69	199.9	4.72	0.48	BW62
10W234	BW70	315	7.1	30	16	7	262	56	34	18	66	0.69	148.6	2.99	0.34	BW70
10W234	BW6	269	6.4	5	155	2	107	42	96	28	4	0.04	0.4	1.78	0.17	BW6

Table 5. *Continued*

10W235	TRAY:	6	POS: 20		LOCAL	SPECIFICITY: B35,B15.3,BW53,B51										
10W235	BW53	531	8.0	7	0	212	312	78	0	96	100	0.14	10.1	6.25	1.00	BW53
10W235	B35	524	7.8	45	0	167	312	77	0	78	100	0.37	72.4	8.52	0.98	B35
10W235	BW75	479	7.8	55	1	112	311	73	1	67	99	0.48	112.1	7.89	0.88	BW75
10W235	BW77	423	8.0	10	1	102	310	65	9	91	91	0.24	24.1	5.69	0.83	BW77
10W235	B15.3	412	7.8	8	1	94	309	61	11	92	89	0.22	20.3	5.24	0.80	B15.3
10W235	B51	403	7.7	27	5	67	304	59	15	71	85	0.42	72.4	5.25	0.63	B51
10W235	BW52	371	7.1	7	6	60	298	47	46	89	54	0.18	11.7	2.80	0.40	BW52
10W235	BW62	358	6.5	55	34	5	264	45	38	8	62	0.69	172.2	3.39	0.36	BW62
10W235	BW70	269	7.2	5	40	0	224	80	88	0	12	0.31	25.4	0.77	0.09	BW70
10W236	TRAY:	6	POS: 37		LOCAL	SPECIFICITY: BW62,B15.3,SH.7										
10W236	BW75	530	7.9	59	0	150	321	92	0	71	100	0.44	102.0	9.17	0.98	BW75
10W236	BW62	471	7.9	115	0	35	321	90	0	23	100	0.83	325.6	9.79	0.97	BW62
10W236	BW77	356	7.8	11	0	24	321	77	0	68	100	0.54	104.1	6.71	0.95	BW77
10W236	BW76	345	7.2	13	0	11	321	70	0	45	100	0.72	180.7	6.56	0.90	BW76
10W236	B15.3	332	8.0	5	2	6	319	72	28	54	72	0.56	103.6	4.05	0.63	B15.3
10W236	BW6	325	6.8	5	221	1	98	50	97	16	3	0.04	0.5	0.61	0.06	BW6
10W237	TRAY:	6	POS: 41		LOCAL	SPECIFICITY: B15										
10W237	BW77	523	8.0	12	0	179	332	83	0	93	100	0.20	21.3	6.18	0.85	BW77
10W237	BW75	511	7.8	60	0	119	332	82	0	66	100	0.50	126.1	8.00	0.85	BW75
10W237	B15.3	451	7.0	8	0	111	332	78	0	93	100	0.22	22.7	5.43	0.81	B15.3
10W237	BW62	443	7.5	99	7	12	325	79	6	10	94	0.88	346.6	6.52	0.65	BW62
10W237	B35	337	6.4	5	29	7	296	41	85	58	15	0.20	13.7	0.47	0.05	B35
10W238	TRAY:	6	POS: 50		LOCAL	SPECIFICITY: B15										
10W238	BW76	530	8.0	18	0	195	317	93	0	91	100	0.23	27.7	7.75	0.99	BW76
10W238	BW77	512	8.0	12	0	183	317	92	0	93	100	0.20	20.0	7.18	0.99	BW77
10W238	BW75	500	7.9	55	1	128	316	92	1	69	99	0.45	103.2	9.00	0.97	BW75
10W238	BW62	444	8.0	108	5	20	311	91	4	15	96	0.86	329.1	8.08	0.80	BW62
10W238	BW63	331	7.1	9	1	11	310	55	10	55	90	0.62	128.0	5.32	0.76	BW63
10W238	B15.3	321	7.0	6	1	5	309	45	14	45	86	0.68	146.4	4.86	0.75	B15.3
10W238	BW57	314	6.0	4	13	1	296	20	76	20	24	0.42	55.2	1.23	0.16	BW57
10W239	TRAY:	6	POS: 36		LOCAL	SPECIFICITY: B15										
10W239	BW62	529	6.3	80	35	26	388	36	30	24	70	0.65	225.0	3.42	0.33	BW62
10W239	BW75	414	6.0	21	37	5	351	30	63	19	37	0.50	102.6	1.18	0.13	BW75
10W240	TRAY:	6	POS: 40		LOCAL	SPECIFICITY: B15										
10W240	BW75	531	7.9	60	0	139	332	93	0	69	100	0.46	112.9	9.27	0.98	BW75
10W240	BW62	471	7.9	115	0	24	332	92	0	17	100	0.88	363.4	10.02	0.98	BW62
10W240	B15.3	356	7.7	7	0	17	332	79	0	70	100	0.53	98.8	6.29	0.95	B15.3
10W240	BW77	349	8.0	10	1	7	331	76	9	41	91	0.72	181.4	5.89	0.82	BW77
10W240	BW63	338	7.2	5	5	2	326	42	50	28	50	0.59	116.7	3.22	0.45	BW63
10W241	TRAY:	6	POS: 46		LOCAL	SPECIFICITY: B15										
10W241	BW77	531	8.0	12	0	191	328	83	0	94	100	0.19	19.8	7.19	0.99	BW77
10W241	BW75	519	7.8	60	0	131	328	82	0	68	100	0.47	116.5	9.07	0.96	BW75
10W241	BW62	459	7.6	112	2	19	326	79	1	14	99	0.89	361.3	9.65	0.95	BW62
10W241	B15.3	345	7.2	5	2	14	324	47	28	73	72	0.42	59.7	4.04	0.61	B15.3
10W241	BW63	338	7.4	7	3	7	321	42	30	50	70	0.58	112.6	3.97	0.56	BW63
10W241	BW57	328	5.6	5	12	2	309	14	70	28	30	0.44	63.9	1.15	0.16	BW57

10W242	TRAY:	6 POS: 47			LOCAL SPECIFICITY: B15											
10W242	BW75	531	8.0	60	0	180	291	93	0	75	100	0.39	82.0	7.84	0.84	BW75
10W242	BW62	471	7.8	115	0	65	291	91	0	36	100	0.72	246.0	8.34	0.83	BW62
10W242	BW76	356	8.0	13	0	52	291	92	0	80	100	0.41	60.4	6.00	0.83	BW76
10W242	BW77	343	8.0	11	0	41	291	90	0	78	100	0.43	63.6	5.80	0.82	BW77
10W242	BW63	332	7.4	10	0	31	291	87	0	75	100	0.47	73.2	5.54	0.80	BW63
10W242	B15.3	322	8.0	5	2	26	289	93	28	83	72	0.31	31.4	3.32	0.52	B15.3
10W242	BW46	315	7.9	22	19	4	270	92	46	15	54	0.64	128.3	2.17	0.25	BW46
10W243	TRAY:	6 POS: 58			LOCAL SPECIFICITY: B15											
10W243	BW62	530	8.0	117	0	100	313	95	0	46	100	0.64	216.6	9.98	0.99	BW62
10W243	BW77	413	8.0	11	0	89	313	91	0	89	100	0.29	35.4	6.90	0.98	BW77
10W243	BW75	402	7.9	57	0	32	313	89	0	35	100	0.76	233.6	9.04	0.98	BW75
10W243	BW63	345	7.8	9	1	23	312	78	10	71	90	0.48	79.7	5.46	0.79	BW63
10W243	B15.3	335	7.7	6	1	17	311	73	14	73	86	0.46	69.5	4.87	0.76	B15.3
10W243	BW57	328	7.7	6	11	11	300	70	64	64	36	0.32	33.1	1.83	0.24	BW57
10W243	BW46	311	7.2	10	32	1	268	63	76	9	24	0.43	58.5	0.98	0.11	BW46
10W245	TRAY:	6 POS: 48			LOCAL SPECIFICITY: BW62,BW63											
10W245	BW75	493	7.9	59	0	128	306	93	0	68	100	0.47	109.7	7.79	0.84	BW75
10W245	BW77	434	8.0	12	0	116	306	92	0	90	100	0.26	29.5	6.00	0.84	BW77
10W245	BW62	422	7.8	97	2	19	304	91	2	16	98	0.87	322.5	8.25	0.84	BW62
10W245	B15.3	323	7.7	6	1	13	303	89	14	68	86	0.50	82.4	4.27	0.66	B15.3
10W245	BW63	316	8.0	6	3	7	300	92	33	53	67	0.54	91.9	3.78	0.56	BW63
10W245	BW4	307	8.0	5	191	2	109	85	97	28	3	0.02	0.2	1.37	0.25	BW4
10W246	TRAY:	6 POS: 38			LOCAL SPECIFICITY: BW62											
10W246	BW62	531	7.9	117	0	87	327	91	0	42	100	0.67	240.5	10.02	0.98	BW62
10W246	BW75	414	7.9	58	0	29	327	83	0	33	100	0.78	253.5	9.12	0.97	BW75
10W246	BW76	356	5.8	13	0	16	327	62	0	55	100	0.65	152.1	6.61	0.89	BW76
10W246	BW77	343	7.8	10	1	6	326	81	9	37	91	0.74	190.1	5.65	0.78	BW77
10W246	B15.3	332	8.0	4	3	2	323	66	42	33	58	0.61	123.4	3.91	0.59	B15.3
10W247	TRAY:	6 POS: 49			LOCAL SPECIFICITY: BW62											
10W247	BW76	530	8.0	18	0	206	306	89	0	91	100	0.22	25.5	6.94	0.89	BW76
10W247	BW77	512	8.0	12	0	194	306	88	0	94	100	0.19	18.3	6.38	0.88	BW77
10W247	BW75	500	7.8	56	0	138	306	87	0	71	100	0.45	99.5	8.11	0.87	BW75
10W247	B15.3	444	7.5	8	1	130	305	86	11	94	89	0.18	14.3	5.67	0.83	B15.3
10W247	BW62	435	7.9	110	1	20	304	86	0	15	100	0.89	340.7	8.31	0.82	BW62
10W247	BW63	324	7.2	5	5	15	299	35	50	75	50	0.32	34.2	3.76	0.54	BW63
10W247	BW57	314	6.7	6	11	9	288	26	64	60	36	0.34	36.8	1.42	0.18	BW57
10W248	TRAY:	6 POS: 33			LOCAL SPECIFICITY: BW62											
10W248	BW76	523	7.7	18	0	156	349	70	0	89	100	0.27	37.4	7.21	0.91	BW76
10W248	BW62	505	7.8	108	6	48	343	68	5	30	95	0.75	281.1	7.72	0.75	BW62
10W248	BW75	391	5.5	42	12	6	331	22	22	12	78	0.80	249.6	4.16	0.45	BW75
10W248	BW6	337	5.2	5	225	1	106	0	97	16	3	0.04	0.6	1.08	0.10	BW6
10W249	TRAY:	6 POS: 39			LOCAL SPECIFICITY: BW62											
10W249	BW75	529	7.4	59	0	117	353	65	0	66	100	0.50	133.2	8.84	0.93	BW75
10W249	BW77	470	5.8	12	0	105	353	60	0	89	100	0.28	37.2	6.55	0.89	BW77
10W249	BW62	458	6.9	100	13	5	340	62	11	4	89	0.89	365.0	6.58	0.64	BW62
10W249	B15.3	345	6.5	4	3	1	337	40	42	20	58	0.67	155.2	3.21	0.48	B15.3

Table 5. *Continued*

10W250	TRAY:	6	POS: 34		LOCAL SPECIFICITY: BW62											
10W250	BW76	531	7.8	18	0	88	425	60	0	83	100	0.38	74.7	7.85	0.97	BW76
10W250	BW62	513	6.8	88	28	0	397	54	24	0	76	0.84	363.5	4.76	0.46	BW62
10W252	TRAY:	6	POS: 35		LOCAL SPECIFICITY: BW62											
10W252	BW76	531	7.8	18	0	112	401	76	0	86	100	0.33	57.5	7.71	0.95	BW76
10W252	BW62	513	7.5	107	9	5	392	74	7	4	93	0.92	435.4	7.37	0.71	BW62
10W253	TRAY:	5	POS: 48		LOCAL SPECIFICITY: BW62											
10W253	BW76	534	7.2	18	0	159	357	63	0	89	100	0.27	37.6	5.65	0.72	BW76
10W253	BW62	516	7.1	100	16	59	341	63	13	37	87	0.65	215.4	4.63	0.45	BW62
10W253	BW77	400	7.8	9	3	50	338	59	25	84	75	0.30	35.7	3.18	0.43	BW77
10W253	BW75	388	7.0	35	19	15	319	54	35	30	65	0.62	150.7	2.33	0.25	BW75
10W254	TRAY:	6	POS: 57		LOCAL SPECIFICITY: B15,SH.7											
10W254	BW75	531	8.0	60	0	211	260	93	0	77	100	0.35	64.9	8.08	0.88	BW75
10W254	BW76	471	8.0	14	0	197	260	91	0	93	100	0.19	17.8	6.31	0.87	BW76
10W254	BW77	457	8.0	12	0	185	260	90	0	93	100	0.19	16.3	6.12	0.86	BW77
10W254	BW62	445	7.8	113	0	72	260	90	0	38	100	0.69	212.9	8.59	0.86	BW62
10W254	BW63	332	8.0	8	2	64	258	90	20	88	80	0.25	20.6	4.75	0.70	BW63
10W254	B15.3	322	7.7	6	1	58	257	89	14	90	86	0.25	19.5	4.29	0.67	B15.3
10W254	BW46	315	7.8	38	3	20	254	89	7	34	93	0.74	173.1	5.69	0.65	BW46
10W254	BW57	274	8.0	14	2	6	252	85	12	30	88	0.77	161.5	4.78	0.64	BW57
10W255	TRAY:	6	POS: 42		LOCAL SPECIFICITY: B15,SH.7											
10W255	BW75	529	7.9	60	0	79	390	77	0	56	100	0.60	189.9	8.82	0.95	BW75
10W255	BW77	469	7.6	11	1	68	389	65	8	86	92	0.32	49.2	5.52	0.77	BW77
10W255	B15.3	457	7.8	8	1	60	388	61	11	88	89	0.29	39.7	4.97	0.73	B15.3
10W255	B51	448	7.1	23	9	37	379	58	28	61	72	0.48	101.6	4.06	0.49	B51
10W255	BW62	416	6.8	24	78	13	301	51	76	35	24	0.29	35.7	0.70	0.07	BW62
10W255	B35	314	6.8	5	29	8	272	30	85	61	15	0.18	10.7	0.50	0.06	B35
10W256	TRAY:	6	POS: 60		LOCAL SPECIFICITY: BW76,B15											
10W256	BW76	529	8.0	18	0	11	500	72	0	37	100	0.78	321.3	7.06	0.85	BW76
10W257	TRAY:	4	POS: 50		LOCAL SPECIFICITY: BW46											
10W257	BW46	532	7.8	56	1	9	466	81	1	13	99	0.91	440.5	7.85	0.80	BW46
10W258	TRAY:	8	POS: 30		LOCAL SPECIFICITY: BW46											
10W258	BW46	518	7.9	56	1	8	453	84	1	12	99	0.92	436.3	7.89	0.81	BW46
10W259	TRAY:	7	POS: 40		LOCAL SPECIFICITY: BW46,CW1											
10W259	BW63	532	6.5	8	2	245	277	88	20	96	80	0.09	4.3	3.01	0.49	BW63
10W259	BW76	522	8.0	18	0	227	277	90	0	92	100	0.20	21.1	3.40	0.43	BW76
10W259	BW77	504	8.0	13	0	214	277	89	0	94	100	0.18	16.3	3.17	0.43	BW77
10W259	BW75	491	8.0	52	4	162	273	88	7	75	93	0.36	62.4	2.79	0.30	BW75
10W259	B15.3	435	8.0	7	2	155	271	85	22	95	78	0.12	6.5	1.66	0.24	B15.3
10W259	CW1	426	7.9	74	2	81	269	84	2	52	98	0.59	148.6	2.34	0.23	CW1
10W259	BW62	350	7.5	76	17	5	252	75	18	6	82	0.84	244.3	1.94	0.19	BW62

10W260	TRAY: 6	POS: 59			LOCAL SPECIFICITY: BW46,B15,B17,CW3.1											
10W260	BW62	530	8.0	116	0	177	237	97	0	60	100	0.48	120.1	6.28	0.63	BW62
10W260	BW75	414	8.0	58	0	119	237	96	0	67	100	0.47	90.3	5.59	0.61	BW75
10W260	BW76	356	8.0	13	0	106	237	94	0	89	100	0.27	26.9	4.15	0.58	BW76
10W260	B15.3	343	8.0	7	0	99	237	94	0	93	100	0.22	16.0	3.68	0.58	B15.3
10W260	BW77	336	8.0	11	0	88	237	93	0	88	100	0.28	27.2	4.01	0.58	BW77
10W260	BW63	325	8.0	10	0	78	237	93	0	88	100	0.29	27.8	3.93	0.58	BW63
10W260	BW57	315	7.9	16	1	62	236	92	5	79	95	0.38	46.4	3.54	0.47	BW57
10W260	BW46	298	7.9	36	4	26	232	91	10	41	90	0.67	134.3	3.32	0.38	BW46
10W260	CW3	258	7.7	23	37	3	195	84	61	11	39	0.52	68.9	0.95	0.09	CW3
10W261	TRAY: 5	POS: 47			LOCAL SPECIFICITY: B15,BW46,B13											
10W261	BW75	531	7.9	60	0	235	236	89	0	79	100	0.32	54.1	6.94	0.76	BW75
10W261	BW46	471	7.9	52	0	183	236	88	0	77	100	0.35	58.7	6.76	0.76	BW46
10W261	B13	419	7.9	29	0	154	236	85	0	84	100	0.31	40.2	6.11	0.75	B13
10W261	BW77	390	8.0	12	0	142	236	83	0	92	100	0.22	19.0	5.15	0.74	BW77
10W261	BW76	378	7.8	12	0	130	236	82	0	91	100	0.23	20.6	5.01	0.72	BW76
10W261	BW63	366	7.5	8	1	122	235	81	11	93	89	0.18	11.5	4.72	0.71	BW63
10W261	B15.3	357	7.6	9	0	113	235	81	0	92	100	0.22	17.8	4.58	0.69	B15.3
10W261	BW62	348	7.7	95	3	18	232	81	3	15	97	0.86	258.6	5.75	0.59	BW62
10W261	BW57	250	6.0	5	11	13	221	38	68	72	32	0.24	14.8	0.85	0.11	BW57
10W261	CW3	234	6.8	10	42	3	179	38	80	23	20	0.32	23.8	0.82	0.07	CW3
10W262	TRAY: 6	POS: 54			LOCAL SPECIFICITY: BW62,BW57,BW63,BW46											
10W262	BW62	526	8.0	114	0	165	247	96	0	59	100	0.49	128.9	8.26	0.84	BW62
10W262	BW75	412	8.0	58	0	107	247	93	0	64	100	0.50	101.0	7.49	0.83	BW75
10W262	BW76	354	8.0	13	0	94	247	89	0	87	100	0.30	31.2	5.67	0.81	BW76
10W262	BW77	341	8.0	11	0	83	247	88	0	88	100	0.30	29.9	5.47	0.80	BW77
10W262	BW63	330	8.0	10	0	73	247	86	0	87	100	0.30	30.7	5.36	0.80	BW63
10W262	B15.3	320	8.0	7	0	66	247	84	0	90	100	0.28	24.2	4.96	0.80	B15.3
10W262	BW57	313	8.0	16	1	50	246	83	5	75	95	0.43	57.6	5.05	0.68	BW57
10W262	BW46	296	7.8	38	2	12	244	78	5	24	95	0.82	201.0	5.51	0.64	BW46
10W262	B35	256	5.6	5	26	7	218	41	83	58	17	0.20	10.3	0.49	0.06	B35
10W263	TRAY: 3	POS: 37			LOCAL SPECIFICITY: B8,BW46											
10W263	B8	532	7.9	31	1	53	447	88	3	63	97	0.56	168.4	7.71	0.86	B8
10W263	BW46	500	7.8	50	6	3	441	84	10	5	90	0.91	412.0	6.50	0.67	BW46
10W264	TRAY: 6	POS: 53			LOCAL SPECIFICITY: BW62,BW63,BW57,BW46											
10W264	BW62	530	7.9	117	0	137	276	92	0	53	100	0.55	163.1	8.76	0.88	BW62
10W264	BW75	413	8.0	58	0	79	276	89	0	57	100	0.57	135.9	7.99	0.87	BW75
10W264	BW76	355	8.0	13	0	66	276	82	0	83	100	0.36	47.1	6.22	0.86	BW76
10W264	B15.3	342	8.0	7	0	59	276	78	0	89	100	0.30	29.9	5.51	0.86	B15.3
10W264	BW77	335	8.0	11	0	48	276	76	0	81	100	0.40	53.2	6.02	0.86	BW77
10W264	BW63	324	8.0	10	0	38	276	70	0	79	100	0.43	59.3	5.90	0.86	BW63
10W264	BW57	314	7.5	16	1	22	275	63	5	57	95	0.60	113.7	5.37	0.71	BW57
10W264	BW46	297	6.8	16	23	6	252	50	58	27	42	0.50	74.0	1.54	0.18	BW46
10W266	TRAY: 6	POS: 43			LOCAL SPECIFICITY: B17,BW63											
10W266	BW58	531	8.0	52	0	54	425	91	0	50	100	0.66	231.1	7.28	0.77	BW58
10W266	BW57	479	7.9	24	0	30	425	85	0	55	100	0.64	198.9	6.39	0.76	BW57
10W266	BW77	455	7.0	8	4	22	421	76	33	73	67	0.40	72.2	4.54	0.61	BW77
10W266	BW63	443	8.0	6	2	16	419	77	25	72	75	0.44	84.7	3.35	0.48	BW63

Table 5. *Continued*

10W267	TRAY:	6	POS: 44			LOCAL	SPECIFICITY: BW63,BW57,BW58									
10W267	BW57		530	7.8	26	0	51	453	71	0	66	100	0.55	160.9	7.40	0.85 BW57
10W267	BW77		504	6.7	9	3	42	450	62	25	82	75	0.34	56.9	5.85	0.77 BW77
10W267	BW58		492	7.1	35	15	7	435	61	30	16	70	0.74	269.3	3.55	0.37 BW58
10W323	TRAY:	4	POS: 57			LOCAL	SPECIFICITY: BW50,BW41,BW62,BW70									
10W323	BW75		530	7.2	59	1	127	343	66	1	68	99	0.47	118.8	7.42	0.81 BW75
10W323	BW62		470	7.5	88	26	39	317	63	22	30	78	0.64	192.1	4.89	0.50 BW62
10W323	BW41		356	4.5	4	3	35	314	30	42	89	58	0.21	15.6	2.96	0.47 BW41
10W323	BW70		349	6.0	19	28	16	286	34	59	45	41	0.40	55.6	1.64	0.19 BW70
10W323	B35		302	6.7	9	25	7	261	43	73	43	27	0.34	34.2	0.85	0.10 B35
10W323	BW60		268	6.5	4	24	3	237	28	85	42	15	0.25	16.8	0.59	0.07 BW60
10W416	TRAY:	3	POS: 54			LOCAL	SPECIFICITY: BW65									
10W416	BW77		532	6.7	11	2	47	472	50	15	81	85	0.37	74.5	2.94	0.38 BW77
10W416	CW8		519	7.1	31	25	16	447	46	44	34	56	0.56	163.4	2.71	0.27 CW8
10W418	TRAY:	4	POS: 48			LOCAL	SPECIFICITY: B12									
10W418	B44		524	7.9	77	0	38	409	92	0	33	100	0.78	321.0	8.71	0.87 B44
10W418	BW76		447	8.0	16	0	22	409	84	0	57	100	0.63	178.6	6.87	0.86 BW76
10W418	B45		431	7.9	14	0	8	409	72	0	36	100	0.79	269.0	6.57	0.84 B45
10W422	TRAY:	8	POS: 42			LOCAL	SPECIFICITY: B12,B44									
10W422	BW76		534	7.9	18	0	113	403	82	0	86	100	0.33	57.3	6.90	0.87 BW76
10W422	B45		516	7.8	16	0	97	403	80	0	85	100	0.34	58.9	6.60	0.85 B45
10W422	B44		500	7.9	76	1	21	402	79	1	21	99	0.86	366.1	7.97	0.81 B44
10W422	BW57		423	4.8	5	18	16	384	23	78	76	22	0.19	14.5	0.73	0.09 BW57
10W428	TRAY:	4	POS: 44			LOCAL	SPECIFICITY: B45									
10W428	BW76		531	7.9	18	0	8	505	80	0	30	100	0.83	361.9	7.58	0.91 BW76
10W428	B45		513	7.2	5	11	3	494	50	68	37	32	0.43	94.8	2.06	0.25 B45
10W432	TRAY:	4	POS: 45			LOCAL	SPECIFICITY: B45									
10W432	BW76		532	8.0	18	0	18	496	88	0	50	100	0.69	256.7	7.53	0.90 BW76
10W432	B45		514	7.7	14	2	4	494	77	12	22	88	0.82	344.8	5.42	0.66 B45
10W478	TRAY:	6	POS: 45			LOCAL	SPECIFICITY: BW57,BW58,BW63,BW62									
10W478	BW58		530	7.9	52	0	60	418	83	0	53	100	0.64	215.2	8.98	0.97 BW58
10W478	BW57		478	7.9	24	0	36	418	71	0	60	100	0.61	176.0	7.84	0.95 BW57
10W478	BW77		454	6.7	12	0	24	418	55	0	66	100	0.56	143.1	6.34	0.87 BW77
10W478	BW63		442	8.0	7	1	17	417	50	12	70	88	0.49	106.9	5.08	0.75 BW63
10W478	BW6		434	6.1	17	365	0	52	29	95	0	5	0.07	2.4	0.91	0.09 BW6
10W493	TRAY:	3	POS: 14			LOCAL	SPECIFICITY: BW4									
10W493	B27		526	7.5	26	2	266	232	84	7	91	93	0.18	16.7	2.74	0.33 B27
10W493	B38		498	7.8	24	2	242	230	85	7	90	93	0.18	16.7	2.02	0.25 B38
10W493	BW57		472	7.9	23	1	219	229	84	4	90	96	0.21	20.1	1.93	0.24 BW57
10W493	BW52		448	7.8	13	1	206	228	83	7	94	93	0.16	11.2	1.56	0.22 BW52
10W493	B44		434	7.8	67	4	139	224	83	5	67	95	0.42	74.9	1.96	0.21 B44
10W493	B51		363	7.5	31	3	108	221	78	8	77	92	0.35	44.4	1.56	0.19 B51
10W493	BW4		329	7.5	97	13	11	208	76	11	10	89	0.84	229.6	2.49	0.44 BW4

10W495	TRAY:	3	POS: 15		LOCAL SPECIFICITY: BW4											
10W495	B13	533	7.9	35	1	282	215	92	2	88	98	0.21	22.8	2.60	0.30	B13
10W495	B27	497	7.8	26	0	256	215	92	0	90	100	0.21	20.9	2.43	0.30	B27
10W495	BW57	471	8.0	26	0	230	215	91	0	89	100	0.22	23.1	2.40	0.30	BW57
10W495	BW52	445	7.9	14	0	216	215	90	0	93	100	0.17	13.5	2.06	0.28	BW52
10W495	B44	431	7.9	73	1	143	214	90	1	66	99	0.44	84.2	2.57	0.27	B44
10W495	B51	357	7.9	33	1	110	213	87	2	76	98	0.38	50.9	2.01	0.24	B51
10W495	BW4	323	7.9	95	5	15	208	84	5	13	95	0.86	239.5	4.19	0.75	BW4
10W496	TRAY:	3	POS: 17		LOCAL SPECIFICITY: BW4											
10W496	BW77	533	8.0	13	0	235	285	64	0	94	100	0.17	15.3	3.01	0.40	BW77
10W496	B27	520	7.2	26	1	209	284	62	3	88	97	0.24	30.0	3.14	0.37	B27
10W496	BW63	493	7.8	10	0	199	284	61	0	95	100	0.17	13.9	2.59	0.36	BW63
10W496	BW52	483	6.6	10	3	189	281	59	23	94	77	0.12	7.0	2.45	0.36	BW52
10W496	BW57	470	7.2	23	1	166	280	59	4	87	96	0.26	32.5	2.45	0.29	BW57
10W496	B51	446	7.4	29	4	137	276	58	12	82	88	0.30	39.1	2.31	0.27	B51
10W496	B38	413	7.6	20	3	117	273	55	13	85	87	0.28	31.8	2.10	0.25	B38
10W496	B13	390	7.2	23	9	94	264	51	28	80	72	0.27	29.1	1.89	0.22	B13
10W496	B37	358	6.0	7	2	87	262	46	22	92	78	0.19	12.7	1.29	0.18	B37
10W496	AW33	349	6.7	25	5	62	257	47	16	71	84	0.41	59.8	1.04	0.11	AW33
10W496	A32	319	6.9	7	4	55	253	45	36	88	64	0.21	14.2	0.66	0.09	A32
10W496	BW4	308	6.6	51	45	4	208	43	46	7	54	0.62	118.3	4.41	0.40	BW4
10W497	TRAY:	3	POS: 18		LOCAL SPECIFICITY: BW4											
10W497	B15.3	534	7.6	5	5	68	456	69	50	93	50	0.15	11.4	2.90	0.42	B15.3
10W497	BW63	524	8.0	5	5	63	451	69	50	92	50	0.15	12.4	2.53	0.36	BW63
10W497	B13	514	7.9	21	14	42	437	66	40	66	60	0.39	79.6	2.95	0.34	B13
10W497	B27	479	7.8	8	18	34	419	52	69	80	31	0.19	16.6	1.29	0.16	B27
10W497	BW61	453	5.7	7	18	27	401	44	72	79	28	0.19	16.0	1.06	0.13	BW61
10W497	B51	428	7.2	5	28	22	373	44	84	81	16	0.11	4.7	0.55	0.06	B51
10W497	B44	395	6.8	12	57	10	316	40	82	45	18	0.24	22.2	0.54	0.06	B44
10W498	TRAY:	3	POS: 16		LOCAL SPECIFICITY: BW4											
10W498	B51	532	7.9	40	0	272	220	94	0	87	100	0.24	30.5	2.61	0.29	B51
10W498	B13	492	8.0	35	0	237	220	94	0	87	100	0.25	30.5	2.54	0.29	B13
10W498	B27	457	8.0	23	0	214	220	93	0	90	100	0.22	22.5	2.27	0.28	B27
10W498	B44	434	7.9	73	2	141	218	92	2	65	98	0.44	83.7	2.25	0.23	B44
10W498	B38	359	7.9	21	1	120	217	90	4	85	96	0.29	31.0	1.69	0.21	B38
10W498	BW4	337	7.9	108	9	12	208	90	7	10	93	0.86	251.3	3.92	0.70	BW4
10W499	TRAY:	3	POS: 19		LOCAL SPECIFICITY: BW6											
10W499	BW75	517	8.0	60	0	347	110	92	0	85	100	0.19	18.3	4.13	0.51	BW75
10W499	BW76	457	8.0	14	0	333	110	91	0	95	100	0.10	4.6	2.71	0.43	BW76
10W499	BW62	443	8.0	111	2	222	108	91	1	66	99	0.31	43.2	3.73	0.42	BW62
10W499	B39	330	7.6	16	0	206	108	87	0	92	100	0.16	8.2	2.59	0.40	B39
10W499	BW54	314	8.0	7	0	199	108	88	0	96	100	0.11	3.8	2.13	0.39	BW54
10W499	BW41	307	8.0	6	0	193	108	87	0	96	100	0.10	3.3	2.02	0.39	BW41
10W499	BW48	301	8.0	5	0	188	108	87	0	97	100	0.10	2.8	1.90	0.38	BW48
10W499	BW60	296	7.9	25	1	163	107	87	3	86	97	0.21	13.1	2.57	0.36	BW60
10W499	B8	270	7.9	20	1	143	106	85	4	87	96	0.21	11.6	2.34	0.34	B8
10W499	B35	249	7.9	29	2	114	104	84	6	79	94	0.28	18.9	2.30	0.31	B35
10W499	B7	218	7.8	35	3	79	101	82	7	69	93	0.37	29.2	2.15	0.28	B7
10W499	BW56	180	8.0	6	1	73	100	79	14	92	86	0.17	5.2	1.41	0.26	BW56
10W499	BW70	173	7.9	31	5	42	95	78	13	57	87	0.46	35.9	1.82	0.24	BW70
10W499	BW61	137	6.7	6	2	36	93	64	25	85	75	0.24	7.9	1.16	0.21	BW61
10W499	BW6	129	7.2	22	11	14	82	63	33	38	67	0.51	33.1	3.56	0.34	BW6

Table 5. *Continued*

10W500	TRAY:	3	POS: 20		LOCAL	SPECIFICITY:	BW6								
10W500	BW75	532	8.0	59	0	421	52	95	0	87	100	0.12	7.2	2.14	0.27 BW75
10W500	BW60	473	8.0	38	0	383	52	94	0	90	100	0.10	5.1	1.83	0.25 BW60
10W500	B35	435	8.0	42	0	341	52	94	0	89	100	0.12	6.3	1.80	0.24 B35
10W500	BW46	393	7.9	48	0	293	52	93	0	85	100	0.15	8.3	1.84	0.24 BW46
10W500	B18	345	8.0	23	0	270	52	93	0	92	100	0.11	4.4	1.49	0.22 B18
10W500	BW70	322	8.0	43	1	227	51	93	2	84	98	0.15	7.2	1.56	0.21 BW70
10W500	BW62	278	8.0	84	3	143	48	92	3	62	97	0.26	18.8	1.73	0.21 BW62
10W500	BW42	191	8.0	15	0	128	48	88	0	89	100	0.17	5.5	1.21	0.20 BW42
10W500	B8	176	7.9	15	0	113	48	86	0	88	100	0.19	6.1	1.14	0.18 B8
10W500	BW76	161	8.0	11	0	102	48	85	0	90	100	0.18	5.0	1.04	0.18 BW76
10W500	B39	150	7.7	14	0	88	48	84	0	86	100	0.22	7.3	1.08	0.18 B39
10W500	BW54	136	8.0	5	0	83	48	84	0	94	100	0.14	2.8	0.72	0.15 BW54
10W500	B7	131	8.0	28	2	55	46	83	6	66	94	0.34	15.1	1.03	0.15 B7
10W500	B45	101	8.0	6	1	49	45	74	14	89	86	0.17	3.0	0.42	0.08 B45
10W500	BW6	94	8.0	20	2	29	43	71	9	59	91	0.43	17.3	2.01	0.19 BW6
10W501	TRAY:	3	POS: 23		LOCAL	SPECIFICITY:	BW6								
10W501	B7	528	7.8	59	3	357	109	79	4	85	96	0.15	11.3	5.15	0.65 B7
10W501	BW76	466	8.0	18	0	339	109	77	0	94	100	0.11	5.7	4.06	0.63 BW76
10W501	B35	448	7.4	45	2	294	107	75	4	86	96	0.16	11.5	4.66	0.62 B35
10W501	BW60	401	7.3	46	0	248	107	75	0	84	100	0.22	18.9	4.70	0.62 BW60
10W501	B39	355	7.7	18	0	230	107	75	0	92	100	0.15	8.2	3.92	0.61 B39
10W501	BW54	337	8.0	10	0	220	107	75	0	95	100	0.12	4.8	3.40	0.60 BW54
10W501	BW41	327	8.0	7	0	213	107	74	0	96	100	0.10	3.5	3.04	0.59 BW41
10W501	BW75	320	6.7	35	1	178	106	73	2	83	98	0.23	17.1	3.70	0.51 BW75
10W501	B8	284	7.2	21	1	157	105	75	4	88	96	0.20	11.0	3.26	0.49 B8
10W501	BW46	262	7.3	33	5	124	100	76	13	78	87	0.23	13.4	3.18	0.44 BW46
10W501	BW42	224	8.0	16	2	108	98	76	11	87	89	0.20	8.9	2.79	0.44 BW42
10W501	BW61	206	7.6	14	2	94	96	73	12	87	88	0.20	8.6	2.47	0.40 BW61
10W501	BW62	190	7.3	52	10	42	86	71	16	44	84	0.48	43.6	2.76	0.35 BW62
10W501	BW70	128	7.2	22	8	20	78	69	26	47	74	0.48	29.2	2.08	0.31 BW70
10W501	B18	98	7.3	9	4	11	74	65	30	55	70	0.47	22.0	1.45	0.24 B18
10W501	BW6	85	7.4	7	4	4	70	54	36	36	64	0.58	28.8	2.54	0.24 BW6
10W502	TRAY:	8	POS: 52		LOCAL	SPECIFICITY:	BW6								
10W502	BW62	530	8.0	113	2	351	64	92	1	75	99	0.17	15.5	1.27	0.13 BW62
10W502	B18	415	8.0	28	0	323	64	90	0	92	100	0.11	5.5	0.93	0.12 B18
10W502	B35	387	7.9	37	0	286	64	90	0	88	100	0.14	8.1	0.97	0.12 B35
10W502	B8	350	8.0	25	0	261	64	89	0	91	100	0.13	6.0	0.88	0.12 B8
10W502	BW70	325	8.0	39	1	222	63	88	2	85	98	0.16	8.5	0.83	0.10 BW70
10W502	BW76	285	8.0	17	0	205	63	86	0	92	100	0.13	5.1	0.71	0.10 BW76
10W502	B39	268	7.8	18	0	187	63	85	0	91	100	0.15	5.9	0.69	0.10 B39
10W502	B45	250	8.0	11	0	176	63	85	0	94	100	0.12	3.9	0.56	0.09 B45
10W502	BW60	239	7.9	32	1	144	62	84	3	81	97	0.21	10.7	0.67	0.08 BW60
10W502	BW41	206	8.0	6	0	138	62	81	0	95	100	0.11	2.7	0.45	0.08 BW41
10W502	BW6	200	7.7	96	9	42	53	80	8	30	92	0.51	52.0	1.81	0.17 BW6

10W504	TRAY:	3	POS: 24		LOCAL	SPECIFICITY: BW6										
10W504	BW75	530	7.8	60	0	348	122	84	0	85	100	0.20	20.2	3.35	0.39	BW75
10W504	B18	470	7.9	29	0	319	122	83	0	91	100	0.15	10.8	2.76	0.36	B18
10W504	BW62	441	7.8	105	3	214	119	82	2	67	98	0.32	44.3	3.30	0.35	BW62
10W504	B7	333	8.0	41	2	173	117	77	4	80	96	0.25	20.8	2.61	0.32	B7
10W504	BW60	290	7.5	26	1	147	116	72	3	84	97	0.24	16.6	2.46	0.32	BW60
10W504	BW76	263	7.8	13	0	134	116	70	0	91	100	0.20	10.8	2.06	0.31	BW76
10W504	B8	250	7.7	19	0	115	116	68	0	85	100	0.27	17.8	1.95	0.26	B8
10W504	BW64	231	8.0	4	0	111	116	65	0	96	100	0.13	4.1	1.30	0.25	BW64
10W504	BW67	227	7.0	4	0	107	116	63	0	96	100	0.14	4.3	1.25	0.24	BW67
10W504	BW56	223	6.9	7	1	100	115	63	12	93	88	0.15	5.2	1.04	0.17	BW56
10W504	BW41	215	7.6	5	1	95	114	63	16	95	84	0.13	3.4	0.83	0.15	BW41
10W504	BW42	209	7.7	14	4	81	110	62	22	85	78	0.20	8.3	0.95	0.13	BW42
10W504	BW54	191	7.5	4	1	77	109	56	20	95	80	0.12	3.0	0.73	0.13	BW54
10W504	BW61	186	6.7	12	3	65	106	55	20	84	80	0.23	10.0	0.74	0.11	BW61
10W504	B39	171	7.8	9	4	56	102	55	30	86	70	0.18	5.8	0.59	0.09	B39
10W504	BW6	158	6.7	45	26	11	76	50	36	19	64	0.53	44.0	2.80	0.27	BW6
10W507	TRAY:	3	POS: 21		LOCAL	SPECIFICITY: BW6										
10W507	BW75	529	8.0	60	0	345	124	87	0	85	100	0.20	20.7	3.92	0.47	BW75
10W507	B35	469	7.9	43	0	302	124	85	0	87	100	0.19	17.0	3.57	0.45	B35
10W507	BW76	426	7.0	14	0	288	124	84	0	95	100	0.12	5.9	2.79	0.43	BW76
10W507	BW67	412	8.0	5	0	283	124	84	0	98	100	0.07	2.2	2.06	0.40	BW67
10W507	BW62	407	8.0	103	2	180	122	84	1	63	99	0.37	54.5	3.53	0.39	BW62
10W507	BW46	302	7.5	37	3	143	119	76	7	79	93	0.26	20.7	2.71	0.36	BW46
10W507	BW60	262	7.7	26	1	117	118	74	3	81	97	0.28	21.1	2.51	0.34	BW60
10W507	BW70	235	7.9	43	2	74	116	71	4	63	96	0.45	46.6	2.61	0.33	BW70
10W507	BW41	190	6.4	5	1	69	115	59	16	93	84	0.16	5.1	1.73	0.32	BW41
10W507	B8	184	6.9	11	5	58	110	59	31	84	69	0.20	7.3	1.11	0.17	B8
10W507	BW63	168	6.0	4	4	54	106	62	50	93	50	0.07	0.9	0.93	0.16	BW63
10W507	B39	160	6.5	8	5	46	101	64	38	85	62	0.17	4.9	0.97	0.15	B39
10W507	B7	147	6.5	17	15	29	86	69	46	63	54	0.25	9.1	0.62	0.08	B7
10W507	BW6	115	7.5	21	15	8	71	75	41	27	59	0.51	30.5	2.82	0.27	BW6
10W509	TRAY:	3	POS: 22		LOCAL	SPECIFICITY: BW6,AN										
10W509	BW62	527	8.0	116	0	365	46	95	0	75	100	0.16	14.2	0.72	0.07	BW62
10W509	B7	411	8.0	47	0	318	46	94	0	87	100	0.13	6.7	0.53	0.06	B7
10W509	B35	364	8.0	38	0	280	46	93	0	88	100	0.13	6.1	0.48	0.06	B35
10W509	B37	326	7.6	10	1	270	45	92	9	96	91	0.03	0.2	0.35	0.05	B37
10W509	BW60	315	7.9	38	0	232	45	92	0	85	100	0.15	7.2	0.42	0.05	BW60
10W509	BW6	277	7.9	169	6	63	39	91	3	27	97	0.46	57.4	1.44	0.14	BW6
10W524	TRAY:	2	POS: 47		LOCAL	SPECIFICITY: CW3										
10W524	CW3	533	8.0	219	1	25	288	98	0	10	100	0.90	436.3	6.46	0.59	CW3
10W524	BW46	313	8.0	19	1	6	287	92	5	24	95	0.84	220.1	2.18	0.22	BW46
10W525	TRAY:	2	POS: 48		LOCAL	SPECIFICITY: CW3										
10W525	CW3	534	7.8	219	2	28	285	91	0	11	100	0.89	423.5	6.91	0.63	CW3
10W525	BW46	313	7.9	19	1	9	284	82	5	32	95	0.79	194.2	2.16	0.22	BW46
10W526	TRAY:	2	POS: 49		LOCAL	SPECIFICITY: CW3										
10W526	CW3	530	7.6	211	6	58	255	78	2	21	98	0.77	317.6	5.20	0.49	CW3
10W526	BW46	313	7.6	15	5	43	250	55	25	74	75	0.38	45.1	0.99	0.10	BW46
10W526	CW1	293	7.2	5	21	38	229	44	80	88	20	0.04	0.5	0.72	0.08	CW1
10W526	BW77	267	6.0	4	7	34	222	42	63	89	37	0.13	4.6	0.47	0.06	BW77

Table 5. *Continued*

10W527	TRAY: 2	POS: 50			LOCAL SPECIFICITY: CW3											
10W527	CW3	533	7.7	218	3	64	248	84	1	22	99	0.77	316.9	5.37	0.49	CW3
10W527	BW46	312	7.6	18	2	46	246	65	10	71	90	0.45	63.3	2.61	0.28	BW46
10W527	BW77	292	7.6	10	1	36	245	58	9	78	91	0.41	48.6	1.78	0.25	BW77
10W527	BW62	281	5.7	6	16	30	229	52	72	83	28	0.13	4.5	1.66	0.17	BW62
10W527	BW75	259	7.5	16	11	14	218	56	40	46	60	0.51	66.9	1.18	0.13	BW75
10W528	TRAY: 2	POS: 51			LOCAL SPECIFICITY: CW3											
10W528	CW3	532	7.9	212	8	33	279	90	3	13	97	0.85	382.2	4.98	0.46	CW3
10W528	BW46	312	7.0	10	10	23	269	51	50	69	50	0.34	35.1	0.52	0.05	BW46
10W544	TRAY: 7	POS: 49			LOCAL SPECIFICITY: BW4,CW6,TEC10											
10W544	BW53	531	8.0	7	0	242	282	83	0	97	100	0.12	8.0	4.14	0.64	BW53
10W544	B15.3	524	7.5	8	2	234	280	82	20	96	80	0.09	4.7	4.23	0.62	B15.3
10W544	B51	514	7.9	36	1	198	279	82	2	84	98	0.29	43.1	5.19	0.60	B51
10W544	BW76	477	6.4	15	0	183	279	80	0	92	100	0.21	21.8	4.40	0.59	BW76
10W544	B44	462	7.9	72	3	111	276	83	4	60	96	0.51	119.0	5.20	0.55	B44
10W544	B45	387	7.8	13	1	98	275	74	7	88	93	0.27	29.2	3.84	0.52	B45
10W544	B35	373	7.9	35	6	63	269	72	14	64	86	0.47	83.0	3.53	0.40	B35
10W544	BW52	332	7.8	8	6	55	263	60	42	87	58	0.20	13.8	1.63	0.22	BW52
10W544	BW75	318	7.4	29	18	26	245	56	38	47	62	0.49	76.0	1.66	0.19	BW75
10W544	CW4	271	6.5	11	13	15	232	34	54	57	46	0.38	39.9	0.56	0.06	CW4
10W544	BW4	247	6.6	10	109	5	123	33	91	33	9	0.09	2.2	1.04	0.19	BW4
10W2064	TRAY: 15	POS: 26			LOCAL SPECIFICITY: A24,BW4											
10W2064	B44	404	8.0	56	0	232	116	96	0	80	100	0.25	26.2	2.18	0.24	B44
10W2064	B27	348	8.0	25	0	207	116	96	0	89	100	0.20	13.5	1.77	0.22	B27
10W2064	B38	323	8.0	23	0	184	116	96	0	88	100	0.21	13.9	1.73	0.22	B38
10W2064	BW58	300	7.9	16	0	168	116	95	0	91	100	0.19	10.7	1.60	0.22	BW58
10W2064	BW77	284	8.0	10	0	158	116	95	0	94	100	0.16	7.2	1.44	0.22	BW77
10W2064	BW52	274	8.0	13	0	145	116	95	0	91	100	0.19	10.0	1.51	0.22	BW52
10W2064	B51	261	8.0	26	1	119	115	95	3	82	97	0.28	20.2	1.36	0.17	B51
10W2064	A23	234	8.0	5	0	114	115	94	0	95	100	0.15	4.9	1.12	0.17	A23
10W2064	B13	229	7.9	21	1	93	114	93	4	81	96	0.30	20.3	1.26	0.16	B13
10W2064	A24	207	7.9	65	6	28	108	93	8	30	92	0.68	94.9	0.99	0.11	A24
10W2064	BW4	136	7.8	23	1	5	107	92	4	17	96	0.86	100.9	4.01	0.75	BW4
10W2065	TRAY: 15	POS: 28			LOCAL SPECIFICITY: A24,BW4											
10W2065	B38	403	7.9	26	0	256	121	96	0	90	100	0.17	11.9	2.08	0.26	B38
10W2065	BW53	377	8.0	6	0	250	121	96	0	97	100	0.09	2.9	1.54	0.26	BW53
10W2065	BW58	371	8.0	16	0	234	121	96	0	93	100	0.15	8.1	1.85	0.26	BW58
10W2065	BW52	355	8.0	14	0	220	121	96	0	94	100	0.15	7.5	1.79	0.25	BW52
10W2065	BW77	341	8.0	13	0	207	121	96	0	94	100	0.15	7.4	1.76	0.25	BW77
10W2065	B51	328	8.0	29	1	178	120	96	3	85	97	0.22	16.0	1.70	0.21	B51
10W2065	B13	298	8.0	26	1	152	119	95	3	85	97	0.24	16.5	1.63	0.21	B13
10W2065	B44	271	8.0	44	2	108	117	94	4	71	96	0.36	35.2	1.65	0.19	B44
10W2065	B27	225	8.0	16	1	92	116	93	5	85	95	0.26	15.7	1.36	0.19	B27
10W2065	BW57	208	8.0	13	1	79	115	92	7	85	93	0.26	14.4	1.26	0.18	BW57
10W2065	A24	194	7.9	63	8	16	107	91	11	20	89	0.74	106.9	0.89	0.10	A24
10W2065	A23	123	8.0	2	0	14	107	81	0	87	100	0.33	13.6	0.51	0.08	A23
10W2065	BW4	121	7.6	9	0	5	107	78	0	35	100	0.78	74.3	3.87	0.72	BW4

10W2097	TRAY: 16	POS: 11			LOCAL SPECIFICITY: B44,B45,B51,BW52											
10W2097	BW76	395	8.0	18	0	64	313	71	0	78	100	0.43	72.0	4.98	0.65	BW76
10W2097	B45	377	6.0	4	2	60	311	64	33	93	67	0.17	10.7	2.37	0.38	B45
10W2097	BW77	371	8.0	6	7	54	304	66	53	90	47	0.16	8.9	1.25	0.17	BW77
10W2097	B44	358	6.6	22	24	32	280	62	52	59	48	0.35	44.2	1.17	0.13	B44
10W2097	BW52	312	7.0	4	10	28	270	65	71	87	29	0.13	5.3	0.78	0.11	BW52
10W2097	B51	298	7.3	8	21	20	249	67	72	71	28	0.20	12.5	0.48	0.06	B51
10W2101	TRAY: 16	POS: 21			LOCAL SPECIFICITY: BW4											
10W2101	BW77	396	7.0	12	1	64	319	68	7	84	93	0.34	46.3	4.62	0.64	BW77
10W2101	BW63	383	7.3	8	1	56	318	68	11	87	89	0.30	34.5	4.22	0.63	BW63
10W2101	B15.3	374	7.5	4	4	52	314	67	50	92	50	0.15	7.9	3.31	0.50	B15.3
10W2101	BW57	366	7.4	7	8	45	306	67	53	86	47	0.19	13.5	2.12	0.29	BW57
10W2101	B13	351	7.7	15	13	30	293	64	46	66	54	0.36	45.2	2.34	0.28	B13
10W2101	B37	323	7.6	5	7	25	286	50	58	83	42	0.22	15.5	1.73	0.24	B37
10W2101	B27	311	6.3	8	14	17	272	44	63	68	37	0.29	25.7	1.06	0.14	B27
10W2101	B51	289	6.8	8	19	9	253	47	70	52	30	0.32	30.3	0.93	0.11	B51
10W2101	BW4	262	5.2	5	89	4	164	33	94	44	6	0.08	1.6	0.96	0.09	BW4
10W2102	TRAY: 16	POS: 12			LOCAL SPECIFICITY: BW4											
10W2102	B44	395	8.0	51	0	178	166	96	0	77	100	0.33	42.5	5.47	0.69	B44
10W2102	B27	344	8.0	25	0	153	166	95	0	85	100	0.27	25.1	4.70	0.67	B27
10W2102	BW58	319	8.0	16	0	137	166	94	0	89	100	0.24	18.3	4.23	0.66	BW58
10W2102	BW57	303	8.0	15	0	122	166	94	0	89	100	0.25	19.1	4.16	0.66	BW57
10W2102	BW52	288	8.0	13	0	109	166	93	0	89	100	0.25	18.5	4.01	0.66	BW52
10W2102	BW77	275	8.0	10	0	99	166	92	0	90	100	0.24	15.8	3.75	0.65	BW77
10W2102	B37	265	8.0	10	0	89	166	91	0	89	100	0.26	17.4	3.75	0.65	B37
10W2102	BW63	255	8.0	5	0	84	166	91	0	94	100	0.19	9.5	3.17	0.64	BW63
10W2102	BW53	250	8.0	5	0	79	166	90	0	94	100	0.20	10.1	3.17	0.64	BW53
10W2102	B51	245	8.0	24	2	55	164	89	7	69	93	0.44	48.0	3.93	0.56	B51
10W2102	B38	219	7.9	14	1	41	163	85	6	74	94	0.43	39.8	2.83	0.44	B38
10W2102	B13	204	7.9	18	3	23	160	82	14	56	86	0.55	62.8	2.77	0.41	B13
10W2102	BW4	183	8.0	10	6	13	154	73	37	56	63	0.47	39.8	4.93	0.48	BW4
10W2103	TRAY: 16	POS: 22			LOCAL SPECIFICITY: BW6											
10W2103	B7	395	7.7	31	3	237	124	81	8	88	92	0.15	9.3	1.34	0.17	B7
10W2103	B8	361	8.0	18	0	219	124	80	0	92	100	0.17	9.9	1.15	0.16	B8
10W2103	BW76	343	7.8	17	1	202	123	78	5	92	95	0.15	7.7	0.72	0.10	BW76
10W2103	B35	325	7.5	34	5	168	118	77	12	83	88	0.19	11.8	0.74	0.09	B35
10W2103	BW54	286	8.0	9	2	159	116	77	18	94	82	0.09	2.5	0.49	0.07	BW54
10W2103	CW8	275	7.5	44	7	115	109	76	13	72	87	0.27	20.8	0.63	0.07	CW8
10W2103	AW34	224	6.7	9	2	106	107	73	18	92	82	0.14	4.3	0.49	0.07	AW34
10W2103	AW33	213	7.0	21	4	85	103	75	16	80	84	0.25	13.3	0.53	0.06	AW33
10W2103	CW1	188	7.6	34	15	51	88	76	30	60	70	0.29	15.6	0.48	0.05	CW1
10W2105	TRAY: 16	POS: 24			LOCAL SPECIFICITY: BW6											
10W2105	BW70	393	6.4	20	3	219	151	71	13	91	87	0.13	7.0	2.00	0.30	BW70
10W2105	BW76	370	7.8	18	0	201	151	75	0	91	100	0.19	13.0	1.83	0.28	BW76
10W2105	BW77	352	7.8	12	1	189	150	73	7	94	93	0.14	6.8	1.42	0.23	BW77
10W2105	B38	339	6.6	13	7	176	143	72	35	93	65	0.05	0.7	0.78	0.13	B38
10W2105	B8	319	6.2	9	5	167	138	73	35	94	65	0.04	0.5	0.67	0.12	B8
10W2105	CW8	305	7.7	38	7	129	131	76	15	77	85	0.25	18.8	1.05	0.12	CW8
10W2105	BW61	260	7.5	8	3	121	128	72	27	93	73	0.10	2.5	0.52	0.08	BW61
10W2105	BW46	249	7.6	36	10	85	118	71	21	70	79	0.28	19.9	0.59	0.07	BW46
10W2105	B37	203	7.7	6	2	79	116	68	25	92	75	0.14	3.8	0.44	0.07	B37
10W2105	A31	195	7.6	10	1	69	115	67	9	87	91	0.25	12.3	0.46	0.06	A31

Table 5. *Continued*

10W2106	TRAY:	16	POS: 14			LOCAL SPECIFICITY: BW6											
10W2106	B18		396	7.9	26	0	280	90	94	0	91	100	0.14	8.2	3.28	0.46	B18
10W2106	BW76		370	8.0	18	0	262	90	94	0	93	100	0.13	6.1	2.98	0.45	BW76
10W2106	BW70		352	8.0	21	1	241	89	94	4	91	96	0.12	5.4	3.10	0.45	BW70
10W2106	BW67		330	8.0	5	0	236	89	93	0	97	100	0.08	1.9	2.23	0.45	BW67
10W2106	B8		325	8.0	14	0	222	89	93	0	94	100	0.13	5.5	2.80	0.45	B8
10W2106	BW56		311	8.0	9	0	213	89	93	0	95	100	0.11	3.7	2.51	0.44	BW56
10W2106	B35		302	7.7	35	1	178	88	92	2	83	98	0.22	14.0	3.02	0.40	B35
10W2106	BW60		266	8.0	39	1	139	87	94	2	78	98	0.27	19.9	3.02	0.40	BW60
10W2106	B7		226	7.8	30	1	109	86	92	3	78	97	0.29	18.9	2.79	0.38	B7
10W2106	BW61		195	7.9	14	1	95	85	92	6	87	94	0.22	9.2	2.20	0.35	BW61
10W2106	BW62		180	7.6	51	6	44	79	92	10	46	90	0.50	45.1	2.11	0.26	BW62
10W2106	BW75		123	7.9	25	4	19	75	97	13	43	87	0.58	42.0	1.77	0.25	BW75
10W2106	BW6		94	8.0	17	11	2	64	100	39	10	61	0.66	40.6	3.81	0.39	BW6
10W2107	TRAY:	16	POS: 15			LOCAL SPECIFICITY: BW6											
10W2107	BW56		395	7.8	9	0	325	61	95	0	97	100	0.07	1.7	1.03	0.18	BW56
10W2107	B7		386	8.0	34	0	291	61	96	0	89	100	0.13	7.0	1.28	0.17	B7
10W2107	BW70		352	8.0	24	0	267	61	95	0	91	100	0.12	5.4	1.14	0.16	BW70
10W2107	B18		328	8.0	22	0	245	61	95	0	91	100	0.13	5.4	1.11	0.16	B18
10W2107	BW76		306	8.0	18	0	227	61	94	0	92	100	0.12	4.8	1.04	0.15	BW76
10W2107	BW54		288	8.0	11	0	216	61	94	0	95	100	0.10	3.1	0.93	0.15	BW54
10W2107	B39		277	8.0	14	0	202	61	93	0	93	100	0.12	4.2	0.98	0.15	B39
10W2107	B8		263	8.0	14	0	188	61	93	0	93	100	0.13	4.5	0.98	0.15	B8
10W2107	BW60		249	7.9	36	1	152	60	93	2	80	98	0.21	11.2	0.98	0.13	BW60
10W2107	BW75		212	7.9	32	1	120	59	92	3	78	97	0.24	12.3	0.94	0.12	BW75
10W2107	B35		179	7.9	29	2	91	57	92	6	75	94	0.26	11.9	0.71	0.09	B35
10W2107	CW2		148	7.7	7	0	84	57	92	0	92	100	0.18	4.6	0.53	0.07	CW2
10W2107	BW6		141	7.8	70	6	14	51	92	7	16	93	0.72	72.4	2.86	0.30	BW6

Table 6. Seratyping of Antigen Society Sera Defining B15 and Its Subgroups

SERUM	ANTIGEN	NO. REAC	AVE	++	MISS +-	EXTR -+	--	STR	PERCENTAGE MISS	PERCENTAGE EXTR	PERCENTAGE INCL	R	CHI	QSCORE	QNORM	ANTIGEN
	TRAY: 280	POS: 3			LOCAL SPECIFICITY: B17,BW63											
10W265	BW57	680	7.5	22	8	64	586	67	26	74	74	0.39	104.6	3.16	0.35	BW57
10W265	BW58	650	7.4	36	18	28	568	60	33	43	67	0.57	214.2	2.76	0.28	BW58
10W265	BW63	596	5.7	6	22	22	546	39	78	78	22	0.18	18.4	0.50	0.06	BW63
	TRAY: 280	POS: 4			LOCAL SPECIFICITY: B15											
10W4486	BW77	684	8.0	13	0	319	352	73	0	96	100	0.14	14.1	5.69	0.78	BW77
10W4486	BW75	671	7.8	61	1	258	351	72	1	80	99	0.32	70.8	6.72	0.72	BW75
10W4486	BW76	609	5.7	14	0	244	351	67	0	94	100	0.18	19.5	5.20	0.70	BW76
10W4486	BW62	595	7.6	157	8	87	343	70	4	35	96	0.68	276.7	6.23	0.59	BW62
10W4486	B15.3	430	6.6	10	1	77	342	48	9	88	91	0.29	34.9	3.99	0.56	B15.3
10W4486	BW63	419	7.4	22	6	55	336	48	21	71	79	0.42	72.5	3.77	0.47	BW63
10W4486	BW57	391	6.7	14	6	41	330	38	30	74	70	0.37	54.6	2.64	0.33	BW57
10W4486	BW46	371	5.7	29	50	12	280	31	63	29	37	0.43	67.2	1.14	0.12	BW46
	TRAY: 280	POS: 5			LOCAL SPECIFICITY: B15											
10W4487	BW76	682	8.0	18	0	15	649	75	0	45	100	0.73	363.6	6.65	0.77	BW76
	TRAY: 280	POS: 6			LOCAL SPECIFICITY: BW62,BW63											
10W4488	BW75	681	7.8	62	0	218	401	81	0	77	100	0.38	97.7	8.07	0.83	BW75
10W4488	BW77	619	8.0	13	0	205	401	78	0	94	100	0.20	24.4	6.32	0.83	BW77
10W4488	B15.3	606	7.4	13	0	192	401	77	0	93	100	0.21	26.0	5.92	0.77	B15.3
10W4488	BW62	593	7.7	151	11	41	390	78	6	21	94	0.80	376.7	6.11	0.56	BW62
10W4488	BW63	431	7.3	24	4	17	386	43	14	41	86	0.68	202.0	4.50	0.52	BW63
10W4488	A31	399	4.9	7	25	9	358	18	78	56	22	0.27	28.8	0.59	0.06	A31
	TRAY: 280	POS: 7			LOCAL SPECIFICITY: BW62											
10W4489	BW76	684	8.0	18	0	263	403	83	0	93	100	0.20	26.5	6.51	0.84	BW76
10W4489	BW77	666	8.0	13	0	250	403	82	0	95	100	0.17	20.3	6.08	0.83	BW77
10W4489	B15.3	653	7.3	9	3	241	400	82	25	96	75	0.10	7.0	5.06	0.78	B15.3
10W4489	BW75	641	7.8	52	6	189	394	82	10	78	90	0.34	73.7	5.79	0.63	BW75
10W4489	BW62	583	7.7	146	17	43	377	78	10	22	90	0.76	337.3	5.79	0.55	BW62
10W4489	BW63	420	6.7	9	19	34	358	41	67	79	33	0.19	15.7	1.01	0.12	BW63
10W4489	BW57	392	6.3	6	15	28	343	41	71	82	29	0.17	11.1	0.72	0.09	BW57
10W4489	BW46	371	5.9	22	57	6	286	39	72	21	28	0.40	59.3	0.70	0.07	BW46
	TRAY: 280	POS: 8			LOCAL SPECIFICITY: BW62											
10W4490	BW77	683	8.0	13	0	226	444	84	0	94	100	0.19	24.6	6.15	0.83	BW77
10W4490	BW76	670	7.8	18	0	208	444	83	0	92	100	0.23	36.3	6.46	0.82	BW76
10W4490	BW75	652	7.8	50	8	158	436	83	13	75	87	0.36	86.4	5.19	0.56	BW75
10W4490	BW62	594	7.6	138	27	20	409	79	16	12	84	0.80	380.7	4.89	0.46	BW62
10W4490	B15.3	429	6.7	6	5	14	404	45	45	70	55	0.38	63.2	2.60	0.41	B15.3
10W4490	BW63	418	6.4	5	23	9	381	42	82	64	18	0.22	19.5	0.67	0.08	BW63
10W4491	TRAY: 280	POS: 9			LOCAL SPECIFICITY: BW62											
10W4491	BW77	653	8.0	13	0	221	419	75	0	94	100	0.19	23.8	6.56	0.89	BW77
10W4491	BW76	640	6.6	18	0	203	419	73	0	91	100	0.23	35.1	6.15	0.79	BW76
10W4491	BW75	622	7.9	50	6	153	413	75	10	75	90	0.38	89.8	6.10	0.66	BW75
10W4491	BW62	566	7.4	129	29	24	384	69	18	15	82	0.77	331.4	4.61	0.44	BW62
10W4491	B15.3	408	6.7	6	5	18	379	33	45	75	55	0.34	48.4	2.57	0.36	B15.3
10W4491	BW63	397	6.0	4	17	14	362	22	80	77	20	0.16	10.8	0.71	0.09	BW63
10W4491	CW6	303	6.0	4	73	6	220	20	94	60	6	0.06	1.2	0.56	0.06	CW6
10W4491	BW6	299	5.8	8	215	2	74	20	96	20	4	0.02	0.2	0.76	0.10	BW6

Table 6. *Continued*

	TRAY: 280	POS: 10			LOCAL SPECIFICITY: BW62											
10W4492	BW77	684	8.0	13	0	240	431	80	0	94	100	0.18	22.6	6.84	0.90	BW77
10W4492	BW75	671	7.8	60	1	180	430	79	1	75	99	0.41	114.4	8.63	0.89	BW75
10W4492	BW76	610	4.9	11	3	169	427	74	21	93	79	0.16	16.6	6.13	0.79	BW76
10W4492	BW62	596	7.5	152	13	17	414	78	7	10	93	0.88	456.7	6.69	0.62	BW62
10W4492	B15.3	431	7.0	6	5	11	409	47	45	64	55	0.42	76.3	2.81	0.43	B15.3
10W4492	BW63	420	6.5	8	20	3	389	36	71	27	29	0.43	79.2	1.45	0.17	BW63
	TRAY: 280	POS: 11			LOCAL SPECIFICITY: BW62											
10W4493	BW76	658	7.7	18	0	161	479	82	0	89	100	0.27	49.5	7.68	0.96	BW76
10W4493	BW62	640	7.6	133	34	28	445	81	20	17	80	0.75	356.3	5.41	0.50	BW62
10W4493	BW75	473	7.5	19	36	9	409	60	65	32	35	0.44	91.6	1.56	0.17	BW75
10W4493	BW60	414	5.6	5	32	4	373	22	86	44	14	0.24	24.6	0.60	0.07	BW60
10W4493	BW6	381	6.0	4	239	0	138	25	98	0	2	0.08	2.3	0.88	0.08	BW6
	TRAY: 280	POS: 12			LOCAL SPECIFICITY: BW62,B15.3											
10W4494	BW76	683	8.0	18	0	271	394	85	0	93	100	0.19	25.2	5.81	0.75	BW76
10W4494	BW75	665	8.0	55	3	216	391	84	5	79	95	0.34	77.0	5.61	0.60	BW75
10W4494	BW77	607	7.7	12	1	204	390	80	7	94	93	0.18	18.6	4.29	0.59	BW77
10W4494	BW62	594	7.8	155	10	49	380	79	6	24	94	0.78	359.9	5.53	0.53	BW62
10W4494	B15.3	429	7.1	7	4	42	376	38	36	85	64	0.27	30.4	2.58	0.41	B15.3
10W4494	B27	414	6.4	5	16	36	357	31	76	87	24	0.11	4.8	0.43	0.06	B27
10W4494	B13	393	5.6	5	29	31	328	30	85	86	15	0.06	1.4	0.47	0.05	B13
10W4494	BW6	363	5.9	26	231	6	100	31	89	18	11	0.07	1.9	1.45	0.20	BW6
	TRAY: 280	POS: 13			LOCAL SPECIFICITY: BW62,B15.3											
10W4495	BW77	686	7.0	12	1	170	503	67	7	93	93	0.21	29.4	5.76	0.74	BW77
10W4495	BW75	673	7.4	49	13	121	490	67	20	71	80	0.39	104.6	5.24	0.54	BW75
10W4495	BW62	611	7.0	113	54	8	436	61	32	6	68	0.74	331.5	4.02	0.37	BW62
	TRAY: 280	POS: 14			LOCAL SPECIFICITY: B51											
10W4496	B51	682	7.6	38	13	8	623	76	25	17	75	0.77	402.4	4.66	0.47	B51
10W4496	BW6	631	5.8	8	503	0	120	37	98	0	2	0.05	1.9	0.64	0.09	BW6
	TRAY: 280	POS: 15			LOCAL SPECIFICITY: BW62,BW63,BW66											
10W4497	BW77	685	8.0	12	1	139	533	70	7	92	93	0.24	38.1	6.42	0.84	BW77
10W4497	BW76	672	7.2	18	0	121	533	67	0	87	100	0.32	70.9	6.58	0.81	BW76
10W4497	BW75	654	7.3	46	12	75	521	66	20	61	80	0.49	156.1	4.44	0.47	BW75
10W4497	BW62	596	6.8	62	104	13	417	57	62	17	38	0.46	128.3	1.39	0.13	BW62
10W4497	BW63	430	5.5	4	23	9	394	30	85	69	15	0.18	13.7	0.56	0.06	BW63
	TRAY: 280	POS: 16			LOCAL SPECIFICITY: BW62,B13											
10W4498	BW76	685	8.0	18	0	194	473	68	0	91	100	0.25	41.2	6.95	0.90	BW76
10W4498	BW77	667	7.8	13	0	181	473	65	0	93	100	0.22	32.3	6.43	0.88	BW77
10W4498	BW75	654	7.7	46	12	135	461	63	20	74	80	0.36	84.8	4.66	0.51	BW75
10W4498	B13	586	7.0	23	19	109	435	56	45	82	55	0.21	26.9	2.36	0.27	B13
10W4498	BW62	554	6.6	90	69	22	373	52	43	19	57	0.57	183.1	2.49	0.24	BW62
10W4498	BW57	395	6.7	9	14	13	359	45	60	59	40	0.36	52.3	1.62	0.20	BW57
10W4498	BW63	372	6.0	5	19	8	340	30	79	61	21	0.25	22.9	0.76	0.10	BW63

	TRAY: 280	POS: 17			LOCAL	SPECIFICITY: BW62,B35,B50,BW77										
10W4499	BW75	666	7.9	61	0	287	318	75	0	82	100	0.30	61.4	7.14	0.75	BW75
10W4499	BW77	605	7.7	13	0	274	318	71	0	95	100	0.16	14.7	5.03	0.68	BW77
10W4499	BW70	592	7.1	48	2	226	316	70	4	82	96	0.30	54.3	6.07	0.67	BW70
10W4499	B15.3	542	6.7	11	1	215	315	73	8	95	92	0.15	12.6	4.34	0.59	B15.3
10W4499	BW62	530	7.6	148	11	67	304	73	6	31	94	0.70	259.8	5.74	0.54	BW62
10W4499	BW56	367	7.3	8	1	57	301	46	11	87	89	0.30	32.1	3.40	0.49	BW56
10W4499	B35	362	7.1	33	7	26	296	42	17	44	83	0.63	144.5	3.56	0.41	B35
	TRAY: 280	POS: 18			LOCAL	SPECIFICITY: BW62,CW4,CW7										
10W4500	BW76	679	8.0	18	0	348	313	84	0	95	100	0.15	15.8	3.89	0.51	BW76
10W4500	B15.3	661	7.1	14	0	334	313	83	0	95	100	0.14	12.9	3.47	0.48	B15.3
10W4500	BW75	647	7.8	56	1	278	312	84	1	83	99	0.29	54.4	4.23	0.47	BW75
10W4500	BW62	590	7.9	159	5	119	307	82	3	42	97	0.62	226.4	4.68	0.45	BW62
10W4500	BW77	426	7.6	10	1	109	306	67	9	91	91	0.23	22.2	2.55	0.37	BW77
10W4500	B35	415	7.6	36	6	73	300	65	14	66	86	0.45	85.3	2.39	0.27	B35
10W4500	CW4	371	7.7	26	3	47	295	54	10	64	90	0.51	97.5	2.26	0.22	CW4
10W4500	BW70	344	7.4	14	31	33	266	38	68	70	32	0.20	13.4	0.53	0.06	BW70
10W4500	BW46	299	4.6	16	61	17	205	21	79	51	21	0.18	10.0	0.53	0.06	BW46
10W4500	BW6	222	5.8	13	114	4	91	41	89	23	11	0.11	2.8	0.85	0.08	BW6
	TRAY: 280	POS: 19			LOCAL	SPECIFICITY: BW4,BW62										
10W4501	BW62	680	7.9	166	3	380	131	85	1	69	99	0.26	45.7	3.23	0.33	BW62
10W4501	B38	511	7.9	26	0	354	131	81	0	93	100	0.14	9.4	2.31	0.31	B38
10W4501	BW76	485	8.0	15	0	339	131	80	0	95	100	0.11	5.7	1.97	0.29	BW76
10W4501	BW58	470	8.0	52	1	287	130	79	1	84	99	0.21	20.1	2.44	0.29	BW58
10W4501	BW77	417	8.0	12	0	275	130	75	0	95	100	0.12	5.6	1.80	0.28	BW77
10W4501	BW52	405	7.6	18	0	257	130	74	0	93	100	0.15	8.9	1.88	0.27	BW52
10W4501	BW57	387	8.0	21	1	236	129	74	4	91	96	0.15	8.8	1.75	0.24	BW57
10W4501	BW75	365	7.9	40	2	196	127	72	4	83	96	0.23	19.4	1.86	0.23	BW75
10W4501	BW63	323	7.8	18	1	178	126	67	5	90	95	0.17	9.8	1.56	0.22	BW63
10W4501	B51	304	7.6	24	1	154	125	64	4	86	96	0.23	15.7	1.59	0.21	B51
10W4501	B15.3	279	7.7	7	1	147	124	61	12	95	88	0.11	3.5	0.90	0.15	B15.3
10W4501	B27	270	6.9	13	2	133	122	60	13	91	87	0.16	6.8	0.91	0.13	B27
10W4501	B44	256	7.2	38	13	96	109	60	25	71	75	0.22	12.5	1.04	0.13	B44
10W4501	AW34	145	8.0	7	0	58	80	56	0	89	100	0.25	9.1	0.63	0.09	AW34
10W4501	A10	138	8.0	9	1	49	79	51	10	84	90	0.27	10.2	0.60	0.07	A10
10W4501	AW33	174	6.5	8	3	64	99	45	27	88	73	0.17	4.8	0.44	0.05	AW33
10W4501	BW4	177	6.8	28	17	44	88	47	37	61	63	0.26	11.6	2.78	0.40	BW4
	TRAY: 280	POS: 20			LOCAL	SPECIFICITY: B49,BW51,BW52,BW63,8W66										
10W4502	BW52	678	7.4	23	2	81	572	75	8	77	92	0.42	117.5	6.30	0.73	BW52
10W4502	B51	653	7.9	47	4	34	568	75	7	41	93	0.70	323.8	5.72	0.58	B51
10W4502	BW77	602	7.1	7	2	27	566	50	22	79	78	0.38	89.2	3.59	0.48	BW77
10W4502	BW63	593	7.1	9	12	18	554	48	57	66	43	0.35	73.5	2.03	0.23	BW63
10W4502	BW4	572	6.4	14	249	4	305	44	94	22	6	0.12	7.6	0.97	0.08	BW4
	TRAY: 280	POS: 21			LOCAL	SPECIFICITY: B17,BW63										
10W4504	BW57	686	7.9	29	2	80	575	84	6	73	94	0.46	146.5	6.16	0.68	BW57
10W4504	BW58	655	8.0	52	2	28	573	80	3	35	97	0.77	388.1	6.32	0.63	BW58
10W4504	BW63	601	6.8	10	18	18	555	42	64	64	36	0.33	63.8	1.06	0.12	BW63
10W4504	AW33	551	5.3	6	42	12	491	33	87	66	13	0.16	14.2	0.50	0.05	AW33
10W4504	BW4	525	5.8	8	223	4	290	41	96	33	4	0.07	2.6	0.83	0.07	BW4

Table 6. *Continued*

10W4505	TRAY: 280	POS: 22			LOCAL	SPECIFICITY: B17,BW63,B49,BN										
10W4505	BW77	685	7.8	9	4	87	585	64	30	90	70	0.22	33.5	3.88	0.49	BW77
10W4505	BW57	672	7.0	25	6	62	579	62	19	71	81	0.44	132.2	4.28	0.47	BW57
10W4505	BW58	641	7.2	38	15	24	564	59	28	38	72	0.63	254.4	3.61	0.37	BW58
10W4505	BW63	588	6.8	8	20	16	544	37	71	66	29	0.28	45.0	1.20	0.13	BW63
10W4505	B51	560	4.8	8	33	8	511	31	80	50	20	0.28	44.2	0.57	0.06	B51
10W4505	BW4	519	7.0	6	204	2	307	50	97	25	3	0.09	4.0	0.87	0.08	BW4
	TRAY: 280	POS: 23			LOCAL	SPECIFICITY: B5,B49,BW63										
10W4506	BW52	681	7.6	25	0	126	530	80	0	83	100	0.37	91.1	6.14	0.70	BW52
10W4506	BW77	656	8.0	13	0	113	530	80	0	89	100	0.29	55.8	5.50	0.70	BW77
10W4506	BW53	643	7.7	13	1	100	529	78	7	88	93	0.30	56.0	5.48	0.69	BW53
10W4506	B51	629	8.0	45	1	55	528	78	2	55	98	0.63	249.1	6.14	0.64	B51
10W4506	BW63	583	7.8	17	5	38	523	61	22	69	78	0.46	123.1	3.63	0.42	BW63
10W4506	BW57	561	6.0	7	20	31	503	50	74	81	26	0.17	16.5	0.76	0.09	BW57
10W4506	BW4	534	6.8	25	201	6	302	54	88	19	12	0.19	19.8	0.62	0.09	BW4
	TRAY: 280	POS: 24			LOCAL	SPECIFICITY: B15,BW46,CW3,B17										
10W4507	BW77	680	8.0	13	0	402	265	91	0	96	100	0.11	8.5	4.55	0.65	BW77
10W4507	B15.3	667	7.8	13	0	389	265	90	0	96	100	0.11	8.7	4.52	0.64	B15.3
10W4507	BW76	654	7.4	18	0	371	265	90	0	95	100	0.14	12.6	4.78	0.64	BW76
10W4507	BW46	636	7.9	89	3	282	262	91	3	76	97	0.32	65.3	5.89	0.62	BW46
10W4507	BW62	544	8.0	155	2	127	260	90	1	45	99	0.60	194.3	6.13	0.60	BW62
10W4507	BW75	387	8.0	50	1	77	259	80	1	60	99	0.54	113.3	5.29	0.60	BW75
10W4507	BW63	336	7.3	25	2	52	257	68	7	67	93	0.49	80.7	3.79	0.47	BW63
10W4507	B38	309	6.5	8	8	44	249	65	50	84	50	0.21	13.3	1.15	0.16	B38
10W4507	BW61	293	7.4	7	9	37	240	68	56	84	44	0.19	10.9	1.08	0.16	BW61
10W4507	A31	273	7.5	4	5	32	232	63	55	88	45	0.17	7.9	0.91	0.10	A31
10W4507	B51	268	7.3	8	12	25	223	63	60	75	40	0.24	15.3	0.63	0.09	B51
10W4507	B35	248	6.8	8	21	17	202	60	72	68	28	0.21	11.1	0.58	0.07	B35
	TRAY: 280	POS: 25			LOCAL	SPECIFICITY: B15,BW57,BW46										
10W4508	BW76	677	8.0	18	0	365	294	92	0	95	100	0.14	14.2	6.25	0.82	BW76
10W4508	BW77	659	8.0	13	0	352	294	91	0	96	100	0.13	10.7	5.86	0.82	BW77
10W4508	B15.3	646	7.7	13	0	339	294	91	0	96	100	0.13	11.1	5.77	0.81	B15.3
10W4508	BW62	633	8.0	164	1	175	293	91	0	51	100	0.55	188.5	8.06	0.78	BW62
10W4508	BW75	468	7.9	55	1	120	292	84	1	68	99	0.46	100.5	6.76	0.75	BW75
10W4508	BW63	412	7.7	27	1	93	291	78	3	77	97	0.40	65.9	5.75	0.70	BW63
10W4508	BW57	384	8.0	19	2	74	289	75	9	79	91	0.37	53.1	4.88	0.63	BW57
10W4508	BW46	363	7.4	62	16	12	273	68	20	16	80	0.77	213.8	3.96	0.42	BW46
	TRAY: 280	POS: 26			LOCAL	SPECIFICITY: BW46,CW1,CW3										
10W4509	CW1	670	7.8	95	20	184	371	74	17	65	83	0.38	95.9	4.10	0.40	CW1
10W4509	CW3	555	7.3	159	52	25	319	65	24	13	76	0.70	273.6	3.34	0.30	CW3
10W4509	BW46	354	7.6	5	2	27	320	37	28	84	72	0.31	33.8	2.96	0.29	BW46
10W4509	BW62	347	5.8	16	33	11	287	29	67	40	33	0.38	49.2	0.98	0.09	BW62
10W4509	BW4	298	5.5	8	182	3	105	18	95	27	5	0.04	0.4	1.07	0.16	BW4
	TRAY: 280	POS: 27			LOCAL	SPECIFICITY: B15,BW22,CW3										
10W4510	CW3	675	7.7	225	37	24	389	85	14	9	86	0.81	441.4	4.24	0.37	CW3
10W4510	BW62	423	7.3	12	49	20	342	62	80	62	20	0.19	14.9	0.73	0.07	BW62
10W4510	AW33	344	6.3	7	21	13	303	55	75	65	25	0.24	20.5	0.52	0.05	AW33

TRAY: 280 POS: 28 LOCAL SPECIFICITY: B5,B35,BW53,B18,B15.3,BW62,BW70

10W4511	BW52	664	8.0	21	0	398	245	89	0	94	100	0.14	12.7	5.54	0.75	BW52
10W4511	B15.3	643	8.0	13	0	385	245	89	0	96	100	0.11	8.2	4.96	0.74	B15.3
10W4511	BW53	630	8.0	13	0	372	245	89	0	96	100	0.12	8.4	4.96	0.74	BW53
10W4511	BW75	617	8.0	59	1	313	244	88	1	84	99	0.26	40.2	6.05	0.69	BW75
10W4511	B35	557	8.0	50	1	263	243	86	1	84	99	0.27	39.9	5.85	0.69	B35
10W4511	B51	506	7.9	45	1	218	242	84	2	82	98	0.29	42.6	5.65	0.67	B51
10W4511	B18	460	7.5	32	1	186	241	81	3	85	97	0.28	35.0	4.93	0.62	B18
10W4511	BW62	427	7.8	119	6	67	235	81	4	36	96	0.67	191.7	5.26	0.55	BW62
10W4511	BW77	302	8.0	4	1	63	234	59	20	94	80	0.18	9.8	2.82	0.51	BW77
10W4511	BW70	297	7.2	33	14	30	220	57	29	47	71	0.52	80.2	2.31	0.28	BW70
10W4511	B38	250	7.3	8	9	22	211	50	52	73	48	0.29	21.2	1.78	0.27	B38

TRAY: 280 POS: 29 LOCAL SPECIFICITY: B15,B35,B18,B5,4C

10W4512	B51	683	8.0	51	0	338	294	78	0	86	100	0.25	41.7	5.05	0.57	B51
10W4512	B35	632	7.9	60	0	278	294	74	0	82	100	0.30	57.7	5.02	0.55	B35
10W4512	BW52	572	8.0	24	0	254	294	70	0	91	100	0.22	26.5	4.22	0.54	BW52
10W4512	BW53	548	8.0	10	0	244	294	68	0	96	100	0.15	11.8	3.28	0.49	BW53
10W4512	BW76	538	7.1	15	0	229	294	66	0	93	100	0.19	18.6	3.46	0.48	BW76
10W4512	B37	516	7.2	12	0	213	291	66	0	94	100	0.18	15.9	3.26	0.47	B37
10W4512	B18	511	7.9	36	1	181	293	65	2	83	98	0.31	49.1	3.93	0.47	B18
10W4512	B15.3	474	6.9	9	2	172	291	59	18	95	82	0.14	9.1	2.75	0.46	B15.3
10W4512	BW75	463	7.8	40	7	132	284	59	14	76	86	0.33	51.5	3.13	0.36	BW75
10W4512	B21	410	6.4	5	1	124	280	50	16	96	84	0.14	7.6	1.89	0.31	B21
10W4512	B38	410	7.1	13	2	114	281	50	13	89	87	0.23	22.6	2.17	0.30	B38
10W4512	BW41	389	6.8	5	2	106	276	49	28	95	72	0.13	6.4	1.42	0.23	BW41
10W4512	B13	382	7.1	23	12	83	264	49	34	78	66	0.27	27.7	1.35	0.17	B13
10W4512	CW8	347	7.6	10	3	73	261	42	23	87	77	0.25	20.9	1.58	0.17	CW8
10W4512	B39	334	5.8	8	9	65	252	35	52	89	48	0.14	6.7	0.83	0.12	B39
10W4512	CW4	317	6.8	17	18	48	234	36	51	73	49	0.24	19.0	0.55	0.06	CW4
10W4512	BW6	288	5.8	46	185	5	52	29	80	9	20	0.12	3.9	0.59	0.09	BW6

TRAY: 280 POS: 30 LOCAL SPECIFICITY: B15,B35,4C

10W4513	BW76	680	8.0	18	0	400	262	75	0	95	100	0.13	11.6	5.57	0.79	BW76
10W4513	B21	653	7.7	6	0	387	260	74	0	98	100	0.08	4.0	4.33	0.77	B21
10W4513	BW62	656	7.7	163	4	231	258	74	2	58	98	0.45	131.6	6.66	0.68	BW62
10W4513	BW75	489	7.9	53	3	178	255	65	5	77	95	0.34	57.0	5.19	0.61	BW75
10W4513	BW77	433	7.5	11	1	167	254	56	8	93	92	0.17	13.0	3.91	0.60	BW77
10W4513	BW41	417	7.0	6	1	159	251	53	14	96	86	0.12	6.3	3.24	0.55	BW41
10W4513	B15.3	414	8.0	8	3	153	250	54	27	95	73	0.11	5.4	3.29	0.51	B15.3
10W4513	BW70	403	6.6	43	8	110	242	51	15	71	85	0.36	53.3	3.64	0.44	BW70
10W4513	B35	352	7.0	30	11	80	231	51	26	72	74	0.33	38.0	2.76	0.34	B35
10W4513	BW53	311	7.0	4	5	76	226	47	55	95	45	0.07	1.7	1.84	0.30	BW53
10W4513	B45	296	6.3	6	6	70	214	46	50	92	50	0.11	3.9	1.70	0.26	B45
10W4513	BW60	288	6.3	16	16	54	202	47	50	77	50	0.21	12.9	1.52	0.20	BW60
10W4513	B44	258	7.0	26	19	28	185	50	42	51	58	0.42	44.7	1.69	0.20	B44
10W4513	BW61	213	6.5	4	7	24	178	35	63	85	37	0.16	5.5	1.13	0.18	BW61
10W4513	B51	202	5.5	8	16	16	162	33	66	66	34	0.24	12.0	0.97	0.14	B51
10W4513	BW57	178	6.5	4	14	12	148	37	77	75	23	0.16	4.3	0.57	0.08	BW57
10W4513	BW6	160	5.4	10	84	2	64	33	89	16	11	0.14	3.2	1.35	0.19	BW6

Table 6. *Continued*

ID	Name															
	TRAY: 280	POS: 31			LOCAL SPECIFICITY: B15,B17											
10W4514	BW76	680	8.0	18	0	448	214	91	0	96	100	0.11	8.5	5.91	0.84	BW76
10W4514	B15.3	662	8.0	14	0	434	214	91	0	96	100	0.10	6.8	5.60	0.84	B15.3
10W4514	BW77	648	8.0	12	0	422	214	91	0	97	100	0.10	6.0	5.41	0.83	BW77
10W4514	BW62	636	8.0	163	1	259	213	90	0	61	100	0.41	108.0	7.78	0.80	BW62
10W4514	BW75	472	8.0	55	1	204	212	85	1	78	99	0.32	48.2	6.50	0.76	BW75
10W4514	BW57	416	8.0	22	1	182	211	81	4	89	96	0.23	21.2	5.32	0.72	BW57
10W4514	BW63	393	7.9	24	2	158	209	79	7	86	93	0.25	23.7	4.82	0.64	BW63
10W4514	BW46	367	7.5	68	8	90	201	76	10	56	90	0.48	84.2	4.99	0.57	BW46
10W4514	BW70	291	7.1	41	8	49	193	72	16	54	84	0.51	76.7	4.38	0.53	BW70
10W4514	B35	242	7.5	30	8	19	185	75	21	38	79	0.63	96.2	3.85	0.49	B35
10W4514	B51	204	7.3	8	14	11	171	68	63	57	37	0.32	21.4	1.07	0.15	B51
10W4514	BW4	182	6.8	10	109	1	62	63	91	9	9	0.14	3.4	1.32	0.20	BW4
	TRAY: 280	POS: 32			LOCAL SPECIFICITY: B17,B15											
10W4515	BW76	682	8.0	18	0	249	415	85	0	93	100	0.21	28.7	6.95	0.90	BW76
10W4515	BW77	664	8.0	13	0	236	415	84	0	94	100	0.18	22.1	6.50	0.89	BW77
10W4515	B15.3	651	7.3	11	2	225	413	83	15	95	85	0.14	13.4	5.04	0.69	B15.3
10W4515	BW62	638	7.8	141	24	84	389	83	14	37	86	0.62	245.6	5.46	0.52	BW62
10W4515	BW75	473	7.6	39	16	45	373	67	29	53	71	0.50	120.4	3.76	0.41	BW75
10W4515	BW63	418	6.5	12	16	33	357	51	57	73	43	0.28	32.2	2.00	0.25	BW63
10W4515	BW57	390	6.4	5	15	28	342	51	75	84	25	0.14	7.4	0.97	0.12	BW57
10W4515	BW46	370	6.6	21	57	7	285	53	73	25	27	0.38	52.9	0.81	0.08	BW46
10W4515	BW6	292	6.0	6	209	1	76	28	97	14	3	0.04	0.5	0.83	0.11	BW6
	TRAY: 280	POS: 33			LOCAL SPECIFICITY: BW62											
10W9228	BW77	527	8.0	13	0	246	268	76	0	94	100	0.16	13.8	5.56	0.84	BW77
10W9228	BW76	514	6.7	18	0	228	268	75	0	92	100	0.20	20.3	5.53	0.79	BW76
10W9228	B15.3	496	6.4	11	0	217	268	76	0	95	100	0.16	13.2	4.80	0.75	B15.3
10W9228	BW62	485	7.6	133	17	84	251	77	11	38	89	0.59	169.5	5.01	0.52	BW62
10W9228	BW75	335	7.8	36	12	48	239	64	25	57	75	0.47	74.3	3.37	0.41	BW75
10W9228	BW63	287	7.1	13	14	35	225	43	51	72	49	0.27	21.1	1.45	0.20	BW63
10W9228	BW57	260	6.8	5	9	30	216	37	64	85	36	0.16	6.3	1.11	0.17	BW57
10W9228	BW46	246	6.2	18	51	12	165	33	73	40	27	0.27	17.3	0.46	0.05	BW46
	TRAY: 280	POS: 34			LOCAL SPECIFICITY: BW62,BW62S,BW76,B45CR											
10W9229	BW76	521	8.0	18	0	180	323	80	0	90	100	0.24	30.4	6.37	0.82	BW76
10W9229	BW62	503	7.7	133	18	47	305	78	11	26	89	0.71	256.8	5.23	0.50	BW62
10W9229	BW75	352	6.7	34	14	13	291	48	29	27	71	0.67	158.7	2.91	0.32	BW75
10W9229	BW6	304	5.3	11	192	2	99	23	94	15	6	0.08	1.9	0.65	0.11	BW6
	TRAY: 281	POS: 3			LOCAL SPECIFICITY: B5											
10W9230	BW52	562	7.7	23	3	61	475	82	11	72	89	0.45	115.9	6.60	0.79	BW52
10W9230	B51	536	7.6	47	2	14	473	78	4	22	96	0.84	382.2	6.97	0.74	B51
10W9230	BW61	487	7.0	4	30	10	443	50	88	71	12	0.15	10.3	0.57	0.06	BW61
10W9230	BW6	453	6.7	9	370	1	73	50	97	10	3	0.03	0.3	0.99	0.16	BW6
	TRAY: 281	POS: 4			LOCAL SPECIFICITY: B5,B35											
10W9231	B51	562	7.8	45	4	45	468	63	8	50	92	0.64	229.5	7.09	0.76	B51
10W9231	BW52	513	6.6	19	7	26	461	33	26	57	74	0.53	141.5	4.12	0.50	BW52
10W9231	BW53	487	6.3	6	7	20	454	23	53	76	47	0.30	44.0	3.26	0.43	BW53
10W9231	B35	474	5.4	13	38	7	416	25	74	35	26	0.37	64.0	0.92	0.10	B35
10W9231	BW61	423	6.4	5	26	2	390	57	83	28	17	0.32	43.1	0.77	0.09	BW61

	TRAY: 281	POS: 5			LOCAL SPECIFICITY: B5											
10W9232	B51	561	7.6	41	7	8	505	77	14	16	86	0.83	387.2	5.22	0.54	B51
	TRAY: 281	POS: 6			LOCAL SPECIFICITY: B5											
10W9234	BW53	561	7.4	10	3	126	422	70	23	92	77	0.19	20.1	2.61	0.35	BW53
10W9234	B51	548	7.6	37	10	89	412	70	21	70	79	0.41	90.2	2.37	0.26	B51
10W9234	BW52	501	6.8	18	9	71	403	64	33	79	67	0.31	46.7	1.73	0.21	BW52
10W9234	BW54	474	6.4	5	8	66	395	63	61	92	39	0.11	5.8	0.77	0.10	BW54
10W9234	B38	461	7.3	6	17	60	378	63	73	90	27	0.08	2.7	0.51	0.06	B38
10W9234	A30	287	7.6	5	6	45	231	66	54	90	46	0.15	6.2	0.40	0.06	A30
	TRAY: 281	POS: 7			LOCAL SPECIFICITY: B15,B35											
10W9235	BW75	563	8.0	47	0	184	332	69	0	79	100	0.36	73.7	5.01	0.57	BW75
10W9235	BW76	516	6.0	14	0	170	332	61	0	92	100	0.22	26.0	3.47	0.49	BW76
10W9235	B35	502	7.8	50	5	120	327	64	9	70	91	0.42	89.7	3.52	0.39	B35
10W9235	BW54	447	6.9	7	6	113	321	52	46	94	54	0.11	5.0	0.89	0.14	BW54
10W9235	B39	434	7.0	8	9	105	312	52	52	92	48	0.10	4.1	0.91	0.12	B39
10W9235	B15.3	417	5.6	5	6	100	306	52	54	95	46	0.08	2.5	0.70	0.10	B15.3
10W9235	A28	406	6.6	20	12	80	294	53	37	80	63	0.26	26.8	0.80	0.09	A28
10W9235	BW60	374	5.5	16	22	64	272	51	57	80	43	0.17	10.8	0.61	0.07	BW60
10W9235	BW6	336	6.8	59	175	5	97	57	74	7	26	0.24	19.0	1.20	0.11	BW6
	TRAY: 281	POS: 8			LOCAL SPECIFICITY: B5,B35											
10W9236	B51	563	7.9	29	20	67	447	47	40	69	60	0.35	67.4	2.23	0.24	B51
10W9236	BW53	514	6.0	6	7	61	440	28	53	91	47	0.16	12.9	1.39	0.20	BW53
10W9236	A30	330	5.0	4	7	37	282	39	63	90	37	0.13	6.0	0.75	0.10	A30
10W9236	BW52	490	6.4	9	17	48	416	29	65	84	35	0.17	14.1	0.61	0.07	BW52
10W9236	A23	464	8.0	4	16	44	400	29	80	91	20	0.07	2.1	0.54	0.07	A23
10W9236	B35	444	5.9	18	34	26	366	22	65	59	35	0.30	40.3	0.54	0.06	B35
10W9236	BW4	392	5.7	12	154	14	212	15	92	53	8	0.02	0.2	0.82	0.15	BW4
	TRAY: 281	POS: 9			LOCAL SPECIFICITY: B5,B35											
10W9237	B51	564	7.7	45	4	31	484	69	8	40	92	0.71	282.6	6.90	0.73	B51
10W9237	BW52	515	6.7	12	15	19	469	38	55	61	45	0.38	74.4	2.27	0.27	BW52
10W9237	B35	488	5.6	15	40	4	429	26	72	21	28	0.43	90.5	1.02	0.11	B35
	TRAY: 281	POS: 10			LOCAL SPECIFICITY: B5,B35											
10W9238	B51	564	7.0	36	13	23	492	59	26	38	74	0.64	227.4	3.55	0.37	B51
10W9238	BW52	515	7.0	10	17	13	475	43	62	56	38	0.37	70.8	1.43	0.16	BW52
	TRAY: 281	POS: 11			LOCAL SPECIFICITY: B5,B35											
10W9239	BW53	557	7.3	12	2	60	483	75	14	83	86	0.35	67.6	4.55	0.58	BW53
10W9239	B35	543	7.7	32	20	28	463	76	38	46	62	0.52	149.1	3.19	0.34	B35
10W9239	B51	491	7.3	21	27	7	436	64	56	25	44	0.54	143.2	2.09	0.22	B51
	TRAY: 281	POS: 12			LOCAL SPECIFICITY: B15											
10W9240	BW76	562	8.0	18	0	274	270	80	0	93	100	0.17	17.2	4.29	0.60	BW76
10W9240	BW77	544	8.0	13	0	261	270	78	0	95	100	0.16	13.1	3.93	0.58	BW77
10W9240	BW75	531	7.9	41	2	220	268	77	4	84	96	0.27	40.0	4.39	0.53	BW75
10W9240	B15.3	488	7.6	10	1	210	267	74	9	95	91	0.14	9.5	2.77	0.43	B15.3
10W9240	BW62	477	7.6	138	21	72	246	73	13	34	87	0.61	177.0	3.15	0.32	BW62
10W9240	BW63	318	6.5	16	11	56	235	54	40	77	60	0.27	22.6	1.40	0.19	BW63
10W9240	B39	291	6.3	6	7	50	228	58	53	89	47	0.15	6.3	0.76	0.11	B39

Table 6. *Continued*

	TRAY: 281	POS: 13			LOCAL SPECIFICITY: B15											
10W9241	BW75	560	7.3	45	2	138	375	54	4	75	96	0.41	92.8	6.13	0.71	BW75
10W9241	BW76	513	6.1	14	0	124	375	47	0	89	100	0.28	39.1	4.73	0.66	BW76
10W9241	B21	486	7.0	4	1	119	362	46	20	96	80	0.13	8.0	2.75	0.47	B21
10W9241	B35	494	6.9	33	22	87	352	45	40	72	60	0.29	42.9	1.80	0.20	B35
10W9241	B18	439	5.7	13	16	74	336	41	55	85	45	0.17	12.2	1.07	0.14	B18
10W9241	CW2	252	5.7	6	10	52	184	48	62	89	38	0.09	2.0	0.86	0.13	CW2
10W9241	B39	394	6.4	5	11	63	315	50	68	92	32	0.08	2.3	0.78	0.11	B39
10W9241	BW62	378	6.6	45	86	18	229	49	65	28	35	0.35	45.1	0.78	0.08	BW62
10W9241	BW6	247	6.3	16	128	2	101	44	88	11	12	0.17	7.5	1.08	0.10	BW6
	TRAY: 281	POS: 14			LOCAL SPECIFICITY: B15											
10W9242	BW76	563	5.5	13	5	34	511	27	27	72	73	0.42	99.2	3.05	0.38	BW76
10W9242	BW6	545	5.6	30	399	4	112	26	93	11	7	0.06	2.0	0.72	0.12	BW6
	TRAY: 281	POS: 15			LOCAL SPECIFICITY: B15											
10W9243	BW76	553	8.0	18	0	123	412	59	0	87	100	0.31	54.4	6.34	0.86	BW76
10W9243	B45	496	8.0	8	0	110	378	53	0	93	100	0.23	26.0	5.20	0.82	B45
10W9243	BW55	527	7.0	4	5	111	407	50	55	96	45	0.07	2.7	2.01	0.31	BW55
10W9243	BW75	518	7.6	23	19	88	388	50	45	79	55	0.24	30.2	2.11	0.25	BW75
10W9243	B39	476	8.0	5	12	83	376	42	70	94	30	0.05	1.4	1.38	0.19	B39
10W9243	BW54	459	6.0	5	8	78	368	38	61	93	39	0.09	3.8	1.14	0.16	BW54
10W9243	BW62	446	6.2	53	98	25	270	38	64	32	36	0.33	49.1	0.84	0.08	BW62
10W9243	BW6	295	5.6	21	170	4	100	28	89	16	11	0.12	4.4	0.53	0.05	BW6
	TRAY: 281	POS: 16			LOCAL SPECIFICITY: B15											
10W9244	BW76	562	7.7	18	0	360	184	76	0	95	100	0.13	9.1	2.28	0.32	BW76
10W9244	B35	544	7.8	53	4	307	180	76	7	85	93	0.19	20.4	1.90	0.22	B35
10W9244	BW55	487	6.2	9	1	298	179	73	10	97	90	0.08	3.2	1.34	0.21	BW55
10W9244	B39	477	7.0	14	4	284	175	73	22	95	78	0.06	1.9	1.38	0.20	B39
10W9244	BW70	459	7.3	27	5	257	170	75	15	90	85	0.13	7.4	1.33	0.17	BW70
10W9244	BW62	427	7.5	121	21	136	149	75	14	52	86	0.36	55.6	1.54	0.16	BW62
10W9244	BW56	285	7.0	8	1	128	148	70	11	94	89	0.15	6.3	0.97	0.15	BW56
10W9244	B8	276	8.0	13	5	115	143	71	27	89	73	0.14	5.2	0.68	0.10	B8
10W9244	BW60	258	7.6	23	10	92	133	67	30	80	70	0.19	9.7	0.73	0.09	BW60
10W9244	CW8	225	7.8	34	3	58	130	65	8	63	92	0.46	47.7	0.90	0.09	CW8
10W9244	B7	188	5.7	13	10	45	120	50	43	77	57	0.21	8.1	0.61	0.08	B7
10W9244	BW61	165	7.4	13	7	32	113	57	35	71	65	0.31	16.3	0.55	0.08	BW61
10W9244	AW34	80	6.7	3	1	17	59	55	25	85	75	0.26	5.6	0.30	0.06	AW34
	TRAY: 281	POS: 17			LOCAL SPECIFICITY: BW76											
10W9245	BW76	564	7.2	18	0	24	522	45	0	57	100	0.64	231.1	6.99	0.88	BW76

TRAY: 281 POS: 18 LOCAL SPECIFICITY: BW46,B17

10W9246	BW76	563	7.7	18	0	339	206	62	0	94	100	0.14	10.7	1.91	0.27	BW76
10W9246	BW77	545	7.2	13	0	326	206	61	0	96	100	0.12	8.1	1.61	0.25	BW77
10W9246	BW46	532	7.7	88	11	238	195	60	11	73	89	0.27	39.1	1.92	0.21	BW46
10W9246	BW75	433	7.4	31	8	207	187	50	20	86	80	0.16	10.4	1.01	0.13	BW75
10W9246	BW56	394	6.4	5	2	202	185	46	28	97	72	0.05	1.0	0.70	0.12	BW56
10W9246	B13	387	6.8	19	8	183	177	46	29	90	71	0.10	3.8	0.87	0.12	B13
10W9246	B18	360	5.6	17	9	166	168	44	34	90	66	0.08	2.4	0.53	0.07	B18
10W9246	AW34	219	6.0	3	1	106	109	44	25	97	75	0.07	1.0	0.43	0.07	AW34
10W9246	BW60	330	7.1	19	9	144	158	46	32	88	68	0.11	4.2	0.50	0.07	BW60
10W9246	BW58	302	6.9	7	5	137	153	44	41	95	59	0.04	0.6	0.44	0.07	BW58
10W9246	BW54	290	7.7	6	3	131	150	43	33	95	67	0.07	1.4	0.40	0.07	BW54
10W9246	CW8	281	6.4	9	7	122	143	41	43	93	57	0.05	0.6	0.58	0.06	CW8
10W9246	CW1	265	7.3	8	9	114	134	41	52	93	48	0.01	0.0	0.56	0.06	CW1
10W9246	A29	165	5.4	7	2	73	83	40	22	91	78	0.14	3.3	0.34	0.06	A29

TRAY: 281 POS: 19 LOCAL SPECIFICITY: CW3,CW1.3

10W9247	BW77	557	8.0	13	0	222	322	73	0	94	100	0.18	18.2	3.88	0.54	BW77
10W9247	B51	544	7.9	41	3	181	319	71	6	81	94	0.32	54.4	3.60	0.41	B51
10W9247	BW63	500	7.9	22	4	159	315	66	15	87	85	0.24	27.8	2.87	0.35	BW63
10W9247	BW52	474	7.8	19	4	140	311	62	17	88	83	0.23	26.1	2.47	0.31	BW52
10W9247	BW53	451	7.0	6	4	134	307	58	40	95	60	0.09	4.0	1.33	0.22	BW53
10W9247	BW57	441	6.2	12	6	122	301	58	33	91	67	0.16	11.7	1.33	0.17	BW57
10W9247	BW76	423	7.6	9	6	113	295	59	40	92	60	0.13	7.4	1.16	0.16	BW76
10W9247	BW46	408	7.0	60	30	53	265	58	33	46	67	0.46	87.6	1.42	0.15	BW46
10W9247	BW58	318	7.0	6	7	47	258	49	53	88	47	0.16	8.5	0.91	0.13	BW58
10W9247	A32	220	6.4	5	4	33	178	52	44	86	56	0.21	9.6	0.48	0.06	A32

TRAY: 281 POS: 20 LOCAL SPECIFICITY: CW3,CW1.3

10W9248	CW3	564	7.9	228	7	78	251	84	2	25	98	0.73	296.9	4.75	0.44	CW3
10W9248	BW57	329	6.2	9	6	69	245	57	40	88	60	0.19	11.4	1.44	0.19	BW57
10W9248	BW46	314	7.0	33	18	36	227	62	35	52	65	0.45	64.8	1.24	0.13	BW46
10W9248	BW62	263	7.1	11	38	25	189	61	77	69	23	0.12	3.9	0.82	0.08	BW62
10W9248	BW54	214	8.0	4	3	21	186	60	42	84	58	0.26	14.5	0.53	0.08	BW54
10W9248	B8	207	5.2	5	14	16	172	52	73	76	27	0.17	6.0	0.55	0.07	B8

TRAY: 281 POS: 21 LOCAL SPECIFICITY: CW3,CW1.3

10W9249	CW1	562	7.9	92	19	185	266	80	17	66	83	0.33	62.5	3.22	0.33	CW1
10W9249	CW3	451	7.4	157	30	28	236	72	16	15	84	0.73	243.4	3.41	0.33	CW3
10W9249	BW46	264	7.1	9	3	19	233	42	25	67	75	0.46	55.0	2.27	0.23	BW46
10W9249	BW62	252	5.1	7	37	12	196	26	84	63	16	0.15	5.4	0.66	0.06	BW62

TRAY: 281 POS: 22 LOCAL SPECIFICITY: CW1,CW3,CW3.1

10W9250	CW3	561	7.9	230	3	128	200	92	1	35	99	0.61	210.2	4.06	0.39	CW3
10W9250	BW46	328	8.0	51	0	77	200	82	0	60	100	0.54	94.4	3.35	0.35	BW46
10W9250	CW1	277	7.7	20	5	57	195	72	20	74	80	0.37	37.3	2.79	0.31	CW1
10W9250	BW56	252	4.0	1	0	56	195	66	0	98	100	0.12	3.4	1.87	0.31	BW56
10W9250	B51	251	7.3	18	10	38	185	67	35	67	65	0.36	32.0	0.86	0.10	B51
10W9250	BW62	223	6.8	10	28	28	157	65	73	73	27	0.11	2.8	0.83	0.08	BW62
10W9250	BW75	185	6.9	9	14	19	143	75	60	67	40	0.25	11.8	0.43	0.05	BW75

Table 6. *Continued*

	TRAY: 281	POS: 23			LOCAL SPECIFICITY: B15,BW46											
10W9251	BW76	563	6.9	18	0	291	254	79	0	94	100	0.16	15.3	5.25	0.72	BW76
10W9251	BW77	545	8.0	13	0	278	254	81	0	95	100	0.15	11.6	4.92	0.72	BW77
10W9251	BW46	532	8.0	97	2	181	252	80	2	65	98	0.44	101.9	5.94	0.63	BW46
10W9251	BW75	433	7.8	36	3	145	249	69	7	80	93	0.32	44.9	4.32	0.52	BW75
10W9251	BW63	394	7.4	19	9	126	240	63	32	86	68	0.18	12.5	2.97	0.39	BW63
10W9251	BW54	366	6.9	7	4	119	236	61	36	94	64	0.11	4.3	2.53	0.38	BW54
10W9251	B15.3	355	8.0	6	5	113	231	63	45	94	55	0.08	2.3	2.21	0.33	B15.3
10W9251	BW62	344	6.9	96	48	17	183	61	33	15	67	0.61	128.4	2.58	0.26	BW62
10W9251	BW57	200	7.3	6	8	11	175	70	57	64	43	0.34	22.8	1.40	0.20	BW57
10W9251	CW1	186	7.0	4	11	7	164	72	73	63	27	0.26	12.6	0.83	0.08	CW1
	TRAY: 281	POS: 24			LOCAL SPECIFICITY: BW76H											
10W9252	BW75	561	6.6	37	10	107	407	47	21	74	79	0.37	75.7	3.13	0.37	BW75
10W9252	BW55	514	5.0	4	5	103	402	43	55	96	45	0.08	3.1	1.89	0.30	BW55
10W9252	BW54	505	6.3	7	6	96	396	45	46	93	54	0.13	9.2	1.67	0.24	BW54
10W9252	B35	492	6.7	29	26	67	370	45	47	69	53	0.30	43.5	1.84	0.21	B35
10W9252	B51	437	6.5	13	35	54	335	43	72	80	28	0.11	5.7	0.64	0.08	B51
	TRAY: 281	POS: 25			LOCAL SPECIFICITY: CW1,CW1.3											
10W9253	BW41	560	6.5	4	2	192	362	59	33	97	67	0.07	2.7	1.91	0.43	BW41
10W9253	BW54	554	7.0	14	2	178	360	59	12	92	88	0.19	20.3	2.31	0.32	BW54
10W9253	BW77	538	5.8	11	2	167	358	60	15	93	85	0.17	16.0	2.06	0.30	BW77
10W9253	CW1	525	7.6	83	17	84	341	62	17	50	83	0.53	149.2	2.24	0.22	CW1
10W9253	BW46	425	6.2	9	14	75	327	39	60	89	40	0.12	5.8	1.72	0.18	BW46
10W9253	AW34	280	7.0	6	5	48	221	38	45	88	55	0.18	9.1	0.55	0.08	AW34
10W9253	B13	391	6.3	8	14	61	308	37	63	88	37	0.12	5.6	0.57	0.07	B13
10W9253	BW6	369	6.1	51	257	10	51	37	83	16	17	0.00	0.0	2.16	0.35	BW6
	TRAY: 281	POS: 26			LOCAL SPECIFICITY: TS1,B35,BW62											
10W9254	B35	566	7.9	54	3	266	243	81	5	83	95	0.26	37.6	4.52	0.53	B35
10W9254	BW75	509	8.0	47	3	219	240	78	6	82	94	0.28	38.7	4.31	0.51	BW75
10W9254	BW62	459	7.7	138	10	81	230	73	6	36	94	0.63	181.5	4.52	0.47	BW62
10W9254	BW70	311	6.8	26	5	55	225	51	16	67	84	0.44	59.8	3.50	0.46	BW70
10W9254	BW76	280	6.8	10	3	45	222	49	23	81	77	0.32	28.3	2.32	0.35	BW76
10W9254	B15.3	267	6.4	5	4	40	218	46	44	88	56	0.19	10.0	1.63	0.27	B15.3
10W9254	BW60	258	5.7	13	17	27	201	45	56	67	44	0.28	20.1	0.95	0.12	BW60
10W9254	B39	228	6.5	4	9	23	192	59	69	85	31	0.14	4.7	0.70	0.11	B39
10W9254	BW6	215	7.3	16	94	7	98	60	85	30	15	0.13	3.5	1.69	0.16	BW6
	TRAY: 281	POS: 27			LOCAL SPECIFICITY: TS1,B35,BW46,B51,BW52,B15											
10W9255	B35	511	7.9	51	2	286	172	86	3	84	97	0.22	24.1	5.63	0.72	B35
10W9255	BW52	458	8.0	24	0	262	172	84	0	91	100	0.18	15.2	5.09	0.71	BW52
10W9255	BW77	434	8.0	12	0	250	172	83	0	95	100	0.14	8.1	4.34	0.70	BW77
10W9255	BW53	422	7.6	10	0	240	172	82	0	96	100	0.13	7.0	4.13	0.67	BW53
10W9255	B18	412	8.0	26	2	214	170	82	7	89	93	0.19	14.8	4.14	0.60	B18
10W9255	B51	384	7.8	40	2	174	168	80	4	81	96	0.28	29.8	4.11	0.54	B51
10W9255	B15.3	342	8.0	7	0	167	168	77	0	95	100	0.14	6.9	3.15	0.54	B15.3
10W9255	BW75	335	7.9	41	2	126	166	76	4	75	96	0.35	40.8	4.12	0.52	BW75
10W9255	BW76	292	6.9	7	3	119	163	70	30	94	70	0.10	3.0	2.11	0.36	BW76
10W9255	BW63	282	6.7	9	4	110	159	70	30	92	70	0.12	4.1	1.90	0.31	BW63
10W9255	BW62	269	7.2	66	33	44	126	71	33	40	67	0.40	43.1	1.72	0.20	BW62
10W9255	BW46	170	7.5	36	35	8	91	75	49	18	51	0.48	39.2	0.96	0.12	BW46

	TRAY: 281	POS: 28			LOCAL	SPECIFICITY: TS1,B35,B51,BW52,B15										
10W9256	B35	571	8.0	55	3	243	270	82	5	81	95	0.29	47.0	5.53	0.63	B35
10W9256	BW52	513	7.8	25	0	218	270	78	0	89	100	0.24	29.2	4.71	0.59	BW52
10W9256	BW53	488	7.6	11	0	207	270	77	0	94	100	0.17	13.9	4.04	0.57	BW53
10W9256	B15.3	477	7.6	9	1	198	269	76	10	95	90	0.14	9.0	3.69	0.55	B15.3
10W9256	BW75	467	8.0	47	2	151	267	75	4	76	96	0.37	64.2	4.10	0.46	BW75
10W9256	BW77	418	8.0	10	1	141	266	68	9	93	91	0.19	14.7	3.02	0.44	BW77
10W9256	B51	407	7.9	38	4	103	262	65	9	73	91	0.40	64.5	3.73	0.43	B51
10W9256	B18	365	7.3	19	7	84	255	54	26	81	74	0.28	27.8	3.03	0.39	B18
10W9256	BW62	339	6.7	54	65	30	190	50	54	35	46	0.35	41.7	0.74	0.08	BW62
	TRAY: 281	POS: 29			LOCAL	SPECIFICITY: TS1,B35,B51,BW52,B15										
10W9257	BW76	551	8.0	18	0	324	209	82	0	94	100	0.14	11.4	4.49	0.70	BW76
10W9257	BW77	533	8.0	13	0	311	209	81	0	95	100	0.13	8.6	4.15	0.69	BW77
10W9257	B15.3	520	7.6	9	0	302	209	80	0	97	100	0.11	6.2	3.79	0.68	B15.3
10W9257	BW53	511	7.5	11	2	291	207	80	15	96	85	0.08	3.6	4.03	0.67	BW53
10W9257	BW75	498	7.9	45	4	246	203	80	8	84	92	0.22	25.0	4.30	0.56	BW75
10W9257	B35	449	8.0	43	5	203	198	77	10	82	90	0.24	26.3	3.70	0.48	B35
10W9257	BW62	401	7.6	124	14	79	184	72	10	38	90	0.57	129.6	3.61	0.41	BW62
10W9257	BW70	263	6.5	19	11	60	173	53	36	75	64	0.26	17.9	2.41	0.35	BW70
10W9257	BW52	233	6.7	12	6	48	167	58	33	80	67	0.27	17.1	1.69	0.26	BW52
10W9257	B51	215	6.8	17	9	31	158	58	34	64	66	0.38	31.6	1.69	0.25	B51
10W9257	BW57	189	6.0	5	8	26	150	54	61	83	39	0.16	5.0	0.88	0.15	BW57
10W9257	BW63	176	6.0	4	9	22	141	57	69	84	31	0.13	2.9	0.68	0.11	BW63
	TRAY: 281	POS: 30			LOCAL	SPECIFICITY: TS1,BW46										
10W9258	BW76	568	6.7	18	0	196	354	63	0	91	100	0.23	30.8	4.15	0.58	BW76
10W9258	BW77	550	7.3	12	1	184	353	64	7	93	93	0.18	18.6	3.79	0.56	BW77
10W9258	BW46	537	7.6	86	17	98	336	63	16	53	84	0.51	137.1	3.11	0.33	BW46
10W9258	BW55	434	6.0	4	3	94	333	44	42	95	58	0.11	4.9	1.41	0.24	BW55
10W9258	BW58	427	6.0	8	6	86	327	44	42	91	58	0.16	10.4	1.40	0.21	BW58
10W9258	BW54	413	7.0	8	4	78	323	45	33	90	67	0.20	15.8	1.38	0.20	BW54
10W9258	BW75	401	7.1	20	18	58	305	43	47	74	53	0.27	29.5	1.04	0.13	BW75
10W9258	CW1	363	6.6	10	13	48	292	36	56	82	44	0.20	13.8	1.30	0.12	CW1
	TRAY: 281	POS: 31			LOCAL	SPECIFICITY: TS1,BW62										
10W9259	B7	558	6.3	28	19	45	466	38	40	61	60	0.42	97.6	2.44	0.27	B7
10W9259	BW60	511	4.9	19	28	26	438	37	59	57	41	0.36	64.4	1.77	0.20	BW60
10W9259	BW75	464	7.5	8	35	18	403	50	81	69	19	0.18	15.1	0.51	0.06	BW75
	TRAY: 281	POS: 32			LOCAL	SPECIFICITY: BW62,BW63										
10W9260	BW76	569	7.8	18	0	190	361	70	0	91	100	0.24	32.3	6.58	0.90	BW76
10W9260	BW77	551	7.2	12	1	178	360	68	7	93	93	0.19	19.7	5.05	0.73	BW77
10W9260	BW75	538	7.0	38	11	140	349	67	22	78	78	0.30	48.1	4.06	0.48	BW75
10W9260	BW62	489	7.2	119	40	21	309	67	25	15	75	0.71	246.3	3.84	0.38	BW62
10W9260	BW63	330	5.3	6	19	15	290	23	76	71	24	0.21	14.1	0.76	0.10	BW63
10W9260	BW60	305	5.3	6	26	9	264	26	81	60	19	0.22	14.6	0.70	0.09	BW60
	TRAY: 281	POS: 33			LOCAL	SPECIFICITY: BW62										
10W9261	BW76	567	8.0	18	0	183	366	68	0	91	100	0.24	33.9	6.11	0.78	BW76
10W9261	BW77	549	6.8	10	3	173	363	65	23	94	77	0.14	11.4	4.74	0.70	BW77
10W9261	BW75	536	7.0	39	10	134	353	65	20	77	80	0.32	55.2	3.91	0.43	BW75
10W9261	BW62	487	7.2	115	42	19	311	65	26	14	74	0.71	243.0	3.38	0.32	BW62

Table 6. *Continued*

	TRAY: 281	POS: 34			LOCAL	SPECIFICITY: TS1,B35,B51,BW52										
10W9262	BW77	568	8.0	12	1	179	376	81	7	93	93	0.19	20.5	5.35	0.75	BW77
10W9262	B51	555	8.0	41	4	138	372	79	8	77	92	0.37	77.6	5.78	0.66	B51
10W9262	B35	510	7.8	47	8	91	364	74	14	65	86	0.46	106.5	5.40	0.61	B35
10W9262	BW52	455	7.6	21	4	70	360	64	16	76	84	0.39	67.7	4.84	0.61	BW52
10W9262	BW53	430	7.1	7	2	63	358	60	22	90	78	0.24	25.5	3.75	0.57	BW53
10W9262	BW75	421	7.2	39	9	24	349	60	18	38	82	0.67	187.1	4.47	0.51	BW75
10W9262	B15.3	373	8.0	4	5	20	344	45	55	83	45	0.24	22.1	3.14	0.47	B15.3
10W9262	CW4	339	7.7	6	25	13	295	36	80	68	20	0.19	12.2	0.56	0.06	CW4
	TRAY: 281	POS: 35			LOCAL	SPECIFICITY: B15										
10W9263	BW76	566	8.0	18	0	181	367	77	0	90	100	0.25	34.3	6.10	0.81	BW76
10W9263	BW77	548	7.1	13	0	168	367	75	0	92	100	0.22	27.0	5.58	0.78	BW77
10W9263	BW62	535	7.5	118	40	50	327	76	25	29	75	0.60	195.0	3.58	0.35	BW62
10W9263	BW75	377	7.2	33	16	17	311	60	32	34	68	0.62	143.2	2.75	0.31	BW75
	TRAY: 281	POS: 36			LOCAL	SPECIFICITY: CW1,CX46										
10W9264	CW1	529	7.2	78	26	24	401	66	25	23	75	0.70	258.2	2.88	0.28	CW1
10W9264	BW46	425	6.3	6	13	18	388	45	68	75	32	0.24	25.1	2.34	0.23	BW46
	TRAY: 282	POS: 3			LOCAL	SPECIFICITY: B51,BW52,BW4										
10W9265	B44	559	7.6	72	5	37	445	72	6	33	94	0.75	311.6	5.47	0.54	B44
10W9265	BW77	482	6.0	7	4	30	441	48	36	81	64	0.32	49.7	2.25	0.30	BW77
10W9265	BW58	471	7.0	8	9	22	432	50	52	73	48	0.32	49.0	1.59	0.20	BW58
10W9265	AW33	419	6.2	9	21	10	379	47	70	52	30	0.34	48.4	1.19	0.12	AW33
10W9265	BW4	424	6.4	11	169	2	242	53	93	15	7	0.15	9.8	1.11	0.10	BW4
	TRAY: 282	POS: 4			LOCAL	SPECIFICITY: B51,BW52										
10W9266	BW52	574	6.0	4	25	25	520	44	86	86	14	0.09	4.9	0.40	0.05	BW52
10W9266	BW4	545	6.2	17	255	8	265	44	93	32	7	0.08	3.4	0.40	0.07	BW4
	TRAY: 282	POS: 5			LOCAL	SPECIFICITY: TS1										
10W9267	BW46	547	5.3	6	96	6	439	16	94	50	6	0.12	8.0	0.14	0.01	BW46
	TRAY: 282	POS: 6			LOCAL	SPECIFICITY: BW46										
10W9268	BW46	568	5.8	11	90	10	457	38	89	47	11	0.18	17.9	0.18	0.02	BW46
	TRAY: 282	POS: 7			LOCAL	SPECIFICITY: BW55,B35										
10W9269	BW52	576	7.7	24	5	112	435	68	17	82	83	0.32	59.2	3.71	0.43	BW52
10W9269	B44	547	7.2	55	21	57	414	64	27	50	73	0.52	146.0	3.17	0.33	B44
10W9269	B51	471	7.2	26	23	31	391	56	46	54	54	0.43	86.3	1.77	0.19	B51
10W9269	A23	420	6.0	4	9	27	380	38	69	87	31	0.16	10.7	1.02	0.12	A23
10W9269	B38	409	6.3	7	13	20	369	40	65	74	35	0.26	27.5	0.92	0.11	B38
10W9269	BW4	389	6.5	13	110	7	259	45	89	35	11	0.17	10.9	1.49	0.13	BW4
	TRAY: 282	POS: 8			LOCAL	SPECIFICITY: BW46										
10W9270	CW4	531	7.0	4	82	14	431	44	95	77	5	0.03	0.5	0.07	0.01	CW4

	TRAY: 282	POS: 9			LOCAL SPECIFICITY: B5,B35,BW60,BW62											
10W9271	B18	582	7.8	35	0	404	143	91	0	92	100	0.14	12.1	4.33	0.58	B18
10W9271	BW76	547	7.7	18	0	386	143	90	0	95	100	0.11	6.6	3.80	0.58	BW76
10W9271	B15.3	529	8.0	11	0	375	143	91	0	97	100	0.09	4.2	3.27	0.55	B15.3
10W9271	BW53	518	7.6	14	0	361	143	90	0	96	100	0.10	5.5	3.44	0.55	BW53
10W9271	B35	504	8.0	53	2	308	141	91	3	85	97	0.19	18.6	4.09	0.52	B35
10W9271	BW75	449	8.0	41	1	267	140	89	2	86	98	0.20	18.1	3.95	0.51	BW75
10W9271	BW70	407	7.4	27	1	240	139	88	3	89	97	0.18	12.7	3.51	0.49	BW70
10W9271	BW52	379	7.9	25	0	215	139	90	0	89	100	0.20	15.5	3.50	0.48	BW52
10W9271	B51	354	7.8	39	4	176	135	89	9	81	91	0.23	18.4	3.38	0.45	B51
10W9271	BW62	311	7.9	116	9	60	126	88	7	34	93	0.60	111.5	3.62	0.41	BW62
10W9271	BW77	186	8.0	4	1	56	125	71	20	93	80	0.17	5.4	1.92	0.37	BW77
10W9271	BW58	181	8.0	7	2	49	123	69	22	87	78	0.23	9.7	1.73	0.30	BW58
10W9271	BW46	172	6.8	34	38	15	85	65	52	30	48	0.35	21.3	0.57	0.07	BW46
	TRAY: 282	POS: 10			LOCAL SPECIFICITY: B5,B35											
10W9272	AW33	517	5.7	7	57	12	441	52	89	63	11	0.15	10.9	0.38	0.04	AW33
	TRAY: 282	POS: 11			LOCAL SPECIFICITY: B51,BW52,B35,TS1,BW59,B37											
10W9273	A2	530	8.0	278	6	90	156	95	2	24	98	0.66	233.4	3.03	0.28	A2
10W9273	BW53	255	7.6	5	0	91	159	84	0	94	100	0.18	8.4	1.30	0.18	BW53
10W9273	BW63	250	7.9	15	0	76	159	84	0	83	100	0.33	27.9	1.27	0.16	BW63
10W9273	BW77	235	8.0	2	0	74	159	82	0	97	100	0.13	4.2	1.11	0.15	BW77
10W9273	B51	233	8.0	20	1	54	158	82	4	72	96	0.43	42.9	1.04	0.12	B51
10W9273	BW58	212	8.0	6	1	48	157	75	14	88	86	0.26	13.8	0.77	0.10	BW58
10W9273	BW57	205	7.3	9	2	39	155	72	18	81	82	0.33	22.1	0.80	0.10	BW57
10W9273	BW52	194	8.0	12	2	27	153	71	14	69	86	0.46	40.4	0.63	0.08	BW52
10W9273	BW70	180	6.0	5	11	22	142	59	68	81	32	0.14	3.6	0.45	0.05	BW70
	TRAY: 282	POS: 12			LOCAL SPECIFICITY: B44,B45,B49,B50,BW63											
10W9274	BW76	512	7.4	18	0	99	395	79	0	84	100	0.35	63.0	4.79	0.60	BW76
10W9274	B44	494	8.0	59	6	40	389	79	9	40	91	0.69	233.7	4.30	0.44	B44
10W9274	B21	429	8.0	4	1	36	388	50	20	90	80	0.26	29.9	2.69	0.42	B21
	TRAY: 282	POS: 13			LOCAL SPECIFICITY: B5,B35,BW53,BW63											
10W9275	BW77	553	7.2	13	0	89	451	86	0	87	100	0.33	58.9	6.27	0.83	BW77
10W9275	BW53	540	7.6	10	4	79	447	88	28	88	72	0.24	31.5	3.62	0.47	BW53
10W9275	B35	526	7.9	31	19	48	428	89	38	60	62	0.43	95.5	3.14	0.34	B35
10W9275	B51	476	7.9	24	17	24	411	85	41	50	59	0.49	116.2	2.81	0.31	B51
10W9275	BW52	435	7.6	9	14	15	397	75	60	62	40	0.35	52.6	1.74	0.21	BW52
10W9275	BW63	412	7.7	6	12	9	385	66	66	60	34	0.34	47.3	1.62	0.20	BW63
10W9275	B18	394	7.2	5	25	4	360	55	83	44	17	0.28	30.1	0.68	0.08	B18
	TRAY: 282	POS: 14			LOCAL SPECIFICITY: BW62											
10W9276	BW77	574	8.0	13	0	251	310	69	0	95	100	0.16	15.6	5.58	0.82	BW77
10W9276	BW76	561	6.8	18	0	233	310	67	0	92	100	0.20	23.0	5.63	0.78	BW76
10W9276	BW75	543	7.8	47	2	186	308	69	4	79	96	0.34	61.8	5.75	0.67	BW75
10W9276	BW62	494	7.4	136	25	50	283	63	15	26	85	0.67	223.0	4.25	0.43	BW62
10W9276	BW46	333	5.8	33	57	17	226	32	63	34	37	0.37	45.3	1.00	0.11	BW46
10W9276	BW52	243	6.5	4	13	13	213	29	76	76	24	0.18	7.7	0.77	0.11	BW52
10W9276	B51	226	5.6	5	21	8	192	23	80	61	20	0.21	9.8	0.56	0.07	B51
10W9276	B35	200	5.2	5	27	3	165	25	84	37	16	0.26	13.4	0.54	0.07	B35

Table 6. *Continued*

	TRAY: 282	POS: 15			LOCAL SPECIFICITY: BW62											
10W9277	BW46	498	5.7	6	90	4	398	40	93	40	7	0.15	10.9	0.11	0.01	BW46
10W9277	BW6	402	6.5	4	324	0	74	50	98	0	2	0.05	0.9	1.29	0.20	BW6
	TRAY: 282	POS: 16			LOCAL SPECIFICITY: B35											
10W9278	CW4	533	7.7	85	1	30	417	75	1	26	99	0.82	361.8	6.34	0.62	CW4
10W9278	B35	483	7.5	4	14	36	429	55	77	90	23	0.10	4.8	2.02	0.23	B35
10W9278	BW6	465	6.3	31	327	5	102	52	91	13	9	0.06	1.8	0.36	0.05	BW6
	TRAY: 282	POS: 17			LOCAL SPECIFICITY: CW3,B35,A2											
10W9279	A2	378	7.9	127	52	26	173	93	29	16	71	0.59	131.1	1.88	0.19	A2
10W9279	A28	199	6.5	11	19	15	154	84	63	57	37	0.30	17.3	0.84	0.11	A28
10W9279	B35	169	8.0	5	7	10	147	100	58	66	42	0.32	17.2	0.49	0.06	B35
	TRAY: 282	POS: 18			LOCAL SPECIFICITY: CW3,A2											
10W9280	A2	573	8.0	295	3	178	97	89	1	37	99	0.45	116.6	2.62	0.27	A2
10W9280	A28	275	8.0	36	2	142	95	74	5	79	95	0.25	17.4	1.58	0.22	A28
10W9280	AW34	169	8.0	11	0	109	49	71	0	90	100	0.17	4.8	1.09	0.19	AW34
10W9280	A10	158	8.0	21	1	88	48	68	4	80	96	0.23	8.4	1.20	0.18	A10
10W9280	BW60	204	6.7	11	0	99	94	59	0	90	100	0.22	9.9	0.55	0.06	BW60
10W9280	BW55	193	8.0	4	3	95	91	59	42	95	58	0.02	0.1	0.37	0.05	BW55
	TRAY: 282	POS: 19			LOCAL SPECIFICITY: A2,BW62											
10W9281	A2	570	7.9	279	17	68	206	84	5	19	95	0.71	288.1	3.78	0.35	A2
10W9281	BW41	274	6.0	5	1	63	205	41	16	92	84	0.20	11.3	1.46	0.23	BW41
10W9281	BW60	268	6.8	10	3	53	202	41	23	84	77	0.28	21.7	1.39	0.16	BW60
10W9281	B18	255	5.4	7	12	46	190	37	63	86	37	0.11	3.2	0.51	0.06	B18
10W9281	B7	236	6.8	18	9	28	181	41	33	60	67	0.43	43.2	0.52	0.06	B7
	TRAY: 282	POS: 20			LOCAL SPECIFICITY: B5											
10W9282	BW77	569	7.7	7	6	69	487	51	46	90	54	0.18	18.8	1.90	0.27	BW77
10W9282	B37	556	6.0	4	6	65	481	47	60	94	40	0.11	7.1	1.44	0.21	B37
10W9282	B51	546	7.0	22	23	43	458	47	51	66	49	0.34	64.0	1.41	0.16	B51
10W9282	BW52	501	6.3	6	20	37	438	41	76	86	24	0.12	7.3	0.40	0.05	BW52
	TRAY: 282	POS: 21			LOCAL SPECIFICITY: BW6											
10W9283	BW76	572	7.6	18	0	301	253	66	0	94	100	0.16	14.7	3.99	0.58	BW76
10W9283	B7	554	8.0	46	2	255	251	65	4	84	96	0.26	36.5	4.20	0.51	B7
10W9283	BW55	506	6.4	9	1	246	250	59	10	96	90	0.11	6.4	2.48	0.40	BW55
10W9283	BW75	496	7.3	43	4	203	246	59	8	82	92	0.27	36.4	3.11	0.38	BW75
10W9283	BW60	449	7.7	37	5	166	241	56	11	81	89	0.28	34.4	2.93	0.36	BW60
10W9283	B8	407	6.3	21	8	145	233	50	27	87	73	0.18	12.9	2.11	0.29	B8
10W9283	BW61	378	7.2	25	7	120	226	51	21	82	79	0.25	23.4	1.98	0.26	BW61
10W9283	B39	346	7.0	10	5	110	221	46	33	91	67	0.14	7.1	1.50	0.22	B39
10W9283	B18	331	6.4	15	10	95	211	46	40	86	60	0.16	8.7	1.21	0.17	B18
10W9283	BW54	306	7.0	8	6	87	205	45	42	91	58	0.12	4.7	1.03	0.16	BW54
10W9283	BW62	292	6.3	49	52	38	153	43	51	43	49	0.30	25.9	0.66	0.07	BW62
10W9283	B35	191	6.0	13	17	25	136	44	56	65	44	0.25	12.3	0.50	0.07	B35
10W9283	BW6	161	6.5	20	41	5	95	48	67	20	33	0.37	22.3	1.02	0.15	BW6
	TRAY: 282	POS: 22			LOCAL SPECIFICITY: CW3											
10W9284	CW4	533	5.0	4	82	7	440	18	95	63	5	0.08	3.4	0.17	0.02	CW4

	TRAY: 282	POS: 23			LOCAL SPECIFICITY: B5,B21											
10W9285	BW77	549	8.0	5	8	32	504	54	61	86	39	0.20	21.3	1.79	0.25	BW77
10W9285	BW52	536	6.4	5	18	27	486	46	78	84	22	0.14	10.6	0.82	0.10	BW52
10W9285	B51	513	6.8	5	38	22	448	48	88	81	12	0.09	3.8	0.52	0.06	B51
10W9285	BW6	470	6.3	19	377	3	71	50	95	13	5	0.01	0.1	0.57	0.08	BW6
	TRAY: 282	POS: 24			LOCAL SPECIFICITY: BW4											
10W9286	BW57	370	7.6	9	6	64	291	61	40	87	60	0.21	16.0	1.78	0.24	BW57
10W9286	BW53	355	7.2	5	5	59	286	59	50	92	50	0.14	7.1	1.46	0.21	BW53
10W9286	BW52	345	7.8	9	8	50	278	59	47	84	53	0.22	16.2	1.50	0.20	BW52
10W9286	B44	328	7.4	19	16	31	262	54	45	62	55	0.38	46.2	1.68	0.20	B44
10W9286	B51	293	6.4	9	14	22	248	45	60	70	40	0.27	21.5	0.69	0.09	B51
10W9286	BW4	270	6.5	15	51	7	197	40	77	31	23	0.30	24.8	1.54	0.15	BW4
	TRAY: 282	POS: 25			LOCAL SPECIFICITY: B5,B18,B35,BW62											
10W9287	B35	571	8.0	58	0	390	123	91	0	87	100	0.18	17.7	5.35	0.68	B35
10W9287	BW75	513	8.0	50	0	340	123	90	0	87	100	0.18	17.5	5.19	0.67	BW75
10W9287	BW52	463	8.0	24	0	316	123	88	0	92	100	0.14	9.2	4.35	0.64	BW52
10W9287	B18	439	7.9	29	0	287	123	87	0	90	100	0.17	12.1	4.34	0.62	B18
10W9287	BW53	410	8.0	11	0	276	123	87	0	96	100	0.11	4.8	3.47	0.61	BW53
10W9287	B15.3	399	8.0	11	0	265	123	86	0	96	100	0.11	5.0	3.47	0.61	B15.3
10W9287	BW77	388	8.0	10	0	255	123	86	0	96	100	0.11	4.8	3.37	0.60	BW77
10W9287	BW76	378	7.1	14	0	241	123	85	0	94	100	0.14	7.0	3.62	0.60	BW76
10W9287	B38	364	7.7	15	0	226	123	85	0	93	100	0.15	8.0	3.54	0.58	B38
10W9287	BW62	349	7.9	122	4	104	119	85	3	46	97	0.50	88.9	4.31	0.50	BW62
10W9287	B51	223	7.9	23	2	81	117	75	8	77	92	0.32	23.3	3.57	0.49	B51
10W9287	BW70	198	7.6	28	1	53	116	69	3	65	97	0.47	43.5	3.43	0.49	BW70
10W9287	B37	169	8.0	4	1	49	115	62	20	92	80	0.18	5.7	1.79	0.37	B37
10W9287	B39	164	7.3	9	4	40	111	59	30	81	70	0.25	10.4	1.36	0.23	B39
	TRAY: 282	POS: 26			LOCAL SPECIFICITY: B35											
10W9288	BW77	570	6.8	12	1	105	452	54	7	89	93	0.27	42.0	5.60	0.75	BW77
10W9288	BW53	557	6.8	8	6	97	446	54	42	92	58	0.16	13.8	2.41	0.34	BW53
10W9288	B15.3	543	6.8	5	6	92	440	54	54	94	46	0.10	5.8	1.80	0.28	B15.3
10W9288	B35	532	6.7	32	22	60	418	54	40	65	60	0.37	74.0	2.37	0.26	B35
10W9288	BW75	478	6.9	28	22	32	396	51	44	53	56	0.45	96.0	2.10	0.23	BW75
10W9288	BW54	428	5.7	6	8	26	388	46	57	81	43	0.25	26.2	1.57	0.22	BW54
10W9288	BW52	414	6.0	5	18	21	370	50	78	80	22	0.15	9.9	0.74	0.09	BW52
10W9288	B51	391	7.1	11	29	10	341	52	72	47	28	0.33	42.9	0.71	0.08	B51
	TRAY: 282	POS: 27			LOCAL SPECIFICITY: B5,BW63											
10W9289	BW77	573	8.0	13	0	103	457	66	0	88	100	0.30	52.4	5.34	0.69	BW77
10W9289	B51	560	7.4	43	3	60	454	62	6	58	94	0.58	188.2	4.89	0.52	B51
10W9289	BW53	514	7.3	8	5	52	449	50	38	86	62	0.25	32.2	2.68	0.35	BW53
10W9289	BW52	501	6.8	19	8	33	441	48	29	63	71	0.47	110.4	2.93	0.35	BW52
10W9289	BW63	474	6.8	5	16	28	425	45	76	84	24	0.14	9.6	1.17	0.15	BW63
10W9289	BW57	453	6.0	6	13	22	412	42	68	78	32	0.22	22.1	0.73	0.09	BW57
	TRAY: 282	POS: 28			LOCAL SPECIFICITY: B5,B35											
10W9290	B51	572	7.5	42	8	35	487	68	16	45	84	0.64	234.0	4.93	0.52	B51
10W9290	BW52	522	7.2	21	6	14	481	54	22	40	78	0.66	229.9	3.78	0.43	BW52
10W9290	BW6	495	5.7	12	406	2	75	28	97	14	3	0.01	0.0	0.51	0.07	BW6
	TRAY: 282	POS: 29			LOCAL SPECIFICITY: B35											
10W9291	B51	562	6.3	6	44	22	490	28	88	78	12	0.10	5.7	0.33	0.04	B51

Table 6. *Continued*

	TRAY: 282	POS: 30			LOCAL SPECIFICITY: B5											
10W9292	BW77	572	7.1	13	0	106	453	47	0	89	100	0.30	50.6	4.82	0.65	BW77
10W9292	BW76	559	6.6	13	5	93	448	44	27	87	73	0.25	34.3	2.41	0.31	BW76
10W9292	BW58	541	7.3	8	9	85	439	43	52	91	48	0.14	11.0	1.71	0.22	BW58
10W9292	BW57	524	7.3	11	10	74	429	41	47	87	53	0.20	21.0	1.60	0.20	BW57
10W9292	B27	503	6.2	9	12	65	417	36	57	87	43	0.17	13.8	1.12	0.14	B27
10W9292	B44	482	6.5	26	40	39	377	33	60	60	40	0.30	44.0	1.07	0.11	B44
10W9292	B38	416	7.0	4	12	35	365	25	75	89	25	0.11	4.8	0.73	0.10	B38
10W9292	BW4	400	5.5	23	115	12	250	22	83	34	17	0.20	16.5	0.85	0.08	BW4
	TRAY: 282	POS: 31			LOCAL SPECIFICITY: B5											
10W9293	BW77	515	7.7	12	0	65	438	79	0	84	100	0.37	69.9	5.49	0.76	BW77
10W9293	BW52	503	7.9	16	7	49	431	78	30	75	70	0.37	68.7	2.94	0.36	BW52
10W9293	B51	480	7.8	26	10	23	421	73	27	46	73	0.58	163.3	3.12	0.36	B51
10W9293	BW53	444	7.5	4	7	19	414	52	63	82	37	0.22	22.3	1.51	0.21	BW53
10W9293	BW4	433	6.8	13	156	6	258	47	92	31	8	0.13	7.2	0.91	0.08	BW4
	TRAY: 282	POS: 32			LOCAL SPECIFICITY: BW4											
10W9294	B44	572	8.0	73	0	246	253	84	0	77	100	0.34	66.4	3.62	0.39	B44
10W9294	BW76	499	7.8	16	0	230	253	79	0	93	100	0.18	17.0	2.29	0.31	BW76
10W9294	BW77	483	8.0	11	0	219	253	78	0	95	100	0.16	12.4	2.10	0.31	BW77
10W9294	BW53	472	7.7	14	0	205	253	77	0	93	100	0.19	16.7	2.15	0.30	BW53
10W9294	B27	458	7.8	18	2	187	251	76	10	91	90	0.19	17.3	1.95	0.26	B27
10W9294	BW57	438	7.8	18	1	169	250	74	5	90	95	0.22	22.0	1.79	0.24	BW57
10W9294	BW52	419	8.0	23	2	146	248	73	8	86	92	0.27	29.5	1.80	0.23	BW52
10W9294	BW55	394	5.6	5	4	141	244	69	44	96	56	0.06	1.4	1.13	0.17	BW55
10W9294	B51	385	7.3	32	6	109	238	70	15	77	85	0.33	41.1	1.11	0.13	B51
10W9294	A23	347	8.0	6	0	103	238	67	0	94	100	0.20	13.3	0.77	0.10	A23
10W9294	BW4	341	7.3	72	15	31	223	66	17	30	83	0.67	153.0	3.04	0.47	BW4
	TRAY: 282	POS: 33			LOCAL SPECIFICITY: B5,B49											
10W9295	BW77	461	8.0	4	4	39	414	67	50	90	50	0.19	15.9	2.25	0.33	BW77
10W9295	BW52	453	7.4	13	11	26	403	64	45	66	55	0.38	66.8	2.11	0.25	BW52
10W9295	BW57	429	7.0	4	13	22	390	57	76	84	24	0.15	9.5	1.15	0.15	BW57
10W9295	B51	412	7.4	7	24	15	366	59	77	68	23	0.22	19.7	0.80	0.09	B51
	TRAY: 282	POS: 34			LOCAL SPECIFICITY: BW6											
10W9296	BW76	529	7.4	18	0	237	274	59	0	92	100	0.19	20.0	3.98	0.64	BW76
10W9296	BW48	316	7.0	4	0	169	143	61	0	97	100	0.10	3.3	2.65	0.60	BW48
10W9296	BW75	507	7.2	41	7	192	267	57	14	82	86	0.26	33.2	2.58	0.34	BW75
10W9296	B8	459	7.2	24	8	168	259	54	25	87	75	0.18	15.6	2.10	0.31	B8
10W9296	B7	427	6.9	34	9	134	250	52	20	79	80	0.27	31.6	2.08	0.28	B7
10W9296	B18	384	6.8	17	9	117	241	50	34	87	66	0.17	11.4	1.52	0.24	B18
10W9296	BW54	358	6.7	9	5	108	236	47	35	92	65	0.14	6.6	1.32	0.22	BW54
10W9296	BW60	344	7.1	22	12	86	224	47	35	79	65	0.24	19.4	1.51	0.21	BW60
10W9296	BW56	310	6.7	6	5	80	219	45	45	93	55	0.11	4.1	1.05	0.19	BW56
10W9296	B39	299	5.8	8	8	72	211	43	50	90	50	0.12	4.7	0.92	0.15	B39
10W9296	B35	283	6.2	23	21	49	190	44	47	68	53	0.26	19.8	0.87	0.12	B35
10W9296	A29	131	6.0	4	1	32	94	50	20	88	80	0.23	7.2	0.47	0.07	A29
10W9296	BW61	234	7.0	10	17	35	172	46	62	77	38	0.16	6.2	0.40	0.06	BW61
10W9296	BW6	207	6.5	34	86	1	86	42	71	2	29	0.36	26.5	1.02	0.16	BW6

	TRAY: 282	POS: 35			LOCAL SPECIFICITY: BW4											
10W9297	B44	573	8.0	72	1	398	102	86	1	84	99	0.17	15.6	1.00	0.11	B44
10W9297	B51	500	7.9	48	0	350	102	83	0	87	100	0.16	13.6	0.92	0.11	B51
10W9297	BW52	452	7.8	27	0	323	102	82	0	92	100	0.14	8.4	0.65	0.08	BW52
10W9297	BW4	425	7.8	147	4	176	98	81	2	54	98	0.37	58.5	3.78	0.58	BW4
	TRAY: 282	POS: 36			LOCAL SPECIFICITY: B5,B18,B35,BW53											
10W9298	B18	570	7.9	34	1	282	253	81	2	89	98	0.21	26.3	5.35	0.71	B18
10W9298	B15.3	535	8.0	10	1	272	252	79	9	96	91	0.11	6.6	4.16	0.69	B15.3
10W9298	BW77	524	7.6	11	0	261	252	79	0	95	100	0.14	10.4	4.12	0.68	BW77
10W9298	BW65	333	7.0	4	0	175	154	83	0	97	100	0.10	3.5	3.29	0.68	BW65
10W9298	B35	509	7.9	54	1	203	251	78	1	78	99	0.33	56.1	5.20	0.64	B35
10W9298	B51	454	7.9	40	2	163	249	74	4	80	96	0.32	47.8	4.86	0.62	B51
10W9298	BW52	412	7.9	24	1	139	248	69	4	85	96	0.29	35.5	4.28	0.60	BW52
10W9298	BW75	387	7.7	42	2	97	246	64	4	69	96	0.44	76.5	4.30	0.55	BW75
10W9298	B37	343	8.0	7	0	90	246	53	0	92	100	0.23	18.1	3.08	0.53	B37
10W9298	BW53	336	8.0	8	1	82	245	50	11	91	89	0.23	18.2	3.08	0.53	BW53
10W9298	BW62	327	6.5	62	52	20	193	45	45	24	55	0.49	80.0	1.24	0.14	BW62
10W9298	BW70	213	5.5	4	25	16	168	35	86	80	14	0.06	0.8	0.48	0.07	BW70
10W9298	BW4	184	5.8	9	92	7	76	37	91	43	9	0.01	0.0	1.05	0.16	BW4
	TRAY: 283	POS: 3			LOCAL SPECIFICITY: A2,BW62											
10W9299	A2	567	7.2	191	104	3	269	74	35	1	65	0.67	254.7	3.60	0.33	A2
	TRAY: 283	POS: 4			LOCAL SPECIFICITY: CW3											
10W9300	CW3	565	7.5	212	27	24	302	79	11	10	89	0.81	375.1	5.08	0.47	CW3
10W9300	CW1	326	6.3	19	44	5	258	54	69	20	31	0.43	59.5	0.74	0.08	CW1
	TRAY: 283	POS: 5			LOCAL SPECIFICITY: CW3											
10W9301	CW3	568	7.6	232	8	25	303	82	3	9	97	0.88	443.6	5.84	0.53	CW3
	TRAY: 283	POS: 6			LOCAL SPECIFICITY: BW4											
10W9302	A29	376	7.3	14	1	303	58	90	6	95	94	0.05	1.0	0.68	0.11	A29
10W9302	BW57	551	8.0	19	0	406	126	88	0	95	100	0.10	5.8	0.59	0.07	BW57
10W9302	B27	532	7.8	26	0	380	126	87	0	93	100	0.13	8.5	0.61	0.07	B27
10W9302	A24	324	7.7	103	9	163	49	89	8	61	92	0.19	11.3	0.59	0.07	A24
10W9302	B44	394	8.0	51	2	226	115	86	3	81	97	0.22	19.7	0.63	0.07	B44
10W9302	CW2	170	7.2	5	2	116	47	88	28	95	72	0.00	0.0	0.44	0.06	CW2
10W9302	BW4	334	7.9	111	17	110	96	84	13	49	87	0.34	39.2	1.09	0.10	BW4
	TRAY: 283	POS: 7			LOCAL SPECIFICITY: B5											
10W9303	BW77	567	7.5	13	0	95	459	57	0	87	100	0.32	56.5	4.95	0.66	BW77
10W9303	BW63	554	6.6	19	10	76	449	53	34	80	66	0.30	50.4	2.63	0.31	BW63
10W9303	B15.3	525	7.3	6	5	70	444	53	45	92	55	0.17	14.6	1.85	0.25	B15.3
10W9303	B37	513	6.0	5	5	64	439	52	50	92	50	0.15	11.7	1.71	0.24	B37
10W9303	BW52	504	6.5	12	11	53	428	53	47	81	53	0.26	33.1	1.76	0.22	BW52
10W9303	BW58	481	7.0	6	8	47	420	56	57	88	43	0.18	14.9	1.39	0.18	BW58
10W9303	B51	467	6.6	20	18	27	402	55	47	57	53	0.42	82.8	1.59	0.18	B51
10W9303	BW57	429	6.8	5	12	22	390	51	70	81	30	0.19	16.0	1.04	0.13	BW57
	TRAY: 283	POS: 8			LOCAL SPECIFICITY: B35											
10W9304	B35	567	7.1	46	12	33	476	59	20	41	80	0.64	230.3	4.56	0.48	B35
10W9304	BW53	509	7.3	6	5	27	471	48	45	81	55	0.29	42.8	3.50	0.45	BW53
10W9304	B15.3	498	7.2	5	7	22	464	44	58	81	42	0.25	31.5	2.33	0.31	B15.3
10W9304	B51	486	6.0	10	36	12	428	40	78	54	22	0.27	34.8	0.70	0.08	B51

Table 6. *Continued*

	TRAY: 283	POS: 9			LOCAL SPECIFICITY: BW6											
10W9305	BW60	567	7.7	51	1	381	134	83	1	88	99	0.16	15.1	3.78	0.45	BW60
10W9305	BW76	515	7.9	18	0	363	134	82	0	95	100	0.11	6.6	3.01	0.43	BW76
10W9305	BW56	497	7.5	11	0	352	134	81	0	96	100	0.09	4.2	2.63	0.41	BW56
10W9305	BW75	486	7.8	40	1	312	133	82	2	88	98	0.17	14.2	3.04	0.38	BW75
10W9305	B7	445	7.6	42	3	270	130	80	6	86	94	0.17	12.9	2.99	0.37	B7
10W9305	B8	400	7.9	27	1	243	129	80	3	90	97	0.17	11.5	2.76	0.36	B8
10W9305	BW70	372	7.3	23	1	220	128	79	4	90	96	0.17	10.5	2.50	0.34	BW70
10W9305	B18	348	7.3	21	1	199	127	80	4	90	96	0.17	10.5	2.45	0.34	B18
10W9305	BW55	326	7.0	8	1	191	126	79	11	95	89	0.10	3.0	1.78	0.29	BW55
10W9305	BW61	317	7.9	26	4	165	122	80	13	86	87	0.17	9.7	2.13	0.28	BW61
10W9305	BW62	287	7.5	95	8	70	114	78	7	42	93	0.53	79.3	2.34	0.25	BW62
10W9305	B39	184	7.6	11	2	59	112	72	15	84	85	0.26	12.9	1.58	0.24	B39
10W9305	B35	171	7.4	25	4	34	108	71	13	57	87	0.49	41.3	1.70	0.22	B35
10W9305	CW2	82	8.0	1	0	15	66	93	0	93	100	0.23	4.2	0.45	0.06	CW2
10W9305	BW6	141	7.4	26	14	7	94	69	35	21	65	0.62	53.9	3.12	0.46	BW6
	TRAY: 283	POS: 10			LOCAL SPECIFICITY: B15											
10W9306	BW75	558	8.0	47	0	272	239	86	0	85	100	0.26	38.5	5.70	0.66	BW75
10W9306	BW76	511	8.0	14	0	258	239	84	0	94	100	0.16	12.6	4.36	0.62	BW76
10W9306	BW62	497	7.9	153	6	105	233	83	3	40	97	0.61	183.9	5.41	0.54	BW62
10W9306	BW77	338	7.1	11	1	94	232	66	8	89	92	0.25	21.3	3.24	0.47	BW77
10W9306	B15.3	326	7.7	6	3	88	229	65	33	93	67	0.14	6.5	2.23	0.34	B15.3
10W9306	BW63	317	6.6	18	8	70	221	64	30	79	70	0.28	24.3	2.51	0.33	BW63
10W9306	BW46	291	7.3	51	35	19	186	71	40	27	60	0.53	83.0	1.39	0.15	BW46
10W9306	CW3	205	7.3	12	36	7	150	63	75	36	25	0.30	18.4	0.78	0.07	CW3
10W9306	BW6	157	6.8	5	106	2	44	57	95	28	5	0.00	0.0	0.65	0.06	BW6
	TRAY: 283	POS: 11			LOCAL SPECIFICITY: CW1,CX46											
10W9307	CW1	534	7.7	100	10	15	409	76	9	13	91	0.86	394.6	6.49	0.63	CW1
10W9307	BW46	424	5.0	4	17	11	392	13	80	73	20	0.19	15.6	2.22	0.21	BW46
	TRAY: 283	POS: 12			LOCAL SPECIFICITY: CW1,CX46											
10W9308	CW1	516	7.7	89	16	40	371	76	15	31	85	0.70	251.1	3.38	0.33	CW1
10W9308	BW77	411	7.6	5	7	35	364	52	58	87	42	0.19	14.3	0.42	0.06	BW77
	TRAY: 283	POS: 13			LOCAL SPECIFICITY: BW62											
10W9309	BW76	559	6.0	18	0	152	389	54	0	89	100	0.28	42.6	6.32	0.81	BW76
10W9309	BW62	541	7.1	113	46	39	343	56	28	25	72	0.62	205.8	3.89	0.37	BW62
10W9309	BW75	382	5.6	27	16	12	327	25	37	30	63	0.62	146.1	2.66	0.30	BW75
10W9309	BW60	339	5.2	5	27	7	300	16	84	58	16	0.21	15.1	0.50	0.06	BW60
	TRAY: 283	POS: 14			LOCAL SPECIFICITY: B15,BW57,BW46											
10W9310	BW62	558	8.0	160	0	181	217	91	0	53	100	0.51	142.8	8.69	0.87	BW62
10W9310	BW76	398	8.0	17	0	164	217	84	0	90	100	0.23	21.3	6.09	0.84	BW76
10W9310	BW77	381	8.0	12	0	152	217	82	0	92	100	0.21	16.4	5.66	0.84	BW77
10W9310	BW57	369	7.9	19	0	133	217	81	0	87	100	0.28	28.6	6.16	0.84	BW57
10W9310	BW63	350	7.5	26	0	107	217	79	0	80	100	0.36	45.8	6.46	0.83	BW63
10W9310	B15.3	324	7.8	9	0	98	217	80	0	91	100	0.24	18.8	5.26	0.83	B15.3
10W9310	BW75	315	7.9	39	1	59	216	79	2	60	98	0.55	94.2	6.28	0.75	BW75
10W9310	BW46	275	7.0	54	31	5	185	67	36	8	64	0.69	129.2	2.57	0.28	BW46

	TRAY: 283	POS: 15			LOCAL SPECIFICITY: BW62,BW63,8W66											
10W9311	BW77	559	8.0	13	0	262	284	81	0	95	100	0.16	13.7	6.77	0.98	BW77
10W9311	BW75	546	7.8	47	0	215	284	80	0	82	100	0.32	55.7	8.21	0.95	BW75
10W9311	B15.3	499	7.1	11	0	204	284	78	0	94	100	0.17	14.9	6.06	0.91	B15.3
10W9311	BW76	488	5.9	14	0	190	284	78	0	93	100	0.20	20.1	5.95	0.85	BW76
10W9311	BW62	474	7.8	151	5	39	279	81	3	20	97	0.81	311.4	8.39	0.84	BW62
10W9311	BW63	318	6.0	15	12	24	267	41	44	61	56	0.40	51.4	2.47	0.33	BW63
10W9311	BW57	291	7.5	4	11	20	256	41	73	83	27	0.16	7.1	1.50	0.21	BW57
10W9311	BW46	276	5.9	18	67	2	189	35	78	10	22	0.36	35.5	0.55	0.06	BW46
	TRAY: 283	POS: 16			LOCAL SPECIFICITY: BW62,8W66,BW74											
10W9312	BW76	557	7.9	18	0	187	352	81	0	91	100	0.24	31.9	6.30	0.84	BW76
10W9312	BW62	539	7.7	140	18	47	334	80	11	25	89	0.73	286.8	5.45	0.54	BW62
10W9312	BW75	381	6.5	33	10	14	324	51	23	29	77	0.70	185.9	3.30	0.38	BW75
10W9312	BW63	338	5.5	4	23	10	301	35	85	71	15	0.16	8.4	0.49	0.06	BW63
	TRAY: 283	POS: 17			LOCAL SPECIFICITY: BW62,BW76											
10W9313	BW76	554	8.0	18	0	143	393	81	0	88	100	0.29	45.4	6.90	0.87	BW76
10W9313	BW62	536	7.7	128	28	15	365	79	17	10	83	0.80	344.9	4.94	0.47	BW62
10W9313	BW6	380	5.3	14	254	1	111	13	94	6	6	0.10	3.9	0.86	0.08	BW6
	TRAY: 283	POS: 18			LOCAL SPECIFICITY: B15											
10W9314	BW62	558	8.0	159	0	213	186	94	0	57	100	0.45	111.2	7.20	0.72	BW62
10W9314	BW75	399	8.0	47	0	166	186	90	0	77	100	0.34	46.5	5.90	0.69	BW75
10W9314	BW63	352	7.9	27	0	139	186	87	0	83	100	0.31	32.8	5.19	0.66	BW63
10W9314	BW76	325	8.0	13	0	126	186	85	0	90	100	0.24	18.1	4.40	0.64	BW76
10W9314	BW77	312	8.0	12	0	114	186	84	0	90	100	0.24	18.4	4.31	0.64	BW77
10W9314	BW57	300	7.7	15	0	99	186	82	0	86	100	0.29	25.8	4.48	0.63	BW57
10W9314	B15.3	285	7.8	9	0	90	186	80	0	90	100	0.25	17.5	3.93	0.62	B15.3
10W9314	BW46	276	7.6	74	10	16	176	80	11	17	89	0.78	169.2	3.78	0.41	BW46
10W9314	A31	192	7.5	4	7	12	169	56	63	75	37	0.25	12.0	0.57	0.06	A31
10W9314	BW4	181	6.4	10	100	2	69	50	90	16	10	0.12	2.7	1.19	0.19	BW4
	TRAY: 283	POS: 19			LOCAL SPECIFICITY: B15,A32,BW76											
10W9315	BW76	557	8.0	18	0	196	343	71	0	91	100	0.23	29.8	5.18	0.71	BW76
10W9315	BW75	539	7.1	36	7	160	336	68	16	81	84	0.29	45.3	3.94	0.47	BW75
10W9315	BW62	496	7.3	125	33	35	303	68	20	21	80	0.69	232.9	3.27	0.33	BW62
10W9315	B39	338	6.6	7	8	28	295	45	53	80	47	0.26	22.3	1.25	0.18	B39
10W9315	BW4	323	6.5	19	180	9	115	42	90	32	10	0.04	0.5	0.48	0.08	BW4
	TRAY: 283	POS: 20			LOCAL SPECIFICITY: BW62,B15											
10W9316	BW77	556	7.7	13	0	232	311	69	0	94	100	0.17	16.9	5.78	0.83	BW77
10W9316	BW76	543	6.2	18	0	214	311	68	0	92	100	0.21	25.0	5.93	0.80	BW76
10W9316	BW75	525	7.5	41	2	173	309	69	4	80	96	0.33	57.8	5.72	0.67	BW75
10W9316	BW62	482	7.4	135	21	38	288	65	13	21	87	0.73	257.1	4.96	0.49	BW62
10W9316	BW63	326	5.9	18	9	20	279	31	33	52	67	0.52	86.5	3.68	0.48	BW63
10W9316	B15.3	299	6.8	5	4	15	275	45	44	75	56	0 34	35.5	2.48	0.45	B15.3
	TRAY: 283	POS: 21			LOCAL SPECIFICITY: BW62											
10W9317	BW76	551	7.7	18	0	182	351	68	0	91	100	0.24	32.7	5.39	0.72	BW76
10W9317	BW75	533	7.2	35	8	147	343	67	18	80	82	0.30	46.4	4.28	0.50	BW75
10W9317	BW77	490	7.4	10	2	137	341	65	16	93	84	0.18	16.7	3.29	0.47	BW77
10W9317	BW62	478	7.1	107	49	30	292	64	31	21	69	0.61	180.6	2.70	0.27	BW62
10W9317	B39	322	6.5	4	11	26	281	46	73	86	27	0.13	5.6	1.15	0.16	B39

Table 6. *Continued*

	TRAY: 283	POS: 22			LOCAL SPECIFICITY: B15,8W66,TS1,BW57,BW46											
10W9318	BW75	558	7.9	47	0	352	159	90	0	88	100	0.19	20.5	5.61	0.69	BW75
10W9318	BW76	511	8.0	14	0	338	159	90	0	96	100	0.11	6.5	4.29	0.66	BW76
10W9318	BW77	497	8.0	13	0	325	159	89	0	96	100	0.11	6.3	4.21	0.66	BW77
10W9318	BW63	484	7.8	29	0	296	159	89	0	91	100	0.18	15.1	4.88	0.65	BW63
10W9318	B15.3	455	7.8	11	0	285	159	89	0	96	100	0.12	6.1	3.90	0.63	B15.3
10W9318	BW62	444	8.0	152	2	133	157	89	1	46	99	0.52	122.2	5.79	0.61	BW62
10W9318	BW57	290	8.0	14	1	119	156	78	6	89	94	0.22	14.4	3.52	0.53	BW57
10W9318	BW46	275	7.8	79	5	40	151	76	5	33	95	0.68	127.0	4.15	0.47	BW46
10W9318	BW53	191	6.3	6	3	34	148	45	33	85	67	0.25	11.9	2.07	0.41	BW53
10W9318	BW70	182	6.7	9	18	25	130	44	66	73	34	0.16	4.5	1.48	0.21	BW70
	TRAY: 283	POS: 23			LOCAL SPECIFICITY: B15											
10W9319	BW77	513	8.0	12	0	256	245	88	0	95	100	0.15	11.2	5.51	0.82	BW77
10W9319	BW76	501	7.8	18	0	238	245	87	0	92	100	0.19	17.9	5.92	0.82	BW76
10W9319	BW75	483	8.0	36	1	202	244	87	2	84	98	0.28	37.0	6.11	0.74	BW75
10W9319	BW62	446	7.9	144	5	58	239	85	3	28	97	0.73	238.1	6.39	0.65	BW62
10W9319	B15.3	297	7.4	7	1	51	238	58	12	87	88	0.29	24.2	3.59	0.58	B15.3
10W9319	BW57	289	7.6	9	4	42	234	56	30	82	70	0.29	24.9	2.80	0.41	BW57
10W9319	BW63	276	6.8	16	9	26	225	50	36	61	64	0.43	50.7	2.14	0.28	BW63
	TRAY: 283	POS: 24			LOCAL SPECIFICITY: B15,BW57											
10W9320	BW75	556	8.0	47	0	271	238	85	0	85	100	0.26	38.4	7.35	0.85	BW75
10W9320	B15.3	509	8.0	11	1	260	237	83	8	95	92	0.12	7.3	5.57	0.82	B15.3
10W9320	BW76	497	7.8	13	0	247	237	82	0	95	100	0.16	12.2	5.56	0.81	BW76
10W9320	BW77	484	7.8	12	0	235	237	81	0	95	100	0.16	11.8	5.46	0.80	BW77
10W9320	BW62	472	7.8	144	11	91	226	81	7	38	93	0.60	171.6	5.78	0.58	BW62
10W9320	BW63	317	7.0	24	3	67	223	63	11	73	89	0.41	52.2	4.10	0.52	BW63
10W9320	BW57	290	6.6	10	5	57	218	64	33	85	67	0.24	16.9	2.77	0.39	BW57
10W9320	BW46	275	7.4	50	34	7	184	70	40	12	60	0.63	110.8	2.12	0.23	BW46
	TRAY: 283	POS: 25			LOCAL SPECIFICITY: B15,8W66,TS1,BW57											
10W9321	BW75	557	8.0	47	0	371	139	90	0	88	100	0.18	17.1	5.83	0.76	BW75
10W9321	BW63	510	7.8	28	0	343	139	89	0	92	100	0.15	11.1	5.12	0.73	BW63
10W9321	BW57	482	8.0	16	0	327	139	89	0	95	100	0.12	6.7	4.56	0.73	BW57
10W9321	BW76	466	8.0	14	0	313	139	88	0	95	100	0.11	6.1	4.41	0.73	BW76
10W9321	BW77	452	8.0	13	0	300	139	88	0	95	100	0.11	5.9	4.32	0.72	BW77
10W9321	B15.3	439	8.0	11	0	289	139	87	0	96	100	0.11	5.2	4.13	0.72	B15.3
10W9321	BW62	428	7.9	152	1	137	138	87	0	47	100	0.51	110.0	6.40	0.71	BW62
10W9321	BW46	275	7.8	83	2	54	136	75	2	39	98	0.64	112.6	5.23	0.63	BW46
10W9321	BW53	190	6.0	6	3	48	133	50	33	88	67	0.19	6.8	2.20	0.48	BW53
10W9321	BW52	181	6.9	7	5	41	128	52	41	85	59	0.19	6.7	1.69	0.29	BW52
10W9321	BW70	169	5.8	13	14	28	114	53	51	68	49	0.24	10.0	1.86	0.28	BW70
10W9321	B13	142	6.5	8	10	20	104	67	55	71	45	0.24	8.0	1.11	0.17	B13
10W9321	B35	124	7.2	10	14	10	90	70	58	50	42	0.34	14.3	0.83	0.12	B35
10W9321	B51	100	6.7	6	13	4	77	60	68	40	32	0.35	12.1	0.61	0.09	B51

	TRAY: 283	POS: 26			LOCAL SPECIFICITY: B15,BW70,B35											
10W9322	BW77	540	8.0	13	0	320	207	85	0	96	100	0.12	8.3	4.81	0.77	BW77
10W9322	BW76	527	7.9	18	0	302	207	85	0	94	100	0.15	12.1	5.08	0.76	BW76
10W9322	B15.3	509	7.8	10	1	292	206	84	9	96	91	0.10	4.6	3.55	0.59	B15.3
10W9322	BW75	498	8.0	39	4	253	202	84	9	86	91	0.20	19.9	4.51	0.58	BW75
10W9322	BW62	455	7.8	142	11	111	191	81	7	43	93	0.53	129.3	4.69	0.50	BW62
10W9322	BW57	302	7.5	11	5	100	186	65	31	90	69	0.16	7.4	2.38	0.36	BW57
10W9322	B35	286	7.2	26	11	74	175	65	29	74	71	0.29	23.3	2.59	0.35	B35
10W9322	BW63	249	8.0	12	11	62	164	63	47	83	53	0.16	6.1	1.51	0.21	BW63
10W9322	BW46	226	7.0	42	32	20	132	56	43	32	57	0.46	47.5	1.49	0.18	BW46
10W9322	BW70	152	5.3	9	17	11	115	35	65	55	35	0.29	12.6	0.87	0.13	BW70
10W9322	B51	126	6.3	6	14	5	101	54	70	45	30	0.33	13.5	0.52	0.08	B51
	TRAY: 283	POS: 27			LOCAL SPECIFICITY: B15											
10W9323	BW76	541	5.8	16	2	213	310	69	11	93	89	0.17	16.5	5.74	0.78	BW76
10W9323	BW77	523	8.0	13	0	200	310	71	0	93	100	0.19	19.4	5.07	0.73	BW77
10W9323	B15.3	510	7.3	11	0	189	310	70	0	94	100	0.18	17.4	4.77	0.71	B15.3
10W9323	BW75	499	7.7	42	1	147	309	69	2	77	98	0.38	71.5	5.48	0.64	BW75
10W9323	BW63	456	6.8	16	11	131	298	63	40	89	60	0.15	9.6	2.08	0.27	BW63
10W9323	BW62	429	7.1	99	47	32	251	64	32	24	68	0.58	144.9	2.31	0.23	BW62
10W9323	BW46	283	6.3	24	59	8	192	56	71	25	29	0.36	36.3	0.66	0.07	BW46
	TRAY: 283	POS: 28			LOCAL SPECIFICITY: B15											
10W9324	BW77	559	8.0	13	0	292	254	83	0	95	100	0.14	11.1	5.00	0.76	BW77
10W9324	BW76	546	6.7	18	0	274	254	82	0	93	100	0.17	16.2	5.02	0.72	BW76
10W9324	B13	528	8.0	35	1	239	253	83	2	87	98	0.25	31.8	5.46	0.69	B13
10W9324	BW75	492	7.8	39	2	200	251	81	4	83	96	0.28	38.8	5.35	0.66	BW75
10W9324	B15.3	451	7.6	9	2	191	249	79	18	95	82	0.12	6.4	3.51	0.55	B15.3
10W9324	BW62	440	7.8	141	10	50	239	79	6	26	94	0.73	233.7	5.29	0.55	BW62
10W9324	BW57	289	7.3	8	8	42	231	40	50	84	50	0.21	12.7	1.79	0.26	BW57
10W9324	BW63	273	7.4	10	14	32	217	35	58	76	42	0.23	14.0	1.44	0.20	BW63
10W9324	B38	249	5.5	4	8	28	209	21	66	87	34	0.14	4.7	0.73	0.13	B38
10W9324	BW60	237	5.1	9	19	19	190	21	67	67	33	0.23	12.6	0.66	0.09	BW60
	TRAY: 283	POS: 29			LOCAL SPECIFICITY: B35,B51,BW53											
10W9325	BW53	563	7.0	10	3	75	475	56	23	88	77	0.27	39.7	4.85	0.70	BW53
10W9325	B35	550	6.7	31	22	44	453	56	41	58	59	0.43	100.2	2.95	0.32	B35
10W9325	B51	497	7.0	25	24	19	429	54	48	43	52	0.49	119.8	2.61	0.28	B51
10W9325	B13	448	8.0	4	29	15	400	31	87	78	13	0.11	5.4	0.58	0.07	B13
10W9325	BW60	415	5.0	8	41	7	359	13	83	46	17	0.25	25.8	0.48	0.05	BW60
	TRAY: 283	POS: 30			LOCAL SPECIFICITY: BW62S											
10W9326	BW46	511	5.5	12	90	8	401	20	88	40	12	0.20	20.9	0.19	0.02	BW46
	TRAY: 283	POS: 31			LOCAL SPECIFICITY: CW3											
10W9327	CW3	566	7.5	222	17	40	287	74	7	15	93	0.80	361.3	3.97	0.36	CW3
10W9327	BW77	327	7.7	6	6	34	281	57	50	85	50	0.22	16.6	0.39	0.05	BW77
10W9327	BW6	315	6.9	26	214	8	67	52	89	23	11	0.00	0.0	0.99	0.15	BW6

Table 6. *Continued*

	TRAY: 283	POS: 32			LOCAL	SPECIFICITY:		CW3								
10W9328	CW3	567	7.8	232	7	77	251	82	2	24	98	0.73	302.0	4.89	0.45	CW3
10W9328	BW77	328	7.5	11	1	66	250	49	8	85	92	0.31	32.2	1.67	0.24	BW77
10W9328	BW46	316	7.0	30	19	36	231	43	38	54	62	0.43	57.1	1.08	0.11	BW46
10W9328	BW62	267	6.0	8	40	28	191	36	83	77	17	0.04	0.5	0.81	0.08	BW62
10W9328	BW75	219	7.1	9	18	19	173	35	66	67	34	0.23	11.7	0.54	0.06	BW75
10W9328	CW1	192	5.5	4	13	15	160	26	76	78	24	0.14	3.9	0.57	0.06	CW1
10W9328	BW63	175	6.0	8	18	7	142	26	69	46	31	0.33	19.2	0.41	0.05	BW63
	TRAY: 283	POS: 33			LOCAL	SPECIFICITY:		CW3,BW46								
10W9329	CW3	557	7.9	229	4	62	262	87	1	21	99	0.78	340.3	6.25	0.58	CW3
10W9329	CW1	324	7.0	41	21	21	241	59	33	33	67	0.58	109.4	1.99	0.22	CW1
10W9329	BW46	262	6.0	4	8	17	233	42	66	80	34	0.20	10.9	1.36	0.14	BW46
10W9329	BW62	250	5.6	5	37	12	196	47	88	70	12	0.09	2.1	0.64	0.06	BW62
	TRAY: 283	POS: 34			LOCAL	SPECIFICITY:		B5,BW49,BW63								
10W9330	BW77	564	8.0	13	0	145	406	72	0	91	100	0.25	34.2	5.28	0.70	BW77
10W9330	BW53	551	7.4	13	1	132	405	69	7	91	93	0.24	32.8	5.23	0.69	BW53
10W9330	B51	537	7.6	43	2	89	403	68	4	67	96	0.50	133.5	5.15	0.56	B51
10W9330	BW52	492	7.6	23	4	66	399	60	14	74	86	0.42	86.8	3.70	0.43	BW52
10W9330	BW58	465	7.5	11	4	55	395	53	26	83	74	0.31	44.5	2.87	0.37	BW58
10W9330	BW63	450	6.4	15	6	40	389	49	28	72	72	0.40	72.0	2.76	0.34	BW63
10W9330	BW57	429	7.1	9	8	31	381	52	47	77	53	0.30	39.8	1.78	0.24	BW57
10W9330	B35	412	6.0	13	36	18	345	48	73	58	27	0.26	28.9	0.63	0.07	B35
10W9330	BW4	363	7.2	15	122	3	223	61	89	16	11	0.21	16.8	1.69	0.15	BW4
	TRAY: 283	POS: 35			LOCAL	SPECIFICITY:		B5,BW49,BW58,BW63								
10W9331	BW52	564	8.0	27	0	147	390	82	0	84	100	0.34	63.6	5.35	0.62	BW52
10W9331	BW53	537	8.0	14	0	133	390	78	0	90	100	0.27	38.1	4.57	0.59	BW53
10W9331	BW77	523	8.0	13	0	120	390	76	0	90	100	0.27	39.1	4.49	0.59	BW77
10W9331	B51	510	7.9	43	2	77	388	74	4	64	96	0.53	142.3	4.62	0.49	B51
10W9331	BW57	465	7.8	16	3	61	385	62	15	79	85	0.38	65.6	2.96	0.36	BW57
10W9331	BW58	446	7.3	11	3	50	382	55	21	81	79	0.34	51.6	2.42	0.31	BW58
10W9331	BW63	432	6.6	14	5	36	377	52	26	72	74	0.42	74.9	2.38	0.31	BW63
10W9331	BW4	413	7.0	23	126	13	251	50	84	36	16	0.18	13.2	2.36	0.21	BW4
	TRAY: 283	POS: 36			LOCAL	SPECIFICITY:		BW62								
10W9332	BW77	558	8.0	13	0	317	228	84	0	96	100	0.13	9.2	5.72	0.84	BW77
10W9332	BW57	545	7.4	20	1	297	227	83	4	93	96	0.15	12.3	5.92	0.79	BW57
10W9332	BW62	524	7.9	153	1	144	226	84	0	48	100	0.56	161.7	7.81	0.78	BW62
10W9332	BW76	370	6.6	17	0	127	226	71	0	88	100	0.27	28.0	5.64	0.78	BW76
10W9332	BW75	353	8.0	43	1	84	225	73	2	66	98	0.49	83.2	6.44	0.76	BW75
10W9332	B15.3	309	7.0	8	1	76	224	59	11	90	89	0.24	17.8	3.96	0.62	B15.3
10W9332	BW63	300	7.3	23	2	53	222	59	8	69	92	0.46	64.1	4.69	0.61	BW63
10W9332	BW46	275	6.6	46	42	7	180	52	47	13	53	0.57	90.6	1.67	0.18	BW46

References

1. Albert ED, Mickey MR, McNicholas AC, Terasaki PI. Seven new HL-A specificities and their distribution in three races. In: Terasaki PI (ed): Histocompatibility Testing 1970. Munksgaard, Copenhagen, 1970;221-230.
2. Alonso A, Ollier W, Doyle P, Williams E, Festenstein H. Family investigation demonstrating the association of HLA-Bw62 with HLA-Bw4. Tissue Antigens 1983;22:32-36.
3. Alonso A, Doyle P, Williams E, Festenstein H. Further splits of HLA-B15. Ninth Workshop Newsletter No. 5, 1984.
4. Alonso A, Doyle P, Williams E, Gill D, Ollier W, Jaraquemada D, et al. HLA-B15 heterogeneity in different populations. Tissue Antigens 1985;25:33-37.
5. Cambon-Thomsen A, Chandanayingyong D, Thomsen M, Hammond MG. Antigen Report: HLA-Bw63 and Other Bw4-Associated Variants of B15. In: Albert ED, et al. (eds): Histocompatibility Testing 1984. Springer-Verlag, Berlin, Heidelberg, 1984;171-174.
6. Campell EM, Taljaard DG, Du Toit ED. B15 Kemp/B15 short Thai. Ninth Workshop Newsletter No. 2, 1983.
7. Chan SH, Naito S. Antigen Report: HLA-Bw46. In: Albert ED, et al. (eds): Histocompatibility Testing 1984. Springer-Verlag, Berlin, Heidelberg, 1984;153-154.
8. Chan SH. Antigen Report: Bw46. In: Aizawa M (ed): HLA in Asia-Oceania 1986. Hokkaido University Press, Sapporo, Japan, 1986;106-107.
9. Chandanayingyong D. A possible new Thai variant of B15. Ninth Workshop Newsletter No. 2, 1983.
10. Chandanayingyong D, Cambon-Thomsen A, Hammond MG. Antigen Report: HLA-Bw62 and other Bw6-associated variants of B15. In: Albert ED, et al. (eds): Histocompatibility Testing 1984. Springer-Verlag, Berlin, Heidelberg, 1984;169-171.
11. Chiewsilp P, Chandanayingyong D, Sujirachato K. B15 short Thai. Ninth Workshop Newsletter No. 2, 1983.
12. Coates E, Stratton A, Dewar PJ. Definition of Bw62, Bw63, and B15.3. Ninth Workshop Newsletter No. 4, 1983.
13. Danilovs J, Pollock C. Joint report: 8w66. In: Terasaki PI (eds): Histocompatibility Testing 1980. UCLA Tissue Typing Laboratory, Los Angeles, 1980;477-479.
14. Dejour G, Fauchet R, Jalais E, Bouhallier O, Genetet B. B15.3 (Chinese) specificity. Ninth Workshop Newsletter No. 3, 1983.
15. Dick HB. HLA-A, B and C serology and antigen reports. In: Boder WF, et al. (eds): Histocompatibility Testing 1977. Munksgaard, Copenhagen, 1977;157.
16. Fan LA, Choo SY, Beatty PG, Hansen JA. HLA-Bw62 variants identified by cytotoxic lymphocytes and isoelectric focusing gel electrophoresis. Tenth Workshop Newsletter No. 3, 1987.
17. Hammond MG. HLA-B15 complex in South African Indians. Ninth Workshop Newsletter No. 8, 1984.
18. Joysey VC, Roger JH, Bland C, et al. Further studies on a Malay population. In: Kissmeyer-Nielsen F (ed): Histocompatibility Testing 1975. Munksgaard, Copenhagen, 1975;251.
19. Kennedy LJ, Marsh SGE, Bodmer JG. Some preliminary thoughts on our results using the core typing sets. Tenth Workshop Newsletter No. 1, 1987.
20. Loon J, Bernoco D, Terasaki P, Takemura S. Report of the 72nd International Cell Exchange. UCLA Tissue Typing Laboratory, Los Angeles, 1981.
21. Loon J, Terasaki P, Takemura S. Report of the 73rd Cell and lst Serum Exchange. UCLA Tissue Typing Laboratory, Los Angeles, 1981.
22. Parham P, et al. In: Dupont B. (ed): Immunobiology of HLA, Volume I: Histocompatibility Testing 1987. New York: Springer-Verlag, 1989.
23. Richiardi P, Castagneto M, D'Amaro J, et al. Four new HLA allelic factors subtypic to HLA-12 and w15. Their correlation with w4 and w6. J Immunogenet 1974;1:323.
24. Saueracker G, Christiansen FT, Dawkins RL. Antigen Report: ABC B15. In: Simons MJ, Tait BD (eds): Proceedings of the Second Asia and Oceania Histocompatibility Workshop Conference. Immunopublishing, Australia, 1981;80.
25. Shiraki T, Akasa T, Asami M, Ueda Y, Morishima Y, Hasegawa I. A study of a new HLA-B locus antigen TS-1 with 9th workshop sera. Ninth Workshop Newsletter No. 3, 1983.
26. Sierp G, Albert ED, Cambon-Thomsen A, Ohayon E. Data analysis of the HLA-typing for "Provinces Francaises," genetique des populations humaines. Colloque INSERM 1986;142:177-196.
27. Singal DP, Lung P. Bw62. In: Terasaki PI (ed): Histocompatibility Testing 1980. UCLA Tissue Typing Laboratory, Los Angeles, 1980;462-464.
28. Taljaard DJ, Martell RW, du Toit ED. HLA Bw62, Bw63 and other B15 variants found in Cape Town. Tenth Workshop Newsletter No. 6, 1987.
29. Thorsby E, Kissmeyer-Nielsen F. New HL-A alleles. Identification by planned immunization. Vox Sang, Basel, 1970;18:138-148.
30. Thorsby E, Kissmeyer-Nielsen F, Svejgaard A. New alleles of the HLA system: Serological and genetic studies. In: Terasaki PI (ed): Histocompatibility Testing 1980. UCLA Tissue Typing Laboratory, Los Angles, 1980;137.
31. Troup GM. Joint Report: Bw63. In: Terasaki PI (ed): Histocompatibility Testing 1980. UCLA Tissue Typing Laboratory, Los Angeles, 1980;465-486.
32. Zhao TM, Lee TD, Bu KJ, Zhang GL, Rubinstein P. Definition of the HLA-B antigen SH7, identical to TS-1 in the Chinese population. Ninth Workshop Newsletter No. 7, 1984.
33. Zhao TM, Shiraki T. Antigen report: HLA-Bw62, Bw63, Bw70, Bw62.1 and B15 variants. In: Aizawa M (ed): HLA in Asia-Oceania 1986. Hokkaido University Press, Sapporo, Japan, 1986;92-95.
34. Zhao TM, Lee TD. Equivalence of SH7, B15SL1 and other Bw6-associated variants of B15. Tenth Workshop Newsletter No. 4, 1987.

Antigen Society #10 Report (Bw57 and Bw58)

S.F. Goldmann,[1] E. Ivaskova,[2] H.J. von Keyserlingk,[3] H. Mervart,[4] E. Yang,[4] V. Lenhard,[5] S.H. Chan,[6] M. Minev,[7] E. Gyodi,[8] G. Petranyi,[8] S. Nyulassy(Bratislava), C. Brautbar,[9] C. Jersild,[10] S. Chen,[11] and G. Dunston[12]

History

Data obtained in the Seventh International Histocompatibility Workshop suggested the existence of two alleles subtypic to HLA-B17 (1,6). The subdivision of HLA-B17 was confirmed during The Eighth and Ninth International Histocompatibility Workshops, and the two subtypic alleles were designated HLA-Bw57 and HLA-Bw58 (2–5). These two alleles are both included in HLA-Bw4 (1–6).

Serology

Of the 60 sera included in the antigen society no. 10 of the Tenth International Histocompatibility Workshop, 14 showed a high r value (> 0.8) for reactivity with both Bw57 and Bw58 cells. These sera are listed in Table 1, and are 10w471, 472, 474, 476, 486, 4524, 4526, 4529, 4532, 4533, 4534, 4543, 4545, and 4555 reacting only with Bw57 or Bw58 positive cells. Sera 10w 266 and 267 were positive also with Bw63 cells. Serum 10w266 reacted additionally with Bw77 cells.

Seven additional antisera, 10w 4517, 4520, 4521, 4522, 4523, 4527, and 4528 (not included in Table 1) showed a significant r value between 0.7 and 0.79 in reacting only with Bw57 and Bw58 cells. Table 1 includes five sera (10w 480, 483, 4540, 4551, and 4553) reacting only with Bw57 cells but not with cells bearing Bw58 or other HLA-antigens. The r values of these monospecific Bw57 sera are between 0.5 and 0.79 (Table 1). Neither a monospecific Bw58 serum nor a duospecific or oligospecific serum detecting Bw58 but not Bw57 was recognized. Thus, Bw57 can be defined with monospecific sera, but detection of Bw58 is based on negative reactions with the monospecific Bw57 antisera.

Conclusions

Bw57 can be clearly defined with the specific sera 10w 480, 483, 4540, 4551, and 4553. Bw58 can only be determined on cells not possessing the Bw57 antigen.

Reporting Laboratories: GERGOL,[1] EAEIVS,[2] GERKEY,[3] NCYMRV,[4] GERLEN[5]
Participating Laboratories: ANZCSH,[6] EAEMIN,[7] EAEPTR,[8] GERBRA,[9] SCAJER,[10] US5FTW,[11] US8DUN[12]

Table 1. Reactivity* of Tenth International Workshop Sera Defining HLA-Bw57 and HLA-Bw58 Cells

		10wsera	Bw57	Bw58	Bw63	Bw77
Core	#	471	++++	++++	–	–
	#	472	++++	++++	–	–
	#	474	++++	++++	–	–
	#	476	++++	++++	–	–
	#	486	++++	++++	–	–
Ag Soc	#	4524	++++	++++	–	–
	#	4526	++++	++++	–	–
	#	4529	++++	++++	–	–
	#	4532	++++	++++	–	–
	#	4533	++++	++++	–	–
	#	4534	++++	++++	–	–
	#	4543	++++	++++	–	–
	#	4545	++++	++++	–	–
	#	4555	++++	++++	–	–
Core	#	266	++++	++++	++++	++++
	#	267	++++	++++	++++	–
	#	478	+++	+++	+++	+++
	#	480	++	–	–	–
	#	483	++	–	–	–
Ag Soc	#	4540	+++	–	–	–
	#	4551	++	–	–	–
	#	4553	++	–	–	–

*r values: ∓ 0.8 = ++++; 0.7–0.79 = +++; 0.5–0.69 = ++.

Immunobiology of HLA
Volume I

Data on frequencies in different populations and on linkage disequilibrium could not be obtained due to cell selection criteria of the Tenth Workshop. However, these data were reported in previous Workshops (2–5). The present analysis provided no evidence for a third B17 subtypic allele.

Acknowledgments. We acknowledge Dr. S.H. Chan, Singapore; Dr. M. Minev, Sofia, Bulgaria; Dr. E. Gyodi and Dr. G. Petranyi, Budapest, Hungary; Dr. S. Nyulassy, Bratislava, Czechoslovakia; Dr. C. Brautbar, Jerusalem, Israel; Dr. C. Jersild, Aalborg, Denmark; Dr. S. Chen, Ft. Wayne, Indiana, USA; Dr. G. Dunston, Washington, USA, for active cooperation in this study.

References

1. Dick HM. Joint report. HLA-A, B, and C serology and antigen reports. In: Bodmer WF, Batchelor JR, et al. (eds): Histocompatibility Testing 1977. Munksgaard, Copenhagen, 1977;157–204.
2. Cambon-Thomsen A, Festenstein H, Alonso A, Thomsen M, Williams E. Antigen report: HLA-Bw57. In: Albert ED, Mayr W, et al. (eds): Histocompatibility Testing 1984. Springer-Verlag, Berlin, 1984;163–164.
3. Cambon-Thomsen A, Alonso A, Festenstein H, Thomsen M, Williams E. Antigen report: HLA-Bw58. In: Albert ED, Mayr W, et al. (eds): Histocompatibility Testing 1984. Springer-Verlag, Berlin, 1984;164–165.
4. Goldmann SF, Bittighofer PM. Joint report Bw57. In: Terasaki PI (ed): Histocompatibility Testing 1980. UCLA Tissue Typing Laboratory, Los Angeles, 1980;444–446.
5. Gyodi E, Petranyi GG. Joint report Bw58. In: Terasaki PI (ed): Histocompatibility Testing 1980. UCLA Tissue Typing Laboratory, Los Angeles, 1980;447–449.
6. Westphal E, Mueller-Eckhardt G, Goldmann SF. HLA-Bw15 and Bw17 in the German population. Tissue Antigens 1977;10:173.

Antigen Society #11 Report (B37, B38, B39, and Bw67)

G.M. Troup,[1] L. Fainboim,[2] S. Naito,[3] A.H. Mota,[2] and E.A. Palavecino[2]

History

The antigen HLA-B37 (TY) was first described by Gelsthorpe and Doughty (1). HLA-B16 (U18, Te 64, Ao81, Da31) was first defined in the Fifth International Workshop (2). This antigen was split into B38 and B39 in the Sixth Workshop (3). B38 was found to be associated with Bw4 while B39 was associated with Bw6 (4). HLA-Bw67 (SN-2, BF, Te90) was described during the Sixth and Seventh Workshops (5,6). This antigen often is defined in sera belonging to the B16 and the B22 cross-reacting groups (7).

Serology

Antigen Society 11 analyzed the above specificities with 26 Tenth Workshop Core sera and 22 Antigen Society sera. A total of 11 laboratories contributed typing for 512 cells: 299 from Caucasians, 134 from blacks, 23 from Orientals, and 56 from other population groups, the majority of which were American Indians.

Tables 1 and 2 present the best sera in the Core set and Antigen Society set for the specificities B37, B16, B38, B39, and Bw67. HLA-B37, Table 1, is well defined by 6 sera, most with R values above 0.500. All HLA-B16 sera, Table 2, detected B38 and B39 with R values for the two antigens combined ranging from 0.688 to 0.896. All B16 sera except 10w458 also contained anti-Bw67 reactivity. HLA-B38 is defined by 10w460, 461, 462, and 4562. Only 10w460 is specific for B38; the other three have weak and/or short B39 activity. HLA-B39 is defined by 9 sera that are non-reactive with B38. However, all these sera except 10w465 and 10w467 react with the majority of Bw67 cells. HLA-Bw67 is defined by extra reactions in the majority of B16 and B39 sera. In addition, 10w344, 10w323, and 10w345 react with the majority of Bw67 cells. Both 10w344 and 10w323 have antiB42 and/or B7. Serum 10w345 reacts with 21 of 25 Bw67 cells and is devoid of another specificity by chi-square analysis. Of interest is that sera 10w467 and 4562 are non-reactive with a subset of B39 cells which are reactive with all other B16/B39 antisera. Thus, this subset may define a further split of the HLA-B16 cross-reacting group.

Table 1. Tenth Workshop Sera for HLA-B37

Antigen	W.S. No.	% Inclusion	R value	Other
HLA-B37	491	97	0.787	–
	488	95	0.775	–
	492	95	0.759	–
	490	95	0.634	–
	489	50	0.554	–
	4573	75	0.726	–
	4574	93	0.595	A29
	4577	67	0.244	Cw4,6
	4575	64	0.582	–
	4572	46	0.538	–
	4571	43	0.331	–

Reporting Laboratories: US6TRP,[1] LATFAI,[2] JAPNAI[3]
Participating Laboratories: GERGAZ, LATLAY, LATPRI, NCYMRV, NCYNIK, SAFDUT, SAFHAM, US4SND

Table 2. Tenth Workshop Sera for HLA-B16, B38, B39, and Bw67

Antigen	W.S. No.	% Inclusion	R value	Other
HLA-B38+B39 (B16)				
	458	96	0.864	–
	459	91	0.688	Bw67
	455	89	0.852	Bw67
	453	85	0.826	Bw67
	452	81	0.748	Bw67
	4561	94	0.896	Bw67
	4559	85	0.857	Bw67
	4558	82	0.840	Bw67
HLA-B38				
	460	74	0.748	–
	461	92	0.708	Weak B39
	462	86	0.583	Weak B39
	4562	92	0.684	Short B39
HLA-B39				
	465	71	0.663	–
	467	80	0.741	Short B39
	466	88	0.772	Bw67
	470	81	0.773	Bw67
	469	90	0.745	Bw67
	4567	90	0.793	Bw67
	4568	74	0.717	Bw67
	4569	69	0.701	Bw67
	4570	84	0.700	Bw67
HLA-Bw67				
	344	88	0.311	Bw42
	323	89	0.410	Bw42,B7
	345	84	0.481	–

Antisera that do not appear in Tables 1 and 2 because of low inclusion percentages, low R values, and/or multispecificity are the following. For HLA-B37, 10w4576 is multispecific. For HLA-B16, B38, and B39, sera 10w446, 454, 457, 468, 4556, 4557, 4560, 4563, 4564, 4565, and 4566 have inclusion of less than 60% and/or R values of less than 0.5. The monoclonal antibody 10w2090 contains HLA-A2, Bw54, Bw41, B14, and Cw8 as well as B16(B38+B39). The R value for the latter is 0.153.

Table 3 presents a summary of the reaction patterns of sera for the above specificities. HLA-B37, B38, B39, and B16 (B38+B39) are well defined by Tenth Workshop sera. HLA-Bw67 is defined by the extra reactions in B16, B39, B42, and/or B7 sera. Serum 10w345, although weak, may be specific for Bw67. With regard to further splits of B16 (i.e., SH6) as reported by Zhao, et al. (8) and Chandanayingyong, et al. (9), the subset of cells, generally from Oriental populations, in the Core and Antigen Society databases which have positive reactions to all Workshop B16/B39 antisera except for sera 10w467 and 4562 may define SH6 (ST-16). Further studies will be necessary in mixed ethnic populations to decide if ST-16 as originally described by Ness et al. in Mexican-Americans (10) is exactly the same as the Oriental SH6.

References

1. Gelsthorpe K, Doughty RW. A new HL-A antigen: TY. Tissue Antigens 1973;3:316.
2. W.H.O. terminology report. In: Dausset J, Colombani J (eds): Histocompatibility Testing 1972. Munksgaard, Copenhagen, 1972;3.
3. Bodmer J. The ABC of HLA-Bw16. In: Kissmeyer-Nielsen F (ed): Histocompatibility Testing 1975. Munksgaard, Copenhagen, 1975; p 53.
4. Schreuder I, D'Amero J, Sandberg L, van Rood JJ. Partition of the antigen w16 into subtypic determinants. Tissue Antigens 1975;5:142.
5. Naito S, Saito S, Toyoda K, Yamamoto H, Okochi K. A study of anti-w22 complex sera in the Japanese population and the characterization of a possible new antigen of the second segregant series, called SN-2. In: Kissmeyer-Nielsen F (ed): Histocompatibility Testing 1975. Munksgaard, Copenhagen, 1975;289.
6. Ivaskoba E, Mackintosh P, Radvany R. Bw16 (Bw38, Bw39). In: Bodmer WF, Batchelor JR, Bodmer JG, Festenstein H, Morris PJ (eds): Histocompatibility Testing 1977. Munksgaard, Copenhagen, 1977;175.
7. Wolf E. 8w57. In: Terasaki PI (ed): Histocompatibility Testing 1980. UCLA Tissue Typing Laboratory, Los Angeles, 1980;469.
8. Zao TM, Lee TD. Definition of antigen ST-16 identical to SH6, in Chinese. Tenth Workshop Newsletter No. 1, p 23.
9. Chandanayingyong D, Bejrachandra S, Rungroung E, Hiranyavasit W. An unusual B16 antigen. Tenth Workshop Newsletter No. 1, p 4.
10. Ness DB, Watson AK, Grumet FC. ST-16: A new Bw16 subtype present in Mexican-Americans. Hum Immunol 1982;4:55.

Table 3. Summary of the Reaction Patterns for B37, B38, B39, and Bw67 Antisera

Sera	488	491	492	490	4573	4574	4575	460	462	461	4562	467	465	466	470	469	4567	4568	4569	4570	344	323	345
Antigen																							
B37	+	+	+	+	+	+	+																
B38								+	+	+	+	–	–	–	–	–	–	–	–	–			
B39								–	w	w	s	s	+	+	+	+	+	+	+	+			
Bw67								–	–	–	–	–	–	+	+	+	+	+	+	+	+	+	s
Other																					42	42 7	

w = weak; s = short

Antigen Society #12 Report (Bw54, Bw55, Bw56 and Bw42)

S. Sekiguchi,[1] H. Neumeyer,[2] N. Kashiwagi,[3] K. Tsuji,[4] K. Kobayashi,[1] Y. Konoeda,[1] M. Ohkubo, M. Atoh, K. Tokunaga,[5] A. Yagita, H. Inoko,[4] R. Fong,[6] H. Mervart,[6] Y. Paik,[8] P. Reekers,[9] M. Hammond,[10] E. Du Toit,[11] and E. Call.[11]

History

The antigen Bw54 was officially designated as a split of Bw22 at the Seventh International Histocompatibility Workshop. This antigen was primarily defined in the Japanese population by several laboratories independently, and designated as J-1 by Juji et al. (1), as SAP-1 by Nakayama et al. (2) and as SN-1 by Saito et al. (3). Splits of Bw22 in Caucasians were recognized mainly by the association with either Cw1 or Cw3. The antigens Bw55 and Bw56, splits of Bw22 in Caucasians that were primarily called BT22 (4) and Te22 (5), Da30 (6) and AA-AJ (7), were confirmed at the Eighth International Workshop in 1980. The antigen Bw55 is associated with Cw3 in Caucasians and with Cw1 in Japanese. The antigen Bw56 is associated with Cw1 in Caucasians and with Cw4 in Japanese. A new Bw22 antigen as a split of Bw56 was primarily described by Tokunaga in 1983 at the Eighth Japanese Histocompatibility Workshop, but was not clearly defined. The antigen Bw42, previously called MWA (8), was defined at the Fifth International Workshop in 1972. This antigen was primarily found in Negroid populations.

Reporting Laboratories: JAPSEK,[1] GERNEU,[2] JAPKSH,[3] JAPTSU,[4] JAPTOK,[5] ANZFON,[6] NCYMRV,[7] US5PAK,[8] BENREE[9]
Participating Laboratories: SAFHAM,[10] SAFDUT[11]

Serology

At this Workshop, 43 antisera and 3 extra antisera were submitted for this society set. Many of the sera were polyspecific (B7+Bw42, B7+Bw55, Bw42+Bw55, and so on). Several sera, however, were confirmed as monospecific for Bw55 and B7. The following sera formed clusters, identifying the Bw55 specificity (4592, 4590, 4595, 4581, and 4578), and the B7 (4607 and 4613). Other sera were polyspecific. The monospecific sera for Bw54, Bw56, and Bw42 were not obtained in this society set. Twenty-one sera (4602, 4608, 4605, 4610, 4586, 4593, 4600, 4601, 4612, 4614, 4615, 4582, 4583, 4613, 4607, 4578, 4581, 4590, 4592, 4588, and 4589) had R values greater than 0.7 with Bw22 antigen group. The reactivities of these sera are given in Table 1. The reaction patterns of these antigens are shown in Table 2. It seems that four sera (4584, 4593, 4617, and 4614) could split Bw56 into two distinct portions (long

Table 1. Analysis of the Correlation Between Serum Activity and Antigenic Specificity of the Sera

Serum Identity	Specificities	R value	% False pos	% False neg	Extras
4582	Bw22	0.89	6.7	2.6	
4583	Bw22	0.77	3.5	22.8	
4592	Bw55	0.82	3.4	8.8	Bw42
4590	Bw55	0.82	2.9	12.1	Bw42
4595	Bw55	0.76	4.2	15.5	Bw42
4581	Bw55	0.75	7.1	5.2	Bw54,Bw67
4578	Bw55	0.85	3.9	7.1	Bw42
4607	B7	0.87	0.9	15.9	
4613	B7	0.82	3.8	10.0	Bw42
4588	Bw55+w42	0.82	5.6	7.9	
4589	Bw55+w42	0.76	1.7	27.8	
4593	Bw22+w42	0.78	5.4	14.8	
4586	Bw22+w42	0.77	9.2	49.0	B7
4602	B7+w42+w55	0.84	7.7	7.5	Bw56
4608	B7+w42+w55	0.83	5.6	11.3	
4610	B7+w42+w55	0.79	4.1	16.8	Bw56
4614	B7+w42+w55	0.75	3.1	25.6	Bw56,Bw67
4606	B7+w42+w55+56	0.91	4.9	3.4	Bw54
4615	B7+w42	0.88	2.1	10.7	
4612	B7+w42	0.87	6.0	1.9	
4601	B7+w42	0.84	7.7	2.0	Bw55
4600	B7+w42	0.77	3.7	22.1	

Table 2. Reaction Pattern of the Sera

	Serum No.																				
HLA	4590	4595	4592	4578	4581	4583	4582	4586	4593	4584	4617	4614	4606	4610	4608	4615	4612	4601	4600	4607	4613
Bw54	–	–	–	–	*	+	+	+	–	–	–	–	–	*	–	–	–	–	–	–	–
Bw55	+	+	+	+	+	*	+	+	+	+	+	*	+	+	+	–	–	–	–	–	–
Bw56	–	–	–	+	–	–	+	+	*	*	*	*	+	+	–	–	–	–	–	–	–
Bw42	–	–	–	–	–	–	–	+	+	+	+	+	+	+	+	+	+	+	+	–	–
B7	–	–	–	–	–	–	–	–	–	–	+	+	+	+	+	+	+	+	+	+	+

pattern and short pattern, Table 3). Most of the Japanese are positive for the cluster sera, but all Caucasians except one are negative. Two families were available for the segregation of Bw56 short, but the family having Bw56 long was not found. There were a few false positives or false negatives, but the segregations were confirmed in these families. The cells of family 1 were the same cells that were used at the Eighth Japanese Workshop. No distinctive patterns suggestive of a split were detected in the 52 positive cells of B7, 20 positive cells of Bw42, 57 positive cells of Bw54, and 48 positive cells of Bw55 (Table 4).

A cluster of sera appears to split the Bw56 antigen into two portions, Bw56.1 and Bw56.2. The "short" Bw56.1 antigen, seen predominantly in Caucasians, reacts only with the sera 4578, 4582, 4605, 4606, and 4610, but does not react with the sera 4584, 4617, 4614, and 4593. The "long" Bw56.1 + Bw56.2 antigen seen predominantly in Japanese reacts with both groups of the tested sera.

In the Core serum set, Bw54 can be defined by three sera, and one serum of Bw42 monospecific sera was found from 30 sera submitted as containing Bw54, Bw55, Bw56, and/or Bw42. The reactivities of these sera are given in Table 5. Sera 303, 304, and 305 as Bw54, and 320 as Bw42 were confirmed as monospecific sera respectively. There were sera in the Core set showing concomitant specificities of Bw56, B35, and Bw62 but not with Bw55 or Bw54. This may be of some significance in that antigen Bw56 is different from the other Bw22 complex antigen in its epitope.

Table 3. The Reaction Pattern of Bw56 Antigen

					Serum No.									
Lab	ID	Ethnic	C	Locus	4578	4582	4605	4606	4610	4584	4617	4614	4593	
TSU	412	Jap	4	–	+	+	+	+	+	+	+	+	+	Long pattern
TSU	401	Jap	1	3	+	+	+	+	+	+	+	+	+	
KSH	6015	Jap	4	–	+	+	+	+	+	+	+	+	+	
KSH	6021	Jap	4	–	+	+	+	+	+	+	+	+	+	
TOK	2515	Jap	1	3	+	+	+	+	+	+	+	+	+	
PAK	122	Jap	1	3	+	+	+	+	+	+	+	–	+	
SEK	3697	Jap	1	7	+	+	+	+	+	+	0	+	+	
MRV	846	Cau	1	7	–	+	+	+	+	–	+	+	+	
NEU	8	Cau	1	7	+	+	+	+	+	–	+	–	–	Short pattern
NEU	9	Cau	1	6	+	+	+	+	+	–	–	–	–	
NEU	12	Cau	1	4	+	+	+	+	+	–	–	–	–	
NEU	53	Cau	1	–	+	+	+	+	+	–	–	–	–	
NEU	4	?	1	–	+	+	+	+	+	–	–	–	–	
TOK	1072	Jap	1	7	+	+	+	0	+	–	–	–	–	Family 1 F
TOK	1853	Jap	1	5	+	+	+	+	+	–	–	–	–	M
TOK	1854	Jap	1	3	+	+	+	+	+	–	–	–	–	C
NEU	1	Cau	1	4	+	+	+	+	+	–	–	–	–	Family 2 F
NEU	2	Cau	1	7	–	–	+	+	+	–	–	–	–	M
NEU	3	Cau	1	1	+	+	+	+	+	–	–	–	–	C1
NEU	4	Cau	1	7	+	+	+	+	+	–	–	–	–	C2
NEU	5	Cau	1	4	+	+	+	+	0	–	–	–	–	C3
NEU	6	Cau	1	7	+	–	–	+	+	–	–	–	–	C4

Table 4. Pattern Analysis in Antigens of Antigen Society #12 Positive Cells

B7 N=52	Serum No. 4602	4608	4610	4606	4615	4612	4601	4600	4614	4607	4613	n=	Freq.
pattern 1	+	+	+	+	+	+	+	+	+	+	+	40	76.9%
pattern 2	+	+	+	+	+	+	+	+	+	−	+	3	5.8
pattern 3	+	−	+	+	+	+	+	+	+	−	+	2	3.8

Bw42 N=20	Serum No. 4593	4586	4602	4608	4610	4606	4615	4612	4601	4600	4614	n=	Freq.
pattern 1	+	+	+	+	+	+	+	+	+	+	+	9	45.0%
pattern 2	−	−	+	+	+	+	+	+	+	+	+	2	10.0
pattern 3	+	−	+	+	+	+	+	+	+	+	+	2	10.0
pattern 4	+	−	+	+	+	+	−	+	+	+	+	2	10.0

Bw54 N=57	Serum No. 4579	4582	4583	4586	4610	n=	Freq.
pattern 1	−	+	+	+	+	17	29.8%
pattern 2	−	+	+	+	−	17	29.8
pattern 3	+	+	+	+	+	11	20.4
pattern 4	−	−	−	−	−	7	13.0

Bw55 N=48	Serum No. 4592	4590	4595	4581	4578	4579	4582	4593	4586	4602	4608	4610	4606	n=	Freq.
pattern 1	+	+	+	+	+	+	+	+	+	+	+	+	+	28	58.3%

Bw56 N=13	Serum No. 4578	4579	4580	4582	4593	4586	4600	4610	4606	n=	Freq.
pattern 1	+	+	+	+	+	+	+	+	+	3	23.1%
pattern 2	+	−	+	+	−	−	−	+	+	2	15.4
pattern 3	+	−	−	+	−	−	−	0	+	2	15.4

Table 5. Analysis of Correlation Between Serum Activity and Antigenic Specificity of the Sera of Core Serum Set

Serum Identity	Specificities	++	+−	−+	−−	Total	R value
303	Bw54	41	1	20	312	374	0.78
304	Bw54	39	4	14	319	376	0.78
303	Bw54	33	9	10	320	372	0.74
320	Bw42	33	3	15	318	369	0.76
325	B7+Bw42	76	13	13	260	362	0.80
324	B7+Bw42	89	3	20	256	368	0.84

Conclusion

The following sera formed clusters defining Bw54 (10W303, 304, and 305), Bw55 (10W4578, 4581, 4590, 4592, and 4595), broad Bw22 (10W296, 4582, and 4583), and Bw42 (10W320). Bw54 and Bw42 could be defined well with the tenth Core serum set and Bw55 with the sera of Antigen Society set 12. Bw56, however, was not defined as a monospecific serum. It was possible to detect Bw55 and Bw56 based on reactivity with several polyspecific sera of the Core serum set containing Bw22 antigen group. We can split Bw56 into two portions, Bw56.1 and Bw56.2, by a cluster of sera. Most Japanese belong to Bw56.1+56.2 (long Bw56), Caucasian, Bw56.1 (short Bw56). The (short) Bw56.1 was seen in a Caucasian family, a Japanese family, and a random Caucasian population. The report by E.L. Milford et al. on The Serologic Exercises of the 10th International Histocompatibility Workshop gives additional detail on the serology of Antigen Society #12.

References

1. Juji T, Hagino Y, Tamaru M, et al. A possible new Japanese specific HL-A antigen (J-1). Jap Exp Med 1973;43:447.
2. Nakayama E, Itakura K, Yakura H, et al. Sa-1, a possible new HL-A specificity found in the Japanese population. Vox Sang, Basel, 1974;27:134.
3. Saito S, Naito S, Toyoda K, et al. A study on HLA system in Japanese. Tissue Antigens 1975;5:217.
4. Batchelor JR, Chapman BA. Genetic background and transplantation antigens. J Clin Path 1967;30:415.
5. Albert ED, Mickey MR, McNicolas, et al. In: Terasaki PI (ed): Histocompatibility Testing 1970. Munksgaard, Copenhagen, 1970;221.
6. Dausset J, Colombani J, Legrand L, et al. Population and family studies in a French population with special reference to non-HLA antibodies. In: Dausset J, Colombani J (eds): Histocompatibility Testing 1972. Munksgaard, Copenhagen, 1973;107.
7. Thorsby E, Lidholm A, Sandberg L, et al. The HL-A system. Further evidence of antigenic heterogeneity. Vox Sang, Basel, 1971;21:68.

Antigen Society #13 Report (B7, B27, Bw47, Bw73)

Danny A. Youngs,[1] Sharon A. Alosco,[2] Leo P. de Waal[3]

HLA-B7 and HLA-B27 are well defined by monospecific Tenth International Workshop sera listed in Table 1 (Core sera) and Table 2 (Antigen Society 13 sera). There are no monospecific Workshop sera to define BW47, but this specificity is easily defined using B27 + Bw47 sera and B40 + B13 + Bw47 sera (Tables 1 and 2). Bw73 was originally described by Mayr and Kirnbauer as "ka" in 1977 (1), and was given WHO nomenclature after the Ninth Workshop, based on the reactivity of three sera (9w245, 246, 247), which also reacted with other B7-CREG (cross reactive group) antigens (2). In the Tenth Workshop there are five Core sera plus seven Antigen Society 13 sera that define Bw73 in the absence of other B7-CREG antigens. In addition, there are two Core sera, 10w336 (9w440) and 10w338, that were submitted as monospecific Bw73 sera. Although these two sera were not strictly monospecific, having extra reactivity toward Bw46 and Cw7, they were not reactive with other B7-CREG antigens. The Cw7 reactivity in these two sera does not appear to be an artifact of Cw7 linkage with B7 and Bw73. A "new" B7-like antigen, pot, previously described by Reekers et al. (3), is identified by three Core sera and five sera from Antigen Society 13. These eight sera are positive with other B7-CREG antigens, so assignment of B pot is possible only in the absence of other B7-CREG antigens.

Table 1. Best Core Sera

	Percent Inclusion					
Serum	B7	B27	Bw47	Bw73	B pot	Others
276	96					
278	89					
335	86					
275	95					Bw42(65)
325	98					Bw42(83)
280		96				
282		98				
283		88				
292		91	92			
288		94	56			
368			90			Bw60(94),Bw61(92), B13(89)
353			99			Bw60(98),Bw61(98), B13(96),Bw48(91)
375			97			Bw60(98),Bw61(98), B13(97),Bw48(95)
370			87			Bw60(95),Bw61(94), B13(90)
339	98	89		68		Bw42(65),Bw67(50)
341	98	97		76		
343	98	95		59		Bw48(68)
340	98	92		77	100	Bw42(93),Bw48(77), Bw60(69),Bw55(74)
333	96	21		86	67	Bw42(47)
336				70		Bw46(55),Cw7(32)
338				70		Bw46(23),Cw7(31)
381	97	90			100	Bw42(81),Bw48(78), Bw67(57)

Reporting Laboratories: US7HAN,[1] NCYYUN,[2] BENENG.[3]

Table 2. Best Society 13 Sera

Serum	Percent Inclusion					Others
	B7	B27	Bw47	Bw73	B pot	
4622	90					
4624	90					
4619	81					
4648		96				
4650		96				
4649		94				
4647		98				
4656		90				
4660		92	72			B13(65)
4671			84			B44(93),B45(94), B13(65)
4672			67			Bw60(94),Bw61(88), B13(81)
4668			57			Bw60(94),Bw61(88), B13(69)
4637	77			73		
4659		94		77		Cw2(73)
4633	94			87	100	Bw48(75)
4645	98	83		86		Bw42(66)
4673	97	98	22	100	96	Bw42(85),Bw60(93), Bw61(100)
4674	84	92		93		Bw42(76)
4680	95	85		85		A29(84),Bw22(92), B40(62)
4642	96				100	Bw42(80),A11(43)
4644	66				86	Bw42(54)
4675	86				83	Bw60(94),Bw48(90)

Subtypes of B7 and B27, defined by cytotoxic T-lymphocytes, have been described (4,5). Moreover, it was shown that CTL-defined B27 subtypes can be recognized by antibody as well (6). Testing of Tenth Workshop sera against B7 and B27 subtypes revealed that these sera can partially identify such subtype molecules (Table 3).

Conclusion

B7 and B27 are antigens well defined by monospecific Workshop antisera. Bw47, Bw73, and a "new" specificity, B pot, can be adequately defined in the absence of cross-reacting antigens by use of multispecific Workshop antisera. Subtypes of B7 and B27, known from T-cell clone and biochemistry studies, can be partially discriminated serologically by some Tenth Workshop antisera.

Table 3. Recognition of CTL-defined B7 and B27 Subtypes by Tenth Workshop Sera

Core Set Serum	CTL-defined		B7 subtypes	
	B7.1	B7.2	B7.3	B7.4
278	+	–	–	–
Antigen-Society-13				
4636	+	–	–	–
4637	+	–	–	–
4655	–	+	+	+
4669	–	+	+	+
4685	–	+	+	+

Core Set Serum	CTL-defined		B27 subtypes	
	B27.5	B27.2	B27.4	B27.6
292	+	–	+	+
Antigen Society 13				
284	+	–	+	+
4661	+	–	+	+
9333	–	–	+	+

References

1. Mayr WR, Kirnbauer M. Rare HLA-A,B,C antigens in caucasoids. In: Steinberg C, Lefkovits L (eds): The immune system, vol 2. Karger, Basel, 1981;217.
2. Kissmeyer-Nielsen F, Graugaard B, Kuhnl P, Mayr WR. Antigen reports: HLA-Bw73. In: Albert ED, Bauer MP, Mayr WR (eds): Histocompatibility Testing 1984. Springer-Verlag, Berlin, 1984;175.
3. Reekers P, van der Beveken M, Andrien M, Beelen JM, Müller-Eckhardt G, Schreuder GMTh. HLA B pot: A new HLA antigen within the B7 cross-reactive group. Tissue Antigens 28:182.
4. Breur-Vriesendorp BS, Neefjes JC, Huis B, van Seventer GA, Ploegh HL, Ivanyi P. Identification of new B27 subtypes (B27C and B27D) prevalent in Oriental populations. Hum Immunol 1986;16:163.
5. Seventer van GA, van der Horst AR, Semeijn J, Lardy NM, Krom FEJM, Breur-Vriesendorp BS. Six subtypes of HLA-B7 defined by histotopes, epitopes and ID-IEF. In: Dupont B (ed): Immunobiology of HLA, Volume I: Histocompatibility Testing 1987, New York: Springer-Verlag, 1989.
6. De Waal LP, Krom FEJM, Breur-Vriesendorp BS, Engelfriet CP, Lopez de Castro JA, Ivanyi P. Conventional alloantisera can recognize the same HLA-B27 polymorphism as detected by cytotoxic T-lymphocytes. Hum Immunol 1987;20:265.

Antigen Society #13 Report (B7, B27, Bw47, Bw73): Public Determinants of the HLA-B7 CREG Defined by Antibodies and T-Cell Clones*

G.A. van Seventer,[1] A.R. van der Horst,[2] L.P. de Waal, P. Reekers,[3] and P. Ivanyi[1]

We have previously described a fine-specificity analysis of a series of HLA-B7 allospecific CTL clones (*1). Besides an HLA-B7-specific reactivity, these CTL clones exerted distinct "extra reactions," i.e., lysis of some non-B7 target cells. These extra reactivities showed a remarkable resemblance with the well-known compartments of the cross-reacting group of HLA-B7 (B7-CREG) as defined by antibodies. We therefore extended the fine-specificity analysis of our B7-specific CTL clones by studying the reactivity patterns of the clones on a combined panel of cell lines from the Workshop (WS) and of local source. The CTL clones KOR-1, KOR-18, KOR-132, HG-31, HG-61, and LIV-8 were isolated by limiting dilution procedures from an MLC stimulation. All clones were derived from stimulation with cells that were either HLA-B7 homozygous or heterozygous HLA-B7, B44, while the clones themselves all express HLA-B60 and B44 (*1). Thus, there may be a bias in the reaction patterns of the clones because of the self-tolerance for a part of the B7-CREG, which includes B60.

The CTL clones can differentiate among the HLA-B7⁺ panel members and define at least six B7-subtypes (described in another report in this volume by van Seventer et al.). The results obtained (see Table 1) indicated that the HLA-B7 expressing cell lines from the WS panel SA, SCHU, WT8, BM14, SAVC, HHKB, EA, and HO104 were of the B7.1.1 subtype, while the cell line LD2B is of the B7.1.2 subtype.

We then further tested the reactivity of the CTL clones on a panel of cell lines expressing HLA antigens belonging to the CREG of HLA-B7 (Table 1), including the seven subtypes of HLA-B27 (between brackets are synonyms given, as used in literature) B27.1 (27W, 27a), B27.2 (27K, 27e), B27.3 (27C, 27b), B27.4 (27D), B27.5 (27f), B27.6 (27J, 27d) and B27c. The clone KOR-132 recognized all B27.1⁺ cells (the WS cell lines BTB, JESTHOM, HOM-2, and WT24 are all of the B27.1 subtype). A B27.2 expressing cell line from local

*This study was supported by a grant from The Netherlands Kidney Foundation (grant no. 84502).
Reporting Laboratories: BENIVA,[1] BENENG,[2] BENREE[3]

Table 1. HLA-B7 Allospecific CTL Clones

HLA (subtype)	KOR-132	KOR-1	KOR-18	LIV-8	HG-31	HG-61
B7.1.1	+	+	+	+	+	−
B7.1.2	+	+	+	+	−	−
B7.2	+	+	+	+	+	+
B7.3	+*	−	−	−	−	−
B7.4	+	+	+	−	−	−
B7.5 (BPot)	+	+*	−	−	−	−
B27.1 (W)	+	−	−	−	−	−
B27.2 (K)	−	−	−	−	−	+
B27.3 (C)	−	−	−	−	−	−
B27.4 (D)	−	−	−	−	−	−
B27.5 (f)	−	−	−	−	−	+
B27.6 (d,J)	−	−	−	−	−	−
B27c	−	−	−	+	+	−
B60 (A)	−	−	−	−	−	−
B60 (B)	−	−	−	−	−	+
B61	−	−	−	−	−	−
B54 nt						
B55	−	+	−	−	−	−
B56	−	−	−	−	−	−
B42	+*	+	−	−	−	−
B47	−	−	−	−	+	−
B13, B41, B48, and B73 all negative.						

The ⁵¹Chromium release assay was used according to the Workshop procedures and was therefore performed in a 4-hour assay with an effector target cell ratio of 10:1. Specific lysis on the stimulator target cells was always above 30%. Positive reaction (+) is considered above 25%, weak reaction (+*) is between 20–25%, negative reaction (−) less than 10%.
Not tested = nt. In all experiments the mean of six-fold determinations was taken.

Immunobiology of HLA
Volume I

Table 2. Protein Sequence From the First Three Extracellular Domains of the CREG of HLA-B7

Alpha 1 Domain.

	10 20 30 40 50 60 70 80 90
HLA-A2.1	GSHSMRYFFTSVSRPGRGEPRFIAVGYVDDTQFVRFDSDAASQRMEPRAPWIEQEGPEYWDGETRKVKAHSQTHRVDLGTLRGYYNQSEA
HLA-B7.2	--------Y--------------S------------------P-E----------------RN-QIY--QA--D-ES-RN----------
B7.3	--------Y--------------S------------------P-E----------------RN-QIY--QA--D-ES-RN----------
HLA-B27.1	--------H--------------T-------L----------P-E----------------R--QIC--KA--D-E--R--LR-------
B27.2	--------H--------------T-------L----------P-E----------------R--QIC--KA--D-EN-RIALR-------
B27.3	--------H--------------T-------L----------P-E----------------R--QIC--KA--D-ES-R--LR-------
B27.4	--------H--------------T-------L----------P-E----------------R--QIC--KA--D-ES-R--LR-------
B27.f	--------H--------------T-------L----------P-E----------------R--QIC--KA--Y-EN-R-ALR-------
B27.d	--------H--------------T-------L----------P-E-------------H--R--QIC--KA--D-E--R--LR-------
HLA-B40*	--------H--------------T-------L--------T-P-K----------------R--QIS-TNT--Y-ES-RN----------
Bw60	--------H-AM-----------T-------L--------T-P-K----------------R--QIS-TNT--Y-ES-RN----------
HLA-B13	--------Y-AM-----------T----------------T-P--A---------------R--QIS-TNT--Y-EN-R-ALR-------
HLA-B41	--------H-AM-----------T-------L--------T-P-K----------------R--QIS-TNT--Y-ES-RN----------
HLA-B42	--------Y--------------S------------------P-E----------------RN-QIY--QA--D-ES-RN----------
HLA-B47	--------Y-AM-----------T-------L--------T-P-K----------------R--QIS-TNT--Y-E--R--LR-------

Alpha 2 Domain.

	100 110 120 130 140 150 160 170 180
HLA-A2.1	GSHTVQRMYGCDVGSDWRFLRGYHQYAYDGKDYIALKEDLRSWTAADMAAQTTKHKWEAAHVAEQLRAYLEGTCVEWLRRYLENGKETLQRT
HLA-B7.2	----L-S-------P-G-L---HD------------N----------T---I-QR-----RE---R------E-------------DK-E-A
B7.3	----L-S-------P-G-L---HD-X----------N----------T---I-QR-----RE---R------E-------------DK-E-A
HLA-B27.1	----L-N-------P-G-L------D----------N---S------T---I-QR-----R-----------E------------------A
B27.2	----L-N-------P-G-L------D----------N---S------T---I-QR-----R-----------E------------------A
B27.3	----L-N-------P-G-L------D----------N---S------T---I-QR-----RE----------E------------------A
B27.4	----L-N-------P-G-L----D------------N---S------T---I-QR-----RE----------E------------------A
B27.f	----L-N-------P-G-L------D----------N---S------T---I-QR-----R-----------E------------------A
B27.d	----L-N-------P-G-L------D----------N---S------T---I-QR-----R-----------E------------------A
HLA-B40*	----L-S-------P-G-L---HN------------N----------T---I-QR-----R-----------E------------------A
Bw60	----L---------P-G-L---HN------------N----------T---ISQR-L---R-----------E-------------DK-E-A
HLA-B13	----W-T-----L-P-G-L---HN-L----------N---S------T---I-QL-----R-----------E------------------A
HLA-B41	----W---------P-G-L---HN------------N----------T---I-QR-----R-------------------------D--E-A
HLA-B42	----L-S-------P-G-L---HN------------N----------T---I-QR-----R----D--------------------D--E-A
HLA-B47	----I---F-----P-G-L------D----------N---S------T---I-QR-----R-----------E------------------A

Alpha 3 Domain.

	190 200 210 220 230 240 250 260 270
HLA-A2.1	DAPKTHMTHHAVSDHEATLRCWALSFYPAEITLTWQRDGEDQTQDTELVETRPAGDGTFQKWAAVVVPSGQEQRYTCHVQHEGLPKPLT
HLA-B7.2	-P----V---PI------------G----------------------C--------R--E----------E------------------

All B-locus products depicted in this table are homologous in the alpha 3 domain with HLA-B7.2 (or the identical B7.3), with the exception that they all have a Q at AA residue 242 instead of an E. Furthermore HLA-B13 and B47 also have two additional AA changes at AA residue 230, an L instead of a C, and at AA residue 239, a G instead of an R. Protein sequence publications of HLA-B7, referred to as B7.1 and B7.2 in the literature by other research groups, are according to our nomenclature, respectively B7.2 and B7.3. The original cell lines used for AA sequence analysis of the HLA-B7 and HLA-B27 group of subtypes were included in our target cell panel. The AA sequences of B40* and B47 were derived from the WS cell lines SWEIG007 (B40*) and PLH (B47). The sequence data are a compilation from a large series of publications from the research groups of Dr. J.A. Lopez de Castro, Dr. J. Strominger, Dr. P. Parham, and Dr. B. Dupont.

source (a previous study showed that all our B27.2 expressing cell lines from local panel, n=11, were recognized by clone HG-61) and the one B27.5 (LIH) expressing cell line were lysed by the clone HG-61. The clones LIV-8 and HG-31 reacted positively with the only available B27c expressing cell line (ETO). The HLA-B42 positive cell line RSH from the WS panel was lysed by the clones KOR-1 and KOR-132. The WS cell line PLH ($B47^+$) was only recognized by clone HG-31. Not included in the WS panel but available from local source were HLA-B55 and B56 expressing cell lines, of which only the B55 expressing cell lines were recognized by the clone KOR-1.

The clone HG-61 reacted with only a small group of HLA-B60 positive cells [B60(B)] including the WS cell lines PE117 and MT14B, whereas the $B40^+$ WS cell lines SWEIG007, SLE005, MADURA, DKB, and EMJ were not recognized. This is intriguing because the clone itself is also typed HLA-B60, indicating that there are at least two subtypes of B60. A further analysis of a very limited number of $B60^+$ cells revealed no clear correlation with the two serologically defined B60 subtypes (with sera from the Core set, e.g., WS 356, 371 vs WS 357, 360, 363, and 384) and the two B60 variants found in 1D-IEF (*2).

A comparison of the reactivity patterns of the antibodies from the Core set, the Antigen Society 13 (B27-B7), and 14 (B40) with that of the CTL clones showed that the reaction patterns of the CTL clones are all included in the specificity patterns of broad reactive alloimmune sera from the WS, although neither of them was completely identical.

Correlations made between stretches of amino-acid residues from the known protein sequences from the first three extracellular domains of the CREG of HLA-B7 (Table 2) with the reaction patterns of the CTL clones were not conclusive. The data indeed indicate that the variable regions in the alpha helices of the HLA molecules are important (e.g., the AA residues spanning position 63-83), and they provide also the warning that one cannot just assign isolated AA residues out of the context of the whole MHC molecule to be the target structures for T-cell recognition.

As an example, the HG-61 clone's lack of reactivity with the B7.3 subtype indicates that AA residue 116 is of importance, because this position is the only difference between the subtypes B7.2 and B7.3. From the B27 subtypes, only the B27.2 and B27.5 are recognized, which suggests that the AA residue on position 81 is involved in the reaction with this clone. The first domain of B42 is identical with B7.2 indicating that the lack of recognition must be caused by the second domain of B42, which is different from B7.2 at AA residues positions 114, 152, 156, 163, and 178. Furthermore, the differential lysis on the B40 expressing cell lines may possibly indicate that the AA residues on position 11 and 12 are also of relevance (the cell line SWEIG007 is not recognized and was the original cell line for the B40* sequence in Table 2). The complexity and difficulties in assigning the relevant AA residues is well illustrated by this example.

In conclusion, the fine-specificity analysis of the HLA-B7-allospecific CTL clones indicates a further heterogeneity of several HLA antigens of the HLA-B7 CREG, e.g., splits of B27 and B60. Furthermore, various CTL-defined public determinants (histotopes) can be included in the serologically defined CREG of HLA-B7.

References

1. Van Seventer GA, Huis B, Melief CJM, Ivanyi P. Fine specificity of human HLA-B7-specific cytotoxic T-lymphocyte clones. I. Identification of HLA-B7 subtypes and histotopes of the HLA-B7 cross-reacting group. Hum Immunol 1986; 16:375.
2. Neefjes JJ, Breur-Vriesendorp BS, Van Seventer GA, Ivanyi P, Ploegh HL. An improved biochemical method for the analysis of HLA class-I antigens. Definition of new HLA class-I subtypes. Hum Immunol 1986;16:169.

Antigen Society #13 Report (B7, B27, Bw47, Bw73): Six Subtypes of HLA-B7 Defined by Histotypes, Epitopes and 1D-IEF*

G.A. van Seventer,[1] A.R. van der Horst,[2] J. Semeijn,[1] N.M. Lardy,[1] F.E.J.M. Krom,[1] B.S. Breur-Vriesendorp,[1] L.P. de Waal,[2] and P. Ivanyi[1]

The HLA-B7 antigen was one of the best serologically defined HLA antigens in the Ninth Histocompatibility Workshop in 1984. Nevertheless, we could describe at least four different subtypes of the HLA-B7 antigen (*1). We now present data indicating a further heterogeneity of the HLA-B7 antigen as analyzed by HLA-B7 allospecific CTL clones, conventional pregnancy sera, and 1D-IEF.

A panel of well-defined human HLA-B7 allospecific CTL clones was tested in a ^{51}Cr-release assay on a selected panel of HLA-B7 subtyped EBV-transformed B-cell lines. The reaction pattern of the clones on the various target cells, summarized in Table 1, indicated that six different subtypes of HLA-B7 could be defined. The previously designated subtype B7.1 was found to include two subtypes, B7.1.1 and B7.1.2, whereas the formerly described BPot antigen (*2) is here considered as an HLA-B7 subtype, namely B7.5.

One-dimensional IEF analysis (Fig. 1 and Table 1) revealed only two variants, an acidic variant (A) and a more basic variant (B). The B7.3 subtype correlated with the basic variant and has been studied by a comparative tryptic peptide analysis (*3). A tyrosine residue at position 116 of the heavy chain was reported to be substituted by an unknown residue when compared with the primary structure of the B7, expressed by the widely used B-cell line JY (HLA-A2,2:B7,7:Cw-,-:DR4,w6). However, a human fibroblast cell line (M1) and a human rhabdomyosarcoma cell line (RD) both transfected with identical B7 genes (as determined by DNA-sequence analysis) isolated from this cell line JY, were differentially recognized by our panel of CTL clones (Table 1). This heterogeneity was not caused by a difference in avidity between the CTL clones toward the transfected cells, because M1 cells transfected with an HLA-B27K gene were recognized by the clone HG-61. This CTL clone is specific for B7 and cross-reacts with B27K^{+} cells (see another report in this volume by Van Seventer et al.). This latter reaction is of a weaker avidity than the B7-specific reaction (as determined by blocking with CD8 specific MoAb). These results indicate that the lack of recognition of M1/B7 cells is not caused by a too low avidity interaction or by the absence of critical adhesion molecules on the transfected cell line. Further detailed analysis suggested that the cell line JY is heterozygous at the B-locus and expresses the B7.2 and the B7.4 subtypes (manuscript submitted). This finding may have implications for most of the biochemical studies performed on the B7 antigens expressed by JY.

We also performed a serologic analysis of the B7 antigen with sera obtained from the Tenth Workshop, including the Antigen Society 13 (B7-B27), and the Core set. On a panel of cells from B7-subtyped healthy blood donors, four distinct reaction patterns (A,B,C, and D in Table 1) could be observed, by which the B7 antigen can be divided into three clusters of epitopes.

Reporting Laboratories: BENIVA,[1] BENENG[2]

*This study was supported by a grant from The Netherlands Kidney Foundation (grant no. 84502).

Table 1. HLA-B7 Specific CTL Clones[a]

HLA-B7 Subtypes	KOR-132	KOR-1	KOR-18	LIV-8	HG-31	HG-61	1D-IEF	Serology Patterns A	B	C	D	Epitope Cluster
B7.1.1	+	+	+	+	+	−	A	+	−	+	+	I
B7.1.2	+	+	+	+	−	−	A	+	−	+	+	I
B7.2	+	+	+	+	+	+	A	−	+	+	+	II
B7.3	−/+	−	−	−	−	−	B	−	+	+	+	II
B7.4	+	+	+	−	−	−	A	−	+	+	+	II
B7.5	+	−/+	−	−	−	−	A	−	−	−	+	III
JY	+	+	+	+	+	+	A					
M1	−	−	−	−	−	−	na					
M1/B7(JY150)	+	+	+	−	−	−	A					
RD	−	−	−	−	nt	−	na					
RD/B7(JY158)	+	+	+	−	nt	−	A					

[a]+ = positive reaction, above 25% specific ^{51}Cr-release at E:T of 5:1, −/+ = only positive above E:T of 10:1, − = negative reaction, less than 5% specific ^{51}Cr-release at E:T of 5:1. All data as determined in a 4-h assay. nt = not tested. na = not applicable.

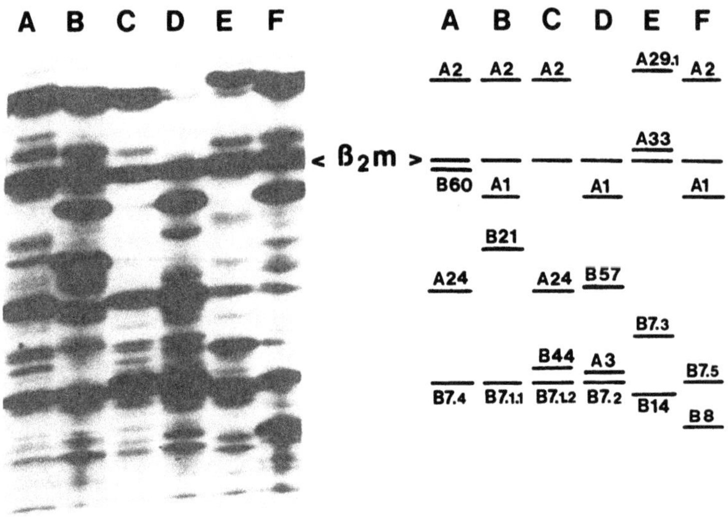

Figure 1. Iso-electric focusing gel comparing the different HLA-B7 subtypes. Iso-electric focusing was performed according to the WS procedures for 1D-IEF for HLA class-I antigens. The panel members were EBV-transformed cell lines A) O-T (HLA-A2,24;B60,7.4), B) KRO (HLA-A1,2;B21,7.1.1), C) HEM (HLA-2,24;B44,7.1.2), D) H-N (HLA-A1,3;B57,7.2), E) CF (HLA-A29[1],33;B14,7.3), and F) Bpot (HLA-A1,2; B8,7.5). For antigen designation the prefix W was omitted.

Cluster I correlates with B7.1.1 and B7.1.2, cluster II includes B7.2, B7.3, and B7.4, and cluster III defines B7.5. There were also sera reactive with all B7 subtypes. WS sera grouped in these reaction patterns were: reaction A–278, 4636, and 4637; reaction B–4669 and 4685; reaction C–275, 276, 277, 325, 331, 4632, 4643, and 4645; reaction D–333, 340, 381, 4633, 4642, and 4677.

Preliminary subtype frequencies of $B7^+$ individuals (n=50) indicated that the majority of the B7 antigens (80%) in the Caucasian population is of the B7.1.1 or B7.1.2 subtype, while the other subtypes are rarely found (each less than 5%). The B7.3 subtype is strongly associated with an 1D-IEF variant of HLA-A29 (4 out of 4).

In conclusion, evidence was obtained by CTL clone reactivity patterns, pregnancy sera, and biochemical analysis in 1D-IEF, that at least six subtypes of the HLA-B7 antigen could be defined.

References

1. Van Seventer GA, Huis B, Melief CJM, Ivanyi P. Fine specificity of human HLA-B7-specific cytotoxic T-lymphocyte clones. I. Identification of HLA-B7 subtypes and histotopes of the HLA-B7 cross-reacting group. Hum Immunol 1986;1375.
2. Reekers P, Van der Beucken M, Andrien M, Beelen J, Mueller-Eckhardt G, Schreuder GMTh. HLA-BPot: A new HLA antigen within the B7-cross-reactive group. Tissue Antigens 1986;28:182.
3. Taketani S, Krangel MS, Spits H, De Vries JE, Strominger JL. Structural analysis of an HLA-B7 antigen variant detected by cytotoxic T-lymphocytes. J Immunol 1984;133:816.

Antigen Society #14 Report (B40 CREG, BW60, BW61, BW41, BW48, B13)

R.M. Radvany,[1] J. van Munster,[2] S. Nakata,[3] A.A. Biegel,[4] Y.K. Paik,[5] D. Middleton,[6] K. Tokunaga,[7] E. Gyodi,[8] N. Nagao,[9] N. Straachan,[10] N. Tokashige,[5] P. Reekers,[2] H. Amemiya,[11] S.H. Chan,[12] P. Chiewsilp,[13] R. Fong,[14] T. Frazer,[10] T. Fukunishi,[16] R. Garrido,[17] K. Gelsthorpe,[18] P. Holland,[19] S. Hsia,[20] X. Jingjuan,[21] M. Kreisler,[22] D. McCluskey,[23] H. Mervart,[24] K. Omori(Japan Sakvra), H. Saji(Kyoto), K. Sakagami(Okayama Japan), M. Tada(Japan Osaka), K. Taniwaki(Nishinomiya Japan), L. deWaal,[25] L.J. Kennedy,[26] G. Woodfield,[27] and Y. Sun[28]

Introduction

In the Tenth International Histocompatibility Workshop the B40 CREG was studied by Antigen Society No. 14. Thirty laboratories analyzed 854 cells against 89 Core and 81 Antigen Society antisera. The cells were of Chinese, Japanese, Polynesian, Moluccan, Maori, Thai, Gypsy, South African and Caucasoid origin. Initial antigen assignments of cells were based on the typing with sera from the Ninth International Histocompatibility Workshop (9IHW) and of local origin.

History

The HLA-B40 (BB,W10) antigen was first described by Svejgaard et al. in 1970 (1). The heterogeneity of HLA-B40 was revealed in the Fifth International Histocompatibility Workshop (2) and lead to the recognition of BW41 (3), BW47 (4), and BW48 (5), as well as BW60 and BW61 (6) specificities. The two splits of B40 (BW60 and BW61) were first reported in an African Indian population by Hammond et al. (7), but the Seventh and Eighth International Histocompatibility Workshops revealed their presence also in Caucasoids and Japanese (4,8). The B13 antigen was described in the fourth International Histocompatibility Workshop (9), and has always been considered a well defined antigen, although many anti-B13 antisera contain additional antibodies directed to antigens of the B40 CREG.

Analysis

Preworkshop analysis of the data was based on 2 × 2 contingency tables from the participating centers. Final analysis was performed utilizing ten IHW programs. All sera were analyzed against all cells in the file using the tail analysis program. Sera with the highest Q scores for the BW60, BW61, BW41, BW48, and B13 specificities were used to check HLA-types assigned locally prior to the Workshop. If correction of preworkshop HLA types was necessary, this was done after extensive inspection of the observed reaction patterns. Final analysis was based on a corrected file of 854 cells. In a separate analysis using cells from individuals representing the different ethnic groups (Negroid, Oriental and Caucasoid), pattern programs and S x C tables were utilized to help demonstrate the existence of possible splits. Reaction patterns of "splits" were based on reactions of antisera with specificities recorded in the cell file, and could not always be verified by demonstrating segregation within families because family data were not available.

Results

The Core Serum set and the Antigen Society serum set contained operationally monospecific antisera for B13, BW41, and BW60 antigens. By contrast, all antisera for BW61 and BW48 and most of the BW60 antisera also cross-reacted with at least one additional antigen of the B40 complex. In addition, many sera contained antibodies to BW47 and/or B7 as well as to other specificities.

BW60, BW61, BW41, BW48, B13 "specific" sera. As shown in Table 1, serum reaction patterns against cells from all available ethnic groups identified five antisera (358, 357, 4742, 356, 4744, and 340) that reacted only with BW60 positive cells but not with W61 positive cells. Two of these sera (356 and 4744) were relatively weak, and serum 340 was broadly reactive with B7 CREG. Serum 363 reacted strongly with BW60 positive but weakly with some BW61 positive cells.

Monospecific BW61 sera were not found (Table 1) and BW61 was defined by subtraction. Serum 367 was the only serum that reacted more strongly with BW61 than with BW60 positive cells. Thus, with the available Workshop sera, it was not possible to positively identify BW60, BW61 heterozygotes in the absence of family segregation data.

There were four operationally monospecific BW41 antisera (390, 9008, 387, and 9659; see Table 1), that reacted with BW41 positive cells from all ethnic groups tested. BW48 was best identified with sera 381, 4743 (weak serum), 382, and 360 (Table 1) irrespective of the ethnic group tested. Numerous excellent monospecific B13 antisera were identified. Table 2 lists antisera reacting with all B13 cells giving their R values and Q scores.

Reporting Laboratories: US4RAD,[1] BENREE,[2] JAPMAT,[3] US1IND,[4] US5PAK,[5] UKIMID,[6] JAPTOK,[7] Budapest,[8] Japan,[9] ANZFRA[10]
Participating Laboratories: JAPAMY,[11] ANZCSH,[12] ANZCHI,[13] ANZFON,[14] JAPTKB,[16] FRAGRA,[17] UKIGEL,[18] NCYSAC,[19] US4SND,[20] CHICHE,[21] FRAKRE,[22] UKIFES,[23] NCYMRV,[24] BENENG,[25] UKIBOD,[26] ANZWOO,[27] CHIBCH[28]

Table 1. Reaction Patterns of Best Sera for Classical B40 CREG: BW60, BW61, BW41, BW48, B13

Serum			4			4							4	4		9	4		9		9		4				4
ID	3	3	7	3	3	7	3	3	3	3	3	3	7	7	3	6	7	3	0	3	6	3	7	3	3	3	7
	5	5	4	6	5	4	4	6	6	5	7	5	2	3	7	6	3	9	0	8	5	8	4	8	6	9	1
Specificity	8	7	2	3	6	4	0	7	5	1	1	5	5	0	4	1	1	0	8	7	9	1	3	2	0	1	1 ....
BW60	+	+	+	+	±	±	+	±	+	+	+	+	+	+	+	+	+	−	−	−	−	−	±	±	±	−	−
BW61	−	−	−	?	−	−	−	+	+	+	+	+	+	+	+	+	+	−	−	−	−	−	−	−	±	−	−
BW41	−	−	−	−	−	−	±	−	−	−	−	−	−	−	−	−	−	+	+	+	+	−	−	−	−	−	−
BW48	−	−	+	+	−	−	+	−	−	−	−	−	+	+	−	−	−	−	−	−	−	+	+	+	+	−	−
B13	−	−	−	−	−	−	−	−	−	−	−	−	−	−	±	+	+	−	−	−	−	−	−	−	−	+	+
additional reactivity																											
B7	−	−	+	−	−	−	+	−	−	−	−	−	−	+	−	−	−	−	−	−	−	+	+	+	−	−	−
BW42	−	−	−	−	−	−	+	−	−	−	−	−	−	−	−	−	−	−	−	−	−	+	−	+	−	−	−
BW47	−	−	−	−	−	−	+	−	−	−	−	−	−	−	−	−	−	−	−	−	−	−	−	−	−	−	−
BW73	−	−	−	−	−	−	+	−	−	−	−	−	−	−	−	−	−	−	−	−	−	−	−	−	−	−	−
extra	−	−	A1	−	−	−	BW22	−	−	−	−	−	−	−	−	−	−	−	−	−	−	−	A3	−	−	−	−
	BW60							BW61 + BW60										BW41				BW48				B13*	

*Many more excellent B13 sera were tested (see Table 2).

Table 2. Reactivity of B13 Antisera With Cells of All Ethnic Groups Represented

Serum #	R value	Q score	Serum #	R value	Q score
4711	0.95	10.29	393	0.93	7.58
391	0.92	10.22	4696	0.92	8.13
4689	0.97	9.93	4691	0.89	8.23
396	0.96	9.23	4698	0.90	7.25
395	0.94	8.81	4708	0.88	8.24
4690	0.93	8.38	4706	0.85	7.33
273 (+B18)	0.81	6.9	394 (+BW41)	0.72	3.68

BW60, BW61, BW41, BW48, B13: B40 CREG broadly reactive sera. Table 3 lists sera that reacted with (almost) all cells bearing one of the B40 CREG-antigens irrespective of the ethnic group tested. The broadly reactive antisera were used in combination with the "specific" antisera to correct HLA-types of cells in the cell file and to define serological splits.

Serological variants of BW60 and BW61. Because of difference in cell files used by the "pattern" program to analyze Core sera and Antigen Society sera, these two groups of sera were analyzed separately. Table 4 demonstrates the long (patterns BW60a and BW61a) and short (patterns BW60b and BW61b) reaction patterns. It appeared that two BW60 (382, 383) and two BW61 antisera (366, 373) could be detecting the shorter variety of BW60 and BW61.

These patterns, however, need to be confirmed by demonstrating their segregation in families and thereby eliminating the possibility that they are the result of weak serum reactions.

Variants of BW41. A short variety of BW41 described in the Ninth International Workshop (15) by the lack of reactivity with serum, 9W241 (EO243 = 91HWS 10W386 = 10W9658), was confirmed in this Workshop in the same Brazilian families as well as in Hungarian Gypsies (data not shown) and eight additional cells of random individuals (Table 5). Sera 326, 385, 386 and probably also 394 showed the same type of short reaction pattern as 9658. The other sera reacting with BW41 positive cells did not consistently follow any of the two reaction patterns and may have weak B41 antibodies.

Table 3. Reaction Patterns of B40 CREG Broadly Reactive Antisera

Serum ID Specificity	384	353	370	375	4747	4739	9662
BW60	+	+	+	+	+	+	+
BW61	+	+	+	+	+	+	+
BW41	+	+	+	±	+	+	+
BW48	+	+	±	+	+	+	+
B13	−	+	+	+	+	+	+
additional reactivity							
BW47	−	+	+	+	−	+	±
B7	±	±	−	−	−	+	+
BW73	−				−	+	+
extra	+					+	+

Variants of BW48. A short variety of BW48 ("FU") was described in the Japanese population by Kamura et al. (14), and segregation of the "FU" antigen was demonstrated in one Japanese family. The Workshop analysis showed that cells bearing the "FU" antigen reacted only with the broadly reactive sera 4730, 4738, 4739, 4746, 384, and 2096, but not with sera such as 381, 360, 382, and 4743, which were more restricted in their reactivity (Table 6). In addition, Workshop data suggested the existence of another short variant of BW48 characterized by negative reactions with sera 343, 4725, and 4729. However, family studies showing segregation of the short BW48 reaction patterns and biochemical studies have not been performed. It was noted that Antigen Society serum 4727 (not shown) reacted only with cells typed by the Chinese laboratories. It was also of interest that the long variety of BW48 was present in all racial groups including Maoris.

Variants of B13. As shown in Table 7, a short variety of B13 originally suggested by Yank et al., Mills et al., and Chiewsilp et al. (10,11,12), was observed in the Oriental population after analysis of B13 sera. Cells representing the short variants of B13 did not react with several monoclonal antisera thereby forming two loosely defined clusters: B13b Oriental (most cells of Chinese origin) and B13c Oriental (most cells of Thai origin). Cells representing both of these short varieties of B13, however, were positive with the B13 alloantisera listed in Table 2, and the three serum patterns shown in Table 7 are based on differences between reactions of monoclonal antisera, which did not always correlate with

Table 4. Reaction Patterns of BW60 and BW61 Antisera

Serum ID Specificity	358	357	382	383	366	373	365	351
BW60a	+	+	+	+	+	+	+	+
BW60b	+	+	−	−	+	+	+	+
BW61a	−	−	±	±	+	+	+	+
BW61b	−	−	−	±	−	−	+	+

Table 5. Variants of Bw41

Serum			9	9		9				9					9	9								4	4	4	9	2
ID	3	3	0	6	3	6	3	3	3	0	3	3	2	3	0	0	3	3	3	3	3	4	3	7	7	7	6	0
	9	8	0	5	8	5	2	8	9	0	8	8	3	6	0	0	4	6	5	7	8	5	9	3	4	4	6	9
Specificity	0	7	8	9	6	8	6	5	4	7	8	9	2	8	5	6	9	8	4	5	4	1	0	9	7	8	2	8
BW41.1	+	+	+	+	+	+	+	+	+	±	±	±	±	±	±	±	±	±	±	±	±	±	±	+	+	+	±	±
BW41.2	+	+	+	+	−	−	−	−	−	+	±	+	±	±	±	±	±	±	±	±	±	±	±	+	+	+	±	±

Table 6. Variants of Bw48

Serum ID Specificity	4730	4738	4739	4746	384	2096	381	4743	382	360	363	376	4742	4745	340	368	383	343	4725	4729
BW48	+	+	+	+	+	+	+	+	+	+	+	+	+	+	±	±	±	+	+	+
BW48 short	+	+	+	+	+	+	+	+	+	+	+	+	+	+	±	±	±	−	−	−
"FU"	+	+	+	+	+	+	−	−	−	−	−	−	−	−	−	−	−	−	−	−

reaction patterns of B13 alloantisera with Caucasoid B13 positive cells. Therefore, it remains to be confirmed whether the weak clusters B13b and B13c represent real variants of an Oriental B13 antigen or are caused by weak serum reactivities. The other B13 alloantisera tested, not listed in Tables 2 and 7, reacted either weakly or showed patterns that could not be correlated with any of the three patterns shown in Table 7. Although Oriental B13 gene products have an isoelectric point (IEP) that is different from the Caucasoid B13 (13), reactivity patterns of monoclonal antibodies listed in Table 7 did not correlate well with the IEPs observed. Thus, more investigations are needed to clarify the possible correlations between serological and biochemical epitopes in Orientals.

New Antigens within the B40 CREG Antigen Group. Based on previous publications, the low frequency antigens BRI (16,17) and BST40 (17,18) were investigated using B40-CREG Antigen Society sera. Those sera that reacted with the 8 BRI and 3 BST40 cells are listed in Table 8. Specific antisera directed to these new determinants were not found.

Class I Gene Biochemistry of B40 CREG. Distinct pH values for neuraminidase-treated immunoprecipitates with 4E/W6-32 were observed for gene products of BW61 (pH ~6.46) and BW48 (double band, upper band pH ~6.04). For BW60, the bands precipitated at about the same pI as β_2 microglobulin. For B13 different bands have been observed in Orientals (pH ~6.16) and Caucasoids (pH ~6.00). BW41 has two charged variants pH ~6.38 and pH ~6.20 respectively).

In the ELISA assay described elsewhere in this publication, monoclonal antibodies detecting B40 CREG antigens were not observed.

Acknowledgments. We wish to thank Drs. A. Amororo (Torino, Italy), A. Moraes (Rio de Janeiro, Brazil), F. Kissmeyer-Nielsen (Aarhus, Denmark), and C. Jersild (Aalborg, Denmark) for their contributions of cells from families with Bw41 variants and Bw48 antigens. We also wish to thank Drs. J. van Rood (Leiden, The Netherlands), J. Beelen (Groningen, The Netherlands), and A. Moraes (Rio de Janeiro, Brazil) for contributing additional interesting sera identifying variants of B40 CREG.

Table 7. Variants of B13

Serum ID Specificity	4711	391	396	2092	2080	2094	2101	2054	2081
B13	+	+	+	±	±	±	±	±	±
B13b (Oriental)	+	+	+	±	±	±	−	−	−
B13c (Oriental)	+	+	+	±	−	−	−	−	−

Table 8. New B40 CREG Determinants: "BRI," B "ST40"

BRI	Reacts with 232, 357, 353, 370, 375 (Core) and 4738, 4743, 4739 (Antigen Society)
ST40	Reacts with Core sera 273, 287, 288, 292, 326, 351, 357, 375, 394

References

1. Svejgaard A, Kissmeyer-Nielsen F, Thorsby E. HLA typing of platelets. In: Terasaki PI (ed): Histocompatibility Testing 1970. Munksgaard, Copenhagen, 1970;153–164.
2. Thorsby E. W10 (Workshop Report). In: Dausset J, Colombani J (eds): Histocompatibility Testing 1972. Munksgaard, Copenhagen, 1972;639–640.
3. Bodmer J. The ABC of HLA. In: Kissmeyer-Nielsen F (ed): Histocompatibility Testing 1974. Munksgaard, Copenhagen, 1974;50–51.
4. Dick HE, Bodmer WF. Joint Report 4: HLA-A, -B and -C serology. In: Bodmer WR, Batchelor JP, Bodmer JG, Festenstein H, Morris PJ (eds): Histocompatibility Testing 1977. Munksgaard, Copenhagen, 1977;178–180.
5. Kissmeyer-Nielsen F, Andersen H, Hause M, Kjerbye KE, Morgensen B, Svejgaard A. HLA types in Danish Eskimos from Greenland. Tissue Antigens 1971;1:74.
6. Dawkins RL, Houliston JB. Joint Report: BW60. In: Terasaki PI (ed): Histocompatibility Testing 1980. UCLA Tissue Typing Laboratory, Los Angeles, 1980;456–457.
7. Hammond MG, Appado B, Brain P. HLA Antigens and Antibodies in South African Indians. Tissue Antigens 1972;2:389.
8. Thompson JP. Joint Report BW60. In: Terasaki PI (ed): Histocompatibility Testing 1980. UCLA Tissue Typing Laboratory, Los Angeles 1980;460–461.
9. Allen FH, Amos DB, Batchelor JR. WHO terminology report. In: Terasaki PI (ed): Histocompatibility Testing 1970. Munksgaard, Copenhagen, 1970;49.
10. Yang E, Wong KH, Mervart H. Recognition of an HLA-B13 variant. Eleventh Workshop Newsletter No. 1, p 5.
11. Mills T, Street J, Darke C. Identification of an HLA-B13 variant. Eleventh Workshop Newsletter No. 2, p 8.
12. Chiewsilp P, Sujirachato K. Evidence of HLA-B13 variant. Eleventh Workshop Newsletter No. 1, p. 10.

13. Yang SY, Chang A, Olivero R, Relias V, Yunis EJ. IEF patterns of HLA-B13 antigens from Orientals and Caucasians. Immunogenetics 1986;21:125–134.
14. Kamura H, Kohara S, Shimizu K, Akaza T. Antigen report: B13, B40(Bw60,Bw61), Bw41, B247, Bw48, Bw73, FU. In: Aizawa M (ed): HLA in Asia-Oceania. Hokkaido University Press, Sapporo, Japan, 1986;120–122.
15. Mervart H, Taylor C, Ting A. Antigen report: HLA-Bw41. In: Albert ED, Baur MP, Mayr WR (eds): Histocompatibility Testing 1984. Springer-Verlag, Berlin, 1984;149.
16. Mueller-Eckhardt G, Schrender I, Mayr WR, Westphal E, Mueller-Eckhardt C. A new Bw6-associated B locus antigen "RI." Ninth Workshop Newsletter No. 5, pp 24–26.
17. Reekers P, Beucken vander M, Andrien M, Beelen J, Mueller-Eckhardt G, Schreuder GMTh. HLA-BPot: A new HLA antigen within the B7 cross-reactive group. Tissue Antigens 1986;28:182–189.
18. Watson K, Ness D, Gantan Z, Perkins H, Grumet FC. ST40: A new Bw40 subtype identified by alloantisera and a monoclonal antibody. Hum Immunol 1982;5:170–179.

Antigen Society #15 Report (Bw4 and Bw6)

A. Arnaiz-Villena,[1] M. Belvedere,[2] F. Decary,[3] M. Fotino,[4] E. Heise,[5] V. Hogan,[6] M. Martinetti,[2] C. Muller,[7] P. Richiardi,[8] J.L. Vicario,[1] M. Barbanti[(Italy)], J. Bruyere,[9] C. Caruso[(Italy)], C. Conighi,[10] K. Gelsthorpe, M. Hammond,[12] C. Lopez-Larrea,[1] H. Mervart,[6] D. Peruccio[(Italy)], J.R. Regueiro,[1] and I. Schreuder[13]

During the Tenth Workshop, 27 anti-sera in the Antigen Society and 22 anti-sera in the Core serology serum set were provided for the serologic analysis of the supertypic HLA-B locus specificities Bw4 and Bw6. The serum sets included 12 monoclonal antibodies with Bw4 or Bw6 specificity. For each of the submitted sera, Q score, R values, % extra, and missed reactions, as well as "tail" antigens derived by 2x2 comparison of their reactivity on Caucasians, Negroes, and Orientals, are shown in Tables 1 and 2.

The Bw4 specificity was best defined with the alloantisera 493, 495, 498 and the monoclonals 2064, 2065, 2102 of the Core serum set, as well as the alloantisera 4750, 4751, 4752, 4758, 4759, 17017II MUE of the Antigen Society with highly concordant positive reaction patterns in 87% of the Bw4 positive cells in all analyzed different ethnic groups. Among the anti-Bw6 anti-sera reagents, 4765, 4766, 4768, 4769 and the monoclonal antibody 2BC WES of the Antigen Society, as well as the sera 499, 500 and the monoclonals 2106, 2107, showed the highest Q scores and R values in all analyzed races. Pattern analysis revealed that in 96% of Bw6 positive cells of all ethnic groups there was a positive reaction with at least six of the nine best Bw6 sera.

Three of the anti-Bw4 alloantisera (4750, 4751, 4752) appeared to be "monospecific" for HLA-Bw4 in all ethnic subpopulations. As in previous Workshops, a larger group (4758, 4759, 17017II MUE, 271MUC, 493, 498, 2064, 2065, 2102) of anti-Bw4 alloantisera and monoclonals were found to recognize also determinants on HLA-A locus antigens, in particular A23, A24, A32, A2. In tail and pattern analysis most of these anti-Bw4 sera showed extra-reactions with A24, A23, A32 in Caucasians (Table 3), whereas in Negroes and Orientals only the monoclonals 2064 and 2065 detected these HLA-A locus antigens frequently. Two anti-Bw4 alloantisera (495, 260 MUC) revealed extra-reactivity with subgroups of the Bw6-associated HLA-B locus antigens B50 and B35 in Caucasians. Monoclonal antibodies 2101 and 2198 appeared to have Bw4 reactivity, but were frequently negative with Bw4 and Bw6 positive cells. In serum-to-serum correlations, 16 of the 25 submitted anti-Bw4 anti-sera were included in one cluster with three serum subgroups of highest correlations (Table 4).

In contrast, all anti-Bw6 anti-sera except the alloantiserum 4764 were operationally monospecific and had high correlations among each other (Table 4). Serum

---

Reporting Laboratories: FRAARN,[1] IT2FER,[2] US8DEC,[3] US8FOT,[4] US1HEI,[5] NCYMRV,[6] GERMUC,[7] IT1CEP[8]
Participating Laboratories: BENBRU,[9] IT1CNG,[10] UKIGEL,[11] SAFHAM,[12] BENROO[13]

Table 1. Tenth Workshop Core Sera With Anti-Bw4, Bw6 Reactivity

				Q score			R value			% extras			% misses		
10th WS No.	Origin	Specificity	C	N	O	C	N	O	C	N	O	C	N	O	Other Specificities
493	LAW	Bw4	4.1	5.4	3.8	0.77	0.83	0.79	4	3	4	4	4	6	A32
495	GAN	Bw4	5.3	5.2	6.4	0.79	0.84	0.90	4	5	3	3	1	2	B50
498	ILE	Bw4	5.6	5.8	7.0	0.83	0.88	0.92	5	3	3	2	2	1	A32
496	MBC	Bw4	3.7	3.8	4.4	0.70	0.67	0.76	3	2	2	12	14	9	
497	DPT	Bw4	0.8	0.7	0.4	0.30	0.26	0.23	1	3	3	43	55	41	
2057*	MUC	Bw4	4.2	4.0	3.1	0.63	0.73	0.67	0	20	2	20	11	13	A32, A23, A24
2064*	TSU	Bw4	8.3	5.7	5.5	0.97	0.89	0.83	0	2	1	1	3	6	A32, A23, A24
2065*	TSU	Bw4	7.3	5.2	5.6	0.94	0.86	0.83	1	4	1	1	3	6	A32, A23, A24
2102*	GEL	Bw4	8.6	5.4	4.7	0.92	0.86	0.81	3	2	7	0	1	2	A32
2098*	WAN	Bw4	1.9	1.6	1.3	0.55	0.47	0.48	2	0	4	23	28	25	
2101*	MUC	Bw4	1.1	0.4	0.3	0.41	0.27	0.21	1	11	1	35	53	45	
499	BRN	Bw6	6.0	5.2	3.3	0.81	0.80	0.62	0	9	2	7	7	11	
500	MUE	Bw6	5.9	5.0	2.5	0.83	0.81	0.56	5	4	1	0	1	7	
507	FAU	Bw6	2.9	3.4	2.3	0.58	0.64	0.51	4	3	2	12	8	13	
509	ROO	Bw6	1.3	1.9	1.2	0.40	0.56	0.41	12	8	8	6	2	2	
501	RIC	Bw6	3.5	3.4	2.4	0.59	0.59	0.52	1	1	2	18	16	14	
504	BER	Bw6	4.1	3.1	1.6	0.70	0.55	0.37	1	2	2	12	17	28	
502	DPT	Bw6	3.6	2.4	1.8	0.68	0.69	0.45	6	4	6	3	4	7	
2106*	KRE	Bw6	6.0	3.6	2.1	0.86	0.72	0.50	2	2	3	2	8	12	
2107*	SFR	Bw6	5.6	4.5	2.4	0.89	0.80	0.54	1	3	4	2	4	8	
2103*	GEL	Bw6	1.3	0.2	–	0.42	0.07	–	5	11	–	21	22	–	
2105*	THP	Bw6	0.1	0.1	–	0.07	0.13	–	13	–	–	23	27	–	
427	HAM	B44	7.2	5.3	7.4	0.81	0.77	0.87	30	4	2	0	1	0	Bw4
166	EAE	B51	4.6	1.7	5.1	0.66	0.46	0.71	3	1	2	0	10	0	Bw4

*Monoclonal antibody; C Caucasians (N=410); N Negros (N=589); O Orientals (N=1001)

Table 2. Tenth Workshop Antigen Society Sera With Anti-Bw6, -Bw6 Reactivity

			Q score		R value		% extras		% misses		
10th WS No.	Origin	Specificity	C	N	C	N	C	N	C	N	Other Specificities
4750	YAG	Bw4	4.7	1.6	0.76	0.57	1	6	11	16	
4751	BRN	Bw4	3.5	3.4	0.74	0.78	5	6	8	3	
4752	ROO	Bw4	8.2	1.3	0.89	0.51	2	8	3	16	
4753	BRU	Bw4	2.4	2.4	0.62	0.26	6	7	13	32	
4755	ROO	Bw4	3.9	1.2	0.64	0.49	0	5	21	22	
4756	KWC	Bw4	2.8	1.9	0.58	0.58	20	18	1	3	
4757	MBC	Bw4	1.5	1.4	0.43	0.53	6	16	21	6	
4758	CEP	Bw4	5.4	1.7	0.84	0.59	2	16	3	3	A32, A23, A24
4759	DPT	Bw4	4.5	1.7	0.73	0.50	6	24	2	1	A32, A23, A24
4760	DPT	Bw4	4.1	3.6	0.67	0.68	15	16	1	0	
17023II#	MUE	Bw4	1.4	–	0.51	–	12	–	5	–	A23, A24
17017II#	MUE	Bw4	3.7	–	0.74	–	6	–	3	–	A23, A24
260#	MUC	Bw4	4.4	–	0.71	–	14	–	0	–	B35
271#	MUC	Bw4	1.6	–	0.53		16	–	6	–	A2
4761	OHA	Bw6	2.7	1.4	0.53	0.40	2	10	22	12	
4762	BAC	Bw6	2.7	1.0	0.59	0.42	3	5	16	22	
4763	BAC	Bw6	0.4	0.7	0.15	0.33	2	4	60	33	
4764	ROO	Bw6	3.5	1.7	0.71	0.55	1	6	7	11	A3
4765	BRU	Bw6	7.3	3.0	0.93	0.74	1	6	1	2	
4766	BRU	Bw6	5.4	2.2	0.84	0.64	4	7	2	5	
4767	MAC	Bw6	7.4	2.5	0.94	0.67	1	6	1	5	
4768	PNW	Bw6	6.3	3.6	0.87	0.78	2	6	3	1	
4769	ROO	Bw6	6.6	2.6	0.81	0.70	1	16	7	3	
4770	HFP	Bw6	4.6	2.2	0.78	0.61	1	5	9	10	
4771	ROO	Bw6	5.4	1.9	0.79	0.59	2	6	6	8	
307#	MUC	Bw6	3.4	–	0.67	–	3	–	10	–	
2BC4*#	WES	Bw6	4.9	–	0.80	–	4	–	3	–	

= Local assignment; * = monoclonal antibody; C = Caucasians (N=585); N = Negroes (N=141)

Table 3. Serologic Patterns of Selected Anti-Bw4 Sera on A23⁺, A24⁺ or A32+, Bw4⁻ Cells

		10th WS Sera							
	%	493	498	2064	2065	2105	2057	4758	4759
In 50 Caucasians	32.0	–	–	+	+	+/–	+/–	+	+
	26.0	+	+/–	+	+	+/–	+/–	+	+
	12.0	+	+	+	+	+	+	+	+
	12.0	–	+	+	+	+/–	+/–	+	+
	6.0	–	–	+	+/–	–	+	–	+
	4.0	+/–	–	+	+	–	–	–	+
	2.0	+	+	+	+	+	+	–	–
	2.0	+	+	+	+	–	+	–	+
	2.0	–	–	–	–	+	+	–	+
	2.0	–	–	+	+	–	+	–	+
In 161 Negroes and Orientals	56.5	–	–	+	+	–	+	nt	nt
	8.6	–	–	–	–	–	–	nt	nt
	7.4	–	–	+	+	–	–	nt	nt
	7.4	+/–	–	+	+	+	+	nt	nt
	6.2	+	+/–	+	+	–	+	nt	nt
	4.9	–	–	–	–	–	+	nt	nt
	4.8	–	–	+	–	–	–	nt	nt
	1.8	–	+	+	+	–	+	nt	nt
	1.8	+/–	+/–	+/–	+/–	–	+/–	nt	nt

Table 4. Internal Correlations of Selected Anti-Bw4 and Bw6 Sera

a) Anti-Bw4 Sera

493												
495	.8											
498	.8	.8										
2102	.7	.7	.8									
2064	.7	.7	.7	.7								
2065	.7	.7	.7	.7	.9							
4756	.6	.6	.6	.6	.6	.7						
4760	.7	.6	.7	.7	.7	.7	.8					
4758	.6	.6	.6	.6	.8	.8	.8	.8				
4759	.5	.5	.5	.5	.7	.7	.7	.8	.8			
4750	.6	.6	.7	.6	.6	.6	.6	.7	.7	.6		
4752	.6	.6	.7	.7	.6	.6	.7	.7	.7	.6	.8	
4751	.7	.7	.7	.6	.6	.6	.7	.7	.7	.6	.8	.8

b) Anti-Bw6 Sera

499			
500	.7		
2106	.7	.8	
2107	.7	.8	.9

4764 showed extra-reactions with A3 in Caucasians. Two monoclonal antibodies 2106 and 2107 were excellent Bw6 reagents in all races. A further monoclonal antibody 2103 with Bw6 reactivity missed many Bw6 positive cells.

In summary, inclusion analysis of the Bw4 and Bw6 specificities revealed no significant deviation with previous reports. The anti-Bw4 sera submitted to the Tenth Workshop were more heterogenous in their reactivity with HLA-A locus antigens than the anti-Bw6 anti-sera and showed differences in this reaction in various ethnic groups. All anti-Bw4 monoclonal reagents with high performance in this serologic analysis also were found to cross-react with HLA-A locus antigens and thus did not lead to a further differentiation of the HLA-Bw4 specificity.

Antigen Society #16 Report (Cw1, Cw3, Cw9, Cw10, Cw11)

P. Chiewsilp,[1] K. Sujirachato,[1] W.R. Mayr,[2] P. Perrier,[3] H. Hasekura,[4] M. Ota,[4] M. Aizawa,[5] A. Wakisaka,[5] H. Ikeda,[5] S. Naito,[6] T. Akaza,[7] M. Jeannet,[8] A. Zachary,[9] and W. Braun[9]

History

HLA-Cw1 and HLA-Cw3 were first described in the Fourth and the Fifth International Histocompatibility Workshops, respectively (1,2). The three subtypic factors of Cw3, namely Cw3.1, Cw3.2, and Cw3.3, were demonstrated in the Ninth International Histocompatibility Workshop (3). The expression of Cw1 and Cw3 on the same haplotype in Orientals was observed by Payne et al. in 1975 (4) and was later found to be strongly associated with Bw46. CwB had been proposed as a new C locus antigen for this Cw1-Cw3 co-segregating phenotype in the Second Asia and Oceania Histocompatibility (AOH) Workshop (5). However, CX46 was more recently used to designate this C locus antigen in the Third AOH Workshop (6). In the present Workshop, the new assignment of HLA specificities is given by the WHO nomenclature committee (7). This included antigens in C locus, i.e., Cw9 (Cw3.1), Cw10 (Cw3.2), and Cw11 (CX46, Cw1+3, Cw1X3, C-Bangkok, CSH1).

Linkage Disequilibrium

There are strong associations of Cw1 with B27 in Caucasoids; with Bw54, Bw55, and Bw59 in Japanese, and with Bw54, Bw55, and B40 in Chinese and Thais (Table 1). Further strong associations were observed for Cw9 with Bw55 and Bw62; Cw10 with Bw62, Bw60 in Caucasians; Cw9 with B35, Bw55; Cw10 with B40; Cw11 with Bw46 in Japanese; and Cw9 with B15; Cw10 with Bw58 and B40; Cw11 with Bw46 in Chinese and Thais.

Serology

Cw1. Eight sera were submitted as anti-Cw1, six from the Core set and two from the Antigen Society (Table 2 and Table 3). The Cw1 antigen was well defined by the Tenth Workshop anti-sera: 508, 512, 513, 514, 259, 510, and 4773. Pattern analysis among Caucasians, Japanese, and Thais indicated that they were strongly correlated with one another.

Cw3, Cw9, Cw10, and Cw11. Twenty-nine sera were submitted as anti-Cw3, 10 from the Core set and 19 from the Antigen Society (Table 2 and Table 3). There appeared to be seven possible subtypes of Cw3, all of which were identified by patterns. It was confirmed that Cw3 could be divided into Cw9 and Cw10 by the sera 524, 525, 526, 527, 529, 530, 260, 528, 9064, 532, 4777, 4778, 4779, 4783, 4785, 4786, 4787, 4788, 4780, 4781, 4793, and 9035 (Table 4). Cw9 reacted with all of the above anti-Cw3 sera, while Cw10 gave negative reactions with sera 9064 and 532.

Only a few Thai individuals were found to be Cw9 short (Cw9S). Approximately 45 cells of several ethnic origins appeared to have the Cw9S pattern as displayed in the CXS Workshop data base. Cw9S can be differen-

Reporting Laboratory: ANZCHI[1]
Participating Laboratories: EAEMYR,[2] FRAPRR,[3] JAPHAS,[4] JAPAIZ,[5] JAPNAI,[6] JAPJUJ,[7] FRAJEA,[8] US6BRN[9]

Table 1. Associations of Cwl and Splits of Cw3 With B Locus Specificities

Caucasoids	
Cwl	– B27
Cw9	– Bw55
Cw9	– Bw62
Cw10	– Bw60
Cw10	– Bw62
Japanese	
Cwl	– Bw54
Cwl	– Bw55
Cwl	– Bw59
Cw9	– B35
Cw9	– Bw55
Cw10	– B40
Cw11	– Bw46
Chinese, Thais	
Cwl	– Bw54
Cwl	– Bw55
Cwl	– B40
Cw9	– B15
Cw10	– Bw58
Cw10	– B40
Cw11	– Bw46

tiated from Cw9 by a negative reaction with serum 532. The pattern of Cw9S showed segregation within family CHI FO4 (Table 5). In the Third AOH Workshop Cw9 was identified as Cw3.11, and Cw9S as Cw3.12.

Cw11 was defined by positive reactions with some Cw3 sera (524, 525, 526, 527, 529, 530, 4777, 4778, and 4785) but also by positive reactions with Cwl sera (508, 512, 513, 514, 259, 510, and 4773). The important negative reactions for Cw3 were with sera 528, 532, and 9064. Cw11 reactivity pattern was clearly segregated within family CHI FO3 (Table 5). As previously observed in the Third AOH Workshop (8), it was found in the Tenth WS Serology Program Analysis of all 177 random Bw46 positive cells that four Thais (Cw11 var1) from DCH Lab (DCH2, DCH13, DCH51 and DCH55) and 1 Chinese (Cw11 var3) from ZHA Lab (ZHA78) were reactive only with Cw3 and Cwl sera, respectively (Table 4). In addition, there were further evidences of other 2 Cw11 variants, Cw11 var2 in one Japanese from MAT Lab (MAT27), which gave negative reaction with all of Tenth WS anti-Cwl but reacted only with some anti-Cw3 sera 524, 529, and 530. Cw11 var4 pattern of reactivity was found among three Japanese from DOH, SAS, and MAT Laboratories (DOH10, SAS5914 and MAT53). The segregation pattern of Cw11 var4 was observed in a Negro family from ROO Lab (Table 4). Because the Cw11 specificity includes a portion of both Cw3 and Cwl, it may explain the apparent occurrence of "triplets" with Cwl/Cw3 present on a single haplotype (4). Furthermore, this division of Cwl may presage an eventual split in Cwl (see also Class I Biochemistry for the Cw locus).

An interesting monoclonal antibody, 9035, by virtue of its reactivity (separate data) with all Cwl and Cw3 positive cells, as well as with some Cw7, Cw11/Cw-, and Cw-/Cw- cells, may define an epitope and possible specificity existing on some Cw blank cells and representing the entity shared by Cwl and Cw3 as found in the Cw11 pattern.

References

1. Sandberg L, Thorsby E, Kissmeyer-Nielsen F, Lindsholm A. Evidence of a third sublocus within the HLA chromosomal region. In: Terasaki PI (ed): Histocompatibility Testing 1970. Munksgaard, Copenhagen, 1970; p 165.
2. Svejgaard A, Staub-Nielsen L, Ryder LP, Kissmeyer-Nielsen F, Sandberg L, Lindholm A, et al. Subdivisions of HLA-A antigens. Evidence of a "new" segregant series. In: Dausset J, Colombani J (eds): Histocompatibility Testing 1972. Munksgaard, Copenhagen, 1972; p 465.
3. Campbell EM, du Toit ED, Tongio MM, Orlacher A, Mayer S, Mayr WR. Antigen Report: HLA-CW3. In: Albert ED (ed): Histocompatibility Testing 1984. Springer-Verlag, Berlin, Heidelberg, 1984; pp 179.
4. Payne R, Radvany R, Grumet FC, Feldmann M, Cann H: Two third-series antigens transmitted together. A possible fourth

Table 3. Analysis of 10WS Antigen Society Sera for Cwl and Cw3 Specificities

Sera	Specificity	R	Sera	Specificity	R
10WS4772	Cwl	0.136	10WS4784	Cw3	0.629
10WS4773	Cwl	0.728	10WS4785	Cw3	0.921
10WS4776	Cw3	0.174	10WS4786	Cw3	0.899
10WS4777	Cw3	0.644	10WS4787	Cw3	0.966
10WS4778	Cw3	0.885	10WS4788	Cw3	0.899
10WS4779	Cw3	0.659	10WS4789	Cw3	0.628
10WS4780	Cw3	0.888	10WS4790	Cw3	0.576
10WS4781	Cw3	0.791	10WS4791	Cw3	0.798
10WS4782	Cw3	0.655	10WS4792	Cw3	0.753
10WS4783	Cw3	0.920	10WS4793	Cw3	0.795
				A28	0.733
				Aw68	1.000

Table 2. Analysis of 10WS Core Sera for Cw1 and Cw3 Specificities

Serum	Specificity	Strength	R	Q score	% Miss	% Extra	Residuals
10WS508	Cw1	96	0.828	5.394	9.02	2.77	A2 = 10%
	Bw46	97	0.694	6.793	6.93	1.33	Cw3 = 8%
10WS510	Cw1	66	0.718	3.678	27.32	2.51	A2 = 8%
	Bw46	78	0.587	4.546	24.51	1.35	Cw7 = 7%
10WS512	Cw1	89	0.792	5.707	6.00	4.41	B44 = 11%
	Bw46	89	0.538	6.211	7.92	3.01	A2 = 6%
							Cw7 = 6%
10WS513	Cw1	94	0.806	5.740	5.79	4.01	A2 = 10%
	Bw46	92	0.577	6.448	5.88	2.58	
10WS514	Cw1	78	0.796	4.972	12.10	3.12	A2 = 11%
	Bw46	89	0.608	6.241	12.74	1.79	Cw7 = 7%
10WS259	Cw1	86	0.544	3.058	11.34	14.27	
	Bw62	76	0.576	2.614	23.45	7.75	
10WS9064	Cw3	60	0.422	0.928	60	5.44	A2 = 9%
	Cw3.1	70	0.723	5.942	8.5	2.31	Cw4 = 6%
10WS524	Cw3	95	0.751	3.931	8.1	9.85	A2 = 7%
	Cw3.1	99	0.569	6.301	1.5	6.60	
	Cw3.2	96	0.619	7.054	1.96	4.01	
	Bw46	96	0.598	7.291	2.38	2.53	
10WS525	Cw3	83	0.754	3.942	9.44	9.03	A2 = 10%
	Cw3.1	95	0.591	6.697	1.98	5.79	
	Cw3.2	93	0.644	6.337	4.58	3.26	
	Bw46	82	0.670	7.307	2.35	1.75	
10WS526	Cw3	79	0.600	2.647	12.68	18.05	Cw4 = 7%
	Cw3.1	89	0.390	4.445	3.45	15.18	A2 = 6%
	Cw3.2	83	0.345	3.539	11.11	13.09	Cw6 = 6%
	Bw46	66	0.260	2.606	17.65	11.98	
10WS527	Cw3	82	0.670	3.239	10.35	14.05	A2 = 8%
	Cw3.1	95	0.466	5.812	1.47	10.94	Cw4 = 6%
	Cw3.2	84	0.461	5.439	3.90	8.50	
	Bw46	70	0.311	2.908	22.62	7.41	
10WS528	Cw3	91	0.711	3.811	7.47	12.76	A2 = 8%
	Cw3.1	96	0.495	6.428	0.98	9.60	Cw7 = 6%
	Cw3.2	95	0.481	5.391	6.49	7.20	
	Cw1,3	73	0.243	3.255	20	6.66	
10WS529	Cw3	92	0.737	3.972	6.67	11.31	A2 = 9%
	Cw3.1	98	0.523	5.841	2	8.14	
	Cw3.2	97	0.559	6.517	1.95	5.53	
	Bw46	79	0.437	6.024	3.17	4.44	
	Cw1	47	0.201	0.356	83.24	3.05	
10WS530	Cw3	90	0.752	4.155	6.08	10.74	A2 = 10%
	Cw3.1	97	0.541	6.120	1.48	7.54	
	Cw3.2	95	0.570	6.470	3.25	4.99	
	Bw46	98	0.534	6.995	2.38	3.52	
	Cw1	40	0.173	0.252	87.06	2.46	
10WS532	Cw3	76	0.455	1.126	63.33	5.31	A2 = 8%
	Cw3.1	88	0.783	7.648	2.94	1.92	Cw7 = 7%
10WS260	Bw62	98	0.506	4.291	1.80	21.59	A2 = 10%
	Bw57	88	0.400	2.402	14.95	17.32	Cw7 = 6%
	Cw3	72	0.388	0.975	49.85	10.74	
	Cw3.1	90	0.417	4.089	7.32	8.69	
	Bw63	95	0.417	4.785	5.49	7.08	
	Bw46	82	0.453	4.705	6.90	5.51	

Table 4. Reaction Patterns of Cw1 and the Splits of Cw3

	Cw1							Cw3																					
	0508	0512	0513	0514	0259	0510	4773	0524	0525	0526	0527	0529	0530	0260	0528	9064	0532	4777	4778	4779	4783	4785	4786	4787	4788	4780	4781	4793	9035
Cw1	+	+	+	+	+	+	+	−	−	−	−	−	−	−	−	−	−	−	−	−	−	−	−	−	−	−	−	−	−
Cw11	+	+	+	+	+	+	+	+	+	+	+	+	+	+	−	−	−	+	+	+	+/−	+	W	+	+	−	−	−	+
Cw11 var1 (T)	−	−	−	−	−	−	0	+	+	+	+	+	+	+	0	−	−	0	0	0	0	0	0	0	0	0	0	0	0
Cw11 var2 (J)	−	−	−	−	−	−	0	+	−	−	−	+	+	+	0	−	−	0	0	0	0	0	0	0	0	0	0	0	0
Cw11 var3 (C)	+	+	+	+	+	−	0	−	−	−	−	−	+	+	0	−	−	0	0	0	0	0	0	0	0	0	0	0	0
Cw11 var4 (J,N)	+	+	+	+	+/−	−	0	+	+/−	−	−	+/−	+	+/−	0	−	−	0	0	0	0	0	0	0	0	0	0	0	0
Cw9	−	−	−	−	−	−	−	+	+	+	+	+	+	+	+	+	+	+	+	+	+	+	+	+	+	+	+	+	+
Cw9S	−	−	−	−	−	−	−	+	+	+	+	+	+	+	+	+	−	+	+	+	+	+	+	+	+	+	+	+	+
Cw10	−	−	−	−	−	−	−	+	+	+	+	+	+	+	+	−	−	+	+	+	+	+	+	+	+	+	+	+	+

J = Japanese
T = Thais
C = Chinese
N = Other Negro (exclude African Negro and American Negro)

Table 5. Segregations of Cw9, Cw10 and Cw11 in Families Using 10WS Sera

Fam	I.D.	Rela	C locus	CoAb										AgSo											
				0524	0525	0526	0527	0529	0530	0260	0528	9064	0532	4777	4778	4779	4783	4785	4786	4787	4788	4780	4781	4793	9035
	Cw9			+	+	+	+	+	+	+	+	+	+	+	+	+	+	+	+	+	+	+	+	+	+
	Cw9S			+	+	+	+	+	+	+	+	+	−	+	+	+	+	+	+	+	+	+	+	+	+
F04	CHI15	F	10,7	8	6	8	8	8	8	1	8	1	1	8	8	8	8	8	8	8	8	8	6	8	8
	CHI95	M	10,9S	8	8	4	8	8	8	2	8	6	1	8	8	8	8	8	8	8	8	8	8	8	8
	CHI99	C1	10,9S	8	8	8	8	8	8	4	8	6	1	0	0	0	0	0	0	0	0	0	0	0	0
	CHI68	C2	10,9S	8	8	1	8	8	8	6	8	6	1	8	8	8	8	8	8	8	8	8	8	8	8
	CHI70	C3	7,10	8	8	8	8	8	8	1	8	1	1	8	8	8	8	8	8	8	8	8	4	8	8
	Cw10			+	+	+	+	+	+	+	+	−	−	+	+	+	+	+	+	+	+	+	+	+	+
F01	CHI77	F	10,11	8	8	6	8	8	8	8	8	1	1	8	8	8	8	8	8	8	8	8	8	8	8
	CHI78	M	11,8	8	8	4	2	8	8	4	1	1	1	8	8	6	6	8	4	8	8	2	4	1	8
	CHI82	C1	10,8	8	8	8	8	8	8	1	8	1	1	8	8	8	8	8	8	8	8	8	8	8	8
	CHI89	C2	10,8	8	8	6	8	8	8	1	8	1	1	8	8	8	8	8	8	6	8	8	6	8	8
	CHI26	C3	10,8	8	8	8	8	8	8	1	8	1	1	8	8	8	8	8	8	8	8	8	6	8	1
	CHI87	C4	10,11	8	8	8	8	8	8	8	8	1	1	8	8	8	8	8	8	8	8	8	8	6	8
	Cw11			+	+	+	+	+	+	+	−	−	−	+	+	+	+/−	+	W	+	+	−	−	−	+
F03	CHI07	F	11,−	8	8	8	8	8	8	8	2	1	1	8	8	8	1	8	6	8	6	1	1	4	8
	CHI57	C1	11,−	8	8	8	6	8	8	8	4	1	1	8	6	1	2	6	2	6	6	1	1	1	8
	CHI64	C4	11,−	8	8	4	6	6	8	4	4	1	1	8	6	4	2	8	1	6	8	1	1	1	8
	Cw11 var 4			+	+/−	−	−	+/−	+	+/−	0	−	−	0	0	0	0	0	0	0	0	0	0	0	0
F	R0029	M	11v4,11v4	8	6	6	1	6	8	8	0	1	1	0	0	0	0	0	0	0	0	0	0	0	0
	R0030	C1	11v4,−	8	6	4	1	4	8	6	0	1	1	0	0	0	0	0	0	0	0	0	0	0	0
	R0031	C2	11v4,7	8	4	6	1	1	8	4	0	1	1	0	0	0	0	0	0	0	0	0	0	0	0
	R0032	C3	11v4,−	8	4	1	1	1	6	1	0	1	1	0	0	0	0	0	0	0	0	0	0	0	0
	R0033	C4	11v4,−	8	1	4	1	6	8	1	0	1	1	0	0	0	0	0	0	0	0	0	0	0	0
	R0034	C5	11v4,−	8	6	1	1	1	6	4	0	1	1	0	0	0	0	0	0	0	0	0	0	0	0

SD locus? In: Kissmeyer-Nielsen F (ed): Histocompatibility Testing 1975. Munksgaard, Copenhagen, 1975; p 343.
5. Chandanayingyong D. Bw46 report. In: Simons MJ, Tait BD (eds): Proceedings of the Second Asia and Oceania Histocompatibility Workshop Conference. Melbourne, 1981; p 127.
6. Shimbo M, Mitani T, Ikeda H, Sekiguchi S: Analysis of "CX46" specificity. In: Aizawa W (ed): Proceedings of the Third Asia and Oceania Histocompatibility Workshop Conference. Sapporo, Japan, p 52.
7. Bodmer WF, Albert E, Bodmer JG, Dupont B, Mach B, Mayr W, et al. Nomenclature for factors of the HLA system.
8. Sun Y, Shimbo M, Mitani T, Ikeda H, Sekiguchi S. HLA-Cw1, Cw3 Report. In: Aizawa M (ed): Proceedings of the Third Asia and Oceania Histocompatibility Workshop Conference. Sapporo, Japan, p 52.

Antigen Society #17 Report (Cw5 and Cw8)

W.R. Mayr,[1] L. Contu,[2] M. Kirnbauer,[1] and H. Mervart[3]

As in the previous Workshops, the definition of Cw5 and Cw8 was rather difficult because of the cross-reactivity between Cw5 and Cw8 and because of the strong linkage disequilibrium between Cw5 and B44 as well as between Cw5 and B14.

Cw5 and Cw8 were nearly absent in non-Caucasian populations; therefore, the reactivity of the anti-sera used to define these factors was analyzed by taking into account only the cells of Caucasians tested in Antigen Society 17 (n = 164).

The best sera for the definition of Cw5 and Cw8 are listed in Table 1. Serum 10w552 contains, besides an anti-Cw5 + Cw8, a strong anti-B44.

There was no good indication for the existence of splits of Cw5 or Cw8.

In spite of the fact that one good anti-Cw5 (serum 10w551) and one anti-Cw8 of reasonable quality (serum 10w554) have been found, difficulties remain with regard to the definition of Cw5 and Cw8.

Participating Laboratories: EAEMYR,[1] IT1CON,[2] NCYMRV[3]

Table 1. Reactivity of the 10w Anti-Cw5 and Cw8 Sera

10w serum	antigen	ave.str	%fp	%fn	r	Qsc
551 #	Cw5	7.5	0	0	0.98	8.96
550 #	Cw5	5.3	0	30	0.79	5.63
554 #	Cw8	6.3	25	10	0.78	5.21
9004 #	Cw5	7.2	41	0	0.72	5.91
	Cw8	2.5	29	77	0.35	1.51
553 #	Cw5	5.1	37	26	0.61	4.58
	Cw8	8.0	10	0	0.90	7.48
4794 ##	Cw5	3.0	60	65	0.28	1.01
	Cw8	6.4	55	10	0.59	5.21

serum of the Core set
serum of Antigen Society 17
ave.str = average strength
%fp = percentage of false positive reactions
%fn = percentage of false negative reactions
r = correlation coefficient
Qsc = quality score

Antigen Society #18 Report (Cw4 and Cw6)

C. Conighi,[1] L. Contu,[2] M.T. Grappa,[3] E. Du Toit,[4] M. Hammond,[5] P. Lulli,[6] W.R. Mayr,[7] A. Menicucci,[8] H. Mervart,[9] and M. Pupura[10]

HLA-Cw4 (formerly T4, RH315) and HLA-Cw6 (formerly T7) obtained the formal HLA designation after the Sixth and Seventh International Histocompatibility Workshops respectively (1,2). Since then, it has been observed that some troubles arise in identifying Cw6 when Cw4 is present: The difficulty is due to the fact that while bispecific Cw4+Cw6 and monospecific Cw4 sera are comparatively frequent, Cw6 sera not recognizing Cw4 are quite rare (3,4).

Reporting Laboratory: IT1CNG[1]
Participating Laboratories: IT1CON,[2] IT2FER,[3] SAFDUT,[4] SAFHAM,[5] IT2GAN,[6] EAEMYR,[7] IT1MTT,[8] NCYMRV,[9] IT1PUR[10]

This work was partly supported by "Regione Emilia-Romagna: Progetti di ricerca sanitaria finalizzata" and partly by Lagitre S.R.L., Milano (One Lambda, Los Angeles).

Table 1. Core Set Sera Analysis: Blacks

		STR	SPEC	AVE	INCL	N	TP	FN	FP	TN	Q	R	CHI²
533	BEN ENG CLB105												
	CW4	90	99	7.5	96	357	137	5	3	212	7.735	0.948	320.593
	RES	14	CW6 = 20%		A1 = 10%		A30 = 10%		B27 = 10%		BW57 = 10%		
536	US5 CRO Martinez2.21.83												
	CW4	76	99	6.6	87	357	122	19	3	213	6.023	0.867	268.529
	RES	14	A2 = 10%		BW46 = 10%		B– = 10%		CW1 = 10%		A11 = 10%		
537	UK1 BRS 5956.CBT												
	CW4	93	95	7.5	94	356	134	8	11	203	5.806	0.884	278.261
+	B7	47	97	2.1	19	214	6	26	5	177	0.544	0.262	14.708
+	BW57	78	99	2.0	14	182	3	18	2	159	0.694	0.262	12.534
	OR	92	99	6.0	73	356	143	52	2	159	4.752	0.726	187.595
	RES	14	CW6 = 20%		A3 = 10%		AW19 = 10%		BW65 = 10%		BW70 = 10%		
539	EAE ZAR MZ467												
	CW4	75	93	6.7	90	337	121	14	14	188	4.475	0.822	227.680
+	BW50	33	94	5.3	75	202	3	1	11	187	3.669	0.380	29.240
	OR	74	94	6.6	89	337	124	15	11	187	4.588	0.835	235.124
	RES	52	CW6 = 10%		A2 = 9%		B7 = 7%	CW2 = 7%					
541	NCY MRV 22166.0												
	CW4	76	100	7.0	94	355	133	9	1	212	7.457	0.936	311.197
	RES	33											
543	US8 JLE Billing												
	CW4	78	77	7.0	92	349	127	11	49	162	2.895	0.669	156.128
	CW6	55	98	6.2	88	211	45	6	4	156	5.839	0.859	155.515
	OR	72	98	6.7	91	349	172	17	4	156	6.045	0.877	268.406
	RES	27	A30 = 11%		BW42 = 11%								
545	US5 TER TER.C4610B												
	CW4	93	70	7.7	98	358	139	3	65	151	3.668	0.666	158.673
	CW6	51	88	6.0	87	216	45	7	20	144	3.728	0.686	101.511
+	CW5	54	90	5.6	71	164	5	2	15	142	3.021	0.380	23.650
*	TEC10	33	90	0.0	0	157	0	0	15	142	12.532	0.106	1.751
	OR	82	90	7.2	94	358	189	12	15	142	5.122	0.842	253.736
	RES	33	A2 = 7%		CW2 = 7%		A1 = 6%	A28 = 6%					
540	FRA DDC Pierson												
	CW4	95	70	7.6	96	356	134	6	64	152	2.648	0.646	148.446
	CW6	86	90	7.1	90	216	47	5	17	147	3.382	0.741	118.589
	OR	93	90	7.4	94	356	181	11	17	147	4.070	0.837	249.426
	RES	84	A30 = 7%		CW3 = 7%		B7 = 6%						
547	GER MUE MUE23928												
	CW6	74	92	6.5	84	359	53	10	23	273	3.939	0.706	178.731
	RES	42	CW4 = 17%		BW53 = 12%		AW36 = 7%		B35 = 7%		A2 = 6%		
558	FRA DDC Devinat												
	CW7	71	78	4.9	61	358	43	27	64	224	0.777	0.338	40.946
+	B39	38	79	4.3	71	388	5	2	59	222	1.493	0.187	10.087
+	AW34	44	81	3.3	42	281	11	15	48	207	0.484	0.170	8.117
+	BW60	45	82	4.7	67	255	4	2	44	205	1.489	0.193	9.461
+	A24	47	84	3.1	43	249	6	8	38	197	0.661	0.167	6.942
+	A26	65	86	3.3	35	235	7	13	31	184	0.620	0.162	6.164
+	BW4	77	96	2.2	18	215	29	129	2	55	0.259	0.178	6.794
	OR	67	96	3.1	35	358	105	196	2	55	0.705	0.246	21.607
	RES	43	A30 = 22%		A29 = 11%		B45 = 11%		BW72 = 11%		CW6 = 11%		
205	GER BRA 918B												
	B35	78	77	5.7	74	360	48	17	67	228	1.825	0.419	63.304
+	B39	6	79	3.7	70	295	7	3	60	225	1.254	0.211	13.181
+	AW34	33	82	3.4	48	285	12	13	48	212	0.585	0.207	12.204
+	BW70	24	85	2.9	42	260	13	18	35	194	0.452	0.225	13.122
+	A1	33	87	2.8	35	229	9	17	26	177	0.498	0.197	8.851
+	BW71	38	89	3.5	42	203	5	7	21	170	1.173	0.224	10.204
+	A29	9	91	2.5	36	191	4	7	17	163	0.942	0.211	8.542
	OR	51	91	4.0	54	360	98	82	17	163	1.199	0.479	82.771
	RES	24	CW4 = 12%		A2 = 10%		BW53 = 7%		A30 = 6%				

Table 1. *Continued*

		STR	SPEC	AVE	INCL	N	TP	FN	FP	TN	Q	R	CHI^2
192	US5 PNW Juli												
	B35	64	92	4.8	62	354	40	24	24	266	1.804	0.539	102.800
+	AW34	20	93	2.0	23	290	6	20	18	246	0.386	0.177	9.082
+	CW4	45	97	1.8	15	264	13	72	5	174	0.314	0.230	13.918
*	BW53	20	98	1.3	7	179	1	14	4	160	0.283	0.105	1.989
	OR	55	97	2.9	34	354	59	116	5	174	0.946	0.398	56.067
	RES	23	CW3 = 17%		A28 = 12%		A30 = 8%		BW70 = 8%		BW71 = 8%		
191	US2 MBC BC.CY.CHAD												
	BW53	70	79	6.3	85	358	73	13	56	216	2.297	0.569	115.706
	B35	75	96	6.2	82	272	47	10	9	206	3.866	0.780	165.463
	OR	73	96	6.3	84	358	120	23	9	206	4.214	0.809	234.089
	RES	43	CW4 = 10%		CW6 = 7%		BW70 = 7%						
562	IT1 PUR CNTS.105												
	CW4	76	99	6.4	84	357	120	23	2	212	5.200	0.852	259.267
	RES	20	A26 = 20%		A30 = 20%		BW42 = 20%		BW58 = 20%		CW3 = 20%		
535	EAE MYR Furlinger												
	CW4	82	98	7.2	94	322	118	7	3	194	6.365	0.928	277.584
	RES	71	BW53 = 22%		CW6 = 22%		A2 = 11%		A11 = 11%		BW61 = 11%		
347	US5 MIT B5612.4												
	B7	93	71	7.9	100	354	42	0	90	222	4.185	0.471	78.543
	BW42	64	82	7.0	98	312	40	1	50	221	4.588	0.584	106.483
	CW4	47	97	3.1	38	271	46	75	4	146	1.338	0.448	54.432
*	BW67	33	97	0.0	0	150	0	0	4	146	17.600	0.213	6.813
	OR	68	97	4.9	63	354	128	76	4	146	2.750	0.610	131.694
	RES	27	A2 = 11%		BW73 = 11%		A1 = 11%		B8 = 11%		B45 = 6%		
196	FRA BET E1218												
+	CW4	98	79	7.8	97	359	139	4	45	171	4.519	0.744	198.518
	BW53	58	85	6.1	83	216	15	3	30	168	3.227	0.458	45.308
	B35	61	91	5.5	74	198	14	5	16	163	2.766	0.525	54.574
+	B51	27	93	3.4	50	179	4	4	12	159	2.054	0.318	18.149
+	CW2	26	97	2.0	19	171	8	34	4	125	0.375	0.267	12.219
	OR	87	97	6.2	78	359	180	50	4	125	4.795	0.717	184.646
	RES	9	A1 = 10%		A30 = 10%		B7 = 10%		BW71 = 10%				
214	US2 WOL NW14002												
	CW4	95	73	7.4	93	358	133	10	59	156	3.033	0.640	146.695
	BW53	62	79	7.1	100	215	17	0	42	156	3.709	0.468	47.054
	B35	84	86	6.9	89	198	17	2	25	154	3.253	0.536	56.781
	B51	58	90	6.5	100	179	8	0	17	154	5.065	0.522	48.828
+	BW58	33	92	4.8	75	171	3	1	14	153	2.799	0.337	19.390
	BW70	20	94	2.1	22	167	6	21	8	132	0.356	0.225	8.430
*	BW52	20	95	3.5	50	140	1	1	7	131	3.186	0.263	9.703
*	TE79	33	94	0.0	0	140	0	0	8	132	14.100	0.151	3.189
	OR	87	94	6.6	84	358	184	34	8	132	3.938	0.766	209.840
	RES	37	CW3 = 10%		BW71 = 8%								
538	BEN BOU 64307 6/7/84												
	CW4	63	98	5.9	79	349	111	30	4	204	4.781	0.797	221.594
	RES	43	A30 = 10%		B35 = 10%		BW42 = 10%		CW7 = 10%		A23 = 10%		
204	IT1 CON CA181												
	CW4	88	91	7.0	89	356	125	16	19	196	4.098	0.791	222.621
+	A32	33	92	4.0	60	215	3	2	16	194	1.909	0.286	17.627
	OR	87	92	6.9	88	356	128	18	16	194	4.068	0.798	226.503
	RES	37	A2 = 7%		CW7 = 7%								
534	UK1 FES Langlais												
	CW4	49	100	4.3	56	355	79	62	1	213	2.872	0.646	148.193
	RES	33											
2101	GER MUC TU109												
*	BW4	52	95	2.7	29	347	70	175	5	97	0.533	0.258	23.043
	RES	23	A24 = 8%		B7 = 8%		A2 = 8%		B45 = 8%		BW70 = 8%		
2103	UK1 GEL 103.1.51												
*	BW6	76	35	5.6	72	351	199	78	48	26	0.104	0.063	1.145
	RES	68	CW4 = 13%		BW53 = 11%		AW33 = 6%						

Table 2. Core Set Sera Analysis: Caucasians

		STR	SPEC	AVE	INCL	N	TP	FN	FP	TN	Q	R	CHI^2
533	BEN ENG	CLB105											
	CW4	90	100	7.8	100	382	98	0	1	283	10.199	0.986	371.724
	RES	33											
536	US5 CRO	Martinez2.21.83											
	CW4	85	100	7.3	95	382	93	5	1	283	8.857	0.952	346.275
	RES	33											
537	UK1 BRS	5956.CBT											
	CW4	91	98	7.7	99	381	96	1	7	277	9.334	0.940	336.847
	RES	38	CW6 = 15%		A24 = 12%		B13 = 8%		A1 = 8%	BW60 = 8%			
539	EAE ZAR	MZ467											
	CW4	75	97	7.2	98	378	95	2	9	272	6.798	0.920	320.215
+	BW50	54	99	3.5	45	281	5	6	4	266	2.192	0.482	65.314
	OR	74	99	6.8	93	378	100	8	4	266	6.246	0.916	317.000
	RES	27	A1 = 12%		A24 = 12%		BW57 = 12%		CW6 = 12%		A2 = 6%		
541	NCY MRV	22166.0											
	CW4	71	99	6.5	88	382	86	12	2	282	7.280	0.897	307.135
	RES	20	A2 = 17%		A31 = 17%		BW50 = 17%		B44 = 17%		CW6 = 17%		
543	US8 JLE	Billing											
	CW4	78	71	7.2	98	381	95	2	81	203	3.447	0.603	138.355
	CW6	74	99	6.9	94	284	78	5	3	198	8.465	0.924	242.343
	OR	77	99	7.1	96	381	173	7	3	198	8.152	0.943	338.464
	RES	78	A2 = 13%		A3 = 7%	B7 = 7%		B13 = 7%		CW2 = 7%			
545	US5 TER	TER.C4610B											
	CW4	90	64	7.7	99	382	97	1	101	183	3.794	0.550	115.722
	CW6	74	89	7.0	95	284	79	4	22	179	5.229	0.794	179.039
+	B5	64	91	4.0	50	201	4	4	18	175	1.317	0.263	13.880
*	TEC10	33	91	0.0	0	193	0	0	18	175	13.381	0.097	1.833
	OR	83	91	7.2	95	382	180	9	18	175	5.196	0.855	279.357
	RES	44	A2 = 11%		CW7 = 8%		B7 = 7%		CW5 = 7%				
540	FRA DDC	Pierson											
	CW4	98	69	8.0	100	382	98	0	89	195	4.191	0.596	135.617
	CW6	92	97	7.8	100	284	83	0	6	195	8.886	0.943	252.745
	OR	96	97	7.9	100	382	181	0	6	195	9.758	0.964	354.972
	RES	33	A2 = 7%		A23 = 7%	B49 = 7%		B51 = 7%		A1 = 7%			
547	GER MUE	MUE23928											
	CW6	79	94	6.8	90	380	81	9	16	274	4.278	0.818	254.447
	RES	61	B35 = 17%		CW4 = 17%		A3 = 11%		A2 = 7%				
558	FRA DDC	Devinat											
	CW7	81	68	6.8	89	381	119	15	78	169	1.840	0.544	112.608
+	CW7L	81	71	7.6	100	247	9	0	69	169	2.850	0.278	19.101
	OR	81	71	6.9	90	381	128	15	69	169	2.092	0.583	129.543
	RES	56	CW4 =9%		B35 = 7%		A3 = 6%		A1 = 6%	A2 = 6%			
205	GER BRA	918B											
	B35	95	95	7.9	100	382	78	0	16	288	8.081	0.880	295.958
+	CW7L	33	96	2.7	30	304	3	7	13	281	1.207	0.221	14.901
+	B39	56	96	2.4	21	294	3	11	10	270	0.799	0.202	12.054
	OR	92	96	6.6	82	382	84	18	10	270	4.504	0.804	246.907
	RES	30	CW7 = 9%		A1 = 7%		CW6 = 7%						
192	US5 PNW	Juli											
	B35	73	95	6.9	94	382	73	5	14	290	6.377	0.849	275.409
+	B51	20	97	1.7	15	304	6	33	8	257	0.275	0.204	12.629
*	BW53	60	97	2.0	14	265	1	6	7	251	0.958	0.150	5.944
	OR	69	97	5.2	68	382	79	38	8	257	2.998	0.704	189.591
	RES	26	A1 = 10%		CW7 = 10%		B8 = 8%		CW6 = 8%	A24 = 8%			
191	US2 MBC	BC.CY.CHAD											
	B35	61	96	5.9	81	382	63	15	13	291	4.412	0.767	224.624
	BW53	69	97	5.4	71	304	5	2	8	289	4.329	0.503	76.985
+	B5	11	98	3.1	38	297	3	5	5	284	3.053	0.370	40.642
+	CW4	33	99	1.8	16	289	3	16	2	268	0.888	0.295	25.173
	OR	59	99	5.0	66	382	74	38	2	268	4.263	0.740	209.121
	RES	14	A2 = 18%		B44 = 18%		CW5 = 18%		B51 = 9%		CW2 = 9%		

Table 2. *Continued*

		STR	SPEC	AVE	INCL	N	TP	FN	FP	TN	Q	R	CHI²
562	IT1 PUR CNTS.105												
	CW4	86	100	6.3	78	376	76	21	1	278	5.818	0.840	265.000
	RES	33											
535	EAE MYR Furlinger												
	CW4	79	99	6.9	93	373	87	7	3	276	6.511	0.922	316.906
	RES	14	A26 = 10%		A11 = 10%		B8 = 10%	B39 = 10%		CW7L = 10%			
347	US5 MIT B5612.4												
	B7	96	80	7.9	100	374	74	0	60	240	5.861	0.660	162.918
	CW4	45	97	4.2	59	300	53	37	7	203	2.455	0.631	119.551
+	BW60	27	98	2.0	22	210	4	14	3	189	0.810	0.328	22.570
*	BW67	33	98	0.0	0	192	0	0	3	189	19.330	0.243	11.321
*	BW42	60	99	8.0	100	192	1	0	2	189	9.370	0.524	52.690
	OR	74	98	5.5	72	374	131	51	3	189	3.604	0.730	199.194
	RES	56	A11 = 12%		A2 = 12%		CW7 = 12%		A25 = 6%		B51 = 6%		
196	FRA BET E1218												
+	CW4	99	95	7.9	99	382	97	1	13	271	7.207	0.905	312.595
	B35	45	97	7.0	100	284	4	0	9	271	6.542	0.526	78.619
*	BW53	71	97	8.0	100	280	2	0	7	271	7.075	0.447	56.037
	OR	97	97	7.9	99	382	101	1	9	271	7.518	0.930	330.413
	RES	62	A2 = 10%		CW7 = 10%		A24 = 6%		B5 = 6%		A11 = 6%		
214	US2 WOL NW14002												
	CW4	96	87	7.7	97	382	95	3	38	246	4.843	0.761	221.408
	B51	57	94	5.4	71	284	24	10	14	236	3.635	0.613	106.827
+	B5	79	98	6.8	89	250	8	1	6	235	4.877	0.679	115.309
	B35	78	99	8.0	100	241	3	0	3	235	7.421	0.654	103.234
*	BW52	33	99	1.3	0	238	0	12	3	223	1.611	0.041	0.395
*	BW53	71	100	8.0	100	238	2	0	1	235	9.444	0.718	122.600
*	BW70	33	99	1.0	0	238	0	6	3	229	0.674	0.074	1.308
*	TE79	33	99	0.0	0	238	0	0	3	235	19.862	0.244	14.195
	OR	89	99	7.1	90	382	130	14	3	235	6.219	0.900	309.756
	RES	56	CW6 = 18%		A29 = 12%		BW53 = 12%		A25 = 6%		A30 = 6%		
538	BEN BOU 64307 6/7/84												
	CW4	62	98	5.9	83	382	81	17	6	278	4.709	0.833	265.101
	RES	45	A24 = 10%										
204	IT1 CON CA181												
	CW4	87	99	7.4	95	380	93	5	4	278	6.754	0.932	329.811
	RES	33	A2 = 12%		B7 = 12%		A24 = 12%		B27 = 12%		CW2 = 12%		
534	UK1 FES Langlais												
	CW4	56	100	4.3	57	380	55	42	1	282	3.419	0.688	179.707
	RES	33											
2101	GER MUC TU109												
+	CW6	66	86	4.5	56	359	48	37	37	237	1.131	0.428	65.637
+	B5	71	89	6.3	82	274	9	2	28	235	3.186	0.405	44.866
*	BW4	71	98	2.1	17	263	26	124	2	111	0.443	0.244	15.600
	OR	69	98	3.1	34	359	83	163	2	111	1.104	0.345	42.716
	RES	43	CW8 = 17%		A3 = 8%		AW33 = 8%		BW64 = 8%		B35 = 8%		
2103	UK1 GEL 103.1.51												
	BW6	74	81	5.3	68	360	205	97	11	47	1.220	0.364	47.776
+	CW4	38	88	3.8	50	58	5	5	6	42	1.078	0.355	7.324
	OR	73	88	5.3	67	360	210	102	6	42	1.537	0.377	51.144
	RES	47	B44 = 14%		A2 = 10%		A32 = 7%		CW5 = 7%		CW7 = 7%		

Table 3. Core Set Sera Analysis: Orientals

		STR	SPEC	AVE	INCL	N	TP	FN	FP	TN	Q	R	CHI²
533	BEN ENG CLB105												
	CW4	86	99	7.1	91	586	50	5	5	526	7.209	0.891	465.682
	RES	9	A26 = 12%	BW61 = 12%	A24 = 12%	CW1 = 12%	CW3 = 8%						
536	US5 CRO Martinez2.21.83												
	CW4	79	92	6.8	87	567	47	7	42	471	3.936	0.632	226.767
+	BW46	40	95	2.8	31	513	20	44	22	427	0.856	0.319	52.070
+	CW1	57	100	2.0	18	449	22	103	0	324	0.812	0.361	58.538
	OR	66	100	3.3	37	567	89	154	0	324	2.015	0.495	139.153
	RES	33											
537	UK1 BRS 5956.CBT												
	CW4	92	96	7.1	87	584	48	7	20	509	6.259	0.755	332.629
	RES	22	A24 = 13%	A26 = 10%	CW1 = 9%	A2 = 7%	BW61 = 7%						
539	EAE ZAR MZ467												
	CW4	77	99	6.5	85	572	47	8	5	512	5.973	0.858	421.570
	RES	27	A11 = 16%	CW6 = 11%	A2 = 11%								
541	NCY MRV 22166.0												
	CW4	78	99	6.7	87	584	48	7	4	525	6.345	0.879	451.257
	RES	14	A24 = 17%	A2 = 8%	B38 = 8%	BW54 = 8%	CW1 = 8%						
543	US8 JLE Billing												
	CW4	78	97	6.7	89	579	49	6	18	506	4.829	0.779	351.478
	CW6	54	100	5.1	67	524	16	8	2	498	4.198	0.747	292.310
	OR	73	100	6.2	82	579	65	14	2	498	5.560	0.872	440.527
	RES	43	A32 = 10%	B44 = 10%	A2 = 10%	BW63 = 10%	CW3 = 10%						
545	US5 TER TER.C4610B												
	CW4	83	94	7.2	93	583	51	4	31	497	5.281	0.725	306.637
	CW6	54	97	5.0	67	528	16	8	15	489	3.068	0.560	165.657
+	BW59	45	98	2.3	22	504	4	14	11	475	0.713	0.232	27.045
*	TEC10	20	98	6.0	100	486	1	0	10	475	8.491	0.300	43.878
	OR	75	98	5.7	73	583	71	26	11	475	3.673	0.756	332.831
	RES	36	A24 = 16%	CW3 = 11%	A11 = 7%	B7 = 7%	AW33 = 7%						
540	FRA DDC Pierson												
	CW4	93	94	7.7	98	584	54	1	31	498	6.096	0.759	336.583
	CW6	69	99	7.1	100	529	24	0	7	498	8.153	0.859	390.401
	OR	86	99	7.5	99	584	78	1	7	498	8.264	0.938	513.480
	RES	41	A24 = 17%	BW62 = 10%	AW33 = 7%	B44 = 7%	A26 = 7%						
547	GER MUE MUE23928												
	CW6	63	96	4.1	52	561	12	11	20	518	1.957	0.415	96.718
+	CW4	60	98	2.4	22	538	11	40	9	478	0.753	0.308	50.953
+	CW1,												
	3	11	99	2.0	27	487	3	8	6	470	1.402	0.304	44.989
	OR	56	99	2.8	31	561	26	59	6	470	1.183	0.451	114.130
	RES	20	A2 = 10%	AW33 = 10%	A11 = 10%	B44 = 7%	CW1 = 7%						
558	FRA DDC Devinat												
	CW7	73	89	6.1	81	586	112	27	50	397	2.682	0.657	253.217
+	A11	52	94	2.7	29	447	30	74	20	323	0.492	0.308	42.402
+	B7	64	95	3.4	40	343	4	6	16	317	1.377	0.264	23.962
	OR	69	95	4.6	58	586	146	107	16	317	1.944	0.584	199.587
	RES	31	A2 = 12%	A24 = 10%	CW1 = 8%	CW3 = 6%							
205	GER BRA 918B												
	B35	69	85	6.4	87	580	60	9	76	435	3.088	0.548	174.250
+	BW54	47	88	2.7	31	511	21	46	55	389	0.263	0.182	16.883
+	B27	69	89	4.9	62	444	5	3	50	386	1.747	0.210	19.635
+	B15	56	89	4.8	60	436	3	2	47	384	1.792	0.172	12.831
+	CW8	60	90	2.8	30	431	6	14	41	370	0.479	0.143	8.797
+	TS1	33	91	3.4	43	411	3	4	38	366	1.162	0.156	10.021
+	CW4	24	92	2.2	23	404	9	30	29	336	0.256	0.159	10.173
+	BW58	33	93	2.6	38	365	3	5	26	331	1.004	0.178	11.532
	OR	60	93	3.9	49	580	110	113	26	331	1.178	0.481	134.119
	RES	27	A2 = 11%	A24 = 10%	CW3 = 8%	A11 = 6%							

Table 3. *Continued*

		STR	SPEC	AVE	INCL	N	TP	FN	FP	TN	Q	R	CHI²
192	US5 PNW Juli												
	B35	58	95	6.1	85	562	58	10	24	470	4.072	0.738	306.432
+	B15S-LI	60	96	6.3	86	494	6	1	18	469	6.522	0.446	98.270
+	CW8	9	97	1.8	21	487	4	15	14	454	0.658	0.199	19.269
*	BW53	33	97	1.0	0	468	0	1	14	453	4.704	0.081	3.079
	OR	55	97	5.2	72	562	68	26	14	454	3.344	0.729	298.730
	RES	29	A2 = 12%		A24 = 10%		CW3 = 10%		BW62 = 7%		BW52 = 6%		
191	US2 MBC BC.CY.CHAD												
	B35	68	96	6.0	81	572	55	13	21	483	3.781	0.727	302.233
+	B51	28	98	2.1	20	504	11	43	10	440	0.563	0.284	40.536
*	BW53	60	98	8.0	100	450	1	0	9	440	8.689	0.314	44.355
	OR	61	98	4.3	54	572	66	56	10	440	2.138	0.623	221.938
	RES	39	A24 = 19%		A2 = 6%		BW60 = 6%		BW52 = 6%		CW3 = 6%		
562	IT1 PUR CNTS.105												
	CW4	86	97	7.0	89	565	47	6	13	499	5.679	0.808	369.230
+	BW62	38	98	1.5	8	512	5	55	8	444	0.157	0.142	10.304
	OR	81	98	4.0	46	565	52	61	8	444	2.027	0.571	184.480
	RES	26	A24 = 13%		A31 = 10%		A2 = 10%		B44 = 6%		CW3 = 6%		
535	EAE MYR Furlinger												
	CW4	90	98	6.9	87	575	47	7	12	509	5.276	0.808	375.230
	RES	52	A24 = 12%		A2 = 8%		CW3 = 8%		B44 = 6%		A26 = 6%		
347	US5 MIT B5612.4												
	B7	89	88	7.6	98	583	61	1	63	458	5.320	0.646	243.489
	CW4	73	93	4.4	54	521	28	24	35	434	1.709	0.426	94.381
	BW67	7	94	2.7	46	469	6	7	29	427	1.277	0.256	30.617
+	BW56	11	94	3.2	60	456	3	2	26	425	2.396	0.241	26.517
+	BW48	38	95	2.0	16	451	5	26	21	399	0.460	0.131	7.709
+	BW55	27	96	1.9	15	420	4	22	17	377	0.731	0.135	7.621
+	B40	27	96	1.8	14	394	4	25	13	352	0.403	0.144	8.135
+	BW60	23	97	1.4	8	365	5	55	8	297	0.166	0.121	5.366
*	BW42	33	97	1.0	0	305	0	1	8	296	4.502	0.106	3.423
	OR	69	97	3.6	42	583	116	162	8	297	1.545	0.475	131.515
	RES	26	A2 = 15%		CW3 = 12%		A26 = 8%		BW61 = 8%				
196	FRA BET E1218												
+	CW4	96	84	7.7	96	575	53	2	81	439	4.411	0.559	179.453
	B35	73	92	6.0	77	520	43	13	38	426	2.969	0.583	176.826
+	B51	14	95	2.5	31	464	17	37	21	389	0.856	0.310	44.522
+	B27	33	96	3.1	43	410	3	4	18	385	1.596	0.240	23.697
*	BW53	20	96	6.0	100	403	1	0	17	385	7.650	0.234	22.111
	OR	74	96	5.3	67	575	116	56	18	385	2.640	0.679	265.296
	RES	38	A24 = 13%		BW60 = 8%								
214	US2 WOL NW14002												
	CW4	91	77	7.6	96	584	53	2	123	406	3.386	0.463	124.928
	B35	69	84	5.8	78	529	47	13	76	393	1.887	0.464	114.004
	B51	77	90	5.0	62	469	33	20	43	373	1.681	0.445	92.706
+	B15S-LI	54	91	5.6	71	416	5	2	38	371	3.492	0.265	29.216
	BW52	35	94	2.5	28	409	17	44	21	327	0.547	0.269	29.700
+	B5	56	95	5.8	75	348	3	1	18	326	3.860	0.316	34.757
*	BW53	20	95	6.0	100	344	1	0	17	326	8.618	0.233	18.619
*	BW70	33	95	1.0	0	344	0	3	18	323	1.272	0.033	0.381
*	TE79	33	95	0.0	0	344	0	0	18	326	17.346	0.105	3.806
	OR	75	95	5.2	66	584	158	82	18	326	2.354	0.647	244.739
	RES	49	BW62 = 11%		CW3 = 11%		A2 = 9%		A24 = 9%		BW46 = 6%		
538	BEN BOU 64307 6/7/84												
	CW4	69	98	5.8	76	579	41	13	10	515	4.866	0.753	328.587
+	BW60	38	99	1.3	7	525	5	69	5	446	0.171	0.149	11.731
	OR	66	99	3.2	36	579	46	82	5	446	1.801	0.507	148.884
	RES	8	A24 = 18%		A2 = 9%		B7 = 9%		CW1 = 9%				

Table 3. *Continued*

	STR	SPEC	AVE	INCL	N	TP	FN	FP	TN	Q	R	CHI²
204 ITI CON CA181												
CW4	79	93	6.5	85	580	46	8	36	490	4.437	0.649	244.548
+ CW3.2	4	95	1.9	22	526	10	35	26	455	0.451	0.191	19.276
+ B27	56	95	4.2	50	481	3	3	23	452	2.411	0.234	26.434
OR	65	95	4.4	56	580	59	46	23	452	2.162	0.565	185.191
RES	14	A24 = 8%		CW1 = 8%		BW61 = 8%		A11 = 8%		CW3 = 8%		
534 UKI FES Langlais												
CW4	61	99	5.1	67	581	37	18	6	520	4.597	0.733	312.372
RES	8	A2 = 13%		BW61 = 9%		A24 = 9%						
2101 GER MUC TU109												
+ B27	87	91	6.4	75	570	6	2	50	512	3.761	0.263	39.364
BW4	37	98	1.9	16	562	43	224	7	288	0.352	0.239	31.971
OR	45	98	2.0	18	570	49	226	7	288	0.409	0.257	37.626
RES	41	A26 = 11%		BW60 = 9%		A24 = 9%		A31 = 9%		A2 = 6%		
2103 UKI GEL 103.1.51												
* BW6	75	17	6.6	88	573	450	60	52	11	0.064	0.058	1.944
RES	79	A24 = 14%		A2 = 8%		B44 = 7%		AW33 = 7%		B51 = 7%		

Core Set Serology

Results of the Core set sera analysis in the three main ethnic groups (Blacks, Caucasians, Orientals) are reported in Tables 1, 2, and 3.

The analysis includes the sera submitted as recognizing Cw4 and/or Cw6 as well as the sera whose reactions have been observed to be associated with these antigens in the previous central data analysis.

The high number and the good quality of the sera submitted allow a very good definition of Cw4. The situation regarding definition of the Cw6 antigen with the Tenth Workshop Core set sera is not satisfactory: when Cw4 is absent, Cw6 may be assigned because of the positivity of the cluster of Cw4+Cw6 sera (10W 543, 10W540, and 10W545): serum 10W547, the only Cw6 monospecific serum, is negative in 14% of Cw4-ve, Cw6+ve cells and positive in 15% of Cw4+ve,Cw6−ve cells. This problem is clearly shown in Table 4, in which the reaction patterns of some selected sera in the three main ethnic groups are reported.

Antigen Society Serology

The results of the Antigen Society sera analysis are reported in Tables 5 and 6: Analysis has not been performed in Orientals because of the very low number (18) of cells tested. Among the 15 sera submitted, two recognize both Cw4 and Cw6, ten Cw4, and two Cw6. The reaction patterns of some selected sera are shown in Table 7: The problems in the definition of Cw6 appear to be similar to those observed with Core set sera.

Table 4. Frequency (×100) of the Reaction Patterns of Selected Core Set Sera (Patterns Observed in Only One Cell Have Not Been Reported)

Phenotype		Tenth W No.							Population		
Cw4	Cw6	543 a	540 a	545 a	547 b	533 c	541 c	535 c	Black	Cauc	Ori
+	−								(N=118)	(N=73)	(N=56)
		+	+	+	−	+	+	+	63	75	57
		+	+	+	+	+	+	+	14	4	16
		+	+	+	−	+	−	+	−	8	−
		+	+	+	−	+	+	−	−	4	7
		+	+	+	+	+	−	+	−	3	−
		−	+	+	−	+	+	+	−	−	5
		+	+	−	−	+	−	+	−	−	4
		+	+	+	+	+	+	−	2	−	−
		+	−	+	+	+	+	+	2	−	−
−	+								(N=54)	(N=84)	(N=26)
		+	+	+	+	−	−	−	50	74	46
		+	+	+	−	−	−	−	9	8	15
		−	+	+	+	−	−	−	−	6	−
		+	+	−	+	−	−	−	11	5	8
		−	+	+	−	−	−	−	−	−	8
		+	−	+	+	−	−	−	11	−	−
−	−								(N=174)	(n=198)	(n=498)
		−	−	−	−	−	−	−	98	87	91
		−	−	+	−	−	−	−	6	8	2
		−	+	−	−	−	−	−	6	−	1
		−	−	−	+	−	−	−	2	−	1
		−	+	+	−	−	−	−	2	1	−

a = Cw4+Cw6; b = Cw6; c = Cw4

Table 5. Antigen Society Sera Analysis: Blacks

		STR	SPEC	AVE	INCL	N	TP	FN	FP	TN	Q	R	CHI²
4795	US5 MIT C5902.4												
	CW4	66	88	5.2	66	200	19	10	20	151	1.642	0.474	44.873
+	A1	7	91	3.4	55	171	6	5	14	146	1.818	0.351	21.108
+	CW6	57	97	2.4	24	160	10	32	4	114	0.569	0.314	15.813
+	BW42	45	100	1.6	11	118	4	33	0	81	0.312	0.261	8.068
	OR	53	100	2.9	33	200	39	80	0	81	1.129	0.398	31.699
	RES	33											
4796	NCY KAP CWR.PLA.241												
+	B35	65	94	7.4	100	200	7	0	11	182	5.736	0.587	68.853
+	A26	9	96	1.7	17	193	4	20	7	162	0.576	0.189	6.897
+	A23	33	97	1.8	14	169	3	19	4	143	0.487	0.197	6.541
*	CW4	71	98	1.7	10	147	2	19	2	124	0.280	0.185	5.035
	OR	42	97	2.5	26	200	14	39	4	143	0.733	0.361	26.064
	RES	82	A29 = 13%		B44 = 13%		BW42 = 13%		B13 = 13%		BW70 = 13%		
4797	US5 MIT D4155.1												
	CW4	28	95	4.3	62	200	18	11	9	162	2.239	0.578	66.768
+	B35	56	96	7.3	100	171	3	0	6	162	6.478	0.539	49.628
+	A26	45	99	2.1	19	168	4	17	2	145	0.700	0.317	16.882
	OR	36	99	3.6	47	200	25	28	2	145	1.869	0.583	68.019
	RES	14	A2 = 17%		A29 = 17%		B13 = 17%		B39 = 17%		CW2 = 17%		
4798	GER GAZ T.1758												
	CW4	97	89	8.0	100	200	29	0	19	152	6.247	0.722	104.182
+	A1	33	94	5.3	82	171	9	2	10	150	3.559	0.577	56.976
	OR	82	94	7.3	95	200	38	2	10	150	4.961	0.820	134.457
	RES	57	A30 = 12%		A26 = 8%		CW6 = 8%		A29 = 8%				

Table 5. *Continued*

		STR	SPEC	AVE	INCL	N	TP	FN	FP	TN	Q	R	CHI²
4799	FRA BET 1268												
	CW4	43	91	5.0	79	200	23	6	16	155	2.681	0.613	75.254
+	A23	39	96	3.0	40	171	10	15	6	140	1.770	0.432	31.873
	OR	42	96	4.1	61	200	33	21	6	140	2.861	0.631	79.623
	RES	7	A30 = 14%		A26 = 7%		B7 = 7%	B15 = 7%		CW3 = 7%			
4800	US5 PNW BORT												
	CW4	30	99	4.1	59	200	17	12	2	169	2.647	0.677	91.727
	RES	33											
4801	IT1 MTT F10113												
	CW4	75	92	7.1	93	199	26	2	14	157	5.001	0.723	104.112
+	A1	26	96	4.6	73	171	8	3	6	154	3.264	0.603	62.195
+	A30	7	100	1.4	10	160	6	53	0	101	0.287	0.246	9.655
	OR	54	100	3.4	41	199	40	58	0	101	1.643	0.501	50.044
	RES	33											
4802	UK1 CXH Lopian												
	CW4	90	95	7.9	100	200	29	0	8	163	7.126	0.850	144.433
+	B44	9	97	1.8	16	171	4	21	4	142	0.763	0.229	8.987
+	A29	82	100	3.0	29	146	4	10	0	132	1.645	0.495	35.750
	OR	82	100	4.6	54	200	37	31	0	132	2.909	0.655	85.790
	RES	33											
4803	EAE ZAR MZ435												
	CW4	49	87	4.7	67	198	18	9	22	149	2.104	0.455	41.073
+	B13	26	91	5.4	80	171	8	2	14	147	3.447	0.491	41.229
+	AW68	5	95	2.1	24	161	8	26	6	121	0.476	0.273	11.987
+	B45	9	98	1.8	27	127	4	11	2	110	1.376	0.377	18.054
	OR	29	98	3.2	44	198	38	48	2	110	1.505	0.516	52.774
	RES	14	A30 = 33%		B44 = 33%		BW58 = 33%		B15 = 33%		CW3 = 33%		
4804	FRA PRR SCH1423												
	CW4	88	93	6.8	85	198	23	4	12	159	4.510	0.692	94.920
+	A1	7	96	3.4	55	171	6	5	6	154	2.738	0.483	39.914
+	B44	45	99	1.9	17	160	4	19	2	135	0.592	0.296	14.013
	OR	68	99	4.3	54	198	33	28	2	135	2.580	0.629	78.248
	RES	14	A30 = 20%		A26 = 20%		B7 = 20%	B15 = 20%		CW3 = 20%			
4805	BEN DPT SMEETS.80												
	CW4	96	91	8.0	100	198	27	0	16	155	6.668	0.742	109.106
+	A1	7	94	3.4	55	171	6	5	10	150	2.680	0.407	28.279
+	A30	47	98	1.7	14	160	8	51	2	99	0.268	0.225	8.092
	OR	74	98	3.6	42	198	41	56	2	99	1.232	0.481	45.896
	RES	71	AW68 = 20%		A29 = 20%		BW57 = 20%		BW70 = 20%		CW6 = 20%		
4803	EAE ZAR MZ435												
	CW4	49	87	4.7	67	198	18	9	22	149	2.104	0.455	41.073
+	B13	26	91	5.4	80	171	8	2	14	147	3.447	0.491	41.229
+	AW68	5	95	2.1	24	161	8	26	6	121	0.476	0.273	11.987
+	B45	9	98	1.8	27	127	4	11	2	110	1.376	0.377	18.054
	OR	29	98	3.2	44	198	38	48	2	110	1.505	0.516	52.774
	RES	14	A30 = 33%		B44 = 33%		BW58 = 33%		B15 = 33%		CW3 = 33%		
4804	FRA PRR SCH1423												
	CW4	88	93	6.8	85	198	23	4	12	159	4.510	0.692	94.920
+	A1	7	96	3.4	55	171	6	5	6	154	2.738	0.483	39.914
+	B44	45	99	1.9	17	160	4	19	2	135	0.592	0.296	14.013
	OR	68	99	4.3	54	198	33	28	2	135	2.580	0.629	78.248
	RES	14	A30 = 20%		A26 = 20%		B7 = 20%	B15 = 20%		CW3 = 20%			
4805	BEN DPT SMEETS.80												
	CW4	96	91	8.0	100	198	27	0	16	155	6.668	0.742	109.106
+	A1	7	94	3.4	55	171	6	5	10	150	2.680	0.407	28.279
+	A30	47	98	1.7	14	160	8	51	299	0.			
										268	0.225	8.092	
	OR	74	98	3.6	42	198	41	56	2	99	1.232	0.481	45.896
	RES	71	AW68 = 20%		A29 = 20%		BW57 = 20%		BW70 = 20%		CW6 = 20%		

Table 5. *Continued*

		STR	SPEC	AVE	INCL	N	TP	FN	FP	TN	Q	R	CHI²
4806	UKI GEL VAU20802												
	CW4	84	94	7.6	100	200	29	0	10	161	7.504	0.823	135.468
+	A1	33	98	3.4	55	171	6	5	4	156	2.434	0.535	48.926
+	B44	9	100	1.5	17	160	4	19	0	137	0.870	0.378	22.824
	OR	68	100	4.6	62	200	39	24	0	137	3.343	0.716	102.621
	RES	33											
4807	BEN ROD VR61809												
	CW4	74	80	6.6	86	198	25	4	34	135	2.377	0.504	50.332
	CW6	45	93	4.0	52	169	26	24	8	111	1.278	0.509	43.761
	OR	60	93	5.0	65	198	51	28	8	111	2.014	0.612	74.275
	RES	26	A30 = 12%		BW58 = 12%		A1 = 6%		B7 = 6%		BW65 = 6%		
4808	US5 SIN 13782												
	CW6	67	91	4.9	62	200	32	20	14	134	1.778	0.537	57.701
+	B39	45	93	5.0	67	148	4	2	10	132	2.674	0.400	23.722
	OR	65	93	4.9	62	200	36	22	10	132	2.069	0.587	68.854
	RES	22	BW58 = 14%		B44 = 14%		A30 = 14%		AW68 = 10%		BW70 = 10%		
4809	UKI AST DB5424												
	CW6	71	80	6.3	83	200	43	9	30	118	1.937	0.563	63.284
+	B39	33	83	6.7	100	148	6	0	24	118	4.012	0.394	23.028
+	CW4	74	88	3.6	40	142	10	15	14	103	0.655	0.285	11.560
+	A3	45	90	3.0	33	117	4	8	10	95	0.621	0.232	6.310
	OR	67	90	5.2	66	200	63	32	10	95	1.718	0.583	67.961
	RES	39	A30 = 18%		BW70 = 12%		CW2 = 12%		A26 = 6%		B7 = 6%		

Table 6. Antigen Society Sera Analysis: Caucasians

		STR	SPEC	AVE	INCL	N	TP	FN	FP	TN	Q	R	CHI²
4795	US5 MIT C5902.4												
	CW4	61	95	5.3	71	348	63	26	14	245	2.931	0.682	162.059
+	BW57	20	97	2.7	29	259	6	15	8	230	1.224	0.310	24.927
+	B37	33	98	2.5	23	238	3	10	5	220	1.163	0.278	18.445
	OR	56	98	4.5	59	348	72	51	5	220	2.595	0.644	144.328
	RES	38	A1 = 11%		CW6 = 11%		A24 = 7%		B44 = 7%		CW3 = 7%		
4796	NCY KAP CWR.PLA.241												
+	B35	61	97	5.1	69	346	47	21	9	269	3.132	0.705	171.945
+	BW53	54	99	5.1	71	278	5	2	4	267	3.922	0.602	100.862
+	B51	33	100	1.5	8	271	3	35	1	232	0.466	0.217	12.817
*	CW4	33	100	1.2	0	233	0	13	1	219	1.073	0.075	1.294
	OR	59	100	3.9	49	346	55	58	1	232	2.470	0.609	128.438
	RES	60	A2 = 20%		A24 = 20%		B7 = 20%		B44 = 20%		CW7 = 20%		
4797	US5 MIT D4155.1												
	CW4	74	92	6.3	83	349	74	15	22	238	3.634	0.724	182.982
+	B51	27	97	3.2	44	260	15	19	7	219	1.530	0.493	63.114
+	B35	33	98	5.3	100	226	3	0	4	219	7.817	0.610	84.090
	OR	65	98	5.4	73	349	92	34	4	219	3.610	0.761	202.141
	RES	64	A2 = 14%		A24 = 9%		B7 = 9%		CW7 = 9%		A29 = 9%		
4798	GER GAZ T.1758												
	CW4	93	91	7.7	98	348	87	2	24	235	5.534	0.823	235.494
+	B51	35	96	3.3	41	259	14	20	10	215	2.071	0.426	46.979
+	B35	78	97	8.0	100	225	3	0	7	215	6.424	0.516	59.796
	OR	86	97	6.5	83	348	104	22	7	215	4.322	0.813	230.287
	RES	65	CW7 = 11%		A2 = 8%		B7 = 8%		CW6 = 8%				
4799	FRA BET 1268												
	CW4	61	98	5.8	80	347	71	18	4	254	4.841	0.824	235.405
	RES	33	A2 = 13%		A24 = 13%		B51 = 13%		B7 = 7%		B44 = 7%		
4800	US5 PNW BORT												
	CW4	54	98	5.6	81	349	72	17	4	256	4.633	0.832	241.471
	RES	33	CW6 = 12%		B44 = 12%		A26 = 6%		A30 = 6%		B13 = 6%		

Table 6. *Continued*

		STR	SPEC	AVE	INCL	N	TP	FN	FP	TN	Q	R	CHI²
4801	ITl MTT F10113												
	CW4	80	91	6.5	84	346	74	14	22	236	3.747	0.730	184.366
+	B35	56	93	7.3	100	258	3	0	19	236	5.469	0.346	30.928
+	CW3	47	95	2.4	24	255	8	25	11	211	0.443	0.251	16.061
	OR	77	95	5.5	69	346	85	39	11	211	2.768	0.677	158.442
	RES	36	CW7 = 14%		A2 = 12%		B7 = 10%		A1 = 9%				
4802	UKl CXH Lopian												
	CW4	86	99	7.4	96	349	85	4	3	257	6.945	0.940	308.401
	RES	60	A2 = 20%		A24 = 20%		B7 = 20%		B44 = 20%		CW7 = 20%		
4803	EAE ZAR MZ435												
	CW4	78	90	5.7	72	347	63	24	26	234	2.275	0.615	131.455
+	CW6	58	97	2.5	26	260	20	58	6	176	0.616	0.338	29.718
	OR	74	97	4.2	50	347	83	82	6	176	1.746	0.534	98.832
	RES	47	B7 = 12%		A2 = 9%		CW2 = 6%		CW7 = 6%		A24 = 6%		
4804	FRA PRR SCH1423												
	CW4	87	87	7.4	97	348	86	3	34	225	4.568	0.761	201.685
+	A23	85	88	4.9	56	259	5	4	29	221	1.326	0.244	15.371
	OR	88	88	7.2	93	348	91	7	29	221	3.854	0.764	203.098
	RES	74	A1 = 9%		A2 = 9%		CW6 = 9%						
4805	BEN DPT SMEETS.80												
	CW4	94	95	7.7	98	319	80	2	11	226	7.739	0.892	254.021
+	B49	11	96	1.8	27	237	3	8	8	218	1.005	0.254	15.317
+	CW5	11	97	1.4	10	226	3	27	5	191	0.232	0.150	5.057
	OR	87	97	5.6	70	319	86	37	5	191	3.259	0.721	165.869
	RES	38	CW6 = 15%		A1 = 12%		A2 = 12%		A24 = 8%		A29 = 8%		
4806	UKl GEL VAU20802												
	CW4	92	98	7.7	99	349	88	1	6	254	8.136	0.942	309.646
	RES	23	CW6 = 16%		A2 = 12%		B44 = 12%		A1 = 8%		CW5 = 8%		
4807	BEN ROO VR61809												
	CW4	66	76	6.6	92	346	81	7	62	196	2.584	0.597	123.490
	CW6	51	98	5.3	76	258	58	18	4	178	4.322	0.783	158.309
	OR	60	98	6.0	85	346	139	25	4	178	4.915	0.832	239.641
	RES	64	A2 = 15%		CW7 = 15%		B7 = 10%		A24 = 10%				
4808	US5 SIN 13782												
	CW6	38	98	3.5	44	340	37	47	6	250	2.321	0.537	97.973
	RES	54	A3 = 12%		CW4 = 12%		A2 = 12%		B51 = 8%		CW7 = 8%		
4809	UKl AST DB5424												
	CW6	73	91	6.0	80	350	69	17	25	239	2.788	0.683	163.163
+	AW68	20	92	3.2	46	264	6	7	19	232	1.324	0.291	22.344
+	BW73	33	93	3.2	50	251	3	3	16	229	1.955	0.263	17.384
	OR	67	93	5.5	74	350	78	27	16	229	2.537	0.696	169.558
	RES	60	A2 = 13%		CW4 = 11%		B35 = 9%		A3 = 7%		B51 = 7%		

When there are taken into account the cells typed both with the Core set and the Antigen Society set sera (Table 8) the definition of Cw6 appears improved, although in some Cw4−ve,Cw6+ve cells the Cw6 sera react as in the Cw4+ve,Cw6−ve cells: This improvement is clearly due to the increase in the number of the monospecific Cw6 sera.

Segregation Data

Some interesting segregation data (S.A. Blacks and Cape colored) have been kindly supplied by Dr. E. Du Toit. In Table 9 the segregation pattern of some Core set sera in family Dut 228 is shown. The four children receive from the father the pattern 10W543+, 10W540+, 10W545+: 10W537 and 10W539 segregate together and in repulsion with 10W547 (the monospecific Cw6 serum); in this family the sera 10W537 and 10W539 segregate with the haplotype Aw68,Cw6,Bw72. Similar segregation patterns have been observed in families Dut 10 and Dut 20, associated respectively with B15K and with Bw57.

Population Data

In Cw4−ve unrelated Black population 10W537 and 10W534 are associated (r=0.54); there are eight cells 10W537+,10W539+: seven of them are Cw6+ve, and five out of them are Bw57+ve. Among the 10 cells Cw4-, Cw6-, Bw57+ none has been observed to be

Table 7. Frequency (×100) of the Reaction Patterns of Selected Antigen Society Sera (Patterns Observed in Only One Cell Have Not Been Reported)

Phenotype		Tenth W No.						Population	
Cw4	Cw6	4807 a	4809 b	4808 b	4802 c	4805 c	4806 c	Black	Cauc
+	−							(N=25)	(N=66)
		+	−	−	+	+	+	36	70
		+	+	−	+	+	+	40	5
		−	−	−	+	+	+	16	−
		+	−	+	+	+	+	8	−
−	+							(N=50)	(N=67)
		+	+	−	−	−	−	12	30
		+	+	+	−	−	−	24	30
		−	+	−	−	−	−	12	7
		−	−	−	−	−	−	4	6
		+	−	−	−	−	−	−	4
		−	+	+	−	−	−	20	3
−	−							(N=119)	(N=158)
		−	−	−	−	−	−	75	91
		−	+	−	−	−	−	7	3
		−	−	−	−	+	−	−	3

a = Cw4 + Cw6; b = Cw6; c = Cw4

10W537+, 10W539+; among the 38 cells Cw4−, Cw6+, Bw57−, only two are positive with both sera.

These family and population data may suggest that 10W537 and 10W539 recognize a split of Cw6 or, perhaps, a new HLA-C antigen cross-reacting with Cw4 and Cw6 and in linkage disequilibrium with Bw57 in Blacks.

Table 9. Segregation Patterns of Some Core Set Sera in Family Dut 228

	Tenth W No.									
	543	540	545	547	533	541	535	537	539	
F	+	+	+	+	−	−	−	+	+	ab
M	−	−	−	−	−	−	−	−	−	cd
C1	+	+	+	−	−	−	−	+	+	ac
C2	+	+	+	−	−	−	−	+	+	ad
C3	+	+	+	+	−	−	−	−	−	bd
C4	+	+	+	+	0	−	−	−	−	bd

Haplotypes: a = Aw68, Cw6, Bw72; b = A3, Cw6, Bw58; c = A30, Cw2, Bw71; d = Aw34, Cw7, B8

Acknowledgment. We gratefully acknowledge the helpful support of Dr. Fiorenza Quoghi.

References

1. Nomenclature for Factors of the HLA System. In: Kissmeyer-Nielsen F (ed): Histocompatibility Testing 1975. Munksgaard, Copenhagen, 1975; p 5.
2. Nomenclature for Factors of the HLA System 1977. In: Bodmer WF, et al. (eds): Histocompatibility Testing 1977. Munksgaard, Copenhagen, 1977; p 14.
3. Tiilikainen A. Cw6. In: Terasaki PI (ed): Histocompatibility Testing 1980. UCLA, Los Angeles, 1980; p 496.
4. Scholz S, Wakisaka A. HLA-Cw6. In: Albert ED, et al. (eds): Histocompatibility Testing 1984. Springer-Verlag, Berlin, Heidelberg, 1984; p 181.

Table 8. Frequency (×100) of the Reaction Patterns of Selected Core and Antigen Society Sera (Patterns Observed in Only One Cell Have Not Been Reported)

Phenotype		Tenth W No.													Population	
Cw4	Cw6	543	540	545	4807	547	4809	4808	4806	533	541	535	4802	4805	Black	Cauc
		a	a	a	a	b	b	b	c	c	c	c	c	c		
+	–														(N=13)	(N=63)
		+	+	+	+	–	–	–	+	+	+	+	+	+	15	67
		+	+	+	+	–	+	–	+	+	+	+	+	+	8	6
		+	+	+	+	–	–	–	+	+	–	+	+	+	–	6
		+	+	+	+	–	–	–	+	+	+	–	+	+	–	3
		+	+	+	+	+	–	–	+	+	–	+	+	+	–	3
–	+														(N=26)	(N=67)
		+	+	+	+	+	+	+	–	–	–	–	–	–	4	29
		+	+	+	+	+	+	–	–	–	–	–	–	–	4	23
		+	+	+	+	–	+	–	–	–	–	–	–	–	–	6
		+	+	+	–	+	+	–	–	–	–	–	–	–	–	4
		+	+	+	+	+	+	+	–	–	–	–	–	+	8	3
		–	+	+	–	+	–	–	–	–	–	–	–	–	–	3
		+	+	+	–	–	–	–	–	–	–	–	–	–	–	3
		+	+	+	–	+	+	+	–	–	–	–	–	–	4	3
		+	+	+	+	+	+	+	+	–	–	–	–	–	–	3
		+	+	–	+	+	+	+	–	–	–	–	–	–	15	–
		+	+	–	–	+	+	+	–	–	–	–	–	–	15	–
		+	+	–	+	+	+	–	–	–	–	–	–	–	8	–
–	–														(N=59)	(N=156)
		–	–	–	–	–	–	–	–	–	–	–	–	–	68	86
		–	–	+	–	–	–	–	–	–	–	–	–	–	–	4
		–	–	+	–	–	+	–	–	–	–	–	–	–	–	3
		–	–	–	–	–	–	–	–	–	–	–	–	+	–	2
		–	+	–	–	–	–	–	–	–	–	–	–	+	–	1
		–	–	–	–	–	+	–	–	–	–	–	–	–	3	–

a = Cw4 + Cw6; b = Cw6; c = Cw4

Antigen Society #19 Report (Cw7)

C. Raffoux,[1] S. Goldmann,[2] and W.R. Mayr[3]

History

The antigen Cw7 was described by Dehay et al. as CVE (1). During the Eighth and Ninth Workshops, Cw7 was defined by two sera with good segregation in the families. The gene frequency was 0.163 in Caucasoids. The linkage disequilibrium with B7 and B8 was confirmed. The most frequent haplotype found in Caucasians was B7, Cw7, DR2 (= 102) but this one was not found in Orientals. A split into two parts was suggested.

Serology

To the Core set, six sera were submitted as defined Cw7 = 10WS 555, 556, 557, 558, 559, and 561. To the Antigen Society n° 19, only one serum was submitted, WS = 4810, and six additional sera have been investigated. The best sera are 10WS 559 – 555 and 558. The R values were between 0.81 and 0.54 in the Core set and between 0.82 and 0.66 in the Antigen Society. The percentage of missed and extra reactions are shown in Table 1. Two other sera, 10WS 336 and 338, submitted as Bw46 and Bw73, are also defining Cw7 were R values 0.40 – 0.45. They both reacted parallel with the same cells. The segregation of Cw7 in families was good except for 10WS 557.

Linkage Disequilibrium

The known linkage disequilibrium of Cw7 with B7 and B8 is confirmed in Caucasoids and with Bw58 in Orientals.

Split of Cw7

During the Eighth and Ninth Workshops, a split of Cw7 was suggested (2–3). To answer the question whether sera are defining a possible Cw7 subspecificity, Cw7 segregation in two families was analyzed. The subdivision was not confirmed in either the Core set or the Antigen Society set. Therefore, 556, 557, 336, and 338 are weak sera.

Conclusions

Cw7 is a well defined antigen in Caucasoid and Japanese populations. The subdivision suggested in the last Workshop has not been confirmed.

References

1. Dehay C, Schmid M, Benajam A, Legrand L, Dausset J. Seventh allele of the HLA-C series (Cve). Transplant Proc 1979;11:613.
2. Sasazuki T, Dausset J, Tokunaga K, Sato H, Okochi K, Naito S. Cw7 joint report. In: Terasaki PI (ed): Histocompatibility Testing 1980. UCLA Tissue Typing Laboratory, Los Angeles, 1980; p 501.
3. Baur MP, Neugebauer M, Albert ED. In: Albert ED, Bauer MP, Mayr WR (eds): Histocompatibility Testing 1984. Springer-Verlag, 1984; p 677.

Reporting Laboratories: FRACOH,[1] GERGOL,[2] EAEMYR[3]

Table 1. Analysis of Core and Antigen Society Sera Defining Cw7

Sera Identity	Strength	% of Missed Reactions	% of Extra Reactions	R value
10WS 555	70	3	10	0.70
10WS 556	54	8	5	0.66
10WS 557	49	10	4	0.62
10WS 558	71	5	12	0.60
10WS 559	69	3	10	0.69
10WS 561	47	14	8	0.43
10WS 4810	75	8	11	0.76
FRA DDC P 0366		31	14	0.44
FRA DDC Q 3153		28	14	0.56
FRA DDC O 6907		25	1	0.59
FRA DDC R 6784		28	0	0.53
FRA MRV 9372.0		25	2	0.53
FRA MRV 18525.1 1B		21	1	0.62

Immunobiology of HLA
Volume I

Antigen Society #20 Report (Cw2)

M.M. Tongio,[1] C. Jersild,[2] H.C. von Keyserlingk,[3] S. Wegener,[4] H. Mervart,[5] and S. Mayer[1]

The first serum describing the Cw2 (previously called T2) specificity was serum 170 (1). At the Eighth Histocompatibility Workshop, the Cw2 antigen frequency was 9.8% in Caucasians, 20.3% in blacks, and less than 1.5% in Asiatics (2). Cw2 is in linkage disequilibrium with B27, Bw61, B44, and in Caucasians the following haplotypes have been found: Cw2, Bw61, DR5; A2, Cw2, B51; and A32, Cw2, Bw61 (2,3).

At the Tenth Histocompatibility Workshop, 11 Cw2 sera were included in the Core cell analysis, and 13 in the Cw2 Antigen Society analysis.

The Core sera were tested by all the laboratories involved in the Tenth Workshop serologic studies with 691 cells carrying the Cw2 antigen. The Antigen Society sera were tested by the antigen society participants with only 114 cells carrying Cw2 (93 Caucasians, 21 blacks, no Asiatics).

In order to perform a comparative study of the quality of the Core sera and the Antigen Society sera, the analysis was realized with the same panel of cells, i.e., those that were used for the Antigen Society studies. A reagent was considered good when its R values was ≥ .82, its strength value ≥ 60, and its average strength ≥ 6.

The results obtained with cells from Caucasians differed from those obtained with cells from Blacks.

When tested with cells from Caucasians, the Cw2 Core sera behaved as excellent (4 + reactions) or very good (3 + reactions) reagents. Most of them were "monospecific" (Table 1) and two contained additional antibodies: 326 contained Bw41 + Bw42 antibodies and 293 contained B27 + B47 antibodies. The quality of the sera included in the Cw2 antigen society was more heterogeneous. Only one half (listed in Table 2) fulfilled the criteria necessary to be considered good, whereas the others were too weak to be useful for Cw2 typing (nos. 4814, 4815, 4816, 4812, 4822, 4821). The most common patterns obtained with the Cw2 anti-sera having an R values ≥ .82 are given in Table 3.

When tested with cells from Blacks, the results were quite different. Only two of the Core sera (no. 517 and no. 520) fulfilled the criteria necessary to be considered good. The other Core sera as well as the Antigen Society sera had a quality score < 3.5 and showed frequent miss reactions. Furthermore, sera 4821, 4822, and 4823 were tested with too few cells to be included in the analysis.

Preliminary data from South Africa (du Toit, personal communication) obtained with cells not included in our

Reporting Laboratories: FRAMYE,[1] SCAJER,[2] GERKEY,[3] EAEWEG,[4] NCYMRV[5]

Table 1. Monospecific Cw2 Core Sera in Caucasians

Serum No	Antigen Soc. Assessment*	R value	Strength Value	Average Strength	Quality Scores
518 NCY KAP	++++	.89	74	6.8	6.3
517 LAT LAY	++++	.87	86	7.4	6.0
520 IT1 CEP	++++	.94	94	7.8	7.7
515 EAE MYR	++++	.94	83	7.6	8.3
522 US2 MBC	+++	.92	75	6.9	6.4
519 US3 GRU	+++	.92	77	7.2	6.8
521 US5 FTW	+++	.89	67	6.8	6.3
563 IT1 PUR	+++	.86	90	7.7	6.9
523 US6 BRN	+++	.78	60	5.8	4.9

*Assessment given in crosses: ++++ excellent; +++ very good.

Table 2. Best Cw2 Antigen Society Sera in Caucasians

Serum No	Antigen Soc. Assessment*	R value	Strength Value	Average Strength	Quality Scores
4823 EAE BRT	++++	.90	83	7.3	8.3
4811 NCY TOM	++++	.92	68	6.7	7.3
4818 US7 DUP	+++	.92	76	7.1	8.4
4813 US1 HJH	+++	.86	60	6.2	5.0
4820 US1 NWO	+++	.96	83	7.7	8.6
4819 SCA SVE	+++	.83	76	7.3	6.5
4917 BEN BOU	+++	.83	74	6.7	4.3

Table 3. Most Common Patterns Found With the Best Cw2 Sera (R value ≥ .82) in Caucasians

Total Cells : 228
Total Matching Selection : 228

	CW2	518	517	520	515	522	519	521	563	4823	4811	4818	4813	4820	4817	4819	Count	Percent	Percent of All Cells
Selection																	228	100.0	100.0
	−	−	−	−	−	−	−	−	−	−	−	−	−	−	−	−	101	44.3	44.3
	+	+	+	+	+	+	+	+	+	+	+	+	+	+	+	+	48	21.1	21.1
	−	−	−	−	−	−	−	−	−	−	−	−	−	−	−	+	10	4.4	4.4
	−	−	−	−	−	−	−	−	+	−	−	−	−	−	−	−	9	3.9	3.9
	+	+	+	+	+	+	+	+	+	+	+	+	−	+	+	+	6	2.6	2.6
	+	+	+	+	+	+	+	+	+	+	+	+	+	+	−	+	4	1.8	1.8
	−	−	−	−	−	−	−	−	−	−	−	−	−	−	+	−	3	1.3	1.3
	−	−	+	−	−	−	−	−	−	−	−	−	−	−	−	−	3	1.3	1.3
	−	−	−	−	−	−	−	−	−	+	−	−	−	−	−	−	3	1.3	1.3
	−	−	−	−	−	−	−	+	−	−	−	−	−	−	−	+	2	0.9	0.9
	+	−	+	+	+	+	+	+	+	+	+	+	+	+	+	+	2	0.9	0.9
	−	−	−	−	+	−	−	−	−	−	−	−	−	−	−	+	2	0.9	0.9
	−	+	+	+	−	−	−	−	+	−	−	+	−	+	+	+	1	0.4	0.4

analysis may indicate that sera 522 and 521 could allow the determination of splits in the South African population. The data obtained with the 21 Black cells tested with both the Core and Antigen Society sera could not confirm this point.

Additional Cw2 sera (not included in the Workshop computer analysis) were also studied by the participating laboratories. It was agreed that some of these reagents tested on Caucasian cells were even better (excellent or very good) than the official Cw2 Antigen Society sera, the best being SCA SVE D5454/73; LO483 H.H. (donated by Dr. H. Hansen, Copenhagen); FRA MYE Aliser; SA 532.4 (donated by Dr. L. Sandberg, Göteborg) and SCA SVE D4307/83.

References

1. Svejgaard A, Staub-Nielsen L, Ryder LP, Kissmeyer-Nielsen F, Sandberg L, Lindholm A, et al. Subdivisions of HL-A antigens. Evidence of a "new" segregant series. In: Dausset J, Colombani J (eds): Histocompatibility Testing 1972. Munksgaard, Copenhagen, Denmark, 1972; p 465.
2. Dick HM. Cw2. In: Terasaki PI (ed): Histocompatibility Testing 1980. UCLA Tissue Typing Laboratory, Los Angeles, 1980; p 485.
3. Grange D, Tongio MM, Hauptmann G, Uring-Lambert B, North ML, Mayer S. In: Albert ED, Baur MP, Mayr WR (eds): Histocompatibility Testing 1984. Springer-Verlag, Berlin, Heidelberg, 1984; p 326.

Antigen Society #21 Report (DR1, DRw10)

L.U. Lamm,[1] C. Brautbar,[2] R. Fauchet,[3] T. Lee,[4] E. Jalais,[3] T. Zhao,[5] S. Battat,[2] J. Hunter,[6] G. Dunston,[4] F. Garrido,[7] R.J. Winchester,[8] H. Mervart,[9] and A. Nunez-Roldan[8]

DR1 has been known as a well defined antigen since the Seventh International Histocompatibility Workshop (1). At the Ninth Workshop the antigen frequency was determined to be 14.7% in Caucasians and 9.45% in Negroids (2). The DRw10 antigen was described by Jakobsen et al. in 1979 (3), then reporting a new DR antigen associated with B37. In the Ninth Workshop (2), the frequency was found to be 2.1% in Caucasians and 3.5% in Negroes. At the Third Asia-Oceania Workshop, this antigen was found to be especially frequent in the Asian-Indian populations (up to 10%). Recent biochemical and DNA sequence studies have provided evidence that the DR1 and the DRw10 β1 chains are quite different (4).

Reporting Laboratories: SCAKIS,[1] GERBRA,[2] FRAFAU,[3] US8JLE,[4] CHIZHA,[5] US2MBC,[6] FRAGRA,[7] US8RJW,[8] NCYMRV[9]

Analysis of Workshop Data

Relevant reagents for this antigen society were 30 Core sera plus nine MAbs together with 53 Antigen Society sera. SERAN analyses were carried out separately for

Table 1. Summary of Tail Analysis on MAbs Relevant for DRl and DRWl0

WS10 No.	Specificity in Tail Order (R Value)				
3008	DR1	(0.735)	DR4	(0.475)	OR (0.716)
3009	DR1	(0.489)	DR4	(0.543)	OR (0.770)
	+DRw9	(0.428)	+DRw14	(0.419)	
3010	DR1	(0.540)	DR7	(0.502)	
	+DRw9	(0.542)			OR (0.720)
3011	DR2	(0.460)	DR1	(0.478)	
	+DR7	(0.395)	+DRw9	(0.504)	OR (0.695)
3028	+DRw52	(0.638)	+DRw10	(0.553)	
	+DRw9	(0.686)			OR (0.813)
3044	DR4	(0.355)	DRw52	(0.550)	OR (0.597)
3045	+DRw53	(0.340)	DRw10	(0.386)	OR (0.464)
3051	+DRw53	(0.396)	DRw10	(0.465)	
	+DQw2	(0.393)			OR (0.569)
3046	No significant association.				

+ indicates not submitted specificity

Antigen Society and core reagents as well as the two combined. These gave very similar results, and the results cited below refer to the combined analyses, i.e., restricted to cells typed for both sets. Furthermore, since preliminary analyses failed to show major ethnic differences, the main analysis was based on the major ethnic groups combined. Only non-flagged cells were included, thus reducing the data from initially more than 1000 individuals tested to 562 analyzed.

Sixteen sera formed one tight cluster with cluster R values above 0.8 clearly defining DR1. The R values for association with DR1 for the best ten of these sera range from 0.948 to 0.897 and in order of decreasing values, the sera are 1006, 1003, 1010, 5009, 5022, 5031, 5012, 1011, 1008, and 5045. Three sera, 5011, 5018, and 5015, show mutual strong association but fail to react with 16 to 17% of DR1 positives. This indicates the existence of a "DR1 short" specificity as was suggested in the Ninth Workshop by Fauchet et al. (2) and is also supported by T-cell and RFLP data from this Workshop although it has not been possible to correlate the various parts of the Workshop for DR1.

Four DRw10 sera formed a tight cluster with r values for serum versus antigen from 0.818 to 0.735. These sera are 1016, 5037, 1020, and 1021. Only one of these, 1016, showed sufficient quality as a stand-alone reagent.

There is a large group of sera defining DR1+w10. Three of these, 1007, 1013, and 1012, showed a particular tight cluster. The specificities identified in the MAbs are summarized in Table 1. None of them was found to be monospecific for DR1 or DRw10 or DR1+DRw10. None of them showed mutual strong correlation. Two sera, 1143 and 1163, were correlated with each other (r=0.8) and to DQw1 (r values approx. 0.8) and could be useful for the definition of this specificity. Specificities for sera not mentioned in text are presented in Table 2.

Table 2. Specificities for Sera Not Mentioned in the Text (R Values)

DRl sera in main cluster	5021	(0.875 + DR2)	5024	(0.868)
(includes "short" sera)	5029	(0.869 + DRw10)	5023	(0.880)
	5030	(0.802)	5016	(0.734)
	1005	(0.815)	5014	(0.742)
	9038	(0.767)	5020	(0.701)
	9037	(0.672)	1009	(0.718)
DRl sera outside main cluster	1004	(0.806)	9041	(0.615)
	5040	(0.535)	5013	(0.369)
	5042	(0.396)	5007	(0.464)
	5010	(0.528)		
DRl + w10 sera in main cluster	5008	5032	5026	5028
	5035	5033	1019	1014
	1015	1018	1017	
DRl + w10 sera outside main cluster	5004	5027	5024	9042
		1136		
DRl + 4 sera	1055	1057	5047	9040
	9039	3009		
DQwl sera in main cluster	5050	(0.801)	5003	(0.602)

A considerable number of reagents show specificity for DR1 + DR4. Most of these were not associated with each other. Exceptions are, however, four reagents falling into two groups, 3008 + 1093 and 5046 + 5048, with somewhat weaker intergroup association. They may define the "MC1" specificity.

In summary, this Workshop has not contributed significantly to definition of these specificities. Monospecific DRw10 anti-sera are still rare. There is no conclusive evidence for a split of DR1. No MAbs specific for DR1 and DRw10 exist.

Linkage Disequilibrium

In addition to the previously known linkage disequilibria, DRw10 has been shown in this Workshop to be highly associated with the A29, Cw7, B7 haplotype in the Jewish population, in whom the linkage with B37 is not as significant as in the non-Jewish population.

References

1. Bodmer JG, Pickbourne P, Richards S. In: Bodmer WF (ed): Histocompatibility Testing 1977. Munksgaard, Copenhagen, 1977; pp 35–84.
2. Fauchet R, van Leuwen A, Yunis EJ, Schreuder GMTh. Antigen Report: HLA-DR1. In: Albert E, et al. (eds): Histocompatibility Testing 1984. Springer-Verlag, Heidelberg, 1984; pp 184–185.
3. Jakobsen B.K., Svejgaard A. A private DR antigen associated with HLA-B37. Tissue Antigens 1979;14:361.
4. Matsuyama T, Winchester R, Lee S, Shookster L, Nunez-Roldan A. Identification of the DRw10 DRβI chain allele as encoding a polymorphic class II MHC epitope otherwise restricted to DRβ2 molecules of the DRw53 haplotype. J Immunol, 1988;140:537.

Antigen Society #22 Report (DR2)

J.R. Batchelor,[1] S. Curtoni,[2] H. Mervart,[3] H. Noreen,[4] N.J. Davey,[1] J.H. Moses,[1] and A. Amoroso[2]

At the Ninth International Histocompatibility Workshop, the Nomenclature Committee did not feel that DR2 could be subdivided serologically (1). In this Tenth Workshop, sufficient evidence was presented to justify splitting DR2 into two major subtypes (DRw15 and DRw16), and also indicated the existence of further heterogeneity. The analysis reported here was based on results obtained with the reagents summarized in Table 1.

The best antisera (those with R values > 0.6) are listed in Table 2. These coefficients are based on the results of all cells tested, irrespective of their ethnic origin and DR2 subtype.

Serological Patterns

A number of reaction patterns were observed with the best quality antisera. Two patterns in particular were quite distinct, coinciding with DR2 (DRw15) and DR2 short (DRw16). The reaction pattern defining DRw16 was established using Tenth Workshop cell line 9016 (RML), AZH and 57 short or variant DR2+ve cells tested with Core and Antigen Society sera (Table 3). Examples of DRw16 were found in all ethnic groups (Table 1). DR2.3 is at present a tentative definition and appears to occur more frequently in Oriental populations (2). However, the number of Oriental cells tested with Tenth Workshop Core and Antigen Society sets was too small for definitive conclusions.

DR2.4 (LUM) has been identified in independent studies by du Toit and co-workers (3) as a variant found among South African blacks and Cape coloreds. However, sufficient cells from this population were not available for testing with Tenth Workshop sera to allow confirmation of this variant.

Table 1. Reagents Used for the Analysis of DR2

Antibodies			
	29 "Core Set" Allosera		
	3 "Core Set" MAbs		
	53 "Antigen Society" Allosera		
Cells			
	DR2 +ve		DRw –ve
	DRw15	DR216	
Caucasoids	159	37	225
Blacks	48	5	181
Orientals	22	4	25
Others*	45	11	20
Totals	274	57	451

*This group contains other ethnic groups, EBV transformed cell lines, CLL's, and other cells.

Reporting Laboratories: UKIBAT,[1] ITICEP,[2] NCYMRV,[3] US2BAC[4]
Participating Laboratories: JAPAMY, US4BRN, ANZBSH, US8DEC, SAFDUT, SAFHAM, FRAJEA, US5MLS, US5SIN, FRAFAU

Table 2. Tenth Workshop Best DR2 Antisera (Data From All Cells)

Workshop Number	Serum Identification: Laboratory Origin	Local Name	AgSoc/CORE	R value	% Strong Reactions Score 6–8
1031*	MRV	19315	CORE	0.86	80%
1032	BRN	CC440.2	CORE	0.84	86%
9046	MRV	27348	AgSoc	0.84	81%
5056	AMY	JR36.348PA	AgSoc	0.80	82%
5095*	MRV	19315	AgSoc	0.80	82%
5053	ROO	62074	AgSoc	0.80	84%
1027	FAU	PACAULT	CORE	0.77	86%
5072	GOR	GM327	AgSoc	0.77	75%
1030	BAC	OLSON.T.79	CORE	0.76	89%
5084	TER	TER.DR2	AgSoc	0.76	70%
9044	MRV	MRV19676	AgSoc	0.75	59%
1029	YUN	ZIAKIS	CORE	0.75	78%
1025	BRN	CCB1010.2	CORE	0.74	86%
1024	MRV	9887.2	CORE	0.74	72%
5065	MRV	26110	AgSoc	0.73	72%
5071	MTT	F10196	AgSoc	0.73	75%
5078	BET	E991	AgSoc	0.72	80%
1026	FAU	COTTAIS	CORE	0.71	71%
5077	FAU	E1105	AgSoc	0.70	67%
5073	JLE	JOYCE	AgSoc	0.69	88%
5094	MRV	10692	AgSoc	0.69	59%
5064	MRV	21483	AgSoc	0.68	69%
5060	PRR	SAL1137	AgSoc	0.68	83%
1023	MRV	9839.1	CORE	0.64	56%

*MRV 19315 was included in both the Core and Antigen Society No. 22 sets of sera (WS No. 1031 and 5095 respectively); the samples were tested and analyzed independently, and showed 5% discordant reactions differing by a score of 4 or greater.

Table 3. Tenth Workshop DR2 Reaction Patterns (Best Antisera)

	5055	1027	1030	1025	5065	5078	5077	5073	5064	5060	1026	1023	1024	1029	5071	1031	1032	9046	5095	5053	5072	5084	9044	5094
Cell Type: DRw15	+	+	+	+	+	+	+	+	+	+	+	+	+	+	+	+	+	+	+	+	+	+	+	+
Cell Type: DRw16	+	+	+	+	+	+	+	+	+	+	+	+	+	+	+	−	−	−	−	−	−	−	−	−
DR2.3	+	+	+	+	+	+	+	+	+	+	+	+	+	−	−	+	+	+	+	+	+	+	+	+
DR2.4(LUM)	+	+	+	+	+	+	+	+	+	+	−	−	−	−	−	+	+	+	+	+	+	+	+	+

N.B. DR2.3 and 2.4 are at present tentative definitions (see text).

RFLP analysis of TaqI-digested genomic DNA using a 530 base pair, DRB-specific probe clearly distinguishes between DRw15 and DRw16. DRw15 yields two distinct bands of approximately 13 kb and 1.3 kb while DRw16 gives a band of approximately 15 kb. A common band estimated as 1.6 kb is seen in both variants.

Finally, the separate identities of DRw15 and DRw16 were confirmed by the different class II polymorphisms associated with them (see Tables 6 to 8 inclusive of the Tenth Workshop Nomenclature Committee Report).

Conclusions

Two major subtypes of HLA-DR2 have been clearly demonstrated by serologic typing: DRw15 (previously DR2 long or 2.1) and DRw16 (previously DR2 short or 2.2). The existence of these variants is confirmed by analysis of RFLP patterns, by the reactions of T cell clones, by Dw typing, and by different linkage disequilibria.

Two further subtypes have been independently reported. DR2.3 is a variant occurring most commonly in Oriental populations; DR2.4 (LUM) is a variant seen in South African blacks and Cape coloreds. These reports are based on data from one or only a few laboratories, and sufficient Tenth Workshop data on the relevant ethnic groups were not available for confirmation of these subtypes.

References

1. Batchelor JR, Hows J, Dodi IA, Betuel H, Borelli I. Antigen Report: HLA-DR2. Histocompatibility Testing 1984; p 186.
2. Mervart H, Mody M, Wong KH. HLA-DR2 Subtypes. Tenth Workshop Newsletter No. 1, p 6.
3. Oudshoorn M, du Toit ED, Taljaard DG. The Heterogeneity of HLA-DR2. Tenth Workshop Newsletter No. 1, p 7.

Antigen Society #23 Report (DR4, DR9): Part 1: Analysis of Core and Antigen Society #23 Panel Data

B. Nisperos,[1] N. Sakahara,[2] J. Awad,[6] V. Drummond,[7] B. Larsen,[8] H. Patel,[5] D. Stewart,[9] and J. Williamson[3]

Report prepared by B. Nisperos and N. Sakahara

Introduction

HLA-DR4 was first defined in the Seventh International Histocompatibility Workshop (7WS) by reactivity with two clusters of alloantisera known as WIA4 and WIA4x7 (1). The WIA4 group of sera was cross-reactive with DR5, and the WIA4x7 group was cross-reactive with DR7. Operationally monospecific anti-sera were available in the 8WS, and thus the definition of DR4 was relatively clear (2). Splits of DR4 have been frequently postulated, particularly in association with the T cell-defined HLA-D subtypes Dw4, Dw10, Dw13, Dw14, and Dw15 (3,4,5). However, there has been no convincing evidence, supported by segregation studies, demonstrating a definite serologic split of DR4.

HLA-DR9 was first clearly defined during the 8WS (6,7). The sera were complex, but a distinct reactivity pattern in the WIA4x7 group of reagents was recognized through family studies. Sera available in the 9WS gave the first clear definition of DRw9 (8). Comparative racial studies were very informative. A shorter DR9 reactivity pattern was observed in Japanese by family study (7) and DR9 was shown to be associated with either DQw3 or DQw2 (8).

Results and Discussion

DR4. A total of 153 alloantisera and 14 monoclonal antibodies from the Core and Antigen Society 23 were tested in this Workshop. Twenty-eight DR4-specific alloantisera and two DR4-specific monoclonal antibodies were identified. Fourteen of the sera had R values greater than 0.8 in the combined population, and the remainder had R values between 0.7 and 0.8 (Table 1). Some of these DR4 sera reacted variably in different racial groups. In Orientals, the lower R values of some sera reflect cross-reactivity with DR9. Some sera that had high DR4 correlations in Caucasoids and Orientals, such as 1056, 1112, 5114, 5104, and 5117, had relatively low correlations in blacks. Interestingly, the anti-DR monoclonal antibodies 3008 and 3009 showed stronger reactions and stronger DR4 correlations in blacks, while many of the DR4 anti-sera reacted relatively weakly in blacks. The tail analysis for the reagents

Reporting Laboratories: US7HAN,[1] US3PER,[2] ANZDAW,[3] ANZDCH,[4] UKIJAM,[5] UKIFES,[6] UKICXH,[7] NCYMAR,[8] US7STE[9]

Participating Laboratories: ANZCRS, ANZCHA, ANZTAI, ANZBCH, CHIYEG, EAEKAS, FRAPRR, JAPAIZ, JAPKSH, NCYMRV, UKISTB, UKIDYR, UKIWOR, US2BAC, US8RJW

Table 1. Reactivity of Selected DR4 Alloantisera and Monoclonal Antibodies in Different Populations

			Combined n = 1059		Caucasoid n = 622		Oriental n = 120		Negro n = 162		Others n = 155	
10WS#	Lab	Local I.D.	r	QSC	r	QSC	r	QSC	r	QSC	r	QSC
1048	MRV	7407.7	0.90	7.2	0.87	5.9	0.90	6.8	0.95	8.2	0.92	6.8
1056	JUJ	1.1908	0.84	5.3	0.82	4.8	0.87	5.7	0.56	6.3	0.92	6.0
1047	FES	Hughes	0.92	7.1	0.90	6.0	0.97	8.9	0.76	7.0	0.93	7.1
1051	BRN	CCB 1064.4	0.86	5.5	0.83	5.2	0.88	5.4	0.72	4.0	0.94	6.8
1046	GOE	Goeken	0.85	5.2	0.82	4.8	0.89	6.0	0.74	5.7	0.82	4.3
1112	BET	E1208	0.83	5.0	0.85	5.5	0.89	5.8	0.39	1.0	0.82	4.8
5098	MRV	9254	0.83	4.2	0.81	3.8	0.67	2.3	0.79	4.4	0.86	5.1
5099	MRV	9697	0.85	5.0	0.83	4.8	0.74	3.3	0.67	4.1	0.95	6.8
5109	MRV	7407	0.88	5.3	0.85	4.9	0.90	5.7	0.93	6.9	0.84	4.3
5112	MID	BEL5759D	0.82	4.8	0.81	4.4	0.72	4.1	0.62	2.0	0.86	5.6
5114	MTT	F103401	0.81	4.3	0.85	4.6	0.57	2.1	0.44	1.4	0.95	7.6
5118	MOR	Lee C.	0.88	5.5	0.87	5.1	0.82	5.2	0.69	6.3	0.94	6.6
5120	ZAR	MO1978	0.83	4.4	0.82	3.9	0.84	5.3	0.80	7.1	0.59	2.4
5127	JLE	Vale	0.84	4.4	0.83	4.1	0.66	2.6	0.78	6.7	0.90	5.4
1044	MRV	5094.1	0.77	5.0	0.74	4.7	0.70	3.3	0.70	4.6	0.80	4.3
1045	MRV	9671	0.71	3.6	0.68	3.8	0.81	4.4	0.67	3.9	0.57	2.6
1191	MBC	BC.Jowauk	0.73	3.2	0.70	3.0	0.87	5.8	0.50	2.1	0.60	1.9
5103	BIG	F51	0.79	4.4	0.80	4.3	0.62	2.8	0.58	2.0	0.79	5.7
5104	PER	Lipski	0.73	3.3	0.70	3.0	0.82	4.3	0.43	1.5	0.75	3.3
5107	MRV	22907	0.75	3.4	0.73	3.3	0.73	3.1	0.60	2.3	0.76	3.7
5111	MID	BEL235D	0.78	3.7	0.72	3.0	0.74	4.8	0.82	5.6	0.85	4.7
5117	DUQ	Kingsbury	0.72	3.2	0.74	3.0	0.74	3.4	0.49	1.4	0.74	5.0
5119	DAW	Matthew	0.72	3.3	0.65	2.6	0.84	4.9	0.72	5.0	0.77	4.6
5121	ZAR	MO1984	0.77	3.6	0.75	3.3	0.68	3.6	0.77	4.7	+DR9	
5122	GEL	MUR20955	0.73	3.3	0.68	3.1	0.64	2.1	0.78	4.5	0.76	3.1
5129	DUP	Vesey3340	0.80	4.0	0.77	4.1	0.84	4.5	0.66	3.1	0.78	4.3
5130	TER	TER.DR4A	0.70	4.0	0.67	3.9	0.62	2.8	0.67	4.0	0.72	3.1
5132	TER	TER.DR4C	0.78	3.7	0.72	3.1	0.83	4.5	0.82	5.1	0.86	5.7
3009	CEP	ACI.59	0.393	0.8	0.68	2.5	0.23	1.0	0.81	4.4	0.26	1.0
3008	BRN	CC.CL20	0.49	1.4	0.50	1.8	0.46	1.0	0.89	6.3	0.46	1.9

Table 2. Matrix of Correlations Between Selected DR4 Sera

10WS #	
1048	
5109	9
1051	98
1046	989
1047	9898
5118	88999
5099	888888
5120	8888888
5114	88888888
5111	888888888
5129	8888888877
5098	88888888778
5112	888888877878
1056	8888887777887
1191	87888777777777
1112	878887877777777
5132	8888788778787877
5127	88878888888777777
1044	887877777778786787
5103	8877778878788767777
5119	87787877777877678776
5121	787878787777777777777
5104	7777787777777767777777
1045	77787777777778677686877
5130	777777767777766868768777
5107	7777788777777766777676766
5122	77777777777777666766777666
5117	777777777777776676667676766
3009	4544444444444444444444443344
3008	44444444444445444443544444345

	15111555555551115515555155553 3
	01000101111010111101111011110 0
	40544192112915913240120430210 0
	89167890419826122743914507279 8

in Table 1 was uninformative except for the two monoclonal antibodies: MAb 3008 reacted with some DR1-positive cells (r = .40); and MAb 3009 reacted with some DR9-positive cells (r = .40). The serum-serum correlations between the 28 DR4 alloantisera were generally good (r > 6), and were very good for 9 sera (r > 8) (Table 2). The correlations between the monoclonal antibodies and the alloantisera, however, were consistently lower (r ≤ 5).

Table 5. Reactivity Patterns of Selected DR9 Sera

10WS #	DR9	DR4	DR7	DRw10
1062	+	−	−	−
1063	+	−	−	−
1058	+	−	−	−
1060	+	−	−	+
1067	+	−	+	−
1066	+	−	+	−

The reactivity of selected anti-DR4 reagents with DR4-associated HLA-D specificities is summarized in Table 3. Reagents 3008, 5119, 3009, 1046, and 1045 reacted with Dw4, Dw13, Dw14, and Dw15 but not with Dw10 positive cells. Furthermore, reagents 3008 and 5119 did not react with Dw13. Monoclonal antibody 3009 reacted strongly with Dw4, Dw13, Dw14, and Dw15, but did not react with Dw10. This pattern, identified in the 9WS and designated DR"4.1" (3), thus appears to represent a distinct DR4.

DR9. The DR9 antigen was well defined in the 10WS (Table 4). Three of the submitted sera were found to be DR9-specific: 1062, 1063, and 1058 (Table 5). There was no consistent evidence of any DR9 split. A tail analysis revealed the presence of additional antibodies in some of these sera: DR7 in 1066 and 1067; DRw10 in 1060.

Summary

HLA-DR4 and DR9 were well defined by alloantisera present in the Antigen Society 23 reagent set. Two monoclonal antibodies were also identified that reacted with DR4, but extra reactions were noted with either DR1 or DRw10. There is consistent evidence for a DR4 split designated DR"4.1" that includes the T cell-defined specificities Dw4, Dw13, Dw14, and Dw15. The Dw10 specificity is DR"4.1" negative.

References

1. Bodmer JG, Pickbourne P, Richards S. Joint Report. 2. Ia. Serology. In: Bodmer WE (ed): Histocompabitility Testing 1977. Munksgaard, Copenhagen, 1978; p 35.

Table 4. Reactivity of Selected DR9 Alloantisera in Different Populations

			Combined (n=1059)		Caucasoid (n=622)		Oriental (n=120)		Negro (n=162)		Others (n=155)	
Reagent #	Lab	I.D.	r	QSC	r	QSC	r	QSC	r	QSC	r	QSC
1062	HJH	252C	0.87	8.9	0.71	7.1	0.85	8.5	0.95	8.5	0.93	7.7
1063	NAI	1014	0.75	6.0	0.51	5.1	0.87	5.5	0.79	7.8	0.97	9.0
1058	MRV	9254	0.86	6.8	0.76	6.8	0.73	4.3	0.95	8.0	0.94	7.4
1060	FAU	BRULE	0.79	8.1	0.75	5.9	0.79	5.9	0.70	6.1	0.85	6.3
1067	FAU	JEGU	0.58	3.8	0.62	2.1	0.67	2.6	0.42	3.4	0.63	3.8
1066	FAU	ANTIN	0.84	7.5	0.85	5.1	0.81	5.4	0.70	6.7	0.67	4.5

Table 3. Reaction Pattern of Selected Sera and Monoclonal Antibodies With the DR4-Associated HLA-D Specificities

10WS Reagent #	Lab	Dw10 n=11	Dw13 n=12	Dw4 n=20	Dw14 n=18	Dw15 n=12	Additional Specificity
3008	BRN	–	–	+/–	+	+	
5119	DAW	–	–	+	+	+	
3009	CEP	–	+	+	+	+	
1046	GOE	–	+	+	+	+	
1045	MRV	–	W	+	+	+	
1056	JUJ	+/–	+/–	+	+	+	
5109	MRV	W	+	+	+	+	
5098	MRV	W	W	+	+	+	
1044	MRV	W	W	+	+	+	
5112	MID	W	+	+	+	+	
1048	MRV	W	+	+	+	+	
1051	BRN	+	+	+	+	+	
1047	FES	+	+	+	+	+	
1112	BET	+	+	+	+	+	
5099	MRV	+	+	+	+	+	
5114	MTT	+	+	+	+	+	
1191	MBC	+	+	+	+	+	
5111	MID	+	+	W	W	+	
Other Informative Sera:							
9417	STE	–	–	–	–	+	DR5
5145	BRN	–	–	–	–	+	DR9,7

"+" indicates positive reaction; "–" indicates negative reaction, and "W" indicates weak reaction.

2. Festenstein H. Joint Report, DR4. In: Terasaki PI (ed): Histocompatibility Testing 1980. UCLA Tissue Typing Laboratory, Los Angeles, 1980; pp 515–517.
3. Williamson J, Tait B, Richiardi P, Awad J, Fauchet R, Curtoni ES. Antigen Report: HLA-DR4. In: Albert ED, Baur MP, Mayr WR (eds): Histocompatibility Testing 1984. Springer-Verlag, Berlin, 1984; pp 188–189.
4. Holbeck SL, Kim SJ, Silver J, Hansen JA, Nepom GT. HLA DR4-associated haplotypes are genetically diverse within HLA. J Immunol 135:637–641.
5. Reinsmoen NL, Bach FH. Five HLA-D clusters associated with HLA-DR4. Hum Immunol 1982;4:249–258.
6. Payne R, Colombe B. On the Trail of the 4x7 B Lymphocyte Antigen. In: Terasaki PI (ed): Histocompatibility Testing 1980. UCLA Tissue Typing Laboratory, Los Angeles, 1980; pp 829–830.
7. Colombe B, Payne R. Joint Report 8W12 (DRw9+10). In: Terasaki PI (ed): Histocompatibility Testing 1980. UCLA Tissue Typing Laboratory, Los Angeles, 1980; pp 543–545.
8. Williamson J, McNicholas A. Antigen Report HLA-DRw9. In: Albert ED, Baur MP, Mayr WR (eds): Histocompatibility Testing 1984. Springer-Verlag, Berlin, 1984; p 200.

Antigen Society #23 Report (DR4, DR9): Part 2: Serologic Typing of B-Lymphoblastoid Cell Lines: DR4 and DR9

J. Williamson,[3] B. Horton,[3] F.T. Christiansen,[3] R.L. Dawkins,[3] D. Chandanayingyong,[4] D.C.O. James,[5] J.A. Hansen,[1] and H. Perkins[2]

In the Ninth Workshop, the DR4 sera revealed some heterogeneity, and it appeared that some of the patterns correlated with B alleles and/or DW splits (1). There was also some difficulty distinguishing well characterized lymphoblastoid cell lines (LCLs) which would have been very useful, but few were available at the time. Furthermore, many laboratories experienced difficulty with DR and DQ typing on LCLs largely because of extra reactions.

Under these circumstances, we undertook a comparison of peripheral blood mononuclears and lymphoblastoid cell lines (LCLs) from the same donors and developed methods that would reduce the artifacts seen with LCLs. Here, we show that LCLS can be used without difficulty but, with one exception, splits of DR4 cannot be recognized with available sera.

See footnote at the beginning of Part 1 for a listing of the reporting laboratories.

Table 1. Analysis of Selected DR4 Sera Using Lymphoblastoid Cell Lines Typed in Laboratory DAW

Serum No	Total No of Cells	TP	FN	FP	TN	R	Extras
1046	82	24	6	3	49	0.76	
5149	84	23	7	6	48	0.66	
1050	84	18	12	2	52	0.63	
5098	86	16	15	0	55	0.64	
5119	85	15	16	0	54	0.61	
1048	86	12	19	0	55	0.54	
1044	86	6	25	0	55	0.37	
1045	80	7	21	8	44	0.12	
5101	86	31	0	11	44	0.77	DR1
9428	81	30	0	17	34	0.65	DRw9 +
1056	84	31	0	1	52	0.98	
9459	69	23	0	1	45	0.97	
5132	86	29	2	0	55	0.95	
9423	86	28	3	6	49	0.78	
5099	86	30	1	0	55	0.95	
1194	78	30	0	4	44	0.90	
5120	86	30	1	3	52	0.90	
5113	86	30	1	2	53	0.93	
5130	85	28	2	0	55	0.95	
9450	83	29	2	2	52	0.90	
5106	84	28	3	0	53	0.93	
1051	81	26	4	0	51	0.90	
5138	84	26	4	1	53	0.87	
5112	86	26	5	3	52	0.80	
9439	69	21	3	6	39	0.72	± DRw9
5121	86	29	2	4	51	0.85	± DRw9
5140	74	26	1	4	43	0.86	± DRw9
1191	84	28	2	12	42	0.68	DR5
1053	81	28	0	10	43	0.77	DRw9
5107	85	29	2	10	44	0.73	± DR7
5125	84	26	5	11	42	0.61	± DRw9 +
5118	85	24	7	0	54	0.83	
5114	79	17	7	0	55	0.79	
5116	70	14	7	1	37	0.37	
1049	80	19	10	14	37	0.37	± DRw9
1047	84	23	8	2	51	0.74	
5127	85	14	17	0	54	0.59	
5109	86	9	22	3	52	0.34	

Methods

In an earlier study (2), we showed that many of the spurious reactions with LCLs could be avoided by absorption of the complement and by typing viable LCLs 24 hours after subculture.

Complement Absorption. *Class I Typing.* A pool of 10 random LCLs was prepared and mixed with complement at a concentration of 10 × 10⁶ cells/ml. The mixture was rotated for 3 hours at 4°C, and centrifuged at 1000 *g* for 15 minutes. The complement was then passed through a 0.8 μ filter to ensure complete removal of LCLs. Aliquots were stored in liquid nitrogen.

Class II Typing. On the day of testing, complement was mixed with 10×10^6 autologous cells per ml and rotated on a Matburn mixer for 3 hours at 4°C. The complement was then centrifuged at 1000 *g* for 15 minutes, passed through a 0.8 μ filter, and kept at 4°C until ready for use.

Preparation of Cells. The viability of LCLs was maximized by subculturing every 2 to 3 days and harvesting within 16 to 24 hours of subculture. Cultures were centrifuged at 500 *g* for 10 minutes, washed once in RPMI with 5% FCS, and resuspended at 2×10^6 cells/ml. Cells were spotted onto the trays as soon as possible.

Incubation Times. Typing was by the standard microcytotoxicity assay. Briefly, 2 μl of LCLs were mixed with 1 μl of serum under liquid paraffin.

Class I Typing. The plates were incubated at 22°C, with 5 μl of complement being added after 30 minutes. After a further 60 minutes, 5 μl of eosin and 5 μl of formalin were added.

Table 2. Putative DR4.1 and DR4.2 Summary of Reaction Patterns on LCLs

	1046	5149	1050	5098	5119	1048	1044	1045	5101	9428	1056	9459	5132
DR4.1	+	+	±	±	±	∓	∓	∓	+	+	+	+	+
DR4.2	–	–	–	–	–	–	–	–	+	+	+	+	+
Extras									DR1	DRw9			
										+			
	9423	**5099**	**1194**	**5120**	**5113**	**5130**	**9450**	**5106**	**1051**	**5138**	**5112**	**9439**	**5121**
DR4.1	+	+	+	+	+	+	+	±	±	±	±	+	+
DR4.2	+	+	+	+	+	+	+	±	±	+	±	+	+
												∓	∓
Extras												DRw9	DRw9
	5140	**1191**	**1053**	**5107**	**5125**	**5188**	**5114**	**5116**	**1049**	**1047**	**5127**	**5109**	
DR4.1	+	+	+	+	±	±	±	±	±	±	+	+	
DR4.2	+	+	+	+	±	±	NT	+	∓	∓	∓	∓	
	±			±	±				±				
Extras	DRw9	DR5	DR9	DR7	DRw9				DRw9				

Table 3. DRw9 Reaction Pattern of Selected Sera with LCLs

Sera	1060	5110	5128	9438	1057	1068	1066	1067	1062	1018	1063	1058	1059
DR9	+	+	+	+	∓	+	+	+	+	+	±	±	∓
DR4	±	+	+	+	+	–	–	–	–	–	–	–	–
DR7	–	–	–	–	–	+	+	+	–	–	–	–	–
DR2	–	–	–	–	–	–	–	–	±	–	–	–	–
Others	–	–	–	–	–	–	–	+ + +	±	+ +	–	–	+

Class II Typing. The initial incubation was for 60 minutes at 37°C, after which time 5 μl of autologous absorbed complement was added. Eosin and formalin were added after a further 120 minutes at 22°C.

Results

The sera that reacted with LCLs expressing DR4 are listed in Table 1. It can be seen that a minority of the sera had false-positive reactions generally with cells containing DRw9. The remaining sera had few if any extras. The reactivity patterns are summarized in Table 2. As shown in Figure 1, some of the short sera were weak, but others appeared to be identifying splits of DR4. Many of these sera were negative with cells 1245, 709, 9026, and 1246. Since one of these four cells (9026) has the B38, Dw10 split of DR4, we conclude that this shorter pattern represents 4.2 (1). When additional cells are available, it is likely that further splits will be identified.

DRw9 was well defined (see Table 3). Reproducibility data are shown in Table 4.

Table 4. Summary of Results of 10W Hidden Duplicate Antisera Against All LCLs Tested in the DAW Laboratory

	Workshop Numbers				
	9051 9052	9053 9054	9055 9056	9057 9058	Total
Specificity*	DQ7	A2+A28	DQ1+8+9	DQ7+8+9	
Both duplicates positive	28	48	87	73	236
Both duplicates negative	101	71	35	58	265
Discordant results†	0	0	6	1	7

*Specificity derived from local LCL panel; †all discordant results were false-negative results.

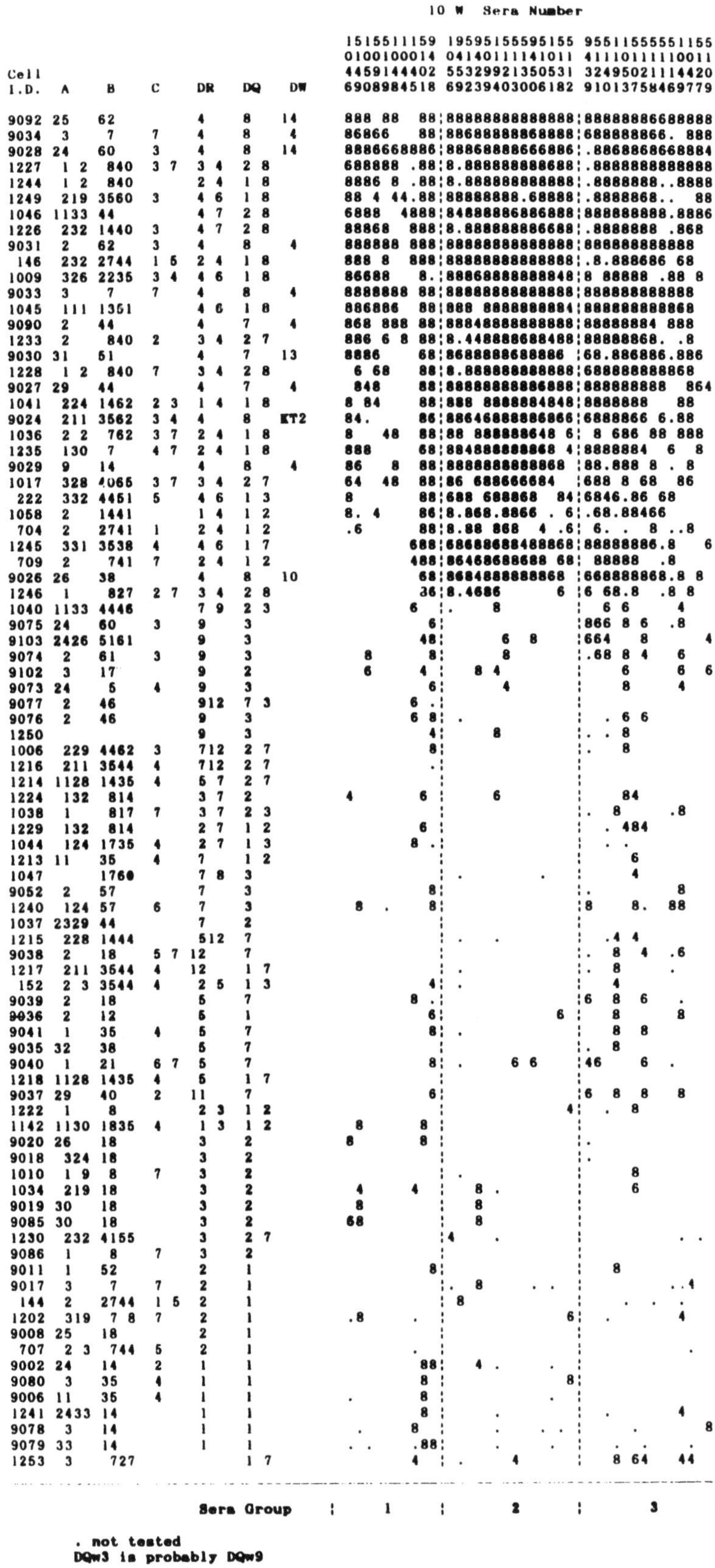

Figure 1. Serological patterns of selected DR4 sera using LCLs.

Conclusion

With some variation of standard techniques, it is possible to achieve excellent serologic typing of LCLs. DR4 remains heterogeneous with one obvious split. Undoubtedly, other splits exist, but the various patterns cannot be classified until additional well characterized LCLs are available.

References

1. Williamson J, Tait B, Richiardi P, et al. Antigen report: HLA-DR4. In: Albert ED, Baur MP, Mayr WR (eds): Histocompatibility Testing 1984. Springer-Verlag, Berlin, 1984:188–189.
2. Grimsley G, Homes P, Townsend D, Williamson J, Dawkins RL. HLA typing of lymphoblastoid cell lines (LCL). In: Aizawa M (ed): HLA in Asia-Oceania 1986. Sapporo, Japan: Hokkaido University Press, 1986:909–913.

Antigen Society #24 Report (DRw11, DRw12, DRw8)

P. Stastny,[1] Z. Layrisse,[2] D.P. Singal,[3] A. Svejgaard,[4] E. van den Berg-Loonen,[5] K. Dohi,[6] L. Caraballo,[7] P. Chiewsilp,[8] B. Colombe,[9] R. Fauchet,[10] E. Haas,[11] M.G. Hammond,[12] B.K. Jakobsen,[4] S. Knight,[13] J. Lee,[14] H. Mervart,[15] G.M.T. Schreuder,[16] and K. Sullivan[17]

HLA-DR5

Groups of sera defining a DR5 specificity correlated with HLA-Dw5 were first observed in the Seventh Workshop. However, sera of excellent quality recognizing a distinct DR5 antigen have been rare. In the Ninth Workshop, it was discovered that most DR5 sera contained antibodies against related DQ-specificities, and shorter patterns (DRw11 and DRw12) were defined.

In the present Workshop no reagents for definition of DR5 have been found. However, DRw11 and DRw12 can be seen quite clearly.

Reporting Laboratories: US4STA,[1] LATLAY,[2] US5SIN,[3] SCASVE,[4] BENBER,[5] JAPDOH[6]
Participating Laboratories: LATCOL,[7] ANZCHI,[8] US3GVY,[9] FRAFAU,[10] LATHAA,[11] SAFHAM,[12] UKIKNT,[13] US8JLE,[14] NCYMRV,[15] BENROO,[16] US1TUL[17]

HLS-DRw11

A large group of sera defining DRw11 was observed. While many of these do have some contaminating antibodies (especially DQw3 and DQw7), enough very good reagents could be selected that reacted in block and established DRw11 typing quite unequivocally (see Tables 1 and 2).

HLA-DRw12

Three reagents of the Workshop (10W, 9999, 9050, and 3068) recognized DRw12 monospecifically. These were three different aliquots of the same monoclonal antibody (DN1), which clustered with high correlation values. Coincident reaction of the replicates provided excellent definition for this antigen in most of the participating laboratories. The continuity of the definition of this

Table 1. Tenth Workshop Sera Used for the Analysis of DRw11, DRw12, and DRw8

HLA Specificity	Core Set Alloantibodies	Monoclonal	Ag. Soc. 24
DRw11	1096, 1103, 1113, 1114	3036	5154, 5157, 5170 5172, 5175, 5181 9059, 9069, 9070 9075
DRw12		9999, 9050, 3068	
DRw8	1085, 1086, 1087, 1089, 1091		5190, 9093
DRw52	1152, 1198, 1199, 1201, 1204, 5384		
DRw52-short		3025, 3062, 3063	9059
DQw1	1153, 1155, 1159	3091	
DQw3	1116, 1176, 1179	3066, 3111	
DQw7		3119, 3120, 3121	9051
DQw4		3101	
DQw5	1136		
DQw6	1141		
IIB3	1217	3086, 3088, 3090 3122	9055, 9056

Table 2. The Most Common Haplotype Associations With HLA-DRw11, DRw12, and DRw8

Observed HLA-DR/DQ	HLA-DR							HLA-DQ						
	w11	w12	w8	w13	w14	w52	w52s	w3	w7(w3)	w4	w1	w6(w1)	w5(w1)	11B3
Haplotypes	1 1 3 9 9 5 5	9 9 3	1 1 1 1 5 9	1 1 3	9 1	1 1 1	3 9	3 3	3 3 3 9	3	1 1	1	1	3 3 1
	0 1 0 0 0 1 1	9 0 0	0 0 0 0 1 0	1 1 0	0 1	1 1 2	0 0	0 1	1 1 1 0	1	1 1	1	1	0 0 2
	9 1 3 5 6 8 5	9 5 6	8 8 8 9 9 9	3 2 2	6 1	9 9 0	6 5	6 1	2 2 1 5	0	5 5	4	3	8 8 1
	6 4 6 9 9 1 4	9 0 8	5 9 7 1 0 3	4 6 1	0 1	9 8 1	3 9	6 1	0 1 9 1	1	3 5	1	6	8 6 7
DRw11 DQw3(w7)	+ + + + + + +	– – –	– – – – – –	– w –	– +	+ + +	+ +	+ +	+ + + +	–	– –	–	–	– – –
DRw11 DQw1	+ + + + + + +	– – –	– – – – – –	– w –	– +	+ + +	+ +	– –	– – – –	–	+ +	+	–	+ + +
DRw12 DQw3(w7)	– – – – – – –	+ + +	– – – – –	– w –	– w	+ + +	– –	+ +	+ + + +	–	– +	–	–	– – –
DRw12 DQw1	– – – – – – –	+ + +	– – – – –	– w +	– w	+ + +	– –	– –	– – – –	–	+ +	–	+	+ + +
DRw8 DQw4	– – – – – – –	– – –	+ + + + +	– – –	– –	+ + +	– –	– –	– – – –	+	– –	–	–	+ + +
DRw8 DQw3(w7)	– – – – – – –	– – –	+ + + + +	– – –	– –	+ + +	– –	+ +	+ + + +	–	– –	–	–	– – –
DRw8 DQw3(w7N)	– – – – – – –	– – –	+ + + + +	– – –	– –	+ + +	– –	+ +	– – – –	–	– –	–	–	+ + +
DRw8 DQw1	– – – – – – –	– – –	+ + + + +	– – –	– –	+ + +	– –	– –	– – – –	–	+ +	?	w	+ + +

w = weak or variable reactions
w7N = DQw7-negative
? = not enough information available

antigen with the previous definition in the Ninth Workshop was demonstrated by typing of the cell line Herluf and several other cells and members of families that had been tested in the previous Workshop. In some laboratories occasional reactivity of DN1 with DRw8 cells was observed. It was brought to our attention that a T-cell clone from M. Thomsen reacted uniquely with DRw12-positive cell lines.

DRw8

DRw8 antigen was defined in this Workshop by several good sera (Table 1). In addition, many other sera on the Antigen Society tray appeared to react with DRw8 cells and to contain antibodies against the new DQw4 specificity.

Haplotype Associations of HLA-DRw11, HLA-DRw12, and DRw8 With DRw52 and DQ Alleles

The haplotype associations are summarized in Table 2. DRw11, DRw12, and DRw8 all reacted with broad DR sera defining DRw52. However, a subset of reagents among those clustering with DRw52 gave shorter patterns, some of which are observed to be negative with DRw12-positive cells and/or DRw8-positive cells.

DQ haplotype associations observed with DRw11 and DRw12 were similar. The most common haplotypes were with DQw3, and these were almost always positive for DQw7. While a few DRw11, DQw3, DQw7-negative cells were present, their number was sufficiently small to likely represent a residual of typings with technical problems.

Other haplotypes observed with DRw11, DRw12, and DRw8 carried DQw1. Such cells reacted either with 10W1141, defining DQw6, or with 10W1136, an antibody correlated with LY1327 (4), reported to define DQw5. Most DRw11 and DRw8, DQw1 cells were positive with 10W1141 (DQw6); the DRw12, DQw1 cells were 10W1141-negative and sometimes reacted with 10W1136, suggesting they were positive for DQw5.

The frequency of haplotype associations with DRw8 and DQ antigens varies in different ethnic groups. In Caucasoids the most common haplotype was DRw8, DQw4. In Orientals, DRw8, DQw1 was more commonly observed. Other haplotypes observed were DRw8, DQw7, and less frequently, DQw3-positive, DQw7-negative.

A cluster of antibodies defining the broad DQ specificity 11B3 reacted with DRw11, DRw12, and DRw8, DQw1-associated cells. Also included were the DRw8, DQw4, and DRw8, DQw3, DQw7-negative haplotypes.

Relationship with DRw13 and DRw14

Cells having DRw11 frequently gave patterns of weak reactivity with sera used to define DRw13. This was also true of cells that were coded as DRw12. DRw14 was defined in the Core set by only one reagent (10W9060). In addition, 10W1111 was a duospecific serum reacting with both DRw11 and DRw14.

References

1. Bodmer JG, Pickbourne P, Richards. Joint Report: 1a Serology. In: Bodmer WF, et al. (eds): Histocompatibility Testing 1977. Munksgaard, Copenhagen, 1978; pp 35–84.
2. Betuel H, Gebuhrer L, Schreuder GMT, Goldmann SF, Arnaiz Villena A, Layrisse Z. Antigen Report: HLA-DR5 and its subtypes HLA-DRw11 and HLA-DRw12. In: Albert ED, Baur M, Mayr W (eds): Histocompatibility Testing 1984. Springer-Verlag, Berlin, 1984; pp 190–192.
3. Park MS, et al. In: Terasaki P (ed): Histocompatibility Testing 1980. UCLA Tissue Typing Laboratory, Los Angeles, 1980; p 548.
4. Gebuhrer L, Betuel H, Freidel AC, Farre A. A division of HLA-DQw1 associated with DR1, DR1x-DQw1, DR2 short, DRw10, and DRw14. 1. Definition of allo-antiserum LY1327. Hum Immunol 1987;18:235–245.

Antigen Society #25 Report (DRw6, DRw13, DRw14)

G.M.Th. Schreuder,[1] L. Gebuhrer,[2] V. Lepage,[3] S.L. Pask,[4] P. de Lange,[1] H. Beutel,[2] L. Degos,[3] J. Awad,[5] E. Du Toit,[6] and F.C. Grumet[4]

History

Since the Ninth International Histocompatibility Workshop 1984, two splits of HLA-DRw6 have been recognized: DRw13 and DRw14 (1). Both specificities were included in DRw52 and highly associated with DQw1. At that time, few monospecific DRw14 sera were available, and DRw13 could only be reliably defined in the absence of DRw11 because all DRw13 sera included DRw11 activity. DRw13 and DRw14 were observed also with DQw3. However, the discrimination between the relatively rare DRw13 DQw3 and the more commonly observed DRw12 DQw3 was extremely difficult.

Antigen Society 25 (AGS-25)

Twenty laboratories collaborated in AGS-25 (Table 1). In addition to the 16 sera provided by the Workshop Organization, 51 other sera were exchanged and tested by the participants. Twelve laboratories also provided their Core serology data so that we were able to perform an analysis of the combined core plus AGS-25 dataset. All data were transferred to the Leiden University mainframe computer. Local software was used for the screening and cluster analyses.

DRw6

Sera that were reported as informative for DRw6-related specificities in the Workshop Newsletters, plus additional sera that our analyses showed were well correlated with those specificities and that gave concordant results in the majority of participating laboratories, were selected from the AGS-25 and Core sets. Unfortunately, the correlation coefficients were fairly low for these serum clusters; not more than 0.5 for DRw13 and 0.4 for DRw14. The most useful sera and MAbs are summarized in Table 2.

DRw13

The DRw13 specificity was clearly distinguished (Table 3). One serum, 10w1134, which is the same as VR64034, reacted almost monospecifically with all DRw13 positive cells, although it was weak in some laboratories. On the majority of DRw13 positive cells, DRw13 was associated with the DQw6 subtype of DQw1 recognized with MAb IIB3 = 10w3088 (2). Cells that were reported as Dw18 or Dw19 all had this DRw13, DQw6 pattern. DRw13 was also observed in association with the DQw7 split of DQw3, estimated to be 2% of Caucasoid DRw13s and reported as DHAG = Dw1802. Serum LYGUE reacted positively only with DRw13+ DQw7+ DHAG+ cells, but not with DRw13+ DHAG-cells.

All sera that reacted with only a fraction of the DRw13 DQw1 positive cells were checked for associations with Dw18 and Dw19. Serum 10w3021 was significantly correlated with Dw18 ($p = 0.005$). The correlation between sera 5206 and 5208 and Dw18 was of borderline significance.

DRw14

The DRw14 specificity was identified with some AGS-25 sera and one monospecific but weak serum (10w9060) in the Core set (Tables 2 and 3). In the majority of cells, DRw14 is associated with the DQw5 split of DQw1, recognized by two sera, LY1327 and 10w1136 (3). Cells that were reported as Dw9 all had this DRw14 pattern. Only a few cells were submitted as

Table 1. AGS25, Laboratories and Submitted Cells

Lab Code	N total	N DRw6	N DRw13	N DRw14
BCH	13	9		
BER	84	36	14	9
BET	93	47	37	10
DDC	68	41	30	9
DPT	80	58	15	6
DUT	250	84	63	10
FES	51	23	12	11
GRU	50	49	15	7
HAN	36	20		5
JEA	73	26	8	3
JLE	53	41		
JON	15	15	9	6
LAY	32	24	4	8
MRV	145	41	26	9
PTR	10	9	5	2
ROO	112	38	25	12
STA	28	24	7	5
TAK	21	16	13	3
THP	23	23	22	1
WAN	28	18	16	2
Total	1301	642	321	118

Reporting Laboratories: BENROO,[1] FRABET,[2] FRADDC,[3] US3GRU,[4] UKIFES,[5] SAFDUT[6]

This work was supported in part by the Dutch Foundation for Medical and Health Research (Medigon); the J.A. Cohen Institute for Radiopathology and Radiation Protection (IRS); and The Netherlands Reference Centre for Histocompatibility Testing.

Table 2. Sera Used for the Identification of DRw13, DRw14, DRw11, DRw12, and DRw8

Core	Sera	AS25 Sera		Specificity
1134	DRw13	= ROO	VR 64034	DRw13
1133	DRw13+w11	= ROO	VR 62008	DRw13+w11
1126	DRw13+w11		10w5209	DRw13+w11+w8+DQw7
3021	DRw13 (Dw18)	HAN	SEAL1209	DRw13+w11+w12+w8
9060	DRw14	MRV	26350	DRw14
3068	DRw12	MRV	24873	DRw14+w11+w12
9050	DRw12	MRV	21152	DRw14+DR4
9999	DRw12	MRV	21604	DRw14+w13+3
3036	DRw11		10w5206	DRw14+w11+w12
1113	DRw11	ROO	VR71200	DRw11
1114	DRw11	BET	LYGUE	DRw11+w8(+w13DHAG)
1087	DRw8	LAY	PLIV	DRw8+
1085	DRw8+w12	JLE	3.80.2.29.1	DRw6+3+w11+w12
1089	DRw8+w12	DDC	P1.4262	DRw6+w11+3
3038	DRw12+w11+w8			
3063	DRw6+w11+3	ROO	VR43821	DRw52
3025	DRw6+w11+w12+3	= ROO	7.3.19.1	DRw6+w11+w12+3
5384	DRw52		10w5194	DQw1
1152	DRw52		10w5197	DQw1
1136	DQw5	BET	LY1327	DQw5+DR1
3088	DQw6+DQw8+DQw9	= ROO	IIB3	DQw6+DQw8+DQw9
3119	DQw7			
3120	DQw7			
1107	DQw7			

= Same serum or MAb

DRw14 associated with DQw7. The reaction pattern of those cells was not conclusive for DRw14, which could be due to the weak reactivity of the anti-DRw14 sera. Several of the DRw14 cells were submitted as Dw16, including HTC 10W9064, which is identical to 9W1701. We have to conclude that DRw14 in association with DQw7 could not be assigned with confidence with these Workshop sera.

Laboratories FES and DUT reported reaction patterns (DR8I and DR8x14 respectively) that included DRw14 and DRw8 sera and that are associated with DQw5. Whether or not these represent one or more new specificities requires further study.

Discrimination of DRw12 from DRw13 (see also the DR5 report)

Three sera in the Core set recognized DRw12 monospecifically. However, these were replicate samples of the same monoclonal antibody. Several of the broader AGS- 25 sera included DRw12. Together with negative reactions obtained with broader DRw52-like sera, the

Table 3. DRw6 Related Patterns of AGS 25 and Core Sera

		DRw8	DRw11	DRw13	DRw12	DRw14	<DRw52	DRw52	DQw5	DQw6	DQw7	
			V	S V		M M M M	3	V		+		
			R	E R R		R R R R	P 8	R	Y			
			7 L	A 6 6		2 2 2	1 0 7	4	Y			
		1 1 1	3 1 1 1 Y 3	1 5 2 1 1 4 1	3 9 9	9 6 4 1 1	5 3 4 2 3 3	3 5 1	1 1	3 I	3 3 1	
		0 0 0	0 1 1 2 G 0	2 2 0 1 1 0 1	0 9 0	0 3 8 1 6	2 0 2 2 0 1	8 3 1	3 1	0 I	1 1 1	
		8 8 8	3 1 1 0 U 3	0 0 0 3 2 3 3	6 9 5	6 5 7 5 0	0 6 6 9 2 9	2 8 5	2 3	8 B	1 2 0	
		5 7 9	6 3 4 0 E 8	9 9 8 3 6 4 4	8 9 0	0 0 3 2 4	6 3 2 1 5 1	1 4 2	7 6	8 3	9 0 7	N*
DRw13	DQw1	− − −	− − − − − −	+ + + + + + +	w − −	− − − − w	w + + + + +	+ + +	− −	+ +	− − −	239
DRw13	DQw3	− − w	w − − w + +	+ + + + + + +	w − −	− − − − w	− + + + + +	+ + +	− −	− −	+ + +	32
DRw14	DQw1	− − −	− − − − − w	− − − − − − −	w − −	w + + w +	+ + + + + +	+ + +	+ +	− −	− − −	82
DRw14	DQw3	− − −	− − − − −	− + − − − − −	− − −	− − − + +	− + + + + +	+ + +	− −	− −	+ + +	15
DRw12	DQw3	+ w w	− − − − − +	+ + − − − − −	+ + +	− − w − −	+ − − w + +	+ + +	− −	− −	+ + +	53
DRw12	DQw1	+ − w	− − − − − +	+ + − − − − −	+ + +	− − + − −	+ − − w + +	+ + +	+ +	− −	− − −	16
DRw11	DQw3	− − −	+ + + + + w	+ + + + w − −	− − −	− − w − −	+ + + w + +	+ + +	− −	− −	+ + +	120
DRw11	DQw1	− − −	+ + + + + w	+ + + + w − −	− − −	− − w − −	+ + + w + +	+ + +	− −	+ +	− − −	16
DRw8		+ + +	− − − w + w	+ + − − − − −	w − −	− − − − −	− − + − − −	+ + +				72
DR3		− − −	− − − − − −	− − w − − − −	− − −	− − − − +	− + + + + +	+ + +				

*N = number of cells analyzed; w = reactivity unreliable

DRw12 specificity could be clearly distinguished from DRw13 (Table 3). DRw12 was predominantly observed with DQw7. Some of these cells were reported as DB6 (D Herluf).

Laboratory DUT originally reported some cells as DR6x12 but the analysis suggests that they could be DRw12 in association with DQw5 (Table 3). This pattern has thus far only been observed in South African blacks.

References

1. Schreuder GMT, Kennedy LJ, Gebuhrer L, Awad J, Betuel H, Degos L, et al. Antigen report: HLA-DRw6 and its subgroups HLA-DRw13 and HLA-DRw14. In: Albert ED, Baur MP, Mayr WR (eds): Histocompatibility Testing 1984. Springer-Verlag, Berlin, Heidelberg, 1984; p 192.
2. Schreuder GMT, Maeda H, Koning F, D'Amaro J. TA10 and 2B3, two new alleles in the HLA-DQ region recognized by monoclonal antibodies. Hum Immunol 1986; 16:127.
3. Gebuhrer L, Betuel H, Freidel AC, Farre A. A division of HLA-DQw1 associated with DR1, DR1x-DQw1, DR2 short, DRw10, and DRw14. I. Definition by alloantiserum LY1327. Hum Immunol 1987;18:235.

Antigen Society #26 Report (DR3, DR7, DQw2): Part 1

A. Cambon-Thomsen,[1] M. Calot,[1] E. Sommer,[1] E. Ohayon,[1] N. Goeken,[2] C. Kaplan,[3] P.L. Mattiuz,[4] A. Menicucci,[4] R. Cross,[5] B. Tait,[6] M. Buc,[7] M. Jeannet,[8] C. Irle,[8] S. Mayer,[9] M.M. Tongio,[9] L. Contu,[10] M. Purpura,[11] A. Nikaein,[12] H. Mervart,[13] K. Sullivan,[14] R. Schweizer,[15] J.A. Hansen,[16] E. Du Toit,[17] and M.G. Hammond[18]

General Introduction

There were 1,019 cells analyzed in the Antigen Society (795 Caucasians, 202 Negroes, 22 Orientals); 1,011 were tested both on Core serology and Antigen Society reagents. The total number of sera and monoclonal antibodies (MAbs) in the antigen group was 182, distributed as shown in Table 1.

The list of these reagents is given in Table 2 with their identity and lab of origin. It must be noted that due to a heterogeneity in the tray layout between the different labs, the central data analysis team cut down the results of 15 anti-DR7 sera from the Antigen Society as indicated between the two horizontal lines in Table 2. The results concerning these sera could not be analyzed in detail.

Table 1. Type and Number of Serologic Reagents Toward the Specificities Studied in Antigen Group 26 : DR3, DR7, DQw2

Specificity	Core Serum	Core MAb	Antigen Society
DR3 (mainly)	14	5	41
Related to DR3 (DRw52...)	10	3	/
DR7 (mainly)	12	4	59
Related to DR7 (DRw53, DRw9...)	6	2	/
DQw2	12	3	11

Reporting Laboratories: FRAOHA,[1] NCYGOE,[2] FRAMUL,[3] IT1MTT[4]
Participating Laboratories: ANZCRS,[5] ANZTAI,[6] EAEBUC,[7] FRAJEA,[8] FRAMYE,[9] IT1CON,[10] IT1PUR,[11] NCYNIK,[12] NCYMRV,[13] US1TUL,[14] US5SWE,[15] US7HAN,[16] SAFDUT,[17] SAFHAM [18]

Table 2. List and Identification of Serologic Reagents for Antigen Group 26

Origin	10WID Original ID
Order = 1 US6 BRN	1039 CC434.1
Specs: DR3	
Order = 2 UK1 DRK	1043 WILLIAMS.773
Specs: DR3	
Order = 3 SAF HAM	1040 N1165
Specs: DR3	
Order = 4 SAF HAM	1037 N1164
Specs: DR3	
Order = 5 ANZ BSH	1068 SPARKS
Specs: DR7	
Order = 6 ANZ DAW	1070 SIMMONDS.P6.5590
Specs: DR7	
Order = 7 IT1 MRA	1071 PV35
Specs: DR7	
Order = 8 IT2 GAN	1075 CO812
Specs: DR7	
Order = 9 UK1 JOY	1152 HAYES
Specs: DR3 DRw6 DRw8	
Order = 10 NCY MRV	1201 18660
Specs: DRw52	
Order = 11 SAF HAM	1069 N1101
Specs: DR7	
Order = 12 UK1 AST	1072 DB2410
Specs: DR7	
Order = 13 US5 TER	1182 TER.DQ3A
Specs: DQw3	
Order = 14 IT1 MTT	1169 F102406
Specs: DQw2	
Order = 15 FRA FAU	1173 CHEVRIER
Specs: DQw2	
Order = 16 US6 BRN	1168 CC413.1
Specs: DQw2	
Order = 17 FRA FAU	1067 JEGU
Specs: DR7 DRw9	
Order = 18 US8 RUB	1038 NYBC002
Specs: DR3	
Order = 19 NCY MRV	1035 10466
Specs: DR3	
Order = 20 NCY MRV	1034 11552
Specs: DR3	
Order = 21 UK1 FES	1036 ALLEN
Specs: DR3	
Order = 22 US2 BAC	1150 REYNOLDS
Specs: DR3 DRw6	
Order = 23 GER NEU	1077 GO810468B
Specs: DR7	
Order = 24 FRA BET	1079 E915
Specs: DR7	
Order = 25 US6 BRN	1080 CC327.5
Specs: DR7	
Order = 26 GER GOL	1076 924.4
Specs: DR7	
Order = 27 US1 THP	1042 EUINK
Specs: DR3	
Order = 28 NCY GOE	1202 PL855
Specs: DRw52	
Order = 29 US5 SIN	1205 15886
Specs: DRw52	
Order = 30 FRA DDC	5384 P6465
Specs: DRw52	
Order = 31 US8 JLE	1216 H181
Specs: DRw53	
Order = 32 IT2 FER	1213 FE200
Specs: DRw53	
Order = 33 GER BRA	1170 422B
Specs: DQw2	
Order = 34 IT1 MTT	1174 F103295
Specs: DQw2	
Order = 35 US2 MBC	1175 BC.CA.MAGN
Specs: DQw2	
Order = 36 BEN BER	1172 MSD20
Specs: DQw2	
Order = 37 US5 TER	1117 TER.DP1
Specs: DPw1	
Order = 38 NCY YUN	1144 BLOUDETTE
Specs: DR3 DR6	
Order = 39 UK1 TAT	1041 WINTER.SOH.487
Specs: DR3	
Order = 40 FRA PRR	1064 VEL1557
Specs: DRw9	
Order = 41 ANZ DAW	1195 R5.6295
Specs: DR4 DR5	
Order = 42 NCY MRV	1212 24784
Specs: DRw53	
Order = 43 UK1 DRK	1078 OWEN.812
Specs: DR7	
Order = 44 NCY MRV	1167 11028
Specs: DQw2	
Order = 45 FRA FAU	1066 ANTIN
Specs: DR7 DRw9	
Order = 46	9060
Specs: UNK	
Order = 47 FRA JEA	1151 DROZ
Specs: DR3 DRw6	
Order = 48 FRA MYE	1147 LECOINTRE
Specs: DR3 DRw6	
Order = 49 BEN BER	1135 MSD6
Specs: DRw13 DRw14	
Order = 50 IT1 MTT	1180 F101404
Specs: DQw3	
Order = 51 IT2 FER	1186 FE94
Specs: DQw3	
Order = 52 US1 DUQ	1171 QUAGLIER
Specs: DQw2	
Order = 53 UK1 DRK	1208 PRATT.80%
Specs: DR3 DR5 DRw6 DRw52	
Order = 54 IT1 MTT	1200 F102422
Specs: DRw52	
Order = 55 GER WAN	3005 C5C5
Specs: DR1 DR3 DR4 DRw8	
Order = 56 FRA DDC	3010 CHE153
Specs: DR1 DR7	
Order = 57 JAP AIZ	3011 HU30
Specs: DR2 DR1	
Order = 58 GER WAN	3020 M4F11
Specs: DR3 DRw13	
Order = 59 FRA DDC	3023 CHE41.2
Specs: DR3 DRw6	
Order = 60 UK1 FES	3031 JA1
Specs: DR3	
Order = 61 JAP JUJ	3048 PLM3
Specs: DR7 DRw9 DRw12	
Order = 62 UK1 BOD	3049 17.3.3
Specs: DR7	

Origin	10WID Original ID
Order = 63 NCY SFR	3050 SFR16.DR7M
Specs: DR7	
Order = 64 US7 GSC	3051 GSP65.1
Specs: DR7 DRw10	
Order = 65 GER WAN	3062 M4G8
Specs: DRw52	
Order = 66	3105
Specs: DQw2	
Order = 67 IT2 FER	3106 MP1
Specs: DR2 DR3 DR4 DR5 DRw6	
Order = 68 IT2 GAN	3107 XIII358
Specs: DQw2	
Order = 69 US7 GSC	3113 GSP91.1
Specs: DQw3	
Order = 70	3025
Specs: DR3 DR5 DRw6	
Order = 71 GER WAN	3046 C6E2
Specs: DR3 DR4 DR7	
Order = 72 IT1 CEP	1081 TORP1017
Specs: DR7	
Order = 73 FRA PRR	5006 PUY.A.217
Specs: DR7	
Order = 74 UK1 BRS	5210 10726.LCS
Specs: DR3	
Order = 75 NCY MRV	5211 27026
Specs: DR3	
Order = 76 NCY MRV	5212 18835
Specs: DR3	
Order = 77 NCY MRV	5213 20571
Specs: DR3	
Order = 78 NCY MRV	5214 2551.6
Specs: DR3	
Order = 79 US2 MBC	5215 BC.HE.BURJ
Specs: DR3	
Order = 80 US2 MBC	5216 BC.JA.SADO
Specs: DR3	
Order = 81 US2 MBC	5217 BC.VE.BREE
Specs: DR3	
Order = 82 IT1 MIT	5218 F101972
Specs: DR3	
Order = 83 FRA BIG	5219 H12
Specs: DR3	
Order = 84 FRA DDC	5220 P2983
Specs: DR3	
Order = 85 ANZ CRS	5221 PAGE
Specs: DR3	
Order = 86 US7 HAN	5222 SEA1250
Specs: DR3	
Order = 87 ANZ DAW	5223 SLEEPY.F29026
Specs: DR3	
Order = 88 NCY MRV	5224 3953.5
Specs: DR3	
Order = 89 NCY MRV	5225 9134.2
Specs: DR3	
Order = 90 US5 ABS	5226 ABS.17
Specs: DR3	
Order = 91 UK1 MID	5227 BEL553D
Specs: DR3	
Order = 92 NCY CAR	5228 BOS/BAT
Specs: DR3	
Order = 93 FRA FAU	5229 BRIAND
Specs: DR3	

Origin	10WID Original ID
Order = 94 US6 BRN	5230 CC519.2
Specs: DR3	
Order = 95 US6 BRN	5231 CCB1060.1
Specs: DR3	
Order = 96 UK1 JOY	5232 DENNING
Specs: DR3	
Order = 97 EAE RIC	5233 H1195.1
Specs: DR3	
Order = 98 NCY GOE	5234 J18571
Specs: DR3	
Order = 99 EAE KAS	5235 KAS7292
Specs: DR3	
Order = 100 US8 JLE	5236 M325
Specs: DR3	
Order = 101 FRA OHA	5237 MARCHE
Specs: DR3	
Order = 102 BEN BER	5238 MSD8
Specs: DR3	
Order = 103 SAF HAM	5239 N1163
Specs: DR3	
Order = 104 SAF HAM	5240 N1233
Specs: DR3	
Order = 105 US7 DUP	5241 NJ4050
Specs: DR3	
Order = 106 FRA PRR	5242 ROU.A212
Specs: DR3	
Order = 107 US4 SND	5243 SI5L178
Specs: DR3	
Order = 108 US5 TER	5244 TER.DR3
Specs: DR3	
Order = 109 UK1 JOY	5245 WATERHOUSE
Specs: DR3	
Order = 110 US3 PER	5246 WOODARDP0109A
Specs: DR3	
Order = 111 US1 DUQ	5247 Y.CARROLL
Specs: DR3	
Order = 112 UK1 TAT	5248 DURNFORD. SOH.517
Specs: DR3	
Order = 113 UK1 DRK	5249 MARSHALL.1070
Specs: DR3	
Order = 114 US1 DUQ	5250 YURUS
Specs: DR3	
Order = 115 US5 SIN	5251 15452
Specs: DR7	
Order = 116 NCY KAP	5252 CWRP1518
Specs: DR7	
Order = 117 US7 POL	5253 045
Specs: DR7	
Order = 118 NCY MRV	5254 11542
Specs: DR7	
Order = 119 GER GOL	5255 15924.4
Specs: DR7	
Order = 120 EAE SHA	5256 SHA.8
Specs: DR7	
Order = 121 GER GOL	5257 16738
Specs: DR7	
Order = 122 NCY MRV	5258 19330
Specs: DR7	
Order = 123 NCY MRV	5259 20387
Specs: DR7	
Order = 124 NCY MRV	5260 20502
Specs: DR7	
Order = 125 NCY MRV	5261 21420
Specs: DR7	

Origin	10WID Original ID
Order = 126 NCY MRV Specs: DR7	5262 25552
Order = 127 BEN BOU Specs: DR7	5263 64352 9/11/82
Order = 128 NCY MRV Specs: DR7	5264 7128
Order = 129 NCY MRV Specs: DR7	5266 9235
Order = 130 GER BRA Specs: DR7	5267 950
Order = 131 US5 ABS Specs: DR7	5268 ABS.20
Order = 132 IT1 GAB Specs: DR7	5269 AN127
Order = 133 UK1 JOY Specs: DR7	5270 ANDREWS
Order = 134 IT1 MTT Specs: DR7	5271 B08.3
Order = 135 IT1 CEP Specs: DR7	5272 CDS9031
Order = 136 UK1 MID Specs: DR7	5273 BEL213D
Order = 137 UK1 MID Specs: DR7	5274 BEL522D
Order = 138 FRA OHA Specs: DR7	5275 BERNON
Order = 139 EAE RIC Specs: DR7	5276 BH3747
Order = 140 UK1 GEL Specs: DR7	5277 CLA22677
Order = 141 IT2 GAN Specs: DR7	5278 CO868
Order = 142 US5 SWE Specs: DR7	5279 DELMASTRO
Order = 143 FRA FAU Specs: DR7	5280 ESNAULT
Order = 144 IT1 MTT Specs: DR7	5281 F101577
Order = 145 IT2 FER Specs: DR7	5282 FE208
Order = 146 IT2 FER Specs: DR7	5283 FE216
Order = 147 ANZ CRS Specs: DR7	5284 HENNING
Order = 148 US6 GAT Specs: DR7	5285 JH
Order = 149 US4 STA Specs: DR7	5286 KWI
Order = 150 US1 DUQ Specs: DR7	5287 MCCLOSKEY12.84
Order = 151 FRA KRE Specs: DR7	5288 MTC8122
Order = 152 FRA KRE Specs: DR7	5289 MTC8337
Order = 153 SAF HAM Specs: DR7	5290 N1188
Order = 154 SAF HAM Specs: DR7	5291 N1201

Origin	10WID Original ID
Order = 155 SAF HAM Specs: DR7	5292 N1234
Order = 156 SAF HAM Specs: DR7	5293 N945
Order = 157 FRA PRR Specs: DR7	5294 PHI1466
Order = 158 UK1 JOY Specs: DR7	5295 RICHARDSON
Order = 159 US8 FOT Specs: DR7	5297 STRATIS
Order = 160 US5 TER Specs: DR7	5298 TER.DR7
Order = 161 FRA MYE Specs: DR7	5299 THOMAS
Order = 162 IT1 CNG Specs: DR7	5300 TSRP170
Order = 163 IT1 CNG Specs: DR7	5301 TSRP307
Order = 164 US3 PER Specs: DR7	5302 ULRICHP0019C
Order = 165 US8 JLE Specs: DR7	5303 W283
Order = 166 UK1 DRK Specs: DR7	5304 JONES. GERRARD.807
Order = 167 UK1 LAW Specs: DR7	5305 WILLIAMS
Order = 168 NCY KAP Specs: DR7	5306 CWR.PLA.1355
Order = 169 NCY KAP Specs: DR7	5307 CWR.NIE.3
Order = 170 NCY KAP Specs: DR7 DRw9 DR3	5308 CWR.PLA.1007
Order = 171 UK1 GEL Specs: DR7	5309 DAY20558
Order = 172 US6 BRN Specs: DR3 DR7	5310 CCB.1045.2
Order = 173 IT2 GAN Specs: DR3 DR7	5311 CO1178
Order = 174 UK1 GLA Specs: DR7 DR3 DQw2	5312 HUTTON
Order = 175 NCY KAP Specs: DR3 DR7 DRw9	5313 CWR.PRIM
Order = 176 US6 BRN Specs: DQw2	5314 CCB.1035.1
Order = 177 US2 BAC Specs: DQw2	5315 M.LARSEN
Order = 178 NCY GOE Specs: DQw2	9104 PL1758
Order = 179 NCY GOE Specs: DQw2	9105 PL2051
Order = 180 NCY GOE Specs: DQw2	9106 PL2282
Order = 181 NCY GOE Specs: DQw2	9107 PL2441
Order = 182 NCY GOE Specs: DQw2	9108 PL2451

Antigen Society #26 Report (DR3, DR7, DQw2): Part 2: HLA-DR3

C. Kaplan,[3] A. Cambon-Thomsen,[1] N. Goeken,[2] P.L. Mattiuz,[4] and A. Menicucci[4]

HLA-DR3, originally described in 1973 by the Leyden group (1), was clearly defined at the Seventh, Eighth, and Ninth International Histocompatibility Workshops. During the Ninth Workshop, there was a suggestion of a split ("short") in DR3 in that three sera had a high percentage of DR3 misses. A segregation of this DR3 short was seen in several families. Because of the weakness of the reactions, the split was not confirmed (2). As expected, a strong linkage disequilibrium with HLA B8, Dw3, DRw52, and DQw2 was found.

Core Set Serology Analysis (Table 1)

During the Tenth Workshop, eight alloantisera and one monoclonal antibody had an anti-DR3 activity with a Q score value over 4.5. These sera are highly correlated with cluster together in the serum x serum analysis.

As suggested in the Ninth Workshop and documented in the Sixth Newsletter of the Tenth Workshop by Oudshoorn et al. (2), a split of DR3 exists. It is defined by positive reactions with sera 1043, 1039, 1040, (1037) and negative with sera 1036, 1034, 1035, and 1038. It is neither associated with Dw3 nor with DQw2; in blacks, it is associated with DQwa (DQw4, monoclonal antibody 3101 positive) and with Dw "RSH" (typing reaction with cell 9021). This antigen specificity is common in blacks and very rare in Caucasians. The segregation of this antigen has been consistently observed in families.

See footnote at beginning of Part 1 for a listing of the reporting laboratories.

Table 1. Core Set Serology (All Cells Non-Flagged)

10thWS ID#	Q Score	r Value	% Extras	% Score 8	Residual	
1043	8.09	0.91	12.6	91.5	DR5	19%
					DQw3	15%
					DQw1	13%
1039	6.94	0.91	5.4	76.3	DQw1	17%
					DRw6	13%
					DR5	13%
1037	6.63	0.91	1.4	69.1		
1040	6.55	0.88	14	80.8	DR5	19%
					DQw1	19%
					DRw6	10%
1035	5.49	0.84	4.6	57.4	DQw3	21%
					DR4	14%
					DR5	14%
					DQw1	14%
1038	4.94	0.82	1.3	38.7		
1036	4.86	0.80	1.3	61.7	DR5	25%
					DRw8	25%
					DQw3	25%
					DR4	25%
1034	4.64	0.77	4.1	60.7	DR2	23%
					DQw1	23%
					DRw6	15%
					DR5	15%
3031	4.04	0.78	15.5	56.4	DR5	13%

Antigen Society Serology (Table 2)

We analyzed 41 anti-HLA DR3 antisera on 1,019 cells (795 Caucasians, 202 Negroes, 22 Orientals) from the collaborating laboratories (see list at the beginning of the report). Twelve sera had a Q score over 5 and were shown to be operationally monospecific with a high correlation between each other. The following five sera had a Q score over 7, when considering the Caucasoid cells: 5213, 5221, 5232, 5246, 5248. The serum 5225 defines the short split when considering the black cells. No difference was found in the DR3 serologic definition between the B18- and the B8-associated DR3.

Table 2. Antigen Society Serology (All Cells Non-Flagged)

10thWS ID#	Q Score	r Value	% Extras	% Score 8	Residual	
5213	7.80	0.96	1.2	77.9		
5221	7.53	0.93	8.2	80.0	DR5	20%
					DQw1	14%
					DQw3	12%
5245	6.88	0.93	4.6	64.8	DR5	31%
					DQw3	19%
					DQw1	12%
					DR4	12%
5248	6.66	0.89	11.0	79.6	DR2	20%
					DR4	20%
					DR5	20%
					DQw1	20%
					DQw3	20%
5247	5.96	0.83	20.4	71.2	DR5	14%
					DQw1	13%
5227	5.83	0.89	9.4	70.5	DR5	16%
					DQw1	12%
					DQw3	10%
5250	5.69	0.83	22.2	81.9	DR5	22%
					DQw3	13%
					DQw1	13%
5229	5.66	0.81	27.5	84.0	DR5	19%
					DQw1	18%
					DQw3	10%
5232	5.45	0.81	24.8	83.2	DR5	22%
					DQw1	19%
					DRw6	12%
5246	5.35	0.80	23.3	75.5	DR5	20%
					DQw1	16%
5230	5.32	0.83	11.2	61.5	DR5	16%
					DQw3	14%
					DQw1	13%
5218	5.23	0.83	4.5	47.3	DR5	19%
					DQw1	19%
					DR1	12%

Conclusion

HLA-DR3 was well defined by alloantisera, both in the Core set and in the Antigen Society; a short split is defined by Core set sera and particularly frequent in Negroes; the definition of DR3 by monoclonal antibodies was more difficult.

The new provisional HLA-DR specificities are DRw17 and DRw18. DRw17 includes most of the originally defined DR3 specificity usually associated with DQw2. DRw18 is a short DR3 specificity found mainly in black populations, and associated with Bw42 and DQw4 ("wa"). This specificity is also defined by T cell-clones but is not assigned a Dw specificity.

Acknowledgments. We would like to thank L. Hallé and F. Castellano for their technical assistance.

References

1. Van Leeuwen A, Schuit HRE, Van Rood JJ. Typing for MLC (LD). II. The selection of non-stimulator cells by MLC inhibition tests using SD identical stimulator cells (MISIS) and fluorescent antibody studies. Transplant Proc 1984;5:1539.
2. Fuller TC, Park MS, Thorsby E. Antigen report HLA-DR3. In Albert ED, Baur MP, Mayr WR (eds): Histocompatibility Testing 1984. Springer Verlag, Berlin, Heidelberg, 1984; p 187.
3. Oudshoorn M, Du Toit ED, Taljaard DG. A possible new Negro HLA-DR3 variant. Tenth Workshop Newsletter No. 1, p 26.

Antigen Society #26 Report (DR3, DR7, DQw2): Part 3: Serologic Definition of HLA-DR7

A. Cambon-Thomsen,[1] M. Calot,[1] N. Goeken,[2] C. Kaplan,[3] P.L. Mattiuz,[4] and E. Ohayon[1]

History

Since 1977, HLA-DR7 has been one of the best defined antigens. During the Eighth and Ninth International Histocompatibility Workshops, excellent anti-DR7 typing reagents were selected (1,2). No DR7 split was suggested. Some monoclonal antibodies were found to react with this antigen and some others were reacting with all HLA-DR specificities except DR7 (3). During the Ninth Workshop, information about associations with Dw, DP, and DQ specificities were obtained in various ethnic groups; in Caucasians, DR7 is associated with Dw7, Dw17 (DQw2), and Dw11 (DQw8 positive, often Bw57); in Negroes and Orientals, DR7 is associated with DQw2 (4). In Caucasians, significant linkage, disequilibria were found with A29, A30, B13, B44, Bw57, Bw64, Cw6, Cw8, Dw7, Dw11, and DQw2; in Negroes with Bw57, Bw64, Cw2, Cw8, DQw2; and in Orientals with B13, Cw6 and DQw2 (5).

See footnote at beginning of Part 1 for a listing of the reporting laboratories.

Table 1. Serological Definition of DR7 with Core Alloantisera (general analysis)

Sera	Q Score	Comment
1068	7.76	
1079	7.58	
1070	7.58	
1077	6.64	
1075	6.55	
1078	5.64	+DQw2
1072	5.31	+DRw53
1080	5.24	
1173	5.09	+DQw2
1071	4.88	
1066	4.85	+DRw9
1175	4.65	+DQw2
1167	4.55	+DQw2
1067	4.30	+DR3

Tenth Workshop DR7 Serology

In the Tenth Workshop Core set of serologic reagents, according to the analysis performed on about 2,400 cells from all ethnic groups and with well contrasted reactions (mean reaction profile > or = 0.75), 15 alloantisera had an anti-HLA-DR7 activity with a Q score value over 4; seven of them were operationally monospecific (Table 1). In addition, four monoclonal antibodies had a Q score value for DR7 between 2 and 4 (3048 to 3051), two of them being monospecific (3049 and 3050). The first five sera of the table were highly correlated to each other in the serum x serum analysis. Reactions defining splits could not be found in the available data.

Combined Core Set and Antigen Society Serology

To better characterize the serologic reagents of Core set and Antigen Society trays, we performed a tail analysis on the high quality Caucasian cells tested both by Core and Antigen Society reagents (N = 284) with the following selection criteria: non-flagged cells; minimum inclusion ratio, X1 = 25; minimum number of positively reacting cells, X2 = 5; minimum chi-squared value for repeated iterations, X3 = 25; minimum number of positive cells for repeated iterations on a serum, X4 = 5; type of cells considered: fresh or frozen PBLs, fresh or frozen B cells, B-cell lines and lectin activated T cells.

Table 2. Quality of 10W Serologic Reagents for the Definition of DR7

10W ID	Type of Reagent	Origin	Name	r Value	Q Score	% Inclusion	Mean Strength	
1068	Core serum	ANZ BSH	SPARKS	0.99	10.6	100	7.8	
1079	Core serum	FRA BET	E915	0.99	10.4	100	7.7	
1077	Core serum	GER NEU	GO810468B	0.94	9.0	97	7.5	
1075	Core serum	IT2 GAN	CO812	0.96	8.9	98	7.6	
5294	Ag. Soc.	FRA PRR	PHI1466	0.85	9.2	95	7.2	
5279	Ag. Soc.	US5 SWE	DELMASTRO	0.95	8.7	97	7.4	
1070	Core serum	ANZ DAW	SIMMONDS.P6.5590	0.96	8.5	95	7.5	
5304	Ag. Soc.	UK1 DRK	JONES.GERRARD.B07	0.90	8.4	95	7.3	
5292	Ag. Soc.	SAF HAM	N1234	0.94	8.0	98	7.7	
5260	Ag. Soc.	NCY MRV	20502	0.92	8.0	96	7.4	
5255	Ag. Soc.	GER GOL	15924.4	0.86	8.0	98	7.6	
1080	Core serum	US6 BRN	CC327.5	0.94	7.9	93	6.7	
5259	Ag. Soc.	NCY MRV	20387	0.93	7.9	98	7.1	
5287	Ag. Soc.	US1 DUQ	MCCLOSKEY12.84	0.95	7.7	98	7.6	
5283	Ag. Soc.	IT2 FER	FE216	0.88	7.6	97	7.2	
5006	Ag. Soc.	FRA PRR	PUY.A.217	0.94	7.5	92	7.1	
5291	Ag. Soc.	SAF HAM	N1201	0.91	7.3	98	7.7	
5298	Ag. Soc.	US5 TER	TER.DR7	0.92	7.2	93	7.2	
5257	Ag. Soc.	GER GOL	167381	0.90	7.2	88	6.6	
5280	Ag. Soc.	FRA FAU	ESNAULT	0.85	7.2	98	7.5	
5262	Ag. Soc.	NCY MRV	25552	0.91	7.0	88	6.8	
1076	Core serum	GER GOL	924.4	0.82	7.0	98	7.7	
1072	Core serum	UK1 AST	DB2410	0.78	6.8	100	7.9	
1066	Core serum	FRA FAU	ANTIN	0.83	6.9	100	7.9	+DRw9
5254	Ag. Soc.	NCY MRV	11542	0.90	6.7	88	6.6	
1069	Core serum	SAF HAM	N1101	0.83	6.6	100	7.2	+DRw9
5258	Ag. Soc.	NCY MRV	19330	0.89	6.3	87	6.2	
3050	MAb	NCY SFR	SFR16.DR7M	0.87	6.1	96	7.5	
5282	Ag. Soc.	IT2 FER	FE208	0.92	7.2	93	7.2	
1081	Ag. Soc.	IT1 CEP	TORP1017	0.85	6.0	79	5.9	
5303	Ag. Soc.	US8 JLE	W283	0.87	6.2	85	6.7	
5251	Ag. Soc.	US5 SIN	15452	0.85	5.9	95	7.2	
5252	Ag. Soc.	NCY KAP	CWRP1518	0.85	5.7	82	5.8	
1071	Core serum	IT1 MRA	PV35	0.80	5.7	98	7.6	
5306	Ag. Soc.	NCY KAP	CWR.PLA.1355	0.79	5.7	97	7.7	
5293	Ag. Soc.	SAF HAM	N945	0.83	5.6	93	7.3	
5301	Ag. Soc.	IT1 CNG	TSRP307	0.86	5.6	95	7.3	
5261	Ag. Soc.	NCY MRV	21420	0.86	5.5	99	6.3	
1078	Core serum	UK1 DRK	OWEN.812	0.82	5.4	97	7.4	
5299	Ag. Soc.	FRA MYE	THOMAS	0.86	5.2	96	7.6	+DR2
5315	Ag. Soc.	US2 BAC	M.LARSEN	0.75	5.1	95	7.3	+DRw9
5300	Ag. Soc.	IT1 CNG	TSRP170	0.81	4.9	77	5.6	
5284	Ag. Soc.	ANZ CRS	HENNING	0.67	4.7	93	7.4	
5290	Ag. Soc.	SAF HAM	N1188	0.59	4.7	100	7.7	
5286	Ag. Soc.	US4 STA	KWI	0.70	4.6	97	7.3	
5307	Ag. Soc.	NCY KAP	CWR.NIE.3	0.77	4.6	94	7.1	
5295	Ag. Soc.	UK1 JOY	RICHARDSON	0.79	4.5	82	6.7	
9108	Ag. Soc.	NCY GOE	PL2451	0.69	4.3	97	7.3	+DRw9 submitted as DQw2
3048	MAb	JAP JUJ	PLM3	0.58	4.4	100	7.8	
3049	MAb	UK1 BOD	17.3.3	0.51	2.4	96	7.5	
3010	MAb	FRA DDC	CHE153	0.42	1.4	88	6.4	+DR1

The results presented are given by the tail analysis of 284 high quality (non-flagged) Caucasoid cells tested by both Core set and Antigen Society reagents.

Table 3. Correlation Between Anti-DR7 Reagents

10WID		1111555115555555553515551155555555555151555115593 0000220002222222203022300322232222230302220023 10 6777870789659569550758078055608599907078886681 04 8905796702298701401653481328102134562714569605 88
1068 DR7		999999999999999999899989999988989888887778 89876
1079 DR7		9 999999999999999998999899999898898888877788 8876
1070 DR7		99 999999999999999988899888889888888788887678 88776
1075 DR7		999 99999999999998888998888898888888887876788 8776
5287 DR7		9999 99999999999988888988888988888888888667888776
5279 DR7		99999 999999999999888998998888888888888777888776
5006 DR7		999999 99999999998888888998888888887777767888775
1077 DR7		9999999 9999999998888898888899888898878766788 8765
1080 DR7		99999999 99999998888889888889988888888777667888775
5292 DR7		999999999 9999999888889888888888888887776678887 75
5262 DR7		9999999999 9998888888888888898888888877776678887 65
5259 DR7		99999999999 98998888888888898888888888777766788877 5
5298 DR7		999999999999 8998888888888999888888787786668888 65
5257 DR7		9999999999988 9998878887888898788888777766788877 5
5260 DR7		99999999998999 99888899888889878888877876678887 75
5291 DR7		999999999989999 88888988888888888887877776788877 5
5254 DR7		9999999989888998 8878888888888788888777766688876 6
3050 DR7		99988988888888888 87888787787788786777766787776 5
5301 DR7		998888888888888888 78887787887777777777766787776 5
1076 DR7		8888888888888788777 88887777777777766777766877765
5255 DR7		99888888888888888888 888887877777877877667877776
5283 DR7		999989888888889988888 88887887878877777677878765
5304 DR7		9999998999888898888888 8888888888888777777888765
1078 DR7		88888888888887888778888 7778777777677666667777 65
1081 DR7		998889988888888888778887 98888777887777666777765
5303 DR7		9988899888899888878788879 888877777777766677776 5
5252 DR7		99888888888888888777778788 98878878777766677776 5
5258 DR7		999998899898999888878888889 8888888777766688877 5
5261 DR7		9988888998888888878778878888 887887777766677776 5
5300 DR7		88888888888887787777778788888 7777777676567776 5
5282 DR7		898888888888888888877788777 7887 78777677666777665
5251 DR7		9888888888888888888877778777 88777 7777777656777765
5293 DR7		88888888888888888877778877 7888787 887777678777655
5294 DR7		998888898888888888877888787 7887778 8777667777776 5
5295 DR7		88788888888887887867677868 788777788 7767567767654
5306 DR7		888888788877787787776778777 77777777 676567777654
1072 DR7		8888887777777777777787777777 77677776 77666877765
5307 DR7		888788787777778777777776777 7767777677 6666777765
1071 DR7		88888877777787777777777677 777777767676 655767664
5284 DR7		777767766666666676667667666 6666666655666 66666654
5285 DR7		7766676666666666666667766666 65657766665 6 7665544
5286 DR7		777777777777677767767776666 666666877766567 766654
1069 DR7	DRw9	88888888888888888888888887777 8777777777877667 88876
1066 DR7	DRw9	8888888888888888887777787777 787777776777666 68 8887
5280 DR7		98888888888888888877778877778 7777777777765688 887
5315 DR7	DRw9	8877777777778777777777777777 77767676677665688 8 86
9108 DR7	DRw9	7777777677676777666676666667 6666565566654578 88 7

The principal results of this analysis are reported in Table 2, where the reagents are ordered by decreasing Q score value. Characteristics of sera with Q score over 4 and of monoclonal antibodies are summarized in this table. The Core sera 1068 and 1069 were perfect (Q score > 10); 21 sera (5 from Core set and 16 from Antigen Society) were excellent strongly reacting reagents with R value between 0.82 and 0.96, Q score between 7.0 and 9.2 and percent of inclusion between 92 and 98. We performed the same analysis on cells from Negroes and from Orientals; no difference of definition appeared between populations. In general, the r values and Q scores were lower on Negroid cells due to a lower number of cells being tested. The results with monoclonal antibodies were disappointing, the best one being 3050 with a Q score of 6.1. Among the Antigen Society sera, which could not be fully analyzed, the best sera were 5277, 5274, and 5278. The sera were highly correlated

Table 4. Characteristic Patterns for Definition of DR7

DR7	DR3	DRw9	DQw2	DQw3	1068	1079	1077	1075	5294	5279	5304	5260	3050	3048	1066	5307	1043	1034	3107	3106
+	−	−	+	−	+	+	+	+	+	+	+	+	+	+	+	+	−	−	+	+
+	−	−	−	+	+	+	+	+	+	+	+	+	+	+	+	+	−	−	−	−
−	+	−	+		−	−	−	−	−	−	−	−	−	−	−	−	+	+	+	+
−	−	+	+	−	−	−	−	−	−	−	−	−	−	+	+	(+)	−	−	+	+
−	−	+	−	+	−	−	−	−	−	−	−	−	−	+	+	+	(+)	−	−	−

to each other (Table 3). All attempts to find specific patterns in relation to the associations with B locus antigens, Dw or DQ specificities, failed and no evidence of split could be found. The major patterns of reaction distinguishing DR7 from other related DR, DQ specificities are shown in Table 4.

In conclusion, DR7 was a very well defined specificity by Core and Antigen Society sera and no evidence of serologic split could be found. No monoclonal antibody reached the same quality level as alloantisera for DR7 definition.

References

1. Albert A. DR7. In: Terasaki PI (ed): Histocompatibility Testing 1980. UCLA Tissue Typing Laboratory, Los Angeles, 1980; p 537.
2. Jakobsen BK, Terasaki PI, Svejgaard A. Antigen report: HLA-DR7. In: Albert ED, et al. (eds): Histocompatibility Testing 1984. Springer-Verlag, Berlin, Heidelberg, 1984; p 198.
3. Bodmer JG, Kennedy LJ, Aizawa M, Dawkins RL, Lepage V, Mazzilli MC. HLA-D region monoclonal antibodies. In: Albert ED, et al. (eds): Histocompatibility Testing 1984. Springer-Verlag, Berlin, Heidelberg, 1984; pp 217–236.
4. Schreuder GMT, Doxiadis I, Parlevliet J, Grosse-Wilde H. HLA-DR, DQ, LB, and TA10 specificities of Ninth Workshop homozygous typing cells. In: Albert ED, et al. (eds): Histocompatibility Testing 1984. Springer-Verlag, Berlin, Heidelberg, 1984; pp 243–248.
5. Baur MP, Neugebauer M, Albert ED. Reference tables of two-locus haplotype frequencies for a II MHC marker loci. In: Albert ED, et al. (eds): Histocompatibility Testing 1984. Springer-Verlag, Berlin, Heidelberg, 1984; pp 677–678.

Antigen Society #26 Report (DR3, DR7, DQw2): Part 4: HLA-DQw2

P.L. Mattiuz,[4] A. Cambon-Thomsen,[1] N. Goeken,[2] C. Kaplan,[3] and M. Calot[1]

History

HLA-DQw2 was first described in 1979 by Duquesnoy et al. (1). This determinant was similar to Te 24 and to DC3 defined in 1980 and 1982 respectively (2,3). In the Eighth Workshop, DQw2 could be defined by a cluster of good reagents (4), most of them containing, however, anti-DR7 antibodies. It could then be fully confirmed that DQw2 belongs to a series of molecules distinct from DR, based on biochemistry data (5,6). In the Ninth Workshop more than 10 sera were shown to be good reagents for this specificity and the corresponding patterns had a good family segregation (7); but there was no monoclonal antibody with such a specificity (8). There was a strong association of DQw2 with DR3, Dw3 and DR7, Dw7 or Dw17. In contrast, some DR3 positive, Dw3 negative cells were associated with either DQw3, TA10 (DQw7), or with DQ blank, the latter being mostly found in Negroes, and corresponding to DRw18 associated with DQw4. The cells DQw2 negative but DR7 positive were generally DQw3 positive, TA10 negative (DQw8), Dw11 positive and often Bw57 (9). In addition, significant positive linkage disequilibrium was found between DQw2 and A1, A29, A30, B8, B13, B44, Bw50, Bw57, Bw64, Cw6, and Cw8 in Caucasians, with A1, B8, B13, Bw41 and Bw64 in Negroes, and with A1, A30, B13, and Cw6 in Orientals (10).

Tenth Serologic Definition of DQw2

In the general analysis available for the Tenth Workshop Core Set of Serologic reagents, there were eight sera with a Q score value over 3 for DQw2, and three monoclonal antibodies with a Q score value over 4 for this specificity (Table 1). Among these reagents we can recognize three clusters in the serum x serum analysis (Table 2): The first shows a strong correlation with the

Table 1. Serologic Definition of DQw2 (core set)

Sera	Q Score	Comment	Cluster
1173	6.6	+DR7	III
1170	6.6	+DR3	I
1169	6.3	+DR7	III
1167	4.3	+DR7	III
1174	4.0	+DR3	I
1175	3.6	+DR7	III
1172	3.2	+DR3	I
1168	3.1	+DR7	III
Monoclonal Antibodies			
3106	6.0	) miss some	II
3107	5.7	) DR3, or DR7	II
3105	4.6	) DQw2-positive	II
		) cells	

Immunobiology of HLA
Volume I

Table 2. Correlation Between Serologic Reagents of Core Set for Antigen Group 26: DR3, DR7, DQw2

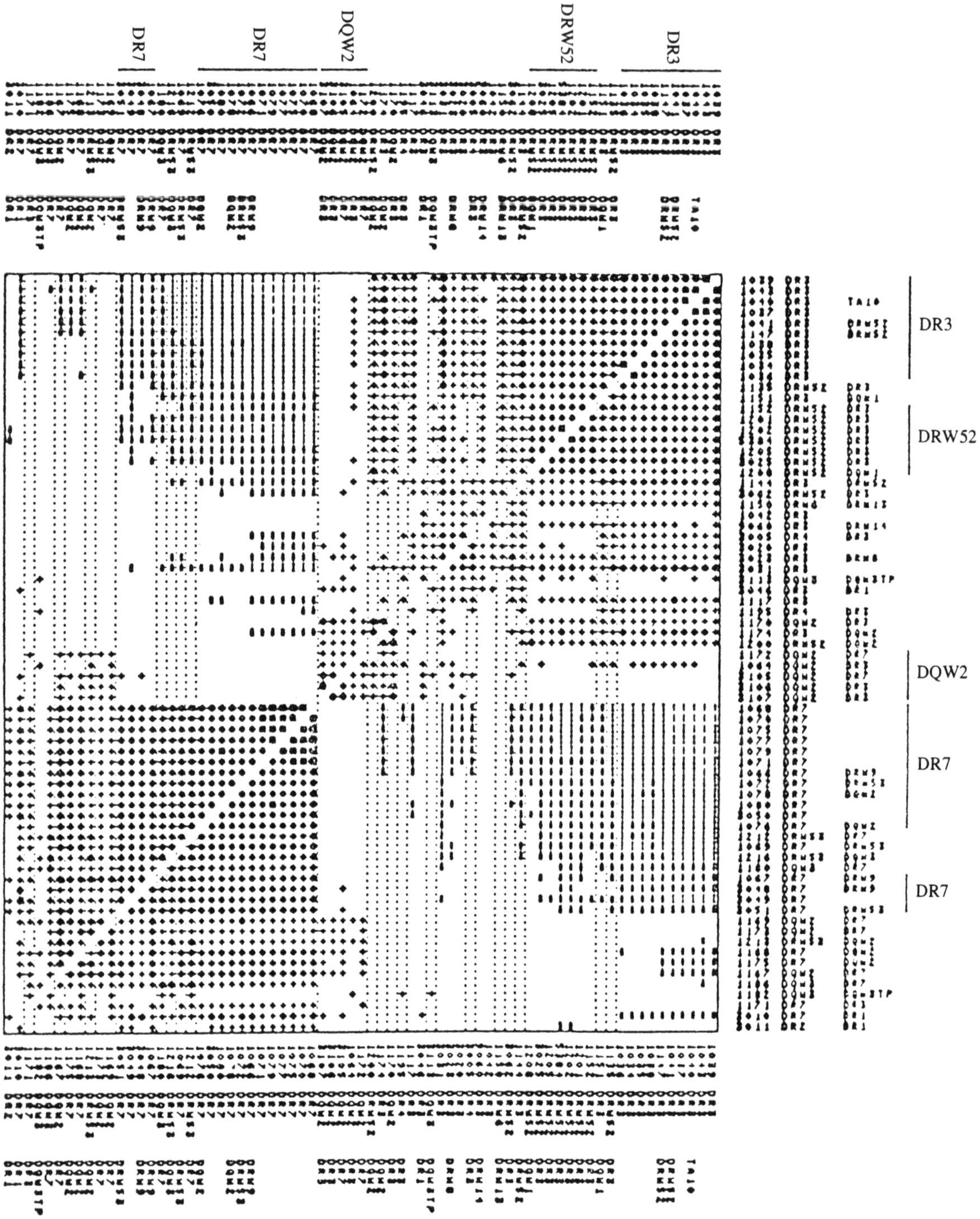

Table 3. Quality of 10W Serologic Reagents for the Definition of DQw2

10W ID	Type of Reagent	Origin	Name	r Value	Q Score	% Inclusion	Mean Strength	
1167	Core serum	NCY MRV	11028	0.90	6.2	90	6.6	+DR7
1168	Core serum	US6 BRN	CC413.1	0.83	5.5	81	5.8	+DR7
1175	Core serum	US2 MBC	BC.CA.MAGN	0.83	5.5	79	5.9	+DR7
1173	Core serum	FRA FAU	CHEVRIER	0.79	4.9	94	7.2	+DR7
3107	MAb	IT2 GAN	XIII358	0.82	4.8	83	6.0	
3106	MAb	IT2 FER	MP1	0.82	4.2	84	6.4	Human MAb
1169	Core serum	IT1 MTT	F102406	0.80	4.1	91	6.9	+DR7
5311	Ag. Soc.	IT2 GAN	CO1178	0.68	3.7	94	7.1	Submitted as DR3+DR7
1170	Core serum	GER BRA	422B	0.70	3.3	93	7.1	+DR3
1172	Core serum	BEN BER	MSD20	0.60	2.8	50	3.6	+DR7
1067	Core serum	FRA FAU	JEGU	0.71	2.8	86	6.7	+DRw9 Submitted as DR7
5314	Ag. Soc.	US6 BRN	CCB.1035.1	0.66	2.7	67	5.0	Submitted as DR7
5305	Ag. Soc.	UK1 LAW	WILLIAMS	0.63	2.6	59	4.5	Submitted as DR7
9107	Ag. Soc.	NCY GOE	PL2441	0.62	2.6	89	6.7	+DRw53
5297	Ag. Soc.	US8 FOT	STRATIS	0.69	2.5	80	6.1	Submitted as DR7
3105	MAb	GER WAN	3105	0.65	2.3	73	5.3	
1174	Core serum	IT1 MTT	F103295	0.58	2.2	82	6.3	+DR3
1213	Core serum	IT2 FER	FE200	0.57	2.1	90	6.9	Submitted as DRw53
1064	Core serum	FRA PRR	VEL1557	0.56	2.1	55	4.1	Submitted as DRw9
9105	Ag. Soc.	NCY GOE	PL2051	0.61	1.9	83	6.1	
9104	Ag. Soc.	NCY GOE	PL1758	0.53	1.6	77	5.7	
1171	Core serum	US1 DUQ	QUAGLIER	0.43	0.8	71	4.9	Weak

anti-DR3 sera, the third with anti-DR7 sera while being negatively associated with anti-DR3 sera; the second cluster is formed by three monoclonal antibodies and is positively correlated with part of the anti-DR7 sera and with the sera of the third cluster. By studying DR3 positive, DQw2 negative cells, and DR7 positive, DQw2 negative cells, it appeared that sera of the first cluster contained anti-DR3 antibodies (both long DR3 (DRw17) and short split (DRw18) and that sera of the third cluster contained anti-DR7 antibodies. The rare DRw9 and DQw2 positive cells react with reagents of all three clusters. Thus, it appears that a probable subdivision of DQw2 may be suggested on the basis of the reactivity of three monoclonal antibodies that miss part of the DR7 or DR3, DQw2 positive cells. However, we are still missing alloantisera containing DQw2 reactivity without anti-DR antibodies. The antigen DQw2 was well defined with the reagents of the Core set when DR3 and DR7 were not present at the same time on the same cells.

Combined Core Set and Antigen Society Serology

In order to better characterize the definition of DQw2 by Core and Antigen Society reagents, we performed a tail analysis with the same selection criteria used for the DR7 analysis on Caucasian, Negroid, and Oriental cells independently. The main results on Caucasian cells are indicated in Table 3. The best anti-DQw2 reagents in all ethnic groups were Core set sera that all contained anti-DR7 antibodies, as demonstrated by pattern on DR7 positive, DQw2 negative cells. Then the two monoclonal antibodies (3107 and 3106) gave rather good results, but they missed a number of DQw2 positive cells. These negative cells did not correspond to any defined subgroup and could be B8, DR3 cells as well as B18, DR3,

Table 4. Correlation Between Anti-DQw2 Reagents

			11111153913331115599 11
			11111053121111113311 01
			67676691010007771000 67
			75839771737650424554 41
1167	DQw2	DR7	8888776658767566665 54
1175	DQw2	DR7	8 77777665766646655555
1168	DR7	DQw2	87 7776565666656665554
1173	DQw2	DR7	877 866665765655556554
1169	DQw2	DR7	8778 66766666655565554
1067	DR7	DQw2	77766 6665665535555453
5297	DQw2	DRw9	776666 555554535554443
5311	DQw2	DRw53	6656765 65655555455543
9107	DQw2	DRw53	66666656 6654444455443
1213	DRw53	DQw2	555565556 543444443343
3107	DQw2		8767665665 87766666553
3106	DQw2	DR3	76666655548 7765666553
3105	DQw2		666565454377 655555453
1170	DQw2	DR3	7666655544776 75556553
1174	DR3	DQw2	54555335446657 5554643
1172	DQw2		666555554465555 665564
5314	DQw2		6665555444665556 65544
5305	DQw2	DRw11	65656555546655566 5554
9105	DQw2		655655455366564555 543
9104	DQw2	DRw11	5555544543554565555 43
1064	DQw2	DPw3	55555544445555464544 3
1171	DR7	DR3	454443333333333444333

Table 5. Characteristic Pattern of Reactions of Anti-DQw2 Serologic Reagents

Cell specificity						Major pattern of reaction													
DR3	DR7	DRw9	DQw2	DQw3	DQw4	1173	1169	1174	1170	3107	3106	3105	5311	5314	9107	5297	1213	9105	9104
+	−	−	+	−	−	+	+	+	+	+	+	+	+	−	+	+	+	±	±
+	−	−	−	−	+	−	−	+	+	−	−	−	−	−	−	−	−	+	−
−	+	−	+	−	−	+	+	+	+	+	+	+	+	+	+	+	+	+	+
−	+	−	−	+	−	+	+	−	−	−	−	−	+	+	+	+	+	−	−
−	−	+	+	−	−	+	+	+	+	+	+	+	+	+	+	+	+	+	+
−	−	+	−	+	−	+	+	−	−	−	−	−	+	−	+	+	+	−	−

or B44, DR7. It is interesting to note that 3106 is a human monoclonal antibody (11). The DR3-contaminated anti-DQw2 Core set reagents were less useful for the definition of this antigen. The best reagents in the Antigen Society were 5311, 5314, 9107, 5305, and 5297, with Q scores between 2.5 and 4. The correlation between anti-DQw2 reagents is presented in Table 4, and characteristic patterns of reactions are given in Table 5.

Conclusion

We still miss good alloantisera for the definition of DQw2. The definition of this antigen was better with the Core set than with the Antigen Society reagents in all ethnic groups. The most useful reagents were the two monoclonal antibodies 3107 and 3106 and to a lesser extent 3105. At the present stage of serology, it is possible to clearly define splits of DQw2 although this is suggested by low percentage of inclusion for several reagents, and as such, subdivisions have been demonstrated by other techniques.

References

1. Duquesnoy RJ, Marrari M, Annen K. Identification of an HLA-DR associated system of a B-cell alloantigens. Transplant Proc 1979;11:1757.
2. Park MS, Terasaki PI. Second DR locus. In: Terasaki PI (ed): Histocompatibility Testing 1980. UCLA Tissue Typing Laboratory, Los Angeles, 1980; p 578.
3. Tanigaki N, Tosi R. The genetic control of human Ia alloantigens: A three-loci model derived from the immunochemical analysis of "supertypic" specificities. Immunol Rev 1982;66:5.
4. Duquesnoy RJ, Marrari M. MB system joint report. In: Terasaki PI (ed): Histocompatibility Testing 1980. UCLA Tissue Typing Laboratory, Los Angeles, 1980; p 552.
5. Tanigaki N, Tosi R, Duquesnoy RJ, Ferrara GB. Three IA species with different structures and alloantigenic determinants in an HLA-homozygous cell line. J Exp Med 1983;157:23.
6. Shackelford A, Strominger JL. Identification of three class II antigens, DR7, MB2, and MT3, from a homozygous human cell line. Hum Immunol 1984;9:159.
7. Marrari M, Duquesnoy RJ. Antigen report: HLA-DQw2. In: Albert ED, et al. (eds): Histocompatibility Testing 1984. Springer-Verlag, Berlin, Heidelberg, 1984; p 207.
8. Bodmer JG, Kennedy LJ, Aizawa M, Dawkins RL, Lepage V, Mazzilli MC. HLA-D region monoclonal antibodies. In: Albert ED, et al. (eds): Histocompatibility Testing 1984. Springer-Verlag, Berlin, Heidelberg, 1984; p 217.
9. Schreuder GMT, Doxiadis I, Parlevliet J, Grosse-Wilde H. HLA-DR, DQ, LB, and TA10 specificities on Ninth Workshop homozygous typing cells. In: Albert ED, et al. (eds): Histocompatibility Testing 1984. Springer-Verlag, Berlin, Heidelberg, 1984; p 243.
10. Baur MP, Neugebauer M, Albert ED. Reference tables of two-locus haplotype frequencies for all MHC marker loci. In: Albert ED, et al. (eds): Histocompatibility Testing 1984. Springer-Verlag, Berlin, Heidelberg, 1984; p 677.
11. Pistillo MP, Hämmerling U, Dupont B, Ferrara GB. In vitro production of a human HLA alloantibody of restricted specificity (DQw2) via Epstein-Barr virus transformation. Hum Immunol 1986;15:109.

Antigen Society #27 Report (DQw1)

J.A. Falk,[1] A.M. Willshire,[1] J.G. Bodmer,[2] S.G.E. Marsh,[2] L.J. Kennedy,[2] T. Juji,[3] K. Matsuki,[3] R.J. Duquesnoy,[4] M. Marrari,[4] A.H. Johnson,[5] and H. Mervart[6]

History

The supertypic specificity now designated DQw1 (DC1, MB1, MT1 or LB12), was first described in the Seventh International Workshop (1). DQw1 was originally described as associated with DR1 + 2 + w6 and, later, DRw10 was included (2). Studies shoed that DQw1 epitope(s) resided on a subset of class II molecules (3), differing from the DR subset in both the and the subunits (4,5).

The Eighth and Ninth International Workshops described three DQw1-included specificities showing patterns of reactivity with:

1) Part of DR1 and DR2 (6)
2) DR1 + DR10 (2)
3) DR2 + DRw6 (2)

Reporting Laboratories: NCYFAL,[1] UKIBOD,[2] JAPJUJ,[3] USIDUQ,[4] USIHJH,[5] NCYMRV[6]

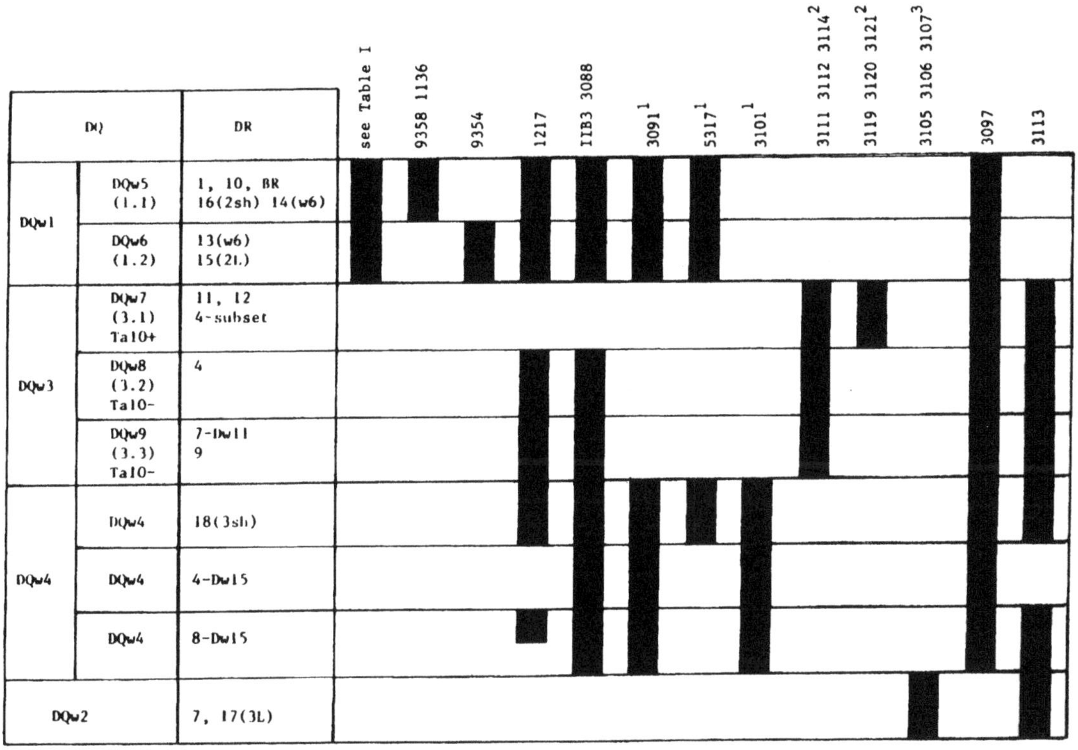

1 See report for AgSoc #35
2 See report for AgSoc #28
3 See report for AgSoc #26

Figure 1. Tenth Workshop DQ patterns.

However, it was deemed inconclusive as to whether these observations showed heterogeneity of DQw1 or were a result of some weakly reacting DQw1 antisera.

Serology (Fig. 1)

Best Sera for Definition. Many good DQw1 antisera were submitted to both the Core and Antigen Society sets, and Table 1 summarizes the best DQw1 antisera.

Splits of DQw1. In the Ninth International Workshop, the "DR1 + DRw10" pattern was defined by several antisera: 9W504, 509, 510, 511, 515, 516, 518, 521, and 522; the "DR2 + DRw6" pattern was defined by four antisera, 9W706, 776, 705, 778 (2). In this Workshop, two groups of antisera gave reaction patterns defining probable splits of DQw1:

DQw5: DR1 + DRw16 + DRw14 + DRw10 (i.e., "DR1 + DRw10")
DQw6: DRw15 + DRw13 (i.e., "DR2 + DRw6")

Our analysis was based on antigen designation for cells as submitted to the Workshop by the participating laboratories. There is still considerable difficulty in defining the antigens DRw16 (DR2 short) and the DRw14 split of DRw6 due to the low frequency of DRw16 in Caucasians and to the lack of specific antisera to DRw16 and DRw14. Therefore, reassignment of the DR specificities, based on the analysis and report of both the DR2 Antigen Society (AGSoc 22) and the DRw6 Antigen

Table 1. Best DQw1 Antisera

10WS#	Core r*	10WS#	AgSOC r*	Extras
1153	0.95	5327	0.95	DR2
1162	0.93	5342	0.91	
1154	0.92	5334	0.89	
1159	0.91	5331	0.88	
1160	0.91	5320	0.88	DR1,DR2?+
1161	0.86	5330	0.88	
3092	0.86	5328	0.88	DR2
1165	0.86	5333	0.87	DR2?
3091	0.86	5336	0.87	
1155	0.85	5335	0.85	
1143	0.85	9342	0.85	DQw4
		9352	0.83	DR3-DQw4
		5329	0.82	DR2?

*Based on Caucasian data only.

Society (AGSoc 25), is essential for full analysis of DQw1 cells.

DQw5. Two sera best defined this specificity, 10W9358 and 10W1136. 10W9358 (LY-1327) was previously described as directed against part of DQw1, a split including DR1, DR1x, DR2sh (DRw16), DRw10, and DRw14 (7). In this Workshop, the reactivity of this serum is concordant with DQw5, both when tested on the PBLs and on a panel of 100 homozygous typing cells, including 34 Tenth Workshop Homozygous Typing Cells (HTC). Extra reactions were detectable on HTCs expressing DR4-Dw14-DQw4, and DR9. (HTCs were tested at the Bodmer Laboratory, London, U.K. and the Falk Laboratory, Toronto, Canada.) 10W9358 was included in a group of sera described in the Tenth Workshop Newsletter, which, from data on testing a panel of CLL cells, fall into a subset predominately associated with DR1, DRw14, and DRw10 (8).

10W1136 was concordant with DQw5 when tested on HTCs and reacted with DRw16, DRw10, and DRw14 but only weakly with DR1 PBLs. When analyzed on our HTC panels, weak extra reactions with some DR7 positive HTCs were found. This serum was nonreactive with DRw14-DW16 HTCs, but was positive with DRw14-DW9 HTCs.

Two monoclonal antibodies not submitted to the Tenth Workshop, but tested in the Ninth Workshop–9W924 (Hu33) and 9W963 (HPA 49.11.100)–were found, when reanalyzed, to have reaction patterns typical of DQw5 (9,10).

DQw6. 10W9354, 10W5343, and 10W5337, on the basis of the Antigen Society data, defined the specificity DQw6. Extra weak reactions on HTCs suggest the possibility that one or more of these may be weak DQw1 reagents. One of these sera (10W9354) was included in a group of sera described in the Tenth Workshop Newsletter, which, from data on testing a panel of CLL cells, define a subset predominately associated with DR2 and DRw13 (8).

Three reagents, S5 (10W3087), NDS 15.38 (10W3012), and T3516, either excluded or not submitted to the Tenth Workshop, have been described in published data as having reactivity that defines DQw6 (11–13).

The heterogeneity of DQw1 is confirmed by Southern blot analyses in this Workshop. Tenth Workshop studies describe six DQB RFLP clusters that include a cluster (B) corresponding to DQw6 (DRw15, DRw13) and a cluster (J) corresponding to DQw5 (DR1, DRw14, DRw16). An independent study by Carlsson et al. describes unique DQ RFLP patterns supporting and extending the Workshop studies (14).

Further Associations

During analysis of DQw1 reagents, patterns of reactivity broader than DQw1 and including DQw3 and DQw4 determinants were identified (see Figure 1). The polymorphism defined by monoclonal antibodies IIB3 and Ta10 have been previously studied and assigned to DQB (15).

References

1. Bodmer JG. Ia Serology. In: Bodmer WF, et al. (eds): Histocompatibility Testing 1977. Munksgaard, Copenhagen, 1977; pp 351–357.
2. Ferrara GB, Park MS, Tanigaki N, Terasaki PI. Antigen Report: HLA-DQw1. In: Albert ED, Baur MP, Mayr WR (eds): Histocompatibility Testing 1984. Munich, Vienna, 1984; pp 205–207.
3. Tosi R, Tanigaki N, Centis D, Ferrara GB, Pressman D. Immunological dissection of human Ia molecules. J Exp Med 1978;148:1592.
4. Corte G, Calabi F, Damiani G, Bargellesi A, Tosi R, Sorrentino R. Human Ia molecules carrying DC1 determinants differ both in alpha and beta subunits from Ia molecules carrying DR determinants. Nature 1981;292: 357.
5. Shackelford DA, Mann DL, Van Rood JJ, Ferrara GB, Strominger JL. Demonstration of structural polymorphism among HLA-DR light chains by two-dimensional gel electrophoresis. Proc Natl Acad Sci USA 1984;78:4566.
6. Park MS, Terasaki PI. Second DR locus. In: Terasaki PI (ed): Histocompatibility Testing 1980. UCLA Tissue Typing Laboratory, Los Angeles, 1980; pp 578–582.
7. Gebuhrer L, Betuel H, Lambert J, Freidel AC, Farre A. A division of HLA-DQw1 associated with DR1, DR1x-DQw1, DR2 short, DRw10 and DRw14. I. Definition of alloantiserum LY1327. Hum Immunol 1987;18:235.
8. Taylor CJ, Bunce M, Ting A. Serological heterogeneity of HLA-DQw1: Definition using B cells from chronic lymphatic leukemia (CLL) patients. In: Dupont B, Barletta M, Flomenberg (eds): Tenth Workshop Newsletter No. 1, pp 6–7.
9. Kasahara M, et al. Serological dissection of HLA-D specificities by the use of monoclonal antibodies. Immunogenetics 1983;18:525.
10. Blankenhorn EP, Symington FW, Pickard A, McKearn TJ. Monoclonal antibodies that distinguish rat Ia antigens and haplotypes. Transplant Proc 1983;15:1569.
11. Johnson JP, Wank R. Biochemical dissection of DRw6-related epitopes using monoclonal antibodies. Hum Immunol 1986;16:69.
12. Fuggle SV, Kirkley J, Ting A, Morris PJ. A monoclonal antibody recognizing HLA-DR2 on malignant and activated cells. Immunogenetics 1980;18:415.
13. Satake M, Matsuki K, Yabe T, Suzuki M, Juji T. A possible split of antigen DQw1 identified by antiserum T3516. Tissue Antigens 1986;27:155.
14. Carlsson B, Wallin J, Bohme J, Moller E. HLA-DR-DQ haplotypes defined by restriction fragment analysis: Correlation to serology. Hum Immunol 1987;20:95.
15. Tilanus MGJ, Morolli B, van Eggermond MCJA, Schreuder GMTh, de Vries RRP, Giphart MJ. Dissection of HLA class II haplotypes in HLA-DR4 homozygous individuals. Immunogenetics 1986;23:333.

Antigen Society #28 Report (DQw3)

B.D. Tait,[1] M.S. Park,[2] A. Barbetti,[2] J. Langley,[3] P.I. Teraski,[2] and M. Aizawa[4]

DQw3 (formerly MB3) is a DQ locus specificity that is found in strong linkage disequilibrium with DR4 and DR5. It is also found on some haplotypes, including DRw6, DRw8, DR2S (DB9), DR9, and the Bw57-DR7 haplotype. At the Seventh International Histocompatibility Workshop in Oxford (1) where the DR series of antigens were serologically defined, a cluster of sera was described, termed WIA4, which was used to aid in the definition of DR4. Most of the sera in this cluster also reacted with DR5 cells and, in retrospect, seemed to define the antigen we now know as DQw3. However, at the time, the pattern of reactivity was attributed to multispecificity of the sera, or cross-reactivity, and the concept of a separate locus was not agreed upon.

Serologic evidence for a second B-cell locus was provided by Park et al. (2) who described the specificity Te22, which was later recognized as a subtype of DQw3, defined by a monoclonal reagent and termed TA10 (3). The TA10 subtype of DQw3 occurs on nearly all Caucasian DR5 haplotypes and approximately 40% of DR4 haplotypes.

Three alleles of the DQ series were described at the Eighth International Histocompatibility Workshop and termed MB1-MB3 (4). Four sera comprised one cluster, which appeared to define MB3 (DQw3). Subsequent immunochemical data supported the view that the antigen MB3 was the product of a locus separate from DR (5). The notation MB remained in use until the nomenclature committee officially named the series DQ, subsequent to the Ninth Histocompatibility Workshop (6).

In addition to the TA10 subtype of DQw3, another specificity termed LBE 13 has been described (7), which is included within DQw3. LBE 13 is a serologically short subtype of DQw3 that can be discriminated from the broad DQw3 sera by the lack of reactivity with DR7+, DQw3+, and DR9+, DQw3+ cells.

Subsequent to the Tenth International Workshop, the subtypes of DQw3 have been given official designations by the W.H.O. Nomenclature Committee (this volume). The TA10 positive and negative subtypes of DQw3 are now known as DQw7 and DQw8 respectively while the DQw3 allele on DR7 and DR9 haplotypes has been given the DQw9 designation. The LBE 13 serologic pattern therefore represents sera that contain both DQw7 and DQw8 antibodies.

The subtypes of DQw3 have significance in type I diabetes where it has been demonstrated both at the DNA level (8) and serologically (9) that the DR4 haplotypes in Caucasian DR3/4 patients almost exclusively include DQw8.

Reporting Laboratories: ANZTAI,[1] US5TER,[2] ANZCRS,[3] JAPAIZ[4]

Serology

Core Sera and Monoclonals. There were several good reagents, both allosera and monoclonals, that appeared to define DQw3 and its subtypes. These reagents are listed in Table 1.

Antigen Society Sera. Twenty-two sera were submitted to the DQw3 antigen society. Table 3 lists these sera together with their submitted and analyzed specificities. Antigen Society sera used in the definition of DQw3 and its subtypes are also listed in Table 1.

The sera and monoclonals listed in Table 1 fall into several groups based on their pattern of reactivity. The key sera for each group are highlighted with an asterisk.

Group I. This group includes the broad DQw3 reagents, i.e., those sera and monoclonals that define DQw7, DQw8, and DQw9. In addition to six Core Sera and four Antigen Society sera, there are four monoclonals that define DQw3 with high R values and good strength of reactivity. These monoclonals are 3066, 3111, 3112, and 3115.

Group II. This group of four Core sera, four Antigen Society sera, and one monoclonal show lack of reactivity with DR7-DQw3 and DR9-DQw3 cells (DQw9). They therefore appear to define DQw7 and DQw8. The reactivity of the monoclonal 3114 suggests that there is an epitope common to DQw7 and DQw8 that is absent on DQw9 molecules.

Group III. These are a group of allosera and four monoclonals that appear to define the TA10 subtype of DQw3; i.e., they are monospecific for DQw7.

Group IV. These are two allosera and two monoclonals that have broad DQw3 reactivity, but in addition react with cells that are DQw4 (DQWa). The two monoclonals 3035 and 3088 and the Antigen Society serum 5346 also contain reactivity to DQw1. This pattern of reactivity is different in IIB3, which appears to define an epitope common to DQw1 and DQw8 (10).

Group V. This group of six anti-sera appear to react with DQw3, but contain additional DRw53 activity. It is not possible to establish whether these sera contain DQw9 activity in addition to DQw7 and DQw8 since all DR7-DQw3 and DR9-DQw3 cells are positive with these sera by virtue of the DRw53 activity. Sera 1183 contains DR1 activity as well as DQw3 and DRw53, while serum 1186 contains DQw2 as an additional specificity. The fact that no monoclonal was found that reacted with DQw3 and DRw53 may indicate that the

Table 1. Allosera and Monoclonals that Define DQw3 and Subtypes DQw7, DQw8, and DQw9

	Group I														Group II								
	Core Sera						Monoclonals				Society Sera												
	*1176	*1179	*1184	1116	1188	1193	*3066	*3111	*3112	*3115	*5350	*5352	*5437	*5356	1118	1187	1177	1190	*3114	*5355	*5364	*5353	5357
DQw7	+	+	+	+	+	+	+	+	+	+	+	+	+	+	+	+	+	+	+	+	+	+	+
DQw8	+	+	+	+	+	+	+	+	+	+	+	+	+	+	+	+	+	+	+	+	+	+	+
DQw9	+	+	+	+	+	+	+	+	+	+	+	+	+	+	–	–	–	–	–	–	–	–	–
DQw4	–	–	–	–	–	–	–	–	–	–	–	–	–	–	–	–	–	–	–	–	–	–	–
DRw53	–	–	–	–	–	–	–	–	–	–	–	–	–	–	–	–	–	–	–	–	–	–	–
Extra Reactions																					DR5		

	Group III									Group IV				Group V					
	*1054	*1106	1107	*3119	*3120	*3121	3118	*5359	*5351	1108	3035	3088	5346	1216	1183	1180	1209	1186	1182
DQw7	+	+	+	+	+	+	+	+	+	+	+	+	+	+	+	+	+	+	+
DQw8	–	–	–	–	–	–	–	–	–	+	+	+	+	+	+	+	+	+	+
DQw9	–	–	–	–	–	–	–	–	–	+	+	+	+	(+)	(+)	(+)	(+)	(+)	(+)
DQw4	–	–	–	–	–	–	–	–	–	+	+	+	+	–	–	–	–	(+)	–
DRw53	–	–	–	–	–	–	–	–	–	–	–	–	–	+	+	+	+	+	+
Extra Reactions								DR5			DQw1	DQw1	DQw1		DR1			DQw2	

Table 2. Possible New Subtypes of DQw3

	?DQw7 Subtype	? New DQw3 Subtype	DQw7	DQw3	DQw7, DQw8
	5 5	5 5 5	5 5	5 5 5 5	5 5 5 5
	3 3	3 3 3	3 3	3 3 4 3	3 3 3 3
	6 6	6 4 6	5 5	5 5 3 5	5 5 6 5
	0 1	2 9 3	9 1	0 2 7 6	3 5 4 7
DR4/DQw3	− −	− − −	− −	+ + + +	+ + + +
	+ +	+ + +	+ +	+ + + +	+ + + +
	− −	+ + +	+ +	+ + + +	+ + + +
	− −	+ + +	− −	+ + + +	+ + + +
DR5/DQw3	+ +	+ + +	+ +	+ + + +	+ + + +
	− −	+ + +	+ +	+ + + +	+ + + +
DR7/DQw3	− −	+ + +	− −	+ + + +	− − − −
DR5/DQw3 Negative	− −	+ − −	+ −	− − − −	− − + +
DR7/DQw3 Negative	− −	− + −	− −	− − − −	− − − −

combination of these specificities in allosera may reflect immunization to two distinct epitopes on DQw3-DRw53 haplotypes rather than a shared epitope between these two specificities.

Possible Further Subdivision of DQw3

Sera 5360 and 5361 appear to further subtype DQw7. These sera react with the majority of DR5 DQw7 cells but give clearly negative results with a subpopulation (approximately 10 to 15%). The sera also react with approximately 20% of DR4, DQw7 cells. They do not contain DR5 activity as demonstrated by their clearly negative reactivity with DR5+ DQw3− cells, of which there are several examples in the American black population. These sera therefore appear to further subdivide DQw7, reacting preferentially with DR5 DQw7 cells.

Serum 5362 is a short DQw3 serum when analyzed against DR4 cells. All the DR4 DQw7 cells are included within its reactivity but in addition, it reacts to some DQw8 cells while being clearly negative with others. This serum in addition has DR5 activity. Sera 5349 and 5363 appear to react with the same subpopulation of DR4 DQw3 cells but do not contain DR5 activity. They do react however with the majority of DR5 cells by virtue of their DQw7 activity. Serum 5349 also contains DR7 antibody activity. The patterns generated with these sera are shown in Tables 2 and 3.

Linkage Disequilibrium

Caucasian. Nearly all Caucasian DR5 haplotypes are associated with DQw7 (formerly called TA10), which also occurs on approximately 40% of DR4 haplotypes. In particular, DQw7 appears to be in linkage disequilibrium with B44-DR4. The remainder of the DR4 haplotypes are associated with DQw8, which demonstrates linkage disequilibria with Bw60-DR4 and Bw62-DR4. In this Workshop there were several examples of DR5+, DQw3− cells but no convincing examples of DR4+, DQw3− could be found.

A population of DR7 haplotypes, namely Bw57-DR7, is associated with DQw9, which is also present on DR9 haplotypes.

Japanese. Among DR5 individuals approximately 90% are DQw7, the remainder appearing to be DQw8. Many Japanese DR4 haplotypes are associated with DQw4 (formerly DQWa). Among those that are DQw3, approximately 25% are DQw7, the remainder being DQw8.

American Blacks. Approximately 50% of DR5 individuals are DQw3, the other 50% are DQw1. Nearly all the DR5/DQw3 individuals are DQw7. Among DRw8 individuals approximately 50% are DQw3, the majority being DQw7. Insufficient DR4 cells were submitted for frequency evaluation. Examples of DR3/DQw3 and DRw6/DQw3 were also found in this racial group.

Correlation of Serology with RFLP and Biochemistry Studies

The three subtypes of DQw3 that can be defined serologically, i.e., DQw7, DQw8, DQw9, are associated with subtype specific RFLP fragments using a DQ probe. These associations are summarized in Table 4.

There also appears to be RFLP differences within DQw7, depending on whether it is found on a DR5 or DR4 haplotype (Table 4). There are two fragments, Kpn 1 12.8 Kb and Eco RV 14.5 Kb, which are seen with a DQ β probe and are present with DR4, DQw7 cells but not DR5, DQw7 cells. Examination of RFLPs obtained

Table 3. DQw3 Antigen Society Sera Analysis

Serum ID	Average Reaction	% Inclusion DQw3	Submitted Specificity	Analyzed Specificity
5344-GM316	6.5	–	DQw3	?DR4
5345-S15M21	1.6	–	DR2,DQw3	?LOW Frequency
5346-1265	7.7	94	DQw3	DQw3,DR5,DQw4,DQw1
5347-AUSTIN	7.4	–	DQw3	DQw1
5348-BC.DO.WAGN	5.2	73	DQw3	DQw3 (Weak)
5349-F103525	6.4	84	DQw3	?SPLIT,DQw3,DR7
*5350-MSD7	7.6	90	DQw3	DQw3
#5351-N1272	5.4	71	DQw3	DQw7
*5352-VR281051	7.8	93	DQw3,LBE13	DQw3
+5353-VR12505	7.5	91	DQw3,LBE13	DQw7,DQw8,DR5
5354-P19363	3.8	40	DQw3	DQw3 Weak
+5355-BIDGOOD.763	7.5	89	DQw3,DR4	DQw7,DQw8
*5356-E1075	7.6	79	DQw3	DQw3
5357.R5.6284	6.8	77	DR4,DR5	DQw7,DQw8
5358-R5-6346	1.9	10	DR4,DR5	?LOW Frequency
#5359-TER DR 22C	6.0	72	DR4,DR5,TA10	DQw7,DR5
5360-TER DR 22B	6.9	–	DR4,DR5,TA10	?DQw7 Subtype
5361-TER DR 22A	6.4	–	DR4,DR5,TA10	?DQw7 Subtype
5362-TORP718	7.3	91	DR4,DR5	?SPLIT,DQw3,DR5
5363-CWRP1381	6.9	71	DR4,DR5	?SPLIT DQw3
+5364-CHURCHILL 39	7.4	89	DR5,DQw3	DQw7,DQw8,DR5
*5437-TUA	7.2	89	DQw3	DQw3

* Key Sera DQw3; # Key Sera DQw7; + Key Sera Defining DQw9 (i.e., DQw7 plus DQw8).

on the same cells with a DR β probe suggests they are not the result of cross-hybridization of the DQ β probe with DR4 specific fragments.

The T-cell clone EM.C3 appeared to have specificity for the same population of DR4/DQw7 cells but in the absence of reactivity with DR5/DQw7, showing correlation therefore with the RFLP data (E. Mickleson, this volume). Both of these observations add support for the interpretation of a subtype of DQw7 based on the reactivity of sera 5360 and 5361. These sera appear to react preferentially with DQw7 on DR5 haplotypes.

Table 4. DQw3 Associated RFLP Patterns

RFLP	DQw7		DQw8			
Enzyme/Size Kb	7.1	7.2	8.1	8.2	DQw9	DQw4
Pst I 4.9	+	+	+	+	+	+
Taq 1 4.6	+	+	–	–	–	–
Bgl II 4.1	+	+	–	–	–	–
Bam HI 3.6	+	+	–	–	–	–
Kpn I 1.2	+	+	–	–	–	–
Kpn I 5.3	+	+	–	–	–	–
Pvu II 3.2	+	+	–	–	–	–
Hind III 3.4	+	+	–	–	–	–
Taq I 2.6	–	–	+	+	+	+
Bgl II 5.6	–	–	+	+	+	+
Taq I 1.9	–	–	+	+	+	+
Kpn I 19.1	–	–	+	+	+	+
Bam HI 10.3	–	–	+	+	+	+
Kpn I 12.8	–	+	–	+	±	–
Eco RV 14.5	–	+	+	+	–	–
Pst I 11.2	–	–	–	+	–	–
Msp I 2.0	–	–	–	+	–	–

The 11.2 Kb Pst 1 fragment and the Msp I 2.0 Kb fragment, while appearing to indicate further polymorphism within DQw8, may in fact indicate DX polymorphism. However, they are included in Table 4 for completeness.

The serologically well defined subtypes of DQw3, i.e., DQw7, DQw8, and DQw9, are all defined biochemically, each allelic product giving unique variant patterns (see Biochemistry report).

Summary

The DQ specificity originally defined as DQw3 consists of at least three subtypes that have been designated DQw7, DQw8, and DQw9. These subtypes can be defined serologically in the Workshop and, in addition, there are RFLP fragments and biochemical variants that are unique to each subtype. DQw7 and DQw8 correspond to the previously described TA10 positive and TA10 negative subtypes of DQw3 respectively. DQw9 is the DQ allele found on DR9 and some DR7 haplotypes. There is some serologic and RFLP evidence suggestive of further heterogeneity within DQw3.

Acknowledgments. The antigen society coleaders wish to acknowledge the expertise of all HLA technical staff who contributed data to the Tenth Workshop DQw3 study.

References

1. Bodmer JG, Pickbourne P, Richards S. Ia serology. In: Bodmer WF, et al. (eds): Histocompatibility Testing 1977. Munksgaard, Copenhagen, 1977; p 35.

2. Park MS, Terasaki PI, Bernoco D, Iwaki Y. Evidence for a second B-cell locus separate from the DR locus. Transplant Proc 1978;10:823.
3. Maeda H. Mouse monoclonal antibody detects a new polymorphic Ia determinant other than the HLA-DR antigen: A possible allele of DC-1. Tissue Antigens 1984;23:163.
4. Duquesnoy RJ, Marrari M. MB systems. In: Terasaki PI (eds): Histocompatibility Testing 1980. UCLA Tissue Typing Laboratory, Los Angeles, 1980; p 552.
5. Karr RW, Kannapell CG, Stein JA, Gebel HM, Mann DL, Duquesnoy RJ, et al. Molecular relationships of the human B-cell alloantigens MT2, MB3, MT4, and DR5. J Immunol 1982;128:1809.
6. Nomenclature for factors of the HLA system. In: Albert ED, et al. (eds): Histocompatibility Testing 1984. Springer-Verlag, Berlin, Heidelberg 1984; p 4.
7. Termijtelen A, Boettcher B, Bradley BA, D'Amaro J, Van Leeuwen A, Van Rood JJ. DR typing in Australian aborigines. An indication for a second locus in the HLA-D region defined by serology. Tissue Antigens 1980;16:140.
8. Serjeantson SW, Tait BD, Chiewsilp P, Gazit E, Hawkins BR, Kohonen-Corish MRJ. Insulin dependent diabetes mellitus. In: Aizawa M (ed): HLA in Asia Oceania 1986. Hokkaido University Press, Sapporo, Japan 1986; p 380.
9. Tait BD, Boyle A. DR4 and susceptibility to type I diabetes mellitus: Discrimination of high-risk and low-risk DR4 haplotypes on the basis of TA10 typing. Tissue Antigens 1986;28:65.
10. Schreuder GMTh, Maeda H, Koning F, D'Amaro J. TA10 and 2B3, two new alleles in the HLA-DQ region recognized by monoclonal antibodies. Hum Immunol 1986;16:127.

Antigen Society #29 Report (DRw52)

C.J. Taylor,[1] L. Ugozzoli,[2] N. Tanigaki,[3] R. Tosi,[4] M. Bunce,[1] A. Ting,[1] and G.B. Ferrara[2]

History

DRw52 was first recognized in the Seventh Workshop as a serologically defined group, seemingly cross-reactive with DR3, DR5, and Te50. In the Eighth Workshop this supertypic specificity, then called MT2, was shown to completely include DR3, 5, w6, w8, and DRw12 (then 8W13). The Ninth Workshop concluded that MT2 must be a member of the HLA-DR subset, although it might be on molecules other than those controlling the then designated HLA-DR specificities. MT2 was thought to be a supertypic antigen encoded by the DRB3 gene. On some haplotypes the gene product was expressed as a separate molecule to that of the private HLA-DR antigen encoded by the DRB1 gene. Thus, the MT2 specificity was formally designated DRw52. The high number was chosen to separate it from the subtypic HLA-DR specificities. More recently, heterogeneity of DRw52 has been defined by molecular,[1] biochemical,[2] cellular,[1] and serologic[3] analyses. It is now clear that DRw52 represents multiple epitopes that can be carried on the DRB1 and/or DRB3 gene products (Table 1).

Reporting Laboratories: UKIMOR,[1] IT2FER,[2] NCYRSP,[3] IT2TOS[4]

Results

The serologic profiles of the Society sera are shown in Table 2. This table gives the percentage of cells of each HLA-DR type to which the sera reacted. All cells that

Table 1. Haplotype Distribution, Gene Localization and Expected Serologic Profiles of DRw52

	Serologic Profiles of DRw52							
	HLA-DR Haplotype						DRB	
Specificity*	3	11	12	13	14	8	Gene	Serologic Profile
DRw52a (LB-Q4)	67†	0	50	32	20	0	III	<DR3,w12,w13,w14
DRw52b (LB-Q1,TR22)	33	100	50	45	80	0	III	DRw11,<3,w12,w13,w14
DRw52c	0	0	0	23	0	0	III	<DRw13 (DRw13-Dw19)
TR81	100	0	0	38	67	0	III	(DRB1 on DR3) DR3,<w13,w14
DRw52/R1	+	+	+	+	+	+	I & III	DR3,w11,w12,w6,w8
DRw52/R2	+	+	−	+	+	(+)	I	DR3,w11,w13,w14,(w8)
DRw52/R3	+	+	−	+	+	−	I & III	DR3,w11,w13,w14

*DRw52a,b,c defined by oligonucleotide typing: LB-Q4, LB-Q1 defined by cellular typing; TR81, TR22, DRw52/R1, R2, R3 defined by radioimmunoassay typing. DRw52b, LB-Q1, and TR22 have been found to have perfect correlation in the haplotype distribution, whereas DRw52a, LB-Q4, and TR81 show very similar but not identical distribution. DRw52/R1 conforms to DRw52-defined serologically in cytotoxicity assays.
†The values represent a provisional analysis of the percentage of cells positive (Termijtelen et al. Tenth International Histocompatibility Conference, 1987). DRB1 on DR3,11,w13,w14 and DRB3 on DRw52a-DR3,w6. DRw52b negative.

Table 2. Serologic Profiles of the DRw52 Antigen Society Sera

10WS Serum No.	HLA-DR Type DR3	DRw11	DRw12	DRw13	DRw14	DRw8
Group 1						
5372	4*	4	4	4	4	4
5373	4	4	4	4	4	4
5382	4	4	4	4	4	4
5371	4	4	4	4	4	4
5376	4	4	4	4	4	4
5379	4	4	4	4	4	4
Group 2						
5386	4	4	4	4	4	3
5377	4	4	4	3	4	3
5380	4	4	4	3	4	3
5366	4	4	3	3	3	3
Group 3						
5378	4	4	2	4	2	4
5369	4	4	3	4	3	4
5383	4	4	4	4	2	4
5374	4	4	2	3	3	3
Group 4						
5388	4	2	1	1	1	1
5389	4	2	1	1	1	1
Group 5						
5368	3	4	4	3	4	4
5385	3	4	4	3	3	4
5370	3	4	4	4	2	4
5381	3	4	4	4	4	4
Group 6						
5387	2	4	2	1	1	2
5365	2	4	2	1	1	1
5367	2	3	3	3	2	3
5390	2	2	2	1	1	1
Total No. of cells	110	39	28	64	20	32

*4 = 80–100% of cells positive, 3 = 50–79% of cells positive, 2 = 25–49% of cells positive, 1 = 0–24% of cells positive.

were DRw52 heterozygous or gave weak HLA-DR typing were excluded from this analysis. Group 1 (10WS 5372, 5373, 5382, 5371, 5376, and 5379) represent the best sera for the definition of the classical DRw52 (DRw52/R1, Table 1), being reactive with HLA-DR3, w11, w12, w13, w14, and w8 positive cells. Group 2 (10WS 5386, 5377, 5380, and 5366) sera did not react with all HLA-DRw13 and w8 cells. Group 3 (10WS 5378, 5369, 5383, and 5374) were negative with some HLA-DRw14 cells and, in addition, 5378, 5369, and 5374 were negative with most DRw12 cells. Group 4 (10WS 5388 and 5389) reacted strongly with DR3 cells and were predominantly negative with DRw11, w12, w13, w14, and w8 cells. These sera cluster together in a distinct group from the other sera in each ethnic group. In the Oriental population they had the highest correlation coefficient (r=0.9). It is possible that these sera react primarily with DR3 with some DRw52 reactivity, although no correlation with DRw52a, b or c was found when tested against a selected panel of Workshop cell lines defining these subtypes. Group 5 (10WS 5368, 5385, 5370, and 5381) showed strong reactivity with DRw11, w12, and w8 cells, and negative with some DR3 cells. Group 6 (10WS 5387, 5365, and 5367) showed strong reactivity with DRw11 cells, but gave lower percentage reactivity with DR3, w12, w13, w14, and w8 cells. These sera may represent the profiles expected for DRw52b, but no clear correlation with DRw52b was observed when tested against a selected panel of Workshop cell lines defining DRw52a, b and c.

Serologic analysis of the DRw52 Antigen Society sera using the direct binding assay with I^{125}-labeled preparations of HLA-DR molecules purified from Class II transfectants revealed distinct antibody populations defining DRw52a (TR81) and DRw52b (TR22) in several sera. The best sera for TR81 are 10w5372, 5373, 5382, and 5388, whereas those for TR22 are 10w5377, 5380, 5385, and 5390. However, these sera are not monospecific and contain antibodies for the broad DRw52 specificity as seen in Table 2. These sera could be made specific by appropriate absorption and then could be used in the cytotoxicity assay.

Core Serology and Monoclonal Antibodies

Forty-two reagents in the Core Serology demonstrated anti-DRw52 reactivity. The best reagents for the definition of the classical DRw52 were 10WS 1198, 1199, 1201, 1205, 1207, and 5384. 10WS 1202 was a good DRw52 reagent, but was weak or negative with most DRw8 cells. Many varied profiles could be seen with the Core DRw52-associated reagents, particularly the monoclonal antibodies. However, these profiles could not be correlated with the definition of DRw52 subgroups.

Two monoclonal antibodies were of particular interest. 10WS 3063 defined the DRw52/R3 antigen (Table 2), being negative for DRw8 and w12. This antibody (UK8.1), described in the Ninth Workshop (Bodmer et al.), has been shown previously to react with a DRw52-like epitope on the DRβI molecule, which is absent on the DRw8 and w12 haplotypes (Termijtelen, 10WS Newsletter, Vol 1; No. 6, 5). 10WS 3048 was specific for DR7, w9, w12 and, in addition, a subset of DRw52 cells. Analysis on a panel of Workshop cell lines typed for LB-Q1 (DRw52b) and LB-Q4 (DRw52a) revealed a high correlation with LB-Q4, although there were exceptions (Table 3). The antibody was positive with B8-DR3 (DRw52a) cells but negative with B18-DR3 (DRw52b) cells. These haplotypes have been shown to be in linkage disequilibrium with DRw52a and DRw52b respectively. The additional reactivity could be correlated with DRw52c, being associated with DRw13-Dw19.

Conclusion

DRw52 was clearly defined on both the Antigen Society and the Core serology typing trays. Many patterns of reactivity were observed with DRw52-related reagents, although their definition as splits could not be sub-

Table 3. Reactivity of Monoclonal Antibody 10WS3048 Against a Selected Panel of Workshop Cell Lines

10WS 3048 Negative Cell Lines						10WS 3048 Positive Cell Lines					
	HLA Type						HLA Type				
10WS ID.	DR	DQ	Dw	LB-Q†	DRB3	10WS ID.	DR	DQ	Dw	LB-Q	DRB3
9019	3	2	3	1	1	9021	3	4		4	3
9018	3	2	3	1	1	9022	3	2	3	4	3
9020	3	2	3	1	1	9023	3	2	3	4	3
9085	3	2	3	1		9086	3	2		4	
9037	11	3	5	1	1 + 2	9087	3	2	3	4	
9042	11	3	?	1	1	9088	3	2		4	
9039	11	3	?	1		9099	14	3	16	4	
9036	11	1	DB2	1	1	9064	14	3	16	4	3
9044	11	3	5	1	1 + 2	9065	13	1	18	4	3
9057	14	1			1	9058	13	1	18	4	3
9071	8	4	8.2	–		9097	13	1	19	–	
9068	8	4	8.1	–		9059	13	1	19	–	2
9100	8	4	8.2	–		9063	13	1	19		2
9101	8	4	8.2	–		9055	13	1	?	–	2
						9053	13	1	?		
						9056	14(13)	1		1	1 + 2
						9038	12	3	DB6	1	1
						9045	12(11)	3		1	1

†LB-Q1 and LB-Q4 in DRw52 Positive B-LCL. A. Termijtelen, W. Schroeyers, V. Jacobs. Reported in the Tenth Workshop Newsletter Vol. 1, No. 6, p 3–4. *DRw52 Subtypes defined by the Tenth Workshop Class II Biochemistry Study, reported in Princeton, November 1987.

stantiated in this Workshop. The direct binding assay demonstrated that most DRw52 alloantisera appear to consist of multiple antibody populations specific for the different DRw52 subtypes. The monoclonal antibody 10WS 3048 (PLM3, Japan-Juji) did define the "a" and possibly 'c' subtypes of DRw52 and, in addition, reacted with all DR7, w9, w12 cells.

References

1. Termijtelen A, Tilanus MGJ, Engelen I, Koning F, van-Rood JJ. Molecular localization of LB-Q1, a DRw52-like T-cell recognition epitope and identification at the genomic level of associated shared hybridizing fragments. Hum Immunol 1987;19:255.
2. Fuggle SV, Carter C, Watts F, Kirkley J, Morris PJ. Monoclonal antibody definition of multiple polymorphic epitopes on HLA-DRw52. Hum Immunol 1987;20:249.
3. Berte CC, Tanigaki N, Tosi R, Gorski J, Mach B. Serological recognition of HLA-DR allodeterminant corresponding to DNA sequence involved in gene conversion. Immunogenetics 1988;27:167.

Antigen Society #30 Report (DRw53)

A. Nuñez-Roldan,[1,2] F. Sanchez,[1] B. Sanchez,[1] H. Mervart,[3] and R. Winchester[2]

The HLA-DRw53 specificity, previously known as WiA4x7 and MT3, was originally considered to be expressed on DR4, DR7, DRw9, and possibly DRw10 cells (1). Based on prior biochemical studies of the class II molecules, the ideal concept of the DRw53 specificity is that it would be only expressed on the DR molecule containing the DRB4 chain encoded by, for example, DR4, 7, and w9 haplotypes (2,3). However, the serologic reality has thus far proven to be somewhat more complex than this ideal goal, as demonstrated by the results of this antigen society analysis.

In addition to the Core serum set, five sera were submitted for analysis by the Antigen Society, which was performed on 200 cells. The tails analysis reveals that four alloantisera and one monoclonal reagent give high correlations with DRw53 and among themselves, Tables 1 and 2. The percentages of inclusion of DRw53 are generally comparable among the ethnic groups, with the interesting exception that serum 1212 does not react with 62% of Black DRw53 individuals. This observation raises the question of whether there is a polymorphism

Reporting Laboratories: FRANRD,[1], US8RJW,[2] NCYMRV[3]

Table 1. The Most Specific Sera for the Definition of DRw53 Grouped by Correlation Scores[a]

	1209	1216	5394	3047
1212	6	7	7	6
1209	–	6	4	5
1216		–	6	6
5394			–	5
3047				–

[a]The scores are rounded values with 4, 5, 6, and 7, respectively, representing correlation coefficients with values of (0.36 to 0.45), (0.46 to 0.55), (0.56 to 0.65), and (0.66 to 0.75).

of DRw53 among w53-positive Blacks characterized by the absence of epitopes recognized by serum 1212.

Table 3 illustrates the percentage of inclusions of the best DRw53 sera with the different haplotypes associated with the DRw53 specificity in an attempt to examine the influence of potential ethnic differences in haplotypic organization. It demonstrates that the lack of reactivity with DRw53 among Blacks shown by serum 1212 is specifically accounted for by a lack of reactivity with DR4 haplotypes and not DR7 haplotypes, suggesting that there may be a polymorphism of the *DRB4* gene in certain DR4 haplotypes. In general, the sera did not distinguish among the serologic subtypes of the DR4 specificity with the possible exception of serum 1212, which reacted with only 29% of Black DR4.1 individuals, as opposed to 50% of Black DR4 individuals. More attention to this point is required.

Also shown in Table 3 is that individual sera preferentially react with either DR4 or DR7 haplotypes, as indicated by the ratios of the inclusion for DR4 and DR7 cells. Values of the DR4/DR7 inclusion percentage less than 1.0 suggest the presence of specificities selectively directed to determinants encoded by the DR7 haplotype, while ratios greater than 1.0 suggest additional specific reactivity with determinants encoded by the DR4 haplotype.

Another possible approach to defining the biochemical specificity of serologic reagents is illustrated in Table 4, which depicts reactivity on cells with the phenotype of HLA A1, Bw57, and DR7. It is presently an unresolved question whether these cells have the DR7 haplotype associated with Dw11 and characterized by the absence of an expressed DRw53 gene product.

Table 3. Percentage of Inclusion of Nominal DRw53 Sera with DR4, DR7, or DRw9 Cells from Different Ethnic Groups

	Serum 1212			Serum 1216			Serum 1209		
	C	B	O[a]	C	B	O	C	B	O
DR4	80	38	56	97	83	64	83	97	92
DR7	88	86	100	100	97	67	82	85	67
DRw9	100	76	92	50	82	89	50	88	97
Ratio									
DR4/DR7[b]	0.91	0.44	0.56	0.97	0.86	0.96	1.01	1.14	1.37

[a]C = Caucasian; B = Black; O = Oriental. The numbers of cells studied for DR4, DR7, and DRw9, respectively, in each ethnic group are C: 30, 17, 2; B: 60, 66, 17; O: 36, 3, 37.
[b]Ratio of inclusion percentages for DR4 cells to that for DR7 cells. Values less than 1.0 suggest preferential reactivity with DR7 cells, greater than 1.0 suggest preferential reactivity with DR4 cells. Note numbers in some comparisons (e.g., Oriental) are small.

Assuming that this is so, the results shown in Table 4 also emphasize the potential complexity of these sera. Serum 5394 is nonreactive with each of the cells, sera 1212 and 1216 are uniformly reactive, and serum 1209 and antibody 3047 give a parallel complex pattern. In the light of this interpretation, antibody 3047, as studied by cytotoxicity, appears to be analogous to antibody PL3 in that the epitope recognized by it may be present on the DR7 B1 chain. Supporting the absence of specificities in serum 5394 directed to the *DR7* haplotype *DRB1* gene product is the finding of no reactivity with 10/17 DR7 cells. Reciprocally, the presence of separate specificities for the *DR4* DRB1 gene product in this serum is suggested by reactivity with all but 2 of 30 DR4 cells.

The analysis revealed that there was no single "perfect" anti-DRw53 reagent in the Workshop, if by this one aspired to a reagent that reacted only with, for example, the product of the *DRB4*0101* gene. All appeared to have additional major serologic specificities to other gene products of, for example, either DR4 or DR7 haplotypes. In most this appeared to be the *DRB1* chain gene product. However, certain combinations of sera were shown to give perfect concordance as shown in Table 5 for Caucasians. In the Antigen Society study, no DRw53 cell was identified that was negative with all three sera and no DRw53 negative cell was positive with both 1212 and 5394. The use of cells transfected with single DR β chain genes has been shown to be a useful method to

Table 2. Best Quality DRw53 Reagents: Correlation Coefficient and Percentage of Inclusion with DRw53 Cells in Different Ethnic Groups

	Caucasian		Black		Oriental	
	r	% Incl.	r	% Incl.	r	% Incl.
1212	0.83	87	0.45	38	0.60	76
1209	0.62	84	0.44	86	0.72	97
1216	0.72	97	0.52	89	0.52	86
3047	0.65	89	0.27	64	0.47	80
5394	0.77	81%		–[a]		–[a]

[a]Not the principal specificity revealed upon tails analysis.

Table 4. Use of Cells Having the A1 Bw57 DR7 Phenotype, Which Presumably Are Dw11, DQw3, and Lack a Functional *DRB4* Gene Product, to Detect the Presence of Specificities Directed to Other Linked Gene Products

Cell Identification	Serum Number and Reactivity				
	1212	1216	1209	3047	5394
MRV 792	+	+	+	+	–
MRV 865	+	+	+	+	–
WAR 59	+	+	–	–	–
DCH 67	+	+	+	+	–

Table 5. Use of Combinations of Reagent Sera to Define DRw53 in Caucasians

Cell Designation	Serum Reactivity Pattern			No. of Cells Identified
DRw53	1212	5394	1216	
+	–	–	–	0
+	+	+	+	4
–	+	+	–	0
–	+	+	+	1

There are 60 DR53-negative cells and 51 DR53-positive cells.

demonstrate the presence of specificities to other than the β chain bearing the DRw53 specificity (2).

In addition to this complexity, it is likely that other factors interfere with a simple interpretation of the results. The previously described distribution of DRw53-related epitopes recognized by monoclonal antibodies is intricate, and serves as a model for interpretation. For example, each of three monoclonal antibodies that detect DRw53-like specificities has a distinct structural basis as has been shown in several laboratories (see reference 2): PL3 recognizes an epitope found on the common B4 chain present on haplotypes DR4, 7 and 9 as well as the B1 chain of the DR7 haplotype. Similarly, the Ab109d6 epitope is found on the same B4 chain, is also encoded by the B1 chain of DRw9 and DRw10, but is not found on the B1 chain of DR4 or DR7. The Ab17.3.3S epitope is present on the DR7 and DR4 B1 chain, but is absent from DRw9. Moreover, while *DR4*, *DR7*, and *DRw9* haplotypes all have the DRw53 B4 chain, biochemical analysis of the DRw10 haplotype does not reveal a B4 chain. The presumed explanation for the controversial inclusion of DRw10 among the DRw53-related specificities may rest on the fact that the third diversity region of DRw10 B1 is identical in sequence to the third diversity region of the DRw53 B4 (3).

In summary, some evidence suggesting the possibility of a polymorphism of DRw53 among Blacks was obtained, defined by the lack of reactivity with serum 1212. In general, no single alloantiserum exhibited a "pure" DRw53 specificity, with separate specificities for either DR4, DR7, or DQw3 being found in each. Several approaches to defining these additional specificities were illustrated. The use of combinations of alloantisera appeared to be a reliable approach to typing for DRw53.

References

1. Nakata S, Park MS, Omori K. MT3. Joint report. In: Terasaki PI (ed): Histocompatibility Testing 1980. UCLA Press, Los Angeles 1980; p 569.
2. Nuñez-Roldan A, Gregersen PK, Winchester RJ, Silver J. Analysis of class II genes from a DR5, DQw1 haplotype: implications for haplotype evolution. ASHI, 1986.
3. Merryman P, Gregersen PK, Lee S, Silver J, Nuñez-Roldan A, Crapper R. Nucleotide sequence of DRw10 β chain gene; identity of the third diversity region with that of the DRw53 allele of the β2 locus and as the probable site encoding a polymorphic MHC class II epitope. J Immunol 1988;140:2447.

Antigen Society #31 Report (Qa and TL-Like Antigens) Part 1: The Leiden "Class IV" or "Class I-like" Workshop*

A. Van Leeuwen,[1] J. D'Amaro,[1] R. Fauchet,[2] A. Longo,[3] E. Gazit,[4] C. Navarrete,[5] P. Richiardi,[6] M.M. Tongio,[7] E.J. Yunis,[8] and J.J. Van Rood[1]

Introduction

The aims of the miniworkshop were the following: 1) to compare two PHA blast preparation techniques with the T gamma (T_G) separation technique and to submit a proposal for a common technique to be used at the Tenth International Histocompatibility Workshop, and 2) to identify and compare the specificities of anti-PHA blast sera (1–4) submitted for the Tenth International Histocompatibility Workshop and the Leiden anti-TCA sera (5,6). The teams from Boston, Genoa, Leiden, London, Rennes, Strasbourg, Tel-Aviv, and Turin met in Leiden to test the 57 anti-blast sera submitted for the Qa-Tla Antigen Working group, Antigen Society 31. These sera and several anti-T cell subset sera (7,8), including the anti-TCA1 and anti-TCA2 sera were tested with two different PHA blast preparations and a T_G-enriched suspension from the same individual. A total of 25 donors were tested. All donors were typed for HLA-A, B, C, DR and for TCA.

Results

In the original protocol, Dr. Fauchet, Dr. Gazit, and Dr. van Leeuwen offered to culture PHA blasts from the Leiden donors, according to the methods described by Fauchet et al. (9) and Gazit et al. (10). Unfortunately, the PHA blasts prepared by Dr. Gazit did not survive the

Reporting Laboratories: BENROO,[1] FRAFAU,[2] IT2FER,[3] GERGAZ,[4] UKIFES,[5] IT1CEP,[6] FRAMYE[7]
Participating Laboratory: NCYYUN[8]

*A miniworkshop for the Antigen Society 31 held in Leiden, November 20–23, 1986.

Immunobiology of HLA
Volume I

Table 1. Work Plan of the Eight Teams

Teams	Team Code	Tested Cells: F-blasts	G-blasts	T_G	PBL
Rennes	BF/FF	X			
Genoa	EF	X			
Turin	RF	X			
Strasbourg	TF	X			
Leiden	AG/AG	X	X	X	X
Boston	YG		X		
Tel-Aviv	GG		X		
London	LG		X		

BF = tests performed in Rennes; FF = tests performed in Leiden.

overseas flight, so only the blasts cultured and frozen by Dr. Fauchet in Rennes and those prepared by Dr. van Leeuwen in Leiden, according to the Gazit protocol, were used in this Workshop. Five teams tested the 57 anti-blast sera and the 14 anti-T cell subset sera, i.e., seven anti-TCA sera and seven anti-T cell subset sera, against the "Fauchet" (F) blasts and four teams tested the same anti-sera against the "Gazit" (G) blasts (Table 1). Finally, all sera were tested against T_G lymphocytes by the Leiden team. Each team used his own technique. A comprehensive report will be published in the January 1988 issue of *Tissue Antigens*. We have used the term F-HA to label the serum clusters because of the high degree of concordance among the results obtained with Fauchet's (F) blasts. The HLA-A like specificities are called HT (3) and HA (9) by Gazit and Fauchet, respectively.

Five clusters of sera were strongly associated with the HLA-A1 (three sera), -A2 (six sera), -A3 (two sera), -A9 (four sera), and -A10 (five sera). Table 2 shows the composition of these serum clusters. All sera were tested with PHA blasts, peripheral blood (PBL) and T_G-enriched lymphocyte suspensions. For the blasts specificity the results of the teams that produced the highest r values were used for this table. It is clear that some sera contain HLA-A antibodies when tested with PBL and T_G cells, such as sera 4866 and 4867 in cluster F-HA2. These sera were probably not sufficiently absorbed before they were submitted to the International Workshop.

However, the majority of the clustered sera do not show significant values for the respective HLA-A like specificities when tested with PBL and T_G cells. Nevertheless, we are still facing the question: Are we really working with antibodies different from the classical HLA-A antibodies? Gazit et al. (3) described an alloantiserum SF 48 (serum no. 4844, this Workshop) with HLA-A3 like reactivity. After exhaustive absorptions with HLA-A3 positive platelets, they still were able to show reactivity with HLA-A3 positive blasts. Fauchet et al. (9) showed that even after using sequential immunoprecipitation by a monoclonal anti-HLA-A3 antibody (six cycles) followed by precipitation with serum 101RE (serum no. 4890, this Workshop), she could still show 41K and 12K molecules. Serum 4859 also helps to answer the question. This serum was produced by Ferrara by planned immunization. Both, the cell donor and recipient carried HLA-A2. These three

Table 3. Comparison of the Reactions Obtained with Anti-T Cell Subset Sera

Cells	TCA1	TCA2	Fe110/13	Fe132/3	Fe133/7	Fe134/7	Fe140/7
1	+	–	–	–	–	–	–
2	+	–	–	–	–	–	–
3	+	–	+	+	+	–	–
4	+	–	–	–	–	+	–
5	+	–	–	–	–	–	–
6	+	+	+	+	–	+	+
7	+	+	+	+	–	–	–
8	+	+	+	–	–	–	–
9	+	+	+	–	–	–	–
10	+	+	+	–	–	–	–
11	+	+	+	+	–	–	–
12	+	+	–	–	–	–	–
13	+	+	–	–	–	–	–
14	+	+	–	–	+	+	–
15	+	+	+	+	–	–	–
16	+	+	–	–	–	–	–
17	+	+	+	+	–	–	–
18	+	+	+	–	–	+	–
19	+	+	–	–	–	–	–
20	–	+	+	+	–	–	–
21	–	+	+	+	–	–	–
22	–	+	+	+	+	+	+
23	–	+	+	+	+	–	–
24	–	+	–	–	–	–	–
25	–	+	–	–	–	–	–

Table 2. Clusters of Anti-Blast Sera that Show HLA-A Specificities by Testing Them with PHA Blasts

Cluster	10th IHWS Serum No.	Blasts Specificity	r	PBL Specificity	r	T_G Specificity	r
F-HA1	4863	A1(BF)[a]	0.614	DRw11	0.572	A1	0.020
				A1	0.000		
	4877	A1(TF)	0.745	A29	0.316	Negative	
				A1	0.000		
	4879	A1(TF)	0.672	DRw11	0.239	DRw13	0.231
				A1	0.000	A1	0.000
F-HA2	4859	A2(FF)	0.817	Aw34	0.239	Bw4	0.448
				A2	0.000	A2	0.091
	4866	A2(BF)	0.834	DR7 1e IT	0.563	A2	0.740
				A2 2e IT	0.427		
	4867	A2(FF)	0.902	A2	0.625	A2	0.375
	4868	A2(BF)	0.913	Negative		A2	0.241
	4888	A2(AF)	0.822	Bw22	0.239	Negative	
				A2	0.000		
	4889	A2(TF)	0.739	Negative		Bw62	0.317
						A2	0.261
F-HA3	4844	A3(TF)	0.739	Negative	DR7	0.262	
						A3	0.162
	4890	A3(BF)	0.797	Negative		A3	0.352
F-HA9	4870[b]	A9(BF)	0.875	Bw22	0.239	DR1	0.407
				A9	0.153	A9	0.000
	4871[b]	A9(BF)	0.875	Negative		Bw58	0.316
						A9	0.037
	4892	A9(FF)	0.808	Bw52	0.479	B39	0.479
				A9	0.000	A9	0.000
	4900	A9(AF)	0.794	Negative		DR4	0.200
						A9	0.191
F-HA10	4880	A25(BF)	0.572	Bw58	0.479	A25	0.316
				A10	0.239		
	4881	A25(TF)	0.719	Negative		A25	0.405
	4883	A25(TF)	0.717	A9	0.352	Negative	
				A10	0.000		
	4884[b]	A25(FF)	0.563	DR5	0.352	Negative	
				A10	0.000		
	5429	A25(BF)	0.572	DR7	0.405	A10	0.432
				A10	0.000	A25	0.118

[a]( ) indicate the team that performed the test (see Table 1). [b]Sera contain strong anti-PHA reactivity as described by Nunez-Roldan and Winchester (11). IT = Iteration.

All r values in this table are the highest values for the particular sera. Values for r are also given for the corresponding HLA-A specificity, using PBL and T_G lymphocytes. Tests with PHA blasts revealed five serum clusters with highly significant associations with different HLA-A specificities. Cluster F-HA1 (three sera) has a good correlation with HLA-A1. However, tests with PBL and T_G-enriched suspensions did not reveal that specificity. Cluster F-HA2 (six sera) is highly associated with HLA-A2. Note that in this cluster some sera (4866 and 4867) also show anti-HLA-A2 specificity with PBL and/or T_G lymphocytes. Both sera in cluster F-HA3 were negative when tested with PBLs. Serum 4890, tested with T_G lymphocytes reacted weakly with two of six HLA-A3-positive panel cells. Cluster F-HA9 (four sera) shows strong associations with HLA-A9 when tested with HLA-A9-positive PHA blasts. Using PBLs only one out of five HLA-A9-positive panel cells yielded a positive reaction with serum 4870. In the F-HA10 cluster (five sera), only sera 4881 and 4883 show strong correlation with HLA-A25, using PHA blasts. Sera 4880, 4881 and 5429 show some HLA-A10 reactivity when tested with PBL and/or T_G cells. However, the reactions were not significantly correlated with HLA-A10.

examples demonstrate that it is very unlikely that these anti-blast sera show classical anti-HLA reactivity.

One of the aims of this Workshop was the comparison between the anti-blast sera and the anti-TCA sera. The anti-TCA sera react mainly with large granular lymphocytes (LGLs) or T_G lymphocytes. In normal individuals, these anti-TCA sera define a di-allelic system, different from any HLA loci. Because it was already known that these anti-TCA do not react similarly with PHA blasts, these sera were not submitted for the "anti-blast" Antigen Society 31 of the International Histocompatibility Workshop.

Anti-TCA sera and some of Ferrara's anti-T cell subset sera were tested against PHA blasts and T_G-enriched lymphocyte suspensions. Tests with PHA blasts revealed that no sera appeared to recognize the TCA deter-

Table 4. Expression (Detection) of HT/HA and TCA by Different Cell Preparations

	HT/HA[a]	TCA
PBL	Negative	Positive (T_G cells)
T-PHA	Positive	System not detectable
B cells lines	Positive	Negative
Thymocytes	Negative	Positive
Increased expression in leukemias		
B-cell malignancies	Yes	No
T-cell malignancies	Yes	Yes
Myeloid malignancies	Yes	Yes

[a]HT and HA represent Gazit's and Fauchet's nomenclature for the HLA-A-like determinants.

minants. Consequently, it is concluded once again that the anti-blast sera were different from the anti-TCA sera. Ferrara's anti-T cell subset sera also did not show any significant correlations with the anti-blast sera.

A comparison of Ferrara's anti-T cell subset sera and TCA1 and TCA2 sera is set out in Table 3. Note the very striking similarity of Fe110/13 and Fe132/3 and their inclusion (except cell no. 3) in TCA2. Table 4 summarizes the different expression of HT/HA and TCA in different cell preparations. The expression of the determinants from one system, i.e., HT/HA seems to exclude the expression of those from the other system, TCA, except for leukemic cells.

The conclusions of this miniworkshop were the following:

1. HT as described by Gazit et al. (3) and HA from Fauchet et al. (9) are probably similar and may be the human homologue of Qa in the mouse.
2. The HLA-A like blast antigens are different from the TCA antigens (5,6).
3. TCA is very likely the homologue of the mouse TL locus (van Leeuwen et al., manuscript in preparation).

Acknowledgments. We would like to thank all participants for their enthusiasm and we are very grateful to Mr. P. de Lange for computer analyses and to Mrs. I. Curiël for typing the manuscript.

This workshop is supported in part by the Dutch Organisation for Pure Research (ZWO), the J.A. Cohen Institute for Radiopathology and Radiation Protection (IRS) (grantnr. 3.3.1), the following Foundations: Jo Keur (MFL), Macropa, Leiden University Alumni Fund (LUF), Developmental Research IHB (SOFI), Holland Biotechnology B.V., and the National Reference Centre for Histocompatibility Testing.

References

1. Crepaldi T, Amoroso A, Curtoni SE, Richiardi P. Alloantisera with apparent HLA-A specificity reacting only against PHA activated cells. Ricerca Clin Lab 1986;16: 449–456.
2. Fauchet R, Charron DJ, Bouhallier O, et al. Class I MHC-linked determinants expressed on activated T cells. In: Albert ED, Baur MR, Mayr WP (eds): Histocompatibility Testing 1984. Berlin, Heidelberg: Springer-Verlag 1984:450–454.
3. Gazit E, Terhorst C, Yunis EJ. The human "T" genetic region of the HLA linkage group is a polymorphism detected on lectin activated lymphocytes. Nature 1980;284:275–277.
4. Mitsuishi Y, Falkenrodt A, Urlacher A, Tongio MM, Mayer S. New human MCH class I antigens segregating with HLA-A antigens detected on some lymphocyte subpopulations. Human Immunol 1986;15:175–197.
5. Van Leeuwen A, Festenstein H, Van Rood JJ. Di-allelic alloantigenic systems on subsets of T cells. Human Immunol 1982;4:109–121.
6. Van Leeuwen A, Giphart MJ, De Groot G, et al. Two different T-cell systems in humans, one of which is probably equivalent to Qa or Tla in mice. Human Immunol 1985; 12:235–246.
7. Ferrara GB, Strelkauscas AJ, Longo A, McDowell J, Yunis EJ, Schlossman SF. Markers of human T cell subsets identified by alloantisera. J Immunol 1979;123:1272–1277.
8. Ferrara GB, Longo A, Tosi R, Biagini A, Mennucci M, Carminati G. HLA segregation of human T cell subpopulation markers. Human Immunol 1981;2:213–223.
9. Fauchet R, Boscher M, Bouhallier O, et al. New class I in man: serological and molecular characterization. Human Immunol 1986;17:3–20.
10. Gazit E, Gothelp Y, Gil R, et al. Alloantibodies to PHA-activated lymphocytes detect human Qa-like antigens. J Immunol 1984;132:165–169.
11. Nunez-Roldan A, Winchester R. PHA-specific antibodies in Antigen Society 31 (Qa-T1) antisera. XIHWS Newsletter 1987, no. 4.

Antigen Society #31 Report Part 2: Antigen Society #31 Report

E. Gazit,[1] R. Fauchet,[2] M. Jones,[3] A. Van Leeuwen,[4] A. Longo,[5] R. Mahoney,[6] C. Navarrete,[7] P. Richiardi,[8] M.M. Tongo,[9] R. Altshuler,[1] O. Bouhallier,[2] S. Balboni,[5] M. Belvedere,[10] S. Cappelacci,[11] T. Crepaldi,[8] G.B. Ferrara,[5] M. Hammond,[12] P. Lulli,[11] M. Martinetti,[10] M. Savi,[5] J. D'Amaro,[4] E.J. Yunis,[13] and J.J. Van Rood[4]

History

The discovery that alloactivated T lymphocytes express new class II antigens (1,2) stimulated experiments in which activated lymphocytes were studied for the expression of new antigens not detected at the quiescent stage. Other workers have shown that resting T lymphocytes could be subdivided by the use of sera recovered from patients with juvenile rheumatoid arthritis (3) or from alloimmunized volunteers (4,5). Later (6), TCA-TCB system expressed on T gamma-enriched cells was also identified by alloantisera. When PHA-activated lymphocytes were used in the screening of pregnancy sera, it appeared that some sera reacted exclusively with the lectin-activated lymphocytes, but not with the resting T or B lymphocytes separated from the same individual (7). Cross-absorption experiments indicated that, indeed, these determinants were not shared by the resting autologous lymphocytes. In blocking experiments, it was shown that the new determinants are associated with B-2 microglobulin, which classified them into the class-I gene product family. Their absence from thymocytes suggested that they may be the human counterpart of the murine Qa-like gene products (8). Family studies have shown that the reactivity segregated with HLA, which maps it to chromosome VI (7). When the lymphocytes used for screening were assigned their HLA phenotype, linkage disequilibrium with HLA was observed, specifically with locus A gene products. Thus, some sera exhibited linkage disequilibrium with HLA-A3,A10,A2,A9 or HLA-A1 (7–10). It was originally suggested by Gazit et al. to term them HT (human T) because of their similarity with the mouse T-region gene products (8) or later H-A by Fauchet et al. (9) because they were expressed by PHA or alloactivated on T cells and B-cell lines. In biochemical experiments, it was shown that the antigenic determinant which was precipitated by the specific alloantibody was a 41-12 K dimer distinct from the HLA class I 44-12 K antigen (7). Sequential immune precipitation with the w6/32 or HLA-A3 monoclonal antibodies did not remove the antigenic reactivity of this determinant, indicating that it is different from HLA-ABC antigens. Taken together, the reports published so far indicate that PHA activation, alloactivation, or beta interferon stimulation (11) induces the expression of new non-HLA class I antigens.

This system is distinct from HLA for the following reasons. The reactivity is not absorbed by either platelets or resting T or B lymphocytes. Lysostripping with HLA alloantibody does not remove the reactivity (7). There is a linkage disequilibrium with HLA; however, there are cells that do not express the linked HLA antigen, but do react with the serum. The molecular weight of the heavy chain is 41 to 42 K and not 44 K, which is typical for the heavy chain of HLA. To date, all efforts have failed to produce a murine monoclonal antibody that specifically reacts with the PHA-activated lymphocyte and precipitates the antigen.

Reporting Laboratories: GERGAZ,[1] FRAFAU,[2] UKILAW,[3] BENROO,[4] IT2FER,[5] NCYMAH,[6] UKIFES,[7] IT1CEP,[8] FRAMYE[9]

Participating Laboratories: IT1MRA,[10] IT2GAN,[11] SAFDUT,[12] NCYYUN[13]

Table 1. Cellular Targets Tested in the Workshop

Name of Target	No. of Cells
1) Peripheral blood leucocytes	70
2) Nylon wool purified T-cells	324
3) Nylon wool purified B-cells	270
4) PHA activated lymphocytes	423
5) Alloactivated lymphocytes	30
6) EBV transformed B cell lines	76
7) Leukemia T-cell lines (HPB, Jurkat, MOLT-4, 8402, PEFR, HSB-2)	6
8) Leukemia lymphoblasts	
T-cell acute lymphoblastic (ALL)	11
Common acute lymphoblastic (C-ALL)	22
Acute Myeloid (AML)	33
Chronic lymphoid (CLL)	15
Unclassified	7

Results

This is the first international HLA workshop in which these novel antigens have been studied. The objectives

Table 2. PHA Activation Protocol

Gazit (8)	Fauchet (9)
PBL + PHA (purified)	PBL + PHA (crude)
3 days	2 days
IL-2	IL-2
4 days	2 days
Harvest + freeze	Harvest + freeze
Cytotoxicity Testing	
1 + 2 hours	1 + 1 hours

Table 3. Patterns of Reactivities of the Alloantibodies

	Group 1	Group 2	Group 3	Group 4	Group 5	Group 6	Group 7
Peripheral blood lymphocytes	−	+	−	−	−	−	−
T lymphocytes	−	+	−	−	−	−	−
B lymphocytes	−	+	+	−	−	+	+
PHA T cells	−	+	+	+	+	+	−
Allogenic T cells	−	+	+	+	+	+	−
EBV cell lines	−	+	+	+	−	−	−
Number of sera	5	13	25	14	6	1	1
	↓	↓	↓				
Possible interpretation	negative	anti-HLA-A, -B	anti-HLA-DR, -DQ		New Class 1 markers ?		B cell

were to compare the different protocols of PHA T-cell preparations to define the linkage with the classic HLA antigens and to identify the serologic clusters. Fifteen Laboratories took active part in these studies. The protocol required a study of 20 to 30 unrelated selected cells and two or three families including different lymphocyte targets: PBL or T cells, PHA T cells, B cells, and EBV cell lines. Additional targets were selected as an optional study. They included thymocytes, pathologic cells, leukemia T-cell lines, and leukemia lymphoblasts. Sixty-nine platelets, allosera, and six monoclonal antibodies were submitted by eight laboratories. Classic class I and class II sera were added as control (Fe1-Fa7: HLA-A2, HLA-DR4,Fe3-Fa9: HA10, Fe2-Fa8: negative sera). Table 1 summarizes the lymphocyte targets that were used. Two protocols were followed for the preparation of the activated lymphocytes (Table 2). The difference between Gazit's and Fauchet's protocols lies in the use of crude PHA by Fauchet and length of the tissue culture (a total of 7 days in the first protocol and 4 days in the second). From preliminary experiments (5), it appeared that the crude PHA is superior for the preparation of blasts, but it has not been decided whether short- or long-term culture is superior. In Table 3, the overall pattern of the reactivities is summarized. It is clear that several sera reacted like HLA ABC or DR antibodies. They were subsequently identified and removed from the analysis. The serum by serum (SxS) analysis resulted in 10 clusters, 4 of which were found to be the classic HLA. The remaining six clusters are summarized in Table 4. It is worth mentioning that some sera that reacted as classic HLA antibodies clustered with the corresponding sera, which showed linkage disequilibrium with a particular HLA antigen. In four clusters, there is a significant linkage disequilibrium with HLA, and the r value is given in Table 4. There is

Table 4. PHA Activated Lymphocytes Form Clusters of Reactivity*

Cluster No.	Serum No.	HLA Association	R value	Reacting with EBV	Reacting with Thymocytes	Reacting with T-cell Lines
	4875	A1	0.26	−	+	+
1	4877	A1	−	−	+	−
	4879	A1	0.29	−	−	−
2	4888	A−2	0.45	+	+	−
	4889	A−2	0.38	+	+	+
	4844	A−3	0.39	++	−	+
3	4890	A−3	0.55	+	−	+
	9378(BRAN)	A−3	0.27	+	−	+
	4880	A−10	0.29	+	−	+
4	4881	A−10	0.25		−	−
	4887	A+10	0.22		−	−
	4889	A−10	−		−	−
	4850	−			++	−
5	9379(SCHN)	−			+	+
	9380(SPIE)	−			+	+
	4880	−			+	+
	4838	−			−	+
6	4839	−			−	+
	4840	−			+	+
	4842					
	4843					

*ALL sera listed in this table reacted positively with PHA-activated lymphocytes.

Table 5. The Reactivity of Sera from Cluster 3 with Leukemia Lymphoblasts

Serum Number	Leukemia Type T-ALL (n = 12)	AML (n = 28)	c-ALL (n = 16)	B-CLL (n = 14)
4844	NS	NS	0.027	0.054
4890	NS	0.017	0.027	0.011
BRAN	NS	NS	NS	NS

The number in the table is the p value, which was calculated for the numbers of cells reacting with the sera in the cluster and having HLA-A3. NS = nonsignificant.

a cluster that is associated with HLA-A1, A2, A3, and A10. The sera in the clusters were analyzed by their reactivities with B-cell lines, thymocytes, T cell lines, segregation in families and reactivities with leukemia lymphoblasts (Tables 4 and 5). Two groups of sera from clusters 2 and 3 were found to react with B-cell lines, and cluster 5 reacted with thymocytes (Table 4).

Leukemia T-cell lines were found to react with sera in clusters 3, 5, and 6. The segregation of the reactivities was studied in several families, and clusters 1 to 4 were found to segregate with HLA. ALL, which classified as c-ALL and were HLA-A3-positive, reacted with sera in cluster 3 (HT-3,), i.e., the reactivity was associated with the presence of the HLA-antigen, and not with the leukemia per se.

Discussion

The experiments performed in this Workshop successfully identified and defined six clusters of non-HLA alloantibodies, four of which were in linkage disequilibrium with HLA. Sera in the remaining two clusters also reacted with thymocytes, T-cell lines, and some T-ALL lymphoblasts. Thus, it is possible to divide the clusters into two groups: Qa-like and the TL-like antigens. In family studies, the reactivities of most clusters segregated with HLA. Clusters 2 and 3 also reacted with EBV-transformed B-cell lines, and cluster 5 reacted with T-cell lines and thymocytes. The reactivity of these non-HLA antibodies with human thymocytes extends the findings in early reports (12–14). Common ALL lymphoblasts in HLA-A3-positive individuals reacted with sera in cluster 3.

These results indicated that the serologic aspects of the new system are almost resolved, even though the reproducibility of the tests have not, as yet, been worked out. Unfortunately, the biochemistry experiments failed. Therefore, no biochemistry data were presented in this Workshop. Also, Qa-Tla-like probes were not assayed in the molecular biology experiments of this workshop.

It is hoped that immune precipitation and Southern blot analysis will be studied in future workshops using specific monoclonal antibodies in biochemistry experiments and non-HLA class I probes in DNA experiments.

References

1. Evans RL, et al. J Exp Med 1978;148:1440.
2. DeWolf WC, Schlossman SF, Yunis EJ. J Immunol 1979;122:1780.
3. Strelkauskas AJ, et al. J Immunol 1978;120:1278.
4. Ferrara GB, et al. J Immunol 1979;123:1272.
5. Van Leeuwen A, et al. Tissue Antigens 1979;14:437.
6. Van Leeuwen A, et al. Tissue Antigens 1987;31:33.
7. Gazit E, Terhorst C, Yunis EJ. Nature 1980;284:275.
8. Gazit E, et al. J Immunol 1984;132:165.
9. Fauchet R, et al. Human Immunol 1986;17:3.
10. Mitsuishi Y, et al. Human Immunol 1986;15:175.
11. Gazit E, et al. Human Immunol 1987;18:53.
12. Ferrara GB, et al. Clin Immunol Immunopathol 1982; 222:428.
13. Mahoney RJ, et al. J Immunol 1982;129:263.
14. Park S, et al. Human Immunol 1988;22:151.

Antigen Society #31 Report Part 3: Leukemic Blasts Express New HLA Class I-Like Alloantigens

D. Peruccio,[1] T. Crepaldi,[1] C. Castagnoli,[1] M. Lecchi,[1] E. Lovisone,[1] P. Saracco,[1] E. Olivetti,[1] and P. Richiardi[1]

New HLA, β2m-associated alloantigens, undetectable on resting T and B cells and platelets, are detected on PHA-activated lymphocytes (PHA-L) using platelet-absorbed alloantisera (1–4). The expression of these new HLA class I-like specificities was examined in newly diagnosed acute leukemias. Bone marrow blasts obtained at onset and peripheral blood lymphocytes during remission were tested by the complement-dependent lymphocytotoxicity technique using platelet-absorbed alloantisera. All leukemia samples (26 cALL, 6 T-ALL, 28 ANLL) were tested with locally selected alloantisera (1), while only a limited number (8 cALL, 3 T-ALL, 19 ANLL) was examined with the alloantisera submitted

ITICEP[1]

Table 1. Correlation Between Reactivity of Anti PHA-L Sera and HLA-A Antigens in a Panel of PHA-L and ANLL

Serum	Correlated HLA Antigen	PHA-L		ANLL					
		r	p	++	+−	−+	−−	r	p
XW4875		0.468	**	1	3	2	6		n.s.
XW4877	A1	0.620	***	1	3	1	12		n.s.
XW4879		0.716	*	0	4	2	11		n.s.
TORP2222		0.717	***	1	4	0	17		n.s.
XW4866		0.679	***	8	1	1	9	0.786	**
XW4867		0.791	***	6	3	0	10	0.716	**
XW4868		0.708	***	6	3	1	9	0.584	**
XW4888	A2	0.432	*	4	5	0	10	0.544	**
XW4889		0.314	*	5	3	0	10	0.693	**
TORP1504		0.557	***	11	4	1	11	0.650	**
TORP902		0.558	***	10	6	0	12	0.645	**
TORP1952		0.786	***	8	8	0	12	0.548	**
XW4844	A3	0.615	**	1	1	3	13		n.s.
XW4890		0.880	***	2	1	0	15	0.791	*
XW4870		0.664	***	6	3	0	10	0.716	**
XW4871	A9	0.632	**	7	2	0	10	0.805	**
XW4892		0.556	*	6	3	0	10	0.716	**
XW4900		0.756	**	7	2	1	9	0.685	**
XW5429	A10	0.827	*	no HLA-A10+ cells					
TORP820	A10	0.894	***	0	3	0	25		n.s.

n.s. = not significant. * = $< 10^{-2}$. ** = $< 10^{-3}$. *** = $< 10^{-4}$. +− = missing reaction. −+ = extra reaction.

Table 2. Expression of HLA-A2 and HLA-A2 Correlated Class I-like Antigens on Different A2+ Leukemias Evaluated by Absorption Experiments

Absorbing Sample Cell	Cells Type	Cytotoxic Reactivity with Class I-like Antisera	[a]$Abs_{50}(x10^{-6})$ of Sera		
			[b]11.53	[c]TORP 902	[c]TORP 1952
10	B-CLL	−	200	> 400	> 400
5	B-CLL	−	120	> 400	> 400
4	B-CLL	−	100	n.d.	> 400
20	ANLL	−	20	200	> 400
30	cALL	−	30	> 400	> 400
39	cALL	−	100	> 400	> 400
37	cALL	−	n.d.	200	> 400
16	ANLL	+	45	> 400	n.d.
62	ANLL	+	30	n.d.	180
24	ANLL	+	40	60	150
34	cALL	+	25	75	100
128	cALL	+	10	35	65
33	cALL	+	n.d.	75	50
01	PBL	−	25	> 400	> 400
01	PBL	−	20	> 400	> 400
01	PHA-L	+	15	70	90
02	PHA-L	+	12	70	90
A2 neg.	ANLL	−	> 400	> 400	> 400

[a]Abs_{50} = Dose of cells used to reduce cytotoxic titer of 1 ml of serum to 50% of the original titer (mean of test against 5 positive target cells)
[b]Anti HLA-A2 antiserum.
[c]HLA-A2 correlated class I-like alloantisera preabsorbed with 100×10^9 platelets/ml serum.
n.d. = not done.

for the Tenth Histocompatibility Workshop (Antigen Society #31). When tested against a panel of PHA-L (n = 70) the reactivity of the sera listed in Table 1 was significantly correlated with HLA-A specificities.

The following results were obtained against ANLL blasts (Table 1). The majority of the tested sera were significantly correlated with the same specificities identified against the PHA-L panel. No such correlation was found for the Workshop alloantisera when tested against cALL and T-ALL samples, probably because of the small number of cell samples. However, the reactivity of the locally selected alloantisera examined against the entire panel of cALL and T-ALL was significantly correlated with the expected HLA-A antigen. Altogether, approximately 50% of the acute leukemias carrying the correlated HLA-A specificity reacted positively with the selected antisera independently of the cell type or the differentiation stage (FAB classification); all the sera correlated with the same HLA-A specificity showed, with few exceptions, a concordant (positive or negative) pattern against each leukemia sample. Six cases of B-CLL, including A2+, A9+ and A10+ samples, were never found positive.

These results were confirmed by absorption studies. Two HLA-A2 correlated class I-like antisera were absorbed using increasing amount of different A2+ leukemia samples (Table 2). The residual reactivity of the sera was tested against 3 A2+ leukemias and 2 A2+ PHA-L. Only ANLL and ALL blasts that expressed the new class I-like specificities could absorb the reactivity of the platelet-absorbed alloantisera both against leukemic blasts and PHA-L; thus, class I-like allospecificities detected on leukemic blasts are the same or very similar to those expressed on PHA-L. On the contrary, the reactivity was not abrogated by blasts that were negative with anti-PHA-L antisera by the lymphocytotoxicity technique, even using high amounts of absorbing cells (4×10^8 blasts/ml of serum).

Quantitative absorption experiments of a typing anti-HLA-A2 alloantiserum (11.53) were performed to test whether the ability to react with class I-like antisera was correlated to a high amount of HLA-A molecules on the cell surface (Table 2). No such correlation was found. In fact, leukemic blasts that were unreactive with platelet-absorbed anti-class I-like antibodies were nonetheless capable to completely remove the anti-HLA-A2 antibodies using the same absorbing dose (3 to 5×10^7/ml serum) as class I-like positive blasts. Thus, the expression of class I-like specificities appears to be independently regulated from that of HLA-A molecules.

The new class I-like specificities were not detectable on peripheral blood lymphocytes of the leukemic patients during remission. However, PHA-L from all the patients, including those whose leukemic blasts did not express class I-like specificities, were capable of reacting with anti-class I-like antisera. Thus, the apparent absence of class I-like specificities on the cell surface of some leukemic blast samples can be explained by a particular differentiation stage or activation state of the leukemic clone, rather than by a genetic defect (5).

The clinical relevance of these results is, for the moment, unclear. It is suggested that β2m-associated HLA molecules may represent markers of leukemic blast activation and/or maturation state.

Acknowledgments. This work was supported by MPI 60 and by CNR, Progetto Finalizato Tecnologie Biomediche. D. Peruccio, M. Lecchi, and E. Lovisone are recipients of a fellowship from "Comitato Ghirotti." The excellent technical assistance of Mr. D. Marsico is gratefully acknowledged.

References

1. Crepaldi T, Peruccio D, Lecchi M, et al. New HLA class I-like alloantigens expressed on blast cells. J Immunogenetics 1987;14:219.
2. Fauchet R, Boscher M, Bouhallier O, et al. New class I in man: serological and molecular characterization. Human Immunol 1986;17:3.
3. Gazit E, Terhorst C, Yunis E. The human "T" genetic region of the HLA linkage group is a polymorphism detected on lectin-activated lymphocytes. Nature 1980;284:275.
4. Mitsuishi Y, Falkenrodt A, Urlacher A, Tongio MM, Mayer S. New human MHC class I antigens segregating with HLA-A antigens detected on some lymphocyte subpopulations. Human Immunol 1986;15:175.
5. Peruccio D, Crepaldi T, Lovisone E, et al. HLA class I-like antigen expression on human leukemic cells. Tissue Antigens 1987;30:76.

Antigen Society #32 Report (DP)

A.H. Johnson,[1] M.S. Park,[2] J.G. Bodmer,[3] L. Kennedy,[3] S. Marsh,[3] G.B. Ferrara,[4] A. Bratlie,[5] S. Goldmann,[6] B.K. Jakobson,[7] H. Mervart,[8] G. Mueller-Eckhardt,[9] A. Svejgaard,[7] E. Thorsby,[5] and K. Tsuji[10]

HLA-DP is defined conventionally using primed lymphocyte typing (PLT) cells. These PLT cells are of limited availability. Six DP alleles are recognized by the WHO Nomenclature Committee, and account only for approximately 70% of the DP antigen frequency. Four to six additional DP alleles have been defined by PLT and/or DNA sequence analysis, but have not received W.H.O. nomenclature recognition (1–4). Recently, several groups have used restriction fragment length polymorphism (RFLP) to define DP alleles (5–9). Using three restriction enzymes, MspI, TaqI, and BglII and DP beta cDNA probes, DPw1, 2, 4, and 5 can be defined with a single specific fragment (5). Serologic definition of HLA-DP would be of enormous benefit for investigating the biological significance of HLA-DP in transplantation and disease association. Several alloantisera that have a high correlation coefficient with an HLA-DP allele have been described (10–13); however, the frequency of alloantisera which contain DP reactivity is very low. The first DP antisera to be defined were the result of planned immunization (10,11). These antisera were shown to react with the DP molecule using the "lysostripping" technique. Cytotoxic reactivity was abolished when a DP monomorphic monoclonal antibody, but not when a DR or a DQ monoclonal antibody, was used to "lysostrip" molecules from the cell surface. Subsequently, multiparous antisera that correlate with HLA-DP have been reported (12,13). Although two apparently monospecific alloantisera have been identified [SOW=DPw1 (10); FE.LD3bis=DPw3 (11)], the majority of these sera appear to react with supertypic DP determinants. These supertypic apperns correlate with DPw3+w6 (12), DPw2+w4 (13), and DPw1+w2+w3 (13). In addition, one monomorphic DP monoclonal antibody B7\21 (14), and three DP polymorphic monoclonal antibodies have been described. Monoclonal antibody ILR-1 reacts with DPw2, DPw3, DPw6, and DR5 (15). Monoclonal antibody DP11.1 was raised by immunization of a mouse with a DP-transfected L-cell and reacts with DPw2 and DPw4 (16). Monoclonal antibody SP4 reacts with DPw2+DPw4 and with DPw1+DPw5 at a lower titer (12).

The purpose of this Antigen Society was to identify additional HLA-DP antisera and determine the reproducibility of HLA-DP antisera reactivity. Forty antisera were submitted to Antigen Society #32. In addition, the following controls were included: 1) a negative control, 2) a positive control (DP monomorphic monoclonal antibody B7\21), 3) hidden negative control, 4) hidden duplicate antiserum (FE134/6), and 5) a hidden DR antiserum control. Three antisera that had been shown previously to correlate with HLA-DP and to react with the DP molecule using the "lysostripping" technique were included in the set (5398, 5399, and 9114). Many of the antisera that were included in this Antigen Society were submitted to the Workshop as DR new or unknown. The remaining antisera were included because they had unusual reaction patterns which did not correlate with HLA-DR or DQ specificities. The serum analysis was performed on approximately 300 cells submitted to the Workshop, which were typed for DP and run on both the Antigen Society antiserum set and on the core antiserum set. Table 1 summarizes the results from Antigen Society #32 antisera. Two new antisera (5400, 5401) were identified in addition to those antisera that were known already to correlate with HLA-DP (5398, 5399, 9114). Antiserum 5400 (VR70311) correlates with DPw3 and DRw11. Antiserum 5401 (VR67864) correlates with DPw3, with many of the residual reactions being DRw8. Analysis of data from individual laboratories participating in Antigen Society #32 confirmed these data and found a stronger correlation of 5401 with DPw3 and DRw8. The reactivity of 5398 (FE 134/6) correlated with DPw3 and DRw11, as had been determined previously (5) Antiserum 5399 (FE LD3bis) also correlated

Reporting Laboratories: US1HJH,[1] US5TER,[2] UKIBOD,[3] IT2FER,[4] SCATSB,[5] GERGOL,[6] SCASVE,[7] NCYMRV,[8] GERMUE,[9] JAPTSU[10]

Table 1. Antigen Society Antisera: Correlation with HLA-DP

WS#	REG	LAB	I.D.	Specificity	Correlation with DP r	Correlation with DP χ^2
5400	BEN	ROO	VR70311	DR5(w11) + DPw3	0.21	10.86
5401	BEN	ROO	VR67864	DPw3 + > DRw8 > DR5	0.16	7.62
5398	IT2	FER	FE.134\6	DR5(w11) + DPw3	0.31	25.27
5399	IT2	FER	FE.LD3.bis	DPw3 + > DR5	0.30	27.29
9114	SCA	TSB	SOW	> DPw1	0.47	59.96

Immunobiology of HLA
Volume I

Table 2. HLA-DP Serum × Serum Correlation

	5398	9117	5400	5401
5398		6	6	4
9117[a]	6	5	3	
5400	6	5	4	
5401	4	3	4	

[a]9117 hidden duplicate of 5398.

Table 3. Core Antisera: Correlation with HLA-DP

WS#	REG	LAB	Specificity
3109	US	TER	DPw4, DPw2+
9001	IT2	FER	DPw3, DPw6
9002	IT2	FER	DPw3, DPw6, DR5
1103	US	WAL	DPw3, DPw6, DR5

with DPw3, as had been determined previously (5). Antiserum 5399 had the fewest false-negative reactions (5/86) with the putative DP specificity. In addition, in the Workshop, antiserum 5399 reacted with a low percentage of DR5 cells. As would be expected, the above antisera form a cluster on serum by serum analysis (Table 2). The cluster includes the duplicate sera 5398 and 9117. Antiserum 9114 (SOW), which is known to be a DPw1 antiserum, was weak but still gave a correlation with DPw1. Two additional antisera could potentially contain DPw2 reactivity (5410 and 5419). The primary reactivity of these sera is DRw52, DQw1, DR2; however, they both contain DPw2 in the "tail" analysis. Due to the high frequency of reactivity of sera 5410 and 5417, it is difficult to draw any definite conclusion about the potential DPw2 reactivity.

Prescreening data on core antiserum using B-lymphoblastoid cell lines from Dr. Min Park's laboratory are summarized in Table 3. Two additional DP antisera could be identified in the core antiserum set. Antiserum 3109 correlated with DPw4 and DPw2. Antiserum 1103 correlated with DPw3, DPw6, and DR5. Antisera 9001 and 9002 are identical to 5399 and 5398. Note that Dr. Park detected additional DPw6 activity that was not detected in the entire Workshop cell analysis. This is most likely explained by inadequate assignment of DPw6 to cells in the entire workshop analysis.

In conclusion, this Workshop demonstrated that HLA-DP can be detected serologically. However, antisera are difficult to identify and much of the time these antisera also contain DR reactivity. Interestingly, a majority of antisera that correlate with DP react with DPw3 (? + DPw6) and DRw11. The reason for this is not clear. One possibility is that antisera against DPw3 may be found more frequently because DPw3 may share an epitope(s) with DRw11. Support for a shared epitope between DRw11 and DPw3 comes from absorption data with antiserum FE134/6. Both the DR and the DP reactivity from antiserum FE134 \ 6 can be absorbed with either a DRw11+,DPw3− or a DRw11−,DPw3+ cell (Johnson, unpublished observations). A correlary of this observation could be that many DRw11 antisera contain some degree of DPw3 reactivity, which could account for the relative difficulty in finding DRw11 monospecific antisera. Monoclonal antibody ILR-1 also reacts with DR5 in addition to DPw2, 3, and 6. Further support for shared epitopes between DP and DR is that more monomorphic monoclonal antibodies react with DR+DP than either DR+DQ or DQ+DP. One might hypothesize that DR5(w11) has acted as the donor gene for gene conversion-like events in the generation of DP polymorphism.

Among the multiple reasons that more HLA-DP antisera have not been identified are the following: 1) HLA-DP alleles share epitopes with each other. All monoclonal antibodies and most alloantisera, to date, react with two or more DP alleles. Thus, one should expect to find supertypic DP antisera with a relatively high frequency. The proposed evolution of DP beta based on RFLP indicates three related DP groups: DPw2+DPw4, DPw3+DPw5+DPw6, and DPw1. DPw1 is more closely related to DPw3+DPw5+DPw6 (5). 2) HLA-DP is more polymorphic than defined by PLT. DNA sequence analysis of amplified DNA has identified subtypes of DPw4 and DPw2 (4). In addition, two sequences have been described for DP alpha (4), which introduces another level of potential polymorphism within DP (17). 3) HLA-DP is present in a much lower quantity on the cell surface relative to other class II molecules and thus is less immunogenic. 4) There is a large percentage of DP "blank" alleles. 5) Many laboratories that currently screen for HLA antisera do not have HLA-DP-typed panels.

Future strategies for HLA-DP serology should include production of monoclonal antibodies using cell lines transfected with DP genes as demonstrated by Heyes et al. (16). In addition, because planned immunization has produced the most reliable antibodies perhaps one might consider using DP-transfected cell lines to immunize chimpanzees since chimpanzees have been shown to produce high quality DR antisera (18).

References

1. Nose Y, Nakatsuji T, Tsuji K, Reinsmoen NL. A new HLA-DP (SB) specificity (Cp63) defined in Japanese and Caucasian cell populations by cloned PLT cells. Human Immunol 1985;13:145.
2. Termijtelen A, Naipal-Van Berge S, Suwandi-Thung L, Van Rood JJ. Recognition of DP determinants with typing reagents prepared with lymphocytes from Dutch unrelated individuals. Tissue Antigens 1985;26:234.
3. Odum N, Hartzman RJ, Jakobsen BK, et al. The HLA-DP polymorphism in Denmark investigated by local and international PLT reagents. Human Immunol 1986;28:105.
4. Bugawan T, Horn G, Hansen J, et al. Analysis of HLA-DP allelic sequences polymorphism using the in vitro enzymatic DNA amplification of DP and DP loci. In: Dupont B (ed): Immunobiology of HLA, Volume I: Histocompatibility Testing 1987. New York: Springer-Verlag, 1989.
5. Hyldig-Neilsen JJ, Morling N, Odum N, et al. Restriction fragment length polymorphism of the HLA-DP subregion and correlations to HLA-DP phenotypes. PNAS 1987; 84:1644.
6. Bodmer J, Bodmer W, Heyes J, et al. Identification of HLA-

DP polymorphism with DP and DP probes and monoclonal antibodies: correlation with primed lymphocyte typing. PNAS 1987;84:4596.
7. Mitsuishi Y, Urlacher A, Tongio MM, Mayer S. Restriction fragment length polymorphism of DP genes defines three new DP alleles. Immunogenetics 1987;26:383.
8. Baas EJ, Bontrop RE, Otting N, Giphart MJ. Tissue Antigens 1987;31:5.
9. Maeda M, Inoko H, Ando A, Uryu N, Nagata Y, Tsuji K. HLA-DP typing by analysis of DNA restriction fragment length polymorphisms in the HLA-DP subregion. Human Immunol 1988;21:239.
10. Johnson AH, Thorsby E, Nakatsuji T, Moen T, Hartzman RJ. Recognition of a polymorphic DPw1 alloantiserum raised by planned immunization. Human Immunol 1986; 17:21.
11. Johnson AH, Tang T, Nunez G, et al. Identification and molecular characterization of HLA-DP alloantisera. In: Aizawa M (ed): The Proceedings of the Third Asia-Oceania Histocompatibility Workshop Conference. Sapporo: Hokkaido University Press, 1986:612.
12. Park MS, Takenoushi T, Terasaki P, Tonai R, Barbertti A. HLA-DP region complexity by CDC, RFLP and cellular assays. In: Dupont B (ed): Immunobiology of HLA, Volume I: Histocompatibility Testing 1987. New York: Springer-Verlag, 1989.
13. Tanigaki N, Tosi R, Parodi B, Sorrentino R, Ferrara GB, Strominger JL. Detection of HLA-DP serological allodeterminants by the use of radioiodinated DP molecules. Eur J Immunol 1987;17:743.
14. Royston I, Omary MB, Trowbridge IS. Monoclonal antibodies to a human T-cell antigen and Ia-like antigen in the characterization of lymphoid leukemia. Trans Proc 1981; 13:761.
15. Shaw S, DeMars R, Schlossman SF, Smith PL, Lampson LA, Nadler LM. Serologic identification of the human secondary B cell (SB) antigens: correlations between function, genetics and structure. J Exp Med 1982;156:731.
16. Heyes J, Austin P, Bodmer J, et al. Monoclonal antibodies to HLA-DP transfected mouse cells. PNAS 1986;83:3417.
17. Ward FE, Johnson AH, Metzgar RS, Strong DM, Cresswell P. Chimpanzee antibodies to purifies human Ia-like B-cell antigen. Transplantation 1981;32:45.
18. Lotteau V, Teyton MM, Soulier A, Thomsen M, Sasportes M, Charron D. Biochemical polymorphism of the HLA-DP heavy chain. Immunogenetics 1987;25:403.

Antigen Society #33 Report (Public Epitopes)

G. Rodey,[1] P. Terasaki,[2] M.S. Park,[2] T. Fuller,[3] D. Kostyu,[4] D. Phelans,[5] A. Fuller,[3] M. Schroeder,[6] T. Zhao,[7] D. Chandanayingyong,[8] R. Duquesnoy,[9] J. Mararri,[9] A. Cambon-Thomsen,[10] M. Calot,[10] H. Gebel,[11] and J. Hunter [12]

The purpose of this study was to identify operationally monospecific antibodies detecting alloepitopes shared by more than one HLA class I gene product (public epitope). In this Workshop, public epitopes are defined as highly cross-reactive homologous private alloepitopes or other shared alloepitopes that are spatially distinct from the previously recognized private alloepitopes.

Alloantibodies that react with public or cross-reactive epitopes often manifest the CYNAP phenomenon, that is, not all of the specificities of the antibody can be detected in standard complement-dependent cytotoxicity (CDC) procedures. A more complete analysis of the two binding characteristics of such alloantibodies is only possible using indirect binding assays (such as flow cytometry, ELISA, or RIA) or by antiglobulin-augmented CDC (AHG-CDC), a procedure previously shown to minimize or override CYNAP.

Reporting Laboratories: NCYROD,[1] US5TER,[2] NCYFUL[3]
Participating Laboratories: US1AMO,[4] NCYSTL,[5] US5MLS,[6] CHIZHA,[7] ANZDCH,[8] US1DUQ,[9] FRAOHA,[10] US2RSH,[11] US1MBC[12]

Methods

Two hundred and six alloantisera and 32 complement-fixing monoclonal antibodies were submitted to this Antigen Society. Four laboratories tested the serum set by standard CDC, and eight laboratories tested the sera by both CDC and AHG-CDC. Total cells used for testing were 750 (CDC) and 450 (CDC + AHG-CDC). Approximately 25% of the sera were differentially adsorbed or were submitted as operationally monospecific reagents. Computer analysis of serum reactions was kindly performed by Dr. Ray Mickey (Los Angeles) using an iterative tail analysis program previously developed for the Eighth International Histocompatibility Workshop.

Results

In the subset of cells simultaneously tested by both CDC and AHG-CDC, cell panel reactivity was increased in alloantisera an average of 40% and in monoclonal antibodies an average of 10%. AHG-CDC monoclonal data are not accurately interpretable because the antiglobulin reagent used was a goat anti-human reagent. Most of the increased reactivity of alloantibodies tested by AHG-CDC could be attributed either to additional specifi-

Immunobiology of HLA
Volume I

cities within the CREG groups or to the detection of additional "public" antibodies.

Most of the antibodies could be assigned to five major cross-reactive groups (CREGs). For analysis, splits and variants were recombined into larger groups (e.g., A25, A26, Aw34, and Aw66=A10). The 5 CREG groups were:

A1C (1,3,11,10,W19,9,28)
A2C (2,28,9,17,10,33)
B5C (5,53,35,18,15,17,70,49)
B7C (7,27,22,42,48,40,41,13,47)
B8C (8,14,16,22,52)
12C (12,21,40,41,13)

In addition, several monoclonal antibodies defined distinctive patterns for which there were no alloantibody counterparts. These included:

A9,25,32
A2,3,w19,28 and
A32, B27

A1C. Several patterns were identified in this group, based upon reactivities with A1,3,10 and 11. (The reactions with the Aw19 group were highly variable and could not be used for cluster assignment) 1,3,11; 1,3,10, 11; 10,11; and 1,10,11. To summarize, the A1C group of molecules appears to contain several sets of public epitopes. A1, 10,11 was defined by six sera, including one monoclonal. A1,3,10,11 was defined by four sera. The remaining patterns were defined by two or less sera.

A2C. This group contained four discrete clusters, each defined by 2 to 5 sera: A2,28; A2,28,9; A2, B17; and A28,33,10. One monoclonal defined a distinct group of cells that included A2, 28,33,3,29,30,31 (not 32 or 33).

B5C. Four general clusters of sera were seen in this group of sera, but the reaction patterns with several specificities, namely B18, 70 and 49, varied within clusters. Major clusters were defined as follows: B5,53,35, 62, B49 positive, B17 negative (11 sera); the same clusters plus reactivity with B17 (13 sera); B5, 53, 35 positive, B70, 49; and B15, 17 negative (7 sera); Finally a fourth group was positive for B5, 53 (but negative for 35, 18) and variable positive with Bw4-associated gene products. This latter group appears to be a Bw4-related specificity rather than a B5C public.

B7C. The B7 CREG had two well defined clusters and an additional two possible clusters. Well defined were: B7, 27, 22, 42 (14 antibodies which included 3 monoclonals); B7, 42,48,40, B41, 13, 47 positive and B27, 22 negative (10 sera). Two possible additional clusters were: B7, 22, 42, B48, 40,41,13 positive and B27 negative; and B7, 27, B48, 40,41,13,47 positive and B22, 42 negative. To summarize, B7 appears to have two and possibly four distinct public epitopes.

B8C. Only one cluster was defined in this group (2 sera): B8, 14,16 and 18. One of the reagents is a monoclonal antibody. Other sera (11) reacted variably with the specificities and in addition, with B22 and 52.

B12C. Sera in this group appears to define a relatively homogenous public epitope shared by: B12, 21,40,41, and 13 (4 sera). An additional 11 sera reacted variably with these same specificities but not with additional specificities.

Conclusions and Recommendations

1. Reasonably well defined clusters of sera were formed which appear to detect public specificities shared by the following gene products:

   a) A2, 28
   b) A2, 28, 9
   c) A2, 17
   d) A28, 33, 34
   e) B7, 27, 22, 42
   f) B7, 42, 40, 13, 41
   g) B8, 14, 16, 18
   h) B12, 21, 40, 41, 13

   Other clusters were suggestive of additional publics, but were not sufficiently defined in this Workshop serum set on preliminary analysis.
2. We recommend that the W.H.O. Nomenclature Committee consider a nomenclature that can be applied to epitopes, specifically public epitopes.
3. Because CYNAP is a common finding in alloantisera used to define public epitopes, we recommend that techniques that override or minimize CYNAP, such as AHG-CDC, flow cytometry or other complement-independent procedures, be utilized in future workshops that define public epitopes.
4. Antisera that are selected for standards that define public epitopes must be differentially absorbed or be used as eluates in order to document that only a single operationally monospecific antibody is present.
5. We recommend that the data obtained in this Antigen Society be entered into the central computer facility so that correlations between putative public epitope groups can be made with other parameters of data such as T-cell specificities and RFLP patterns.

Antigen Society #35 Report (DQw4 [DQ "WA"])

A. Wakisaka,[1] M.S. Park,[2] A. Barbetti,[2] T. Juji,[3] P.I. Terasaki,[2] and T. Takenouchi[2]

DQw4(DQ"WA") is defined by a monoclonal antibody HU-46 (10W3101), which was established by immunizing C3H mice with EBV-WA line (A24-Bw54-Cw1-DR4-DRw53-Dw15-DQ blank homozygous).

By two-dimensional (2D) gel analyses, immunoprecipitates of HU-46 appeared in the same position as DQ-beta 1, but in a different position from DR beta. Thus, HU-46 is likely to react to DQ molecules (1).

The results of cytotoxicity testing shown in Figure 1 indicate that in this Workshop almost 90% of DQw4 cells tested in a random population were DQ blank. The remaining cells that are presumably heterozygous are also listed. The black bars indicate the small portion of cells that were triplets. All of the data from the Workshop were used in this analysis and none, including questionable data, were excluded.

The DR association in two ethnic groups is shown in Figure 2. The association with DR4.1 in Japanese and with DR3 in African Blacks was observed in the 3rd Asia-Oceania Workshop (AOH) report (2). DR8 was reported to be significantly associated with DQw4 in Caucasians.

A similar report was written from the Eighth International Workshop data utilizing a similar alloantibody submitted by our laboratory. This antibody (AIZ1094) in Caucasians reacted to most DRw8 and a very small fraction of DR4; however, in Japanese it reacted to DR4 (presumably Dw15) and not DRw8 (3).

Although family study data were not available in this Workshop, using data from the 3rd AOH Workshop, proper segregation with HLA haplotypes was observed.

Among the core cell lines, the positive cells were:

Reporting Laboratory: JAPAIZ[1]
Participating Laboratories: US5TER,[2] JAPJUJ[3]

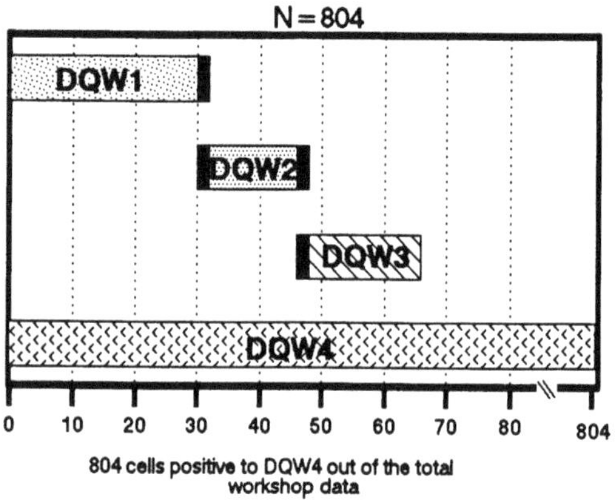

Figure 1. Allelism of DQw 1,2,3 and 4.

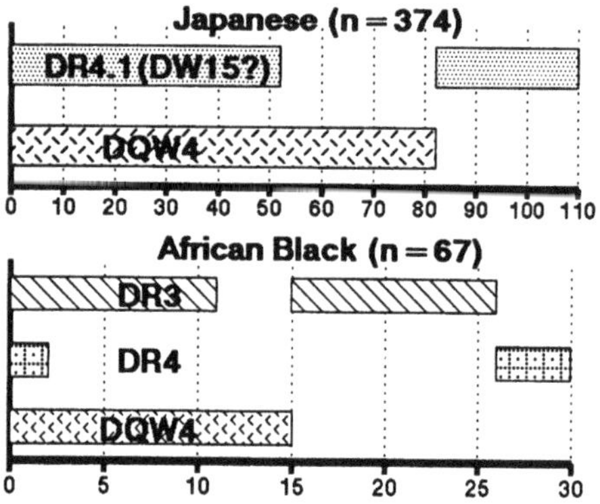

Figure 2. Association to DR antigen (3rd AOH workshop).

9021 (DR3-Dw new, African Black), 9107 (DR4-Dw15, Japanese), and 9067,9068,9069,9071,9072 (DR8-Dw8.1, Caucasian and DR8-Dw8.2, South American Indian). All five of the preceding DR8-DQw4 haplotypes were positive to DRA-72, DQA1-84, DQB2-XB1, and DQA2-XA1 which were defined by RFLP (4). There was no RFLP-associated band to DQw4 alone. Table 1 shows DQw4 associated bands. Two RFLP banding patterns corresponding to DQw4,DQw8, DQw9 and DQw4,DQw8,DQw9,DQw7 were observed. Thus, it appears that the monoclonal reagent provides the best definition of DQw4.

Since the central data did not have the DQw4 specificity, Figure 3 shows a summary of UCLA 250 cell lines. The antibodies, including DQw4-related antibodies, are listed along with their specificities. The 10W1092, an alloantibody, reacts somewhat similarly to the panel as the monoclonal antibody (HU-46) DQw4. The other sera listed in Figure 3 contain additional specificities. For example, 10W3091 is reactive to DQw1 and DQw4 and 10W1181 reacts with DQw4+DQw3+DQw2. One of the T-cell clones (6058), not shown in the figure, was reactive and it is still not clear whether it recognized DQw4 or the DQA1-A84 antigen (see Joint Report).

Table 1. RFLP Band-Related DQw4

Probe	Enzyme/Kb	Specificity
DQB	BAM−10.2	DQw4, DQw8, DQw9
DQB	KPN−19.1	DQw4, DQw8, DQw9
DQB	TAQ−1.9	DQw4, DQw8, DQw9
DQB	BGL−5.6	DQw4, DQw8, DQw9
DQB	TAQ−2.5	DQw4, DQw8, DQw9
DQB	PVU−3.4	DQw4, DQw8, DQw9, DQw5
DQB	HID−3.3	DQw4, DQw8, DQw9, DQw6, DQw5.2
DRB	PST−4.9	DQw4, DQw8, DQw9, DQw7
DQB	MSP−3.6	DQw4, DQw8, DQw9, DQw2, DRw15

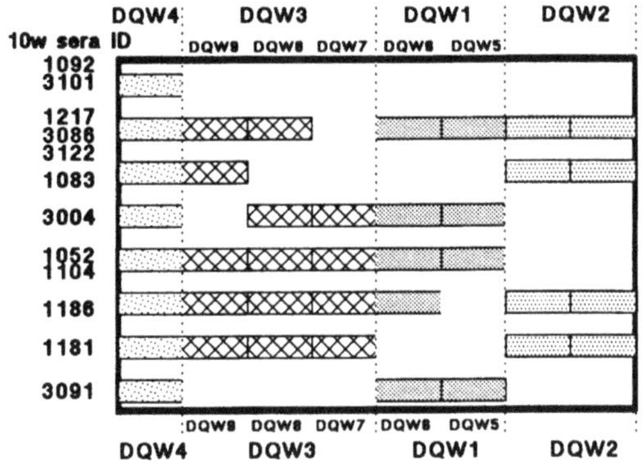

Figure 3. DQw4 (10w 3101) and DQw4 related antibodies.

Figure 4 shows the gene frequency of DQB1 antigens in Caucasians based on 170 random individuals. DQw4 was approximately 4% in this panel. The antigen frequencies from the 3rd Asia-Oceania Workshop vary according to the ethnic group (e.g., Caucasian 14%, African Black 24%, Australian Aborigine 30%, Japanese 22%).

Table 2 represents the Hardy-Weinberg equilibrium for DQ antigens. It should be noted that the observed and expected values fit the formula.

In conclusion, DQw4, which is clearly defined by a monoclonal reagent in this Workshop, is a DQ molecule by immunoprecipitation and is mainly reactive to DQ blank by cytotoxicity. It is allelic to DQw1(w5,w6),w2, and w3(w7,w8,w9). DQw4 has different associations to DR in three different ethnic groups (DR8 in Caucasians, DR4 in Japanese, and DR3 in African Blacks). DQw4 segregates with HLA haplotypes. This new WHO recognized antigen fits into DQB1 locus and fills most of the DQ blanks.

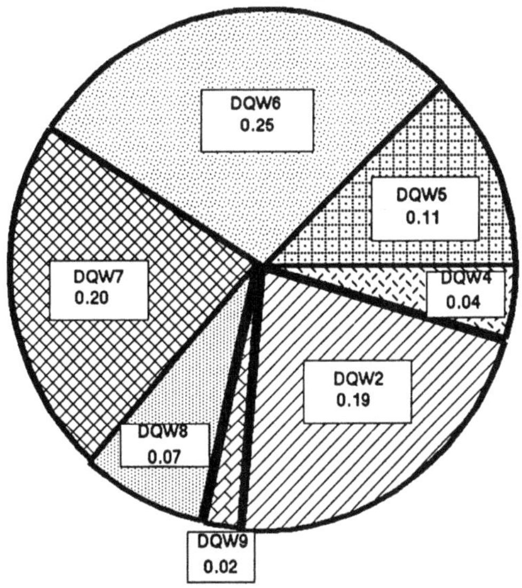

Figure 4. Gene frequency of DQBl in Caucasians.

Table 2. Hardy-Weinberg Equilibrium of DQB1 Antigens

Alleles	Gene Frequency	AG Frequency
DQw5	0.115	0.217
DQw6	0.248	0.435
DQw7	0.202	0.364
DQw8	0.066	0.129
DQw9	0.020	0.041
DQw2	0.191	0.347
DQw4	0.042	0.082
DQ Blank	0.111	0.000

Phenotype		Observed	Expected	Chi-Square
	DQw5	2.000	6.636	3.239
DQw5	DQw6	12.000	9.759	0.514
DQw5	DQw7	7.000	7.969	0.117
DQw5	DQw8	4.000	2.628	0.715
DQw5	DQw9	2.000	0.816	1.713
DQw5	DQw2	8.000	7.537	0.028
DQw5	DQw4	2.000	1.651	0.073
	DQw6	22.000	19.902	0.221
DQw6	DQw7	12.000	17.149	1.546
DQw6	DQw8	7.000	5.657	0.318
DQw6	DQw9	3.000	1.758	0.877
DQw6	DQw2	16.000	16.220	0.002
DQw6	DQw4	2.000	3.554	0.679
	DQw7	24.000	14.679	5.918
DQw7	DQw8	0.000	4.619	4.619
DQw7	DQw9	0.000	1.435	1.435
DQw7	DQw2	17.000	13.245	1.064
DQw7	DQw4	2.000	2.902	0.280
	DQw8	4.000	3.294	0.151
DQw8	DQw9	0.000	0.473	0.473
DQw8	DQw2	6.000	4.369	0.608
DQw8	DQw4	1.000	0.957	0.001
	DQw9	1.000	0.860	0.022
DQw9	DQw2	1.000	1.357	0.094
DQw9	DQw4	0.000	0.297	0.297
	DQw2	8.000	13.525	2.257
DQw2	DQw4	3.000	2.745	0.023
	DQw4	4.000	1.891	2.348
--Blank--		0.000	2.104	2.104
Sum		170.000	169.999	31.750

Chi-Square/DF = 31.750/(29 − 8) = 1.5.

References

1. Ishikawa N, Wakisaka A, Aizawa M, et al. Detection of a novel HLA-DQ specificity: serological and immunochemical analyses by a monoclonal antibody. Immunogenetics 1987;26:143.
2. Ikeda H, Shimbo M, Mitani T, Sekiguchi S. DQwA HLA in Asia-Oceania. In: Aizawa M (ed): HLA in Asia-Oceania 1986. Sapporo, Japan: Hokkaido University Press, 1986:188.
3. Nakata S, Terasaki PI, Omori K. Interracial crossreactivity. In: Terasaki PI (ed): Histocompatibility Testing 1980. Los Angeles: UCLA Tissue Typing Laboratory, 1980:590.
4. Park MS, Barbetti A, Takenouchi T, Terasaki PI. Analysis of Class II RFLP patterns and T cell clone reactions. In: Dupont B (ed): Immunobiology of HLA, Volume I: Histocompatibility Testing 1987. New York: Springer-Verlag, 1989.

Micro EIA and Monoclonal Antibodies

K. Nelson,[1] J. Bodmer,[2] A. Martin,[3] C. Navarette,[4] and D.M. Strong[5]

Binding of monoclonal antibodies (MAb) to HLA antigens on human leukocytes was evaluated using Micro EIA, a modified ELISA assay in Terasaki trays. One goal of this section of the serology exercises was to provide binding data to correlate with data from the biochemistry and T-cell exercises. Another goal was to compare the reactivities of MAb to Micro EIA with those in microcytotoxicity.

Design of the Study

Micro EIA used an indirect ELISA technique to determine MAb binding to cells in standard tissue typing trays. A microdensitometer developed by Genetic Systems allowed an objective reading of the results. The Micro EIA reader, software, and standardized reagents were made available to the 49 laboratories participating in these exercises. Table 1 lists the participating laboratories. Training workshops were given to familiarize the Micro EIA participants with the assay and the system.

Selection of MAb from those submitted to the Workshop was based on titration and specificity through screening in both Micro EIA and microcytotoxicity. Priority was given to MAb that worked well in both assays. Appropriate dilutions of the MAb were arranged on four plates in a pattern designed to detect position effects and carryover. Positive controls included MAb to nonpolymorphic determinants on class I heavy chains or DR molecules.

Analysis of the Data

Data analysis was performed using the SERAN program developed for the Workshop by Dr. Jean-Marc Lalouel and colleagues. Mark Ferguson wrote the programs for Micro EIA data analysis. The analysis was conducted in four stages. First, data submitted on floppy discs was edited to insure conformity of the written information in each file. Deviations from standard maps for each plate were also encoded this time. Deviations most often were due to missing sera.

In the second stage of analysis, the number of plates with acceptable data was determined for each laboratory as were the mean and variance of values for control MAb. Acceptability of plates was determined by comparison of the values of positive and negative control MAb on each plate with established limits. For positive controls, two of three values needed to be greater than 20 optical density (OD) units for analysis of peripheral blood leukocytes (PBL) or greater than 30 OD units for analysis of B-cell lines (BLCL). Negative control limits required four of six values to be less than 10 OD units for PBL or 15 OD units for BLCL. A laboratories' dataset was included in the analysis if greater than 70% of their plates were accepted and if the coefficient of variance of their positive controls was less than 50%. Table 2 presents a summary of the number of laboratories and plates included in the final analysis.

In the third stage of data analysis, the data from these laboratories was translated from OD units encoded by the Micro EIA reader into cytotoxicity scores. This involved three steps. First, the OD unit score of each MAb in each test was normalized using the positive and negative controls for that test. Second, these normalized values were compared to the cutoff established for that MAb to determine if the reaction was positive or negative. Third, the set of positive values was ranked in descending order. The top third was given a score of 8, the next third given 6 and the lowest third 4. The negative values were similarly ranked, the top half was given a score of 2, the bottom half a score of 1. In the fourth and final stage, the translated data were analyzed by the SERAN program.

The most difficult part of this analysis was the determination of cutoffs to define negative and positive reactions. These assignments were based on data from 11

Reporting Laboratories: US7GSC,[1] UKIBOD,[2] UKIBRS,[3] UKIFES,[4] US7NTC[5]

Table 1. Laboratories Participating in Micro EIA Study

ANZDAW	IT1CNG	LATGOR	NCYCAR	US5MIT
BENENG	IT1MTT	SAFDUT	NCYFAL	US5MLS
BENREE	IT2FER	SAFHAM	NCYMRV	US5TER
FRACOH	IT2GAN	SCAKIS	NCYSTL	US6BRN
FRADDC	JAPJUJ	SCASVE	NCYYUN	US7DUP
FRAFRA	JAPNAI	UKIBOD	US1DUQ	US7GSC
FRAJEA	JAPSEK	UKIBRS	US1HJH	US7HAN
FRAMYE	JAPTSU	UKIFES	US2BAC	US7POL
GERGOL	LATLAY	UKIKNT	US4RAD	US8RJW
GERMUC	LATJRM	UKILAW	US4STA	

Table 2. Micro EIA Data Analysis Summary

	Number	Percent
Plates run	9318	
Plates acceptable by evaluation of controls	7239	78
Plates used in analysis	4864	
Percent of plates run		52
Percent of acceptable plates		67
Laboratories submitting data	38	
Laboratories included in analysis	24	63

Immunobiology of HLA
Volume I

Table 3. Monoclonal Antibodies to HLA-A

10WS#	Local Name	Specificity	r Value	Comments	Microcytotoxicity
2005	GAP.A3	A3	***		Same
2009	XI.23	A3	**	Weak on PBL	Same
2013	GSP55.2	A3	***		Same
2020	76.3.10	A2,28	***		Same
2021	NFLD.M2	A2,28	***		Same
2025	76.3.21	A2,28	*	Weak on PBL	A23 also on BLCL
2026	CBRI.M1.20.86	A2,28	***		Same
2027	CBRI.M3.63.14	A2,28	***		Same
2029	XV.17	A2,28	***	Very weak on PBL	Same
2030	MAD.148	A2	**	Very weak on PBL	A3,29,31, also on BLCL
2032	4B	A2,28	***	Weak on PBL	Same, misses some Aw68
2033	TU101	A2,28	***		Same
2035	TU160	A2,28	*	Very weak on PBL	Not tested
2037	BB7.2	A2,w69	***		Same
2039	GSP20.1	A2,w69	***		A31,w34,w68 also on BLCL
2040	MTC301338	A2,28	**		Same
2041	AC9.51	A2,3,28,29,30,31,w33	**	Weak on PBL	Same
2042	GSP5.4	A2,3,28,29,30,31,w33	***		Same
2048	DCH184.3	A29	**	Weak on PBL	Same
2050	TER.M.A3031	A30,31	*	Extra reactions on BLCL	A30,31 only
2056	TER.M.A2532	A25,32	*	Extra reactions	A24,23,24,B27 also on BLCL
2060	HA9	A9	*	Very weak on PBL	Same
2061	HU49	A9	***		Same
2063	GS114.1	Aw24	***		Same
2066	TER.M.A10	A26	*	Extra reactions	A25,26,w34,w43

r value *** = 0.8–1.0; ** = 0.6–0.8; * = 0.4–0.6.

laboratories. For many MAb, the cutoff was obvious as there was a biphasic distribution of values with a clear separation between positive and negative. However, the majority of MAb were assigned standard cutoffs based upon the locus encoding the antigen and the type of cells tested. For BLCL, cutoffs varied from 20% to 50% of the positive control signals. For PBL, cutoffs varied from 50% of the class I positive control signal for MAb to A locus antigens to 10% of the class II positive control signal for MAb to DQ antigens.

Analysis of MAb to Class I Antigens

Tables 3 and 4 present summaries of the data from class I analysis. The specificity assigned to each MAb is presented as is the r value of that assignment. A MAb was included if the r value of any assignment was greater than 0.4. Variation in binding and r values due to type of cell assayed is indicated in the "Comments" column. Specificity of each MAb in microcytotoxicity is presented for comparison. These data are extracted from the report, "Cytotoxic Monoclonal Antibodies" in this volume.

Twenty-three MAb were reactive with A locus antigens: nine for epitopes present on a single specificity, 16 for public determinants. Twenty MAb were reactive for B locus antigens: eight with epitopes on single specificities, and twelve with public determinants. The latter included four MAb that reacted with the AxB public determinant found on A9 and A32 and on B alleles expressing Bw4.

The conclusions from this part of the study were that six A locus alleles and five B locus alleles as defined by alloantibodies were also definable by MAb and Micro EIA. Several public antigen epitopes were also defined and showed consistent correlation with clusters of class I alleles.

Analysis of MAb to Class II Antigens

Tables 5 and 6 present data from the class II analysis. The format is identical to that of Tables 3 and 4. The classification of a specificity as DR or DQ is based solely on serologic analysis in Micro EIA or microcytotoxicity and does not reflect a biochemical assignment of an epitope to the product of a particular locus.

Forty-two MAb were assigned specificities in this study. Twenty-one were specific for DR antigens, 20 for DQ antigens. R values were significantly lower when DQ MAb were tested on resting B cells rather than on BLCL. This may be due to lower signals in the test itself; lower signals reflecting the lower amount of DQ antigen due either to lower expression on resting cells or lower expression of a single specificity on heterozygous PBL versus consanguinous BLCL. Another source of higher r values on BLCL could be the clearer assignment of DQ specificities to the Core BLCL which were typed by many laboratories. The r values for MAb to DQw4 or DQw7 were based on BLCL data, as many PBL panels were not typed for these specificities.

The conclusions from this part of the study were that eight DR alleles could be defined by MAb and Micro

Table 4. Monoclonal Antibodies to HLA-B

10WS#	Local Name	Specificity	r Value	Comments	Microcytotoxicity
1220	SY1	B13	**		
1250	VE5	B13	*	Very weak on PBL	Same
2054	GSP56.1	A32,B13, 27	*	Weak on PBL	A9, other B locus on BLCL
2057	TU48	A9,32, Bw4	*	Very weak on PBL	Same
2064	TOK45.3A	A9,32, Bw4	***		Same
2065	TOK45.3B	A9,32, Bw4	***		Same
2070	KS.3	B7,w22,27	***		Also B39,w42,w64,w65
2075	GS145.2	B27	**	Extra reactions B7,w52	Not tested
2076	KS.4	B7,14,w22,27	*	Weak on PBL	Also Bw42
2080	BD.7	A9,32, short Bw4	**	Weak on PBL	Also A25
2083	GSP7.1	B7,17,w22,27,w42, w63	***	Also B14 on BLCL	Not tested
2086	TER.M.B8	B8	*	Weak on PBL	Also Bw59
2088	XIII397	B8	***		Also B14 on BLCL
2089	GSP8.1	B8	***		Also Bw59 on BLCL
2091	XXIII.23	B5,8,14,18	**	Also B38 on BLCL	Broadly reactive
2092	TER.M.B13	B13	*	Extra reactions on BLCL	B13 only
2094	UL.39.10	B13	**		Same
2097	KT4	B5,12,17	*	Weak on PBL	B5,12 on PBL, more on BLCL
2101	TU109	Bw4	*	Very weak on PBL	A23 weak Bw4
2106	MTC29306	Bw6	**	Weak on PBL	Same

r value *** = 0.8–1.0, ** = 0.6–0.8, * = 0.4–0.6.

Table 5. Monoclonal Antibodies to DR

10WS#	Local Name	Specificity	r Value	Comments	Microcytotoxicity
3010	CHE153	DR1,7,9	**	Too weak on PBL	Same BLCL and PBL
3011	HU30	DR2,1	**	Extra reactions on BLCL	DR1,2,7,9,
3012	NDS15.38	DR2(Dw2)	**	BLCL only	Not tested
3025	7.3.19.1	DRw52	**	Not DRw8 or DRw12	Same
3028	NDS12	DRw52	**	Weak on PBL	Also DR9,w10
3031	JS1	DR3	**	Extra reactions, DR4,w6	Same
3036	GSP88.2	DRw11	***	A few extra reactions on PBL	Same
3044	HU48	DR4,w11	*	Weak on PBL	DRw52,4,w10
3045	GSP97.1	DR9	**	Extra reactions on BLCL	DRw53,w10,w14
3048	PLM3	DR7,9,w12	**	DR7 only on PBL	DR7,9,w52
3049	17.3.3	DR7	***	A few extra reactions on BLCL	DR7,4(Dw15),12
3050	SFR16.DR7M	DR7	***		Same
3051	GSP65.1	DRw53	**		Same
3054	50D6	All but DR7	**		Same
3055	GSP14.1	All but DR7	**		Same
3056	HU20	All but DR7	***		Same
3057	HU26	All but DR7	**		Same
3059	PD3	DR1,4,9	**		Most cells positive
3063	8.1	DRw52	*	Not DRw8 or DRw12	Same
3065	MAJA51	DR2,w8,w14	*	Weak on PBL	MS
3068	TER.M.DN1	DRw12	**	Extra reactions on PBL	Same

r value *** = 0.8–1.0; ** = 0.6–0.8; * = 0.4–0.6.

Table 6. Monoclonal Antibodies to DQ

10WS#	Local Name	Specificity	r Value	Comments	Microcytotoxicity
1217	77.34	DQw1	*	Extra reactions DRw53	Same
3086	R1	DQw1	*	BLCL only and extra reactions	DQw1,w4,w8,w9
3090	II18B	DQw1,w4	**	BLCL only	DQw1,w4,w8,w9
3091	KS.11	DQw1,w4	***		Same
3094	GENOX3.53	DQw1	***	Weak on PBL	Not tested
3095	G2A5	DQw1	***	Weak on PBL	DQw1
3096	GS148.1	DQw1	**	Weak on PBL	DQw1, DR7
3100	1.2	DQw1	***	Very weak on PBL	Not tested
3101	HU46	DQw4	***	BLCL only	Same
3105	C1A2	DQw2	**	Very weak on PBL	Same
3107	XII358	DQw2	***		Same
3108	IVD12	DQw3	**	Weak on PBL	Not tested
3111	HU18	DQw3	***		Same
3112	PLM12	DQw3	***		Same
3114	GS100.1	DQw3	***	Weaker on PBL	Same
3115	17.15	DQw3	**	Very weak on PBL	DQw3, DR7
3118	AC60	DQw3	*	Weak on PBL	Same
3119	A10/13	DQw7	*		Same
3120	PLM2	DQw7	***	Weak on PBL	Same
3121	HU23	DQw7	**		Same
3064	DP11.1	DPw4,w2	**	BLCL only	Same

r value *** = 0.8–1.0; ** = 0.6–0.8; * = 0.4–0.6.

EIA, seven of these on resting B cells. Several public class II epitopes were also defined although their assignment to DR or DQ locus products is not known. MAb also defined DQw1, DQw2, DQw3, DQw4, and DQw7. Clear assignments could be made for DR specificities on B cells. DQ specificities were also assigned but lower r values (0.5 to 0.7) may reflect limits on the sensitivity of this version of the assay.

Class II MAb and Core BLCL

One goal of this exercise was analysis of the binding of MAb to aid in interpretation of biochemical analysis and T-cell blocking studies. The SERAN program allowed presentation of a cluster analysis in its CxS format. In Figures 1, 2, and 3, data are presented for clusters of class II MAb reacting on Core BLCL. Tenth Workshop numbers of Core BLCL are given down the left-hand side and the number of each MAb across the top of each figure as in the CxS display. Positive reactions are indicated by a numeral, negative reactions (1 or 2) are left blank. Each data point is the average of duplicate or triplicate assays. BLCL that had less than two valid assays were not included.

In Figure 1, clusters corresponding to DRw52-like reactions are presented. These antibodies clustered closely to a MAb reacting with DRw11, 3036. In Figures 2 and 3, two clusters are shown. The first consists of MAb that bound to DR1 and DR4 plus other DR alleles. They cluster just to the left of MAb that bound to DQw1 and DQw1 plus other cells which expressed DR4, DR7, or DRw8. Figure 2 presents results on cells that type as DQw1; Figure 3 presents results with the other BLCL. In Figure 4, there are three distinct clusters, DQw2, DR7, and DQw3. The MAb, which define DQw7, 3120, and 3121, are bracketed within the DQw3 cluster.

Class I MAb and Core BLCL

One set of clusters is presented from the class I analysis. Figure 5 shows a set of MAb that bound to cells expressing Bw4 and/or various A-locus alleles, A9,10,32. The data are of interest in sorting out the different patterns of binding shown by MAb to this public determinant.

Analysis of Splits of DRw52 by MAb and Micro EIA

DRw52 is a complex serologic specificity found on haplotypes expressing DR3, DR5, DRw6, and DRw8. DRw52 antigens expressed on Core BLCL were analyzed using T-cell clones and two-dimensional gel electrophoresis; the DNA from these cells was analyzed by RFLP techniques. Results are given in detail in the report from the DRw52 Antigen Society. Data presented in Princeton indicated that there were at least three distinct clusters of DRw52 epitopes detected by T cells and that these clusters correlated with clusters of DRβ chains and distinct DRB RFLP. These same BLCL were also studied with MAb and Micro EIA. Seven MAb were particularly informative. Figure 6 presents the results of these MAb tested on BLCL used in the other exercises for analysis of DRw52. Five subgroups of DRw52 were defined. Their relationship to subgroups defined by T-cell clones and to assignments of DRβ chain electro-

WSNO	DR	DQ	3023	3062	3061	3076	3028	3063	3025	3027	3026	3036	3041
9018	3, w52	w2	6	4			6	6	6	4	4		
9019	3, w52	w2	6	6	4		6	6	8	6	8		8
9020	3, w52	w2	8	6	4		8	8	8	8	8		6
9021	3, w52	w4							8				
9023	3, w52	w2	4	4			6	6	8	4			4
9085	3, w52	w2	6	4			8	6	8	4	4		6
9086	3, w52	w2	6	4			8	6	6	4			6
9087	3, w52	w2	4	6			6	6	8	4			6
9088	3, w52	w2	4	6	6	6	6	8	8	4			8
9035	w11, w52						6	8	6	6	4	8	6
9036	w11, w52	w7					8	8	8	8	8	8	6
9037	w11, w52	w7						8	4	8	4	8	4
9039	w11, w52	w7					4	6	4		4	6	4
9040	w11, w52	w7						6	4		4	6	
9041	w11, w52	w7		4		4		8	4	8	4	8	6
9042	w11, w52	w7					4	4	6	8	6	6	6
9043	w11, w52	w7						4	4	6	4	6	4
9044	w11. w52	w7						6	4	6	4	6	4
9038	w12, w52	w7					4		6	6	8		
9045	w11, w12, w52, w7						4	4	4	8	4	8	4
9105	w11, w52	w1					4	4	4	8	4	8	4
9054	w14, w52	w5	6	8	6		8	6		8	8		8
9055	w6, w52	w1	4	4	6	8	6	6	6	4			6
9056	w14, w52	w5		4			4	6	6	6	4		6
9057	w13, w52	w6	6	6			6		6	6	6		6
9058	w13, w52	w6	6	6		8	6	8	8	4			6
9059	w13, w52	w6	4	6	4		8	8	4	4			
9060	w13, w52	w6	6	4	4		4	6	6	4	6		8
9061	w14, w52	w5	6	4	4	6	8		4	8	6		8
9063	w13, w52	w6	4				4	6	6				
9064	w14, w52	w7		4	4		4	8	4				6
9097	w13, w52	w6		6	6		6	6					4
9099	w14,w52	w7	4	6	6		6	6		4			6

Figure 1. C×S analysis of class II MAb and Core BLCL. Clusters of MAb with DRw52-like specificity.

WSNO	DR	DQ	3031	3008	3059	3006	3009	3044	3065	3011	3012	3088	3085	3086	3089	1217	3090	3091	3094	3095	3096	3100
9001	1	w5	6	8	6	6				4			4	4	6	6	4	6	8	6		6
9002	1	w5	6	8						6				4	8	6	6	8	8	6	6	6
9003	1	w5	6	8	8	4	4		4			8	4	8	6	6	6	8	8	6	6	6
9006	1	w5	6	4	4				4	6						8	4	4	8	6	4	
9078	1	w5	6	8	8		6			4			4	8	6	6	8	8	8	8	6	8
9079	1	w5	6	4	6		4								6	4	6	4	4			6
9080	1	w5	6	6	8	6	4							6	6	4	8	6	8	6	8	4
9007	w16, 4, w53	w5, w3	4		4			4	6	4			4		4	4	4	4	6	4	4	4
9008	w15	w6				4			6	8		4	6	6	8	6	6	6	8	8	4	8
9009	w15	w6				4			8	8				6	6	8	6	4	6	6		8
9010	w15	w6				8	8		6	6	6	6		8	4	6	6	6	6	8	4	6
9011	w15	w6				8	8			8				4		4	4	4	6	6		
9012	w16	w5							6	8				4	4	6	4	6	8	6	4	6
9013	w15	w6							6	6	6	6	6	6	6	6	6	8	8	6	8	6
9014	w15	w6							8	8	8	8	8	4	6	8	6	8	8	8	8	6
9015	w15	w6			4	4	4		8	6		4	8	8		6	8	6	8	8	4	6
9016	w15	w3			4				6	4				6								
9017	w15	w6							8	8	8	8	8	8	8	8	8		8	8	8	8
9082		w6							6	8	8		4			8	6	6	8	6	6	6
9083	w15	w6		8	4				8	4	4		6	6	6	8	6	8	8	6	8	8
9084	w15	w1							6	8				4	6	4			6		6	4
9066	w8, w52	w1							8				6	4		8	6	8	8	6	6	4
9036	w11,w52	w1	4		4			6	4				4	4	4	8	8	6	8	6	6	8
9105	w11,w52	w1											4			8	4	4	8	8		8
9055	w6, w52	w1							6				6	4		8	6	4	6	6	4	4
9057	w13, w52	w6			4			4	4					6	6	6	4	6	6	6	4	6
9058	w13, w52	w6	4						8		6	4	4	6	8	6	6	8	8	8	6	4
9059	w13, w52	w6							6				6	6	8	8	4	8	6	4	6	4
9060	w13, w52	w6		4					8		4	6	6	6	6	6	4	6	8	8	8	6
9063	w13, w52	w6			4	4			6			4	6	6	6	6	6	8	8	6	4	6
9097	w13, w52	w6							6				6	4	6	6	4	6	8	6	6	6
9054	w14, w52	w5	4	6	4		4	6					6	8	6	6	4	6	8	8	8	6
9056	w14, w52	w5							4				4	4	6	8	6	6	6	4	4	6
9061	w14, w52	w5		4	6			4					4	6	6	6	6	6	6	6	6	4

Figure 2. C×S analysis of class II MAb and Core BLCL. Clusters of MAb with DQw1-like reactivity assayed on cells expressing DQw1.

WSNO	DR	DQ	3031	3008	3059	3006	3009	3044	3065	3011	3012	3088	3085	3086	3089	1217	3090	3091	3094	3095	3096	3100
9047	7, w53								4													
9048	7, w53	w2							4	4					6							
9050	7,w53																					
9052	7, w53	w9							4				6	4	8	6	6					
9094	7, w53	w2							4	4					4	4						
9095	7, w53	w9							4				4	4			6		4			
9096	7, w53																					
9106	7, w53	w9							4	6			4	6	6	4	6					
9074	w9, w53	w9			4		6						4	4	4	4	6					
9075	w9, w53				4										8	4	4					
9076	w9, w53						4		6			6		6			8					
9077	w9, w53	w9	6	6			8		8			8		8			8					
9102	w9, w53				8				4													
9103	w9, w53				6		6		4			4	4	4	6	4	6					
9024	4, w53	w3	8		8	8	8	6	4				8	6	8	6	6					
9025	4, w53	w7	6	4	4	6	4	4										8	4			
9026	4, w53	w8						4	6				4	4	4	8	6					
9027	4, w53	w3	6		8		4	8									6					
9028	4, w53	w3	8	6	6		4	8	4				4	4	8	4						
9029	4, w53	w3		6	6	4		6						4		8	4					
9030	4, w53	w7			8	6	6	6					4									
9031	4, w53	w3	6		8							6	8	8	8	6	6					
9032	4, w53	w8	6	8	6	6	8					6	6	8	8	4	6					
9090	4, w53		6		6		6	8	4													
9091	4, w53	w7	4	8		6	4		4													
9098	4, w53	w3	4	8		6	8	4				6		4	6	6						
9107	4, w53	w4	4	4	6			6							4		4					
9067	w8, w52	w4		6		4			8								4	4				
9068	w8, w52	w4			4	4	6	4	6						4		4	6		6		
9069	w8, w52						4	6	4					6			4	6				
9071	w8, w52	w4			6		4	6	6			6	4	6			6	6				
9100	w8, w52	w4							6			4					4	6	4			
9101	w8, w52	w4							4					4			4	4				
9018	3, w52	w2	8		4												4					
9019	3, w52	w2	6	4	4				4													
9020	3, w52	w2	6		4				4													
9021	3, w52	w2	4		4	4											4	4	4			
9022	3, w52	w2	6																			
9023	3, w52	w2	4		4				4								4					
9085	3, w52	w2	8						4													
9086	3, w52	w2	8		4																	
9087	3, w52	w2	4		4				4													
9088	3, w52	w2	6		6				4								4					

Figure 3. C×S analysis of class II MAb and Core BLCL. Clusters of MAb with DQw1-like reactivity assayed on cells not expressing DQw1.

WSNO	DR	DQ	3103	3010	3047	3021	3068	3033	3107	3105	3049	3050	3048	3051	3023	3066	3113	3117	3120	3121	3112	3111	3114	3108	3118	3115
9018	3, w52	w2							4	6																
9019	3, w52	w2							8	6				4	8											
9020	3, w52	w2						8	6	8				4	6		4									
9021	3, w52	w4												4	6		6									
9022	3, w52	w2							6	4			4	6				4								
9023	3, w52	w2							6	4			4													
9085	3, w52	w2							6	8					4		6									
9086	3, w52	w2							6	4				4	4		4									
9087	3, w52	w2							6	6			4	4	4		4	4								
9088	3, w52	w2							8	8			6	6	6		6									
9046	7, w53	w2							8	8	8	6	4	4	6	6										
9047	7, w53	w2		4		4			4	6	2	4	4	6				4						4		4
9048	7, w53	w2	8	6					8	8	8	8	6	4												
9049	7, w53	w2							8	4		8	8		4											
9050	7, w53	w2							8	6	6	8	6	4	8											
9051	7, w53	w2							8	8	6	8	6	4	8											
9052	7, w53	w9									6	6	6	6	4		4				6	6	6	6	6	8
9094	7, w53	w2		6					8	8	6	6	6	8	4										4	
9095	7, w53	w2, w9							6	4	6	6	4	6							4		4	4	6	6
9096	7, w53	w2							8	8	4	6	6		6											
9106	7, w53	w2, w9		8	4				6	6	8	8	6	8	4						6	6	4	6	6	4
9074	9, w53	w9		6									4	4	4			4			6	4	4	4	4	6
9075	9, w53	w9		4						4			4	4		4	4	6			6	6	4		4	4
9076	9, w53	w3		4	8					4			4	6	4		4	8		4	8	8	8		6	8
9077	9,w12	w3		8	8	8	8	6		4		4	8	6		6	8	8	8	6	6	6	4	4	4	8
9102	9	w2							4	4																
9103	9	w3		4	4	6				4				6	4		6				6	6	4		4	4
9024	4. w53	w3												6	8	6	4			4	6	6	6	4	6	6
9025	4, w53	w7						4							4		4		6	4	6		6		4	4
9026	4, w53	w8													8	4	4				6	6	6	4	8	8
9027	4, w53	w7								4		4		6	8	8	6	8	8	6	8	8	8	6	8	8
9028	4, w53	w3	8											4				4			6	6	4	6	4	4
9029	4, w53	w3													4			4			6	4	6	8	6	8
9030	4, w53	w7																4	6	4	6	6	6	6	4	6
9031	4, w53	w3													6		6				6	4	6	4	6	6
9032	4, w53	w8														4	8				6	4	6	4	6	6
9089	4, w53	w7													4	6	6	6	6	6	6	6	6	4	4	6
9090	4, w53	w7												4	6	6	4	6	8	6	6	6	6	6	6	4
9091	4, w53	w7												4	6	4	8	6	6	8	8	4	6	6	8	8
9098	4, w53	w3														4	4	6			8	6	6	6	6	8
9035	w11, w52	w7													4		4	4	6	8	8	6	6	8	4	8
9037	w11, w52	w7								4		6	4		8	6	6	8	8	8			8	6	6	8
9039	w11, w52	w7														4	6	6	6		6	6	4	4	6	6
9040	w11, w52	w7															4	6	8		8	8	4	6	4	6
9041	w11, w52	w7														8	6	4	6	8	8	6	6	4	8	8
9042	w11, w52	w7													4	8	6	6	8	4	8	8	6	8	6	6
9043	w11, w52	w7																4	4		6	6	4	4	4	4
9044	w11, w52	w7													4		4	6	8	4	6	6	4	6	6	4
9038	w12, w52	w7													4		6	4	8	6	8	4	6	6	6	6
9045	w11,w12	w7											6				4	6	6	4	6	6	4	6	4	4
9064	w14, w52	w7													6	6	6	6	8	6	6	8	6	6	6	8
9016	w2	w7														4	6	6	4	6	6	6	8	6	8	

Figure 4. C×S analysis of class II MAb and Core BLCL. Clusters of MAb with DQw2, DR7, or DQw3-like reactivity.

WS #	A	B	2067	2057	2064	2065	2080	2098	2101	2055	2062	2058	2047	2110	2100	2108	2054	2052
9001	24	5		8	8	8	6			8	6	4		4	4	6		
9002	24	14			6	4	4								6	6		
9003	9	5		6	6	6	6	8	4	6		4		6	6	4		
9018	3, 24	18			6	4	4							4				
9028	24	60		6	6	6	4			4								
9044	9	5		4	6	4	6	4	6	4					6	4		
9075	24	60		4	6	6	4			4					6	6		
9103	24, 26	51, 61	6		6	8	4			4	4				6	4		
9107	24	55		4	6	6	6	4		4	6	4	4	4	6	6	4	
9008	25	18	8				8	4		8	8							
9035	32	38		8	8	8	8	6		8		8		4	8		8	4
9063	32	44		6	6	8	8	8		8	4		4	6	8	6	8	
9010	28	5	4	6	6	4		6		4				6	6	4		
9011	1	52			8	6		8	6	4					6			
9016	2	51	4	6	6	6	6	6	8	4	4				6	4		
9030	31	51		4	6	6	4	6	8	6	6			8	6	8	6	
9045	2	51			6	6		4		4					6	4		
9070	2	51		4	6	6	4	6	6	6	4			4	8	8	8	
9027	29	44		6	6	6		6					8	4	6	4		
9036	2	12		6	8	4	8								8			
9050	29	44		4	6	6	6	6					8		4			
9051	29	44		4	6	6	6	6					8	6	6	4		
9053	33	44		4	6	6	4	4					4	4	4	4		
9054	2	12		6	6	6	4	4							6	4		
9090	2	44		4	6	6	4	4							4	4		
9040	1	21		4	8	4	4	6	6	6	4				8			
9012	2	57		4	6	6	6	6	6	4	6			6		6	4	
9052	2	57		6	8	8	6	4	6	6	4	4		8	6	8	4	
9102	3	17		4	6	4	4						4	4	4	4		
9046	2	13		4	6	6	6	8		4			4	8	8	4	6	4
9048	30	13		4				4						4	8	4		
9093	2	13		4	6	6	6	4	8						6	4		
9047	3	47		8	8	6									4	4		
9004	2	27		4	6	6	8			6			4		6	6	6	
9005	3	27		4	6	6	6			6	4		4	8	8	6	8	
9015	2	27	8	8	8	8	8			8	6		6	6	8	6	6	
9067	2	27		4	6	6	4					6			8	6		
9026	26	38	8	6	6	8	6	4		4	4			6	8	8		
9057	26	38	4	8	6	8	6	4		4	4			4	6	4		

Figure 5. C×S analysis of class I MAb and Core BLCL. Clusters of MAb that react with cells expressing Bw4.

DR	3	3	3	13	3	3	13	14	11	11	11	11	12	13	13	14	8	8	8	8	8	8
Dw	3	3	N	18	3	3	18	9	5	5	N	N	B6	19	19	16	8.1	8.1	8.2	8.2	8.3	8.3
C	9	9	9	9	9	9	9	9	9	9	9	9	9	9	9	9	9	9	9	9	9	9
E	0	0	0	0	0	0	0	0	0	0	0	0	0	0	0	0	0	0	0	1	0	0
L	2	2	2	5	1	2	6	6	3	4	3	4	3	6	5	6	6	6	7	0	7	6
MAb L	2	3	1	8	9	0	0	1	7	4	9	2	8	3	9	4	8	0	1	0	0	6
PLM3 3048	+	+	+	+	-	-	-	-	-	-	-	-	-	-	-	-	-	-	-	-	-	-
NDS10 3026	-	-	-	-	+	+	+	+	+	+	+	+	+	-	-	-	-	-	-	-	-	-
7.3.19 3025	+	+	+	+	+	+	+	+	+	+	+	+	+	+	+	+	-	-	-	-	-	-
8.1 3063	+	+	+	+	+	+	+	-	-	+	+	+	+	+	+	+	-	-	-	-	-	-
NDS13 3027	+	+	-	+	+	+	+	+	+	+	+	+	+	-	+	-	+	+	+	+	+	+
M4G-8 3062	+	+	+	+	+	+	+	+	-	-	-	-	-	-	+	+	+	+	+	+	+	+
R3 3061	+	-	-	-	+	+	+	+	-	-	-	-	-	-	+	+	+	+	+	+	+	+
T CELL CLUSTER	Dw24				Dw25				Dw25					Dw26	Dw24							
2-D GEL DR ßIII	3				1				1	,2	1			2	2	3						

Figure 6. MAb define splits of DRw52 that correlate with those defined by T-cell clones and by biochemical analysis.

phoretic mobility is indicated. MAb 3048 bound to cells typed as Dw24 with the exception of the cell expressing DRw14, Dw16. MAb 3026 defined the same cluster as T-cell clones defining Dw25. However, this cluster was further split by reactivity to MAb 3061 and 3062. The DRw13-related haplotypes were somewhat more complex. Although T cells and MAb agreed on cluster of cells expressing DRw18, they disagreed on DRw19 cells splitting the Dw26 cluster. The T-cell clones and four of the MAb did not react with cells expressing DRw8. Of the three MAb that did bind to DRw8-positive cells, three were negative on cells with DR5-related haplotypes.

Thus, MAb defined splits of DRw52 that corresponded closely to those defined by T-cell clones. They may aid in mapping the relevant epitopes to the DRBl or the DRB4 chain in each haplotype.

Conclusions

1. A substantial portion of MAb showed good agreement with antigen specificity in Micro EIA and microcytotoxicity.
2. DQ specificities (DQw1, DQw2, DQw3, DQw4, and DQw7) were well defined by MAb.
3. DR specificities defined by alloantisera (DR3, DR7, DR9 DRw11, DRw12, DRw52, and DRw53) were also defined by MAb. Other MAb detected complex but reproducible class II specificities.
4. There was good definition of several A- and B-locus alleles.
5. Public epitopes common to several specificities were defined, including interlocus (AxB) determinants.

Cytotoxic Monoclonal Antibodies

L.J. Kennedy,[1] S.G.E. Marsh,[1] and J. Bodmer[1]

A total of 82 ABC and 74 D-region cytotoxic monoclonal antibodies (MAbs) were tested in the Tenth Workshop. Special tray layouts were used for the MAbs, which involved putting a variable number of wells containing diluent between each MAb. This allowed carryover to occur into the diluent wells and prevented contamination of the next MAb. Analysis of Ninth Workshop MAb data showed extensive carryover with both class I and class II antibodies. The addition of diluent wells into the Tenth Workshop tray layouts appears to have successfully solved the carryover problem, and the overall computer analyses were as informative as those for the alloantisera.

Tables 1 and 2, respectively, list all the cytotoxic ABC and D-region MAb tested in the Tenth Workshop giving their Tenth Workshop number, Ninth Workshop number (if appropriate), local name, laboratory of origin, speci-

Reporting Laboratory: UKIBOD[1]

Table 1. Cytotoxic ABC Monoclonal Antibodies

10WS#	9WS#	Local Name	Lab of Origin		Specificity on PBL and Lymphoid Lines	Extra Specificities Found on Lymphoid Lines Only
1220	1134	SY1	US7	DUP	B13	
1250		VE5	GER	WAN	B13	
2001		POSITIVE	US7	DUP		
2002		NEGATIVE	US7	DUP		
2003		TER.M.A1	US5	TER	Multispecific	
2004		GSP16.1	US7	GSC	A1,11,23,24,25,26,w43,w36	Aw34,29,30,31,w33,w68
2005		GAP.A3	US1	WAR	A3	
2009		XI.23	FRA	FAU	A3	
2010	1131	X15.4	IT2	GAN	A3	
2013		GSP55.2	US7	GSC	A3	
2014	1126	AUF5.13	FRA	DDC	A3,11,w43,w36	A29,32,w34
2017		TER.M.A11	US5	TER	A3,11	A32,B7,w42,8,w60,w61, 41
2020		76.3.10	UKI	GEL	A2,w68,w69	
2021		NFLD.M2	NCY	MAR	A2,w68,w69	
2025		76.3.21	UKI	GEL	A2,w68,w69	A23
2026	1142	CBRI.M1.20.86	UKI	LAW	A2,w68,w69	
2027		CBRI.M3.63.14	UKI	LAW	A2,w68,w69	
2028		CBRI.M3.70.2	UKI	LAW		A2,w68
2029		XV.17	FRA	FAU	A2,w68,w69	
2030		MAD148	FRA	DDC	A2	A3,29,31
2032		4B	US7	DUP	A2,w68,w69 misses some Aw68	
2033	1156	TU101	GER	MUC	A2,w68,w69	
2034		TU152	GER	MUC	A2,w68,w69	
2036		UL.39.24	GER	GOL	A2,26	Aw68,w69,25,w34,w43,29
2037	1139	BB7.2	UKI	BOD	A2,w69	
2039		GSP20.1	US7	GSC	A2,w69	A31,w34,w68
2040		MTC301338	FRA	KRE	A2,w68,w69	
2041		AC9.51	IT1	CEP	A2,3,w68,w69,29,30,31,w33	
2042		GSP5.3	US7	GSC	A2,3,w68,w69,29,30,31,w33	
2044		129.2.47	UKI	GEL	A3,25,w34,26*,31*,w33*,11*,30*	A2,w68,29
2045		HU31	JAP	AIZ	??	
2046		KT6	GER	WAN	A2,29,3*	A31,w33
2047		H281	FRA	FAU	??	
2048	1121	OCH.184.3M	FRA	DDC	A29	
2049		H287	FRA	FAU	??	
2050		TER.M.A3031	US5	TER	A30,31	
2052		TER.M.A32	US5	TER	A32	A25,23,24,B27
2054		GSP56.1	US7	GSC	A32,B13*,27*	A23,24,B37,49,17,w47*
2055	1125	CATA1	FRA	VIV	A24,25,32,B51,52	A23,B49,13,17,27,37,w63
2056		TER.M.A2532	US5	TER	A25,32	A23,24,B27
2057	1155	TU48	GER	MUC	A23,24,32,Bw4	
2058		BS112.1	US7	GSC	A23,24,25,32,B27,13*	B5,8,17,49,w63,?44
2060		HA9	JAP	MRI	A23,24	
2061	926	HU49	JAP	AIZ	A23,24	
2062		TER.M.A910	US5	TER	A23,24,25,32,Bw4*	
2063		GS114.1	US7	GSC	A24	
2064		TOK45.3A	JAP	TSU	A23,24,32,Bw4	
2065		TOK45.3B	JAP	TSU	A23,24,32,Bw4	
2066		TER.M.A10	US5	TER	A25,26,w34,w43	
2067		MA10	JAP	MRI	A25,26,w34,w68*,w69*	w33*
2069		MP5	IT2	FER	Bw60,w61,w47,27,13,7	Bw48
2070		KS.3	NCY	FRN	B7,w42,27,w22,w64,39,?w73	Bw65
2071	1128	B27M1	US3	GRU	B27,7*,14*,?w73	
2072		IB.134	FRA	FAU	B7,27,w42,14,w47,16,w22,49,18+	B12,40,5++++++
2073		POSITIVE	US7	DUP		
2074		NEGATIVE	US7	DUP		
2076		KS.4	NCY	FRN	B7,w42,27,w22,w64,?w67,?w73	Bw65
2077	1129	B27M2	US3	GRU	B27,w47,12*,13*,14*,15*,40*	B41,49,38,37
2078		TER.M.B27	US5	TER	B27	B14,13,37,w47,49,?A32

Table 1. *Continued*

10WS#	9WS#	Local Name	Lab of Origin		Specificity on PBL and Lymphoid Lines	Extra Specificities Found on Lymphoid Lines Only
2079		MAD120	FRA	DDC	B7,8,41,40,A32*	Aw33,Bw42,13,27,15, w52,w48,?A30,31
2080		BD.7	FRA	FAU	A32,B27,37,13	A23,24,25,29,B17,38, 44,w47,49,?A26
2081		TER.M.B37	US5	TER	B13,27,37	A32,Bw47,44
2086	1154	TER.M.B8	US5	TER	B8,w59	
2087		NDS.8	UKI	MOR	B8	B14, ?Bw42
2089		GSP8.1	US7	GSC	B8	Bw59
2090	1132	AB10.58	IT1	CEP	A2,B8,14,16,18,41	Bw59,37,w22,w42
2091	1117	XX111.23	FRA	FAU	B8,14,5,18,38*,49*,A25*,w33*	A26,w68,B15,35,37
2092	1153	TER.M.B13	US5	TER	B13	
2094		UL.39.10	GER	GOL	B13	
2096		BER3.74	FRA	DDC	A32,B7,8,w60,w61,41,w48,44,45	A23,25,29,30,w33,B14, 15,35,?A31
2097		KT4	GER	WAN	B5,44,45	B8,w59,13,14,15,17,18, 49,w54
2098		S4	GER	WAN	B51,w52,13,57,w58,49,38*,44*	Bw53,w47,A23,24,25,26, 32,?B37
2101	1157	TU109	GER	MUC	A23, weak Bw4	
2102		116.5.28	UKI	GEL	A32,25,Bw4	A23,24
2103		103.1.51	UKI	GEL	??	
2105		T11H10.1A	US1	THP	??	
2106		MTC29306	FRA	KRE	Bw6	
2107	1143	SFR8.B6	NCY	SFR	Bw6	
2109		YU3.20A	JAP	TSU	POSITIVE	
2111		HU34	JAP	AIZ	Multispecific	
2116		BOS29.5	US7	DUP	A25	
3017		M2D2	GER	WAN	B8*,40*,A30*,31*,w33*	B7,w42,A29,32 ?A23,B44,14

ficity on peripheral blood lymphocytes (PBL), plus any extra reactions found with B-lymphoblastoid cell lines (LYL). However, it should be noted that many of the LYL tested were homozygous typing cells. Many of these MAb are also discussed in the appropriate Antigen Society Reports.

ABC Monoclonal Antibodies

There were few specific class I MAb in the Tenth Workshop. Most detected epitopes present on several different antigens, which can be locus specific or interlocus. Many of the MAb had additional reactions when tested on LYL compared with PBL and they can be divided into four categories on this basis: 1) MAb with identical patterns on both PBL and LYL, 2) MAb that react only on LYL, 3) MAb that show an extended reaction pattern on LYL, and 4) MAb that show an extended reaction pattern on LYL, but which may also be seen on PBL occasionally, i.e., if the cell is homozygous for a particular antigen, or has two or more of the antigens included in the extended reaction pattern. These antigens are indicated with an asterisk in Table 1 and Table 2. Some of these reaction patterns may be a direct result of the concentration at which the antibody was used in the Workshop.

All the antibodies submitted as Bw4 were shown to contain other reactivities, these being A9 (23,24) and A32. Sequence data available (Parham et al., these proceedings) indicate that the Bw4/Bw6 epitope has been localized to three amino acid residues in positions 79,80, and 83 The same sequence of amino acids at these positions have been found in both A9 (23,24) and A32. It is probable, therefore, that it will be impossible to obtain an antibody that would recognize Bw4 alone. There are also a number of MAb that react to A25 as well as A9,A32 and Bw4. Presumably there is a minor sequence difference in the A25 antigen that accounts for this reactivity.

D-Region Monoclonal Antibodies

There were not many specific DR MAb in the Tenth Workshop, echoing the results of the Ninth Workshop, although those few that were monospecific were exceptional reagents. These included 3021-DRw13, 3036-DRw11, 3068-DRw12, and 3050-DR7. It is particularly interesting to see a MAb against DRw13, as a serum with the same specificity has yet to be identified.

Various MAb reacted with all or a subset of DRw52 (see the report for Antigen Society 29 for further details) but the variants that have been detected biochemically and with T-cell clones could not be identified using these DRw52 MAb. No good DRw53 MAb was found, which is interesting, since good DRw53 sera are also extremely rare.

Table 2. Cytotoxic D-region Monoclonal Antibodies

10WS#	9WS#	Local Name	Lab of Origin			Specificity on B Cells and Lymphoid Lines	Extra Reactions Found on Lymphoid Lines Only
1217	943	77.34	NCY	FAL	*	DQw1 + < DR4	DRw9,7(Dw11), < DQw4
3002		MCANEGCON	US7	DUP		Negative control	
3004		NDS14	UK1	MOR		DR5++	All but DR3,w9
3005		C5C5	GER	WAN		< DR3,4	DR1,w8
3008		CC.CL20	US6	BRN		DR1,4 +++	DR3++
3009		AC1.59	IT1	CEP		DR1,4,w9,w14,w8	MS
3010		CHE153	FRA	DDC		DR1,7,w9 ++	MS
3011	906	HU30	JAP	AIZ		DR1,2,7,w9 ++	
3016		M1H5	GER	WAN		DR2,3,4 ++	
3019	929	16.23	GER	WAN		Weak DR3,w13(Dw18)	
3020		M4F11	GER	WAN		DR3, < w6	
3021		S2	GER	WAN	*	DRw13	
3023		CHE41.2	FRA	DDC		DRw52 (not DR5)	
3025	930	7.3.19.1	BEN	ROO	*	DRw52 (not DRw8)	?+DR7
3028		NDS12	UK1	MOR		DRw52 + DRw9,w10	DR7
3031		JS1	UK1	FES		DR3++	DR4,w6
3034	954	CHE249.2	FRA	DDC		All positive	
3035		MP3	IT2	FER		ABC activity	
3036		GSP88.2	US7	GSC	*	DRw11	
3038		TER.M.D58	US5	TER		DR5.8,++	DR1,4, w14
3041		MAD209.6	FRA	DDC		MS	All positive
3042		MAD88	FRA	DDC		Drw8 +++	DRw12 ++
3044	923	HU48	JAP	AIZ		DRw52,DR4,w10	
3045		GSP97.1	US7	GSC		DRw53,DRw10,w14	DR1
3046		C6E2	GER	WAN		DR1,3,4,7,w10,w14	All but most DR2,5,w8
3047		NFLD.M3	NCY	MAR		DRw53 +++	DR3+
3048		PLM3	JAP	JUJ		DR7,w9,DRw52	
3049	949	17.3.3	UK1	BOD		DR7 +DR4(Dw15)	DRw12
3050		SFR16.DR7M	NCY	SFR	*	DR7	
3051		GSP65.1	US7	GSC		DRw53+++	
3052		TER.M.DR9	US5	TER		All but DR3,7,w13	
3053		AB7.AE9	NCY	FRN		All negative	
3054		50D6	NCY	FAL		All but DR7	
3055		GSP14.1	US7	GSC		All but DR7	
3056		HU20	JAP	AIZ		All but DR7	
3057	904	HU26	JAP	AIZ		All but DR7	
3059		PD3	BEN	ROO		Nearly all positive	All but Dw2,DRw10, w13,Br
3060		1B6	BEN	ROO		Nearly all positive	Like 3059
3061		R3	GER	WAN		Weak DRw52	
3062		M4G8	GER	WAN		Weak DRw52 (not DR3)	
3063	946	8.1	UK1	BOD	*	DRw52 (not DRw8,w12)	
3064		DP11.1	UK1	BOD		Weak DPw2,w4	*DPw2,w4
3065	955	MAJA51	FRA	DDC		MS	
3066		XIV.466	IT2	GAN		DQw3 ++	
3068		TER.M.DN1	US5	TER	*	DRw12	
3075		E1	BEN	ENG		All negative	
3076		M3H5	BEN	END		All negative	
3079		HKB.1	SCA	TSB		All positive	
3086		R1	GER	WAN	*	DQw1,4 +DR4 =2B3	DR7(Dw11),w9
3088		IIB3	BEN	ROO	*	2B3	
3090		II18B	BEN	ROO	*	2B3	
3091		KS.11	NCY	FRN	*	DQw1,4	
3092	945	4.1	UK1	BOD	*	DQw1	
3093		NFLD.H1	NCY	MAR		MS	
3095		G2A5	UK1	BOD	*	DQw1	DRw53
3096		GS148.1	US7	GSC	*	DQw1	DR7
3097	913	HU11	JAP	AIZ	*	DQw1 ++	All but DQw2
3101	921	HU46	JAP	AIZ	*	DQw4	
3105		C1A2	GER	WAN	*	DQw2	
3106		MP1	IT2	FER	*	DQw2	
3107		XIII358	IT2	GAN	*	DQw2	

Table 2. *(Continued)*

10WS#	9WS#	Local Name	Lab of Origin		Specificity on B Cells and Lymphoid Lines	Extra Reactions Found on Lymphoid Lines Only
3110		MA10	JAP	MRI	Nearly all positive	
3111	915	HU18	JAP	AIZ	* DQw3	
3112		PLM12	JAP	JUJ	* DQw3	
3113		GSP91.1	US7	GSC	DQw3	DR3,7,w8
3114		GS100.1	US7	GSC	* DQw3	
3115	969	17.15	US7	HAN	DQw3	DR7
3117		SFR20.DQ4	NCY	SFR	All positive	
3118		AC60	NCY	FRN	* DQw3	DRw13
3119	790	A10/13	JAP	JUJ	* DQw7 (TA10)	
3120		PLM2	JAP	JUJ	* DQw7 (TA10)	
3121	909	HU23	JAP	AIZ	* DQw7 (TA10)	
3122		PLM13	JAP	JUJ	MS	All but DQw2,w7
7220		Pos. control	US7	DUP	Positive control	

* = specificities detected preferentially on LYL (see text).

Many of the other DR MAb had clean reaction patterns (with strong positives and few weak reactions) but they did not correlate with known patterns of cross-reaction. Such antibodies are presumably detecting epitopes nor yet defined in conventional serology or by biochemical analysis, although some correlation with RFLP banding patterns has been seen in local data. (See also the RFLP-MAb correlation study reported elsewhere in this volume.)

The success of the Tenth Workshop MAb study, however, has been the identification of a number of extremely good DQ typing reagents defining DQw1, DQw2, DQw3, DQw4, DQw7, and 2B3. Again it is interesting to note that there is still only one "pure" DQw1 MAb, i.e., 3092: all the others are either DQw1+DQw4, DQw1+DR7, or 2B3. These MAb are particularly useful in the cytotoxicity assay because they cannot be contaminated with any associated DR antibodies, and therefore make the identification of unusual DR-DQ combinations fairly straightforward.

Overall, the results with the MAb were very encouraging, with a good number of useful ABC and D-region antibodies being identified. Many of the MAb were also tested by ELISA with equally good results. (See the ELISA Joint Report for more details.) As the use of MAb becomes more common and the MAb themselves become more reliable as we learn how to handle and store them, tissue typing will improve, especially in the serologic definition of DQ.

Joint Reports: HLA Class I Polymorphism Detected by One-Dimensional Isoelectric Focusing

Population Analysis of Class I HLA Antigens by One-Dimensional Isoelectric Focusing Gel Electrophoresis: Workshop Summary Report

Soo Young Yang[1]

A total of 918 cells obtained from 58 serology laboratories were analyzed by 14 laboratories using one-dimensional isoelectric focusing gel electrophoresis (1D-IEF). This cell panel represented all ethnic groups designated by the Tenth Workshop and included 19 A locus, 35 B locus, and 8 C locus serologically defined alleles. To facilitate antigen assignment for each IEF band, two monoclonal antibodies (4E and W6/32) were used in sequential immunoprecipitations. This study demonstrated that the 1D-IEF method could define most of the serologically defined alleles and was very effective in detecting serologically indistinguishable variants or subtypes for the class I HLA antigens. The Biochemistry Class I Workshop has identified a total of 34 alleles for A locus, 52 alleles for B locus, and 11 alleles for C locus detected by 1D-IEF from this Workshop exercise. The study also demonstrated that there is an allelic consistency across different races (i.e., the same 1D-IEF alleles were frequently detected in several ethnic groups), indicating that HLA polymorphism predates the racial diversification. However, ethnically restricted specificities were observed for many alleles, suggesting that generation of new polymorphic determinants may have occurred recently. Those ethnic-specific determinants may have occurred by chance or by natural selection in adaptation to different environmental conditions.

The present Workshop Summary Report contains the description of the HLA-A and HLA-B class I alleles detected by 1D-IEF. The description of the HLA-C IEF alleles is presented in the report prepared by Hajek-Rosenmayr and Doxiadis (this volume).

A summary of the number of IEF alleles detected for the serologically defined HLA-A and HLA-B alleles is shown schematically in Figures 1A, B. This figure also demonstrates the relative pI position for each of the IEF-detectable alleles. The present report describes many examples of genetic linkage disequilibrium between IEF alleles of the HLA-A and HLA-B loci or, HLA-B and HLA-C loci, as summarized in Table 5. Serologic identification of IEF variants could be made in a few instances. It is very likely that many other IEF variants will be serologically defined in the future. Serologic data obtained with the Workshop core serum set were, however, only available for 386 of the 918 cells studied in the Biochemistry Class I Workshop component, and only very few cells were also tested with the supplemental antisera from the appropriate Antigen Societies. The serologic data base for correlation analysis between serology and IEF specificities was, therefore, insufficient to allow an in-depth analysis.

The present collaborative Workshop study allowed for the first time a comprehensive determination of HLA class I 1D-IEF variants on a large panel of unrelated cell donors representing all ethnic groups. This was made possible because of the very helpful and dedicated cooperation of 58 HLA serology laboratories that arranged for the shipment of representative cells to the 14 laboratories performing the biochemical studies. The present data analysis report demonstrated a large number of new IEF variants and, in particular, allowed for identification of the ethnic distribution of the different HLA alleles. Further characterization of these alleles, including nucleotide sequence analysis and development of monospecific cellular or serologic reagents, will be necessary for determination of the biological significance of IEF-detectable differences within the individual serologically defined alleles.

Introduction

The extensive polymorphism of class I HLA antigens was originally demonstrated through the use of serologic reagents. More recently, even greater diversity of class I antigens has been revealed by cell-mediated lympholysis and by molecular and biochemical analysis. One-dimensional IEF has been particularly informative in detecting variant forms of serologically established alleles (1–3). Taking advantage of relatively simple 1D-IEF procedures and the collaboration within the International Workshop, the Class I Biochemistry component was undertaken to evaluate the extent of class I HLA diversity by 1D-IEF at the population level.

The aims of this study were: 1) to evaluate the serologic definition of each allelic product to provide a biochemical definition for each antigen by 1D-IEF, 2) to identify subtypes or variants for serologically defined allelic products, 3) to evaluate ethnic differences for serologically defined alleles that are shared by different ethnic groups, and 4) to detect nonclassic HLA class I molecules.

Cell Panel

Panel cells were collected from 58 Antigen Society leaders; these cells plus the core cell lines were analyzed by 14 laboratories (Table 1). The serologically defined

---

Participating Laboratory: US7SYY[1]

The Workshop Summary Report was prepared by the chairperson on behalf of the participants of the Class I Biochemistry component: A.M. Little (UKIAST), A. Madrigal (UKIBOD), P. Klouda (UKIBRS), P. Dyer (UKIDYR), L.P. de Waal (BENENG), Y. Choo (US7HAN), S. Ellis (UKIMOR), A. Hajek-Rosenmayr (EAEMYR), K. Wong (NCYMRV), M.J. Giphart (BENROO), B. Duffy (NCYSTL), R. Tonai (US5TER), and P. Reekers (BENREE).

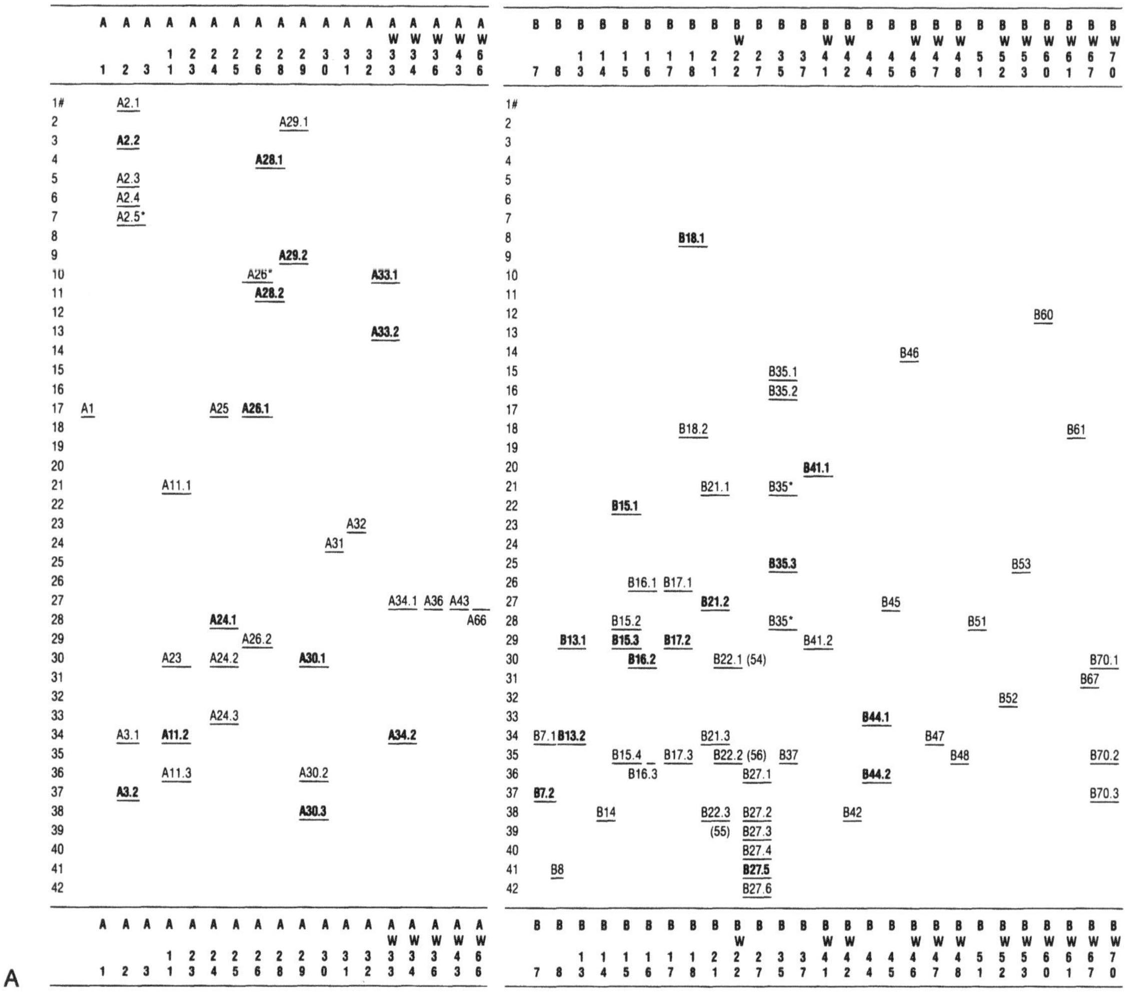

Figure 1. IEF band patterns for HLA-A (A) and HLA-B locus (B) gene products.

Table 1. Participating Laboratories

Biochemistry Laboratory	Serology Laboratory
AST	AST, MUE, DRK, DCH
BOD	BOD
BRS	BRS
DYR	DYR
ENG	ENG
HAN	HAN
MOR	MOR
MYR	MYR
MRV	MRV
ROO	ROO
STL	STL, CRO, MRV
TER	TER, DCH
REE	REE
SYY	ABS, ARN, BRA, BRN, CAR, CEP, CHI, DCH, DUP, DUT, FAI, FAL, FAU, FRA, FTW, GAN, GOE, GOR, GRU, JEA, JER, JRM, KAS, KAP, KYL, KNT, LAW, MIT, MLS, MUE, MUL, OHA, POL, PRR, SAC, SND, TOM, TRP, WOL, WSH, YAG, ZHA

HLA specificities were determined by the serology laboratory donating the cells using the Tenth Workshop sera. Cells that were typed with Ninth Workshop sera were also included in the IEF analysis in many instances. A total of 918 panel cells were used for the final analysis, and the composition of ethnic groups is listed in Table 2.

Methods

A standard IEF protocol was designed for the Biochemical Class I Workshop component using a modification of the original O'Farrell's two-dimensional gel method (4) as well as the Triton X-114 solubilization of membrane (5), which was first applied for class I HLA by Neefjes et al. (6). During the Workshop, it was realized that some of the HLA antigens were poorly bound to the Triton X-114, making data interpretation difficult. It was, however, found that Triton X-114-lysed cell lysates gave cleaner background when compared with that of Nonidet P-40-lysed cell extract. Cell lysis steps in the recommended protocol have been revised. A detailed 1D-IEF protocol is presented in this volume (in the report Method for Class I HLA Analysis by One-Dimensional Isoelectric Focusing (IEF) Gel Electrophoresis). The Workshop protocol utilized two monoclonal antibodies – 4E (3) and W6/32 (7) – for immunoprecipitation. The 4E, which recognizes a HLA-B and -Aw19 common epitope, discriminates HLA-A antigens except the Aw19 from the HLA-B antigens. The W6/32 identifies all the remaining class I antigens in the cell lysates. To eliminate heterogeneity produced by sialic acid residues of the carbohydrate moiety of HLA heavy chain, immunoprecipitates were treated with neuraminidase.

Data Management

Data were collected on floppy diskettes provided by the Data Analysis Group, Salt Lake City. The diskettes contained programs for gel lay out and contained cell information and the interpretation of IEF bands. The central standardization program was prepared by Jean Marc Lalouel; however, too many overlapping IEF bands made the central standardization by computer impractical. Therefore, interpretation of each band was made manually utilizing commonly occurring antigens as standard bands on each gel.

Table 2. Ethnic Codes and Number of Cells Studied

Ethnic Codes	# Studied	Ethnic Codes	# Studied
Negro (N)	69	Caucasian (C)	502
10: African Negro	22	510: Canadian Caucasian	2
20: American Negro	41	515: U.S. Caucasian	168
30: Cape Colored	6	516: Amish	1
Oriental (O)	54	517: U.S. Hispanic	10
110: Chines	18	520: Australian	18
120: Japanese	25	525: Austrian	43
130: Other Oriental	11	535: Dutch	9
American Indian (AI)	29	540: English	96
210: N. Am. Indian	15	541: Scot	5
220: S. Am. Indian	9	545: French	25
230: Mexican Indian	5	550: German	21
Southeast Asian (SEA)	42	555: Hungarian	1
310: Thai	33	560: Italian	26
320: Malaysian	1	565: Ashk. Jew	10
330: Indonesian	2	566: Non-Ashk. Jew	15
399: Other S.E. Asian	6	575: Scandinavian	16
Unknown	222	580: South African	4
00: No information	222	585: Swiss	4
		595: Yugoslavia	1
		599: Other Caucasian	37

Designation of IEF Bands

The names for each IEF band were decided on by the participants in the Biochemistry Class I component at the Workshop meeting in Princeton, New Jersey. Sequential decimal numbers from the top of the gel (cathode) were adopted after the serologic specificity numbers for antigens with more than one band. For example, three bands that correspond to the A11 antigen are designated as A11.1, A11.2, and A11.3 starting from the basic (top) to the acidic (bottom) part of gel. In the case of a serologic split with no correlation with IEF bands, supertypic serologic specificities were used. For example, a total of three IEF bands corresponding to the B17 antigen were observed, but no correlation could be made between the Bw57 or Bw58 antigens and the three IEF bands. These three B17 bands were named B17.1, B17.2, and B17.3, starting from the top of the gels.

The IEF assignment of the 918 panel cells used for the IEF analysis is listed in Table 3. To keep the original cell specificities, IEF subtype designation was added to the original cell specificities with decimal numbers as described above. For example, Bw57.2 cells and Bw58.2 cells have the same B17.2 IEF band.

Results

The report of the final analysis is individually presented for each of the Workshop Antigen Groups (Antigen Group/Society #1 to #14 as designated for the serology workshop exercises) these reports are given below. Within the collaborative Workshop exercises, it was not possible to make accurate assignments for the HLA-C locus antigens not only because of poor cell surface expression, but also as a result of multiple bands for each Cw-allelic product, which were masked by predominant A or B antigens on IEF gels. Accurate analysis of the IEF studies for HLA-C locus products at the Workshop level must wait until HLA-C-specific monoclonal antibodies are available. Reports for the HLA-C locus products are presented in this volume as independent studies. Non-classic HLA Class I antigens also failed to be identified in this Workshop study. It has been shown that the class I products encoded by some non-HLA-A, -B, -C genes also react with monoclonal antibody W6/32 (8), although these antigens are the most divergent class I genes from the classic HLA genes (9,10). If there was any cell surface expression, the nonclassic HLA antigens seem to have been masked on the 1D-IEF gels by the HLA-A, -B, -C antigens. Many uninterpretable minor bands were, however, precipitated by 4E or W6/32 and may belong to the nonclassic HLA Class I antigens.

Summary of Data

Table 4 lists the alleles detected by IEF for HLA-A and HLA-B gene products. The IEF method has detected a considerable number of novel HLA serologically indistinguishable subtypes. HLA-A2, the most basic, and B27, the most acidic protein, showed the largest number of IEF variants. Figure 1 illustrates schematically the IEF band patterns for all HLA-A and B alleles detected by IEF. In five cases (A25/A26, Aw68/Aw69, Bw64/Bw65, Bw57/Bw58, and B49/Bw50), IEF did not detect the serologically defined splits. However, the IEF method was particularly useful in discriminating between many of the serologically cross-reactive antigens (A30/31, B38/B39, B7/Bw70, B35/Bw70, Bw71/Bw72, etc.; see Fig. 1 and individual reports for each Antigen Group). One of the most interesting observations from this study was the allelic consistency across different ethnic groups. The majority of serologically defined common HLA alleles found in different ethnic populations also had the same IEF patterns. This suggests that the serologically detected HLA allelic epitopes predate racial diversification. Available DNA sequence data for serologically indistinguishable IEF subtypes found in different ethnic groups show that very few, if any, nucleotide differences exist in the coding regions of those genes, suggesting that HLA class I alleles only very rarely undergo new mutations. HLA crystal structure (11) has now revealed the amino acid sequences for antigenic sites, the T-cell recognition site, and the foreign antigen binding site (12). When nucleotide sequences for more HLA alleles become available, it may be possible to predict if nucleotide substitutions in different regions of the molecule are equally important in clinical transplantation. It is, for example, presently unknown if mismatches for serologically indistinguishable HLA class I alleles represent a major HLA class I antigen mismatch.

While IEF typing for HLA antigens is very effective in detecting variants or subtypes, this method does have its own disadvantages. Interpretation of IEF patterns is entirely dependent on the serologic definitions. Wrongly assigned serologic specificities will result in misleading IEF band assignment. Too many overlapping bands with different HLA antigens make the IEF method impractical to substitute for HLA typing. However, serology and IEF together can provide a better definition for each allelic product. The use of cells defined by these two methods for characterization of tissue typing reagents will, in turn, facilitate an accurate assignment of HLA class I antigens.

Antigen Group Reports

Antigen Group #1 (A1, A3, A11, Aw36, A9, A23, and A24). The common IEF bands of the HLA-A1, A3, Aw36, and A11 antigens were readily distinguishable from each other (Fig. 2). A single rare subtype for A3 and two variants for A11 were observed, whereas no IEF variants were identified for the A1 and Aw36 antigens. The A11 variants (A11.1 and A11.3) were found only in Orientals (Table 6). Five of the nine A11.1-positive cells and all three A11.3-positive cells were typed as Bw46. The A3.1 and A11.2 comigrate; A3.2 and A11.3 also share the same position on an IEF gel.

The A9 Antigen Group subdivides into A23 and A24. There is serologic evidence that the A9 group may have

Table 3. Cell Panel for 1D-IEF

BIO	SER	ID	ETH	HLA-A		HLA-B		HLA-C	
AST	AST	828	540	A1.1	A3.2	B8	Bw64	Cw7	
AST	AST	435	540	A2.2		B27.5	Bw64	Cw1	
AST	AST	902	540	A2.2	A28.2	B44.1	Bw64		
AST	AST	1013	540	A3.2	A29	B44.1	Bw65	Cw8	
AST	AST	880	540	A2.2	A3.2	B7.2	Bw64		
AST	AST	1056	540	A1	A2.2	B8	Bw64	Cw7	
AST	AST	457	540	A1	A3.2	B7.2	Bw65	Cw6	
AST	AST	971	540	A3.2		B44.1	Bw65	Cw8	
AST	AST	1049	540	A1	A3.2	B44.1	Bw65	Cw8	
AST	AST	1008	540	A1	A2.2	B8	Bw65		
AST	AST	1017	540	A11.2	A30.1	B18.1	Bw65	Cw5	
AST	AST	764	540	A3.2	Aw33.1	Bw62.3	Bw65	Cw3	
AST	MUE	94	0	A28.1	A2.2	Bw60	Bw65	Cw3	Cw8
AST	MUE	38	0	A11.2	A24.1	B35.3	Bw65	Cw4	Cw8
AST	MUE	10	0	A2.2		Bw57.2	Bw65	Cw6	Cw8
AST	DRK	8998	0	A3.2	A31	Bw62.3	Bw65	Cw3	Cw8
AST	DRK	8995	0	A24.1	A32	Bw60	Bw65	Cw3	Cw8
AST	DRK	58	0	A2.2	A28.1	Bw62.3	Bw65		
AST	MUE	76	0	A3.2		B18	Bw64	Cw5	Cw8
AST	MUE	33	0	A1	A26.1	B8	Bw64	Cw8	
AST	MUE	70	0	A2.2	A24.1	B35.3	Bw64	Cw4	
AST	DRK	8999	0	A24.1	A28.2	B44.2	Bw64	Cw7	Cw8
AST	DRK	4816	0	A3.2	Aw33.1	Bw62.3	Bw65	Cw3	Cw8
AST	DRK	8997	0	A3.2	A32	B7.2	Bw64	Cw7	Cw8
AST	AST	91	540	A2.2	A3.2	B37	Bw65	Cw6	
AST	AST	876	540	A28.1	A30.3	B44.2	Bw65	Cw5	
AST	AST	900	540	A1	A3.2	Bw62	Bw65	Cw1	Cw6
AST	AST	1050	540	A1	A2.2	B44.2	Bw65	Cw5	Cw8
AST	DRK	3530	0	A3.2	A28.2	B44.2	Bw65	Cw7	Cw8
AST	AST	342	540	A30.1	A32	Bw50.1	Bw64	Cw4	
AST	DRK	323	0	A2.2	A3.2	B7.2	Bw64	Cw7	Cw8
AST	DRK	4964	0	A2.1	A32	Bw58.2	Bw64	Cw7	Cw8
AST	AST	910	566	A3.2	A24.1	B15.1	B35.2	Cw4	
AST	AST	8225	540	A32		Bw63.1	B44.2		
AST	AST	8226	540	A1	A11.2	Bw63.1	B51		
AST	AST	8236	0	A2.1	A28.2	Bw63.1	Bw65		
AST	AST	771	540	A32	A29	Bw63.1	B44.2	Cw5	
AST	AST	45	540	A2.2	A30.1	B51	Bw62.3	Cw1	Cw3
AST	AST	334	540	A1	A31	B44.2	Bw62.3	Cw3	Cw5
AST	AST	351	540	A2.2	A24.1	B18	Bw62.3	Cw3	
AST	AST	397	540	A2.2	A28.1	Bw55.3	Bw62.3	Cw1	Cw4
AST	AST	434	540	A1	A2.2	B8	Bw62.3	Cw3	
AST	AST	439	540	A1	A2.2	B17.2	Bw62.3	Cw3	
AST	DCH	2	310	A24.3	A11.1	Bw62.2	Bw46	Cw3	Cw8
AST	DCH	34	310	A24	A2	Bw62.2	Bw46	Cw1	Cw8
AST	DCH	40	310	A2.2	A24.3	Bw62.2	Bw46	Cw1	Cw8
AST	DCH	68	310	A24.1		B15.3	Bw46	Cw1	Cw3
AST	DCH	90	310	A24.1	A2.3	B27.6	B15.3	Cw3	Cw3
AST	DCH	99	310	A2.3		Bw46	B15.3	Cw1	Cw3
AST	AST	868	540	A31	Aw33.2	Bw60	Bw62.3	Cw3	
AST	AST	831	540	A1	A11.2	B7.2	Bw62.3		
AST	AST	994	540	A2.2	A24.1	B44.2	Bw62.3	Cw3	Cw5
AST	AST	905	540	A2.2	A28.2	B44.2	Bw62.3		
AST	AST	1057	540	A2.2		Bw60	Bw62.3	Cw3	
AST	AST	1062	540	A1	A2.2	B17.2	Bw62.3	Cw3	Cw6
AST	DCH	13	0	A11.2	A11.2	B7.1	Bw46	Cw3	Cw7
AST	DCH	16	0	A2.2	A2.3	Bw60	Bw46	Cw1	
AST	DCH	36	0	A2.2	A29	B7.1	Bw46	Cw1	
AST	DCH	17	0	A2.2	A31	B35	Bw46	Cw1	Cw4
AST	DCH	23	0	A11.3	A11.1	B18	Bw46	Cw1	Cw7
AST	DCH	64	0	A2.2	Aw33.2	Bw58.2	Bw46	Cw1	
AST	DCH	11	0	A2.2	A11.2	B15.2	B51	Cw8	
AST	DCH	12	0	A2.3	A11.2	B15.2	Bw46	Cw1	Cw8
AST	DCH	31	0	A2.3	A11.2	B15.2	Bw62.4	Cw8	
AST	DCH	29	0	A2.3	A11.2	Bw46	Bw62.4	Cw1	
AST	DCH	30	0	A11.2	A11.2	Bw46	Bw62.4	Cw1	
AST	DCH	45	0	A11.2	A11.2	Bw46	Bw62.3	Cw1	Cw4
AST	DCH	44	0	A2.2	A11.2	Bw62.3	Bw46	Cw1	Cw4
AST	DCH	46	0	A2.1	Aw34.1	Bw46	Bw56.1	Cw1	Cw3
AST	DCH	79	0	A24.2	Aw33.2	B44.1	Bw63.1	Cw7	
AST	DCH	85	0	A2.3	A24.1	B27.6	Bw46	Cw1	Cw3
AST	AST	8241	540	A1	A30.1	B18.1	B15.1	Cw5	
AST	BRS	208	0	A11.2	A24.1	Bw62.3	B14	Cw3	Cw8
BOD	BOD	307	540	A1		B8	Bw57.2	Cw6	
BOD	BOD	310	599	A1	Aw69.1	B8	B35.1		
BOD	BOD	309	599	A1	Aw69.1	Bw57.2	B35.1	Cw5	
BOD	BOD	308	599	A29.1	Aw69.1	B35.1		Cw4	
BOD	BOD	311	599	A1	A29.1	B8	B35.3	Cw4	
BOD	BOD	459	565	A26.2	Aw69.1	B38.1	B18.1		
BOD	BOD	252	110	A2.3	A24.1	B13.1	B27.4?	Cw3	
BOD	BOD	2	540	A2.2	A24.1	B39.2	Bw55.3	Cw3	
BOD	BOD	23	540	A2.2	A26.1	B38.1	B35.1	Cw6	
BOD	BOD	52	10	A2.1	Aw33	B51	B35.3	Cw4	
BOD	BOD	67	540	A2.2	A29.2	B44.1	B39.4	Cw7	
BOD	BOD	30	599	A2.2	A3.2	Bw62.3	Bw56.3	Cw1	Cw7
BOD	BOD	8	540	Aw68.2	A32	Bw55.3	Bw61	Cw2	Cw3
BOD	BOD	9	599	A11	Aw69.2	Bw52			
BOD	BOD	26	540	A1	Aw68.2	B44.2	Bw57.2	Cw6	Cw7
BOD	BOD	27	540	A25	Aw68.2	B18.1	Bw65	Cw8	
BOD	BOD	33	540	A3.2	Aw68.2	B7.2	Bw47	Cw6	Cw7
BOD	BOD	45	515	A3.3	Aw68.2	Bw65		Cw8	
BOD	BOD	80	540	A2.2	Aw33	Bw41.1	Bw65	Cw8	
BRS	BRS	132	540	A23	A29.2	B45	Bw72.3	Cw2	Cw6
BRS	BRS	103	540	A2.2	A26.1	B44.2	Bw72.3	Cw7	
BRS	BRS	341	540	A2.2	A30.3	B44.2	Bw72.3	Cw2	Cw5
BRS	BRS	348	540	A2.2		B44.2	Bw72.2	Cw5	Cw7
BRS	BRS	70	540	A23	A32	B44.1	Bw72.1		
BRS	BRS	353	540	A26.1	A11.2	B45	Bw72.1	Cw7	
BRS	BRS	103	540	A2.2	A26.1	B44.2	Bw72.2	Cw7	
BRS	BRS	614	540	A3.2	Aw68.2	B18.1	Bw71.2	Cw5	
BRS	BRS	345	540	A1	A11.2	B18.1	Bw72.2	Cw7	
BRS	BRS	343	540	A2.2	A23	B44.2	Bw71.2	Cw5	Cw7
BRS	BRS	347	540	A23	A24.1	B49.2	Bw71.1	Cw7	
BRS	BRS	348	540	A2.2		B44.2	Bw72.2	Cw5	Cw7
BRS	BRS	344	540	A1	A2.2	B44.2	B35.3	Cw4	Cw5
BRS	BRS	82	540	A2.2	A30.1	B18.1	Bw56.2	Cw5	
BRS	BRS	605	599	Aw68.2	A32	B18.1	B51	Cw7	
BRS	BRS	47	540	A2.2	Aw69.2	B8	B18.1	Cw7	
BRS	BRS	163	540	A23	A30.1	B18.1	Bw41.1	Cw2	Cw5
BRS	BRS	271	540	A2.2	A25	B18.1	B27.5	Cw1	
BRS	BRS	600	0	A2.2	A11.2	B18.1	B49.2	Cw7	
BRS	BRS	25	540	A3.2	A24.1	B18.1	B27.5	Cw2	Cw5
DYR	DYR	1	540	A2.2	A29.2	B14	Bw52	Cw6	Cw8
DYR	DYR	69	540	A2.2	A29.2	B44.1	B51		
DYR	DYR	200	540	A3.2	A29.2	B44.1	Bw47		
DYR	DYR	12	540	A2.2	A29.1	B7.1	B8		
DYR	DYR	52	540	A-	A29.2	B7.2	B44.1	Cw7	
DYR	DYR	53	540	A1	A32	B27.5	B44.1	Cw1	
DYR	DYR	2	540	A1	A32	B44.2	Bw58.1	Cw5	Cw6
DYR	DYR	201	540	A24.1	A32	B27.5	B44.2		
DYR	DYR	49	540	A2.3	A32	Bw65	B49	Cw6	
DYR	DYR	46	540	A26.1	A32	B13.2	B38	Cw6	
DYR	DYR	202	540	A11.2	A32	B27.5	Bw57		
DYR	DYR	204	540	A3.2	A32	B39.4	B44.2		
DYR	DYR	209	540	A2.2	A30.3	B13.2	B44.2		
DYR	DYR	66	540	A2.1	A30.3	B13.2	Bw50.1	Cw6	
DYR	DYR	210	540	A2.2	A30.3	B27.5	B-		
DYR	DYR	211	541	A28.2	A30.1	B7.2	B18.1		
DYR	DYR	212	599	A1	A30.3	B13.2	B35.3		
DYR	DYR	72	540	A11.2	A29.2	B35.3	B44.1	Cw5	
DYR	DYR	6	541	AX	A30.1	B13.2	B18.1	Cw5	Cw6
DYR	DYR	71	541	A3.2	A31	B7.2	B51*		
DYR	DYR	213	540	A1	A32	B37	B44.2		
DYR	DYR	214	540	A2.2	Aw33.2	B44.2	B44.1		
DYR	DYR	215	540	A25	A30.3	B18.1	B35.3		
DYR	DYR	28	540	A1	A30.3	B8	B-	Cw7	

BIO:Biochemistry Laboratory
SER:Serology Laboratory
ID :Cell Identification
ETH:Ethnic Code (see Table 2)
* :Rare Variant

Table 3. *Continued*

BIO	SER	ID	ETH	HLA-A		HLA-B		HLA-C	
DYR	DYR	14	540	A3.2	A30.3	B8	Bw47	Cw6	Cw7
DYR	DYR	29	540	A3.2	A30.3	B7.2	B8	Cw7	
DYR	DYR	6	541	A30.1	A30.3	B13.2	B18.1	Cw5	Cw6
DYR	DYR	205	540	A1	A2.2	B18.1	B37		
DYR	DYR	206	540	A1	A25	B18.1	Bw57		
DYR	DYR	208	540	A2.2	A24.1	B7.2	B18.1		
DYR	DYR	63	540	A28.2	A30.1	B18.1	B44.2	Cw5	Cw7
DYR	DYR	216	540	A3.2	A31	B7.2	Bw60		
DYR	DYR	37	540	A2.2	A31	B44.2	B49	Cw5	Cw7
DYR	DYR	71	540	A3.2	A31	B7.2	B51		
ENG	ENG	90	0	A1	A2.2	B49.2	B7.2		
ENG	ENG	91	0	A1	A3.2	Bw57.2	B7.2	Cw6	Cw7
ENG	ENG	94	0	A29.1	Aw33.2	B14	B7.1	Cw5	Cw8
ENG	ENG	94	0	A29.1	Aw33.2	B14	B7.1	Cw5	Cw8
ENG	ENG	95	0	A29.1	Aw33.2	B44.1	B7.1	Cw4	
ENG	ENG	9067	0	A2.2		B27.5			
ENG	ENG	89	0	A2.2	A24.1	Bw60	B7.2	Cw3	
ENG	ENG	100	0	A2.2	A3.2	B35.3	B7.2	Cw3	Cw4
ENG	ENG	96	0	A2.2	A32	B51	B7.2	Cw3	
ENG	ENG	125	0	A30.1		Bw72.1	Bw42		
ENG	ENG	124	0	A1	A30.1	B8	Bw42		
HAN	HAN	301	515	A2.2	A3.2	B27.5	B44.2	Cw3	Cw5
HAN	HAN	351	110	A11.1	A11.2	B27.5	Bw48	Cw6	
HAN	HAN	421	515	A11.2	A31	B27.3	B35.3		
HAN	HAN	521	210	A2.2	A24.1	B27.5	B35.3	Cw2	Cw4
HAN	HAN	311	210	A24.1		B27.5	Bw61	Cw2	Cw3
HAN	HAN	361	110	A3.2	A11.1	B27.5	B38.1		
HAN	HAN	431	0	A2.2	A11.2	B27.2	B35.3	Cw2	Cw4
HAN	HAN	531	210	A2.2	A24.1	B27.5	B35.3	Cw3	Cw7
HAN	HAN	321	210	A24.1		B27.5	B8	Cw3	Cw7
HAN	HAN	371	0	A11.2	A24.1	B27.5	Bw62.3	Cw6	
HAN	HAN	441	515	A3.2	A26.1	B27.2	B35.3	Cw4	Cw5
HAN	HAN	541	515	A1	A2.2	B27.5	B8	Cw2	Cw7
HAN	HAN	331	515	A24.1	A30.3	B27.5	B7.2	Cw2	
HAN	HAN	381	399	A2.3	A24.1	B27.5	Bw60	Cw4	Cw8
HAN	HAN	451	0	A2.2	A24.1	B27.2	B18.1	Cw2	Cw5
HAN	HAN	551	515	A1	A3.2	B27.5	B8	Cw2	Cw7
HAN	HAN	341	210	A2.2	A28.2	B27.5	B51	Cw2	Cw3
HAN	HAN	391	399	A2.2	A11.2	B27.4	B38.1		
HAN	HAN	461	515	A3.2	A25	B27.2	B35.3	Cw2	Cw4
HAN	HAN	481	599	A11.2	A24.1	B27.1			
HAN	HAN	501	20	A23		B27.5	B8		
HAN	HAN	401	20	A31	A32	B27.3	B18.1	Cw2	Cw7
HAN	HAN	471	515	A24.1		B27.1	B8	Cw1	Cw7
HAN	HAN	491	330	A2.2	A11.2	B27.6	B5		
HAN	HAN	511	130	A2.2	Aw33.2	B27.5	B44.1	Cw1	
HAN	HAN	411	515	A2.2		B27.2	B44.2	Cw2	Cw5
HAN	HAN	321	210	A24		27.5	B8		
MOR	MOR	1	540	A3.2	A30.1	B14	B18.1		
MOR	MOR	4	0	A2.2		B18.1	BX		
MOR	MOR	6	0	A2.1	A3.2	B7.2	Bw57		
MOR	MOR	7	0	A1	A2.2	B27.5	Bw57		
MRV	MRV	26	20	A23	Aw33	Bw58	Bw66		
MYR	MYR	34	525	A28.2	A32	B44.2	Bw55.3	Cw3	Cw7
MYR	MYR	48	525	A2.2	A3.2	Bw60	Bw62.3	Cw3	
MYR	MYR	17	525	A2.2	A28.2	B13.2	Bw62.3	Cw3	Cw6
MYR	MYR	8	525	A3.2		B7.2	Bw62.3	Cw3	Cw7
MYR	MYR	23	525	A1	A24.1	B7.2	B13.2	Cw6	Cw7
MYR	MYR	31	525	A2.2	A30.3	B7.2	B13.2	Cw6	Cw7
MYR	MYR	10	525	A1	A24.1	B13.2	B37	Cw6	
MYR	MYR	35	525	A2.2	A29.2	B44.1	Bw57.2	Cw6	
MYR	MYR	7	525	A3.2	A29.2	B51	B45	Cw6	Cw7
MYR	MYR	12	525	A2.2	A11.2	B51	Bw47	Cw6	
MYR	MYR	1	525	A3.2		B7.2	Bw65	Cw2	Cw7
MYR	MYR	58	525	A3.2	A11.2	B7.2	Bw61	Cw2	Cw7
MYR	MYR	69	525	A2.2	A11.2	Bw62.4	B27.5	Cw2	Cw3
MYR	MYR	24	525	A2.2	A23	B44.1		Cw2	Cw4
MYR	MYR	47	525	A1	A24.1	B37	Bw61	Cw2	Cw6
MYR	MYR	77	525	A2.2	A3	B8	B35.3	Cw4	Cw7
MYR	MYR	19	525	A11.2	A24.1	B18.2	B35.3	Cw4	Cw7
MYR	MYR	46	525	A25	A32	B18.1	B35.3	Cw4	
MYR	MYR	25	525	A2.2		B39.2	B35.3	Cw4	
MYR	MYR	16	525	A3.2	A23	B7.2	B44.1	Cw4	Cw7
MYR	MYR	72	525	A26.1		B7.2	Bw55.3	Cw3	Cw7
MYR	MYR	60	525	A2.2		Bw62	B44.2	Cw5	Cw3
MYR	MYR	74	525	A2.2	A3.2	B7.2	B44.2	Cw5	
MYR	MYR	2	525	A1	A2	B44	Bw57	Cw5	Cw6
MYR	MYR	83	525	A2	A31	B44	B35	Cw4	Cw5
MYR	MYR	21	525	A1	A32	B18	B35	Cw4	Cw5
MYR	MYR	61	525	A2.2	A3.2	B18.1	B35.3	Cw4	Cw5
MYR	MYR	85	525	A3.2		B14	Bw47	Cw6	Cw8
MYR	MYR	63	525	A1	A24.1	B14	Bw57.2	Cw6	Cw8
MYR	MYR	67	525	A24.1		B44.2	B14	Cw7	Cw8
MYR	MYR	65	525	A2.2	A26.1	B51	Bw62.3	Cw3	Cw8
MYR	MYR	6	525	A2.2	Aw34.2	B8	B14	Cw7	Cw8
MYR	MYR	71	525	A1	Aw34.2	B8	B14	Cw7	Cw8
MYR	MYR	29	525	A1	A2.2	B51	B7.2	Cw1	Cw7
MYR	MYR	57	525	A2.2	A26.1	B7	Bw56	Vw1	Cw7
MYR	MYR	56	525	A11.2	A25	B51	Bw55.3	Cw1	Cw3
MYR	MYR	76	525	A2.2	A23	B51	Bw62.3	Cw1	Cw3
MYR	MYR	86	525	A3.2	A28.2	B35.3	Bw55.3	Cw1	Cw3
MYR	MYR	18	525	A2.2	A30.3	B51	B7.2	Cw7	
MYR	MYR	70	525	A3.2	A31	B7.2	B60	Cw3	Cw7
MYR	MYR	78	525	A2.2		B18.1	Bw60	Cw3	Cw7
MYR	MYR	81	525	A3.2		B7.2	Bw60	Cw3	Cw7
MYR	MYR	79	525	A1	A23	B49.2		Cw7	
REE	REE	17	0	A24.1	A31	Bw55.3	Bw60		
REE	REE	19	0	A2.2	A3.2	B35.3	Bw60		
REE	REE	9	0	A24.1	A66	Bw60	BN		
REE	REE	18	0	A2.2	A3.2	Bw60	B35.3		
REE	REE	11	0	A2.2	A23	Bw60	Bw63.1		
REE	REE	15	0	A2.2	A11.2	B37	Bw60		
REE	REE	14	0	A2.2	A31	B39.2	Bw60		
REE	REE	8	0	A3.2	Aw66	B35.3	Bw60		
REE	REE	26	0	A3.2	A24.1	B35.3	B37		
REE	REE	22	0	A3.2		B18.1	Bw61		
REE	REE	21	0	A1	A32	B8	Bw61		
REE	REE	20	0	A3.2	A32	Bw61	Bw65		
REE	REE	25	0	A24.1	A29.2	B37	Bw61		
REE	REE	10	0	A11.1	A24.1	Bw52	Bw60		
REE	REE	79	0	A2.2	A28.1	B13.1	Bw61		
REE	REE	82	0	A2.2	A11.2	Bw61			
REE	REE	80	0	A11.2	Aw33.2	Bw58.2	Bw61		
REE	REE	162	0	A2.3		Bw52	Bw61		
REE	REE	34	0	A2.2	A3.1	Bw41.1	B44.2		
REE	REE	35	0	A3.2	A25	B44.1	B44.2		
REE	REE	29	0	A3.2	A32	B14	Bw41.2		
REE	REE	36	0	A2.2	A24.1	Bw41.1	Bw55.3		
REE	REE	28	0	A23	A32	Bw41.2	B49.1		
REE	REE	32	0	A1	A32	Bw41.1	Bw52		
REE	REE	30	0	A1	A28.1	Bw41.2	Bw53		
REE	REE	31	0	A1	A2.2	B27.3	Bw41.2		
REE	REE	27	0	A1		B8	Bw41.2		
REE	REE	46	0	A29.2	A2.2	B13.2	B45		
REE	REE	42	0	A31	AN	B13.2	Bw70.3		
REE	REE	1111	0	A2.2	A10	B13.2	B35.3		
REE	REE	39	0	A2.2	A11.1	B13.1	Bw46		
REE	REE	43	0	A2.2	Aw66	B13.1	B38.1		
REE	REE	40	0	A2.2		Bw46*			
REE	REE	38	0	A3.2		B13.2	B35.3		
REE	REE	1112	0	A2.2	Aw66	B13.1	B35.3		
REE	REE	41	0	A24*	A29.2	B13.2	B35.3		
REE	REE	65	0	A29.1	Aw36	Bw42	B44.1		
REE	REE	49	0	A24.1		B15.1	Bw48		
REE	REE	48	0	A24		Bw48	Bw62		
REE	REE	50	0	A24.1	Aw34.1	B18.1	Bw48		

BIO:Biochemistry Laboratory
SER:Serology Laboratory
ID :Cell Identification
ETH:Ethnic Code (see Table 2)
* :Rare Variant

Table 3. *Continued*

BIO	SER	ID	ETH	HLA-A		HLA-B		HLA-C	
REE	REE	51	0	A24.1	Aw34.1	B35.3	Bw61		
REE	REE	90	0	A24.1	Aw34.1	Bw48	Bw54		
REE	REE	53	0	A24.1	AN	Bw48			
REE	REE	52	0	A24.1		B5*	Bw48		
REE	REE	85	0	A2.2	A11.2	Bw61	Bw62.1		
REE	REE	73	0	A2.2	A3.2	B7	Bw62.3		
REE	REE	70	0	A2.2	A3.2	B7.2	B44.2		
REE	REE	1	0	A2.2	A24.1	B44.2	BN		
REE	REE	69	0	A2.2	A24.1	Bw62.2	BN		
REE	REE	2	0	A2.2	A24.1	Bw62.2	BN		
REE	REE	68	0	A2.2	A24.1	B44.2	BN		
REE	REE	7	0	A3.2	A24.1	B7.2	BN		
REE	REE	6	0	A3.2	A24.1	B35.3	BN		
REE	REE	1154	0	A24.1	Aw34.1	Bw60	BN		
REE	REE	1113	0	A24.1	Aw34.1	Bw60	BN		
REE	REE	130	0	A2.2	A3.2	B7.2	BN		
REE	REE	127	0	A1	A2.2	B8	BN		
REE	REE	125	0	A1	A2.2	B44.2	BN		
REE	REE	126	0	A1	A2.2	B44.2	BN		
REE	REE	3	0	A1	A2.2	B8	BN		
REE	REE	129	0	A2.2		B62.1	BN		
REE	REE	128	0	A2.2	A3.2	B7.2	BN		
REE	REE	1114	0	A2.2	A32	B51	BN		
REE	REE	1115	0	A2.2	A3.2	B35	BN		
REE	REE	1116	0	A2.2	A32	Bw62.1	BN		
REE	REE	13	0	A1	A2.2	B44.2	Bw73		
REE	REE	269	0	A2.2	A23	Bw60	Bw73		
REE	REE	10	0	A11.1	A24.1	Bw60	Bw52		
REE	REE	145	0	A23	A24.1	B7.2	Bw72.3		
REE	REE	264	0	A1	A28	B8	Bw71.2		
REE	REE	144	0	A2.2	A11.2	B51	B35.3		
REE	REE	262	0	A3.1	A24.1	Bw41.1	Bw72.1		
REE	REE	141	0	A2.2	A10	B15.2	Bw72.1		
REE	REE	136	0	A1	A11.2	Bw52	Bw71.2		
REE	REE	134	0	A1	A28.2	B8	Bw71.2		
REE	REE	135	0	A1	Aw33.2	B44.1	Bw71.2		
REE	REE	138	0	A24	A32	Bw62.3	Bw72.3		
REE	REE	142	0	A1	A11.2	B37	Bw72.1		
REE	REE	1117	0	A1	A3.2	B17.2	Bw47		
REE	REE	116	0	A1	A26.1	B7.2	Bw47		
REE	REE	113	0	A2.2	A26.1	Bw57.2	Bw47		
REE	REE	1118	0	A3.2	A11.2	B35	Bw47		
REE	REE	1119	0	A2.2	A3.2	B27.5	Bw47		
REE	REE	1120	0	A1		B8	Bw47		
REE	REE	108	0	A2.2	A3.2	B35*	Bw47		
REE	REE	104	0	A24	A32	B15.2	B27.5		
REE	REE	1121	0	A2	A28	B15.2	B27.6		
REE	REE	1122	0	A23	A31	Bw52	B27.2		
REE	REE	189	0	A24.1	Aw33.2	Bw55.3	Bw64		
REE	REE	190	0	A3.2		B27.5	Bw55.3		
REE	REE	1123	0	A2.2	A29.2	B51	Bw55.3		
REE	REE	187	0	A11.2	A23	B44.1	Bw55.3		
REE	REE	12	0		Aw33.2	B51	Bw54.1		
REE	REE	1124	0	A11	Aw33.2	B51	Bw54.1		
REE	REE	183	0	A2.2	A24.2	B15.2	Bw54.1		
REE	REE	193	0	A2.2		B27.6	Bw55.3		
REE	REE	1125	0				BN		
REE	REE	202	0	A30.3		Bw42	Bw72.1		
REE	REE	1126	0	A1	A30.3	B8	Bw42*		
REE	REE	200	0	A1	A30.3	B8	Bw42*		
REE	REE	1110	0	A28.1			Bw42		
REE	REE	4	0	A2.2	A31	B51	Bw56.2		
REE	REE	1127	0	A1	A28.2	Bw53	Bw56.2		
REE	REE	198	0	A31	A28.1	B18.1	Bw56.3		
REE	REE	1128	0	A23	A29.2	B44.1	Bw56.2		
REE	REE	1129	0	A1	A2.2	B51	Bw56.2		
REE	REE	1130	0	A28.1	A29.1	B7.2	B39.1		
REE	REE	1131	0	A2.3	A32	Bw61	Bw52		

BIO	SER	ID	ETH	HLA-A		HLA-B		HLA-C	
REE	REE	1132	0	A11.2	A32	Bw41.1	Bw55.3		
REE	REE	1133	0	A11.2	A31	Bw61	Bw67		
REE	REE	1134	0		A31		Bw48		
REE	REE	1135	0	A3.2	A24.1	B44.1	B35.3		
REE	REE	1137	0	A11.2		B13.1	Bw62.2		
REE	REE	1138	0	A11.2	Aw33	B13.2	Bw55.3		
REE	REE	1139	0	A2.2	A29.2	B44.1	Bw55.3		
REE	REE	1140	0	A1	A11.2	B8	Bw56.2		
REE	REE	61	0	A30.2		Bw41.1			
REE	REE	60	0	A2.2	A28.1	Bw41.1	Bw58.2		
REE	REE	62	0	A2.1	A29.1	B38	Bw41.1		
REE	REE	1141	0	A24.1	A31	Bw41.1	Bw63.1		
REE	REE	1142	0	A2.2	A30.3	Bw42	Bw52		
REE	REE	64	0	A30.1	Aw36	B7.2	Bw42		
REE	REE	1143	0	A2.2		B13.2	B14		
REE	REE	1144	0	A2.2	A24.1	Bw61	BN		
REE	REE	1145	0	A31		Bw48			
REE	REE	1146	0	A23	Aw66	B49.2	Bw41.2		
REE	REE	1147	0	A2.2	A24.1	Bw61	Bw72.3		
REE	REE	1148	0	A24.1	A30.1	B18.1	Bw61		
REE	REE	1149	0	A3.2	A24.1	BN	BN		
REE	REE	1150	0	A3.2	A25	B39.2	BN		
REE	REE	63	0	A2.1		B7.1	Bw41.1		
REE	REE	1151	0	A2.2	AX	B49.1	Bw61		
REE	REE	1152	0	A1	Aw66	B35.3	Bw41.2		
REE	REE	1153	0	Aw33.2		B44.1	Bw61		
REE	REE	1154	0	AN	A24.1	B27.1	Bw63.2		
REE	REE	1155	0	A24.1	A11.2	Bw52	Bw54.1		
REE	REE	1156	0	A1	A23	B49.2	Bw53		
REE	REE	1157	0	A3.2	A26.1	B38.1	Bw41.1		
REE	REE	1159	0	A1	A2.1	Bw41.1			
REE	REE	65	0	A29.1	Aw36	B44.1	Bw42		
REE	REE	1161	0	A1	Aw66	B35.2	Bw41.2		
REE	REE	1162	0	A2.2	Aw34	B13.1	B15.1		
REE	REE	1163	0	A2.2	A29	B7.2	Bw41.1		
REE	REE	1164	0	A1	A23	B35.2	B49.2		
REE	REE	1166	0	A3.2	A24.1	B7	Bw73.3		
REE	REE	1167	0	A24.1	A31	B7.2	Bw60		
REE	REE	1169	0	A2.2	A30.1	B35.3	B40		
REE	REE	1170	0	A1	A31	B8	Bw60		
REE	REE	1172	0	A24.1	Aw33.2	B44.1	Bw60		
REE	REE	1173	0	A24.1	A32	B7.2	Bw61		
REE	REE	1175	0	A1	A3.2	B8	Bw61		
REE	REE	1176	0	A2.2	A29.2	B44.1	BN		
REE	REE	187	0	A11.2	A23	B44.1	Bw55.3		
REE	REE	1178	0	A2.2	A11.2	Bw55.3	Bw62		
REE	REE	63	0	A2.1		B7.2	Bw41.1		
REE	REE	65	0	A29.2	Aw36	Bw42	B44.1		
REE	REE	144	0	A2.2	A11.2	B35.3	B51		
REE	REE	1182	0	A2.2	A3.2	B51	Bw41.1		
REE	REE	46	0	A2.2	A29.2	B13.2	B45		
REE	REE	185	0	A2.2	A29.2	B13.2	B45		
REE	REE	43	0	A2.2	Aw66	B13.1	B38.1		
REE	REE	1112	0	A2.2	Aw66	B13.1	B35.3		
REE	REE	1188	0	A2.2		B15.2	B38.1		
REE	REE	1189	0	A2.2	Aw66	B15.2	B38.1		
REE	REE	1190	0	A.2	Aw66	B15.2			
REE	REE	1191	0	A1	A23	B35.2	B49.1		
REE	REE	1192	0	A1	Aw66	B35.2	Bw41.2		
REE	REE	1193	0	A23	Aw66	Bw41.2	B49.2		
REE	REE	1194	0	A1	Aw66	B35	Bw41.2		
TER	TER	1262	515	A1	A25	B8	B18.1		
TER	TER	1263	515	A1	A2.2	B7.2	B18.1	Cw7	
TER	TER	1264	515	A1	A30.1	B18.1	Bw57.2	Cw5	
TER	TER	1265	515	A11.2	A25	B18.1	Bw62		
TER	TER	1261	515	A3.2	A28.2	B18.2	Bw60	Cw3	Cw7
TER	TER	1260	515	A28.2	A30.1	B18.1	Bw60	Cw3	Cw5
TER	TER	1266	515	A2.2		B18.1	Bw57.2		

BIO:Biochemistry Laboratory
SER:Serology Laboratory
ID :Cell Identification
ETH:Ethnic Code (see Table 2)
* :Rare Variant

Table 3. *Continued*

BIO	SER	ID	ETH	HLA-A		HLA-B		HLA-C	
TER	TER	1252	515	A2.2	A25	B7.2	B18.1	Cw2	Cw6
TER	TER	1254	515	A1	A24.1	B18.1		Cw7	
TER	TER	1256	515	A2.2	A32	B13.2	B18.1	Cw6	Cw7
TER	TER	1257	515	A2.2	A26.2	B18.1	Bw58.3		
TER	TER	1259	515	A2.2	A25	B18.1	B27.5		
TER	TER	1318	515	A24.1	A29.2	B35.3	B45	Cw4	
TER	TER	1319	515	A1	A11.2	B8	B35.2	Cw4	
TER	TER	1320	20	A30.3	Aw36	B35.3	Bw42	Cw4	Cw7
TER	TER	1321	515	A3.2	A24.1	B35.3	B49.1	Cw4	
TER	TER	1322	515	A2.2	A24.1	B35.2	B44.2	Cw4	Cw5
TER	TER	1311	515	A23	A28.1	B35.3	B44.1	Cw4	
TER	TER	1323	120	A2.2	A11.2	B7.2	B35.3	Cw3	Cw7
TER	TER	1324	515	A2.2	A3.2	B35.3	Bw60	Cw3	Cw4
TER	TER	1325	515	A3.2	A24.1	B27.2	B35.3	Cw2	Cw4
TER	TER	1388	517	A24.1	A24.1	B35.3	B51	Cw4	Cw7
TER	TER	1309	515	A3.2		B35.3		Cw4	
TER	TER	1303	515	A3.2	A24.1	B35.3	Bw58	Cw3	
TER	TER	1244	20	A23	Aw36	Bw53		Cw4	
TER	TER	1331	20	A30.1	Aw34.2	B44.1	Bw53	Cw4	
TER	TER	1246	20	A1	A26.2	Bw53	Bw70.3	Cw4	
TER	TER	1332	20	A23	Aw34.2	B44.1	Bw53	Cw4	
TER	TER	1335	20	A2.1	Aw34.2	B7.1	Bw53	Cw4	
TER	TER	1336	20	A30.3	Aw33	Bw42	Bw53	Cw7	
TER	TER	1337	20	A2.2	A30.1	Bw53	Bw57.2	Cw4	
TER	TER	1330	20	A30.1	Aw34.2	B44.1	Bw53	Cw4	
TER	TER	1333	20	A3.2		Bw53		Cw4	
TER	TER	1334	20	A2	A23	Bw53		Cw2	Cw4
TER	TER	1128	20	A1	A30.3	Bw42*	Bw53	Cw4	
TER	TER	1251	20	Aw19*	A28.1	Bw53	Bw57.1	Cw4	
TER	TER	1282	515	A1	A3.2	B14	B51	Cw1	Cw8
TER	TER	1283	515	A2		B8	B51	Cw2	
TER	TER	1284	517	A2	A29	B44	B51	Cw3	
TER	TER	1270	515	A2	A3	B44	B51	Cw1	
TER	TER	1275	515	A2.2	A32	B38.1	B51		
TER	TER	1281	515	A2.2	A3.2	B8	B51		
TER	TER	1272	515	A1	A2.2	B7.2	B51	Cw1	Cw7
TER	TER	1276	515	A2.2	A3.2	Bw48	B51	Cw6	Cw8
TER	TER	1286	399	A2	A24.1	B51	Bw60	Cw3	
TER	TER	1285	120	A2.2	A3.2	B44.2	B51	Cw1	Cw5
TER	TER	1278	120	A24.1	A31	B51	B59	Cw1	Cw6
TER	TER	1269	515	A2.2		B8	B51		
TER	TER	1204	20	A1	A32	Bw52		Cw2	
TER	TER	1212	120	A2.2	A24.1	Bw52	Bw54.1	Cw1	
TER	TER	1215	515	A23	A32	B49.3	Bw52	Cw7	
TER	TER	1093	120	A11.2	A24.1	Bw52	Bw67	Cw7	
TER	TER	1144	515	A3.2	A11.2	B8	Bw52		
TER	TER	1208	130	A2.2	A11.2	B37	Bw52		
TER	TER	1287	515	A1		B8	Bw52		
TER	TER	1247	20	A2.2	Aw34.2	Bw52	Bw62.2	Cw2	
TER	TER	1288	120	A24.1	Aw33.2	B44.1	Bw52		
TER	TER	1291	515	A23	Aw33.1	B14	Bw52		
TER	TER	1245	120	A24	Aw33	B44	Bw52		
TER	TER	1293	515	A2.2	A31	B51	Bw60	Cw3	
TER	TER	1308	120	A2	A24	B35	Bw46	Cw1	
TER	DCH	12	310	A2.3	A11.2	B15.2	Bw46	Cw3	Cw8
TER	DCH	13	310	A11.2		B7	Bw46	Cw3	Cw7
TER	DCH	14	310	A2.2		B27	Bw46	Cw1	Cw3
TER	DCH	15	310	A2.3		B13.1	Bw46	Cw1	Cw3
TER	DCH	16	310	A2.2	A2.3	Bw46	Bw60	Cw1	
TER	TER	1383	110	A2.2	Aw33.2	B44.1	Bw46	Cw1	Cw3
TER	TER	1384	120	A2.2	A31	Bw46		Cw1	Cw3
TER	TER	1386	130	A2	A11.3	Bw46		Cw1	
TER	TER	1081	110	A11.2	A26.1	Bw46	Bw60	Cw1	Cw7
TER	TER	1387	120	A2.2	Aw33.2	Bw46	Bw58.2		
TER	TER	1385	110	A2.2	A11.1	Bw46	Bw60	Cw7	Cw8
TER	TER	1343	515	A1	A2.2	Bw57.2	Bw62.3	Cw3	
TER	TER	1348	515	A1	A32	B8	Bw62.3	Cw4	Cw7
TER	TER	1306	515	A2.2	A3.2	B35.3	Bw62.3	Cw3	Cw4
TER	TER	1350	515	A24.1	A29.2	B44.1	Bw62.3	Cw3	
TER	TER	1273	515	A2.2	A3.2	B51	Bw62.3	Cw1	Cw3
TER	TER	1373	515	A2.2	A3.2	B7.2	Bw62.3	Cw3	Cw7
TER	TER	1372	515	A1	A2.2	B7.2	Bw62.3	Cw3	Cw7
TER	TER	1265	515	A11.2	A25	B18.1	Bw62.3		
TER	TER	1341	515	A1	A2.2	Bw61	Bw62.3	Cw2	Cw3
TER	TER	1250	130	A24	Aw34	B27.5	B15.2		
TER	TER	1340	20	A11.2	A23	Bw58.2	Bw62.3	Cw2	Cw6
TER	TER	1294	515	A2.2	A11.2	B35.3	Bw62.3		
TER	TER	1126	515	A1	A3.2	B7.2	Bw63.1	Cw2	Cw7
TER	TER	1243	515	A2.2	A24.1	B8	Bw63.1		
TER	TER	1146	515	A24.1	A31	B35.3	Bw63.1	Cw4	Cw7
TER	TER	1381	20	Aw19*	A23	B14	Bw63.1	Cw7	Cw8
TER	TER	1045	515	A2.2	A23	B8	Bw63.1	Cw4	
TER	TER	1386	130	A2.2	A11.3	Bw46		Cw1	
TER	TER	1267	515	A2.2	A24.1	B18.1	Bw63.3	Cw3	Cw7
TER	TER	1375	20	A2.2	A26.1	B7.2	Bw63.1		
TER	TER	1376	20	A2.2	A26.1	B7.2	Bw63.1	Cw2	
TER	TER	1377	20	A28	A30	Bw57.1	Bw63.4	Cw3	
TER	TER	1382	515	A2.2	A2.1	B14	Bw63.1	Cw7	
TER	TER	1379	515	A23	Aw34.2	B14	Bw63.3	Cw7	
SYY	CHI	75	110	A2.2	A11.2	Bw46	Bw56.1	Cw11	
SYY	CHI	24	110	A2.2	A29.2	Bw46	Bw62.2	Cw4	Cw11
SYY	DCH	1	310	A24.1	A11.2	Bw62.2	Bw60	Cw4	Cw8
SYY	DCH	2	310	A24.1	A11.1	Bw62.2	Bw46	Cw11	Cw8
SYY	DCH	3	310	A24.1	Aw33.2	Bw62.2	Bw58.2	Cw3	Cw8
SYY	DCH	4	310	A2.2	A11.2	Bw62.2	B17.2		Cw3
SYY	DCH	5	310	A11.1	A11.2	A18.2	Bw62.2	Cw7	Cw8
SYY	DCH	6	310	A2.2	A11.2	B15.3	B51	Cw3	
SYY	DCH	7	310	A2.2	Aw33.2	B15.3	B44.1	Cw3	Cw7
SYY	DCH	8	310	A24.1	A11.2	B15.2	Bw62.3	Cw3	Cw8
SYY	DCH	9	310	A11.1	A24.1	B15.2	Bw62.3		Cw8
SYY	DCH	10	310	A2.2	A24.1	B44.1	B15.2	Cw7	Cw8
SYY	DCH	11	310	A2.2	A11.2	B15.2	B51	Cw8	
SYY	DCH	12	310	A11.2	A2.3	Bw46	B15.2	Cw8	Cw11
SYY	DCH	13	310	A11.2		B7.2	Bw46	Cw3	Cw11
SYY	DCH	14	310	A2.2		Bw46	B27.4	Cw11	
SYY	DCH	15	310	A2.3		Bw46	B13.1	Cw11	Cw3
SYY	DCH	16	310	A2.2	A2.3	Bw46	Bw60	Cw11	
SYY	DCH	17	310	A2.2	A31	Bw46	B35.3	Cw11	Cw4
SYY	DCH	18	310	A2.3	Aw33.2	B39.2	B44.1	Cw7	
SYY	DCH	19	310	A2.3	A11.2	B39.4	Bw61	Cw3	
SYY	DCH	20	310	A11.2	Aw33.2	B7.2	Bw62.1	Cw1	Cw7
SYY	DCH	21	310	A24.1	A11.2	Bw62.1	B18.2	Cw7	Cw8
SYY	DCH	22	310	A2.2	A11.2	B15.2	B51	Cw3	
SYY	FRA	2	599	A2.2	A24.1	B44.1	Bw61	Cw2	
SYY	FRA	4	520	A24.1	A26.1	B39.4	Bw60	Cw3	Cw7
SYY	FRA	6	520	A24.1	A32	B7.2	Bw61	Cw2	
SYY	FRA	7	320	A2.2	A11.2	B37	Bw60	Cw3	Cw6
SYY	FRA	20	520	A1	A32	Bw57.2	Bw60	Cw3	
SYY	FRA	21	520	A1	A31	B8	Bw60	Cw3	
SYY	FRA	22	520	A2.2	A26.1	B44.1	Bw61		
SYY	FRA	32	520	A2.2		B13.2	B14		
SYY	FRA	53	520	A2.2	A24.1	Bw55.3	Bw60	Cw1	Cw3
SYY	FRA	54	520	A2.2	A30.1	Bw62.1	Bw60	Cw3	
SYY	FRA	55	520	A1	A30.3	B8	B13.2		
SYY	ZHA	4	110	A2.2	A26.1	Bw61	B15.2	Cw3	Cw8
SYY	ZHA	5	110	A2.2	A11*	B15.2	B35.3	Cw1	Cw8
SYY	ZHA	9	110	A2.2	A2.3	B38.1	B15.2	Cw3	Cw7
SYY	ZHA	11	110	A11.2	Aw33.2	Bw57	B7	Cw3	Cw7
SYY	ZHA	13	110	A11.2	X	B15.2	B39.2	Cw1	Cw7
SYY	ZHA	18	110	Aw33.2	A11.2	B15.1	Bw57.2	Cw1	Cw3
SYY	KAS	531	599	A2.2	Aw33.1	B44.2	Bw62.3	Cw4	Cw7
SYY	KAS	541	599	A2.2	A32	B27.2	B35.3	Cw2	Cw4
SYY	KAS	551	599	A2.2	A31	B8	B44.2	Cw5	Cw7
SYY	KAS	561	599	A2.2	A31	B8	Bw60	Cw3	Cw7
SYY	KAS	571	599	A2.2	A30.3	B27.5	B13.2	Cw2	Cw6
SYY	KAS	581	599	A2.2	Aw33.1	B38.1	B51		

BIO:Biochemistry Laboratory
SER:Serology Laboratory
ID :Cell Identification
ETH:Ethnic Code (see Table 2)
* :Rare Variant

Table 3. *Continued*

BIO	SER	ID	ETH	HLA-A		HLA-B		HLA-C	
SYY	KAS	601	599	A1	A32	B8	B41.1	Cw2	Cw7
SYY	KAS	611	599	A2.2	A29.2	B44.1	Bw60	Cw3	
SYY	ARN	21	590	A2.2	A32	B44.1	Bw60	Cw2	Cw3
SYY	JEA	22	585	A3.2	A24.1	B27.5	B44.1	Cw1	Cw4
SYY	JEA	360	585	A1	A3.2	B7.2	B8	Cw3	Cw7
SYY	JEA	1794	585	A2.2	A25	B8	B27.5	Cw1	Cw7
SYY	JEA	2398	585	A1	A24.1	B51	B8	Cw7	Cw1
SYY	MUL	9800	545	A2.1	A2.2	B41.1	B50.2		
SYY	MUL	9808	545	A2.2		Bw62.3	B38.1	Cw3	Cw6
SYY	MUL	9819	545	A23	A28.1	B51	B65	Cw6	Cw8
SYY	MUL	9821	545	A2.2	A30.1	B18.1		Cw5	
SYY	MUL	9857	545	A1	A28.2	B8	B27.2	Cw2	Cw7
SYY	OHA	1	545	A24.1	A10	Bw60	B39.2		
SYY	OHA	2	399	Aw33.2	A2.2	Bw58.2	Bw46	Cw11	
SYY	OHA	3	399	Aw33.2	A2.2	Bw58.2	Bw46	Cw11	
SYY	FAU	191	545	A24.1		B14	Bw62.3	Cw2	Cw3
SYY	FAU	193	545	A31	A24.1	Bw56.2	B14	Cw1	Cw2
SYY	FAU	189	545	A24.1	A28.2	B14	B44.2	Cw2	Cw5
SYY	FAU	295	545	A28.2	A23	Bw71.2	B49.2	Cw7	
SYY	FAU	333	10	A2.1	A3.2	B42	B44.1		
SYY	PRR	10	10	A11.2	A26.2	B39.2	B27.5	Cw1	
SYY	PRR	37	545	A26.1	A68.2	B27.5	Bw64	Cw1	Cw8
SYY	PRR	42	545	A2.2	A11.2	B35.3	Bw62.3	Cw3	Cw4
SYY	BRA	2337	565	A29.1	A28.1	B7.1	B38.1		
SYY	BRA	2354	565	A2.2	A26.1	B35.3	B38.1	Cw4	
SYY	BRA	2355	565	A3.2		B7.2	Bw65	Cw7	Cw8
SYY	BRA	2432	565	A3.2.A	30.1	B14	B38.1	Cw8	
SYY	BRA	2448	566	A1	A2.1	B41.1		Cw2	
SYY	KEY	8951	550	A3.2	A30.1	B44.1	Bw57.2	Cw5	Cw6
SYY	KEY	8952	550	A26.1	Aw33.2	B39.2	Bw58.2	Cw3	
SYY	KEY	8953	550	A2.2		Bw57.2	Bw60	Cw3	Cw6
SYY	KEY	8954	550	A3.2	A32	Bw58.2	B65	Cw3	
SYY	KEY	8955	550	A2.2	A31	Bw57.2	Bw62.3	Cw3	Cw6
SYY	KEY	8956	550	A1	A2.2	B7.2	Bw58.2	Cw3	Cw7
SYY	MUE	10	550	A2.2		Bw65	Bw57.2	Cw6	
SYY	MUE	18	550	A1		B8			
SYY	MUE	33	550	A1	A26.1	B8	Bw64	Cw7	Cw8
SYY	MUE	77	550	A2.1	A2.2	Bw50.2	Bw65	Cw6	Cw7
SYY	CEP	13	560	A2.2	A24.1	B51	Bw55.3	Cw3	
SYY	CEP	14	560	A31	A26.2	B27.5	B35.1	Cw1	
SYY	CEP	17	560	A3.2	Aw68.1	B39.2	Bw61		
SYY	CEP	19	560	A24.1	A1	B51	Bw52		
SYY	CEP	64	560	A3.2	A2.2	B51		Cw1	
SYY	CEP	64	560	A1	A2.2	B8	Bw62.3	Cw3	Cw7
SYY	GAN	1429	560	A1	A28.2	B7.2	Bw70.3	Cw2	Cw7
SYY	GAN	3914	560	A24.1	A28.2	B44.1	Bw70.1		
SYY	GAN	3922	560	A2.1	A29.1	B7.2	B14	Cw7	Cw8
SYY	GAN	3927	560	A2.2	A26.1	B27.1	B38.1	Cw1	
SYY	GAN	3928	560	A1	A2.1	B8	Bw50.2	Cw6	Cw7
SYY	MAT	1	120	A24.1	A11.2	B52	Bw54.1	Cw1	
SYY	MAT	13	120	A24.1	Aw33.2	Bw60	B44.1	Cw3	
SYY	MAT	14	120	A2.2	A26*	Bw61	B38.1	Cw7	
SYY	FAI	301	565	A26.1		B38.1			
SYY	FAI	302	599	A2.2	A31	B39.4	B51	Cw1	Cw7
SYY	FAI	306	565	A1	A26.1	B38.1	B57.2	Cw6	
SYY	FAI	307	599	A29.2	A32	B37	B14	Cw8	
SYY	FAI	309	566	A2.2	A3.2	B51	B39.4	Cw7	
SYY	GOR	3	230	A2.2	A23	B39.1	B18.1	Cw7	
SYY	GOR	6	230	A2.2	A24.1	B7.2	B52	Cw3	Cw7
SYY	GOR	7	230	A2.2		B35.3	B49.3	Cw7	Cw4
SYY	GOR	8	230	A2.2	A24.1	B52		Cw3	
SYY	GOR	10	230	A2.2	A26.1	B35.2	B44.1	Cw4	Cw3
SYY	GOR	49	599	A11.2	A32	B52	B18.2	Cw7	
SYY	GOR	51	599	A34.1	A26.1	A38.1	B35.2	Cw4	
SYY	JRM	3	20	A29.2	Aw68.2	B35.3	Bw62.1		Cw4
SYY	JRM	4	20	A3.1	A30.2	B35.3	Bw42	Cw4	
SYY	JRM	8	220	A25	Aw33.1	Bw64	B38.1	Cw7	
SYY	JRM	14	20	A29.2	A31	B35.3	X	Cw4	

BIO	SER	ID	ETH	HLA-A		HLA-B		HLA-C	
SYY	JRM	18	599	A3.2	A30.3	B7.2	B35.3	Cw7	
SYY	JRM	27	20	A1	Aw33.2	B35.3	X	Cw3	
SYY	JRM	30	599	A11.2	A29.1	B7.2		Cw7	
SYY	JRM	32	220	A25	Aw33.1	Bw65	B18.1	Cw8	
SYY	JRM	35	220	A1	A30.3	B44.1	Bw42		
SYY	JRM	36	599	A3.2	A32	B7.2	Bw64	Cw7	Cw8
SYY	CAR	1	515	A2.2	A32	B44.2	Bw62.3	Cw3	
SYY	CAR	5	20	Aw33.1		B35.3	B14	Cw4	Cw8
SYY	CAR	11	515	A2.2	Aw33.1	B27.5	B38.1	Cw1	
SYY	CAR	12	515	A24.1	A31	B35.3	B27.5	Cw3	Cw4
SYY	CAR	20	515	A1	A29.2	B8	B44.1		
SYY	CAR	25	515	A1	A30.1	B14	B35.3	Cw4	Cw8
SYY	CAR	42	515	A31	Aw33.2	B14	Bw61	Cw2	
SYY	GOE	25	515	A2.2	A31	B8	Bw62.2	Cw1	Cw7
SYY	GOE	26	20	A28.2	A34.2	B17.2	B35.3	Cw6	
SYY	GOE	200	515	A11.2	Aw33.1	B13.2	Bw22.3	Cw3	
SYY	SAC	1002	517	A24.1	A29.2	B44.1	Bw61		
SYY	SAC	1010	599	A28.2	A32	B44.2	B48		
SYY	SAC	4038	110	A11.2		B13.1	Bw62.2		
SYY	SAC	6022	210	A24.1	A31	B7.2	Bw60		
SYY	TOM	4	517	A2.2	A29.2	B51	Bw62.3	Cw3	
SYY	TOM	5	517	A1	A24.1	B8	Bw62.3		Cw3
SYY	TOM	6	515	A1	A26.1	B13.2	Bw60	Cw3	
SYY	TOM	4922	515	A29.2		B27.5	B45	Cw1	
SYY	TOM	2006	515	A1	A3.2	B8	B27.5	Cw1	
SYY	YAG	3	515	A3.2	A25	B7.2	B39.2	Cw7	
SYY	YAG	8	515	A24.1	A29.2	B38.1	B44.1		
SYY	DUT	57	30	A29.1	Aw34.1	B7.1	B15.2	Cw4	
SYY	DUT	123	30	A2.3	A11.2	Bw60	B15.3	Cw7	Cw8
SYY	DUT	603	30	A1	A25	B18.1	Bw57.2	Cw6	
SYY	DUT	604	10	A26.2	Aw33.1	Bw42	Bw58.2	Cw2	Cw6
SYY	DUT	605	10	A26.1	A29.2	Bw42	B51	Cw7	
SYY	DUT	606	10	A23	Aw34.2	Bw42	Bw70.3	Cw2	Cw4
SYY	DUT	607	10	A2.1	A26.2	Bw58.2	Bw63.1	Cw6	
SYY	DUT	608	10	A29.2	Aw43	B18.1	Bw42	Cw2	Cw7
SYY	DUT	609	10	Aw43	A30.1	B45	Bw70.2	Cw4	
SYY	DUT	610	10	A28.1	Aw34.2	B7.2	B44.1	Cw7	
SYY	DUT	611	10	A24.1	Aw34.2	B7.2	B8	Cw7	
SYY	DUT	612	580	A2.1	A25	B18.1	Bw50.2	Cw6	
SYY	DUT	613	10	A28.1	X	B7.2	Bw58.2	Cw2	Cw4
SYY	DUT	614	10	A3.2	X	B35.3	Bw58.2	Cw4	Cw6
SYY	DUT	615	580	A24.1	A31	B14	B27.5	Cw2	Cw8
SYY	DUT	616	10	A28.1	Aw33.2	B7.2	Bw71.2	Cw3	Cw7
SYY	DUT	617	10	A23		Bw72.3	B39.2	Cw2	
SYY	DUT	618	10	A2.1	A29.2	Bw72.2	B8	Cw2	
SYY	DUT	619	10	A2.2	A29.2	Bw42	Bw49.2	Cw2	Cw7
SYY	DUT	620	10	A3.2	A30.3	Bw42	Bw58.2	Cw7	
SYY	DUT	621	30	A2.2	A24.1	Bw62.1	Bw46	Cw3	Cw11
SYY	DUT	622	30	A2.2	A28.2	Bw46	Bw53	Cw4	Cw11
SYY	DUT	623	10	A26.2	Aw33.1	Bw58.2	Bw42	Cw2	Cw6
SYY	DUT	624	30	Aw68.1	A31	Bw58.1	Bw72.3	Cw6	
SYY	JER	101	575	A1	A3.2	B35.3	Bw57.2	Cw4	
SYY	JER	102	575	A11.2	A28.2	B38.1	Bw57.2	Cw6	
SYY	JER	103	575	A3.2		Bw57.2	B51	Cw3	Cw6
SYY	JER	104	575	A11.2	A25	B35.3	Bw58.2	Cw4	Cw7
SYY	JER	105	575	A2.2	Aw33.2	Bw58.2	B51	Cw3	Cw6
SYY	JER	106	575	A2.2	Aw33.2	Bw58.2	B51	Cw3	
SYY	JER	202	575	A2.2	Aw68.2	B27.5	B44.2	Cw2	Cw5
SYY	JER	2022	575	A1	A2.2	B7.2	Bw61	Cw2	
SYY	LAW	11	540	A24.1	Aw68.2	Bw55.3	Bw71.2	Cw3	Cw7
SYY	KNT	22	540	A11.2	Aw68.2	B7.2	B44.2	Cw7	
SYY	KNT	23	540	A2.2	A28.2	B7.2	Bw57.2	Cw6	Cw7
SYY	BIE	261	515	A1	A3.2	B8	Bw61	Cw2	Cw7
SYY	KYL	1	515	A1	A2.2	B8		Cw3	Cw4
SYY	KYL	2	515	A1	A31	B8	B51	Cw7	
SYY	KYL	4	20	A2.1	Aw36	B35.3	Bw53	Cw3	Cw4
SYY	KYL	36	515	A1	A2.1	B8	Bw53	Cw4	Cw7
SYY	KYL	44	515	A2.1	A3.2	B44.2	Bw62.3	Cw3	Cw5

BIO:Biochemistry Laboratory
SER:Serology Laboratory
ID :Cell Identification
ETH:Ethnic Code (see Table 2)
* :Rare Variant

Table 3. *Continued*

BIO	SER	ID	ETH	HLA-A		HLA-B		HLA-C	
SYY	KYL	48	515	A1	A24.1	B7.2	Bw49.2	Cw7	
SYY	KYL	49	515	A2.2	A3.2	B18.1	Bw61	Cw2	Cw7
SYY	KYL	50	515	A1	A3.2	B8	Bw62.3	Cw3	
SYY	KYL	51	515	A3.2	A11.2	Bw62.3	B35.3	Cw3	Cw4
SYY	KYL	52	515	A3.2	A24.1	Bw55.3	Bw60	Cw3	
SYY	KYL	53	515	A23	A30.3	B13.2	Bw57.2	Cw4	
SYY	KYL	56	515	A24.1	A31	B7.2	Bw62.3	Cw3	
SYY	KYL	57	20	A1		B35.3	Bw58.1	Cw4	
SYY	KYL	58	515	A1	A11.2	B7.2	Bw57.2	Cw4	Cw7
SYY	KYL	59	515	A2.2	A24.1	B44.1	B27.5		
SYY	KYL	64	20	A23	A2.2	B44.2	Bw70.2		
SYY	KYL	66	20	A23	A29.2	Bw70.2	Bw49.2		
SYY	KYL	67	20	A23	A36	B44.1	X		
SYY	KYL	29	515	A2.2	A11.2	Bw55.3	Bw60	Cw3	
SYY	WOL	1	515	A2.2	A3.2	B51	B7.2	Cw7	
SYY	WOL	2	515	A2.2	A25	B51	B13.2	Cw5	Cw6
SYY	WOL	3	515	A3.2	A11.2	B18.1	Bw38.1		
SYY	WOL	4	515	A3.2	A24.1	B35.3	B39.2		
SYY	WOL	5	20	A28.2	Aw19*	Bw52	B35.3		
SYY	WOL	6	515	A11.2	A24.1	B18.1	Bw53	Cw4	
SYY	WOL	7	515	A11.2	A26.1	B7.2	Bw53	Cw4	
SYY	WOL	8	515	A1*	A11.2	B35.3	Bw49.2	Cw4	
SYY	WOL	9	515	A25	A28.2	B51	B18.1		
SYY	WOL	10	20	A28.1	A36	B14	Bw53	Cw4	
SYY	WOL	11	20	Aw34.2	Aw36	B35.3	Bw58.2	Cw4	
SYY	WOL	12	515	A2.2	A26.1	B18.1	Bw38.1	Cw7	
SYY	WOL	13	20	A28.1	A11.2	B44.2	Bw53	Cw4	Cw5
SYY	WOL	14	515	A1	A11.2	B7.2			
SYY	WOL	16	599	A24.1	A30.3	B13.2	B15.3	Cw6	
SYY	WOL	17	20	Aw33.1	Aw34.2	Bw58.2	B17	Cw7	
SYY	WOL	18	120	A11.2	A31	Bw61	Bw67	Cw3	
SYY	GRU	15	517	A1	A2.2	Bw58.2	Bw39.2	Cw3	
SYY	SND	3	517	A2.2	A30.1	B35.3	Bw60	Cw3	Cw6
SYY	SND	35	517	A24.1		Bw39.4	B35.3	Cw4	
SYY	SND	40	517	A28.2	A30.1	B14	Bw39.1	Cw7	Cw8
SYY	ABS	2838	210	A2.2	A24.1	B21.2	B51	Cw3	
SYY	ABS	2841	210	A24.1	A31	B21.3	B35.3	Cw3	Cw4
SYY	FTW	10	515	A2.1	A2.2	B44.2	Bw50.2		Cw6
SYY	FTW	14	515	A1	A2.2	B17.2	B44.2	Cw6	
SYY	FTW	15	20	A30.1	Aw33.1	B45	Bw57.3	Cw4	Cw6
SYY	FTW	16	515	A2.2	A11.2	B44.2	Bw56.2	Cw1	Cw5
SYY	FTW	21	515	A24.1	A28.2	B35.3	B44.1	Cw2	Cw4
SYY	FTW	27	515	A1	A2.2	B8	B44.2		
SYY	FTW	28	515	A28.2	Aw33.1	B14	B44.2		
SYY	FTW	31	515	A2.2	A23	B44.2	Bw50.2		
SYY	FTW	37	515	A1		Bw49.2	Bw55.3	Cw3	Cw5
SYY	FTW	40	515	A1	A2.2	B8	Bw49.2		Cw7
SYY	FTW	42	515	A2.2	A29.2	B13.2	B45	Cw6	
SYY	FTW	46	515	A3.2	A32	B35.3	B45	Cw4	
SYY	FTW	51	515	A1	A3.2	B27.5	B44.1	Cw1	Cw3
SYY	FTW	53	515	A2.2	A29.2	B44.1	B51		
SYY	FTW	57	515	A2.2	A25	B18.1	Bw49.2		
SYY	FTW	60	20	A2.1	A30.1	B45	Bw63.1		
SYY	FTW	65	515	A26.1	A29.2	Bw61	B44.1	Cw2	
SYY	MIT	3	515	A1	A3.2	B8	B35.3	Cw4	
SYY	MIT	4	515	A1	A2.2	B37	B44.2	Cw5	Cw6
SYY	MIT	5	515	A2.2		B44.2	Bw62.3	Cw5	Cw3
SYY	MIT	6	515	A2.2	A28.1	B44.2		Cw3	
SYY	MIT	4420	0	A24.1	A26.1	B44.2	Bw58.2		
SYY	MIT	4543	515	A23	A26.1	Bw49.2	Bw62.3	Cw7	Cw3
SYY	MLS	37	545	A2.2	A3.2	B44.2	Bw64	Cw7	Cw8
SYY	MLS	45	510	A1	A11.2	B8	B35.3	Cw4	
SYY	MLS	46	541	A1	A2.2	B51	B8	Cw7	
SYY	KAP	4	550	A1	A31	Bw57.2	B27.5	Cw1	Cw6
SYY	KAP	5	550	A2.2	A3.2	B51	B7.2	Cw1	Cw7
SYY	KAP	16	550	A11.2	A1	Bw60	B35.2	Cw3	Cw4
SYY	BRN	20	110	A2.2	A26.1	Bw60	Bw46	Cw11	Cw7
SYY	TRP	1	515	A2.2	A24.1	B38.1	B51		
SYY	TRP	2	515	A2.2	A31	B38.1	B35.2	Cw5	Cw6
SYY	TRP	3	515	A26.1	A28.2	Bw22*	B38.1	Cw3	
SYY	TRP	4	515	A2.2	A11.2	Bw22.3	B39.2	Cw3	Cw7
SYY	TRP	5	515	A2.2	A25	B39.4	Bw41.2	Cw7	
SYY	TRP	6	0	A2.2	X	B39.2	B44.2	Cw3	
SYY	TRP	7	210	A2.2	A31	B35.3	B39.2	Cw4	Cw7
SYY	TRP	8	210	A2.2		B39.2		Cw7	
SYY	TRP	9	515	A2.1	A31	Bw50.2	Bw60	Cw3	Cw6
SYY	TRP	10	515	A24.1	A31	Bw60	B35.2	Cw3	Cw4
SYY	TRP	11	515	A11.2	A30.3	B13.2	B35.3	Cw1	Cw4
SYY	TRP	12	210	A2.2	A31	B21.1	Bw48	Cw7	Cw8
SYY	TRP	13	210	A31		Bw48		Cw8	
SYY	TRP	14	210	A31		Bw48		Cw8	
SYY	POL	107	330	A2.5	A32	Bw61	Bw52	Cw6	
SYY	POL	277	515	A24.1		B35.3	B39.4		
SYY	POL	2496	10	Aw33.1	A30.3	B35.3	Bw57	Cw3	Cw6
SYY	POL	3025	565	A3.2	A26.1	B38.1	Bw41.1		
SYY	POL	8457	515	A3.2	A24.1	B35.3	B44.1	Cw2	Cw4
SYY	DUP	7398	515	A1		B13.2	B14	Cw6	
SYY	DUP	4161	515	A1	A24.1	B8	B38.1	Cw7	
SYY	DUP	61	555	A2.1	A30.1	Bw58.2	Bw50.2	Cw3	Cw6
SYY	DUP	229	110	A26.1	Aw33.2	Bw57.2	Bw60	Cw3	Cw7
SYY	DUP	6578	120	A2.2	A11.2	B35.3	Bw55.3		
SYY	DUP	7335	130	A11.2	A28.2	B35.3		Cw3	Cw4
SYY	DUP	6678	130	A2.2	A24.1	B37	Bw60	Cw3	Cw6
SYY	DUP	9145	515	A2.2	A28.2	B35.3	B51		
SYY	DUP	4004	130	A24.1		B13.1	B51*	Cw3	
SYY	DUP	6652	120	A24.1	A26.1	Bw61	Bw62.4	Cw3	Cw8
SYY	DUP	9324	515	A1		B44.1	Bw57.2		Cw4
SYY	DUP	4274	515	A2.1	A29.1	B44.1	B41.1		
SYY	DUP	7724	399	A24.1	Aw34.2	B35.3	Bw61	Cw4	
SYY	DUP	2786	515	A2.2	A24.1	B18.1	B51		
SYY	DUP	7102	515	A1	A3.2	B38.1	Bw52		
SYY	DUP	9357	560	A3.2	A28.2	B35.2		Cw4	
SYY	DUP	9352	515	A1	A2.2	B8	B18.1	Cw7	
SYY	DUP	6946	515	A2.2		B44.2	Bw61	Cw2	
SYY	DUP	6674	515	A3.2	A11.2	B35.3	Bw52	Cw4	
SYY	DUP	9356	515	A11.2		Bw62.3	B16.1	Cw1	
SYY	DUP	2108	566	A26.2	A28.1	B14	Bw41.2	Cw8	
SYY	DUP	3475	515	A24.1	Aw34.2	Bw42	Bw53	Cw6	
SYY	DUP	9368	515	A2.2	A3.2	B7.2	Bw57.2	Cw6	
SYY	DUP	9366	110	A2.3	A11.2	Bw62.2		Cw1	Cw3
SYY	DUP	2684	515	A2.2	A3.2	B7.2	Bw60	Cw3	Cw7
SYY	DUP	2319	515	A1	A3.2	B7.2	B8	Cw7	
SYY	DUP	7338	130	A3.2	A11.2	B39.2	Bw62.3		
SYY	DUP	9365	120	A24.1	A2.2	Bw52	Bw61		
SYY	DUP	7963	515	A2.1	A2.2	B7.2	Bw70.3		
SYY	DUP	9367	515	A2.2	A25	B18.1	B35.2	Cw4	
SYY	DUP	6482	515	A30.3	A24.1	Bw61	B13.2	Cw4	Cw3
SYY	DUP	5657	515	A2.3	A28.2	Bw49.2	B18.1	Cw7	Cw5
SYY	DUP	5658	515	A24.1	A2.1	B7.2	Bw49.2	Cw7	
SYY	DUP	5263	515	A1	A26.1	Bw57.2	B38.1	Cw6	Cw7
SYY	DUP	5264	515	A24.1		B38.4	Bw53		
SYY	DUP	3919	515	A24.1	A29.2	B44.1	Bw62.3	Cw3	
SYY	DUP	5143	515	Aw33.1	A11.2	B7.2	B18.1	Cw7	
SYY	DUP	5209	515	Aw33.1	A30.2	B7.2	Bw62.3	Cw7	Cw3
SYY	DUP	5234	515	Aw33.1	A2.1	B7.2	Bw50.2	Cw7	
SYY	DUP	5010	515	A29.2	A3.2	B44.1	B38.1		
SYY	DUP	5013	515	A3.1	A3.2	B35.2	B38.1	Cw4	
SYY	DUP	8039	515	A24.1	A28.2	B35.2	Bw62.3	Cw4	Cw3
SYY	DUP	1910	515	A24.1		Bw55.3	B35.3		
SYY	DUP	7341	515	A30.1	A31	B35.3	B57.4	Cw3	Cw7
SYY	DUP	7017	515	A24.1		Bw63.2		Cw4	

BIO:Biochemistry Laboratory
SER:Serology Laboratory
ID :Cell Identification
ETH:Ethnic Code (see Table 2)
* :Rare Variant

Table 4A. HLA-A Alleles Detected by 1D-IEF

Serologic Specificity		IEG Allele Designation
A1		A1, A1*
A2		A2.1, **A2.2**, A2.3, A2.4, A2.5
A28		**A28.1, A28.2**
A3		A3.1, **A3.2**
A9:	A23	A23
	A24	**A24.1**, A24.2, A24.3, A24*
A10:	A25	A25
	A26	**A26.1**, A26.2, A26*
	Aw34	Aw34.1, **Aw34.2**
	Aw66	Aw66
A11		A11.1, **A11.2**, A11.3
Aw19:	A29	A29.1, **A29.2**
	A30	**A30.1**, A30.2, **A30.3**
	A31	A31
	A32	A32
	Aw33	**Aw33.1, Aw33.2**
Aw36		Aw36
Aw43		Aw43

*Rare variants.
IEF alleles, shown in **bold type**, represent common subtypes.

other subtypes. A total of 214 donor cells in this Antigen Group were tested for IEF analysis. A23 and A24 are clearly distinguishable on IEF gel (Fig. 2). The A23 band had 100% correlation with the serologic typing of 47 panel cells tested. Three IEF bands corresponding to the A24 antigen were detected. The A24.1 band, the common A24 type, occurred in 97% of all A24 panel cells tested. The rare variant A24.2, which migrates very closely to the A23 band, was found in two cells and the A24.3 variant was also found in two donors. Both of the rare A24 variants were found only in Southeast Asians. Another variant for A24, which migrates with the B17.2 band, has been observed in one case.

Antigen Group #2 (A10, A25, A26, Aw34, Aw66, and Aw43). The subdivision of A10 into A25 and A26 has long been established. Aw34 and Aw66 also belong to the A10 group. Aw43 is a serologically related antigen and has been primarily found in South Africans.

The A10 (A25 and A26) band is clearly distinguishable from the rest of the antigens (Aw34, Aw66, and Aw43). The A26.2, a variant of A26, was observed in Blacks and Caucasians, but not in Orientals; two cases of the other A26 variant (A26*) were observed in Orientals (Fig. 3) and also carried Bw61. Three out of the five black A26.2+ cells were also found as Bw58.2. These associations need to be confirmed by family studies. The Aw34.1 (a variant of Aw34), Aw43, and Aw66 bands are difficult to distinguish from each other, even on the same IEF gel. The common Aw34.2 comigrates with the common A11.2 band. The ethnic distribution of the IEF alleles for this Antigen Group is shown in Table 7.

Antigen Group #3 (A29, A30, A31, A32, and Aw33). Aw19 is a complex Antigen Group that includes A29,

Table 4B. HLA-B Alleles Detected by 1D-IEF

Serologic Specificity		IEF Allele Designation
B5:	B51	B51, B51*
	Bw52	Bw52
	Bw53	Bw53
B7		B7.1, **B7.2**
B12:	B44	**B44.1, B44.2**
	B45	B45
B13		**B13.1, B13.2**
B14:	Bw64	B14
	Bw65	B14
B15:	Bw63	B15.1
	Bw62	**B15.2, B15.3** B15.4
B16:	B38	B16.1
	B39	**B16.2,** B16.3
B17:		B17.1, B17.2, B17.3
	Bw57	**B17.2**
	Bw58	**B17.2**
B18		**B18.1,** B18.2
B21:		B21.1, B21.3
	B49	**B21.2**
	Bw50	**B21.2**
Bw22:	Bw54	Bw22.1
	Bw55	Bw22.3
	Bw56	Bw22.2
B27		B27.1, B27.2, B27.3, B27.4, **B27.5,** B27.6, B27*
B35		B35.1, B35.2, **B35.3,** B35*
B37		B37
B40:	Bw60	Bw60
	Bw61	Bw61
Bw41		**Bw41.1,** Bw41.2
Bw42		Bw42, Bw42*
Bw46		Bw46, Bw46*
Bw47		Bw47
Bw48		Bw48
Bw59		Bw59
Bw67		Bw67
Bw70:	Bw71	**Bw70.2**
	Bw72	**Bw70.1, Bw70.3**

*Rare variants.
IEF alleles, shown in **bold type**, represent common subtypes.

Table 5. Two Locus HLA Class I IEF Haplotypes

HLA-A	HLA-B	HLA-C
A2	B44.2	
A28	B44.2	
A2.1	Bw50	
A11.1	Bw46	
A11.3	Bw46	
A29.1	B7.1	
A29.2	B44.1	
A30.1	B18.1	
A30.3	B13.2	
Aw33.1	B14	
Aw33.2	B44.1	
Aw33.2	B17.2	
	B15.2	Cw8
	B15.3	Cw3
	B27.2	Cw2

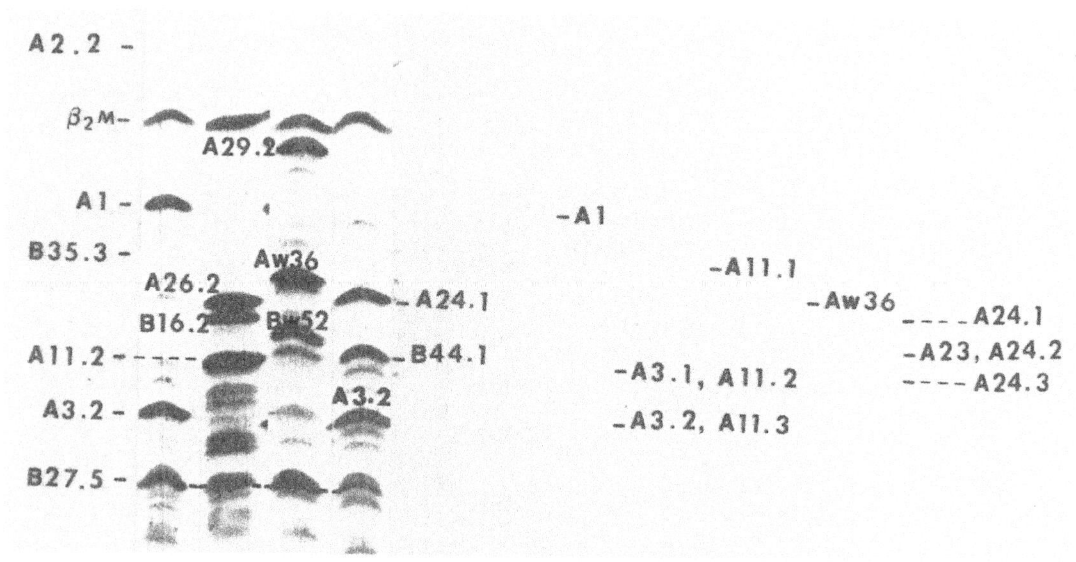

Figure 2. IEF band patterns and a schematic illustration of Antigen Society #1 (A1, A3, A11, Aw36, A9, A23, and A24). A11.1 is positioned slightly above the B35.3 band (see Fig. 10). A23 and A24.2 are located between B16.2 and Bw52 and A24.3 bands migrate very closely to the B44.1 band.

A30, A31, A32, Aw33, and a new split, Aw74 (Th). One of the unique characteristics of this Aw19 Antigen Group is its reactivity with the 4E antibody that recognizes a B locus-specific epitope.

A total of nine IEF bands that belong to the Aw19 Antigen Group were identified (Fig. 4). The A29.2 is a common band; A29.1 (25% of a total of 60 A29 cells on the panel) is a variant form that migrates slightly basic to the common A2.2 bands. This A29.1 variant can be identified by its reactivity with the 4E monoclonal antibody that does not react with A2. Six individuals out of 15 A29.1 cells (40%) had the rare B7.1 variant.

Table 6. Number of Donor Cells Tested for Antigen Society #1

	Ethnic Group						
Allele	N	O	AI	SEA	C	U	Total
A1	7	1	1	0	138	55	202
A3.1	1	0	0	0	1	3	5
A3.2	4	3	0	0	108	37	152
A11.1	0	3	0	6	0	0	9
A11.2	4	17	0	17	44	30	112
A11.3	0	2	0	1	0	0	3
Aw36	4	–	–	–	–	4	8
A23	3	–	1	–	20	13	47
A24.1	2	15	10	10	77	48	162
A24.2	–	–	–	1	–	1	2
A24.3	–	–	–	2	–	–	2

IEF alleles, shown in **bold type**, represent common subtypes. Ethnic codes: N = Negro; O = Oriental; AI = American Indian; SEA = Southeast Asian; C = Caucasian; U = unknown.

Two subtypes, A30.1 and A30.3, were observed with approximately equal frequency in Blacks and Caucasians, but were not observed in Orientals (Table 8). The A30.1 subtype was associated preferentially with B18, whereas the A30.3 was associated with the Caucasian B13.3 subtype and with Bw42 in Negroes. The A30.3 band was almost indistinguishable from the B14, Bw42, Bw55, and B27.2 bands. The A30.2 is a rare variant form of A30 and was observed in only three individuals.

Two IEF bands were identified for the Aw33 antigen from a total of 57 Aw33+ panel cells. The Aw33.1 subtype was found primarily in Caucasians and was associated with the B14 antigen. The Aw33.2 subtype was observed frequently in Orientals and was associated with Bw44.1 and B17.2. The Aw33.1 band is located between the A29.2 and A28.2 bands, and the Aw33.2

Table 7. Number of Donor Cells Tested for Antigen Society #2

	Ethnic Group						
Allele	N	O	AI	SEA	C	U	Total
A25	1	–	2	–	35	4	42
A26.1	3	5	1	–	30	5	44
A26.2	5	–	–	–	5	–	10
A26v*	–	2	–	–	–	–	2
Aw34.1	1	–	–	–	–	6	7
Aw34.2	10	–	–	1	4	–	15

*Rare variant.
IEF alleles, shown in **bold type**, represent common subtypes. Ethnic codes: N = Negro; O = Oriental; AI = American Indian; SEA = Southeast Asian; C = Caucasian; U = unknown.

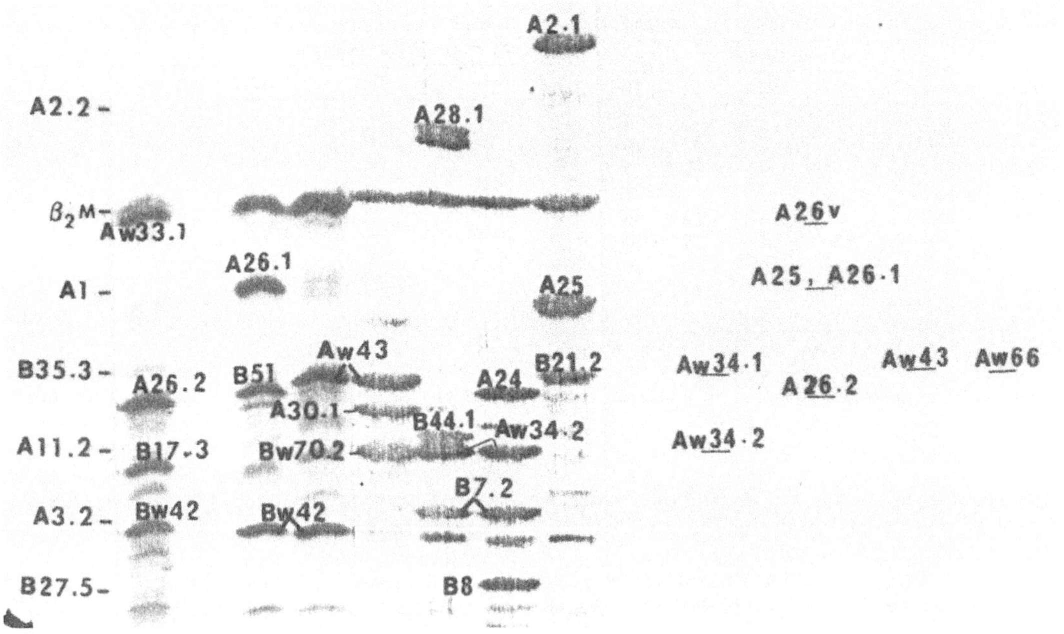

Figure 3. IEF band patterns and a schematic illustration for Antigen Society #2 (A10, A25, A26, Aw34, Aw66, and Aw43). A25 and A26.1 are indistinguishable on IEF gel. Aw34.1, Aw43, and Aw66 are also indistinguishable on IEF gel. Aw34.2 comigrates with A11.2.

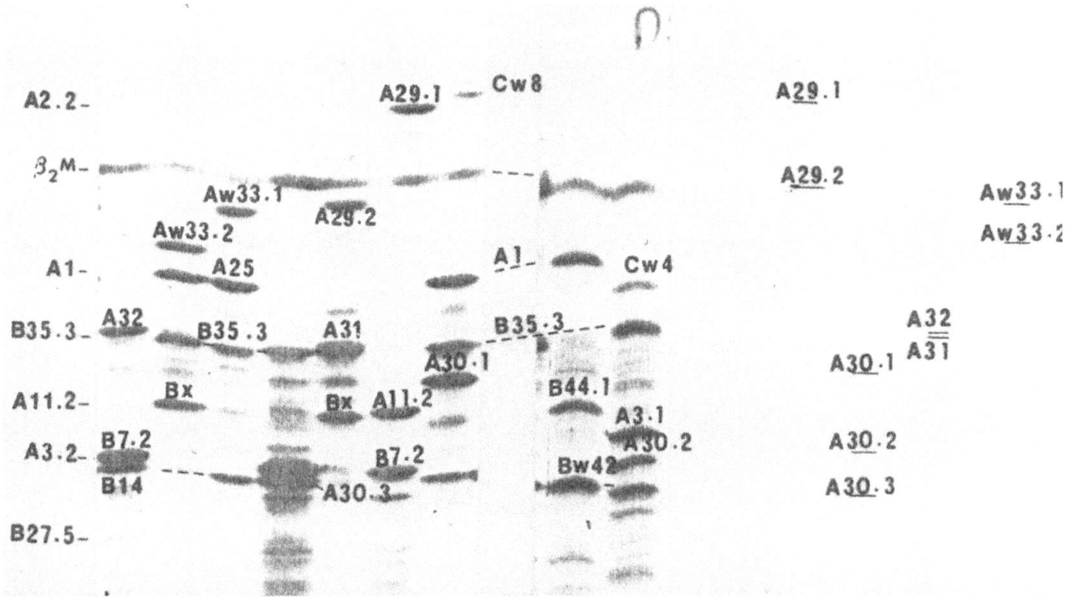

Figure 4. IEF band patterns and a schematic illustration for Antigen Society #3 (Aw19, A29, A30, A31, A32, and Aw33).

Table 8. Number of Donor Cells Tested for Antigen Society #3

	Ethnic Group						
Allele	N	O	AI	SEA	C	U	Total
A29.1	1	–	–	–	7	7	15
A29.2	7	1	–	–	27	10	45
A30.1	6	–	–	–	25	5	36
A30.2	1	–	–	–	1	1	3
A30.3	5	–	1	–	20	4	30
A31	3	4	9	1	35	14	66
A32	2	–	–	1	36	16	55
Aw33.1	6	–	2	–	13	1	22
Aw33.2	2	9	–	8	6	10	35

IEF alleles, shown in **bold type**, represent common subtypes. Ethnic codes: N = Negro; O = Oriental; AI = American Indian; SEA = Southeast Asian; C = Caucasian; U = unknown.

band migrates just below the Bw60 band, which is located close to the band of Bw46. The two Aw33 IEF subtypes have been shown previously (13).

No variants were observed for A31 and A32 antigens. These two antigens can be distinguished only on well resolved IEF gel.

No information is available for Aw74 (Th).

Antigen Group #4 (A2, AW68, AW69, and A28). A2 was one of the first recognized HLA antigens and was subdivided early into A2 and A28. A28 was split into Aw68 and Aw69. Aw68 was further subdivided into Aw68.1 and Aw68.2 (14). Exon-shuffling experiments between the A2 and A28 genes have revealed that Aw69 allele may have been generated by reciprocal recombination of Aw68.1 and A2 (15).

Table 9. Reaction Patterns for Selected A2-Specific Core Sera on HLA-A2 IEF Subtypes

Sera	A2.1	A2.2	A2.3	A2.4
26	+	+	+	–
32	+	+	+	–
22	+	+	+	–
2032	+	+	–	–
2030	–	+	+	+
2046	–	+	+	+
2033	+	+	–	+

A total of seven bands were identified for the antigens belonging to this Antigen Group; five bands for A2 and two bands for A28. This group of antigens represents the most basic proteins among HLA-A and -B antigens identified thus far (Fig. 5).

A2.2 is the most common form of the A2 group of antigens, and is found in all ethnic groups tested. A2.1 is the most basic protein and is not observed in Orientals. Five of 11 A2.1 cells had Bw50. A2.3 is an Asian variant primarily found in Southeast Asians; A2.4 variant was observed in American Indians. The A2.3 and A2.4 bands migrate very closely together on the IEF gel and are, therefore, difficult to assign if samples are run on different gels. A2.5 was observed in only one cell, and the donor was also of Southeast Asian origin. Some

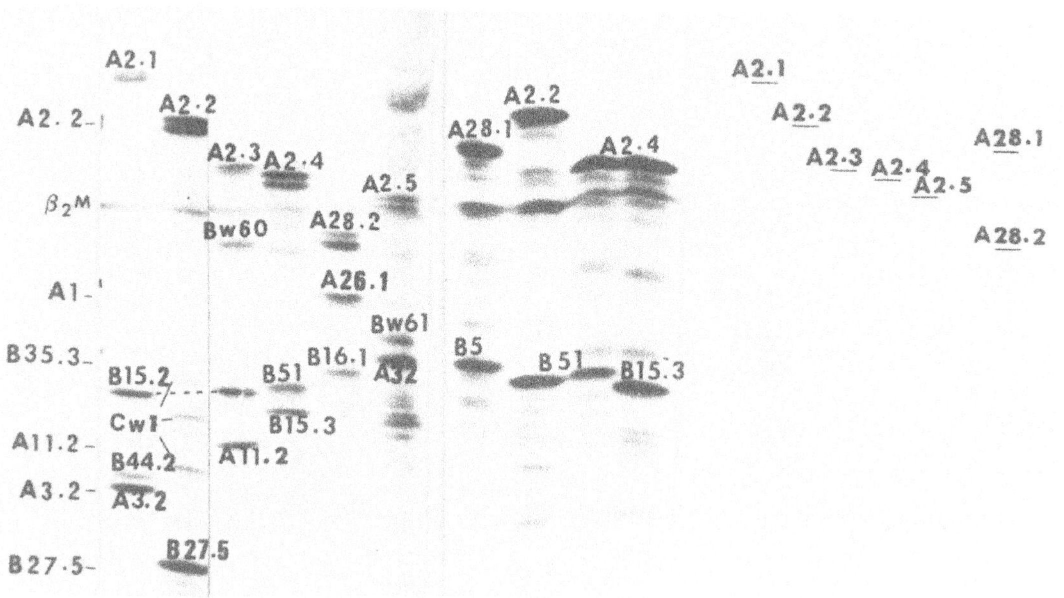

Figure 5. IEF band patterns and a schematic illustration for Antigen Society #4 (A2, Aw68, Aw69, and A28). Well resolved IEF gels show a doublet for A2/A28 family of antigens.

of the core sera that can distinguish the different A2 subtypes are listed in Table 9.

After typing a total of 72 donor cells with A28, Aw68, and Aw69 specificities, two IEF subtypes were observed. The basic form A28.1 was observed in four cells typed as Aw69 and two cells typed as Aw68, and the remaining cells were typed as A28. The acidic subtype A28.2 was found in 11 donors with Aw68, 2 donors with Aw69, and the remaining cells were typed as A28. It has been shown previously that Aw68 has two subtypes with charge differences designated as Aw68.1 and Aw68.2 (16). Aw68.2, which differs from Aw69 by a single substitution from valine to methionine at position 12, may correspond to the A28.1 IEF subtype. It appears that the A28.2 IEF band is identical to the Aw68.1 subtypes, whereas the A28.1 IEF band corresponds to both Aw68.2 and Aw69 as designated by Holmes et al. (16). The charge difference generated by a change from valine to methionine was not large enough to be detected by the pH gradient used. Both A28.1 and A28.2 IEF subtypes were observed with B14 with equal frequency. The ethnic distribution of these IEF alleles is shown in Table 10.

Table 10. Number of Donors Tested for Antigen Society #4

	Ethnic Group						
Allele	N	O	AI	SEA	C	U	Total
A2.1	7	–	–	–	17	8	32
A2.2	8	20	12	8	199	85	332
A2.3	1	3	–	16	4	3	27
A2.4	–	–	3	–	–	–	3
A2.5	–	–	–	1	–	–	1
A28.1	7	–	–	–	8	8	23
A28.2	1		1	–	38	5	45

IEF alleles, shown in **bold type**, represented common subtypes. Ethnic codes: N = Negro; O = Oriental; AI = American Indian; SEA = Southeast Asian; C = Caucasian; U = unknown.

Antigen Group #5 (B5, B51, Bw52, Bw53, B35, and B18). Subdivision of B5 into B51 and Bw52 along with the Bw53 specificity was officially recognized in the 1977 Workshop. Although B35 and Bw53 are well recognized alleles, serologic distinction between these two alleles is, however, problematic due to a lack of monospecific sera.

B51, Bw52, and three B35 subtypes are readily identifiable on an IEF gel (Fig. 6). However, the Bw53 band is indistinguishable from the B35.3 band, the common type of B35. No variants for B51, Bw52, and Bw53 were found; however, a few donor cells typed as B5 showed a band migrating close to Bw53. Serologic specificities for these cells need to be confirmed to distinguish the B5 subtype from the Bw53 antigen.

Two relatively common variants, B35.1 (3% of B35 panel) and B35.2 (17% of B35 panel), of the B35 antigen were identified only in Caucasians (Table 11). These two variants are difficult to distinguish unless they are compared on the same gel (Fig. 6). No apparent association was observed between a particular HLA-A specificity and the B35 variants. An additional variant (marked as B35v in Fig. 6) of the B35 antigen was observed in one laboratory.

B18 is a well defined antigen, and no evidence has been presented to indicate the presence of splits. Two IEF bands were recognized from 74 B18+ panel cells. The IEF variant B18.2 was observed with 6% of B18-positive donors. The common B18.1 type was

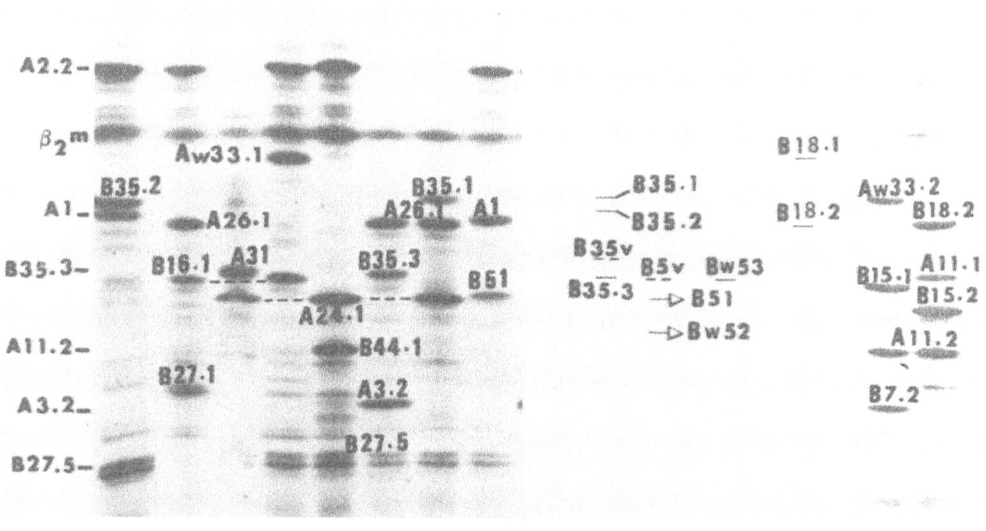

Figure 6. IEF band patterns and a schematic illustration for Antigen Society #5 (B5, B51, Bw52, Bw53, B35, and B18). B18.1 is the most basic HLA-B antigen among those tested and is located slightly above the Aw33.1 band, whereas B18.2 is just below the A1 band.

Table 11. Number of Donor Cells Tested for Antigen Society #5

	Ethnic Group						
Allele	N	O	AI	SEA	C	U	Total
B51	2	6	3	5	56	12	84
Bw52	3	6	–	1	11	9	30
Bw53	18	–	–	–	8	3	29
B35.1	–	–	–	–	5	–	5
B35.2	–	–	–	–	14	4	18
B35.3	14	6	5	2	54	20	101
B35v	–	–	–	–	–	3	3
B18.1	3	2	–	–	59	6	70
B18.2	–	–	–	1	3	–	4

IEF alleles, shown in **bold type**, represent common subtypes. Ethnic codes: N = Negro; O = Oriental; AI = American Indian; SEA = Southeast Asian; C = Caucasian; U = unknown.

Table 12. Number of Donor Cells Tested for Antigen Society #6

	Ethnic Group						
Allele	N	O	AI	SEA	C	U	Total
B44.1	6	5	2	3	40	15	71
B44.2	2	1	–	–	53	12	68
B45	5	–	–	–	8	3	16
B21.1	–	–	1	–	2	–	3
B21.2	2	–	1	–	14	–	17
B21.3	–	–	1	–	–	1	2

IEF alleles, shown in **bold type**, represent common subtypes. Ethnic codes: N = Negro; O = Oriental; AI = American Indian; SEA = Southeast Asian; C = Caucasian; U = unknown.

associated with the A30 antigen in 17 of 70 donors. Most of the A30 cells in association with B18 were of the A30.1 subtype. In two cases, they were found as A30.3.

Antigen Group #6 (B12, B44, B45, B21, B49, and Bw50). B44 and B45, the split of B12, are well defined antigens. A further split of B44 has been demonstrated by cell-mediated lympholysis as well as by IEF analysis.

Three major and equally spaced IEF bands that belong to the B12 group were identified (Fig. 7). The most basic band belongs to the B45, and the other two belong to the B44 antigens named B44.1 and B44.2. These two B44 subtypes occurred with approximately equal frequency. The B44.2 subtype was found predominantly in Caucasians, whereas the B44.1 subtypes were found in all ethnic groups tested; the Oriental B44 was included in this B44.1 subtype (Table 12). Among 71 donors with B44.1, 25 (77.5%) had the common A29.2 subtype, while only a single individual (1%) from 68 B44.2 donors had the A29 antigen. The B44.2 subtype was preferentially associated with the A2 (65%) or A28 (18%) antigens. Linkage disequilibrium within HLA-linked genes with B44 subtypes has been shown previously; B44.1 with complotype FC31 and DR7, and B44.2 with SC30-DR4

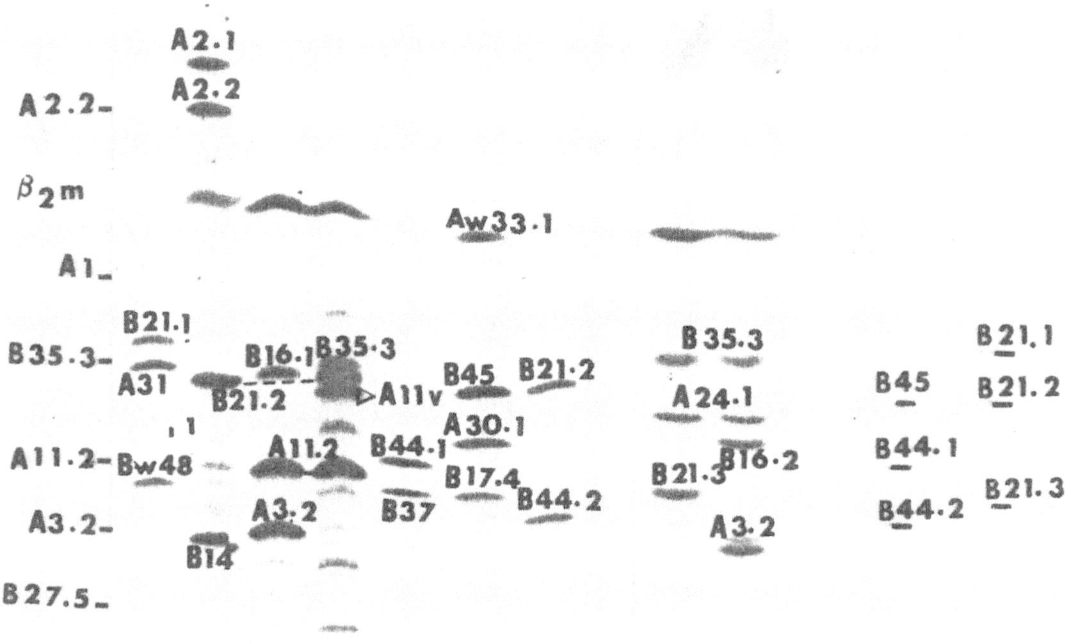

Figure 7. IEF band patterns and a schematic illustration for Antigen Society #6 (B12, B44, B45, B21, B49, and Bw50).

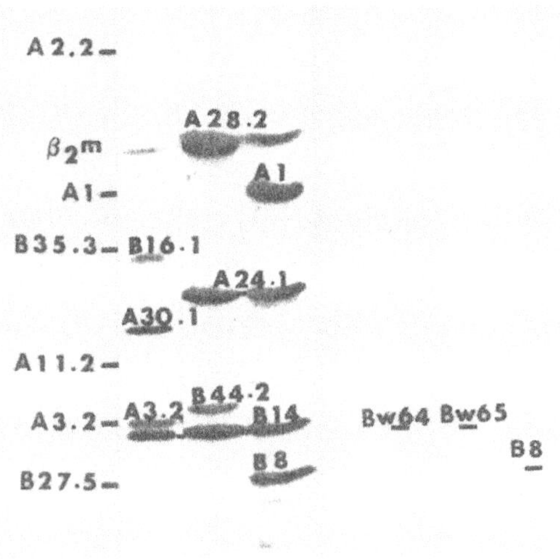

Figure 8. IEF band patterns and a schematic illustration for Antigen Society #7 (B8, B14, Bw64, and Bw65).

(17). No clear correlation was observed for B44.1 and B44.2 with core sera, although there were sera reaction patterns within B44 donor cells. Serology for defining the two B44 subtypes needs to be established.

B21 is composed of two serologic specificities– B49 and Bw50. Three IEF bands have been identified for this Antigen Group (Fig. 7). B21.2 is the most commonly occurring band (81% of the total B21 panel) and was typed as B49 (38%) as well as Bw50 (85%). Since B21.1 and B21.3 subtypes are rare, the B21.2 band may represent both Bw49 and Bw50 serologic subtypes. This Antigen Group needs further characterization.

Antigen Group #7 (B8, B14, Bw64, Bw65, and Bw59). B8 and B14 are clearly defined antigens. B14 has been subdivided into Bw64 and Bw65. B14 subtypes have strong linkage disequilibrium with complotypes; Bw64 with BfS-C2C-C4A3B1 and Bw65 with BfS-C2C-C4A2B2 or BfF-C2C-C4A3B1.

This Antigen Group is the most homogeneous and showed the simplest IEF band patterns (Fig. 8). The two bands corresponding to B14 and B8 are identified in the very acidic region. Although there is evidence for serologic splits of the B14 antigen, an IEF could not distinguish the B14 subtypes. The B14 panel (Table 13) consisted of 16 cells with B14, 19 cells with Bw64, and 56 cells with Bw65.

No IEF pattern is available for Bw59.

Antigen Group #8 (Bw70, Bw71, and Bw72). Bw70 has two serologic splits, Bw71 and Bw72, originally known as Bu and Sv, respectively. These antigens are known to be cross-reactive with B5, B35, B18, B15, and B21. IEF patterns for the Bw70 group can be distinguished from those for the cross-reactive antigens. The three IEF bands corresponding to this Antigen Group are named Bw70.1, Bw70.2, and Bw70.3 (Fig. 9) and have

Table 13. Number of Donor Cells Tested for Antigen Society #7

	Ethnic Group						
Allele	N	O	AI	SEA	C	U	Total
B8	3	–	2	–	68	16	89
B14	4	–	3	–	51	24	82

Ethnic codes: N = Negro; O = Oriental; AI = American Indian; SEA = Southeast Asian; C = Caucasian; U = unknown.

relative frequencies of 26%, 50%, and 24%, respectively (Table 14). Bw70.1 and B70w.3 correlate with Bw72 specificity, and Bw70.2 correlates with Bw71 specificity. Further correlation of IEF bands with specific Workshop sera will be described elsewhere in this volume.

Antigen Group #9 (B15, Bw62, Bw63, and Bw46). B15 was originally subdivided into Bw62 and Bw63, with their association with Bw6 and Bw4, respectively. Bw62 was further subdivided based on serologic reaction patterns. IEF analysis of a total of 155 panel cells with B15 revealed four distinct bands, B15.1, B15.2, B15.3, and B15.4 (Fig. 10). Among the 22 Bw63-positive cells tested, 17 donor cells (77%) had the B15.1 band, suggesting that the B15.1 band corresponds to the Bw63 antigen. This B15.1 subtype is found in all major ethnic groups, but was more frequent in Negroes and Caucasians (Table 15).

The B15.2 band was shown by 40 individuals (26%) and primarily observed in Orientals. This subtype includes B15T (Thai), Kemp, SH7, and Te79. B15.2 cells were associated with Cw8 (64%), but this was observed only in Orientals. Cw8 (labeled as Cw8.2 in Fig. 10) associated with B15.2 may have a different IEF pattern from that of Cw8 associated with B14 in Caucasians. This B15.2 band comigrates with A24 or B51. To distinguish B15.2 from A24, sequential immunoprecipitation with 4E and W6/32 antibodies is helpful since A24 is the most common antigen in Orientals.

The B15.3 subtype is most common in Caucasians, but is also rarely found in other ethnic groups tested. B15S showed this B15.3 subtype. Of the B15.3 donors, 83% also carried Cw3 antigens.

B15.4 was only observed in five cells (4%) in the B15 panel cells tested and was found primarily in Orientals.

The WHO Nomenclature Committee (1987) has designated Bw62 subtypes as Bw75 (Bw62.1), Bw76 (15S), and Bw77 (15T). Designation of the new WHO specificities to the corresponding IEF bands requires further studies. Cells submitted as 8w66 showed variable IEF patterns; therefore, 8w66+ cells need further serologic as well as IEF studies.

Antigen Group #10 (B17, Bw57, and Bw58). B17 has two subtypes, Bw57 and Bw58, and both alleles are included in Bw4. Three well separated IEF bands were identified from the analysis of 66 donors with the B17 Antigen Group (Fig. 11). The B17.2 band was the most

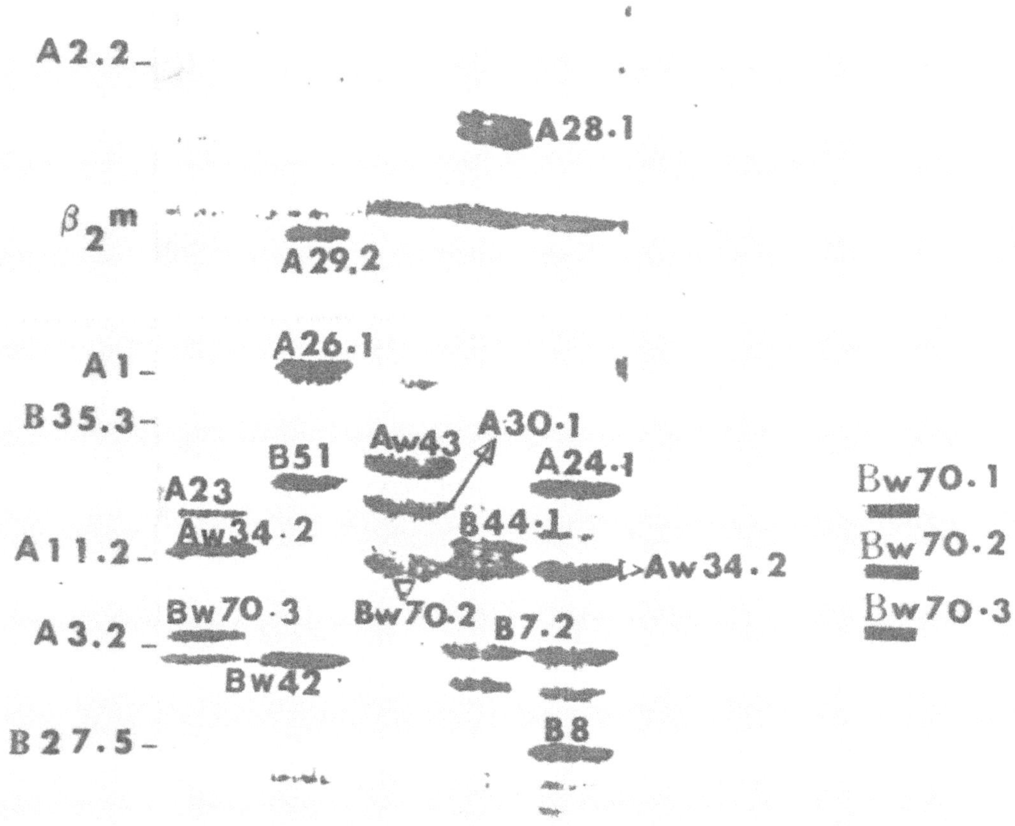

Figure 9. IEF band patterns and a schematic illustration for Antigen Society #8 (Bw70, Bw71, and Bw72). Bw70.1 band is located just below A24.1 band and very close to the B15.3 band.

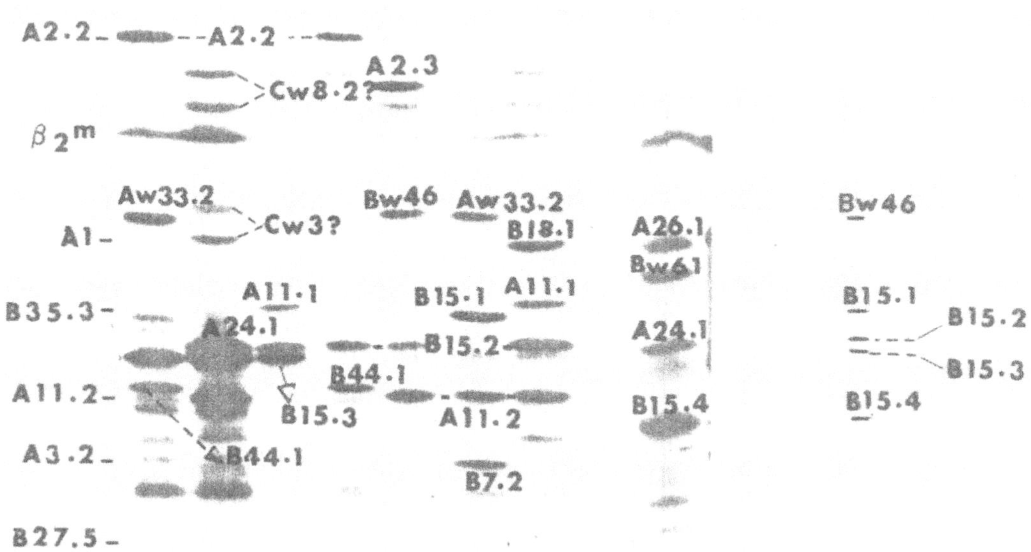

Figure 10. IEF band patterns and a schematic illustration for Antigen Society #9 (B15, Bw62, Bw63, and Bw46). B15.2 and B15.3 bands run very closely on IEF gel. A24.1 band is the best marker to distinguish these two subtypes; B15.2 comigrates with A24.1 and B15.3 band just below the A24.1 band. Bw46 band can be located just below Bw60 band (see Fig. 15).

Table 14. Number of Donor Cells Tested for Antigen Society #8

	Ethnic Group						
Allele	N	O	AI	SEA	C	U	Total
Bw70.1	–	–	–	–	4	5	9
Bw70.2	5	–	–	–	8	4	17
Bw70.3	2	–	–	–	3	3	8

Ethnic codes: N = Negro; O = Oriental; AI = American Indian; SEA = Southeast Asian; C = Caucasian; U = unknown.

Table 15. Number of Donor Cells Tested for Antigen Society #9

	Ethnic Group						
Allele	N	O	AI	SEA	C	U	Total
B15.1	7	2	–	2	12	7	30
B15.2	2	7	–	18	2	11	40
B15.3	2	3	5	7	53	9	79
B15.4	1	1	1	3	1	–	7

IEF alleles, shown in **bold type**, represent common subtypes. Ethnic codes: N = Negro; O = Oriental; AI = American Indian; SEA = Southeast Asian; C = Caucasian; U = unknown.

common (89% of B17 panel), and was found in both Bw57 donors (56% of B17.2 positive cells) and Bw58 donors (46% of 17.2$ cells). Two patterns of core sera were noted within this B17.2 group, as shown in Table 16.

Two different sera reactivity patterns for the B17.2 subtype indicate that IEF typing cannot distinguish

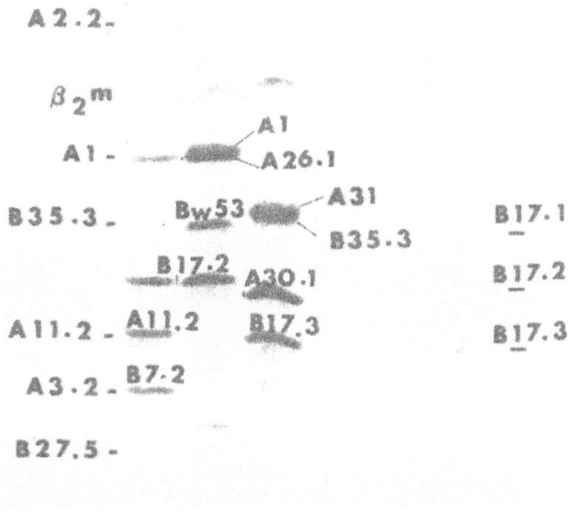

Figure 11. IEF band patterns and a schematic illustration for Antigen Society #10 (B17, Bw57, and Bw58). B17.1 band is located just below B35.3 marker, B17.2 comigrates with B15.3 band, and B17.3 runs just below A11.2 band.

Table 16. Core Sera Reaction Patterns for B17 IEF Subtypes

	264	220	260	262	98
B17.1	+	+	+	+	+
B17.2	+	+	+	+	+
B17.2	–	–	–	–	–
B17.3	–	–	–	–	–

between the two serologically defined subtypes Bw57 and Bw58.

Both the B17.1 and B17.3 IEF subtypes were rare and were observed in Negroes and Caucasians. Core sera reactivity for these subtypes are also shown in Table 16. The Antigen Group #10 panel is shown in Table 17.

Antigen Group #11 (B16, B38, B39, B37, and Bw67). B16 is subdivided into B38 and B39 serologic specificities. B38 is associated with Bw4, and B39 is associated with Bw6. Three equally spaced bands, B16.1, B16.2, and B16.3, were identified from studying a total of 68 cells with B16 specificity (Fig. 12, Table 18). B16.1 correlated 100% with the cell specificity B38, indicating that the B16.1 IEF subtype is B38.

B39 splits into two IEF subtypes – B16.2 (69%) and B16.3 (31%). No core sera that can distinguish between the two B39 subtypes B16.2 and B16.3 were observed. From the data, the following conclusions can be made:

B16.1 = B38
B16.2 = B39.1
B16.3 = B39.2

No variant form of the B37 antigen was identified from the 13 individuals with Oriental and Caucasian backgrounds.

Antigen Group #12 (Bw22, Bw54, Bw55, Bw56, and Bw42). The Bw22 serologic splits are Bw54, Bw55, and Bw56. Bw55 and Bw56 can be distinguished mainly by the C locus association – Bw55 and Cw3, and Bw56 with Cw1.

A total of 48 panel cells expressing the Bw22 antigens produced three distinctive IEF bands designated as Bw22.1, Bw22.2 and Bw22.3 (Fig. 13). The B21.1 band was found only in Orientals, whereas Bw22.2 and Bw22.3 were found in Caucasians (Table 19). The IEF

Table 17. Number of Donor Cells Tested for Antigen Society #10

	Ethnic Group						
Allele	N	O	AI	SEA	C	U	Total
B17.1	4	–	–	–	1	–	5
B17.2	11	3	–	4	34	7	59
B17.3	1	–	–	–	3	–	4
Bw67	–	2	–	–	–	1	3

IEF alleles, shown in **bold type**, represent common subtypes. Ethnic codes: N = Negro; O = Oriental; AI = American Indian; SEA = Southeast Asian; C = Caucasian; U = unknown.

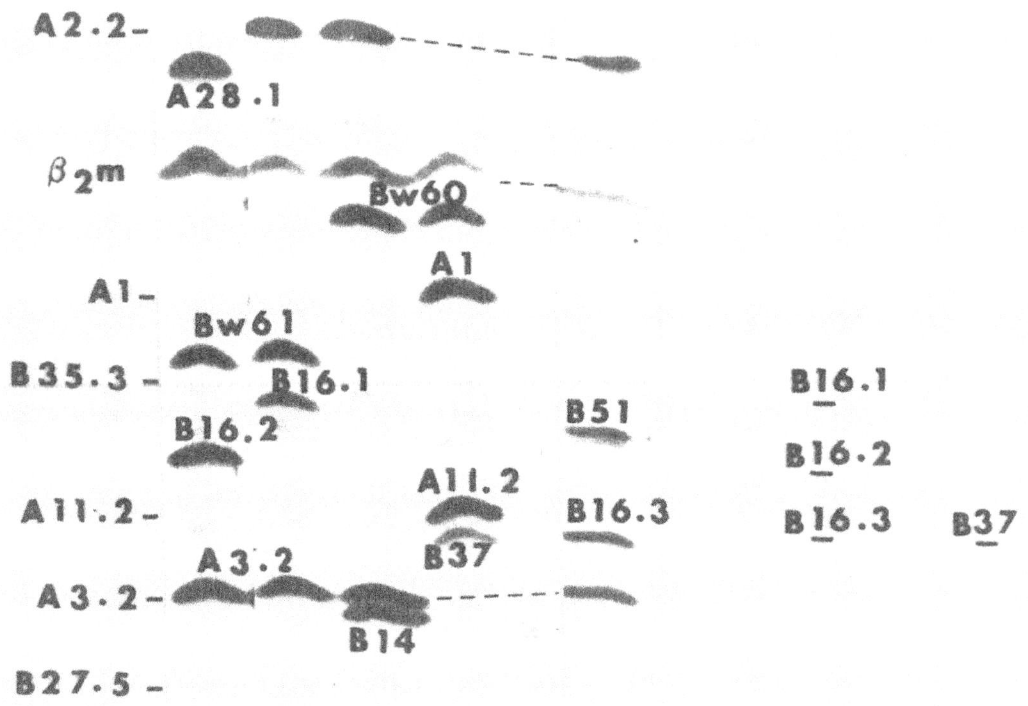

Figure 12. IEF band patterns and a schematic illustration for Antigen Society #11 (B16, B38, B39, and B37). B16.1 band runs just below B35.3, B16.2 band just below B17.2, and B16.3 just below A11.2 band.

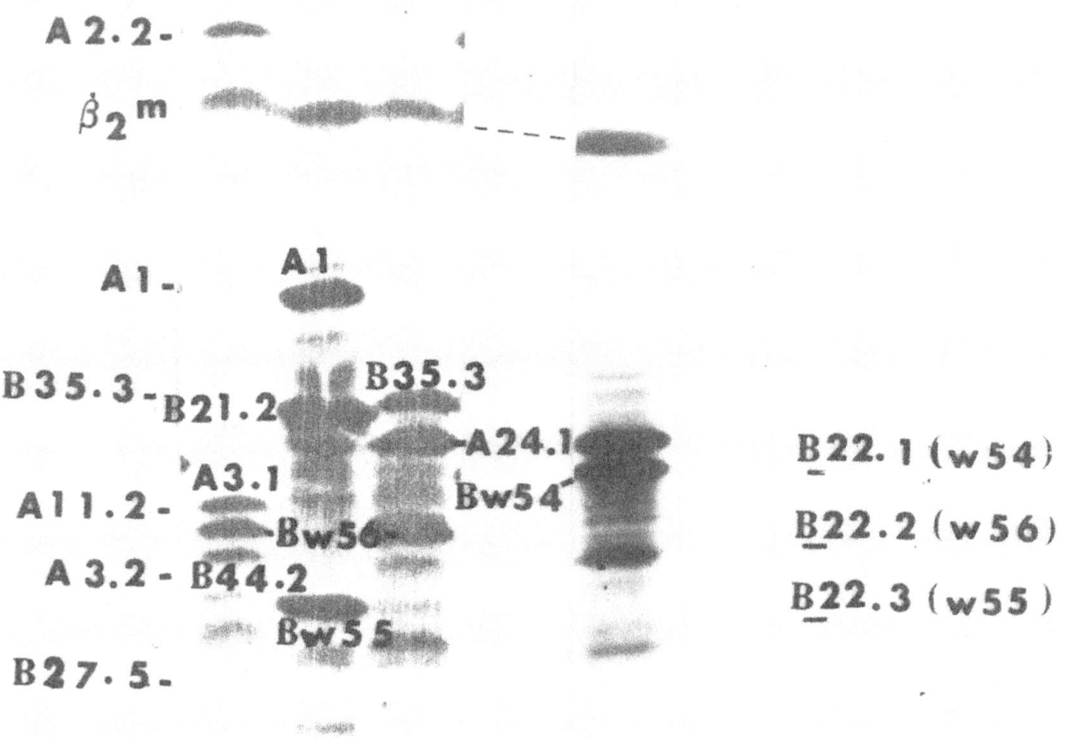

Figure 13. IEF band patterns and a schematic illustration for Antigen Society #12 (Bw22, Bw54, Bw55, Bw56, Bw42). Bw42 band is shown together with Antigen Group #2 in Figure 3.

Table 18. Number of Donor Cells Tested for Antigen Society #11

	Ethnic Group						
Allele	N	O	AI	SEA	C	U	Total
B16.1	–	3	1	1	27	5	37
B16.2	2	2	2	1	8	3	18
B16.3	–	–	–	1	7	–	8
B37	–	3	–	1	6	3	13

IEF alleles, shown in **bold type**, represent common subtypes. Ethnic codes: N = Negro; O = Oriental; AI = American Indian; SEA = Southeast Asian; C = Caucasian; U = unknown.

Table 19. Number of Donor Cells Tested for Antigen Society #12

	Ethnic Group						
Allele	N	O	AI	SEA	C	U	Total
Bw22.1 (Bw54)	–	4	–	–	–	5	9
Bw22.2 (Bw56)	–	–	–	–	3	5	8
Bw22.3 (Bw55)	–	1	–	–	17	13	31
Bw42	14	–	1	–	1	11	27

Ethnic codes: N = Negro; O = Oriental; AI = American Indian; SEA = Southeast Asian; C = Caucasian; U = unknown.

subtypes were correlated with sera specificities—Bw22.1 with Bw54, Bw22.2 with Bw56, and Bw22.3 with Bw55.

The Bw42 antigen is primarily found in Negroes. No apparent subtype of the Bw42 antigen was identified from 27 donor panel cells tested. However, one laboratory has identified a variant form of the Bw42 antigen that needs to be confirmed.

Antigen Group #13 (B7, B27, Bw47, and Bw73). The B7 antigen is considered as a classic specificity and is known to cross-react with B27 and Bw22. The IEF band patterns for the antigens belonging to this group are shown in Figure 14. The common B7 (B7.2) band is easily distinguishable from the B27 common band (B27.5) and the Bw54, Bw55, and Bw56 bands. B7.2 comigrates with A3.2, the common A3 band. The rare variant B7.1 is identified with a relative frequency of 10% (10 cases out of 107 B7 panel cells). The B7.1 variant is expressed together with A29.1 (7 of 10) or with the rare A2.1 subtype (2 cases out of 10).

B27 is a highly heterogeneous group of antigens and comprise the most acidic class I HLA antigens identified

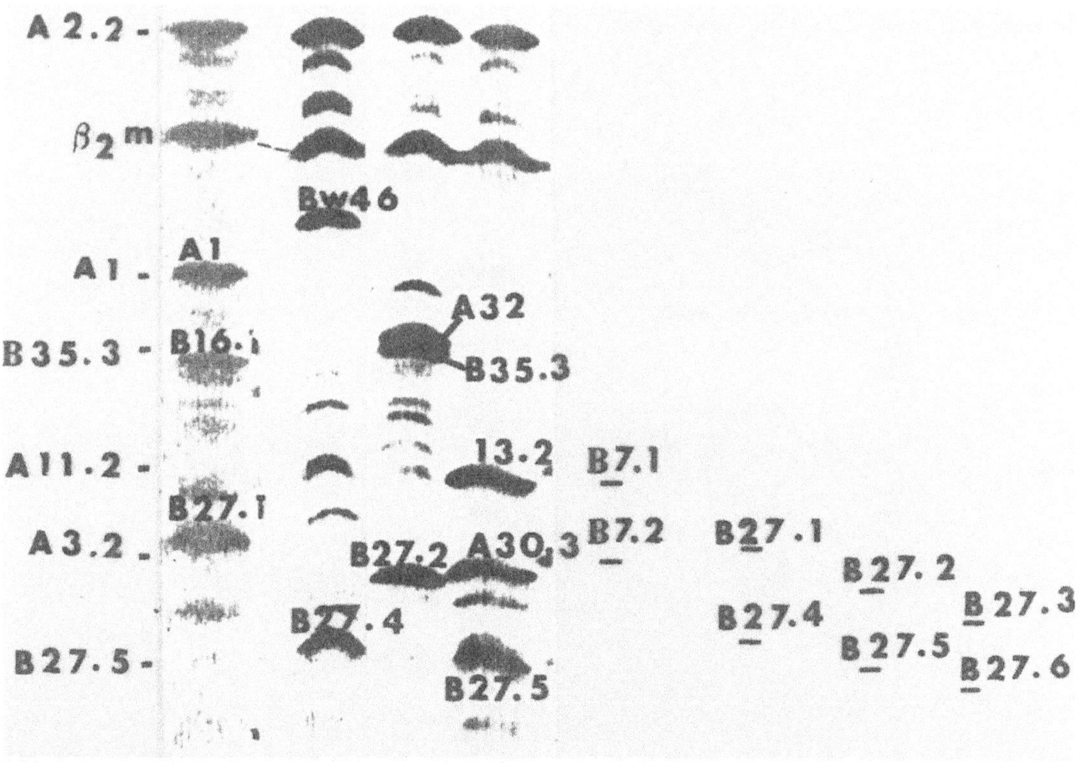

Figure 14. IEF band patterns and a schematic illustration for Antigen Society 13 (B7, B27, and Bw47). Bw47 band is shown together with Antigen Group #14 in Figure 15.

Table 20. B27 Designations

10th WS	Choo et al. (19)	Breur-Vriesendorp et al. (18)	Racial Distribution
B27.1	B27f	–	Caucasian
B27.2	B27e	B27K	Caucasian
B27.3	B27d	B27J	Caucasian/Negro
B27.4	B27b	B27C	Oriental
B27.5	B27a	B27W	All races
B27.6	–	B27D	Oriental

so far. A total of six IEF band patterns were identified for B27. The B27 variants have been extensively studied previously by IEF (1,18). Different designations are listed in Table 20.

The B27.5 subtype is the most common form of B27 and has been found in all races examined. B27.2 is the most common variant among the five and is associated with Cw2. Small numbers of the other variants did not allow us to arrive at any conclusions. B27.1, B27.2, and B27.6 variants can be easily identified on an IEF gel with the Workshop protocol. The other variants may require a special pH gradient, in addition to having them run on the same gel, to distinguish them from each other. Combined structural data on B27 variants indicate that they differ by one to four amino acids (19).

The Bw47 band can be clearly distinguished from B7.2 or B27.2 and their variants, but can be located in the same position as B7.1 and B13.2 (see Fig. 14). The ethnic distribution of the IEF alleles in Antigen Group #13 is shown in Table 21.

Table 21. Number of Donor Cells Tested for Antigen Society #13

	Ethnic Group						
Allele	N	O	AI	SEA	C	U	Total
B7.1	2	–	–	2	2	4	10
B7.2	6	2	2	2	66	20	98
Bw47	–	–	–	–	6	8	14
B27.1	–	–	–	–	3	1	4
B27.2	–	–	–	–	6	3	9
B27.3	1	–	–	–	1	1	3
B27.4	–	1	–	3	–	–	4
B27.5	2	4	5	2	31	9	53
B27.6	–	–	–	3	–	1	4

IEF alleles, shown in **bold type**, represent common subtypes. Ethnic codes: N = Negro; O = Oriental; AI = American Indian; SEA = Southeast Asian; C = Caucasian; U = unknown.

Antigen Group #14 (B40, Bw60, Bw61, B13, Bw41, and Bw48). The Bw40 splits, Bw60 and Bw61, have been clearly defined and distinguished serologically from the cross-reacting antigens B13, Bw41, Bw48, and Bw47. IEF analysis of the Bw60 and Bw61 antigens was also consistent with serology wherein two distinct bands were identified corresponding to each Bw40 subtype. Many cells submitted as Bw40 were found to be Bw61,

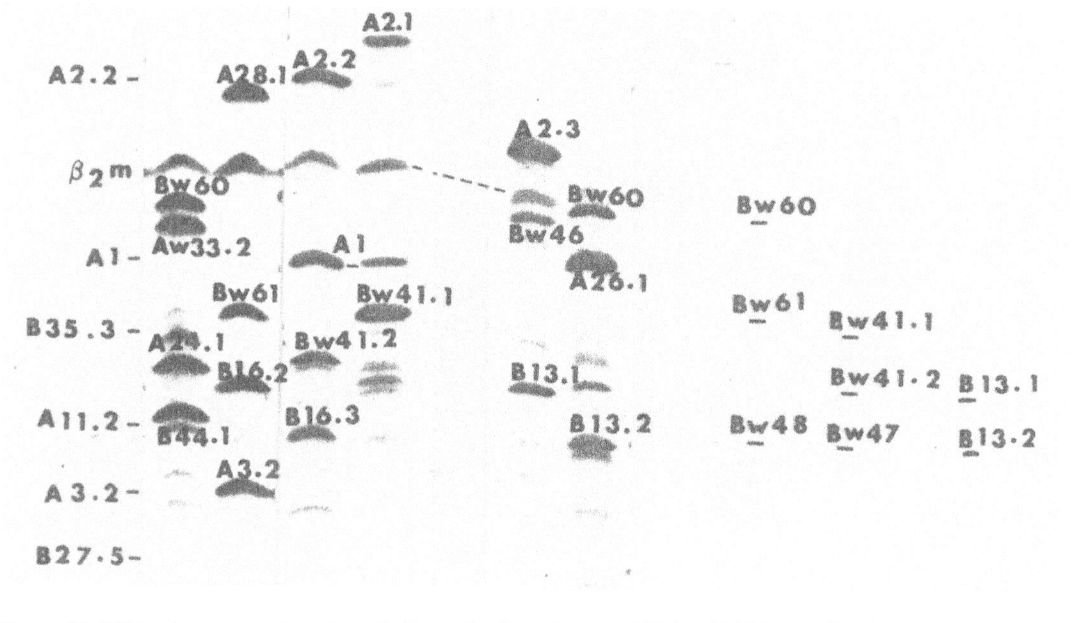

Figure 15. IEF band patterns and a schematic illustration for Antigen Society #14 (B40, Bw60, Bw61, B13, Bw41, and Bw48). Bw41.1 band runs just below Bw61, and Bw41.2 band between A24.1 and B16.2 (note that the gel pattern is a composite of three different runs).

Table 22. Number of Donor Cells Tested for Antigen Society #14

	Ethnic Group						
Allele	N	O	AI	SEA	C	U	Total
Bw60	1	7	2	6	36	20	72
Bw61	–	7	2	4	20	19	52
B13.1	–	3	–	2	–	8	13
B13.2	–	–	–	–	27	9	36
Bw41.1	–	–	–	–	3	15	18
Bw41.2	–	–	–	–	2	11	13
Bw48	–	1	3	–	1	8	13

IEF alleles, shown in **bold type**, represent common subtypes. Ethnic codes: N = Negro; O = Oriental; AI = American Indian; SEA = Southeast Asian; C = Caucasian; U = unknown.

indicating that serologic typing for Bw61 is still problematic. Although no major variants were identified (Fig. 15), variable bands were observed in cells typed as the Bw40 antigen. The Bw40 Antigen Group deserves further study.

B13 is a serologically well defined antigen, and splits have been biochemically identified previously. This Workshop also made similar observations. Only two previously recognized IEF B13 subtypes were observed. The B13.1 subtype was seen only in Orientals, whereas B13.2 was observed only in Caucasians (Table 22). This is one of the rare cases in which a major allele is an ethnic-specific prototype. The Caucasian-specific B13.2 bands overlap with Bw48 as well as Bw47 (Fig. 15).

Bw41 is also a well defined antigen. The Bw41 split was recognized by IEF for the first time. These two subtypes, Bw41.1 (58%) and Bw41.2 (42%), have a similar frequency in the panels tested. The majority of the cells were not identified for their racial background.

The Bw48 antigen is found in 12% of Eskimos, but is very rare in Caucasians. All 13 Bw48 donor cells showed the same band, which is indistinguishable from B13.2, Bw47, and A11.2.

References

1. Choo SY, Antonelli P, Nisperos B, Nepom GT, Hansen JA. Six variants of HLA-B27 identified by isoelectric focusing. Immunogenetics 1986;23:24.
2. Hahn AWA, Schendel DJ, Hansen PW, Ploegh HL. Biochemical and cellular identification of HLA-Bw44 subtypes. Human Immunol 1984;11:69.
3. Yang SY, Morishima Y, Collins N, et al. Comparison of one-dimensional IEF patterns for serologically detectable HLA-A and HLA-B allotypes. Immunogenetics 1984; 19:217.
4. O'Farrell PH. High resolution two-dimensional electrophoresis of proteins. J Biol Chem 1975;250:4007.
5. Bordier L. Phase separation of integral membrane proteins in Triton X-114 solution. J Biol Chem 1981;256:1604.
6. Neefjes JJ, Breur-Vriesendorp BS, van Seventer GA, Ivanyi P, Ploegh HL. An improved biochemical method for the analysis of HLA Class I antigens: definition of new HLA Class I subtypes. Human Immunol 1986;16:169.
7. Parham P, Barnstable CJ, Bodmer WF. Use of a monoclonal antibody (W6/32) in structural studies of HLA-A,B,C antigens. J Immunol 1979;123:324.
8. Shimizu Y, Geraghty DE, Koller BH, Orr HT, DeMars R. Transfer and expression of three cloned human non-HLA-A,B,C Class I major histocompatibility complex genes in mutant lymphoblastoid cells. Proc Natl Acad Sci USA 1988;85:227.
9. Koller BH, Geraghty DE, DeMars R, Orr HT. HLA-E: a novel HLA class I gene expressed in resting T lymphocytes. J Immunol 1988;141:897.
10. Mizuno S, Trapani JA, Koller BH, Dupont B, Yang SY. Isolation and nucleotide sequence of a cDNA clone encoding a novel HLA Class I gene. J Immunol 1988;140:4024.
11. Bjorkman PJ, Saper MA, Samraoui B, Bennett WS, Strominger JL, Wiley DC. Structure of the human class I histocompatibility antigen, HLA-A2. Nature 1987;329: 506.
12. Bjorkman PJ, Saper MA, Samraoui B, Bennett WS, Strominger JL, Wiley DC. The foreign antigen binding site and T cell recognition regions of class I histocompatibility antigens. Nature 1987;329:512.
13. de Waal LP, Mesman HJ, Engelfriet CP. Serological and biochemical identification of HLA-Aw33 subtypes. Immunogenetics 1987;26:211.
14. Holmes DJ, Miller WV, Rodey GE. The detection of two antigenic determinants of the A28 molecule different from but serologically distinct HLA-A28 specificities. Tissue Antigens 1982;19:161.
15. Holmes N, Parham P. Exon shuffling in vivo can generate novel class I MHC antigens. EMBO J 1985;4:2849.
16. Holms N, Ennis P, Wan AM, Denney DW, Parham P. Multiple genetic mechanisms have contributed to the generation of the HLA-A2/A28 family of class I MHC molecules. J Immunol 1987;139:936.
17. Kruskall MS, Eynon EE, Awdeh Z, Alper CA, Yunis EJ. Identification of HLA-B44 subtypes associated with extended MHC haplotypes. Immunogenetics 1987;26:216.
18. Breur-Vriesendorp BS, Neefjes JC, Huis B, van Seventer G, Ploegh HL, Ivanyi P. Identification of new B27 subtypes (B27C and B27D) prevalent in Oriental populations. Human Immunol 1986;16:163.
19. Choo SY, St. John T, Orr HT, Hansen JA. Molecular analysis of the variant alloantigen HLA-B27d (HLA-B*2703) identifies a unique single amino acid substitution. Human Immunol 1988;21:209.

A Standardized Method for Detection of HLA-A and HLA-B Alleles by One-Dimensional Isoelectric Focusing (IEF) Gel Electrophoresis

Soo Young Yang[1]

Introduction and Brief Description

One-dimensional isoelectric focusing (1D-IEF) gel electrophoresis was used for the study of class I HLA molecules in the Class I HLA Biochemistry component of the Workshop. The method used is a modified version of the two-dimensional gel electrophoresis technique originally developed by O'Farrell (1). The Workshop technique utilized a vertical slab gel apparatus and IEF samples obtained from metabolically labeled cell lysates by immunoprecipitation using monoclonal antibodies. The immunoprecipitated HLA antigens are desialated by neuraminidase treatment to obtain a single, strongly expressed IEF band for each antigen.

Typing by 1D-IEF of HLA class I antigens is very effective in detecting variants or subtypes for HLA-A and HLA-B antigens, but the method has distinct limitations. Interpretation of IEF patterns is entirely dependent on the serologic antigen definitions. Wrongly assigned serologic specificities will result in misleading IEF band assignment. Many overlapping bands for different HLA antigens make the IEF method impractical to substitute for HLA typing. However, serology and IEF together can provide a better definition for each allelic product. The use of cells defined by these two methods for characterization of tissue typing reagents will, in turn, facilitate an accurate assignment of HLA class I antigens.

Typing for HLA-A and HLA-B IEF alleles by the method described in this paper utilize the two monoclonal antibodies 4E (2) and w6/32 (3), which are used in sequential immunoprecipitations. 4E precipitates all B antigens as well as the Aw19 splits (A29, A30, A31, A32, Aw33), and the w6/32 antibody precipitates the remaining class I HLA antigens from each cell lysate. Figure 1 shows an example of the IEF gel layout used in the Workshop. Each gel includes a lane with four IEF bands (A2.2, B35.2, A1, and B27.5) marking the most acidic end (B27.5) and the most basic end (A2.2) of each gel, which allows us to evaluate the quality of the pH gradient. Although the relative position of an IEF band for each antigen remained constant, considerable differences in the resolution of bands were observed between gels

Participating Laboratory: US7SYY[1]

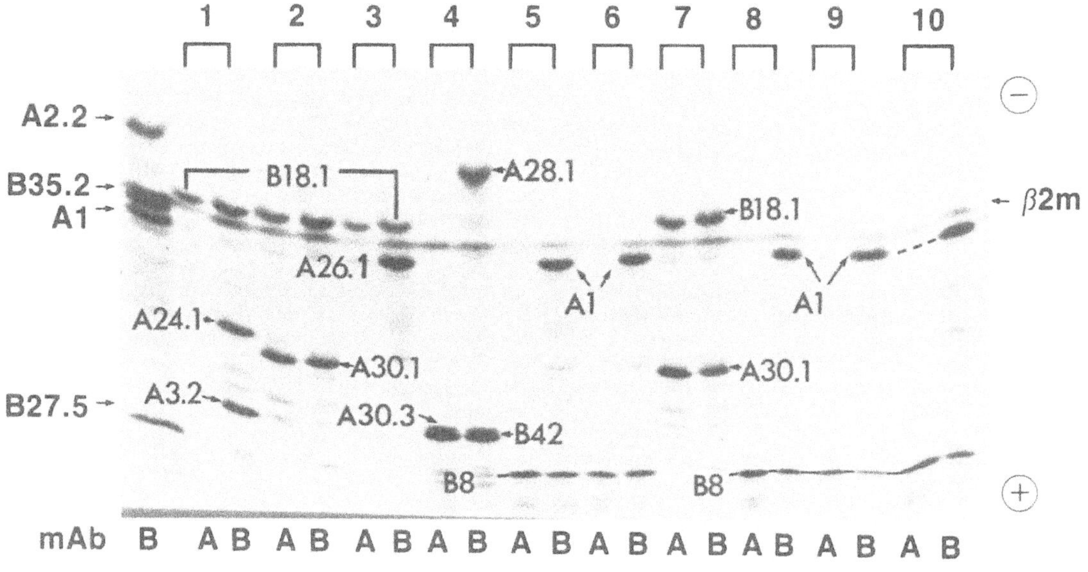

Figure 1. IEF patterns for HLA-A and B antigens isolated from 10 core cell lines. 1 = WSH 9018 (LOO81785, A3, A24; B18; Cw5), 2 = WSH 9019 (DUCAF, A30; B18; Cw5), 3 = WSH 9020 (QBL, A26; B18; Cw5), 4 = WSH 9021 (RSH, Aw68, A30; Bw42; Cw2), 5 = WSH 9022 (COX, A1; B8; Cw7), 6 = WSH 9023 (VAVY, A1; B8; Cw7), 7 = WSH 9085 (EJ32B, A30; B18; CW5), 8 = WSH 9086 (LO541265, A1; B8; Cw7); 9 = WSH 9087 (STEINLIN, A1; B8; Cw-), 10 = WSH 9088 (PFO4015, A1; B8; Cw7). Monoclonal antibodies (mAb) A = 4E, B = w6/32.

and also between laboratories. The bands for the common antigens were used as internal markers to assign specificity for each band. It is an advantage to have a large gel plate with more lanes and longer gels to achieve better resolution and easier interpretation of IEF bands. The pH gradient used in the Workshop protocol provided a wide range resolution; however, modification of the pH gradient may be necessary for a fine resolution of a particular region of the IEF gel.

It should be noted that the recommended technique utilized sequential immunoprecipitation with the monoclonal antibodies 4E followed by W6/32. The method is designed with a single 4E precipitation followed by the W6/32 precipitation. The HLA class I antigens reacting with the 4E (e.g., HLA-B antigens) are not completely removed by a single immunoprecipitation. Therefore, precipitation with W6/32 (which reacts with all class I antigens) will also provide IEF bands corresponding to the 4E-precipitated bands. Data interpretation is accordingly made by comparison of Lane A (4E precipitation) with Lane B (W6/32 precipitation).

It is strongly recommended that the IEF typing gel layout in addition to the markers Lane with A2.2, B35.2, A1 and B27.5 also include reference cell lines representing the serologic antigens for which IEF types are tested on the test cells.

It is not recommended to attempt to make an antigen specificity assignment based on the specific pI values for each HLA-A and HLA-B IEF allele. The present method provides, as stated above, a relative position of an IEF band for each antigen relative to any other antigen. The large number of overlapping or closely spaced bands contributes to serious limitations for data interpretation if this is made based on a pI value. The relative pI position for all IEF HLA-A and HLA-B alleles detected by this technique during the Tenth International Histocompatibility Workshop is summarized in Table 5 in the preceding paper (Population Analysis of Class I HLA Antigens by One-Dimensional Isoelectric Focusing Gel Electrophoresis).

Methods

Preparation of the Radiolabeled Cell Extract

Equipment
- Table top centrifuge
- Microcentrifuge (at least 13,000 × *g*)
- CO_2 incubator

Reagents
- Methionine free RPMI 1640 containing nonessential amino acids supplemented with 10% FCS, penicillin (100 U/ml) and streptomycin (100 μg/ml)
- L-^{35}S-methionine, translation grade >500 Ci/mM
- PMSF stock: 100 mM PMSF (phenylmethylsulfonylfluoride, 100x) in absolute ethanol
- PBS (phosphate buffered saline)
- TE buffer: 50 mM Tris, pH 7.4 to 7.6, 5 mM EDTA
- Lysis buffer: 0.5% Triton X-114 (Sigma) in TE buffer and 0.5% Nonidet P-40 (NP-40) in TE buffer

Procedure
1. Prepare 3–5 × 10^6 EBV-transformed B-cell line (B-LCL) or 5 to 10 × 10^6 PBL (fresh or cryopreserved) with good viability.
2. Wash the cells with PBS and resuspend the cells in 0.5 ml methionine-free RPMI supplemented with 10% FCS, and antibiotics in a flat-bottomed, 2 ml tissue culture plate or 10 ml round-bottomed tissue culture tube. Incubate the cells at 37°C in a CO_2 incubator for 30 min to deplete endogenous methionine.
3. Add 50 μCi ^{35}S-met to the cell culture and continue to culture for a minimum of 3 hours (particularly for nonsterile cells) or a maximum of overnight in a CO_2 incubator.
4. Spin the cells, decant the supernatant, and remove the labeling culture medium by blotting the tube onto a paper towel. There is no need to wash the radiolabeled cells.
5. Before adding lysis buffer, resuspend the cell pellet by tapping the tube (with no liquid). Then add 0.5 ml of cold 0.5% Triton X-114 lysis buffer-2 mM PMSF (always add fresh PMSF). Mix gently with pipet, transfer the cell lysate with the same pipet to a 1.5 ml eppendorf tube, and incubate for exactly 30 min on ice.
6. Centrifuge for 5 min at 4°C using microcentrifuge (at least 13,000 × *g*) and transfer the supernatant (cell lysate) into a new eppendorf tube. Add 0.2 ml 0.5% NP-40 lysis buffer to the cell lysate. The NP-40 prevents phase separation of Triton X-114.
7. Check the labeling efficiency: take 5 μl of cell lysate and mix with 4 ml aquasol or equivalent and count radioactivity using the ^{14}C-channel of the scintillation counter.

Immunoprecipitation

Reagents
- TNEN stock solution (20 mM Tris-HCl, pH 7.6, 0.5% NP-40, 10 mM EDTA, 0.1 M NaCl).
- 1st washing solution (0.1% SDS, 0.5% DOC [Na-deoxycholate]) and 1% BSA in TNEN.
- 2nd washing solution (0.5 M NaCl in 1/10 dilution of TNEN.
- IEF sample buffer (IFSB):

  For 25 ml IFSB:

Urea (ultrapure grade)	14.26 g
10% NP-40 stock	5.0 ml
Ampholine pH 3.5–10	1.25 ml
Dithiothreitol	0.38 g
(or 2-mercaptoethanol, 1.25 ml)	
1% bromophenol blue stock	0.25 ml
Add dd H_2O to total 25 ml	

  Note: Make 1 to 2 ml aliquots of the IFSB and keep them frozen.
- Prewashed 10% Staph-A cells (The Enzyme Center, Boston, MA.) Wash 3x with PBS and make a 10%

suspension in the 1st washing solution; make aliquots and keep it frozen).
- RAM (rabbit anti-mouse Ig) serum or purified RAM Ig.
- Monoclonal antibodies to class I HLA antigens
  4E: HLA-Aw19 splits, HLA-B, and HLA-C antigens.
  w6/32: Class I HLA framework.

Procedure. Preclearing the Cell Lysate
1. Spin 250 µl Staph-A cells in an eppendorf tube for 2 min using a microcentrifuge. Remove supernatant and suspend the pellet by scraping the eppendorf tube on a metal rack.
2. Transfer the cell lysate to the tube with the Staph-A pellet, and add 5 µl hyperimmunized RAM serum or 50 µg RAM Ig to the mixture.
3. Incubate the above mixture on ice or rotate at least 30 min at 4°C and spin 10 min in cold.
4. Preclear once more using 0.25 ml Staph-A cells without RAM (repeat step #1 and #2).
5. Check the background of the precleared cell lysate.
   a. Add 50 µl Staph-A cells to the cell lysate.
   b. Rotate 15 min, spin and save the cell lysate for immunoprecipitation.
   c. Wash the pellet.
      - Unclump the pellet (scrape the tube on a metal rack) and add 1 ml of the 1st washing solution. Mix, spin, and save the pellet.
      - Repeat once more. Wash the pellet once more using the 2nd washing solution, and remove the supernatant completely by blotting the tube on a paper towel.
      - Add 50 µl IEF sample buffer. Resuspend and let it sit for 30 min at room temperature.
      - Spin and count 5 µl of the supernatant mixed with 4 ml of scintillation fluid using the ^{14}C-channel of scintillation counter. Acceptable if < 1000 cpm.

Note: If the background cpm of your precleared cell lysate is consistently low, then you may skin step #5.

Sequential Immunoprecipitation of HLA-B and HLA-A Antigens Using MAbs 4E (IgG2a) and w6/32 (IgG2b)
1. Add 5 µg of MAb 4E (purified, ascites, or hybridoma supernatant) to each precleared radiolabeled cell lysate. Mix and add 50 µl prewashed Staph-A cell suspension.
2. Rotate exactly 15 min in cold and spin 5 min.
3. Transfer the supernatant to a new tube, add 5 µg of mAb w6/32 and 50 µl of Staph-A to the extract, and rotate 15 min. Keep the 4E-Staph-A pellet in cold while w6/32 is processed.
4. Wash the pellets (4E and w6/32) once with 1st washing solution and once with 2nd washing solution.
5. Neuraminidase treatment of the pellets.
   a. Add 20 µl neuraminidase (Type VIII, Sigma 10 U/ml dissolved in 0.05 M EDTA, pH 6.5 to 7.0) containing 1 mM PMSF to the washed pellets. Resuspend (scrape the tubes on a metal rack).
   b. Incubate for 3 hours at 37°C in a shaking water bath or shake the tubes occasionally during the incubation time.
   c. Add 1 ml of TNEN, mix, spin and remove the supernatant.
6. Add 50 µl IEF sample buffer to the pellet. Resuspend and let sit for 30 min at room temperature.
7. Spin and measure the radioactivity in the sample. Carefully take out 2.5 µl of the sample supernatant and mix with scintillation fluid, as described above.

Elcctrofocusing

Apparatus
- Vertical slab gel box
- Glass plates, at least 20 cm long
- Spacers with 0.75 to 1 mm thickness
- Comb
- IEF power supply
- Gel drier
- X-ray film cassette

Reagents
- 30% acryamide-bis stock solution:
  29.2 g acrylamide
  0.8 g N,N′-methylene-bis-acrylamide
  dd H_2O to 100 ml
  Note: Filter the acrylamide stock solution through Whatman #1 filter paper.
- 10% NP-40 stock solution in dd H_2O
- Urea (ultrapure grade)
- TEMED (N,N,N′,N′-tetramethyl ethylene diamide)
- 10% ammonium persulfate (APS) (make weekly)
- Ampholines; make aliquots and freeze at −70°C
- 1.5% agarose in dd H_2O
- 20 mM NaOH (degas thoroughly for more than 1 hour)
- 15 mM H_3PO_4 (1 ml con. H_3PO_4 in 1000 ml dd H_2O)

Procedure. Casting IEF Gel.
1. Prepare IEF gel mixtures. For 10 ml IEF gel mix (measure the volume of your gel and multiply):

30% acrylamide stock	1.5 ml
Urea	5.7 g
10% NP-40 stock	2.0 ml
dd H_2O	2.5 ml

   Dissolve the urea carefully; do not make bubbles; degas thoroughly (at least 30 min).
2. While degassing the solution, assemble IEF gel plate using hot soft wax or hot 1.5% agarose.
3. To the degassed gel mix, add ampholines:

pH 3.5 to 10	0.2 ml
pH 5.0 to 7	0.15 ml
pH 6.0 to 8	0.15 ml

4. Mix the solution gently.
5. Add 25 µl of 10% APS and mix.
6. Add 10 µl TEMED and mix.
7. Pipet the gel mix into the glass plate and insert comb.
8. Takes approximately 1 hour for complete polymerization. *Note*: APS and TEMED are added just before casting gel; the solution polymerizes very quickly.

Electrophoresis.

1. Carefully remove the comb and bottom spacer.
2. Assemble the gel plate onto the gel box.
3. Seal space between plate and gel box using hot 1.5% agarose, and add additional hot agarose in both corners of the gel and the spacer junction to prevent gaps.
4. Load; 25 μl of the sample.
5. Overlay the wells with 25 μl of overlay buffer (dilute IFSB 1:2 with dd water).
6. Add H_3PO_4 in the low chamber and NaOH in the upper chamber without disturbing the samples in the wells.
7. Remove air bubbles from bottom of the gel with a bent needle with a syringe.
8. Run the gel at 10 mA with setting the maximum voltage at 500 for <30 ml gel plate, at 800 for >30 ml but <50 ml gel plate, and 1000 for >50 ml gel plate. Amperage will drop to 3 to 5.
9. Running time:
   Minimum 13 hours
   Maximum 16 hours
10. Fix gel in 15% acetic acid (500 ml) for 30 min on a platform rocker and treat the gel with beta-enhancer (Amplifier from Amersham, Autofluor from National Diagnostics, or DMSO/PPO as directed in manufacturer's instructions).
11. Dry the gel and expose to X-ray film overnight or longer.

Summary

The 1D-IEF method described here was used for the study of class I HLA antigens in the Tenth Workshop. This method can be applied to any type of viable cells (e.g., PBLs, B-cell lines, tumor cell lines, transfectants, sterile or nonsterile cells, and simple enough to use as a routine procedure). The Workshop studies demonstrated that 1D-IEF is an effective method for detecting serologically indistinguishable subtypes or variants of HLA-A and HLA-B antigens, and can distinguish between serologically cross-reactive antigens. In addition, this method is useful in identifying homozygosity or heterozygosity at the HLA-A locus and HLA-B locus. The method can also be applied for monitoring engraftment for patients receiving HLA incompatible bone marrow grafts. However, this method by itself is not suitable for HLA typing because of the fact that many alleles have very similar pI values.

Accurate IEF band assignment in this Workshop was achieved by the use of 4E (HLA-B locus plus Aw19 splits), W6/32 (HLA framework) monoclonal antibodies, and commonly occurring antigens as standard markers. The relative positions of IEF bands which were established during this Workshop will provide additional guidelines for the assignment of class I HLA alleles in the IEF gels. Information on the relative pI values for the class I HLA-A and HLA-B alleles is provided in the Workshop Summary Report (Population Analysis of Class I HLA Antigens by One-Dimensional Isoelectric Focusing Gel Electrophoresis) (this volume).

An alternative approach to the Workshop method would be to use pI markers similar to those applied in the Workshop two-dimensional gel analysis of the HLA class II alpha- and beta-chains. Because of the many closely spaced IEF bands seen from the HLA class I alleles, it is, however, not feasible to design a standardized technique for the purpose of assigning an absolute pI value to each HLA class I allele.

Compared with HLA-A and B antigens, HLA-C antigens are weakly expressed on the cell surface, and the HLA-C genes produce multiple IEF bands. Therefore, a routine 1D-IEF technique for determination of HLA-C alleles must wait for more information on HLA-C gene products and development of HLA-C locus-specific monoclonal antibodies.

References

1. O'Farrell PH. High resolution two-dimensional electrophoresis of proteins. J Biol Chem 1975;250:4007.
2. Yang SY, Morishima Y, Collins NH, et al. Comparison of one dimensional IEF patterns for serologically detectable HLA-A and HLA-B allotypes. Immunogenetics 1984;19: 217.
3. Parham P, Barnstable CJ, Bodmer WF. Use of a monoclonal antibody (w6/32) in structural studies of HLA-A,B,C antigens. J Immunol 1979;123:324.

Detection of HLA-Cw3 Subtypes and a New HLA-C Antigen by One-Dimensional Isoelectric Focusing (1D-IEF)

Ilias Doxiadis[1] and Agathe Hajek-Rosenmayr[2]

The Ninth International Histocompatibility Workshop (9WS) had brought about some indications for a serologic subdivision of HLA-Cw3 (1). During the 10WS, the two subtypes, Cw3.1 and Cw3.2, were termed Cw9 and Cw10. HLA-Cw9 and Cw10 are found in association with HLA-Bw55 and Bw60, respectively.

A subdivision of Cw3 into Cw3.1 and Cw3.2 was also possible by biochemical analysis in 1D-IEF gels. The biochemical variant Cw3.1 is in linkage disequilibrium with HLA-Bw55, whereas Cw3.2 is associated with Bw60. The third HLA-Cw3-associated HLA-B antigen, Bw62, can be found with both biochemical variants of HLA-Cw3.

The two splits of Cw3 are shown in Figure 1, in which seven HLA-Cw3-positive individuals were analyzed by immunoprecipitation with monoclonal antibody (Moab) W6/32 and 1D-IEF. In lanes 1, 2, 3, and 4, the pattern for Cw3.1 can be seen mainly in association with Bw55. HLA-Cw3.2 is seen in association with Bw62 and Bw60 is presented in lanes 4, 5, 6, and 7. Individual 4, in lane 4, is heterozygous for Cw3.1 and Cw3.2.

A panel of 13 individuals typed for HLA-Cw3 by 10WS serology has been biochemically typed for HLA-C in 10WS 1D-IEF gels. A comparison between the two methods results in a perfect correlation between serologically and biochemically defined splits of HLA-Cw3

Participating Laboratories: GERGSW,[1] EAEMYR[2]

Table 1. Comparison of Serologically and Biochemically Defined Splits of HLA-Cw3

		Biochemistry		
		Cw3.1	Cw3.2	
Serology	Cw9	6	0	
	Cw10	0	7	
				13

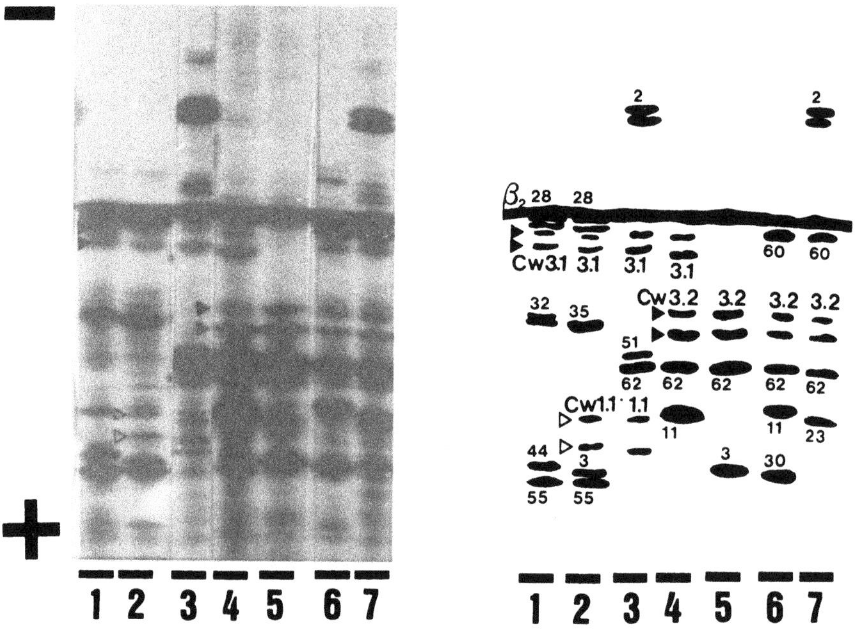

Figure 1. An autoradiogram of an 1D-IEF gel. Results presented in the left panel are explained by schematic drawings in the right panel. In this autoradiogram, four Cw3.1-positive and four Cw3.2-positive individuals can be seen; one is Cw3.1/Cw3.2 heterozygous. HLA types of individuals tested are as follows: 1=A28,30; B44,w55;Cw3.1; 2=A3,28; B35,w55; Cw1.1,w3.1; 3=A2; B51,w62; Cw1.1,w3.1; 4=A11,Bw62; Cw3.1,w3.2; 5=A3; Bw62; Cw3.2; 6=A11,30; Bw60,w62; Cw3.2; 7=A2,23; Bw60,w62; Cw3.2.

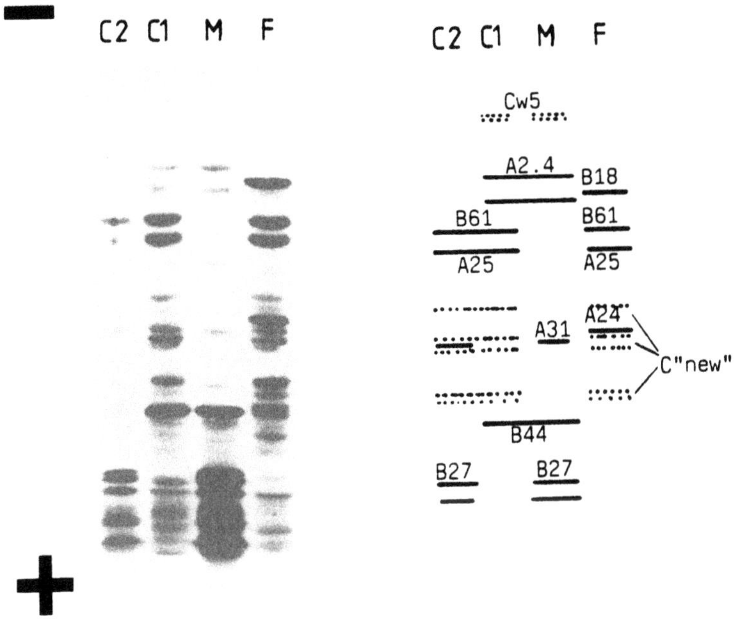

Figure 2. Immunoblot results presented in the left panel are explained by schematic drawings in the right panel. In this autoradiogram, four Cw3.1-positive and four Cw3.2-positive individuals can be seen; one is Cw3.1/Cw3.2 heterozygous. In B, a family, where a HLA-C "blank" allele is segregating, is presented. HLA types of individuals tested are as follows: F=A24,25; B18,w61; C-; M=A2,31; B27,44; Cw5; C1=A2,25; B44,w61; Cw5, C-; C2=A25,31; B27,w61; C-.

(Table 1). Thus, similar subtypes obviously are detected both by serology and biochemistry.

In addition to splits of HLA-C specificities (2), a new biochemical HLA-C variant with no serologic equivalent was found in a family by biochemical typing. Class I 1D-IEF patterns of family E, investigated by the immunoblotting technique (3), are shown in Figure 2. A typical, but so far unknown HLA-C banding pattern, can be seen in F, C1, and C2, segregating with a serologic C "blank" in the Bw61 haplotype.

These results demonstrate that both classic 1D-IEF and immunoblotting are useful techniques for a more exact definition of HLA-C locus antigens.

References

1. Campbell EM, du Toit ED, Tongio MM, Urlacher A, Mayer, S, Mayr WR. Antigen report: HLA-Cw3. In: Albert ED, Baur M, Mayr WR (eds): Histocompatibility Testing 1984. Berlin, Heidelberg, New York, Tokyo: Springer-Verlag 1984:179.
2. Hajek-Rosenmayr A, Doxiadis I. Identification of HLA-C gene products by one dimensional isoelectro focusing (1D-IEF). In: Dupont B (ed): Immunobiology of HLA, Volume I: Histocompatibility Testing 1987. New York: Springer-Verlag.
3. Neefjes JJ, Doxiadis I, Stam JN, Beckers CJ, Ploegh HL. An analysis of class I antigens of man and other species by one dimensional IEF and immunoblotting. Immunogenetics 1986;28:164–171.

Identification of HLA-C Locus Products by One-Dimensional Isoelectric Focusing (1D-IEF)

Agathe Hajek-Rosenmayr[1] and Ilias Doxiadis[2]

Historically, the HLA-C locus was accepted during the Seventh International Histocompatibility Workshop (7WS) as being the third independent class I gene of the human major histocompatibility complex (1). The 9WS defined eight HLA-C specificities, named Cw1 to Cw8 (2). A high linkage disequilibrium of these antigens to various HLA-B gene products is well known. Although C locus antigens have been intensively studied, there is still a high proportion of HLA-C "blank" haplotypes with no serologically defined HLA-C antigen.

Before 10WS, only a few reports on the biochemical definition of HLA-C locus products can be found. One of them has shown that a monoclonal antibody (Moab) reactive with the β_2 microglobulin-free HLA class I heavy chain, Moab HC10, precipitates HLA-C locus products very well (3).

During the present Workshop, a collaborative analysis including a comparison of immunoprecipitation–the Workshop technique–with the immunoblotting technique (4) was performed. It has brought about the identification of banding patterns corresponding to all so far known HLA-C antigens, and has shown that both Moab W6/32 and Moab HC10 are able to recognize HLA-C locus antigens.

Participating Laboratories: EAEMYR,[1] GERGSW[2]

Table 1. Serologic and Biochemical Evaluation of 10WS HLA-C Specificities

Serology	Biochemistry	++	+−	−+	−−	X^2	r
Cw1	Cw1.1	5	2	0	51	39.86	0.82
Cw1	Cw1.2	6	1	0	51	48.75	0.91
Cw2	Cw2	6	1	0	51	48.75	0.91
Cw3	Cw3.1	10	7	0	41	29.14	0.70
Cw3	Cw3.2	7	10	0	41	19.19	0.57
Cw4	Cw4	10	0	0	48	58.00	1.00
Cw5	Cw5	6	0	0	52	58.00	1.00
Cw6	Cw6	11	2	0	45	46.98	0.90
Cw7	Cw7.1	19	9	0	30	30.27	0.72
Cw7	Cw7.2	24	4	0	30	43.86	0.86
Cw8	Cw8	8	0	0	50	58.00	1.00

The results of the analysis are shown in Figures 1 and 2. In Figure 1, isoelectric focusing has been carried out after immunoprecipitation with Moab W6/32. The resulting autoradiogram (Fig. 1), containing all HLA-C locus banding patterns from Cw1 to Cw8, is presented (left

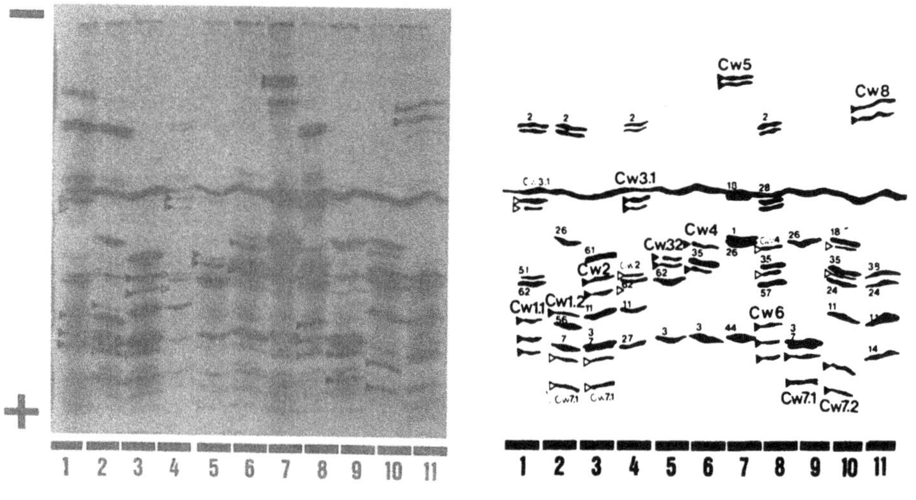

Figure 1. An autoradiogram of an isoelectric focusing gel. Eleven individuals, each of them presenting clearly at least one HLA-C locus banding pattern, have been analyzed in lanes 1 through 11. Results presented in the left panel are explained by schematic drawings in the right panel. HLA-C locus banding patterns are marked with a ► or a ▷ and are identified as Cw . . .; other HLA class I bands are designated only with the appropriate antigen number. HLA types of individuals 1 to 11 are as follows: 1=A2; B51,w62; Cw1.1,w3.1; 2=A2,26; B7,w56; Cw1.2,w7.1; 3=A3,11; B7,w61; Cw2,w7.1; 4=A2,11; B27,w62; Cw2,w3.1; 5=A3; Bw62; Cw3.2; 6=A3; B35, Cw4; 7=A1,26; B18,44; Cw5; 8=A2,28; B35,w57; Cw4,w6; 9=A3,26; B7; Cw7.1; 10=A11,24; B18,35; Cw4, w7.2; 11=A24, w11, B14,38; Cw8.

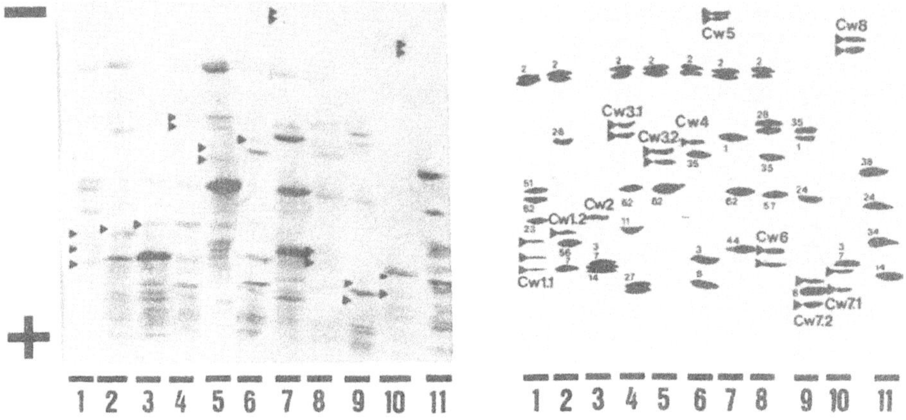

Figure 2. An immunoblot. Eleven individuals, each of them presenting clearly at least one HLA-C locus banding pattern, have been analyzed in lanes 1 through 11. Results presented in the left panel are explained by schematic drawings in the right panel. HLA-C locus banding patterns are marked with a ► or a ▷ and are identified as Cw . . . ; other HLA class I bands are designated only with the appropriate antigen number. HLA types of the individuals 1 to 11 are as follows: 1=A2,23; B51,w62; Cw1.1, w3.1; 2=A2,26; B7,w56; Cw1.2,w7.1; 3=A3; B7,14; Cw2,w7.1; 4=A2,11; B27,w62; Cw2,w3.1; 5=A2; Bw62; Cw3.1,w3.2; 6=A2,3; B8,35; Cw4,w7.2; 7=A1,26; B18,44; Cw5; 8=A2,28; B35,57; Cw4,w6; 9=A1,24, B8,35; Cw7.2; 10=A3, B7; Cw7.1; 11=A24,w34; B14,38; Cw8.

panel) and explained in a schematical diagram (right panel). The left panel of Figure 2 shows an immunoblot, again presenting all HLA-C locus banding patterns which are explained in the right panel. Immunoblotting has been carried out with Moab HC10 and a rabbit anti-heavy chain serum according to a technique described in the same volume (4).

The use of immunoprecipitation as well as immunoblotting allowed us to identify well known HLA-C specificities, and define further variants in this restricted polymorphism. We were able to identify variants of Cw1, (two variants), Cw3 (two variants), and Cw7 (two variants). The variants of Cw1 and Cw7 seem to segregate with HLA-Bw51 for Cw1.1, Bw56 for HLA-Cw1.2, B7 for Cw7.1, and B8 and B49 for Cw7.2.

The comparative analysis of 58 unrelated individuals is given in Table 1.

In summary, the methodologies presented in this communication will help in the future to identify "new" or serologically so far unidentified specificities.

References

1. Sandberg L, Thorsby E, Kissmeyer-Nielsen F, Lindsholm A. Evidence of a third sublocus within the HLA-chromosomal region. In: Terasaki PI (ed). Histocompatibility Testing 1970, Copenhagen: Munksgaard, 1970:165.
2. WHO Nomenclature Report. Nomenclature for the factors of the HLA system 1984. Tissue Antigens 1984;24:73.
3. Stam NJ, Spits H, Ploegh HL. Monoclonal antibodies raised against denaturated HLA-B locus heavy chains permit biochemical characterisation of certain HLA-C locus products. J Immunol 1986;137:2299.
4. Frenz G, Doxiadis I, Vögeler U, Doxiadis G, Grosse-Wilde H. Segregation analysis of HLA class I determinants by biochemistry. In: Dupont B (ed). Immunobiology of HLA, Volume II: Immunogenetics and Histocompatibility. New York: Springer-Verlag.

Serological Definition of HLA-A2 Variants

Lorna J. Kennedy,[1] Steven G.E. Marsh,[1] J. Alejandro Madrigal,[1] Genevieve J. Rabiasz,[1] and Julia G. Bodmer[1]

Variants of HLA-A2 have previously been defined by 1) using the highly specific ability of cytotoxic T lymphocytes to recognize HLA class I antigens on virus infected target cells (1), 2) differences in isoelectric focusing (IEF) points (2), 3) peptide mapping (3), and more recently, 4) different patterns of reactivity with monoclonal antibodies (MAb) (4).

We studied a number of A2 variants plus a large panel of A2 donors with the Tenth Histocompatibility Workshop core sera and MAb, and performed ELISA and isoelectric focusing on most of those cells with A2 variants previously defined using cytotoxic T lymphocytes. The A2 variant cells used included JY, the common variant A2.2 (previously A2.1); AM, a basic IEF variant A2.1Y (previously A2.2Y and A2.2-AM); RT, a basic IEF variant A2.1F (previously A2.2F and A2.2-RT) (5); DK1, an acidic IEF variant A2.3; TROG, with the same IEF as A2.2 but distinguishable from A2.2 cells using cytotoxic T cells (previously A2.4); DEW-ND, an acidic IEF variant A2.3 locally defined as A2.3-DW; KCH-ND, an acidic IEF variant A2.3, locally defined as A2.3-KC; and XLI-ND with an IEF identical with A2.2, locally defined as A2.2-Lee. Interestingly, DK1, DEW-ND, KCH-ND, and XLI-ND were all from Oriental donors. Most of these cells were tested in duplicate and as both peripheral blood lymphocytes and Epstein-Barr virus-transformed B-lymphoblastoid cell lines.

Participating Laboratory: UK1BOD[1]

The serological patterns for these variants are shown in Table 1, together with the patterns for the related antigens Aw68 and Aw69 for comparison. Table 1 shows that the Oriental A2 variants, A2.3-DW, A2.3-KC, and A2.2-Lee can be clearly distinguished from all the other variants using the MAb (Workshop numbers greater than 10W2000). Interestingly, no MAb appeared to differentiate the A2.1 subtypes, although several alloantisera may distinguish A2.1Y and possibly A2.1F from the other A2 variants. Similar results using the MAb were obtained in ELISA, although some of the antibodies were less consistent than others. Several other reagents (not shown in Table 1) (for example, 10W26 [serum], 10W2026 [MAb], and 10W2027 [MAb]) react with all A2 variants and both splits of A28. Table 1 also indicates some of the problems that are encountered when these reagents are used to type B-lymphoblastoid cell lines: several reagents are negative with peripheral blood lymphocytes, but positive with the cell line from the same individual. When using alloantisera, this is usually ascribed to contaminating class II antibodies, but this explanation cannot be applied to (cloned) MAb which must contain only one antibody. Careful studies testing several of these MAb in both microcytotoxicity and ELISA at many different dilutions suggest that the extra reactivity of the B-cell lines is probably due to the

Table 1. Serological Patterns for A2 Variants, in the Tenth Workshop Core Reagents

Post 10WS Antigen Name	Previous Antigen Name	10WS Reagents: 20	32	35	29	16	18	30	31	2037	2036	22	33	37	19	2021	2020	2040	2029	2033	2025	2034	2032	40	38	39	36	42
Aw68		−	−	−	−	−	−	−	−	−	−	−	−	−	−	+	+	+	+	+	+	+	±	+	+	+	+	+
Aw69		−	−	−	−	−	+	+	+	+	−	−	−	−	−	+	+	+	+	+	+	+	+	+	+	+	+	+
A2.2	A2.1	+	+	+	+	+	+	+	+	+	+	+	+	+	+	+	+	+	+	+	+	+	±	−	−	−	−	−
A2.1F	A2.2F/RT	+	+	+	+	+	+	+	+	+	+	+	+	+	−*	+	+	+	+	+	+	+	−*	−	−	−	−	−
A2.1Y	A2.2Y/AM	+	+	+	+	+	+	+	+	+	+	−*	−	−*	−	+	+	+	+	+	+	+	−*	−	−	−	−	−
A2.2-Lee	A2-Lee	+	+	+	+	+	+	+	+	−	−*	−	−	+	−	−*	+	+	+	+	+	+	+	−	−	−	−	−
A2.3-DW	A2-DW	+	+	+	+	+	+	+	+	+	+	+	+	+	−	−*	−*	−*	−	−*	−*	−	−	−	−	−	−	−
A2.3-KC	A2-KC	+	+	+	+	+	+	+	+	+	+	+	−	+	−	+	+	+	−	−*	−*	−	−	−	−	−	−	−
A2.3	A2.3	+	+	+	+	+	+	+	+	+	+	+	+	+	−	+	+	+	+	+	+	−	−	−	−	−	−	−
A2.2-TROG	A2.4	+	+	+	+	+	+	+	+	+	+	+	+	+	+	+	+	+	+	+	+	+	+	−	−	−	−	−

*These reagents were negative with peripheral blood lymphocytes of that specificity, but positive with the B-lymphoblastoid cell line from the same individuals.
± These reagents may be positive or negative.
All reagents numbered above 10W2000 are MAb, and were also tested by ELISA on these cells.
Identical results were obtained for all the above MAb in both microcytotoxicity and ELISA, except 10W2021, 10W2020 and 10W2040, which gave less consistent results.

Immunobiology of HLA
Volume I

Table 2. Numbers of A2 Variants Found in Different Ethnic Groups

Ethnic Group	Number of A2s Tested	A2.2 Lee	A2.3 DW	A2.3 KC	A2.1 Y
Black	239	6	–	4	–
Chinese	168	1	6	10	–
Japanese	301	7	–	2	1
Thai/Malay	165	–	5	22	–
Other Oriental	110	1	4	–	1
Caucasoid	233	1	–	1	3
Total	1216	16	15	39	5

increased amount of antigen on the cell surface of cell lines. It is, therefore, probably necessary to determine different working dilutions for MAb, depending on the cell type to be tested.

A2.3-DW and A2.3-KC may represent splits of A2.3, (as AM and RT represent splits of A2.1), because although they clearly have different reaction patterns with the MAb, there is some recognition of KCH-ND by effector T cells from DEW-ND, and they both have an identical acidic isoelectric point (4).

A search through the appropriate Tenth Workshop data files was undertaken to ascertain whether A2 cells with any of these patterns existed in data from other laboratories. The results of this search are shown in Table 2. Cells with a particular pattern were only included in Table 2 when the majority of other A2 cells from that laboratory had the full reaction pattern (i.e., A2.2). Although IEF and cytotoxic T-cell data are not available for these cells, so that absolute identity with the A2 variants from this laboratory cannot be proven, there are sufficient numbers of the three Oriental A2 variants (A2.2-Lee, A2.3-DW and A2.3-KC) to suggest that these reaction patterns can be used to identify these variants with a fair amount of reliability.

It would, therefore, seem possible to identify most A2 variants using a panel of sera and MAb in the standard microcytotoxicity assay.

References

1. McMichael AJ, Ting A, Zwerink HJ, Askonas BA. HLA restriction of cell-mediated lysis of influenza virus-infected human cells. Nature 1977;270:524.
2. Van der Poel JJ, Molders A, Ploegh HL. Definition of four HLA-A2 subtypes by CML typing and biochemical analysis. Immunogenetics 1983;17:609.
3. Krangel MS, Taketani S, Biddison WE, Strong DM, Strominger JL. Comparative structural analysis of HLA A2 antigens distinguishable by cytotoxic T lymphocytes: variants M7 and DR1. Biochemistry 1982;21:6313.
4. Kennedy LJ, Wallace LE, Madrigal AJ, Rickinson AB, Bodmer JG. New HLA-A2 variants defined by monoclonal antibodies and cytotoxic T lymphocytes. Immunogenetics 1987; 26:155.
5. Parham P, Lomen CE, Lawlor DA, et al. Nature of polymorphism in HLA-A, -B, and -C molecules. Proc Natl Acad Sci USA 1988;85:4005.

Variants of HLA-Aw68 Recognized by Isoelectric Focusing

J. Alejandro Madrigal,[1] Lorna J. Kennedy,[1] Steven G.E. Marsh,[1] and Julia G. Bodmer[1]

The family of related molecules called HLA-A2/28 includes all the subtypes of A2 and the two splits of A28, namely Aw68 and Aw69 (1). Moreover, Aw68 has been split serologically and by DNA sequencing into two further groups (2,3).

In recent years, the HLA A2/28 family has been analyzed with cytotoxic T cells, monoclonal antibodies (MAb,) restriction enzymes, DNA sequencing, and more recently, a crystallographic analysis of the class I HLA-A2 antigen has been published (3,4). The A2 subtypes (A2.1, A2.2, A2.3, and A2.4) are easily identified in one-dimensional isoelectric focusing gels (1D-IEF) using monomorphic as well as polymorphic monoclonal antibodies (5).

In this study, a panel of Aw68 cell lines, (Table 1), were studied by one-dimensional isoelectric focusing gels (1D-IEF). It was demonstrated that Aw68 antigens can be subdivided into subsets with different isoelectric points (IEP). Aw68a has an acidic IEP just below β2m, whereas Aw68b has a more basic IEP (Fig. 1).

Sequencing data from Holmes et al. (3) have shown that the two variants in that study of Aw68 (Aw68.1 and Aw68.2) differ from each other by six nucleotide substitutions. In exon 2, one single change produces a V–M substitution at position 12 in the alpha-1 domain. The changes in exon 3 lead to four amino acid (AA) substitutions in the alpha-2 domain (Fig. 2). The amino acid substitutions at AA 12, 114 and 105 do not produce a charge effect, although the substitution at AA 97 from methionine to arginine (positive charge) and at AA 116 from aspartic acid (negative charge) to tyrosine (positive

Participating Laboratory: UKIBOD[1]

Table 1. Class I Types of the Cell Lines Analyzed in This Study

Cell	HLA-A	HLA-B	HLA-C
AMR	2.1, 32	49, w60	w10, w7
DK1	2.3, w33	44, w60	w10
LB-S	w68	40	w2, w3
MRH-TC	w68, 32	51, 27	w2
GRH-TC	2.2, w68	13, 51	nt
KHE-TC	29, w68	7, 51	nt
GMR	3, w68	7, 8	w3, w4
BREL	w68, 30	7, 39	nt
IARC9	w68, 30	13, 37	w6
HFF	1, w68	14, 37	w6, w8
HAL-NF	29, w69	35	w4
MAL-NF	1, w69	57, 35	w6
LLICRF	1, w69	8, 35	w7

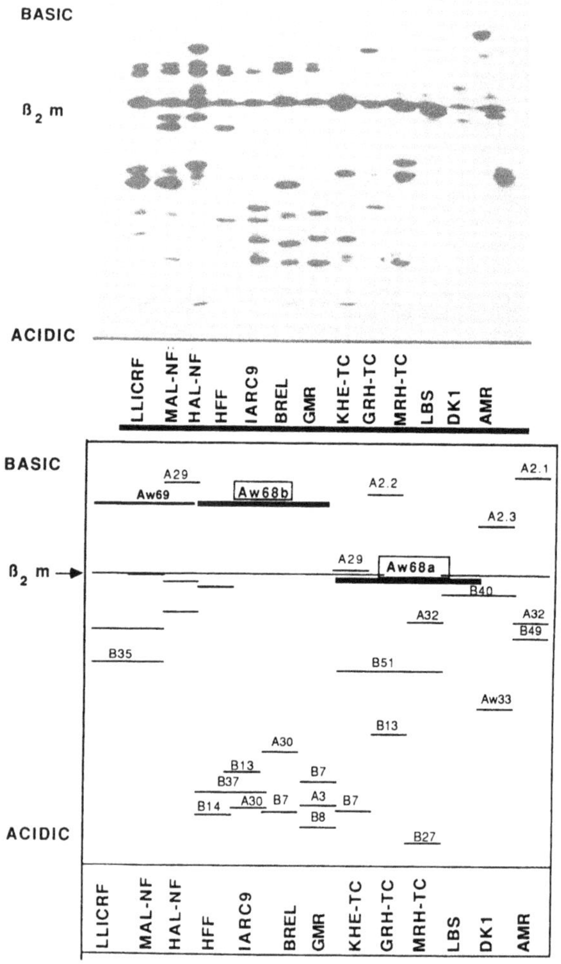

Figure 1. A large panel of Aw68 individuals was analyzed by 1D-IEF using the MAb W6/32. This figure demonstrates the two different IEP positions of Aw68: Aw68a just below the β2m (KHE-TC, GRH-TC, MRH-TC, LB-S,) and Aw68b above the β2m (HFF, IARC9, BREL, GMR). This figure also demonstrates that Aw69 (LLICRF, MAL-NF, HAL-NF) has the same IEP as Aw68b.

POSITION	SUBSTITUTION	Aw68.1	Aw68.2	CHARGE EFFECT
12	V---M	Valine	Methionine	none
97	M---R	Methionine	Arginine+	+1
105	S---P	Serine	Proline	none
114	R---H	Arginine +	Histidine+	none
116	D---Y	Aspartic Ac. -	Tyrosine	+1
			Total net charge	=+2

Figure 2. Amino acid differences between Aw68.1 and Aw68.2.

POSITION	SUBSTITUTION	Aw69	Aw68.2	CHARGE EFFECT
12	V---M	Valine	Methionine	none
95	V---I	Valine	Isoleucine	none
105	S---P	Serine	Proline	none
107	W---G	Tryptophan	Glycine	none
156	L---W	Leucine	Tryptophan	none
205	A---V	Alanine	Valine	none

Figure 3. Amino acid differences between Aw69 and Aw68.2.

charge) produces a net positive charge. The substitutions found in this Aw68.2 variant lead to a more basic protein, which may account for the differences found in our IEF analysis, and it is possible that our Aw68b variant corresponds to the same gene product as Aw68.2.

Aw69 is a hybrid molecule of A2 and Aw68 (6). Thus, Aw69 has an identical sequence to Aw68 in the first extracellular domain and is identical to A2.2 in the second and third domains. IEF analysis of Aw69 shows it to have the same IEP as Aw68b (Fig. 1). Holmes and co-workers (3) had previously shown that there are six amino acid differences between Aw68.2 and Aw69 (Fig. 3). However, none of these differences produce a charge effect in the protein and, consequently, Aw69 focuses at the same IEP as Aw68b.

Serologically, it is possible to distinguish the splits of A28, Aw69 and Aw68, using the 10WS core sera and monoclonals (7). However, there is no evidence to suggest that the Aw68a and Aw68b splits can be defined by serology using 10WS reagents.

References

1. Biddison WE, Kostyu DD, Strominger JL, Krangel MS. Delineation of immunologically and biochemically distinct HLA-A2 antigens. J Immunol 1982;129:730.
2. Holmes DJ, Miller WV, Rodey GE. The detection of two serologically distinct HLA-A28 specificities. Tissue Antigens 1982;19:161.
3. Holmes N, Ennis P, Wan AM, Denney DW, Parham P. Multiple genetic mechanisms have contributed to the generation of the HLA-A2/28 family of class I MHC molecules. J Immunol 1987;139:936.

4. Bjorkman PJ, Saper MA, Samraoui B, Bennett WS, Strominger JL, Wiley DC. Structure of human class I histocompatibility antigen, HLA-A2. Nature 1987;329:506.
5. Kennedy LJ, Wallace LE, Madrigal JA, Rickinson AB, Bodmer JG. New HLA-A2 variants defined by monoclonal antibodies and cytotoxic T lymphocytes. Immunogenetics 1987;26:155.
6. Holmes N, Parham P. Exon shuffing in vivo can generate novel class I MHC molecules. EMBO J 1985;4:5175.
7. Kennedy LJ, Marsh SGE, Madrigal JA, Rabiasz GJ, Bodmer JG. Serological definition of HLA-A2 Variants. In: Dupont B (ed): Immunobiology of HLA, Volume I: Histocompatibility Testing 1987. New York: Springer-Verlag, 1989.

Biochemical Variants of the Bw70 Antigen (Bw71, Bw72)

M.G. Guttridge,[1] G.J. Laundy,[1] and P.T. Klouda[1]

The Bw70 antigen splits Bw71 (Bu) and w72 (Sv) were first described serologically in 1977 as Bw6-associated antigens reacting with some multispecific antisera, identifying the highly cross-reactive antigens B5, 35, 15, 18, 21. There are no monospecific sera recognizing Bw71 or Bw72, and the serologic definition relies on reactions with multispecific reagents.

A biochemical investigation of 30 Bw70-positive individuals (14 Bw71, 16 Bw72) has identified three variants of Bw70 with distinct isoelectric points. These have been designated Bw70.1, Bw70.2, and Bw70.3 in order of increasing isoelectric point acidity. The IEF band Bw70.1 focused slightly acidic to A24, the Bw70.2 band focused slightly acidic to A11, and the Bw70.3 band focused acidic to B44 in the region of A3 and B7 (Fig. 1).

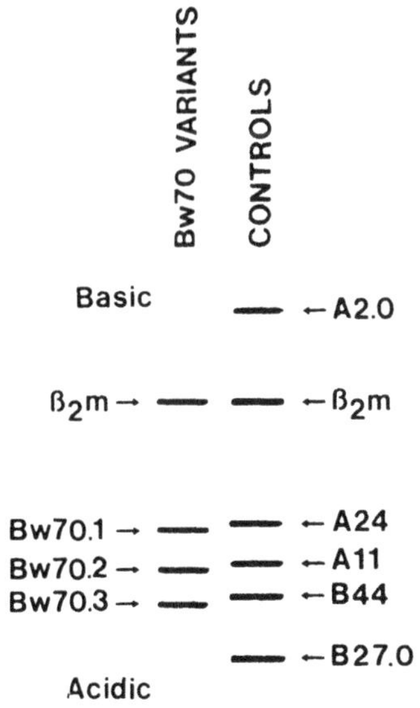

Figure 1. Schematic 1D-IEF gel patterns showing three variants of the Bw70 antigen.

Table 1. Comparison of Serologic and Biochemical Variants of Bw72. Five Cells Not Split Serologically

	Biochemistry		
Serology	Bw70.1	Bw70.2	Bw70.3
Bw72.1	1	1	5
Bw72.2	2	1	1

A serologic and biochemical comparison showed that the Bw70.2 IEF band was present in 13 of the 14 Bw71-positive individuals; 1 Bw71 individual had the Bw70.1 IEF band (Fig. 2).

Analysis of the 16 Bw72-positive donors revealed that 10 (62%) had the Bw70.3 IEF band, 4 had the Bw70.1 IEF band, and 2 had the Bw70.2 IEF band (Fig. 2).

The Bw71 antigen, with one exception is, homogeneous by biochemistry, and there is no evidence of splits by serology. The Bw72 antigen is heterogeneous by biochemistry, with Bw70.1 and Bw70.3 being the main variants. Tenth Workshop serology also shows possible splits of Bw72, which are described in the Bw70 Antigen Society Report (in this volume). These serologic splits appear not to correlate with the biochemically defined variants (Table 1). While the biochemical variants of Bw72 are well defined, the serology is problematic with few participating laboratories having discriminating data. This may account for the lack of correlation between the biochemical and serological definitions.

Participating Laboratory: UKIBID[1]

CELL IDENTITY	HLA* -B	-B	-A	-A	1D-IEF BAND 70.1	70.2	70.3
GAN 3914	71	44	24	68		+	
DUP 7335	71	35	11	28		+	
FAU 8295	71	49	23	28		+	
DUT 616	71	7	33	68		+	
LAW 8020	71	55	24	68		+	
OHA 8295	71	49	23	28		+	
REE 264	71	8	1	28		+	
REE 136	71	52	1	11		+	
REE 135	71	44	1	33		+	
BRS 614	71	18	3	68		+	
BRS 343	71	44	2	23		+	
BRS 345	71	18	1	11		+	
DUT 618	71	8	2	29		+	
BRS 347	71	49	23	24	+		
BRS 70	72	44	23	32	+		
BRS 353	72	45	11	26	+		
REE 142	72	37	1	11	+		
REE 202	72	42	30		+		
DUT 609	72	45	30	43		+	
BRS 348	72	44	2			+	
GAN 1429	72	7	1	68			+
DUT 624	72	58	31	68			+
DUT 617	72	39	23	24			+
ENG 125	72	42	30				+
REE 145	72	7	23	24			+
REE 138	72	62	24	32			+
REE 1147	72	61	2	24			+
BRS 132	72	45	23	29			+
BRS 103	72	44	2	26			+
BRS 341	72	44	2	30			+

Figure 2. Comparison of serologic and biochemical definitions of the Bw70 antigen.
* prefix w omitted.

B "SNA": HLA-B5 Split Associated with Bw6 Studied by Serology and 1D-IEF

M. Andrien,[1] P. Reekers,[2] and E. Dupont[1]

During the Ninth International Histocompatibility Workshop, a new B locus antigen SNA was described by Andrien et al. (1). B "SNA" is a Bw6-associated antigen belonging to the B5-CREG antigens: B51, Bw52, Bw53, B35, Bw70, Bw71, and Bw72.

Originally, the SNA antigen was found in a North African Moroccan family of Berber ethnic origin. In the present workshop (Table 1) we studied an additional family (KHA) and a patient (BOU) from the same ethnic background with core serology and 1D-IEF after immunoprecipitation with Mab 4E and W6/32. Most of the

Table 1. 10WS Study Results

Haplotypes in family SNA		
Father	: gh	a : A3, B7, Cw7, DRw8
Mother (004)	: ac	c : A28, *SNA*, DRw13
Children(009→012)	: gc	g : A3, B7, Cw7, DRw8
		h : A24, B44, Cw5
Haplotypes in family KHA		
Father (001)	: ab	a : Aw36, *SNA*, DRw13
Mother	: cd	b : Aw33, Bw65, Cw8, DRw8
Child (003)	: ac	c : A30, B7, Cw7, DRw9
		d : A29, B39, DRw13
Phenotype in patient BOU		
A24 Aw33 B8 *SNA* Cw7 DR3 DRN13		

Participating Laboratories: BENDPT,[1] BENREE[2]

Immunobiology of HLA
Volume I

Table 2. Serology of B "SNA" Compared with B5-CREG Antigens

Families	B51+ SNA+	B51+ SNA+	B52	B5+B49+ SNA−	B5+B49+ SNA+	B53+B35 SNA−	B53+B35 SNA+	BW70+ SNA−	BW70+ SNA+
SNA:OO4, 009→012	1	8	1	1	8	1	8	1	8
KHA:001, 003	1	8	1	1	8	1	8	1	8
BOU	1	8	1	1	8	1	8	1	8
10WS sera	174	178	182	222	224	191	208	220	198
	177	179	183	223		196	188	221	202
	180		442	226			193	239	228

10WS sera, detecting B51, Bw52, Bw53, B35, and the BW70 antigens, were negative with SNA-positive cells.

SNA reactivity was found predominantly in multispecific anti-B5 sera, as listed in Table 2. This contrasts with the reactivity pattern of sera 10W222, 223 and 226, which provide a negative reaction with SNA-positive cells, confirmed by absorption studies.

In addition, informative sera for identification of SNA antigen include 10W178 and 179 antisera. They are known to recognize B51 antigen and also to detect SNA, whereas monospecific sera defining antigens belonging to the various B5-CREG are negative with SNA-positive cells. Broad Mab detecting the B5 antigen including 10W2091 and 2098 are reactive with SNA-positive cells. Despite the lack of monospecific alloantisera and monoclonal antibodies, SNA antigen can easily be defined by a combination of B51 and B5 sera from the 10WS. Recently, other antigens (Te 76 and BX1) have been described (2,3) that might be similar to SNA due to BW6 association and African ethnic background. The possible identity of SNA and the antigens Te 76/BX1 remains to be investigated.

After 35S-methionine labeling of cells from the families SNA (EBV-transformed B-LCL) and KHA (PHA-blast), class I molecules were immunoprecipitated sequentially with Mab 4E and W6/32, and analyzed after neuraminidase-treatment on 4.5% acrylamide 1D-IEF gels according to the 10WS protocol. The SNA-gene

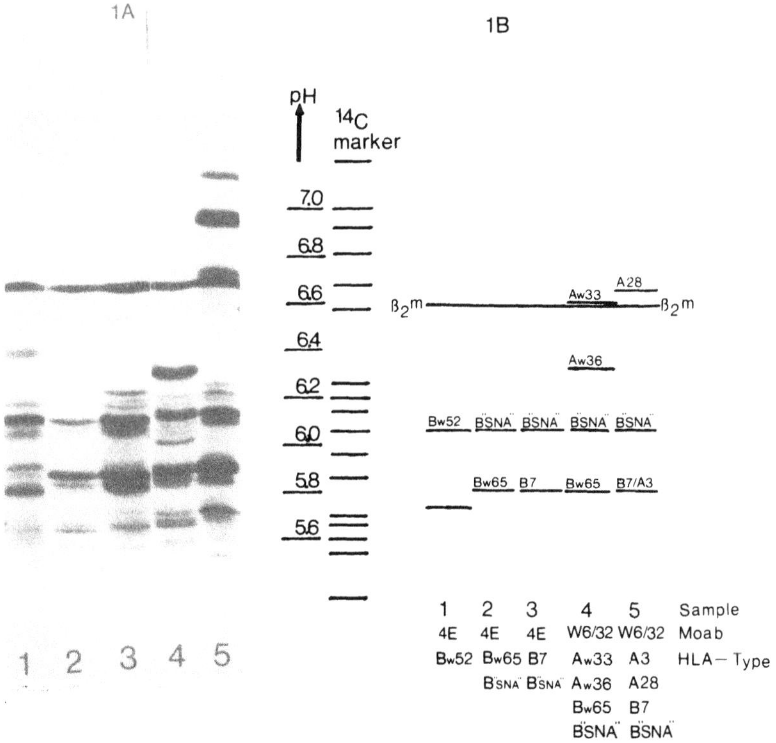

Figure 1. 1D−IEF pattern of B"SNA" positive cells.

product precipitated as most B locus products with Mab 4E and showed haplotype segregation in three additional family members tested. The isoelectric point (pH = 6.08, Fig. 1) was identical to that of Bw52 and one of the charge variants of B51, but differed from other B5 CREG molecules.

References

1. Andrien M, Vandercruys M, Hoven C, Dupont E. Analysis of a Bw6 associated B locus antigen belonging to the B5 cross group (CREG) by family and absorption studies. 9th Histocompatibility Workshop Newsletter 1984;9:3.
2. Clark BD, Park MS, Tonai R, Terasaki PI. Serologic and biochemical analysis of HLA B15 and B5 complexes Human Immunol 1987;20:47–57.
3. Ward FF, Stewart AD, et al. Model for the evolution of diversification of epitopes within the B5 CREG 13th ASHI Meeting, 1987; Abstract #38.

B40 CREG Variants Detected by Serology, IEF, and CTL

P. Reekers,[1] G. van Seventer,[2] L. de Waal,[2] A. Amoroso,[3] P. Chiewsilp,[4] and J.R. Moraes[5]

The antigenic variations within the B40 cross-reactive group (B40 CREG: Bw60, Bw61, Bw41, B13, and Bw48) have been investigated with 10WS core serum and Antigen Society serum sets. The serologic data have been related to 1D-IEF charge variants of neuraminidase-treated HLA gene products, as well as to response of cytotoxic T lymphocytes (CTL). Some of the families positive for Bw41 or B13 have been tested during the 9WS in other centers (Torino, Bangkok, and Rio de Janeiro), and were selected for this additional study based on their aberrant serologic pattern with 9WS reagents. For the CTL-study, the 10WS B-LCL reference cell lines have been used.

Participating Laboratories: BENREE,[1] BENENG,[2] ITICEP,[3] ANZCHI,[4] LATJRM[5]

Serologic Data

During the 9WS, serologic splits of Bw41 have been proposed based on the reactivity with serum 9W241 (ED234, contributed by Moraes) (1). In the 10WS Antigen Society #14, serum 9W241 has been retested; ED234 formed a cluster with the sera 10W386, 10W387, and 10W451 that did not react with part of the Bw41-positive cells, including the Bw41-short families tested also in the 9WS. Thus, the previously proposed existence of two serologic variants – Bw41-long (Bw41A) and Bw41-short (Bw41B) – could be confirmed (see also Antigen Society #14 Report).

Based on the lower cytotoxicity score for Mab 10W2094 with Oriental B13 cells compared with Caucasoid B13-positive cells, the B13 antigen might be serologically different between various ethnic groups.

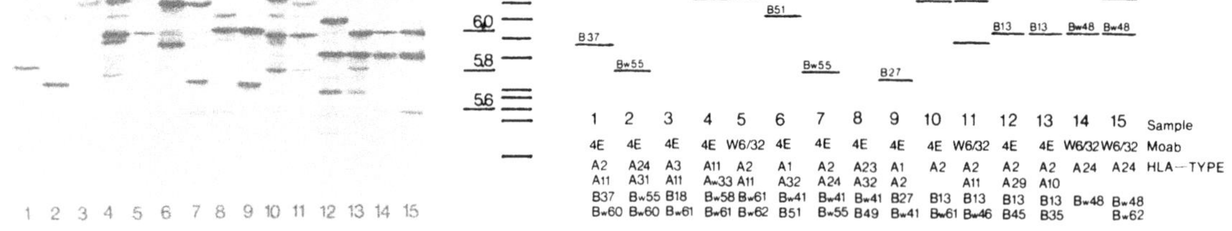

Figure 1. IEF pattern of B40-CREG gene products. As ^{14}C-marker, the reference markers from the Class II Biochemistry Workshop have been used.

However, more sera are needed to confirm these serological variations (see also Antigen Society #14 Report). For the Bw60; Bw61; and Bw48-positive cells, no serologic variants have been observed in our local study.

IEF Data

In Figure 1A, charge variants of neuraminidase-treated HLA class I B locus gene product(s) are given together with the approximate pH at which the IEF-band(s) were observed. For immunoprecipitation with Mab's 4E and W6/32, the Workshop protocol was used. Charge variants were not found for Bw61 (sample #3, 4, 5, and 10) and Bw48 (sample #14 and 15). Bw41-positive cells gave two distinct banding patterns segregating in families, either at approximate pH 6.38 (sample #6 and 7) or pH 6.20 (sample #8 and 9). No correlation was found between the serologic variants Bw41A/Bw41B and the Bw41 charge variants. B13-positive cells of Oriental origin precipitated at approximate pH 6.16 (sample #10 and 11), whereas Caucasian B13 gene products precipitated at pH 6.0 (sample #12 and 13). Thus, the charge variants of HLA-B13 are probably due to exchanges in charged amino acids, conserved within ethnic groups, that might have also influenced serologic epitopes possibly detected with Mab 10W2094. For Bw60 (sample #1 and 2), there are some indications that except for the gene product of approximate pH 6.6, another charge variant does exist. More family data are necessary to confirm this observation.

CTL Data

CTL reagents against B7 variants showed a heterogeneity in CTL response for Bw60- and Bw41-positive Workshop B-LCLs (2). The B7-related CTLs were not reactive against the Bw61 and B13 cells tested. Studies are in progress to correlate variation in CTL reactivity with the serologic and charge variants of the B40 CREG within families.

We conclude that until now, serologic splits detected with pregnancy alloantisera correlate poorly with charge variants of these gene products. Only one anti-B13 monoclonal antibody (10W2094) has been observed that might detect epitopes that correlate between serologic and IEF charge variants.

Acknowledgments. We would like to thank Dr. F. Kissmeyer-Nielsen (Aarhus) and Dr. C. Jersild (Aalborg), who provided Bw48 positive cells from Eskimo origin.

References

1. Mervart H, Taylor C, Ting A. Antigen report: HLA-BW41. In: Albert E, et al. (eds): Histocompatibility Testing 1984. New York: Springer-Verlag, 1984:149.
2. van Seventer GA, et al. Six subtypes of HLA-B7 defined by histotopes, epitopes and 1D-IEF. In: Dupont B (ed): Immunobiology of HLA, Volume II: Immunogenetics and Histocompatibility. New York: Springer-Verlag.

Serology and IEF Analysis of New HLA-B Locus Gene Products B"POT," B"RI," B"ST40," and B"LEV"

P. Reekers,[1] H. Tijssen,[1] G. Mueller-Eckhardt,[2] F.C. Grumet,[3] and G. Leverenz[4]

In the recent years, various new HLA-B locus antigens have been postulated based on serologic studies within families. The B"POT" antigen (three families) has been detected with broadly reactive B7 CREG antisera (1); B"RI" (two families) and B"ST40" (two families) belonged to the B40 CREG (2,3) and B"LEV" (one family) was a recently proposed serologic blank of the HLA-B locus (4). Within the framework of the 10 IHWS, cells from these informative families have been tested with the core serum set and reagents from Antigen Societies #12, 13, and 14. In Table 1, informative core sera for each of these new serologic specificities have been listed. Many broadly reactive anti–B7 sera contained (sometimes weak) anti B"POT"; the strongest sera were 10W381, 10W382, 10W340, and 10W333. In the Antigen Society serum set, serum #9995 was the only serum that was positive with B"POT," but negative with B7.

The specificity B"RI" reacted strongly with sera 10W232 and 10W287; weak reactions were observed with some other sera such as 10W340 and 10W333. The specificity B"ST40" reacted with many B40 CREG antisera; sera 10W353 and 10W357 are examples of such reactivity patterns. No core set or Antigen Societies #12, 13, 14 sera were found that reacted with the proposed serologic blank B"LEV." Thus, since none of the sera were specific for the new determinants, they can only be detected in combination with alloantisera monospecific for the WHO-recognized B locus antigens belonging to the same crossreactive group.

Participating Laboratories: BENREE,[1] GERMUE,[2] US5GRU,[3] EAERIC[4]

Table 1. 10 1HWS Serologic Reaction Patterns of the New B Locus Specificities: B"POT," B"RI," B"ST40," and B"LEV"

Specificity	9295	381	382	340	333	276	275	331	347	372	275	331	347	372	332	387	53	57
B"POT"	+	+	+	+	+	±	±	±	±	±	±	±	±	±	−	−	−	−
B"RI"	−	−	−	±	±	−	−	−	−	−	−	−	−	−	+	+	−	−
B"ST40"	+	−	−	−	−	−	−	−	−	−	−	−	−	−	−	±	+	+
B"LEV"	−	−	−	−	−	−	−	−	−	−	−	−	−	−	−	−	−	−
B7	−	+	+	+	+	+	+	+	+	+	+	+	+	+	−	−	±	−
B40	+	−	−	−	−	−	−	−	−	−	−	−	−	+	−	−	+	+
BW41	−	±	−	−	−	−	−	−	−	−	±	±	−	−	+	−	±	±
BW42	−	+	−	−	+	−	−	−	−	−	+	+	+	−	−	−	−	−
BW48	−	+	+	−	−	±	−	−	−	−	−	−	+	+	−	−	+	±
BW73	−	−	−	−	+	−	−	−	−	−	−	+	−	+	−	−	−	−

In Figure 1, the isoelectric points have been given for the bands obtained after immunoprecipitation with Mabs 4E and W6/32 according to the 10 IHWS Class I Biochemistry protocol. The gene product of HLA-B"POT" (sample #1, 2, and 3) precipitates at the same pH (approximately 5.84) as the most common HLA-B7. The gene product of B"POT" can be shown to be different from B7 by the use of other monoclonals such as 38D4 and by RFLP-analysis (5).

The gene product of B"RI" (sample #6 and 7) precipitates as a double band exactly like the most common B*2705 in Caucasians (approximate pH 5.66). Under standard conditions, B"ST40" could not be precipitated with Mab 4E, but with W6/32 a dense band was observed at approximate pH 6.06 (sample #4 and 5).

B"LEV" was only weakly reactive with 4E as well as W6/32; the weak specific band observed was located close to B7 at approximate pH 5.80 (sample #8 and 9). It needs to be investigated whether B"LEV" is a new B locus antigen associated with Bw4 that can not yet be detected with alloantisera or that weak expression of a B locus gene product causes serologic and biochemical aberrant reactivity.

1B

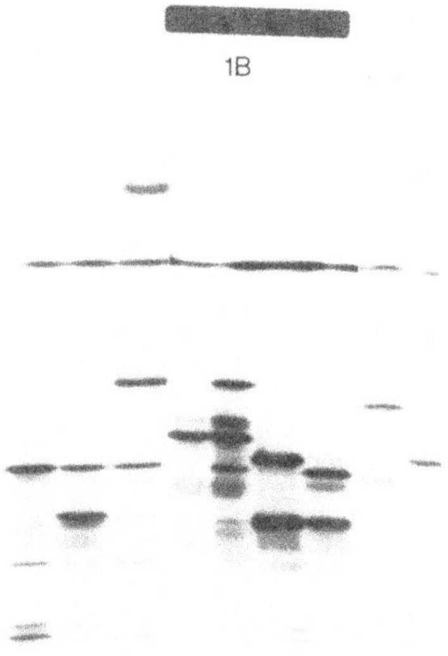

1A

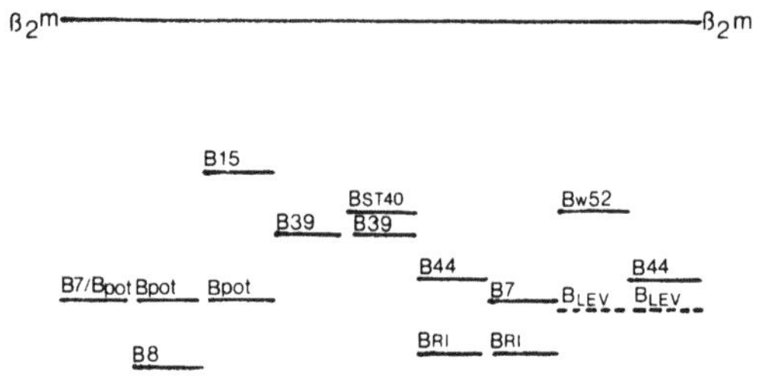

Sample	1	2	3	4	5	6	7	8	9
Moab	4E	4E	W6/32	4E	W6/32	4E	4E	4E	4E
HLA type	A2	A1	A2	A3	A3	A2	A3	A1	A28
	A3	A2		A9	A9	A24	A24	A31	A31
	B7	B8	B15	B39	B39	B44	B7	Bw52	B44
	Bpot	Bpot	Bpot	BST40	BST40	BRI	BRI	BLEV	BLEV

Figure 1. IEF Patterns of gene products from some new B Locus determinants. As ^{14}C-markers, the reference markers from the Class II Biochemistry Workshop have been used.

References

1. Reekers P, van der Beucken M, Andrien M, Beelen J, Mueller Eckhardt G, Schreuder GMTh. HLA-B POT: a new HLA antigen within the B7-cross reactive group. Tissue Antigens 1986;28:182–189.
2. Mueller-Eckhard G, Schreuder I, Mayr WR, Westphal E, Mueller Eckhardt C. A new BW6 associated B locus antigen "RI." 9th Workshop Newletter 1983;5:24–26.
3. Watson K, Ness D, Gantan Z, Perkins H, Grumet FC. ST-40: a new BW40 subtype identified by alloantsera and as monoclonal antibody. Human Immunol 1982;5:178–179.
4. Leverenz G. 1987, personal communication.
5. Reekers P, Harpprecht J, de Hoog W. HLA-B POT: a new gene product from B7-cross reactive group. Human Immunol 1988 (submitted for publication).

HLA Class I Typing by Computer Analysis of 1D-IEF Data

T.F. Rinke de Wit,[1] J. D'Amaro,[1] P. de Lange,[1] P. Reekers,[2] M. Giphart,[1] and P. v.d. Elsen[1]

Abstract. We may summarize our results as follows: 1) We have developed a rapid and reproducible method for storage of 1D-IEF data, which may be applied to any polymorphic protein system. 2) Significant correlations between 1D-IEF bands (pI values) and serologic typing data have been established. 3) 1D-IEF computer analysis may be used to type the most common HLA class I specificities.

Introduction

The polymorphism of the human leukocyte antigen (HLA) class I system is reflected by the charge differences of its products. One-dimensional isoelectric focusing (1D-IEF) is a powerful technique to separate proteins on the basis of this feature, and can even be applied to subtype cells that are serologically indistinguishable.

The aim of this study was to develop a rapid and reproducible method for analyzing large amounts of 1D-IEF data on HLA class I antigens. The application of this method was tested on the cell panels from two BENROO and BENREE laboratories participating in the Tenth International Histocompatibility Workshop. Typical results are presented, and the use of our computer analysis for HLA class I typing is discussed.

Materials and Methods

The 1D-IEF analysis was performed according to the protocol established at the Tenth International Histocompatibility Workshop. A sonic digitizer (model GP-7, Science Accessories Corporation) was used for measuring the autoradiographs. Data were collected on a personal computer (IBM-AT) and analyzed on the Leiden University mainframe computer.

Participating Laboratories: BENROO,[1] BENREE[2]
This work was supported in part by the Dutch Foundation for Medical and Health Research (MEDIGON) and the J.A. Cohen Institute for Radiopathology and Radiation Protection (IRS).

Figure 1. 1D-IEF patterns from HLA-A1-positive cells (W6/32 precipitates).

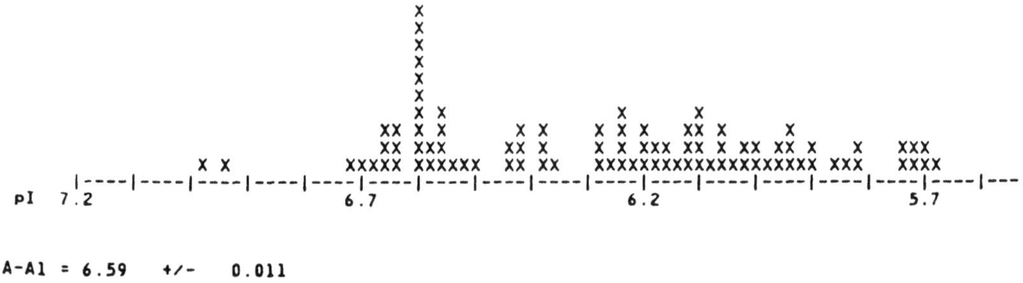

Figure 2. Frequency distribution of 1D IEF patterns from HLA-A1-positive cells (W6/32 precipitates).

Results

A total of 187 human peripheral blood lymphocytes (PBLs), typed with the Workshop Serology Core Set, was metabolically labeled with ^{35}S-methionine. Immunoprecipitates using the HLA class I-specific monoclonal antibodies 4E and W6/32 were separated on 1D-IEF gels according to the Workshop protocol. Each gel contained one lane of marker proteins: HLA-A1, −A2, −B27, and −B35. The relative positions of class I proteins on the remaining lanes were measured on the autoradiographs using a sonic digitizer connected to a PC. Beta-2-microglobulin (β2m) was assumed to be 0% (basic), and HLA-B27 was assumed to be 100% (acidic).

The linearity of the pH gradient was checked by elution of the focused ampholines and pH measurement. A linear pH gradient was found between −30 and 100%. Because almost all class I proteins focus in this region (exceptions are the very basic splits of HLA-A2 and −A28), a nearly complete comparative analysis could be performed, eliminating intra- and intergel differences.

As an initial step, the 1D-IEF patterns of all cells typed positive for a certain class I specificity were selected and printed. Figure 1 presents the results for HLA-A1. The patterns obtained were translated to a graphic presentation, which plots the frequency distribution of specific protein bands across the linear pH scale (Fig. 2). For each peak, the best fitting vertical line was calculated, which represents a certain isoelectric point (pI) value. These lines were subjected to two different analyses. In the first approach, the line with the lowest standard deviation was taken to represent the relevant serologic specificity. The second approach involved the use of a serologic screening program to correlate the absence or presence of pI lines with the absence or presence of a certain HLA class I phenotype. A correlation coefficient was calculated, and all correlations less than or equal to sign 0.3 were taken to be significant (Table 1). This analysis has been performed for all HLA specificities encountered in our panels. Figure 3 represents an overview of the pI values that were significantly correlated with the antigen definition based on the Workshop Core Set results.

Table 1. Correlation of pI 6.59 with HLA-A1, calculated by the Screenings Program (W6/32 precipitates)

pI 6.59/HLA-A1	+/+	+/−	−/+	−/−
	26	9	18	109

n = 162; r = 0.539.

Discussion

The BENREE panel contained 162 cells from many different ethnic groups, and represented 62 serologically defined specificities. On analysis, 19 specificities could be significantly correlated with pI values. Naturally, a larger panel would make it possible to define even more specificities. However, some limitations will remain that can not be resolved by this methodology. 1) Using 1D-IEF, it is impossible to separate HLA class I proteins that do not differ in overall net charge. 2) The ampholine mix prescribed in the Workshop protocol restricts the linear separation region of IEF gels to approximately 5 cm, covering about 1.75 pH units. This range is clearly insufficient for assigning more than 100 different HLA class I proteins a certain pI value. 3) The use of a sonic digitizer introduces a standard deviation of 2 to 3% in the measurements of IEF patterns.

Concerning data interpretation, for some specificities, the graphic presentations of the IEF patterns show several peaks. This may be due to linkage disequilibrium. Alternatively, the extra peaks may represent antigenic splits, which are not yet recognized by the alloantisera. To discriminate between these possibilities, it is possible to perform a negative selection using the computer program. In that way, if IEF patterns of HLA-A2 positives/-B7 negatives are selected, the second peak will disappear if it represents the HLA-B7 specificity. If it is a split of HLA-A2, the peak will still be evident.

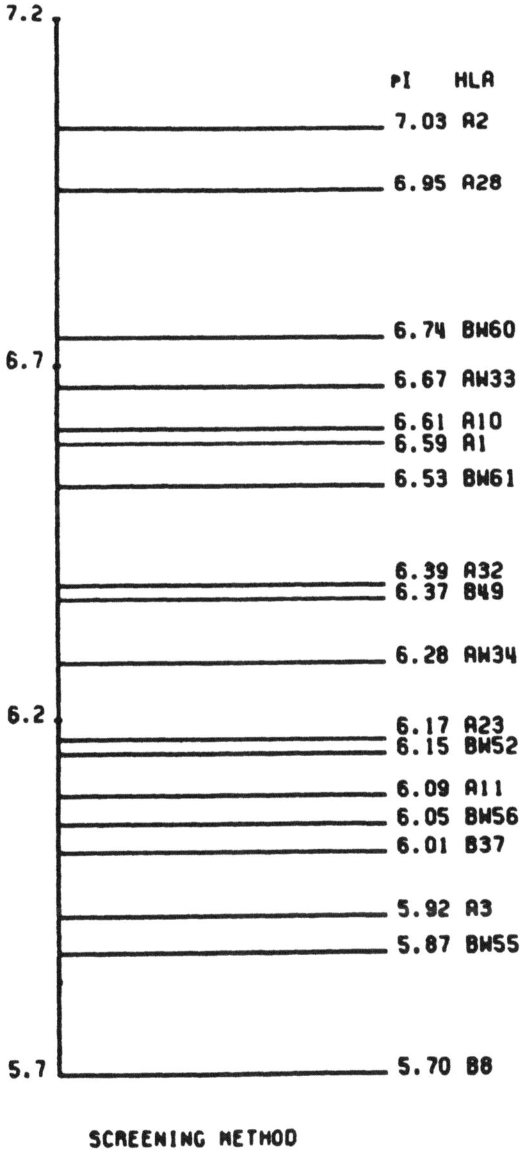

Figure 3. pI values that correlate significantly with Workshop core set HLA class I antigen definitions.
◄

Data for HLA-C typing are not presented here as a result of the poor precipitation performance of W6/32 on these molecules under the conditions used.

To determine whether HLA class I typing by this 1D-IEF computer analysis program could be performed, the data from the BENREE panel were used as the basis for the analysis of the 25 cells from the BENROO panel. Only the most common specificities were identified (HLA-A1, −A2, −A3, −A28, and −B8). When more data on cells with rare specificities become available, this typing performance may be improved. A description of the computer programs will be published elsewhere (1).

Reference

1. D'Amaro J, Rinke de Wit TF, Van den Elsen P, De Lange P. Manuscript in preparation.

Segregation Analysis of HLA Class I Determinants by Biochemistry

Gabriele Frenz,[1] Ilias Doxiadis,[1] Udo Vögeler,[1] Gabriele Doxiadis,[1] and Hans Grosse-Wilde[1]

To date, the HLA class I alloantigenic determinants have been defined by classic serology. Methods such as one dimensional isoelectric focusing (1D-IEF) were used predominantly for the biochemical definition of these antigens from B-lymphoblastoid cell lines (B-LCL) by immunoprecipitation with respective monoclonal antibodies (1). The data defined HLA-A,B products by comparative analyses. Non-A and non-B locus products have not yet been identified by gel electrophoresis.

For the analysis of the HLA class I gene products presented here, we have used 5×10^6 lymphocytes of family members (43 families, with a total number of 238 individuals). The cells were lysed in Triton-X-114 buffer directly after isolation, and the class I gene products were enriched by phase separation. The resulting detergent phase was subjected to neuraminidase treatment, the samples denaturated and applied onto an 1D-IEF gel. After the run and respective washes, the banding pattern was transferred onto a solid phase and subjected to immunoblotting using a rabbit anti-human heavy chain serum (2). The resulting bands were then correlated to HLA typing.

By analyzing families, it was possible to calculate respective antigen frequencies of the subtypes. These results are given in Table 1. The data obtained can be summarized as follows:

Table 1. Gene Frequencies (gf) of Biochemically Defined HLA Specificities

HLA	gf	HLA	gf	HLA	gf
A1	22.0	B7	14.0	Bw55	2.8
A2.1	4.7	B8	13.6	Bw56	1.4
A2.2	8.9	B13	2.8	Bw57	2.3
A2.3	11.2	B18	4.2	Bw58	0.5
A2.4	4.2	B27.1	1.9	Bw60.1	2.3
A3.1	0.5	B27.2	0.5	Bw60.2	0.5
A3.2	12.2	B27.3	0.5	Bw61	0.5
A11.1	0.9	B35.1	2.3	Bw62.1	3.7
A11.2	3.7	B35.2	3.3	Bw62.2	5.6
A23	2.8	B37	2.3	Bw62.3	0.5
A24.1	3.3	B38	1.4	Bw63	0.5
A24.2	5.6	B39.1	1.4	Bw65	2.3
A25	4.2	B39.2	0.9		
A26	1.9	Bw41	0.5		
A28.1	0.5	B44.1	12.1		
A28.2	1.9	B44.2	2.8		
A28.3	2.3	B45	0.5		
A28.4	0.9	B49	0.9		
A29	1.4	Bw50.1	0.5		
A30	0.5	Bw50.2	1.4		
A31	1.4	B51.1	1.4		
A32	3.7	B51.2	6.5		
Aw33	1.4	Bw53	1.4		

The most basic variant of a given determinant has the lowest index.

1. Every single serologically defined antigen represents a distinct band on the gel, which segregates with the respective HLA specificity in families.
2. Among the known subtypes, new ones, we could identify which also segregated in the families analyzed.
3. Not only HLA-A and HLA-B locus products can be identified in the immunoblots, but especially banding patterns, segregating with the respective haplotypes in the analyzed families and with a high association to HLA-C locus products.
4. Some individuals and patients expressed more bands, not segregating, but per definition, due to the antibody used, being of class I nature.

In summary, the method presented here allowed us to analyze lymphocytes of families directly after isolation without the use of radioactive labeling. The results obtained are comparable with those presented by others (1,2). In this respect, we could also identify HLA-C locus-associated products.

References

1. Yang SY, Morishima Y, Collins NY, et al. Comparison of one-dimensional IEF patterns for serologically detectable HLA-A and B allotypes. Immunogenetics 1984;19:217.
2. Neefjes JJ, Doxiadis I, Stam NJ, Beckers CJ, Ploegh HL. An analysis of class I antigens of man and other species by one-dimensional IEF and immunoblotting. Immunogenetics 1986;23:164.

Participating Laboratory: GERGSW[1]
This work was supported in part by the Deutsche Forschungsgemeinschaft through Gr 608-4, Do 313-1, and SFB102 TPE4.

Charge Variants of HLA-A and B Gene Products

P. Reekers,[1] D. Coates,[2] I. Doxiadis,[3] S. Ellis,[4] P. Klouda,[5] A. Hajek-Rosenmayr,[6] A. van der Horst,[7] M. Little,[8] A. Madrigal,[9] and T. Rinke de Wit[10]

Ten European centers have jointly analyzed their 10 IHWS one-dimensional isoelectric focusing (1D-IEF) patterns of class I HLA-A and B locus gene products. Seven centers have used the proposed Workshop protocol for ^{35}S-methionine labeling and electrophoresis of neuraminidase-treated immunoprecipitates (with Mabs 4E and W6/32) on 4.5% acrylamide IEF gels. In center #7, an ampholine concentration different from the Workshop protocol was used; center #3 performed immunoblotting with rabbit anti-heavy chain antiserum (HC-10) and center #10 used ^{125}I surface labeling according to the Workshop protocol for Antigen Society #31 (HLA class I-like proteins). For comparison between the various gels and the various centers, band positions were related to the following reference bands: A2, B18, β_2m, B35, A1, BW61, A24, A11, A3, B8, B27. In some centers, ^{14}C-labeled reference proteins from the Biochemistry Class II Workshop component have been used in addition to the reference cell lines. For determination of the isoelectric points (IEP) of the bands, the pH gradient of the gel has been measured and found to be linear between pH 5.5–7.0 (1).

The 10 centers contributed 71 families (195 Caucasian and 35 non-Caucasian individuals) and 273 selected donors (212 Caucasians and 61 non-Caucasians). All have been analyzed and typed serologically either with 10WS antisera (centers #1, 5, 6, 7, 8, 9, and 10) or with 9WS/standardized local sera (centers #2, 3, and 4). In Figure 1, the relative positions of the various HLA-A,B locus gene product bands (defined by the serologic nomenclature) have been ranked from the basic (upper) to the acidic (lower) pH of the gel. The most basic band observed was a HLA-A2 variant; the most acidic was a B27 charge variant. Biochemical variants of a serologically defined HLA antigen were considered to be well established if at least one of the following criteria were fulfilled: 1) expected segregation of the band was documented in informative families, or 2) the band occurred in at least three selected unrelated donors with different HLA types except for the antigen under investigation. In Tables 1 and 2, the well established A and B locus gene products are listed with their corresponding

Participating Laboratories: BENREE,[1] UKINEW,[2] GERGSW,[3] UKIMCM,[4] UKIBID,[5] EAEMYR,[6] BENENG,[7] UKIDYR,[8] UKIBOD,[9] BENROO[10]

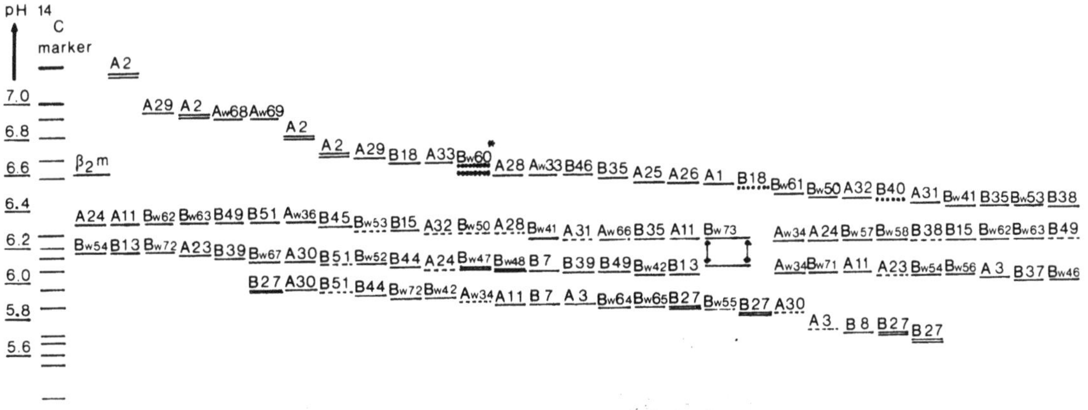

Figure 1. Sequential 1D-IEF band position of HLA-A, B locus gene products. . . . . .=Band observed in single center; *=BW60 band either above or under β2M.

Immunobiology of HLA
Volume I

Table 1. Charge Variants of Neuraminidase-Treated HLA-A Locus Gene Products

Serotype	IEP§	Serotype	IEP§
A1	6.50	A2	
AW36	6.34	A28 unspecified	
		AW68	
A10 unspecified	6.36	AW69	
A25	6.50		
A26	6.50	A3	5.84,† 5.68
AW34	6.18, 6.00, 5.86*	A11	6.33,‡ (6.20), 6.00,† 5.86
AW66	6.20		
		A29	6.94, 6.66
A32	6.44		
AW33	6.64,† 6.58‡	A30	6.10, 5.94
		A31	6.42, 6.21*
A9 unspecified	6.04		
A23	6.14,† 5.98*		
A24	6.32, 6.18, 6.04*		

*Single center observations.
†Most frequently observed gene product in Caucasians.
‡Most frequently observed gene product in non Caucasians.
§IEP must be considered as approximate values.

Table 2. Charge variants of Neuraminidase-Treated HLA-B Locus Gene Products

Serotype	Approximate IEP	Serotype	Approximate IEP
B51	6.24, 6.08,† 5.92	B7	6.00, 5.84
BW52	6.08	BW42	6.00, 5.86
BW53	6.38, 6.23†	BW22-BW54	6.16, 5.98
		BW55	5.80
		BW56	6.00
B18	6.64, 6.48	BW73	6.20 + 6.06
B35	6.56, 6.38, 6.22	B27:Caucasian	5.94, 5.80, 5.66‡
B21-B50	6.42, 6.24†	B27:non-Caucasian	5.77, 5.68, 5.57§
BW70-BW72	6.16, 5.86	BW47	6.06
BW71	6.00		
		B37	5.94
B15-B62	6.30, 6.18		
BW63	6.30, 6.18	B40 unspecified	6.41†
B15.3	6.24	BW60*	6.62/6.58
B15.T	6.24	BW61	6.46
B15.S	6.18	BW41	6.38, 6.20
		BW48	6.04
B17-BW57	6.18	B13	6.16,§ 6.00‡
		B12-B44	6.06, 5.88
BW46	6.56, 5.94	B45	6.26
BW67	6.10	B8	5.66
B16-B38	6.36, 6.18†	B14-BW64	5.83
B39	6.12, 6.00	-BW65	5.83

*In duplicate testings, BW60 band can be found either slightly above or slightly under β_2m band.
†Single center observations.
‡Most frequently observed gene product in Caucasians.
§Most frequently observed gene product in non-Caucasians.

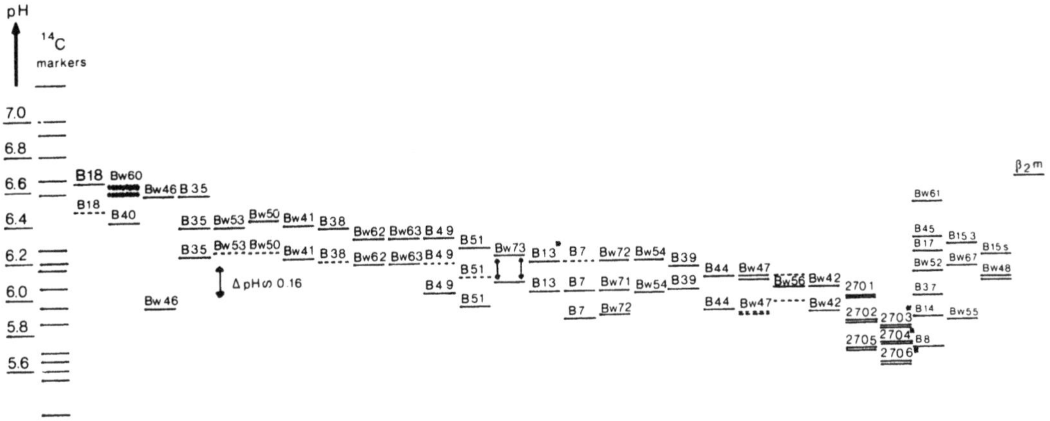

Figure 2. Charge differences between HLA-B locus gene product variants are approximately constant. – – =Band observed in single center; *=Charge variants specific for non-Caucasian ethnic groups.

pH estimated from the pH plot. The most common variants in various ethnic groups are highlighted. For each serologic HLA-B type, if the gene products are ordered according to their IEP position, an almost constant pH difference of approximately 0.16 is apparent between a large proportion of the charge variants identified (Fig. 2). From the data in Table 1, a plot for HLA-A locus gene products with similar charge differences can be constructed (not shown in this report). A possible explanation for this observation is that these almost constant charge differences represent exchanges of charged amino acids (e.g., Glu, Asp, Arg) in those parts of the class I heavy chain that do not considerably alter the structure of the serologically recognized epitope.

Reference

1. Rinke de Wit TF. HLA class I typing by computer analysis of 1D IEF data. In: Dupont B, (ed): Immunobiology of HLA, Volume I: Histocompatibility Testing 1987. New York: Springer-Verlag.

Joint Reports: T-Cell Recognition of HLA Class I Molecules

T-Cell Recognition of HLA Class I Molecules

Patrick G. Beatty,[1] Zulay Layrisse,[2] Gijsbert A. Van Seventer,[3] Li-an Fan,[4] Cristina Magli,[2] Pavlo Ivanyi,[3] James Jenkin,[1] and John A. Hansen[1]

Introduction and Methods

The purpose of the Class I T-Cell Clone component of the Tenth Workshop (10WS) was to determine the extent to which HLA class I molecules express polymorphisms recognizable by alloreactive or antigen-specific class I-restricted cloned T cells. A standard protocol for testing available proliferative and cytolytic T-cell clones for reactivity against the 10WS core cell line panel was developed in parallel to the Class II T-Cell Clone component, with the exception that monoclonal antibody-blocking studies were not performed with the anti-class I T-cell clones. The same method for data reduction (whereby a specific proliferative or cytolytic activity was converted into simple numerical scores ("1"=negative; "6" or "8"=strong positive) was used.

Results

Data for 13 anti-class I T-cell clones were submitted from three laboratories: BEN/IVA, LAT/LAY, and US7/HAN (Table 1). Twelve of the clones were prepared by priming with alloantigen, and one clone was selected for HLA class I-restricted reactivity to EBV-association antigen. Eleven of the clones were CD8+, one was CD4+, and one was not tested. Ten of the clones were cytolytic (CTL) and the reactivity of three were assayed by proliferation. Each of the 13 clones was tested for reactivity with the 10WS core cell line panel (Fig. 1). Eight of the T-cell clones formed three distinct reactivity patterns or clusters. Cluster 1 was defined by clones HAN 1, HAN 2, and IVA 4. Cluster 2 was defined by clones IVA 4, IVA 1, IVA 3, and IVA 5. Cluster 3 was defined by clones HAN 4 and LAY 1. Three additional clones, IVA 2, LAY 2, and LAY 3, appeared to individually define distinct specificities. One clone, HAN 3, reacted with three of the targets in the 10WS cell line panel, but no distinct specificity was apparent. Finally, one clone, LAY 3, did not react with any of the targets in the 10WS cell line panel. In some instances, the 10WS cell line panel data were supplemented by testing with additional local target cells (Table 2).

Cluster 1: T-Cell Clones Defining HLA-B27

The three T-cell clones that formed cluster 1, HAN 1, HAN 2, and IVA 4, reacted with four of four B27+ target cells present in the 10WS core cell line panel. Each of these four B27+ cells was shown to express the B*2705 variant as defined by IEF (see IEF component, 10WS). Two laboratories supplemented the 10WS cell line panel with local target cells representative of additional B27 IEF variants. T-cell clone HAN 1 reacted only with the B*2705–positive targets present in the 10WS core and supplemental cell line panel and with none of the other B27 variants tested in the supplemental panel

Participating Laboratories: US7HAN,[1] LATLAY,[2] BENIVA,[3] and CHIZHA[4]

Table 1. Anti-HLA Class I T-Cell Clones Tested in the Tenth Workshop

T-Cell Clones		Local Assigned		
10WS #	Local #	Specificity*	Phenotype	Effector Function†
HAN 1	4-47	B*2705, Bw67	CD8+	Allo CTL
HAN 2	1-73	B*2702,2703,2705,2701s	CD8+	Allo CTL
HAN 3	11-10	Unknown	CD8+	Allo CTL
HAN 4	No. 6	Bw62	CD8+	Anti-EBV CTL
IVA 1	KOR-1	B7s, Bw55	CD8+	Allo CTL
IVA 2	KOR-4	A2s	CD8+	Allo CTL
IVA 3	KOR-18	B7s	CD8+	Allo CTL
IVA 4	KOR-132	B7s, B*2705, BW42	CD8+	Allo CTL
IVA 5	HG-31	B7s, B56s, B*2704	CD8+	Allo CTL
IVA 6	HG-61	B7s, B*2701, B*2702, Bw60s	CD8+	Allo CTL
LAY 1	0-7	Bw62	CD4+	Allo PTL
LAY 2	0-8	A29, A31	CD8+	Allo PTL
LAY 3	F-12	A23	NT	Allo PTL

*s = short: a reactivity pattern suggestive of a split.
†Allo CTL = alloreactive cytolytic T cell; Anti-EBV CTL = cytolytic T cell specific for EBV-associated antigen and class I restricted; Allo PTL = alloreactive proliferative T cell.

Tenth Workshop Target Cell Panel

T Cell Clone	9005	9004	9067	9109	9015	9013	9065	9082	9083	9001	9017	9033	9034	9081	9021	9031	9060	9064	9024	9032	9071	9072	9112	9113	9059	9069	9074	9090	9111	9037
HAN 1	8	8	8	8	8	1	1	/	/	1	1	1	1	/	1	1	1	1	1	1	1	1	1	1	1	1	1	/	1	1
HAN 2	8	8	8	8	8	1	1	/	/	1	1	1	1	/	1	1	1	1	1	1	1	1	1	1	1	1	1	/	1	1
IVA 4	8	8	8	8	8	8	8	8	8	8	8	8	8	8	6	1	1	1	1	1	1	1	1	8	1	1	1	1	1	1
IVA 1	1	1	1	1	1	8	8	8	8	8	8	8	8	8	8	1	1	1	1	1	1	1	1	1	1	1	1	1	1	1
IVA 3	1	1	1	1	1	8	8	8	8	8	8	8	8	8	1	1	1	1	1	1	1	1	1	1	1	1	1	1	1	1
IVA 5	1	1	1	1	1	8	6	8	4	1	6	8	6	1	1	1	1	1	1	1	1	1	1	1	1	1	1	1	1	1
HAN 4	1	1	1	1	1	1	1	/	/	1	1	1	1	/	1	8	8	8	8	8	8	8	8	8	1	1	1	/	1	1
LAY 1	1	1	1	8	1	1	1	1	1	1	1	1	1	1	1	8	8	8	1	1	/	1	1	1	1	1	1	1	4	1
IVA 2	1	8	8	8	1	1	1	1	1	1	1	1	1	1	1	1	1	1	1	1	1	1	1	1	8	8	8	8	8	1
IVA 6	1	1	1	1	1	1	1	1	1	1	1	1	1	1	1	1	1	1	1	1	1	1	1	1	1	1	1	1	1	1
LAY 2	1	1	1	1	1	1	1	1	1	1	1	1	1	1	1	1	1	1	1	1	1	1	1	1	1	1	1	1	6	8
LAY 3	1	1	1	1	1	1	1	1	1	1	1	1	1	1	1	1	1	1	1	1	1	1	1	1	1	1	1	1	1	1
HAN 3	1	1	1	1	8	1	1	/	/	1	1	1	1	/	1	1	1	1	1	1	1	1	1	1	1	1	1	/	1	8

T Cell Clone	9002	9047	9107	9028	9098	9030	9114	9048	9050	9097	9110	9003	9006	9007	9008	9009	9010	9011	9012	9014	9016	9018	9019	9020	9022	9023	9025	9026	9027	9029
HAN 1	1	1	/	1	/	1	1	1	1	/	1	1	1	1	1	1	1	1	1	1	1	1	1	1	1	1	1	1	1	1
HAN 2	1	1	/	1	/	1	1	1	1	/	1	1	1	1	1	1	1	1	1	1	1	1	1	1	1	1	1	1	1	1
IVA 4	1	1	1	1	1	1	1	1	1	1	1	1	1	1	1	1	1	1	1	1	1	1	1	1	1	1	1	1	1	1
IVA 1	1	1	1	1	1	1	1	1	1	8	1	1	1	1	1	1	1	1	1	1	1	1	1	1	1	1	1	1	1	1
IVA 3	1	1	1	1	1	1	1	1	1	1	1	1	1	1	1	1	1	1	1	1	1	1	1	1	1	1	1	1	1	1
IVA 5	1	8	4	1	1	1	1	1	1	1	1	1	1	1	1	1	1	1	1	1	1	1	1	1	1	1	1	1	1	1
HAN 4	1	1	/	1	/	1	1	1	1	/	1	1	1	1	1	1	1	1	1	1	1	1	1	1	1	1	1	1	1	1
LAY 1	1	1	1	1	1	4	6	1	1	1	1	1	1	1	1	1	1	1	1	1	1	1	1	1	1	1	1	1	1	1
IVA 2	1	1	1	1	1	1	1	1	1	1	1	1	1	1	1	1	1	1	1	1	1	1	1	1	1	1	1	1	1	1
IVA 6	1	1	1	8	8	1	1	1	1	1	1	1	1	1	1	1	1	1	1	1	1	1	1	1	1	1	1	1	1	1
LAY 2	1	1	1	1	1	4	1	4	8	1	1	1	1	1	1	1	1	1	1	1	1	1	1	1	1	1	1	1	1	1
LAY 3	1	1	1	1	1	1	1	1	1	1	1	1	1	1	1	1	1	1	1	1	1	1	1	1	1	1	1	1	1	1
HAN 3	8	1	/	1	/	1	1	1	1	/	1	1	1	1	1	1	1	1	1	1	1	1	1	1	1	1	1	1	1	1

T Cell Clone	9035	9036	9038	9039	9040	9041	9042	9043	9044	9045	9046	9049	9051	9052	9053	9054	9055	9056	9057	9058	9061	9062	9063	9066	9068	9070	9073	9075	9108	9115
HAN 1	1	1	1	1	1	1	1	1	1	1	1	1	1	1	1	1	1	1	1	1	1	1	1	1	1	1	1	1	1	1
HAN 2	1	1	1	1	1	1	1	1	1	1	1	1	1	1	1	1	1	1	1	1	1	1	1	1	1	1	1	1	1	1
IVA 4	1	1	1	1	1	1	1	1	1	1	1	1	1	1	1	1	1	1	1	1	1	1	1	1	1	1	1	1	1	1
IVA 1	1	1	1	1	1	1	1	1	1	1	1	1	1	1	1	1	1	1	1	1	1	1	1	1	1	1	1	1	1	1
IVA 3	1	1	1	1	1	1	1	1	1	1	1	1	1	1	1	1	1	1	1	1	1	1	1	1	1	1	1	1	1	1
IVA 5	1	1	1	1	1	1	1	1	1	1	1	1	1	1	1	1	1	1	1	1	1	1	1	1	1	1	1	1	1	1
HAN 4	1	1	1	1	1	1	1	1	1	1	1	1	1	1	1	1	1	1	1	1	1	1	1	1	1	1	1	1	1	1
LAY 1	1	1	1	1	1	1	1	1	1	1	1	1	1	1	1	1	1	1	1	1	1	1	1	1	1	1	1	1	1	1
IVA 2	1	1	1	1	1	1	1	1	1	1	1	1	1	1	1	1	1	1	1	1	1	1	1	1	1	1	1	1	1	1
IVA 6	1	1	1	1	1	1	1	1	1	1	1	1	1	1	1	1	1	1	1	1	1	1	1	1	1	1	1	1	1	1
LAY 2	1	1	1	1	1	1	1	1	1	1	1	1	1	1	1	1	1	1	1	1	1	1	1	1	1	1	1	1	1	1
LAY 3	1	1	1	1	1	1	1	1	1	1	1	1	1	1	1	1	1	1	1	1	1	1	1	1	1	1	1	1	1	1
HAN 3	1	1	1	1	1	1	1	1	1	1	1	1	1	1	1	1	1	1	1	1	1	1	1	1	1	1	1	1	1	1

Figure 1. Reactivity of 13 T-cell clones versus 90 B-lymphoblastoid cell line targets. A score of 8 represents specific lysis greater than 35%; 6=15 to 35%; 4=10 to 15%; 2=5 to 10%; 1= less than 5%; and /=not tested. Positive reactivity was clustered into three patterns (from upper lefthand corner): cluster 1=B27; cluster 2=B7; cluster 3=Bw62.

Table 2. Additional Target Cell Lines Tested with T-Cell Clones

Tenth Workshop Cell Line ID #	Tenth Workshop Serologic ID #	Local ID	HLA Specificity
9138, 9128	HAN 0471, ENG 0112	LIH	B*2701
9139	HAN 0481	HS	B*2701
9116	ENG 0111	APO	B*2702
9133	HAN 0421	PG	B*2702
9134	HAN 0431	KRIJ	B*2702
9135	HAN 0441	DMA	B*2702
9136	HAN 0451	R-34	B*2702
9137	HAN 0461	KS	B*2702
9132	HAN 0401	CHI	B*2703
9118	ENG 0110	ETO	B*2704
9127	HAN 0351	AS-4	B*2704
9128	HAN 0361	SMF	B*2704
9131	HAN 0391	SIAO	B*2704
9140	HAN 0491	LIE	B*2706
9144	HAN 0643	BB	Bw62.1
nd	HAN 0151	JAH	Bw62.2
9129	HAN 0371	WEWAK I	Bw62.3
9146	HAN 0645	SB	Bw62.4
9127	ENG 0090	KROON	7.1.1
9139	ENG 0080	ZEXL	7.1.1
9123	ENG 0199	HEMMING	7.1.2
9125	ENG 0092	JY	7.2
9122	ENG 0091	HEES	7.2
9119	ENG 0093	FIGDOR	7.3
9134	ENG 0089	OOSTER	7.4
9135	ENG 0194	POTZE	BPot

(Table 3). Clone HAN 2 reacted with B*2702, B*2703, and B*2705-positive target cells, but did not react with cells expressing B*2704 or B*2706. Two B*2701-positive target cells were tested with clone HAN 2, but only one of the two was lysed by clone HAN 2, suggesting that there may be more than one unique B27 allele within the IEF-defined B*2701 group. Clone IVA 6 did not react with any of the B*2705-positive cells included in the 10WS core cell line panel, but was shown in local experiments to lyse target cells expressing B*2701 and B*2702, in addition to cells expressing B7 and Bw60. Clone IVA 4 reacted with B7, B42, and B*2705-positive target cells, but not with B*2701, B*2702, B*2703, B*2704, or B*2706-positive cells. Clone IVA 5 reacted only with one B*2704-positive cell and none of the other B*27 variants. Three local T-cell clones, SEA 1, SEA 11, and SEA 65, as well as a local cell panel, were also tested against the 10WS cell line panel. Clones SEA 1 and SEA 11 reacted with B*2704 and B*2706-positive cells, and clone SEA 65 reacted only with B*2704-positive cells. No workshop sera were found which correlated with the reaction patterns of these T-cell clones.

Table 3. Summary of T-Cell Clone Reactivity with HLA-B27 Alleles*

T Cell Clones†	B*2701‡ (n = 3)	B*2702 (n = 6)	B*2703 (n = 1)	B*2704 (n = 5)	B*2705 (n = 5)	B*2706 (n = 1)	Other Antigens
HAN 1	–	–	–	–	4/4	–	Bw67
HAN 2	1/2	5/5	1/1	–	4/4	–	–
IVA 6	1/1	1/1	–	–	–	–	B7,B60
IVA 4	–	–	–	–	4/4	–	B7,B42
IVA 5	–	–	–	1/1	–	–	B47
SEA 1	–	–	–	4/4	–	1/1	–
SEA 11	–	–	–	4/4	–	1/1	–
SEA 65	–	–	–	4/4	–	–	–

*HLA-B27 alleles defined according to the *WHO HLA Nomenclature Committee Report*, 1987.
†T-cell clones SEA 1, SEA 11, and SEA 65 are local clones that were not directly tested in the 10WS protocol experiments.
‡Target cell lines tested included 10WS core cell line panel and local cell lines: 1) B*2701 – HAN 0471, HAN 0481, ENG 0112; 2) B*2702 – HAN 0421, HAN 0431, HAN 0441, HAN 0451, HAN 0461, ENG 0111; 3) B*2703 – HAN 0401; 4) B*2704 – HAN 0351, HAN 0361, HAN 0381, HAN 0391, ENG 0110; 5) B*2705 – 10WS 9004, 9005, 9015, 9067, Supplemental 9109; and 6) B*2706 – HAN 0491.

Table 4. Summary of T-Cell Clone Sensitivity with HLA-B7 Alleles*

T-Cell Clones	7.1.1[†]	7.1.2	7.2	7.3	7.4	BPOT	Other
IVA 1	2/2	1/1	2/2	–	1/1	1/1	Bw55
IVA 3	2/2	1/1	2/2	–	1/1	–	–
IVA 4	2/2	1/1	1/1	1/1	1/1	1/1	B*2705,Bw42
IVA 5	2/2	–	1/1	–	–	–	B*2704,Bw47
IVA 6	–	–	1/1	–	–	–	B*2701,B*2702,Bw60

*B7 alleles defined by ref. 1.
[†]Target cell lines tested included 10WS core cell line panel and local cell lines: 1) B7.1.1 – ENG 0090; ENG 0080; 2) B7.1.2 – ENG 0199; 3) 7.2 – ENG 0092, ENG 0091; 4) 7.3 – ENG 0094; 5) 7.4 – ENG 0089; and 6) BPOT – ENG 0199.

Two 10WS sera, however, were identified that appeared to split HLA-B27. Serum 284 reacted with B*2704, B*2705, and B*2706, and serum 4661 reacted with B*2703, B*2704, B*2705, and B*2706.

Cluster 2: T-Cell Clones Defining HLA-B7

The five T-cell clones that formed cluster 2, IVA 4, IVA 1, IVA 3, IVA 5, and IVA 6, reacted with B7-positive target cells. IVA 1, IVA 3, IVA 4, and IVA 5 lysed B7-positive cells from the 10WS panel. Clone IVA 6 lysed a local B7-positive cell. In addition, clone IVA 4 lysed B*2705-positive cells, clone IVA 5 lysed B*2704-positive cells, and clone IVA 6 lysed B*2701 and B*2702-positive cells. These studies, using local target cells (Table 4), demonstrated that these CTL appeared to define six variants of B7 (1,2). IEF analysis, however, identified only two B7 variants. All clones except IVA 3 lysed cells expressing other antigens as indicated. Four serum clusters subdivided the B7 antigen; however, none of the serum patterns correlated with the CTL lysis patterns.

Cluster 3: T-Cell Clones Defining HLA-Bw62

One cytolytic clone, HAN 4, and one proliferative clone, LAY 1, reacted with all 10WS core cell lines expressing Bw62. In addition, clone HAN 4 was tested on selected local cells expressing four IEF-defined Bw62 variants (Table 5). Two of these, Bw62.2 and Bw62.4, were lysed by clone HAN 4 (3). No workshop sera were found whose reaction patterns correlated with the reaction patterns of these clones.

Table 5. Summary of T-Cell Clone Reactivity with HLA-Bw62 Variants*

T-Cell Clone	Bw62.1	Bw62.2	Bw62.3	Bw62.4
HAN 4	–	+	–	+
LAY 1	–	+	nt	nt

*HLA-Bw62 variants defined by isoelectric focusing gel electrophoresis (see IEF Component, 10WS).
Target cell lines tested included 10WS core cell line panel and local cell lines: 1) Bw62.1 - HAN 0643; 2) Bw62.2 - 10WS 9024, 9031, 9032, 9060, 9064, 9071, 9072 and HAN 0151; 3) Bw62.3 - HAN 0371; and 4) Bw62.4 - HAN 0045.

Individual T-Cell Clones Defining HLA Class I Determinants

Three T-cell clones reacted with different HLA-A locus antigens (Table 1, Fig. 1). Clone IVA 2 reacted with some but not all A2-positive cells, and thus appeared to identify an A2 variant. Clone LAY 2 reacted with a determinant shared by A29 and A31-positive cells. Clone LAY 3 reacted with A23-positive cells.

Summary

The results of this component of the Workshop demonstrate that it is biologically possible and technically feasible to generate T-cell clones capable of recognizing distinct epitopes expressed on class I molecules. It is also evident that different laboratories, employing different cloning strategies, can isolate cloned reagents that have similar, if not identical, lytic profiles. Although the scope of these Workshop studies was constrained by current technical problems with the growth and long-term maintenance of cytolytic T-cell clones specific for class I determinants, it is clear that these reagents allow precise definition of unique, functionally relevant, class I alleles.

References

1. van Seventer GA, van der Horst AR, et al. Six subtypes of HLA-B7 defined by histotopes, epitopes and 1D-IEF. In: Dupont B (ed): Immunobiology of HLA, Volume II: Immunogenetics and Histocompatibility. New York: Springer-Verlag, 1989.
2. van Seventer GA, van der Horst AR, de Waal LP, Reekers R, Ivanyi P: Public determinants of the HLA-B7 CREG defined by antibodies and T cell clones. In: Dupont B (ed): Immunobiology of HLA, Volume II: Immunogenetics and Histocompatibility. New York: Springer-Verlag, 1989.
3. Fan L, Choo SY, Beatty PG, Hansen JA: HLA-Bw62 variants identified by a cytotoxic T lymphocyte clone and isoelectric focusing gel electrophoresis. In: Dupont B (ed): Immunobiology of HLA, Volume II: Immunogenetics and Histocompatibility. New York: Springer-Verlag, 1989.

Joint Reports: Structural Polymorphism of the HLA Class II Alpha and Beta Chains

Structural Polymorphism of the HLA Class II α and β Chains: Summary of the 10th Workshop 2-D Gel Analysis

Robert W. Knowles[1]

A comprehensive analysis of structural polymorphism in the HLA class II antigens has been performed using a standardized two-dimensional (2-D) gel electrophoresis system. Polymorphic patterns have been defined for the α and β chains of the DR and DQ molecules isolated from 71 of the 10th Workshop core B-cell lines representing 34 functionally distinct HLA class II haplotypes. For the DR molecules, a total of 19 DRβ chain patterns were found associated with the DRB1 locus, with 5 additional patterns associated with a second expressed DRB gene. For the DQ molecules, 10 β chain patterns were found along with 9 DQα chain patterns. For the DP molecules, two DPα chain patterns were found, but the DPβ chains were not resolved. This corresponds well to the extent of polymorphism found in the first domain sequences of cloned DR and DQ genes from these haplotypes. While some of the known sequence variation does not result in distinct 2-D gel patterns, additional polymorphism has been found in this study, where first domain sequences are identical. Although there is extensive linkage disequilibrium between these genes, it is also clear from this study that recombination events within the class II region have played a significant role in the generation of functionally distinct haplotypes.

Goals of the Study. The major goal of this study was to determine the extent of structural polymorphism in the α and β chains of the HLA class II antigens, detectable as distinctive 2-D gel patterns. Since this technique detects differences in isoelectric point or molecular weight of individual subunits following immunoprecipitation, these structural polymorphisms can be localized to the α or β chains of the class II dimers. Because only the expressed gene products are examined, this technique avoids one of the limitations of Southern blot analysis, which commonly detects additional polymorphism in the introns and flanking regions of the class II genes or their closely related pseudogenes. Serologic and cellular analysis of class II polymorphism examines epitopes, which are often small segments of a polypeptide chain and can therefore be supertypic or shared by several distinct alleles. These epitopes can also be dimer specific, created by unique combinations of polymorphic determinants located on both the α and the β subunits of a class II dimer. This is expected in the DQ and DP molecules, but not in the DR molecules, where the DR α chain is not polymorphic. While serology and cellular typing are important approaches to the complete understanding of functional polymorphism of the class II antigens, the use of 2-D gels is closest to amino acid sequence analysis in defining structural polymorphisms of each expressed α or β chain gene. Therefore, this study serves to link specific α and β chain polymorphisms with the serologic and cellular polymorphisms defined in the other studies performed with a common set of Workshop cell lines.

Another important goal has been the analysis of HLA class II haplotypes, including the HLA-DR and DQ loci, which are in strong linkage disequilibrium in the human population. The Workshop core B-cell line panel represents most of the previously identified functionally distinct haplotypes defined by Dw typing (1). Because each of the gene products in the DR and DQ subregions was examined in the 2-D gel analysis, it has been possible to identify the specific combinations of alleles that define the haplotypes represented by each of these cell lines.

In the tradition of the International Histocompatibility Workshop studies, considerable effort was made to develop a standardized technique that could be used in a large number of participating laboratories. Although 2-D gels have been used for the analysis of class II polymorphism since the production of the first anti-class II monoclonal antibodies (2), it has not previously been possible to standardize the technique sufficiently to compare the results obtained in different laboratories (3). Therefore, in this study, every aspect of the technique including reagents and equipment has been standardized (4). The two major innovations involved the development of radiolabeled protein standards, which were added to each sample as internal markers for the class II patterns on the autoradiograms, and the development of premixed liquid buffers formulated for maximum stability during long-term storage. Through these efforts in standardization, it has been possible for many laboratories with little or no previous experience to contribute significantly to this study. Because of the large number of monoclonal antibodies required for a comprehensive analysis of the 75 Workshop cell lines, it was essential to the success of the project to have the maximum number of participating laboratories.

Experimental Design. The analysis was restricted to the Workshop core cell lines (1) to provide a broad range of distinct haplotypes that would also be analyzed in other Workshop studies. Generally the participants studied the DR haplotypes of their choice (Table 1), but the assignment of specific cell lines was made in such a way that every cell line would be analyzed independently in two laboratories. The standardized Workshop techniques were used by all participants, which required the purchase of common equipment and reagents by each laboratory (4). The monoclonal antibodies and additional

---

Participating Laboratory: US7KNO[1]

Table 1. Laboratories in Each Antigen Group Participating in the 2-D Gel Analysis of Workshop Cell Lines

Antigen Group	Laboratories					
DR1	FRAOHA[a] (3),[b]	JAPTSU (6),	UKIFES (2),	US6BRN (1)		
DR2	UKIMOR[a] (5),	FRACHR (2),	JAPTSU (2),	NCYMAV (2),	NCYMRV (4),	US6BRN (2)
DR3	UKIMOR[a] (3),	FRAMUL (3),	SAFDUT (3),	US7GSC (3)		
DR4	UKIWEL[a] (3),	FRAPRR (3)	NCYMAR (3),	US1ZMI (5),	US6BRN (2),	US7GSC (1),
		US7VMC (4),	US7KNO (1)			
DR5	UKIBOD[a] (7),	BENROO (2),	FRAMUL (2),	GERGOL (2),	IT2GAN (3),	
		NCYMRV (3),	US7VMC (2)			
DRw6	FRACHR[a] (3),	BENROO (5),	GERGOL (3),	SAFDUT (2),	US6IND (5),	US5SWE (5)
DR7	US8RJW[a] (2),	FRADDC (5),	IT2GAN (2),	UKIFES (3),	US7KNO (3)	
DRw8	US7SYY[a] (4),	BENROO (3),	JAPJUJ (5)			
DR9	FRAFAU[a] (3),	US7KNO (2),	US8RJW (1)			
DRw52	BENROO,[a] and others					
DRw53	US7KNO,[a] and others					
DQw1	FRAFAU,[a] and others					
DQw2	US7GSC,[a] and others					
DQw3	IT2GAN,[a] and others					
DQw4	US7SYY,[a] and others					

[a]Laboratories responsible for writing the antigen group reports, which follow this summary report.
[b]Number of cell lines in the DR antigen groups examined by each participating laboratory.

reagents for immunoprecipitation were prepared centrally and distributed with the radioactive standard proteins to each laboratory. Two samples were prepared from each immunoprecipitate for analysis on two separate pH gradients. The α chains were analyzed on a relatively acidic gradient, with 80% of the ampholytes in the narrow pH range of 4 to 6, whereas a more basic gradient with 80% of the ampholytes in the narrow pH range of 6 to 8 was used for the β chains. A panel of mAbs (Table 2) selected for this study included several that were monomorphic, but DR, DQ, or DP locus specific, and these antibodies were studied on all cell lines. Additional polymorphic antibodies were chosen for each cell line based on a serologic binding analysis (7) and previous studies. Those with broad supertypic DQ-related specificities (DQw1, DQw1+3, DQw3) or broad supertypic DRB3- and DRB4-related specificities (DRw52 and DRw53) were given highest priority to maximize the number of cell lines examined with each antibody. This allowed the assignment of groups of spots within each 2-D gel pattern to a specific class II gene product for each cell line. Comparisons were made between cell lines of the same DR type to identify common or distinct allelic patterns for the DRB1 and DQ molecules within each DR antigen group. The patterns of related molecules from different DR antigen groups were then compared for cell lines with shared supertypic DRw52, DRw53, or DQ specificities. Identification of the DR and DQ β chain patterns was performed by the participants at the Workshop meeting in Princeton, and the criteria for their provisional assignment as the gene product of a given class II locus are described in the individual antigen reports. The identification of the DQ and DP α chain patterns was performed centrally.

Table 2. The mAbs Used for the Workshop 2-D Gel Study

Specificity[a]	mAbs
Broad	TAL-1B5, Tü 39, 2D6
DR	GSP 4.1, PL8, L243
DQ	1a3, Tü 22
DP	B7/21, PL15
DQw1	UK 4.1, NDS 15.38, 77.34, KS6
DQw1,3	Leu10, PLM 13, GS4AA7, KS7
DQw3	17.15, 1VD12, PLM 12, PLM 2
DQ polym.	C1, GSP 17.1, IIB3
DRw52	NDS 13, UK 8.1, 7.3.19.1, NDS 10
DRw53	PL3
DR polym.	IIIE3, CC CL 20, JS1, NFLD M1, 16.23, CHE249.2, GSP88.2, SFR3 DR5, SFR16-DR7M, MAD 88
DP polym.	DP11.1

[a]Only the approximate specificities are listed, with further details and the list of contributors given in Knowles et al. (7).

Results and Discussion

Resolution and Reproducibility. The Workshop 2-D gel system provided excellent resolution for most of the β chains in the study with the exception of the very basic DQw5(w1) associated molecules. Reproducibility of the patterns within one laboratory was often excellent as shown in Figure 1, where patterns obtained with four different antibodies are compared. Although it often appears clear by visual inspection which of the spots are shared between the various patterns, a simple comparison of the absolute (x,y) coordinates of each spot in the pattern was generally not reliable due to slight variations in both the first-dimension pH gradient and the second-dimension SDS size separation. This was more pronounced when patterns from different laboratories were compared. This variability was not uniform over the entire gel but was generally localized to small regions, making simple corrections for stretching the patterns impossible. This was anticipated from previous studies and the use of radioactive carbamylated standard proteins, creatine phosphokinase and carbonic anhydrase (CPK and CA in Fig. 1), made it possible to correct for much of this variability across the pH gradient. Since the molecular weight of the standards placed them above and below most of the class II spots, interpolation of the position of each class II spot relative to the closest spots in the patterns of the protein standards was possible. Other proteins, such as actin, were found in most immunoprecipitates and could also be used, whereas proteins such as the invariant chain (I in Fig. 1) were not found consistently with every antibody used for the immunoprecipitations.

Nomenclature. Groups of spots that appear to represent glycosylated forms of a single polypeptide were given 2-D gel pattern numbers. Therefore, the provisional 10th Workshop 2-D gel nomenclature indicates the conclusion of the Workshop participants in each antigen group as to whether the group of spots found in different cell lines have identical or distinct pI and molecular weight. For example, DRβI-7 was assigned to the distinct DRβ chain pattern found in all serologically DR7 positive cell lines by immunoprecipitation with the DR7 specific monoclonal antibody SFR16-DR7M. It was clearly distinct from a second DRβ chain pattern, DRβIV-1, found in DR7 cells as well as DR4 and DR9 cells with DRw53-related antibodies. This Workshop nomenclature should be considered provisional, since analysis of the 2-D patterns in this study alone does not formally demonstrate that a given pattern, such as DRβI-7, represents the DRB1 gene product. In this particular example, however, additional studies performed in this laboratory have demonstrated that an identical DRβI-7 pattern is found for the DRβ chain expressed by murine L cells transfected with the DR7 allele of the DRB1 gene isolated from a Workshop cell line MOU (9050). (The transfectants were kindly provided by D. Wilkinson and J. Trowsdale.)

A major goal of this study has been to define the 2-D gel patterns that correspond to the alleles for each class II locus. These alleles will eventually be given official WHO nomenclature based on confirmed cDNA sequences (8). Therefore, the 10th Workshop nomenclature for the 2-D gel patterns shown in Table 3 will be provisional until the corresponding alleles are given official WHO designations.

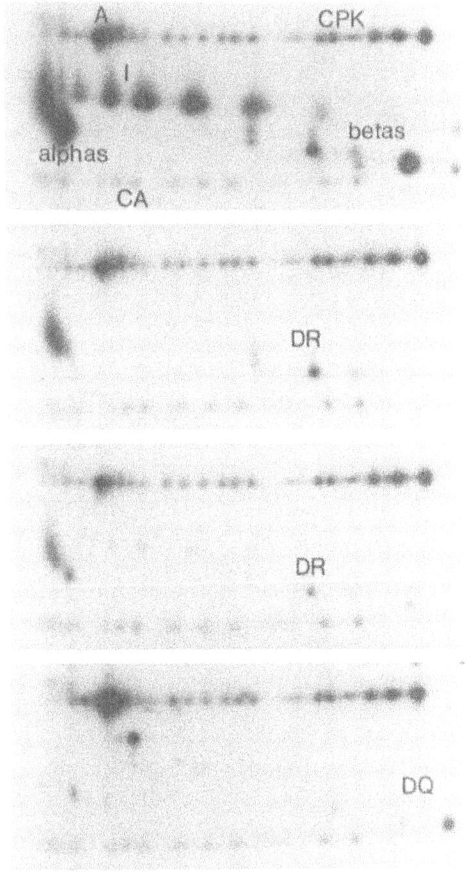

Figure 1. HLA class II molecules analyzed using the standardized Workshop 2-D gel system. Class II molecules were immunoprecipitated from an NP-40 lysate of the ^{35}S-methionine-labeled DR4, Dw4 cell line SAVC (9034) using the broadly reactive mAb Tu39 (*top panel*), the DR specific mAbs GSP 4.1 and L243 (*second and third panels*), and the DQ specific mAb Tu22 (*bottom panel*). They were then separated using the Workshop system with ampholytes selected for optimal β chain separation (80% narrow range pH 6 to 8, 20% broad range pH 3.5 to 10). The position of the ^{14}C-labeled standard proteins, CPK (40 kDa) and CA (30 kDa), the ^{35}S-labeled class II α and β chains, and the coprecipitating cellular proteins, actin (A) and the invariant chain (I), are indicated in the top panel. This cell line expresses the DRβI-4.1 and the DQβ-3.2 chains, as discussed in the text.

Associations Between Alleles Defined by 2-D Gel Patterns and DNA Sequences. A total of 45 distinct 2-D gel patterns were identified in this study for the polymorphic DR and DQ α and β chain genes. There is an excellent correlation between these patterns and the large number of distinct alleles identified by DNA sequence analysis. A complete listing of the 2-D gel patterns assigned to specific class II DR and DQ loci is shown in Table 3 with the DNA sequence of the most closely associated allele. Detailed descriptions of the 2-D patterns may be found in the individual antigen reports that follow this summary report. The specific associations with sequenced alleles should be considered provisional, since in many cases the sequences were derived from non-Workshop cell lines that appear to have the same haplotype (9). Of the 33 known alleles of the various DR β genes and 14 alleles of the DQ β genes, 36 of these alleles have been distinguished by 2-D gel patterns, including 2 that have not been defined by distinct

Table 3. Complete Listing of Allelic 2-D Gel Patterns and Their Associated First-Domain Amino Acid Sequences

DQβ 2-D Patterns	DQB1 Sequences[a]	DQα 2-D Patterns	DQA1 Sequences	DRβI 2-D Patterns	DRB1 Sequences	DRβ-[b] 2-D Patterns	DRβ-[b] Sequences
DQβ-(1.1*)[c]	DR1 DQw1.1	DQα-1.1	DR1 DQw1.1[d]	DRβI-1.1	DR1 βI	DQβ-b-2.1[e]	Dw2 βIII
DQβ-1.2	DR2 DQw1.2	DQα-1.2	DR2 DQw1.2	DRβI-1.1	DR1 NASC	DQβ-b-2.1[e]	Dw12 βIII
DQβ-(1.9*)[c]	DRw6 DQw1.9	DQα-1.14	DRw6 DQw1.9[d]	DRβ-a-2.2[e]	DR2 Dw2βI	DQβ-b-2.1[e]	AZH βIII
DQαb-1.12	DR2 DQw1.12	DQα-1.18	DRw6 DQw1.18	DRβ-a-2.12[e]	DR2 Dw12βI	DRβIII-1	DRw52bβIII
DQβ-1.18	DRw6 DQw1.18	DQα-2	DR5 DQw3.1	DRβ-a-2.16[e]	DR2 AZHβI	DQβIII-2	DRw52cβIII
DQβ-1.19	DRw6 DQw1.19	DQα-3.1	DRw9 DQw3[f]	DRβI-3	DR3 βI	DRβIII-3	DRw52aβIII
DQβ-(1.21*)[c]	DR2 DQw1.AZH	DQα-3.2	DR4 DQw3.2[f]	DRβI-3	DR3 DQWaβI	DQβIV-1	DR4 DRw53βI
DQβ-2	DR3 DQw2	DQα-3.7	DR7 DQw2 and w3	DRβI-4.1	DR4 Dw14βI		
DQβ-3.1	DR4 DQw3.1	DQα-3.8	DRw8 ARC (WA)	DRβI-4.1	DR4 Dw4βI		
DQβ-3.2	DR4 DQw3.2	DQα-3.8	DRw8 LUY (DQw3)	DRβI-4.2	DR4 Dw13βI		
DQβ-3.3	DRw9 DQw3.3			DRβI-4.2	DR4 KT2βI		
DQβ-Wa	DRw8 DQ WA			DRβI-4.3	DR4 Dw10βI		
DQβ-(Wa15*)[c]	DR4 DQ WA			DRβI-(4.4*)[g]	DR4 Dw15βI[g]		
DQβ-1.12*	–			DRβI-7	DR7 βI		
				DRβI-8.1	DRw8 βI		
				DRβI-8.2	DRw8 SPLβI		
				DRβI-8.1	DRw8 TABβI		
				DRβI-9	DRw9 βI		
				DRβI-11.1	DR5 Dw5βI		
				DRβI-11.2	DR5 JVMβI		
				DRβI-12	–[h]		
				DRβI-13	DRw6aβI		
				DRβI-14.1	DRw6bβI		
				DRβI-14.2	DRw6 AMALAβI		
				DRβ-[g]	DR1 CETUS[g]		
				DRβ-[g]	DRw10βI[g]		

[a]First-domain DNA sequences are named according to Gregersen et al (9). Official WHO nomenclature has not been established for these alleles. Association of 2-D patterns with specific first-domain sequences is based on the criteria discussed in the text.
[b]DRβ- is used to designate the second expressed DRβ chain from the second expressed DRB gene, which can be the unnamed DRB gene in DR2 cells, the DRB3 gene, or the DRB4 gene, depending on the haplotype.
[c]DQβ-1.1, DQβ-1.21, DQβ-1.9, and DQβ-Wa15 were not characterized by Workshop 2-D gel analysis.
[d]DR1 DQw1.1 and DRw6 DQw1.9 have identical DQA1 first-domain DNA sequences.
[e]The designations "DRβ-a" and "DRβ-b" are used since these patterns have not been assigned to DRB1 or a second DRB- locus.
[f]DR4 DQw3.2 and DRw9 DQw3 have identical DQA1 first-domain DNA sequences.
[g]No DR1 CETUS, DR4,DW15, or DRw10 cell lines were included in the Workshop.
[h]No sequence is known to be associated with this pattern.

first-domain DNA sequences. Similarly, of the 10 known alleles of DQα chains, 9 can be distinguished by 2-D gels, including 2 that have not been defined by distinct first-domain DNA sequence analysis.

Since the ability to distinguish alleles by 2-D gel analysis is dependent on the isoelectric point of the individual chain, not all amino acid substitutions are detectable with this technique. Therefore, the contribution of the first-domain amino acids to the net charge of the molecule has been calculated from the reported sequences of the associated alleles (9), as shown in Tables 4 and 5. This calculation is based on the simple assumption that each aspartic acid (D) and glutamic acid (E) will add one negative charge, whereas each arginine (R), histidine (H), and lysine (K) will add one positive charge. The minor variations in isoelectric point contributed by other amino acid substitutions have been ignored in this calculation. In several cases it is clear that the amino acid substitutions observed in the first-domain sequences would not be expected to lead to a distinct 2-D gel pattern (Table 4). As expected, the two distinct DRB1 alleles found in DR1 cells give rise to a single DRβI-1.1 pattern and two DR3 associated alleles give the DRβI-3 pattern. Two of the DR4 associated alleles give the DRβI-4.1 pattern, whereas two other alleles give the DRβI-4.2 pattern. Similarly, two DR8 alleles give the DRβI-8.1 pattern, whereas three alleles of the "DRβ-II" or "DRβ-b" gene in DR2 cells give the DRβ-b-2.1 pattern. The ability to distinguish closely related DQA1 and DQB1 alleles, which have the same predicted net charge from

Table 4. Net Charge of DRβ First-Domain Amino Acid Sequences

DRβ Sequence[a]	DRβ 2-D[b]	+/−[c] NET	1–10	11–20	21–30	31–40	41–50	51–60	61–70	71–80	81–90	91–94
				S1[d]	S2	S3	S4	H1		H2	H3	
			- (+) (+)	(± ±) (± + + ±) (+)	- + (+) (±)+(±)	(+ +) - -(-) +	- - - +	(+)- + (- -)-(+)	+ -	-(±) (±)+ (±) -	+ (+) -	+ +
			GDTRPRFLWQ	LKFECHFFNG	TERVRLLERC	IYNQEESVRF	DSDVGEYRAV	TELGRPDAEY	WNSQKDLLEQ	RRAAVDTYCR	HNYGVGESFT	VQRR
DR1 w1 βI	(1.1)	+1	----------	----------	----------	----------	----------	----------	----------	----------	----------	----
DR1 NASC βI	(1.1)	+1	----------	----------	----------	----------	----------	----------	----------	----------	----AV----	----
DR2 Dw2 βI	(a-2.1)	+2	--------Q-	D-Y-------	-----F-H-D	------FL--	----------	----------	------F--D	----------	----------	----
DR2 Dw12 βI	(a-2.12)	+4	--------Q-	D-Y-------	-----F-H-G	------N---	----------	----------	------F--D	----------	----------	----
DR2 ASH βI	(a-2.16)	+3	-----C--Q-	D-Y-------	-----F-H-G	------N---	----------	----------	------I---	A---------	----AV----	----
Dw2 βII	(b-2.1)	+4	----------	P-R-------	-----F-D-Y	F---------	------F---	----------	------I---	A---------	-----V----	----
Dw12 βII	(b-2.1)	+4	----------	P-R-------	-----F-D-Y	F---------	------F---	----------	------I---	A---------	------	
ASH βII	(b-2.1)	+4	----------	P-R-------	-----F-D-Y	F---------	----------	----------	------F--D	----------	----------	----
DRw6a βI	(13)	0	--------EY	STS-------	-----F-D-Y	FH----N---	------F---	----------	------I--D	E---------	-----V----	----
DRw6b βI	(14.1)	+5	--------EY	STS-------	-----F-D-Y	FH----F---	----------	------A--H	---------R	---E------	-----V----	----
DRw6 AMALA βI	(14,2)	+3	--------EY	STS-------	-----F---Y	FH----N---	----------	----------	----------	----------	----------	----
DRw52a βIII	(-3)	+7	--------EL	R-S-------	-----Y-D-Y	FH----FL--	----------	------V--S	----------	K-GR--N---	----------	----
DRw52b βIII	(-1)	+6	--------EL	--S-------	-----F---H	FH----YA--	----------	R---------	----------	K-GQ--N---	-----V----	----
DRw52c βIII	(-2)	+5	--------EL	--S-------	-----F---Y	FH----LA--	----------	------V--S	----------	K-GQ--N---	----------	----
DR3 βI	(3)	+4	--------EY	STS-------	-----Y-D-Y	FH----N---	------F---	----------	----------	K-GR--N---	-----V----	----
DR3 DQw βI	(3)	+4	--------EY	STS-------	-----F-E-Y	FH----N---	----------	----------	----------	K-GR--N---	----------	----
DR5 Dw5 βI	(11.1)	0	--------EY	STS-------	-----F-D-Y	F-----Y---	------F---	-------E--	------F--D	----------	----------	----
DR5 JVM βI	(11.2)	-2	--------EY	STS-------	-----F-D-Y	F-----Y---	------F---	-------E--	------I--D	E---------	-----V----	----
DRw8 βI	(8.1)	+1	--------EY	STG--Y----	-----F-D-Y	F-----Y---	----------	------S---	------F--D	---L------	------	
DRw8 TAB βI	(8.1)	+1	--------EY	STG--Y----	-----F-D-Y	F-----Y---	----------	------S---	------I--D	---L------		
DRw8 SPL βI	(8.2)	0	--------ET	STG--Y----	-----F-D-Y	F-----Y---	----------	----------	------F--D	--GL------		
DR4 Dw4 βI	(4.1)	+5	--------E-	V-H-------	-----F-D-Y	F-H---Y---	----------	----------	----------	K---------	----------	----
DR4 Dw14 βI	(4.1)	+5	--------E-	V-H-------	-----F-D-Y	F-H---Y---	----------	----------	----------	----------	-----V----	----
DR4 Dw15 βI	(4.4)	+6	--------E-	V-H-------	-----F-D-Y	F-H---Y---	----------	------S---	----------	----------	----------	----
DR4 Dw10 βI	(4.3)	+3	--------E-	V-H-------	-----F-D-Y	F-H---Y---	----------	----------	------I--D	E---------	-----V----	----
DR4 Dw13 βI	(4.2)	+4	--------E-	V-H-------	-----F-D-Y	F-H---Y---	----------	----------	----------	---E------	-----V----	----
DR4 KT2 βI	(4.2)	+4	--------E-	V-H-------	-----F-D-Y	F-H-------	----------	----------	----------	---E------	-----V----	----
DR7 βI	(7)	+4	---Q------	G-YK------	----QF---L	F-----F---	----------	------V--S	------I--D	--GQ---V--	----------	----
DRw9 βI	(9)	+7	---Q----K-	D---------	-----Y-H-G	------N---	----------	------V--S	------F--R	---E---V--	----------	----
DRw53 βIV		+1	---Q----E-	A-C----L--	----MN-I-Y	------YA-Y	N--L---Q--	----------	---------R	---E------	Y----V----	----
			- (+) (+)	(± ±) (± + + ±) (+)	- + (+) (±)+(±)	(+ +) - -(-) +	- - - +	(+)- + (- -)-(+)	+ -	-(±) (±)+ (±) -	+ (+) -	+ +

[a]First-domain sequences are named according to Gregersen et al. (9). Official WHO nomenclature has not been established for these alleles. The predicted amino acid sequence from the DR1 Dw1 β1 gene is shown, using the single-letter code. Only the amino acids that differ from this sequence in other genes are indicated.

[b]The Workshop DRβ 2-D gel pattern most closely associated with each sequence is indicated as in Table 3.

[c]Positions in the sequence where negatively charged amino acids (D and E) or positively charged amino acids (R, H, and K) occur as indicated "−" or "+." If the charge at a given position is variable for this group of sequences, it is indicated as "(−)," "(+)," or "(±)." The net charge of the first domain amino acid is indicated for each sequence, assuming a charge of −1 for D and E and +1 for R, H, and K.

[d]The amino acids predicted to form strands of β-pleated sheets or α-helical structures in class II molecules are indicated by S1, S2, S3, S4 and H1, H2, H3 according to the model of Brown et al (10).

Table 5a. Net Charge of DQα First Domain Amino Acid Sequences

Secondary structure[d]: S1 (positions 10–18), S2, S3, S4 (to 40), H1 (50–), H2 (60–80)

DR	DQ	DQα 2-D[b]	+/− NET[c]	1–10	11–20	21–30	31–40	41–50	51–60	61–70	71–80	81–
				-- -+		+- -	--(-) - (-)	++- (+) (-)	(+ +) (+ +) -	(+) +	+ ++	
				EDIVADHVAS	CGVNLYQFYG	PSGQYTHEFD	GDEEFYVDLE	RKETAWRWPE	FSKFGGFDPQ	GALRNMAVAK	HNLNIMIKRY	NSTAATN
DR1	DQw1.1	(DQα-1.1)	−2	----------	----------	----------	----------	----------	----------	----------	----------	-------
DRw6	DQw1.9	(DQα-1.1)	−2		---	----------	----------	----------	----------	----------	----------	-------
DR2	DQw1.2	(DQα-1.2)	−1	----------	----------	----------	---Q------	----------	----------	----------	----------	-------
DR2	DQw1.AZH	(DQα-1.2)	−1	----------	----------	----------	---Q------	----------	----------	----------	----------	-------
DRw6	DQw1.19	(DQα-1.2)	−1		---	----------	---Q------	----------	----------	----------	----------	-------
DRw6	DQw1.18	(DQα-1.18)	−1		---	----F-----	---Q------	K---------	----------	----------	----------	
DRw8	DQw1	(DQα-1.18	−1		---	----F-----	---Q------	K---------	----------	----------	----------	
DR5	DQw3.1	(DQα-2)	0	----------	Y------S--	----------	---Q-----G	----V-CL-V	LRQ-R[ ]----	F--T-I--L-	----SL---S	
DRw8	ARC (NA)	(DQα-3.8)	0	----------	Y------S--	----------	---Q-----G	----V-CL-V	LRQ-R[ ]----	F--T-I--T-	-----L---S	-------
DRw8	LUY	(DQw3)	0		S--	----F-----	---Q-----G	----V-CL-V	LRQ-R[ ]----	F--T-I--T-	-----L---S	
DR4	DQw3.2	(DQα-3.2)	0	----------	Y------S--	-----S----	----------	----V-QL-L	-RR-RR----	F--T-I--L-	-----V---S	-------
DRw9	DQw3	(DQα-3.1)	0	----------	Y------S--	-----S----	----------	----V-QL-L	-RR-RR----	F--T-I--L-	-----V---S	-------
DR7	DQw2,w3	(DQα-3.7)	0	----------	Y------S--	----F-----	----------	----V-KL-L	-HRLR[ ]----	F--T-I--L-	-----L---S	
				-- -+		+- -	--(-) - (-)	++- (+) (-)	(+ +) (+ +) -	(+) +	+ ++	

Table 5b. Net Charge of DQβ First Domain Amino Acid Sequences

Secondary structure[d]: S1 (10–), S2 (to 30), S3 (40), S4 (to 50), H1 (60), H2 (70), H3 (80–90)

DR	DQ	DQβ (2-D)[b]	+/− NET[c]	1–10	11–20	21–30	31–40	41–50	51–60	61–70	71–80	81–90	91–
				+- --	+	-(+) + +(+)	+--(-) +	- - (- -) +	(+) (-) -	+- -(±)	(-)+ (-) -(+) +	+ (-) (-) +	++
				RDSPEDFVYQ	FKGLCYFTNG	TERVRGVTRH	IYNREEYVRF	DSDVGVYRAV	TPQGRPVAEY	WNSQKEVLEG	ARASVDRVCR	HNYEVAYRGI	LQRR
DR1	DQw1.1	(DQβ-1.1*)	+3	----------	----------	----------	----------	----------	----------	----------	----------	----------	----
DR2	DQw1.21	(DQβ-1.21*)	+3	----------	----------	----------	----------	----------	------S---	----------	----------	----------	----
DRw6	DQw1.9	(DQβ-1.9*)	+2			---------	----------	----------	------D---	----------	---------		
DR2	DQw1.2	(DQβ-1.2)	−1	--------F-	---M------	-----L---Y	-------A--	----------	------D---	----------	T--EL-T---	------F---	----
DRw6	DQw1.18	(DQβ-1.18)	0			---L-----	-------A--	----------	------D---	----------	T--EL-T-		
DRw6	DQw1.19	(DQβ-1.19)	+2	----------	---M------	-----L----	-------A--	----------	----------	---------R	T--EL-T---	-----G----	----
DR2	DQw1.12	(DQβ-1.12)	−1	--P-----L-	--AM------	-----Y---Y	------D---	----------	------D---	-----DI--R	T--EL-T---	------F---	----
DR3	DQw2	(DQβ-2)	+2	----------	---M------	-----L-S-S	------I---	-----EF---	-LL-L-A---	-----DI--R	K--A------	---QLEL-TT	----
DR4	DQ NA	(DQβ-Na*)	−3	--------F-	---M------	--L------Y	-------A--	----------	--L--LD---	-----DI--E	D-----T---	---QLEL-TT	----
DRw8	DQ NA	(DQβ-Na)	−2	--------F-	---M------	---------Y	-------A--	----------	--L--LD---	-----DI--E	D-----T---	---QLEL-TT	----
DR4	DQw3.1	(DQβ-3.1)	−2	----------	--AM------	-----Y---Y	-------A--	----E-----	--L-P-D---	---------R	T--EL-T---	---QLEL-TT	----
DR4	DQw3.2	(DQβ-3.2)	0	----------	---M------	-----L---Y	-------A--	----------	--L-P-A---	---------R	T--EL-T---	---QLEL-TT	----
DRw9	DQw3.3	(DQβ-3.3)	−1	----------	---M------	-----L---Y	-------A--	----------	--L-P-D---	---------R	T--EL-T---	---QLEL-TT	----
				+- --	+	-(+) + +(+)	+--(-) +	- - (- -) +	(+) (-) -	+- -(±)	(-)+ (-) -(+) +	+ (-) (-) +	++

[a]First-domain sequences are named according to Gregersen et al (9). Official WHO nomenclature has not been established for these alleles. The predicted amino acid sequences from the DQα and DQβ genes from a DR1 DQw1.1 haplotype are shown, using the single-letter code. Only the amino acids that differ in other genes are indicated.

[b]The Workshop DQα and DQβ 2-D gel pattern most closely associated with each sequence is indicated as in Table 3.

[c]Positions in these sequences where charged amino acids occur and the net charge of the first-domain amino acids are indicated as in Table 4.

[d]The amino acids predicted to form strands of β-pleated sheets or α-helical structures in class II molecules are indicated as in Table 4.

first-domain sequence alone, suggests that there may be an additional contribution from second-domain sequences (Table 5).

In addition to the previously known class II β chain alleles, the DRβI-12 chain pattern from the DRw12 cell line and an apparently new DQw1-related DQβ-1.12* chain pattern from a DQw1-DR5 cell line have been characterized. DNA sequences from these haplotypes have not been previously reported. In the DQα chain analysis two new alleles were identified. A DQw5 associated DQα-1.14 chain found in the DRw14,Dw9 cell lines was shown to be distinct from the DQα-1.1 chain found in DR1 cell lines, although there is no difference in the reported first-domain DQA1 sequences from these haplotypes (Table 5). This suggests that the distinct 2-D gel patterns arise from differences in the second domain or C-terminal regions of the molecule.

Similarly, in the DQw3 cells two DQα chains have been distinguished for the first time. These are the DQα-3.1 and DQα-3.2 chains. These two α chains are found in linkage disequilibrium with the DQβ-3.1 (DQw7) and DQβ-3.2 (DQw8) chains, respectively. Since the DQα-3.2 chain is only found associated with the DQβ-3.2 chain in these DR4 haplotypes, it may contribute to the disease susceptibility associated with the DQw8 specificity. One DR4 Workshop cell line was found to have the DQα-3.1 chain associated with the DQβ-3.2 chain, suggesting a recombination event may have occurred between these genes, as discussed below. The DQα-3.1 chain is also found associated with the DQβ-3.3 in DR9 cells. No differences have been observed in any of the first-domain sequences of DQα chains isolated from DR4 (DQw7 or DQw8) cells or from DR9 (DQw9) cells. As with the DQw5 α chains described above, this suggests differences outside the first domain that need to be analyzed at the sequence level.

Table 6. DRβ Chain 2-D Gel Patterns and Their Associated DR and Dw Specificities Observed in Workshop Cell Lines

DRβ 2-D Patterns	DR Typing[a]	Dw Typing[a]
DRβI-1.1	DR1	Dw1,Dw20
DRβ-a-2.2[b]	DRw15(2)	Dw2
DRβ-a-2.12[b]	DRw15(2)	Dw12
DRβ-a-2.16[b]	DRw16(2)	Dw21,Dw22
DRβI-3	DRw17(3),DRw18(3)	Dw3,Dw"new"
DRβI-4.1	DR4	Dw4,Dw14
DRβI-4.2	DR4	Dw13,DwKT2
DRβI-4.3	DR4	Dw10
DRβI-(4.4*)[c]	DR4	Dw15
DRβI-7	DR7	Dw17,Dw11,DB1
DRβI-8.1	DRw8	Dw8.1,Dw8.3
DRβI-8.2	DRw8	Dw8.2
DRβI-9	DR9	Dw23
DRβI-11.1	DRw11(5)	Dw5,DB2
DRβI-11.2	DRw11(5)	Dw5,Dw"new"
DRβI-12	DRw12(5)	DB6
DRβI-13	DRw13	Dw18,Dw19
DRβI-14.1	DRw14	Dw9
DRβI-14.2	DRw14	Dw16
DRβ-b-2.1[b]	DRw15(2),DRw16(2)	Dw2,Dw12,Dw21, Dw22
DRβIII-1	DRw52	Dw25
DRβIII-2	DRw52	Dw26
DRβIII-3	DRw52	Dw24
DRβIV-1	DRw53	–

[a]Association of the 2-D patterns defined in this study with DR and Dw specificities (8) is based on the results of 10th Workshop serologic (11) and cellular (12) typing on these cell lines.
[b]The designations "DRβ-a" and "DRβ-b" are used since these patterns have not been assigned to DRB1 or a second DRB-locus.
[c]DR4,Dw15 cell lines were included in the Workshop.

Association Between the DRβ Chain Patterns and DR and Dw Typing. The observed association between specific DRβ chain patterns and the DR and Dw typing of the Workshop cell lines is shown in Table 6. The serologic and cellular specificities listed are from the WHO Nomenclature (8) based on the 10th Workshop studies (11). Of the 24 distinct DRβ chain patterns, 11 were associated with more than one DR or Dw specificity. In some cases the distinct DRB1 amino acid sequences (9) may explain the serologic and cellular differences described above. These include DR1 cells, where Dw1 and Dw20 differ by two conservative first domain amino acid substitutions [the DRB1 sequence "DR1 NASCβ1" (9) has been associated with Dw20 (J. Gorski and D. Eckels, personal communication)]; DR3 cells, where DRw17 and DRw18 differ by four conservative amino acid substitutions; DR4 cells, where DRw4 and DRw14 differ by two conservative first-domain amino acid substitutions, and where Dw13 and DwKT2 differ by one conservative first-domain amino acid substitution; and DRw8 cells, where Dw8.1 and one Dw8.3 cell line (TABO) differ by one conservative first-domain amino acid substitution.

In other cases, it would appear that Dw differences are due to polymorphism in the DRB3 gene or the DQ genes. In DR7 cells the DRB1 gene is identical in the three haplotypes distinguished as Dw17, Dw11, and DB1 (9). The Dw11 cells lack the expression of their DRB4 gene and have a distinct DQB1 gene, which may account for the Dw11 cellular response. However, no differences have been found to distinguish the DB1 haplotype from the Dw17 haplotype. In the DRw13 cells, the difference between Dw18 and Dw19 is apparently due to a different DRB3 allele or to the different DQA1 and DQB1 alleles found in these haplotypes. In the DRw11 cells, the difference between Dw5 and DB2 cells may be due to the DQw3–DQw1 difference. Similarly, in the DRw16 cells the difference between Dw21 and Dw22 appears to be due to the DQw5–DQw3 difference. In contrast to the DRw8 cells discussed above, the Dw8.3 cell line (LUY) has an identical DRB1 sequence but differs only at DQ. In each case where the DRB3 or DQ molecules may lead to Dw differences, recombination events appear to have created these functionally distinct haplotypes, as discussed below.

Table 7. DQ α-β Dimers and Their Association with DQ and Dw Specificities Observed in the Workshop Cell Lines

DQ α-β Dimers		Associated Serologic DQ and Cellular Dw Specificities[a]	
DQβ Chain	DQα Chain	DQ Typing	Dw Typing
DQβ-(1.1*)[b]	DQα-1.1	DQw5 (w1)	Dw1,Dw20
DQβ-(1.9*)[b]	DQα-1.14	DQw5 (w1)	Dw9
DQβ-(1.21*)[b]	DQα-1.2	DQw5 (w1)	Dw21
DQβ-1.2	DQα-1.2	DQw6 (w1)	Dw2
DQβ-1.12	DQα-1.18	DQw6 (w1)	Dw12
DQβ-1.12*	DQα-1.2	DQw1	DB2
DQβ-1.18	DQα-1.18	DQw6 (w1)	Dw18
DQβ1.19	DQα-1.2	DQw6 (w1)	Dw19
DQβ-2	DQα-2	DQw2	Dw3
DQβ-2	DQα-3.7	DQw2	Dw17, DB1
DQβ-3.1	DQα-3.1	DQw7 (w3)	Dw4, Dw13
DQβ-3.1	DQα-2	DQw7 (w3)	Dw5, DB6, Dw"new5," Dw16, Dw22
DQβ-3.1	DQα-3.8	DQw7 (w3)	Dw8.3(L)
DQβ-3.2	DQα-3.2	DQw8 (w3)	Dw4, Dw14, DwKT2
DQβ-3.2	DQα-3.1	DQw8 (w3)	Dw4
DQβ-3.3	DQα-3.1	DQw9 (w3)	Dw23
DQβ-3.3	DQα-3.7	DQw9 (w3)	Dw11
DQβ-Wa	DQα-3.8	DQw4	Dw8.1, Dw8.2, Dw"new3"
DQβ-(Wa15*)[b,c]	DQα-(3.1*)[c]	DQw4	Dw15

[a]Association of 2-D patterns with DQ and Dw specificities is based on the results of 10th Workshop serologic (11) and cellular (12) typing on these cell lines.
[b]DQβ-1.1, DQβ-1.21, DQβ-1.9, and DQβ-Wa15 allele designations are based on first-domain DNA sequences (9), but were not characterized by Workshop 2-D gel analysis.
[c]No Dw15 cell lines were included in the Workshop panel.

Determining the associations between 2-D gel patterns and the serologic and cellular polymorphisms defined by mAbs or T-cell clones is complicated by the fact that these epitopes can correspond to small segments of the polypeptide chain that are shared between alleles or even between genes of different loci. One of the clearest examples of epitope mapping with mAbs in the 2-D gel study is found with monoclonal antibodies PL3 and 109d6, which bind to distinct DRw53-related determinants. The Workshop 2-D gel analysis has extended previous studies by characterizing the binding specificity of each antibody for the DRβIV or the DRβI chain in the Workshop DR4, DR7, or DR9 cell line (13). These results demonstrate that both antibodies bind to the DRβIV chain, but show differential binding to allelic forms of the polymorphic DRβI chain. Neither antibody binds to the DRβI chain from DR4 cells; PL3, but not 109d6, binds to the DRβI chain from DR7 cells; and both PL3 and 109d6 bind to the DRβI chain from DR9 cells. It is now possible to suggest polymorphic sequences on the DRβIV chain that are shared with specific DRβI chains (Table 4) and may therefore represent the epitopes responsible for the distinct binding specificities of these antibodies. For example, glutamine is found at position 4 on every chain that binds PL3: DRβI-7, DRβI-9 and DRβIV. Similarly, a group of amino acids shared between DRβIV and DRβI-9, including (arg-arg-//-glu) at polymorphic positions (70-71-//-74), are likely candidates for the 109d6 epitope. A similar analysis of the molecular specificity of each of the polymorphic mAbs, with each of the Workshop cell lines, has not been accomplished in this study. More comprehensive epitope mapping is an important goal for future studies.

The same mAbs were used for both the 2-D gel study and for blocking studies in the analysis of alloreactive T-cell clones. This was designed to allow the molecular localization of T-cell defined epitopes. T-cell blocking studies were generally successful at localizing determinants to the DR or DQ region molecules (6). However, testing a larger panel of antibodies to determine the fine specificity for each clone was not generally possible due to the limited number of cells available. As with the mAb epitope mapping, further mapping of the T-cell defined epitopes will be important in future studies.

Linkage Disequilibrium Between DQ α and β Alleles in the Workshop Cell Lines and the Association Between Specific α–β Dimers and DQ Typing. Only 19 distinct DQ α–β dimers are observed between the 14 DQβ alleles and the 10 DQα alleles found in the Workshop cell lines, as shown in Table 7. This strong linkage disequilibrium between the DQB1 and DQA1 genes may simply reflect the strong linkage disequilibrium generally found within the class II region. However, the associations between α and β chain alleles are not absolute, and several of the DQw2, DQw7, DQw8, and DQw9 haplotypes appear to have resulted from recombination events between the DQB1 and DQA1 genes. These apparent recombination events are discussed in more detail below, but the resulting haplotypes serve to clearly demonstrate that the serologic determinants identified by Workshop DQ typing reagents (11), as summarized in the WHO nomenclature (8), are only associated with the DQβ chain alleles, regardless of the α chain allele (Table 7). In contrast, there are numerous examples of DQ specific T-cell clones that appear to be specific for determinants on the DQ α chain or for determinants found only on unique DQ α–β dimers (14).

In the DQw1 haplotypes, it appears that serologic reagents detect supertypic specificities, since additional polymorphism is detected at the protein level in both the DQ α and β chains. The DQα-1.2 chain is shared between the DQw5 associated haplotype (Dw21) and the DQw6 haplotypes (Dw2, Dw19), suggesting that DQβ chains are also important in distinguishing the DQw5 and DQw6 specificities, although the DQ typing of the Dw21 cell lines with the Workshop core allosera was not definitive (11). Two distinct DQw4 haplotypes are defined by different alleles of both the DQ α and β chains associated with either DR4 or DRw8 (9). In the case of the DR4 haplotype, the DQ α chain is not unique, being shared with DQw7 and DQw9 haplotypes, suggesting that the supertypic serologic specificity (Wa) common to all DQw4 molecules is localized on the DQ β chain.

Haplotypes Defined by Unique Combinations of DQβ, DQα, DRβI, and DRβ- Chains in the Workshop Cell Lines. A total of 34 different combinations of DQβ, DQα, DRβI, and a second DRβ chain were found in the Workshop homozygous cell lines. The diagram in Figure 2 demonstrates the interrelationship between distinct haplotypes that share one or more of the specific alleles defined in this study (Table 8). The details of the results listed in Table 8 are presented by the participants in the antigen group reports that follow this summary. All but two of the functionally distinct haplotypes were also distinguished by their 2-D gel patterns, clearly demonstrating the power of the Workshop 2-D gel system.

Of these 34 haplotypes, 27 were distinguished by MLC (Dw) typing (12) and the remaining 7 were distinguished with T-cell clones (14). Two haplotypes were defined in the DR1 cell lines with the use of T-cell clones that recognize the Dw20 specificity, and T cell clones also split the DR5,Dw5 haplotype (14). The two DR3,Dw3 and two DRw13,Dw18 haplotypes were distinguished by several clones recognizing DRB3 associated determinants, Dw24 and Dw25 (15). The two DRw8,Dw8.3 haplotypes (TABO-DQw6 and LUY-DQw7) were distinguished by the broad DQw3 clone (6022). There are also three DR4,Dw4 haplotypes: DQw7 (DQβ-3.1/DQα-3.1), DQw8 (DQβ-3.2/DQα-3.2), and the recombinant haplotype SAVC (9034) (DQβ-3.2/DQα-3.1). The first two can be distinguished by a combination of a T-cell clone (EM.C3) recognizing a DQw7 associated determinant specific for the (DQβ-3.1/DQα-3.1) dimer and a clone (6094) recognizing a DQw8 determinant associated with the DQβ-3.2 β chain allele (14). The recombinant cell line SAVC can be distinguished by the lack of reactivity with an additional clone (6095) reported to recognize a DQw8-like determinant associated with the DQα-3.2 α chain allele (16). Thus, every haplotype identified with a distinct combination of DQβ, DQα, DRβI, and the second DRβ chain is functionally distinct.

Recombination Events in the Class II Region. Despite extensive linkage disequilibrium in the class II region, several recombination events appear to have resulted in unusual combinations of alleles. This is clear from the diagram in Figure 2, which indicates the interrelationship between alleles that constitute the 34 distinct haplotypes identified in this study of the Workshop cell lines. Table 9 lists the possible recombination events between the DQB1-DQA1, DQA1-DRBI, and DRBI-DRB3 (or B4) genes suggested by the results of the Workshop 2-D gel study for DQ and DR genes, as summarized in Table 8.

DQB1-DQA1 recombination events. As discussed above, only 19 distinct DQ α–β dimers are observed between the 14 DQβ alleles and the 10 DQα alleles found in the Workshop cell lines (Table 7). This strong linkage disequilibrium between the DQB1 and DQA1 genes reflects, in part, the strong linkage disequilibrium generally found within the class II region. For example, there are three possible recombination events suggested by the associations between the DQB1 and DQA1 alleles in the DQw1 group of haplotypes (Table 9), but there is only minimal DQα chain polymorphism at the sequence level (Table 5). Strong linkage disequilibrium in this region probably accounts for the continued association of the α chains with new DQw1 associated β chain variants.

In the DQw2 and DQw3 haplotypes, recombination events seem to be a more likely explanation of the observed DQ α–β chain associations. The DQβ-2 chain, which determines the DQw2 serologic specificity, is associated with either the DQα-2 chain (in DR3 cells) or the DQα-3.7 chain (in DR7,Dw17 cells). Since the DQα-3.7 is also found with the DQβ-3.3 chain (in the DR7,Dw11 haplotype) and the DQβ-3.3 chain is also found with the DQα-3.1 chain (in DR9 cells), there appear to have been several recombination events between DQB1 and DQA1 to create these DQw2 and DQw9 haplotypes as indicated in Table 9. In the DR4 haplotypes, two DQα chains, DQα-3.1 and DQα-3.2, have been distinguished for the first time by 2-D gel analysis. The DQβ-3.1 (DQw7) chain is in linkage disequilibrium with the DQα-3.1 chain, whereas the DQα3.2 chain is in linkage disequilibrium with the DQβ-3.2 (DQw8) chains. Analysis of the cell line SAVC has suggested that a recombination event may have

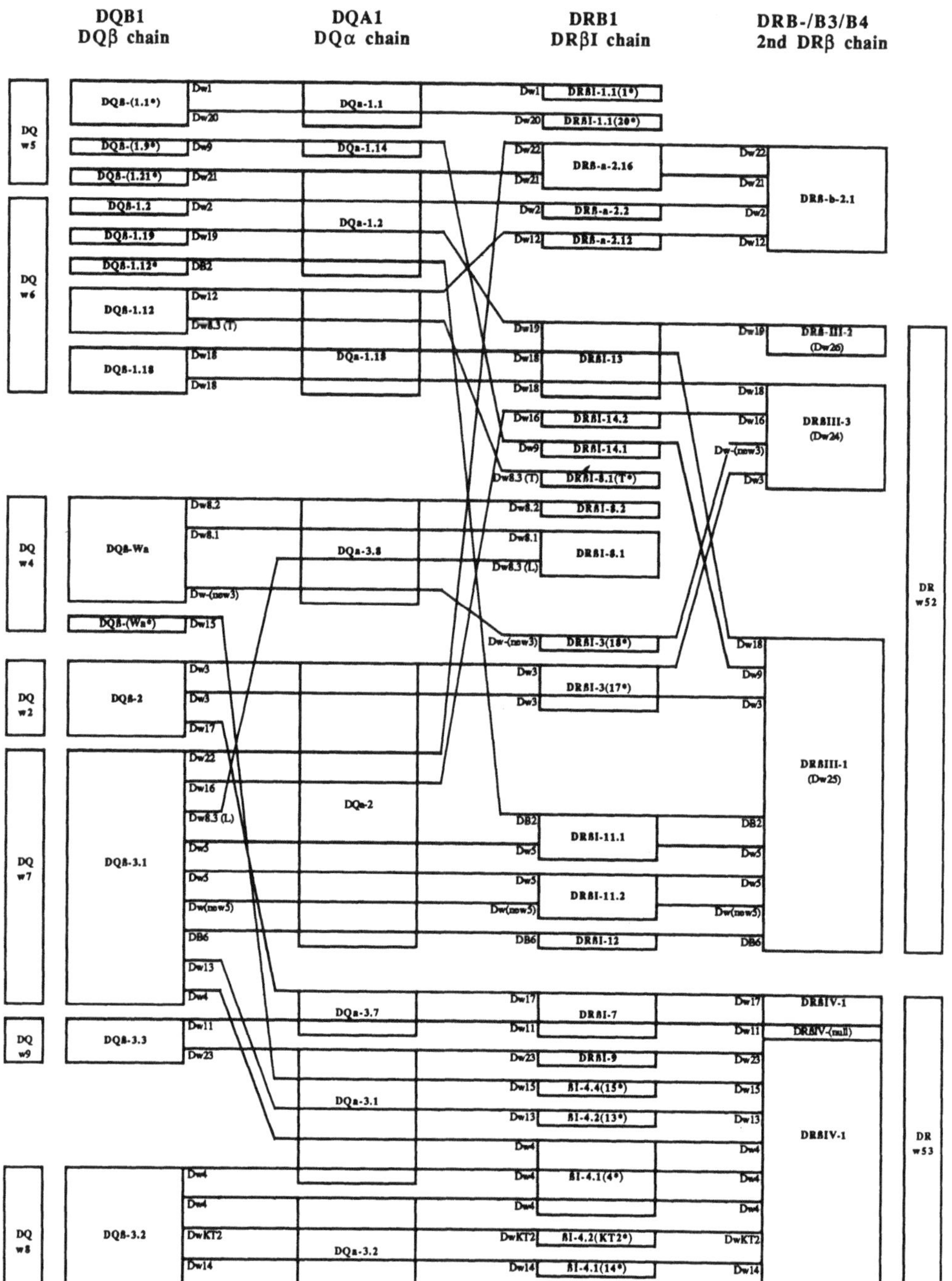

Figure 2. Haplotypes defined by different combinations of polymorphic DQβ, DQα, and DRβ chain 2-D gel patterns found in the Workshop core cell lines. A complete listing of the allelic 2-D gel patterns, provisionally assigned to the DQB1, DQA1, DRB1, and the second expressed DRB gene for each core cell line, may be found in Table 8. Details regarding the nomenclature may be found in Tables 3, 6, and 7. Individual haplotypes are indicated as lines connecting the boxes containing the appropriate class II alleles at each of the four loci. The Dw typing (12) of each haplotype is indicated at several positions along the line. DQ typing and DRw52/DRw53 typing are also indicated (in boxes on either side of the figure), as defined in the Workshop serologic analysis (11), using the WHO nomenclature (8). Possible recombination events, indicated by diagonal lines in this diagram, are suggested when an allele at one locus is shared by two distinct haplotypes that both show strong linkage disequilibrium in this region, as discussed in the text.

Table 8. Complete Listing of 2-D Gel Patterns for the Workshop Core Cell Lines[a]

DR	CELL #	CELL NAME	Serol. DQ	2-D DQ beta	2-D DQ alpha	2-D DR beta	2-D DR beta	Serol. DR	Serol. DRw52/53	T cell Dw	T cell DRw52
1	9003	KAS116	DQw5	- (1.1*)	1.1	I-1.1	-	DR1		Dw1	
1	9004	JESTHOM	DQw5	- (1.1*)	-	I-1.1	-	DR1		Dw1	
1	9005	HOM2	DQw5	- (1.1*)	1.1	I-1.1	-	DR1		Dw1	
1	9006	WT1OOBIS	DQw5	- (1.1*)	1.1	I-1.1	-	DR1		Dw1	
1	9001	SA	DQw5*	- (1.1*)	1.1	I-1.1	-	DR1		Dw1	
1	9002	MZO7O782	DQw5	- (1.1*)	-	I-1.1	-	DR1		Dw20	
2	9010	AMAI	DQw6	1.2	1.2	a-2.2	(b-*)	DRw15		Dw2	
2	9008	DO2O8915	DQw6	1.2	1.2	a-2.2	b-2.1	DRw15		Dw2	
2	9013	SCHU	DQw6	1.2	1.2	a-2.2	b-2.1	DRw15		Dw2	
2	9014	MGAR	DQw6	1.2	1.2	a-2.2	b-2.1	DRw15		Dw2	
2	9017	WT8	DQw6*	1.2	1.2	a-2.2	b-2.1	DRw15		Dw2	
2	9011	E4181324	DQw6	1.12	1.18	a-2.12	b-2.1	DRw15		Dw12	
2	9012	WJRO76	DQw5*	- (1.21*)	1.2	a-2.16	b-2.1	DRw16		Dw21	
2	9009	KASO11	DQw5*	- (1.21*)	1.2	a-2.16	b-2.1	DRw16		Dw21	
2	9015	WT24	DQw5	- (1.21*)	-	a-2.16	b-2.1	DRw16		Dw21	
2	9016	RML	DQw7	3.1	2	a-2.16	b-2.1	DRw16		Dw22	
w6	9055	HO3O1	DQw6	1.19·	1.2	I-13	III-2	DRw13	DRw52	Dw"new"	Dw26
w6	9053	HOR	DQw6	-	1.2	I-13	-	DRw13	DRw52	Dw19	Dw26
w6	9059	SLEOO5	DQw6	1.19	1.2	I-13	III-2	DRw13	DRw52	Dw19	Dw26
w6	9063	WT47	DQw6	1.19	1.2	I-13	III-2	DRw13	DRw52	Dw19	Dw26
w6	9058	OMW	DQw6	1.18	1.18	I-13	III-3	DRw13	DRw52	Dw18	Dw24
w6	9062	WDV	DQw6	1.18	1.18	I-13	III-3	DRw13	DRw52	Dw18	Dw24
w6	9065	HHKB	DQw6	1.18	1.18	I-13	III-3	DRw13	DRw52	Dw18	Dw24
w6	9060	CB6B	DQw6*	1.18	1.18	I-13	III-1	DRw13	DRw52	Dw18	Dw25
w6	9064	AMALA	DQw7	3.1	2	I-14.2	III-3	DRw14	DRw52	Dw16	Dw24
w6	9054	EK	DQw5	- (1.9*)	1.14	I-14.1	III-1	DRw14	DRw52	Dw9	Dw25
w6	9057	TEM	DQw5*	- (1.9*)	1.14	I-14.1	III-1	DRw14	DRw52	Dw9	Dw25
w6	9061	31227ABO	DQw5	- (1.9*)	1.14	I-14.1	III-1	DRw14	DRw52	Dw9	Dw25
w6	9056	KOSE	DQw5+6*	-	1.2+1.14	I-(13+14.1)	III-1+2	DRw13+14	DRw52	Dw9+19	Dw25+26
3	9018	LOO81785	DQw2	2	-	I-3	III-1	DRw17*	DRw52	Dw3	Dw25
3	9019	DUCAF	DQw2	2	2	I-3	III-1	DRw17*	DRw52	Dw3	Dw25
3	9020	QBL	DQw2	-	-	I-3	III-1	DRw17*	DRw52	Dw3	Dw25
3	9022	COX	DQw2	-	2	I-3	III-3	DRw17*	DRw52	Dw3	Dw24
3	9023	VAVY	DQw2	2	2	I-3	III-3	DRw17*	DRw52	Dw3	Dw24
3	9021	RSH	DQw4	Wa	3.8	I-3	III-3	DRw18*	DRw52	Dw"new"	Dw24
5	9036	SPOO1O	DQw5	1.12*	1.2	I-11.1	III-1	DRw11	DRw52	DB2	Dw25
5	9039	JVM	DQw7	-	-	I-11.2	III-1	DRw11	DRw52	Dw"new"	Dw25
5	9040	BM15	DQw7	3.1	-	I-11.2	III-1	DRw11	DRw52	Dw5	Dw25
5	9042	TISI	DQw7	3.1	2	I-11.2	III-1	DRw11	DRw52	Dw"new"	Dw25
5	9037	SWEIGOO7	DQw7	3.1	2	I-11.1	III-1	DRw11	DRw52	Dw5	Dw25
5	9043	BM21	DQw7	3.1	2	I-11.1	III-1	DRw11	DRw52	Dw5	Dw25
5	9044	BRIP	DQw7	3.1	2	I-11.1	III-1	DRw11	DRw52	Dw5	Dw25
5	9035	JBUSH	DQw7	3.1	2	I-11.1	III-1	DRw11	DRw52	Dw5	Dw25
5	9045	TUBO	DQw7	3.1	-	I-11.1+12	III-1	DRw11+12	DRw52	Dw5	Dw25
5	9038	BM16	DQw7	3.1	2	I-12	III-1	DRw12	DRw52	DB6	Dw25
7	9050	MOU	DQw2	2	3.7	I-7	IV-1	DR7	DRw53	Dw17	
7	9051	PITOUT	DQw2	2	3.7	I-7	IV-1	DR7	DRw53	Dw17	
7	9049	IBW9	DQw2	2	-	I-7	IV-1	DR7	DRw53	Dw17	
7	9047	PLH	DQw2	2	3.7	I-7	IV-1	DR7	DRw53	DB1	
7	9046	BH	DQw2	2	-	I-7	IV-1	DR7	DRw53	DB1	
7	9048	LBUF	DQw2	2	3.7	I-7	IV-1	DR7	DRw53	DB1	
7	9052	DBB	DQw9	3.3	3.7	I-7	IV-null	DR7	DRw53	Dw11	
w9	9073	KT12	DQw9	3.3	3.1	I-9	IV-1	DR9	DRw53	Dw23	
w9	9075	DKB	DQw9	3.3	3.1	I-9	IV-1	DR9	DRw53	Dw23	
w9	9074	HID	DQw9+w1*	3.3+1.12*	3.1	I-9	IV-1	DR9	DRw53	Dw23	
4	9028	PE117	DQw8	3.2	3.2	I-4.1	IV-1	DR4	DRw53	Dw14	
4	9029	WT51	DQw8	3.2	3.2	I-4.1	IV-1	DR4	DRw53	Dw4	
4	9031	BOLETH	DQw8	3.2	3.2	I-4.1	IV-1	DR4	DRw53	Dw4	
4	9032	BSM	DQw8	3.2	3.2	I-4.1	IV-1	DR4	DRw53	Dw4	
4	9033	BM14	DQw8	3.2	3.2	I-4.1	IV-1	DR4	DRw53	Dw4	
4	9034	SAVC	DQw8	3.2	3.1	I-4.1	IV-1	DR4	DRw53	Dw4	
4	9025	DEU	DQw7	3.1	3.1	I-4.1	IV-1	DR4	DRw53	Dw4	
4	9027	PF97387	DQw7	3.1	3.1	I-4.1	IV-1	DR4	DRw53	Dw4	
4	9030	JHAF	DQw7	3.1	3.1	I-4.2	IV-1	DR4	DRw53	Dw13	
4	9024	KT17	DQw8	3.2	3.2	I-4.2	IV-1	DR4	DRw53	KT2	
4	9026	YAR	DQw8	3.2	3.2	I-4.3	IV-1	DR4	DRw53	Dw10	
w8	9070	LUY	DQw7	3.1	3.8	I-8.1	-	DRw8	DRw52	Dw8.3(L)	
w8	9068	BM9	DQw4	Wa	3.8	I-8.1	-	DRw8	DRw52	Dw8.1	
w8	9069	MADURA	DQw4	Wa	3.8	I-8.1	-	DRw8	DRw52	Dw8.1	
w8	9071	OLGA	DQw4	Wa	3.8	I-8.2	-	DRw8	DRw52	Dw8.2	
w8	9066	TABO89	DQw6	1.12	1.18	I-8.1	-	DRw8	DRw52	Dw8.3(T)	

[a]Details regarding the nomenclature for the 2-D gel patterns defined in this study may be found in the text. The WHO nomenclature for DR, DQ, and Dw specificities (8) is based on the results of 10th Workshop serologic (11) and cellular (12) typing of these cell lines, except where indicated (*).

Table 9. Possible Recombination Events Between the DQB1 and DQA1, DQA1 and DRB1, or DRB1 and The Other Expressed DRB Genes (DRB–)[a]

DQB1	DQA1	DQA1	DRB1	DRB1	DRB-
DQβ-(1.21*)	DQα-1.2	DQα-1.2	DRβ-a-2.16	DRβI-13	DRβIII-2
DQβ-1.2	DQα-1.2	DQα-1.2	DRβ-a-2.2		
DQβ-1.19	DQα-1.2	DQα-1.2	DRβI-13	DRβI-13	DRβIII-1
		DQα-1.2	DRβI-11.1	DRβI-14.1	DRβIII-1
DQβ-1.12	DQα-1.18			DRβI-3	DRβIII-1
DQβ-1.18	DQα-1.18	DQα-1.18	DRβ-a-2.12	DRβI-11.1	DRβIII-1
		DQα-1.18	DRβI-13	DRβI-11.2	DRβIII-1
DQβ-2	DQα-2	DQα-1.18	DRβI-8.1	DRβI-12	DRβIII-1
DQβ-2	DQα-3.7				
		DQα-2	DRβ-a-2.16*	DRβI-13	DRβIII-3
DQβ-3.3	DQα-3.7	DQα-2	DRβI-14.2	DRβI-14.2	DRβIII-3
DQβ-3.3	DQα-3.1	DQα-2	DRβI-3	DRβI-3	DRβIII-3
		DQα-2	DRβI-11.1		
DQβ-3.1	DQα-2	DQα-2	DRβI-12	DRβI-7	DRβIV-1
DQβ-3.1	DQα-3.8			DRβI-9	DRβIV-1
DQβ-3.1	DQα-3.1	DQα-3.1	DRβI-9	DRβI-4.1	DRβIV-1
		DQα-3.1	DRβI-4.1	DRβI-4.2	DRβIV-1
DQβ-3.2	DQα-3.1	DQα-3.1	DRβI-4.2	DRβI-4.3	DRβIV-1
DQβ-3.2	DQα-3.2				
		DQα-3.2	DRβI-4.1		
		DQα-3.2	DRβI-4.2		
		DQα-3.8	DRβI-8.1		
		DQα-3.8	DRβI-3		

[a]Details regarding the nomenclature for the 2-D patterns may be found in Tables 3, 6, and 7. Possible recombination events are suggested when an allele at one locus is shared by two or more haplotypes with apparently unrelated alleles at their other loci, as discussed in the text.

occurred between these genes, resulting in a unique haplotype in which the DQα-3.1 chain is paired with the DQβ-3.2 (DQw8) chain. Another recombination event involving the DQα-3.1 chain is suggested by its association with the DQβ-3.3 (DQw9) chain in DRw9 cells. As discussed above, there are no differences in the first-domain sequences of DQα chains isolated from DR4 (DQw7 or DQw8) cells or from DR9 (DQw9) cells. In the DR5 haplotypes, the DQβ-3.1 (DQw7) chain is always found with the DQα-2 chain. Again, this represents a recombination event between the DQB1 and DQA1 genes (perhaps from a DR3 and a DR4 haplotype).

In the haplotypes where recombination has been suggested (DR3,DQw2; DR7,DQw2; DR4,DQw7; DR5,DQw7; DR7,DQw9; DR9,DQw9), each DQβ chain allele has been found in association with more than one DQα chain. However, the suggestion that a "hot spot" of recombination exists between DQB1 and DQA1 does not seem appropriate, since linkage disequilibrium appears to be maintained after these recombination events. Furthermore, no recombination between any of these haplotypes and the DQw1-related haplotypes has been observed, suggesting that the mechanism responsible for these recombination events may be restricted to related DQB1-DQA1 haplotypes. Alternatively, the hybrid DQ α–β dimers formed between DQw1 chains and DQw2 or DQw3 chains may have a selective disadvantage for individuals with these recombinant haplotypes. This selective disadvantage may also account for the fact that several combinations of the other possible DQ α–β chains within the DQw2 and DQw3 haplotypes are not observed. The DQw8 (DQα-3.2 and DQβ-3.2) chains are never seen in recombinant haplotypes (except the SAVC cell line discussed above). It may be noteworthy that of the DQw3 haplotypes, the DR4,DQw8 haplotype is highly associated with autoimmune diabetes mellitus (17). In light of this new α chain polymorphism, it may be interesting to examine patient populations for the presence of the DQα-3.2 allele by 2-D gel analysis, or by oligonucleotide typing once the altered DQA1 sequence has been identified.

DQA1-DRB1 recombination events. Although there is considerable linkage disequilibrium between the DQA1 and DRB1 genes, several recombination events are suggested by the combination of alleles found in the Workshop haplotypes. In particular, the haplotypes that involve apparent recombinations between DQw1 haplotypes and DQw2 or DQw3 haplotypes are only seen as recombination events between DQA1 and DRB1 and are never seen between DQB1 and DQA1, as noted above. For example, the apparent recombinant DQw1-DRw11 (Dw-DB2) haplotype represented by the cell line SP0010 (9036) includes the DQα-1.2 allele as well as the apparently unique DQβ-1.12* chain. Similarly, the DQw1-DRw8 (Dw8.3T) haplotype represented by the cell line TAB089 (9066) includes both the DQβ-1.12 and DQα-1.18 alleles normally found in the DR2,Dw12 haplotype. Both the DQw7-DR2 (Dw22) haplotype, represented by the cell line RML (9016), and the DQw7-DRw14 (Dw16) haplotype, represented by the cell line AMALA (9064), include both the DQβ-3.1 and DQα-2 alleles found in all DR5 haplotypes. The DQw4-DRw18 (Dw "new3") haplotype represented by the cell RSH (9021) includes both the DQβ-Wa and DQα-3.8 alleles found in DRw8 cells.

The remaining haplotypes with shared DQα chain alleles, shown in Table 9, could simply reflect the strong linkage disequilibrium between these genes and the common evolution of closely related haplotypes. These include sharing of DQα-1.2 between DR2,Dw2 and DR2,Dw16. In contrast, the association of the same DQα-1.2 allele with the DRw13 (Dw19) haplotype may be due to recombination. Similarly, recombination appears to have resulted in the sharing of the DQα-1.18 between DRw13 and Dw12 haplotypes. Sharing of DQα-2 between DRw11.1, DRw11.2, and DRw12 haplotypes and sharing of DQα-3.1 between DR9 and DR4 haplotypes are probably due to common evolution.

DRB1-DRB3 recombination events. The three alleles of the DRB3 gene are only found in the DR3, DR5, and DRw6 haplotypes, whereas the DRB4 gene is only found in the DR4, DR7, and DR9 haplotypes. Similarly, the two DRB genes found in DR2 cells are not seen other haplotypes. While recombination events between the DRB1 gene and the DRB3 gene appear to have played a role in the evolution of the DR3, DR5, and DRw6 haplotypes, as described below, no recombination events have been observed between the DRB3 gene and the DRB1 gene of any other haplotype. This is likely to be due to the significantly different organization of genes and pseudogenes associated with the DRB3 haplotypes compared, for example, to that in the DRB4 haplotypes.

Evidence for recombination between the DRB1 and DRB3 genes is seen most clearly in the DR3 haplotypes. Two extended haplotypes have been characterized with apparently identical DQB1, DQA1, and DRB1 alleles. However, they differ in their DRB3 alleles and each of the other MHC genes from the putative recombination site (between DRB1 and DRB3) to the HLA-B locus, as shown in Figure 3.

Other recombination events appear to have occurred in the DRw6 cell lines, where strong linkage disequilibrium is also observed between the DRB1 and DRB3 genes in the various haplotypes. For example, the Dw18 and Dw19 haplotypes share the DRβI-13 chains, but Dw19 is exclusively associated with the DRβIII-2 allele, whereas the DRw18 haplotype is associated with the DRβIII-3 or the DRβIII-1 allele, but not with DRβIII-2. Similarly, the two alleles of DRw14 are associated with different DRβIII alleles. Dw9 cells express DRβI-14.1 and DRβIII-1, whereas Dw16 cells express DRβI-14.2 and DRβIII-3. In contrast, all DR5 cells express the DRβIII-1 allele regardless of which DRBI allele is expressed.

MHC genes	DQB1 --	DQA1 --	DRB1 --------	DRB3	-- C4-B C4-A	-- Bf	-- B
	DQβ-2 --	DQα-2 --	DRβI-3 ---//--	DRβIII-1	-- C4(BQ0, A3)	-- Bf(F1)	-- B18
	DQβ-2 --	DQα-2 --	DRβI-3 ---//--	DRβIII-3	-- C4(B1, AQ0)	-- Bf(S)	-- B8

Figure 3. Recombination between DRB1 and DRB3 in two DR3 haplotypes suggested by the combination of alleles. Two DR3 haplotypes showing strong linkage disequilibrium between the DQB1 and HLA-B loci appear to result from a recombination event (---//--) between the DRB1 and DRB3 genes, as discussed in the text.

Two DRB genes are also found in the DR2 haplotypes, although the gene order is not known. In this study three alleles of the DRβ-a chains were distinguished, but the DRβ-b chains had a common 2-D gel pattern. These DRβ-b chains correspond to three reported DNA sequences of the gene called "DRβII," which define three distinct alleles, each associated with a distinct allele of the DRβ-a chain, DRβ-a-2.2, DRβ-a-2.12, or DRβ-a-2.16 (9). Analysis of possible recombination events between the DRβ-a and DRβ-b genes in the four DR2 haplotypes (Dw2, Dw12, Dw21, and Dw22) will require oligonucleotide typing or sequence analysis of these cells.

Recombination between DRB1 and DRB4 in the DRw53-related haplotypes could not be analyzed in this study due to the lack of polymorphism in the 2-D gel pattern of the DRβIV chain found in all DR4, DR7, and DR9 cell lines with the exception of the DR7,Dw11 cell line, DBB (9052). DBB lacks the expression of the DRβIV-1 chain, although Southern blot analysis demonstrates that this gene is present (18) and an abnormal DRB4-specific mRNA has been found in these cells (19). Since no protein product has been identified, this can be considered a null allele of the DRB4 gene. The MHC genes of this haplotype are in very strong linkage disequilibrium from DQB1 to the HLA-B locus, and the null allele has not been found in other DRw53-related haplotypes.

Corresponding null alleles of the DRB3 gene or the DRB- gene in DR2 haplotypes have not been found in this study. The DR1 and DRw8 cell lines expressed only one DRB1 gene, detectable by 2-D gel analysis, but additional null alleles or pseudogenes may be found in these haplotypes in further studies at the genomic level.

DPα Chain Polymorphism Associated with Cellular DP Typing. The DPα chain polymorphism was limited to two patterns, DPα-1.1 and DPα-1.2, as described in the DP report (20). There was an apparent correlation between the expression of the less common chain, DPα-1.2, and the DP1 and DP5 specificities (12) for the Workshop cell lines, suggesting linkage disequilibrium between DPA1 and DPB1. However, there was no apparent correlation with DQB1 alleles. Therefore, the linkage disequilibrium seen between the DR and DQ genes is not found between DQ and DP genes, as demonstrated previously by serologic DQ typing and cellular DP typing.

Unfortunately, the DPβ chains were not evaluated in this study due to technical difficulties, apparently in the solubilization of the β chains following immunoprecipitation. This was surprising, since the same urea-containing sample buffer was effective for solubilization of the DR and DQ molecules, as well as the DPα chains. Alternatively, the pH range may not have been optimal for focusing these chains in the Workshop system. Although this problem has been encountered in previous studies, the biochemical basis for this apparent solubility problem is not understood and will be the focus for future studies of the DP molecules.

Future Aims

Computer Matching. An effort was made to computerize the scanning and matching of the 2-D gel patterns produced in the study. Scanning of more than 1000 Workshop films was performed and the (x,y) position and intensity of each spot in the 2-D patterns was determined using the BioImage system (BioImage Corp., Ann Arbor, MI) (5). A program was also developed to assign a relative pI and molecular weight to each spot based on the pattern of internal protein standards. This allowed the development of composite maps in a standard coordinate system, which were useful in the visual analysis of these films performed by the participants at the Princeton meeting. A fully automated matching program is currently being developed and a preliminary analysis of a small number of Workshop films has been performed (5). One of the future goals of this approach will be the development of additional programs to allow an objective comparison of unknown 2-D gel patterns with the pattern of each class II allele defined in this Workshop study.

Molecular Localization of mAb and T-Cell Epitopes. From the analysis of epitopes defined by the monomorphic antibodies, it was clear that significant cross-reactivities were observed in the immunoprecipitates. These monomorphic antibodies showed a preference for DR, DQ, or DP, but in many cases the specificity was not absolute and only a quantitative preference for one subregion product was observed in the 2-D gel patterns. For example, of the three mAbs with specificity for DR molecules, GSP4.1 was the most specific, but detectable amounts of DQ molecules were also found in the immunoprecipitates with the other DR "specific" mAbs, PL8 and L243. Similarly, the DQ specific mAb Tu22 showed more cross-reactivity with DR molecules than 1a3. In both cases, the antibodies with the most reactivity for precursor molecules have the most cross-reactivity with products of the other locus. These precursors are seen as the most basic spots in each pattern representing the least glycosylated forms, lowest in the negatively charged sugar sialic acid. These molecules are unlikely to be on the cell surface, but they are seen in the immunoprecipitates since ^{35}S-methionine used in the study will label all molecules synthesized during the 4-hour labeling procedure. Since the amount of the precursors in each lysate appeared to vary considerably, a complete analysis of the specificity and cross-reactivity of each mAb on each of the Workshop cell lines has not been accomplished. Relative specificity for the DRβIII or DRβIV chains was observed for some of the DRw52 and DRw53 antibodies along with additional reactivity for the DRβI chains, depending on the haplotype, as discussed above. This has not been analyzed systematically, although a considerable amount of quantitative data are available from the computerized scanning, which can be analyzed further with the availability of additional programs.

Although T-cell blocking studies were generally successful at localizing determinants to the DR or DQ region molecules (6), it may be possible to better characterize their fine specificity, in future studies, if each clone could be tested with a larger panel of antibodies.

Dimers Between Products of Different Class II Subregions in the Same Haplotype. In this Workshop study the DR α chain was found in almost every immunoprecipitate obtained using anti-DQ antibodies. This may be due to weak cross-reactivity of DQ antibodies for DR molecules, as discussed above, or to dimer formation between the DR α chain and β chain of the DQ subregion. Evidence for the existence of this type of hybrid molecule has been reported recently (21). To distinguish between these two explanations for DR α chains in the DQ precipitates in this study, it is necessary to demonstrate that DR β chains are not present in the same immunoprecipitate. Since most DQ "specific" antibodies in the Workshop study also immunoprecipitated a small amount of DR β chain, cross-reactivity appears to be common. By choosing the appropriate combination of monoclonal antibodies for preclearing and immunoprecipitation, it is theoretically possible to identify each of the interlocus dimer molecules in homozygous cell lines, DRα-DQβ, DQα-DRβI, and DQα-DRβIII or -DRβIV chains. However, since most antibodies were cross-reactive and preclearing studies were not performed, the comprehensive analysis of interlocus dimers in Workshop cell lines was beyond the scope of the present study.

Hybrid DQ Molecules in Heterozygotes. One of the important questions that could not be examined in the present study is the formation of hybrid DQ molecules in heterozygotes. Since a number of recombinant haplotypes in the homozygous Workshop cell lines demonstrated that some DQα alleles can be expressed with different DQβ alleles encoded on the same chromosome, there is every reason to expect that these dimers will also form in heterozygotes when they are encoded on separate chromosomes. For example, DR3,DQw2/DR4,DQw7 heterozygotes would be expected to express hybrid DQ molecule DQβ-3.1/DQα-2 dimers, which are identical to the DQ dimers seen in DR5 homozygotes. They may also express DQα-3.1/DQβ-2 dimers, which have not been observed in the homozygous Workshop cell lines. Evidence for the existence of hybrid DQ molecules has been reported previously (22). As discussed above for the interlocus hybrids, it may be possible to choose the appropriate combination of polymorphic allele specific monoclonal antibodies for preclearing and immunoprecipitation to identify each of the hybrid dimer molecules. In the heterozygous situation there may be competition between DQβ chains for a given α chain, and differential hybrid formation may be found depending on the specific alleles. Therefore, a quantitative analysis would be necessary with computerized scanning of the films.

Alternative Techniques. Several alternative techniques should be examined in future studies. The pH gradient needs to be extended at the basic end of the β chain gels to include DQβ-1.1, DQβ-1.9, DQβ-1.21, and any other chains that were not resolved in this system. The use of a one-dimensional IEF system should also be considered when the specificities of the monoclonal antibodies are known in advance. Solubility of the DPβ chains needs to be improved, presumably through an alternative procedure, perhaps involving the use of SDS before the urea sample solution used for the first-dimension IEF analysis. Surface labeling procedures such as ^{125}I labeling with lactoperoxidase should be compared with the current ^{35}S-methionine labeling procedure to determine which of the glycosylated forms of each class II α and β chains is expressed on the cell surface and which of the precursor forms remain in the cytoplasm. This may also reduce the amount of cross-reactivity of the antibodies, as seen, for example, in the apparent cross-reactivity of DQ mAbs for the precursor forms of DR molecules and the cross-reactivity of DR mAbs for the precursor forms of DQ molecules. Associations with serologic and cellular typing would ideally be made with the polymorphisms seen in the surface structures. This is particularly important when the 2-D gel patterns are being used to localize T-cell epitopes blocked with these mAbs.

Despite these limitations, the 2-D gel system developed for the Workshop study has been highly successful in characterizing the extensive structural polymorphism of the HLA class II α and β chains.

Acknowledgments. The results summarized in this report were contributed by the participating laboratories listed in Table 1 and are described in detail in the following reports. These reports have been edited to provide the maximum consistency in style and nomenclature. The high level of cooperation between the laboratories and the willingness to follow standardized techniques was greatly appreciated and essential to the success of the project. Dominique Charron provided valuable advice and assistance to many of the participants who were less experienced with these techniques. He also organized several important meetings for planning and data analysis and assisted greatly in the writing and editing of the antigen group reports. David Sammons provided many helpful discussions regarding computerized 2-D gel analysis, which were important in the planning stages of the study, and performed all of the scanning and matching of films in preparation for the Princeton Meeting. Nancy Kendrick made important suggestions that helped in the development of the protein standards. The enthusiastic support of Dennis Klenk, Robert Vigna, and Robert Anderson of Health Products Inc. and Pierce Chemical Co. was critical to the development of the reagents and equipment used in the study. Bernadette Kienzle played an important role in the development and testing of new techniques and trained many of the Workshop participants. Several other members of this laboratory provided technical assistance, including Vishwa Adluri, Serena Kilaru, and Alexandra Marshall. Janet Lee, Kimberly Stranick, Mark Geisberg, and Vivien Sutton provided helpful discussions and comments on the manuscript. Bo Dupont and Neal Flomenberg provided advice and encouragement throughout the study, and MaryAnn Barletta, Halena Korsun, and Sally Krell provided expert organizational skill and patience. In addition, I am particularly grateful for the objective advice and constant support provided by Trudy Small. Financial support for this work was provided, in part, by grants AI22916 and CA22507 from the NIH and a grant from the Society of Memorial Sloan-Kettering Cancer Center.

References

1. Yang SY, Milford E, Hammerling V, Dupont B. Description of the reference panel of B-lymphoblastoid cell lines for factors of the HLA system. In: Dupont B (ed): Immunobiology of HLA, Volume I: Histocompatibility Testing 1987. New York: Springer-Verlag, 1989.
2. Charron DJ, McDevitt HO. Analysis of HLA-D region-associated molecules with monoclonal antibody. Proc Natl Acad Sci USA 1979;76:6567.
3. Crumpton MJ, Bodmer JG, Bodmer WF, Heyes JM, Lindsay J, Rudd CE. Biochemistry of class II antigens: Workshop report, in: Albert ED, Baur MP, Mayr WR (eds): Histocompatibility Testing 1984. Berlin: Springer-Verlag, 1984; p 29.
4. Knowles RW, Kienzle B. Standardized equipment, reagents and techniques developed for the 10th Workshop 2-D gel study of HLA class II antigens. In: Dupont B (ed): Immunobiology of HLA, Volume I: Histocompatibility Testing 1987. New York: Springer-Verlag, 1989.
5. Sammons DW, Pan N, Kwan KH, Ko W, Lin NH, Fong YM, et al. Computerized analysis of the 2-D gel patterns for the DRβI and DQβ chains from HLA-DR4 cell lines. In: Dupont B (ed): Immunobiology of HLA, Volume I: Histocompabitility Testing 1987. New York: Springer-Verlag, 1989.
6. Flomenberg N, Plätke R, Zemmour J, Small TN, Klein JS. Immunogenetic analysis of HLA class II cellular data. In: Dupont B (ed): Immunobiology of HLA, Volume I: Histocompatibility Testing 1987. New York: Springer-Verlag, 1989.
7. Knowles RW, Adluri V, Kilaru S, Sirotina A, Small T. Binding patterns of the monoclonal antibodies in the 10th Workshop 2-D gel analysis of class II analysis. In: Dupont B (ed): Immunobiology of HLA, Volume I: Histocompatibility Testing 1987. New York: Springer-Verlag, 1989.

8. Bodmer WF, Albert E, Bodmer JG, Dupont B, Mach B, Mayr W, et al. Nomenclature for factors of the HLA system, 1987. In: Dupont B (ed): Immunobiology of HLA, Volume I: Histocompatibility Testing 1987. New York: Springer-Verlag, 1989.
9. Gregersen PK, Todd JA, Erlich HA, Long E, Servenius B, Choi E, et al. First domain sequence diversity of DR and DQ subregion alleles. In: Dupont B (ed): Immunobiology of HLA, Volume I: Histocompatibility Testing 1987. New York: Springer-Verlag, 1989.
10. Brown JH, Jardetzky T, Saper MA, Samraoui B, Bjorkman PJ, Wiley DC. A hypothetical model of the foreign antigen binding site of class II histocompatibility molecules. Nature 1988;332:845.
11. Milford EL, Kennedy LJ, Yang SY, Dupont B, Lalouel J-M, Yunis EJ. Serologic characterization of the reference panel of B-lymphoblastoid cell lines for factors of the HLA system. In: Dupont B (ed): Immunobiology of HLA, Volume I: Histocompatibility Testing 1987. New York: Springer-Verlag, 1989.
12. Mickelson E, Reinsmoen N, Robbins FM, Hartzman R, Odum N, Svejgaard A, et al. HLA-Dw and HLA-DP typing of the reference panel of B-lymphoblastoid cell lines. In: Dupont B (ed): Immunobiology of HLA, Volume I: Histocompatibility Testing 1987. New York: Springer-Verlag, 1989.
13. Knowles RW, Kienzle B, Winchester R, Mathews E, Bouhallier O, Fauchet R, et al. DRw53: 2-D gel patterns. In: Dupont B (ed): Immunobiology of HLA, Volume I: Histocompatibility Testing 1987. New York: Springer-Verlag, 1989.
14. Flomenberg N. Functional polymorphisms of the HLA class II gene products detected by T lymphocyte clones: Summary of the 10th International Histocompatibility Workshop cellular studies. In: Dupont B (ed): Immunobiology of HLA, Volume I: Histocompatibility Testing 1987. New York: Springer-Verlag, 1989.
15. Ihrle C, Mickelson E, Masaewicz S, Hansen JA, Thomsen M, Termijtelen AM. DRw52: Antigen report of the cellular studies of the 10th International Histocompatibility Workshop. In: Dupont B (ed): Immunobiology of HLA, Volume I: Histocompatibility Testing 1987. New York: Springer-Verlag, 1989.
16. Sharrock C, Lundin K, Zier K, Zeevi A, Thomsen M, Eiermann T, et al. DR4 and DQw3: Antigen report of the cellular studies of the 10th International Histocompatibility Workshop. In: Dupont B (ed): Immunobiology of HLA, Volume I: Histocompatibility Testing 1987. New York: Springer-Verlag, 1989.
17. Kim SJ, Holbeck S, Nisperos B, Hansen JA, Maeda H, Nepom GT. Identification of a polymorphic variant associated with HLA-DQw3 and characterized by specific restriction sites within the DQβ chain gene. Proc Natl Acad Sci USA 1985;82:8139.
18. Sutton VR, Knowles RW. Mapping of a restriction fragment length polymorphism associated with defective DR β4 chain expression to the HLA-DRB1 gene. Hum Immunol 1988;22:123.
19. Sutton VR, Gorski J, Knowles RW. An abnormal DRB4 (DRw53) mRNA is produced by DR7 individuals who fail to express the gene product. (These proceedings, vol. II, 1989.)
20. Knowles RW, Kienzle B. DPα chains: 2-D gel patterns. In: Dupont B (ed): Immunobiology of HLA, Volume I: Histocompatibility Testing 1987. New York: Springer-Verlag, 1989.
21. Lotteau V, Teyton L, Burroughs D, Charron DJ. A novel HLA class II molecule (DRalpha-DQbeta) created by mismatched isotype pairing. Nature 1987;329:339.
22. Charron DJ, Lotteau V, Turmel P. Hybrid HLA-DC antigens provide molecular evidence for gene trans-complementation. Nature 1984;312:157.

DR1: 2-D Gel Patterns

Jean Tkaczuk,[1] Nelson Fernandez,[2] Takako Nakatsuji,[3] Corazon Dejelo,[4] Ronald Bontrop,[5] and Dominique Charron[6]

Six HLA-DR1 lymphoblastoid cell lines, listed in Table 1, were studied with ten monomorphic monoclonal antibodies (mAb) and ten polymorphic mAbs. The analysis was performed by comparison of the 2-D gel patterns from each of the Workshop DR1 cell lines with the pattern of spots obtained by scanning the films and plotting the composite map for the JESTHOM cell line.

Results

DRβ Chain. The DRβI-1.1 2-D gel pattern was observed on the autoradiograms obtained with the mAbs L243, GSP 4.1, and PL8. An identical set of four distinct DRβ chain spots was found in each of the six cell lines tested (Fig. 1). The most basic spot was located just under the most basic spot of the invariant chain pattern, and the three other spots were located at equal intervals from each other towards the acidic end. Although the cell line MZ070782 (9002) was described during the 10th Workshop as a DR1 variant (Dw20) through the use of T-cell clones and RFLP analysis, the Workshop 2-D gel study failed to detect any polymorphism in the pattern of the DRβI-1.1 chains from any of the DR1 cell lines. Recent DNA sequencing of the β1 exon of the DRB1 gene from MZ070782, however, has identified two conservative amino acid substitutions (-V-G- to -A-V-) at positions 85 and 86 that may be responsible for the Dw20 specificity, but would not change the pI of the DRβI-1.1 pattern (J. Gorski and D. Eckels, personal communication).

DQβ Chain. No distinct DQ pattern was obtained with the monomorphic DQ specific antibodies Tü22 and 1a3.

Participating Laboratories: FRAOHA,[1] UKIFES,[2] JAPTSU,[3] US6BRN,[4] BENROO,[5] FRACHA[6]

Table 1. 2-D Gel Patterns from DR1 Cell Lines[a]

DR	CELL #	CELL NAME	Serol. DQ	2-D DQ beta	2-D DQ alpha	2-D DR beta	2-D DR beta	Serol. DR	T cell Dw
1	9003	KAS116	DQw5	- (1.1*)	1.1	I-1.1	-	DR1	Dw1
1	9004	JESTHOM·	DQw5	- (1.1*)	-	I-1.1	-	DR1	Dw1
1	9005	HOM2	DQw5	- (1.1*)	1.1	I-1.1	-	DR1	Dw1
1	9006	WT1OOBIS	DQw5	- (1.1*)	1.1	I-1.1	-	DR1	Dw1
1	9001	SA	DQw5*	- (1.1*)	1.1	I-1.1	-	DR1	Dw1
1	9002	MZO7O782	DQw5	- (1.1*)	-	I-1.1	-	DR1	Dw20

[a]DR, DQ, and Dw nomenclature (1) is based on the 10th Workshop studies of these cell lines (2,3) except where indicated (DQw5*). The DQβ-(1.1*) cell line was not obtained with the Workshop 2-D gel system.

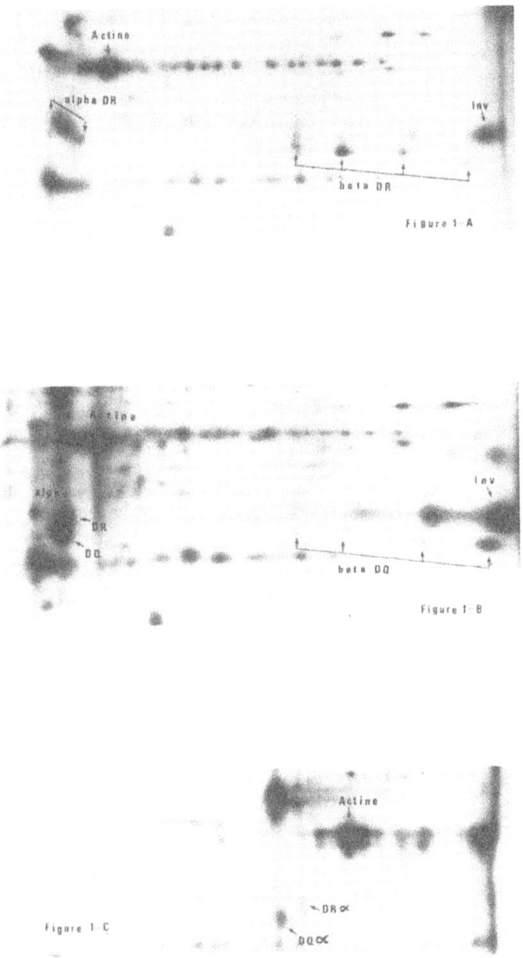

Figure 1. DR and DQ α and β chain patterns from DR1 cell lines. (*A*) DRβ chain pattern precipitated by anti-DR mAb, JS1, from WT100BIS (9006) (J. Tkaczuk, FRAOHA). (*B*) Pattern precipitated by anti-DQ mAb UK4.1 from WT100BIS (9006) (J. Tkaczuk, FRAOHA). (*C*) DQα pattern of KAS-116 (9003) cell line precipitated by anti-DQ mAb UK4.1 (J. Tkaczuk, FRAOHA).

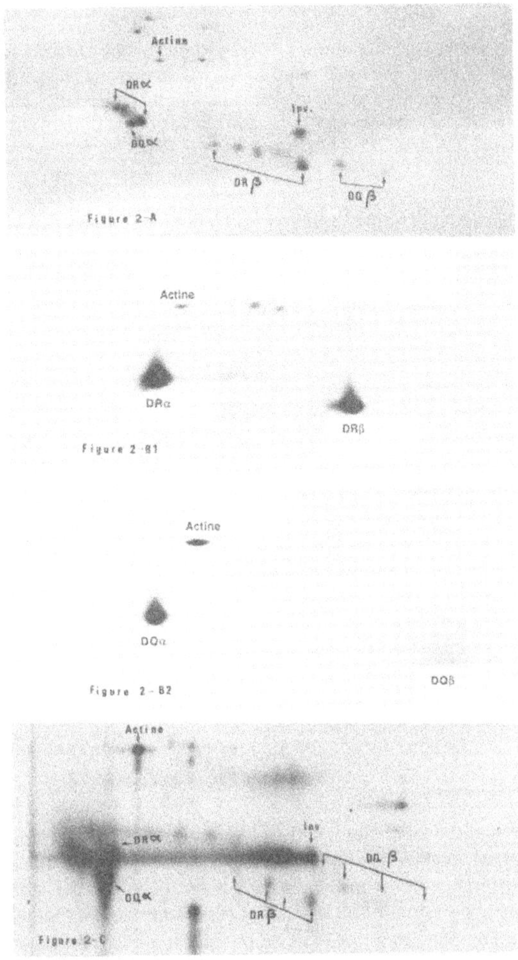

Figure 2. DQ β chain patterns displayed by DR1 cell lines in additional studies using nonequilibrium 2-D gel electrophoresis. (*A*) NEPHGE 2-D analysis of DR and DQ antigens from a DR1 cell line (D. Charron, FRACHR). (*B*) 2-D gel analysis of DRβ molecules of the cell line OOS (DR1,DQw1) immunoprecipitated with anti-DR mAb B8.11.2 (R. Bontrop, BENROO). [*Note*: In (*B*) and (*C*), all the molecules were treated with neuraminidase; for other technical details, see Charron and McDevitt (4).] (*C*) DQβ molecules from the same cell line, OOS, immunoprecipitated with the anti-DQ mAb SPV-L3 and analyzed as in (*B*) (R. Bontrol, BENROO). (*D*) The workshop cell line WT100 (9006) was immunoprecipitated with the anti-DQ mAb Leu10 (Jean Tkaczuk, FRAOHA).

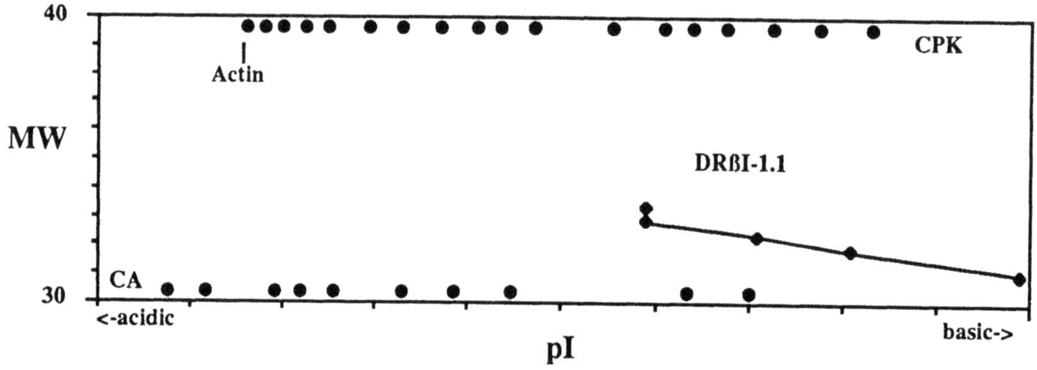

Figure 3. Diagram of the 2-D gel pattern for the DRβI chain isolated from DR1 cell lines. DRβI chains were labeled with ^{35}S-methionine and isolated by immunoprecipitation. Each sample was mixed with ^{14}C-labeled carbamylated protein standards (CA and CPK) and separated by 2-D gel electrophoresis using the standardized Workshop technique with the β chain mixture of ampholytes (80% pH 6 to 8, 20% pH 3.5 to 10) and detected by autoradiography. Spots of the DRβI-1.1 chain pattern (♦) are indicated, relative to the spots of the CA and CPK standard proteins and actin (●).

All six cell lines displayed four spots indistinguishable from the DRβI-1.1 chain pattern. When the polymorphic mAbs Leu10, UK4.1, PLM13, GS4 AA7, KS7, and 77-34 were used, the same four spots of the DRβI-1.1 pattern were observed with only minor differences in their relative intensities. Using a mixture of ampholytes covering a more basic pH range (3), the WT100BIS cell line was found to have a very basic DQβ chain pattern (Fig. 2) identical to that reported previously for DR1 cells (3,4).

DR and DQα Chain. Monomorphic anti-DR and anti-DQ as well as polymorphic anti-DQ mAbs displayed clear and reproducible DR and DQα chain patterns. The initial and most intense spot of this DQα pattern, called DQα-1.1, was positioned directly under the second DRα spot from the basic end. In addition to this major DQα spot, fainter spots were also observed, aligned in a row extending the initial DRα and/or the initial DQα spot. No polymorphism was found for the DRα or the DQα-1.1 pattern from these cell lines.

Discussion

When the DRβ chain and DQβ chain patterns were compared, as shown in Figure 1, it was impossible to see a clear difference between them. They both included four spots with the same initial basic spot (diagrammatically shown in Fig. 3). From the Workshop study it appeared that the DQ β pattern is identical to the DR β pattern. This may be due to coprecipitation of the DRβ chains with both the monomorphic and polymorphic anti-DQ mAbs. In additional studies using a NEPHGE procedure (4), the DQβ chain of DR1 cell lines migrates in a more basic area, with a pH range from 7 to 8. This basic pH range was not obtained with the ampholytes used in the present Workshop study, as illustrated in Figure 2 (5). This result further supports the conclusion that the DRβI-1.1 pattern found with the DQ mAbs in the Workshop pH gradient is simply due to coprecipitation of the DRβI-1.1 molecule.

In conclusion, no variants were detectable by analysis of the DRβ chain 2-D gel patterns for the six DR1 cells studied. Moreover, the data suggest that only one DR β gene, encoding the DRβ-1.1 chain, is expressed by the DR1 cells tested in this Workshop study. Similarly, a single DQα pattern, DQα-1.1, was found in these cells. Although the DQβ chain was not resolved using the Workshop ampholytes, a basic DQβ chain pattern could be detected with the NEPHGE procedures.

Acknowledgments. We gratefully acknowledge the expert technical assistance of N. Woloszyn, C. Offer, M. Sorbara, and D. Clement (FRAOHA). This work was supported, in part, by grant INSERM C.A.R. #487007 (FRACHR and FRAOHA).

References

1. Bodmer WF, Albert E, Bodmer JG, Dupont B, Mach B, Mayr W, et al. Nomenclature for factors of the HLA system, 1987. In: Dupont B (ed): Immunobiology of HLA, Volume I: Histocompatibility Testing 1987. New York: Springer-Verlag, 1989.
2. Milford EL, Kennedy LJ, Yang SY, Dupont B, Lalouel J-M, Yunis EJ. Serologic characterization of the reference panel of B-lymphoblastoid cell lines for factors of the HLA system. In: Dupont B (ed): Immunobiology of HLA, Volume I: Histocompatibility Testing 1987. New York: Springer-Verlag, 1989.
3. Mickelson E, Reinsmoen N, Robbins FM, Hartzman R, et al. HLA-Dw and HLA-DP typing for the reference panel of B-lymphoblastoid cell lines. In: Dupont B (ed): Immunobiology of HLA, Volume I: Histocompatibility Testing 1987. New York: Springer-Verlag, 1989.
4. Charron DJ, and McDevitt HO. Characterization of HLA-DR antigens by two dimensional gel electrophoresis: Molecular genotyping. J Exp Med 1980;152:18.
5. Bontrop RE, Baas EJ, Otting N, Schreuder GMT, Giphart MJ. Molecular diversity of HLA-DQ:DQ alpha and beta chain isoelectric point differences and their relation to serologically defined HLA-DQ allospecificities. Immunogenetics 1987;25:305.

DR2 (DRw15, 16): 2-D Gel Patterns

Nigel Rust,[1] Christine Carter,[1] Susan Fuggle,[1] Steven Sacks,[1] Pierre Hermans,[2] Dominique Charron,[2] Takako Nakatsuji,[3] Kim Wong,[4] Cora Dejelo,[5] and William H. Marshall[6]

The results of the 9th Workshop identified two serologic splits within the DR2 haplotype (long and short). Heterogeneity was also defined by cellular typing; three forms were identified and described as Dw2 (common Caucasian), Dw12 (Japanese variant), and "FJO" ("LD2s"). A comparative analysis of the results indicated that the cellular type "LD2s" was recognized by sera as DR2 short (1). Other studies by RFLP (2) and 2-D gel (3,4) analysis described further polymorphism. In this Workshop, these three subtypes have been confirmed and given WHO recognition (5). DR2 long is now called DRw15 and DR2 short is DRw16. DRw16 can be associated with either DQw1 or DQw3, which have been given separate Dw specificities, Dw21 (LD2s) and Dw22 (LD5a), respectively. Two subtypes of DQw1 have been serologically defined, namely, DQw5 (DQw1.1) and DQw6 (DQw1.2). The association of these new specificities with the Workshop 2-D gel patterns is shown in Table 1.

Results and Discussion

In the Workshop 2-D gel analysis, ten DR2 typed cell lines (Table 1) were studied with ten monomorphic antibodies and a number of polymorphic antibodies. Because little is known about the genomic organization of the DR2 haplotypes, their DRβ chains cannot be assigned to either the DRB1 or another DRβ gene locus. Therefore, the terminology DRβ-a-2.2, DRβ-a-2.12, and DRβ-a-2.16 or DRβ-b-2.1 will be used in this report to indicate the members of two allelic series of DRβ chains.

One DRα chain family and one DRβ chain family (designated DRβ-b-2.1) were consistently found (Fig. 1) in all cell lines except AMAI. In the Workshop DRβ chain analysis AMAI was not assigned a specificity for the DRβ-b chains, since there was a preliminary indication that the pattern was not identical to the DRβ-b-2.1 pattern. Further studies since the Workshop indicate that this cell line expresses a variant form that is slightly more acidic than the common form of the DRβ-b chain, DRβ-b-2.1 (N. Rust, unpublished results).

Another DRβ chain family was also detected for which the 2-D gel pattern varied between cell lines. This DRβ chain family (DRβ-a) subdivided the DR2 cell lines into three separate subtypes corresponding to the specificities Dw2, Dw12, and DRw16 (Dw21 and Dw22). These families were designated DRβ-a-2.2, DRβ-a-2.12, and DRβ-a-2.16 and were clearly distinct as demonstrated in Figure 1. All of the monomorphic DR antibodies precipitated both the DRβ-a and DRβ-b chains, and none of the polymorphic antibodies was specific for either of these DRβ chain families. It was also noted that an additional β chain-like molecule was immunoprecipitated from three of the DRw16 cell lines. In a previous study this molecule was found to have approximately the same pI value as the DRβ-a-2.16 chain, but a lower molecular weight (8). It is unclear whether this chain originates from the same gene as the DRβ-a-2.16 chain through an alternative splicing event or is the product of a different gene. In another study, this chain was found to be specifically recognized by a human monoclonal antibody (9).

In the analysis of DQβ chain polymorphism within DR2 cell lines, two forms of DQw1 β chains and one DQw3 β chain (from RML, discussed below) were identified. One form of DQw1 was associated with Dw2 typed cells (DQβ-1.2) and one form with Dw12 typed cells (DQβ-1.12). Further studies were performed after the Workshop study to compare the DQβ-1.12 pattern from the E4181324 Workshop cell line with those from

Participating Laboratories: UK1MOR,[1] FRACHA,[2] JAPTSU,[3] NCYMRV,[4] US6BRN,[5] NCYMAR[6]

Table 1. DR2 2-D Gel Patterns[a]

DR	CELL #	CELL NAME	Serol. DQ	2-D DQ beta	2-D DQ alpha	2-D DR beta	2-D DR beta	Serol. DR	T cell Dw
2	9010	AMAI	DQw6	1.2	1.2	a-2.2	(b-*)	DRw15	Dw2
2	9008	DO208915	DQw6	1.2	1.2	a-2.2	b-2.1	DRw15	Dw2
2	9013	SCHU	DQw6	1.2	1.2	a-2.2	b-2.1	DRw15	Dw2
2	9014	MGAR	DQw6	1.2	1.2	a-2.2	b-2.1	DRw15	Dw2
2	9017	WT8	DQw6*	1.2	1.2	a-2.2	b-2.1	DRw15	Dw2
2	9011	E4181324	DQw6	1.12	1.18	a-2.12	b-2.1	DRw15	Dw12
2	9012	WJR076	DQw5*	- (1.21*)	1.2	a-2.16	b-2.1	DRw16	Dw21
2	9009	KASO11	DQw5*	- (1.21*)	1.2	a-2.16*	b-2.1	DRw16	Dw21
2	9015	WT24	DQw5	- (1.21*)	-	a-2.16*	b-2.1	DRw16	Dw21
2	9016	RML	DQw7	3.1	2	a-2.16*	b-2.1	DRw16	Dw22

[a]DR, DQ, and Dw nomenclature (5) is based on the 10th Workshop studies of these cell lines (6,7) except where indicated (*).

Immunobiology of HLA
Volume I

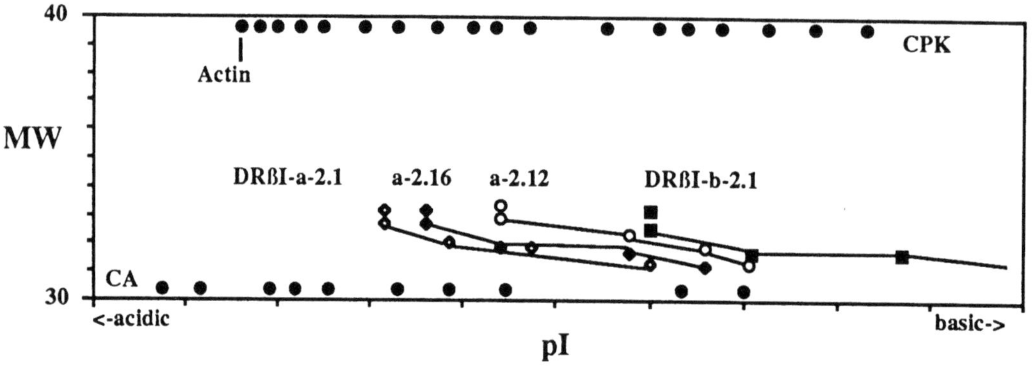

Figure 1. Diagram of the 2-D gel patterns for the DRβ chains isolated from DR2 cell lines. DRβ chains were labeled with ^{35}S-methionine and isolated by immunoprecipitation. Each sample was mixed with ^{14}C-labeled carbamylated protein standards (CA and CPK) and separated by 2-D gel electrophoresis using the standardized Workshop technique with the β chain mixture of ampholytes (80% pH 6 to 8, 20% pH 3.5 to 10) and detected by autoradiography. Spots of the DRβ-a-2.1 chain pattern (◇), DRβ-a-2.12 chain pattern (O), DRβ-a-2.16 chain pattern (♦), and the DRβ-b-2.1 chain pattern (■) are indicated, relative to the spots of the CA and CPK standard proteins and actin (●).

two previously studied Dw12 cell lines, AKIBA and BGE (3). The AKIBA and BGE patterns were found to be identical, but the E4181324 pattern was more basic (N. Rust, unpublished results). In the Workshop 2-D gel pH gradient, it was not possible to identify a DQβ chain in the DR2,Dw21 cell lines. This chain is related to the very basic DQβ chains of other DQw5(DQw1.1) positive cells. Analysis of the β1 domain DNA sequences of the DQβ genes from DR1,DQw1.1 and DR2,Dw21 (DQw1.AZH) cells indicates that only a single amino acid substitution at position 57 distinguishes these alleles (10). Further description of the DQw1 2-D gel patterns can be found in the DQw1 report (11).

Two DQw1 associated DQα chains were also identified in these DR2 cell lines. The common form, DQα-1.2, found in all of the Dw2 and two of the three Dw21 cell lines examined, is also found in DRw6 cells with the DRw13,Dw19 haplotype. The DQα-1.18 chain, found in the Dw12 cell line, is also found in DRw6 cells with the DRw13,Dw18 haplotype. This suggests that recombination events between the DQB1 and DQA1 genes as well as recombination events between the DQA1 and DRB1 genes have played a significant role in the creation of these distinct DQw1-related haplotypes.

Unlike the other DR2,DQw1 cell lines, the Dw22 cell line, RML, was typed serologically as the DQw7 subtype of DQw3. Comparison of its DQβ chain pattern with the DQβ chains from DR4 and DR5 cell lines showed that it expresses the DQβ-3.1 chain, which defines DQw7. RML also expresses the same DQα-2 chain found in DR3 and DR5 cell lines. This suggests both DQA1 and DQB1 genes are derived from a DR5 haplotype by a single recombination event between the DQA1 gene of the DR5 haplotype and the DRB1 gene of a DR2,Dw21 haplotype.

The analysis of the DQ core antibodies revealed different reactivity patterns for Tu22 and 1a3. Their reactivity was dependent on the maturation state of the DQ molecule, Tu22 reacting with the immature forms and 1a3 with the more mature forms of the DQβ families. The polymorphic DQw1 and DQw(1+3) antibodies included in this study precipitated the DQ molecules. The polymorphic antibody NDS15.38, originally submitted to the Workshop as a DR2 specific antibody, was found to react exclusively with the DQ molecules expressed on the DR2 positive cell lines in the Workshop study and in further local analysis (12).

Acknowledgments. We gratefully acknowledge the expert technical assistance of Gareth Plant (UKIMOR) and Yasmin Wasi (NCYMAR). This work was supported, in part, by grants from the Medical Research Council U.K. (UKIMOR), National Kidney Research Fund U.K. (UKIMOR), Wellcome Foundation (UKIMOR), and Arthritis Society of Canada (NCYMAR).

References

1. Batchelor JR, Hors J, Dopi IA, Beuteul H, Borelli I. Antigen report: HLA-DR2, in Albert ED, Baur MP, Mayr WR (eds): Histocompatibility Testing 1984. Berlin, Springer-Verlag, 1984, p 186.
2. Bell JI, Denny D, MacMurray A, Foster L, Watung D, McDevitt HO. Molecular mapping of class II polymorphism in the human histocompatibility complex. I. DRβ. J Immunol 1987;139:562.
3. Lee BSM, Rust NA, McMichael AJ, McDevitt HO. HLA-DR subtypes from an additional supertypic family of DRβ alleles. Proc Natl Acad Sci USA 1987;84:4591.
4. Nishimura Y, Tsukamoto K, Sone T, Hirayama K, Takenouchi T, Ogasawara K, et al. Molecular and functional analysis of class II molecules characteristic to HLA-Dw2 and Dw12, in Albert ED, Baur MP, Mayr WR (eds): Histocompatibility Testing 1984. Berlin, Springer-Verlag, 1984, p 504.
5. Bodmer WF, Albert E, Bodmer JG, Dupont B, Mach B, Mayr W, et al. Nomenclature for factors of the HLA sys-

tem, 1987. In: Dupont B (ed): Immunobiology of HLA, Volume I: Histocompatibility Testing 1987. New York: Springer-Verlag, 1989.
6. Milford EL, Kennedy LJ, Yang SY, Dupont B, Lalouel J-M, Yunis EJ. Serologic characterization of the reference panel of B-lymphoblastoid cell lines for factors of the HLA system. In: Dupont B (ed): Immunobiology of HLA, Volume I: Histocompatibility Testing 1987. New York: Springer-Verlag, 1989.
7. Mickelson E, Reinsmoen N, Robbins FM, Hartzman R, et al. HLA-Dw and HLA-DP typing for the reference panel of B-lymphoblastoid cell lines. In: Dupont B (ed): Immunobiology of HLA, Volume I: Histocompatibility Testing 1987. New York: Springer-Verlag, 1989.
8. Charron DJ, Haziot A, Lotteau V, Neel D, Merlu B, Teyton L. Biochemical diversity of the human class II antigens, in Feldman M, McMichael A (eds): Regulation of Immune Gene Expression. Clifton, NJ: Humana Press, 1986, p 61.
9. Marshall WH, Wasi Y, Drover S. A human antibody binds to a subset of HLA-DR molecules on DR2/Dw1 cell line cells. In: Dupont B (ed): Immunobiology of HLA, Volume II: Immunogenetics and Histocompatibility. New York: Springer-Verlag, 1989.
10. Gregersen PK, Todd JA, Erlich HA, Long E, Servenius B, Choi E, et al. First-domain sequence diversity of DR and DQ subregion alleles. In: Dupont B (ed): Immunobiology of HLA, Volume I: Histocompatibility Testing 1987. New York: Springer-Verlag, 1989.
11. Kaplan C, Morel M-C, Hermans P, Tkaczuk J, Naskatsuji T, Fernandez N, et al. DQw1: 2-D gel patterns. (These proceedings, vol. I, 1989.)
12. Fuggle SV, Rust N, Carter C, McMichael AJ, Morris PJ. Definition of polymorphism within DR2-associated DQw1 molecules. In: Dupont B (ed): Immunobiology of HLA, Volume II: Immunogenetics and Histocompatibility. New York: Springer-Verlag, 1989.

(DR3 (DRw17, DRw18): 2-D Gel Patterns

Susan Fuggle,[1] Christine Carter,[1] Steven Sacks,[1] Susan Radka,[2] Barbara Nepom,[2,3] Cecile Kaplan,[4] Marie-Christine Morel,[4] Ernette du Toit,[5] Stephanie Crafford,[5] and Nigel Rust[1]

Six DR3 cell lines were included in the Workshop study (Table 1). Two-dimensional (2-D) gel patterns for both α chains and β chains were obtained using the panel of core monomorphic and selected polymorphic monoclonal antibodies. Each cell line was studied in two laboratories. The results obtained in different laboratories were generally concordant, and minor differences could be explained on the basis of differences in the efficiency of incorporation of labeled methionine. The complex pattern of class II β chains precipitated by the broadly reactive monomorphic antibodies could be dissected by using a combination of the results from the locus specific monomorphic and polymorphic antibodies. In this way, a specificity was assigned to each family of spots. In all six cell lines two distinct DRβ gene products could be identified, and a DQβ gene product was found in four of the cell lines.

Results and Discussion

The DRB1 gene product, designated DRβI-3, represented the major family of spots precipitated by the monomorphic antibodies, whereas the DRB3 gene product, DRβ-III, was more weakly represented with these antibodies. The DRβI chain had an identical 2-D profile in all six cell lines (Fig. 1). Five of the cell lines had the conventional DR3 long antigen, DRw17, whereas the cell line RSH had the DR3 short antigen, DRw18. Thus, the DRw17 and DRw18 DRβI chains could not be distinguished in the present 2-D gel study. This finding is consistent with sequence data, where conservative substitutions have been demonstrated in the DRB1 gene product of this allele isolated from RSH (4) and another DR3,DQw4 haplotype (C.K. Hurley and A.H. Johnson, personal communication).

The DRB3 gene product, the DRβIII chain, was often weakly present in the profiles using the monomorphic DR antibodies. In contrast, DRβIII was the major product precipitated by DRw52 reactive antibodies, NDS 10, NDS 13, 7.3.19.1, and UK 8.1. Comparison of the DRβIII profiles from the six cell lines studied revealed two distinct forms of DRβIII expressed in linkage disequilibrium with the two DR3 extended haplotypes included in the study. DRβIII-1 was present on three cell lines with the DR3–DRβIII-1–B18 extended haplotype (L0081785, DUCAF and QBL); DRβIII-3 was present on two cell lines with the DR3–DRβIII-3–B8 extended haplotype (COX and VAVY); and DRβIII-3 was present DQw4,DR3 cell line (RSH) (Table 1). The monoclonal antibody NDS 10 differentiated the two alleles, reacting only with DRβIII-1 positive cells (5), whereas the other DRw52 reactive antibodies were positive with all cells. While DRβIII-1 had an acidic pI, DRβIII-3 was more basic, and in some gels only the more mature forms of the product were resolved in the Workshop system (see DRw52 report). The migration of these products relative to the DRβI-3 chain was consistent with previous reports (6–8).

DQβ chain spots of these DQw2 positive DR3 cell lines appeared to be identical to those detected on DQw2,DR7 cells (9). They were precipitated by the DQ antibody 1a3 in cell lines L0081785,DUCAF, and

Participating Laboratories: UK1MOR,[1] US7GSC,[2] US7VMC,[4] FRAMUL,[4] SAFDUT[5]

Table 1. 2-D Patterns from DR3 Cell Lines[a]

DR	CELL #	CELL NAME	Serol. DQ	2-D DQ beta	2-D DQ alpha	2-D DR beta	2-D DR beta	Serol. DR	Serol. DRw52/53	T cell Dw	T cell DRw52
3	9018	LOO81785	DQw2	2	-	I-3	III-1	DRw17*	DRw52	Dw3	Dw25
3	9019	DUCAF	DQw2	2	2	I-3	III-1	DRw17*	DRw52	Dw3	Dw25
3	9020	QBL	DQw2	-	-	I-3	III-1	DRw17*	DRw52	Dw3	Dw25
3	9022	COX	DQw2	-	2	I-3	III-3	DRw17*	DRw52	Dw3	Dw24
3	9023	VAVY	DQw2	2	2	I-3	III-3	DRw17*	DRw52	Dw3	Dw24
3	9021	RSH	DQw4	Wa	3.8	I-3	III-3	DRw18*	DRw52	Dw"new"	Dw24

[a]DR, DQ, and Dw nomenclature (1) is based on the 10th Workshop studies of these cell lines (2,3) except where indicated (*).

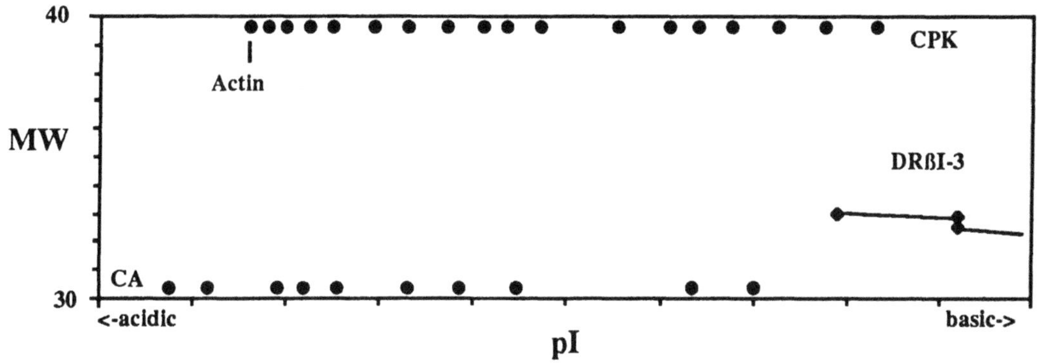

Figure 1. Diagram of the 2-D gel pattern for the DRβI chain isolated from DR3 cell lines. DRβI chains were labeled with ^{35}S-methionine and isolated by immunoprecipitation. Each sample was mixed with ^{14}C-labeled carbamylated protein standards (CA and CPK) and separated by 2-D gel electrophoresis using the standardized Workshop technique with the β chain mixture of ampholytes (80% pH 6 to 8, 20% pH 3.5 to 10) and detected by autoradiography. Spots of the DRβI-3 chain pattern (♦) are indicated, relative to the spots of the CA and CPK standard proteins and actin (●).

VAVY. The DQw4 cell line RSH gave a distinctly different DQβ profile (10), being more acidic and identical to the DQβ-Wa chain of the DQw4 molecule detected on DRw8 cells (Table 1). Both Tu22 and 1a3 were reactive with this product. Tu22 appeared to have preferential reactivity with the more immature forms of the molecule, whereas 1a3 reacted with the more mature forms.

The α chain analysis revealed an identical DRα pattern in all cells. The DQw2α chain, the most basic α chain identified in the study, was only detected in DUCAF, COX, and VAVY (9). In RSH a DQα chain was found that corresponded to the DQα-3.8 chain found in all DRw8,DQw4 cell lines (10). Another α chain spot, similar in position to DQα-1.1, was also found with RSH; it is possible that RSH is heterozygous for DQα, but it seems more likely that this additional chain corresponds to the DPα-1.2 chain, which has pI and molecular weight identical to those of the DQα-1.1 chain. It thus appears that this DQw4–DR3 haplotype (RSH) resulted from a recombination event between the DQA1 gene from a DQw4–DRw8 cell line and the DRB1 gene from a DQw2–DR3 cell line.

Acknowledgments. We gratefully acknowledge the expert technical assistance of Gareth Plant (UK1MOR), Janet Johnston (US7GSC), and David A. Schwarz (US7GSC). This work was supported, in part, by grants from the Medical Research Council U.K. (UK1MOR), National Kidney Research Fund U.K. (UK1MOR), Wellcome Foundation (UK1MOR), and INSERM C.A.R. #487007 (FRAMUL).

References

1. Bodmer WF, Albert E, Albert E, Bodmer JG, Dupont B, Mach B, Mayr W, et al. Nomenclature for factors of the HLA system, 1987. In: Dupont B (ed): Immunobiology of HLA, Volume I: Histocompatibility Testing 1987. New York: Springer-Verlag, 1989.
2. Milford EL, Kennedy LJ, Yang SY, Dupont B, Lalouel J-M, Yunis EJ. Serologic characterization of the referencepanel of B-lymphoblastoid cell lines for factors of the HLA system. In: Dupont B (ed): Immunobiology of HLA, Volume I: Histocompatibility Testing 1987. New York: Springer-Verlag, 1989.
3. Mickelson E, Reinsmoen N, Robbins FM, Hartzman R, et al. HLA-Dw and HLA-DP typing for the reference panel of B-lymphoblastoid cell lines. In: Dupont B (ed): Immunobiology of HLA, Volume I: Histocompatibility Testing 1987. New York: Springer-Verlag, 1989.
4. Gregersen PK, Todd JA, Erlich HA, Long E, Servenius B, Choi E, et al. First-domain sequence diversity of DR and DQ subregion alleles. In: Dupont B (ed): Immunobiology of HLA, Volume I: Histocompatibility Testing 1987. New York: Springer-Verlag, 1989.
5. Fuggle SV, Carter C, Bell JI, Morris PJ. Monoclonal antibody definition of polymorphism within DRw52. In: Dupont B (ed): Immunobiology of HLA, Volume I: Histocompatibility Testing 1987. New York: Springer-Verlag, 1989.
6. Bontrop RM, Titanus M, Mikulski M, van Eggermond M, Termijtelen A, Giphart M. HLA-DR polymorphism detected at the protein and DNA levels are reflected by T-cell recognition. Immunogenetics 1987;23:401.
7. Sacks SH, Bushell A, Rust NA, Karagiannis JA, Jewell DP, Ledingham JGG, et al. Functional and biochemical subtypes of the haplotype HLA-DR3 in patients with celiac disease or idiopathic membranous nephropathy. Hum Immunol 1987;20:175.
8. Fuggle SV, Carter C, Watts F, Kirkley J, Morris PJ. Monoclonal antibody definition of multiple polymorphic epitopes on HLA-DRw52. Hum Immunol 1987;20:249.
9. Radka S, Fuggle S, Mazzilli MC, Carter C, Rust N, Winchester R, et al. DQw2: 2D gel patterns. In: Dupont B (ed): Immunobiology of HLA, Volume I: Histocompatibility Testing 1987. New York: Springer-Verlag, 1989.
10. Kulova L, Maeda H, Bontrop R, Rust N, Fuggle S, Carter C, et al. DQw4 (Wa): 2-D gel patterns. In: Dupont B (ed): Immunobiology of HLA, Volume I: Histocompatibility Testing 1987. New York: Springer-Verlag, 1989.

DR4: 2-D Gel Patterns

Jerry S.S. Lanchbury,[1] Gerald T. Nepom,[2] Susan F. Radka,[3] Cora Dejelo,[4] William H. Marshall,[5] Pascale Perrier,[6] Ann-Marie Henry,[6] Nan Knitter-Jack,[2] Barbara S. Nepom,[2,3] Malek Kamoun,[7] Chester M. Zmijewski,[7] William E. Braun,[4] Robert W. Knowles,[8] and Dimitri Monos[7]

HLA-DR4 is a public allospecificity carried by a series of polymorphic DRβI chains encoded by alleles at the DRB1 locus. DR4 is in strong linkage disequilibrium with the DRw53 antigen in most racial groups and with DQw3 in Caucasoids (1). Included within DR4 are the mixed lymphocyte culture (MLC) types Dw4, Dw10, Dw13, Dw14, Dw15, and DwKT2 (1). Functional MLC defined polymorphism correlates with molecular polymorphism of the DRB1 alleles (2–4).

Results and Discussion

Homozygous cell lines covering the defined Dw types were studied for DR and DQ polymorphisms (Table 1). The Workshop B-cell panel included five DR4 associated MLC specificities. A representative Dw15 cell line, SJAH, was also studied in one laboratory (UKIWEL). Although it was not included in the 10th Workshop as a core cell line, SJAH was studied previously in the 8th Workshop (8W#612, P. Goodfellow and R. Payne) and was typed as Dw4 with 10th Workshop serologic reagents (J. Bodmer, personal communication). By two-dimensional (2-D) gel electrophoresis, four allelic DRβI chain patterns were distinguished. These patterns are shown diagramatically in Figure 1. The most basic allele, DRβI-4.4, is associated with the Dw15 specificity (SJAH). The DRβI-4.1 pattern, associated with the Dw4 and Dw14 specificities, is more acidic than DRβI-4.4 and is found in Dw4 and Dw14 cell lines represented by BOLETH and PE117. It has not been possible to distinguish the Dw4 from Dw14 associated alleles with 2-D gels. A third electrophoretic allele, DRβI-4.2, more acidic than DRβI-4.1, is associated with the Dw13 and DwKT2 MLC types, represented by JHAF and KT17. The most acidic DR4 allele is expressed by Dw10 cell lines represented by YAR. Table 1 lists the allele designations, which are distinguished by their 2-D patterns. As expected, no polymorphism was detected in the DRα pattern. Recent cDNA nucleotide sequencing data have clarified the full extent of DR4 DRβI chain polymorphism (3–5). It is now clear that DRβI chains from Dw4 and Dw14 positive HTCs differ by two amino acids at positions 71 and 86 in the first domain. Although the protein charge difference is insufficient to enable discrimination by 2-D gels, it is sufficient to alter T-cell recognition in the MLC. Similarly, a sequence difference has been shown at position 37 between DRβI chains associated with Dw13 and KT2 MLC specificities, although this micropolymorphism cannot be resolved by 2-D gels (8).

In DR4 positive haplotypes the DRw53 antigenic determinants are carried by the DRβIV chain, which is the product of the DRB4 gene. For the cell lines in the study, the quantity of the DRβIV molecule expressed is considerably less than the quantity of the DRβI molecule. No allelic polymorphism of the DRβIV chain could be detected.

Analysis of DQβ polymorphism among the DR4, DQw3 cell lines revealed the existence of two allelic

Participating Laboratories: UKIWEL,[1] US7VMC,[2] US7GSC,[3] US6BRN,[4] NCYMAR,[5] FRAPRR,[6] US1ZMI,[7] US7KNO[8]

Table 1. 2-D Gel Patterns from DR4 Cells[a]

DR	CELL #	CELL NAME	Serol. DQ	2-D DQ beta	2-D DQ alpha	2-D DR beta	2-D DR beta	Serol. DR	Serol. DRw52/53	T cell Dw
4	9028	PE117	DQw8	3.2	3.2	I-4.1	IV-1	DR4	DRw53	Dw14
4	9029	WT51	DQw8	3.2	3.2	I-4.1	IV-1	DR4	DRw53	Dw4
4	9031	BOLETH	DQw8	3.2	3.2	I-4.1	IV-1	DR4	DRw53	Dw4
4	9032	BSM	DQw8	3.2	3.2	I-4.1	IV-1	DR4	DRw53	Dw4
4	9033	BM14	DQw8	3.2	3.2	I-4.1	IV-1	DR4	DRw53	Dw4
4	9034	SAVC	DQw8	3.2	3.1	I-4.1	IV-1	DR4	DRw53	Dw4
4	9025	DEU	DQw7	3.1	3.1	I-4.1	IV-1	DR4	DRw53	Dw4
4	9027	PF97387	DQw7	3.1	3.1	I-4.1	IV-1	DR4	DRw53	Dw4
4	9030	JHAF	DQw7	3.1	3.1	I-4.2	IV-1	DR4	DRw53	Dw13
4	9024	KT17	DQw8	3.2	3.2	I-4.2	IV-1	DR4	DRw53	KT2
4	9026	YAR	DQw8	3.2	3.2	I-4.3	IV-1	DR4	DRw53	Dw10

[a]DR, DQ, and Dw nomenclature (5) is based on the 10th Workshop studies of these cell lines (6,7).

Immunobiology of HLA
Volume I

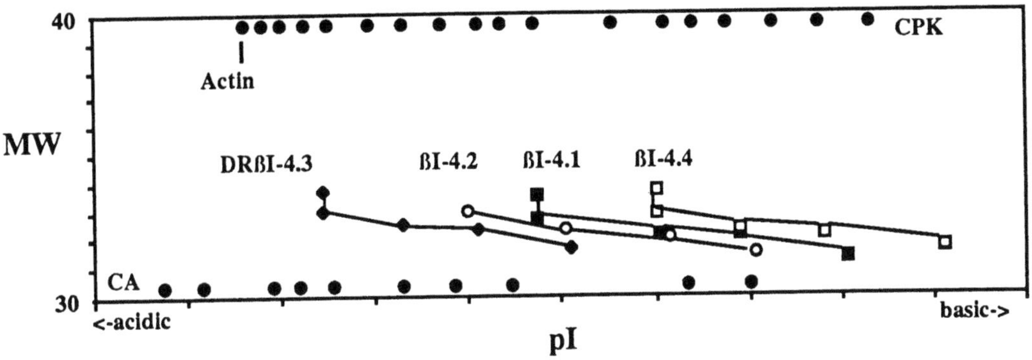

Figure 1. Diagram of the 2-D gel patterns for the DRβI chains isolated from DR4 cell lines. DRβI chains were labeled with ^{35}S-methionine and isolated by immunoprecipitation. Each sample was mixed with ^{14}C-labeled carbamylated protein standards (CA and CPK) and separated by 2-D gel electrophoresis using the standardized Workshop technique with the β chain mixture of ampholytes (80% pH 6 to 8, 20% pH 3.5 to 10) and detected by autoradiography. Spots of the DRβI-4.1 chain pattern (■), DRβI-4.2 chain pattern (○), DRβI-4.3 chain pattern (♦), and DRβI-4.4 chain pattern (□) are indicated, relative to the spots of the CA and CPK standard proteins and actin (●).

DQβ chain patterns, designated DQβ-3.1 and DQβ-3.2. All 11 workshop lines were positive for one or the other of these alleles, and none was heterozygous. The results for each cell line are given in Table 1. As described previously, the presence of DQβ-3.1 correlates with the DQw7(TA10) specificity, whereas DQβ-3.2 is associated with DQw8 and reactivity for the mAb IIB3 (9). The Dw15 cell line included for comparison is characteristically DQw4 positive, but the DQ molecules were not analyzed by 2-D gels.

The DR4 DQw3 associated DQα chain also exhibited molecular polymorphism detected by the Workshop 2-D gel system. Although the intensity of the DQα chain pattern was reduced relative to the DQβ pattern, an acidic electrophoretic variant, DQα-3.1, was distinguished from a more basic form, DQα-3.2. Molecular weight polymorphism was also detected, with DQα-3.1 being the heavier chain. Each cell line expresses either DQα-3.1 or α-3.2, and these polymorphic molecules thus appear to behave as alleles (Table 1). No nucleotide sequence difference that might account for the observed 2-D gel heterogeneity has yet been reported among DR4 associated DQα alleles. Haplotype associations were also found between these DQα and DQβ alleles. Among the cell lines studied, all three that are positive for DQα-3.1 also express DQβ-3.1. Seven of the eight cell lines that carry DQα-3.2 are also positive for DQβ-3.2, whereas SAVC expresses the unusual DQα-3.1/DQβ-3.2 combination. It is possible, therefore, that SAVC represents a haplotype generated as the result of an intra-DQ region recombination.

The relationship between Dw specificity and DQβ polymorphism also deserves comment. While Dw4 and Dw13 lines may be either DQβ-3.1 or DQβ-3.2 positive, the data gathered here, together with previously reported data (10), demonstrate that Dw10 and Dw14 types are exclusively associated with DQβ-3.2. The DwKT2 specificity is usually also associated with DQβ-3.2.

In conclusion, DR4 positive haplotypes express polymorphic DRβI, DQβ, and DQα molecules together with a nonpolymorphic DRβIV chain. Specific combinations of alleles are inherited as haplotypes and are associated with distinct MLC specificities.

Acknowledgments. We gratefully acknowledge the expert technical assistance of Yasim Wasi (NCYMAR), Peter Lo (US1ZMI), and Bernadette Kienzle (US7KNO). This work was supported, in part, by grants from the Arthritis and Rheumatism Council U.K. (UKIWEL), INSERM C.A.R. #487007 (FRAPRR), the Arthritis Society of Canada (NCYMAR), JDF 19525 (US1ZMI), NIH RR-240 (US1ZMI), and AI22916 (US7KNO).

References

1. Schreuder GMT, Doxiadis I, Parlevliet J, Grosse-Wilde H. HLA-DR, DQ, LB, and TA10 specificities of Ninth Workshop homozygous typing cells, in Albert ED, Baur MP, Mayr WR (eds): Histocompatibility Testing 1984. Berlin, Springer-Verlag, 1984, p 243.
2. Nepom BS, Nepom GT, Mickelson E, Antonelli P, Hansen JA. Electrophoretic analysis of human HLA-DR antigens from HLA-DR4 homozygous cell lines: Correlation between β-chain diversity and HLA-D. Proc Natl Acad Sci USA 1983;80:6962.
3. Cairns JS, Curtsinger JM, Dahl CA, Freeman S, Alter BJ, Bach FH. Sequence polymorphism of HLA DRβ alleles relating to T-cell-recognized determinants. Nature 1985; 317:166.
4. Gregersen PK, Shen M, Song Q-L, Merryman P, Degar S, Seki T, et al. Molecular diversity of HLA-DR4 haplotypes. Proc Natl Acad Sci USA 1986;83:2642.
5. Bodmer WF, Albert E, Bodmer JG, Dupont B, Mach B, Mayr W, et al. Nomenclature for factors of the HLA system, 1987. In: Dupont B (ed): Immunobiology of HLA, Volume I: Histocompatibility Testing 1987. New York: Springer-Verlag, 1989.

6. Milford EL, Kennedy LJ, Yang SY, Dupont B, Lalouel J-M, Yunis EJ. Serologic characterization of the reference panel of B-lymphoblastoid cell lines for factors of the HLA system. In: Dupont B (ed): Immunobiology of HLA, Volume I: Histocompatibility Testing 1987. New York: Springer-Verlag, 1989.
7. Mickelson E, Reinsmoen N, Robbins FM, Hartzman R, et al. HLA-Dw and HLA-DP typing for the reference panel of B-lymphoblastoid cell lines. In: Dupont B (ed): Immunobiology of HLA, Volume I: Histocompatibility Testing 1987. New York: Springer-Verlag, 1989.
8. Gregersen PK, Goyert SM, Song Q-L, Silver J. Microheterogeneity of HLA-DR4 haplotypes: DNA sequence analysis of LD"KT2" and LD"TAS" haplotypes. Hum Immunol 1987;19:287.
9. Bontrop RE, Schreuder GMT, Mikulski EMA, van Miltenburg RT, Giphart MJ. Polymorphisms within the HLA-DR4 haplotypes. Various DQ subtypes detected with monoclonal antibodies. Tissue Antigens 1986;27:22.
10. Kim SJ, Holbeck S, Nisperos B, Hansen JA, Maeda H, Nepom GT. Identification of a polymorphic variant associated with HLA-DQw3 and characterized by specific restriction sites within the DQβ chain gene. Proc Natl Acad Sci USA 1985;82:8139.

DR5 (DRw11, DRw12): 2-D Gel Patterns

Alejandro Madrigal,[1] Ronald Bontrop,[2] Maria Cristina Mazzilli,[3] Cecile Kaplan,[4] Kim Wong,[5] Nan Knitter-Jack,[6] Gerald Nepom,[6] Marie-Christine Morel,[4] Joan Lindsay,[1] Dominique Charron,[7] and Manfred Ballas[8]

HLA-DR5 was subdivided after the 9th Workshop into two specificities, HLA-DRw11 and HLA-DRw12 (1). Despite some discussion in the 10th Workshop (2), it was finally agreed to retain both DRw11 and DRw12 as splits of DR5 (3). The biochemical characteristics of the class II molecules expressed by the DRw11 and DRw12 cell lines are described in this report.

Results

Polymorphic variation within DRw11 and between DRw11 and DRw12 cell lines has been identified using a large panel of monoclonal antibodies with the Workshop two-dimensional (2-D) gel technique. Four different DRβ 2-D gel patterns were detected in a panel of 11 Workshop DRw11 and DRw12 cell lines (Table 1).

Pattern A was found in the cell lines, JVM, BM15 and TISI that are DRw11 homozygous. It consists of one DRB1 gene product, DRβI-11.2, found in some DRw11 cell lines, and one DRB3 gene product, DRβIII-1, that was found in all of the cells studied and corresponds to the DRw52b allele defined previously (6,7).

Pattern B was found in the BM16 cell line, known to be homozygous for DRw12. This cell line expresses one DRB1 gene product, DRβI-12, that is different from the DRw11 cell lines and appears to be associated with the DRw12 specificity. The DRB3 gene product expressed

Participating Laboratories: UK1BOD,[1] BENROO,[2] IT2GAN,[3] FRAMUL,[4] NCYMRV,[5] US7VMC,[6] FRACHA,[7] GERGOL[8]

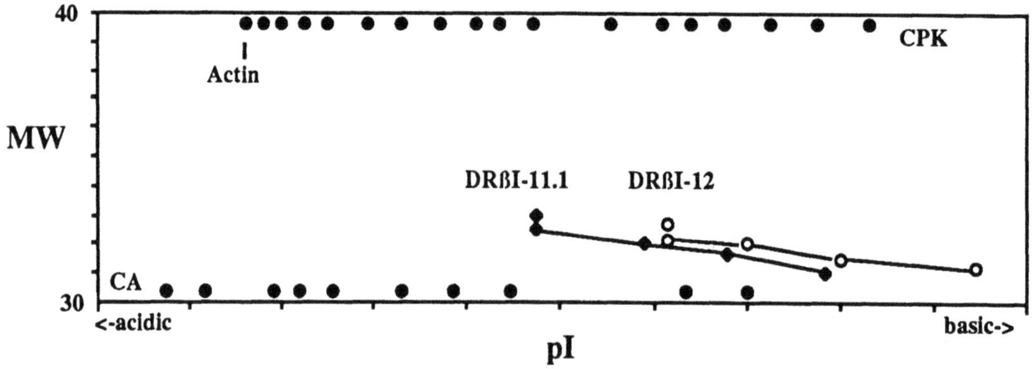

Figure 1. Diagram of the 2-D gel patterns for the DRβI chains isolated from DR5 cell lines. DRβI chains were labeled with ^{35}S-methionine and isolated by immunoprecipitation. Each sample was mixed with ^{14}C-labeled carbamylated protein standards (CA and CPK) and separated by 2-D gel electrophoresis using the standardized Workshop technique with the β chain mixture of ampholytes (80% pH 6 to 8, 20% pH 3.5 to 10) and detected by autoradiography. Spots of the DRβI-11.1 chain pattern (♦) and the DRβI-12 chain pattern (○) are indicated, relative to the spots of the CA and CPK standard proteins and actin (●).

Table 1. 2-D Gel Patterns from DR5 Cell Lines[a]

DR	CELL #	CELL NAME	Serol. DQ	2-D DQ beta	2-D DQ alpha	2-D DR beta	2-D DR beta	Serol. DR	Serol. DRw52/53	T cell Dw	T cell DRw52
5	9036	SPOO1O	DQw5	1.12*	1.2	I-11.1	III-1	DRw11	DRw52	DB2	Dw25
5	9039	JVM	DQw7	-	-	I-11.2	III-1	DRw11	DRw52	"new"	Dw25
5	9040	BM15	DQw7	3.1	-	I-11.2	III-1	DRw11	DRw52	Dw5	Dw25
5	9042	TISI	DQw7	3.1	2	I-11.2	III-1	DRw11	DRw52	"new"	Dw25
5	9037	SWEIGOO7	DQw7	3.1	2	I-11.1	III-1	DRw11	DRw52	Dw5	Dw25
5	9043	BM21	DQw7	3.1	2	I-11.1	III-1	DRw11	DRw52	Dw5	Dw25
5	9044	BRIP	DQw7	3.1	2	I-11.1	III-1	DRw11	DRw52	Dw5	Dw25
5	9035	JBUSH	DQw7	3.1	2	I-11.1	III-1	DRw11	DRw52	Dw5	Dw25
5	9045	TUBO	DQw7	3.1	-	I-11.1+12	III-1	DRw11+12	DRw52	Dw5	Dw25
5	9038	BM16	DQw7	3.1	2	I-12	III-1	DRw12	DRw52	DB6	Dw25

[a]DR, DQ, and Dw nomenclature (3) is based on the 10th Workshop studies of these cell lines (4,5). The DQβ-1.12* is similar to the DQβ-1.12 pattern, but appears to be a distinct allele.

by BM16 is identical to the DRβIII-1 molecule described in pattern A.

Pattern C was found in the cell lines SWEIG007, BM21, and BRIP. These cell lines share both the DRB1 and the DRB3 gene product, DRβI-11.1, DRβIII-1.

Pattern D is simply a combination of patterns B and C and was found in the heterozygous cell line, TUBO. This cell appears to express both the DRB1 gene products of the DRw11 and DRw12 haplotypes (Fig. 1).

Analysis of the results from three cell lines, JVM, BM15, and TISI were difficult to interpret based on the Workshop films alone. However, β1 domain DNA sequence data for the DRβI gene from JVM (8) identified three amino acid differences, including arg(+) to glu(−) at position 71. This predicts a more acidic pI for the JVM DRβI 2-D gel pattern, which was observed in some Workshop films and studies using other 2-D gel systems (9), and is indicated as DRβI-11.2 in Table 1. Two of these cell lines were also distinguished from the other DRw11 cell lines by Dw typing (9) and in the Workshop by T-cell clones (10).

In the DQβ chain analysis, all the DR5 cell lines, except SPOO10, shared the DQβ-3.1 chain pattern associated with the DQw7 subtype of the DQw3 specificity (formerly DQw"3.1"). In the DQα chain analysis, the DR5,DQw7 cell lines were found to express the DQα-2 chain pattern, which is also found in DR3,DQw2 cell lines. No polymorphism was found in the pattern of DRα chains.

The remaining DR5 cell line, SPOO10, is serologically DQw5 and expresses a DQβ chain with the DQβ-1.12* pattern, which, although similar to the DQβ chain expressed by the DR2,Dw12 cell line in this study, appears to be distinct. Unlike the DR2,DQw6,Dw12 cell line, this DR5 cell line SPOO10 expresses the DQα-1.2 chain found in the other DR2,DQw6 cell lines, which are Dw2. A recombination event between the DQA1 and DRB1 genes may be responsible for the generation of this haplotype.

Discussion

The DRw11 and DRw12 subtypes of DR5 haplotypes express one DQβ chain and at least two different DRβ chains that are products of the corresponding DRB1 and DRB3 genes, respectively. The DRw11 and DRw12 molecules appear to be allelic products of the DRB1 gene that can be readily distinguished by their 2-D gel patterns, DRβI-11.1 and DRβI-12. Another DRβI allele (DRβI-11.2), identified in the JVM cell line in previous DNA sequence (8) and IEF studies (9), was found in three cell lines.

The DRB3 gene of the DR5 haplotypes encodes the DRβIII-1 molecule corresponding to the DRw52b allele (6,7) of the DRB3 gene. This chain can also be recognized by T-cell clones with the newly defined Dw25 specificity (10,11). The DRβIII-1 chain was present in all the cell lines studied, indicating that this allele is in close linkage with both DRw11 and DRw12.

Acknowledgments. We gratefully acknowledge the expert technical assistance of Nel Otting (BENROO) and thank Steven G.E. Marsh (UK1BOD) for his helpful discussions. This work was supported, in part by grants CNR87.000864.51 (IT2GAN) and INSERM C.A.R. #487007 (FRAMUL,FRACHR).

References

1. Betuel H, Gebuhrer L, Schreuder GMT, Goldman SF, Villena A, Layrisse Z. DR5 and its subtypes HLA-DRw11 and HLA-DRw12, in Albert ED, Baur MP, Mayr WR (eds): Histocompatibility Testing 1984. Berlin, Springer-Verlag, 1984, p 190.
2. Park MS, Tonai R, Kaszubowski K, Terasaki PI. The orphan antigens: DRw12 and DRw14. Tenth Int Histocompatibility Workshop Newsl 1987;3:18.
3. Bodmer WF, Albert E, Bodmer JG, Dupont B, Mach B, Mayr W, et al. Nomenclature for factors of the HLA system, 1987. In: Dupont B (ed): Immunobiology of HLA, Volume I: Histocompatibility Testing 1987. New York: Springer-Verlag, 1989.
4. Milford EL, Kennedy LJ, Yang SY, Dupont B, Lalouel J-M, Yunis EJ. Serologic characterization of the reference panel of B-lymphoblastoid cell lines for factors of the HLA system. In: Dupont B (ed): Immunobiology of HLA, Volume I: Histocompatibility Testing 1987. New York: Springer-Verlag, 1989.
5. Mickelson E, Reinsmoen N, Robbins FM, Hartzman R, et al. HLA-Dw and HLA-DP typing for the reference panel of B-lymphoblastoid cell lines. In: Dupont B (ed): Immunobiology of HLA, Volume I: Histocompatibility Testing 1987. New York: Springer-Verlag, 1989.
6. Didier DK, Schiffbauer J, Schuman S, Abrazzini LF, Gorski J, Watling DL, et al. Characterization of two distinct β-chain alleles at the βIII locus of DR5 haplotype: βIII alleles are highly conserved. J Immunol 1986;137: 2627.
7. Gorski J, Mach B. Polymorphism of human Ia antigens: Gene conversion between two DRβ loci results in a new HLA-D/DR specificity. Nature 1986;322:67.
8. Gregersen PK, Todd JA, Erlich HA, Long E, Servenius B, Choi E, et al. First-domain sequence diversity of DR and DQ subregion alleles. In: Dupont B (ed): Immunobiology of HLA, Volume I: Histocompatibility Testing 1987. New York: Springer-Verlag, 1989.
9. Bontrop RE, Tilanus MGJ, Mikulski MMA, Elferink DG, Termijtelen A, de Vries RRP, et al. Polymorphism and complexity of HLA-DR: Evidence for intra-HLA-DR region crossing over events. Immunogenetics 1988;27:40.
10. Mickelson E, Masaewicz S, Thomsen M, Reinsmoen N, Sterkers G, Urlacher A, et al. DR5: Antigen report of the cellular studies of the 10th International Histocompatibility Workshop. In: Dupont B (ed): Immunobiology of HLA, Volume I: Histocompatibility Testing 1987. New York: Springer-Verlag, 1989.
11. Ihrle C, Mickelson E, Masaewicz S, Hansen JA, Thomsen M, Termijtelen AM. DRw52: Antigen report of the cellular studies of the 10th International Histocompatibility Workshop. In: Dupont B (ed): Immunobiology of HLA, Volume I: Histocompatibility Testing 1987. New York: Springer-Verlag, 1989.

DRw6 (DRw13, DRw14): 2-D Gel Patterns

Dominique Charron,[1] Pierre Hermans,[1] Manfred Ballas,[2] Zacharie Brahmi,[3] Ernette du Toit,[4] Robert Martell,[4] Laurine Bow,[5] and Ronald Bontrop[6]

DRw6 (DRw13,DRw14) haplotypes were previously considered to be the most enigmatic group of HLA class II specificities. This was largely due to the lack of monospecific anti-DRw6 alloantisera. Through the combined efforts of serology, cellular typing, and biochemistry, it has been possible over the past few years to recognize individual molecular variants of DRβ and DQβ chains and their associations, thus defining several DRw6 haplotypes (1,2). A common feature of all the DRw6 (DRw13,DRw14) haplotypes is the presence of at least one DRβ product belonging to the DRw52 supertypic specificities, thus ranking the DRw6 haplotypes among the DRw52 associated haplotypes with DR3 and 5.

Results and Discussion

DRβ Chains. Every DRw6 cell line studied in the workshop possessed two distinct DRβ chains that were distinguishable by two-dimensional (2-D) gels (Table 1). However, in contrast to the other DR specificities, no common DRβ chain was found associated with all DRw6 haplotypes. Only the DRw13 haplotypes share the same DRβ chain (DRβI-13), which is rather acidic, whereas the DRw14 haplotypes can be split into two, each displaying a unique DRβI chain (DRβI-14.1 or DRβI-14.2). Therefore, it is not surprising that monospecific DRw6 alloantisera have not been found.

The DRw13 group can be divided into three 2-D gel patterns corresponding to three distinct haplotypes. The DRβI-13 chain is common to all DRw13 haplotypes. This common chain is alternatively associated with one of three types of second DRβ chain that are most probably three allelic products of the DRβIII locus. The DRβIII-1 chain is the most acidic, the DRβIII-2 has an intermediate electrophoretic mobility, and the DRβIII-3 is slightly more basic. The DRβIII-2 and DRβIII-3 patterns are quite similar, but were distinguishable from each other in this study.

The DRw14 haplotypes include two distinct biochemical patterns. The DRβI-14.1 chain is very acidic and is only found associated with the DRβIII-1 chain. The DRβI-14.2 chain, which is more basic, is only found associated with the DRβIII-3 chain. Thus, five DRw6 groups are recognized from these DRβ chain patterns (Fig. 1).

DQβ Chains. A common DQβ pattern (designated DQβ-1.18) was found in all DRw13,Dw18 cells (with DRβIII-1 or DRβIII-3) haplotypes. It has an electrophoretic mobility more basic than that of DQβ-1.2 chain found in DR2,Dw2 cells, which is consistent with the single amino acid change at position 30 from Tyr (uncharged) in the DQβ-1.2 chain to His (positive charge) in the DQβ-1.18 chain (6). Another pattern, the DQβ-1.19 chain, was associated with the DRw13, Dw19 haplotype. This chain is more basic than the DQβ1.18 chain. The characterization of the even more basic DQw1 β chain in DRw14,Dw9 was difficult due to the pH range selected for the Workshop study. However, from previously published and unpublished data using a NEPHGE (7) or IEF (8) procedure with a more basic ampholyte mixture, a more complete picture can be proposed of the DQβ products. The DRw14,Dw9 haplotype (DRβI-14.1 + DRβIII-1) possesses a very basic DQβ chain that can only be found using the NEPHGE procedure. It appears similar in electrophoretic mobility to the DQw1 β chain of the DR1 haplotypes, DQβ1.1, which was also too basic to be resolved in the Workshop system (9). The DNA sequence of the DQβ chain from Dw9 cells, called DQβ-1.9, has been found to differ from the DRβ-1.1 sequence by one amino acid in the β1 domain (6).

The DRw14,DQw3 haplotype (DRβI-14.2 + DRβIII-3) expresses DQβ-3.1, which corresponds to the most acidic chain of the DQw3 associated DQβ family (associated with DQw7), also found in most DR4 and all DR5 haplotypes.

DR and DQ α Chains. A DRα chain 2-D gel pattern consisting of four spots was found to be common to all cell lines. The DR α pattern can be used as an internal reference marker in the gel for analysis of the polymorphisms of the DQ and DP α chains.

The Workshop protocol was highly efficient in resolving the DQα chain polymorphisms. Among the DRw6 cell lines, three distinct DQα patterns were identified. The most acidic of the three was found in cell lines of the DRw13,Dw19 haplotype. It is similar to the DQα chain found in DR2,Dw2 haplotypes and was thus designated DQα-1.2. An intermediate pattern located just below the initial major spot of the DRα chains was found in all three cell lines with the DRw14,Dw9 haplotype and was designated DQα-1.14. The third and most basic DQ α pattern was found in all four cell lines of the DRw13,Dw18 haplotype. Designated DQα-1.18, this chain appears identical to the DQα chain found in the DR2,Dw12 haplotype. The DQα chain isolated from the DRw14,DQw3 cell line had the DQα-2 pattern seen in all DR3 and DR5 cell lines. Thus, the combination of DQα-2 and DQβ-3.1 is only found in DR5 cells, suggesting that this haplotype resulted from a recombination

---

Participating Laboratories: FRACHA,[1] GERGOL,[2] US11ND,[3] SAFDUT,[4] US5SWE,[5] BENROO[6]

Table 1. 2-D Gel Patterns from DRw6 Cell Lines[a]

DR	CELL #	CELL NAME	Serol. DQ	2-D DQ beta	2-D DQ alpha	2-D DR beta	2-D DR beta	Serol. DR	Serol. DRw52/53	T cell Dw	T cell DRw52
w6	9058	OMW	DQw6	1.18	1.18	I-13	III-3	DRw13	DRw52	Dw18	Dw24
w6	9062	WDV	DQw6	1.18	1.18	I-13	III-3	DRw13	DRw52	Dw18	Dw24
w6	9065	HHKB	DQw6	1.18	1.18	I-13	III-3	DRw13	DRw52	Dw18	Dw24
w6	9060	CB6B	DQw6*	1.18	1.18	I-13	III-1	DRw13	DRw52	Dw18	Dw25
w6	9055	HO3O1	DQw6	1.19	1.2	I-13	III-2	DRw13	DRw52	Dw"new"	Dw26
w6	9053	HOR	DQw6	-	1.2	I-13	-	DRw13	DRw52	Dw19	Dw26
w6	9059	SLEOO5	DQw6	1.19	1.2	I-13	III-2	DRw13	DRw52	Dw19	Dw26
w6	9063	WT47	DQw6	1.19	1.2	I-13	III-2	DRw13	DRw52	Dw19	Dw26
w6	9056	KOSE	DQw5+6*	-	1.2+1.14	I-(13+14.1)	III-(1+2)	DRw13+14	DRw52	Dw9+19	Dw25+26
w6	9054	EK	DQw5	- (1.9*)	1.14	I-14.1	III-1	DRw14	DRw52	Dw9	Dw25
w6	9057	TEM	DQw5*	- (1.9*)	1.14	I-14.1	III-1	DRw14	DRw52	Dw9	Dw25
w6	9061	31227ABO	DQw5	- (1.9*)	1.14	I-14.1	III-1	DRw14	DRw52	Dw9	Dw25
w6	9064	AMALA	DQw7	3.1	2	I-14.2	III-3	DRw14	DRw52	Dw16	Dw24

[a]DR, DQ, and Dw nomenclature (3) is based on the 10th Workshop studies of these cell lines (4,5) except where indicated (*). The DQβ pattern DQβ-(1.9*) was not determined by 2-D gel analysis in this study, as discussed in the text.

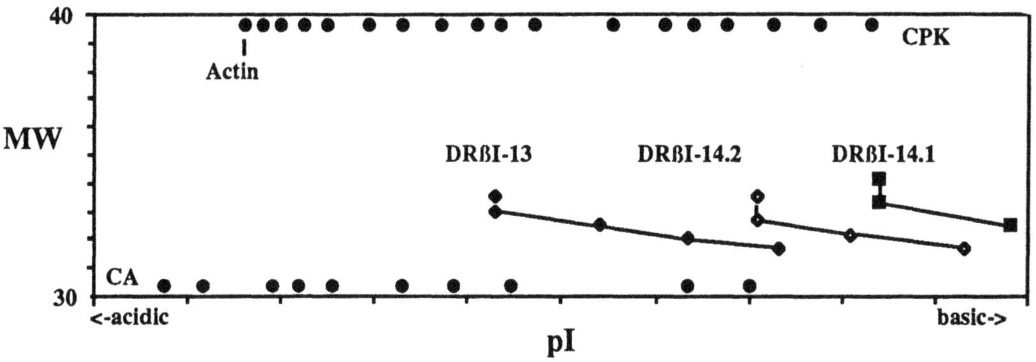

Figure 1. Diagram of the 2-D gel patterns for the DRβI chains isolated from DRw6 cell lines. DRβI chains were labeled with ^{35}S-methionine and isolated by immunoprecipitation. Each sample was mixed with ^{14}C-labeled carbamylated protein standards (CA and CPK) and separated by 2-D gel electrophoresis using the standardized Workshop technique with the β chain mixture of ampholytes (80% pH 6 to 8, 20% pH 3.5 to 10) and detected by autoradiography. Spots of the DRβI-13 chain pattern (♦), the DRβI-14.1 chain pattern (■) and the DRβI-14.2 chain pattern (◇) are indicated, relative to the spots of the CA and CPK standard proteins and actin (●).

Table 2. DRw6 Haplotypes Characterized in This Study

		DRβI	2nd DRβ	DQβ	DQα
DRw6	DRw13	I.13	III.1	1.18	1.18
	DRw13	I.13	III.3	1.18	1.18
	DRw13	I.13	III.2	1.19	1.2
	DRw14	I.14.1	III.1	1.9*[a]	1.14
	DRw14	I.14.2	III.3	3.1	2

[a]The DQβ pattern DQβ-(1.9*) was not determined by 2-D gel analysis in this study, as discussed in the text.

event between the DQA1 gene of a DR5 haplotype and the DRB1 gene of a DRw6 haplotype.

In conclusion, Table 2 lists the DRw6 haplotypes that have been characterized biochemically.

Acknowledgments. This work was supported in part by grants INSERM 487007 and 862002 (FRACHR).

References

1. Haziot A, Lepage V, Freidel AC, Betuel H, Degos L, Charron DJ. Biochemistry of HLA-DRw6: Evidence for seven distinct haplotypes. J Immunol 1986;136:3767.
2. Bosch MR, Termijtelen A, Gerrets R, Schreuder GMT, Bontrop RE, Giphart M. Polymorphisms within the HLA-DRw6 haplotype: Protein charge heterogeneity reflects MLC subtyping of HLA-DRw6 homozygous cells. Immunogenetics 1985;22:23.
3. Bodmer WF, Albert E, Bodmer JG, Dupont B, Mach B, Mayr W, et al. Nomenclature for factors of the HLA system, 1987. In: Dupont B (ed): Immunobiology of HLA, Volume I: Histocompatibility Testing 1987. New York: Springer-Verlag, 1989.
4. Milford EL, Kennedy LJ, Yang SY, Dupont B, Lalouel J-M, Yunis EJ. Serologic characterization of the reference panel of B-lymphoblastoid cell lines for factors of the HLA system. In: Dupont B (ed): Immunobiology of HLA, Volume I: Histocompatibility Testing 1987. New York: Springer-Verlag, 1989.
5. Mickelson E, Reinsmoen N, Robbins FM, Hartzman R, et al. HLA-Dw and HLA-DP typing for the reference panel of B-lymphoblastoid cell lines. In: Dupont B (ed): Immunobiology of HLA, Volume I: Histocompatibility Testing 1987. New York: Springer-Verlag, 1989.
6. Gregersen PK, Todd JA, Erlich HA, Long E, Servenius B, Choi E, et al. First-domain sequence diversity of DR and DQ subregion alleles. In: Dupont B (ed): Immunobiology of HLA, Volume I: Histocompatibility Testing 1987. New York: Springer-Verlag, 1989.
7. Charron DJ, McDevitt HO. Characterization of HLA-DR antigens by two dimensional gel electrophoresis: Molecular genotyping. J Exp Med 1980;152:18.
8. Bontrop R, Van Miltenburg R, Konig F, Gerrets R, Hyland C, Giphart M. Divergent and invariant HLA class II beta chain isoelectric points. Hum Immunol 1986;16:38.
9. Tkaczuk J, Fernandez N, Nakatsuji T, DeJelo C, Bontrop R, Charron D. DR1: 2-D gel patterns. In: Dupont B (ed): Immunobiology of HLA, Volume I: Histocompatibility Testing 1987. New York: Springer-Verlag, 1989.

DR7: 2-D Gel Patterns

Robert Winchester,[1] Erin Matthews,[1] Francois Amesland,[2] Maria Christina Mazzilli,[3] Nelson Fernandez,[4] Bernadette Kienzle,[5] and Robert W. Knowles[5]

Results

The two-dimensional (2-D) gel analysis of the Workshop DR7 cell lines formed two distinctive patterns according to whether the cells were Dw17,DQw2 or Dw11,DQw9 (Table I). The DRβI chains of all DR7 cells, including both haplotypes, were indistinguishable with no evidence of detectable polymorphism in this analysis. This DRβI chain, designated DRβ I-7 exhibited a characteristic mobility that distinguished it from other DRβI chains (Fig. 1).

Cells with the more common Dw17,DQw2 haplotype also expressed a second DRβ chain, designated DRβIV-1. This chain was identical in all properties to the DRβIV chains expressed by DR4 and DR9 haplotypes. In the case of the Dw11,DQw9 cell line DBB (9052), no identifiable DRB4 gene product was detected by any workshop monoclonal reagent, confirming the reports of a defect in the expression of this chain in cells of this haplotype (4,5).

Among all of the DR7 cells examined, the DQ α chains gave an identical 2-D gel pattern. They were designated DQα-3.7 by virtue of their distinctive isoelectric point. In contrast, the two groups of DR7 cells were characterized by distinctive DQβ chains. The Dw17,DQw2 cells had DQβ chains indistinguishable from those of DR3,DQw2 cells and were designated DQβ-2 whereas the Dw11,DQw9 cell had a DQβ chain designated DQβ-3.3, also found in DR9 cells.

Discussion

These findings support and extend the results reported previously on the characteristics of the DR7 gene products. In particular, there was a close relationship among serology, cellular typing, and the 2-D gel patterns. The Dw17,DQw2 group of cells were all found to be identical by this analysis despite the earlier evidence of a MLC distinction (Table 1). As a group they differed from the Dw11,DQw9 cell line by the absence of the DRβIV chain and a distinctive DQβ chain. These changes emphasize that at least two separate events occurred in the generation of this cell type, one involving an alteration in the DRβIV gene that prevents its expression (6) and a recombination event between the DQA1 and DQB1 genes, most probably with the DQβ chain gene and centromeric portion of the class II region derived from a DR9 haplotype.

This heterogeneity within DR7 cells involving the presence or absence of the DRβIV-1 chain allows mapping of particular serologic specificities to the DRβI or DRβIV chain. While certain DRw53 alloreagents react with DBB (9052), cells of this haplotype are not recognized by other DRw53 alloreagents (7). Indeed there is strong support for the identification of the DRβIV-1 chain as the molecule that encodes the critical determinants of the serologic DRw53 allospecificity, whereas some of these determinants are also expressed on the DRβI chain in DR7 cells (4,5). The further complexity of this supertypic specificity is described in the DRw53 report (8).

Acknowledgments. This work was supported in part by grants CNR 87.000864.51 (IT2GAN) and AI22916 (US7KNO).

References

1. Bodmer WF, Albert E, Bodmer JG, Dupont B, Mach B, Mayr W, et al. Nomenclature for factors of the HLA system, 1987. In: Dupont B (ed): Immunobiology of HLA, Volume I: Histocompatibility Testing 1987. New York: Springer-Verlag, 1989.

Participating Laboratories: US8RJW,[1] FRADDC,[2] IT2GAN,[3] UKIFES,[4] US7KNO[5]

Table 1. 2-D Gel Patterns from DR7 Cell Lines[a]

DR	CELL #	CELL NAME	Serol. DQ	2-D DQ beta	2-D DQ alpha	2-D DR beta	2-D DR beta	Serol. DR	Serol. DRw52/53	T cell Dw
7	9050	MOU	DQw2	2	3.7	I-7	IV-1	DR7	DRw53	Dw1[illegible]
7	9051	PITOUT	DQw2	2	3.7	I-7	IV-1	DR7	DRw53	Dw1[illegible]
7	9049	IBW9	DQw2	2	-	I-7	IV-1	DR7	DRw53	Dw1[illegible]
7	9047	PLH	DQw2	2	3.7	I-7	IV-1	DR7	DRw53	DB1
7	9046	BH	DQw2	2	-	I-7	IV-1	DR7	DRw53	DB1
7	9048	LBUF	DQw2	2	3.7	I-7	IV-1	DR7	DRw53	DB1
7	9052	DBB	DQw9	3.3	3.7	I-7	-	DR7	DRw53	Dw1[illegible]

[a]DR, DQ, and Dw nomenclature (1) is based on the 10th Workshop studies of these cell lines (2,3).

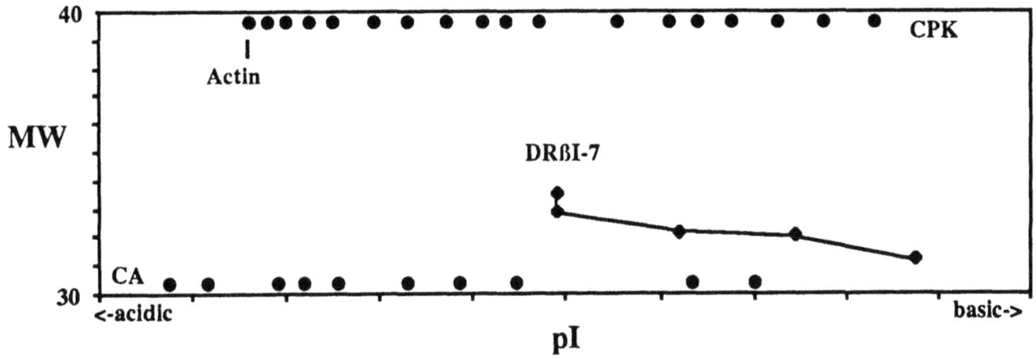

Figure 1. Diagram of the 2-D gel pattern for the DRβI chain isolated from DR7 cell lines. DRβI chains were labeled with ^{35}S-methionine and isolated by immunoprecipitation. Each sample was mixed with ^{14}C-labeled carbamylated protein standards (CA and CPK) and separated by 2-D gel electrophoresis using the standardized Workshop technique with the β chain mixture of ampholytes (80% pH 6 to 8, 20% pH 3.5 to 10) and detected by autoradiography. Spots of the DRβI-7 chain pattern (♦), are indicated, relative to the spots of the CA and CPK standard proteins and actin (●).

2. Milford EL, Kennedy LJ, Yang SY, Dupont B, Lalouel J-M, Yunis EJ. Serologic characterization of the reference panel of B-lymphoblastoid cell lines for factors of the HLA system. In: Dupont B (ed): Immunobiology of HLA, Volume I: Histocompatibility Testing 1987. New York: Springer-Verlag, 1989.
3. Mickelson E, Reinsmoen N, Robbins FM, Hartzman R, et al. HLA-Dw and HLA-DP typing for the reference panel of B-lymphoblastoid cell lines. In: Dupont B (ed): Immunobiology of HLA, Volume I: Histocompatibility Testing 1987. New York: Springer-Verlag, 1989.
4. Knowles RW, Flomenberg N, Horibe K, Winchester R, Radka SF, Dupont B. Complexity of the supertypic HLA-DRw53 specificity: Two distinct epitopes differentially expressed on one or all of the DRβ chains depending on the HLA-DR allotype. J Immunol 1986;137:2618.
5. Matsuyama T, Schwenzer J, Silver J, Winchester R. Structural relationships between the DRβ1 and DRβ2 subunits in DR4, 7 and w9 haplotypes and the DRw53(MT3) specificity. J Immunol 1986;137:934.
6. Sutton VR, Gorski J, Knowles RW. An abnormal DRB4 (DRw53) mRNA is produced by DR7 individuals who fail to express the gene product. (These proceedings, vol. II, 1989.)
7. Nunez-Roldan A, Sanchez F, Sanchez B, Mervart H, Winchester R. DRw53 Antigen Society Joint Report. In: Dupont B (ed): Immunobiology of HLA, Volume I: Histocompatibility Testing 1987. New York: Springer-Verlag, 1989.
8. Knowles RW, Kienzle B, Winchester R, Mathews E, Bouhallier O, Fauchet R, et al. DRw53: 2-D gel patterns. In: Dupont B (ed): Immunobiology of HLA, Volume I: Histocompatibility Testing 1987. New York: Springer-Verlag, 1989.

DRw8: 2-D Gel Patterns

Lydia Kulova,[1] Ronald Bontrop,[2] Soo Young Yang,[1] and Hiroo Maeda[3]

Results and Discussion

Five DRw8,Dw8 cell lines (Table 1) that have been previously characterized serologically and cellularly (4,5) were analyzed in this study. The biochemical analysis of their class II antigens was performed using the Workshop two-dimensional (2-D) gel procedure (IEF-SDS), following immunoprecipitation with monomorphic and polymorphic monoclonal antibodies.

The 2-D gel patterns reveal that all DR8 cells express a single β chain product of the DRB1 gene as reported previously (6,7). This pattern contains a family of four to five consecutive spots, indicating the various stages of intracellular processing of the antigen. Two electrophoretically distinct variants were observed (Fig. 1). The more basic chain, DRβI-8.1, was found in TABO089, BM9, MADURAQ, and LUY, whereas the more acidic DRβI-8.2 chain was expressed in OLGA. These two distinct β chains form heterodimers with one invariant form of DRα chain.

Although the DRw8 specificity is included in the DRw52 supertypic group of haplotypes, not all of the submitted Workshop DRw52 monoclonal antibodies react (e.g., 7.3.19.1 is negative). After immunoprecipitation with NDS.13, the 2-D gel pattern consists of one or two

Participating Laboratories: US7SYY,[1] BENROO,[2] JAPJUJ[3]

Table 1. 2-D Gel Patterns from DRw8 Cell Lines[a]

DR	CELL #	CELL NAME	Serol. DQ	2-D DQ beta	2-D DQ alpha	2-D DR beta	2-D DR beta	Serol. DR	Serol. DRw52/53	T cell Dw
w8	9070	LUY	DQw7	3.1	3.8	I-8.1	-	DRw8	DRw52	Dw8.3(L)
w8	9068	BM9	DQw4	Wa	3.8	I-8.1	-	DRw8	DRw52	Dw8.1
w8	9069	MADURA	DQw4	Wa	3.8	I-8.1	-	DRw8	DRw52	Dw8.1
w8	9071	OLGA	DQw4	Wa	3.8	I-8.2	-	DRw8	DRw52	Dw8.2
w8	9066	TABO89	DQw6	1.12	1.18	I-8.1	-	DRw8	DRw52	Dw8.3(T)

[a]DR, DQ, and Dw nomenclature (1) is based on the 10th Workshop studies of these cell lines (2,3).

spots comigrating with the main spots of the DRβI chain pattern. It would thus appear that DRw52-like epitopes recognized by alloantisera and some monoclonal antibodies are encoded by the DRB1 gene in DRw8 haplotypes. These findings are consistent with RFLP analysis (7), indicating that DRw8 individuals do not have a DRB3 gene. It was also shown recently that the DRw8 haplotypes are negative when probed with any of the three DRw52 allele specific oligonucleotides (8).

In previous serologic studies, DRw8 has been found associated with DQ-blank ("Wa"), DQw1, or DQw3 (4). The five DR8 cell lines in the present study included haplotypes with all three DQ serologic types, which gave DQβ 2-D gel patterns corresponding to the DQβ-Wa, DQβ-1.12, and DQβ-3.1 patterns. The DQβ-Wa pattern was the most acidic, represented by a set of four spots. This DQβ chain is associated with one molecular variant of DQ α chains, the DQα-3.8. This α chain is difficult to distinguish from the DQα-3.1 pattern, but has a slightly lower molecular weight as predicted from its sequence (9). The same DQ α and β chains are found for the DQ heterodimer of the DRw18,Dw24 cell line, RSH, which is serologically "WA" positive. The same DQα-3.8 polypeptide is also present in the DRw8,DQw7 haplotype (LUY), paired with the DQβ-3.1 allele, also found in the DQw7 positive DR4, DRw11, and DRw12 cell lines. It can be speculated that the DRw8,DQw7 haplotype is generated through recombination between the DQA1 gene of a DRw8,DQw4 haplotype and the DQB1 gene of a DQw7 encoding haplotype. From the biochemical data presented, it can also be concluded that the DQw4 and DQw7 specificities are determined by the DQβ polypeptide.

The 2-D gel patterns for DQβ and α chains in TABO89 (DQβ-1.12 and DQα-1.18) comigrate with those found for the DRw15,Dw12,DQw6 cell line, E 4181324. However, the DQβ chain in TABO89 is expressed as three doublets of spots. The spots in each doublet have similar charges, but different molecular weights. The finding that the DQB1 gene from the DRw15,Dw12 haplotype consists of six exons and generates two distinct mRNAs, one of which utilizes exon 5, while the other DQB1 genes do not (10), suggests that this specific 2-D gel profile in TABO89 is probably the result of alternative RNA splicing. The DQα-1.13 chain is more acidic than the DQα-3.8 chain. These immunochemical data suggest that the DRw8,DQw6 haplotype in TABO89 may have been generated by a recombinatorial event between DRw8,Dw8,DQw4 and DRw15,Dw12,DQw6.

A clear correlation has been found between cellularly

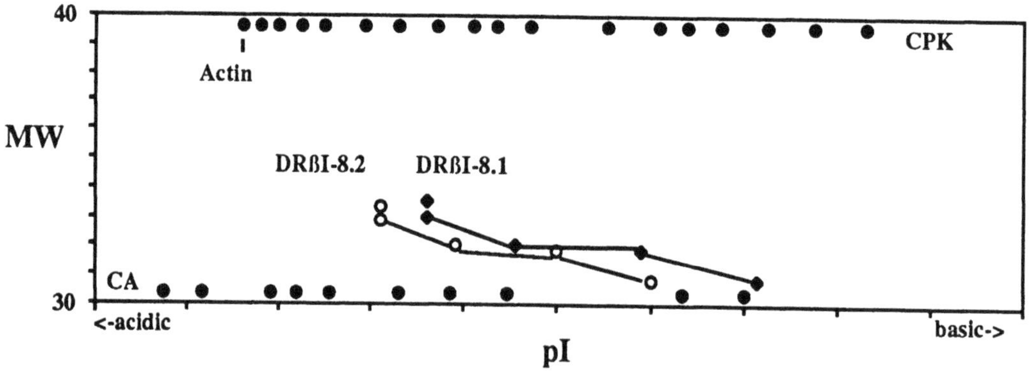

Figure 1. Diagram of the 2-D gel patterns for the DRβI chains isolated from DRw8 cell lines. DRβI chains were labeled with ^{35}S-methionine and isolated by immunoprecipitation. Each sample was mixed with ^{14}C-labeled carbamylated protein standards (CA and CPK) and separated by 2-D gel electrophoresis using the standardized Workshop technique with the β chain mixture of ampholytes (80% pH 6 to 8, 20% pH 3.5 to 10) and detected by autoradiography. Spots of the DRβI-8.1 chain pattern (♦) and the DRβI-8.2 chain pattern (○) are indicated, relative to the spots of the CA and CPK standard proteins and actin (●).

defined Dw8 subtypes and protein charge polymorphism in the 2-D gel patterns, if both DRB1 and DQB1 variants are considered together (Table 1). This is consistent with the evidence that both DR and DQ associated determinants contribute to the definition of Dw clusters in other haplotypes (11). However, CTL clones were described recently that discriminate subtle differences in DRw8 molecules (12), and further biochemical variations have been reported (6). Therefore, additional complexity may exist in the DRw8 haplotypes that is not detected in the 2-D gel patterns described here.

Acknowledgment. This work was supported in part by grant CA22507 (US7SYY).

References

1. Bodmer WF, Albert E, Bodmer JG, Dupont B, Mach B, Mayr W, et al. Nomenclature for factors of the HLA system, 1987. In: Dupont B (ed): Immunobiology of HLA, Volume I: Histocompatibility Testing 1987. New York: Springer-Verlag, 1989.
2. Milford EL, Kennedy LJ, Yang SY, Dupont B, Lalouel J-M, Yunis EJ. Serologic characterization of the reference panel of B-lymphoblastoid cell lines for factors of the HLA system. In: Dupont B (ed): Immunobiology of HLA, Volume I: Histocompatibility Testing 1987. New York: Springer-Verlag, 1989.
3. Mickelson E, Reinsmoen N, Robbins FM, Hartzman R, et al. HLA-Dw and HLA-DP typing for the reference panel of B-lymphoblastoid cell lines. In: Dupont B (ed): Immunobiology of HLA, Volume I: Histocompatibility Testing 1987. New York: Springer-Verlag, 1989.
4. Betuel H, Gebuhrer L, Schreuder GMT, Layrisse Z, Villena AA, Goldmann SF. HLA-DRw8, in Albert ED, Baur MP, Mayr WR (eds): Histocompatibility Testing 1984. Berlin, Springer-Verlag, 1984; p 198.
5. Layrisse Z, Carreno B, Svejgaard A, Platz P, Ryder LP, Kristensen T. First level testing of Dw8 HTC, in Albert ED, Baur MP, Mayr WR (eds): Histocompatibility Testing 1984. Berlin, Springer-Verlag, 1984; p 278.
6. Baldwin GC, Michelson EM, Hansen JA, Nisperos B, Antonelli P, Nepom BS, et al. Electrophoretic variation between class II molecules expressed on HLA-DRw8 homozygous cells reveals multiple distinct haplotypes. Immunogenetics 1985;21:49.
7. Bohme S, Andersson M, Andersson G, Moller E, Peterson PA, Rask L. HLA-DR genes vary in number between different DR specificities whereas the number of DQ genes is constant. J Immunol 1985;135:2149.
8. Tiercy JM, Gorsky J, Jeannet M, Mach B. Identification and distribution of three serologically unrelated alleles of HLA-DR by oligonucleotide–DNA typing analysis. Proc Natl Acad Sci USA 1988;85:198.
9. Gregersen PK, Todd JA, Erlich HA, Long E, Servenius B, Choi E, et al. First-domain sequence diversity of DR and DQ subregion alleles. In: Dupont B (ed): Immunobiology of HLA, Volume I: Histocompatibility Testing 1987. New York: Springer-Verlag, 1989.
10. Tsukamoto K, Yasunami M, Kimura M, Inoko H, Ando A, Hirose, T, et al. DQw1 gene from HLA-DR2/Dw12 consists of 6 exons and expresses multiple DQβ polypeptides through alternative splicing. Immunogenetics 1987;25:343.
11. Bach F. The HLA class II genes and products: The HLA-D region, 1985. Immunol Today 1985;6:89.
12. Tajima Y, Kojima H, Ishikawa N, Wakisaka A, Aizava M. New DRw8 associated class II specificities defined by cytotoxic T lymphocyte clones. Tissue Antigens 1987;30:204.

DR9: 2-D Gel Patterns

Renée Fauchet,[1] Odile Bouhallier,[1] Dominique J. Charron,[2] Robert Winchester,[3] Erin Matthews,[3] Bernadette Kienzle,[4] and Robert W. Knowles[4]

Three DR9 cell lines were examined in the Workshop study: DKB, HID, and KT12. Serologic typing in the Workshop indicates that only DKB is completely homozygous throughout the HLA region. KT12 is clearly not consanguinous, being heterozygous at the A and B loci, but appears to be homozygous in the D region, as indicated in Table 1. HID appears to be heterozygous at the DP locus by cellular typing and at the C locus by Workshop serology and showed additional DQ reactivity with several DQw1-related mAbs, including 77.34 and KS6 and KS7, which were negative with the DKB and KT12 cell lines in the immunofluorescence analysis (4). All

Participating Laboratories: FRAFAU,[1] FRACHA,[2] US8RJW,[3] US7KNO[4]

Table 1. Serologic and Cellular HLA Typing of the DR9 Cell Lines [a]

	A	C	B		Dw	DR	DRw53	DQ	DP
9073 DKB	A24	Cw10	Bw60	Bw6	Dw23	DR9	DRw53	DQw9	DP4
9074 HID	A2	Cw8,Cw10	Bw61	Bw6	Dw23	DR9	DRw53	DQw9	DP2,DP5
9075 KT12	A24,A31	Cw4	B35,Bw52	Bw4,Bw6	Dw23	DR9	DRw53	DQw9	DP5

[a]Serologic and cellular nomenclature (1) is based on the 10th Workshop studies of these cells lines (2,3).

Table 1. 2-D Gel Patterns from DR9 Cell Lines[a]

DR	CELL #	CELL NAME	Serol. DQ	2-D DQ beta	2-D DQ alpha	2-D DR beta	2-D DR beta	Serol. DR	Serol. DRw52/53	T cell Dw
w9	9073	KT12	DQw9	3.3	3.1	I-9	IV-1	DR9	DRw53	Dw23
w9	9075	DKB	DQw9	3.3	3.1	I-9	IV-1	DR9	DRw53	Dw23
w9	9074	HID	DQw9+w1*	3.3+1.12*	3.1	I-9	IV-1	DR9	DRw53	Dw23

[a]DR, DQ, and Dw nomenclature (1) is based on the 10th Workshop studies of these cell lines (2,3). HID expresses a DQβ-1.12* chain t[hat] is similar to the DQβ-1.12 pattern seen in a DR2,Dw12 cell line.

three cells lines were serologically DQw9 in the Workshop serology analysis and reacted with the broad DQw3-related mAbs 17.15, IVD12, and PLM12, as well as IIB3, which does not react with DQw7(TA10+) cells in the immunofluorescence analysis. Together, these preliminary serologic results suggested that either HID was not homozygous in the HLA-DQ region or it carried a unique DQ allele reactive with all of these antibodies.

Results and Discussion

A summary of the two-dimensional (2-D) gel patterns found with these three cell lines is shown in Table 2. Two distinct DRβ chains were immunoprecipitated with anti-DR mAbs (Fig. 1). One of the chains, called the DRβI-9, has a pattern with a rather basic pI, the initial most basic spot being located just under the most basic Ii chain spot. This first spot migrates at the same position as the first spot of the DRβI-1.1 chain (5). A more acidic and fainter chain, DRβIV-1, migrates with a pattern identical to that of the DRβIV-1 chain found in all DR4 cells and in the DR7,Dw17 haplotype. This chain corresponds to the previously described DRw53 β chain found in DR9 cells (6,7). Both the DRβI-9 and the DRβIV-1 chains are immunoprecipitated with the DRw53-related mAbs PL3 and 109d6, as discussed further in the DRw53 report (8).

The DQβ chain, called DQβ-3.3, was immunoprecipitated from all three DR9 cell lines. This is the most basic of the three DQw3 β chains and is clearly distinct from the DQβ-3.1 chain (associated with some DR4 and all DR5 cells) and the DQβ-3.2 chain (associated with the remaining DR4 cells). It is also found on the DR7,Dw11 Workshop cell line DBB, which also types serologically as DQw9. An additional DQβ chain was found in the HID cell line with a pattern that appears similar but not identical to that of the DQβ-1.12 chain found in a DR2,Dw12 cell line. This suggests HID is heterozygous for DQB1 and may account for the additional reactivity of DQw1-related mAbs (4).

Analysis of the α chains identified four spots of the nonpolymorphic DRα chain. One common DQβ spot was found in KT12, HID, and DKB using Tü22, IIB3, IVD12. It represents the initial spot of the DQα-3.1 chain also found in DR4 cells that express the DQw7 associated DQβ-3.1 chain. Two DPα chains were detected with Tü39 and 2D6: DPα-1.1, found under the third spot of the DRα pattern, and DPα-1.2, which was slightly more basic. Spots for both DPα chains were found for KT12 and HID, whereas only the DPα-1.1 spot was found for DKB.

[*Note*: In one laboratory additional α chain spots were observed in the more basic end of the α chain gradient for DKB using both monomorphic and polymorphic DQ antibodies, which may represent a second DQα

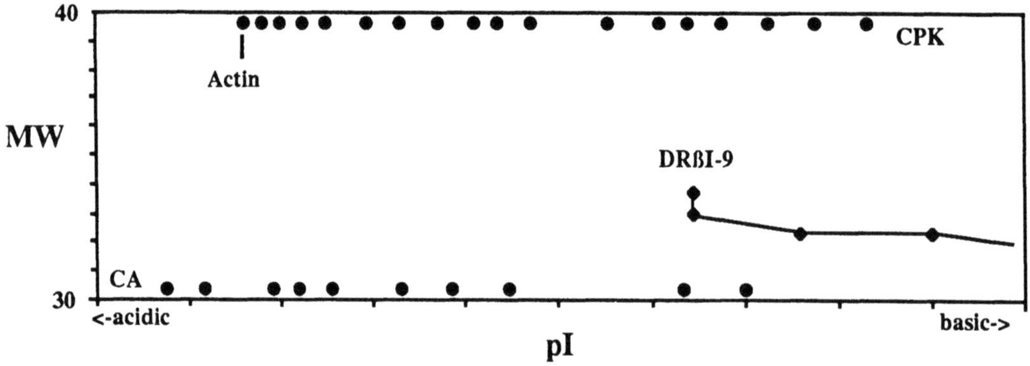

Figure 1. Diagram of the 2-D gel pattern for the DRβI chain isolated from DR9 cell lines. DRβI chains were labeled with ^{35}S-methionine and isolated by immunoprecipitation. Each sample was mixed with ^{14}C-labeled carbamylated protein standards (CA and CPK) and separated by 2-D gel electrophoresis using the standardized Workshop technique with the β chain mixture of ampholytes (80% pH 6 to 8, 20% pH 3.5 to 10) and detected by autoradiography. Spots of the DRβI-9 chain pattern (♦), are indicated, relative to the spots of the CA and CPK standard proteins and actin (●).

chain or the most acidic form of the DQβ-3.3 chain (FRAFAU).]

Acknowledgments. This work was supported, in part, by grants INSERM C.A.R. 4870007 (FRAFAU and FRACHR) and AI22916 (US7KNO).

References

1. Bodmer WF, Albert E, Bodmer JG, Dupont B, Mach B, Mayr W, et al. Nomenclature for factors of the HLA system, 1987. In: Dupont B (ed): Immunobiology of HLA, Volume I: Histocompatibility Testing 1987. New York: Springer-Verlag, 1989.
2. Milford EL, Kennedy LJ, Yang SY, Dupont B, Lalouel J-M, Yunis EJ. Serologic characterization of the reference panel of B-lymphoblastoid cell lines for factors of the HLA system. In: Dupont B (ed): Immunobiology of HLA, Volume I: Histocompatibility Testing 1987. New York: Springer-Verlag, 1989.
3. Mickelson E, Reinsmoen N, Robbins FM, Hartzman R, et al. HLA-Dw and HLA-DP typing for the reference panel of B-lymphoblastoid cell lines. In: Dupont B (ed): Immunobiology of HLA, Volume I: Histocompatibility Testing 1987. New York: Springer-Verlag, 1989.
4. Knowles RW, Knowles R, Adluri V, Kilaru S, Sirotina A, Small T. Binding patterns of the monoclonal antibodies in the 10th Workshop 2-D gel study of class II antigens. In: Dupont B (ed): Immunobiology of HLA, Volume I: Histocompatibility Testing 1987. New York: Springer-Verlag, 1989.
5. Fauchet R, Bouhallier O, Gautier A, Semana G, Lotteau V, Charron DJ. Biochemical studies of the DRw9 specificity in Caucasians. In: Dupont B (ed): Immunobiology of HLA, Volume I: Histocompatibility Testing 1987. New York: Springer-Verlag, 1989.
6. Yabe T, Suzuki M, Mukai R, Juji T, Hamaguchi H. Analysis of the DRw9 antigen using two dimensional gel electrophoresis and allosera. Jpn J Hum Genet 1984;29:17.
7. Maeda H, Hirata R, Thompson A, Mukai R. Molecular characterization of three HLA class II molecules on DR4 and DRw9 haplotypes: Serologic and structural relationship at the polypeptides level. Hum Immunol 1986;15:1.
8. Knowles RW, Kienzle B, Winchester R, Mathews E, Bouhallier O, Fauchet R, et al. DRw53: 2-D gel patterns. In: Dupont B (ed): Immunobiology of HLA, Volume I: Histocompatibility Testing 1987. New York: Springer-Verlag, 1989.

DRw52: 2-D Gel Patterns

Ronald Bontrop,[1] Manfred Ballas,[2] Alejandro Madrigal,[3] Pierre Hermans,[4] Susan Fuggle,[5] Nel Otting,[1] Christine Carter,[5] Steven Sacks,[5] Cecile Kaplan,[6] Marie-Christine Morel,[6] Susan F. Radka,[7] Barbara Nepom,[7,10] Laurine Bow,[8] Ernette Du Toit,[9] Stephanie Crawford,[9] Nan Knitter-Jack,[10] Gerald Nepom,[10] Maria Christina Mazzilli,[11] Zacharie Brahmi,[12] and Dominique Charron[4]

The HLA-DRw52 specificity comprises a group of supertypic determinants, normally expressed on DR3, DR5, DRw6, and DRw8 positive cells, which can be detected by means of alloantisera (1), monoclonal antibodies (2,3), and T-cell reagents (4–6). DR3, DR5, DRw6 cell lines express two DRβ chains, which have been called DRβI and DRβIII (7). This report summarizes the structural polymorphisms detected in the DRβIII chains that express DRw52-related determinants.

Results

A panel of 28 HLA-DR3, DR5, and DRw6 positive cell lines were examined in the Workshop study (Table 1). Three different two-dimensional (2-D) gel patterns were observed for DRβIII chains, designated DRβIII-1 (formerly 52b), DRβIII-2 (formerly 52c), and DRβIII-3 (formerly 52a). The DRβIII-1 chain has the most acidic isoelectric point, the DRβIII-2 chain is intermediate in pI, and the DRβIII-3 chain has the most basic isoelectric point (Fig. 1).

The DR3 positive Workshop cell lines express either the DRβIII-1 or DRβIII-3 chain from the DRB3 locus in addition to a common DRβI-3 chain from the DRB1 locus (Table 1). The DRβIII-1 chain allele is in linkage disequilibrium with the extended haplotype DR3–DRβIII-1–B18, whereas the DRβIII-3 chain allele is in linkage disequilibrium with the extended haplotype DR3– DRβIII-3–B8. The only DR3,DQw4 cell line included in the study expressed the DRβIII-3 chain. The DR5 positive Workshop cell lines were found to express the DRβIII-1 molecule, whereas the DRβIII-2 and DRβIII-3 chain alleles are in strong negative linkage disquilibrium with DR5 haplotypes (Table 1). Within the HLA-DRw6 associated haplotypes, all three DRβIII chains were observed, but strong linkage disequilibrium was seen between DRB3 alleles and some of the DRw6 extended haplotypes (Table 1). Only the DRβIII-2 chain was expressed by the DRw13,Dw19 cell lines tested, whereas the DRw13,Dw18 cell lines expressed either the DRβIII-1 or DRβIII-3 chains. Only the DRβIII-1 chain was expressed by the DRw14,Dw9 cell lines, whereas the only DRw14,Dw16 cell line in the study

---

Participating Laboratories: BENROO,[1] GERGOL,[2] UKIBOD,[3] FRACHA,[4] UKIMOR,[5] FRAMUL,[6] US7GSC,[7] US5SWE,[8] SAFDUT,[9] US7VMC,[10] IT2GAN,[11] USIIND[12]

Table 1. 2-D Gel Patterns from DRw52 Cell Lines[a]

DR	CELL #	CELL NAME	Serol. DQ	2-D DQ beta	2-D DQ alpha	2-D DR beta	2-D DR beta	Serol. DR	Serol. DRw52/53	T cell Dw	T cell DRw52
w6	9055	HO3O1	DQw6	1.19	1.2	I-13	III-2	DRw13	DRw52	Dw"new"	Dw26
w6	9053	HOR	DQw6	-	1.2	I-13	-	DRw13	DRw52	Dw19	Dw26
w6	9059	SLEOO5	DQw6	1.19	1.2	I-13	III-2	DRw13	DRw52	Dw19	Dw26
w6	9063	WT47	DQw6	1.19	1.2	I-13	III-2	DRw13	DRw52	Dw19	Dw26
w6	9058	OMW	DQw6	1.18	1.18	I-13	III-3	DRw13	DRw52	Dw18	Dw24
w6	9062	WDV	DQw6	1.18	1.18	I-13	III-3	DRw13	DRw52	Dw18	Dw24
w6	9065	HHKB	DQw6	1.18	1.18	I-13	III-3	DRw13	DRw52	Dw18	Dw24
w6	9060	CB6B	DQw6*	1.18	1.18	I-13	III-1	DRw13	DRw52	Dw18	Dw25
w6	9064	AMALA	DQw7	3.1	2	I-14.2	III-3	DRw14	DRw52	Dw16	Dw24
w6	9054	EK	DQw5	- (1.9*)	1.14	I-14.1	III-1	DRw14	DRw52	Dw9	Dw25
w6	9057	TEM	DQw5*	- (1.9*)	1.14	I-14.1	III-1	DRw14	DRw52	Dw9	Dw25
w6	9061	31227ABO	DQw5	- (1.9*)	1.14	I-14.1	III-1	DRw14	DRw52	Dw9	Dw25
w6	9056	KOSE	DQw5+6*	-	1.2+1.14	I-(13+14.1)	III-(1+2)	DRw13+14	DRw52	Dw9+19	Dw25+26
3	9018	LOO81785	DQw2	2	-	I-3	III-1	DRw17*	DRw52	Dw3	Dw25
3	9019	DUCAF	DQw2	2	2	I-3	III-1	DRw17*	DRw52	Dw3	Dw25
3	9020	QBL	DQw2	-	-	I-3	III-1	DRw17*	DRw52	Dw3	Dw25
3	9022	COX	DQw2	-	2	I-3	III-3	DRw17*	DRw52	Dw3	Dw24
3	9023	VAVY	DQw2	2	2	I-3	III-3	DRw17*	DRw52	Dw3	Dw24
3	9021	RSH	DQw4	Wa	3.8	I-3	III-3	DRw18*	DRw52	Dw"new"	Dw24
5	9036	SPOO1O	DQw5	1.12*	1.2	I-11.1	III-1	DRw11	DRw52	DB2	Dw25
5	9039	JVM	DQw7	-	-	I-11.2	III-1	DRw11	DRw52	Dw"new"	Dw25
5	9040	BM15	DQw7	3.1	-	I-11.2	III-1	DRw11	DRw52	Dw5	Dw25
5	9042	TISI	DQw7	3.1	2	I-11.2	III-1	DRw11	DRw52	Dw"new"	Dw25
5	9037	SWEIGOO7	DQw7	3.1	2	I-11.1	III-1	DRw11	DRw52	Dw5	Dw25
5	9043	BM21	DQw7	3.1	2	I-11.1	III-1	DRw11	DRw52	Dw5	Dw25
5	9044	BRIP	DQw7	3.1	2	I-11.1	III-1	DRw11	DRw52	Dw5	Dw25
5	9035	JBUSH	DQw7	3.1	2	I-11.1	III-1	DRw11	DRw52	Dw5	Dw25
5	9045	TUBO	DQw7	3.1	-	I-11.1+12	III-1	DRw11+12	DRw52	Dw5	Dw25
5	9038	BM16	DQw7	3.1	2	I-12	III-1	DRw12	DRw52	DB6	Dw25
w8	9070	LUY	DQw7	3.1	3.8	I-8.1	-	DRw8	DRw52	Dw8.3(L)	
w8	9068	BM9	DQw4	Wa	3.8	I-8.1	-	DRw8	DRw52	Dw8.1	
w8	9069	MADURA	DQw4	Wa	3.8	I-8.1	-	DRw8	DRw52	Dw8.1	
w8	9071	OLGA	DQw4	Wa	3.8	I-8.2	-	DRw8	DRw52	Dw8.2	
w8	9066	TABO89	DQw6	1.12	1.18	I-8.1	-	DRw8	DRw52	Dw8.3(T)	

[a]DR, DQ, and Dw nomenclature (8) is based on the 10th Workshop studies of these cell lines (9,10) except where indicated (*). The DQβ-1.12* pattern is similar to the DQβ-1.12 pattern, but appears to be a distinct allele.

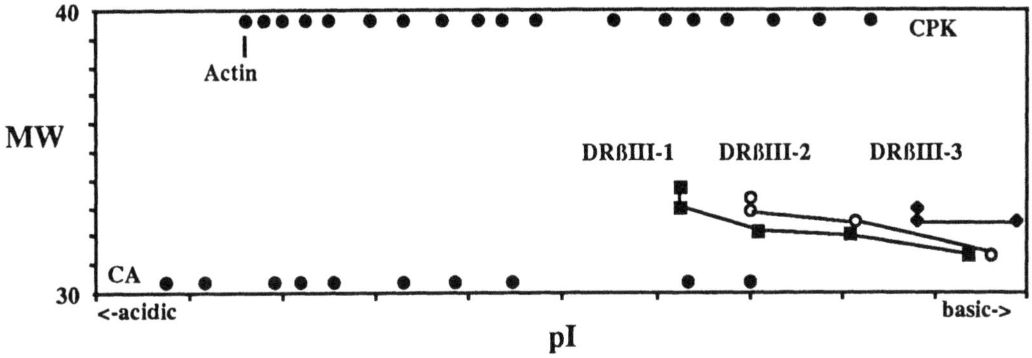

Figure 1. Diagram of the 2-D gel patterns for the DRβIII chains isolated from DRw52 cell lines. DRβIII chains were labeled with ^{35}S-methionine and isolated by immunoprecipitation. Each sample was mixed with ^{14}C-labeled carbamylated protein standards (CA and CPK) and separated by 2-D gel electrophoresis using the standardized Workshop technique with the β chain mixture of ampholytes (80% pH 6 to 8, 20% pH 3.5 to 10) and detected by autoradiography. Spots of the DRIII-1 chain pattern (■), the DRβIII-2 chain pattern (○),and the DRβIII-3 chain pattern (♦) are indicated, relative to the spots of the CA and CPK standard proteins and actin (●).

expressed the DRβIII-3 chain. The cell lines KOSE (9056) appeared to be heterozygous at each HLA class II locus examined in this study, and the heterozygosity of the donor has since been serologically confirmed in a family study (M. Tongio, personal communication).

Discussion

Three different DRβIII chain alleles were detected using 2-D gel electrophoresis in the Workshop cell lines with the family of HLA-DRw52 associated haplotypes. The same DRB3 gene polymorphism can also be detected at the genomic level by oligonucleotide typing (7) and at the cell surface by alloreactive T cells (6,8). T-cell clones have been used in the 10th Workshop T-cell study to define new Dw specificities corresponding to the alleles Dw24 (DRβIII-3, 52a), Dw25 (DRβIII-1, 52b), and Dw26 (DRβIII-2, 52c) (6,8). This is in sharp contrast to the DRB4 locus, which exhibits no polymorphism for the Workshop cell lines with HLA-DRw53 associated haplotypes (11). Extensive linkage disequilibrium was found between the DRB3 alleles and several DRB1 alleles. However, two of the DRB3 alleles are found with more than one DRB1 allele, suggesting that recombination events between these two DRβ genes has been a common mechanism in generating new extended haplotypes, as suggested previously (12).

In DR3, DR5, and DRw6 positive cells, DRw52 or DRw52-like determinants can be located on either or both of the DRB1 and DRB3 gene products (2), whereas in DRw8 positive cells such determinants are exclusively located on DRB1 gene products since DRw8 cells do not express any DRB3 gene products. Therefore, the polymorphisms that contribute to the DRw52 specificity are more complex than the polymorphisms that define the three alleles of the DRB3 gene. A similar situation is found for the DRw53 supertypic specificities (11).

Acknowledgments. We gratefully acknowledge the expert technical assistance of Gareth Plant (UKIMOR), Janet Johnston (US7GSC), and David A. Schwarz (US7GSC). This work was supported, in part, by MRC Programme Grants (UKIMOR), Welcome Grant 16065/1.5/PAJB/LEC/JJ (UKIMOR), a National Kidney Research Grant (UKIMOR), INSERM C.A.R. #487007 (FRAMUL, FRACHR), and CNR 87.000864.51 (IT2GAN).

References

1. Conighi C, Mattiuz PL, Schreuder GMT. Antigen report: HLA-DRw52, in Albert ED, Baur MP, Mayr WR (eds): Histocompatibility Testing 1984. Berlin, Springer-Verlag, 1984; p 202.
2. Karr RW, Lipman LK, Alber C, Koning F, Goeken NE. Ia molecular localization of the DRw52 allodeterminant and the DRw52-like determinants defined by monoclonal antibodies. J Immunol 1985;135:2642–2646.
3. Fuggle SV, Carter C, Watts F, Kirkley J, Morris PJ. Monoclonal antibody definition of multiple polymorphic epitopes on HLA-DRw52. Hum Immunol 1987;20:249.
4. Termijtelen A, Tilanus MGJ, Engelen I, Koning F, van Rood JJ. Molecular localization of LB-Q1, a DRw52-like T-cell recognition epitope and identification at the genomic level of associated shared hybridizing fragments. Hum Immunol 1987;19:255–276.
5. Sheehy MJ, Rowe JR, Koning F, Jorgensen L. Functional polymorphism of the HLA-DR beta III chain. Hum Immunol 1988;21:49–62.
6. Ihrle C, Mickelson E, Masaewicz S, Hansen JA, Thomsen M, Termijtelen AM. DRw52: Antigen report of the cellular studies of the 10th International Histocompatibility Workshop. In: Dupont B (ed): Immunobiology of HLA, Volume I: Histocompatibility Testing 1987. New York: Springer-Verlag, 1989.
7. Tiercy JM, Gorski M, Jeannet M, Mach B. Identification and distribution of three serologically undetected alleles of HLA-DR by oligonucleotide-DNA typing analysis. Proc Natl Acad Sci USA 1988;85:198–202.
8. Bodmer WF, Albert E, Bodmer JG, Dupont B, Mach B,

Mayr W, et al. Nomenclature for factors of the HLA system, 1987. In: Dupont B (ed): Immunobiology of HLA, Volume I: Histocompatibility Testing 1987. New York: Springer-Verlag, 1989.
9. Milford EL, Kennedy LJ, Yang SY, Dupont B, Lalouel J-M, Yunis EJ. Serologic characterization of the reference panel of B-lymphoblastoid cell lines for factors of the HLA system. In: Dupont B (ed): Immunobiology of HLA, Volume I: Histocompatibility Testing 1987. New York: Springer-Verlag, 1989.
10. Mickelson E, Reinsmoen N, Robbins FM, Hartzman R, et al. HLA-Dw and HLA-DP typing for the reference panel of B-lymphoblastoid cell lines. In: Dupont B (ed): Immunobiology of HLA, Volume I: Histocompatibility Testing 1987. New York: Springer-Verlag, 1989.
11. Knowles RW, Kienzle B, Winchester R, Mathews E, Bouhallier O, Fauchet R, et al. DRw53: 2-D gel patterns. In: Dupont B (ed): Immunobiology of HLA, Volume I: Histocompatibility Testing 1987. New York: Springer-Verlag, 1989.
12. Bontrop RE, Tilanus MGJ, Mikulski MMA, Elferink DG, Termijtelen A, de Vries RRP, et al. Polymorphism and complexity of HLA-DR: Evidence for intra-HLA-DR region crossing over events. Immunogenetics 1988;27: 40–45.

DRw53: 2-D Gel Patterns

Robert W. Knowles,[1] Bernadette Kienzle,[1] Robert Winchester,[2] Erin Matthews,[2] Odile Bouhallier,[3] Renee Fauchet,[3] William H. Marshall,[4] Dimitri Monos,[5] Francoise Amesland,[6] Maria Christina Mazzilli,[7] Marie-Anne Henry,[8] Pascale Perrier,[8] Cora Dejelo,[9] Nelson Fernandez,[10] Barbara Nepom,[11,12] Nan Knitter-Jack,[12] Gerald Nepom,[12] Malek Kamoun,[5] Chester Zmijewski,[5] and Jerry Lanchbury[13]

The supertypic DRw53 specificity is defined by alloantisera reacting with HLA-DR4, DR7, and DR9 positive cells (1). Biochemical studies of the class II molecules recognized by DRw53 alloantisera have identified the DRBIV chain, which is distinct from the DRβI chain carrying the DR4, DR7, or DR9 allotypic determinants (2). Analysis with monoclonal antibodies has demonstrated additional complexity in the expression of multiple distinct DRw53-related epitopes depending on the allotype (3,4).

Participating Laboratories: US7KNO,[1] US8RJW,[2] FRAFAU,[3] NCYMAR,[4] US1ZMI,[5] FRADDC,[6] IT2GAN,[7] FRAPRR,[8] US6BRN,[9] UKIFES,[10] US7GSC,[11] US7VMA,[12] UKIWEL[13]

Table 1. 2-D Gel Patterns Expressed by DRw53 Cell Lines[a]

DR	CELL #	CELL NAME	Serol. DQ	2-D DQ beta	2-D DQ alpha	2-D DR beta	2-D DR beta	Serol. DR	Serol. DRw52/53	T cell Dw
4	9028	PE117	DQw8	3.2	3.2	I-4.1	IV-1	DR4	DRw53	Dw14
4	9029	WT51	DQw8	3.2	3.2	I-4.1	IV-1	DR4	DRw53	Dw4
4	9031	BOLETH	DQw8	3.2	3.2	I-4.1	IV-1	DR4	DRw53	Dw4
4	9032	BSM	DQw8	3.2	3.2	I-4.1	IV-1	DR4	DRw53	Dw4
4	9033	BM14	DQw8	3.2	3.2	I-4.1	IV-1	DR4	DRw53	Dw4
4	9034	SAVC	DQw8	3.2	3.1	I-4.1	IV-1	DR4	DRw53	Dw4
4	9025	DEU	DQw7	3.1	3.1	I-4.1	IV-1	DR4	DRw53	Dw4
4	9027	PF97387	DQw7	3.1	3.1	I-4.1	IV-1	DR4	DRw53	Dw4
4	9030	JHAF	DQw7	3.1	3.1	I-4.2	IV-1	DR4	DRw53	Dw13
4	9024	KT17	DQw8	3.2	3.2	I-4.2	IV-1	DR4	DRw53	KT2
4	9026	YAR	DQw8	3.2	3.2	I-4.3	IV-1	DR4	DRw53	Dw10
7	9050	MOU	DQw2	2	3.7	I-7	IV-1	DR7	DRw53	Dw17
7	9051	PITOUT	DQw2	2	3.7	I-7	IV-1	DR7	DRw53	Dw17
7	9049	IBW9	DQw2	2	-	I-7	IV-1	DR7	DRw53	Dw17
7	9047	PLH	DQw2	2	3.7	I-7	IV-1	DR7	DRw53	DB1
7	9046	BH	DQw2	2	-	I-7	IV-1	DR7	DRw53	DB1
7	9048	LBUF	DQw2	2	3.7	I-7	IV-1	DR7	DRw53	DB1
7	9052	DBB	DQw9	3.3	3.7	I-7	-	DR7	DRw53	Dw11
w9	9073	KT12	DQw9	3.3	3.1	I-9	IV-1	DR9	DRw53	Dw23
w9	9075	DKB	DQw9	3.3	3.1	I-9	IV-1	DR9	DRw53	Dw23
w9	9074	HID	DQw9+w1*	3.3+1.12*	3.1	I-9	IV-1	DR9	DRw53	Dw23

[a]DR, DQ, and Dw nomenclature (5) is based on the 10th Workshop studies of these cell lines (6,7) except where indicated (*).

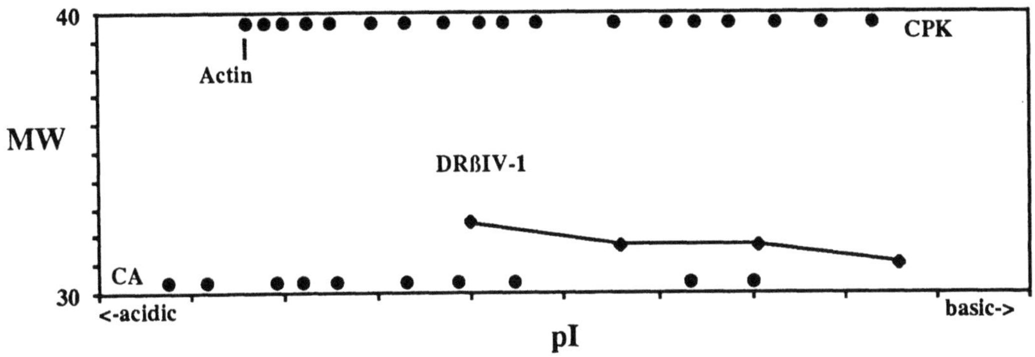

Figure 1. Diagram of the 2-D gel pattern for the DRβIV chain isolated from DRw53 cell lines. DRβIV chains were labeled with ^{35}S-metionine and isolated by immunoprecipitation. Each sample was mixed with ^{14}C-labeled carbamylated protein standards (CA and CPK) and separated by 2-D gel electrophoresis using the standardized Workshop technique with the β chain mixture of ampholytes (80% pH 6 to 8, 20% pH 3.5 to 10) and detected by autoradiography. Spots of the DRβIV-1 chain pattern (♦), are indicated, relative to the spots of the CA and CPK standard proteins and actin (●).

Results

Analysis of the 21 DRw53 positive Workshop cell lines (Table 1) clearly identified the DRβIV chain on every DR4, DR7, and DR9 haplotype except the DR7,Dw11 cell line DBB. This DRβIV chain was immunoprecipitated with the mAb PL3 (DRw53 related) and has a characteristic two-dimensional (2-D) gel pattern (Fig. 1). No polymorphism was detected in this pattern for any of the Workshop DRw53 cell lines.

Several cell lines were also examined with another DRw53-related mAb, 109d6, in parallel with PL3 in one laboratory (US7KNO). These results demonstrate that both antibodies bind to the DRβIV chain, but they show differential binding to allelic forms of the polymorphic DRβI chain. Neither antibody binds to the DRβI chain from DR4 cells; PL3, but not 109d6, binds to the DRβI chain from DR7 cells; and both PL3 and 109d6 bind to the DRβI chain from DR9 cells (Table 2).

Table 2. Expression of Two DRw53-Related Determinants on DRβI and DRβIV Chains[a]

Serol. allo-DR	T cell Dw	2-D gel pattern DRßI	Serol. allo-DRw53	2-D gel pattern DRßIV	mAb binding (2-D gel patterns): DRßI chain PL3	DRßI chain 109d6	DRßIV chain PL3	DRßIV chain 109d6
DR4	Dw4	ßI-4.1(4*)	DRw53	ßIV-1	-	-	+	+
DR4	Dw10	ßI-4.3	DRw53	ßIV-1	-	-	+	+
DR4	Dw13	ßI-4.2(13*)	DRw53	ßIV-1	-	-	+	+
DR4	DwKT2	ßI-4.2(KT2*)	DRw53	ßIV-1	-	-	+	+
DR4	Dw14	ßI-4.1(14*)	DRw53	ßIV-1	-	-	+	+
DR4	Dw15	ßI-4.4	DRw53	ßIV-1	-	-	+	+
DR7	Dw17	ßI-7	DRw53	ßIV-1	+	-	+	+
DR7	Dw11	ßI-7	DRw53	-	+	-	(not expressed)	
DR9	Dw23	ßI-9	DRw53	ßIV-1	+	+	+	+

[a]Expression of the DRw53 epitopes, PL3 and 109d6, on the DRβI chain or DRβIV chain of each haplotypes was determined by analysis of the 2-D gel patterns found by immunoprecipitation with each antibody. The serologic DR allotyping and T-cell Dw nomenclature (5) for these haplotypes is based on the 10th Workshop studies of the core cell lines (6,7). Asterisks indicate the 2-D gel patterns for each haplotype based on Workshop analysis of the core cell lines (Table 1). DRβI-4.1 from Dw4 cells cannot be distinguished by 2-D gels from DRβI-4.1 from Dw14 cells, and DRβI-4.2 from Dw13 cells cannot be distinguished by 2-D gels from DRβI-4.2 from DwKT2 cells. No DRβI-4.4 (Dw15) cell lines were included in the core cell line panel; however, this haplotype was analyzed previously (3).

Discussion

Four DRβ genes have been characterized with genomic clones isolated from a DR4 positive cell line. These include the DRB1 gene, encoding the DRβI chain; the DRB4 gene, encoding the DRβIV chain; and two pseudogenes, DRβΨI and DRβΨII (8). Southern blot analysis suggests that DR7 cells have the same DRβ genomic organization, with two expressed genes and two pseudogenes (9). Isolation of DRβ cDNA clones from a variety of DR4, DR7, and DR9 cells demonstrated a high degree of polymorphism in the expressed DRB1 gene, but no polymorphism in the expressed DRB4 gene (10).

Previous studies using the murine monoclonal antibodies PL3 and 109d6, recognizing distinct DRw53-related determinants, have demonstrated the molecular complexity of the DRw53 specificity (3,4). In DR4 positive cells both mAbs bind to the nonpolymorphic DRβIV chain and neither mAb binds to the DRβI chain, whereas in DR7 cells PL3, but not 109d6, also binds to the DRβI chain (3). The present Workshop study confirms and extends the previous analysis by demonstrating the binding of both PL3 and 109d6 to the DRβI chain of the DR9 haplotype. It is now possible to suggest polymorphic sequences on the DRβIV chain that are shared with specific DRβI chains and may represent the epitopes responsible for the distinct binding specificities of these antibodies. For example, glutamine is found at position 4 on every chain that binds PL3: DRβI-7, DRβI-9, and DRβIV. Similarly, a group of amino acids shared between DRβIV and DRβI-9, including (arg-arg-//-glu) at polymorphic positions (70-71-//-74), are likely candidates for the 109d6 epitope.

Cells of one DR7 haplotype (DR7,Dw11) fail to express the DRβIV chain, but remain positive with some DRw53 alloantisera and with PL3, the DRw53-related mAb that also binds to a determinant on the DRβI chain of all DR7 cells (3,4). No alternative allelic form of the DRβIV chain has been found, and these cells are completely negative with the DRw53-related mAb 109d6, which is specific for the DRβIV chain in DR7 cells (3,4). This haplotype exhibits strong linkage disequilibrium across the extended haplotype DQw9–DR7–Bw57 and has several unique features, including an RFLP near the DRB1 gene (9) and an apparent recombination event between the DQB1 from a DR9 haplotype and the DQA1 gene from a DR7 haplotype, resulting in the expression of a DQβ-3.3/DQα-3.7 dimer that carries the DQw9 specificity (Table 1). These genetic events do not, however, appear to be responsible for defective DRβIV chain expression, since Southern blot analysis detects no obvious genetic rearrangements in the DRB4 gene (9). However, an abnormal DRB4 mRNA species has been identified (11), suggesting that the molecular basis for defective DRβIV chain expression will be found in the nucleotide sequence of this gene.

Acknowledgments. We gratefully acknowledge the expert technical assistance of Peter Lo (US1ZMI), Yasmin Wasi (NCYMAR), Janet Johnston (US7GSC), and David A. Schwarz (US7GSC). This work was supported, in part, by grants AI22916 and CA22507 (US7KNO), the Arthritis Society of Canada (NCYMAR), JDF 19525 (US1ZMI), NIHRR-240 (US1ZMI), and CNR 86.00094.51 (IT2GAN).

References

1. Lepage V, Grumet FC, Busson M, Colombani J, Ness D, Degos L. Antigen report: HLA-DRw53, in Albert ED, Baur MP, Mayr WR (eds): Histocompatibility Testing 1984. Berlin, Springer-Verlag, 1984, p 203.
2. Mukai R, Suzuki, M, Yabe T, Hamaguchi H, Maeda H. Identification of the MT3 molecule from HLA-DR4, DR7, and w9 homozygous cell lines. J Immunol 1984;33:3211.
3. Knowles RW, Flomenberg N, Horibe K, Winchester R, Radka SF, Dupont B. Complexity of the supertypic HLA-DRw53 specificity: Two distinct epitopes differentially expressed on one or all of the DRβ chains depending on the HLA-DR allotype. J Immunol 1986;137:2618.
4. Matsuyama T, Schwenzer J, Silver J, Winchester R. Structural relationships between the DRβ1 and DRβ2 subunits in DR4, 7 and w9 haplotypes and the DRw53(MT3) specificity. J Immunol 1986;137:934.
5. Bodmer WF, Albert E, Bodmer JG, Dupont B, Mach B, Mayr W, et al. Nomenclature for factors of the HLA system, 1987. In: Dupont B (ed): Immunobiology of HLA, Volume I: Histocompatibility Testing 1987. New York: Springer-Verlag, 1989.
6. Milford EL, Kennedy LJ, Yang SY, Dupont B, Lalouel J-M, Yunis EJ. Serologic characterization of the reference panel of B-lymphoblastoid cell lines for factors of the HLA system. In: Dupont B (ed): Immunobiology of HLA, Volume I: Histocompatibility Testing 1987. New York: Springer-Verlag, 1989.
7. Mickelson E, Reinsmoen N, Robbins FM, Hartzman R, et al. HLA-Dw and HLA-DP typing for the reference panel of B-lymphoblastoid cell lines. In: Dupont B (ed): Immunobiology of HLA, Volume I: Histocompatibility Testing 1987. New York: Springer-Verlag, 1989.
8. Andersson G, Larhammar D, Widmark E, Servenius B, Peterson PA, Rask L. Class II genes of the major histocompatibility complex. Organization and evolutionary relationship of the DRβ genes. J Biol Chem 1987;262:8748.
9. Sutton VR, Knowles RW. Mapping of a restriction fragment length polymorphism associated with defective DR β4 chain expression to the HLA-DRB1 gene. Hum Immunol 1988;22:123.
10. Gregersen PK, Moriuchi T, Karr RW, Obata F, Maccari J, Goldberg D, Winchester RJ, Silver J. Polymorphism of HLA-DR β chains in DR4, -7, and -9 haplotypes: Implications for the mechanisms of variation. Proc Natl Acad Sci USA 1986;83:9149.
11. Knowles RW. Structural polymorphism of the HLA class II alpha and beta chains: Summary of the 10th Workshop 2-D gel analysis. In: Dupont B (ed): Immunobiology of HLA, Volume I: Histocompatibility Testing 1987. New York: Springer-Verlag, 1989.
12. Sutton VR, Gorski J, Knowles RW. An abnormal DRB4 (DRw53) mRNA is produced by DR7 individuals who fail to express the gene product. In: Dupont B (ed): Immunobiology of HLA, Volume I: Histocompatibility Testing 1987. New York: Springer-Verlag, 1989.

DQw1 (DQw5, DQw6): 2-D Gel Patterns

Cecile Kaplan,[1] Marie-Christine Morel,[1] Pierre Hermans,[2] Jean Tkaczuk,[3] Takako Naskatsuji,[4] Nelson Fernandez,[5] Nigel Rust,[6] Christine Carter,[6] Susan Fuggle,[6] Corazon Dejelo,[7] William H. Marshall,[8] Kim Wong,[9] Manfred Ballas,[10] Zacharie Brahmi,[11] Ernette Du Toit,[12] Robert Martell,[12] Laurine Bow,[13] Lydia Kulova,[14] Robert Winchester,[15] Ronald Bontrop,[16] and Dominique Charron[2]

The HLA-DQw1 specificity was the first supertypic specificity recognized and was described at the 7th Workshop (1). Structural differences have been demonstrated between DQ molecules associated with DR1, DR2, and DRw6 (2). These differences have also been distinguished in vitro by CTL lines (3) and clones (4) and monoclonal antibodies such as Genox 3.53 (5).

Results and Discussion

DQβ Chains. Distinct DQβ patterns were found to be associated with the new provisional DQ specificities: DQw5 and DQw6 (Table 1). The newly defined DQw5 antigens (formerly DQw1.1) include the very basic DQβ chains expressed in DR1 cells (DQβ-1.1), DR2,Dw21 cells (DQβ-1.21), and DRw14,Dw9 cells (DQβ-1.9). The β1 domain DNA sequences of the genes encoding the DQβ chains from DQw5 haplotypes is consistent with their very basic isoelectric points compared to the DQw6 associated β chains (9). These DQβ chains were not resolved in the Workshop system and need to be analyzed with a more basic pH gradient (10).

In contrast, several distinct DQβ chains associated with the DQw6 antigens (formerly DQw1.2) were well resolved in the Workshop pH gradient (Fig. 1). These include, from the most basic to the most acidic in isoelectric point: DQβ-1.19, associated with DRw13,Dw19 cell lines; DQβ-1.18, associated with DRw13,Dw18 cell lines; DQβ-1.12, associated with DR2,Dw12 and DRw8,DQw1 cell lines; and DQβ-1.2, associated with DR2,Dw2 cell lines. The relative isoelectric points of the DQβ-1.19, β-1.18, and β-1.2 patterns are consistent with the relative charge of the molecules expected from the amino acid substitutions predicted from the nucleotide sequence of their β1 domains (9). The DQβ-1.12 pattern is more basic than expected based on the β1 domain sequences alone, suggesting that additional β2 domain substitutions have also occurred. The difference found in the DQβ chain pattern between DRw13,Dw19 cell lines (DQβ1.19) and DRw13,Dw18 cell lines (DQβ-1.18) is also associated with distinct DQβ gene RFLP patterns (11).

Analysis of the DQβ-1.12 pattern was somewhat difficult using the Workshop celllines. The DQβ pattern obtained from the DR2,Dw12 Workshop cell line, E4181324, is similar to the pattern found in the DRw8, DQw1 cell line, TAB089. While this DQβ-1.12 pattern has a pI similar to that of the DQβ-1.18 pattern, the DQβ-1.12 pattern contained the distinctive pattern of vertical doublets, reported previously, which may result from differential RNA splicing (12). Sequence data from the β1 exon of the DQβ allele found in TAB089 (DRw8) indicate that it is identical to the DQβ chain of the DR2,Dw12 cell line BGE (13). This sequence predicts ten amino acid substitutions compared to the DQβ-1.2 chain, but approximately the same pI. Two other DQw1 positive cell lines, SPOO10 (DR5) and HID (DR9), had DQβ patterns containing a single spot that migrates near one spot of the DQβ-1.12 pattern. Therefore, they have been called DQβ-1.12*, simply to indicate that they may be related. The Workshop data were not conclusive on this point, however, and the DQβ-1.12* patterns may represent new alleles. This DR9, DQw1 cell line, HID, encodes two DQβ chains: the DQβ-3.3 chain normally associated with the other DR9,DQw9 cell lines, and the DQβ-1.12*-related chain. This may account for the unusual serologic binding patterns with some of the DQw1-related mAbs observed in the Workshop analysis (14). Analysis of the DQβ-1.12 chain pattern is further complicated by more recent evidence, discussed in the DR2 report (15), suggesting that the E4181324 DQβ chain is not identical to that found in the DR2,Dw12 cell line BGE (N. Rust, unpublished observations).

DQα Chains. Four DQα chain patterns have been identified (Fig. 1), from the most basic to the most acidic pI: DQα-1.18, associated with DR2,Dw12; DRw13,Dw18 and DRw8,DQw6 haplotypes; DQα-1.14, associated with the DRw14,Dw9 haplotype; DQα-1.2, associated with DRw13,Dw19; DR2,Dw2; DR2,Dw21; DRw11, DQw1 haplotypes; and DQα-1.1, associated with the DR1 haplotype. The Workshop two-dimensional (2-D) gel system resolved some differences in the DQα chain that have not been identified by analysis of the nucleotide sequence of their β1 domains (9). For example, the DQα chain from a DR14,Dw9 cell line is predicted to have the same amino acid sequence as the DQα chain from DR1 cells, suggesting that additional β2 domain substitutions have also occurred. While the more acidic DQα-1.2 pattern is consistent with the single glu (−) to gln (uncharged) substitution found in the β1 domain, the DQα-1.18 chain sequence only differs from the DQα-1.2 pattern by a tyr (uncharged) to phe (uncharged) substitution and an arg (+) to lys (+) substitution, suggesting that the subtle differences in the pK of these

Participating Laboratories: FRAMUL[1], FRACHA[2], FRAOHA[3], JAPTSU[4], UKIFES[5], UKIMOR[6], US6BRN[7], NCYMAR[8], NCYMRV[9], GERGOL[10], US1IND[11], SAFDUT[12], US5SWE[13], US7SYY[14], US8RJW[15], BENROO[16]

Table 1. 2-D Gel Patterns from DQw1 Cell Lines[a]

DR	CELL #	CELL NAME	Serol. DQ	2-D DQ beta	2-D DQ alpha	2-D DR beta	2-D DR beta	Serol. DR	Serol. DRw52/53	T cell Dw	T cell DRw52
1	9003	KAS116	DQw5	- (1.1*)	1.1	I-1.1	-	DR1		Dw1	
1	9004	JESTHOM	DQw5	- (1.1*)	-	I-1.1	-	DR1		Dw1	
1	9005	HOM2	DQw5	- (1.1*)	1.1	I-1.1	-	DR1		Dw1	
1	9006	WT1OOBIS	DQw5	- (1.1*)	1.1	I-1.1	-	DR1		Dw1	
1	9001	SA	DQw5*	- (1.1*)	1.1	I-1.1	-	DR1		Dw1	
1	9002	MZO7O782	DQw5	- (1.1*)	-	I-1.1	-	DR1		Dw20	
2	9012	WJRO76	DQw5*	- (1.21*)	1.2	a-2.16	b-2.1	DRw16		Dw21	
2	9009	KASO11	DQw5*	- (1.21*)	1.2	a-2.16	b-2.1	DRw16		Dw21	
2	9015	WT24	DQw5	- (1.21*)	-	a-2.16	b-2.1	DRw16		Dw21	
w6	9054	EK	DQw5	- (1.9*)	1.14	I-14.1	III-1	DRw14	DRw52	Dw9	Dw25
w6	9057	TEM	DQw5*	- (1.9*)	1.14	I-14.1	III-1	DRw14	DRw52	Dw9	Dw25
w6	9061	31227ABO	DQw5	- (1.9*)	1.14	I-14.1	III-1	DRw14	DRw52	Dw9	Dw25
w6	9056	KOSE	DQw5+6*	-	1.2+1.14	I-(13+14.1)	III-(1+2)	DRw13+14	DRw52	Dw9+19	Dw25+26
2	9010	AMAI	DQw6	1.2	1.2	a-2.2	(b-*)	DRw15		Dw2	
2	9008	DO2O8915	DQw6	1.2	1.2	a-2.2	b-2.1	DRw15		Dw2	
2	9013	SCHU	DQw6	1.2	1.2	a-2.2	b-2.1	DRw15		Dw2	
2	9014	MGAR	DQw6	1.2	1.2	a-2.2	b-2.1	DRw15		Dw2	
2	9017	WT8	DQw6*	1.2	1.2	a-2.2	b-2.1	DRw15		Dw2	
w6	9055	HO3O1	DQw6	1.19	1.2	I-13	III-2	DRw13	DRw52	Dw"new"	Dw26
w6	9053	HOR	DQw6	-	1.2	I-13	-	DRw13	DRw52	Dw19	Dw26
w6	9059	SLEOO5	DQw6	1.19	1.2	I-13	III-2	DRw13	DRw52	Dw19	Dw26
w6	9063	WT47	DQw6	1.19	1.2	I-13	III-2	DRw13	DRw52	Dw19	Dw26
w6	9058	OMW	DQw6	1.18	1.18	I-13	III-3	DRw13	DRw52	Dw18	Dw24
w6	9062	WDV	DQw6	1.18	1.18	I-13	III-3	DRw13	DRw52	Dw18	Dw24
w6	9065	HHKB	DQw6	1.18	1.18	I-13	III-3	DRw13	DRw52	Dw18	Dw24
w6	9060	CB6B	DQw6*	1.18	1.18	I-13	III-1	DRw13	DRw52	Dw18	Dw25
2	9011	E4181324	DQw6	1.12	1.18	a-2.12	b-2.1	DRw15		Dw12	
w8	9066	TABO89	DQw6	1.12	1.18	I-8.1	-	DRw8	DRw52	Dw8.3(T)	
5	9036	SPOO1O	DQw1*	1.12*	1.2	I-11.1	III-1	DRw11	DRw52	DB2	Dw25
w9	9074	HID	DQw9+w1*	3.3+1.12*	3.1	I-9	IV-1	DR9	DRw53	Dw23	

[a]DR, DQ, and Dw nomenclature (6) is based on the 10th Workshop studies of these cell lines (7,8) except where indicated (*). DQβ-(1.1*), DQβ-(1.21*), and DQβ-(1.9*) patterns were not obtained in this study. The DQβ-1.12* pattern appears to be similar to the DQβ-1.12 pattern, but may be distinct.

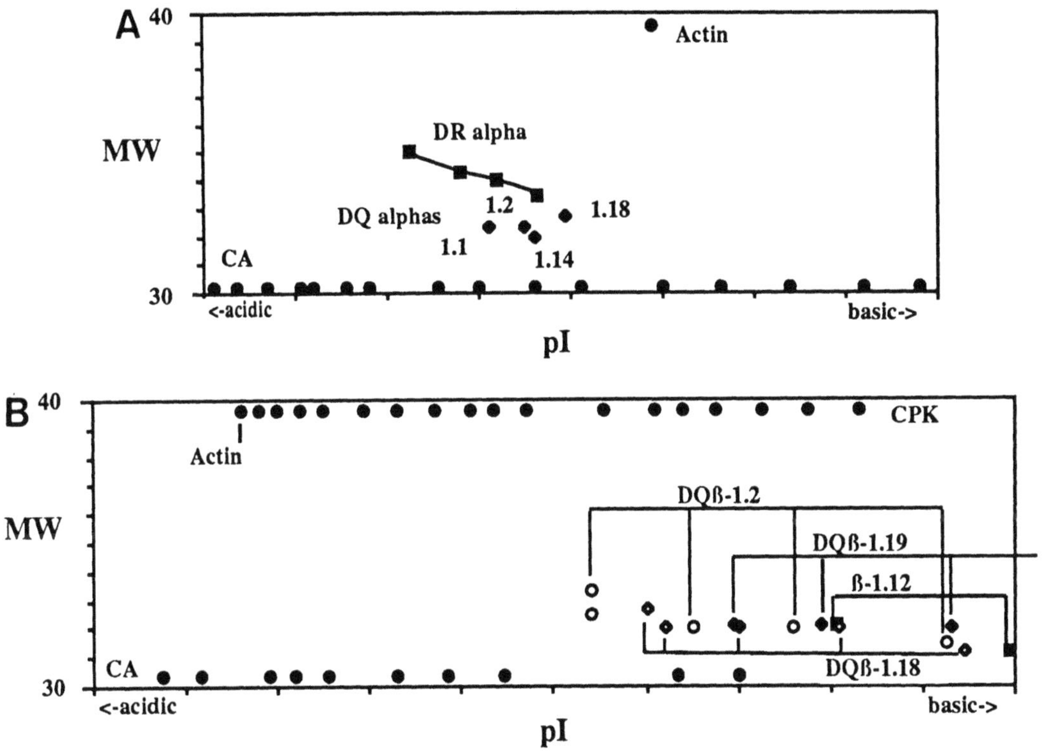

Figure 1. (*A*) Diagram of the 2-D gel patterns for the DQα chains isolated from DQw1 cell lines. DQα chains were labeled with ^{35}S-methionine and isolated by immunoprecipitation. Each sample was mixed with ^{14}C-labeled carbamylated protein standards (CA and CPK) and separated by 2-D gel electrophoresis using the standardized Workshop technique with the α chain mixture of ampholytes (80% pH 4 to 6, 20% pH 3.5 to 10) and detected by autoradiography. The major spots of the DQα-1.1, DQα-1.2, DQα-1.14, and DQα-1.18 chain patterns are indicated (♦), relative to the spots of the DRα chain (■), and the CA standard proteins and actin (●). (*B*) Diagram of the 2-D gel patterns for the DQβ chains isolated from DQw1 cell lines. DQβ chains were labeled with ^{35}S-methionine and isolated by immunoprecipitation. Each sample was mixed with ^{14}C-labeled carbamylated protein standards (CA and CPK) and separated by 2-D gel electrophoresis using the standardized Workshop technique with the β chain mixture of ampholytes (80% pH 6 to 8, 20% pH 3.5 to 10) and detected by autoradiography. Spots of the DQβ-1.2 chain pattern (○), The DQβ-1.12 chain pattern (■), the DQβ-1.18 chain pattern (◇), and the DQβ-1.19 chain pattern (♦) are indicated, relative to the spots of the CA and CPK standard proteins and actin (●).

amino acid side groups or additional amino acid substitutions outside the β1 domain are being detected in the isoelectric point of their 2-D gel patterns.

This study provides evidence for structural polymorphisms in both the DQα and DQβ chains of DQw1 positive haplotypes. The DQw5 and DQw6 splits of DQw1, identified by monoclonal antibodies and alloantisera, had been previously characterized, in part, by the 2-D gel patterns of their DQβ chains, whereas some of these patterns had not been resolved prior to this study. Considerable polymorphism in the DQα chains has also been identified by their 2-D gel patterns. Furthermore, the analysis shows that the DQα-1.2 and DQα-1.18 chains can associate with different DQβ chains, providing further evidence for naturally occurring hybrid DQ molecules, as reviewed elsewhere (16). In heterozygous individuals, the occurrence of hybrid DQ molecules, formed by the association of DQα and DQβ chains from different haplotypes, has been demonstrated (17). The present study indicates that a large number of different hybrid class II molecules could be formed in heterozygous individuals.

Acknowledgments. We gratefully acknowledge the expert technical assistance of Gareth Plant (UKIMOR) and Yasmin Wasi (NCYMAR). This work was supported, in part, by a grant from INSERM C.A.R. #487007 (FRAMUL, FRAOHA, and FRACHR), MRC Programme Grants (UKIMOR), a National Kidney Research Grant (UKIMOR), and the Arthritis Society of Canada (NCYMAR).

References

1. Bodmer JG. Ia serology, in Bodmer WF, et al. (eds): Histocompatibility Testing 1977. Copenhagen, Munksgaard, 1977, p 351.

2. Shakelford DA, Eibl B, Strominger JL. Structural polymorphism of the DC1 light chains from DR1, DR2 and DRw6 positive cell lines. Immunogenetics 1983;18:625.
3. Spits H, Borst J, Giphart M, Coligan J, Terhorst C, DeVries JE. HLA-DC antigens can serve as recognition elements for cytotoxic T lymphocytes. Eur J Immunol 1984;14:299.
4. Fraser JD, Strominger JL. A human cytotoxic lymphocyte subdividing, the DQ1 antigen. Immunogenetics 1987;25: 347.
5. Shakelford DA, Mann DL, van Rood JJ, Ferrara GB, Strominger JL. Human B cell alloantigens DC1, MT1, and LB12 are identical to each other but distinct from the HLA-DR antigen. Proc Natl Acad Sci USA 1981;78:4566.
6. Bodmer WF, Albert E, Bodmer JG, Dupont B, Mach B, Mayr W, et al. Nomenclature for factors of the HLA system, 1987. In: Dupont B (ed): Immunobiology of HLA, Volume I: Histocompatibility Testing 1987. New York: Springer-Verlag, 1989.
7. Milford EL, Kennedy LJ, Yang SY, Dupont B, Lalouel J-M, Yunis EJ. Serologic characterization of the reference panel of B-lymphoblastoid cell lines for factors of the HLA system. In: Dupont B (ed): Immunobiology of HLA, Volume I: Histocompatibility Testing 1987. New York: Springer-Verlag, 1989.
8. Mickelson E, Reinsmoen N, Robbins FM, Hartzman R, et al. HLA-Dw and HLA-DP typing for the reference panel of B-lymphoblastoid cell lines. In: Dupont B (ed): Immunobiology of HLA, Volume I: Histocompatibility Testing 1987. New York: Springer-Verlag, 1989.
9. Gregersen PK, Todd JA, Erlich HA, Long E, Servenius B, Choi E, et al. First-domain sequence diversity of DR and DQ subregion alleles. In: Dupont B (ed): Immunobiology of HLA, Volume I: Histocompatibility Testing 1987. New York: Springer-Verlag, 1989.
10. Tkaczuk J. DR1: 2-D gel patterns. (These proceedings, vol. I, 1989.)
11. Bosch ML, Fei H, Bontrop RE, Gerrets R, Tilanus MGJ, Termijtelen A, et al. Polymorphisms within the HLA-DRw6 haplotype. III. DQα and DQβ polymorphism associated with HLA-D. Hum Immunol 1987;19:91.
12. Tsukamoto K, Yasunami M, Kimura M, Inoko H, Ando A, Hirose T, et al. DQw1 gene from HLA-DR2/Dw12 consists of 6 exons and expresses multiple DQβ polypeptides through alternative splicing, Immunogenetics 1987;25:343.
13. Horn GT, Bugawan TL, Long CM, Erlich HA. Allelic sequence variation of the HLA-DQ loci: Relationship to serology and to insulin-dependent diabetes susceptibility. Proc Natl Acad Sci USA 1988;85:6012.
14. Knowles RW, Adluri V, Kilaru S, Sirotina A, Small TN. Binding patterns of the monoclonal antibodies in the 10th Workshop 2-D gel analysis of class II antigens. In: Dupont B (ed): Immunobiology of HLA, Volume I: Histocompatibility Testing 1987. New York: Springer-Verlag, 1989.
15. Rust N, Carter C, Fuggle S, Sacks S, Hermans P, Charron D, et al. DR2 (DRw15,16): 2-D gel patterns. In: Dupont B (ed): Immunobiology of HLA, Volume I: Histocompatibility Testing 1987. New York: Springer-Verlag, 1989.
16. Charron DJ, Lotteau V, Hermans P, Burroughs D, Faille A, Ju LY, et al. Hybrid HLA class II molecules: Expression and regulation, in Silver J (ed): Molecular Biology of Class II antigens. Boca Raton, FL, CRC Press (in press).
17. Charron DJ, Lotteau V, Turmel P. Hybrid HLA-DC antigens provide molecular evidence for gene trans-complementation. Nature 1984;312:157.

DQw2: 2-D Gel Patterns

Susan Radka,[1] Susan Fuggle,[2] Maria Christina Mazzilli,[3] Christine Carter,[2] Nigel Rust,[2] Robert Winchester,[4] Erin Matthews,[4] Cecile Kaplan,[5] Barbara Nepom,[1,6] Francoise Amesland,[7] Nelson Fernandez,[8] Ernette Du Toit,[9] Stephanie Crafford,[9] Marie-Christine Morel,[5] and Robert W. Knowles[10]

Results

The pH gradient used in this study was not optimal for the two-dimensional (2-D) gel analysis of the DQβ chains from DQw2 cell lines. Therefore, only the most acidic, highly glycosylated form of this extremely basic molecule was observed. Nevertheless, whenever this DQβ spot was found in any of the DR3,DQw2 or DR7,DQw2 cell lines listed in Table 1, its pI and molecular weight were identical relative to the internal standards (Fig. 1). This result is consistent with the published cDNA sequences for DQw2 β chains from both DR3 and DR7 haplotypes, which show that the DR3,DQw2 and DR7,DQw2 haplotypes share the same allele of the DQB1 gene (see Discussion).

In contrast, the α chains were well resolved in the pH gradient used to investigate DQα chain variability. Although the DQβ chains are identical in all DQw2 haplotypes, the DR3 and DR7 cell lines carry distinct DQα chains. Those carried by DR3,DQw2 cells (DQα-2) possess the most basic pI of the DQα chains, whereas those carried by DR7 cells (DQα-3.7) are considerably more acidic (Fig. 1). The pIs of the DQα chains from DR3 cells are identical to those from DR5 cell lines (both DRw11 and DRw12), although the DR5 cells express the DQw7 subtype of the DQw3 serologic specificity associated with their DQβ-3.1 chain. The pI of the DQα chains carried by DR7,DQw2 cells (DQα-3.7) is identical to that of DR7,DQw3 cells, indicating that these two DR7 positive haplotypes carry an identical DQA1 allele.

Participating Laboratories: US7GSC[1], UKIMOR[2], ITGAN[3], US8RJW[4], FRAMUL[5], US7VMC[6], FRADDC[7], UKIFES[8], SAFDUT[9], US7KNO[10]

Immunobiology of HLA
Volume I

Table 1. 2-D Gel Patterns from DQw2 Cell Lines[a]

DR	CELL #	CELL NAME	Serol. DQ	2-D DQ beta	2-D DQ alpha	2-D DR beta	2-D DR beta	Serol. DR	Serol. DRw52/53	T cell Dw	T cell DRw52
3	9018	LOO81785	DQw2	2	-	I-3	III-1	DRw17*	DRw52	Dw3	Dw25
3	9019	DUCAF	DQw2	2	2	I-3	III-1	DRw17*	DRw52	Dw3	Dw25
3	9020	QBL	DQw2	-	-	I-3	III-1	DRw17*	DRw52	Dw3	Dw25
3	9022	COX	DQw2	-	2	I-3	III-3	DRw17*	DRw52	Dw3	Dw24
3	9023	VAVY	DQw2	2	2	I-3	III-3	DRw17*	DRw52	Dw3	Dw24
7	9050	MOU	DQw2	2	3.7	I-7	IV-1	DR7	DRw53	Dw17	
7	9051	PITOUT	DQw2	2	3.7	I-7	IV-1	DR7	DRw53	Dw17	
7	9049	IBW9	DQw2	2	-	I-7	IV-1	DR7	DRw53	Dw17	
7	9047	PLH	DQw2	2	3.7	I-7	IV-1	DR7	DRw53	DB1	
7	9046	BH	DQw2	2	-	I-7	IV-1	DR7	DRw53	DB1	
7	9048	LBUF	DQw2	2	3.7	I-7	IV-1	DR7	DRw53	DB1	

[a]DR, DQ, and Dw nomenclature (1) is based on the 10th Workshop studies of these cell lines (2,3) except where indicated (*).

Discussion

A direct comparison of the entire 2-D gel pattern of the DQw2 associated DQβ chains from the DR3 and DR7 haplotypes was not possible with the Workshop pH gradient, due to their extremely basic pI. However, a comparison of the published DQβ cDNA sequences obtained from DR3 DQw2 and DR7 DQw2 (4–6) shows them to be identical. Since the most acidic spot of their 2-D gel pattern appears to be identical, it seems clear that these two haplotypes share the same DQB1 allele.

In contrast, the DQα chains carried by each of these haplotypes have distinct pIs. This is supported by additional biochemical studies of DQα polymorphism presented during the Workshop Conference (7) and previously published nucleotide sequence analysis (6, 8–10). The DQα chains carried by DR3 haplotypes are identical in electrophoretic mobility (7) and nucleotide sequence (8) to those of DR5 haplotypes (9). The DQα chain carried by DR7,DQw2 cells is identical to that of DR7,DQw3 haplotypes (10) and is distinct from all other DQα chains so far examined.

The biochemical and sequence data strongly suggest that the DQw2 serologic specificity is a DQβ polymorphism and have some implications for linkage relationships among alleles at the DRB1, DQA1, and DQB1 loci. In the DQw2 haplotypes, recombination seems to have occurred between DQA1 and DQB1 genes. This results in a stronger association between the DRB1 alleles (which carry the DR specificity) and the DQA1 alleles than between DQB1 and DRB1 alleles. These results are consistent with previously published molecular data, which have determined the gene order within the class II region to be centromere-DQB1-DQA1-DRB1 (11).

Acknowledgments. We gratefully acknowledge the expert technical assistance of Janet Johnson (US7GSC), David Schwarz (US7GSC), and Bernadette Kienezle (US7KNO). This work was supported, in part, by an MRC Programme Grant (UKIMOR), a National Kidney

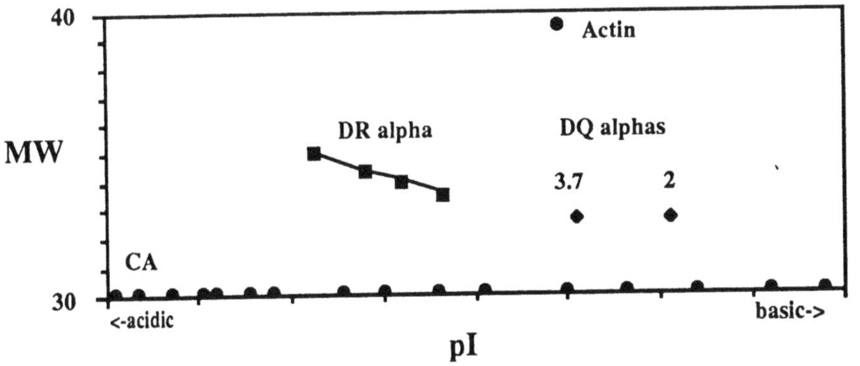

Figure 1. Diagram of the 2-D gel patterns for the DQα chains isolated from DQw2 cell lines. DQα chains were labeled with ^{35}S-methionine and isolated by immunoprecipitation. Each sample was mixed with ^{14}C-labeled carbamylated protein standards (CA and CPK) and separated by 2-D gel electrophoresis using the standardized Workshop technique with the α chain mixture of ampholytes (80% pH 4 to 6, 20% pH 3.5 to 10) and detected by autoradiography. The major spot of the DQα-2 and DQα-3.7 chain patterns are indicated (♦), relative to the spots of the DRα chain (■), the CA standard proteins and actin (●).

Research Grant (UKIMOR), INSERM C.A.R. #487007 (FRAMUL), CNR 86.00094.51 (IT2GAN), and AI22916 (US7KNO).

References

1. Bodmer WF, Albert E, Bodmer JG, Dupont B, Mach B, Mayr W, et al. Nomenclature for factors of the HLA system, 1987. In: Dupont B (ed): Immunobiology of HLA, Volume I: Histocompatibility Testing 1987. New York: Springer-Verlag, 1989.
2. Milford EL, Kennedy LJ, Yang SY, Dupont B, Lalouel J-M, Yunis EJ. Serologic characterization of the reference panel of B-lymphoblastoid cell lines for factors of the HLA system. In: Dupont B (ed): Immunobiology of HLA, Volume I: Histocompatibility Testing 1987. New York: Springer-Verlag, 1989.
3. Mickelson E, Reinsmoen N, Robbins FM, Hartzman R, et al. HLA-Dw and HLA-DP typing for the reference panel of B-lymphoblastoid cell lines. In: Dupont B (ed): Immunobiology of HLA, Volume I: Histocompatibility Testing 1987. New York: Springer-Verlag, 1989.
4. Boss JM, Strominger JL. Cloning and sequencing of the human major histocompatibility complex gene DC3β. Proc Natl Acad Sci USA 1984;81:5199.
5. Song Q-L, Gregersen PK, Karr RW, Silver J. Recombination between DQα and DQβ genes generates human histocompatibility leukocyte antigen class II haplotype diversity. J Immunol 1987;139:2993.
6. Gregersen PK, Todd JA, Erlich HA, Long E, Servenius B, Choi E, et al. First-domain sequence diversity of DR and DQ subregion alleles. In: Dupont B (ed): Immunobiology of HLA, Volume I: Histocompatibility Testing 1987. New York: Springer-Verlag, 1989.
7. Radka SF, Schwarz D, Johnston JV, Nepom BS. Biochemical analysis of DQα polymorphisms defined by a monoclonal antibody. In: Dupont B (ed): Immunobiology of HLA, Volume II: Immunogenetics and Histocompatibility. New York: Springer-Verlag, 1989.
8. Schenning L, Larhammar D, Bill P, Wiman K, Jonsson A-K, Rask L, et al. Both α and β chains of HLA-DC class II histocompatibility antigens display extensive polymorphism in their amino-terminal domains. EMBO J 1984;3:447.
9. Schiffenbauer J, Didier DK, Klearman M, Rice K, Shuman S, Tieber VL, et al. Complete sequence of the DQα and DQβ cDNA from a DR5/DQw3 cell line. J Immunol 1987;139:228.
10. Moriuchi J, Moriuchi T, Silver J. Nucleotide sequence of an HLA-DQα chain derived from a DRw9 cell line: Genetic and evolutionary implications. Proc Natl Acad Sci USA 1985;82:3420.
11. Hardy DA, Bell JI, Long EO, Lindsten T, McDevitt HO. Mapping of the class II region of the human major histocompatibility complex by pulsed-field gel electrophoresis. Nature 1986;323:453.

DQw3 (DQw7, DQw8, DQw9): 2-D Gel Patterns

Christina Mazzilli,[1] Enrico Gandini,[1] Jerry Lanchbury,[2] Alejandro Madrigal,[3] Manfred Ballas,[4] Susan Fuggle,[5] Christine Carter,[5] Nigel Rust,[5] Marie-Anne Henry,[6] Pascale Perrier,[6] William H. Marshall,[7] Dimitri Monos,[8] Cora Dejelo,[9] Cecile Kaplan,[10] Marie-Christine Morel,[10] Kim Wong,[11] Pierre Hermans,[12] Robert Winchester,[13] Erin Matthews,[13] Francoise Amesland,[14] Odile Bouhallier,[15] Renee Fauchet,[15] Barbara Nepom,[16,17] Nan Knitter-Jack,[17] Gerald Nepom,[17] Malek Kamoun,[8] Chester Zmijewski,[8] Ronald Bontrop,[18] Robert W. Knowles,[19] and Dominique Charron[12]

Results and Discussion

DQw3 polymorphism has been described previously at the serologic, biochemical, and genomic level (1–8). In the present study, HLA class II α and β chain polymorphism was investigated by two-dimensional (2-D) gel electrophoresis of immunoprecipitated molecules in a panel of 27 HLA-DQw3 lymphoblastoid cell lines (Table 1).

Three different DQβ chain patterns were observed, named DQβ-3.1, DQβ-3.2, and DQβ-3.3 from the most acidic to the most basic side of the gel (Fig. 1). The DQβ-3.1 pattern was observed in all of the DR5 cell lines examined, 3 of 11 DR4 cells, RML (DR2,DQw3), LUY (DRw8,DQw3), and AMALA (DRw14,DQw3). The DQβ-3.2 pattern was observed in the remaining eight DR4 cell lines, whereas the DQβ-3.3 pattern was found in the DR9 cells and in DBB (DR7,DQw3) (Table 1). This DQβ chain polymorphism perfectly overlaps the new DQw3 serologic specificities: DQw7, DQw8, and DQw9.

Five different DQ α chains, named DQα-3.1, DQα-3.2, DQα-3.7, DQα-3.8, and DQα-2, were observed to pair with the three DQw3 β chains. Seven DQ dimers have been observed: DQα-3.1 has been found with DQβ-3.1 in some DR4 haplotypes, DQβ-3.3 in DR9 haplotypes, and DQβ-3.2 in SAVC, a DR4 cell line; DQα-3.2 is found associated with DQβ-3.2 in DR4 haplotypes; DQα-3.7 has been found associated with DQβ-3.3 in the DR7,DQw3 cell line DBB; DQα-3.8 has been found associated with DQβ-3.1 in the DR8,DQw3 cell line LUY; and DQα-2 is associated with DQβ-3.1 in

---

Participating Laboratories: IT2GAN[1], UKIWEL[2], UKIBOD[3], GERGOL[4], UKIMOR[5], FRAPRR[6], NCYMAR[7], US1ZMI[8], US6BRN[9], FRAMUL[10], NCYMRV[11], FRACHA[12], US8RJW[13], FRADDC[14], FRAFAU[15], US7GSC[16], US7VMC[17], BENROO[18], US7KNO[19]

Table 1. 2-D Gel Patterns from DQw3 Cells[a]

DR	CELL #	CELL NAME	Serol. DQ	2-D DQ beta	2-D DQ alpha	2-D DR beta	2-D DR beta	Serol. DR	Serol. DRw52/53	T cell Dw	T cell DRw52
2	9016	RML	DQw7	3.1	2	a-2.16	b-2.1	DRw16		Dw22	
w6	9064	AMALA	DQw7	3.1	2	I-14.2	III-3	DRw14	DRw52	Dw16	Dw24
5	9039	JVM	DQw7	-	-	I-11.2	III-1	DRw11	DRw52	Dw"new"	Dw25
5	9040	BM15	DQw7	3.1	-	I-11.2	III-1	DRw11	DRw52	Dw5	Dw25
5	9042	TISI	DQw7	3.1	2	I-11.2	III-1	DRw11	DRw52	Dw"new"	Dw25
5	9037	SWEIGOO7	DQw7	3.1	2	I-11.2	III-1	DRw11	DRw52	Dw5	Dw25
5	9043	BM21	DQw7	3.1	2	I-11.2	III-1	DRw11	DRw52	Dw5	Dw25
5	9044	BRIP	DQw7	3.1	2	I-11.2	III-1	DRw11	DRw52	Dw5	Dw25
5	9035	JBUSH	DQw7	3.1	2	I-11.2	III-1	DRw11	DRw52	Dw5	Dw25
5	9045	TUBO	DQw7	3.1	-	I-11.1+12	III-1	DRw11+12	DRw52	Dw5	Dw25
5	9038	BM16	DQw7	3.1	2	I-12	III-1	DRw12	DRw52	DB6	Dw25
w8	9070	LUY	DQw7	3.1	3.8	I-8.1	-	DRw8	DRw52	Dw8.3(L)	
4	9025	DEU	DQw7	3.1	3.1	I-4.1	IV-1	DR4	DRw53	Dw4	
4	9027	PF97387	DQw7	3.1	3.1	I-4.1	IV-1	DR4	DRw53	Dw4	
4	9030	JHAF	DQw7	3.1	3.1	I-4.2	IV-1	DR4	DRw53	Dw13	
4	9034	SAVC	DQw8	3.2	3.1	I-4.1	IV-1	DR4	DRw53	Dw4	
4	9028	PE117	DQw8	3.2	3.2	I-4.1	IV-1	DR4	DRw53	Dw14	
4	9029	WT51	DQw8	3.2	3.2	I-4.1	IV-1	DR4	DRw53	Dw4	
4	9031	BOLETH	DQw8	3.2	3.2	I-4.1	IV-1	DR4	DRw53	Dw4	
4	9032	BSM	DQw8	3.2	3.2	I-4.1	IV-1	DR4	DRw53	Dw4	
4	9033	BM14	DQw8	3.2	3.2	I-4.1	IV-1	DR4	DRw53	Dw4	
4	9024	KT17	DQw8	3.2	3.2	I-4.2	IV-1	DR4	DRw53	KT2	
4	9026	YAR	DQw8	3.2	3.2	I-4.3	IV-1	DR4	DRw53	Dw10	
7	9052	DBB	DQw9	3.3	3.7	I-7	-	DR7	DRw53	Dw11	
w9	9073	KT12	DQw9	3.3	3.1	I-9	IV-1	DR9	DRw53	Dw23	
w9	9075	DKB	DQw9	3.3	3.1	I-9	IV-1	DR9	DRw53	Dw23	
w9	9074	HID	DQw9+w1*	3.3+1.12*	3.1	I-9	IV-1	DR9	DRw53	Dw23	

[a]DR, DQ, and Dw nomenclature (9) is based on the 10th Workshop studies of these cell lines (10,11) except where indicated (*). The DQβ-1.12* pattern is similar to the DQβ-1.12 pattern of DR2,Dw12 cells, but may be distinct. The DRβ-a- and DRβ-b- patterns have not been assigned to specific DRB genes.

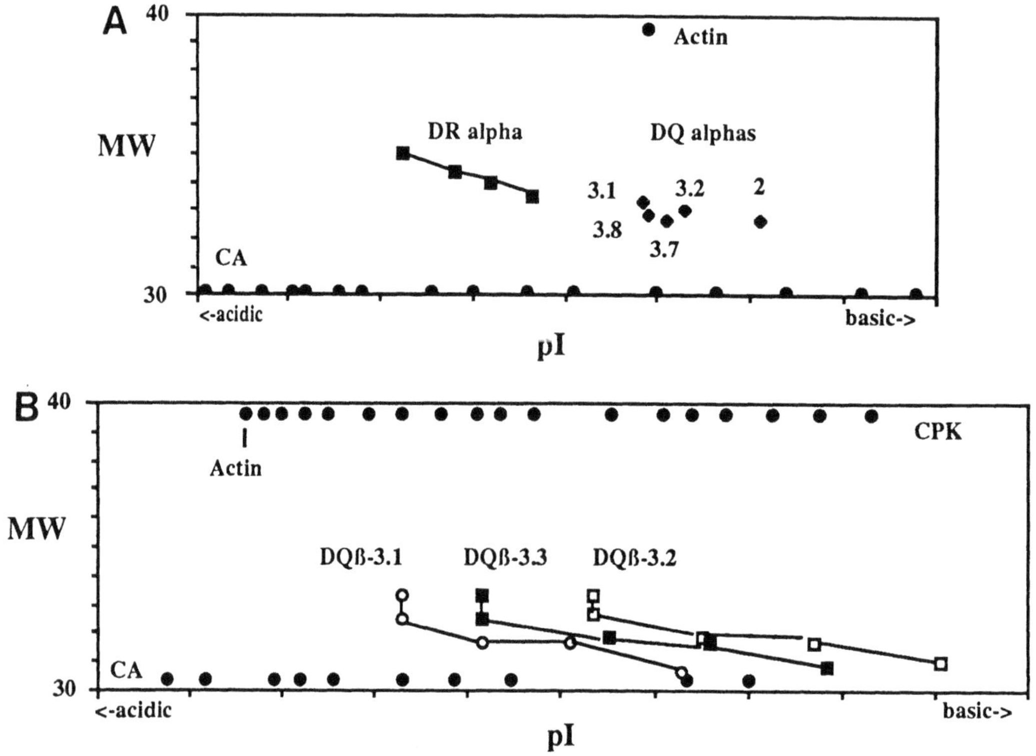

Figure 1. (*A*) Diagram of the 2-D gel patterns for the DQα chains isolated from DQw3 cell lines. DQα chains were labeled with ^{35}S-methionine and isolated by immunoprecipitation. Each sample was mixed with ^{14}C-labeled carbamylated protein standards (CA and CPK) and separated by 2-D gel electrophoresis using the standardized Workshop technique with the α chain mixture of ampholytes (80% pH 4 to 6, 20% pH 3.5 to 10) and detected by autoradiography. The major spot of the DQα-2, DQα-3.1, DQα-3.2, DQα-3.7, and DQα-3.8 chain patterns are indicated (♦), relative to the spots of the DRα chain (■), the CA standard proteins and actin (●). (*B*) Diagram of the 2-D gel patterns for the DQβ chains isolated from DQw3 cell lines. DQβ chains were labeled with ^{35}S-methionine and isolated by immunoprecipitation. Each sample was mixed with ^{14}C-labeled carbamylated protein standards (CA and CPK) and separated by 2-D gel electrophoresis using the standardized Workshop technique with the β chain mixture of ampholytes (80% pH 6 to 8, 20% pH 3.5 to 10) and detected by autoradiography. Spots of the DQβ-3.1 chain pattern (○), the DQβ-3.2 chain pattern (□), and the DQβ-3.3 chain pattern (■) are indicated, relative to the spots of the CA and CPK standard proteins and actin (●).

all DR5 haplotypes, in the DR2,DQw3 cell line RML (DR2), and in the DRw14,DQw3 cell line AMALA.

Although it is difficult to distinguish DQα-3.1 from DQα-3.8 in the Workshop 2-D gel system, the DQα-3.8 chain appears slightly smaller than the DQα-3.1 chain. DNA sequence data have demonstrated that the LUY cell line (12) expresses a DQα chain with an α1 domain sequence that is identical to the DQα chain of other DRw8,DQw4 cell lines, including MADURA (except for the conservative change F for Y at position 25), and very different from the DQα-3.1 chain found in the DR9 and DR4 cells (12, 13). The following discussion is based on the assumption that these α chains are distinct.

From these results, it is clear that the DQw3 serologic specificity is associated with the DQβ chain (Table 1). For example, DR5,DQw7 cells share the same DQα-2 chain with DR3 cells, but the DR3 cells carry a DQβ-2 chain and are DQw2 positive. The DR7,DQw9 cell line DBB shares the same DQα-3.7 chain with the other DR7 haplotype, which carries the DQβ-2 chain and types as DQw2. The DR9,DQw9 cell lines share the same DQα-3.1 chain with the several DR4 haplotypes, which carry the DQβ-3.1 chain and type as DQw7. In the DR4 cells there are two DQα chains, DQα-3.1 and DQα-3.2. Within DR4 cells the linkage disequilibrium between the DQα-3.1/β-3.1 and DQα-3.2/β-3.2 alleles is strong, but not complete, SAVC being the exception. As noted above, the serologic specificity is again associated with the DQβ chain, since SAVC carries the DQβ-3.2 chain with the DQα-3.1 chain and types as DQw8 serologically.

In conclusion, 2-D gel analysis of DQw3 products shows three DQβ alleles, coincident with the serologic specificities, that can associate with five different α chains. Thus, two types of hybrid DQα–β dimer molecules are observed: trans-hybrid molecules in heterozygous individuals (14), and the naturally occurring cis-hybrid molecules seen in this study when rare combinations of DQ α and β chains are found on the same haplotype, as reviewed elsewhere (15). Within the observed DQα–β combinations, however, some appear to be preferentially associated, suggesting DQA1-DQB1 linkage disequilibrium. Linkage disequilibrium also involves DRβ alleles, but is not complete. In some cases DQα is found with different DR haplotypes within the same DQβ specificity, whereas in other cases the same DQα chain can be shared by different DR and DQ serotypes. Only 7 of the 15 possible DQα–β chain dimers have been observed in this study. While this may simply be due to linkage disequilibrium or the fact that only 27 DQw3 cell lines have been tested, it remains possible that certain dimers are "forbidden" (4).

Acknowledgments. We gratefully acknowledge the expert technical assistance of Gareth Plant (UK1MOR), Yasmin Wasi (NCYMAR), David A. Schwarz (US7GSC), Peter Lo (US1ZMI), and Bernadette Kienzle (US7KNO). This work was supported, in part, by grant CNR 86.00094.51 (IT2GAN), a grant from the Arthritis and Rheumatism Council of U.K. (UK1WEL), MRC Programme Grants (UK1MOR), a National Kidney Research Grant (UK1-MOR), the Arthritis Society of Canada (NCYMAR), INSERM C.A.R. #4870007 (FRAMUL and FRACHR), JDF 19525 (US1ZMI), NIHRR-240 (US1ZMI), and AI22916 (US7KNO).

References

1. Schreuder GMT, Maeda H, Koning F, D'Amaro J. TA10 and 2B3, two new alleles in the HLA-DQ region recognized by monoclonal antibodies. Hum Immunol 1986;16:127.
2. Radka SF, Scott RG, Stewart SJ. Molecular complexity of HLA-DQw3: The TA 10 determinant is located on a subset of DQw3 β chains. Hum Immunol 1987;18:287.
3. Mickelson EM, Nepom GT, Nisperos B, Hansen JA. DQw3 variants defined by cloned alloreactive T cells. Hum Immunol 1988;21:63.
4. Bontrop RE, Baas EJ, Otting N, Schreuder GMT, Giphart MJ. Molecular diversity of HLA-DQ: DQ α and β chain isoelectric point differences and their relation to serologically defined HLA-DQ allospecificities. Immunogenetics 1987;25:305.
5. Rodriguez de Cordoba S, Nunez-Roldan A, Winchester R, Marshall P, Carrier C, Mollen N, et al. Molecular characterization by high-resolution isoelectric focusing of the products encoded by the class II region loci of the major histocompatibility complex in humans. I. DR and DQ gene variants. Hum Immunol 1987;10:71–93.
6. Tilanus MGJ, van Eggermond MCJA, van de Bijl M, D'Amaro J, Schreuder GMT, De Vries RRP, et al. An overview of the restriction fragment length polymorphism of the HLA-D region: Its application to individual D-, DR-typing by computerized analyses. Tissue Antigens 1986;28:218.
7. Amar A, Radka SF, Holbeck SL, Kim SJ, Nepom BS, Nelson K, et al. Characterization of specific HLA-DQβ allospecificities by genomic, biochemical, and serological analysis. J Immunol 1987;138:3986.
8. Holbeck SL, Nepom GT. Exon-specific oligonucleotide probes localize HLA-DQβ allelic polymorphisms. Immunogenetics 1986;24:251.
9. Bodmer WF, Albert E, Bodmer JG, Dupont B, Mach B, Mayr W, et al. Nomenclature for factors of the HLA system, 1987. In: Dupont B (ed): Immunobiology of HLA, Volume I: Histocompatibility Testing 1987. New York: Springer-Verlag, 1989.
10. Milford EL, Kennedy LJ, Yang SY, Dupont B, Lalouel J-M, Yunis EJ. Serologic characterization of the reference panel of B-lymphoblastoid cell lines for factors of the HLA system. In: Dupont B (ed): Immunobiology of HLA, Volume I: Histocompatibility Testing 1987. New York: Springer-Verlag, 1989.
11. Mickelson E, Reinsmoen N, Robbins FM, Hartzman R, et al. HLA-Dw and HLA-DP typing for the reference panel of B-lymphoblastoid cell lines. In: Dupont B (ed): Immunobiology of HLA, Volume I: Histocompatibility Testing 1987. New York: Springer-Verlag, 1989.
12. Horn GT, Bugawan TL, Long CM, Erlich HA. Allelic sequence variation of the HLA-DQ loci: Relationship to serology and insulin-dependent diabetes susceptibility. Proc Natl Acad Sci USA 1988;85:6012.
13. Todd JA, Bell JI, McDevitt HO. HLA-DQβ gene contributes to susceptibility and resistance to insulin-dependent diabetes mellitus. Nature 1987;329:599.
14. Charron DJ, Lotteau V, Turmel P. Hybrid HLA-DC antigens provide molecular evidence for gene transcomplementation. Nature 1984;312:157.
15. Charron DJ, Lotteau V, Hermans P, Burroughs D, Faille A, Ju LY, et al. Hybrid HLA class II molecules: Expression and regulation, in Silver J (ed): Molecular biology of class II antigens. Boca Raton, FL, CRC Press (in press).

DQw4 (Wa): 2-D Gel Patterns

Lydia Kulova,[1] Hiroo Maeda,[2] Ronald Bontrop,[3] Nigel Rust,[4] Susan Fuggle,[4] Christine Carter,[4] Ernette Du Toit,[5] Soo Young Yang,[1] and Robert W. Knowles[6]

Results and Discussion

DQw4 (Wa) is a class II specificity defined in the 10th Workshop as a fourth allele at the DQ locus. It is revealed serologically with the mouse monoclonal antibody Hu-46, which reacts with some DR4 and DRw8 haplotypes (1) previously designated as DQ "blank" (2). The DQw4 antigen shows linkage disequilibrium with DR4,Dw15 in Japanese with a gene frequency of 20% and with DRw8/Dw8 in Caucasians (1). Cytotoxic T-cell clones

Participating Laboratories: US7SYY[1], JAPJUJ[2], BENROO[3], UKIMOR[4], SAFDUT[5], US7KNO[6]

Table 1. 2-D Gel Patterns from DQw4 Cell Lines[a]

DR	CELL #	CELL NAME	Serol. DQ	2-D DQ beta	2-D DQ alpha	2-D DR beta	2-D DR beta	Serol. DR	Serol. DRw52/53	T cell Dw	T cell DRw
3	9021	RSH	DQw4	Wa	3.8	I-3	III-3	DRw18*	DRw52	Dw"new"	Dw2
w8	9068	BM9	DQw4	Wa	3.8	I-8.1	-	DRw8	DRw52	Dw8.1	
w8	9069	MADURA	DQw4	Wa	3.8	I-8.1	-	DRw8	DRw52	Dw8.1	
w8	9071	OLGA	DQw4	Wa	3.8	I-8.2	-	DRw8	DRw52	Dw8.2	

[a]DR, DQ,and Dw nomenclature (3) is based on the 10th Workshop studies of these cell lines (4,5) except where indicated (*).

can also be generated that recognize DQw4 ("Wa") in accordance with the Hu-46 reactivity pattern (2).

During the 10th Workshop only a limited variety of haplotypes were available to study the "Wa" antigen. One DR3,DRw18,Dw24 cell line (RSH) and three DRw8,Dw8 cell lines (BM9, MADURA, OLGA), were studied (Table 1). The 2-D gel analysis revealed a specific DQ α–β heterodimer in each of the DQw4 cell lines. The DQβ chain pattern DQβ-Wa contains a series of three to four spots, indicating different stages of intracellular glycosylation (Fig. 1). DNA sequence analysis has demonstrated that the distinct DQβ-Wa chains from DRw8,DQw4 cells and DR3,DQw4 cells are identical in the β1 domain and only differ from the DQβ-Wa chains from DR4,Dw15,DQw4 cells by a singe amino acid substitution in the β1 domain (6–8). This would be expected to be detectable as a shift

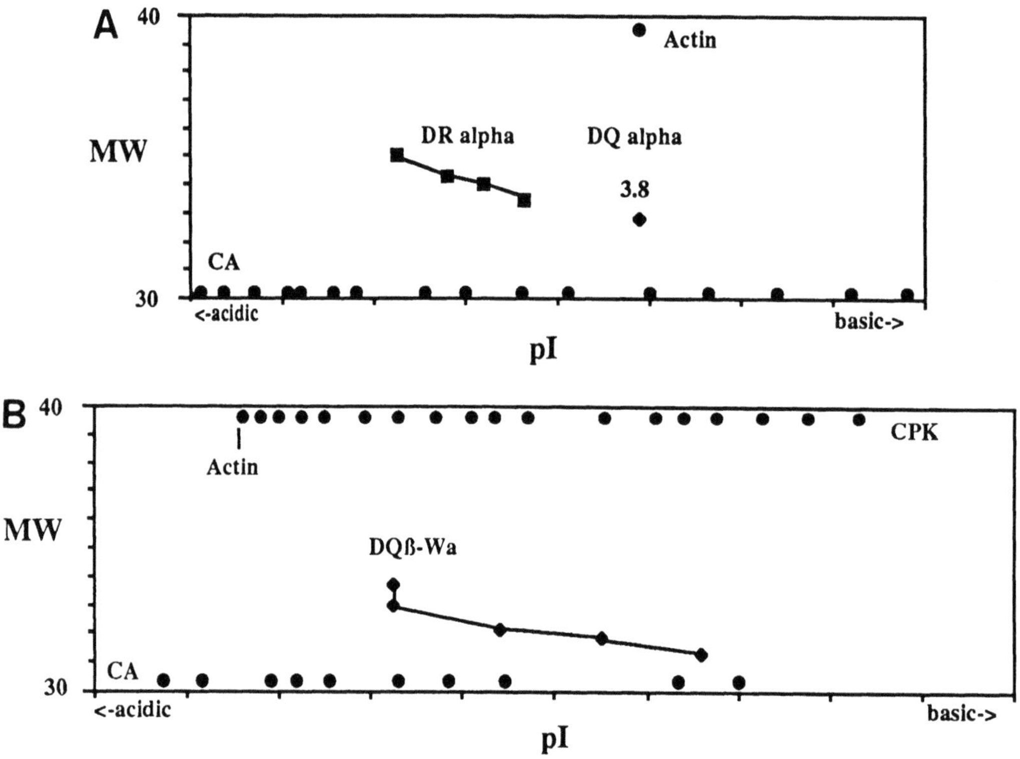

Figure 1. (*A*) Diagram of the 2-D gel pattern for the DQα chain isolated from DQw4 cell lines. DQα chains were labeled with ^{35}S-methionine and isolated by immunoprecipitation. Each sample was mixed with ^{14}C-labeled carbamylated protein standards (CA and CPK) and separated by 2-D gel electrophoresis using the standardized Workshop technique with the α chain mixture of ampholytes (80% pH 4 to 6, 20% pH 3.5 to 10) and detected by autoradiography. The major spot of the DQα-3.8 chain pattern is indicated (♦), relative to the spots of the DRα chain (■), the CA standard proteins and actin (●). (*B*) Diagram of the 2-D gel pattern for the DQβ chain isolated from DQw4 cell lines. DQβ chains were labeled with ^{35}S-methionine and isolated by immunoprecipitation. Each sample was mixed with ^{14}C-labeled carbamylated protein standards (CA and CPK) and separated by 2-D gel electrophoresis using the standardized Workshop technique with the β chain mixture of ampholytes (80% pH 6 to 8, 20% pH 3.5 to 10) and detected by autoradiography. Spots of the DQβ-Wa chain pattern (♦) are indicated, relative to the spots of the CA and CPK standard proteins and actin (●).

in isoelectric point on the two-dimensional (2-D) gels, but no DR4,Dw15 cells were included in the Workshop study. However, heterogeneity within the DQw4 "Wa" positive polypeptides in different ethnic groups was reported previously (1).

The DQα chain associated with DQw4, DQα-3.8, has a molecular weight slightly lower than that of the DQα-3.1 form of the DQw3 α chain seen in some DR4 and all DR9 cell lines. Although it is difficult to distinguish these two α chain patterns, DNA sequence analysis of the α1 domains of the DQα chains from DRw8 cells and the DR3, DQw4 cell line, RSH has demonstrated that they are distinct from the DQα chains found on DR4 and DR9 cells (6–8). In agreement with the identity of the 2-D gel patterns from DRw8,DQw4 and DR3,DQw4 (RSH) in this study, similar DQA1 and DQB1 RFLP patterns have been observed between DRw8,DQw4 cells and other DR3,DQw4 individuals (8).

One Workshop DRw8 cell line, LUY, expresses the DQβ-3.1 chain and is serologically typed as DQw7, but also expresses the DQα-3.8 chain, detected by its 2-D gel patterns as described in the DRw8 report (9). DNA sequence analysis shows it has one conservative amino acid substitution compared the α chain found on another Workshop DRw8,DQw4 cell line, MADURA, and is distinct from the α chain found on DR4,DQw3.1 cells (6). Since this DQα-3.8 chain is found in both DRw8, DQw4 and DRw8,DQw7 individuals, it is clear that the DQw4 specificity is determined by DQβ chain.

Acknowledgments. We are grateful for the technical assistance of Gareth Plant (UKIMOR) and Robert Martell (SAFDUT). This work was supported in part, by MRC Programme Grants (UKIMOR), a National Kidney Research Grant (UKIMOR), CA22507 (US7SYY, US7KNO), and AI22916 (US7KNO).

References

1. Ishikawa N, Kojima H, Nakayama T, Kunikane H, Hawkin S, Fukasawa Y, et al. Detection of a novel HLA-DR specificity; serological and immunochemical analysis by a monoclonal antibody. Immunogenetics 1987;26:143.
2. Baur MP, Neugebauer M, Deppe H, Sigmund M, Luton T, Mayr WR, et al. Population analysis on the basis of deduced haplotypes from random families, in Albert ED, Baur MP, Mayr WR (eds): Histocompatibility Testing 1984. Berlin, Springer-Verlag, 1984, p 333.
3. Bodmer WF, Albert E, Bodmer JG, Dupont B, Mach B, Mayr W, et al. Nomenclature for factors of the HLA system, 1987. In: Dupont B (ed): Immunobiology of HLA, Volume I: Histocompatibility Testing 1987. New York: Springer-Verlag, 1989.
4. Milford EL, Kennedy LJ, Yang SY, Dupont B, Lalouel J-M, Yunis EJ. Serologic characterization of the reference panel of B-lymphoblastoid cell lines for factors of the HLA system. In: Dupont B (ed): Immunobiology of HLA, Volume I: Histocompatibility Testing 1987. New York: Springer-Verlag, 1989.
5. Mickelson E, Reinsmoen N, Robbins FM, Hartzman R, et al. HLA-Dw and HLA-DP typing for the reference panel of B-lymphoblastoid cell lines. In: Dupont B (ed): Immunobiology of HLA, Volume I: Histocompatibility Testing 1987. New York: Springer-Verlag, 1989.
6. Gregersen PK, Todd JA, Erlich HA, Long E, Servenius B, Choi E, et al. First-domain sequence diversity of DR and DQ subregion alleles. In: Dupont B (ed): Immunobiology of HLA, Volume I: Histocompatibility Testing 1987. New York: Springer-Verlag, 1989.
7. Gregersen PK, Shen M, Song Q, Merryman P, Degas S, Seki T, et al. Molecular diversity of HLA-DR4 haplotypes. Proc Natl Acad Sci USA 1986;83:2642.
8. Hurley C, Gregersen P, Steiner N, Bell J, Hartzman R, Nepom G, et al. Polymorphism of the HLA-D region in American blacks: A DR3 haplotype generated by recombination. J Immunol 1988;140:885.
9. Kulova L, Bontrop R, Yang SY, Maeda H. DRw8: 2-D gel patterns. In: Dupont B (ed): Immunobiology of HLA, Volume I: Histocompatibility Testing 1987. New York: Springer-Verlag, 1989.

DPα Chains: 2-D Gel Patterns

Robert W. Knowles[1] and Bernadette Kienzle[1]

The two-dimensional (2-D) gel patterns of the DPα chains for the Workshop core cell lines were examined centrally on behalf of the participants from their submitted autoradiograms. Since the DPβ chain analysis was not successful using the Workshop β chain system, the immunoprecipitates from the DP specific antibodies were not generally run on the α chain gradient. Most laboratories did, however, submit films from immunoprecipitates with the broadly reactive antibodies, including Tü39, which contained DPα chains as part of a more complex pattern including DRα chains and in some cases DQα chains. The DPα chains were also found in precipitates with other antibodies, although this was not examined systematically for all of the submitted films. The purpose of this analysis was to determine the extent of DPα chain polymorphism and whether there were significant associations with the 10th Workshop cellular DP typing of the core cell lines (1).

Results

The DPα 2-D gel chain patterns found for 59 of the 73 cell lines are listed in Table 1 with the DP cellular typing

---

Participating Laboratory: US7KNO[1]

Table 1. DPα Chain 2-D Gel Patterns Found on the Workshop Core Cell Lines and Their Associations with Cellular DP Typing

10W 2-D gel patterns for DP alpha chains (by cell line)

DR TYPE	10W #	CELL LINE	2-D DP alpha (1.1 or 1.2)	10W-DP TYPE
1	9001	SA	1.1+*	4
1	9002	MZ070782	(-)+*	4
1	9003	KAS116	(-)+*	-
1	9004	JESTHOM	1.1+*	4
1	9005	HOM2	1.1+*	-
1	9006	WT1OOBIS	1.1+*	1
2	9008	DO2O8915	1.1	2+4
2	9009	KASO11	1.1	3+4
2	9010	AMAI	1.1	-
2	9011	E4181324	1.1	2+4
2	9012	WJRO76	1.1	2
2	9013	SCHU	1.1	-
2	9014	MGAR	1.1	4
2	9015	WT24	-	3
2	9016	RML	1.1	-
2	9017	WT8	-	2+4
3	9018	LOO81785	1.1	3
3	9019	DUCAF	1.1	-
3	9020	QBL	1.1	2
3	9021	RSH	1.1+1.2	1
3	9022	COX	1.1	3
3	9023	VAVY	1.2	1
4	9024	KT17	1.2	5
4	9025	DEU	-	4
4	9026	YAR	1.1	4
4	9027	PF97387	1.1	4
4	9028	PE117	1.1	4
4	9029	WT51	1.1	2
4	9030	JHAF	1.1	3
4	9031	BOLETH	-	4
4	9032	BSM	1.1	2
4	9033	BM14	1.1	4
4	9034	SAVC	1.2	-
5	9035	JBUSH	1.1	4
5	9036	SPOO1O	1.1	2
5	9037	SWEIGOO7	1.1	-
5	9038	BM16	1.1	2
5	9039	JVM	-	2
5	9040	BM15	-	3
5	9041	JO528239	1.1+1.2	-
5	9042	TISI	1.1	-
5	9043	BM21	1.2	-
5	9044	BRIP	1.1	2
5	9045	TUBO	1.1	2+4
7	9046	BH	-	4
7	9047	PLH	1.1	-
7	9048	LBUF	1.2	-
7	9049	IBW9	-	1
7	9050	MOU	1.1	2
7	9051	PITOUT	1.1	4
7	9052	DBB	1.1	4
w6	9053	HOR	1.1	4
w6	9054	EK	1.1	-
w6	9055	HO3O1	-	5
w6	9056	KOSE	1.1	2+N
w6	9057	TEM	-	4
w6	9058	OMW	1.2	1
w6	9059	SLEOO5	1.1	3
w6	9060	CB6B	1.2	-
w6	9061	31227ABO	1.1	4
w6	9062	WDV	-	4
w6	9063	WT47	-	2
w6	9064	AMALA	1.1	-
w6	9065	HHKB	1.1	4
w8	9066	TABO89	1.2	-
w8	9067	BTB	-	4
w8	9068	BM9	1.1	2
w8	9069	MADURA	-	4
w8	9070	LUY	1.1+1.2	1+4
w8	9071	OLGA	1.1	3
w9	9073	KT12	1.1+1.2	5
w9	9074	HID	1.1+1.2	2+5
w9	9075	DKB	1.1	4

Associations between 2-D gel patterns and DP typing

DR TYPE	10W #	CELL LINE	2-D DP alpha (1.1 or 1.2)	10W-DP TYPE
2	9012	WJRO76	1.1	2
3	9020	QBL	1.1	2
4	9029	WT51	1.1	2
4	9032	BSM	1.1	2
5	9036	SPOO1O	1.1	2
5	9038	BM16	1.1	2
5	9044	BRIP	1.1	2
7	9050	MOU	1.1	2
w8	9068	BM9	1.1	2
2	9008	DO2O8915	1.1	2+4
2	9011	E4181324	1.1	2+4
5	9045	TUBO	1.1	2+4
w6	9056	KOSE	1.1	2+N
3	9018	LOO81785	1.1	3
3	9022	COX	1.1	3
4	9030	JHAF	1.1	3
w6	9059	SLEOO5	1.1	3
w8	9071	OLGA	1.1	3
2	9009	KASO11	1.1	3+4
2	9014	MGAR	1.1	4
4	9026	YAR	1.1	4
4	9027	PF97387	1.1	4
4	9028	PE117	1.1	4
4	9033	BM14	1.1	4
5	9035	JBUSH	1.1	4
7	9051	PITOUT	1.1	4
7	9052	DBB	1.1	4
w6	9053	HOR	1.1	4
w6	9061	31227ABO	1.1	4
w6	9065	HHKB	1.1	4
w9	9075	DKB	1.1	4
3	9021	RSH	1.1+1.2	1
w8	9070	LUY	1.1+1.2	1+4
w9	9074	HID	1.1+1.2	2+5
w9	9073	KT12	1.1+1.2	5
3	9023	VAVY	1.2	1
w6	9058	OMW	1.2	1
4	9024	KT17	1.2	5
2	9010	AMAI	1.1	-
2	9013	SCHU	1.1	-
2	9016	RML	1.1	-
3	9019	DUCAF	1.1	-
5	9037	SWEIGOO7	1.1	-
5	9042	TISI	1.1	-
7	9047	PLH	1.1	-
w6	9054	EK	1.1	-
w6	9064	AMALA	1.1	-
5	9041	JO528239	1.1+1.2	-
4	9034	SAVC	1.2	-
5	9043	BM21	1.2	-
7	9048	LBUF	1.2	-
w6	9060	CB6B	1.2	-
w8	9066	TABO89	1.2	-
1	9006	WT1OOBIS	1.1+*	1
1	9001	SA	1.1+*	4
1	9004	JESTHOM	1.1+*	4
1	9005	HOM2	1.1+*	-
1	9002	MZ070782	(-)+*	4
1	9003	KAS116	(-)+*	-
7	9049	IBW9	-	1
5	9039	JVM	-	2
w6	9063	WT47	-	2
2	9015	WT24	-	3
5	9040	BM15	-	3
4	9025	DEU	-	4
4	9031	BOLETH	-	4
7	9046	BH	-	4
w6	9057	TEM	-	4
w6	9062	WDV	-	4
w8	9067	BTB	-	4
w8	9069	MADURA	-	4
w6	9055	HO3O1	-	5
2	9017	WT8	-	2+4

[a]Presence of the 2-D gel pattern, DPα-1.1 or DPα-1.2, is indicated, whereas " – " indicates that no films were available for analysis.

[b]DP typing is based on the Workshop analysis of these cell lines, where DP-blank is indicated as " – " and a new DP type is indicated as "N"; DP6 and DP7 were not assigned (1).

[c]The DR1 cell lines all express DQα-1.1, which comigrates with the DPα-1.2 pattern. These cell lines could only be analyzed for the presence or absence of the DPα-1.1 pattern, as indicated by "1.1+*" or "(−)+*, where the asterisk indicates the presence of DQα-1.1 with or without DPα-1.2.

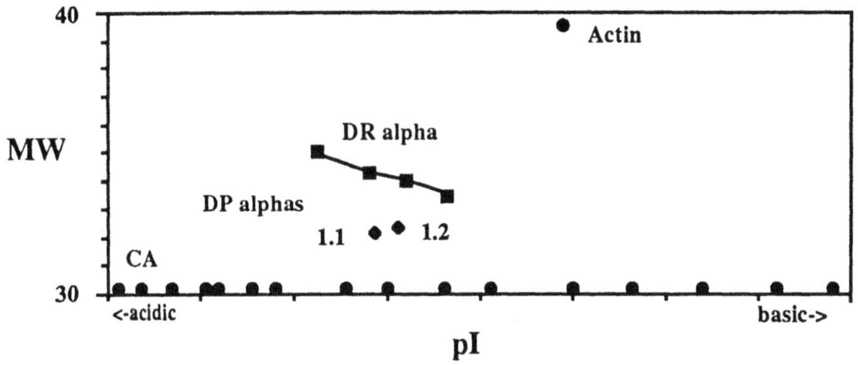

Figure 1. Diagram of the 2-D gel patterns for the DPα chains isolated from Workshop cell lines. DPα chains were labeled with ^{35}S-methionine and isolated by immunoprecipitation. Each sample was mixed with ^{14}C-labeled carbamylated protein standards (CA and CPK) and separated by 2-D gel electrophoresis using the standardized Workshop technique with the α chain mixture of ampholytes (80% pH 4 to 6, 20% pH 3.5 to 10) and detected by autoradiography. The major spots of the DPα-1.1 and DPα-1.2 chain patterns are indicated (♦), relative to the spots of the DRα chain (■), the CA standard proteins and actin (●).

(1) for each cell line. Analysis of DPα chain polymorphism in this study showed two chains that were easily distinguishable (Fig. 1). In the 53 cell lines with clear results, the DPα-1.1 chain pattern was found in 40 cell lines, the DPα-1.2 chain pattern in 8 cell lines, and both patterns in 5 cell lines. Six DR1 cell lines could only be evaluated for the presence or absence of the DPα-1.1 chain, since the DPα-1.2 chain comigrates with DQα-1.1, which was present on all of these DR1 cell lines. For the remaining cell lines no films were available to evaluate the DPα chains.

Table 1 also shows the association between 2-D gel patterns and DP cellular typing. When the presence of the DPα-1.2 chain is compared with the presence of either DP1 or DP5 typing for these cell lines a clear correlation is observed, as shown in Table 2. Expression of the DPα-1.2 chain is negatively associated with DP2, DP3, and DP4 typing when the 17 DP-blank cell lines are removed from the analysis. DP-blank cell lines were heterogeneous with regard to these DPα chain patterns, although further associations with DP6 or DP7 may be found when these specificities are assigned.

Discussion

Two forms of DPα chain were distinguished by the Workshop 2-D gel system. A similar result was reported at the 10th Workshop conference in a study of 94 haplotypes in 137 members of 25 insulin-dependent diabetes mellitus (IDDM) families, performed using a one-dimensional isoelectric focusing technique (2). In that study the frequency of the acidic form (designated here as DPα-1.1) was 85%, in close agreement with the results in this study. Of the five cell lines in the Workshop 2-D gel study expressing both DPα chains, it is likely that this is due to heterozygosity at the DPA1 locus. RSH is heterozygous at the HLA-A locus and is not consanguinous, LUY was heterozygous by DP typing and is not consanguinous, HID is heterozygous at the HLA-C locus and by DP typing, KT12 is heterozygous at both the HLA-A and -B locus and is not consanguinous, and the consanguinity of J0528239 is unknown. There is no evidence from this study to suggest that the second DPA2 gene is expressed.

Most of the correlations found in the present study between 2-D gel patterns and DP typing are consistent with those found in the IDDM family study (2). The acidic form (designated here as DPα-1.1) was also associated with DP3 and DP4, whereas the basic form (here DPα-1.2) was associated with DP5; however, DP1 and DP2 specificities were heterogeneous with regard to the DPα chain patterns. These differences may be due to the relatively small sample number and the different selection criteria for the haplotypes in the two studies. A very strong linkage disequilibrium observed in the IDDM study between "affected" DR7,DQw2 haplotypes and the basic form (here DPα-1.2) was not seen with the "unaffected" DR7,DQw2 haplotypes (2). This association is not seen in the present study, more closely reflecting the results expected for "unaffected" haplotypes.

Table 2. Correlation Between the Expression of the DPα-1.2 Chain and DP1 and DP5 Cellular Typing on 10th Workshop Core Cell Lines (Based on the Data in Table 1)

	DPα 2-D Gel Patterns[a]	
	DPα-1.2$^+$	DPα-1.2$^-$
DP(1 or 5)$^+$	7	0
DP(1 or 5)$^-$	0	31
DP-blank	6	9

[a]Workshop core cell lines typed for DPα 2-D gel patterns (either DPα-1.1 or DPα-1.2), including heterozygotes, but excluding cell lines untyped by 2-D gels and DR1 cells (n = 6) because they all contain the DQα-1.1 chain, which comigrates with DPα-1.2.

In conclusion, the Workshop 2-D gel system was successful in the analysis of two DPα chain alleles, which showed clear correlations with DP typing specificities. These correlations suggest that the DQα chains contribute part of the polymorphisms defined by DP typing, DP(1+5) and DP(2+3+4), much as supertypic determinants contribute to DR and DQ serologic specificities. They may also contribute to the DP specificities by only forming dimers with a restricted number of the polymorphic β chains, as suggested in a parallel study (2). These two alleles have also been defined by DNA sequence analysis of the DP α1 domains reported at the Workshop conference (3).

Acknowledgments. The α chain films were contributed by the participating laboratories in the 10th Workshop Biochemistry Study of HLA Class II Antigens. A complete listing of participants may be found in the summary report (4).

References

1. Mickelson E, Reinsmoen N, Robbins FM, Hartzman R, et al. HLA-Dw and HLA-DP typing for the reference panel of B-lymphoblastoid cell lines. In: Dupont B (ed): Immunobiology of HLA, Volume I: Histocompatibility Testing 1987. New York: Springer-Verlag, 1989.
2. Rodriguez de Cordoba S, Marshall P, Ginsberg-Fellner F, Rubinstein P. Biochemical polymorphism of the DP alpha and DP beta chains. In: Dupont B (ed): Immunobiology of HLA, Volume II: Immunogenetics and Histocompatibility. New York: Springer-Verlag, 1989.
3. Bugawan TL, Horn GT, Hansen JA, Mickelson EM, Angelini G, Ferrara GB, et al. Analysis of HLA-DP allelic sequence polymorphism using the in vitro enzymatic amplification of DP alpha and DP beta loci. In: Dupont B (ed): Immunobiology of HLA, Volume II: Immunogenetics and Histocompatibility. New York: Springer-Verlag, 1989.
4. Knowles RW. Structural polymorphism of the HLA class II α and β chains: Summary of the 10th Workshop 2-D gel analysis. In: Dupont B (ed): Immunobiology of HLA, Volume I: Histocompatibility Testing 1987. New York: Springer-Verlag, 1989.

Binding Patterns of the Monoclonal Antibodies in the 10th Workshop 2-D Gel Analysis of Class II Antigens

R.W. Knowles,[1] V. Adluri,[1] S. Kilaru,[1] A. Sirotina,[2] and T.N. Small[2]

The monoclonal antibodies used for the Workshop two-dimensional (2-D) gel study were selected to satisfy several criteria. In order to use them for the 2-D gel study and the study of alloreactive T-cell clones, mAbs with demonstrated ability to immunoprecipitate class II molecules and block alloreactive T-cell clones were given highest priority. Additional antibodies were selected that had unique polymorphic specificities or unique α or β chain specificity. A listing of antibodies with the laboratory of origin is given in Table 1. Their reported specificity at the time of submission is indicated, but in many cases the results of serologic and biochemical analysis during this Workshop 2-D gel study suggested that they are broader or more limited than was expected (as indicated by the asterisk). Additional polymorphic antibodies were submitted, but these preparations were found to have weak reactivity in the serologic screening and were not analyzed further. In order to set the priority for choosing mAbs to use for immunoprecipitation in the 2-D gel study, they were examined for binding to the Workshop core cell lines using a quantitative indirect immunofluorescence assay.

Participating Laboratories: US7KNO[1], US7NFL[2]

Results and Discussion

Cell lines were tested with a 1:100 dilution of ascites fluid of the antibodies listed in Table 1, in parallel with one negative control mAb (H65, anti-CD5), no mAb (wash buffer, PBS with FCS, without any mAb), and one positive control mAb (W6/32, anti-HLA-A,B,C). Binding was measured by a standard indirect immunofluorescence technique using FITC-conjugated goat antimouse IgG (TAGO, Burlingame, CA, USA) at a saturating concentration and analyzed using the Coulter EPICS-C. The photomultiplier gain and background threshold settings were selected for each cell line to maximize the intensity of the positive control sample, W6/32, while reducing the intensity of the negative control samples, in order to obtain patterns with less than 10% of the cells above the threshold value. Since W6/32 recognizes the HLA-A,B,C antigens, it was strongly bound to these B-cell lines. A maximum value was expected. In some cases, however, the W6/32 pattern was weaker than expected. This generally occurred when the background binding of the FITC-goat anti-mouse Ig reagent was high in the negative control samples, generally due to a relatively high number of dead cells for a particular cell line. This required turning down the photomultiplier gain setting to keep the background below 10% in the negative con-

Table 1. Monoclonal Antibodies Submitted for the Workshop 2-D Gel Study of Class II Antigens

mAb Name	Specificity	REGLAB
TAL-1B5	BROAD (alphas)	UKIBOD
Tü 39	BROAD	A. Ziegler
2D6	BROAD	US7POL
GSP 4.1	.DR	US7GSC
PL8	DR	US7KNO
L243	DR	Becton-Dickinson
1a3	DQ	US8RJW
Tü 22	DQ	A. Ziegler
B7/21	DP	Becton-Dickinson
PL15	DP	US7KNO
UK 4.1	DQwl	UKIBOD
NDS 15.38	DQwl*	UKIMOR
77.34	DQwl	NCYFAL
KS6	DQwl	US7SFE
Leu10	DQw1,3	Becton-Dickinson
PLM 13	DQw1,3*	JAPJUJ
GS4AA7	DQw1,3*	US7GSC
KS7	DQw1,3*	US7SFE
17.15	DQw3	US7GSC
1VD12	DQw3	US4STE
PLM 12	DQw3	JAPJUJ
PLM 2	DQw3*	JAPJUJ
W6/32,(+) control	A,B,C	US7KNO (UKIBOD)
C1	DQw2*	GERWAN
GSP 17.1	DQw2,3*	US7GSC
IIB3	DQ*	BENROO
NDS 13	DQw52	UKIMOR
UK 8.1	DRw52	UKIBOD
7.3.19.1	DRw52	BENROO
NDS 10	DRw52	UKIMOR
IIIE3	DRw52,53*	US4STE
PL3	DRw53	US7KNO
CC CL 20	DR 1,4*	US6BRN
JS1	DR 3*	UKIFES
NFLD M1	DR 4*	NCYMAR
16.23	DR 3,13,18*	GERWAN
CHE249.2	DR 5*	FRADDC
GSP88.2	DR 5*	US7GSC
SFR3 DR5	DR 5*	NCYSFR
SFR16-DR7M	DR 7*	NCYSFR
MAD 88	DRw 8*	FRADDC
DP11.1	DP-alpha(polym)	UKIBOD
Tü 35	DR (wk)	A. Ziegler
AB8.25	DR 2,5*(broad)	IT1CEP
3G5	DR 1 (wk)	BENENG
UL65	DR 3 (wk)	GERGOL
AB7AE9	DRw 9 (wk)	NCYFRN

trols. Therefore, if the W6/32 pattern is not 8 for a given cell line, then the intensities of all the monoclonal antibodies for that cell line are relatively lower than would be expected. The results are shown in Table 2. For each cell line, antibodies were tested together on the same day under the same conditions, so the most valid intensity comparisons in these tables can be made vertically, for all of the antibodies binding to one cell line, rather than horizontally, for one antibody across all of the cell lines. Because of the large number of antibodies and cell lines in this study, immunofluorescence for each mAb–cell line combination was only performed once, and therefore the results should not be considered definitive.

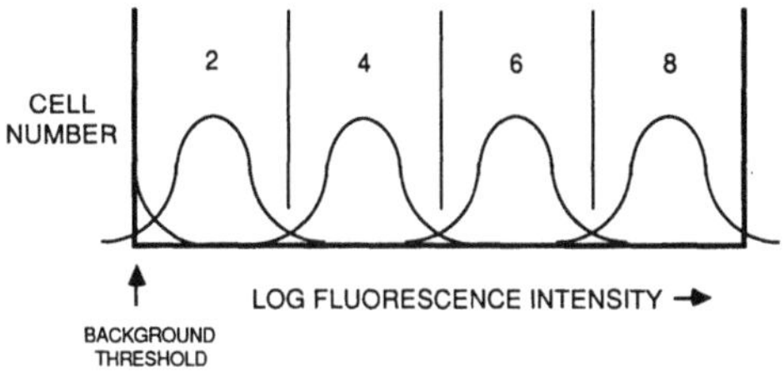

Figure 1. Criteria for scoring the indirect immunofluorescence profiles. Binding patterns of cell number versus log fluorescence intensity were examined and scored as shown, based on the position of the peak channel of fluorescence intensity, by dividing the fluorescence intensity scale into four equal segments. Further details are provided in the text.

Binding patterns of cell number versus fluorescence intensity were examined and scored as shown in Figure 1, based on the position of the peak channel of fluorescence intensity, by dividing the log scale into four equal segments: —=negative; 2=first segment; 4=second segment; 6=third segment; and 8=fourth segment.

If no clear peak was observed, due to a bimodal or more complex distribution, the two segments containing the majority of the cells were used to assign intermediate scores: 1=—/2, 3=2/4, 5=4/6, and 7=6/8. Some antibodies were not tested on all cell lines, as indicated by the blank spaces in Table 2.

The goal of the 2-D gel study was to examine 10 polymorphic antibodies on each cell line in addition to the 10 monomorphic antibodies. Therefore, a priority was established for selection of the polymorphic mAbs to be tested in the 2-D gel study for each cell line (Table 3). The priorities were based on their predominant binding patterns on the panel of core cell lines. These decisions were made in collaboration with the T-cell study in an effort to find antibodies that could be used on more than one DR type to maximize the overlap in the study. The supertypic determinants related to the allodeterminants of the DQ genes (DQw1, w2, w3 and DQw1,3) and the DR genes (DRw52 and DRw53) were given highest priority. A few of the other polymorphic antibodies were also chosen, especially for cell lines when only a small number of supertypic antibodies were found to bind.

Limitations of This Analysis

Because each mAb–cell line combination was only performed once, the results are considered preliminary and were only used to assist in the assignment of priorities for further use. Some false-negative results are indicated by the failure of monomorphic antibodies to bind in some experiments and would therefore be expected with the polymorphic antibodies as well (Table 2). Weak

Table 2a. Binding Patterns of Polymorphic and Monomorphic mAbs Used in the Workshop 2-D Gel Analysis of Class II Antigens[a]

Specificity*	mAb	DR1>					Dw20	Dw2>					Dw12	Dw21>			Dw22	Dw19>				Dw18>			Dw16	Dw9>			Dw9+19	DR3>					DRw18	DB2	DR5>							DRw12	DR7> Dw17>						Dw11	DR9>			DR4> DQw8>				3.2/3.1	DQw7>			w8>		8.3(L)	DRw8>			8.3(T)
	10W CELL LINES	9003	9004	9005	9006	9001	9002	9008	9010	9013	9014	9017	9011	9012	9009	9015	9016	9055	9053	9059	9063	9058	9065	9060	9064	9054	9057	9061	9056	9018	9019	9020	9022	9023	9021	9036	9040	9042	9037	9043	9044	9035	9045	9038	9050	9051	9049	9047	9046	9048	9052	9073	9075	9074	9028	9029	9031	9033	9034	9025	9027	9030	9024	9026	9070	9068	9069	9071	9066
DQw1	UK 4.1	2	2	2	3	2	4	2	-	2	2	2	2	2	3	6	1	6	6	8	4	2	2	2	-	2	2	2	4	-	-	-	2	-	-	2	-	2	2	2	2	-	-	2	-	-	-	2	2	-	-	2	2	-	-	-	2	-	3	2	-	2	2	-	-	2	-	2	4
DQw1*	NDS 15.38	-	-	-	-	-	-	2	2	6	2	6	2	-	2	2	1	6	4	4	2	2	2	-	-	2	-	2	4	-	-	-	2	-	2	2	2	3	2	2	2	-	-	-	-	-	2	2	2	-	2	3	3	2	-	2	3	2	5	2	-	2	2	2	-	-	-	2	4
DQw1	77.34	2	2	2	3	2	2	2	4	2	2	7	2	2	4	2	2	5	5	6	2	2	5	2	-	3	2	2	5	-	-	-	2	-	2	2	-	-	2	2	2	-	-	2	-	-	-	2	2	-	3	2	2	6	2	2	2	6	4	2	-	3	2	3	-	-	2	2	2
DQw1	KS6	-	4	2	-	-	-	-	6	2	2	7	3	2	2	2	2	3	2	4	2	2	6	3	-	2	2	2	2	2	-	-	-	-	-	2	-	2	2	2	2	-	-	-	-	-	-	2	2	-	4	2	2	7	2	2	2	6	3	2	-	2	2	-	-	2	2	2	2
DQw1,3	Leu10									6		7		6	7		6	8	6	8		4			6	7	5	3	6				-					8	8	6	6			8			-	-	-	-		6	6	7			7	6	7	7			6			4			6
DQw1,3*	PLM 13	2	2	2	2	2	6	3	2	6	3	6	2	6	6	6	1	7	6	7	4	4	5	2	-	6	4	3	6	-	-	-	2	-	6	6	-	2	2	2	2	-	-	-	-	-	-	-	-	-	3	5	4	6	2	4	8	5	6	2	-	2	5	3	-	2	2	6	6
DQw1,3*	GS4AA7	3	2	2	6	3	6	3	3	6	5	7	2	6	6	-	1	7	7	8	5	6	5	2	-	6	3	2	6	-	-	-	-	-	6	6	-	2	2	-	2	-	-	-	-	-	-	2	2	-	-	2	2	-	-	-	2	-	3	2	-	2	2	-	-	2	3	6	6
DQw1,3*	KS7	2	3	2	2	2	2	2	6	4	2	8	3	4	6	3	2	3	5	6	2	2	5	4	-	4	2	2	5	-	-	-	-	-	2	2	-	2	-	-	2	-	-	-	-	-	-	2	-	-	4	2	2	7	2	2	2	6	5	-	-	2	2	4	-	4	2	3	5
DQw3	17.15									-		-		-	-		3	-	-	-		2			3	-	-	-	-				2					6	6	4	5			6			2	2	2	3		4	3	6			6	4	6	5			4			-			-
DQw3	1VD12									-		-		-	-		5	-	-	-		2			5	-	-	-	-				-					6	6	4	6			6			-	-	-	-		6	5	6			6	6	6	6			6			-			-
DQw3	PLM 12	-	-	-	-	-	2	-	-	-	-	-	-	-	2	-	6	2	2	2	-	2	2	-	6	2	-	-	2	3	-	-	2	-	-	-	6	7	8	6	6	6	6	7	-	-	-	2	-	-	4	6	5	7	2	5	6	6	7	6	2	6	6	5	6	-	-	-	2
DQw3*	PLM 2	-	-	-	-	-	-	-	-	-	-	-	-	2	2	-	6	2	2	2	-	2	-	-	6	2	-	2	-	4	-	-	2	-	-	-	6	7	8	6	6	5	6	8	-	-	-	-	-	-	-	2	2	-	-	2	-	-	2	6	2	6	2	-	5	-	-	-	2
DQw2*	C1	2	-	-	-	2	2	2	-	8	3	-	-	4	5	2	2	6	6	6	4	4	2	-	2	3	2	2	6	-	2	2	6	2	2	2	2	4	3	3	4	-	-	2	2	3	5	6	6	5	2	5	5	-	-	2	2	-	6	4	-	2	4	2	-	2	-	2	6
DQw2,3*	GSP 17.1	-	-	-	-	-	-	-	-	2	-	-	-	2	2	-	2	2	3	3	-	2	-	-	2	2	-	-	2	2	2	2	4	2	2	3	2	6	6	4	2	2		6	2	2	2	2	2	2	2	2	2	2	-	2	3	2	2	2	-	2	2	2	2	2	2	4	2
DQ*	IIB3	2	-	-	2	2	2	4	4	6	6	7	2	3	6	2	-	8	7	8	5	6	6	2	-	2	2	2	6	-	-	-	2	-	4	3	-	2	2	2	2	-	-	2	-	-	-	2	2	-	4	6	6	6	2	5	7	6	7	2	-	2	6	5	-	2	2	4	6
DRw52	NDS 13	2	-	-	-	-	-	2	2	2	2	2	-	4	6	4	2	8	7	8	5	6	5	4	6	8	7	7	7	4	6	6	7	5	7	7	6	8	8	8	8	6	6	7	2	4	6	6	6	2	2	2	2	-	-	2	3	3	5	3	-	2	2	2	6	6	5	6	7
DRw52	UK 8.1	-	-	-	-	-	2	2	-	2	2	-	-	4	3	2	-	8	8	8	5	7	6	2	6	7	6	6	7	4	6	4	8	3	6	6	6	8	8	7	7	5	6	2	-	-	-	2	-	-	-	2	2	-	-	-	-	-	2	2	-	2	2	-	-	2	-	-	2
DRw52	7.3.19.1	2	-	-	2	2	2	2	2	3	2	2	-	5	6	4	-	8	8	8	6	6	5	3	6	7	6	6	7	3	6	6	8	6	8	6	5	7	6	-	6	4	6	7	-	-	2	2	2	-	2	4	3	-	2	2	2	-	6	2	-	2	3	2	-	-	-	3	5
DRw52	NDS 10	2	-	-	2	2	2	-	2	2	-	-	-	2	3	2	2	5	5	6	2	2	2	3	3	8	6	7	6	3	4	5	4	2	2	7	5	7	7	6	6	4	6	6	-	-	-	2	2	-	-	2	2	-	-	2	2	-	6	2	-	2	2	-	-	-	-	2	2
DRw53+*	IIE3	2	2	-	2	2	3	-	-	2	-	-	2	2	2	-	2	4	2	4	2	2	-	-	2	8	6	6	6	-	2	3	6	2	6	3	2	4	4	3	4	-	-	2	3	6	6	7	3	6	-	8	7	8	2	3	5	4	6	4	2	4	5	2	-	2	-	2	4
DRw53	PL3	-	-	-	-	-	-	-	-	2	-	-	-	2	2	-	-	2	2	2	-	-	-	-	-	2	-	-	2	-	-	-	-	-	-	-	-	2	2	2	2	-	-	-	6	4	4	6	6	8	5	6	5	8	2	2	2	2	3	2	2	5	2	2	-	-	-	2	-
DR 1,4*	CC CL 20	2	3	2	2	2	4	-	2	2	2	-	-	2	5	2	1	6	6	6	2	4	2	-	4	4	2	3	6	-	-	2	6	2	-	3	2	4	2	4	4	2	2	3	-	-	-	2	2	-	-	2	2	-	2	3	3	2	6	4	-	2	3	2	-	-	-	2	2
DR 3*	JS1	2	2	2	2	2	5	2	2	3	2	-	-	3	6	3	2	6	6	6	3	5	2	2	6	6	4	6	6	2	6	6	7	6	8	6	2	6	6	5	6	2	2	5	-	2	2	2	2	-	2	2	2	2	2	4	2	6	6	6	2	2	5	2	2	2	2	2	4
DR 4*	NFLD M1	2	2	-	2	2	2	2	2	4	2	4	2	4	6	-	2	6	6	8	2	2	2	2	2	6	6	4	5	2	2	2	4	2	2	6	4	6	6	2	4	2	2	4	-	-	-	2	2	-	-	2	2	-	2	3	6	7	8	6	2	4	7	2	2	5	2	4	6
DR3,13,w18*	16.23									-		-		-	-		-	-	-	-		3			2	-	-	-	-	5	5	3	6	5	6			-	-	-	-			-			-	-	-	-		-	-	-			-	-	-	-			-			-			-
DR5*	CHE249.2	-	-	-	-	-	-	-	-	2	-	2	-	2	3	-	-	4	5	4	-	3	-	-	2	6	2	3	4	-	-	2	4	2	2	4	2	5	4	3	5	-	-	2	-	-	2	3	4	2	-	2	2	2	-	-	2	2	4	2	-	2	2	-	-	2	-	-	4
DR5*	GSP88.2	-	-	-	-	-	-	-	-	3	2	-	-	2	2	2	-	4	4	4	2	2	-	-	-	2	2	2	3	4	-	-	2	-	-	6	7	8	7	6	6	5	5	2	-	-	-	2	2	-	-	3	2	-	-	2	2	-	5	2	-	2	2	-	-	2	-	2	4
DR5*	SFR3 DR5	-	-	-	-	-	-	-	-	-	-	-	-	2	2	2	-	3	2	2	-	-	-	-	2	2	2	2	3	2	-	-	2	-	-	5	3	5	6	5	4	3	2	2	-	-	-	2	2	-	-	2	2	-	-	-	-	-	2	2	-	-	2	-	-	2	-	-	2
DR7*	SFR16-DR7									-		-		-	2		-	-	-	-		-			-	-	-	-	-				-					-	-	2	2			-			6	8	7	5		5	4	3			-	-	-	-			-			-			-
DRw8*	MAD 88	2	-	-	-	2	-	2	2	2	2	2	2	2	4	3	2	4	4	6	2	2	2	2	2	5	2	3	5	2	-	2	6	2	4	4	5	6	2	4	6	3	3	4	-	-	-	2	2	3	2	4	4	7	-	2	3	2	5	4	2	6	2	2	2	2	2	6	3
A,B,C	W6/32	6	4	4	6	6	7	5	6	7	6	8	4	8	8	6	6	7	8	8	6	6	6	4	6	8	6	8	8	3	6	4	8	4	5	6	7	7	8	6	7	6	6	6	5	6	6	8	8	8	6	8	8	8	4	6	7	8	8	8	4	7	8	6	6	8	5	7	6

[a]The scoring system (1 through 8) for the indirect immunofluorescence profiles is described in the text. –, negative binding; blank space, not tested.

Table 2b. Binding Patterns of Monomorphic mAbs Used in the Workshop 2-D Gel Analysis of Class II Antigens[a]

Specificity*	mAb	9003	9004	9005	9006	9001	9002	9008	9010	9013	9014	9017	9011	9012	9009	9015	9016	9055	9053	9059	9063	9058	9065	9060	9064	9054	9057	9061	9056	9018	9019	9020	9022	9023	9021	9036	9040	9042	9037	9043	9044	9035	9045	9038	9050	9051	9049	9047	9046	9048	9052	9073	9075	9074	9028	9029	9031	9033	9034	9025	9027	9030	9024	9026	9070	9068	9069	9071	9066
	10W CELL LINES	DR1>					Dw20	Dw2>					Dw12	Dw21>			Dw22	Dw19>				Dw18>			Dw16	Dw9>			Dw9+19	DR3>					DRw18	DB2	DR5>							DRw12	DR7> Dw17>						Dw11	DR9>			DR4> DQw8>				3.2/3.1	DQw7>			w8>		8.3(L)	DRw8>			8.3(T)
DR	GSP 4.1	6	6	4	6	6	7	6	6	8	7	8	6	8	8	8	7	8	8	8	5	8	6	6	6	8	7	8	8	5	6	6	8	6	6	8	8	8	8	8	8	7	6	8	6	8	7	8	8	8	6	8	8	8	6	7	8	8	8	8	5	8	8	6	6	8	6	8	8
DR	PL8	6	5	5	6	6	8	6	6	8	7	8	6	8	8	8	6	8	8	8	7	8	6	6	6	8	7	8	6	6	6	6	8	6	8	8	8	8	8	6	8	6	7	8	6	8	8	8	8	8	6	8	8	8	5	8	8	8	8	8	3	8	8	7	6	8	6	8	8
DR	L243	6	6	5	6	6	6	6	6	8	6	8	6	8	8	7	7	7	8	8	6	7	6	5	6	8	6	7	7	6	6	6	8	6	7	8	8	4	8	8	7	6	6	6	5	7	7	8	6	8	6	8	8	8	3	6	7	8	8	8	3	8	8	6	6	6	5	6	6
DQ	IA3	2	2	2	2	2	4	3	4	6	5	7	2	6	6	5	5	7	6	6	3	4	2	-	6	6	4	3	6	3	7	2	6	2	5	5	4	7	7	7	6	5	6	8	2	4	6	6	6	5	3	6	5	6	2	4	7	6	7	7	2	6	6	3	5	4	2	6	6
DQ	TU 22	2	4	2	5	3	2	4	6	7	6	8	2	7	8	6	6	4	7	8	6	6	6	4	6	7	6	6	7	5	3	3	6	2	6	6	5	8	8	6	6	6	5	8	2	5	6	6	5	6	5	6	6	8	2	6	8	6	8	8	2	8	7	6	6	4	2	-	7
DP	B7/21	4	5	2	4	4	4	3	4	7	6	6	2	7	8	6	6	8	8	8	5	6	4	2	6	7	6	6	7	3	2	2	6	3	6	6	4	8	8	6	6	4	4	6	2	4	6	7	5	6	4	6	6	6	2	4	7	6	8	7	2	6	6	4	4	4	2	6	7
DP	PL15	2	2	2	3	2	-	2	3	6	4	6	2	6	6	4	4	6	6	6	2	4	3	2	3	6	4	6	5	2	2	2	6	2		6	2	6	6	4	6	2	3	5	2	2	6	6	5	5	4	6	5	6	2	2	6	6	6	5	2	5	6	2	2	3	2	4	6
BROAD	TU 39	2	6	4	2	2	-	2	6	3	3	8	4	3	5	4	6	5	4	5	2	2	8	6	2	3	2	2	5	6	7	-	2	2	-	4	-	4	3	4	4	-	-	2	2	-	2	3	2	7	2	3	3	8	3	2	2	6	6	3	2	2	2	-	-	6	4	2	3
BROAD	2D6	-	6	2	2	-	-	-	6	2	2	8	6	3	3	2	7	3	3	4	-	2	6	6	2	3	2	2	4	6	2	-	2	3	4	2	-	3	2	2	2	-	-	2	3	-	2	4	2	6	5	2	2	8	2	-	2	7	4	2	2	4	2	2	-	6	6	-	3
A,B,C	W6/32	6	[4]	[4]	6	6	7	[5]	6	7	6	8	[4]	8	8	6	6	7	8	8	6	6	6	[4]	6	8	6	8	8	[3]	6	[4]	8	[4]	[5]	6	7	7	8	6	7	6	6	6	[5]	6	6	8	8	8	6	8	8	8	[4]	6	7	8	8	8	[4]	7	8	6	6	8	[5]	7	6

[a]The scoring system (1 through 8) for the indirect immunofluorescence profiles is described in the text. –, negative binding; blank space, not tested.

Table 3. Priority for Use of Polymorphic mAbs in the Workshop 2-D Gel Analysis[a]

Specificity*	mAb	DR1>					Dw20	Dw2>					Dw12	Dw21>			Dw22	Dw19>				Dw18>				Dw16	Dw9>			Dw9+19	DR3>					DRw18
	10W CELL LINES	9003	9004	9005	9006	9001	9002	9008	9010	9013	9014	9017	9011	9012	9009	9015	9016	9055	9053	9059	9063	9058	9062	9065	9060	9064	9054	9057	9061	9056	9018	9019	9020	9022	9023	9021
DQw1	UK 4.1	1	1	1	1	1	1	1	·	1	1	1	1	1	1	1	6	1	1	1	1	1	1	1	1	·	1	1	1	1						
DQw1	77.34	2	2	2	2	2	2	2	1	2	2	2	2	2	2	2	7	2	2	2	2	2	2	2	2	·	2	2	2	2						
DQw1	KS6	·	3	3		·	·	·	2	3	3	3	3	3	3	3	8	3	3	3	3	3	3	3	3	·	3	3	3	3						
DQw1,3	Leu10	3	4	4	3	3	3	3	3	4	4	4	4	4	4	4	1	4	4	4	4	4	4	4	4	1	4	4	4	4						13
DQw1,3*	PLM 13	4	5	5	4	4	4	4	4	5	5	5	5	5	5	5	9	5	5	5	5	5	5	5	5	·	5	5	5	5						10
DQw1,3*	GS4AA7	5	6	6	5	5	5	5	5	6	6	6	6	6	6	·	10	6	6	6	6	6	6	6	6	·	6	6	6	6						11
DQw1,3*	KS7	6	7	7	6	6	6	6	6	7	7	7	7	7	7	6	11	7	7	7	7	7	7	7	7	·	7	7	7	7						12
DQw3	17.15																2									2					10			10		
DQw3	1VD12																3									3					11					
DQw3	PLM 12																4		21	21		21				4				21	8			11		
DQw3*	PLM 2																5		22	22		22				5					9			12		
DQw2*	C1							14		17	17			16	17	15	18	20	20	20	19	20	20	18		16	20	19	20	20	·	1	1	1	1	1
DQw2,3*	GSP 17.1																														1	2	2	2	2	2
DQ*	IIB3							13	14	16	16	12	10	15	16	14		19	19	19	18	19	19	17	15		19	18	19	19						13
DRw52	NDS 13							10	11	12	12	10		11	12	10	16	8	8	8	8	8	8	8	8	6	8	8	8	8	2	3	3	3	3	3
DRw52	UK 8.1							11	·	13	13	·		12	13	11	·	9	9	9	9	9	9	9	9	7	9	9	9	9	3	4	4	4	4	4
DRw52	7.3.19.1	11			11	11	11	12	12	14	14	11		13	14	12	·	10	10	10	10	10	10	10	10	8	10	10	10	10	4	5	5	5	5	5
DRw52	NDS 10	12			12	12	12		13	15	15	·		14	15	13	17	11	11	11	11	11	11	11	11	9	11	11	11	11	5	6	6	6	6	6
DRw53+*	IIE3	10	11	·	10	10	10											12	12	12	12	12	12	·		10	12	12	12	12	·	7	7	7	7	7
DRw53	PL3																																			
DR 1,4*	CC CL 20	7	8	8	7	7	7	·	8	9	9	·	·	8	9	8	13	14	14	14	14	14	14	13		11	14	13	14	14						
DR 3*	JS1	8	9	9	8	8	8	8	9	10	10	·	·	9	10	9	14	15	15	15	15	15	15	14	12	12	15	14	15	15	7	9	9	9	9	9
DR 4*	NFLD M1	9	10	·	9	9	9	9	10	11	11	9	9	10	11	·	15	16	16	16	16	16	16	15	13	13	16	15	16	16						
DQw1*	NDS 15.38							7	7	8	8	8	8	·	8	7	12	13	13	13	13	13	13	12			13	·	13	13						
DR 3,13,18	16.23																														6	8	8	8	8	8
DR 5	CHE249.2																	18	18	18	·	18	18	·		14	18	17	18	18						
DR 5	GSP88.2																														12					
DR 5	SFR3 DR5																																			
DR 7	SFR16-DR7																																			
DRw 8	MAD 88																	17	17	17	17	17	17	16	14	15	17	16	17	17						

Specificity*	mAb	DB2	DR5>							DRw12	DR7> Dw17>						Dw11	DR9>			DR4> DQw8>					3.2/3.1	DQw7>			w8>		8.3(L)	DRw8>			8.3(T)
	10W CELL LINES	9036	9040	9042	9037	9043	9044	9035	9045	9038	9050	9051	9049	9047	9046	9048	9052	9073	9075	9074	9028	9029	9031	9032	9033	9034	9025	9027	9030	9024	9026	9070	9068	9069	9071	9066
DQw1	UK 4.1																·									15										
DQw1	77.34																10			12	13				14	16			16		14					
DQw1	KS6																11			13					15	17										
DQw1,3	Leu10	1	5	5	5	5	5	5	5	5							7	5	5	4	4	5	4		4	5	5	5	5	5	4	1	1	1	1	1
DQw1,3*	PLM 13	2															8	6	6	5	5	6	5	5	5	6	6	·	6	6	5	·	2	2	2	2
DQw1,3*	GS4AA7	3															·	12	13		·		6	6		19	7	·	7	16	·	·	3	3	3	3
DQw1,3*	KS7																9	13	14	6	6	7	7	7	6	7	·	·	8	7	6	·	4	4	4	4
DQw3	17.15	·	1	1	1	1	1	1	1	1	·	·	7	7	7	7	4	2	2	1	1	1	1	1	1	1	1	1	1	1	1	2				
DQw3	1VD12	·	2	2	2	2	2	2	2	2							5	3	3	2	2	2	2	2	2	2	2	2	2	2	2	3				
DQw3	PLM 12	·	3	3	3	3	3	3	3	3							6	4	4	3	3	3	3	3	3	3	3	3	3	3	3	4				
DQw3*	PLM 2	·	4	4	4	4	4	4	4	4							·	14	15	·	·	4	·	4	·	4	4	4	4	4	·	5				
DQw2*	C1										1	1	1	1	1	1	1	1	1	·						20	16									10
DQw2,3*	GSP 17.1	4	6	6	6	6	6	6	6	6	6	6	6	6	6	6	13						17												10	
DQ*	IIB3																14	11	11	11	12	15	15	15	16	18	15	·	17	15	15	·	9	9	9	9
DRw52	NDS 13	5	7	7	7	7	7	7	7	7	5	5	5	5	5	5	12					16	16	16	17	21	17			16	16	6	5	5	5	5
DRw52	UK 8.1	6	8	8	8	8	8	8	8	8																										
DRw52	7.3.19.1	7	9	9	9	·	9	9	9	9								15	12							22									11	11
DRw52	NDS 10	8	10	10	10	9	10	10	10	10																23										
DRw53+*	IIE3										3	3	3	3	3	3		8	8	8	7	8	8	8	7	8	8	6	9	8	7					12
DRw53	PL3										4	4	4	4	4	4	3	9	9	9	8	9	9	9	8	9	9	7	10	9	8					
DR 1,4*	CC CL 20																				10	11	11	11	10	11	11	·	12	11	10					
DR 3*	JS1	13	15	15	15	14	15	14	14	15											11	12	12	12	11	12	12	9	13	12	11	7	6	6	6	6
DR 4*	NFLD M1	14	16	16	16	15	16	15	15	16											9	10	10	10	9	10	10	8	11	10	9	9	8	8	8	8
DQw1*	NDS 15.38																	16	16	14	·	13	13	13	12	13	13	·	14	13	12					
DR 3,13,18	16.23																																			
DR 5	CHE249.2	9	11	11	11	10	11	·	·	11			8	7	8	8										21										13
DR 5	GSP88.2	10	12	12	12	11	12	11	11	12								17								22										14
DR 5	SFR3 DR5	11	13	13	13	12	13	12	12	13																										15
DR 7	SFR16-DR7										2	2	2	2	2	2	2	7	7	7																
DRw 8	MAD 88	12	14	14	14	13	14	13	13	14								10	10	10	·	14	14	14	13	14	14	10	15	14	13	8	7	7	7	7

[a]The criteria for assigning the priority for the use of each mAb with each cell line are discussed in the text.

binding is difficult to interpret, although significant quantitative differences have been observed for supertypic determinants expressed on different alleles (1). An independent immunofluorescence analysis of these antibodies was also performed as part of the Workshop T-cell study (2), and some of these antibodies were studied in the Workshop serology study.

Because most immunoprecipitations for any given combination of mAb and cell line were only performed once per laboratory and many of the polymorphic antibodies were only studied in one laboratory, it has not been possible to describe the molecular specificity of each antibody from this study alone. In one case, the apparent DR2 specificity of NDS 15.38 was clearly localized to the DQ molecules (3). Antibodies with DRw52- and DRw53-related specificities have also been discussed in some detail (4,5).

Acknowledgments. The Workshop participants gratefully acknowledge the generous contributions of antibodies from the individuals listed in Table 1, the Becton-Dickinson Corporation (Mountain View, CA, USA), and the Genetic Systems Corporation (Seattle, WA, USA).

References

1. Horibe K, Knowles RW, Dupont B. A new supertypic class II determinant expressed with DR7,DRw11(5), DR14(w6), DR3 and DR2 and biochemically localized to DR7 molecules. Hum Immunol 1986;16:14.
2. Flomenberg N, Plätke R, Zemmour J, Small TN, Klein JS. Immunogenetic analysis of HLA class II cellular data. In: Dupont B (ed): Immunobiology of HLA, Volume I: Histocompatibility Testing 1987. New York: Springer-Verlag, 1989.
3. Fuggle SV, Rust N, Carter C, McMichael AJ, Morris PJ. Definition of polymorphism within DR2-associated DQw1 molecules. In: Dupont B (ed): Immunobiology of HLA, Volume II: Immunogenetics and Histocompatibility. New York: Springer-Verlag, 1989.
4. Bontrop R, Ballas M, Madrigal A, Hermans P, Fuggle S, Otting N, et al. DRw52: 2-D gel patterns. In: Dupont B (ed): Immunobiology of HLA, Volume I: Histocompatibility Testing 1987. New York: Springer-Verlag, 1989.
5. Knowles RW, Kienzle B, Winchester R, Mathews E, Bouhallier O, Fauchet R, et al. DRw53: 2-D gel patterns. In: Dupont B (ed): Immunobiology of HLA, Volume I: Histocompatibility Testing 1987. New York: Springer-Verlag, 1989.

Standardized Equipment, Reagents, and Techniques Developed for the 10th Workshop 2-D Gel Study of HLA Class II Antigens

Robert W. Knowles[1] and Bernadette Kienzle[1]

The goal of this study was the identification of structural polymorphism in the primary amino acid sequence of the HLA class II α and β chains using two-dimensional (2-D) gel electrophoresis. This required the ability to distinguish slight shifts in the pI of individual spots in the 2-D patterns representing various glycosylated forms of polypeptides encoded by closely related genes. Many of the HLA class II molecules were known or expected to have only minor differences in their net charge and little or no detectable difference in the molecular weight of the polypeptide backbone. Therefore, it was essential to be able to compare 2-D gel patterns produced in different laboratories. Based on a preliminary study involving eight laboratories, it was decided that complete standardization of every aspect of the 2-D gel techniques would be important to the success of the study. The major innovations of the 10th Workshop 2-D gel system were the development of protein standards (^{14}C-carbamylated proteins), which were added to each sample to serve as internal markers for each 2-D gel pattern on the autoradiograms, and the development of liquid buffers, premixed in large batches, which were formulated and aliquoted for long-term storage either at 4 °C or frozen at −20 °C. This report provides a brief description of the system.

Standardized Equipment

IEF Tube Gel Apparatus. Standardization of the first-dimension equipment were considered essential to the reproducibility of the pH gradient. The IEF tube gel apparatus chosen for the study (MEGA ISO-20, Health Products Inc., Rockford, IL, USA) provides the ability to pour 20 tube gels simultaneously to a predefined level from a common reservoir at the bottom of the tubes. This was a considerable improvement over systems in which each tube is filled sequentially from the top over several minutes. It is based, in part, on a previous design (1).

SDS Slab Gel Apparatus. Several designs of SDS slab gel equipment were used by the participants before this study. Nevertheless, standardizing this equipment was considered important for reproducibility of the molecular weight gradient for comparisons between molecules with subtle differences in the size of the polypeptide backbone or carbohydrate structures. The equipment chosen for

Participating Laboratory: US7KNO[1]

the study (MEGA DALT4, Health Products Inc., Rockford, IL, USA) allowed simultaneous pouring of four slab gels to the same predefined level and running of all four slab gels in a common buffer tank sharing the same electrodes. It is based, in part, on a previous design (2).

Other Equipment. Although it may have been preferable to standardize other equipment, including the power supplies to ensure that the same voltages were used in all laboratories for the first- and second-dimension electrophoresis, the costs were considered prohibitive, and this was probably not a major variable in the study. Instead, the second dimension was standardized by measuring the distance of migration of the bromophenol blue dye.

Standardized Reagents

Monoclonal Antibodies. Monoclonal antibodies (4) were diluted 1:5 in PBS with 0.1% sodium azide, filter sterilized through 0.22-μm GV filters (Millipore, Corpo, Bedford, MA, USA), and stored at 4°C.

Rabbit Anti-(Mouse Ig) Coated Staphylococcus aureus (RAM-Staph). Formalin-fixed protein A containing *S. aureus* (IgGsorb, Enzyme Center, Boston, MA, USA) was washed in PBS, passed through a 22-gauge needle, and washed in lysis buffer (8.76 g/L NaCl, 1.46 g/L Na_2EDTA, 6 g/L Tris, 0.5% NP-40). The RAM-Staph were then resuspended in lysis buffer as a 10% solution and coated with rabbit antibodies recognizing mouse immunoglobulins, using 50 μl of serum per milliliter of 10% RAM-Staph, for 30 minutes at 4°C. The rabbit serum was prepared locally by immunizing rabbits with purified murine IgG. This RAM-Staph reagent was washed in lysis buffer three times, resuspended at 10%, and stored at −70°C. The rabbit anti-(mouse Ig) antibodies were used to ensure that all murine mAbs would bind to the protein A–*S. aureus* during the immunoprecipitation procedure, including those of the IgG1 subclass.

Protein Standards (Radiolabeled and Carbamylated Internal Markers for the 2-D Gels). Protein standards, carbonic anhydrase and creatine phosphokinase (Sigma, St. Louis, MO, USA), were radiolabeled with ^{14}C-formaldehyde (5) and carbamylated (6) to provide internal markers for the pI and relative molecular weights of spots in each pattern of HLA class II proteins analyzed in the study. Alternatively, unlabeled carbamylated carbonic anhydrase and creatine phosphokinase were purchased (Carbamylytes, Pharmacia, Piscataway, NJ, USA) and radiolabeled with ^{14}C-formaldehyde. The protein standards were produced specifically for this study at the Kendrick Laboratory (Madison, WI, USA).

Premixed Liquid 2-D Gel Buffers. Liquid buffers were prepared for each part of the 2-D gel procedure. Stock solutions of these reagents were premixed in large batches, filtered under nitrogen, and formulated for long-term storage either at 4°C or frozen at −20°C. Each reagent was aliquoted for daily or weekly use to provide standardization of the reagents over the 6 to 9 months of the study. The 2-D gel reagents were produced specifically for this study by Pierce Chemical Company (Rockford, IL, USA). They are based, in part, on previously described reagents (1–3).

Fluorography Reagent. Because the study used ^{35}S-methionine to label the class II molecules, fluorography was used to increase the sensitivity of the x-ray film to the labeled molecules. The study recommended the use of a commercial reagent (Amplify, Amersham Radiochemicals, Arlington Heights, IL, USA), but other fluors were also used with longer exposure times.

Standardized Techniques

^{35}S-Methionine Labeling and Preparation of NP-40 Cell Lysates. Briefly, 40 × 10^6 cells were washed and resuspended in 1 ml of methionine free medium without fetal calf serum for 30 minutes at 37°C. Then, 2 mCi of ^{35}S-methionine was added and the cells were incubated at 37°C for 4 hours. Cells were then washed three times in 50 ml of PBS at 4°C and lysed in 0.5 ml of an NP-40 lysis buffer (8.76 g/L NaCl, 1.46 g/L Na_2EDTA, 6 g/L Tris, 0.5% NP-40), with PMSF (1 mM) added fresh before use. After 30 minutes at 4°C the nuclei were removed by centrifugation for 2 minutes at 4°C, and the lysate was stored at −70°C.

Preclearing the Lysates and Immunoprecipitation. The lysate was cleared for 16 hours at 4°C by adding 0.5 ml of 20% RAM-Staph solution to 0.5 ml of lysate. The RAM-Staph was removed by centrifugation. The lysate was cleared a second time by adding an additional 0.5 ml of 20% RAM-Staph for at least 1 hour at 4°C. Immunoprecipitation was performed with a 0.1-ml aliquot of cleared lysate and 5 μl of mAb (1:5 dilution of ascites) incubated at 4°C for 30 minutes followed by 50 μl of 10% RAM-Staph at 4°C for 30 minutes. The RAM-Staph immunoprecipitates were washed three times in lysis buffer and split to form two pellets (one for the α and one for the β chain analysis). Pellets were stored at −70°C.

First-Dimension Gels—IEF. Tube gels were prepared using a solution containing A (12 g of urea, 0.45 g of NP-40, 9.05 g H_2O), B (3 ml of 30% acrylamide, 1.8% bis-acrylamide), C [1.25 ml of ampholyte solution (80% narrow range, 20% wide range)], D (75 μl of 10% ammonium persulfate), and 20 μl of TEMED. The narrow-range ampholytes used for the HLA class II α chains were in the pH range of 4 to 6, whereas those used for the β chains were pH 6 to 8. The wide-range ampholytes were pH 3.5 to 10. The apparatus was assembled with the lower buffer chamber containing 5 ml of 34% phosphoric acid in 2 L of H_2O and the upper buffer chamber containing 5 ml of 2.4% NaOH in 150 ml of H_2O. Frozen pellets were resuspended in 25 μl of IEF sample buffer containing E (0.3 ml of 9 M urea, 4% NP-40, 2% 2-mercaptoethanol), F (25 μl of ampholyte mixture used in the IEF gels), and G (0.2 ml of standard proteins ^{14}C-CA and ^{14}C-CPK in 9 M urea, 4% NP-40, 2% 2-mercaptoethanol). The labeled class II molecules were eluted at 50°C for 30 minutes and the Staph was removed by centrifugation. Samples were loaded to the top of the tube gels. Electrophoresis was performed at 700 V for 17 hours. Gels were removed by gentle water pressure from a syringe fitted with a 1-ml pipet tip and stored at −70°C in freezing vials.

Second Dimension–SDS. Slab gels were poured from a solution containing A (124 ml of 30% acrylamide and 0.8% bis-acrylamide), B (2.7 ml of 10% ammonium persulfate), C (180 ml of 0.78 M Tris, pH 8.6), D (3.1 ml of 10% SDS), and 20 μl of TEMED. Slab gels were poured to within 0.5 cm from the top of the plates. A hot agarose overlay solution (0.1 g in 20 ml of the tank buffer with 50 μl of a 1% bromophenol blue solution) was added to 0.25 cm from the top, and the tube gel was layered onto the agarose surface and covered with the agarose solution to the top of the plates. The tank buffer was diluted from a 10× stock buffer (containing 504 g of glycine, 112 g of Tris, 35 g of SDS, and 3.5 L of H_2O). The slab gels were run at 75 V until the dye reached 2 cm from the bottom of the slab, about 18 hours.

Autoradiography. Gels were fixed in 7% acetic acid and 30% methanol overnight and treated with 125 ml of Amplify for 15 minutes immediately before slow drying at 60°C, followed by exposure to XAR-5 x-ray film (Kodak, Rochester, NY, USA).

Acknowledgments. Several Workshop participants, including Dominique Charron, Susan Fuggle, Christine Carter, Nigel Rust, and Andrew McMichael, provided helpful advice in the design of the study. Dominique Charron and the members of his laboratory also sponsored an important meeting of Workshop participants in August 1986. This allowed a critical discussion of the need to standardize specific reagents and techniques. Support from each of the participating laboratories for the difficult decision to standardize both the equipment and reagents is gratefully acknowledged. David Sammons (University of Arizona, Tucson) and Dennis Klenk (Pierce Chemical Co., Rockford, IL) provided advice on the equipment and reagents for the 2-D gel system. Nancy Kendrick (Kendrick Laboratories, Madison, WI) provided advice on the protein labeling technique. Andrew McMichael and Nigel Rust, Nelson Fernandez and Hilliard Festenstein, Dominique Charron, Gerald Nepom, Takako Nakatsuji and Kimyosi Tsuji, Takehiko Sasazuki, Sanna Goyert, and Alejandro Madrigal participated in preliminary standardization studies. Bo Dupont provided encouragement throughout the study. This work was supported, in part, by grants from the U.S. Public Health Service (AI22916, CA22507) and the Society of Memorial Sloan-Kettering Cancer Center.

References

1. Anderson NG, Anderson NL. Analytical techniques for cell fractions. XXII. Two-dimensional analysis of serum and tissue proteins: Multiple isoelectric focusing. Anal Biochem 1978;85:331.
2. Anderson NG, Anderson NL. Analytical techniques for cell fractions. XXII. Two-dimensional analysis of serum and tissue proteins: Multiple gradient-slab gel electrophoresis. Anal Biochem 1978;85:341.
3. O'Farrell PH. High resolution two-dimensional electrophoresis of proteins. J Biol Chem 1975;250:4007.
4. Knowles RW, Adluri V, Kilaru S, Sirotina A, Small TN. Binding patterns of the monoclonal antibodies in the 10th Workshop 2-D gel analysis of HLA class II antigens. In: Dupont B (ed): Immunobiology of HLA, Volume I: Histocompatibility Testing 1987. New York: Springer-Verlag, 1989.
5. Jentoft N, Dearborn DG. Protein labeling by reductive alkylation. Methods Enzymol 1983;91:570.
6. Anderson NL, Hickman BJ. Analytical techniques for cell fractions. XXIV. Isoelectric point standards for two-dimensional electrophoresis. Anal Biochem 1979;93:312.

Computerized Analysis of 2-D Gel Patterns of the HLA Class II Antigens

D.W. Sammons,[1] N. Pan,[1] K.H. Kwan,[1] W.J. Ko,[1] N.H. Lin,[1] Y.M. Fong,[1] J.S.S. Lanchbury,[2] and R.W. Knowles[3]

One goal of the 10th Workshop 2-D gel analysis of HLA class II antigens was to evaluate whether computer programs could be used to compare patterns more objectively than by visual analysis alone. In a previous 2-D gel study of HLA class II antigens, computerized analysis was found to be limited by the lack of internal reference points in the patterns (1). These reference points are necessary to convert the x,y coordinates of each pattern into a standardized coordinate system before the comparisons between patterns can be made. Therefore, in the 10th Workshop study, radioactively labeled and carbamylated protein standards were developed to serve as internal markers for the isoelectric point (pI) and molecular weight (MW) of spots in the patterns (2). Preliminary studies demonstrated that the BioImage

Participating Laboratory: US7DWS,[1] UKIWEL,[2] US7KNO[3]

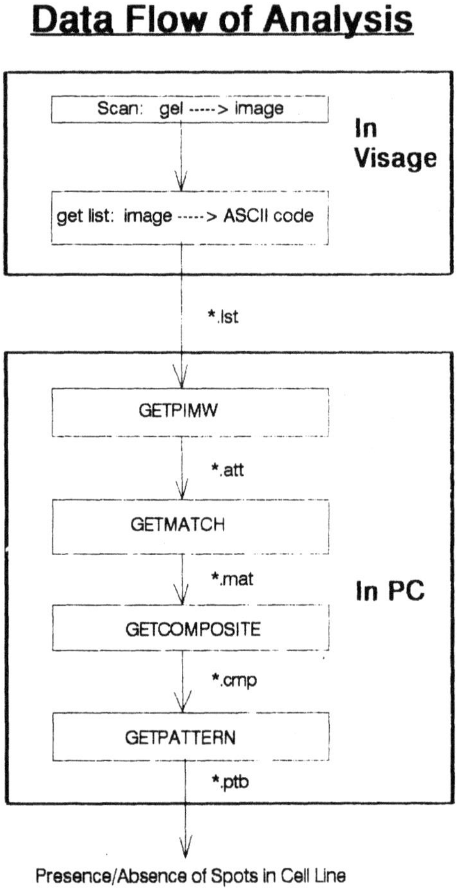

Figure 1. Flow chart for the data analysis. The analysis began with scanning of films and the generation of spot lists using the BioImage Visage system software, called EQ. The spot lists were transported as ASCII code to the IBM-compatible personal computer and then transformed into pI and MW values. The data were then manipulated through GETPIMW, GETMATCH, GETCOMPOSITE, and GETPATTERN with the indicated file names.

Visage System (BioImage Corporation, Ann Arbor MI) could reliably scan the 2-D gel patterns from the Workshop x-ray films, digitize the images, and determine the x,y coordinates and intensities for each spot. However, none of the commercially available computer systems included the capacity to automatically transform the coordinates and create a database that would be adequate for the large number of Workshop 2-D gel patterns. Algorithms for the transformation of the x,y coordinates into standardized pI and MW values had been developed previously at the University of Arizona (3,4). In preliminary studies, modifications of these algorithms were found to be well suited for use with the Workshop internal protein standards. However, new programs were also needed to develop a database and make comparisons between the 2-D gel patterns in the study.

The Workshop x-ray films contained the 2-D patterns of the immunoprecipitated ^{35}S-labeled class II antigens, mixed with the ^{14}C-labeled protein standards and separated using the standardized equipment, reagents and procedures developed for this study (2). Films were visually inspected, and those with adequate resolution of the class II molecules and standard proteins were selected for computerized analysis. Each selected film was scanned, the image was digitized, and a "spot list" of the x,y coordinates and intensities was generated for each film. A map of this pattern was plotted from the spot lists for each film. Spot lists were then converted into a standardized coordinate system using the analysis program called MATCHWARE™ (1986, 1987, 1988, © University of Arizona, all rights reserved). A relative isoelectric point and molecular weight was automatically assigned to each spot in approximately 1000 Workshop films and a database was developed. For each cell line, patterns obtained with different mAbs were compared and composite tables, with corresponding composite maps, were generated and distributed to the participants in the November 1987 Princeton Workshop meeting, to assist in the visual interpretation of the 2-D patterns.

After the Princeton meeting, additional analysis was performed on a smaller number of films, from the HLA-DR4 group of cell lines. The class II molecules isolated from cell lines expressing known subgroups of functionally distinct HLA-DR4 associated molecules were chosen for this study to evaluate the use of computerized analysis in the study of genetic polymorphism. The goal was to identify spots that correspond to the DRβ and DQβ chain 2-D gel patterns for each cell line, and to correlate any structural polymorphism in these patterns with the serologic (DR and DQ) or cellular (Dw) subtypes of these cell lines.

Materials and Methods

Computer Hardware. The gel scanning, digitization, and the generation of spot lists were accomplished with a BioImage Visage System, including a Mass Comp minicomputer with a 68010 microprocessor. Further analysis of the patterns was performed with the MATCHWARE™ group of programs using an IBM-compatible personal computer. The IBM-compatible personal computer was also used to select gels for database formatting and to drive the plotter.

Computer Software. The Visage EQ program (BioImage Corporation) was used to scan the films, digitize the patterns, and generate the spot lists including the location (x,y coordinates) and the quantitative integrated intensity (II) of each spot. The gel scanning was standardized by initializing the scan at the same position on each film relative to the internal standard patterns. Several parameters of contrast and background were optimized for the study and were kept constant. The EQ program generated spot lists in a file format that could be used on other computers for further analysis. From each digitized film, the centers of the spots were automati-

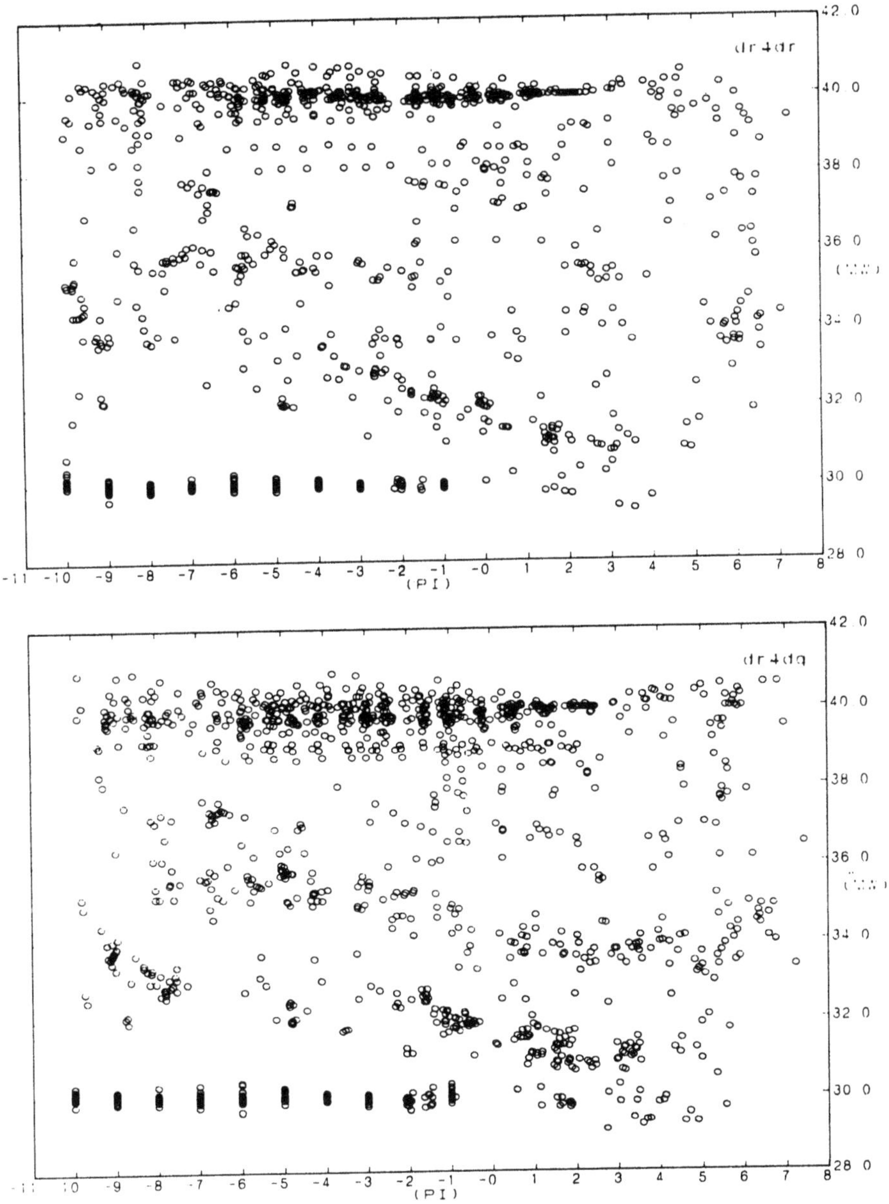

Figure 2. (*A*) Maps of DRβ chain patterns obtained with DR-specific mAbs (from *.ATT files). The map is an aggregate of all spots located in the 23 DR films included in the study and is generated by transforming x,y coordinates into pI and MW values with the GETPIMW program. Spots that are located between 30 kD and 40 kD MW and 8 pI and −11 pI are included. (*B*) Maps of DQβ chain patterns obtained with DQ-specific mAbs (from *.ATT files). The map is derived from in the same way as in Figure 2A except the map is an aggregate of 29 DQ films.

cally assigned by the EQ software and these positions were listed as x,y coordinates in the *.1st files (Fig. 1). The intensity values were also determined with the EQ software, and are listed as integrated intensities (II) although they were not considered in the maps of Figures 2A and B.

MATCHWARE™ programs were used for data transformation and analysis. This group of programs (GETPIMW™, SPOTMATCH™, MATCHBASE™, and GETPATTERN™) was developed at the University of Arizona (1986, 1987, 1988. © University of Arizona, all rights reserved) (1,3). Figure 1 illustrates the flow of work through the various hardware and software systems. The first stage in the analysis of the spot lists was

Table 1. Correlation of the Presence or Absence of Individual DR Spots by Cell Line

SID	pI	MW	II	BSM Dw4	WT51 Dw4	BM14 Dw4	SAVC Dw4	DEU Dw4	MTF Dw14	YAR Dw10	JHAF Dw13	KT17 DwKT2	KT2 DwKT2	SJAF Dw15
1	-5.73	33.02	5.26	NA	NA	-	-	-	-	+	-	-	-	-
2	-5.68	33.85	0.87	NA	NA	-	+	-	-	+	-	-	-	-
3	-5.42	32.51	0.10	NA	NA	-	-	-	+	-	-	-	-	-
4	-5.25	33.85	0.08	-	-	-	+	NA	NA	NA	NA	NA	NA	NA
5	-4.76	32.47	2.17	-	-	+	+	-	-	+	+	-	-	-
6	-4.45	33.33	0.01	-	-	-	+	NA	NA	NA	NA	NA	NA	NA
7	-3.89	33.54	0.07	NA	NA	NA	+	NA	NA	-	NA	-	NA	NA
8	-3.57	33.18	3.18	NA	NA	-	+	-	+	+	+	+	+	-
9	-3.06	32.58	0.14	-	-	+	-	-	+	-	-	-	-	-
10	-2.81	31.24	0.07	NA	NA	-	-	-	-	-	-	-	+	-
11	-2.66	33.60	0.22	NA	NA	-	+	-	NA	NA	NA	-	NA	NA
12	-2.47	32.84	0.82	+	-	+	+	-	-	-	-	-	+	-
13	-2.32	33.77	0.27	-	-	+	+	+	+	-	-	-	-	-
14	-1.91	32.40	4.77	NA	NA	-	-	+	+	+	+	+	-	-
15	-1.20	33.79	0.20	-	-	+	+	-	-	-	-	-	-	+
16	-1.19	32.17	5.17	+	+	+	+	+	+	+	+	+	+	+
17	-0.93	31.08	0.01	NA	NA	+	-	-	NA	NA	NA	-	NA	NA
18	-0.65	33.67	0.04	-	-	+	-	NA	NA	NA	NA	NA	NA	NA
19	-0.49	31.79	0.33	-	+	-	-	-	-	-	-	-	+	-
20	-0.05	31.93	1.89	+	-	+	+	+	+	+	+	-	-	+
21	0.19	33.77	0.20	-	-	+	-	NA	NA	NA	NA	NA	NA	NA
22	0.50	31.43	4.34	NA	NA	-	-	-	-	-	-	+	-	-
23	0.76	33.48	0.15	-	-	+	+	-	-	-	-	-	-	-
24	1.25	31.90	1.78	-	-	+	-	-	-	-	-	-	-	+
25	1.55	31.30	1.68	+	-	+	+	+	+	-	-	+	+	-
26	1.95	33.81	0.01	-	-	+	-	NA	NA	NA	NA	NA	NA	NA
27	2.03	31.12	0.02	NA	NA	-	-	+	-	-	-	-	-	+
28	2.52	31.08	0.35	NA	NA	-	-	-	-	+	-	-	-	-
29	2.78	32.61	0.36	NA	NA	+	-	-	-	-	-	-	+	-
30	3.22	31.41	1.46	-	-	-	+	-	-	+	-	-	-	+
31	3.54	33.64	0.05	NA	NA	-	-	-	-	-	-	+	-	-
32	3.59	31.02	0.03	NA	NA	NA	-	NA	NA	-	NA	+	NA	NA

The table lists spots in DR films that were matched and assigned to the exact match, cluster match, or unique category. The ∗.CMP file was generated within the GETCOMPOSITE algorithm and presence (+) or absence (−) is indicated. NA indicates that films were not included in the study to allow determinations.

the transformation of the x,y coordinates of each spot into their relative isoelectric point and molecular weight values using the GETPIMW program (Figs. 2A and B). Each 2-D gel included two sets of radioactively labeled carbamylated internal protein standards that form patterns above and below the HLA class II antigens in the autoradiogram. The x,y coordinates of each class II spot were compared with the internal standards and transformed into relative pI and MW values. The pattern of the larger internal protein standard, creatine phosphokinase (CPK), was assigned the molecular weight value of 40 kDa, while the smaller internal protein standard, carbonic anhydrase (CA), was assigned the molecular weight value of 30 kDa. Because most of the films included additional spots from the immunoprecipitated samples at the same molecular weight as the CPK marker proteins, this pattern could not be used in the pI calculations. In contrast, there were no spots from the immunoprecipitated class II molecules within the CA protein standard pattern at 30 kDa. In addition, the CA standard formed a simple pattern of regularly spaced spots making it ideal for this analysis. One of the CA marker spots was, however, consistently weaker than the other spots or absent, preventing an accurate pI assignment for these gels. Therefore, a special feature was created for the automatic assignment of the missing CA spot so that films with this defect could be included. Some of the class II antigens were more basic than the most basic CA marker, and it was also necessary to use the algorithm to extend the CA markers toward the basic direction by extrapolation. An arbitrary pI numbering system was established with both positive and negative numbers. The most basic spot of the CA pattern was designated pI = 0, and each spot in the carbamylated pattern was numbered −1, −2, −3, etc., decreasing towards the acidic end. The extrapolated spots toward the basic end were given positive numbers 1, 2, 3, etc., increasing towards the basic end. This transformation to standardized pI and MW coordinates was required for the next step of the analysis. The x,y coordinates from films with insufficient exposure, skewed patterns, or missing internal marker spots could not be transformed

Table 2. Correlation of the Presence or Absence of Individual DQ Spots by Cell Line

SID	pI	MW	II	BSM w8	WT51 w8	BM14 w8	SAVC w8	AS w8	MTF w8	KT17 w8	PF97 w7
1	-5.91	32.75	0.02	NA	NA	-	-	NA	NA	NA	+
2	-5.50	33.32	0.09	+	-	+	+	-	-	-	-
3	-5.18	32.04	0.92	-	-	+	-	NA	NA	-	NA
4	-4.77	32.17	0.30	-	-	+	+	-	-	-	+
5	-4.44	32.26	0.03	-	-	+	-	NA	NA	NA	NA
6	-4.18	33.28	0.04	-	NA	+	-	-	+	NA	-
7	-3.46	31.93	1.06	+	-	-	-	NA	NA	-	+
8	-3.24	33.68	0.01	+	NA	-	NA	NA	NA	-	-
9	-2.81	32.81	0.05	+	-	+	-	NA	NA	-	-
10	-2.22	32.45	0.30	+	+	+	-	NA	NA	-	-
11	-2.02	31.15	1.10	-	NA	-	-	NA	NA	-	+
12	-1.56	32.47	0.27	+	+	+	+	-	-	-	-
13	-1.17	31.95	0.78	NA	NA	+	-	-	-	NA	NA
14	-0.94	31.95	0.57	+	+	+	+	+	+	-	+
15	-0.57	33.53	0.27	-	-	+	+	NA	NA	-	+
16	-0.51	31.83	0.74	+	-	+	+	+	+	+	-
17	0.09	31.33	0.21	-	-	-	-	NA	NA	+	-
18	0.22	32.52	0.00	NA	NA	+	-	-	-	NA	NA
19	0.58	31.49	0.39	-	+	+	-	NA	NA	-	-
20	0.62	33.63	1.01	-	-	+	+	NA	NA	-	+
21	0.78	32.48	0.00	NA	NA	+	-	-	-	NA	NA
22	0.92	31.61	0.42	-	+	+	+	-	-	-	-
23	1.14	31.19	0.66	-	-	+	+	NA	NA	-	-
24	1.19	33.63	0.12	NA	NA	+	-	NA	NA	NA	-
25	1.53	33.72	1.26	NA	-	+	-	NA	NA	-	+
26	1.66	31.41	1.29	+	-	+	+	+	+	-	-
27	1.97	32.62	0.00	NA	NA	+	-	-	-	NA	NA
28	2.24	33.55	0.71	-	+	+	-	-	-	-	+
29	2.91	33.70	6.35	-	NA	+	-	NA	NA	-	+
30	2.96	31.33	0.52	-	NA	-	+	NA	NA	NA	-
31	3.30	31.20	5.32	+	-	-	+	-	+	+	+
32	3.30	33.72	0.54	+	-	+	+	+	-	-	+
33	3.89	33.83	4.15	-	NA	-	-	NA	NA	-	+

Spots in DQ films are listed in the same format as used in Table 1.

and were not included in the subsequent computerized comparative analysis. However, x,y coordinates and intensity values were generated for all the films which were scanned. These were shared with the workshop participants and were useful in the visual analysis.

Data reduction to composite maps and tables was accomplished with the SPOTMATCH algorithm, designed to match all of the spots within a defined area from a series of films with different 2-D gel patterns. The input file contained an *.ATT spot list with the transformed coordinates generated with the GETPIMW program (see Fig. 1). In comparing two films, the GETMATCH algorithm took each spot on film 1 and identified the spots on film 2 with the closest coordinates within a preset window of tolerance in pI and MW values, and generated a table of matched spots and unique spots. The limits of the window of tolerance were set after examination of the spotlists and estimation of the average distances between spots in the patterns of the CA and CPK standard proteins. When more than one spot was found within a given window on film 2, the closest of these spots was selected as the exact match, and the other spots were declared unique. The window of tolerance was selected so that the windows generally contained only one spot. This matching process was extended to each pattern in a series of gels. Whenever two spots were located near each other in several gels, there appeared to be a clustering of the spots across two window boundaries and the algorithm created another classification of match called a cluster. Thus, SPOTMATCH Tables (*.MAT file, see Fig. 1) were created with all the spots found in the DRβ and DQβ groups of matched gels.

The GETCOMPOSITE program was used to generate the *.CMP file and table (Tables 1 and 2, Figs. 4 and 5). The data were summarized according to specific cell lines, groups of cell lines, monoclonal antibodies, and groups of monoclonal antibodies. In the early part of the study, a database program, called Accent R, was used with the VAX 8650 mainframe computer to group the 2-D gel patterns and their transformed data into tables that could be analyzed further according to specific cell lines and specific monoclonal antibodies. These tables were used to create a single table and composite map for each cell line which were used by the participants of the

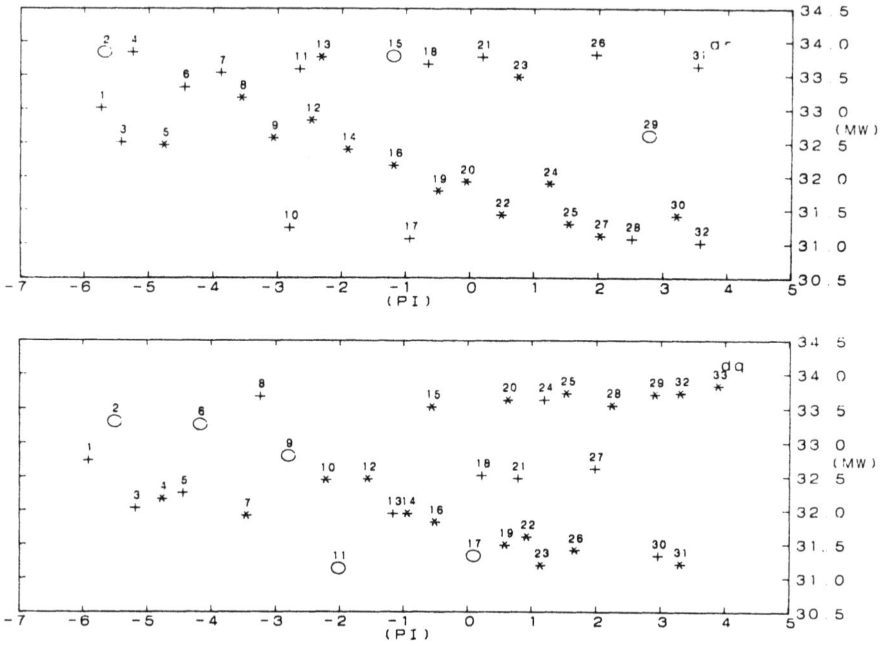

Figure 3. (*A*) Composite map of DRβ chain patterns obtained with DR-specific mAbs. The map is a composite generated with GETMATCH and GETCOMPOSITE algorithms. Spots in a window of 31 to 34 kD and pI of 4 to −6 are included (-O-O-) Exact match spots fall within a tolerance setting of pI=0.3 and MW=0.8;. (- * - * -) Cluster match spots are derived from the cluster program that determines a match of those spots that fall within the tolerance settings, but more than one spot within a gel matches with a spot outside the tolerance boundary. (−+−+−) Unique spots are found only in one gel pattern. (*B*) Composite map of DQβ chain patterns. Map was generated as in Figure 3A except that the patterns are obtained with DQ-specific mAbs.

10th Workshop Meeting in Princeton (November 1987). After the meeting, the IBM-compatible package used in the HLA-DR4 study presented herein was developed with the same features as the program operating on the mainframe computer, except that fewer gels could be analyzed together as a group.

The GETPATTERN program was used to generate a composite spot list table and map for groups of cell lines based on their Dw subtype. Six Dw subtypes of DR4 were included in the study and the DRβ chain patterns were grouped by Dw subtype to generate the composite maps. Two DQw3 subtypes, DQw7 and DQw8, were found in these cells, and the DQβ chain patterns were grouped by DQ type. The presence or absence of the composite protein patterns was derived from the GETPIMW program's *.ATT file, GETMATCH program's *.MAT file, and the GETCOMPOSITE program's *.CMP file.

Cell Lines and Monoclonal Antibodies. Most of the 2-D gel patterns analyzed in this report were from films submitted to the Workshop by participating laboratories using DR4 cell lines from the Workshop Reference Panel described in the DR4 Report (5). These included BSM (9032), WT51 (9029), BM14 (9033), SAVC (9034), DEU (9025), JHAF (9030), KT17 (9024), and PF97387 (9027). Additional patterns were also analyzed from films produced by J.S.S.L. using non-Workshop DR4 cell lines, kindly provided by Dr. Julia Bodmer. These included MTF (derived from Dw14 homozygous typing cells, originally provided by Dr. H. Festenstein); KT2 (a DwKT2, DQw8 cell line, originally provided by Dr. Kashiwagi, and studied in the 8th Workshop); SJAF (a Dw15 cell line, originally provided by P. Goodfellow and R. Payne); and AS (Dw4, DQw8 cell line, also called Gerber, originally provided by Dr. R. Ceppelini). The cell lines included had been previously shown to be positive by immunofluorescence with the DR- and DQ-specific mAbs. In addition, the cell lines had been typed serologically for DR and DQ polymorphisms and by Dw typing for the DR4 related subtypes. The mAb, cell line, and Dw type information about each film were entered into the MATCHWARE™ database and associated with each film's unique scan number.

The DR-specific monoclonal antibodies (L243, GSP4.1, PL8, and NFLD M1) or the DQ-specific antibodies (TU22, PLM12, IVD12, KS7, 17.15, and LEU10) were used in this analysis. These antibodies were selected because they were previously characterized to be relatively monospecific for either DR or DQ cell surface antigens.

Results

The present study is limited to DRβ and DQβ class II antigens of DR4 cell lines. Figure 2A includes scans of

23 β chain films from DR immunoprecipitates from 11 cell lines (Table 1) and Figure 2B includes scans of 29 β chain films from DQ immunoprecipitates from 8 cell lines (Table 2). Each circle represents a single spot from a single film. In addition to the class II antigens, these patterns also include additional spots from proteins which are coprecipitated. The intensities are not reflected in the size of the spots in these composite maps but they were listed in Tables 1 and 2.

All of the matched spots in the 2-D gel patterns of DRB or DQB chains immunoprecipitated using DR- or DQ-specific antibodies found in the DR4 cell lines are shown in the composite maps in Figures 3A and B. The composite patterns of all proteins in the DRβ and DQβ groups are illustrated from a window corresponding to a pI range of +4 to −6 and to a MW range of 31 to 34 kDa. The maps are the result of the GETMATCH algorithm that assigns spots into three categories:

1) Exact matches are spots that fall within the tolerance setting and have no close neighbors that crossmatch.
2) Cluster matches are a group of spots on several films that are related because each spot forms matches with at least one, but not all of the other spots in the cluster.
3) Unique spots occur on only one film.

Table 1 lists the composite proteins from DR and DQ films according to increasing pI. A spot identification number (SID) is listed in increasing order to correspond with the increasing order of the pI in each DRβ and DQβ grouping. The presence (+) or absence (−) of a spot is indicated for each cell line. Scoring of the presence/absence of a protein on a cell's surface was accomplished by using the Workshop mAbs. Often, it was impossible to conclusively score the presence or absence of a spot to a cell line and, therefore, it was listed as not available (NA). For example, though a spot might have been absent on a film pattern generated by a mAb/cell line combination, it was scored NA if the spot in question did not appear with at least one of the mAbs in the cell line being tested.

The complete list of 32 spots in the DRβ chain patterns found in 11 DR4 cell lines by immunoprecipitation with the mAbs is presented in Table 1. Table 2 shows the complete list of 35 spots in the DQβ chain patterns found in the 8 DR4 cell lines by immunoprecipitation. Comparison of composite spots of Tables 1 and 2 shows that several spots are found in both the DRβ and DQβ sets of films. The fact that the DRβ and DQβ patterns have individual spots in common is presumably due to cross-reactivity of the monoclonal antibodies, which are not always specific for either DR or DQ molecules. In addition, it appears that some of the spots in the DRβ and DQβ chain patterns are distinct, but simply have a similar charge and size. To trace a given protein throughout the analysis, a spot identification number (SID) was assigned to each composite (see Tables 1 and 2). This number is listed about each spot within Figures 3A and B and also identifies chain pattern spots in Figures 4 and 5.

In Tables 1 and 2, the cellular Dw typing or serologic DQ typing of the cell lines is also shown. Composite tables and maps were produced using smaller groups of films from the individual Dw subgroups of cell lines, (Fig. 4). These represent the Dw4, Dw14, Dw10, Dw13, DwKT2, and Dw15 subgroups of DR4, which have been well characterized previously as discussed in the DR4 report (5). Indicated by number on the composite maps in Figure 4 are the individual spots found for each Dw or DQ subtype. Listed with each map is a data table that has the pI, MW, and II for these spots. For cross-reference, the SID numbers of Tables 1 and 2 are listed. Comparison of the patterns from the Dw4 (spots 20, 16, 11, 8 and 9, 7 and 10, Fig. 4A) and Dw14 cell lines (spots 8, 7, 6, 5, and 4, Fig. 4B) suggests that their DRβ patterns are identical. Spots 8 and 9 and 7 and 10 were split by the algorithm because of conductivity gaps in the 2-D gels. This pattern has been called DRβI-4.1 in the Workshop nomenclature (5). Similarly, the patterns from the Dw13 (spots 5, 4, and 3, Fig. 4C) and DwKT2 cells (spots 8, 6, and 5, Fig. 4D) appear identical and have been called DRβI-4.3 (5). Two additional patterns from the Dw10 (spots 5, 3, 1, and 2, Fig. 4E) and Dw15 (spots 6, 4, 3, 2, and 1, Fig. 4F) cell lines appear to be distinct and have been called DRβI-4.2 and DRβI-4.4 in the Workshop studies (5). The DRB4 gene product is only weakly expressed on DR4 cell lines and is difficult to detect without DRw53-specific antibodies (5).

The DQβ chain patterns were analyzed according to the serologic DQ typing of the cell lines, but only one DQw7 cell line was included in this study. Figures 5A and B show the composite tables and maps for the spots found in the DQw7 and DQw8 cell lines, respectively. The DQβ-3.1 chain pattern corresponds to spots 4, 3, and 2 in Figure 5A. The first spot of this pattern can be seen immediately over the CA pI standard at position pI = −1 in Figure 2A. Because of the MW window that was selected, it does not appear in Figure 5A. In Figure 5B, the consensus pattern of DQβ-3.2 corresponds to spots 32, 22, 15, and 10. These DQβ-3.1 and DQβ-3.2 chain patterns are clearly distinct as discussed in the DR4 report (5).

Discussion

One of the objectives of this study was to determine the feasibility of analyzing 2-D patterns from gels run in different laboratories. This was quite successful since the films were contributed by a variety of laboratories. The study demonstrates the feasibility of using a computerized approach for the matching of the HLA class II patterns. There was an excellent correlation between the computer-generated composite patterns for the DRβ and DQβ chains and the consensus patterns determined by visual analysis in the DR4 Workshop report (5).

There are, however, a few remaining limitations in the analysis. Most of the monoclonal antibodies are not completely gene-specific. For example, in Figure 5B, it appears that spots 20 or 23 and 13 may be due to coprecipitation of the DR antigen by the DQ mAbs.

Spots mapping between dw4.cmp and dr.cmp:

dw4.cmp SID	PI	MW	II	<--->	dr.cmp SID	PI	MW	II
1	-5.62	33.72	0.01	<--->	2	-5.60	33.85	0.87
				<--->	1	-5.73	33.02	5.26
2	-5.25	33.85	0.08	<--->	4	-5.25	33.85	0.08
3	-4.76	32.10	0.04	<--->	5	-4.76	32.47	2.17
4	-4.45	33.33	0.01	<--->	6	-4.45	33.33	0.01
5	-3.87	33.56	0.10	<--->	7	-3.89	33.54	0.07
				<--->	8	-3.57	33.18	3.18
6	-2.98	32.92	0.05	<--->	9	-3.06	32.58	0.14
7	-2.52	33.76	0.23	<--->	13	-2.32	33.77	0.27
				<--->	11	-2.66	33.60	0.22
8	-2.49	32.92	0.48	<--->	12	-2.47	32.84	0.82
				<--->	11	-2.66	33.60	0.22
9	-2.07	32.77	0.12	<--->	14	-1.91	32.40	4.77
10	-2.06	33.64	0.14	<--->	13	-2.32	33.77	0.27
11	-1.25	32.21	3.28	<--->	16	-1.19	32.17	5.17
12	-1.16	33.84	0.07	<--->	15	-1.20	33.79	0.20
13	-0.93	31.08	0.01	<--->	17	-0.93	31.08	0.01
14	-0.65	33.67	0.04	<--->	18	-0.65	33.67	0.04
15	-0.58	31.69	0.06	<--->	19	-0.49	31.79	0.33
16	-0.04	31.97	0.55	<--->	20	-0.05	31.93	1.89
17	0.19	33.77	0.20	<--->	21	0.19	33.77	0.20
18	0.76	33.48	0.15	<--->	23	0.76	33.48	0.15
19	1.14	31.59	0.02	<--->	24	1.25	31.90	1.78
20	1.51	31.25	1.33	<--->	25	1.55	31.30	1.68
				<--->	24	1.25	31.90	1.78
21	1.95	33.81	0.01	<--->	26	1.95	33.81	0.01
22	2.00	31.19	0.02	<--->	27	2.03	31.12	0.02
23	2.91	32.75	0.71	<--->	29	2.78	32.61	0.36
24	3.40	31.19	0.45	<--->	32	3.59	31.02	0.03
				<--->	30	3.22	31.41	1.46

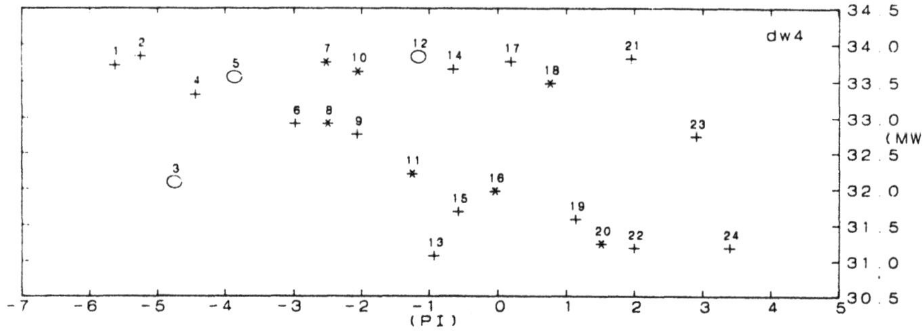

A

Spots mapping between dw14.cmp and dr.cmp:

dw14.cmp SID	PI	MW	II	<--->	dr.cmp SID	PI	MW	II
1	-5.42	32.51	0.10	<--->	3	-5.42	32.51	0.10
2	-3.66	33.85	0.30	<--->	7	-3.89	33.54	0.07
				<--->	8	-3.57	33.18	3.18
3	-3.14	32.25	0.23	<--->	9	-3.06	32.58	0.14
4	-2.07	33.72	0.54	<--->	13	-2.32	33.77	0.27
5	-1.97	32.68	11.43	<--->	14	-1.91	32.40	4.77
6	-1.12	32.29	52.72	<--->	16	-1.19	32.17	5.17
7	0.13	32.03	6.97	<--->	20	-0.05	31.93	1.89
8	1.77	31.47	6.07	<--->	25	1.55	31.30	1.68
				<--->	27	2.03	31.12	0.02

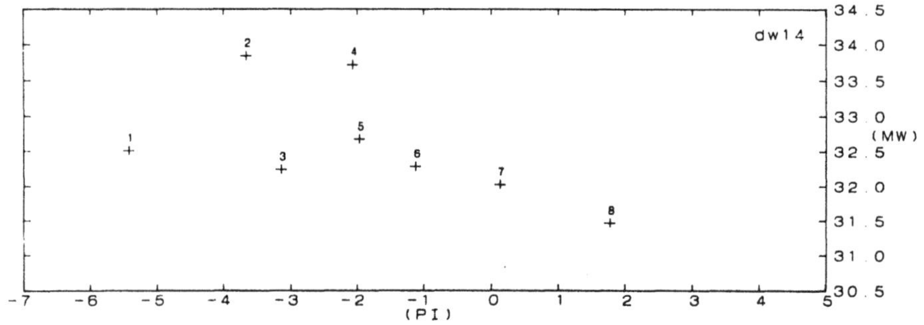

B

Spots mapping between dw13.cmp and dr.cmp:

dw13.cmp SID	PI	MW	II	<--->	dr.cmp SID	PI	MW	II
1	-4.92	33.21	0.23	<--->	5	-4.76	32.47	2.17
2	-3.59	33.30	6.24	<--->	8	-3.57	33.18	3.18
3	-1.97	32.46	13.12	<--->	14	-1.91	32.40	4.77
4	-1.11	32.14	5.09	<--->	16	-1.19	32.17	5.17
5	0.10	31.56	0.69	<--->	20	-0.05	31.93	1.89

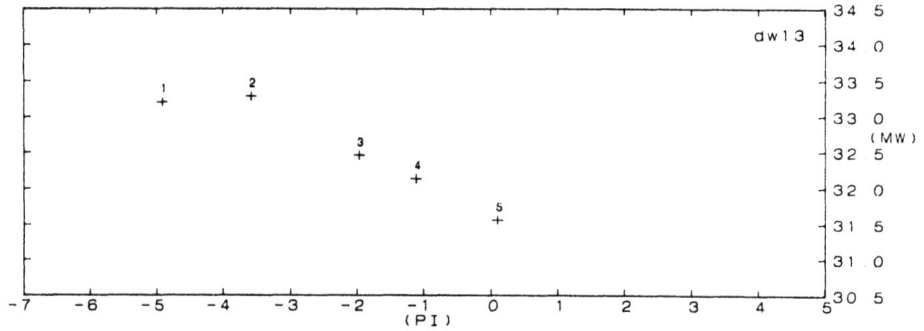

C

Figure 4. Composite maps of DRβ chain for the Dw subtypes of DR4 cell lines. Composite maps were generated from DR films from subgroups of cell lines as described in Figure 3A. (*A*) Dw4, (*B*) Dw14, (*C*) Dw13, (*D*) DwKT2, (*E*) Dw10, and (*F*) Dw15. Spots in each Dw map are numbered sequentially. These spots are related in each section's table and correlated to the composite SID number of Table 1 and Figure 3A.

Spots mapping between dwkt2.cmp and dr.cmp:

dwkt2.cmp				<--->	dr.cmp			
SID	PI	MW	II	<--->	SID	PI	MW	II
1	-3.76	32.76	0.72	<--->	8	-3.57	33.18	3.18
				<--->	7	-3.89	33.54	0.07
2	-3.32	33.03	0.09	<--->	8	-3.57	33.18	3.18
				<--->	9	-3.06	32.98	0.14
3	-2.81	31.24	0.07	<--->	10	-2.81	31.24	0.07
4	-2.31	32.15	3.88	<--->	12	-2.47	32.84	0.82
5	-1.76	32.35	2.04	<--->	14	-1.91	32.40	4.77
6	-1.02	31.97	0.60	<--->	16	-1.19	32.17	5.17
7	-0.40	31.89	0.60	<--->	19	-0.49	31.79	0.33
8	0.50	31.43	4.34	<--->	22	0.50	31.43	4.34
9	1.52	31.20	0.20	<--->	25	1.55	31.30	1.68
				<--->	24	1.25	31.90	1.78
10	2.65	32.47	0.01	<--->	29	2.78	32.61	0.36
11	3.54	33.64	0.05	<--->	31	3.54	33.64	0.05
12	3.59	31.02	0.03	<--->	32	3.59	31.02	0.03

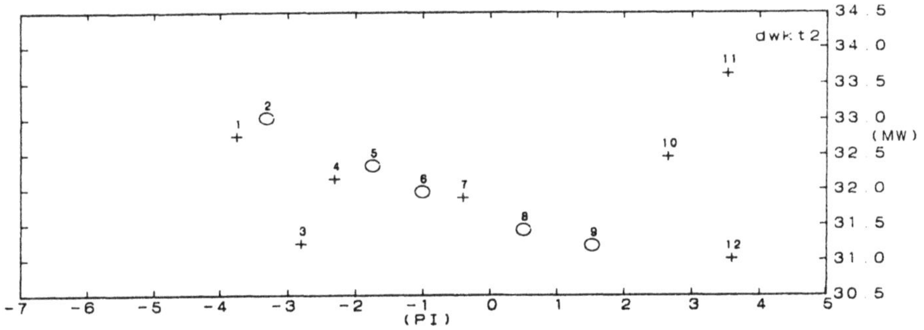

D

Spots mapping between dw10.cmp and dr.cmp:

dw10.cmp				<--->	dr.cmp			
SID	PI	MW	II	<--->	SID	PI	MW	II
1	-5.73	33.02	5.26	<--->	1	-5.73	33.02	5.26
2	-5.73	33.97	1.72	<--->	2	-5.68	33.85	0.87
3	-4.74	32.38	8.53	<--->	5	-4.76	32.47	2.17
4	-3.48	32.72	14.69	<--->	8	-3.57	33.18	3.18
5	-2.06	31.85	2.60	<--->	14	-1.91	32.40	4.77
6	-1.14	31.72	0.04	<--->	16	-1.19	32.17	5.17
				<--->	17	-0.93	31.08	0.01
7	0.00	31.64	0.08	<--->	20	-0.05	31.93	1.89
8	2.52	31.08	0.35	<--->	28	2.52	31.08	0.35
9	3.19	31.34	0.02	<--->	30	3.22	31.41	1.46

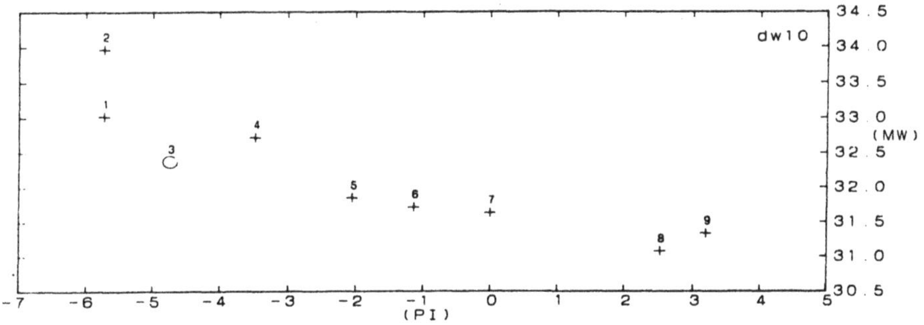

E

Spots mapping between dw15.cmp and dr.cmp:

dw15.cmp				<--->	dr.cmp			
SID	PI	MW	II	<--->	SID	PI	MW	II
1	-1.28	33.70	0.48	<--->	15	-1.20	33.79	0.20
2	-1.26	32.80	3.68	<--->	16	-1.19	32.17	5.17
3	-0.13	32.24	11.82	<--->	20	-0.05	31.93	1.89
4	1.35	32.20	3.54	<--->	24	1.25	31.90	1.78
5	2.06	31.06	0.02	<--->	27	2.03	31.12	0.02
6	3.06	31.71	3.90	<--->	30	3.22	31.41	1.46

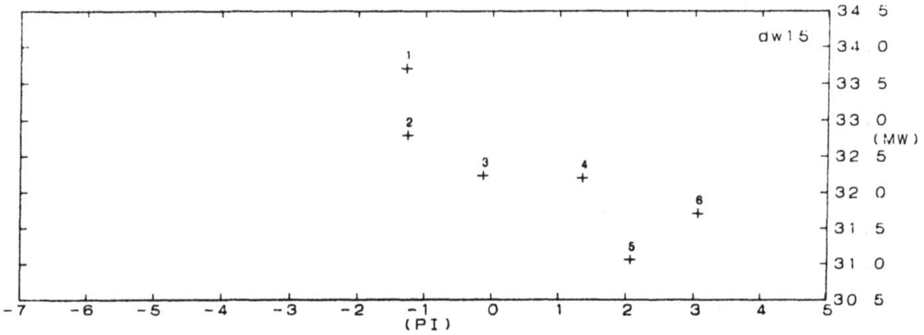

F

Spots mapping between dqb7.cmp and dq.cmp:

dqb7.cmp SID	PI	MW	II	<--->	dq.cmp SID	PI	MW	II
1	-5.91	32.75	0.02	<--->	1	-5.91	32.75	0.02
2	-4.82	32.45	0.26	<--->	4	-4.77	32.17	0.30
3	-3.54	31.73	1.42	<--->	7	-3.46	31.93	1.06
4	-2.02	31.15	1.10	<--->	11	-2.02	31.15	1.10
5	-1.20	31.59	0.11	<--->	13	-1.17	31.95	0.78
				<--->	14	-0.94	31.95	0.57
6	-0.55	33.77	0.51	<--->	15	-0.57	33.53	0.27
7	0.76	33.65	3.71	<--->	20	0.62	33.63	1.01
8	1.69	33.60	5.77	<--->	25	1.53	33.72	1.26
9	2.05	33.40	1.42	<--->	28	2.24	33.55	0.71
				<--->	27	1.97	32.62	0.00
10	2.88	33.67	9.13	<--->	29	2.91	33.70	6.35
11	3.41	31.20	0.00	<--->	31	3.30	31.20	5.32
12	3.62	33.62	2.27	<--->	33	3.89	33.83	4.15
13	4.00	33.94	2.40	<--->	33	3.89	33.83	4.15

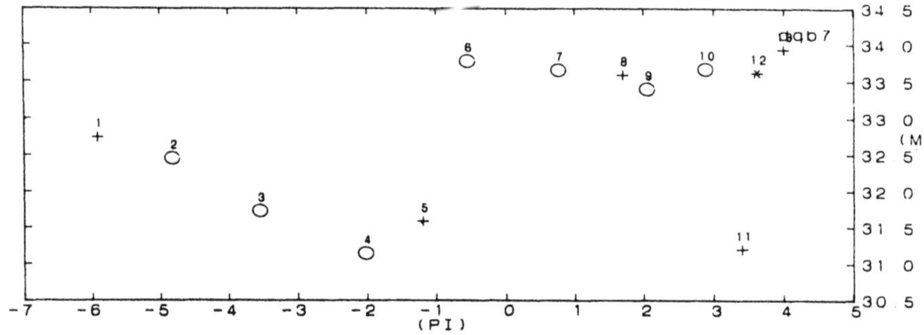

Spots mapping between dqb8.cmp and dq.cmp:

dqb8.cmp SID	PI	MW	II	<--->	dq.cmp SID	PI	MW	II
1	-5.50	33.32	0.09	<--->	2	-5.50	33.32	0.09
2	-5.18	32.04	0.92	<--->	3	-5.18	32.04	0.92
3	-4.75	32.07	0.32	<--->	4	-4.77	32.17	0.30
4	-4.44	32.26	0.03	<--->	5	-4.44	32 26	0.03
5	-4.18	33.28	0.04	<--->	6	-4.18	33.28	0.04
6	-3.24	33.68	0.01	<--->	8	-3.24	33.68	0.01
7	-3.21	32.55	0.00	<--->	7	-3.46	31.93	1.06
8	-2.81	32.81	0.05	<--->	9	-2.81	32.81	0.05
9	-2.22	32.45	0.30	<--->	10	-2.22	32.45	0.30
10	-1.62	32.59	0.34	<--->	12	-1.56	32.47	0.27
11	-1.28	32.16	0.26	<--->	13	-1.17	31.95	0.78
				<--->	12	-1.56	32.47	0.27
12	-0.81	33.31	0.01	<--->	15	-0.57	33.53	0.27
13	-0.76	31.91	0.62	<--->	14	-0.94	31.95	0.87
				<--->	16	-0.51	31.83	0.74
14	-0.44	31.04	0.04	<--->	16	-0.51	31.83	0.74
15	-0.39	31.91	1.37	<--->	16	-0.51	31.83	0.74
16	-0.36	33.27	0.04	<--->	15	-0.57	33.53	0.27
17	0.09	31.33	0.21	<--->	17	0.09	31.33	0.21
18	0.22	32.52	0.00	<--->	18	0.22	32.52	0.00
19	0.57	33.62	0.10	<--->	20	0.62	33.63	1.01
20	0.58	31.49	0.39	<--->	19	0.58	31.49	0.39
21	0.78	32.48	0.00	<--->	21	0.78	32.48	0.00
22	0.92	31.61	0.42	<--->	22	0.92	31.61	0.42
				<--->	23	1.14	31.19	0.66
23	1.14	31.19	0.66	<--->	23	1.14	31.19	0,66
				<--->	22	0.92	31.61	0.42
24	1.19	33.63	0.12	<--->	24	1.19	33.63	0.12
25	1.49	33.75	0.14	<--->	25	1.53	33.72	1.26
				<--->	24	1.19	33.63	0.12
26	1.66	31.41	1.29	<--->	26	1.66	31.41	1.29
27	1.97	32.62	0.00	<--->	27	1.97	32.62	0.00
28	2.30	33.60	0.48	<--->	28	2.24	33.55	0.71
29	2.96	31.33	0.52	<--->	30	2.96	31.33	0.52
30	2,97	33.75	0.78	<--->	29	2.91	33.70	6.35
31	3.28	33.72	0.51	<--->	32	3.30	33.72	0.54
32	3.29	31.20	5.85	<--->	31	3.30	31.20	5.32

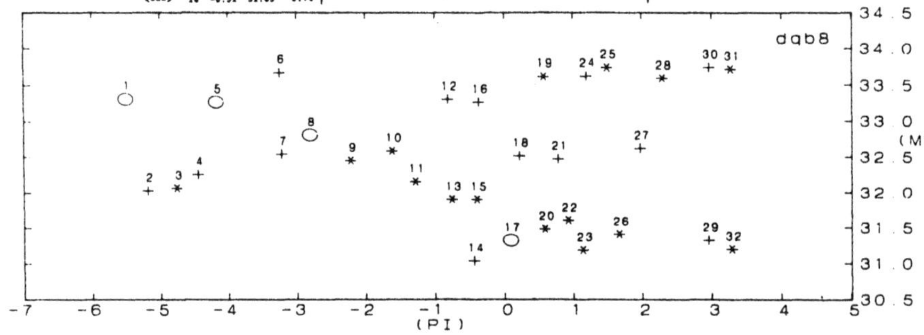

Figure 5. Composite maps of DQβ chain spots for the DQ subtypes of DR4 cell lines. Composite maps were generated from DQ films from subgroups of cell lines as described in Figure 4. The proteins present in Table 2 were correlated with Dw type and listed in the format described in Figure 3A. These DQ spots are also correlated to the composite SID number of Table 2 and Figure 3B. (*A*) DQw7 and (*B*) DQw8.

Likewise, the coprecipitation of DQ chains by the DR mAbs was also observed. Usually, these can be distinguished by examination of the relative integrated intensities, but future studies will be directed toward developing algorithms that can use integrated intensity to eliminate cross-reactive contaminants. Another problem involved a conductivity gap in the ampholyte mixture used for the β chain isoelectric focusing separation. The spot at a relative pI position of −2 was often distorted and could not be used in the algorithm. Other spots in this area often appeared to be split into two clusters (e.g., 9 and 10 of Fig. 5B and 7 and 10 and 8 and 9 of Fig. 4A). This problem should be easily remedied using a different batch of ampholytes. A further limitation of the study is the small number of films included within some of the DR4 subgroups. The window of tolerance used in the matching algorithm is also a potentially important variable, and the values selected (0.3 pI units and 0.8 kDa MW units) may not have been optimum. These tolerances can be adjusted in future studies to optimize the discrimination of spots that are closely related by size and charge. Because the class II antigens are closely related in size, it may be important to improve resolution in the MW dimension. In future studies, it may be possible to increase the resolution using different acrylamide gel concentrations or gradient gels. Reducing the window of tolerance may also improve the accuracy of the matching analysis.

In summary, it is clearly possible to computerize the analysis of HLA class II antigens. This study illustrates that with careful standardization of the 2-D gel techniques, sufficient reproducibility can be obtained to allow the matching of patterns from different laboratories. Application of this approach to the other HLA-DR antigen groups should result in the generation of a database that can be used as a standard reference for future studies. Furthermore, the programs developed for this analysis have been run on a microcomputer, suggesting that large and expensive systems are not required. These programs have general applications beyond the analysis of the HLA class II antigens, and should prove useful in a variety of applications where 2-D gel electrophoresis is being used.

Acknowledgments. We thank Dr. Julia Bodmer for providing several non-Workshop DR4 cell lines. This work was supported, in part, by NASA Grants NAG 8-716 and NAGW-693; gifts from American Cyanamid Company, Pierce Chemical Company, and BioImage Corporation (US7DWS); funding from the Arthritis and Rheumatism Council of Great Britain (UKIWEL); and NIH grant AI22916 (US7KNO).

References

1. Crumpton MJ, Bodmer JG, Bodmer WF, Heyes JM, Lindsay J, Rudd CE. Biochemistry of class II antigens: Workshop report. In: Albert ED, Baur MP, Mayr WR (eds): Histocompatibility Testing 1984. New York: Springer-Verlag, 1984: 29–37.
2. Knowles RW, Kienzle B. Standardized reagents and techniques developed for the 10th annual workshop: 2-D gel study of Class II antigens. In: Dupont B (ed): Immunobiology of HLA, Volume I: Histocompatibility Testing 1987. New York: Springer-Verlag, 1989.
3. Dwyer VG, Benson B, Kwan KH, et al. A computer-enhanced comparative study of brain region polypeptides and proteins by 2-D gel electrophoresis. Pharmaceut Biomed Anal (special edition) 1988;6(2):185.
4. Sammons DW, Humphreys RC, Lin NH, et al. A computerized strategy for estimation of protein purity. Proceedings of Frontiers in BioProcessing Conference, Boulder, Colorado, Boca Raton, FL: CRC Press, 1989: Chapter 24.
5. Lanchbury JSS, Nepom GT, Radka SF, et al. DR4: 2-D gel patterns. In: Dupont B (ed): Immunobiology of HLA, Volume I: Histocompatibility Testing 1987. New York: Springer-Verlag, 1989.

Joint Reports: T-Cell Recognition of HLA Class II Molecules

"Who would have imagined . . .

. . . that so few T-cell clones . . .

. . . could cause so much grief . . .

. . . but would be so worthwhile."

Margo Honeyman
Spring 1987

Preface

The Tenth International Histocompatibility Workshop component concerning T-cell recognition of HLA class II molecules drew its strength from the hard work, diligence, and selfless spirit of the 23 participating laboratories. The enthusiasm and camaraderie exhibited by the participants in Princeton during November 1987 bear testimony to the caliber of the individuals involved.

Numerous individuals, both inside and outside of the Organizing Committee, contributed significantly to the success of this component of the Workshop. We thank David Eckels, Adrianna Zeevi, Nancy Reinsmoen, and Eric Mickelson for their continued advice, encouragement, and hard work throughout this endeavor. We thank Deborah Richardson for her help during the early stages of this work, and Fran Berman for her help in preparing the report. MaryAnn Barletta, Sally Krell, and Halina Korsun provided invaluable help with a multitude of organizational and operational issues. Bo Dupont and Bob Knowles provided sound advice, helpful discussions, and continued support. John Hansen, Jean Marc Lalouel, and the other members of the Organizing Committee made valuable contributions. Rosemarie Plätke's enthusiasm, diligence, and statistical expertise were vital ingredients in this undertaking. John Klein was always ready to step in and assume whatever responsibility was necessary to keep the work going. Above all, I owe special and personal thanks to my family, whose patience, support, and understanding helped to sustain me over the last 2 years.

Neal Flomenberg
June 1988

Background, Goals, and Experimental Design

N. Flomenberg,[1] J.A. Hansen,[2] J.S. Klein,[1] and B. Dupont[3]

Introduction

The Tenth International Histocompatibility Workshop component concerning T-cell recognition of HLA class II molecules was designed with careful attention to the overall Workshop structure. Our goal was to design a collaborative study which could stand as an independent and self-sufficient scientific investigation, but which, at the same time, could be integrated with the other Workshop efforts. The Workshop B-Lymphoblastoid Cell Panel was an obvious connecting link to the other HLA class II components around which such a study could be built. We hoped that cellular studies of the cell panel might indeed point out functionally relevant polymorphisms between the cell lines, which would complement the information gained from the Southern blot, cDNA sequencing, and protein chemistry components.

Along with other advances in technology during the last 5 to 10 years, it has become feasible to dissect the allocytotoxic response and secondary alloproliferative response into their component parts mediated by individual T-cell clones. It was our belief that application of this new technology might allow us to extend and expand the knowledge obtained from cellular studies performed in previous Workshops and individual laboratories. In designing this section of the Workshop, our challenge was to introduce the use of T-cell cloning technology into a large-scale international collaboration, including laboratories with both extensive, as well as limited, expertise in this field.

To begin to systematically dissect cellular recognition of HLA class II molecules into its clonal components, we were faced with two specific goals. These were: 1) to produce a common core definition of the specificity of each clone studied in the Workshop, in a manner that would allow clones derived in different laboratories and tested in different assay systems (i.e., proliferation vs. cytotoxicity) to be compared one to another; and 2) to map the epitope recognized each T-cell clone to a specific HLA class II family (i.e., DR, DQ, or DP) and, when possible, to map the epitope to specific alpha-beta chain pairs. Additionally, it was necessary to reduce the results of the cellular studies to a form that could be compared and correlated with results obtained in the other Workshop components.

Study Design

The development of the Workshop protocol was influenced, in part, by a small-scale workshop involving five participating laboratories sponsored by the American Society for Histocompatibility and Immunogenetics in 1984 (1). This study utilized 12 HLA homozygous cell lines for panel analyses and 30 monoclonal antibodies for inhibition assays. The consensus of the participants was that the approach was worthwhile. Scoring of the panel results and the blocking data were based on individual local criteria (with the studies being performed in a blinded fashion). Despite the subjective scoring system, the patterns of allorecognition and blocking were sufficiently clear-cut to validate the study. It was the feeling of the participants that useful information would be gained "by studying a substantially large number of T-cell clones, preferably with overlapping specificities and different functions." The Committee also noted that "such a project will succeed to the degree that well characterized T-cell clones and monoclonal antibodies are used, and that appropriate statistical procedures are developed."

The problem to be solved in developing a successful protocol was, therefore, to balance two opposing needs. First, we had to be sufficiently extensive and intensive in these studies to produce meaningful results, both with the panel analyses and the antibody blocking assays. It was also crucial that the data be produced in a way that would allow for comparisons from clone to clone, from laboratory to laboratory, and from experiment to experiment in an objective fashion. At the same time, we were confronted with the reality that the growth of T-cell clones remains as much an art as a science. Therefore, it was necessary to simultaneously be frugal in our design to allow this information to be obtained from realistic numbers of cloned T-cells and from experiments of manageable size. Given these considerations, it seemed totally impractical to expect that clones could be grown to sufficient number to distribute and test in a variety of laboratories. Our goal, therefore, focused on standardizing the assays and analyses so that the results of each individual laboratory could be synthesized into a larger whole.

Since the techniques utilized by successful T-cell cloning laboratories are quite diverse, it was decided that the method for priming and expansion of the clones would be left to the individual laboratories. Pitfalls to be avoided were summarized for laboratories with less experience. In particular, the need for vigilant mycoplasma surveillance and prophylaxis was emphasized. In contrast to the paucity of restrictions in clonal generation and growth, the assays of alloreactivity (or MHC restriction) and antibody blocking were carefully standardized. (The details of the experimental methods are included in the Appendix of this chapter.)

Workshop Definition of Clonal Specificity: Selection and Analysis of the B-Lymphoblastoid Cell Line Panel. The total HLA class II cell panel utilized for the cellular

---

Participating Laboratories: US7NFL,[1] US7HAN,[2] US7DUP[3]

Immunobiology of HLA
Volume I:

Table 1. Priority List for Cellular Studies

Priority Number	WS-ID #	Cell ID
1	9006	WT 100-BIS
2	9017	WT 8
3	9023	VAVY
4	9031	BOLETH
5	9045	TUBO
6	9060	CBG-IB (CB6B)
7	9050	MOU
8	9068	BM 9
9	9075	DKB
10	9004	JES THOM
11	9011	E 4181324
12	9019	DUCAF
13	9030	JHAF
14	9038	BM 16
15	9064	AMALA
16	9048	LBUF
17	9066	TAB 089
18	9028	PE 117
19	9039	JVM
20	9059	SLE-005
21	9002	MZ 070782
22	9016	RML
23	9021	RSH
24	9027	PF 97387
25	9042	TISI
26	9065	HHKB
27	9052	DBB
28	9069	MADURA
29	9001	SA
30	9026	YAR
31	9037	SWEIG 007
32	9061	ABO
33	9067	BTB
34	9062	WDV
35	9003	KAS 116
36	9009	KAS-011
37	9022	COX
38	9034	SAVC
39	9036	SPO-010
40	9056	KOSE
41	9049	IBW 9
42	9071	OLGA
43	9074	HID
44	9012	WJR 076
45	9014	M GAR
46	9025	DEU
47	9054	EK/OH
48	9008	DO 208915
49	9029	WT51
50	9040	BM 15
51	9063	WT 47
52	9005	HOM-2
53	9010	AMAI
54	9020	QBL
55	9032	BSM
56	9035	J BUSH
57	9055	HO 301
58	9046	BM (BH)
59	9070	LUY
60	9073	KT12
61	9007	DEM
62	9018	LO 081785

Table 1. *Continued*

Priority Number	WS-ID #	Cell ID	
63	9033	BM 14	
64	9044	BRI-P	
65	9058	OMW	
66	9047	PLH	
67	9072	S PACHEO	
68	9013	SCHU	
69	9024	KT17	
70	9043	BM 21	
71	9057	TEM	
72	9015	WT 24	
73	9041	JO 528239	
74	9051	PITOUT	
75	9053	HOR	
Blinded Duplicates			
C-A	9108	DBB	(9052)
C-B	9109	JES THOM	(9004)
C-C	9110	E 4181324	(9011)
C-D	9111	BM 9	(9068)
C-E	9112	BOLETH	(9031)
C-F	9113	CBG-IB (CB6B)	(9060)
C-G	9114	BM 16	(9038)
C-H	9115	VAVY	(9023)

studies of the Workshop included 75 core cell lines and 8 blinded duplicates. These cell lines were selected from the overall Workshop Cell Line Panel based on the availability of prior detailed immunogenetic information, with particular emphasis on HLA-D typing and HLA-DP typing. The selection was performed in concert with Dr. Robert Knowles to produce maximal overlap between the results of the cellular and biochemical analyses. The remaining cell lines were provided as a supplemental set for testing when, and if, possible.

The core cell lines and eight blinded duplicates were given priority numbers to indicate the sequence in which the cell lines were to be tested. This sequence was designed to mix the cell lines that would be tested according to their DR type and, if known, their Dw and DP types. This would ensure that if a laboratory could only test a particular clone or clones against half the Workshop Cell Panel, that the clone would not end up being tested against all the cell lines expressing DR1 through DR4, and none of the cell lines expressing DR5 through DRw9. Similarly, this approach would ensure that the cell lines tested in various laboratories would overlap as much as possible. The Workshop cell lines are listed in order of their assigned priority score in Table 1.

Workshop cell lines were to be tested together with a local positive and negative control cell line in each experiment. Aside from ensuring that the clone was behaving appropriately in a particular experiment, the positive and negative controls were included in order to allow the data to be normalized. In this way the entire Workshop Cell Panel did not have to be tested at one time, but could be tested in several sections in experi-

Table 2. Monoclonal Antibodies Used in HLA Class II Cellular Studies

Antibody Name	Submitted Specificity	Core	Priority #	Source Reg	Source Lab
PL8-C (A-1)	DR	Y	1	US7	KNO
1A3-C (A-2)	DQ	Y	2	US8	RJW
B7/21-C (A-3)	DP	Y	3		BD
Tu39	DP, DR	Y	4	GER	MUC
w6/32	HLA-A,B,C – CONTROL	Y	5	US7	SYY
M101	CONTROL	Y	6	US7	KNO
Tu35	DP,DR	Y	7	GER	MUC
GSP 4.1	DR	Y	8	US7	GSP
1A3	DQ	Y	9	US8	RJW
B7/21	DP	Y	10	US2	BAC
PL8	DR	Y	11	US7	DUP
Tu22	DQ	Y	12	GER	MUC
PL15	DP	Y	13	US7	DUP
L243	DR	Y	14		BD
2D6	DP,DR	Y	15	US7	POL
Leu 10	DQw1,DQw3	Y	16		BD
3G5	DR1			BEN	ENG
7.3.19.1	DRw52			BEN	ROO
16.23	DRw3,w13,w18			GER	WAN
17.15	DQw3			US7	GSP
77-34	DQw1			NCY	FAL
AB7.AE9	DRw9			US7	SFE
AB8.25	DR2,5			IT1	CEP
C1	DQw2			GER	WAN
CC-CL 20	DR1,4			US6	BRN
CHE249.2	DR5			FRA	DDC
DP11.1	DP-ALPHA			UKI	BOD
GS 4AA7	DQw1,w3 SUB,BLANK			US7	GSP
GSP 17.1	DQw2,w3,BLANK			US7	GSP
GSP 88.2	DR5			US7	GSP
IIB3	DQ SPLIT, NOT TA10			BEN	ROO
IIIE3	DRw52,w53			US4	STA
IVD12	DQw3			US4	STA
JS1	DR3			UKI	FES
KS-6	DQw1-SUB			US7	SFE
KS-7	DQw1,w3-SUB			US7	SFE
MAD88	DRw8			FRA	DDC
NDS10	DRw52,3,5,w6			UKI	MOR
NDS13	DRw52,3,5,w6,w8			UKI	MOR
NDS15.38	DR2			UKI	MOR
NFLD.M1	DR4			NCY	MAR
PL3	DRw53			US7	DUP
PLM2	DQw3-TA10			JAP	JUJ
PLM12	DQw3			JAP	JUJ
PLM13	DQw1,w3 SUB			JAP	JUJ
UK 4.1	DQw1			UKI	BOD
UK 8.1	DRw52			UKI	BOD
UL-65	DR3			GER	GOL
SFR3-DR5	DR5			NCY	SFR
SFR16-DR7M	DR7			NCY	SFR

A-1, A-2, A-3 are the duplicate controls. SUB = subset BD = Becton Dickinson

ments of smaller size. The positive and negative cell lines would allow us to control for variations in how briskly the clones proliferated, or how strongly they killed on a day-to-day basis. In assays of proliferation, pool-PLT or PHA-PLT reagents were to be developed by the local laboratories and tested against each cell line to ensure that the line was capable of inducing a proliferative response on the day of the assay.

Determination of Molecular Specificity: The Antibody Panel. Antibodies were selected for use in the cellular and protein chemistry studies in tandem to

Table 3. Participants in the HLA Class II Cellular Studies

Laboratory	No. of Clones Submitted	Clone Type
AIZ	5	Cytotoxic
BAC	14	Cytotoxic
DDC	1	Cytotoxic
KUR	5	Cyto/Prolif
NFL	9	Cytotoxic
BAT	6	Proliferative
BSH	3	Proliferative
DUQ	4	Proliferative
FER	3	Proliferative
FES	2	Proliferative
GOL/HEI	5	Proliferative
HAN	11	Proliferative
JEA	2	Proliferative
KSH	2	Proliferative
LEV	6	Proliferative
MBC	18	Proliferative
MYE	6	Proliferative
NIK	5	Proliferative
OHA	6	Proliferative
ROO	7	Proliferative
TSB	6	Proliferative
ZAP	1	Proliferative
ZIE	3	Proliferative

facilitate integration of the results. Careful attention was placed on selecting critical core reagents with a high likelihood of mapping the epitope in question to either the DR, DQ, or DP family. Selection of the DR reagents was guided by the ASHI Workshop, in which three antibodies (GSP4.1, L243, and PL8) produced blocking that was limited to T-cell clones specific for HLA DR gene products. Monoclonal antibody B7/21 was selected as the "gold standard" for DP-specific reagents. The selection of reagents specific for HLA-DQ was more difficult in that several of the best reagents, such as Tu-22 and Leu-10, bind with lesser affinity (or do not bind at all) to DQw2 gene products. This problem was diminished greatly with the availability of monoclonal antibody 1A3, which appeared to represent a pan HLA-DQ reagent. Additionally, several irrelevant control antibodies were included in the core panel, as well as a blind duplicate of a DR-, DQ-, and DP-specific reagent. These core antibodies were given priority numbers for reasons analogous to those discussed above for the cell panel. In addition to the core antibodies, a large variety of polymorphic antibodies were selected, with the hope that they might assist in the fine mapping of the various epitopes. These were distributed as a supplemental antibody set. Antibodies utilized in this section of the Workshop are listed in order of their assigned priority numbers in Table 2.

Experimental Sequence. Each clone was to be tested sequentially as follows:

1. Panel analysis of the reactivity of the clone against the first 34 core cell lines and the 8 blinded duplicates.
2. Antibody inhibition studies using the core antibodies, including the blinded duplicates.
3. Panel analysis of the reactivity of the clone against the remaining core cell lines and as many relevant supplemental cell lines as possible.
4. Antibody inhibition studies using relevant supplemental antibodies.

It should be reemphasized that experiments were to be performed in duplicate. Panel results that were not performed in duplicate were not accepted for the central analysis, since we felt there was no way in which we could accurately gauge the significance of unexpected patterns of reactivity from a single testing. We did attempt to analyze antibody blocking data from a single testing. Here, for example, we felt that inhibition by several different anti-DQ reagents (each only tested once) would together provide reasonable assurance that the clone recognized a DQ gene product. Additional panel data might have been incorporated had we applied the same concept to the panel studies. However, we were quite concerned that this would lead to the inclusion of potentially erroneous data or that, alternatively, we would be tempted to include results that conformed to preconceived ideas and reject results that did not. For this reason, the only objective and uniformly fair approach was to reject all panel data when results from two concordant experiments were not provided.

Cell lines and antibodies were distributed to 35 laboratories that had indicated an interest in this component. Twenty-three laboratories returned data to us, consisting of analyses of 130 T-cell clones. The laboratories that submitted data are listed in Table 3. The statistical approaches used for the reduction and analysis of this data are described in the next two papers, while the immunogenetic findings resulting from the analysis are discussed in the remaining papers in this section.

Reference

1. Eckels DD, Zeevi A. ASHI workshop summary report of the science and education subcommittee: Structural and functional relationships of human class II MHC molecules. Hum Immunol 1986;15:68–74.

Appendix: Methods

A. Thawing Cell Lines.

1. Vials are warmed in a 37°C waterbath, with constant shaking until only a small ice crystal remains.
2. The contents of the vial are removed and placed in a sterile tube.
3. Ten ml of sterile growth medium (see below) are added to each vial, one drop at a time, with gentle shaking.
4. Cells are centrifuged 1400 RPM for 8 minutes.
5. Supernatant is removed, pellets are resuspended by shaking, and cells are transferred to flasks in 10 ml of growth medium.
6. Cells should be in culture for at least 48 hours and

should have viabilities of at least 80% prior to use in functional studies.

Growth medium = RPMI-1640 supplemented with:
Penicillin
Streptomycin
Tylocine
Glutamine
10% fetal calf serum

Tylocine (Anti-PPLO Agent) (100X) 6000 μg/ml (GIBCO Cat. #600-5220)

Penicillin	(100X)	10,000 U/ml
Streptomycin	(100X)	10,000 μg/ml
L-Glutamine	(100X)	29.2 ng/ml

B. Experimental Hierarchy (Suggested).

1. The minimum data set that will be analyzed centrally consists of panel analysis of core cell lines 1 through 34 and the eight blinded duplicate cell lines (C-A through C-H), as well as analysis of core antibodies 1 through 11.

a. Test 23 core cell lines (CA-CH, 1–13, and the local positive and negative controls) in the proliferative or cytotoxicity assays as described below. Repeat the entire analysis once. Resolve any bad data points or discordant data in a third experiment.

b. Test core cell lines 14 to 34 plus the two local controls in the same manner as described above.

c. Select the cell line to be used in the blocking experiments for each clone. This cell line will be the first Workshop cell line (i.e., priority number closest to 1 which (a) matches the DR type of the specific stimulator cell, and (b) strongly stimulates or is strongly lysed by the clone being tested. (% specific lysis ≥ 30% or proliferation ≥ 8-fold above the background proliferation of the clone and cell line in question.

d. Perform blocking study (see below) at 1:40 dilution only with core antibodies (1 to 16). Repeat analysis of antibodies which fail to block (<20% inhibition) at 1:40 dilution only. Repeat analysis of blocking antibodies (≥ 20% inhibition) at four concentrations (1:40, 1:200, 1:1000, 1:5000). Repeat titration analysis of blocking antibodies. Resolve discordant results if necessary.

2. Proceed to test the rest of the cell panel.

a. Test core cell lines 35 to 55 plus the two local controls as described in section la above.

b. Test cell lines 56 to 75 plus the two local controls as described in section la above.

c. Test the appropriate supplemental cell lines plus the two local controls as described in Section la above. Example: If the clone recognizes only DR1 cell lines, test the supplemental DR1 cells. If the clone recognizes DQw1 cell lines, test the supplemental DR1, DR2, and DRw6 cell lines. Test every clone with the DR-supplemental cell lines.

3. Test the supplementary antibody set.

a. Utilize the same cell line for each clone as selected in section lc above.

b. Perform blocking studies with all antibodies that bind to cell lines of interest as described in section ld above. (The list of binding patterns will be forwarded to you as soon a possible.)

4. **Please Note**

All functional assays are to be performed in:

RPMI-1640 supplemented with:
Penicillin
Streptomycin (or gentamicin)
Tylocine
Glutamine
10% pooled normal human serum

Tylocine (Anti-PPLO agent) (100X) 6,000 μg/ml (GIBCO Cat. #600-5520)

Penicillin	(100X)	10,000 U/ml
Streptomycin	(100X)	10,000 μg/ml
L-Glutamine	(100X)	29.2 ng/ml

C. Proliferation Assay.

1. All experimental points are tested in triplicate in round bottom plates.
2. LCLs are irradiated 10,000 rads, resuspended at a concentration of 2.5×10^5 cells per ml in RPMI-1640 + 10% human serum. Aliquots of 100 μl are dispensed to each well. Three microwells are plated for each clone to be tested plus three additional wells to measure background proliferation (media control).
3. Clones are resuspended in RPMI-1640 + 10% human serum at a concentration of 1×10^5 cells per ml. Aliquots of 100 μl are dispensed. Three additional microwells are plated to measure background proliferation (media control).
4. One hundred microliters of RPMI + 10% human serum is added to each of the media control microwells.
5. Plates are incubated for 48 hours at 37°C in a 5% CO_2, humidified atmosphere. Plates are labeled with ^{3}H-TdR or ^{14}C-TdR for 18 hours, after which they are refrigerated until harvested and counted. Median values are selected from each triplicate.
6. PHA-PLT or Pool-PLT reagents must be included in each study to ensure that the cell lines are capable of effective stimulation on the date tested.

D. Cytotoxic Assay.

1. All experimental points are tested in triplicate in round bottom plates.
2. 2 to 5×10^6 target cells are labeled with ^{51}Cr for 60 to 90 minutes at 37°C, washed twice, and resuspended at 3×10^4 cells per ml. One hundred microliter aliquots are dispensed. Six additional wells are plated for spontaneous ^{51}Cr release and six for maximum ^{51}Cr release.
3. Cloned cells are utilized at an effector to target ratio which produces at least 25% specific lysis of a known relevant target cell. One hundred microliter aliquots are dispensed.
4. One hundred microliters of RPMI-1640 + 10% human serum are added to the spontaneous release microwells and 100 μl of 5% Triton-X detergent are added to maximum release microwells.

5. Plates are spun momentarily at 2000 RPM and incubated for 4 hours at 37°C in a humidified, 5% CO_2 atmosphere. Plates are then spun at 2000 RPM for 5 minutes, and supernatants are harvested for counting.

E. Blocking Assays.

1. All experimental parts are tested in triplicate.
2. Monoclonal antibodies are supplied as ascites fluids. These should be aliquoted (50 to 100 μl) and frozen.
3. Antibodies which are thawed for use are spun at 15,000 *g* to remove aggregates.
4. Antibodies are diluted 1:10, 1:50, 1:250, and 1:1250 in RPMI 1640 + 10% human serum.
5. Stimulator cells or target cells are resuspended at twice their usual concentration in RPMI-1640 + 10% human serum. Fifty microliter aliquots of antibody and stimulator/target cells are distributed. Cells and antibody are permitted to incubate for 30 minutes at room temperature prior to addition of clones. Final antibody concentrations are therefore 1:40, 1:200, 1:1000, and 1:5000.
6. Six wells are plated as an unblocked control, receiving 50 μl of medium alone in place of antibody.
7. Control points (background proliferation, spontaneous ^{51}Cr release, and maximum ^{51}Cr release) are all prepared as detailed above.

F. Preparation of PHA-PLT Reagent.

1. Isolate peripheral blood mononuclear cells by ficoll-hypaque density gradient centrifugation.
2. Resuspend cells in media consisting of RPMI-1640 supplemented with penicillin, tylocine, glutamine, 10% pooled normal serum, and 0.1% v/v PHA-P at a concentration of 2×10^5 cells per ml. Incubate in a humidified, 5% CO_2 atmosphere at 37°C.
3. Replace media when cultures become acidic (yellow). Harvest cells after 10 days.
4. Cryopreserve cells with 5 to 10×10^6 per vial.
5. PLA-PLT cells are thawed on the day clones are to be tested and are adjusted to a concentration of 2×10^5 per ml in RPMI-1640 + 10% human serum. One hundred microliter aliquots are dispensed. This reagent is utilized to test the ability of each B-LCL to serve as an allogeneic stimulator cell on the date in question. The PHA-PLT reagent cells are also cultured in the absence of any B-LCL to determine the background proliferation of this reagent. One may alternatively utilize a Pool-PLT reagent (described below).

G. Preparation of Pool-PLT Reagent.

1. Draw 60 ml of heparinized blood from two normal donors who will serve as responders.
2. For the stimulator pool, draw 40 ml of heparinized blood from three normal donors. Select individuals disparate for as many HLA class II antigens as possible.
3. Isolate mononuclear cells by ficoll-hypaque centrifugation.
4. Irradiate stimulator cells for 3000 rads. Pool the stimulator cells.
5. Culture 8 to 10×10^6 responder and 8 to 10×10^6 pooled stimulator cells in 15 ml of RPMI-1640 + 20% pooled normal human serum in 50 ml flasks. Feed each PLT on days 2, 4, 6, and 8 with 2 ml of medium.
6. Harvest the two PLTs on day 10. Combine the cells and cryopreserve at 5 to 10×10^6 per vial.
7. Pool-PLT cells are thawed on the day clones are to be tested and are adjusted to a concentration of 2×10^5 per ml in RPMI-1640 + 10% human serum. One hundred microliter aliquots are dispensed. This reagent is utilized to test the ability of each B-LCL to serve as an allogenic stimulator cell on the date in question. The Pool-PLT cells are also cultured in the absence of any B-LCL to determine the background proliferation of this reagent. One may alternatively utilize a PHA-PLT reagent (described above).

Collection and Processing of HLA Class II Cellular Data

R. Plätke,[1] M. Ferguson,[1] J.S. Klein,[2] J. Zemmour,[3] N. Flomenberg,[2] and J.M. Lalouel[1]

Introduction

Data collection and processing consisted of three steps:

1. Collecting the data from the participating laboratories through the use of computer programs for data entry.
2. Checking the triplicate measurements for errors or inconsistencies, and reducing the raw data to indices of cytotoxicity, proliferation, and blocking.
3. Discarding data when the indices of cytotoxicity, proliferation, and blocking obtained from the duplicate experiments were discordant.

In the following sections, we describe each of these steps separately.

Data Collection

All laboratories participating in the study received four computer programs for data entry. The first program was designed to uniquely define each T-cell clone for the Workshop and record relevant information about that clone. The second program was used to enter the actual experimental results. The other two programs were used to print the information entered using programs 1 and 2. All programs were written in Pascal for use on IBM PCs or compatible computers.

Before entering data, each T-cell clone had to be defined using program 1, in order to provide a unique Workshop identification. Basic information on the clone, including the laboratory of origin, the local name of the clone, the sensitizing HLA antigens (and/or nominal antigens), the donor's HLA phenotype (or genotype, if known), the type of clone (antigen-specific MHC-restricted or allospecific, cytotoxic or proliferative), etc. were recorded.

The experimental data entry system provided in program 2 allowed the user to enter and view the data directly on the computer screen. For each pair of duplicate experiments, the user was provided with a blank table on the screen to be filled out with the relevant

Participating Laboratories: US7LAL,[1] US7NFL,[2] US7DUP[3]

CYTOTOXICITY

PANEL

ORDER	LAB	TCELL	BCELL	BCELLNAME	EXPTYPE	FILEPOS	DNUM
1		1	9006		1	1	1

	FIRST EXP NO. : 1			SECOND EXP NO. : 2		
POS CTRL CELL :						
SPONT.REL	130	121	158	117	94	118
MAX.REL	942	949	1015	1002	955	1033
EXP.REL	412	418	409	429	381	403
NEG CTRL CELL:						
SPONT.REL	171	177	213	112	126	131
MAX.REL	1715	1795	1864	1108	1132	1173
EXP.REL	184	219	268	167	185	150
EXPERIMENTAL CELL LINE						
SPONT.REL	114	106	96	138	74	113
MAX.REL	903	970	1019	1191	1013	1167
EXP.REL	89	99	102	110	126	108

BLOCKING

ORDER	LAB	TCELL	BCELL	BCELLNAME	EXPTYPE	FILEPOS	DNUM
3		1	9027		2	200	3

	FIRST EXP NO. : 1			SECOND EXP NO. : 2		
SPONT.REL	237	218	226	322	340	389
MAX.REL	1292	1252	1283	1442	1397	1450
UNBLOCKED	624	664	678	679	676	720
UNBLOCKED	0	0	0	723	657	711
1:40	422	423	485	398	414	424
1:200	570	583	597	567	589	610
1:1000	598	603	615	650	656	682
1:5000	650	689	703	711	698	6870

PROLIFERATION

PANEL

ORDER	LAB	TCELL	BCELL	BCELLNAME	EXPTYPE	FILEPOS	DNUM
1		1	9020		1	1	1

	FIRST EXP NO. : 1			SECOND EXP NO. : 2		
POS. B-LCL:						
MEDIA	769	623	713	1005	657	1029
EXP	7774	9416	7811	24093	23380	22646
NEG. B-LCL:						
MEDIA	589	652	665	357	458	309
EXP	180	158	194	272	219	252
MED. CONTROL T-CLONE	187	125	113	267	113	200
PHA/POOLED PLT: MEDIA	449	420	484	872	881	957
EXP	1200	1239	1440	2078	2245	2908
MEDIA CONTROL B-LCL	769	623	713	1005	657	1029
EXPERIMENTAL VALUE	7774	9416	7811	24093	23380	22646

BLOCKING

ORDER	LAB	TCELL	BCELL	BCELLNAME	EXPTYPE	FILEPOS	DNUM
20		4	9065		2	700	2

	FIRST EXP NO. : 473			SECOND EXP NO. : 20		
MEDIA CONTROL B-LCL	665	660	500	669	720	655
MEDIA CONTROL T-CLN	315	165	125	153	186	168
UNBLOCKED EXP	14950	18990	19200	12200	11210	12365
UNBLOCKED EXP	21965	17635	18960	0	0	0
1:40	1522	1523	1585	3456	3226	3459
1:200	3098	3126	3280	5418	5075	5266
1:1000	6099	6212	6125	7689	7543	7559
1:5000	9985	10027	9876	9876	9854	9543

Figure 1. Tables as they appeared on the computer screen for data entry. *Upper left*, Cytotoxicity panel data. *Upper right*, Cytotoxicity antibody blocking data. *Lower left*, Proliferation panel data. *Lower right*, Proliferation antibody blocking data.

Immunobiology of HLA
Volume I

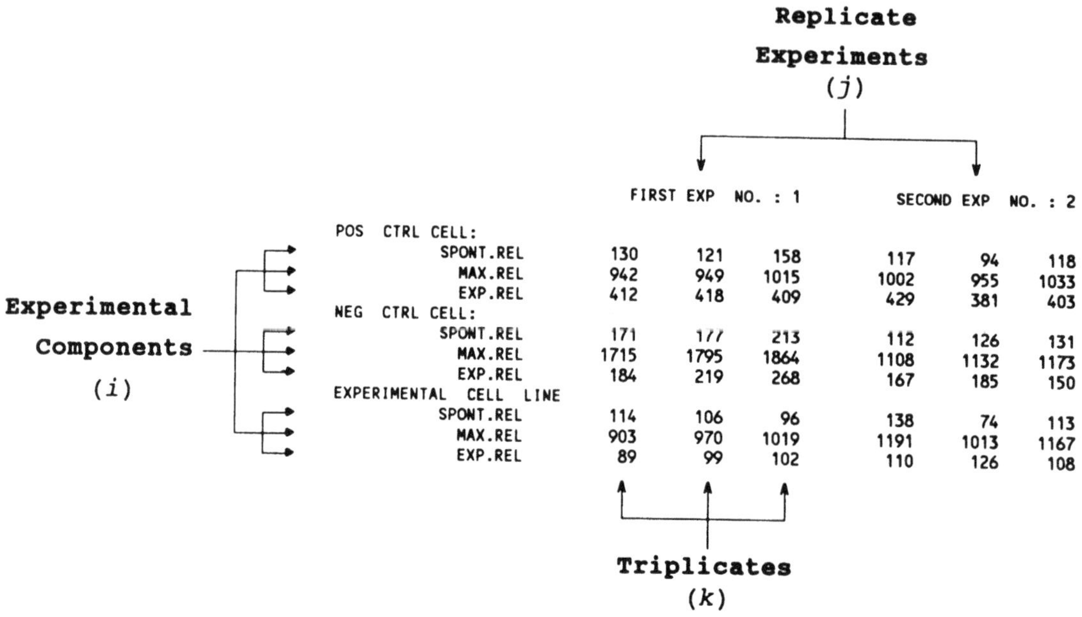

Figure 2. Definitions of experimental component (*i*), replicate experiment (*j*), and triplicate (*k*) are illustrated using the cytotoxicity panel data entry form.

experimental data (see Fig. 1). The program also allowed for the recall of data, addition of new information, and correction of mistakes. The information concerning T-cell clone definitions and experimental data were stored in separate binary files.

After receiving the diskettes from the participating laboratories, the data files were transferred to a VAX 8650 computer, merged, and converted to new record files, which could be accessed by the VAX. We analyzed all data on the VAX, using a Pascal program to identify and eliminate outliers (spurious points), calculate basic variables for the final analysis, and print text files containing the intermediate results. Statistical analysis was performed using the statistical software package BMDP. Bartlett's test and the Kolmogorov-Smirnov test, however, were performed using Pascal programs.

Checking and Reducing Repeated Measurements

The Method. Each group of three (repeated) measurements within an experiment is hereafter referred to as a "triplicate." In the cytotoxicity and proliferation panel studies, there were thus nine different triplicates for each experiment and, therefore, a total of 18 triplicates for the two replicate experiments (Fig. 2). For the antibody inhibition studies, there were, at most, eight triplicates per experiment, and a total of 16 for the two replicate experiments (see Fig. 1). Each row of data in the tables in Figures 1 and 2 thus reflects a different set of measurements (e.g., maximum release of the positive control cell line, spontaneous release for the positive control cell line, etc.) as indicated by the labels to the left of the tables. Hereafter, each of these nine measurements is referred to as an "*experimental component*" (see Fig. 2). For the detection and elimination of outliers, each of the nine experimental components was analyzed separately. In the calculations that follow, each experimental component is referred to by the letter *i*, corresponding to the number of its row in the table. Each of the two replicate experiments is referred to by the letter *j*, and each individual value within a triplicate is referred to by the letter *k* (see Fig. 2). To find and eliminate outliers (i.e., "bad" values within triplicates) prior to subsequent stages of the analysis, we adopted the following method.

First, all values were transformed to the logarithmic scale. The mean of the log transformed triplicates (x_{1ij}, x_{2ij}, and x_{3ij}) was calculated as:

$$\bar{x}_{ij} = (1/3) \sum_{k=1}^{3} x_{kij}$$

Next, we calculated the squared difference $(x_{kij}-\bar{x}_{ij})^2$ between each value and its corresponding mean. Finally, we summed all of the squared differences for each experimental component within each laboratory and calculated the following standard deviation for each experimental component *i* as:

$$s(i) = ((1/3n) \sum_{j=1}^{n} \sum_{k=1}^{3} (x_{kij}-\bar{x}_{ij})^2)^{1/2}$$

where *n* is the number of experiments performed in one laboratory.

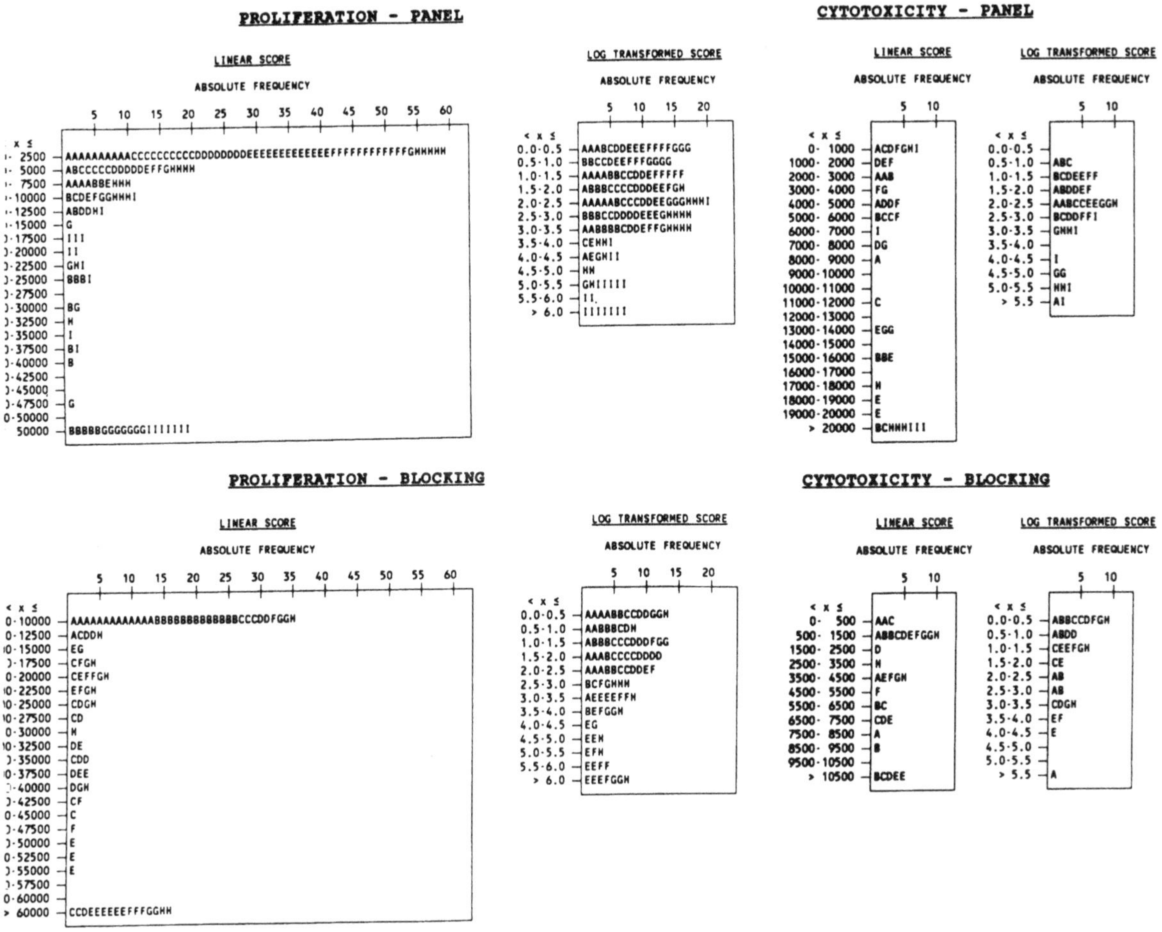

Figure 3. Distribution of the range of the first triplicate values and of their logs for all experimental components. Each letter illustrates the range of one experimental component from one laboratory. For example, 'A' represents the range of the first experimental component (e.g., spontaneous release of the positive control cell line in the panel studies of the cytotoxicity section), 'B' of the second experimental component, etc. The letters are in alphabetical order, corresponding to rows 1 to 9 in the tables illustrated in Figure 1. Further details are found in the Appendix.

A value was marked as an outlier when its deviation from the mean was larger than three times the corresponding standard deviation (3s(i)). Triplicates containing outliers were checked by the central and local laboratories, and typographic errors were eliminated.

Triplicates of each experiment were reduced for the final analysis as follows: If a triplicate contained no outlier, the mean of the logs was calculated and retransformed to the original scale. If one of the three values was an outlier, the mean of the logs of the two remaining values was calculated and retransformed. When two members of a triplicate were outliers, the two values were identified whose difference (in the log scale) was smaller than 3s(i) (i.e., the two values closest together). The mean of these two log values was calculated and retransformed to a linear value. If all three values were outliers, the median was taken.

Reflection on the Method. The range of results from experiments carried out at different times and in different laboratories within the cytotoxicity and proliferation studies differed from each other very substantially. In the cytotoxicity section, for example, the values reported for one laboratory for the spontaneous release of the positive control cell line varied from below 100 to over 4000. In another laboratory, the same values ranged between 2950 and 5472. In the proliferation section, the media control of the positive control cell line ranged between 385 and 11850 counts in one laboratory (see also Fig. 3). Thus, we are dealing with samples from populations which differ in their statistics of location and, therefore, also in their variances (i.e., heterogeneity of variances) (1). In practical terms, this means that using a linear scale, the sensitivity of detecting outliers among the triplicates varies, depending on the

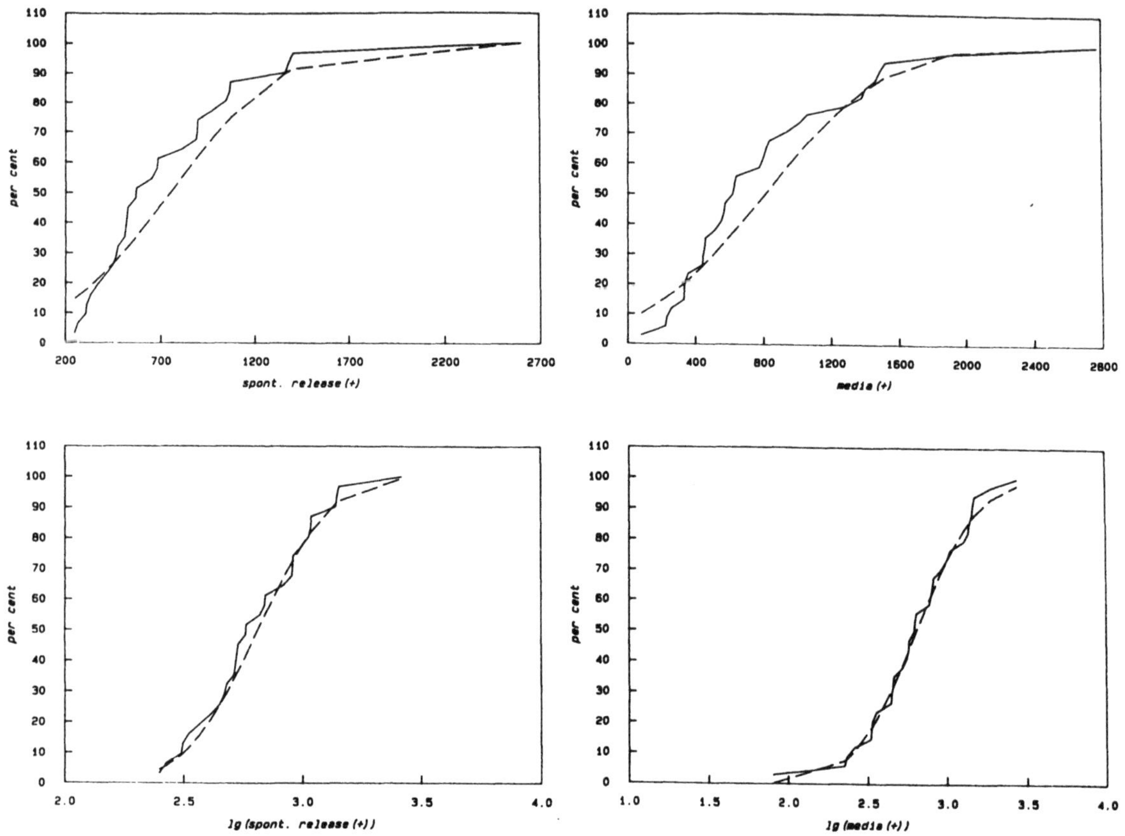

Figure 4. Cumulative frequency distributions of the first triplicate values and the corresponding logs from the experimental components "media control of positive control cell line" (proliferation panel) and "spontaneous release of positive control cell line" (cytotoxicity panel) from two laboratories. The curve of the expected normal distribution is illustrated by the dashed lines.

magnitude of the response. This is clearly very undesirable. To eliminate this heterogeneity of variances, we transformed the data before doing the analysis of outliers. We chose a logarithmic transformation because it has a stabilizing effect on the variance (i.e., the logarithms of large values change very slowly over a wide linear range) (1).

To detect and eliminate outliers, we used standard deviations and means (of logarithms). Thus, we indirectly assumed that the logarithms of the data are normally distributed. As indicated, we calculated the standard deviation for each experimental component by using all data points of that component from that particular laboratory. The deviation of each score from its corresponding mean (i.e., the mean of the triplicate, to which it belongs) was squared and summed for this calculation. Hence, we actually calculated the variance within groups assuming that each triplicate forms a group. When analyzing samples consisting of groups, one can calculate a number of variances that provide different information about the variation of the data (e.g., total variance, variance between groups, variances within groups, and variance of each group). In our case, the variance of a group is the variance calculated using only the three triplicate values. The variance within groups is then the average of the variances of triplicates. Thus,

$$\text{Variance within groups} = (1/3n) \sum_{j=1}^{n} \sum_{k=1}^{3} (x_{kij} - \bar{x}_{ij})^2$$

If we now assume that the variances of all triplicates are identical (let us call this variance s^2), then it easily follows that the variance within groups is equal to s^2.

$$(1/3n) \sum_{j=1}^{n} \sum_{k=1}^{3} (x_{kij} - \bar{x}_{ij})^2 = (1/n) \sum_{j=1}^{n} ((1/3) \sum_{k=1}^{3} (x_{kij} - \bar{x}_{ij})^2) = s^2$$

Thus, when calculating the standard deviation of each experimental component, we get an estimate for the standard deviation s for each individual triplicate. Moreover, we use much more information from the data for this estimation, resulting in a much better estimate

Table 1. Skewness and Kurtosis of the Distribution of the First Triplicate Values and Their Logarithms (panel studies)

Experimental Component	n	Skewness Linear	Skewness Log	Kurtosis Linear	Kurtosis Log
Proliferation – Panel					
Pos. Control – B-LCL					
Media Prolif.	34	1.43*	0.15	2.15*	−0.85
Exper. Prolif.	34	1.11*	−0.26	0.68	−0.65
Neg. Control – B-LCL					
Media Prolif.	34	0.99*	0.0	−0.02	−1.11
Exper. Prolif.	34	1.21*	−0.07	0.93	−0.86
Media Control – T-Clone	34	1.66*	0.58	1.93*	−0.46
Workshop B-LCL					
Media Prolif.	997	2.74*	0.07	10.26*	0.16
Exper. Prolif.	997	3.61*	0.99*	15.49*	0.37*
Cytotoxicity – Panel					
Pos. Control – B-LCL					
Spont. Release	31	1.89*	0.31	4.61*	−0.48
Max. Release	31	0.29	−0.29	−1.16	−1.06
Exp. Release	31	0.84	0.19	−0.44	−1.28
Neg. Control – B-LCL					
Spont. Release	31	2.32*	0.19	6.91*	−0.60
Max. Release	31	1.57*	0.08	2.97*	−0.41
Exp. Release	31	3.04*	0.38	11.22*	0.11
Workshop B-LCL					
Spont. Release	734	8.20*	0.36*	115.46*	0.57*
Max. Release	734	12.85*	0.82*	225.89*	2.83*
Exp. Release	734	11.90*	0.57*	223.56*	0.14

Data are derived from one laboratory. n = number of cases analyzed. A star indicates that the parameter is significantly different from zero ($2p < 0.05$), indicating that the distribution deviates from a normal distribution.

than if we were to use the data from each triplicate separately.

Certainly, the two assumptions mentioned above are valid for the data in the proliferation section. Since this section measures the growth of cells, we may assume that the variables measured here follow the law of exponential growth. Thus, the logs of these values have a normal distribution. It is not as immediately obvious that logarithmically transformed values of the counts obtained from the cytotoxicity experiments are normally distributed.

To check whether the above described method for finding and eliminating outliers is appropriate for our data, we first analyzed distributions of the first values of triplicates for each experimental component of the panel data from two laboratories. We also looked at coefficients of variation and standard deviations from all laboratories after having corrected the data.

Looking at the cumulative frequency distribution of the first triplicate values, we always observed that logarithmic transformed values seemed to fit a normal distribution much better than non transformed data (Fig. 4). We therefore calculated the skewness and kurtosis of this set of data and tested their deviations from zero (2). While skewness measures the asymmetry of a distribution, kurtosis describes its peakedness or flatness (e.g., see reference 2). For normally distributed variables both parameters are equal to zero. In the proliferation section, all skewness parameters of nontransformed data are significantly different from zero. In the cytotoxicity section, only the skewness of the distributions of maximum release and experimental release of the positive cell lines are not significantly different from zero. In all other experiments both parameters are significantly different from zero. In both sections, skewness and kurtosis of all log transformed data are not significantly different from zero when positive or negative control·cell lines were tested (Table 1). In general, logarithmic transformation seems to produce normally distributed variables in the cytotoxicity as well as in the proliferation section.

Figures 3 and 5 show the effect of the logarithmic transformation on the range and the coefficient of variation after correction of the data. We looked at the first triplicate value for each experimental component in all laboratories.

Taking into consideration the experimental design, we expect large variation in experimental components which deal with the testing of different cell lines (i.e., the Workshop cell lines and not the control cell lines), because we are dealing with the testing of a heterogeneous sample. Figures 3 and 5, however, show that even after data correction in experimental components testing

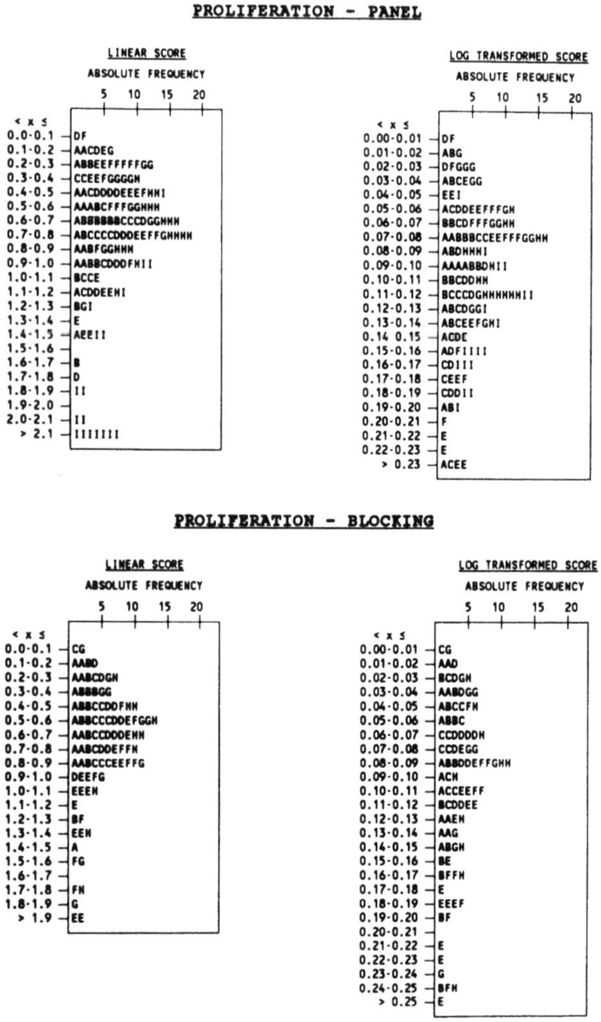

Figure 5. Distribution of the coefficients of variation of the first triplicate values for all experimental components and laboratories. Notation is the same as in Figure 3.

just one cell line (i.e., positive or negative controls), data from different times and laboratories vary widely before transformation. Even in these experimental components, the coefficient of variation is larger than 1, indicating a large variability of the scores. (In normally distributed variables one expects a coefficient of variation of about 0.33.) The effect of the log transformation is obvious: all coefficients of variation are now smaller than 0.15 in the cytotoxicity section and 0.3 in the proliferation section, respectively.

To check the stabilizing effect of the log transformation, we compared standard deviations (within groups) for all experimental components and laboratories. Certainly, this method is not theoretically correct for answering this specific question. We actually should compare standard deviations from different triplicates of each experimental component and laboratory. However, since the standard deviation from each triplicate is calculated from only three values, we find it inappropriate to utilize these figures for this analysis.

As Figure 6 and Table 2 show, standard deviations of different laboratories vary widely in each experimental component after log transformation. While the range of the corresponding standard deviations is not too large in the cytotoxicity section, standard deviations differ substantially in the proliferation section. In order to compare standard deviations from different laboratories we used Bartlett's test of homogeneity of variances (2). We compared the standard deviations of different laboratories for each experimental component. All comparisons were significant ($2p < 0.01$). Also standard deviations from different experimental components were significantly different ($2p < 0.01$) in each laboratory. Thus, these comparisons indicate that the stabilizing effect of the logarithmic transformation on the magnitude of the standard deviations does not seem to be fully established.

Figure 5. *Continued*

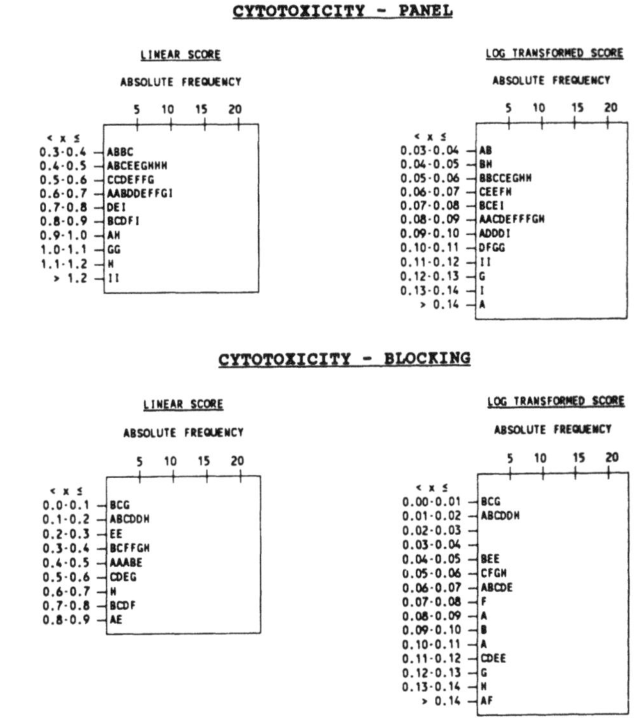

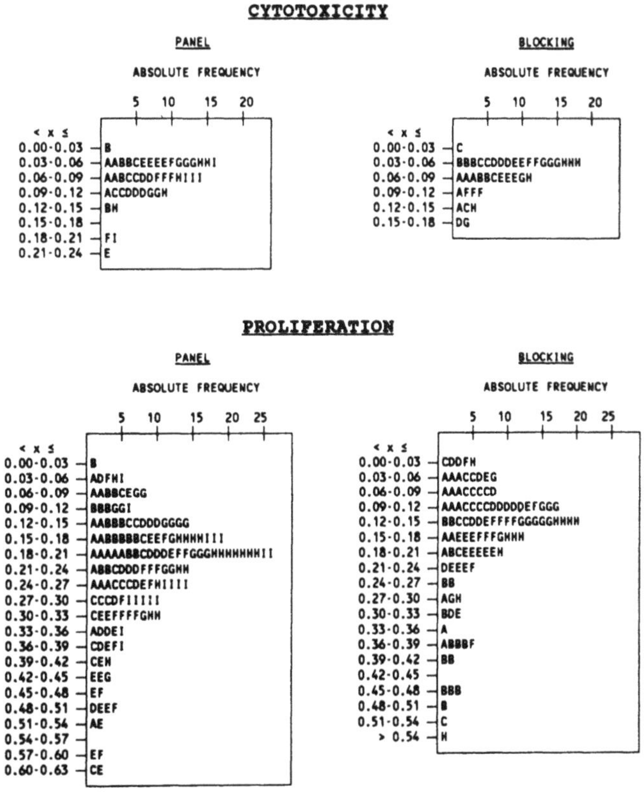

Figure 6. Distribution of the standard deviations for each experimental component (i.e., row) and each laboratory for panel and blocking experiments. Data were transformed to the logarithmic scale. Notation is the same as in Figure 3.

Table 2. Minimum and Maximum of Standard Deviations in Each Experimental Component in the Cytotoxicity and Proliferation Section (all laboratories)

Experimental Component	Smallest s	Largest s
Proliferation – Panel		
Pos. Control – B-LCL		
Media Prolif.	0.047	0.520
Exper. Prolif.	0.019	0.233
Neg. Control – B-LCL		
Media Prolif.	0.070	0.612
Exper. Prolif.	0.049	0.491
Media Control – T-Clone	0.086	0.622
PHA-PLT		
Media Prolif.	0.038	0.576
Exper. Prolif.	0.063	0.429
Workshop B-LCL		
Media Prolif.	0.053	0.391
Exper. Prolif.	0.042	0.373
Proliferation – Blocking		
Media – B-LCL	0.042	0.377
Media – T-Clone	0.138	0.503
Unblocked Exp. 1	0.014	0.531
Unblocked Exp. 2	0.006	0.319
Dilution 1:40	0.031	0.327
Dilution 1:200	0.027	0.376
Dilution 1:1000	0.034	0.276
Dilution 1:5000	0.019	0.607
Cytotoxicity – Panel		
Pos. Control – B-LCL		
Spont. Release	0.047	0.097
Max. Release	0.030	0.124
Exp. Release	0.053	0.102
Neg. Control – B-LCL		
Spont. Release	0.076	0.113
Max. Release	0.038	0.232
Exp. Release	0.049	0.182
Workshop B-LCL		
Spont. Release	0.040	0.114
Max. Release	0.035	0.131
Exp. Release	0.057	0.183
Cytotoxicity – Blocking		
Spont. Release	0.069	0.134
Max. Release	0.036	0.067
Unblocked Exp. 1	0.027	0.133
Unblocked Exp. 2	0.034	0.166
Dilution 1:40	0.046	0.086
Dilution 1:200	0.051	0.119
Dilution 1:1000	0.046	0.152
Dilution 1:5000	0.045	0.145

Data were transformed to the logarithmic scale.

The large variation among these standard deviations may be due to a variety of factors. A few extremely large values, perhaps due to typographic errors, may still be present in the data despite our extensive efforts at data correction. Moreover, laboratories tested the Workshop cell panel in experiments of widely different size, ranging from the entire panel at once to 10 to 15 cell lines at a time. As a result, some laboratories have a very small number of values (less than five) for the experimental components involving the positive and negative control cell lines. This can also affect the magnitude of standard deviations. We must also consider that the quality of the technical performance of experiments has an effect on the magnitude of these standard deviations.

Conclusions. Transforming the data to the logarithmic scale reduces the variances and produces normal distributions for (nearly) all experimental components. However, logarithmic transformation does not seem to completely stabilize the variances. Unfortunately, the experimental design does not allow for definitive testing of the stabilizing effect of the logarithmic transformation.

Since we did not have prior information about statistical properties of our data, we had to use the data for estimating means and standard deviations, which were, in turn, then used for screening the same data. It is certainly a disadvantage to use the data for estimating parameters which are subsequently used for detecting outliers in the same data. Despite this problem, when screening the data, we were left with the clear impression that this method was successful in detecting outliers in a manner comparable to those which would have been flagged subjectively. In contrast to some nonparametric methods of detecting outliers (3), this method also offered the advantage of being able to use as much information from the data as possible for this purpose.

Criteria for Discarding Discordant Experiments

After outliers had been removed from each of the triplicate measurements and the triplicates reduced to single values, we calculated simple indices of cytotoxicity and proliferation, which were to be used in quality controlling the database. We had requested that each Workshop cell line be tested twice, and we were interested in retaining only those data in which the two experiments led to the same conclusion. Specifically, we wished to eliminate any data in which one testing of a given clone and cell line indicated that the clone strongly recognized the cell line, while the second testing suggested no such recognition.

Cytotoxicity. The criteria for discarding discordant data in cytotoxicity experiments were as follows. Divisions were established based on the percent cytotoxicity. Values less than 10% were considered negative, those from 10 to 15% uncertain, 15 to 20% probably positive, 20 to 30% positive and larger than 30% strongly positive. Based on this system, data were eliminated when one experiment suggested a positive result (i.e., kill) while the second experiment suggested a negative result (i.e., no kill). Practically, this meant that the results of the two replicate testings were separated by more than two of the divisions defined above.

The above criteria served to retain experiments where replicate testings led to the same conclusions. When this was true, the data were utilized for the subsequent analysis. In most cases, the mean of the two replicate experiments was used. In some cases, one of the replicates (though consistent with its partner) was technically suboptimal. In cytotoxicity assays, this usually reflected a small difference between the lysis of the positive and negative control cell lines. Such an observation suggested that, amongst other possibilities, the clone might not be functioning optimally on the day of testing, compromising the discrimination between positive and negative results. If the difference in kill of the positive and negative control cell lines was less than 15%, then that replicate testing was eliminated and not averaged with its matched partner. Based on these criteria, there were 290 instances in the cytotoxicity panel experiments in which one of the replicate testings was eliminated from the final analysis.

A similar approach was utilized for the antibody inhibition experiments. Here, gradations of <25% inhibition, 25 to 50% inhibition, and >50% inhibition were established. Data were eliminated when the two replicate testings led to the opposite conclusions. While data from the highest antibody concentration were used primarily, data from the titration studies were also considered in this judgment when available. Similarly if the percent lysis of the cell line in the absence of antibody was <20%, then that replicate was not averaged with its matched pair. Using this criterion, there were 93 instances in which one of the replicate testings was eliminated.

Proliferation. The criteria for the proliferation section were similar to those used for the cytotoxicity section. However, for these experiments the criteria relied more heavily on the positive controls for each experiment. First, divisions were established based on the amounts of thymidine incorporation (counts per minute, CPM). In experiments in which the positive control was < 10000 CPM, the divisions were as follows: < 1000 were considered negative, 1000 to 2500 probably negative, 2500 to 5000 uncertain/weakly positive, 5000 to 10000 positive, and > 10000 strongly positive. The values used for establishing these ranges were scaled to reflect the magnitude of the positive control.

In addition to the above criterion, special cases were considered. If the experimental value was > 10 times the negative control, the experimental value was then considered to be (at least) weakly positive in strength. If the experimental value was >50% of the positive control, the experimental value was considered to be (at least) positive in strength. Finally, if the experimental value was greater than the positive control or other positive values in the same experiment, the experimental value was considered to be strongly positive. Experiments were again discarded when the replicate testings led one to reach opposite conclusions.

In analogy to the cytotoxicity data, one of the two replicate experiments would be excluded from the final analysis when the difference between proliferation toward the positive and negative control cell lines was <2400 CPM. Here again, we felt the ability to discriminate between positive and negative results might be compromised. A total of 1404 single experiments were eliminated using this criterion. For the blocking studies, single experiments were eliminated when the unblocked response was <2900 CPM. A total of 167 single experiments were discarded.

Table 3. Number of Replicate Experiments That Were Excluded From or Retained for the Final Analysis

Type of Experiment	Number of Replicates		
	Received	Eliminated	Analyzed
Cytotoxicity			
Panel	2250	114 (5.1%)	2136 (94.9%)
Blocking	957	288 (30.1%)	669 (69.9%)
Proliferation			
Panel	8788	2253 (25.6%)	6535 (74.4%)
Blocking	2200	835 (38.0%)	1365 (62.0%)
Total	14195	3490 (24.6%)	10705 (75.4%)

Number of Discarded Experiments. Table 3 shows the number and proportion of experiments that were submitted to the Workshop, but which were excluded from the final analysis. In the cytotoxicity section, 5.1% of the panel and 30.1% of the blocking experiments were eliminated, while in the proliferation studies, 25.6% of the panel and 38% of the blocking experiments were removed. Thus, 75% of the data records originally submitted were retained for subsequent analysis. This suggested that the criteria used to screen the data were not overly demanding, while still allowing us to feel comfortable that the data retained had been demonstrated to be reproducible.

Appendix

Definition of letters used in Figures 3, 5, and 6

Letter	Proliferation		Cytotoxicity	
	Panel	Blocking	Panel	Blocking
A	$Media^{+}$	$Media^{B}$	$Spont^{+}$	Spont. Rel.
B	Exp^{+}	$Media^{T}$	Max^{+}	Max. Rel.
C	$Media^{-}$	1. Unbl. Exp.	Exp^{+}	1. Unbl. Exp.
D	Exp^{-}	2. Unbl. Exp.	$Spont^{-}$	2. Unbl. Exp.
E	$Media^{T}$	Dil. 1:40	Max^{-}	Dil. 1:40
F	$Media^{Pha}$	Dil. 1:200	Exp^{-}	Dil. 1:200
G	Exp^{Pha}	Dil. 1:1000	$Spont^{Exp}$	Dil. 1:1000
H	$Media^{Exp}$	Dil. 1:5000	Max^{Exp}	Dil. 1:5000
I	Exp^{Exp}		Exp^{Exp}	

References

1. Lienert GA. Verteilungsfreie Methoden in der Biostatistik, Volume 1. Anton Hain, Meisenheim, 1986.
2. Sokal RR, Rohlf FJ. Biometry. W.H. Freeman and Company, New York, 1981.
3. Barnett V, Lewis T. Outliers in Statistical Data. John Wiley and Sons, New York, 1984.

Statistical Analysis of HLA Class II Cellular Data

R. Plätke,[1] J.M. Lalouel,[1] and N. Flomenberg[2]

Introduction

In analyzing the capacity of a population of T-cells to recognize MHC gene products (as evidenced by a proliferative or cytotoxic response), results are frequently reduced to a qualitative scale (i.e., either the stimulator/target T-cell in question elicits a response or it does not). Although the final interpretation is qualitative, the assays of T-cell proliferation and cytotoxicity produce results which are not only quantitative, but also continuous. While cloned populations of T-cells are more likely than bulk culture (i.e., uncloned) populations to produce responses that are either present or absent, intermediate responses can be seen even with clonal responder populations. Biologically, weak or intermediate responses produced by T-cell clones can be understood as:

1. a less intense response to a "cross-reactive" antigen,
2. a less intense response due to quantitative differences in the amount of antigen present,
3. a less intense response due to some other underlying characteristic of the stimulator/target T-cells (quantity of other cell surface molecules necessary for conjugate formation), mycoplasma contamination, etc., and
4. a combination of these and/or other factors.

Thus, proliferative responses can vary continuously from background levels up to the maximum proliferative capacity of the clone (or bulk T-cell population) in question. Cytolytic activity can vary similarly. For this reason, objective statistical methods are required for the interpretation of T-cell clone reactivity, just as they are required for analyses of bulk T-cell populations.

In addition, the experimental design of the cellular Workshop studies introduced additional requirements for the statistical methods utilized for the analysis of the data generated within this component. Specifically, these relate to the fact that the Workshop cell panel was tested in sections over several experiments (at least in most laboratories), rather than in one large experiment. The variables utilized for the analysis additionally had to be capable of reconciling differences of scale introduced by the fact that the magnitude of responses of T-cell clones will vary widely between different laboratories, or when measured at different times in the same laboratory. Thus, the first statistical problem to be solved was to define variables which are able to measure cytotoxicity, proliferation, or inhibition (antibody blocking) in an objective manner, which can provide reproducible results, and which allow data from multiple laboratories and multiple experiments to be compared with one another.

Participating Laboratories: US7LAL,[1] US7NFL,[2] US7DUP[3]

Since the results from this component were to be compared with data from other Workshop components, it was also necessary to choose a common scoring system to facilitate cross-comparisons. The traditional HLA serologic scores were chosen for this purpose. Thus, our second problem was to define a scale consisting only of the scores 1, 2, 4, 6, and 8 based on the continuous variables. (Hereafter, we refer to these as serologic scores, despite the fact that they are not defined with serologic techniques.)

In this report, we describe the way in which variables were selected and their continuous scales reduced to an ordinal scale. We also illustrate how we checked as to whether these newly defined variables were capable of standardizing results from cytotoxicity, proliferation, or antibody blocking experiments. Finally, we discuss how these approaches were utilized to analyze data from this component.

Criteria and Methods for Selecting the Variables

Sample Complexity. Assays of the reactivity of T-cell clones require the interactions of T-cells and (in these experiments) B-cells, leading to cytolytic or proliferative responses by the T-cells. Strictly speaking, the elements of the samples we are dealing with generally consist of pairs of certain T- and B-cells. (Exceptions are those samples used in measuring proliferation in media alone or spontaneous ^{51}Cr release in cytotoxicity assays.) Moreover, in blocking experiments, the elements of the samples are triples: T-cells, B-cells, and antibodies. We often refer to the pairs of B- and T-cells as "responder or nonresponder combinations," depending on whether the interactions of those pairs result in the cytotoxicity or proliferation of T-cells or not. For simplification, however, we often disregard the complex structure of the sample and mention only the properties of the T-cells.

Continuous Variables. Four criteria were used for determining the most appropriate variables capable of measuring cytotoxicity, proliferation, or inhibition. Such a variable must:

1. have a simple biological interpretation,
2. be standardized to allow comparisons across different experiments,
3. clearly demonstrate that its distribution of experimental results is a mixture of distributions due to the presence of both responder and nonresponder combinations, and
4. be (approximately) normally distributed, when one considers only one component of this mixture (i.e., responders or nonresponders).

Regarding criterion 1: When defining variables, we always utilized a biologic model through which the formula could be explained. This allowed us to avoid guessing in the definition of formulae.

Regarding criterion 2: Certainly, "standardization" is always required in order to perform a proper statistical analysis. It is the most important requirement we proposed.

Since the experimental design included blind duplicate controls, we first check whether a variable is standardized by comparing the results of experiments and their hidden controls in scattergrams. If a variable allows comparisons across experiments and the results of experiments are reliable, T-cell clones must be scored nearly identically in these matched experiments. Consequently, the points in the scattergram must lie on, or at least nearby, the straight line of identity $y = x$. In order to assess the deviation of the points from the straight line of identity, the correlation coefficient as well as the regression line are calculated, and statistical tests are performed which check whether the regression coefficient and the intercept of the straight line with the vertical axis are significantly different from 1 and 0, respectively. (We use the t-test for both parameters) (1). However, the magnitude of the correlation coefficient and the results of the statistical tests are affected by the heterogeneity of the sample. Therefore, we mainly interpret the scattergram and use the results of the statistical test as an additional descriptive tool. Although correlation coefficients are presented, they are not often used in interpreting the scattergrams.

One laboratory tested its clones in assays of cytotoxicity as well as proliferation. In this case, we can compare the results from both sections in a scattergram. This allows us to assess whether the criteria used for assigning ordinal values are comparable for proliferative and cytotoxic experiments.

Regarding criterion 3: Based on the experimental design, it is clear that we are dealing with heterogeneous samples of responder and nonresponder B-cell lines and T-cell clones. Therefore, the third criterion is unquestionably as important as the second. We cannot choose a formula which has a nice biological interpretation, but cannot distinguish between responder and nonresponder combinations.

We check this property through descriptive analyses. Specifically, we look at frequency polygons, cumulative frequency distributions, and parameters as local and dispersion measurements. If a distribution is a mixture of (equal) distributions, it often has two or more modes and a large standard deviation. Therefore, we will always mention these parameters in this discussion. As Figure 1 shows, curves of cumulative frequency distributions show characteristics which are unique for certain types of distributions and which become obvious when compared with their expected normal distribution. Thus, we often use cumulative frequency distributions instead of frequency polygons. Figure 1 may be a useful guideline when examining the other figures in this paper.

Regarding criterion 4: This criterion is not as important as the preceding ones. Therefore, it did not often decide the matter. We used it as one of our requirements, since we wanted to have the option of using statistical methods which presume normally distributed data.

We generally analyze the results of experiments carried out with positive or negative cell lines or unblocked controls to obtain information about the type of distribution of the variables. We then assume that experiments carried out only with responder or nonresponder combinations will produce similar results.

In order to determine whether a distribution is normal, we analyze frequency polygons, compare cumulative frequency distributions with those of the expected normal distributions, and use the Kolmogorov-Smirnov test (2,3).

Serologic Score. Our goal was to find simple rules for defining the serologic scores from the continuous variables. The rules could then be implemented by computer and the scores determined automatically and objectively.

In order to establish simple rules for reducing cytotoxicity data into serologic scores, we first looked at a variety of cytotoxicity variables together with the raw data in one laboratory's experiments. For all laboratories, we performed a cluster analysis (k-means algorithm) (4) for each clone separately and used the results when the cytotoxicity index fell near a boundary value.

For the proliferation section, we used a similar approach. We again compared various proliferation indices with the original data of one laboratory in order to find a simple rule. After establishing these rules, we did not find it necessary to additionally use the results of cluster analysis. The rules were directly implemented by the computer, and the data did not require further corrections.

We check whether rules for defining serologic scores provide reproducible results by comparing scores from experiments and their hidden matched controls in contingency tables and determining the number and percentage of discordant experiments. Matched experiments are called discordant when a T-cell clone was scored differently in the two experiments. We also compared serologic scores of clones tested in both the cytotoxicity and proliferation sections.

Since a subjective method was used for defining the serologic scale, we will try to demonstrate in two different ways that our decisions were not arbitrary. We asked four laboratories (two having done experiments in the proliferation section and the other two in the cytotoxicity section) to score their experiments based on their own local criteria. We refer to these locally derived scores as "subjective scores," while the serologic score defined by us is named "Workshop score." We compared these subjective scores with our assignments and also determined the number and percentage of discordant experiments.

Each laboratory used a different approach to determine its subjective scores. Two laboratories used subjective approaches based on analysis of the original data

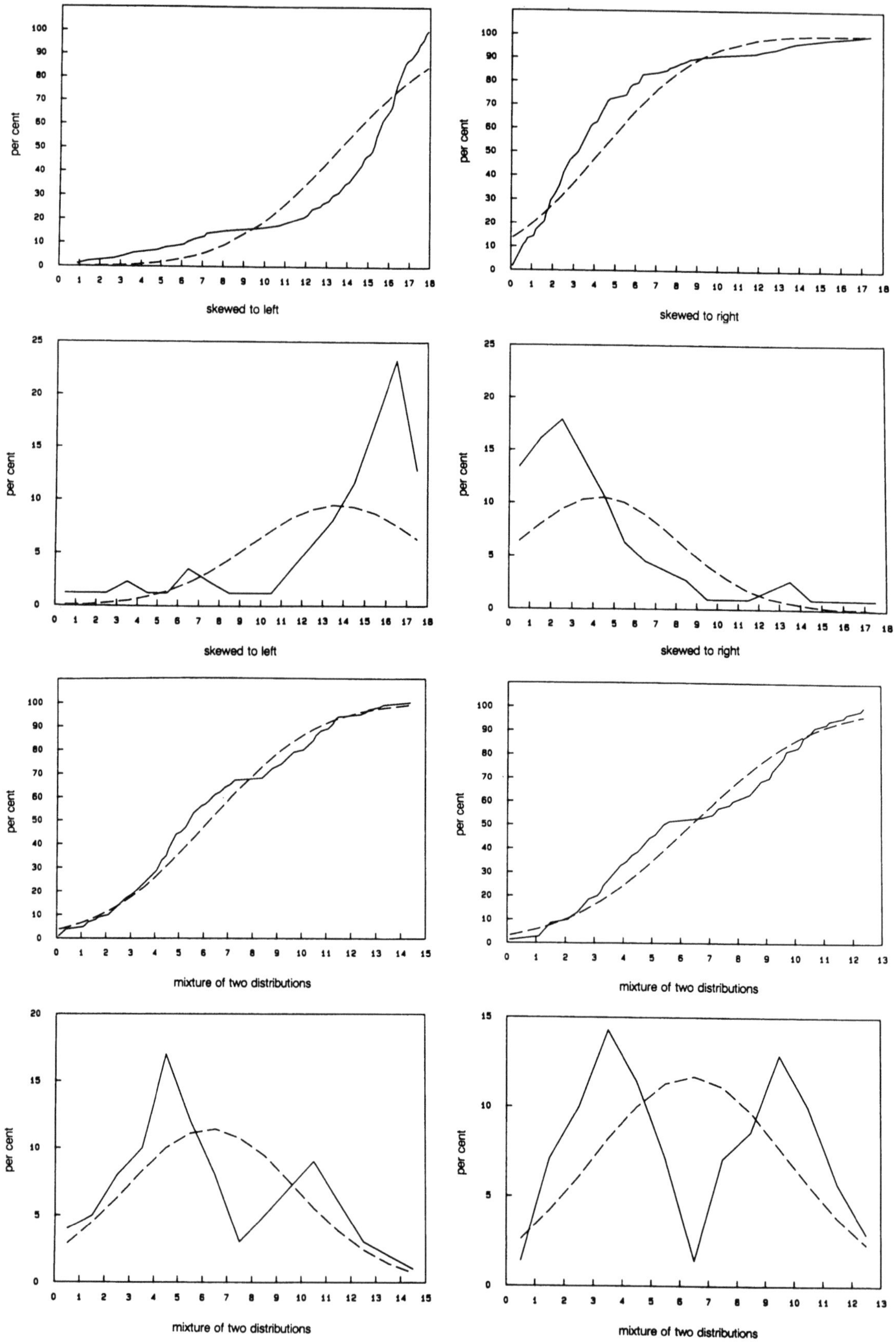

Figure 1. Histograms and cumulative frequency distributions of some fictitious data. The curves of the expected normal distributions are always dashed.

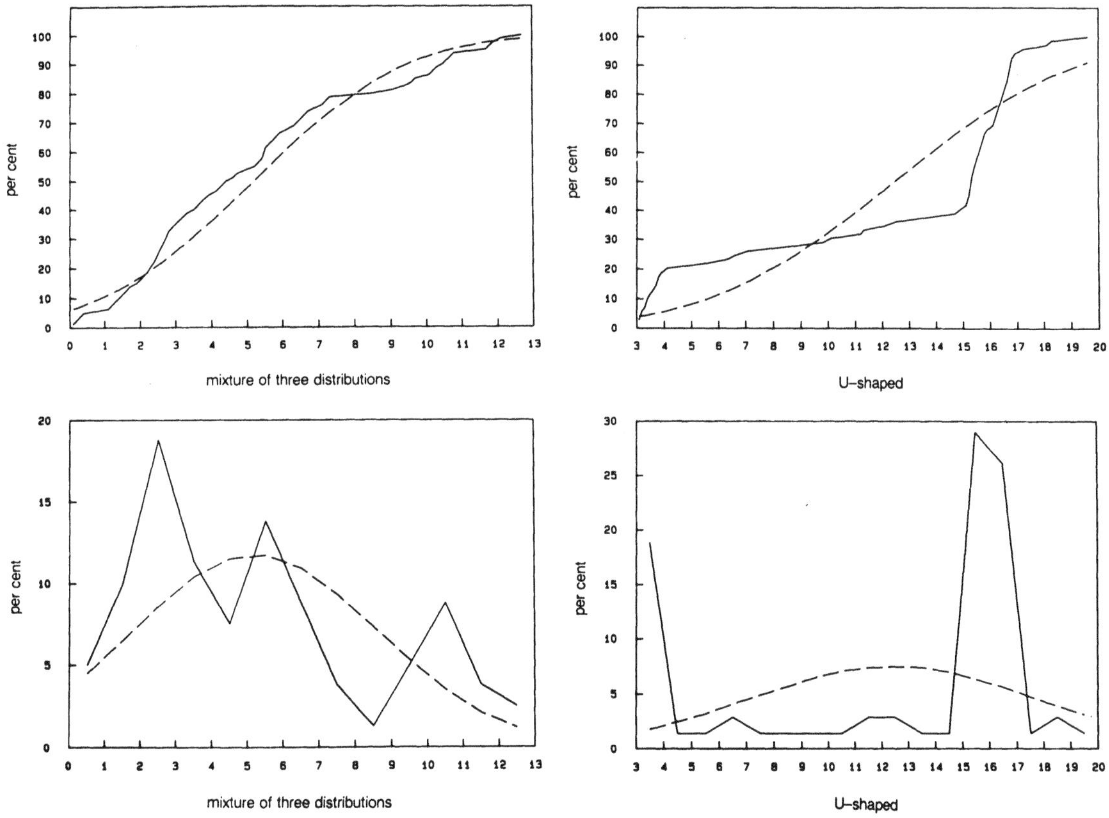

Figure 1. *Continued.*

including results of the positive and negative control cell lines. The other laboratories used computer programs performing either centroid cluster analysis (5) or T-max (6) analysis for scoring the data.

The second approach toward evaluating our assignments was based upon linear discriminant analysis (7,8). This statistical method is generally used for variable selection when the membership of each element of the sample to a certain group is known previously. It analyzes whether the groups can be distinguished on the basis of one or more continuous variables and determines a discriminating function. This function is then used for allocating each member of the sample into one of the previously defined groups. Finally, the number of misclassifications is determined by comparing the assignment of the discriminant analysis with the (true) membership of each element.

The serologic score was used as the variable defining the membership of each T-cell/B-cell combination to a certain group. Since the scores assigned are 1, 2, 4, 6, or 8, there are five different groups. For convenience, let us refer to these groups by these numbers. As the discriminating variable, we use one of the continuous variables measuring cytotoxicity, proliferation, or inhibition. This approach serves to demonstrate that the heterogeneity of the samples is well detected by discriminant analysis using the variables we defined, and that the assignments of the clones via discriminant analysis do not differ greatly from our assignments. Thus, discriminant analysis serves as an additional (descriptive) tool for evaluating our approach. We have, therefore, only presented the classification matrix of the analysis. Strictly speaking, discriminant analysis classifies pairs of B- and T-cells in the experiments (or triples of B, T, and antibodies in blocking experiments). For simplification, we often describe the "misclassification of *experiments*" when presenting the results of the discriminant analysis, hoping that this simplification makes the text easier to read.

Decisions for Each Section

Cytotoxicity–Panel. The cytotoxicity index for the panel data was based on the following model. Cell death is not an exponential process. Thus, in this case, transforming the original data does not seem to be necessary. On the other hand, the amount of ^{51}Cr released by the cells can vary between zero and any positive number. The magnitude of the maximum positive number depends on the amount of ^{51}Cr that is originally incorporated by a cell. Thus, we must expect that the original scale of

Table 1. Criteria for Serologic Scores in the Experiments of the Cytotoxicity Section (Panel)

Cytotoxicity Index	Serologic Scale
$cyt^{exp} < 0.05$	1
$0.05 \leq cyt^{exp} < 0.10$	2
$0.10 \leq cyt^{exp} < 0.15$	4
$0.15 \leq cyt^{exp} < 0.35$	6
$0.35 \leq cyt^{exp}$	8

counts per minute (CPM) is skewed to the right. Certainly, we could have tried to eliminate this skewness by using the log transformation. In the cytotoxicity data, log transformation appeared to unnecessarily scale down differences between responders and nonresponders. Therefore, it was not utilized.

Let us define $spont^+$, max^+, and exp^+ as the spontaneous, maximum and experimental ^{51}Cr release measured in experiments with positive control cell lines. $Spont^-$, max^-, and exp^- are defined similarly for experiments with negative control cell lines. $Spont^{exp}$, max^{exp}, and exp^{exp} indicate the results using the Workshop (i.e., experimental) cell lines. We then calculate an index of cytolytic activity, cyt^i (where i = +, −, or exp) as follows:

$$cyt^i = \frac{exp^i - spont^i}{max^i - spont^i}$$

Thus for each Workshop cell line tested, we calculate three indices, cyt^{exp}, cyt^+, and cyt^-, reflecting cytolytic activity toward the Workshop cell line itself and toward the positive and negative control cell lines tested in the same experiment. It was our original intention to calculate a normalized experimental cytotoxicity index using cyt^+ and cyt^- to correct for day to day variations in lytic activity of the clones. This was to have been done by multiplying cyt^{exp} by a correction factor defined as:

$$\text{correction factor} = \frac{\text{mean}(cyt^+) - \text{mean}(cyt^-)}{cyt^+ - cyt^-}$$

However, analysis of the data revealed that in many cases, there were large variations in cyt^+. This resulted in a very large correction factor when cyt^+ was appreciably smaller than mean(cyt^+). As a consequence, weakly lysed cell lines with cyt^{exp} of 3 to 5% began to appear as if appreciably lysed. This distortion in the data seemed undesirable and normalization was, therefore, eliminated. Thus, the results of the experiments with positive and negative control cell lines were used only to eliminate certain experiments from the final analysis. (As described in the preceding report by Plätke et al., we excluded experiments when the difference $cyt^+ - cyt^-$ was ≤ 0.15.)

The definition of the cytotoxicity index is the traditional formula for percent specific lysis that is utilized by most laboratories for analysis of cytotoxicity data. This

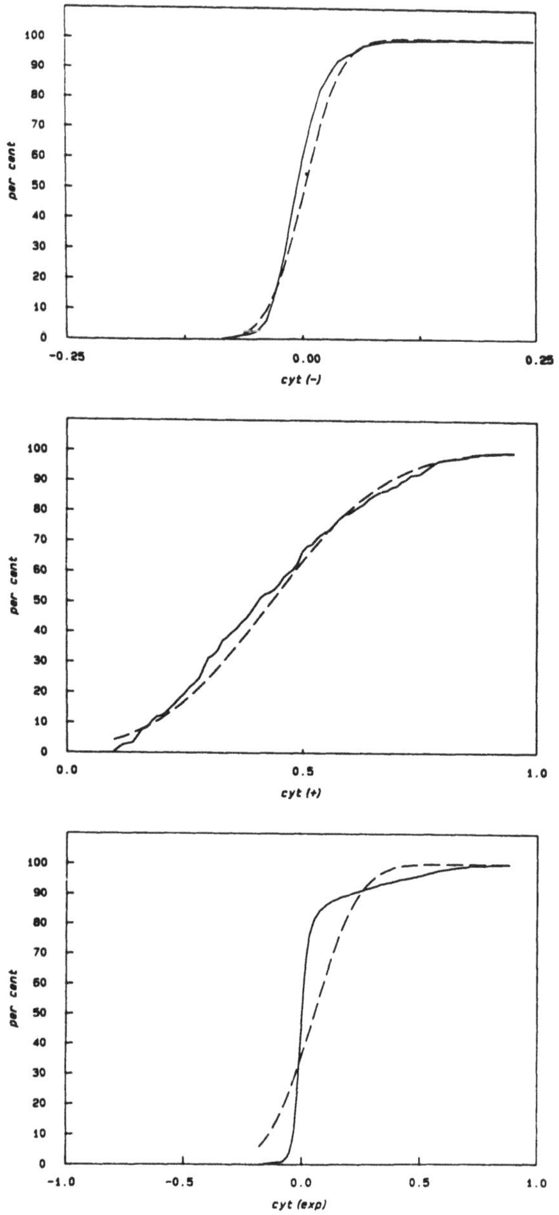

Figure 2. Cumulative frequency distribution of cyt^+, cyt^-, and cyt^{exp}.

formula allows for a simple biological interpretation; the cytotoxicity of a T-cell clone is expressed as a quotient of experimental over maximum release after subtracting spontaneous release (experimental background). Thus, the cytotoxicity index can be interpreted as a deviation of experimental release from maximum release in percent. A cytotoxicity index close to zero indicates a noncytotoxic cell, whereas a number nearby one (or 100%) shows high cytotoxicity.

Figure 3. Frequency distribution of cyt^{exp}.

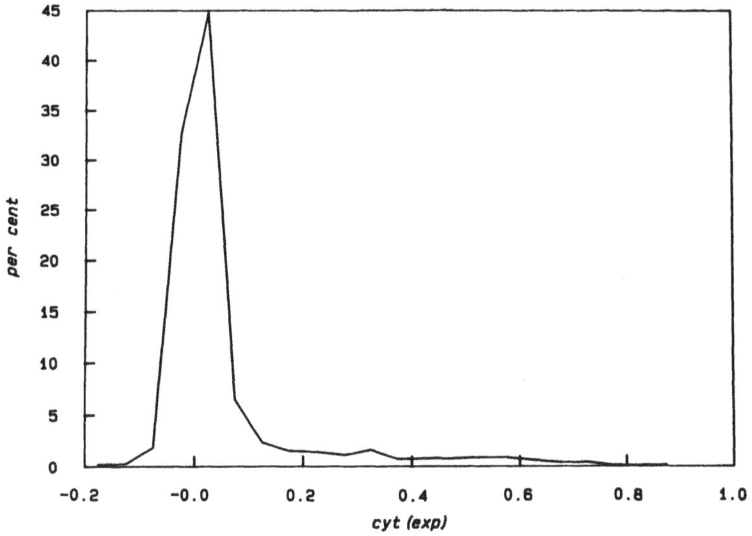

The criteria shown in Table 1 were utilized for converting cyt^{exp} into the serologic scale. As previously mentioned, if the cytotoxicity index fell near a boundary value, a cluster analysis was also used. Thus, a borderline value was assigned the lower score if it was clustered with values falling below the boundary, while it received the higher score if it was clustered with values above the boundary. We used this additional criterion in 68 (3.2%) of 2136 cases.

Results of the Analysis

Distribution of cyt^+ and cyt^-. Figure 2 shows the observed and expected (assuming normally distributed variables) cumulative frequency distributions of the cytotoxicity index in the experiments with positive, negative, or experimental cell lines. Although the cumulative frequency distributions of positive and negative experiments do not seem to deviate from that of a normal distribution, the Kolmogorov-Smirnov test is significant in both cases ($2p < 0.05$, $n=466$). Therefore, the distribution of the log transformed variables cyt^+ and cyt^- were also tested. In both cases, however, the Kolmogorov-Smirnov test remains significant ($2p < 0.05$). The magnitude of the mode, median, and arithmetic mean (mode = 0.5, mean = 0.43, median = 0.41) as well as the magnitude of skewness (0.40) indicate that the frequency distribution of the cytotoxicity index for experiments testing the positive control cell line is skewed to the right (i.e., the distribution is asymmetric and its right tail is drawn out). The same statistical measurements (mean = 0.002, median = 0.0, mode = −0.01; skewness = 1.77) for cyt^- show that its distribution is also skewed to the right. Obviously, in both cases, the measurements of localization do not differ tremendously.

Distribution of cyt^{exp}. Figures 2 and 3 clearly show that the frequency distribution of cyt^{exp} is a mixture of distribution (Kolmogorov-Smirnov test: $2p < 0.05$, $n = 2136$; $s = 0.144$), although a comparison of the measurements of localization (mode = median = 0, mean = 0.049) does not indicate this. This is certainly due to the difference in the number of experiments belonging to each group (i.e., there are many non-responders and a small group of responder combinations).

Reproducibility of Results. There are 227 experiments that have a matched (hidden) control. Figure 4 compares the cytotoxicity indices in those experiments. The scattergram shows the heterogeneity of our sample. It consists of two nearly separate groups of responders and nonresponders. There are only two pairs of experiments whose scores differ extremely. One experiment is scored 0.09, while the score is 0.28 for its hidden control. In the second pair of experiments the scores are 0.4 and 0.0. The correlation coefficient of the scattergram is 0.921. The regression line $y = -0.007+0.973x$ does not significantly deviate from the straight line $y = x$. (For both coefficients, $2p > 0.05$).

The concordance of the results of these experiments can also be observed when comparing the serologic scores (Table 2). Eleven percent (25) of the experiments were scored differently from their matched controls, a percentage which is acceptable. If we define a coarser scale consisting only of the three values – 1 plus 2 (negative), 4 (weak or uncertain), and 6 plus 8 (positive) – there are only 12 matched experiments (5.3%) that would be considered discordant. This indicates that the scores differ slightly in about half of the discordant experiments without affecting the interpretation of the results. (Hereafter, the serologic scale consisting of five scores is referred to as the "finer" scale,

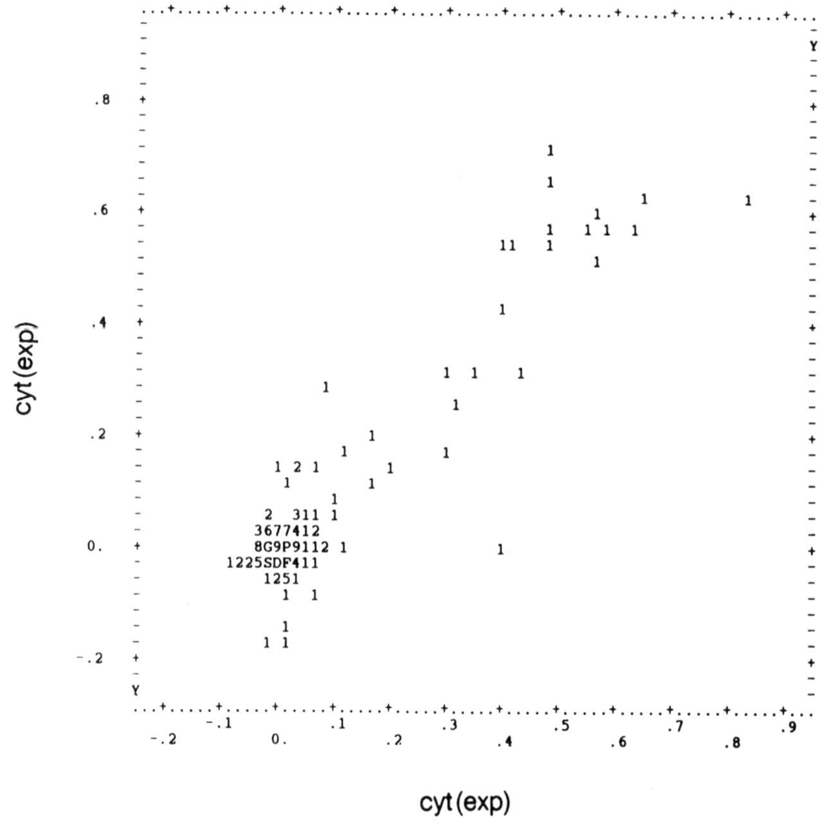

Figure 4. Concordance of cyt^{exp} values in experiments having hidden controls. Numbers in the plot are the frequencies of points having the same coordinates. For counts greater than 9, letters are used. A denotes 10, B denotes 11, etc.

while the scale with three scores is referred to as the "coarser" scale.)

As already mentioned above, two laboratories subjectively scored some of their experiments. Since they used different methods of scoring (Table 3), we present their results separately (Table 4). The first laboratory calculated the cytotoxicity index by using cyt^{exp}, but used local criteria for converting to a serologic scale. Consideration was also given to the results of cyt^{+} and cyt^{-}. The second laboratory used only the results of the experimental cell line (without considering the positive and negative control cell lines) in establishing its local scoring.

Table 2. Comparison of Serologic Scores from Experiments with Hidden Controls in the Cytotoxicity Section (Panel)

1st	2nd Score					
Score	1	2	4	6	8	Total
1	182	7	1	0	1	191
2	4	2	1	0	0	7
4	4	1	0	2	0	7
6	0	1	1	4	2	8
8	0	0	0	0	14	4
Total	190	11	3	6	17	227

First Laboratory. Ten (6.7%) of the experiments differed in their scores. If we do not distinguish between

Table 3. Local Criteria for Scoring Clones in Two Laboratories (Cytotoxicity)

Percent Cytotoxicity	Subjective Score	Score in Table 4
	First Laboratory	
0 - 9.9	Negative	1
10 - 14.9	Probably Negative	2
15 - 19.9	Probably Positive	4
20 - 34.9	Positive	6
≥ 35	Strongly Positive	8
	Second Laboratory	
≤ 8.0		1
8.1 - 14.0		4
≥ 14.1		8

Table 4. Comparison of Subjective and Workshop Scores

	First Laboratory					
Workshop Score	Subjective Score 1	2	4	6	8	Total
1	115	0	0	0	0	115
2	7	0	0	0	0	7
4	0	1	0	0	0	1
6	0	0	0	2	1	3
8	0	0	0	1	22	23
Total	122	1	0	3	23	149
	Second Laboratory					
1	122		0		0	122
2	0		0		0	0
4	0		0		0	0
6	0		1		10	11
8	0		0		9	9
Total	122		1		19	142

scores of 1 and 2 (negative) or between scores of 6 and 8 (positive), then just one experiment (0.7%) is scored differently.

Second Laboratory. This laboratory scored its data in a three division scale (negative = 1, weak = 4, and positive = 8). Eleven (7.7%) of the experiments received local scores that differed from the Workshop score. However, if we do not distinguish between Workshop scores of 1 and 2 (negative) or between scores of 6 and 8 (positive), then just 1 experiment (0.7%) is scored differently.

The serologic score assigns experimental results into different groups. A discriminant analysis was performed to assess whether this statistical method can discriminate different types of T-cell responses on the basis of cyt^{exp}. Table 5 compares the Workshop assignments with the assignments of the discriminant analysis. The discriminant analysis nearly always classified 90% or more of the experiments in the same way as the Workshop. Eighty-two percent of the experiments having a Workshop score of 6 got the same score in the discriminant analysis, representing the smallest number of correctly classified experiments. Overall, 150 (7%) of 2136 experiments were classified by discriminant analysis into different groups. If we consider the coarser scale, there are only 1.2% (26) misclassifications. These results support the criteria used for converting cyt^{exp} results into the serologic scores. They also demonstrate that a discriminant analysis is capable of classifying T-cell/B-cell combinations on the base of cyt^{exp}.

Other Variables. As mentioned above, we wished to avoid transforming the original data to a logarithmic or other scale, since we suspected that such a transformation might unnecessarily scale down differences between responders and non-responders. To assess whether this assumption was correct, we analyzed frequency distributions of two variables defined using logarithmic transformation. A discriminant analysis was also performed

Table 5. Classification Matrix of Cyt^{Exp} Produced by Discriminant Analysis

Workshop Score	Percent Correct	Score of Discriminant Analysis 1	2	4	6	8	Total
1	93.6	1644	112	0	0	0	1756
2	94.5	0	86	5	0	0	91
4	98.1	0	1	51	0	0	52
6	81.7	0	0	20	89	0	109
8	90.6	0	0	0	12	116	128
Total	93.0	1644	199	76	101	116	2136

in order to check the capability of these two alternate variables to discriminate the previously defined groups.

The two alternate variables, cyt^{diff} and cyt^{quot}, were defined in the following way:

$$cyt^{quot} = \frac{lg(exp^{exp} - spont^{exp})}{lg(max^{exp} - spont^{exp})}$$

and

$$cyt^{diff} = lg(exp^{exp} - spont^{exp}) - lg(max^{exp} - spont^{exp}) + 1$$

In this example, and in several subsequent examples, we added 1 in the formula to produce a scale where non-responder combinations are close to zero and responder combinations close to 1.

Experiments having a response value ≤ 0 were not included as the logarithm of a negative number is undefined. Therefore, 2114 experiments were analyzed. Table 6 shows the frequency distribution of the two alternate variables, taking into consideration the classification of results by the Workshop score.

Both of these variables seem to detect the heterogeneity of the sample. There are some experimental results (cyt^{diff} at least 4; cyt^{quot} at least 2) which are misclassified when using these variables for assigning the T-cell clones into groups. A discriminant analysis was performed with each variable separately. In evaluating the power of cyt^{diff} and cyt^{quot} to discriminate the groups, we have assumed (based on comparisons of the local and Workshop assignments) that the Workshop score represents a good approximation for the true recognition patterns of the clones. Table 7 shows the results of both analyses. The percentage of correct allocation for cyt^{quot} and cyt^{diff} is not as high as for cyt^{exp}. Using cyt^{diff} the percentage of correct classification is over 90% only for those experiments having a score of 4 or 8. Using cyt^{quot}, the percentage of correct classification is larger than 90% only in the group of experiments having a score of 8. There were 395 (18.7%) misclassifications for cyt^{diff} and 428 (20.2%) for cyt^{quot}. Thus, neither of these variables seems to detect the structure of the sample as well as cyt^{exp}. However, the large percentage of misclassification disappears when we use the coarser serologic scale (1 plus 2, 4, and 6 plus 8). In this case, cyt^{diff} misclassifies 1.5% and cyt^{quot} 2.2% of the experiments.

Table 6. Frequency Distribution of Cyt^{diff} and Cyt^{quot}

$<$ Cyt^{diff} $\leq$		Serologic Score 1	2	4	6	8	Total
	−3.0	50					50
−3.0	−2.5	269					269
−2.5	−2.0	352					352
−2.0	−1.5	452				1	453
−1.5	−1.0	267	2				269
−1.0	−0.5	271	5				276
−0.5	0.0	85	81	9		1	176
0.0	0.1		2	21			23
0.1	0.2			14	5		19
0.2	0.3			1	24		25
0.3	0.4				26	1	27
0.4	0.5				33		33
0.5	0.6				14	15	29
0.6	0.7			1		32	33
0.7	0.8	1			1	48	50
0.8	0.9					23	23
0.9	1.0					7	7
Total		1747	90	46	103	128	2114
$<$ Cyt^{quot} $\leq$							
	0.0	537					537
0.0	0.1	38					38
0.1	0.2	171				1	172
0.2	0.3	429					429
0.3	0.4	196	6				202
0.4	0.5	168	2				170
0.5	0.6	170	26	1			197
0.6	0.7	37	51	16	1		105
0.7	0.8		5	28	34	1	68
0.8	0.9	1		1	67	41	110
0.9	1.0				1	85	86
Total		1747	90	46	103	128	2114

Table 7. Classification Matrix of Cyt^{diff} and Cyt^{quot} Produced by Discriminant Analysis

Workshop Score	Percent Correct	Score of Discriminant Analysis 1	2	4	6	8	Total
Cyt^{diff}							
1	79.6	1391	355	0	0	1	1747
2	82.2	2	74	14	0	0	90
4	93.5	0	2	43	0	1	46
6	85.4	0	0	11	88	4	103
8	96.1	1	0	1	3	123	128
Total	81.3	1394	431	69	91	129	2114
Cyt^{quot}							
1	79.0	1380	364	2	0	1	1747
2	70.0	6	63	21	0	0	90
4	82.6	0	5	38	2	1	46
6	82.5	0	0	12	85	6	103
8	93.8	1	0	1	6	120	128
Total	79.8	1387	432	74	93	128	2114

Table 8. Criteria for the Serologic Score in Blocking Experiments of the Cytotoxicity Section

Inhibition 1:40	Serologic Scale
Inh < 0.5	1
0.5 ≤ Inh < 0.6	4
0.6 ≤ Inh < 0.8	6
0.8 ≤ Inh	8

Cytotoxicity–Blocking. For the blocking experiments, the variables which measure inhibition by monoclonal antibodies were defined in an analogous fashion as in the panel section. The same theoretical model was also used for the blocking experiments. Let us define rel^{max} and rel^{spont} as maximum and spontaneous ^{51}Cr release in the blocking experiments. Rel^{i} is defined as the ^{51}Cr release of the unblocked experiment (i=u) or of the experiments with monoclonal antibody dilutions of 1:i (where i=40, 200, 1000, 5000).

For each experiment i, we define lysis and inhibition as follows:

$$lys^{i} = \frac{rel^{i} - rel^{spont}}{rel^{max} - rel^{spont}}$$

$$inh^{i} = \frac{lys^{u} - lys^{i}}{lys^{u}} = 1 - \frac{lys^{i}}{lys^{u}} = \frac{rel^{u} - rel^{i}}{rel^{u} - rel^{spont}}$$

The variable inh^{i} measures the percentage of inhibition of T cell-mediated cytotoxicity in the unblocked experiments after eliminating the amount of spontaneous ^{51}Cr release. Thus, a low value of inh^{i} means little or no inhibition while a value nearby or larger than one indicates strong inhibition.

Criteria for finding the serologic scale by using inh^{40} are shown in Table 8. Scores were revised upwards if we found comparable blocking in experiments performed with a 1:200 dilution of antibody. Scores were revised downwards if we observed strong blocking at the 1:40 dilution, but little or no blocking at the 1:200 dilution.

Results of the Analysis

Distribution of lys Variables. Figure 5 shows the cumulative frequency distributions of the variables lys^{u}, lys^{40}, and lys^{200}. The deviation of the cumulative curve of lys^{u} from the expected normal curve is small but significant (Kolmogorov-Smirnov test: $2p < 0.05$, $n = 669$). The curve is slightly skewed to the right (skewness = 0.26). Mean and median are nearly identical (mean = 0.48, median = 0.47, s = 0.18), while the mode (0.27) is smaller. This again indicates the slight skewness of lys^{u}.

The deviations of all other variables lys^{i} from the expected normal distribution (i = 40, 200, 1000, 5000) are also small but significant ($2p < 0.05$, n(40) = 669, n(200) = 172, n(1000) = 168, n(5000) = 168). For lys^{40}, the mode is not unique indicating that lys^{40} is a

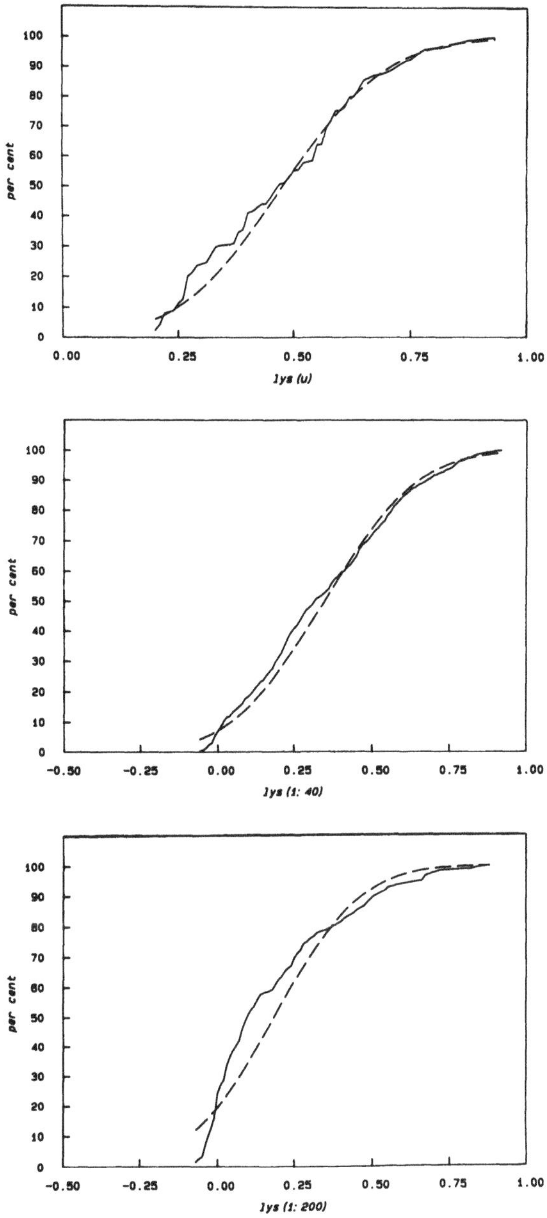

Figure 5. Cumulative frequency distribution of lys^u, lys^{40}, and lys^{200}

mixture of distributions (mean = 0.35, median = 0.32, s = 0.24), although the deviation from its expected normal curve is relatively small.

Distribution of inh Variables. The cumulative frequency distributions of inh^{40} and inh^{200} clearly show that they are mixtures of distributions (Fig. 6). Their deviation from their expected normal distribution is obvious even from inspection of the plots. The Kolmogorov-

Table 9. Results of the Regression Analysis with Experiments Having Hidden Controls (Cytotoxicity-Blocking)

		Regression Line $y = a + bx$			
Variable	n	a	b	Result of Stat. Tests	r
Lys^u	92	0.047	0.895	$a = 0 \quad b \neq 1$	0.90
Lys^{40}	92	0.011	0.931	$a = 0 \quad b = 1$	0.94
Inh^{40}	92	0.046	0.89	$a \neq 0 \quad b \neq 1$	0.92

Tests were carried with a significance level of 2p = 0.05.

Smirnov test is significant ($2p < 0.05$) for all inh^i variables (i = u, 40, 200, 1000, 5000). For inh^{40} the mode is not unique (mean = 0.24, median = 0.10, s = 0.4), while for inh^{200} the mode is 1 (mean = 0.61, median = 0.69, s = 0.41).

Reproducibility of Results. In the blocking section 92 experiments have a blind control for inh^{40}. In the scattergram, the dispersion of the points seem to be larger than for cyt^{exp} (Fig. 7). Although the lys^u and lys^{40} data points do not scatter very much in the plot, there are five pairs of experiments for which inh^{40} differs substantially (−0.26/0.14, −0.17/0.55, 0.29/0.72, 0/0.86, 0.88/0.37). As demonstrated in Table 9, the regression line for inh^{40} deviates from the straight line y = x significantly.

Although the analysis of the scattergrams is not as positive as for the panel data, we nevertheless obtain good results when comparing the experiments using the serologic score rather than the raw data (Table 10). Only 5.4% (5) of the experiments were scored differently. Using the coarser serologic scale, there are 4.3% (4) discordant experiments.

The results of the discriminant analysis confirm that inh^{40} is not as reliable (for the blocking data) as cyt^{exp} (for the panel data) (Table 11). In particular, experiments scored as 4 or 6 are misclassified by discriminant analysis. Of all experiments having a score of 6 or 4 only 35.7% or 73.7%, respectively, are allocated into the right group by discriminant analysis. Some 11.2% (75) of all experiments are misclassified by discriminant analysis. Even when scores of 6 and 8 are grouped together, 7.9% (53) of the experiments are still assigned scores by the discriminant analysis which differ from their Workshop score.

Table 10. Comparison of Serologic Scores in Experiments and Their Hidden Controls in Cytotoxicity (Blocking)

	2nd Score				
1st Score	1	4	6	8	Total
1	59	0	0	1	60
4	0	1	0	1	2
6	0	0	0	0	0
8	2	0	1	27	30
Total	61	1	1	29	92

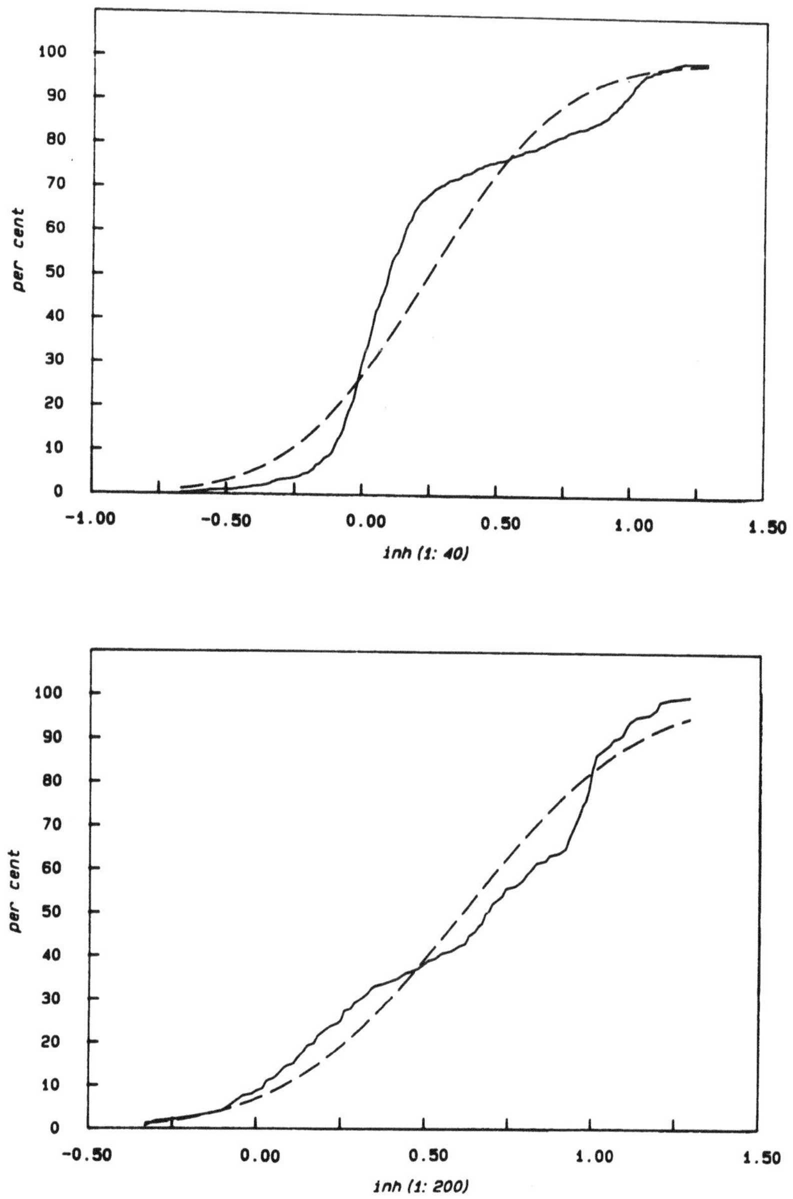

Figure 6. Cumulative frequency distribution of inh^{40} and inh^{200} (cytotoxicity–blocking).

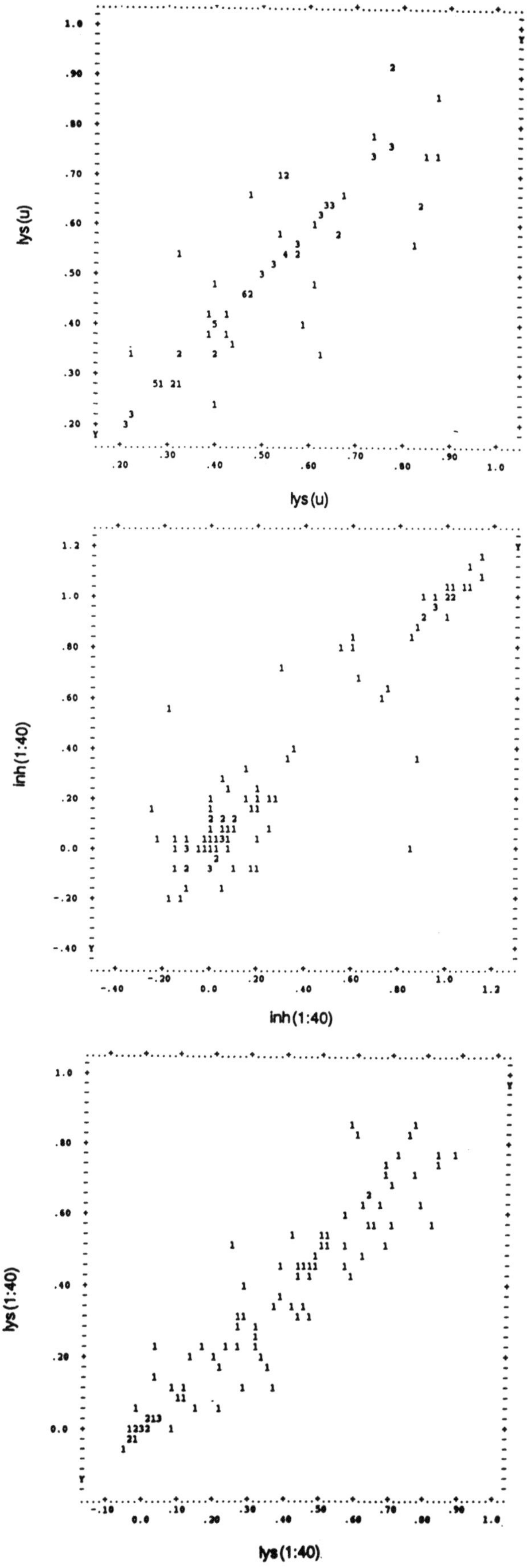

Figure 7. Concordance of inh^{40} (cytotoxicity), $\mathrm{lys}^{\mathrm{u}}$, and lys^{40} values in experiments having hidden controls. Numbers in the plot are the frequencies of points having the same coordinates.

Table 11. Classification Matrix of inh^{40} Produced by Discriminant Analysis (Cytotoxicity–Blocking)

Workshop Score	Percent Correct	Score of Discriminant Analysis 1	4	6	8	Total
1	93.8	484	32	0	0	516
4	73.7	0	14	5	0	19
6	35.7	0	14	15	13	42
8	88.0	0	2	9	81	92
Total	88.8	484	62	29	94	669

Table 12. Criteria for the Serologic Score in the Panel Experiments of the Proliferation Section

Proliferation Index	Serologic Scale
$\text{Prol} < 0$	1
$0 \leq \text{Prol} < 0.398$	2
$0.398 \leq \text{Prol} < 0.699$	4
$0.699 \leq \text{Prol} < 0.875$	6
$0.875 \leq \text{Prol}$	8

Proliferation–Panel. To define a variable measuring proliferation of T-cells, we utilize a model in which two processes contribute to the growth of T-cells during the experiment. 1) Background proliferation is random and not exponential. We therefore assume that the growth of B-cells (which have been irradiated) and of T-cells (cultured in the absence of exogenous IL-2) follows a normal distribution in the media control subcomponents of the experiment. 2) The growth of T-cells follows exponential law after stimulation with appropriate B-cell lines. We additionally assume that the results that we observe are the sum of the results of the two different growth processes. Based on these assumptions, we first calculate an intermediate response value by subtracting the backgrounds. Since the resultant variable is presumed to be lognormally distributed, we next transform the response values to the logarithmic scale to obtain standardized variables.

Thus, in experiments with positive, negative, and experimental cell lines, the response values, res^i ($i = +, -, exp$), are calculated in the following way:

$$res^i = exp^i - media^i\,B - media^i\,T$$

For convenience, let us call those three variables positive, negative, and experimental response. Using the logarithms of res^i, we define two variables which may serve as indices of proliferation.

$$prol^{quot} = \frac{lg(res^{exp}) - lg(res^-)}{lg(res^+) - lg(res^-)}$$

and

$$prol^{diff} = lg(res^{exp}) - lg(res^+) + 1 = lg\frac{res^{exp}}{res^+} + 1$$

For the purposes of these calculations, values ≤ 1.0 were set to 1.0 (since $\log x \leq 0$ for $0 < x \leq 1$ and $\log x$ is not defined for $x \leq 0$). We again added 1 to $prol^{diff}$ to produce a scale in which nonresponders are close to 0 and responders are close to 1.

Both variables allow for a simple biologic interpretation. $Prol^{diff}$ measures the logarithm of the percentage of experimental response over positive control response, while $prol^{quot}$ is the discrepancy in percent of the logarithms of experimental and positive response after subtracting the negative control response. $Prol^{diff}$ and $prol^{quot}$ have small values when T-cells do not proliferate. Their values both increase when proliferation occurs in the experiments, reaching values of 1 when $res^{exp} = res^+$.

The variable that was ultimately used for measuring proliferation (prol) and for determining the serologic score is the minimum of $prol^{quot}$ and $prol^{diff}$

$$prol = \min(prol^{diff}, prol^{quot})$$

Table 12 lists the criteria for the serologic scale. In our final analysis 666 (10.2%) experiments were scored using $prol^{quot}$. In 24 experiments, both variables were numerically identical. In all other cases (5869), $prol^{diff}$ was chosen. The rationale for choosing the minimum of the two variables is explained below.

Results of the Analysis

Distribution of res^+ and res^-. Res^- and res^+ are not normally distributed (Fig. 8). The Kolmogorov-Smirnov test is significant in both cases ($2p < 0.05$, $n = 356$). While res^- has a distribution skewed to the left (skewness $= -1.62$, mean $= -182.9$, median $= -71.3$, mode is not unique, $s = 1095.3$), the distribution of res^+ is skewed to the right (skewness $= 1.57$, mean $= 24266.2$, median $= 16410$, mode $= 36326.9$, $s = 22006.2$). The distributions of the logarithmic transformed variables do not significantly deviate from their expected normal distribution ($2p > 0.05$). This fact seems to support the model of T-cell proliferation described above.

Distribution of $prol^{diff}$ and $prol^{quot}$. As illustrated in Figures 9 and 10, $prol^{diff}$ is obviously a mixture of three unequal distributions (mean $= -1.68$, median $= -2.08$, mode $= -2.55$, $s = 1.50$). In contrast, the distribution of $prol^{quot}$ (Fig. 10) is skewed to the left (skewness: -2.63, mean $= -0.06$, median $= 0$, mode $= 0$, $s = 0.84$). $Prol^{quot}$ does not seem to recognize the heterogeneity of the sample as clearly as $prol^{diff}$.

The variable prol again demonstrates the heterogeneity of the sample quite clearly. The distributions of all three variables deviate significantly from the expected normal distribution ($2p < 0.05$, $n = 6535$).

Certainly, it is unusual to take the minimum of two differently defined variables which have different frequency distributions and to subsequently use the minimum for a statistical analysis. However, one must take

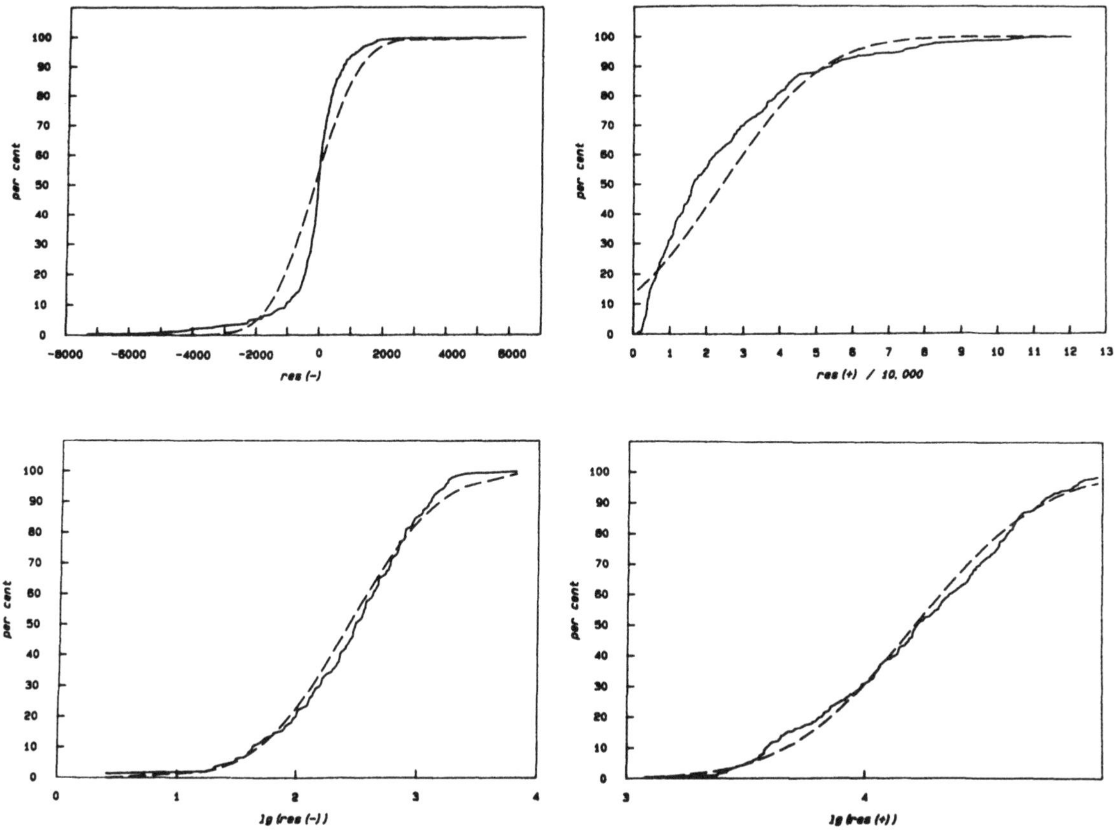

Figure 8. Cumulative frequency distributions of res⁺, res⁻, and their logarithmic transformations.

into consideration that by using prol as the variable measuring proliferation, we were conservative in our evaluation of the experimental results. When comparing the results of $prol^{diff}$ or $prol^{quot}$ with the original data of each experiment, we recognized that the variable $prol^{diff}$ usually works fine when there is a large difference between res^+ and res^- or when res^- is small. However, for experiments with a very high negative and/or a small positive control, the variable $prol^{quot}$ seems to characterize the proliferative response much better than $prol^{diff}$. This phenomenon can probably be explained using our model. In experiments having a high negative and (to a lesser degree) a low positive control, conditions for T-cell growth are probably not optimal. Therefore, in these experiments the growth of cells is most likely to deviate from the exponential law.

Figure 11 illustrates that in certain cases, either $prol^{quot}$ or $prol^{diff}$ will more reliably score the raw data. This figure contains three sections from the Workshop database in which the differential utility of $prol^{quot}$ or $prol^{diff}$ can be demonstrated. In the upper portion of the figure, there are two examples where res^{exp}, res^+, and res^- are all > 1000 and < 10000 (i.e., they fall within a fairly small numerical range). Because $prol^{diff}$ does not consider res^-, it would tend to score these responses as more strongly positive than they really are. In one case, res^{exp} is actually smaller than res^-. $Prol^{quot}$ provides a more accurate interpretation in these cases. In the middle and lower portion of the figure, there are several examples where res^{exp} is less than 10% of res^+, yet (because of a negligible, often negative, res^-) would be scored as significant responses using $prol^{quot}$. $Prol^{diff}$ provides the more accurate interpretation in these cases. Using both variables together thus tends to provide for an accurate interpretation even in the extreme cases, where one or the other variable alone is misleading. Consequently, we chose the minimum of $prol^{quot}$ and $prol^{diff}$ as the final variable, which provided satisfactory results in most cases.

The difference $res^+ - res^-$ is an indicator for the quality of experiments. Figure 12 shows the correlation between this difference and the experimental response, and demonstrates that in experiments where $res^{exp} > res^+ - res^-$, $prol^{quot}$ is taken, otherwise $prol^{diff}$ is taken. This indicates that $prol^{quot}$ actually characterizes experiments with low res^+ and/or high res^-. It should be noted that the use of $prol^{quot}$ requires that $res^+ > res^-$. If this is not the case, then $prol^{quot}$ will be negative irre-

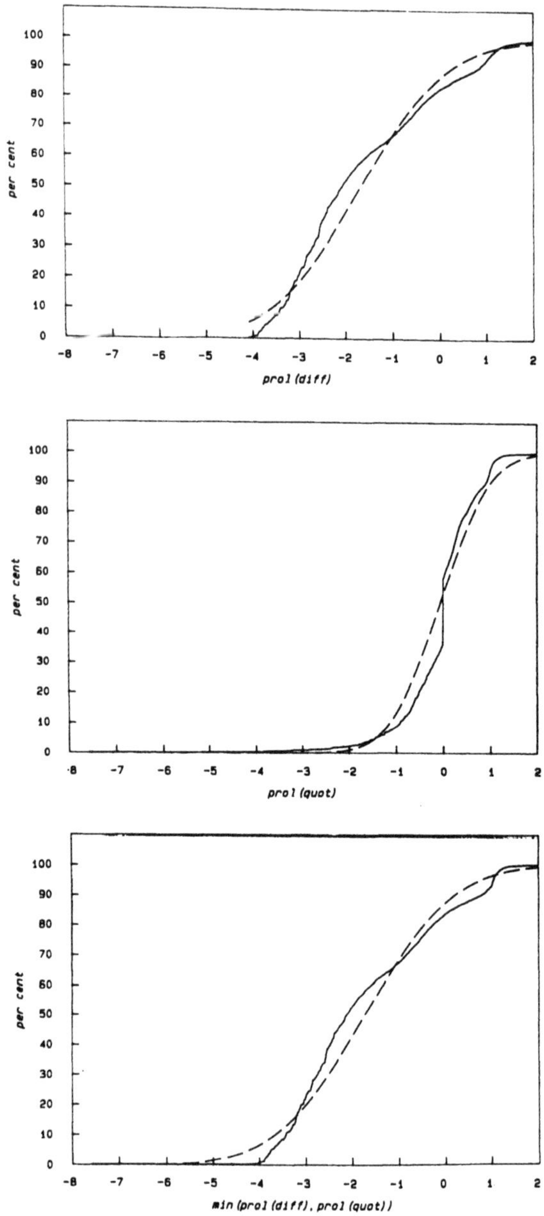

Figure 9. Cumulative frequency distributions of $prol^{diff}$, $prol^{quot}$, and prol.

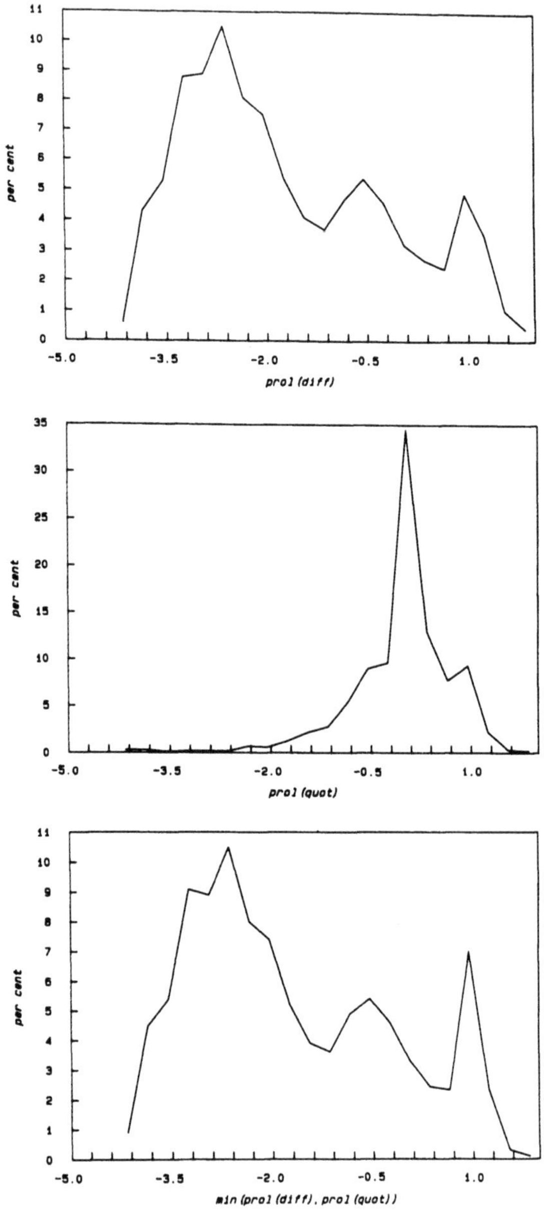

Figure 10. Frequency distribution of $prol^{diff}$, $prol^{quot}$, and prol.

11/07/87 WORKSHOP PANEL · PROLIFERATION Page 9

Fp	TC	BC	Res^-	Res^+	Res^{exp}	Quot	Diff	Score
381	2	9110	1104	2873	5545	1.69	1.29	8
343	2	9012	1730	6074	4345	.73	.85	6
315	2	9064	1788	6728	2527	.26	.57	2
321	2	9016	1730	6074	2370	.25	.59	2
384	2	9113	1104	2873	1668	.43	.76	4
360	2	9007	1788	6728	1959	.07	.46	2
368	2	9024	361	6127	2572	.69	.62	4
326	2	9052	1104	2873	875	-.24	.48	1
348	2	9029	361	6127	1566	.52	.41	4
363	2	9044	361	6127	2178	.63	.55	4

$Prol^{Quot}$ preferred (Fp 360, 326)

11/07/87 WORKSHOP PANEL · PROLIFERATION Page 100

Fp	TC	BC	Res^-	Res^+	Res^{exp}	Quot	Diff	Score
816	3	9030	-342	4333	57833	1.31	2.13	8
869	3	9015	-3276	15660	73604	1.16	1.67	8
841	3	9071	-342	4333	52900	1.30	2.09	8
873	3	9097	-3276	15660	80494	1.17	1.71	8
858	3	9007	-3276	15660	71301	1.16	1.66	8
836	3	9022	-342	4333	53798	1.30	2.09	8
848	3	9040	-342	4333	54069	1.30	2.10	8
819	3	9059	-342	4333	60043	1.31	2.14	8
859	3	9018	-5574	44329	87367	1.06	1.29	8
821	3	9016	-342	4333	4172	1.00	.98	8
824	3	9042	-342	4333	1445	.87	.52	4
822	3	9021	-342	4333	2242	.92	.71	6
835	3	9009	-342	4333	1863	.90	.63	4
808	3	9045	-342	4333	1079	.83	.40	4
823	3	9027	-342	4333	1404	.87	.51	4
834	3	9003	-342	4333	326	.69	-.12	1
812	3	9075	-342	4333	1266	.85	.47	4
799	3	9110	-342	4333	605	.76	.15	2
826	3	9052	-342	4333	166	.61	-.42	1

$Prol^{Diff}$ preferred (Fp 834, 799, 826)

11/07/87 WORKSHOP PANEL · PROLIFERATION Page 136

Fp	TC	BC	Res^-	Res^+	Res^{exp}	Quot	Diff	Score
781	2	9112	-1037	65575	141794	1.07	1.33	8
786	2	9034	-400	13906	44467	1.12	1.50	8
797	2	9029	-400	13906	42537	1.12	1.49	8
803	2	9032	-400	13906	28347	1.07	1.31	8
744	2	9031	-1037	65575	119520	1.05	1.26	8
764	2	9027	-1037	65575	131726	1.06	1.30	8
794	2	9025	-400	13906	1281	.75	-.04	1
765	2	9042	-1037	65575	8273	.81	.10	2

$Prol^{Diff}$ preferred (Fp 794, 765)

Figure 11. Examples showing that in certain circumstances $prol^{diff}$ or $prol^{quot}$ alone provide a less accurate assessment than when they are used together.

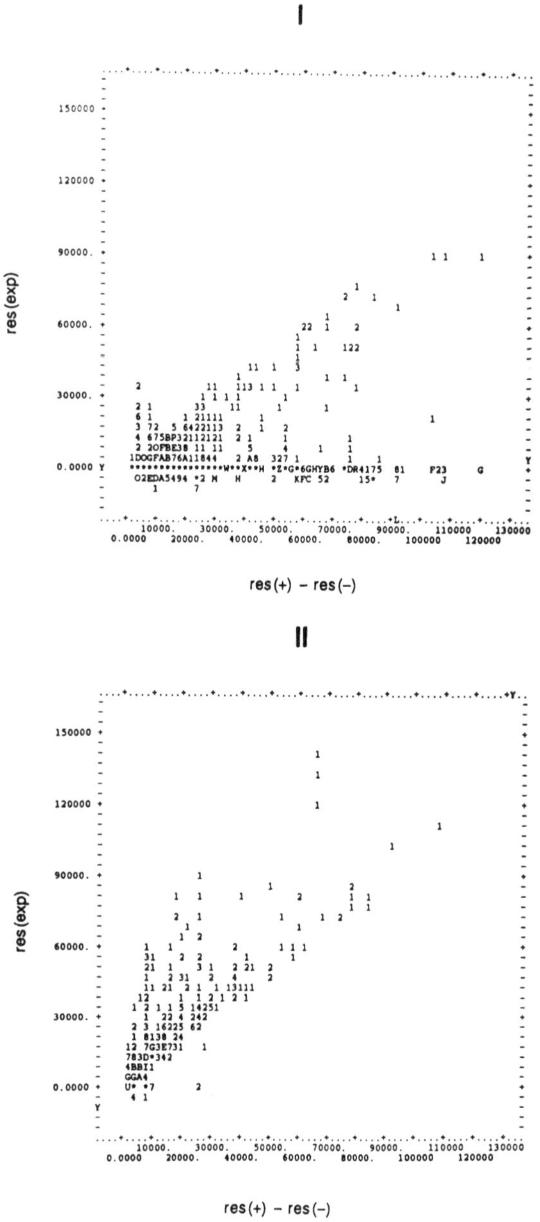

Figure 12. Correlation between the difference $res^- - res^+$ and res^{exp} in experiments having $prol^{quot} > prol^{diff}$ (I) or $prol^{quot} < prol^{diff}$ (II). For counts greater than 9, letters are used. A denotes 10, B denotes 11, etc. An asterisk indicates more than 36 points having the same position in the plot.

spective of the size of res^{exp} (unless res^{exp} is also smaller than res^-) or will be undefined when $res^+ = res^-$. Thus, $prol^{quot}$ cannot be used unless the experiment is technically satisfactory and the minimal criterion $res^+ > res^-$ is true.

Figure 13 shows that when $prol^{diff}$ and $prol^{quot}$ are less than zero (i.e., no proliferation), $prol^{quot}$ is often much larger than $prol^{diff}$. For all experiments having positive values for $prol^{diff}$ and $prol^{quot}$, the deviation of the points from the straight line of identity, $y = x$, is not very large. Thus, Figure 13 illustrates that when positive, $prol^{quot}$ and $prol^{diff}$ do not differ appreciably. This indicates that $prol^{diff}$ alone is also a good candidate for measuring proliferation.

If $prol^{diff}$ is actually also a good variable for characterizing proliferation, the serologic score defined by using $prol^{diff}$ must not differ from the Workshop score appreciably. To test this assumption, we looked at all (642) results of experiments having $prol^{quot}$ smaller than $prol^{diff}$ and defined a serologic score $score^{diff}$ using the criteria from Table 12, but based upon $prol^{diff}$ alone. Table 13 shows the deviation of $score^{diff}$ and the Workshop score. The scores differ in 48 (7.5%) experiments. Although the combination of the two variables deals more accurately with certain experimental extremes, using $prol^{diff}$ alone does not misclassify a significant number of experiments.

Table 13. Comparison of Workshop Score and $Score^{diff}$

Workshop Score	$Score^{diff}$ 1	2	4	6	8	Total
1	161	25	0	0	0	186
2	0	26	17	1	0	44
4	0	0	17	4	0	21
6	0	0	0	3	1	4
8	0	0	0	0	381	387
Total	161	51	34	8	388	642

$Score^{diff}$ was determined using the criteria in Table 12.

Reproducibility of Results. Five hundred ninety-five experiments have a matched, hidden control. Their concordance regarding $prol^{diff}$, $prol^{quot}$, and prol are shown in Figure 14.

In all three plots, the points deviate appreciably from the straight line of identity. Consequently, the regression lines also deviate significantly from the straight line of identity ($2p < 0.05$, see Table 14).

Converting results to the serologic score reduces the discrepancies in the matched experiments (Table 15). Some 9.4% (56) of the experiments received different serologic scores. Using the coarser serologic scale, there are 28 experiments (4.7%) which are scored differently.

Discriminant analysis was carried out for each variable separately (Table 16). The concordance of the assignments of the discriminant analysis and our classification is not as good as in the panel section of cytotoxicity. In the discriminant analysis which utilized prol, 17.8% of experiments were classified differently. However, using the coarser scale, the number of misclassifications is only 0.4%. The misclassification of experiments is 18.5% (coarser scale: 0.6%) for $prol^{diff}$, and 26.3% (coarser scale: 4.8%) for $prol^{quot}$. The lowest percentages of correct classifications for a given score were 79.1% for $prol^{diff}$ (group 8), 32.9% for $prol^{quot}$

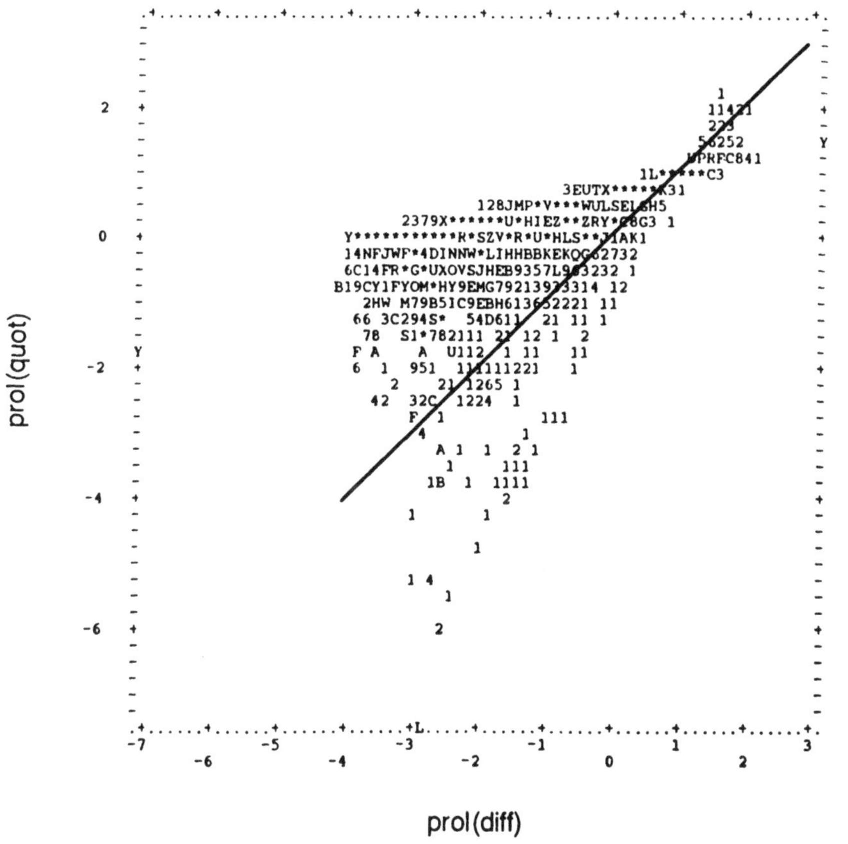

Figure 13. Relationship between $prol^{diff}$ and $prol^{quot}$. For counts greater than 9, letters are used. A denotes 10, B denotes 11, etc. An asterisk indicates more than 36 points having the same position in the plot.

(group 4), and 80.5% for prol (group 1). Using prol or $prol^{diff}$, classification of experiments is best for those experiments scored as 2, 4, or 6. With regard to the total percentage of misclassifications, there is very little difference between the results of discriminant analyses using prol or $prol^{diff}$.

As mentioned above, one laboratory tested its clones in experiments of cytotoxicity as well as proliferation. Scattergrams, showing the relationship of prol or $prol^{diff}$ and cyt^{exp}, do not differ substantially (Fig. 15). The relationship is not linear in either case. However, prol (or $prol^{diff}$) and cyt^{exp} can clearly discriminate two groups of experimental combinations (i.e., responders and nonresponders). This indicates that they must be powerful variables for classifying T-cell responses when used simultaneously.

Table 17 compares the serologic scores based on prol and cyt^{exp}. Even when the results of the matched

Table 14. Results of the Regression Analysis Performed with Those Experiments Having Hidden Controls (Proliferation-Panel)

Variable	n	Regression Line $y = a + bx$: a	b	Result of Stat. Tests	r
$Prol^{diff}$	595	−0.536	0.708	$a \neq 0$ $b \neq 1$	0.72
$Prol^{quot}$		0.029	0.709	$a = 0$ $b \neq 1$	0.56
Prol		−0.580	0.709	$a \neq 0$ $b \neq 1$	0.69

Tests were carried with a significance level of $2p = 0.05$.

Table 15. Serologic Scores from Experiments and Their Hidden Controls in Proliferation (Panel)

1st Score	2nd Score: 1	2	4	6	8	Total
1	484	4	2	0	3	493
2	14	5	2	0	1	22
4	3	5	4	0	2	14
6	2	3	2	3	4	14
8	2	0	1	6	43	52
Total	505	17	11	9	53	595

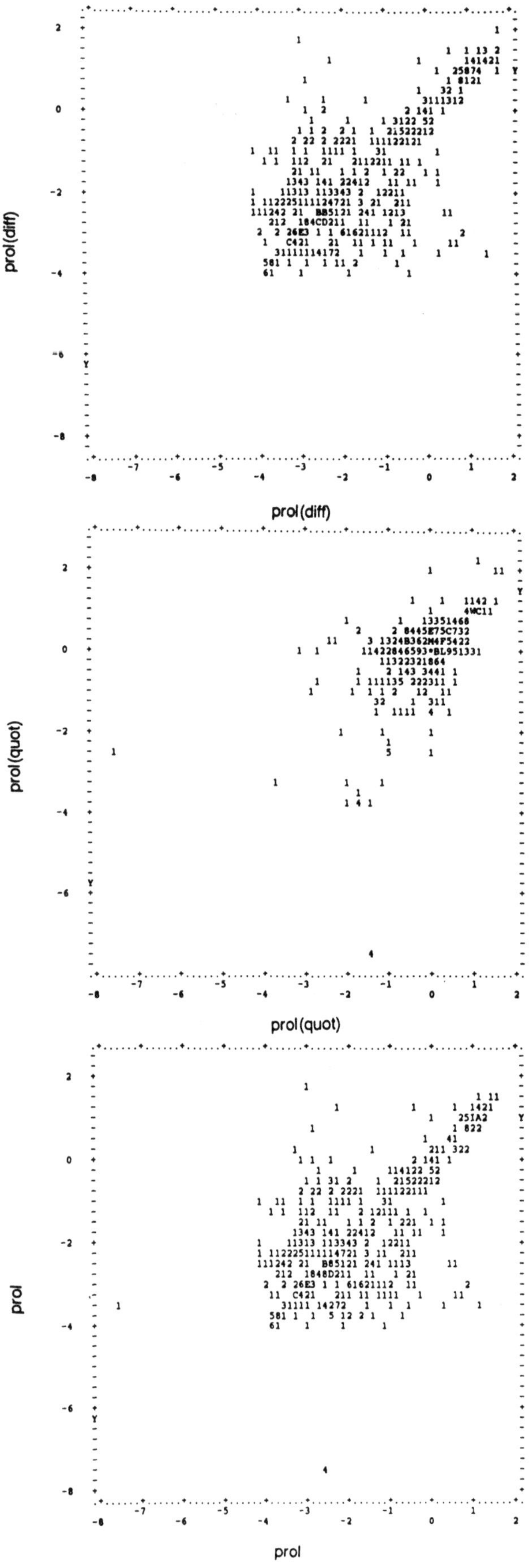

Figure 14. Concordance of $prol^{diff}$, $prol^{quot}$, and prol in experiments having a hidden control. For counts greater than 9, letters are used. A denotes 10, B denotes 11, etc. An asterisk indicates more than 36 points having the same position in the plot.

Table 16. Classification Matrix of Prol^{diff}, Prol^{quot}, and Prol Produced by Discriminant Analysis

Workshop Score	Percent Correct	Score of Discriminant Analysis 1	2	4	6	8	Total
		Prol^{diff}					
1	80.8	4413	1052	0	0	0	5465
2	89.8	0	212	23	1	0	236
4	91.8	0	2	134	10	0	146
6	100.0	0	0	0	110	0	110
8	79.1	0	0	0	121	457	578
Total	81.5	4413	1266	157	242	457	6535
		Prol^{quot}					
1	76.1	4161	1191	113	0	0	5465
2	48.7	21	115	97	3	0	236
4	32.9	0	34	48	64	0	146
6	96.4	0	0	4	106	0	110
8	66.6	0	0	0	193	385	578
Total	73.7	4182	1340	262	366	385	6535
		Prol					
1	80.5	4399	1066	0	0	0	5465
2	94.5	0	223	13	0	0	236
4	91.1	0	0	133	13	0	146
6	100.0	0	0	0	110	0	110
8	87.7	0	0	0	71	507	578
Total	82.2	4399	1289	146	194	507	6535

experiments differ in the proliferation section, the percentage of misclassification is not large. Approximately 9.8% (18) of the experiments are scored differently in the two sections. Using the coarser scale, only five (2.7%) of the T-cell/B-cell combinations are not concordant.

Two laboratories voluntarily scored some of the T-cell clones which were analyzed in the Workshop using their own criteria. The first laboratory determined its scores using centroid cluster analysis, producing a scale of 0, 1, 2, 6, and 8. The second laboratory used a T-max analysis (8), which produces a dichotomous scale consisting of the two elements "no proliferation" or "proliferation." Table 18 explains in more detail the scores that were used by these laboratories. Again, the number of discrepancies between the Workshop scores and subjective scores is small (Table 19).

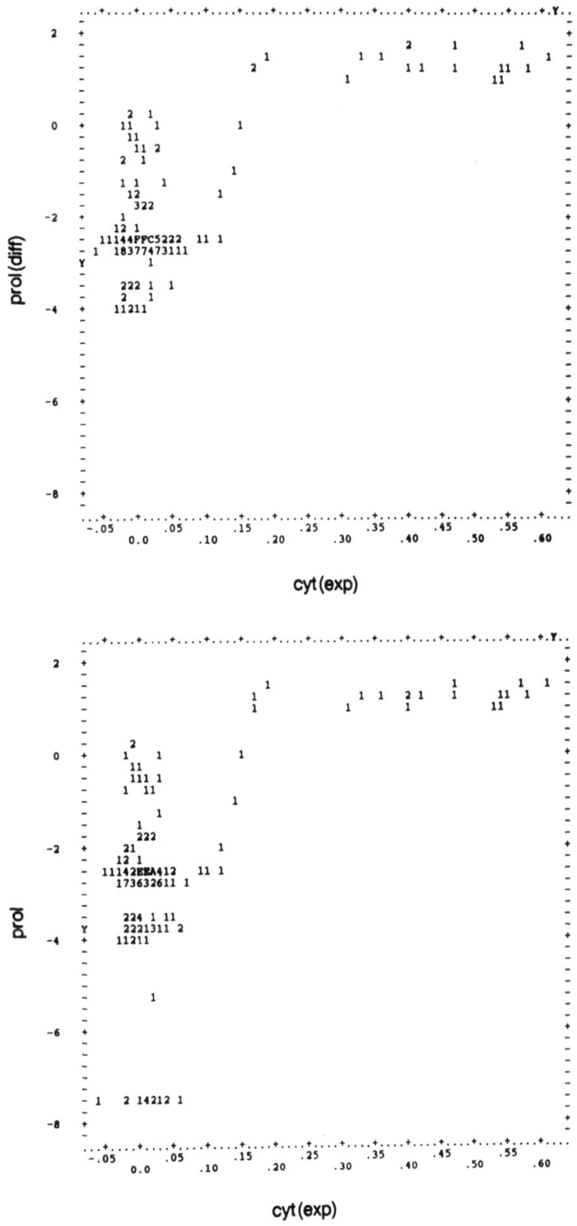

Figure 15. Comparisons of cyt^{exp} with prol and cyt^{exp} with prol^{diff}. For counts greater than 9, letters are used. A denotes 10, B denotes 11, etc. An asterisk indicates more than 36 points having the same position in the plot.

Table 17. Serologic Scores from Experiments in Cytotoxicity and Proliferation (Panel)

Prol	Cyt^{exp} 1	2	4	6	8	Total
1	152	5	5	0	0	162
2	3	0	0	0	0	3
4	0	0	0	0	0	0
6	0	0	0	0	0	0
8	0	0	0	5	14	19
Total	152	5	5	5	14	184

First Laboratory. A total of 9.8% (17) of the experiments are scored differently. Using the coarser scale, there are 6.9% (12) experiments that are scored differently. If we omit the three experiments that were scored as 0 locally, then 8.2% and 5.3% of the experiments received different scores using the finer and coarser scales, respectively.

Table 18. Criteria for Scoring Clones in Two Laboratories (Proliferation)

Subj. Score	Score in Table 19
First Laboratory	
Stimulator LCL	0
Did Not Work	
Definitive Negative	1
Probably Negative	2
Probably Positive	6
Definitive Positive	8
Second Laboratory	
Not Scored	0
No Proliferation	1
Proliferation	8

Second Laboratory. This laboratory scored each of its two replicate experiments separately. Therefore, we compared the scoring of each experiment with the Workshop score. Since this laboratory used a dichotomous scale, we determined the percentage of discordant experiments using the coarser scale only. In this case, 3.7% (44) of the experiments received discordant scores.

Other Variables. In our model, we assumed that proliferation in the media control subcomponents of the experiments is not exponential. If we alternately assume that these subcomponents also exhibit exponential growth of T-cells, we can generate another group of variables for analysis. These variables are generated in a similar fashion to those described above; however, the results are transformed to the logarithmic scale before subtracting out the background. Therefore, we define the following alternate response values and variables measuring proliferation.

Table 19. Comparison of Subjective and Workshop Scores (Proliferation – Panel)

First Laboratory

Workshop Score	Subjective Score 0	1	2	6	8	Total
1	3	117	1	0	0	121
2	0	3	0	1	0	4
4	0	1	2	5	0	8
6	0	0	0	2	1	3
8	0	0	0	0	0	37
Total	3	121	3	8	38	173

Second Laboratory

Workshop Score	Subjective Score 00	01	10	11	08	18	81	88	Total
1	1	12	2	925	0	1	0	0	941
2	0	1	0	78	0	1	0	0	80
4	0	0	0	25	0	2	5	4	36
6	0	0	0	3	0	0	0	18	21
8	0	0	0	0	1	2	0	123	126
Total	1	13	2	1031	1	6	5	145	1204

Table 20. Classification Matrix of $Prol^{lgdiff}$ and $Prol^{lgquot}$ Produced by Discriminant Analysis

Workshop Score	Percent Correct	Score of Discriminant Analysis 1	2	4	6	8	Total
$Prol^{lgdiff}$							
1	77.1	4212	1174	65	14	0	5465
2	61.0	27	144	60	4	1	236
4	52.1	0	25	76	45	0	146
6	57.3	1	11	13	63	22	110
8	74.9	0	10	21	114	433	578
Total	75.4	4240	1364	235	240	456	6535
$Prol^{lgquot}$							
1	76.9	4204	1152	107	2	0	5465
2	44.1	56	104	72	3	1	236
4	47.9	5	21	70	50	0	146
6	73.6	0	3	19	81	7	110
8	65.6	0	0	16	183	379	578
Total	74.0	4265	1280	284	319	387	6535

$$\mathrm{lgres}^i = \lg(\exp^i) - \lg(\mathrm{media}^i\,B) - \lg(\mathrm{media}^i\,T),$$

$$\mathrm{prol}^{lgquot} = \frac{\lg(\mathrm{res}^{exp}) - \lg(\mathrm{res}^-)}{\lg(\mathrm{res}^+) - \lg(\mathrm{res}^-)}$$

$$\mathrm{prol}^{lgdiff} = \lg(\mathrm{res}^{exp}) - \lg(\mathrm{res}^+)$$

Though $prol^{lgquot}$ and $prol^{lgdiff}$ have distributions that are similar to $prol^{diff}$ and $prol^{quot}$ (Fig. 16), we could not identify any simple rules for defining the serologic scale for these variables, either alone or together.

The discriminant analysis based on $prol^{lgdiff}$ misclassified 24.6% of the experiments. The percentage of misclassification was 26% when $prol^{lgquot}$ was used (Table 20). Using the coarser serologic scale, about 4% of experiments were misclassified ($prol^{lgdiff}$: 4.1%, $prol^{lgquot}$: 4.6%). The percentage of correct classifications of experiments assigned to a particular group was generally low. The highest percentage of correct classification was 77.1% for $prol^{lgdiff}$ (group 1) and 76.9% for $prol^{lgquot}$ (group 1). Therefore, these variables seemed less useful for the analysis than $prol^{diff}$ and $prol^{quot}$

Proliferation – Blocking. For the blocking experiments in the proliferation section, we assumed the same model as for the panel data. For reasons that will be described below, however, we do not transform the response values before calculating inhibition.

For each experimental subcomponent i (i = u for unblocked experiments, 40 for experiments with dilution 1:40, etc.), we define a variable measuring response analogously to that used for analysis of the panel data.

$$\mathrm{res}^i = \exp^i - \mathrm{media}^i\,B - \mathrm{media}^i\,T.$$

The variable measuring inhibition is defined in each experiment i as follows (i = u, 40, 200, 1000, 5000):

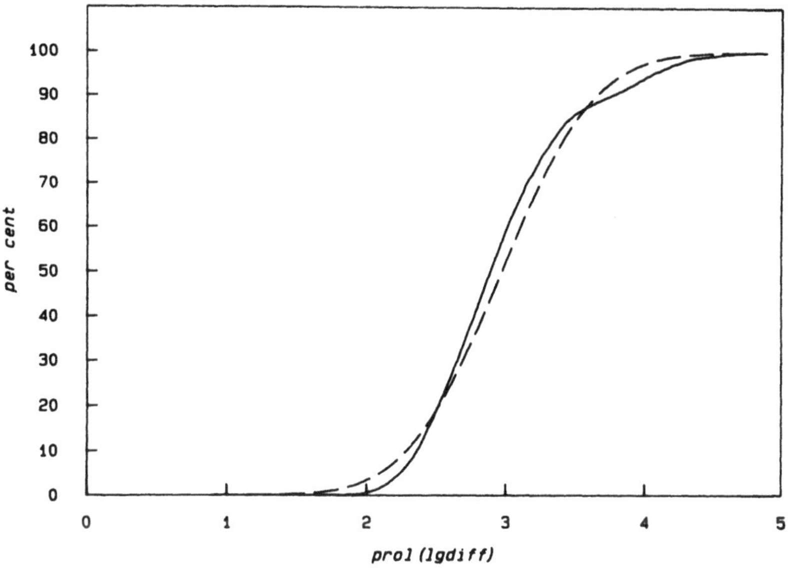

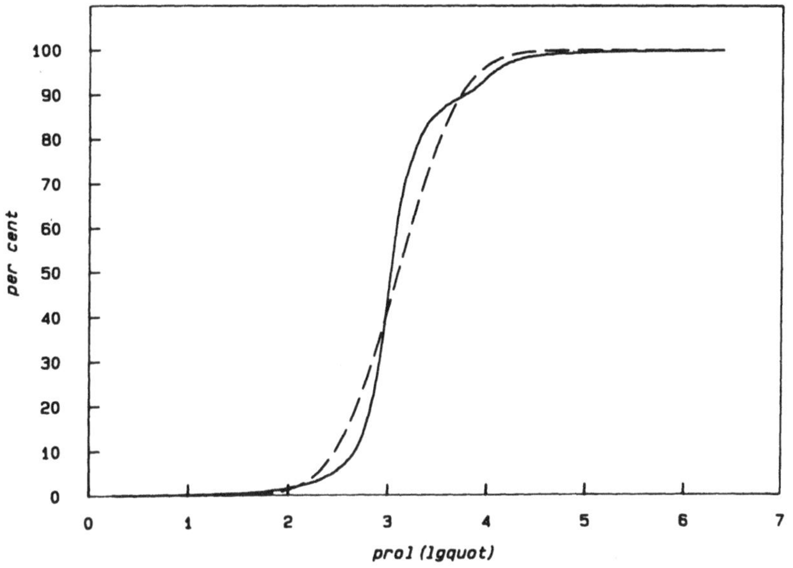

Figure 16. Cumulative frequency distribution of $prol^{lgdiff}$ and $prol^{lgquot}$

$$inh^{i} = \frac{res^{u} - res^{i}}{res^{u}}$$

Inh^{i} measures the percentage of deviation of res^{i} from res^{u}. Thus, a value of inh^{i} which lies near one indicates strong inhibition, while a value less than zero or near zero indicates no inhibition. Res values are not transformed to logarithms for calculation of inh, since this results in tremendous scaling down of differences among the data. For example, changes in proliferation producing 50 to 90% inhibition using untransformed data produces scores between 0.1 and 0.39 after transformation. This is the reason for the apparent inconsistency between the analyses of the panel and antibody blocking experiments.

Table 21 shows the criteria used for converting the blocking data to the serologic scale. One laboratory did not perform experiments at the 1:40 dilution, but began at 1:200 instead. In this case, the results from experi-

Table 21. Criteria for the Serologic Score in the Blocking Experiments (Proliferation)

Inhibition 1:40	Serologic Scale
Inh < 0.5	1
0.5 ≤ Inh < 0.6	4
0.6 ≤ Inh < 0.8	6
0.8 ≤ Inh	8

ments performed at the 1:200 dilution were used to determine the serologic scores. When analyzing the variable inh^{40}, however, these results were excluded from the analysis. For this reason, there are different numbers of T-cell clones analyzed in some of the scattergrams and contingency tables.

Results of the Analysis

Distribution of res^i. The frequency distribution of res^u is skewed to the right (skewness: 1.67, see Fig. 17). It significantly deviates from a normal distribution ($2p < 0.05$, $n = 184$; mean = 20186, median = 18138, $s = 14467$) and has two modes that lie close together. The heterogeneity within this sample (i.e., responders and nonresponders) is very well detected in the distribution of logarithmic transformed values, which also deviates significantly from the expected normal distribution.

The distributions of all other response variables are unequal mixtures of distributions (res^{40}: mean = 9899.4, median = 5371.5, mode is not unique, $s = 12906.6$; res^{200}: mean = 11745, median = 5377, mode is not unique, $s = 17245.6$). The Kolmogorov-Smirnov test is significant for all distributions ($2p < 0.05$).

Distribution of inh^{40} and inh^{200}. Figure 18 shows that inh^{40} and inh^{200} are mixtures of unequal distributions (inh^{40}: mean = 0.42, median = 0.46, mode = 1.0, $s = 0.75$; inh^{200}: mean = −0.23, median = 0.74, mode = 1.0, $s = 1.65$). Since inh^i measures the relative difference of res^u and res^i, the distribution is truncated at one. Therefore, it always deviates from a normal distribution (Kolmogorov-Smirnov test is significant: $2p < 0.05$, $n = 1365$).

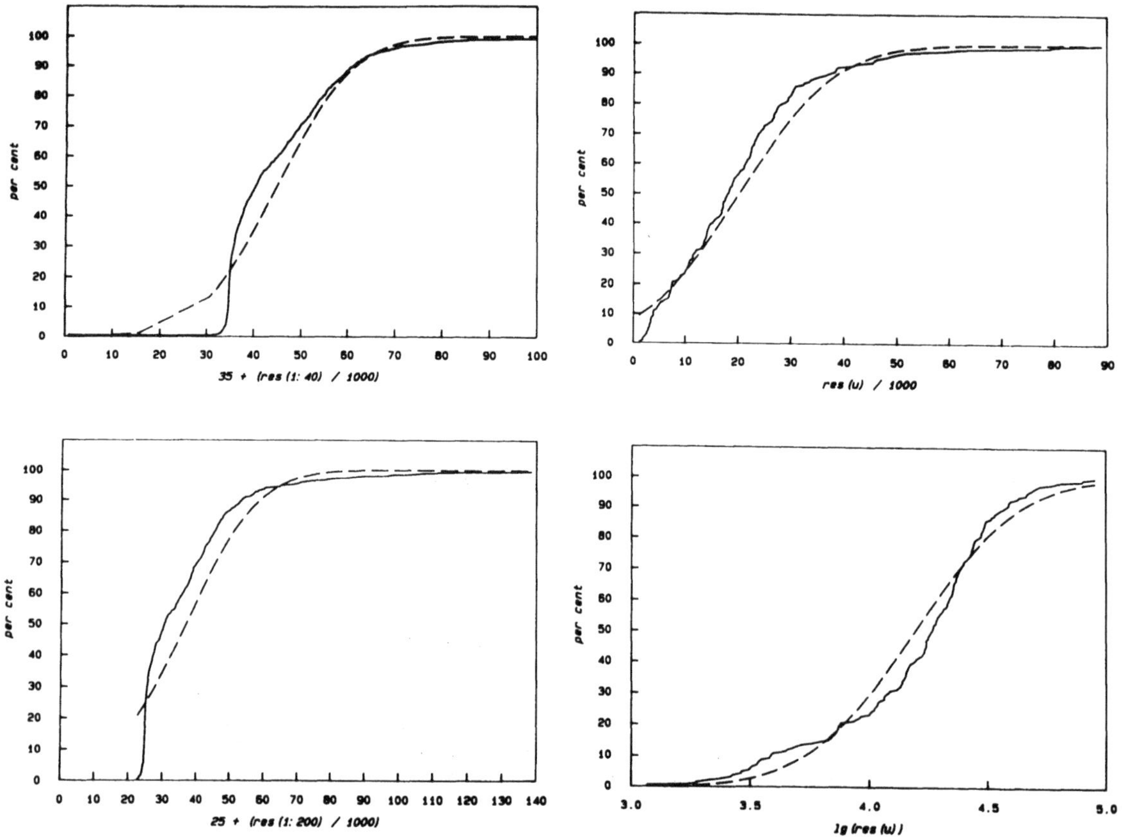

Figure 17. Cumulative frequency distribution of res^u, its logarithmic transformation, res^{40}, and res^{200}.

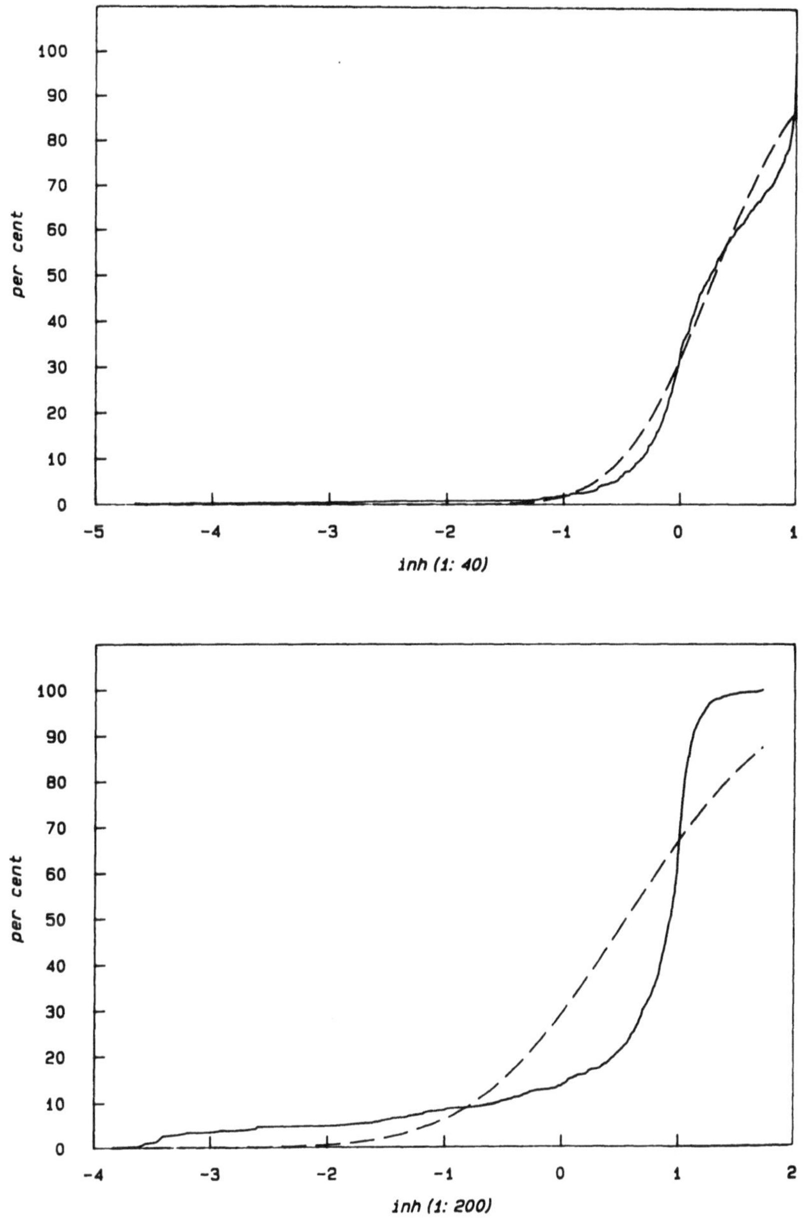

Figure 18. Cumulative frequency distribution of inh^{40} and inh^{200} (proliferation– blocking).

Reproducibility of Results. A total of 230 experiments has blind duplicates. Figure 19 compares the results of inh^{40} for these experiments. Although the values do not appear to scatter much, scores from the matched experiments often differ (e.g., −1/0.1, −0.8/0.6, −0.05/0.13, 0.54/0.2, 0.52/0.04). Therefore, the regression line deviates significantly from the straight line of identity. The regression coefficient and the interception with the y-axis are both significantly different from 1 and 0, respectively (regression line: $y = 0.809x + 0.103$, $r = 0.91$).

Regarding the serologic score, 11.3% (26) of the matched experiments have different scores. If we do not distinguish between 6 and 8, we still have 24 pairs of experiments (10.4%) which are discordant (Table 22).

There are only 39 instances where antibody inhibition experiments were performed with the same clones in both the cytotoxicity and proliferation sections. The

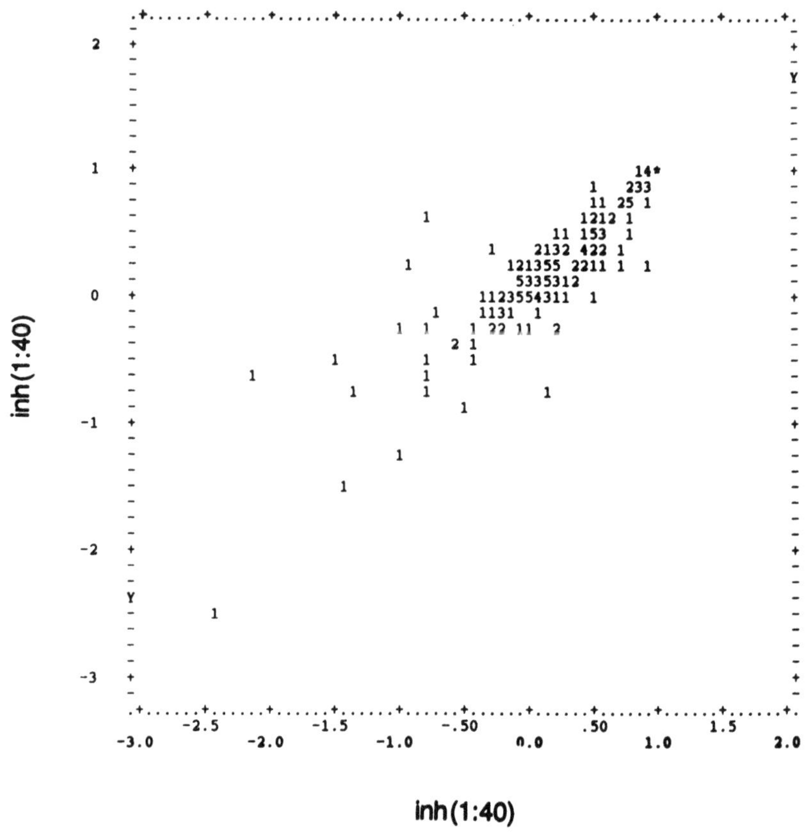

Figure 19. Concordance of inh^{40} (proliferation) in experiments having hidden controls. For counts greater than 9, letters are used. A denotes 10, B denotes 11, etc. An asterisk indicates more than 36 points having the same position in the plot.

comparison of the serologic scores of those data shows that seven experiments were scored differently in the proliferation and cytotoxicity section (Table 23).

Unfortunately, the total number of analyzed experiments is too small to allow for conclusions. Clearly, however, the data in this section of the cellular studies were far less satisfactory than for the previous three sections. It was for this reason that only extremely strong inhibition of proliferation was scored as significant. Nevertheless, the discriminant analysis based on inh^{40} still yields good results (Table 24), especially considering that inh^{40} has a truncated distribution. A total of 15.7% (214) of the experiments are classified differently by discriminant analysis. Using the coarser scale results in 14.3% (195) misclassifications.

Other Variables. We defined another group of variables for the blocking section by transforming response values to the logarithmic scale before calculating the variables measuring inhibition. Since we calculate and

Table 22. Serologic Scores from Experiments Having Hidden Controls (Proliferation – Blocking)

	2nd Score				
1st Score	1	4	6	8	Total
1	121	10	3	1	135
4	1	3	2	1	7
6	2	3	7	2	14
8	1	0	0	73	74
Total	125	16	12	77	230

Table 23. Comparison of the Serologic Scores in Cytotoxicity and Proliferation (Blocking) Experiments

Score in	Score in Cytotoxicity				
Proliferation	1	4	6	8	Total
1	24	0	0	0	24
4	2	0	1	0	3
6	2	0	0	0	2
8	1	1	0	8	10
Total	29	1	1	8	39

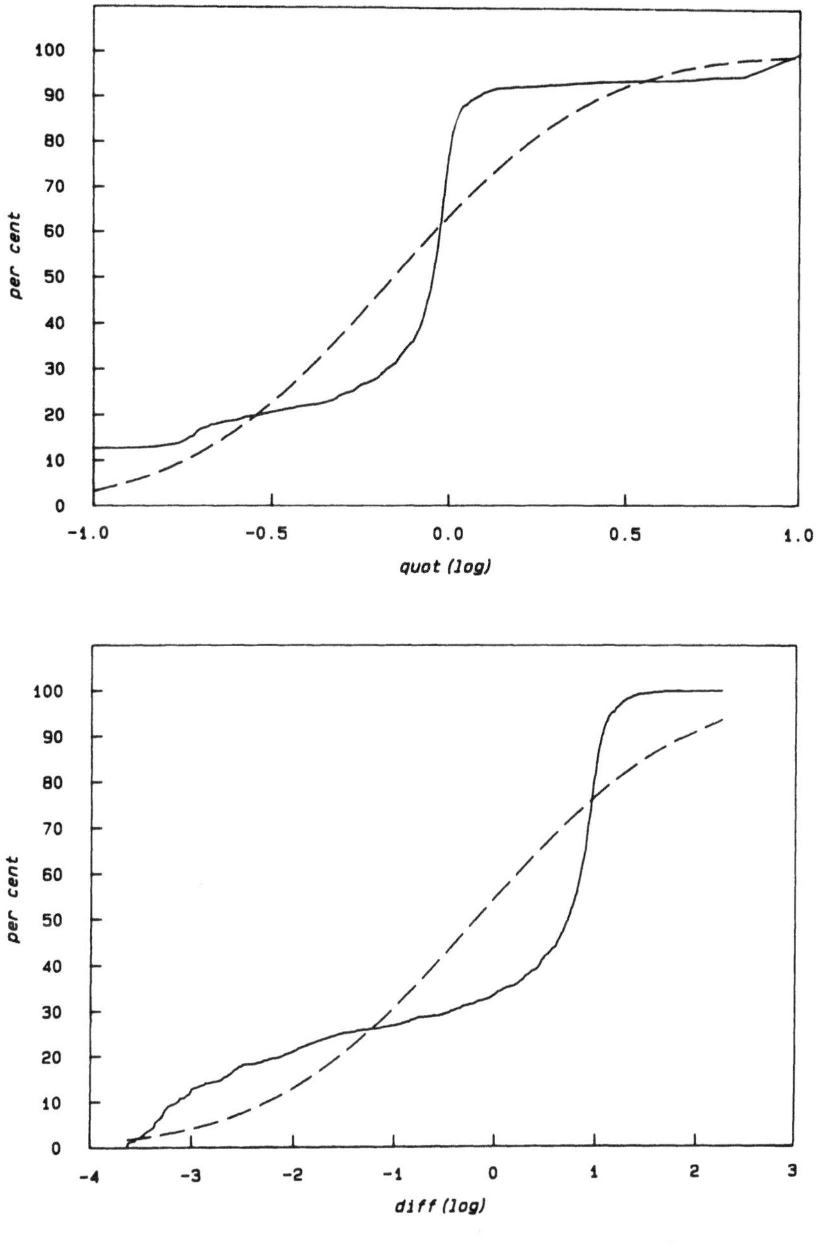

Figure 20. Cumulative frequency distribution of $quot^{log}$ and $diff^{log}$.

Table 24. Classification Matrix of Inh^{40} Produced by Discriminant Analysis

Workshop Score	Percent Correct	Score of Discriminant Analysis 1	4	6	8	Total
1	73.6	520	187	0	0	707
4	100.0	0	60	0	0	60
6	92.1	0	8	93	0	101
8	96.6	0	0	19	478	497
Total	84.3	520	255	112	478	1365

analyze these variables only for the experiments performed at the 1:40 dilution, we omit the index 40 in their name.

$$diff^{log} = lg(res^{40}) - lg(res^{u}) + 1$$

$$quot^{log} = \frac{lg(res^{40}) - lg(res^{u})}{lg(res^{u})}$$

The distributions of both variables also show the heterogeneity of the sample (Fig. 20). They tend to be u-shaped (i.e., they have more items in their tails than in the intermediate regions) ($diff^{log}$: mean = −0.19, median = 0.73, mode = 0.97, s = 1.61; $quot^{log}$: mean = 0.75, median = 0.91, mode = 1.0, s = 0.33). Although these variables are candidates for measuring inhibition, their u-shaped distribution makes their statistical analysis very difficult. Moreover, it was not possible to define simple rules for the serologic scale using one or both of the variables.

Discussion

We chose four criteria to guide us in the selection of variables measuring proliferation, cytotoxicity, or inhibition. We required that the variables have a simple biological interpretation, allow for comparisons across experiments carried out at different times and/or locations, clearly show the heterogeneity of the sample, and exhibit a normal distribution.

The presentations of the results of our analysis demonstrate that all but the last condition could be fully established. All variables which were utilized deviate from their expected normal distribution, with the single exception being $prol^{diff}$. Even when deviations from normal distributions seemed to be very small in the figures, the Kolmogorov-Smirnov test was always significant.

Table 25 summarizes the results of our analyses. If we accept 11% as an upper limit for the number of misclassified or discordant experiments, only the variables for the cytotoxicity section are optimal. While the percentage of discordant experiments is under (or equal) 11% in the experiments of the proliferation section, misclassification of experiments is often larger than 15% using discriminant analysis. To evaluate the type of misclassification or discordance, we also calculated the percentage for the coarse scale (1 plus 2, 4, and 6 plus 8). Analyzing these results, one can see that in the panel studies most of the misclassifications disappear when the coarse scale is used. In contrast, a more substantial number of misclassifications remain in both blocking sections, even using the coarser scale. The concordance of the serologic scores of those T-cell clones which were tested in cytotoxicity as well as in proliferation is excellent. In the panel section, 9.8% of T-cell clones are scored differently, with only 2.7% scored differently using the coarse scale. Unfortunately, the number of blocking experiments that can be compared between the cytotoxicity and proliferation sections is small (39) in this laboratory. Thus, it is impossible to evaluate the true concordance of the variables measuring inhibition in these experiments.

Comparing the Workshop scores with scores assigned locally by four laboratories produced similar high levels of concordance in all four cases. The percentage of discordant experiments was below 6% in all cases, ranging from 0.7 to 5.3% (coarser scale). The high percentage of concordant assignments in all four laboratories provides

Table 25. Summary of the Analysis of Variables

Variable	Analysis of Matching Experiments: Regression Line Sign. Different from Y=X	Analysis of Matching Experiments: Discordant %	Discriminant Analysis: Misclassif. %	Comparison with Local Assignments: Discordant %
Cyt^{exp}	No	11.0 / 5.3	7.0 / 1.2	7/0.7; 0.7
Inh^{40} (Cyt)	Yes	5.4 / 4.3	11.2 / 7.9	
Prol	Yes	9.4 / 4.7	17.8 / 0.4	8.2/5.3; 3.7
$Prol^{diff}$	Yes		18.5 / 0.6	
Inh^{40} (Prol)	Yes	11.3 / 10.4	15.7 / 14.3	
Comparison of Results from Cytotoxicity and Proliferation	Panel : 9.8 / 2.7 Blocking : 7 Experiments from 39			

The first number always shows the percentage for the scale 1, 2, 4, 6, and 8, while the second number shows the percentage for the reduced scale 1 plus 2, 4, 6 plus 8.

extremely strong support for the variables utilized in the panel sections.

While comparisons of serologic scores often show high concordance, direct comparisons of the variables by analyzing the scattergrams are often not as good. In the panel experiments of the cytotoxicity section, the differences of the results of the two matched experiments are so small that the points in the plot fall quite close to the line of identity, $y = x$. In all other sections, points often deviate substantially from this line. However, one has to take into consideration the fact that the scattering of the points in the plot is not solely caused by the limitations of the variable utilized. Scattering will increase if the quality of the experiments is suboptimal, but it can also be seen when both the variables and the data are optimal. As an example of the last possibility, one can consider the situation when two matched experiments produce proliferation indices of 0.85 and 2.0. This point would not fall close to the line of identity, but both experiments nevertheless indicate a very strong, unequivocally positive proliferative response. It is for this reason that we make frequent use of the coarse serologic scale, since in this form, we can actually make a biologic interpretation from the data.

Our approach for defining the variable for the panel data analysis in the proliferation section was clearly unusual in that we selected a composite variable. The variable works fine for our purpose of defining a serologic scale and it allows us to retain the results of 666 experiments which actually were not optimal because of the small differences between their positive and negative control responses. For experiments which are technically optimal, $prol^{diff}$ appears to be an excellent variable for measuring proliferative responses. Its distribution clearly shows the heterogeneity of the sample. Moreover, $prol^{diff}$ has some nice statistical properties, which neither $prol^{quot}$ nor prol possess. As already shown, res^{+} and res^{-} are lognormally distributed. We can assume that res^{exp} has the same distribution if the analysis is limited to either responder or nonresponder combinations. In this case, $prol^{diff}$ must also have a normal distribution (because the difference of two normally distributed, independent variables is again normally distributed (9)). Consequently, $prol^{quot}$ must have a Cauchy distribution (because the quotient of two independent, normally distributed random variables has a Cauchy distribution) whose mean, standard deviation, and other moments cannot be estimated. Unfortunately, the experimental design does not allow to tėst these assumptions directly. Therefore, our approach is limited. We can only analyze the distribution of $prol^{diff}$ within the five groups of experiments (each defined by the serologic group). The distribution of $prol^{diff}$ does not deviate significantly from a normal distribution in the groups scored 2, 4, and 6 ($n(2) = 236$, $n(4) = 146$, $n(6) = 110$, Kolmogorov-Smirnov test: $2p > 0.05$). In groups 1 and 8, however, it deviates significantly from the expected normal distribution ($n(1) = 5465$, $n(8) = 578$, $2p < 0.05$). Figure 21 clearly shows that the distribution of $prol^{diff}$ is still a mixture of distributions in group 1, while in group 8 it is skewed to the right (skewness: 1.22). We cannot unequivocally explain the heterogeneity of T-cell clones which are scored 1. The asymmetry of $prol^{diff}$ may indicate that the threshold value for assigning T-cell clones into group 8 was too high. On the whole, however, this analysis seems to support our assumptions and our model.

Certainly, when analyzing data using $prol^{diff}$ alone, there is the disadvantage that the quality of an experiment (i.e., the difference between the positive and negative control responses) is not taken into consideration, since this variable compares only the magnitude of the experimental and positive control responses. Therefore, it seems prudent to additionally use $prol^{quot}$ and/or res^{-} and res^{+} for screening data.

When analyzing the data of blocking experiments in proliferation, we found it hard to define and choose among variables. One reason for having this problem is that one cannot easily use an analogous or similar variable to that in the panel section. When combining results from different experiments of the panel sections, we use the results of experiments that were carried out in a similar way and that differed only in the type of cell line which was used. In the blocking experiments, however, we actually deal with two different types of experiments. The first group of experiments (unblocked experiments) simply measures the response of the T-cell clones, while the second group measures the blocking effect of the antibodies. One can certainly assume that there is exponential growth in the unblocked experiments, but it is questionable if cell growth also follows exponential law in experiments with antibodies. Thus, when calculating variables measuring inhibition in the proliferation section, we use the results of two subtly different types of experiments simultaneously. This should be taken into consideration when searching for the "optimal" variable measuring inhibition. It also should be restated that the blocking data for the proliferation section were the least technically satisfactory data in this component of the Workshop. For this reason, only extremely strong blocking effects were scored as significant. It may be that the variables would have behaved better from a statistical viewpoint had the data been better overall.

Our approach to variable selection was indeed unusual. We calculated continuous variables which conformed to certain rules or models. Next, we defined another variable having a more limited (serologic) scale and used it for checking the quality of the continuous variables. Defining the serologic scale was done in a subjective and, thus, unusual way through comparisons of the original data with the newly defined variables. After defining the serologic scores, we used linear discriminant analysis to show that the pattern we detected in the data could also be recognized by an objective statistical method. Linear discriminant analysis uses means and variances of groups to determine the discriminant function. Therefore, normal distribution of the data is a necessary condition in order to obtain reliable results. However, the variables we analyzed often

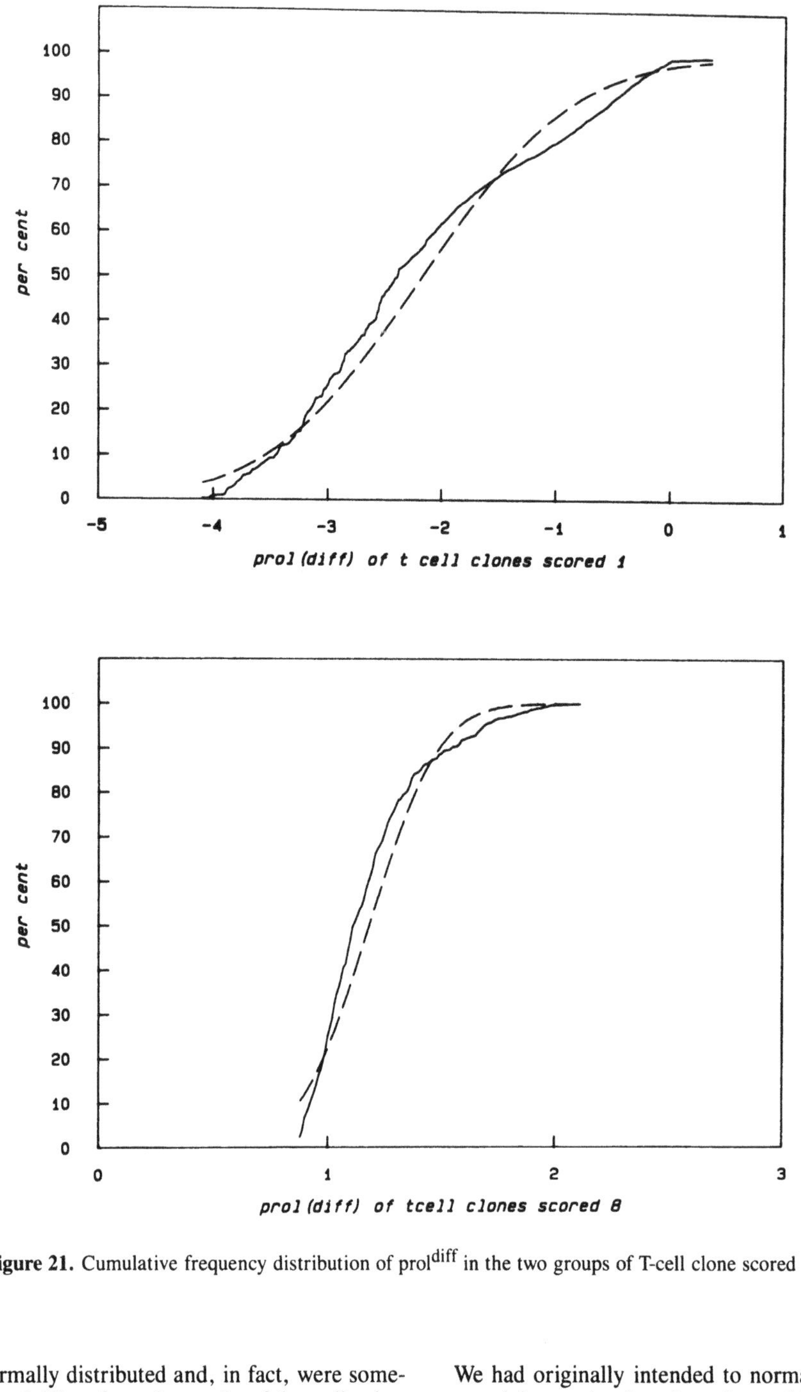

Figure 21. Cumulative frequency distribution of $prol^{diff}$ in the two groups of T-cell clone scored 1 or 8.

were not normally distributed and, in fact, were sometimes truncated. Therefore, the results of these discriminant analysis must be interpreted carefully with regard to the high percentage of misclassifications. Despite all the limitations, the percentage of misclassification by discriminant analysis remains under 20% for all variables chosen. This reflects the robustness of the discriminant analysis with regard to deviations of the data from normal distributions.

We had originally intended to normalize data in the cytotoxicity section based on the positive and negative control cell lines. Because of the wide variation in the strength of lysis of the positive control cell lines (ranging from 15 to 60% for certain clones), this approach introduced distortion into the results, and it was therefore avoided. The variable cyt^{exp} was used which represents the traditional cytotoxicity index used in most laboratories. Despite the lack of normalization, cyt^{exp} was the best of the variables used (based on statistical criteria). This may reflect that this variable itself (without normalization) produces a scale in which results tend to fall between 0 and 1. This contrasts with the clear need to use some measure of the positive control response in the proliferative experiments. Here, a raw score of 5,000 for res^{exp} would be highly significant if res^{+} were 7000, but probably insignificant if res^{+} was 70,000, a level of variation which was indeed seen for several clones within the Workshop.

We also performed cluster analyses (k-means algorithm) with the variable cyt^{exp}, but used the results in a very limited way. Because the k-means algorithm searches for a partition so that the members of each cluster have a minimum distance to the mean of the cluster, data must be normally distributed. Since our variables are not normally distributed, the results of the cluster analysis had to be carefully interpreted and were used only to slightly modify the serologic scores established with our other methods.

Certainly, the comparison of Workshop scores and "subjective" scores of the four laboratories is the most powerful approach for validating the variables. Additionally, it should be pointed out that there was a consensus among the participants in Princeton that the Workshop scores accurately reflected the reactivity of their clones, suggesting that agreement was not limited to these laboratories alone. Retrospectively, it is unfortunate that we did not solicit such "subjective" scoring at the beginning of our statistical analysis. This would have allowed us to pursue alternative and perhaps more effective approaches. At the least, we would have added the additional criterion for variable selection that an optimal variable should not misclassify more than 10% of the experiments in discriminant analysis. Discriminant analysis could then have been applied at the beginning of the process for the purpose of screening the variables. This approach would have also allowed for a more careful and direct analysis of the distributions of the selected variables.

Despite these limitations, it is clear that these variables allowed the T-cell clones to be scored on the basis of quantitative objective methods, with good agreement between the local and Workshop interpretations. Moreover, in the case of the panel data, these results also validate the concept that the serologic scores for cytotoxicity and proliferation can be integrated and analyzed together. Thus, the results presented here in many ways lend credibility and validity to the immunogenetic analyses performed with these results.

References

1. Sokal RR, Rohlf FJ. Biometry. W.H. Freeman and Company, New York, 1981.
2. Zar JH. Biostatistical Analysis. Prentice Hall, Englewood Cliffs, NJ, 1974.
3. Buening H, Trenkler G. Nichtparametrische Statistische Methoden. Walter de Gruyter, Berlin and New York, 1978.
4. Hartigan JA. Clustering Algorithms. John Wiley & Sons, New York, 1975.
5. Dixon WJ (chief editor). BMDP Statistical Software. University of California Press, Berkeley, 1985.
6. Mendell NR, Lee KL, Reinsmoen N, Yunis E, Amos DB, Emme L. Statistical methods for the evaluation responses in HLA-D typing. Transplant Proc 1977;9:99–106.
7. Lindeman RH, Merenda PF, Gold RZ. Introduction to Bivariate and Multivariate Analysis. Scott, Foresman and Company, Glenview, Illinois, 1980.
8. Kendall M, Stuart A, Ord JK. The Advanced Theory of Statistics, Vol. 3. Charles Griffin & Co., London, 1983.
9. Mood AM, Graybill FA, Boes DC. Introduction to the Theory of Statistics. McGraw-Hill Inc., Duesseldorf, 1974.

Immunogenetic Analysis of HLA Class II Cellular Data

N. Flomenberg,[1] R. Plätke,[2] J. Zemmour,[3] T.N. Small,[1] and J.S. Klein[1]

Summary of Data Processing and Statistical Approaches

As described in detail in the report "Collection and Processing of HLA class II Cellular Data" by Plätke et al. in this section, data was collected on floppy diskettes, and files were merged in the central computer at the Howard Hughes Medical Institute in Salt Lake City, Utah. The initial database was over 20 megabytes in size. The triplicate measurements were analyzed using the approach described in that report, which allowed spurious points within the triplicates to be identified and eliminated (1). Initial indices of cytotoxicity and proliferation were calculated, which were reviewed together with the raw data on a record by record basis. Records in which the two replicate experiments did not lead to the same experimental interpretation were flagged and the raw data was printed out, inspected and returned to the laboratory of origin for correction or clarification. If clarification could not be provided, these records were eliminated from the database. It should be restated that we required duplicate testings for all panel studies. Cell lines that were tested only a single time with a given clone were eliminated, in that we did not feel we could ensure reproducibility of results. We also established certain minimal criteria for the inclusion of particular experiments. For example, in cytotoxicity experiments the lysis of the positive control cell line had to exceed the lysis of the negative control cell line by at least 15%. Similarly, in proliferation experiments, proliferation in response to the positive control cell line had to exceed the response to the negative control cell line by at least 2400 CPM. Details of the criteria utilized are discussed in the preceding report (Statistical Analysis of HLA Class II Cellular Data by Plätke et al.) of this section (2).

After the database had been corrected, the data was subjected to the statistical analyses also described in the preceding report. Indices of cytotoxicity, proliferation, or antibody inhibition (blocking) were calculated and converted into a single digit score (8 = strongly positive, 6 = probably positive, 4 = weak or uncertain, 2 = probably negative, and 1 = negative). Thus, each of these resultant single digit numbers (reflecting the reactivity of a given T-cell clone with a particular B-cell line) represented the analysis of two replicate experiments which, including their controls, had required 54 raw CPM values to be entered by the local laboratories. As a result of this tremendous data reduction, the original 20 megabytes of data can be reduced to fit on a single floppy disk of 360 kilobytes. The scores used for this analysis were based on the scores of traditional HLA serology, and were adopted as a means of facilitating cross-comparisons with other Workshop components. Based on the results of the statistical analysis (2), we felt comfortable that: 1) the results of our interpretation of the data matched nicely (95% agreement) with local interpretation, and 2) the scales used for scoring cytotoxic and proliferative clones were comparable, such that these two groups of clones could therefore be analyzed together. It should also be stated that we eliminated clones from the central analysis when there was more than one instance in which the blind duplicates provided information that differed from their matched counterparts.

The results of small portions of the database are listed in graphic form in Figures 1 and 2. Figure 1 demonstrates the results of selected panel studies, testing the reactivity of the clones versus the Workshop cell lines. Figure 2 illustrates the results of the selected antibody blocking studies. (The composition of the Workshop cell panel is described in detail elsewhere in this volume (3)).

Inspection of the results indicates that the patterns of reactivity can be clustered into distinct groups. In the panel analyses, the most obvious initial observation is that there are clusters that can be correlated with conventionally defined HLA class II specificities, as well as with subsets and supersets of these specificities. These immediate observations were quite important in that they demonstrated that our statistical efforts had indeed yielded immunogenetically interpretable data. Detailed analysis of the patterns of clonal reactivities resulted in the establishment of seven new W.H.O. recognized specificities, as well as over 30 additional cellularly defined polymorphisms. These findings are described in the antigen reports and summary chapters that follow.

Analysis of the antibody blocking data is illustrated in Figure 2. It should be noted that the subjective assessment of the data (by those who screened it for errors) was that the studies of antibody blocking in the proliferation section were not optimal. It was only by setting extremely stringent criteria for scoring these results that the antibody panel began to demonstrate differential inhibition of proliferative responses (2). Nevertheless, once these stringent requirements were set, meaningful patterns began to emerge for the majority of the clones analyzed.

As illustrated in Figure 2, there are clusters of clones that are inhibited primarily by anti-HLA-DR antibodies, by anti-HLA-DQ specific antibodies or by anti-HLA-DP reagents. Additionally, two groups of clones were identified that were either not inhibited by any of the antibodies, or whose patterns of inhibition could not be deciphered. Our interpretations of the antibody blocking data are listed in the right hand column of Figure 2, and

Participating Laboratories: US7NFL,[1] US7LAL,[2] US7DUP[3]

Immunobiology of HLA
Volume I

T-Cell Clones — Panel Analysis

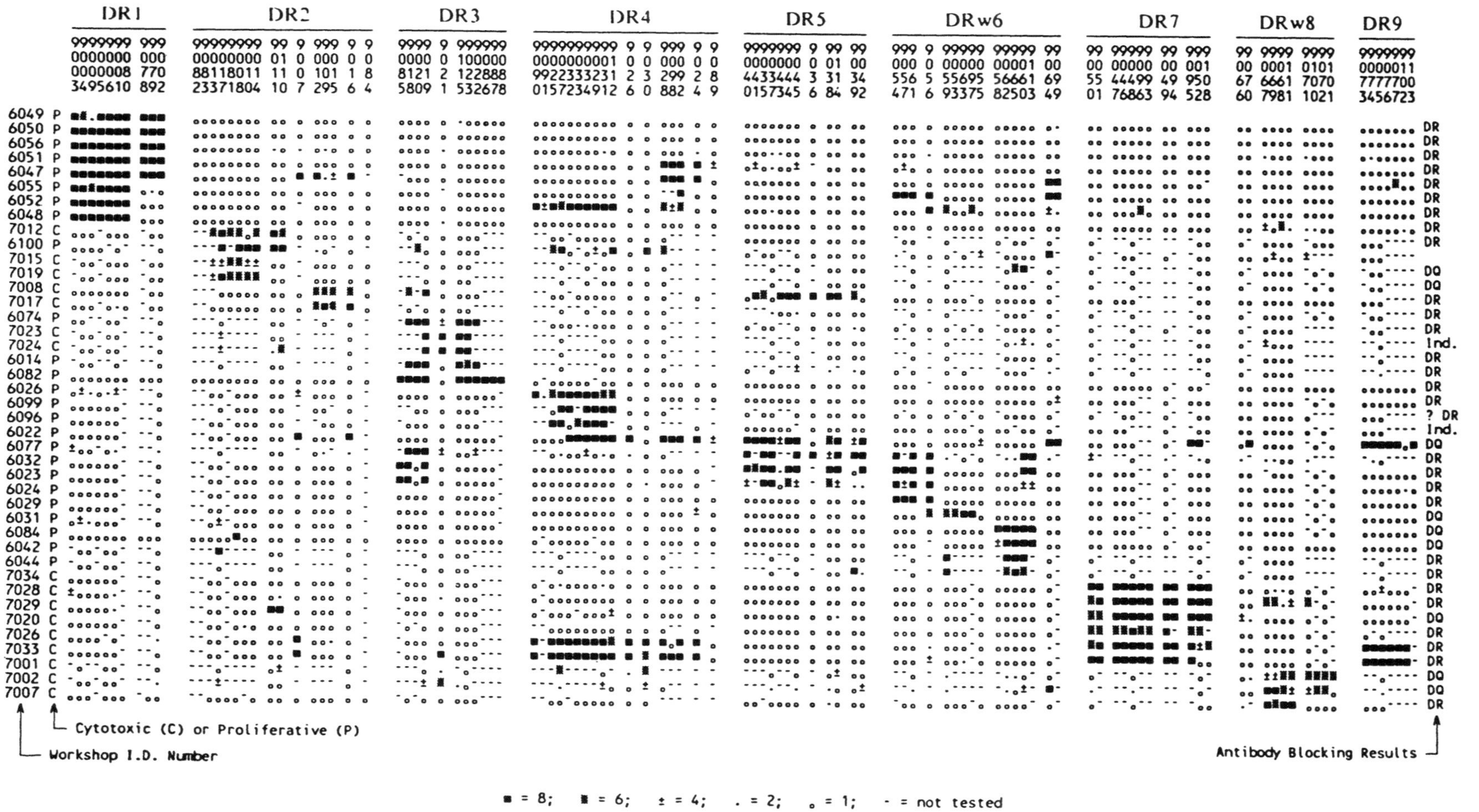

Figure 1. A portion of the cellular panel data is illustrated. The Workshop cell lines have been grouped according to their DR and Dw types. The Workshop number assigned to each clone is listed in the left column. The functional assay in which the clone was tested (cytotoxicity = C or proliferation = P is listed in the second column. The HLA class II family recognized by each clone (as inferred from the monoclonal antibody blocking data) is listed in the right column. The symbols in the figure reflect the following scores: solid square = 8 = positive; shaded rectangle = 6 = probably positive; ± =4 = weak or uncertain; closed dot = 2 = probably negative; open circle = 1 = negative; − = not tested.

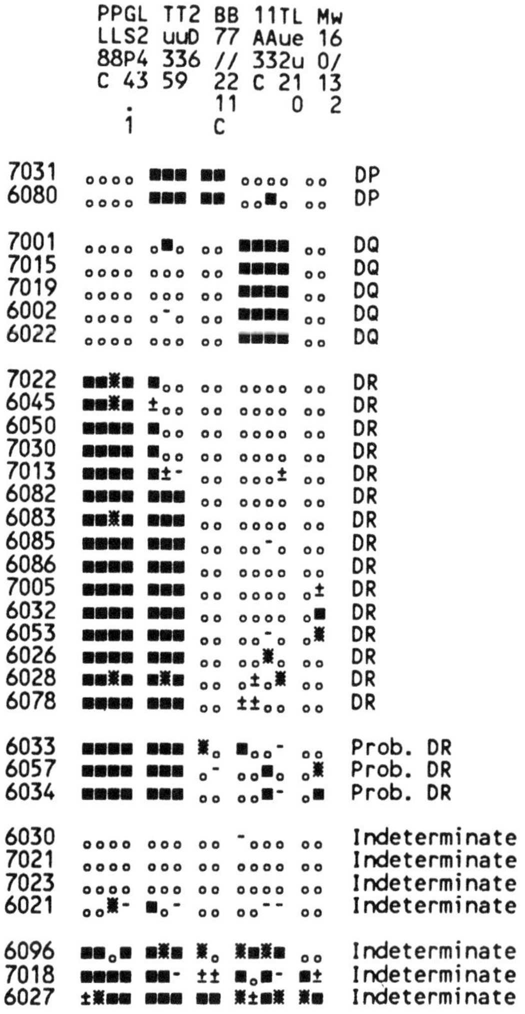

■ = 8; ✱ = 6; ± = 4; . = 2; ₒ = 1; - = not tested

Figure 2. A portion of the antibody inhibition data is illustrated. The Workshop number assigned to each clone is listed in the left column. Antibodies have been separated into five groups representing (from left to right): HLA-DR-specific antibodies, broadly reactive anti-HLA class II antibodies recognizing more than a single class II family, HLA-DP-specific antibodies, HLA-DQ-specific antibodies, and control antibodies that do not bind to HLA class II molecules. The term "specific" indicates that an antibody reacts predominantly, though not necessarily exclusively with products of that HLA class II region. The symbols in the figure reflect the following scores: solid square = 8 = positive; shaded rectangle = 6 = probably positive; ± = 4 = weak or uncertain; closed dot = 2 = probably negative; open circle = 1 = negative; – = not tested. The interpretations of the patterns observed are listed in the right column of the figure.

are also listed in the right hand column of Figure 1. "Probable" blocking assignments were made when there was blocking activity within two or more antibody groups (i.e., DR and DQ), but where there were clearly more blocking responses or stronger blocking responses within one group. Otherwise the patterns were scored as indeterminate.

In these figures and in the antigen reports and summary chapters that follow, cytotoxic clones have received 7000 numbers, while proliferative clones received 6000 numbers. Those who attended the Workshop should note that the numbers for proliferative clones in this report are different than those used during the meeting in Princeton. This is due to the inclusion of several additional clones. Clones with other types of designations represent additional local clones, included for illustrative purposes at the discretion of the individuals preparing the reports. These clones were tested with the Workshop cell

Table 1. Workshop Clone Designations

Lab/Clone	Old Designation	New Designation
BAT 1	6001	6001
BAT 2	6002	6002
BAT 3	6003	6003
BAT 4	6004	6004
BAT 5	6005	6005
BAT 6	6006	6006
BSH 2	6007	6007
BSH 1	6008	6008
BSH 3	6009	6009
DUQ 1	6010	6010
DUQ 2	6011	6011
DUQ 3	6012	6012
DUQ 4	6013	6013
FER 1	6014	6014
FER 2	6015	6015
FER 3	6016	6016
FES 1		6017
FES 2		6018
HEI 1	6017	6019
HEI 2	6018	6021
HAN 1	6019	6022
HAN 2	6020	6023
HAN 3	6021	6024
HAN 4	6022	6025
HAN 5	6023	6026
HAN 6	6024	6027
HAN 7	6025	6028
HAN 8	6026	6029
HAN 9	6027	6030
HAN 10	6028	6031
HAN 11	6029	6032
JEA 1	6030	6033
JEA 2	6031	6034
KUR 5	6035	6035
KUR 2	6032	6036
KUR 3	6033	6037
KUR 4	6034	6038
KUR 6	6036	6039
LEV 2	6037	6040
LEV 3	6038	6041
LEV 4	6039	6042
LEV 5	6040	6043
LEV 6	6041	6044
LEV 7	6042	6045
MBC 1	6043	6046
MBC 2	6044	6047
MBC 3	6045	6048
MBC 4	6046	6049
MBC 5	6047	6050
MBC 6	6048	6051
MBC 7	6049	6052
MBC 8	6050	6053
MBC 9	6051	6054
MBC 10	6052	6055
MBC 11	6053	6056
MBC 12	6054	6057
MBC 13	6055	6058
MBC 14	6056	6059
MBC 15	6057	6060
MBC 16	6058	6061
MBC 17	6059	6062
MBC 18	6060	6063
MYE 1	6061	6064
MYE 2	6062	6065
MYE 3	6063	6066
MYE 4	6064	6067
MYE 5	6065	6068
MYE 6	6066	6069
NIK 2	6067	6070
NIK 3	6068	6071
NIK 4	6069	6072
NIK 5	6070	6073
OHA 1		6074
OHA 2		6075
OHA 3		6076
OHA 4		6077
OHA 5		6078
OHA 6		6079
ROO 1	6071	6080
ROO 2	6072	6081
ROO 3	6073	6082
ROO 4	6074	6083
ROO 5	6075	6084
ROO 6		6085
ROO 7		6086
TSB 2	6076	6093
TSB 1	6077	6094
TSB 3	6078	6095
TSB 4	6079	6096
TSB 6	6080	6097
ZIE 2	6081	6099
ZIE 3	6082	6100
AIZ 1	7001	7001
AIZ 2	7002	7002
AIZ 3	7003	7003
AIZ 4	7004	7004
AIZ 5	7005	7005
BAC 1	7006	7006
BAC 2	7007	7007
BAC 3	7008	7008
BAC 4	7009	7009
BAC 5	7010	7010
BAC 6	7011	7011
BAC 7	7012	7012
BAC 8	7013	7013
BAC 9	7014	7014
BAC 10	7015	7015
BAC 11	7016	7016
BAC 12	7017	7017
BAC 13	7018	7018
BAC 14	7019	7019
DDC 1	7020	7020
KUR 2	7021	7021
KUR 3	7022	7022
KUR 4	7023	7023
KUR 5	7024	7024
KUR 6	7025	7025
NFL 1	7026	7026
NFL 2	7027	7027
NFL 3	7028	7028
NFL 4	7029	7029
NFL 5	7030	7030
NFL 6	7031	7031
NFL 7	7032	7032
NFL 8	7033	7033
NFL 9	7034	7034

Table 2A. Quantitative Immunofluorescent Analysis of Workshop Cell Panel – Part A

		Monoclonal Antibodies																									
Cell #	Cell Name	Neg	W6/32	A-1	A-2	A-3	Tu 39	M101	Tu 35	GSP 4.1	1A3	B7/21	PL 8	Tu 22	PL 15	L243	2D6	Leu 10	16.23	17.15	3G5	7.3.19.1	77.34	AB7.AE9	AB8.25	C1	CC-CL20	CHE249.2
9001	SA	6	100	100	96	98	98	92	99	100	95	99	99	98	99	99	99	99	22	12	83	20	98	22	97	35	96	88
9002	MZ 070782	9	99	94	93	94	87	61	91	97	93	93	97	94	93	89	92	96	9	13	18	12	94	20	79	15	89	54
9003	KAS-116	8	99	99	95	98	98	87	99	99	95	98	99	97	92	98	98	98	25	37	66	23	96	28	93	34	96	83
9004	JES THOM	3	97	77	95	95	92	57	93	98	98	96	98	98	97	93	94	99	8	6	7	10	99	14	94	7	98	81
9005	HOM-2	7	99	99	98	99	98	100	99	98	98	99	99	98	99	99	96	99	17	21	57	8	98	19	97	9	97	84
9006	WT 100-BIS	4	99	98	88	93	95	81	93	98	89	91	97	93	92	97	97	93	8	8	20	5	93	11	85	19	91	67
9007	DEM	5	99	99	98	98	99	88	99	98	97	98	99	99	99	97	99	98	25	95	67	60	97	27	97	13	71	73
9008	DO 208915	6	99	95	89	91	91	45	90	95	91	90	95	92	92	94	90	95	9	7	29	26	94	18	87	18	5	56
9009	KAS-011	6	100	98	93	96	95	81	96	97	92	96	98	96	89	98	98	97	21	28	50	34	93	25	94	35	24	83
9010	AMAI	6	99	96	92	97	96	80	96	97	93	15	97	96	97	95	98	97	19	18	41	67	96	27	96	21	10	74
9011	E 4181324	3	98	94	92	89	92	22	93	96	93	90	95	93	89	85	94	91	5	6	11	41	93	7	78	8	4	47
9012	WJR 076	10	99	98	96	97	98	79	98	98	97	97	99	98	98	98	98	99	12	18	54	25	97	31	96	30	13	89
9013	SCHU	4	97	98	97	97	94	55	88	99	97	96	99	97	97	97	96	98	4	3	14	8	97	8	96	5	3	40
9014	M GAR	3	99	94	92	95	90	78	87	91	91	96	93	92	95	83	96	95	9	14	28	73	93	9	86	21	13	77
9015	WT 24	10	98	99	98	99	90	36	97	99	98	97	100	98	95	99	96	99	6	5	8	9	99	9	98	8	10	20
9016	RML	6	98	99	99	99	97	96	97	99	99	99	99	99	99	98	97	99	6	98	10	15	11	16	99	18	12	81
9017	WT 8	9	98	97	97	97	96	16	95	97	97	96	98	97	97	98	95	97	6	8	16	25	97	20	94	12	8	52
9018	LO 081785	4	99	99	99	100	99	77	99	99	99	100	100	100	99	100	99	11	99	72	34	99	16	10	98	97	75	88
9019	DUCAF	5	98	97	91	87	93	60	90	95	89	85	97	86	84	78	91	32	93	27	32	97	29	26	91	88	55	84
9021	RSH	6	99	99	97	99	83	73	93	99	97	97	98	96	98	96	94	98	98	10	33	99	86	12	95	14	24	69
9022	COX	5	99	98	93	98	98	92	98	98	95	98	99	96	97	98	98	20	99	57	49	99	22	18	95	98	58	92
9023	VAVY	5	98	93	60	89	87	35	88	90	83	84	93	76	81	82	88	10	83	8	11	93	11	9	97	72	11	75
9024	KT17	4	96	95	97	98	95	64	88	94	97	98	97	97	95	97	96	99	4	95	17	33	98	17	93	7	26	61
9025	DEU	5	96	97	96	98	93	47	93	97	92	90	98	96	94	96	91	97	11	90	19	14	18	33	88	34	51	68
9026	YAR	5	100	99	99	100	98	92	98	99	99	99	100	100	99	99	99	99	29	100	63	86	100	20	99	28	21	98
9027	PF 97387	6	96	91	9	86	83	65	80	92	94	85	91	92	85	87	81	95	9	90	20	9	10	17	59	15	51	59
9028	PE 117	9	95	97	88	89	91	72	93	94	90	88	98	89	89	93	87	93	8	86	16	34	85	16	78	20	86	61
9029	WT 51	8	96	99	99	99	98	89	99	99	99	99	99	100	99	100	99	9	99	65	66	69	98	19	98	16	84	96
9030	JHAF	4	99	99	95	98	97	60	95	98	95	96	98	99	100	98	79	96	10	92	11	9	100	100	92	22	52	80
9031	BOLETH	3	99	99	95	97	95	52	94	98	96	97	99	97	97	98	89	98	3	96	25	21	97	6	96	5	76	57
9032	BSM	4	99	99	98	98	97	79	99	99	99	96	99	99	97	99	98	99	13	97	61	37	97	25	97	22	71	84
9033	BM 14	6	99	99	96	98	98	91	98	99	98	98	99	97	98	99	99	98	46	97	52	82	97	16	97	19	79	78
9034	SAVC	7	99	98	95	98	98	92	97	98	96	98	98	98	97	98	97	99	19	97	56	79	96	28	92	19	74	89
9035	J BUSH	3	99	99	97	97	98	72	98	98	97	96	99	97	98	99	98	97	12	96	35	98	7	8	97	12	20	78
9036	SPO-010	3	93	94	98	97	72	96	93	99	98	94	99	96	96	95	95	98	14	7	44	98	98	5	96	12	35	69
9037	SWEIG-007	3	100	100	99	99	98	90	98	100	100	99	100	99	99	100	98	100	7	99	32	99	9	4	99	6	10	93
9038	BM 16	3	99	96	87	93	92	28	89	94	90	87	94	91	84	96	91	95	5	89	20	95	6	6	86	8	14	51

Cell #	Cell Name	Neg	W6/32	A-1	A-2	A-3	Tu 39	M101	Tu 35	GSP 4.1	1A3	B7/21	PL 8	Tu 22	PL 15	L243	2D6	Leu 10	16.23	17.15	3G5	7.3.19.1	77.34	AB7.AE9	AB8.25	C1	CC-CL20	CHE249.2
9039	JVM	7	97	95	94	80	93	59	94	95	94	75	94	92	77	97	93	96	10	88	17	94	15	13	81	14	16	64
9040	BM 15	7	99	99	99	99	99	84	98	99	98	99	99	99	98	99	99	99	10	98	60	100	16	19	97	18	27	90
9041	JO 528239	3	99	99	96	97	96	69	95	99	97	98	99	95	93	97	95	97	4	96	16	97	5	4	97	5	10	42
9042	TISI	10	100	100	100	100	100	68	100	100	100	99	100	100	100	100	100	100	18	100	83	100	21	22	100	19	65	86
9043	BM 21	3	99	98	98	98	96	54	98	98	97	97	99	97	70	98	95	99	2	97	21	98	3	4	96	5	17	42
9044	BRI-P	5	99	99	95	97	98	78	97	99	96	4	99	97	96	99	98	98	11	93	44	98	14	9	95	13	14	85
9045	TUBO	3	99	97	96	95	91	55	90	97	95	95	96	78	94	94	65	94	8	93	33	97	11	10	88	15	48	55
9046	BH	6	99	99	98	98	98	99	98	100	98	99	99	99	99	99	98	32	18	88	43	81	30	42	98	98	17	79
9047	PLH	6	98	96	98	95	88	75	91	90	99	96	97	86	97	95	95	15	6	80	21	48	12	18	92	83	8	55
9048	LBUF	6	99	98	96	97	97	98	93	98	96	95	97	94	62	97	95	18	5	65	62	91	24	29	94	93	35	30
9049	IBW 9	8	99	98	97	98	99	89	98	99	98	98	98	99	97	98	99	99	25	21	59	98	99	20	96	23	47	78
9050	MOU	4	99	84	87	84	88	29	73	90	76	57	93	85	71	89	84	39	5	52	8	35	11	19	80	87	5	54
9051	PITOUT	4	99	94	84	81	88	59	85	89	79	77	92	72	79	90	86	6	6	40	16	14	6	6	76	80	7	36
9052	DBB	3	98	96	94	94	92	46	72	96	94	92	96	93	93	89	87	96	2	90	3	67	93	4	84	6	4	60
9053	HOR	2	99	96	91	92	92	47	94	96	92	89	95	94	91	94	90	94	3	4	21	96	94	4	88	8	19	50
9054	EK/OH	7	99	98	97	97	98	78	98	99	98	97	99	98	99	96	98	98	11	12	46	99	98	10	94	14	40	63
9055	HO 301	5	99	99	97	99	99	85	99	98	98	99	99	99	97	98	99	99	25	20	43	99	98	23	91	17	28	80
9056	KOSE	5	99	97	97	96	94	74	96	98	95	96	97	97	95	95	97	62	7	54	45	88	58	17	94	61	36	73
9057	TEM	8	96	95	96	96	93	55	85	95	95	95	96	96	95	95	95	97	11	15	21	97	96	18	88	20	62	68
9058	OMW	8	93	96	93	83	93	52	95	95	93	94	94	91	95	91	90	95	87	12	13	95	94	18	85	15	7	53
9059	SLE-005	6	99	97	98	99	94	86	95	99	98	99	99	98	98	97	98	99	9	8	20	99	98	12	96	13	23	82
9060	CBG-1B	6	98	90	89	89	88	83	83	90	89	89	89	87	79	82	82	89	11	8	37	89	90	11	73	14	37	76
9061	ABO	7	98	97	88	96	98	85	97	98	88	97	98	96	96	97	97	96	23	14	40	98	96	20	94	18	37	80
9062	WDV	6	98	98	99	99	98	97	98	99	99	99	99	98	99	98	98	98	99	14	84	99	99	12	95	39	29	95
9063	WT 47	7	99	98	94	94	93	98	94	97	94	94	97	97	83	97	97	97	23	21	80	97	96	16	91	17	38	79
9064	AMALA	7	97	98	98	96	91	84	93	98	97	94	97	97	96	92	93	98	87	93	79	98	12	14	89	14	95	76
9065	HHKB	4	99	99	97	98	98	84	97	100	97	97	99	98	98	98	98	98	95	12	51	99	95	10	92	20	14	89
9066	TAB 089	6	97	89	74	68	87	70	81	92	76	67	89	76	62	82	85	90	9	9	16	31	85	12	74	11	10	57
9067	BTB	7	100	100	98	99	97	97	97	99	98	97	100	96	98	99	98	99	6	5	38	30	85	7	96	8	5	83
9068	BM 9	7	99	94	90	83	92	55	92	96	89	84	96	90	83	91	89	89	11	14	15	19	69	19	82	19	10	71
9069	MADURA	5	100	100	99	100	98	46	98	100	100	99	100	99	99	99	99	100	11	10	53	41	95	23	99	10	8	89
9070	LUY	6	99	100	100	100	99	89	99	99	100	100	98	100	100	99	97	100	18	99	76	11	24	22	99	17	12	95
9071	OLGA	8	95	98	98	99	100	93	99	98	97	99	99	97	99	99	99	98	49	41	93	75	95	36	96	44	35	98
9072	S PACHEO	5	94	93	89	88	88	65	92	93	90	89	96	91	90	90	64	91	8	10	22	12	72	12	76	14	9	52
9073	KT 12	3	98	98	96	97	97	74	98	98	96	97	98	96	96	98	97	98	9	97	47	42	97	18	99	15	5	75
9074	HID	8	99	99	99	99	99	90	98	99	98	99	100	99	98	100	99	99	13	98	54	68	98	34	98	26	11	83
9075	DKB	6	99	99	88	90	98	40	97	98	91	90	99	98	95	95	97	97	5	82	11	42	91	11	91	10	6	66

Quantitative scores reflect uncorrected values without subtraction of the negative control results.

Table 2B. Quantitative Immunofluorescent Analysis of Workshop Cell Panel – Part B

		Monoclonal Antibodies																								
Cell #	Cell Name	W6/32	DP11.1	GS4AA7	GSP 17.1	GSP 88.2	IIB3	IIIE3	IVD1.2	JS1	KS-6	KS-7	MAD 88	NDS 10	NDS 13	NDS 15.38	NFLD1	PL 3	PLM 12	PLM 13	PLM 2	SFR16-DR7	SFR-DR5	UK-4.1	UK-8.1	UL-65
9001	SA	100	98	98	18	22	88	98	25	99	98	98	77	85	65	19	87	17	10	96	24	51	76	98	16	19
9002	MZ 070782	99	93	94	12	12	73	90	16	95	94	95	66	41	28	12	71	14	7	84	9	15	12	92	11	17
9003	KAS-116	99	18	97	26	39	86	96	32	96	98	97	86	94	70	32	89	29	17	98	23	60	69	96	20	26
9004	JES THOM	97	99	99	4	6	97	98	9	99	99	98	77	17	15	7	63	9	4	98	11	12	8	99	8	8
9005	HOM-2	99	98	99	15	19	94	97	21	98	98	98	97	90	64	17	89	31	8	98	11	36	22	98	14	32
9006	WT 100-BIS	99	6	94	10	7	67	89	11	91	93	91	60	52	16	10	57	13	10	88	5	10	11	89	8	10
9007	DEM	99	97	96	79	36	97	97	98	97	97	98	98	86	98	61	99	96	98	98	8	26	42	75	24	23
9008	DO 208915	99	90	91	9	14	93	25	15	14	94	96	66	46	57	81	82	14	6	92	9	22	14	76	16	12
9009	KAS-011	100	81	93	20	34	100	100	99	98	97	98	96	95	94	54	84	43	16	93	27	62	66	81	19	32
9010	AMAI	99	94	94	19	20	94	71	30	71	96	98	77	67	83	92	90	38	11	96	16	37	25	87	39	19
9011	E 4181324	98	88	95	4	7	96	76	8	13	95	96	78	20	52	64	82	7	3	89	5	9	5	75	6	4
9012	WJR 076	99	97	1	18	19	85	58	25	67	98	98	78	88	97	32	84	22	13	97	13	32	24	87	22	19
9013	SCHU	97	96	98	12	6	98	17	4	28	96	97	89	20	65	94	59	5	2	91	3	9	5	82	6	7
9014	M GAR	99	97	94	13	30	91	87	22	27	96	95	76	78	84	84	71	18	8	92	10	46	50	78	10	11
9015	WT 24	98	5	99	7	51	92	25	5	77	99	98	98	14	87	5	62	9	5	97	5	10	6	84	10	9
9016	RML	98	99	10	98	11	9	59	97	92	95	53	96	70	98	17	81	22	99	14	99	16	17	15	12	17
9017	WT 8	98	97	97	15	11	96	43	8	32	97	97	98	33	69	94	78	11	8	96	7	22	9	89	14	15
9018	LO 081785	99	10	12	97	83	9	99	12	100	10	30	100	100	100	19	99	16	6	8	7	14	48	9	100	98
9019	DUCAF	98	87	17	90	13	24	95	24	83	26	58	69	97	97	30	93	50	10	14	12	20	27	19	96	85
9021	RSH	99	97	98	88	8	95	98	11	93	97	98	95	65	98	31	90	8	5	92	4	8	10	9	98	94
9022	COX	99	17	26	90	57	17	97	27	98	12	36	96	98	99	24	96	19	7	10	9	14	52	19	98	90
9023	VAVY	98	13	8	62	13	25	73	7	87	7	30	38	86	90	14	64	15	23	7	5	7	11	8	91	62
9024	KT17	96	8	5	92	7	98	97	97	96	98	97	96	25	53	71	97	96	98	91	4	6	8	6	12	7
9025	DEU	96	88	20	66	14	11	81	95	89	79	15	99	53	73	16	96	75	95	15	95	20	22	18	24	16
9026	YAR	100	99	53	97	50	98	99	99	81	99	99	98	92	96	95	99	100	99	96	27	36	73	19	21	38
9027	PF 97387	96	85	11	69	7	11	76	85	88	70	13	85	44	67	18	92	68	96	10	94	10	14	16	10	12
9028	PE 117	95	90	9	69	13	91	90	85	90	92	90	63	44	70	53	97	83	92	86	9	12	13	14	12	14
9029	WT 51	96	100	33	88	21	99	99	100	99	99	100	96	89	99	77	100	99	100	99	8	22	45	13	23	9
9030	JHAF	99	14	13	81	45	12	95	91	82	78	10	88	40	77	8	97	89	96	12	96	7	10	7	8	8
9031	BOLETH	99	97	5	90	5	98	95	97	97	97	97	83	33	79	67	98	89	97	97	3	4	6	5	6	6
9032	BSM	99	96	31	86	23	99	98	99	99	99	99	92	85	94	69	99	98	100	99	9	30	31	17	32	26
9033	BM 14	99	98	26	85	15	98	98	98	98	97	97	90	72	94	76	98	93	97	97	10	13	27	15	14	12
9034	SAVC	99	20	24	91	26	98	96	98	96	98	99	93	88	95	78	98	93	96	97	14	25	52	25	20	11
9035	J BUSH	99	97	8	89	98	9	57	97	91	84	10	96	100	97	8	93	7	97	10	97	77	93	5	98	95
9036	SPO-010	93	96	98	13	91	94	34	6	87	98	98	93	98	97	17	95	11	5	92	6	78	94	97	99	98
9037	SWEIG-007	100	99	12	98	99	10	59	98	94	88	4	99	100	100	8	99	4	99	8	99	98	98	5	100	98
9038	BM 16	99	83	3	86	3	6	20	92	77	74	5	83	96	95	5	89	15	92	9	92	5	4	4	9	4

Cell #	Cell Name	W6/32	DP11.1	GS4AA7	GSP 17.1	GSP 88.2	IIB3	IIIE3	IVD1.2	JS1	KS-6	KS-7	MAD 88	NDS 10	NDS 13	NDS 15.38	NFLD1	PL 3	PLM 12	PLM 13	PLM 2	SFR16-DR7	SFR-DR5	UK-4.1	UK-8.1	UL-65
9039	JVM	97	77	9	83	86	11	38	83	69	65	12	88	97	97	28	94	14	95	10	91	50	89	14	97	74
9040	BM 15	99	14	26	96	99	20	74	98	93	85	16	99	100	100	24	97	20	99	17	99	82	97	17	99	81
9041	JO 528239	99	39	5	95	96	4	37	96	89	87	97	94	99	98	7	75	8	97	8	96	52	96	5	99	96
9042	TISI	100	100	21	100	99	15	69	100	100	97	13	100	100	100	25	99	24	100	40	100	97	100	17	100	100
9043	BM 21	99	2	4	89	96	3	38	96	88	88	5	89	99	99	6	79	5	97	5	97	46	86	5	99	95
9044	BRI-P	99	92	14	88	97	15	54	93	84	74	11	92	100	99	13	97	14	97	11	96	84	98	9	99	89
9045	TUBO	99	97	11	90	89	9	70	81	93	88	11	89	98	96	12	94	12	94	21	93	63	92	12	96	91
9046	BH	99	97	28	92	26	25	96	38	57	28	85	98	88	99	41	17	99	12	18	17	100	74	26	25	26
9047	PLH	98	98	10	88	14	13	95	8	88	12	70	91	51	94	14	10	94	5	11	11	97	16	10	14	13
9048	LBUF	99	25	31	47	10	31	96	16	51	30	51	96	51	95	34	31	96	15	18	21	96	27	37	26	38
9049	IBW 9	99	79	98	21	20	98	98	24	92	99	99	92	99	98	53	93	20	7	98	16	20	35	97	98	41
9050	MOU	99	80	8	60	8	7	87	16	17	11	30	21	57	91	20	24	93	4	8	5	90	25	9	13	17
9051	PITOUT	99	75	7	48	34	5	84	10	8	6	27	9	32	76	7	4	90	5	4	9	86	11	9	7	5
9052	DBB	98	93	4	55	4	96	37	93	3	96	95	62	16	94	54	2	93	95	93	2	95	11	5	3	3
9053	HOR	99	92	94	9	52	93	48	3	76	95	91	78	69	96	33	92	5	3	89	4	4	6	93	94	42
9054	EK/OH	99	96	95	11	10	77	98	14	92	99	99	90	100	99	12	94	13	4	96	7	10	15	88	98	30
9055	NO 301	99	14	98	14	16	99	89	26	87	99	99	81	85	98	80	95	21	6	97	11	16	24	86	97	51
9056	KOSE	99	38	58	61	16	60	97	14	57	59	88	91	75	99	21	52	71	5	54	7	75	26	62	62	25
9057	TEM	96	97	97	16	11	88	71	12	95	97	97	96	97	97	20	93	18	11	95	15	17	13	82	97	69
9058	OMW	93	13	94	8	8	96	50	15	93	96	94	86	44	96	84	79	14	7	89	8	11	12	74	96	41
9059	SLE-005	99	12	98	12	53	99	65	13	97	99	98	93	62	98	47	97	13	7	97	8	8	17	98	99	58
9060	CBG-1B	98	15	87	17	12	91	42	15	81	89	88	80	94	93	75	87	12	8	85	10	11	16	77	90	46
9061	ABO	98	96	89	20	23	66	97	30	83	97	97	83	99	99	23	87	21	8	94	11	20	31	64	96	30
9062	WDV	98	99	99	73	22	98	99	29	99	99	98	97	96	99	98	98	14	13	97	29	58	92	99	98	89
9063	WT 47	99	14	96	17	19	96	85	31	89	98	97	89	82	97	65	91	17	9	96	11	22	29	93	96	48
9064	AMALA	97	94	14	93	12	13	96	95	97	78	21	92	79	97	25	84	15	98	18	97	12	19	15	98	68
9065	HHKB	99	99	98	19	12	98	85	25	97	99	99	96	92	99	86	93	17	14	96	17	43	66	94	99	60
9066	TAB 089	97	61	88	19	12	87	63	10	42	87	89	78	51	92	51	87	15	7	82	9	10	13	58	11	51
9067	BTB	100	98	98	91	9	92	60	9	25	97	98	95	52	100	54	96	9	5	95	5	8	10	7	5	77
9068	BM 9	99	84	91	72	12	80	39	15	19	90	91	86	42	96	35	92	17	9	86	10	37	18	13	17	69
9069	MADURA	100	99	100	97	9	98	12	56	48	99	100	98	66	100	62	99	11	5	98	7	11	12	9	8	92
9070	LUY	99	98	15	98	15	15	88	99	78	95	86	99	85	100	21	99	18	100	19	99	13	31	19	15	90
9071	OLGA	95	96	98	94	53	97	91	58	81	97	98	96	97	99	81	97	47	13	98	20	46	73	58	42	90
9072	S PACHEO	94	87	93	80	14	87	29	8	35	88	89	82	61	92	35	90	19	7	83	10	12	9	10	14	65
9073	KT 12	98	85	12	82	13	98	98	99	44	98	97	96	45	29	70	8	98	96	97	6	95	20	12	12	12
9074	HID	99	96	20	94	18	99	100	99	58	99	99	99	77	45	85	10	99	99	99	13	98	46	14	26	14
9075	DKB	99	94	6	56	6	95	98	98	22	96	94	94	33	23	43	5	99	94	89	3	86	6	3	5	6

Quantitative scores reflect uncorrected values without subtraction of the negative control results.

Table 3A. Qualitative Immunofluorescent Analysis of Workshop Cell Panel – Part A

Monoclonal Antibodies

Cell #	Cell Name	Neg Cont	W6/32	A-1	A-2	A-3	Tu 39	M101	Tu 35	GSP 4.1	1A3	B7/21	PL 8	Tu 22
9001	SA	N	H	H	L/I	L/I	I/H	W	I/H	I/H	L	L/I	I/H	L/I
9002	MZ 070782	N	H	I	L	L	L	W	L/I	I	L	L	I	I
9003	KAS-116	N	I	I	W	L	L/I	W	I	I	W	W/L	I	I
9004	JES THOM	N	I/H	I/H	W/L	W/L	L/I	W	I	I/H	L	L	H	I/H
9005	HOM-2	N	I	I	L	L	I	W	I	I	L	L	I	I
9006	WT 100-BIS	N	I/H	I/H	W/L	L/I	L/I	W	I	I/H	W/I	L/I	I/H	I
9007	DEM	N	I/H	I/H	W/L	W/L	I/H	W	I/H	I	W/L	W/L	I	I
9008	DO 208915	N	I	L/I	W	W	W/L	W	L/I	L	W	W	L/I	W/L
9009	KAS-011	N	I/H	I	L/I	W/L	L/I	W	L/I	I	W/L	W/L	I	I
9010	AMAI	N	I	L/I	W/L	W	L/I	W	L/I	L/I	W/L	N	L/I	L
9011	E 4181324	N	I	I/H	W/L	W	I	W	I/H	I	L	W	I	L/I
9012	WJR 076	N	I	L/I	W	W	L/I	W	L/I	I	W	W	I	L
9013	SCHU	N	I/H	I/H	I	L/I	I/H	W	I/H	I/H	L/I	L	I/H	I/H
9014	M GAR	N	I	L/I	W	W	W	W	W	L	W	W	L	W/L
9015	WT 24	N	H	H	I	I	I/H	W	I/H	I/H	I	I	H	L/I
9016	RML	N	H	H	I/H	I	I/H	W/L	I/H	H	I/H	I	H	I/H
9017	WT 8	N	I	I/H	I	L	I/H	W	I	I/H	L/I	L	I/H	I/H
9018	LO 081785	N	I	I	W/L	L	I/H	W	I	I	L	L	I	I
9019	DUCAF	N	H	H	L	W/L	H	W	I/H	H	W/L	W/L	I	I
9021	RSH	N	H	H	L/I	L/I	I/H	W	I/H	H	L/I	L/I	H	I
9022	COX	N	I/H	I	W/L	L	L/I	W	I.H	I	L	L	I/H	L/I
9023	VAVY	N	I/H	I	W	W/L	W/L	W	L	I	W	W/L	I	L
9024	KT17	N	H	H	L/I	I	I/H	W	I/H	I	L	I	H	I/H
9025	DEU	N	L	L/I	W/L	W	W/L	W	W/L	W/L	W	W	L/I	W/L
9026	YAR	N	I/H	H	L/I	L	I	W	H	I	L/I	L	H	I
9027	PF 97387	N	I/H	I/H	L	W	W/L	W	W/L	I	L	W	I/H	I
9028	PE 117	N	H	I/H	L	L	I/H	W	I/H	I/H	L	L	H	L/I
9029	WT 51	N	I	I	W/L	W/L	L/I	W	I	I	W/L	I	L/I	L
9030	JHAF	N	H	H	L	I	H	W	I/H	H	L/I	L/I	H	I/H
9031	BOLETH	N	I/H	H	L	L	I/H	W	L/I	I/H	L/I	L	I/H	I
9032	BSM	N	I	L/I	W/L	W	L/I	W	L/I	I	W/L	W	I	L
9033	BM 14	N	I	I/H	L	L	I/H	W	I/H	I	L	L	I	I
9034	SAVC	N	I	I	W/L	W/L	I	W	L/I	I	W/L	W/L	L/I	L/I
9035	J BUSH	N	I	I	L	W/L	I	W	I	I	L	W/L	I	I
9036	SPO-010	N	H	H	L/I	L/I	L	L	I/H	H	L/I	L	H	H
9037	SWEIG-007	N	I/H	H	I	L	I/H	W	I/H	I/H	I	L	H	I
9038	BM 16	N	I/H	I	L/I	L	I	W	I	I	L/I	W	I/H	I

Cell #	Cell Name	PL 15	L243	2D6	Leu 10	16.23	17.15	3G5	7.3.19.1	77.34	AB7.AE9	AB8.25	C1	CC-CL20	CHE249.2
9001	SA	L/I	I/H	I	I	W	N	W	W	I	W	I	W	L	W
9002	MZ 070782	W/L	I	I	L/I	N	N	W	W	L/I	W	W/L	W	L	W
9003	KAS-116	W	I	I	L	W	W	W	W	L	W	L	W	W/L	W
9004	JES THOM	L	L/I	I	I	W	N	N	W	I	N	I	N	I	W
9005	HOM-2	L	I	L/I	L/I	N	N	W	N	L/I	W	L/I	N	L	W
9006	WT 100-BIS	L	I/H	I/H	I	W	W	W	N	I	N	W/L	W	W/L	W
9007	DEM	W/L	I/H	I/H	L/I	W	W	W	W	L	W	L/I	N	W	W
9008	DO 208915	W	L/I	W/L	W/L	N	N	W	W	W	N	L	W	N	W
9009	KAS-011	W	I	L/I	L/I	N	W	W	W	L	W	L/I	W	N	W
9010	AMAI	W	L/I	I/H	L	N	N	W	W	W/L	W	L	N	N	W
9011	E 4181324	W	L/I	I	L/I	N	N	W	W	L	N	L	N	N	W
9012	WJR 076	W	I	L/I	L	N	N	W	N	L	W	L	N	N	W
9013	SCHU	L	H	I/H	I	N	N	W	N	I	N	L/I	N	N	W
9014	M GAR	W	W	W	L	N	N	W	W	L	N	L	N	N	W
9015	WT 24	H	H	I/H	H	N	N	W	W	I/H	N	I	N	W	W
9016	RML	I	H	H	I	N	L/I	N	N	N	N	I	N	N	W
9017	WT 8	L	I/H	I/H	I	N	N	W	W	I	W	L/I	W	N	W
9018	LO 081785	L	I	I	N	L/I	W	W	I	N	N	L	W/L	W	W
9019	DUCAF	W/L	I/H	I/H	W	I	W	W	N	W	W	L/I	W/L	W	W
9021	RSH	L/I	I/H	I/H	I	I	N	W	N	W/L	N	L/I	N	W	W
9022	COX	L	I	L/I	N	L/I	W	W	I/H	N	N	L/I	W/L	W	W
9023	VAVY	W/L	L/I	W/L	N	L	N	W	I	W	W	W/L	W	W	W/L
9024	KT17	L	H	I/H	I	N	W/L	W	W	L/I	N	I	N	W	W
9025	DEU	W	L	W/L	L	N	W	N	N	N	N	W/L	W	W	W
9026	YAR	L/I	I/H	H	I	W	L	W	W	I	W	I	W	W	W
9027	PF 97387	W	I	W/L	L/I	N	W	W	N	N	W	W	W	W	W
9028	PE 117	L	L/I	L/I	L/I	N	W/L	N	W	L	N	L/I	N	L/I	N
9029	WT 51	I	L/I	L	N	W/L	W	W	W	L	N	L	N	W	W
9030	JHAF	I	H	W	I	W	L	W	N	I	L	L/I	W	W	W
9031	BOLETH	L/I	I/H	I	I	N	W/L	W	W	L/I	N	I	N	W	W
9032	BSM	W	I	L/I	W/L	N	W	W	W	W	W	W/L	N	W	W
9033	BM 14	L	I	I	L/I	W	W	W	W	L	N	L/I	N	W	W
9034	SAVC	W/L	I	L	L/I	N	W	W	W	L	W	W/L	N	W	W
9035	J BUSH	W/L	I	I	L/I	N	W	W	W/L	N	N	W/L	N	W	W
9036	SPO-010	L/I	H	H	I	W	N	W	L/I	L/I	N	I	N	W	W
9037	SWEIG-007	L	H	I/H	I	N	L	W	L/I	N	N	I	N	N	W
9038	BM 16	W/L	I/H	I	I	N	L	W	I	N	N	L	N	N	W

Cell #	Cell Name	Neg Cont	W6/32	A-1	A-2	A-3	Tu 39	M101	Tu 35	GSP 4.1	1A3	B7/21	PL 8	Tu 22	PL 15	L243	2D6	Leu 10	16.23	17.15	3G5	7.3.19.1	77.34	AB7.AE9	AB8.25	C1	CC-CL20	CHE249.2
9039	JVM	N	I/H	I	L/I	W	I/H	W	I	I	L	W	I	L/I	W	L/I	I/H	L/I	N	W/L	N	L	N	N	L	N	N	W
9040	BM 15	N	I	I	L	W/L	L/I	W	L/I	I	L	W/L	I	L/I	W/L	I	I	L	N	W/L	W	L	N	N	L/I	N	W	W
9041	JO 528239	N	H	H	I	I	I/H	W	I/H	H	I	I	H	H	L	H	I/H	I	N	L	W	L/I	N	N	I	N	N	W
9042	TISI	N	I/H	I/H	I	L/I	I/H	W	I/H	H	I	L/I	H	I/H	L/I	I/H	H	I	W	L/I	W	I	W	W	L/I	W	W	W
9043	BM 21	N	H	H	I	I/H	I/H	W	I/H	I/H	H	I	H	H	W	H	H	H	N	I	W	I	N	N	I	N	W	W
9044	BRI-P	N	I	I	W/L	W	L/I	W	L/I	I	W/L	N	I	L/I	W	I	L/I	L	N	W	W	W/L	N	N	W/L	N	N	W
9045	TUBO	N	I/H	I/H	I	L	I	W	I	H	I	L	H	L/I	L	I/H	W/L	I	W	L	W	I	W	N	L/I	W	W	W
9046	BH	N	I	I	L	L	I	L/I	I	I	L	L	I	L/I	W	I	I	W	N	W	W	W	W	W	L/I	L	N	W
9047	PLH	N	I/H	H	L/I	L/I	I/H	W	I/H	I/H	L/I	L/I	H	I/H	I	I/H	I/H	N	N	W	W	W	N	W	L/I	L	N	W
9048	LBUF	N	I/H	H	I	L	I/H	W/L	L/I	I/H	I	L	I	I	W	I/H	I/H	N	N	W	W	L	W	W	L	L	W	W
9049	IBW 9	N	I	I	W	W	I	W	I	I	W	W	I	L/I	W	I	I	L	N	N	W	L	L	N	L/I	N	W	W
9050	MOU	N	I/H	I/H	L	W/L	L/I	W	W/L	I	W/L	W	I/H	I	W	I	L	W	N	W	N	W	W	W	L	W/L	N	W
9051	PITOUT	N	H	I	L	W/L	L	W	I	I	L	W	I	L	W/L	I/H	W/L	N	N	W	W	W	N	N	L/I	L	N	W
9052	DBB	N	H	H	L	L	I/H	W	I	H	L	L	H	I/H	L	H	H	L/I	N	W	N	W	L	N	I	N	N	W
9053	HOR	N	H	I/H	L	L	I/H	W	I/H	I/H	L	L	I/H	I	L	I/H	I/H	L/I	N	N	W	I	L/I	N	L	N	W	W
9054	EK/OH	N	I	I	W/L	W/L	L/I	W	L/I	L/I	W/L	W/L	L/I	L/I	W/L	I	L/I	L	N	N	W	L	L	N	W/L	N	W	W
9055	HO 301	N	I	I	W	W/L	L/I	W	L/I	I	W	W	I	L/I	W	I	I	L	W	N	W	L	W/L	N	L	N	W	W
9056	KOSE	N	I	I	W/L	L	L/I	W	L/I	I	W/L	W/L	I	I	W/L	I	I	W/L	N	W	W	W/L	W/L	W	L	W/L	W	W
9057	TEM	N	I	I/H	L/I	L	I/H	W	L/I	I/H	L	L	H	I	L	I/H	I	I	N	N	W	I	I	N	L/I	W	W	W
9058	OMW	N	I/H	H	W/L	L	H	W	H	I/H	L	I	H	I/H	L	L/I	I/H	I	W	W	W	I	I	W	L/I	W	N	W
9059	SLE-005	N	H	H	L	L/I	I/H	W	I/H	H	L	L/I	H	I/H	L/I	H	H	I	N	N	W	I	I	N	I	N	W	W
9060	CBG-1B	N	I/H	I/H	L/I	L	I	W	L/I	I	L/I	L	I	I	W/L	L/I	L/I	I	N	N	W	I	I	N	L	N	W	W
9061	ABO	N	I	L	W	W	L	W	W/L	L	W	W	L/I	W/L	W	W/L	W/L	W/L	W	N	W	L	W/L	W	L	N	W	W
9062	WDV	N	I/H	I/H	L/I	L/I	I	L	I	I	I	L	H	I	L/I	I/H	I/H	I	L	N	W	I	I	N	L/I	W	W	W
9063	WT 47	N	I	I	W	W	L/I	W	I	I	W	W	I	I	W	I	I	L/I	N	N	W	L/I	L/I	N	L/I	N	W	W
9064	AMALA	N	I/H	H	I	L	I/H	W	I/H	H	I	L	H	H	L/I	I/H	I/H	I	W	L	W	I	N	N	I	N	L/I	W
9065	HHKB	N	I/H	I/H	I	L/I	I/H	W	I/H	H	L/I	L/I	I/H	I	L/I	I/H	I/H	I	L	N	W	I	I	N	L/I	W	N	W
9066	TAB 089	N	I/H	I	W	W	W/L	W	W/L	I	W	W	I	W/L	W	W/L	L	W/L	N	N	W	W	W	N	W	N	N	W
9067	BTB	N	I/H	I/H	I	L/I	I/H	W	I/H	I/H	L/I	L	H	I	L	H	I/H	I	N	N	W	W	W	N	L/I	N	N	W
9068	BM 9	N	I/H	I/H	W/L	W	L/I	W	I	I/H	L	W	I/H	I	W	I	L/I	W/L	W	W	W	W	W	W	L	W	N	W
9069	MADURA	N	H	H	I	L/I	H	W	H	H	I	L/I	H	I	L/I	H	I/H	I	N	N	W	W	L	W	I	N	N	W
9070	LUY	N	I	I	L	L	H	W	I/H	I	L	L	I	I	L	I	I/H	L/I	N	L	W	N	W	W	I	N	N	W
9071	OLGA	N	I	H	L/I	L/I	I	W	I	I/H	L/I	L/I	H	I	L/I	I	I	I	W	W	W	W	L	W	I	W	W	W
9072	S PACHEO	N	H	H	L	L	I/H	W	I/H	I/H	L	L	H	L/I	L	I/H	I/H	L/I	N	N	W	N	W	N	W/L	W	N	W
9073	KT 12	N	I	I	W	W/L	L/I	W	L/I	L/I	W	W	I	L	W	I	L/I	W/L	N	W	W	W	W/L	N	W/L	N	N	W
9074	HID	N	I	I	L	L	I	W	I	I	W/L	W/L	I	I	W/L	I	L/I	L	N	W	W	W	L	W	L	W	N	W
9075	DKB	N	H	H	W	W	I	W	I	I/H	W	W	I	L/I	W	I	I	L	N	W	N	W	W/L	N	L	N	N	W

Qualitative scores were assigned based on where along the x-axis the peak channel of the fluorescence histogram occurred. For this purpose, the x-axis was divided into four equal partitions scored as (from left to right) weak (W), low (L), intermediate (I), or high (H). Histograms with peak channels along the y-axis were scored as negative (N).

Table 3B. Qualitative Immunofluorescent Analysis of Workshop Cell Panel – Part B

Monoclonal Antibodies

	Cell Name	W6/32	DP 11.1	GS4AA7	GSP 17.1	GSP 88.2	IIB3	IIIE3	IVD 1.2	JS1	KS-6	KS-7
9001	SA	H	L	I	W	W	W/L	L/I	W	L/I	I	I
9002	MZ 070782	H	W/L	L	N	N	W	W/L	W	L/I	L/I	L/I
9003	KAS-116	I	N	L	W	W	W	W/L	W	L/I	L/I	L
9004	JES THOM	I/H	L/I	I	N	N	L	L/I	W	I	I	I
9005	HOM-2	I	L	L/I	N	N	W/L	L	W	L	I	L/I
9006	WT 100-BIS	I/H	N	I	W	W	W	W/L	W	L	I	I
9007	DEM	I/H	W	W	W	W	W/L	W	W/L	W/L	L/I	L/I
9008	DO 208915	I	W	W	N	N	W/L	W	N	N	W/L	W/L
9009	KAS-011	I/H	W	L	N	W	I	L	W	W	L/I	L
9010	AMAI	I	W	W/L	N	N	W/L	W	W	W	L	L
9011	E 4181324	I	W	L	N	N	L/I	W	N	W	L/I	L/I
9012	WJR 076	I	W	N	N	N	W	W	N	W	L	L
9013	SCHU	I/H	W/L	I	W	N	I	W	N	W	I	I
9014	M GAR	I	W	W	N	W	L	W	N	N	L	L
9015	WT 24	H	N	I	W	L	L/I	W	N	L	H	I
9016	RML	H	L/I	N	I	N	N	W	I/H	W/L	L/I	W
9017	WT 8	I	L	I	W	W	I	W	N	W	I	I
9018	LO 081785	I	N	N	W/L	W	N	L/I	N	I	N	W
9019	DUCAF	H	W/L	W	W/L	W	W	L/I	W	I/H	W	W
9021	RSH	H	L	I	W/L	N	L	I	N	I/H	L/I	L/I
9022	COX	I/H	N	N	W	W	N	L	N	L/I	N	W
9023	VAVY	I/H	W	N	W	W	W	W/L	N	L	N	W
9024	KT17	H	N	N	W/L	N	L/I	L/I	I	I	I	I
9025	DEU	L	W	N	W	N	N	W	W/L	W	W	N
9026	YAR	I/H	L/I	W	L	W	I	L	I	W	I	L/I
9027	PF 97387	I/H	W	N	W	N	N	W	W/L	W/L	W	W
9028	PE 117	H	L	N	W	N	L	L	L	L/I	L	L/I
9029	WT 51	I	W/L	W	W	N	L	W	L	L	L	L
9030	JHAF	H	W	W	W	W	N	L	I	W	W/L	N
9031	BOLETH	I/H	L	N	W/L	N	I	W/L	L/I	L/I	L/I	L/I
9032	BSM	I	W	W	W	W	W/L	W	L	W/L	W/L	W/L
9033	BM 14	I	L	W	W	N	L	W	L	L/I	L/I	L
9034	SAVC	I	N	W	W	N	L/I	W	L	L	L/I	L
9035	J BUSH	I	W/L	N	W	L	N	W	L/I	W	W	N
9036	SPO-010	H	L	I	W	I	L	W	N	W	I	I
9037	SWEIG-007	I/H	L	N	L	I	N	W	I	W	L	N
9038	BM 16	I/H	L	N	L/I	N	N	W	I	W	W	N

	Cell Name	MAD 88	NDS 10	NDS 13	NDS 15.38	NFLD1	PL 3	PLM 12	PLM 13	PLM 2	SFR16DR7	SFR3DR5	UK-4.1	UK-8.1	UL-65
9001	SA	W	W	W	W	W	N	N	L/I	W	W	W	I	N	W
9002	MZ 070782	W	W	W	N	W	W	N	W/L	N	W	W	L	N	W
9003	KAS-116	W	W	W	W	W	W	N	L	W	W	W	L	N	W
9004	JES THOM	W	W	W	N	W	N	N	L/I	W	W	N	I	N	N
9005	HOM-2	L	W	W	N	W	W	N	L	N	W	W	L/I	N	W
9006	WT 100-BIS	W	W	W	W	W	W	W	W/L	W	W	W	L/I	W	W
9007	DEM	L	W	W	W	L/I	W	W/L	L	N	N	W	W	N	N
9008	DO 208915	W	W	W	W	W	N	N	W	N	W	N	W	N	N
9009	KAS-011	W	W	W	W	W	W	N	W/L	N	W	W	W	N	W
9010	AMAI	W	W	W	W	W	W	N	W	N	W	W	W	W	N
9011	E 4181324	W	E	W	W	W	N	N	W/L	N	N	N	W	N	N
9012	WJR 076	W	W	W	W	W	N	N	W	N	W	N	W	N	N
9013	SCHU	W/L	W	W	L	W	N	N	L	N	N	N	W	N	N
9014	M GAR	W	W	W	W	W	N	N	L	N	W	W	W	N	N
9015	WT 24	W	W	L	N	L	W	N	I	N	W	W	L	W	W
9016	RML	I	W	L	N	W	W	I/H	N	I/H	N	N	N	N	N
9017	WT 8	L/I	W	W	L/I	W/L	N	N	L	N	W	N	W/L	N	W
9018	LO 081785	L	L	L/I	N	L	N	N	N	N	N	W	N	I	W
9019	DUCAF	W	I	I	W	L/I	W	N	W	W	W	W	W	I/H	W/L
9021	RSH	I	W	I	W	L	N	N	W/L	N	N	N	N	I	W/L
9022	COX	L	W	L/I	N	W	N	N	N	N	N	W	N	I	W
9023	VAVY	W	W	L	W	W	W	W	N	N	N	W	N	L/I	W/L
9024	KT17	I	W	W	W	I	W/L	I	W/L	N	N	N	N	N	N
9025	DEU	W/L	W	W	N	L	W	L	N	W/L	N	N	N	N	N
9026	YAR	L/I	W	W	W	I	L	I	L	W	W	W	N	N	W
9027	PF 97387	W/L	W	W	W	I	W	L/I	N	L/I	N	W	W	N	N
9028	PE 117	W	W	W	W	I	W	L/I	L	N	N	N	N	N	N
9029	WT 51	W	W	W	W	I	W	L	W/L	N	N	W	N	N	N
9030	JHAF	I	W	W	N	I	W	I	N	I	N	N	N	N	N
9031	BOLETH	W/L	W	W	W	I	W	L/I	L	N	N	N	N	N	N
9032	BSM	W	W	W	W	L/I	W	W/L	W	N	W	W	N	W	W
9033	BM 14	W	W	W	W	I	W	L/I	W/L	N	N	W	N	N	N
9034	SAVC	W/L	W	W	W	I	W	L	W/L	N	W	W	W	W	N
9035	J BUSH	L	L	L/I	N	W/L	N	L/I	N	L/I	W	W	N	I	W
9036	SPO-010	W/L	L/I	I	W	L	N	N	I	N	W	L	L	H	I
9037	SWEIG-007	L/I	L	I	N	L	N	I	N	I	W	L	N	I	L
9038	BM 16	L/I	I	I	N	L	N	I	N	I	N	N	N	N	N

Cell #	Cell Name	W6/32	DP 11.1	GS4AA7	GSP 17.1	GSP 88.2	IIB3	IIIE3	IVD 1.2	JS1	KS-6	KS-7	MAD 88	NDS 10	NDS 13	NDS 15.38	NFLD1	PL 3	PLM 12	PLM 13	PLM 2	SFR16DR7	SFR3DR5	UK-4.1	UK-8.1	UL-65
9039	JVM	I/H	W	N	W	L	N	W	L	W	W	N	W/L	L	I	W	L	N	I	N	I	W	W/L	N	I/H	W
9040	BM 15	I	N	W	W	L/I	N	W	L/I	W	W	N	L/I	L	I	N	W/L	N	L	N	L	W	W/L	N	I	W
9041	JO 528239	H	W	N	L	I	N	W	I	L	L	L/I	I	L	I	N	W	N	I	N	I	W	L	N	I/H	L
9042	TISI	I/H	I	W	L/I	L/I	N	W	I	L	L	N	I	I	I	W	L/I	W	I	W	I	W	L	N	H	L
9043	BM 21	H	N	N	I	I	N	W	I/H	L	L/I	N	I/H	I	H	W	L	N	H	N	H	W	L	N	H	L/I
9044	BRI-P	I	W	N	W	L	N	W	W/L	W	W	N	W	W	L/I	N	W	W	L	N	W/L	W	W	N	I	W
9045	TUBO	I/H	L	W	L	L/I	N	W	L/I	W/L	L	W	L/I	I	I	W	I	W	I/H	W	I	W	W/L	W	I/H	W/L
9046	BH	I	W	W	W	W	W	W/L	W	W	W	W	W	W	L	W	N	I	N	N	N	L/I	W	W	W	W
9047	PLH	I/H	I	N	L	N	N	I	N	W	N	W	L	W	L	N	N	I	N	N	N	L/I	W	N	N	N
9048	LBUF	I/H	W	W	W	N	W	I	N	W	W	W	L	W	L	W	W	I	N	N	W	L	W	W	W	W
9049	IBW 9	I	W	L	N	N	W/L	L/I	N	W	L	L	L	W/L	L/I	W	W	N	N	L	N	N	W	L	I	W
9050	MOU	I/H	W/L	N	W	W	W	L	W	W	W	W	W	W	L	W	W	L/I	N	W	N	L	W	W	W	W
9051	PITOUT	H	W	N	W	W	N	L	N	N	N	W	N	W	W	N	N	L/I	N	N	N	L	N	N	N	N
9052	DBB	H	L	N	W	N	L/I	W	L	N	L/I	L/I	W	W	L	W	N	H	L	L	N	I	W	N	N	N
9053	HOR	H	L	L/I	N	W	L/I	W	N	W	L/I	L/I	W/L	W	I	W	L	N	N	L	N	N	N	L/I	I/H	W
9054	EK/OH	I	W/L	W/L	N	N	W	L	N	W	L	L	W	L	L/I	N	W/L	N	N	W/L	N	N	N	W	L	W
9055	HO 301	I	N	L	N	N	L	W	N	W	L	W/L	W	W	L	W	W/L	N	N	W/L	N	N	N	W	I	W
9056	KOSE	I	W	W/L	W	N	W	L/I	N	W	W/L	W	W	W	L	W	W	I	N	W	N	L/I	W	W	W/L	W
9057	TEM	I	L	L/I	W	N	W	L	N	L	I	I	I	I	I	W	L	W	N	L/I	N	N	N	W	I	W
9058	OMW	I/H	N	L/I	N	N	L/I	W	W	L	I	I	L/I	W	L/I	W/L	W	N	N	L/I	N	N	N	W	I/H	W
9059	SLE-005	H	N	I	N	W	I	W	N	L	I	I	I	W	I	W	L/I	N	N	L/I	N	N	N	I	H	W
9060	CBG-1B	I/H	W	I	W	W	I	W	W	W/L	L/I	L/I	L	I	I	L	L	W	N	L/I	W	W	W	L	I/H	W
9061	ABO	I	W	W	N	W	W	W/L	W	W	W/L	W/L	W	L	L/I	N	W	N	N	W	N	N	W	W	W/L	W
9062	WDV	I/H	L/I	I	W	W	I	W	W	L/I	I	I	L/I	W	I	L	L	N	N	L	W	W	W	L	I/H	W
9063	WT 47	I	N	L	N	N	L/I	W	N	W	L/I	L	L	W	L	W	L	N	N	L	N	N	N	L	I	W
9064	AMALA	I/H	L	N	L	N	N	L	I	I	L	W	L/I	W	I	W	W	N	I	N	I	N	N	N	H	W
9065	HHKB	I/H	L/I	I	W	N	I	W	W	L/I	I	I	L/I	W	L/I	W/L	W/L	N	N	I	N	W	W	L	I/H	W
9066	TAB 089	I/H	W	W	W	N	W	W	N	W	W/L	W/L	W/L	W	L/I	W	W/L	N	N	W	W	N	N	W	N	W
9067	BTB	I/H	L	L/I	L	N	L	W	N	W	I	L/I	I	W	I	W	L	N	N	L	N	N	N	N	N	W
9068	BM 9	I/H	W/L	L	W	W	W/L	W	W	W	L	W/L	L	W	I	W	L	W	N	W/L	N	W	W	W	W	W
9069	MADURA	H	L	I	L	N	L	N	W	W	I	I	I/H	W	I	W	L/I	N	N	L/I	N	N	N	N	N	L
9070	LUY	I	W	N	W	N	N	W	L/I	W	W/L	W	I	W	L/I	N	L/I	N	L	N	L/I	N	W	N	N	W
9071	OLGA	I	L	I	L	W	L	W	W	W	I	I	L/I	W	I	W	I	W	N	L	N	W	W	W	W	L
9072	S PACHEO	H	W	I	W	W	L	W	N	W	L	L	L	W	L/I	W	L	W	N	L	N	N	N	N	N	W
9073	KT 12	I	W	N	W	N	W/L	L/I	L	W	L	W/L	L/I	W	W	W	N	I	W/L	W	N	W	W	N	N	N
9074	HID	I	W	N	W	N	L	I	L/I	W	L	L	I	W	W	W	N	I	L	W/L	N	W	W	N	N	N
9075	DKB	H	W	N	W	N	L	H	L	W	L	L	L/I	W	W	W	N	I	L	W	N	L	N	N	N	N

Qualitative scores were assigned based on where a long the x-axis the peak channel of the fluorescence histogram occurred. For this purpose, the x-axis was divided into four equal partitions scored as (from left to right) weak (W), low (L), intermediate (I), or high (H). Histograms with peak channels along the y-axis were scored as negative (N).

Table 4. Titration of Workshop Monoclonal Antibodies by Quantitative Immunofluorescent Analysis

				Dilution of Monoclonal Antibody							
Ab Code	Ab Name	Pos	Neg	1:40	1:200	1:1000	1:5000	1:25000	1:125000	1:625000	Cell Line
1	A-1	100	4	99	99	100	97	85	73	27	9001
2	A-2	100	4	94	93	92	89	54	14	7	9001
3	A-3	100	4	98	98	96	97	90	65	33	9001
4	TU 39	100	4	97	98	98	97	88	43	13	9001
5	W6/32	100	4	99	99	91	73	34	18	8	9001
7	TU 35	100	4	97	97	98	98	90	55	10	9001
8	GSP 4.1	100	4	99	99	99	96	85	32	7	9001
9	1A3	99	5	99	100	99	98	83	37	35	9043
10	B7/21	100	4	97	97	98	96	82	54	9	9001
11	PL 8	99	6	99	99	98	98	87	44	35	9004
12	TU 22	99	6	98	98	97	95	93	80	17	9004
13	PL 15	99	8	97	97	95	96	84	70	78	9047
14	L243	99	5	100	100	99	100	90	48	27	9043
15	2D6	99	5	100	100	99	99	79	38	9	9043
16	LEU 10	99	5	100	99	100	100	98	87	39	9043
17	16.23	100	5	90	74	33	9	9	11	9	9018
19	3G5	100	4	36	36	45	27	27	27	5	9001
20	7.3.19.1	99	8	99	99	99	99	85	51	31	9019
21	77.34	95	8	94	74	22	8	6	3	3	9015
22	AB7.AE9	99	8	20	14	13	18	17	15	12	9030
23	AB8.25	100	4	90	88	88	74	36	9	3	9001
24	C1	99	8	95	89	60	27	13	11	11	9047
25	CC-CL 20	99	5	90	92	76	63	47	18	5	9028
26	CHE249.2	100	4	75	35	12	7	4	4	4	9001
27	DP 11.1	99	8	97	96	81	68	35	20	16	9047
28	GS4AA7	100	4	96	97	95	91	64	21	7	9001
29	GSP 17.1	98	6	89	85	84	72	37	11	7	9016
30	GSP 88.2	99	5	100	99	99	97	88	62	31	9043
32	IIIE3	99	8	98	97	97	93	72	18	20	9047
33	IVD12	98	6	92	88	84	54	11	6	8	9016
34	JS1	99	8	98	98	97	93	93	52	35	9019
35	KS-6	100	4	97	98	96	97	95	79	33	9001
36	KS-7	100	4	96	95	97	96	95	92	79	9001
37	MAD 88	98	6	92	87	70	29	10	7	8	9016
38	NDS 10	99	8	100	99	98	96	81	52	19	9019
39	NDS 13	99	5	100	100	100	100	95	72	21	9043
41	NFLD,M1	99	8	92	67	53	27	14	11	10	9019
42	PL 3	99	8	99	98	93	91	72	39	21	9047
43	PLM 12	98	6	93	93	90	85	46	12	7	9016
44	PLM 13	95	8	92	60	19	10	4	6	4	9015
45	PLM 2	99	5	99	99	99	88	46	17	6	9043
46	SFR16-DR7	99	8	98	98	97	91	77	47	25	9047
47	SFR3-DR5	99	5	99	99	94	58	14	5	4	9043
48	UK 4.1	100	4	96	97	95	91	82	49	15	9001
49	UK 8.1	99	5	98	99	99	96	76	30	12	9043
50	UL-65	97	6	95	87	49	9	7	7	6	9036

Quantitative scores reflect uncorrected values without subtraction of the negative control results. "Pos" refers to the positive control antibody (w6/32), while "Neg" refers to the negative control (mouse immunoglobulin). "Cell line" refers to the number of the Workshop cell line on which the titration analysis was performed.

Table 5. Titration of Workshop Monoclonal Antibodies by Qualtitative Immunofluorescent Analysis

Ab Code	Ab Name	Pos	Neg	Dilution of Monoclonal Antibody 1:40	1:200	1:1000	1:5000	1:25000	1:125000	1:625000	Cell Line
1	A-1	I	N	I	L/I	L/I	W/L	W	W	W	9001
2	A-2	I	N	W/L	W/L	W/L	W	W	N	N	9001
3	A-3	I	N	W/L	W/L	W/L	W/L	W	W	W	9001
4	TU 39	I	N	L/I	L/I	L/I	W/L	W	W	N	9001
5	W6/32	I	N	I	L/I	W	W	W	N	N	9001
7	TU 35	I	N	W/L	W/L	W/L	W/L	W	W	N	9001
8	GSP 4.1	I	N	I	I	I	W/L	W	W	N	9001
9	1A3	I	N	I	I	L/I	W/L	W	W	W	9043
10	B7/21	I	N	W/L	W/L	W/L	W/L	W	W	N	9001
11	PL 8	L/I	N	I	I	L/I	W/L	W	W	W	9004
12	TU 22	L/I	N	I	L/I	L/I	L	W	W	N	9004
13	PL 15	I	N	W/L	W/L	W	W	W	W	W	9047
14	L243	I	N	I/H	I/H	L/I	W/L	W	W	W	9043
15	2D6	I	N	H	I/H	L/I	W	W	W	N	9043
16	LEU 10	I	N	I	I	I	L	W	W	W	9043
17	16.23	I	N	W	W	W	N	N	N	N	9018
18	17.15	I	N	W/L	W/L	W/L	W	W	N	N	9016
19	3G5	I	N	W	W	W	W	W	W	N	9001
20	7.3.19.1	I/H	N	I	I	I	L/I	W	W	W	9019
21	77.34	W/L	N	W	W	W	N	N	N	N	9015
22	AB7.AE9	L/I	N	N	N	N	N	N	N	N	9030
23	AB8.25	I	N	W/L	W/L	W/L	W	W	N	N	9001
24	C1	I	N	W/L	W	W	W	N	N	N	9047
25	CC-CL 20	I	N	W	W	W	W	W	N	N	9028
26	CHE249.2	I	N	W	W	N	N	N	N	N	9001
27	DP 11.1	I	N	W	W	W	W	W	N	N	9047
28	GS4AA7	I	N	L	L	W/L	W/L	W	W	N	9001
29	GSP 17.1	I	N	W	W	W	W	W	N	N	9016
30	GSP 88.2	I	N	I	I	L/I	W/L	W	W	W	9043
32	IIIE3	I	N	L	W/L	W/L	W	W	N	N	9047
33	IVD12	I	N	L/I	W/L	W	W	N	N	N	9016
34	JS1	I/H	N	L/I	L/I	L	W	W	W	W	9019
35	KS-6	I	N	L	L	W/L	W/L	W	W	W	9001
36	KS-7	I	N	W/L	W/L	W/L	W/L	W	W	W	9001
37	MAD 88	I	N	L	W/L	W	W	N	N	N	9016
38	NDS 10	I/H	N	L	L	L	W	W	W	N	9019
39	NDS 13	I	N	I	I	L/I	W/L	W	W	W	9043
41	NFLD,M1	I/H	N	W/L	W	W	W	N	N	N	9019
42	PL 3	I	N	L/I	L/I	W/L	W/L	W	W	W	9047
43	PLM 12	I	N	L/I	L/I	W/L	W	W	N	N	9016
44	PLM 13	W/L	N	W	W	N	N	N	N	N	9015
45	PLM 2	I	N	I	I	L/I	W	W	N	N	9043
46	SFR16-DR7	I	N	L	L	W/L	W	W	W	W	9047
47	SFR3-DR5	I	N	L	W/L	W	W	N	N	N	9043
48	UK 4.1	I	N	L	L	W/L	W/L	W	N	N	9001
49	UK-8.1	I	N	I	I	L	W	W	W	N	9043
50	UL-65	W/L	M	W	W	W	N	N	N	N	9036

Qualitative scores were assigned based on where along the x-axis the peak channel of the fluorescence histogram occurred. For this purpose, the x-axis was divided into four equal partitions scored as (from left to right) weak (W), low (L), intermediate (I), or high (H). Histograms with peak channels along the y-axis were scored as negative (N). "Pos" refers to the positive control antibody (w6/32), while "Neg" refers to the negative control (mouse immunoglobulin). "Cell line" refers to the number of the Workshop cell line on which the titration analysis was performed.

panel and antibodies, but were not formally submitted for central analysis. In view of the 95% correlation between Workshop and local scoring, these supplemental clones have been included when appropriate. The designations for the clones included in the central analysis are listed in Table 1.

Workshop Antibody Panel

Interpretation of the antibody blocking data required knowledge of antibody titer and binding specificity versus the Workshop cell panel. Upon arrival in New York, antibodies were dialyzed to remove sodium azide. They were tested for their ability to inhibit the growth of B-lymphoblastoid cell lines (B-LCL) or T-cell clones as a means of insuring that they had been rendered azide-free. Their binding profiles toward the entire core set of 75 reagents were determined, and antibody titer was measured against a cell line to which the antibody had exhibited its strongest degree of binding. These analyses were performed by indirect immunofluorescence using an EPICS-C cell sorter. The cell sorter's sensitivity was adjusted to maximize the separation between the positive and negative controls. The positive control was the w6/32 monoclonal antibody recognizing HLA class I molecules, while the negative control antibody was mouse immunoglobulin.

The binding profiles of the antibodies are listed in quantitative and qualitative form in Tables 2 and 3. Table 2 lists the percentage of cells scored as positive with a given antibody (used at a saturating concentration). Table 3 provides an idea as to the relative amounts of antigen detected by the antibodies as estimated by the brightness of the fluorescence profiles. A fluorescence histogram was assigned a qualitative score according to where along the x-axis its peak channel occurred. For this purpose, the x-axis was divided into four equal partitions scored as (from left to right) weak (W), low (L), intermediate (I), or high (H). Histograms with peak channels along the y-axis were scored as negative (N). The results of the titration assays are listed in Tables 4 and 5. Table 4 lists the change in the percentage of positively stained cells as the antibodies were diluted. Table 5 lists titration results based on the leftward shift of the immunofluorescent profiles. Usually the shift in the fluorescence profiles (indicating that not all accessible antigen was bound) preceded the change in the number of cells stained as the antibodies were progressively diluted.

Acknowledgment. We wish to thank the laboratories contributing monoclonal antibodies for their generosity and for the hard work they undertook in order to provide these reagents for use in the Workshop.

References

1. Plätke R, Ferguson M, Klein JS, Zemmour J, Flomenberg N, Lalouel J-M. Collection and processing of HLA class II cellular data. In: Dupont B (ed): Immunobiology of HLA, Volume I: Histocompatibility Testing 1987. New York: Springer-Verlag, 1989.
2. Plätke R, Lalouel J-M, Flomenberg N. Statistical analysis of HLA class II cellular data. In: Dupont B (ed): Immunobiology of HLA, Volume I: Histocompatibility Testing 1987. New York: Springer-Verlag, 1989.
3. Milford EL, Kennedy LJ, Yang SY, Dupont B, Lalouel J-M, Yunis EJ. Serologic characterization of the reference panel of B-lymphoblastoid cell lines for factors of the HLA system. In: Dupont B (ed): Immunobiology of HLA, Volume I: Histocompatibility Testing 1987. New York: Springer-Verlag, 1989.

DR1: Antigen Report of the Cellular Studies of the Tenth International Histocompatibility Workshop

D. Eckels,[1] T. Sell,[1] T. Eiermann,[2] and A. Nikaein[3]

The Workshop B-lymphoblastoid cell lines (B-LCLs) expressing HLA-DR1 could be split into two groups based on the reactivity of T-cell clones. The patterns of reactivity of 22 clones are shown in Figure 1, while 10 are shown in more detail in Figure 2. Seven of the clones in Figure 2, derived from three different responder and stimulator combinations, recognized all of the HLA-DR1+ B-LCLs, and responded to these cell lines exclusively. However, three clones, again derived from three different donors, did not respond to cell lines 9002, 9078, and 9079, though they responded quite clearly to the other DR1+ Workshop B-LCLs. Cell lines 9002, 9078, and 9079, all of caucasoid origin, are all HLA-B14+ and express a new HLA-Dw specificity, designated Dw20, possibly identical to the "LVA" specificity (1). A B14-associated DR1 "split" has also been described (2), which seems to be derived from an α-β heterodimer that is resistant to dissociation by conventional methods leading to a dominant form of suppressed T-cell recognition. It is not clear whether the Workshop findings might similarly reflect in part a DR α chain polymorphism of the type described in that report.

Interestingly, Workshop DRβ restriction fragment length polymorphisms (RFLP) also correlated with the Dw1-Dw20 split. Using Pst I, Dw1 cell lines revealed a

Participating Laboratories: US2MBC,[1] GERGOL,[2] NCYNIK[3]

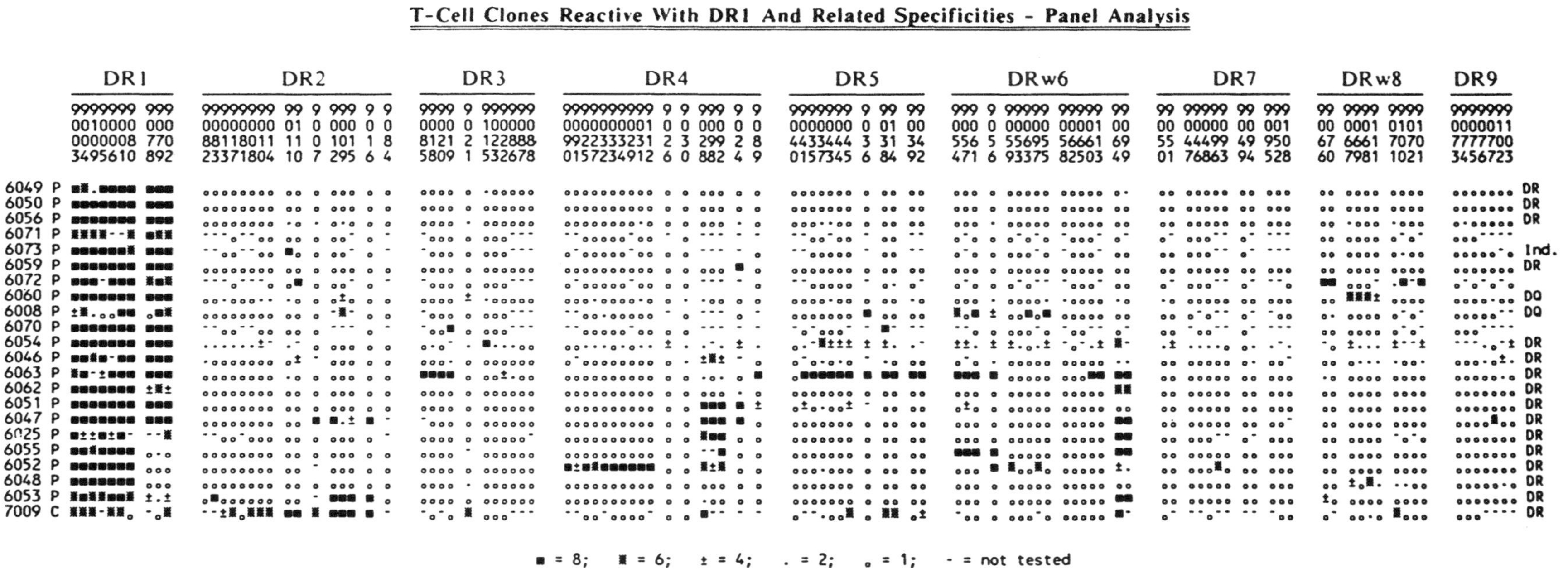

Figure 1. A portion of the cellular panel data relevant to HLA-DR1 is illustrated. The Workshop cell lines have been grouped according to their DR and Dw types. The Workshop number assigned to each clone is listed in the left column. The functional assay in which the clone was tested (cytotoxicity–C or proliferation–P) is listed in the second column. The HLA class II family recognized by each clone (as inferred from the monoclonal antibody blocking data) is listed in the right column. The symbols in the figure reflect the following scores: solid square = 8 = positive; shaded rectangle = 6 = probably positive; ± = 4 = weak or uncertain; closed dot = 2 = probably negative; open circle = 1 = negative; – = not tested.

B-LCL	DR	Dw	6052	6055	6048	6073	6059	6049	6050	6056	6062	6071	Pst I 3.73	Pst I 1.42
			DR1 SUBTYPES											
9080	1 (B14-)	?	+	+	+	+	+	+	+	+	+	+		
9001	1 (B14-)	w1	+	+	+	+	+	+	+	+	+	nt		
9004	1 (B14-)	w1	+	+	+	+	+	+	+	+	+	+	-	+
9006	1 (B14-)	w1	+	+	+	+	+	+	+	+	+	nt	-	+
9005	1 (B14-)	w1	+	+	+	+	+	+	+	+	+	+	-	+
9109	1 (B14-)	w1	+	+	+	+	+	+	+	+	+	+		
9003	1 (B14-)	w1	+	+	+	+	+	+	+	+	+	+	-	+
9078	1 (B14+)	w20	-	-	-	+	+	+	+	+	+	+		
9079	1 (B14+)	w20	-	-	-	+	+	+	+	+	+	+		
9002	1 (B14+)	w20	-	-	-	+	+	+	+	+	+	+	+	-

Figure 2. The reactivities of 10 T-cell clones toward 10 HLA-DR1 + Workshop cell lines are illustrated. "+" denotes recognition of the cell line by the clone (scores of 6 or 8), while "−" denotes lack of recognition (scores of 1 or 2). "nt" indicates combinations which were not tested or excluded from the central analysis. RFLP patterns of four Dw1 and one Dw20 cell lines are illustrated in the right columns of the figure. "+" and "−" refer to the presence or absence of 3.73 and 1.42 kb fragments using the Workshop DRβ probe and Pst I enzyme.

1.42 kb fragment, while the Dw20 cell line 9002 reciprocally demonstrated a 3.73 kb fragment. Using EcoRI and Pvu II restriction enzyme digests, Dw1 LCLs yielded 3.84 kb and 10.35 kb bands, respectively, that are missing from the Dw20 B-LCL 9002. (The Dw20 cell lines 9078 and 9079 were not tested in the Workshop RFLP analysis.) These data thus suggest that the Dw1 and Dw20 can be distinguished by DRβ restriction fragment length polymorphisms, although these findings require confirmation with additional Dw20 cell lines. In this regard, a variant DR1 sequence has been described (3) in cells of American Black origin that is localized to amino acids 85 (V → A) and 86 (G → V). More recently, these same substitutions have been observed in the DRβ chains of the three Dw20 Workshop cell lines (J. Gorski and D. Eckels, manuscript in preparation). To date, an associated α chain polymorphism has not been definitively ruled out.

References

1. Cohen N, Friedmann A, Szafer F, Amar A, Cohen D, Brautbar C. Polymorphism of the HLA DR1 haplotype in the Israeli population investigated at the serological, cellular, and genomic levels. Immunogenetics 1986;23:252–259.
2. Davis JE, Rich RR, Van M, Le HV, Pollack MS, Cook RG. Defective antigen presentation and novel structural properties of DR1 from an HLA haplotype associated with 21-hydroxylase deficiency. J Clin Invest 1987;80:998–1008.
3. Hurley CK, Ziff BL, Silver J, Gregersen PK, Hartzman R, Johnson A. Polymorphism of the HLA-DR1 haplotype in American blacks. J Immunol 1988;140:4019–4023.

DR2: Antigen Report of the Cellular Studies of the Tenth International Histocompatibility Workshop

N. Reinsmoen,[1] D. Eckels,[2] K. Zier,[3] D. Elferink,[4] R.P. de Vries,[4] A. Zeevi,[5] and N. Flomenberg[6]

Among the serologically defined DR2 (DRw15) cluster, two distinct haplotypes, Dw2 and Dw12, have been recognized since the Eighth International Workshop. These two haplotypes differ by at least three amino acids in one of their DRβ molecules and by at least one amino acid in the second DRβ chain. Although Dw2 and Dw12 are both associated with the DQw6 subtype of DQw1, they nevertheless exhibit multiple amino acid differences in their DQβ chains (1) and have distinct DQα chains (2). A third DR2-associated specificity, which is distinct from Dw2 and Dw12, has been designated LD-2s since cells with this specificity have been associated with the "DR2 short" serologic specificity, now designated DRw16. This specificity has been designated MN2, FJO, AZH, and WT24 in various reports (3) and is now designated Dw21. LD-5a is a related specificity which is also associated with the DR2 short–DRw16 specificity (4) and is now designated Dw22. Sequence analysis has demonstrated that one DRβ chain is identical in cells typing as Dw21 and Dw22 (but differs by at least five amino acids from each of the DRw15 haplotypes) (1,5). The second DRβ chain differs only by a single amino acid between Dw21 and Dw22 (which does not alter electrophoretic mobility) (6). Here again there are at least four or five amino acid differences versus the DRw15 haplotypes (1,5). Despite their striking similarities in the DR region, Dw21 and Dw22 can be readily differentiated based on the DQw1 (DQw5) association of Dw21 versus the DQw3 (DQw7) association of Dw22. Dw21 and Dw22 can also be distinguished from the DRw15-associated haplotypes based on a series of restriction fragment length polymorphisms (7,8). Within the Workshop cell panel, three cell lines typed as Dw21 (9009, 9015, and 9012), while one (9016) typed as Dw22. (An additional DR2 variant present in the black population DR2 LUMCT has been reported, but was not known

Participating Laboratories: US1BAC,[1] US2MBC,[2] NCYZIE,[3] BENROO,[4] US1DUQ,[5] US7NFL[6]

T-Cell Clones Reacting With DR2 And Related Specificities – Panel Analysis

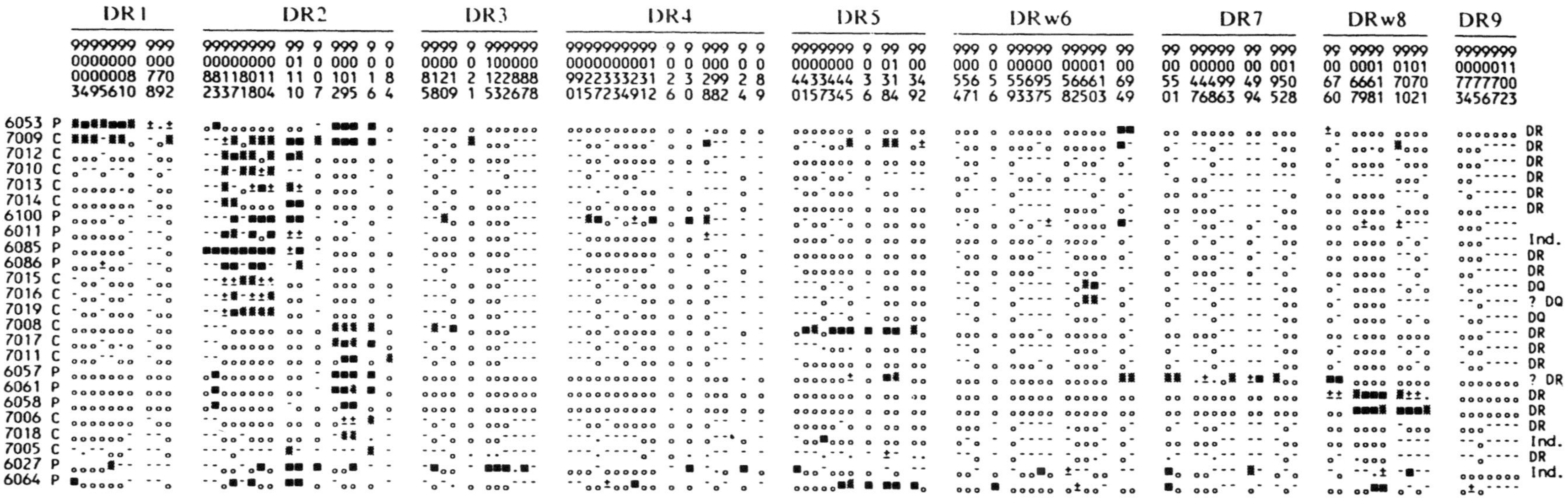

■ = 8; ▪ = 6; ± = 4; . = 2; ∘ = 1; - = not tested

Figure 1. A portion of the cellular panel data relevant to HLA-DR2 is illustrated. The Workshop cell lines have been grouped according to their DR and Dw types. The Workshop number assigned to each clone is listed in the left column. The functional assay in which the clone was tested (cytotoxicity = C or proliferation = P) is listed in the second column. The HLA class II family recognized by each clone (as inferred from the monoclonal antibody blocking data) is listed in t right column. The symbols in the figure reflect the following scores: solid square = 8 = positiv shaded rectangle = 6 = probably positive; ± = 4 = weak or uncertain; closed dot = 2 = probab negative; open circle = 1 = negative; − = not tested.

DR2 SUBTYPES

B-LCL	DR	Dw	7009	7008	6061	6057	7017	7006	7011	7018	6058	7010	7013	7012	6085	2D-Gel Patterns: DR-Beta			DQ-Beta
9009	2 (w16)	21	+	+	+	+	+	+	+	+	+					2.1	2.16	X	1.21
9015	2 (w16)	21	+	+	+	+	+	+	+	+	+					2.1	2.16	X	1.21
9012	2 (w16)	21	+	+	+	+	+			nt						2.1	2.16		1.21
9016	2 (w16)	22	+	+	+	+	+	+				nt				2.1	2.16	X	3.1
9011	2 (w15)	12	+									+	+	+		2.1	2.12		1.12
9017	2 (w15)	2	+									nt	nt	+	+	2.1	2.2		1.2
9013	2 (w15)	2										+	+	+	+	2.1	2.2		1.2
9008	2 (w15)	2	+									+	+	+		2.1	2.2		1.2
9010	2 (w15)	2	+									+	+		+		2.2		1.2
9014	2 (w15)	2	+									+	+	+	+	2.1	2.2		1.2

Figure 2. The reactivities of 13 T-cell clones toward 10 HLA-DR2+ Workshop cell lines are illustrated. "+" denotes recognition of the cell line by the clone (scores of 6 or 8), while a blank space indicates lack of recognition (scores of 1 or 2). "nt" indicates combinations which were not tested or excluded from the central analysis. The two-dimensional gel patterns, as recognized in the Protein Chemistry component, are also shown. "X" denotes an additional DRβ product, presumably related to DRβ-2.16 found in three of the four DRw16 cell lines.

to be represented in the Workshop cell panel [9]).

Twenty clones analyzed within the Workshop demonstrated appreciable reactivity with DR2+ cell lines, although many of these clones also exhibited reactivity outside the DR2 cluster as well. The reactivities of these clones are shown in Figure 1, while 13 are illustrated in more detail in Figure 2. At least three of the 20 clones recognized HLA-DQ gene products and are discussed separately. It is noteworthy that only a single clone exhibited broad reactivity within the DR2 cluster, recognizing cells expressing either Dw2, Dw12, Dw21, or Dw22 (see Fig. 1). The majority of DR-specific clones separated into two groups, recognizing either the DR2 long (DRw15)-associated specificities (Dw2 plus Dw12) or the DRw16-associated clusters (Dw21 plus Dw22). It is also noteworthy that none of the DR specific Workshop clones distinguished between Dw2 and Dw12, although clones of this type have been described previously (10). (In contrast, several HLA-DQ specific clones submitted to the Workshop clearly distinguished between Dw2 and Dw12.)

Within the DRw15 and DRw16 clusters, clear examples of microheterogeneity were seen. Clone 7014, for example, reacted with Dw12 plus a subset of Dw2. Both of the cells recognized expressed the extended A3;B7;DR2 haplotype, while the cells missed (with the exception of 9081) expressed other DR2-associated haplotypes (Fig. 1). Several DRw15-specific clones did not recognize cell line 9010 (AMAI) which has been shown to have a more acidic DRβ chain in the protein chemistry studies (Fig. 1) (6). Several clones detected heterogeneity within DRw16, which in several cases could be correlated with either protein chemistry or DNA sequence data. Several clones recognized DRw21, but did not respond to cell line 9012, which interestingly appears to lack a DRβ chain product found in other DRw16 cell lines (shown as X in Fig. 2, see also [6]). Two of these clones also failed to recognize cell line 9016 (Dw22), suggesting that they may be able to recognize the single amino acid difference present in the second DRβ chains of these haplotypes.

Of additional interest is cell line 9084. This cell line was typed as DRw16 in the Tenth Workshop, and was tested with seven clones recognizing DRw15 and six clones recognizing DRw16 (Fig. 1). Only a single clone recognizing a subset of Dw21 reacted with this cell line, suggesting it may belong to a distinct DR2-associated family. Cell line 9083 was unfortunately tested with only a single clone recognizing DRw15 (and was recognized by this clone). Surprisingly, it was recognized by three clones otherwise reactive with DRw16 and not DRw15 (Fig. 1). This would suggest that despite its homozygous expression of the A3;B7;DR2 haplotype, this cell line may either be heterozygous for DRw15/DRw16 or express some other unusual DR2-associated haplotype. Finally, the DR2/DR4 heterozygous cell line DEM (9007) was recognized only by the single broadly DR2 reactive clone, suggesting that its DR2 haplotype is also distinctive in some way (Fig. 1).

In summary, these Workshop analyses of HLA-DR2 haplotypes support the presence of two cellularly defined clusters associated with DRw15 and DRw16. Within these large clusters, further heterogeneity clearly exists as presently defined by the Dw2, Dw12, Dw21, and Dw22 subtypes. Although not conclusive, there is suggestive evidence of additional heterogeneity within Dw2, strengthened in one instance by the finding of a more acidic DRβ chain. In the case of Dw21, the functional heterogeneity detected in the present studies is well correlated with the protein chemistry studies, in which a DRβ chain product is absent from at least one of the members of this cluster.

References

1. Gregersen PK, Todd JA, Erlich HA, Long E, Servenius B, Choi E, et al. First domain sequence diversity of DR and DQ subregion alleles. In: Dupont B (ed): Immunobiology of HLA, Volume I: Histocompatibility Testing 1987. New York: Springer-Verlag, 1989.
2. Kaplan C, Morel MC, Hermans P, Tkaczuk J, Naskatsuji T, Fernandez N, et al. DQw1 (DQw5, DQw;6): 2-D gel patterns. In: Dupont B (ed): Immunobiology of HLA, Volume I: Histocompatibility Testing 1987. New York: Springer-Verlag, 1989.
3. Reinsmoen NL, Layrisse Z, Betuel H, Freidel AC, Bach FH. A study of HLA-DR2 associated HLA-Dw/LD specificities. Hum Immunol 1984;11:105–115.
4. Layrisse Z, Simoney N, Park MS, Terasaki PI. HLA-D and DRw determinants in an American indigenous isolate. Transplant Proc 1979;11:1788–1791.
5. Liu CP, Wu S, Bach FH. Molecular studies of the rare DR2/LD51/DQw3 haplotype: Implications for mechanisms of generating HLA class II polymorphisms. In: Dupont B (ed): Immunobiology of HLA, Volume II: Immunogenetics and Histocompatibility. New York: Springer-Verlag, 1989.
6. Rust N, Carter C, Fuggle S, Sacks S, Hermans P, Charron D, et al. DR2 (DRw15,16): 2-D gel patterns. In: Dupont B (ed): Immunobiology of HLA, Volume I: Histocompatibility Testing 1987. New York: Springer-Verlag, 1989.
7. Bidwell JL, Bidwell EA, Savage D, Middleton D, Cullen C, Klouda PT, et al. Identification of Dw2, Dw12 and short DR2 splits with sequential exon-specific DRbeta, DQbeta and DQalpha cDNA probes. In: Dupont B (ed): Immunobiology of HLA, Volume II: Immunogenetics and Histocompatibility. New York: Springer-Verlag, 1989.
8. Bignon JD, Semana G, Tiercy JM, Simons M, Lalouel JM, Cohen D. DNA-RFLP analysis: HLA-DR beta workshop report. In: Dupont B (ed): Immunobiology of HLA, Volume I: Histocompatibility Testing 1987. New York: Springer-Verlag, 1989.
9. Oudshoorn M, Du Toit ED, Martell RW, Taljaard DG. The heterogeneity of HLA-DR2. In: Dupont B (ed): Immunobiology of HLA, Volume II: Immunogenetics and Histocompatibility. New York: Springer-Verlag, 1989.
10. Reinsmoen NL, Bach FH. T-cell clonal analysis of HLA-DR2 haplotypes. Hum Immunol 1987;20:13–32.

DR3: Antigen Report of the Cellular Studies of the Tenth International Histocompatibility Workshop

J. Kurnick,[1] M. Thomsen,[2] C. Irle,[3] D. DeTotero,[4] G.B. Ferrara,[4] C. Navarette,[5] H. Festenstein,[5] and A. Termijtelen[6]

There were 11 clones evaluated which appeared to react dominantly or exclusively with DR3-positive target cells. The clones assayed included alloproliferative and allocytotoxic clones as well as a tetanus toxoid specific clone. The reactivities of these clones are illustrated in Figure 1, while nine are illustrated in more detail in Figure 2. The combined data from these clones indicate a split in DR3 recognition. Four clones (three allospecific and one tetanus toxoid-specific in the context of DR3) recognized DR3 haplotypes associated with both B8 and B18, but failed to recognize a DR3-positive cell line associated with Bw42 from a Black African (RSH, Workshop B-LCL 9021). A distinct cluster of clones recognized all of the DR3-positive cell lines, including the RSH (9021) target/stimulator cell. Many of these latter clones expressed both proliferative and cytotoxic allo-directed functions, indicating that the same DR recognition events could lead to both proliferative and/or cytotoxic function by the same T-cell clone. These T-cell-defined DR3 subsets are in agreement with cDNA sequencing studies (1) in which the DRB1 gene of HLA-DR3 haplotypes associated with B8 and B18 were found to be identical. In contrast, the Bw42-associated DR3 haplotype has four conservative amino acid substitutions in its DRB1 gene sequence (1,2). (Although not included in the Workshop analysis, Hurley and Johnson

Participating Laboratories: NCYKUR,[1] FRAOHA,[2] FRAJEA,[3] IT2FER,[4] UKIFES,[5] BENROO[6]

T-Cell Clones Reactive With DR3 And Related Specificities - Panel Analysis

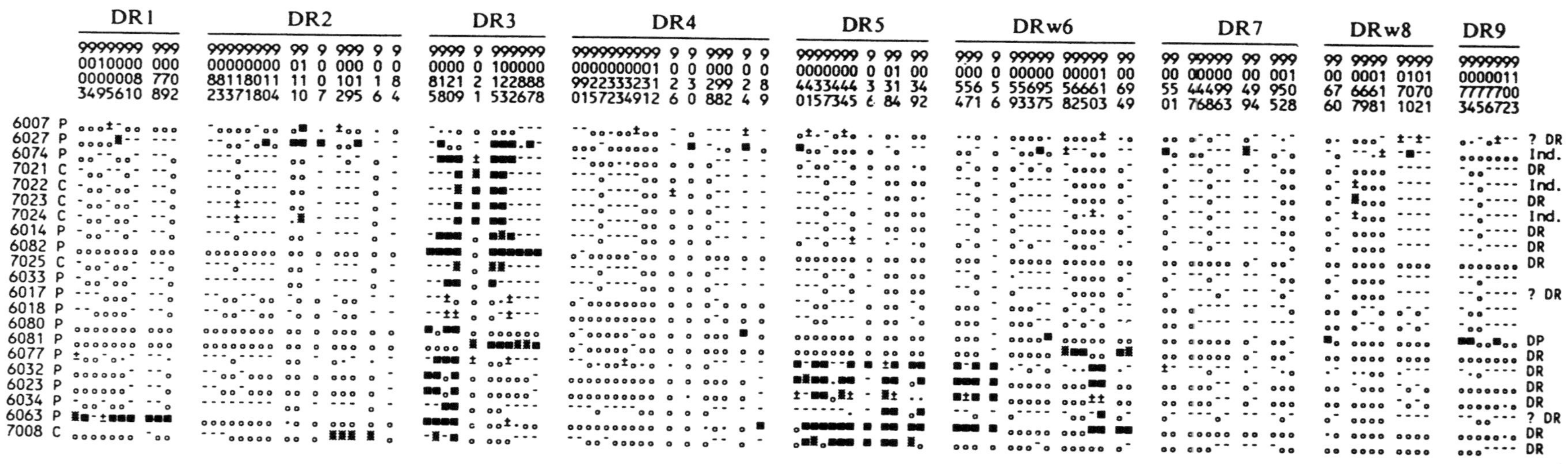

■ = 8; ▪ = 6; ± = 4; . = 2; ∘ = 1; - = not tested

Figure 1. A portion of the cellular panel data relevant to HLA-DR3 is illustrated. The Workshop cell lines have been grouped according to their DR and Dw types. The Workshop number assigned to each clone is listed in the left column. The functional assay in which the clone was tested (cytotoxicity = C or proliferation = P) is listed in the second column. The HLA class II family recognized by each clone (as inferred from the monoclonal antibody blocking data) is listed in the right column. The symbols in the figure reflect the following scores: solid square = 8 = positive; shaded rectangle = 6 = probably positive; ± = 4 = weak or uncertain; closed dot = 2 = probably negative; open circle = 1 = negative; − = not tested.

	DR3 WS - B-LCL					
	B8 - DR3		B18 - DR3			B42-DR3
CLONE	9023	9022	9018	9019	9020	9021
6082	+	+	+	+	+	-
6033	+	+	+	+	+	-
6014	+	+	+	+	+	-
6036/7021	+	nt	nt	+	nt	+
6037/7022	+	nt	nt	+	nt	+
6038/7023	+	nt	nt	+	nt	+
6035/7024	+	nt	nt	+	nt	+
6039	+	nt	nt	+	nt	+
6074	+	+	+	+	+	+

Figure 2. The reactivities of nine T-cell clones toward six HLA-DR3+ Workshop cell lines are illustrated. "+" denotes recognition of the cell line by the clone (scores of 6 or 8), while "−" indicates lack of recognition (scores of 1 or 2). "nt" indicates combinations which were not tested or excluded from the central analysis. Clones with dual designations were tested in both the cytotoxicity and proliferation sections.

also have isolated clones which recognize RSH (9021), but not the Caucasian DR3 [2]). This cellular polymorphism corresponds to the serologic polymorphism of DR3 defined in the Tenth Workshop studies, where the HLA-Bw42-associated haplotype behaves as a "short" DR3 (DRw18), in contrast to the HLA-B8- and B18-associated haplotypes (DRw17). Taken together, these studies indicate that specific areas of the DRβ-I molecule can act as restricting elements for antigen presentation or for allorecognition by functional T lymphocytes. These data confirm the functional role of DR epitopes in regulating T-cell recognition of conventional antigens such as tetanus toxoid, as well as serving as alloantigenic determinants. Evidence for additional DR3-bearing haplotypes in Black and Chinese populations is presented elsewhere in this book (2,3). The heterogeneity of the DR β-III product in DR3-associated haplotypes is discussed in the report on DRw52 (4).

References

1. Gregersen PK, Todd JA, Erlich HA, Long E, Servenius B, Choi E, et al. First domain sequence diversity of DR and DQ subregion alleles. In: Dupont B (ed): Immunobiology of HLA, Volume I: Histocompatibility Testing 1987. New York: Springer-Verlag, 1989.
2. Johnson AH, Tang TF, Robbins FM, Rosen-Bronson S, Steiner N, Hartzman RJ, et al. DR3 Heterogeneity in American blacks. In: Dupont B (ed): Immunobiology of HLA, Volume II: Immunogenetics and Histocompatibility. New York: Springer-Verlag, 1989.
3. Cotner T, Mellins E, Pious DA. DR3 beta chain heterogeneity. In: Dupont B (ed): Immunobiology of HLA, Volume II: Immunogenetics and Histocompatibility. New York: Springer-Verlag, 1989.
4. Irle C, Mickelson E, Masaewicz S, Hansen JA, Thomsen M, Termijtelen A. DRw52: Antigen report of the cellular studies of the Tenth International Histocompatibility Workshop. In: Dupont B (ed): Immunobiology of HLA, Volume I: Histocompatibility Testing 1987. New York: Springer-Verlag, 1989.

DR4 and DQw3: Antigen Report of the Cellular Studies of the Tenth International Histocompatibility Workshop

C. Sharrock,[1] K. Lundin,[2] K. Zier,[3] A. Zeevi,[4] M. Thomsen,[5] T. Eiermann,[6] J.A. Hansen,[7] E. Mickelson,[7] and N. Flomenberg[8]

HLA-DR4 is one of the most heterogeneous of the serologically defined HLA class II specificities. Sequence analyses (1) of the HLA-D subtypes of DR4 indicate that the amino acid sequences of Dw4, Dw10, Dw13, Dw14, Dw15, and DKT2 are identical from residues 1 through 56 of the DR β-I molecule, with the single exception of a Y → S substitution at position 37 in DKT2 (Fig. 1). Eight T-cell clones were submitted to the Tenth International Workshop which recognized polymorphisms of HLA-DR4, and these clustered into four groups: 1) those recognizing Dw4 (clones 6026, 6096, and 6099), 2) those recognizing Dw13 and DKT2 (clones 6001 and 6006), 3) those recognizing Dw14 (clones 6010 and 6078), and 4) a single clone (6097) recognizing Dw10. Additionally, clone 6021 recognized cells typing as Dw4, Dw10, Dw13, and Dw14. The reactivities of these clones are illustrated in Figure 1.

Clones 6026, 6096, and 6099 have apparent specificity for the DR4;Dw4 β-I molecule. These clones recognized Dw4 cell lines irrespective of their DQw3 subtype, and they failed to recognize cells expressing other polymorphisms of DR4 expressed within the Workshop panel. In addition, monoclonal antibody blocking studies indicate that clones 6026 and 6099 recognize

Participating Laboratories: UKIBAT,[1] SCATSB,[2] NCYZIE,[3] US1MBC,[4] FRAOHA,[5] GERGOL,[6] US7HAN,[7] US7NFL[8]

T-Cell Clones Reacting With HLA-DR4 And Related Specificities - Panel Analysis

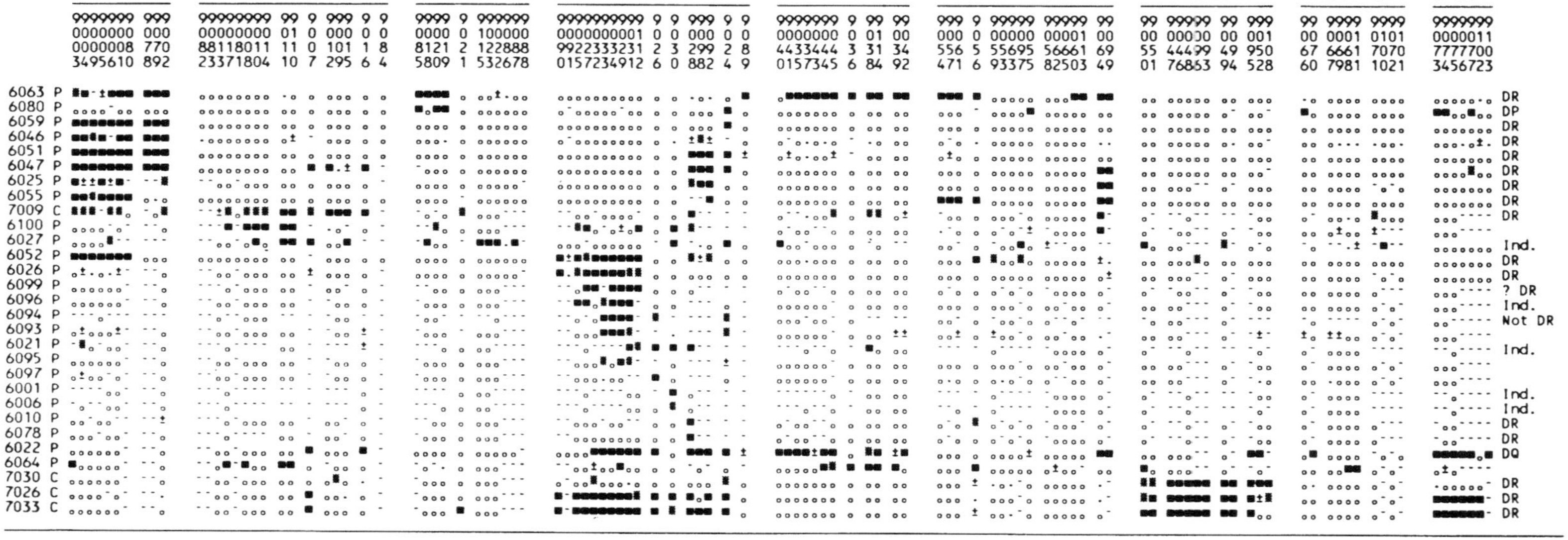

DR Beta I First Domain Amino Acid Sequence

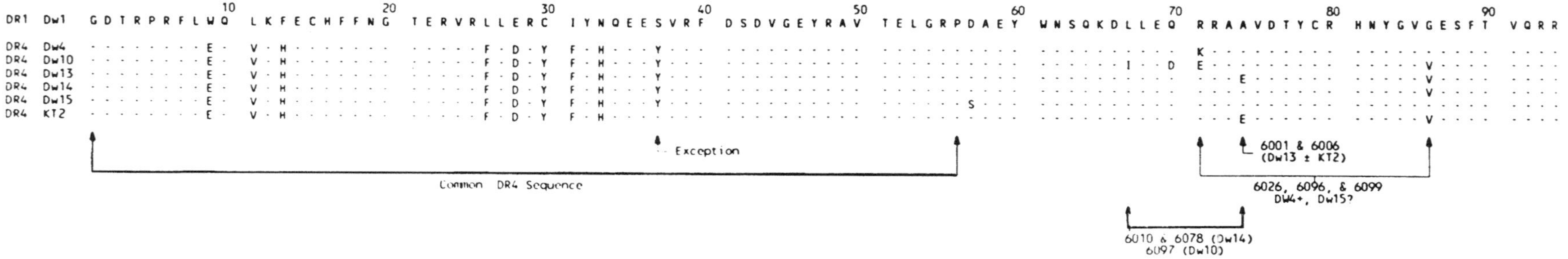

HLA-DR molecules. Analyses of the sequences of the DR4 subtypes would indicate that residues 71 and 86 are important for the restricted pattern of recognition of these clones. Dw4 differs from the other DR4 subtypes by expressing K (vs. R or E) at residue 71. Dw4 and Dw15 express G (vs. V) at residue 86; however, Dw15 was not represented within the Workshop cell panel. Clone 6026 was known from local studies to be nonreactive with cells typing as Dw15, and so must recognize an epitope including residue 71. In the case of clones 6096 and 6099, it is not possible to assess whether these clones also recognized Dw15. Thus, it is possible that residues 71 and/or 86 influence the specificity of these two clones.

Clones 6001 and 6006 recognized only one cell line in the core panel, JHA (9030), which expresses the Dw13 subtype of DR4. The most probable molecular specificity for these T-cell clones was DR β-I, although the antibody blocking studies were not definitive. Dw13 and DKT2 exhibit the same amino acid sequence from residues 38 through 94. They are unique amongst the DR4 subtypes in their expression of E versus A at residue 74. Within the first domain, the only difference between Dw13 and DKT2 is the expression of Y versus S at residue 37. Both clones were thought to react with DKT2 in local studies, though the studies of cell line 9024 (the only DKT2 cell line in the Workshop panel) were not included in the central analysis. It is likely, therefore, that residue 74 contributes significantly to the epitopes recognized by these clones (Fig. 1). Clones 6010 and 6078 recognizing Dw14 were inhibited by anti-DR monoclonal antibodies, and thus are also likely to recognize the DR β-I molecule, particularly the region between residues 67 and 74, which differentiates Dw14 from other DR4 subtypes. Dw14 cell lines were also recognized by several clones which also recognized DR1 (which exhibits the identical sequence throughout this region). In some instances, cell lines typing as Dw4 or DKT2 were recognized as well (2). Clone 6097, recognizing Dw10, is also likely to recognize sequences falling between residues 67 and 74. It should also be noted that all DR4 cell lines, with the exception of 9089 (BOB), were recognized by DRw53 specific T-cell clones. This cell line also failed to bind the DRw53-specific monoclonal antibody 109d6 (3) and was subsequently serologically retyped as DRw11;DRw52 (4). The cellular analyses are consistent with these assignments.

Table 1. Associations Between DR Specificities and DQw Subtypes

DR	DQw7	DQw8	DQw9
DR4;Dw4	X	X	
DR4;Dw10		X	
DR4;Dw13	X	X	
DR4;Dw14		X	
DRw11;Dw5	X		
DRw12;DB6	X		
DRw14;Dw16	X		
DR7;Dw11			X
Dw8.3 (Dutch)	X		
DR9			X

Of the DR4 haplotypes, Dw4, Dw10, Dw13, and Dw14 type as DQw3, while Dw15 haplotypes express DQw4. After the Tenth Workshop, three subtypes of DQw3 (DQw7, DQw8, and DQw9) were recognized. The DR-DQw3 subtype associations of DR4 and other DQw3-associated DR types are illustrated in Table 1. Several clones analyzed in the Tenth Workshop were relevant to the analysis of DQw3, of which four are illustrated in Figure 2. All were inhibited by HLA-DQ-specific antibodies. Clone 6022 recognized all DQw3 cell lines, except the subset of DQw7 cell lines that were also DR4-positive. Clone EM.C3 identified the reciprocal subset, recognizing only the DQw7 cells within the DR4 cluster. Clone 6094 recognized DQw8 (as did clones 6093 and 6095). The serologically defined DQw3 subtypes correlate well with the type of DQ β chain detected by 2D-gel analysis (Fig. 3 and Mazzilli et al., [5]). The specific recognition of DR4, DQw7 by clone EM.C3 suggests that a new epitope may result from the pairing of this particular α-β dimer (α-3.1, β-3.1), since the DQ α chain in these cell lines appears similar to that found in DRw9 cells, while the DQ β chain is similar to that found in Dw22, DR5, and the subsets of DRw8 and DRw14 cell lines expressing DQw3 (5,6). This requirement for both the DQ α and β chains in the formation of this epitope has been confirmed in transfection experiments (E. Mickelson and G. Nepom, personal communication). Conversely, this particular α-β combination

◄

Figure 1. *Upper*: A portion of the cellular panel data relevant to DR4 and DQw3 is illustrated. The Workshop cell lines have been grouped according to their DR and Dw types. The Workshop number assigned to each clone is listed in the left column. The functional assay in which the clone was tested (cytotoxicity = C or proliferation = P) is listed in the second column. The HLA class II family recognized by each clone (as inferred from the monoclonal antibody blocking data) is listed in the right column. The symbols in the figure reflect the following scores: solid square = 8 = positive; shaded rectangle = 6 = probably positive; ± = 4 = weak or uncertain; closed dot = 2 = probably negative; open circle = 1 = negative; – = not tested. *Lower*: The first domain sequences of six DR4 subtypes are illustrated using the single letter amino acid codes. The DR1 sequence is listed as the reference against which the other sequences are compared. "–" denotes the identical amino acid as is found in the DR1. All six DR4 subtypes exhibit a common sequence from residues 1 to 56 as indicated, with the exception of LDKT2 at residue 37 (shown as "exception"). The substitutions which are relevant to the specificities of particular clones are shown on the right side of the figure. A heavy line denotes localization of the specificity to a particular region of the molecule, while a light line denotes localization to one or two relevant amino acides. Adapted from Gregersen et al. (1).

DQw3 Subtypes

<u>10 WS B-LCL</u>

	DR2	DRw11						DRw12	DRw14		DR7		DRw8	DRw9			DR4								
	DQw7	DQw7						DQw7	DQw7		DQw9		DQw7	DQw9			DQw8						DQw7		
T Cell Clones	9016	9035	9037	9040	9041	9043	9044	9038	9064	9099	9052	9095	9070	9073	9074	9075	9029	9031	9033	9024	9026	9034	9025	9027	9030
6022	+	+	+	+	+	+	+	+	+	+	+	+	+	+	+	+	+	+	+	+	+	+	-	-	-
EM.C3	-	-	-	-	-	-	-	nt	-	-	-	-	-	-	-	-	-	-	-	-	-	-	+	+	+
6094	-	-	-	-	-	-	-	-	-	nt	-	nt	-	-	-	-	+	+	+	+	+	+	-	-	-
6095	-	-	-	-	-	-	-	-	-	nt	-	nt	nt	-	-	-	+	+	+	+	+	-	-	-	-
DQ-α	2	2						2	2		3.7		3.8	3.1			3.2					3.1			
DQ-β	3.1	3.1						3.1	3.1		3.3		3.1	3.3			3.2						3.1		

(shaded) = α 3.1 & 3.2; **Identical AA Sequence in First Domain**
(open box) = β 3.2 & 3.3; **One AA Difference in First Domain**

Figure 2. The reactivities of four T-cell clones toward 25 HLA-DQw3+ Workshop cell lines are illustrated. "+" denotes recognition of the cell line by the clone (scores of 6 or 8), while "−" indicates lack of recognition (scores of 1 or 2). "nt" indicates combinations which were not tested or excluded from the central analysis. The results of the biochemical analysis of the HLA DQ α and β chains are shown as well. As illustrated, the serologic DQ specificities and the specificity of clone 6094 can be correlated with the type of DQ β chain detected in the biochemical studies. The specificities of clones 6022, EM.C3, and 6095 can be correlated with the presence of particular α-β dimers. Shaded areas illustrate similarities between the DQ products found in DQw8 cell lines and those found in other subtypes of DQw3.

must prevent the formation of another conformational epitope (recognized by clone 6022), which is found when these chains are present separately on other haplotypes, but which cannot form when they are present together.

Particularly informative in the analysis of DQw7 and DQw8 is cell line 9034, which uniquely (among DR4 cell lines) expresses a DQα-3.1 chain and DQβ-3.2 chain. It types as DQw8 serologically, and behaves as DQw8 in its cellular recognition pattern with clones 6022, 6094, 6093, and EM.C3. This would suggest that clone 6094 and 6093 recognize a DQw8-specific epitope present on the DQβ-3.2 chain. Most likely, this epitope would include residue 57 (A in DQw8, D in DQw7 and DQw9) (1,7). This is the only residue in the first domain of DQ β which distinguishes DQβ-3.2 from DQβ-3.3. This residue also differentiates DQβ-3.2 from 3.1, although additional differences also exist between DQβ-3.1 and DQβ-3.2. (Additional residues must also contribute to this epitope as well, since DQw2 β chains also exhibit A at residue 57.) Of interest is an additional clone (6095) which recognizes all DQw8 cell lines with the exception of cell line 9034 which expresses the DQα-3.1 chain (7). Analysis of the patterns of reactivity exhibited toward DR4 and DR9 cell lines (Fig. 2) demonstrates that the specificity of the DQw8 directed clones is exquisitely precise. The broadly reactive DQw8 clones (6093 and 6094) would appear to be sensitive to the single amino acid substitution presently recognized as differentiating DQβ-3.2 and 3.3. (The first domains of the DQ α chains in these cell lines appear to differ by only a single silent nucleotide substitution [1]). Thus, while recognizing a DQα-3.1–DQβ-3.2 heterodimer in cell line 9034, these two clones fail to recognize a DQα-3.1–DQβ-3.3 heterodimer in the DR9 cell lines. Clone 6095 must recognize an epitope to which *both* DQα-3.2 and DQβ-3.2 contribute. This is an especially interesting observation since DQα-3.1 and DQα-3.2 (despite their electrophoretic differences) are identical in their first domain sequences (J. Lee, personal communication, [1]). The biochemical and functional difference between these two α chains must be due to differences in the more proximal portions of the molecule. Thus, clones 6093 and 6094 recognize the DQβ-3.2 chain, paired with either a DQα-3.1 or a DQα-3.2 chain, while 6095 recognizes only the DQβ-3.2, DQα-3.2 dimer.

In summary, DR4-directed clones may exhibit specificity for subsets of DR4 cell lines defined previously by HTC typing. In some cases, the patterns of reactivity can be correlated with specific amino acid substitutions within the DR β chain. The DQw3 subtypes defined by serology are best correlated with the type of DQ β chain expressed, while further subdivisions are possible based on the biochemical and cellular data. The DQw8 specificity may be largely dependent on the presence of A (versus D) at residue 57 of the DQ β molecule, although in some cases, the α chain may contribute to the epitopes recognized as well. Several additional examples are presented of the creation and suppression of combinatorial epitopes to which both the α and β chains contribute. Overall there is excellent agreement between the patterns of cellular reactivity observed and the information provided by the protein chemistry and DNA sequence analyses.

References

1. Gregersen PK, Todd JA, Erlich HA, Long E, Servenius B, Choi E, et al. First domain sequence diversity of DR and DQ subregion alleles. In: Dupont B (ed): Immunobiology of HLA, Volume I: Histocompatibility Testing 1987. New York: Springer-Verlag, 1989.
2. Eckels D, Sell T, Zier K, Mickelson E, Hansen JA, Flomenberg N, et al. T-cell clones recognizing shared or cross-reactive epitopes: Antigen report of the Tenth International Histocompatibility Workshop. In: Dupont B (ed): Immunobiology of HLA, Volume I: Histocompatibility Testing 1987. New York: Springer-Verlag, 1989.
3. Sanchez B, Winchester R, Nunez-Roldan A. Workshop B-cell reference panel reactivity with monoclonal antibody 109d6. Tenth Workshop Newsletter, No. 4, pp 19–20, 1987.
4. Milford EL, Kennedy LJ, Yang SY, Dupont B, Lalouel J-M, Yunis EJ. Serologic characterization of the reference panel of B-lymphoblastoid cell lines for factors of the HLA system. In: Dupont B (ed): Immunobiology of HLA, Volume I: Histocompatibility Testing 1987. New York: Springer-Verlag, 1989.
5. Mazzilli C, Gandini E, Lanchbury J, Madrigal A, Ballas M, Fuggle S, et al. DQw3 (DQw6, DQw9): 2-D gel patterns. In: Dupont B (ed): Immunobiology of HLA, Volume I: Histocompatibility Testing 1987. New York: Springer-Verlag, 1989.
6. Mickelson E, Nepom GT, Nisperos B, Hansen JA. DQw3 variants defined by cloned alloreactive T cells. Hum Immunol 1988;21:63–73.
7. Lundin EA, Gaudernack G, Paulsen G, Sollid LM, Thorsby E. Epitopes on HLA-DQw3 molecules recognized by T-lymphocyte clones. In: Dupont B (ed): Immunobiology of HLA, Volume II: Immunogenetics and Histocompatibility. New York: Springer-Verlag, 1989.

DR5: Antigen Report of the Cellular Studies of the Tenth International Histocompatibility Workshop

E. Mickelson,[1] S. Masaewicz,[1] M. Thomsen,[2] J.A. Hansen,[1] N. Reinsmoen,[3] G. Sterkers,[4] A. Urlacher,[5] M. Tongio,[5] A. Zeevi,[6] and N. Flomenberg[7]

Within the DR5 cluster several distinct subtypes have been identified. The classic Dw5 specificity encompasses a heterogeneous group of cells, including many of the DR5 cell lines represented in the Workshop cell panel. Additional DR5 specificities that can be serologically recognized include the DB2 group (associated with DQw1 rather than DQw3) and the DB6 group (associated with DRw12 rather than DRw11). Two Workshop cell lines, 9039(JVM) and 9042(TISI), represent distinct DR5-associated specificities. These cells can be distinguished from Dw5 in cellular studies, although they cannot presently be serologically distinguished from other DRw11 haplotypes. Evidence for additional cellularly defined polymorphism within DR5 exists as well.

Several T-cell clones reacted broadly with the DR5 cluster, although the majority of these clones recognized cells from other DR clusters as well (Fig. 1). Clone 6077 recognizing the Dw25 subtype of DRw52 responded to all the DR5 cell lines against which it was tested, including all of the DR5 subtypes represented within the Workshop panel. Two additional clones recognizing Dw25 (6032 and 6023) recognized most DR5 cell lines, although neither of the clones responded to cell lines 9043 (BM21) or 9039 (JVM). It is noteworthy that Dw25 expression among DR5 cell lines appears to be heterogeneous, based on the reactivity of some T-cell clones (1,2) and DNA sequence heterogeneity (J. Gorski, personal communication).

Heterogeneity can also be detected in the DR β-I molecule. The most clear cut split differentiates DRw11 and DRw12 cells. Clone 6075 reacted with the DRw12;DB6 cell line, 9038. In local studies, this clone responded to the Herluf cell characterized in the Ninth International Workshop (9w1503). It showed no reactivity with cells from the DRw11 or DRw8 clusters. DRw12 could also be recognized by the DN1 monoclonal antibody characterized in the Workshop serologic analysis (3), while the DRw11 cluster was reciprocally

Participating Laboratories: US7HAN,[1] FRAOHA,[2] US2BAC,[3] FRALEV,[4] FRAMYE,[5] US1DUQ,[6] US7NFL[7]

T-Cell Clones Reactive With DR5 And Related Specificities - Panel Analysis

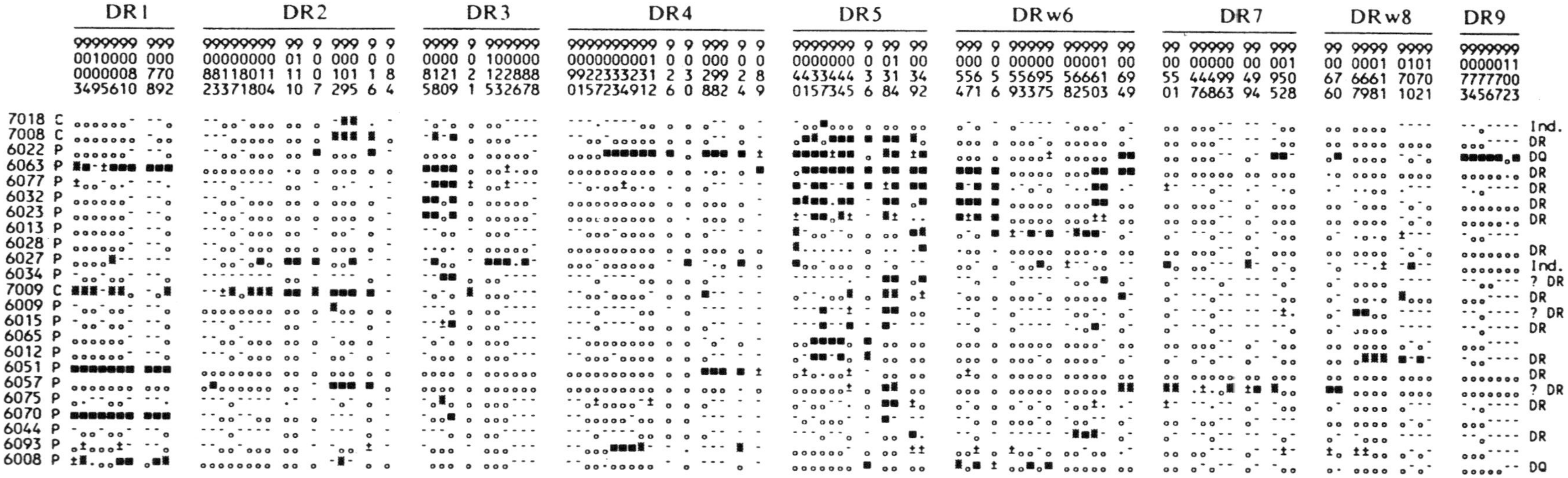

■ = 8; ▪ = 6; ± = 4; . = 2; ∘ = 1; - = not tested

Figure 1. A portion of the cellular panel data relevant to HLA-DR5 is illustrated. The Workshop cell lines have been grouped according to their DR and Dw types. The Workshop number assigned to each clone is listed in the left column. The functional assay in which the clone was tested (cytotoxicity = C or proliferation = P) is listed in the second column. The HLA class II family recognized by each clone (as inferred from the monoclonal antibody blocking data) is listed in th right column. The symbols in the figure reflect the following scores: solid square = 8 = positive shaded rectangle = 6 = probably positive, ± = 4 = weak or uncertain; closed dot = 2 = probabl negative; open circle = 1 = negative; — = not tested.

recognized by monoclonal antibodies CHE249.2, GSP 88.2, and SFR-DR5, tested within this component (4). The two cell lines typed as DwN (9039 and 9042) could be separated from one another by several clones, including 7008 and 6044 which recognized only 9039 (JVM), while clones 6032 and 6028 recognized only cell line 9042 (TISI). It is noteworthy that cell lines 9040 and 9042 behave concordantly in many instances and may represent a distinct DR5 subset. JVM has already been shown to contain a distinct DRB1 gene (5). Clones (6065 and 6012) responded only to a discrete subset of Dw5 (cell lines 9035, 9037, 9043, and 9044) and to DB2, but not to other Dw5 cells, nor to cells typed as DB6 or DwN. Additional microheterogeneity within the Dw5 subtype was also detected. Several clones (7008, 6065, and 6012) failed to recognize cell line 9040 (BM15). However, this cell line was recognized by clones 6027, 6013, and 6028, which failed to detect any of the other Dw5 cell lines. Clone 7008 also missed the Dw5 cell line 9037 (SWEIG), which was, reciprocally, the only DR5 cell recognized by clone 7018. Interestingly several of the clones reactive with DR5 were also reactive with DRw16, including clones 7008 (DRw16) and 7018 (Dw21 subset). Cross-reactivity with DRw8 cell lines (especially Dw8.1 and Dw8.2) was also a frequent observation. The basis of this cross-reactivity is discussed elsewhere in this section (see Eckels et al. [6]). The DR5 cell lines could also be split on the basis of their associated DQ types. Clone 6022 recognizing a split of DQw3 responded to all DR5, DQw7 cell lines, but did not proliferate in response to the DR5;DQw6;DB2 cell line (9036).

Overall, DR5 appears to be quite complex. Although the serologic relationships between DR5, DRw11, DRw12, and DR8 are the subject of debate, DRw12 can be recognized as a specificity distinct from either DRw11 or DRw8 by T-cell clones, by unique RFLP fragments (7), and by monoclonal antibody reactivity (3). Although not conclusive, there is suggestive evidence for further heterogeneity in the DRw11 and DRw52 products expressed by these cell lines. A discrete DR5 subset is defined by clones 6065 and 6012. Cell lines (TISI) and 9039 (JVM) also appear to represent distinct DR5 subtypes. In some cases, DQ products may contribute to the heterogeneity of the haplotypes, particularly DB2, which is associated with DQw1. It is also noteworthy that clone 6064, recognizing DPw2, is reactive with a large subset of the DR5 Workshop cell lines (consistent with their known DP types).

Additional factors may contribute to the heterogeneity of DR5 haplotypes. Analysis of the protein chemistry of the DR β chain products of these cell lines indicates that two DR β-I products may be expressed (8). Cell line 9045 (TUBO) expressed a second DR β-I chain which appears identical to that seen in DRw12 cell lines. TUBO was typed as a DRw11/DRw12 heterozygote serologically (9). This cell line was not recognized by the DRw12-specific clone 6075, although it was recognized by two more broadly reactive clones (7009 and 6057) whose reactivity within the DR5 cluster was otherwise limited to DRw12. The simplest explanation is that this cell line is a true heterozygote. Alternatively, it is conceivable, though less likely, that some of the complexity of DR5 may be due to expansion and contraction of the MHC within this haplotype, as well as to sequence heterogeneity of the genes which are present.

References

1. Mickelson EM, Masaewicz SA, Hansen JA. Variants of HLA-DRw52 defined by proliferative T-lymphocyte clones. In: Dupont B (ed): Immunobiology of HLA, Volume II: Immunogenetics and Histocompatibility. New York: Springer-Verlag, 1989.
2. Irle C, Mickelson E, Masaewicz S, Hansen JA, Thomsen M, Termijtelen A. DRw52: Antigen report of the cellular studies of the Tenth International Histocompatibility Workshop. In: Dupont B (ed): Immunobiology of HLA, Volume I: Histocompatibility Testing 1987. New York: Springer-Verlag, 1989.
3. Stastny P, Layrisse Z, Singal DP, Svejgaard A, Van den Berg-Loonan E, Dohi K, et al. Antigen Society #24 report (DRw11, DRw12, DRw8). In: Dupont B (ed): Immunobiology of HLA, Volume I: Histocompatibility Testing 1987. New York: Springer-Verlag, 1989.
4. Flomenberg N, Platke R, Zemmour J, Small TN, Klein JS. Immunogenetic analysis of HLA class II cellular data. Antigen report of the cellular studies of the Tenth International Histocompatibility Workshop. In: Dupont B (ed): Immunobiology of HLA, Volume I: Histocompatibility Testing 1987. New York: Springer-Verlag, 1989.
5. Gregersen PK, Todd JA, Erlich HA, Long E, Servenius B, Choi E, et al. First domain sequence diversity of DR and DQ subregion alleles. In: Dupont B (ed): Immunobiology of HLA, Volume I: Histocompatibility Testing 1987. New York: Springer-Verlag, 1989.
6. Eckels D, Sell T, Zier K, Mickelson E, Hansen JA, Flomenberg N, et al. T-cell clones recognizing shared or cross-reactive epitopes. Antigen report of the Tenth International Histocompatibility Workshop. In: Dupont B (ed): Immunobiology of HLA, Volume I: Histocompatibility Testing 1987. New York: Springer-Verlag, 1989.
7. Bignon JD, Semana G, Tiercy JM, Simons M, Lalouel JM, Cohen D. DNA-RFLP analysis: HLA-DR beta workshop report. In: Dupont B (ed): Immunobiology of HLA, Volume I: Histocompatibility Testing 1987. New York: Springer-Verlag, 1989.
8. Madrigal A, Bontrop R, Mazzilli MC, Kaplan C, Wong K, Knitter-Jack N, et al. DR5 (DRw11, DRw12): 2-D gel patterns. In: Dupont B (ed): Immunobiology of HLA, Volume I: Histocompatibility Testing 1987. New York: Springer-Verlag, 1989.
9. Milford EL, Kennedy LJ, Yang SY, Dupont B, Lalouel J-M, Yunis EJ. Serologic characterization of the reference panel of B-lymphoblastoid cell lines for factors of the HLA system. In: Dupont B (ed): Immunobiology of HLA, Volume I: Histocompatibility Testing 1987. New York: Springer-Verlag, 1989.

DRw6: Antigen Report of the Cellular Studies of the Tenth International Histocompatibility Workshop

E. Mickelson,[1] G. Sterkers,[2] A. Zeevi,[3] A. Termijtelen,[4] C. Farrell,[5] M. Honeyman,[5] J.A. Hansen,[1] D. Eckels,[6] N. Reinsmoen,[7] and N. Flomenberg[8]

At present, DRw6 has been subdivided into four cellularly defined specificities, Dw18 plus Dw19 (associated with DRw13) and Dw9 plus Dw16 (associated with DRw14). Each of the DRw6-associated DR and Dw subtypes (DRw13, DRw14, Dw9, Dw16, Dw18, and Dw19) could be distinguished using the Tenth Workshop T-cell clones (Fig. 1). The entire serologically defined DRw14 group was recognized by clone 6055, which was inhibited by anti-DR antibodies. This clone also exhibited reactivity with Dw1 cell lines and one DR4 cell line (9092) which clustered with the Dw14 cell lines in these cellular studies. The DRw14;Dw9 subgroup was recognized by clone 6024 (HLA-DR specific), which was limited in its reactivity solely to this Dw type, being nonreactive with cell lines from the DRw14;Dw16 cluster. None of the Workshop clones exhibited patterns of reactivity exclusively limited to the Dw16 subgroup. The Dw16 cell lines could readily be separated from the other DRw6 cell lines, however, by T-cell clones with broader patterns of reactivity. Several clones recognized DR1 and/or DR2 together with Dw16 cell lines, often with recognition of DR4 subsets as well. Based on the antibody blocking studies, these clones appear to recognize DR rather than DQ molecules. These patterns of reactivity are in keeping with protein chemistry which demonstrates that the DR β-I molecules of Dw9 and Dw16 can be electrophoretically distinguished, despite the fact that both express DRw14 serologic determinants (1). Moreover, sequence analysis indicates that these two DRw14 subtypes exhibit multiple differences in their DRB1 sequences (2). The Dw16 cell lines are also differentiated from the other DRw6 cell lines on the basis of their serologic typing as DQw3 (DQw7) and their recognition by clones specific for DQw3 determinants (3). Dw9 and Dw16 can, additionally, be distinguished from one another on the basis of their reactivity with DRw52 specific reagents—Dw16 cells typing as Dw24 and Dw9 cells typing as Dw25 (4).

The DRw13 haplotypes were recognized by several DR-specific clones, including 6041, 6042, 6044, and 6013. Clone 6013 and, to a lesser extent, clone 6044 also exhibited reactivity within the DR5;DwN subgroup. The two subtypes of DRw13 could not be separated from one another by DR specific clones, but were readily split by the DQ specific clones 6029 (Dw19) and 6031, 6084 (Dw18) (5). These findings are supported by the fact that the DR β-I chains of Dw18 and Dw19 cannot be distinguished on the basis of protein chemistry (1), although they exhibit a single conservative amino acid substitution (residue 86 valine → glycine) in their DRB1 genes (6). In contrast, the DQ products are readily distinguishable (7).

Heterogeneity within the Dw19 cluster was evident from the patterns of reactivity of several clones, including 6008, 6052, 6027, and 6029. Heterogeneity was also observed within the Dw18 cluster, where cell line 9060 and its blind duplicate control (9113) behaved distinctively from other Dw18 cell lines in several cases. This cell line could be distinguished from other Dw18 cell lines in serologically (but not cellularly or biochemically) typing as DQw5 rather than DQw6 and by several RFLPs (7–9). Several of the clones that recognized this cell line, but not the other Dw18 cell lines, had specificity for the newly defined DRw52 split, Dw25. Unlike the other Dw18 cell lines, 9060 was not recognized by clones specific for Dw24. As discussed in the DRw52 report, the protein chemistry and Southern blot analysis of this cell line are consistent with expression of a Dw25, rather than a Dw24 product. DRw13;Dw19 cell lines express a third DRw52 allele, Dw26 (4). It should be noted that cell line 9060 is not unique among Dw18 homzygous typing cells in its expression of Dw25. Other Dw18 HTCs may also type as Dw25, which may explain the "noise" seen in this Dw cluster in primary MLC (Mickelson et al., unpublished observations). The DQw5 association is less certain in that these cell lines have not been readily distinguished by Dw18;DQw6-specific clones (5), nor has cell line 9060 demonstrated a distinctive DQ pattern in 2D gel analysis (7).

In summary the two DRw13 haplotypes (Dw18 and Dw19) appear to exhibit very similar DR β-I products, but distinct DR β-III and DQ products. Dw18 cell lines may exhibit either the Dw24 or Dw25 alleles of DRw52, while Dw19 cell lines express Dw26. The Dw18;Dw24 subset expresses DQw6. The Dw18;Dw25 cell line 9060 was serologically typed as DQw5, although this was not confirmed in the cellular or protein studies, and requires analysis of additional cell lines. In contrast, the two DRw14 haplotypes Dw9 and Dw16 can be distinguished on the basis of their DR β-I, DR β-III, and DQ products. Thus five potential DRw6-associated haplotypes can be identified among the Tenth Workshop cell lines, representing Dw9, Dw16, Dw19, and two Dw18 haplotypes. Additional polymorphisms may exist in other

---

Participating Laboratories: US7HAN,[1] FRALEV,[2] US1DUQ,[3] BENROO,[4] ANZBSH,[5] US2MBC,[6] US2BAC,[7] US7NFL[8]

T-Cell Clones Reactive With DRw6 And Related Specificities - Panel Analysis

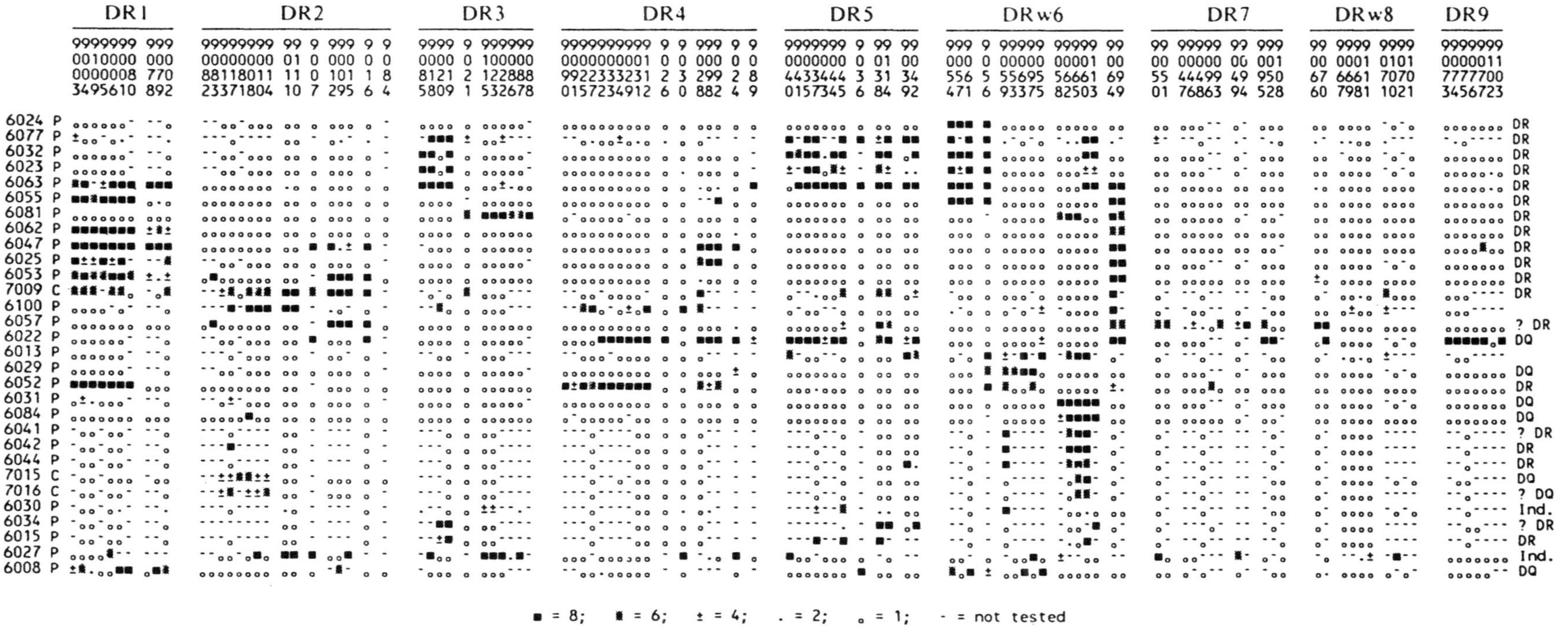

Figure 1. A portion of the cellular panel data relevant to HLA-DRw6 is illustrated. The Workshop cell lines have been grouped according to their DR and Dw types. The Workshop number assigned to each clone is listed in the left column. The functional assay in which the clone was tested (cytotoxicity = C or proliferation = P) is listed in the second column. The HLA class II family recognized by each clone (as inferred from the monoclonal antibody blocking data) is listed in the right column. The symbols in the figure reflect the following scores: solid square = 8 = positive; shaded rectangle = 6 = probably positive; ± = 4 = weak or uncertain; closed dot = 2 = probably negative; open circle = 1 = negative; — = not tested.

ethnic groups as well (10,11). There is good overall agreement between patterns of clonal reactivity, protein chemistry, and DNA sequence, which helps to explain the heterogeneity of DRw6-associated haplotypes.

References

1. Charron D, Hermans P, Ballas M, Brahmi Z, du Toit E, Martell R, et al. DRw6 (DRw13, DRw14): 2-D gel patterns. In: Dupont B (ed): Immunobiology of HLA, Volume I: Histocompatibility Testing 1987. New York: Springer-Verlag, 1989.
2. Gregersen PK, Todd JA, Erlich HA, Long E, Servenius B, Choi E, et al. First domain sequence diversity of DR and DQ subregion alleles. In: Dupont B (ed): Immunobiology of HLA, Volume I: Histocompatibility Testing 1987. New York: Springer-Verlag, 1989.
3. Sharrock C, Lundin K, Zier K, Zeevi A, Thomsen M, Eiermann T, et al. DR4 and DQw3: Antigen report of the cellular studies of the Tenth International Histocompatibility Workshop. In: Dupont B (ed): Immunobiology of HLA, Volume I: Histocompatibility Testing 1987. New York: Springer-Verlag, 1989.
4. Irle C, Jaques D, Tiercy JM, Jeannet M, Mach B. Functional and structural analysis of HLA-DRw52. In: Dupont B (ed): Immunobiology of HLA, Volume II: Immunogenetics and Histocompatibility. New York: Springer-Verlag, 1989.
5. Reinsmoen N, Eckels D, Mickelson E, Honeyman M, Farrel C, Hansen JA, et al. DQw1 report: Antigen report of the cellular studies of the Tenth International Histocompatibility Workshop. In: Dupont B (ed): Immunobiology of HLA, Volume I: Histocompatibility Testing 1987. New York: Springer-Verlag, 1989.
6. Gorski J, Eckels DD, Tiercy J, Ucla C, Mach B. Sequence analysis of the DRw13 beta chain genes: The Dw19 specificity may be encoded by the DR beta III locus. In: Dupont B (ed): Immunobiology of HLA, Volume I: Histocompatibility Testing 1987. New York: Springer-Verlag, 1989.
7. Kaplan C, Morel MC, Herman P, Tkaczuk J, Naskatsuji T, Fernandez N, et al. DQw1 (DQw5, DQw6): 2-D gel patterns. In: Dupont B (ed): Immunobiology of HLA, Volume I: Histocompatibility Testing 1987. New York: Springer-Verlag, 1989.
8. Bignon JD, Semana G, Tiercy JM, Simons M, Lalouel JM, Cohen D. DNA-RFLP Analysis: HLA-DR beta workshop report. In: Dupont B (ed): Immunobiology of HLA, Volume I: Histocompatibility Testing 1987. New York: Springer-Verlag, 1989.
9. Milford EL, Kennedy LJ, Yang SY, Dupont B, Lalouel J-M, Yunis EJ. Serologic characterization of the reference panel of B-lymphoblastoid cell lines for factors of the HLA system. In: Dupont B (ed): Immunobiology of HLA, Volume I: Histocompatibility Testing 1987. New York: Springer-Verlag, 1989.
10. du Toit E, Oudshoorn M, Martell RW, MacGregor KJ. DRw6 and its complexity. In: Dupont B (ed): Immunobiology of HLA, Volume I: Histocompatibility Testing 1987. New York: Springer-Verlag, 1989.
11. Lee K, Johnson AH, Tang T, Steiner N, Robbins FM, Hartzman R, et al. DRw6 polymorphism in American blacks. In: Dupont B (ed): Immunobiology of HLA, Volume II: Immunogenetics and Histocompatibility. New York: Springer-Verlag, 1989.

DR7, DQw2, and DRw53: Antigen Report of the Cellular Studies of the Tenth International Histocompatibility Workshop

N. Flomenberg,[1] M. Sasportes,[2] and J. Klein[1]

HLA-DR7 has been subdivided into two official and one unofficial subtypes. Dw11 was the first subtype split off from the large Dw7 group. This haplotype is in linkage disequilibrium with Bw57 and DQw3. Quite often, though not invariably, the DRB4 (DRw53) gene product is not expressed by cells of this type (1,2). HLA-Dw17 was defined as the complementary split of Dw7 (versus Dw11). Cells of this type are frequently associated with Bw44 and express DQw2 and DRw53 products. DB1, often associated with HLA-B13, would technically be considered as part of Dw17. Cells typed as DB1 or Dw17 appear to express identical DR and DQ products (3), although they can apparently be distinguished in HTC experiments (E. Mickelson and N. Reinsmoen, personal communications).

Eleven T-cell clones were submitted to the Workshop, recognizing DR7 or the DR7-associated specificities, DRw53 and DQw2 (Figs. 1 and 2). Four clones (7034, 7028, 7030, and 7020) reacted broadly across the cluster, responding to cells expressing all DR7-associated Dw specificities including Dw11, Dw17, and the DB1 clusters. These clones were blocked by anti-DR antibodies and presumably recognize products of the DRB1 gene. The broad reactivity of these clones with all subtypes of DR7 is not surprising, given the lack of polymorphism of the DR β chains on DR7 haplotypes (3–5). One of these four clones also exhibited appreciable reactivity with DRw8 cell lines, including cells expressing the Dw8.1 and Dw8.2 specificities (7028). An additional clone (6057) recognized a subset of DR7 cell lines. This clone also recognized cells expressing DRw16, DRw12, Dw16, and Dw8.3. The molecular basis for this pattern of cross-reactivities cannot presently be deduced from nucleotide sequence analysis of these haplotypes.

Participating Laboratories: US7NFL,[1] FRADDC[2]

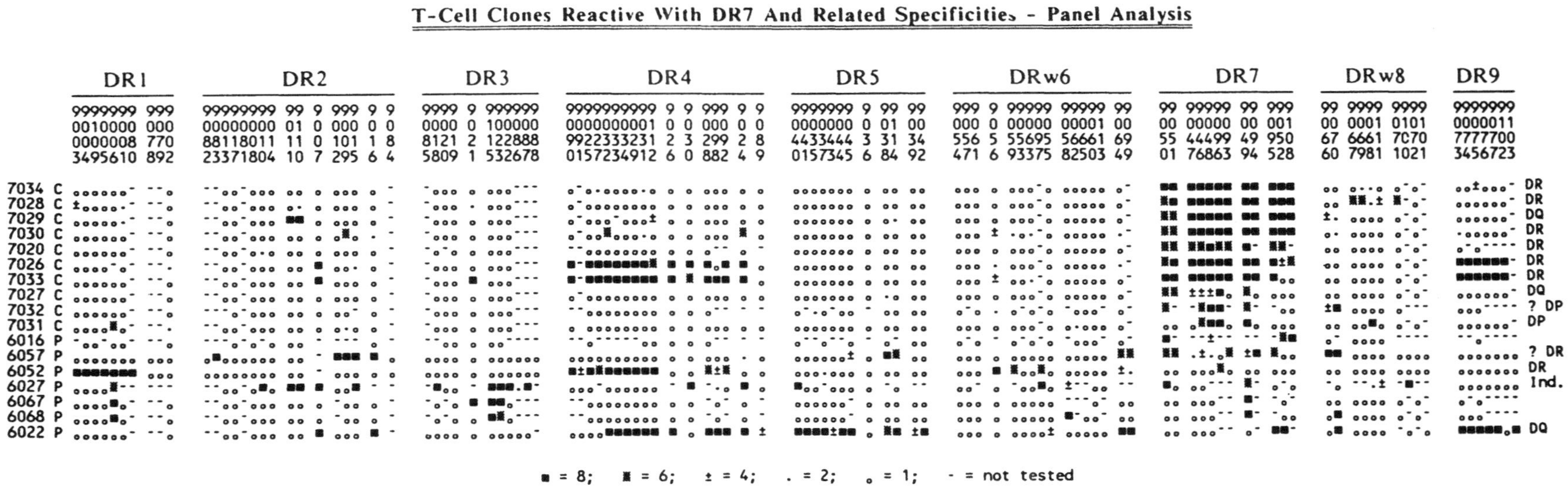

Figure 1. A portion of the cellular panel data relevant to HLA-DR7 is illustrated. The Workshop cell lines have been grouped according to their DR and Dw types. The Workshop number assigned to each clone is listed in the left column. The functional assay in which the clone was tested (cytotoxicity = C or proliferation = P) is listed in the second column. The HLA class II family recognized by each clone (as inferred from the monoclonal antibody blocking data) is listed in the right column. The symbols in the figure reflect the following scores: solid square = 8 = positive; shaded rectangle = 6 = probably positive; ± = 4 = weak or uncertain; closed dot = 2 = probably negative; open circle = 1 = negative; — = not tested.

DR7 Directed T-Cell Clones

		DR				?DP	DQ	
		7020	7034	7030	7028	7032	7027	7029
9046	DB1	+	+	+	+	+	+	+
9048	DB1	+	+	+	+	+	+	+
9047	DB1	+	+	+	+	nt	+	+
9049	Dw17	+	+	+	+	+	+	+
9050	Dw17	+	+	+	+	+	+	+
9051	Dw17	+	+	+	+	nt	+	+
9052	Dw11	+	+	+	+	-	-	+
9108	Dw11	nt	+	+	+	-	-	+
					8.1 8.2	8.3		Dw12 9066 DQ1.12

Figure 2. The reactivities of seven T-cell clones toward eight HLA-DR7+ Workshop cell lines are illustrated. "+" denotes recognition of the cell line by the clone (scores of 6 or 8), while "–" indicates lack of recognition (scores of 1 or 2). "nt" indicates combinations which were not tested or excluded from the central analysis. The molecular families recognized, as inferred from the antibody blocking studies, are shown above the clones. Other DR types recognized by the clones are listed at the bottom of the figure.

Two clones (7027 and 7029) recognize DQ specificities associated with HLA-DR7. Neither clone exhibits appreciable reactivity with DR3 cell lines, which also type as DQw2. One of the clones (7029) reacted with all DR7 lines and exhibited an interesting cross reactivity with Dw12. This clone was clearly inhibited by anti-DQ but not anti-DR monoclonal antibodies. The clone reacts equally well with cells typing as DR7;DQw2 or DR7;DQw3;Dw11. This would suggest that clone 7029 recognizes an epitope dominantly influenced by the DQ α chain, which appears to be similar, if not identical, for all DR7-associated subspecificities, in contrast to the DQ β chain which is clearly different between the DR7;DQw2 and DR7;DQw3 haplotypes (3,6). The second DQ-specific clone (7027) exhibited a narrower pattern of reactivity. This clone specifically failed to recognize those DR7 cell lines expressing DQw3, suggesting it recognized an epitope to which the DQ β chain also contributes. However, it also failed to recognize two DR7 cell lines typing as DQw2 and all HLA-DR3;DQw2 cell lines. Since the DQ β chains from DR3 and DR7 haplotypes exhibit identical sequences (3), this clone most likely recognizes a combinatorial epitope produced by the unique DQ α chain of DR7 haplotypes, together with the DQw2 β chain. The DR7;DQw3;Dw11 cell lines were appropriately recognized by DQw3 specific clones (7).

Two additional clones also detected DR7-related specificities. One clone (7032) reacted with cells typing as either Dw17 or DB1, but not Dw11. This clone also responded to the two Workshop cells typing as Dw8.3. An additional clone (7031) reacted primarily with cells of the DB1 subtype. Interestingly, these clones exhibited unusual patterns of blocking by monoclonal antibodies, being inhibited primarily by broadly reactive anti-HLA class II antibodies or by anti-DP reagents. Several explanations exist for this observation. Although these patterns could not be clearly correlated with the DP types of the cell lines, DNA sequence analyses have identified DP polymorphisms extending well beyond the presently defined DP alleles (8). Moreover, Rodriguez de Cordoba and associates (9) have identified surprisingly strong linkage disequilibrium between DPβb-DPαb and DR7; DQw2 ($p = 0.00017$). Thus, DP polymorphisms might contribute to the heterogeneity of the DR7 cell lines observed here. Alternatively, these clones may be recognizing DR or DQ peptide fragments being presented by DP molecules, or DQ-DP (or less likely DR-DP) heterodimers. In this context, it is noteworthy that the existence of DRα–DQβ hybrid products has been demonstrated. The expression of these heterodimers is dependent on the ratios of the various α and β transcripts (10). Thus, a precedent for this type of heterodimer in human B-LCLs has indeed been demonstrated. Taken together, these findings suggest that heterogeneity exists within DR7 haplotypes, which cannot be readily explained on the basis of DR β or DQ β sequence variation and which, therefore, forces us to consider novel models for an explanation.

HLA-DRw53 determinants were recognized by two Workshop clones (7026 and 7033) (Fig. 1). These cells reacted broadly with cells expressing HLA-DR4, although they both failed to recognize cell line 9089 (BOB). This cell line was not studied in either the Workshop RFLP or protein chemistry sections, but failed to react with a DRw53-like monoclonal antibody (109d6) in a preworkshop analysis (11), and was subsequently retyped as DRw11;DRw52. These two clones also recognized all six DRw9 cell lines against which they were tested. The broad reactivity of these clones is consistent with the relative lack of sequence variation in the DR β-IV product (although a variant sequence has recently been identified [J. Gorski, personal communication]). These clones do, however, differ in their fine specificity. Clone 7033 failed to lyse cell line 9052/9108 expressing the Bw57;DR7;Dw11;DQw3 haplotype. This haplotype frequently fails to express a DR β-IV molecule and fails to react with monoclonal antibodies such as 109d6 (1,11). It can be recognized by a RFLP detected using β chain probes, which is actually 3′ to the DRB1 gene, rather than associated with the DRB4 gene itself (2). Other Bw57;DR7 homozygous cell lines appear to express a DRβ-IV molecule and are heterozygous with respect to this RFLP (V. Sutton and R. Knowles, personal communication). In contrast, clone 7026 recognizes cell line 9052/9108, although with weaker lytic activity than exhibited toward the other DRw53-associated haplotypes. Since this cell line is known to lack expression of a DR β-IV molecule, this would suggest that clone 7026 may additionally interact with a cross-reactive epitope expressed on the DR β-I molecule. This is consistent with the observation that some DRw53 monoclonal antibodies react with both the DR β-I and the DR β-IV chains on DR7 cell lines (1). This reactivity may be influenced by the R → Q substitution at residue 4, common to DR7, DRw9, and DRw53 β chains.

In summary, heterogeneity exists with the DR7 cluster, which cannot be explained on the basis of known sequence variation. A variety of novel mechanisms, therefore, require consideration to explain the heterogeneity of DR7 associated haplotypes.

References

1. Knowles RW, Flomenberg N, Horibe K, Winchester R, Radka SF, Dupont B. Complexity of the supertypic HLA-DRw53 specificity. J Immunol 1986;137:2618–2826.
2. Sutton VR, Knowles RW. Mapping of a restriction fragment length polymorphism associated with defective DRβ4 chain expression to the HLA-DRB1 gene. Hum Immunol 1988;22:123–134.
3. Gregersen PK, Todd JA, Erlich HA, Long E, Servenius B, Choi E, et al. First domain sequence diversity of DR and DQ subregion alleles. In: Dupont B (ed): Immunobiology of HLA, Volume I: Histocompatibility Testing 1987. New York: Springer-Verlag, 1989.
4. Erlich HA, Scharf SJ, Long CM, Horn GT. Analysis of isotypic and allotypic sequence variation in the HLA-DR beta region using the in vitro enzymatic amplification of specific DNA segments. In: Dupont B (ed): Immunobiology of HLA, Volume I: Histocompatibility Testing 1987. New York: Springer-Verlag, 1989.
5. Winchester R, Matthews E, Amesland F, Mazzilli MC, Fernandez N, Kienzle B, et al. DR7: 2-D gel patterns. In: Dupont B (ed): Immunobiology of HLA, Volume I: Histocompatibility Testing 1987. New York: Springer-Verlag, 1989.
6. Radka S, Fuggle S, Mazzilli MC, Carter C, Rust N, Winchester R, et al. DQw2: 2-D gel patterns. In: Dupont B (ed): Immunobiology of HLA, Volume I: Histocompatibility Testing 1987. New York: Springer-Verlag, 1989.
7. Sharrock C, Lundin K, Zier K, Zeevi A, Thomsen M, Eiermann T, et al. DR4 and DQw3: Antigen report of the cellular studies of the Tenth International Histocompatibility Workshop. In: Dupont B (ed): Immunobiology of HLA, Volume I: Histocompatibility Testing 1987. New York: Springer-Verlag, 1989.
8. Bugawan TL, Horn GT, Hansen JA, Mickelson EM, Angelini G, Ferrara GB, et al. Analysis of HLA-DP allelic sequence polymorphism using the in vitro enzymatic amplification of DPα and DPβ loci. In: Dupont B (ed): Immunobiology of HLA, Volume I: Histocompatibility Testing 1987. New York: Springer-Verlag, 1989.
9. Rodriguez de Cordoba S, Marshall P, Ginsberg-Fellner F, Rubinstein P. Biochemical polymorphism of the DPα and DPβ chains. In: Dupont B (ed): Immunobiology of HLA, Volume II: Immunogenetics and Histocompatibility. New York: Springer-Verlag, 1989.
10. Lotteau V, Teyton L, Turmel P, Charron D. DR alpha-DQ beta: A novel class II molecule whose expression is influenced by the ratio of alpha-beta chains transcripts. In: Dupont B (ed): Immunobiology of HLA, Volume II: Immunogenetics and Histocompatibility. New York: Springer-Verlag, 1989.
11. Sanchez B, Winchester R, Nunez-Roldan A. Workshop B-cell reference panel reactivity with monoclonal antibody 109d6. Tenth Workshop Newsletter, No. 4, pp 19–20, 1987.

DRw8 and DQw4: Antigen Report of the Cellular Studies of the Tenth International Histocompatibility Workshop

Y. Tajima,[1] M. Aizawa,[1] D. Eckels,[2] N. Reinsmoen,[3] A. Nikaein,[4] A. Zeevi,[5] M. Honeyman,[6] C. Farrell,[6] and N. Flomenberg[7]

DRw8 was recognized in its entirety by a single Workshop clone (6061), a DR-specific clone, which also recognizes DRw16 (Fig. 1). The Dw8.1, 8.2, and 8.3 subsets could each be defined using combinations of Tenth Workshop clones. The Dw8.1 subgroup was recognized by two clones (7007 and 6040) that exhibited reactivity only with Dw8.1 cell lines. (Clone 6040 was not tested with Dw8.2 cell lines.) These clones were blocked by anti-DR antibodies, and did not react with Dw8.3 cell lines. The DR β chains of Dw8.1 and Dw8.3 are indistinguishable by two-dimensional gel analysis (1), but differ by a single amino acid (2). Dw8.1 may contain at least two subtypes based on differential recognition of cell line 9068 (BM9) versus cell lines 9067 (BTB) and 9069 (MADURA). Several clones including 6009, 7004, and 7028 (many of which are specific for HLA-DR molecules) differentially recognized these two Dw8.1 groups. In other instances, Dw8.1 appeared to split in a different manner. Clones 7022 and 6012, both blocked by anti-DR antibodies recognized 9067 (BTB) versus 9068 (BM9) and 9069 (MADURA), respectively.

Dw8.2 cell lines were frequently recognized together with other DRw8 subsets. Clones 6058 and 6012 (both anti-HLA-DR specific) recognized Dw8.1 + Dw8.2. Clone 6072 recognized Dw8.2 + Dw8.3. These patterns were surprising since the DR β chains from the Dw8.2 subset were readily distinguished from their Dw8.1 and Dw8.3 counterparts (1).

Dw8.3 (but not Dw8.1 or Dw8.2) was recognized by two clones, 6057 and 7032. Both of these clones exhibited appreciable cross-reactivity with DR7 cell lines, and clone 6057 also recognized Dw16, DB6, and DRw16. The two Dw8.3 cell lines were differentially recognized by several clones, including 6093, 6022, 6053, and 7029. Several of these clones recognize HLA-DQ molecules, which may explain some of the disparities between these two Dw8.3 cell lines, since 9066 (TAB089) was typed as DQw7 in the Workshop while 9070 (LUY) was typed as DQw6.

Clones 7001 and 7002 recognized HLA-DQ molecules on cell lines typing as Dw8.1 or 8.2 and thus appear to recognize DQw4. In local studies, these clones recognize cells typing as DRw8;DQw4, but fail to lyse cell lines typing as DR4;Dw15;DQw4. Clone 7002 (but not 7001) also recognized cell line 9021, which expresses the DR3;DRw18;DQw4 haplotype. The first domain of the DQ α chain of the DR3;DQw4 cell line RSH is identical to that seen in other DR3 and DR5 cell lines, with two exceptions (residue 69 = threonine/leucine, and residue 75 = isoleucine/serine). The DQ β chain sequence is identical to that seen in Dw15 cell lines, except for an arginine/leucine substitution at position 23. An identical heterodimer was identified in a DRw8 cell line (3). Clones 7001 and 7002, which distinguish between DR4;Dw15;DQw4 and DRw8;DQw4, may recognize the substitution at position 23 and/or the contribution of the α chain, which in Dw15 cell lines is comparable to that seen in DQw7 cell lines. DQw4 was further split by the DQ-specific clone 6060, which proliferated when stimulated with DR1 cell lines, with DRw8;DQw4 cell lines typing as Dw8.1 but not Dw8.2, and weakly with the DR3;DRw18;DQw4 cell line (9021, RSH). Additional clones of this type have been reported (4). DQ α and β gene products from Dw8.1 and Dw8.2 cell lines appear to exhibit identical patterns in two-dimensional electrophoresis in the Tenth Workshop protein chemistry studies (5). Differential recognition of DQw4 products between the Dw8.1 or Dw8.2 subtypes is, therefore, likely to reflect either 1) conservative amino acid substitutions undetected in the protein chemistry studies, 2) presentation of DR β peptide fragments (for which Dw8.1 and 8.2 cell lines differ) by the DQ molecules, or 3) recognition of DRβ–DQα heterodimers analagous to the findings of Lotteau and associates (6).

Participating Laboratories: JAPAIZ,[1] US2MBC,[2] US2BAC,[3] NCYNIK,[4] USIDUQ,[5] ANZBSH,[6] US7NFL[7]

Immunobiology of HLA
Volume I

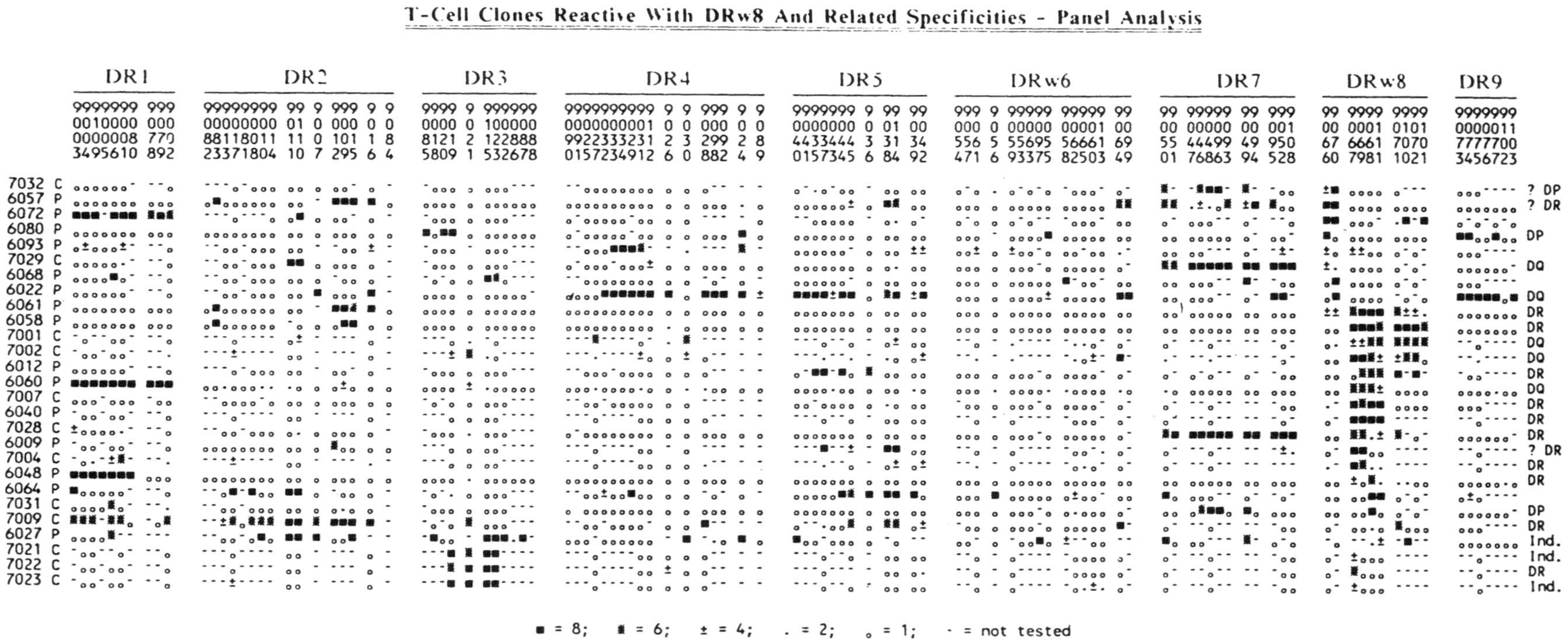

Figure 1. A portion of the cellular panel data relevant to HLA-DRw8 and DQw4 is illustrated. The Workshop cell lines have been grouped according to their DR and Dw types. The Workshop number assigned to each clone is listed in the left column. The functional assay in which the clone is tested (cytotoxicity = C or proliferation = P) is lited in the second column. The HLA class II family recognized by each clone (as inferred from the monoclonal antibody blocking data) is listed in th right column. The symbols in the figure reflect the following scores: solid square = 8 = positiv shaded rectangle = 6 = probably positive; ± = 4 = weak or uncertain; closed dot = 2 = probab negative; open circle = 1 = negative; — = not tested.

References

1. Kulova L, Bontrop R, Yang SY, Maeda H. DRw8: 2-D gel patterns. In: Dupont B (ed). Immunobiology of HLA, Volume I: Immunogenetics and Histocompatibility. New York: Springer-Verlag, 1989.
2. Gregersen PK, Todd JA, Erlich HA, Long E, Servenius B, Choi E, et al. First domain sequence diversity of DR and DQ subregion alleles. In: Dupont B (ed): Immunobiology of HLA, Volume I: Immunogenetics and Histocompatibility, New York: Springer-Verlag, 1989.
3. Hurley CK, Steiner N, Gregersen PK, Silver J, Gorski J, Hartzman R, et al. Evolution of DQw2-related haplotypes. In: Dupont B (ed). Evolution of DQw2-related haplotypes. Immunobiology of HLA, Volume II: Immunogenetics and Histocompatibility. New York: Springer-Verlag, 1989.
4. Tajima Y, Kojima H, Ishikawa N, Hawkin S, Fukasawa Y, Takada A, et al. Cellular analyses of DQ antigens. In: Dupont B (ed). Immunobiology of HLA, Volume II: Immunogenetics and Histocompatibility. New York: Springer-Verlag, 1989.
5. Kulova L, Maeda H, Bontrop R, Rust N, Fuggle S, Carter C, et al. DQw4 (Wa): 2-D gel patterns. In: Dupont B (ed): Immunobiology of HLA, Volume I: Immunogenetics and Histocompatibility. New York: Springer-Verlag, 1989.
6. Lotteau V, Teyton L, Turmel P, Charron D. DR alpha-DQ beta: A novel class II molecule whose expression is influenced by the ratio of alpha-beta chains transcripts. In: Dupont B (ed). Immunobiology of HLA, Volume II: Immunogenetics and Histocompatibility. New York: Springer-Verlag, 1989.

DRw52: Antigen Report of the Cellular Studies of the Tenth International Histocompatibility Workshop

C. Irle,[1] E. Mickelson,[2] S. Masaewicz,[3] J.A. Hansen,[2] M. Thomsen,[3] and A. Termijtelen[4]

The serologically defined DRw52 specificity can be subdivided into four distinct subtypes of reactivity using allospecific and antigen-specific DRw52-restricted T-cell clones and Tenth Workshop homozygous B-lymphoblastoid cell lines as stimulator/antigen presenting cells. Among these four patterns, three are positively identified by the T-cell clones (Fig. 1). The first cluster referred to as LB-Q4 or DR52a (1–4) is associated with some DR3 and DRw6 (DRw13;Dw18 and DRw14;Dw16) haplotypes, and has been officially named Dw24. The second cluster, Dw25, was previously known as LB-Q1 or DR52b (2,4,5). Dw25 is associated with other DR3, all DR5, and other DRw6 (DRw13;Dw18 and DRw14;Dw9) haplotypes. In two instances, B-LCL 9039 and 9061 did not present antigen to the tetanus-specific clone 6034. This was not due to a true failure to react with the APC in the case of B-LCL 9039, since peripheral blood mononuclear cells from the same donor presented antigen effectively to this clone (2). The third cluster, Dw26, was previously designated DRw52c. It has, thus far, been found in some DR5(w11), and DRw6(w13);Dw19 haplotypes (4,6,7). The DRw8-associated DRw52 was not identified by any of the T-cell clones tested, and is therefore distinct from the previous three DRw52 subgroups.

Clone 6030 that defines a DRw6(14),Dw19-associated reactivity referred to as Dw26 (DRw52c) also reacted to two DR5 B-LCL (9037 and 9045), indicating the presence of a Dw26 associated epitope on these cells. Two-dimensional gel analysis of the DR β-III molecules of the DR3, 5, and w6 B-LCL tested with the T-cell clones showed concordance between the expression of a DRβIII-3 chain and Dw24 (LB-Q4/DRw52a) reactivity. There was similar concordance between expression of the DRβIII-2 chain and Dw26 (DRw52c) reactivity. The DRβIII-1 chain correlates well with Dw25 (LB-Q1/DRw52b) (8). The reactivity of clone 6030 with the Dw25 cell lines 9037 and 9045 may reflect additional microheterogeneity within Dw25 which has been observed at the DNA level (6, J. Gorski, personal communication).

Polymorphism within DRw52 was detected by DRβ RFLP studies and was previously correlated to the Dw25 (LB-Q1) subgroup of DRw52 (5). More recently, the Dw24-, Dw25-, and Dw26-associated subgroups were shown to correlate respectively with different DR β-III alleles (9,10) that can now also be identified using locus and allele specific oligonucleotide probes (9,11). These probes fail to hybridize with DRw8 haplotypes, which may relate to the presence of only one DR β gene in DRw8 cells, as previously suggested (12). Similarly, the Workshop RFLP analysis of the B-LCLs that were tested with the T-cell clones showed agreement between two RFLP fragments (a 272 base fragment using the restriction enzyme KPN and a 107 base fragment with Pst) and Dw24 T-cell clone reactivity. One RFLP fragment (an 1148 base TaqI fragment) correlated with Dw25. The relation between RFLP fragments and Dw26 is less clear cut, since the three fragments (270 base fragment of Hind III, 1375 base fragment of Sst I, and 164 base fragment of MspI) are associated with both Dw26 and Dw19. They may more specifically be associated with Dw19 than with Dw26.

Informative blocking results were obtained using monomorphic anti-DR antibodies, except for clone 6030

Participating Laboratories: FRAJEA,[1] US7HAN,[2] FRAOHA,[3] BENROO[4]

Immunobiology of HLA
Volume I

DRw52 T-Cell Clones

DR	3			13		14		3		11				12	13	14	13		8		
Dw	3	3	"N"	18		16		3	3	5			"N"	B6	18	9	19		8.1	8.2	8.3
WS#	9022	9023	9021	9062	9065	9064	9099	9019	9020	9037	9045	9039	9042	9038	9060	9061	9063	9059	9068	9071	9066
Clone																					
6081	+	+	+	+	+	+	+	-	-	-	-	-	-	-	-	-	-	-	-	-	-
C6	+	+	+	+	+	+	+	-	-	-	-	-	-	-	-	-	-	-	-	-	-
16	+	+	nt	+	+	nt	nt	-	-	nt	nt	nt	nt	nt	nt	nt	nt	nt	nt	nt	nt
6032	-	-	-	-	-	-	-	+	-	+	+	-	+	+	+	+	-	-	-	-	-
FC40	-	-	-	-	-	w	-	+	+	+	+	+	+	+	+	+	+	w	-	-	-
6077	-	w	w	-	-	-	-	+	+	+	+	+	+	+	+	+	-	-	-	-	-
6034	-	-	-	-	-	-	nt	+	+	+	+	w	+	+	+	-	nt	-	-	-	-
6015	-	-	-	-	-	-	nt	+	w	+	+	nt	nt	+	+	-	nt	-	-	-	-
6045	nt	-	-	-	-	-	nt	-	nt	-	-	-	-	-	-	-	nt	+	-	-	-
6030	-	w	-	-	-	-	-	-	nt	w	+	-	-	-	-	-	+	+	-	-	-

RFLP

Kpn I 272 nt
Pst I 107 nt
Taq I 1148
Hind III 270
Sst I 1375
MspI 164

2-D Gel Pattern

III - 3
III - 1
III - 2

Figure 1. The reactivities of 10 T-cell clones toward 21 HLA-DRw52+ Workshop cell lines are illustrated. "+" denotes recognition of the cell line by the clone (scores of 6 or 8). "w" indicates a weak or uncertain response (score of 4), while "–" denotes lack of recognition (scores lf 1 or 2). "nt" indicates combinations that were not tested or excluded from the central analysis. The three cellularly defined subclusters are indicated by the boxes. In the middle section of the figure, RFLP patterns of the cell lines using the Workshop DRβ probe are shown, which correlate with the cellularly defined clusters. In the bottom portion of the figure, the results of the protein chemistry studies are shown, which also correlate with the cellular findings. Several DR5 cell lines exhibit DR βIII-1 and DR βIII-2 products, consistent with their recognition by Dw25-specific clones as well as the Dw26-specific clone 6030.

which was not inhibited (Fig. 2). Neither DQ nor DP mAb inhibited the clones significantly (i.e, 80% of inhibition) at any dilution of ascites. Among the polymorphic monoclonal antibodies, the DR3 directed antibody JS1 inhibited the DR3, but not the DRw52 directed clones. Similarly, UK8.1 (a DRw52-like monoclonal antibody) blocked DR3, but not the DRw52 reactive clones, suggesting that this antibody binds to a DRw52 epitope on the DR β-I molecule. In contrast, the DRw52-directed monoclonal antibody NDS10 and NDS13 inhibited only the Dw25-specific clones, while the DRw52 antibody 7.3.19.1 blocked both DR3 and DRw52-specific clones, suggesting it may bind to both DR β-I and DR β-III molecules. Together, the Workshop results show a correlation between the DRw52 series defined by T-cell clones, two-dimensional gel analysis, and the RFLP studies, and map the epitopes defined by the clones of this Workshop component to a DRB3 gene product.

References

1. Irle C, Jaques D, Tiercy JM, Jeannet M, Mach B. Functional and structural analysis of HLA-DRw52. In: Dupont B (ed): Immunobiology of HLA, Volume II: Immunogenetics and Histocompatibility. New York: Springer-Verlag, 1989.
2. Irle C, Jaques D, Tiercy JM, Fuggle SV, Gorski J, Termijtelen A, et al. Functional polymorphism of each of the two HLA-DR beta-chain loci demonstrated with antigen-specific DR3 and DRw52 restricted T-cell clones. J Exp Med 1988;167:853.

T Cell Clone Inhibition by Moabs

	DR				DQ	DP	DRw52				DR3	
	GSP41	L243	PL8	A1	A2	A3	73191	NDS10	NDS13	UK81	JS1	
6081	+	+	+	+	-	-	+	-	-	-	-	Dw24
C6	+	+	+	+	-	-	+	-	-	-	-	
C16	+	nt	nt	nt	-	-	+	-	nt	nt	nt	
6032	+	+	+	+	-	-	+	+	+	-	nt	Dw25
6034	+	+	+	+	-	-	+	+	+	-	-	
6015	+	nt	+	nt	nt	-	+	nt	nt	nt	nt	
6045	-	+	+	+	-	-	nt	nt	nt	nt	nt	Dw26
6030	-	-	-	-	-	-	+	-	-	-	nt	
6082	+	+	+	+	-	-	+	-	-	+	+	DR3
6033	+	+	+	+	-	-	+	-	-	+	+	
6014	+	+	+	+	-	-	+	-	-	+	+	

Figure 2. The patterns of inhibition of eleven T-cell clones by 11 monoclonal antibodies are shown. The specificities of the antibodies are shown along the top of the figure, while the specificities of the clones are shown in the right column. "+" denotes inhibition of the clone by the antibody, while "–" denotes lack of inhibition. "nt" indicates combinations which were not tested or excluded from the central analysis.

3. Termijtelen AM, Robbins FM, Schroeijers WEM, Jacobs VPG, Gorski J, Tosi R. Subtypes of DRw52: Different typing techniques reveal very similar but distinct typing results. In: Dupont B (ed): Immunobiology of HLA, Volume II: Immunogenetics and Histocompatibility. New York: Springer-Verlag, 1989.
4. Sheehy MJ, Rowe JR, Konig F, Jorgensen L. Functional polymorphism of the HLA-DR beta III chain. Hum Immunol 1988;21:49–61.
5. Termijtelen A, Tilanus MGJ, Engelen I, Koning F, van Rood JJ. Molecular localization of LB-Q1, a DRw52-like recognition epitopic and identification at the genomic level of associated shared hybridizing fragments. Hum Immunol 1987;19:255.
6. Mickelson EM, Masewicz SA, Hansen JA. Variants of HLA-DRw52 defined by proliferative T-lymphocyte clones. In: Dupont B (ed): Immunobiology of HLA, Volume II: Immunogenetics and Histocompatibility. New York: Springer-Verlag, 1989.
7. Michon JM, Henin Y, Sterkers G, Freidel C, Gebuhrer H, Betuel H, et al. Highly polymorphic products of both HLA-DR and HLA-DQ genes contribute to the polymorphism of the HLA-DRw13 haplotype. Immunogenetics 1986;24:177.
8. Bontrop R, Ballas M, Madrigal A, Hermans P, Fuggle S, Otting N, et al. DRw52: 2-D Gel Patterns. In: Dupont B (ed): Immunobiology of HLA, Volume I: Histocompatibility Testing 1987. New York: Springer-Verlag, 1989.
9. Gorski J, Mach B. Polymorphisms of human Ia antigens: Gene conversion between two DR beta loci result in a new HLA-D/DR specificity. Nature 1986;322:67.
10. Gorski J, Eckels DD, Tiercy JM, Ucla C, Mach B. Sequence analysis of the DRw13 beta-chain genes: the Dw19 specificity may be encoded by the DR alpha beta III locus. In: Dupont B (ed): Immunobiology of HLA, Volume II: Immunogenetics and Histocompatibility. New York: Springer-Verlag, 1989.
11. Tiercy JM, Gorski J, Jeannet M, Mach B. Identification of three serologically undetected alleles of HLA-DR by oligonucleotide DNA typing analysis. Proc Natl Acad Sci 1988;85:198.
12. Bohme J, Andersson M, Andersson G, Moller E, Peterson PA, Rask L. J Immunol 1985;135:2149.

DQw1: Antigen Report of the Cellular Studies of the Tenth International Histocompatibility Workshop

N. Reinsmoen,[1] D. Eckels,[2] E. Mickelson,[3] M. Honeyman,[4] C. Farrell,[4] J.A. Hansen,[3] A. Termijtelen,[5] and N. Flomenberg[6]

In the course of the Tenth Workshop, DQw1 was split into two specificities, DQw5 and DQw6 distinguished by alloantisera (Nomenclature for Factors of the HLA System, elsewhere in this volume). DQw5 is associated with DR1, DRw16;Dw21, DRw14;Dw9, and DRw10. Cells typing as DRw13;Dw18;*Dw25* (but not Dw24) may also express DQw5, although this requires further confirmation on a larger number of cells. The single cell line typing as DR5;DB2 typed as DQw6, but appeared more like DQw5 biochemically and cellularly. DQw6 is associated with DRw15;Dw2, DRw13;Dw18;*Dw24*, DRw13; Dw19; Dw26, and probably with DRw15;Dw12. The single DRw8;Dw8.3;DQw1 cell line typed as DQw6 as well (1,2).

Cellular analyses, RFLP, protein chemistry, and cDNA sequence have revealed DQw1 to be more complex than this two-part split. Nine DQw1-related patterns of reactivity were demonstrated by the Tenth International Workshop clones. The reactivities of nine Workshop clones and three additional clones are shown in Figure 1. All of these clones were inhibited by anti-HLA-DQ, but not anti-HLA-DR antibodies. None of these clones recognized the entire DQw1 specificity. In some cases, the clones recognized subsets of DQw1, which were concordant in their distribution with specific DR or Dw types. These included clones specific for DR1;Dw1+Dw20 (clone 6060), DR2(DRw15);Dw2 (clone 7019), DR2 (DRw15);Dw2+Dw12 (clone 6011), DRw6(DRw13); Dw18 (clones 6031 and 6084), and DRw6(DRw13); Dw19 (clone 6029). The reactivity of clone 6029 could be correlated with a unique set of DQβ RFLPs in Dw19 cell lines, designated DQB1.6 (3). Clones recognizing HLA-DQw1-specific determinants whose expression was limited to DR2(DRw15);Dw12, DR2(DRw16); Dw21, DRw6(DRw14);Dw9, and DR5(DRw11);DB2 haplotypes were not identified within the Workshop. These haplotypes were recognized, however, by more broadly reactive clones, as illustrated in Figure 1. It should also be noted that the clones 6084 and 6031 reacted with all Dw18 cell lines, including cell line 9060 which, unlike the others, was serologically typed as DQw5 rather than DQw6.

Clone 6008 exhibited a pattern of recognition similar to the serologically defined specificity DQw5 in that it primarily recognized cells typing as DR1, DRw16, and DRw14;Dw9. It also recognized the single DRw11;DB2 cell line and two cell lines typing as DRw13;DQw6. These DQw5 haplotypes exhibit identical DQ β sequences from residues 58 through 78 and a common set of RFLPs designated DQB1.3 (3,4). This clone failed to react with several DQw5 cell lines in the DR1 group and exhibited some reactivity outside the cluster, raising the question of microheterogeneity within these DQw1 subtypes. Several clones identified a subset of DQw6 in that they recognized most cell lines typing as DR2(DRw15) and DR13;Dw18, but did not react with DRw13;Dw19 cell lines. This pattern can be correlated with an identical DQ β sequence from residues 31 through 78 in Dw2 and Dw18 haplotypes as well as a common set of RFLPs designated DQB1.5 (3,4). Clone EMC4 recognized a broadly reactive DQw1-associated specificity including parts of both DQw5 and DQw6. This clone reacted with cells expressing DR1;DQw5, DR2(DRw15);Dw2;DQw6, DR13;Dw18;DQw6, and DRw14;Dw9;DQw5. It failed to recognize cells expressing DR2(DRw16);Dw21, DRw13;Dw19;DQw6 and the single Workshop cell lines expressing either DR2 (DRw15);Dw12;DQw6 or DRw11;DB2.

In the case of HLA-DQ molecules, both α and β chain polymorphisms can contribute to the determinants recognized by T cells. Given the detection of at least five different α and seven different β chains by two-dimensional gel analysis (1), it is not surprising that such a large number of broadly distributed, as well as private epitopes, are recognized by the T-cell clones. The haplotype-specific DQw1 subsets likely reflect recognition of combinatorial epitopes produced by particular α-β dimers or, alternatively, HLA-DQ presentation of DR β peptide fragments. Analyses of the patterns of reactivity of the more broadly reactive clones can in some cases be correlated with the presence or absence of particular RFLPs and/or with regions of shared amino acid sequence. In other cases, similar correlations are not readily apparent. These latter patterns of reactivity are most easily explained through recognition of combinatorial epitopes which can be produced by subsets of DQw1 α-β heterodimers.

Participating Laboratories: US2BAC,[1] US2MBC,[2] US7HAN,[3] ANZBSH,[4] BENROO,[5] US7NFL[6]

DQw1 Subtypes

WS#	DR	Dw	7019	6011	EME5	7016	7015	EMC4	6098	NR91	6660	6684	6031	6029	DQ	DQβ	DQα
			T-Cell Clones														
9017	2	2	+	+	+	+	+	+	-	-	-	-	-	-			
9018	2	2	+	+	+	+	+	+	-	-	-	+	-	-			
9010	2	2	+	-	+	+	+	+	-	-	-	-	-	-	w6	1.2	1.2
9013	2	2	+	+	+	+	+	+	-	-	-	-	-	-			
9014	2	2	+	+	nt	+	+	nt	-	-	-	-	-	-			
9011	2	12	-	+	+	-	-	-	-	-	-	-	-	-	w6	1.12	1.18
9065	13	18	-	-	+	+	+	+	-	-	-	+	+	-			
9062	13	18	-	-	+	-	-	+	-	nt	-	+	+	-	w6	1.18	1.18
9060	13	18	-	-	+	+	+	+	-	-	-	+	+	-			
9058	13	18	-	-	-	-	-	-	-	nt	-	+	+	-			
9001	1	1	-	-	-	-	-	+	+	+	+	-	-	-			
9002	1	20	-	-	-	-	-	+	+	nt	+	-	-	-			
9003	1	1	-	-	-	-	-	-	w	+	+	-	-	-	w5	1.1	1.1
9004	1	1	-	-	-	-	-	+	+	+	+	-	-	-			
9005	1	1	-	-	-	-	-	+	-	nt	+	-	-	-			
9006	1	1	-	-	-	-	-	+	-	nt	+	-	-	-			
9009	2	21	-	-	-	-	-	-	+	+	-	-	-	-			
9012	2	21	-	-	nt	-	-	-	+	+	-	-	-	-	w5	1.21	1.2
9015	2	21	-	-	nt	-	-	-	+	+	-	-	-	-			
9059	13	19	-	-	-	-	-	-	-	nt	-	-	-	+			
9097	13	19	-	-	-	-	-	-	-	nt	-	-	-	+	w6	1.19	1.2
9063	13	19	-	-	-	-	-	-	+	nt	-	-	-	+			
9053	13	19	-	-	-	-	-	-	-	nt	-	-	-	+			
9055	13	19v	-	-	-	-	-	-	+	nt	-	-	-	-			
9054	14	9	-	-	-	-	-	+	+	nt	-	-	-	-			
9061	14	9	-	-	-	-	-	-	+	+	-	-	-	-	w5	1.9	1.14
9057	14	9	-	-	w	-	-	+	-	nt	-	-	-	-			
9036	11	DB2	-	-	+	-	-	-	+	nt	-	-	-	-	w1	1.12*	1.2

Figure 1. A portion of the cellular panel data relevant to HLA-DQw1 is illustrated. The Workshop cell lines have been grouped according to their DR and Dw types. The serologic DQw1 subtypes (i.e., DQw5 or DQw6) assigned to the cell lines and the results of the biochemical analyses of the DQ α and DQ β chains are listed along the right side of the figure. "+" denotes recognition of the cell line by the clone (scores of 6 or 8). "w" indicates a weak or uncertain response (score of 4), while "−" denotes lack of recognition (scores of 1 or 2). "nt" indicates combinations that were not tested or excluded from the central analysis.

References

1. Kaplan C, Morel MC, Hermans P, Tkaczuk J, Naskatsuji T, Fernandez N, et al. DQw1 (DQw5, DQw;6): 2-D gel patterns. In: Dupont B (ed): Immunobiology of HLA, Volume I: Histocompatibility Testing 1987. New York: Springer-Verlag, 1989.
2. Milford EL, Kennedy LJ, Yang SY, Dupont B, Lalouel J-M, Yunis EJ. Serologic characterization of the reference panel of B-lymphoblastoid cell lines for factors of the HLA system. In: Dupont B (ed): Immunobiology of HLA, Volume I: Histocompatibility Testing 1987. New York: Springer-Verlag, 1989.
3. Trucco M, Ball E. RFLP analysis of DQ-beta chain gene: Workshop report. In: Dupont B (ed): Immunobiology of HLA, Volume I: Histocompatibility Testing 1987. New York: Springer-Verlag, 1989.
4. Gregersen PK, Todd JA, Erlich HA, Long E, Servenius B, Choi E, et al. First domain sequence diversity of DR and DQ subregion alleles. In: Dupont B (ed): Immunobiology of HLA, Volume I: Histocompatibility Testing 1987. New York: Springer-Verlag, 1989.

DP: Antigen Report of the Cellular Studies of the Tenth International Histocompatibility Workshop

A. Urlacher,[1] Y. Mitsuishi,[1] and M. Tongio[1]

Several clones recognizing HLA-DP specificities were included in the Tenth Workshop cellular studies. Five clones recognized specificities associated with DPw1 (6067 and 6068), DPw2 (6064), and DPw3 (6066 and 6069). Overall, these clones exhibited good concordance with the typings established by bulk-cultured reagents. In most of the cell lines missed by these clones, either DPα or DPβ specific RFLPs were detected capable of differentiating the atypical cell lines from other cell lines of similar DP type. Since the polymorphism of DP molecules extends well beyond the presently defined specificities (1,2), structural explanations for these DP splits are likely to be forthcoming. The details of this analysis are presented in detail by Urlacher et al. (3) in Volume II of this book.

Participating Laboratory: FRAMYE[1]

References

1. Bugawan TL, Horn GT, Hansen JA, Mickelson EM, Angelini G, Ferrara GB, et al. Analysis of HLA-DP allelic sequence polymorphism using the in vitro enzymatic amplification of DPα and DPβ loci. In: Dupont B (ed): Immunobiology of HLA, Volume I: Histocompatibility Testing 1987. New York: Springer-Verlag, 1989.
2. Rodriguez de Cordoba S, Marshall P, Ginsberg-Fellner F, Rubinstein P. Biochemical polymorphism of the DPa and DPb chains. In: Dupont B (ed): Immunobiology of HLA, Volume II: Immunogenetics and Histocompatibility. New York: Springer-Verlag, 1989.
3. Urlacher A, Mitsuishi Y, Tongio M. T-cell clones show polymorphism within DPw1, DPw2 and DPw3 specificities. In: Dupont B (ed): Immunobiology of HLA, Volume II: Immunogenetics and Histocompatibility. New York: Springer-Verlag, 1989.

T-cell Clones Recognizing Shared or Cross-Reactive Epitopes: Antigen Report of the Cellular Studies of the Tenth International Histocompatibility Workshop

D. Eckels,[1] T. Sell,[1] K. Zier,[2] E. Mickelson,[3] J.A. Hansen,[3] N. Flomenberg,[4] and A. Nikaein[5]

Interesting T-cell reactivity was observed with specificities shared by DR1-, DR4-, and DRw6-associated gene products (Fig. 1). Clone 6025 was primed against Dw14, yet recognized all of the HLA-DR1 + Workshop B-LCLs against which it was tested and two Dw16 cell lines (9064 and 9099). The determinant recognized by 6025 correlated with an oligonucleotide probe corresponding to amino acids 68 to 74 of HLA-DR4 β-I (1). An independently derived T-cell clone (6046) produced in a DR1-specific priming, recognized the LCL panel with a similar specificity pattern as 6025, although the reactivity with Dw16 was weaker. A total of six Workshop clones exhibited cross-reactivity between DR1 and Dw14, while seven exhibited cross-reactivity between DR1 and Dw16. Therefore, it appears that epitopes are shared by DR1, Dw14, and Dw16 class II molecules, which T-cells of independent origin are capable of recognizing with appreciable frequency. Additional patterns within this group include clone 6051, which recognized all DR1 cell lines and DR4 cell lines expressing either the Dw14 or DKT2 specificity. Clone 6052 reacted with DR1;Dw1 LCLs, DR4;Dw4 + Dw14, and subsets of Dw16 and Dw19.

The sequence of the Dw16 β chain is identical to that of DR1 from residues 38 through 94 (2), which would theoretically incorporate the third and fourth strands of β pleated sheet and the entire α helix. This region is also identical, with one exception (G → V) at residue 86, with the DRB1 sequence in Dw14 cell lines. Thus, from residues 38 through 85, all three sequences are identical. Between residues 38 and 94, Dw14 and Dw16 show fewer differences from Dw1 than does the related DR1; Dw20 haplotype. It is significant that within the third hypervariable region of DRB1, DR1 and Dw14 are identical, whereas Dw4 and Dw13 have single amino acid substitutions at positions 71 (R → Q) and 74 (A → E), respectively, suggesting that the epitope recognized by these clones is closely associated with the third hypervariable region of HLA-DR1, Dw14, and Dw16 β-I chains.

Participating Laboratories: US2MBC,[1] NCYZIE,[2] US7HAN,[3] US7NFL,[4] NCYNIK[5]

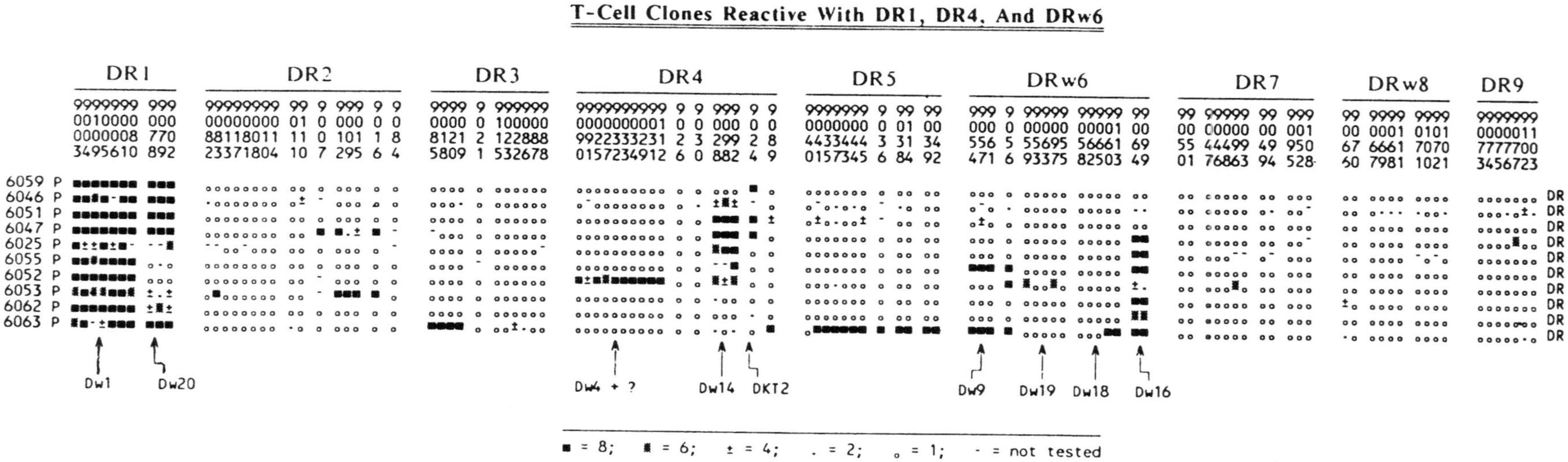

Figure 1. A portion of the cellular panel data relevant to DR1, DR4, and DRw6 cross-reactive specificities is illustrated. The Workshop cell lines have been grouped according to their DR and Dw types. The Workshop number assigned to each clone is listed in the left column. The functional assay in which the clone was tested (cytotoxicity = C or proliferation = P) is listed in the second column. The HLA class II family recognized by each clone (as inferred from the monoclonal antibody blocking data) is listed in the right column. The symbols in the figure reflect the following scores: solid square = 8 = positive; shaded rectangle = 6 = probably positive; ± = 4 = weak or uncertain; closed dot = 2 = probably negative; open circle = 1 = negative; − = not tested.

T-Cell Clones Recognizing DR1 and DRw8

B-LCL	DR	Dw	6048	6072
9080	1	?	+	+
9001	1	1	+	+
9004	1	1	+	+
9006	1	1	+	+
9005	1	1	+	+
9109	1	1	+	+
9003	1	1	+	+
9078	1	20	-	+
9079	1	20	-	+
9002	1	20	-	+
9069	8	8.1	-	-
9067	8	8.1	w	-
9068	8	8.1	+	-
9111	8	8.1	-	w
9100	8	8.2	-	+
9101	8	8.2	-	+
9066	8	8.3	-	+
9070	8	8.3	-	+

Figure 2. The reactivities of two T-cell clones toward 18 HLA-DR1+ or DRw8+ Workshop cell lines are illustrated. "+" denotes recognition of the cell line by the clone (scores of 6 or 8). "w" indicates a weak or uncertain response (score of 4), while "−" denotes lack of recognition (scores of 1 or 2).

The MC1 serologic specificity has been described as common to subsets of DR1 and DR4 (3). T-cell clones with similar specificity associations suggest a high degree of heterogeneity within this serologically defined cross-reactive group. However, this must be stated cautiously since the reagents described here are clonal, while alloantisera may consist of multiple diverse antibody subpopulations, some of which might individually recapitulate the clonal specificities described here.

Other examples of shared or cross-reactive epitopes are illustrated by the Workshop clones. Clones recognizing epitopes shared between DR2, DR5, and DRw8 cell lines are described in detail by Reinsmoen and Bach (4). Their findings represent additional examples where "cross-reactivities" can be explained by shared sequences (DR βI residues 26 to 31 and 67 to 74). Indeed, it may be more proper to think of shared rather than cross reactive epitopes in these instances.

A specificity common to DR1 and DRw8 was observed with two clones derived independently (6048 and 6072) (Fig. 2). Clone 6048 recognized DR1;Dw20 cell lines and 2 of 4 Dw8.1 B-LCLs (9067 and 9068), whereas clone 6072 recognized all DR1-positive B-LCLs and Dw8.1, 8.2 and 8.3 exclusive of those recognized by clone 6048. One exception was noted in that neither clone reacted with cell line 9069. These data suggest that epitopes are shared by DR1 and DRw8 molecules similar to DR1 and DR4. Other interesting examples include clone 6063, which recognizes a DR1 + Dw25 + Dw16-like specificity, and clone 6100, which recognizes DRw15 and a subset of DR4. In these latter cases, however, shared sequences cannot readily explain the patterns of reactivity observed. These clones are thus more likely to illustrate true examples where two distinct epitopes are likely to be cross-reactive rather than shared between two molecules.

References

1. Seyfried CE, Mickelson E, Hansen JA, Nepom GT. A specific nucleotide sequence defines a functional T-cell recognition epitope shared by diverse HLA-DR specificities. Hum Immunol 1988;21:289.
2. Gregersen PK, Todd JA, Erlich HA, Long E, Servenius B, Choi E, et al. First domain sequence diversity of DR and DQ subregion alleles. In: Dupont B (ed): Immunobiology of HLA, Volume I: Histocompatibility Testing 1987. New York: Springer-Verlag, 1989.
3. Duquesnoy RJ, Marrari M, Hackbarth S, Zeevi A. Definition of MC1, a novel HLA determinant associated with DR1 and DR4. In: Histocompatibility Testing, 1984;426.
4. Reinsmoen NL, Bach FH. Correlation of primary sequence and functional determinants of HLA class II molecules. In: Dupont B (ed): Immunobiology of HLA, Volume II: Immunogenetics and Histocompatibility. New York: Springer-Verlag, 1989.

Functional Polymorphisms of HLA Class II Gene Products Detected by T-Lymphocyte Clones: Summary of the Tenth International Histocompatibility Workshop Cellular Studies

N. Flomenberg[1]

Introduction

During the course of a normal immune response, much of the attention of the T-cell is focused on the MHC molecules expressed on the surfaces of other cells with which the T-cell interacts. As a consequence of this fundamental immunologic property, T-cells are exquisitely sensitive to polymorphisms exhibited by MHC molecules. Cellular reagents in the form of homozygous typing cells and/or primed lymphocyte reagents have been utilized in the Sixth through the Ninth Workshops, as well as in local studies, for the definition of cellularly recognizable polymorphisms of HLA class II molecules. The structural basis for many of these polymorphisms has become apparent as more sophisticated biochemical and molecular genetic tools have been applied to the study of the major histocompatibility complex (MHC).

Along with these other advances in technology, it has become feasible to dissect the process of cellular recognition into its component parts, mediated by individual T-cell clones. In the Tenth International Histocompatibility Workshop, a comprehensive investigation of HLA class II polymorphisms detectable in functional assays was undertaken using T-lymphocyte clones. Together with data from other Workshop components, these analyses have helped to define new HLA class II specificities, and have additionally helped to refine our concepts concerning many of the HLA haplotypes defined in earlier investigations. As a consequence, a more integrated functional picture of the HLA class II region has begun to emerge.

Goals of the Study

The investigations undertaken had two major goals:

1. Immunogenetic Analyses

Definition of HLA class II polymorphisms by T-cell clones. The definition of a cellularly detectable polymorphism has two components. First, the distribution of the specificity in question is characterized either in a population study or through analysis of an appropriately selected cell panel. Second, the specificity must be mapped to the specific HLA class II family (DR, DQ, DP) on which it is expressed and, if possible, to the specific α-β dimers. Mapping is usually performed through monoclonal antibody blocking experiments.

Analysis of cellularly defined polymorphisms in the context of polymorphisms detected by serology, protein chemistry, Southern blot analysis, and cDNA sequencing. Each of these approaches for investigation of HLA class II polymorphism has its own strengths and limitations. Serology may fail to detect polymorphisms recognized by T-cells. Protein chemistry may be insensitive to amino acid substitutions that do not alter charge. Southern blots may detect polymorphisms associated with pseudogenes or flanking sequences, yet may fail to detect the subtle but functionally significant sequence variations which differentiate certain haplotypes from one another. Cellular analyses require the testing of a large number of reagents to ensure that all potential polymorphisms can be recognized. When applied together, these various approaches tend to offset each other's weaknesses and emphasize their mutual strengths.

Utilizing the composite information to reconstruct the structural basis for the various class II haplotypes as they exist within the population. Clearly, epitopes do not exist in isolation, but must be viewed in the context of the entire class II molecule. Moreover, the entire haplotype, consisting of a variety of HLA class II products must be considered in interpreting cellular reactivities. Together, the combination of clonal reactivities, protein chemistry, Southern blot, and cDNA sequencing are extremely valuable in constructing composite pictures of the various class II haplotypes as they exist within the population.

2. Statistical Standardization

Analysis of patterns of clonal reactivity is typically performed within single laboratories within limited numbers of experiments. The analyses of large numbers of cell lines as performed in the Workshop studies posed an additional statistical problem since the testing of clones was split over multiple experiments in most laboratories. These experiments had to be integrated together, and this data, in turn, had to be integrated with data from other clones and other laboratories using a common scale. Developing objective statistical approaches for this purpose was clearly essential for a meaningful analysis.

Experimental Design

T-cell clones were tested in a hierarchical fashion against the Workshop cell line panel and eight blind duplicate

---

Participating Laboratory: US7NFL.[1]

This work was supported in part by grants from the National Institutes of Health, CA-22507, CA-44322, CA-23766, and CA-33050.

Immunobiology of HLA
Volume I

cell lines. Seventy-five of the Workshop cell lines were designated as core cell lines and were assigned priority numbers to ensure maximum overlap in the cell lines tested in various laboratories. Duplicate testing was required in all cases to ensure reproducibility of results. Local positive and negative control cells were required in each experiment as a means of normalizing the data and providing an independent quality control assessment. The blind duplicate cell lines also provided a means for quality control assessment. A panel of selected monoclonal antibodies was distributed for blocking studies, through which the epitope in question could be mapped to a specific class II family or α-β dimer. The cell lines and antibodies utilized in the cellular studies are described in detail earlier in this volume (1,2).

Results

Statistical Standardization. The detailed description of the data reduction and statistical analyses are presented in the accompanying papers by Plätke et al. (3,4). Briefly, triplicates were screened for outliers, which were removed prior to subsequent calculations. In proliferative assays, medium control values for the B-cell line and T-cell clone of interest were calculated. After subtraction of the media controls, the resultant values were transformed to a log scale and two indices of proliferation were calculated as follows:

$$\mathrm{Prol}^{\mathrm{quot}} = \frac{\lg(\mathrm{res}^{\mathrm{exp}}) - \lg(\mathrm{res}^{-})}{\lg(\mathrm{res}^{+}) - \lg(\mathrm{res}^{-})}$$

$$\mathrm{Prol}^{\mathrm{diff}} = \lg(\mathrm{res}^{\mathrm{exp}}) - \lg(\mathrm{res}^{+}) + 1$$

where "res" = corrected counts per minute (CPM) values after subtraction of the media controls, "exp" = experimental cell line, "+" = positive control cell line, and "−" = negative control cell line.

Although $\mathrm{Prol}^{\mathrm{diff}}$ provided accurate assessments of the proliferative response in most cases, under certain extremes of experimental conditions, $\mathrm{Prol}^{\mathrm{quot}}$ proved more appropriate (see Fig. 11 in Plätke et al. [4]). Using the minimum of $\mathrm{Prol}^{\mathrm{diff}}$ and $\mathrm{Prol}^{\mathrm{quot}}$ provided subjectively accurate interpretation of the data in virtually all cases. The minimum value of $\mathrm{Prol}^{\mathrm{diff}}$ and $\mathrm{Prol}^{\mathrm{quot}}$, termed "Prol," was thus selected and transformed to a five-part scale, where 8 = positive, 6 = probably positive, 4 = weak or uncertain, 2 = probably negative, and 1 = negative. In most cases, scores of 6 and 8 or 1 and 2 were considered together producing a three-part scale, positive, uncertain, or negative.

For cytotoxicity experiments, specific lysis was calculated as:

$$\text{specific lysis} = \frac{\text{experimental release} - \text{spontaneous release}}{\text{maximum release} - \text{spontaneous release}}$$

Attempts at further correcting this cytotoxicity index based on the positive and negative control cell lines introduced excessive distortion into the results. These corrections were therefore eliminated. Similar indices were calculated for the antibody blocking experiments. All were converted to a scale of 8, 6, 4, 2, or 1, as described above. The details of this analysis and other statistical variables that were considered and discarded are described in Plätke et al. (4).

Three important points emerged from the statistical efforts:

1. There was a high concordance (95%) in scoring between the blind duplicate cell lines and their matched controls, validating both the data and the scoring system used for its analysis.
2. For those clones tested in both cytotoxicity and proliferation, there was 97% concordance in scoring, indicating that the scales established for cytotoxicity and proliferation were comparable.
3. There was 95 to 99% concordance between scoring performed using these Workshop criteria and the assignments made in four local laboratories. Additionally, there was a consensus agreement by the Workshop participants in Princeton that the centralized scores accurately reflected their data.

Together, these results indicated that it was indeed possible to develop objective means for combining together data obtained over the course of multiple experiments and for combining studies performed in multiple laboratories. Moreover, all Workshop clones (both cytotoxic and proliferative) could be analyzed jointly.

Definition of Cellular Polymorphisms: Overview. The panel studies for 125 centrally analyzed clones are listed in Figure 1, while antibody blocking data is listed for 100 clones in Figure 2. As illustrated in Figure 2, there is a large cluster of clones (n=61) blocked by anti-DR antibodies. An additional eight clones were strongly blocked by anti-DR antibodies, with more scattered or weaker blocking by anti-DQ antibodies. These clones were felt to most likely recognize DR gene products. As detailed in the summary of the class II protein studies, some of the anti-DQ reagents exhibited weak cross reactivity with DR products (5), which might explain these observations in part. Fifteen clones were blocked primarily by anti-DQ monoclonal antibodies, while three were blocked primarily by broadly reactive or DP-specific reagents. The patterns of blocking exhibited by 11 clones could not be deciphered because either none of the antibodies produced inhibition or because virtually all of the antibodies were inhibitory. Thus, based on the blocking patterns that could be interpreted with some certainty, 77% of the clones recognized DR products, 18% recognized DQ products, while only 3% recognized DP products. The interpretations of the blocking studies are listed in the right columns of Figures 1 and 2.

With regard to the 125 clones analyzed in the panel analysis (Fig. 1), there are clear clusters that segregate with each of the serologically defined DR specificities. Within these large groups, there are a large number of

T-Cell Clones - Panel Analysis - Part A

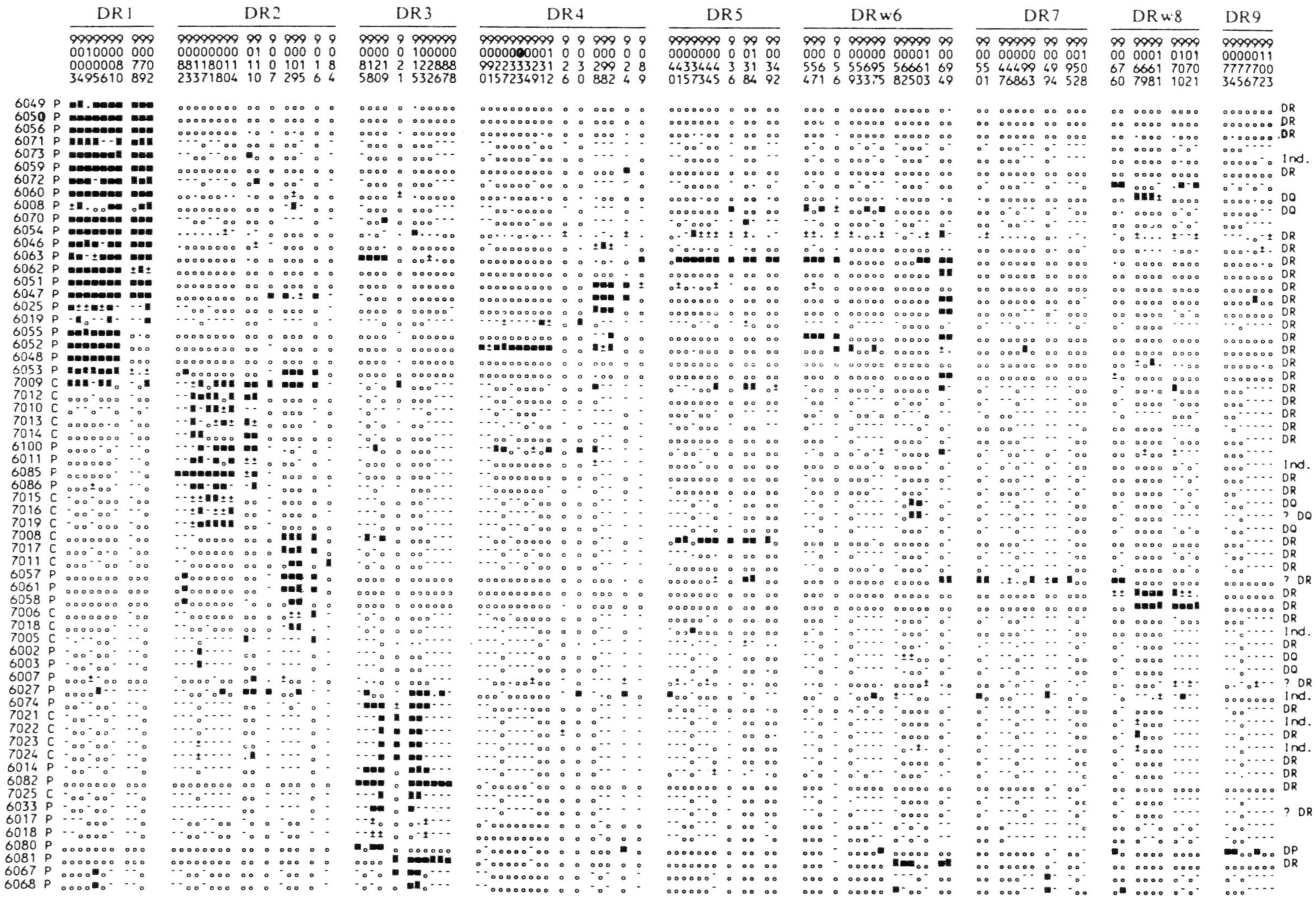

■ = 8; ▮ = 6; ± = 4; . = 2; ∘ = 1; - = not tested

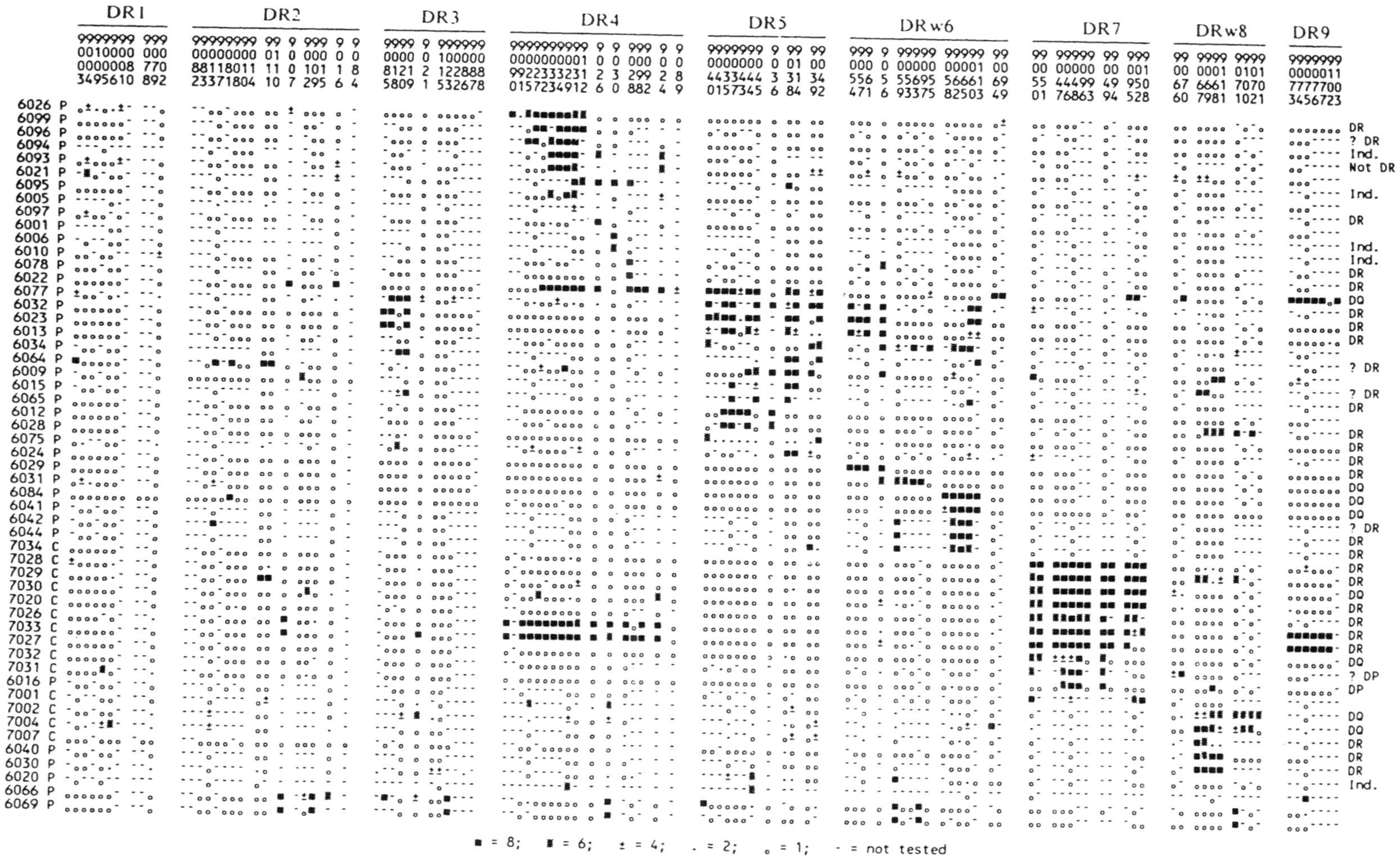

Figure 1. Cellular panel data for 125 centrally analyzed clones are illustrated. The Workshop cell lines have been grouped according to their DR and Dw types. The Workshop number assigned to each clone is listed in the left column. The functional assay in which the clone was tested (cytotoxicity = C or proliferation = P) is listed in the second column. The HLA class II family recognized by each clone (as inferred from the monoclonal antibody blocking data) is listed in the right column. The symbols in the figure reflect the following scores: solid square = 8 = positive; shaded rectangle = 6 = probably positive; ± = 4 = weak or uncertain; closed dot = 2 = probably negative; open circle = 1 = negative; − = not tested.

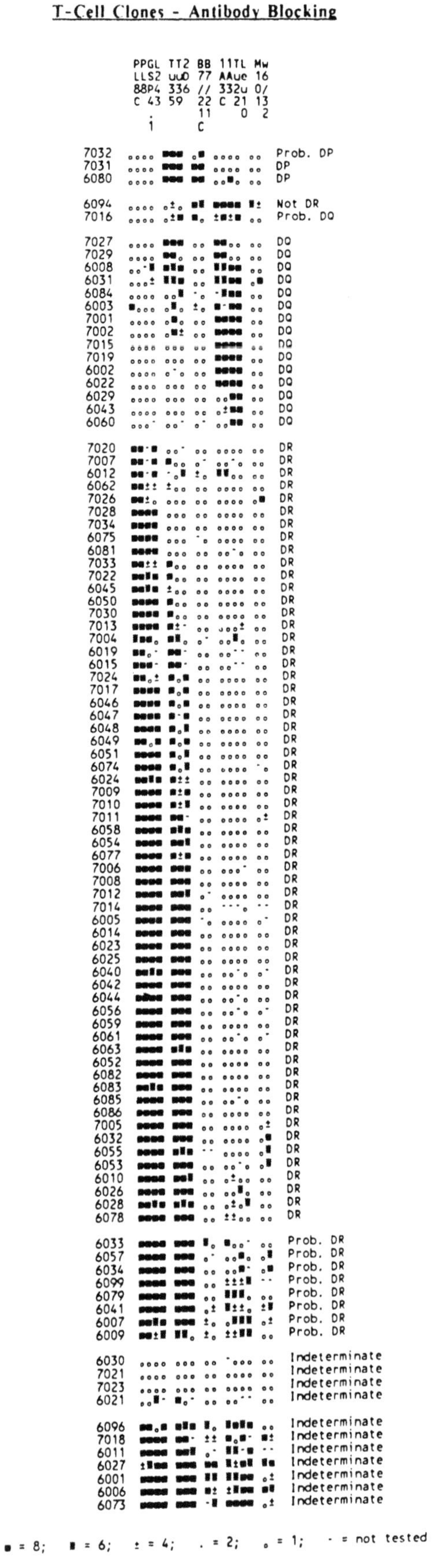

Figure 2. Antibody inhibition data for 100 centrally analyzed clones are illustrated. The Workshop number assigned to each clone is listed in the left column. Antibodies have been separated into five groups representing (from left to right): HLA-DR specific antibodies, broadly reactive anti-HLA class II antibodies recognizing more than a single class II family, HLA-DP-specific antibodies, HLA-DQ-specific antibodies, and control antibodies that do not bind to HLA class II molecules. The term "specific" indicates that an antibody reacts predominantly, though not necessarily exclusively with products of that HLA class II region. The symbols in the figure reflect the following scores: solid square = 8 = positive; shaded rectangle = 6 = probably positive; ± = 4 = weak or uncertain; closed dot = 2 = probably negative; open circle = 1 = negative; − = not tested. The interpretations of the patterns observed are listed in the right column of the figure.

subclusters. Indeed, it is probably the exception rather than the rule when several clones exhibit exactly the same pattern of reaction across the entire cell panel. Moreover, it is often the subtle differences which the clones exhibit which prove to be most informative. The specificities identified through this analysis are described in detail in the sections that follow. In these sections, we have utilized the biochemistry nomenclature and the RFLP nomenclature described in other Workshop reports (5,6).

WHO Recognized Specificities. As a result of these investigations, the Workshop serologic studies, and cDNA sequence analysis, new HLA class II specificities were recognized by the World Health Organization Nomenclature Committee. When serologic reagents were capable of recognizing the products, a DR (or DQ) designation was given, irrespective of the independent recognition of the product by cellular reagents. When a serologic equivalent was not available, a Dw designation was assigned. Thus, Dw designations continue to be used for cellularly defined epitopes, although prior to the Tenth Workshop, they have reflected cellular epitopes stimulating strong responses in primary mixed lymphocyte cultures (see Nomenclature for Factors of the HLA System, elsewhere in this volume).

Dw20. This specificity represents a variant of DR1 which can be mapped to the DR product. In the present analysis, Dw20 was detected as a "short" DR1 specificity, which was missed by a group of clones recognizing DR1;Dw1. Dw20 is in linkage disequilibrium with B14 in the Jewish population. Subsequent to the Workshop, the three Dw20 cell lines (9002, 9078, 9079) were demonstrated to exhibit two conservative amino acid substitutions at residues 85 and 86 of their DR β chains (Gorski and Eckels, manuscript in preparation). The identical DR1 β chain sequence has been identified in American Blacks (7), although in this case there is no linkage disequilibrium with an HLA-B locus product (8).

Dw21 and Dw22. These specificities represent distinct haplotypes associated with a '"short" DR2 serologic specificity, which has been designated DRw16. These two haplotypes are distinguishable in primary MLC (9), by DR-specific T-cell clones (10,11), and on the basis of their associated DQ molecules - DQw1(w5) in the case of Dw21 and DQw3(w7) in the case of Dw22.

One of the two DR β chains encoded by these haplotypes is identical in Dw21 and Dw22 cells, while the second differs by a single conservative amino acid substitution (12). Both chains differ from those expressed in Dw2 or Dw12 cell lines. As discussed below, there is additional heterogeneity within the DRw16 cluster. The previously defined Dw2 and Dw12 haplotypes are both associated with the traditional ("complete") DR2 serologic specificity, now designated DRw15. DRw15 and DRw16 can also be recognized by DR-specific T-cell clones (10,11).

DRw17 and DRw18. These specificities represent subtypes of DR3 that can be defined either cellularly or serologically. DRw17 represents the traditional DR3 specificity in linkage disequilibrium with HLA-B8 and B18. DRw18 represents a serologically short DR3, present in the Black population in linkage disequilibrium with Bw42. In the Workshop, DRw18 also behaves as a "short" cellular specificity in that it is missed by several DR specific clones reactive with the DRw17 subtype of DR3. This specificity can also be distinguished from the DRw17 haplotypes on the basis of its association with DQw4, rather than DQw2. Sequence analysis of DRw18 reveals four conservative amino acid substitutions within the first domain of the DR β-I chain (13,14).

Dw24, Dw25, and Dw26. These three specificities represent cellularly defined subtypes of DRw52. They can be defined by clones reactive with DR products, and in most cases, can be blocked by DRw52-specific antibodies (such as 7.3.19.1) as well (15). These three molecules have additionally been differentiated by protein chemistry (16) and cDNA sequence (17,18). Dw24 (formerly DRw52a or LB-Q4) is associated with B8;DR3 haplotypes, Bw42;DR3(DRw18), some DRw13;Dw18 haplotypes, and DRw14;Dw16. Dw25 (formerly DRw52b or LB-Q1) is associated with B18;DR3 haplotypes, DR5 haplotypes, some DRw13;Dw18 haplotypes, and DRw14; Dw9. Dw26 (formerly DRw52c) is associated with DRw13;Dw19 haplotypes, although some DR5 cell lines are recognized by Dw26 as well as Dw25 reactive clones. DRw8 cells, though DRw52-positive, are not included in any of these DRw52 subsets (15).

Specificities Identified in the Cellular Analyses that Have Not Been Incorporated into WHO Nomenclature. Evidence for other class II splits or subtypes was obtained from the Workshop studies. The following nomenclature is unofficial and is intended solely for the purpose of facilitating future communication. The designations are preceded by "TCD" (T-lymphocyte clone-defined or T-cell defined) to differentiate this nomenclature from official WHO specificities. The designations incorporate the most closely related DR, DQ, or Dw specificity. Specificities that correlate precisely with serologic equivalents are, in general, not redefined, except in those cases where they are differentiated from closely related but distinct specificities. The decimal addition ".0" is used in these rare instances to indicate complete agreement with the serologic counterpart. Other decimal extensions (".1", ".2", ".3", etc.) indicate a cellularly defined subtype. In the case of DQ products, some of the subtypes identified exhibit concordance with particular DR or Dw specificities, in which case the DR or Dw type is included in parenthesis. In one instance, a DQ specificity exhibited concordance with the presence of a particular DQ α chain, irrespective of the serologic DQ type. In this case, the designation TCD-DQA was utilized to reflect the DQ α specificity. These specificities are described in detail below and are summarized in Tables 1 to 3.

Subtypes of DR2. *TCD-DRw15.1* represents a specificity limited to cells typing as Dw12 and to most cells expressing the extended A3;B7;DR2;Dw2 haplotype.

Table 1. HLA-DR Associated Specificities Recognized in the Tenth International Histocompatibility Workshop Cellular Studies

Cluster Designation	Specificity and/or Prototypic Cell Line
TCD-DRw15.1	Dw12 + subset of Dw2
TCD-DRw15.2	DRw15;Dw2 variant – DRβ also exhibts more acidic pattern – Cell line 9010 (AMAI)
TCD-DRw16.1	DRw16 subset – Associated with an additional DRβ product in cells of either Dw21 or Dw22 type
TCD-DRw16.2	DRw16 subset – Associated with an additional DRβ product in cells of Dw21 type only
TCD-DR2.1	DR2 variant – Cell line 9084 (CALOGERO)
TCD-DR2.2	DR2 variant – Cell line 9007 (DEM)
TCD-DRw18.0	Cellular equivalent of DRw18 – Cell line 9021 (RSH)
TCD-DRw11.1	DR5(w11) variant – Cell line 9042 (TISI)
TCD-DRw11.2	DR5(w11) variant – Cell line 9039 (JVM)
TCD-DRw11.3	DB2 + Dw5 subset
TCD-DRw11.4	DR5 subset (TCD-DRw11.1–, TCD-DRw11.2–, TCD-DRw11.3–)
TCD-Dw11.1	Dw11 cells lacking DRB4 product
TCD-Dw11.2	Dw11 cells expressing DRB4 product
TCD-DRw8.1, TCD-DRw8.2, and TCD-DRw8.3	DR specificities associated with 3 subtypes of Dw8. Additional heterogeneity in the TCD-DRw8.1 group.
TCD-Dw25.1	Dw25 (DRw52b) variant – Cell line 9043 (BM21)
TCD-DRw53.1	DRw53-like epitope expressed on the DRB1 gene product of DR7 cells

This specificity was defined by clone 7014 and seems clearly associated with a DR product. Cells expressing other (non-HLA-B7 associated) Dw2 haplotypes are not recognized.

TCD-DRw15.2 represents another short DRw15 specificity associated with cell line 9010 (AMAI). This cell line was recognized by nine DRw15 reactive clones, but was missed by three others. This cell line was noted to have a more acidic DR β pattern in the protein chemistry studies, independently demonstrating its unique characteristics (19).

TCD-DRw16.1 and TCD-DRw16.2. TCD-DRw16.1 is a specificity limited to Workshop cell lines 9009 and 9015, both Dw21, but lacking in cell line 9012 (also Dw21). This pattern of reactivity was detected by clones 7011 and 6058. TCD-DRw16.2 is similar to TCD-DRw16.1, differing only in that the Dw22 cell line (9016) is also recognized (clone 7006). TCD-DRw16.1 and 16.2 can be correlated with the presence of an extra DR β product most likely related to the DRβ-2.16 chain (19). TCD-DRw16.2 can be correlated with the presence of this extra product in any DRw16 cell line, while TCD-DRw16.1 correlates with its presence in cells typing as Dw21. In this latter case, the clones appear to recognize

Table 2. HLA-DQ-Associated Specificities Recognized in the Tenth International Histocompatibility Workshop Cellular Studies

Cluster Designation	Specificity and/or Prototypic Cell Line
TCD-DQw5	Similar to serologically defined DQw5 Associated with DQβ1.3 RFLP Common sequence (residues 58-78 of DQβ)
TCD-DQw5(Dw1+Dw20)	DQw5 subset limited to DR1 cells
TCD-DQw6(Dw2)	DQw6 subset limited to Dw2 cells
TCD-DQw6(Dw2+Dw12)	DQw6 subset limited to Dw2 and Dw12 cells
TCD-DQw6(Dw18)	DQw6 subset limited to Dw18 cells
TCD-DQw6(Dw19)	DQw6 subset limited to Dw19 cells Associated with DQβ1.6 RFLP
TCD-DQw6(Dw2+Dw18)	DQw6 subset limited to Dw2 and Dw18 cells Associated with DQβ1.5 RFLP Common sequence (residues 31-78 of DQβ)
TCD-DQw2(DR7)	DQw2 subset limited to DR7;DQw2 cells
TCD-DQA(DR7)	DQ specificity common to all DR7 cells irrespective of DQw2 or DQw3 typing
TCD-DQw7.1	DQw3 subset limited to cells of DR4;DQw7 type
TCD-DQw3.1	DQw3 subset including all cells but DR4;DQw7 homozygotes
TCD-DQw8.0	Equivalent to serologically defined DQw8 Dependent upon presence of DQβ-3.2 chain only
TCD-DQw8.1	DQw8 subset – expressed only in presence of both DQβ-3.2 and DQα-3.2 chains
TCD-DQw4.1	DQw4 subset common to DRw8 and DRw18 (lacking in Dw15)
TCD-DQw4.2	DQw4 subset limited to DRw8
TCD-DQw4.3	DQw4 subset limited to DW8.1 and DRw18

Table 3. Supertypic Specificities Recognized in the Tenth International Histocompatibility Workshop Cellular Studies

Cluster Designation	Specificity and/or Prototypic Cell Line
TCD-DR81	Supertypic specificity associated with DR1, DR4, DRw6, ± DRw16
TCD-DR81.1	Supertypic specificity associated with DR1 + Dw14 + Dw16 Common sequence (38-86 of DRβ)
TCD-DR82	Supertypic specificity associated with DR2 (particularly DRw16), DR5, and DRw8
TCD-DR83	Supertypic specificity associated with DR1 and DRw8
TCD-DR84	DR1 + Dw25 + Dw16

the unique amino acid substitution which differentiates the DR products in these two haplotypes (10,12).

TCD-DR2.1 is represented by cell line 9084 (Calogero). This cell line was serologically typed as DRw16. It was tested with seven clones recognizing DRw15 and six recognizing DRw16-associated specificities. Only a single clone recognizing a subset of Dw21 reacted with this cell line, indicating that its DR2 haplotype is distinct.

TCD-DR2.2 is represented by the DR2/4 heterozygous cell line 9007 (DEM). This clone was recognized only by one broadly reactive DR2-directed clone, suggesting that its DR2-associated haplotype is also distinct in some way.

Subtypes of DR3. *TCD-DRw18.0* is the cellular equivalent of the serologically defined DRw18 specificity.

Subtypes of DR5. *TCD-DRw11.1.* This specificity is represented by cell line 9042 (TISI), which frequently behaves as an outlier in its recognition by DR5 reactive T-cell clones and which does not type as Dw5 in HTC experiments (20,21). Cell line 9040 is frequently concordant in its pattern of reactivity with 9042 and may reflect a similar DR5 subtype.

TCD-DRw11.2 is represented by cell line 9039 (JVM) which is known to have a distinct DRB1 gene compared to other DRw11 cell lines (18). This cell line also behaves distinctly with a variety of clones reactive with DR5 haplotypes and does not type as Dw5 in HTC experiments (20,21).

TCD-DRw11.3 is identified by two clones, 6012 and 6065. This cluster includes cell lines 9035, 9037, 9043, 9044 (Dw5) and 9036 (DB2). Specifically missed were the DRw12 cell line (9038), the TCD-DRw11.1 cell lines (9040 and 9042), the TCD-DRw11.2 cell line (9039), and two additional cell lines (9041 and 9045).

Within the TCD-DRw11.3 cluster, cell line 9037 exhibited a distinct pattern of recognition with two additional clones (7018 and 7008, see Fig. 1). One of these clones (7008) also recognizes DR3;Dw25 cell lines, raising the possibility that the additional heterogeneity exhibited by cell line 9037 may be due to variation in the DR β-III molecule, rather than DR β-I variation.

TCD-DRw11.4. As noted above, cell line 9045 was missed by the two clones reactive with TCD-DRw11.3. It was recognized (although in one case weakly) by two clones (7009 and 6057), which also recognize DRw16 and DRw12. Although typed as heterozygous for DRw11 and DRw12 serologically, this cell line was not recognized by the DRw12 specific clone 6075. This pattern suggests that its DRB1 gene products may be distinct from classic DRw11 and DRw12. This cell line as well as cell line 9041 (which is also recognized by clone 6057) have been grouped as TCD-DRw11.4. It should be emphasized that this group is defined more by exclusion (TCD-DRw11.1−, TCD-DRw11.2−, TCD-DRw11.3−) than by positive reactivity (20).

Subtypes of DRw6. There is evidence for heterogeneity in DRw6 haplotypes, particularly within Dw18 and Dw19. However, distinct subclusters cannot be clearly defined with the possible exception of cell line 9062, which was not recognized by clones 7015 and 7016.

Subtypes of DR7. *TCD-Dw11.1 and TCD-Dw11.2.* Dw11 haplotypes are an unusual subtype of DR7 in linkage disequilibrium with Bw57 and DQw3. In many cases, this haplotype is associated with absence of a DR β-IV gene product. An associated RFLP can be used to detect this haplotype, although the RFLP actually reflects a change 3' to the DRB1 gene rather than associated with the defective DRB4 gene (22). Some of these cell lines express DRw53 and are heterozygous for the associated RFLP (Sutton and Knowles, personal communication). Dw11 cell lines can thus be split on the basis of their reactivity with DRw53-specific T-cell clones and monoclonal antibodies. Cell line 9052 clearly lacks a DRw53 product and is designated TCD-Dw11.1. Cell line 9095 expresses a DRw53 product on the basis of its reactivity with the DRw53-specific clone 7033 (23) and monoclonal antibody 109d6 (24). Although this cell line may be a Dw11/Dw17 heterozygote, other Dw11 cell lines have also been demonstrated to express a DRβ-IV product. Those Dw11 cells that express a DR β-IV product are designated TCD-Dw11.2.

Among the DR7;DQw2 cell lines, two subclusters could be identified on the basis of the reactivity of clones 7031 and 6057. Clone 7031 recognized cell lines 9046, 9048, 9049, and 9096, while clone 6057 recognized cell lines 9050, 9051, 9093, and 9094. Clone 7031 is unusual in that it was blocked only by broadly reactive class II monoclonal antibodies and the DP-specific antibody B7/21. Its reactivity within the panel cannot be correlated with DP type, although Rodriguez de Cordoba et al. (25) have identified linkage disequilibrium between DR7;DQw2 and a particular DP product (DPαb, DPβb).

Conceivably, clone 7031 recognizes this unique DP product, or alternatively, may recognize a DPα-DQβ heterodimer analogous to the DRα–DQβ heterodimer described by Lotteau et al. (26). Clearly, however, the blocking studies indicate that it does not recognize a DR product. Polymorphism of the DR or DQ products expressed by DR7;DQw2 cell lines has not been clearly demonstrated to date (18,27,28). Although the DR7 cell lines can be split in this way, these subtypes have not been assigned names, since a clear structural explanation for this subdivision is lacking at present.

Subtypes of DRw8. *TCD-DRw8.1 and TCD-DRw8.3.* Electrophoretically DR molecules in Dw8.1 and Dw8.3 are indistinguishable (29). However, a single conservative amino acid substitution has been identified in the Dw8.3 cell line (TABO) versus the DR β chain in Dw8.1 cell lines (18). Clones 7007 and 6040 (which recognize DR products expressed by Dw8.1, but not by Dw8.3 cell lines) presumably recognize this difference. We, therefore, refer to the DR β chain of Dw8.1 as TCD-DRw8.1 to differentiate it from the reciprocal TCD-DRw8.3 product recognized by clones 6057 and 7032. There is additional evidence (clones 6009, 7028) to suggest that cell line 9068 is distinct from cell lines 9067 and 9069 (TCD-DRw8.1a).

TCD-DRw8.2 was not recognized independently, but was recognized together with either TCD-DRw8.1 or TCD-DRw8.3. This molecule is clearly distinct from its other Dw8 counterparts on the basis of its two-dimensional gel pattern and cDNA sequence (29,30).

Subtypes of DRw52. *TCD-Dw25.1.* Although defined at the time of the Tenth Workshop, evidence for functional heterogeneity within Dw25 exists (15,20,31), as do suggestions of DNA sequence variability (J. Gorski, personal communication). Cell line 9043 (BM21) was missed by several Dw25-specific clones (6023 and 6032), suggesting it carries a Dw25 variant. Additional heterogeneity may exist within Dw25, as previously described in the section on subtypes of DR5 (TCD-DRw11.3) above.

Subtypes of DRw53. *TCD-DRw53.1.* DRw53 has been associated with the DRB4 gene product in cells typing as DR4, DR7, and DRw9. As described above, a subset of Dw11 cell lines (TCD-Dw11.1) lack expression of a DRB4 product. Cell line 9052 is known to be such an example (32). Despite this fact, cell line 9052 and its blind duplicate were recognized (although less vigorously) by clone 7026. This suggests that this clone may recognize the DR β-I product in this cell line, albeit with lower affinity. This observation parallels the finding that the DRw53 monoclonal antibody PL3 binds only to the DR β-IV product on DR4 haplotypes, but to both DR β-I and DR β-IV products on DR7 haplotypes (32). Similar types of observations have been made with DRw52 monoclonal antibodies (15). This may reflect recognition of an epitope involving residue 4 of the DR β chain where a R → Q substitution is found on DR7, DRw9, and DRw53 β chains (18). TCD-DRw53.1 is thus a DRw53-like epitope expressed on the DRB1 gene product of DR7 cells.

Recently, a variant DRw53 sequence has been identified (J. Gorski, personal communication). Cell line 9098 (DR4;Dw14) was repeatedly missed by the DRw53-specific clone 7026, although it was recognized by another DRw53-specific clone (7033). This cell line may conceivably contain such a DRw53 variant gene.

Subtypes of DQw1. *TCD-DQw5* is defined by clones 6008 and NR91, recognizing DQ products on DR1, Dw21, and Dw9 cell lines. This pattern is correlated with the DQB1.3 RFLP (6) and with shared sequences between residues 58 through 78 of the DQ β chains (18). Clone 6008 missed a few Dw1 cell lines, raising the possibility of some heterogeneity within Dw1-associated DQ products.

TCD-DQw5(Dw1+Dw20) is defined by clone 6060, recognizing DQ products expressed only on cell lines typing as Dw1 or Dw20. This clone presumably recognizes the DQα-1.1, DQβ-1.1 dimer (33). The DQ products expressed by Dw20 cell lines of Black origin exhibit identical first domain sequences compared with their Dw1 counterparts (7). The DQ products of Dw20 cell lines of Jewish origin have not been sequenced to date.

TCD-DQw6(Dw2) is defined by clone 7019, recognizing DQ products expressed only on cell lines typing as Dw2. Presumably these clones recognize the DQα-1.2, DQβ-1.2 dimer, which is unique to these cell lines (33).

TCD-DQw6(Dw2+Dw12) is defined by clone 6011, recognizing DQ products expressed on cell lines typing as either Dw2 or Dw12.

TCD-DQw6(Dw18) is defined by clones 6084 and 6031, presumably recognizing the DQα-1.18, DQβ-1.18 dimer limited to cells typing as Dw18 (33).

TCD-DQw6(Dw19) is defined by clone 6029, reactive with cells typing as Dw19 and associated with the DQB1.6 RFLP pattern described in the Southern blot report (6). This clone presumably reacts with a DQα-1.2, DQβ-1.19 dimer (33). This clone specifically misses the Dw19 variant cell line 9055.

TCD-DQw6(Dw2+Dw18) is defined by clones 7015 and 7016. This pattern of reactivity is associated with the DQB1.5 RFLP pattern (6) and with shared sequences between residues 31 to 78 of the DQ β molecules (18,34).

Subtypes of DQw2. *TCD-DQw2(DR7).* This specificity is defined by clone 7027, which recognized HLA-DQ products expressed by DR7;DQw2 cell lines. This clone presumably recognizes the DQα-3.7, DQβ-2 dimer since DQ products from DR3;DQw2 cell lines (which expresses DQα-3.1, DQβ-2 dimers) and DR7;DQw3 cell lines (which express DQα-3.7, DQβ-3.3 dimers) are not recognized (35).

TCD-DQA(DR7). This DQ specificity is common to all DR7 cell lines, irrespective of their DQw2 or DQw3 type. Presumably it reflects recognition of the DR7-associated DQ α chain (DQα-3.7) which is common throughout the DR7 cluster, while the DQ β chains differ depending on the DQw2 or DQw3 association. It is labeled DQA because it appears to reflect an α chain polymorphism which is independent of serologically defined DQ types (23,35).

Subtypes of DQw3. *TCD-DQw7.1*. This specificity is defined by clone EM.C3 and recognizes DQw7 products on DR4 cell lines. Presumably this reflects recognition of the DQα-3.1, DQβ-3.1 dimer unique to this subgroup of DQw7 cells (36,37).

TCD-DQw3.1 is broad DQw3 specificity defined by clone 6022. All DQw3 cell lines are recognized with the exception of those defined as TCD-DQw7.1. It appears that this clone can recognize any DQw3 α-β dimer except the DQα-3.1, DQβ-3.1 pair. Thus with regard to homozygous cell lines, TCD-DQw3.1 and TCD-DQw7.1 are reciprocal specificities (36,37).

TCD-DQw8.0 and TCD-DQw8.1. TCD-DQw8.0 is defined by clones 6094 and 6093, which recognize all DQw8 cell lines. In this respect, this specificity parallels serologically defined DQw8, being expressed whenever the DQβ-3.2 chain is present. It is defined to illustrate the distinction from TCD-DQw8.1, which requires the DQα-3.2 chain as well. Clone 6095 defines TCD-DQw8.1. It recognizes all DQα-3.2, DQβ-3.2 cell lines, but repeatedly fails to recognize the DQα-3.1, DQβ-3.2 cell line 9034 (36,38). The contribution of the DQ α and β chains to these two epitopes are extremely well defined molecularly. These epitopes are dependent on single amino acid substitutions in the DQ β chain (TCD-DQw8.0) and additionally to changes outside of the first domain of the DQ α chain (TCD-DQw8.1) (39).

Subtypes of DQw4. *TCD-DQw4.1 and TCD-DQw4.2* are defined by clones 7001 and 7002. In local studies, both clones recognized DQw4-associated with DRw8, but not with Dw15 (40). This differential recognition may be based on both the DQ α and β chains present in these haplotypes. The DQ α chains are quite distinct between the Dw15 and DRw8 haplotypes (41), while (in the first domain) the β chains differ only at residue 23 (arginine → leucine). The DQ α and β chains in DRw18; DQw4 and DRw8;DQw4 cell lines were shown to be identical in their first domain sequences (18,42). TCD-DQw4.1 represents a specificity common to DQw4 in both DRw18 and DRw8 haplotypes (but apparently lacking in Dw15 haplotypes), as recognized by clone 7002. Clone 7001 recognized all DRw8;DQw4 cell lines, but not the DRw18;DQw4 cell line. TCD-DQw4.2 is thus a DQw4 specificity associated with DRw8, but not DRw18 or Dw15. This specificity may be generated by conservative amino acid substitutions outside of the first domain or may reflect recognition of DRw8 peptide fragments presented by DQw4 molecules.

TCD-DQw4.3 was defined by clone 6060, which recognized DQ products associated with DR1 and Dw8.1, but not Dw8.2 cell lines. This would suggest the presence of an, as yet unidentified, difference in the DQw4 molecules present in Dw8.1 and Dw8.2 haplotypes. This clone also recognized the DRw18;DQw4 cell line (9021) (30).

Supertypic Specificities. In defining nomenclature for supertypic specificities, we have followed the W.H.O. example of DRw52 and DRw53, using higher numbers for the supertypic determinants. Since W.H.O. may choose to add a DRw54 specificity at some future time, we have begun the numbering system in the 80s to avoid future confusion.

TCD-DR81 and TCD-DR81.1. This designation is used to represent the cross-reactivities detected between DR1, DR4, and DRw6. Occasionally, DRw16 is included as well. The prototypic example DR1 + Dw14 + Dw16 probably reflects the large stretch of common amino acid sequence between residues 38 through 86 of these DR β-I molecules (18,43). Because of its strong structural bases, this example is given the specific designation TCD-DR81.1, while the entire group (containing several different patterns of DR1, DR4, DRw6 ± DRw16 cross-reactivity) is referred to as TCD-DR81.

TCD-DR82. This refers to the cross-reactive group formed between DR2 (particularly DRw16), DR5, and DRw8. These cross-reactivities may be based on common sequences between residues 26-31 and 67-74 of these DR β-I molecules (11,18).

TCD-DR83. This represents the DR1 + DRw8 cross-reactive group (44).

TCD-DR84 is defined by clone 6063, recognizing DR1 + Dw25 + and Dw16 (44).

Clonal Specificity at the Molecular Level. In some instances, it is possible to find very specific structural bases for the reactivities of the clones. Analysis of the first domain sequences of the DR β chains in certain haplotypes, such as DR4, leads one to the generalization that the serologically defined DR type is heavily influenced by the amino acid sequence from residues 1 to approximately 35 to 40. Additional polymorphisms through the rest of the first domain contribute to the cellularly defined subtypes of these specificities. This is less clearly the case in certain other DR allotypes such as DR3 (18). In the case of DQ α and β, similar generalizations are not possible. The significance of even single amino acid substitutions in dictating the cellularly defined subtype is perhaps best indicated in the specificity of clone 6026 recognizing Dw4, but no other DR4 subtypes. This clone clearly must recognize an epitope to which the presence of K at residue 71 contributes, as this is the unique difference between Dw4 and the other DR4 subtypes. Similarly, clones 6001 and 6006 reactive with Dw13 and DKT2 must recognize an epitope dependent on the presence of E at residue 74, as this uniquely differentiates these two DR4 subtypes from the others (39).

In the case of DQ reactive clones, similar changes in structure can alter specificity. The DQw8 serologic specificity (and the TCD-DQw8.0 cellular specificity) must be largely dependent on the presence of A at residue 57 of the DQw8 β chain, which uniquely differentiates the first domain of this β chain from the β chains of DQw7 and DQw9. (Other residues must contribute as well, since DQw2 β chains also contain A at

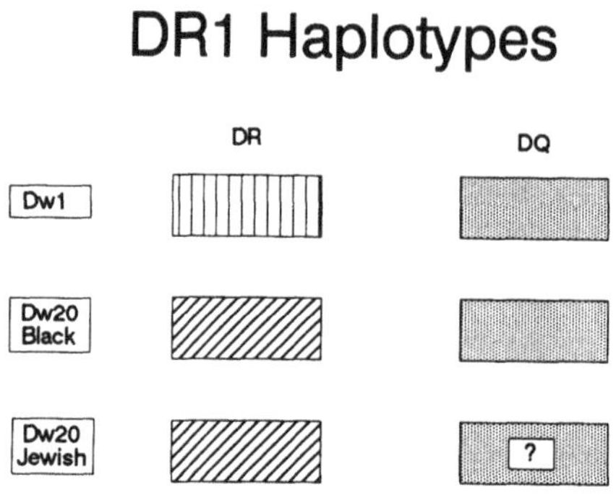

Figure 3. The DR and DQ products expressed by HLA-DR1+ cells are graphically represented. Different symbols are used to represent documented differences between the molecules. "?" refers to the fact that the DQ products in Dw20 cells of Jewish origin appear identical to those in Dw1 cells and Dw20 cells of Black origin, although sequence analyses have not yet provided formal proof.

residue 57 [18,38].) In the case of DQ products, one or both chains may contribute to the specificity recognized. TCD-DQA(DR7) and TCD-DQw8.0 are examples where the epitopes appear to be dependent only upon one chain (the DQ α and the DQ β chains respectively), and are recognized regardless of the other chain with which they are paired. In contrast, specificities TCD-DQw8.1, TCD-DQw2(DR7), TCD-DQw7.1, and several of the cellularly defined subtypes of DQw1 are dependent on the presence of specific pairs of α and β chains. These specificities are not detected when either chain is present separately, even when paired with a very closely related but distinct partner. Perhaps the best example is TCD-DQw8.1 where the DQα-3.1 chain cannot substitute for the DQα-3.2 chain, despite the fact that these two α chains are identical to one another within the first domain of the molecule.

Structural explanations for shared epitopes can also be deduced from the first domain sequences. Cross-reactivity between DR1 (including both Dw1 and Dw20), Dw14, and Dw16 are more easily understood when one considers the long stretches of shared sequence between these molecules. Taken together, these various examples illustrate the extreme precision with which T-cell clones can differentiate between subtle differences in MHC molecules.

Cellularly Defined Haplotypes. Although the definition of HLA-D was changed somewhat by the W.H.O. Nomenclature Committee after the last Workshop, it is perhaps appropriate to reask a question posed during the Ninth Workshop – specifically, "What is HLA-Dw?" The answer to this question is probably clearer now than before, though still capable of provoking a variety of passionate arguments. Prior to making any generalizations, it is probably best to consider each of the DR types individually. In this discussion, we consider only officially recognized or well established specificities.

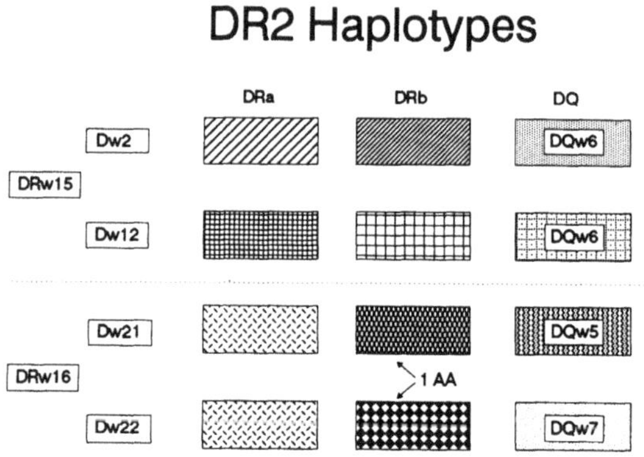

Figure 4. The DR and DQ products expressed by HLA-DR2+ cells are graphically represented. DRa and DRb designations are utilized to reflect the fact that the DR products in these haplotypes have not been formally assigned to the DRB1, DRB3, DRB4, or other genes as has been done in other haplotypes. Different symbols are used to represent documented differences between the molecules. "1 AA" refers to the fact that only a single amino acid difference has been identified. Serologically defined DR and DQ subtypes are indicated.

DR1. DR1 can be subdivided into two subtypes Dw1 and Dw20 (Fig. 3). Both Dw1 and Dw20 appear to express a single DR and DQ product. The DR β chain in Dw20 cell lines differs from that in Dw1 cell lines by two conservative amino acid substitutions at residues 85 and 86 of the first domain. The identical Dw20 sequence has been identified in DR1 of both American Blacks and Jews. The first domain sequence of the DQ α and β chains from Dw20 cells of Black origin and from Dw1 cells are identical (7). This is presumably true for Dw20 cells of Jewish origin as well, in that the molecules are indistinguishable on two-dimensional gels (33), though it has not been formally proven by sequence analysis.

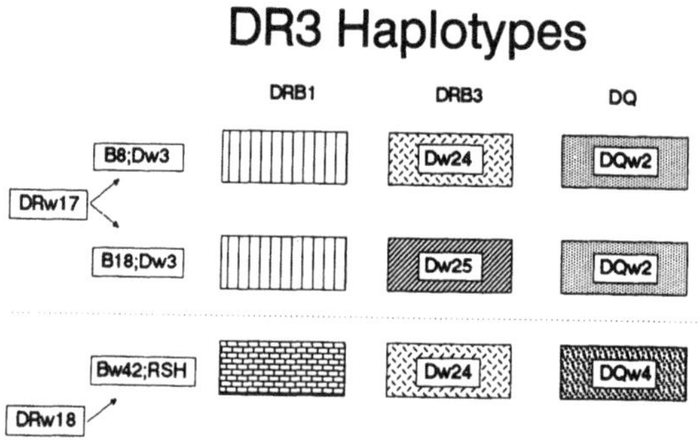

Figure 5. The DR and DQ products expressed by HLA-DR3+ cells are graphically represented. Different symbols are used to represent documented differences between the molecules. Serologically defined DR and DQ subtypes are indicated, as are cellularly defined subtypes of DRw52.

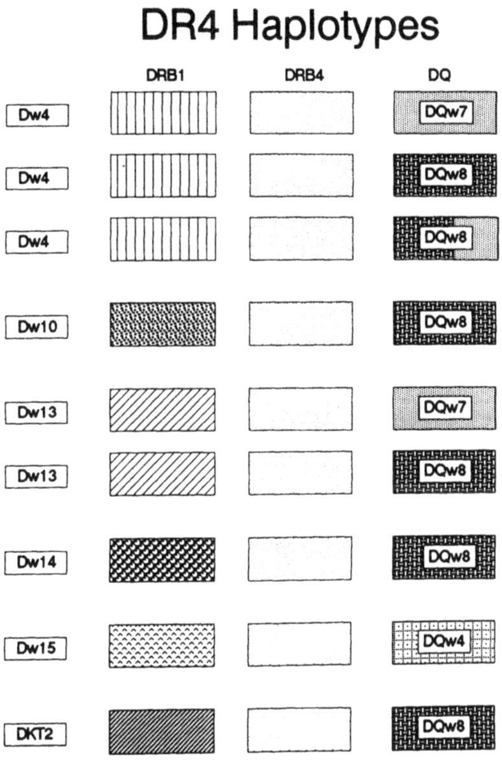

Figure 6. The DR and DQ products expressed by HLA-DR4+ cells are graphically represented. Different symbols are used to represent documented differences between the molecules. Serologically defined DQ subtypes are indicated. The composite symbol used for the third Dw4 haplotype reflects identification of a cell line which expresses the DQ α chain normally associated with DR4;DQw7 haplotypes together with the DQ β chain normally associated with DR4;DQw8 haplotypes.

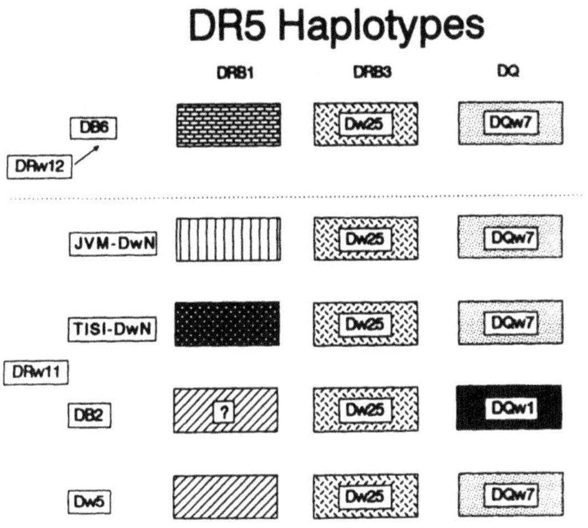

Figure 7. The DR and DQ products expressed by HLA-DR5+ cells are graphically represented. Different symbols are used to represent documented differences between the molecules. In the case of "TISI-DwN," the molecule appears functionally distinct, though a structural difference has not yet been identified. "?" refers to products which appear functionally or biochemically similar to those expressed by other members of the DR5 cluster but which have not been proven to be identical by sequence analysis. Serologically defined DR and DQ subtypes are indicated, as are cellularly defined subtypes of DRw52.

DR2. The DR2 specificity is subdivided into two subgroups, DRw15 (classic DR2) and DRw16 (DR2-short) (Fig. 4). DRw15 is associated with Dw2 and Dw12. DRw16 is associated with Dw21 and Dw22. Two DR β chains are expressed by each of these haplotypes, as well as a single DQ α and DQ β product. The DR β chains in DR2 haplotypes have not been formally assigned to the DRB1, DRB3, or DRB4 loci, as has been done in most other haplotypes. The DR β chains and the DQ α and β products in Dw2 and Dw12 all exhibit either distinct two-dimensional gel patterns (5) or distinct sequences (18). In Dw21 and Dw22, one pair of DR chains is identical, while the second pair differs by a single conservative amino acid substitution (12,18). The DQ product in Dw21 contains a unique DQ β chain (33) and types as DQw5. In contrast, Dw2 and Dw12 type as DQw6 (1,33). Dw22 cells express a DQw3(w7) product unlike the other DR2 haplotypes.

DR3. DR3 has been split into two subtypes DRw17 and DRw18. Two DRw17 haplotypes have been recognized (Fig. 5) in linkage disequilibrium with B8 and B18, respectively. These haplotypes express identical DR β-I chains and identical DQ products. They differ in their DR β-III products, the B8 haplotype expressing Dw24, and the B18 expressing Dw25 (16). Since these two haplotypes may exhibit low level reactivity in primary MLC, we suggest that the unofficial designations Dw3.24 and Dw3.25 be used to describe these Dw types. The DRw18 haplotype expresses a distinct DR β-I product with four conservative amino acid substitutions, a Dw24 β-III product, and distinct DQ α and β chains associated with the DQw4 specificity (13).

DR4. At least six subtypes of DR4 are recognized, including Dw4, Dw10, Dw13, Dw14, Dw15 and DKT2 (Fig. 6). All exhibit distinct DR β-I molecules and a common DR β-IV molecule (18,45). Dw10, Dw14, and DKT2 are associated with DQw8. Dw15 is associated with DQw4. Dw4 and Dw13 may be associated with either DQw7 or DQw8. In at least one instance, a Dw4 cell line has been found to express a DQw8 β chain together with the α chain normally associated with DQw7. (See the discussion of TCD-DQw8.1 above and [36].)

DR5. DR5 haplotypes are probably more heterogeneous than previously recognized (20,46). The classic Dw5 specificity is associated with a DRw11 β-I product, a Dw25 β-III product, and a DQw3(w7) product (Fig. 7). The DB2 haplotype is associated with serologically identical DR β-I and β-III products, together with a DQw1 product. (Sequence analyses comparing the DB2-associated DR products with those found in Dw5 cells have not yet been performed.) The DB6 haplotype

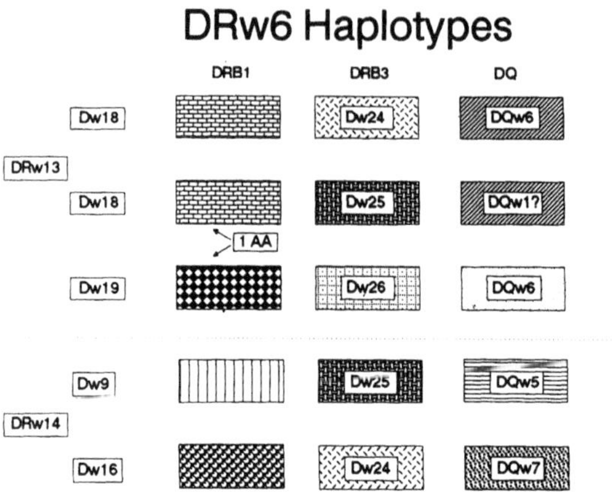

Figure 8. The DR and DQ products expressed by HLA-DRw6+ cells are graphically represented. Different symbols are used to represent documented differences between the molecules. "1 AA" refers to the fact that only a single amino acid difference has been identified. Serologically defined DR and DQ subtypes are indicated, as are cellularly defined subtypes of DRw52. The DQ products in the two Dw18 haplotypes appear biochemically and functionally identical (and thus are designated with the same symbol). However, the single Dw18;Dw25 cell line tested in the workshop was serologically typed as DQw5. Because of this inconsistency, the designation "DQw1?" is utilized.

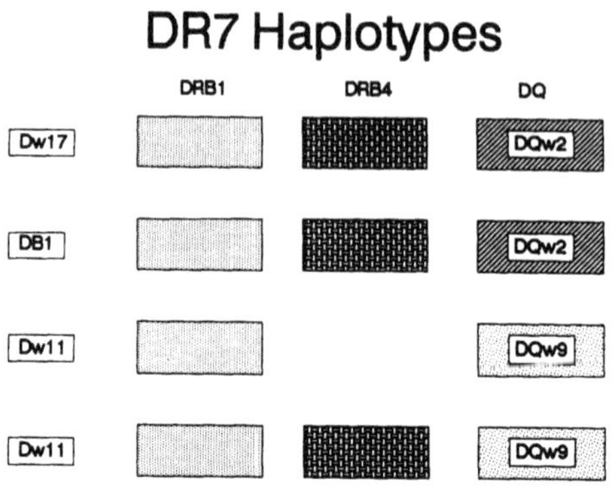

Figure 9. The DR and DQ products expressed by HLA-DR7+ cells are graphically represented. Different symbols are used to represent documented differences between the molecules. In some Dw11 cell lines, products of the DRB4 gene are not detectable on the cell surface. Serologically defined DQ subtypes are indicated.

is associated with a DRw12 β-I product, a Dw25 β-III product, and a DQw3(w7) product. At least two additional haplotypes defined by the JVM (9039) and TISI (9042) cell lines can also be distinguished. JVM expresses a DRw11 molecule with a distinct sequence (18). There is evidence for additional heterogeneity within Dw5 which may reflect microheterogeneity within both the DR β-I and DR β-III products (20).

DRw6. DRw6 can be split into two serologic subtypes DRw13 and DRw14. At least three haplotypes are associated with DRw13, two typing as Dw18 and one as Dw19 (Fig. 8). The two Dw18 haplotypes express identical DR β-I products (47). Their DQw1 products appear identical biochemically and cellularly (DQw6), although they may be distinguishable serologically (1). These two Dw18 haplotypes clearly differ, however, in their DR β-III products (either Dw24 or Dw25) (15,16). This observation helps to explain some of the low level reactivity among Dw18 HTCs. Analogous to Dw3, we suggest referring to these as Dw18.24 and Dw18.25. The DRw13 product of Dw19 differs only by an extremely conservative amino acid substitution (valine/glycine) at residue 86 (48). Dw19 is distinct in expressing a Dw26 β-III product (15,16,48). Although the DQ molecules of Dw19 cells type as DQw6 similarly to most Dw18 cells, both the DQ α and β chains differ between these two Dw types (33), and these distinct DQw6 subtypes are readily differentiated by T-cell clones (34). The Dw9 haplotype is associated with a DRw14 β-I product, a Dw25 DR β-III product, and a DQw5 product. Dw16 is associated with a DR β-I molecule which, although serologically cross-reactive with that found in Dw9, exhibits a very distinct nucleotide sequence (18). Dw16 is associated with a Dw24 DR β-III product, and a DQw3(w7) product (34,47).

DR7. This set of haplotypes will be discussed in more detail subsequently. Identical DR β-I and β-IV products are present in all DR7 subtypes, although in some but not all Dw11 cells, a DR β-IV product is not expressed (32). Dw11 is also distinct in its expression of DQw3(w9), while the remaining haplotypes express DQw2 (Fig. 9).

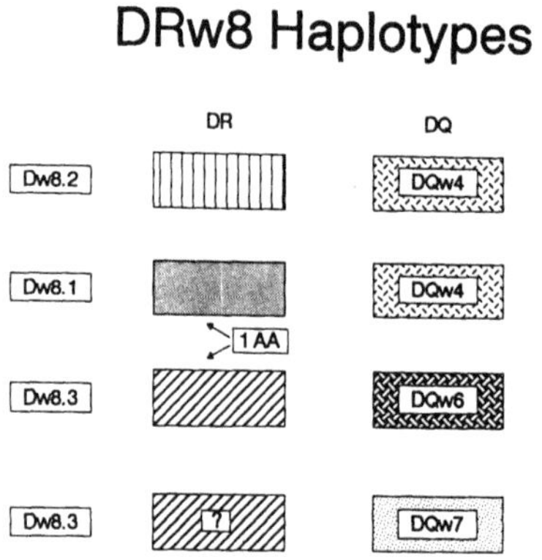

Figure 10. The DR and DQ products expressed by HLA-DRw8+ cells are graphically represented. Different symbols are used to represent documented differences between the molecules. "?" refers to products which appear functionally or biochemically similar to those expressed by other members of the DRw8 cluster but which have not been proven to be identical by sequence analysis. "1 AA" refers to the fact that only a single amino acid difference has been identified. Serologically defined DQ subtypes are indicated.

Table 4. Summary of HLA Class II Haplotypes

HLA-DR		HLA-Dw	HLA-DRw52/53/Dw	HLA-DQ	HLA-B	Ethnic	Comments
HLA-DR1		Dw1		DQw5 (DQw1)	B35		
		Dw20		DQw5 (DQw1)	B14	Jewish	
						Black	
HLA-DR2*	DRw15	Dw2		DRw6 (DQw1)	B7		
	DRw15	DW12		DQw6 (DQw1)	Bw52	Japanese	
	DRw16	Dw21		DQw5 (DQw1)			MN2, FJO, AZH
	DRw16	Dw22		DQw7 (DQw3)		Warao Indian	tb24, LD-5a
HLA-DR3	DRw17	Dw3	DRw52a, Dw24	DQw2	B8		
	DRw17	Dw3 variant	DRw52b, Dw25	DQw2	B18		
	DRw18	Dw new	DRw52a, Dw24	DQw4	Bw42	Black	
HLA-DR4		Dw4	DRw53	DQw7 (DQw3)	B44(12)		Dw4.1
				DQw8 (DQw3)	Bw62(15)		Dw4.2
		Dw10	DRw53	DQw8 (DQw3)	B38	Jewish	
		Dw13	DRw53	DQw7 (DQw3)			
				DQw8 (DQw3)			
		Dw14	DRw53	DQw8 (DQw3)	Bw60(w40)		
		Dw15	DRw53	DQw4		Japanese	
HLA-DR5*	DRw11	Dw5	DRw52b, Dw25	DQw7 (DQw3)			
	DRw11	DB2	DRw52b, Dw25	DQw1			
	DRw12	DB6	DRw52b, Dw25	DQw7 (DQw3)			
HLA-DRw6	DRw13	Dw18	DRw52a, Dw24	DQw6 (DQw1)			
	DRw13	Dw18	DRw52b, Dw25	DQw1			
	DRw13	Dw19	DRw52c, Dw26	DQw6 (DQw1)			
	DRw14	Dw9	DRw52b, Dw25	DQw5 (DQw1)			
	DRw14	Dw16	DRw52a, Dw24	DQw7 (DQw3)	Bw62	American Indian	
HLA-DR7		Dw7	DRw53	DQw2			
		Dw17	DRw53	DQw2	B44		
		Dw11	blank	DQw9 (DQw3)	Bw57		
		Dw11	DRw53	DQw9 (DQw3)	Bw57		
		DB1	DRw53	DQw2	B13		
HLA-DRw8		Dw8.1		DQw4		Caucasian	
		Dw8.2		DQw4		American Indian	
		Dw8.3		DQw1		Oriental	
		Dw8.3		DQw7 (DQw3)		Dutch	LUY
HLA-DR9		Dw23	DRw53	DQw9 (DQw3)		Oriental	
HLA-DRw10		Dw new		DQw5 (DQw1)		Jewish	SHY
HLA-DR Bon		Dw new		DQw1		French	27-CH3

*Additional haplotypes described in the text.

DRw8. DRw8 haplotypes exhibit a single DR β chain (Fig. 10). These DR molecules differ in each of the Dw8.1, 8.2, and 8.3 subtypes (18), although the differences between Dw8.1 and 8.3 cannot be distinguished by two-dimensional gel analysis (29). Dw8.1 and 8.2 express DQw4 products, while Dw8.3 has been associated with either DQw1 or DQw3(w7). As discussed above, there is cellular evidence for heterogeneity of the DR products within the Dw8.1 group. The DQw4 products in the Dw8.1 and Dw8.2 products may also be differentiated by T-cell clones (30).

Other DR Specificities. DR9, DRw10, and DR-Bon (49,50) have been associated, thus far, with only single haplotypes. They are included in the summary table (Table 4), but will not be discussed in detail here.

Cellularly Defined Haplotypes – Correlating Structure and Function. In considering the Dw specificitie as a whole, several points can be made. First, in virtually every case, variability is seen in the DR β-I molecule. This may be limited to a single amino acid difference, but there are clear examples where this difference in the DR β-I molecule is sufficient by itself to explain the Dw type. Dw4 serves as a case in point. Here, a single amino acid substitution in the DR β-I molecule, without any change in the DR β-IV or DQ molecules, is sufficient for recognition of the cell as Dw4, as opposed to another DR4 subtype. Other more subtle changes, such as the single, very conservative, substitution in the DRw13 β-I molecules of Dw18 and Dw19, may or may not be sufficient for distinguishing between these two haplotypes. Here, however, the distinctions between the DR β-III products Dw24/25 (in Dw18) versus Dw26 (in Dw19) may be responsible for the reactivity seen in primary MLC. In this case, DQ differences could theoretically contribute as well. Indeed, in the case of Dw3 and Dw18, controlling for the type of the β-III molecule present may reduce the low level reactivity seen in MLC within these Dw clusters. In these haplotypes, it might be appropriate to refer to Dw3.24 and Dw3.25 or Dw18.24 and Dw18.25 in order to distinguish the subtypes.

The only clear example where a DR sequence variation of at least one amino acid is not present in two distinct Dw types are the Dw11, Dw17, and DB1 subtypes of DR7. Dw11 was the first split of Dw7. This specificity is in linkage disequilibrium with Bw57 and DQw3. As discussed above, it is sometimes, though not invariably, associated with absence of a DR β-IV product. Dw17 was defined as the reciprocal split of Dw7. However, in common usage, Dw17 is frequently associated with the Bw44 or B14-associated DR7;DQw2 haplotypes, while DB1 is associated with the B13;DR7;DQw2 haplotype. DB1 is perhaps the easiest DR subtype to distinguish in primary MLC (E. Mickelson and N. Reinsmoen, personal communications). At present, sequence variations have not been identified in the DR products of these haplotypes. It is difficult to know what role the absence of the DR β-IV molecule may have played in splitting off Dw11, since it is not known which of the homozygous typing cells used were actually homozygous for absence of DRw53 expression and which did express a DRw53 product. Despite the DQw2-DQw3 difference, the primary MLC response between Dw17 and Dw11 cells can be small. In these DR7 haplotypes one must consider alternate possibilities to explain the stronger MLC reactivity seen toward the DB1 cluster. Are B13 peptide fragments presented by DR7 molecules strongly stimulatory? Does B13 generate an unusual high frequency of class I-directed proliferative clones? Does the DP linkage disequilibrium described by Rodriguez de Cordoba and associates (25) make an important contribution? Do unusual DQ-DR DP-DR or similar hybrid products form as suggested by Lotteau and associates (26)? For the DR7 haplotypes, definitive answers cannot be given, but in most Dw types, one can envision a hierarchy of responses: DR β-I > DR β-III/IV > DQ > DP > class I. Clearly DR β-I and probably DR β-III contribute to the primary MLC response. It is possible that the other products may contribute to a lesser degree in some but not all cases.

Conclusions and Future Directions

The studies of cellular recognition of HLA-class II molecules proved to be a productive enterprise, reflecting the enthusiastic participation of a number of laboratories. The studies succeeded in certain of the traditional aspects of HLA Workshops in that, together with data from other components, they provided information which could be used for better defining HLA-molecules. Much of the traditional emphasis of Workshops has been placed on separating antigens into subtypes, allowing for more precise definitions of the allospecificities. To that end, these studies contributed to the definitions of Dw20, Dw21, Dw22, DRw18, Dw24, Dw25, and Dw26. The studies have also pointed out numerous additional complexities, of which 39 were assigned Workshop (TCD) designations. Some of these latter specificities may be important for clinical typing purposes and will find their way into the official nomenclature in the future. Others will not necessarily be important for clinical typing, but may, nonetheless, help to unravel the puzzle as to how the T-cell receptor and MHC molecule interact.

The recognition of MHC products by T cells in primary MLC led to the definition of HLA-Dw specificities beginning in the Sixth International Workshop. The HLA-Dw subtypes of DR4 were recognized for many years, before the structural basis for their existence could be delineated. As analyses of the MHC incorporate clonal cellular reagents, new complexities have been, and will continue to be, uncovered. This is particularly likely to be true in analyses of the DQ family, where both α and β chains may contribute to the polymorphism detected by T-cells. The various cellularly defined subtypes of DQw1, extending well beyond DQw5 and DQw6, illustrate the complexity waiting to be resolved. The subdivisions of DQw2 and even DQw7 described in this report provide further examples. In addition to defining new specificities, past Workshops have additionally served to confirm the specificities defined previously. Perhaps the most beleaguered specificity in the Tenth Workshop was DRw12, which fortuitously was recognized by one of the Tenth Workshop clones without any cross-reactivity with DRw11 or DRw8. A summary of the class II haplotypes and cellularly defined specificities, as they are currently recognized, is provided in Tables 4 and 5. Most important is that for many haplotypes, we are beginning to develop a clear picture as to how the cellular subdivisions, protein chemistry, and cDNA sequence interrelate in very subtle and intricate ways.

A fascinating example of the intricacies revealed by the combination of cellular, protein, and cDNA studies is the specificity defined as TCD-DQw8.1. The cellular studies identified the DQw8+ cell line 9034 as unusual. The protein chemistry identified an unexpected DQ α product. The sequence analysis demonstrated that the α chain's difference fell outside of the first domain. Thus, the biochemistry explains the cellular pattern of reactivity, and the cellular reactivity illustrates the functional significance of the biochemical and sequencing observations.

Perhaps the most intriguing observations of this component deal with the numerous examples of "supertypic" clones recognized. While some conform to known DRw52/53 or DQ determinants, others are clearly quite distinct. One of the nicest examples is the cross-reactivity between DR1, Dw14, and Dw16 (TCD-DR81). On the surface, this would appear to represent a very unusual cross-reaction between cells of quite distinct evolutionary backgrounds. However, these cells exhibit identical sequences in the third hypervariable region of the DR β-I molecule. Thus, a clearcut and substantial basis for this cross-reactivity exists. In this case, it may be more proper to think of the epitopes as shared rather than cross-reactive. In other cases, the primary sequence may not offer such a simple explanation, and it may be more proper to think of those epitopes as cross-reactive although physically distinct. Here again are examples

Table 5. T Cell Defined HLA-Dw Subtypes of HLA-DRw52

HLA-Dw	HLA-DRw52	Associations		
HLA-Dw24	DRw52a	DR3	Dw3	(B8-associated)
		DR3	Dw new	(Bw42-associated)
		DRw13	Dw18	
		DRw14	Dw16	
HLA-Dw25	DRw52b	DR3	Dw3 variant	(B18-associated)
		DRw11	Dw5	
		DRw11	DB2	
		DRw12	DB6	
		DRw13	Dw18	
		DRw14	Dw9	
HLA-Dw26	DRw52c	DRw13	Dw19	

where the structural similarities explain the cellular patterns of reactivity, and the cellular reactivities in turn emphasize the relevance of the structural observations. Although these sorts of observations did not originate within the confines of the Tenth Workshop, they nevertheless gained substantial credibility due to the numerous examples which were demonstrated in the Workshop exercises and from the large panel of cell lines against which the clones were tested. All who were involved with the analysis of this data were struck by the number of cross-reactions present, but were even more impressed by the exquisite immunogenetic specificity which they exhibited in many cases. The incorporation of clonal reagents into these studies thus forces one to rethink the relationships between MHC gene products. Clearly, the analysis must consider the unexpected (and expected) similarities between MHC gene products and must not be focused solely on their differences.

One of the most exciting aspects of these studies is that, in many cases, sequence data have been readily available to explain and independently validate the observations. In some cases, such as the analysis of Dw20, these cellular investigations may have served as the impetus and guide for the sequence analysis. In most cases, sequence variations occur in defined regions of class II molecules which are recognized as functionally significant. In some cases, the sequence variations may either fall outside of the typical regions of variability (i.e., DQα-3.1 and DQα-3.2) or may be quite conservative in nature. In these cases, analyses such as those described here can clearly demonstrate the biologic relevance of these polymorphisms. More than ever, cellular analyses require correlation with other investigations into MHC polymorphisms.

One of the major questions in immunology has been how MHC molecules, nominal antigen, and T-cell receptors interact as a ternary complex. The elucidation of the crystal structure for HLA-A2 (51,52) and its predicted similarities to class II structure (53) have provided major insights in this area and have illustrated clearly how the widely distributed variable regions of MHC molecules are brought into close proximity in the mature protein. The groove present in the class I surface and the presence of copurified peptides within that groove have provided a straightforward physical model as to how MHC molecules may present processed antigens to T-cells (51,52).

The nature of alloreactivity is less clear. Does alloreactivity represent the interaction of T-cells with naked MHC molecules? Are MHC peptides or other peptide fragments present in the molecular cleft? Do both of these processes contribute to alloreactivity? Final answers are not available, yet, at least in some cases, peptides appear to play a role (54,55). As the targets of alloreactivity become defined with increased precision, understanding the mechanisms of allorecognition and tolerance to self becomes increasingly important.

The question as to whether to pursue further efforts of this type in future Workshops will (fortunately) have to be answered by those organizing the future studies. Such questions are never easy to address, and answers must be based upon the state of the art of both the biology of the MHC and the technology available for its analysis at the time of the next Workshop. The organizers of the ASHI T-Cell Clone Workshop (56) were wise in their predictions that:

1. useful information would be gained "*by studying a substantially larger number of T-cell clones, preferably with overlapping specificities and different functions.*"
2. "*such a project will succeed to the degree that well characterized T-cell clones and monoclonal antibodies are used, and that appropriate statistical procedures are developed.*"

Clearly, subtleties of T-cell recognition were most obvious for the haplotypes against which the largest number of clones were tested and for the haplotypes which contained the broadest and most varied representation within the workshop cell panel. The statistical approaches developed in the course of the present Workshop have clearly allowed data from multiple laboratories and multiple experiments to be integrated into a meaningful database, capable of then being subjected to a detailed immunogenetic analysis. With appropriate refinements of the statistics and larger numbers of clones tested, future studies should prove better still. The Tenth Workshop cell panel represented an excellent start, though one would ideally like to see additional specificities such as Dw15 included in the future. It is gratifying that at least one laboratory has begun to employ T-cell clones in a clinical transplantation setting for the characterization of HLA class II products expressed by unrelated donor-recipient pairs (E. Mickelson, personal communication).

The primary residual obstacle to these studies is that the growth of T-cell clones remains as much an art as a science in 1989. While the testing of clones was long and laborious in this project, the primary difficulty (for most labs) was growing the T-cell clones to large enough cell numbers to test against the entire panel. There were several instances in which interesting immunogenetic questions could not be definitively answered because the clones had not been tested against certain informative cell lines. For this particular problem, the future may indeed provide important help. It is conceivable that as our understanding of T-cell physiology evolves, it may be possible to introduce either oncogenes or genes producing appropriate cytokines into clones to stabilize and facilitate their growth. Alternatively, murine T-cell clones derived from transgenic mice (expressing human MHC products) may prove more hardy, allowing both clones and cell lines to be internationally distributed for analysis. Even relying on technology as it presently exists, it is clearly possible to obtain a great deal of useful information provided that the participating laboratories exhibit the same degree of enthusiasm and hard work as they exhibited in 1987.

References

1. Milford EL, Kennedy LJ, Yang SY, Dupont B, Lalouel J-M, Yunis EJ. Serologic characterization of the reference panel of B-lymphoblastoid cell lines for factors of the HLA system. In: Dupont B (ed): Immunobiology of HLA, Volume I: Histocompatibility Testing 1987. New York: Springer-Verlag, 1989.
2. Flomenberg N, Platke R, Zemmour J, Small TN, Klein JS. Immunogenetic analysis of HLA class II cellular data: Antigen report of the cellular studies of the Tenth International Histocompatibility Workshop. In: Dupont B (ed): Immunobiology of HLA, Volume I; Histocompatibility Testing 1987. New York: Springer-Verlag, 1989.
3. Plätke R, Ferguson M, Klein JS, Zemmour J, Flomenberg N, Lalouel JM. Collection and processing of HLA class II cellular data. In: Dupont B (ed): Immunobiology of HLA, Volume I; Histocompatibility Testing 1987. New York: Springer-Verlag, 1989.
4. Plätke R, Lalouel JM, Flomenberg N. Statistical analysis of HLA class II cellular data. In: Dupont B (ed): Immunobiology of HLA, Volume I; Histocompatibility Testing 1987. New York: Springer-Verlag, 1989.
5. Knowles RW. Structural polymorphism of the HLA class II α and β chains: Summary of the 10th Workshop 2-D gel analysis. In: Dupont B (ed): Immunobiology of HLA, Volume I; Histocompatibility Testing 1987. New York: Springer-Verlag, 1989.
6. Trucco M, Ball E. RFLP analysis of DQ-beta chain gene: Workshop report. In: Dupont B (ed): Immunobiology of HLA, Volume I; Histocompatibility Testing 1987. New York: Springer-Verlag, 1989.
7. Hurley CK, Ziff BL, Silver J, Gregersen PK, Hartzman R, Johnson A. Polymorphism of the HLA-DR1 haplotype in American blacks. J Immunol 1988;140:4019–4023.
8. Eckels D, Sell T, Eiermann T, Nikaein A. DR1: Antigen report of the Tenth International Histocompatibility Workshop. In: Dupont B (ed): Immunobiology of HLA, Volume I; Histocompatibility Testing 1987. New York: Springer-Verlag, 1989.
9. Reinsmoen NL, Layrisse Z, Betuel H, Freidel AC, Bach FH. A study of HLA-DR2 associated HLA-Dw/LD specificities. Hum Immunol 1984;11:105–115.
10. Reinsmoen N, Eckels D, Zier K, Elferinu D, deVries RP, Zeevi A, et al. DR2: Antigen report of the cellular studies of the Tenth International Histocompatibility Workshop. In: Dupont B (ed): Immunobiology of HLA, Volume I; Histocompatibility Testing 1987. New York: Springer-Verlag, 1989.
11. Reinsmoen NL, Bach FH. Correlation of primary sequence and functional determinants of HLA class II molecules. In: Dupont B (ed): Immunobiology of HLA, Volume II; Immunogenetics and Histocompatibility. New York: Springer-Verlag, 1989.
12. Liu C, Wu S, Bach FH. Molecular studies of the rare DR2/LD-5a/DQw3 haplotype: Implications for mechanisms of generating HLA class II polymorphisms. In: Dupont B (ed): Immunobiology of HLA, Volume II; Immunogenetics and Histocompatibility. New York: Springer-Verlag, 1989.
13. Johnson AH, Tang TF, Robbins FM, Rosen-Bronson S, Steiner N, Hartzman RJ, et al. DR3 heterogeneity in American blacks. In: Dupont B (ed): Immunobiology of HLA, Volume II; Immunogenetics and Histocompatibility. New York: Springer-Verlag, 1989.
14. Kurnick J, Thomsen M, Irle C, DeTotero D, Ferrara GB. DR3: Antigen report of the cellular studies of the Tenth International Histocompatibility Workshop. In: Dupont B (ed): Immunobiology of HLA, Volume I; Histocompatibility Testing 1987. New York: Springer-Verlag, 1989.
15. Irle C, Mickelson E, Masaewicz S, Hansen JA, Thomsen M, Termijtelen A. DRw52: Antigen report of the cellular studies of the Tenth International Histocompatibility Workshop. In: Dupont B (ed): Immunobiology of HLA, Volume I; Histocompatibility Testing 1987. New York: Springer-Verlag, 1989.
16. Bontrop R, Ballas M, Madrigal A, Hermans P, Fuggle S, Otting N, et al. DRw52: 2-D gel patterns. In: Dupont B (ed): Immunobiology of HLA, Volume I; Histocompatibility Testing 1987. New York: Springer-Verlag, 1989.
17. Tiercy JM, Gorski J, Jeannet M, Mach B. Identification of three serologically undetected alleles of HLA-DR by oligonucleotide DNA typing analysis. Proc Natl Acad Sci 1988; 85:198.
18. Gregersen PK, Todd JA, Erlich HA, Long E, Servenius B, Choi E, et al. First domain sequence diversity of DR and DQ subregion alleles. In: Dupont B (ed): Immunobiology of HLA, Volume I; Histocompatibility Testing 1987. New York: Springer-Verlag, 1989.
19. Rust N, Carter C, Fuggle S, Sacks S, Hermans P, Charron D, et al. DR2 (DRw15,16): 2-D gel patterns. In: Dupont B (ed): Immunobiology of HLA, Volume II; Immunogenetics and Histocompatibility. New York: Springer-Verlag, 1989.
20. Mickelson EM, Masaewicz S, Thomsen M, Hansen JA, Reinsmoen N, Sterkers G, et al. DR5: Antigen report of the cellular studies of the Tenth International Histocompatibility Workshop. In: Dupont B (ed): Immunobiology of HLA, Volume I; Histocompatibility Testing 1987. New York: Springer-Verlag, 1989.

21. Mickelson E, Reinsmoen N, Robbins FM, Hartzman R, Odum N, Svejgaard A, et al. HLA-Dw and HLA-DP typing of the reference panel of B-lymphoblastoid cell lines. In: Dupont B (ed): Immunobiology of HLA, Volume I; Histocompatibility Testing 1987. New York: Springer-Verlag, 1989.
22. Sutton VR, Knowles RW. Mapping of a restriction fragment length polymorphism associated with defective DRβ4 chain expression to the HLA-DRB1 gene. Hum Immunol 1988;22:123.
23. Flomenberg N, Sasport M, Klein J. DR7, DQw2 and DRw53: Antigen report of the cellular studies of the Tenth International Histocompatibility Workshop. In: Immunobiology of HLA, Volume I; Histocompatibility Testing 1987. New York: Springer-Verlag, 1989.
24. Sanchez B, Winchester R, Nunez-Roldan A. Workshop B-cell reference panel reactivity with monoclonal antibody 109d6. Tenth Workshop Newsletter, No. 4, pp 19–20, 1987.
25. Rodriguez de Cordoba S, Marshall P, Ginsberg-Fellner F, Rubinstein P. Biochemical polymorphism of the DPa and DPb chains. In: Dupont B (ed): Immunobiology of HLA, Volume II; Immunogenetics and Histocompatibility. New York: Springer-Verlag, 1989.
26. Lotteau V, Teyton L, Turmel P, Charron D. DR alpha-DQ beta: A novel class II molecule whose expression is influenced by the ratio of alpha-beta chains transcripts. In: Dupont B (ed): Immunobiology of HLA, Volume II; Immunogenetics and Histocompatibility. New York: Springer-Verlag, 1989.
27. Erlich HA, Scharf SJ, Long CM, Horn GT. Analysis of isotypic and allotypic sequence variation in the HLA-DR beta region using the in vitro enzymatic amplification of specific DNA segments. In: Dupont B (ed): Immunobiology of HLA, Volume I; Histocompatibility Testing 1987. New York: Springer-Verlag, 1989.
28. Winchester R, Matthews E, Amesland F, Mazzilli MC, Fernandez N, Kienzle B, et al. DR7: 2-D gel patterns. In: Dupont B (ed): Immunobiology of HLA, Volume I; Histocompatibility Testing 1987. New York: Springer-Verlag, 1989.
29. Kulova L, Bontrop, Yang SY, Maeda H. DRw8: 2-D gel patterns. In: Dupont B (ed): Immunobiology of HLA, Volume I; Histocompatibility Testing 1987. New York: Springer-Verlag, 1989.
30. Tajima Y, Aizawa M, Eckels D, Reinsmoen N, Nikaen A. DRw8 and DQw4: Antigen report of the cellular studies of the Tenth International Histocompatibility Workshop. In: Dupont B (ed): Immunobiology of HLA, Volume I; Histocompatibility Testing 1987. New York: Springer-Verlag, 1989.
31. Mickelson EM, Masewicz SA, Hansen JA. Variants of HLA-DRw52 defined by proliferative T-lymphocyte clones. In: Dupont B (ed): Immunobiology of HLA, Volume II; Immunogenetics and Histocompatibility. New York: Springer-Verlag, 1989.
32. Knowles RW, Flomenberg N, Horibe K, Winchester R, Radka SF, Dupont B. Complexity of the supertypic HLA-DRw53 specificity. J Immunol 1986;137:2618–2826.
33. Kaplan C, Morel MC, Hermans P, Tkaczuk J, Naskatsuji T, Fernandez N, et al. DQw1 (DQw5, DQw6): 2-D gel patterns. In: Dupont B (ed): Immunobiology of HLA, Volume I; Histocompatibility Testing 1987. New York: Springer-Verlag, 1989.
34. Reinsmoen NL, Eckels D, Mickelson E, Honeyman M, Farrel C, Hansen JA, et al. DQw1 report: Antigen report of the cellular studies of the Tenth International Histocompatibility Workshop. In: Dupont B (ed): Immunobiology of HLA, Volume I; Histocompatibility Testing 1987. New York: Springer-Verlag, 1989.
35. Radka S, Fuggle S, Mazzilli MC, Carter C, Rust N, Winchester R, et al. DQw2: 2-D gel patterns. In: Dupont B (ed): Immunobiology of HLA, Volume I; Histocompatibility Testing 1987. New York: Springer-Verlag, 1989.
36. Mazzilli C, Gandini E, Lanchbury J, Madrigal A, Ballas M, Fuggle S, et al. DQw3 (DQw6, DQw9): 2-D gel patterns. In: Dupont B (ed): Immunobiology of HLA, Volume I; Histocompatibility Testing 1987. New York: Springer-Verlag, 1989.
37. Mickelson E, Nepom GT, Nisperos B, Hansen JA. DQw3 variants defined by cloned alloreactive T cells. Hum Immunol 1988;21:63.
38. Lundin EA, Gaudernack G, Paulsen G, Sollid LM, Thorsby E. Epitopes on HLA-DQw3 molecules recognized by T-lymphocyte clones. In: Dupont B (ed): Immunobiology of HLA, Volume II; Immunogenetics and Histocompatibility. New York: Springer-Verlag, 1989.
39. Sharrock C, Lundin K, Zier K, Zeevi A, Thomsen M, Eiermann T, et al. DR4 and DQw3: Antigen report of the cellular studies of the Tenth International Histocompatibility Workshop. In: Dupont B (ed): Immunobiology of HLA, Volume I; Histocompatibility Testing 1987. New York: Springer-Verlag, 1989.
40. Tajima Y, Kojima H, Ishikawa N, Hawkin S, Fukasawa Y, Takada A, et al. Cellular analyses of DQ antigens. In: Dupont B (ed): Immunobiology of HLA, Volume II; Immunogenetics and Histocompatibility. New York: Springer-Verlag, 1989.
41. Kulova L, Maeda H, Bontrop R, Rust N, Fuggle S, Carter C, et al. DQw4 (Wa): 2-D gel patterns. In: Dupont B (ed): Immunobiology of HLA, Volume I; Histocompatibility Testing 1987. New York: Springer-Verlag, 1989.
42. Hurley CK, Steiner N, Gregersen PK, Silver J, Gorski J, Hartzman R, et al. Evolution of DQw2-related haplotypes. In: Dupont B (ed): Immunobiology of HLA, Volume II; Immunogenetics and Histocompatibility. New York: Springer-Verlag, 1989.
43. Seyfried CE, Mickelson E, Hansen JA, Nepom GT. A specific nucleotide sequence defines a functional T-cell recognition epitope shared by diverse HLA-DR specificities. Hum Immunol 1988;21:289.
44. Eckels D, Sell T, Zier K, Mickelson E, Hansen JA, Flomenberg N, et al. T-cell clones recognizing shared or cross-reactive epitopes: antigen report of the Tenth International Histocompatibility Workshop. In: Dupont B (ed): Immunobiology of HLA, Volume I; Histocompatibility Testing 1987. New York: Springer-Verlag, 1989.
45. Knowles RW, Kienzle B, Winchester R, Mathews E, Bouhallier O, Fauchet R, et al. DRw53: 2-D gel patterns. In: Dupont B (ed): Immunobiology of HLA, Volume I; Histocompatibility Testing 1987. New York: Springer-Verlag, 1989.
46. Madrigal A, Bontrop R, Mazzilli MC, Kaplan C, Wong K, Knitter-Jack N, et al. DR5 (DRw11, DRw12): 2-D gel patterns. In: Dupont B (ed): Immunobiology of HLA, Volume I; Histocompatibility Testing 1987. New York: Springer-Verlag, 1989.
47. Charron D, Hermans P, Ballas M, Brahmi Z, du Toit E, Martell R, et al. DRw6 (DRw13, DRw14): 2-D gel patterns. In: Dupont B (ed): Immunobiology of HLA, Volume I; Histocompatibility Testing 1987. New York: Springer-Verlag, 1989.
48. Goiski J, Eckels DD, Tiercy J, Ucla C, Mach B. Sequence analysis of the DRw13 beta chain genes: the Dw19 speci-

ficity may be encoded by the DR beta III locus. In: Dupont B (ed): Immunobiology of HLA, Volume I; Histocompatibility Testing 1987. New York: Springer-Verlag, 1989.
49. Coppin H, Huchenq A, Garner JM, dePreval C. Cloning and sequence analysis of the DRβ gene coding for the specificity HLA-D/DR-BON. In: Dupont B (ed): Immunobiology of HLA, Volume II; Immunogenetics and Histocompatibility. New York: Springer-Verlag, 1989.
50. Tkaczuk J, Thomsen M, Calot M, Ohayon E, Cambon-Thomsen A. Structure of the HLA-D/DR-BON antigen. In: Dupont B (ed): Immunobiology of HLA, Volume II; Immunogenetics and Histocompatibility. New York: Springer-Verlag, 1989.
51. Bjorkman PJ, Saper MA, Samraoui B, Bennett WS, Strominger JL, Wiley DC. Structure of the human class I histocompatibility antigen, HLA-A2. Nature 1987;329:506.
52. Bjorkman PJ, Saper MA, Samraoui B, Bennett WS, Strominger JL, Wiley DC. The foreign antigen binding site and T-cell recognition regions of class I histocompatibility antigens. Nature 1987;329:512-518.
53. Brown JH, Jardetzky T, Saper MA, Samraoui B, Bjorkman PJ, Wiley DC. A hypothetical model of the foreign antigen binding site of class II histocompatibility molecules. Nature 1988;332:845.
54. Marrack P, Kappler J. T cells can distinguish between allogeneic major histocompatibility complex products on different cell types. Nature 1988;332:840.
55. Song ES, Linsk R, Olson CA, McMillan M, Goodenow S. Allospecific cytotoxic T-lymphocytes recognize an H-2 peptide in the context of a murine major histocompatibility complex class I molecule. Proc Natl Acad Sci 1988;85:1927.
56. Eckels D, Zeevi A. ASHI workshop summary report of the science and education subcommittee: Structural and functional relationships of human class II MHC molecules. Human Immunol 1986;15:68-74.

Joint Reports: Restriction Fragment Length Polymorphism (RFLP) with HLA Region Probes (Southern Blot)

Standardized Southern Blot Workshop Technique

A. Marcadet,[1] P. O'Connell,[2] and D. Cohen[1]

Establishing a Reference Protocol

For the Southern blot section of the Tenth Histocompatibility Workshop, it was necessary to establish a suitable reference protocol with reproducible, viable, and simple techniques. This is the first workshop with a Southern blot section. Moreover, this protocol had to take account of:

1) peculiarities of the HLA system, and
2) organization and amount of work of each laboratory,
3) necessity to compare: a) the different results (films) from the same laboratory, b) the results of different labs using the same DNAs, enzyme and probe, and c) the results from core cells and from families.

These data were analyzed as follows:

1) The HLA system contains many genes, the products of which are highly polymorphic; the sequences of these genes are very homologous. This means that hybridization with an HLA probe of restriction digests from different genomic DNAs results, most often in *many* fragments with different molecular weights, *from 100bp to 20–30Kb*. To distinguish between fragments (long or small) very close to each other, it was necessary to perform for the same DNAs two electrophoresis procedures (S and L) with different agarose gel concentrations, voltages, and migration times. It was also necessary to use many molecular weight markers from 500 bp to 30 Kb (digests of λ phage DNA with HindIII or KpnI or SmaI or BsteII).

2) Each laboratory had to hybridize each time at least six membranes and perform at least six successive hybridizations. We decided to choose the labeling procedure of DNA with random primers (cheaper than "nick-translation"), and check the quality of the results obtained when plastic boxes (Tupperware) were used for prehybridization, hybridization, washing, and dehybridization. The efficiency for rehybridization of the chosen membrane was tested (BIOTRACE).

3) To be able to compare the films obtained with the same enzyme and probe, we decided to use a technique where the molecular weight markers were detected on each membrane after each hybridization (even several weeks after the first hybridization). We established and checked a new method where a small quantity of the marker mix was loaded on the gels (ng amounts), transferred to the membrane, and then hybridized with radioactive complementary DNA. The molecular weight markers (M1 and M2) were composed of a mix of digests of λ phage DNA, and the specific probe was total λ phage DNA. This DNA was labeled (pg amounts of DNA is sufficient for a signal) at the same time as the HLA probe used for hybridization.

As the molecular weight markers are not visible under UV, and because electrophoresis of DNA with identical conditions of agarose concentration, voltage, and running time often give different results in migration of the fragments, we decided to add to the gels a marker visible under UV, a large quantity of linearized plasmid, PAT 153.

Additional peculiarities have been introduced to this protocol to facilitate the work, such as special combs for 30 wells. All times of experiment and quantities of products have been calculated and tested several times with the particular conditions used for this workshop (quantity of loaded DNA-genomic or markers, enzymes, etc.). Twenty laboratories have tested with us different parameters such as buffers for transfer, gel loading, and dehybridization, binding of DNA to membranes by UV treatment, baking or dry air, etc.

DNA Extraction Protocol

Buffers and Preparation of Phenol

White cells lysis buffer: WCLB (filtered stock solution)
- Tris pH 7.6   10 mM
- EDTA pH 8   10 mM
- NaCl   50 mM
- SDS   0.2%
- Proteinase K   200 μg/ml

Red cells lysis buffer: RCLB (filtered stock solution)
- Tris pH 7.6   10 mM
- $MgCl_2$ 5 mM
- NaCl   10 mM

Phenol: prepared on the day of use or previous day (and stored in a dark place at 4°C):
- We use phenol from Carlo Erba ref:451287 (no need to distill it)
- Mix 3 volumes of melted phenol with 1 volume of chloroform-isoamyl alcohol (24/1), referred to hereafter as PCI
- Add 40 ml of PCI to 10 ml Tris 1M pH 8, in 50 ml Falcon tube mix
- Spin 10 min at 1500 rpm/min
- Remove upper aqueous phase
- Add 2ml Tris 0.1M pH 8, 0.2% mercaptoethanol; mix
- Spin 10 min at 1500 rpm/min
- Eliminate upper aqueous phase

DNA Extraction

a) *Lysis*

1) *From fresh peripheral blood*:
Collect 30 ml of blood from one individual and mix it

Participating Laboratories: [1]FRACOH, [2]US7LAL

Immunobiology of HLA
Volume I

with 5 ml of EDTA 5% (and not with heparin, which inhibits restriction enzymes)
On the day of collection, spin the mix 5 min at 1500 rpm/min
Remove the upper phase of plasma without touching the leukocytes
Add red cell lysis buffer (RCLB) up to 50 ml
Mix carefully
Spin 10 min at 1500 rpm
Remove supernatant of lysed erythrocytes
Again add RCLB, up to 50 ml
Spin 10 min at 1500 rpm
Remove the supernatant
Resuspend the pellet in 2 ml of RCLB
Quickly add the white cell lysis buffer (WCLB), up to 15 ml
Gently agitate, horizontally, in a rotary shaker (50 rpm) overnight at 42°C
Then use the protocol for the phenol-chloroform extraction

2) *From frozen peripheral blood*:
Sampling on EDTA (no heparin, since it inhibits restriction enzymes)
Freeze the white blood cell pellet, obtained after two red cell lysis cycles at −80°C. Extract DNA within 1 month
Thawing must be done quickly: place tubes in hot water (50°C) and leave until thawed
Quickly add WCLB, up to 15 ml, and proceed as in a1)

3) *From cell lines and frozen lymphocytes*:
Pellet 300 to 350 × 10^6 cells in a 50 ml Falcon tube
Resuspend in 2 ml of RPMI or PBS
Quickly add WCLB, up to 15 ml
Gently agitate, horizontally, in a rotary shaker (50 rpm) overnight at 42°C

b) *Phenol chloroform extraction*
Add 15 ml of PCI (1 vol)
Shake 10 min in a tridimensional shaker or shake gently by hand (a complete emulsion must be obtained)
Spin 3 min at 1500 rpm/min
Discard the lower phase of PCI with a long metal needle (18 or 20 gauge) attached to a 50 ml glass syringe; leave the interphase
Extract a second time with PCI, agitating for 5 min only
Extract two times, 5 min each, with chloroform-isoamyl alcohol 24/1 (discard the lower phase of chloroform after each extraction)

c) *Isopropanol precipitation*
Isopropanol allows the use of one volume of alcohol, instead of two, with ethanol. Moreover, it does not precipitate RNA (precipitated RNA can lead to an overestimated amount of DNA when checked by OD)
Add 300 µl of NaCl (3 M) to obtain a final concentration of 60 mM, followed by up to 30 ml (1 vol) of cold 100% isopropanol
Agitate gently until the precipitate has formed; precipitation will have been achieved when the precipitate floats
Put the precipitate into a 5 ml sterile Falcon tube
Rinse three times with 4 ml *70%* ethanol (do not use 100% ethanol or isopropanol)
Dry 5 min in a vacuum

d) *Resuspension of precipitated DNA*
1) It is essential to wash the precipitate with 70% ethanol
2) Next, resuspend it in Tris 1 mM, EDTA 0.1 mM (filtered)
3) It has recently been established that resuspension is delayed when microbubbles are present between the precipitate and the aqueous solution.
These microbubbles can be eliminated when resuspension is carried out in an unheated speed vacuum. Under these conditions, the DNA is resuspended in a shorter time
4) If you do not have access to a speed vacuum, resuspend precipitate in a rotary shaker (slowly) for several hours (better at 37°C)
5) The optimal final concentration is 200 µg/ml. It is strongly recommended that you obtain *equal concentrations of all samples* to assure equal loading in the gel. This equal loading is essential in family studies to discriminate between homozygosity and heterozygosity. For this purpose, it is recommended that you load the gel only when you are sure that homogeneity has been obtained. Clearly, the *initial estimate of DNA concentration* (before digest) is the *decisive step* in achieving this goal. We generally check the OD of each sample (every 24 hours) by three successive measurements of the sample, diluted at approximately 200 µg/ml. This is followed by a visual check of the DNA size by electrophoresis in 0.7% agarose gel. You must obtain high molecular weight DNA, greater or equal to λ phage, with equal amounts for each sample. A second check is carried out after the digest during control (see below). There is a final check after migration: a photograph must be taken of each migration and sent to the standardization center.

e) *Storage*
Make aliquots of 50 µg
Store at −80°C (for several years)
Keep one aliquot for use, stored for a few months at 4°C

Preparation of Probes. Most of the probes used in this study were cDNAs in the plasmids PBR 322 or PAT 153. Cloning was performed by classic techniques (e.g., insertion into a restriction site of the plasmid, transformation of competent bacteria, selection of recombinants on appropriate media with antibiotics).

Amplification of the plasmids was performed according to classical techniques, e.g.,

Culture of transformed bacteria
Addition of chloramphenicol for amplification
Lysis of bacteria (lysozyme; SDS)

CsCl gradient and purification of plasmids DNAs by phenol-chloroform extractions
Ethanol precipitation then resuspension in TE

We preferred to send to participating laboratories cDNA inserts rather than total plasmids. These inserts were prepared from total plasmids also by classical methods, e.g.,

Digestion of a large quantity of plamids DNA
Separation of inserts from the vectors by electrophoresis
Excision from the gel of inserts, followed by electro-elution
Purification of inserts by extraction with phenol-chloroform, DEAE chromatography column, ethanol precipitation and suspension in TE 1-0.1 (concentration: 5 ng/μl

Note: It is not always necessary to excise inserts from plamids in order to label the probe. The recombinant plasmid can be easily labeled and directly used to hybridize. Another way to label an insert can be performed as follows:

Digest a small quantity of plasmid DNA
Separate the insert from the vector in an agarose-gel (Low Melting Point Agarose)
Excise the insert in a slice of agarose (Eppendorf tube)
Melt the gel at 65°C in a water bath for 5 min (gel must be completely melted)
Aliquot it: 25 ng of DNA/tube and store at −20°C
At the moment of use, thaw the sample then perform labeling of 25 ng of insert by the random primer method (III-VII).

Southern Technique

Genomic DNA Endonuclease Treatment. (For all steps up to loading, use the same pipet but change the tip for each new DNA). For core DNA, it is recommended that you digest *40 μg* per sample:

In Eppendorf tubes (1.4 ml)
Final enzyme concentration: 5 u/1 μg of DNA
Final DNA concentration: 40 μg/300 μl

Temperature and buffer according to manufacturer's recommendations (given in "Specific Recommendations"). The temperature is 37°C for all enzymes except TaqI (65°C). The buffer must be prepared on the day of use from filtered stock solutions. Do not use BSA (bovine serum albumin) even when indicated by the manufacturer (unnecessary). The easiest buffer to prepare is the 10X.

Time: overnight (except for TaqI).
We advise you to proceed according to the following protocol, paying close attention to the order:

- Always resuspend gently DNA that has just been thawed
- Put the DNA into Eppendorf tubes
- Add the distilled water (see table, next column)
- Add the T10X buffer
- Then add spermidine *systematically* at a final concentration of 2mM (store solution at −20°C: 200 mM); the spermidine improves restriction endonuclease digests
- Finally, add the enzyme. Always resuspend the enzyme solution before pipeting it. The enzyme should be stored at −20°C. It must be kept on ice during use, and restored at −20°C immediately after use. The enzyme is added in three stages, in three equal parts (keep it at −20°C between each stage): First, add one third of the total amount necessary to digest all the DNA. Mix gently. Place in a water bath at 37°C. One hour later, add second amount (one-third) of enzyme. Place at 37°C. Add the final quantity (one-third) of enzyme and leave the digestion overnight at 37°C.
- Special protocol for TaqI digest: This digestion is done at 65°C. Digestion is completed in only 3 hours. You should proceed in the same way as for the other enzymes, except that the waterbath should be at 65°C. After 1 hour of digestion, spin the tubes in a microfuge for 1 min, then add the second quantity (one-third) of enzyme. After one further hour, spin again before adding the last quantity of enzyme (one-third). Leave the digestion at 65°C for another hour, then spin and control as indicated below.

Example

	DNA	Buffer	Spermidine	Enzyme	H_2O
Initial concentration	200 μg/ml	10X	200 mM	10 or 20 u/μl	
Volume used	200 μl	30 μl	3 μl	In 3 stages 20 or 10 μl (for 40 μg DNA)	Up to 300 μl
Final concentration	40 μg/300 μl	1X	2 mM	5 μ/μg DNA	

Digestion Control

Spin each tube for 1 min in a microfuge.
Run 1 μg of each digest (7.5 μl) on a large 0.7% agarose gel for at least 2 hours at 70 volts.

a) In most cases, a smear is obtained: The reaction is stopped by adding EDTA, 400 mM (11.5 μl), to obtain a final concentration of 15 mM. Digests are then stored at 4°C, if loaded the same day, or at −80°C.
b) In very rare cases (< 1%), a smear is not obtained and the DNA is not (well) hydrolyzed: Add one-third of the total amount of enzyme previously used, plus 2 mM final concentration of spermidine

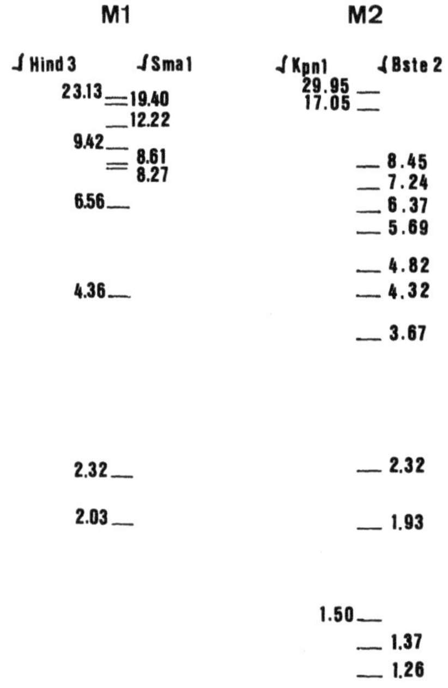

Figure 1. S conditions.

and incubate for 2 hours. Check, using the same procedure. If there is still no smear, and if the enzyme quality has been checked, this DNA sample is considered unsuitable for endonuclease cleavage. The main reasons for this are: 1) DNA is not well resuspended, or 2) is not clean enough (very rare); in the latter case, check OD ratio 260/280 – it must be between 1.7 and 2. If it is less, the DNA is not clean enough, in which case, do another phenol extraction and precipitate it again.

c) On rare occasions, over-digestion can occur. If this happens, incubate for 3 to 4 hours only.

Sample Concentration. This can be done in two ways:

speed vacuum – the simplest method
ethanol precipitation, if speed vacuum is not available.

1) *In a speed vacuum centrifuge*:
   Place, in an Eppendorf tube, a volume of reaction containing a sufficient amount of DNA for one well only: 7 µg (for a single DNA sample, as many tubes as loadings). After control and stop reaction, you will have 39 µg in 304 µl; 7 µg corresponds to 55 µl.
   Concentrate the samples in open tubes in a speed vacuum centrifuge (Savant) connected to a lyophilizer with the heater turned on (5 to 10 min).
   Check that the volume expected (16 µl) has been obtained. If the volume obtained is less than 16 µl, dilute with distilled water; if it is greater than 16 µl, leave it to concentrate for a longer period.
   The high salt concentration of these samples does not affect migration.

2) *Ethanol precipitation* (ref Focus, 1985; 7: no. 4)
   In an Eppendorf tube, place the volume of reaction containing the desired amount of DNA (all reactions can be precipitated in one tube)
   Add Na acetate (1 M) to obtain a final concentration of 0.3 M followed by two volumes of 100% ethanol
   Shake gently
   Keep on *ice* (not dry ice) for 10 min
   Spin *30 min* in a microfuge (speed: 12,000 *g* at least)
   Remove ethanol
   Dry under vacuum
   Resuspend in TE 1 to 0.1 (wash the walls of the tube carefully) at a final concentration of 7 µg/16 µl
   Allow 30 min for complete resuspension, at 37°C
   Spin ten seconds in a microfuge to bring down the condensate, then resuspend

Loading. Seven µg genomic DNA per well (for special combs with 30 teeth):

Loading buffer 5X: Tris 50 mM, pH 7.6, EDTA 50 mM, SDS 0.5%, BP blue 0.1%, sucrose 15% (Stored solutions at 4°C: Tris pH 7.6 67 mM, EDTA 67 mM, SDS 0.67%, BP blue 0.13% and sucrose 60%)
Add 4µl of loading buffer 5X for 16 µl of DNA
Heat samples for 5 min at 65°C *including the λ markers* (see molecular weight markers)
Spin in a microfuge for 1 min, then resuspend
Load, *very slowly,* 20 µl per well (for the WS combs)
Following *loading order* indicated in "Rules and Nomenclature"

Molecular Weight Markers. We have provided you with three different sets of markers labeled M1, M2 and P, respectively. M1 contains a mixture of single digests of λ DNA digested with HindIII and SmaI: 1 µl of M1 contains 5 ng of λ HindIII and 5 ng of λ SmaI. M2 contains a mixture of single digests of λ DNA digested with KpnI and Bste II: 1 µl of M2 contains 5 ng of λ KpnI and 5 ng of λ BsteII. P is the linearized plasmid, PAT 153, at a concentration of 50 ng/µl. A list of molecular weights is given in Figures 1 and 2. P (500 ng) will be loaded into a well at one extremity of each gel and will serve as a visual check of the length of migration, since it can be detected with ethidium bromide.

Only 100 ng of M1 and M2 will be loaded and it will, therefore, not be visible under UV. These λ fragments will be detected by hybridization with radiolabeled λ DNA mixed with the probe.

A small amount (1%) of λ DNA is added to the labeling mixture (see "Labeling"). Thus, at each cycle of hybridization, the transferred λ fragments will be lightened. This will allow markers to be made visible on every autoradiogram, even after several cycles of hybridization. The ratio of the amount of λ DNA fragments loaded onto the gel and the amount of λ DNA labeled have been carefully calculated and checked several times in Paris. It is important to respect this ratio and *not to forget λ DNA in the labeling reaction*.

Thus, you will have to load, per well:

For P: A sample of 20 μl containing:
4 μl loading buffer 5X
10 μl P (500 ng)
6 μl H_2O
For M1 or M2: Samples of 20 μl containing
4 μl loading buffer
10 μl M1 or M2 mixture
6 μl H_2O

Note: To prepare the molecular weight markers, individually perform one digest of λ phage DNA with KpnI, one with SmaI, one with HindIII, and one with BsteII, using the same protocol as for genomic DNA (times can be reduced to 3 hours). Control the digestions, and then for M1, mix equal quantities of λ HindIII and λ SmaI (for M2, equal quantities of λ KpnI and λ BsteII). Precipitate with ethanol as indicated in the protocol and resuspend in TE 10-1 at a concentration of 5 ng of each digest in 1 μl. Linearize double-stranded PAT 153, for example, with Eco RI; control the digest; precipitate with ethanol then resuspend in TE 10-1 at a concentration of 50 ng/μl.

Gel Electrophoresis (BRL H4)

Buffer

Two litres of TEA 1X × 125 μl of ethidium bromide, 8 mg/ml (final concentration: 0.5 μg/ml):
TEA 50x:

Tris 40 mM × 50
Glacial acetic acid 20 mM × 50
EDTA 2 mM × 50

Adjust to pH 8 with acetic acid and filter.

Gel dimensions: 20 cm × 25 cm × 0.6 cm = 300 ml (BRL H4).

Gel concentration: For resolution of large fragments (L), prepare 0.6% agarose gel. For resolution of small fragments (S), prepare 0.9% agarose gel.

- Mix agarose (Sigma type II), 6 ml of TEA 50X (TEA 1X final), and water
- Boil
- Cool, then add 18 μl of ethidium bromide (8 mg/ml) before pouring (final concentration: 0.5 μg/ml)

Before pouring the gel, check that the teeth of the comb are 1 mm from the plate (you must be able to slide one layer of Whatmann 3MM between the teeth and the plate).
Combs: teeth are 5.5 mm × 1 mm. Load 7 μg of DNA per well.

Migration:

- *Allow the buffer to recirculate* with a peristaltic pump (this is very important for long periods of electrophoresis)
- For resolution of large fragments (L), 0.6% agarose gels are run 43 hours at 40 volts (the first half hour after loading, they should be run at 30 volts, then 40 volts)

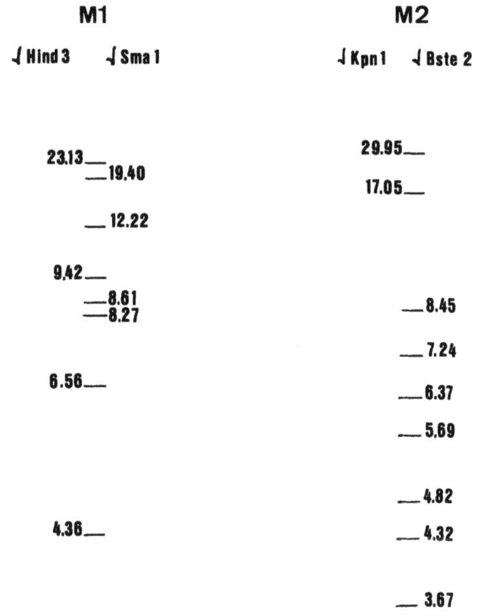

Figure 2. L conditions.

- For resolution of small fragments (S), 0.9% agarose gels are run 48 hours at 25 volts
- Set the current (constant voltage)
- Check the migration of PAT 153 several hours before the end of running time. Check it every half hour when P has reached 160 mm for L or 60 mm for S. *Stop migration only when P has reached*:

173 mm for L (approximately 43 hours)
70 mm for S (approximately 48 hours)

Very important: Some plastic gel trays recently produced by BRL transmit UV light poorly. In this case, it can be very difficult to detect P. Please test your trays before starting the work. If you are unable to change these defective trays, try to obtain a portable UV light to check the P migration. (It is dangerous to pull the gel out of the tray.)

Advice:

For L conditions
Set current at end of the afternoon
Check after 41 hours in the morning
For S conditions
Set current at end of the morning
Check after 46 hours

- If two or more gels are run on the same power supply, *check each gel separately* as considerable differences can be observed.

Photograph the final migration showing the wells and all the samples.

Transfer onto Biotrace Membranes (Alkaline Transfer)

Gel treatment

Under UV cut the gel at the level of the wells.

Discard the track corresponding to PAT 153, which must not be transferred since it would be detected strongly by probes contaminated by plasmids.

Then, at room temperature, with gentle agitation:

1) Soak the gel in 400 ml 0.15 M HCL, for precisely 10 min (*no longer and no less*). Briefly rinse the gel with distilled water to remove excess acid.
2. Soak the gel in 400 ml of 0.4 N NaOH for 30 min. While the gel is equilibriating in the NaOH solution, prewet the nylon membrane (Bio-Trace) in distilled water for 30 min before starting transfer.

The transfer system will include (from the bottom to the top):

- Plexiglass plate across a baking dish containing 0.4 N NaOH.
- Two prewetted (in 0.4 N NaOH) Whatmann bridges, each consisting of one layer of Whatmann 3MM paper.
- Gel upside-down, surrounded by a frame of film or equivalent. To get your gel upside-down, cover it with one layer of Whatmann. When this layer is wet, cover with a second layer. Then press one hand on top of the two sheets of Whatmann paper and invert the dish with the gel. The gel now rests in your hand, upside-down. Dispatch the gel with the two layers of Whatmann on the bridges. This step is best performed with two people, one to hold the gel and the other to remove the dish. This procedure is especially important for 0.6% gel.
- The prewetted (in distilled water) membrane filter (Bio-Trace: 21 × 23 cm). Check that the membrane is in no contact whatsoever with the Whatmann surrounding the gel (cut the membrane a little if necessary). Dispatch it to have one edge of the membrane exactly at the place of the wells.
- Four layers of prewetted (in 0.4 N NaOH) Whatmann 3MM paper.
- Ten cm of paper towels.
- A glass plate.
- 0.5 kg weight only.

After 1 or 2 hours, the wet layers of paper towels are replaced by dry paper towels. Allow DNA to transfer overnight with a sufficient quantity of 0.4 N NaOH solution as transfer buffer.

After transfer, rinse off any adhering agarose by rinsing 5 to 10 min in 2X SSPE (cf Solutions in "Hybridization"). Then place the washed membrane on dry 3MM filter paper and allow it to air dry at room temperature. *Do not bake.*

Check under UV that DNA is transferred onto the membrane. For the 0.9% gel, you should be able to clearly see the DNA on the membrane; for the 0.6% gel, it may be more difficult but still possible. If not, do not use the membrane.

Store the membrane in a plastic bag or Tupperware box in prehybridization mix at 4°C prior to use.

Radiolabeling DNA to High Specific Activity with Random Primer or Multiprime. (Ref: Feinburg and Vogelstein, Anal. Biochem. 1983;132:6–13, Feinburg and Vogelstein, Anal Biochem 1987;137:266–267):

5X oligomer labeling buffer: OLB 5X

- Solution "0":
  1.25 M Tris pH 8.0
  0.125 M $MgCl_2$
  Store at 4°C
- Solution "A":
  1 ml solution "0"
  18 μl β2 mercaptoethanol 3M
  5 μl dATP each triphosphate previously
  5 μl dTTP dissolved at 0.1 M in 3 mM Tris pH 7.0
  5 μl dGTP 0.2 mM EDTA
  Store at −20°C
- Solution "B":
  2 M Hepes, titrated to pH 6.6 with 4 M NaOH
  Store at 4°C
- Solution "C":
  Hexadeoxynucleotides (Pharmacia: $pd(N)_6$ No. 27-2166-01; 50 units) suspended in 3 mM Tris pH 7.0, 0.2 mM EDTA at 90 OD units/ml (555 μl for 50 OD units).
  Store at −20°C

5X OLB:

Mix A:B:C in a ratio of 100:250:150.

OLB is stored at −20°C and is not adversely affected by repeated freeze/thaw.

You have been provided with 1.85 ml 5X OLB and 50 OD units of hexadeoxynucleotides (Pharmacia: $pd(N)_6$ No. 127-2166-01; 50 units). The multiprime kit from Amersham can be used.

Labeling Procedure:

Use tubes with perforated caps

For one 21 cm × 23 cm membrane:

- Mix 25 ng of probe (5 μl of probe in TE 1-0.1 at 5ng/μl) with *250 pg of λ DNA* (1 μl of λ DNA in TE 1-0.1 at 250 pg/μl).
- Boil for 2 min, then put on ice for 5 min
- Spin 10 sec in microfuge
- Add 5 μl OLB 5X
- Then 8.7 μl H_2O
- Then 4 μl αP32 dCTP at 10 μCi/μl (> 3,000 Ci/mMole)
- Then finally, 1.3 μl Klenow (initial concentration = 2 u/μl)
- Follow this order; the final volume is 25 μl
- Label at room temperature from 3 hours to overnight
- Eliminate nonincorporated nucleotides by passing the mixture through a 9 ml Sephadex G200 column (in a 10 ml plastic pipet)
- Count an aliquot for calculation of specific activity

- *Remark*: Labeling of larger quantities of probe can be done in a single tube if proportions are respected (Max = 6).

Do Not Forget λ DNA

Hybridization.

Solutions:

- Formamide deionized on the day of use with 5 g/100 ml of mixed bed resin AG 501-X8 (Biorad, cat ref. 142 64 25) for 30 min

Eliminate the resin by passing through a coffee filter

- SSPE 30X: (agitate overnight at room temperature)
  NaCl 5.4 M
  NaH_2PO_4 0.3 M
  EDTA 30 mM

If necessary, adjust with NaOH to pH 7.7

- "MAGIC" Denhart. For 200 ml:

1) Prepare 100 ml of 2% BSA (Fraction V, Boehringer) according to the following protocol:
   2 g of BSA in 80 ml H_2O
   Adjust pH to 3 with HC1 2N

Boil for 15 min, then put on ice for 10 min
   Adjust pH to 7.5 with NaOH 2 M on ice
   Add H_2O up to 100 ml

2) Prepare 100 ml of 2% PVP 40 and 2% Ficoll (400); autoclave it for 30 min at 100°C
3) Mix the 100 ml of 2% PVP–Ficoll and the 100 ml of 2% BSA; MAGIC Denhart can be aliquoted and stored at −20°C

- Salmon sperm DNA: ss DNA (10 mg/ml)
  Sonicated for several minutes (ss sonicated DNA can be aliquoted and stored at −20°C)
  If necessary, denature on the time of use by boiling for 10 min, then putting on ice for 10 min
- Dextran sulphate 50%: dissolve it by overnight agitation at 4°C; can be aliquoted and stored at −20°C
- Pre-hybridization mix (for Bio-Trace membrane):
  Formamide 50%
  Denhart 0.1%
  SSPE 5X
  SDS 1%
  Dextran sulphate 5%
  Salmon sperm sonicated and denatured DNA 200 μg/ml
- Hybridization mix for 20 ml (for Bio-Trace membrane):

1) Prepare in a Falcon tube a mixture (Mix 1) containing:

Final		Stock	Volume
Formamide	50%	100%	10 ml
Denhart	0.1%	1%	2 ml
SSPE	5X	30X	3.3 ml
Dextran	5%	50%	2 ml
SDS	1%	10%	2 ml

2) Mix in an Eppendorf tube (with perforated cap):
   25 ng of radio-labeled probe (containing 250 pg of radiolabeled λ phage)
   500 μl of ss sonicated DNA (10 mg/ml)
   Boil for 10 min, then put on ice for 5 min
   Mix gently and carefully with mix 1 (19.3 ml) in a Falcon tube

Hybridization of a single membrane (21 cm × 23 cm):

- Prehybridization:
  Place the membrane in a plastic bag with 30 ml of prehybridization mix and let it prehybridize from 4 hours to overnight at 42°C with agitation.
- Hybridization:
  Replace the prehybridization mix with 20 ml of hybridization mix
  Massage for homogenization
  Let it hybridize 40 h at +42°C with agitation
- Washing:
  With gentle agitation:
  Twice, 5 min each, with 2X SSPE at room temperature
  Once, 15 min, with 2X SSPE, 0.5% SDS at +65°C
  Once, 15 min with 0.5X SSPE at +65°C
  For Any Probe

Hybridization of several membranes (21 cm + 23 cm) with the same probe:

- Prehybridization and hybridization are performed in one single closed Tupperware box (20 cm × 20 cm) containing the 6 or more (up to 15) membranes. The first membrane is placed on the bottom of the box, recovered with the mix: 30 ml for prehybridization and 20 ml for hybridization. The second membrane is placed on the first one (avoid bubbles). Pour mix, again 30 ml for prehybridization and 20 ml for hybridization, etc. Let the membranes prehybridize or hybridize in the same way as for the single membrane.
- Washes are also performed in a single closed Tupperware box (20 cm × 20 cm) containing 160 to 200 ml of washing solution per membrane (wash conditions as previously described). No more than six membranes per box with 1000 ml of washing solution.

Exposure

Dry the membrane on air (on a sheet of Whatmann paper)

Place it on a Whatmann 3 MM sheet, then recover it with Saran wrap

Expose it overnight on an X-OMAT-AR film (Kodak) between two screens Li+ *overnight* at −80°C

Reexpose on a new film for a longer period, i.e., 3 to 6 days depending on the results of the overnight exposure

Dehybridization

In a water bath at 42°C with agitation:

- 30 min in NaOH 0.4 M
- 30 min in Tris 0.2 M pH 7.5, SDS 0.5%, SSPE 0.1X

Check de-hybridization by overnight exposure

The membrane must be stored at 4°C in a Tupperware box or sealed in a plastic bag, in prehybridization mix.

- Several membranes can be dehybridized together in a single closed Tupperware box.

Criticisms of the Reference Protocol

It must be repeated that this protocol has been established to study the HLA system. To study other systems, many of the conditions applied here are not necessary: for example, the molecular weight markers, the concentration of agarose-gel, times, and voltages used for electrophoresis, the quantity of loaded DNA, etc.

From comments of the different participating laboratories, we can add the following.

1) The *suspension of genomic DNA* was often a *very long step*, even when microbubbles had been eliminated. Because the concentration must be appropriate to perform an efficient digestion, and to be able to distinguish between homozygotes and heterozygotes, we agree that it is necessary to achieve a stable DNA concentration before digestion with a restriction enzyme, perhaps as long as 1 to 2 weeks.

2) For several participating laboratories, the times of migration for PAT 153 to reach the distances indicated in the protocol for S and for L gels was different and sometimes *very different* from that indicated. This means that, unhappily, even when the same voltage and a same concentration of agarose-gel is used, DNA migrations can be different in two electrophoresis systems. So, we advise, first, perform a migration with only molecular weight markers, with the conditions of the desired migration and observe the time necessary for PAT 153 to reach the desired distance. Then perform the migration with genomic DNA and check it under UV several hours before the time indicated by migrating the markers.

3) It was observed that when a Bio-Trace membrane has been used several times, it shrinks. This may lead to some difficulty in comparing autoradiograms of repeated hybridizations of same membrane.

4) It appeared that the depurination conditions of S and L gels have to be different. The efficiency of transfer of DNA using L conditions seemed to be inferior to that of DNA using S conditions. Time and/or HC1 concentration must be changed to improve the result of the "L transfer."

Acknowledgments. We wish to thank D. Beaud'huy-Lancelin, I. Le Gall, M. Martell, and A. Piscone for their help for technical problems, and E. May for secretarial assistance.

Organization and Design of the Southern Blot Component of the Tenth Histocompatibility Workshop

A. Marcadet,[1] B. Dupont,[2] and D. Cohen[1]

The Southern blot section of this Workshop had two main objectives: first, to introduce the Southern blot technology to the HLA scientific community, and second to define a standard set of polymorphic restriction fragments associated with the MHC genes.

The Pre-Workshop

The first effort was to elaborate a robust Southern blot protocol (from September 1985 to September 1986). This pre-workshop consisted in testing several key parameters of this technique. Twenty participating laboratories were involved, the different values or aspects being tested in parallel by groups of four laboratories. The main parameters tested were the conditions of migration, transfer, probe-labeling and dehybridization.

For the migration, the purpose was to find standardized migration conditions (i.e., for a given agarose concentration) to evaluate the electric field and migration time ensuring the best resolution. It appeared that from one laboratory to another, important variations occurred, even with the same migration parameters (voltage, amperage, time, and agarose concentration). Thus, it was decided that all laboratories would use the same electrophoretic conditions, but would stop the run when a visible marker (pAT153) had migrated a certain distance (see reference protocol).

Transfer conditions were carefully studied. Nylon membranes from several manufacturers as well as different transfer protocols were tested. Two more or less equivalent protocols emerged: one using MSI (Hybond) membrane, and the other using Bio-Trace membrane with alkaline transfer; the former was finally adopted for its greater simplicity. The labeling system selected was the hexamer priming method for two reasons: higher specific activity and use of a reduced amount of probe (5 to 10 times less than for nick-translation).

Participating Laboratories: FRACOH,[1] US7DUP[2]

The Workshop

The work comprised the following: first, the study of 70 homozygous cell lines of which 2 were duplicated (the core cell line work), then family studies, and finally, optional studies of disease associations, recombinant families, and rare haplotypes.

The Core Cell Line Work. DNA from the 70 cell lines was extracted at CEPH in Paris and sent to all participants with batch controlled restriction enzymes, probes, labeling kits, and DNA markers (see Appendix I). The 70 cell lines have been studied blind with 13 HLA probes (see Appendix II) and 12 restriction enzymes (see Appendix II). Each laboratory used a single enzyme. Each digested DNA was systematically subjected to two different migration conditions: L for optimal resolution of large fragments, and S for optimal resolution of small fragments. A comb with 30 wells was specially made for this Workshop and used by all participants. The order of loading was the same in all laboratories. Thus, each laboratory prepared six membranes, 3L and 3S, each containing markers in four lanes and 24 samples of digested DNA. All the details are described in the "rules and nomenclature" (see Appendix III). The six membranes were hybridized and washed together in a single Tupperware box (also distributed to all participants). The set of six membranes were hybridized successively with nine probes in each laboratory. Eighty laboratories were involved; they were assembled into 58 units, each unit being composed of 1 to 3 laboratories. The 58 units were organized into 12 groups, 10 of 5 units and 2 of 4 units. Each group was dedicated to the study of DNA from the 72 cell lines, a single enzyme and the 13 probes. In each group, each unit had to produce six membranes (3L and 3S) and hybridize with nine probes (see Appendix IV). The five (or four) units used in parallel the same five probes, in the same order: DRB, DQB, DPB, DQA, and class I. Three of five units (or two of four) then used the four class III region probes (C4, 210H, C2, and BF), while the remaining two used the four other class II probes (DRA, DOB, DPA, and DNA) (previous name DZ alpha) (see Appendix IV). Thus, the 156 enzyme/probe combinations also called *systems* were studied independently two to five times.

For standardization purposes, each unit input local data on floppy disks through an IBM-PC program LSD. The data included for each autoradiogram: markers mobilities, fragment mobilities, presence/absence assignment on the 72 cell lines (see Appendix V), and locus assignment. This locus assignment was indispensable for class II beta probes since a given beta probe can also hybridize with genes at other beta loci (i.e., DRB with DQB, DPB or DOB sequences), but giving fainter signals than the corresponding probe (see Appendix VI).

All the diskettes and autoradiograms converged towards 12 standardization centers. Each center was responsible for 13 systems comprising the four class II beta probes with a single enzyme and the nine remaining probes with another enzyme (see Appendix IV). A standardization program allowed the center to easily compare data from each unit and to assign definitively for each fragment a standard molecular weight (mean value from 2 to 30 measurements), definitive presence/absence assignment, and locus specificity. Floppy disks containing the standardized data then converged towards the analysis center, which will be described elsewhere.

The Family Work. Each unit had to contribute two families essentially selected with these conditions: two parents and four HLA different and nonrecombinant children. These two families had to be studied locally with six probes (DRB, DQB, DPB, class I, DQA, and C4) and six enzymes (TaqI, Eco RI, Eco RV, Bam HI, PvuII, HindIII) (i.e., 36 systems) under L and S conditions. Thus, in each unit, six membranes were produced and hybridized successively with six probes (see Appendix IV). Local diskettes and autoradiograms converged through a complex network towards standardization centers. The principle was that a given standardization center received all family data relevant to the system with which it was dealing during the core cell line RFLP standardization (see Appendix IV). Each standardization center first had to check the input of each unit, then assign each fragment to a standard core fragment and indicate new fragments (by comparing visually family blots with core cell lines autoradiograms). Standardization data then converged towards the analysis center.

The Optional Work. This included studies of disease associations, recombinant families, and rare haplotypes. The only condition was to use the Workshop probe/enzyme combinations as well as the reference protocol.

The Meeting in Princeton. The meeting in Princeton was dedicated only to the core cell line data. One hundred participants received initial core data analysis, which consisted of fragment to fragment and fragment to cell cluster analysis. Fragment to antigen correlation tables were also available. The first interpretation of the data occurred during three full days and was performed by several specialized groups. In each group, a secretary was responsible for presenting the data to all other Workshop participants.

After Princeton. The core cell line data were checked by standardization centers before the final analysis that was sent to each one of the secretaries responsible for analysis reports.

Acknowledgments. We would like to thank M.F. Legrand, D. Leleu, F. Lethrosne, C. Massart, M. Masset, P. Pasturaud for the quality controls of restriction enzymes and for the preparation of all products (except probes, markers, and restriction enzymes, which were produced by Appligene Society, Strasbourg, France), N. Lallemand for informatic help, and E. May, M. Marceau, and M. Verna for secretarial assistance.

Appendix I: Tenth Histocompatibility Workshop–Southern Blot–Shipment Content

- 88 DNA samples at the concentration 200 μg/ml, 1 ml each.
- Restriction enzymes
  The core enzyme: 6 tubes of 4000 units at 10 or 20 u/μl
  The 6 family enzymes: 1 tube of 4000 units per enzyme at 10 or 20 u/μl
- Kleenow enzyme
  1 tube of 1500 u (750 μl at 2 u/μl)
- OLB 5X (labeling mix)
  1 tube of 1.85 ml
- Lyophilized hexamer PL
  1 flask of 50 OD units
- Probe inserts at 5 ng/μl: 13 tubes
  Class I = pHLA 2 : 6 μg
  DRB . . . . .3 μg (3 μg will be sent to you later)
  DQB . . . . .6 μg
  DPB . . . . .6 μg
  DQA . . . . .6 μg
  C4 . . . . .6 μg
  DRA . . . . .1.5 μg (1.5 μg will be sent to you later)
  DPA . . . . .3 μg
  DOB . . . . .3 μg

Table I-1. Probe References

Name	Reference	Vector	Antibiotic	Digest	Size bp
Class I	Sood et al. Proc Nat Acad Sci USA 1981;616:78	PSP 65	Amp	Pst I	1400
DRB	Long et al. Embo J 1983;2:389	PBR 322	Tet	HindIII-Sal I	786
DRA	Gustaffson et al. Scand J Immunol 1984;19:91–97	PBR 322	Tet	Pst I	2 × 700
DQB	Sheldon et al. Proc Nat Acad Sci USA 1986;83:9085	M13 mp10	–	EcoRI	400 + 800
DQA	Auffray et al. Immunogenetics 1987;26:63	PUC 9	Amp	BamHI-HindIII	1600
DOB	Tonnelle et al. Embo J 1985;4:2839	PUC 12	Amp	BamHI	1400
DNA(DZα)	Trowsdale et al. Embo J 1985;4:2231	pAT 153	Amp	PstI	1600
DPB	Roux et al. Proc Nat Acad Sci USA 1983;80:6036	PUC 13	Amp	EcoRI HindIII	1000
DPA	Erlich et al. Fed Proc 1984;43:3025	PBR 328	Amp	EcoRI	1150
C4	Belt et al. Cell 1984;36:907	pAT153/PvuII/8	Amp	BamHI-KpnI	500
21-OH	White et al. Proc Nat Acad Sci USA 1986;83:5111	PBR 322	Amp	BamHI	1700
C2	Bentley et al. Immunogenetics 1985;22:377	pAT 153	Amp	KpnI-BamHI	350
BF	Morley & Campbell Embo J 1984;3:153	pAT 153	Amp	ClaI-BamHI	1600 + 660

DNA . . . . .3 μg
(previous name DZ alpha)
C2 . . . . .3 μg
BF . . . . .3 μg
210H . . . . .3 μg

- Bacterial strains in 50% glycerol for 12 probes (all except DQB). It is recommended to grow them in selective medium as soon as possible. Inoculate 1 ml LB+AB with 10 μl of bacterial (+ glycerol) suspension. Plate after incubation (with selective AB).
- DQB in M13 mp10: 1.4 μg RF in 1 tube
  Transform JM103
  To amplify RF, incubate transformed bacteria at 30°C in LB without antibiotic (AB), then purify like a plasmid
- Markers
  M1 = λ phage-HindIII + λ phage-SmaI
  2 tubes of 400 μl at 10 ng/μl
  M2 = λ phage-KpnI =λ phage-BstE II
  2 tubes of 400 μl at 10 ng/μl
  Ex: 1 μl of M1 contains 5 ng of λ phage-HindIII and 5 ng of λ phage-SmaI
  pAT 153:
  1 tube of 400 μl at 50 ng/μl
  λ phage total DNA (for labeling)
  1 tube of 600 μl at 250 pg/μl

Appendix II: Tenth Histocompatibility Workshop–Southern Blot

Table II-1.

	Probes (Full Length)	Restriction Enzymes
Cell Line Studies	DRB, DQB, DPB, DOB, DRA, DQA, DPA, DNA(DZα), Class I, BF, C2, C4, 210H	EcoRI, MspI, BamHI, PstI, BglII, HindIII, TaqI, SstI, EcoRV, PvuII, KpnI, HincII
Family Studies	DRB, DQB, DPB, DQA, Class I, C4	EcoRI, BamHI, HindIII, TaqI, EcoRV, PvuII

Appendix III Tenth Histocompatibility Workshop–Southern Blot–Rules and Nomenclature

1. Core cell lines genomic DNA nomenclature:
   Genomic DNAs have been prepared from 93 cell lines and numbered sequentially from 1 to 93.
   Genomic DNAs from 88 homozygous cell lines have been sent to you.
   Five other cell line DNAs will be sent to you later.
   Only the 72 DNA samples numbered from 1 to 72 should be used in the core part. The 21 remaining samples can be used for any purpose you decide. The HLA types of the 72 core cell lines are given, but the loading number corresponding to each HLA type (or cell) is not indicated. These 72 samples will be analyzed blind by each participant and each standardization center. The corresponding HLA types will be delivered to each participant at the beginning of the central analysis.
2. Family DNA digest nomenclatures:
   You will have prepared DNA from 2 families of 6 members with:
   father (a/b), mother (c/d), sibling 1 (a/c), sibling 2 (a/d), sibling 3 (b/c), sibling 4 (b/d).
   These two families will be studied with EcoRI, TaqI, BamHI, EcoRV, PvuII, and HindIII. Each *digest* will have an identification number in accordance with the rules outlined in Table 1.
3. Loading order and membrane nomenclature:
   Core part
   Each gel will be loaded with:
   24 genomic DNA digests
   2 M1 samples, one on each side
   2 M2 samples, one on each side
   1 P sample at one extremity
   It is essential for standardization that the gels be loaded in the same way in each laboratory. Blots not in keeping with this rule will be rejected.
   The loading order is indicated in the figure below.

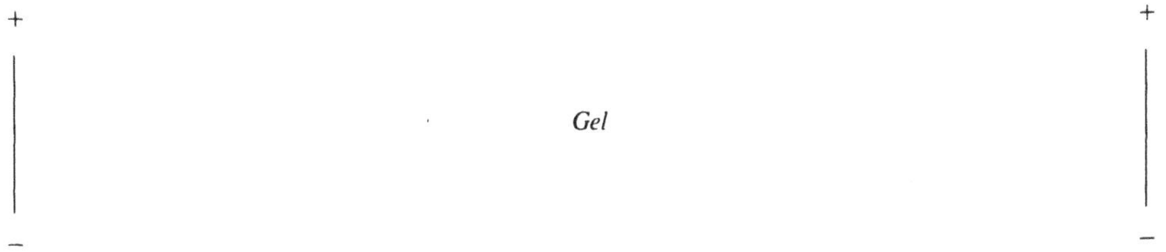

P M1 M2 1 2 3 4 5 6 7 8 9 10 11 12 13 14 15 16 17 18 19 20 21 22 23 24 M1 M2 - A
P M1 M2 25 26 27 28 29 30 31 32 33 34 35 36 37 38 39 40 41 42 43 44 45 46 47 48 M1 M2 - B
P M1 M2 49 50 51 52 53 54 55 56 57 58 59 60 61 62 63 64 65 66 67 68 69 70 71 72 M1 M2 - C

Gel A contains core DNA from 1 to 24; Gel B contains core DNA from 25 to 48; Gel C contains core DNA from 49 to 72

AL, BL and CL correspond to gels A, B, and C migrated under L conditions
AS, BS and CS correspond to gels A, B, and C migrated under S conditions
Family study:
All gels will be similar to the core part gels (Fig. 1). They will comprise:
Markers in the same order as core part gels
24 genomic DNA samples, also in the same order
Gel D contains family DNA from digest 1 to 24
Gel E contains family DNA from digests 25 to 48
Gel F contains family DNA from digests 49 to 72
Family DNA digests are coded according to Table 1
DL, EL and FL correspond to gels D, E and F, migrated under L conditions
DS, ES and FS correspond to gels D, E and F, migrated under S conditions

Table III-1. Identification Number Attributed to Each DNA Digest in the Family Study

		EcoRI	TaqI	BamHI	EcoRV	PvuII	HindIII
F	F	1	13	25	37	49	61
A	M	2	14	26	38	50	62
M	S1	3	15	27	39	51	63
I	S2	4	16	28	40	52	64
L	S3	5	17	29	41	53	65
Y	S4	6	18	30	42	54	66
I							
F	F	7	19	31	43	55	67
A	M	8	20	32	44	56	68
M	S1	9	21	33	45	57	69
I	S2	10	22	34	46	58	70
L	S3	11	23	35	47	59	71
Y	S4	12	24	36	48	60	72
II							

Appendix IV: Tenth Histocompatibility Workshop–Southern Blot–Example of Specific Recommendations

The 58 laboratories participating in the Southern blot component constitute 12 groups comprising:

10 groups of 5 laboratories
2 groups of 4 laboratories
(cf list of participating laboratories)

A number from 1 to 12 is attributed to each group. All groups will study the DNA of core cell lines with a single enzyme. Each group will be divided in two subgroups: subgroup 1 will use probe set 1, while subgroup 2 will use probe set 2. Subgroup 1 will comprise either three laboratories (for groups 1 to 9 and 12) or two laboratories (for groups 10 and 11). Subgroup 2 will comprise two laboratories.

The code attributed to each laboratory is composed of the Workshop lab code followed by a number. The last digit of this number corresponds to the probe set (subgroup 1 or 2); the first digit corresponds to the group (enzymes 1 to 12).

Example. Your Southern blot code is: JAP/SEK 2.2. You belong to group 2, which is responsible for TaqI, and you will use probe set 2.

This means that:

1. For the core part you will need to:
Digest DNA of the 72 cell lines with TaqI according to the protocol.
Produce six membranes: AL, BL, CL, AS, BS, and CS (see Rules and Nomenclature)
Hybridize these six membranes with probe set 2, which comprises nine probes that must be used successively in the following order:
DRB, DQB, DPB, DQA, Class I, C4, 210H, C2, BF
Send the autoradiograms and diskettes to Standardization Center N° 2 (indicated in Table 2)
The beta autoradiograms are AL, BL, CL, AS, BS, CS, hybridized with
DRB, DQB, DPB (see Rules and Nomenclature)
Send your non-beta autoradiograms and diskettes to Standardization center N° 8 (indicated in Table 2)
The non-beta autoradiograms are AL, BL, CL, AS, BS, CS, hybridized with: class I, C4, 210H, C2, BF, DQA

2. For the family part, the work is the same for each laboratory, i.e.:
Digest six DNAs from two families with the enzymes EcoRI, TaqI, BamHI, EcoRV, PvuII, HindIII
Produce six membranes, DL, EL, FL, DS, ES, FS (see Rules and Nomenclature)
Hybridize these six membranes with six probes, which must be used successively in the following order (see Rules and Nomenclature):
DRB, DQB, DPB, Class I, DQA, and C4
Send your beta and non-beta autoradiograms (see Rules and Nomenclature) and diskettes to the standardization centers indicated in Tables 1 and 2.

Table IV-1. Standardization Centers

Autoradiograms	Enzyme	Standardization Center
1/2 DL,1/2 DS: DRB,DQB,DPB	EcoRI	1
1/2 DL,1/2 DS: ClassI,DQA,C4	EcoRI	7
1/2 DL,1/2 DS: DRB,DQB,DPB	TaqI	2
1/2 DL,1/2 DS: ClassI,DQA,C4	TaqI	8
1/2 EL,1/2 ES: DRB,DQB,DPB	BamHI	3
1/2 EL,1/2 ES: ClassI,DQA,C4	BamHI	9
1/2 EL,1/2 ES: DRB,DQB,DPB	EcoRV	4
1/2 EL,1/2 ES: ClassI,DQA,C4	EcoRV	10
1/2 FL,1/2 FS: DRB,DQB,DPB	PvuII	5
1/2 FL,1/2 FS: ClassI,DQA,C4	PvuII	11
1/2 FL,1/2 FS: DRB,DQB,DPB	HindIII	6
1/2 FL,1/2 FS: ClassI,DQA,C4	HindIII	12

Remark: "1/2" means half autoradiograms (see Rules and Nomenclature). Half of a blot corresponds to one enzyme.

Table IV-2. Names of Standardization Centers

Standardization Center N°	Probes	Enzyme	Name
1	Beta	EcoRI	Bodmer
	Non-beta	MspI	
2	Beta	TaqI	Cohen
	Non-beta	BglII	
3	Beta	BamHI	Svejgaard
	Non-beta	PstI	
4	Beta	EcoRV	Terasaki
	Non-beta	SstI	
5	Beta	PvuII	Dupont
	Non-beta	HincII	
6	Beta	HindIII	Thorsby
	Non-beta	KpnI	
7	Beta	MspI	Bach
	Non-beta	EcoRI	
8	Beta	BglII	Van Rood
	Non-beta	TaqI	
9	Beta	PstI	Albert
	Non-beta	BamHI	
10	Beta	SstI	Tsuji
	Non-beta	EcoRV	
11	Beta	HincII	Ferrara
	Non-beta	PvuII	
12	Beta	KpnI	Festentein
	Non-beta	HindIII	

Table IV-3. List of Participating Laboratories

Country	Southern Code		Participants
Argentina	LAT/FAI	8.2	Fainboim
Australia	ANZ/BSH	4.1	Geczy
Australia	ANZ/DAW	11.2	Dawkins/Christiansen/Tait
Canada	NCY/FAL	12.2	Falk
Canada	US5/SIN	3.1	Singal
China	CHI/BCH	1.1	Sun
Denmark	SCA/SVE	3.1	Svejgaard/Kissmeyer-Nielsen
Finland	SCA/KOS	5.2	Koskimies
France	FRA/BET	1.2	Betuel
France	FRA/BIG	2.1	Bignon
France	FRA/COH	2.1	Cohen
France	FRA/FAU	6.1	Fauchet
France	FRA/OHA	9.2	Cambon-Thomsen
France	FRA/MYE	4.2	Mayer/Tongio
Germany	GER/ALB	9.2	Albert
Germany	GER/BOE	3.1	Boehm
Germany	GER/GRZ	12.2	Grzeschik
Germany	GER/MUC	7.2	Muller
Israel	GER/BRA	12.1	Brautbar/Kalil
Italy	IT1/CEP	11.1	Ceppellini
Italy	IT1/CON	9.1	Contu
Italy	IT2/FER	5.1	Ferrara
Japan	JAP/AIZ	9.1	Aizawa
Japan	JAP/JUJ	5.1	Juji
Japan	JAP/KSH	7.2	Kashiwagi
Japan	JAP/NAI	1.1	Naito
Japan	JAP/SAS	10.2	Sasazuki
Japan	JAP/SEK	2.2	Sekiguchi
Japan	JAP/TAK	8.1	Takata
Japan	JAP/TSU	10.1	Tsuji

Table IV-3. *Continued*

Country	Southern Code		Participants
Netherlands	BEN/REE	10.2	Reekers
Netherlands	BEN/ROO	8.1	Van Rood/Termijtelen
Norway	SCA/TSB	6.1	Thorsby
S. Africa	SAF/DUT	12.1	Du Toit
S. Africa	SAF/HAM	6.2	Hammond
Spain	FRA/ARN	7.1	Arnaiz-Villena
Sweden	SCA/LOW	7.1	Low
Switzerland	FRA/JEA	6.1	Jeannet
Thailand	ANZ/DCH	11.1	Chandanayingyong/Chiewsilp
U.K.	UKI/BOD	1.1	Bodmer/Batchelor/Bidwell/Barr
U.K.	UKI/FES	12.1	Festenstein/James/Welch
U.K.	UKI/MOR	3.2	Morris/Gels/Darke/Middleton
U.S.A.	US2/BAC	7.1	Bach/Segall
U.S.A.	US5/UAB	6.2	Acton/Barger
U.S.A.	US6/BRN	11.1	Braun
U.S.A.	NCY/CAR	2.1	Carpenter
U.S.A.	US7/DUP	5.2	Dupont/Yang
U.S.A.	US3/GVY	1.2	Garovoy
U.S.A.	US7/HAN	5.1	Hansen
U.S.A.	US5/PAK	4.2	Paik
U.S.A.	US8/RUB	8.2	Rubinstein/Gorodezky
U.S.A.	US4/STA	8.1	Stastny
U.S.A.	US5/TER	4.1	Terasaki
U.S.A.	US6/TRP	10.1	Troup
U.S.A.	USl/DUQ	9.1	Trucco/Duquesnoy
U.S.A.	US5/WAL	2.2	Walford/Gatti
U.S.A.	NCY/YUN	3.2	Yunis/Siddiqi
U.S.A.	US8/RJW	4.1	Winchester

Appendix V: Tenth Histocompatibility Workshop–Southern Blot–Using the DNA Fragment/Presence Program

The codes indicate whether a fragment is present or absent in a given lane. In addition, a rough quantification of intensity may be performed. The allowable codes are as follows:

1 = Single copy intensity (presumed)
2 = Double copy intensity (presumed)
3 = Intensity greater than two-fold above assumed single copy intensity
8 = Faint fragment present
9 = Unable to read or not done (default value)
Blank = Fragment absent

Appendix VI: Tenth Histocompatibility Workshop–Southern Blot–Locus Assignment

For class II beta probes, much cross-hybridization occurred. For a fragment that is observed with at least two different probes, a code has been used to compare the intensities of hybridization. This **locus assignment** has been established by standardization centers, which tried to reflect the **most often observed result** (taking into account the results of six films in two to five laboratories for each enzyme).

Shown below is the guideline for BETA system locus assignment.

DR	DQ	DP	DO	
1	0	0	0	DR-specific
0	1	0	0	DQ-specific
0	0	1	0	DP-specific
0	0	0	1	DO-specific
1	1	0	0	DR and DQ detected; same intensity
1	2	0	0	DR and DQ detected; DR more intense
1	2	2	0	DR, DQ, and DP detected; DR most intense; DQ and DP same intensity (DR > DQ=DP)
1	1	2	0	DR, DQ, and DP detected; DR and DQ most intense (DR=DQ > DP)
1	2	2	3	DR, DQ, DP, and DO detected; DR most intense; DQ and DP same intensity; DO least intense (DR > DQ=DP > DO)
2	1	3	4	DR, DQ, DP and DO detected; DQ > DR > DP > DO

Southern Blot Experiments: Data Processing

M. Ferguson,[1] D. Cohen,[2] B. Dupont,[3] and J-M. Lalouel[1]

The Southern blot component of the Workshop primarily focused on the core Workshop cell lines. Ancillary subcomponents included the analysis of families for haplotype studies, case-control studies for selected HLA- associated diseases, recombinant families and studies of rare haplotypes. Computer programs were provided for data entry by each participating laboratory.

Processing of Core Cell Lines

Design Overview. Seventy-two select cells were characterized for a combination of 12 HLA probes and 12 restriction enzymes. The design of the experiments provided for significant redundancy among participating laboratories, so that up to five replicates from distinct laboratories were expected for any given probe-enzyme combination (referred to as "system" throughout). Characterization of a given system required autoradiography of six filters: two distinct conditions of migration were used, referred to as long (L) and short (S) migration respectively; for any migration condition, the 72 DNA samples were electrophoresed in three separate experiments, leading to three filters each including 24 lanes of experimental data. Each filter included two additional lanes of reference markers to be used for the determination of molecular weights of DNA fragments detected in the experimental lanes.

The rationale for designing data processing was as follows. Short and/or long migration data would be used to infer the presence or absence of a given DNA fragment. Because mobilities of DNA fragments cannot be compared from one gel to another, standardization had to operate on estimated molecular weights derived from the observed mobilities. The marker lanes would be used to estimate the parameters of a function relating measured mobilities to known molecular weights of the markers. That function would in turn be used to estimate the molecular weights of DNA fragments in experimental lanes. Because replicates available from distinct laboratories would inevitably reveal inconsistencies, a round of standardization would be necessary to reduce the replicates to a set of standard DNA fragments and an array of individual scores indicating the presence or absence of each standard fragment in each core cell. Moreover, because of cross-hybridization between class II beta-chain sequence, an attempt could be made to assign each standard fragment to any one of four loci, DR, DQ, DP, and DO.

Local Data Entry. The programs for data entry consisted of a series of six procedures, each implementing a particular function selected through a master menu (Fig. 1). Having selected a given probe-enzyme combination, the user would first enter the observed mobilities of the

Participating Laboratories: US7LAL,[1] FRACOH,[2] US7DUP[3]

```
Southern Blot Data Base  (C) 1987 Howard Hughes Medical Institute for the
               Tenth International Histocompatibility Workshop

                        CORE CELL LINE PANEL STUDIES


Region Code : US7                               Probe Name : DRB
Lab Code    : DUP                               Enzyme Name: PVUII

                        1 - Marker Mobility

                        2 - Fragment Mobility

                        3 - Data on Individual Target Cells

                        4 - Locus Assignment


                        6 - Filter Data Initialization

                  Enter Selection (hit return to exit):
```

Figure 1. Menu of the package for local entry of Southern blot data.

Marker Mobility

Probe Name: DRB Enzyme Name: PVUII

Frag	ASm1	ASm2	BSm1	BSm2	CSm1	CSm2	ALm1	ALm2	BLm1	BLm2	CLm1	CLm2
1	0	100	0	105	0	80	0	445	0	435	0	440
2	110	0	120	0	85	0	495	0	480	0	485	0
3	125	0	135	0	100	0	540	0	525	0	532	0
4	0	140	0	145	0	120	0	580	0	560	0	568
5	195	0	200	0	175	0	735	0	705	0	722	0
6	265	0	265	0	245	0	890	0	855	0	880	0
7	290	0	295	0	275	0	955	0	920	0	940	0
8	0	295	0	300	0	280	0	965	0	932	0	952
9	305	0	310	0	295	0	990	0	950	0	975	0
10	0	350	0	355	0	335	0	1090	0	1055	0	1078
11	395	0	395	0	380	0	1175	0	1135	0	1160	0
12	0	405	0	405	0	390	0	1200	0	1165	0	1185
13	0	455	0	455	0	435	0	1290	0	1262	0	1280
14	0	540	0	540	0	525	0	1465	0	1445	0	1455
15	0	590	0	590	0	570	0	1550	0	1536	0	1545
16	595	0	595	0	575	0	1555	0	1540	0	1550	0
17	0	690	0	690	0	665	0	1725	0	1733	0	1728
18	0	950	0	955	0	920	0	0	0	2220	0	2185

Page Up = Top Page Dn = Bottom Ctrl/E = Exit to Menu

Figure 2. Definition of mobilities in marker lanes.

bands observed in the marker lanes for each of six filters (Fig. 2). He or she would then define the mobilities of the DNA fragments observed across all experimental lanes (Fig. 3). The entry of individual scores, indicating the presence or absence of each DNA fragment in each cell, could then be performed (Fig. 4). A semiquantitative code was used, with ' ' for absent fragment, '1' for present in single copy, '2' present in two copies, '3' present in more than two copies, '8' for a faint fragment, and '9' for 'not done' or 'inconclusive.' For class II beta probes, locus assignment of each DNA fragment could be attempted. However, this turned out to be very impractical in local entry, and locus assignment was performed at a later stage (regional standardization). Two additional functions were the definition of the filter layouts, for subcomponents other than the core cell lines, and file initialization. All functions provided were implemented as specialized full screen editors, allowing the user to review and revise his data. A structured file was produced for each probe-enzyme combination.

Fragment Mobility

Probe Name: DRB Enzyme Name: PVUII

Frag	AS	BS	CS	AL	BL	CL
1	0	0	0	600	595	580
2	175	180	145	625	0	585
3	180	215	160	660	640	630
4	220	230	190	735	710	705
5	225	250	215	775	755	755
6	265	0	240	855	820	830
7	0	365	0	975	945	950
8	385	390	0	1105	1075	1080
9	400	395	0	1145	1115	1110
10	410	420	390	1165	1120	1140
11	425	460	405	1215	1185	1195
12	440	505	425	1270	0	1225
13	0	0	0	1280	0	1235
14	460	530	445	1295	1265	1260
15	510	555	0	1375	0	1340
16	512	565	500	1380	1355	1365
17	525	570	0	1400	1375	1375
18	545	580	0	1430	1400	1405

Page Up = Top Page Dn = Bottom Ctrl/E = Exit to Menu

Figure 3. Definition of mobilities of experimental fragments.

```
               DNA Fragment Presence/Absence in Target Cells
  Probe Name: DRB           Enzyme Name: PVUII
     000000000111111111122222222223333333333444444444455555555556666666666777
Frag 123456789012345678901234567890123456789012345678901234567890123456789012
   1 88888888889 988888888888 8888888889888888888888888888888888888888888888
   2            2      1                               2             1
   3    2   2         2 222    2 22        2  2 2     2 2 2      2 2   12 2 2
   4 11 11111 11  1   1111 11111111  1191 1118111   1 1 1111111  111111111111
   5   1     1  11 111 1  1        11    1        111   1      11
   6 2 2  2   2  222   2      2 2  2 22  22 22   222       2 2      22     2
   7 88   8    88888888888888 888888888988888888888888      88
   8 888888888888888888888888888888888888888888888888888888888888888888888888
   9  1  1   1                                 1                  1   1
  10         2      22                 1                       2
  11    2   1         2 222    2 22        2  2 2   2 2 2 2      2 2   82 2 2
  12 2                                                       2
  13  8        88888888888888                          8 8888   8    8   88
  14    2   2                  2 22        2  2     2 2 2   1    2 2    2 2 2
  15           888 8                                   1
  16    2   2         2 222    2 22        2  2 2   2 2 2 2      2 2   12 2 2
  17         8    8    8        1  1 11   1 1            1 118  1   11     1
  18          888888   8   8    1  1 11   1 1          1   1 8      11     1

        Page Up = Top       Page Dn = Bottom       Ctrl/E = Exit to Menu
```

Figure 4. Entry of individual experimental data.

The data so entered were mailed to specific standardization centers. Each of 12 centers assumed the responsibility of producing consensus data sets for selected probe-enzyme combinations.

Regional Standardization. Another series of procedures were developed to assist the regional standardization centers, again implementing specific functions driven through a general menu. For a given system, the standardization task first involved merging relevant files from each of up to five laboratories. Once these data were pooled in a new file, the two marker lanes of each filter could be used to fit a function relating observed mobilities to the known molecular weights of the markers; thereafter, this function was automatically applied to estimate the molecular weights of the DNA fragments observed in the experimental lanes from their reported mobilities. The definition of standard fragments and consensus individual scores could then be performed.

Sizing of DNA Fragments. For a given probe-enzyme combination, an experimenter has generated data from three sets of filters, A, B and C, which include a filter for each of two conditions of migration, short (S) and long (M) migration. Each of these six filters corresponds to a particular experiment. Therefore, different parameters will relate experimental mobilities to estimated molecular weights for each of six filters, and sizing must be performed within a filter. Thereafter, to derive a single estimate of molecular weight for a given experimental fragment across all filters where it is observed, some averaging becomes necessary.

The SIZING Programs Proceeds as Follows

1. An input file is selected, which contains core cells data for a given probe-enzyme combination for one of up to five laboratory replicates.
2. Each laboratory, data can be submitted in turn to the analytical steps (3) and (4).
3. For each filter in this data file, a function is fitted to the marker mobilities, and the estimated parameters that define this function are stored in the corresponding laboratory data record.
4. Once all available filters have been subjected to the previous analysis, sizing of the experimental fragments is performed across all available filters; the molecular weights thus generated are stored in the corresponding laboratory records.

The accuracy of the estimation of molecular weights of DNA molecules in Southern experiments depends on the experimental protocol, the quality of markers, mathematical functions, and numerical methods used in curve fitting. Because it is well known that extrapolation of molecular weights is not reliable, the markers used in the Workshop have been chosen so that they are likely to bracket the size range of almost all fragments that might be detected in experimental lanes. Twenty-five markers are used, covering a range from 29.9 to 0.6 Kb.

Of the various functions that have been proposed, the logarithmic relationship:

$$\ln(L) = a + b * m$$

between length (L), and mobility (m) is the most commonly used. This function, however, yields fair estimates of molecular weights only on a restricted range of fragment lengths, of the order of 8 kb or less. When the range of molecular weights cover the whole area of a Southern gel, as can be expected of the various fragments detected in the Tenth HLA Workshop Southern experiments, the error in estimation of molecular weight

Started at : 13:17:00 , ended at : 13:17:22
Filter: A Migration: L PTG : -0.00000151

Weighted least squares (x100) : 0.13016891

Parameters : 3.712 -9.788 35.497 0.373

band	true kb	fitted kb	band	true kb	fitted kb
1	29.946	29.965	14	4.822	4.773
2	23.130	22.997	15	4.324	4.373
3	19.397	19.420	16	4.361	4.350
4	17.053	17.196	17	3.675	3.661
5	12.220	12.154	18	2.323	0.000
6	9.416	9.437	19	2.322	0.000
7	8.614	8.610	20	2.027	0.000
8	8.454	8.494	21	1.929	0.000
9	8.271	8.215	22	1.503	0.000
10	7.242	7.238	23	1.371	0.000
11	6.557	6.543	24	1.264	0.000
12	6.369	6.358	25	0.564	0.000
13	5.686	5.750	26	0.000	0.000

Figure 5. Sample output of the SIZING program.

can exceed 2 Kb in the range 8 to 2 kb; it is yet much larger for higher molecular weights.

The inadequacy of the logarithmic function was recognized by Southern (1). He pointed out that the elution time of a DNA fragment from preparative gels is directly proportional to its size over a large molecular weight range, and that elution time from a preparative gel is inversely proportional to mobility at the voltage gradients used in the Workshop conditions. Therefore, he introduced a function, further refined in Elder and Southern (2), which makes molecular weight inversely proportional to mobility:

$$L = L0 + c/(m - m0)**p,$$

where L0 and m0 are two asymptotes, and p is a power term. Therefore, four parameters (m0, L0, c and p) need to be estimated to describe the relationships between observed mobilities and known molecular weights for the markers, while only two parameters have to be estimated with the log function. Estimation is performed by weighted least squares, minimizing the function:

$$f = 100 \text{ SIGMAi } w[i] (m[i] - E[mi])**2 / \text{ SIGMAi } w[i],$$

where, from the above function, the expected mobilities are given by:

$$E[mi] = m0 + \exp(\ln(c/(Kb[i] - L0))/p).$$

The choice of the weights is empirical. We chose the known molecular weights of the marker fragments, for it gave better relative weighting over the whole range of the gel. Minimizing f involves the estimation of four parameters mo, L0, c and p by an iterative process. Because m0 and L0 are asymptotes and p is a power term, accurate numerical estimation is difficult. For this purpose, a powerful, general optimization algorithm has been used (GEMINI [3]), which estimates first derivatives numerically and constructs a recursive approximation of second derivatives. Moreover, calculations are performed in double precision. Parameter estimates have to remain within bounds during the iterative process; thus, constraints are applied to the parameters during the estimation process so that they remain within bounds.

The SIZING Program Estimates m0, L0, c, and p as Follows

1. Initial starting values are selected for the four parameters. Default values are provided by the program, distinct for each condition of migration, which were usually satisfactory.
2. If x is the vector of initial values for the four parameters, the program obtains improved values x'

$$x' = x + t * p$$

where p is a vector of derivatives which are scaled by an approximation to the second derivatives, and t is a step size parameter computed by the program so that

... sizing experimental fragments ...
DQA.Pvu lab: 1 : MOLECULAR WEIGHTS OF FRAGMENTS SEEN ON SIX FILTERS

i	AV. KBs	st.dev.	KBs-AS	KBs-AL	KBs-BS	KBs-BL	KBs-CS	KBs-CL
1	8.250	0.132	8.224	8.437	8.072	8.258	8.076	8.333
2	7.301	0.127	7.185	7.464	7.140	7.409	7.171	7.338
3	4.959	0.028	4.955	5.003	4.950	4.959	4.892	4.960
4	2.857	0.032	2.899	2.882	2.861	2.855	2.861	2.798
5	2.806	0.037	2.851	2.838	2.790	2.804	2.836	2.743
6	1.503	0.037	1.553	0.000	1.495	0.000	1.450	0.000

Figure 6. Averaging molecular weights of experimental fragments across filters.

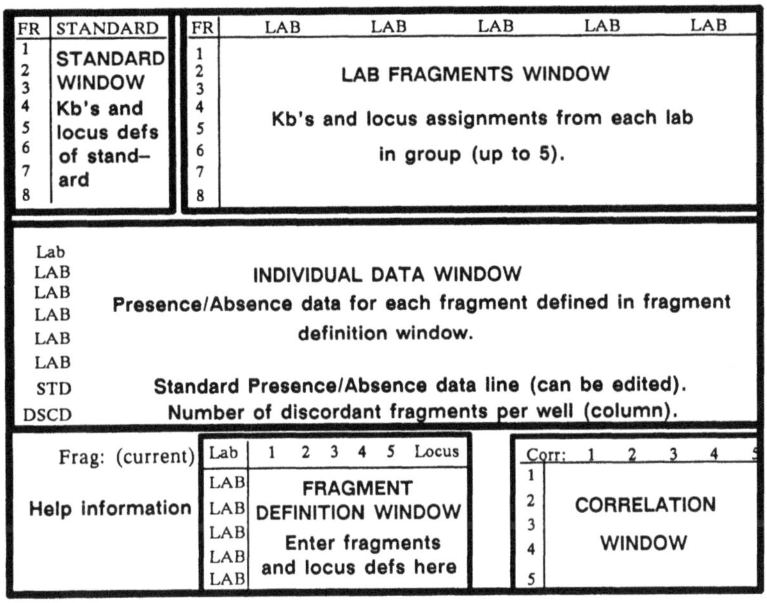

Figure 7. Schematic description of standardization screen.

the function to minimize decreases optimally from one iteration to the next.

3. Convergence is reached when PTG, the momentum of the iterative process downhill towards the true solution, a function of the derivatives, becomes smaller in absolute value than a specified tolerance, set to 0.00001 in the program.

Figure 5 gives a typical example of the final output tables that appear on the screen as the size of the fragments originating from one laboratory are determined.

Sizing experimental fragments is the last step of the estimation procedure for the data arising from a given laboratory. Using the parameter estimates obtained for each filter, the program computes estimated molecular weights for each of the experimental fragments seen on the corresponding filter. Hence, for each experimental fragment, we may have as many as six different estimates of molecular weight, which now have to be combined into a final estimate across all six filters. This is done by weighted averaging of the estimates obtained on the six filters. Weights have been chosen so that they take into account the final value of the loss function: the smaller the loss function on a filter, the larger the weight attributed to the corresponding filter. Moreover, the actual mobility of a fragment also affects the weight, so that long migration has greater weight in determining the molecular weights of the large fragments. The final table summarizing the estimation of molecular weights across the six filters contributed by a laboratory includes these estimates as well as their standard errors across filters. These estimates are subjected to standardization across laboratories in another program. A typical output is presented in Figure 6.

Once molecular weights have been estimated for each experimental fragment defined by each of the five laboratory replicates, standard fragments and consensus individual scores can be defined. Computer assistance was implemented through multiple windowing on a screen, as described in Figures 7 and 8. The former is a schematic description of the functions offered on the screen, while the latter is a typical screen at a particular stage of the standardization process. The "standard window" displays a partial list of the currently defined standards and their associated molecular weights. The "lab fragment window" displays the list of molecular weights of the experimental fragments defined in up to five replicates. These two windows can be scrolled independently of one another. "Frag: (current)" identifies the standard fragment currently being defined. The "fragment definition window" allows the user to establish a correspondence between the sequential code of a standard and the sequential position of experimental fragments in the replicates. As this correspondence is established, the correlation window displays the correlation in the distribution of experimental fragments so assumed to be equivalent among all pairs of laboratories; a low value would indicate that the equivalence assumed is erroneous. Simultaneously, the individual scores observed for all cells in all replicates are displayed in the 'individual data window'. The row labeled "STD" in that window gives a consensus interpretation which can be edited by the user, while the row labeled "DSCD" identify the lanes in which discordant results are noted (Fig. 8). These functions, together with a review of the original

```
FR  STANDARD      FR     DUP          KOS           HAN           FER
 1 14.83 0000      1 16.19 0000  14.92 0000  14.75 0000  24.81 0000   0.00 0000
 2 13.50 1200      2 14.83 0000  13.74 0000  13.29 0000  14.53 0000   0.00 0000
 3 11.43 0000      3 13.48 0000  10.45 0000  11.13 0000  12.38 0000   0.00 0000
 4 10.35 0000      4 11.73 0000   9.81 0000   9.83 0000  11.50 0000   0.00 0000
 5  9.80 0000      5 10.88 0000   6.60 0000   9.48 0000  10.05 0000   0.00 0000
 6  6.57 0000      6  9.84 0000   6.28 0000   6.52 0000   6.66 0000   0.00 0000
 7  6.16 0000      7  8.03 0000   6.11 0000   6.12 0000   6.30 0000   0.00 0000
 8  5.92 0000      8  6.95 0000   5.56 0000   5.92 0000   6.11 0000   0.00 0000
     000000000111111111122222222223333333333444444444455555555556666666666777
 Lab 123456789012345678901234567890123456789012345678901234567890123456789012
 DUP    2   2         2 222    2 22        2  2 2     2 2 2       2 2  12 2 2
 KOS    1   1 9      918111    1 11 9   9  1  191   1999999999999999999999999
 HAN    1   1         1 111    1 18        8  8 8   8 1 1 2       1 1  81 1 8
 FER

 STD    2   2         2 222    2 22        2  2 2   1 2 2 2       2 2  12 2 2
DSCD 000000000000000000100000000000000000000000000001000000000000000000000000
  Standard Frag:  2   Lab  1  2  3  4  5  Locus  Corr: 1    2    3    4    5
 Page Dn = Next Frag  DUP  3  0  0  0  0   1200      10.00 0.95 0.99 0.71 0.00
 Page Up = Prev Frag  KOS  2  0  0  0  0             20.98 0.00 0.98 0.67 0.00
 Ctrl/U/D = LFW Up/Dn HAN  2  0  0  0  0             30.92 0.88 0.00 0.69 0.00
 Ctrl/K = Compute Kb  FER  0  0  0  0  0             40.71 0.70 0.79 0.00 0.00
    Ctrl/E = End           0  0  0  0  0             50.00 0.00 0.00 0.00 0.00
```

Figure 8. Sample screen of standardization in progress.

autoradiograms, led to the eventual generation of a list of standard fragments and their estimated molecular weights as well as the consensus scores in the experimental lanes.

Thus were generated the data that are analyzed in various sections of these proceedings. While we did our best to facilitate the work of the local laboratories and the standardization centers, we must extend our deep appreciation for the extraordinary quality and quantity of the work performed by all to process such data, particularly the individuals who performed the standardization.

References

1. Southern EM. Measurement of DNA length by gel electrophoresis. Anal Biochem 1979;100:319.
2. Elder JK, Southern EM. Measurement of DNA length by gel electrophoresis. II. Comparison of methods for relating mobility to fragment length. Anal Biochem 1983;128: 227.
3. Lalouel JM. GEMINI: A Computer Program for Optimization of Nonlinear Functions. Department of Biophysics and Medical Computing Technical Report No. 14, University of Utah, Salt Lake City, Utah, 1979.

RFLP Standardization Reports

RFLP Standardization Report for DR Beta/SstI

A. Ando,[1] T. Sasazuki,[2] P. Reekers,[3] and K. Tsuji[1]

The DR beta/SstI system identified the 28 fragments as shown in Figure 1 and Table 1. Some close fragments from St. fragment no. 8 to no. 14 were difficult to analyze even in the long migration condition. St. fragment no. 4, no. 22, and no. 25 were found only in a single laboratory. Two fragments (no. 22 and no. 25) are DQ-dominant. Cross-hybridizing bands are shown in Table 2. Of the 28 fragments, 18 are dominantly DR-specific. St. fragment no. 16 and no. 17 are DR-DO codominant. Six fragments are DQ-dominant and four are DP-dominant. Fine bands (St. fragment no. 1, no. 6, no. 7, no. 14, and no. 28) showed no cross-hybridization with other beta probes. The fragments absence/presence data on each cell line with the DRβ probe are shown in Table 3. St. fragment no. 12 (11.59 kb) and no. 16 (6.15 kb) are highly correlated with DR1 and DRw15, respectively. St. fragment no. 2 (21.12 kb), no. 14 (10.95 kb), and no. 17 (5.19 kb) are shown the common banding pattern in each of the 72 core cell lines. Twenty-two cell lines showed these three bands, including all but one of the DRw53 cells.

Participating Laboratories: JAPTSU,[1] JAPSAS,[2] BENREE[3]

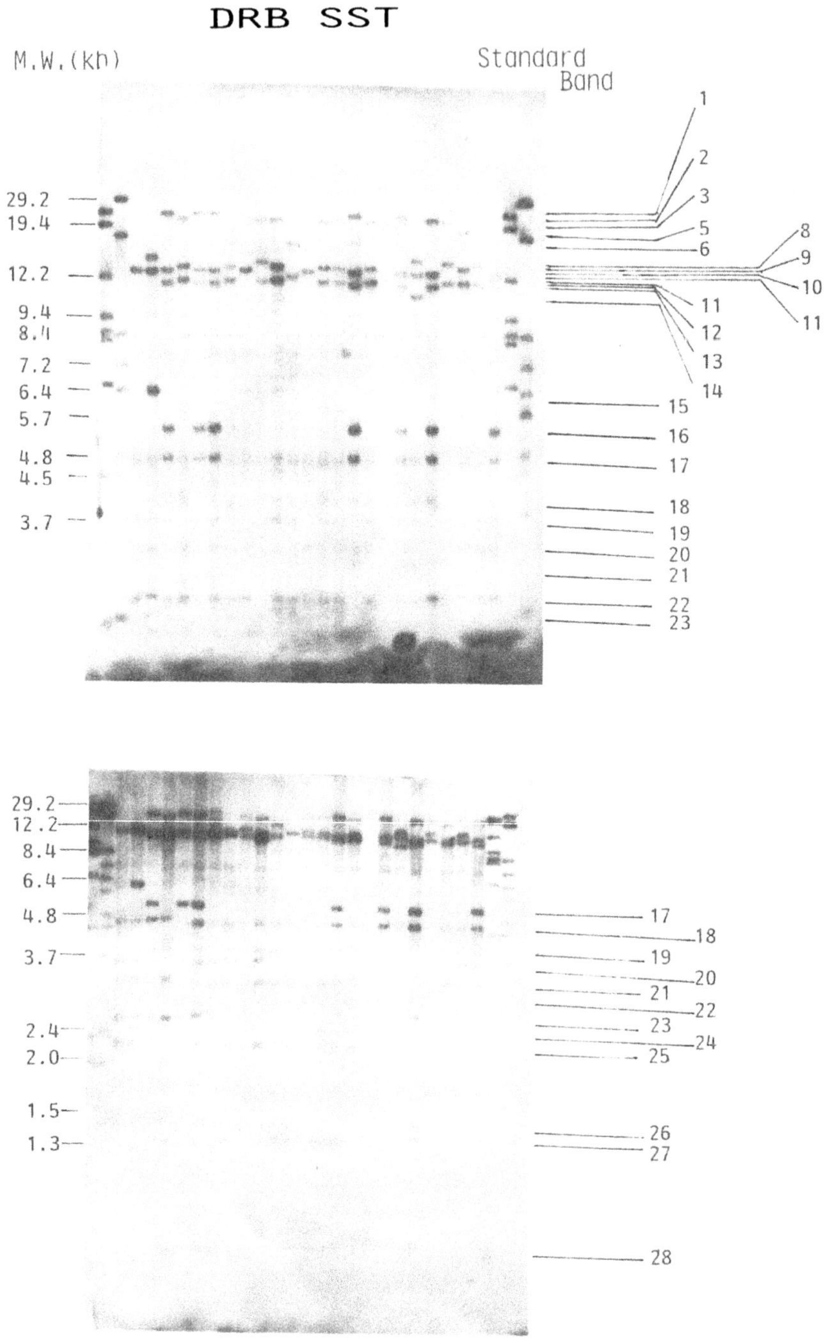

Figure 1. Fragments identified with the DR beta/SstI system.

Table 1. Fragments Identified by the DR Beta/SstI System

Fragment No.	kb	DRB SST TSU Locus	Freq	% Faint	56*	32*
1	23.96	1 0 0 0	0.043	0.000		
2	21.12	1 3 2 0	0.304	0.000		
3	19.53	1 0 2 0	0.261	0.053		
4	18.69	1 0 0 0	0.000	0.000		
5	17.02	2 2 1 0	0.072	0.400		
6	14.97	2 2 1 0	0.246	0.667	f	
7	13.75	1 0 0 0	0.072	0.333		
8	13.19	1 3 2 0	0.159	0.000		
9	12.87	1 3 2 0	0.058	0.000		
10	12.38	1 3 2 0	0.319	0.000		
11	12.01	1 3 2 0	0.522	0.027		
12	11.59	1 3 2 0	0.072	0.000		
13	11.25	1 3 2 0	0.377	0.000		
14	10.95	1 0 2 0	0.304	0.000		
15	10.30	1 0 0 0	0.043	0.000		
16	6.15	1 3 2 1	0.116	0.000		
17	5.19	1 2 2 1	0.304	0.000		
18	4.57	1 3 2 0	0.986	0.214		
19	3.82	2 1 0 0	0.261	1.000		f
20	3.46	2 3 1 0	0.696	0.280		
21	3.02	2 1 0 0	0.333	1.000	f	
22	3.00	2 1 0 0	0.000	0.000		
23	2.38	3 2 1 0	0.275	1.000		
24	2.23	2 1 0 0	0.400	0.448		
25	2.20	3 1 2 0	0.000	0.000		
26	1.29	3 2 1 0	0.029	1.000		
27	1.20	3 1 2 0	0.343	1.000	f	f
28	0.44	1 0 0 0	0.100	1.000		f

*Discrepancies in faint bands (f) or strong bands (s) of hidden duplicates (WS ID #9056 and 9032).

Table 2. Cross-Hybridization Bands

DRB	kb	DQB	kb	DPB	kb	DOB	kb
2	21.12	1	21.66	1	20.83		
3	19.53			2	19.59		
5	17.02	2	17.91	3	17.42		
6	14.97	3	15.94	4	15.34		
8	13.19	6	13.82	7	13.14		
9	12.87	7	13.40	8	12.38		
10	12.38	8	12.68	9	12.32		
11	12.01	9	12.23	10	11.96		
12	11.59	10	11.89	11	11.54		
13	11.25	11	11.47	12	11.18		
16	6.15	17	6.08	17	6.12	1	6.26
17	5.19	19	5.20	18	5.17	2	5.19
18	4.57	21	4.56	19	4.56		
		22	4.35	20	4.34		
19	3.82	23	3.78				
20	3.46	24	3.44	22	3.46		
21	3.02	25	3.07				
23	2.38	27	2.42	25	2.40		
24	2.23	28	2.28				
		30	2.01	27	2.05		
		31	1.33	29	1.29		
		32	1.09	30	1.14		
		33	0.34	31	0.47		

Underline denotes primary locus assignment.

Table 3. RFLP Pattern of DR Beta/SstI System

	Cell	Lines	1	2	3	4	5	6	7	8	9	10	11	12	13	14	15	16	17	18	19	20	21	22	23	24	25	26	27	28	DW	DR	DW2	DRw	DQw	DPw
	Fragment size (kb)		23.96	21.12	19.53	18.69	17.02	14.97	13.75	13.19	12.87	12.38	12.01	11.59	11.25	10.95	10.30	6.15	5.19	4.57	3.82	3.46	3.02	3.00	2.38	2.23	2.20	1.29	1.20	0.44						
59	9003	KAS116						1						2						1			8			8			8		1	1			5	
9	9004	JESTHOM						1						1						8		8				8					1	1			5	1
16	9006	WT100BIS						2						2						1						8					1	1			5	1
35	9002	MZ070782						1						1						1		1				1					20	1			5	1
17	9005	HOM2						2						2						1		1				8					1	1			5	
19	9009	KAS011								1			1				1			8		8	8			8			8		21	2			1	3,4
2	9014	MGAR								2			2					2		8	8	1	8		8	1					2	15			6	4
58	9008	DO208915								1			1					2		8		1				1			8		2	15			6	2,4
7	9013	SCHU								1			1					1	1	8	8	8	8		8	8					2	15			6	
66	9082	HO104								2			2					2		1	8	1	8		8	1			8			15			6	4
24	9010	AMAI								2			1					2		1		1	8			1					2	15			6	
62	9011	E4181324								2			2					2		1	8	1							8		12	15			6	2,4
5	9017	WT8								1			1					1		8		1				8					2	15			1	2,4
43	9016	RML								2			2				2			1		1									22	16			7	
13	9022	COX	1					8				2			2					1						8					3	3	24	52	2	3
28	9019	DUCAF			1							2			2					8			8		8						3	3	25	52	2	
15	9023	VAVY	1					8				2			2					1											3	3	24	52	2	1
31	9088	PF04015	1					8				1			1					8	8	8	8		8				8	8		3	24	52	2	1,4

Table 3. *Continued*

	Fragment No.		1	2	3	4	5	6	7	8	9	10	11	12	13	14	15	16	17	18	19	20	21	22	23	24	25	26	27	28						
	Fragment size (kb)		23.96	21.12	19.53	18.69	17.02	14.97	13.75	13.19	12.87	12.38	12.01	11.59	11.25	10.95	10.30	6.15	5.19	4.57	3.82	3.46	3.02	3.00	2.38	2.23	2.20	1.29	1.20	0.44						
	Cell	Lines																													DW	DR	DW2	DRw	DQw	DPw
71	9020	QBL			1							2			2					1									8		3	3	25	52	2	2
16	9021	RSH					1	8				2			2					1		1	8								N	3	24	52	4	1
10	9018	L0081785			1			8				2			2					1	8										3	3	25	52	2	3
61	9029	WT51		2									2			2			2	1	8	8							8		4	4		53	8	2
50	9092	BM92		1					8				2			1			2	1	8	8	8		8				8	8	14	4		53	8	
63	9026	YAR		2									2			2			2	1		1							8		10	4		53	8	1
27	9028	PE117		2									2			2			2	1		1			8						14	4		53	8	4
4	9107	LKT3		1									1			1			1	1											15	4		53	4	5
8	9033	BM14		1									1			1				1		1									4	4		53	8	4
18	9032	BSM		1									1			1			2	1		1	8		8						4	4		53	8	2
52	9034	SAVC		1									1			1			2	1	8		8						8	8	4	4		53	8	
29	9025	DEU		2									2			1			2	1	8	1			8						4	4		53	7	4
12	9091	MLF		1									2			2			2	1		1			8						4	4		53	7	
68	9032	BSM		1									2			2			2	1	8	1	8		8				8	8	4	4		53	8	2
30	9031	BOLETH		1									2			2			2	1		1	8		8				8		4	4		53	8	4
39	9030	JHAF		1				8					1			1			2	1											13	4		53	7	3
72	9007	DEM		1						1			2			1	1		2	1		8							8			16		53	5,8	3,4
64	9039	JVM			1							2			2					8	8	8							8		N	11	25	52	7	2
20	9104	DHIF		1									1			1			1	1											5	11		52	7	N,T
23	9040	BM15			1			1				1			2					1											5	11	25	52	7	3
34	9036	SPOO10			1							2			2					1		1				1					DB2	11	25	52	5	2
26	9037	SWEIG007								2			2					2		8		1				1					5	11	25	52	7	
14	9043	BM21			1			8				2			2					1						1					5	11	25	52	7	
6	9042	T1S1			1							1			2					1		1									N	11	25	52	7	
41	9105	FPAF	9	9	9		9	9	9	9	9	9	9	9	9	9	9	9	9	9	9	9	9		9						FS	11		52	6	2
1	9045	TUBO			8		8					1			2					8	8	1			8	8					5	11,12	25	52	7	2,4
57	9038	BM1G					1								1					8		1							8		DB6	12		52	7	2
11	9055	HO301							1			2			2					1	8		8			1					N	6	26	52	6	5
17	9062	WDV			1						2				2					1		1	8		8	1					18	13	24	52	6	4
15	9065	HHKB			1						2				2					1		1	8		8	1					18	13	24	52	6	4
33	9060	CB6B			1						1				1					1						1		8			18	13	25	52	5	
3	9058	OMW			1			8			1				1					1	8		8			1					18	13	24	52	6	1
12	9063	WT47							1			2			2					1	8		8			1					19	13	26	52	6	2
51	9059	SLE005						8	1			2			2					8	8					8			8		19	13	26	52	6	3
19	9056	KOSE			1				1			2			2					8		8				8					9,19	13,14	25	52	5	2,N
65	9056	KOSE			1			8	8			2			2					1		8	8			8			8		9,19	13,14		52	5	2,N
37	9064	AMALA			1							1			1					1		1			8						16	14	24	52	7	
55	9061	31227ABO			1							2			2					1		1	8			8			8		9	14	25	52	5	4
40	9054	EK			1							2			2					1		1				1					9	14	25	52	5	
38	9057	TEM			1							2			2					1		1	8			1					9	14	25	52	1	4
18	9050	MOU		1									2			1			2	1		1			8						17	7		53	2	2
22	9047	PLH		1			8						2			2			2	1											DB1	7		53	2	
70	9052	DBB		2			2									2			2	1	8	1	8		8				8	8	11	7		53	9	4
21	9051	PITOUT		1									2			2			2	1		1									17	7		53	2	4
67	9106	MANIKA		1									8			1			1	1		8							8			7		53	2?3	N,T
14	9048	LBUF		1									2			2			2	1								8			DB1	7		53	2	
32	9066	TAB089											2							1			8		8						8.3	8		52	6	
53	9068	BM9											2							1		1							8		8.1	8		52	4	2
25	9069	MADURA											1								8	8									8.1	8		52	4	4
56	9067	BTB											2							1		1							8	8	8.1	8		52	4	4
36	9071	OLGA						8					2							1		1	8		8						8.2	8		52	4	3
60	9070	LUY						8					2							1		8							8	8	8.3	8		52	7	1,4
69	9072	SPACH										1								8		8							8	8	8.2	8		52	4	4
54	9075	DXB		1									2			2			2	1		1							8		23	9		53	9	4

RFLP Standardization Report for DR Beta/PstI

E. Keller,[1] A. Andreas-Zietz,[1] E.D. Albert,[1] H. Coppin,[2] L. Contu,[3] A. Takada,[4] and M. Trucco[5]

The DR beta/PstI system identified 35 fragments on the core cell lines as shown in Table 1 and Figures 1 and 2. Table 1 shows the presence/absence pattern of the standardized fragments in all 72 cell lines, the corresponding molecular weight, as well as the locus assignment and probable associations with DR/DQ antigens. The intensity of bands is indicated by numbers. "1," "2," "3," means increasing intensity, whereas "8" means faint fragment, and "9" unreadable.

In this complex system the intensity does not necessarily indicate the gene dosage. "1" especially does not generally mean heterozygosity.

Participating Laboratories: GERALB,[1] FRAOHA,[2] IT1CON,[3] JAPAIZ,[4] US1DUQ[5]

This study was supported by SFB 217.

Table 1.

Probe Name: DRB Enzyme Name: PSTI

										kb	locus	associated with:
	99999	999999999	9999999	99999999999999	9999999999	9999999999999	999999	9999999	9	WS-ID number		
	00000	000000000	0000000	10000000000000	0100000001	0000000000000	000010	0000000	0			
	00000	011118011	2282112	02399233322330	3044334430	6656656555565	554405	6676767	7			
	65234	947302816	2381890	75012831496227	6523970584	0485253966417	017862	8917062	5			
	11350	400026564	1134127	02345203566467	3401622052	3304411516453	122467	5235636	5	position on blots		
	67599	925746823	3516081	49920780213882	4164463170	3735712195058	812470	3566029	4			
DR		11111111			1111111111	11111 1111111						
	11111	255555556	3333333	44444444444444	1111111121	3433363333444	777777	8888888	9			
DR				1	1	11						
				6	2	44						
DQ	55555	161666667	2224222	47778888888888	5677777777	5766666655551	222229	444476-	9			
DQ				5			3			kb	locus	associated with:
1		888888		888 88888	89	8 88 8				10.96	2 1 3 0	
2		111111		8	8 9	8 111 8				10.57	2 1 3 0	part DQw 1
3		88								9.40	1 0 0 0	DRw 16
4		8	888		9	8 8 8				8.55	2 1 0 0	
5		8	888 888		8988898 8	8888888888888				7.60	1 0 0 0	part DR3+5+6
6					9	11188				6.32	2 1 3 0	
7	11111	111111111	1111111	11111111111111	1111111112	1111111111111	222222	1111111	2	5.96	1 2 2 0*	
8		111111			9 9	22211				5.37	1 2 2 0*	DRw15+ part DRw13
9					21111	1				5.38	1 0 0 0	part DRw 11
10		1	122 222		9 1		11121			5.18	2 1 3 0	DQw 2
11			1		111	1111 11111		1		5.11	1 2 0 0	part DR5+6+3
12		8	1	12212212222221	1189111	1	12	21111 8	2	4.93	2 1 3 0	DQw 3 + 4
13					2		2112			4.90	1 0 2 0	DR 7 - DQw 2
14	11188	8			1	18888888				4.44	2 1 2 0	part DQw 1
15	88188	88 9 888	8888	888888	1 88 88888	8 8 88 8	8888 8	888881	8	4.19	2 1 2 0	
16				11111111111111						4.11	1 2 2 0	DR 4
17	2									3.73	1 2 2 0	
18	22222	222222222		2	9					3.66	2 0 0 1	DR 1 + 2
19				11111111111111						3.55	1 0 0 0	DR 4
20				11111111111111	8 1	8	111111		1	2.87	2 2 1 0	DR 4 + 7 + 9
21					22	22222				2.73	1 2 2 0	DRW 12 + 14
22			2222222	11111111111111	2122292221	21 22222222	111111		1	2.73	1 2 2 0	neg. with DR1+2+8
23	11118	111221111	1112121	1811181 1 888	2121121118	1112211111211	88 1 8	1111111	8	2.37	2 3 1 0	
24					3		333323		3	2.15	1 0 0 0	DR 7 + 9
25				23323323333333						2.05	1 2 3 0*	DR 4
26	88888	888888888	8888888	88888888888888	8988888888	8888888888888	888888	8888888	8	1.89	3 2 1 0	
27	8888	888888888	8888 88	88888888888888	8988888888	888888 888888	888888	8888888	8	1.74	2 2 1 0	
28	11 11									1.42	1 2 2 0	part DR 1
29	11811	8			1	111				1.33	2 1 3 0	DQw 5
30	22121	232332332	1222332	23333333333333	3233233333	3111232222333	333323	2212121	3	1.24	1 2 3 0	
31			2222			2222				1.07	1 0 0 0	part DR 3 + 6
32	88888	888888888	8888888	88888888888888	8888888888	8888888888888	888888	8888888	8	0.87	2 3 1 0	
33	1	81111111	1111111		111111111	1111111111111				0.57	1 0 0 2	DR 2+3+5+6
34	1111	1		1						0.55	1 0 0 0	DR 1 + 16
35		888888888	8888	8	88888888	8888888888		8888888		0.23	1 0 2 0	

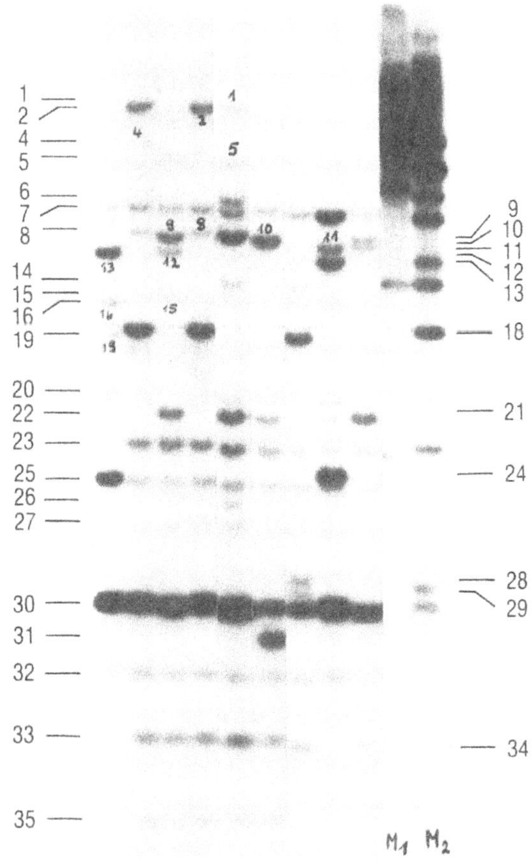

Figure 1. DR beta/PstI core fragments (gel 0.9%). For fragments no. 3 and 17 see Fig. 2.

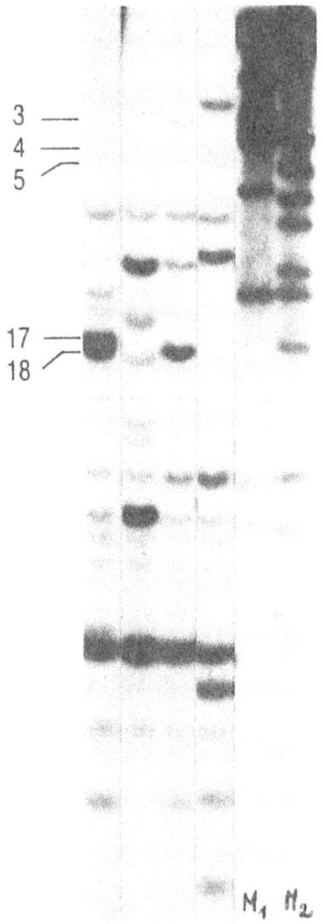

Figure 2. DR beta/PstI core fragments no. 3 and 17 (gel 0.9%).

Core cell no. 26 showed two different banding patterns in different laboratories. Therefore, those fragments not identical in all laboratories were coded as "9." Because the different banding patterns were confirmed in three laboratories, the most likely explanation for this discrepancy is a mixup of DNA.

Table 2 shows the cross-hybridizing bands in all four beta/PstI systems. It should be mentioned that the assignment of standard fragments nos. 8, 9, 10, and 11 was rather difficult under Workshop conditions, because these fragments are poorly separated. There are some DR beta/PstI fragments (e.g., nos. 7, 8, and 25) where the locus assignment is not identical for all cell lines positive for a band in that position (i.e., the first locus, in this case DR). This band is identical for all cell lines, but the cross hybridization with DQ and DP, respectively, is different.

Table 2. Cross-Hybridization Table–Enzyme: Pst–Probe: HLA Class II Beta

DRB	kb	DQB	kb	DPB	kb	DOB	kb
1	10.96	1	11.22	2	10.86		
2	10.57	2	10.68	3	10.42		
4	8.55	3	8.33				
6	6.32	6	6.43	4	6.39		
7	5.96	8	6.07	5	5.96		
8	5.37	9	5.51	6	5.37		
10	5.18	10	5.23	7	5.20		
11	5.11	11	5.08				
12	4.93	12	4.93	9	4.94		
13	4.90			10	4.82		
14	4.44	13	4.42	11	4.43		
15	4.19	14	4.17	12	4.20		
16	4.11	15	3.96	13	4.00		
17	3.73	16	3.81	14	3.70		
18	3.66					3	3.60
20	2.87	18	2.84	15	2.85		
21	2.73	19	2.66	16	2.71		
22	2.73	19	2.66	17	2.67		
23	2.37	20	2.35	18	2.39		
25	2.05	21	2.05	19	2.02		
26	1.89	22	1.91	20	1.91		
27	1.74	23	1.73	21	1.72		
28	1.42	24	1.45	22	1.41		
29	1.33	25	1.36	23	1.32		
30	1.24	26	1.26	24	1.25		
32	0.87	27	0.85	25	0.87		
33	0.57					4	0.59
35	0.23			26	0.24		

Underline denotes primary locus assignment.

Fragment no. 7. This fragment is present in all core cells but with different intensity; all strong bands of intensity "2" show the locus assignment "1220." All other bands of intensity "1" have the locus assignment "1000" or "1200" (DR, DQ, DP, DO, respectively). We have noted that the locus assignment "1220" includes all DR7-positive cells and a few others (see also core fragment no. 8 DQ beta/PstI).

Fragment no. 8. Cell line nos. 11, 12, 14, 51, 58, and 65 are positive for this fragment and show the locus assignment "1000" (five cell lines with DRw6 and one with DR2), while others are coded as "1220" (all five positive cells have DR2). This indicates that there are two different fragments in the same position.

Fragment no. 25. The results from one laboratory showed that there are perhaps two bands in the same position. One very strong band, for which only a few cells are positive (all DR4 and one DR2), is clearly seen on the films of all participating laboratories and has the locus assignment "1230" (DR, DQ, DP, DO). There is also one fainter band, which is present in all cells with the locus assignment "1120." Because only the strong band is clearly visible in all films, we decided to include only this one in the DR beta/PstI standardization (see also core fragment no. 21 DQ beta/PstI and core fragment no. 19 DP beta/PstI).

RFLP Standardization Report for DR Beta/KpnI

P.I. Biro,[1] S. Cutbush,[1] P. Cassell,[1] H. Avakian,[1] R. Vaughan,[1] M. Oudshoorn,[2] R. Martell,[2] and K. Siminovitch[3]

Fifteen standard bands have been defined in this system (Table 1, Fig. 1). Seven are relatively DR-specific and polymorphic (Tables 2 and 3). DNA samples differ in both number and intensity of bands; for example; DR5 DNAs show no characteristic, intense bands, whereas DR 4 DNAs are associated with bands 1, 4, and 8. Table 4 shows the correlation between the std. bands and DR-types. Table 3 shows that a few DR types can be assigned on the basis of RFLPs; for example, bands 2 and 6 unambiguously define DR types 1 and 2, respectively. However, some DR specificities share patterns (DR 4 and DR 7 DNAs share bands 1, 4, and 6) and others have more than one pattern; for example, not all DR 3 DNAs have bands 11 and 14. Kpn polymorphisms may have limited applicability in DR typing by RFLP analysis.

Participating Laboratories: UKIFES,[1] SAFDUT,[2] NCYFAL[3]

Immunobiology of HLA
Volume I

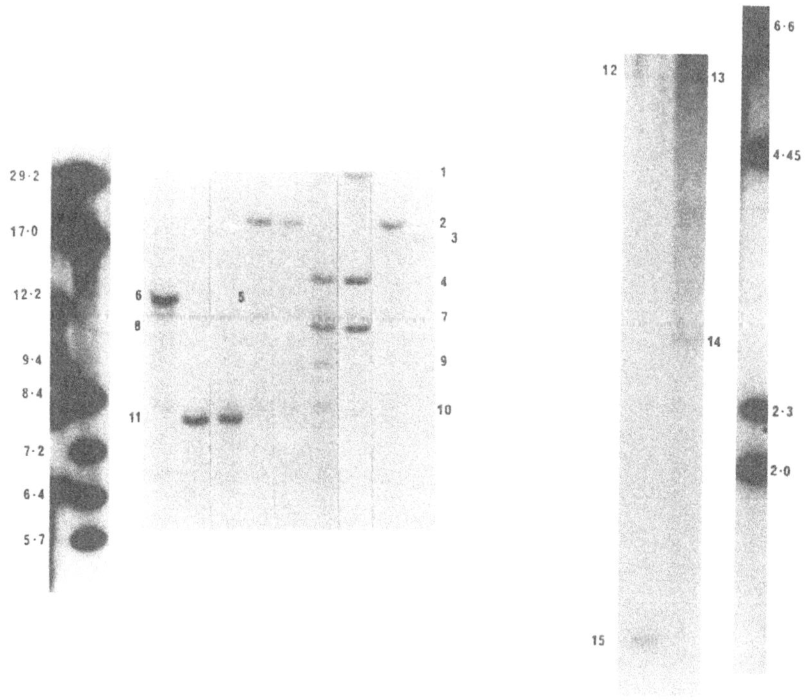

Figure 1. Standard bands produced by the KpnI/DRbeta enzyme/probe combination.

Table 1. Sizes and Frequencies of Standard Bands Produced by the KpnI/DRB Enzyme/Probe Combination

	kb	Locus	Frequency	% Faint
1	37.30	1 0 1 0	0.229	0.563
2	17.95	1 2 2 0	0.114	0.500
3	15.99	1 2 3 0	0.129	0.333
4	12.80	1 2 3 0	0.329	0.042
5	12.50	2 1 0 0	0.029	0.000
6	11.84	1 2 3 0	0.200	0.143
7	11.05	2 2 1 0	1.000	0.611
8	10.49	1 2 3 0	0.362	0.037
9	9.43	2 1 2 0	0.072	0.000
10	7.94	2 1 3 0	1.000	1.000
11	7.69	1 2 2 0	0.116	0.000
12	5.53	2 1 2 0	0.174	0.583
13	5.39	2 1 2 0	0.357	0.654
14	2.72	1 0 0 0	0.116	0.125
15	1.22	2 1 3 0	0.174	0.250

Table 2. Cross Hybridization Table—Enzyme: Kpn—Probe: HLA Class II Beta

DRB	kb	DQB	kb	DPB	kb
1	37.30			1	38.04
2	17.95	3	17.75	4	18.64
3	15.99	4	15.90		
4	12.80	6	12.81		
5	12.50	7	12.03		
6	11.84	8	11.71		
7	11.05	11	11.04	9	11.11
8	10.49	12	10.68	10	10.83
9	9.43	14	9.20	11	9.36
10	7.94	17	7.98		
11	7.69	16	7.73	13	7.68
12	5.53	20	5.33	14	5.59
13	5.39	21	5.15	15	5.44
15	1.22	23	1.20	17	1.23
		1	25.91	2	24.44

Underline denotes primary locus assignment.

Table 3. RFLP Pattern of DRB/KPNI System

Cell Lines		DR Type	Fragment No. 111111 1232456789012345
MZ070782	9002	1	1 81 8 1
KAS116	9003	1	8 8 8 8
JESTHOM	9004	1	2 8 8 8
HOM2	9005	1	2 1 8 8 8
WT100BIS	9006	1	2 1 8 8 8
DEM	9007	2/4	8 1 181 8 8
D0208915	9008	2	8 9999 99
KAS011	9009	2	18 8
AMAI	9010	2	28 8 8
E4181324	9011	2	28 8
SCHU	9013	2	18 8 8
MGAR	9014	2	28 8 8
RML	9016	2	21 8 1 1
WT8	9017	2	18 8 8
L0081785	9018	3	8 8
DUCAF	9019	3	1 8
QBL	9020	3	8 8
RSH	9021	3	1 81 1
COX	9022	3	8 82 1
VAVY	9023	3	8 82 1
DEU	9025	4	2 3 13 8 1 1
YAR	9026	4	8 2 81 8
PE117	9028	4	2 3 13 8
WT51	9029	4	8 1 11 8
JHAF	9030	4	1 2 12 8 1 1
BOLETH	9031	4	1 2 12 8 8
BSM	9032	4	1 81 8
BM14	9033	4	1 81 8
SAVC	9034	4	1 81 8
SP0010	9036	5	1 1 8 81
SWEIG007	9037	5	1 8 1 1
BM16	9038	5	8 8 8
JVM	9039	5	8 8 8 1
BM15	9040	5	8 8 8 1
TISI	9042	5	8 8 8 8
BM21	9043	5	8 8 8 1
TUBO	9045	5	8 8
PLH	9047	7	3 8318
LBUF	9048	7	1 82 1218
MOU	9050	7	8 3 8318
PITOUT	9051	7	8 3 8318
DBB	9052	7	8 82 81 8
EK	9054	6	8 1 8 1
H0301	9055	6	83 8 8
KOSE	9056	6	83 8 8

Table 3. *Continued*

Cell Lines		DR Type	Fragment No. 111111 123456789012345
TEM	9057	6	1 8 1
OMW	9058	6	88 82 81
SLE005	9059	6	81 8 8
CB6B	9060	6	11 8 1
31227ABO	9061	6	1 8 8
WDV	9062	6	11 81 11
WT47	9063	6	83 8 8
AMALA	9064	6	1 828 81
HHKB	9065	6	11 81 11
TAB089	9066	8	8 1 8 1
BTB	9067	8	1 1 8
BM9	9068	8	1 8 8
MADURA	9069	8	1 1 8
LUY	9070	8	2 8 8 8
OLGA	9071	8	1 1 8
SPACH	9072	8	1 8 8
DKB	9075	9	2 2 12 8
H0104	9082	?	28 8 8
PF04015	9088	3	1 82 1
MLF	9091	4	1 81 11 8 1 1
BM92	9092	4	8 1 81 8
DHIF	9104	11	88 1 8818
FPAF	9105	11	19 8 1
MANIKA	9106	7	8 1 81 8
LKT3	9107	4	1 82 8

Table 4. Relative Associations Between Standard Bands and DR Type

Std Band	DR 1	2	3	4	5	6	7	8	9	11
1				8/14			6/6		1/1	
2	5/5									
3								6/7		
4				14/14			6/6		1/1	1/1
6		9/9								
8				14/14		5/13	6/6		1/1	1/1
11			4/7			4/13				
14			4/7			4/13				
No strong band			3/7		6/6	4/13		1/7		

RFLP Standardization Report for DR Beta/EcoRV

T. Takenouchi,[1] R. Tonai,[1] P. Terasaki,[1] S. Mayer,[2] S. Bashir,[3] R. Winchester,[4] and Y. Paik[5]

DRbeta/EcoRV data were received from five laboratories and standardized using the following criteria:

1. Discrepancies were resolved by majority consensus.
2. Weak bands were accepted as standard fragments when they were detected by most labs.
3. Apparent cases of misassignment or mismeasurement of fragments were reevaluated.
4. When sizing, all measurements of the heavy fragments in the S membranes were ignored.
5. Intensity and locus assignment of fragments were all reevaluated.

Through the standardization of this system, we found several experimental problems. First, many of the DNA were nicked before digesting with enzyme. Therefore, with this system which produces heavy bands, some fragments were difficult to observe. Second, presumably the DNA of sample #26 was incorrectly provided to some laboratories. In those cases, we did not accept the apparently incorrect data. Third, although fragments that show cross-hybridization should have the same fragment size through all systems, some discrepancies were found due to the limitation of the standardization program.

In this system, we found 20 standard fragments (Table 1, Fig. 1). Figure 2 shows the molecular weight of markers utilized in this study. Of the 20 fragments, 14 are dominantly DR-specific. One fragment is DR-DQ codominant. Three fragments are DQ- and two are DP-dominant. In the case of cross-hybridization, we show which standard fragment number of one system corresponds to the others in Table 2. There are some discrepancies in the patterns of the hidden duplicates in the beta probe systems; however, these discrepancies could be explained by weakness of the cross-hybridization band signals.

Determination of the number of alleles is of major interest using this one enzyme-one probe combination. In Table 3, we show the banding pattern of each of the 72 core cell lines. Ten patterns were found looking at DR-specific bands. Cells KOSE, EK, 3122, BM16, and DEM appear to be heterozygous in this system.

Participating Laboratories: US5TER,[1] FRAMYE,[2] ANZBSH,[3] US8RJW,[4] US5PAK[5]

Table 1. Fragments Found Using the DR Beta/EcoRV System

		DRB	ECV	TER		
	kb	Locus	Frequency	% Faint	56	32
1	49.17	1 0 0 0	0.371	0.962		
2	25.05	1 2 2 0	0.371	0.630		f
3	22.98	1 2 2 0	0.186	0.000		
4	21.99	1 2 0 0	0.429	0.645		
5	20.47	1 1 0 0	0.086	0.000		
6	20.04	1 0 0 0	0.229	0.294		
7	18.24	1 2 2 0	0.500	0.111		
8	16.48	1 0 2 0	0.300	0.000		
9	15.34	1 0 0 0	0.386	0.000		
10	10.96	1 2 2 0	0.057	0.000		
11	10.02	2 2 1 0	0.343	0.840		
12	7.52	1 2 0 0	0.414	0.033		
13	6.88	2 2 1 0	0.771	0.929		
14	6.32	1 0 0 0	1.000	1.000		
15	6.30	1 2 0 0	0.071	0.000		
16	5.26	2 1 3 0	0.200	0.933	f	
17	2.71	2 1 2 0	0.157	1.000	f	
18	2.61	2 1 2 0	0.157	1.000	f	
19	1.83	1 2 2 0	0.071	0.600		
20	0.96	1 0 0 0	1.000	1.000		

Table 2. Cross-Hybridization Bands

Standard Fragment					
DRB	kb	DQB	kb	DPB	kb
3	22.98	2	22.70		
5	20.47	3	20.01	2	20.18
7	18.24			3	18.39
		5	15.14	4	14.94
		6	14.52	5	14.46
11	10.02	10	9.79	7	10.10
13	6.88	11	6.89	8	6.89
16	5.26	13	5.24	9	5.24
17	2.71	15	2.71		
18	2.61	16	2.62	12	2.59
19	1.83	17	1.84	13	1.83

Underline shows dominancy.

Table 3. Banding Pattern of the DRβ/EcoRV System in Each Cell Line

	Fragment No.													
Cell	1	2	3	4	5	6	7	8	9	10	12	15	19	20
9014 M GAR	8	1												8
9013 SCHU	8	1												8
9010 AMAI	8	1												8
9017 WT8		1												8
9008 DO 208915		1												8
9082 HO 104		2												8
9011 E 4181324		1												8
9016 RML		1												8
9009 KAS-011		1												8
9007 DEM	8	1	1	8				1						8
9034 SAVC	8	8	1	8				1						8
9030 JHAF	8	8	1	8				2						8
9025 DEU	8	8	1	8				2						8
9031 BOLETH	8	8	1	8				2						8
9028 PE 117	1	8	1	8				2						8
9092 BM92	8	8	1	8				1						8
9029 WT51	8	8	1	8				1						8
9026 YAR	8	8	1	8				1						8
9107 LKT 3	8		1	8				1						8
9032 BSM			1	8				1						8
9032 BSM			1	8				2						8
9091 MLF			1	8				2						8
9033 BM14			1	8				1						8
9048 LBUF	8			8	2			2						8
9104 DHIF	8			8	1			1						8
9050 MOU	8			8	1			1						8
9051 PITOUT	8			8	1			1						8
9047 PLH	8			8	1			1						8
9075 DKB	8			8	1			1						8
9106 MANIKA	8			1				1						8
9052 DBB	8			1				1						8
9071 OLGA				1			1							8
9068 BM9				1			1							8
9067 BTB				1			1							8
9072 S PACHEO				1			1							8
9069 MADURA				1			8							8
9038 BM16				1		1	1		1		2			8
9054 EK				1		1	1		1		2			8
9061 3-12-27 AOB				1		1	1		1		2			8
9036 SPO 010						1	2		1		2			8
9057 TEM						1	2		1		2			8
9020 QBL						1	2		1		2			8
9039 JVM						1	2		1		2			8
9037 SWEIG 007						1	2		1		2			8
9019 DUCAF						1	2		1		2			8
9042 TISI						8	2		1		2			8
9018 LO 081785						1	2		1		2			8
9043 BM21						1	2		1		2			8
9040 BM15						8	2		1		2			8
9045 TUBO				8		8	1		1		2			8
9105 FPAF						8	1		1		2			8
9056 KOSE						1	2		1	1	2			8
9056 KOSE						1	2		1	1	2			8
9059 SLE-005							2		1	1	2			8
9055 HO 301							2		1	1	2			8
9063 WT47							2		1	1	2			8
9062 WDV							1		1		2			8
9065 HHKB	8						1		1		2			8
9064 AMALA	8						1		1		2			8

Table 3. *Continued*

Cell	Fragment No.													
	1	2	3	4	5	6	7	8	9	10	12	15	19	20
9060 CBG-IB						8	1		1		2			8
9021 RSH	8						1		1		2			8
9088 PF 04015							1		1		2			8
9058 OMW							1		1		2			8
9022 COX							1		1		2			8
9023 VAVY							1		1		2			8
9066 TAB 089							2							8
9070 LUY							2							8
9003 KAS-116												2	1	8
9006 WT100 BIS												2	8	8
9004 JES THOM												2	8	8
9005 HOM-2												2	8	8
9002 MZ 070782												2	1	8

1 = single intensity; 2 = double intensity; 8 = faint.

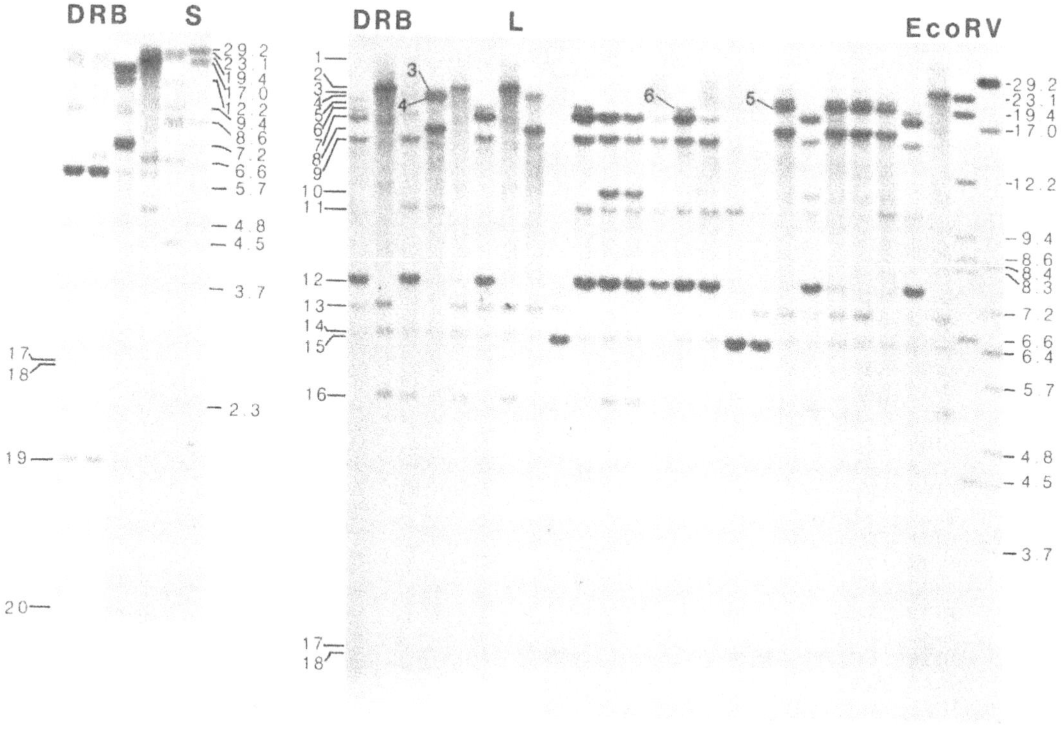

Figure 1. Twenty standard fragments found in the DRβ/EcoRV system.

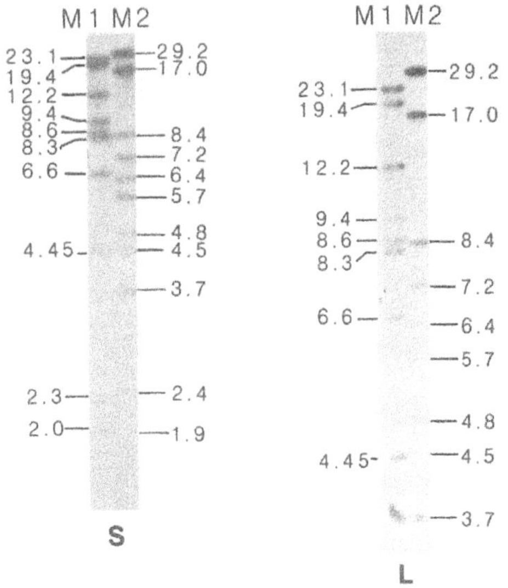

Figure 2. Molecular weights of standard markers. ►

RFLP Standardization Report for DR Beta/TaqI

A. Marcadet,[1] I. LeGall,[1] D. Cohen,[1] J.D. Bignon,[2] S. Sekiguchi,[3] C.B. Carpenter,[4] and R.L. Walford[5]

Twenty-seven fragments for the DR beta/TaqI system were standardized on the core cell lines (Table 1). The number of fragments revealed by the DRβ probe was so great that some participating laboratories interpreted more than 40 fragments. Even the removal of uncommon fragments necessitated standardizing more than 40 bands. Therefore, for the DRβ system, we standardized only specific DRβ bands and those cross-hybridizing class II bands which were more intense with DRβ probe. Of the remaining 27 standardized fragments, no constant fragments were found. Standard "fragment" no. 13 represents several different fragments (probably four that were too close to each other to be differentiated under these migration conditions. Eight bands of weak intensity were revealed—this weak intensity was not due to cross-hybridization. The locus assignment for each band reflects the *most frequently observed result* (obtained from 30 films from five different laboratories). One fragment (no. 5) gave the same intensity with DRβ and DQβ probes (1 1 0 0). The relative intensity of the cross-hybridizing bands obtained with DQ, DP, and DOβ probes (more intense with DRβ) are indicated in Table 1. Those cross-hybridizing fragments revealed with the DRβ probe, which proved more intense in another class II system, were standardized in this latter system. Where possible, these fragments have been indicated directly

Participating Laboratories: FRACOH,[1] FRABIG,[2] JAPSEK,[3] NCYCAR,[4] US5WAL[5]

Table 1. Standard Bands in the DR BETA/TaqI System

		DRB	Taq	COH			
	kb	Locus		Frequenty	% Faint	56	32
1	13.81	1 0 2 0		0.329	0.000		
2	11.48	1 2 2 0		0.229	0.000		
3	11.11	1 0 0 0		0.309	0.636		
4	9.79	1 2 2 0		0.257	0.263		
5	9.37	1 1 0 0		0.235	1.000		
6	8.66	1 0 2 0		0.100	0.000		
7	7.23	1 0 0 0		0.043	0.000		
8	7.00	1 2 0 0		0.257	0.000		
9	6.08	1 0 0 0		0.171	0.333		
10	5.80	1 0 0 0		0.086	0.000		
11	5.22	1 2 2 0		0.186	0.000		
12	4.40	1 0 0 0		0.143	0.000		
13	4.12	1 0 0 0		0.500	0.000		
14	3.15	1 0 0 0		0.246	1.000		
15	2.84	1 0 0 0		0.300	1.000		
16	2.69	1 0 0 0		0.257	0.000		
17	2.60	1 0 0 0		0.043	0.000		
18	2.12	1 0 2 0		0.123	1.000		
19	2.08	1 0 2 0		0.174	0.000		
20	1.72	1 0 0 0		0.426	0.633	f	
21	1.69	1 0 0 0		0.043	0.000		
22	1.42	1 0 0 0		0.114	0.000		
23	1.23	1 0 0 0		0.143	1.000		
24	1.18	1 0 0 0		0.406	1.000		
25	1.08	1 0 2 0		0.420	1.000	f	
26	0.48	1 0 0 0		0.130	1.000		
27	5.74	1 0 0 0		0.257	0.000		

Table 2. Cross-Hybridization Table–Enzyme: TaqI–Probe: HLA Class II Beta

DRB	kb	DQB	kb	DPB	kb	DOB	Kb
1	13.81			Sa			
2	11.48	Sa		NR			
4	9.79	Sb		NR			
5	9.37	Sc					
6	8.66			NR			
8	7.00	Sf					
Sa		2	5.34	NR			
11	5.22	NR		NR			
Sb				3	4.80		
Sb				4	4.64		
Sb		5	4.44	NR			
Sb				5	4.39		
Sc		NR		6	4.04		
				7	2.92	1	2.89
Sf		8	2.87				
Sg		9	2.61	NR			
NR		12	2.35				
Sh		13	2.19				
18	2.12			NR			
19	2.08			NR			
Si		14	1.98				
Si		15	1.94				
Si		16	1.90				
NR				10	1.82		
Si				11	1.73		
NR		17	1.58				
25	1.08			NR			

NR = Not represented. Underline denotes primary locus assignment (see text).

on the photographs. A cross-hybridization table (Table 2) summarizes all class II, TaqI cross-hybridizing fragments, giving their standardization number, molecular weight (in the system in which they showed the greatest intensity), and "name" on the figures (for systems in which they were not standardized).*

*Note: In comparing your results with those of the Workshop (WS), you may encounter problems with fragments that do not appear to be standardized in the system studied. Fragments giving a strong signal (on *L and S* migrations) may in fact be new fragments correlating with specificities not used in the WS. For other fragments, you will have to check whether the faint intensity is due to cross-hybridization. This can be done by comparing hybridization of the different class II probes with the same DNAs. The locus assignment established for the WS reflects the most frequently observed result, and not the result of every blot from every laboratory. This means that your locus assignment for one band can differ slightly from the standard locus assignment. However, each cross-hybridizing fragment has been standardized at least once, and you should be able to find each of your cross-hybridizing bands in one of the TaqI systems.

Table 3 indicates the distribution of standard bands in the core cell lines. Five specific DR fragments can be observed: no. 10 correlates with DR1, no. 11 = DR4, no. 21 = DRw16, no. 22 = DRw15, and no. 6 = DRw8. Fragment no. 1 correlates with DRw16+DR4+DR7 (+DRw9?); no. 13 with DR3+DR5 (except DR5 DB2) +DRw6+DR7 (+DRw9?); no. 3 with DRw15+DR4+DR7–$DQw2^-$ (faint intensity for DR4 and DR7); and no. 8 with B8–DR3+B18–, DR3+DRw6. DR3–$B18^+$ and DR3–$B18^-$ can be discriminated against in fragments no. 2 and no. 4. If we consider only bands giving an intense signal, it is possible to distinguish the specific patterns of bands for DR specificities: DR1 to DRw8.

The figure for S (Fig. 1) is a photograph taken from a film from laboratory FRA/COH and the figure for L (Fig. 2) is from a photograph from laboratory FRA/BIG. All standardized fragments are indicated at least once for S and/or L migration. The presence of a standardized fragment is indicated by a hyphen. Cross-hybridizing bands are indicated by lower case letters.

Comments on S migration. Molecular weight and locus assignment of the cross-hybridizing bands indicated on the photographs are given in the cross-hybridization table.

The fragment identified as "a" represents the standardized fragment DQβ TaqI no. 2.

Fragments identified as "b" represent several standardized fragments: –DPβ TaqI no. 3, DPβ TaqI no. 4, DPβ TaqI no. 5 and DQβ TaqI no. 5. For the others:

"c" = DPβ TaqI no. 6
"e" = DQβ TaqI no. 7 (exceptionally light on this blot)
"f" = DQβ TaqI no. 8
"g" = DQβ TaqI no. 9
"h" = DQβ TaqI no. 13

Fragments identified as "i" represent several standardized fragments: –DQβ TaqI no. 14, DQβ TaqI no. 15, DQβ TaqI no. 16 and DPβ TaqI no. 11. The fragment identified as "d," only light on S migration, has not been standardized.

In other systems where a cross-hybridizing band has been standardized in several systems, you will note that the molecular weights may differ (by less than 100 pb) although the locus assignments are of course the same. In addition, the interpretations may be different. Obviously, if a fragment is detected by cross-hybridization (most often with faint intensity), it will be difficult to interpret, but when it gives a strong signal, it is easier to analyze. Therefore, if you are sure that *your faint fragment* is due to cross-hybridization, we advise that you *analyze it in the system in which it gives the strongest signal*. You must then *refer to the system in which the fragment gave the strongest signal when analyzed by the standardization center, in order to obtain its true frequency, molecular weight, correlation, etc.* If you have hybridized with only one class II probe, which is not

Table 3. RFLP DRβ/TaqI System. Distribution of Standard Bands in 10th WS Core Cell Lines

Standard Fragments N =

Cell Lines	HLA DR	1	2	3	4	5	6	7	8	9	10	27	11	12	13	14	15	16	17	18	19	20	21	22	23	24	25	26
9022	1										2			2						9	2							
9004	1										2			2							2							
9005	1										2			2							2							
9006	1										2			2							2	8						
9003	1										2			2							2	8				8		
9007	w16/4	2		9		9						2	2			9	8	2		9		2	2		8			
9009	2	2																		9	9	2	2		8			
9016	w16	2																		9		2	2					8
9008	w15			2																8	2	2		2	8			8
9010	w15			2										2						8	2	2		2	8			8
9013	w15			2										2						8	2	2		2	8			8
9014	w15			2										2						8		9		2	8			8
9017	w15			2										2						8	2	2		2	8			8
9011	w15			2										2						8		2		2	8			8
9082	w15			2																8	2	2		2	8			8
9019	3°		2						2						2											8	8	
9020	3°		2						2						2											8	8	
9018	3°		2						2						2							8				8	8	
9088	3^{+}				2				2						2							9				8	8	
9023	3^{+}				2				2						2							8					8	
9022	3^{+}				2				2						2							8					8	
9021	3				2					2					2							8				8	8	
9025	4	2		8		8						2	2			8	8	2										
9026	4	2		8		8						2	2			8	8	2										
9028	4	2		8		8						2	2			8	8	2										
9029	4	2		8		8						2	2			8	8	2										
9031	4	2		8		8						2	2			8	8	2										
9032	4	2		8		8						2	2			8	8	2										
9032	4	2		8		8						2	2			8	8	2										
9033	4	2		8		8						2	2			8	8	2										
9091	4	2		8		8						2	2			8	8	2										
9092	4	2		8		8						2	2			8	8	2										
9107	4	2		8		8						2	2			8	8	2				8						
9030	4	2		8		8						2	2			8	8	2				8						
9034	4	2		8		8						2	2			8	8	2				8						
9105	w11		1							2					2					9		9				9	9	9
9040	w11		1							2					2							8				8	8	
9037	w11			2																8	2	2		2	8			8
9039	w11		2							2					2											8	8	
9036	w11		2							2					2											8	8	
9042	w11		2							2					2											8	8	
9043	w11		2							2					2							8					8	
9045	w11/12		2							2					2						2	9				8	8	
9038	w12		2												2											8		
9104	w11	2			8	9				8		2			2	8	8	2										
9060	w13		2						2						2							8				8	8	
9061	w14		2						2						2											8	8	
9054	w14		2						2						2											8	8	
9057	w14		2		8				2						2											8	8	
9056	w13/14		2		8				2						2											8		
9056	w13/14		2		2				2						2							8				8	8	
9055	w6				2				2						2							8				8	8	
9058	w13				2				2						2											8	8	
9062	w13				2				2						2											8	8	
9064	w14				2				2						2											8	8	
9065	w13				2				2						2											8	8	
9059	w13				2				2						2							8				8		
9063	w13				2				2						2							8				8		
9047	7*	2			8			2		8					2		8		2			8						
9050	7*	2			8			2		8					2		8		2									
9051	7*	2			8			2		8					2		8		2									
9048	7*	2			1	8						2			2	8	8	2										
9052	7	2		8		8						2			2	8	8	2										
9106	7	2		9		8						9			2	8	8	2										
9071	w8						2															8				8	8	
9066	w8						2																			8	8	
9068	w8						2																			8	8	
9069	w8						2																			8	8	
9070	w8						2																				8	
9072	w8						2																				8	
9067	w8						2																				8	
9075	w9	2			1	8						2			2	8	8											

° = DR3-B18; + = DR3-B8; * = DR7-DQw2.

listed in WS reports as giving the strongest signal, you must assume that your result is inaccurate, in which case you should dehybridize then rehybridize with the most homologous probe. In the TaqI system, if you do not find a standardization number and molecular weight for a faint fragment you have detected, you can assume that this fragment is due to cross-hybridization, and interpret the fragment in another system (with a lower margin of error).

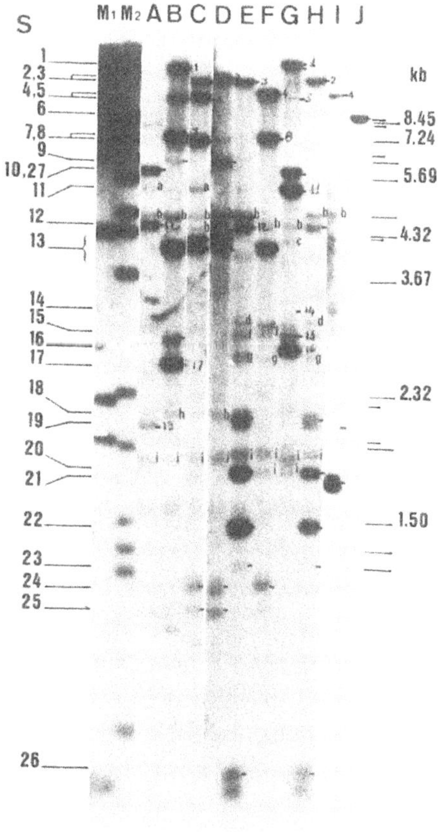

Figure 1. Standardized fragments for S migration.

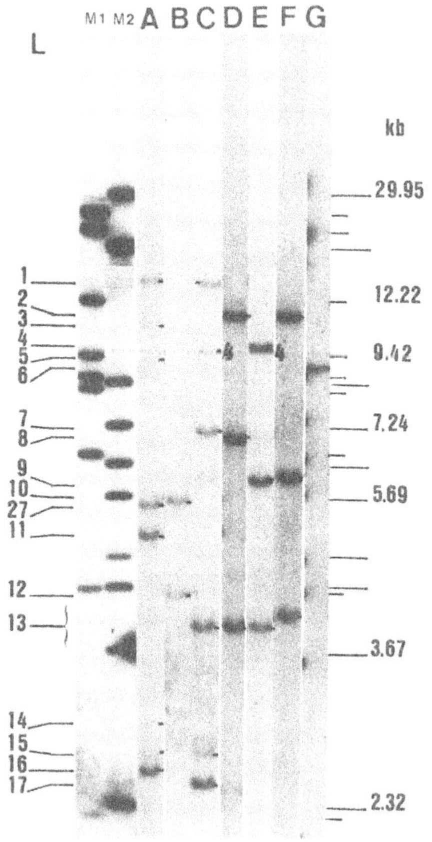

Figure 2. Standardized fragments for L migration.

RFLP Standardization Report for DR Beta/BglII

R.E. Bontrop,[1] M. Van Eggermond,[1] H. Takata,[2] P. Rubenstein,[3] L. Fainboim,[4] and P. Stastny[5]

Data were received from five different laboratories, and standardization was performed based on the following two criteria:

1) weak hybridization signals were accepted as standard fragments when they were detected by most laboratories
2) discrepancies were resolved by majority consensus

For this system, 14 different standard fragments were observed (Table 1, Fig. 1). Standard fragment 1 has a molecular weight of 17.76 kb, but is also detected with a DP beta probe (Table 2). Standard fragment 2 appears to be DR locus-specific and has a molecular weight of 16.70 kb and a panel frequency of 18.6%. Standard fragment 3 appears to be DR-specific, and this fragment has a molecular weight of 12.55 kb and a panel frequency of 10.0%. Standard fragment 4 is always seen in combination with standard fragment 6. Standard fragment 4 has a molecular weight of 11.74 kb, whereas standard fragment 6 has a molecular weight of 10.27 kb. These two standard fragments are detected for most HLA-DR4-positive cell lines. Standard fragment 5 has a molecular weight of 11.14 kb and a panel frequency 21.4%. This fragment is most often detected on BglII digests of HLA-DR1 positive cell lines. Standard fragment 7 has a molecular weight of 9.77 kb and a panel frequency of 34.3%. This fragment appears to be HLA-DR-specific and is most often detected in the example of HLA-DRw8-positive cell lines. Fragment 8 has a molecular weight of 9.34 kb and a panel frequency of 50%. This fragment is also detected by use of DQ beta and DP beta probes (Table 2). Fragment 8 is often seen in combination with fragment 9; however, exceptions to the rule do exist, which explains its panel frequency of 40%. Fragment 10 is detected by hybridization with DR, DQ, and DP beta probes and has a molecular weight of 6.96 kb. Fragment 11 has a molecular weight of 5.64 kb, but like fragment 10, is also detected with DQ and DP beta probes. Fragments 12, 13, and 14 do not appear to be HLA-DR-specific. Standard fragment 11 is often seen in combination with standard fragment 12 (Fig. 1, lanes 39, 44, and 48).

Participating Laboratories: BENROO,[1] JAPTAK,[2] US8RUB,[3] LATFAI,[4] US4STA[5]

Table 1. Standard Fragments Observed with the DR Beta/BglII System

Fragment No.	kb	Locus	Frequency	% Faint	56	32
1	17.76	2010	0.290	0.650	f	
2	16.70	1000	0.186	0.786		s
3	12.55	1000	0.100	0.143		
4	11.74	1020	0.386	0.071		
5	11.14	1000	0.214	0.067		
6	10.27	1020	0.186	0.071		
7	9.77	1000	0.343	0.231	s	
8	9.34	1320	0.500	0.000		
9	9.16	1020	0.400	0.034		
10	6.96	2130	0.086	0.000		
11	5.64	1120	0.300	0.000		
12	3.96	1020	0.429	0.097		
13	3.09	2130	0.171	0.167		
14	2.60	2210	0.129	0.000	s	

Table 2. Cross-Hybridization Table–Enzyme: BglII–Probe: HLA Class II Beta

DRB	kb	DQB	kb	DPB	kb	DOB	kb
1	17.76			2	17.49		
4	11.74			4	11.58		
6	10.27			5	10.18		
8	9.34	1	9.38	6	9.38		
9	9.16			7	8.99		
10	6.96	2	6.93	9	6.97		
		3	5.87			4	5.17
11	5.64	4	5.61	11	5.93		
12	3.96			15	3.89		
13	3.09	7	3.27	16	3.22		
14	2.60	9	2.78	17	2.80		

Underline denotes primary locus assignment.

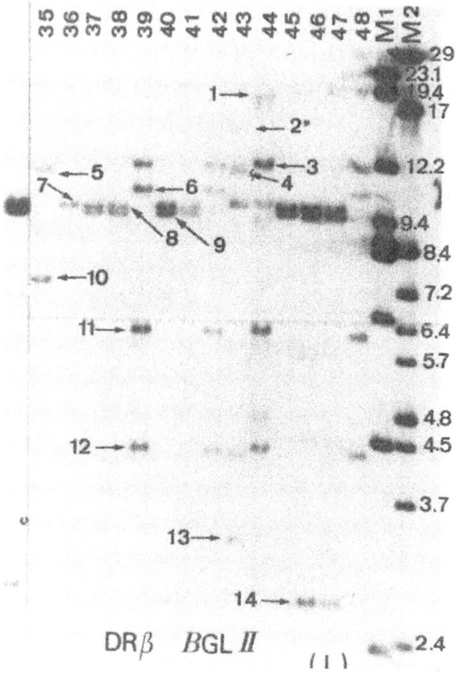

Figure 1. Standard fragments observed with the DR beta/ BglII system.
◄

Table 3. Fragment Presence/Absence Pattern of the DR Beta/BglII System in Each Cell Line

	Fragment No.													
Cell	1	2	3	4	5	6	7	8	9	10	11	12	13	14
9055	8							2	2					
9040	8							2	2					
9022	1							2	2					
9043	1							2	2					
9023	1							2	2					
9018	8	8						2	2					
9063	8	8						2	2					
9105								1	8					
9037								1	2					
9059								1	2					
9060								1	1					
9042								2	2					
9019								2	2					
9088								2	2					
9036								2	2					
9061								2	2					
9058	8							2	2				2	2
9056	8							2	2				1	1
9045								2	2				1	1
9046								2	2				1	1
9057								2	2				1	1
9054								2	2				1	1
9065								2	2				1	1
9021								2	2				1	1
9062								2	2				1	1
9056								1	2					
9020							2	1	2					
9039							2	1	2		8			

	Fragment No.													
Cell	1	2	3	4	5	6	7	8	9	10	11	12	13	14
9038							1	1	1					
9066							1	2						
9010	8					1		1	1					
9014						1		1	2				2	
9016			1		1		1	2						
9069								2						
9068								2						
9067								2						
9071							1							
9072							1							
9082							2				1			
9009			1		1		1					1		
9104				2						2	2			
9051				2						2	2			
9047	8			2						2	1			
9050	1	8		2						2	2			
9048	1	8		2						2	2			
9032				1		8	1			2	1			
9107	8			1		1				2	2			
9030	8			1		1				2	2			
9028				1		1				2	1			
9031				1		1				2	1			
9091				1		1				2	1			
9092				1		1				2	1			
9033				1		1				1	2			
9034		8		1		1	8			2	1			
9029		8		1		1	8			2	1			
9026		1		1		1	8			1	8			

Table 3. *Continued*

Cell	1	2	3	4	5	6	7	8	9	10	11	12	13	14
	Fragment No.													
9032		1		1		1	8			1	8			
9052		8		2		8				1	2			
9106		8		1			8			1	8			
9007		8		1	1	1	1			1	1			
9003		8		1	1				2					
9075		1	1	2			1				2	1		
9011			1	2	1		1				1			
9013	1		1	8	1		1					1		

Cell	1	2	3	4	5	6	7	8	9	10	11	12	13	14
	Fragment No.													
9008	8		1	1	1		1				1			
9017	8		8	8	1		1					1		
9025				1		1				1	2	1		
9004					1					2				
9006					1					2				
9002					1					2				
9004					1					2				
9005					8					2				

RFLP Standardization Report for DR Beta/PvuII

K. Kato,[1] A. Smith,[2] J. Partanen,[3] P. Piccioli,[4] B. Dupont,[1] and S.Y. Yang[1]

Twenty four bands were standardized with the DR beta/PvuII system as shown in Table 1. It is demonstrated in Figure 1 that several additional faint bands could be seen, but only in a single laboratory. These faint bands were not counted as standard bands and are probably generated by slightly less stringent conditions than indicated in the Workshop protocol. The region between standard fragment nos. 14 to 16 is the most compressed zone, and any bands migrating within this zone were separated into three standard bands. All bands except no. 24 (0.27 kb) were polymorphic. Four bands – no. 1 (14.83 kb), no. 3 (11.43 kb), no. 4 (10.35 kb), and no. 6 (6.57 kb) showed no cross-hybridization with other beta probes, and should be considered DR beta-unique fragments. Cross-hybridizing bands are shown in Table 2. Nine of 24 standard bands are highly correlated with individual DR specificities ($r > 0.85$) – fragment no. 6 (6.57 kb) with DR1, fragment no. 13 (4.47 kb) with DR2, fragment no. 9 (5.65 kb) with DR4, fragment no. 11 (4.75 kb) with DR7 and fragment no. 2 (13.50 kb), fragment no. 7 (6.16 kb), fragment no. 10 (5.17 kb), fragment no. 12 (4.60 kb), and fragment no. 17 (3.88 kb) with DRw53. These correlations for bands nos. 6, 9, 10, and 12 are consistent with a previous report in which the data were observed with the 3′ = DR beta probe (1). The RFLP pattern observed with this system is shown in Figure 2.

Table 1. Standard Bands of DR Beta/PvuII

	kb	DRB Locus	PVU Frequency	DUP % Faint	56	32
1	14.83	1 0 0 0	0.057	0.800		
2	13.50	1 2 0 0	0.300	0.000		
3	11.43	1 0 0 0	0.725	0.019		
4	10.35	1 0 0 0	0.243	0.000	s	
5	9.80	1 2 0 0	0.357	0.000		
6	6.57	1 0 0 0	0.071	0.000		
7	6.16	1 2 2 0	0.300	0.045		
8	5.92	1 0 2 0	0.029	0.000		
9	5.65	1 1 1 0	0.186	0.000		
10	5.17	1 1 1 0	0.300	0.000		
11	4.75	1 2 0 0	0.100	0.000		
12	4.60	2 1 2 0	0.286	0.000		
13	4.47	1 2 0 0	0.143	0.000		
14	4.28	2 1 2 0	0.143	0.000		
15	4.23	1 0 2 0	0.414	0.000	s	
16	4.15	1 1 1 0	0.214	0.000		
17	3.88	1 2 0 0	0.300	0.000		
18	3.65	2 1 3 0	0.743	0.259		f
19	3.39	2 1 2 0	0.400	1.000	f	
20	2.37	3 1 2 0	0.429	1.000		
21	2.15	2 1 0 0	0.171	0.333		
22	1.74	1 2 1 0	0.986	0.239		
23	1.01	2 1 3 0	0.943	1.000	f	f
24	0.27	1 1 1 0	1.000	1.000		

From left to right, band no., band kb, locus assignment of DR, DQ, DP, and DO, band frequency, frequency of faint bands, and hidden duplications.

Participating Laboratories: US7DUP,[1] US7HAN,[2] SCAKOS,[3] IT2FER [4]

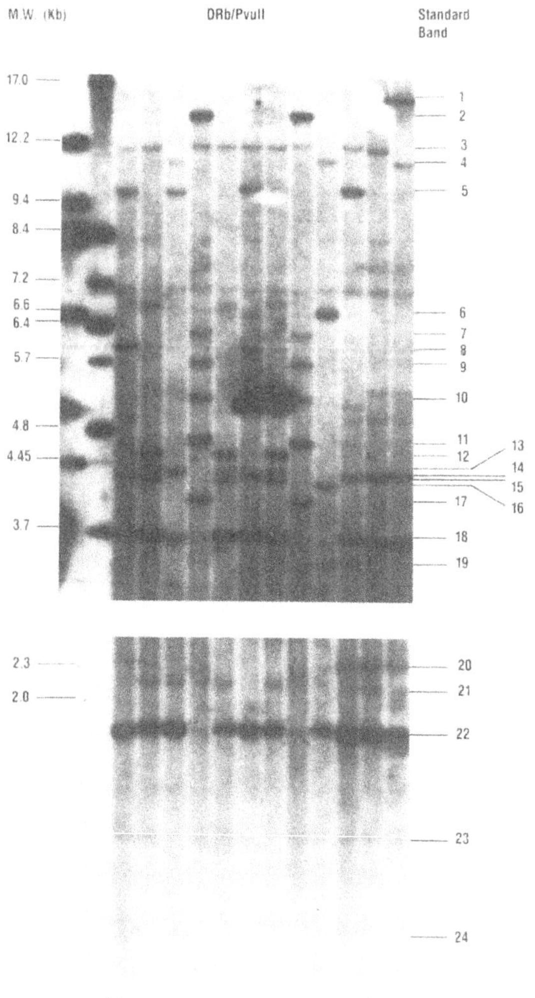

Figure 1. DR beta/PvuII top: long migration, bottom: short migration.

Table 2. Cross-Hybridization Table – Enzyme: PvuII – Probe: HLA Class II Beta

DRB	kb	DQB	kb	DPB	kb
2	13.50	1	13.49		
5	9.80	2	9.58		
7	6.16	5	6.05	3	6.11
8	5.92			4	5.80
9	5.65	6	5.75	5	5.68
10	5.17	7	5.21	6	5.17
11	4.75	8	4.78		
12	4.60	9	4.60	7	4.54
13	4.47	10	4.45		
14	4.28	11	4.31	8	4.26
15	4.23			9	4.25
16	4.15	12	4.20	10	4.15
17	3.88	13	4.01		
18	3.65	15	3.60	12	3.59
19	3.39	16	3.40	13	3.38
20	2.37	17	2.38	14	2.39
21	2.15	18	2.17		
22	1.74	20	1.77	16	1.75
23	1.01	22	1.02	17	1.03
24	0.27	25	0.53	18	0.63

Underline denotes primary locus assignment.

Reference

1. Bell JI, Denney J, MacMurray A, Foster L, Watling D, McDevitt HO. Molecular mapping of class II polymorphisms in the human major histocompatibility complex. I. DRβ. J Immunol 1987;139:562.

Cell lines	1	2	3	4	5	6	7	8	9	10	11	12	13	14	15	16	17	18	19	20	21	22	23	24
9107 LKT 3		2	1				2		2	2		2					2		8			8	8	8
9038 BM16		2					1		2	2		2					2		8			8	8	8
9028 PE 117		2	1				2		2	2		2					2	1	8			1	8	9
9092 BM92		2	1				2		2	2		2					2		8			8	8	9
9031 BOLETH		2	1				2		2	2		2					2		8	8		8	8	8
9025 DEU		2	1				2		2	2		2					2			8		8	8	9
9030 JHAF		2	1				2		2	2		2					2			8		8	8	9
9091 MLF		2	1				2		2	2		2					2			8		8	8	8
9034 SAVC		2	1				2		2	2		2					2	8	8			8	8	9
9029 WT51		2	1				2		2	2		2					2	8	8			8	8	9
9026 YAR		2	1				2		2	2		2					2	8	8			8	8	9
9032 BSM		2	1				2		2	2		2					2	8	8			1	8	9
9075 DKB		2	1				2			2		2					2	8	8			8	8	9
9052 DBB		2	1				2			2	2	2					2	8	8			8		9
9050 MOU		2	1				2			2	2	2					2			8		8	8	8
9104 DHIF		2	1				2			2	2	2					2			8		1	8	8
9051 PITOUT		2	1				2			2	2	2					2			8		1	8	8
9048 LBUF		2	1				2			2	2	2					2			8		8	8	9
9047 PLH		2		1			2			2	2	2					2			8		8	8	8
9017 WT8			1										2	2		2		2			1	3	8	8
9013 SCHU			1										2	2		2		2			1	3	8	8
9064 AMAI			1										2	2		2		2			1	2	8	8
9016 RML			1										2	2		2		2		8		8	8	9
9011 E 4181324			1										2	2		2		2		8		1	8	9
9082 HO 104			1										2	2		2		2			8	1	8	9
9014 M GAR			1										2			2		2			1	3	8	8
9008 DO 208915			1										2			2		2			8	2	8	9
9042 TISI			1		2										2			2		8		3	8	8
9018 LO 081785			1		2										2			2		8	1	3	8	8
9043 BM21			1		2										2			2		8		3	8	8
9040 BM15			1		2										2			2		8		3	8	8
9037 SWEIG 007			1		2										2			2		8		3	8	8
9019 DUCAF			1		2										2			2		8		3	8	8
9039 JVM			1		2										2			2		8		3	8	9
9020 QBL			1		2										2			2		8		3		9
9060 CB6B			1		2										2			2			8	3	8	8
9056 KOSE	1		1	1	2										2			2	8	8		3	8	8
9105 FPAF			8		2										2			2			8	3	8	8
9056 KOSE	8		1		2													2		8		3		9
9021 RSH				1	2													2	8			3	8	8
9062 WDV				1	2													2			1	3	8	8
9065 HHKB					2													2			1	3	8	8
9069 MADURA			1												2			2	8			2	8	8
9068 BM9			1												2			8	8			1	8	9
9067 BTB			1												2			8	8			3	8	8
9072 S PACHEO			1												2			8	8			1	8	9
9071 ORGA			1												2			8	8			2	8	8
9066 TAB 089				1											2					8		3	8	8
9070 LUY				1											2					8		2	8	9
9055 HO 301	8		1												2			2		8		3	8	8
9063 WT47	8			1										2				2		8		3	8	8
9059 SLE-005	8			1														2		8		3	8	9
9006 WT100 BIS				1		1										2		8	8			2	8	8
9005 HOM-2				1		1										2		8	8			2	8	8
9004 JES THOM				1		1										2			8			2	8	8
9003 KAS-116				1		1										2		8	8			3	8	9
9002 MZ 070782			9			1										1						2	8	9
9009 KAS-011													2	2		2		2	8			1	8	9
9106 MANIKA		1	1				8			1	2						2							9
9032 BSM		1	1				2		2	2		2					2		8			8	8	9
9007 DEM		2	1				2		2	2		2	2	2		2	2	2	8			1		9
9036 SPO 010			1		2										2			2	8			3	8	8
9057 TEM			1		2										2			2	8			3	8	8
9054 EK			1		2										2			2	8			3	8	8
9061 3-12-27 AOB			1		2										2			2	8			3	8	9
9045 TUBO			1		2			2						2	2			2		8		3	8	8
9038 BM16			1		2			2										8		8		3	8	8
9058 OMW				1	2										2			2			1	3	8	8
9022 COX				1	2										2			2		8		3	8	8
9023 VAVY				1	2										2			2		8		3	8	8
9088 PF 04015				1	2										2			2		8		3	8	8
9064 AMALA				1	2										2			2				2	8	8

Figure 2. RFLP pattern of DR beta/PvuII system.

RFLP Standardization Report for DR Beta/EcoRI

S. Tonks,[1] V. Groves,[1] H. Betuel,[2] J.L. Bidwell,[3] M.L. Garavoy,[4] R. Lechler,[5] S. Naito,[6] J. Bodmer,[1] and J. Trowsdale[1]

A high degree of polymorphism shown in the DRbeta/EcoRI system that identified 20 fragments as shown in Figure 1 (band number 20 not shown). Only nine of these fragments were unique to DRB – 5, 6, 9, 10, 11, 12, 14, 17, and 20. The others were also found in the DQB and DPB systems (Table 1). The fragment sizes and patterns appear to be consistent with those previously reported, as shown in Table 2. From Table 2, cell numbers 9066 and 9069 appear not to have any fragments. This is because only fragment 7 is present in these two cell lines, and Table 2 shows only those fragments that are DR-unique or dominant.

Participating Laboratories: UKIBOD,[1] FRABET,[2] UKIBID,[3] US4GVY,[4] UKIBAT,[5] JAPNAI[6]

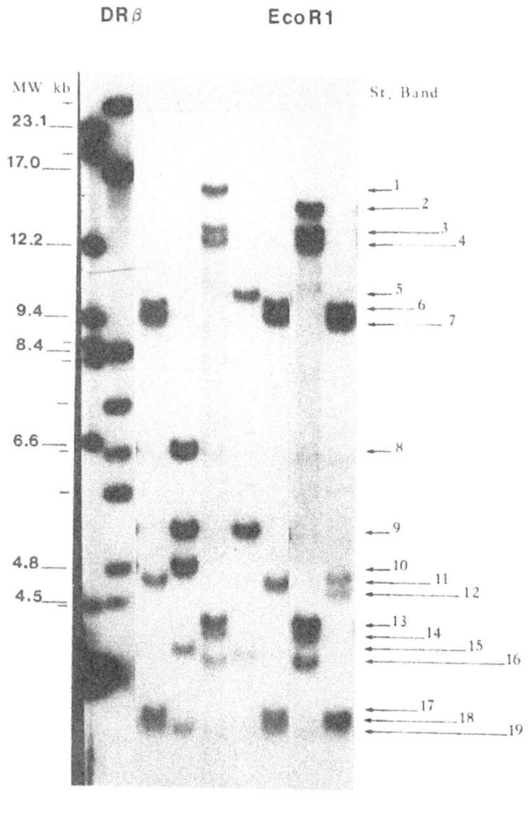

Figure 1. Fragments identified with the DR beta/EcoRI system.

Table 2. Cross-Hybridization Table – Enzyme: EcoRI – Probe: HLA Class II Beta

DRB	kb	DQB	kb	DPB	kb
1	15.13			1	19.47
2	14.13			3	14.75
3	12.55			4	12.93
		7	10.20	5	10.24
7	9.15	8	9.22		
8	6.28	9	6.21	8	6.17
13	4.10			9	4.02
15	3.84	12	3.83	10	3.88
16	3.73			11	3.77
18	3.29			13	3.33
19	3.19			14	3.25

Underline denotes primary locus assignment.

There were a number of good correlations between DR types and both individual bands and groups of bands in this system. DR4 was found to correlate well with fragment 1 (15.13kb; $\chi^2 = 53.009$) and in association with DR7 shared a number of fragments; 3 (12.55 kb), 13 (4.10 kb), 14 (3.98 kb), and 16 (3.73 kb); all had a $\chi^2 = 59.933$. DR1 was correlated with fragment 5 (10.06 kb) and with fragments 9 (5.13 kb) and 15 (3.98 kb) in association with DR2, which also had unique correlations with fragments 8 (6.28 kb) and 10 (4.71 kb). DR11 and three shared fragments 11 (4.56 kb) and 18 (3.29 kb). Fragment 12 (4.40 kb) showed good correlation with DRw14.

Discrepancies between duplicates in this system occurred in several fragments – 9, 12, and 17 – and were caused by poor resolution. There were some differences in the computed sizes of the higher molecular weight fragments, probably due to inaccuracies in the measurement of mobilities. Slight differences in the running conditions of the gels may also have contributed to these discrepancies.

Table 1. Banding Pattern of the DRβ/EcoRI System in Each Cell Line

WS		Fragment No.																
ID No.	Cell	1	2	3	5	6	9	10	11	12	13	14	15	16	17	18	19	20
9104	DHIF				2		3											
9004	JESTHOM				1		3						1					
9006	WT100BIS				2		3						1					
9005	HOM2				2		3						1					
9004	JESTHOM				1		1						1					
9058	OMW					1			1									
9505	FPAF					3			8									
9060	CB6B					3			1									
9059	SLE005					3			2						2			8
9039	JVM					2			1						1	8		
9019	DUCAF					3			2						2	1		
9088	PFO4015					3			1						8	8		
9036	SPOO10					3			2						1	8		
9040	BM15					2			1						1	1	8	
9020	QBL					2			2						2	2	1	
9043	BM21					1			1						1	1	8	8
9037	SWEIG007					8			2						1		8	
9045	TUBO					1			1							1	8	
9042	TISI					1			1							1	8	
9018	LOO81785					1			1							1	8	
9023	VAVY					1			1							1		
9022	COX					1			1							1		
9064	AMALA					3				8								
9054	EK					3				1								
9038	BM16					3				1								
9065	HHKB					3				1								
9021	RSH					3				1								
9061	31227ABO					3				1								
9057	TEM					3				8								
9062	WDV					3				1								
9056	KOSE					2		1	1						2	1	8	
9055	HO301					1		1								1	8	8
9063	WT47					1		1										
9082	HO104						1	1		1								
9016	RML						3	3					1					
9013	SCHU						3	3					1				8	
9010	AMAL						3	2					1				8	
9008	DO208915						1	1					1				8	
9011	E418324						2	2					1				8	
9017	WT8						3	3					1		1		8	
9014	MGAR						3	3				1	1				8	8
9007	DEM	1		1			1	2			2	1	1	1			1	
9032	BSM	1		1			1				1	1		1			8	
9092	BM92	8		2				1			2	8		1				
9091	MLF	2		1							3	3		1				
9029	WT51	1		1							1	8		8				
9026	YAR	1		1							1	8		8				
9033	BM14	1		1							1	1		1				
9034	SAVC	8		1							2	8		1		2		
9107	LKT3	1		2							1			1				
9028	PE117	3		2							2			1				
9025	DEU	3		2							3			1				
9031	BOLETH	3		2							3			1				
9052	DBB	1		2							2	2		1				
9030	JHAF		2	2							3	3		1				
9050	MOU		1	2							2	2		2			8	
9051	PITOUT		1	2							1	1		1			8	
9047	PLH		1	2							1	1		1			8	
9104	DHIF		1	2							1	1		1			8	
9048	LBUF		1	1							3	3		1	1		8	
9075	DKB		1	1							2	8		1				
9106	MANIKA			1							1	8	1	8				

Table 1. *Continued*

WS		Fragment No.																
ID No.	Cell	1	2	3	5	6	9	10	11	12	13	14	15	16	17	18	19	20
9067	BTB																8	
9068	BM9																8	
9071	OLGA																8	
9070	LUY													8				8
9072	SPACH	8																
9066	TAB089																	
9069	MADURA																	

1 = Single intensity; 2 = double intensity; 8 = faint.

RFLP Standardization Report for DR Beta/HindIII

G. Paulsen,[1] G. Markussen,[1] R.T. Acton,[2] J.M. Tiercy,[3] M.G. Hammond,[4] and R. Fauchet[2]

Twenty-eight bands were identified by RFLP in the DRB/HID system, as shown in Table 1 and Figure 1. The distribution among core cell lines of 25 bands with high DRB specificity is shown in Table 2.

Participating Laboratories: SCATSB,[1] US5UAB,[2] FRAFAU,[3] SAFHAM,[4] FRAJEA[3]

A cross-hybridization table for the HLA class II/HID systems is indicated in Table 3. Bands 13 and 14 had almost identical migration, but were decided to be two distinct fragments. The same decision was made for bands 18 and 19. This made the band number assignment of these bands difficult.

Regarding WS. ID. 9037 (RFLP gel load number 26), DNA from two different cell lines seems to have been

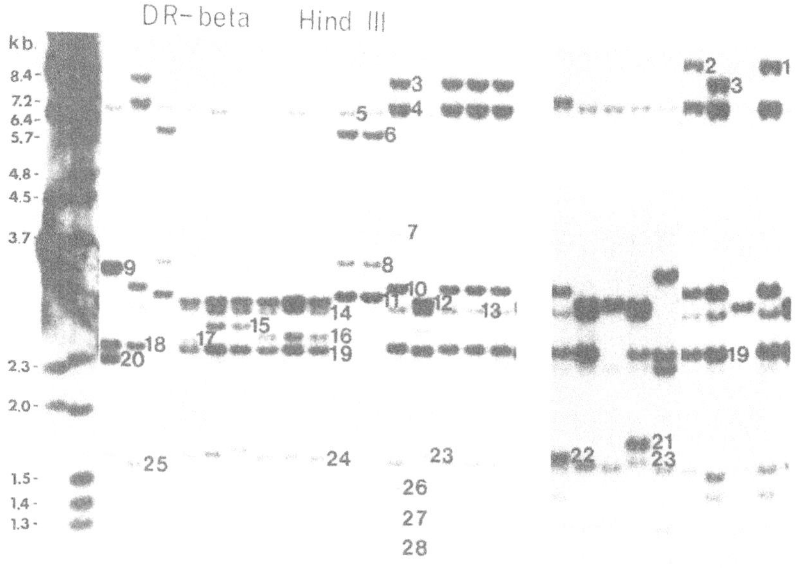

Figure 1. RFLP of 25 DNA samples run in 0.9% agarose gel. Representative core bands are shown on the right. The WS. ID. of the core cell lines is shown from the left following marker 1 and marker 2: 9013, 9033, 9004, 9018, 9055, 9063, 9022, 9043, 9023, 9006, 9005, 9050, 9056, 9104, 9051, 9047, open lane, 9075, 9061, 9067, 9038, 9008, 9106, 9032, 9072, and 9052.

Immunobiology of HLA
Volume I

Table 1. Standard Bands in the DRBeta/HID System

Band	kb	Locus[a]	Frequency[b]	% Faint[c]	56[d]	32[d]
1	9.66	1 2 2 0	0.029	0.000		
2	9.65	1 0 2 0	0.014	0.000		
3	8.44	1 0 2 0	0.257	0.000		
4	7.17	1 2 0 0	0.300	0.000		
5	6.94	2 3 1 0	1.000	1.000		
6	6.10	1 0 0 0	0.071	0.000		
7	3.84	1 0 0 0	1.000	1.000		
8	3.43	1 2 1 0	0.071	0.000		
9	3.32	1 0 2 0	0.143	0.000		
10	3.13	1 0 0 0	0.300	0.000		
11	3.02	1 2 3 0	0.071	0.000		
12	2.96	1 0 2 0	0.500	0.000		
13	2.89	1 0 1 0	0.529	0.000	s	
14	2.85	1 0 2 0	0.171	0.000	s	
15	2.70	1 0 0 0	0.057	0.000		
16	2.60	1 2 0 0	0.129	0.000		
17	2.55	1 0 0 0	0.057	0.000		
18	2.51	1 3 2 0	0.429	0.000		
19	2.49	1 3 2 0	0.443	0.000		
20	2.36	1 0 0 0	1.000	0.861		
21	1.83	1 2 0 0	0.029	0.000		
22	1.72	1 0 0 0	0.014	0.000		
23	1.68	1 2 1 0	0.343	0.000		
24	1.63	1 2 1 0	0.414	0.000		
25	1.58	1 2 1 0	0.171	0.000		
26	1.44	1 2 1 0	0.300	0.000		
27	1.30	2 1 2 0	0.086	1.000		
28	1.15	2 1 2 0	1.000	1.000		

[a]Locus assignment in the order DRB, DQB, DPB, DOB. 1 indicates high specificity; 2 and 3 indicate lower specificity.
[b]Band frequency in the core cell lines.
[c]Frequency of faint bands
[d]Discrepancies in faint bands (f) or strong bands (s) of the hidden duplicates (WS. ID. 9056 and 9032).

Table 3. Cross-Hybridization Table–Enzyme: HID–Probe: HLA Class II Beta

DRB	kb	DQB	kb	DPB	kb
		1	<u>12.80</u>	1	13.07
		2	<u>11.42</u>	3	11.66
1	<u>9.66</u>	3	9.76	4	9.76
2	<u>9.65</u>			5	9.64
3	<u>8.44</u>			6	8.59
4	<u>7.17</u>	7	7.24		
5	6.94	8	7.08	7	<u>7.05</u>
		10	5.48	9	<u>5.43</u>
		11	5.13	10	<u>5.06</u>
8	<u>3.43</u>	15	3.47	14	<u>3.50</u>
		16	<u>3.41</u>	15	3.42
9	<u>3.32</u>			16	3.37
		17	<u>3.30</u>	17	3.31
		19	<u>3.20</u>	18	3.42
11	<u>3.02</u>	20	3.09	19	3.09
12	<u>2.96</u>			20	2.99
13	<u>2.89</u>			21	<u>2.90</u>
14	<u>2.85</u>			22	2.83
16	<u>2.60</u>	25	2.62		
18	<u>2.51</u>	26	2.52	23	2.51
19	<u>2.49</u>	27	2.50	24	2.50
21	<u>1.83</u>	28	1.79		
23	<u>1.68</u>	29	1.70	27	<u>1.68</u>
24	<u>1.63</u>	30	1.66	28	<u>1.65</u>
25	<u>1.58</u>	31	1.61	29	<u>1.60</u>
26	<u>1.44</u>	32	1.51	30	<u>1.48</u>
27	1.30	33	<u>1.35</u>	31	1.32
28	1.15	34	<u>1.19</u>	32	1.17

Underline denotes primary locus assignment.

Table 2. Distribution of Bands With High DR Beta Specificity in the DRB/HID System

		Band number																								
WS ID	DR	01	02	03	04	06	07	08	09	10	11	12	13	14	15	16	17	18	19	20	21	22	23	24	25	26
9106	7	2	2		2		8			2			1						2	8				1		1
9052	7	2			2		8			2			1					2		8				1		1
9047	7			2	2		8			2			1					2		8				1		1
9048	7			2	2		8			2			1					2		8				1		1
9050	7			2	2		8			2			1					2		8				1		1
9051	7			2	2		8			2			1					2		8				1		1
9104	11			2	2		8			2			1					2		8				1		1
9007	4/16			2	2		8		1	2			1					2		2				1		1
9025	4			2	2		8			2			1					2		8					1	1
9026	4			2	2		8			2			1					2		8					1	1
9028	4			2	2		8			2			1					2		8					1	1
9029	4			2	2		8			2			1					2		8					1	1
9030	4			2	2		8			2			1					2		8					1	1
9031	4			2	2		8			2			1					2		8					1	1
9033	4			2	2		8			2			1					2		8					1	1
9034	4			2	2		8			2			1					2		8					1	1
9091	4			2	2		8			2			1					2		8					1	1
9092	4			2	2		8			2			1					2		8					1	1
9107	4			2	2		8			2			1					2		8					1	1

Table 2. *Continued*

		Band number																								
WS ID	DR	01	02	03	04	06	07	08	09	10	11	12	13	14	15	16	17	18	19	20	21	22	23	24	25	26
9032	4			2	2		8			2			1						2	8					1	1
9032	4			2	2		8			2			1						2	8					1	1
9075	9				2		8			2			1					2		8		2	1			1
9002	1					2	8	1			3									8						1
9003	1					2	8	1			3									8						
9004	1					2	8	1			3									8						
9005	1					2	8	1			3									8						
9006	1					2	8	1			3									8						
9010	15						8		3									2		2			1			
9011	15						8		3									2		2			1			
9082	15						8		3									2		2			1			
9017	15						8		3									2		2				1		
9013	15						8		3									2		2				1		
9014	15						8		3									2		2				1		
9008	15						8		3									2		2				1		
9009	2						8		1									2		2				2		
9016	16						8		1									2		2				2		
9019	3						8					3	3				2		2	8			1			
9020	3						8					3	2				2		2	8			1			
9018	3						8					3	2				2		2	8				1		
9022	3						8					3		2		2			2	8				1		
9023	3						8					3		2		2			2	8				1		
9088	3						8					3		2		2			2	8				1		
9021	3						8					3		2					2	8				1		
9036	11						8					3	3			2			2	8			1			
9037	11						8					3	2			2			2	8			1			
9039	11						8					3	3			1			2	8			1			
9042	11						8					3	2			2			2	8			1			
9043	11						8					3	3			2			2	8			1			
9060	13						8					3	3			2			2	8			1			
9105	11						8					3	3			1			2	8				1		
9045	11/12						8					3	2				2		2	8	1			1		
9038	12						8					3	3						2	8	2		1			
9040	11						8					3	2						2	8			1			
9054	14						8					3	3						2	8			1			
9061	14						8					3	3						2	8			1			
9057	14						8					3	3						2	8			1			
9058	13						8					3		2					2	8			1			
9062	13						8					3		2					2	8			1			
9065	13						8					3		2					2	8			1			
9064	14						8					3		2					2	8				1		
9055	6						8					3		2	2				2	8			1			
9059	13						8					3		2	2				2	8			1			
9063	13						8					3		2	2				2	8			1			
9056	13/14						8					3		2	2				2	8			1			
9056	13/14						8					3	2		2				2	8			1			
9066	8						8					3								8				1		
9067	8						8					3								8				1		
9068	8						8					3								8				1		
9069	8						8					3								8				1		
9070	8						8					3								8				1		
9071	8						8					3								8				1		
9072	8						8					3								8				1		
		01	02	03	04	06	07	08	09	10	11	12	13	14	15	16	17	18	19	20	21	22	23	24	25	26

Bands designation: 1-3 = increasing intensity; 8 = faint.

distributed among the participating laboratories. The "variant" analyzed by FAU and TSB has been included in the report.

The DRw53 association of two fragments in accordance with bands 3 and 4 in the present report has previously been described. A fragment corresponding to standard band 3 was associated with a HLA restriction element, which seemed to be more narrow than the serologically determined DRw53 specificity (1). This is in agreement with the present standardization where band 3 was missing in two of seven HLA-DR7 core cell lines.

Reference

1. Paulsen G, Qvigstad E, Gaudernack G, Rask L, Winchester R, Thorsby E. Identification, at the genomic level, of an HLA-DR restriction element for cloned antigen-specific T4 cells. J Exp Med 1985;161:1569.

RFLP Standardization Report for DR Beta/MsPI

M. Segall,[1] L. Schluender,[1] A. Arnaiz-Villena,[2] N. Kashiwagi,[3] and C. Muller[4]

DRbeta/MspI blots were received from four laboratories. In spite of the standardized technique, the overall intensity of the blots and the number of fragments detected were somewhat variable. Marker bands were remeasured where necessary and the data were pooled as follows:

1. Faint bands were listed as standard fragments when they were identified by at least two of the four laboratories.
2. Faint bands seen by only one laboratory, sometimes in only one lane, were not included. This was seen most often with bands of high kb and may have been due to incomplete digestion.
3. Discrepancies between blots were resolved in favor of the majority, or in favor of the positive identification when two labs recorded a band and two did not.
4. Faint bands in particular were sometimes resolved as a doublet in one lab's blots, but appeared as a singlet in others. Such bands were considered as two standard fragments when they could be consistently resolved on the blots from at least two labs; otherwise they were considered as one band.
5. Because of small differences in measurement from blot to blot and from lab to lab, the standardized molecular sizes of cross-hybridizing bands were not always exactly the same. In addition, the pattern of positives and negatives of the same band with different cross-hybridizing probes was not always identical. Therefore, cross-hybridizing bands were identified by overlaying the pairs of autoradiographs of blots hybridized with the relevant probes.

The DRB/MspI system identified 27 fragments on the core cell lines, as shown in Table 1 and the Figure 1. Nine of these were uniquely seen with the DRB probe, and eight additional ones were strongest with DRB; one (fragment 6) was equally strong with DRB and DQB. In Figure 1, the DRB-unique or "dominant" fragments are indicated by a dot next to the fragment number.

Table 2 shows all cross-hybridizing bands for MspI with the class II beta probes; the "primary" assignment for each fragment is underlined (e.g., DRB fragment 1 was strongest with the DQB probe and was also seen with the DRB probe, but not with the DPB or DOB probes). It should be noted that the intensity of many fragments varied from lane to lane within a single blot in patterns that were not due to loading of different amounts of DNA in different lanes; hence, for example, a fragment with the strongest hybridization to the DRB probe would still be faint for certain lanes and might have a stronger signal with the DQB probe in certain lanes. Fragments detected with only one probe are not shown in Table 2.

Discrepancies for the "hidden duplicates" (9056 and 9032) were due to faint bands or to difficulties in reading fainter blots where a fragment was not identified by all four labs. As mentioned above, some of the fragments could not be identified or doublets resolved on all blots, largely due to relatively dark background in some blots. Fragments 14a, 21a, and 22a (see Figure 1) were clear on one lab's blots, very faint on a second lab's, and could not be identified on the other two. Fragment 27 was classified by the computer analysis as freq 0.000,% faint 0.000, although this fragment was identified by two labs in exactly the same manner as Fragments 23, 24, 25, and 26; it has the same presence/absence pattern as Fragment 26.

Participating Laboratories: US2BAC,[1] FRAARN,[2] JAPKSH,[3] GERMUC[4]
This is publication #498 from the Immunobiology Research Center, University of Minnesota, Minneapolis, Minnesota 55455, USA. The work was supported in part by Juvenile Diabetes Foundation Grant #186175 and March of Dimes Birth Defects Foundation Grant #6-496.

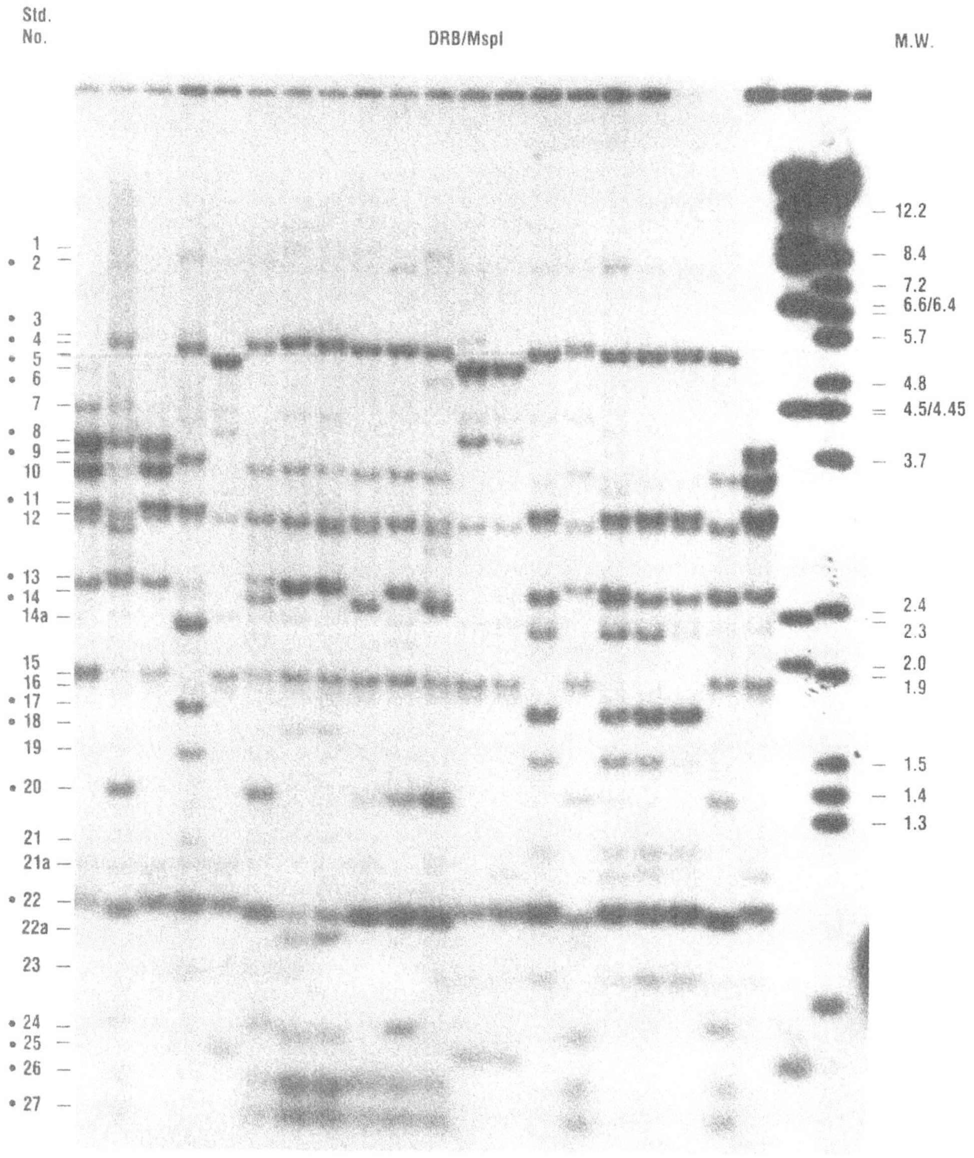

Figure 1. Fragments identified with the DR beta/MspI system.

Presence/absence data for the polymorphic fragments classified as DR beta fragments are given in Table 3; in this table, cells are arranged according to DR, Dw (when available), and DQ specificity.

Several fragments were unique to a single specificity. Fragment 5 was positive with 5/5 DR1 and 1/1 DRw16-Dw22; fragment 8 with 7/7 DRw15 (including 1 DRw15-Dw12); fragment 9 with 13/13 DR4; fragment 18 with 2/2 DRw13-Dw19-DQw6 and 1/1 DRw6-Dw"new"-DQw6. Fragment 17 was positive with 20/20 DRw53 plus cell 9104, which although listed as a DRw11-Dw5, had RFLP patterns characteristic for a DR7.

Other fragments identified several specificities or had additional scattered positive cells which did not fit a pattern. Fragment 6 was positive with 3/3 DRw16 as well as scattered other positives; fragment 14 was positive with 7/7 DR3, 4/4 DR7-DQw2 (cell 9106, which was negative for this fragment, was listed as questionable DQw2), 9104, and scattered other positives which were not DQw2, including a DR7-DQw9. Fragment 11 was positive with 10/10 DR2 (both DRw15 and DRw16),

Table 1. DRBeta/MspI Standard Fragments

Fragment	kb	Locus Assignment	Frequency	% Faint	Discrepant for 9056	Discrepant for 9032
1	8.93	2 1 0 0	0.086	0.000		
2	8.08	1 2 0 0	0.286	0.000	s	
3	5.62	1 0 0 0	0.157	0.083		
4	5.24	1 2 0 0	0.571	0.000		
5	4.93	1 2 0 0	0.086	0.000		
6	4.99	1 1 0 0	0.100	0.143		
7	4.17	2 1 2 0	0.314	0.043		
8	3.81	1 0 0 0	0.100	0.000		
9	3.68	1 0 2 0	0.186	0.000		
10	3.51	2 1 2 0	0.600	0.070		
11	3.15	1 0 2 0	0.514	0.162		
12	3.07	2 3 1 0	0.971	0.129		
13	2.53	1 0 2 0	0.429	0.000		
14	2.43	1 0 0 0	0.286	0.200		
15	1.93	2 0 1 0	0.557	0.050		
16	1.87	2 1 2 0	0.257	0.389		
17	1.76	1 2 2 0	0.300	0.000		
18	1.64	1 0 0 0	0.043	0.000		
19	1.52	2 0 0 1	0.129	0.000		
20	1.36	1 0 0 0	0.143	0.200	s	
21	1.15	2 0 0 1	0.300	0.636		
22	0.91	1 2 2 0	0.900	0.246		
23	0.67	2 2 1 0	0.257	0.667		f
24	0.55	1 0 0 0	0.229	0.813	f	
25	0.50	1 0 0 0	0.157	0.636		
26	0.41	1 0 0 0	0.414	1.000		
27	0.33	1 0 0 0	0.000	0.000		

Table 2. Cross-Hybridizing Fragments MspI, HLA Class II Beta Probes

DRB	kb	DQB	kb	DPB	kb	DOB	kb
1	8.93	3	9.23				
2	8.08	4	8.17				
4	5.24	6	5.56				
5	4.93	7	5.34				
6	4.99	8	4.96				
7	4.17	9	4.37	4	4.38		
9	3.68			7	3.73		
10	3.51	10	3.62	9	3.46		
11	3.15			10	3.26		
12	3.07	11	3.15	11	3.11		
13	2.53			12	2.60		
15	1.93			14	1.97		
16	1.87	16	1.85	15	1.91		
17	1.76	17	1.78	16	1.79		
19	1.52					7	1.55
21	1.15					9	1.16
22	0.91	20	0.86	19	0.95		
23	0.67	22	0.67	20	0.71		

13/13 DR4, and 5/6 DR7 (negative with 9106), as well as scattered other positives. Other fragments showed similar complex patterns (see Table 3). A number of fragments had no obvious correlation with any specificity or group of specificities.

Table 3. Distribution of Polymorphic Fragments in Core Cell Lines, DRB/MspI Standard Fragment Number

Cell No.	Name	2	3	4	5	6	8	9	11	13	14	17	18	20	22	24	25	26	27
9002	MZO70782				1										1		8		
9003	KAS116				2										1		1		
9004	JESTHOM				2										1		8		
9005	HOM2	1			2										8		8		
9006	WT100BIS	1			2	1									1				
9008	D0208915						1		1	1					8				
9010	AMAL						1		2	1	8				8				
9013	SCHU						1		2	1					1				
9014	MGAR	1					1		2	1					8				
9017	WT8						1		2	1					1				
9011	E4181324						1		2	1					1				
9082	HO104						1		2	1					1				
9009	KAS011					1			1		1				1				
9016	RML	1			1	1			2	1					1				
9007	DEM			1		1		1	1	1	1	1			1				
9018	L0081785			1							1				1	8		8	
9019	DUCAF			1							1			1	1	8	8		
9020	QBL			1							1				1			8	
9022	COX			1							1			8	1			8	
9023	VAVY			1		1					1			1	1			8	
9088	PFO4015		1			8			8		8			1	1			8	
9021	RSH	1		1					8		8				1				
9025	DEU	1		1				1	1			1			8				
9091	MLF	1		1				1	1			1			1				
9029	WT51			1				1	1			1			1				
9031	BOLETH			1				1	2			1			8				
9032	BSM			1					1			1			2				

Table 3. *Continued*

Cell No.	Name	2	3	4	5	6	7	9	11	13	14	17	18	20	22	24	25	26	27
9032	BSM			1				1	1			1			1				
9033	BM14			1				1	1			1			8				
9034	SAVC			1				1	1			1			2				
9026	YAR			1				1	1			1			8				
9030	JHAF			1				1	2			1			1				
9028	PE117			1				1	2			2			8				
9092	BM92			1				1	1			1			8				
9107	LKT3	1		1				1	1			1			1				
9037	SWEIG007		1							1					1				
9040	BM15			1						1	1			1	1	8		8	
9043	BM21	1		1						1	1			1	1	8		8	
9104	DHIF	1		1					1	1	1	1			8				
9039	JVM		1							1					1	1		8	
9042	TISI	1	8	1						1					1	8		8	
9036	SP0010		1							1				1	1	8		8	
9105	FPAF		1							1					1				
9045	TUBO	1		1						1					1	8		8	
9038	BM16			1										8	2	1			
9052	DBB			1					1		1	2			1				
9050	MOU	1		1					1	1	1	1			1				
9051	PITOUT	1		1					1		1	1			8				
9047	PLH	1		1					1		1	1			1				
9048	LBUF	1		1					1		1	1			1				
9106	MANIKA			1						1		1			1				
9058	OMW	1		1		1				1				1	1			8	
9065	HHKB			1					1	1					1				
9062	WDV	1		1					8	1					1				
9060	CB6B		1							1					1	8		8	
9059	SLE005		1							1			1				1	8	
9063	WT47		1							1			1			8		8	
9054	EK			1											1	8		8	
9061	31227ABO			1											8	1		8	
9057	TEM			1											1	8		8	
9064	AMALA			1					8	1					1			8	
9056	KOSE		1	1						1					1			8	
9056	KOSE	1	1	1						1				1	1	8		8	
9055	HO301	1	1							1			1			8		8	
9067	BTB																1	8	
9068	BM9																1	8	
9069	MADURA														8		8	8	
9071	OLGA														8		8	8	
9072	SPACH																8	8	
9066	TAB089								8						8		8	8	
9070	LUV								8									8	
9075	DKB			1					1	1		1			1				

RFLP Standardization Report for DR Beta/BamHI

J.J. Hyldig-Nielsen,[1] B.O. Boehm,[2] K.J. Wood,[3] A.R. Bushell,[3] D.P. Singal,[4] and A. Svejgaard[1]

The DR beta/BamHI system identified the 17 fragments listed in Table 1. Four of these (St. fragment nos. 2, 6, 13, and 16) were not included in the analysis (no locus assignment) as they were very faint and seen only on a few films. Possible reasons for these fragments could be either a very weak cross-hybridization to another (new?) locus or a partial digestion of some DNA preparations. Of the remaining 13 fragments, only 9 hybridized most strongly with the DRβ probe (Table 2). Three of the remaining fragments (nos. 3,4, and 17) hybridized more strongly with the DQβ probe and one (no. 5) with the DPβ probe. All the DR-specific bands were polymorphic. The DR associations of these bands are described in the joint report.

Table 3 shows the 9 banding patterns obtained with this system, of which 3 patterns are identical (St. fragments 7, 10, and 11). St. fragment no. 12 clearly consists of fragments with "normal" intensity and very strongly hybridizing fragments (Fig. 1). Comparing the distribution of this strongly hybridizing fragment with the assignments made in a preliminary report from the Biochemistry Study Group showed an almost perfect correlation to the III-1 and III-2 assignments of DRβ-III. This corresponds to the Dw25 and DW26 assignments from the New Provisional HLA-D (T cell-defined) specificities. Furthermore, all cell lines possessing this strongly hybridizing fragment have a lower number of hybridizing fragments, which supports the assumption that the heavily labeled fragments are caused by the cross-hybridization to the DRB3 genes.

Participating Laboratories: SCASVE,[1] GERBOE,[2] UKIMOR,[3] US5SIN[4]

Table 1. Fragments Identified with the DR Beta/BamHI System

		DRB	BAM	SVE		
	kb	Locus	Frequency	% Faint	56	32
1	11.33	1 0 0 0	0.400	0.000		
2	10.38	0 0 0 0	0.000	0.000		
3	10.87	3 1 2 0	0.000	0.000		
4	9.45	3 1 2 0	0.000	0.000		
5	8.16	2 3 1 0	1.000	1.000		
6	7.18	0 0 0 0	0.000	0.000		
7	6.59	1 0 0 0	0.300	0.000		
8	5.26	1 3 2 0	0.186	0.000		
9	4.78	1 0 0 0	0.229	0.000		
10	4.63	1 0 0 0	0.300	0.000		
11	4.28	1 0 0 0	0.300	0.000		
12	4.17	1 3 2 0	0.600	0.000		
13	4.05	0 0 0 0	0.143	0.000		
14	3.92	1 0 2 0	0.386	0.000		
15	3.80	1 3 2 0	0.300	1.000		
16	3.54	0 0 0 0	0.000	0.000		
17	2.78	2 1 0 0	0.171	0.750		

Table 2. Cross-Hybridization Table–Enzyme: BamHI–Probe: HLA Class II Beta

DRB	kb	DQB	kb	DPB	kb
3	10.87	5	10.26	2	10.79
4	9.45	6	9.40	3	9.36
5	8.16	8	8.09	5	8.08
8	5.26	11	5.27	7	5.23
		13	4.57	8	4.50
12	4.17	15	4.19	9	4.13
14	3.92			10	3.85
15	3.80	17	3.83	11	3.74
16	3.54	18	3.55	12	3.47
		19	3.03	13	3.01
17	2.78	20	2.89	14	2.87

Underline denotes primary locus assignment.

Table 3. Band Patterns Obtained with the DR Beta/BamHI System

WS ID#	Name	HLA-DR	HLA-DQ	HLA-DP	Standard Fragment No.								
					1	7	8	9	10	11	12	14	15
9002	MZ070782	1	5	4				6				6	
9003	KAS116	1	5					6				6	
9004	JESTHOM	1	5	4				6				6	
9005	HOM2	1	5					6				6	
9006	WT100BIS	1	5	1				6				6	
9009	KAS011	2	1	3,4				6			6	6	
9018	LO081785	3	2	3	6						8		
9019	DUCAF	3	2		6						8		4
9020	QBL	3	2	2	6						8		4
9021	RSH	3	4	1	6						6	6	
9022	COX	3	2	3	6						6	6	
9023	VAVY	3	2	1	6						6	6	
9088	PF04015	3	2	1,4	6						6	6	
9025	DEU	4	7	4		6	6		6	6			4
9026	YAR	4	8	4		6	6		6	6			4
9028	PE117	4	8	4	6	6	6		6	6			4
9029	WT51	4	8	2		6	6		6	6			4
9030	JHAF	4	7	3		6	6		6	6			4
9031	BOLETH	4	8	4		6	6		6	6			4
9032	BSM	4	8	2		6	6		6	6			4
9032	BSM	4	8	2		6	6		6	6			4
9033	BM14	4	8	4		6	6		6	6			4
9034	SAVC	4	8			6	6		6	6		6	4
9091	MLF	4	7			6	6		6	6			4
9092	BM92	4	8			6	6		6	6		6	4
9107	LKT3	4	4	5		6	6		6	6			4
9055	HO301	6	6	5	6						8		
9047	PLH	7	2			6			6	6			4
9048	LBUF	7	2			6			6	6			4
9050	MOU	7	2	2		6			6	6			4
9051	PITOUT	7	4	4		6			6	6			4
9052	DBB	7	9	4		6			6	6			4
9106	MANIKA	7	2,3	NT		6			6	6			4
9066	TABO89	8	6								6		
9067	BTB	8	4	4							6		
9068	BM9	8	4	2							6		
9069	MADURA	8	4	4									
9072	SPACH	8	4										
9070	LUY	8	7	1,4							6		
9071	OLGA	8	4	3							6		
9075	DKB	9	9	4		6			6	6		6	
9036	SPO010	11	5	2	6						8		
9037	SWEIG007	11	7					6			6	6	
9039	JVM	11	7	2	6						8		
9040	BM15	11	7	3	6						8		
9042	TISI	11	7		6						8		
9043	BM21	11	7		6						8		
9104	DHIF	11	7	NT		6			6	6			4
9105	FPAF	11	6	2	6						8		
9045	TUBO	11,12	7	2,4	6						8		
9038	BM16	12	7	2	6						6		
9058	OMW	13	6	1	6						6	6	
9059	SLE005	13	6	3	6						8		
9060	CB6B	13	5		6						8		
9062	WDV	13	6	4	6						6	6	
9063	WT47	13	6	2	6						8		
9065	HHKB	13	6	4	6						6	6	
9056	KOSE	13,14	5	2	6						8		
9056	KOSE	13,14	5	2	6						6		
9054	EK	14	5		6						6		

Table 3. *Continued*

WS ID#	Name	HLA-DR	HLA-DQ	HLA-DP	Standard Fragment No. 1	7	8	9	10	11	12	14	15
9057	TEM	14	1	4	6						6		
9061	31227ABO	14	5	4	6						6		
9064	AMALA	14	7		6						6	6	
9008	DO208915	15	6	2,4				6			6	6	
9010	AMAI	15	6					6			6	6	
9011	E4181324	15	6	2,4				6			6	6	
9013	SCHU	15	6					6			6	6	
9014	MGAR	15	6	4				6			6	6	
9017	WT8	15	1	2,4				6			6	6	
9082	HO104	15	6	4				6			6	6	
9016	RML	16	7					6			6	6	
9007	DEM	16,4	5,8	3,4		6	6	6	6	6		6	

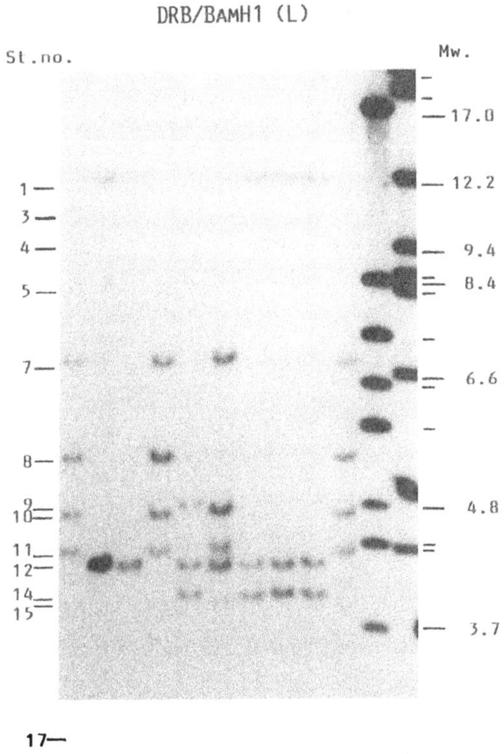

Figure 1. Fragments identified with the DR beta/BamHI system.

RFLP Standardization Report for: DR Beta/HincII

P. Piccioli,[1] D. Chandanayingyong,[2] M. Vijod,[3] and G.B. Ferrara[1]

In this system, eight fragments were identified. However, because of the poor quality of the autoradiographs and the weak signal the interpretation of the data, the assignment of a specific fragment to each cell line was very difficult and imprecise. Indeed, it was possible to calculate only the sizing for each fragment.

Table 1 shows the kb of the fragments detected in this system. In Table 2 the fragments for which we detected a cross-hybridization between DR, DQ, and DP locus are summarized. For the reasons mentioned above, Table 2 should be viewed as an "hypothetic model." Locus assignment for DR locus was determined by superimposing autoradiographs (the same filter rehybridized with different beta probes), looking only for the presence/absence of each specific band (not for the band intensity) and checking if there was a correlation between molecular weights. Both parameters were kept in consideration for DQ and DP locus.

Participating Laboratories: IT2FER,[1] ANZDCH,[2] ANZDAW[3]
P. Piccioli was recipient of a fellowship from A.I.R.C. (Associazione Italiana per la Ricerca sul Cancro). This work was supported in part by CNR Grants 86.0265.51 and 86.00659.

Table 1. kb of Fragments Identified with the DR Beta/HincII System

Fragment	kb
1	10.77
2	8.31
3	7.04
4	4.11
5	3.63
6	3.45
7	2.77
8	2.20

Table 2. Cross-Hybridization Table–Enzyme HincII–Probe: HLA Class II Beta

DRB	kb	DQB	kb	DPB	kb
		2	12.54	1	12.65
1	10.77	3	10.61		
		4	10.18	2	10.17
2	8.31	6	8.50	3	8.69
		7	7.66	5	7.53
3	7.04	8	6.91	6	7.07
		9	3.81	10	3.77
5	3.63	10	3.66	11	3.67
6	3.45	11	3.32	12	3.41
		12	2.93	13	2.92
7	2.77	13	2.69	14	2.72
8	2.28	15	2.21		
		16	2.07	15	2.09

Underline denotes primary locus assignment.

RFLP Standardization Report for DQ Beta/SstI

A. Ando,[1] T. Sasazuki,[2] P. Reekers,[3] and K. Tsuji[1]

The DQ beta/SstI system identified the 33 fragments as shown in Figure 1 and Table 1. All bands were polymorphic except three bands–St. fragment no. 23 (3.78 kb), no. 30 (2.01 kb), and no. 32 (1.09 kb) (Table 2). These three nonpolymorphic fragments hybridized only with the DQ beta probe. Cross hybridizing bands are shown in Table 3. There are some discrepancies in fragment size through all systems. The fragments between St. fragments no. 6 to no. 9, which show cross-hybridization, show considerable discrepancies in fragment size through all systems. Seven fragments–no. 12, no. 15, no. 16, no. 20, no. 25, no. 26, and no. 28–are polymorphic and are dominantly DQ-specific. St. fragment no. 26 (2.86 kb) and no. 29 (2.22 kb) are highly associated with DQw7 (DQ"α1", TA10+, IIB3−). St. fragment no. 28 (2.28 kb) is highly correlated with DQw1, DQw5, and DQw6.

Participating Laboratories: JAPTSU,[1] JAPSAS,[2] BENREE[3]

Immunobiology of HLA
Volume I

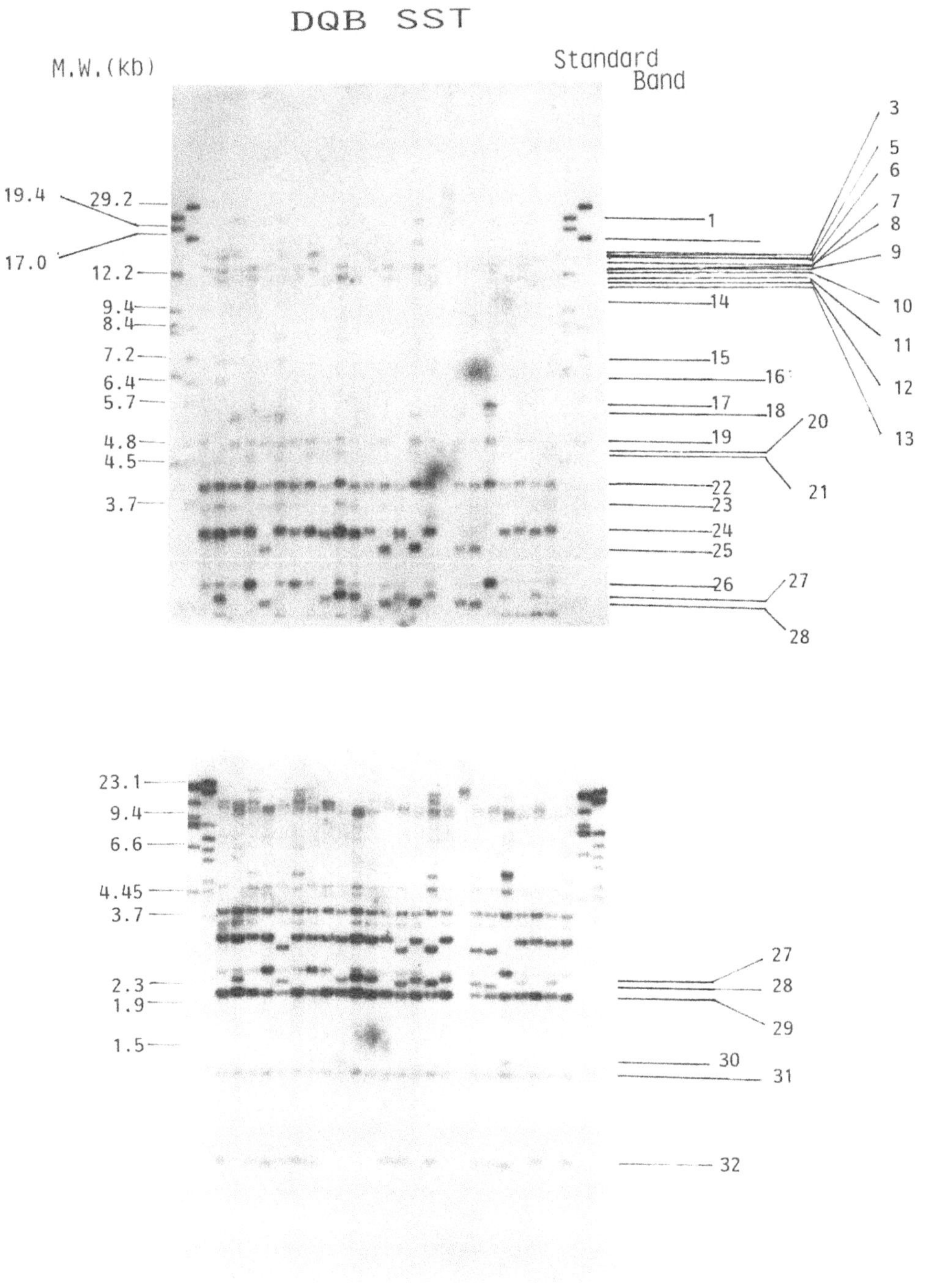

Figure 1. Fragments observed with the DQ beta/SstI system.

Table 1. Fragments Identified with the DQ Beta/SstI System

	kb	DQB Locus	SstI Frequency	TSU % Faint	56	32
1	21.66	1 3 2 0	0.290	1.000		f
2	17.91	2 2 1 0	0.029	1.000		
3	15.94	2 2 1 0	0.319	1.000	f	
4	14.16	1 3 2 0	0.000	0.000		
5	14.14	0 1 0 0	0.261	0.944		
6	13.82	1 3 2 0	0.200	1.000		
7	13.40	1 3 2 0	0.057	1.000		
8	12.68	1 3 2 0	0.406	1.000		
9	12.23	1 3 2 0	0.329	1.000		f
10	11.89	1 3 2 0	0.072	1.000		
11	11.47	1 3 2 0	0.449	1.000		
12	11.22	0 1 0 0	0.086	1.000		f
13	10.79	0 1 0 0	0.000	0.000		
14	10.09	1 2 0 0	0.000	0.000		
15	9.37	0 1 0 0	0.043	1.000		
16	6.75	0 1 0 0	0.261	1.000		
17	6.08	1 3 2 1	0.071	1.000		
18	5.40	0 1 0 0	0.072	0.000		
19	5.20	1 2 2 1	0.275	1.000		
20	4.64	0 1 0 0	0.739	1.000		
21	4.56	1 3 2 0	0.186	1.000		
22	4.35	0 2 1 1	0.217	1.000	f	f
23	3.78	2 1 0 0	1.000	0.000		
24	3.44	2 3 1 0	0.638	1.000		
25	3.07	2 1 0 0	0.754	0.000		
26	2.86	0 1 0 0	0.188	0.077		
27	2.42	3 2 1 0	0.739	0.788	f	
28	2.28	2 1 0 0	0.377	0.000		
29	2.22	0 1 0 0	0.174	0.000		
30	2.01	0 1 2 0	1.000	0.000		
31	1.33	0 2 1 0	0.043	1.000		
32	1.09	0 1 1 0	1.000	1.000		
33	0.34	0 1 1 0	0.594	1.000		

Table 3. RFLP Pattern of DQ Beta/SstI

		Fragment No.	1	2	3	4	5	6	7	8	9	10	11	12	13	14	15	16	17	18	19	20	21	22	23	24	25	26	27	28	29	30	31	32	33						
		Fragment size (kb)	21.66	17.91	15.94	14.16	14.14	13.82	13.40	12.68	12.23	11.89	11.47	11.22	10.79	10.09	19.37	6.75	6.08	5.40	5.20	4.64	4.56	4.35	3.78	3.44	3.07	2.86	2.42	2.28	2.22	2.01	1.33	1.09	0.34						
	Cell	Lines																																		DW	DR	DW2	DRw	DQw	DPw
67	9106	MANIKA																							1		1		8			1		8	8		7		53	2?3	N,T
72	9007	DEM			8													8				8			1		2			1		2		8	8	—	16,4		53	5,8	3,4
49	9009	KAS011																				8			1		2			1		2		8	8	21	2			1	3,4
5	9017	WT8						8			8							8	8			8		8	1	8	2		8	1		2		8		2	15			1	2,4
38	9057	TEM	8				8			8			8					8							1	8	2		8	1		2		8		9	14	25	52	1	4
17	9005	HOM2										8	8									8			1	8	2		8	2		2		8		1	1			5	—
19	9056	KOSE			8					8			8									8		8	1	8	2		8	1		2		8		9,19	13,14	25	52	5	2,N
65	9056	KOSE								8			8									8			1	8	2			1		2		8		9,19	13,14		52	5	2,N
59	9003	KAS116			8							8										8			1		2			2		2		8		1	1			5	—
35	9002	MZ070782			8							8										8			1	8	1		8	2		2		8		20	1			5	4
34	9036	SP0010	8							8			8									8		8	1	8	2		8	2		2		8		DB2	11	25	52	5	2
16	9006	WT100BIS			8							8	8									8			1		2			2		2		8		1	1			5	1
9	9004	JESTHOM										8	8									8			1	8	2		8	1		2		8		1	1			5	4
40	9054	EK								8			8									8			1	8	2		8	1		2		8		9	14	25	52	5	—
55	9061	31227ABO								8			8									8			1	8	2		8	2		2		8		9	14	25	52	5	4
33	9060	CB6B							8				8									8		8	1		1			1		2	8	8		18	13	25	52	5	—
62	9011	E4181324					8	8			8											8			1	8	2		8			2		8		12	15			6	2,4
3	9058	OMW			8				8				8					8							1		2			1		2		8		18	13	24	52	6	1
41	9105	FPAF	9	9	9		9			9		9	9				9	9		9	9	9		9	9	9	9	9	9	9	9	9		9	9	FS	11		52	6	2
66	9082	H0104					8	8			8							8	8						1	8	2		8	1		2		8			15			6	4
47	9062	WDV							8				8									8			1	8	1		8	1		1		8		18	13	24	52	6	4
58	9008	D0208915						8			8														1	8	1			1		2		8		2	15			6	2,4
24	9010	AMAI																8	8						1	8	2		8	1		2		8		2	15			6	—
2	9014	MGAR						8			8		8						8			8			1	8	2		8	1		2		8		2	15			6	4
32	9066	TAB089					1				8											8		8	1		1					2		8		8.3	8		52	6	—
7	9013	SCHU						8			8		8					8	8						1	8	1		8	1		2		8		2	15			6	—
12	9063	WT47			8					8			8					8						8	1		2			1		2		8		19	13	26	52	6	2
45	9065	HHKB	8				8		8				8									8			1	8	1		8	1		1		8		18	13	24	52	6	4
11	9055	H0301			8					8			8									8			1		2			1		2		8		N	6	26	52	6	5
51	9059	SLE005			8					8			8					8							1		2			1		2		8		19	13	26	52	6	3
10	9018	L0081785			8					8			8									8		8	1		1		2			1		8	8	3	3	25	52	2	3
31	9088	PF04015			8		8			8			8									8			1	8	1		1			1		8	8		3	24	52	2	1,4
18	9050	MOU	8							8										1	8	8		8	1	8			2			1		8	8	17	7		53	2	2
22	9047	PLH	8	8						8										1	8	8	8		1				2			1		8	8	DB1	7		53	2	—
28	9019	DUCAF	8							8			8									8		8	1		1		2			1		8	8	3	3	25	52	2	—
44	9048	LBUF	8								8			8						1	8	8	8		1				2			2	8	8	8	DB1	7		53	2	—

Table 3. *Continued*

		Fragment No.	1	2	3	4	5	6	7	8	9	10	11	12	13	14	15	16	17	18	19	20	21	22	23	24	25	26	27	28	29	30	31	32	33						
		Fragment size (kb)	21.66	17.91	15.94	14.16	14.14	13.82	13.40	12.68	12.23	11.89	11.47	11.22	10.79	10.09	19.37	6.75	6.08	5.40	5.20	4.64	4.56	4.35	3.78	3.44	3.07	2.86	2.42	2.28	2.22	2.01	1.33	1.09	0.34						
	Cell	Lines																																		DW	DR	DW2	DRw	DQw	DPw
21	9051	PITOUT	8							8	8			8						1	8	8	8		1	8			2			1		8	8	17	7		53	2	4
13	9022	COX			8					8			8									8		8	1		1		1			1		8	8	3	3	24	52	2	3
15	9023	VAVY			8					8			8									8			1		1		1			1		8	8	3	3	24	52	2	1
71	9020	QBL								8			8									8			1		1		2			1		8	8	3	3	25	52	2	2
37	9064	AMALA	8				8			8			8									8			1	8		1	8		1	1		8	8	16	14	24	42	7	–
1	9045	TUBO								8			8									8		8	1	8		1	8		1	1		8	8	5	11,12	25	52	7	2,4
42	9091	MLF	8							8	8										8	8			1	8		1	8		1	1		8	8	4	4		53	7	–
20	9104	DHIF	8																	1	8	8	8		1			8	1			1	8	8	8	5	11		52	7	N,T
57	9038	BMIG											8									8			1	8		1	8		1	1		8	8	DB6	12		52	7	2
6	9042	TISI								8			8				8					8			1	8		1	8		1	1		8	8	N	11	25	52	7	–
23	9040	BM15			8					8			8									8			1			1			1	1		8	8	5	11	25	52	7	3
14	9043	BM21			8					8			8				8					8			1			1			1	1		8	8	5	11	25	52	7	–
39	9030	JHAF	8		8						8										8	8	8	8	1			1			1	1		8	8	13	4		53	7	3
26	9037	SWEIG007					8											8							1	8	2	8	1			2		8		5	11	25	52	7	–
29	9025	DEU	8				8			8	8										8	8	8		1	8		1	8		1	1		8	8	4	4		53	7	4
64	9039	JVM								8			8					8							1	8		1	8		1	1		8	8	N	11	25	52	7	2
60	9070	LUY			8		8			8												8			1			1			1	1		8	8	8.3	8		52	7	1,4
43	9016	RML						8		8	8											8			1	8		1	8		1	1		8	8	22	16			7	–
52	9034	SAVC	8		8			8										8			8				1		1					2		8	8	4	4		53	8	–
68	9032	BSM	8															8			8		8		1	8	1		8			2		8	8	4	4		53	8	2
8	9033	BM14						8										8			8		8	8	1	8	1		8			2		8	8	4	4		53	8	4
61	9029	WT51						8			8							8			8		8		1	8	1		8			2		8	8	4	4		53	8	2
27	9028	PE117	8				8				8			8				8			8	8	8		1	8	1		8			2		8	8	14	4		53	8	4
63	9026	YAR															8	8			8				1	8	1		8			2		8	8	10	4		53	8	4
30	9031	BOLETH	8				8				8			8				8			8		8		1	8	1		8			2		8	8	4	4		53	8	4
50	9092	BM92	8		8			8			8			8							8	8			1	8	1		8			2		8	8	14	4		53	8	–
70	9052	DBB	8	8			8														8	8	8		1	8	1		8			2		8	8	11	7		53	8	4
54	9075	DXB	8				8	8			8										8	8			1	8	1		8			2		8	8	23	9		53	8	4
48	9032	BSM									8			8				8			8		8	8	1	8	1		8			2		8	8	4	4		53	8	2
69	9072	SPACH																							1	8	1					2		8	8	8.2	8		52	4	4
25	9069	MADURA					8				8											8			1	8	1		8			2		8	8	8.1	8		52	4	4
56	9067	BTH					8	8			8											8			1	8	1		8			2		8	8	8.1	8		52	4	4
4	9107	LKT3	8		8		8														8	8	8		1		1					2		8		15	4		53	4	5
46	9021	RSH			8					8			8									8			1	8	1		8			2		8	8	N	3	24	52	4	1
36	9071	OLGA			8		8				8											8		8	1	8	1		8			2		8	8	8.2	8		52	4	3
53	9068	BM9						8			8											8			1	8	1		8			2		8	8	8.1	8		52	4	2

RFLP Standardization Report for DQ Beta/PstI

E. Keller,[1] A. Andreas-Zietz,[1] E.D. Albert,[1] H. Coppin,[2] L. Contu,[3] A. Takada,[4] and M. Trucco[5]

The DQ Beta/PstI system identified 28 fragments on the core cell lines as shown in Table 1 and Figures 1 and 2. Table 1 shows the presence/absence pattern of all 72 cell lines for the 28 standardized DQ beta/PstI fragments, the corresponding molecular weight, as well as the locus assignment and probable associations with DR/DQ antigens. The intensity of bands is indicated by numbers. "1," "2," "3" means increasing intensity, whereas "8" means a faint fragment and "9" means unreadable. In this complex system, the intensity does not necessarily indicate the gene dosage. Especially "1" generally does not mean heterozygosity (see also the report on DR beta/PstI). Core cell no. 26 showed two different banding patterns in different laboratories. Therefore, "9" was assigned to all fragments not identical in all laboratories. As the

Participating Laboratories: GERALB,[1] FRAOHA,[2] IT1CON,[3] JAPAIZ,[4] US1DUQ[5]
This work was supported by SFB 217.

Table 1. Banding Pattern of All 72 Core Cell Lines, as Standardized in the DQ Beta/PstI System

```
Probe Name: DQB            Enzyme Name: PSTI

    |99999999999999 999999999999999 99999999999 99999999999999 999999999 99 9999999| WS-ID number
    |00000000000000 000000000100000 00000000001 10000000000000 000000000 00 0000001|
    |00000556530056 150181616061565 22811255440 04443343293716 333322329 75 6667720|
    |36452661769740 088327542561539 23889001786 42307958510064 312496282 52 9872117|
    |.............. ............... ........... .............. ......... .. .......| ...............
    |51013165334743 205060404436115 11312712246 20122605243643 034566625 57 2556340| position on
    |96975955849203 438765527122121 35108181247 06436417929037 808213870 40 5369664| blots
    |.............. ............... ........... .............. ......... .. .......| ...............
DR  |     11111  11 1111111111 1 11             11111111    11                     |
    |11111334412443 535555353185633 33333377777 11111112444864 444444444 97 8888834|
    |.............. ............... ........... .............. ......... .. .......|
DR  |     11    1                                     1                            |
    |     44    6                                     2                            |
    |.............. ............... ........... .............. ......... .. .......|
DQ  |55555555151555 666661666666666 22222222222 77777777777777 888888888 99 444-444|
    |.............. ............... ........... .............. ......... .. .......|
DQ  |           8                             3                                    | kb     locus      associated with:
    |.............. ............... ........... .............. ......... .. .......| ..............   ...............
  1 |        2  1   111111     2112                 92         12122221            | 11.22  2 1 3 0
  2 |             2 2222222222                      9                              | 10.68  2 1 3 0   part DQw 1
  3 |            11      81         111       8                                    |  8.33  2 1 0 0
  4 |           8                                              818888818           |  6.84  0 1 0 0   DQw 8
  5 |88 88  8 888          8888        8118881  88889 88888888        88 81 8888818|  6.39  0 1 0 0
  6 |     81                    111                                                |  6.43  2 1 3 0   part DQw 6
  7 |                        1                                                     |  6.26  0 1 0 0
  8 |         111                   88181188818 8                        81        |  6.07  1 2 2 0*  DQw 2 + 9 +
  9 |                  888 8                                                 88    |  5.51  1 2 2 0
 10 |                 8             22223222221 2                        11 8118   |  5.23  2 1 3 0*  DQw 2 + 9 +
 11 |111118111   1            11                                                  1|  5.08  1 2 0 0*  DQw 5
 12 |           2                             1  2229222323222 233222222 23 2222232|  4.93  2 1 3 0   DQw 3
 13 |23233223232 3                 322      8888  8888988888818                    |  4.42  2 1 2 0*
 14 |3323332313221     2    22232         23222221 22229 22222222        22 22 2222222|  4.17  2 1 2 0
 15 |                        9                                 888888888          8|  3.96  1 2 2 0   DR 4
 16 |    8                                                                         |  3.81  1 2 2 0
 17 |            22      12         222       1                                    |  3.01  0 1 0 0
 18 |88   88   88     8          888 88 8    8     88       88      8             888|  2.84  2 2 1 0
 19 |     81111  1               888 88 811    8889 88                           8 |  2.66  1 2 2 0
 20 |  88888888888  8 8888888988       8 8888 88 88  888888 888 888 88888 88 888888 |  2.35  2 3 1 0
 21 |88888888888888 888888888988888 88888888888 88888888888888 888888888 88 8888888|  2.05  1 2 3 0*
 22 |88888888888888 888888888988888 88888888888 88888888888888 888888888 88 8888888|  1.91  3 2 1 0
 23 |88888888888888 888888888988888 88888888888 88888888888888 888888888 88 8888888|  1.73  2 2 1 0
 24 |8888                    9                                                     |  1.45  1 2 2 0
 25 |2222212222222                                                                 |  1.36  2 1 3 0   DQw 5
 26 |88888888888888 888888888988888 88888888888 88888888888888 888888888 88 8888888|  1.26  1 2 3 0
 27 |88888888888888 888888888988888 88888888888 88888888888888 888888888 88 8888888|  0.85  2 3 1 0
 28 |     88  888 8 888888888988888 88888888888 88888888888888 888888888 88 8888888|  0.36  0 1 0 0
```

Cells are listed according to DQ and DR antigens and banding patterns. Numbers indicate intensity (1,2,3 = increasing intensity; 8 = faint; 9 = unreadable).
* = For some cells different locus assignment, see text.

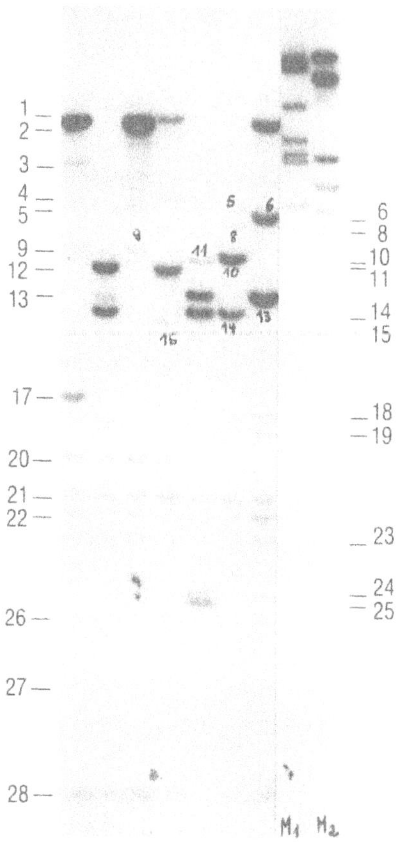

Figure 1. DQ beta/PstI core fragments (gel 0.9%).

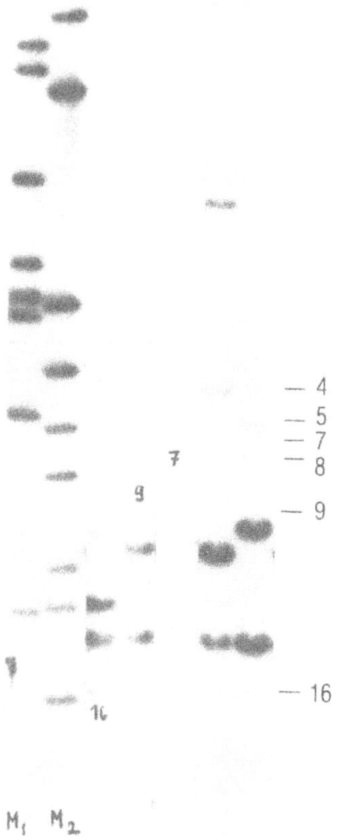

Figure 2. DQ beta/PstI core fragments no. 7 and 16 (gel 0.6%).

different banding patterns were confirmed in three laboratories, the most likely explanation is a mixup of DNA (see also report on DR beta/PstI and DP beta/PstI). For cross-hybridization with DR; DP and DO beta probes, see Table 2 in the Workshop report on DR beta/PstI. It should be mentioned that the assignment of standard fragments no. 8, 9, 10, and 11 was rather difficult under Workshop conditions, also leading to difficulties in locus assignment.

Fragment no. 11. This actually consists of two separate fragments not distinguishable under Workshop conditions. Some cells positive for this band clearly show the locus assignment "0100" (all DR1 and one DRw14), the others have the locus assignment "1200" (DR, DQ, DP, DO, respectively).

Fragments no. 10 and 13. There are different cross-hybridization patterns for DR and DP, while there is agreement in the primary locus assignment to DQ. A few cells (no. 53 DRw8, no. 54 DRw9, no. 56 DRw8, no. 69 DRw8, no. 70 DRw8) have *fragment no. 10* with the locus assignment "0100," indicating that this cross-hybridization pattern is due to genetic variation rather than technical artifacts. The other cells positive for fragment no. 10 show the locus assignment "2130". Concerning *fragment no. 13*, the locus assignment "2120" is only valid for the very strong bands. The faint ones show the locus assignment "0100" (DR, DQ, DP, DO, respectively).

Fragments no. 8 and 21. There are different cross-hybridization patterns for DQ and DP, while there is agreement in the primary locus assignment to DR. There are some cells (no. 10, 13, 15, 28, 31, 34, 49, 71, 72, six of them DR3-positive, two DR2 and one DR5) positive for *fragment no. 8* with the locus assignment "1200," while the other positive cells are coded as "1220" (see also comments on core fragment no. 7 of the DR beta/PstI system).

Fragment DQ Beta/PstI no. 21 is present in all cells. The locus assignment is "1120" except in those cells that show a very strong signal for *DR* beta/PstI fragment no. 25 (DR4-associated) where the locus assignment is "1230" (see also comments on core fragments no. 25 DR beta/PstI and no. 19 DP beta/PstI).

RFLP Standardization Report for DQ Beta/KpnI

P.I. Biro,[1] B. Herbut,[1] S. Cutbush,[1] P. Cassell,[1] H. Avakian,[1] R. Vaughan,[1] M. Oudshoorn,[2] R. Martell,[2] and K. Siminovitch[3]

Twenty-three fragments are defined by the DQ beta probe (Fig. 1), the majority of which are faint, polymorphic, and relatively DQ-specific (Tables 1 and 2). Among the nonpolymorphic bands, st. bands 11, 13, and 22 are faint and no. 16 is intense. Most DNA samples display one or two relatively intense polymorphic bands, which show limited correlation with DR type (Table 3). For example, nearly all DR 3 and some DR 8 samples contain st. band 1, but not bands 20 or 21, and most DR 8 and some DR 4 samples contain band 2. Band 6 is shared by DR 4 and DR 7 DNAs, and band 8 is found mainly in DR 2 DNAs. DQ beta/KpnI RFLPs may be of some value in DQ or DR typing, but other enzymes may show better correlations.

Table 1. Sizes and Frequencies of Standard Bands Produced by the DQ Beta/KpnI Enzyme/Probe Combination

	kb	Locus	Frequency	% Faint
1	25.91	0 1 2 0	0.129	0.333
2	19.06	3 1 2 0	0.271	0.150
3	17.75	1 2 2 0	0.074	0.000
4	15.44	1 2 3 0	0.072	0.000
5	15.90	1 2 3 0	0.043	1.000
6	12.81	1 2 3 0	0.271	0.000
7	12.03	2 1 0 0	0.101	0.000
8	11.71	1 2 3 0	0.171	0.250
9	11.82	0 1 0 0	0.143	0.500
10	11.75	0 1 0 0	0.044	0.000
11	11.04	2 2 1 0	1.000	0.070
12	10.68	1 2 2 0	0.101	0.429
13	9.58	2 1 0 0	1.000	0.000
14	9.20	2 1 2 0	0.087	0.000
15	8.36	0 1 0 0	0.058	0.500
16	7.89	2 1 2 0	1.000	0.000
17	7.73	1 2 2 0	0.087	0.833
18	6.85	0 1 0 0	0.074	0.600
19	6.06	0 1 0 0	0.015	0.000
20	5.33	2 1 2 0	0.186	0.077
21	5.15	2 1 2 0	0.414	0.000
22	1.54	0 1 0 0	1.000	1.000
23	1.20	2 1 3 0	0.174	0.000

Table 2. Cross Hybridization Table – Enzyme: KpnI – Probe: HLA Class II Beta

DRB	kb	DQB	kb	DPB	kb
1	37.30			1	38.04
2	17.95	3	17.75	4	18.64
3	15.99	4	15.90		
4	12.80	6	12.81		
5	12.50	7	12.03		
6	11.84	8	11.71		
7	11.05	11	11.04	9	11.11
8	10.49	12	10.68	10	10.83
9	9.43	14	9.20	11	9.36
10	7.94	17	7.98		
11	7.69	16	7.73	13	7.68
12	5.53	20	5.33	14	5.59
13	5.39	21	5.15	15	5.44
15	1.22	23	1.20	17	1.23
		1	25.91	2	24.44

Underline denotes primary locus assignment.

Participating Laboratories: UKIFES,[1] SAFDUT,[2] NCYFAL [3]

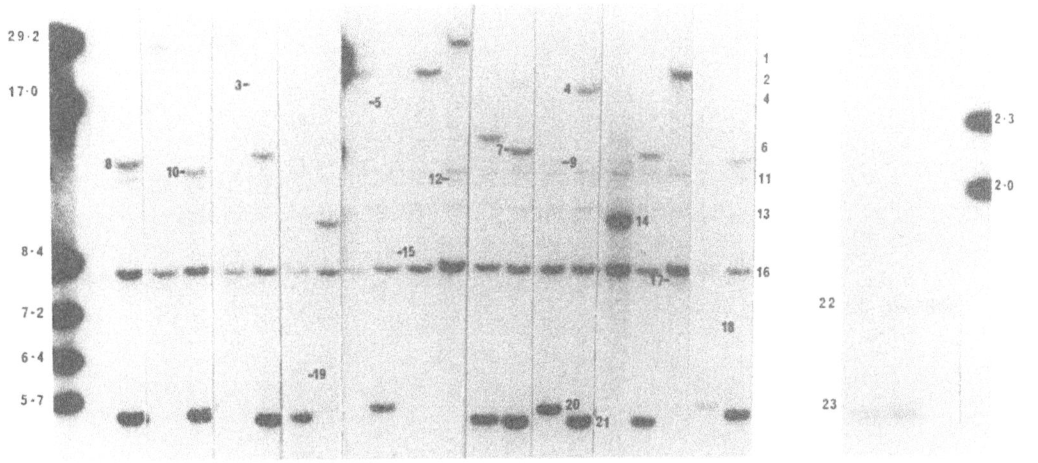

Figure 1. Standard bands produced by the DQ beta/KpnI enzyme/probe combination.

Table 3. RFLP Pattern of DQB/KpnI System

Cell Lines		DR Type	Fragment No. 1	2	3	4	5	6	7	8	9	10	11	12	13	14	15	16	17	18	19	20	21	22	23
MZ070782	9002	1			1				1				1		1			1					1	8	
KAS116	9003	1			1				1				1		1			1					1	8	
JESTHOM	9004	1			1				1				1		1			1					1	8	
HOM2	9005	1			1				1				1		1			1					1	8	
WT100BIS	9006	1			1				1				1		1			1					1	9	
DEM	9007	2/4		1				2					1		1			1					1	8	
DO208915	9008	2			9	9	9		9			9	9	9	9	9	9	1	9	9	9		1	9	9
KAS011	9009	2						2		1			1		1			1					1	8	
AMAI	9010	2								1			1		1			1					1	8	
E4181324	9011	2						2		8			1		1			1					1	8	
SCHU	9013	2								1			1		1			1					1	8	
MGAR	9014	2								1			1		1			1					1	8	
RML	9016	2								1			1		1			1				1		8	1
WT8	9017	2			9					1			1		1			1					1	8	
LO081785	9018	3	1										1		1			1						8	
DUCAF	9019	3	1										1	1	1			1						8	
QBL	9020	3	1										1		1			1						8	
RSH	9021	3		8									1		1			1						8	
COX	9022	3	1										1		1			1	8					8	
VAVY	9023	3	1										1		1			1	8					8	
DEU	9025	4						1			1		1		1			1				1		8	1
YAR	9026	4		8				1					1		1			1						8	
PE117	9028	4		1				1					1		1			1						8	
WT51	9029	4		1				1					1		1			1						8	
JHAF	9030	4						1			1		1	8	1			1				1		8	1
BOLETH	9031	4		1				1					1		1			1						8	
BSM	9032	4		1				1					1		1			1						8	
BM14	9033	4		1									1		1			1						8	
SAVC	9034	4	8	8									8		1			1						8	
SP0010	9036	5						2					1		1			1					1	8	
SWEI6007	9037	5											1		1		1	1				1		8	1
BM16	9038	5											8		1			1		8		1		8	1
JVM	9039	5											1		1		8	1				1		8	1
BM15	9040	5									8		1		1			1		8		1		8	1
TISI	9042	5											8		1			1		1		1		9	1

Cell Lines		DR Type	Fragment No. 1	2	3	4	5	6	7	8	9	10	11	12	13	14	15	16	17	18	19	20	21	22
BM21	9043	5											1		1		1	1				1		
TUBO	9045	5											1		1			1		1			1	
PLH	9047	7						1			1		1		1	1		1						
LBUF	9048	7						1			8		1		1	1		1						
MOU	9050	7						1			8		1		1	1		1						
PITOUT	9051	7						1			1		1		1	1		1						
DBB	9052	7		1				1					1		1			1						
EK	9054	6				1							1		1			1					1	
HO301	9055	6										1	1	1	1			1					1	
KOSE	9056	6				1							1		1			1			1		1	
TEM	9057	6				1							1		1			1					1	
OMW	9058	6								8			1		1			1	8				1	
SLE005	9059	6										1	1	8	1			1					1	
CB6B	9060	6								8			1		1			1					1	
31227ABO	9061	6				1							1		1			1					1	
WDV	9062	6							1	1			1		1			1	8				1	
WT47	9063	6										1	1	1	1			1					1	
AMALA	9064	6									8		1		1			1				1		
HHKB	9065	6							1				1		1			1	1				1	
TAB089	9066	8						1					1		1			1					1	
BTB	9067	8		1									1		1			1						
BM9	9068	8	8	1									1		1			1						
MADURA	9069	8		1			8						1		1			1						
LUY	9070	8					8						1		1			1		8		1		
OLGA	9071	8		1			8						1		1			1						
SPACH	9072	8		1									8		1			1						
DKB	9075	9		1									1		1			1						
HO104	9082	?				1				1			1		1			1						
PF04015	9088	3	1										1	1	1			1	8					
MLF	9091	4						1			1		1	8	1			1				1		
BM92	9092	4	8	1									1		1			1						
DHIF	9104	11						1			8		1		1	1		1				8		
FPAF	9105	11								1		9	1		1			1		9	9		1	
MANIKA	9106	7		1									8		9	1		1						
LKT3	9107	4		1									1		1			1						

RFLP Standardization Report for DQ Beta/EcoRV

T. Takenouchi,[1] R. Tonai,[1] P. Terasaki,[1] S. Mayer,[2] S. Bashir,[3] R. Winchester,[4] and Y. Paik[5]

DQ beta/EcoRV data were received from five laboratories and standardized using the same criteria as the DR/EcoRV system. In this system, the same experimental problems as the DR/EcoRV system were encountered and treated similarly. For DQ beta/EcoRV, 18 standard fragments (Table 1 and Fig. 1) were assigned. Refer to Figure 2 in the RFLP standardization report for DR beta EcoRV earlier in this section for the molecular weight of markers seen on the two lanes on the right side of the gel. Of the 18 fragments, 13 are dominantly DQ-specific. One fragment is DR-DQ codominant. Two fragments are DR-specific and two are DP-dominant. In the case of cross-hybridization see Table 2 in DR beta/EcoRV system.

The data of the one enzyme-one probe combination DQ beta/EcoRV are shown in Table 2. Looking at DQ-specific bands, eight banding patterns are observed (Fig. 2). This number of clusters is greater than those detected by serology. The specificity of the clusters will be decided when core cell line serology has been completed. Furthermore, cDNA of the DQ beta probe is thought to hybridize with DQB1 and DQB2 genes. Therefore, the nonpolymorphic or nonallelic DQ loci-dominant fragments observed in this system may show DQB2 specificity. Cells KOSE and DEM appear to be heterozygous in this system.

Participating Laboratories: US5TER,[1] FRAMYE,[2] ANZBSH,[3] US7RJW,[4] US5PAK[5]

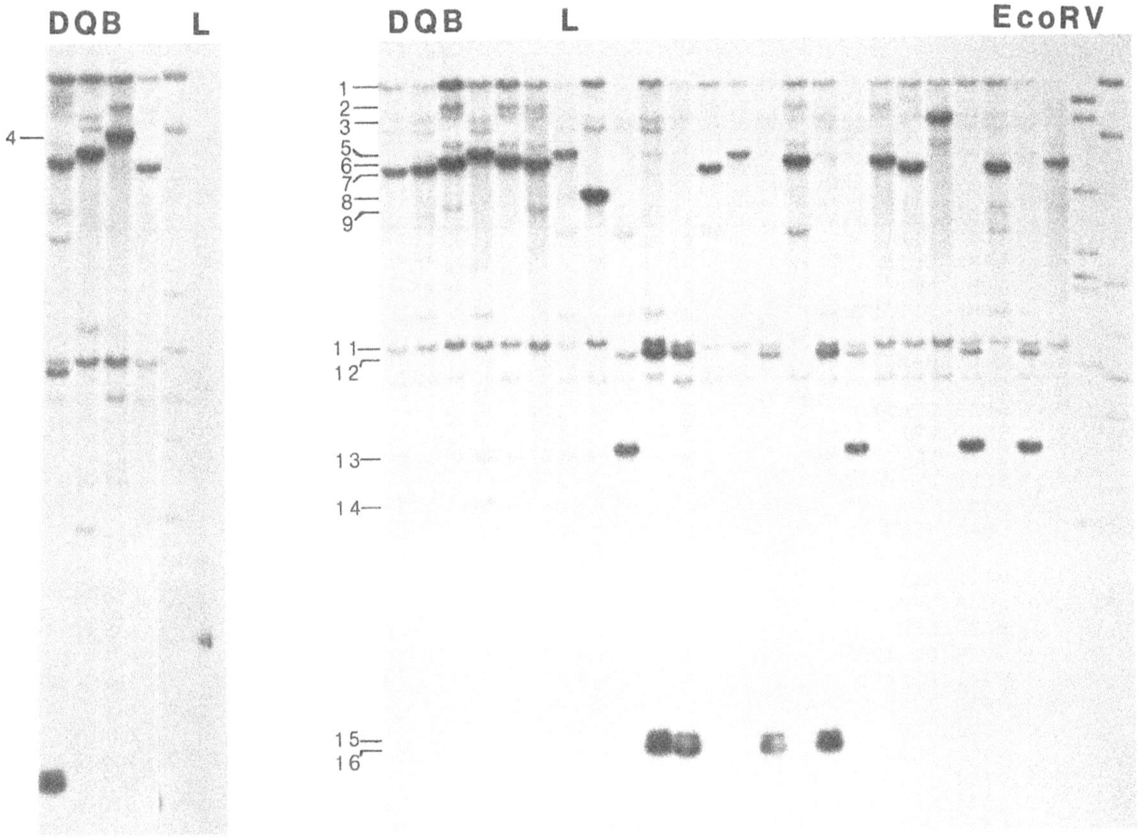

Figure 1. Fragments identified with the DQ beta/EcoRV system.

Table 1. DQ Beta/EcoRV Data

		DQB	ECV	TER		
	kb	Locus	Frequency	% Faint	56	32
1	29.90	0 1 2 0	1.000	1.000		
2	22.70	1 2 2 0	0.200	1.000		
3	20.01	1 1 0 0	0.086	0.000		
4	16.43	0 1 2 0	0.029	0.000		
5	15.14	0 1 2 0	0.100	0.000		
6	14.52	2 1 2 0	0.200	0.000		
7	13.91	2 1 2 0	0.229	0.063		
8	11.82	2 1 2 0	0.029	0.000		
9	10.97	0 1 0 0	0.243	1.000		
10	9.79	2 2 1 0	0.329	1.000		
11	6.89	2 2 1 0	0.771	0.339		
12	6.73	0 1 0 0	0.371	0.000		
13	5.24	2 1 3 0	0.214	0.000		
14	4.63	0.1 0 0	0.129	1.000	f	
15	2.71	2 1 2 0	0.171	0.000		
16	2.62	2 1 2 0	0.171	0.000		
17	1.84	1 2 2 0	0.071	1.000		
18	1.77	0 1 0 0	1.000	1.000		

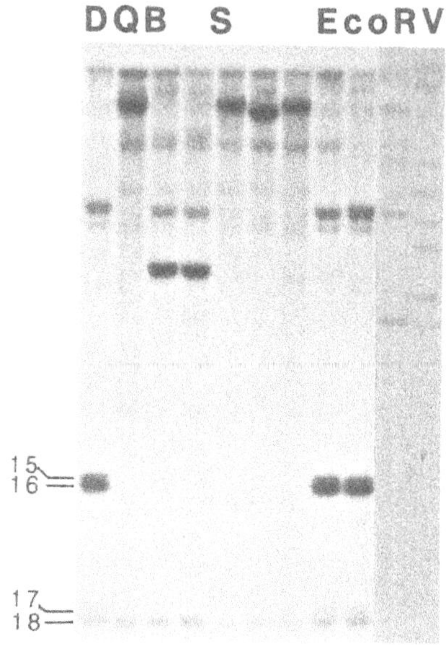

Figure 2. Banding patterns in DQ-specific bands.

Table 2. Banding Pattern of the DQ Beta/EcoRV System in Each Cell Line

	Fragment No.													
Cell	1	3	4	5	6	7	8	9	12	13	14	15	16	18
9047 PLH	8	2									8			8
9104 DHIF	8	2				8								8
9050 MOU	8	2									8			8
9051 PITOUT	8	2									8			8
9048 LBUF	8	2												8
9106 MANIKA	8		1											8
9052 DBB	8		2											8
9019 DUCAF	8			2							8			8
9020 QBL	8			2							8			8
9018 LO 081785	8			2										8
9088 PF 04015	8			2										8
9022 COX	8			2										8
9023 VAVY	8			2										8
9064 AMALA	8			2										8
9091 MLF	8				2									8
9032 BSM	8				2			8						8
9032 BSM	8				2			8						8
9028 PE 117	8				2			8						8
9031 BOLETH	8				2			8						8
9030 JHAF	8				2									8
9025 DEU	8				2									8
9007 DEM	8				2			8	1			1	2	8
9107 LKT 3	8				2			8						8
9034 SAVC	8				2			8						8
9029 WT51	8				2			8						8
9026 YAR	8				2			8						8
9092 BM92	8				2									8
9033 BM14	8				2			8						8
9075 DKB	8	1			2									8
9071 OLGA	8					2								8
9069 MADURA	8					2								8

Table 2. *Continued*

		Fragment No.										
9068	BM9	8		2								8
9067	BTB	8		2								8
9072	S PACHEO	8		2								8
9038	BM16	8		2					8			8
9039	JVM	8		2		8						8
9043	BM21	8		2								8
9037	SWEIG 007	8		2								8
9040	BM15	8		2								8
9045	TUBO	8		2					8			8
9042	TISI	8		2								8
9021	RSH	8		2		8						8
9070	LUY	8		2								8
9016	RML	8		2								8
9066	TAB 089	8			2							8
9011	E 4181324	8			2							8
9082	HO 104	8				8	1	2				8
9008	DO 208915	8				8	1	2				8
9013	SCHU	8				8	1	2				8
9014	M GAR	8					1	2				8
9010	AMAI	8					1	2				8
9017	WT8	8					1	2				8
9105	FPAF	8					1	2				8
9065	HHKB	8					1	2				8
9062	WDV	8					1	2				8
9058	OMW	8					1	2				8
9060	CBG-IB	8					1	2				8
9063	WT47	8				8	1	2				8
9059	SLE-005	8				8	1	2				8
9055	HO301	8					1	2				8
9056	KOSE	8					1·	2	8	1	2	8
9056	KOSE	8					1	2	8	1	2	8
9009	KAS-011	8					1			1	2	8
9054	EK	8	8				1			1	2	8
9061	3-12-27 AOB	8					1			1	2	8
9057	TEM	8				8	1			1	2	8
9036	SPO 010	8					1			1	2	8
9003	KAS-116	8					1		8	1	2	8
9006	WT100 BIS	8					1			1	2	8
9004	JES THOM	8					1			1	2	8
9005	HOM-2	8					1			1	2	8
9002	MZ 070782	8					1			1	2	8

1 = single intensity; 2 = double intensity; 8 = faint.

RFLP Standardization Report for DQ Beta/TaqI

A. Marcadet,[1] M. Martell,[1] D. Cohen,[1] J.D. Bignon,[2] K. Kobayashi,[3] C.B. Carpenter,[4] and R.L. Walford[5]

The DQ beta/TaqI system was standardized as the DR beta/TaqI system. When considering only specific DQβ bands and those cross-hybridizing class II bands that were more intense with DQβ probe, 22 fragments were standardized on the core cell lines (Table 1). Only one (no. 5) was constant. Ten bands of weak intensity were revealed; this weak intensity was not due to cross-hybridization. Important remarks about class II cross-hybridizing fragments have been reported in DR beta/TaqI system, and you must refer to this report to analyze DQβ blots. In the DR beta/TaqI report, a cross-hybridization table summarizes all class II, TaqI cross-hybridizing fragments, giving their standardization number and molecular weight (and "name" on the figures).

Table 2 indicates the distribution of standard bands in the core cell lines. No specific fragment for DQw specificities can be observed. Two fragments (no. 10 and no. 16) can be correlated with DQw8+ DQw9+ DQw4. One fragment no. 3 can be correlated with DQw2-DR3 +DQw7. Respectively, DQw2-DR7 and DQW2-DR3 can be discriminated against in fragments no. 1 and 3. If we consider only *bands giving an intense signal*, it is possible to distinguish the specific patterns of bands for DQw1 and for DQw2-DR7, but DQw8 cannot be distinguished from DQw7. But when taking account of DR and DQβ bands, discrimination between DQw-DRw haplotypes is very efficient.

The figure for S (Fig. 1) is a photograph taken from a film from laboratory NCY/CAR and the figure for L (Fig. 1) is from a photograph from FRA/COH. All standardized fragments are indicated at least once for S and/or L migration. The presence of a standardized fragment is indicated by a hyphen. Cross-hybridizing bands are indicated by lower case letters (cf DR Beta/TaqI cross-hybridization table).

Comments on S Migration. The fragment identified as "a" represents the standardized fragment DRβ TaqI no. 2. For the others:

"b" = DRβ TaqI no. 4
"c" = DRβ TaqI no. 5
"f" = DRβ TaqI no. 8

The fragments identified as "d," "e," and "g," only light on S migration, have not been standardized.

Comments on L Migration. The fragment identified as "a" represents the standardized fragment DRβ TaqI no. 12, which is exceptionally light on this blot. For the others:

"b" = DRβ TaqI no. 13
"c" = DRβ TaqI no. 15

Fragments "a" and "b" are exceptionally light on this blot.

Table 1. Standard Bands in the DQ Beta/TaqI System

		DQB	TAQ	COH		
	kb	Locus	Frequency	% Faint	56	32
1	6.69	0 1 0 0	0.121	0.000		
2	5.34	2 1 2 0	0.391	0.429		
3	4.62	0 1 0 0	0.275	0.000		
4	4.50	0 1 0 0	0.029	0.000		
5	4.44	2 1 2 0	1.000	0.000		
6	3.41	0 1 0 0	0.910	0.000		
7	2.99	0 1 0 0	0.424	1.000		
8	2.87	2 1 0 0	0.171	0.000		
9	2.61	2 1 2 0	0.171	0.000		
10	2.55	0 1 0 0	0.243	0.000		
11	2.46	0 1 0 0	0.119	1.000		
12	2.35	2 1 0 0	0.229	1.000		
13	2.19	2 1 0 0	0.776	1.000		
14	1.98	2 1 0 0	0.101	0.000		
15	1.94	2 1 0 0	0.265	0.000		
16	1.90	2 1 0 0	0.271	0.000		
17	1.58	2 1 0 0	0.136	1.000		
18	1.48	0 1 0 0	0.108	1.000		
19	1.12	0 1 0 0	0.250	1.000		
20	1.00	0 1 0 0	0.159	1.000		
21	0.87	0 1 0 0	0.152	1.000		
22	0.59	0 1 0 0	0.200	1.000		

Participating Laboratories: FRACOH,[1] FRABIG,[2] JAPSEK,[3] NCYCAR,[4] US5WAL[5]

Table 2. RFLP DQ Beta/TaqI System: Distribution of Standard Bands in Tenth Workshop Core Cell Lines

Cell Lines	HLA DQw	1	2	3	4	5	6	7	8	9	10	11	12	13	14	15	16	17	18	19	20	21	22
9007	5/8		2			2	2	8			2			8		2	2						8
9054	5		2			2	2							8	2					9			
9061	5		2			2	2							8	2					9			
9036	5		2			2	2							8	2					9			
9002	5		2			2	2							8	2								
9004	5		2			2	2							8	2								
9003	5		2			2	2	8						8	2			9					
9057	1		2			2	2	8							9	9				9			
9059	6		2			2	2	8								2				9			
9063	6		2			2	2	8								2				8			
9055	6		2			2	2	8						8		2				8			
9006	5		2			2	2	8						8		2							
9005	5		2			2	2							8		2							
9056	5		2			2	2							8		2				8			
9056	5		2			2	2							8		2				9			
9009	1		2			2	2							8		2							
9060	5		8			2	2		2	2				8		2							
9065	6		8			2	2		2	2				8		2				9			
9062	6		8			2	2		2	2				8		2				9			
9008	6		8			2	2	8	2	2				8		2				9			
9105	6		8			2	9	9	2	2				9		9		9	9	9			9
9058	6		8			2	2	8	2	2						2				8			
9017	1		8			2	2	8	2	2				8		2							
9082	6		8			2	2	8	2	2						2							
9010	6		8			2	2	8	2	2						2							
9013	6		8			2	2	8	2	2						2							
9014	6		8			2	2	9	2	2				8		2							
9011	6	2			2	2	2							8								8	
9066	6	2			2	2	2							8						8		8	
9016	2?3	2				2	9							9			2						
9047	2	2				2		9						8				8				8	
9050	2	2				2								8				8				8	
9051	2	2				2								8				8				8	
9048	2	2				2	2							8				9				9	
9020	2			2		2	9							8				9				9	
9019	2			2		2								8				8		8		8	
9088	2			2		2	2							8				8		8		8	
9018	2			2		2	2	8						8				8		8		8	
9022	2			2		2	2	8						8				8		8		8	
9023	2			2		2	2	8						8				8		8		8	

Cell Lines	HLA DQw	1	2	3	4	5	6	7	8	9	10	11	12	13	14	15	16	17	18	19	20	21	22
9104	7	2				2	2							8				8				9	
9037	7		9	2		2	2	8	2	2				9	2								
9030	7	9		2		2	2	8				8		8							8		
9039	7			2		2	2	8				8		8					8	9	8		
9043	7			2		2	2	8				8		8					8		8		
9040	7			2		2	2	9				9		8					8		8		
9064	7		9	2		2	2								8					9	9	8	
9038	7			2		2	2						8		8							8	
9016	7			2		2	2						9		8					9		9	
9025	7		9	2		2	2						9		8					9		8	
9091	7		9	2		2	2						8		8					8		8	
9070	7			2		2	2						8		8					8		8	
9042	7			2		2	2						8		8					8	9	8	
9045	7			2		2	2						8		8					8	8	8	9
9026	8					2	2	8			2		8				2		9				8
9028	8					2	2	8			2		8	8			2						8
9029	8					2	2	8			2		8				2						8
9031	8					2	2	8			2		8				2						8
9032	8					2	2	8			2		8				2						9
9032	8					2	2	8			2		8				2						8
9033	8					2	2	8			2		8				2						8
9034	8					2	2	8			2		8				2						8
9075	9					2	2				2		8	8			2						9
9052	9			9		2	2				2		8	8			2						
9092	8					2					2		8	8			2						9
9021	4					2		8			2		8				2			9			8
9069	4					2	2						8	8			2			8			8
9067	4					2	2				2		8	8			2			8			8
9071	4					2	2				2		8	8			2			8			8
9068	4					2	2				2			8			2			9			9
9107	4		8			2	2				2		8	8			2						8
9072	4					2	2				2		8	8			2						8

Standard Fragments N =

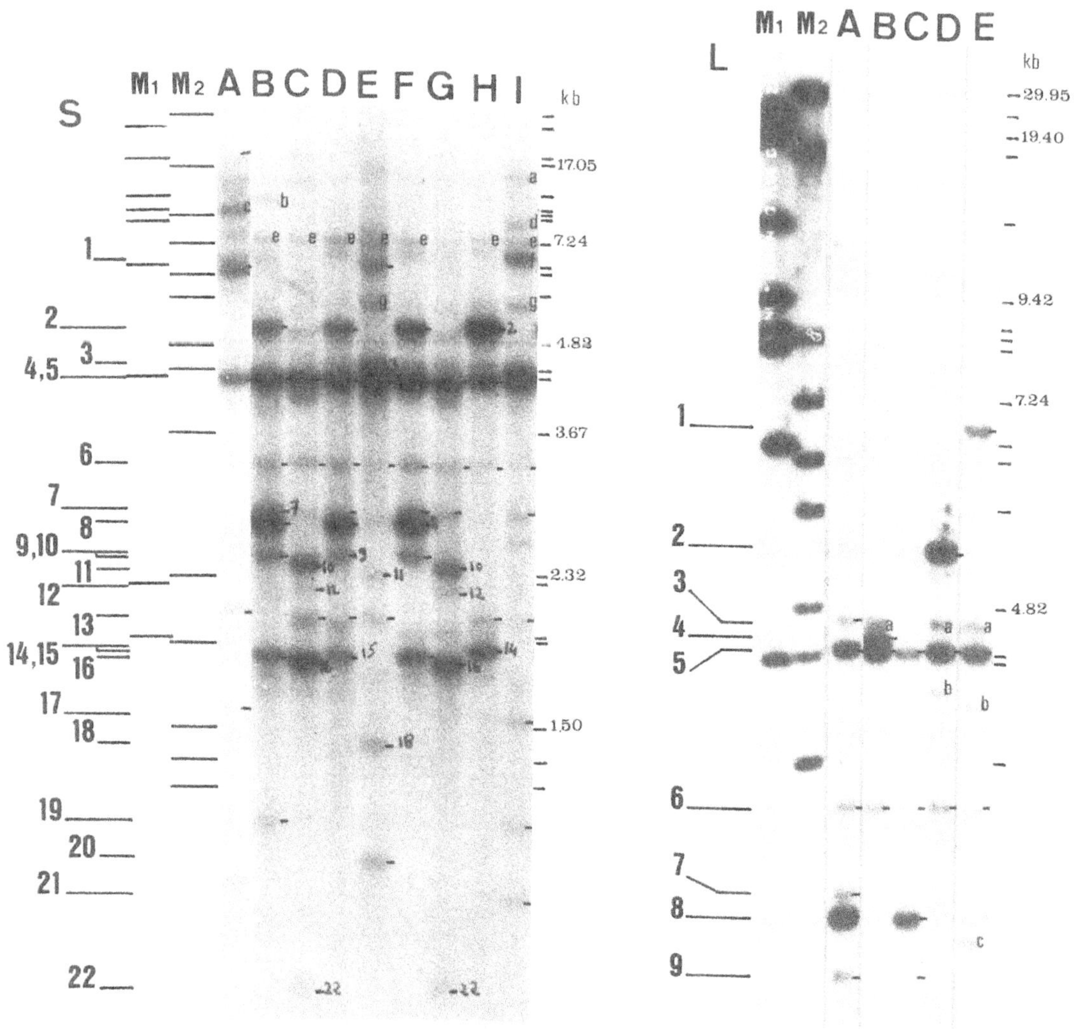

Figure 1. Fragments identified with the DQ beta/TaqI system.

RFLP Standardization Report for DQ Beta/BglII

R.E. Bontrop,[1] M. Van Eggermond,[1] L. Fainboim,[2] P. Stastny,[3] H. Takata,[4] and P. Rubenstein[5]

Data were received from five different laboratories and standardization was performed based on the following criteria:

1. weak hybridization signals were accepted as standard fragments when they were detected by most laboratories
2. discrepancies were resolved by majority consensus

For this system 10 different standard fragments were observed (Table 1 and Fig. 1). The first fragment has a molecular weight of 9.38 kb and a frequency of 48.6%. In most examples (74.3%), this fragment is detected as a weak hybridization signal. This phenomenon is probably due to cross-hybridization since this fragment is also detected as standard fragment 8 in the DR beta/BglII and as standard fragment 6 in the DP beta/BglII systems (see Table 2 in the DR beta/BglII report). The same observation is made for standard fragment 2 (Table 2 in the DR beta/BglII report), which has a molecular weight of 6.93 kb. This fragment, with a frequency of 34.8% is observed as a faint band (Table 1); however, it appears to be more specific for HLA-DQ than for HLA-DR and -DP. Standard fragment 3 has a molecular weight of 5.87 kb and is observed for all cells tested. This fragment is identical to standard fragment 4, which has been detected in the DO beta/BglII system. Standard fragment 4 has a molecular weight of 5.61 kb and is identical to standard fragment 11 in the DR beta/BglII system and to standard fragment 11 in the DP beta/BglII systems (see Table 2 in the DR beta/BglII report). Standard fragment 5 appears to be DQ-specific (Table 1). Also, standard fragment 6 with a molecular weight of 4.12 kb and a frequency of 20% appears to be DQ beta-specific. This fragment seems to be associated with the presence of the serologically defined HLA-DQw7 (TA10) determinant or with the DQ beta chain type 3.1, which has been described in the Biochemistry Section of this Workshop. Standard fragment 7 with a molecular weight of 3.27 kb and a frequency of 92.9% is also detectable with DR- and DQ beta region probes (see Table 2 in the DR beta/BglII report). This fragment has a frequency of 38.8% and is observed as a faint band. Standard fragment 8 is apparently DQ beta-specific and has a molecular weight of 3.10 kb and a panel frequency of 28.6%. When this fragment is observed in a high staining intensity, fragment 7 is usually also present in a high staining intensity (Fig. 1, lanes 51 and 58). Standard fragment 9, with a molecular weight of 2.78 kb, is probably more HLA-DP-specific, but is possibly detected due to cross-hybridization (see Table 2 in the DR beta/BglII report). Standard fragment 10 has a molecular weight of 1.29 kb and seems to be DQ-specific. Its panel frequency appears to be 16.6%, and 67.6% of these fragments are detected as faint signals (Table 1).

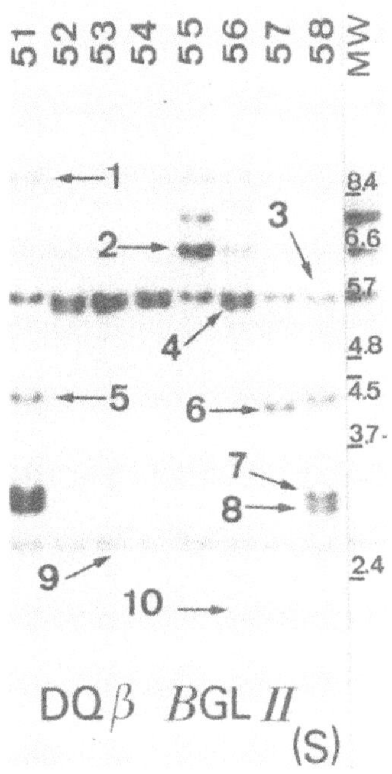

Figure 1. Fragments identified with the DQ beta/BglII system.

Participating Laboratories: BENROO,[1] LATFAI,[2] US4STA,[3] JAPTAK,[4] US8RUB[5]

Table 1. DQ Beta/BglII System Data

Fragment	kb	Locus	Frequency	% Faint	56	32
1	9.38	1320	0.486	0.743		
2	6.93	2130	0.300	0.348		
3	5.87	0101	1.000	0.125		
4	5.61	1120	0.271	0.150		
5	4.23	0100	0.200	0.533		
6	4.12	0100	0.243	0.118		
7	3.27	2130	0.929	0.388		
8	3.10	0100	0.286	0.238		
9	2.78	2210	1.000	1.000		
10	1.29	0100	0.166	0.676		

Table 2. Fragment Presence/Absence Patterns for the DQ Beta/BglII System as Defined on 72 Core Cell Lines

	Fragments											Fragments									
Cell	1	2	3	4	5	6	7	8	9	10	Cell	1	2	3	4	5	6	7	8	9	10
9045	8		2			2			1		9056	1	2	2				2	1	8	
9042	8		2			2			1		9036	8	2	1				1	8	1	
9043	1		2			2	8		1		9057	8	1	8						1	
9023	1		2				8		1		9054		1	8						1	
9040	1		2			1	8		1		9009		2	1						1	
9037	8		1			1			1		9061	8	2	2						1	
9091	8		1			1			1	8	9003		2	2						1	
9038	8		2			2			1	8	9056	8	2	2		8		1	1	1	
9070	8		2			2			1	1	9002		2	1				8		1	
9039	8		2			2			1	8	9018	1		2						1	8
9025			1			1			1	1	9104			2						1	
9064	8		1			1	8		1	1	9051			2				8	8	8	1
9030			1			1			1	1	9047			2				8		1	1
9107			2	2					1		9088	8		1						1	1
9033			2	2				8	1		9066	8		1					8	1	
9069			2	1	8				1		9105			8						1	
9028	8		1	1					1		9016			8						1	
9031			1	1					1		9019			1						1	1
9071	8		1	1					1		9014	1		2		8		2	2	1	
9021	8		1	1				8	1		9058	1		2				2	2	1	
9032			1	1					1		9017	8		2		1		2	2	1	
9092			2	2					1		9013	8		2		1		2	2	1	
9034			2	2					1		9055	1		2		1		2	2	1	
9068	8		2	2					1		9063	1		2		1		2	2	1	
9075		8	2	2					1		9022	1		2				8	8	1	
9067	8	1	2	2					1		9050			2				8		1	8
9029		1	2	2					1		9048			2				8		1	
9026		1	2	2					1		9020	8		2				8		1	
9106		1	2	1					1		9010			2				2	1	1	
9032		1	2	2					1		9060	8		1				1	8	1	
9072	8	1	2	2					1		9065	8		1				1	8	1	
9052	8		2	2					1		9062			8				1	8	1	
9007		2	2	2					1		9059	8		2		8		2	2	1	
9004		2	2						1		9008			2		1		2	2	1	
9006		2	2				8		1		9011	8		2		8			1	1	
9005		2	2				8		1		9082	8		2				2	2	1	

RFLP Standardization Report for DQ Beta/PvuII

K. Kato,[1] A. Smith,[2] J. Partanen,[3] P. Piccioli,[4] B. Dupont,[1] and S.Y. Yang[1]

Twenty-six bands were standardized with DQ beta/PvuII system as shown in Table 1. All bands present in Figure 1 were counted as standard bands since each band could clearly be assigned in more than one laboratory. Four bands – no. 4 (6.98kb), no. 20 (1.77kb), no. 22 (1.02kb), and no. 25 (0.53kb) – were nonpolymorphic. MacMurray et al.[1] have previously reported three nonpolymorphic bands of 7.3 kb, 1.1 kb, and 0.7 kb obtained with the 5′ DQ beta probe which contains the untranslated region to the majority of the second domain. We consistently observed, however, both the 1.02 kb fragment and 1.27 kb fragment. Eight bands – no. 3 (8.19 kb), no. 4 (6.98 kb), no. 14 (3.72 kb), no. 19 (2.02 kb), no. 21 (1.66 kb), no. 23 (0.88 kb), no. 24 (0.63 kb), and no. 26 (0.34 kb) – showed no cross-hybridization with other beta probes (Fig. 2). The DQ beta cross-hybridizing bands are shown in Table 2 in the DR beta/PvuII report. Seven of 26 bands are correlated with specific DQw types. These bands are no. 19 (2.02 kb) and no. 18 (2.17 kb) with DQw6 subtypes and no. 24 (0.63 kb), no. 21 (1.66 kb) and no. 8 (4.78 kb) with DQw2 subtypes and no. 14 (3.72 kb), and no. 26 (0.34 kb) with DQw7.

Participating Laboratories: US7DUP,[1] US7HAN,[2] SCAKOS,[3] IT2FER[4]

Table 1. Standard Bands of DQ Beta/PVUII

	kb	DQB Locus	PVU Frequency	DUP % Faint	56	32
1	13.49	1 2 0 0	0.300	0.455		
2	9.58	1 2 0 0	0.229	0.813	f	
3	8.19	0 1 0 0	0.729	1.000		
4	6.98	0 1 0 0	1.000	0.000		
5	6.05	1 2 2 0	0.229	1.000		f
6	5.75	1 1 1 0	0.186	0.286		
7	5.21	1 1 1 0	0.286	1.000		
8	4.78	1 2 0 0	0.086	1.000		
9	4.60	2 1 2 0	0.286	0.429		
10	4.45	1 2 0 0	0.100	0.857		
11	4.31	2 1 2 0	0.600	0.651		
12	4.20	1 1 1 0	0.200	0.571		
13	4.01	1 2 0 0	0.271	0.700		
14	3.72	0 1 0 0	0.200	0.000		
15	3.60	2 1 3 0	0.729	0.019		
16	3.40	2 1 2 0	0.443	0.030		
17	2.38	3 1 2 0	0.457	0.061		
18	2.17	2 1 0 0	0.157	0.000		
19	2.02	0 1 0 0	0.029	0.000		
20	1.77	1 2 1 0	1.000	0.028		
21	1.66	0 1 0 0	0.200	0.071		
22	1.02	2 1 3 0	1.000	0.042		
23	0.88	0 1 0 0	0.557	0.825		
24	0.63	0 1 0 0	0.157	0.000		
25	0.53	1 1 1 0	1.000	1.000		
26	0.34	0 1 0 0	0.186	0.846		

From left to right: band no., band kb, locus assignment of DR, DQ, DP, and DO, band frequency, frequency of faint bands, and hidden duplications.

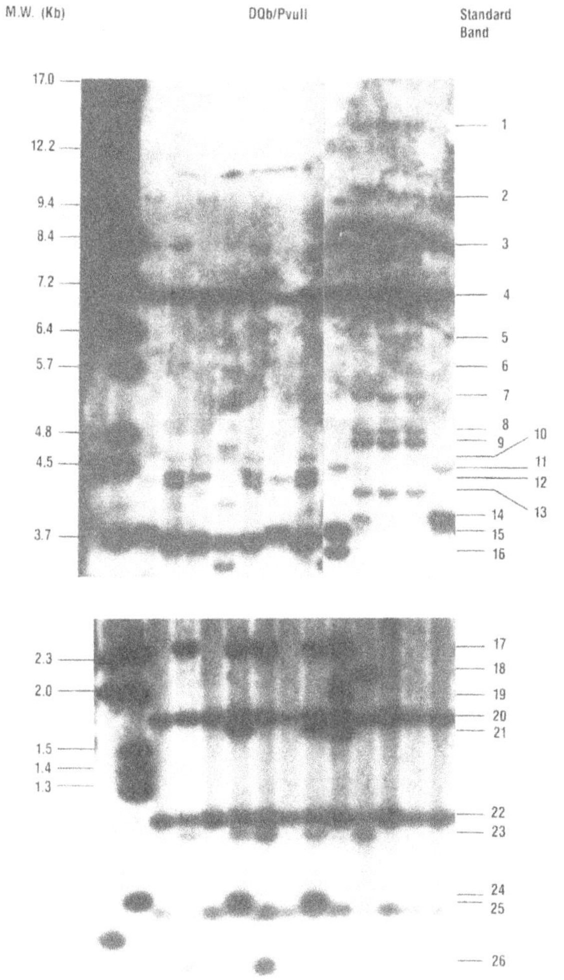

Figure 1. DQ beta/PvuII top:long migration, bottom:short migration.

Reference

1. MacMurray AJ, Bell JI, Denney J, Watling D, Foster LS, McDevitt HO. Molecular mapping of class II polymorphism in the human major histocompatibility complex. II. DQb. J Immunol 1987;139:574.

Immunobiology of HLA
Volume I

```
                              Fragment #
                              11111111112222222
Cell lines           12345678901234567890123456
-----------------------------------------------
9045 TUBO             882         2  1  1 11 8
9038 BM16             882         2  2  1 18 81
9042 TISI               82      8  2  2  1 18 88
9043 BM21             882      8  2  2  2 18 88
9040 BM15             882      8  2  2  1 18 88
9037 SWEIG 007          82      1  2  2  2 18 88
9064 AMALA               2      1  2  2  1 18 88
9070 LUY                82      8  2  2  1 18 88
9039 JVM                 2      8  2  2  1 88 88
9030 JHAF            1 82818 8   82  2  1 18 98
9025 DEU             2 82818 1   12  2  2 21 81
9091 MLF             1 82818 8   82  2  1 18 88
9016 RML                82     888 2  2  2 18 88
9065 HHKB             882      1   2  2 1 88 8
9062 WDV              882      1   2  2 2 18 8
9060 CB6B               82      1   2  2 2 11 9
9105 FPAF               82      1   2  2 2 18 8
9013 SCHU                2     811  2  2 2 11 8
9017 WT8                82     881  2  2 1 18 8
9082 HO 104              2      81  2  2 1 18 8
9014 M GAR              82     881  2  2 1 11 88
9058 OMW              8 2      8   2  2 1 18 8
9064 AMAI                2     188  2  2 1 18 8
9008 DO 208915          82     888  2  2 1 18 8
9004 JES THOM           82       1  22   1 1  8
9006 WT100 BIS          82       8  22   1 1  8
9005 HOM-2              82       8  22   1 1  8
9036 SPO 010          182      1   22   2 2  8
9054 EK               182      1   22   2 1  8
9057 TEM                 2      1   22   2 1  8
9021 RSH              8 2      1   22   2 1  8
9071 OLGA               82      1   22   1 1  8
9002 MZ 070782          82      8   22   1 1  8
9009 KAS-011             2      88  22   1 1  8
9069 MADURA             82          22   2 2  8
9068 BM9                82          22   1 1  8
```

```
Fragment #
                              11111111112222222
Cell lines           12345678901234567890123456
-----------------------------------------------
9075 DKB             1 828 8 1   1 22   1 1  8
9034 SAVC            1  2818 1   8 22   1 1  8
9028 PE 117          2 82818 1   1 22   2 2  8
9026 YAR             8  2818 1   1 228  1 1  8
9052 DBB             8 828 881   8 22   1 1  8
9092 BM92            8 82888 8   8 22   1 1  8
9031 BOLETH          1  2818 8   8 22   1 1  8
9029 WT51            8  2888 8   8 22   8 1  8
9032 BSM             8  2818 8   8 22   1 1  8
9107 LKT 3           8 82 88 8   8 22   1 1  8
9038 BM16            8  2 18 8   8 21   1 1  8
9007 DEM                2 88 8  8  22   1 1  8
9003 KAS-116           82       8  22   1 1  8
9061 3-12-27 AOB      882      8   22   1 1  8
9067 BTB               82      8   22   1 1  8
9072 S PACHEO          82      8   22   1 1  8
9032 BSM             8    18 1   1 22   1 1  8
9055 HO 301          8 82      8   2 2  2 11 8
9063 WT47            8  2      8   2 2  2 11 8
9059 SLE-005            2      8   2 2  1 18 8
9018 LO 081785          2      8   2 2  121818
9022 COX              882      8   1 2  121818
9020 QBL              882      8   1 2  221829
9023 VAVY             882      8   1 2  221818
9019 DUCAF            182      1   1 2  222828
9088 PF 04015          82      1   1 2  221829
9104 DHIF            1 828 881   81  2  121818
9050 MOU             1 828 881   8   2  121818
9051 PITOUT          1 828 881   8   2  121818
9047 PLH             1 828 881   8   2  121818
9048 LBUF            1 828 881   1   2  121828
9011 E 4181324         82     881  8 2 1121  8
9056 KOSE              82      8   222  1 18 8
9056 KOSE             882      8   222  1 18 8
9106 MANIKA            82          118  888  8
9066 TAB 089           82      1   8222222  8
```

Figure 2. RFLP pattern of the DQ beta/PvuII system.

RFLP Standardization Report for DQ Beta/EcoRI

S. Tonks,[1] V. Groves,[1] H. Betuel,[2] J.L. Bidwell,[3] M.L. Garavoy,[4] R. Lechler,[5] S. Naito,[6] J. Bodmer,[1] and J. Trowsdale[1]

The DQ beta/EcoRI system identified 13 fragments in the core cell lines (Fig. 1). Locus assignments in this system (Table 1) showed fragments 1,2,4,5,6,10, & 13 to be DQB specific, the remaining fragments cross-hybridizing with both DRB and DPB (Table 3 in the DR Beta/EcoRI report). A number of different banding patterns were revealed in this system, standard fragment 6 being a faint constant band. All the other unique bands were found to be polymorphic.

Participating Laboratories: UKIBOD,[1] FRABET,[2] UKIBID,[3] US4GVY,[4] UKIBAT,[5] JAPNAI[6]

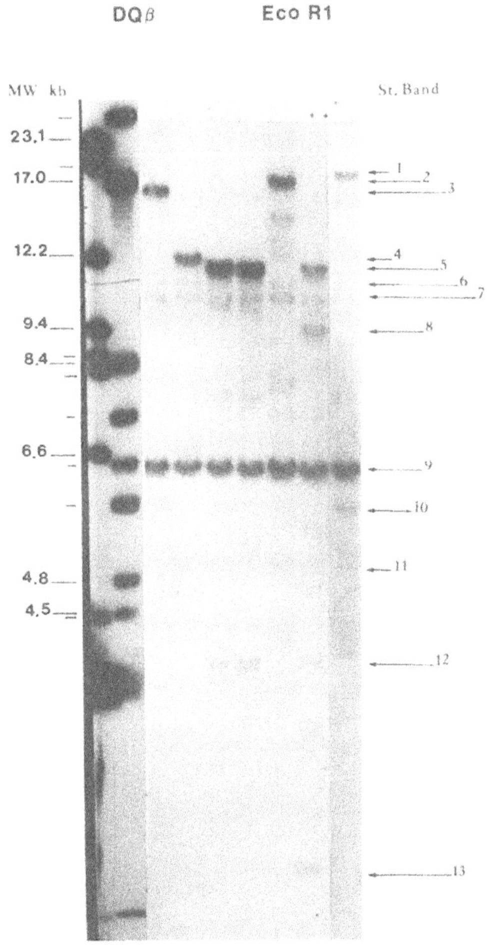

►

Figure 1. Fragments observed with the DQ beta/EcoRI system.

Table 1. Banding Pattern of the DQ Beta/EcoRI System in Each Cell Line

WS ID		Fragment No.							
No.	Cell	1	2	4	5	6	8	10	13
9107	LKT3	1				8		8	
9033	BM14	1				8		8	
9069	MADURA	1				8		8	
9028	PE117	1				8		8	
9071	OLGA	1				8		8	
9021	RSH	1				8		8	
9034	SAVC	1				8		8	
9068	BM9	1				8		8	
9067	BTB	1				8		8	
9029	WT51	1				8		8	
9026	YAR	1				8		8	
9072	SPACH	1				8		8	
9071	OLG	1				8			
9092	BM92	1				8		8	2
9007	DEM	1			3	8		8	
9050	MOU		1			8			
9104	DHIF		1			8			
9051	PITOUT		1			8			
9047	PLH		1			8			
9106	MANIKA		1			8			
9048	LBUF		1			8			1
9045	TUBO			1		8		8	1
9018	LOO181785			2		8			
9022	COX			2		8			
9023	VAVY			2		8			
9040	BM14			3		8			
9019	DUCAF			3		8			
9025	DEU			3		8			
9088	PFO4015			3		8			
9064	AMALA			2		8			
9009	KASO11			1		8			
9036	SPOO10			3	3	8			
9020	QBL			1	3	8			
9030	JHAF			2		8		8	
9091	MLF			3		8		8	
9075	DKB			3		8		8	
9038	BM16			3		8		8	
9070	LUY			3		8		1	
9016	RML			3		8		8	1
9004	JESTHOM				2	8			
9006	WT100B1S				2	8			
9005	HOM2				2	8			
9057	TEM				3	8			
9054	EK				3	8			
9061	31277ABO				3	8			
9003	KAS116				3	8			
9066	TAB089					8			
9002	MZ070782					8			
9032	BSM					8			
9011	E4181324					8			
9042	TISI					8		1	
9043	BM21					8		1	
9037	SWE1G007					8		8	
9026	JVM					8		8	
9052	DBB					8		8	
9014	MGAR					8	1		1
9058	OMV					8	2		1
9017	WT8					8	2		1
9013	SCHU					8	2		1
9055	HO301					8	2		1
9063	WT47					8	2		1

Table 1. *Continued*

WS ID		Fragment No.							
No.	Cell	1	2	4	5	6	8	10	13
9056	KOSE					8	2		1
9010	AMAL					8	2		1
9060	CB6B					8	8		8
9062	WDV					8	8		1
9008	DO208915					8	8		1
9067	BTB					8	1		1
9105	FPAF					8	8		
9065	HHKB					8	1		
9059	SLE005					8	1		

1 = single intensity; 2 = double intensity; 8 = faint.

RFLP Standardization Report for DQ Beta/HindIII

G. Paulsen,[1] G. Markussen,[1] J. Demopulos,[3] J.M. Tiercy,[4] S. van Tonder,[5] and F. Quillivic[5]

Thirty-four bands were identified in the DQB/HID system (Table 1 and Fig. 1). The distribution of nineteen bands showing high DQB specificity is shown in Table 2.

Considering bands with intensities scored from 1 to 3, five band clusters (A, B, C1, C2, D, and E) can be observed. All DQ2 cell lines associated with DR7 in addition to WS ID 9104-typed DQ7 made cluster A (bands 1 and 19). Cluster B (band 2 and 19) includes all DQ2-associated with DR3. Cluster C1 and C2 (bands 9, 17 or 18 and 19) includes all DQ1, DQ5, and DQ6. Cluster D (band 16 and 19) includes DQ7, and cluster E (bands 17 and 19) includes DQ4, DQ8 and DQ9. Although bands 17 and 18 had very similar migration, they were determined to be two distinct fragments. A cross-hybridization table for the HLA-class II/HID systems can be found in the DR Beta/HID report.

Participating Laboratories: SCATSB,[1] FRAFAU,[2] US5UAB,[3] FRAJEA,[4] SAFHAM[5]

Table 1. Standard Bands in the DQBeta/HID System

Band	kb	Locus[a]	Frequency[b]	% Faint[c]	56[d]	32[d]
1	12.80	0 1 2 0	0.086	0.000		
2	11.42	0 1 2 0	0.086	0.000		
3	9.76	1 2 2 0	0.014	1.000		
4	9.55	0 1 0 0	0.243	1.000	f	
5	8.60	0 1 0 0	0.829	1.000		
6	8.10	0 1 0 0	0.029	1.000		
7	7.24	1 2 0 0	0.471	0.765		
8	7.03	2 3 1 0	1.000	1.000		
9	6.44	0 1 0 0	0.400	0.000		
10	5.48	0 2 1 0	0.514	1.000		
11	5.13	0 2 1 0	0.429	1.000	f	
12	4.78	0 1 0 0	0.271	1.000		
13	3.77	0 1 0 0	0.286	1.000		
14	3.51	0 1 0 0	0.499	1.000		
15	3.47	1 2 1 0	0.071	1.000		
16	3.41	0 1 2 0	0.186	0.000		
17	3.30	0 1 2 0	0.529	0.000		
18	3.27	0 1 0 0	0.129	0.000		
19	3.20	0 1 2 0	1.000	0.000		
20	3.09	1 2 3 0	0.071	1.000		
21	3.01	0 1 0 0	0.071	1.000		
22	2.92	0 1 0 0	0.329	1.000		
23	2.86	0 1 0 0	1.000	1.000		
24	2.68	0 1 0 0	0.029	1.000	f	
25	2.62	1 2 0 0	0.129	1.000		
26	2.52	1 3 2 0	0.385	1.000		
27	2.50	1 3 2 0	0.443	1.000		
28	1.79	1 2 0 0	0.029	1.000		
29	1.70	1 2 1 0	0.342	1.000		
30	1.66	1 2 1 0	0.414	1.000		
31	1.61	1 2 1 0	0.171	1.000		
32	1.51	1 2 1 0	0.300	1.000		
33	1.35	2 1 2 0	1.000	1.000		
34	1.19	2 1 2 0	1.000	1.000		

[a]Locus assignment in the order DRB, DQB, DPB, DOB. 1 indicates high specificity; 2 and 3 indicate low specificity.
[b]Band frequency in the core cell lines.
[c]Frequency of faint bands.
[d]Discrepancies in strong bands (s) and faint bands (f) of hidden duplicates (WS ID 9056 and 9032) in the standardization of RFLP bands.

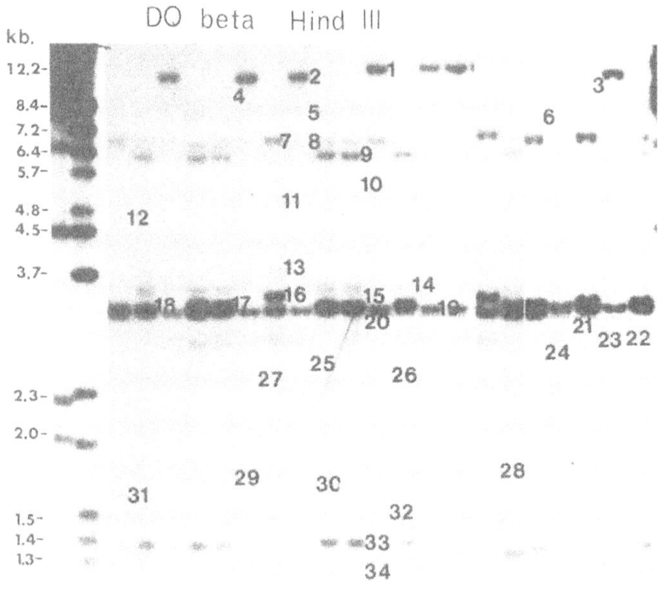

Figure 1. RFLP in the DQB/HID of 21 DNA samples run in 0.9% agarose gel. Representative core bands are shown in the figure numbered on their right side. The WS ID of the core cell lines are from the left following marker 1 and marker 2: 9033, 9004, 9018, 9055, 9063, 9022, 9043, 9023, 9006, 9005, 9050, 9056, 9104, 9051, 9038, 9008, 9032, 9072, 9052, 9020, and 9007.

Table 2. Distribution of Bands with High DQBeta Specificity in the DQB/HID System

WS ID No.	DR	DQ	Band No. 01	02	04	05	06	09	12	13	14	16	17	18	19	21	22	23	24	33	34	
9047	7	2	2			8				8					2			8		8	8	
9048	7	2	2			8				8					2			8		8	8	
9050	7	2	2			8				8					3		8	8		8	8	A
9051	7	2	2			8				8					2			8		8	8	
9104	11	7	2			8				8		8			2			8		8	8	
9106	7	2?3	1			8			8				1		2			8		8	8	—
9018	3	2		2		8					8				2		8	8		8	8	
9019	3	2		2		8					8				3		8	8		8	8	
9020	3	2		2		8					8				2		8	8		8	8	
9022	3	2		2		8					8				2			8		8	8	B
9023	3	2		2		8					8				2			8		8	8	
9088	3	2		2		8					8				2			8		8	8	—
9003	1	5				8		2			8		3		3			8		8	8	
9036	11	5				8		2			8		3		3			8		8	8	
9060	13	5				8		1			8		3		2			8		8	8	
9054	14	5				8		1			8		3		3		8	8		8	8	
9061	14	5				8		1			8		3		3		8	8		8	8	
9007	4/16	5/8			8	8		1	8		8		2		3			8		8	8	
9055	6	6				8		1			8		3		3		8	8	8	8	8	
9066	8	6				8		1		8			3		2			8		8	8	
9105	11	6			8	8		1					2		2		8	8		8	8	
9058	13	6			8			1			8		3		3			8		8	8	
9059	13	6			8			1			8		3		3			8		8	8	
9062	13	6				8		1			8		2		2			8		8	8	C1
9063	13	6			8			1			8		3		3		8	8	8	8	8	
9065	13	6				8		1			8		2		2			8		8	8	
9008	15	6				8		1			8		2		2			8		8	8	
9010	15	6			8			1			8		3		2			8		8	8	
9011	15	6				8		1		8			3		3			8		8	8	
9013	15	6			8			1			8		3		3			8		8	8	
9014	15	6				8		1			8		3		3			8		8	8	
9082	15	6			8			1			8		3		3			8		8	8	

Table 2. *Continued*

WS ID No.	DR	DQ	Band No. 01	02	04	05	06	09	12	13	14	16	17	18	19	21	22	23	24	33	34	
9017	15	1			8	8		1			8		3		3			8		8	8	—
9009	2	1				8		1			8			2	3			8		8	8	
9057	14	1			8			1			8			3	2		8	8		8	8	
9002	1	5				8		1			8			3	3	8	8	8		8	8	
9004	1	5				8		1			8			3	3			8		8	8	C2
9005	1	5				8		1			8			3	3			8		8	8	
9006	1	5				8		1			8			3	3			8		8	8	
9056	13/14	5				8		1			8			2	3		8	8		8	8	
9056	13/14	5				8		1			8			2	2		8	8		8	8	
9016	16	7				8				8		2			2			8		8	8	—
9025	4	7				8				8		2			2			8		8	8	
9030	4	7				8				8		2			2		8	8		8	8	
9037	11	7				8				8		2			2			8		8	8	
9038	12	7				8				8		2			2		8	8		8	8	
9039	11	7			8					8		2			2		8	8		8	8	D
9040	11	7				8				8		2			2		8	8		8	8	
9042	11	7				8				8		2			2	8		8		8	8	
9043	11	7				8				8		2			2		8	8		8	8	
9045	11/12	7								8		2			2			8		8	8	
9064	14	7				8	8			8		2			2			8		8	8	
9070	8	7				8				8		2			2	8		8		8	8	
9091	4	7				8				8		2			2			8		8	8	—
9021	3	4				8			8				2		2			8		8	8	
9067	8	4				8			8				3		3	8		8		8	8	
9068	8	4				8			8				3		3	8		8		8	8	
9069	8	4				8			8				2		2			8		8	8	
9071	8	4							8				2		2			8		8	8	
9072	8	4				8			8				2		2	8		8		8	8	
9107	4	4				8			8				3		3			8		8	8	
9026	4	8			8	8			8				2		2		8	8		8	8	
9028	4	8				8			8				2		2			8		8	8	E
9029	4	8			8	8			8				2		2		8	8		8	8	
9031	4	8			8		8		8				3		3			8		8	8	
9032	4	8							8				3		3			8		8	8	
9032	4	8			8		8		8				2		2			8	8	8	8	
8033	4	8			8	8			8				3		2			8		8	8	
9034	4	8			8	8			8				2		2		8	8		8	8	
9092	4	8				8			8				2		3			8		8	8	
9052	7	9			8	8			8				3		3		8	8		8	8	
9075	9	9				8			8				2		2		8	8		8	8	—
			01	02	04	05	06	09	12	13	14	16	17	18	19	21	22	23	24	33	34	

Bands designation: 1 to 3 increasing intensity, 8 = faint.

RFLP Standardization Report for DQ Beta/MspI

M. Segall,[1]* L. Schluender,[1]* A. Arnaiz-Villena,[2] N. Kashiwagi,[3] and C. Muller.[4]

DQ beta/MspI blots were received from four laboratories and were analyzed as detailed in the report on DR beta/MspI. The DQ beta/MspI system identified 22 fragments on the core cell lines (Table 1 and Figs. 1 and 2). Ten of these were unique to DQ beta, four were cross-hybridizing but strongest with DQ beta, and one (fragment 8) was equally strong with DR beta and DQ beta, as shown in Table 2 of the report on DR beta/MspI. Fragments detected uniquely with one probe are not shown in that table. In Figures 1 and 2, DQ beta-dominant or unique fragments are indicated by a dot next to the fragment number.

Discrepancies for the "hidden duplicates" (9056 and 9032) were generally due to faint bands in cases where a fragment was not identified by all four labs. Some doublets could not be resolved on all blots; fragment 10 was not uniformly subdivided and hence was assigned a single fragment number, as was fragment 12.

The presence/absence of the DQ beta polymorphic fragments in the individual core cells is given in Table 2, in which cells are arranged according to DR, Dw (when

Participating Laboratories: US2BAC,[1] FRAARN,[2] JAPKSH,[3] GERMUC[4]

This is publication #497 from the Immunobiology Research Center, University of Minnesota, Minneapolis, Minnesota 55455, USA.
*Supported in part by Juvenile Diabetes Foundation Grant #186175 and March of Dimes Birth Defects Foundation Grant #6-496.

Table 1. DQ Beta/MspI Standard Fragments

Fragment	kb	Locus Assignment	Frequency	% Faint	Discrepant for 9056	Discrepant for 9032
1	13.30	0 1 0 0	0.114	0.250		s
2	10.45	0 1 0 0	0.014	0.000		
3	9.23	2 1 0 0	0.143	0.100		
4	8.17	1 2 0 0	0.629	0.089		
5	6.30	0 1 0 0	0.214	1.000		
6	5.56	1 2 0 0	0.229	0.563	s	
7	5.34	1 2 0 0	0.257	0.111		
8	4.96	1 1 0 0	0.157	0.364		
9	4.37	2 1 2 0	0.543	0.325		
10	3.62	2 1 2 0	0.700	0.000		
11	3.15	2 3 1 0	0.514	0.514		
12	2.62	0 1 0 0	0.686	0.020		
13	2.34	0 1 0 0	0.100	0.000		
14	2.13	0 1 0 0	0.171	0.000	s	
15	1.96	0 1 0 0	0.357	0.154		
16	1.85	2 1 2 0	1.000	0.000		
17	1.78	1 2 2 0	0.257	1.000		
18	1.30	0 1 0 0	0.429	1.000		
19	2.01	0 1 0 0	0.271	1.000		
20	0.86	1 2 2 0	0.243	0.706		
21	0.80	0 1 0 0	0.429	0.633	f	f
22	0.67	2 2 1 0	0.271	1.000		

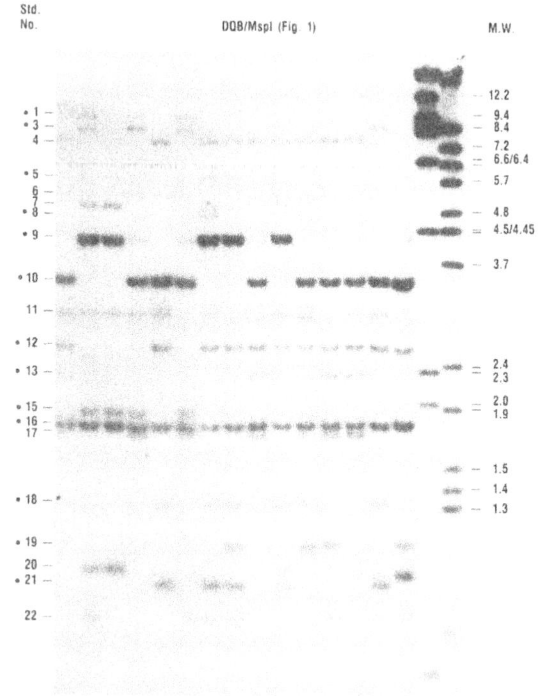

Figure 1. Fragments identified with the DQ beta/MspI system.

Immunobiology of HLA
Volume I

Figure 1. Fragments identified with the DQ beta/MspI system. ►

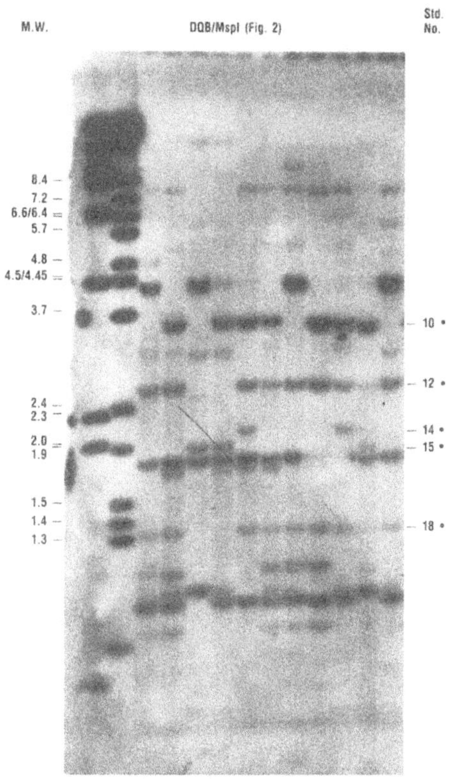

available), and DQ specificity. Most fragments were found in cells of more than one DR or DQ specificity. For example, fragment 5 was positive with 5/5 DR1 and 5/6 DRw15-Dw2, and also showed scattered other positives. Fragment 10 was negative for 5/5 DR1, 1/1 DRw15-Dw12, 1/1 DR2-Dw21, 1/1 DRw11-DB5, 1/1 DRw11-Dw"FS", 1/1 DRw8-Dw8.3-DQw6, and all DRw6 except one DRw14-Dw16; it was positive for all other cells, including all DQw2 and DQw3. Because this fragment may be a doublet, there may be greater specificity of the individual fragments. Fragments 3 and 13 distinguished DR3-DW3-Dw24 from DR3-Dw3-Dw25, but showed other positives in which reactivity was not correlated with the DRw52 subtypes.

Table 2. Distribution of Polymorphic Fragments in Core Cell Lines, DQ Beta/MspI

Cell #	Name	1	2	3	5	8	9	10	12	13	14	15	16	18	19	21
9002	MZO70782			1	8		2		1				1	8	8	8
9003	KAS116				8		2		1				1	8		1
9004	JESTHOM				8		2		1				1		8	
9005	HOM2				8		2		1				1	8	8	8
9006	WT100BIS				8	1	2		1				1	8		8
9008	D0208915							2	1			1	1			
9010	AMAL				8		8	2	1				1		8	
9013	SCHU				8		1	2				1	1		8	
9014	MGAR				8			2	1				1	8		
9017	WT8	1		1	8		1	2			1	1	1			
9011	E4181324						2				1	1	1			
9082	HO104	1			8		8	2				1	1			
9009	KAS011					1	2		1				1			1
9016	RML							2	1				1	8	8	8
9007	DEM					1	2	1	1				1	8		
9018	L0081785							2	1	1			1			
9019	DUCAF						8	2	1	1	1		1			
9020	QBL							2	1	1	1	8	1			
9022	COX			1			8	2				1	1			
9023	VAVY			1		1	8	2				1	1		8	
9088	PFO4015			1		1	8	2				1	1			
9021	RSH					1		2	1	1			1		8	8
9025	DEU							2	1				1	8		8
9091	MLF							2	1				1	8	8	8
9029	WT51	1					8	2			1	1	1			1
9031	BOLETH						1	2				1	1			8
9032	BSM						8	1			1	1	1			
9032	BSM	1					8	2			1	1	1			8
9033	BM14						1	2				1	1		8	
9034	SAVC	1					8	2				1	1			
9026	YAR	8					8	2				1	1			1
9030	JHAF							2	1				1	8		8
9028	PE117						8	2	1			8	1	8		
9092	BM92							2	1				1	8		1
9107	LKT3							2	1				1	8		8
9037	SWEIG007							2	1				1			
9040	BM15			1				2	1				1	8		8
9043	BM21							2	1				1	8		8
9104	DHIF					8		2	1				1	8	8	
9039	JVM	8					8	2			1	1	1			1
9042	TISI							2	1				1		8	
9036	SP0010					8	2		1		1		1	8		8
9105	FPAF						1		1				1			
9045	TUBO							1	1				1			8
9038	BM16							2	1		1		1	8		1
9052	DBB							2	2				1	8		
9050	MOU					8		2	1	1	1		1			
9051	PITOUT							2	1	1			1	8		
9047	PLH					8		2	1	1			1			
9048	LBUF							2	1				1	8	8	
9106	MANIKA							1	8				1			
9058	OMW				8	1	2					1	1			
9065	HHKB		1	8	8		2					1	1		8	
9062	WDV						2		1				1	8		
9060	CB6B			1			2					8	1			
9059	SLE005	1			8		2					1	1			
9063	WT47				8		2					1	1			
9054	EK			1			2					1	1		8	8
9061	31227ABO						2		1				1	8		1
9057	TEM						2					8	1		8	8
9064	AMALA							2	1				1		8	1
9056	KOSE						2		1		1		1	8		
9056	KOSE						2		1				1	8		8
9055	HO301			1	8		2					1	1			
9067	BTB							2	1				1	8		1
9068	BM9							2	1		1		1	8		
9069	MADURA							2	1				1	8		
9071	OLGA							2	1				1	8		8
9072	SPACH							1	1				1	8		
9066	TAB089						2		1		1		1	8		
9070	LUY							2	1				1			1
9075	DKB							2	1				1	8		

RFLP Standardization Report for DQ Beta/BamHI

J.J. Hyldig-Nielsen,[1] B.O. Boehm,[2] K.J. Wood,[3] A.R. Bushell,[3] D.P. Singal,[4] and A. Svejgaard[1]

The DQ-beta/BamHI system identified 22 fragments (Table 1). Eight of these fragments – St. fragment nos. 1, 2, 3, 4, 12, 14, 16, and 21 – were not included in the analysis (no locus assignments) as they were very faint and seen only on a few films. Possible reasons for the occurrence of these fragments could be either a very weak cross-hybridization to another locus or a partial digestion of some DNA preparations. Of the remaining 14 fragments, eight hybridized exclusively or most strongly with the DQ beta probe (see Table 1), while one fragment (no. 6) hybridized equally strongly with both the DQ beta and the DP beta probes (see also the report for the DP-beta/BamHI system). Three fragments hybridized most strongly with the DR beta probe and another three hybridized most strongly with the DP beta probe (see Table 2 in the DR-beta/BamHI report).

Table 2 indicates the seven different banding patterns obtained with this system. St. fragment no. 9 clearly consists of several (2 to 3) fragments with very small size differences, some also hybridizing more strongly than others (Fig. 1). Because a clear distinction between these fragments was not possible, these fragments have been registered as one fragment (St. fragment no. 9).

Participating Laboratories: SCASVE,[1] GERBOE,[2] UKIMOR,[3] US5SIN[4]

Immunobiology of HLA
Volume I

Table 1. DQ Beta/BamHI Data

	kb	DQB Locus	BAM Frequency	SVE % Faint	56	32
1	28.94	0 0 0 0	0.000	0.000		
2	22.60	0 0 0 0	0.000	0.000		
3	16.30	0 0 0 0	0.000	0.000		
4	12.41	0 0 0 0	0.000	0.000		
5	10.26	3 1 2 0	0.271	0.000		
6	9.40	0 1 1 0	1.000	0.000		
7	8.72	0 1 0 0	0.029	0.000		
8	8.09	2 3 1 0	1.000	1.000		
9	6.36	0 1 0 0	0.700	0.314		f
10	5.76	0 1 0 0	0.171	0.000		
11	5.27	1 3 2 0	0.186	0.000		
12	4.62	0 0 0 0	0.243	1.000		
13	4.57	0 2 1 0	1.000	1.000		
14	4.22	0 0 0 0	0.000	0.000		
15	4.19	1 3 2 0	0.743	0.019		
16	3.88	0 0 0 0	0.522	0.973		f
17	3.83	1 3 2 0	0.171	0.000		
18	3.55	0 1 2 0	0.171	0.000		
19	3.03	0 1 2 0	0.257	0.000		
20	2.89	2 1 0 0	0.171	0.000		
21	1.34	0 0 0 0	0.000	0.000		
22	0.25	0 1 0 0	0.214	0.000		

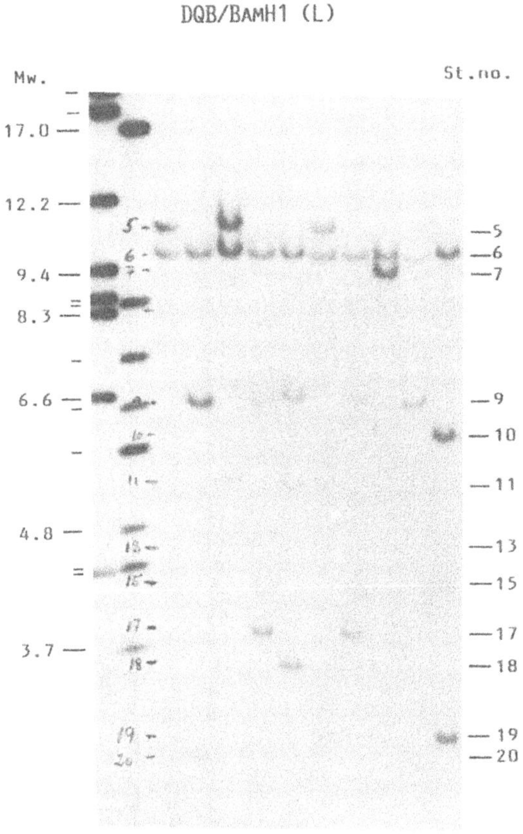

Figure 1.

Table 2. Banding Patterns Obtained with the DQ Beta/BamHI System

WS ID No.	Name	HLA-DR	HLA-DQ	HLA-DP	Standard Fragment No. 5	7	10	19	18	9	20
9009	KAS011	2	1	3,4			6	8			
9057	TEM	14	1	4			6	6			
9017	WT8	15	1	2,4						8	6
9018	LO081785	3	2	3						6	
9019	DUCAF	3	2							6	
9020	QBL	3	2	2						6	
9022	COX	3	2	3						6	
9023	VAVY	3	2	1						6	
9088	PF04015	3	2	1,4						6	
9047	PLH	7	2					6		4	
9048	LBUF	7	2					6		4	
9050	MOU	7	2	2				6		4	
9106	MANIKA	7	2,3	NT	6			6		4	
9021	RSH	3	4	1	6						
9107	LKT3	4	4	5	6						
9051	PITOUT	7	4	4				6		4	
9067	BTB	8	4	4	6						
9068	BM9	8	4	2	6						
9069	MADURA	8	4	4	6						
9071	OLGA	8	4	3	6						
9072	SPACH	8	4		6						
9002	MZ070782	1	5	4			8	6			
9003	KAS116	1	5				8	8			
9004	JESTHOM	1	5	4			6	6			
9005	HOM2	1	5				8	6			

Table 2. *Continued*

WS ID No.	Name	HLA-DR	HLA-DQ	HLA-DP	Standard Fragment No.						
					5	7	10	19	18	9	20
9006	WT100BIS	1	5	1			8	6			
9036	SPO010	11	5	2			8	6			
9060	CB6B	13	5							6	6
9056	KOSE	13,14	5	2			6	6			
9056	KOSE	13,14	5	2			6	6			
9054	EK	14	5				8	6			
9061	31227ABO	14	5	4			8	8			
9007	DEM	16,4	5,8	3,4	6		6	6			
9066	TABO89	8	6			6					
9011	E4181324	15	6	2,4		6					
9055	HO301	6	6	5						8	
9105	FPAF	11	6	2						6	6
9058	OMW	13	6	1						8	6
9059	SLE005	13	6	3						8	
9062	WDV	13	6	4						6	6
9063	WT47	13	6	2						8	
9065	HHKB	13	6	4						8	6
9008	DO208915	15	6	2,4						8	6
9010	AMAI	15	6							8	6
9013	SCHU	15	6							8	6
9014	MGAR	15	6	4						8	6
9082	HO104	15	6	4						8	6
9025	DEU	4	7	4					6	6	
9030	JHAF	4	7	3					6	8	
9091	MLF	4	7						6	6	
9070	LUY	8	7	1,4					8	8	
9037	SWEIG007	11	7							8	6
9039	JVM	11	7	2					8	6	
9040	BM15	11	7	3					6	6	
9042	TISI	11	7						6	6	
9043	BM21	11	7						8	6	
9104	DHIF	11	7	NT			6		4		
9045	TUBO	11,12	7	2,4					6	6	
9038	BM16	12	7	2					6	6	
9064	AMALA	14	7						6	6	
9016	RML	16	7						6	6	
9026	YAR	4	8	4	6						
9028	PE117	4	8	4	6						
9029	WT51	4	8	2	6						
9031	BOLETH	4	8	4	6						
9032	BSM	4	8	2	6						
9032	BSM	4	8	2	6						
9033	BM14	4	8	4	6						
9034	SAVC	4	8		6						
9092	BM92	4	8		6						
9052	DBB	7	9	4	6						
9075	DKB	9	9	4	6						

RFLP Standardization Report for DQ Beta/HincII

P. Piccioli,[1] D. Chandanayingyong,[2] M. Vijod,[3] and G.B. Ferrara[1]

The DQ beta-HincII system has identified 17 fragments (Table 1) on the core cell lines, but none specific for a DQw specificity (with perhaps one exception represented by fragment no. 6). As for all the beta-HincII systems, the interpretation of the results was very difficult both for the presence of several weak fragments and for some cross-hybridization. Moreover, this study was performed using only short migration conditions, and only 24 of 72 cell lines were analyzed from all three laboratories.

All the DQw1 and DQw5 specificities are characterized by the presence of 8.50 kb fragment that is also shared by 12 of 14 DQw6 cell lines tested. The other two cell lines carrying DQw6 specificity present the fragment 1 (15.36 kb). Indeed, we can look at the fragments 1 and 6 as two alleles for DQw6. Fragment 6 and its allele are specific for DQw1-5 and DQw6 with one exception represented by a DQw7,DR11 specificity (Table 2). Within DQw2 specificity, it is possible to discriminate the DR specificity for a positive correlation with a 7.66 kb fragment found in all DR7 cell lines. No specific fragment seems to be correlated with DR3-B8 and -B18. Also, for DQw7,w8,w9 few significant data emerge. In particular, among DQw7 lines, the lack of fragment 3 seems to be peculiar to all DQw7,DR11. Some fragments show a wide distribution (for instance fragment 5 is present in all DQw3 and new defined splits, except for sample 9028, in all DQw1, w5, w6, except for sample 9060, and in six of seven DQw4 cell lines). Also, the 2.69 kb fragment is absent in 14% of DQw7, 12% of DQw8, one of two DQw9, and 22% of DQw4. No specific correlation with Dw specificity has been found that can explain the lack of this fragment. Figures 1 and 2 represent an example of Southern blot obtained after digestion with HincII and hybridization with the DQ beta-specific probe.

Table 1. DQ Beta/HincII Data

Fragment No.	kb	Locus[a]	Frequency[b]	% Faint[c]
1	15.36	0 1 0 0	0.029	0.000
2	12.54	0 1 2 0	0.414	0.724
3	10.61	2 1 0 0	0.300	0.238
4	10.18	0 1 2 0	0.442	0.129
5	9.50	0 1 0 0	0.843	0.068
6	8.50	2 1 2 0	0.386	0.000
7	7.66	0 1 1 0	0.100	0.571
8	6.91	2 1 2 0	0.329	0.957
9	3.81	0 2 1 0	0.614	0.907
10	3.66	2 2 1 0	0.600	0.880
11	3.32	2 1 1 0	0.057	1.000
12	2.93	0 1 1 0	0.186	0.923
13	2.69	2 1 1 0	0.757	1.000
14	2.39	0 1 0 0	0.100	1.000
15	2.21	2 1 0 0	0.157	0.909
16	2.07	0 1 0 0	0.129	1.000
17	1.85	0 1 0 0	0.029	0.500

[a] DR, DQ, DP, DO locus assignment, for instance: 0 1 0 0 means DQ-specific; 0 1 2 0 means DQ- and DP-specific, DQ more intense; 2 1 2 0 means DR, DQ-, DP-specific, DQ most intense, DR and DP same intensity.
[b] Band frequency has been calculated as number of cell lines presenting a specific fragment versus number of cell lines tested.
[c] Frequency of faint band has been calculated as number of cell lines presenting a faint band versus number of cell lines presenting the same fragment (i.e., frequency 1.000 and % faint 0.000 means that all the samples tested share the same band and it is a strong band; on the other hand, 1.000 and 1.000 means that it is a constant and a faint band).

Participating Laboratories: IT2FER,[1] ANZDCH,[2] ANZDAW.[3]
P. Piccioli was recipient of a fellowship from A.I.R.C. (Associazione Italiana per la Ricerca sul Cancro). This work was in part supported by CNR Grants 86.0265.51 and 86.00659.

Table 2. Banding Pattern of the DQB/HincII System in Each Cell Line

Cell	DQ	DR	Fragment No. 1	2	3	4	5	6	7	8	9	10	11	12	13	14	15	16	17
9009 KAS 011	1	2					8	2							8				
9017 WT 8	1	15					1	2			8			8	8				
9057 TEM	1	14		8			1	2		8	8	8			8		8		
9002 MZ 070782	5	1		8		8	1	2		8	8		8	1	8				
9003 KAS 116	5	1					1	2		8		8		8	8	8			
9004 JESTHOM	5	1					1	1											
9005 HOM 2	5	1					1	3		8	8			8					
9006 WT100 BIS	5	1						3		8			8	8	8	8			
9036 SPO 010	5	11		8		8	1	2			1	1			8	8			
9060 CB6B	5	13		8				1			8	8			8				
9054 EK	5	14		8	8		1	2		8	8	8			8		8		
9061 31227 ABO	5	14					1	2			8	8			8		8	8	
9056 KOSE	5	13/14					1	3				8			8		8		
9056 KOSE	5	13/14					1	2				8			8		8		
9007 DEM	5/8	4/16				1		2			8			8	8				
9058 OMW	6	13					1	2				8							
9059 SLE 005	6	13					1	2				8			8			8	
9062 WDV	6	13				1	1	2			8	8			8		8		
9063 WT 47	6	13		8	8		1	1				8			8				
9065 HHKB	6	13		8			1	2			8				8		8		
9008 DO 208915	6	15					1	2				8		8	8				
9010 AMAI	6	15					1	2			8			8	8				
9011 E4181324	6	15	1				1				8			8	8				1
9013 SCHU	6	15					1	2				8		8	8				
9014 MGAR	6	15					1	2			8			8	8				
9082 HO104	6	15					1	2		8	8	8		8	8				
9105 FPAF	6	11		8			8	1			8	8			8				
9055 HO 301	6	6		8	8		1	2	3	8		8			8				
9066 TAB 089	6	8	1		8		1			1	8	8			8				8
9018 LO 081785	2	3		1	2		1					8			8	8			
9019 DUCAF	2	3		1	1						8	8			8	8		8	
9020 QBL	2	3		1	1						8	8			8				
9022 COX	2	3			1		1					8							
9023 VAVY	2	3				1	1			8		8	8		8	8			
9088 PF 04015	2	3		8	1		1			8	8	8			8				
9047 PLH	2	7		2					1			8			8		1		
9048 LBUF	2	7		8	8		1		8		8	8			8			8	
9050 MOU	2	7		2					8		8				8		8		

9051	PITOUT	2	7	1					1	8	8				8		8	
9106	MANIKA	2?3	7	8		8	8				8				8			
9016	RML	7	16	8	1		1			8	8	8		8	8			
9025	DEU	7	4	8	1	1	1			8	8				8			8
9030	JHAF	7	4	8	1	1	1			8	1				8			
9091	MLF	7	4	8	1		1			8	1				8			
9037	SWEIG 007	7	11	8	1		1	1		8	8	1	8		8			
9039	JVM	7	11			1	1				8	8			8			
9040	BM 15	7	11			2	2					8			8			
9042	TISI	7	11			1	1					8						
9043	BM 21	7	11			2	2								8	8		
9104	DHIF	7	11	1			1		8		8				8		8	
9045	TUBO	7	11/12			1	1											
9038	BM 16	7	12		1		1					8						
9064	AMALA	7	14	8	1	1	1			8	1	8			8			
9070	LUY	7	8		1		1					8						
9026	YAR	8	4			1	1				8				8			8
9028	PE 117	8	4	8		1				8	8				8			8
9029	WT 51	8	4			1	1								8			
9031	BOLETH	8	4	8	1	1	1			8	8				8			
9032	BSM	8	4			1	1			8					8			
9032	BSM	8	4			1	1			8					8			
9033	BM 14	8	4			1	1					8						
9034	SAVC	8	4			1	1					8			8			8
9092	BM 92	8	4			1	1				8							
9052	DBB	9	7			1	1		8	8	8				8			
9075	DKB	9	9			1	1				8							8
9021	RSH	4	3	1						8		8			8		8	
9107	LKT3	4	4			1	1											
9067	BTB	4	8			1	1				8	8						
9068	BM 9	4	8			1	1				8	8						
9069	MADURA	4	8		1	1	1				8	1						
9071	OLGA	4	8	8	1	1	1			8	1	8			8			
9072	SPACH	4	8			8	8					8			8			

1 = Single intensity; 2 = double intensity; 3 = more than double intensity; 8 = faint.

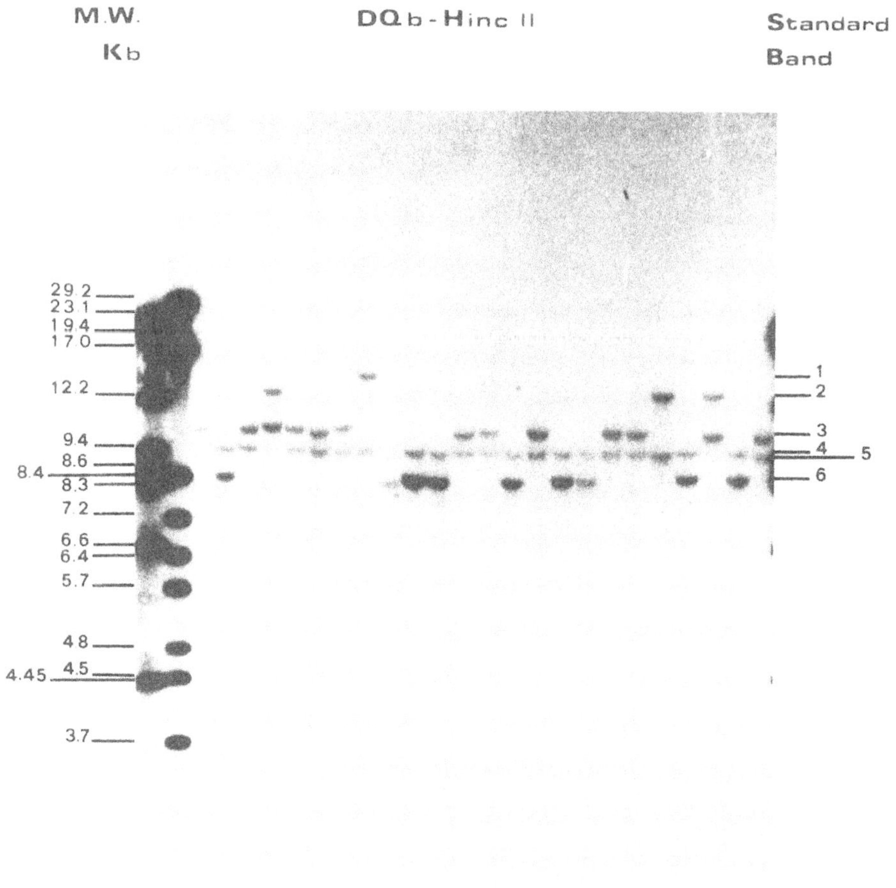

Figure 1. Southern blot of 24 of 72 selected cell lines digested with HincII and hybridized with HLA-DQB probe. Long migration conditions were used to run the gel represented. From left to right: M1, M2, ID 9069, 9037, 9028, 9019, 9025, 9031, 9088, 9066, 9060, 9036, 9002, 9071, 9064, 9057, 9030, 9054, 9005, 9091, 9016, 9048, 9065, 9021, 9062, and 9032.

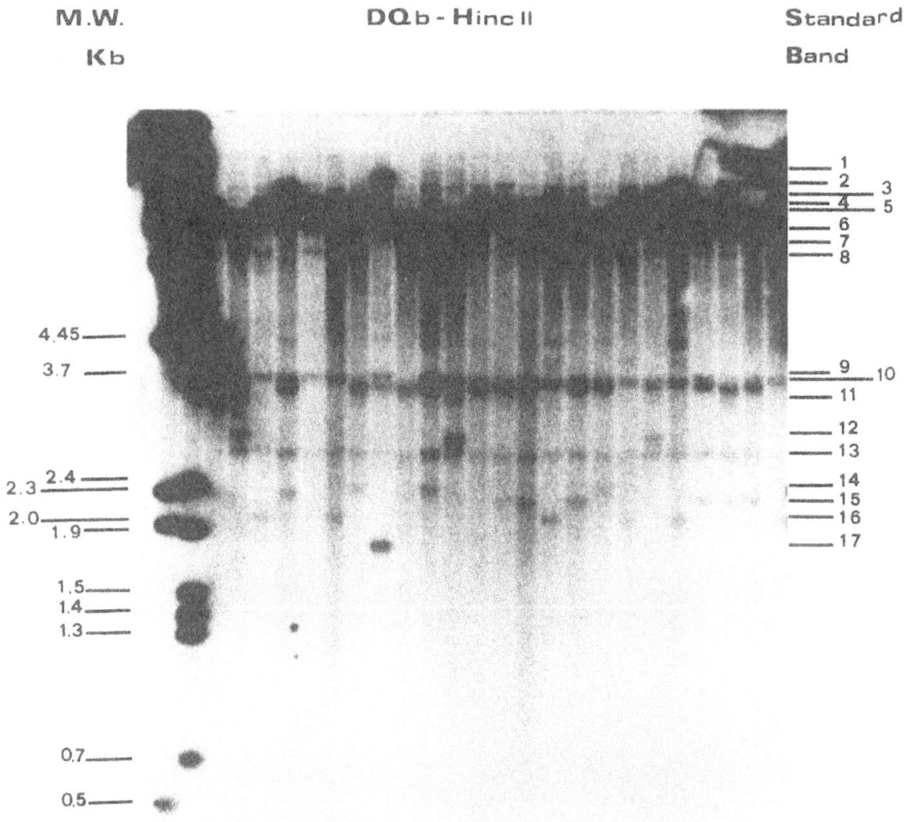

Figure 2. Standard fragments of HLA-DQB HincII system. The samples represented are the same shown in Figure 1, but under different migration conditions.

RFLP Standardization Report for DP Beta/SstI

A. Ando,[1] T. Sasazuki,[2] P. Reekers,[3] and K. Tsuji[1]

The DP beta/SstI system identified the 31 fragments (Table 1 and Figs. 1 and 2). All bands were polymorphic except four bands–St. fragment no. 16 (7.68 kb), no. 26 (2.20 kb), no. 28 (1.54 kb), and no. 30 (1.14 kb) (Table 2). St fragment no. 16, no. 26, and no. 28 hybridized only with the DPB probe. Cross-hybridizing bands are shown in Table 3. All the DP specific bands are not correlated with individual DR and DQ specificities. St. fragment no. 14 (10.03 kb) and no. 22 (3.46 kb) are associated with DPw2 or DPw4 specificities.

Participating Laboratories: JAPTSU,[1] JAPSAS,[2] BENREE[3]

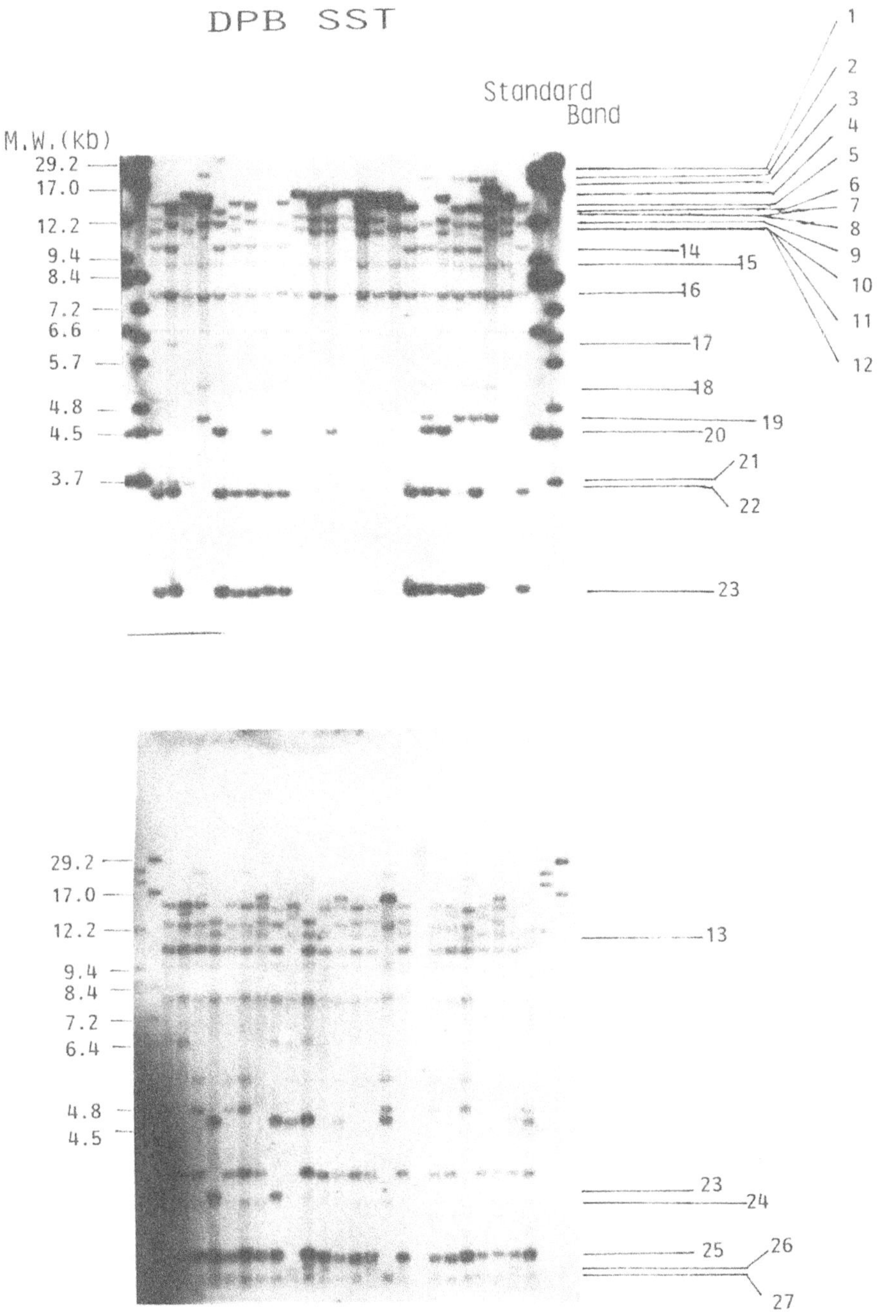

Figure 1. Fragments observed with the DP beta/SstI system.

Table 1. DP Beta/SstI Data

		DPB	SST	TSU		
	kb	Locus	Frequency	% Faint	56	32
1	20.83	1 3 2 0	0.174	0.667		
2	19.59	1 0 2 0	0.086	1.000		
3	17.42	2 2 1 0	0.029	0.000		
4	15.34	2 2 1 0	0.319	0.000		
5	14.02	0 0 1 0	0.565	0.026		
6	13.65	0 0 1 0	0.159	0.091		
7	13.14	1 3 2 0	0.145	0.300		
8	12.38	1 3 2 0	0.057	1.000		
9	12.32	1 3 2 0	0.319	0.087		
10	11.96	1 3 2 0	0.478	0.353		f
11	11.54	1 3 2 0	0.072	0.000		
12	11.18	1 3 2 0	0.377	0.111		
13	10.80	1 0 2 0	0.000	0.000		
14	10.03	0 0 1 0	0.768	0.000		
15	9.13	0 0 1 0	1.000	1.000		
16	7.68	0 0 1 0	1.000	0.042		
17	6.12	1 3 2 1	0.130	0.556		
18	5.17	1 2 2 1	0.275	1.000		
19	4.56	1 3 2 0	0.304	0.091		
20	4.34	0 2 1 1	0.391	0.000		
21	3.65	1 0 1 0	0.000	0.000		
22	3.46	2 3 1 0	0.696	0.000		
23	3.18	0 0 1 1	0.043	0.000		
24	3.17	2 0 1 0	0.000	0.000		
25	2.40	3 2 1 0	0.812	0.052		
26	2.20	0 0 1 0	1.000	0.225		
27	2.05	0 1 2 0	0.913	0.908		
28	1.54	0 0 1 0	1.000	1.000		
29	1.29	0 2 1 0	0.043	0.000		
30	1.14	0 1 1 0	1.000	1.000		
31	0.47	0 1 1 0	0.507	1.000		

Table 3. Cross-Hybridization Bands

DRB	kb	DQB	kb	DPB	kb	DOB	kb
2	21.12	1	21.66	1	20.83		
3	19.53			2	19.59		
5	17.02	2	17.91	3	17.42		
6	14.97	3	15.94	4	15.34		
8	13.19	6	13.82	7	13.14		
9	12.87	7	13.40	8	12.38		
10	12.38	8	12.68	9	12.32		
11	12.01	9	12.23	10	11.96		
12	11.59	10	11.89	11	11.54		
13	11.25	11	11.47	12	11.18		
16	6.15	17	6.08	17	6.12	1	6.26
17	5.19	19	5.20	18	5.17	2	5.19
18	4.57	21	4.56	19	4.56		
		22	4.35	20	4.34		
19	3.82	12	3.78				
20	3.46	24	3.44	22	3.46		
21	3.02	25	3.07				
23	2.38	27	2.42	25	2.40		
24	2.23	28	2.28				
		30	2.01	27	2.05		
		31	1.33	29	1.29		
		32	1.09	30	1.14		
		33	0.34	31	0.47		

Underline denotes primary locus assignment.

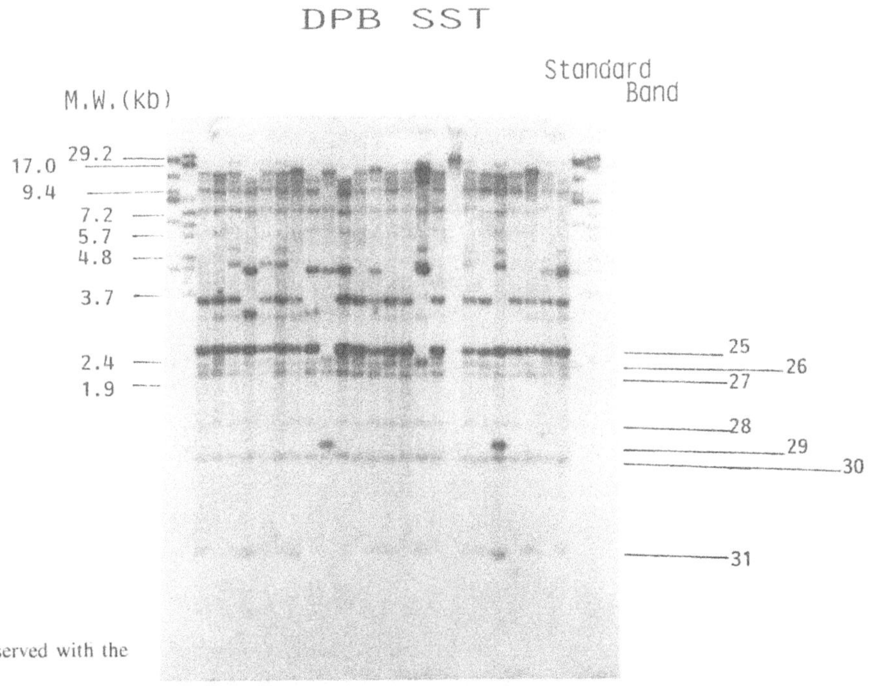

Figure 2. Fragments observed with the DP beta/SstI system.

Table 2. RFLP Pattern of DP Beta/Sstl

			1	2	3	4	5	6	7	8	9	10	11	12	13	14	15	16	17	18	19	20	21	22	23	24	25	26	27	28	29	30	31						
Fragment No.												1	1	1	1	1	1	1	1	1	1	2	2	2	2	2	2	2	2	2	2	3	3						
			1	2	3	4	5	6	7	8	9	0	1	2	3	4	5	6	7	8	9	0	1	2	3	4	5	6	7	8	9	0	1						
			2	1	1	1	1	1	1	1	1	1	1	1	1	1																							
Fragment size (kb)			0	9	7	5	4	3	3	2	1	1	1	0	0	9	7	6	5	4	4	3	3	3	3	3	2	2	2	1	1	1	0						
			.	.	.	.	.	.	.	.	.	.	.	.	.	.	.	.	.	.	.	.	.	.	.	.	.	.	.	.	.	.	.						
			8	5	4	3	0	6	1	3	3	9	5	1	8	0	1	6	1	1	5	3	6	4	1	1	4	2	0	5	2	1	4						
			3	9	2	4	2	5	4	8	2	6	5	1	8	0	1	6	1	1	5	3	5	6	8	7	0	0	5	4	9	4	7						
	Cell	Lines																																DW	DR	DW2	DRw	DQw	DPw
64	9039	JVM									1			1		1	8	1				1		2			3	1	8	8		8	8	N	11	25	52	7	2
18	9050	MOU		8								1				1	8	1		8	1	2		3			3	1	8	8		8	8	17	7		53	2	2
48	9032	BSM														1	8	1		8	1	2		2			3	1	1	8		8	8	4	4		53	8	2
53	9068	BM9										1				1	8	1				2		2			3	1	8	8		8	8	8.1	8		52	4	2
71	9020	QBL									1			1		1	8	1				2			1		3	1	8	8		8	8	3	3	25	52	2	2
12	9063	WT47				2					1			1			8	1				2						1	8	8		8		19	13	26	52	6	2
41	9105	FPAF	9		9	9	9	9	9		9	9	9	9		9	9	9	9	9	9	9		9	9		9	9	9		9	9	9	FS	11		52	6	2
61	9029	WT51	8									8				1	8	1		8	1	1		2			3	1	8	8		8	8	4	4		53	8	2
34	9036	SPOO10									1			1		1	8	1				2		2			3	1	8	8		8		DB2	11	25	52	5	2
68	9032	BSM										8				1	8	1		8	1	2		2			3	1	8	8		8	8	4	4		53	8	2
57	9038	BM16												8		1	8	1				1		1			2	8	8	8		8	8	DB6	12		52	7	2
19	9056	KOSE				1					1			1		1	8	1				2		2			2	8	8	8		8		9,19	13,14	25	52	5	2,N
65	9056	KOSE				2					1			1		1	8	1				1		1			2	8	8	8		8		9,19	13,14		52	5	2,N
62	9011	E4181324					1	1	1			8				1	8	1	1			1		2			3	1	8	8		8		12	15			6	6
58	9008	DO208915							8			8				1	8	1	8			1		2			3	1	8	8		8		2	15			6	6
5	9017	WT8							8			8				1	8	1	8			2		2			3	8	8	8		8		2	15			1	6
1	9045	TUBO					1				8			8		1	8	8				1		1			3	1		8		8		5	11,12	25	52	7	6
45	9065	HHKB					1			8				1		1	8	1						2			3	8	8	8		8		18	13	24	52	6	4
63	9026	YAR	8									8				1	8	1		8	1	1		2			3	1	8	8		8	8	10	4		53	8	4
54	9075	DXB		8				1				1				1	8	1		8	1			2			3	8	8	8		8	8	23	9		53	9	4
25	9069	MADURA					1									1	8	1						1			2	8	8	8		8	8	8.1	8		52	4	4
47	9062	WDV					1			8				1		1	8	1				1		2			3	8	8	8		8		18	13	24	52	6	4
66	9082	H0104					1		1			1				1	8	1	1					2			3	1	8	8		8				15		6	4
69	9072	SPACH					1				8					1		8						1			2	8	8	8		8	8	8.2	8		52	4	4
55	9061	31227ABO						1			1			1		1	8	1						2			3	1	8	8		8		9	14	25	52	5	4
9	9004	JESTHOM					1						1			1	8	1						2			3	1	8	8		8		1	1			5	4
35	9002	MZ070782					1						1			1	8	1						2			3	1	8	8		8		20	1			5	4
2	9014	MGAR					1		1			8				1	8	1	1					2			3	1	8	8		8		2	15			6	4
56	9067	BTB					1					1				1	8	1						2			3	1	8	8		8	8	8.1	8		52	4	4
27	9028	PE117	1				2					1				1	8	1		8	1			2			3	1	8	8		8	8	14	4		53	8	4

29	9025	DEU	1				2					1			1	8	1		8	1		2		3	1	8	8		8	8	4	4		53	7	4
70	9052	DBB	1		1		1								1	8	1		8	1		2		3	1	8	8		8	8	11	7		53	9	4
8	9033	BM14	8									8			1	8	1		8	1	2	2		3	1	8	8		8	8	4	4		53	8	4
30	9031	BOLETH	1				2					1			1	8	1		8	1		2		3	1	8	8		8	8	4	4		53	8	4
38	9057	TEM					1				1			1	1	8	1					2		3	1	8	8		8		9	14	25	52	1	4
21	9051	PITOUT		8			1					1			1	8	1		8	1		2		3	1	8	8		8	8	17	7		53	2	4
60	9070	LUY				2	1					1			1	8	1					1		2	1	8	8		8		8.3	8		52	7	7
31	9088	PF04015				1	1				1			1	1	8	1					1		2	8	8	8		8	8		3	24	52	2	7
49	9009	KASO11				1		8				1			1	8	1					1		1	8	8	8		8		21	2			1	4,3
72	9007	DEM	8			2	8		1			1			1	8	1			8	1	1		2	1	8	8		8	8	–	16,4		53	5,8	4,3
15	9023	VAVY				2	1				1			1		8	1							8	1	8	8		8	8	3	3	24	52	2	1
46	9021	RSH				2	1				1			1	1	8	1					1		2	1	1	8		8	8	N	3	24	52	4	1
16	9006	WT100B1S				2	1						1			8	1								1	1	8		8		1	1			5	1
3	9058	OMW				2	1			8				1		8	1								1	8	8		8		18	13	24	52	6	1
23	9040	BM15				2	1				1			1		8	1								1	8	8		8		5	11	25	52	7	3
36	9071	OLGA				2	1					1			1	8	1				1	2		2	1	8	8		8	8	8.2	8		52	4	3
10	9018	LOO81785				2					1			1		8	1				2				1	8	8		8	8	3	3	25	52	2	3
13	9022	COX				2					1			1		8	1				2			8	1	8	8		8	8	3	3	24	52	2	3
39	9030	JHAF				1						8				8	1		8	1	2				1		8		8		13	4		53	7	3
51	9059	SLE005				2					1			1		8	1				2				1	8	8		8	8	19	13	26	52	6	3
11	9055	HO301				2	1				1			1		8	1								8	8	8		8		N	6	26	52	6	5
4	9107	LKT3	8			2	1					1				8	1		8	1					1	8	8		8		15	4		53	4	5
20	9104	DH1F		8								8			1	8	1		8	1				2	8	8	8	1	8	8	5	11		52	7	N,T
67	9106	MANIKA				1									1	8	1			1	1	1		2	1	8	8		8	8		7		53	2?3	N,T
52	9034	SAVC	8			2		1				1				8	1		8	1					1	8	8		8	8	4	4		53	8	–
50	9092	BM92	8				1					8			1	8	1	8	8	8		2		2	1	8	8		8	8	14	4		53	8	–
6	9012	T1S1					1				1			1	1	8	1					2		3	1	8	8		8		N	11	25	52	7	–
42	9091	MLF	8				1					1			1	8	1	8				2		3	1		8		8	8	4	4		53	7	–
26	9037	SWEIG007					1		1			1			1	8	1	8				2		3	1		8		8		5	11	25	52	7	–
37	9064	AMALA					1				1			1	1	8	1					2		3	1	8	8		8		16	14	24	52	7	–
14	9043	BM21				2		1			1			1		8	1								1	8	8		8	8	5	11	25	52	7	–
43	9016	RML					1		1			1			1	8	1					2		3	1		8		8		22	16			7	–
32	9066	TAB089						1				1			1	8	1				2		2	3	1	8	8		8		8.3	8		52	6	–
7	9013	SCHU					1	1	8			8			1	8	8	8				1		1	8	8	8		8		2	15			6	–
24	9010	AMAI					1		1			1			1	8	1	1				2		3	1	1	8		8		2	15			6	–
33	9060	CB6B					2			8				8		8	1				2				1	8	8	1	8		18	13	25	52	5	–
17	9005	HOM2					1						1		1	8	1					2		3	1	1	8		8		1	1			5	–
40	9054	EK					1				1			1	1	8	1					2		3	1	1	8		8		9	14	25	52	5	–
59	9003	KAS116				2		1					1			8	1								8	8	8		8		1	1			5	–
28	9019	DUCAF									1			1	1	8	1				2		2	3	1	8	8		8	8	3	3	25	52	2	–
44	9048	LBUF		8				1				1			1	8	1		8	1				3	8		8	1	8	8	DB1	7		53	2	–
22	9047	PLH		8	2			1				1				8	1		8	1				8	1	8	8		8	8	DB1	7		53	2	–

RFLP Standardization Report for DP Beta/PstI

E. Keller,[1] A. Andreas-Zietz,[1] E.D. Albert,[1] H. Coppin,[2] L. Contu,[3] A. Takada,[4] and M. Trucco[5]

The DP beta/PstI system identified 26 fragments on the core cell lines (Table 1 and Figs. 1 and 2). Table 1 shows the positive and negative reactions of all 72 core cells for the 26 standardized fragments, the corresponding molecular weight as well as the locus assignment, and the probable association with DP antigens. The intensity of bands is indicated by numbers. Scores of "1," "2," and "3" mean increasing intensity, "8" means a faint fragment and "9" is unreadable. The intensity, however, does not necessarily indicate the gene dosage. Because of the obvious contrast of strong and weak bands, seen especially in the L gels (0.6%) of most laboratories, we decided to assign the intensities "1," "2," and "3" only to the strong real DP beta/PstI bands. To all the others, we assigned "8" (faint).

Core cell no. 26 showed two different banding patterns in different laboratories. Therefore, "9" was assigned to all fragments not identical in all laboratories. As the different banding patterns were confirmed, the most likely explanation for this discrepancy is a mixup of DNA (see also report on DR beta/PstI and DQ beta/PstI).

For cross-hybridization with DR, DQ, and DO-beta probes, see Table 2 in the report on DR beta/PstI.

Fragment DP beta/PstI no. 19. This is present in all cells. The locus assignment is "1120" except in those cells that also show a very strong signal for DR beta/PstI fragment no. 25 (DR4-associated) where the locus assignment is "1230" (see also comments on core fragments no. 25 DR beta/PstI and no. 21 DQ beta/PstI).

Participating Laboratories: GERALB,[1] FRAOHA,[2] IT1CON,[3] JAPAIZ,[4] US1DUQ[5]
Supported by SFB 217.

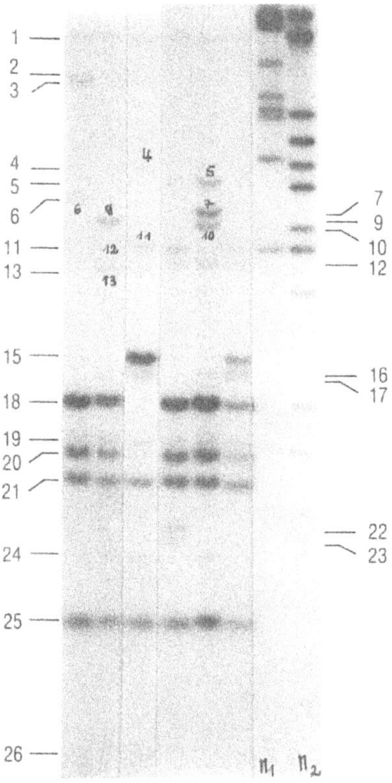

Figure 1. DP beta/PstI core fragments (gel 0.9%).

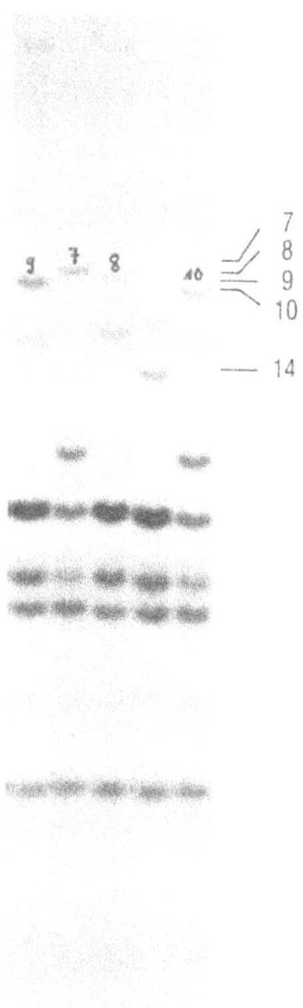

Figure 2. DP beta/PstI core fragments no. 8 and 14 (gel 0.9%).

Immunobiology of HLA
Volume I

Table 1. Banding Pattern of All 72 Core Cell Lines, as Standardized in the DP Beta/PstI System

```
Probe Name: DPB          Enzyme Name: PSTI

    |9999999999999999 999999999 99999999999999999999999999999999999999999999999| WS-ID number
    |0000000001000000 000000001 00000100000000000000000000000000000000000100000|
    |0256123450503464 872550070 33223033561014001823232666555667011119934056467|
    |6383820097534307 801669716 22096598087815424253816521712975530691272444862|
    |---------------- --------- -----------------------------------------------| -------------
    |1101113250155132 364164736 46763465150560030620236445327255102424520243436| position on
    |6532039314192432 106959267 88114147835821952698703575810564774382066007429| blots
    |---------------- --------- -----------------------------------------------| -------------
DP  |11123333355----- 11122333- 222222222222224444444444444444-----------------|
    |---------------- --------- -----------------------------------------------|
DP  |                 44   44            4444                                  | kb    locus      associated with:
    |---------------- --------- -----------------------------------------------| ------------     ----------------
   1|8888888888888888 888888888 88888988888888888888888888888888888888898888888| 16.16  0 0 1 0
   2|  88   8 8 8            9  88 8 98     8    8 8888   8      88    9       | 10.86  2 1 3 0
   3| 8           8              9    88    88     88           88    9       | 10.42  2 1 3 0
   4|  8    8 8                                                                |  6.39  2 1 3 0
   5|             8          8         8                   88  8        8  8   |  5.96  1 2 2 0
   6|                                   8    88                   88            |  5.37  1 2 2 0
   7| 8  88        8 8       8   8     8                     8     8     8  8   |  5.20  2 1 3 0
   8|                       8              8                                8  |  5.03  2 1 2 0
   9|     88 8  88    88     88 88 8  88 8   8      88888     8888   8 8898  8 8|  4.94  2 1 3 0
  10|             8                     8                    8          8  8   |  4.82  1 0 2 0
  11|8  8    8 88        8888      8          88          88     8        8    |  4.43  2 1 2 0
  12|8   8 88 8 8 8 8   88888999 898889988888888889888 8 88 888888  8888888 9889|  4.20  2 1 2 0
  13|     8  8  8            8   88 8                   88888         88        |  4.00  1 2 2 0
  14|                                         8                                 |  3.70  1 2 2 0
  15|2222223222222222 212111121                                                 |  2.85  2 2 1 0   DP 1 + 3 + 5
  16|                   88             8      8            88            8      |  2.71  1 2 2 0
  17| 8 888 8888 8898 8 888    9 88888 888     8     98898     88        888888  8|  2.67  1 2 2 0
  18|                 111111111 22222222222222222222222222222222222222222222222|  2.39  2 3 1 0   DP 2 + 4
  19|888888888888888 8 888888888 88888888888888888888888888888888888888888888888|  2.02  1 2 3 0*
  20|                 111111111 22222222222222222222222222222222222222222222222|  1.91  3 2 1 0   DP 2 + 4
  21|2222222222222222 222222222 22222222222222222222222222222222222222222222222|  1.72  2 2 1 0
  22|8           8                          8                   8              |  1.41  1 2 2 0
  23|8           8       9898        8      88          88      8        8     |  1.32  2 1 3 0
  24|8888888888888888 888888888 88888888888888888888888888888888888888888888888|  1.25  1 2 3 0
  25|2222222222222222 222222222 22222222222222222222222222222222222222222222222|  0.87  2 3 1 0
  26|8888888888888888 888888888 88888888888888888888888888888888888888888888888|  0.24  1 0 2 0
```

Includes kb values, locus assignment and probable DP associations. Cells are listed according to banding patterns and DP types. Numbers indicate intensity (1,2,3 = increasing intensity, 8 = faint, 9 = unreadable).
* = For some cells different locus assignment, see text.

RFLP Standardization Report for DP Beta/KpnI

S. Cutbush,[1] P.I. Biro,[1] P. Cassell,[1] H. Avakian,[1] R. Vaughan,[1] M. Oudshoorn,[2] R. Martell,[2] and K. Siminovitch[3]

DP beta gives a complex pattern with KpnI (Fig. 1). Seventeen fragments have been defined (Table 1), the majority of which are polymorphic and cross-hybridize with other D-region probes (Table 2), but do not correspond to the DP genes themselves. However, four bands (no. 5, 6, 9, and 16) hybridize more strongly with DPB than with the other beta probes. Bands 9 and 16 are relatively intense, nonpolymorphic, and probably contain the DPB genes. Nos. 5 and 6 are faint and show polymorphism. The presence or absence of band 6 correlates weakly with DP type (1) (Table 3) and can be used to group DP types into two clusters (Table 4) (DP types 1 and 3, and, in the second cluster, DP types 4, 2, 2B, and possibly 4B) which correspond to the III_β and IV_β clusters defined by Mitsuishi et al. (2).

References

1. Bodmer J, Bodmer W, Heyes J, et al. Identification of HLA-DP polymorphism with DPα and DPβ probes and monoclonal antibodies: correlation with primed lymphocyte typing. Proc Natl Acad Sci USA 1987;84:4596.
2. Mitsuishi Y, Urlacher A, Tongio M-M, Mayer S. Restriction fragment length polymorphism of DP genes defines three new DP alleles. Immunogenetics 1987;26:383.

Participating Laboratories: UKIFES,[1] SAFDUT,[2] NCYFAL[3]

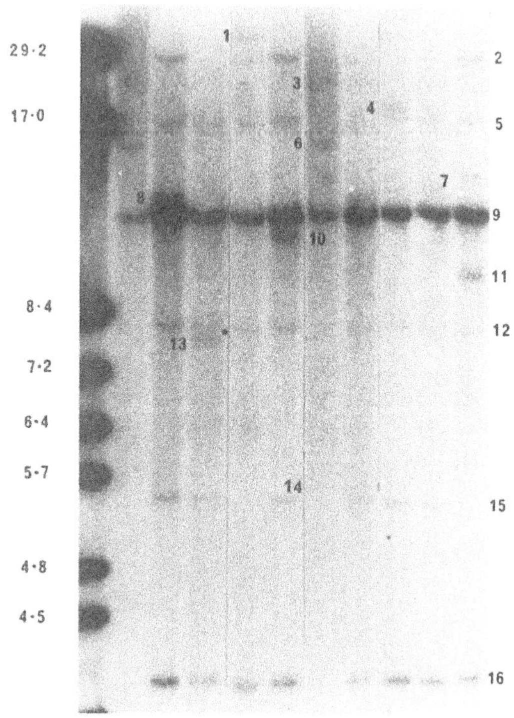

Figure 1. Standard bands produced by the DP beta/KpnI enzyme/probe combination. St. band 17 is not shown.

Table 1. Sizes and Frequencies of Standard Bands Produced by the DP Beta/KpnI Enzyme/Probe Combination

	kb	Locus	Frequency	% Faint
1	38.04	2 0 2 0	0.275	0.263
2	24.44	0 1 2 0	0.478	0.171
3	19.16	3 1 2 0	0.206	0.800
4	18.64	1 2 2 0	0.075	0.200
5	16.80	0 0 1 0	0.986	1.000
6	15.50	0 0 1 0	0.309	0.636
7	13.10	1 0 2 0	0.377	0.444
8	11.94	1 1 2 0	0.086	1.000
9	11.11	2 2 1 0	1.000	0.000
10	10.83	1 2 2 0	0.043	0.333
11	9.36	2 1 2 0	0.125	0.375
12	7.97	2 1 3 0	1.000	0.000
13	7.68	1 3 3 0	0.132	0.100
14	5.59	2 1 2 0	0.171	0.250
15	5.44	2 1 2 0	0.397	0.036
16	3.90	0 0 1 0	1.000	0.000
17	1.23	2 1 3 0	0.132	1.000

Table 2. Cross-Hybridization Table–Enzyme: Kpn–Probe: HLA Class II Beta

DRB	kb	DQB	kb	DPB	kb
1	37.30			1	38.04
2	17.95	3	17.75	4	18.64
3	15.99	4	15.90		
4	12.80	6	12.81		
5	12.50	7	12.03		
6	11.84	8	11.71		
7	11.05	11	11.04	9	11.11
8	10.49	12	10.68	10	10.83
9	9.43	14	9.20	11	9.36
10	7.94	17	7.98		
11	7.69	16	7.73	13	7.68
12	5.53	20	5.33	14	5.59
13	5.39	21	5.15	15	5.44
15	1.22	23	1.20	17	1.23
		1	25.91	2	24.44

Underline denotes primary locus assignment.

Immunobiology of HLA
Volume I

Table 3. RFLP Pattern of DP Beta/KpnI System

Cell Lines		DR Type	Fragment No. 1	2	3	4	5	6	7	8	9	10	11	12	13	14	15	16	17
MZ070782	9002	1				1	8				2			1			1	1	
KAS116	9003	1				8	8				2			1			1	1	
JESTHOM	9004	1				1	8				2			1			1	1	
HOM2	9005	1				1	8				2			1			1	1	
WT100BIS	9006	1				1	8				2			1			1	1	
DEM	9007	2/4	1				8	1	1		2			1			1	1	
D0208915	9008	2		9	9	9	8	9			2	9	9	1	9		9	1	9
KAS011	9009	2					8		8		2			1			1	1	
AMAI	9010	2		1			8			8	2		9	1			1	1	
E41B1324	9011	2		1			8		1	8	2			1			1	1	
SCHU	9013	2		1			8			8	2			1			1	1	
MGAR	9014	2		1			8	8		8	2			1			1	1	
RML	9016	2		1			8			8	2			1		1		1	8
WT8	9017	2		1			8	8			2			1			1	1	
L0081785	9018	3	8	1			8				2			1				1	
DUCAF	9019	3		8			8				2		9	1				1	
QBL	9020	3		1			8				2			1				1	
RSH	9021	3		1	1		8				2			1	1			1	
COX	9022	3					8				2			1	1			1	8
VAVY	9023	3					8	1			2			1	1			1	
DEU	9025	4	1				8		1		2			1		1		1	8
YAR	9026	4		1			8		1		2			1				1	
PE117	9028	4	1		1		8		1		2		9	1				1	
WT51	9029	4		1			8		1		2			1				1	
JHAF	9030	4	1				8		1		2			1		1		1	8
BOLETH	9031	4	8		8		8		8		2			1				1	
BSM	9032	4		8			8		1		2			1				1	
BM14	9033	4					8	8	8		2			1				1	
SAVC	9034	4			8		8	8	8		2			1				1	
SP0010	9036	5		1			8		8		2			1			1	1	
SWEI6007	9037	5		1			8				2		9	1		1		1	8
BM16	9038	5		1	8		8	8			2			1		8		1	
JVM	9039	5		1			8				2			1		8		1	8
BM15	9040	5					8	1			2			1		1		1	
TISI	9042	5						1	8		2			1		1		1	9
BM21	9043	5	8				8				2			1	8	1		1	
TUBO	9045	5	8			9	8	1			2			9			9	1	
PLH	9047	7					8		1		2		1	1				1	8
LBUF	9048	7		1			8		1		2		1	1				1	
MOU	9050	7		1			8		1		2		1	1				1	
PITOUT	9051	7	1				8		1		2		1	1				1	
DBB	9052	7	1				8	1	1		2			1				1	
EK	9054	6		1			8				2			1			1	1	
H0301	9055	6		1			8				2	1		1			1	1	
KOSE	9056	6		1			8	1			2		9	1			1	1	
TEM	9057	6	1				8				2			1			1	1	
OMW	9058	6					8				2			1	1		1	1	
SLE005	9059	6	1		8		8				2	8		1			1	1	
CB6B	9060	6					8				2			1			1	1	
31227ABO	9061	6	1		8		8	8			2			1			1	1	
WDV	9062	6		8			8				2			1	1		1	1	
WT47	9063	6					8				2	1		1			1	1	
AMALA	9064	6		1			8				2			1	1	1		1	
HHKB	9065	6	1				8				2			1	1		1	1	
TABO89	9066	8		1			8				2			1			1	1	
BTB	9067	8	1		8		8	8			2		8	1				1	
BM9	9068	8		1	8		8	8			2		8	1				1	
MADURA	9069	8					8	8			2			1				1	
LUY	9070	8	1	8	8		8	8			2			1		8		1	8
OLGA	9071	8		8			8	8			2			1				1	
SPACH	9072	8		1			8		8		2			1				1	
DKB	9075	9	1		8		8	8	1		2		8	1				1	
H0104	9082	?	1		8		8	8		8	2			1			1	1	
PF04015	9088	3					8				2			1	1			1	
MLF	9091	4		1			8		8		2			1		1		1	8
BM92	9092	4		1	8		8	1	8		2			1				1	
DHIF	9104	11		1			8		1		2		1	1				1	
FPAF	9105	11	9		9		8	9	9		2			1			1	1	
MANIKA	9106	7					9		8		2			1				1	
LKT3	9107	4		1	1		8		8		2			1				1	

Table 4. Correlation of DP Beta/KpnI St. Band 6 with DP Type

DP	4	2	2B	4B	1	3
%	36	33	33	22	17	17

% of each DP type displaying st. band 6.

RFLP Standardization Report for DP Beta/EcoRV

T. Takenouchi,[1] R. Tonai,[1] P. Terasaki,[1] S. Mayer,[2] S. Bashir,[3] R. Winchester,[4] and Y. Paik[5]

Data from five laboratories were used for standardization according to the criteria outlined in the DR/EcoRV system. The same experimental problems found in the DR/EcoRV system were observed in this system and the data received similar treatment. For DP beta/EcoRV, 13 standard fragments were assigned (Table 1 and Fig. 1). The molecular weights of markers are shown in Figure 1. Of the 13 fragments, 6 are dominantly DP-specific. Three fragments are DR-specific and four are DQ-specific. In the case of cross-hybridization, see Table 2 in DR Beta/EcoRV System report. Considering only the DP-specific bands two allelic banding patterns are observed. There is no obvious association of these cluster patterns with the others.

Participating Laboratories: US5TER,[1] FRAMYE,[2] ANZBSH,[3] US8RJW,[4] US5PAK[5]

Table 1. DP Beta/EcoRV Data

		DPB	ECV	TER		
	kb	Locus	Frequency	% Faint	56	32
1	23.93	1 2 2 0	0.343	1.000		f
2	20.18	0 2 1 0	1.000	1.000		
3	18.39	1 2 2 0	0.486	1.000		
4	14.94	2 1 2 0	0.100	1.000		
5	14.46	2 1 2 0	0.171	1.000		
6	12.38	0 0 1 0	1.000	1.000		
7	10.10	2 2 1 0	0.343	0.000		
8	6.89	2 2 1 0	0.771	0.000		
9	5.24	2 1 3 0	0.214	1.000		
10	3.36	0 0 1 0	0.771	0.125		
11	3.27	0 0 1 0	1.000	1.000		
12	2.59	2 1 2 0	0.143	1.000	f	
13	1.83	1 2 2 0	0.071	1.000		

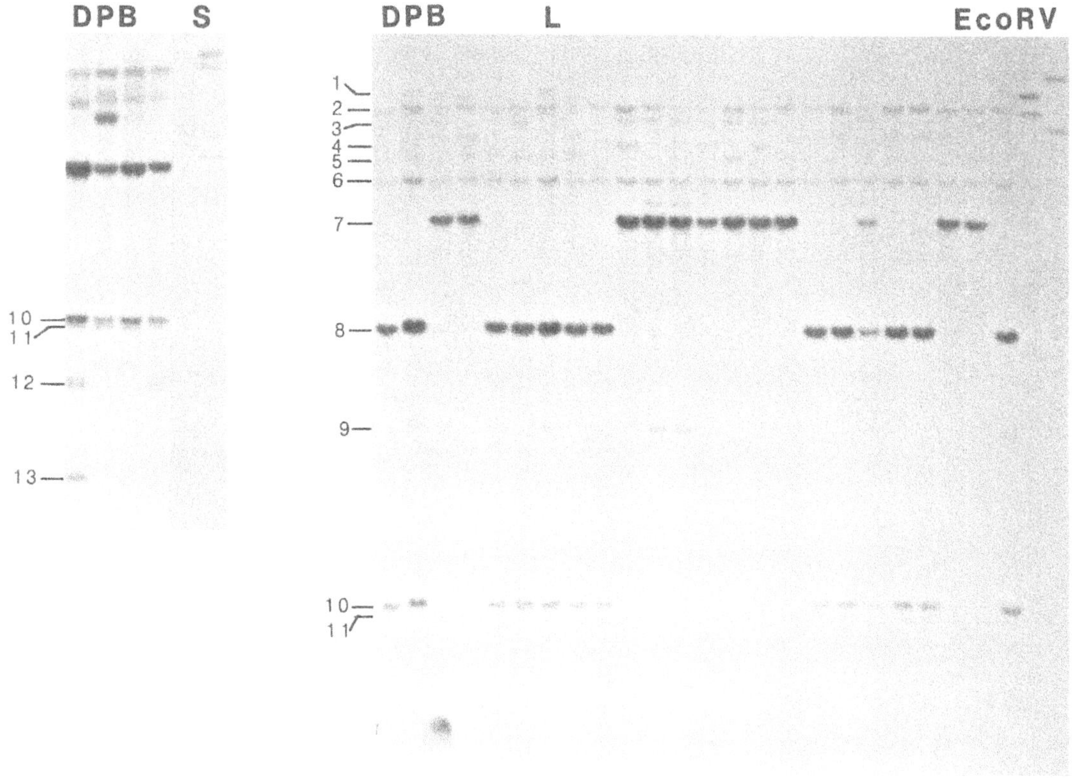

Figure 1. Fragments observed with the DP beta/EcoRV system.

Table 2. Banding Pattern of the DP Beta/EcoRV System in Each Cell Line

	Fragment No.					
Cell	2	6	7	8	10	11
9030 JHAF	8	8	2			8
9034 SAVC	8	8	2			8
9107 LKT 3	8	8	2			8
9018 LO 081785	8	8	2			8
9022 COX	8	8	2			8
9023 VAVY	8	8	2			8
9040 BM15	8	8	2			8
9043 BM21	8	8	2			8
9058 OMW	8	8	2			8
9055 HO 301	8	8	2			8
9063 WT47	8	8	2			8
9060 CBG-IB	8	8	2			8
9059 SLE 005	8	8	2			8
9006 WT10	8	8	2			8
9003 KAS-116	8	8	2			8
9047 PLH	8	8	2			8
9056 KOSE	8	8	1	1	8	8
9056 KOSE	8	8	1	1	1	8
9088 PF 04015	8	8	1	1	1	8
9071 OLGA	8	8	1	1	8	8
9021 RSH	8	8	1	1	8	8
9070 LUY	8	8	1	1	8	8
9007 DEM	8	8	1	1	8	8
9009 KAS-011	8	8	1	1	8	8
9106 MANIKA	8	8	1	1	8	8
9014 M GAR	8	8		2	1	8
9105 FPAF	8	8		2	1	8
9017 WT8	8	8		2	1	8
9013 SCHU	8	8		2	1	8
9010 AMAI	8	8		2	1	8
9008 DO 208915	8	8		2	1	8
9082 HO 104	8	8		2	1	8
9065 HHKB	8	8		2	1	8
9062 WDV	8	8		2	1	8
9045 TUBO	8	8		2	1	8
9028 PE 117	8	8		2	1	8
9025 DEU	8	8		2	1	8
9092 BM92	8	8		2	1	8
9075 DKB	8	8		2	1	8
9029 WT51	8	8		2	1	8
9026 YAR	8	8		2	1	8
9032 BSM	8	8		2	1	8
9032 BSM	8	8		2	1	8
9011 E 4181324	8	8		2	1	8
9052 DBB	8	8		2	1	8
9050 MOU	8	8		2	1	8
9104 DHIF	8	8		2	1	8
9051 PITOUT	8	8		2	1	8
9031 BOLETH	8	8		2	1	8
9016 RML	8	8		2	1	8
9048 LBUF	8	8		2	1	8
9019 DUCAF	8	8		2	1	8
9064 AMALA	8	8		2	1	8
9020 QBL	8	8		2	1	8
9037 SWEIG 007	8	8		2	1	8
9066 TAB 089	8	8		2	1	8
9068 BM9	8	8		2	1	8
9067 BTB	8	8		2	1	8
9038 BM16	8	8		2	1	8
9039 JVM	8	8		2	1	8
9072 S PACHEO	8	8		2	1	8

Table 2. *Continued*

Cell	Fragment No. 2	6	7	8	10	11
9042 TISI	8	8		2	1	8
9069 MADURA	8	8		2	1	8
9036 SPO 010	8	8		2	1	8
9057 TEM	8	8		2	1	8
9054 EK	8	8		2	1	8
9061 3-12-27 AOB	8	8		2	1	8
9004 JES THOM	8	8		2	1	8
9005 HOM-2	8	8		2	1	8
9002 MZ 070782	8	8		2	1	8
9091 MLF	8	8		2	1	8
9033 BM14	8	8		2	1	8

1 = Single intensity; 2 = double intensity; 8 = faint.

RFLP Standardization Report for DP Beta/TaqI

A. Marcadet,[1] A. Piscone,[1] D. Cohen,[1] J.D. Bignon,[2] S. Sekiguchi,[3] C.B. Carpenter,[4] and R.L. Walford[5]

The DP beta/TaqI system was standardized as the DR beta/TaqI system. When considering only specific DPB bands and those cross-hybridizing class II bands that were more intense with DPB probe, 11 fragments were standardized on the core cell lines (Table 1). Seven fragments (nos. 1, 2, 3, 4, 5, 7, and 10) were constant. Important remarks about class II cross-hybridizing fragments have been reported in DR beta/TaqI system, and the reader must refer to this report to analyze DPB blots. In DR beta/TaqI report, a cross-hybridization table summarizes all class II, TaqI cross-hybridizing fragments giving their standardization number and molecular weight (and "name" on the figures).

Table 2 indicates the distribution of the standardized bands on the core cell lines. The seven constant fragments were not listed. According to previous HLA typing, at least 10 cells were not homozygous for HLA-DP. Moreover HLA-DP typing was incomplete for 21 cells. This DP beta/TaqI system does not seem to be an efficient system for analyzing DPw specificities because 1) no specific fragment for DPw specificities can be observed; 2) fragments 4 and 5 are present for all cells, although the intensities of signal differ (the faint intensity can be correlated with DPw1 +DPw3 +DPw5 for fragment 4 and with DPw2 + DPw4 for fragment no. 5; 3) we can not observe specific patterns of standardized bands to distinguish DPw2 from DPw4, and DPw1 from DPw3 and from DPw5; and 4) standard fragment 8 is present for all cells except one (DP−); this cell may be a variant which presents the standardized fragment 9.

Table 1. Standard Bands in the DP Beta/TaqI System

	kb	DPB Locus	TAQ Frequency	COH % Faint	56	32
1	6.37	0 0 1 0	1.000	0.000		
2	5.24	0 0 1 0	1.000	0.000		
3	4.80	2 0 1 0	1.000	0.000		
4	4.64	2 0 1 0	1.000	0.250		
5	4.39	2 0 1 0	1.000	0.686		
6	4.04	2 2 1 0	0.183	0.000		
7	2.92	0 0 1 2	1.000	0.000		
8	2.72	0 0 1 0	0.986	0.000		
9	2.64	0 0 1 0	0.014	0.000		
10	1.82	2 0 1 0	1.000	0.000		
11	1.73	2 0 1 0	0.329	0.000		

Figure 1 is a photograph taken from two films from laboratory FRA/BIG. All standardized fragments are indicated at least once for S and/or L migration. The presence of a standardized fragment is indicated by a hyphen. Cross-hybridizing bands are indicated by lower case letters.

Participating Laboratories: FRACOH,[1] FRABIG,[2] JAPSEK,[3] NCYCAR,[4] US5WAL[5]

Immunobiology of HLA
Volume I

Table 2. RFLP DP Beta/TaqI System: Distribution of Standard Bands in Tenth Workshop Core Cell Lines

Cell lines	HLA DPw	n = 4	5	6	8	9	11
9058	1	8	2		2		2
9023	1	8	2		2		2
9006	1	8	2		2		2
9021	1	2	2		2		2
9088	1/4	2	2		2		2
9070	1/4	2	2		2		2
9063	2	8	2		2		2
9036	2	2	8		2		
9038	2	2	8		2		
9039	2	2	8		2		
9068	2	2	8		2		
9020	2	2	8		2		
9029	2	2	8	9	2		
9105	2	2	8	9	2		
9032	2	2	8	9	2		
9032	2	2	8	2	2		
9050	2	2	8	2	2		
9045	2/4	2	8		2		
9011	2/4	2	8		2		
9008	2/4	2	8		2		
9017	2/4	2	8		2		
9056	2/N	2	8		2		2
9056	2/N	2	9		2		2
9030	3	8	2	9	2		2
9040	3	8	2		2		2
9018	3	8	2		2		2
9022	3	8	2		2		2
9071	3	8	2		2		2
9059	3	8	2		2		2
9007	3/4	2	2		2		2
9009	3/4	2	2		2		2

Cell lines	HLA DPw	n = 4	5	6	8	9	11
9057	4	2	8		2		
9061	4	2	8		2		
9062	4	2	8		2		
9065	4	2	8		2		
9002	4	2	8		2		
9004	4	2	8		2		
9014	4	2	8		2		
9082	4	2	8		2		
9069	4	2	8		2		
9028	4	2	8	9	2		
9033	4	2	8	9	2		
9067	4	2	8	9	2		
9025	4	2	8	2	2		
9075	4	2	8	2	2		
9031	4	2	8	2	2		
9026	4	2	8	2	2		
9051	4	2	8	2	2		
9052	4	2	8	2	2		
9107	5	8	2	9	2		2
9055	5	8	2		2		2
9106	NT	8	9		2		
9005	–	2	8		2		
9010	–	2	8		2		
9013	–	2	8		2		
9016	–	2	8		2		
9037	–	2	8		2		
9042	–	2	8		2		
9054	–	2	8		2		
9064	–	2	8		2		
9066	–	2	8		2		
9072	?	2	8		2		
9019	–	2	8		2		
9091	–	2	8	9	2		
9092	–	2	8	9	2		
9104	NT	2	8	2	2		
9048	–	2	8	2	2		
9034	–	8	2	2	2		2
9047	–	8	2	2	2		2
9060	–	8	2			2	2
9043	–	8	2		2		2
9003	–	8	2		2		2

N = New?

Standardized fragment DPB TaqI 7 (molecular weight = 2.92 kb) is the same fragment as standardized fragment DO beta TaqI 1 (molecular weight = 2.89 kb; locus assignment = 0 0 1 2). Standardized fragment 9 has been positioned on this figure of S migration, but is not present for these four DNA samples (cf L migration). The fragment identified as "a" on S migration represents the standardized fragment DRB TaqI no. 1. The fragment identified as "b" on S migration and "a" on L migration is due to partial digest of genomic DNA.

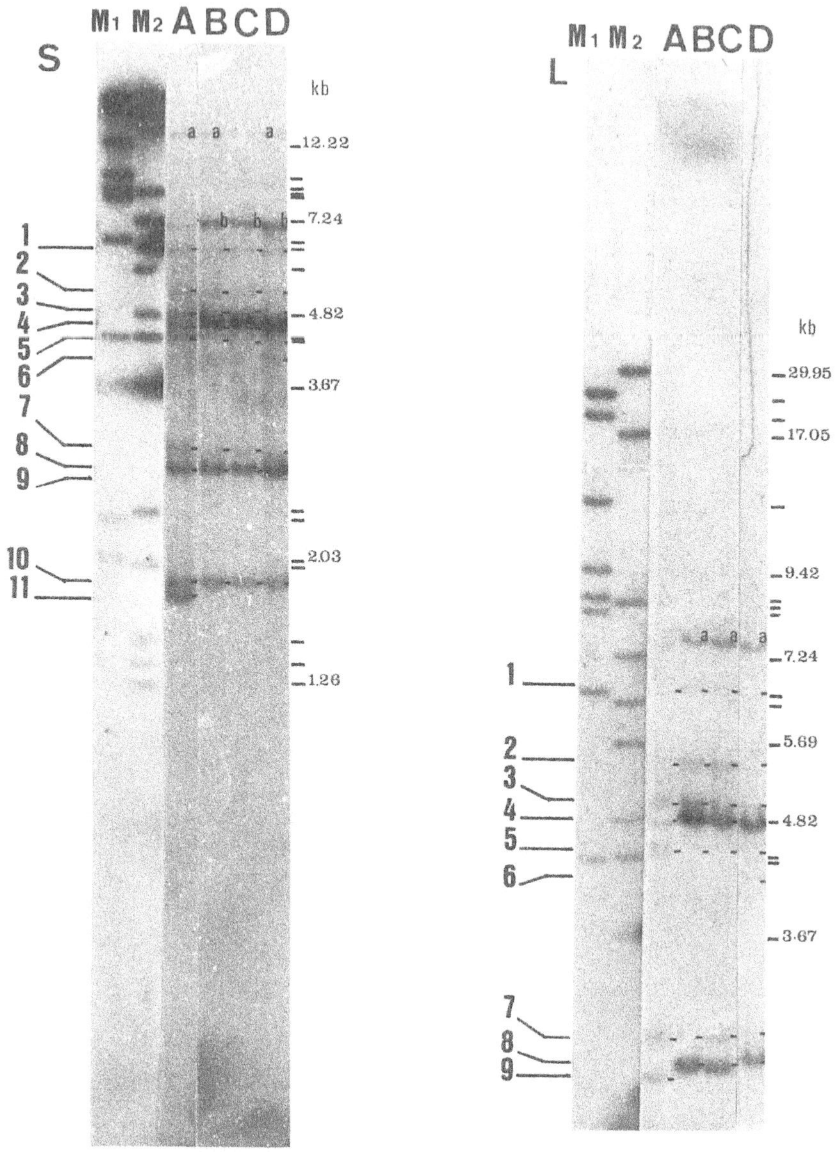

Figure 1. Fragments identified with the DP beta/TaqI system.

RFLP Standardization Report for DP Beta/BglII

R.E. Bontrop,[1] M. Van Eggermond,[1] P. Stastny,[2] H. Takata,[3] P. Rubenstein,[4] and L. Fainboim[5]

Data were received from five different laboratories and standardization was performed based on the following criteria: 1) weak hybridization signals were accepted as standard fragments when they were detected by most laboratories, and 2) discrepancies were resolved by majority consensus. For this system, 18 different standard fragments were observed (Table 1 and in Fig. 1). The first standard fragment has a molecular weight of 20.50 kb and a panel frequency of 100%. This fragment appears to be HLA-DP-specific. The second standard fragment has a molecular weight of 17.49 kb and is observed in 71.4% of the cells tested. This fragment is also detected in the DR beta/BGlII system as standard fragment 1. Standard fragment 3 appears to be HLA-DP-specific and is observed with a panel frequency of 78.6%. This fragment is always present in an abundant staining intensity (Fig. 1). Standard fragments 4 and 5 are also detected with a DR beta probe (see DR Beta/BglII report Table 2) and have a molecular weight of 11.58 and 10.18 kb, respectively. Standard fragment 6 has a molecular weight of 9.38 kb and is also detected with DR and DQ beta probes (see DR Beta/BglII report Table 2). In contrast to this, standard fragment 7 with a molecular weight of 8.99 kb is detected both with DR and DP beta probes, but not with a DQ beta probe (see DR Beta/BglII report Table 2). Fragment 8 seems to be HLA-DP specific and has a molecular weight of 7.28 kb (Table 1). The same situation that is observed for standard fragment 6 is also noticed for standard fragment 9 (Table 2). Standard fragment 10 appears to be DP-specific and has a molecular weight of 6.56 kb and a panel frequency of 17.1%. Again, fragment 11, like fragments 6, 9, 16, and 17, are also detected within the DR beta/BglII and DQ beta/BglII systems (see DR beta/BglII report Table 2). Fragment 12 has a molecular weight of 5.55 kb and a panel frequency of 92.9% (Table 1). Standard fragments 13 and 14 are HLA-DP-specific and have molecular weight of 5.15 and 5.05 kb respectively (Table 1). Fragment 15, which has a molecular weight of 3.89 kb, can also be detected with a DR beta probe (see DR Beta/BglII report Table 2). Fragment 17 has a molecular weight of 2.80 kb and a panel frequency of 22%. Fragment 18 appears to be HLA-DP-specific and is always present when fragment 3 is observed, which is also HLA-DP-specific (Table 1 and Fig. 1).

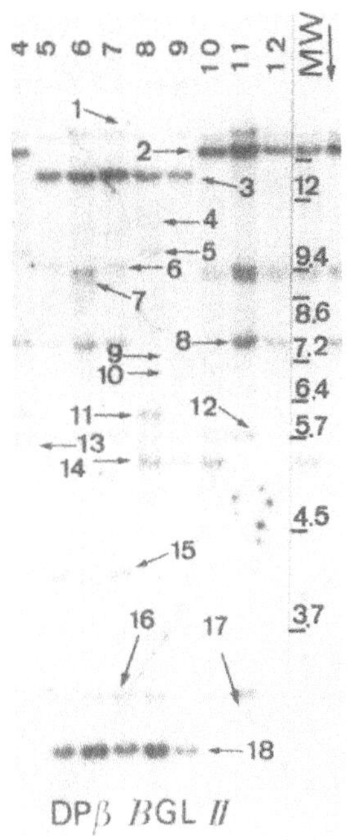

Figure 1. Fragments observed with the DP beta/BglII system.

Participating Laboratories: BENROO,[1] US4STA,[2] JAPTAK,[3] US8RUB,[4] LATFAI[5]

Table 1. DP Beta/BglII Data

Fragment	kb	Locus	Frequency	% Faint	56	32
1	20.50	0010	1.000	0.000		
2	17.49	2010	0.714	0.538		
3	15.05	0010	0.786	0.000		
4	11.58	1020	0.314	0.174		
5	10.18	1020	0.500	0.676		f
6	9.38	1320	0.586	0.048		
7	8.99	1020	0.371	0.000		
8	7.28	0010	0.571	0.548	f	
9	6.97	2130	0.143	0.182		
10	6.56	0010	0.171	0.492		
11	5.93	1120	0.600	0.000		
12	5.55	0010	0.929	0.239		
13	5.15	0010	0.250	0.316		f
14	5.05	0010	0.529	0.350		
15	3.89	1020	0.286	0.350		
16	3.22	2130	1.000	0.000		
17	2.80	2130	0.222	0.069		
18	2.23	0010	0.786	0.289		

Table 2. Fragment Presence/Absence Patterns for the DPBeta/BglII System as Defined on 72 Core Cell Lines

	Fragment No.																	
Cell	1	2	3	4	5	6	7	8	9	10	11	12	13	14	15	16	17	18
9045	1		1			1	1	1							8	1	1	2
9042	1		2			1	1	1							8	1	1	2
9038	1	8	1		8	1	1	1			1	1	1			8	1	1
9070	1	1	1			1		8			1	1	1	1		1		8
9039	1	8	1		8	1	1	8			1	1	1			1		1
9007	1	1	1		8				1	1	1	1	1	1	8	1		1
9014	1		2			1								1	1	1	1	2
9017	1	8	1		8	1		1						1	1	1	1	1
9013	1	8	1		8	1		1							1	1	1	1
9004	1	1	1		8	8	1	1		8						1		1
9010	1		1			1		1								1		2
9037	1		1			1	1	8						8	1	1		1
9068	1		1		8	1		8		1	1	1		1		1		1
9088	1	1	1			1	1							8		1		8
9066	1	8	1		8	1		8	8		1	1		8		1		8
9060	1	8	1		8	1	1									1		8
9016	1	8	1		8	1		8			1	1	1		1	1		8
9065	1	8	1			1								8		1		8
9021	1	1	1			1	1	8								1		8
9075	1	8	1	1						8	1	1	8			1		1
9061	1	8	1			1			1		1	1		1		1		1
9067	1	8	1			1			8		1	1		1		1		1
9008	1	8	1		8	1		8			1	1	1	1	8	1		1
9029	1	8	1	1	1			8			1	1				1		1
9011	1	8	1	8	8	1		8			1	1	1	1	8	1	1	1
9026	1	8	1	8	8						1	1		1		1		1
9056	1	1	1		8	1	1	8	1	8	1	1				1	1	1
9082	1	8	1			1					1	1	8	1	8	2	1	1
9106	1	1	1	1	8						1	1		1		1		8
9032	1	8	1	1	1				8	1	1	1		1		1		1
9072	1		1	8	8			8			1	1		1		1		1
9052	1		1	1						1	1	1		1		1		1
9020	1		1		8	1	1	8			1	1				1		1
9033	1		1	1	1						1	1		1		1		1
9004	1		2						1	1				1		1		2
9005	1		2						1	1				1		1	1	1

	Fragment No.																	
Cell	1	2	3	4	5	6	7	8	9	10	11	12	13	14	15	16	17	18
9050	1		8	2	1			1			1	1				1		2
9104	1		2	1				1			1	1				1		1
9051	1	8	2	8							1	1		1		1		1
9069	1		1								1	1		8		1	8	8
9028	1		1	1	1						1	1	1	1	1	1		1
9019	1	8	1		8	1	1	8	8		1	1				1		1
9025	1	8	1	1	1						1	1	8	8	1	1		8
9031	1	8	1	1	1						1	1		8		1		8
9036	1	8	1		8	1	1	8	1	8	1	1				1	8	1
9002	1	8	1						1		1	1		8		1		1
9071	1	1	1		8	1	1			1	1			8	1	1		8
9064	1	8	1		8	1	1	8					1		1	1		1
9057	1		1		8	1					1	1		8		1		8
9054	1	8	1		8	1	1	8		8	1	1				1		1
9105	1		1		8	1	1				1	1				1		8
9091	1	8	1		1			8					1			1		8
9048	1	8	1	1	8			8					1			1		1
9062	1		1		1	1	8				1	1		8		1		8
9032	1		1	1	1		1	8			1	1	1		1	1		
9009	1	1	1	1							1	1	1	1	8	1		
9092	1		1	1	1			1			1	1				1		
9107	1	1		1	1			1			1	1	1	1		1		
9047	1	1	1		1						1	1		1		1		
9030	1	1		1	1						1	1	8	8	1	8		
9059	1	1				1				1	1	1		1		1	8	
9034	1	1		1	1			8			1	1				1		
9058	1	1				1	1	1								1	8	
9018	1	1				1	1							1	1	1		
9055	1	2				1	1	1								1	1	
9063	1	2				1	1	1								1		
9022	1	2				1	1	1						1		1		
9043	1	2				1	1	1							1	1		
9023	1	1				8	1	1								1		
9006	1	2						1	1	1						1		
9003	1	1						1	1	8	1	1				8		

RFLP Standardization Report for DP Beta/PvuII

K. Kato,[1] A. Smith,[2] J. Partanen,[3] P. Piccioli,[4] B. Dupont,[1] and S.Y. Yang[1]

Eighteen standard bands were standardized with the DP beta/PvuII system as shown in Table 1. All bands present in Figure 1 are counted as standard bands since they were consistently demonstrated. Seven bands—no. 1 (10.64 kb), no. 2 (6.33 kb), no. 11 (3.80 kb), no. 15 (2.05 kb), no. 16 (1.75 kb), no. 17 (1.03 kb) and no. 18 (0.63 kb)—were nonpolymorphic. Four bands—no. 1 (10.64 kb), 2 (6.33 kb), no. 11 (3.80 kb), and no. 15 (2.05 kb)—showed no cross-hybridization with other beta probes and should be considered DP beta-specific. Cross-hybridizing bands are shown in Table 2 in the DR Beta/PvuII report. No clear correlations between any of these fragments with the DP specificities were observed. No previous reports have obtained RFLP polymorphism at the DP beta locus obtained with PvuII. The RFLP pattern of the DP beta PvuII system is illustrated in Figure 2.

Participating Laboratories: US7DUP,[1] US7HAN,[2] SCAKOS,[3] IT2FER[4]

Table 1. Standard Bands of DP Beta/PvuII

		DPB	PVU	DUP		
	kb	Locus	Frequency	% Faint	56	32
1	10.64	0 0 1 0	1.000	0.000		
2	6.33	0 0 1 0	1.000	0.000		
3	6.11	1 2 2 0	0.286	0.095		
4	5.80	1 0 2 0	0.029	0.500		
5	5.68	1 1 1 0	0.143	0.000		s
6	5.17	1 1 1 0	0.300	0.091		
7	4.54	2 1 2 0	0.114	0.125		
8	4.26	2 1 2 0	0.129	0.111		
9	4.25	1 0 2 0	0.486	0.029		
10	4.15	1 1 1 0	0.200	0.429		
11	3.80	0 0 1 0	1.000	0.028		
12	3.59	2 1 3 0	0.914	0.152		
13	3.38	2 1 2 0	0.371	1.000	f	
14	2.39	3 1 2 0	0.286	0.333	f	
15	2.05	0 0 1 0	1.000	0.000		
16	1.75	1 2 1 0	1.000	0.014		
17	1.03	2 1 3 0	1.000	1.000		
18	0.63	1 1 1 0	1.000	1.000		

From left to right, Band no., band Kb, locus assignment of DR, DQ, DP, and DO, band frequency, frequency of faint bands, and hidden duplications.

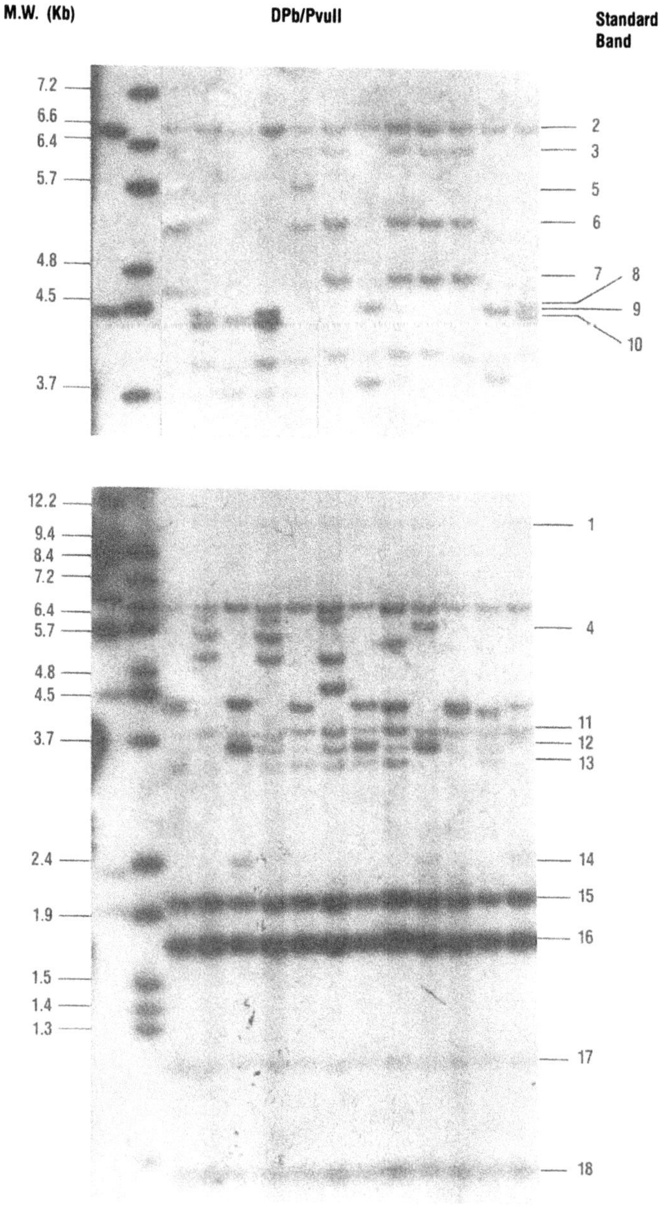

Figure 1. DP beta/PvuII top:long migration, bottom:short migration.

Cell lines	1	2	3	4	5	6	7	8	9	10	11	12	13	14	15	16	17	18
9107 LKT 3	2	2	2		2	2					2	2	8		2	2	8	8
9028 PE 117	2	2	2		2	2					2	2			2	2	8	8
9025 DEU	2	2	2		2	2					2	2		8	2	2	8	8
9031 BOLETH	2	2	2		2	2					2	2			2	2	8	8
9032 BSM	2	2	2		2	2					2	2	8		2	2	8	8
9034 SAVC	2	2	2		2	2					2	2	8		2	2	8	8
9030 JHAF	2	2	2		2	2					2				2	2	8	8
9091 MLF	2	2	2		2	2			2		2				2	2	8	8
9092 BM92	2	2	2		2	2					2	8	8		2	2	8	8
9038 BM16	2	2	8		2	2					2	8	8		2	2	8	8
9050 MOU	2	2	2			2	2				2			1	2	2	8	8
9048 LBUF	2	2	2			2	2				2			1	2	2	8	8
9104 DHIF	2	2	2			2	2				2	8		1	2	2	8	8
9051 PITOUT	2	2	2			2	2				2	8		1	2	2	8	8
9047 PLH	2	2	2			2	2				2	8		1	2	2	8	8
9075 DKB	2	2	2			2	1				2	2	8		2	2	8	8
9052 DBB	2	2	2			2	1				2	2	8		2	2	8	8
9032 BSM	2	2	2			2					2	2	8		2	2	8	8
9029 WT51	2	2	2			2					2	2	8		2	2	8	8
9026 YAR	2	2	2			2					2	2	8		2	2	8	8
9058 OMW	2	2							2		2	2			2	2	8	8
9042 TISI	2	2							2		2	2			2	2	8	8
9043 BM21	2	2							2		2	2			2	2	8	8
9056 KOSE	2	2							2		2	2			2	2	8	8
9040 BM15	2	2							2		2	2			2	2	8	8
9037 SWEIG 007	2	2							2		2	2			2	2	8	8
9060 CB6B	2	2							2		2	2			2	2	8	8
9064 AMALA	2	2							2		2	2			2	2	8	8
9039 JVM	2	2							2		2	2			2	2	8	8
9065 HHKB	2	2							2		2	2			2	2	8	8
9062 WDV	2	2							2		2	2			2	2	8	8
9059 SLE-005	2	2							2		2	2		8	2	2	8	8
9070 LUY	2	2							2		2	2		8	2	2	8	8
9056 KOSE	2	2							2		2	2	8	8	2	2	8	8
9069 MADURA	2	2							2		2	2	8		2	2	8	8
9021 RSH	2	2							2		2	2	8		2	2	8	8
9068 BM9	2	2							2		2	2	8		2	2	8	8
9061 3-12-27 AOB	2	2							2		2	2	8		2	2	8	8
9067 BTB	2	2							2		2	2	8		2	2	8	8
9057 TEM	2	2							2		2	2	8		2	2	8	8
9036 SPO 010	2	2							2		2	2	8		2	2	8	8
9071 OLGA	2	2							2		2	2	8		2	2	8	8
9018 LO 081785	2	2							2		2	2		1	2	2	8	8
9055 HO 301	2	2							2		2	2		1	2	2	8	8
9063 WT47	2	2							2		2	2		1	2	2	8	8
9022 COX	2	2							2		2	2		1	2	2	8	8
9023 VAVY	2	2							2		2	2		1	2	2	8	8
9019 DUCAF	2	2							2		2	2		1	2	2	8	8
9088 PF 04015	2	2							2		2	2		1	2	2	8	8
9020 QBL	2	2							2		2	2		1	2	2	8	8
9066 TAB 089	2	2							2		2			1	2	2	8	8
9016 RML	2	2						2	2	2	2				2	2	8	8
9014 M GAR	2	2						2		2	2	2			2	2	8	8
9013 SCHU	2	2						2		2	2	2			2	2	8	8
9064 AMAI	2	2						2		2	2	2			2	2	8	8
9017 WT8	2	2						2		2	2	2			2	2	8	8
9006 WT100 BIS	2	2								2	2	2	8		2	2	8	8
9005 HOM-2	2	2								2	2	2	8		2	2	8	8
9004 JES THOM	2	2								2	2	8	8		2	2	8	8
9008 DO 208915	2	2						1		8	2	2			2	2	8	8
9082 HO 104	2	2						1		8	2	2			2	2	8	8
9045 TUBO	2	2		8					8		2	2			2	2	8	8
9054 EK	2	2									2	2	8		2	2	8	8
9105 FPAF	2	2									2	2			2	2	8	8
9038 BM16	2	2		2							2	2		8	2	2	8	8
9106 MANIKA	2	2	8			8	8				2	2		8	2	8	8	8
9003 KAS-116	2	2								8	2	2	8		2	2	8	8
9011 E 4181324	2	2						1		8	2	8		8	2	2	8	8
9007 DEM	2	2				8					2	8	8		2	2	8	8
9002 MZ 070782	2	2								8	8	2	8		2	2	8	8
9009 KAS-011	2	2						8		8	8	8	8		2	2	8	8
9072 S PACHEO	2	2							2		2	8	8		2	2	8	8

Figure 2. RFLP pattern of DP beta/PvuII system.

RFLP Standardization Report for DP Beta/EcoRI

S. Tonks,[1] V. Groves,[1] H. Betuel,[2] J.L. Bidwell,[3] M.L. Garavoy,[4] R. Lechler,[5] S. Naito,[6] J. Bodmer,[1] and J. Trowsdale[1]

The majority of the 14 fragments identified in the DP beta/EcoRI system were cross-hybridizing with DRB and DQB (Table 3 in the DR Beta/EcoRI report). The DP beta-specific bands were numbers 2,6,7, and 12 (Table 1 and Fig. 1), the latter being a constant band. Six different banding patterns were also identified. Only fragment number 7 (7.20 kb) was significantly correlated with a specific DP type (DPw2; $\chi^2=15.166$). EcoRI appears to have limited value in DP typing using RFLP. The discordance between hidden duplicates in this system (fragment 13) was caused by lack of resolution of close bands.

Participating Laboratories: UKIBOD,[1] FRABET,[2] UKIBID,[3] US4GVY,[4] UKIBAT,[5] JAPNAI[6]

Table 1. Banding Pattern of the DP Beta/EcoRI System in Each Cell Line

WS ID No.	Cell	Fragment No. 2	6	7	12
9091	MLF	1	1		1
9092	BM92	1	1		1
9052	DBB	1	2		1
9107	LKT3	1	2		1
9042	TISI	8	2		1
9028	PE117	2	2		1
9025	DEU	1	2		1
9031	BOLETH	1	1		1
9030	JHAF	1		1	1
9029	WT51	8		1	1
9033	BM14	1		1	1
9043	BM21	1		1	1
9057	TEM		1		1
9054	EK		1		1
9016	RML		1		1
9065	HHKB		1		1
9021	RSH		2		1
9070	LUY		1		1
9082	H0104		1		1
9072	SPACH		1		1
9004	JESTHOM		1		1
9055	H0301		1		1
9023	VAVY		1		1
9006	WT100B/S		1		2
9005	HOM2		1		1
9051	PITOUT		2		2
9040	BM15		2		1
9010	AMAL		2		1
9069	MADURA		1		1
9037	SWE1G007		1		1
9088	PFO4015		1		1
9002	MZ070782		1		1
9064	AMALA		1		1
9105	FPAF			1	1
9048	LBUF			1	1
9032	BSM			1	1
9059	SLE005			1	1
9034	SAVC			1	1
9068	BM9			1	1
9075	DKB			1	1

Table 1. *Continued*

WS ID No.	Cell	Fragment No. 2	6	7	12
9038	BM16			1	1
9008	D0208915			1	1
9003	KAS116			1	1
9029	WT51			1	1
9011	E4181324			1	1
9039	JVM			1	1
9020	QBL			1	1
9017	WT8			1	1
9018	LO081785			1	1
9063	WT47			1	1
9022	COX			1	1
9050	MOU			1	1
9056	KOSE			1	1
9104	DHIF			1	1
9047	PLH			1	2
9019	DUCAF			1	1
9066	TAB089			1	1
9060	CB6B			1	1
9036	SPO010			1	1
9062	WDV		1	1	1
9009	KAS011		8	8	1
9061	31277ABO		1	1	1
9026	YAR		1	1	1
9106	MANIKA		1	1	1
9007	DEM		2	1	1
9045	TUBO		1	1	1
9013	SCHU		1	1	1
9071	OLGA		1	1	1
9029	WT51	8		1	1
9043	BM21	1		1	1

1 = Single intensity; 2 = double intensity; 8 = faint.

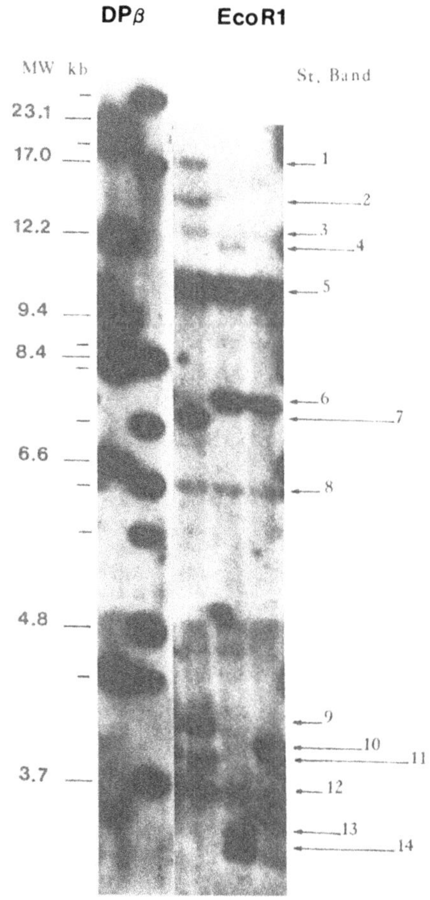

Figure 1. ◄

RFLP Standardization Report for DP Beta/HindIII

G. Paulsen,[1] G. Markussen,[1] B.O. Barger,[3] R. Fauchet,[2] M.G. Hammond,[5] and J.M. Tiercy[4]

Thirty-two bands were identified by RFLP in the DP beta/HID system as shown in Table 1 and in Figure 1. The distribution among core cell lines of 16 bands with high DPB specificity is shown in Table 2. A cross-hybridization table for the HLA class II/HID system can be found in the DR Beta/HID report.

Participating Laboratories: SCATSB,[1] FRAFAU,[2] US5UAB,[3] FRAJEA,[4] SAFHAM[5]

Immunobiology of HLA
Volume I

Table 1. Standard Bands in the DP Beta/HID System

Band	kb	Locus[a]	Frequency[b]	% Faint[c]	56[d]	32[d]
1	13.07	0 1 2 0	0.086	1.000		
2	12.57	0 0 1 0	1.000	1.000		
3	11.66	0 1 2 0	0.086	1.000		
4	9.76	1 2 2 0	0.014	1.000		
5	9.64	1 0 2 0	0.014	1.000		
6	8.59	1 0 2 0	0.257	1.000		
7	7.05	2 3 1 0	1.000	0.000		
8	5.47	0 0 1 0	1.000	0.000		
9	5.43	0 2 1 0	0.586	0.000		s
10	5.06	0 2 1 0	0.543	0.000		
11	4.36	0 0 1 0	1.000	1.000		
12	4.10	0 0 1 0	0.014	1.000		
13	4.00	0 0 1 0	0.986	1.000		
14	3.50	1 2 1 0	0.071	0.000		
15	3.42	0 1 2 0	0.171	1.000		
16	3.37	1 0 2 0	0.114	0.000		
17	3.31	0 1 2 0	0.643	1.000		
18	3.24	0 1 2 0	1.000	1.000		
19	3.09	1 2 3 0	0.071	1.000		
20	2.99	1 0 2 0	0.514	1.000		
21	2.90	1 0 1 0	0.514	1.000	f	
22	2.83	1 0 2 0	0.171	0.769	f	
23	2.51	1 3 2 0	0.385	1.000		
24	2.50	1 3 2 0	0.442	1.000		
25	1.86	0 0 1 0	0.986	0.000		
26	1.81	0 0 1 0	0.014	0.000		
27	1.68	1 2 1 0	0.342	0.000		
28	1.65	1 2 1 0	0.414	0.000		
29	1.60	1 2 1 0	0.171	0.000		
30	1.48	1 2 1 0	0.300	0.000		
31	1.32	2 1 2 0	1.000	1.000		
32	1.17	2 1 2 0	1.000	1.000		

[a]Locus assignment in the order DRB, DQB, DPB, DOB; 1 indicates high, 2 and 3 lower specificity.
[b]Band frequency in the core cell lines.
[c]Frequency of faint bands.
[d]Discrepancies in faint bands (f) or strong bands (s) of the hidden duplicates (WS ID 9056 and 9032).

Table 2. Distribution of Bands With High DPB Specificity in the DP Beta/HID System

WS ID	DR	DP	Band No. 02	07	08	09	10	11	12	13	14	21	25	26	27	28	29	30
9006	1	1	8	2	2	2		8		8	2		2					
9004	1	4	8	2	2	2		8		8	2		2					
9005	1	–	8	2	2	2		8		8	2		2					
9002	1	4	8	2	2	2		8		8	2		2					
9058	13	1	8	2	2	2		8		8			2		1			
9040	11	3	8	2	2	2		8		8		8	2		1			
9082	15	4	8	2	2	2		8		8			2		1			
9065	13	4	8	2	2	2		8		8			2		1			
9055	6	5	8	2	2	2		8		8			2		1			
9010	15	–	8	2	2	2		8		8			2		1			
9037	11	–	8	2	2	2		8		8		8	2		1			
9042	11	–	8	2	2	2		8		8		8	2		1			
9054	14	–	8	2	2	2		8		8		8	2		1			
9021	3	1	8	2	2	2		8		8			2			1		
9023	3	1	8	2	2	2		8		8			2			1		
9014	15	4	8	2	2	2		8		8			2			1		
9067	8	4	8	2	2	2		8		8			2			1		
9069	8	4	8	2	2	1		8		8			2			1		
9088	3	1,4	8	2	2	2		8		8			2			1		
9070	8	1,4	8	2	2	2		8		8			2			1		
9016	16	–	8	2	2	2		8		8			2			1		
9051	7	4	8	2	2	2		8		8		8	2			1		1
9052	7	4	8	2	2	2		8		8		8	2			1		1
9025	4	4	8	2	2	1		8		8		8	2				1	1
9028	4	4	8	2	2	2		8		8		8	2				1	1
9031	4	4	8	2	2	2		8		8		8	2				1	1
9107	4	5	8	2	2	2		8		8		8	2				1	1
9091	4	–	8	2	2	2		8		8		8	2				1	1
9092	4	–	8	2	2	2		8		8		8	2				1	1
9064	14	–	8	2	2	2		8		8			2			1		
9057	14	4	8	2	2	2		8		8		8	2		1			
9072	8		8	2	2	1		8		8		1	2			1		
9062	13	4	8	2	2	1	2	8		8			2		1			
9011	15	2,4	8	2	2	2	2	8		8			2		1			
9045	11/12	2,4	8	2	2	1	2	8		8		8	2			1		
9009	2	3,4	8	2	2	1	2	8		8			2			1		
9013	15	–	8	2	2	2	2	8		8			2			1		
9007	4/16	3,4	8	2	2	2	2	8		8		8	2			1		1
9106	7	–	8	2	2	1	2	8		8		8	2			1		1
9032	4	2	8	2	2	1	2	8		8		8	2				1	1
9032	4	2	8	2	2		2	8		8		8	2				1	1
9020	3	2	8	2	2		2	8		8		8	2		1			
9036	11	2	8	2	2		2	8		8		8	2		1			
9038	12	2	8	2	2		2	8		8		8	2		1			
9039	11	2	8	2	2		2	8		8		8	2		1			
9063	13	2	8	2	2		2	8		8			2		1			
9059	13	3	8	2	2		2	8		8			2		1			
9061	14	4	8	2	2		2	8		8		8	2		1			
9019	3	–	8	2	2		2	8		8		8	2		1			
9043	11	–	8	2	2		2	8		8		8	2		1			
9056	13/14		8	2	2		2	8		8			2		1			
9056	13/14		8	2	2		2	8		8		8	2		1			
9075	9	4	8	2	2		2	8		8		8	2		1			1
9068	8	2	8	2	2		2	8		8			2			1		
9018	3	3	8	2	2		2	8		8		8	2			1		
9022	3	3	8	2	2		2	8		8			2			1		
9105	11	2	8	2	2		2	8	8			8	2			1		
9071	8	3	8	2	2		2	8		8			2			1		
9008	15	2,4	8	2	2		2	8		8			2			1		
9017	15	2,4	8	2	2		2	8		8			2			1		
9066	8	–	8	2	2		2	8		8			2			1		
9104	11		8	2	2		2	8		8		8	2			1		1
9047	7	–	8	2	2		2	8		8		8	2			1		1
9048	7	–	8	2	2		2	8		8		8	2			1		1
9050	7	2	8	2	2		2	8		8		8	2			1		1
9029	4	2	8	2	2		2	8		8		8	2				1	1
9030	4	3	8	2	2		2	8		8		8	2				1	1
9026	4	4	8	2	2		2	8		8		8	2				1	1
8033	4	4	8	2	2		2	8		8		8	2				1	1
9034	4	–	8	2	2		2	8		8		8	2				1	1
9003	1	–	8	2	2		2	8		8	2		2					
9060	13	–	8	2	2		2	8		8		8		1	1			
			02	07	08	09	10	11	12	13	14	21	25	26	27	28	29	30

Bands designation: 1–2 = increasing intensity, 8 = faint.

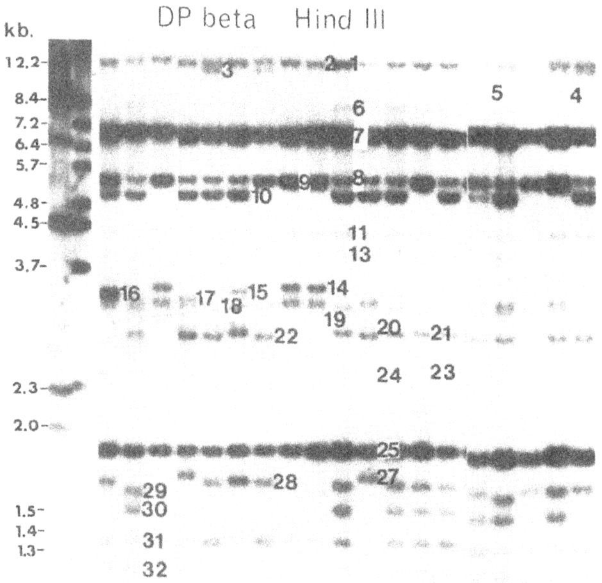

Figure 1. RFLP in the DP beta/HID system of 19 DNA samples run in 0.9% agarose gel. Representative core bands are shown in the figure numbered on their right side. The WS ID of the core cell lines are from the left following marker 1 and marker 2: 9013, 9033, 9004, 9063, 9022, 9043, 9023, 9006, 9005, 9050, 9056, 9104, 9051, 9047, 9106, 9032, 9072, 9052, and 9020.

RFLP Standardization Report for DP Beta/MspI

M. Segall,[1]* L. Schluender,[1]* A. Arnaiz-Villena,[2] N. Kashiwagi,[3] and C. Muller[4]

DP beta/MspI blots were received from four laboratories and were analyzed as detailed in the report on DR Beta/MspI. The DP beta/MspI system identified 20 fragments on the core cell lines (Figs. 1 and 2). Nine of these fragments were unique to DPB, and three were cross-hybridizing but strongest with DPB, as shown in Table 2 of the report on DR Beta/MspI. Unique fragments are are not shown in that table. DPB-dominant or unique fragments are indicated in Figures 1 and 2 with a dot next to the fragment number.

Discrepancies for the "hidden duplicates" (9056 and 9032) were found in bands identified by only two or three of the participating laboratories and generally involved faint bands. Fragments 17 and 18, which were both unique to DPB, were separable on one lab's blots and are certainly a doublet, but could not always be clearly distinguished on the blots of other labs.

Presence/absence data for the polymorphic DPB fragments are shown in Table 1; the cells are arranged in order by DPw specificity. Fragments 2 and 3 both were positive in 6/8 DPw1 (both were negative with cell #9022); fragment 2 also was positive with 1/2 DPw5. Fragment 13 was positive with 16/16 DPw2 and 1 DPw4.

Participating Laboratories: US2BAC,[1] FRAARN,[2] JAPKSH,[3] GERMUC[4]

*This is publication #496 from the Immunobiology Research Center, University of Minnesota, Minneapolis, Minnesota 55455, USA. Supported in part by Juvenile Diabetes Foundation Grant #186175 and March of Dimes Birth Defects Foundation Grant #6-496.

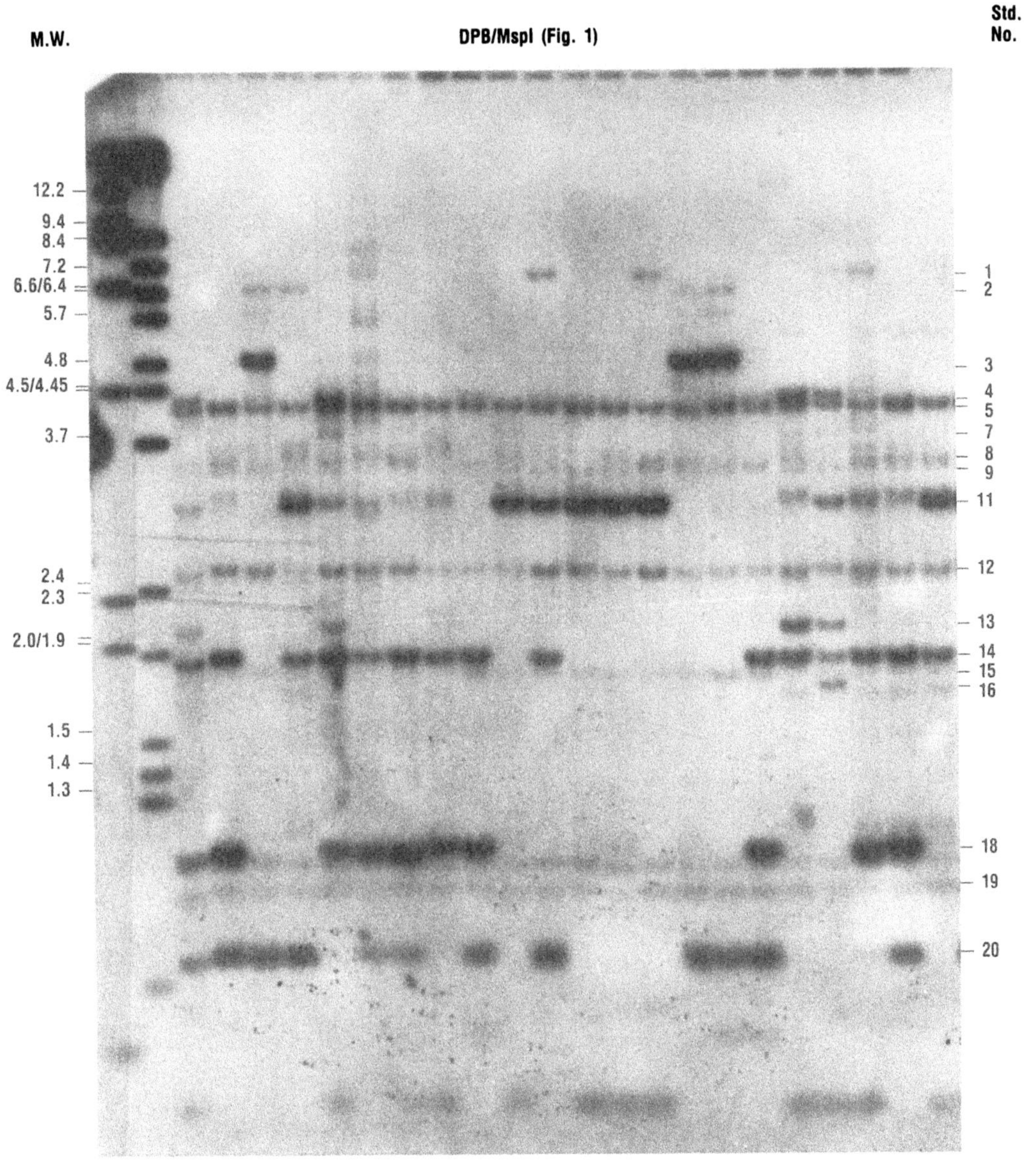

Figure 1. Fragments identified with the DP beta/MspI system.

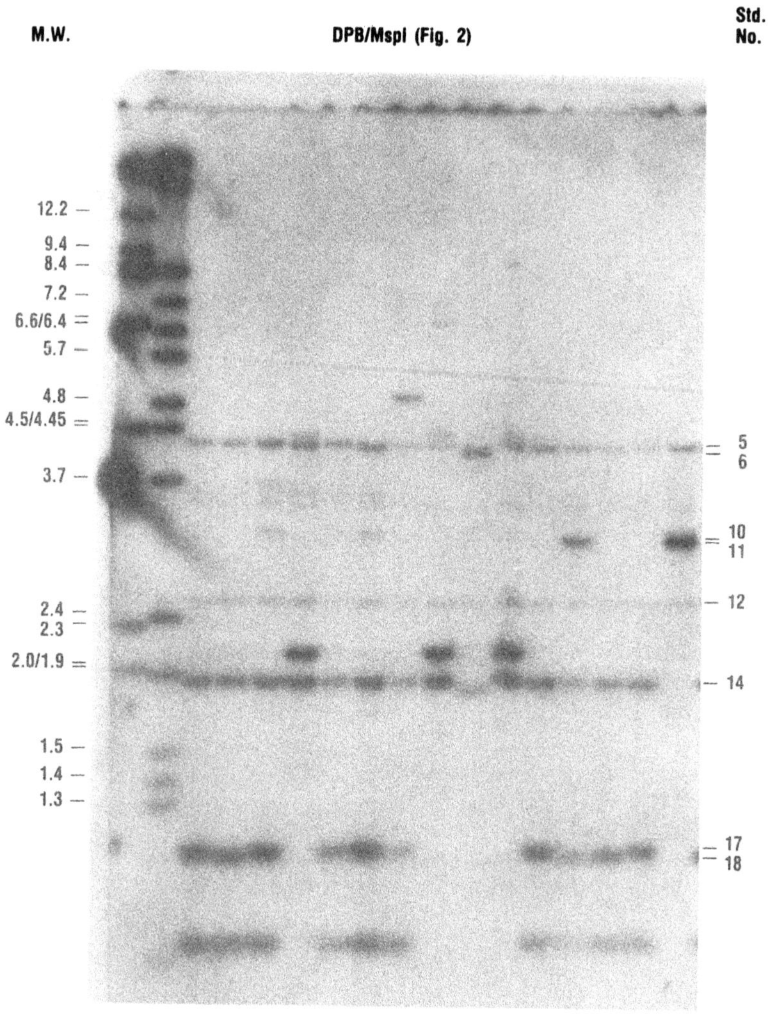

Figure 2. Fragments identified with the DP beta/MspI system.

Table 1. Distribution of Polymorphic Fragments in Core Cell Lines, DPBeta/MspI System

		Standard Fragment Number											
Cell #	Name	1	2	3	5	6	8	11	13	14	17	18	20
9034	SAVC	1			1		8	2					
9047	PLH				1			2		2			
9064	AMALA				1					1		1	1
9066	TAB089		1		1			8	2	1			
9092	BM92				1		8	8		2		2	2
9005	HOM2				1					2		2	2
9016	RML				1			8		1		1	1
9019	DUCAF				1		8		2	1			
9010	AMAL				1		8			2		2	2
9003	KAS116	1			1			2				1	
9091	MLF				1			8		1		1	1
9037	SWEIG007				1		8			1		2	1
9048	LBUF	1			1			8		1		2	
9054	EK				1					1		1	1
9043	BM21	1			1		8	2					
9021	RSH		1	2	1		8			1	1	1	1
9023	VAVY		1	3	1		8						2
9022	COX				1		8	2					
9006	WT100BIS		1	3	1								2
9058	OMW		1	3	1								2
9060	CB6B		8		8	2							
9070	LUY		1	1	1					1		1	1
9088	PFO4015			2	1					1	1	1	1
9029	WT51				1				2	1			
9038	BM16				8		8		2	1		2	
9039	JVM				1				2	1			
9020	QBL				8				2	1			
9050	MOU				2			1	2	2			
9036	SP0010				1				2	1			
9063	WT47				1			2					
9068	BM9				1		8		2	1			
9105	FPAF	1			8				1	1			
9032	BSM				1				2	1			
9032	BSM				1			8	2	2			
9011	E4181324				1				2	1		1	1
9017	WT8				1				2	2		2	

		Standard Fragment Number											
Cell #	Name	1	2	3	5	6	8	11	13	14	17	18	20
9008	D0208915				1				1	1		1	
9045	TUBO				1				2	1		1	2
9056	KOSE				1			1	1	2			
9056	KOSE				2			1	2	1			
9071	OLGA				1			1		1		1	1
9042	TISI				1		1			2		2	2
9040	BM15				1			2					2
9072	SPACH				1					1		1	1
9018	L0081785				1		8	2					
9059	SLE005				1			2					
9030	JHAF				1			3				8	
9009	KAS011				1			1		1		1	1
9007	DEM				1			1		1		1	1
9067	BTB				1		8			1		2	2
9075	DKB				1		8			1		2	2
9025	DEU				1		8	1		1	1	2	1
9026	YAR				1		8	8		1		1	
9028	PE117				1		8	1		2	1	2	1
9082	HO104				1					1		1	1
9014	MGAR				1		8			2		2	2
9013	SCHU				1					2		2	2
9069	MADURA				1					1		2	1
9004	JESTHOM				1					2		2	2
9002	MZO70782				1					1		2	1
9065	HHKB				1					1	1	2	1
9062	WDV				1				1	1	1	1	1
9052	DBB				1			1		2		2	1
9031	BOLETH				1			1		1	1	2	1
9061	31227ABO				1					1		2	2
9057	TEM				1					1	1	1	1
9033	BM14				1					2		2	
9051	PITOUT	1								2		2	2
9107	LKT3		1		1			2		2			2
9055	HO301	1			1		8	2		2			2
9104	DHIF	1			1		8	1		2		2	
9106	MANIKA				1			1		1		8	1

RFLP Standardization Report for DP Beta/BamHI

J.J. Hyldig-Nielsen,[1] B.O. Boehm,[2] K.J. Wood,[3] A.R. Bushell,[3] D.P. Singal,[4] and A. Svejgaard[1]

The DP beta/BamHI system identified the 15 fragments (Table 1). Two of these fragments, (nos. 14 and 15) were not included in the analysis (no locus assignment) as they were very faint and seen only on a few films. A possible reason for these fragments could be weak cross-hybridization to another locus. Of the remaining 13 fragments, five hybridized exclusively or most strongly with the DP beta probe, while one fragment was equally strong with both the DPB and the DQB probes (see Table 2 of the DR beta/BamHI report). Four fragments hybridized most strongly with the DRB probe and three fragments most strongly with the DQB probe (see Table 1).

Table 2 gives the banding patterns obtained with the DP beta/BamHI System. St. fragment no. 3, which is one of the two polymorphic DPB fragments, has the same size as a nonpolymorphic, cross-hybridizing DQB fragment (DQB St. fragment 6). The weak fragments seen in lanes 1, 2, 3, and 5 in Figure 1 are due to this cross-hybridization. Although this fragment has been excluded from the further analysis of the results, we have indicated its distribution in the Workshop cell lines.

Participating Laboratories: SCASVE,[1] GERBOE,[2] UKIMOR,[3] US5SIN[4]

Table 1. Fragments Identified with the DP Beta/BamHI System

		DPB	BAM	SVE		
	kb	Locus	Frequency	% Faint	56	32
1	11.13	0 0 1 0	1.000	0.000		
2	10.79	3 1 2 0	0.257	0.526		
3	9.36	0 1 1 0	1.000	0.429		
4	9.00	0 0 1 0	0.529	0.000		
5	8.08	2 3 1 0	1.000	0.000		
6	7.20	0 0 1 0	1.000	1.000		
7	5.23	1 3 2 0	0.186	0.000		
8	4.50	0 2 1 0	1.000	0.000		
9	4.13	1 3 2 0	0.743	0.000		
10	3.85	1 0 2 0	0.357	0.000		
11	3.74	1 3 2 0	0.357	0.000		
12	3.47	0 1 2 0	0.171	0.000		
13	3.01	0 1 2 0	0.286	1.000	f	
14	2.87	0 0 0 0	0.000	0.000		
15	2.38	0 0 0 0	0.000	0.000		

Table 2.

WS ID #	Name	HLA-DR	HLA-DQ	HLA-DP	Standard fragment # 1	3	4	5	6	8
9002	MZ070782	1	5	4	6	6		6	6	6
9003	KAS116	1	5		6		6	6	6	6
9004	JESTHOM	1	5	4	6	6		6	6	6
9005	HOM2	1	5		6	6		6	6	6
9006	WT100BIS	1	5	1	6	6		6	6	6
9007	DEM	16,4	5,8	3,4	6	6	6	6	6	6
9008	DO208915	15	6	2,4	6		6	6	6	6
9009	KAS011	2	1	3,4	6	6	6	6	6	6
9010	AMAI	15	6		6	6		6	6	6
9011	E4181324	15	6	2,4	6	6	6	6	6	6
9013	SCHU	15	6		6	6	6	6	6	6
9014	MGAR	15	6	4	6	6		6	6	6
9016	RML	16	7		6	6		6	6	6
9017	WTB	15	1	2,4	6		6	6	6	6
9018	LO081785	3	2	3	6		6	6	6	6
9019	DUCAF	3	2		6		6	6	6	6
9020	QBL	3	2	2	6		6	6	6	6
9021	RSH	3	4	1	6	6		6	6	6
9022	COX	3	2	3	6		6	6	6	6
9023	VAVY	3	2	1	6	6		6	6	6
9025	DEU	4	7	4	6	6		6	6	6
9026	YAR	4	8	4	6		6	6	6	6
9028	PE117	4	8	4	6	6		6	6	6
9029	WT51	4	8	2	6		6	6	6	6

Table 2. *Continued*

WS ID #	Name	HLA-DR	HLA-DQ	HLA-DP	Standard fragment # 1	3	4	5	6	8
9030	JHAF	4	7	3	6		6	6	6	6
9031	BOLETH	4	8	4	6	6		6	6	6
9032	BSM	4	8	2	6		6	6	6	6
9032	BSM	4	8	2	6		6	6	6	6
9033	BM14	4	8	4	6		6	6	6	6
9034	SAVC	4	8		6		6	6	6	6
9036	SPO010	11	5	2	6		6	6	6	6
9037	SWEIG007	11	7	6	6	6		6	6	6
9038	BM16	12	7	2	6		6	6	6	6
9039	JVM	11	7	2	6		6	6	6	6
9040	BM15	11	7	3	6	6		6	6	6
9042	TISI	11	7		6	6		6	6	6
9043	BM21	11	7		6		6	6	6	6
9045	TUBO	11,12	7	2,4	6	6	6	6	6	6
9047	PLH	7	2		6		6	6	6	6
9048	LBUF	7	2		6		6	6	6	6
9050	MOU	7	2	2	6		6	6	6	6
9051	PITOUT	7	4	4	6	6		6	6	6
9052	DBB	7	9	4	6	6		6	6	6
9054	EK	14	5		6	6		6	6	6
9055	HO301	6	6	5	6	6		6	6	6
9056	KOSE	13,14	5	2	6		6	6	6	6
9056	KOSE	13,14	5	2	6		6	6	6	6
9057	TEM	14	1	4	6	6		6	6	6
9058	OMW	13	6	1	6	6		6	6	6
9059	SLE005	13	6	3	6		6	6	6	6
9060	CB6B	13	5		6		6	6	6	6
9061	31227ABO	14	5	4	6		6	6	6	6
9062	WDV	13	6	4	6	6	6	6	6	6
9063	WT47	13	6	2	6		6	6	6	6
9064	AMALA	14	7		6	6		6	6	6
9065	HHKB	13	6	4	6	6		6	6	6
9066	TABO89	8	6		6		6	6	6	6
9067	BTB	8	4	4	6	6		6	6	6
9068	BM9	8	4	2	6		6	6	6	6
9069	MADURA	8	4	4	6	6		6	6	6
9070	LUY	8	7	1,4	6	6		6	6	6
9071	OLGA	8	4	3	6	6	6	6	6	6
9072	SPACH	8	4		6	6		6	6	6
9075	DKB	9	9	4	6		6	6	6	6
9082	HO104	15	6	4	6	6		6	6	6
9088	PF04015	3	2	1,4	6	6		6	6	6
9091	MLF	4	7		6	6		6	6	6
9092	BM92	4	8		6	6		6	6	6
9104	DHIF	11	7	NT	6		6	6	6	6
9105	FPAF	11	6	2	6		6	6	6	6
9106	MANIKA	7	2,3	NT	6	6		6	6	6
9107	LKT3	4	4	5	6	6		6	6	6

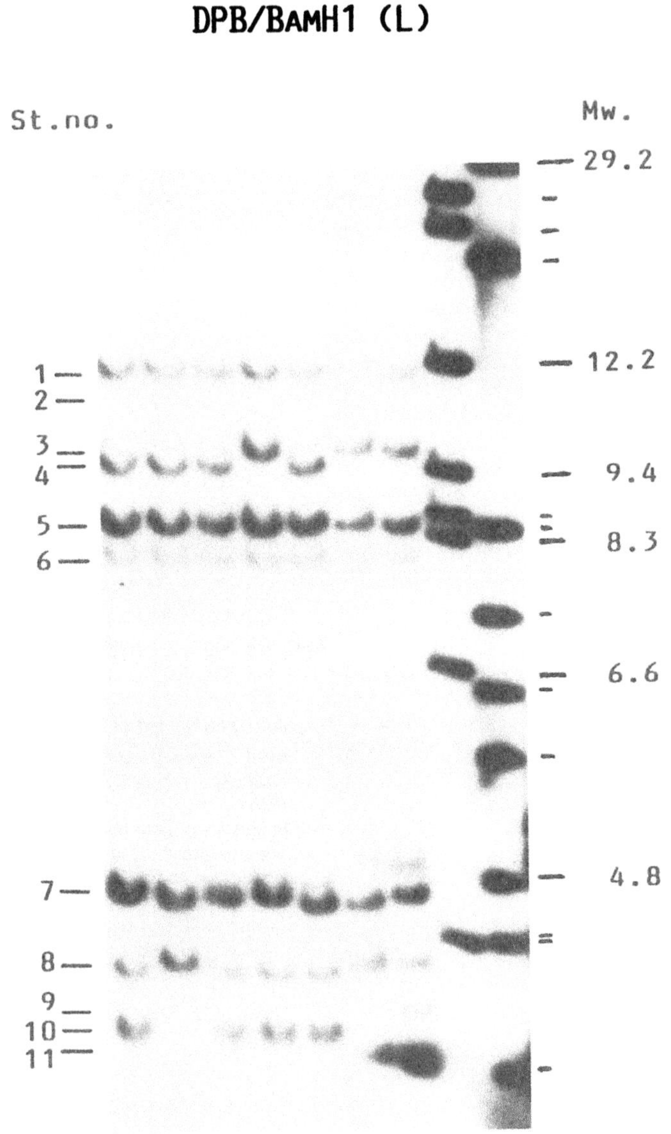

Figure 1. Fragments identified with the DP beta/BamHI system.

RFLP Standardization Report for: DP Beta/HincII

P. Piccioli,[1]* D. Chandanayingyong,[2] G. Angelini,[1] and G.B. Ferrara[1]

DP beta/HincII system has identified 15 fragments on the core cell lines (Table 1). The interpretation of the results was very difficult both for the presence of several weak fragments (including some DPw-specific fragments) and for some cross-hybridization. Morever, this study was performed using only short migration conditions, and only 24 of 72 cell lines were analyzed from all the labs. In spite of the problems mentioned above, all the DPw1 cell lines tested were identified for the presence of a 3.41 kb unique fragment and the lack of the 9th and 10th fragments. Even if a DPw2-correlated specific fragment was not found, all the cell lines carrying this specificity showed the presence of the ninth fragment. Concerning DPw3, DPw4, and DPw5 specificities, no specific band or patterns could be detected. Figure 1 and Table 2 show the fragments and their distribution on the core cell lines.

Participating Laboratories: IT2FER,[1] ANZDCH,[2] ANZDAW[3]
*P. Piccioli was recipient of a fellowship from A.I.R.C. (Associazione Italiana per la Ricerca sul Cancro) This work was in part supported by CNR grants n.86.0265.51 and 86.00659.

Table 1. Fragments Identified with the DP Beta/HincII System

Fragment	kb	Locus[a]	Frequency[b]	% Faint[c]
1	12.65	0 1 2 0	0.986	0.085
2	10.17	0 1 2 0	0.271	0.900
3	8.69	2 1 2 0	0.214	1.000
4	7.85	0 0 1 0	0.157	1.000
5	7.53	0 1 1 0	0.100	1.000
6	7.07	2 1 2 0	0.471	0.061
7	6.59	0 0 1 0	0.043	0.500
8	4.75	0 0 1 0	0.414	1.000
9	4.39	0 0 1 0	0.400	0.367
10	3.77	0 2 1 0	0.729	0.000
11	3.67	2 2 1 0	0.257	0.056
12	3.41	2 1 1 0	0.086	0.333
13	2.92	0 1 1 0	0.100	0.571
14	2.72	2 1 1 0	0.214	1.000
15	2.09	0 1 1 0	0.186	0.846

[a]DR, DQ, DP, DO locus assignment, for instance:
0 0 1 0 means DP-specific
0 1 2 0 means DQ- and DP-specific, DQ more intense
2 2 1 0 means DR-, DQ-, DP-specific, DP most intense, DR and DQ same intensity.

[b]Band frequency has been calculated as number of cell lines presenting a specific fragment versus number of cell lines tested.

[c]Frequency of faint band has been calculated as number of cell lines presenting a faint band versus number of cell lines presenting the same fragment (i.e., frequency 1.000 and % faint 0.000 means that all samples the tested share the same band and it is a strong band; on the other hand, 1.000 and 1.000 means that it is a constant and a faint band).

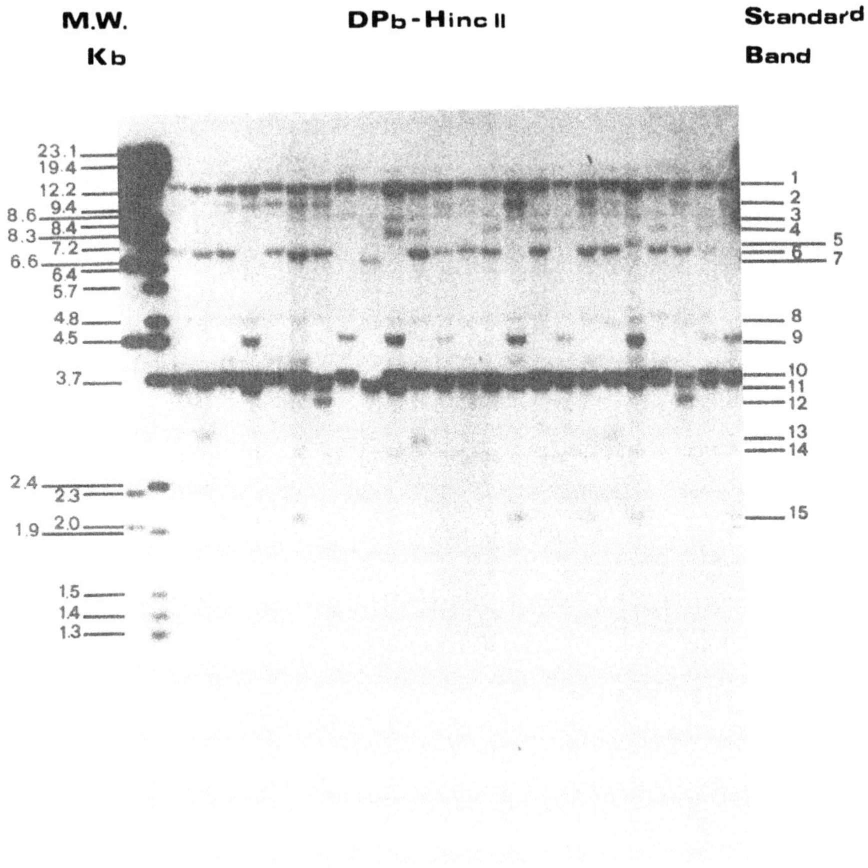

Figure 1. Southern blot of 24 of 72 selected cell lines hybridized with a HLA-DPB probe. The samples were digested with HincII endonuclease, and the gel was run using short migration conditions. From left to right: M1, M2, ID 9069, 9037, 9028, 9019, 9025, 9031, 9088, 9066, 9060, 9036, 9002, 9071, 9064, 9057, 9030, 9054, 9005, 9091, 9016, 9048, 9065, 9021, 9062, and 9032.

Table 2. Banding Pattern of the DP Beta/HincII System in Each Cell Line

Cell	DP	Fragment No.														
		1	2	3	4	5	6	7	8	9	10	11	12	13	14	15
9058 OMW	1	1					1						1			
9023 VAVY	1	1	8	8		8	1		8			1	2		8	
9006 WT100 BIS	1	1		8	8		1		8				1			
9021 RSH	1	1					1				1	8	1		8	8
9063 WT 47	2	1							8	1		1				
9050 MOU	2	1				8			8	1	2					8
9036 SPO 010	2	2		8	8				8	1	2				8	
9105 FPAF	2	1		8	8				8	8	2				8	
9032 BSM	2	1								8	2					
9032 BSM	2	1	8							1						
9068 BM 9	2	1								8	2					
9038 BM 16	2	1								8	1					
9029 WT 51	2	1								8	1					
9039 JVM	2	1								1	2					
9020 QBL	2	1	8							8	2					
9056 KOSE	2,N	1						8	8	1		2				
9056 KOSE	2,N	1						1		8		2				
9018 LO 081785	3	1	8						8	1		2				
9022 COX	3	1	8						8	1		1		8		
9040 BM 15	3	8					1					2				
9071 OLGA	3	1					1		8		2					
9030 JHAF	3	1	1	8	8				8	1	1	1			8	8
9059 SLE 005	3	1	8							1		1				
9014 MGAR	4	1		8	8		1				2					
9033 BM 14	4	8								1		2				
9004 JESTHOM	4	8									1					
9051 PITOUT	4	1				8	1				2					8
9069 MADURA	4	1					8				2					
9028 PE 117	4	1	8				1		8		2				8	8
9025 DEU	4	1	8				1		8		2					8
9031 BOLETH	4	1	8	8			1		8		2					8
9002 MZ 070782	4	1		8	8		1		8		2			1		
9057 TEM	4	1					1		8		2				8	
9065 HHKB	4	1			8		1		8		2				8	
9062 WDV	4	1					8		8		2					

9075	DKB	4	1									2					1
9061	31227 ABO	4	9									2					1
9067	BTB	4	1					1				2					
9026	YAR	4	1						1			2					
9082	HO 104	4	1					1				2			8		
9052	DBB	4	1				8	1				2					
9107	LKT3	5	1	8				1									
9055	HO 301	5	1		8	8		1		8			2				
9042	TISI		1					1					2				
9013	SCHU		1		8	8		1			1		2		8		
9043	BM 21		1	8	8					8	1					8	
9005	HOM 2		1			8		1		8		2					
9047	PLH		1				8						2				8
9010	AMAI		1					1				2					
9037	SWEIG 007		1					1		8		2			1	8	
9019	DUCAF		1	8						8	1	2				8	
9066	TAB 089		1		8						1	2					
9060	CB6B		8						8				1				
9064	AMALA		2	8				1		8		2					
9054	EK		1	8	8			1		8		2				8	8
9091	MLF		1	1		8		1		8		2				8	8
9016	RML		1					1		8		2			1		
9048	LBUF		1	8	8		8			8	1	2				8	8
9092	BM 92		1	8				1				2					
9034	SAVC		1	8							8		1				
9003	KAS 116		1								1		1				
9072	SPACH		1									1					
9104	DHIF	NT	1				8			8	1	2					8
9106	MANIKA	NT	8									1					
9088	PF 04015	1/4	1		8			1				1	1	8			
9070	LUY	1/4	1					1				1		8			
9045	TUBO	2/4	1									1					
9017	WT 8	2/4	1								1	2			8	8	
9008	DO 208915	2/4	1									1					
9011	E 4181324	2/4	1								8	2					
9009	KAS 011	3/4	1									2					
9007	DEM	3/4	1	8							8	2					

1 = Single intensity; 2 = double intensity; 8 = faint; 9 = unable to read.

RFLP Standardization Report for DO Beta/SstI

A. Ando[1] and K. Tsuji[1]

The DO beta/SstI system identified the five fragments (Table 1 and Fig. 1). The results are based on the data from one laboratory. Cross-hybridization bands are shown in Table 2. All the bands were polymorphic except St. Fragment no. 5.

Two bands, St. fragment no. 1 (6.26 kb) and no. 2 (5.19 kb), are DO-DR codominant. The other two bands, St. fragment no. 3 (4.10 kb) and no. 4 (2.94 kb), are DO-DP codominant. St. fragment no. 5 (0.45 kb) is dominantly DO-specific. The banding pattern of each of the 72 core cell lines is shown in Table 3. Four patterns are found in Table 3. These patterns are not associated with any of the class II antigen specificities.

Participating Laboratory: JAPTSU[1]

Table 1. Fragments Identified with the DO Beta/SstI System

		DOB	SST	TSU		
	kb	Locus	Frequency	% Faint	56	32
1	6.26	1 3 2 1	0.414	0.000		
2	5.19	1 2 2 1	0.343	0.042		
3	4.10	0 2 1 1	0.671	0.041		
4	2.94	0 0 1 1	0.657	0.000		
5	0.45	0 0 0 1	0.971	1.000		

Table 2. Cross-Hybridization Bands

DRB	kb	DQB	kb	DPB	kb	DOB	kb
2	21.12	1	21.66	1	20.83		
3	19.53			2	19.59		
5	17.02	2	17.91	3	17.42		
6	14.97	3	15.94	4	15.34		
8	13.19	6	13.82	7	13.14		
9	12.87	7	13.40	8	12.38		
10	12.38	8	12.68	9	12.32		
11	12.01	9	12.23	10	11.96		
12	11.59	10	11.89	11	11.54		
13	11.25	11	11.47	12	11.18		
16	6.15	17	6.08	17	6.12	1	6.26
17	5.19	19	5.20	18	5.17	2	5.19
18	4.57	21	4.56	19	4.56		
		22	4.35	20	4.34		
19	3.82	23	3.78				
20	3.46	24	3.44	22	3.46		
21	3.02	25	3.07				
23	2.38	27	2.42	25	2.40		
24	2.23	28	2.28				
		30	2.01	27	2.05		
		31	1.33	29	1.29		
		32	1.09	30	1.14		
		33	0.34	31	0.47		

Underline denotes primary locus assignment.

Figure 1. Fragments identified with the DO beta/SstI system.

DOB SST

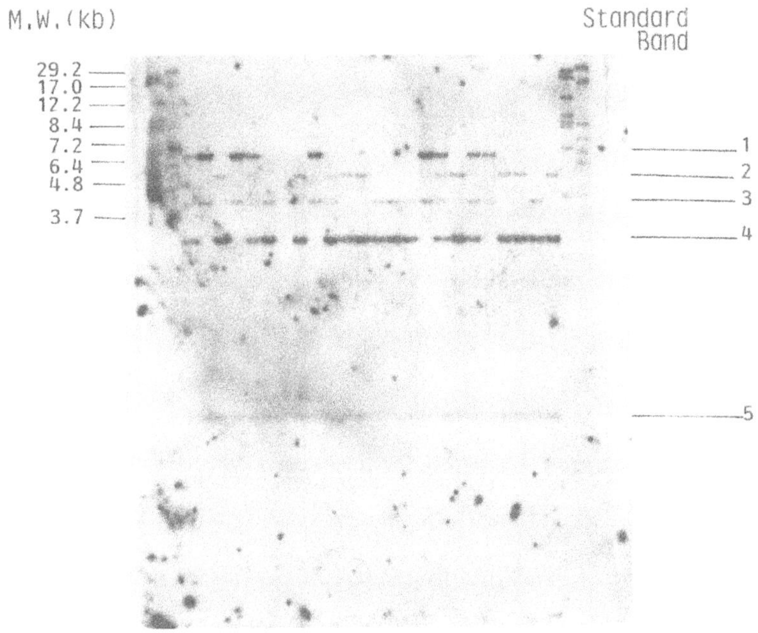

Table 3. RFLP Pattern of DO Beta/SstI System

			1	2	3	4	5							
Fragment No.			1	2	3	4	5							
Fragment Size (kb)			6.26	5.19	4.10	2.94	0.45							
Cell Lines								DW	DR	DW2	DRw	DQw	DPw	GLO
17	9005	HOM2	2	1	1	2	8	1	1			5	–	
5	9017	WT8	2	8	8	2	8	2	15			1	2,4	
2	9014	MGAR	3		1		8	2	15			6	4	1,4
4	9107	LKT3	3		1		8	15	4		53	4	5	
9	9004	JESTHOM	3		1		8	1	1			5	4	
16	9006	WT100BIS	3		1		8	1	1			5	1	
32	9066	TAB089	3		1		8	8.3	8		52	6	–	2
33	9060	CB6B	3		1		8	18	13	25	52	5	–	
34	9036	SPO010	3		1		8	DB2	11	25	52	5	2	
35	9002	MZ070782	3		1		8	20	1			5	4	1,2
40	9054	EK	3		1		8	9	14	25	52	5	–	
43	9016	RML	3		1		8	22	16			7	–	
44	9048	LBUF	3		1		8	DB1	7		53	2	–	
45	9065	HHKB	3		1		8	18	13	24	52	6	4	
53	9068	BM9	3		1		8	8.1	8		52	4	2	
60	9070	LUY	3		1		8	8.3	8		52	7	1,4	
70	9052	DBB	3		1		8	11	7		53	9	4	1
20	9104	DHIF	2		1		8	5	11		52	7	NT	
49	9009	KASO11	2		1		8	21	2			1	3,4	
50	9092	BM92	2		1		8	14	4		53	8	–	1
57	9038	DM16	2		1		8	DB6	12		52	7	2	
67	9106	MANIKA	2		1		8		7		53	2?3	NT	
69	9072	SPACH	2		1		8	8.2	8		52	4	4	
72	9007	DEM	2		1	3	8	–	16,4		53	5,8	3,4	1,2

Table 3. *Continued*

		Cell Lines	1	2	3	4	5	DW	DR	DW2	DRw	DQw	DPw	GLO
		Fragment No.	1	2	3	4	5							
		Fragment Size (kb)	6.26	5.19	4.10	2.94	0.45							
36	9071	OLGA	2		1	3	8	8.2	8		52	4	3	
27	9028	PE117	2		1	2	8	14	4		53	8	4	
19	9056	KOSE	2		1	2	8	9,19	13,14	25	52	5	2,N	
65	9056	KOSE	2		1	2	8	9,19	13,14		52	5	2,N	
1	9045	TUBO	1		8	1	8	5	11,12	25	52	7	2,4	
25	9069	MADURA	1				8	8.1	8		52	4	4	
10	9018	LO081785		1	1	3	8	3	3	25	52	2	3	
62	9011	E4181324		1	1	3	8	12	15			6	2,4	
6	9042	TISI			1	3	8	N	11	25	52	7	–	2
13	9022	COX			1	3	8	3	3	24	52	2	3	2
14	9043	BM21			1	3	8	5	11	25	52	7	–	
15	9023	VAVY			1	3	8	3	3	24	52	2	1	1
23	9040	BM15			1	3	8	5	11	25	52	7	3	
29	9025	DEU			1	3	8	4	4		53	7	4	
30	9031	BOLETH			1	3	8	4	4		53	8	4	
31	9088	PF04015			1	3	8		3	24	52	2	1,4	
37	9064	AMALA			1	3	8	16	14	24	52	7	–	
39	9030	JHAF			1	3	8	13	4		53	7	3	
46	9021	RSH			1	3	8	N	3	24	52	4	1	1
47	9062	WDV			1	3	8	18	13	24	52	6	4	
48	9032	BSM			1	3	8	4	4		53	8	2	
68	9032	BSM			1	3	8	4	4		53	8	2	
56	9067	BTB			1	3	8	8.1	8		52	4	4	
55	9061	31227ABO			1	3	8	9	14	25	52	5	4	
59	9003	KAS116			1	3	8	1	1			5	–	
52	9034	SAVC			1	3	8	4	4		53	8	–	
3	9058	OMW		1		3	8	18	13	24	52	6	1	2
8	9033	BM14		1		3	8	4	4		53	8	4	
11	9055	HO301		1		3	8	N	6	26	52	6	5	
12	9063	WT47		1		3	8	19	13	26	52	6	2	
18	9050	MOU		1		3	8	17	7		53	2	2	
21	9051	PITOUT		1		3	8	17	7		53	2	4	2
22	9047	PLH		1		3	8	DB1	7		53	2	–	
24	9010	AMAI		1		3	8	2	15			6	–	
26	9037	SWEIG007		1		3	8	5	11	25	52	7	–	
28	9019	DUCAF		1		3	8	3	3	25	52	2	–	2
38	9057	TEM		1		3	8	9	14	25	52	1	4	
42	9091	MLF		1		3	8	4	4		53	7	–	
51	9059	SLE005		1		3	8	19	13	26	52	6	3	
54	9075	DXB		1		3	8	23	9		53	9	4	
58	9008	DO208915		1		3	8	2	15			6	2,4	
61	9029	WT51		1		3	8	4	4		53	8	2	
63	9026	YAR		1		3	8	10	4		53	8	4	1,2
64	9039	JVM		1		3	8	N	11	25	52	7	2	
66	9082	H0104		1		3	8		15			6	4	
71	9020	QBL		1		3	8	3	3	25	52	2	2	
7	9013	SCHU						2	15			6	–	
41	9105	FPAF						FS	11		52	6	2	

RFLP Standardization Report for DO Beta/PstI

E. Keller,[1] A. Andreas-Zietz,[1] E.D. Albert,[1] L. Contu,[2] A. Takada,[3] and M. Trucco[4]

The DO beta/PstI system identified 4 fragments on the core cell lines (Table 1 and Figs. 1 and 2). Table 1 shows the presence/absence pattern of all 72 core cell lines for the DO beta/PstI core fragments, the corresponding molecular weight and the locus assignment. The intensity of bands is indicated by numbers. For cross-hybridization with DR, DQ, and DP beta probes see Table 2 in the Workshop report on DR beta/PstI.

Core fragment no. 3, present in all cells, except no. 21 and core fragment no. 2, only present in cell no. 21 are the main DO beta/PstI fragments. As fragment no. 1 was only seen in one laboratory, family analysis would be necessary to find out whether it is a real DO beta/PstI fragment.

The locus assignment for *fragment no. 3* is "0001" (DR,DQ,DP, and DO, respectively) except for those cells which show in addition a strong signal for *DR* beta/PstI fragment no. 18 (DR1 and DR2 associated), where the locus assignment is "2001."

The same is true for the faint DO beta/PstI *fragment no. 4* which has the locus assignment "0001" except for those cells and show in addition a stronger signal for *DR* beta/PstI fragment no. 33, where the locus assignment is "1002." Perhaps in both cases, there are two separate fragments in the same position, not distinguishable under Workshop conditions.

Participating Laboratories: GERALB,[1] IT1CON,[2] JAPAIZ,[3] US1DUQ[4]

Table 1. Banding Pattern of All 72 Core Cell Lines, as Standardized in the DO Beta/PstI System, Including kb Values and Locus Assignment

```
 Probe Name: DOB                Enzyme Name: PSTI

    |9999999999999999999999999999999999999999999999999999999999999999 9 999999 9| WS-ID number
    |0000000000000000000100000000000000010000000000000000000100000000 0 000000 1|
    |0000011110181122182023923932233034404334355666565566544067666777 5 023555 0|
    |6523974308126823981750181249622762340795858352907414078691876025 1 403662 5|
    ................................................................ . ...... . |...............
    |1135400025664111234023423556646730122260510144533453124623553665 2 070167 4| position on
    |6759952748263035816499270021388246403641713257138057824756362094 1 918950 1| blots
    ................................................................ . ...... . |...............
DR  |     11111111                   111111111 1111111111                   11  1|
    |1111255555556333333444444444444411111111263333333444477778888889 7 134337 1|
    ................................................................ . ...... . |
DR  |                                1       1                              11   |
    |                                6       2                              44   |
    ................................................................ . ...... . |
DQ  |5555116666667222224477788888888557777777766666651557222244446 7-9 2 528559 6|
    ................................................................ . ...... . |
DQ  |                                8                       3                   |  kb    locus
    ................................................................ . ...... . |...............
   1|                                                                   888888 9|  5.51  0 0 0 1
   2|                                                                 2        9|  4.35  0 0 0 1
   3|2222222222222222222222222222222222222222222222222222222222222222   222222 9|  3.60  2 0 0 1*
   4|8888888888888888888888888888888888888888888888888888888888888888   888888 9|  0.59  1 0 0 2*
```

Numbers indicate intensity (1,2,3 = increasing intensity; 8 = faint , 9 = unreadable).
* = For some cells different locus assignment, see text.

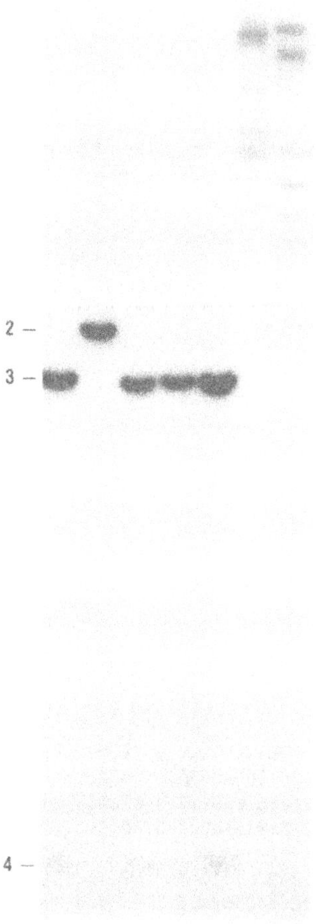

Figure 1. DO beta/PstI core fragment nos. 2 to 4 (gel 0.9%).

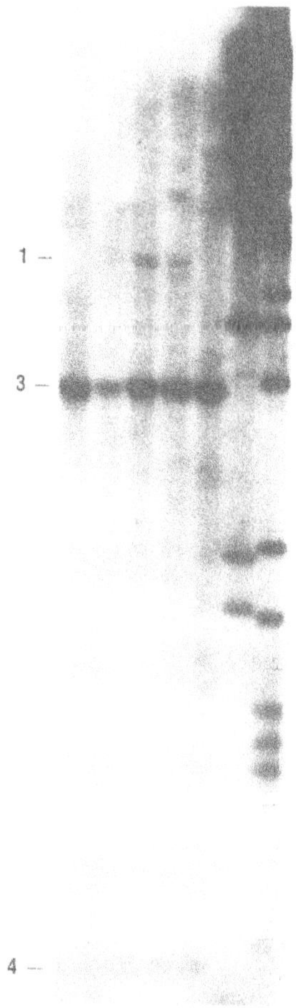

Figure 2. DO beta/PstI core fragment no. 1 (gel 0.9%).

RFLP Standardization Report for DO Beta/KpnI

P.I. Biro,[1] S. Cutbush,[1] P. Cassell,[1] H. Avakian,[1] R. Vaughan,[1] M. Oudshoorn,[2] and R. Martell[2]

Two nonpolymorphic, very high molecular weight, DOB-specific bands are produced with this enzyme/probe combination (Table 1 and Fig. 1). Their high molecular weight makes them indistinct in most blots and not useful for further studies.

Participating Laboratories: UKIFES,[1] SAFDUT[2]

Table 1. Sizes and Frequencies of Standard Bands Produced by the DOβ/KpnI Enzyme/Probe Combination

	kb	Locus	Frequency	% Faint
1	48.31	0 0 0 1	1.000	1.000
2	38.02	0 0 0 1	1.000	1.000

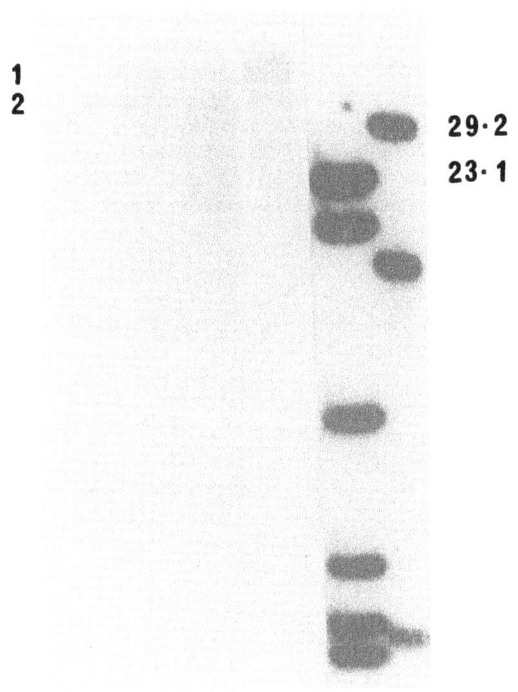

Figure 1. Standard bands produced by the DO beta/KpnI enzyme/probe combination.

RFLP Standardization Report for DR Beta/EcoRV

T. Takenouchi,[1] R. Tonai,[1] P. Terasaki,[1] S. Mayer,[2] and R. Winchester[3]

Data from three laboratories were used for standardization, the criteria for data selection is outlined in the DR/EcoRV system. For DO beta/EcoRV, four standard fragments were assigned (Table 1 and Fig. 1). Marker molecular weights are given in Fig. 1. All bands are specific for DO and no cross-hybridization appeared. We show in Table 2, the DO beta/EcoRV enzyme-probe combination bands for each of the 72 core cell lines. Two allelic patterns are apparent. Only four cell lines showed heterozygotic patterns.

Participating Laboratories: US5TER,[1] FRAMYE,[2] US8RJW[3]

Table 1. Fragments Identified with the DR Beta/EcoRV System

		DOB	ECV	TER		
	kb	Locus	Frequency	% Faint	56	32
1	9.26	0 0 0 1	1.000	0.000		
2	3.73	0 0 0 1	0.357	0.000		
3	2.65	0 0 0 1	0.700	0.000		
4	0.20	0 0 0 1	1.000	1.000		

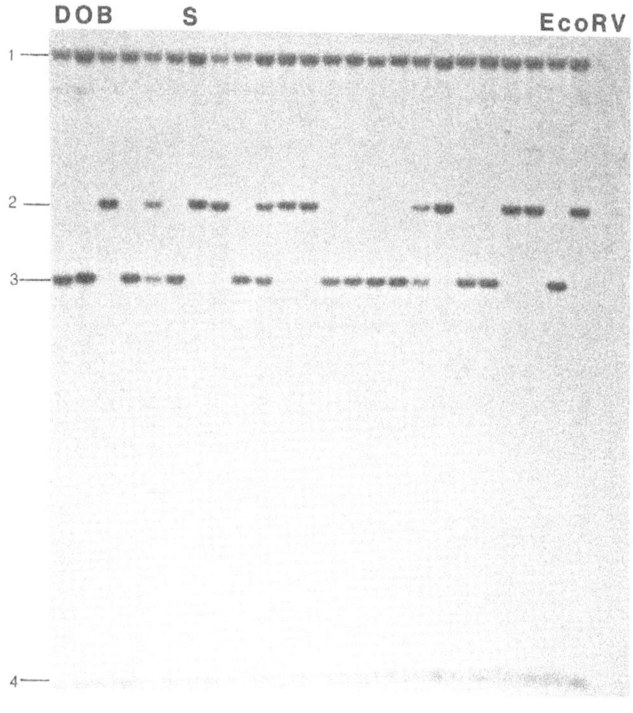

Figure 1. Fragments identified with the DR beta/EcoRV system.

Table 2. Banding Pattern of the DO Beta/EcoRV System in Each Cell Line

Cell	Fragment No. 1	2	3	4
9058 OMW	1	2		8
9013 SCHU	1	2		8
9033 BM14	1	2		8
9055 HO 301	1	2		8
9063 WT47	1	2		8
9050 MOU	1	2		8
9051 PITOUT	1	2		8
9047 PLH	1	2		8
9010 AMAI	1	2		8
9019 DUCAF	1	2		8
9057 TEM	1	2		8
9105 FPAF	1	2		8
9091 MLF	1	2		8
9059 SLE 005	1	2		8
9075 DKB	1	2		8
9008 DO 208915	1	2		8
9029 WT51	1	2		8
9026 YAR	1	2		8
9039 JVM	1	2		8
9082 HO 104	1	2		8
9020 QBL	1	2		8
9017 WT8	1	1	1	8
9018 LO 081785	1	1	1	8
9005 HOM-2	1	1	1	8
9011 E 4181324	1	1	1	8
9045 TUBO	1		2	8
9014 M GAR	1		2	8
9107 LKT 3	1		2	8
9042 TISI	1		2	8
9004 JES THOM	1		2	8
9022 COX	1		2	8
9043 BM21	1		2	8
9023 VAVY	1		2	8
9006 WT100 BIS	1		2	8
9056 KOSE	1		2	8
9056 KOSE	1		2	8
9104 DHIF	1		2	8
9040 BM15	1		2	8
9069 MADURA	1		2	8
9037 SWEIG 007	1		2	8
9028 PE 117	1		2	8
9025 DEU	1		2	8
9031 BOLETH	1		2	8
9088 PF 04015	1		2	8
9066 TAB 089	1		2	8
9060 CBG-IB	1		2	8
9036 SPO 010	1		2	8
9002 MZ 070782	1		2	8
9071 OLGA	1		2	8
9064 AMALA	1		2	8
9030 JHAF	1		2	8
9054 EK	1		2	8
9016 RML	1		2	8
9048 LBUF	1		2	8
9065 HHKB	1		2	8
9021 RSH	1		2	8
9062 WDV	1		2	8
9032 BSM	1		2	8
9032 BSM	1		2	8
9009 KAS-011	1		2	8

Table 2. *Continued*

Cell	Fragment No. 1	2	3	4
9092 BM92	1		2	8
9034 SAVC	1		2	8
9068 BM9	1		2	8
9061 3-12-27 AOB	1		2	8
9067 BTB	1		2	8
9038 BM16	1		2	8
9003 KAS-116	1		2	8
9070 LUY	1		2	8
9106 MANIKA	1		2	8
9072 S PACHEO	1		2	8
9052 DBB	1		2	8
9007 DEM	1		2	8

1 = Single intensity; 2 = double intensity; 8 = faint.

RFLP Standardization Report for DO Beta/TaqI

A. Marcadet,[1] C. Bellane,[1] D. Cohen,[1] J.D. Bignon,[2] and C.B. Carpenter[3]

The DO beta/TaqI system identified six fragments on the core cell lines (Table 1). Three fragments were constant. Three fragments of weak intensity were revealed; this weak intensity was not due to cross-hybridization.

Table 2 indicates the distribution of the standardized bands on the core cell lines. The three constant fragments were not listed. Fragment nos. 1 and 2 have the same distribution in the core cell lines. The fragment no. 4 seems to be allelic with the two others bands. Only two different patterns of standardized bands can be observed—positive for fragment nos. 1 and 2 or positive for fragment no. 4.

The figure is taken from a film from laboratory NCY/CAR. All standardized fragments are indicated at

Participating Laboratories: FRACOH,[1] FRABIG,[2] NCYCAR[3]

Table 1. Standard Bands in the DO Beta/TaqI System

		DOB	TAQ	COH		
	kb	Locus	Frequency	% Faint	56	32
1	2.89	0 0 1 2	0.956	1.000		
2	1.55	0 0 0 1	0.928	1.000		
3	0.80	0 0 0 1	1.000	0.000		
4	0.65	0 0 0 1	0.058	1.000		
5	0.39	0 0 0 1	1.000	0.000		
6	0.26	0 0 0 1	1.000	0.000		

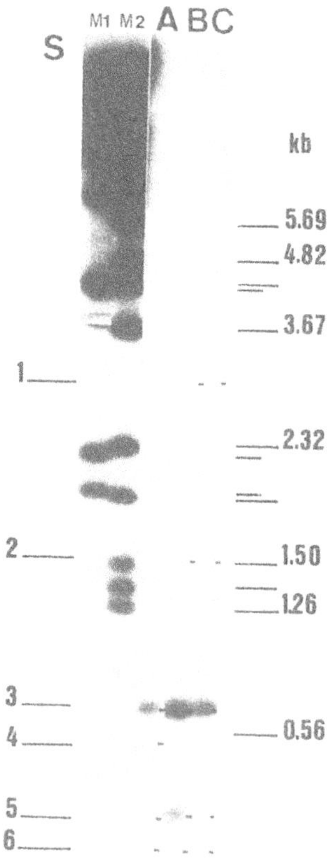

Figure 1. Fragments identified with the DO beta/TaqI system.

Immunobiology of HLA
Volume I

Table 2. RFLP DO Beta/TaqI System: Distribution of Standard Bands in Tenth Workshop Core Cell Lines

		Standard Bands					Standard Bands		
			n =					n =	
Cell Lines	HLA-DR	1	2	4	Cell Lines	HLA-DR	1	2	4
9002	1	8	8		9039	w11	8	8	
9003	1	8	8		9042	w11	8	8	
9004	1	8	8		9043	w11	8	8	
9005	1	8	8		9104	w11	8	8	
9006	1	8	8		9055	w6	8	8	
9007	w16/4	8	8		9058	w13	8	8	
9008	w15	8	8		9059	w13	8	8	
9010	w15	8	8		9060	w13	8	8	
9011	w15	8	8		9062	w13	8	8	
9013	w15	8	8		9063	w13	8	8	
9016	w16	8	8		9065	w13	8	8	
9017	w15	8	8		9054	w14	8	8	
9082	w15	8	8		9056	w13/14	8	8	
9018	3	8	8		9056	w13/14	8	8	
9019	3	8	8		9057	w14	8	8	
9020	3	8	8		9061	w14	8	8	
9021	3	8	8		9064	w14	8	8	
9022	3	8	8		9047	7	8	8	
9023	3	8	8		9048	7	8	8	
9088	3	8	8		9050	7	8	8	
9025	4	8	8		9051	7	8	8	
9026	4	8	8		9052	7	8	8	
9028	4	8	8		9106	7	8	8	
9029	4	8	8		9066	w8	8	8	
9030	4	8	8		9067	w8	8	8	
9031	4	8	8		9068	w8	8	8	
9032	4	8	8		9069	w8	8	8	
9032	4	8	8		9070	w8	8	8	
9033	4	8	8		9071	w8	8	8	
9034	4	8	8		9072	w8	8	8	
9091	4	8	8		9075	w9	8	8	
9092	4	8	8		9045	w11/12	9		9
9107	4	8	8		9014	w15	9		8
9040	w11	8	9		9009	2			8
9105	w11	8	8		9036	w11			8
9037	w11	8	8		9038	w12			8

least once for S and/or L migration. The presence of a standardized fragment is indicated by a hyphen. Standardized fragment DOβ TaqI no. 1 (Molecular weight = 2.89 kb) is the same fragment as standardized fragment DPβ TaqI no. 7 (Molecular weight = 2.92 kb; locus assignment = 0 0 1 2). The fragment identified as "a," light with a weak intensity, has not been standardized.

RFLP Standardization Report for DO Beta/BglII

R.E. Bontrop,[1] P. Rubenstein,[2] M. Van Eggermond,[1] L. Fainboim,[3] P. Stastny,[4] and H. Takata[5]

Data were received from five different laboratories and standardization was performed based on the following criteria: 1) weak hybridization signals were accepted as standard fragments when they were detected by most laboratories, and 2) discrepancies were resolved by majority consensus.

For this system, four different standard fragments were observed (Table 1 and Fig. 1). Standard fragment 1 has a molecular weight of 8.27 kb and a frequency of 37.1%. This fragment appears to be DO specific (Table 1). Standard fragment 2 has a molecular weight of 7.70 kb and a frequency 34.3% and this fragment appears to be DO region-specific. Besides this, about 8% of the total number of fragment 2 is observed as a weak hybridization signal. Standard fragment 3 has a molecular weight of 7.37 kb and a panel frequency of 40%. This fragment is always observed as a strong hybridization signal. Some of the cell lines analyzed appear to be "heterozygous" (e.g., for fragments 2 and 3) (Table 2).

Standard fragment 4 has a molecular weight of 5.17 kb and a panel frequency of 100% (Fig. 1). This fragment is also detected in the DQ beta/BglII system as fragment 3 (see Table 2 in the DR Beta/BglII report).

Computer scanning of the reported sequence of hte DO beta gene (1) revealed the absence of BglII restriction sites within the coding sequences itself. Therefore, the polymorphism observed with the DO beta probe and the restriction enzyme BglII probably reflects the presence of polymorphic sequences surrounding the gene or the presence of such sequences within the introns of the DO beta gene.

Participating Laboratories: BENROO,[1] US8RUB,[2] LATFAI,[3] US4STA,[4] JAPTAK[5]

Table 1. Fragments Identified with the DO Beta/BglII System

Fragment	kb	Locus	Frequency	% Faint	56	32
1	8.27	0.001	0.371	0.000		
2	7.70	0.001	0.343	0.080		
3	7.37	0.001	0.400	0.000		
4	5.17	0.101	1.000	0.000		

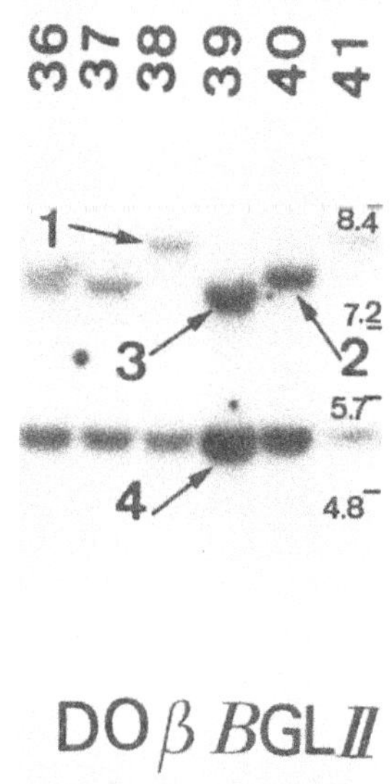

Figure 1. Fragments observed with the DO beta/BglII system.

Reference

1. Tonelle C, DeMars R., Long EO. DO beta: a new beta chain gene in HLA-D with a distinct regulation of expression. EMBO J 1985;4:2839.

Table 2. Fragment Presence/Absence Pattern of the DO Beta/BglII System in Each Core Cell Line

	Fragment					Fragment			
Cell	1	2	3	4	Cell	1	2	3	4
9058	1			1	9106		1		1
9013	1			1	9072		1		1
9033	1			1	9052		1		1
9055	1			1	9045			1	1
9063	1			1	9014			1	1
9050	1			1	9042			1	1
9051	1			1	9022			1	1
9047	1			1	9043			1	1
9010	1			1	9023			1	1
9069	1			1	9040			1	1
9057	1			1	9025			1	1
9105	1			1	9031			1	1
9091	1			1	9088			1	1
9059	1			1	9036			1	1
9075	1			1	9064			1	1
9008	1			1	9030			1	1
9029	1			1	9021			1	1
9026	1			1	9062			1	1
9039	1			1	9032			1	1
9082	1			1	9009			1	1
9020	1			1	9034			1	1
9107		1		1	9061			1	1
9004		1		1	9067			1	1
9006		1		1	9038			1	1
9104		1		1	9003			1	1
9069		1		1	9032			1	1
9066		1		1	9017	1	1		1
9060		1		1	9005	1	1		1
9002		1		1	9018	1		1	1
9054		1		1	9011	1		1	1
9016		1		1	9056		1	1	1
9048		1		1	9037		1	1	1
9065		1		1	9028		1	1	1
9092		1		1	9071		1	1	1
9068		1		1	9056		1	1	1
9070		1		1	9007		1	1	1

RFLP Standardization Report for DO Beta/PvuII

K. Kato,[1] A. Smith,[2] B. Dupont,[1] and S.Y. Yang[3]

Two bands were standardized with the DO beta/PvuII system as shown in Table 1. These two bands were non-polymorphic as shown in Figure 1. Even with the difference of band intensity, no faint bands were observed in this study. These two bands showed no cross-hybridization with other beta probes and should be considered DO beta-specific.

Participating Laboratories: US7DUP,[1] US7HAN,[2] US7SYY[3]

Table 1.

		DOB	PvuII	DUP		
	kb	Locus	Frequency	% Faint	56	32
1	4.38	0 0 0 1	1.000	0.000		
2	2.66	0 0 0 1	1.000	0.000		

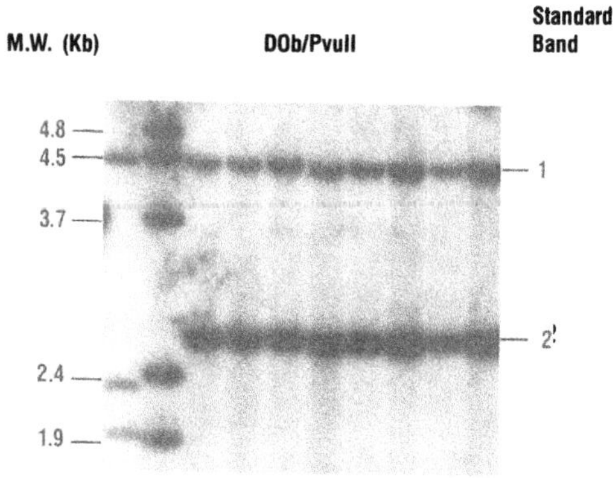

Figure 1. DO beta/PvuII short migration.

Cell line	Fragment # 12
9063 WT47	22
9055 HO 301	22
9013 SCHU	22
9050 MOU	22
9019 DUCAF	22
9031 BOLETH	22
9060 CB6B	22
9002 MZ 070782	22
9054 EK	22
9067 BTB	22
9026 YAR	22
9052 DBB	22
9020 QBL	22
9045 TUBO	12
9014 M GAR	12
9058 OMW	12
9107 LKT 3	12
9017 WT8	12
9042 TISI	12
9038 BM16	12
9004 JES THOM	12
9022 COX	12
9043 BM21	12
9023 VAVY	12
9006 WT100 BIS	12
9005 HOM-2	12
9056 KOSE	12
9104 DHIF	12
9051 PITOUT	12
9047 PLH	12
9040 BM15	12
9064 AMAI	12
9069 MADURA	12
9037 SWEIG 007	12
9028 PE 117	12
9025 DEU	12

Cell line	Fragment # 12
9088 PF 04015	12
9066 TAB 089	12
9036 SPO 010	12
9071 OLGA	12
9064 AMALA	12
9057 TEM	12
9030 JHAF	12
9105 FPAF	12
9091 MLF	12
9016 RML	12
9048 LBUF	12
9065 HHKB	12
9021 RSH	12
9062 WDV	12
9032 BSM	12
9009 KAS-011	12
9092 BM92	12
9059 SLE-005	12
9034 SAVC	12
9068 BM9	12
9075 DKB	12
9061 3-12-27 AOB	12
9038 BM16	12
9008 DO 208915	12
9003 KAS-116	12
9070 LUY	12
9029 WT51	12
9011 E 4181324	12
9039 JVM	12
9056 KOSE	12
9082 HO104	12
9106 MANIKA	12
9032 BSM	12
9072 S PACHEO	12
9007 DEM	12

Figure 2. RFLP pattern of DO beta/PvuII system.

RFLP Standardization Report for DO Beta/EcoRI

S. Tonks,[1] V. Groves,[1] H. Betuel,[2] J.L. Bidwell,[3] M.L. Garavoy,[4] R. Lechler,[5] S. Naito,[6] J. Bodmer,[1] and J. Trowsdale[1]

The DOB/EcoRI system was the least polymorphic of the beta-probe combinations. Only three fragments were identified; however, all these were DO beta-unique (Table 1, Fig. 1) (i.e., did not cross-hybridize with other class 2 probes). The banding pattern for this system was simple despite the fact that all three fragments were polymorphic.

Participating Laboratories: UKIBOD,[1] FRABET,[2] UKIBID,[3] US3GVY,[4] UKIBAT,[5] JAPNAI[6]

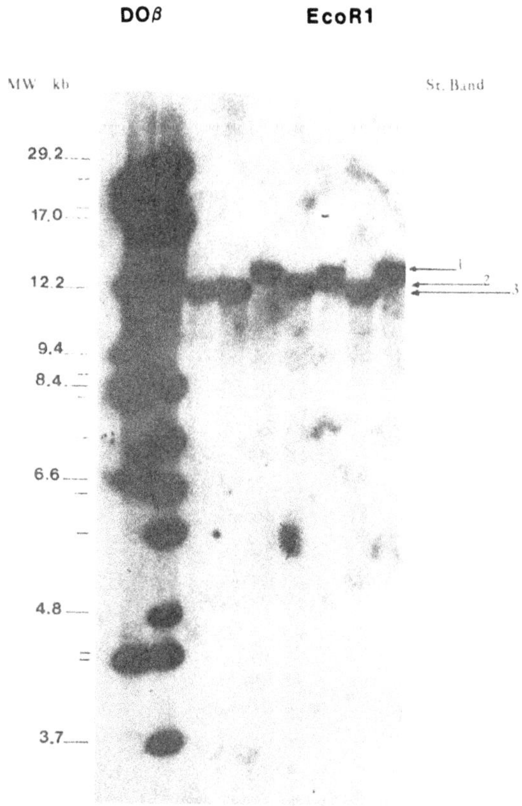

Figure 1.

Table 1. Banding Pattern of the DO Beta/EcoRI System in Each Cell Line

WS ID No.	Cell	Fragment No. 1	2	3
9058	OMW	3		
9013	SCHU	3		
9033	BM14	3		
9055	H0301	3		
9050	MOU	3		
9051	PITOUT	3		
9047	PLH	3		
9010	AMAL	3		
9019	DUCAF	3		
9057	TEM	3		
9105	FPAF	3		
9091	MLF	3		
9059	SLE005	2		
9075	DKB	1		
9008	D0208915	8		
9029	WT51	3		
9026	YAR	3		

Immunobiology of HLA
Volume I

Table 1. *Continued*

WS ID No.	Cell	Fragment No. 1	2	3
9039	JVM	3		
9082	H0104	3		
9020	QBL	3		
9017	WT8	3	1	
9006	WT100BIS	3	1	
9005	HOM2	3	1	
9018	LO081785	3		1
9011	E4181324	3		8
9107	LKT3		3	
9004	JESTHOM		3	
9056	KOSE		3	
9104	DHIF		3	
9069	MADURA		3	
9037	SEE1G007		3	
9028	PE117		3	
9066	TAB089		3	
9060	CB6B		3	
9002	MZ070782		3	
9071	OLGA		3	
9064	AMALA		3	
9030	JHAF		3	
9054	EK		3	
9016	RML		3	
9048	LBUF		3	
9065	HHKB		3	
9021	RSH		3	
9062	WDV		3	
9032	BSM		1	
9092	BM92		1	
9068	BM9		1	
9070	LUY		3	
9106	MANIKA		3	
9072	SPACH		3	
9052	DBB		2	
9007	DEM		3	
9045	TUBO			1
9014	MGAR			1
9042	TISI			1
9022	COX			1
9043	BM21			1
9023	VAVY			1
9040	BM15			1
9025	DEU			1
9031	BOLETH			1
9088	PF04015			1
9036	SPO010			1
9009	KAS011			1
9034	SAVC			1
9061	31227ABO			1
9067	BTB			1
9038	BM16			8
9003	KAS116			

1 = Single intensity; 2 = double intensity; 8 = faint.

RFLP Standardization Report for DO Beta/HindIII

G. Paulsen,[1] G. Markussen,[1] J.M. Tiercy,[3] and R. Fauchet[2]

In the DO beta/HID system, one single band was detected in all the core cell lines (Table 1 and Fig. 1).

Participating Laboratories: SCATSB,[1] FRAFAU,[2] FRAJEA[3]

Table 1. Standards Bands in the DO Beta/HID System

Band	kb	Locus[a]	Frequency[b]	% Faint[c]	56[d]	32[d]
1	7.73	0 0 0 1	1.000	0.000		

[a]Locus assignment in the order DRB, DQB, DPB, DOB. 1 indicates high specificity.
[b]Band frequency in the core cell lines.
[c]Frequency of faint bands.
[d]Discrepancies in strong bands (s) and in faint bands (f) of hidden duplicates (WS.ID. 9056 and 9032).

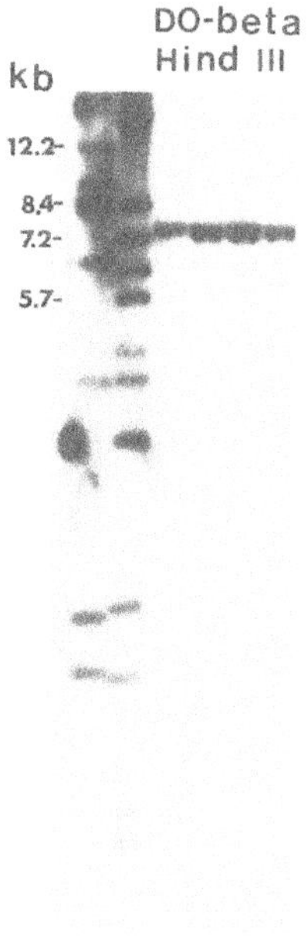

Figure 1. RFLP bands in the DO beta/HID system of four core cell lines run in 0.9% agarose gel. The WS.IDs. of these core cell lines are from the left following marker 1 and marker 2: 9009, 9092, 9059, and 9034.

RFLP Standardization Report for DO Beta/MspI

M. Segall,[1]* L. Schluender,[1]* and A. Arnaiz-Villena[2]

DO beta/MspI blots were received from two laboratories. Fragments well identified on at least one lab's blots were included as standard fragments.

The DO beta/MspI system identified nine fragments on the core cell lines (Table 1 and Fig. 1). The two bands seen

Participating Laboratories: US2BAC,[1] FRAARN[2]
*Supported in part by Juvenile Diabetes Foundation Grant #186175 and March of Dimes Birth Defects Foundation Grant #6-496.
This is publication #495 from the Immunobiology Research Center, University of Minnesota, Minneapolis, Minnesota 55455, USA.

Table 1. DO Beta/MspI Standard Fragments

Fragment	kb	Locus Assignment	Frequency	% Faint	Discrepant for 9056	Discrepant for 9032
1	6.01	0 0 0 1	0.357	0.000		
2	4.82	0 0 0 1	0.029	1.000		
3	4.62	0 0 0 1	0.857	0.000		
4	3.44	0 0 0 1	0.229	0.000		
5	2.75	0 0 0 1	0.300	0.333	f	
6	2.45	0 0 0 1	0.014	0.000		
7	1.55	2 0 0 1	1.000	0.000		
8	1.42	0 0 0 1	0.700	0.000		
9	1.16	2 0 0 1	1.000	0.000		

Immunobiology of HLA
Volume I

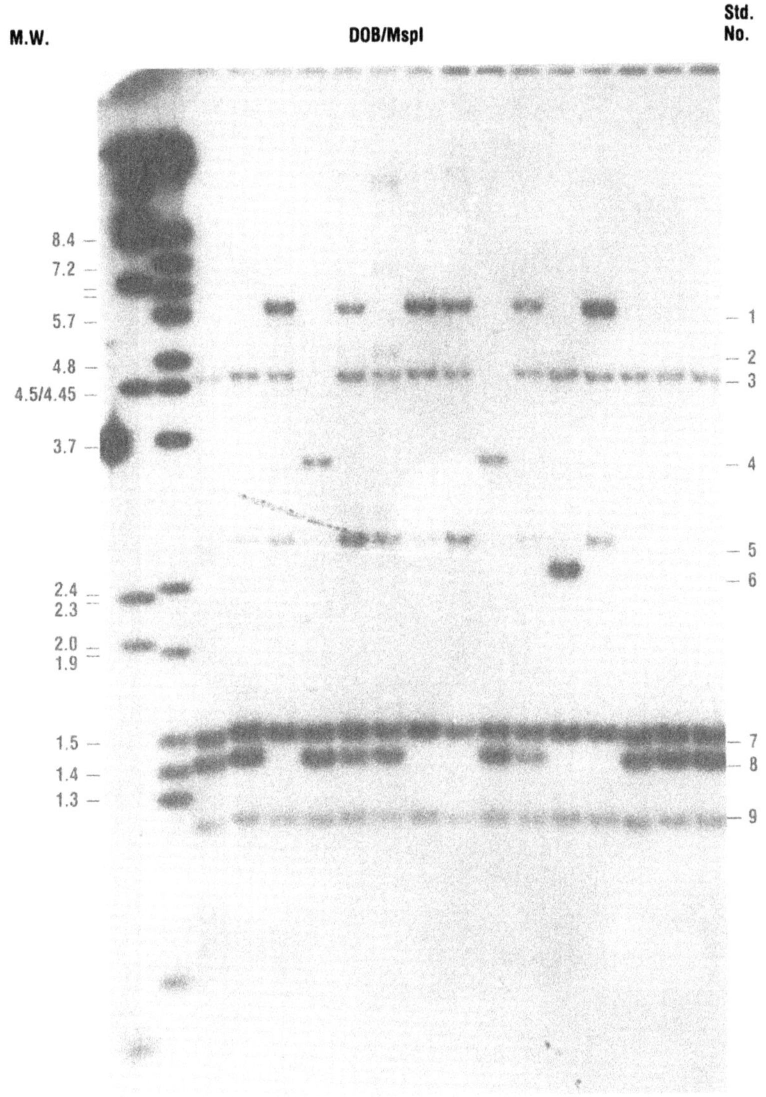

Figure 1. Fragments identified with the DO beta/MspI system.

above fragment 1 in one lane of Figure 1 appeared on only one blot of one laboratory and were therefore not assigned as core fragments. Fragment 5 was identified by only one of the two labs and was extremely variable in intensity; the discrepancy for the duplicates of 9056 involved a faint band in one lane that was not seen in the other. All other fragments were clearly identified by both labs. Fragments 7 and 9 showed cross-hybridization with the DR beta probe (see Table 3 of the DR Beta/MspI Report).

The distribution of fragments in the core cell lines studied is given in Table 2. Table 2 is arranged by pattern of fragments.

Table 2. Distribution of Polymorphic Fragments in Core Cell Lines, DOBeta/MspI

		St. Fragment Number									St. Fragment Number						
Cell #	Name	1	2	3	4	5	6	8	Cell #	Name	1	2	3	4	5	6	8
9048	LBUF				2			2	9105	FPAF	1		1				
9072	SPACH				1			2	9091	MLF	2		1				
9002	MZO70782				2			2	9020	QBL	2		1				
9107	LKT3				1			2	9059	SLE005	2		1				
9070	LUY				1			1	9047	PLH	2		1				
9004	JESTHOM				1			2	9029	WT51	2		1				
9016	RML				2			2	9075	DKB	2		1				
9052	DBB				1			2	9008	D0208915	2		1				
9006	WT100BIS		8		1			2	9082	HO104	2		1				
9104	DHIF				1	1		2	9057	TEM	2		1				
9030	JHAF			1				2	9039	JVM	2		1				
9032	BSM			1				2	9033	BM14	1		1		1		
9054	EK			1				2	9063	WT47	2		1		1		
9061	31227ABO			1				2	9050	MOU	2		1		8		
9009	KAS011			1				2	9019	DUCAF	2		1		1		
9021	RSH			1				2	9010	AMAL	2		1		8		
9025	DEU			1				2	9058	OMW	2		1		1		
9028	PE117			1				2	9013	SCHU	2		1		8		
9023	VAVY			1				2	9051	PITOUT	2		1		1		
9022	COX			1				2	9011	E4181324	2		1				1
9003	KAS116			1				2	9005	HOM2	1		1	1			1
9065	HHKB			1				2	9037	SWEIG007	2		1	1	1		1
9068	BM9			1				2	9017	WT8	1		1		2		2
9092	BM92			1				2	9018	L0081785	1		1		8		1
9067	BTB			1				2	9056	KOSE			1	1			2
9064	AMALA			1				2	9007	DEM			1	1			2
9062	WDV			1				2	9106	MANIKA			1	1			2
9060	CB6B			1				2	9042	TISI		8	1		1		2
9045	TUBO			1				2	9071	OLGA			1	1	1		2
9043	BM21			1				2	9056	KOSE			1	1	8		2
9034	SAVC			1				2	9069	MADURA			1		1		2
9038	BM16			1				2	9014	MGAR			1		1		2
9032	BSM			1				2	9088	PFO4015			1		8		2
9031	BOLETH			1				2	9066	TAB089			1		1		2
9055	HO301			1			2		9036	SP0010			1		1		2
9026	YAR	2		1					9040	BM15			1		8		2

RFLP Standardization Report for DO Beta/BamHI

J.J. Hyldig-Nielsen,[1] B.O. Boehm,[2] K.J. Wood,[3] A.R. Bushell,[3] D.P. Singal,[4] and A. Svejgaard[1]

The DO beta/BamHI system identified the four fragments shown in Table 1. Three of these four fragments were polymorphic (nos. 2,3, and 4). Fragment 3 was only found in one of the 70 different cell lines investigated, while fragment 4 was about twice as frequent as fragment 2. Using the enzyme BamHI, it was not possible within the core cells to detect any linkage disequilibrium between the DO and the DR/DQ subregions.

Table 1. Fragments Observed with the DO Beta/BamHI System

		DOB	BAM	SVE		
	kb	Locus	Frequency	% Faint	56	32
1	13.82	0 0 0 1	1.000	1.000		
2	7.29	0 0 0 1	0.357	0.000		
3	6.10	0 0 0 1	0.014	0.000		
4	5.09	0 0 0 1	0.686	0.000		

Participating Laboratories: SCASVE,[1] GERBOE,[2] UKIMOR,[3] US5SIN[4]

Immunobiology of HLA
Volume I

Figure 1. Fragments identified with the DO beta/BamHI system.

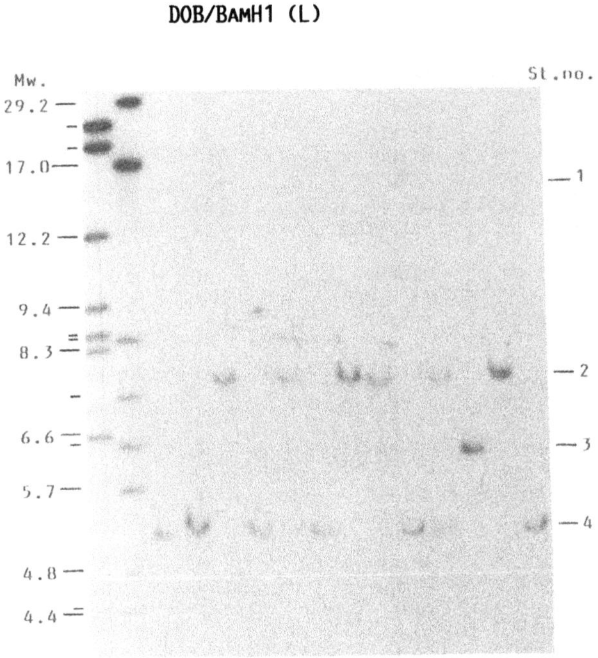

Also the DO beta probe did not cross-hybridize to any of the other beta chains (see Table 1). A smaller fifth fragment (2.36 kb) was seen in a single case but was most likely an artifact due to overdigestion of that particular DNA preparation. Table 2 indicates the banding patterns obtained with this system.

Table 2. Banding Patterns Obtained with the DO Beta/BamHI System

					Standard Fragment No.		
WS ID	Name	HLA-DR	HLA-DQ	HLA-DP	2	3	4
9009	KAS011	2	1	3,4			6
9057	TEM	14	1	4	6		
9017	WT8	15	1	2,4	6		
9018	LO081785	3	2	3	6		
9019	DUCAF	3	2		8		
9020	QBL	3	2	2	8		
9022	COX	3	2	3			6
9023	VAVY	3	2	1			6
9088	PF04015	3	2	1,4			6
9047	PLH	7	2		8		
9048	LBUF	7	2				6
9050	MOU	7	2	2	8		
9106	MANIKA	7	2,3	NT			6
9021	RSH	3	4	1			6
9107	LKT3	4	4	5			6
9051	PITOUT	7	4	4	8		
9067	BTB	8	4	4			8
9068	BM9	8	4	2			8
9069	MADURA	8	4	4			8
9071	OLGA	8	4	3			6
9072	SPACH	8	4				6
9002	MZ070782	1	5	4			6
9003	KAS116	1	5				8
9004	JESTHOM	1	5	4			6
9005	HOM2	1	5		6		
9006	WT100BIS	1	5	1			6

Table 2. *Continued*

WS ID	Name	HLA-DR	HLA-DQ	HLA-DP	Standard Fragment No.		
					2	3	4
9036	SPO010	11	5	2			6
9060	CB6B	13	5				6
9056	KOSE	13,14	5	2			8
9056	KOSE	13,14	5	2			8
9054	EK	14	5				6
9061	31227ABO	14	5	4			8
9007	DEM	16,4	5,8	3,4			8
9055	HO301	6	6	5		6	
9066	TABO89	8	6				6
9105	FPAF	11	6	2	6		
9058	OMW	13	6	1	8		
9059	SLE005	13	6	3	8		
9062	WDV	13	6	4			6
9063	WT47	13	6	2	6		
9065	HHKB	13	6	4			6
9008	DO208915	15	6	2,4	8		
9010	AMAI	15	6		6		
9011	E4181324	15	6	2,4	6		
9013	SCHU	15	6		6		
9014	MGAR	15	6	4			8
9082	HO104	15	6	4	8		
9025	DEU	4	7	4			8
9030	JHAF	4	7	3			6
9091	MLF	4	7		8		
9070	LUY	8	7	1,4			8
9037	SWEIG007	11	7		8		
9039	JVM	11	7	2	8		
9040	BM15	11	7	3			8
9042	TISI	11	7				6
9043	BM21	11	7				6
9104	DHIF	11	7	NT			8
9045	TUBO	11,12	7	2,4			6
9038	BM16	12	7	2			6
9064	AMALA	14	7				6
9016	RML	16	7				6
9026	YAR	4	8	4	8		
9028	PE117	4	8	4			8
9029	WT51	4	8	2	8		
9031	BOLETH	4	8	4			6
9032	BSM	4	8	2			8
9032	BSM	4	8	2			6
9033	BM14	4	8	4	6		
9034	SAVC	4	8				6
9092	BM92	4	8				6
9052	DBB	7	9	4			8
9075	DKB	9	9	4	8		

RFLP Standardization Report for Class I/SstI

T. Takenouchi,[1] M. Kanatani,[1] P. Terasaki,[1] T. Sasazuki,[2] K. Tsuji,[3] and P. Reekers[4]

We standardized data from three laboratories and encountered several problems for this enzyme-probe system. First, the DNA for sample #41 was not digested in all laboratories. Second, there were many discrepancies in the data in fragment assignment due to the difficulty, at times, of achieving complete digestion with this particular enzyme. Third, because of the small number of participating laboratories, it was difficult to achieve a consensus in certain samples of DNA which showed problems in more than one laboratory. To resolve these problems, we reevaluated the data or reassigned the fragments as much as possible. However, we regretfully had to delete the data in those cases we could not make any meaningful interpretation. For the most part, we were able to standardize the data in the same manner we used for the beta systems. But, we accepted measurements in the S filters because this system produced relatively small fragments.

In this system, 23 standard fragments ranging from ca. 1 kb to ca. 14 kb were assigned (Table 1 and Fig. 1). In this system, approximately 13 fragments were observed for each A,B,C, or unknown locus by sorting analysis. Some of these bands appear to be allelic; however, this system does not contain full sets of alleles and no further interpretation or analysis is thought to be useful.

Table 1. Fragments Identified with the Class I/SstI System

		Class I	SST	TER		
	kb	Locus	Frequency	% Faint	56	32
1	13.34	0 0 0 0	0.246	1.000		
2	10.52	0 0 0 0	0.986	1.000		
3	9.57	0 0 0 0	0.130	1.000		
4	7.97	0 0 0 0	0.058	0.000		
5	6.92	0 0 0 0	1.000	1.000		
6	5.92	0 0 0 0	0.116	0.000		
7	5.28	0 0 0 0	0.043	0.000		
8	5.06	0 0 0 0	0.899	0.000		
9	4.21	0 0 0 0	0.130	1.000		
10	3.76	0 0 0 0	1.000	1.000		
11	3.43	0 0 0 0	1.000	0.000		
12	3.17	0 0 0 0	0.059	0.000		
13	3.14	0 0 0 0	0.200	0.000	s	
14	3.11	0 0 0 0	1.000	0.000		
15	2.97	0 0 0 0	0.203	0.000		
16	2.78	0 0 0 0	0.043	0.000		
17	2.56	0 0 0 0	1.000	0.000		
18	2.39	0 0 0 0	0.101	0.000		
19	2.24	0 0 0 0	0.836	0.000		
20	2.10	0 0 0 0	1.000	0.000		
21	1.62	0 0 0 0	0.088	0.000		
22	1.07	0 0 0 0	0.676	0.000		
23	0.84	0 0 0 0	1.000	0.000		

Participating Laboratories: US5TER,[1] JAPSAS,[2] JAPTSU,[3] BENREE[4]

Figure 1. Fragments identified with the class I/SstI system.

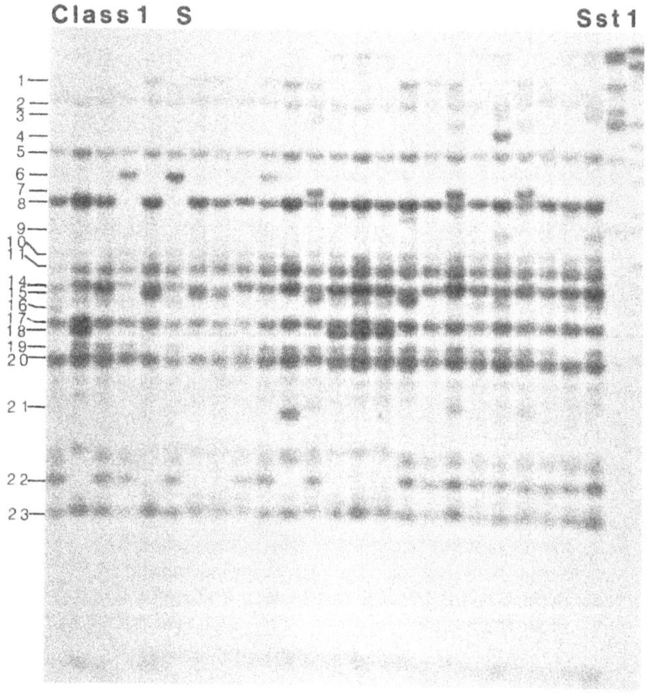

Fragment #12 and #13 are not seen in the S filter.

RFLP Standardization Report for Class I/PstI

J.J. Hyldig-Nielsen,[1] M. Trucco,[2] H. Coppin,[3] A. Andreas-Zietz,[4] E. Keller,[4] C. Carcassi,[5] A. Takada,[6] and A. Svejgaard[1]

The class I/PstI system identified a total of 30 fragments (Table 1 and Fig. 1). Five of these (St. fragment 7,15, 17,29, and 30) have been excluded from the analysis because they were very weak fragments seen only on a few films, which also made the assignments uncertain. St. fragment 12 is a weakly hybridizing fragment, the assignment of which is somewhat uncertain because it is very close to the nonpolymorphic St. fragment 11. The same problem concerns the distinction between fragments 13 and 14. St. fragment 13 might even consist of two fragments. St. fragments 18 and 26 are not seen in the blot in Figure 1.

PstI might not be a good enzyme to use in the distinction between the different class I genes, because many of the fragments are positioned so close that the assignments often are uncertain. Table 2 shows the banding patterns for the 19 polymorphic fragments.

Participating Laboratories: SCASVE,[1] US1DUQ,[2] FRAOHA,[3] GERALB,[4] IT1CON,[5] JAPAIZ[6]

Table 1. Fragments Observed with the Class I/PstI System

	kb	CL1 Locus	PST Frequency	SVE % Faint	56	32
1	10.76	0 0 0 0	0.914	0.000		
2	7.21	0 0 0 0	0.814	0.000		
3	6.72	0 0 0 0	0.171	0.000		
4	6.42	0 0 0 0	0.771	0.000		
5	5.73	0 0 0 0	0.243	0.000		
6	5.43	0 0 0 0	1.000	0.000		
7	5.22	0 0 0 0	0.000	0.000		
8	4.30	0 0 0 0	0.086	0.000		
9	4.05	0 0 0 0	0.971	0.000		
10	3.34	0 0 0 0	0.612	1.000		
11	3.00	0 0 0 0	1.000	0.000		
12	2.95	0 0 0 0	0.603	1.000		
13	2.67	0 0 0 0	0.806	0.885		
14	2.57	0 0 0 0	1.000	0.000		
15	3.04	0 0 0 0	0.000	0.000		
16	2.28	0 0 0 0	0.129	1.000		
17	2.13	0 0 0 0	0.000	0.000		
18	1.72	0 0 0 0	0.014	0.000		
19	1.49	0 0 0 0	0.881	1.000		
20	1.43	0 0 0 0	0.057	0.000		
21	1.31	0 0 0 0	0.971	0.000		
22	1.13	0 0 0 0	0.786	0.000		
23	1.01	0 0 0 0	0.200	0.000		
24	1.01	0 0 0 0	1.000	1.000		
25	0.81	0 0 0 0	0.900	0.000		
26	0.79	0 0 0 0	0.043	0.667		
27	0.66	0 0 0 0	1.000	0.000		
28	0.61	0 0 0 0	1.000	1.000		
29	0.58	0 0 0 0	0.000	0.000		
30	0.50	0 0 0 0	0.000	0.000		

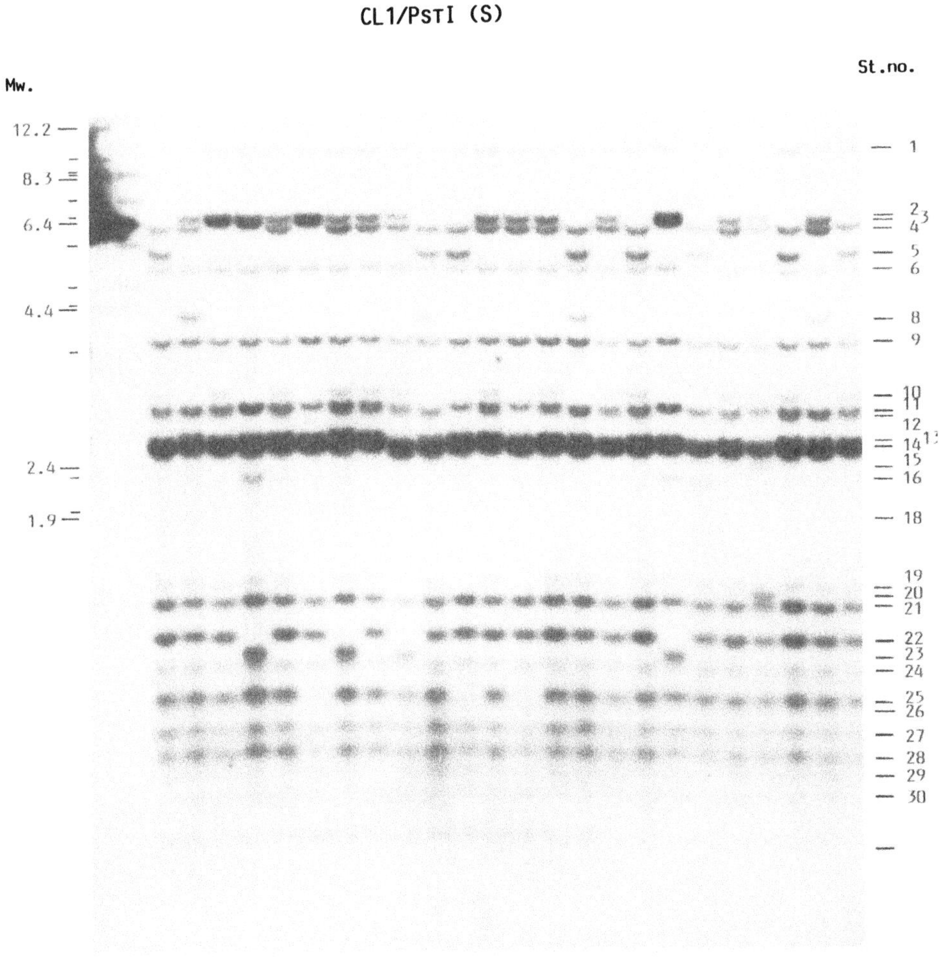

Figure 1. Fragments identified with the class I/PstI system.

Table 2. Class I/PstI Banding Patterns for 19 Polymorphic Fragments

WS ID#	Name	HLA-DR	A	C	B	Stardard Fragment No.																		
						18	20	3	23	5	6	26	16	10	12	13	19	4	22	2	25	21	1	9
9013	SCHU	15	3	7	7			8	6				4			4	4			6	6	6	6	6
9017	WTB	15	3	7	7			8	6				4				4			6	6	6	6	6
9033	BM14	4	3	7	7			8	6				4			4	4			6	6	6	6	6
9034	SAVC	4	3	7	7			6	6								4			6	6	6		6
9065	HHKB	13	3	7	7			6	6				4				4			8	6	6	6	6
9082	HO104	15	3	7	7											4				8	6	6	6	6
9022	COX	3	1	7	8				6	6				4	4		4			6	6	6	6	6
9023	VAVY	3	1	7	8				6	6				4	4		4			6	6	6	6	6
9088	PF04015	3	1		8				6	6				4	4		4			6	6	6	6	6
9014	MGAR	15	26	7	8				6	6	6				4	6	4			6	6	6	6	6
9048	LBUF	7	30	6	13					6			4			4	4	6	6		6	6	6	6
9055	HO301	6	3	8	14								4			4	4	6	6	6	6	6	6	6
9002	MZ070782	1	24	2,8	14											4		6	6	6		6	6	6
9039	JVM	11	2	5	18									4	4	4		6	6	6	6	6	6	6
9038	BM16	12	2	7	18				6					4	4		4	6		6	6	6	6	6
9061	31227ABO	14	2	7	18				6					4	4	4	4	6		6	6	6	6	6
9018	LO081785	3	3	5	18											4	4	6	6	6	6	6	6	6
9008	DO208915	15	25		18					6	6					4	4	6	6		6	6	6	6
9020	QBL	3	26	5	18						6				4	4		6	6	6	6	6	6	6
9019	DUCAF	3	30	5	18										4	4	4	6	6	6	6	6	6	6
9004	JESTHOM	1	2	1	27									4		4		6	6	6	6	6	6	6
9067	BTB	8	2	1	27									4	4	4	4	6	6	6	6	6	6	6
9005	HOM2	1	3	1	27								4				4	6	6	6	6	6	6	6
9105	FPAF	11	1	4	35										4		4	6	6	6	6	6	6	
9068	BM9	8	2	4	35									4	4	4	4	6	6	6	6	6	6	6
9056	KOSE	13,14	2		35					6				4	4	4	4	6	6		6	6	6	6
9056	KOSE	13,14	2		35					6				4	4	4		6	6		6	6	6	6
9006	WT100BIS	1	11	4	35								4		4		4	6	6	6	6	6	6	6
9042	TISI	11	24	4	35											4	4	6	6	6		6	6	6
9025	DEU	4	31	4	35		6							4		4		6	6	6	6	6		6
9009	KAS011	2	1	6	37					6				4		6	4	6	6		6	6	6	6
9062	WDV	13	2		38			6	6	6				4	4	4	4	6	6		6	6	6	6
9026	YAR	4	26		38					6	6				4	4	4	6	6		6	6	6	6

9057	TEM	14	26		38					6	6				4	4	4	6	6		6	6	6	6
9104	DHIF	11	31		38					6				4		4	4	6	6		6	6	6	6
9043	BM21	11	1		41			6				6		4	4		4			6	6		6	6
9021	RSH	3	68,30	2	42			6				4					4			6	6		6	6
9036	SP0010	11	2	5	44									4	4	4	4	6	6	6	6	6	6	6
9054	EK	14	2	5	44									4	4	4	4	6	6	6	6	6	6	6
9063	WT47	13	2	5	44							4		4	4		4	6	6	6	6	6	6	6
9050	MOU	7	29		44									4	4	6	4	6	6	6	6	6	6	
9051	PITOUT	7	29		44									4	4	6	4	6	6	6	6	6	6	
9058	OMW	13	2		45									4	4	4	4	6	6	6	6	6	6	6
9066	TAB089	8	2	11	46									4	4	4	4	6	6	6	6	6	6	6
9047	PLH	7	3	6	47	6				6			4			4	4		6		6	6	6	6
9040	BM15	11	1	7	49				6					4	4		4	6		6	6	6	6	6
9106	MANIKA	7	3		50					6						4		6	6		6	6		6
9016	RML	16	2		51									4	4	4	4	6	6	6	6	6	6	6
9070	LUY	8	2		51									4	4	4		6	6	6	6	6	6	6
9045	TUBC	11,12	2,3	7	51				6								4	6	6	6	6	6	6	6
9003	KA5116	1	24		51					6						4		6	6			6	6	6
9092	BM92	4	25	1	51						6			4	4	6	4	6	6	6	6	6	6	6
9030	JHAF	4	31		51		6							4				6	6	6	6	6		6
9011	E41B1324	15	11		52									4	4	4	4	6	6	6	6	6	6	6
9010	AMAI	15	28	4	53									4	4	6	4	6	6	6	6	6	6	6
9107	LKT3	4	24	1	54											4		6	6	6		6	6	6
9007	DEM	16,4	2	6	57					6				4	4	4		6	6		6	6	6	6
9052	DBB	7	2	6	57					6				4	4	4		6	6		6	6	6	6
9059	SLE005	13	2	10	60			6						4	4		4		6	6	6	6	6	6
9069	MADURA	8	2	10	60			6						4	4		4		6	6	6	6	6	6
9028	PE117	4	24	10	60			6								4		6	6	6		6	6	6
9075	DKB	9	24	10	60			6								4			6	6		6	6	6
9037	SWEIG007	11	29	2	61									4	4	4	4	6	6	6	6	6	6	
9060	CB6B	13	1	9	62									4	4		4	6	6	6	6	6	6	6
9032	BSM	4	2	9	62									4	4	4		6	6	6	6	6	6	6
9032	BSM	4	2	9	62									4	4	4	4	6	6	6	6	6	6	6
9064	AMALA	14	2	9	62									4	4	4	4	6	6	6	6	6	6	6
9091	MLF	4	2	9	62									4	4	4	4	6	6	6	6	6	6	6
9031	BOLETH	4	2	10	62									4	4	4	4	6	6	6	6	6	6	6
9072	SPACH	8	31	1	62		6							4		4		6	6	6	6	6		6
9071	OLGA	8	31	1,10	62		6							4		4		6	6	6	6	6		6
9029	WT51	4	23	8	65													6	6	6		6	6	6

RFLP Standardization Report for Class I/KpnI

G. Paulsen,[1] M. Oudshoorn,[2] S. Cutbush,[3] and K. Siminovitch[4]

Thirty-two bands were identified by the RFLP analysis in the Class I/KPN system as shown in Table 1. The RFLP patterns of some of the core cell lines are shown in Figure 1. Band nos. 29, 30, 31, and 32 are not present in Figure 1. Distribution of bands in the core cell lines is shown in Table 2.

Participating Laboratories: SCATSB,[1] SAFDUT,[2] UKIFES,[3] NCYFAL[4]

Table 1. Standard Bands in the Class I/KPN System

Band	kb	Frequency[a]	% Faint[b]	56[c]	32[c]
1	32.99	0.821	0.281		
2	26.44	0.045	0.667		
3	21.18	0.900	0.077		
4	18.71	0.100	0.000		
5	16.41	1.000	0.056		
6	16.13	0.091	0.000		
7	15.70	0.015	0.000		
8	14.92	0.586	0.238		
9	14.08	0.090	0.000		
10	13.37	0.750	0.057		
11	12.70	0.057	0.500		
12	12.04	0.984	0.935		
13	10.95	1.000	0.069		
14	9.99	0.700	0.102		
15	9.27	0.304	0.571		f
16	8.47	1.000	0.069		
17	7.97	1.000	0.139		
18	7.46	1.000	0.083		
19	6.67	0.614	0.023		
20	6.25	1.000	0.014		
21	5.85	0.286	0.200		f
22	5.62	0.114	0.250		
23	5.43	1.000	0.028		
24	5.12	0.086	0.000		
25	4.84	0.119	0.000		
26	4.45	0.586	0.000		
27	3.48	0.800	0.017		
28	3.23	0.229	0.000		
29	2.79	0.043	0.667		
30	1.36	1.000	0.000		
31	0.97	1.000	1.000		
32	0.82	0.557	0.872		

[a]Band frequency in the core cell lines.
[b]Frequency of faint bands.
[c]Discrepancies in strong bands (s) or faint bands (f) of the hidden duplicates (WS.ID 9056 and 9032).

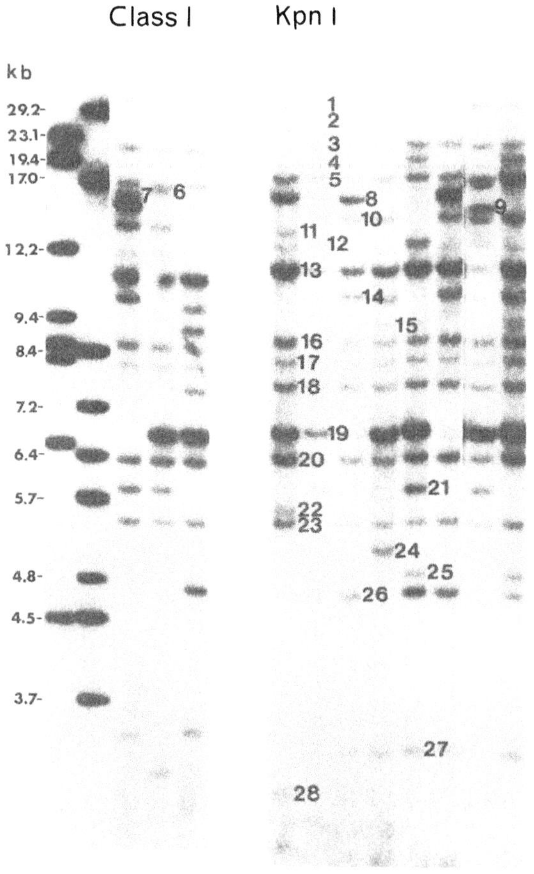

Figure 1. RFLP bands in the class I/KPN system of 11 core cell lines run in 0.6% agarose gel. Representative core bands are shown numbered on their right side. The WS.ID. of these core cell lines are from the left following marker 1 and marker 2: 9006, 9005, 9050, open lane, 9002, 9071, 9064, 9057, 9030, 9054, 9065, and 9021.

Immunobiology of HLA
Volume I

Table 2. Distribution of Bands in Core Cell Lines in the Class I/KPN System

		Band No.																															
WS ID	HLA-B	01	02	03	04	05	06	07	08	09	10	11	12	13	14	15	16	17	18	19	20	21	22	23	24	25	26	27	28	29	30	31	32
9025	35	1		1	1	1			2				8	2		8	1	1	1		1	1		1		1	1	1			1	8	8
9010	53	8		1	1	1			2		1		8	2	1	1	1	1	1		1			1		1	1	1		8	1	8	
9055	14	1		1		1	1		2		1		8	2			1	1	1		1	1		1					1		1	8	8
9006	35	1		1		1		1	2		1		8	2	1		1	1	8		1	1		1				1			1	8	
9032	62	8		1		1			2		1		8	2	1	8	1	1	1		1	8		1			1	1			1	8	
9014	8	1		1		1			2		1		8	2	1	1	1	1	1		1			1	2			1			1	8	
9020	18	1		1		1			2		1		8	1	1	8	1	1	1		1			1	1			1			1	8	
9022	8	8		1		1			2		1		8	2	1	1	1	1	1		1			1			1	1			1	8	8
9023	8	1		1		1			2		1		9	2	1	8	1	1	1		1			1			1	1			1	8	8
9036	44	1		1		1			2		1		8	2	1		1	1	1		1			1			1	1			1	8	
9054	44	1		1		1			2		1		8	2	1		1	1	1		1			1			1	1			1	8	
9070	51	1		1		1			2		1		1	8	1		1	1	1		1			1			1	1			1	8	
9088	8	8		1		1			2		1		8	2	1		1	1	1		1			1			1	1			1	8	8
9019	18	1		1		1			2		1		8	2	1		1	1	1		1			1				1			1	8	8
9039	18	1		1		1			2		1		8	2	1		1	1	1		1			1			1	1			1	8	
9068	35	1		1		1			2		1		8	2	1		1	8	1		1			1			1	1			1	8	
9060	62	8		1		1			2		1		8	2	1		1	1	1		1			1			1	1			1	8	8
9064	62	1		1		1			2		1		8	2	1		1	1	1	8	1			1			1	1			1	8	
9091	62	1		1		1			2		1		8	2	1		1	1	1		1			1			1	1			1	8	
9032	62	1		1		8			2		1		8	2	1		1	1	1		1			1			1	1			1	8	
9031	62	8		1		1			2		1		8	2	1		1	1	1		1			1			1	1			1	8	8
9063	44	8		1		1			2		1		8	2	1		1	1	1		1			1					1		1	8	
9059	60	1		1		1			2		1		8	2	1		1	1	1		1			1			1	1			1	8	
9069	60	8	8	1		1			2		1		9	2	1		1	1	1		1			1			1	1			1	8	
9028	60	1	1			1			2			1	8	2			1	1	1	2	1		1	1					1		1	8	8
9018	18	1		1		1			2		8		9	2			1	1	1		1	1		1				1	1		1	8	8
9002	14	1		8		1			2			1	8	2	8		1	1	1	2	1		1	1					1		1	8	8
9075	60	1		8		1			2					2			1	8	1		1		1	1				1			1	8	8
9071	62	1	8	1	1	1			8		8		8	2			1	1	1	2	1	1			1		1	1	1		1	8	
9021	42	1		1	1	1			8		1		8	2	1	1	1	1	1	2	1			1		1	1	1			1	8	8
9030	51	1		1	1	1							1	2		8	1	1	1	2	1	1		1		1	1	1			1	8	8
9072	62	1		1	1	1							1	1			1	1	1	2	1	1		1		1	1	1			1	8	1
9047	47	8		1		1	1				1		8	2			1	1	1	2	1	1		1					1		1	8	8
9017	7	8		1		1	1		8	1	1		9	1			1	1	1	2	1	1		1					1		1	8	8
9065	7	1		1		1	1		8	1	1		8	1			1	1	1	2	1	1		1					1		1	8	8
9013	7	8		1		1	1			1	1		8	1			1	1	1	3	1	1		1					1		1	8	8
9040	49	1		1		1			8	1	1		8	1	1		1	1	1	2	1			1			1	1			1	8	8
9061	18	1		1		1				1	1		8	1	1		1	8	1	2	1			1			1	1			1	8	
9082	7	1		1		1				1	1		1	2			1	1	1	2	1	1		1					1		1	8	1
9057	38	1		1		1					1		8	2	1	8	1	1	1	2	1	8		1	1			1			1	8	
9026	38	1		1		1					1		8	2	1	8	1	1	1	2	1			1	1			1			1	8	
9011	52	1		1		1					1		8	2	1	8	1	1	1	2	1			1			1	1			1	8	1
9052	57	1		1		1					1		8	2	1	8	1	1	1	2	1			1			1	1			1	8	
9007	57	1		1		1					1		8	2	1	8	1	1	1	2	1			1			1	1			1	8	
9043	41	1		1		1					1		8	2	1	1	1	1	1	3	1			1			1	1			1	8	8
9016	51	1		1		1					1		8	2	1		1	1	1	2	1			1			1	1			1	8	
9056	35	1		1		1					1		8	2	1		1	1	1	2	1			1			1	1			1	8	
9056	35	1		1		1					1		8	2	1		1	8	1	2	1			1			1	1			1	8	
9004	27	8		1		1					1		8	2	1		1	1	1	2	1			1			1	1			1	8	
9062	38	8		1		1					1		8	2	1		1	1	1	2	1			1			1	1			1	8	
9106	50	1		1		1					1		8	2			1	1	1	2	1	1		1					1		1	8	1
9050	44	1		1		1							8	2	1	1	1	1	1	2	1			1			1	1			1	8	8
9051	44	1		1		1							8	2	1	1	1	1	1	2	1			1			1	1			1	8	8
9033	7	8		1		1							8	1			1	1	1	2	1	1		1					1		1	8	8
9029	65	1				1			2			8	9	2			1	1	1		1		1	1				1			1	8	1
9003	51	8				1						8	8	2			1	1	1	2	1		1	1				1			1	8	8
9107	54	8				1							8	2			1	1	1	2	1		1	1					1		1	8	8
9037	61		8	1		1			8				8	2	1	1	1	1	1	2	1			1			1	1			1	8	8
9048	13			1	1	1	9				1		8	2	1	1	1	1	1	2	1	8	8	1		1		1			1	8	8
9005	27			1		1	1				1		8	2			1	1	1	2	1	1		1					1	8	1	8	8
9067	27			1		1			1		1		8	2	1		1	8	1	2	1			1			1	1			1	8	

Table 2. *Continued*

		Band No.																													
WS ID	HLA-B	01	02	03	04	05	06	07	08	09	10	11	12	13	14	15	16	17	18	19	20	21	22	23	24	25	26	27	28	29	30	31	32
9092	51			1		1			8		8		8	1	1	8	1	1	1	1	1			1	1			1			1	8	
9038	18			1		1			1				8	8	1		8	8	8	1	1			1			1	1			1	8	
9034	7			1		1			8				8	8	8		8	8	8	1	1	1		1					1	1	1	8	
9104	38			1		1					1		8	2	1		1	1	1	2	1			1		1		1			1	8	8
9009	37			1		1					1		8	2	1		1	1	1	2	1			1			1	1			1	8	8
9058	45			8		1					1		8	2	1		1	1	1	2	1			1			1	1			1	8	
9066	46			1		1					1		8	2	1		1	1	1	2	1			1			1	1			1	8	
9045	51	9	9	8		8	9	9	8	9			9	8	8	9	8	8	8	1	1	8		1		9	1	8	1		1	8	8
9008	18	9	9	8		8	9	9		9	9		9	1	8	8	8	8	8	1	1			8	1	9		1			1	8	8
9105	35	9	9			8	9	9	8	9	9		9	8	8		8	8	8		8			8		9	1	1			1	8	8
9042	35					1			1				9	1			1	1	1		1		8	1				1			1	8	8
		01	02	03	04	05	06	07	08	09	10	11	12	13	14	15	16	17	18	19	20	21	22	23	24	25	26	27	28	29	30	31	32

Band designation: 1 = single intensity, 2 = double intensity, 8 = low intensity, 9 = unreadable.

RFLP Standardization Report for Class I/EcoRV

A. Ando,[1] T. Takenouchi,[2] Y.K. Paik,[3] E. Dwyer,[4] Y. Mitsuishi,[5] J.A. Trejant,[6] and K. Tsuji[1]

The class I/EcoRV system identified the 23 fragments (Fig. 1). The presence/absence pattern of all 72 core cell lines is shown in Table 1. Four bands out of the 23 standard bands are highly associated with individual class I specificities (r>0.80): St. fragment no. 4 (17.20 kb) and St. fragment no. 7 (13.55 kb) with A2, St. fragment no. 11 (10.96 kb) with Cw4, St. fragment no. 15 (7.86 kb) with B8.

Participating Laboratories: JAPTSU,[1] US5TER,[2] US5PAK,[3] US8RJW,[4] FRAMYE,[5] ANZBSH[6]

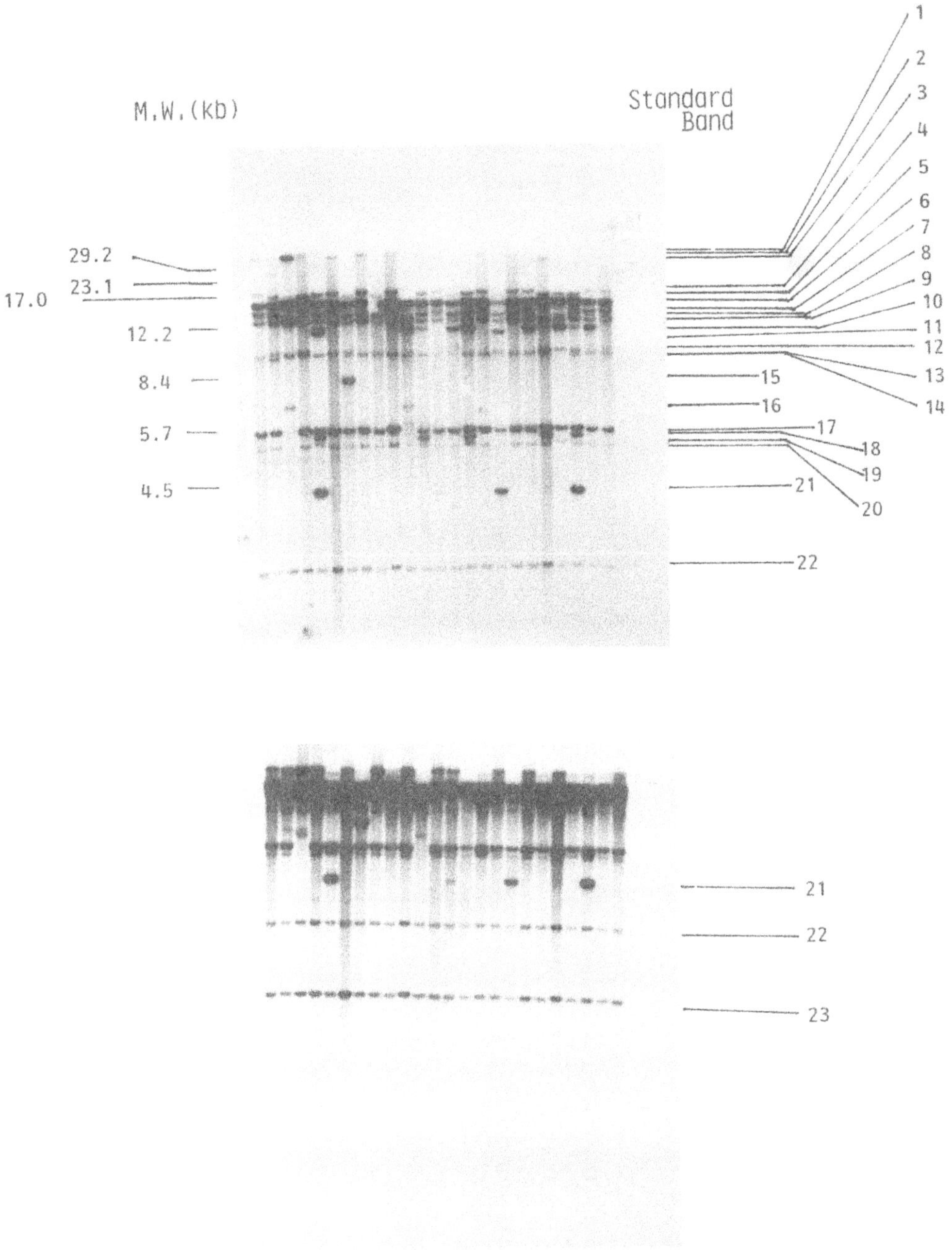

Figure 1. Fragments identified with the class I/EcoRV system.

Table 1. RFLP Pattern of Class I/EcoRV

	Fragment No.		1	2	3	4	5	6	7	8	9	10	11	12	13	14	15	16	17	18	19	20	21	22	23				
	Fragment size (kb)		70.81	38.95	26.71	17.20	16.31	14.16	13.55	12.93	11.93	11.16	10.98	9.46	9.20	8.98	7.86	6.64	5.97	5.73	5.56	5.46	5.40	4.20	3.93				
	Cell	Lines																								A	C	B	Bw
43	9016	RML				8		2	1		1	2		3	8				2					8	1	2	--	51	4
37	9064	AMALA				8		2	1		1				8				2					8	1	2	9	62	6
64	9039	JVM	8			8		2	1		1			8	8				2					8	1	2	5	18	6
42	9091	MLF		8		8		2	1		1			8	8				2					8	1	2	9	62	6
65	9056	KOSE				8		2	1		1	2		2	8				2					8	1	2	--	35	6
60	9070	LUY				8		2	1		1			1	8		2		2					8	1	2	-	51	4
56	9067	BTB	1			8		2	1	3	1			1	8				2					8	1	2	1	27	4
51	9059	SLE005	8			8		2	1		1				8				2					8	1	2	10	60	6
9	9004	JESTHOM	1			8		2	1		1				8				2					8	1	2	1	27	4
68	9032	BSM				8		2	1		1			1					2					8	1	2	9	62	6
57	9038	BM16		1		8		2	1		1			8					2					8	1	2	7	18	6
40	9054	EK	8			8		2	1		1				8				2					8	1	2	5	44	4
55	9061	31227ABO		1		8		2	1		1			8					2					8	1	2	7	18	6
70	9052	DBB		1		8		2	1		1			1	8				2					8	1	2	6	57	4
19	9056	KOSE				8		2	1		8	2			8				2					8	1	2	--	35	6
30	9031	BOLETH		1		8		2	1		1			8	2				2					8	1	2	10	62	6
32	9066	TABO89	8			8		2	1		1			1	8				2					8	1	2	11	46	6
34	9036	SPOO10	8			8		2	1		1			1	8				2					8	1	2	5	44	4
53	9068	BM9				8		2	1		1		2	1	8				2				2	8	1	2	4	35	6
72	9007	DEM		1		8		2		8	1				8				2					8	1	2	6	57	4
46	9021	RSH					8	2		1	1			8	8				2	1			2	8	1	68,30	2	42	6
24	9010	AMAI					8	2		1	1		2	8	8				2	1			2	8	1	28	4	53	4
44	9048	LBUF		8			8	2			1			8	8				2	1				8	1	30	6	13	4
20	9104	DHIF		1			8	2			1	8		8	8				2	1				8	1	31		38	
29	9025	DEU						2		1	1	8	2		8				2	1			2	8	1	31	4	35	6
39	9030	JHAF						2		1	1	2		1	8				2	1				8	1	31	–	51	4
69	9072	SPACH	8					2		1	1			1	1				2	1				8	1	31	1,3.1	62	
36	9071	OLGA	8					2		1	1			8	8				2	1				8	1	31	1,10	62	6
6	9042	TIS1						2			1		2		8								2	8	1	24	4	35	6
16	9006	WT100BIS					8	2			1		2	8	8				2				2	8	1	11	4	35	6

41	9105	FPAF				2		1	1	2	1	8				2	2	8	1	1	4	35	6
15	9023	VAVY			8	2		1	1			8	8	2		2		8	1	1	7	8	6
2	9014	MGAR			1	2		2	1			8		2		2		8	1	26	7	8	6
31	9088	PF04015				2		2	8			8	8	2		2		8	1	1		8	6
13	9022	COX			8	2		1	8			8	8	2		2		8	1	1	7	8	6
71	9020	QBL	8		8	2		1	1			1	8			2		8	1	26	?5	18	6
58	9008	DO208915	8		8	2		1	1			8				2		8	1	25	-	18	6
28	9019	DUCAF	8		8	2			1			8	8			2		8	1	30	5	18	6
17	9005	HOM2	1		8	2			1			8				2		8	1	3	1	27	4
10	9018	LOO81785	1		8	2			1			8	8			2		8	1	3	5	18	6
12	9063	WT47	1		8	2		1	1				8			2		8	1	32	5	44	4
54	9075	DXB	1			2			2			8	8					8	1	24	10	60	6
3	9058	OMW		8		2	1		8				8			2		8	1	2	-	45	6
14	9043	BM21		2	8	2		1	1			8	8			2		8	1	1	-	41	6
38	9057	TEM			8	2		1	1	2			8			2		8	1	26	-	38	4
63	9026	YAR			8	2		1	1	2		8	2			2		8	1	26	-	38	4
18	9050	MOU		2	8	2		1	1			8	8			2		8	1	29	-	44	4
67	9106	MANIKA		1		2			1			8				2		8	1	3	-	50	6
21	9051	PITOUT		2	8	2		1	1			8	8			2		8	1	29	-	44	4
47	9062	WDV				2	1		1	2		8	8			2		8	1	2	-	38	4
62	9011	E4181324			8	2		1	1			8	8			2	2	8	1	1	-	52	4
59	9003	KAS116				2			2	2		8	8					8	1	24	-	51	4
35	9002	MZ070782				2			1	2		8	8		8			8	1	24	2,8	14	6
4	9107	LKT3				2			1				8					8	1	24	1	54	6
50	9092	BM92			8	2		1	1	2			8			2		8	1	25	1	51	4
26	9037	SWEIGOO7				2		1	1			8	8			2		8	1	29	2	61	6
22	9047	PLH		1	8	2			1			8				2		8	1	3	6	47	4
49	9009	KASO11			8	2		1	1				8			2		8	1	1	6	37	4
66	9082	HO104			8	2			1	2		8	8			2		8	1	3	7	7	6
23	9040	BM15		1	8	2		1	1			8	8			2		8	1	1	7	49	4
8	9033	BM14				2			8	2		8				2		8	1	3	7	7	6
7	9013	SCHU			8	2			8	2		8				2		8	1	3	7	7	6
1	9045	TUBO				2	8		8				8			2	2	8	1	2,3	7	51	4
45	9065	HHKB			8	2		8	1	2		8				2		8	1	3	7	7	6
5	9017	WT8			8	2			8	2		8				2		8	1	3	7	7	6
52	9034	SAVC			8	2		1	1	2		8				2		8	1	3	7	7	6
11	9055	HO301			8	2			1	2		8				2		8	1	3	8	14	6
61	9029	WT51				2			2	2		1	8					8	1	23	8	65	6
48	9032	BSM				2	1		1				8			2		8	1	2	9	62	6
33	9060	CB6B			8	2		1	8			8	8			2		8	1	1	9	62	6
25	9069	MADURA		1		2			1				8			2		8	1	2	10	60	6
27	9028	PE117		1		2			2				8		8			8	1	24	10	60	6

RFLP Standardization Report for Class I/TaqI

R.E. Bontrop,[1] C.B. Carpenter,[2] R. Walford,[3] J.D. Bignon,[4] D. Cohen,[5] and S. Sekiguchi[6]

Data were received from five different laboratories and standardization was performed on the following criteria: 1) weak hybridization signals were accepted as standard fragments when they were detected by most laboratories, and 2) discrepancies were resolved by majority consensus.

For this system, 24 different standard fragments were detected (Table 1 and Fig. 1). The first fragment has a molecular weight of 12.08 kb, whereas the second fragment has a molecular weight of 11.12 kb. Standard fragment 3 has a molecular weight of 7.33 kb and a panel frequency of 84.3%. Some of these fragments are observed as a faint signal (Table 1). Standard fragment 4 has a molecular weight of 6.80 kb and appears to be relatively rare. Standard fragments 5 and 6 have molecular weights of 6.20 and 5.76 kb and panel frequencies of 72.2%, respectively. Some of these bands are observed as weak hybridization signals. Standard fragment 7 is also seen for 72.2% of the panel cells tested (Table 1) and 25% of of these bands are seen as a weak signal. In most examples, fragments 6 and 7 are observed together. Standard fragments 8 and 9 have molecular weights of 4.39 and 3.75 kb, whereas fragments 10 and 11 have molecular weights of 3.24 and 2.80 kb, respectively. Most of these bands are sometimes observed as faint signals. Standard fragment 12 has a molecular weight of 2.74 kb and a panel frequency of 100%. In contrast to this, standard fragment 13, with a molecular weight of 2.50 kb, is seen for 78.6% of the cells. Fragment 14 has a molecular weight of 2.40 kb and is observed in 55.3% of the cells tested. Standard fragments 15 and 16 have molecular weights of 2.14 and 2.06 kb, respectively. Standard fragment 17 has a molecular weight of 1.83 kb and a panel distribution of 40.0%. Standard fragments 18 and 19 have molecular weights of 1.76 and 1.34 kb, whereas fragments 20 and 21 have molecular weights of 1.23 and 1.11 kb, respectively. Standard fragments 22 and 23 have molecular weights of 0.98 and 0.96 kb and panel frequencies of 100%, respectively. Standard fragment 24 has a molecular weight of 0.62 kb and a panel frequency of 100%.

Participating Laboratories: BENROO,[1] NCYCAR,[2] US5WAL,[3] FRABIG,[4] FRACOH,[5] JAPSEK[6]

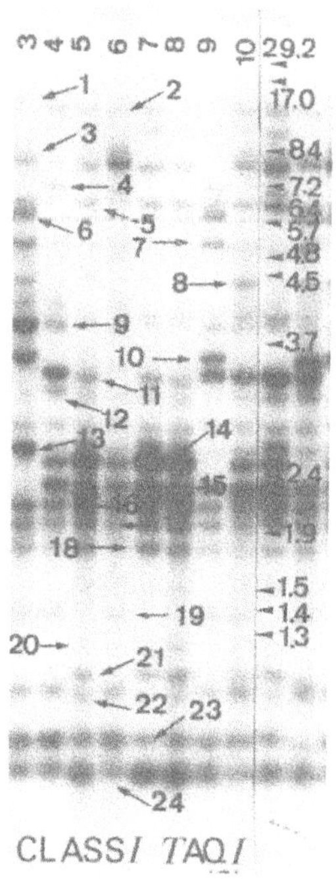

Figure 1. Fragments observed with the class I/TaqI system.

Table 1. Fragments Identified with the Class I/TaqI System

Fragment	kb	Frequency	% Faint	56	32
1	12.08	0.388	0.000		
2	11.12	0.652	0.000		
3	7.33	0.843	0.148	s	
4	6.80	0.416	0.067		
5	6.20	0.722	0.256		
6	5.76	0.722	0.184		
7	5.20	0.722	0.250		
8	4.38	0.543	0.156	s	
9	3.75	0.613	0.362		
10	3.24	0.643	0.070		
11	2.80	0.514	0.075		
12	2.74	1.000	0.623		f
13	2.50	0.786	0.167		
14	2.40	0.553	0.167	s	s
15	2.14	0.843	0.043		
16	2.06	0.957	0.145		
17	1.83	0.400	0.345		
18	1.76	1.000	0.433		
19	1.34	1.000	0.682		s
20	1.23	0.166	0.300		
21	1.11	0.157	0.273		
22	0.98	1.000	0.259		
23	0.96	1.000	0.157		
24	0.62	1.000	0.286		

Table 2. DNA Fragment Presence/Absence Patterns in 72 Core Cell Line Target Cells: Class I/TaqI

	Fragment No.																							
Cell	1	2	3	4	5	6	7	8	9	10	11	12	13	14	15	16	17	18	19	20	21	22	23	24
9058	8				1	1	1	2	1	1		1	2		1	1	1	1	1			1	1	1
9004	8			1	1	1	1			1	2	1			1	1	1	1	1			1	1	1
9056	8				1	1	2			1	2	1			1	1	1	1	1			1		1
9010	8		2		1	1	1			1		1			1	1	1	1	1			1	1	1
9069	8		1		1	8	8	8		1		1			1	1	1	1	1			1	1	1
9037	8		1			1	1			1		1			1	1	1	1	1			1	1	1
9028	8		1					8				1		2	1	1	1	1	1			1	1	1
9031	8		1		1	1	1	2		1		1			1	1	1	1	1			1	1	1
9036	8				2	1	1	1		1	2	1			1	1	1	1	1			1	1	1
9066	8				2	1	1	1	1	1	2	1			1	1	1	1	1			1	1	1
9064	8		1		2	8	1	1		1		1			1	1	1	1	1			1	1	1
9054	8				2	8	1	2		1	2	1			1	1	1	1	1			1	1	1
9105	8		1	8	2	8	1			1	1	1		8	1	1	1	1	1			1	1	1
9091	8		1		2	8	1	1		1		1			1	1	1	1	1			1	1	1
9016	8		8		2	1	1		1	1		1			1	1	1	1	1			1	1	1
9021	8		8	1	8	8	1	2		1		1	2	2	1	1	1	1	1			1	1	1
9062	8			2	2	1	1			1	2	1			1	1	1	1	1			1	1	1
9032	8		1		2	8	1	1		1		1			1	1	1	1	1			1	1	1
9059	8		1	1	2	1	1	1		1		1	8	2	1	1	1	1	1			1	1	1
9068	8		1	2	2	1	1	1		1		1	2		1	1	1	1				1	1	1
9061	8		8	1	2	1	1	1		1	2	1			1	1	1	1	1			1	1	1
9038	8			1	2	1	1	1		1	2	1	8		1	1	1	1	1			1	1	1
9070	8				2	1	1	1		1	2	1	1		1	1	1	1	1		1	1	1	1
9039	8		8	1	2	1	1	1	1	1	2	1	1		1	1	1	1	1			1	1	1
9056	8		1	2	2	1	1			1	2	1	1		1	1	1	1	1			1	1	1
9032	8		2	1	2	1	1	2		1		1	1		1	1	1	1	1			1	1	1
9052	8			2	2	1	1			1		1			1	1	1	1	1			1	1	1
9007	8			2	2	1	1			1		1			1	1	1	1	1			1	1	1
9045		8	1		8	8	8	1		1	8	1	1	2	1	1	1	1	1	1	8	1	1	1
9017		8			8						8	1	2	1	1	1	1	1		1	1	1	1	1
9013		8		1	8						8	1	2	1	1	1	1	1		1	1	1	1	1
9033		8			8						8	1	2	1	1	1	1	1		1	1	1	1	1
9018		8	1		8				1		2	1		1	1	1	1	1	1	1	1	1	1	1
9055		8	1		1						2	1	1	1	1	1	1	1		1	1	1	1	1
9005		8		8	8						2	1	1	1	1	1	1	1		1	1	1	1	1
9047		8			8				1		8	1	1	1	1	1	1	1		1	1	1	1	1
9065		8		1	2						2	1	1	2	1	1	1	1	1	2	8	1	1	1
9034		8		1	2	1		1			8	1	2	2	1	1	1	1		1	1	1	1	1
9082		8		2	2						8	1	2	2	1	1	1	1		1	1	1	1	1
9106		8			2							1		2	1	1	1	1		1	1	1	1	1
9009		8	8			1		2		1		1		2	1	1	1	1	1			1	1	1
9067		8	1	2	2	2	1	2		1	2	1	8		1	1	1	1	1			1	1	1
9014		8	2			1	1			1		1		2	1	1	1	1	1			1	1	1
9042		8	1	1					1			1		1	1	1	1	1	1			1	1	1
9063		8	1			1	1	1		1	2	1	1		1	1	1	1	1			1	1	1
9022		8	2			1	1	8		1		1		1	1	1	1	1	1			1	1	1
9043		8	8			1	1	2		1		1		1	1	1	1	1	1			1	1	1
9023		8	1			1	1	1		1		1		1	1	1	1	1	1			1	1	1
9050		8	1			1	2	2	1	1		1			1	1	1	1	1			1	1	1
9051		8	1			8	2	2	1	1		1				1	1	1	1			1	1	1
9040		8	2			8	8			1		1		1	1	1	1	1	1		1	1	1	1
9025		8	1	2	1	2	1		1	1		1			1	1	1	1	1			1	1	1
9088		8	2		8	1				1		1		2	1	1	1	1	1			1	1	1
9060		8	1			1		1		1		1		2	1	1	1	1	1			1	1	1
9071		8	1		2	1	1	2	1	1	2	1			1	1	1	1	1			1	1	1
9030		8	1		2	1	1	1	1	1		1			1	1	1	1	1			1	1	1
9011		8	8	8	2	1		2		1		1	1	1	1	1	1	1	1			1	1	1
9072		8	8	1	2	1	1	2	1	1	2	1	1		1	1	1	1	1			1	1	1
9075		8						8	1			1	1	2	1	1	1	1	1			1	1	1
9107		8		1					1		2	1		1	1	1	1	1	1			1	1	1
9104		8		1		8	8	2		1	8	1		1	1	1	1	1	1			1	1	1
9002		8							1		2	1	1	2	1	1	1	1	1			1	1	1

Table 2. *Continued*

Cell	Fragment No.																						
	1	2	3	4	5	6	7	8	9	10	11	12	13	14	15	16	17	18	19	20	21	22	23	24
9057		8		1	2	8	1			1	2	1			1	1	1	1	1			1	1	1
9048		8		1				2		1		1		2	1	1	1	1	1			1	1	1
9003		8			1			1	1		2	1	2	2	1	1	1	1	1			1	1	1
9092		8				8	1	1		1	2	1			1	1	1	1				1	1	1
9006		8	1	8	8			1	1	1		1		1	1	1	1	1				1	1	1
9019		8	8	1	8			1	1	1	2	1	1	2	1	1	1	1				1	1	1
9008		8	8	1		1	1	1		1	2	1	8		1	1	1	1				1	1	1
9029		8						1			2	1	1	1	1	1	1	1				1	1	1
9026		8	8	2	2	1	1	1		1	2	1	1		1	1	1	1				1	1	1
9020		8	1	1	2	1	1	2		1	2	1			1	1	1	1				1	1	1

RFLP Standardization Report for Class I/BglII

A. Marcadet,[1] D. Cohen,[1] M.L. Satz,[2] H. Takata,[3] M.C.A. Van Eggermond,[4] C. Gorodezky,[5] and L. Dombrausky[6]

The Class I/BglII system identified 25 fragments on the core cell lines (Table 1). Seven fragments were constant. One fragment (no. 11) of weak intensity was revealed. The pattern of bands observed with the class I probe was very difficult to analyze, especially for large fragments, even on L migration.

Table 2 indicates the distribution of standardized bands in core cell lines. The seven constant fragments

Participating Laboratories: FRACOH,[1] LATFAI,[2] JAPTAK,[3] BENROO,[4] US8RUB,[5] US4STA[6]

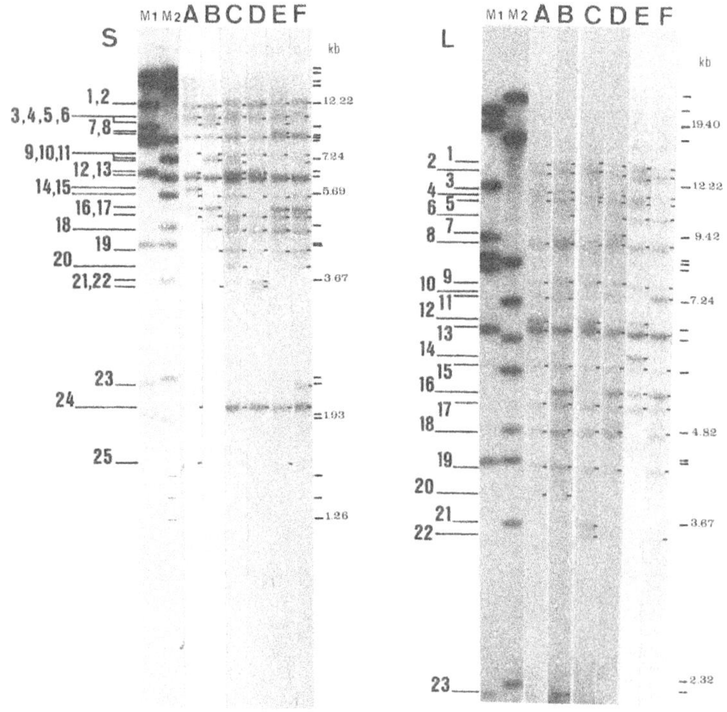

Figure 1. Fragments identified with the class I/BglII system.

Immunobiology of HLA
Volume I

Table 1. Standard Bands in the Class I/BglII System

	kb	CLI Locus	BGL Frequency	COH % Faint	56	32
1	13.97	0 0 0 0	1.000	0.000		
2	13.39	0 0 0 0	0.809	0.000		
3	11.68	0 0 0 0	0.929	0.000		
4	11.23	0 0 0 0	0.014	0.000		
5	10.94	0 0 0 0	1.000	0.000		
6	10.56	0 0 0 0	0.742	0.000		
7	9.48	0 0 0 0	0.029	0.000		
8	9.04	0 0 0 0	1.000	0.000		
9	7.69	0 0 0 0	1.000	0.000		
10	7.47	0 0 0 0	0.043	0.000		
11	7.32	0 0 0 0	0.116	1.000		
12	6.74	0 0 0 0	0.457	0.000		
13	6.49	0 0 0 0	1.000	0.000		
14	6.02	0 0 0 0	0.116	0.000		
15	5.82	0 0 0 0	1.000	0.000		
16	5.37	0 0 0 0	0.586	0.000		
17	5.11	0 0 0 0	1.000	0.000		
18	4.79	0 0 0 0	0.914	0.000		
19	4.44	0 0 0 0	0.914	0.000		
20	3.92	0 0 0 0	0.214	0.000		
21	3.70	0 0 0 0	0.129	0.000		
22	3.59	0 0 0 0	0.449	0.000	s	
23	2.27	0 0 0 0	0.114	0.000		
24	2.07	0 0 0 0	0.929	0.000		
25	1.59	0 0 0 0	0.729	0.925	f	

were not listed. Fragment 4 may be specific for A11 (but the number of cell A11⁺=1). Fragment 7 may be specific for A29-B44 (but n=2). Fragment 20 seems specific for A3(n=11) + Aw30(n=3). Fragment 22 seems to correlate with A2+A25+A26. Fragments 18 and 19 are present for all cells except for A23(n=1) and some A24(4/6). Fragments 2 and 3 are present for all cells except for Aw23(n=1)+A24(n=6)+A29-B44(n=2)+-A32(n=1)+A31(n=3). Fragment 14 may be allelic with fragments 18 and 19. Fragment 10 may be allelic with fragment 24.

Figure 1 is taken from two films from laboratory JAP/TAK. All standardized fragments are indicated at least once for S and/or L migration. The presence of a standardized fragment is indicated by a hyphen.

Comments on L Migration. Standardized fragments 3, 4, and 5, which are very close to each other, are difficult to differentiate, even under these migration conditions:

No. 3 is present for DNA samples A, B, D, E, and F.
No. 4 is present for DNA sample C.
No. 5 is present for DNA samples A, B, C, D, E, and F (constant fragment).

Standardized fragments 10 and 11, which are very close from each other, are difficult to differentiate, even under these migration conditions:

No. 10 is present for DNA sample F (great intensity).
No. 11 is present for DNA samples A and B.

Table 2. RFLP Class I/Bgl II System: Distribution of Standard Bands in Tenth Workshop Core Cell

Cell lines	HLA A	HLA Cw	HLA B	Standard bands N = 23467111111112222222 / 0124689012345
9018	3	5	18	22 2 22 2222 2
9021	w68/30	2	w42	22 2 22222228
9104	31	-	38	22 9 8 2222 22
9048	30	6	13	22 2 8 2222 22
9034	3	7	7	22 2 8 92222 28
9082	3	7	7	22 2 8 2222 28
9013	3	7	7	22 2 8 2222 2
9047	3	6	w47	22 2 8 2222 2
9005	3	1	27	22 2 2222 2
9017	3	7	7	22 2 2222 2
9065	3	7	7	22 2 2222 2
9033	3	7	7	22 9 2222 2
9045	2/3	7	51	22 2 2 2222 2 28
9092	25	1	51	22 2 222 2 28
9008	25	-	18	22 2 222 2 28
9038	2	7	18	22 2 222 2 28
9061	2	7	18	22 2 222 2 28
9058	2	-	45	22 2 222 2 28
9067	2	1	27	22 2 222 2 28
9009	1	6	37	22 2 222 28
9011	1	-	w52	22 2 222 28
9040	1	7	49	22 2 222 28
9043	1	-	w41	22 2 222 28
9106	3	-	w50	99 9 222 28
9004	2	1	27	22 2 222 9 2
9016	2	-	51	22 2 222 2 2
9066	2	11	w46	22 2 2 222 2 8
9019	30	5	18	22 2 82 222 2
9014	26	7	8	22 2 2 22 2 28
9020	26	?5	18	22 2 2 22 2 28
9069	2	10	w60	22 2 2 22 2 22
9031	2	10	w62	22 2 2 22 2 28
9032	2	9	w62	22 2 2 22 2 28
9032	2	9	w62	22 2 92 22 2 28
9091	2	9	w62	22 2 2 22 2 28
9036	2	5	44	22 2 2 22 2 28
9054	2	5	44	22 2 2 22 2 28
9070	2	-	51	22 2 2 22 2 28
9059	2	10	w60	22 2 2 22 2 2
9022	1	7	8	22 2 2 22 28
9023	1	7	8	22 2 2 22 28
9088	1	-	8	22 2 2 22 28
9060	1	9	w62	22 2 2 22 28
9006	11	4	35	222 2 22 22 2
9007	2	6	w57	22 2 222 2 8
9052	2	6	w57	22 2 222 2 8
9010	28	4	w53	22 2 22 2222
9037	29	2	w61	22 2 22 22 28
9039	2	5	18	22 92 22 2 28
9064	2	9	w62	22 2 22 2 28
9068	2	4	35	22 2 22 2 22
9105	1	4	35	22 2 22 28
9055	3	8	14	22 82 222 2
9056	2	-	35	22 222 2
9056	2	-	35	22 222 2 28
9062	2	-	38	22 222 2 28
9026	26	-	38	22 222 2 28
9057	26	-	38	22 222 2 28

Table 2. *Continued*

Cell lines	HLA A	HLA Cw	HLA B	Standard bands N = 2	3	4	6	7	10	11	12	14	16	18	19	20	21	22	23	24	25
9025	31	4	35								2			2	2				2	2	8
9063	32	5	44								2			2	2		2			2	8
9042	24	4	35								2	2					2			2	
9075	24	10	w60				2				2	2					2				8
9029	23	8	w65				2				2	2								2	8
9107	24	1	w54				2					2	2							2	
9003	24	-	51				2					2	2				2			2	8
9028	24	10	w60				2				2	2	2							2	2

Table 2. *Continued*

Cell lines	HLA A	HLA Cw	HLA B	Standard bands N = 2	3	4	6	7	10	11	12	14	16	18	19	20	21	22	23	24	25
9030	31	-	51				2						2	2	2				2	2	8
9071	31	1/10	w62				2						2	2	2				2	2	8
9072	31	1	w62	9	9		9						2	2	2				2	2	8
9002	24	2/8	14				2				2	2	2	2	2					2	
9050	29	-	44				2	2					2	2	2					2	8
9051	29	-	44				2	2					2	2	2					2	8

RFLP Standardization Report for Class I/PvuII

P. Piccioli,[1]* K. Kato,[2] A. Smith,[3] J. Partanen,[4] B. Parodi,[1] and G.B. Ferrara[1]

The class I/PvuII system has identified 34 fragments on the 70 cell lines tested (Tables 1 and 2). (The 6th and 25th bands have been detected by only one laboratory that performed the standardization.) Even though this system is very polymorphic, only few fragments presented a specific association with the A, B, and C specificities. In particular a specific association between B35,Cw4 and the 2.69 kb fragment has been found; the same fragment was also evident in sample 9056 typed as B-,Cw4. The 14.82 kb fragment segregates in the cell lines typed as A3 (with one exception represented by cell lines 9006 carrying the specificity A11), and in all A3,B7,Cw7. Cw7 cell lines on the contrary lack the 17th fragment. A31 positive cell lines present an interesting pattern, because, apparently, 5.49 kb and 4.8 kb fragments represent allelic fragments for this specificity, and 3.22 kb fragment seems to be specifically associated with this specificity. Moreover 3 bands are correlated with individual A or B specificities: 8.81 kb fragment (7th) with A3 and A24; 4.80 kb fragment (15th) with A1; 3.22 kb fragment (20th) with A31. Figures 1 and 2 show the fragments obtained using the class I-specific probe and the enzyme PvuII.

Participating Laboratories: IT2FER,[1] US7DUP,[2] US7HAN,[3] SCAKOS[4]

*P. Piccioli was recipient of a fellowship from A.I.R.C. (Associazione Italiana per la Ricerca sul Cancro) This work was in part supported by CNR grants n.86.0265.51 and 86.00659.

Table 1. Banding Pattern of the Class I/PvuII System in each cell line

Cell		A	C	B	Fragment 1	2	3	4	5	6	7	8	9	10	11	12	13	14	15	16	17	18	19	20	21	22	23	24	25	26	27	28	29	30	31	32	33	34
9105	FPAF	1	4	35	(8)			2				2	8	2			9	2	2		2		2			8	8				8	8	8	8	(8)	8	(8)	1
9022	COX	1	7	8	8			2				2	(8)	2			8	2	2	2		1					1		1		1	1	1	1		8	8	1
9023	VAVY	1	7	8	8			2				2	(8)	2			8	2	2	2		1					1		1		1	1	1	1		8		1
9040	BM 15	1	7	49	8			1				2	1	2			8	2	2			1	1				1		1		1	1	1	1	8	8	8	1
9088	PF 04015	1	-	8	8			2				2	(8)	2			8	2	2	2		8					8		8				8	8		(8)		(1)
9043	BM 21	1	-	41	8			2		8		2	(8)	2			8	2	2	2		1					1		1		1	1	1	1		8		1
9011	E 4181324	1	-	52	8			2				2	(8)	1			8	2	2		1	1	1				8		8		8	8	8	8	9	9	9	(8)
9060	CB 6B	1	9	62	8			2		8		2	8	2			9	2	2	2		8					8		8		(8)	8	8	8		8	8	8
9009	KAS 011	1	6	37	8			8				2	(8)	2			8	2	2	2	2						8		8			8	8	8		8	8	8
9058	OMW	2	-	45	8			2				2	(8)	2				2		2		8			8		8		8	1		8	8	8		8	8	
9036	SPO 010	2	5	44	8			2		8		2	8	2				2		2	2				8		8			1		8	8	1		8		
9054	EK	2	5	44	8			2				2	8	2			8	2		2	2				8		8		8	1	(8)	8	8	8		8	8	8
9004	JESTHOM	2	1	27	8							2	(8)	2			8	2			1		1		8		1		1	1	(1)		1	1	8	8	(8)	1
9067	BTB	2	1	27	8			2		8		2	(8)	2			8	2			2	1	1		1		8		8	1	8	8	8	8	8	8	8	8
905b	KOSE	2	-	35	8			2				2	(8)	2			8	2			1		1		1	1	1		1	1		1	1	1	(8)	(8)		1
9056	KOSE	2	-	35	8			2		8		2	(8)	2			1	2			1	1	1		1	1	8		(8)	1	(8)	8	8	8	8	8	8	(8)
9068	BM9	2	4	35	8			2		8		2	(8)	2			8	2			2		8		8	8	8		8	1	8	8	8	8	8	8	8	8
9062'	WDV	2	-	38	8			2				2	8	2			8	2			2		8		8	8	8		8	1	(8)	8	8	8	(8)	8	(8)	(8)
9016	RML	2	-	51	8			2		8		2	8	2			9	2				1	8		8		8		8	1	(8)	8	8	8	8	8		8
9070	LUY	2	-	51	8			2		8		2	(8)	2			1	2			1	8	1		1		8		8	1	8	8	8	8	8	8	(8)	(1)
9061	31227 ABO	2	-	18	8			2				2	(8)	2			8	2				2	1		1		8		8	1	8	8	8	8	8	8	8	(8)
9038	BM 16	2	7	18	8			2				2	(8)		2		8	2			1	1	1		8		8		8	1	8	8	8	8	8	8	8	8
9039	JVM	2	5	18	8			2		8		2	(8)	2			1	2			1	1	1		8		8		8	1	8	8	8	8	8	8	8	(8)
9052	DBB	2	6	57	8			2		8		2	(8)	2			1	2			1	1	1		1		8		8	1	8	8	8	8	8	8	8	8
9007	DEM	2	6	57	8			2		8		2	(8)	2			1	2			1	1	1		8		1		8	1	8	9	1	9	9	9	9	8
9069	MADURA	2	10	60	(8)			2				2	(8)	2			8	2				1	1		1		8		8	1	8	8	8	8	8	8	8	8
9059	SLE 005	2	10	60	8			2				2	(8)	2			8	2				2	1		1		8		8	1	9	8	8	8	8	8	8	8
9031	BOLETH	2	10	62	(8)			2		8		2	(8)	2			8	2		2		8			8		8		8	1		8	8	8		8	8	(8)
9064	AMALA	2	9	62	(8)			2				2	8	2			8	2		2		8			8		8			1		8	8	8		8	8	8
9032	BSM	2	9	62	8			2		8		2	8	2			9	2		2		8			8		8		8	1	(8)	8	8	8		8	(8)	8
9032	BSM	2	9	62	8			2		8		2	(8)	2			1	2		1		1			1		8		8	1	8	8	8	8		8	(8)	8
9091	MLF	2	9	62	8			2				2	8	2			9	2		2		8			8		8		8	1	8	8	8	8		8	8	8
9066	TAB 089	2	11	46	8			2		8		2	8	2			1	2		2	2				8		8		8	2	(8)	8	8	8		8	8	8
9045	TUBO	2/3	7	51	8						8	2	(8)	1			1	2			1				1		1		1	1	(1)	(1)	1	1	8	8	(8)	1
9017	WT 8	3	7	7	8		2				1	2	(8)				8	2		2		1			1	1	1		1	1	(1)	(1)	1	1		8		1
9013	SCHU	3	7	7	8		2			8	1	2	(8)				1	2		2		1			1		1		1	1	(1)	(1)	1	1		8		1
9033	BM 14	3	7	7	8		2	2			1	2	(8)				8	2		2		1			1		1		1	1	(1)	(1)	1	1		8		1
9065	HHKB	3	7	7	8		2			8	1	2	8				8	2		2		8			8		8		8	1		8	8	8		8		8

Table 1. *Continued*

9034	SAVC	3	7	7	8		2				1	2	(8)				1	2		2		2			1		8		8	1	8	8	8	8	(8)	8	8	8
9082	HO 104	3	7	7	8		2				1	2	(8)				1	2		1		1			1		8		8	1	(8)	8	8	8	(8)	(8)	9	(8)
9018	LO 081785	3	5	18	8		2				1	2	8	2			1	2			1		1		8		1		1	1	1	(1)	1	1	8	8	(8)	1
9005	HOM 2	3	1	27	8		2				1	2	(8)				8	2			1		1		1		1		1	1	1	1	1	1	(8)	(8)	(8)	1
9047	PLH	3	6	47	8		2				1	2	(8)				1	2			1		1		1		1		1	1	1	1	1	1	8	8	8	1
9106	MANIKA	3	-	50	8						1	2	(8)	8			8	2			1	8	1		8		8		8	1	8	8	8	8	8	8	8	8
9055	HO 301	3	8	14	8		2			8	1	2	(8)				8	2			1		1		1		1		1	1	1	1	1	1	8	(8)	(8)	1
9029	WT 51	23	8	65	(8)						1	2	(8)	2				2			1	1	2				8		8		8	8	8	8	8	8	8	(8)
9028	PE 117	24	10	60	(8)						1	2	(8)	2				2				8	1				8				(8)	8	8	8	8	(8)	(8)	(1)
9075	DKB	24	10	60	8					8	1	2	(8)	2				2				2	1				8		8		8	8	8	8	8	8	8	8
9107	LKT3	24	1	54	8						1	2	(8)	2				2			1		1				1		1		(1)	(1)	1	1	8	8		1
9042	TISI	24	4	35	8						1	2	(8)	2				2		2	1						1		1			(1)	1	1		8	(8)	1
9003	KAS 116	24	-	51	8					8	1	2	(8)	2				2			1	1	1				8		8		8	8	8	8	8	8	8	(1)
9002	MZ 070782	24	2/8	14	(8)						1	2	8	2			1	2			2		8				8		8			8	8	8	(8)	8		(8)
9092	BM 92	25	1	51	8			2		8		2	(8)	2			8	2			2		1		1		8		8	1		8	8	8	8	8	8	8
9008	DO 208915	25	-	18	8			2		8		2	(8)	1				2			1	1	1		8		8		8	1	8	8	8	8	8	8	8	8
9057	TEM	26	-	38	(8)			2				2	8	2			9	2			2		8		8		8		8		1	(8)	8	8	8	8	8	8
9026	YAR	26	-	38	8			2		8		2	(8)	2			1	2			1	1	2		1		8		8	1	8	8	8	8	8	8	8	(8)
9014	MGAR	26	7	8	8			2				2	(8)	2			8	2		2		1			1		1		1	1	(1)	(1)	1	1		8	(8)	1
9020	QBL	26	?5	18	8							2	(8)	2			1	2			1	1	1		8		8		8	1	8	8	8	8	8	8	8	8
9006	WT100 BIS	11	4	35	8		2					2	(8)		2		8	2			2		1			1	1		1		(1)	1	1	1	(8)	8	(8)	1
9050	MOU	29	-	44	8							2	(8)	2			8	2			1		1		1		1		1	1	1	1	1	1	(8)	(8)	(8)	1
9051	PITOUT	29	-	44	8			2		8		2	(8)	2			1	2			1		1		1		1		1	1	1	1	1	1	(8)	8	8	1
9037	SWEIG 007	29	2	61	(8)			2				2	(8)	2			8	2			2		8		8		8		8	1	8	8	8	8	8	8		8
9048	LBUF	30	6	13	8	(1)			2	8		2	8	2			9	2	2		2		8	1			8	8	8		(8)	8	8	8	8	8	8	8
9021	RSH	68/30	2	42	8	(1)		2	2	8		2	8	2	2		9	2	2	2		8		1			8	8	8		8	8	8	8		8		1
9019	DUCAF	30	5	18	(8)				1	8		2	(8)	2			8	2		8	2		8		8		8		8	2	8	8	8	8	8	8	8	8
9104	DHIF	31		38	8				2			2	(8)	2			1	2	2		1		1	2			1	1	1	1	1	1	1	1	(8)	(8)	8	1
9030	JHAF	31	-	51	(8)	(1)		2		8		2	8	2		2	2	2				8		1	8		1	(8)	8	1	(8)	8	8	8	8	8	(8)	8
9025	DEU	31	4	35	(8)	1		2		8		2	1	2		2	2	2			2		8	1	8	8	8	(8)	8	1		8	8	8	8	8	8	8
9072	SPACH	31	1	62	8			2				2	(8)	1		1	1	2		1				1	8		(8)	8	8	1	8	8	8	8		8	8	8
9071	OLGA	31	1/10	62	(8)	1		2				2	8	2		2	8	2		2	2			1	8		8	8	8	1	8	8	8	8	(8)	8	8	8
9063	WT 47	32	5	44	8			2				2	(8)		2		1	2		2	1				1		1		1	1	1	1	1	1		8	(8)	1
9010	AMAI	28	4	53	8			2				2	8	2			1	2	2		1		1	2		8	1	(1)	1		1	1	1	1	8	8	8	1

1 = Single intensity; 2 = double intensity; 8 = faint; 9 = unable to read. ( ) = discordance between labs that performed this study.

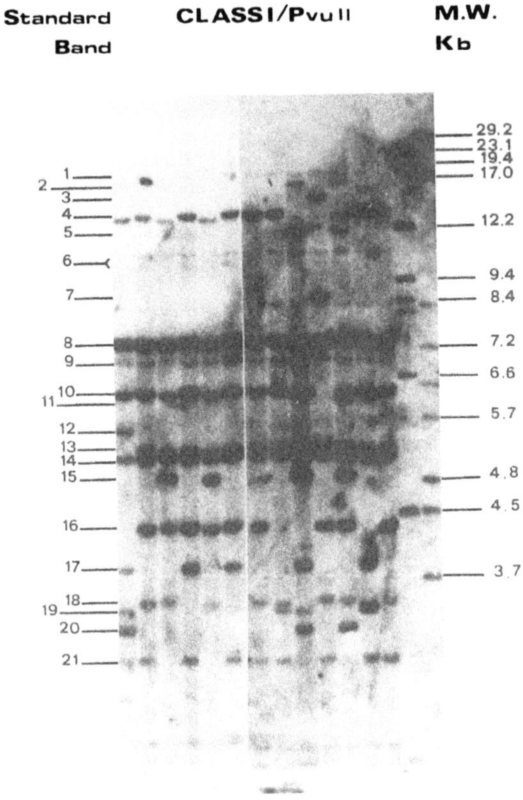

Figure 1. Southern blot analysis of PvuII-digested genomic DNAs from core cell lines, hybridized with a class I probe. The samples were digested with PvuII endonuclease and the gel was run using long migration conditions. From left to right: DEU, BOLETH, PF04015, TAB089, CB6B, SP0010, BM92, RML, LBUF, HHKB, RSH, WDV, BSM, and molecular weight controls.

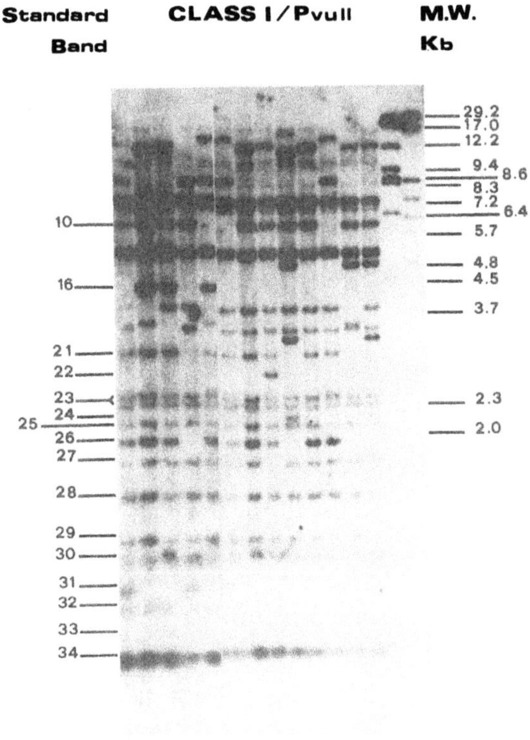

Figure 2. Southern blot analysis of PvuII-digested genomic DNAs from core cell lines, hybridized with a class I probe. The samples were digested with PvuII endonuclease and the gel was run using short migration conditions. From left to right: TUBO, MGAR, OMW, LKT3, WT8, HOM2, MOU, KOSE, DHIF, PITOUT, PLH, BM15, AMAL, and molecular weight controls.

Table 2. Fragments Identified with the Class I/Pvu II System

Fragment	kb	Frequencies[a]	% Faint[b]
1	18.15	0.757	1.000
2	16.19	0.029	0.000
3	14.82	0.157	0.000
4	13.10	0.671	0.021
5	11.81	0.057	0.000
6	10.43	0.000	0.000
7	8.81	0.271	0.057
8	7.45	1.000	0.000
9	7.05	0.314	0.909
10	6.21	0.800	0.018
11	6.05	0.058	0.000
12	5.49	0.058	0.000

Table 2. *Continued*

Fragment	kb	Frequencies[a]	% Faint[b]
13	5.23	0.992	0.508
14	5.10	1.000	0.000
15	4.80	0.186	0.000
16	4.20	0.386	0.037
17	3.79	0.357	0.000
18	3.41	0.057	0.308
19	3.37	0.600	0.238
20	3.22	0.114	0.000
21	2.95	0.814	0.357
22	2.69	0.086	0.500
23	2.37	1.000	0.686
24	2.14	0.057	0.750

Table 2. *Continued*

Fragment	kb	Frequencies[a]	% Faint[b]
25	2.06	0.000	0.000
26	1.90	0.686	0.000
27	1.70	0.528	0.338
28	1.41	0.814	0.728
29	1.08	1.000	0.645
30	0.94	0.927	0.695
31	0.76	0.500	1.000
32	0.64	0.891	1.000
33	0.49	0.514	1.000
34	0.37	0.778	0.541

Table 2. *Continued*

[a]Band frequency has been calculated as number of cell lines presenting a specific fragment versus number of cell lines tested.
[b]Frequency of faint band has been calculated as number of cell lines presenting a faint band versus number of cell lines presenting the same fragment (i.e., frequency 1.000 and % faint 0.000 means that all samples tested share the same band and it is a strong band; on the other hand 1.000 and 1.000 means that it is both a constant and faint band). Fragments 6 and 25 present both frequency and % faint 0.000 because were found in one lab; for this reason STD gave 0.000.

RFLP Standardization Report for Class I/EcoRI

M. Segall,[1]* L. Schluender,[1]* H. Betuel,[2] B. Bradley,[3] M. Garovoy,[4] and S. Naito[5]

Class I/EcoRI blots were received from four laboratories, and were analyzed as described for the DR beta/MspI system. The class I/EcoRI system identified 20 fragments on the core cell lines (Table 1 and Fig. 1). All fragments were readily identifiable except for 13 and 14. Fragments 13 and 14, listed as frequency 0.000, % faint 0.000, were recorded by one of four laboratories participating, as a faint band on one blot and unreadable on the other two. They were visible in some portions of the blots submitted by other laboratories, but could not be uniformly identified. There were no discrepancies for the duplicate DNAs 9032 and 9056.

Because the core cell lines were not selected to represent a panel of class I antigens, analysis of the correlations of the fragments with particular specificities is difficult. Presence/absence of fragments in the core cell lines is given in Table 2, in which the cells are arranged by Workshop number. Fragment 6 is positive with 22/25 A2+ cells; fragment 7 is negative with 2/2 A30, 6/6 A24, 1/7 A1 and positive with all other cells.

Table 1. Standard Fragments, Class I/EcoRI

				Discrepant for	
Fragment	kb	Frequency	% Faint	9056	9032
1	20.93	1.000	0.000		
2	17.71	1.000	0.986		
3	13.81	0.057	0.000		
4	12.99	0.014	0.000		
5	12.43	1.000	0.000		
6	11.79	0.300	0.000		
7	11.46	0.871	0.016		
8	10.83	0.171	0.083		
9	8.75	1.000	0.000		
10	7.96	0.157	0.000		
11	7.45	0.857	0.000		
12	7.20	0.929	0.731		
13	7.11	0.000	0.000		
14	7.04	0.000	0.000		
15	6.65	1.000	0.000		
16	6.47	1.000	0.000		
17	5.98	1.000	0.000		
18	5.76	0.114	0.000		
19	5.18	0.681	0.776		
20	2.87	0.171	0.083		

Fragment 8 is positive with 12/12 A3; fragment 18 is positive with 5/5 A3, 1/1 A28, 1/1 A68 and 1/2 A30; fragment 20 is positive with 6/6 B7 plus scattered other positives. Fragments 10 and 11 show reciprocal patterns but no apparent specificity.

Participating Laboratories: US2BAC,[1] FRABET,[2] UKIBRS,[3] US3GVY,[4] JAPNAI[5]

*This is publication #499 from the Immunobiology Research Center, University of Minnesota, Minneapolis, Minnesota 55455, USA.

Supported in part by Juvenile Diabetes Foundation Grant #186175 and March of Dimes Birth Defects Foundation Grant #6-496.

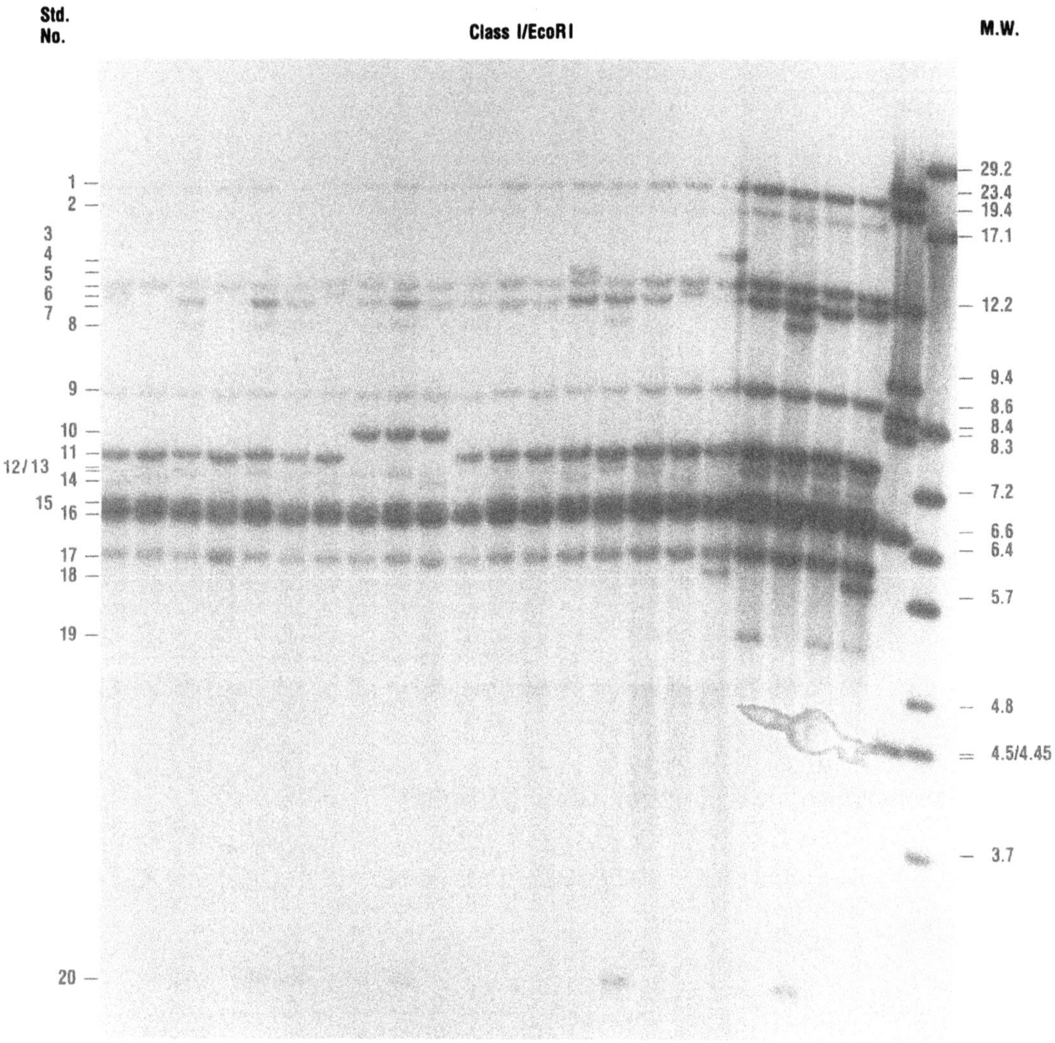

Figure 1. Fragments observed with the class I/EcoRI system.

Table 2. Distribution of Polymorphic Fragments in Core Cell Lines, Class I/EcoRI

		St. Fragmant No.									
Cell No.	Name	3	4	6	7	8	10	11	18	19	20
9002	MZO70782						1	1			
9003	KAS116							1			
9004	JESTHOM			1	1			2		1	
9005	HOM2				2	1		2			1
9006	WT100BIS		2		2			2			
9007	DEM			1	1			2		8	
9008	D0208915				1			1		8	
9009	KAS011				1			1		8	
9010	AMAL				1			2	2	1	
9011	E4181324						2			8	
9013	SCHU				2	1		2			1
9014	MGAR				2			2		1	

Table 2. *Continued*

		St. Fragmant No.									
Cell No.	Name	3	4	6	7	8	10	11	18	19	20
9016	RML			1	1			1		8	
9017	WT8				2	1		2			1
9018	L0081785				1	8	2				8
9019	DUCAF	2					2			8	
9020	QBL				1		2			8	
9021	RSH	1			1			1	2	8	
9022	COX				1			2			
9023	VAVY				1			2		8	
9025	DEU				2			1	2		
9026	YAR				1			2		8	
9028	PE117							2			
9029	WT51				1		2			8	
9030	JHAF				2			1	2	8	

Table 2. *Continued*

Cell No.	Name	St. Fragmant No. 3	4	6	7	8	10	11	18	19	20
9031	BOLETH			1	1			1		8	
9032	BSM				1			1		8	
9032	BSM				1			2		8	
9033	BM14				1	1		2			1
9034	SAVC				1	1		2		9	1
9036	SP0010			1	1		2			1	
9037	SWEIG007				1			1		8	
9038	BM16			1	1			1		8	
9039	JVM			1	1		2			8	
9040	BM15				1			2		8	
9042	TISI							2			
9043	BM21				2			2		1	
9045	TUBO				1	1		2			1
9047	PLH				1	1		2			1
9048	LBUF	2						1	2	8	
9050	MOU				2			2		1	
9051	PITOUT				1			2		1	
9052	DBB			1	1			2		8	
9054	EK			1	1		2			8	
9055	HO301				2	2	2				1
9056	KOSE			1	1			2		1	
9056	KOSE			1	1			1		8	
9057	TEM				1			1		8	
9058	OMW			1	1			2		1	

Table 2. *Continued*

Cell No.	Name	St. Fragmant No. 3	4	6	7	8	10	11	18	19	20
9059	SLE005			1	1			2		8	
9060	CB6B				1			1		8	
9061	31227ABO			1	1			2		8	
9062	WDV			1	1			1		8	
9063	WT47				1		2			1	
9064	AMALA			1	1			1		8	
9065	HHKB				2	1		1			1
9066	TAB089			1	1			1		8	
9067	BTB			1	1			2		8	
9068	BM9			1	1			2		8	
9069	MADURA			1	1			2		8	
9070	LUY			1	1			2		8	
9071	OLGA				2			1	2		
9072	SPACH				2			2	1		
9075	DKB							2			
9082	HO104			2	1		2			1	
9088	PFO4015				1			1		8	
9091	MLF			1	1			1		8	
9092	BM92				1			2		8	
9104	DHIF	1			8			2	2	1	
9105	FPAF				1			1		8	
9106	MANIKA				1	1		2			1
9107	LKT3							2			

RFLP Standardization Report for Class I/HindIII

P.A. Biro,[1] S. Cutbush,[1] J.M. Tiercy,[2] G. Paulsen,[3] J. Demopulos,[4] R.T. Acton,[4] S.V. Van Tonder,[5] G. Semana,[6] and F. Quillivic[6]

Twenty-eight fragments have been defined using the class I probe (Table 1 and Fig. 1). The family studies have revealed the existence of a few additional faint bands that are not detected in every blot. The majority of the fragments are polymorphic, although many of their frequencies are either above 85% or below 15%. Most individual DNA samples display 15 to 18 bands which correlate well both in number and size with those reported (1,2). There is no relationship between RFLP pattern and DR type (Table 2).

Table 1. Sizes and Frequencies of Standard Bands Produced by the Class I/HindIII Enzyme/Probe Combination

	kb	Frequency	% Faint
1	43.87	0.271	0.050
2	31.25	0.757	0.000
3	24.48	0.300	0.000
4	18.91	0.671	0.000

Participating Laboratories: UKIFES,[1] FRAJEA,[2] SCATSB,[3] US5UAB,[4] SAFHAM,[5] FRAFAU[6]

Table 1. *Continued*

	kb	Frequency	% Faint
5	14.91	0.014	0.000
6	9.93	0.057	0.000
7	9.00	1.000	0.000
8	8.89	1.000	1.000
9	7.69	1.000	0.000
10	7.39	0.171	0.000
11	7.27	0.859	0.035
12	7.07	0.071	0.000
13	6.72	0.843	0.000
14	6.07	1.000	0.000
15	5.89	1.000	0.000
16	5.70	1.000	0.000
17	5.57	0.043	0.000
18	5.46	0.929	0.000
19	5.35	0.900	0.000
20	5.17	0.057	0.000
21	5.09	0.686	0.000
22	4.98	0.114	0.000
23	4.86	0.100	0.000
24	4.76	0.157	0.000
25	2.98	1.000	1.000

Table 1. *Continued*

	kb	Frequency	% Faint
26	2.36	0.214	0.000
27	2.12	0.914	0.015
28	1.66	0.571	0.000

References

1. Koller BH, Geraghty D, Orr HT, Shimizu Y, DeMars R. Organization of the human class I major histocompatibility complex genes. Immunol Res 1987;6:1.
2. Shimizu Y, Geraghty DE, Koller B, Orr HT, DeMars R. Transfer and expression of three cloned human non-HLA-A-B-C class I major histocompatibility complex genes in mutant lymphoblastoid cells. Proc Natl Acad Sci USA 1988;85:227.

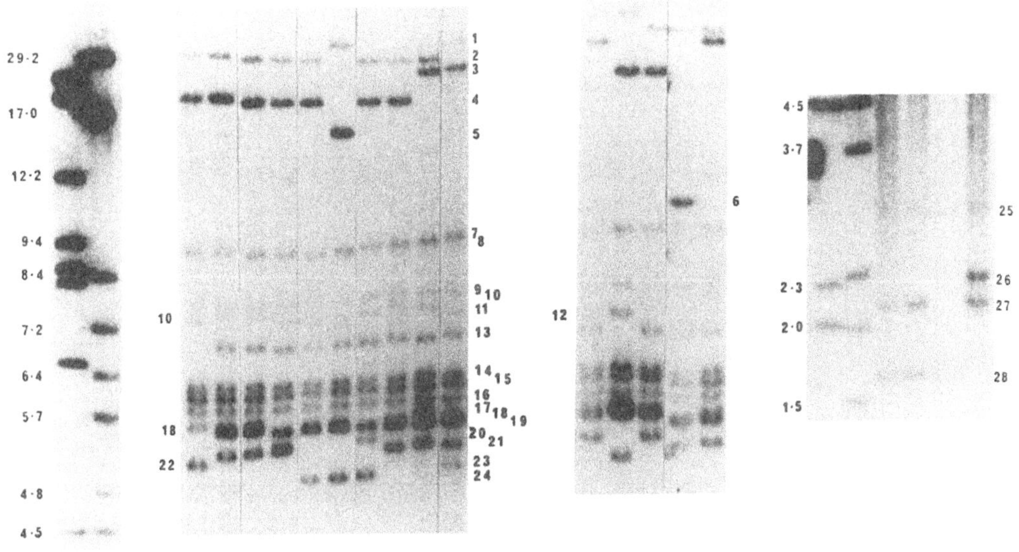

Figure 1. Standard bands produced by the class I/HindIII enzyme/probe combination.

Table 2. RFLP Pattern of the Class I/Hind III System

Cell lines		DR Type	Fragment No. 1111111111222222222 1234567890123456789012345678	Cell lines		DR Type	Fragment No. 1111111111222222222 1234567890123456789012345678
MZ070782	9002	1	1 3 281 1 222 2 2 8 1	BM21	9043	5	1 3 281 2 2222 22 2811
KAS116	9003	1	1 2 281 1 222 2 2 8 1	TUBO	9045	5	13 281 2 2222 22 2 8 11
JESTHOM	9004	1	1 3 281 2 2222 22 2 8 11	PLH	9047	7	1 2 28119 2222 22 2 8 11
HOM2	9005	1	1 3 2811 2222 22 2 8 11	LBUF	9048	7	1 2 281 2 2222 22 2 2 8 1
WT100BIS	9006	1	1 3 281 9 2222 22 28 11	MOU	9050	7	12 2811 2222122 2 8 1
DEM	9007	2/4	1 2 281 2 2222 22 2 8 11	PITOUT	9051	7	12 28119 2222122 2 8 1
DO208915	9008	2	1 2 281 2 2222 22 2 8 11	DBB	9052	7	1 2 281 2 2222 22 2 8 11
KAS011	9009	2	822 281 2 2222 22 28 11	EK	9054	6	1 3 281 2 2222 22 2 8 11
AMAI	9010	2	1 3 281 2 2222 22 228 11	HO301	9055	6	1 3 2811 2222 22 2 8 11
E4181324	9011	2	12 281 2 2222 22 2821	KOSE	9056	6	1 2 281 2 2222 22 2 8 11
SCHU	9013	2	1 3 2811 2222 22 2 8 1	TEM	9057	6	1 2 281 2 2222 22 2 8 11
MGAR	9014	2	1 3 281 2 2222122 2 8 1	OMW	9058	6	12 281 2 2222 22 2 8 1
RML	9016	2	1 3 281 2 2222 22 2 8 11	SLE005	9059	6	1 3281 2 2222 22 2 8 11
WT8	9017	2	1 3 28119 2222 22 2 8 1	CB6B	9060	6	1 3 281 2 2222 22 2811
LO081785	9018	3	1 3 281 2 2222 22 22 818	31227ABO	9061	6	1 3 281 2 2222 22 2 8 11
DUCAF	9019	3	1 3 281 2 2222 22 2 811	WDV	9062	6	1 2 281 2 2222 22 2 8 11
QBL	9020	3	1 3 281 2 2222 22 2 811	WT47	9063	6	1 3 281 2 2222 222 8 11
RSH	9021	3	1 2 281 2 2222 2222 2 811	AMALA	9064	6	1 3 281 2 2222 22 2 811
COX	9022	3	1 3 281 2 2222 22 28 11	HHKB	9065	6	1 3 2811 2222 2 2 8 1
VAVY	9023	3	1 3 281 2 2222 22 28 11	TABO89	9066	8	1 3 281 2 2222 22 2 811
DEU	9025	4	1 3 281 82 222 22 2 8 11	BTB	9067	8	1 3 281 2 2222 22 2 8 11
YAR	9026	4	1 2 281 2 2222 22 2 8 11	BM9	9068	8	1 3 281 2 2222 22 2 8 11
PE117	9028	4	112 3281 1 222 2 2 8 1	MADURA	9069	8	1 3281 2 2222 22 2 8 11
WT51	9029	4	1 3 281 1 222 2 2 8	LUY	9070	8	1 3 281 2 2222 22 2 8 11
JHAF	9030	4	1 3 281 12 222 22 2 8 11	OLGA	9071	8	11 3 281 12 222 22 2 811
BOLETH	9031	4	1 3 281 2 2222 22 2 811	SPACH	9072	8	1 3 281 2 222 22 811
BSM	9032	4	1 3 281 2 2222 22 2 811	DKB	9075	9	1 3281 1 222 2 2 8 11
BM14	9033	4	1 3 2811 2222 22 2 8 1	HO104	9082	?	1 3 2811 2222 22 2 8 1
SAVC	9034	4	1 3 28118 2222 22 2 8 1	PF04015	9088	3	1 3 281 2 2222 22 28 11
SPO010	9036	5	1 3 281 2 2222 22 2 8 11	MLF	9091	4	1 3 281 2 2222 22 2 811
SWEIG007	9037	5	121 281 2 2222 22 2 8 11	BM92	9092	4	1 3 281 2 2222 22 2 8 11
BM16	9038	5	1 3 281 2 2222 222 8 11	DHIF	9104	11	1 2 281 2 2222 22 2 2 8 1
JVM	9039	5	1 3 281 2 2222 22 2 811	FPAF	9105	11	1 3 281 9 2222 2 28 11
BM15	9040	5	1 3 281 2 2222 22 28 1	MANIKA	9106	7	1 2 2811 2222 22 2 8 1
TISI	9042	5	1 3 281 92 222 2 2 8 1	LKT3	9107	4	1 3 281 1 222 2 2 8 1

RFLP Standardization Report for Class I/MspI

S. Tonks,[1] V. Groves,[1] C. Muller,[2] F. Otani,[3] J.R. Regueiro,[4] M. Segall,[5] J. Bodmer,[1] and J. Trowsdale[1]

The class I/MspI system showed extensive polymorphism, identifying 21 fragments on the core cell lines (Table 1 and Fig. 1). However, the large number of bands made interpretation of the data difficult despite the high resolution obtained by running two gel concentrations. The problems of interpretation in turn led to difficulty in assignment of some bands, which caused discrepancies between duplicates in some standard fragments, band number 5, for example. Standard fragment 19, 20, and 21 at the lower end of the gel where faint and indistinct accounting for the discrepancy for 9032 in 19 and for 9056 in 20 and 21.

The majority of bands identified in this system were polymorphic producing complex banding patterns (Table 2, duplicates not included). Only one fragment, 11 (2.62 kb), showed significant correlation with a specific class I type, A30 (χ^2=51.931). MspI polymorphisms have very limited use for class I typing.

Participating Laboratories: UKIBOD,[1] GERMUC,[2] JAPKSH,[3] FRAARN,[4] US2BAC[5]

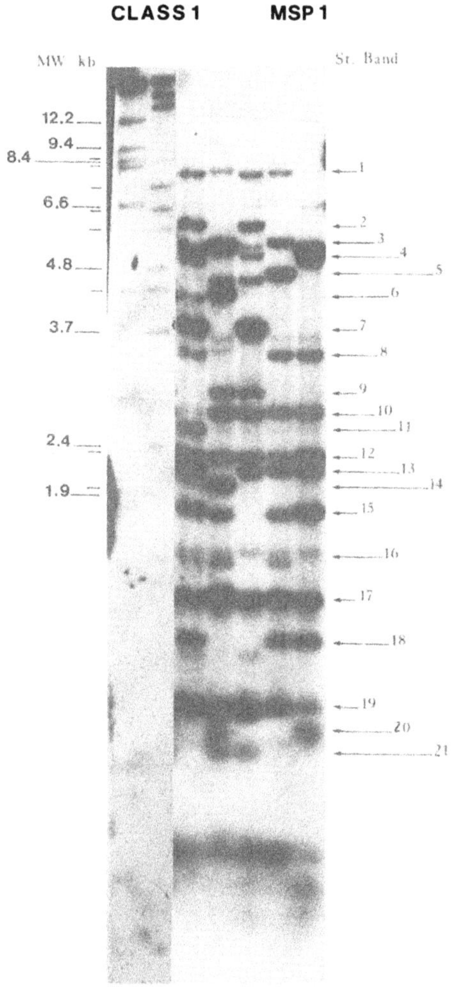

►

Figure 1. Fragments identified with the class I/MspI system.

Table 1. Banding Pattern of the Class I/MspI System in Each Cell Line

		Fragment No.																				
WS. ID.	Cell	1	2	3	4	5	6	7	8	9	10	11	12	13	14	15	16	17	18	19	20	21
9011	E4181324	1		2		1			1		1		1	1		1	9	2	1	1		
9018	LO081785	1			2					1	1		1	2		1	1	2	1	2		
9063	WT47				2					1	1		1	1	1	1	1	2	1	2		
9022	COX	1				2			1		1		1	1		1	1	1	1	1		
9043	BM21	1		1				2	1		2		2	2		2	1	2	2	1		
9023	VAVY	1		1		2			1	1	1		2	1		2	1	2	2	1		
9006	WT100BIS	1			2					1	2		2	2		1	1	2	2	1		
9104	DHIF	1	1				1	2	1	8	2	1	2	2		1	1	1	2	1		
9040	BM15	1		1	2	2					1		2	2		1	1	2	2	2		
9088	PF04015	1		1		2			1	8	1		1	1		1	1	2	1	1		
9002	MZ070782	1			2	2				1			2	1		1	1	2	2	8		
9030	JHAF	1				2				1	1		2		1	1	1	2	2	1	8	
9048	LBUF	1			1			2	1			1	2	2		2	1	2	2	2		
9021	RSH	2	1	2	1				1	8	1	1	2	2		1	1	2	2	1		
9009	KAS011	1		1				2	1		1		1	1		1	9	1	8	8	1	1
9003	KAS116	1					2		1		1		1	1		1	9	1	1	1		
9028	PE117	1			2	2				1			2			1	1	2	1	8	8	
9014	MGAR	1				2					1		1	2	1		1	2	2	2		
9017	WT8	1				2				1	1		1	1			1	1	1	1		
9008	D0208915	1					2				1		2	2	2		9	2	1	2		
9026	YAR	1					2				1		1	1	1		9	1	1	1		
9020	QBL	1	8		2	8					1		1	1			9	2	8	1		
9061	31227ABO	1				2				1	1		2	1			9	1		1		8
9092	BM92	1				2					1		1	1	1		9	1	8			
9105	FPAF	1		1	2				1		1		1	1			1	2	1			
9057	TEM	1					2				1		1		1		1	2	2	2		
9010	AMAI	8			2						1		2	1		2	1	1	1			8
9042	TISI	1			1				1				1	1			1	1	1			
9107	LKT3	1				2				1			2	2		1	1	1				
9050	MOU	1		2			1			1	2		2		2	1	1	2				
9051	PITOUT	1		2			1			1	2		2		1	1	1	2				
9037	SWEIG007					2				1	1		2		1	1	1	2		8	8	
9019	DUCAF				2					1		1	1	1		1	1	2		2		
9025	DEU	1			2					1	1		1	1	1	1	1	2		8	8	8
9062	WDV	1	1				2			1	1		2	1			1	2		1		
9060	CB6B	1		1	2						1		1			1	1	2				
9075	DKB	1		2		8			1				1			1	9	1	8			
9029	WT51	1			2	1							1	1	1	1	9	1		8		
9071	OLGA	1				2				2	2		2	1	1		1	2		8	8	
9055	HO301	1			2				1	1	1		2	2			1	1		2		
9036	SPO010	2		1	2				1	2	2		2	2			1	2		2		
9045	TUBO	1						1			1		2							1		
9047	PLH	1	1	1				2		1	1		1	2			1	1		2		
9106	MANIKA	1		1				2		1	1		1	1			9	1		8		
9052	DBB	1				8		2		1	1		2	2			9	1		1		
9007	DEM	1		1		8		2		1	1			1			9			1		
9056	KOSE	1		1			2			1	2		2	2			1	2		2		
9058	OMW	1		2			1			1	1		2				1	1		1	8	
9013	SCHU	1	1	8		2				1	1		1	2			1	2		2		
9033	BM14	1				2				1	1		1	1			1	1		2		
9004	JESTHOM	1				2				1	1		2	1			1	1			8	
9005	HOM2	1		1		2				1	2		2	2			1	1				
9066	TAB089	1		1		2				1	1		2	2			1	2		8	8	
9016	RML	1		1		2				1	1		2				1	2			8	
9065	HHKB	1	1			2				1	1		1	1			1	2		2		
9059	SLE005	1			2	1				1	1		2				9	1				
9034	SAVC	1				2				1	1		1	1			9	1				
9068	BM9	1		1	2	8				1	1		2	1			9	1				
9067	BTB	1				2				1	1		2	1			9	1		8	8	8
9038	BM16	1				2				1	1		2	1			9	1		1	8	8
9070	LUY	1			2	1				1	1		2	1			9	1		8	8	8
9039	JVM	1			2	1				1	1		2	1			9	1		1		
9082	HO104	1		1		2				1	1		2	2			9	1		1		
9032	BMS	1		1	2					1	1						9	2		1		
9091	MLF	1			2					1	1		2				1	2				
9054	EK/OH	1		1	2					1	1		2	2			1	2		2		
9064	AMALA	1		1	2					1	1		2	1			1	2		1	8	
9031	BOLETH	1			2					1	1		2				1	2		8	8	8
9069	MADURA	1		1						1	1		2				1	2			8	
9072	SPACH	1				2				1	1		1	1			9	1				

1 = Single intensity; 2 = double intensity; 8 = faint; 9 = default.

RFLP Standardization Report for Class I/BamHI

A. Andreas-Zietz,[1] E. Keller,[1] E.D. Albert,[1] B.O. Boehm,[2] J.J. Hyldig-Nielsen,[3] A. Bushell,[4] and D.P. Singal[5]

For the class I/BamHI system, 23 core fragments were first identified on the core cell lines (Table 1 and Figs. 1 and 2). Table 1 shows the reaction pattern of all 72 core cell lines for the class I/BamHI core fragments as well as the corresponding molecular weights and probable association with A,B,C antigens.

The family study of the same system revealed 10 additional fragments, some of which were also detected afterwards on the core cell line films (i.e., a rare band present in only one cell or one band that can be split into two neighboring bands by family analysis; see comments below).

The class I/BamHI system shows a great number of either nonpolymorphic or very frequent bands. The number of fragments which seem to be informative is relatively low (see Table 1).

Participating Laboratories: GERALB,[1] GERBOE,[2] SCASVE,[3] UKIMOR,[4] US5SIN[5]

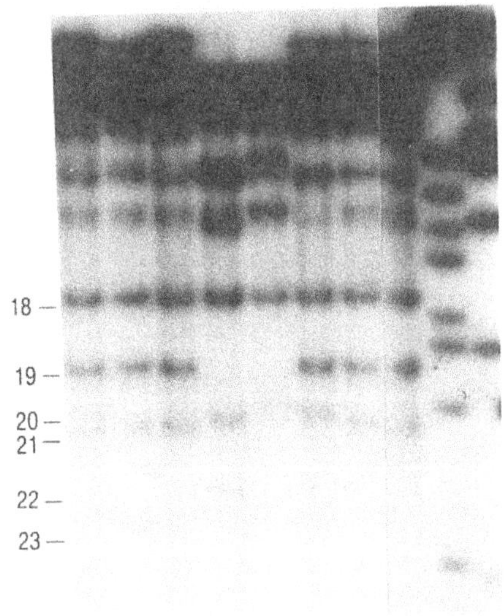

Figure 2. Core bands no. 18–23 of the class I/BamHI system (gel 0.9%).

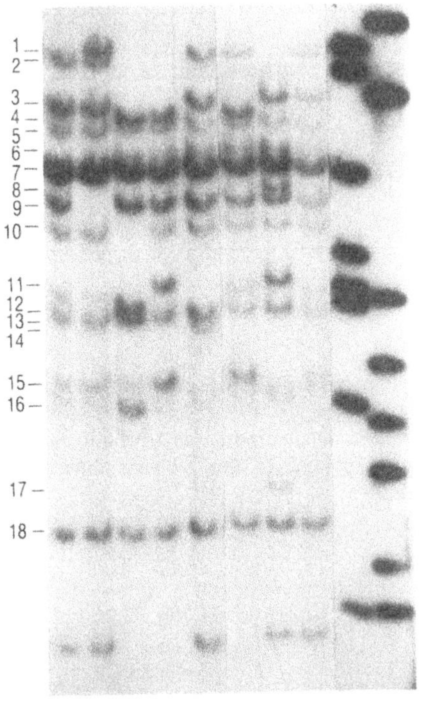

Figure 1. Core bands no. 1 to 18 of the class I/BamHI system (gel 0.6%).

Particular fragments deserve some comments:

1. An additional fragment exists *between core fragment 2 and 3*, which was not reported as a core band, but which is present in core cell no. 60.
2. From family analysis, we now know that *core fragment 3* can be split into two neighboring bands, both of normal intensity. The upper band is more frequent, the lower band is rare. For a certain distinction, especially in unrelated individuals, reference cells are needed.
3. Just below *core fragment 6*, a faint band is seen on the films, but not reported as a core band. It is very frequent, but not monomorphic, as we know from family analysis.
4. Exactly the same situation as in point no. 3 exists for a band just below *core fragment 11*.
5. From family analysis, we now know that *core fragment no. 15* consists of two different fragments located in exactly the same position. One fragment is faint and frequent, but not monomorphic, the other fragment is rare and of stronger intensity (see Table 1 and Fig. 1).
6. As we now know from family analysis, *core fragment no. 16* can be split into two different bands. The original fragment 16 is very faint and probably nonpolymorphic. An additional fragment located just below band 16 is very rare, but of normal intensity and only present in core cell line no. 16.

Table 1. Banding Pattern of All 72 Core Cell Lines, as Standardized in the Class I/BamHI System, Including kb Values and Probable A and C Locus Associations

Probe Name: CLASSI   Enzyme Name: BAMHI

	999999	9999	9999	9999999	999	9999999	99999	99	999999	99	99	9999999	9999	99999	99999999	WS-ID	
	000000	0000	0000	0000000	000	1000000	00001	00	000000	00	01	0000000	0100	00000	00000000	number	
	113638	2281	4502	6331021	060	0556042	06520	42	556355	64	40	4170931	1050	65273	66933377		
	733542	2384	8529	1898809	475	5668625	92764	31	013648	67	06	5603201	0727	99857	04122112		
	000456	1130	4136	5561572	051	4165102	44362	14	121340	32	26	0465536	2077	25252	33446336	position	
	578526	3512	4151	5740818	967	1953669	97830	46	812403	22	37	1309092	4402	51746	37288069	on blots	
A		2	3 22	223		123	223	3	223			223	22	222	33		
	333333	1116	0343	2223560	223	1222141	12661	10	992222	23	13	2224511	8422	22449	12222211		
A								6									
								8				3					
C										1				1111	11		
	77777-	77-7	6888	7755-55	111	4--4444	6----	-2	--555	16	7-	7---1--	4166	00002	9999900-		
B			1116	1111111	222	3333333	33333	44	444444	44	45	5555555	5555	66666	66666666		
	777777	8888	3445	8888888	777	5555555	78888	12	444445	67	90	1111112	3477	00001	22222222		associated with
1		1111														23.56 kb	B 8
2		1111	1	111 111	11	1111	11111	11	111111	1	1	111 1 1	1 11	11 1	111111	21.40	
3		1111	11	111111	11	1111 11	1111	11	111111	1	1	1111111	1111	11111	11111111	16.53	
4	111811		11	1 1	1	1	1	8		1	1	1				15.34	A 3 + 11 + 30
5	888888	8888	8888	8888888	888	8888888	88888	88	888888	88	88	8888888	8888	88888	88888888	14.64	
6	111111			11							1	1				13.43	part Cw7 (B8 ne
7	333333	3333	3333	3333333	333	3333333	33333	33	333333	33	33	3333333	3333	33333	33333333	12.24	
8			1			1	1	1				1	1	1	8	11.41	part Aw 19
9			1111	11111	111	1111111	11111	11	111111	11	1	1111111	1111	11111	11111111	11.00	neg. with Cw 7
10	111111	1111	1111	1111111	111	1111 11	11111	11	111111	11	11	1111111	1111	11111	11111111	10.08	neg. with A 11
11	111111		1	1	1	1				1	1	1 1			11	8.55	A 3 + 31
12						1										8.25	A 11
13	111111	1111	1111	1111111	111	1111111	11111	11	111111	11	11	1111111	1111	11111	11111111	7.91	
14									88							7.78	A 29
15	111811	8881	1188	8881881	881	888888	88881	88	8888	81	88	18888 8	8888	88888	888888	6.86	neg. with A 31
16	888888	8888	8888	8888888	888	8888188	88888	88	888888	88	88	8888888	8888	88888	88888888	6.58	
17						1						1			11	5.36	A 31
18	111111	1111	1111	1111111	111	1111111	11111	11	111111	11	11	1111111	1111	11111	11111111	4.91	
19		1111		111 11	11	1111 1	1111	11	111111	1	1	111 111	1 11	11 1	11111111	3.96	neg. with A 3+1
20	111111		1	1	1					1	1					3.59	A 3
21	888888	1111	1811	1118111	118	1111111	11111	11	111111	18	18	1111111	1111	11111	11111111	3.42	
22	111111		1	1	1	1 1				1	1	1			11	2.78	A 3 + 11 + 31
23								12								2.50	

Cells are listed according to B locus and A locus antigens and banding patterns.
Numbers indicate intensity (1,2,3 = increasing intensity, 8 = faint, 9 = unreadable).

7. *Core fragment no. 20* can be subdivided into two neighboring fragments, both relatively rare. For a certain distinction reference cells are needed.
8. *Core fragment no. 21* can also be split into two neighboring bands. The original fragment 21 seems to be nonpolymorphic, whereas the additional fragment, located just above band 21, is relatively rare. For a certain determination of this band, the segregation pattern in the corresponding families is needed.

RFLP Standardization Report for BF/SstI

T. Takenouchi,[1] R. Tonai,[1] P. Terasaki,[1] T. Sasazuki,[2] and P. Reekers[3]

The data from two participating laboratories showed two nonpolymorphic fragments as shown in Table 1 and Figure 1.

Participating Laboratories: US5TER,[1] JAPSAS,[2] BENREE[3]

Table 1. Fragments Identified with the BF/SstI System

	kb	BF Locus	SST Frequency	TER % Faint	56	32
1	7.34	0 0 0 0	1.000	0.000		
2	6.65	0 0 0 0	1.000	0.000		

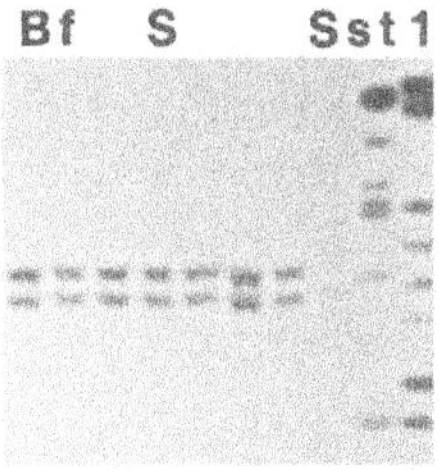

Figure 1.

RFLP Standardization Report for BF/PstI

J.J. Hyldig-Nielsen,[1] M. Trucco,[2] H. Coppin,[3] A. Andreas-Zietz,[4] E. Keller,[4] C. Carcassi,[5] A. Takada,[6] and A. Svejgaard[1]

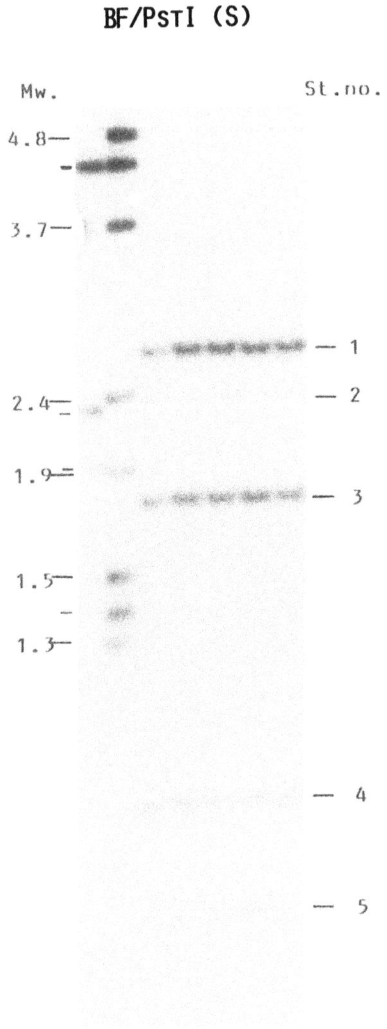

The BF/PstI system identified five nonpolymorphic fragments, as shown in Table 1 and Figure 1.

Participating Laboratories: SCASVE,[1] US1DUQ,[2] FRAOHA,[3] GERALB,[4] IT1CON,[5] JAPAIZ[6]

Table 1. Fragments Identified with the BF/PStI System

		BF	PST	SVE		
	kb	Locus	Frequency	% faint	56	32
1	2.56	0 0 0 0	1.000	0.000		
2	2.39	0 0 0 0	1.000	1.000		
3	1.82	0 0 0 0	1.000	0.000		
4	0.76	0 0 0 0	1.000	0.000		
5	0.49	0 0 0 0	1.000	1.000		

◄
Figure 1.

RFLP Standardization Report for BF/KpnI

G. Paulsen[1] and A.K. Somani[2]

Two bands were identified by the RFLP analysis in the BF/KPN system as shown in Table 1 and in Figure 1. RFLP patterns of the core cell lines are shown in Table 2.

Participating Laboratories: SCATSB,[1] NCYFAL[2]

Table 1. Standard Bands in the BF/KPN System

Band	kb	Frequency[a]	% Faint[b]	56[c]	32[c]
1	6.89	1.000	0.069		
2	1.37	1.000	1.000		

[a]Band frequency in the core cell lines.
[b]Frequency of faint bands.
[c]Discrepancies in strong bands (s) or faint bands (f) of the hidden duplicates (WS.ID. 9056 and 9032).

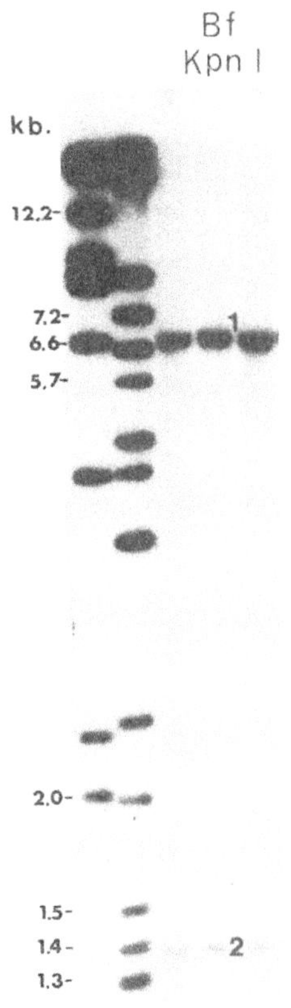

►

Figure 1. RFLP bands in the BF/KPN system of three core cell lines run in 0.9% agarose gel. Representative core bands are shown in the figure numbered on their right side. The WS.ID. of these core cell lines are from the left following marker 1 and marker 2: 9009, 9092, and 9059.

Table 2. Distribution of Bands in the Core Cell Lines in the BF/KPN System

	Band No.			Band No.	
WS ID	1	2	WS ID	1	2
9002	1	8	9043	2	8
9003	3	8	9045	1	8
9004	1	8	9047	2	8
9005	2	8	9048	3	8
9006	2	8	9050	2	8
9007	1	8	9051	2	8
9008	1	8	9052	3	8
9009	2	8	9054	2	8
9010	2	8	9055	8	8
9011	2	8	9056	2	8
9013	2	8	9056	2	8
9014	3	8	9057	1	8
9016	2	8	9058	2	8
9017	2	8	9059	2	8
9018	2	8	9060	2	8
9019	2	8	9061	2	8
9020	2	8	9062	2	8
9021	2	8	9063	8	8
9022	8	8	9064	8	8
9023	2	8	9065	2	8
9025	2	8	9066	1	8
9026	2	8	9067	3	8
9028	2	8	9068	3	8
9029	1	8	9069	1	8
9030	2	8	9070	1	8
9031	2	8	9071	1	8
9032	2	8	9072	1	8
9032	2	8	9075	1	8
9033	1	8	9082	1	8
9034	2	8	9088	1	8
9036	2	8	9091	2	8
9037	2	8	9092	2	8
9038	1	8	9104	8	8
9039	2	8	9105	8	8
9040	2	8	9106	2	8
9042	1	8	9107	2	8

Bands designation: 1 to 3 = increasing intensity, 8 = faint.

RFLP Standardization Report for BF/TaqI

R.E. Bontrop,[1] S. Sekiguchi,[2] and R. Walford[3]

Data for standardization were received from two laboratories. The results obtained from the two laboratories did not show any significant discrepancies in molecular weight or distribution of the fragments. Therefore, standardization was easily performed. For this system, five different standard fragments were observed (Table 1 and in Fig. 1). As can be seen, fragment 1 and fragment 2 show about equal staining intensities (Fig. 1) and have molecular weights of 6.76 and 5.05 kb respectively (Table 1). However, fragment 1 displays a 10% frequency, whereas fragment 2 has a frequency of 95.5% within the panel of cells tested (Table 1). The majority of the panel cells appear either to be positive for fragment 1 or fragment 2. On the other hand, as the frequencies of both fragments already do suggest, some panel cells were found to be "heterozygous" for these two fragments. Standard fragment 3, with a molecular weight of

Participating Laboratories: BENROO,[1] JAPSEK,[2] US5WAL[3]

Table 1. Fragment Observed with the BF/TaqI System

Fragment	kb	Frequency
1	6.76	0.100
2	5.05	0.955
3	3.18	1.000
4	1.84	1.000
5	0.82	1.000

3.18 kb, appears to exhibit the most abundant staining intensity. This fragment and fragments 4 and 5 are observed for all panel cells tested. In conclusion, it will be of interest whether the Bf-polymorphism described with the enzyme TaqI and the Bf probe can be matched with heterogeneity observed at the product level.

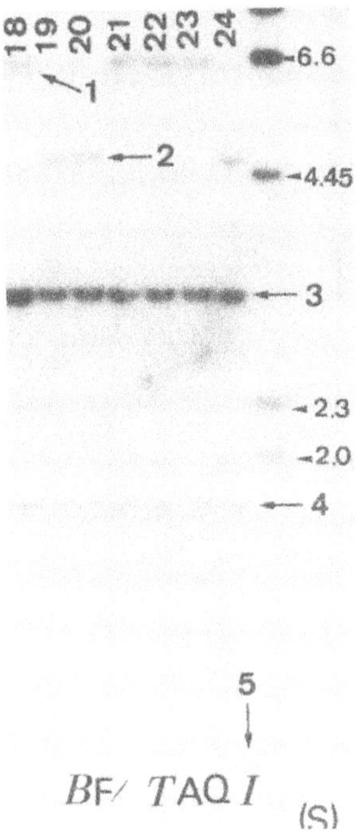

Figure 1. Fragments observed with the Bf/TaqI system. ►

Table 2. Fragment Presence/Absence Patterns for the Bf/TaqI System

	Fragment No.						Fragment No.				
Cell No.	1	2	3	4	5	Cell No.	1	2	3	4	5
9058	1		1	1	1	9057		1	1	1	1
9050	1		1	1	1	9030		1	1	1	1
9051	1		1	1	1	9054		1	1	1	1
9047	1		1	1	1	9105		1	1	1	1
9040	1		1	1	1	9091		1	1	1	1
9060	1		1	1	1	9016		1	1	1	1
9045	1	1	1	1		9048		1	1	1	1
9014		1	1	1	1	9065		1	1	1	1
9107		1	1	1	1	9021		1	1	1	1
9017		1	1	1	1	9062		1	1	1	1
9042		1	1	1	1	9032		1	1	1	1
9013		1	1	1	1	9009		1	1	1	1
9033		1	1	1	1	9092		1	1	1	1
9004		1	1	1	1	9059		1	1	1	1
9018		1	1	1	1	9034		1	1	1	1
9063		1	1	1	1	9068		1	1	1	1
9022		1	1	1	1	9075		1	1	1	1
9043		1	1	1	1	9061		1	1	1	1
9023		1	1	1	1	9067		1	1	1	1
9006		1	1	1	1	9038		1	1	1	1
9005		1	1	1	1	9008		1	1	1	1
9056		1	1	1	1	9003		1	1	1	1
9104		1	1	1	1	9070		1	1	1	1
9010		1	1	1	1	9029		1	1	1	1
9069		1	1	1	1	9011		1	1	1	1
9037		1	1	1	1	9026		1	1	1	1
9028		1	1	1	1	9039		1	1	1	1
9019		1	1	1	1	9056		1	1	1	1
9025		1	1	1	1	9082		1	1	1	1
9031		1	1	1	1	9106		1	1	1	1
9088		1	1	1	1	9032		1	1	1	1
9066		1	1	1	1	9072		1	1	1	1
9036		1	1	1	1	9052		1	1	1	1
9002		1	1	1	1	9020		1	1	1	1
9071		1	1	1	1	9007		1	1	1	1
9064		1	1	1	1	9055	1	1	1	1	1

RFLP Standardization Report for BF/BglII

A. Marcadet,[1] D. Cohen,[1] L.P. Chertkoff,[2] and A. Olivo[3]

The Bf/BglII system identified three fragments on the core cell lines (Table 1). These three fragments are constant for all tested cell lines. Figure is taken from two films from laboratory LAT/FAI. All standardized fragments are indicated at least once for S and/or L migration. The presence of a standardized fragment is indicated by a hyphen.

Table 1. Standard Bands in the Bf/BGL II System

		BF	BGL	COH		
	kb	Locus	Frequency	% Faint	56	32
1	9.92	0 0 0 0	1.000	0.000		
2	5.47	0 0 0 0	1.000	0.000		
3	2.75	0 0 0 0	1.000	0.000		

Participating Laboratories: FRACOH,[1] LATFAI,[2] US8RUB[3]

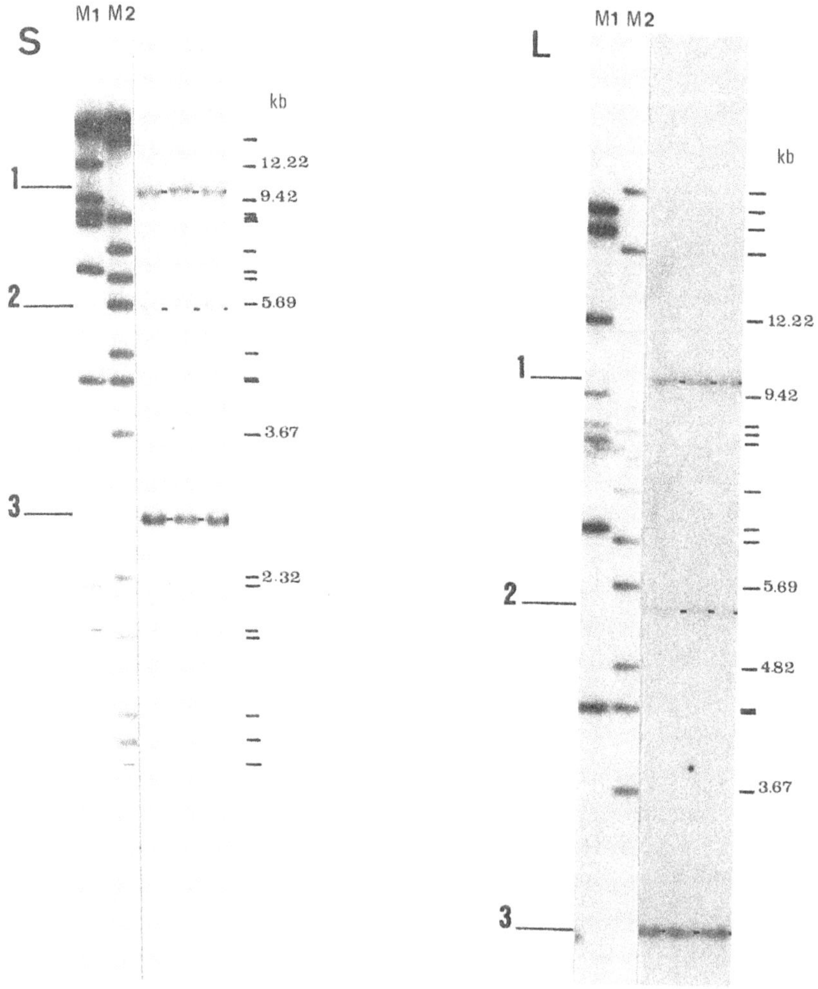

Figure 1. Fragments identified with the Bf/BglII system.

Immunobiology of HLA
Volume I

RFLP Standardization Report for BF/PvuII

P. Piccioli,[1]* K. Kato,[2] J. Partanen,[3] and G.B. Ferrara[1]

The Bf probe detected two fragments that presented an ubiquitous distribution on the core cell lines. These two fragments of 8.17 kb and 0.41 kb hybridized as a strong and a faint band, respectively (Fig. 1).

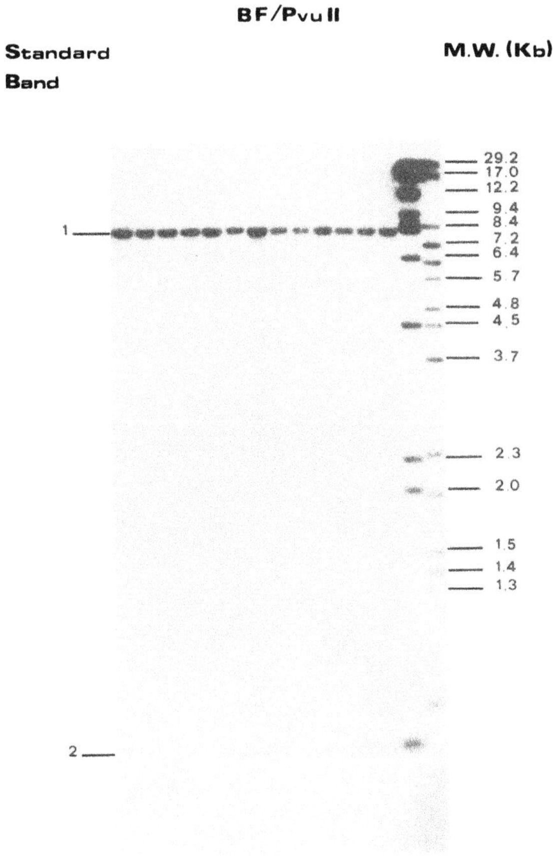

Participating Laboratories: IT2FER,[1] US7DUP,[2] SCAKOS[3]
*P. Piccioli was recipient of a fellowship from A.I.R.C. (Associazione Italiana per la Ricerca sul Cancro).

This work was in part supported by CNR grants n.86.0265.51 and 86.00659.

Figure 1. RFLP pattern of 12 DNAs from core cell lines digested with PvuII endonuclease and hybridized with the probe for BF genes. From left to right: WT47, COX, BM21, VAVY, WT100BIS, HOM2, MOU, KOSE, DHIF, PITOUT, PLF, BM15, AMAL, and molecular weight controls (M1 and M2). The gel was run using short migration conditions.

RFLP Standardization Report for BF/EcoRI

M. Segall,[1]* L. Schluender,[1]* H. Betuel,[2] and M. Garovoy[3]

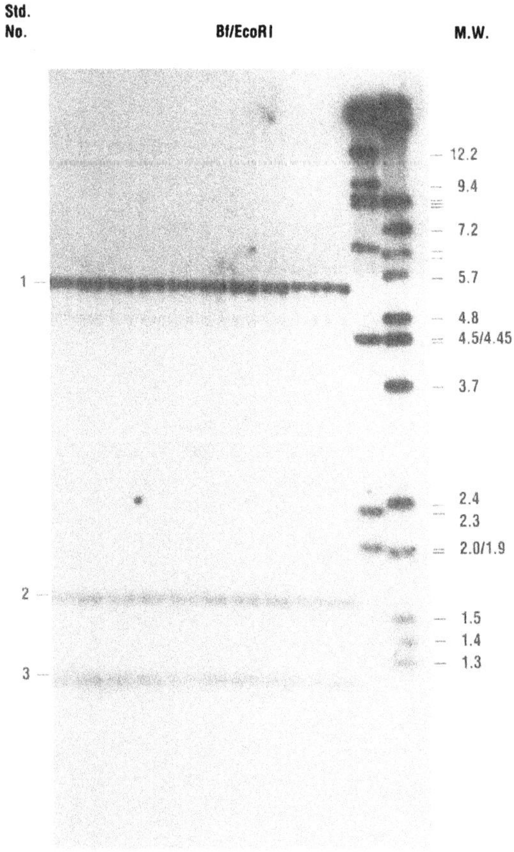

Figure 1. Fragments identified with the Bf/EcoRI system.

Bf/EcoRI blots were received from two laboratories. The Bf/EcoRI system identified three fragments on the core cell lines (Table 1 and Fig. 1). Some of the blots were uniformly rather faint, so that a few lanes were so faint that the presence or absence of the band could not be unequivocally determined. These three fragments are most likely all constant fragments present in all cells and, therefore, no table of polymorphic fragments is given. The apparent band at 4.8 in Figure 1 was seen in only one blot and was, therefore, not included as a standard fragment.

Participating Laboratories: US2BAC,[1] FRABET,[2] US3GVY[3]
*This is publication #491 from the Immunobiology Research Center, University of Minnesota, Minneapolis, Minnesota 55455, USA.
Supported in part by Juvenile Diabetes Foundation Grant #186175 and March of Dimes Birth Defects Foundation Grant #6-496.

Table 1. BF/EcoRI Standard Fragments

				Discrepant for	
Fragment	kb	Frequency	% Faint	9056	9032
1	5.49	1.000	0.000		
2	1.65	0.986	0.338		
3	1.17	0.986	0.338		

RFLP Standardization Report for BF/HindIII

P.A. Biro,[1] S. Cutbush,[1] J. Demopulos,[2] R.T. Acton,[2] and S.V. Van Tonder[3]

Two nonpolymorphic, low molecular weight bands (Table 1) are produced by this enzyme/probe combination (Fig. 1). The 5′ end of the Bf gene is probably contained in the less intense, 2.86 kb band (standard band 2) and the majority of the coding sequence, up to the 3′ end, lies in the 4.52 kb band (1).

Reference

1. Cross SJ, Edwards JH, Bentley DR, Campbell RD. DNA polymorphism of the C2 and factor B genes. Immunogenetics 1985;21:39.

Participating Laboratories: UKIFES,[1] US5UAB,[2] SAFHAM[3]

Table 1. Sizes and Frequencies of Standard Bands Produced by the BF/HindIII Enzyme Probe Combination

	kb	Frequency	% Faint
1	4.52	1.000	0.000
2	2.86	1.000	0.000

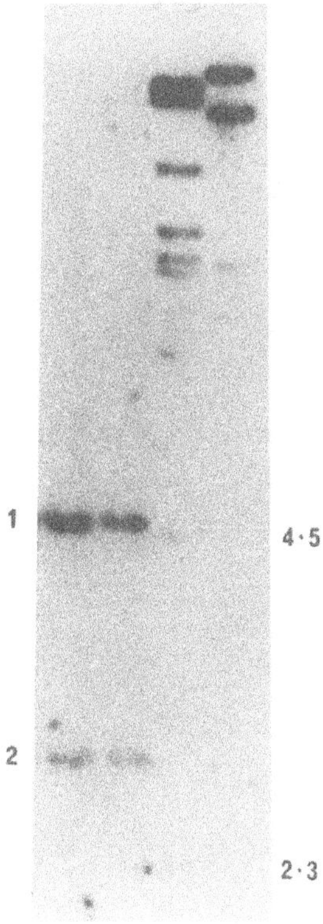

Figure 1. Standard bands produced by the Bf/HindIII enzyme/probe combination.

RFLP Standardization Report for BF/MspI

S. Tonks,[1] V. Groves,[1] C. Muller,[2] F. Otani,[3] J.R. Regueiro,[4] M. Segall,[5] J. Bodmer,[1] and J. Trowsdale[1]

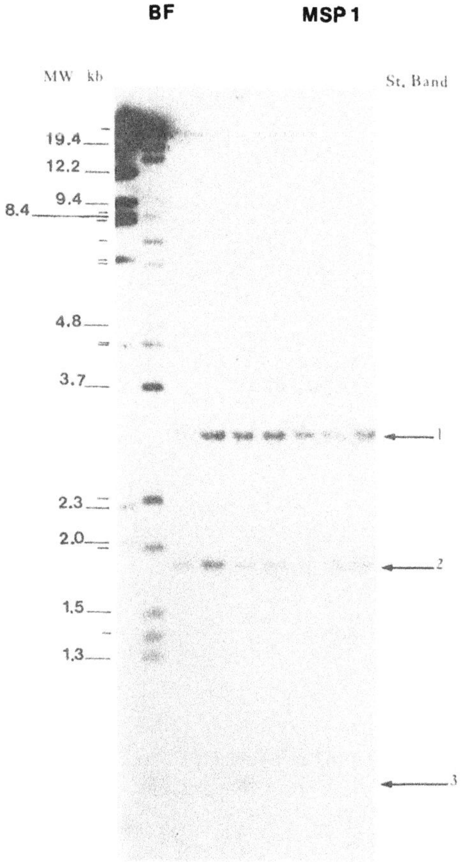

Figure 1. Fragments identified with the Bf/MspI system.

The Bf/MspI system identified a simple pattern of three fragments (Table 1). Standard fragments 1 and 2 were constant throughout the core cell lines. Fragment 3 was found to be polymorphic, but of low molecular weight and faint (Fig. 1), leading to problems in its assignment and thus discordance between duplicates.

Participating Laboratories: UKIBOD,[1] GERMUC,[2] JAPKSH,[3] FRAARN,[4] US2BAC[5]

Table 1. Banding Pattern of the Bf/Mspl System in Each Cell Line

		Fragment No.		
WS ID	Cell	1	2	3
9058	OMW	2	2	8
9055	H0301	2	2	8
9006	WT10081S	2	2	8
9050	MOU	2	2	8
9056	KOSE	2	2	8
9104	DHIF	2	2	8
9051	PITOUT	2	2	8
9047	PLH	2	2	8
9040	BM15	2	2	8
9010	AMAI	2	2	8
9028	PE117	2	2	8
9019	DUCAF	2	2	8
9025	DEU	2	2	8
9031	BOLETH	2	2	8

Table 1. *Continued*

WS ID	Cell	Fragment No. 1	2	3
9060	CB6B	2	2	8
9036	SP0010	2	2	8
9002	MZ070782	2	2	8
9071	OLGA	2	2	8
9064	AMALA	2	2	8
9057	TEM	2	2	8
9030	JHAF	2	2	8
9054	EK/OH	2	2	8
9105	FPAF	2	2	8
9091	MLF	2	2	8
9016	RML	2	2	8
9048	LBUF	2	2	8
9065	HHKB	2	2	8
9021	RSH	2	2	8
9032	BMS	2	2	8
9068	BM9	2	2	8
9045	TUBO	2	2	
9014	MGAR	2	2	
9107	LKT3	2	2	
9017	WT8	2	2	
9042	TISI	2	2	
9013	SCHU	2	2	
9033	BM14	2	2	
9004	JESTHOM	2	2	
9018	L0081785	2	2	
9063	WT47	2	2	
9022	COX	2	2	
9043	BM21	2	2	
9023	VAVY	2	2	
9005	HOM2	2	2	
9069	MADURA	2	2	
9037	SWEIG007	2	2	
9088	PF04015	2	2	
9066	TAB089	2	2	
9062	WDV	2	2	
9009	KAS011	2	2	
9092	BM92	2	2	
9059	SLE005	2	2	
9034	SAVC	2	2	
9075	DKB	2	2	
9061	31227ABO	2	2	
9067	BTB	2	2	
9038	BM16	2	2	
9008	D0208915	2	2	
9003	KAS116	2	2	
9070	LUY	2	2	
9029	WT51	2	2	
9011	E4181324	2	2	
9026	YAR	2	2	
9039	JVM	2	2	
9082	H0104	2	2	
9106	MANIKA	2	2	
9072	SPACH	2	2	
9052	DBB	2	2	
9020	QBL	2	2	
9007	DEM	2	2	

1 = Single intensity; 2 = double intensity; 8 = faint.

RFLP Standardization Report for BF/BamHI

A. Andreas-Zietz,[1] E. Keller,[1] E.D. Albert,[1] A. Bushell,[2] and E.J. Yunis[3]

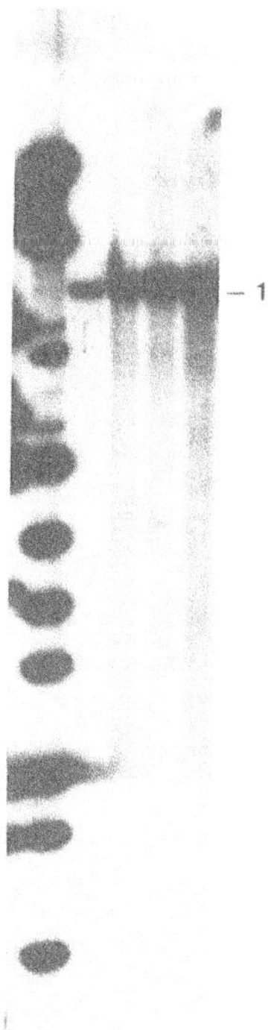

Figure 1. Core bands of the Bf/BamHI system (gel 0.6%).

The Bf/BamHI system identified one fragment on all core cell lines (Table 1 and Fig. 1). This fragment has the same size as the main band of the 210H/BamHI system (fragment no. 4). It may be due to cross-hybridization.

Participating Laboratories: GERALB,[1] UKIMOR,[2] NCYYUN[3]
This study was supported by SFB 217.

Table 1. Banding Pattern of All 72 Core Cell Lines, as Standardized in the Bf/BamHI System, Including kb Value

Probe Name: BF **Enzyme Name: BAMHI**

	999999999999999999999999999999999999999999999999999999999999999999999999	WS-ID
	000100000000000000010000000000000000000010000000000000000000000001000000	number
	415014130156242005505441632123866307653509146263095367663007212358037520	
	548772334853233650641700978951860621470451685122929485178830916962622207	
	....	....
	000000000111111111122222222223333333333444444444455555555556666666666777	position
	123456789012345678901234567890123456789012345678901234567890123456789012	on blots
	....	....
1	222222222222222212222222123333322222999999911111233333333222222222222332	12.76 kb

Cells are listed according to their position on the blots. Numbers indicate intensity)1,2,3 = increasing intensity, 8 = faint, 9 = unreadable.

Immunobiology of HLA
Volume I

RFLP Standardization Report for BF/EcoRV

A. Ando,[1] Y.K. Paik,[2] J.A. Trejant,[3] and K. Tsuji[1]

The Bf/EcoRV system identified the two fragments as shown in Figure 1 and Table 1. All the bands were not polymorphic (Table 1).

Participating Laboratories: JAPTSU,[1] US5PAK,[2] ANZBSH[3]

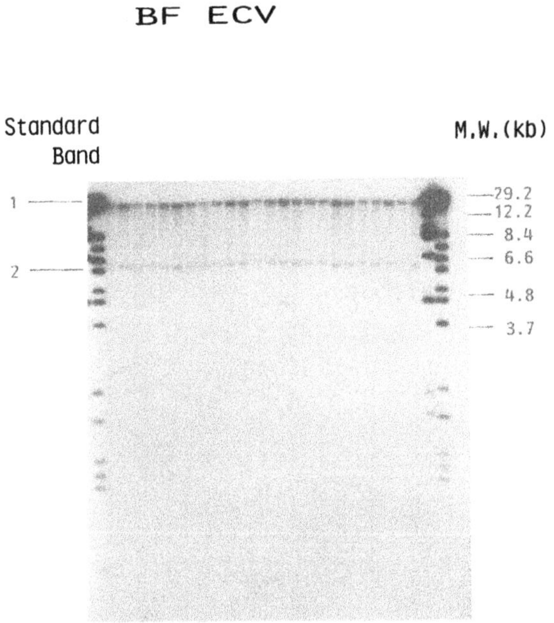

Figure 1. Fragments identified with the Bf/EcoRV system.

Table 1. RFLP Pattern of the Bf/EcoRV System

	Fragment No.	1	2	A	C	B	Bw	Bf	C2	C4A	C4B
	Fragment Size (kb)	24.50	6.10								
	Cell Lines			A	C	B	Bw	Bf	C2	C4A	C4B
28	9019 DUCAF	3	3	30	5	18	6	F1		A3	BQ0
51	9059 SLE005	3	3	2	10	60	6				
60	9070 LUY	3	3	2	-	51	4				
1	9055 HO301	3	2	3	8	14	6				
2	9014 MGAR	3	2	26	7	8	6	S			
5	9017 WT8	3	2	3	7	7	6				
6	9042 TISI	3	2	24	4	35	6	S		A3	B1
13	9022 COX	3	2	1	7	8	6	S	C	AQ0	B1
14	9043 BM21	3	2	1	-	41	6				
15	9023 VAVY	3	2	1	7	8	6	S	C	AQ0	B1
18	9050 MOU	3	2	29	-	44	4			AQ0	B1
19	9056 KOSE	3	2	2	-	35	6	S			
26	9037 SWEIG007	3	2	29	2	61	6				
27	9028 PE117	3	2	24	10	60	6				
30	9031 BOLETH	3	2	2	10	62	6				
32	9066 TAB089	3	2	2	11	46	6	S	C	A4	B2
34	9036 SPOO10	3	2	2	5	44	4				
39	9030 JHAF	3	2	31	-	51	4				
43	9016 RML	3	2	2	-	51	4				
44	9048 LBUF	3	2	30	6	13	4	S		A3	B1
52	9034 SAVC	3	2	3	7	7	6	S		A3	B1
53	9068 BM9	3	2	2	4	35	6				
54	9075 DXB	3	2	24	10	60	6	S	1	A4	B2
55	9061 31227ABO	3	2	2	7	18	6				
56	9067 BTB	3	2	2·	1	27	4				

Table 1. *Continued*

	Fragment No.		1	2								
	Fragment Size (kb)		24.50	6.10								
	Cell Lines				A	C	B	Bw	Bf	C2	C4A	C4B
61	9029	WT51	3	2	23	8	65	6				
66	9082	HO104	3	2	3	7	7	6				
70	9052	DBB	3	2	2	6	57	4	S		A6	B1
71	9020	QBL	3	2	26	?5	18	6	F1	1	A3	BQ0
72	9007	DEM	3	2	2	6	57	4	S		A6	B1
7	9013	SCHU	3	1	3	7	7	6				
10	9018	LO081785	3	1	3	5	18	6	F1		A3	BQ0
16	9006	WT100B1S	3	1	11	4	35	6				
22	9047	PLH	3	1	3	6	47	4				
23	9040	BM15	3	1	1	7	49	4				
24	9010	AMAI	3	1	28	4	53	4				
29	9025	DEU	3	1	31	4	35	6				
35	9002	MZ070782	3	1	24	2,8	14	6	S			
36	9071	OLGA	3	1	31	1,10	62	6				
40	9054	EK	3	1	2	5	44	4				
45	9065	HHKB	3	1	3	7	7	6	S	1	A3	B1
57	9038	BM16	3	1	2	7	18	6				
62	9011	E4181324	3	1	1	–	52	4	S		A3	B1
63	9026	YAR	3	1	26	–	38	4	S			
64	9039	JVM	3	1	2	5	18	6				
65	9056	KOSE	3	1	2	–	35	6				
50	9092	BM92	2	3	25	1	51	4				
25	9069	MADURA	2	2	2	10	60	6				
49	9009	KASO11	2	2	1	6	37	4	S			
59	9003	KAS116	2	2	24	–	51	4				
1	9045	TUBO	2	1	2,3	7	51	4	S			
3	9058	OMW	2	1	2	–	45	6	F	C	A3	B1
4	9107	LKT3	2	1	24	1	54	6				
8	9033	BM14	2	1	3	7	7	6				
9	9004	JESTHOM	2	1	2	1	27	4	S			
12	9063	WT47	2	1	32	5	44	4				
17	9005	HOM2	2	1	3	1	27	4				
20	9104	DHIF	2	1	31		38		S		A3	B1
21	9051	PITOUT	2	1	29	–	44	4	F	C	A3	B1
31	9088	PF04015	2	1	1		8	6	S		AQ0	B1
33	9060	CB6B	2	1	1	9	62	6	F		A3	B1
37	9064	AMALA	2	1	2	9	62	6				
38	9057	TEM	2	1	26	–	38	4				
41	9105	FPAF	2	1	1	4	35	6	S	C	A3	B1
42	9091	MLF	2	1	2	9	62	6				
46	9021	RSH	2	1	68,30	2	42	6	F	C	A1	BQ0
47	9062	WDV	2	1	2	–	38	4		1	A2	B1
67	9106	MANIKA	2	1	3	–	50	6				
58	9008	DO208915	2	1	25	–	18	6	S	Q0	A4	B2
69	9072	SPACH	2	1	31	1,3.1	62					
48	9032	BSM	2	1	2	9	62	6				
68	9032	BSM	2	1	2	9	62	6				

RFLP Standardization Report for C2/SstI

T. Takenouchi,[1] R. Tonai,[1] P. Terasaki,[1] T. Sasazuki,[2] and P. Reekers[3]

Only two laboratories contributed data for standardization of this enzyme-probe system. In instances where data were not consistent in the two laboratories, we reinspected the films and judged the assignment of the standard fragments. Also sample #41 could not be digested by either laboratory.

Participating Laboratories: US5TER,[1] JAPSAS,[2] BENREE[3]

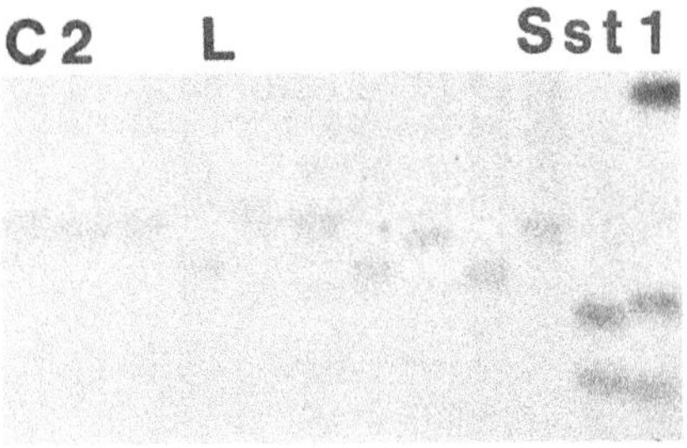

Figure 1. Fragments observed with the C2/SstI system.

Table 1. Fragments Identified with the C2/SstI System

		C2	SST	TER		
	kb	Locus	Frequency	% Faint	56	32
1	2.75	0 0 0 0	0.161	0.000		
2	2.70	0 0 0 0	0.288	0.000		
3	2.64	0 0 0 0	0.409	0.000		
4	2.48	0 0 0 0	0.088	0.000		

Four standardization fragments were assigned (Table 1 and Fig. 1). Four exclusively allelic patterns were distinctively observed (Table 2). There is no significant association between C2 and class II phenotypes.

Table 2. Banding Pattern of the C2/Sstl System in Each Cell

	Fragment No.			
Cell	1	2	3	4
9014 M GAR	1			
9033 BM14	1			
9043 BM21	1			
9056 KOSE	1			
9056 KOSE	1			
9036 SPO 010	1			
9092 BM92	1			
9075 DKB	1			
9008 DO 208915	1			
9029 WT51	1			
9082 HO 104	1			
9017 WT8		1		
9013 SCHU		1		
9018 LO 081785		1		
9022 COX		1		
9023 VAVY		1		
9005 HOM-2		1		
9104 DHIF		1		
9010 AMAI		1		
9019 DUCAF		1		
9088 PF 04015		1		
9066 TAB 089		1		
9002 MZ 070782		1		
9057 TEM		1		
9030 JHAF		1		
9048 LBUF		1		
9065 HHKB		1		
9021 RSH		1		
9045 TUBO			1	
9107 LKT 3			1	
9042 TISI			1	
9004 JES THOM			1	

Table 2. *Continued*

	Fragment No.			
Cell	1	2	3	4
9063 WT47			1	
9006 WT100 BIS			1	
9047 PLH			1	
9069 MADURA			1	
9037 SWEIG 007			1	
9028 PE 117			1	
9025 DEU			1	
9031 BOLETH			1	
9071 OLGA			1	
9064 AMALA			1	
9054 EK			1	
9105 FPAF			1	
9091 MLF			1	
9016 RML			1	
9062 WDV			1	
9032 BSM			1	
9032 BSM			1	
9009 KAS-011			1	
9061 3-12-27 AOB			1	
9067 BTB			1	
9003 KAS-116			1	
9070 LUY			1	
9011 E 4181324			1	
9026 YAR			1	
9072 S PACHEO			1	
9058 OMW				1
9055 HO 301				1
9050 MOU				1
9051 PITOUT				1
9040 BM15				1
9060 CBG-IB				1
9106 MANIKA	9	9	9	9
9059 SLE-005	9	9	9	9
9034 SAVC	9	9	9	9
9068 BM9	9	9	9	9
9038 BM16	9	9	9	9
9039 JVM	9	9	9	9
9052 DBB	9	9	9	9
9020 QBL	9	9	9	9
9007 DEM	9	9	9	9

1 = Single intensity; 9 = uninterpretable.

RFLP Standardization Report for C2/PstI

J.J. Hyldig-Nielsen,[1] M. Trucco,[2] H. Coppin,[3] A. Andreas-Zietz,[4] E. Keller,[4] C. Carcassi,[5] A. Takada,[6] and A. Svejgaard[1]

The C2/PstI system defines the two polymorphic fragments (Table 1 and Fig. 1). The largest fragment of 12.31 kb was much more frequent among the 72 core cell lines that the other fragment of 11.65 kb. The banding patterns of the C2/PstI system are shown in Table 2.

Participating Laboratories: SCASVE,[1] US1DUQ,[2] FRAOHA,[3] GERALB,[4] IT1CON,[5] JAPAIZ[6]

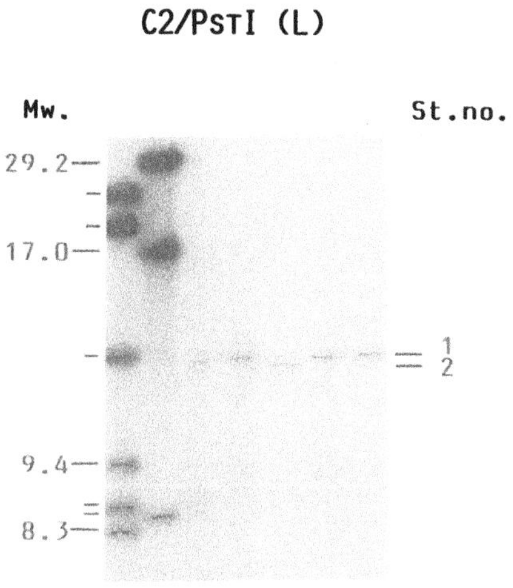

Table 1. Fragments Identified with the C2/PstI System

	kb	C2 Locus	PST Frequency	SVE % Faint	56	32
1	12.31	0 0 0 0	0.929	0.000		
2	11.65	0 0 0 0	0.071	0.000		

Table 2. Banding Pattern of the C2/PstI System

WS ID	Name	HLA-DR	Bf	C2	C4A	C4B	B	Standard Fragment No. 1	Standard Fragment No. 2
9002	MZ070782	1	S				14		8
9003	KAS116	1					51		8
9004	JESTHOM	1	S'				27		8
9005	HOM2	1					27		8
9006	WT100BIS	1					35		8
9007	DEM	16,4	S		A6	B1	57		8
9008	DO208915	15	S	QO	A4	B2	18		8
9009	KAS011	2	S				37		8
9010	AMAI	15					53		8
9011	E4181324	15	S		A3	B1	52		8
9013	SCHU	15					7		8
9014	MGAR	15	S				8		8
9016	RML	16					51		8
9017	WT8	15					7		8
9018	LOO81785	3	F1		A3	BQO	18		8
9019	DUCAF	3	F1		A3	BQO	18		8
9020	QBL	3	F1	1	A3	BQO	18		8
9021	RSH	3	F	C	A1	BQO	42		8
9022	COX	3	S	C	AQO	B1	8		8
9023	VAVY	3	S	C	AQO	B1	8		8
9025	DEU	4					35		8
9026	YAR	4	S				38		8
9028	PE117	4					60		8
9029	WT51	4					65		8
9030	JHAF	4					51		8
9031	BOLETH	4					62		8
9032	BSM	4					62		8
9032	BSM	4					62		8
9033	BM14	4					7		8

Immunobiology of HLA
Volume I

Table 2. *Continued*

WS ID	Name	HLA-DR	Bf	C2	C4A	C4B	B	Standard Fragment No. 1	Standard Fragment No. 2
9034	SAVC	4	S		A3	B1	7		8
9036	SPD010	11					44		8
9037	SWEIG007	11					61		8
9038	BM16	12					18		8
9039	JVM	11					18		8
9040	BM15	11					49	8	
9042	TISI	11	S		A3	B1	35		8
9043	BM21	11					41		8
9045	TUBO	11,12	S				51		8
9047	PLH	7					47		8
9048	LBUF	7	S		A3	B1	13		8
9050	MOU	7			AQ0	B1	44	8	
9051	PITOUT	7	F	C	A3	B1	44	8	
9052	DBB	7	S		A6	B1	57		8
9054	EK	14					44		8
9055	H0301	6					14	8	
9056	KOSE	13,14	S				35		8
9056	KOSE	13,14	S				35		8
9057	TEM	14					38		8
9058	OMW	13	F	C	A3	B1	45	8	
9059	SLE005	13					60		8
9060	CB6B	13	F		A3	B1	62		8
9061	31227ABO	14					18		8
9062	WDV	13		1	A2	B1	38		8
9063	WT47	13					44		8
9064	AMALA	14					62		8
9065	HHKB	13	S	1	A3	B1	7		8
9066	TAB089	8	S	C	A4	B2	46		8
9067	BTB	8					27		8
9068	BM9	8					35		8
9069	MADURA	8					60		8
9070	LUY	8					51		8
9071	OLGA	8					62		8
9072	SPACH	8					62		8
9075	DKB	9	S	1	A4	B2	60		8
9082	H0104	15					7		8
9088	PF04015	3	S		AQ0	B1	8		8
9091	MLF	4					62		8
9092	BM92	4					51		8
9104	DHIF	11	S		A3	B1	38		8
9105	FPAF	11	S	C	A3	B1	35		8
9106	MANIKA	7					50		8
9107	LKT3	4					54		8

RFLP Standardization Report for C2/KpnI

G. Paulsen[1] and P.C. Kwong[2]

Two bands were identified by the RFLP analysis in the C2/KPN system as shown in Table 1 and Figure 1. The distribution of bands among core cell lines is shown in Table 2.

Participating Laboratories: SCATSB,[1] NCYFAL[2]

Table 1. Standard Bands in the C2/KPN System

Band	kb	Frequency[a]	% Faint[b]	56[c]	32[c]
1	10.50	0.912	0.297		
2	9.88	0.072	0.000		

[a]Band frequency in the core cell lines.
[b]Frequency of faint bands.
[c]Discrepancies in strong bands (s) or faint bands (f) of the hidden duplicates (WS.ID. 9056 and 9032).

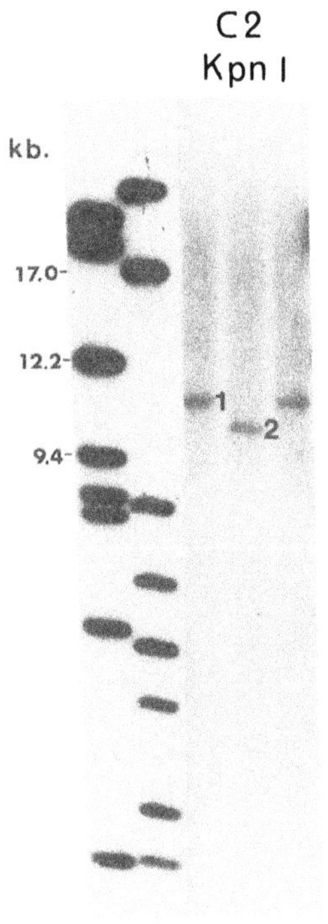

Figure 1. RFLP bands in the C2/KPN system of three core cell lines run in 0.6% agarose gel. Representative core bands are shown in the figure numbered on their right side. The WS.ID. no. of these core cell lines are from the left following marker 1 and marker 2: 9047, 9040, and 9010.

Table 2. Distribution of Standard Bands in Core Cell Lines in the C2/KPN System

		Band No.				Band No.	
WS. ID.	C2	1	2	WS. ID.	C2	1	2
9045		8		9091		1	
9014		1		9016		1	
9107		8		9048		1	
9017		1		9065	1	1	
9042		1		9021	C	1	
9013		1		9062	1	1	
9033		8		9032		1	
9004		1		9009		8	
9018		1		9092		8	
9063		1		9059		8	
9022	C	1		9034		8	
9043		1		9068		1	
9023	C	1		9075	1	1	
9006		1		9061		1	
9005		1		9067		1	
9056		1		9038		8	
9104		9	9	9008	Q0	8	
9047		1		9003		1	
9010		1		9070		1	
9069		1		9029		1	
9037		1		9011		1	
9028		1		9026		1	
9019		1		9039		1	
9025		1		9056		1	
9031		8		9082		1	
9088		8		9106		8	
9066	C	8		9032		8	
9060		8		9072		8	
9036		1		9052		1	
9002		1		9020	1	8	
9071		1		9058	C		1
9064		8		9055			1
9057		8		9050			1
9030		1		9051	C		1
9054		1		9040			1
9105	C	9		9007			1

The bands are designated: 1 = strong, 8 = faint, 9 = unreadable.

RFLP Standardization Report for C2/EcoRV

A. Ando,[1] Y.K. Paik,[2] J.A. Trejant,[3] and K. Tsuji[1]

The C2/EcoRV system identified the two fragments (Table 1 and Fig. 1). Standard fragment no. 1 was not polymorphic (Table 1). Standard fragment no. 2 was polymorphic, but was a faint band in most cell lines. Both St. fragment no. 1 and no. 2 are also found in DO208915, which is known to be associated with C2 deficiency (1).

Reference

1. Fu SM, Stern R, Kunkel HG, et al. Mixed lymphocyte culture determinants and C2 deficiency: LD-7a associated with C2 deficiency in four families. J Exp Med 1975;142:495.

Participating Laboratories: JAPTSU,[1] US5PAK,[2] ANZBSH[3]

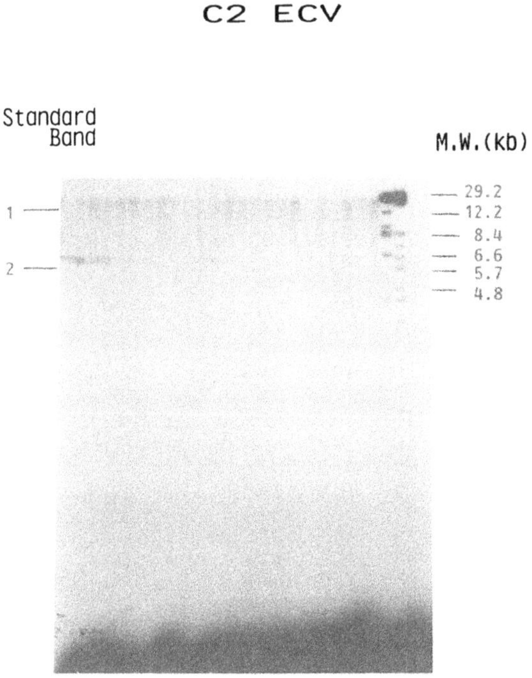

Figure 1. Fragments identified with the C2/EcoRV system.

Table 1. RFLP Pattern of the C2/EcoRV System

	Fragment No.		1	2							
	Fragment Size (kb)		24.04	8.55							
	Cell Lines			A	C	B	Bw	Bf	C2	C4A	C4B
25	9069	MADURA	1	2	10	60	6				
31	9088	PF04015	1	1		8	6	S		AQ0	B1
33	9060	CB6B	1	1	9	62	6	F		A3	B1
36	9071	OLGA	1	31	1,10	62	6				
37	9064	AMALA	1	2	9	62	6				
38	9057	TEM	1	26	-	38	4				
41	9105	FPAF	1	1	4	35	6	S	C	A3	B1
42	9091	MLF	1	2	9	62	6				
45	9065	HHKB	1	3	7	7	6	S	1	A3	B1
46	9021	RSH	1	68,30	2	42	6	F	C	A1	BQ0
47	9062	WDV	1	2	-	38	4		1	A2	B1
62	9011	E4181324	1	1	-	52	4	S		A3	B1
63	9026	YAR	1	26	-	38	4	S			
64	9039	JVM	1	2	5	18	6				
65	9056	KOSE	1	2	-	35	6				
69	9072	SPACH	1	31	1,3.1	62					
71	9020	QBL	1	26	?5	18	6	F1	1	A3	BQ0
26	9037	SWEIG007	2	29	2	61	6				
27	9028	PE117	2	24	10	60	6				

Table 1. *Continued*

		Fragment No.	1	2								
		Fragment Size (kb)	24.04	8.55								
		Cell Lines			A	C	B	Bw	Bf	C2	C4A	C4B
28	9019	DUCAF	2		30	5	18	6	F1		A3	BQ0
29	9025	DEU	2		31	4	35	6				
30	9031	BOLETH	2		2	10	62	6				
32	9066	TAB089	2		2	11	46	6	S	C	A4	B2
34	9036	SP0010	2		2	5	44	4				
35	9002	MZ070782	2		24	2,8	14	6	S			
39	9030	JHAF	2		31	–	51	4				
40	9054	EK	2		2	5	44	4				
43	9016	RML	2		2	–	51	4				
44	9048	LBUF	2		30	6	13	4	S		A3	B1
66	9082	H0104	2		3	7	7	6				
70	9052	DBB	2		2	6	57	4	S		A6	B1
72	9007	DEM	2		2	6	57	4	S		A6	B1
48	9032	BSM	8		2	9	62	6				
68	9032	BSM	8		2	9	62	6				
67	9106	MANIKA	8		3	–	50	6				
50	9092	BM92	1	3	25	1	51	4				
52	9034	SAVC	1	2	3	7	7	6	S		A3	B1
51	9059	SLE005	2	2	2	10	60	6				
56	9067	BTB	2	1	2	1	27	4				
54	9075	DXB	2	1	24	10	60	6	S	1	A4	B2
49	9009	KAS011	8	1	1	6	37	4	S			
55	9061	31227ABO	1	1	2	7	18	6				
53	9068	BM9	2	1	2	4	35	6				
61	9029	WT51	1	1	23	8	65	6				
3	9058	OMW	1	8	2	–	45	6	F	C	A3	B1
4	9107	LKT3	1	8	24	1	54	6				
5	9017	WT8	1	8	3	7	7	6				
7	9013	SCHU	1	8	3	7	7	6				
8	9033	BM14	1	8	3	7	7	6				
9	9004	JESTHOM	1	8	2	1	27	4	S			
12	9063	WT47	1	8	32	5	44	4				
17	9005	HOM2	1	8	3	1	27	4				
20	9104	DHIF	1	8	31		38		S		A3	B1
21	9051	PITOUT	1	8	29	–	44	4	F	C	A3	B1
23	9040	BM15	1	8	1	7	49	4				
24	9010	AMAI	1	8	28	4	53	4				
60	9070	LUY	1	8	2	–	51	4				
2	9014	MGAR	2	8	26	7	8	6	S			
6	9042	TISI	2	8	24	4	35	6	S		A3	B1
10	9018	LO081785	2	8	3	5	18	6	F1		A3	BQ0
11	9055	H0301	2	8	3	8	14	6				
13	9022	COX	2	8	1	7	8	6	S	C	AQ0	B1
14	9043	BM21	2	8	1	–	41	6				
15	9023	VAVY	2	8	1	7	8	6	S	C	AQ0	B1
16	9006	WT100BIS	2	8	11	4	35	6				
18	9050	MOU	2	8	29	–	44	4			AQ0	B1
19	9056	KOSE	2	8	2	–	35	6	S			
22	9047	PLH	2	8	3	6	47	4				
57	9038	BM16	2	8	2	7	18	6				
58	9008	D0208915	2	8	25	–	18	6	S	Q0	A4	B2
1	9045	TUBO	8	8	2,3	7	51	4	S			
59	9003	KAS116	8	8	24	–	51	4				

RFLP Standardization Report for C2/TaqI

R.E. Bontrop,[1] R. Walford,[2] and S. Sekiguchi[3]

Data for standardization were received from two laboratories. This system allowed the detection of two different standard fragments. Standard fragment 1, with a molecular weight of 4.91 kb, has a frequency of 91.4%, whereas standard fragment 2, with a molecular weight of 4.75 kb, displays a frequency of 8.6%. These data seem to suggest a "diallelic" distribution of restriction sites that are detected with the C2 probe and the restriction enzyme TaqI. Careful analysis of the Bf/TaqI and C2/TaqI systems seems to suggest that standard fragment 2 for the Bf/TaqI system is identical to standard fragment 1 detected for the C2/TaqI system. In most panel cells tested it is seen that when a cell is positive for standard fragment 1 in the Bf/TaqI system it is also positive for fragment 2 in the C2/TaqI system. However, exceptions to the rule do occur. For example, panel cell 22 is positive for fragment 1 in the Bf/TaqI system (see Fig. 1 in the Bf/TaqI report), but is also positive for fragment 1 in the C2/TaqI system (Fig. 1). This observation probably explains the apparent difference in frequency of fragment 2 in the Bf/TaqI system (95.5%) and fragment 1 in the C2/TaqI system (91.4%).

Participating Laboratories: BENROO,[1] US5WAL,[2] JAPSEK[3]

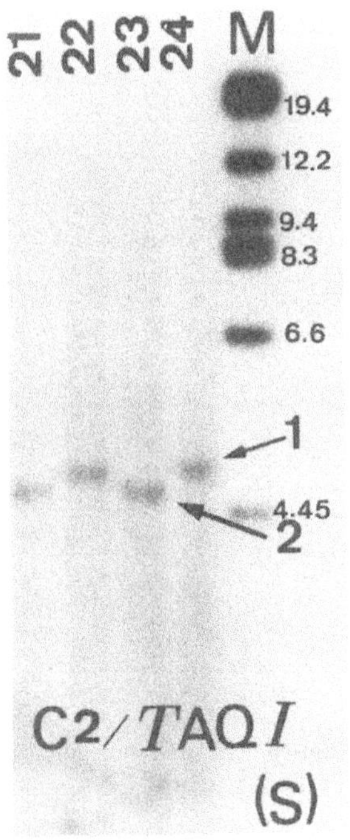

Figure 1. Fragments observed with the C2/TaqI system.

Table 1. Fragments Identified with the C2/TaqI System

Fragment	kb	Frequency
1	4.91	0.914
2	4.75	0.086

Table 2. Fragment Presence/Absence Pattern for the C2/TaqI System on the 72 Core Cell Lines

	Fragment No.			Fragment No.	
Cell	1	2	Cell	1	2
9045	1		9059	1	
9014	1		9034	1	
9107	1		9068	1	
9017	1		9075	1	
9042	1		9061	1	
9013	1		9067	1	
9033	1		9038	1	
9004	1		9008	1	
9018	1		9003	1	
9063	1		9070	1	
9022	1		9029	1	
9043	1		9011	1	
9023	1		9026	1	
9006	1		9039	1	
9005	1		9056	1	
9056	1		9082	1	
9104	1		9106	1	
9047	1		9032	1	
9010	1		9072	1	
9069	1		9052	1	
9037	1		9020	1	
9028	1		9007	1	
9019	1		9058		1
9025	1		9055		1
9031	1		9050		1
9088	1		9051		1
9066	1		9040		1
9036	1		9060		1
9002	1				
9071	1				
9064	1				
9057	1				
9030	1				
9054	1				
9105	1				
9091	1				
9016	1				
9048	1				
9065	1				
9021	1				
9062	1				
9032	1				
9009	1				
9092	1				

Immunobiology of HLA
Volume I

RFLP Standardization Report for C2/BglII

A. Marcadet,[1] D. Cohen,[1] M. Herrero,[2] and S. Rodriguez de Cordoba[3]

The C2/BglII system only identified two constant fragments on the core cell lines (Table 1). Figure 1 is taken from two films from laboratory LAT/FAI. All standardized fragments are indicated at least once for S and/or L migration. The presence of a standardized fragment is indicated by a hyphen.

Participating Laboratories: FRACOH,[1] LATFAI,[2] US8RUB[3]

Table 1. Standard Bands in the C2/BglII System

	kb	C2 Locus	BGL Frequency	COH % Faint	56	32
1	11.33	0 0 0 0	1.000	0.000		
2	8.18	0 0 0 0	1.000	0.000		

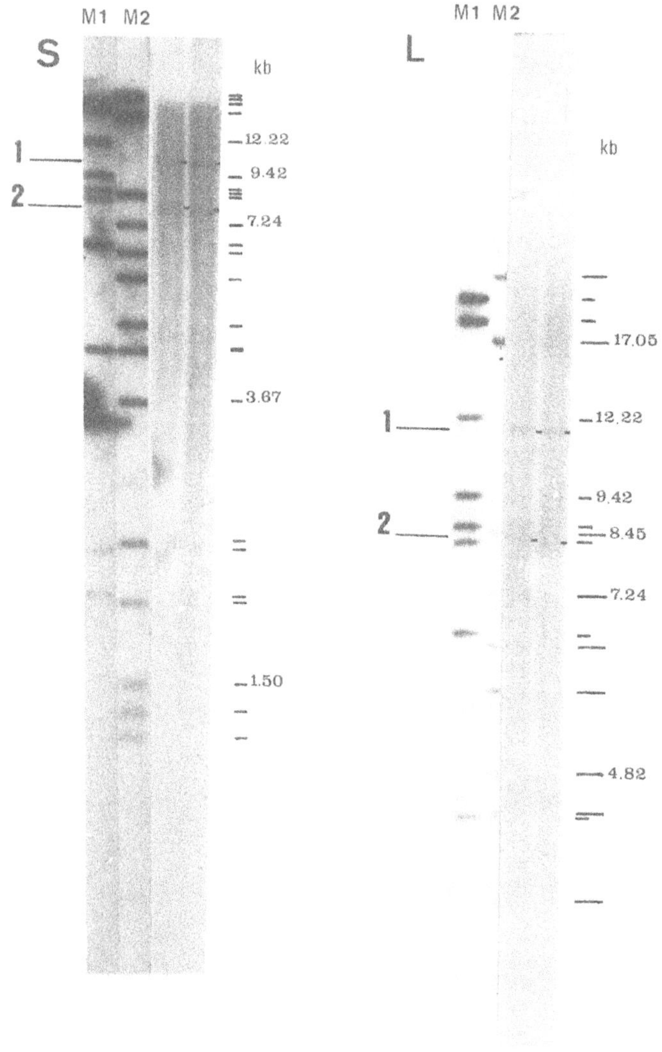

Figure 1. Fragments identified with the C2/BglII system.

RFLP Standardization Report for C2/PvuII

P. Piccioli,[1]* K. Kato,[2] J. Partanen,[3] G. Angelini,[1] and G.B. Ferrara[1]

Seventy-two DNA samples analyzed by Southern blot technique using the restriction enzyme PvuII and probed with a 350 bp BamHI-Kpn I genomic fragment (1) allowed us to observe three strong fragments of 4.40, 4.31, and 4.18 kb. (Table 1 and Fig. 1). Although a real interpretation of the data is difficult because of the few cell lines typed for this complement component, 34 (48%) carried the 4.40 kb fragment, 28 (40%) carried the 4.31 kb fragment, 6 the 4.18 kb fragment and 2 both the first and the second one. In the 48% are included four of a total of seven cell lines carrying the C2-C allele, all the cell lines typed as C2-1 and the only one typed as QO. The last three cell lines typed as C2-C showed the second fragment (sample ID 9105) and the third (ID 9058 and ID 9051, the last one typing as B44-DR7-FC31). Polymorphism of C2 has been demonstrated by isoelectrofocusing (2–4) and by Southern blot with two different enzymes (1,5).

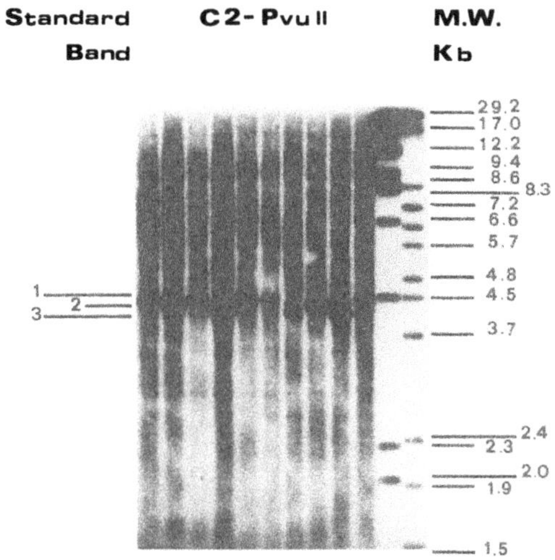

Figure 1. VAVY, WT100BIS, HOM2, MOU, KOSE, DHIF, PITOUT, PLH, BM 15, and AMAL cell lines pattern using the restriction enzyme PvuII and a C2 probe (350 bp KpnI-BamHI genomic probe). On the left are indicated the three standard fragments of 4.40, 4.31 and 4.18 Kb, respectively.

References

1. Bentley DR, Campbell RD, Cross SJ. DNA polymorphism of the C2 locus. Immunogenetics 1985;22:377.
2. Alper CA. Inherited structural polymorphism in C2: evidence for genetic linkage between C2 and Bf. J Exp Med 1976;144:1111.
3. Meo T, Atkinson J, Bernoco M, Bernoco D, Ceppellini R. Mapping of the HLA locus controlling C2 structural variants and linkage disequilibrium between alleles C22 and Bw15. Eur J Immunol 1976;6:916.
4. Meo T, Atkinson J., Bernoco M., Bernoco D. and Ceppellini R. Structural heterogeneity of C2 complement protein and its genetic variants in man: a new polymorphism of the HLA region. Proc Natl Acad Sci USA 1977;74:1672.
5. Woods DE, Edge MD, Colten HR. Isolation of a complementary DNA clone for the human complement protein C2 and its use in the identification of a restriction fragment length polymorphism. J Clin Invest 1984;14:634.

Participating Laboratories: IT2FER,[1] US7DUP,[2] SCAKOS[3]
*P. Piccioli was recipient of a fellowship from A.I.R.C. (Associazione Italiana per la Ricerca sul Cancro).
This work was in part supported by CNR grants n.86.0265.51 and 86.00659.

Table 1. Banding Pattern of the C2/PvuII System in Each Cell Line

Cell	C2	Fragment No. 1	2	3
9002 MZ 070782		2		
9003 KAS 116			2	
9004 JES THOM		2		
9005 HOM 2		2		
9006 WT100 BIS			2	
9007 DEM		2		
9008 DO 208915	QO	2		
9009 KAS 011			2	
9010 AMAL		2	1	
9011 E 4181324			2	
9013 SCHU		2		
9014 M GAR		2		
9016 RML			2	
9017 WT 8		2		

Table 1. *Continued*

Cell	C2	Fragment No. 1	2	3
9018 LO 081785		2		
9019 DUCAF		2		
9020 QBL	1	2		
9021 RSH	C	2		
9022 COX	C	2		
9023 VAVY	C	2		
9025 DEU			2	
9026 YAR			2	
9028 PE 117		2		
9029 WT 51		2		
9030 JHAF		2		
9031 BOLETH			2	
9032 BSM			2	
9032 BSM			2	
9033 BM 14		2		
9034 SAVAC		2		
9036 SPO 010		2		
9037 SWEIG 007			2	
9038 BM 16		2		
9039 JVM		2		
9040 BM 15				2
9042 TISI			2	
9043 BM 21		2		
9045 TUBO			2	
9047 PLH			2	
9048 LBUF		2		
9050 MOU				2
9051 PITOUT	C			2
9052 DBB		2		
9054 EK			2	
9055 HO 301				2
9056 KOSE		2		
9056 KOSE		2		
9057 TEM			2	
9058 OMW	C			2
9059 SLE 005		1	2	
9060 CB6B				2
9061 31227 ABO			2	
9062 WDV			2	
9063 WT 47			2	
9064 AMALA			2	
9065 HHKB	1	2		
9066 TAB 089	C	2		
9067 BTB			2	
9068 BM 9		2		
9069 MADURA			2	
9070 LUY			2	
9071 OLGA			2	
9072 SPACH			1	
9075 DKB	1	2		
9082 HO 104		2		
9088 PF 04015		2		
9091 MLF			2	
9092 BM 92		2		
9104 DHIF		2		
9105 FPAF	C		2	
9106 MANIKA			2	
9107 LKT3			2	

1 = Single intensity; 2 = double intensity.

RFLP Standardization Report for C2/EcoRI

M. Segall,[1]* L. Schluender,[1]* H. Betuel,[2] and M. Garovoy[3]

C2/EcoRI blots were received from two laboratories. The C2/EcoRI system identified two fragments on the core cell lines (Table 1 and Fig. 1). The two fragments were complementary except for one discordance (cell 9060), which had a different fragment in each of the two blots. No other resolution of discrepancies was required. The pattern of the two fragments in the cell lines is given in Table 2.

Table 1. C2/EcoRI Standard Fragments

Fragment No.	kb	Frequency	% Faint	Discrepant for 9056	Discrepant for 9032
1	6.76	0.929	0.000		
2	6.45	0.086	0.000		

Participating Laboratories: US2BAC,[1] FRABET,[2] US3GVY.[3]
*This is publication #488 from the Immunobiology Research Center, University of Minnesota, Minneapolis, Minnesota 55455, USA.
Supported in part by Juvenile Diabetes Foundation Grant #186175 and March of Dimes Birth Defects Foundation Grant #6-496.

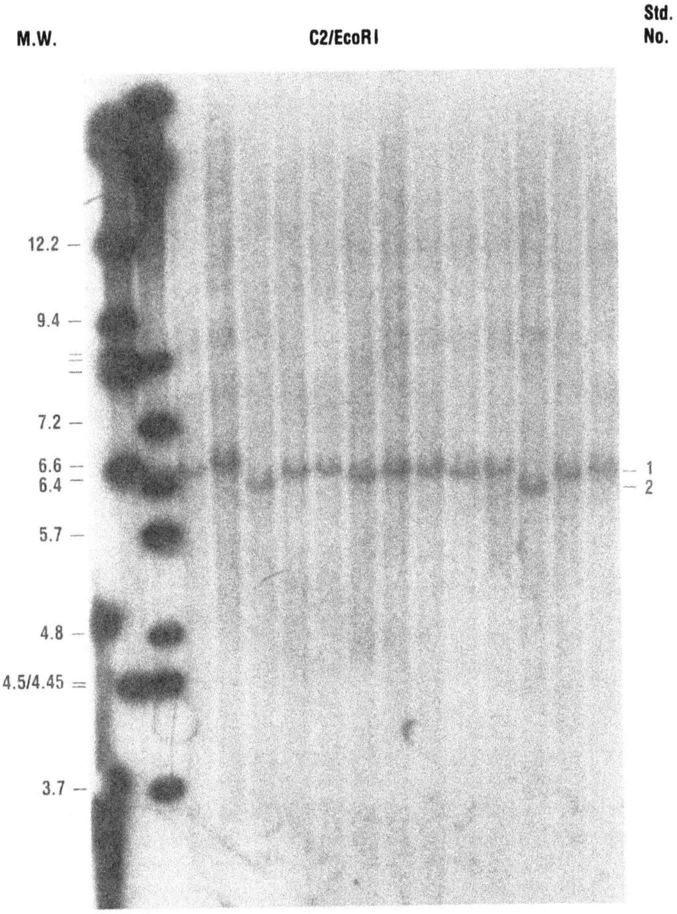

Figure 1. Fragments identified with the C2/EcoRI system.

Table 2. Distribution of Polymorphic Fragments in Core Cell Lines, C2/EcoRI

Cell No.	Name	Std. Fragment No. 1	Std. Fragment No. 2	Cell No.	Name	Std. Fragment No. 1	Std. Fragment No. 2
9020	QBL	1		9017	WT8	1	
9025	DEU	1		9014	MGAR	1	
9028	PE117	1		9011	E4181324	1	
9030	JHAF	1		9009	KAS011	1	
9032	BSM	1		9007	DEM	1	
9034	SAVC	1		9032	BSM	1	
9037	SWEIG007	1		9005	HOM2	1	
9039	JVM	1		9003	KAS116	1	
9056	KOSE	1		9072	SPACH	1	
9043	BM21	1		9106	MANIKA	1	
9047	PLH	1		9104	DHIF	1	
9048	LBUF	1		9091	MLF	1	
9052	DBB	1		9082	HO104	1	
9054	EK	1		9071	OLGA	1	
9057	TEM	1		9069	MADURA	1	
9059	SLE005	1		9067	BTB	1	
9062	WDV	1		9065	HHKB	1	
9064	AMALA	1		9063	WT47	1	
9066	TAB089	1		9061	31227ABO	1	
9068	BM9	1		9045	TUBO	1	
9070	LUY	1		9042	TISI	1	
9075	DKB	1		9038	BM16	1	
9088	PFO4015	1		9036	SP0010	1	
9092	BM92	1		9033	BM14	1	
9105	FPAF	1		9031	BOLETH	1	
9107	LKT3	1		9029	WT51	1	
9002	MZO70782	1		9026	YAR	1	
9004	JESTHOM	1		9023	VAVY	1	
9006	WT100BIS	1		9021	RSH	1	
9056	KOSE	1		9022	COX	1	
9008	D0208915	1		9058	OMW		1
9010	AMAL	1		9051	PITOUT		1
9013	SCHU	1		9055	HO301		1
9016	RML	1		9040	BM15		1
9018	L0081785	1		9050	MOU		1
9019	DUCAF	1		9060	CB6B	1	1

RFLP Standardization Report for C2/HindIII

P.A. Biro,[1] S. Cutbush,[1] J. Demopulos,[2] R.T. Acton,[2] and S.V. Van Tonder[3]

The C2 probe detects a single fragment (Table 1) in each DNA sample. This fragment shows the same allelic micro-heterogeneity (Table 2 and Fig. 1) as previously described (1) for *SstI* digests. However, because of the small size differences among the *HindIII* fragments, it is not possible to assign each fragment to its appropriate allele. Consequently, only two alleles have been defined for this enzyme/probe combination. The 4.46 kb band includes the three longest fragments described by Bentley et al. (1), and the 4.32 kb band is equivalent to the shortest allelic fragment of Bentley.

Participating Laboratories: UKIFES,[1] US5UAB,[2] SAFHAM[3]

Reference

1. Bentley DR, Campbell RD, Cross SJ. DNA polymorphism at the C2 locus. Immunogenetics 1985;22:377.

Table 1. Sizes and Frequencies of Standard Bands Produced by the HindIII/C2 Enzyme/Probe Combination

	kb	Frequency	% Faint
1	4.46	0.929	0.000
2	4.32	0.086	0.000

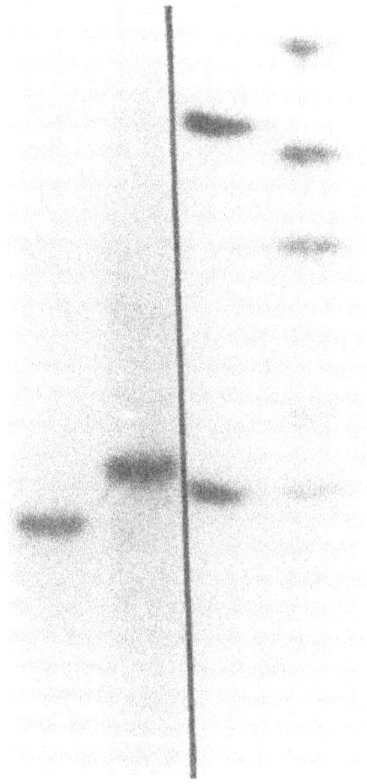

►

Figure 1. Standard bands produced by the C2/HindIII enzyme/probe combination.

Table 2. RLFP Pattern of C2/HindIII System

Cell Lines		DR Type	Fragment No. 1	2	Cell Lines		DR Type	Fragment No. 1	2
MZ070782	9002	1	1		BM21	9043	5	1	
KAS116	9003	1	1		TUBO	9045	5	1	
JESTHOM	9004	1	1		PLH	9047	7	1	
HOM2	9005	1	1		LBUF	9048	7	1	
WT100BIS	9006	1	1		MOU	9050	7		1
DEM	9007	2/4	1		PITOUT	9051	7		1
DO208915	9008	2	1		DBB	9052	7	1	1
KAS011	9009	2	1		EK	9054	6	1	1
AMAI	9010	2	1		HO301	9055	6		1
E4181324	9011	2	1		KOSE	9056	6	1	1
SCHU	9013	2	1		TEM	9057	6	1	1
MGAR	9014	2	1		OMW	9058	6		1
RML	9016	2	1		SLE005	9059	6	1	
WT8	9017	2	1		CB6B	9060	6	1	1
LO081785	9018	3	1		31227ABO	9061	6	1	
DUCAF	9019	3	1		WDV	9062	6	1	
QBL	9020	3	1		WT47	9063	6	1	
RSH	9021	3	1		AMALA	9064	6	1	
COX	9022	3	1		HHKB	9065	6	1	
VAVY	9023	3	1		TAB089	9066	8	1	
DEU	9025	4	1		BTB	9067	8	1	
YAR	9026	4	1		BM9	9068	8	1	
PE117	9028	4	1		MADURA	9069	8	1	
WT51	9029	4	1		LUY	9070	8	1	
JHAF	9030	4	1		OLGA	9071	8	1	
BOLETH	9031	4	1		SPACH	9072	8	1	
BSM	9032	4	1		DKB	9075	9	1	
BM14	9033	4	1		HO104	9082	?	1	
SAVC	9034	4	1		PF04015	9088	3	1	
SPO010	9036	5	1		MLF	9091	4	1	
SWEI6007	9037	5	1		BM92	9092	4	1	
BM16	9038	5	1		DHIF	9104	11	1	
JVM	9039	5	1		FPAF	9105	11	1	
BM15	9040	5		1	MANIKA	9106	7	1	
TISI	9042	5	1		LKT3	9107	4	1	

RFLP Standardization Report for C2/MspI

S. Tonks,[1] V. Groves,[1] C. Muller,[2] F. Otani,[3] J.R. Regueiro,[4] M. Segall,[5] J. Bodmer,[1] and J. Trowsdale[1]

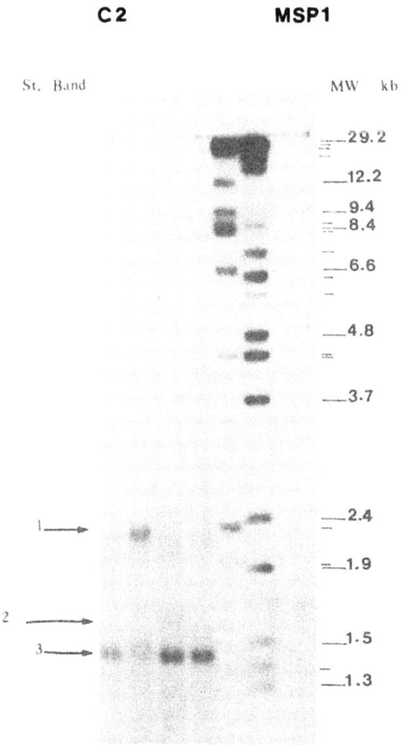

The C2/MspI system identified three standard fragments in the core cell lines (Fig. 1 and Table 1). Fragments 1 (2.30 kb) and 2 (1.79 kb) were found to be polymorphic, while fragment 3 (1.53 kb) was constant. These banding patterns can be seen in Table 1 (duplicate cells not shown).

This system was standardized on data from only two laboratories, only one set of which was easily interpreted. The DNA appears to have been lost from lane 19 in these data, producing a discrepancy in the 9056 set of duplicates.

Participating Laboratories: UKIBOD,[1] GERMUC,[2] JAPKSH,[3] FRAARN,[4] US2BAC[5]

◄
Figure 1. Fragments observed with the C2/MspI system.

Table 1. Banding Pattern of the C2/MspI System in Each Cell Line

		Fragment No.		
WS ID	Cell	1	2	3
9047	PLH	2		2
9043	BM21		1	2
9006	WT100B1S		1	2
9050	MOU		1	2
9056	KOSE		1	2
9051	PITOUT		1	2
9040	BM15		1	2
9069	MADURA		1	2
9028	PE117		1	2
9019	DUCAF		1	2
9025	DEU		1	2
9031	BOLETH		1	2
9088	PF04015		1	2
9066	TAB089		1	2
9060	CB6B		1	2
9036	SPOO10		1	2
9002	MZ070782		1	2
9071	OLGA		1	2
9064	AMALA		1	2
9057	TEM		1	2

Immunobiology of HLA
Volume I

Table 1. *Continued*

WS ID	Cell	Fragment No. 1	2	3
9030	JHAF		1	2
9054	EK/OH		1	2
9105	FPAF		1	2
9091	MLF		1	2
9016	RML		1	2
9048	LBUF		1	2
9065	HHKB		1	2
9021	RSH		1	2
9062	WDV		1	2
9032	BMS		1	2
9045	TUBO			2
9014	MGAR			2
9058	OMW			2
9107	LKT3			2
9017	WT8			2
9042	TISI			2
9013	SCHU			2
9033	BM14			2
9004	JESTHOM			2
9018	LOO81785			2
9055	HO301			2
9063	WT47			2
9022	COX			2
9028	PE117			2
9005	HOM2			2
9104	DHIF			2
9010	AMAI			2
9037	SWEIG007			2
9009	KAS011			2
9092	BM92			2
9059	SLE005			2
9034	SAVC			2
9068	BM9			2
9075	DKB			2
9061	31227ABO			2
9067	BTB			2
9038	BM16			2
9008	DO208915			2
9003	KAS116			2
9070	LUY			2
9029	WT51			2
9011	E4181324			2
9026	YAR			2
9039	JVM			2
9082	HO104			2
9106	MANIKA			2
9072	SPACH			2
9052	DBB			2
9020	QBL			2
9007	DEM			2

1 = Single intensity; 2 = double intensity; 8 = faint.

RFLP Standardization Report for C2/BamHI

A. Andreas-Zietz,[1] E. Keller,[1] E.D. Albert,[1] and E.J. Yunis[2]

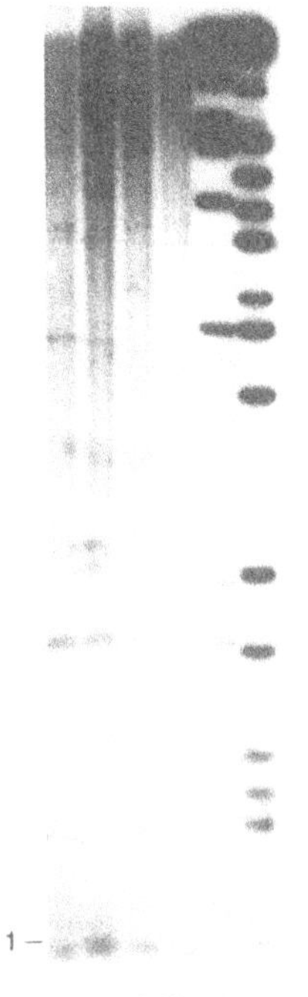

◄ **Figure 1.** Core bands of the C2/BamHI system (gel 0.9%).

The C2/BamHI system identified one fragment on all core cell lines (Table 1 and Fig. 1). The results are based on the data of only one lab.

Cell no. 58 carries the A25,B18,DR2 haplotype, which is known to be associated with C2-deficiency (1). However, cell no. 58 is positive for the C2/BamHI core fragment.

Reference

1. Fu SM, Stern R, Kunkel HG, et al. Mixed lymphocyte culture determinants and C2 deficiency: LD-7a associated with C2 deficiency in four families. J Exp Med 1975;142:495.

Participating Laboratories: GERALB,[1] NCYYUN[2]
This study was supported by SFB 217.

Table 1. Banding Pattern of All 72 Core Cell Lines, as Standardized in the C2/BamHI System, Including kb Value

```
Probe Name: C2              Enzyme Name: BAMHI

 |999999999999999999999999999999999999999999999999999999999999999999999999| WS-ID
 |000100000000000000010000000000000000000000100000000000000000000000100000| number
 |415014130156242005505441632123866307653509146263095367663007212358037520|
 |548772334853233650641700978951860621470451685122929485178830916962622207|
 |------------------------------------------------------------------------|---------
 |000000000111111111122222222223333333333444444444455555555556666666666777| position
 |123456789012345678901234567890123456789012345678901234567890123456789012| on blots
 |------------------------------------------------------------------------|---------
1|111111111111111111111111111111111111111111111111111111111111111111111111|  0.90 kb
```

Cells are listed according to their position on the blots. Numbers indicate intensity (1,2,3 = increasing intensity, 8 = faint, 9 = unreadable).

Immunobiology of HLA
Volume I

RFLP Standardization Report for C4/SstI

T. Takenouchi,[1] R. Tonai,[1] P. Terasaki,[1] T. Sasazuki,[2] and P. Reekers[3]

Standardization of the data from the two participating laboratories was carried out in a similar manner to the handling of the C2/SstI system. Once again, sample #41 could not be digested by either laboratory.

Two standardization fragments were assigned (Table 1 and Fig. 1). Two allelic patterns were observed as shown in Table 2. There is no significant association between C4 and C2 and class II phenotypes.

Participating Laboratories: US5TER,[1] JAPSAS,[2] BENREE[3]

Table 1. Fragments Observed with the C4/SstI System

	kb	C4 Locus	SST Frequency	TER % Faint	56	32
1	4.10	0 0 0 0	0.942	0.000		
2	3.98	0 0 0 0	0.471	0.000		

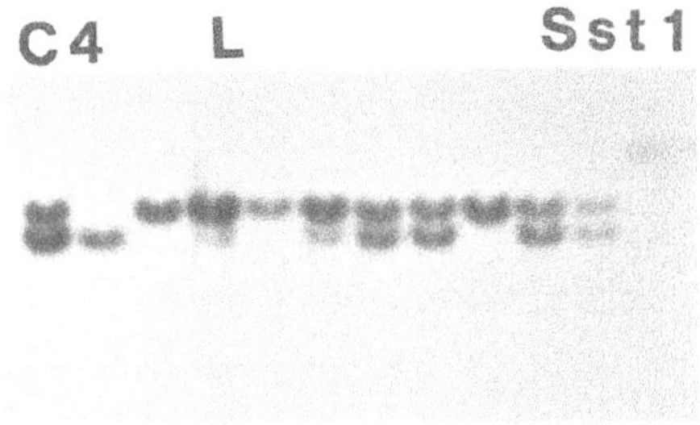

Figure 1. Fragments observed with C4/SstI system.

Table 2. Banding Pattern of the C4/Sstl in Each Cell

Cell	Fragment No. 1	2
9045 TUBO	1	
9014 M GAR	1	
9017 WT8	1	
9042 TISI	1	
9013 SCHU	1	
9004 JES THOM	1	
9018 LO 081785	1	
9063 WT47	1	
9006 WT100 BIS	1	
9050 MOU	1	
9047 PLH	1	
9019 DUCAF	1	
9025 DEU	1	
9031 BOLETH	1	
9010 AMALA	1	
9057 TEM	1	
9030 JHAF	1	
9054 EK	1	
9091 MLF	1	
9016 RML	1	
9065 HHKB	1	
9062 WDV	1	
9032 BSM	1	
9032 BSM	1	
9009 KAS-011	1	
9068 BM9	1	
9061 3-12-27 AOB	1	
9067 BTB	1	
9038 BM16	1	
9003 KAS-116	1	
9070 LUY	1	
9029 WT51	1	

Table 2. *Continued*

Cell	Fragment No. 1	2
9011 E 4181324	1	
9026 YAR	1	
9039 JVM	1	
9082 HO 104	1	
9020 QBL	1	
9071 OLGA	1	9
9058 OMW	1	1
9107 LKT 3	1	1
9033 BM14	1	1
9055 HO 301	1	1
9043 BM21	1	1
9005 HOM-2	1	1
9056 KOSE	1	1
9056 KOSE	1	9
9104 DHIF	1	1
9051 PITOUT	1	1
9040 BM15	1	1
9010 AMAI	1	1
9037 SWEIG 007	1	1
9028 PE 117	1	1
9066 TAB 089	1	1
9060 CBG-IB	1	1
9036 SPO 010	1	1
9002 MZ 070782	1	1
9048 LBUF	1	1
9021 RSH	1	1
9092 BM92	1	1
9059 SLE-005	1	1
9034 SAVC	1	1
9075 DKB	1	1
9008 DO 208915	1	1
9106 MANIKA	1	1
9072 S PACHEO	1	1
9052 DBB	1	1
9007 DEM	1	1
9022 COX		1
9023 VAVY		1
9069 MADURA		1
9088 PF 04015		1
9105 FPAF	9	9

1 = Single intensity; 9 = uninterpretable.

RFLP Standardization Report for C4/PstI

J.J. Hyldig-Nielsen,[1] M. Trucco,[2] H. Coppin,[3] A. Andreas-Zietz,[4] E. Keller,[4] C. Carcassi,[5] A. Takada,[6] and A. Svejgaard[1]

The C4/PstI system identified a single nonpolymorphic fragment of 2.34 kb (Fig. 1).

Participating Laboratories: SCASVE,[1] US1DUQ,[2] FRAOHA,[3] GERALB,[4] IT1CON,[5] JAPAIZ[6]

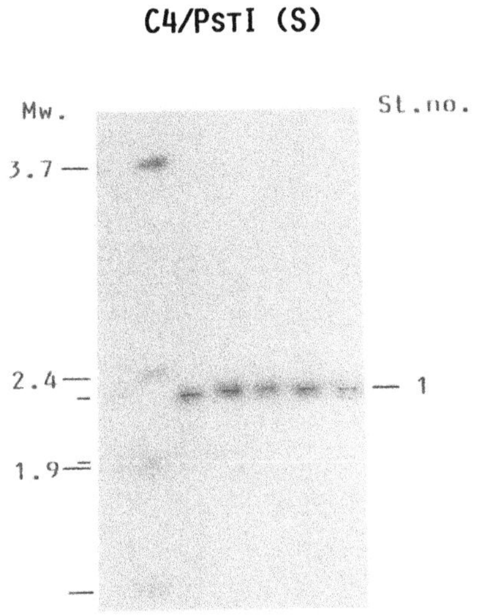

Figure 1. Fragments identified with the C4/PstI system.

RFLP Standardization Report for C4/KpnI

G. Paulsen[1] and K. Siminovitch[2]

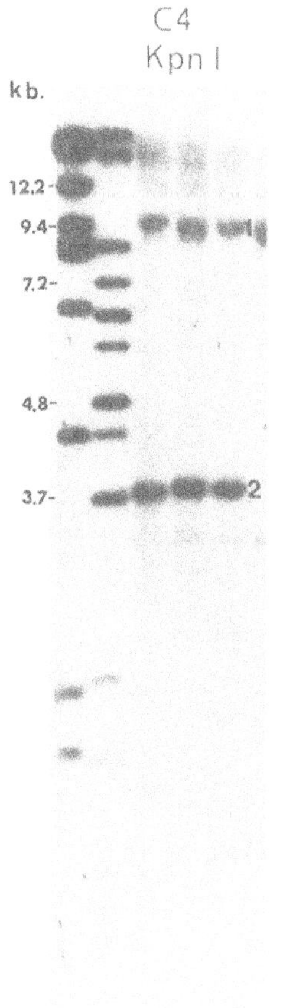

Two bands were identified by the RFLP analysis in the C4/KPN system (Table 1 and Fig. 1). RFLP patterns of the core cell lines are indicated in Table 2. Three different patterns were observed: pattern 1 (band 1), pattern 2 (band 2), and pattern 3 (band 1+2). Pattern 1 occurred in 11 core cell lines, 6 of which bearing the haplotype HLA DR3/B8 or B18; pattern 2 in 1 core cell line (WS.ID. 9105) (Table 2).

Participating Laboratories: SCATSB,[1] NCYFAL[2]

Table 1. Standard Bands in the C4/KPN System

Band	kb	Frequency[a]	% Faint[b]	56[c]	32[c]
1	10.65	0.986	0.043		
2	3.81	0.841	0.017		

[a]Band frequency in the core cell lines.
[b]Frequency of faint bands.
[c]Discrepancies in strong bands (s) or faint bands (f) of the hidden duplicates (WS.ID. 9056 and 9032).

◄
Figure 1. RFLP bands in the C4/KPN system of three core cell lines run in 0.9% agarose gel. Representative core bands are shown in the figure numbered on their right side. The WS.ID. of these core cell lines are from the left following marker 1 and marker 2: 9069, 9037, and 9028.

Table 2. Distribution of Bands in the Core Cell Lines in the C4/KPN System

			Band No.					Band No.	
	C4A	C4B	1	2		C4A	C4B	1	2
9002			2	1	9029			1	1
9003			1	2	9030			2	1
9004			2	2	9031			1	1
9005			2	2	9032			1	2
9007	A6	B1	8	1	9033			2	2
9008	A4	B2	1	1	9034	A3	B1	1	2
9009			1	1	9036			2	1
9010			2	2	9037			2	1
9011	A3	B1	2	2	9038			1	1
9014			2	2	9040			2	2
9013			2	2	9042	A3	B1	2	2
9016			2	2	9043			2	2
9017			1	1	9045			1	2
9025			2	1	9048	A3	B1	2	2
9026			1	2	9051	A3	B1	2	2
9028			2	1	9052	A6	B1	2	2

Immunobiology of HLA
Volume I

Table 2. *Continued*

	C4A	C4B	Band No. 1	Band No. 2		C4A	C4B	Band No. 1	Band No. 2
9054			1	1	9072			8	1
9055			2	2	9075	A4	B2	2	1
9056			2	2	9091			1	1
9056			2	1	9092			2	2
9082			2	2	9106			1	2
9057			1	1	9107			1	1
9058	A3	B1	2	2	9006			2	
9059			2	2	9018	A3	BQ0	2	
9060	A3	B1	2	2	9019	A3	BQ0	2	
9061			2	2	9020	A3	BQ0	1	
9062	A2	B1	1	1	9022	AQ0	B1	2	
9063			2	2	9023	AQ0	B1	2	
9064			8	1	9039			1	
9065	A3	B1	1	1	9047			2	
9066	A4	B2	1	1	9050	AQ0	B1	2	
9067			2	2	9068			2	
9069			1	1	9088	AQ0	B1	1	
9070			1	1	9105	A3	B1		8
9071			2	1	9104	A3	B1	9	9

Bands designation: 1 = single intensity, 2 = double intensity, 8 = faint, 9 = unreadable.

RFLP Standardization Report for C4/EcoRV

A. Ando,[1] Y.K. Paik,[2] and Y. Mitsuishi[3]

C4/EcoRV system identified the four fragments (Table 1 and Fig. 1). The presence/absence pattern of all 72 core cell lines is shown in Table 1. Four patterns are found in Table 1. St. Fragment no. 4 (7.46 kb) correlates with the C4B long gene.

Participating Laboratories: JAPTSU,[1] US5PAK,[2] FRAMYE[3]

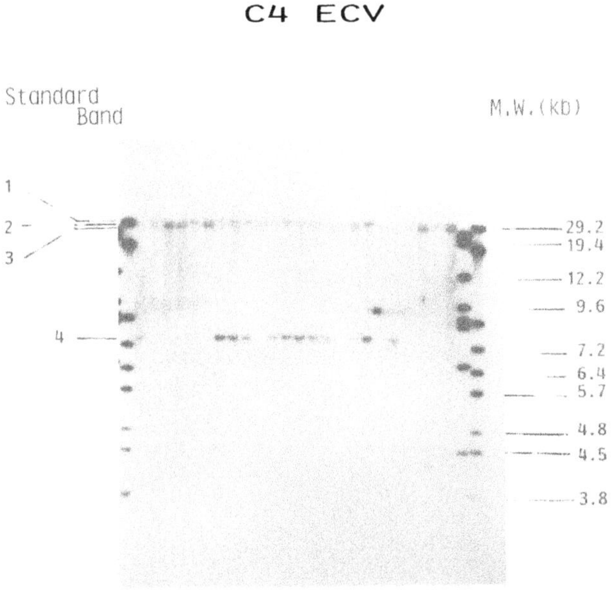

Figure 1. Fragments identified with the C4/EcoRV system.

Immunobiology of HLA
Volume I

Table 1. RFLP Pattern of the C4/EcoRV System

	Fragment No.		1	2	3	4								
	Fragment Size (kb)		67.36	28.50	25.15	7.49								
	Cell Lines						A	C	B	Bw	Bf	C2	C4A	C4B
25	9069	MADURA	1	2	2		2	10	60	6				
19	9056	KOSE	8	2	2	2	2	–	35	6	S			
13	9022	COX	8	2	2		1	7	8	6	S	C	AQ0	B1
15	9023	VAVY	8	2	2		1	7	8	6	S	C	AQ0	B1
28	9019	DUCAF	8	2	2		30	5	18	6	F1		A3	BQ0
31	9088	PF04015	8	2	2		1		8	6	S		AQ0	B1
3	9058	OMW		3	3		2	–	45	6	F	C	A3	B1
4	9107	LKT3		3	3		24	1	54	6				
8	9033	BM14		3	3		3	7	7	6				
11	9055	HO301		3	3		3	8	14	6				
14	9043	BM21		3	3		1	–	41	6				
22	9047	PLH		3	3		3	6	47	4				
24	9010	AMAI		3	3		28	4	53	4				
32	9066	TAB089		3	3		2	11	46	6	S	C	A4	B2
34	9036	SPOO10		3	3		2	5	44	4				
35	9002	MZ070782		3	3		24	2,8	14	6	S			
44	9048	LBUF		3	3		30	6	13	4	S		A3	B1
46	9021	RSH		3	3		68,30	2	42	6	F	C	A1	BQ0
51	9059	SLE005		3	3		2	10	60	6				
52	9034	SAVC		3	3		3	7	7	6	S		A3	B1
54	9075	DXB		3	3		24	10	60	6	S	1	A4	B2
72	9007	DEM		3	3		2	6	57	4	S		A6	B1
70	9052	DBB		3	2		2	6	57	4	S		A6	B1
10	9018	LO081785		2	2		3	5	18	6	F1		A3	BQ0
16	9006	WT100BIS		2	2		11	4	35	6				
18	9050	MOU		2	2		29	–	44	4			AQ0	B1
23	9040	BM15		2	2		1	7	49	4				
50	9092	BM92		2	2		25	1	51	4				
53	9068	BM9		2	2		2	4	35	6				
58	9008	DO208915		2	2		25	–	18	6	S	Q0	A4	B2
64	9039	JVM		2	2		2	5	18	6				
69	9072	SPACH		2	2		31	1,3.1	62					
71	9020	QBL		2	2		26	?5	18	6	F1	1	A3	BQ0
21	9051	PITOUT		1	1		29	–	44	4	F	C	A3	B1
67	9106	MANIKA		1	1		3	–	50	6				
68	9032	BSM		8	8	2	2	9	62	6				
5	9017	WT8		3	3	2	3	7	7	6				
27	9028	PE117		3	3	2	24	10	60	6				
7	9013	SCHU		2	2	2	3	7	7	6				
29	9025	DEU		2	2	2	31	4	35	6				
36	9071	OLGA		2	2	2	31	1,10	62	6				
40	9054	EK		2	2	2	2	5	44	4				
45	9065	HHKB		2	2	2	3	7	7	6	S	1	A3	B1
63	9026	YAR		2	2	2	26	–	38	4	S			
65	9056	KOSE		2	2	2	2	–	35	6				
38	9057	TEM		1	1	2	26	–	38	4				
41	9105	FPAF		1	1	2	1	4	35	6	S	C	A3	B1
42	9091	MLF		1	1	2	2	9	62	6				
47	9062	WDV		1	1	2	2	–	38	4		1	A2	B1
49	9009	KASO11		1	1	2	1	6	37	4	S			
59	9003	KAS116		1	1	2	24	–	51	4				
2	9014	MGAR		3	3	3	26	7	8	6	S			
6	9042	TISI		3	3	3	24	4	35	6	S		A3	B1
66	9082	HO104		3	3	3	3	7	7	6				
9	9004	JESTHOM		2	2	3	2	1	27	4	S			
17	9005	HOM2		2	2	3	3	1	27	4				
26	9037	SWEIG007		2	2	3	29	2	61	6				

Table 1. *Continued*

	Fragment No.		1	2	3	4								
			6	2	2									
	Fragment Size (kb)		7	8	5	7								
			.	.	.	.								
			3	5	1	4								
			6	0	5	9								
	Cell Lines						A	C	B	Bw	Bf	C2	C4A	C4B
30	9031	BOLETH		2	2	3	2	10	62	6				
33	9060	CBGB		2	2	3	1	9	62	6	F		A3	B1
37	9064	AMALA		2	2	3	2	9	62	6				
39	9030	JHAF		2	2	3	31	-	51	4				
43	9016	RML		2	2	3	2	-	51	4				
56	9067	BTB		2	2	3	2	1	27	4				
57	9038	BM16		2	2	3	2	7	18	6				
60	9070	LUY		2	2	3	2	-	51	4				
61	9029	WT51		2	2	3	23	8	65	6				
62	9011	E4181324		2	2	3	1	-	52	4	S		A3	B1
1	9045	TUBO		1	1	3	2,3	7	51	4	S			
12	9063	WT47		1	1	3	32	5	44	4				
48	9032	BSM		1	1	3	2	9	62	6				
55	9061	31227ABO		1	1	3	2	7	18	6				
20	9104	DHIF		2	2	8	31		38		S		A3	B1

RFLP Standardization Report for C4/TaqI

R.E. Bontrop[1] and R. Walford[2]

Data for standardization were only received from one laboratory. The C4/TaqI system allowed the detection of five different standard fragments (Table 1 and Fig. 1). Standard fragment 1 appears to be the most abundant and has a molecular weight of 7.04 kb and a frequency of 94.3%. Standard fragment 2 (6.27 kb) has a frequency of 5.7%. This implies that when one of the panel cells is negative for standard fragment 1, it is positive for fragment 2. Standard fragment 3 has a molecular weight of 5.77 kb and a frequency of 48.6%, whereas fragment 4, with a molecular weight of 4.62 kb, exhibits a frequency of 42.9%. On the other hand, fragment 5, with a molecular weight of 3.82 kb, is the most rare and has a panel frequency of 1.4%. Some panel cells are only positive for fragment 1 (Fig. 1, lane 16) or for fragment 2 (Fig. 1, lanes 13 and 15), or positive for fragment 1 in combination with fragment 3 (Fig. 1, lane 13) or with fragment 4 (Fig. 1, lanes 14, 20 and 21), whereas fragment 1 also can be observed in combination with fragments 3 and 4 (Fig. 1 lane 17) or in combination with fragments 4 and 5 (Fig. 1, lane 19). None of the fragments detected with the C4/TaqI system appeared to be identical to the standard fragments detected in the C2/TaqI or Bf/TaqI systems. Fragment presence/absence patterns for the C4/TaqI system are shown in Table 2. A detailed interpretation of these data is presented in "Workshop Report of RFLP in C4, C2, BF, and 21OH Genes" (Alper et al., this volume).

Table 1. Fragments Identified with the C4/TaqI System

Fragment	kb	Frequency
1	7.04	0.943
2	6.27	0.057
3	5.77	0.486
4	4.62	0.429
5	3.82	0.014

Participating Laboratories: BENROO,[1] US5WAL[2]

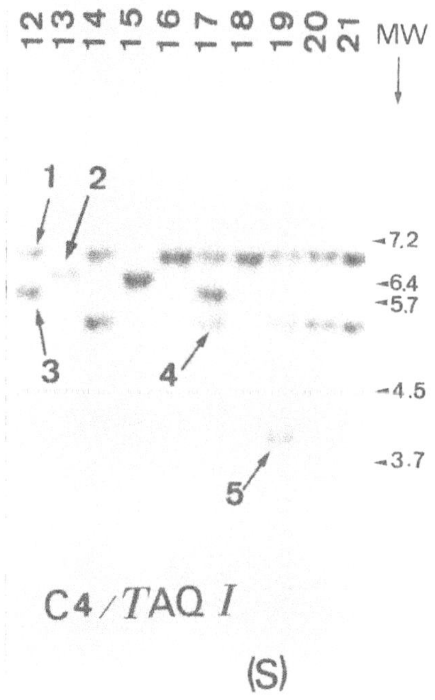

Figure 1. Fragments observed with the C4/TaqI system.

Table 2. Fragment Presence/Absence Pattern for the C4/TaqI System as Defined on 72 Core Cell Lines

	Fragment No.						Fragment No.				
Cell	1	2	3	4	5	Cell	1	2	3	4	5
9018	1					9058	1			1	
9006	2					9107	1			1	
9050	1					9033	1			1	
9047	1					9018	1			1	
9019	1					9043	1			1	
9068	1					9104	1			1	
9020	1					9051	1			1	
9045	1		1			9040	1			1	
9014	1		1			9036	1			1	
9017	1		1			9066	1			1	
9042	1		1			9036	1			1	
9013	1		1			9002	1			1	
9004	1		1			9048	1			1	
9063	1		1			9021	1			1	
9037	1		1			9092	1			1	
9025	1		1			9059	1			1	
9031	1		1			9034	1			1	
9057	1		1			9075	1			1	
9030	1		1			9008	1			1	
9054	1		1			9056	1			1	
9105	1		1			9106	1			1	
9016	1		1			9072	1			1	
9065	1		1			9052	1			1	
9062	1		1			9007	1			1	
9032	1		1			9005	1		1	1	
9009	1		1			9028	1		1	1	
9061	1		1			9060	1		1	1	
9067	1		1			9071	1		1	1	
9038	1		1			9064	1		1	1	
9003	1		1			9022		1			
9070	1		1			9023		1			
9029	1		1			9088		1			
9011	1		1			9069		1		1	
9026	1		1			9056	1			1	1
9082	1		1			9056	1			1	1
9032	1		1								

RFLP Standardization Report for C4/PvuII

P. Piccioli,[1]* K. Kato,[2] and G.B. Ferrara[1]

Figure 1 shows the results of Southern blot analysis of core cell line DNAs using the restriction endonuclease PvuII and C4 probe (1). A 3.69 kb fragment was common to all samples tested; no correlation is evident between RLFP and A and B alleles of the fourth complement component. Analyzing the pattern obtained, 15 of 16 cell lines sharing C4 alleles (excluding C4QO alleles) presented all the fragments detected (3.69, 3.22, and 0.80 kb), whereas 6 out of 8 cell lines carrying either AQ0 or BQ0 lacked the second fragment. In Table 1, the fragment distribution on the core cell lines panel is represented.

Reference

1. Belt KT, Carroll C, Porter RR, The structural basis of the multiple forms of human complement component C4. Cell 1984;36:907.

Participating Laboratories: IT2FER,[1] US7DUP[2]
*P. Piccioli was recipient of a fellowship from A.I.R.C. (Associazione Italiana per la Ricerca sul Cancro).
This work was in part supported by CNR grants n.86.0265.51 and 86.00659.

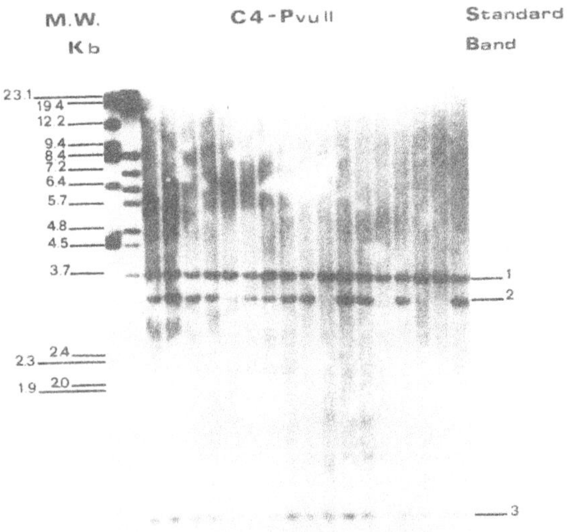

Figure 1. Southern Blot Analysis of DNAs extracted from TUBO, MGAR, OMW, LKT3, WT8, TISI, SCHU, JESTHOM, LO081785, HO301, WT47, COX, BM21, VAVY, WT100BIS and HOM2 cell lines. Three bands of 3.69, 3.22 and 0.80 kb were obtained after digestion with PvuII and hybridization with a C4 probe.

Table 1. Banding Pattern of the C4/PvuII System in Each Cell Line

			Fragment No.		
Cell	C4A	C4B	1	2	3
9002 MZ 070782			2	3	2
9003 KAS 116			2	2	2
9004 JESTHOM			2	2	2
9005 HOM 2			2	2	2
9006 WT100 BIS			2		1
9007 DEM	A6	B1	2	2	2
9008 DO 208915	A4	B2	2	2	2
9009 KAS 011			2	2	2
9010 AMAL			2	2	2
9011 E 4181324	A3	B1	1	1	
9013 SCHU			2	1	1
9014 M GAR			2	2	2
9016 RML			2	3	2
9017 WT 8			2	1	1
9018 LO 081785	A3	BQ0	2		2
9019 DUCAF	A3	BQ0	2		2
9020 QBL	A3	BQ0	2		2

Table 1. *Continued*

Cell	C4A	C4B	Fragment No. 1	2	3
9021 RSH	A1	BQ0	2	2	2
9022 COX	AQ0	B1	2		1
9023 VAVY	AQ0	B1	2		2
9025 DEU			2	2	2
9026 YAR			2	2	2
9028 PE 117			2	2	2
9029 WT 51			2	2	2
9030 JHAF			2	2	2
9031 BOLETH			2	2	2
9032 BSM			2	2	2
9032 BSM			2	2	2
9033 BM 14			2	2	2
9034 SAVAC	A3	B1	2	2	2
9036 SPO 010			1	1	
9037 SWEIG 007			2	2	2
9038 BM 16			2	2	2
9039 JVM			2		
9040 BM 15			2	2	2
9042 TISI	A3	B1	2	2	1
9043 BM 21			2	2	2
9045 TUBO			2	2	2
9047 PLH			2		1
9048 LBUF	A3	B1	2	2	2
9050 MOU	AQ0	B1	2		1
9051 PITOUT	A3	B1	2	2	2
9052 DBB	A6	B1	2	2	2
9054 EK			2	2	2
9055 HO 301			2	2	2
9056 KOSE			2	2	2
9056 KOSE			2	2	2
9057 TEM			1	1	2
9058 OMW	A3	B1	2	2	2
9059 SLE 005			2	2	2
9060 CB 6B	A3	B1	1	3	2
9061 31227 ABO			1	2	2
9062 WDV	A2	B1	2	2	2
9063 WT 47			2	2	2
9064 AMALA			1	2	2
9065 HHKB	A3	B1	2	2	2
9066 TAB 089	A4	B2	2	2	2
9067 BTB			2	2	2
9068 BM 9			2		2
9069 MADURA			2	2	2
9070 LUY			2	2	2
9071 OLGA			2	2	2
9072 SPACH			1	1	2
9075 DKB	A4	B2	2	2	2
9082 HO 104			2	2	1
9088 PF 04015	AQ0	B1	1		
9091 MLF			2	2	2
9092 BM 92			2	2	2
9104 DHIF	A3	B1	2	2	2
9105 FPAF	A3	B1	2	1	2
9106 MANIKA			1	2	2
9107 LKT3			2	2	2

1 = Single intensity; 2 = double intensity; 3 = more than double intensity.

RFLP Standardization Report for C4/EcoRI

M. Segall,[1]* L. Schluender,[1]* H. Betuel,[2] and M. Garovoy[3]

Only one set of blots could be used for this standardization; however, the fragments were clear, as shown in Figures 1 and 2.

The C4/EcoRI system identified six fragments on the core cell lines (Table 1 and Figs. 1 and 2). There were no discrepancies for the "hidden duplicates." The pattern of polymorphic fragments is given in Table 2.

Table 1. C4/EcoRI Standard Fragments

				Discrepant for	
Fragment	kb	Frequency	% Faint	9056	9032
1	15.23	0.014	0.000		
2	13.52	0.414	0.033		
3	12.25	0.486	0.000		
4	8.96	0.014	0.000		
5	7.50	0.057	0.250		
6	6.18	0.943	0.000		

Participating Laboratories: US2BAC,[1] FRABET,[2] US3GVY[3]
*This is publication #489 from the Immunobiology Research Center, University of Minnesota, Minneapolis, Minnesota 55455, USA.
Supported in part by Juvenile Diabetes Foundation Grant #186175 and March of Dimes Birth Defects Foundation Grant #6-496.

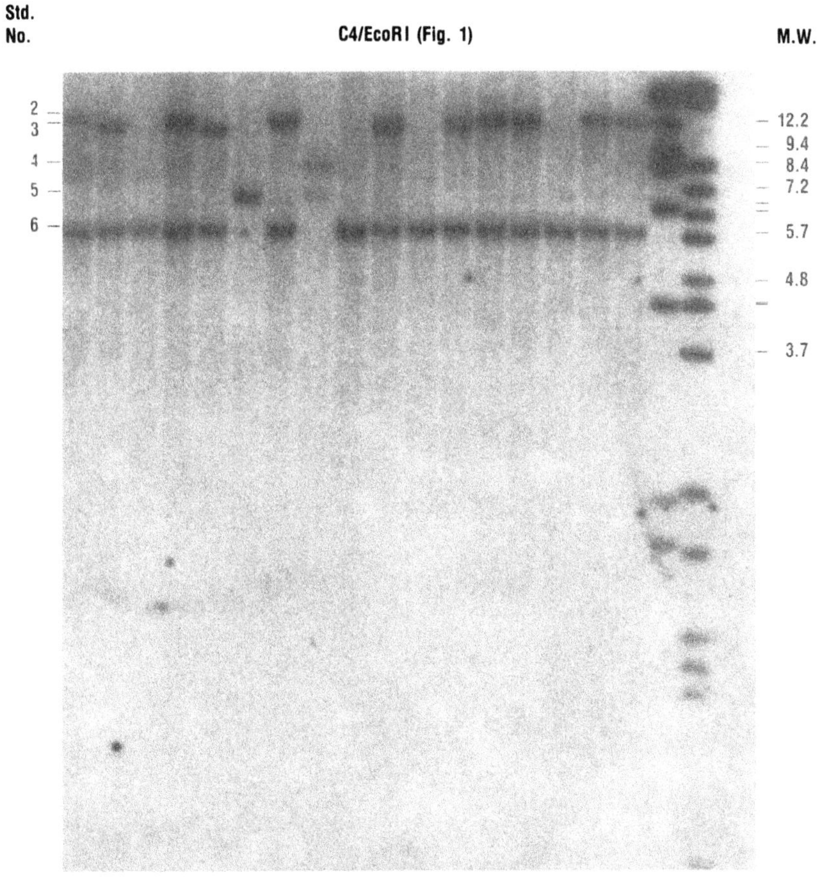

Figure 1. Fragments identified with the C4/EcoRI system.

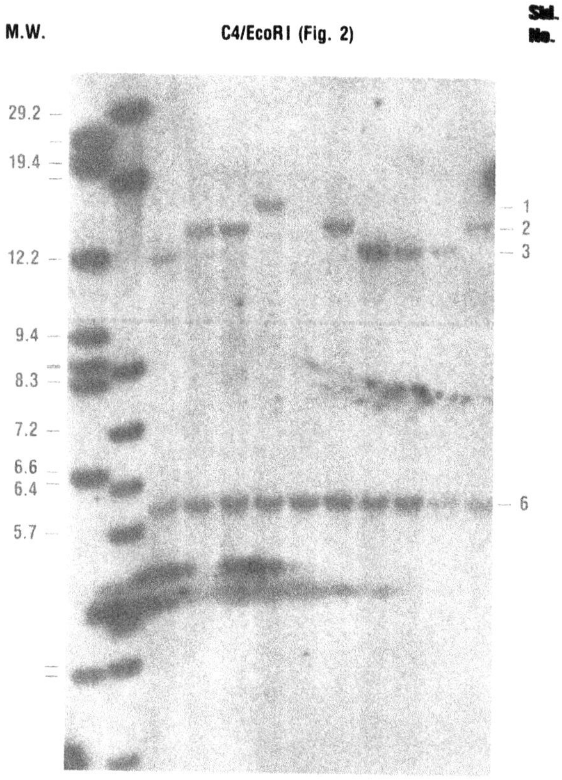

Figure 2. C4/EcoRI system standard fragments.

Table 2. Distribution of Polymorphic Fragments in Core Cell Lines, C4/EcoRI

		Std. Fragment No.								Std. Fragment No.					
Cell No.	Name	1	2	3	4	5	6	Cell No.	Name	1	2	3	4	5	6
9022	COX					1		9058	OMW		1				1
9088	PFO4015					1		9059	SLE005		1				1
9069	MADURA		1			1		9051	PITOUT		1				1
9047	PLH						1	9107	LKT3		1				1
9050	MOU						1	9002	MZO70782		1				1
9039	JVM						1	9072	SPACH		1				1
9006	WT100BIS						1	9008	D0208915		1				1
9018	L0081785						1	9010	AMAL		1				1
9019	DUCAF						1	9021	RSH		1				1
9068	BM9						1	9016	RML			1			1
9020	QBL						1	9017	WT8			1			1
9034	SAVC	1					1	9014	MGAR			1			1
9048	LBUF		1				1	9091	MLF			1			1
9043	BM21		1				1	9082	HO104			1			1
9104	DHIF		1				1	9067	BTB			1			1
9106	MANIKA		1				1	9065	HHKB			1			1
9007	DEM		1				1	9045	TUBO			1			1
9092	BM92		1				1	9042	TISI			1			1
9075	DKB		1				1	9038	BM16			1			1
9066	TAB089		1				1	9030	JHAF			1			1
9033	BM14		1				1	9032	BSM			1			1
9040	BM15		1				1	9057	TEM			1			1
9037	SWEIG007		1				1	9061	31227ABO			1			1
9036	SP0010		1				1	9062	WDV			1			1
9052	DBB		1				1	9004	JESTHOM			1			1
9055	HO301		1				1	9003	KAS116			1			1

Table 2. *Continued*

		Std. Fragment No.								Std. Fragment No.					
Cell No.	Name	1	2	3	4	5	6	Cell No.	Name	1	2	3	4	5	6
9032	BSM			1			1	9029	WT51			1			1
9011	E4181324			1			1	9025	DEU			1			1
9013	SCHU			1			1	9026	YAR			1			1
9009	KAS011			1			1	9028	PE117		1	1			1
9105	FPAF			1			1	9005	HOM2		1	1			1
9070	LUY			1			1	9056	KOSE		1	1			1
9064	AMALA			1			1	9071	OLGA		1	1			1
9063	WT47			1			1	9056	KOSE		1	1			1
9054	EK			1			1	9060	CB6B		8	1			1
9031	BOLETH			1			1	9023	VAVY				1	8	1

RFLP Standardization Report for C4/HindIII

S. Cutbush,[1] P.A. Biro,[1] J. Demopulos,[2] R.T. Acton,[2] and S.V. Van Tonder[3]

Seven standard bands have been defined using the C4 probe (Table 1). Band 3 (19.26 kb) is only found in one sample (WT8, 9017) and is probably a partial digestion product of bands 5 and 7 (Table 2 and Fig. 1). Band 4 is found in all nondeletion haplotypes and constitutes the 3′ half of the C4A gene. In many C4A deletion haplotypes (1), (lanes 13, 15, and 31: WS.ID. 9022, 9023, and 9088, respectively) st. band 4 is replaced by an 8.0 kb band (st. band 6), which is produced as a result of the fusion of the C4A and C4B genes.

Bands 1 and 2 are allelic and correspond to the "long" and "short" forms of the C4B gene (2,3). The C4 probe did not reproducibly detect the 3′ half of the C4B gene, which would have displayed a 30.5 kb fragment of the same size as st. band 1 (see also 210H/HindIII Report). This band was seen in long exposures of some autoradiograms, but was not defined for the standardization analysis.

Lane 25 (WS.ID. 9069) and lanes 4 and 27 (WS.ID. 9107 and 9028, respectively) give nonstandard patterns (bands 5 and 7), which are the result of C4 gene rearrangements. Lane 25 contains a band with the same mobility as band 6, but which may not correspond to the same DNA fragment (this sample also contains a band at position 2); it may be a rare allelic variant of fragment 4. Alternatively, WS.ID. 9069 may contain a complex, recombinant C4 locus, comprising a deleted C4A gene, a "short" C4B gene (giving rise to band 6), and a recombinant 210HA gene. These genes are followed by a second C4B gene and its accompanying 210HB gene, which give rise to band 2 and produce the normal 210H RFLP pattern (D. Campbell, Oxford University, personal communication).

Participating Laboratories: UKIFES,[1] US5UAB,[2] SAFHAM[3]

Table 1. Sizes and Frequencies of Standard Bands Produced by the C4/HindIII Enzyme/Probe Combination

	kb	Frequency	% Faint
1	31.94	0.586	0.093
2	24.77	0.414	0.000
3	19.26	0.014	0.000
4	13.69	0.943	0.000
5	12.02	0.029	0.000
6	8.01	0.043	0.000
7	6.45	0.029	0.000

References

1. Carroll MC, Palsdottir A, Belt KT, Porter RR. Deletion of complement C4 and steroid 21-hydroxylase genes in the HLA class III region. EMBO J. 1985;4:2547.
2. Prentice HL, Schneider PM, Strominger JL. C4B gene polymorphism detected in a human cosmid clone. Immunogenetics 1986;23:274.
3. Palsdottir A, Fossdal R, Arnason A, Edwards JH, Jensson O. Heterogeneity of human C4 gene size. Immunogenetics 1987;25:299.

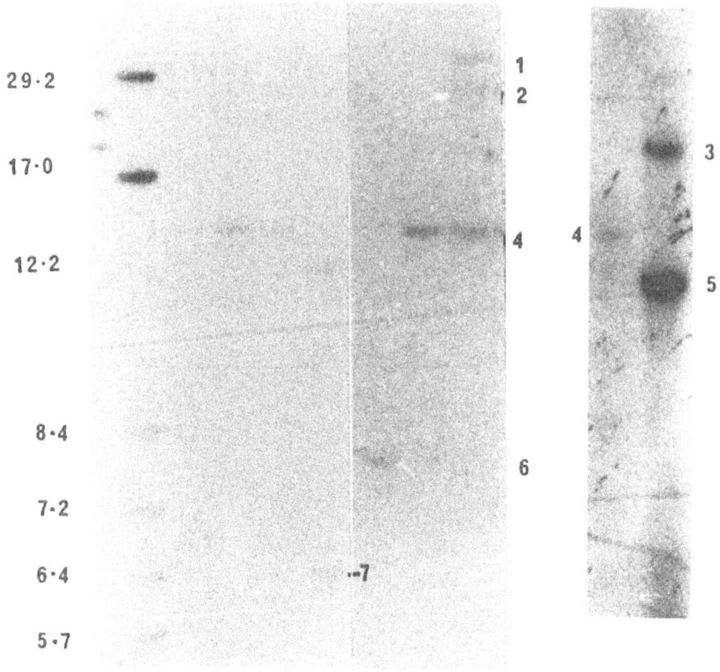

Figure 1. Standard bands produced by the C4/HindIII enzyme/probe combination.

Table 2. RLFP Pattern of C4/HindIII System

Cell Lines		DR Type	Fragment No. 1	2	3	4	5	6	7
MZ070782	9002	1		1		1			
KAS116	9003	1	1			1			
JESTHOM	9004	1	1			1			
HOM2	9005	1	1	1		1			
WT100BIS	9006	1				1			
DEM	9007	2/4	1	1		1			
D0208915	9008	2	8	1		1			
KAS011	9009	2	1			1			
AMAI	9010	2		2		2			
E4181324	9011	2	1			1			
SCHU	9013	2	1			1			
MGAR	9014	2	1			1			
RML	9016	2	1			1			
WT8	9017	2	1			1			
L0081785	9018	3				1			
DUCAF	9019	3				1			
QBL	9020	3	1			1			
RSH	9021	3		1		1			
COX	9022	3						1	
VAVY	9023	3						1	
DEU	9025	4	1			1			
YAR	9026	4	1			1			
PE117	9028	4	1			1	1		1
WT51	9029	4	1			1			
JHAF	9030	4	1			1			
BOLETH	9031	4	1			1			
BSM	9032	4		1		1			
BM14	9033	4		1		1			
SAVC	9034	4		1		1			
SP0010	9036	5		1		1			
SWEI6007	9037	5		1		1			
BM16	9038	5	1			1			
JVM	9039	5	8			1			
BM15	9040	5		2		2			
TISI	9042	5	1			1			
BM21	9043	5		1		1			
TUBO	9045	5				1			
PLH	9047	7				2			
LBUF	9048	7		1		1			
MOU	9050	7				1			
PITOUT	9051	7		1		1			
DBB	9052	7		1		1			
EK	9054	6	1			1			
H0301	9055	6		1		1			
KOSE	9056	6	1	1		1			
TEM	9057	6	1			1			
OMW	9058	6		1		1			
SLE005	9059	6	1	1		1			
CB6B	9060	6	1	1		1			
31227ABO	9061	6	1			1			
WDV	9062	6	1			1			
WT47	9063	6	1			1			
AMALA	9064	6	1			1			
HHKB	9065	6	1			1			
TAB089	9066	8		1		1			
BTB	9067	8	1			1			
BM9	9068	8	1			1			
MADURA	9069	8		1		1			
LUY	9070	8	1			1			
OLGA	9071	8	1	1		1			
SPACH	9072	8	8	1		1			
DKB	9075	9	1	1		1			
H0104	9082	?	1			1			
PF04015	9088	3						1	
MLF	9091	4	1			1			
BM92	9092	4		1		1			
DHIF	9104	11		1		1			
FPAF	9105	11	1			1			
MANIKA	9106	7		1		1			
LKT3	9107	4	8		1		1		1

RFLP Standardization Report for C4/MspI

S. Tonks,[1] V. Groves,[1] C. Muller,[2] F. Otani,[3] J.R. Regueiro,[4] M. Segall,[5]
J. Bodmer,[1] and J. Trowsdale[1]

The panel of core cell lines cut with restriction enzyme MspI and hybridized with the C4 probe revealed a very simple pattern (Fig. 1). Two fragments were identified, of which only the first was polymorphic (1.37 kb) (Table 1).

Participating Laboratories: UKIBOD,[1] GERMUC,[2] JAPKSH,[3] FRAARN,[4] US2BAC[5]

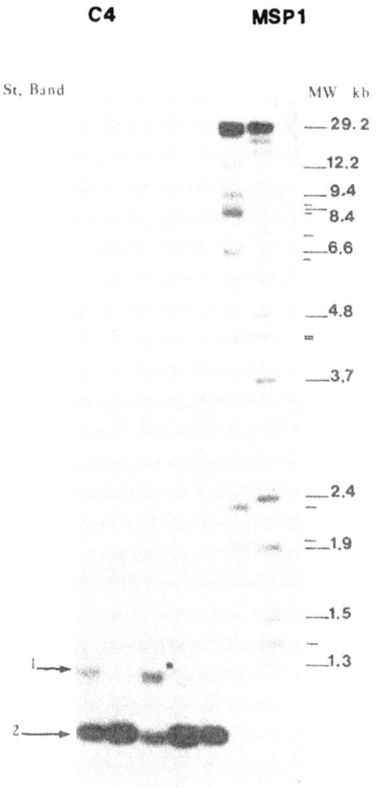

Table 1. Banding Pattern of the C4/MspI System in Each Cell Line

WS. ID.	Cell	Fragment No. 1	Fragment No. 2	WS. ID.	Cell	Fragment No. 1	Fragment No. 2
9045	TUBO	8	2	9006	WT100BIS	8	2
9014	MGAR	8	2	9005	HOM2	8	2
9058	OMW	8	2	9050	MOU	1	2
9107	LKT3	8	2	9056	KOSE	2	2
9017	WT8	8	2	9104	DHIF	1	2
9042	TISI	8	2	9051	PITOUT	8	2
9013	SCHU	8	2	9047	PLH	2	2
9033	BM14	8	2	9040	BM15	8	2
9004	JESTHOM	8	2	9010	AMAL	8	2
9018	LO081785	8	2	9069	MADURA	8	2
9055	H0301	8	2	9037	SWEIG007	8	2
9063	WT47	8	2	9028	PE117	1	2
9022	COX	8	2	9019	DUCAF	8	2
9043	BM21	8	2	9025	DEU	1	2
9023	VAVY	8	2	9031	BOLETH	8	2

Table 1. *Continued*

WS. ID.	Cell	Fragment No. 1	2	WS. ID.	Cell	Fragment No. 1	2
9088	PFO4015	8	2	9059	SLE005	8	2
9066	TAB089	8	2	9034	SAVC	8	2
9060	CB6B	1	2	9068	BM9	8	2
9036	SPOO10	2	2	9075	DKB	8	2
9002	MZ070782	8	2	9061	31227ABO	8	2
9071	OLGA	1	2	9067	BTB	8	2
9064	AMALA	8	2	9038	BM16	8	2
9057	TEM	8	2	9008	D0208915	8	2
9030	JHAF	8	2	9003	KAS116	8	2
9054	EK	8	2	9070	LUY	8	2
9105	FPAF	8	2	9029	WT51	8	2
9091	MLF	8	2	9011	E4181324	8	2
9016	RML	8	2	9026	YAR	8	2
9048	LBUF	8	2	9039	JVM	8	2
9065	HHKB	8	2	9082	H0104	8	2
9021	RSH	8	2	9106	MANIKA	8	2
9062	WDV	8	2	9072	SPACH	8	2
9032	BSM	8	2	9052	DBB	8	2
9009	KAS011	8	2	9020	QBL	8	2
9092	BM92	8	2	9007	DEM	8	2

1 = Single intensity; 2 = double intensity; 8 = faint.

RFLP Standardization Report for C4/BamHI

A. Andreas-Zietz,[1] E. Keller,[1] E.D. Albert,[1] and E.J. Yunis[2]

The C4/BamHI system identified two fragments on the core cell lines (Table 1 and Fig. 1). The results are based on the data from one laboratory.

Table 1 shows the reaction pattern of all 72 core cell lines for the C4/BamHI core fragments, as well as the corresponding molecular weights. It is known from the literature (1,2) that standard fragment no. 1 represents the C4A gene as well as the long type of the C4B-gene. Standard fragment no. 2 correlates with the C4B short gene.

Because of the high frequency of deletions and duplications in the region of C4 and 210H genes, gene dosage effects in this system would be of great interest. However, the fact that standard fragment 1 includes both the C4A gene and the C4B long gene leads to great problems in interpretation—even in families.

Participating Laboratories: GERALB,[1] NCYYUN[2]
This study was supported by SFB 217.

References

1. Carroll MC, Plasdottir A, Belt KT, Porter RR. Deletion of complement C4 and steroid 21-hydroxylase genes in the HLA class III region. Embo J 1985;4:2547.
2. Yu CY, Belt KT, Giles CM, Campbell RD, Porter RR. Structural basis of the polymorphism of human complement components C4A and C4B: gene size, reactivity and antigenicity. Embo J 1986;5:2873.

Table 1. Banding Pattern of All 72 Core Cell Lines, as Standardized in the C4PstI System, kb Values and Probable C4A- and C4B-Locus Associations.

```
Probe Name: C4                 Enzyme Name: BAMHI

 |999 9999999999 99999 999999 999999999 999999999999999999999999999999999999 999| WS-ID
 |000 0000000010 10001 000000 000000000 000000000000001000000000000000000000 000| number
 |226 3845002507 04560 055267 334915057 116112163064020605665544310713369332 832|
 |239 3885281862 43167 566801 460202795 735480918472655897234075063017141229 296|
 |... .......... ..... ...... ......... .................................... ...|.........
 |112 0341354066 21230 161233 532527755 004017255050124543414120345662334466 666| position
 |355 8141586379 04124 759736 243040214 575201857966691398720821939026072881 643| on blots
 |... .......... ..... ...... ......... .................................... ...|.........
A|      3 223  3 3 2 2    2 3    22   2    2 23    213   2 3 2  3 2  2     2   2|
 |112 3103450231 11924 322411 321582224 333636022224111216222932124219222223 326|
 |... .......... ..... ...... ......... .................................... ...|
A|          6                                                                    |
 |          8                                                  3                 |
 |... .......... ..... ...... ......... .................................... ...|
B|  6   11114456 34445 233666  44555566     11111223333333444455555566666666  13|
 |880 7834482502 81464 755022 749137700 777888888877555557884447111112122225 788|                   associated with:
 |... .......... ..... ...... ......... .................................... ...|.........          ................
1|    1111111111 11111 222222 222222222 222222222222222222222222222222222222 111|  4.77 kb          C4 A + C4B long
2|222 2222222222 11111 111111 222222222                                         |  3.28              C4 B short
```

Includes kb values and probable C4A- and C4B-locus associations.
Cells are listed according to banding patterns and HLA-A and HLA-B-locus antigens.
Numbers indicate intensity (1,2,3 = increasing intensity, 8 = faint, 9 = unreadable).

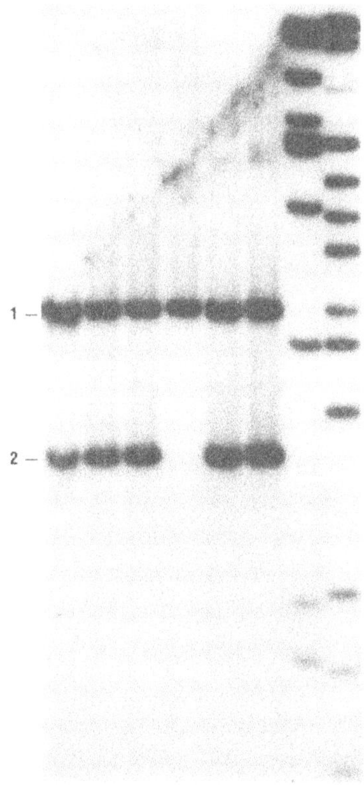

Figure 1. Core bands of the C4/BamHI system (gel 0.9%).

RFLP Standardization Report for 210H/SstI

T. Takenouchi,[1] R. Tonai,[1] P. Terasaki,[1] T. Sasazuki,[2] and P. Reekers[3]

Data from the three participating laboratories revealed two nonpolymorphic fragments, shown in Table 1.

Participating Laboratories: US5TER,[1] JAPSAS,[2] BENREE[3]

Table 1. Nonpolymorphic Fragments Found Using the 21OH/SstI System

		21OH	SST	TER		
	kb	Locus	Frequency	% Faint	56	32
1	7.32	0 0 0 0	1.000	0.000		
2	0.55	0 0 0 0	1.000	1.000		

RFLP Standardization Report for 210H/PstI

J.J. Hyldig-Nielsen,[1] M. Trucco,[2] H. Coppin,[3] A. Andreas-Zietz,[4] E. Keller,[4] C. Carcassi,[5] A. Takada,[6] and A. Svejgaard[1]

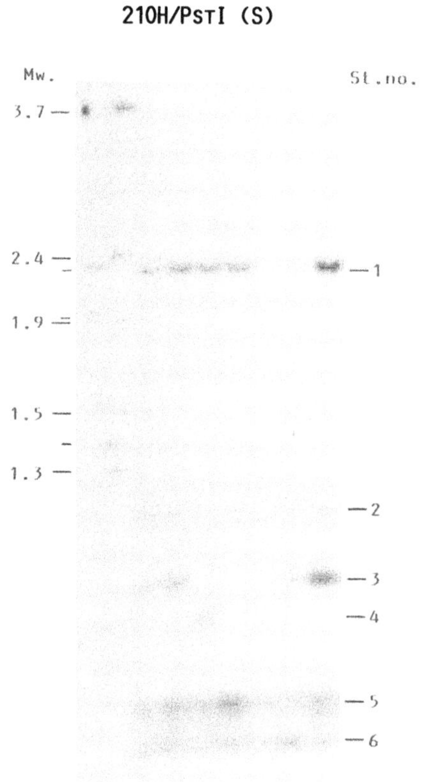

The 210H/PstI system identified the four polymorphic and two nonpolymorphic fragments indicated in Table 1. St. fragment no. 4 was found in only one of the 72 cell-lines and it should be noted that the correct size of this fragment is 0.7 kb (the automatic calculation defined a value of 0.84 kb).

Participating Laboratories: SCASVE,[1] US1DUQ,[2] FRAOHA,[3] GERALB,[4] IT1CON,[5] JAPAIZ[6]

◄ **Figure 1.** Fragments observed with the 210H/PstI system.

Immunobiology of HLA
Volume I

Table 1. Fragments Observed with the 210H/PstI System

WS ID	Name	HLA-DR	Bf	C2	C4A	C4B	B	Standard Fragment No. 4	1	3	6	2	5
9002	MZ070782	1	S				14		6	6	6	6	6
9003	KAS116	1					51		6	6	6	6	6
9004	JESTHOM	1	S				27		6	6	6	6	6
9005	HOM2	1					27		8	6	6	6	6
9006	WT100BIS	1					35				6	6	6
9009	KAS011	2	S				37		6	6	6	6	6
9018	L0081785	3	F1		A3	BQ0	18				6	6	6
9019	DUCAF	3	F1		A3	BQ0	18				6	6	6
9020	QBL	3	F1	1	A3	BQ0	18				6	6	6
9021	RSH	3	F	C	A1	BQ0	42		6		6	6	6
9022	COX	3	S	C	AQ0	B1	8				6	6	6
9023	VAVY	3	S	C	AQ0	B1	8				6	6	6
9088	PF04015	3	S		AQ0	B1	8				6	6	6
9025	DEU	4					35		6		6	6	6
9026	YAR	4	S				38		6	6	6	6	6
9028	PE117	4					60		6	6	6	6	6
9029	WT51	4					65		6	6	6	6	6
9030	JHAF	4					51		6		6	6	6
9031	BOLETH	4					62		8	8	6	6	6
9032	BSM	4					62		6	6	6	6	6
9032	BSM	4					62		6	6	6	6	6
9033	BM14	4					7		6	6	6	6	6
9034	SAVC	4	S		A3	B1	7		6		6	6	6
9091	MLF	4					62		6	6	6	6	6
9092	BM92	4					51		6	6	6	6	6
9107	LKT3	4					54		6	6	6	6	6
9055	H0301	6					14		6		6	6	6
9047	PLH	7					47				6	6	6
9048	LBUF	7	S		A3	B1	13		8	6		6	6
9050	MOU	7			AQ0	B1	44				6	6	6
9051	PITOUT	7	F	C	A3	B1	44		6	6	6	6	6
9052	DBB	7	S		A6	B1	57		8	6		6	6
9106	MANIKA	7					50		6	6	6	6	6
9066	TAB089	8	S	C	A4	B2	46		6		6	6	6
9067	BTB	8					27		6	6	6	6	6
9068	BM9	8					35				6	6	6
9069	MADURA	8					60		6	6	6	6	6
9070	LUY	8					51		6	6	6	6	6
9071	OLGA	8					62		6	6	6	6	6
9072	SPACH	8					62		6	6	6	6	6
9075	DKB	9	S	1	A4	B2	60			6	6	6	6
9036	SP0010	11					44		6	8	6	6	6
9037	SWEIG007	11					61		6		6	6	6
9039	JVM	11					18				6	6	6
9040	BM15	11					49		6	6	6	6	6
9042	TISI	11	S		A3	B1	35		6	6	6	6	6
9043	BM21	11					41		6	6	6	6	6
9104	DHIF	11	S		A3	B1	38		8	6	6	6	6
9105	FPAF	11	S	C	A3	B1	35		6	6	6	6	6
9045	TUBO	11,12	S				51		6	6	6	6	6
9038	BM16	12					18		6	6	6	6	6
9058	OMW	13	F	C	A3	B1	45		6	6	6	6	6
9059	SLE005	13					60	6	6		6	6	6
9060	CB6B	13	F		A3	B1	62		6	6	6	6	6
9062	WDV	13		1	A2	B1	38		6	6	6	6	6
9063	WT47	13					44		6		6	6	6
9065	HHKB	13	S	1	A3	B1	7		6	6	6	6	6
9056	KOSE	13,14	S				35		6	6	6	6	6
9056	KOSE	13,14	S				35		6	6	6	6	6
9054	EK	14					44		6	6	6	6	6
9057	TEM	14					38		6		6	6	6
9061	31227AB0	14					18		8	8	6	6	6
9064	AMALA	14					62		6	6	6	6	6
9008	D0208915	15	S	Q0	A4	B2	18		6	6	6	6	6
9010	AMAI	15					53		6		6	6	6
9011	E4181324	15	S		A3	B1	52		6	6	6	6	6
9013	SCHU	15					7		6	6	6	6	6
9014	MGAR	15	S				8		6	6	6	6	6
9017	WT8	15					7		6	6	6	6	6
9082	H0104	15					7		6	6	6	6	6
9016	RML	16					51		6	6	6	6	6
9007	DEM	16,4	S		A6	B1	57		8	6		6	6

RFLP Standardization Report for 210H/KpnI

G. Paulsen[1] and E. Lui[2]

Three bands were identified by the RFLP analysis in the 210H/KpnI system as shown in Table 1 and in Figure 1. The distribution of the bands in the core cell lines is shown in Table 2.

Three different patterns were seen: pattern 1 (band 1+2+3), pattern 2 (band 2+3), and pattern 3 (band 1+3). Pattern 2 occurred in 11 core cell lines, seven of which were DQw2. Pattern 3 was found in one core cell line also having the DQw2 specificity (Table 2).

Participating Laboratories: SCATSB,[1] NCYFAL[2]

Table 1. Standard Bands in the 21OH/KPN System

Band	kb	Frequency[a]	% Faint[b]	56[c]	32[c]
1	4.16	0.843	0.000		
2	2.98	0.986	0.000		
3	1.38	1.000	0.352		

[a]Band frequency in the core cell lines.
[b]Frequency of faint bands.
[c]Discrepancies in strong bands (s) or faint bands (f) of the hidden duplicates (WS.ID. 9056 and 9032).

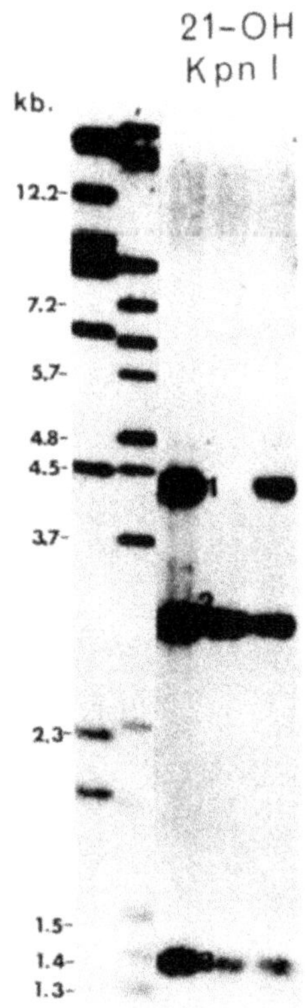

Figure 1. RFLP bands in the 210H/KpnI system of 3 core cell lines run in 0.9% agarose gel. Representative core bands are shown in the figure numbered on their right side. The WS.ID. of these core cell lines are, from the left following marker 1 and marker 2: 9052, 9020, and 9007.

Table 2. Distribution of Standard Bands in the Core Cell Lines in the 21OH/KPN System

			Band No.						Band No.		
WS ID	DR	DQ	1	2	3	WS ID	DR	DQ	1	2	3
9045	11/12	7	3	3	1	9033	4	8	2	2	8
9014	15	6	3	3	1	9004	1	5	2	2	8
9058	13	6	3	3	1	9069	8	4	2	2	8
9107	4	4	3	3	1	9037	11	7	2	2	8
9017	15	1	3	3	1	9028	4	8	2	2	8
9040	11	7	3	3	1	9104	11	7	2	1	8
9010	15	6	3	3	1	9043	11	7	2	1	8
9026	4	8	3	3	1	9002	1	5	2	1	1
9061	14	5	3	2	1	9064	14	7	2	1	1
9003	1	5	3	2	1	9016	16	7	2	1	1
9056	13/14	5	3	2	1	9029	14	5	2	1	1
9060	13	5	3	2	1	9011	15	6	2	1	1
9056	13/14	5	3	2	8	9055	6	6	1	2	8
9051	7	2	3	2	8	9063	13	6	1	2	8
9031	4	8	3	2	8	9091	4	7	1	2	1
9106	7	2?3	2	3	1	9065	13	6	1	2	1
9005	1	5	2	3	8	9071	8	4	1	1	1
9036	11	5	2	2	1	9054	14	5	1	1	1
9057	14	1	2	2	1	9105	11	6	1	1	1
9030	4	7	2	2	1	9075	9	9	1	1	1
9048	7	2	2	2	1	9070	8	7	1	1	1
9021	3	4	2	2	1	9082	15	6	1	1	1
9062	13	6	2	2	1	9007	16/4	5/8	1	1	1
9032	4	8	2	2	1	9066	8	6	1	1	8
9009	2	1	2	2	1	9047	7	2	3		8
9092	4	8	2	2	1	9050	7	2		3	8
9059	13	6	2	2	1	9020	3	2		2	1
9034	4	8	2	2	1	9068	8	4		2	1
9067	8	4	2	2	1	9018	3	2		2	8
9038	12	7	2	2	1	9022	3	2		2	8
9008	15	6	2	2	1	9023	3	2		2	8
9032	4	8	2	2	1	9019	3	2		2	8
9072	8	4	2	2	1	9025	4	7		2	8
9052	7	9	2	2	1	9088	3	2		2	9
9042	11	7	2	2	8	9006	1	5		1	8
9013	15	6	2	2	8	9039	11	7		1	8

Band designation: 1–3 = increasing intensity, 8 = faint, 9 = unreadable.

RFLP Standardization Report for 210H/EcoRV

A. Ando,[1] Y.K. Paik,[2] Y. Mitsuishi,[3] J.A. Trejant,[4] and K. Tsuji[1]

Two bands were standardized with the 210H/EcoRV system as shown in Figure 1. Table 1 shows the presence/absence pattern of all 72 cell lines. St. Fragment no. 2 (24.40 kb) does not present in three cell lines: COX, VAVY and PF04015, which are known to be deletions in the C4/C210H region.

Participating Laboratories: JAPTSU,[1] US5PAK,[2] FRAMYE,[3] ANZBSH[4]

Table 1. RFLP Pattern of the 21OH/EcoRV System

	Cell Lines		Fragment No. 1	Fragment No. 2	A	C	B	Bw	Bf	C2	C4A	C4B
		Fragment Size (kb)	24.63	2.40								
13	9022	COX	1		1	7	8	6	S	C	AQ0	B1
15	9023	VAVY	1		1	7	8	6	S	C	AQ0	B1
19	9056	KOSE		3	2	–	35	6	S			
65	9056	KOSE		3	2	–	35	6				
32	9066	TAB089		3	2	11	46	6	S	C	A4	B2
33	9060	CB6B		3	1	9	62	6	F		A3	B1
34	9036	SP0010		3	2	5	44	4				
5	9017	WT8		3	3	7	7	6				
35	9002	MZ070782		3	24	2,8	14	6	S			
7	9013	SCHU		3	3	7	7	6				
36	9071	OLGA		3	31	1,10	62	6				
9	9004	JESTHOM		3	2	1	27	4	S			
37	9064	AMALA		3	2	9	62	6				
11	9055	HO301		3	3	8	14	6				
39	9030	JHAF		3	31	–	51	4				
14	9043	BM21		3	1	–	41	6				
40	9054	EK		3	2	5	44	4				
17	9005	HOM2		3	3	1	27	4				
43	9016	RML		3	2	–	51	4				
3	9058	OMW		3	2	–	45	6	F	C	A3	B1
44	9048	LBUF		3	30	6	13	4	S		A3	B1
22	9047	PLH		3	3	6	47	4				
51	9059	SLE005		3	2	10	60	6				
24	9010	AMAI		3	28	4	53	4				
52	9034	SAVC		3	3	7	7	6	S		A3	B1
27	9028	PEI17		3	24	10	60	6				
54	9075	DXB		3	24	10	60	6	S	1	A4	B2
30	9031	BOLETH		3	2	10	62	6				
55	9061	31227ABO		3	2	7	18	6				
4	9107	LKT3		3	24	1	54	6				
56	9067	BTB		3	2	1	27	4				
8	9033	BM14		3	3	7	7	6				
57	9038	BM16		3	2	7	18	6				
12	9063	WT47		3	32	5	44	4				
60	9070	LUY		3	2	–	51	4				
18	9050	MOU		3	29	–	44	4			AQ0	B1
61	9029	WT51		3	23	8	65	6				
23	9040	BM15		3	1	7	49	4				
62	9011	E4181324		3	1	–	52	4	S		A3	B1
29	9025	DEU		3	31	4	35	6				
63	9026	YAR		3	26	–	38	4	S			
6	9042	TISI		3	24	4	35	6	S		A3	B1
2	9014	MGAR		3	26	7	8	6	S			
16	9006	WT100BIS		3	11	4	35	6				
66	9082	HO104		3	3	7	7	6				
26	9037	SWEIG007		3	29	2	61	6				
68	9032	BSM		3	2	9	62	6				
10	9018	LO081785		3	3	5	18	6	F1		A3	BQ0
1	9045	TUBO		3	2,3	7	51	4	S			
20	9104	DHIF		3	31		38		S		A3	B1
70	9052	DBB		3	2	6	57	4	S		A6	B1
72	9007	DEM		3	2	6	57	4	S		A6	B1
21	9051	PFTOUT		2	29	–	44	4	F	C	A3	B1
28	9019	DUCAF		2	30	5	18	6	F1		A3	BQ0
42	9091	MLF		2	2	9	62	6				
45	9065	HHKB		2	3	7	7	6	S	1	A3	B1
46	9021	RSH		2	68,30	2	42	6	F	C	A1	BQ0

Table 1. *Continued*

	Cell Lines		Fragment No. 1 (24.63 kb) / 2 (24.0 kb)	A	C	B	Bw	Bf	C2	C4A	C4B
50	9092	BM92	2	25	1	51	4				
53	9068	BM9	2	2	4	35	6				
58	9008	DO208915	2	25	-	18	6	S	Q0	A4	B2
59	9003	KAS116	2	24	-	51	4				
64	9039	JVM	2	2	5	18	6				
67	9106	MANIKA	2/	3	-	50	6				
69	9072	SPACH	2	31	1,3.1	62					
71	9020	QBL	2	26	?5	18	6	F1	1	A3	BQ0
25	9069	MADURA	1	2	10	60	6				
38	9057	TEM	1	26	-	38	4				
41	9105	FPAF	1	1	4	35	6	S	C	A3	B1
47	9062	WDV	1	2	-	38	4		1	A2	B1
48	9032	BSM	1	2	9	62	6				
49	9009	KASO11	1	1	6	37	4	S			
31	9088	PF04015		1		8	6	S		AQ0	B1

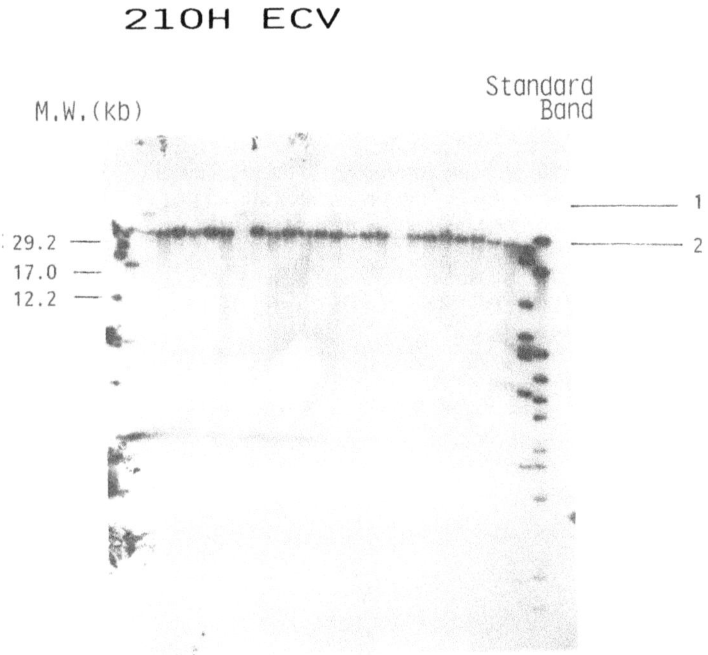

Figure 1. Bands standardized with the 21OH/EcoRV system.

RFLP Standardization Report for 21OH/TaqI

R.E. Bontrop,[1] R. Walford,[2] and S. Sekiguchi[3]

Data for standardization were received from two different laboratories. In the example of the 21OH/TaqI system four different standard fragments were detected (Table 1 and Fig. 1). Standard fragment 1 has a molecular weight of 4.56 kb and a panel frequency of 98.6%. Standard fragment 2 has a molecular weight of 4.06 kb and a frequency of 85.7%. For this system, both fragment 1 and 2 show the most abundant staining intensities. However, some faint hybridization signals were observed in the example of fragment 2. It can clearly be seen that fragments 3 and 4 show remarkably weaker hybridization signals (Fig. 1). Standard fragment 3 has a molecular weight of 3.74 kb and a frequency of 100% whereas standard fragment 4 has a molecular weight of 3.01 kb and a frequency of 90%. The standard fragments detected for the 21OH/TaqI system had no obvious relation to the standard fragments detected in the Bf/TaqI, C2/TaqI and C4/TaqI systems.

TaqI has most frequently been used in previously published reports describing deletions in the 21OHB gene, resulting in 21OH-deficiency. Fragment 1 (4.56 kb) is present in all core cell lines except 9047, which is known to have homozygous deletion of the 21OH gene. The Workshop fragment 1 (4.56 kb) probably corresponds to the 3.7 kb TaqI fragment previously reported while fragment 2 (4.06 kb) probably corresponds to the previously reported 3.2 kb TaqI fragment. Fragments 2 and 4 (3.01 kb) are absent from panel cells with known deletion of 21OHA (for further details see "Workshop Report for RFLP in C4, C2, BF and 21OH Genes," Alper et al. this volume and White, PC "Molecular Genetics of the Class III Region of the HLA Complex," Volume II).

Participating Laboratories: BENROO,[1] US5WAL,[2] JAPSEK[3]

Table 1. Standard Fragments Detected Using the 21OH/TaqI System

Fragment	kb	Frequency	% Faint
1	4.56	0.986	0.000
2	4.06	0.857	0.016
3	3.74	1.000	0.000
4	3.01	0.900	0.000

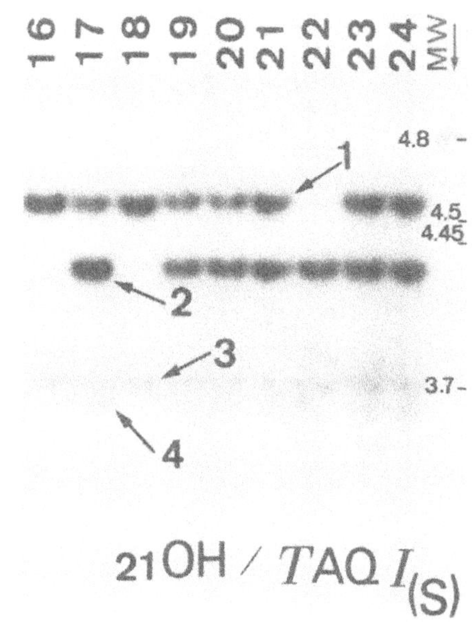

Figure 1. Fragments detected using the 21OH/TaqI system.

Immunobiology of HLA
Volume I

Table 2. Fragment Presence/Absence Patterns for the 21OH/ TaqI System as Defined on the 72 Core Cell Lines

	Fragment No.					Fragment No.					Fragment No.					Fragment No.			
Cell No.	1	2	3	4	Cell No.	1	2	3	4	Cell No.	1	2	3	4	Cell No.	1	2	3	4
9045	1	1	1	1	9021	1	1	1	1	9069	1	1	1	1	9082	1	1	1	1
9014	1	1	1	1	9062	1	1	1	1	9037	1	1	1	1	9106	1	1	1	1
9058	1	1	1	1	9032	1	1	1	1	9028	1	1	1	1	9032	1	1	1	1
9107	1	1	1	1	9009	1	1	1	1	9031	1	1	1	1	9072	1	1	1	1
9017	1	1	1	1	9092	1	1	1	1	9066	1	1	1	1	9052	1	1	1	1
9042	1	1	1	1	9059	1	1	1	1	9060	1	1	1	1	9007	1	1	1	1
9013	1	1	1	1	9034	1	1	1	1	9036	1	1	1	1	9018	1		1	
9033	1	1	1	1	9075	1	1	1	1	9002	1	1	1	1	9022	1		1	
9004	1	1	1	1	9061	1	1	1	1	9071	1	1	1	1	9023	1		1	8
9055	1	1	1	1	9067	1	1	1	1	9064	1	1	1	1	9006	1		1	
9063	1	1	1	1	9038	1	1	1	1	9057	1	1	1	1	9050	1		1	
9043	1	1	1	8	9008	1	1	1	1	9030	1	1	1	1	9019	1		1	
9005	1	1	1	1	9003	1	1	1	1	9054	1	1	1	1	9025	1		1	8
9056	1	1	1	1	9070	1	1	1	1	9105	1	1	1	1	9088	1		1	8
9104	1	1	1	1	9029	1	1	1	1	9091	1	1	1	1	9068	1		1	8
9051	1	1	1	1	9011	1	1	1	1	9016	1	1	1	1	9039	1		1	
9040	1	1	1	1	9026	1	1	1	1	9048	1	1	1	1	9020	1		1	
9010	1	1	1	1	9056	1	1	1	1	9065	1	1	1	1	9047		1	1	1

RFLP Standardization Report for 210H/BglII

A. Marcadet,[1] D. Cohen,[1] M.L. Satz,[2] and A. Olivo[3]

The 210H/BglII system identified two fragments on the core cell lines (Table 1). One fragment was constant (no. 2).

Table 2 indicates the distribution of the standardized band no. 1 in core cell lines.

The figures are photographs taken from two films from laboratory LATFAI. All standardized fragments are indicated at least once for S and/or L migration. The presence of a standardized fragment is indicated by a hyphen.

Table 1. Standard Bands in the 21OH/BglII System

		21OH	BGL	COH		
	kb	Locus	Frequency	% Faint	56	32
1	12.57	0 0 0 0	0.843	0.000		
2	10.94	0 0 0 0	1.000	0.000		

Participating Laboratories: FRACOH,[1] LATFAI,[2] US8RUB[3]

Table 2. RFLP 21OH/BglII System. Distribution of Standard Bands in 10th WS Core Cell Lines

Cell	HLA				Std. Band	Cell	HLA				Std. Band
Lines	C2	C4A	C4B	DR	no. 1	Lines	C2	C4A	C4B	DR	no. 1
9002				1	2	9021		A1	BQO	3	2
9004				1	2	9025				4	2
9005				1	2	9026				4	2
9003				1	2	9028				4	2
9009	QO			2	2	9029				4	2
9007		A6	B1	w16/4	2	9031				4	2
9016				w16	2	9032				4	2
9008		A4	B2	w15	2	9032				4	2
9010				w15	2	9033				4	2
9013				w15	2	9091				4	2
9014				w15	2	9092				4	2
9017	C			w15	2	9107				4	2
9011		A3	B1	w15	2	9030	C			4	2
9082				w15	2	9034		A3	B1	4	2

Table 2. *Continued*

Cell Lines	HLA				Std. Band no. 1	Cell Lines	HLA				Std. Band no. 1
	C2	C4A	C4B	DR			C2	C4A	C4B	DR	
9104		A3	B1	w11	2	9051		A3	B1	7	2
9105		A3	B1	w11	2	9048		A3	B1	7	2
9040				w11	2	9052	C	A6	B1	7	2
9037				w11	2	9106				7	2
9036				w11	2	9071				w8	2
9042		A3	B1	w11	2	9066		A4	B2	w8	2
9043				w11	2	9069				w8	2
9045				w11/12	2	9070	1			w8	2
9038				w12	2	9072		A3	B1	w8	2
9055				w6	2	9067				w8	2
9061				w14	2	9075		A4	B2	w9	2
9054				w14	2	9047				7	
9057				w14	2	9068				w8	
9064				w14	2	9020	C	A3	BQO	3	
9056	C			w13/14	2	9019	C	A3	BQO	3	
9056	1			w13/14	2	9018		A3	BQO	3	
9060	1	A3	B1	w13	2	9023		AQO	B1	3	
9058		A3	B1	w13	2	9022		AQO	B1	3	
9062		A2	B1	w13	2	9088		AQO	B1	3	
9065	C	A3	B1	w13	2	9050		AQO	B1	7	
9059				w13	2	9006				1	
9063				w13	2	9039	1			w11	

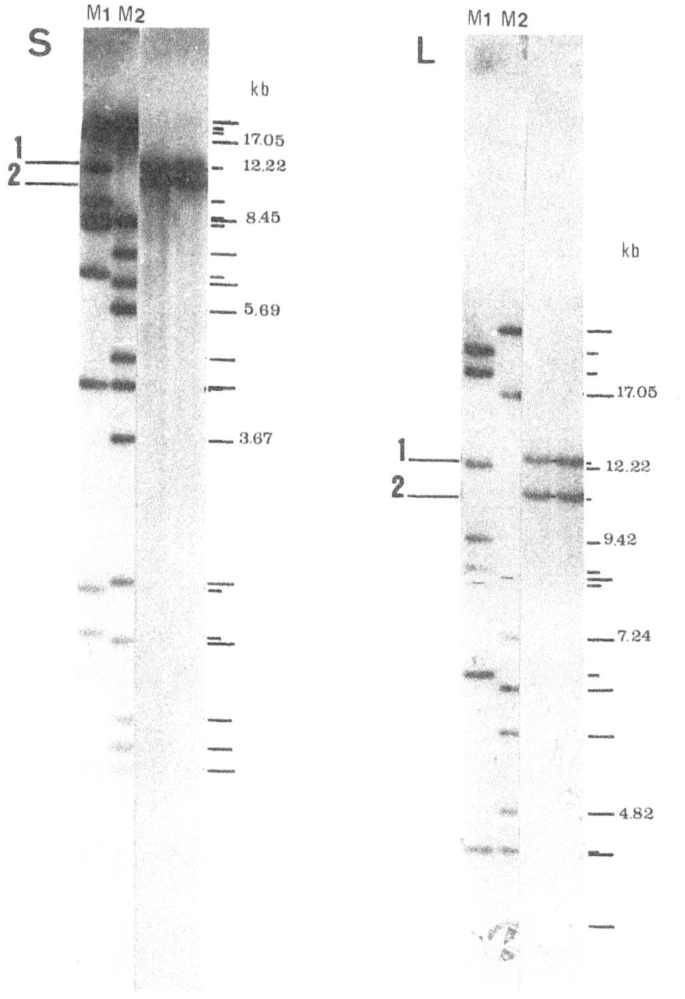

Figure 1. Left: Fragments identified by the 210H/BglII system for S migration. Right: Fragments identified for L migration.

RFLP Standardization Report for 21OH/PvuII

P. Piccioli,[1]* K. Kato,[2] J. Partanen,[3] and G.B. Ferrara[1]

Two 21OH genes have been found in man, each mapping immediately to the 3′ side of C4A and C4B. Of these two 21OH, A gene appears to be a pseudogene, while B gene encodes the enzyme necessary for the metabolic pathway in cortisol synthesis, as it is evident in 21OH deficiency, associated with an homozygous deletion of this gene. The 21OH/Pvu II system identified 10 different fragments on the core cell lines, including two cell lines (ID 9047 and ID 9069) from 21-hydroxylase deficient patients. Neither a specific fragment nor a specific pattern related to this disease was found. Moreover, some fragments showed weak signals and they were not evident on all filters. Indeed, a correlation between a haplotype and a specific band seems difficult with an exception perhaps represented by 1.95 fragment. This fragment appears in 59 out of 70 samples and is absent in some of the HLA haplotypes associated with common complotypes such as all DR3 B18, DR3 B8, DR1 B35, DR7 B44. The results are summarized in Figure 1 and Table 1.

Table 1. Fragments on the Core Cell Lines Identified by the 21OH/PvuII System

Fragment	kb	Frequency[b]	% Faint[c]
1	15.43	0.071	1.000[a]
2	4.05	0.200	0.867[a]
3	3.69	0.371	0.115
4	3.60	0.957	0.000
5	3.50	0.385	0.000
6	1.95	0.843	0.049
7	1.77	0.943	0.121
8	1.36	0.029	0.000
9	0.48	1.000	1.000
10	0.41	1.000	1.000

[a]Discordance between the hidden duplicates 19 and 65 (ID 9056) has been found.
[b]Band frequency has been calculated as number of cell lines presenting a specific fragment versus number of cell lines tested.
[c]Frequency of faint band has been calculated as number of cell lines presenting a faint band versus number of cell lines presenting the same fragment (i.e., frequency 1.000 and % faint 0.000 means that all the samples tested share the same band; on the other hand 1.000 means that it is a constant and a faint band).

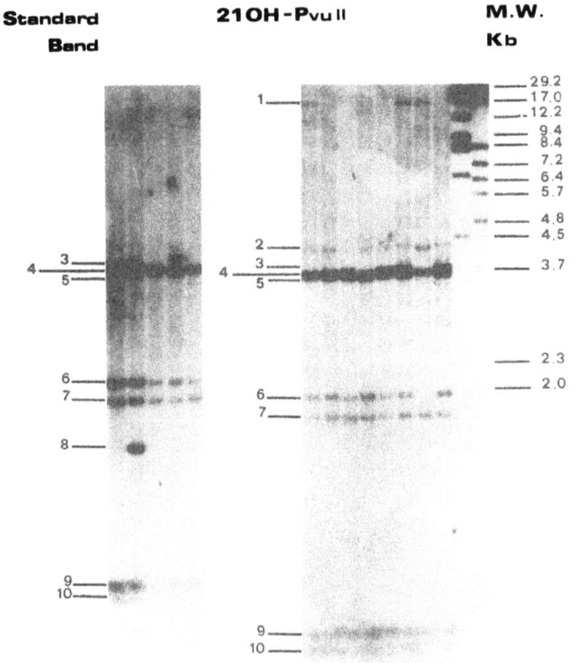

Figure 1. Southern blot analysis of PvuII-digested genomic DNAs from ID 9056, 9082, 9106, 9032, 9072, 9052, 9020, 9007, (on the left) and 9058, 9107, 9017, 9042, 9013 (on the right) core cell Lines hybridized with a 21OH probe.

Participating Laboratories: IT2FER,[1] US7DUP,[2] SCAKOS[3]
*P. Piccioli was recipient of a fellowship from A.I.R.C. (Associazione Italiana per la Ricerca sul Cancro).
This work was in part supported by CNR grants n.86.0265.51 and 86.00659.

RFLP Standardization Report for 210H/EcoRI

M. Segall,[1]* L. Schluender,[1]* H. Betuel,[2] and M. Garovoy[3]

210H/EcoRI blots were received from two laboratories. Fragments well-identified on either laboratories blots were included as standard fragments.

The 210H/EcoRI system identified seven fragments on the core cell lines as shown in Table 1 and the Figures. Fragments 2 and 7 were identified by one of the two laboratories but were quite clear on this laboratories blots and were included as standard fragments. Fragment 5, which shows a discrepancy for the duplicates of 9032, was very variable in intensity across the blots, even in adjacent lanes. Table 2 indicates the distribution of polymorphic fragments in core cell lines identified by the 210H/EcoRI system.

Table 1. 21OH/EcoRI Standard Fragments

				Discrepant for	
Fragment	kb	Frequency	% Faint	9056	9032
1	16.00	1.000	0.042		
2	14.67	0.329	0.435		
3	13.05	0.500	0.028		
4	11.89	0.986	0.000		
5	11.16	0.857	0.262		s
6	8.62	0.971	0.014		
7	0.43	1.000	1.000		

Participating Laboratories: US2BAC,[1] FRABET,[2] US3GVY[3]
*This is publication #490 from the Immunobiology Research Center, University of Minnesota, Minneapolis, Minnesota, 55455, USA.

Supported in part by Juvenile Diabetes Foundation #186175 and March of Dimes Birth Defects Foundation Grant #6-496.

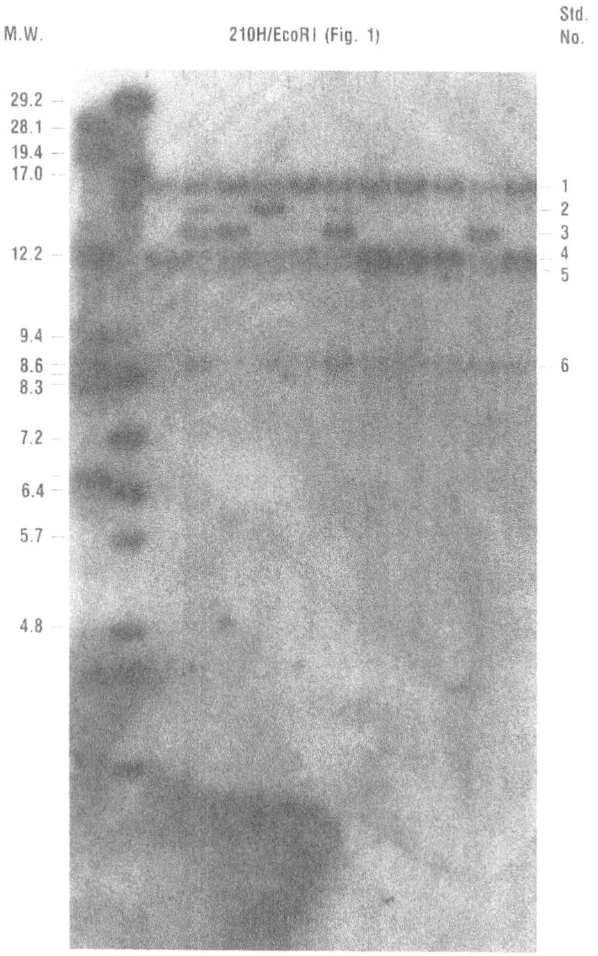

Figure 1. Fragments on the core cell lines identified by use of the 210H/EcoRI system.

Figure 2. Fragments on the core cell lines identified by use of the 210H/EcoRI system.

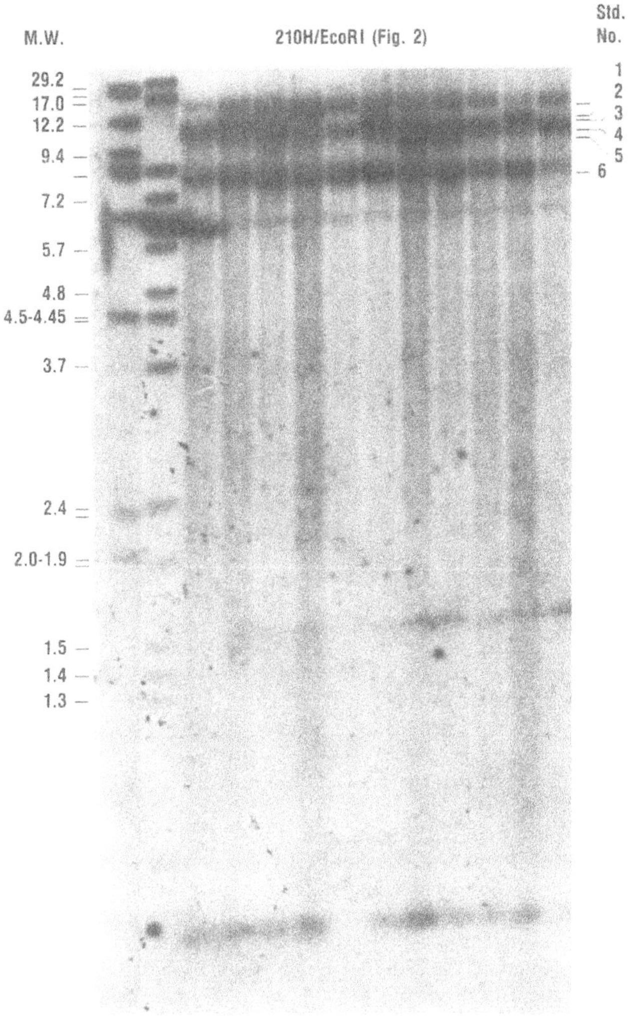

Table 2. Distribution of Polymorphic Fragments in Core Cell Lines, 210H/EcoRI

		Std. Fragment No.							Std. Fragment No.				
Cell No.	Name	2	3	4	5	6	Cell No.	Name	2	3	4	5	6
9047	PLH				1	1	9013	SCHU			2	1	1
9023	VAVY			1			9014	MGAR			3	1	1
9006	WT100BIS			1			9011	E4181324			2	8	8
9105	FPAF			1	1		9003	KAS116			2	1	1
9039	JVM			1		1	9029	WT51			2	1	1
9038	BM16			2		1	9042	TISI			3	1	1
9032	BSM			2		1	9045	TUBO			2	1	1
9018	L0081785			1		1	9054	EK			2	1	1
9019	DUCAF			1		1	9057	TEM			2	1	1
9088	PFO4015			1		1	9062	WDV			2	1	1
9022	COX			1		1	9064	AMALA			2	8	1
9050	MOU		8	1		1	9065	HHKB			2	1	1
9026	YAR		2	2		1	9068	BM9			1	1	1
9061	31227ABO			3	1	1	9070	LUY			2	1	1
9030	JHAF			2	1	1	9063	WT47			3	1	1
9032	BSM			1	1	1	9091	MLF(MCF)			2	1	1
9031	BOLETH			2	1	1	9082	HO104			2	1	1

Table 2. *Continued*

Cell No.	Name	Std. Fragment No. 2	3	4	5	6	Cell No.	Name	Std. Fragment No. 2	3	4	5	6
9005	JESTHOM			2	1	1	9107	LKT3	8	2	1	1	1
9016	RML			2	1	1	9010	AMAL	8	2	1	8	1
9017	WT8			2	1	1	9072	SPACH	2	1	1	1	1
9025	DEU			2	1	1	9005	HOM2	8	1	2	1	1
9009	KAS011		2	1	1	1	9002	MZO70782	1	2	1	1	1
9056	KOSE		1	2	1	1	9106	MANIKA	1	2	1	1	1
9052	DBB		2	1	1	1	9075	DKB	1	2	1	1	1
9058	OMW		1	1	8	1	9071	OLGA	8	1	1	8	1
9067	BTB		2	3	1	1	9066	TAB089	1	2	1	8	1
9059	SLE005		2	1	1	1	9060	CB6B	1	1	2	1	1
9056	KOSE		1	2	1	1	9055	HO301	1	2	1	8	1
9007	DEM		2	1	1	1	9051	PITOUT	8	2	1	8	1
9008	D0208915		2	1	1	1	9048	LBUF(LBF)	1	2	1	8	1
9037	SWEIG007		1	2	8	1	9040	BM15	8	2	1	8	1
9020	QBL		2	1	1	1	9036	SP0010	1	2	1	1	1
9043	BM21	8	2	1	8	1	9021	RSH	8	2	1	8	1
9069	MADURA	1	2	1	8	1	9034	SAVC	2	2	1	8	1
9092	BM92	1	1	1	1	1	9028	PE117	1	1	2	1	1
9104	DHIF	8	1	1	8	1	9033	BM14	8	2	1	8	1

RFLP Standardization Report for 210H/HindIII

P.A. Biro,[1] S. Cutbush,[1] J. Demopulos,[2] R.T. Acton,[2] and S.V. Van Tonder[3]

The 210H probe detects bands of three size classes (Fig. 1, Table 1). Std. band 1 contains one fragment of 30.50 kb encompassing the nonpolymorphic 210HB gene (1) and also, in "long" C4B haplotypes (2), a second fragment containing the 210HA gene. In "short" C4B haplotypes the 210HA gene is detected on a 25.19 kb band (std. band 2). The difference between the two haplotypes is determined by the presence or absence of an intron of ~6.5 kb in the C4B gene. Std. band 3 (19.57 kb) is found in only two DNA samples, lanes 4 and 27 (WS 9107 and 9028, Table 2) and is due to DNA rearrangements in the C4/210H loci (see also C4/HindIII report).

Participating Laboratories: UKIFES,[1] US5UAB,[2] SAFHAM[3]

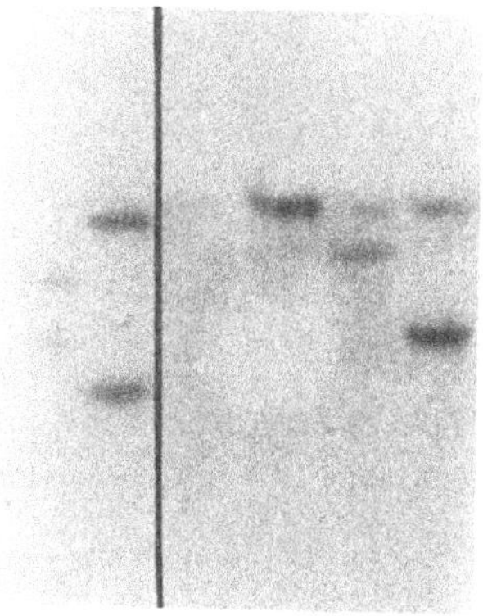

Figure 1. Standard bands produced by the HindIII/21OH enzyme/probe combination.

Table 1. Sizes and Frequencies of Standard Bands Produced by the HindIII/21OH Enzyme/Probe Combination

	kb	Frequency	% Faint
1	30.50	1.000	0.000
2	25.19	0.406	0.000
3	19.57	0.029	0.000

References

1. Prentice HL, Schneider PM, Strominger JL. C4B gene polymorphism detected in a human cosmid clone. Immunogenetics 1986;23:274.
2. Palsdottir A, Fossdal R, Arnason A, Edwards JH, Jensson O. Heterogeneity of human C4 gene size. Immunogenetics 1987;25:299.

Table 2. RFLP Pattern of 21OH/HindIII System

Cell Lines		DR Type	Fragment No. 123	Cell Lines		DR Type	Fragment No. 123
MZ070782	9002	1	11	TUBO	9045	5	1
KAS116	9003	1	1	PLH	9047	5	2
JESTHOM	9004	1	1	LBUF	9048	7	11
HOM2	9005	1	11	MOU	9050	7	1
WT100BIS	9006	1	1	PITOUT	9051	7	11
DO208915	9008	2/4	11	DBB	9052	7	22
KAS011	9009	2	1	EK	9054	7	1
AMAI	9010	2	22	HO301	9055	6	11
E4181324	9011	2	1	KOSE	9056	6	11
SCHU	9013	2	1	TEM	9057	6	1
MGAR	9014	2	1	OMW	9058	6	11
RML	9016	2	1	SLE005	9059	6	11
WT8	9017	2	1	CB6B	9060	6	11
LO081785	9018	2	1	31227ABO	9061	6	29
DUCAF	9019	3	1	WDV	9062	6	1
QBL	9020	3	1	WT47	9063	6	1
RSH	9021	3	11	AMALA	9064	6	1
COX	9022	3	1	HHKB	9065	6	1
VAVY	9023	3	1	TAB089	9066	6	11
DEU	9025	3	2	BTB	9067	8	2
YAR	9026	4	1	BM9	9068	8	1
PE117	9028	4	2 1	MADURA	9069	8	11
WT51	9029	4	1	LUY	9070	8	2
JHAF	9030	4	1	OLGA	9071	8	11
BOLETH	9031	4	1	SPACH	9072	8	11
BSM	9032	4	1	DKB	9075	8	11
BM14	9033	4	11	HO104	9082	9	2
SAVC	9034	4	11	PF04015	9088	?	1
SPOO10	9036	4	11	MLF	9091	3	1
SWEIG007	9037	5	11	BM92	9092	4	11
BM16	9038	5	2	DHIF	9104	4	11
JVM	9039	5	1	FPAF	9105	11	1
BM15	9040	5	22	MANIKA	9106	11	21
TISI	9042	5	1	LKT3	9107	7	1 1
BM21	9043	5	11				

RFLP Standardization Report for 210H/MspI

S. Tonks,[1] V. Groves,[1] C. Muller,[2] F. Otani,[3] J.R. Regueiro,[4] M. Segall,[5] J. Bodmer,[1] and J. Trowsdale[1]

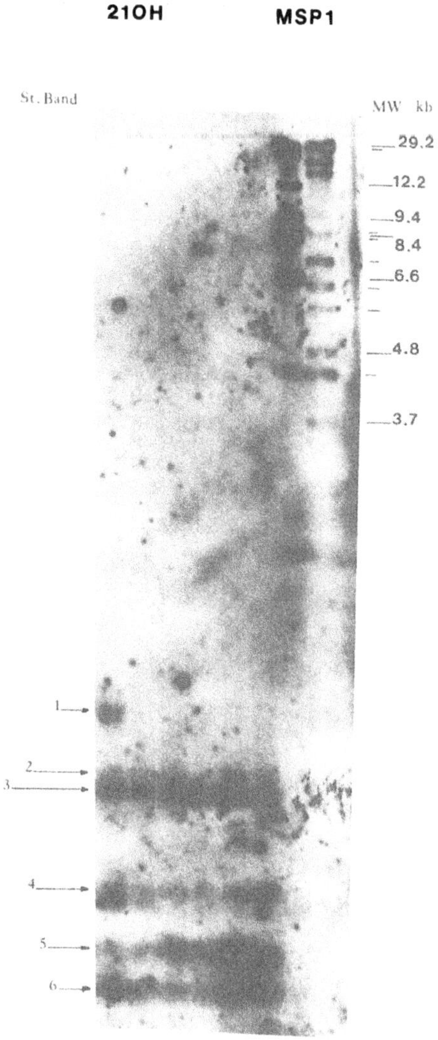

The 210H/MspI system identified six standard fragments in the core cell lines as shown in Table 1 and Figure 1. Much of the data received for this system was difficult to interpret because all the fragments were of low molecular weight and thus produced indistinct bands. High background on several autoradiographs was also a contributory factor in the discrepancy between duplicates in the first fragment. Only fragments 1 (1.27 kb) and 4 (0.49 kb) were found to be polymorphic.

Participating Laboratories: UKIBOD,[1] GERMUC,[2] JAPKSH,[3] FRAARN,[4] US2BAC[5]

◄ **Figure 1.** Standard fragments in the core cell lines identified by the 210H/MspI system.

Table 1. Banding Pattern of the 21OH/Mspl System in Each Cell Line

		Fragment No.					
WS ID	Cell	1	2	3	4	5	6

Immunobiology of HLA
Volume I

Table 1. Banding Pattern of the 21OH/MspI System in Each Cell Line

		Fragment No.					
WS ID	Cell	1	2	3	4	5	6
9048	LBUF	8	1	2	1	1	2
9021	RSH	8	1	2	2	2	2
9045	TUBO		8	1	1	2	2
9058	OMW		8	1	8	2	2
9107	LKT3		8	1	8	2	2
9042	TISI		8	1	1	2	2
9033	BM14		8	1	1	2	2
9004	JESTHOM		8	1	1	2	2
9055	H0301		8	1	1	2	2
9063	WT47		8	1	2	2	2
9022	COX		1	2	2	2	2
9043	BM21		8	1	1	2	2
9023	VAVY		8	1	1	2	2
9006	WT100BIS		1	2	1	2	2
9005	HOM2		1	2	1	2	2
9050	MOU		1	2	2	2	2
9104	DHIF		1	2	1	1	1
9051	PITOUT		8	1	1	2	2
9047	PLH		1	2	1	2	2
9040	BM15		8	2	1	2	2
9010	AMAI		8	1	1	2	2
9069	MADURA		8	1	1	2	2
9037	SWEIG007		8	1	2	2	1
9028	PE117		8	1	2	2	1
9031	BOLETH		1	2	2	2	2
9088	PF04015		1	2	1	2	2
9066	TAB089		1	2	1	2	2
9030	JHAF		1	2	2	1	1
9054	EK/OH		1	2	2	2	2
9105	FPAF		1	2	2	2	2
9091	MLF		1	2	2	2	2
9016	RML		1	2	2	2	2
9065	HHKB		1	2	1	2	1
9047	PLH		1	2	2	2	2
9032	BMS		1	2	2	2	2
9009	KAS011		1	2	1	2	2
9092	BM92		1	2	1	2	2
9034	SAVC		1	2	2	2	2
9068	BM9		1	2	1	2	2
9075	DKB		1	2	1	2	2
9061	31227ABO		1	2	2	2	2
9067	BTB		1	2	2	2	2
9008	DO208915		1	2	1	2	2
9003	KAS116		1	2	2	2	2
9070	LUY		1	2	1	2	1
9029	WT51		1	2	1	2	2
9026	YAR		1	2	2	2	2
9082	HO104		1	2	1	2	2
9106	MANIKA		1	2	2	2	2
9072	SPACH		1	2	8	2	2
9070	LUY		1	1	8	2	2
9071	OLGA		1	2	8	2	1
9007	DEM		8	1	8	2	2
9014	MGAR		8	1		2	2
9017	WT8		8	1		2	2
9013	SCHU		8	1		2	2
9018	LO081785		8	1		2	2
9019	DUCAF		1	1		2	2
9051	PITOUT		1	2		2	2
9038	BM16		1	2		2	2
9011	E4181324		1	2		2	2
9039	JVM		1	2		2	1

1 = Single intensity; 2 = double intensity; 8 = faint.

RFLP Standardization Report for 210H/BamHI

A. Andreas-Zietz,[1] E. Keller,[1] E.D. Albert,[1] A. Bushell,[2] and E.J. Yunis[3]

The 210H/BamHI system identified four fragments on the core cell lines as shown in Table 1 and Figure 1.

Table 1 shows the reaction pattern of all 72 core cell lines for the 210H/BamHI core fragments as well as the corresponding molecular weights.

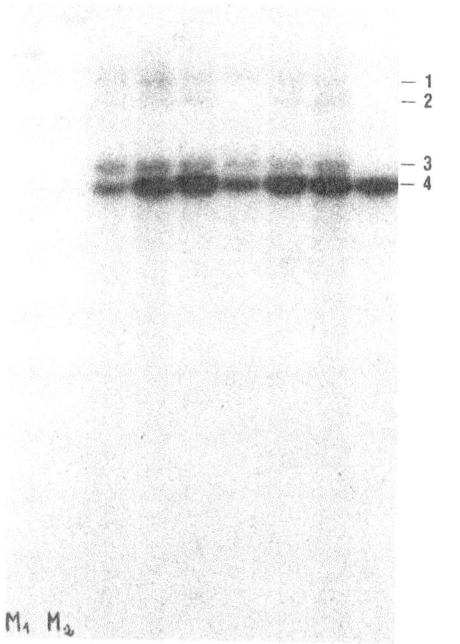

Figure 1. Core bands of the C210H/BamHI system (gel 0.6%).

Fragment no. 4 seems to be identical with the 14 kb main band as described in the literature (1), representing the 210HA gene as well as the 210HB gene. Therefore a fourfold dosage of fragment no. 4 would be expected in normal cases.

For some cell lines, where a deletion in the C4/210H region is known, one observes a double dose rather than the "normal" fourfold dosage (e.g., no. 13 and no. 15[A1, B8, DR3], and no. 22 [A3, Bw47, DR7]). For other cell lines the gene dosage is less clear. As structural variants like deletions and duplications are frequent in the region of C4 and 210H genes it is difficult to judge the intensity of this band. Furthermore this fragment no. 4 has the same size as the core fragment no. 1 of the Bf/BamHI system. This may be due to cross-hybridization.

The faint bands no. 1, 2, and 3 are only detected by one laboratory. Without further testing it is not possible to decide whether these bands are relevant for 210H or represent cross-hybridizations with other genes.

Reference

1. Carroll MC, Plasdottir A, Belt KT, and Porter RR. Deletion of complement C4 and steroid 21-hydroxylase genes in the HLA class III region. Embo J 1985;4:2547.

Participating Laboratories: GERALB,[1] UKIMOR,[2] NCYYUN[3]
This study was supported by SFB 217.

Table 1. Banding Pattern of All 72 Core Cell Lines, as Standardized in the 21OH/BamHI System, Including kb Values

```
Probe Name: 21OH            Enzyme Name: BAMHI


  |999999 9999 9999 9999999 999 9999999 9999999 9999999 9 999999999 9999 99999 99999999| WS-ID
  |000000 0000 0000 0000000 000 1000000 0000100 0000000 0 010000000 0100 00000 00000000| number
  |113638 2281 4502 3211063 060 0065542 0652042 5563556 4 404170931 1050 65273 66933377|
  |733542 2384 8529 9098818 475 5686625 9276431 0136486 7 065603201 0727 99857 04112212|
  |------ ---- ---- ------- --- ------- ------- ------- - --------- ---- ----- --------|---------
  |000456 1130 4136 6721555 051 4151602 4436214 1213403 2 260465536 2077 25252 33434636| position
  |578526 3512 4151 4180857 967 1639569 9783046 8124032 2 371309092 4402 51746 37208869| on blots
  |------ ---- ---- ------- --- ------- ------- ------- - --------- ---- ----- --------|---------
A |          2 3 22  23 2        1   23   223 3 223            223  22     222       33|
  |333333 1116 0343 2603522 223 1122241 1266110 9922222 3 132224511 8422 22449 12222211|
  |------ ---- ---- ------- --- ------- ------- ------- - --------- ---- ----- --------|
A |                                           6                                        |
  |                                           8             3                          |
  |------ ---- ---- ------- --- ------- ------- ------- - --------- ---- ----- --------|
B |            1116 1111111 222 3333333 3333344 4444444 4 455555555 5555 66666 66666666|
  |777777 8888 3445 8888888 777 5555555 7888812 4444456 7 901111112 3477 00001 22222222|
  |------ ---- ---- ------- --- ------- ------- ------- - --------- ---- ----- --------|---------
 1|988888 99 8 8888  188889 888  188888 8888888 8888888 8 889888888 8988 88881 88888888| 19.53 kb
 2|988888 99 8 8888    9989 888   98888 8888888 9888888 9 889888888 8988 88898 88888888| 17.36
 3|911181 88 1 1111  119919 118  891811 1191111 8111891 1 119111181 1911 11191 19111189| 14.53
 4|333333 2233 3333 3222333 333 2233333 3333333 2333333 2 333333333 3333 33333 33333332| 12.72
```

Cells are listed according to HLA-A and HLA-B locus antigens and banding patterns. Numbers indicate intensity (1,2,3 = increasing intensity, 8 = faint, 9 = unreadable).

RFLP Standardization Report for DR Alpha/SstI

T. Takenouchi,[1] R. Tonai,[1] P. Terasaki,[1] and K. Tsuji[2]

For this enzyme-probe system, data from only one laboratory was obtained. One nonpolymorphic 4.78 kb fragment was observed (Table 1).

Participating Laboratories: US5TER,[1] JAPTSU[2]

Table 1. Fragment Found Using DR Alpha/SstI System

		DRA	SST	TER		
	kb	Locus	Frequency	% Faint	56	32
1	4.78	0 0 0 0	1.000	0.000		

RFLP Standardization Report for DR Alpha/PstI

J.J. Hyldig-Nielsen,[1] M. Trucco,[2] H. Coppin,[3] A. Andreas-Zietz,[4] E. Keller,[4] C. Carcassi,[5] A. Takada,[6] and A. Svejgaard[1]

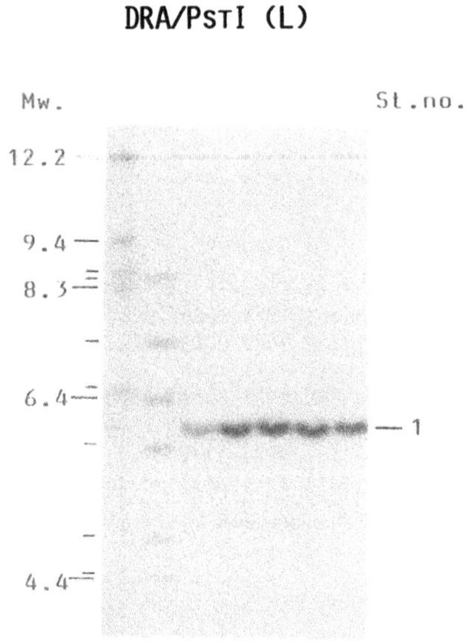

The DR alpha/PstI system identified a single nonpolymorphic fragment of 5.90 kb as shown in Figure 1.

Participating Laboratories: SCASVE,[1] US1DUQ,[2] FRAOHA,[3] GERALB,[4] IT1CON,[5] JAPAIZ[6]

◄ **Figure 1.** The nonpolymorphic fragment identified by the DR alpha/PstI system.

Immunobiology of HLA
Volume I

RFLP Standardization Report for DR Alpha/KpnI

G. Paulsen,[1] H. Avakian,[2] and E. Du Toit[3]

Four bands were identified by the RFLP analysis in the DR alpha/KPN system as shown in Table 1 and in Figure 1. RFLP patterns of the core cell lines are shown in Table 2.

Bands 1 and 3 were present in all 70 core cell lines. Bands 2 and 4 were faint bands readable on one x-ray film only.

Participating Laboratories: SCATSB,[1] UKIFES,[2] SAFDUT[3]

Table 1. Standard Bands in the DR Alpha/KPN System

Band	kb	Frequency[a]	% Faint[b]
1	31.64	1.000	0.000
2	11.49	0.042	1.000
3	8.07	1.000	0.000
4	5.79	0.083	1.000

[a]Band frequency in the Core cell lines.
[b]Frequency of faint bands.

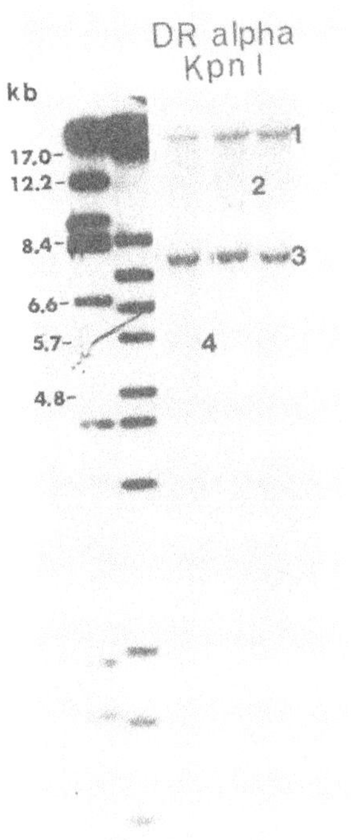

►

Figure 1. RFLP bands in the DR alpha/KPN system of three core cell lines run in 0.9% agarose gel. Representative bands are shown in the figure numbered on their right side. The WS.ID. of these core cell lines are, from the left following marker 1 and marker 2: 9036, 9002, and 9030.

Table 2. Distribution of Bands in the Core Cell Lines in the DR Alpha/KPN System

		Band No.						Band No.			
Cell	DR	1	2	3	4	Cell	DR	1	2	3	4
9045	11/12	1	9	2	9	9050	7	1	9	2	9
9014	15	1	9	2	9	9056	13/14	1	9	2	9
9058	13	1	9	2	9	9104	11	1	9	2	9
9107	4	1	9	2	9	9051	7	1	9	2	9
9017	15	1	9	2	9	9047	7	1	9	2	9
9042	11	1	9	2	9	9040	11	1	9	2	9
9013	15	1	9	2	9	9010	15	1	9	2	9
9033	4	1	9	2	9	9069	8	1		2	
9004	1	1	9	2	9	9037	11	1		2	
9018	3	1	9	2	9	9028	4	1		2	
9055	6	1	9	2	9	9019	3	1		2	
9063	13	1	9	2	9	9025	4	1		2	
9022	3	1	9	2	9	9031	4	1		2	8
9043	11	1	9	2	9	9088	3	1		2	
9023	3	1	9	2	9	9066	8	1		2	
9006	1	1	9	2	9	9060	13	1		2	
9005	1	1	9	2	9	9036	11	1		2	8

Table 2. *Continued*

Cell	DR	Band No. 1	2	3	4	Cell	DR	Band No. 1	2	3	4
9002	1	1		2		9075	9	1	9	2	9
9071	8	1		2		9061	14	1	9	2	9
9064	14	1		2		9067	8	1	9	2	9
9057	14	1		2		9038	12	1	9	2	9
9030	4	1	8	2		9008	15	9	9	9	9
9054	14	1		2		9003	1	1	9	2	9
9105	11	1		2		9070	8	1	9	2	9
9091	4	1		2		9029	14	1	9	2	9
9016	16	1		2		9011	15	1	9	2	9
9048	7	1		2		9026	4	1	9	2	9
9065	13	1		2		9039	11	1	9	2	9
9021	3	1		2		9056	13/14	1	9	2	9
9062	13	1		2		9082	15	1	9	2	9
9032	4	1		2		9106	7	1	9	2	9
9009	2	1	9	2	9	9032	4	1	9	2	9
9092	4	1	9	2	9	9072	8	1	9	2	9
9059	13	1	9	2	9	9052	7	1	9	2	9
9034	4	1	9	2	9	9020	3	1	9	2	9
9068	8	1	9	2	9	9007	16/4	1	9	2	9

Bands designation: 1 = single intensity; 2 = double intensity; 8 = faint; 9 = unreadable.

RFLP Standardization Report for DR alpha/EcoRV

A. Ando,[1] T. Takenouchi,[2] E. Dwyer,[3] and K. Tsuji[1]

Two bands were standardized with the DR alpha/EcoRV system as shown in Table 1 and Figure 1. These two fragments were not polymorphic (Table 1).

Participating Laboratories: JAPTSU,[1] US5TER,[2] US8RJW[3]

Figure 1. Bands standardized using the DR alpha/EcoRV system.

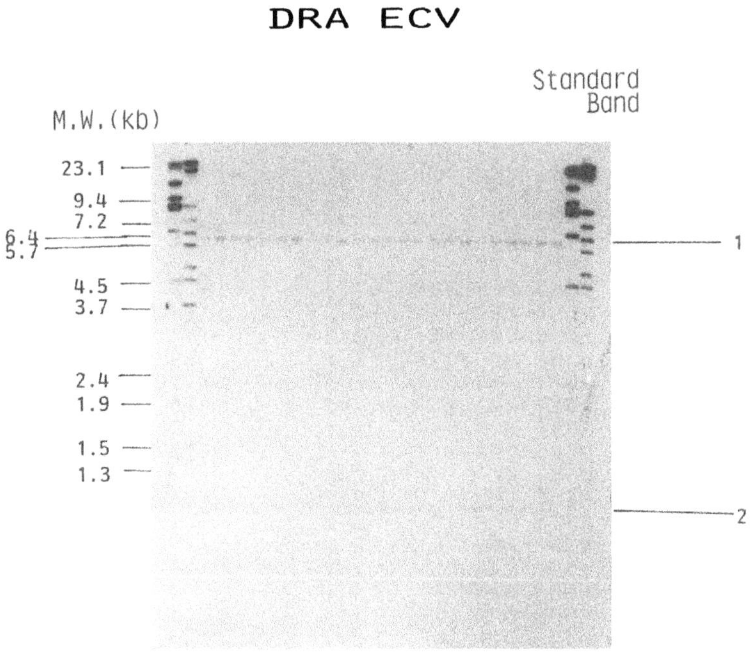

Table 1. RFLP Pattern of DR Alpha/EcoRV

	Fragment No.		1	2							
	Fragment Size (kb)		6.36	1.02							
	Cell Lines				DW	DR	DW2	DRw	DQw	DPw	GLO
1	9045	TUBO	2	2	5	11,12	25	52	7	2,4	
2	9014	MGAR	2	2	2	15			6	4	1,4
3	9058	OMW	2	2	18	13	24	52	6	1	2
4	9107	LKT3	2	2	15	4		53	4	5	
5	9017	WT8	2	2	2	15			1	2,4	
6	9042	TISI	2	2	N	11	25	52	7	-	2
7	9013	SCHU	2	2	2	15			6	-	
8	9033	BM14	2	2	4	4		53	8	4	
9	9004	JESTHOM	2	2	1	1			5	4	
10	9018	LOO81785	2	2	3	3	25	52	2	3	
11	9055	HO301	2	2	N	6	26	52	6	5	
12	9063	WT47	2	2	19	13	26	52	6	2	
13	9022	COX	2	2	3	3	24	52	2	3	2
14	9043	BM21	2	2	5	11	25	52	7	-	
15	9023	VAVY	2	2	3	3	24	52	2	1	1
16	9006	WT100BIS	2	2	1	1			5	1	
17	9005	HOM2	2	2	1	1			5	-	
18	9050	MOU	2	2	17	7		53	2	2	
20	9104	DHIF	2	2	5	11		52	7	N,T	
21	9051	PITOUT	2	2	17	7		53	2	4	2
22	9047	PLH	2	2	DB1	7		53	2	-	
23	9040	BM15	2	2	5	11	25	52	7	3	
24	9010	AMAI	2	2	2	15			6	--	
25	9069	MADURA	2	2	8.1	8		52	4	4	
26	9037	SWEIG007	2	2	5	11	25	52	7	...	
27	9028	PE117	2	2	14	4		53	8	4	
28	9019	DUCAF	2	2	3	3	25	52	2	-	2
29	9025	DEU	2	2	4	4		53	7	4	

Table 1. *Continued*

	Fragment No.		1	2							
	Fragment Size (kb)		6.36	1.02							
	Cell Lines				DW	DR	DW2	DRw	DQw	DPw	GLO
30	9031	BOLETH	2	2	4	4		53	8	4	
31	9088	FF04015	2	2		3	24	52	2	1,4	
32	9066	TAB089	2	2	8.3	8		52	6	–	2
33	9060	CB6B	2	2	18	13	25	52	5	–	
34	9036	SPOO10	2	2	DB2	11	25	52	5	2	
35	9002	MZ070782	2	2	20	1			5	4	1,2
36	9071	OLGA	2	2	8.2	8		52	4	3	
37	9064	AMALA	2	2	16	14	24	52	7	–	
38	9057	TEM	2	2	9	14	25	52	1	4	
39	9030	JHAF	2	2	13	4		53	7	3	
40	9054	EK	2	2	9	14	25	52	5	–	
41	9105	FPAF	2	2	FS	11		52	6	2	
49	9009	KASO11	2	2	21	2			1	3,4	
50	9092	BM92	2	2	14	4		53	8	–	1
51	9059	SLE005	2	2	19	13	26	52	6	3	
52	9034	SAVC	2	2	4	4		53	8	–	
53	9068	BM9	2	2	8.1	8		52	4	2	
54	9075	DXB	2	2	23	9		53	9	4	
55	9061	31227ABO	2	2	9	14	25	52	5	4	
56	9067	BTB	2	2	8.1	8		52	4	4	
57	9038	BM16	2	2	DB6	12		52	7	2	
58	9008	DO208915	2	2	2	15			6	2,4	
59	9003	KAS116	2	2	1	1			5	–	
60	9070	LUY	2	2	8.3	8		52	7	1,4	
61	9029	WT51	2	2	4	4		53	8	2	
62	9011	E4181324	2	2	12	15			6	2,4	
63	9026	YAR	2	2	10	4		53	8	4	1,2
64	9039	JVM	2	2	N	11	25	52	7	2	
66	9082	HO104	2	2		15			6	4	
68	9032	BSM	2	2	4	4		53	8	2	
70	9052	DBB	2	2	11	7		53	9	4	1
71	9020	QBL	2	2	3	3	25	52	2	2	
72	9007	DEM	2	2	–	16,4		53	5,8	3,4	1,2
19	9056	KOSE	2	2	9,19	13,14	25	52	5	2,N	
65	9056	KOSE	2	2	9,19	13,14		52	5	2,N	
67	9106	MANIKA	1	2		7		53	2?3	N,T	
69	9072	SPACH	1	2	8.2	8		52	4	4	
42	9091	MLF	2	8	4	4		53	7	–	
43	9016	RML	2	8	22	16			7	–	
44	9048	LBUF	2	8	DB1	7		53	2	–	
45	9065	HHKB	2	8	18	13	24	52	6	4	
46	9021	RSH	2	8	N	3	24	52	4	1	1
47	9062	WDV	2	8	18	13	24	52	6	4	
48	9032	BSM	2	8	4	4		53	8	2	

RFLP Standardization Report for DR Alpha/TaqI

R.E. Bontrop,[1] J.D. Bignon,[2] C.B. Carpenter,[3] and D. Cohen[4]

Data were received from three different laboratories and standardization was performed based on the following two criteria: 1) weak hybridization signals were accepted as standard fragments when they were detected by most laboratories, and 2) discrepancies were resolved by majority consensus.

In this system, five different standard fragments were detected (Table 1 and Fig. 1). Standard fragment 1 has a molecular weight of 2.72 kb and a panel frequency of 38.6%. This fragment is sometimes observed as a strong hybridization signal, however, weak hybridizations are also observed (Table 1). Standard fragments 2 (molecular weight 1.98 kb) and 3 (molecular weight 1.92 kb) display the most abundant staining intensities (Fig. 1). Fragment 2 has a panel frequency of 43.5% whereas fragment 3 has a panel frequency of 59.4%. Standard fragment 4 is represented by a weak hybridization signal, has a molecular weight of 1.64 kb, and is seen for all panel cells analyzed. Standard fragment 5 is represented by a stronger hybridization signal than fragment 4, has a molecular weight of 1.63 kb, and is seen in all cells tested.

Strangely enough, protein studies and DNA sequence analyses have demonstrated that the DR alpha chain gene itself is not polymorphic. For that reason, the heterogeneity observed with a DR alpha probe detected after digestion with the restriction enzyme TaqI probably reflects the presence of polymorphic restriction sites flanking the coding sequences of the DR alpha gene.

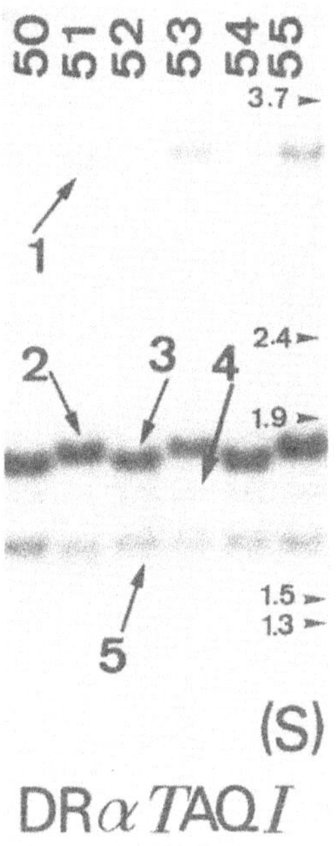

Figure 1. Standard fragments identified by the DR alpha/TaqI system.

Participating Laboratories: BENROO,[1] FRABIG,[2] NCYCAR,[3] FRACOH[4]

Table 1. Standard Fragments Detected by the DR Alpha/TaqI System

Fragment	kb	Frequency	% Faint	56	32
1	2.72	0.386	0.571	f	
2	1.98	0.435	0.000		
3	1.92	0.594	0.000		
4	1.64	1.000	0.000		
5	1.63	1.000	0.556		

Comment. The fragment presence/absence patterns as obtained for the different laboratories did not show any major discrepancies, however, the molecular weights of the standard fragments somehow did. This explains why the molecular weights of the fragments shown in Figure 1 do not exactly match the data presented in Table 1.

Table 2. Fragment Presence/Absence Pattern for the DR Alpha/TaqI System as Defined on 72 Core Cell Lines

	Fragment						Fragment				
Cell	1	2	3	4	5	Cell	1	2	3	4	5
9014	1	2		1		9004			2	1	
9017	1	2		1		9018			2	1	
9055	1	2		1		9043			2	1	
9069	8	2		1		9006			2	1	
9071	8	2		1		9005			2	1	
9059	8	2		1		9050			2	1	
9068	1	2		1		9104			2	1	
9061	8	2		1		9051			2	1	
9067	1	2		1		9047			2	1	
9038	8	2		1		9040			2	1	
9008	8	2		1		9010			2	1	
9070	8	2		1		9037			2	1	
9011	8	2		1		9028			2	1	
9082	8	1		1		9019			2	1	
9052	8	2		1		9025			2	1	
9013		2		1		9031			2	1	
9056		2		1		9060			2	1	
9056		2		1		9036			2	1	
9054		2		1		9057			2	1	
9058	2		1	1		9030			2	1	
9021		2		1	8	9091			2	1	
9063	1	2		1	1	9016			2	1	
9022	1	1		1	8	9048			2	1	
9023	1	2		1	8	9032			2	1	
9088	1	2		1	8	9009			2	1	
9066	8	2		1	1	9092			2	1	
9065	8	2		1	1	9034			2	1	
9062	8	2		1	8	9075			2	1	
9045	1		2	1		9003			2	1	
9032	8		2	1		9029			2	1	
9002	1	2	2	1		9026			2	1	
9106	8	2	9	1		9039			2	1	
9020	8	2	2	1		9007			2	1	
9107			2	1		9105			2	1	1
9042			2	1		9064				8	9
9033			2	1		9072					1

RFLP Standardization Report for DR Alpha/BglII

A. Marcadet,[1] D. Cohen,[1] H. Takata,[2] R. Bontrop,[3] and E.J. Ball[4]

The DR alpha/BglII system identified six fragments on the core cell lines (Table 1). Three constant fragments of weak intensity were revealed (no. 1, 5 and 6).

Table 2 indicates the distribution of standardized bands in core cell lines. The three constant fragments were not listed. In this system, *no specific band* for DRw specificities can be observed. Moreover *no spectific pattern of bands* for DRw specificities can be observed. The

Table 1. Standard Bands in the DR Alpha/BglII System

		DRA	BGL	COH		
	kb	Locus	Frequency	% Faint	56	32
1	7.02	0 0 0 0	1.000	1.000		
2	4.49	0 0 0 0	0.757	0.000		
3	4.22	0 0 0 0	0.186	0.000		
4	3.89	0 0 0 0	0.086	0.000		
5	3.63	0 0 0 0	1.000	1.000		
6	1.93	0 0 0 0	1.000	1.000		

Participating Laboratories: FRACOH,[1] JAPTAK,[2] BENROO,[3] US4STA[4]

Immunobiology of HLA
Volume I

Table 2. RFLP DR Alpha/BglII System. Distribution of Standard Bands in 10th WS Core Cell Lines

		Standard Bands					Standard Bands		
			no.					no.	
Cell Lines	HLA-DR	2	3	4	Cell Lines	HLA-DR	2	3	4
9004	1^d			2	9105	w11	2		
9005	1^d			2	9042	w11	2		
9003	1^d			2	9043	w11	2		
9006	1^d			2	9104	w11	2		
9002	1^e	2			9037	w11	2		
9009	2	2			9039	w11	2		
9016	w16	2			9036	w11	2		
9014	w15	2			9040	w11	2		
9008	w15	2			9045	w11/12	2		
9010	w15	2			9060	w13	2		
9011	w15	2			9061	w14	2		
9082	w15	2			9054	w14	2		
9013	w15	2			9057	w14	2		
9017	w15	2			9055	w6	2		
9007	w16/4	2			9056	w13/14	2	2	
9019	3^a	2			9056	w13/14	2	2	
9020	3^a	2			9058	w13		2	
9018	3^a	2			9062	w13		2	
9088	3		2		9065	w13		2	
9023	3		2		9059	w13		2	
9022	3		2		9063	w13		2	
9021	3		2		9064	w14		2	
9025	4	2			9047	7^b	2		
9026	4	2			9050	7^b	2		
9028	4	2			9051	7^b	2		
9029	4	2			9048	7^b	2		
9091	4	2			9052	7			2
9092	4	2			9106	7	2		2
9034	4	2			9070	w8		2	
9031	4	2			9066	w8		2	
9032	4	2			9068	$w8^c$	2		
9032	4	2			9069	$w8^c$	2		
9030	4	2			9072	$w8^c$	2		
9033	4	2			9067	$w8^c$	2		
9107	4	2			9071	$w8^c$	2		
9038	w12	2			9075	w9	2		

[a] = DR3-B18
[b] = DR7-DQw2
[c] = DRw8-DQw4
[d] = DR1-Dw1
[e] = DR1-Dw20

three fragments 2, 3 and 4 may be allelic (the only exception is the cell no. 9106 HLA-typed DR7). Fragment no. 2 is present for DR1;Dw20 + DR2 + B18;DR3 + DR4 + DR5 + DR7;DQw2 + DRw8;DQw4 + some DRw6 (+ DRw9?). Fragment no. 3 is present for $B18^-$;DR3 + DRw8;$DQw4^-$ + some DRw6 and fragment no. 4 is present for DR1;Dw1 + DR7;$DQw2^-$. To summarize, we can say that this system can be used to discriminate between DR1;$Dw20^+$ and DR1;$Dw20^-$, between DR3;$B18^+$ and DR3;$B18^-$, between DR7;$DQw2^+$ and DR7;$DQw2^-$, and between DRw8;$DQw4^+$ and DRw8;$DQw4^-$.

The figures are photographs taken from two films from laboratory JAPTAK. All standardized fragments are indicated at least once for S and/or L migration. The presence of a standardized fragment is indicated by a hyphen.

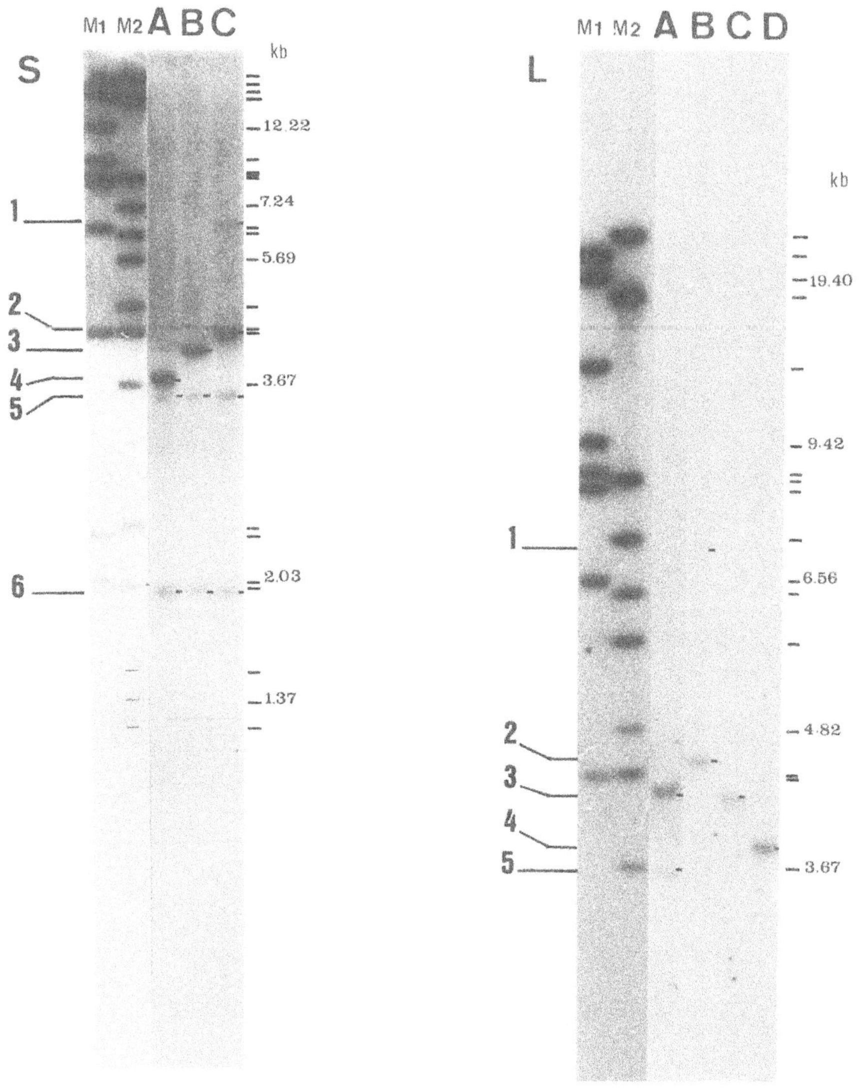

Figure 1. Standardized fragments identified by the DR alpha/BglII system. All fragments are indicated at least once for S and/or L migration.

RFLP Standardization Report for DR Alpha/PvuII

P. Piccioli,[1]* A. Smith,[2] G. Angelini,[1] and G.B. Ferrara[1]

Seventy-two cell lines digested with PvuII and hybridized with a HLA-DR alpha-specific probe identified two fragments as shown in Figure 1.

The first fragment was shared by 55 cell lines, while the second was present in 28 of the 70 (in the core cell line two hidden duplicates have been included). Table 1 shows pattern distribution as analyzed by Southern blot. Discrepancies in the assignment of each of the two fragments sometimes becomes evident when comparing results from the laboratories which performed this RFLP study. Although DR alpha gene is known to be nonpolymorphic (so far only one polymorphism has been detected (1, 2) it was still possible to discriminate some extended haplotypes; indeed the first fragment (11.93 kb) appears to be present in B18-DR3 positive cell lines whereas the 11.01 kb fragment is correlated with B8-DR3. The same association was also evident for HLA-Dw8 specificities defined in MLR (3); DR8-Dw8.1 and DR8-Dw8.2 can be distinguished from DR8-Dw8.3 by the presence of either 11.93 or 11.01 kb specific fragment. Looking at the supertypic specificity DR3B*0101,*0201,*0301 previously defined as DRw52 a,b,c, absence of the first fragment seems to be correlated with DR3B*0101, as seven of the eight cell lines tested share this pattern. The situation appears much more complicated for DR3B*0201; 53% (8/15) of these cell lines present the first fragment and lack the second one, whereas the remaining show both the fragments. Concerning DR3B*0301, any interpretation is a pure speculation because of the low number of cell lines tested carrying this specificity.

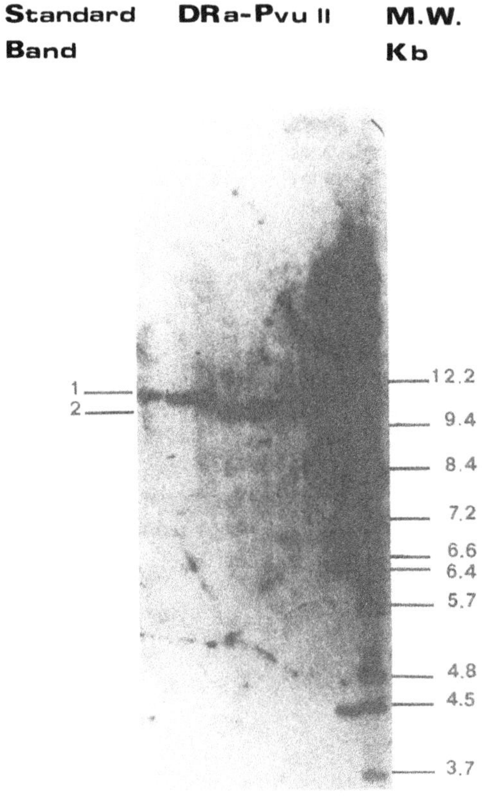

Figure 1. Standard fragments from MLF, RML, LBUF, HHKB, RSH, WDV, and BSM DNAs. On the right and on the left are respectively shown the molecular weights referring to the markers used and the standard bands 1 and 2 of 11.93 and 11.01 kb obtained after digestion with PVU II and hybridization with a specific HLA-DR alpha probe.

References

1. Lee JS, Trowsdale J, Bodmer WF. cDNA clones coding for the heavy chain of human HLA-DR antigen. Proc Natl Acad Sci USA 1987;79:545.
2. Larhammar D, Gustafsson K, Claesson L, Bill P, Wiman K, Schenning L, Sundelin J, Widmark E, Peterson PA, Rask L. Alpha chain of HLA-DR transplantation antigens is a member of the same protein superfamily as the immunoglobulins. Cell 1982;30:153.
3. Mickelson EM, Nisperos B, Layrisse Z, Kim SJ, Thomas ED, Hansen JA. Recognition by HTC typing on three distinct antigen complexes in Caucasian, Native Americans and Orientals. Immunogenetics 1983;17:399.

Participating Laboratories: IT2FER,[1] US7HAN[2]
*P. Piccioli was recipient of a fellowship from A.I.R.C. (Associazione Italiana per la Ricerca sul Cancro).
This work was in part supported by CNR grants n.86.0265.51 and 86.00659.

Table 1. Banding Pattern of the DR Alpha/PvuII System in Each Cell Line

Cell	DR	DRw	Dw	Fragment 1	Fragment 2
9002 MZ 070782	1		1	1	
9003 KAS 116	1		20	(1)	(1)
9004 JESTHOM	1		1		1
9005 HOM 2	1		1		1
9006 WT100 BIS	1		1		1
9009 KAS 011	2		21	1	
9008 DO 208915	15		2	1	
9010 AMAL	15		2	(1)	(1)
9011 E 4181324	15		12	1	
9013 SCHU	15		2	1	
9014 MGAR	15		2	1	(1)
9017 WT 8	15		2	1	
9082 HO 104	15			1	(1)
9016 RML	16		22	1	
9018 LO 081785	3	52	3	1	
9019 DUCAF	3	52	3	(1)	(1)
9020 QBL	3	52	3	1	
9021 RSH	3	52	N		1
9022 COX	3	52	3		1
9023 VAVY	3	52	3		1
9088 PF 04015	3	52	3		1
9025 DEU	4	53	4	2	
9026 YAR	4	53	10	1	
9028 PE 117	4	53	14	(2)	(1)
9029 WT 51	4	53	4	1	
9030 JHAF	4	53	13	1	
9031 BOLETH	4	53	4	(1)	(1)
9032 BSM	4	53	4	1	
9032 BSM	4	53	4	1	
9033 BM 14	4	53	4	1	
9034 SAVC	4	53	4	1	
9091 MLF	4	53	4	(1)	(1)
9092 BM 92	4	53	14	1	
9107 LKT3	4	53	15	1	
9036 SPO 010	11	52	DB2	(1)	(1)
9037 SWEIG 007	11	52	5	(1)	(1)
9039 JVM	11	52	N	1	
9040 BM 15	11	52	5	1	
9042 TISI	11	52	N	1	
9043 BM 21	11	52	5	1	
9104 DHIF	11	52	5	1	
9105 FPAF	11	52	FS	1	
9038 BM 16	12	52	DB6	1	
9055 HO 301	6	52	N	1	
9058 OMW	13	52	18		1
9059 SLE 005	13	52	19	(1)	1
9060 CB6B	13	52	18	(1)	
9062 WDV	13	52	18		1
9063 WT47	13	52	19		1
9065 HHKB	13	52	18		1
9054 EK	14	52	9	1	
9057 TEM	14	52	9	(1)	
9061 31227 ABO	14	52	9	1	
9064 AMALA	14	52	16		(1)
9047 PLH	7	53	DB1		1
9048 LBUF	7	53	DB1	1	

Table 1. *Continued*

Cell	DR	DRw	Dw	Fragment 1	2
9050 MOU	7	53	17	1	
9051 PITOUT	7	53	17	1	
9052 DBB	7	53	11	1	
9106 MANIKA	7	53		1	
9066 TAB 089	8	52	8.3		1
9067 BTB	8	52	8.1	1	
9068 BM 9	8	52	8.1	1	
9069 MADURA	8	52	8.1	(1)	(1)
9070 LUY	8	52	8.3		1
9071 OLGA	8	52	8.2	1	
9072 SPACH	8	52	8.2	1	
9075 DKB	9	53	23	1	
9007 DEM	4/16	53		1	
9045 TUBO	11/12	52	5	1	
9056 KOSE	13/14	52	9/19	1	1
9056 KOSE	13/14	52	9/19	1	(8)

1 = Single intensity; 2 = double intensity; 8 = faint; ( ) = discordance between the two laboratories that performed RFLP study.

RFLP Standardization Report for DR Alpha/EcoRI

M. Segall,[1]* L. Schluender,[1]* B. Bradley,[2] and S. Naito[3]

DR alpha/EcoRI blots were received from two laboratories. Fragments well-identified on at least one laboratory's blots were included as standard fragments.

The DR alpha/EcoRI system identified five fragments on the core cell lines as shown in Table 1. Only one of the two laboratories submitting blots for this probe-enzyme combination identified the 13.43 and 10.88 kb bands, and a portion of one of their gels could not be read, resulting in the discrepancies for duplicates. These two fragments are probably both nonpolymorphic fragments.

The discrepancies for fragment 5 are probably due to difficulties in resolving fragments 4 and 5, as the blots were generally rather faint in this region. However, there are certainly two fragments. They are reciprocal in distribution (Table 2); fragment 5 was found in 8/8 Dw24 cells, 2/3 Dw26 cells (9059 and 9063, but not 9055) and 2/2 Dw8.3 cells, while fragment 4 was found in all other cells. The absence of fragment 5 in 9055 is probably not a result of difficulty in resolving the fragments, as this cell was different from the other two Dw26 cells for other fragments which distinguished Dw24 and Dw26 from other cells (see report on DR alpha fragments).

Table 1. DR Alpha/EcoRI Standard Fragments

Fragment	kb	Frequency	% Faint	Discrepant for 9056	9032
1	13.43	0.871	0.000	s	
2	10.88	0.814	1.000	f	f
3	4.52	1.000	1.000		
4	3.18	0.857	0.000		
5	3.14	0.214	0.000	s	s

Participating Laboratories: US2BAC,[1] UKIBRS,[2] JAPNAI[3]
*This is publication #492 from the Immunobiology Research Center, University of Minnesota, Minneapolis, Minnesota 55455, USA.
Supported in part by Juvenile Diabetes Foundation Grant #196175 and March of Dimes Birth Defects Foundation Grant #6-496.

Table 2. Distribution of Polymorphic Fragments in Core Cell Lines, DR Alpha/EcoRI

Cell No.	Name	Std. Fragment No. 4	5	Cell No.	Name	Std. Fragment No. 4	5
9021	RSH		1	9017	WT8	1	
9032	BSM		1	9014	MGAR	1	
9058	OMW		1	9011	E4181324	1	
9064	AMALA		1	9009	KAS011	1	
9063	WT47		1	9007	DEM	1	
9062	WDV		1	9032	BSM	1	
9059	SLE005		1	9005	HOM2	1	
9023	VAVY		1	9003	KAS116	1	
9066	TAB089		1	9072	SPACH	1	
9070	LUY		1	9106	MANIKA	1	
9088	PFO4015		1	9104	DHIF	1	
9065	HHKB		1	9091	MLF	1	
9022	COX		1	9082	HO104	1	
9030	JHAF	1		9071	OLGA(OLE)	1	
9034	SAVC	1		9069	MADURA	1	
9037	SWEIG007	1		9067	BTB	1	
9039	JVM	1		9052	DBB	1	
9040	BM15	1		9060	CB6B	1	
9043	BM21	1		9061	31227ABO	1	
9047	PLH	1		9055	HO301	1	
9050	MOU	1		9057	TEM	1	
9068	BM9	1		9054	EK	1	
9075	DKB	1		9051	PITOUT	1	
9092	BM92	1		9048	LBUF	1	
9105	FPAF	1		9045	TUBO	1	
9107	LKT3	1		9042	TISI	1	
9002	MZO70782	1		9038	BM16	1	
9004	JESTHOM	1		9036	SP0010	1	
9006	WT100BIS	1		9033	BM14	1	
9056	KOSE	1		9031	BOLETH	1	
9008	D0208915	1		9029	WT51	1	
9010	AMAL	1		9020	QBL	1	
9013	SCHU	1		9028	PE117	1	
9016	RML	1		9025	DEU	1	
9018	L0081785	1		9026	YAR	1	
9019	DUCAF	1		9056	KOSE	1	1

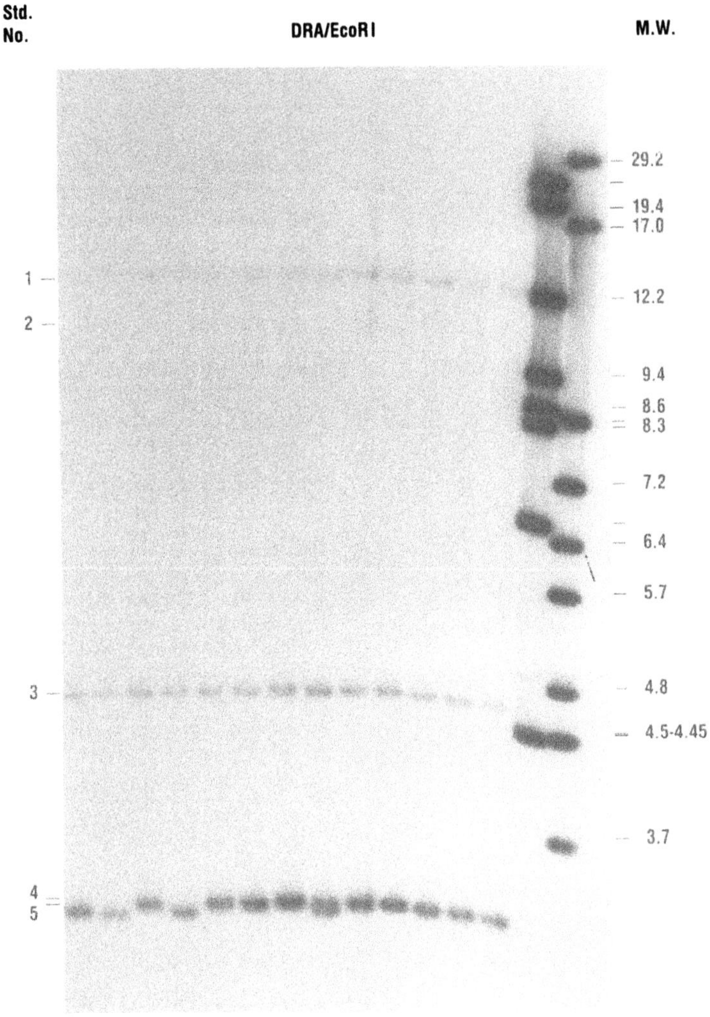

Figure 1. Fragments identified by the DR alpha/EcoRI system.

RFLP Standardization Report for DR Alpha/HindIII

P.A. Biro,[1] S. Cutbush,[1] G. Paulsen,[2] G. Semana,[3] and F. Quillivic[3]

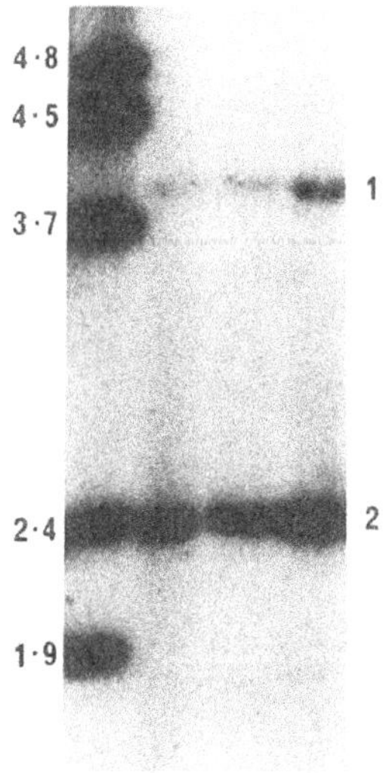

Figure 1. Standard bands produced by the DR alpha/HindIII enzyme-probe combination.

Two nonpolymorphic, low molecular standard weight fragments are produced by this probe-enzyme combination (Fig. 1). Standard band 2 (2.35 kb) is the more intense band and spans the majority of the exons of the gene (nos. 2, 3, and 4) (1).

Table 1. Sizes and Frequencies of Standard Bands Produced by the DR Alpha/HindIII Enzyme/Probe Combination

	kb	Frequency	% Faint
1	3.93	1.000	0.000
2	2.35	1.000	0.000

Reference

1. Das HK, Lawrance SK, Weissman SM. Structure and nucleotide sequence of the heavy chain gene of HLA-DR. Proc Natl Acad Sci USA 1983;80:3543.

Participating Laboratories: UKIFES,[1] SCATSB,[2] FRAFAU[3]

RFLP Standardization Report for DR Alpha/MspI

S. Tonks,[1] V. Groves,[1] C. Muller,[2] F. Otani,[3] J.R. Regueiro,[4] M. Segall,[5] J. Bodmer,[1] and J. Trowsdale[1]

Three standard bands were recognized in the DR alpha/MspI system (Table 1 and Fig. 1). The resolution of fragments 1 and 2 on some autoradiographs was not good enough to allow easy interpretation producing discrepancies in both duplicated cell lines. Despite this, the data are believed to be correct as there is good correlation with other DR alpha polymorphisms reported in the workshop.

Participating Laboratories: UKIBOD,[1] GERMUC,[2] JAPKSH,[3] FRÁARN,[4] US2BAC[5]

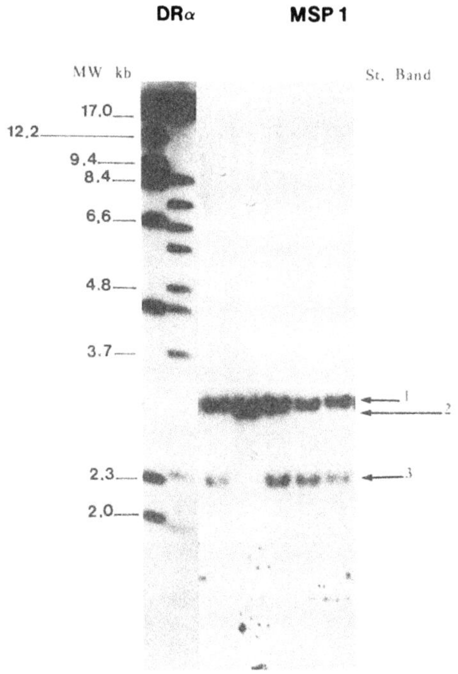

► **Figure 1.** Standard band recognized by the DR alpha/MspI system.

Table 1. Banding Pattern of the DR Alpha/MspI System in Each Cell Line

		Fragment No.		
WS ID	Cell	1	2	3
9014	MGAR	1		
9013	SCHU	1		
9063	WT47	1		
9022	COX	1		
9031	BOLETH	1		
9105	FPAF	1		
9059	SLE005	1		
9082	H0104	1		
9045	TUBO	1		1
9107	LKT3	1		1
9042	TISI	1		1
9033	BM14	1		1
9004	JESTHOM	1		1
9018	LO081785	1		1
9055	H0301	1		1
9043	BM21	1		1
9006	WT100BIS	1		1
9005	HOM2	1		1
9050	MOU	1		1
9104	DHIF	1		1
9051	PITOUT	1		1
9040	BM15	1		1

Table 1. *Continued*

WS ID	Cell	Fragment No. 1	2	3
9010	AMAI	1		1
9069	MADURA	1		1
9037	SWEIG007	1		1
9028	PE117	1		1
9019	DUCAF	1		1
9025	DEU	1		1
9060	CB6B	1		1
9036	SPO010	1		1
9002	MZ070782	1		1
9071	OLGA	1		1
9057	TEM	1		1
9030	JHAF	1		1
9054	EK	1		1
9091	MLF	1		1
9016	RML	1		1
9048	LBUF	1		1
9032	BSM	1		1
9009	KAS011	1		1
9092	BM92	1		1
9034	SAVC	1		1
9068	BM9	1		1
9075	DKB	1		1
9061	31227ABO	1		1
9067	BTB	1		1
9038	BM16	1		1
9003	KAS116	1		1
9029	WT51	1		1
9011	E4181324	1		1
9026	YAR	1		1
9039	JVM	1		1
9106	MANIKA	1		1
9072	SPACH	1		1
9020	QBL	1		1
9007	DEM	1		1
9056	KOSE		1	1
9066	TAB089		1	1
9070	LUY		1	1
9052	DBB		1	1
9058	OMW		1	
9023	VAVY		1	
9088	PF0401S		1	
9064	AMALA		1	
9065	HHKB		1	
9021	RSH		1	
9062	WDV		1	
9008	D0208915		1	
9070	LUY		1	
9052	DBB		1	

1 = Single intensity; 2 = double intensity; 8 = faint.

RFLP Standardization Report for DQ Alpha/SstI

T. Takenouchi,[1] M. Kanatani,[1] P. Terasaki,[1] T. Sasazuki,[2] K. Tsuji,[3] and P. Reekers[4]

Data from three laboratories were standardized for this enzyme-probe system. Also, we encountered the same problems observed with class I/SstI and handled the data and its interpretation in the same manner as described in the class I/SstI system.

In this system, nine standard fragments were assigned (Table 1 and Fig. 1). Standard fragments nos. 6 and 8 appear in only one cell each. At the time of standardization we found this fragment in more than two laboratories, however, we now think these bands are the result of incomplete digestion because these fragments are inconsistent with the total number of fragments appearing in each cell line. In cells which possess fragments nos. 6 and 8, there are a total of four fragments, however in all other cell lines only three fragments appear. In the photograph, fragments nos. 2, 5, 6, and 8 are not shown. Fragment no. 2 migrates to a position just below fragment no. 1 while fragment no. 5 is just below fragment no. 4. From band pattern analysis an apparent five allele pattern was observed and is shown in Table 2.

Participating Laboratories: US5TER,[1] JAPSAS,[2] JAPTSU,[3] BENREE[4]

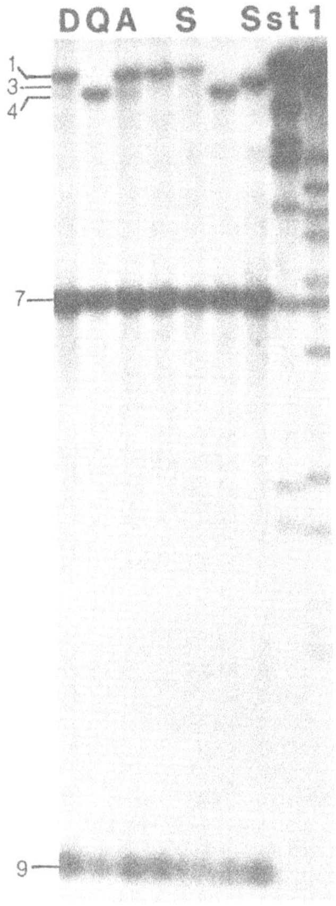

Figure 1. Standard fragments identified by the DQ alpha/SstI system.

Table 1. Standard Fragments Found by the DQ Alpha/SstI System

		DQA	SST	TER		
	kb	Locus	Frequency	% Faint	56	32
1	24.78	0 0 0 0	0.250	0.000		
2	23.55	0 0 0 0	0.029	0.000		
3	19.38	0 0 0 0	0.217	0.000		
4	16.53	0 0 0 0	0.391	0.036		
5	15.53	0 0 0 0	0.057	0.000		
6	9.30	0 0 0 0	0.015	1.000		
7	4.53	0 0 0 0	1.000	0.000		
8	2.05	0 0 0 0	0.014	1.000		
9	0.36	0 0 0 0	1.000	1.000		

Table 2. Banding Pattern of the DQ Alpha/SstI System in Each Cell

	Fragment No.						
Cell	1	2	3	4	5	7	9
9021 RSH	1					2	8
9032 BSM	1					2	8
9032 BSM	9					2	8
9034 SAVC	1					2	8
9028 PE 117	1					2	8
9025 DEU	1					2	8
9031 BOLETH	1					2	8
9030 JHAF	1					2	8
9091 MLF	1					2	8

Table 2. *Continued*

Cell	Fragment No. 1	2	3	4	5	7	9
9107 LKT 3	1					2	8
9033 BM14	1					2	8
9036 SPO 010	1					2	8
9104 DHIF	1					2	8
9050 MOU	1					2	8
9051 PITOUT	1					2	8
9047 PLH	1					2	8
9075 DKB	1					2	8
9048 LBUF		1				2	8
9052 DBB		1				2	8
9014 M GAR			2			2	8
9082 HO 104			1			2	8
9017 WT8			1			2	8
9010 AMAI			1			2	8
9007 DEM	9		1			2	8
9008 DO 208915			1			2	8
9013 SCHU			1			2	8
9011 E 4181324			1			2	8
9037 SWEIG 007			1			2	8
9070 LUY			1			2	8
9069 MADURA			1			2	8
9071 OLGA			1			2	8
9068 BM9			1			2	8
9067 BTB			1			2	8
9072 S PACHEO			1			2	8
9059 SLE-005				8		2	8
9055 HO 301				1		2	8
9056 KOSE				1		2	8
9056 KOSE				1		2	8
9063 WT47				1		2	8
9064 AMALA				1		2	8
9057 TEM				1		2	8
9054 EK				1		2	8
9004 JES THOM				1		2	8
9006 WT100 BIS				1		2	8
9005 HOM-2				1		2	8
9002 MZ 070782				1		2	8
9003 KAS-116				1		2	8
9016 RML				1		2	8
9018 LO 081785				1		2	8
9022 COX				1		2	8
9023 VAVY				1		2	8
9019 DUCAF				1		2	8
9088 PF 04015				1		2	8
9020 QBL				1		2	8
9045 TUBO				1		2	8
9042 TISI				1		2	8
9043 BM21				1		2	8
9040 BM15				1		2	8
9038 BM16				1		2	8
9039 JVM				1		2	8
9066 TAB 089				1		2	8
9061 3-12-27 AOB				1		2	8
9058 OMW					1	2	8
9060 CBG-BI					1	2	8
9065 HHKB					1	2	8
9062 WDV					1	2	8
9009 KAS-011				9		2	8
9105 FPAF				9	9	9	9
9092 BM92	9					2	8
9029 WT51	9					2	8
9026 YAR	9					2	8
9106 MANIKA	9					2	8

1 = Single intensity; 2 = double intensity; 8 = faint; 9 = uninterpretable.

RFLP Standardization Report for DQ Alpha/PstI

J.J. Hyldig-Nielsen,[1] M. Trucco,[2] H. Coppin,[3] A. Andreas-Zietz,[4] E. Keller,[4] C. Carcassi,[5]
A. Takada,[6] and A. Svejgaard[1]

The DQ alpha/PstI system identified seven fragments as shown in Table 1. Standard fragment nos. 5 and 6 are very close, so the distinction between these two fragments might be difficult (Fig. 1). Table 2 indicates the banding patterns obtained with this system. The fragments can be grouped into two sets of complimentary patterns, one including standard fragments 1, 5, 6, 7, and another consisting of standard fragments 2, 3, and 4.

Participating Laboratories: SCASVE,[1] US1DUQ,[2] FRAOHA,[3] GERALB,[4] IT1CON,[5] JAPAIZ[6]

Table 1. Standard Fragments Identified by the DQ Alpha/PstI System

		DQA	PST	SVE		
	kb	Locus	Frequency	% Faint	56	32
1	13.45	0 0 0 0	0.406	0.000		
2	9.12	0 0 0 0	0.029	0.000		
3	7.62	0 0 0 0	0.857	0.000		
4	6.84	0 0 0 0	0.143	0.000		
5	4.31	0 0 0 0	0.072	0.000		
6	4.24	0 0 0 0	0.261	0.056		
7	2.32	0 0 0 0	0.300	0.000		

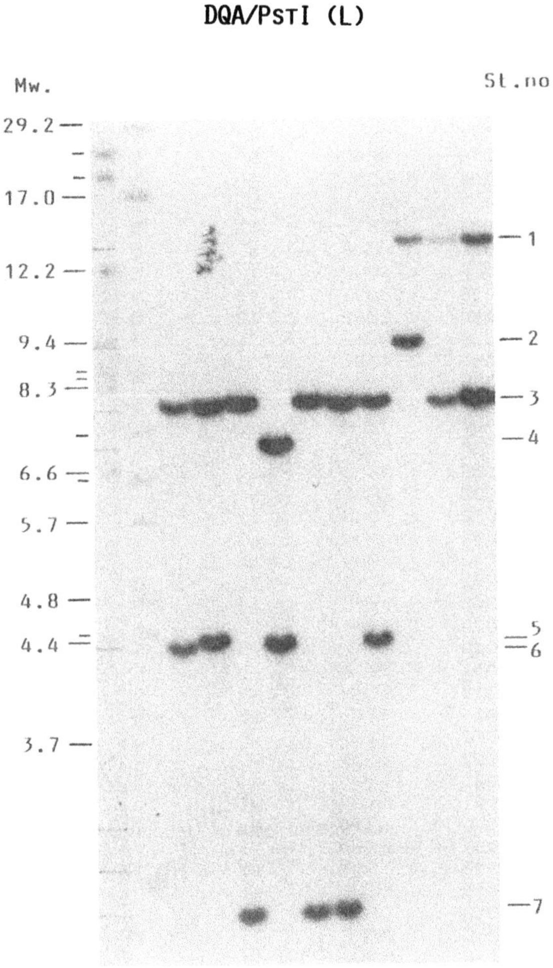

Figure 1. Standard fragments identified by the DQ alpha/PstI system.

Table 2. Banding Patterns Obtained by the DQ Alpha/PstI System

					Standard Fragment No.						
WS ID	Name	HLA-DR	HLA-DQ	HLA-DP	1	5	6	7	2	4	3
9009	KAS011	2	1	3,4	6						8
9057	TEM	14	1	4	6						8
9017	WT8	15	1	2,4	6						8
9018	LOO81785	3	2	3			8				8
9019	DUCAF	3	2				8			8	
9020	QBL	3	2	2			8			8	
9022	COX	3	2	3			8				8
9023	VAVY	3	2	1			8				8
9088	PF04015	3	2	1,4			8				8
9047	PLH	7	2					8		8	

Table 2. *Continued*

WS ID	Name	HLA-DR	HLA-DQ	HLA-DP	Standard Fragment No. 1	5	6	7	2	4	3
9048	LBUF	7	2					8			8
9050	MOU	7	2	2				8		8	
9106	MANIKA	7	2,3	NT				8			8
9021	RSH	3	4	1			8				8
9107	LKT3	4	4	5				8			8
9051	PITOUT	7	4	4				8		8	
9067	BTB	8	4	4			8				8
9068	BM9	8	4	2			8				8
9069	MADURA	8	4	4			8				8
9071	OLGA	8	4	3			8				8
9072	SPACH	8	4				8				8
9002	MZ070782	1	5	4	6						8
9003	KAS116	1	5		6					8	
9004	JESTHOM	1	5	4	6						8
9005	HOM2	1	5		6						8
9006	WT100BIS	1	5	1	6						8
9036	SP0010	11	5	2	6						8
9060	CB6B	13	5		6						8
9056	KOSE	13,14	5	2	6					6	8
9056	KOSE	13,14	5	2	6					6	8
9054	EK	14	5		6						8
9061	31227ABO	14	5	4	6						8
9007	DEM	16,4	5,8	3,4				8			8
9055	HO301	6	6	5	6						8
9066	TAB089	8	6		6				8		
9105	FPAF	11	6	2	6					8	
9058	OMW	13	6	1	6						8
9059	SLE005	13	6	3	6						8
9062	WDV	13	6	4	6						8
9063	WT47	13	6	2	6						8
9065	HHKB	13	6	4	6						8
9008	DO208915	15	6	2,4	6						8
9010	AMAI	15	6		6						8
9011	E4181324	15	6	2,4	6						8
9013	SCHU	15	6		6						8
9014	MGAR	15	6	4	6						8
9082	HO104	15	6	4	6						8
9025	DEU	4	7	4				8			8
9030	JHAF	4	7	3				8			8
9091	MLF	4	7					8			8
9070	LUY	8	7	1,4			8				8
9037	SWEIG007	11	7								8
9039	JVM	11	7	2		8					8
9040	BM15	11	7	3			8				8
9042	TISI	11	7			8	6				8
9043	BM21	11	7			8					8
9104	DHIF	11	7	NT			4	8			8
9045	TUBO	11,12	7	2,4			8			6	8
9038	BM16	12	7	2		8				8	
9064	AMALA	14	7				8				8
9016	RML	16	7			8					8
9026	YAR	4	8	4				8			8
9028	PE117	4	8	4				8			8
9029	WT51	4	8	2				8			8
9031	BOLETH	4	8	4				8			8
9032	BSM	4	8	2				8			8
9032	BSM	4	8	2				8			8
9033	BM14	4	8	4				8			8
9034	SAVC	4	8					8			8
9092	BM92	4	8					8			8
9052	DBB	7	9	4				8			8
9075	DKB	9	9	4				8	6		

RFLP Standardization Report for DQ Alpha/KpnI

G. Paulsen,[1] R. Martell,[2] P. Cassells,[3] and P.C. Kwong[4]

Thirteen bands were identified by the RFLP analysis in the DQ alpha/KPN system as shown in Table 1. RFLP patterns of nine of the core cell lines are visualized in Figure 1. Band nos. 10, 11, 12 and 13 are not present in the figure.

Distribution of bands in the core cell lines is shown in Table 2.

Participating Laboratories: SCATSB,[1] SAFDUT,[2] UKIFES,[3] NCYFAL[4]

Table 1. Standard Bands in the DQ Alpha/KPN System

Band	kb	Frequency[a]	% Faint[b]	56[c]	32[c]
1	25.20	0.086	0.167		
2	18.36	0.043	0.667		
3	17.75	0.271	0.300		
4	16.06	0.257	0.368		
5	14.24	0.286	0.000		
6	11.76	0.300	0.000		
7	10.42	0.129	0.000		
8	8.40	0.157	0.000		
9	8.07	0.871	0.000		
10	7.16	0.483	0.800		f
11	6.60	0.014	1.000		
12	4.65	0.014	1.000		
13	2.32	0.125	1.000		

[a]Band frequency in the core cell lines.
[b]Frequency of faint bands.
[c]Discrepancies in strong bands (s) or faint bands (f) of the hidden duplicates (WS.ID. 9056 and 9032) in the standardization of RFLP bands in the DQ alpha/KPN system.

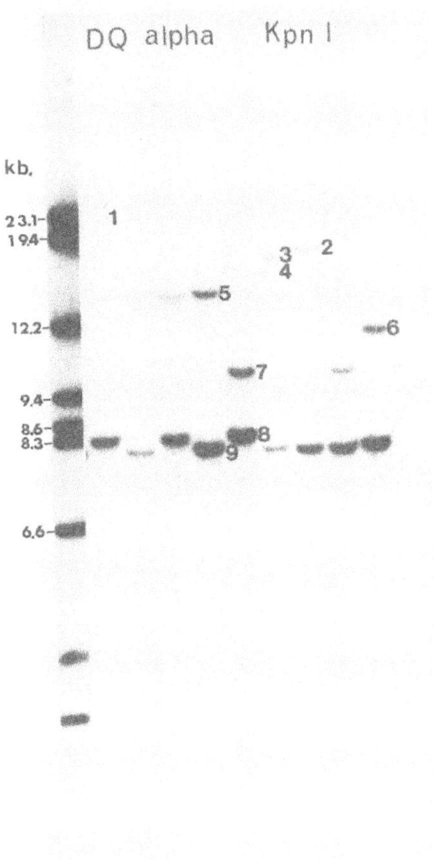

Figure 1. RFLP bands in the DQ alpha/KPN system of nine core cell lines run in 0.6% agarose gel. Representative core bands are shown in the figure numbered on their right side. The WS.ID. of these core cell lines are, from the left following marker 1: 9088, 9066, 9060, 9036, 9002, 9071, 9064, 9057, and 9030.

Table 2. Distribution of Standard Bands in Core Cell Lines in the DQ Alpha/KPN System

			Band No.																Band No.												
	DR	DQ	01	02	03	04	05	06	07	08	09	10	11	12	13		DR	DQ	01	02	03	04	05	06	07	08	09	10	11	12	13
9023	3	2	1			1				2					9	9058	13	6					2				2	8			9
9022	3	2	1							2					9	9063	13	6					2				2	8			9
9088	3	2	8							2						9010	15	6					2				2	1			9
9018	3	2	1								2				9	9009	2	1					1				2	9			9
9019	3	2	1								2	1				9059	13	6					1				2	8			9
9020	3	2	1								2	8			9	9008	15	6					2				2	9			9
9064	14	7		2							2					9011	15	6					1				2	8			9
9052	7	9		8		1		1			2	8			9	9066	8	6					1				2				
9007	16/4	5/8		8		8	1	1			2	9			9	9036	11	5					1				2				
9056	13/14	5			8	1	1		1		2	9			9	9062	13	6					1				2				
9056	13/14	5			8	8	1		1		2	8			9	9105	11	6					1				2				
9082	15	6			8	8	2				2	8			9	9065	13	6					2			2					
9075	9	9			8	1		2			2	8			9	9106	7	2?3						1		1	1	9			9
9061	14	5			8	8			2		2	8			9	9028	4	8						2			2	1			
9040	11	7			1	1				2					9	9048	7	2						1			2	8			8
9045	11/12	7			8	1					1				9	9029	14	5						1			2	8			9
9042	11	7			1	1					2				9	9026	4	8						1			2	8			9
9069	8	4			1	1					2					9032	4	8						1			2	8			9
9068	8	4			1	1					2	9			9	9091	4	7						1			2				
9067	8	4			1	1					2	8			9	9050	7	2						2			2				9
9038	12	7			1	1					2	9			9	9104	11	7						2			2	9	9	9	9
9070	8	7			1	8					2	8			9	9051	7	2						2			2				9
9071	8	4			1	8					2	9				9047	7	2						2			2				9
9043	11	7			1						2	1			9	9025	4	7						2			2				
9037	11	7			1						2	8			8	9031	4	8						2			2				
9039	11	7			1						2	8			9	9107	4	4						1			2				9
9072	8	4			1						2	9			9	9033	4	4						1			2				9
9016	16	7			1						2					9030	4	7						2			2	9			
9021	3	4			1						2					9002	1	5							2	2					
9092	4	8				1		1			2	9			9	9054	14	5							1	2					
9034	4	8				8		1			2	9			9	9005	1	5							1		2	1	8	8	9
9017	15	1					1			1	1				9	9004	1	5							1		2	8			9
9055	6	6					2			2					9	9003	1	5							1		2	8			9
9060	13	5					1			2					9	9006	1	5							1		2	9			9
9013	15	6					2				2	8			9	9057	14	1							1		2	9			
9014	15	6					2				2	1			9																
			01	02	03	04	05	06	07	08	09	10	11	12	13				01	02	03	04	05	06	07	08	09	10	11	12	13

Band designation: 1 = single intensity; 2 = double intensity; 8 = faint; 9 = unreadable.

RFLP Standardization Report for DQ Alpha/EcoRV

A. Ando,[1] T. Takenouchi,[2] Y.K. Paik,[3] E. Dwyer,[4] Y. Mitsuishi,[5] J.A. Trejant,[6] and K. Tsuji[1]

The DQ alpha/EcoRV system identified the two fragments as shown in Figure 1 and Table 1. These two fragments were not polymorphic (Table 1). One additional polymorphic band could be seen in only one of the laboratories. This band (12.46 kb) was not counted as a standard fragment and was probably due to partial digestion.

Participating Laboratories: JAPTSU,[1] US5TER,[2] US5PAK,[3] US8RJW,[4] FRAMYE,[5] ANZBSH[6]

Immunobiology of HLA
Volume I

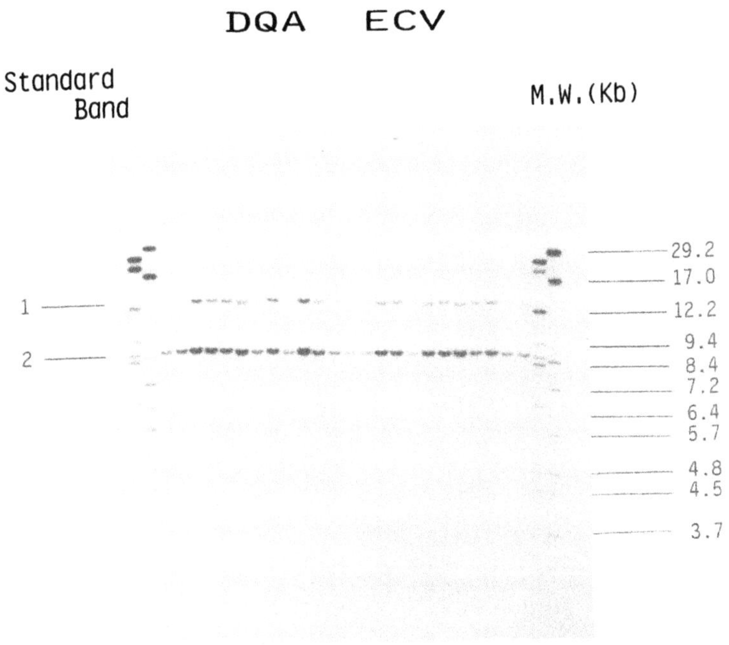

Figure 1. Fragments identified by the DQ alpha/EcoRV system.

Table 1. RFLP Pattern of DQ Alpha/EcoRV

			Fragment No. 1	Fragment No. 2							
		Fragment Size (kb)	12.71	8.61							
		Cell Lines			DW	DR	DW2	DRw	DQw	DPw	GLO
2	9014	MGAR	3	3	2	15			6	4	1,4
11	9055	HO301	3	3	N	6	26	52	6	5	
27	9028	PE117	3	3	14	4		53	8	4	
28	9019	DUCAF	3	3	3	3	25	52	2	-	2
29	9025	DEU	3	3	4	4		53	7	4	
32	9066	TAB089	3	3	8.3	8		52	6	-	2
34	9036	SPO010	3	3	DB2	11	25	52	5	2	
40	9054	EK	3	3	9	14	25	52	5	-	
46	9021	RSH	3	3	N	3	24	52	4	1	1
47	9062	WDV	3	3	18	13	24	52	6	4	
51	9059	SLE005	3	3	19	13	26	52	6	3	
53	9068	BM9	3	3	8.1	8		52	4	2	
54	9075	DXB	3	3	23	9		53	9	4	
56	9067	BTB	3	3	8.1	8		52	4	4	
60	9070	LUY	3	3	8.3	8		52	7	1,4	
62	9011	E4181324	3	3	12	15			6	2,4	
63	9026	YAR	3	3	10	4		53	8	4	1,2
64	9039	JVM	3	3	N	11	25	52	7	2	
65	9056	KOSE	3	3	9,19	13,14		52	5	2,N	
66	9082	HO104	3	3		15			6	4	
70	9052	DBB	3	3	11	7		53	9	4	1
71	9020	QBL	3	3	3	3	25	52	2	2	
72	9007	DEM	3	3	-	16,4		53	5,8	3,4	1,2
5	9017	WT8	2	3	2	15			1	2,4	
6	9042	TISI	2	3	N	11	25	52	7	-	2
7	9013	SCHU	2	3	2	15			6	-	
9	9004	JESTHOM	2	3	1	1			5	4	

Table 1. RFLP Pattern of DQ Alpha/EcoRV

			Fragment No.								
			1	2							
		Fragment Size (kb)	1 2 7 1	8 6 1							
	Cell Lines				DW	DR	DW2	DRw	DQw	DPw	GLO
10	9018	L0081785	2	3	3	3	25	52	2	3	
12	9063	WT47	2	3	19	13	26	52	6	2	
13	9022	COX	2	3	3	3	24	52	2	3	2
14	9043	BM21	2	3	5	11	25	52	7	–	
15	9023	VAVY	2	3	3	3	24	52	2	1	1
16	9006	WT100BIS	2	3	1	1			5	1	
17	9005	HOM2	2	3	1	1			5	–	
18	9050	MOU	2	3	17	7		53	2	2	
19	9056	KOSE	2	3	9,19	13,14	25	52	5	2,N	
22	9047	PLH	2	3	DB1	7		53	2	–	
23	9040	BM15	2	3	5	11	25	52	7	3	
24	9010	AMAI	2	3	2	15			6	–	
26	9037	SWEIG007	2	3	5	11	25	52	7	–	
30	9031	BOLETH	2	3	4	4		53	8	4	
31	9088	PF04015	2	3		3	24	52	2	1,4	
33	9060	CB6B	2	3	18	13	25	52	5	–	
35	9002	MZ070782	2	3	20	1			5	4	1,2
36	9071	OLGA	2	3	8.2	8		52	4	3	
37	9064	AMALA	2	3	16	14	24	52	7	–	
38	9057	TEM	2	3	9	14	25	52	1	4	
39	9030	JHAF	2	3	13	4		53	7	3	
42	9091	MLF	2	3	4	4		53	7	–	
43	9016	RML	2	3	22	16			7	–	
44	9048	LBUF	2	3	DB1	7		53	2	–	
45	9065	HHKB	2	3	18	13	24	52	6	4	
48	9032	BSM	2	3	4	4		53	8	2	
52	9034	SAVC	2	3	4	4		53	8	–	
55	9061	31227ABO	2	3	9	14	25	52	5	4	
59	9003	KAS116	2	3	1	1			5	–	
61	9029	WT51	2	3	4	4		53	8	2	
68	9032	BSM	2	3	4	4		53	8	2	
1	9045	TUBO	1	3	5	11,12	25	52	7	2,4	
3	9058	OMW	1	3	18	13	24	52	6	1	2
4	9107	LKT3	1	3	15	4		53	4	5	
8	9033	BM14	1	3	4	4		53	8	4	
20	9104	DHIF	1	3	5	11		52	7	N,T	
21	9051	PITOUT	1	3	17	7		53	2	4	2
25	9069	MADURA	1	3	8.1	8		52	4	4	
41	9105	FPAF	1	3	FS	11		52	6	2	
50	9092	BM92	1	3	14	4		53	8	–	1
57	9038	BM16	1	3	DB6	12		52	7	2	
58	9008	D0208915	1	3	2	15			6	2,4	
49	9009	KASO11	1	2	21	2			1	3,4	
67	9106	MANIKA	1	2		7		53	2?3	N,T	
69	9072	SPACH	1	2	8.2	8		52	4	4	

RFLP Standardization Report for DQ Alpha/TaqI

R.E. Bontrop,[1] C.B. Carpenter,[2] R. Walford,[2] S. Sekiguchi,[3] J.D. Bignon,[5] and D. Cohen[6]

Data were received from five different laboratories and standardization was performed based on the following criteria: 1) weak hybridization signals were accepted as standard fragments when they were detected by most laboratories, and 2) discrepancies were resolved by majority consensus.

For this system 12 different standard fragments were detected (Table 1 and Fig. 1). Standard fragment 1 is represented by a weak hybridization signal, has a molecular weight of 9.50 kb, and a frequency of 11.4%. This fragment is detected almost exclusively on TaqI digests of HLA-DR5-Dw5-DQw7 positive cells (Fig. 1, lanes 1 and 6). However, standard fragment 1 is always observed in combination with fragment 2, which is also represented by a weak hybridization signal, or in combination with fragment 7, which is in that case represented by a strong hybridization signal (Fig. 1, lanes 1 and 6, and Table 2). It is very well possible that standard fragment 7, which has a molecular weight of 4.56 kb and a panel frequency of 35.6% and represents the presence of the gene that is coding for the DQ alpha chain type 2 which has been reported on by the biochemistry section of the workshop. This DQ alpha chain type is usually expressed by the majority of HLA-DRw17-DQw2, HLA-DRw5-DQw7, HLA-DRw8-DQw7 and possibly by some HLA-DRw4-DQw7 positive cells (1). Standard fragment 2, which has a molecular weight of 8.17 kb and a frequency of 42.9%, is not always observed in combination with fragment 1 (Fig. 1, lane 2). Standard fragment 3 has a molecular weight of 7.60 kb, a frequency of 17.1%, and again this fragment is often detected in combination with standard fragment 7 (Fig. 1, lane 10). In such an example one is often dealing with HLA-DRw17-DQw2 positive

Participating Laboratories: BENROO,[1] NCYCAR,[2] US5WAL,[3] JAPSEK,[4] FRABIG,[5] FRACOH[6]

Table 1. Standard Fragments Detected by the DQ Alpha/TaqI System

Fragment	kb	Frequency	% Faint
1	9.50	0.114	0.000
2	8.17	0.429	0.194
3	7.60	0.171	0.154
4	6.99	0.527	0.243
5	6.19	0.371	0.067
6	5.53	0.171	0.154
7	4.56	0.357	0.000
8	3.76	0.277	0.045
9	2.85	0.512	0.058
10	2.78	0.148	0.000
11	2.45	0.395	0.082
12	2.09	0.605	0.000

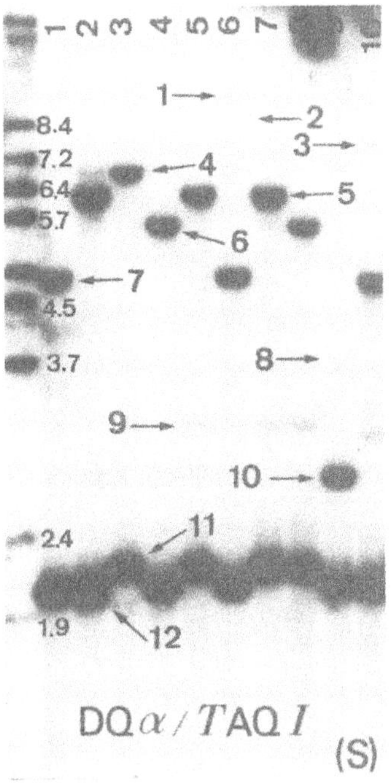

Figure 1. Standard fragments detected by the DQ alpha/TaqI system.

cells. Standard fragment 4 has a molecular weight of 6.99 kb and a frequency of 52.7%. This fragment is normally represented by a weak hybridization signal, however, exceptions to the rule do exist (Fig. 1, lane 4). If this hybridization signal is present in strong intensity then one is almost certainly dealing with HLA-DRw13-Dw24-DQw6 positive cell lines. On the other hand a few HLA-DR2 positive cell lines (for example, AMAI WS ID 9010) do display this fragment. For that reason it is possible that this strong hybridization signal reflects the presence of the DQ alpha 1.12 allele (1) which has been named DQ alpha 1.13 in the biochemistry section of this Workshop.

Fragment 5 has a molecular weight of 6.19 kb and a frequency of 37.1%. When this band is detected as a strong signal then one is almost certainly dealing with HLA-DR2-Dw2-DQw6 positive cells or with HLA-DRw13-Dw19-DQw6 positive cells. This fragment probably represents the DQ alpha 1.2 chain type which is known to be expressed by the latter two haplotypes (1 and Biochemistry Section of this Workshop). Standard fragment 6 has a molecular weight of 5.53 kb and a frequency of 17.1%. This fragment is generally detected in the example of HLA-DR7-DQw2 and some HLA-

Table 2. Fragment Absence/Presence Patterns for the DQ Alpha/TaqI System as Defined on 72 Core Cell Lines

	Fragment No.											
Cell	1	2	3	4	5	6	7	8	9	10	11	12
9045	1	2		8			2					2
9042	2	2		1			2					2
9040	2	1					2					2
9016	1	8		1			1		1			2
9038	2	1		1			2					2
9039	1				1		2	1	8		2	
9037	1	1	1	1			2					2
9014		2		2				8			2	
9006		1		1	8					2		2
9056		1		1	2					2		2
9104		1		1		2	8					2
9025		8		1		2						2
9060		1		2					1		2	
9071		1		1	2							2
9091		8		8		2						2
9048		8		8		2						2
9021		2			2				2			2
9092		2				2						2
9068		2		1	2							2
9075		2		1		2						2
9008		8			2				2		2	2
9003		1							1	2		2
9070		1			2			1	1			2
9056		8			1					1		2
9106		1		8		1		1	8		1	1
9072		8			1			1	1			2
9052		8		8		1						2
9107		8	8			2			1			2
9018		8	2	1			2		1			2
9043		1	2	1			2		2			2
9105		1	1	2								2
9011		8	1	2				8	1			2
9007		8	1	8	1	8		1	1		2	2
9028			1	1		2			2		2	2
9064			1	1			1		8			2
9065			1	2					2		2	

	Fragment No.											
Cell	1	2	3	4	5	6	7	8	9	10	11	12
9062			1	2					8			2
9020			1	8	8	8	2		2			2
9069			1		2							2
9088			1				2		1		2	
9036			1			2			8			2
9058				2					8		2	
9004				1						2		2
9010				2					1		2	
9005				1	8					2		2
9066				1	2							2
9009				1	2							2
9061				1	1					2		2
9067				1	2							2
9082				8	1			1	8		2	
9030				8		1						2
9019				8			1		1			1
9054				8					1	2	2	
9029				8		1		8	8		2	
9026				8		2		1	8		2	
9032				8		2		1	8		2	
9017					2				2		2	
9013					2				2			2
9055					2				2			2
9063					2				2		2	
9057					1				1	1		2
9059					2				2		2	
9033						2			2		2	
9050						2			2			2
9051						2			2			2
9047						2			8			2
9031						2			2		2	
9032						2		8	1		2	
9034						2			1		2	
9022							2		8		2	
9023							2		2		2	
9002							1			2		2

DR4-DQw7 positive cell lines. As a result of this, this fragment possibly corresponds with the presence of the DQ alpha 3.2 type (1). Fragment 7 has been described in detail above. Fragment 8, with a molecular weight of 3.76 kb and a panel frequency of 27.7%, is always present as a weak hybridization signal. However it is often present in combination with standard fragment 10 which is in such examples present in an abundant staining intensity (Fig. 1, lane 9). If this is the case then one is likely to be dealing with HLA-DR1-Dw1-DQw5 or with HLA-DRw14-DW9-DQw5 positive cells. This fragment is consequently representing the DQ alpha 1.1 allele (1). Standard fragment 9 has a molecular weight of 2.85 kb and a frequency of 51.2%. This fragment is always present as a weak signal and usually present in combination with other standard fragments which display a high staining intensity (Fig. 1, lanes 5, 7, 8, and 10). Standard fragment 11 (2.45 kb) and standard fragment 12 (2.09 kb) do represent the DQA2 gene which is known to belong to a diallelic system (2,3).

References

1. Bontrop RE, Baas EJ, Otting N, Schreuder GMTh, Giphart MJ. Molecular diversity of HLA-DQ. DQ alpha and beta chain isoelectric point differences and their relation to serologically defined HLA-DQ allospecificities. Immunogenetics 1987;25:305.
2. Hitman GA, Sachs J, Cassel P, Awad J, Botazzo GF, Toorn AC, Schwartz G, Monson JP, Festenstein H. A DR3 related DX alpha gene polymorphism strongly associates with insulin dependent diabetes melittus. Immunogenetics 1986;23:47.
3. Tilanus MGJ, van Eggermond MCJA, Fei H, D'Amaro J, Schreuder GMTh, Giphart MJ. RFLP of the HLA-DQ alpha region: A diallelic DX alpha polymorphism, not linked to DR and DQ specificities. Tissue Antigens 1987;30:128.

RFLP Standardization Report for DQ Alpha/BglII

A. Marcadet,[1] D. Cohen,[1] M.L. Herrero,[2] H. Takata,[3] M.J. Giphart,[4] P. Rubenstein,[5] and P. Stastny[6]

The DQ alpha/BglII system identified eight fragments on the core cell lines (Table 1). No constant fragment was revealed. Table 2 indicates the distribution of standardized bands in core cell lines.

In comparing DRw-DQw haplotypes of the core cell lines with associations of new provisional DR, DQ specificities listed in "Nomenclature for factors of the HLA system 1987" (Bulletin of the World Health Organization, in press), several "abnormal" DR-DQ associations can be observed. For example, cell no. 9021 was typed DR3-DQw4 instead of expected DR3-DQw2. Most of these cells (but not all) present an "abnormal" pattern (AP) of standardized fragments when comparing with others cells with the same DQw specificity. On Table 2, we have listed separately all cells with "abnormal" association (AA) *and* "abnormal" pattern (n=5). Moreover, in this system, some cells with apparently "normal" DR-DQ associations present "abnormal" patterns of bands, which were not currently revealed for cells sharing the same DQw (n=9).

When considering only cells with "normal" pattern and haplotypes, we notice that *no specific fragment* for DQw specificities can be observed. Moreover, two different patterns of bands can be found for DQw1, three for DQw2, three for DQw3, and only one for DQw4 (but n=3). One of the two DQw1 patterns of fragments is only revealed for DQw1 (specific), but the second pattern of bands, observed for some DQw1, is present for all DQw3-DR4. The two others patterns of bands observed for DQw3 are not found for other specificities (specific). One specific pattern of bands is present for DR7-DQw2, different from one specific pattern for B18-DR3-DQw2, and different from one specific pattern for B8-DR3-DQw2. In addition, one specific pattern of fragments can be correlated with DRw8-DQw4. To summarize this system, we can say that *out of nine* "normal" patterns of standardized bands, *only one is shared* by two different DQw specificities: some DQw1 and some DQw3.

Most of the cells with both "abnormal" pattern and association present the same pattern of bands as other cells with other DR-DQ associations.

Comments on S Migration. The fragment identified as "a," only light on S migration, has not been standardized.

Comments on L Migration. The two fragments identified as "a" for DNA sample B and "b" for DNA sample D, giving a weak signal, have not been standardized.

Participating Laboratories: FRACOH,[1] LATFAI,[2] JAPTAK,[3] BENROO,[4] US8RUB,[5] US4STA[6]

Table 1. Standard Bands in the DQ Alpha/BglII System

		DQA	BGL	COH		
	kb	Locus	Frequency	% Faint	56	32
1	13.12	0 0 0 0	0.200	0.071		
2	8.38	0 0 0 0	0.100	0.000		
3	7.26	0 0 0 0	0.157	0.000		
4	7.01	0 0 0 0	0.729	0.000		
5	6.32	0 0 0 0	0.786	0.000		
6	3.82	0 0 0 0	0.043	0.000		
7	3.78	0 0 0 0	0.271	0.000		
8	1.66	0 0 0 0	0.686	0.000		

Table 2. RFLP DQ Alpha/BglII System. Distribution of Standard Bands in Tenth Workshop Core Cell Lines

Cell	HLA		Standard Bands N =								A	A
Lines	DQw	DR	1	2	3	4	5	6	7	8	A	P
9017	1	w15			1	1	2			2		+
9002	5	1			2		2			2		
9054	5	w14			2		2			2		
9060	5	w13			2		2			2	+	
9065	6	w13			2		2			2		
9055	6	w6			2		2			2		
9009	1	2				2	2			2		
9036	5	w11				2	2			2	+	
9004	5	1				2	2			2		
9003	5	1				2	2			2		
9006	5	1				2	2			2		
9005	5	1				2	2			2		
9061	5	w14				2	2			2		
9057	1	w14				2	2			2		
9056	5	w13/14				2	2			2	+	
9056	5	w13/14				2	2			2	+	
9059	6	w13				2	2			2		
9063	6	w13				2	2			2		
9062	6	w13				2	2			2		
9058	6	w13				2	2			2		
9082	6	w15				2	2			2		
9010	6	w15				2	2			2		
9013	6	w15				2	2			2		
9014	6	w15				2	2			2		
9011	6	w15				2	2			2		
9008	6	w15				2	2			2		
9105	6	w11				2	2			2	+	
9066	6	w8				2	2			2	+	
9007	5/8	w16/4	2			2	2			2		+
9048	2	7				2	2			2		+
9047	2	7		2		2				2		
9050	2	7		2		2				2		
9051	2	7		2		2				2		
9020	2	3^a		2		2			2			
9019	2	3^a		2		2			2			
9018	2	3^a		1		2	2		2			+

Table 2. *Continued*

Cell Lines	HLA DQw	HLA DR	Standard Bands no. 1	2	3	4	5	6	7	8	AA	AP
9088	2	3^b			2		2		2			
9022	2	3^b			2		2		2			
9023	2	3^b			2		2		2			
9106	2?3	7	1		2		2			2		+
9040	7	w11			2		2		2			+
9037	7	w11				2	2	2				
9043	7	w11				2	2	2				
9038	7	w12				2	2		2			
9039	7	w11				2	2		2			
9042	7	w11				2	2		2			
9045	7	w11/12				2	2		2			
9025	7	4				2	2			2		
9091	7	4				2	2			2		
9030	7	4				2	2			2		
9104	7	w11	8			2	2			2		+
9028	8	4	2			2	2			2		+
9026	8	4				2	2			2		
9029	8	4				2	2			2		
9031	8	4				2	2			2		
9032	8	4				2	2			2		
9032	8	4				2	2			2		

Table 2. *Continued*

Cell Lines	HLA DQw	HLA DR	Standard Bands no. 1	2	3	4	5	6	7	8	AA	AP
9033	8	4				2	2			2		
9034	8	4				2	2			2		
9092	8	4				2	2			2		
9075	9	w9				2	2			2		
9052	9	7	2							2		
9071	4	w8	2			2	2		2			+
9069	4	w8	2						2			
9067	4	w8	2						2			
9068	4	w8	2						2			
9072	4	w8	1						2			
9070	7	w8	2						2		+	+
9064	7	w14	2						2		+	+
9016	7	w16	2					2			+	+
9107	4	4	2							2	+	+
9021	4	3		2		2			2		+	+

AA = Abnormal DR-DQ association. AP = Abnormal pattern of bands.
[a] = DR3-B18
[b] = DR3-B8.

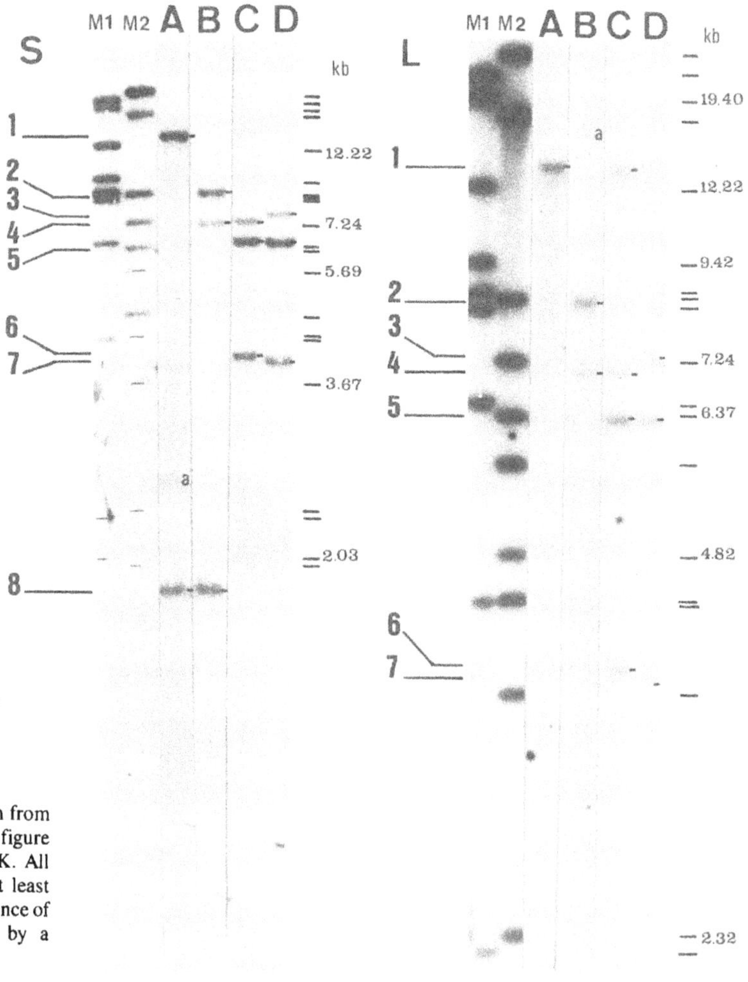

Figure 1. The figure for S (left) is taken from a film from laboratory US8RUB and the figure for L (right) is from laboratory JAPTAK. All standardized fragments are indicated at least once for S and/or L migration. The presence of a standardized fragment is indicated by a hyphen.

RFLP Standardization Report for DQ Alpha/PvuII

P. Piccioli,[1]* K. Kato,[2] A. Smith,[3] J. Partanen,[4] M. Vijod,[1] and G.B. Ferrara[1]

The DQ alpha/PvuII system identified five fragments on the core cell lines (Table 1 and in Fig. 1). Out of these five fragments, one (1.5 kb) was shared by all the cell lines, while the others identified four different RFLP patterns. Due to the high homology between HLA-DQA1 (previously named DQ alpha) and HLA-DQA2 (previously DX alpha) genes (1–2) two fragments are probably assignable to the HLA-DQA2 locus (8.19, 7.22 kb), while the other three belong to the HLA-DQA1 locus (4.92, 2.81 kb, 1.5 kb). From pattern distribution analysis it is evident that the DQw1, DQw5, and DQw6 specificities, identified by patterns II and III, are distinct from all the other DQw specificities. The HLA-DQA1 fragments of these two patterns are identical, while HLA-DQA2 associated fragments 1 and 2 are, respectively, shown in 64% (9/14) DQw6 cell lines and in 76% (8/10) DQw5 homozygous cell lines. A single cell line (ID 9008) shows both first and second fragments, probably due to heterozygosity.

Concerning the HLA-DQA1 fragments, DQW2, DQw7, DQw9, and DQw4 specificities can not be identified by specific patterns, whereas HLA-DQA2 associated fragments allowed us to find out specific correlations. For instance, within DQw2 specificity, six DR3 lines have been tested, and two specific fragments have been detected that distinguish the extended haplotype DR3, B18,F1C30 (pattern I) from DR3,B8,SC01 (pattern IV).

Participating Laboratories: IT2FER,[1] US7DUP,[2] US7HAN,[3] SCAKOS[4]
*P. Piccioli was recipient of a fellowship from A.I.R.C. (Associazione Italiana per la Ricerca sul Cancro)
This work was in part supported by CNR grants n.86.0265.51 and 86.00659.

Table 1. Fragments on the Core Cell Line Identified by the DQ Alpha/PvuII System

Fragment No.	kb	Frequency[a]	% Faint[b]
1	8.19	0.371	0.000
2	7.22	0.643	0.000
3	4.92	0.386	0.000
4	2.81	0.600	0.000
5	1.50	1.000	1.000

[a]Band frequency has been calculated as number of cell lines presenting a specific fragment versus number of cell lines tested.
[b]Frequency of faint band has been calculated as number of cell lines presenting a faint band versus number of cell lines presenting the same fragment (i.e., frequency 1.000 and % faint 0.000 means that all the samples tested share the same band and it is a strong band; on the other hand, 1.000 and 1.000 means that it has a constant and faint band).

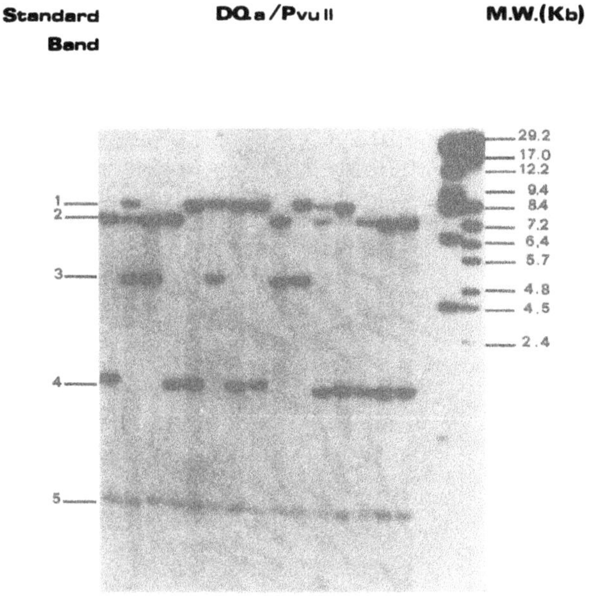

Figure 1. PvuII digestion of DNAs from core cell lines panel hybridized with the HLA-DQA probe. From left to right: BM16, DO208915, KAS116, LUY, WT51, E4181324, YAR, JVM, KOSE, HO104, MANIKA, BSM, SPACH, DBB, QBL, and molecular weight controls (M1 and M2).

Fragment 4 was originally split into two fragments (4a and 4b). Comparing those fragments (4a-4b) from different autoradiographies and laboratories, some discrepancies (due to small difference in M.W.) have been found, so that they were put together as a single fragment.

Similar results have been found by using DQw2-associated specificity, in our hands, all the DR7, DQw2 lines tested showed a common pattern (I). Within DQw3 and associated specificities, the first fragment is present in all DQw8 cell lines, except for sample ID 9092, and in WS ID 9039 (DQw7). A high correlation was also found for DRw8 specificities and fourth fragment (except for sample ID 9066). Patterns and relative DQ associations are summarized in Table 2. Distribution of five HLA-DQA1,HLA-DQA2 associated fragments among 70 cell lines studied is summarized in Table 3. These data are concordant with previous data reported in literature (2).

References

1. Auffray C, Lillie JW, Korman AJ, Boss JM, Frechin N, Guillemot F, Cooper J, Mulligan RC, Strominger JL. Structure and expression of HLA-DQ alpha and -DX alpha genes:

Table 2. RFLP Patterns and Relative DQ Associations Detected by the DQ Alpha/PvuII System

	RFLP Pattern[a]					DQw Specificity[b]									
Fragment	1	2	3	4	5	1	5	6	2	3	7	8	9	4	%[c]
I	−	+	−	+	+				66%		93%	12%	100%	100%	43 %
II	−	+	+	−	+	33%	76%	29%							18.5%
III	+	−	+	−	+	67%	19%	64%							18.5%
IV	+	−	−	+	+				29%		7%	82%			15.8%
V	+	+	+	−	+			7%							1.4%
VI	+	+	+	+	+		4%					6%			1.4%
VII	+	+	−	+	+				5%	100%					1.4%

[a] +/− means fragment presence/absence.
[b] Out of 70 cell lines tested, 3 carry DQw1 specificity, 10.5 DQw5, 14 DQw6, 10.5 DQw2, 0.5 DQw3, 14 DQw7, 8.5 DQw8, 2 DQw9 and 7 DQw4. Heterozygous have been considered as 0.5, homozygous as 1).
[c] % has been calculated as number of total cell lines which share RFLP pattern (i.e. 43% means that 30 out of 70 present the 1st pattern − + − + +).

Table 3. Banding Pattern of the DQ Alpha/PvuII System in Each Cell Line

			Fragment No.				
Cell	DQ	DR	1	2	3	4	5
9009 KAS-011	1	2		2	2		8
9017 WT8	1	15	2		2		8
9057 TEM	1	14	2		2		8
9002 MZ 070782	5	1		2	2		8
9003 KAS-116	5	1		2	2		8
9004 JES THOM	5	1		2	2		8
9006 WT100 BIS	5	1		2	2		8
9005 HOM-2	5	1		2	2		8
9036 SPO 010	5	11		2	2		8
9060 CB6B	5	13	2		2		8
9054 EK	5	14	2		2		8
9061 31227 AOB	5	14		2	2		8
9056 KOSE	5	13/14		2	2		8
9056 KOSE	5	13/14		2	2		8
9007 DEM	5/8	4/16	1	1	1	1	8
9058 OMW	6	13	2		2		8
9059 SLE-005	6	13	2		2		8
9062 WDV	6	13		2	2		8
9063 WT47	6	13	2		2		8
9065 HHKB	6	13	2		2		8
9008 DO 208915	6	15	1	1	2		8
9010 AMAI	6	15	2		2		8
9011 E 4181324	6	15	2		2		8
9013 SCHU	6	15	2		2		8
9014 MGAR	6	15		2	2		8
9082 HO 104	6	15	2		2		8
9105 FPAF	6	11		2	2		8
9055 HO 301	6	6	2		2		8
9066 TAB 089	6	8		2	2		8
9018 LO 081785	2	3		2		2	8
9019 DUCAF	2	3		2		2	8
9020 QBL	2	3		2		2	8
9022 COX	2	3	2			2	8
9023 VAVY	2	3	2			2	8
9088 PF 04015	2	3	2			1	8
9047 PLH	2	7		2		2	8
9048 LBUF	2	7		2		2	8
9050 MOU	2	7		2		2	8
9051 PITOUT	2	7		2		2	8

Table 3. *Continued*

Cell	DQ	DR	Fragment No. 1	2	3	4	5
9106 MANIKA	2?3	7	1	1		1	8
9016 RML	7	16		2		2	8
9025 DEU	7	4		2		2	8
9030 JHAF	7	4		2		2	8
9091 MLF	7	4		2		2	8
9037 SWEIG 007	7	11		2		2	8
9039 JVM	7	11	2			2	8
9040 BM 15	7	11		2		1	8
9042 TISI	7	11		2		2	8
9043 BM21	7	11		2		2	8
9104 DHIF	7	11		2		2	8
9045 TUBO	7	11/12		2		2	8
9038 BM 16	7	12		2		2	8
9064 AMALA	7	14		2		2	8
9070 LUY	7	8		2		2	8
9026 YAR	8	4	2			2	8
9028 PE 117	8	4	2			2	8
9029 WT 51	8	4	2			2	8
9031 BOLETH	8	4	2			2	8
9032 BSM	8	4	2			2	8
9032 BSM	8	4	2			2	8
9033 BM 14	8	4	2			2	8
9034 SAVC	8	4	2			2	8
9092 BM 92	8	4		2		2	8
9052 DBB	9	7		2		2	8
9075 DKB	9	9		2		2	8
9021 RSH	4	3		2		2	8
9107 LKT 3	4	4		2		2	8
9067 BTB	4	8		2		2	8
9068 BM 9	4	8		2		2	8
9069 MADURA	4	8		2		2	8
9071 OLGA	4	8		2		2	8
9072 SPACH	4	8		2		2	8

1 = Single intensity; 2 = double intensity; 8 = faint.

interallelic alternate splicing of the HLA-DQ alpha gene and functional splicing of the HLA-DXalpha gene using a retroviral vector. Immunogenetics 1987;26:63.

2. Le Gall I, Marcadet A, Font MP, Auffray C, Strominger J, Lalouel JM, Dausset J, Cohen D. Exuberant restriction fragment length polymorphism associated with the DQalpha-chain gene and the DXalpha-chain gene. Proc Natl Acad Sci USA 1985;82:5433.
3. Wescott MZ, Awdeh ZL, Yunis EJ, Alper CA. Molecular analysis distinguishes two HLA-DR3-bearing major histocompatibility complex extended haplotypes. Immunogenetics 1987;26:370.

RFLP Standardization Report for DQ Alpha/EcoRI

M. Segall,[1] L. Schluender,[1] H. Betuel,[2] B. Bradley,[3] M. Garovoy,[4] and S. Naito[5]

The DQ alpha/EcoRI system identified six fragments on the core cell lines as shown in Table 1 and Figure 1. All fragments were clearly identified, and there were no discrepancies for the duplicate DNAs 9056 and 9032.

Table 2 shows the presence/absence pattern for the EcoRI DQ alpha fragments; the cells are arranged in order by DR, Dw (where available) and DQ specificity. It can be seen that fragments 1, 2, and 3 form an allelic system. Fragment 1 is positive with all DQw1⁺ cells; fragment 3 with all DRw53⁺ cells; and fragment 2 with the remainder. Fragments 5 and 6 are almost identical in pattern to the TaqI 155 and TaqI 209 fragments which identify a DQA2 polymorphism.

Table 1. Standard Fragments, DQ Alpha/EcoRI

Fragment	kb	Frequency	% Faint	Discrepant for 9056	Discrepant for 9032
1	13.39	0.400	0.000		
2	11.07	0.314	0.000		
3	6.34	0.300	0.000		
4	5.26	1.000	0.000		
5	1.95	0.657	0.021		
6	1.27	0.386	0.036		

Participating Laboratories: US2BAC,[1] FRABET,[2] UKIBRS,[3] US3GVY,[4] JAPNAI[5]

This is publication #487 from the Immunobiology Research Center, University of Minnesota, Minneapolis, Minnesota 55455, USA.

Supported in part by Juvenile Diabetes Foundation Grant #186175 and March of Dimes Birth Defects Foundation Grant #6-496.

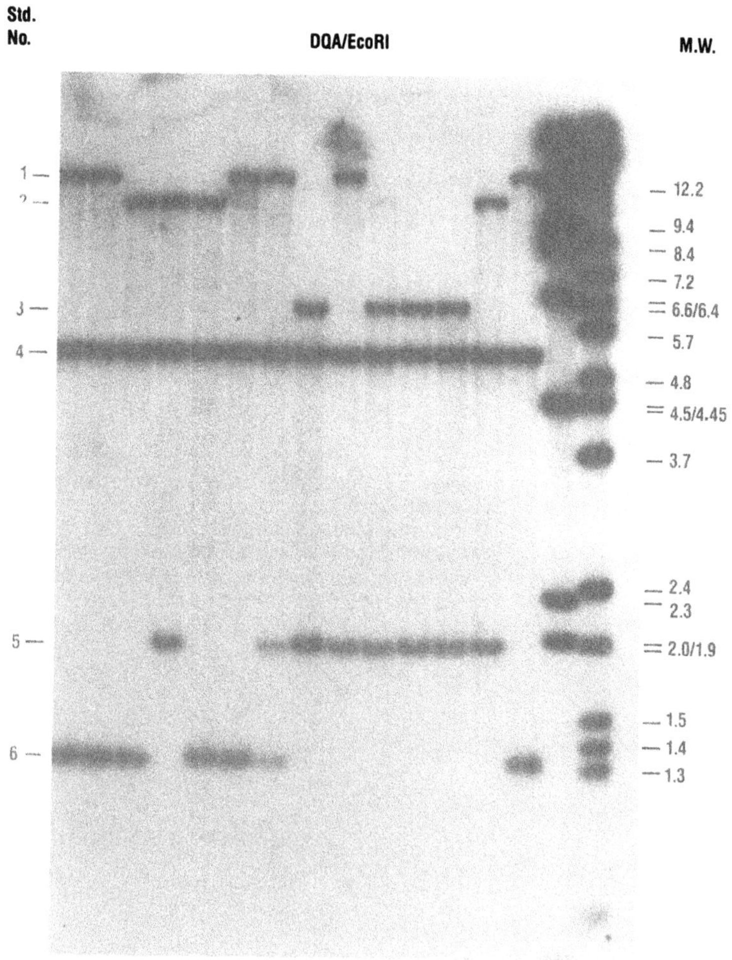

Figure 1. Fragments on the core cell line identified by the DQ alpha/EcoRI system.

Immunobiology of HLA
Volume I

Table 2. Distribution of Polymorphic Fragments in Core Cell Lines, DQ Alpha/EcoRI

		Std. Fragment No.							Std. Fragment No.				
Cell No.	Name	1	2	3	5	6	Cell No.	Name	1	2	3	5	6
9002	MZO70782	1			1		9040	BM15		1		1	
9003	KAS116	1			1		9043	BM21		1		1	
9004	JESTHOM	2				1	9104	DHIF			2	1	
9005	HOM2	1			8	1	9036	SP0010	1			2	
9006	WT100BIS	1				1	9039	JVM		1			1
9008	D0208915	1			1		9042	TISI		1		1	
9010	AMAL	1				1	9105	FPAF	1			1	
9013	SCHU	1				1	9045	TUBO		1		1	
9014	MGAR	2			1		9038	BM16		1		1	
9017	WT8	1				1	9052	DBB			2	1	
9011	E4i81324	1				1	9050	MOU			1	1	
9082	HO104	1				1	9051	PITOUT			2	1	
9009	KAS011	1			1		9047	PLH			2	1	
9016	RML		2		1		9048	LBUF			2	1	
9007	DEM	1		2	1	1	9106	MANIKA			1	1	
9018	L0081785		1		1		9058	OMW	2				1
9019	DUCAF		2		1		9062	WDV	1			1	
9020	QBL		1		1		9065	HHKB	1				1
9022	COX		1			1	9060	CB6B	1				1
9023	VAVY		1			1	9059	SLE005	1				1
9088	PFO4015		2			1	9063	WT47	1				1
9021	RSH		2		1		9054	EK	1				1
9025	DEU			2	1		9057	TEM	1				1
9091	MLF			2	1		9061	31227ABO	1			1	
9029	WT51			1	1		9064	AMALA		2		1	
9031	BOLETH			2		1	9056	KOSE	1			1	
9032	BSM			2		1	9056	KOSE	1			1	
9032	BSM			1		1	9055	HO301	2				1
9033	BM14			1		1	9067	BTB		1		1	
9034	SAVC			1		1	9068	BM9		1		1	
9026	YAR			1		1	9069	MADURA		2		1	
9030	JHAF			2	1		9071	OLGA		2		1	
9028	PE117			2	1	8	9072	SPACH		1		1	
9092	BM92			1	1		9066	TAB089	1			1	
9107	LKT3			2	1		9070	LUY		1		1	
9037	SWEIG007		2		1		9075	DKB			1	1	

RFLP Standardization Report for DQ Alpha/HindIII

P.A. Biro,[1] S. Cutbush,[1] J.M. Tiercy,[2] G. Paulsen,[3] J. Demopulos,[4] R.T. Acton,[4] S.V. Van Tonder,[5] G. Semana,[6] and F. Quillivic[6]

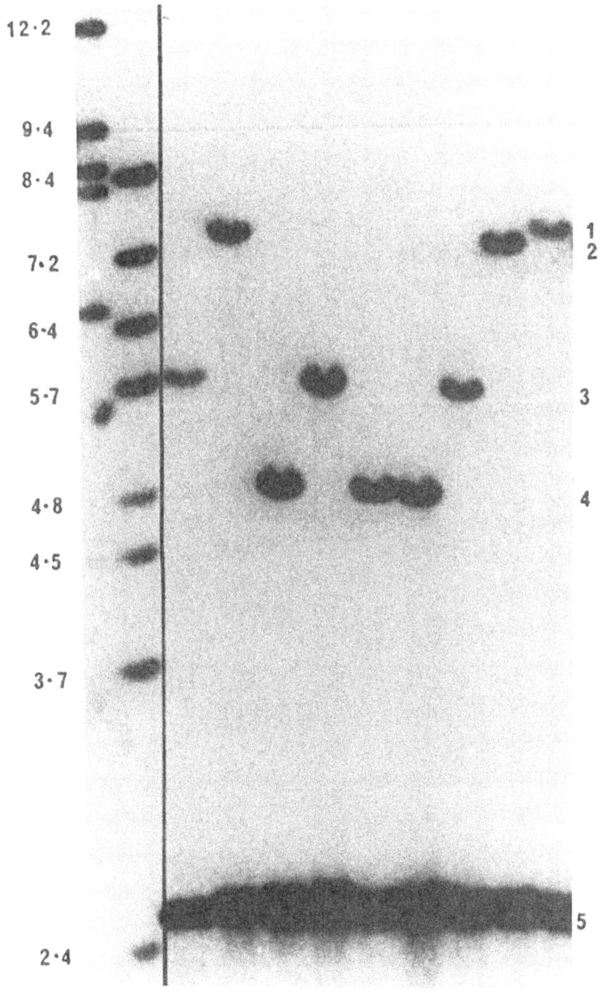

The DQ alpha probe detects five bands. Bands 1–4 are allelic and correlate broadly, but not exclusively, with the DR types of the cells (Table 2). Band 5 is the DQA2 gene (1,2).

Table 1. Sizes and Frequencies of Standard Bands Produced by the HindIII/DQα Enzyme/Probe Combination

	kb	Frequency	% Faint
1	7.68	0.071	0.000
2	7.45	0.343	0.000
3	5.79	0.329	0.043
4	4.83	0.300	0.000
5	2.54	1.000	0.000

References

1. Gardiola J, Maffei A, Carrel S, Accolla RS. Molecular genotyping of the HLA-DQα gene region. *Immunogenetics*, 1988;27:12.
2. Le Gall, Marcadet A, Font M-P, Auffray C, Strominger J, Lalouel J-M, Dausset J, Cohen D. Exuberant restriction fragment length polymorphism associated with DQα-chain gene and the DXα-chain gene. *Proc Natl Acad Sci USA*, 1985;82:5433.

Participating Laboratories: UKIFES,[1] FRAJEA,[2] SCATSB,[3] US5UAB,[4] SAFHAM,[5] FRAFAU[6]

◄ **Figure 1.** Standard bands produced by the HindIII/DQ alpha enzyme-probe combination.

Table 2. RFLP Pattern of DQ Alpha/HindIII System

Cell Lines		DR Type	Fragment No. 12345
MZ070782	9002	1	1 2
KAS116	9003	1	1 2
JESTHOM	9004	1	1 2
HOM2	9005	1	1 2
WT100BIS	9006	1	1 2
DEM	9007	2/4	1 12
DO208915	9008	2	1 2
KAS011	9009	2	1 2
AMAI	9010	2	1 2
E4181324	9011	2	1 2
SCHU	9013	2	1 2
MGAR	9014	2	1 2
RML	9016	2	1 2
WTB	9017	2	1 2
LO081785	9018	3	1 2
DUCAF	9019	3	1 2
QBL	9020	3	1 2
RSH	9021	3	1 2
COX	9022	3	1 2
VAVY	9023	3	1 2
DEU	9025	4	12
YAR	9026	4	12
PE117	9028	4	12
WT51	9029	4	12
JHAF	9030	4	12
BOLETH	9031	4	12
BSM	9032	4	12
BM14	9033	4	12
SAVC	9034	4	12
SP0010	9036	5	1 2
SWEI6007	9037	5	11 2
BM16	9038	5	1 2
JVM	9039	5	1 2
BM15	9040	5	1 2
TISI	9042	5	1 2
BM21	9043	5	1 2
TUBO	9045	5	1 2
PLH	9047	7	12
LBUF	9048	7	12
MOU	9050	7	12
PITOUT	9051	7	12
DBB	9052	7	12
EK	9054	6	1 2
HO301	9055	6	1 2
KOSE	9056	6	1 2
TEM	9057	6	1 2
OMW	9058	6	1 2
SLE005	9059	6	1 2
CB6B	9060	6	1 2
31227ABO	9061	6	1 2
WDV	9062	6	1 2
WT47	9063	6	1 2
AMALA	9064	6	1 2
HHKB	9065	6	1 2
TAB089	9066	8	1 2
BTB	9067	8	1 2
BM9	9068	8	1 2
MADURA	9069	8	1 2
LUY	9070	8	1 2
OLGA	9071	8	1 2
SPACH	9072	8	1 2
DKB	9075	9	12
HO104	9082	?	1 2
PF04015	9088	3	1 2
MLF	9091	4	12
BM92	9092	4	12
DHIF	9104	11	12
FPAF	9105	11	1 2
MANIKA	9106	7	12
LKT3	9107	4	12

RFLP Standardization Report for DQ Alpha/MspI

S. Tonks,[1] V. Groves,[1] C. Muller,[2] F. Otani,[3] J.R. Regueiro,[4] M. Segall,[5] J. Bodmer,[1] and J. Trowsdale[1]

The core cell lines revealed 11 standard fragments in the DQ alpha/MspI system as shown in Figure 1. A fragment between standard bands 4 and 5 (at approximately 5.19 kb) has been reported by some laboratories, however, in these data the band was faint and inconsistent. This fragment may require consideration in some data.

Participating Laboratories: UKIBOD,[1] GERMUC,[2] JAPKSH,[3] FRAARN,[4] US2BAC[5]

The patterns of reaction can be seen in Table 1 (duplicates not shown). Only one fragment showed significant correlation with a specific DQ type; band 8 (2.87 kb) corresponded with DQw1 (χ^2=47.761). However, good correlation was found with a number of DR types, for example, fragment 1 (6.05kb) was associated with DR7 (χ^2=58.498) and fragment 11 (2.15kb) with DR4 (χ^2=57.67). Several fragments were well correlated with DRw8, fragments 7 (3.25 kb), 9 (2.5 kb), 10 (2.31 kb) specifically and fragment 6 (4.73 kb) in association with DR3.

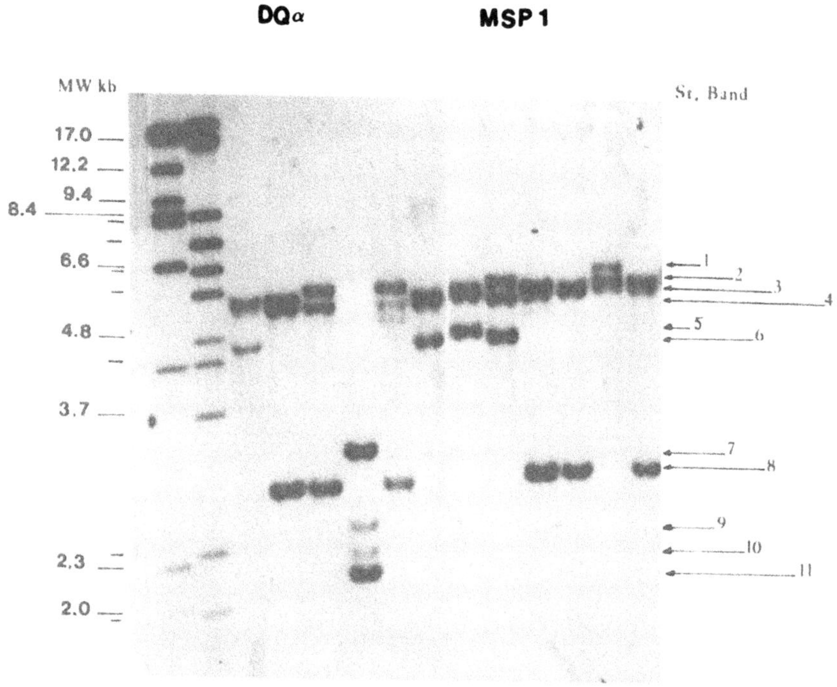

Figure 1. Standard fragments on the core cell lines identified by the DQ alpha/MspI system.

Table 1. Banding Pattern of the DQ Alpha/MspI System in Each Cell Line

		Fragment No.										
WS ID	Cell	1	2	3	4	5	6	7	8	9	10	11
9050	MOU	2		2	1							
9051	PITOUT	2		1	2							
9048	LBUF	1		2	1							
9047	PLH	2		2	1							
9020	QBL			1	2		2					
9045	TUBO			1	2		2					

Immunobiology of HLA
Volume I

Table 1. *Continued*

WS ID	Cell	Fragment No. 1	2	3	4	5	6	7	8	9	10	11
9040	BM15			2	2		2					
9018	LO081785			1	2		2					
9019	DUCAF			2	1		2					
9037	SWEIGOO7				1	1		2			1	
9067	BTB						2	2		1	2	
9070	LUY						2	2		1	1	
9072	SPACH						2	2		1	1	
9068	BM9						2	2		1	2	
9069	MADURA						2	2		1	1	
9016	RML							2		1	1	
9052	DBB	2						2		1	1	
9106	MANIKA	2			2			2		1	1	
9104	DHIF	2		2	1	8		1		8	8	
9107	LKT3							2		1	1	2
9007	DEM				2			2	2	1	1	2
9026	YAR				2							2
9029	WT51				2							2
9025	DEU				1							2
9034	SAVC		1		2							2
9031	BOLETH		1		2							2
9033	BM14		1		2							2
9032	BSM		1		2							2
9075	DKB			1	2							2
9092	BM92			2	1							2
9091	MLF			2	1	1						2
9038	BM16			2	2	1						
9042	TISI			1	2	1						
9030	JHAF			2	1							
9060	CB6B				1				2			
9058	OMW		1		2				2			
9013	SCHU		1		2				2			
9055	H0301		1		2				2			
9063	WT47		1		2				2			
9065	HHKB		1		1				2			
9059	SLE005		1		2				2			
9017	WT8		1		2				2			
9082	H0104		1		2				2			
9088	PFO4015		1		2		2		2			
9023	VAVY		1		2		2					
9010	AMAL		8		2							
9039	JVM		1		2	1						
9014	MGAR			1	2				2			
9004	JESTHOM			1	2				2			
9006	WT100BIS			2	1				2			
9005	HOM2			1	2				2			
9056	KOSE			2	1				2			
9066	TAB089			2	1				2			
9036	SPOO10			2	1				2			
9002	MZ070782			1	1				2			
9105	FPAF			2	1				2			
9062	WDV			1	1				2			
9009	KAS011			1	2				2			
9061	31227ABO			1	2				2			
9011	E4181324			1	2				2			
9003	KAS116			1	2				2			
9057	TEM				2				2			
9054	EK				2				2			2
9008	D0208915				2				2			
9021	RSH				2		2					
9064	AMALA				8		1	2				
9071	OLGA			1	1		2	2			1	1
9022	COX				2		2					

1 = Single intensity; 2 = double intensity; 8 = faint.

RFLP Standardization Report for DQ Alpha/BamHI

E. Keller,[1] A. Andreas-Zietz,[1] E.D. Albert,[1] B.O. Boehm,[2] J.J. Hyldig-Nielsen,[3] A. Bushell,[4] and D.P. Singal[5]

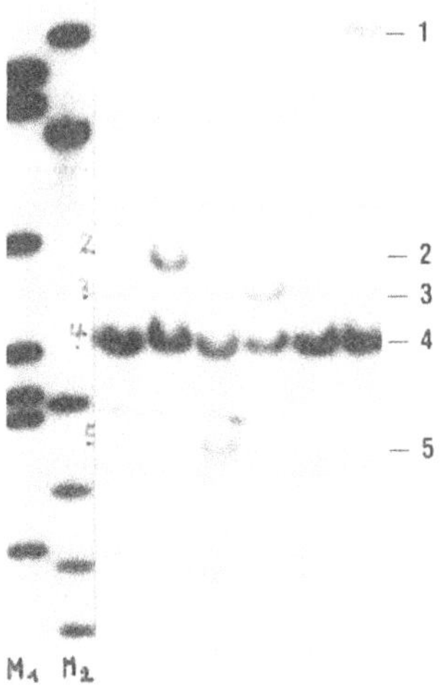

Figure 1. DQ alpha/BamHI core fragments (gel 0.6%).

The DQ alpha/BamHI system identified five fragments on the core cell lines as shown in Table 1 and Figure 1.

Table 1 shows the presence/absence pattern of the standardized fragments in all 72 cell lines, including the molecular weight and probable associations with DQ and DR antigens.

The intensity of the main band (fragment no. 4) is greater than double dose, which is confirmed by family analysis. For the other bands it is not possible to decide about the dosage (i.e. homo- or heterozygous state) without family data for each cell.

Participating Laboratories: GERALB,[1] GERBOE,[2] SCASVE,[3] UKIMOR,[4] US5SIN[5]
This study was supported by SFB 217.

Table 1. Banding Pattern of All 72 Core Cell Lines, as Standardized in the DQ Alpha/BamHI System, Including kb Values and Probable DR/DQ Associations

```
Probe Name: DQA              Enzyme Name: BAMHI

    | 9999999999999 9999999999999999 99999999999 99999999999999 999999999 99 9999999| WS-ID
    | 0000000000000 0000000000000001 00000000001 00000000001000 000000000 00 1000000| number
    | 0000056565530 0111108565566160 12218255440 34443436170239 323932233 75 0676672|
    | 3465241706667 9743082539852165 82398001786 72309584604501 381249622 52 7918721|
    | ------------- ---------------- ----------- -------------- --------- -- -------|---------
    | 5011345331637 4000256115044634 11123712246 20126053462234 023556646 57 0235564| position
    | 9967505839542 9527486121357221 03581181247 66434177300992 870021388 40 4563696| on blots
    | ------------- ---------------- ----------- -------------- --------- -- -------|---------
 DR |      1111111   111111 111111 1             111111111 1                       |
    | 1111144433314 2555555633333581 33333377777 11111124681444 444444444 97 4888883|
    | ------------- ---------------- ----------- -------------- --------- -- -------|
 DR |          11 1                                   1                            |
    |          44 6                                   2                            |
    | ------------- ---------------- ----------- -------------- --------- -- -------|
 DQ | 5555555155555 1166666666666666 22222222222 77777777777777 888888888 99 44444-4|
    | ------------- ---------------- ----------- -------------- --------- -- -------|
 DQ |             8                            3                                   |           associated with:
    | ------------- ---------------- ----------- -------------- --------- -- -------|---------  --------------
   1|          11          111                                                      | 23.93 kb  part DQw 1
   2|            11 1111111                      1                                  | 10.87     part DQw 1
   3| 11111111111            111111                                               | 10.09     part DQw 1
   4| 3333333333333 3333333333333333 33333333333 33333333333333 333333333 33 3333333|  9.18
   5|             1                        11111           1111 111111111 11 1      |  7.63     DR 4 + 7 + 9
```

Cells are listed according to DQ and DR antigens and banding patterns. Numbers indicate intensity (1,2,3 = increasing intensity; 8 = faint; 9 = unreadable).

RFLP Standardization Report for DP Alpha/SstI

T. Takenouchi,[1] M. Kanatani,[1] P. Terasaki,[1] and K. Tsuji[2]

For this enzyme-probe system data from only one laboratory was obtained. Therefore, standardization could not be performed. We reinspected the films, reconfirmed the fragment assignments, and then made these fragments the standard data instead of standardization.

Six fragments were assigned standard fragments shown in Table 1 and Figure 1.

Participating Laboratories: US5TER,[1] JAPTSU[2]

Table 1. Standard Fragments Identified by the DP Alpha/SstI System

		DPA	SST	TER		
	kb	Locus	Frequency	% Faint	56	32
1	18.82	0 0 0 0	0.265	0.105		
2	16.16	0 0 0 0	0.853	0.000		
3	10.42	0 0 0 0	0.735	0.000		
4	6.27	0 0 0 0	0.162	1.000		
5	4.45	0 0 0 0	0.824	0.000		
6	0.99	0 0 0 0	1.000	0.000		

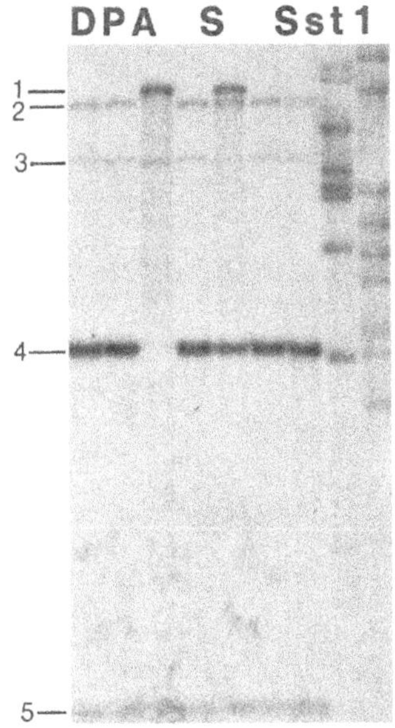

Figure 1. Fragments identified by the DP alpha/SstI system.

RFLP Standardization Report for DP Alpha/PstI

J.J. Hyldig-Nielsen,[1] M. Trucco,[2] H. Coppin,[3] A. Andreas-Zietz,[4] E. Keller,[4] C. Carcassi,[5] A. Takada,[6] and A. Svejgaard[1]

The DP alpha/PstI system identified two polymorphic and four nonpolymorphic fragments (Table 1 and Fig. 1). The two polymorphic fragments divide the DP alleles in two groups. Standard fragment no. 2 correlates with the DPw1, w5, and CDP-HEI alleles, and standard fragment no. 3 correlates with the DPw2, w3, w4, and w6 alleles (1). Table 2 indicates the banding patterns obtained with this system.

Participating Laboratories: SCASVE,[1] US1DUQ,[2] FRAOHA,[3] GERALB,[4] IT1CON,[5] JAPAIZ[6]

Reference

1. Hyldig-Nielsen JJ, Morling N, Odum N, Ryder LP, Platz P, Jakobsen BK, Svejgaard A. DNA Polymorphism of HLA-DP: further studies of the Restriction Fragment Length Polymorphism of DPα and DPβ. In: Dupont B (ed). Immunobiology of HLA: Histocompatibility Testing 1987. New York: Springer-Verlag, 1989.

Table 1. Fragments Identified by the DP Alpha/PstI System

	kb	DPA Locus	PST Frequency	SVE % Faint	56	32
1	3.13	0 0 0 0	1.000	1.000		
2	2.63	0 0 0 0	0.265	0.000		
3	2.43	0 0 0 0	0.824	0.034		
4	2.36	0 0 0 0	1.000	0.000		
5	1.59	0 0 0 0	1.000	0.014		
6	0.51	0 0 0 0	1.000	1.000		

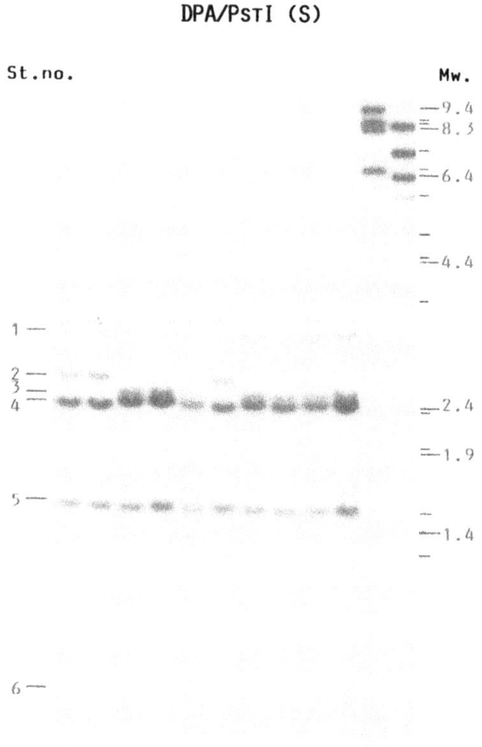

Figure 1. Fragments identified by the DP alpha/PstI system.

Table 2. Banding Patterns Obtained with the DP Alpha/PstI System

WS ID	Name	HLA-DR	HLA-DQ	HLA-DP	Standard Fragment No. 2	3	1	4	5	6
9006	WT100BIS	1	5	1	6		6	6	6	6
9021	RSH	3	4	1	6	6	6	6	6	6
9023	VAVY	3	2	1	6		6	6	6	6
9058	OMW	13	6	1	6		6	6	6	6
9070	LUY	8	7	1,4	6	6	6	6	6	6
9088	PF04015	3	2	1,4	6	4	6	6	6	6
9020	QBL	3	2	2		6	6	6	6	6
9029	WT51	4	8	2		6	6	6	6	6
9032	BSM	4	8	2		6	6	6	6	6
9032	BSM	4	8	2		6	6	6	6	6
9036	SPO010	11	5	2		6	6	6	6	6
9038	BM16	12	7	2		6	6	6	6	6
9039	JVM	11	7	2		6	6	6	6	6
9050	MOU	7	2	2		6	6	6	6	6
9056	KOSE	13,14	5	2	6	6	6	6	6	6
9056	KOSE	13,14	5	2	6	6	6	6	6	6
9063	WT47	13	6	2		6	6	6	6	6
9068	BM9	8	4	2		6	6	6	6	6
9105	FPAF	11	6	2			6	6	6	6
9008	DO208915	15	6	2,4		6	6	6	6	6
9011	E4181324	15	6	2,4		6	6	6	6	6
9017	WT8	15	1	2,4		6	6	6	6	6
9045	TUBO	11,12	7	2,4		6	6	6	6	6
9018	LO081785	3	2	3		6	6	6	6	6
9022	COX	3	2	3		6	6	6	6	6
9030	JHAF	4	7	3		6	6	6	6	6
9040	BM15	11	7	3		6	6	6	6	6
9059	SLE005	13	6	3			6	6	6	6
9071	OLGA	8	4	3		6	6	6	6	6

Table 2. *Continued*

WS ID	Name	HLA-DR	HLA-DQ	HLA-DP	Standard Fragment No. 2	3	1	4	5	6
9007	DEM	16,4	5,8	3,4		6	6	6	6	6
9009	KAS011	2	1	3,4	6	4	6	6	6	6
9002	MZ070782	1	5	4		6	6	6	6	6
9004	JESTHOM	1	5	4		6	6	6	6	6
9014	MGAR	15	6	4		6	6	6	6	6
9025	DEU	4	7	4		6	6	6	6	6
9026	YAR	4	8	4		6	6	6	6	6
9028	PE117	4	8	4		6	6	6	6	6
9031	BOLETH	4	8	4		6	6	6	6	6
9033	BM14	4	8	4		6	6	6	6	6
9051	PITOUT	7	4	4		6	6	6	6	6
9052	DBB	7	9	4		6	6	6	6	6
9057	TEM	14	1	4		6	6	6	6	6
9061	31227ABO	14	5	4		6	6	6	6	6
9062	WDV	13	6	4		6	6	6	6	6
9065	HHKB	13	6	4		6	6	6	6	6
9067	BTB	8	4	4		6	6	6	6	6
9069	MADURA	8	4	4		6	6	6	6	6
9075	DKB	9	9	4		6	6	6	6	6
9082	HO104	15	6	4		6	6	6	6	6
9055	HO301	6	6	5	6		6	6	6	6
9107	LKT3	4	4	5	6		6	6	6	6
9104	DHIF	11	7	NT	6		6	6	6	6
9106	MANIKA	7	2,3	NT	6	6	6	6	6	6
9003	KAS116	1	5		6		6	6	6	6
9005	HOM2	1	5			6	6	6	6	6
9010	AMAI	15	6			6	6	6	6	6
9013	SCHU	15	6			6	6	6	6	6
9016	RML	16	7			6	6	6	6	6
9019	DUCAF	3	2			6	6	6	6	6
9034	SAVC	4	8		6		6	6	6	6
9037	SWEIGOO7	11	7			6	6	6	6	6
9042	TISI	11	7			6	6	6	6	6
9043	BM21	11	7		6		6	6	6	6
9047	PLH	7	2			6	6	6	6	6
9048	LBUF	7	2		6		6	6	6	6
9054	EK	14	5			6	6	6	6	6
9060	CB6B	13	5		6		6	6	6	6
9064	AMALA	14	7			6	6	6	6	6
9066	TAB089	8	6		6		6	6	6	6
9072	SPACH	8	4			6	6	6	6	6
9091	MLF	4	7			6	6	6	6	6
9092	BM92	4	8			6	6	6	6	6

RFLP Standardization Report for DP Alpha/KpnI

G. Paulsen,[1] R. Vaughan,[2] and E.D Du Toit[3]

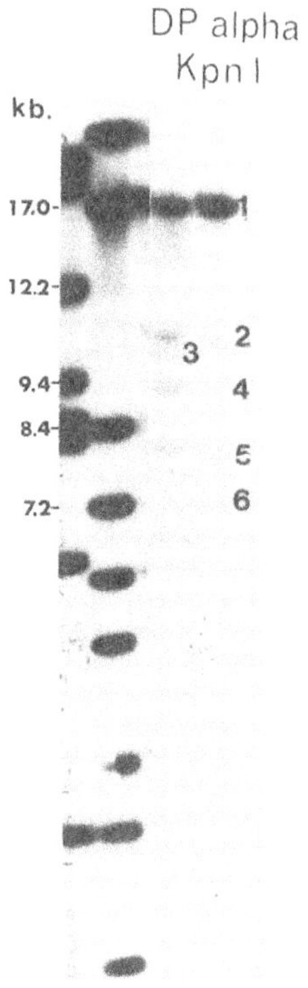

Six bands were identified by the RFLP analysis in the DP alpha/KPN system as shown in Table 1 and in Figure 1. RFLP patterns of the core cell lines are shown in Table 2.

Participating Laboratories: SCATSB,[1] UKIFES,[2] SAFDUT[3]

Table 1. Standard Bands in the DP Alpha/KPN System

Band	kb	Frequency[a]	% Faint[b]	56[c]	32[c]
1	18.17	1.000	0.000		
2	10.78	1.000	0.000		
3	9.99	0.083	1.000		
4	9.34	0.939	0.563		f
5	8.14	1.000	0.507		
6	7.33	0.543	0.579		

[a]Band frequency in the core cell lines.
[b]Frequency of faint bands.
[c]Discrepancies in strong bands (s) or faint bands (f) of the hidden duplicates (WS.ID. 9056 and 9032).

◄

Figure 1. RFLP bands of two core cell lines run in 0.6% agarose gel. Representative core bands are shown in the figure numbered on their right side. The WS.ID. of these core cell lines are from the left, following marker 1 and marker 2: 9002 and 9071.

Table 2. Distribution of Bands in the Core Cell Lines in the DP Alpha/KPN System

		Band No.								Band No.					
WS ID	DP	1	2	3	4	5	6	WS ID	DP	1	2	3	4	5	6
9045	2,4	2	9	9	8	8	9	9064	–	2	1		1	1	1
9014	1,2	2	1	9	8	8	8	9057	4	2	1		1	1	1
9058	1	2	1	9	8	8	8	9030	3	2	1		1	1	
9107	5	2	1	9	8	8	8	9054	–	2	1	8	1	1	
9017	2,4	2	1	9	9	8	9	9105	2	2	1		1	1	
9042	–	2	1	9	9	8	9	9091	–	2	1		1	1	
9013	–	2	1	9	9	8	9	9016	–	2	1			1	1
9033	4	2	1	9	9	8	9	9048	–	2	1			1	1
9004	4	2	1	9	8	8	9	9065	4	2	1		1	1	
9018	3	2	1	9	8	8	9	9021	1	2	1		1	1	
9055	5	2	1	9	8	8	9	9062	4	2	1		1	1	1
9063	2	2	1	9	8	8	9	9032	2	2	1			1	1
9022	3	2	1	9	8	8	9	9009	3,4	2	1	9	8	9	9
9043	–	2	1	9	9	8	9	9092	–	2	1	9	8	1	9
9023	1	2	1	9	9	8	8	9059	3	2	1	9	8	1	9
9006	1	2	1	9	9	8	8	9034	–	2	1	9	9	8	9
9005	–	2	1	9	9	8	9	9068	2	2	1	9	8	1	9
9050	2	2	1	9	9	8	9	9075	4	2	1	9	8	1	9
9056	2,N	2	1	9	9	8	9	9061	4	2	1	9	8	8	8
9104	NT	2	1	9	9	8	8	9067	4	2	1	9	8	1	9
9051	4	2	1	9	9	8	9	9038	2	2	9	9	9	8	9
9047	–	2	1	9	9	8	8	9008	2,4	2	1	9	9	8	9
9040	3	2	1	9	9	8	9	9003	–	2	1	9	8	8	9
9010	–	2	1	9	9	8	8	9070	1,4	2	1	9	9	8	9
9069	4	2	1		1	1		9029	2	2	1	9	9	1	9
9037	–	2	1		1	1		9011	2,4	2	1	9	8	8	9
9028	4	2	1		1	1		9026	1,2	2	1	9	8	1	8
9019	–	2	1		1	1		9039	2	2	1	9	8	1	9
9025	4	2	1		1	1		9056	2,N	2	1	9	8	1	9
9031	4	2	1		1	1		9082	4	2	1	9	9	8	9
9088	2	2	1		1	1	1	9106	NT	2	1	9	8	8	9
9066	–	2	1		1	1		9032	2	2	1	9	8	8	9
9060	–	2	1		1	1		9072		2	1	9	8	8	9
9036	2	2	1		1	1		9052	4	2	1	9	8	1	9
9002	4	2	1	8	1	1		9020	2	2	1	9	8	1	9
9071	3	2	1		1	1	1	9007	3,4	2	1	9	8	8	8
		1	2	3	4	5	6			1	2	3	4	5	6

The bands are designated: 1 = single intensity; 2 = double intensity; 8 = faint; 9 = unreadable.

RFLP Standardization Report for DP Alpha/EcoRV

A. Ando,[1] T. Takenouchi,[2] E. Dwyer,[3] and K. Tsuji[1]

The DP alpha/EcoRV system identified the six fragments as shown in Figure 1 and Table 1. Standard fragments nos. 3, 4, 5, and 6 were polymorphic (Table 1). Standard fragment no. 5 was found with 40 cells, including 28 cells that were either DPw2, DPw3, or DPw4. Twenty of the 72 cells showed standard fragment no. 6, including all the DPw1 and DPw5 cells. These correlations for band no. 6 are consistent with a previous report in which the data were obtained with other DP alpha cDNA probe (1).

Participating Laboratories: JAPTSU,[1] US5TER,[2] US8RJW.[3]

Reference

1. Mitsuishi Y, Urlacher A, Tongio M-M, Mayer S. Restriction fragment length polymorphism of DP genes defines three new DP alleles. Immunogenetics 1987;26:383.

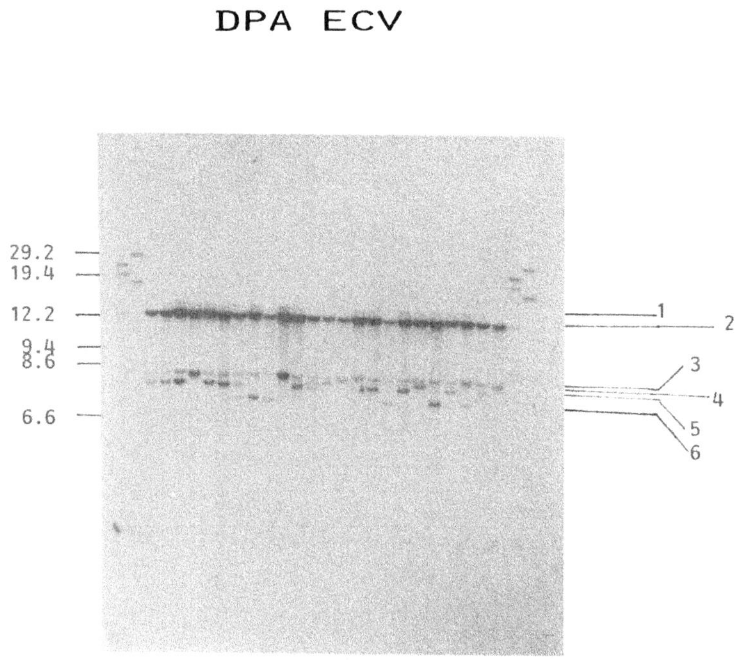

Figure 1. Fragments identified by the DP alpha/EcoRV system.

Immunobiology of HLA
Volume I

Table 1. RFLP Pattern of DP Alpha/EcoRV

Cell Lines			1	2	3	4	5	6	DW	DR	DW2	DRw	DQw	DPw	GLO
Fragment No.			1	2	3	4	5	6							
Fragment Size (kb)			12.47	12.04	7.91	7.61	7.50	6.97							
12	9063	WT47	8	2	1				19	13	26	52	6	2	
18	9050	MOU	8	2	1				17	7		53	2	2	
22	9047	PLH	8	2	1				DB1	7		53	2	–	
24	9010	AMAI	8	2	1				2	15			6	–	
28	9019	DUCAF	8	2	1				3	3	25	52	2	–	2
34	9036	SPOO10	8	2	1				DB2	11	25	52	5	2	
38	9057	TEM	8	2	1				9	14	25	52	1	4	
46	9021	RSH	8	2	1			1	N	3	24	52	4	1	1
53	9068	BM9	8	2	1				8.1	8		52	4	2	
57	9038	BM16	8	2	1				DB6	12		52	7	2	
61	9029	WT51	8	2	1				4	4		53	8	2	
64	9039	JVM	8	2	1				N	11	25	52	7	2	
65	9056	KOSE	8	2	1			8	9,19	13,14		52	5	2,N	
48	9032	BSM	8	2	1				4	4		53	8	2	
68	9032	BSM	8	2	1				4	4		53	8	2	
71	9020	QBL	8	2	1				3	3	25	52	2	2	
37	9064	AMALA	8	2	8	1			16	14	24	52	7	–	
43	9016	RML	8	2	8	1			22	16			7	–	
36	9071	OLGA	8	2	8	8	8		8.2	8		52	4	3	
69	9072	SPACH	8	2	8	8			8.2	8		52	4	4	
62	9011	E4181324	8	2	1		1		12	15			6	2,4	
26	9037	SWEIGOO7	8	2	1		1		5	11	25	52	7	–	
2	9014	MGAR	8	2	8		1		2	15			6	4	1,4
5	9017	WT8	8	2	8		1		2	15			1	2,4	
6	9042	TISI	8	2	8		1		N	11	25	52	7	–	2
7	9013	SCHU	8	2	8		1		2	15			6	–	
8	9033	BM14	8	2	8		1		4	4		53	8	4	
9	9004	JESTHOM	8	2	8		1		1	1			5	4	
10	9018	LO081785	8	2	8		1		3	3	25	52	2	3	
13	9022	COX	8	2	8		1		3	3	24	52	2	3	2
17	9005	HOM2	8	2	8		1		1	1			5	–	
21	9051	PITOUT	8	2	8		1		17	7		53	2	4	2
23	9040	BM15	8	2	8		1		5	11	25	52	7	3	
25	9069	MADURA	8	2	8		1		8.1	8		52	4	4	
27	9028	PE117	8	2	8		1		14	4		53	8	4	
29	9025	DEU	8	2	8		1		4	4		53	7	4	
30	9031	BOLETH	8	2	8		1		4	4		53	8	4	
35	9002	MZ070782	8	2	8		1		20	1			5	4	1,2
39	9030	JHAF	8	2	8		1		13	4		53	7	3	
40	9054	EK	8	2	8		1		9	14	25	52	5	–	
42	9091	MLF	8	2	8		1		4	4		53	7	–	
45	9065	HHKB	8	2	8		1		18	13	24	52	6	4	
50	9092	BM92	8	2	8		1		14	4		53	8	–	1
51	9059	SLE005	8	2	8		1		19	13	26	52	6	3	
54	9075	DXB	8	2	8		1		23	9		53	9	4	
55	9061	31227ABO	8	2	8		1		9	14	25	52	5	4	
56	9067	BTB	8	2	8		1		8.1	8		52	4	4	
63	9026	YAR	8	2	8		1		10	4		53	8	4	1,2
66	9082	HO104	8	2	8		1			15			6	4	
70	9052	DBB	8	2	8		1		11	7		53	9	4	1
72	9007	DEM	8	2	8		1		–	16,4		53	5,8	3,4	1,2
59	9003	KAS116	8	2	8		8	1	1	1			5	–	
31	9088	PF04015	8	2	8		8	8		3	24	52	2	1,4	
49	9009	KASO11	8	2	8		8	8	21	2			1	3,4	
60	9070	LUY	8	2	8		8	8	8.3	8		52	7	1,4	
67	9106	MANIKA	8	2	8		8	8		7		53	2?3	N,T	
1	9045	TUBO	8	2	8		8		5	11,12	25	52	7	2,4	

Table 1. *Continued*

		Fragment No.	1	2	3	4	5	6							
		Fragment Size (kb)	12.47	12.04	7.91	7.61	7.50	6.97							
	Cell Lines								DW	DR	DW2	DRw	DQw	DPw	GLO
47	9062	WDV	8	2	8		8		18	13	24	52	6	4	
58	9008	DO208915	8	2	8		8		2	15			6	2,4	
3	9058	OMW	8	2	8			1	18	13	24	52	6	1	2
4	9107	LKT3	8	2	8			1	15	4		53	4	5	
11	9055	HO301	8	2	8			1	N	6	26	52	6	5	
14	9043	BM21	8	2	8			1	5	11	25	52	7	–	
15	9023	VAVY	8	2	8			1	3	3	24	52	2	1	1
16	9006	WT100BIS	8	2	8			1	1	1			5	1	
19	9056	KOSE	8	2	8			1	9,19	13,14	25	52	5	2,N	
20	9104	DHIF	8	2	8			1	5	11		52	7	N,T	
32	9066	TAB089	8	2	8			1	8.3	8		52	6	–	2
33	9060	CB6B	8	2	8			1	18	13	25	52	5	–	
41	9105	FPAF	8	2	8			1	FS	11		52	6	2	
44	9048	LBUF	8	2	8			1	DB1	7		53	2	–	
52	9034	SAVC	8	2	8			1	4	4		53	8	–	

RFLP Standardization Report for DP Alpha/TaqI

R.E. Bontrop,[1] D. Cohen,[2] C.B. Carpenter,[3] and J.D. Bignon[4]

Data were received from three different laboratories and standardization was performed based on the following criteria: 1) weak hybridization signals were accepted as standard fragments when they were detected by most laboratories, and 2) discrepancies were resolved by majority consensus.

In this system, we detected nine standard fragments which are shown in Table 1 and Figure 1. Standard fragment 1 has a molecular weight of 13.21 kb and a frequency of 62.9%. Standard fragment 2 with a molecular weight of 12.07 kb has a frequency of 27.1% which is lower than that of fragment 1. This fragment is some-

Participating Laboratories: BENROO,[1] FRACOH,[2] NCYCAR,[3] FRABIG[4]

Table 1. Standard Fragments Detected by the DP Alpha/TaqI System

Fragment	kb	Frequency	% Faint	56	32
1	13.21	0.629	0.000		
2	12.07	0.271	0.095		
3	8.40	0.043	0.000		
4	7.66	0.271	0.000		
5	6.45	0.057	0.000		
6	5.47	0.214	0.000		
7	4.66	0.043	0.000		
8	2.99	1.000	0.000		
9	2.49	1.000	0.000		

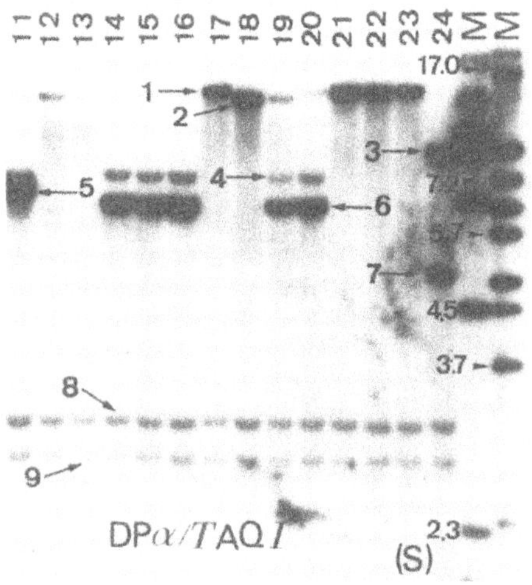

Figure 1. Fragments detected by the DP alpha/TaqI system.

Immunobiology of HLA
Volume I

Table 2. Fragment Presence/Absence Pattern for the DP Alpha/TaqI System as Defined on 72 Core Cell Lines

	Fragment No.										Fragment No.								
Cell	1	2	3	4	5	6	7	8	9	Cell	1	2	3	4	5	6	7	8	9
9014	2							2	2	9008	2	2						2	2
9042	2							2	2	9011	2	2						2	2
9013	2							2	2	9063		2						2	2
9033	2							2	2	9050		2						2	2
9004	2							2	2	9019		2						2	2
9018	2							2	2	9036		2						2	2
9022	2							2	2	9057		2						2	2
9005	2							2	2	9032		2						2	2
9051	2							2	2	9068		2						2	2
9047	2							2	2	9038		2						2	2
9040	2							2	2	9029		2						2	2
9069	2							2	2	9039		2						2	2
9028	2							2	2	9032		2						2	2
9025	2							2	2	9020		2						2	2
9031	2							2	2	9104	1			2		2		2	2
9002	2							2	2	9088	1			1		2		2	2
9071	2							2	2	9070	2			2		2		2	2
9064	2							2	2	9106	2			2		2		2	2
9030	2							2	2	9055		8		2		2		2	2
9054	2							2	2	9056		2		2		2		2	2
9091	2							2	2	9009		2		2		2		2	2
9016	2							2	2	9056		2		2		2		2	2
9065	2							2	2	9058				2		2		2	2
9092	2							2	2	9043				2		2		2	2
9059	2							2	2	9023				2		2		2	2
9075	2							2	2	9006				2		2		2	2
9061	2							2	2	9105				2		2		2	2
9067	2							2	2	9048				2		2		2	2
9026	2							2	2	9034				2		2		2	2
9082	2							2	2	9003				2		2		2	2
9072	2							2	2	9060				2				2	2
9052	2							2	2	9107				2	2			2	2
9007	2							2	2	9066		8		2	2			2	2
9045	2	2						2	2	9021			2	2	2		2	2	2
9017	2	2						2	2	9010			2			2	2	2	2
9062	2	2						2	2	9037			2			2	2	2	2

times seen in an abundant staining intensity (for instance Fig. 1, lane 18) whereas less intense hybridization signals are also observed (Fig. 1, lane 11). Standard fragment 3 appeared to be relatively rare and displayed a panel frequency of 4.3%. Fragment 3 is always seen in combination with standard fragment 7, molecular weight 4.66 kb, and this fragment has the same panel frequency as fragment 3 (Fig. 1, lane 24). Standard fragment 4 has a molecular weight of 7.66 kb and a frequency of 27.1%. This fragment is always seen in combination with standard fragment 5, molecular weight 6.45 kb and frequency 5.7%, (Fig. 1, lane 11) or in combination with standard fragment 6, which has a molecular weight of 5.47 kb and a frequency of 21.4% (Fig. 1, lanes 14, 15, 16, 19, and 20). Standard fragments 8, molecular weight 2.99 kb, and 9, molecular weight 2.49 kb, are seen for all panel cells. The next major hybridization patterns are seen in: 1) fragment 1 in combination with fragment 8 and 9 (Fig. 1 lane 13, 21, 22, and 23), 2) fragment 2 in combination with fragments 8 and 9 (Fig. 1, lanes 12 and 18), 3) fragment 3 in combination with fragments 7, 8, and 9 (Fig. 1, lane 24), 4) fragment 4 and 5 in combination with fragments 8 and 9 (Fig. 1, lane 11) and 5) fragment 4 in combination with fragments 6, 8, and 9 (Fig. 1, lanes 14, 15, and 16). Limited polymorphism for the DP alpha chain has been demonstrated at the product level (1). In this regard it will be of interest to find out whether cell lines such as 19 and 20 are heterozygous for HLA-DP alpha alleles since these cells have the hybridization patterns 2 and 4 and 1 and 4, respectively.

Reference

1. Lotteau V, Teyton L, Tongio MM, Soulier A, Thomsen M, Sasportes M, Charron D. Biochemical polymorphism of the HLA-DP alpha chain. Immunogenetics 1987;25:403.

RFLP Standardization Report for DP Alpha/BglII

A. Marcadet,[1] D. Cohen,[1] H. Takata,[2] M.C.A. Van Eggermond,[3] and P. Stastny[4]

The DP alpha/Bgl II system identified five fragments on the core cell lines (Table 1). Table 2 indicates the distribution of standardized bands in core cell lines. According to previous HLA typing, at least 10 cells were not homozygous for HLA-DP. Moreover HLA-DP typing was incomplete for 21 cells (- or not tested). No *specific fragment* for DPw specificities can be observed. However, some different *specific patterns of bands* can be observed: fragments no. 1 and no. 4 are present for

Table 1. Standard Bands in the DP Alpha/BglII System

		DPA	BGL	COH		
	kb	Locus	Frequency	% Faint	56	32
1	7.24	0 0 0 0	0.643	0.000		
2	5.29	0 0 0 0	0.471	0.000		
3	3.58	0 0 0 0	0.071	0.000		
4	3.41	0 0 0 0	0.200	0.000		
5	2.23	0 0 0 0	0.829	0.017		

Participating Laboratories: FRACOH,[1] JAPTAK,[2] BENROO,[3] US4STA[4]

Table 2. RFLP DP Alpha/BglII System. Distribution of Standard Bands in 10th WS Core Cell Lines

		Standard Bands							Standard Bands				
Cell Lines	HLA DPw	1	2	N = 3	4	5	Cell Lines	HLA DPw	1	2	N = 3	4	5
9021	1	2		2		2	9004	4		2			2
9058	1	2			2		9014	4		2			2
9023	1	2			2		9082	4		2			2
9006	1	2			2		9069	4		2			2
9088	1/4	2	2		2		9028	4		2			2
9070	1/4	2	2		2	2	9033	4		2			2
9105	2	2			2		9067	4		2			2
9063	2	2				2	9025	4		2			2
9036	2	2				2	9075	4		2			2
9038	2	2				2	9031	4		2			2
9039	2	2				2	9026	4		2			2
9068	2	2				2	9051	4		2			2
9020	2	2				2	9052	4		2			2
9029	2	2				2	9107	5	2		2		
9032	2	2				2	9055	5	2		2		
9032	2	2				2	9010	–	2				2
9050	2	2				2	9013	–	2				2
9045	2/4	2				2	9016	–	2				2
9011	2/4	2	2			2	9037	–	2				2
9008	2/4	2	2			2	9042	–	2				2
9017	2/4	2	2			2	9054	–	2				2
9056	2/N	2			2	2	9064	–	2				2
9056	2/N	2			2	2	9072	?	2				2
9040	3	2				2	9019	–	2				2
9071	3	2	2			2	9091	–	2				2
9030	3		2			2	9092	–	2				2
9018	3		2			2	9104	NT	2			2	8
9022	3		2			2	9048	–	2			2	
9059	3		2			2	9034	–	2			2	
9007	3/4		2			2	9043	–	2			2	
9009	3/4	2	2		2	2	9003	–	2			2	
9062	4	2	2			2	9066	–	2		2		
9057	4		2			2	9060	–	2		2		2
9061	4		2			2	9106	NT		2		2	2
9065	4		2				9047	–		2			2
9002	4		2			2	9005	–		2			2

N = New.

Immunobiology of HLA
Volume I

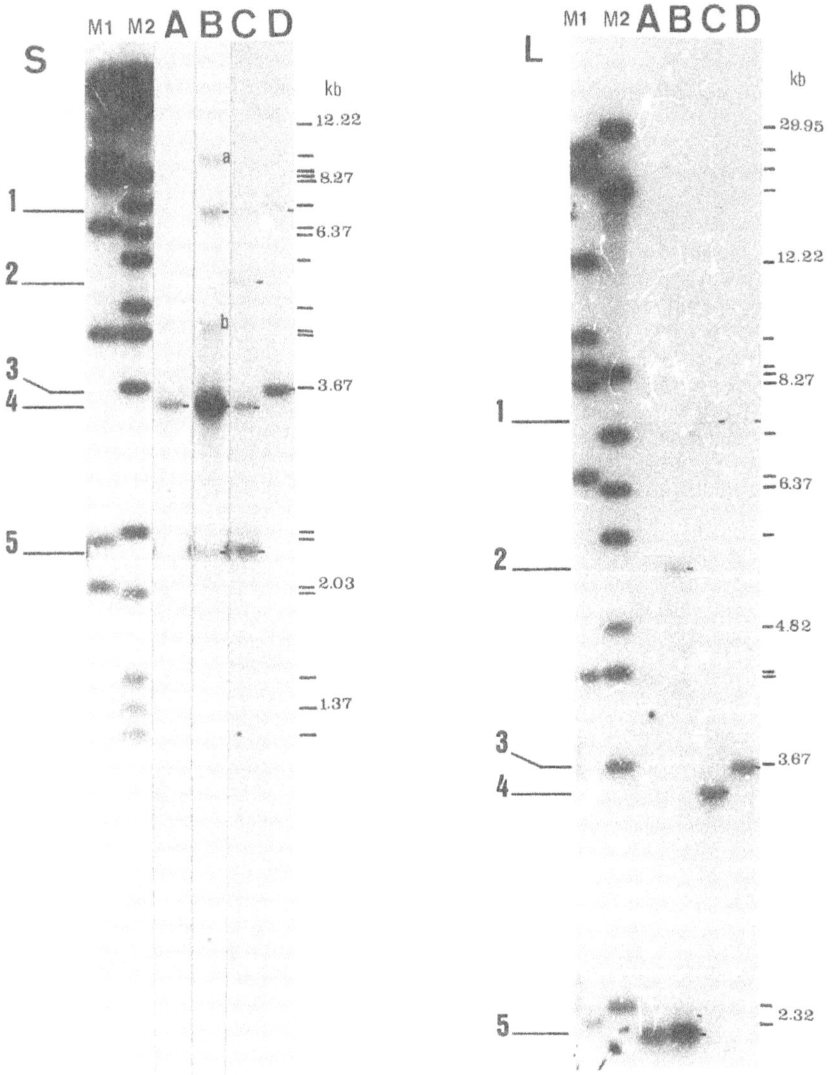

Figure 1. The figures are photographs taken from two films from laboratory JAPTAK. All standardized fragments are indicated at least once for S and/or L migration. The presence of a standardized fragment is indicated by a hyphen.

DPw1; nos. 1 and 5 for Dpw2; nos 1 and 3 for DPw5; and fragments no. 2 and no. 5 are present for DPw3 and DPw4. Correlations between these patterns of fragments and DPw specificities are not very good, as we observed a lot of exceptions. The "new" DPw specificity of cell no. 9056 seems to be positive for fragment no. 4 as DPw1.

Comments on S Migration. The two fragments identified as "a" and "b," only light on S migration, have not been standardized.

RFLP Standardization Report for DP Alpha/PvuII

P. Piccioli,[1] A. Smith,[2] M. Vijod,[1] and G.B. Ferrara[1]

HLA-DPA probe does not seem to detect any significant polymorphism on the DP typed cell lines, although seven different fragments were evident (Table 1). As shown in Table 2, all the lines typed as DPw3 and DPw4 and most of the DPw2 with few exceptions (FPAF,SP0010, and JVM) share the same pattern of fragments, different from the other patterns, where DPw1 and DPw5 specificities can be included. The heterogeneity detected within this system can be also due to the discordance in the data interpretation by different labs. Figure 1 shows the fragments obtained after digestion with PvuII and hybridization with the HLA-DPA probe. Our results seem to correlate with previous studies on this system (1).

Participating Laboratories: IT2FER,[1] US7HAN[2]
*P. Piccioli was recipient of a fellowship from A.I.R.C. (Associazione Italiana per la Ricerca sul Cancro).
This work was in part supported by CNR grants n.86.0265.51 and 86.00659.

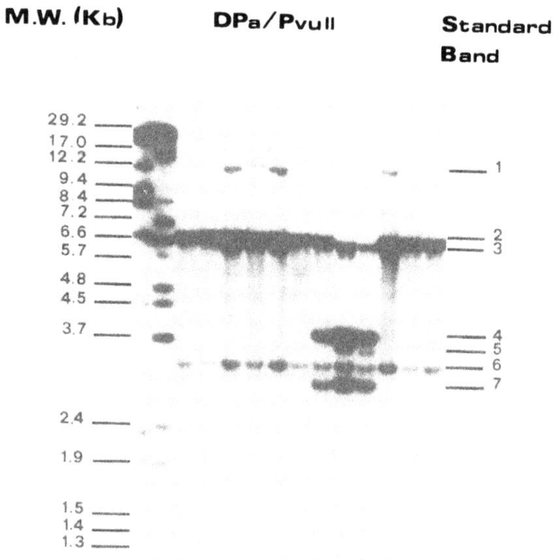

Figure 1. Restriction patterns of MADURA, SWEIG007, PE117, DUCAF, DEU, BOLETH, PF04015, TAB089, CB6B, SP0010, MZ070782, and OLGA cell lines digested with PvuII and hybridized with HLA-DPA probe. Markers are represented on the left. The gel was run using short migration conditions.

Table 1. Fragments Detected by the DP Alpha/PvuII System

Fragment	kb	Frequency[a]	% Faint[b]
1	12.86	0.926	1.000
2	6.58	0.957	0.058
3	6.52	1.000	0.058
4	3.81	0.300	0.143
5	3.42	0.114	1.000
6	3.23	0.917	0.041
7	2.95	0.286	0.175

[a]Band frequency has been calculated as number of cell lines presenting a specific fragment versus number of cell lines tested.
[b]Frequency of faint band has been calculated as number of cell lines presenting a faint band versus number of cell lines presenting the same fragment (i.e. frequency 1.000 and % faint 0.000 means that all the samples tested share the same band and it is a strong band; on the other hand 1.000 and 1.000 means that it has a constant and faint band.

Reference

1. Hylding-Nielsen JJ, Morling N, Odum N, Ryder CP, Platz P, Jakobsen B, Svejgaard A. Restriction fragment length polymorphism of the HLA-DP subregion and correlations to HLA-DP phenotypes. Proc Natl Acad Sci USA, 1987; 84:1644.

Immunobiology of HLA
Volume I

Table 2. Banding Pattern of the DP Alpha/PvuII System in Each Cell Line

		Fragment No.						
Cell	DP	1	2	3	4	5	6	7
9020 QBL	2	8	2	8			8	
9029 WT 51	2	8	2	8			8	
9032 BSM	2	8	2	8			8	
9032 BSM	2	8	2	8			8	
9038 BM 16	2	8	2	8			8	
9050 MOU	2	8	2	8			8	
9063 WT 47	2	8	2	8			8	
9068 BM 9	2	8	2	8			8	
9018 LO 081785	3	8	2	8			8	
9022 COX	3	8	2	8			8	
9030 JHAF	3	8	2	8			8	
9040 BM 15	3	8	2	8			8	
9059 SLE-005	3	8	2	8			8	
9071 OLGA	3	8	2	8			8	
9002 MZ 070782	4	8	2	8			8	
9004 JESTHOM	4	8	2	8			8	
9014 MGAR	4	8	2	8			8	
9025 DEU	4	8	2	8			1	
9026 YAR	4	8	2	8			8	
9028 PE 117	4	8	2	8			8	
9031 BOLETH	4	8	2	8			8	
9033 BM14	4	8	2	8			8	
9051 PITOUT	4	8	2	8			8	
9052 DBB	4	8	2	8			8	
9057 TEM	4	8	2	8			8	
9061 31227 ABO	4	8	2	8			8	
9065 HHKB	4	8	2	8			8	
9067 BTP	4	8	2	8			8	
9069 MADURA	4	8	2	8			8	
9075 DKB	4	8	2	8			8	
9082 HO 104	4	8	2	8			8	
9062 WDV	4	8	2	8			8	
9008 DO 208915	2/4	8	2	8			8	
9011 E 4181324	2/4	8	2	8			8	
9017 WT 8	2/4	8	2	8			8	
9045 TUBO	2/4	8	2	8			8	
9007 DEM	3/4	8	2	8			8	
9037 SWEIG 007		8	2	8			8	
9092 BM92		8	1	8			8	
9005 HOM-2		8	2	8			8	
9010 AMAI		8	2	8			8	
9013 SCHU		8	2	8			8	
9016 RML		8	2	8			8	
9019 DUCAF		8	2	8			8	
9042 TISI		8	2	8			8	
9047 PLH		8	2	8			8	
9054 EK		8	2	8			8	
9072 SPACH		8	1	8			8	
9091 MLF		8	2	8			8	
9006 WT100 BIS	1	8	8	8	1		8	1
9023 VAVY	1	8	8	8	2		8	1
9055 HO 301	5	8	8	8	2		8	2
9070 LUY	1/4	8	1	8	1		8	8
9088 PF 04051	1/4	8	2	8	2		8	8
9056 KOSE	2/N	8	2	8	2		8	1
9056 KOSE	2/N	8	1	8	1		8	8

Table 2. *Continued*

Cell	DP	Fragment No. 1	2	3	4	5	6	7
9104 DHIF	NT	8	1	8	2		8	1
9043 BM 21		8	8	8	2		8	1
9048 LBUF		8	8	8	2		8	1
9058 OMW	1		1	8	2		8	1
9009 KAS-011	3/4		1	8	1		8	1
9107 LKT 3	5		8	8	2	8	8	1
9066 TAB 089			1	8	2	8	1	1
9060 CB6B				8	2	8	8	2
9036 SPO 010	2	8	2	8			1	1
9039 JVM	2	8	1	8	8		8	
9064 AMALA		8	2	8	8		8	
9105 FPAF	2	8		8	2		8	1
9021 RSH	1	8	2	8	2	8	8	1
9003 KAS-116			8	8	2			1
9034 SAVC		8			2		8	1
9106 MANIKA	NT		1	9	8			8

1 = Single intensity; 2 = double intensity; 8 = faint; 9 = unable to read.

RFLP Standardization Report for DP Alpha/EcoRI

M. Segall,[1]* L. Schluender,[1]* B. Bradley,[2] and S. Naito[3]

DP alpha/EcoRI blots were submitted by two laboratories. Fragments defined on either laboratories' blots were included as standard fragments. The DP alpha/EcoRI system identified six fragments on the core cell lines as shown in Table 1 and Figure 1. Portions of the blots submitted showed relatively faint bands or could not be read due to high background. Fragment 2 was identified by one of the two labs reporting this system and was a faint band which did not photograph satisfactorily. For fragments 4, 5 (which are probably non-polymorphic), and 6, portions of the blots could not be read; blot C was especially problematic, so that presence/absence for the DNAs in lanes 49 to 72 is unreliable, and inability to determine presence/absence of a given fragment is indicated with an asterisk (*) in Table 2. The discrepancies for fragments 4, 5, and 6 involved portions of the blots which could not be read.

Table 1. DP Alpha/EcoRI Standard Fragments

Fragment	kb	Frequency	% Faint	Discrepant for 9056	9032
1	10.18	1.000	0.000		
2	5.22	1.000	1.000		
3	4.94	1.000	0.028		
4	4.17	0.694	0.280	s	s
5	3.08	0.639	1.000	f	f
6	2.41	0.139	0.000	s	

Participating Laboratories: US2BAC,[1] UKIBRS,[2] JAPNAI[3]
*This is paper #493 from the Immunobiology Research Center, University of Minnesota, Minneapolis, Minnesota 55455, USA.
Supported in part by Juvenile Diabetes Foundation Grant #186175 and March of Dimes Birth Defects Foundation Grant #6-496.

Figure 1. Fragments on the core cell line identified by the DP alpha/EcoRI system.

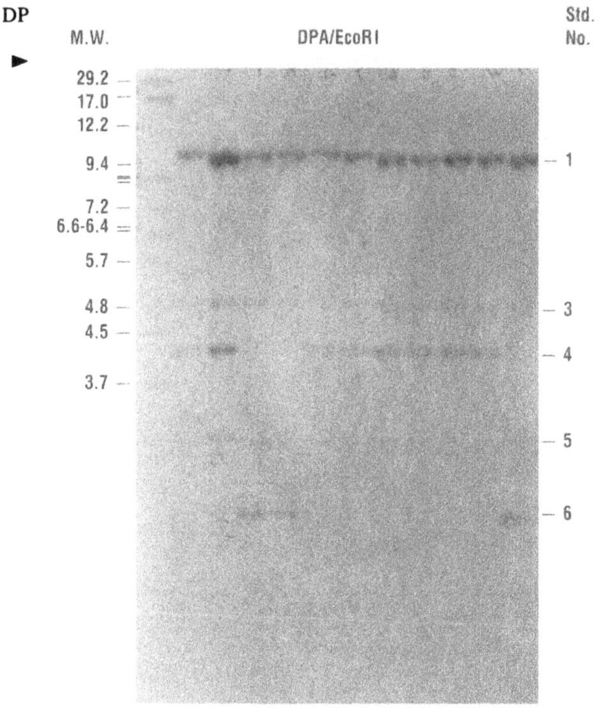

Table 2. Distribution of Polymorphic Fragments in Core Cell Lines, DP Alpha/EcoRI

Cell No.	Name	Std. Fragment No. 4	5	6	Cell No.	Name	Std. Fragment No. 4	5	6
9002	MZO70782	2	8		9043	BM21		8	1
9003	KAS116	*	*	*	9045	TUBO	2	8	
9004	JESTHOM	2	8		9047	PLH	2	8	
9005	HOM2	2	8		9048	LBUF		8	1
9006	WT100BIS		8	1	9050	MOU	2	8	
9007	DEM	8	*	*	9051	PITOUT	2	8	
9008	D0208915	*	*	*	9052	DBB	8	*	*
9009	KAS011	8	*	*	9054	EK	2	8	
9010	AMAL	2	8		9055	HO301	8	8	1
9011	E4181324	*	*	*	9056	KOSE	2	8	1
9013	SCHU	2	8		9056	KOSE	*	*	*
9014	MGAR	2	8		9057	TEM	2	8	
9016	RML	*	*	*	9058	OMW	8	8	1
9017	WT8	2	8		9059	SLE005	8	*	*
9018	L0081785	2	8		9060	CB6B		8	
9019	DUCAF	2	8		9061	31227ABO	8	*	*
9020	QBL	8	*	*	9062	WDV	2	8	
9021	RSH	2	8	1	9063	WT47	2	8	
9022	COX	2	8		9064	AMALA	2	8	
9023	VAVY		8	1	9065	HHKB	2	8	
9025	DEU	2	8		9066	TAB089		8	
9026	YAR	*	*	*	9067	BTB	8	*	*
9028	PE117	2	8		9068	BM9	8	*	*
9029	WT51	*	*	*	9069	MADURA	2	8	
9030	JHAF	2	8		9070	LUV	*	*	*
9031	BOLETH	2	8		9071	OLGA	2	8	
9032	BSM	*	*	*	9072	SPACH	8	*	*
9032	BSM	2	8		9075	DKB	8	*	*
9033	BM14	2	8		9082	HO104	*	*	*
9034	SAVC	*	*	*	9088	PFO4015		8	
9036	SP0010	2	8		9091	MLF	2	8	
9037	SWEIG007	2	8		9092	BM92	8	*	*
9038	BM16	8	*	*	9104	DHIF		8	1
9039	JVM	*	*	*	9105	FPAF	*	*	*
9040	BM15	2	8		9106	MANIKA	*	*	*
9042	TISI	2	8		9107	LKT3	8	8	1

RFLP Standardization Report for DP Alpha/HindIII

P.A. Biro,[1] S. Cutbush,[1] J.M. Tiercy,[2] G. Paulsen,[3] G. Semana,[4] and F. Quillivic[4]

DP alpha (Fig. 1) gives rise to three nonpolymorphic (std. bands 1, 2, and 4; Table 1) and two polymorphic, allelic fragments (std. bands 3 and 5). The allelic bands can be correlated with the DP types of the cells (1,2). Band 3 correlates with DP type 1 (and probably 5, 6, and 7) and band 5 associates with DP types 2, 2B, 3, 4, and 4B (Tables 2 and 3).

Participating Laboratories: UKIFES,[1] FRAJEA,[2] SCATSB,[3] FRAFAU[4]

Table 1. Sizes and Frequencies of Standard Bands Produced by the DP Alpha/HindIII Enzyme/Probe Combination

	kb	Frequency	% Faint
1	5.82	1.000	0.000
2	5.50	1.000	0.000
3	4.90	0.243	0.000
4	4.73	1.000	0.000
5	3.25	0.814	0.000

Table 2. Correlation Between Standard Bands and DP Type

	DP								
Band	1	2	2B	3	4	4B	5	6	7
3	5/6	1/17	1/3		1/27		1/1	1/1	1/1
5	1/6	16/17	2/3	7/7	26/27	9/9			

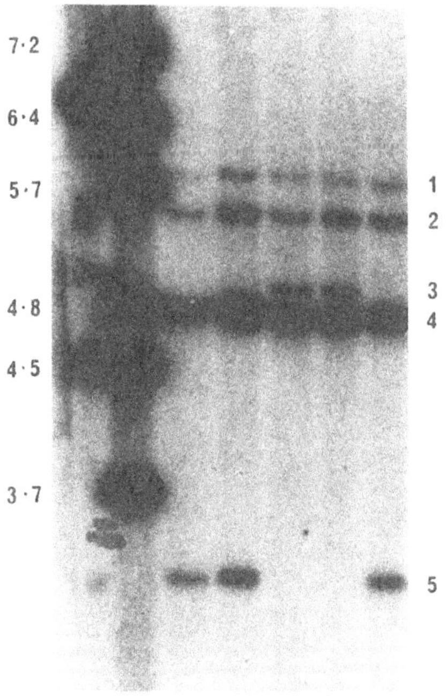

Figure 1. Standard bands produced by the DP alpha/HindIII enzyme-probe combination.

References

1. Bodmer J, Bodmer W, Heyes J, So A, Tonks S, Trowsdale J, Young J. Identification of HLA-DP polymorphism with DPα and DPβ probes and monoclonal antibodies: correlation with primed lymphocyte typing. Proc Natl Acad Sci USA, 1987;84:4596.
2. Mitsuishi Y, Urlacher A, Tongio M-M, Mayer S. Restriction fragment length polymorphism of DP genes defines three new DP alleles. Immunogenetics 1987;26:383.

Table 3. RFLP Pattern of DP Alpha/HindIII System

Cell Lines		DR Type	Fragment No. 12345	Cell Lines		DR Type	Fragment No. 12345
MZ070782	9002	1	11 21	BM21	9043	5	1112
KAS116	9003	1	1112	TUBO	9045	5	11 21
JESTHOM	9004	1	11 21	PLH	9047	7	11 21
HOM2	9005	1	11 21	LBUF	9048	7	1112
WT100BIS	9006	1	1112	MOU	9050	7	11 21
DEM	9007	2/4	11 21	PITOUT	9051	7	11 21
DO208915	9008	2	11 21	DBB	9052	7	11 21
KAS011	9009	2	11 21	EK	9054	6	11 21
AMAI	9010	2	11 21	HO301	9055	6	1112
E4181324	9011	2	11 21	KOSE	9056	6	11121
SCHU	9013	2	11 21	TEM	9057	6	11 21
MGAR	9014	2	11 21	OMW	9058	6	1112
RML	9016	2	11 21	SLE005	9059	6	11 21
WT8	9017	2	11 21	CB6B	9060	6	1112
LO081785	9018	3	11 21	31227ABO	9061	6	11 21
DUCAF	9019	3	11 21	WDV	9062	6	11 21
QBL	9020	3	11 21	WT47	9063	6	11 21
RSH	9021	3	11121	AMALA	9064	6	11 21
COX	9022	3	11 21	HHKB	9065	6	11 21
VAVY	9023	3	1112	TABO89	9066	8	1112
DEU	9025	4	11 21	BTB	9067	8	11 21
YAR	9026	4	11 21	BM9	9068	8	11 21
PE117	9028	4	11 21	MADURA	9069	8	11 21
WT51	9029	4	11 21	LUY	9070	8	11121
JHAF	9030	4	11 21	OLGA	9071	8	11 21
BOLETH	9031	4	11 21	SPACH	9072	8	11 21
BSM	9032	4	11 21	DKB	9075	9	11 21
BM14	9033	4	11 21	HO104	9082	?	11 21
SAVC	9034	4	1112	PF04015	9088	3	11121
SP0010	9036	5	11 21	MLF	9091	4	11 21
SWEIG007	9037	5	11 21	BM92	9092	4	11 21
BM16	9038	5	11 21	DHIF	9104	11	1112
JVM	9039	5	11 21	FPAF	9105	11	1112
BM15	9040	5	11 21	MANIKA	9106	7	11 21
TISI	9042	5	11 21	LKT3	9107	4	1112

RFLP Standardization Report for DP Alpha/MspI

S. Tonks,[1] V. Groves,[1] C. Muller,[2] F. Otani,[3] J.R. Regueiro,[4] M. Segall,[5] J. Bodmer,[1] and J. Trowsdale[1]

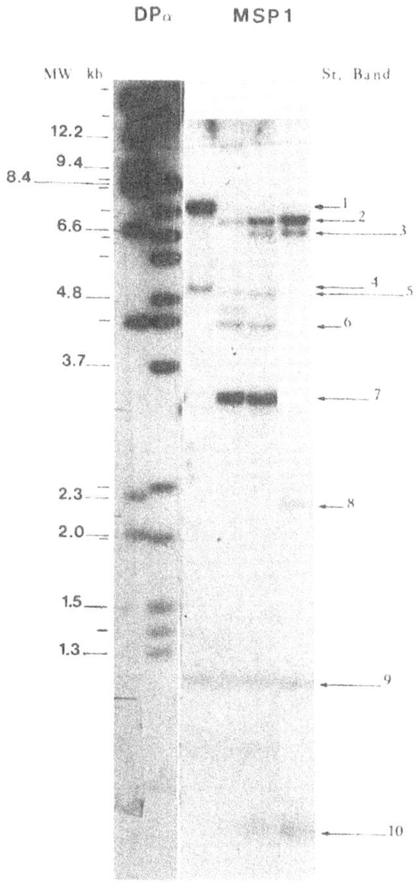

The DP alpha/MspI system was found to be relatively polymorphic, identifying 11 standard fragments in the core cell lines as shown in Figure 1 (band number 11 not shown). Fragment number 9 (1.02 kb) was the only constant band. The banding patterns for the core cell lines (excluding duplicates) can be seen in Table 1. The most common pattern appears to be the presence of bands 5, 6, 7, and 9. Despite the polymorphism shown in this system none of the fragments were found to have a strong correlation with any DP types. The only significant association was fragment 2 (6.71 kb) with DPw1 ($\chi^2=38.699$). MspI has limited application for DP typing using the DP alpha probe.

Participating Laboratories: UKIBOD,[1] GERMUC,[2] JAPKSH,[3] FRAARN,[4] US2BAC[5]

◄ **Figure 1.** Standard fragments in the core cell lines identified by the DP alpha/MspI system.

Table 1. Banding Pattern of the DP Alpha/MspI System in Each Cell Line

		Fragment No.								
WS ID	Cell	1	2	3	4	5	6	7	8	9
9058	OMW	8	1		1					
9107	LKT3	8	1	1						
9066	TAB089	8	1	1					1	1
9010	AMAL	1			1				1	1
9018	LO081785	2			1					1
9105	FPAF	1			1					1
9048	LBUF	1			1					1
9034	SAVC	2			1					1
9003	KAS116	2			1					1
9104	DHIF	1			1		8	8		1
9056	KOSE	1			1	1	1	1		1
9106	MANIKA	1			1	1	1	2		1
9009	KAS011	1			8	8	1	2		1
9031	BOLETH				1	1	1	2		1

Table 1. *Continued*

WS ID	Cell	Fragment No. 1	2	3	4	5	6	7	8	9
9045	TUBO					1	1	2		1
9014	MGAR					1	1	3		1
9017	WT8					1	1	2		1
9004	JESTHOM					1	1	2		1
9063	WT47					1	1	2		1
9022	COX					1	1	2		1
9005	HOM2					1	1	3		1
9050	MOU					1	1	2		1
9051	PITOUT					1	1	3		1
9047	PLH					1	1	3		1
9040	BM15					1	1	3		1
9010	AMAL					1	1	3		1
9069	MADURA					1	1	2		1
9037	SWEIG007					1	1	2		1
9028	PE117					1	1	3		1
9019	DUCAF					1	1	2		1
9025	DEU					1	1	3		1
9002	MZ070782					1	1	3		1
9071	OLGA					1	1	2		1
9064	AMALA					1	1	2		1
9057	TEM					1	1	2		1
9054	EK					1	1	2		1
9091	MLF					1	1	1		1
9016	RML					1	1	2		1
9065	HHKB					1	1	2		1
9062	WDV					1	1	1		1
9032	BSM					1	1	1		1
9092	BM92					1	1	2		1
9059	SLE005					1	1	2		1
9075	DKB					1	1	2		1
9061	31227ABO					1	1	2		1
9067	BTB					1	1	2		1
9038	BM16					1	1	2		1
9008	D0208915					1	1	2		1
9029	WT51					1	1	2		1
9011	E4181324					1	1	2		1
9026	YAR					1	1	2		1
9039	JVM					1	1	1		1
9082	H0104					1	1	3		1
9072	SPACH					1	1	2		1
9052	DBB					1	1	3		1
9020	QBL					1	1	2		1
9007	DEM					1	1	2		1
9042	TISI		2			1	1	2		1
9013	SCHU		1			1	1	2		1
9088	PFO4015		1			1	1	2		1
9021	RSH		1	1		1	1	1		1
9070	LUY		1		1	1	1	1		1
9018	LO081785					1	1	2	1	1
9033	BM14					1	1	2	1	1
9036	SPO010					1	1	3	1	1
9030	JHAF					1	1	2	1	1
9068	BM9					1	1	2	8	1
9060	CB6B		1	1					8	1
9023	VAVY			2	1					1
9006	WT100BIS			2	1					1

1 = Single intensity; 2 = double intensity; 8 = faint.

RFLP Standardization Report for DP Alpha/BamHI

A. Andreas-Zietz,[1] E. Keller,[1] E.D. Albert,[1] B.O. Boehm,[2] J.J. Hyldig-Nielsen,[3] and D.P. Singal[4]

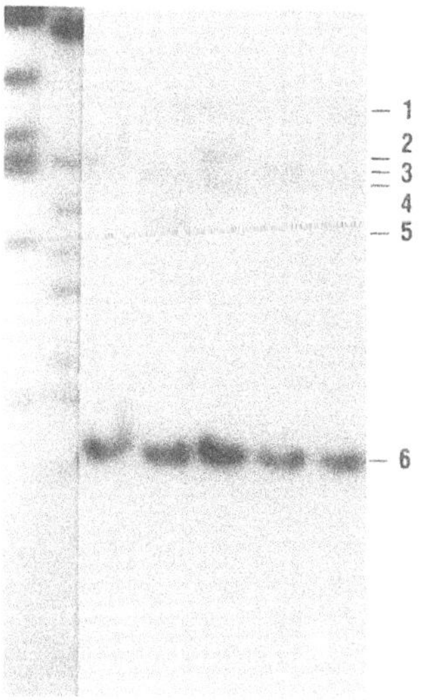

The DP alpha/BamHI system identified six fragments on the core cell lines as shown in Table 1 and Figure 1.

Table 1 shows the reaction pattern of all 72 cell lines for the DP alpha/BamHI core fragments as well as the corresponding molecular weights and probable associations with DP.

The polymorphic pattern in this system (band nos. 2, 3, and 5) allows a clear assignment of homo-and heterozygous state (expressed by numbers "l" and "2," respectively). Intensity of band no. 6 seems to be more than double dose.

Participating Laboratories: GERALB,[1] GERBOE,[2] SCASVE,[3] US5SIN[4]
This study was supported by SFB 217.

◄
Figure 1. Core bands of the DP alpha/BamHI system (gel 0.9%).

Table 1. Banding Pattern of All 72 Core Cell Lines, as Standardized in the DP Alpha/BamHI System, Including kb Values and Probable DP Associations

Probe Name: DPA Enzyme Name: BAMHI

	9999999999999	999	99999999999999999999999999999999	9999	99999999999999999	999	WS-ID	
	0001100100000	001	00000000000000000000000000000000	0000	00000000000000000	000	number	
	5200054066430	870	12473500130562230676628504105997	4116	65363232335413161	255		
	8365753460843	806	82010997434198512551762282354122	5712	30688990227707946	166		
	0114011233455	366	11233547000222233455566750014456	0064	11355667463222234	416	position	
	3561414023429	107	03369192289157905545636086770209	1527	28437141888246873	695	on blots	
DP	111255-------	11-	3333333344444444444444442-------	2224	22222222224------	122		
DP		44	44 4	444				associated with:
1	8888888888888	888	88888888888888888888888888888888	8888	88888888888888888	888	11.12 kb	
2				1111	22222222222222222	111	9.00	DP 2
3		111	22222222222222222222222222222222	1111			8.67	DP 3 + 4
4	2222222222222	222	22222222222222222222222222222222	2222	22222222222222222	222	8.16	
5	2222222222222	111				111	6.92	DP 1 + 5
6	3333333333333	333	33333333333333333333333333333333	3333	33333333333333333	333	3.75	

Cells are listed according to banding pattern and DP typing. Numbers indicate intensity (1,2,3 = increasing intensity; 8 = faint; 9 = unreadable).

RFLP Summary Reports

DNA-RFLP Analysis: HLA-DR Beta Workshop Report

J.D. Bignon,[1] G. Semana,[2] J.M. Tiercy,[3] M. Simons,[4] J.M. Lalouel,[5] and D. Cohen[6]

Introduction

One of the aims of the Tenth International Workshop was to describe the extended HLA polymorphism at the DNA level using a panel of 72 cells (some of them being consanguineous) and a standardized technique based on Southern transfer (1) called "RFLP analysis." This technique was applied to the 72 extracted DNAs (Laboratoire du CEPH, Paris, France, D. Cohen and J. Dausset), and each selected laboratory had to digest these DNAs by 1 of the 12 selected restriction enzymes and hybridize the corresponding blots with one of the two sets of probes (see Workshop DNA protocol). Therefore, DNA of each of the 72 cells was digested by 12 enzymes and hybridized with 13 probes. All data and blots were collected by the standardization centers. Central analysis was performed in Salt Lake City, Utah (J.M. Lalouel). Herein, we report on the synthetic RFLP data obtained with DR beta probe hybridization (2).

To be clear and useful, this report takes into account only unquestionable fragments because either some bands were too faint or missing on some blots to be used for referencing. Moreover, due to high degree of nucleotide homology between beta chain genes of the class II region, we have considered as real DR hybridizing fragments: 1) fragments only revealed with DR beta probe and consequently not observed with DQ beta, DP beta or DO beta probes (= DR beta-unique); and 2) fragments primary DR locus-assigned by using the revised locus assignments in case of cross-hybridization (= DR beta cross-hybridizing).

These two types of DR beta bands (unique and cross-hybridizing) were only selected if they had $<20\%$ scores of <4 in intensity (i.e., $>80\%$ of 6 or 8). In addition, monomorphic DR fragments were excluded in the schematic tables.

Participating Laboratories: FRABIG,[1] FRAFAU,[2] FRAJEA,[3] US5SIM,[4] US7LAL,[5] FRACOH[6]

In conclusion, only DR beta-unique and DR beta cross-hybridizing strong fragments were selected for this report and then could be considered as reference Workshop bands.

Results

Table 1 (from I-1 to I-12 according to restriction enzyme) depicts the distribution of all selected fragments (see above) observed in the 30 cellular types that were defined according to DR, DQW, and DW typing among 66 of 72 cells. DR 1, 2, 4, 7, W8, and 9 displayed clearly allele specific bands: one specific band for DR7, one for DRW 8 and for DR 9, three for DR 2, seven for DR 1, and ten for DR 4.

Two splits of the well defined DR1 pattern were defined correlating with DW1 and DW20 cellular subtypes. Since only one DW20 cell was tested, this RFLP split should be confirmed. DRW 15 and DRW 16 differed by 6.15 kb Sst I and 1.42 kb TaqI fragments present in all DRW 15 cells (not found in DRW 16 cells). Conversely, two bands were present in DRW 16 cells (not found in DRW 15 cells): 13.81 kb TaqI was also surprisingly noted in DRW53 cells, and 1.69 kb TaqI, this latter being a split specific band. Two characteristic patterns of DR3 were revealed; one associated with HLA-B8 haplotype and the second associated with HLA-B18 haplotype, correlating, respectively, with the now called DW 24 and DW 25 cellular specificities. Moreover, a third DR 3 pattern was individualized in the DR 3 BW 42 haplotype of RSH cell (WSNO: 9021).

Concerning DR 4 cells, no clear DQW and/or DW correlating split was noted using DR beta probe hybridization, except with 4.11 kb HincII split-specific fragment present in all DR 4–DQW7 cells and only one of nine DR 4 non-DQW 7 cells.

The serologically DRW 11 and DRW 12 alleles were clearly characterized by a differential RFLP pattern, especially with two DRW 12 specific fragments (1.83 kb HindIII and 5.92 kb PvuII). However this DRW 12

Table 1.1-1.12. Diagrammatic Patterns of DR Beta Hybridizing Fragments According to DR, DQw and Dw Cell Typing (WHO Nomenclature Report)

Thirty cellular types were defined considering all the Workshop cells used in "RFLP typing" except four cells (WSNO: 9007, 9045, 9056, 9104). "n" represents the number of cells corresponding to a defined cellular type.

9082 (DR2 cell) and 9106 (DR7 cell) are the Workshop number (WSNO) of cells with untested Dw type. Three cells are noted with their Workshop name (RSH = 9021; TAB = 9066; LUY = 9070).

A specific band defining a DR allele or a unique split is represented by ■ or by ◤ when, respectively, all cells or only some of them of a given cellular type display this fragment. A nonspecific band is defined by ■ or by ◤ when, respectively, all cells or only some of them of a given cellular type display this fragment.

Table 1-1.

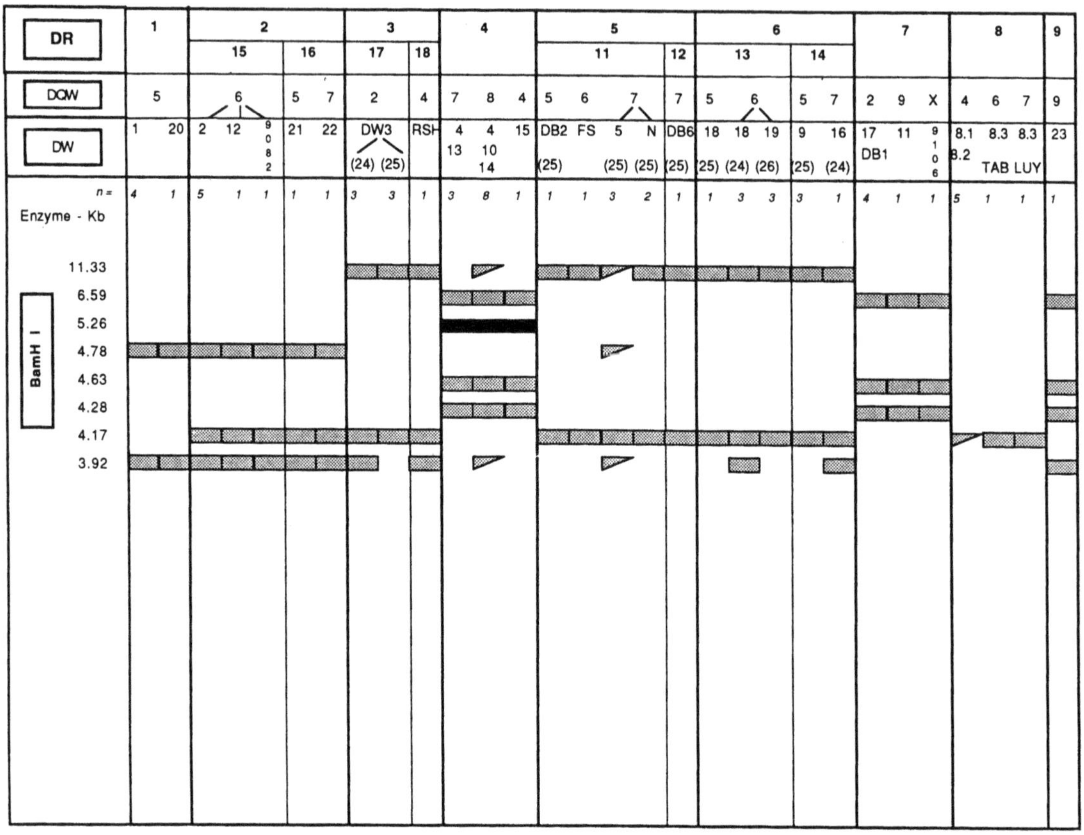

Table 1-2.

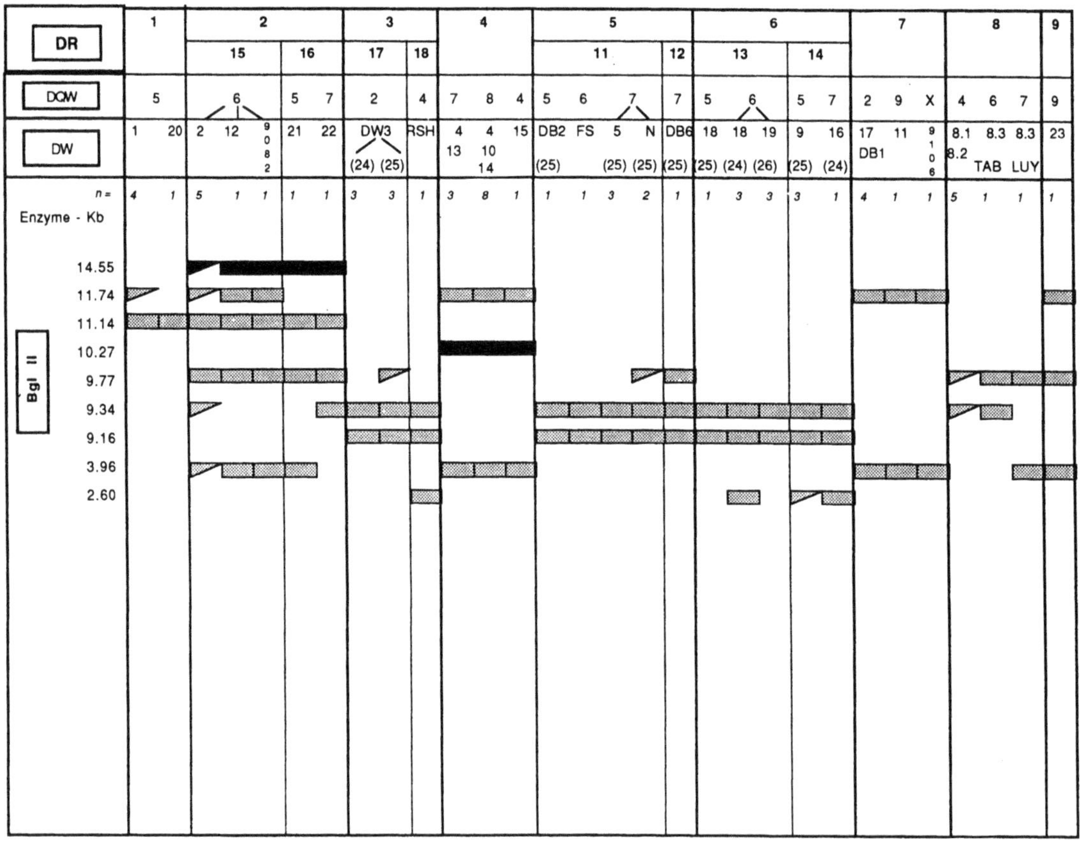

Table 1-3.

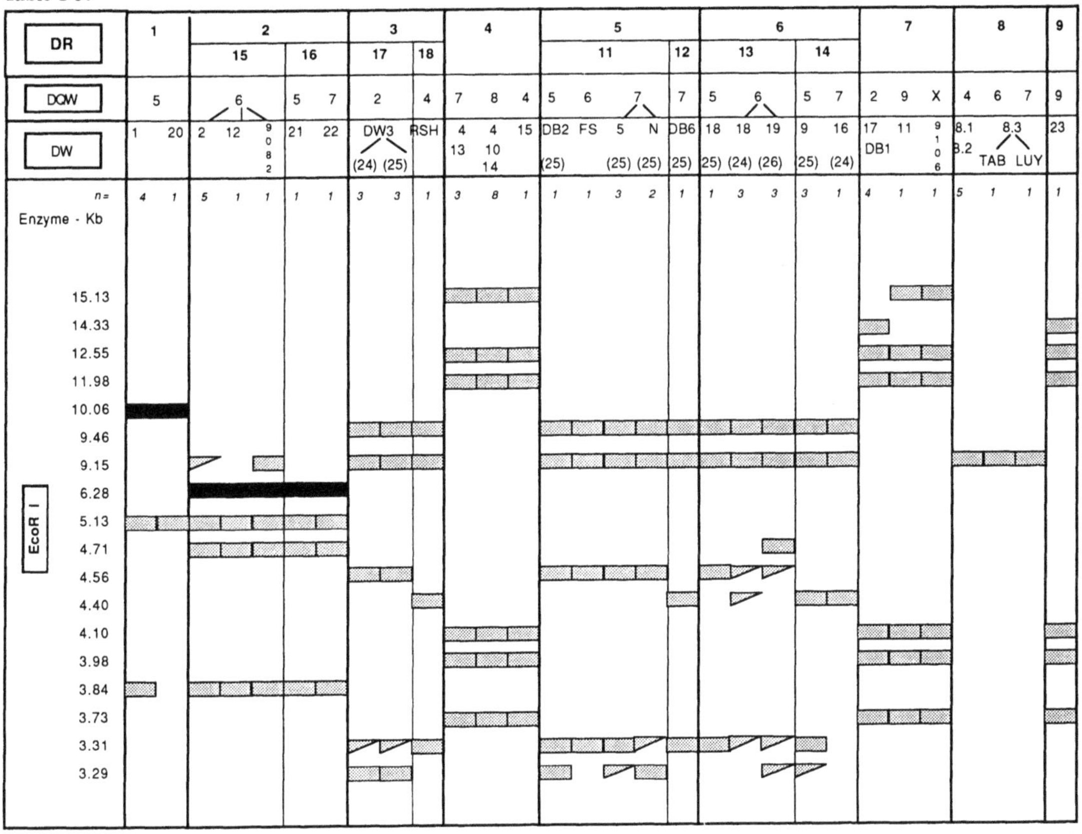

Table 1-4.

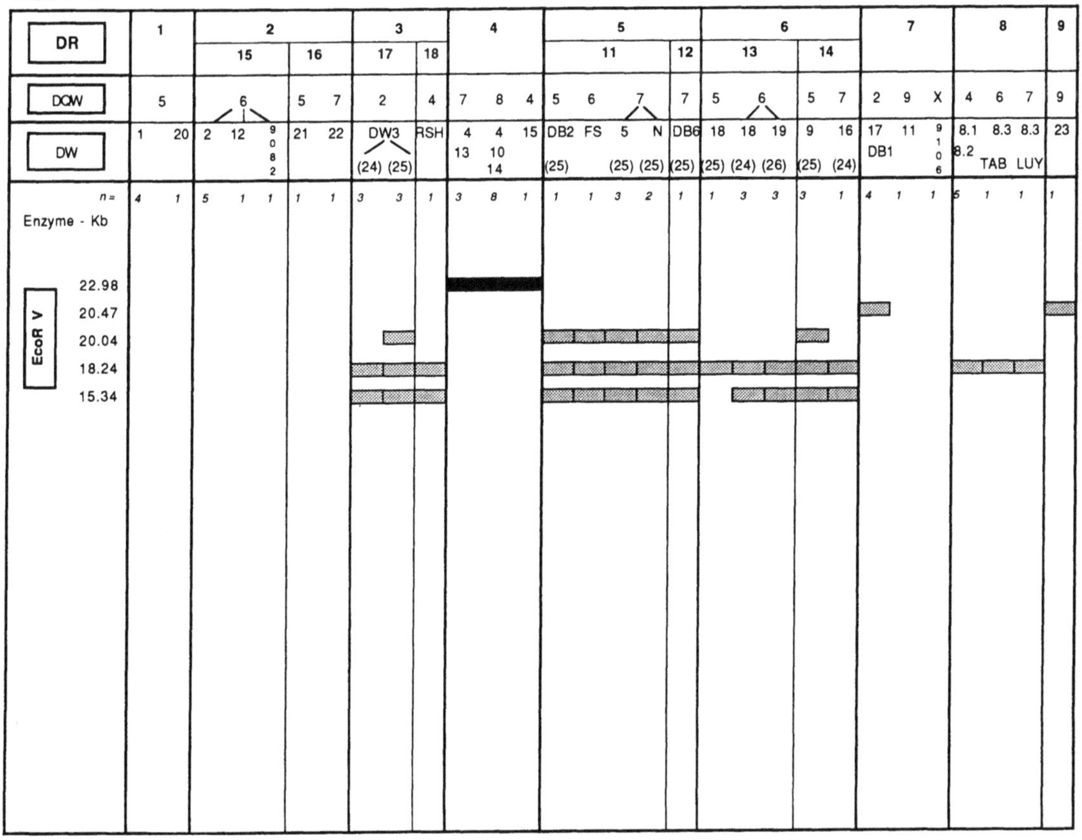

Table 1-5.

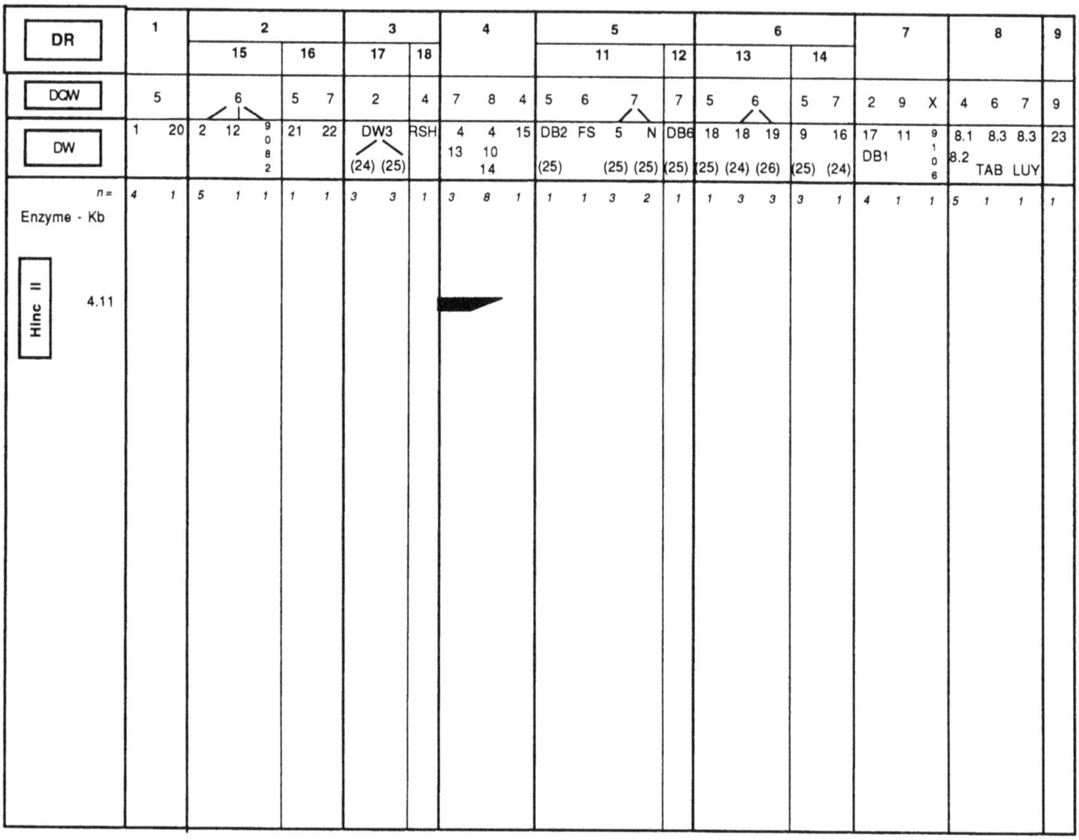

Table 1-6.

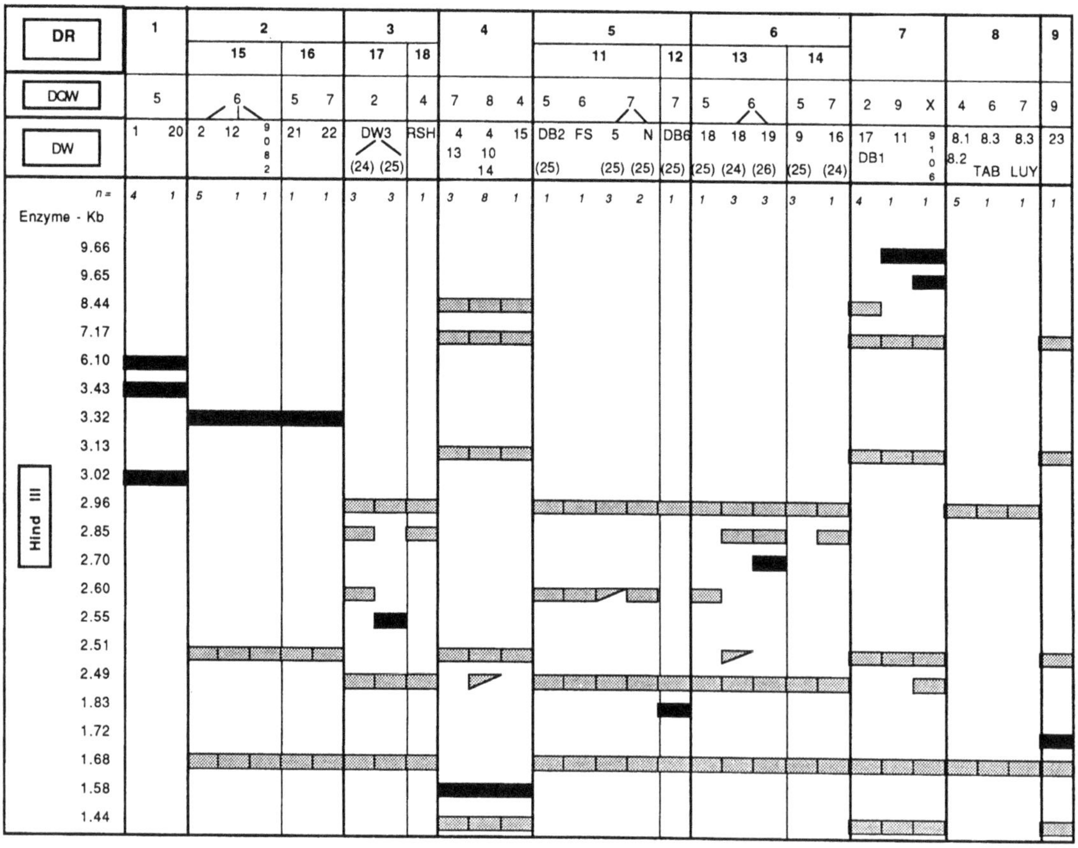

Table 1-7.

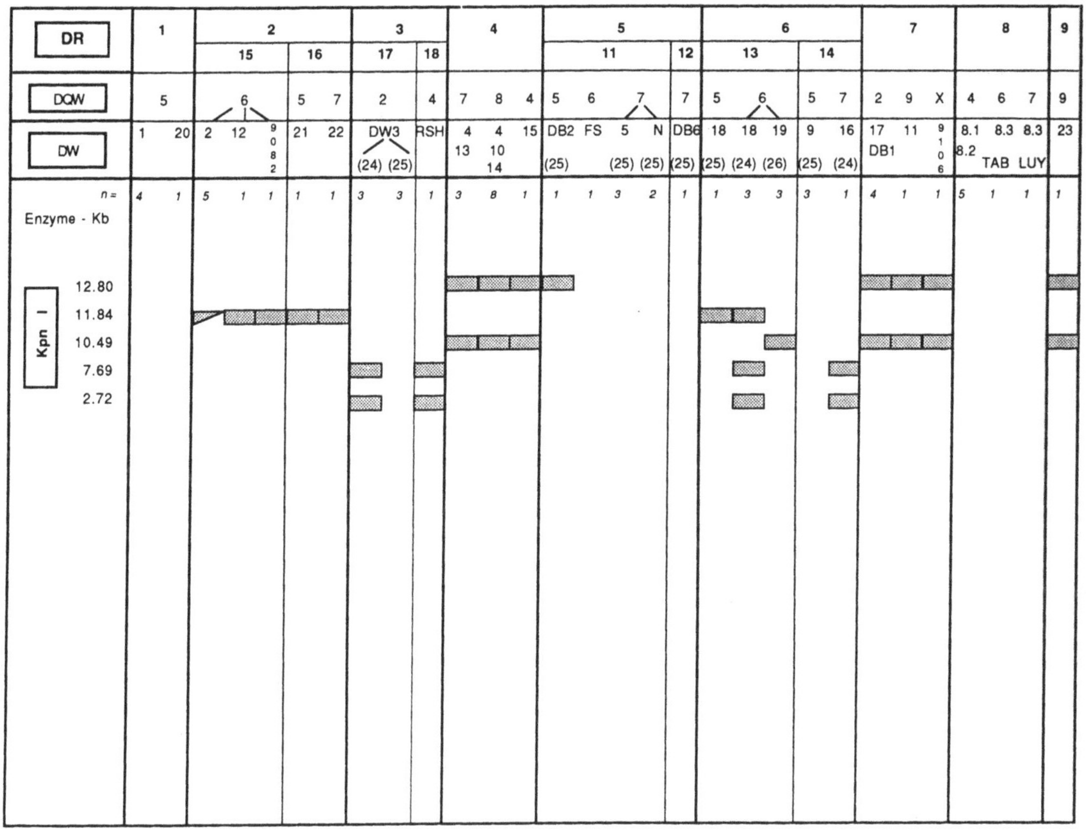

Table 1-8.

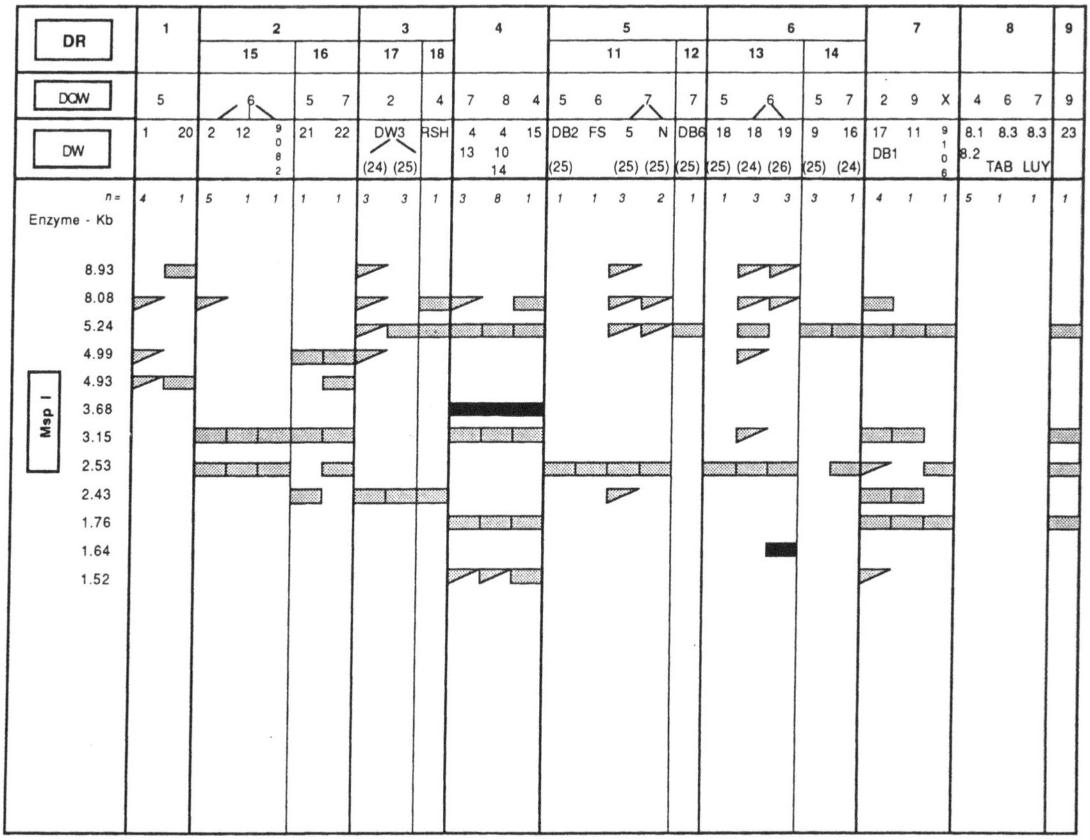

Table 1-9.

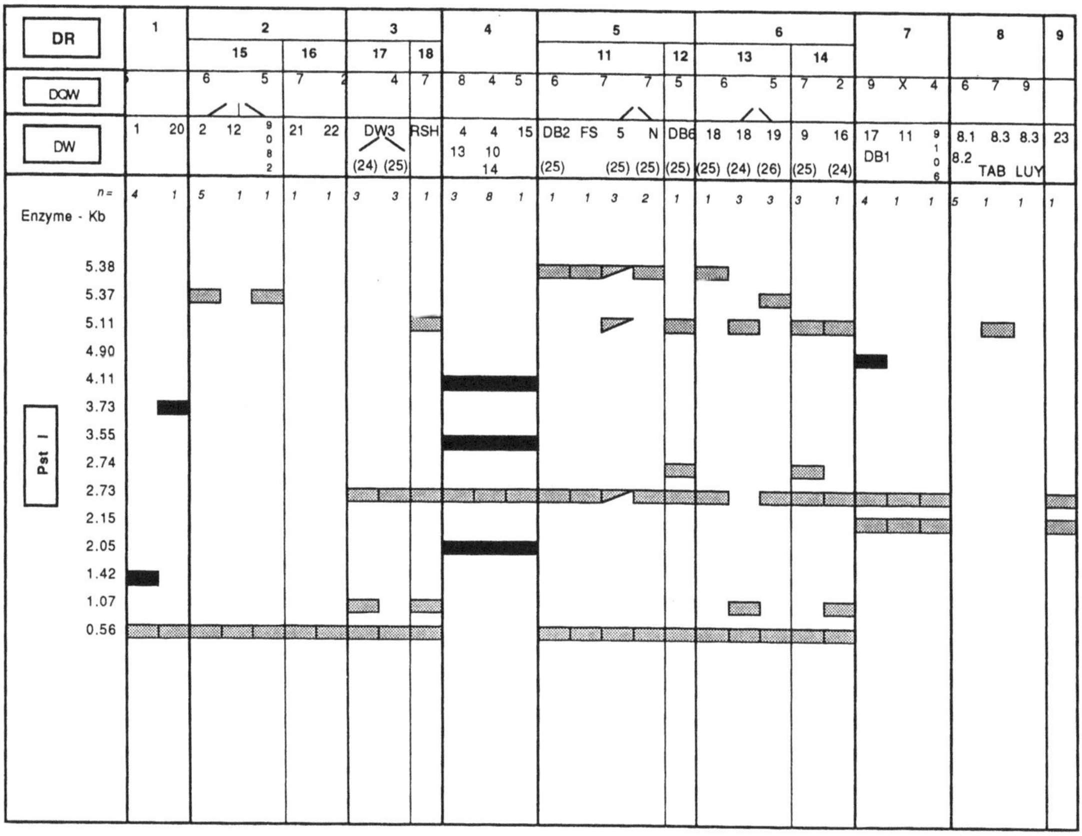

Table 1-10.

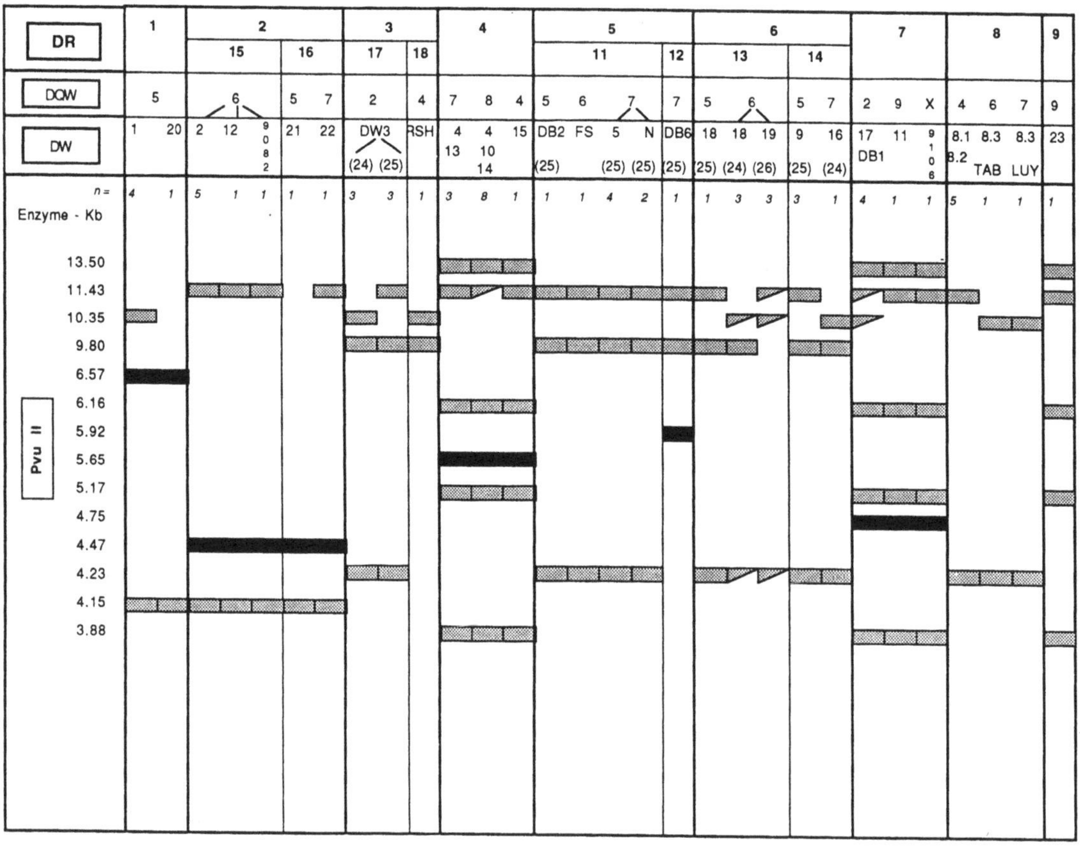

Table 1-11.

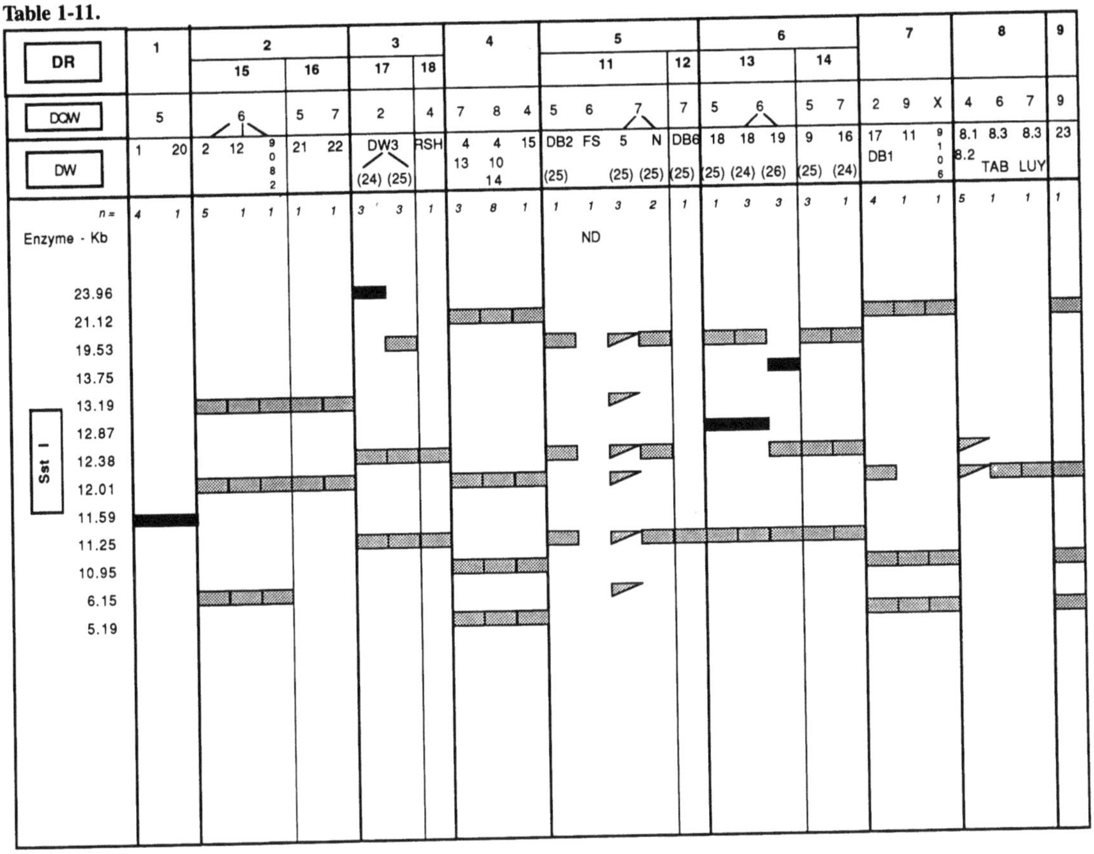

Table 1-12.

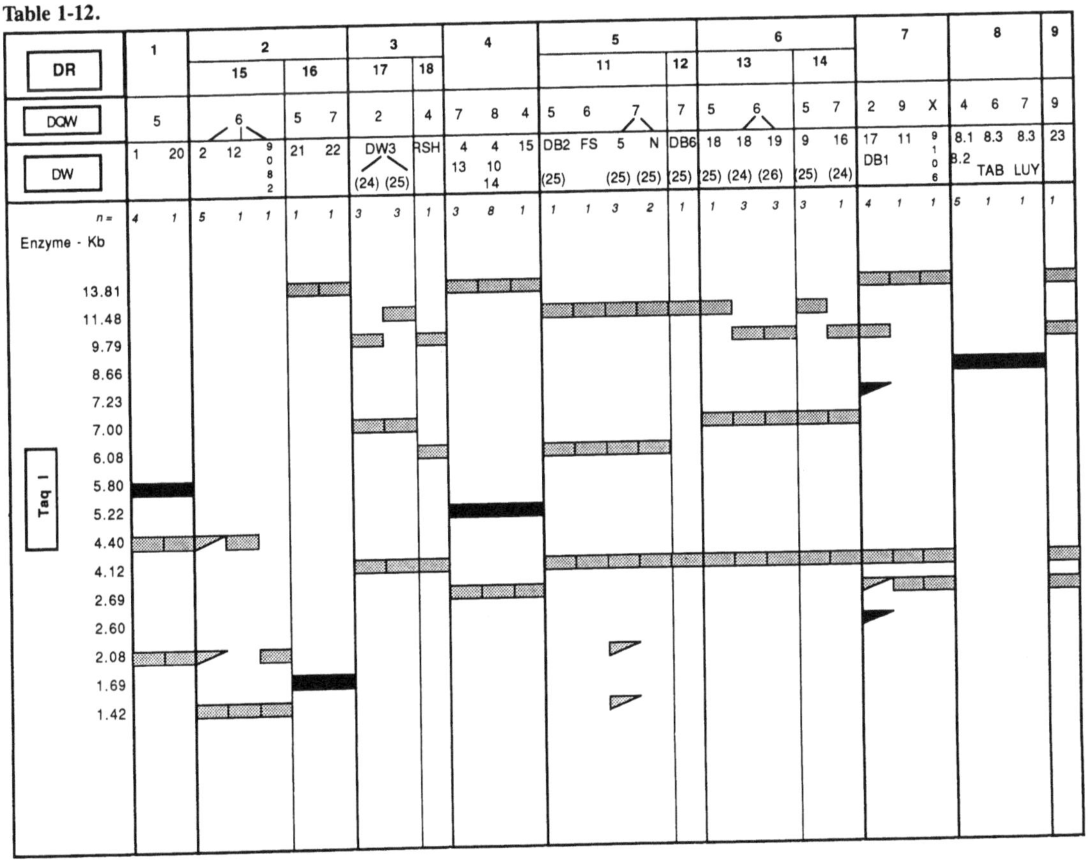

Table 2. DR Alleles and Splits Definition Using Workshop DR Beta Probe Hybridization of DNA from Workshop Cells Digested by 12 Restriction Enzymes

DR4 n = 12 9025,9026, 9028, 9029,9030, 9031, 9032,9033, 9034, 9091,9092, 9107	526 BamH I 1027 Bgl II 2298 EcoR V 158 Hind III 368 Msp I 411 Pst I 355 Pst I 205 Pst I 565 Pvu II 522 Taq I	a: 411 °Hinc II (9025,30,31,91) - see DRW53 for complete data -	a DR4-DQW7 (+ 9031)
DR5 n = 10 9036, 9037, 9038, 9039, 9040, 9042, 9043, 9045, 9104, 9105	no specific band	a: 456 *EcoR I; 253 *Msp I; 423 *Pvu II; 608 *Taq I (9036,37,39,40, 42,43,45,104,105) b: 183 °Hind III; 592 °Pvu II (9045,9038) - see DRW52 for complete data -	a DRW11 b DRW12
DRW6 n = 12 9054, 9055, 9056, 9057, 9058, 9059, 9060, 9061, 9062, 9063, 9064, 9065	no specific band	a: 471 *EcoR I; 270 °Hind III; 1049 *Kpn I; 164 °Msp I; 537 *Pst I; 1375 °Sst I (9055,59,63,56) b: 1184 *Kpn I; 1287 °Sst I (9058,60,62,65) c: 2004 *EcoR V; 274 *Pst I (9054,57,61,56) - see DRW52 for complete data -	a DRW13-DW19 b DRW13-DW18 c DRW14-DW9
DR7 n = 6 9047, 9048, 9050, 9051, 9052, 9106	475 Pvu II	a: 1433 *EcoR I; 844 *Hind III; 808 *Msp I; 490 °Pst I (9047,48,50,51) b: 1513 *EcoR I; 966 °Hind III (9052,106) - see DRW53 for complete data -	a DR7-DQW2 (DW17/DB1) b DR7-DQW9 (DW11) (+ 9106)
DRW8 n = 7 9066, 9067, 9068, 9069, 9070, 9071, 9072	866 Taq I	a: 1143 *Pvu II (9067,68,69,71,72) b: 1035 *Pvu II (9066,70) - see DRW52 for complete data -	a DRW8-DQW4 (DW8.1) b DRW8-DQW6 DRW8-DQW7 (DW8.3)
DR9 n = 1 9075	172 Hind III	Only one cell	

All allele specific fragments are listed (bp = base pairs; Enz = enzyme). The most reliable splits are represented in boxes with the most reliable bands (bp/Enz) allowing their definitions; ° = split-specific band; * = non split-specific band. All cells are referenced with their Workshop numbers (WSNO).

Table 3a. Analysis of DRW52 and DRW53 Supertypic Groups: Identification of Associated DR Alleles

DRW52 (DR3, DR5, DRW6 DRW8)	1824 Eco RV 296 Hind III	a: 946 EcoR I 916 Bgl II (DR3, DR5, DRW6)	a1: 700 Taq I a2: 592 Pvu II, 183 Hind III a3: 608 Taq I b: 866 Taq I	a1 DRW17 ; DRW6 a2 DRW12 a3 DRW11 (+DRW18) b DRW8
DRW 53 (DR4, DR7, DRW9)	428 BamH I 463 BamH I 659 BamH I 1255 EcoR I 410 EcoR I 398 EcoR I 373 EcoR I 717 Hind III 313 Hind III 144 Hind III 176 Msp I 1350 Pvu II 517 Pvu II 388 Pvu II 616 Pvu II 2112 Sst I 1095 Sst I 519 Sst I	DR4	DR7 DR9 - *See DR4, DR7, DRW9 for complete data* -	

Table 3b. Analysis of DRW52 Supertypic Group and Identification of DRW52 Alleles: DW24 (52a), DW25 (52b) and DW26 (52c)

DRW52 (DR3, DR5, DRW6, DRW8)	1824 Eco RV 296 Hind III	= DRW52 a + b a: 980 Pvu II b: 270 Hind III, 1375 Sst I = DRW52 c	= DRW52 b a1: 1148 Taq I a2: 107 Pst I, 272 Kpn I = DRW52 a	a DRW52 a + b DW24-DW25 a1 DRW52 b DW25 a2 DRW52 a DW24 a3 DRW52 c DW26

pattern should be confirmed with a larger number of cells. Here again in DRW 11 cells, no clear DQW and/or DW correlating split could be defined. Among the 11 DRW 6 homozygous cells, DRW 13 - DW 18 - DQW 5; DRW 13 - DW 18 - DQW 6; DRW 13 - DW 19 - DQW 6; DRW 14 - DW 9 - DQW 5; and DRW 14 - DW 16 - DQW 7 could be differentiated using global pattern and split-specific fragments. Only one allele-specific fragment defined DR 7 specificity (4.75 kb PvuII). Inside this general DR 7 pattern, two RFLPs are revealed according to the associated DQW-DW haplotypes (DR 7 -DQW 2 - DW 17 or DB 1 and DR 7 - DQW 9 - DW 11). The DRW 8 haplotype was characterized by an allele-specific band (8.66 kb TaqI), and three different patterns correlating with their DQW typing could be identified. Finally, a DR 9 homozygous cell displayed a 1.72 kb HindIII allele-specific band. However, since only one cell was in the Workshop panel, this pattern should be confirmed.

For each DR allele, Table 2 gives all specific bands and the most reliable splits observed (these splits are defined by specific or nonspecific fragments).

Table 3a shows DRW 52 definition and differentiation of DR 3, DR 5, DRW 6, and DRW 8. DRW 53 specificity is defined by a large number of specific fragments. All of the associated DR alleles (DR 4 - 7 - 9) displayed specific bands. In Table 3b, DRW 52 is shown to be split into three alleles corresponding to DW 24, DW 25, and DW 26 specificities and previously called 52 a, 52 b, and 52c (3,4).

In conclusion, this exhaustive Workshop study using 12 restriction endonucleases describes a large polymorphism with DR beta probe based on RFLP patterns and/or allele- or split-specific fragments. This polymorphism correlates not only with all serologically defined DR alleles and splits, but also with DW typing in some cases. However, we must point out the difficulties in recognizing the DW splits of DR4 and clearly differentiate DR3 and DRW6 with the Workshop DR beta probe. From this point of view, oligonucleotide probes are obviously more useful than full length probes since they identify small variations in nucleotide sequence. Finally, as recently reported (5), the choice of TaqI has to be recommended because of its ability to identify a large polymorphism with DR beta, DQ alpha, and DQ beta hybridization.

Acknowledgments. We would like to thank all the members of Scanticon-Princeton "DRβ group," and A. Vergracht, M.L. Cheneau, F. Quillivic, and V. Gallee for their excellent assistance.

References

1. Southern EM. Detection of specific sequences among DNA fragments separated by gel electrophoresis. J Mol Biol 1975;98:503.
2. Long EO, Wake CT, Gorski J, Mach B. Complete sequence of an HLA-DRβ chain deduced from a cDNA clone and identification of multiple nonallelic DR β chain genes. EMBO J 1983;2:389.
3. Gorski J, Mach B. Polymorphism of human Ia antigens: gene conversion between two DR β loci results in a new HLA-D/DR specificity. Nature 1986;322:67.
4. Tiercy JM, Gorski J, Jeannet M, Mach B. Identification and distribution of three serologically undetected alleles of HLA-DR by oligonucleotide DNA typing analysis. Proc Natl Acad Sci USA 1988;85:198.
5. Bidwell J. DNA-RFLP analysis and genotyping of HLA-DR and DQ antigens. Immunol Today 1988;9:18.

RFLP Analysis of DQ-Beta Chain Gene: Workshop Report

Massimo Trucco[1] and Edward Ball,[2] in collaboration with Carlo Carcassi,[3] Lori Dombrausky,[2] Isabella Cascino,[4] Bernard Boehm,[5] Marie Hoover,[8] Narinder Mehra,[6] Kamala Balakrishnan,[7] and Licinio Contu[3]

Introduction

During the Ninth International Histocompatibility Workshop (1984), several laboratories reported the results of restriction fragment length polymorphism (RFLP) studies that used HLA class II gene probes (1–6). It was apparent from this work and other reports in the literature (7–12) that this new technique could provide a powerful tool for the study of HLA region polymorphism. One of the aims of the Tenth International Histocompatibility Workshop (1987) was to provide a reference standard of RFLP patterns using a battery of restriction enzymes and a panel of well characterized lymphoblastoid B cell lines. These cell lines were chosen because they were recognized to be HLA-D homozygous by the laboratories of origin. Homozygosity was a characteristic that allowed their use in cellular typing and would facilitate the interpretation of the results of the RFLP analysis.

In particular, the specific goals of these exercises were as follows:

1. to verify a standard protocol, developed in preliminary studies, that could be adopted by all laboratories for both:
a. the identification of DQ beta-specific fragments and fragments cross-hybridizing or comigrating with those of other class II genes, and
b. the determination of a consensus list of sizes for the fragments identified by DQ beta probes.
2. to identify of the most informative restriction enzymes for DQ beta allele analysis to determine the minimum number of enzymes required for a complete DQ beta molecular typing.
3. to attempt to correlate DQ beta-specific DNA fragment polymorphisms with DQ specificities identified by serologic and cellular methods to find phenotypic counterparts of the RFLP alleles.

We summarize here the results of this work, which represents the combined efforts of all the participating laboratories. The genomic DNA from 72 cell lines was analyzed for RFLP in each laboratory according to a strict common protocol, with common vendors for key materials. Twelve different endonucleases were used to generate DNA fragments, which were then probed with the same DQ beta-specific cDNA. The organization plan is described in detail in Cohen et al. elsewhere in this volume.

The first step in the analysis of the collected material was the "locus assignment" (i.e., the classification of all the fragments into different categories). The first category included only these fragments that hybridized with the DQ beta probe, called DQ beta "unique" fragments. DQ beta "primary" fragments, included in the second category, were those fragments that gave a strong

Participating Laboratories: US2DUQ,[1] US4STA,[2] IT1CON,[3] IT2TOS,[4] GERBOE,[5] ANZMEH,[6] US5BAL,[7] USIHOO[8]

signal on the autoradiograms when hybridized with the DQ beta probe and a weaker signal when hybridized with the DQ beta probe and a weaker signal when hybridized with other beta probes. The fragments that gave a strong signal on the autoradiograms when probes other than DQ beta were used were included into the third category, the "non"-DQ beta fragments. In this report, only the unique and primary fragments have been subdivided into discrete informative groups.

We used "cluster analysis" to group these fragments; fragments were considered to be part of the same cluster when their correlation coefficients (r values), as calculated from 2 × 2 contingency tables, were > 0.7.

Twenty such clusters were identified, 17 of which were attributed to the DQB1 gene, and three to the DQB2 gene. Of the 17 DQB1 clusters, 14 were found to have a strong association with expressed D-region markers.

The presence and intensity of restriction fragment bands found with each cell line are shown in Tables 1 and 2. The data obtained are presented in a way that stresses the comparison between the serologic and cellular specificities of the cells and fragment clusters. These specificities are indicated in the tables according to the new nomenclature rules (see the Bulletin of the Nomenclature Committee, this volume). The fragment clusters were numbered sequentially for either the DQB1 or the DQB2 locus. The clusters were hereafter referred to as DQB1.1, DQB1.2, DQB1.3, etc. The restriction enzymes used were indicated in the tables with a three-letter code: BAM stands for BamHI, BGL for BglII, ECI

Table 1. DQ Beta RFLP Clusters

<table>
<tr><td colspan="4">SEROLOGY:</td><td colspan="7">DQw1</td><td colspan="3">DQw2</td><td colspan="4">DQw3/w4</td><td></td><td></td><td></td><td></td></tr>
<tr><td colspan="4">CLUSTER #:</td><td>1</td><td>2</td><td>3</td><td>4</td><td>5</td><td>6</td><td>7</td><td>8</td><td>9</td><td>10</td><td>11</td><td>12</td><td>13</td><td>14</td><td>15</td><td>16</td><td></td><td>17</td></tr>
<tr><td colspan="4">RESTRICTION ENZYME :</td><td>HK
IP
DN</td><td>HEHTS
ICIAS
CVCQT</td><td>PBEEBHE
SACCAIC
TMVVMDI
1</td><td>TEE
ACC
QVI</td><td>PBTTP
SAAAV
TMQQU
1</td><td>PK
SP
TN
1</td><td>BTHHP
AAIIV
MQCCU
1</td><td>PTP
SAV
TQU</td><td>HEK
ICP
DVN
112</td><td>HKTEHS
IPACIS
DNQVCT
1 2</td><td>PE
SC
TI</td><td>HHPKKPBTESS
IIVPPVAACSS
DDUNNUMQITT
1</td><td>HBTBTE
IAAGAC
DMQLQI
1 1</td><td>PE
VC
UV
1</td><td>HP
IV
DU</td><td>PT
VA
UQ</td><td>P
V
U</td><td>EST
CSA
ITQ
1</td></tr>
<tr><td colspan="4">SIZE IN KB X 100 :</td><td>65
41
45</td><td>86852
57432
03948</td><td>1522331
3776025
6612372</td><td>152
922
446</td><td>02222
68861
89717</td><td>61
47
35</td><td>84512
75480
20052</td><td>5 1
286
376</td><td>155
419
241</td><td>296075
826064
009160</td><td>45
95
38</td><td>33351 31222
74732350228
71230450326</td><td>401528
729652
860150</td><td>54
75
52</td><td>33
36
00</td><td>24
36
82</td><td>3
4
0</td><td>661
579
158</td></tr>
<tr><td colspan="4">LOCUS ASSIGNED[1] :</td><td>I|</td><td>|I|||</td><td>|I||||I</td><td>||I</td><td>|||||</td><td>|I</td><td>IIIII</td><td>|II</td><td>|||</td><td>||I||I</td><td>|I</td><td>I|I||I|IIII</td><td>I|||II</td><td>||</td><td>||</td><td>|I</td><td>|</td><td>II|</td></tr>
<tr><td colspan="22">HLA-DQw1 Haplotypes</td></tr>
<tr><td>Cell</td><td>DR</td><td>Dw</td><td>DQ</td><td colspan="18">Band Score[2]</td></tr>
<tr><td>9003</td><td>1</td><td>1</td><td>5</td><td>86</td><td>88888</td><td>88688.8</td><td>...</td><td>.....</td><td>..</td><td>.....</td><td>...</td><td>...</td><td>......</td><td>..</td><td>...........</td><td>......</td><td>..</td><td>88</td><td>..</td><td>8</td><td>..8</td></tr>
<tr><td>9004</td><td>1</td><td>1</td><td>5</td><td>66</td><td>66686</td><td>8668688</td><td>...</td><td>.....</td><td>..</td><td>.....</td><td>...</td><td>...</td><td>......</td><td>..</td><td>...........</td><td>......</td><td>..</td><td>.8</td><td>..</td><td>8</td><td>..8</td></tr>
<tr><td>9006</td><td>1</td><td>1</td><td>5</td><td>66</td><td>86888</td><td>8868688</td><td>8..</td><td>.....</td><td>..</td><td>.....</td><td>...</td><td>...</td><td>......</td><td>..</td><td>...........</td><td>......</td><td>..</td><td>.8</td><td>..</td><td>8</td><td>...</td></tr>
<tr><td>9005</td><td>1</td><td>1</td><td>5</td><td>66</td><td>86888</td><td>8868688</td><td>8..</td><td>.....</td><td>..</td><td>.....</td><td>...</td><td>...</td><td>......</td><td>..</td><td>...........</td><td>......</td><td>..</td><td>.8</td><td>..</td><td>8</td><td>...</td></tr>
<tr><td>9002</td><td>1</td><td>20</td><td>5</td><td>66</td><td>86888</td><td>8868688</td><td>...</td><td>.....</td><td>..</td><td>.....</td><td>...</td><td>...</td><td>......</td><td>..</td><td>...........</td><td>......</td><td>..</td><td>.8</td><td>..</td><td>8</td><td>..8</td></tr>
<tr><td>9009</td><td>2</td><td>21</td><td>1</td><td>66</td><td>86886</td><td>866888.</td><td>8..</td><td>.....</td><td>..</td><td>.....</td><td>...</td><td>...</td><td>......</td><td>..</td><td>........6..</td><td>......</td><td>..</td><td>.8</td><td>..</td><td>8</td><td>...</td></tr>
<tr><td>9007</td><td>16/4</td><td></td><td>5/8</td><td>66</td><td>86886</td><td>8668688</td><td>8..</td><td>.....</td><td>..</td><td>.....</td><td>...</td><td>...</td><td>......</td><td>88</td><td>........6..</td><td>468488</td><td>48</td><td>.8</td><td>..</td><td>8</td><td>.4.</td></tr>
<tr><td>9036</td><td>11</td><td>DB2</td><td>5</td><td>86</td><td>86888</td><td>88686..</td><td>...</td><td>.....</td><td>..</td><td>.....</td><td>...</td><td>...</td><td>......</td><td>..</td><td>........6..</td><td>......</td><td>..</td><td>88</td><td>..</td><td>8</td><td>..8</td></tr>
<tr><td>9061</td><td>14</td><td>9</td><td>5</td><td>66</td><td>88888</td><td>88688.8</td><td>...</td><td>.....</td><td>..</td><td>.....</td><td>...</td><td>...</td><td>......</td><td>..</td><td>...........</td><td>......</td><td>..</td><td>88</td><td>..</td><td>8</td><td>..8</td></tr>
<tr><td>9054</td><td>14</td><td>9</td><td>5</td><td>66</td><td>86886</td><td>88686.8</td><td>...</td><td>.....</td><td>..</td><td>.....</td><td>...</td><td>...</td><td>......</td><td>..</td><td>...........</td><td>......</td><td>..</td><td>88</td><td>..</td><td>8</td><td>..8</td></tr>
<tr><td>9056</td><td>14/13</td><td>9/19</td><td>5</td><td>66</td><td>86886</td><td>6668688</td><td>888</td><td>.....</td><td>..</td><td>.....</td><td>...</td><td>...</td><td>......</td><td>..</td><td>...........</td><td>......</td><td>..</td><td>.8</td><td>8.</td><td>8</td><td>...</td></tr>
<tr><td>9056</td><td>14/13</td><td>9/19</td><td>5</td><td>66</td><td>86886</td><td>6668688</td><td>888</td><td>.....</td><td>6.</td><td>.....</td><td>...</td><td>...</td><td>......</td><td>..</td><td>...........</td><td>......</td><td>..</td><td>.8</td><td>8.</td><td>8</td><td>...</td></tr>
<tr><td>9057</td><td>14</td><td>9</td><td>1</td><td>66</td><td>86886</td><td>8668688</td><td>0..</td><td>.....</td><td>..</td><td>.....</td><td>...</td><td>...</td><td>......</td><td>..</td><td>...........</td><td>......</td><td>..</td><td>.8</td><td>..</td><td>8</td><td>.40</td></tr>
<tr><td>9014</td><td>15</td><td>2</td><td>6</td><td>66</td><td>86846</td><td>.......</td><td>888</td><td>86888</td><td>..</td><td>.....</td><td>...</td><td>...</td><td>......</td><td>..</td><td>.....4.....</td><td>......</td><td>..</td><td>88</td><td>..</td><td>.</td><td>...</td></tr>
<tr><td>9008</td><td>15</td><td>2</td><td>6</td><td>66</td><td>86846</td><td>.......</td><td>888</td><td>86888</td><td>.0</td><td>.....</td><td>4..</td><td>...</td><td>.0....</td><td>..</td><td>....0......</td><td>......</td><td>..</td><td>88</td><td>..</td><td>.</td><td>...</td></tr>
<tr><td>9017</td><td>15</td><td>2</td><td>1</td><td>66</td><td>86846</td><td>.......</td><td>888</td><td>86888</td><td>..</td><td>.....</td><td>...</td><td>...</td><td>......</td><td>..</td><td>...........</td><td>......</td><td>..</td><td>88</td><td>..</td><td>.</td><td>.4.</td></tr>
<tr><td>9013</td><td>15</td><td>2</td><td>6</td><td>66</td><td>86846</td><td>.......</td><td>888</td><td>86888</td><td>..</td><td>.....</td><td>...</td><td>...</td><td>......</td><td>..</td><td>...........</td><td>......</td><td>..</td><td>88</td><td>..</td><td>.</td><td>.4.</td></tr>
<tr><td>9010</td><td>15</td><td>2</td><td>6</td><td>66</td><td>86846</td><td>.......</td><td>888</td><td>86888</td><td>..</td><td>.....</td><td>...</td><td>...</td><td>......</td><td>..</td><td>...........</td><td>......</td><td>..</td><td>88</td><td>..</td><td>.</td><td>.4.</td></tr>
<tr><td>9082</td><td>15</td><td>?</td><td>6</td><td>66</td><td>86846</td><td>.......</td><td>888</td><td>86888</td><td>..</td><td>.....</td><td>...</td><td>...</td><td>......</td><td>..</td><td>...........</td><td>......</td><td>..</td><td>88</td><td>..</td><td>.</td><td>.4.</td></tr>
<tr><td>9105</td><td>11</td><td>FS</td><td>6</td><td>66</td><td>66640</td><td>.......</td><td>084</td><td>86888</td><td>.0</td><td>.....</td><td>...</td><td>...</td><td>.....0</td><td>..</td><td>.........00</td><td>......</td><td>..</td><td>88</td><td>..</td><td>.</td><td>.0.</td></tr>
<tr><td>9060</td><td>13</td><td>18</td><td>6</td><td>66</td><td>66646</td><td>.......</td><td>888</td><td>86888</td><td>..</td><td>.....</td><td>...</td><td>...</td><td>......</td><td>..</td><td>...........</td><td>......</td><td>..</td><td>88</td><td>..</td><td>.</td><td>...</td></tr>
<tr><td>9065</td><td>13</td><td>18</td><td>6</td><td>66</td><td>86846</td><td>.......</td><td>888</td><td>86888</td><td>..</td><td>.....</td><td>...</td><td>...</td><td>......</td><td>..</td><td>...........</td><td>......</td><td>..</td><td>88</td><td>..</td><td>.</td><td>...</td></tr>
<tr><td>9062</td><td>13</td><td>18</td><td>6</td><td>66</td><td>86846</td><td>.......</td><td>888</td><td>86888</td><td>..</td><td>.....</td><td>...</td><td>...</td><td>......</td><td>..</td><td>...........</td><td>......</td><td>..</td><td>88</td><td>..</td><td>.</td><td>...</td></tr>
<tr><td>9058</td><td>13</td><td>18</td><td>6</td><td>66</td><td>86846</td><td>.......</td><td>888</td><td>86888</td><td>..</td><td>.....</td><td>...</td><td>...</td><td>......</td><td>..</td><td>...........</td><td>......</td><td>..</td><td>88</td><td>..</td><td>.</td><td>.4.</td></tr>
<tr><td>9055</td><td>6</td><td>N</td><td>6</td><td>66</td><td>86886</td><td>.......</td><td>888</td><td>.....</td><td>66</td><td>.....</td><td>...</td><td>...</td><td>....8.</td><td>..</td><td>...........</td><td>......</td><td>..</td><td>88</td><td>8.</td><td>.</td><td>...</td></tr>
<tr><td>9063</td><td>13</td><td>19</td><td>6</td><td>66</td><td>66686</td><td>.......</td><td>888</td><td>.....</td><td>66</td><td>.....</td><td>...</td><td>...</td><td>......</td><td>..</td><td>...........</td><td>......</td><td>..</td><td>88</td><td>8.</td><td>.</td><td>.4.</td></tr>
<tr><td>9059</td><td>13</td><td>19</td><td>6</td><td>66</td><td>86886</td><td>.......</td><td>888</td><td>.....</td><td>66</td><td>.....</td><td>...</td><td>...</td><td>......</td><td>..</td><td>...........</td><td>......</td><td>..</td><td>88</td><td>8.</td><td>.</td><td>.4.</td></tr>
<tr><td>9066</td><td>8</td><td>8.3</td><td>6</td><td>66</td><td>.....</td><td>.......</td><td>...</td><td>.....</td><td>..</td><td>68648</td><td>.48</td><td>...</td><td>..8...</td><td>..</td><td>4..........</td><td>......</td><td>..</td><td>8.</td><td>8.</td><td>4</td><td>8..</td></tr>
<tr><td>9011</td><td>15</td><td>12</td><td>6</td><td>66</td><td>.....</td><td>.......</td><td>...</td><td>.....</td><td>..</td><td>68666</td><td>.48</td><td>...</td><td>..8...</td><td>..</td><td>4..........</td><td>......</td><td>..</td><td>84</td><td>8.</td><td>.</td><td>8..</td></tr>
<tr><td colspan="22">HLA-DQw2 Haplotypes</td></tr>
<tr><td>Cell</td><td>DR</td><td>Dw</td><td>DQ</td><td colspan="18"></td></tr>
<tr><td>9018</td><td>3</td><td>3</td><td>2</td><td>..</td><td>.....</td><td>.......</td><td>...</td><td>.....</td><td>..</td><td>.....</td><td>848</td><td>886</td><td>......</td><td>..</td><td>........8..</td><td>......</td><td>..</td><td>.8</td><td>88</td><td>.</td><td>...</td></tr>
<tr><td>9019</td><td>3</td><td>3</td><td>2</td><td>..</td><td>.....</td><td>.......</td><td>...</td><td>.....</td><td>..</td><td>.....</td><td>848</td><td>886</td><td>......</td><td>..</td><td>........8..</td><td>......</td><td>..</td><td>.6</td><td>88</td><td>.</td><td>...</td></tr>
<tr><td>9020</td><td>3</td><td>3</td><td>2</td><td>..</td><td>.....</td><td>......8</td><td>...</td><td>.....</td><td>..</td><td>.....</td><td>808</td><td>886</td><td>......</td><td>..</td><td>...........</td><td>......</td><td>..</td><td>.6</td><td>88</td><td>.</td><td>...</td></tr>
<tr><td>9022</td><td>3</td><td>3</td><td>2</td><td>..</td><td>.....</td><td>.......</td><td>...</td><td>.....</td><td>..</td><td>.....</td><td>848</td><td>886</td><td>......</td><td>..</td><td>........8..</td><td>......</td><td>..</td><td>.6</td><td>88</td><td>.</td><td>...</td></tr>
<tr><td>9023</td><td>3</td><td>3</td><td>2</td><td>..</td><td>.....</td><td>.......</td><td>...</td><td>.....</td><td>..</td><td>.....</td><td>848</td><td>886</td><td>......</td><td>..</td><td>........8..</td><td>......</td><td>..</td><td>.6</td><td>88</td><td>.</td><td>...</td></tr>
<tr><td>9088</td><td>3</td><td>?</td><td>2</td><td>..</td><td>.....</td><td>.......</td><td>...</td><td>.....</td><td>..</td><td>.....</td><td>848</td><td>886</td><td>......</td><td>..</td><td>........8..</td><td>......</td><td>..</td><td>.6</td><td>88</td><td>.</td><td>...</td></tr>
<tr><td>9050</td><td>7</td><td>17</td><td>2</td><td>..</td><td>.....</td><td>....6..</td><td>...</td><td>.....</td><td>..</td><td>.....</td><td>848</td><td>...</td><td>868646</td><td>..</td><td>4..........</td><td>.....8</td><td>..</td><td>..</td><td>8.</td><td>.</td><td>...</td></tr>
<tr><td>9051</td><td>7</td><td>17</td><td>2</td><td>..</td><td>.....</td><td>....6..</td><td>...</td><td>.....</td><td>..</td><td>.....</td><td>848</td><td>...</td><td>868666</td><td>..</td><td>4..........</td><td>.....8</td><td>..</td><td>..</td><td>8.</td><td>.</td><td>...</td></tr>
<tr><td>9047</td><td>7</td><td>DB1</td><td>2</td><td>..</td><td>.....</td><td>....6..</td><td>...</td><td>.....</td><td>..</td><td>.....</td><td>848</td><td>...</td><td>868666</td><td>..</td><td>4..........</td><td>.....8</td><td>..</td><td>..</td><td>8.</td><td>.</td><td>...</td></tr>
<tr><td>9048</td><td>7</td><td>DB1</td><td>2</td><td>..</td><td>.....</td><td>....6..</td><td>...</td><td>.....</td><td>..</td><td>.....</td><td>808</td><td>...</td><td>868846</td><td>..</td><td>4..........</td><td>......</td><td>..</td><td>..</td><td>8.</td><td>.</td><td>8..</td></tr>
<tr><td>9104</td><td>11?</td><td>8?</td><td>7?</td><td>..</td><td>.....</td><td>....6..</td><td>...</td><td>.....</td><td>..</td><td>.....</td><td>808</td><td>...</td><td>868646</td><td>..</td><td>4.64......4</td><td>.....8</td><td>..</td><td>..</td><td>8.</td><td>.</td><td>...</td></tr>
</table>

Table 1. *Continued*

SEROLOGY:				|---------DQw1						---------|	|----DQw2		----|	|--------DQw3/w4		-------|			
CLUSTER #:				1	2	3	4	5	6	7	8	9	10	11	12	13	14	15	16		17
RESTRICTION ENZYME:				HK	HEHTS	PBEEBHE	TEE	PBTTP	PK	BTHHP	PTP	HEK	HKTEHS	PE	HHPKKPBTESS	HBTBTE	PE	HP	PT	P	EST
				IP	ICIAS	SACCAIC	ACC	SAAAV	SP	AAIIV	SAV	ICP	IPACIS	SC	IIVPPVAACSS	IAAGAC	VC	IV	VA	V	CSA
				DN	CVCQT	TMVVMDI	QVI	TMQQU	TN	MQCCU	TQU	DVN	DNQVCT	TI	DDUNNUMQITT	DMQLQI	UV	DU	UQ	U	ITQ
SIZE IN KB X 100:						1		1	1	1		112	1 2		1	1 1	1				1
				65	86852	1522331	152	02222	61	84512	5 1	155	296075	45	33351 31222	401528	54	33	24	3	661
				41	57432	3776025	922	68861	47	75480	286	419	826064	95	74732350228	729652	75	36	36	4	579
				45	03948	6612372	446	89717	35	20052	376	241	009160	38	71230450326	860150	52	00	82	0	158
LOCUS ASSIGNED[1]:				I|	|I|||	|I||||I	||I	|||||	|I	IIIII	|II	|||	||I||I	|I	I|I||I|IIII	I|||II	||	||	|I	|	II|
HLA-DQw3 and HLA-DQw4 Haplotypes																					
Cell	DR	Dw	DQ																		
9037	11	5	7	..	6.606	....	...	0688.	..	.....	...	...	......	04	488664.....	......	..	..	88	.	848
9045	11/12	5	7	.6	.....	4	...	.....	..	.....	.0.	...	......	86	488...64666	......	..	..	68	.	6..
9043	11	5	7	..	.....	....	...	.....	..	.....	...	...	......	86	48866464.66	......	..	..	88	.	8..
9040	11	5	7	..	.....	....	...	.....	..	.....	...	...	......	86	48866464866	......	..	..	88	.	...
9042	11	N	7	..	.....	....	...	.....	..	.....	...	...	......	86	48866464.66	......	..	..	88	.	6..
9039	11	N	7	..	.....	....	...	.....	..	.....	...	...	......	84	48866484.66	......	..	..	88	.	.4.
9038	12	DB6	7	..	.....	....	...	.....	..	.....	...	...	......	84	48866664866	......	..	..	88	.	8..
9016	16	22	7	..	.....	....	...	.....	..	.....	...	...	......	84	48866460866	......	..	..	88	.	...
9070	8	8.3	7	..	.....	....	...	.....	..	.....	...	...	......	86	48866484866	......	..	..	88	.	...
9064	14	16	7	..	.....	....	...	.....	..	.....	...	.8.	..0...	8.	48866484866	......	..	..	88	.	...
9025	4	4	7	..	.....	....	...	.....	..	.....	...	...	..0...	86	48866664866	......	68	..	88	.	...
9030	4	13	7	..	.....	....	...	.....	..	.....	...	...	..0...	84	48866464866	......	68	..	88	.	...
9091	4	4	7	..	.....	....	...	.....	..	.....	...	...	..0...	84	48866464866	......	68	..	88	.	...
9032	4	4	8	..	.....	....	...	.....	..	.....	...	...	......	84	....	468688	68	88	..	8	.4.
9032	4	4	8	..	.....	....	...	.....	..	.....	...	...	......	84	....	468688	68	88	..	8	64.
9029	4	4	8	..	.....	....	...	.....	..	.....	...	...	......	84	....	468688	48	88	..	8	.4.
9031	4	4	8	..	.....	....	...	.....	..	.....	...	...	......	84	....	468688	68	88	..	8	.4.
9033	4	4	8	..	.....	....	...	.....	..	.....	...	...	......	84	....	468688	68	88	..	6	.4.
9034	4	4	8	..	.....	....	...	.....	..	.....	...	..4	......	84	....	468688	68	88	..	8	.4.
9026	4	10	8	..	.....	....	...	.....	..	.....	...	...	......	84	....	468488	68	88	4.	8	.4.
9028	4	14	8	..	.....	....	...	.....	..	.....	...	...	......	84	....	468688	68	88	..	8	.4.
9092	4	14	8	..	.....	....	...	.....	..	.....	...	..4	......	84	....	468688	48	88	..	8	...
9075	9	23	9	..	.....	....	...	.....	..	.....	6..	...	...6..	84	........6..	46868.	.8	88	..	8	...
9052	7	11	9	..	.....	....	...	.....	..	.....	6..	...	....4.	84	........6..	46868.	..	88	.0	8	...
9106	7	?	2/3	..	.....	....68.	...	.....	..	.....	6.4	...	668...	6.	....	4686..	..	.6	4.	6	8..
9107	4	15	4	..	...4.	....	...	.....	..	.....	...	...	......	84	....	468688	48	88	..	8	...
9021	3	N	4	..	.....	....	...	.....	..	.....	...	...	......	84	....	468688	..	88	..	8	...
9069	8	8.1	4	..	.....	....	...	.....	..	.....	4..	...	......	84	....	4686.8	..	88	..	8	...
9068	8	8.1	4	..	.....	....	...	.....	..	.....	6..	..4	......	84	....	468688	..	88	..	8	...
9067	8	8.1	4	..	.....	....	...	.....	..	.....	6..	...	......	84	....	468688	..	88	..	8	...
9071	8	8.2	4	..	.....	....	...	.....	..	.....	...	...	......	84	....	468686	..	88	..	8	...
9072	8	8.2	4	..	.....	....	...	.....	..	.....	4..	...	......	84	....	468488	..	88	..	8	...

[1]I = DQ beta "unique" fragment; | = DQ beta "primary" fragment.

[2]Band score is based on intensity, i.e., 8 = double intensity, 6 = single intensity, 4 = low intensity, . = absent, 0 = not determined.

for EcoRI, ECV for EcoRV, HIC for HincII, HID for HindIII, KPN for KpnI, MSP for MspI, PST for PstI, PVU for PvuII, SST for SstI, and TAQ for TaqI.

Results

Table 1 shows DQB1 (and putative DQB1) fragments, while Table 2 shows putative DQB2 fragments. Data for cells 9056 and 9032 are included twice in each table because they represented duplicate DNA samples included in the cell list as blind internal controls. The results obtained with these duplicated samples were nearly identical, indicating the high degree of reproducibility obtained in these exercises.

DQB1 RFLP Clusters

In Table 1, one observes two overlapping series of fragment clusters. The first series contains clusters that correspond well to DQw1 (cluster DQB1.1, 28/28 cases). DQw2 (cluster DQB1.8, 10/10), and DQw3+DQw4 (cluster DQB1.11, 31/31). Duplicated controls and one cell (9104) with uncertain typing (see below) were not considered in these case numbers. The second series results in allelic "splits" of these broader specificities.

DQw1 Splits. DQw1 (cluster DQB1.1) had a number of splits. Cluster DQB1.3 correlated well with DQw5 (12/12), and included haplotypes DRw1,Dw1,DQw5, DRw16,Dw-,DQw5, and DRw14,Dw9,DQw5.

No cluster correlated completely with DQw6, although cluster DQB1.4 was present in all but two of the DQw6 cells (14/16) and represents the DQw6-defining fragments in several reports in the literature (4,7,9). Three apparent splits of DQw6 were observed. Cluster DQB1.5 was associated with DRw15,Dw2,DQw6 and DRw13,Dw18,DQw6 haplotypes (10/10). Cluster DQB1.6 was found in DRw13,Dw19,DQw6 haplotypes (3/4). A third split of DQw6 was shown by cluster DQB1.7 and was found in the DRw8,Dw8.3,DQw6 (1/1) and DRw15,Dw12,DQw6 (1/1) cells.

Cluster DQB1.2 was found in all DQw1 haplotypes except those positive for cluster DQB1.7 (26/28).

In the workshop analysis, cell 9007 turned out to be a DRw16/DR4 and DQw5/DQw8 heterozygote. We hypothesize that the haplotypes present in this cell are DRw16,DQw5,DQB1.3 and DR4,DQw8,DQB1.13.

Cell 9056 which, with the Workshop reagents, was recognized as a heterozygote DRw13/w14 and Dw9/19, by RFLP also would appear to be a heterozygote DQw5/DQw6 (clusters DQB1.3 and DQB1.4). The failure to identify by serology its heterozygosity at the DQ locus (DQw6 allele was not detected) is presumably due to the current paucity of reagents able to define these splits. This is an example where RFLP analysis adds a further level of detail to serologic HLA typing. We attributed the absence of a full DQB1.6 cluster pattern on the DNA of this cell line to its heterozygous status. In heterozygotes, any specific band is expected to be half the intensity of that in homozygotes.

There were two DRw11,DQw1 cells in the data set. One was DQw5,-DQB1.3 (cell 9036) and the second was DQw6,DQB1.5 (cell 9105). It may be that the DRw11, DQw1 cells represent the outcome of two independent cross-over events, one involving a DQw5(w1) donor haplotype (11) and the second a DQw6(w1) donor haplotype. Whether other differences exist between these two haplotypes may be of interest in future work.

DQw2 Splits. DQw2 (cluster DQB1.8) was sharply split into two subsets corresponding to DR3,DQw2 (cluster DQB1.9, 6/6) and DR7,DQw2 (cluster DQB1.10, 5/5) haplotypes.

Cell 9104 was exceptional in that the cell DHIF was reported to be DRw11,DQw7 serologically, but clearly fell into the DR7,DQw2 RFLP pattern, being DQB1.8- and DQB1.10- positive. This DNA sample also appeared

Table 2. DQB2 RFLP Clusters

10WS #	Cell & Name	A	C	B	DR	52/53	DQ	DP	Dw	DQB2 1	DQB2 2	DQB2 3
										PMETHS	PPM	PPMMM
										SSCAIS	SSS	SSSSS
										TPVQDT	TTP	TTPPP
										1 1		
										110296	839	64281
										299957	302	31613
										267955	313	97270
									LOCUS ASSIGNED[1] :	|XXXXX	|X|	X|X|X
		HLA Serology								Band Score[2]		
HLA-DQw1 Haplotypes												
9003	KAS116	24	-	51	1		5	-	1	...4..	...	48664
9004	JESTHOM	2	1	27	1		5	4	1	......	...	.866.
9006	WT100BIS	11	4	35	1		5	1	1	...4..	...	48664
9005	HOM2	3	1	27	1		5	5	1	......	...	48664
9002	MZ070782	24	2,8	14	1		5	4	20	......	..6	486.4
9009	KAS011	1	6	37	2		1	3/4	21	......	...	4864.
9007	DEM	2	6	57	16/4	53	5/8	3/4	X	644444	...	48644
9036	SP0010	2	5	44	11	52	5	2	DB2,25	......	...	48664
9061	31227ABO	2	7	18	14	52	5	4	9,25	......	...	48664
9054	EK	2	5	44	14	52	5	-	9,25	.6....	686	.6...
9056	KOSE	2	-	35	13/14	52	5	2/N	9/19,25	......	...	68664
9056	KOSE	2	-	35	13/14	52	5	2/N	9/19,25	......	...	68664
9057	TEM	26	-	38	14	52	1	4	9,25	844444	...	.6...
9014	MGAR	26	7	8	15		6	4	2	...0..	...	48664
9008	D0208915	25	-	18	15		6	2/4	2	6644..	...	.86..
9017	WT8	3	7	7	15		1	2/4	2	66.444	466	.....
9013	SCHU	3	7	7	15		6	-	2	664444	...	.....
9010	AMAL	28	4	53	15		6	-	2	6..444	...	..6..
9082	H0104	3	7	7	15		6	4		664444	...	.....
9105	FPAF	1	4	35	11	52	6	2	FS	...040	...	4866.
9060	CB6B	1	9	62	13	52	6	-	18,25	.4....	686	.....
9065	HHKB	3	7	7	13	52	6	4	18,24	.6....	684	.....
9062	WDV	2	-	38	13	52	6	4	18,24	......	...	48664
9058	OMW	2	-	45	13	52	6	1	18,24	66.444	...	.....
9055	H0301	3	8	14	6	52	6	5	N,26	66.4..	..6	.....
9063	WT47	32	5	44	13	52	6	2	19,26	664444	...	.....
9059	SLE005	2	10	60	13	52	6	3	19,26	864444	...	.....
9066	TAB089	2	11	46	8	52	6	-	8.3	......	...	48664
9011	E4181324	1	-	52	15		6	2/4	12	86....	...	.8.6.
HLA-DQw2 Haplotypes												
9018	L0081785	3	5	18	3	52	2	3	3,25	...4..	...	4866.
9019	DUCAF	30	5	18	3	52	2	-	3,25	......	...	6866.
9020	QBL	26	?5	18	3	52	2	2	3,25	......	...	6866.
9022	COX	1	7	8	3	52	2	3	3,24	.6.4..	686	.....
9023	VAVY	1	7	8	3	52	2	1	3,24	.6.4..	686	.....
9088	PF04015	1		8	3	52	2	1/4	24	.6....	686	.....
9050	MOU	29	-	44	7	53	2	2	17	......	...	4866.
9051	PITOUT	29	-	44	7	53	2	4	17	......	...	4866.
9047	PLH	3	6	47	7	53	2	-	DB1	...0..	...	4866.
9048	LBUF	30	6	13	7	53	2	-	DB1	......	...	68664
9104	DHIF	31		38	11	52	7	NT	5	......	...	48664

Table 2. *Continued*

10WS Cell # & Name	A	C	B	DR	52/53	DQ	DP	Dw	DQB2 1	DQB2 2	DQB2 3
									PMETHS SSCAIS TPVQDT 1 1 110296 299957 267955	PPM SSS TTP 839 302 313	PPMMM SSSSS TTPPP 64281 31613 97270
LOCUS ASSIGNED[1] :									|IIIII	|I|	I|I|I
	HLA Serology								Band Score[2]		
HLA-DQw3 and HLA-DQw4 Haplotypes											
9037 SWEIG007	29	2	61	11	52	7	-	5,25	0..4.4	...	4666.
9045 TUBO	2/3	7	51	11/12	52	7	2/4	5,25	......	...	4866.
9043 BM21	1	-	41	11	52	7	-	5,25	...4..	...	48664
9040 BM15	1	7	49	11	52	7	3	5,25	...0..	..6	486.4
9042 TISI	24	4	35	11	52	7	-	N,25	......	...	4866.
9039 JVM	2	5	18	11	52	7	2	N,25	864444	...	.....
9038 BM16	2	7	18	12	52	7	2	DB6,25	......	...	48664
9016 RML	2	-	51	16		7	-	22	......	...	48664
9070 LUY	2	-	51	8	52	7	1/4	8.3	......	...	4866.
9064 AMALA	2	9	62	14	52	7	-	16,24	......	...	4864.
9025 DEU	31	4	35	4	53	7	4	4	......	...	48664
9030 JHAF	31	-	51	4	53	7	3	13	...4..	...	48664
9091 MLF	2	9	62	4	53	7	-	4	......	...	48664
9032 BSM	2	9	62	4	53	8	2	4	6644.4	...	.....
9032 BSM	2	9	62	4	53	8	2	4	6644.4	...	.....
9029 WT51	23	8	65	4	53	8	2	4	864444	...	.....
9031 BOLETH	2	10	62	4	53	8	4	4	864444	...	.....
9033 BM14	3	7	7	4	53	8	4	4	664444	...	.....
9034 SAVC	3	7	7	4	53	8	-	4	864444	...	.....
9026 YAR	26	-	38	4	53	8	4	10	864444	...	.....
9028 PE117	24	10	60	4	53	8	4	14	6444.4	...	48664
9092 BM92	25	1	51	4	53	8	-	14	......	...	48664
9075 DKB	24	10	60	9	53	9	4	23	......	...	48664
9052 DBB	2	6	57	7	53	9	4	11	....4.	...	68864
9106 MANIKA	3	-	50	7	53	2?3	NT		......	46.	.644.
9107 LFT3	24	1	54	4	53	4	5	15	..4...	...	48664
9021 RSH	30/68	2	42	3	52	4	1	N,24	..44..	...	6866.
9069 MADURA	2	10	60	8	52	4	4	8.1	......	...	48664
9068 BM9	2	4	35	8	52	4	2	8.1	......	...	48664
9067 BTB	2	1	27	8	52	4	4	8.1	......	...	48664
9071 OLGA	31	1/10	62	8	52	4	3	8.2	......	...	48664
9072 SPL	31	1	62	8	52	4	-	8.2	......	...	486.4

[1]I = DQ beta "unique" fragment; | = DQ beta "primary" fragment.
[2]Band score is based on intensity: 8 = double intensity, 6 = single intensity, 4 = low intensity,   = absent, 0 = not determined.

to be from a DR7,DQw2 haplotype when probed with DQ alpha and DR beta probes (see DQ Alpha and DR Beta RFLP Workshop Reports, this volume). Thus, it seems likely that this represents a sample mixup rather than an unusual D-region haplotype.

DQw3 and DQw4 Splits. Cluster DQB1.11 was shared by DQw3 and DQw4 haplotypes, while cluster DQB1.12 correlated well with the DQw7 split of DQw3 (13/13). An allelic cluster (DQB1.13) was found in all DQw8 (10/10), DQw9 (3/3), and DQw4 (7/7) haplotypes. No fragments were found to uniquely identify the DQw8, DQw9, and DQw4 haplotypes, although certain fragments (Table 1) may be useful in distinguishing these DQ types. Cluster DQB1.14, for instance, appeared in all DR4 haplotypes, regardless of the DQw type of the cell. One must question whether this might represent a fortuitous cross-reaction with DR beta or a real DQ beta fragment. This cluster had to be included here because of the "primary" locus assignment of its two fragments to DQ-beta. Restriction fragment ECI 1820 (included in cluster DQB1.13) may be of use to distinguish DQw4 and DQw8 cells from DQw9 cells. This band was present in all DQw4 (7/7) and DQw8 (9/9) cells, but absent in DQw9 (0/3) cells. It should be noted, however, that this band also appeared in most of DR7,DQw2 cells (3/4).

RFLP Clusters Without Evident Serologic Correlates. A number of DQ beta clusters had no immediately evident correlation with known D-region markers. Many of these clusters were composed of weak bands and are not shown here. However, some such clusters contained intense bands and had "unique" or "primary" locus assignment to DQ-beta. Clusters DQB1.15 to .17 are of this type. These clusters may sometimes be useful in HLA RFLP typing. For example, clusters DQB1.15 and DQB1.16 associate with the DQw7 and DQw8,9,4 splits, respectively. Unfortunately, the allelism of these two clusters is not immediately evident in the DQw1 and DQw2 haplotypes.

DQB2 RFLP Clusters

Table 2 presents the restriction fragments putatively attributed to the DQB2 gene. These clusters have been said to represent DQB2 by a number of investigators because they appear to form an allelic series independent of DQB1 (12–15). In separate experiments, these bands were found to be very weak when hybridized with DQB1 first domain-specific probes, providing more direct evidence that these truly represent DQB2 fragments (16).

The cells are arranged in Table 2 in the same order as in Table 1. Three allelic clusters were identified, DQB2.1, .2, and .3. It became apparent that there is linkage disequilibrium between some DQB2 alleles and other HLA-D-region markers. For this reason, the data are discussed separately for each of the broader DQB1 groups.

DQB2 Alleles in DQw1 Haplotypes. All three DQB2 clusters appeared in DQw1-positive cells. All but two DQw5 haplotypes were DQB2.3 (10/12). The two exceptions were both DRw6–TEM (9057), which was DQB2.1, and EK (9054) DQB2.2. Cell DEM (9007) is a heterozygote, probably DRw16,DQw5,DQB2.3/DR4, DQw8-DQB2.1.

The distribution of DQB2 alleles among DQw6 cells was more complex. Most of the DQw6 cells were DQB2.1 (10/16). All but one of the DRw15,DQw6 cells were DQB2.1 (6/7). Cell MGAR (9014) was DQB2.3. Cells WT8 (9017) and E4181324 (9011) were heterozygous at the DP locus (Table 2), and were found to be DQB2.1/DQB2.2 and DQB2.1/DQB2.3, respectively. Most of the DRw13(w6),DQw6 cells also were DQB2.1 (4/7). Cells HHKB (9065) and CB6B (9060) were both DQB2.2, while cell WDV (9062) was DQB2.3. The distribution of the DQB2 alleles among the DRw6 haplotypes did not appear to be associated with either the HLA-Dw or DRw52 subtypes (Dw24,25,26).

DQB2 Alleles in DQw2 Haplotypes. HLA-DQw2-positive cells were found to be either DQB2.2 or DQB2.3. The data indicate a clear distinction of two extended haplotypes among the DR3,DQw2-positive cells. All of the B8,DR3,DQw2 cells (3/3) were DQB2.2, while all of the B18,DR3,DQw2 cells (3/3) were DQB2.3. Also, all of the DR7,DQw2 cells were DQB2.3 (4/4).

DQB2 Alleles in DQw3 and DQw4 Haplotypes. Only DQB2.1 or DQB2.3 fragment clusters appeared in DQw3 and DQw4 cells. All of the DQw4 cells (7/7) were DQB2.3. Among the DQw3 cells, DQB2 alleles correlated quite well with the DQw7, DQw8, and DQw9 subtypes. Both of the DQw9 cells (9052 and 9075), and all but one of the DQw7 cells (12/13) were DQB2.3. Cell JVM (9039) was DQB2.1. All but one of the DQw8 cells (8/9) were DQB2.1. Cell BM92 (9092) was DQB2.3.

Cell PE117 (9028) may be a DQ heterozygote since it was found to be both DQB2.1 and DQB2.3.

It is of interest that the DQw8 haplotype, which is not distinguishable from DQw4 and DQw9 by DQB1 RFLP clusters (DQB1.13, Table 1), may be differentiated by DQB2 cluster (DQB2.1 for DQw8 versus DQ2.3 for DQw4 and DQw9). This points out the possible utility of DQB2 RFLP analysis in identifying certain DQB1 subtypes.

The data obtained with DQB2 RFLP analysis seem to indicate the existence of linkage disequilibrium between DQB2 and DQB1 alleles, as would be expected by their proximity in the HLA-D-region. This association is not as easily appreciated when one considered only the serological DQw1, 2, and 3 alleles of the DQB1 gene. However, when one considers the more recently described allelic subtypes of DQ and the "extended haplotypes" present for example in the DR3 family (i.e., B8,DR3 versus B18,DR3), the correlations were quite striking (also see ref. 12). These observations may be cause for reevaluation of the proposed recombination "hot spot" between DQB1 and DQB2. We might propose that the particular DQB1/DQB2 allele associations were fixed at (or prior to) the appearance of the DQB1 subtypes.

Concluding Remarks

These exercises were conducted by a large number of laboratories. The goal was to establish a DQ beta RFLP database using B lymphoblastoid cell lines that had been typed for HLA using the best available serologic and cellular reagents. This report, therefore, may be used as a reference to provide DQ beta RFLP patterns for prototypic HLA haplotypes with which individual laboratories may compare their RFLP typings. The enzymes PstI, TaqI, and KpnI seem to be sufficient to completely type for the clusters defined in this report; it is recommended that they be used in parallel.

An important goal of this component of the Workshop was to compare the identified RFLP clusters with the expressed D-region products detected by serology and T-cell typing. One expectation was that RFLP analysis would reveal further polymorphism of the DQ beta gene that cannot be detected by serologic or cellular means. Although additional polymorphisms were seen in many cases, it was clear in other cases the RFLP analysis may have a limited ability to identify DQ beta polymorphisms. Keeping this in mind, the data presented here seem to fall into three categories.

First, some of the RFLP clusters were well correlated with the complexity of the D-region products. For example, the DQB1.12 RFLP cluster in Table 1 correlated perfectly with the serologic specificity of DQw7. Even more precisely, DQB1.12 correlated with the DQ beta nucleotide sequence variability shown in Table 3.

Second, in some instances RFLP clusters identified D-region polymorphism not currently recognized in expressed products. For example, the DR3,DQw2 and DR7,DQw2 haplotypes are distinguished by the DQB1.9 and DQB1.10 clusters (Table 1). As far as is known, the expressed DQw2 beta genes are identical in nucleotide sequence (Table 3). The DQw2 beta RFLP polymorphism may indicate the presence of an unrecognized sequence variation that will be revealed on further nucleotide sequencing. However, the more likely case is that this RFLP indicates variability in noncoding portions (introns) in the vicinity of the DQB1 gene. This suggests the utility of RFLP analysis in searching, for example, for HLA-DQB1 regulatory differences.

Table 3. Correlation of HLA Haplotypes with DQB1 RFLP Clusters and Oligonucleotide Sequences

HLA Type			DQB1	Amino Acid						Amino Acid					
DR	Dw	DQw	RFLP Cluster	24	25	26	27	28	29	54	55	56	57	58	59
				Oligonucleotide[1]						Oligonucleotide					
1	1	5(1)	DQB1.3	GTG	CGG	GGT	GTG	ACC	AGA	GGG	CGG	CCT	GTT	GCC	GAG
w16	21	5(1)	DQB1.3	---	---	---	---	---	---	---	---	---	AGC	---	---
w14	9	5(1)	DQB1.3	---	---	---	---	---	---	---	---	---	-AC	---	---
w15	2	6(1)	DQB1.5	---	--T	CT-	---	---	---	---	---	---	-A-	---	---
w13	18	6(1)	DQB1.5												
w13	19	6(1)	DQB1.6	---	--T	CT-	--A	---	---	---	---	---	---	---	---
w15	12	6(1)	DQB1.7	---	--T	TA-	---	---	---	---	---	---	-AC	---	---
3	3	2	DQB1.9	---	--T	CT-	---	-G-	---	---	-T-	---	-CC	---	---
7		2	DQB1.10	---	--T	CT-	---	-G-	---	---	-T-	---	-CC	---	---
w11	5	7(3)	DQB1.12	---	--T	TA-	---	---	---	---	-C-	---	-AC	---	---
4		7(3)	DQB1.12	---	--T	TA-	---	---	---	---	-C-	---	-AC	---	---
4		8(3)	DQB1.13	---	--T	CT-	---	---	---	---	-C-	---	-CC	---	---
9	23	9(3)	DQB1.13	---	--T	CT-	---	---	---	---	-C-	---	-AC	---	---
w8		4	DQB1.13	---	---	---	---	---	---	---	---	-T-	-AC	---	---

[1]For appropriate references see Janet Lee, this volume.

Third, in several instances, RFLP analysis failed to reveal currently recognized HLA-DQ specificities. One example is the failure of RFLP analysis to distinguish between the DQw4 and DQw8 alleles. This is a well defined serologic split. Furthermore, nucleotide variability has been described between the DQw4, DQw8, and DQw9 allelic DQ beta sequences (Table 3). Another example is the DQw5 allele. As shown in Table 1, RFLP does not recognize variation in this family, whereas DQ beta nucleotide sequences (Table 3) show three allelic forms within the DQw5 group.

These apparent limitations of the RFLP technique may indicate the need to expand the battery of enzymes used for this analysis. On the other hand, since 12 different enzymes were tested, it may be that this exercise has tested the limits of discrimination that this technique has to offer. By all means, however, the data collected in these exercises certainly provided a strong base for the goal of better understanding the serologically and cellularly defined polymorphism of HLA at the molecular genetic level.

Acknowledgments. We would like to thank Drs. Daniel Cohen and Bo Dupont for organizing this component of the Workshop with the aim of providing the community with a reference HLA RFLP database. A study of this size would have been impossible for any single laboratory to complete. In the tradition of the International Histocompatibility Workshops, this successful effort involved the cooperation of a large number of laboratories at the expense of individual gain. Also great thanks are given to Drs. Jean Marc Lalouel and Malcolm Simons for providing the statistical analysis output in a comprehensive form.

References

1. Cohen D, Robbins E, Dausset J. Recombinant DNA technology: its use in the study of HLA. In: Albert ED, Baur MP, Mayr WR (eds). Histocompatibility Testing 1984. Heidelberg: Springer-Verlag, 1984:585.
2. So AKL, Trowsdale J, Bodmer JG, Bodmer WF. DNA polymorphism of HLA-DR-beta and HLA-DP-beta genes. In: Albert ED, Baur MP, Mayr WR (eds). Histocompatibility Testing 1984. Heidelberg: Springer-Verlag, 1984:565.
3. Wallin J, Bohme J, Carlsson B, Moller E, Peterson PA, Rask L. HLA Class II polymorphism: restriction fragment patterns correlated to Ninth Workshop serology and function. In: Albert ED, Baur MP, Mayr WR (eds). Histocompatibility Testing 1984. Heidelberg: Springer-Verlag, 1984:572.
4. Tilanus M, Bosch M, Gerrets R, Giphart MJ. Molecular analysis of HLA: haplotype-specific DNA hybridization patterns using class II cDNA probes. In: Albert ED, Baur MP, Mayr WR (eds). Histocompatibility Testing 1984. Heidelberg: Springer-Verlag, 1984:577.
5. Hui K, Groves A, Jaraquemada D, et al. DNA restriction pattern length polymorphism of the human class II genes as analyzed by Southern blotting technique. In: Albert ED, Baur MP, Mayr WR (eds). Histocompatibility Testing 1984. Heidelberg: Springer-Verlag, 1984:590.
6. Angelini G, De Praval C, Ferrara G, Mach B. Analysis and segregation of HLA-DR by DNA typing: correlation with supertypic specificities. In: Albert ED, Baur MP, Mayr WR (eds). Histocompatibility Testing 1984. Heidelberg: Springer-Verlag, 1984:582.
7. Cohen D, LeGall I, Marcadet A, Font MP, Lalouel JM, Dausset J. Clusters of HLA class II beta restriction fragments describe allelic series. Proc Natl Acad Sci USA 1984;81:787.
8. Andersson M, Bohme J, Anderson G, et al. Genomic hybridization with class II transplantation antigen cDNA probes as a complementary technique in tissue typing. Human Immunology 1984;11:57.
9. Spielman R, Lee J, Bodmer W, Bodmer J, Trowsdale J. Six

HLA-D region alpha chain genes of human chromosome 6: polymorphisms and associations of DC alpha related sequences with DR types. Proc Natl Acad Sci USA 1984; 81:3461.
10. Nepom B, Palmer J, Kim S, Hansen J, Holbeck S, Nepom G. Specific genomic markers for the HLA-DQ subregion discriminate between DR4+ insulin-dependent diabetes mellitus and DR4+ seropositive juvenile rheumatoid arthritis. J Exp Med 1986;164:345.
11. Rosenshine S, Cascino I, Zeevi A, Duquesnoy RJ, Trucco M. DQ-alpha and -beta RFLP reveals the composition of the DQ molecule recognized by T-cell clones. Immunogenetics 1986;23:187.
12. Cascino I, Rosenshine S, Turco E, Marrari M, Duquesnoy RJ, Trucco M. Relationship between DQ alpha and DQ beta RFLP and serologically defined class II HLA antigens, J Immunogenet 1986;13:387.
13. Trowsdale J, Young J, Kelly A, et al. Structure, sequence and polymorphism in the HLA-D region. Immunol Rev 1985;85:1.
14. Korman A, Boss J, Spies T, Sorrentino R, Okada K, Strominger J. Genetic complexity and expression of human class II histocompatibility antigens. Proc Natl Acad Sci USA 1985;79:3687.
15. Trucco M, Rosenshine S, Cascino I, Duquesnoy R. HLA-DQ polymorphisms analyzed by sequential restriction endonuclease DNA digestion. Immunogenetics 1986; 24:184.
16. Turco E, Fritsch R, Trucco M. First domain encoding sequence mediates human class II beta chain gene cross-hybridization. Immunogenetics 1988;28:193.

Workshop Report for the HLA-DP Region

Jens Jorgen Hyldig-Nielsen[1] and Arne Svejgaard[1]

Introduction

Determination of the different DP alleles has become increasingly desirable during the last few years in studies of both graft rejection of transplants (1,2) and of new disease associations (3,4). DP typing has until now been done by the very laborious primed lymphocyte typing (PLT) technique, but lately good correlations between DP types defined by PLT typing and restriction fragment length polymorphism (RFLP) of the DP region has been established (5–7).

The 72 cell lines included in this Workshop were selected to represent as many of the known DR alleles as possible and for being of consanguineous origin. Previous attempts to define the polymorphism of the DP region in such panels have proven unsuccessful, mostly because the distribution of the different DP alleles is distorted, and because some of the rarer alleles may not be represented. DP typing of the Workshop cells by means of PLT has been presented in XIHWS Newsletters (nos. 1 and 3), and there were some discrepancies of the DP assignments, which may be due to differences in the PLT reagents used, but may also reflect true biological differences. In the final DP assignments of the Workshop cell lines, only DPw1-w5 has been assigned, while DPw6 and different local specificities (e.g., DPw7, CDP-HEI, and DPCp63) have not been included. This might be one of the reasons for a rather high number of "blanks" among the 72 cell lines. Another explanation could be that lymphoblastoid cells generally are more difficult to type than are peripheral blood lymphocytes.

Participating Laboratory: SCASVE[1]

General Remarks Concerning the DP Beta Hybridizations

In addition to the Workshop, there have been four other studies dealing with the polymorphism of the DP region (5–8). Hyldig-Nielsen et al., Mitsuishi et al., and the Workshop studies used almost identical cDNA probes (Fig. 1), although of two different origins (9,10), while other investigators used two different genomic probes (11,12). The homology between the two cDNA probes used is about 95% if deletions and insertions are excluded from the analysis. The clone isolated by Roux-Dosseto et al. (10) starts in the first extracellular domain at amino acid (a.a.) 51 and the clone isolated by Gustafsson et al. (9) at a.a. 53. Results obtained with these probes would be expected to be almost identical, which was the case when comparing previous published results.

Comments on the Workshop Registrations Using the DP Beta Probe

BamHI. For the Workshop cells, only a single polymorphic fragment of 9.00 kb has been registered. In two other investigations (7,8), two polymorphic fragments of 10.0/9.5 kb and 10.2/10.0 kb, respectively, have been reported. Hyldig-Nielsen et al. (7) showed that these two fragments have the same distribution in the population as two HindIII fragments previously related to the DP beta pseudogene (13). In the Workshop, the 9.00 kb BamHI fragment has almost the same distribution as a 5.06 kb HindIII fragment. The larger polymorphic BamHI fragment defined in earlier studies was assigned 9.36 kb in the Workshop. Because the DP beta probe cross-hybridized to a nonpolymorphic DQ beta fragment

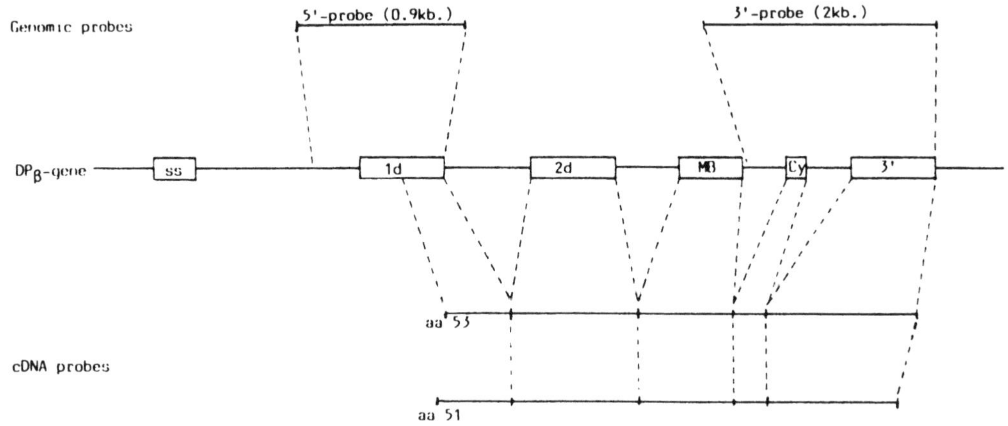

Figure 1. The cDNA and genomic probes used for the elucidation of the RFLP of the DP region. The cDNA amino acid 51 probe was used in the Workshop and in reference 6, cDNA amino acid 53 in references 5 and 7, and the two genomic probes in reference 8.

of the same size, this fragment has been excluded from the analysis. The distribution of the DP beta fragment is shown in Table 2 of the DP Beta/BamHI standardization report. This fragment has almost the same distribution in the workshop cell lines as the 5.43 kb HindIII fragment (see also comments concerning HindIII).

BglII. Results using BglII have previously been reported by Hyldig-Nielsen et al. (5,7) and as is the case for the workshop data, the two smallest fragments have identical (in the Workshop almost identical) distributions in the population. The only discrepancy is the surprisingly big difference in the size of the smallest fragment, which in the Workshop was determined to be 5.05 kb, while Hyldig-Nielsen et al. (5,7) reported this fragment to be 2.8 kb. The explanation for this difference is uncertain.

EcoRI. Before the Workshop, two studies using EcoRI have been published (6,7). Both report two polymorphic fragments of 7.5/7.2 kb and 7.7/7.3 kb, respectively. The 15.83 kb fragment registered for the Workshop data, which is found in a few of the DPw4 and blank specificities, has not been noted earlier. The 2 × 2 tables against the different DR, DQ, and DP specificities did not indicate any significant correlation to a single specificity (data not shown).

EcoRV. There is very good agreement between the results obtained using EcoRV and the DP beta probe in the Workshop and previous published results (6,7).

HincII. The use of the enzyme HincII and the DP beta probe to define the RFLP of the DP region has not been reported previously. For the Workshop cells, 10 polymorphic fragments were defined. Some of these correlated well with the DPw1 specificity and with DP beta clusters IV and V. One fragment of 7.53 kb correlated to DR7, while the remaining six fragments did not correlate to any single DR, DQ, or DP specificity. Neither did these six fragments correlate to any combination of DP alleles.

HindIII. Previous results using the enzyme HindIII and the DP beta probe showed two polymorphic fragments of about 5.0 kb and 5.3 kb (5–7) and three nonpolymorphic fragments (data not shown). Kappes et al. (13) indicated that these fragments arise from the DP beta pseudogene. In the Workshop results, these two fragments have been assigned 5.06 kb and 5.43 kb. One of the fragments previously found to be nonpolymorphic was split for the Workshop cell lines been split into an almost nonpolymorphic fragment of 1.86 kb, and a rather infrequent fragment of 1.81 kb. The smallest fragment was found in a cell line, which in the preliminary assignments, had been typed DPCp63, a newly defined antigen found primarily in the Asian population. Apart from these four fragments, a total of nine additional polymorphic DP beta fragments has been registered for the Workshop cells. However, the 2 × 2 tables show that most, if not all, of these fragments correlate to either the DR or DQ series, and, thus, they are not DP beta fragments.

KpnI. Data using the enzyme KpnI have not been reported earlier. KpnI gave rise to three polymorphic fragments. All three fragments are large (38.04 kb, 16.80 kb, 15.50 kb), giving rise to rather uncertain assignments. This is also seen from the large number of uncertain fragments registered.

MspI. MspI has been used in three other investigations of the polymorphism of the DP region (5–7). In these reports, MspI is clearly the enzyme that gives the best distinction between the different DP alleles, with specific fragments correlating almost perfectly with DPw1, w2, w4, and some of the local specificities (e.g., CDP-HEI). In the Workshop panel, fragments correlating almost absolutely with DPw1 and w2 were also seen.

PstI. In the Workshop cells, it was possible to define three polymorphic fragments using the enzyme PstI. Two of these fragments had the same distribution between the individual cell lines. There is an absolute agreement between these results and those of Hyldig-Nielsen et al. (7). Bodmer et al. (8) also used the enzyme PstI, but found only a single polymorphic fragment of 4.2 kb. This is probably related to the difference in the probes used (see Fig. 1).

PvuII. There is quite a difference between previous results using PvuII (7) and those obtained using the Workshop cell lines. Hyldig-Nielsen et al. (7) found an almost perfect correlation between the two polymorphic PvuII fragments of 2.6 kb and 2.3 kb and the two HindIII fragments related to the DP beta pseudogene. For the Workshop cell lines, three fragments of 5.68 kb, 5.17 kb, and 4.15 kb are registered, and none of these correlated with the HindIII fragments related to the pseudogene. Actually, the two smallest fragments correlates to different DR alleles. It is uncertain why no fragments correspond to the polymorphic fragments previously found.

SstI. SstI has previously been used in this context only by Bodmer et al. (8), who defined two polymorphic fragments of 20.0 kb and 12.0 kb using a 3′-specific DP beta probe. The difference between the probes used makes it almost impossible to relate the results obtained in these two studies. For the Workshop cells, a total of 12 polymorphic fragments were defined, among which were fragments correlating to DP beta clusters II, IV, and VI. Several of the remaining seven fragments are probably related to one of the other class II loci.

TaqI. This enzyme has been used in two previous studies (6,7) revealing three polymorphic fragments of 4.6 kb, 2.9 kb, and 1.7 kb. The large fragment correlates with DPw2 and w4, while the two smaller fragments have identical distributions and correlate with DPw1, w3, w5, w6, and CDP-HEI. Apart from these three fragments, we also found three nonpolymorphic fragments, one of which has a size of 2.7 kb. In the Workshop, two polymorphic fragments of 2.73 kb and 2.64 kb were defined. The 2.73 kb fragment was found in all cell lines except one, which had a size of 2.64 kb instead. The nonpolymorphic nature of the 2.73 kb fragment and the rarity of the smaller fragment indicate that these fragments may be splits of the previously defined nonpolymorphic 2.7 kb fragment. However, it is surprising that none of the other polymorphic fragments seen earlier were detected in the Workshop cells.

Results of the DP Beta Hybridizations for the Workshop Cells

Figure 2 shows the cluster analysis of the DP types against the polymorphic fragments obtained with the 12 enzymes used. Figure 3 gives a more schematic presentation of the clusters that correlate well with the

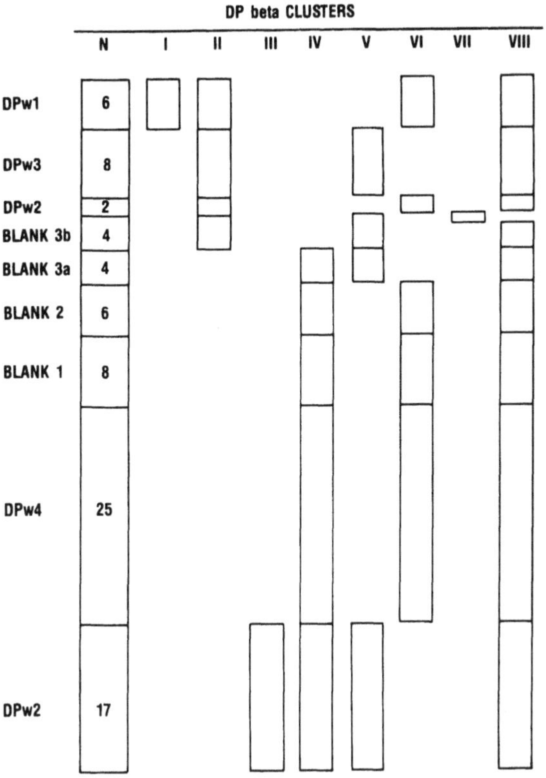

Figure 3. Schematic presentation of DP beta clusters correlating to DP types.

assigned DP specificities. Table 3 presents some of the 2 × 2 tables for the correlation between the presence of the DP antigens and representative fragments included in the clusters defined in Figure 3. To achieve the best possible correlations, we have divided the 22 cell lines with no DP assignments into four groups, as indicated in Table 1.

As can be seen in Figure 3, only fragments specific for DPw1 and DPw2 were defined in the Workshop (clusters I and III in Fig. 3). The correlations between these fragments and the DP specificities were very high (Table 3). Before the Workshop DP beta fragments specific for DPw1, w2, w4, and the local specificity CDP-HEI were defined (6,7). It is surprising that a MspI fragment of about 1.1 kb previously found almost specific for the DPw4 allele is not registered for the Workshop cell lines, as this is a strongly hybridizing fragment not particularly close to any other MspI fragment. Inspection of 2 × 2 tables for fragments assigned to the DR, DQ, and DO series did not identify a fragment wrongly assigned to one of these specificities.

Apart from the fragments correlating to a single DP allele, three sets of complementary clusters were defined by the DP beta probe. These clusters include from 2 to 10 fragments. Several of the earlier RFLP studies of the DP region (5,6,7) clearly defined two major comple-

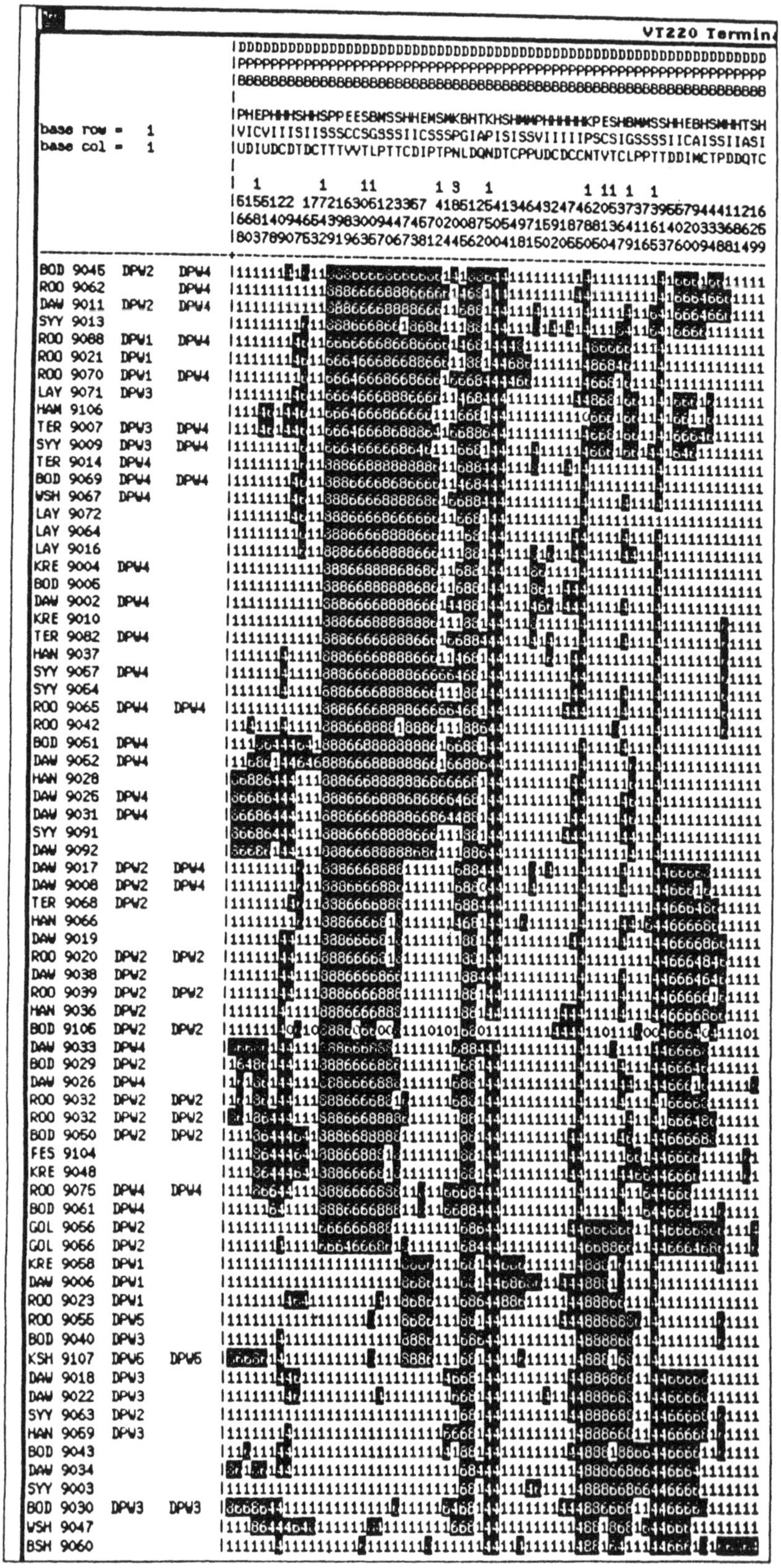

Figure 2. Cluster analysis of polymorphic DP beta fragments defined in the Workshop.

mentary clusters of fragments, one correlating with the DPw2 and w4 determinants, and the other with DPw1, w3, w5, w6, and some local specificities. In the Workshop cell lines, several fragments corresponding to the DPw2 + w4 cluster have also been found (cluster IV in Fig. 3 and Table 3), while only a few fragments representing the other major cluster has been defined (cluster II in Fig. 3 and Table 3). This is especially surprising when the high degree of homology between the probes used is taken into account. One possible explanation could be that some DP-related fragments have been assigned to one of the other class II loci. Inspection of the 2 × 2 tables for the fragment assigned to these other loci identified only three fragments correlating more closely to DP alleles than to any other class II specificity, namely a 5.37 kb PstI fragment, a 14.97 kb SstI fragment, and a 4.96 kb MspI fragment. Both the SstI and the MspI fragments had already been registered as DP-specific fragments (double registrations).

Two complementary clusters (nos. V and VI in Fig. 3) include the two HindIII fragments previously assigned to the DP beta pseudogene (13). The fragments in cluster V correlate to DPw2, w3, and Blank 3 (a + b), while cluster VI correlates to the DPw1, w4, w5, blank 1, and blank 2 specificities. As can be seen in the calculations in Table 3, there are many more exceptions and extras than in the other clusters, which is in agreement with the expectation that these clusters represents the pseudogene.

In addition to the clusters mentioned above, which all had been reported earlier, two complementary clusters each consisting of two fragments were defined for the Workshop cell lines. Cluster no. VIII, which are composed of a 1.86 kb HindIII fragment and a 2.72 kb TaqI fragment, are found in all the cell lines except one (WS no. 9060). This cell line possessed instead a 1.81 kb

Table 1. Cell Lines Included in the Four Groups of Blanks

Blanks	Cell Lines (Workshop No.)
Blank 1	9005, 9013, 9028, 9042, 9054, 9072, 9091, 9092
Blank 2	9010, 9016, 9019, 9037, 9047, 9064
Blank 3a	9048, 9066, 9104, 9106
Blank 3b	9003, 9034, 9043, 9060

Table 2. Associations Between DP Alpha and DP Antigens

	DP Antigen	Cluster	Fragments[a]	++	+−	−+	−−	No.	p
Monospecific fragments	w5	I	BglII 3.58 TaqI 5.47	2	0	3	65	70	4.1×10^{-3}
	w2	V	TaqI 12.07	14	1	5	48	70	1.3×10^{-9}
	w4	VI	BglII 5.29	24	1	9	36	70	2.0×10^{-10}
"Supertypic" fragments	w1 Blank3	II	HindIII 4.90 BglII 3.41 TaqI 4.66	11	3	4	52	70	1.9×10^{-7}
	w1,w5	III	MspI 6.71 MspI 0.19	7	1	2	60	70	2.3×10^{-7}
	w1,w5 Blank3	IV	PvuII 3.81 PvuII 2.95 SstI 18.82 TaqI 6.45 PstI 2.63 EcoRV 6.97 BamHI 6.92 (HindIII 4.90)	16	0	3	51	70	3.9×10^{-13}
	w2,w3,w4 Blank1 Blank2	VII	SstI 16.16 SstI 4.45 HindIII 3.25 PstI 2.43 MspI 4.73 MspI 4.15 MspI 3.14 BglII 2.23 (EcoRI 4.17) (PvuII 6.58)	56	1	1	12	70	1.6×10^{-11}
	w2,Blank2	VIII	BamHI 9.00	19	2	3	46	70	4.5×10^{-12}
	w3,w4 Blank1	IX	BamHI 8.67 EcoRV 7.50 TaqI 13.21	38	1	1	30	70	1.7×10^{-17}

[a]Fragments in brackets indicate fragments not correlating to the same extent as the other fragments in the cluster.

Table 3. Associations Between DP Beta Clusters and DP Antigens

	DP Antigen	Cluster	Fragment[a]	++	+−	−+	−−	No.	p
Monospecific fragments	w1	I	MspI 4.98 HincII 3.41	6	0	0	64	70	7.6×10^{-9}
	w2	III	MspI 4.38	10	5	3	52	70	1.7×10^{-6}
	DPCp63	VII	TaqI 2.64 HindIII 1.81	1	0	0	69	70	.01
"Supertypic" fragments	w2,w4 Blank 1 Blank 2 Blank 3a	IV	PstI 2.39 PstI 1.91 EcoRV 6.89 EcoRV 3.36 SstI 10.03 BglII 15.05 MspI 1.97 SstI 2.40 SstI 3.46 HincII 3.77	52	2	2	14	70	7.0×10^{-11}
	w1,w3,w5 Blank 3b	II	PstI 2.85 EcoRV 10.10 SstI 15.34	20	0	5	45	70	3.3×10^{-13}
	w1,w4,w5 Blank 1 Blank 2	VI	HindIII 5.43 EcoRI 7.58 MspI 0.71 SstI 14.02	37	8	3	22	70	9.2×10^{-9}
	w2,w3 Blank 3	V	HindIII 5.06 EcoRI 7.20 BamHI 9.00 (HincII 4.39)	30	1	8	31	70	2.2×10^{-11}

[a]Fragments in brackets indicate fragments not correlating to the same extent as the other fragments in the cluster.

HindIII fragment and a 2.64 kb TaqI fragment (cluster no. VII). Workshop cell line no. 9060 has no DP assignments in Figure 2. On the other hand, this cell had the preliminary local assignment DPCp63, which is a DP antigen common in the Asiatic population (14). These two fragments, slightly smaller than the commonly found fragments, could be specific markers for the local specificity DPCp63. At the Workshop in Princeton a 1.91 kb MspI fragment was also found to be specific for this allele. This fragment has disappeared during the corrections after the Workshop. We wonder if this is correct because a fragment of about the same size has previously been defined by Hyldig-Nielsen et al. (5) (the 1.8 kb MspI fragment termed marker no 14), although it could not be correlated to any DP specificity.

General Remarks Concerning the DP Alpha Hybridizations

In addition to the Workshop, DP alpha probes have been used in four individual RFLP studies (5–8). Servenius et al. (15) used a genomic probe encoding the entire DPα1 gene, which does not cross-hybridize to the DP alpha pseudogene. Auffray et al. (16) used the same cDNA probe as distributed for the Workshop. Trowsdale et al. (17) also used an almost full length cDNA probe. The homology between these two cDNA probes is about 97%.

Comments to the Workshop Registrations Using the DP Alpha Probe

BamHI. Three polymorphic fragments are reported for the Workshop cell lines. There has not been any previous report of polymorphic fragments using this enzyme/probe combination.

BglII. BglII and the DP alpha probe have been used in two previous studies of the polymorphism of the DP region (5,7). Hyldig-Nielsen et al. (5,7) described four polymorphic fragments, while five fragments were defined in the Workshop. The 1.6 kb fragment found previously and having the same distribution as the 2.3 kb fragment was not found in the Workshop cells. However, two fragments of 7.24 kb and 5.29 kb were found in the Workshop cell lines – the 5.29 kb fragment correlated almost perfectly with the DPw4 allele (Table 4).

EcoRI. Results using the EcoRI/DP alpha system have previously been reported in two individuals studies (6,8). Bodmer et al. (8) defined one fragment of 2.8 kb, while Mitsuishi et al. reported two polymorphic fragments of 3.9 kb and 2.4 kb. The Workshop assignments for this system are rather uncertain (e.g., several of the cell lines have not been registered as positive for any of the three polymorphic fragments identified).

Table 4. Comparison of Fragments Reported for the BglII/DP Alpha and TaqI/DP Alpha Systems

Bodmer et al. (8)		Hyldig-Nielsen et al. (7)		Workshop		Mitsuishi et al. (6)	
kb	Association	Association	kb	kb	Association	Association	kb
				7.24	w1,w2,w3 w5, B1.1, B1.2, B1.3		
				5.29	w4		
		w5	3.8	3.58	w5		
		w1	3.5	3.41	w1, B1.3		
		w2, w3, w4	2.3	2.23	w2, w3, B1.1, B1.2		
		w2, w3, w4	1.6				
		w3, w4, w6	14.0	13.21	w3, w4, B1.1	w3, w4, B1.a	15.0
		w2	13.5	12.07	w2	Some w2	11.5
				7.66	In 3 cell lines		
				6.45	w1, w5, B1.3		
7.6	w5	w5	6.5	5.47	w5	w5	7.4
		w1, CDP-HEI	6.3	4.66	w1, B1.3	w1, B1.b+c	6.9
						w1, w5, B1.b+c	6.3
				3.99	As for 7.66		

EcoRV. Results using EcoRV have previously only been published by Mitsuishi et al. (6). There are good agreement between these results and the results obtained for the Workshop cell lines.

HincII. This enzyme has not been used previously, and no data have been reported for the Workshop cell lines.

HindIII. There is good agreement between the previously reported results (6) and the results registered for the Workshop cell lines using this system.

KpnI. The KpnI/DP alpha system has been used for the first time in the Workshop study. Three polymorphic fragments were reported. None of these fragments correlated with single DP antigens or clusters thereof.

MspI. The MspI/DP alpha system has previously been used in three studies (5–7). There was generally good agreement between these studies and the results obtained in the Workshop, although some extra fragments were found in one or the other study. These extra fragments can be explained either by the differences between the probes used or because these fragments were found only in very few cell lines.

PstI. Three polymorphic fragments have previously been reported using the enzyme PstI (5,7). The two smallest fragments correlate with the two major DP alpha clusters (nos. IV and VII). These two fragments were also defined for the Workshop cell lines. The last fragment of 3.5 kb was found in only a few of the individuals previously investigated.

PvuII. The use of PvuII and the DP alpha probe in defining the polymorphism of the DP region has previously been reported by Hyldig-Nielsen et al. (5,7), who defined three fragments all correlating to one of the two major clusters defined by the DP alpha probe. Six polymorphic fragments were defined for the Workshop cells. Three fragments were rather weak, and two of these were almost nonpolymorphic. The remaining three fragments had nearly the same size as the ones previously defined and correlated to the two major DP alpha clusters, although there were a few exceptions and more extras than were previously found.

TaqI. The enzyme TaqI has been used in all studies, and a detailed comparison of the results is discussed below.

Results of the DP Alpha Hybridizations for the Workshop Cells

Figure 4 shows the cluster analysis of the DP types versus the polymorphic fragments obtained with the 11 enzymes for which data were reported for the Workshop cell lines. Figure 5 is a schematic representation of clusters of fragments correlating well with the assigned DP specificities. Table 2 provides some 2 × 2 tables of the correlations between the presence of certain DP alleles and a representative fragment included in the clusters defined in Figure 5. To obtain the best possible correlations, we have divided the 22 cell lines without DP assignments into four groups, as indicated in Table 1.

As can be seen in Figure 5 and Table 2, three clusters (including a total of five fragments) correlate well with a single DP allele. These are cluster I, which correlates with DPw5, cluster V with DPw2, and cluster VI with DPw4. Before the Workshop, three laboratories (6–8) had investigated the RFLP of the DP region using a DP alpha probe. In those studies, three fragments correlated with single DP specificities, namely DPw1, w2, and w5.

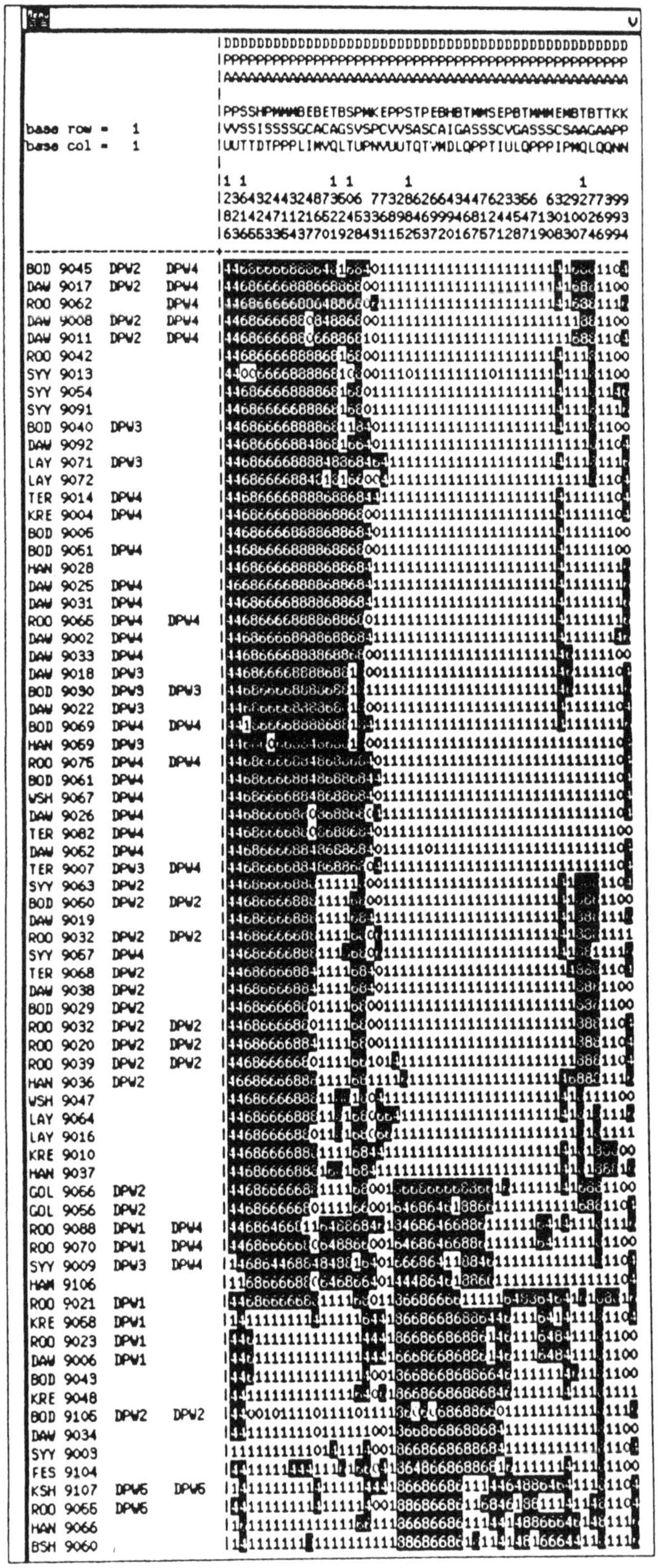

Figure 4. Cluster analysis of polymorphic DP alpha fragments defined in the Workshop.

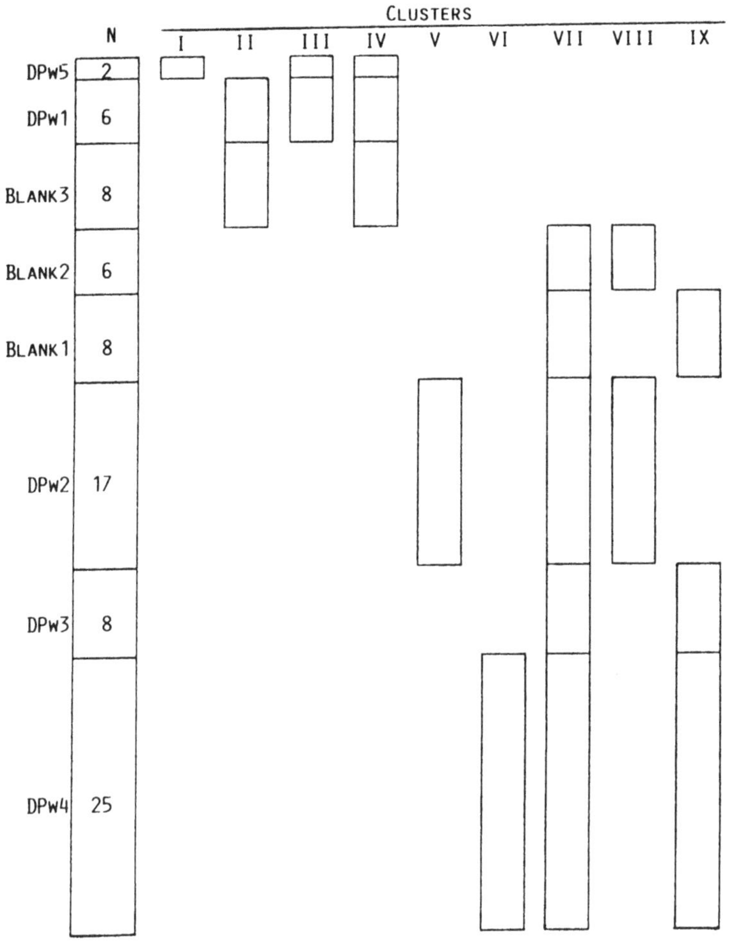

Figure 5. Schematic presentation of DP alpha clusters correlating to DP types.

When using the DP alpha probe, the two enzymes that show the best correlations between PLT-determined DP types and RFLP patterns are BglII and TaqI. Because there are some discrepancies between previously published results and the data obtained at the Workshop, we have compared the results obtained using these two enzymes in Table 5. The differences seen between the results obtained by Hyldig-Nielsen et al. and the Workshop study could be explained by the differences in the probes used for the investigation. For BglII, the main difference is the definition in the Workshop of two fragments of 7.24 kb and 5.29 kb not seen in the previous study. The 5.29 kb fragment is specific for DPw4 (See Table 2).

In the Workshop, TaqI also defined fragments not seen previously, namely three fragments of 7.66 kb, 6.45 kb, and 3.99 kb. None of these fragments correlated to any of the other class II loci. The reason why Hyldig-Nielsen et al. did not see as many fragments as registered in the Workshop could be the difference in the probes used (see Table 4). It is a little surprising that these fragments were not reported by Mitsuishi et al. because they used the same probe as in the Workshop. Moreover, the order of the three smallest fragments seems to be mixed up. Considering the high homology between the probes used by Bodmer et al. and in the Workshop, it is also surprising that they found only one polymorphic band (see Table 4). This could be due to a very uneven distribution of the DP alleles among the cells used in this study. It will take further investigation to explain these discrepancies between the four different studies. They illustrate the problems in comparing the results obtained in different laboratories because small differences in the techniques used are likely to play a role.

For the Workshop cell lines, as well as in previous investigations, two complementary clusters were found (clusters IV and VII in Fig. 5). Within cluster VII, two other complementary clusters were defined—cluster VIII correlating with DPw2 + Blank 2, and cluster IX correlating with DPw3, w4, and blank 1.

Table 5. Specific Fragments for DP Typing by RFLP Analysis

Probe	DP Allele	MspI	BglII	TaqI
	DPw2			12.07 kb
DP alpha	DPw4		5.29 kb	
	DPw5		3.58 kb	5.47 kb
	DPw1	4.98 kb		
	DPw2	4.38 kb		
	DPw3+w6	3.0 kb[a]		
DP beta	DPw4	1.1 kb[b]		
	CDP-HEI	1.8 kb[b]		
	DPCp63			2.64 kb

[a]From reference 7. This fragment correlates with DPw3, w5, w6. Typing for DPw3 + w6 can be achieved when cells are positive for this fragment, but negative for the DPw5-specific DP alpha fragments.
[b]From reference 7.

Concluding Remarks

From the combined RFLP data of the DP region, it can be concluded that the enzyme MspI gives rise to the largest polymorphism when using a DP beta probe, while BglII and TaqI both reveal a high degree of polymorphism when a DP alpha probe is used. Using these three enzymes and the DP alpha and DP beta probes one can find fragments "mono"-specific for DPw1, DPw2, DPw4, DPw5, and the local specificities CDP-HEI and probably also DPCp63 (Table 5). Only DPw3 and DPw6 cannot be separated from each other with these probe/enzyme combinations or by any other combination reported so far. Fortunately, the DPw6 antigen is of low frequency and, thus, the loss of information is quite small. Finally, it should be kept in mind that the number of individuals typed both at the DNA level and on the cell surface for DP is still small, and additional parallel studies are clearly indicated before DNA typing can substitute PLT.

In addition to the fragments and clusters correlating with DP determinants, a number of fragments were seen that did not show such correlations. Some of these fragments most likely define "new" genetic markers of the DP region, which consequently expands our knowledge of this region considerably.

References

1. Odum N, Platz P, Jakobsen BK. HLA-DP and bone marrow transplantation: DP-incompatibility and severe acute graft versus host disease. Tissue Antigens 1987;30:213.
2. Amar A, Nepom GT, Mickelson E, Erlich H, Hansen A. HLA-DP and HLA-DO genes in presumptive HLA-identical siblings: structural and functional identification of allelic variation. J Immunol 1987;138:1947.
3. Niven MJ, Caffrey C, Sachs JA, et al. Susceptibility to coeliac disease involved genes in HLA-DP region. Lancet 1987;ii:805.
4. Oedum N, Hyldig-Nielsen JJ, Morling N, Sandberg-Wollheim M, Platz P, Svejgaard A. HLA-DP antigens are involved in the susceptibility to multiple sclerosis. Tissue Antigens 1988;31:235.
5. Hyldig-Nielsen JJ, Morling N, Odum N, et al. Restriction fragment length polymorphism of the HLA-DP subregion and correlations to HLA-DP phenotypes. Proc Natl Acad Sci USA 1987;84:1644.
6. Mitsuishi Y, Urlacher A, Tongio M-M, Mayer S. Restriction fragment length polymorphism of DP genes defines three new DP alleles. Immunogenetics 1987;26:383.
7. Hyldig-Nielsen JJ, Morling N, Odum N, et al. DNA polymorphism of HLA-DP: further studies of the restriction fragment length polymorphism of DP-alpha and DP-beta. In: Dupont B (ed). Immunobiology of HLA: Immunogenetics and Histocompatibility, Volume II. New York: Springer-Verlag, 1989.
8. Bodmer J, Bodmer W, Heyes J, So A, Tonks STJ, Young J. Identification of HLA-DP polymorphism with DP-alpha and DP-beta probes and monoclonal antibodies: correlation with primed lymphocyte typing. Proc Natl Acad Sci USA 1987;84:4596.
9. Gustafsson K, Emmoth E, Widmark E, Boehme J, Peterson PA, Rask L. Isolation of a cDNA clone coding for an SB beta-chain. Nature 1984;309:76.
10. Roux-Dosseto M, Auffray C, Lillie JW, et al. Genetic mapping of a human class II antigen beta-chain cDNA clone to the SB region of the HLA complex. Proc Natl Acad Sci USA 1983;80:6036.
11. Kelly A, Trowsdale J. Complete nucleotide sequence of a functional HLA-DP-beta gene and the region between the DP-beta 1 and DP-alfa 1 genes: comparison of the 5′ ends of HLA class II genes. Nucleic Acids Res 1985;13:1607.
12. So AKL, Trowsdale J, Bodmer JG, Bodmer WF. DNA Polymorphism of HLA-DR-beta and HLA-DP-beta genes. Histocompatibility Testing 1984:565.
13. Kappes DJ, Arnot D, Okada K, Strominger JL. Structure and polymorphism of the HLA Class II SB light chain genes. EMBO J 1984;3:2985.
14. Nose Y, Nakatsuji T, Tsuji K, Reinsmoen NL. A new HLA-DP (SB) specificity (Cp63) defined in Japanese and Caucasian cell populations by cloned PLT cells. Human Immunol 1985;13:145.
15. Servenius B, Gustafsson K, Widmark E, et al. Molecular map of the human HLA-SB(HLA-DP) region and sequence of an SB[(DP)] pseudogene. EMBO J 1984;3:3209.
16. Auffray C, Lillie JW, Arnot D, Grossberger D, Kappes D, Strominger JL. Isotypic and allotypic variation of human class II histocompatibility antigen α-chain genes. Nature 1984;308:327.
17. Trowsdale J, Young AT, Kelly AP, et al. Structure, Sequence and Polymorphism in the HLA-D Region. Immunol Rev 1985;85:5.

Final Workshop Report: HLA Class I RFLP

J. Kalil,[1] S. Ward,[2] D. Tyan,[3] H. Coppin,[4] L. Satz,[5] L. Degos,[6] and E. Dyer[7]

In this Workshop section, 69 core cell line DNA samples were studied with the class I cDNA probe pHLA$_2$ (1) and 11 restriction enzymes (Table 1). Two hundred and four fragments were detected. Comparison of clustered RFLPs or of individual antigen DNA fragments has been performed. However, cell lines were originally selected for their class II specificities and, therefore, some HLA-A, B, and C locus alleles are poorly represented or absent, making class I analysis tentative in some cases. The great majority of the cell lines were homozygous for HLA-A, B, and C, but a few of them were not. Table 2 shows the HLA class I types represented in the panel. Analysis of the A3 cluster is summarized in Table 3.

A good correlation was observed between HLA-A alleles and particular fragments, although in most

Participating Laboratories: LATKAL,[1] US6TRP,[2] US3GRU,[3] FRAOHA,[4] FRADDC,[5] US8RJW[6]

Table 1. List of the Enzymes Used

Code	Name	Code	Name
BAM	BamHI	MSP	MspI
BGL	BglII	PST	PstI
ECI	EcoRI	PVU	PvuII
ECV	EcoRV	SST	SstI
HID	HindIII	TAQ	TaqI
KPN	KpnI		

Table 2. HLA Specificities Available in the 69 Core Cell Line Panel

HLA-A	HLA-B	HLA-C
A1 = 9	B5 = 1	Cw1 = 5
A2 = 23	B7 = 6	Cw2 = 3
A3 = 12	B8 = 4	Cw4 = 6
A11 = 1	B13 = 1	Cw5 = 7
A23 = 2	B14 = 2	Cw6 = 4
A24 = 6	B18 = 7	Cw7 = 14
A25 = 2	B27 = 3	Cw8 = 2
A26 = 4	B35 = 6	Cw9 = 5
A28 = 1	B37 = 1	Cw10 = 6
A29 = 3	B38 = 4	Cw11 = 1
A30 = 3	Bw41 = 1	CwX = 15
A31 = 4	Bw42 = 1	NT = 2
A32 = 1	B44 = 5	
Aw68 = 1	B45 = 1	Total = 70
	B46 = 1	
Total = 72	Bw47 = 1	
	Bw49 = 1	
	Bw50 = 1	
	B51 = 6	
	Bw52 = 1	
	Bw53 = 1	
	Bw54 = 1	
	Bw57 = 2	
	Bw60 = 4	
	Bw61 = 1	
	Bw62 = 5	
	BX = 1	
	Total = 69	

Table 3. Analysis of "A3 Cluster"

Fragment		A3 Cells with Fragment (Total 12 Cells)	Extras Antigen Present/Total	Preliminary Conclusion
EcoRI	2.87	12	–	A3
EcoRI	10.83	12	–	A3
BamHI	3.59	11[a]	–	A3 homozygous
TaqI	.96	10	–	A3 short
PvuII	14.82	10[a]	A11 1/1	A3 short + A11
BamHI	8.55	12	A31 4/4	A3 + A31
BamHI	2.78	11[a]	A31 4/4, A11 1/1	A3 homozygous + A31 + A11
KpnI	5.85	12	A31 4/4, A11 1/1, 2 others	A3 + A31 + A11 + extras
BamHI	15.34	12	A11 1/1, A30 3/3	A3 + A30 + A11
PstI	2.28	7	A11 1/1, A30 3/3	A3 short + A30 short + A11
BglII	3.92	11	A30 3/3	A3 short + A30
PvuII	8.81	12	A9 7/7	A3 + A9
TaqI	.62	10	A23 1/1	A3 short + A23
HindIII	7.39	10[a]	A29 2/3	A3 short + A29 short
KpnI	3.23	12	A32 1/1, A24 3/6	A3 + A32 + A24 short
SstI	13.34	10[a]	A11 1/3, A19 6/11	A3 short + A11 + A19 short
KpnI	16.13	6[a]	–	A3 short

Analysis of a cluster of 17 fragments correlating with HLA-A3 allele. From the top to the bottom, a decrease in specificity of the fragment is observed. First two are only observed in A3 cells. BamHI 3.59 misses the only heterozygous cell for A3 in the panel.
[a]Negative with the only heterozygous cell for A3.

Immunobiology of HLA
Volume I

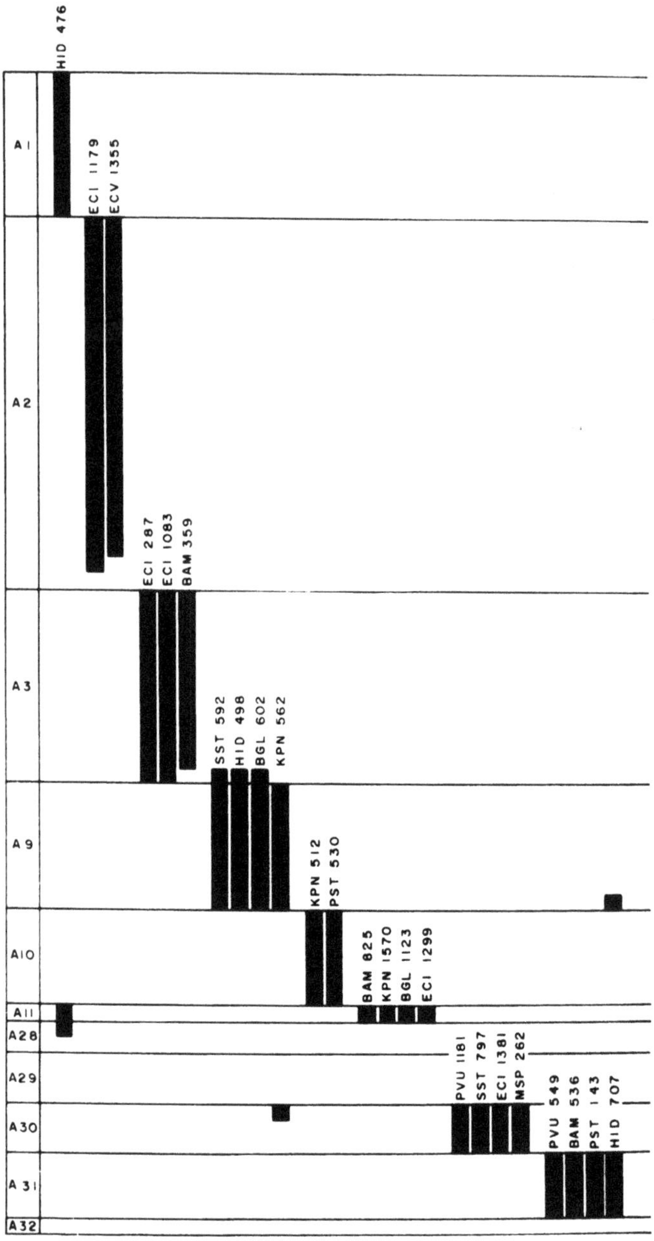

Figure 1. Distribution of RFLPs among HLA-A specificities. The height of the column represents the number of cells with the serologic HLA-A type.

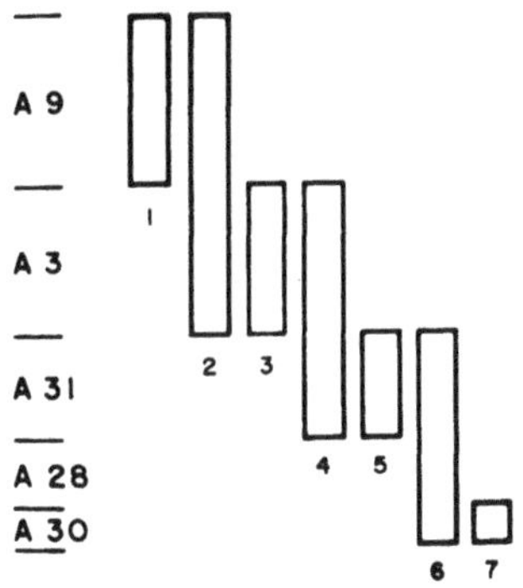

Figure 2. Schematic representation of fragments specific to a certain allele or shared by different ones. Examples of such fragments are as follows: 1: SstI 5.92; 2: PvuII 8.81; 3: BamHI 3.59; 4: BamHI 8.55; 5: BamHI 5.36; 6: KpnI 18.71; 7: SstI 7.97.

instances, splits of classical specificities could not be observed. The results are depicted in Figure 1. Definition of A1, A2, A3, A9, A10, A11, A30, and A31 were possible. Some fragments were shared between two or more specificities. A schematic representation of these observations is presented in Figure 2.

Other fragments had a broad distribution and were characterized by a negative correlation with one or a few well defined specificities. These clusters of fragments are shown in Figure 3. As many as 14 different RFLPs were present in all cells except the ones typed as A9. In several instances, one could observe a specific fragment for a given allele as opposed to another correlating with the remaining specificities. Employing the same A9 example, SstI could define two fragments: SstI 5.92 present in all A9 cells and SstI 5.06 found in the remainder.

Table 4. Fragments Specific to HLA-B or C Alleles

HLA B or C	Fragment		No. positive / No. cells	Extra
B8	BamHI	23.56	4/4	1/6 B51
	EcoRV	7.86	4/4	1/6 B7
B35	PvuII	2.69	5/6	–
B38 + B14	SstI	1.62	6/6	–
B40	KpnI	26.44	3/5	–
B60	HindIII	9.93	4/4	–
B51	SstI	2.78	3/6	–
Cw4	EcoRV	10.96	6/6	2/14 Cw7
	EcoRV	4.44	6/6	1/15 CwX
	PvuII	2.69	4/6	2/14 Cw7, 1/15 CwX
Cw5 + Cw8	EcoRI	7.96	10/10	–
Cw7	PstI	1.01	13/13	1/15 CwX
	BamHI	23.43	10/13	–

Selection of the best fragments found in association with HLA-B and C alleles. Fragment SstI 1.62 is present in B38 and B14 cells. Since all B14 are also Cw8, the specificity assigned could be B38 + Cw8.

Table 5. Possible New Allelic Series for HLA Class I

HindIII	31.25	HindIII	43.87	KpnI	14.92	KpnI	6.67
PstI	7.21	PstI	5.73	BglII	6.74	BglII	5.37
HindIII	18.91	HindIII	24.48	MspI	5.19		
TaqI	1.34			Pvu	6.05	HindIII	5.46
TaqI	0.31			HindIII	5.17		
TaqI	0.27						

Clusters of class I fragments not correlated to defined HLA class I genes. For three of the clusters, an allelic counterpart could be identified. These clusters may indicate new class I genes.

HLA-B and C series did not yield good definition with RFLPs, and a general picture could not be clearly drawn. At least seven fragments were identified as having good correlation with HLA-B specificities and are listed in Table 4. The fragment SstI 1.62 associated with B14 and B38 is again tentative since all B14 cells expressed also Cw8. Thus, this fragment may as well indicate a sharing of B38 plus Cw8. For the HLA-C series, six fragments displayed good association and are also shown in Table 4.

Four new clusters independent of known HLA specificities were found. Three of them had corresponding allelic clusters (Table 5). One of these clusters shows a weak association with Cw6, and the other is complementary. These may be alleles supertypic to the C locus. The other two allelic clusters are new and do not correlate with any known A, B, C, or class III locus specificities.

It was not possible to find the allelic counterpart for the fourth cluster. These clusters may define new class I loci or could determine the equivalents to Qa or Tl in the mouse.

Table 6. Correlation Between Class I and Class III RFLPs

Class III	Class I	Fragment	
C2	Haplotype A29 B44	SstI	5.28
		BamHI	7.78
		BglII	9.48
		SstI	3.17
		HindIII	5.57
	Other than A29 B44	TaqI	1.76
		PstI	4.05
	Bw47 (one cell)	PstI	1.72
C4	B8	SstI	2.39
		BamHI	23.56
	Bw60	KpnI	26.44
		HindIII	9.93
	Other	KpnI	2.79

Clusters between class I and class III RFLPs. Three clusters were observed with C2 polymorphism. One present only with the haplotype A29, B44. Another present in all other cells but one (Bw47). Three clusters were also seen in association with C4. One correlates with B8, another with Bw60, and the last is present in all other cells.

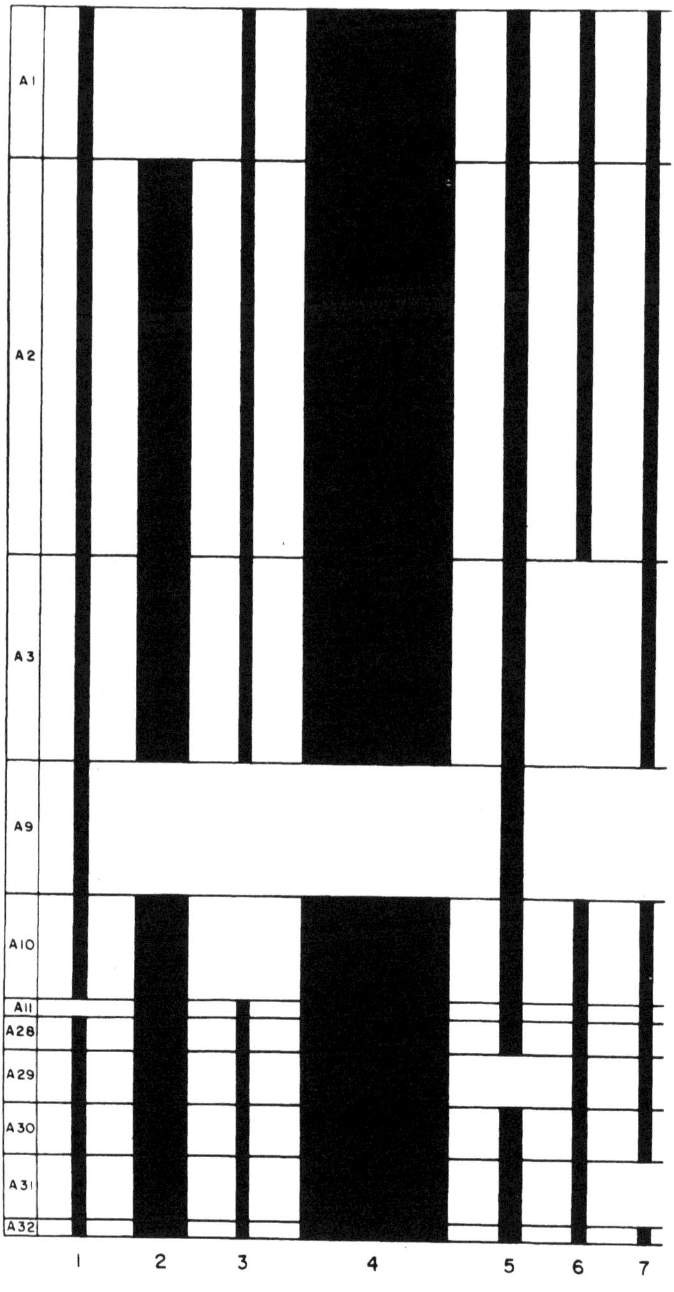

Figure 3. Distribution of RFLPs with broad distribution among HLA-A specificities. They are shared by several alleles, but they are absent in others. The width of the bars represent the number of fragments found in the cluster. Examples of each one are: 1: BamHI 10.08; 2: BglII 13.39; 3: HindIII 5.35; 4: SstI 5.06; 5: PstI 4.05; 6: TaqI 2.72; 7: HindIII 6.72.

Table 7. Average of the Number of Fragments Detected

A1, B8	(n = 3)	= 109
A1, B other	(n = 5)	= 131.6
A2	(n = 24)	= 131.3
A9	(n = 10)	= 110
A10	(n = 10)	= 110
A19	(n = 9)	= 134.9

Average number of fragments detected by the HLA class I probe in different haplotypes according to HLA-A specificity. Range: 96 to 161. Two groups are seen—one with an average of 110 and the other group with about 131 fragments.

Some class I and class III fragments clustered together. For C2, three clusters were seen, possibly correlating with the accompanying haplotype. RFLPs for C4 formed three clusters, two of which correlated with particular B locus alleles. Examples are found in Table 6.

Another interesting feature observed concerns the number of fragments generated from each cell line by all enzymes (Table 7). Fragments ranged from 96 to 161. One group of genes disclosed about 110 fragments (A1 associated with B8, A9, and A10 haplotypes) and the other about 131 fragments (A1 associated with B haplotypes other than B8, A2, and A19 haplotypes). This fact points to a probable expansion of the number of class I genes in the latter group.

In conclusion, although the Workshop design of Southern blot analysis was mainly directed to HLA class II genes, interesting results were obtained with the class I probe. A general picture of HLA-A polymorphism could be discerned, as well as several fragments correlating with HLA-B and C specificities. Fragments correlating with the absence of a given allele can be a new feature of HLA analysis. More interestingly, new allelic series for class I genes not correlated with classic specificities were identified. The notion of duplication of class I genes in certain haplotypes compared with others, as already described for HLA class III genes, was also certified.

Reference

1. Sood AK, Pereira D, Weissman B. Isolation and partial nucleotide sequence of a cDNA clone for human histocompatibility antigen HLA-B by use of an oligodeoxynucleotide primer. Proc Natl Acad Sci USA 1981;78:616.

Workshop Report of RFLP in C4, C2, BF, and 21OH Genes

C. Alper,[1] A. Andreas-Zietz,[2] A. Arnaiz-Villena,[3] R. Dawkins,[4] E. Keller,[2] J. Partanen,[5] and J. Regueiro[3]

The purpose of the Workshop studies of restriction fragment length polymorphisms in class III genes was to correlate specific bands with alleles of C2, BF, C4A, C4B, 210HA, and 210HB. At the time of the Workshop, there were few protein typings available for the Workshop core cells. It was, therefore, decided to guess at complotypes (BF, C2, C4A, and C4B allelic sets occurring in populations and inherited in families as single genetic units) in those panel cells carrying HLA-B and DR specificities known to mark extended haplotypes (1) or supratypes (2). After the Workshop, complement typings became available for 51 core cells (Table 1). The complotype guesses for 14 of the 23 cells with probable extended haplotypes were confirmed in all but one instance. One cell (9027) was guessed to be homozygous SC30 (BFS, C2C, C4A3, C4BQ0), but was in fact heterozygous SC30/SC31. Two other individuals carrying probable nonextended haplotypes were probably heterozygous: C4A5, C4A6 (9095), and BFF, BFS (9080). All cells for which RFLP analyses were performed and for which complement types were available or guessed were used for correlation analyses.

For 210H, two main clusters of restriction fragments were identified with multiple enzymes that corresponded to previously published markers (3,4) for 210HA or 210HB, as shown in Table 2. The enzyme TaqI had been used in most of the published studies with fragment sizes of 3.2 kb and 3.7 kb corresponding to 210HA and 210HB, respectively. In the Workshop, the corresponding fragments apparently were 4.06 (plus 3.01) kb and 4.56 kb. The 210HB fragments were present in all panel cells except for 9047, from a homozygote for the extended haplotype known to be increased among European Caucasians with salt-losing congenital adrenal hyperplasia secondary to deletion of the 210HB gene: [HLA-B47, FC91,0, DR7]. The 210HA gene, on the other hand, was deleted, as expected, on the extended haplotypes [HLA-B8, SC01, DR3], [HLA-B18, F1C30, DR3], and [HLA-B35, FC(3,2)0, DR1]. There appeared to be several RFLP clusters possibly related to minor

Participating Laboratories: NCYCBR,[1] GERALB,[2] FRAARN,[3] ANZDAW,[4] SCAKOS[5]

Table 1. Complement Types of Core Cells

	Cell No.	HLA-A	HLA-C	HLA-B	Complotype[a] BF	C2	C4A	C4B	HLA-DR
Extended haplotypes									
	9002	24	w2,w8	14	S	(C	2	1,2)	1
	9006	11	w4	35	(F	C	3,2	0)	1
	9008	25		18	S	0	4	2	2
	9018	3	w5	18	F1	(C)	3	0	3
	9019	30	w5	18	F1	(C)	3	0	3
	9020	26	?w5	18	F1	C	3	0	3
	9022	1	w7	8	S	C	0	1	3
	9023	1	w7	8	S	C	0	1	3
	9026	26		38	(S	C	2	1)	4
	9027	29		44	S	(C)	3	0/1[b]	4
	9031	2	w10	w62	(S	C	3	3)	4
	9032	2	w9	w62	(S	C	3	3)	4
	9047	3	w6	w47	(F	C	91	0)	7
	9051	29		44	F	C	3	1	7
	9052	2	w6	w57	S	(C)	6	1	7
	9082	3	w7	7	(S	C	3	1)	w15
	9083	3		7	S	(C)	3	1	w15
	9085	30	w5	18	F1	(C)	3	0	3
	9086	1	w7	8	S	(C)	0	1	3
	9088	1		8	S	(C)	0	1	3
	9090	2	w5	44	S	(C)	3	0	4
	9091	2	w9	w62	(S	C	3	3)	4
	9106	3		w50	(S1	C	2	1,18)	7
Nonextended haplotypes									
	9004	2	w1	27	S				1
	9007	2	w6	w57	S		6	1	w16,4
	9009	1	w6	37	S				2
	9011	1		w52	S		3	1	w15
	9012	2	w7	w57	S		6	1	w16
	9021	w68,30	w2	w42	F		1	0	3
	9034	3	w7	7	S		3	1	4
	9035	32		38	S				w11
	9041	1	w4	35	S		3	1	w11
	9042	24	w4	35	S		3	1	w11
	9045	2,3	w7	51	S				w11,w12
	9048	30	w6	13	S		3	1	7
	9050	29		44			0	1	7
	9056	2		35	S				w13,w14
	9058	2		45	F	C	3	1	w13
	9060	1	w9	w62	F		3	1	w13
	9062	2		38		C	2	1	w13
	9065	3	w7	7	S	C	3	1	w13
	9066	2	w11	w46	S	C	4	2	w8
	9075	24	w10	w60	S	C	4	2	9
	9078	3,w33	w8	w65	S	C			1
	9079	w33	w8	14	S				1
	9080	3	w4	35	F/S[b]		3,2	1	1
	9094	2,3	w8	14	S		3	1	7
	9095	1	w6	w57	S		5/6[b]	1	7
	9098	31	w10	w60	S		3	1	4
	9104	31		38	S		3	1	w11
	9105	1	w4	35	S	C	3	1	w11

[a]Alleles and complotypes given in parentheses are guessed. Those without parentheses have been determined.
[b]These individuals are apparently heterozygous at the indicated complement loci. When C4 types are given as two numbers separated by a comma (e.g., SC2(1,2), cell no. 9002), it indicates duplication of that locus on both haplotypes.

Table 2. RFLP in 21OH Genes

I.D. No.	Enzyme-Fragment Size		Concordance	Locus Assignment	Correlation and Comments
1	KPN	2.98	1	21OHB	Cluster 1: Present in all panel cells tested except [Bw47, FC91,0, DR7] (salt-wasting 21OH def)
2	TAQ	4.56	1		
3	ECI	11.89	1		
4	PVU	1.95	1	21OHA	Cluster 2: Present in all panel cells tested except those with known or suspected deletions of 21OHA: [B8, SC01, DR3], [B18, F1C30, DR3], [B35, FC(3,2)0, DR1] and two nonextended haplotypes.
5	KPN	4.16	1		
6	TAQ	4.06	1		
7	BAM	17.36	1[a]		
8	BGL	12.57	1		
9	PST	2.29	1		
10	TAQ	3.01	1		
11	ECI	11.16	2		
12	PST	0.83	1		

[a]Not recommended for use.

polymorphism in 21OH genes, but these could not be further defined.

Five clusters of RFLP fragment patterns were identified with the C4 probe used, as shown in Table 3. The first corresponded to C4A and was positive in all cells except those with deleted C4A: all [HLA-B8, SC01, DR3] and one from a homozygote for HLA-Bw60, DRw8 of unknown complotype. The TaqI fragment size was that previously reported in the literature (3,4). The second cluster was found in all individuals who carried the extended haplotypes [HLA-B7, SC31, DR2], [HLA-B38, SC21, DR4], [HLA-Bw57, SC61, DR7], and [HLA-Bw62, SC33, DR4]. The TaqI fragment size corresponded, more or less, to that published (5) for long C4B genes: 5.77 kb by the Workshop compared with 6.0 kb in the literature. Cluster 3 apparently marked the short C4B gene since it was found on [HLA-B8, SC01, DR3] and [HLA-B44, FC31,DR7] (4,5). The Workshop TaqI fragment of 4.62 was rather different from the published size of 5.4 kb (3). This fragment was also found in the Workshop to characterize [HLA-B14, SC2(1,2), DR1], [HLA-Bw50, S1C2(18,1), DR7], and [HLA-B18, S042, DR2]. A fourth cluster marked the deleted C4A on [HLA-B8, SC01, DR3] (3) and the cell homozygous for HLA-Bw60 and DRw8. The Workshop size of the TaqI fragment was 6.27, corresponding to a published size of 6.4 kb. There was a fifth cluster of absent clusters two and three fragments that seemed to mark deletion of C4B. Patterns with this feature were associated with the extended haplotypes [HLA-B18,

Table 3. RFLP in C4 Genes

I.D. No.	Enzyme-Fragment Size		Concordance	Locus Assignment	Correlation and Comments
1	SST	4.10	1	C4A	Cluster 1: Present in all cells except for those with deleted C4A: [B8, SC01, DR3] and one cell homozygous for B60 and DRw8 of unknown complotype.
2	TAQ	7.04	1		
3	ECI	6.18	1		
4	BAM	4.77	1		
5	HID	13.69	1		
6	HID	31.94	2	C4B	Cluster 2: C4B Long. On [B7, SC31, DR2], [B38, SC21, DR4], [Bw57, SC61, DR7], and [Bw62, SC33, DR4].
7	TAQ	5.77	1		
8	ECI	12.25	1		
9	SST	3.98	2	C4B	Cluster 3: C4B Short. On [B44, FC31, DR7], [B14, SC2(1,2), DR1], [Bw50, S1C2(18,1), DR7], and [B18, S042, DR2].
10	BAM	3.28	1		
11	HID	24.77	2		
12	ECI	13.52	1		
13	TAQ	4.62	1		
14	HID	8.01	1	C4B	Cluster 4: C4A Deletion. Found in all [B8, SC01, DR3] and one cell homozygous for B60 and DRw8 of unknown complotype.
15	TAQ	6.27	1		
16	ECI	7.50	1		
17	ECI	8.96	1		
				C4B	Cluster 5: C4B Deletion. Absence of 6–13 above. Found on [B18, F1C30, DR3], [B35, FC(3,2)0, DR1], and [Bw47, FC91,0].

Table 4. RFLP in C2 Genes

I.D. No.	Enzyme-Fragment Size		Concordance	Locus Assignment	Correlation and Comments
1	PVU	4.40	1	C2	Cluster 1: Common C2C-associated allele. All extended haplotypes tested are positive.
2	KPN	10.50	1		
3	HID	4.46	1		
4	ECI	6.76	1		
5	PST	12.31	1		
6	TAQ	4.91	1		
7	PVU	4.18	1	C2	Cluster 2: Found on 8% of haplotypes in cell panel. "Allelic" to cluster 1, above.
8	KPN	9.88	1		
9	HID	4.32	1		
10	ECI	6.45	1		
11	PST	11.65	1		
12	TAQ	4.75	1		
13	SST	2.48	1		

F1C30, DR3], [HLA-B35, FC(3,2)0, DR1], and [HLA-Bw47, FC91,0, DR7] (with a known deletion of 210HB [6]), as noted in the literature (4).

There are a number of published RFLP in C4 genes that were not detected in the Workshop. For example, the SC61-specific BglII polymorphism (7), the XbaI common 3′ polymorphism (8), and two common polymorphisms in the 5′ region of the C4 genes detected with BglII (9) and HindIII (10).

It is known that C2 and particularly BF are far less polymorphic at both the protein and DNA level than C4. This was also evident in the Workshop studies. Two main clusters were detected in C2 that appeared to define a pair of alleles. The first cluster included all extended haplotypes and over 90% of the Workshop panel. As seen in Table 4, the fragments defining these alleles are produced by a large number of enzymes and the differences in size are fairly constant at about 200 to 500 bp. It is thus clear that the differences relate to those described in the literature (11), which result from size differences in the C2 genes. Thus, it is likely that cluster 1 corresponds to the common form of C2 marked by a 2.7 kb SstI restriction fragment (in the literature, not in the Workshop) on haplotypes with BF S, S1, and F1. The remaining haplotypes, with BF F, mostly carry SstI restriction fragments of 2.4 kb, although a minority carry 2.6 or 2.65 kb. Because the panel cells are homozygous, it is highly unlikely that differences could have been detected between the 2.7, 2.65, and 2.6 kb fragments. The absence of an SstI fragment of around 2.7 kb is thus a puzzle. The Workshop studies did not identify other described genetic polymorphisms, one of which detects the BF F/S point nucleotide difference through the use of MspI (or HaeIII) (12), the other of which distinguishes BF Fb (6.6 kb) from BF Fa (4.5 kb) in TaqI fragments. In most instances, this was because the appropriate enzymes were not used or, if used, the informative cells were not analyzed.

In summary, the Workshop studies defined the RFLP that correspond to deletions in C4A, C4B, 210HA, and 210HB and to polymorphism in the length of C4B genes. However, some published polymorphisms in C4, BF, and C2 genes were not found, largely because the appropriate enzymes were not used for potentially informative cells. It is hoped that in the future, core cell donors will be regularly typed for C2, BF, and C4.

References

1. Awdeh ZL, Raum D, Yunis EJ, Alper CA. Extended HLA/complement allele haplotypes: evidence for T/t-like complex in man. Proc Natl Acad Sci USA 1983;80:259.
2. Dawkins RL, Christiansen FT, Kay PH, et al. Disease associations with complotypes, supratypes and haplotypes. Immunol Rev 1983;70:5.
3. Carroll MC, Palsdottir A, Belt KT, Porter RR. Deletion of complement C4 and steroid 21-hydroxylase genes in the HLA class III region. EMBO J. 1985;4:2547.
4. Schneider PM, Carroll MC, Alper CA, et al. Polymorphism of the human complement C4 and steroid 21-hydroxylase genes. Restriction fragment length polymorphisms revealing structural deletions, homoduplications and size variants. J Clin Invest 1986;78:650.
5. Partanen J, Peltonen L, Koskimies S, Carroll MC. DNA polymorphism of human HLA-linked complement C4 allotypes, including C4 null alleles, in the Finnish population. Hum Hered 1987;37:241.
6. White PC, New MI, Dupont B. *HLA*-linked congenital adrenal hyperplasia results from a defective gene encoding a cytochrome P-450 enzyme specific for steroid 21-hydroxylation. Proc Natl Acad Sci USA 1984;81:7505.
7. Palsdottir A, Cross SJ, Edwards JH, Carroll MC. Correlation between a DNA restriction fragment length polymorphism and C4A6 protein. Nature 1983;306:615.
8. Whitehead AS, Woods DE, Fleischnick E, et al. DNA polymorphism of the C4 genes: a new marker for analysis of the major histocompatibility complex. N Engl J Med 1984; 310:88.
9. Palsdottir A, Fossdal R, Jensson O, Arnason A. BglII polymorphic site upstream to the human complement component C4A gene. Nucleic Acids Res 1987;15:2395.
10. Goldstein R, Gruhn J, Arnett FC, Duvic M. A frequent polymorphism of the complement component C4 gene. Nucleic Acids Res 1986;14:5570.
11. Bentley DR, Campbell RD, Cross SJ. DNA polymorphism of the *C2* locus. Immunogenetics 1985;22:377.
12. Bentley DR, Campbell RD. C2 and factor B: structure and genetics. Biochem Soc Symp 1985;51:7.

Specificity Analysis of DRA Fragments in Core Cell Lines

Miriam Segall[1] with the cooperation of A. Ando,[2] K.W.J. Badenhoop,[3] H. Betuel,[4] H. Festenstein,[5] L. Gebuhrer,[4] C. Gorodezky,[6] R. Heard,[7] R.I. Lechler,[7] E. Casilio-Lonardo,[8] M. Maeda,[2] D.P. Singal,[9] H. Tokata,[2] and A. Zachary[8]

The polymorphic fragments identified in the core cell lines with the DRA probe and the various restriction enzymes are listed in Table 1. They are identified by enzyme, fragment size, and standard fragment number. The concordance (Con) value shows the reproducibility of the presence or absence of the fragment; 1 indicates no discrepancy for the two pairs of "hidden duplicates" among the DNA samples, 2 indicates a discrepancy for one duplicate pair, 3 a discrepancy for both duplicate pairs, 99 that one or more of the duplicates could not be read for technical reasons. Fragments with similar presence/absence patterns are grouped together in clusters, which are discussed below; strong antigen correlations are briefly indicated in Table 1.

Most of the DRA fragments fell into two clusters (I and II) which were negative and positive, respectively, with cells typed as Dw24 or Dw26. Some of these polymorphisms have been previously reported.[1] There were several subclusters within these major clusters, distinguished as follows:

Table 1. DRA Fragments in Core Cell Lines

Enz	kb	Std No.	Con	Clus	Strong Antigen Correlations
PvuII	1193	1	1	IA	Neg for DR1-Dw1, Dw24, Dw26
PvuII	1101	2	1	IIB	Pos for Dw24, Dw26
PvuII	909	3	1		No apparent specificity
BglII	449	2	1	IA	Neg for DR1-Dw1, Dw24, Dw26
BglII	422	3	1	IIA	Pos for Dw24, Dw26
BglII	389	4	1		See text
EcoRI	1343	1	2		Probably nonpolymorphic
EcoRI	1088	2	3		Probably nonpolymorphic
EcoRI	318	4	1	IB	Neg for Dw24, Dw26
EcoRI	314	5	3	IIA	Pos for Dw24, Dw26
MspI	306	1	3	IB	Neg for Dw24, Dw26
MspI	298	2	2	IIA	Pos for Dw24, Dw26
MspI	228	3	1	IC	Neg for Dw24, Dw26, DRw15
TaqI	272	1	2	III	Includes cluster II; see text
TaqI	198	2	1	III	Includes cluster II; see text
TaqI	192	3	1	IC	Neg for Dw24, Dw26, DRw15
TaqI	163	5	1	IIA	Pos for Dw24, Dw26
KpnI	1149	2	99		No apparent specificity
KpnI	579	4	99		No apparent specificity

Cluster IA (*PvuII 1193* and *BglII 449*). These fragments were not found in 4/4 DR1-Dw1, 4/4 DR3-Dw24, 4/4 DRw6-Dw24, and 2/2 DRw8-Dw8.3 cells. Of three DRw6-Dw26 cells, two (9059 and 9063) were negative for these fragments; the third (9055), although typed as Dw26, was consistently positive for cluster I fragments. Cluster IB (*EcoRI 318* and *MspI 306*) had a similar pattern except that 4/4 DR1-Dw1 cells were positive for these fragments. Cluster IC (*MspI 228 and TaqI 192*) had a similar pattern to cluster IB except that 6/7 and 5/7 DRw15 cells, respectively, were also negative for these two fragments (9010 was positive with both). MspI 228 was also negative in 8/8 DRw8 cells, while TaqI 192 was positive in these cells and also in 9021. Most other cells were positive for all cluster I fragments, although a few additional scattered negatives were seen (the presence/absence patterns are given in the specific Workshop probe/enzyme reports).

Cluster IIA (*BglII 422*, *EcoRI 314*, *MspI 298*, and *TaqI 163*). These were essentially the reciprocal of IB and IC, being positive with 4/4 DR3-Dw24, 4/4 DRw6-Dw24, 2/3 DRw6-Dw26 (negative with 9055), and 2/2 DRw8-Dw8.3, as well as a few other scattered positives. Cluster IIB (*PvuII 1101*) was essentially the reciprocal of IA, being positive in addition with 5/5 DR1, as well as 2/2 DR7-DwDB1 and a few other scattered cells. It should be noted that some of the fragments distinguishing these two clusters are very close in size (e.g., EcoRI 318 and 314 and MspI 306 and 298), and the agreement of pattern speaks well for the resolving power of the Workshop Southern blot techniques.

Cluster III (*TaqI 272* and *TaqI 198*). This includes the Dw24 and Dw26-positive cells, as well as all DRw8 cells, 5/7 and 6/7 DRw15 cells, respectively (9010 is negative with these fragments, as well as being positive

Participating Laboratories: US2BAC,[1] JAPTSU,[2] UKISTB,[3] FRABET,[4] UKIFES,[5] LATGOR,[6] UKIBAT,[7] US6BRN,[8] US5SIN[9]

*This is publication #500 from the Immunobiology Research Center, University of Minnesota, Minneapolis, Minnesota 55455, USA.

Supported in part by Juvenile Diabetes Foundation Grant #186175 and March of Dimes Birth Defects Foundation Grant #6-496.

for the fragments of cluster IC), and scattered other positives including some DRw6-Dw25 cells.

Of the additional fragments, *BglII 389* was positive with 4/4 DR1-Dw1, 1/1 DR7-Dw11, and 1/1 DR7-DwX cells. *PvuII 909* had no apparent specificity, and *EcoRI 1343* and *EcoRI 1088* are probably nonpolymorphic fragments. For *KpnI 1149* and *KpnI 579*, many lanes could not be read and therefore no pattern could be established.

Reference

1. Stetler D, Grumet FC, Erlich HA. Polymorphic restriction endonuclease sites linked to the HLA-DR alpha gene: localization and use as genetic markers of insulin-dependent diabetes. Proc Natl Acad Sci USA 1985;82:8100.

Specificity Analysis of DQA Fragments in Core Cell Lines

Miriam Segall[1] with the cooperation of A. Ando,[2] K.W.J. Badenhoop,[3] H. Betuel,[4] H. Festenstein,[5] L. Gebuhrer,[4] C. Gorodezky,[6] R. Heard,[7] R.I. Lechler,[7] E. Casilio-Lonardo,[8] M. Maeda,[2] D.P. Singal,[9] H. Tokata,[2] and A. Zachary[8]

The polymorphic fragments identified in the core cell lines with the DQA probe and the various restriction enzymes are listed in Table 1. They are identified by enzyme, fragment size, and standard fragment number. The concordance (Con) value shows the reproducibility of the presence or absence of the fragment; 1 indicates no discrepancy for the two pairs of "hidden duplicates" among the DNA samples, 2 a discrepancy for one duplicate pair, 3 a discrepancy for both duplicate pairs, 99 that one or more of the duplicates could not be read for technical reasons. Fragments with similar presence/absence patterns are grouped together in clusters, which are discussed below. Strong antigen correlations are briefly indicated in Table 1; however, the fragment may also be positive (or negative) with DNA of cells not expressing the relevant specificity.

Cluster IA *PvuII 492, EcoRI 1339, PstI 1345*) is positive with 28/28 DQw1⁺ cells. Fragment *PvuII 281* is the apparent reciprocal of this cluster, being negative with 28/28 DQw1⁺. **Cluster IB** includes a group of related fragments: *MspI 287* is positive with 28/28 DQw1 and #9037. *HindIII 745* is positive with 23/28 DQw1; it is negative with 4/4 DRw13-Dw18 and with #9105, but positive with #9037. *KpnI 1424* is positive with 20/28 DQw1; it is negative with 5/5 DR1 and 3/3 DRw14-Dw9.

Cluster II *(BamHI 763, KpnI 1176, HindII 483, EcoRI 634, PstI 232*) is positive with 20/20 DRw53-positive cells plus #9104, which is listed as a DRw11-Dw5 but has DR7 RFLP patterns. *TaqI 456* has the same pattern plus scattered other positives, and *MspI 215* is positive for 13/13 DR4 and 1/1 DRw9.

Cluster III *(SstI 1553* and *HindIII 768*) is positive with 4/4 DRw13-Dw18; HindIII 768 is also positive with 1/1 DRw11-Dw"FS".

Cluster IV *(KpnI 2520 and TaqI 699)* is positive with 3/3 B8-DR3 and 3/3 B18-DR3, all of which are probably Dw3; TaqI 699 also shows scattered other positives.

Cluster V *(KpnI 840 and BglII 726)* is positive with 3/4 DR3-Dw24 cells (#9021 is negative), 1/1 DR1-Dw20, plus scattered other positives.

Cluster VI *(BglII 838 and TaqI 278)* is positive for 3/3 DR3-Dw25, #9021, and 2/2 DR7-Dw17, plus one other positive for each fragment.

Cluster VII *(KpnI 807 and BglII 701)* is positive with 4/4 DR1-Dw1, 10/10 DR2, 3/3 DR3-Dw25, #9021, 7/7 DRw8, and 13/13 DR4. Both fragments show numerous other positives with BglII 701 being slightly "shorter."

Cluster VIII *(BamHI 1009, KpnI 1042, TaqI 245)* is positive with 5/5 DR1 and 4/4 DRw14-Dw9. In addition, BamHI is positive with 4/4 DRw13-Dw18, and both BamHI 1009 and TaqI 245 show scattered other positives.

Cluster IX includes a series of fragments all of which are positive with 7/7 DR3-DQw2, 6/6 DRw8 which are not DQw1, and 1/1 DRw14-Dw16-DQw7. In addition, *HindIII 579* and *EcoRI 1107* are positive with 1/1 DRw16-Dw22 and with 8/8 and 7/8 DRw5-DQw7, respectively. *BglII 378*, *PstI 424*, and *MspI 474* are positive with 5/8, 4/8, and 2/8 DRw5-DQw7, respectively. Clusters I, II and IX form an allelic series, identifying DQw1⁺, DRw53⁺, and all remaining cells, respectively.

Participating Laboratories: US2BAC,[1] JAPTSU,[2] UKISTB,[3] FRABET,[4] UKIFES,[5] LATGOR,[6] UKIBAT,[7] US6BRN,[8] US5SIN[9]

*This is publication #501 from the Immunobiology Research Center, University of Minnesota, Minneapolis, Minnesota 55455, USA.

Supported in part by Juvenile Diabetes Foundation Grant #186175 and March of Dimes Birth Defects Foundation Grant #6-496.

Table 1. DQA Fragments in Core Cell Lines

Enz	kb	Std No.	Con	Clus	Strong Antigen Correlations
KpnI	2520	1	1	IV	DR3-Dw3; see text
KpnI	1836	2	1		No apparent specificity
KpnI	1726	3	1		DQw4, DQw7 not DRw53
KpnI	1424	4	1	IB	DQw1
KpnI	1176	5	1	II	DRw53
KpnI	1042	6	1	VIII	DR1, DRw14-Dw9
KpnI	840	7	1	V	DR3-Dw24
KpnI	807	8	1	VII	Several; see text
KpnI	716	9	2		DR1-Dw1, DRw18-Dw19, DRw15
KpnI	660	10	1		No apparent specificity
KpnI	465	11	1		No apparent specificity
KpnI	232	12	99		Many lanes not readable
PstI	1345	1	1	IA	DQw1
PstI	912	2	1		No apparent specificity
PstI	762	3	1	DXA	See text
PstI	684	4	1	DXA	See text
PstI	431	5	1		No apparent specificity
PstI	424	6	1	IX	Several; see text
PstI	232	7	1	II	DRw53
HindIII	768	1	1	III	DRw13-Dw18
HindIII	745	2	1	IB	DQw1
HindIII	579	3	1	IX	Several; see text
HindIII	483	4	1	II	DRw53
BglII	1312	1	1	X	DRw8-not DQw1, DRw16
BglII	838	2	1	VI	Several; see text
BglII	726	3	1	V	DR3-Dw24
BglII	701	4	1	VII	Several; see text
BglII	632	5	1		No apparent specificity
BglII	382	6	1		No apparent specificity
BglII	378	7	1	IX	Several; see text
BglII	166	8	1		Positive for all of Clusters I and II
BamHI	2393	1	1		No apparent specificity
BamHI	1087	2	1		DRw15-Dw2; DRw16-Dw21
BamHI	1009	3	1	VIII	DR1, DRw14-Dw9, DRw13-Dw18
BamHI	763	5	1	II	DRw53
PvuII	819	1	1	DXA	Several; see text
PvuII	722	2	1	DXA	See text
PvuII	492	3	1	IA	DQw1
PvuII	281	4	1	1	Negative with DQw1

Enz	kb	Std No.	Con	Clus	Strong Antigen Correlations
EcoRI	1339	1	1	IA	DQw1
EcoRI	1107	2	1	IX	Several; see text
EcoRI	634	3	1	II	DRw53
EcoRI	195	5	1	DXA	See text
EcoRI	127	6	1	DXA	Several; see text
MspI	605	1	1		6/6 DR7 + 9104
MspI	591	2	1		No apparent specificity
MspI	568	3	1		No apparent specificity; may be related to DXA
MspI	548	4	1		No apparent specificity
MspI	474	5	1	IX	Several; see text
MspI	473	6	1		No apparent specificity
MspI	325	7	1	X	DRw8-not DQw1, DRw16
MspI	287	8	1	IB	DQw1
MspI	250	9	1	X	DRw8-not DQw1, DRw16
MspI	231	10	1	X	DRw8-not DQw1, DRw16
MspI	215	11	1	II	DR4, DRw9
TaqI	950	1	1		No apparent specificity
TaqI	817	2	1		No apparent specificity
TaqI	699	3	2	IV	DR3-Dw3; see text
TaqI	619	4	3		No apparent specificity, may be related to DXA
TaqI	553	5	2		No apparent specificity
TaqI	456	6	2	II	DRw53
TaqI	376	8	3		No apparent specificity
TaqI	285	10	2		No apparent specificity
TaqI	278	9	2	VI	Several; see text
TaqI	245	11	99	VIII	DR1, DRw14-Dw9
TaqI	209	12	1	DXA	See text
TaqI	155	13	1	DXA	See text
SstI	2478	1	99		Most DRw53 (related to Cluster II)
SstI	2355	2	1		No apparent specificity
SstI	1938	3	1		DRw15, DRw8-not DQw1
SstI	1653	4	1		No apparent specificity
SstI	1553	5	1	III	DRw13-Dw18
SstI	930	6	1		No apparent specificity
SstI	205	8	1		DR4-Dw15

Cluster X *(BglII 1312, MspI 325, MspI 231, and MspI 250)* are positive for 2/2 DRw16, 6/6 DRw8 which are not DQw1, and scattered other positives.

Cluster DQA2 includes several series of fragments. *PvuII 722*, *EcoRI 195*, and *TaqI 155* make up a cluster identifying one of the two allelic DX alpha fragments (the TaqI 1.9 fragment) (1). The pattern of the cluster consisting of *TaqI 209*, *PvuII 819*, and *EcoRI 127*, identifying the TaqI 2.1 fragment, is essentially reciprocal with the first cluster, with four heterozygotes (#9007, 9008, 9028, and 9106). These two alleles show high correlations with two different DQ beta fragment clusters: TaqI 209 with a cluster including DQB PstI 1122, MspI 196, and SstI 675; TaqI 155 with a cluster including DQB PstI 417, MspI 262, and MspI 817. DQA *Pst 762* is positive with all cells positive for TaqI 209 and also with a subset of cells positive for the TaqI 155 cluster, while DQA *PstI 684* is positive with 8/10 members of that subset of the TaqI 155 cluster which is negative with Pst 762, suggesting a "split" of the TaqI 155 cluster.

Specificities of the remaining fragments which did not fall into clusters but which showed a specificity are indicated in the Table, as are those fragments which showed no apparent specificity.

Reference

1. Spielman RS, Lee J, Bodmer WF, Bodmer JG, Trowsdale J. Six HLA-D region alpha-chain genes on human chromosome 6: polymorphisms and associations of DC alpha-related sequences with DR types. Proc Natl Acad Sci USA 1984;81:3461.

Unusual DR-DQ Haplotypes

H. Betuel,[1] E.D. Du Toit,[2] L. Gebuhrer,[1] M.P. LaBonne,[1] M. Oudshoorn,[2] L. Dombrausky,[3] C. Gorodezky,[4] and D. Cohen[5]

The linkage disequilibrium existing between HLA-DR and DQ antigens is so strong that associations that do not follow the pattern appear as exceptions. In Caucasoids, the associations are well known: DQw1 is linked with DR1, DR2, DRw6, DRw10, DQw2 with DR3 and DR7, DQw3 with DR4, DR5, DR9, DQw4 (or WA) with DRw8. Different combinations are found when other ethnies are investigated, for example, in South African Negroes, DR3 is sometimes associated with DQw3 or DQ blank (1).

Such unusual DR-DQ associations were analyzed as an optional study in the Southern blot component of the Workshop. The aim of the study was to investigate whether these rare DR-DQ combinations correlated with particular DNA fragments hybridizing with DR or DQ probes or whether they followed the regular DNA polymorphic pattern of DR and DQ antigens.

Material and Methods

Families, unrelated genotyped individuals, and homozygous typing cells (HTCs) were analyzed. The data corresponding to nongenotyped individuals were not considered. The main ethnies involved were Caucasoids from Europe, North America and South Africa, Cape Coloured, South African Negroes and Mestizo. Fifteen different unusual haplotypes were studied (Table 1A). DRw13 DQw7 is recognized cellularly by HTC HAG (2) and by an alloserum (3); however, some individuals termed HAG- are not recognized by this HTC (4) nor by the alloserum.

Besides, seven other combinations were also examined. Details are given in Table 1B. HTC DRw11 TIS (10W9042) is a new non-Dw5 specificity. It seemed worthy of interest to study DRw10, which was not analyzed in the cell lines of the Workshop. DR COR is an unusual family from Bodmer's laboratory typing as DR blank DQw7.

The RFLP was obtained with the technical procedures recommended for the Southern component on core lymphoblastoid cell lines and on families. It was agreed among the participants to use the following six restriction enzymes: BamHI (BAM), EcoRI (ECI), EcoRV (ECV), HindIII (HID), PvuII (PVU), TaqI (TAQ). The blots were hybridized successively with DRβ, DQβ, and DQα probes provided by the central organization. The weight in kilobases (kb) of the fragments studied were compared to that of controls with the common DR-DQ associations as well as with the data on the core cell lines compiled by the central analysis (this volume).

Results

DRβ Probe. Four specificities correlated with single fragments: DRw14 DQw7 with ECV 8.65 kb, DR7 DQw9 with HID 9.66, 9.65 kb, COR DQw7 with ECI 9.9 kb, DRw10 with TAQ 13.0 kb. The other unusual DR-DQ haplotypes always shared one or more fragments with the usual DR-DQ associations and sometimes with other specificities as well. For instance, ECI 4.4 kb

---

Participating Laboratories: FRABET,[1] SAFDUT,[2] US4STA,[3] LATGOR,[4] FRACOH,[5] US8RUB

Immunobiology of HLA
Volume I

Table 1. Unusual DR-DQ Haplotypes Investigated

DR	DQw	Dw	Ethny[a]	(Fam)[b]	No. Family Members	No. Unrelated	No. Total
w14	7	NT	Cauc.			2	2
w13	7	HAG+	Cauc.	(1)	3	2	5
w13	7	HAG−	Cauc.			3	3
w12×6	7	JOH	C.Col.	(1)	4		4
w11	1	Bl.	N.SA				
		NT	Cauc.			3	3
w8	4	8	C.Col.				
		NT	Cauc.	(1)	4	1	5
w8	BI*	NT	Mes.	(1)	2		2
w8	7	8	N.SA			1	1
w8	7	Bl	C.Col.	(1)	2		2
w8×14	1	NT	C.Col.			2	2
7	9	11	Cauc.			1(HTC)	1
7	9	NT	Cauc.	(1)	2	3	5
7	9	7	Cauc.	(1)	3		3
3	4	Bl			↗1		
		RSH	N.SA	(1)	4		4
					↘3		
3	3	NT	Mes.	(1)	3		3
Total 15 ≠ Haplotypes				(9)	27	18	45
Other Combinations							
w11	7	TIS	Cauc.			1(HTC)	1
w10	1	NT	Cauc.	(1)	3		3
4	7	NT	Cauc.	(1)	3		3
4	8	NT	Cauc.	(2)	5		5
		Bl.	N.SA	(1)	5	/	5
4	8	4	Cauc.	(2)	6		6
4	8	13	C.Col.	(1)	3		3
COR	7	NT	Cauc.	(1)	3		3
Total 7 ≠ Haplotypes				9	28	1	29

[a]Cauc. = Caucasoid; C.Col. = Cape Coloured (S. Africa); N.SA = South African Negroes (Xhosa); Mes. = Mestizo (Mexican).
[b]Fam. = family; Bl = Blank.

correlated with DRw13 DQw3, DRw14 DQw3, DR3 DQw4, and with DRw14 DQw1. Fragment TAQ 11.48 kb was found with DRw13 DQw3, DRw12 DQw3, DRw11 DQw1, DRw11 DQw7, DR3 DQw2, and DR3 DQw4.

Table 2. Patterns of Specificities Correlating with DQw7

Pattern	Specificities	Fragments (kb)
a	Some DR4, DRw11, DRw12, DRw13, DQw7	HID 3.41, BAM 3.55
	DRw8 DQw7, COR DQw7	PVU 3.72.
b	"a" plus DQw2, DRw13 DQw6 Dw19	PVU 2.38.
c	"a" plus DR3 DQw2, neg. with DR3 DQw4	TAQ 4.62.
d	"a" plus DR7 DQw17, DRw15, Dw12[a] DQw6 DRw8 DQw6[a]	HID 3.77
e	DR7 DQw7 Dw11, DRw8 DQw7, COR DQw7, DRw15 DQw6 Dw12,[a] DRw8 DQw6[a]	ECl 16.51

[a]The two cells in patterns d and e DRw15 Dw12 (10W 9011), DRw8 DQw6 (10W 9066) have the same DQ beta chain DQβ 1.12.

DQβ Probe. The distinction between the subtypes of DQw1: DQw5, DQw6 previously recognized by serology (5) and by RFLP (6) has been largely confirmed in the present Workshop and has also been documented by T-cell clones and biochemistry (this volume). These subdivisions, thus, will not be considered in this report. However, it can be stressed that distinct DNA fragments can differentiate Dw18, Dw19, and Dw9.

With the enzymes and probes studied, none of the rare haplotypes was individualized by a fragment or cluster of fragments. The fragments correlating with DQw7 and with other specificities form a complex pattern as shown in Table 2. The specificities associated with DQw4, DQw8, DQw9, and their fragments are given in Table 3.

Table 3. Fragments and Specificities Associated with DQw4, DQw8, DQw9

Specificities	Fragments (kb)
DR4 DQw8, DR7 DQw9, DR3 DQw4, DRw8 DQw4	TAQ 1.90, BAM 10.26
DR4 DQw8, DR3 DQw4, DRw8 DQw4	ECl 18.20.

Table 4. Rare and Usual Haplotypes Revealed by DQα Probe

Specificities	Fragments (kb)
DQw1-like cluster	
Usual DQw1 DRw11 DQw6	PVU 4.92
Usual DQw1 DRw11 DQw5	ECl 13.39
Usual DQw1 (except Dw18) DRw11 DQw5	HID 7.45
DRw13 Dw18, DRw11 DQw6	HID 7.68
DR1, DwBON, DRw10, Dw9, DRw8×14	TAQ 2.45
DRw15, DRw8 DQw1, DRw11, DQw5, Dw19	TAQ 5.53
Dw19	BAM 23.93
DR1, DwBON, DRw10, Dw9, DRw8×14	BAM 10.09
Dw18, DRw11, DQw6, DRw8 DQw6 DR2, DRw11, DQw5	BAM 10.87
DRw52-like cluster	
DR3, DRw11, DQw7, DRw12 DQw7, DRw6, DQw7 DRw8 DQw4, DRw8 DQw7	HID 5.79, ECl 5.07
DR3, DRw11, DQw7, DRw6 DQw7, COR DQw7 cell FRA BET 16 (DRw13 DQw7 Dw HAG−) negative	TAQ 3.76
DRw53-like cluster	
DR4 DQw3, DR7 DQw2, DR7 DQw8, DRw9 DQw9	BAM 7.63, ECl 6.34
cell FRA BET 16 positive	HID 4.83, TAQ 4.56

It is worthy to note that fragment ECV 15.14 kb correlating with DRw14 DQw7, DRw13 DQw7 Dw HAG-reacted also with one Dw5 HTC, but marked as well DR3 DQw2 cells and remained negative on DR3 DQw4 cells.

DQα Probe. The rare haplotypes could not be distinguished from the usual DR-DQ combinations, both could be arranged in three main clusters (Table 4). Concerning DQw1-like cluster, two points can be made: 1) a distinction can be made between Dw specificities linked with DRw6 namely Dw18, Dw19 and Dw9, and 2) DRw11 DQw1 cells are heterogeneous; some are DQw5 and some type as DQw6. In cluster DRw52-like are present all cells typing as DRw52 and DQw3, as well as those which are DQw2, DQw4. The salient point is that cells that type as DRw52 and DQW1 such as DRw6 DQw1, DRw8 DQw1, DRw11 DQw1 are not found in this group. Another interesting observation concerns family FRA BET 16. The 'c' haplotype in the mother and two children type as DRw13 DQw7 DRw52, but there are strong extra reactions with DRw53 sera.

These three individuals are not in the DRw52 cluster, but have all the DNA fragments encountered in DRw53 cells.

Conclusion

The DNA pattern of 15 different rare HLA DR-DQ haplotypes and of seven other combinations were investigated in, respectively, 45 and 29 genotyped individuals of various ethnical origins. Three probes were used: DRβ, DQβ, DQα. Hybridization with DRβ probe individualized four specificities with single fragments: DRw14 DQw7, DR7 DQw9, COR DQw7, and DRw10. The other rare haplotypes, even with DQβ and DQα probes, fitted in the various patterns observed with usual DR-DQ combinations. The use of DQα and DQβ probes allowed to distinguish between Dw18, Dw19, Dw9 and showed an heterogeneity in DRw11-DQw1 cells. Hybridization with DQα revealed that cells typing as DRw52 and DQw1 are not found in the DRw52-like cluster.

References

1. Oudshoorn M, Schreuder GMT, Campbell EM, du Toit ED. Segregation of DR and DQ: 'exception to the rules.' In: Albert ED, Baur MP, Mayr WR eds. Histocompatibility Testing 1984. Berlin: Springer-Verlag, 1984:419.
2. Pawelec G, Muller C, Rehbein A, Balko I, Schunter F, Wernet P. The new HLA-DRw6 and -8 associated Dw HAG specificity defined by homozygous typing cell 9W 1802. Tissue Antigens 1984;24:292.
3. Schreuder GMT, Gebuhrer L. et al. HLA-DRw6 and its subgroups HLA-DRw13 and DRw14. In: Dupont B (ed). Immunobiology of HLA, Volume I: Histocompatibility Testing 1987. New York: Springer-Verlag, 1989.
4. Gebuhrer L, Labonne MP, Freidel AC, Betuel H. Personal observation, 1987.
5. Gebuhrer L, Betuel H, Lambert J, Freidel AC, Farre A. A division of HLA. DQw1 associated with DR1, DR1x-DQw1, DR2 short, DRw10 and DRw14. I. Definition by alloantiserum LY 1327. Hum Immunol 1987;18:235.
6. Gebuhrer L, Font MP, Freidel AC, Labonne MP, Betuel H. Recognition of DR and Dw alleles by DNA typing associated with serologic and cellular techniques. Touraine JL, et al, eds. In: Transplantation and Clinical Immunology, XIX. Amsterdam: Excerpta Medica, 1987:77.

Linkage Disequilibrium, Interlocus Association and Ancestral Haplotypes

R.L. Dawkins,[1] E. Martin,[1] A. Andreas-Zietz,[2] E. Keller,[2] J. Partanen,[3] A. Arnaiz-Villena,[4] J.L. Vicario,[4] and C. Alper[5]

The accumulated data in the cell line panel made it possible to examine the associations between alleles at widely separated loci within the MHC. The calculation of 2- or 3-point delta values was clearly inappropriate given the limited size and the selection of the panel. Rather, we could determine whether particular "supratypes" or "extended haplotypes" could "split" apparently homogeneous alleles at multiple loci.

It was already known that the 1,8,3 and 18,F1,3 supratypes differ in the DR and DQ regions, even though both contain DR3 and DQw2. For example, several probe-enzyme combinations yield supratype specific patterns (1). Furthermore, serologic reactions distinguish 1,8,3 and 18,F1,3 apparently by splitting DRw52 in DRB III (Williamson et al., in preparation). Since three examples of each of the DR3, DQw2 containing supratypes were available, we could make meaningful comparisons of the RFLP patterns.

We examined the output for all class II probes. As shown in Figure 1, there are complementary differences at DRA, DRB, DQA, and DQB. Since so many enzymes yield supratype specific patterns, we conclude that there must be substantial structural differences between the two supratypes, even though the serologic patterns are similar at DRβ_1. It should be emphasized that the two supratypes begin to "fragment" between DQ and DP (2).

Next, we asked whether there were differences between two supratypes containing DR2 and DQw1 (Fig. 2). Again there must be major structural variations although only one example of 25, 18, 2 was available.

Finally, we could compare three supratypes with DR7. In striking contrast to the two situations described above, no D region differences are apparent in those sharing DR7 and DQ2 (Fig. 3). The contrasting DQw associations between these DR7-containing supratypes have been reported elsewhere (3).

Given the extremely frequent differences between supratypes (Fig. 4), it is apparent that it will be necessary to characterize complete chromosome segments of > 1 Mb. With such characterization, further heterogeneity of serologic specificities can be expected. The functional significance of these differences should be examined. Undoubtedly there has been striking conservation of at least some ancestral haplotypes.

Participating Laboratories: ANZDAW,[1] GERALB,[2] SCATSB,[3] FRAARN,[4] NCYCBR[5]

References

1. Martin E, Gin W, Kay PH, Christiansen FT, Dawkins RL. Complotypes and Class III gene arrangements can be assigned from supratype specific restriction fragment length polymorphisms. 10th International Histocompatibility Workshop Newsletter 1987;1(3):13.
2. Saueracker GC, Martin E, Dawkins RL. Some supratypes extend to DQ. In: Dupont B (ed). Immunobiology of HLA. Volume II: Immunogenetics and Histocompatibility. New York: Springer-Verlag, 1989.
3. Kay PH, Dawkins RL et al. Coexistence of an MHC chromosomal segment marked by HLA B17, BfS, C4A6, B1, DR7, and DQw9 in different ethnic groups. Human Immunol 1988;23:27.

DRA

CELL ID	B-DR-DQW	Bgl 422	Taq 163	EcI 314	Msp 298	Pvu1 101
9022	8 3 2	+	+	+	+	+
9023	8 3 2	+	+	+	+	+
9088	8 3 2	+	+	+	+	+
9018	18 3 2	-	-	-	-	-
9019	18 3 2	-	-	-	-	-
9020	18 3 2	-	-	-	-	-

DRB

CELL ID	B-DR-DQW	Bam 392	Hid 200	Pst 107	Kpn 769	Kpn 272	Hid 285	Taq 979	Pvu1 035	Hid 163	Bgl 977	Taq1 148	EcV2 004	Msp 55	EcI 331	Hid 289
9022	8 3 2	+	+	+	+	+	+	+	+	+	-	-	-	-	-	-
9023	8 3 2	+	+	+	+	+	+	+	+	+	-	-	-	-	-	-
9088	8 3 2	+	+	+	+	+	+	+	+	+	-	-	-	-	-	-
9018	18 3 2	-	-	-	-	-	-	-	-	-	+	+	+	+	+	+
9019	18 3 2	-	-	-	-	-	-	-	-	-	+	+	+	+	+	+
9020	18 3 2	-	-	-	-	-	-	-	-	-	+	+	+	+	+	+

DQA

CELL ID	B-DR-DQW	Pvu 819	EcI 127	Kpn 840	Bgl 726	Bgl 838
9022	8 3 2	+	+	+	+	-
9023	8 3 2	+	+	+	+	-
9088	8 3 2	+	+	+	+	-
9018	18 3 2	-	-	-	-	+
9019	18 3 2	-	-	-	-	+
9020	18 3 2	-	-	-	-	+

DQB

CELL ID	B-DR-DQW	Msp 196	Bam 383	Pst 833	Pst 301	Msp 773	Taq 341	Hic 924	Hid 165	Hic1 160	Hid 292	Hid 167	Pst 639	Pst 417	Msp 262	Msp 817	EcV 463	Msp 239	Msp 213
9022	8 3 2	+	+	+	+	+	+	+	+	-	-	-	-	-	-	-	-	-	-
9023	8 3 2	+	+	+	+	+	+	+	+	-	-	-	-	-	-	-	-	-	-
9088	8 3 2	+	+	+	+	+	+	+	+	-	-	-	-	-	-	-	-	-	-
9018	18 3 2	-	-	-	-	-	-	-	-	+	+	+	+	+	+	+	+	+	+
9019	18 3 2	-	-	-	-	-	-	-	-	+	+	+	+	+	+	+	+	+	+
9020	18 3 2	-	-	-	-	-	-	-	-	+	+	+	+	+	+	+	+	+	+

Figure 1. The RFLPs obtained after digestion with different enzymes and probing for DRA, DRB, DQA, and DQB are shown. The informative fragments and their sizes (base pairs) are listed across the top, and the cells tested (10W Nos.) down the side. Cells 9022, 9023, and 9088 are homozygous for the B8,DR3,DQw2 supratype, while cells 9018, 9019, and 9020 are homozygous for the B18,DR3,DQw2 supratype. (+) represents fragment present, and (−) represents fragment absent. It can be seen that the patterns are supratype specific and the two supratypes, although serologically similar at class II, are different at DRA, DRB, DQA, and DQB at the molecular level.

CELL ID.	B-DR-DQW	DRB Pvu1 428	DRB Kpn 184	DRB EcI 915	DRB H*id 249	DRB H*id 249	DRB Hic 364	DRB Pst 419	DQA EcI 127	DQA Pvu 722	DQA EcI 195	DQB Hid 942	DQB Kpn1 175	DQB Hid 332	DQB Msp 630	DQB Msp 437	DQB Pst 551	DQB Taq 310	DQB Pst 417	DQB Msp 262	DQB Pst 523
9017	7 2 1	+	+	+	+	-	-	-	+	-	-	+	+	+	+	+	+	-	-	-	-
9013	7 2 1	+	+	+	+	-	-	-	+	-	-	+	+	+	+	+	+	-	-	-	-
9082	7 2 1	+	+	+	+	-	-	-	+	-	-	+	+	+	+	+	+	-	-	-	-
9008	18 2 1	-	-	-	-	+	+	+	-	+	+	-	-	-	-	-	-	+	+	+	+

(* = Mistake in computer assignment: 2 different fragments are called Hid 249.)

Figure 2. The RFLPs obtained after digestion with different enzymes and probing for DRB, DQA, and DQB are shown. The informative fragments and their sizes (base pairs) are listed across the top, and the cells tested (10W Nos.) down the side. 9017, 9013 and 9082 are homozygous for the B7,DR2,DQw1 supratype and cell 9008 is homozygous for the B18,DR2,DQw1 supratype. It can be seen that, once again, the RFLPs are supratype-specific and the two supratypes are different at DRB, DQA and DQB. The DRB/HID 2.49 fragment on these cells is: −/−/−/+ (see Simons et al., Figure 25A, this volume).

DRA:

CELL ID.	B-DR-DQW	Taq 192	Bgl 398	Taq 277	Taq 198
9051	44 7 2	+	-	-	-
9047	47 7 2	+	-	-	-
9052	57 7 2	-	+	+	+

DRB:

CELL ID.	Hid 130	Pvu 101	Taq 979	Pst 518	Pvu 237	Pst 490	Kpn 943	EcV2 047	EcI1 533	Taq 723	Taq 260	Msp 152	Taq 608	EcI 319	Msp 808	Hid 844	Bgl 977	Pvu 365	Kpn1 599	Pst 493	EcI 558	Pvu 339	Hid 133	EcI1 513	Taq 937	Taq 315	Taq 269
9051	+	+	+	+	+	+	+	+	+	+	+	+	+	+	+	+	-	-	-	-	-	-	-	-	-	-	-
9047	+	+	+	+	+	+	+	+	+	+	+	+	+	+	+	+	-	-	-	-	-	-	-	-	-	-	-
9052	-	-	-	-	-	-	-	-	-	-	-	-	-	-	-	-	+	+	+	+	+	+	+	+	+	+	+

DQA:

CELL ID.	Msp 548	Taq 155	Bgl 701	Sst2 478	Pst 762	Pvu 284	Taq 950	Msp 215	Sst2 355	Taq 619	Kpn1 836	Bgl1 312	Msp 325
9051	+	+	+	+	-	-	-	-	-	-	-	-	-
9047	+	+	+	+	-	-	-	-	-	-	-	-	-
9052	-	-	-	-	+	+	+	+	+	+	+	+	+

DQB:

CELL ID.	Hic 230	Msp 496	Hid1 280	Kpn 920	Taq 669	EcV2 001	Kpn1 182	Hic1 160	Pvu 88	Hid 377	Pvu 238	Msp 315	Bam 383	Hic 388	Bgl 938	Bgl 693	Pvu 640	Taq 341	Hid 392	Hid 278	Bam1 026	Kpn1 906	Taq 190	Bgl 161	Taq 565	Hic1 229	Hid 076	EcV1 943	Msp 130
9051	+	+	+	+	+	+	+	+	+	+	+	+	-	-	-	-	-	-	-	-	-	-	-	-	-	-	-	-	-
9047	+	+	+	+	+	+	+	+	+	+	+	+	-	-	-	-	-	-	-	-	-	-	-	-	-	-	-	-	-
9052	-	-	-	-	-	-	-	-	-	-	-	-	+	+	+	+	+	+	+	+	+	+	+	+	+	+	+	+	+

Figure 3. The RFLPs obtained after digestion with different enzymes and probing for DRA, DRB, DQA, and DQB are shown in the same format as that seen in Figures 1 and 2. Cells 9051 and 9047 are homozygous for the B44,DR7,DQw2 and B47,DR7,DQw2 supratypes, respectively. Cell 9052 is homozygous for the B57,DR7,DQw3 supratype. In contrast to the previous comparisons, the two DR7,DQw2 containing supratypes have similar RFLPs and are different from the B57,DR7,DQw3 supratype.

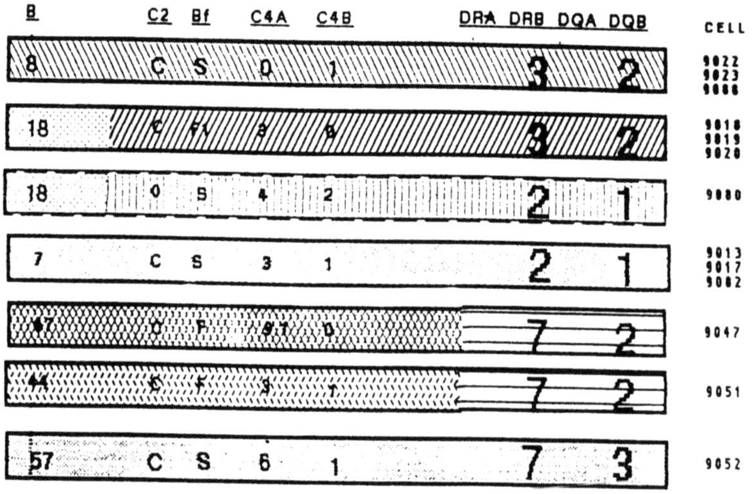

Figure 4. Schematic summary of results suggesting that ancestral haplotypes are more or less unique over extensive segments of the MHC. The examples given are only representative; other comparisons with larger numbers of cell lines are required. To date the similarities between cell lines with the same supratypes have been very striking. At the same time, the differences between different supratypes are far greater than might have been expected. At present we presume cells 9051 and 9047 are similar if not identical in the D region, possibly suggesting that each is a recombinant.

The Genomic Structure of Ancestral Haplotypes Revealed by Pulsed Field Gel Electrophoresis (PFGE)

R.L. Dawkins,[1] P.H. Kay, E. Martin,[1] and F.T. Christiansen[1]

For several years, it has been obvious that certain sets of major histocompatibility complex (MHC) alleles (supratypes) mark haplotypes that have been highly conserved (1). In these instances, we might predict that genomic structure would be essentially supratype-specific viz. apparently unrelated subjects with the same supratype would have similar or identical RFLP patterns when probed for widely separated regions within the MHC. There is now abundant supporting evidence. As only one example, there is extensive genomic polymorphism within and around the 21-hydroxylase genes (2), and the RFLP pattern is predictable in the presence of certain sets of HLA-B, C4A, C4B, and DR alleles (3). See also the Interlocus Report (Dawkins et al., this volume).

It should be emphasized that specificity only relates to certain combinations of alleles, and there has been some difficulty distinguishing between such sets and others that are not specific, presumably because they reflect relatively recent recombinations. This problem has led to a proliferation of terms such as fixed supratypes, extended haplotypes, preferential allelic associations, population haplotypes, super-haplotypes, etc. None of these terms is very clearly defined. Before it was possible to demonstrate similarities and differences in genomic structure, it was generally necessary to depend on the presence of particular complotypes (4) or the presence of very rare and specific alleles such as Bw47 and Bw65. At the present time, there are very few, if any, DR and DQ alleles with the same specificity.

Table 1 provides a partial listing of supratypes that define conserved ancestral haplotypes. These are identified by the B locus allele since this is generally the best single marker, although some haplotypes appear to share the same B locus allele. Pending further splitting of such alleles, we have used the designation 35.1, 35.2, 35.3 to distinguish between different haplotypes. Clearly, a list of this type can only be provisional at this time, but several points should be emphasized. First, although some may be defined by class I and class II alleles alone, this approach can be misleading as, for example, with 35.2, 35.3, 62.1, and 62.2. It should also be emphasized that these combinations cannot be identified on the basis of frequency alone. For example, several combinations that include A2 are relatively frequent, reflecting the frequency and heterogeneity of A2 and the frequency of historical A/B recombination. It should be stated that most of the haplotypes listed are basically Caucasoid, although some (e.g., 57.1) are common to two or more racial groups (5). The list does not include haplotypes that appear to be restricted to other ethnic groups.

Table 1. Ancestral Haplotypes

	A	C	B	C2	Bf	C4A	C4B	DR	DQ
7.1	3	7	7	C	S	3	1	15	6
8.1	1	7	8	C	S	0	1	3	2
13.1	30	6	13	C	S	3	1	7	2
18.1	25	–	18	0	S	4	2	15	6
18.2	?	5	18	C	F1	3	0	3	2
18.3	?	?	18	?	S	3	1	11	7
35.1	?	4	35	?	S	3	1	11	?
35.2	3	4	35	C	F	3+2	0	1	5
35.3	11	4	35	?	S	3	0	1	5
37.1	1	6	37	?	F	?	?	10	?
38.1	26	6	38	?	S	2	1	4	8
44.1	2	5	44	C	S	3	0	4	7
44.2	?	4	44	C	F	3	1	7	2
47.1	3	6	47	C	F	1	0	7	2
50.1	?	?	50	C	S1	2	1+1.5	7	?
51.1	?	–	51	C	F	3	QO	4	3
55.1	?	3	55	B	S	4	5	14	?
57.1	1	6	57	C	S	6	1	7	9
60.1	?	3	60	C	S	3	1	4	3
60.2	?	3	60	C	S	3	0	8	4
60.3	2	3	60	C	S	0	2	13	?
62.1	2	3	62	C	S	3	3	4	8
62.2	?	3	62	B	S	4	2	4	3
63.1	?	?	63	C	F	3	1	6	?
64.1	?	8	64	C	F	3	1	13	6
65.1	?	8	65	C	S	2	1+2	1	5

The Tenth International Workshop provided an excellent opportunity for examining the genomic structure of ancestral haplotypes carried by subjects from different geographic regions of the world. We could assume that identity would be due to conservation rather than recent descent. Therefore, we examined selected cell lines by PFGE since this allows the separation of large fragments of DNA spanning most of the MHC (Fig. 1). Using restriction enzymes that cut at infrequent but conserved sites, we asked whether supratypes carried by unrelated subjects would show a specific pattern in terms of length and gene copy number. The Epstein-Barr virus (EBV)-transformed cell lines were cultured until it was possible to prepare plugs containing 6 + 10^5 cells. After treatment with proteinase K, washing in tris-EDTA and digestion, the plugs were equilibrated with STOP and loaded into gel slots. Digested DNA was run on 1.5% agarose at 20°C using 120 V, 35 H, 90 s pulses for large fragments and 90 V, 65 H, 60 s pulses for small fragments. The gels were then depurinated and denatured

Participating Laboratory: ANZDAW[1]
This is publication no. 8811 of the Departments of Clinical Immunology, Royal Perth Hospital, Queen Elizabeth II Medical Centre and the University of Western Australia, Perth, Western Australia.

Immunobiology of HLA
Volume I

Ancestral Haplotype	Probe	DPA			DPA		DOB		DQB		DRB DQB		DRB					C4	
	FRAGMENT	Mlu	Mlu	Mlu	Hic	Hic	Hic	Hic	Hic	Hic	Mlu	Mlu	Taq	Pvu	Pvu	Pvu	Taq	Taq	Taq
	T(kb)	900[a]	310	150	12	10	7	6	15	10	680	850	12	500	800	200	10	7	6.4
18.2	9085	+	−	−	−	+	+	−	+	−	+	−	+	+	−	−	−	+	−
	9020	+	−	−	+	−	+	−	+	−	−	−	+	+	−	−	−	+	−
	9019	+	−	+	+	−	+	−	+	−	−	+[b]	+	+	−	−	−	+	−
	9018	+	+	−	−	+	+	+	+	+	+	−	+	+	−	−	−	+	−
8.1	9022	−	+	−	−	+	−	+	−	+	+	−	−	−	+	+	+	−	+
	9023	−	+	−	−	−	−	+	−	+	+	−	−	−	+	+	+	−	+
	9088	−	+	−	−	−	−	+	−	+	+	−	−	−	+	+	+	−	+
	9086	−	+	−	−	+	−	+	−	+	+	−	−	−	+	+	+	−	+

[a]This fragment (>900kb) also hybridizes with DRB and DQB.
[b]This large fragment probably indicates differences centromeric of DO.

Figure 1. The RFLP patterns for C4, DRB, DQB, DOB, DPA, and DPB seen by PFGE and conventional gel electrophoresis for Tenth Workshop examples of the 18.2 and 8.1 ancestral haplotypes. Mlu and Pvu indicate MluI and PvuI PFGE-generated fragments. Hic and Taq indicate HincII and TaqI fragments generated using the Workshop Southern blotting protocol. It can be seen that the RFLP patterns are entirely ancestral haplotype-specific and complementary at C4 and DRB. Centromeric of DRB, it can be seen that these patterns begin to fragment and heterozygosity is demonstrated.

prior to Southern transfer. Membranes were prehybridized, hybridized, and exposed essentially according to Tenth International Workshop methods. After stripping and checking, the membranes were reprobed in various orders for DP A, DQ A, DR A, DR B, 21A plus B, Bf, TNF, and class I.

The results are summarized in Table 2, which shows the cell lines that contained patterns for the ancestral haplotypes listed in Table 1. It should be emphasized that there were substantial differences in length and gene copy number. For example, the B-DQ segment of 57.1 is longer than 8.1 by approximately 200 kb. Furthermore, 8.1 contains only one C4 and one 21 gene per haplotype, whereas 65.1 contains three of each. Unfortunately, there were only two supratypes (8.1 and 18.2) carried by a sufficient number of haplotypes to determine whether the haplotypes were similar in genomic structure, but in these two instances, the similarity between B and DR was striking. In one cell line (9018), the DQ and DO regions appeared to be heterozygous for the two different ancestral haplotypes. Several cells (9019, 9018, and 9023) show differences at DP A and/or DP B loci.

Although many additional cell lines need to be characterized, the present results provide yet further evidence for conservation of ancestral haplotypes, at least between HLA B and DR. We therefore use the term "ancestral haplotypes" to describe those haplotypes which are at least very similar at coding and noncoding regions between B and DR.

The term "haplotype" was introduced by Ceppelini et al. (6) to describe patterns carried by some members of a family. Unfortunately, the same term is now used in many different ways (multilocus specificity, half the genome, particular supratypes). In the present context, a haplotype refers to any combination of alleles which can be shown to segregate within a family. The qualification "ancestral" is intended to indicate that the haplotype has been maintained through many generations so as to distinguish between conserved haplotypes and those that are a consequence of more recent recombination. There is certainly no doubt that some ancestral haplotypes do recombine. It remains to be determined whether such recombinants are simply a function of the population frequencies of the ancestral haplotypes or whether there is preferential recombination between some rather than others. The fact that there are such substantial differences in length and gene copy number between different ancestral haplotypes might suggest that there will be different rates of recombination. To approach this problem, it is necessary to define the genomic structure of more ancestral haplotypes. It should then be possible to identify recombinants and the site of recombination. Operationally, an ancestral haplotype can be regarded as a unique structure that cannot be attributed to recombination between other ancestral haplotypes. Ultimately, it may be feasible to type subjects in terms of the ancestral haplotype that they have at each coding region within the MHC. Clearly, there are major implications for disease association, transplantation, paternity testing, etc. In the immediate future, it will be useful to use ancestral haplotypes to predict splits of alleles.

Table 2. Tenth Workshop Cell Lines with Supratypes Apparently Reflecting Ancestral Haplotypes and Their PFGE Patterns

Cell ID	Ancestral Haplotype	PFGE	21A+B NotI	DRB PvuI
9023	8.1	8.1	960	80,350
9086	8.1	8.1	960	80,350
9088	8.1	8.1	960	80,350
9022	8.1	8.1	960	80,350
9018	18.2	18.2[a]	960	430
9085	18.2	18.2	960	430
9020	18.2	18.2	960	430
9019	18.2	18.2[b]	960	430
9002	65.1	65.1	1060	320,390
9078	65.1	65.1	1060	320,390
9079	65.1	65.1	1060	320,390
9052	57.1	57.1[c]	1120	780
9090	44.1	44.1[c]	1120	780
9047	47.1	47.1	1095	780
9076	46.1	46.1[c]	1120	780
9080	35.2	35.2	1050	780
9006	35.3	35.3	1000	780

[a]Heterozygosity at DQ and between DO and DP.
[b]Heterozygosity between DO and DP.
[c]Cannot be distinguished by PFGE using Not I and Pvu I.

References

1. Dawkins RL, Christiansen FT, Kay PH, et al. Disease associations with complotypes, supratypes and haplotypes. Immunol Rev 1983;70:5.
2. Gillet D, Mornet E, Rocca A, Degos L, Cohen D, Pla M. Extensive genomic polymorphism in mouse 21 hydroxylase region. Immunogenetics 1988;27:133.
3. Dawkins RL, Martin E, Kay PH, Garlepp MJ, Wilton AN, Stuckey MS. Heterogeneity of steroid 21-hydroxylase genes in classical congenital adrenal hyperplasia. J Immunogenetics 1987;14:89.
4. Awdeh ZL, Raum D, Yunis EJ, Alper CA. Extended HLA/complement allele haplotypes: evidence for a T/t-like complex in man. Proc Natl Acad Sci USA 1983;80:259.
5. Kay PH, Dawkins RL et al. Coexistence of an MHC chromosomal segment marked by HLA B17,BfS,C4A6,B1,DR7 and DQw9 in different ethnic groups. Human Immunol 1988;23:27.
6. Ceppellini R, Curtoni ES, Mattiuz PL, Miggiano V, Scudeller G, Serra A. Genetics of leukocyte antigens: a family study of segregation and linkage. In: Curtoni ES, et al. (eds). Histocompatibility Testing 1967. Copenhagen: Munksgaard, 1967:149.

Myasthenia Gravis: Joint Report

W. Gin,[1] R.L. Dawkins,[1] E.D. Albert,[2] B.R. Hawkins,[3] C. Muller,[4] K. Tsuji,[5] M. Jeannet,[6] and B. Schalke[7]

Background

Previous studies have demonstrated the importance of immunogenetic factors in myasthenia gravis (MG) (Table 1). In Caucasoids, generalized MG (GMG) is associated with A1,B8,DR3 (1,8,3). This particular supratype is also protective against the development of antistriational antibodies (AStr) and thymoma (8). However, only weak MHC associations have been found in other races such as the Chinese and Japanese. It is also apparent that antigen associations differ between different racial groups. Different MHC associations have also been found for MG due to different inducing agents (e.g., D-penicillamine-induced MG [DPMG] has been found to be associated with B35,DR1) (9). Thus, the development of MG does not seem to be controlled by any specific class I or class II MHC gene. Nevertheless, by examining other races by restriction fragment length polymorphism (RFLP) analysis, it may be possible to determine whether intervening genes are shared.

The association between 1,8,3 and MG in Caucasoids suggests that the relevant genes might lie within the segment of chromosome marked by this combination of alleles. By examining various subsets of Caucasoid patients, it may be possible to identify the location of these genes.

To address these issues, it was proposed that the core component consist of comparisons of RFLP patterns between:

1. GMG, DPMG, restricted ocular MG, congenital MG and healthy subjects;
2. GMG in different racial groups; and
3. mothers and offspring with neonatal MG.

Participating Laboratories: ANZDAW,[1] GERALB,[2] ANZHAW,[3] GERMUC,[4] JAPTSU,[5] FRAJEA,[6] GERWUE[7]

Immunobiology of HLA
Volume I

Table 1. MHC Associations and Relative Risks in Myasthenia Gravis in Different Races

	Caucasoid (1)	American Black (1)	Chinese: Hong Kong (2)	Chinese: Taiwan (3)	Chinese: Shanghai (4)	Asian: Japanese (5)	Asian: Indian (6)	Asian: Thai (7)
A1	2.6	3.3		2.8				5.2
2				2.0				
9					1.4		2.4	
19								
B8	3.0	2.7	4.2*	3.7	1.7		2.4	19
12						15.7		
21				3.8			5.0	
35			1.8[a]					
46			5.6[a]	NT	NT			
DR3	2.9	0.9	NT	0.5	1.3	NT	NT	
DR4			NT		3.3	NT	NT	
DR5	2.6	2.6	NT	1.0		NT	NT	
DR9			NT	3.6		NT	NT	

[a]Juvenile onset MG. References are shown in parentheses. NT = not tested.

RFLP analysis would also allow the assessment of the report that a HincII DQ beta fragment is associated with MG (10).

The variable class I and class II associations indicate that other factors must contribute to the expression of disease. One of these factors might be due to differences in the acetylcholine receptor (AChR) gene itself. AChR polymorphism might be important, for instance, in determining the type of MG (e.g., generalized versus the congenital forms). In the optional, noncore component of the study, it was proposed that AChR gene polymorphism be examined in relation to the different forms of MG.

Materials and Methods

A list of the six participating laboratories and their contributions is shown in Table 2. Most patients were HLA-typed using Tenth Workshop sera, but some had local typing. Most laboratories sent plasma and/or sera to the Perth laboratory for complement allotyping and the measurement of anti-acetylcholine receptor antibodies (anti-AChR). The C4 and Bf complement allotypes were determined by high voltage agarose gel electrophoresis and immunofixation (11). Assignment was based on the use of international standards in accordance with convention (12).

Anti-AChR was measured by radioimmune precipitation using alpha-bungarotoxin-labeled AChR derived from pooled muscle preparations (13). Extracted DNA was also received from some patients for MHC and AChR gene probing.

Results

Serology. HLA. A summary of HLA frequencies in the different groups is shown in Table 3. As expected, A1,B8,DR3 was increased in all Caucasoid groups studied. In the ALB subset, however, some patients had A1,B8 without DR3. Although C4 allotyping was not available, 21-hydroxylase gene probing on DNA samples permitted characterization of the class III genes (see gene localization). There was also a significant increase in DRw6, which was not accompanied by B8 in the ALB subset.

In the Hong Kong Chinese, Bw46 and DRw9 were increased; DR5 and DR7 were significantly decreased, suggesting a possible protective role of these alleles in this race. Three percent of Chinese patients had B8. In the Japanese, A31, Bw52, Bw61, and DRw9 were increased, but the differences were not statistically significant. DRw6 was significantly decreased.

Complement. The results of C4 and Bf allotyping are shown in Table 4. The C4AQ0 alleles were inferred from the densitometric ratio of C4A/C4B products. The frequency of C4AQ0 was increased in all Caucasoid groups.

Table 2. Laboratories Participating in the Tenth Workshop Myasthenia Gravis Joint Study

Lab	Investigator	Ethnic Group	HLA Typing	C4 Allotyping[a]	MHC RFLP	Anti-ACHR[b]
ANZ/DAW	Dawkins	Australian Caucasoid	10WS and local	Yes	Yes	Yes
	Hawkins	Hong Kong Chinese	Local	Yes	No	Yes
FRA/JEA	Jeannet	Swiss Caucasoid	10WS	Yes	Yes	Yes
GER/ALB	Albert	German Caucasoid	Local	No	Yes	No
GER/MUC	Muller	German Caucasoid	10WS	Yes	Yes	Yes
JAP/TSU	Tsuji	Japanese	10WS	Yes	Yes	Yes

[a]Typing performed by the Perth laboratory (DAW). [b]Anti-acetylcholine receptor antibodies (anti-AChR) measured by the Perth laboratory (DAW).

Table 3. Selected HLA (%) Myasthenia Gravis in Different Ethnic Groups

	Caucasoids					Hong Kong Chinese		Japanese	
	MG					MG^{b}		MG	
Antigen	1	2	3	4	$Control^{a}$	1	$Control^{c}$	5	Control
	(24)	(68)	(29)	(11)	(1053)	(166)	(100)	(28)	(645)
A1	41.7	41.2*	55.2*	36	26.7	2	0	3.6	0.8
2	33.3†	42.6	55.2	45	48.9	55.4	53	35.7	41.7
24	8.3	29.4	24.1	27	19.5	30.1	22	60.7	62.9
31	8.3	1.5	0	18	5.8	0	0	21.4	16.6
	(24)	(68)	(29)	(11)	(1053)	(166)	(100)	(28)	(650)
B8	33.3	38.2***	44.8***	45	18.5	3.0	0	0	0
35	20.8	14.7	13.8	9.0	19.9	6.0	6.0	7.1	18.0
46	0	0	0	0	0	47.0*	31.0	7.1	8.3
52	4	0	0	0	3.9	0	0	35.7	22.9
61	4	4.4	0	0	4.1	0	0	28.6	17.5
	(24)	(61)	(30)	(11)	(1026)	(166)	(100)	(28)	(439)
DR2	29.2	27.9	33.3	27	29.0	31.9	38.0	32.1	33.0
3	33.3	29.5	46.7**	45	22.8	5.4	8.0	0	0
5	12.5	27.9	0	0	26.9	5.4†	33	0	0
6	16.7	29.5*	0	0	17.5	16.9	14.0	3.66†††	31.0
7	16.7	9.8	20.0	36	22.3	3.0†	11.0	0	0
9	0	0	0	0	1.7	60.8***	21.0	32.1	26.2

[a]Pooled Caucasoid controls (9WS). [b]Includes some cases of ocular MG. [c]Local Hong Kong controls.
Key: 1 DAW; 2 ALB; 3 MUC; 4 JEA; 5 TSU. Number tested in each group is shown in parenthesis.
Significantly increased: * $p < 0.05$; ** $p < 0.01$; *** $p < 0.001$. Significantly decreased: †$p < 0.05$; ††$p < 0.01$; †††$p < 0.001$.

This was not unexpected since the BfS,C4AQ0,C4B1 is the complotype found with the 1,8,3 supratype.

A C4A3 variant (C4A3 Fast) was encountered in the Hong Kong Chinese, and was difficult to distinguish from C4A3 in the presence of C4A4. Cases in which it was difficult to make this distinction (i.e., between the phenotypes C4A3,C4A4 and C4A3F, C4A4) were designated C4A34, although C4A4 was almost certainly present in most instances. The frequency of C4A4 (including C4A34) was increased in the Chinese as was C4B2. C4A3 was increased in the Japanese patients.

The increase in Bw46,DRw9, C4A4 and C4B2 in the Chinese could be accounted for by the 46,S,4,2,9 supratype that was found in young patients, particularly males under 20 years of age. These patients tended to have absent or low titers of anti-AChR, and AStr was not detected. In contrast to the 1,8,S,0,1,3 supratype that is associated with GMG in adult female Caucasoids, 46,S,4,2,9 appears to be a unique genetic marker for MG occurring in Chinese children. The differences between the two supratypes are shown in Table 5. Further characterization of Chinese pediatric MG has been reported in more detail elsewhere (14). No clear MHC association was evident in the older Chinese patients.

Table 4. C4 and BF Allotypes (%) in MG in Different Ethnic Groups

	Caucasoids				H K Chinese		Japanese	
	MG				MG^{b}		MG	
Allele	1	3	4	$Control^{a}$	1	$Control^{c}$	5	$Control^{d}$
	(21)	(25)	(11)	(558)	(166)	(61)	(32)	(341)
C4AQ0	29	48	45	24	9	12	0	7
C4A3	71	88	91	91	32	61	90^{h}	69
C4A4	25	0	0	12	67^{gi}	20	16	13
C4BQ0	8	4	0	26	7	15	6	16
C4B1	92	92	100	92^{e}	60	42	78	59
C4B2	25	12	9	17^{e}	76^{i}	41	34	17
BfF	17	20	0	37^{f}	20	5	28	20
BfF1	4	0	0	3^{f}	1	0	0	0
BfS	96	100	100	95^{f}	97	94	97	80

[a]Pooled Caucasoid controls (9WS). [b]Includes some cases of ocular MG. [c]Local Hong Kong controls. [d]Pooled Oriental controls (9WS). [e]Controls n = 557. [f]Controls n = 691. [g]Includes C4A34 (see text). [h]$p < 0.01$. [i]$p < 0.001$.
Key: 1 DAW; 3 MUC; 4 JEA; 5 TSU. Number tested in each group is shown in parenthesis.

Table 5. Differences Between Myasthenia Gravis Associated With 1,8,S,0,1,3 and 46,S,4,2,9

Clinical Features	1,8,S,0,1,3	46,S,4,2,9
Race	Caucasoids	Chinese
Sex predominance	Female	Male
Age of onset	Adulthood	Childhood
Anti-AChR titers	Elevated	Absent or low
Antistriational antibodies	May be present	Absent
Thymic hyperplasia	Yes	No

Anti-AChR = Anti-acetylcholine receptor antibodies.

Table 7. 15 kb HincII DQ Beta Fragment Is Not Specific for Generalized Myasthenia Gravis

	GMG		Non-GMG[b]
HincII DQ Beta 15 kb Band	18S013 (n=3)	Not 18S103[a] (n=6)	(n=8)
+	0	0	3[c]
−	3	6	5

[a]Haplotypes that also do not contain DR3 or DR7. [b]Selected supratypes only. [c]Two B18, BfF1, DR3 and one DR7 containing haplotypes.

MHC RFLP Analysis. The three enzymes assigned to the MG disease study consisted of BamHI, HindIII, and TaqI. The enzyme/probe combinations that were undertaken are listed in Table 6. Results from Southern blotting have been reported elsewhere by each individual laboratory.

Bell et al. (10) have previously described an association between a 15 kb HincII DQ beta fragment and Caucasoid patients with MG. Because HLA typing in this report was limited to HLA-DR only, this issue was reexamined in patients with full HLA A,B,DR typing. On the whole, results indicate that this 15 kb fragment is not specific for MG, but suggest that this fragment is specific for the B18, BfF1, C4A3, C4BQ0,DR3 supratype (Table 7). Thus, MHC RFLPs may be specific for a supratype, rather than for a particular disease.

Localization of Disease Susceptibility Gene(s). In Caucasoids, B8 generally occurred in similar frequencies to DR3 (DAW,JEA,MUC) (Table 3). However, in the ALB subset, approximately 10% had B8 without DR3. Thus, if the disease susceptibility gene(s) is (are) closer to B8, then these patients would be expected to have a different class III from those with DR3. On the other hand, if the gene(s) is (are) closer to HLA-D, they should have common class III genes. From Figure 1, it can be seen that virtually all patients with B8 and DR3 also have C4Q0, which is characteristic of this particular supratype. Complement allotyping was not available for the ALB subset, so class III was characterized by 21-hydroxylase gene probing, since the 1,8,3 supratype is known to contain a deletion of the 21-A gene (15). Patients with A1,B8 but not DR3 were found not to have a deletion of 21A and, therefore, lacked all of the more centromeric features of the 1,8,3 supratype (in preparation). This suggests that relevant gene(s) is (are) likely to be located telomeric of class III in this form of MG (i.e., between B8 and C4).

Table 6. Laboratories[a] Contributing to RFLP Analysis

	Probe		
Enzyme	Class I	Class II	Class III
BamHI	5	3,4,5	4,5
HindIII	5	3,4,5	4,5
TaqI	5	3,4,5	1,4,5
HincII	–	1	–

[a]*Key*: 1 DAW; 3 MUC; 4 JEA; 5 TSU.

Are AChR Genes Important? Different structural forms of AChR may be more prevalent in disease states than in healthy states and may, therefore, be important in determining susceptibility to MG. Two allelic forms of the alpha subunit of the AChR gene have been shown in humans using a murine AChR gene probe (Kay, unpublished observations). AChR alpha phenotypes were determined in 62 Caucasoids with GMG and compared with 48 random healthy control subjects. An increase in the heterozygous state was found in patients with MG (Table 8).

Conclusions

Major differences in MHC associations exist between races and even within subpopulations. In view of these differences, it was surprising that 1,8,3 was once again found to be so clearly associated with GMG in Caucasoids. The previously reported HincII DQ beta 15 kb fragment was found not to be specific for MG in Caucasoids. This is not unexpected, particularly in view of the variable class I and class II antigen associations that have been reported. However, the evidence suggests that in patients with MG and the 1,8,3 supratype, the important gene(s) is (are) likely to be located near HLA B8 rather than DR3.

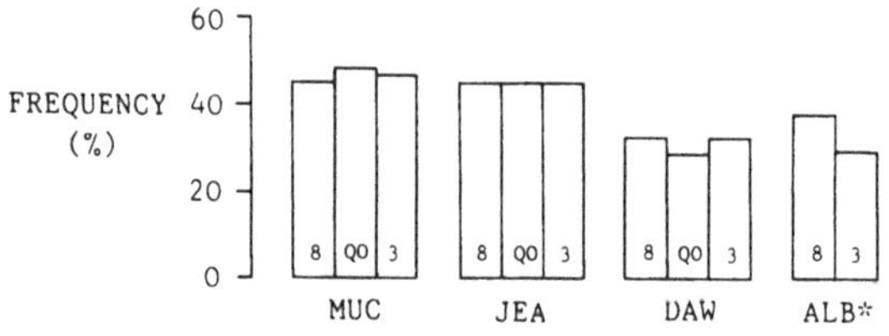

Figure 1. The frequencies of B8, DR3, and C4AQ0 are shown for Caucasoid patients with MG. Where B8 and DR3 occur in approximately equal proportions (MUC, n=29; DAW, n=24; JEA, n=11), C4AQ0 is also found in a similar frequency. *Complement allotyping not available.

Table 8. Comparison of ACHR Alpha Phenotypes Between Caucasoid Patients With MG and Controls

AChR Alpha Phenotype	MG (n=62)	Controls (n=48)	χ^2
1	8	9	0.39[a]
2	13	18	2.38[a]
1,2	41	21	4.18[b]

[a] $p > 0.05$; [b] $p < 0.05$ (Fisher's exact test).

No 1,8,3 equivalent was found in any other race. Pediatric onset MG in the Chinese is marked by the 46,S,4,2,9 supratype, but appears to be a distinct form of MG. Characterization of the 1,8,3 and 46,S,4,2,9 supratypes by pulsed field electrophoresis and Southern blotting has revealed multiple differences (14).

It is still not clear why no MHC association was found in the Japanese or older Chinese patients. Further disease classification within each of these races may reveal important immunogenetic associations.

The existence of heterogeneous anti-AChR antibody profiles in patients with GMG suggests that polymorphic MHC or AChR epitopes may be important in determining antibody specificity. Although data are not sufficient, disease susceptibility may be associated with an increase in AChR gene heterozygosity.

Material for study of the less common forms of MG is still being received.

References

1. Christiansen FT, Pollack MS, Garlepp MJ, Dawkins RL. Myasthenia gravis and HLA antigens in American Blacks and other races. J Neuroimmunol 1984;7:121.
2. Hawkins BR, Chan-Lui WY, Choi EEK, Ho AY. Strong association of HLA Bw46 with juvenile onset myasthenia gravis in Hong Kong Chinese. J Neurol Neurosurg Psychiatry 1984;47:555.
3. Thajeb P, Chee C-Y, Huang C-C. The distribution of HLA A,B,DR antigens in Chinese myasthenia gravis. Tissue Antigens 1987;29:273.
4. Lee TD, Zhao TM, Bu KJ, Lu CZ, O'Donnell M, Sandler SG. Association of HLA-DR4 with myasthenia gravis in the Chinese. Tissue Antigens 1984;23:127.
5. Yoshida T, Tsuchiya M, Ono A, Yoshimatsu H, Satoyoshi E, Tsuji K. HLA antigens and myasthenia gravis in Japan. J Neurol Sci 1977;32:195.
6. Mehra NK, Ahuja GK, Taneja V, Vaidya MC. HLA antigens and myasthenia gravis in North India. J Neurol Neurosurg Psychiatry 1983;46:361.
7. Garlepp MJ, Christiansen FT, Chiewslip P, Dawkins RL. Myasthenia gravis disease report. In: Tait B, Simons M (eds). Proceedings of the Second Asia and Oceania Histocompatibility Conference. Melbourne: Immunopublishing, 1983:343.
8. Dawkins RL, Myasthenia gravis. In: Terasaki PI (ed). Histocompatibility Testing 1980. Los Angeles: UCLA Tissue Typing Laboratory, 1980:662.
9. Garlepp MJ, Dawkins RL. D-penicillamine induced myasthenia gravis. In: Albert ED, Baur MP, Mayr WR (eds). Histocompatibility Testing 1984. Berlin: Springer-Verlag, 1984:384.
10. Bell JS, Smoot S, Newby C, et al. HLA-DQ beta chain polymorphism linked to myasthenia gravis. Lancet 1986;1:1058.
11. Zhang W, Kay PH, Cobain TC, Dawkins RL. C4 allotyping on plasma and serum: applications to routine laboratories. Human Immunology 1988;21:165.
12. Mauff G, Alper CA, Awdeh Z, et al. Statement on the nomenclature of human C4 allotypes. Immunobiol 1983;164:184.
13. Garlepp MJ, Kay PH, Dawkins RL. The diagnostic significance of autoantibodies to the acetylcholine receptor. J Neuroimmunol 1982;3:337.
14. Dawkins RL, Gin W, Hawkins BR, Tokunaga K, Wong VCN, Yu YL. Striking association between paediatric myasthenia gravis and MHC supratype in the Chinese (submitted for publication).
15. Garlepp MG, Wilton AN, Dawkins RL, Kelly H, White P. Rearrangement of 21-hydroxylase genes in disease-associated MHC supratypes. Immunogenetics 1986;23:100.

Analysis of HLA Recombinant Families by Southern Blots

Mogens Thomsen,[1] Carlos Alvarez,[2] and Charles B. Carpenter[2]

In the past, the analysis of recombinations within the HLA system have led to the establishment of the HLA region map as we know it today. Recombination events are of prime importance in the generation of new constellations of alleles in the cis position or new alleles if the cross-over takes place within a structural gene. In addition, unequal cross-overs may result in duplications or deletions of genetic material.

The application of methods of molecular biology has thrown new light on the organization of genes in the HLA region, and the comparison of sequences of different alleles suggests that two additional mechanisms for generation of polymorphism are operating: point mutations and gene conversion. The latter implies that a DNA strand is replaced by another strand from a homologous gene as already described in mouse (1) and man (2).

The technique of pulsed field electrophoresis has facilitated the linkage of the loci of the HLA region together (3–5). Figure 1 shows the physical distances compared with the genetic distances based on the results from "Provinces Francaises" (6). With its about 1400 families, this collaborative study represents the largest existing unbiased collection of families typed for HLA and other genetic markers (7). Previous workshops have accumulated large numbers of recombinant families that have confirmed the mapping order of HLA loci (8–10), but due to selection bias towards submission of recombinant families, the materials have not been used for the estimation of cross-over frequencies. Estimation of cross-over frequencies can also be obtained by the so-called haplotype method (11).

Methods

In the present Workshop, the laboratories participating in the Southern blot component were invited to contribute HLA recombinant families in addition to the compulsory Workshop components.

The purpose of this study was: 1) to establish whether the Tenth Workshop enzyme-probe systems clearly could demonstrate recombinations by restriction fragment length polymorphism (RFLP) patterns and 2) to evaluate the possibility of appearance of variant fragments due to cross-over near the structural genes. The families should ideally have been typed by Workshop core serology, complement allotypes, and glyoxalase-1 (GLO). The following family members should be included: the father, the mother, the recombinant and two siblings who have inherited the same nonrecombinant haplotype, as well as each of the two different haplotypes involved in the recombination event. Not all submitted families met all criteria.

The enzyme probe systems were chosen according to the location of the recombination. The standard Workshop techniques were employed including a long and a short run.

The autoradiograms and the floppy disks were submitted to the two study chairmen who analyzed the segregation patterns of the various restriction fragments and by comparison with the films and the standardized fragment size from the standardization laboratories assigned a size in kilobases (kb). Whenever possible, it was attempted to look for correlations between specific fragments and serologic specificities based on the preliminary Workshop analysis of the homozygous cell lines.

Results

Table 1 shows the HLA haplotypes of the families submitted for the analysis. Two of the families have recombinations in the class II region, one family between HLA-A and -C, two families between HLA-B and

---

Participating Laboratories: FRAOHA,[1] NCYCAR[2]
This work was supported by NATO Grant 0546/87.

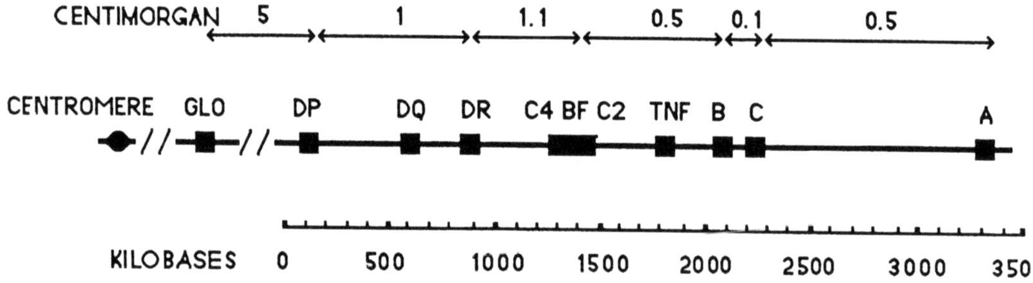

Figure 1. Simplified map of the HLA region. The physical distances are taken from references 3, 4 and 5. The genetic distances are from reference 6. Other family studies (reference 15) show the genetic distance for complement-DR to be 0.3 centimorgans, and for DR-B to be 0.7 centimorgans.

Table 1. HLA Haplotypes in Recombinant Families

		Paternal Haplotypes											Maternal Haplotypes										Offspring				
		A	Cw	B	C2	Bf	C4A,B	DR	DQ	DP	GLO		A	Cw	B	C2	Bf	C4A,B	DR	DQ	DP	GLO	ac	ad	bc	bd	rec
OHA 101	a	2	–	44	.	F	3,1	7	2	4*	2	c	1	4	35	.	S	Q0,1	14	1	4*	2	1	1	1	3	b,c*d
	b	3	–	44	.	F	3,1	–	1	4*	1	d	24	7	7	.	S	4,2	–	1	4*	1					
OHA 102	a	2	4	35	.	S	3,1	5	7	1	1	c	23	4	35	.	S	3,1	11	7	4	1	4	1	1	1	a,d*c
	b	29	4	35	.	F	3,1	1	1	–	1	d	30	6	50	.	S	3,1	10	1	2						
OHA 103	a	2	–	39	.	S	3,1	1	1	4	2	c	2	7	8	.	S	QO,1	3	1	1	1	1	1	0	1	a,c*d
	b	29	–	44	.	F	3,1	7	2	4*	2	d	11	4	35	.	S	3,1	13	1	4	2					
BIG HAN	a	2	5	44	.	S	3,QO	5	.	5	2	c	1	2	51		S	3,1	5	.	4	2	1	1	0	1	a*b,c b,d*c
	b	24	3	62	.	S	3,1	2	.	2	1	d	25	6	57		S	6,1	7	.	3	1					
HAM 902	a	1	–	44	.	.	. .	7	2	.	.	c	1	–	63	.	.	. .	7	4	.	.	1	1	1	1	b*a,d
	b	31	4	62	.	.	. .	6	1	.	.	d	24	–	62s	.	.	. .	6	3	.	.					
BOD 004	a	2	–	52	1	S	3,1	2	6	.	.	c	2	7	63	1	S	2,1+2	1	5	.	.	1	0	2	1	b,c*d
	b	26	–	38	1	S	3,QO	12	7	.	.	d	3	7	8	1	F	3,QO	17	2	.	.					
CAR 001	a	1	–	8	C	S	QO,1	3	.	.	2	c	2	5	44	C	S	3,1	5	.	.	1	1	1	1	1	a,c*d
	b	23	4	44	C	F	3,1	7	.	.	2	d	3	6	57	C	S	6,1	7	.	.	1					
CAR 002	a	26	–	38	C	S	2,1	5	3	.	.	c	2	–	8	C	S	QO,1	3	2	.	.	1	0	0	1	b,c*d
	b	11	–	57	C	F	3,1	6	1	.	.	d	11	4	35	C	S	3,1	5	3	.	.					
FAI 004	a	1	7	8	.	.	. .	9	–	.	.	c	2	7	39	.	.	. .	2	1	.	.	1	1	1	1	b*a,d
	b	24	3	62	.	.	. .	4	3	.	.	d	24	4	35	.	.	. .	8	3	.	.					

The recombinant haplotypes are framed.
 indicates not tested; – indicates a blank; * indicates that presence of allele cannot be excluded.

complement, and the other families have recombinations between HLA-B or complement and DR. The data in Table 1 are as submitted by the laboratories. In the designation of C2 complement, "C" and "1" are equivalent. In the following tables, the segregation patterns of the polymorphic fragments in each of the investigated families will be shown and commented upon. Unless otherwise noted, **bold type** is used to indicate the fragments found in the recombinant individual. If the kb of a fragment is placed in parenthesis, it indicates a nonofficial band which nevertheless segregated in the family.

Family OHA 101 was first studied during the Ninth Workshop. In this family, there is a DR blank allele (DR-BON), which is present in homozygous state in the recombinant (12). The recombination is between B and C4, and BF is not informative (see Table 1). The following enzymes were used: HID, BAM, TAQ, ECV, PVU, and ECI. The probes used were class I, C2, BF, C4, and 21 OH. Due to technical problems, a number of the blots could not be interpreted and only class I and C4 probes showed polymorphism. Table 2 shows that the location of the cross-over between class I and C4 is confirmed as the c haplotype is present in the recombinant with class

Table 2. Family OHA 101

B* Complement Recombination

F	ab	w 44
M	cd	w 45
C1	bc*d	w 49
C2	bc	w 50
C3	ad	w 48

Probe	Haplotypes a	b	c	d	bc
Class I		**HID 24.48**	**TAQ 2.14**	PVU 4.20	**BAM 11.00**
		BAM 8.55*	**ECV 9.46**		**PVU 14.82***
		TAQ 5.20	**ECV 5.56**		**PVU 6.21**
		TAQ 2.50	**PVU 4.80***		
C4			HID 31.94		**ECI 13.52**
			TAQ 6.27		
			ECV 7.49		
			ECI 12,25		

*Fragments correlated with: A3 (BAM 8.55, PVU 14.82); A1 (PVU 4.80).

Table 3. Family OHA 102

B*DR Recombination		
F	ab	W 60
M	cd	W 61
C1	ad*c	W 64
C2	ac	W 51
C3	ad	W 52
C4	bc	W 62
C5	bd	W 63

Probe	Haplotypes a	b	c	d	ac	bd
Class I	**BGL 3.70**		BGL 6.02	**HID 43.87***		
				HID 24.48		
				TAQ 3.24		
				BGL 5.37		
				BGL 3.92*		
				BGL 2.27*		
				ECI 13.81		
C2		HID 4.32				
		TAQ 4.75				
		ECI 16.45				
BF		TAQ 6.76				
C4		HID 24.77				
		BAM 3.28				
		TAQ 4.62				
		ECV 28.50				
		ECI 15.23				
		ECI 13.52				
21OH		HID 26.19				
		ECI 14.67				
		ECI 13.05				
DRB	**TAQ 9.79**	HID 3.32	**HID 2.60**	HID 1.15	**HID 2.85**	HID 6.94
	TAQ 4.12	TAQ 7.23	**TAQ 11.48**	TAQ 13.81	**HID 2.51**	BAM 4.78
	ECV 10.96	TAQ 5.80	**TAQ 4.12**	BGL 3.96	**BAM 11.33**	BGL 10.27
	ECI 4.40	BGL 6.96	**BGL 2.60**		**TAQ 6.08**	ECI 5.13
		ECV 1.83	**ECI 4.56**		**ECV 18.24**	
		ECI 3.73	**ECI 3.31**		**ECV 15.34**	
					ECV 7.52	
					ECI 9.15	

*Fragments correlated with: Cw6 (HID 43.87); A3 (BGL 3.92); A31 (BGL 2.27).
Most of the DRB fragments show correlation with specificities in the Workshop lines that do not correspond to the ones of the haplotypes in this family.

I probes and not with C4 probes. Three of the class I fragments showed a good correlation ($r > 0.75$) with the appropriate A locus specificity.

The OHA 102 family is of arabic origin from North Africa and was studied because of a hematologic malignancy. The patient was found to have a recombination between B and DR. The serologic typing was made in Toulouse, the Southern blots were performed by J.D. Bignon in Nantes, France.

The enzymes used were HID, BAM, TAQ, BGL, ECV and ECI. The probes were class I, C2, BF, C4, 21 OH and DRB. As Table 3 shows, the recombinant clearly has the haplotypes a and d with the class I probe and is ac with DRB. The four class III probes show a characteristic pattern for the b haplotype, but are noninformative for the recombinant material haplotype. The present data, thus, still do not allow us to locate precisely the recombination point. It should be noted that this family by serologic typing had several atypical reactions and by Workshop serology the DR5 of the father could not be assigned as DRw11 or DRw12. It is, thus, not surprising that the RFLP patterns showed a rather poor correlation between the specificity expected from the Workshop cell line RFLPs and that observed on the haplotypes.

The OHA 103 family is a normal family from the study Provinces Francaises (7). It was selected because of a HLA-DR/GLO cross-over, and on a new blood drawing the cross-over was confirmed and was shown to be located between DR and DP. The enzymes used were HID, BAM, TAQ, and ECI. The probes used were DRB, DQA, DQB, DPA, and DPB. The DR and DQ fragments shown in Table 4 are in concordance with the serologic typing, and the recombinant has the haplotypes ac. The DPA probe surprisingly shows that the recombinant is

Table 4. Family OHA 103

DQ*DP Recombination

F	ab	w 54
M	cd	w 55
C1	ac*d	w 57
C2	ac	w 58
C3	ad	w 56
C4	bd	w 59

	Haplotypes					
Probe	a	b	c	d	bc	Other
DRB	**HID 6.10*** **HID 3.02*** **BAM 4.78*** **TAQ 5.80*** **TAQ 4.40*** **ECI 10.06*** **ECI 5.13***	HID 8.44* HID 7.17* HID 3.13* BAM 6.59* BAM 4.62* BAM 4.27* TAQ 13.81* TAQ 7.23 TAQ 2.60 ECI 15.13* ECI 12.55* ECI 11.98*	**TAQ 9.78** **TAQ 4.11*** **ECI 4.56***	BAM 4.16* TAQ 11.48 ECI 4.39		
DQA	**HID 7.45*** **TAQ 3.76**	HID 4.83* BAM 7.63* TAQ 5.53 ECI 6.34*	**HID 5.79*** **TAQ 4.56*** **ECI 11.07**	HID 7.68* TAQ 6.99		
DQB	**BAM 5.76*** **TAQ 5.33** **ECI 11.52***	HID 12.80* HID 3.77 TAQ 6.69* ECI 16.51	**HID 11.40*** **TAQ 4.62*** **ECI 12.33**	BAM 2.89* TAQ 2.87* TAQ 2.60* ECI 9.22 ECI 2.26*		
DPA						BAM 6.92* TAQ 7.66 TAQ 6.45* ECI 2.41
DPB	**TAQ 5.24*** **ECI 3.77**	TAQ 4.04* ECI 19.46 ECI 14.74* ECI 4.02* ECI 3.87 ECI 3.53	**ECI 12.93**		**BAM 4.13*** **BAM 3.47**	TAQ 2.92 TAQ 1.73*

*Fragments correlated with: (DR1 (HID 6.10, HID 3.02, BAM 4.78, TAQ 5.80. TAQ 4.40, ECI 10.06, ECI 5.13, BAM 5.76, ECI 11.52); DRw53 (HID 8.44, HID 7.17, HID 3.13, BAM 6.59, BAM 4.62, BAM 4.27, TAQ 13.81 ECI 12.55, ECI 11.98, HID 4.83, BAM 7.63, ECI 6.34, TAQ 5.24, TAQ 4.04, ECI 4.02); DR7 (ECI 15.13, HID 12.80, TAQ 6.69, ECI 14.74); DRw52 (BAM 4.16, TAQ 4.11, ECI 4.56, HID 5.79, TAQ 4.56); DQw1 (HID 7.45, ECI 2.26); Dw18 (HID 7.68, BAM 2.89, TAQ 2.87, TAQ 2.60); DR3 (HID 11.40, TAQ 4.62); DPw1 (BAM 6.92, TAQ 6.45 TAQ 1.73); DQw2 (BAM 4.13).

unique: for several enzymes ("other" column). In addition, a fragment with DPA is missing in the recombinant, but is present in all other family members. No other polymorphism is seen. The DPB probe shows in some cases a missing fragment in the recombinant as with DPA, while in other cases, there is a segregation like that seen for DR and DQ. Logically, the recombinant should always have fragments which are identical with either those of child 2 or child 3 (i.e., the two siblings who have the paternal haplotype in common with the recombinant and who have each of the two maternal haplotypes that constitute the recombinant haplotype). It should be stressed that the family has been tested for a multitude of non-HLA markers to exclude extrapaternity. The possible explanations at the present time include a deletion of a fragment in the DP region on the recombinant haplotype, a duplication on one of the maternal haplotypes of a DP locus gene, or a cross-over event on the paternal haplotype in child 3. Preliminary data from hybridization with a DO beta probe suggest the possibility of a paternal cross-over between DQ and DO in child 3, but this does not explain the presence of bands specific for the c haplotype in the recombinant with the DPB probe. The possibility of cross-hybridization with other class II genes can, however, not be excluded. The studies of this family are continuing.

Table 5. Family BIG HAN

	Double Recombination			B*DR DR*DP	F ab M cd C1 ac C2 a*bc C3 bd*c C4 bd	
			Haplotypes			
Probe	a	b	c	d	ac	Other
Class I	_TAQ 12.08_ _TAQ 6.20_ _TAQ 4.38_ _TAQ 2.80_ _ECI 11.79*_ _ECI 7.96*_		_HID 4.76*_ _ECV 13.55_	__HID 24.48*__ __TAQ 6.80__		
C4				__HID 24.77__ __TAQ 4.62__ __ECI 13.52__		
DRB		___HID 3.32*___ ___HID 2.36___ ___TAQ 4.40___ ___TAQ 2.60___ ___TAQ 2.08___ ___ECV 7.52*___ ___ECI 5.13___ ___ECI 3.98___		__HID 9.66__ __HID 7.17*__ __HID 3.13*__ __TAQ 13.81__ __TAQ 5.80__ __TAQ 4.12__ __TAQ 2.69__ __ECI 14.13__ __ECI 4.10*__ __ECI 3.73*__	_HID 3.02*_ _HID 2.70*_ _TAQ 11.48_ _TAQ 6.08_ _ECV 15.34*_ _ECI 4.56*_ _ECI 3.31*_	
DQA		___HID 7.45*___ ___TAQ 6.19___ ___TAQ 2.78___ ___TAQ 2.09___ ___ECI 13.79___ ___ECI 1.27___		__HID 4.83*__ __TAQ 4.56__ __ECI 6.34*__	_HID 5.79*_ _TAQ 9.50_ _TAQ 3.76_ _ECI 11.07*_	
DQB		___HID 8.60___ ___HID 6.44*___ ___TAQ 5.38___ ___TAQ 2.87___ ___ECI 2.26___		__HID 4.78__ __ECV 22.70__ __ECI 10.20__	_HID 3.77_ _ECV 20.01*_ _ECI 12.33_	_HID 5.13**_ _TAQ 1.48***_ _ECI 4.72***_
DPA	TAQ 13.21 TAQ 7.66 TAQ 6.45 ECV 4.76 ECI 3.08					
DPB		___ECI 3.88___	HID 5.06*			

Note: DR and DQ on haplotypes a and c are identical; _haplotypes present in C2 are in italics._ **haplotypes present in C3 are in bold**.
*Fragments correlated with A1 (HID 4.76); Cw6 (HID 24.48); A2 (ECI 11.79); Cw5 (ECI 7.96); DR2 (HID 3.32); DRw53 (HID 7.17, HID 3.13, ECI 4.10, ECI 3.73, HID 4.83, ECI 6.34); DR1 (HID 3.02); Dw19 (HID 2.70); DRw52 (ECV 15.34, ECV 7.52, ECI 4.56, ECI 3.31, HID 5.79, ECI 11.07); DQw1 (HID 7 HID 6.44); DR7 (ECV 20.01); DPw2 (HID 5.06).
**faint in parents, strong in C1 and C2; ***absent in parents, present in C1 and C2.

Family BIG HAN is a family submitted by L.U. Lamm in Aarhus, Denmark, and the Southern blots were performed by J.D. Bignon in Nantes, France. The family was studied during the Ninth Workshop (SCA KIS 02). No serology has been performed in this Workshop. The family has two recombinants, one between B and C4, the other between DQ and DP (12). The following enzymes were used: HID, TAQ, ECV, and ECI. The probes used were class I, C4, DRB, DQA, DQB, DPA, and DPB. Table 5 shows that the recombinants have fragment patterns as expected. Child 2 has the haplotypes a and c with the class I probe, and haplotypes

Table 6. Family HAM 902

Intra-Class I Recombination

F	ab	w 97
M	cd	w 98
C1	bc	w 99
C2	bd	w 100
C3	ad	w 103
C4	b*a,d	w 104
C5	ac	w 105

	Haplotypes			
Probe	a	b	d	ac
Class I	PVU 6.05	**BAM 5.36*** **TAQ 5.76** ECV 10.83 **PVU 5.49*** PVU 4.20 PVU 3.79 **PVU 2.95** **PVU 1.90**	**TAQ 4.72** **HID 4.98** **PVU 8.81**	BAM 23.56 HID 6.72 HID 4.76* PVU 4.80

*Strong serology correlations: HID 4.76 (A1), BAM 5.36 (A31), PVU 5.49 (A31).

b and c with class II probes. The C4 probe was not informative for the recombinant. The second recombinant (child 3) is b,d*c with DR, DQA, and DQB and the DP probes confirming the observation of a recombination between DQ and DP, although only one fragment is informative for segregation of maternal haplotypes (HID 5.06 is d haplotype, not present in the recombinant). In this family, a curious finding was the presence of distinctive bands in child 1 and child 2 when hybridized with the DQB probe, whereas the parents were negative for these bands (TAQ 1.48 and ECI 4.72) or faintly positive (HID 5.13). Repeated testings will clarify if these findings are artifacts.

Family HAM 902 was submitted as a well documented HLA-A/C crossover. (Table 6) This paternal recombination occurred between HLA-A and -C by serologic criteria. Segregation of fragments produced by five enzymes was readily discerned in this family, while

Table 7. Family BOD 004

Complement/DR Recombination

F	ab	w 312
M	cd	w 313
C1	ac	w 314
C2	bd	w 315
C3	bc	w 316
C4	b,c*d	w 317
C5	bc	w 318

	Haplotypes					
Probe	a	b	c	d	ab	ac
Class I	ECI 7.96 TAQ 3.75	**TAQ 1.76**		ECI 10.83* ECI 12.87* TAQ (8.30) TAQ 2.74 TAQ 2.50 TAQ 1.23*	**TAQ 4.38**	
DRB	ECI 6.28* ECI 4.71*	**TAQ 9.79** **TAQ 4.12**	ECI 10.06* TAQ 6.08 TAQ 4.40	**ECI 12.55*** **TAQ 11.48** **TAQ 7.00**		ECI 5.13* ECI 3.98*
DQA	TAQ 6.99	**TAQ 6.19**	TAQ 2.78*	**TAQ 4.56**		ECI 13.39*
DQB	ECI 18.20* TAQ 6.69		TAQ 5.34 TAQ 1.94*			
DPB		**TAQ 4.64**	TAQ 5.24	**ECI 7.58***		

*Strong serology correlations: ECI 10.83 (A3), ECI 2.87 (A3), TAQ 1.23 (A3), ECI 6.28 (DR2) ECI 10.06 (DR1), ECI 12.55 (DRw53), ECI 5.13 (DR1, DR2), ECI 4.71 (DR2), ECI 3.98 (DRw53), TAQ 2.78 (DQw5), ECI 13.39 (DQw1), ECI 18.20 (DQw4), TAQ 1.94 (DQw1), ECI 7.58 (DPw4).

Table 8. Family CAR 001

B/Complement Recombination

F ab
M cd
C1 a,c*d
C2 bc
C3 ac
C4 ad

	Haplotypes				
Probe	a	b	c	d	bd
Class I	**BAM 23.56*** **ECV 16.31** **ECV 12.93*** **ECV 7.86** **HID 4.76*** **HID 2.12***	ECV 10.83 HID 4.98	**ECI 7.96*** **ECV 70.81** **ECV 17.20** **ECV 13.55***	ECI 10.83* ECI 2.87* BAM 15.34* BAM 3.59* BAM 2.78* HID 43.87* HID 24.48* HID (12.50)	ECV 38.95
C4	**ECI 8.96** **ECI 7.50** **ECV 67.36** **HID 8.01**	ECI 15.23	ECI 12.25 ECV 7.49 HID 31.94		**ECV 25.15** **HID 24.77**
21OH		ECI 14.67			**ECI 13.05** **HID 25.19***

*Strong serology correlations: BAM 23.56 (B8), ECI 7.96 (Cw5), ECI 10.83 (A3), ECI 2.87 (A3), ECV 12.93 (A1+), BAM 15.34 (A3), ECV 13.55 (A2), BAM 3.59 (A3), HID 4.76 (A1), BAM 2.78 (A3,31) HID 2.12 (A1), HID 43.87 (Cw6), HID 24.48 (Cw6).

EcoRI fragments were monomorphic. All of the a fragments were absent in the recombinant, while five of the eight b haplotype fragments were present. The remaining three fragments may represent HLA-C or -B alleles. The three fragments having strong allelic correlations on the core cell lines match the serology in this family.

Family BOD 004 has a maternal recombination shown by serology and complement typing to be between complement and HLA-DR. (Table 7) Two enzymes only were used in this study. The C4 probe showed polymorphisms for two bands with each of the enzymes, but the segregation patterns a and cd were uninformative for the cross-over. The results show that the recombinant lacked class I d fragments, but had class II d fragments. An interesting anomaly appeared in the nonrecombinant bd child. ECI 3.83, a DQB-hybridizing fragment that cross-hybridizes with DRB, was absent. The fragment, which correlates well with DQw1 on the core cell lines, gave a strong signal with both probes in all other family members. The remainder of the fragment pattern in this individual was unremarkable, except for absence of a TAQ (1.50) fragment with the class I probe, again present in all other family members.

Family CAR 001 has a cross-over between HLA-B and complement (Table 8). This family, provided through the courtesy of E.J. Yunis of the Dana Farber Cancer

Table 9. Family CAR 002

Complement/DR Recombination

F ab
M cd
C1 b,c*d
C2 bd
C3 ac

	Haplotypes			
Probe	a	b	c	d
C4		**ECI 13.52** **ECV 25.15** **HID 24.77**	**ECI 7.50** **ECV 67.36** **HID 8.01**	ECI 12.25 ECV 7.49 HID 31.94
21OH		**ECI 13.05**	**ECV 45.63**	
DRB		**ECV 6.88** **ECV 6.32** **ECV 1.83** **HID 3.84**	BAM 3.92 ECV 49.17	**ECI 3.19**

Table 10. Family FAI 004

		B/DR Recombination	F ab M cd C1 b,d*c C2 ac	w 100 w 101 w 102 w 103
		Haplotypes		
Probe	a	c	d	ac
Class I	PVU 13.10 PVU 4.80	PVU 2.95* PVU 1.90	**PVU 3.79** **PVU 2.69***	PVU 12.78*
DRB	HID 2.49		HID 2.96*	
DPB			HID 5.43* HID 2.99	

*Strong serology correlations: PVU 2.95 (A2), PVU 12.78 (A2,A1), PVU 2.69 (B35, Cw4), HID 2.96 (DRw8, DRw52), HID 5.43 (DPw4).

Institute, Boston, Massachusetts was the first family that showed by this recombination the separate identity of the locus(i) for the MLR. (14) This family was also included in the Eighth Workshop (YUN 06) and the Seventh Workshop (USA7-SCH) (see also 15,16). The enzymes used were ECI, BAM, ECV, and HID, with probes class I, C4, 210H, DRB, and DRA. The recombinant had several class I fragments for haplotypes a and c, and was ad for C4 complement and 210H. Not shown in Table 8, the DRB probe confirmed the presence of the d haplotype in the recombinant.

Family CAR 002 was also contributed by E.J. Yunis. (Table 9). It was selected because it has a recombination between complement and HLA-DR by serology and complement typing. The recombinant in this family shows clearcut inheritance of c haplotype fragments for C4 and 210H, and a d fragment for DRB.

Family FAI 004 (Table 10) was originally selected as a normal family for the Southern blot family studies of the Workshop, and subsequently was found to have a HLA-B/DR recombinant. The family has only two children, but was of interest because of the presence of DRw9 in the father and second child. This antigen was not represented on the core cell lines. We were able to obtain the films for PVU and HID with class I, DRB, DQB, DQA, and DPB probes, and could show that the d haplotype fragments were present for class I, but not for class II, in the recombinant.

Limited studies were submitted on two families with complement/DR cross-overs. In one of these (DUP-AND), in which the C4 probe was used with TAQI, the RFLP pattern was not clearly readable. The other family (DUP-EDM) showed a C2/DR cross-over, marked by a single SstI fragment (2.70 kb) with the C2 probe, and present with the a haplotype, while DRB and DQA probes revealed b haplotype fragments.

Discussion

The number of families available for this analysis was limited, but a number of conclusions can be drawn. The fragments obtained by Southern blotting in most instances allowed confirmation of the location of the cross-over. Even if the class I probe was not locus-specific, some of the fragments correlated well with serologic specificities, and in the one HLA-A/C cross-over on a paternal haplotype (family HAM 902), it was possible to define fragments that followed either the A or the C and B alleles in the recombinant. The class III probes did not give a very polymorphic pattern and were in general less informative than typing of the proteins by usual electrophoretic methods. No information was available concerning correlation between class III fragments and complement allotypes on the Workshop cell lines. The class II beta probes gave a very polymorphic pattern, and cross-hybridization phenomena were sometimes a problem in the interpretation. The class II alpha probes showed a limited polymorphism except for DQ. DPA was informative in some cases, whereas no polymorphism was seen with DRA in the two families tested with this probe (CAR 001 and 002). In many cases, the fragments that have shown a correlation with class II specificities in the Workshop cell lines correlated with the same specificities in the families. In some families, however, the correlation was poor, for instance OHA 102. This may be due to the fact that this family is of North African origin and the serologic specificities often showed atypical reaction patterns. It is, thus, possible that attempts to correlate HLA specificities and RFLP patterns are only valid when applied to defined ethnic subgroups.

The analysis showed unexpected findings in three families. The family BOD 004 with a cross-over between complement, and in one of the nonrecombinant children, DR showed absence of DQA and DQB fragments that normally should be present. Another family BIG HAN showed in two children the appearance of bands that either were very weak or absent in the parents with the DQB probe and several enzymes. Finally, the family OHA 103 with a DQ/DP cross-over showed a segregation pattern for DPA and DPB probes that was difficult to explain. The DPA probe indicated that the recombinant individual missed fragments that were present in all other family members, including the two siblings that were identical with the recombinant for the paternal haplotype and had each of the two maternal haplotypes that were recombined in the recombinant. This might be explained by a recombination within the DPA gene, but the same observation of a missing fragment in the recombinant was made for two of the DPB fragments. For the other DPB fragments, the pattern was as if no recombination had taken place. The patterns of this family, thus, cannot be explained in a simple way and involves either several cross-overs or events of deletion, duplication, or conversion of genes.

In conclusion, this analysis has shown that RFLP analysis in recombinant families gives information on the recombination point with segregation of bands informative for the recombinant. In several families, however, the pattern of fragments suggests that the recombinations are not isolated events, but may be associated with other genetic rearrangements in the HLA region.

References

1. Weiss EH, Mellor A, Golden L, et al. The structure of a mutant H-2 gene suggests that the generation of polymorphism in H-2 genes may occur by gene conversion-like events. Nature 1983;301:671.
2. Gorski J, Mach B. Polymorphism of human Ia antigens: gene conversion between two DR beta loci results in a new HLA-DR specificity. Nature 1986;322:67.
3. Hardy DA, Bell JI, Long EO, Lindsten T, McDevitt HO. Mapping of the class II region of the human major histocompatibility complex by pulsed-field gel electrophoresis. Nature 1986;323:453.
4. Carroll MC, Katzman P, Alicot EM, et al. Linkage map of the human major histocompatibility complex including the tumor necrosis factor genes. Proc Natl Acad Sci USA 1987;84:8535.
5. Dunham I, Sargent CA, Trowsdale J, Duncan Campbell R. Molecular mapping of the human major histocompatibility complex by pulsed-field gel electrophoresis. Proc Natl Acad Sci USA 1987;84:7237.
6. Thomsen M. Preliminary analysis of HLA recombinations in "Provinces Françaises." In: Ohayon E, Cambon-Thomsen A (eds). Colloque INSERM. Human Population Genetics. Volume 142.
7. Cambon-Thomsen A, Thomsen M, Abbal M, Sommer E, Calot M, Ohayon E. A new HLA-D specificity associated with DR blank: D-Bon. Tissue Antigens 1986;27:256.
8. Lamm LU, Cullen P, Edwards JH, et al. Family analysis. In: Bodmer WF, et al. (eds). Histocompatibility Testing 1977. Munksgaard: Copenhagen, 1977.
9. Hawkins BR, Danilovs JA, O'Neill GJ. Analysis of recombinant families. In: Terasaki PL (ed). Histocompatibility Testing 1980. Los Angeles: UCLA Tissue Typing Laboratory, 1980.
10. Baur MP, Sigmund S, Sigmund M, Rittner C. Analysis of MHC recombinant families. In: Albert ED, Baur MP, Mayr WR (eds). Histocompatibility Testing 1984. Berlin: Springer-Verlag, 1984.
11. Olaisen B, Teisberg P, Jonassen R, Thorsby T, Gedde-Dahl T Jr. Gene order and gene distance in the HLA region studied by the haplotype method. Ann Hum Genet 1983;47:285.
12. Cambon-Thomsen A, Ohayon E. 1988. Practical application to population genetics: the genetic survey "Provinces Françaises." In: Mayr W (ed). Advances in Forensic Haemogenetics. Heidelberg: Springer-Verlag, 1988;535.
13. Munck Petersen C, Juncker Sorensen J, Hyldig Nielsen JJ, Lamm LU. Segregation of HLA-DR, DQ and DP phenotypes and restriction fragment length polymorphism fragments in two recombinant families. Tissue Antigens 1988;32:104.
14. Yunis EJ, Amos DB. Three closely linked genetic systems relevant to transplantation. Proc Nat Acad Sci USA 1971;68:3031.
15. Yunis EJ, Awdeh Z, Johnson A, et al. Complotype genetic loci segregate more frequently with HLA-DR than with HLA-B. Immunogenetics 1985;21:25.
16. Raum DD, Awdeh ZL, Glass D, Yunis EJ, Alper CA. The location of C2, C4 and BF relative to HLA-B and HLA-D. Immunogenetics 1981;12:473.

Southern Blot Analysis of Rheumatoid Arthritis

E. Du Toit,[1] J.A. Hansen,[2] N.K. Mehra,[3] M.O. Oudshoorn,[1] A. Smith,[2] T. Takenouchi,[4] and P.I. Terasaki[4]

Introduction and Methods

The methods and materials used in the Southern blot analysis of rheumatoid arthritis (RA) study were identical to the conditions used in the core and family study sections of the Tenth Workshop (10WS) Southern blot component. This allowed a direct comparison of the fragments seen in the study films with the fragments seen on the 10WS standard films. Informative data were available from three laboratories: SAF/DUT, US3/TER, and US7/HAN. The DUT laboratory studied 24 South African Blacks (12 patients and 12 controls). Ten pairs were complete DR matches. DUT used the enzymes BamHI, EcoRI and TaqI, and probes for DRB, DQB, DPB, DQA, and class I. The TER laboratory studied 24 Caucasians (12 patients and 12 controls). Ten pairs were DR-matched. The two patients who were not DR matched with their controls were both typed as DR4, DN1. The HAN laboratory studied 12 patients and 12 controls, all of whom were DR-matched. Both TER and HAN used the enzymes BamHI, HindIII, and TaqI, and probes for DRB and DQB. TER also used DPB while HAN used DQA. When data for the same enzymes and probes were available, the TER and HAN data were combined. The frequencies of DR4 in these patient groups were: DUT, 3/12 (25%); HAN, 10/12 (83%); and TER, 5/12 (42%) (Table 1).

Participating Laboratories: SAFDUT,[1] US7HAN,[2] ANZMEH,[3] US5TER[4]

Results and Discussion

If any fragment observed in these patients or controls could be differentiated from the fragments identified on the standard films, it was designated as a new band. Fragments that were weakly visible with one probe, but present on the standard films hybridized with any other probe, were not designated as new fragments. A total of 38 new fragments were seen in these studies (Table 2). Thirteen were identified with a DRB probe, 6 with a DQB probe, 1 with a DQA probe, 4 with a DPB probe, and 14 with a class I probe. Most of these new bands were found on films from the DUT lab, and thus most of these new bands may represent restriction site polymorphisms that are characteristic for the South African Black population.

DRB fragments that tended to occur with different frequencies in the patient and control groups are listed in Table 3. None of these apparent differences, however, were significant. The distribution of DRB bands in DR4 positive and negative individuals is summarized in Table 4. The DRB TaqI "S11" fragment, which is specific for DR4 on standard films, was present in all $DR4^+$ patients and controls in both the combined TER/HAN and DUT data, and absent in all DR4-individuals (data not shown). The same DRB fragment was sometimes associated with different DR antigens when the TER/HAN and DUT data were compared. TaqI bands "S7" and "S17" were only observed in $DR4^-$, $DR7^+$ individuals in the combined TER/HAN data, but in five of seven $DR4^+$ and two $DR4^-$, $DR9^+$ individuals in the DUT data. DRB TaqI

Table 1. HLA-DR Antigens Present in Rheumatoid Arthritis Patients and Controls

	SAF/DUT		US7/HAN		US3/TER		Combined TER + HAN	
	Pts (n=12)	Controls (n=12)	Pts (n=12)	Controls (n=12)	Pts (n=12)	Controls (n=12)	Pts (n=24)	Controls (n=24)
DR1	1	1	1	1	5	5	6	6
DR2	1	1	2	2	4	4	6	6
DR3	4	4	3	3	1	1	4	4
DR4	3	4	10	10	5	5	15	15
DR5	0	0	0	0	1	2	1	2
DR11	3	2	0	0	0	0	0	0
DRw12	0	1	0	0	0	0	0	1
DRw6	3	0	–	–	1	1	4	1
DR7	0	0	2	2	2	2	4	4
DR8	2	2	–	–	1	1	3	3
DR9	2	0	1	1	1	1	4	2
DRw10	1	1	1	1	0	0	2	2
DRw13(w6)	1	4	0	0	0	0	1	4
DRw14(w6)	0	1	0	0	0	0	0	1
DN1	–	–	–	–	2	0	2	0

Pts = patients

Table 2. New RFLP Fragments Identified in the Rheumatoid Arthritis Study[a]

Probe	Enzyme	Local Band No.	New Band No.	kb	No. of Patients	No. of Controls
DRB	BamHI	D3	N1	7.6	1	0
		D6	N2	5.2	1	1
		D7	N3	5.0	0	1
		H9	N4	4.1	1	1
		D12	N5	4.6	0	2
DRB	EcoRI	D11	N1	4.6	1	0
DRB	HindIII	T1	N1	13	1	0
		TSa	N2	6.4	1	0
		T22,H22	N3	1.3	3	2
DRB	TaqI	D2,H2,T2	N1	15	5	3
		D5	N2	11	1	2
		D14	N3	4.6	2	1
		T13	N4	4.2	2	1
DQB	BamHI	D2a	N1	7.6	1	0
		D3,T4,H4	N2	6.9	11	7
DQB	EcoRI	D8	N3	8.4	2	0
DQB	TaqI	H5	N1	4	2	3
		D4	N2	3.8	2	0
		D7,H9	N3	2.7	7	5
DPB	BamHI	D1	N1	17	1	3
		D7	N2	4.6	6	7
DPB	TaqI	D6	N1	2.4	4	3
		D9	N2	1.4	1	1
DQA	TaqI	D3	N1	6.7	3	6
Class I	BamHI	D19	N1	3.6	2	3
		D20	N2	3.5	12	12
Class I	TaqI	D2	N1	12.1	1	4
		D5	N2	8.3	6	2
		D7	N3	6.8	1	1
		D11	N4	5.2	9	9
		D13	N5	4.2	6	4
		D16	N6	3.4	2	2
		D18,D19	N7	2.8	2	1
		D20	N8	2.7	3	3
		D22	N9	2.6	4	4
		D24	N10	2.5	7	5
		D28	N11	2.2	2	0
		D32	N12	1.6	12	11

[a]"Local Band No." refers to the band numbers assigned in each individual lab: "D", DUT; "H", HAN; and "T", TER.

bands "S16" and "S28" were found with all DR4⁺ and DR4⁻, DR7⁺ individuals in the TER/HAN data, but only in three of seven DR4⁺ individuals in the DUT data. Thus, DRB TaqI fragments "S7/S17" and "S16/S28" correlate with the DRB4 DRw53 specificity present in DR4 and DR7 individuals in the combined TER/HAN Caucasian population, but appear to discriminate between DR4 and DR7 in the DUT South African Black population.

Table 3. DRB RFLP Analysis of RA Patients and DR- Matched Controls[a]

		TER + Han		DUT	
Enzyme	Standard Fragment No.	Patients (n=24)	Controls (n=24)	Patients (n=12)	Controls (n=12)
TaqI	S2	5 (21%)	7 (29%)	5 (42%)	4 (33%)
TaqI	S4	4 (17%)	3 (12%)	3 (25%)	6 (50%)
TaqI HindIII	S8 S16	3 (12%)	5 (21%)	5 (21%)	5 (21%)
TaqI	S7 S17	1 (4%)	3 (13%)	5 (21%)	2 (8%)

[a]Only DRB fragments that showed an apparent difference in the patient and control groups are included; however, none of these were significant.

Table 4. DRB RFLP Analysis of DR4-Positive and DR4-Negative Individuals

	Enzyme	Standard Fragment No.	TER + HAN Patients	TER + HAN Controls	DUT Patients	DUT Controls
DR4-Positive			(n = 15)	(n = 15)	(n = 3)	(n = 4)
	TaqI	S2	3 (20%)	5 (33%)	1 (33%)	1 (2.5%)
	TaqI	S4	2 (13%)	1 (7%)	0	1 (2.5%)
	TaqI HindIII	S8 S16	1 (7%)	3 (20%)	1 (25%)	1 (25%)
	TaqI	S7 S17	0	0	3 (100%)	2 (50%)
	TaqI	S16 S28	15 (100%)	15 (100%)	1 (33%)	2 (50%)
DR4-Negative			(n = 9)	(n = 9)	(n = 9)	(n = 8)
	TaqI	S2	2 (22%)	2 (22%)	4 (24%)	3 (38%)
	TaqI	S4	2 (22%)	2 (22%)	3 (33%)	5 (63%)
	TaqI HindIII	S18 S16	2 (22%)	2 (22%)	4 (44%)	4 (50%)
	TaqI TaqI	S7 S17	1 (11%)	3 (33%)	2 (22%)	0
	TaqI TaqI	S16 S28	2 (22%)	1 (11%)	0	0

DQB fragments that tended to occur with different frequencies in the patient and control groups are listed in Table 5. None of these apparent differences, however, were significant. The distribution of DQB fragments in $DR4^+$ and $DR4^-$ individuals is summarized in Table 6. A series of fragments, BamHI/"N2", BamHI/"S18", HindIII/"S13", TaqI/"S18" and TaqI/"S20" were found in 11 of 15 (73%) $DR4^+$ patients compared to 5 of 15 (33%) $DR4^+$ controls in the combined TER/HAN data. The BamHI/"S18" fragment has been previously shown to identify a DQw3 variant known as DQw7 or DQ3.1(1). The second cluster of DQB fragments, BamHI/"S5" and EcoRI/"S1", identify a second DQw3 variant known as DQ3.2. In the DUT data, the DQ3.1 variant was found in none of three $DR4^+$ patients, while the DQ3.2 variant was found in all three $DR4^+$ patients and all four $DR4^+$ controls. In the combined TER/HAN data, DQ3.2 occurred in some $DR3^+$,$DR4^+$ and $DR8^+$ individuals. In

Table 5. DQB RFLP Analysis of RA Patients and DR Matched Controls[a]

Enzyme	Standard Fragment No.	TER + HAN Patients (n=24)	TER + HAN Controls (n=24)	DUT Patients (n=12)	DUT Controls (n=12)
TaqI	S8	4 (17%)	7 (29%)	4 (33%)	4 (33%)
BamHI TaqI TaqI TaqI BamHI	S3 N3 S17 S21 S17	10 (42%) 5 (21%)	4 (17%) 7 (29%)	4 (33%) 4 (33%)	2 (17%) 1 (8%)
HindIII BamHI EcoRI TaqI TaqI	S12 S5 S1 S10 S16	10 (42%)	14 (58%)	4 (33%)	7 (58%)
BamHI BamHI HindIII TaqI TaqI	N2 S18 S13 S18 S20	12 (50%)	7 (29%)	4 (33%)	5 (42%)

[a]Only DQB fragments that showed an apparent difference in the patient and control groups are included; however, none of these were significant.

Table 6. DQB RFLP Analysis of DR4-Positive and DR4-Negative Individuals

	Enzyme	Standard Fragment No.	TER + HAN Patients	TER + HAN Controls	DUT Patients	DUT Controls
DR4-Positive			(n = 15)	(n = 15)	(n = 3)	(n = 4)
DQw7 (3.1):	BamHI	N2				
	BamHI	S18				
	HindIII	S13				
	TaqI	S18				
	TaqI	S20	11 (73%)	5 (33%)	0	2 (50%)
DQw8 (3.2):	BamHI	S5				
	EcoRI	S1				
	HindIII	S12				
	TaqI	S10				
	TaqI	S16	8 (53%)	12 (80%)	3 (100%)	4 (100%)
	BamHI	S17				
	TaqI	S17				
	TaqI	S21	1	3	0	0
	TaqI	N3	7%	20%		
DR4-Negative			(n = 9)	(n = 9)	(n = 9)	(n = 8)
	BamHI	N2				
	BamHI	S18	1 (11%)	2 (22%)	4 (44%)	3 (33%)
	HindIII	S13				
	TaqI	S18				
	TaqI	S20				
	BamHI	S5				
	EcoRI	S1	2 (22%)	2 (22%)	1 (11%)	3 (33%)
	HindIII	S12				
	TaqI	S10				
	TaqI	S16				
	BamHI	S17				
	TaqI	S17	4 (44%)	4 (44%)	4 (44%)	1 (12%)
	TaqI	S21				
	TaqI	N3				

the South African study the DQ3.1 associated fragments were found in only two of seven DR4$^+$ individuals, but in all DR8$^+$ and some DR3$^+$, DR5$^+$, and DRw6$^+$ individuals. Thus, the RFLP marker for DQ3.1 does not appear to be DR4 associated in South African Blacks.

DQA fragments that tended to occur with different frequencies in patients and controls are listed in Table 7. None of these apparent differences however, were significant. Similarly, no significant differences were noted for the DPB RFLP analysis (Table 8) or the HLA class I RFLP analysis (Table 9).

Summary

This study was designed to determine if RFLP analysis using DRB, DQB, DQA, and DPB probes could detect genomic differences between RA patients and DR-matched normal controls. Although several trends were observed, the total number of patients and controls, 48 Caucasians and 24 South African Blacks, was relatively small and none of the apparent differences were significant. No new RFLP fragments were observed exclusively in RA patients. Many new fragments were found

Table 7. DQA RFLP Analysis of RA Patients and Controls[a]

Enzyme	Standard Fragment No.	TER + HAN Patients	TER + HAN Controls	DUT Patients	DUT Controls
BamHI	S2	1(8%)	3 (25%)	2 (17%)	1 (8%)
TaqI	S5				
BamHI	S3	2 (17%)	1 (8%)	5 (42%)	3 (25%)
TaqI	S11				
EcoRI	S6	8 (67%)	11 (92%)	5 (42%)	7 (58%)
TaqI	S12				

[a]Only the DQA bands that showed an apparent difference in the patient and control groups are included; however, none of these were significant.

Table 8. DPB RFLP Analysis of RA Patient and Controls[a]

		TER		DUT	
Enzyme	Standard Fragment No.	Patients (n=12)	Controls (n=12)	Patients (n=12)	Controls (n=12)
TaqI	S5	11	11	9 (75%)	6 (50%)

[a]Only DPB bands that showed an apparent difference in the patient and control groups are included; however, none of these were significant.

in this study, but some of these occurred in only a single individual. Lacking duplicate testing or family study we have no way of confirming these preliminary findings. Nevertheless, many of these new fragments were observed in more than one individual and most of the latter were South African Blacks. Thus a number of these new fragments may represent differences that are characteristic of South African Blacks, polymorphisms that were probably not well represented among the 10WS core cell lines from which the RFLP standard films were generated. The DR4 specific DRB TaqI "S11" fragment was present in all DR4+ Caucasians and Blacks studied. Finally, the previous observation that DQ3.1 is increased in DR4+ RA patients was confirmed by the analysis of the 24 Caucasian patients and 24 Caucasian controls studied. In the South African Black population, however, this association between RA and DQ3.1 does not appear to exist. In the latter population, DQ3.2-associated fragments were found in all DR4+ patients and controls.

Table 9. HLA Class I RFLP Analysis of RA Patients and Controls[a]

		DUT	
Enzyme	Standard Fragment No.	Patient (n=12)	Controls (n=12)
TaqI	S20	2 (17%)	0
TaqI	N7		
TaqI	N11		
EcoRI	S8		
EcoRI	S2		
BamHI	S2	3 (25%)	0
TaqI	N2	6 (50%)	2 (17%)
TaqI	S14[b] (D24+D25)	7 (58%)	1 (8%)

[a]Only class I bands that showed an apparent difference in the patient and controls groups are included; however, none of these were significant.
[b]The TaqI D24 and D25 fragments appeared to represent a single band on standard films designated "S14"; however, on films from the DUT laboratory, they appeared as discrete bands.

References

1. Kim SC, Holbeck SL, Nisperos B, Hansen JA, Maeda H, Nepom GT. Identification of a polymorphic variant associated with HLA-DQw3 and characterized by specific restriction sites within the DQB chain gene. Proc Natl Acad Sci USA 1985;82:8139.
2. Nepom BS, Palmer J, Kim SJ, Hansen JA, Holbeck SL, Nepom GT. Specific genomic markers for the HLA-DQ subregion discriminate between DR4+ insulin-dependent diabetes mellitus and DR4+ seropositive juvenile rheumatoid arthritis. J Exp Med 1986;164:345.

An RFLP Study of HLA-D Gene Polymorphism in Multiple Sclerosis

R. Heard,[1] W.I. McDonald,[1] J.R. Batchelor,[1] R.I. Lechler,[1] C. Cullen,[2] D. Middleton,[2] B.R. Hawkins,[3] A. Dos Santos,[4] H. Festenstein,[4] G. Semana,[5] R. Fauchet,[5] L. Smeysters,[6] and P. Reekers[6]

Introduction

An association between the demyelinating central nervous system disease, multiple sclerosis (MS), and DR2, DQw1 has been defined in numerous Northern European populations (1,2). The relative risk conferred by possession of DR2 is low, ranging from 2.0 to 3.8 in different studies (3). This study was designed to explore an association between HLA-D-region alleles and MS in more detail, using restriction fragment length polymorphism (RFLP) analysis. The selection of

Reporting Laboratory: UKIBAT[1]
Participating Laboratories: UKIMID,[2] ANZHAW,[3] UKIFES,[4] FRAFAU,[5] BENREE[6]

Table 1. RFLP Characterization of DR and DQ Alleles on the DR2 Haplotype in Patients and Controls

Fragment	Allele	All MS	All Controls	DR2+ MS	DR2+ Control
DQβ: Eco RI 2.2	Dw2	54/112 (48.2)	49/109 (45.0)	54/56 (96.4)	49/53 (92.5)
DQα: MspI 2.87	DQw1	69/79 (87.3)	69/78 (88.5)	51/52 (98.1)	53/53 (100)
DQβ: BglII 4.2	DQw1.2	65/83 (78.3)	38/53 (71.7)	53/56 (94.6)	30/30 (100)

enzyme/probe combinations was based on a recent RFLP study conducted in our laboratory in which we used the pooled DNA approach first described by Arnheim et al. (4). Using pooled DNA samples from DR2$^+$ patients, DR2$^+$ controls, DR2$^-$ patients, and DR2$^-$ controls, we were able to screen 70 enzyme/probe combinations including DRβ, DQβ, DPβ, DQα, and DPα and 14 different restriction enzymes. This analysis highlighted MspI/DQα fragments that occurred at a higher frequency in patients than controls, and provided the basis for a multicenter study.

Patients and Methods

All the individuals included in this study were Northern European Caucasoids and were drawn from five different centers. The patients were all cases of clinically definite MS as defined by the classic Poser criteria, and the controls were healthy, unrelated individuals from the same geographic locations. The frequency of DR2$^+$ and DR2$^-$ individuals in disease and control groups were matched. The representation of DR alleles other than DR2 in the two groups was also very similar.

The enzyme/probe combinations used in this study were:

	DR	DQ	DQ
BamHI	+	+	−
BglII	+	+	−
EcoRI	+	+	−
MspI	+	−	+
TaqI	+	+	+

These were selected on the basis of the results of a previous study conducted in our laboratory as described above. Samples of genomic DNA, prepared from peripheral blood were digested, electrophoresed, and blotted, as per the Workshop Southern blot protocol for "short" gels. Radiolabeled λ KpnI and λ BstEII were used as molecular weight markers. The filters were hybridized and washed at high stringency as per the Workshop protocol. Marker and fragment mobilities were measured, and banding patterns tabulated in the different centers before the data and autoradiographs were sent to the Department of Immunology, RPMS, London for standardization and statistical analysis.

In accumulating data from the different centers, films were "read" independently by three observers, and fragments were only included that were present on films from all of the centers. Where doubt existed about interpreting the films for any particular fragment, those films were excluded. The fragments were scored as either present or absent, and entered onto a computer disc prior to simple statistical analysis. The results are shown as follows.

Results

Further definition of the DR2-containing haplotype was achieved by the detection of three fragments, Eco RI/DQ 2.2 kb, MspI/DQ 2.87 kb, and BglII/DQ 4.2 kb, characteristic of Dw2, DQw1 (Tenth Workshop core cell line analysis), and DQw1.2 (A. Bushell et al., personal communication). The representation of these fragments in DR2$^-$ and DR2$^+$ patients and controls is shown in Table 1. Clearly, the vast majority of the DR2 haplotypes in this study encode the Dw2, DQw1.2 alleles of DR and DQ, as expected in a Caucasoid population.

The most interesting fragments in this study are MspI/DQ 3.25, 2.50, and 2.31 kb as seen in Table 2. The 3.25 kb fragment is present in 22.7% of patients and only 6.4% of controls, irrespective of DR serotype, and in 21.2% of DR2$^+$ patients and only 5.7% of DR2$^+$ controls. The 2.50- and 2.31-kb fragments had a very similar distribution. These results are statistically significant with p values of <0.01 and <0.05. If these were the results of a single study, correction for the number of compari-

Table 2. Significant Differences in the Frequency of MspI/DQα Fragments in Disease Versus Control Groups

Group	Fragment (kb)	MS	Control	χ^2	p
All	3.25	18/79 (22.7)	5/78 (6.4)	8.416	< 0.01
All	2.50	15/53 (28.3)	5/53 (9.4)	4.992	< 0.05
All	2.31	13/53 (24.5)	5/53 (9.4)		NS
DR2+	3.25	11/52 (21.2)	3/53 (5.7)	5.452	< 0.05
DR2+	2.50	9/39 (23.1)	3/41 (7.3)		NS
DR2+	2.31	9/39 (23.1)	3/41 (7.3)		NS
DR2−	3.25	7/27 (25.9)	2/25 (8)		NS
DR2−	2.50	5/14 (35.7)	2/12 (16.7)		NS
DR2−	2.31	4/14 (28.6)	1/12 (8.3)		NS

Table 3. Phenotype of Tenth Workshop Core Cell Lines with the MspI/DQα 3.25-kb Fragment

Name	10th Workshop No.	Phenotype DR	Dw	DQ
RML	9016	2	LD5a	w1,w3
PE117	9028	4	14	w3
LKT3	9107	4	?	–
DBB	9052	7	11	w2,w3
Manika	9106	7	?	w2,w3
BTB	9067	w8	8.1	?
BM9	9068	w8	8.1	?
Madura	9069	w8	8.1	w1
Luy	9070	w8	8.3	w3 TP
Olga	9071	w8	8.2	–
Spach	9072	w8	8.2	–
DHIF	9104	?w11	5	?

sons made would normally be expected. However, the results presented here represent the amalgamation of two consecutive studies, the first conducted by the authors, the second carried out in several centers for the Tenth Workshop. Similar results were obtained in both studies.

Table 4. DR and DQ Phenotype of Patients and Controls with the MspI/DQα 3.25-kb Fragment

Sample No.	DR[a]	DQ[a]
101	2,4	1,3
103	2,4	1,3
104	2,4	1,3
106	2,7	1,–
108	2,2	1,1
118	1,1	1,–
119	1,1	1,–
132	2,2	1,1
133	2,4	1,–
134	1,4	1,–
177	2,7	2,3
190	1,1	3,3
196	2,7	1,3
204	3,7	1,3
208	4,7	NT[b]
235	2,6	NT
236	2,6	NT
502	2,–	NT
505	2,1	NT
518	2,5	NT
527	2,8	NT
529	1,7	NT
530	1,7	NT
539	5,6	NT

[a]Typed using Ninth Workshop sera; NT = not tested.

Discussion

These results raise the possibility that two MHC haplotypes may be contributing to disease susceptibility in at least some patients with MS. The findings in this study support the concept that the DR2-containing haplotype associated with MS in Caucasian populations comprises the Dw2 subtype of DR, and the DQw1.2 allele of DQ that is usually seen in conjunction with Dw2. The MspI/DQα fragments that occur with significantly raised frequency in the patient group appear to be contributed by the non-DR2 haplotype. The 3.25-kb and 2.50-kb fragments described here are seen in the RFLP data acquired from the Workshop analysis of the homozygous core cell lines, and are clustered with all the DRw8, and some of the DR4 and DR7 cells (Table 3). These fragments were never seen in the DQw1 Workshop cells, unlike the MspI/DQα 2.87-kb fragment which was present in all the DQw1 cells. The allele(s) of DQ characterized by the 3.25- and 2.50-kb fragments has yet to be defined; however, the DR phenotypes of the core cell lines in which these fragments were seen is suggestive of a split of DQw3 or DQ blank. This is currently under investigation. The DR and DQ phenotypes of the individuals in this study with the 3.25-kb fragment are presented in Table 4. While the DR alleles include several specificities, the DQ serotypes are all DQw3 or DQ blank.

It is notable that the DQα fragments described above appear to be associated with disease independently of DR2, in that they are equally represented in DR2$^+$ and DR2$^-$ patients. However, of all the patients with these two fragments, 83% also have the MspI/DQα 2.87-kb fragment. This raises the possibility that susceptibility to disease is conferred by the presence of an allele of DQw1 on one haplotype, and a DQ allele characterized by the MspI/DQα 3.25- and 2.50-kb fragments on the other haplotype.

References

1. McDonald WI. Multiple sclerosis: Epidemiology and HLA associations. Ann NY Acad Sci 1984;436:109.
2. Batchelor JR. Immunological and genetic aspects of multiple sclerosis. In: Matthews WB (ed): McAlpines' Multiple Sclerosis. Edinburgh, Churchill Livingstone, 1985, p 281.
3. Tiwari JL, Terasaki PI. HLA and Disease Associations. Springer-Verlag, New York, 1985, p 152.
4. Arnheim N, Strange C, Erlich H. Use of pooled DNA samples to detect linkage disequilibrium of polymorphic restriction fragments in human disease: Studies of the HLA Class II loci. Proc Natl Acad Sci USA 1985;82:6970.

DP RFLP Studies of the 72 Core Cell Lines Selected for the Southern Blot Protocol*

Yoko Mitsuishi,[1] Arlette Urlacher,[1] Simone Mayer,[1] and Marie M. Tongio[1]

DNA was digested with the requested 12 enzymes and hybridized with the DPα and DPβ probes given to the participating laboratories. Only clusters correlating with the PLT-defined DP specificities were retained from the Workshop computer analysis.

Determination of the Main DP Clusters

Most of the fragments obtained with the majority of the enzymes were correlated with broad DP clusters (called main DP clusters). Two main clusters were obtained with the DPα probe (DPIA and DPIIA) and four with the DPβ probe (called DPIB, DPIIB, DPIIIB, DPIVB). The RFLPs allowing their determination as well as the DP alleles associated to these clusters are given in Table 1. Since it is known from the study of Okada et al. (1) that the fragments defining the DPIB and DPIIB main clusters are correlated with the DPB2 pseudogene, these were not taken into consideration for most of the further DP polymorphism analysis. Nevertheless, the data concerning DPIB and DPIIB are given in the tables for the sake of complete information.

Determination of the Main DPA and DPB Cluster Combinations

Each DNA sample was defined by the main clusters DPA (I or II) and B (III or IV) to which it belonged. Three DPA-DPB combinations were found (Table 2): 1) DPA-DPB which was correlated with both DPw1 and DPw5, 2) DPIIA-DPIVB which was correlated with DPw2 and DPw4, and 3) DPIIA-DPIIIB which was correlated with DPw3. Also, DP blank alleles were correlated with one of these different cluster combinations, as will be discussed further.

Determination of the Specific DPw1 to DPw5 Patterns

Knowing the main DP clusters and their DPA-DPB combination, a specific DP pattern could be defined for each DP allele by studying a few fragments obtained with some enzymes inside a given main cluster. In Table 2, arabic numerals show the fragments determining the main clusters, and *italic* numerals indicate the fragments determining the individual DP alleles. For example, DPw1 and DPw5 alleles, defined both by the cluster combination DPIA-DPIIIB, could be differentiated from each other by a few fragments (printed in *italics*) obtained with MspI, BglII, TaqI in the DPIA cluster and MspI in the DPIIIB cluster.

Determination of DPw1 to DPw5 Subdivisions

As shown in Table 3, some of the cells typed as DPw1 to DPw5 by the PLT technique did not have the expected specific DP allelic pattern defined in Table 2. Either major modifications (changes in one of the main clusters), minor modifications (changes in a DP pattern inside a main cluster, or modifications of a few fragments with a few enzymes inside a given DP pattern) occurred, which suggested the existence of several DP subdivisions inside the classic DP specificities. A nomenclature was devised in which the specific DP pattern was called DPx.1 (for example, for DPw1, the specific pattern was DPw1.1) and the following less frequent subdivisions DPx.2, DPx.3, and so forth (for DPw1 these were DPw1.2, DPw1.3). Major modifications were not frequent. They concerned 1) cell no. 9009, typed by the PLT technique as DPw3 which belonged to the DPIA-DPIIIB combination instead of the expected DPIIA-DPIIIB (see Table III subdivision DPw3.3), and 2) cell no. 9063 typed by the PLT test as

Participating Laboratory: FRAMYE[1]

*This paper is an individual laboratory report. The consensus report for DP RFLP is presented by Hyldig-Nielsen and Svejgaard (this volume).

Table 1. RFLPs Defining the Main DP Clusters

Probe	Main DP Clusters	DP Association	RFLP (in kb)
DPA	IA	1, 5, b and c†	Hind III 4.90, Pst I 2.63, Taq I 6.45, Bam HI 6.92, Sst I 18.82, 6.27
	IIA	2, 3, 4 and a	Hind III 3.25, Pst I 2.43, Msp I 4.79, 3.14, Bgl II 2.23, Sst I 16.16, 4.45
DPB	IB*	1, 4, 5 and a	Hind III 5.35, Msp I 0.71, Eco RI 7.58
	IIB*	2, 3, b and c	Bam HI 9.00, Hind III 5.06, Eco RI 7.20
	IIIB	1, 3, 5 and b	Taq I 1.73, Pst I 2.85, Eco RV 10.10
	IVB	2, 4, a and c	Pst I 2.39, 1.91, Eco RV 6.89, 3.36, Bgl II 15.05, 2.80

*The fragments determining the DPIB and DPIIB main clusters are governed by the DPB2 pseudogene; †a, b, c: see reference 2.

Table 2. RFLPs Allowing the Determination of the DPw1 to DPw5 Patterns

Cellular Typing		DPA Hind III	DPA Pst I	DPA Msp I	DPA Bgl II	DPA Taq I	DPA Bam HI	DPA Sst I		DPB1 Taq I	DPB1 Pst I	DPB1 Eco RV	DPB1 Msp I	DPB1 Bgl II		DPB2 Bam HI	DPB2 Hind III	DPB2 Eco RI
		4.90	2.63			6.45	6.92	18.82		1.73	2.85	10.10				–*	5.35	7.58
								6.27										
DPw1				*6.71*	*7.24*								*6.71*	*7.28*				
				6.30									*4.98*					
				4.87	*3.41*	*4.66*												
	IA														IB			
		4.90	2.63			6.45	6.92	18.82		1.73	2.85	10.10				–	5.35	7.58
								6.27										
DPw5				*6.71*	*7.24*				IIIB				*6.71*	*7.28*				
				6.30									*1.97*					
					3.58	*5.47*							*3.11*					
		3.25	2.43	4.79	2.23			16.16		1.73	2.85	10.10				9.00	5.06	7.20
DPw3				3.14				4.45										
					5.29	*13.21*	*8.67*	–					*3.11*	*5.05*				
															IIB			
		3.25	2.43	4.79	2.23			16.16		–	2.39	6.89		15.05		9.00	5.06	7.20
DPw2	IIA			3.14				4.45			1.91	3.36		2.80				
					7.24	*12.07*	*9.00*	*10.42*					*1.97*					
													2.17	*7.28*				
									IVB									
		3.25	2.43	4.79	2.23			16.16		–	2.39	6.89		15.05		–	5.35	7.58
DPw4				3.14				4.45			1.91	3.36		2.80	IB			
					5.29	*13.21*	*8.67*	*10.42*					*1.97*					
													1.07	*5.05*				

*No fragment found with this enzyme. Arabic numberals: RFLPs allowing the determination of the main DP clusters. Italic numerals: RFLPs allowing the determination of the allelic specific DP patterns as well as DP blank alleles (see Table IV). All fragments are given in kb.

Table 3. RFLPs Allowing the Determination of the DPw1 to DPw5 Subdivisions

Cellular Typing	DP Subdivision	Number of Cells*		DPA–RFLP (in kb)		DPB–RFLP (in kb)
	1.1	4		†———		———
DPw1	1.2	1		DPw5 pattern		———
	1.3	1		———		——— Bgl II 7.28 neg,[6], 5.05 pos§
			IA			
DPw5	5.1	1		———		———
	5.2	1		——— Msp I 6.71, 6.30 neg, 4.87 pos		——— Msp I 6.71 neg, 7.16 pos
					IIIB	
	3.1	6	IIA	———		———
DPw3	3.2	1	———	——— Bgl II 5.29 neg, 7.24 pos		——— **Bgl II 5.05 neg, 7.28 pos
	3.3	1	IA	DPw1 pattern, Msp I 6.71, 6.30 neg, 7.15 pos Bam HI 6.92 neg, 8.67 pos		———
	2.1	12		———	IVB	———
DPw2	2.2	2		——— Bst EII 10.0 neg‖		———
	2.3	1		———	IIIB	DPw3 pattern Bgl II 5.05 neg, 7.28 pos
	2.4	1		——— Bam HI 9.00 neg, 8.67 pos		———
			IIA			
	4.1	20		———		———
DPw4	4.2	1		DPw2 pattern, Bgl II 7.24 neg, 5.29 pos	IVB	———
	4.3	6		———		——— (6)
	4.4	1		DPw3 pattern		———

*Total number of cells having the specificity among the 72 core cell lines tested. †Specific DP pattern as defined in Table 2. †§Fragments given as neg were absent, whereas they were usually present in the specific (or indicated) pattern; they were replaced by the fragments mentioned as pos. ‖Result of supplementary work performed in our lab: BstEII 10.0 kb fragment is specific for the DPw2 specificity. **Modification found inside the DPIB or DPIIB main clusters (see text, Determination of the Main DP Clusters).

Table 4. RFLPs Allowing the Determination of Blank Alleles

DP Blank Allele	No. of Cases	DPA Probe			DPB Probe		
			Analogous DP Pattern	RFLPs Differing from the DP-Specific Analogous Pattern		Analogous DP Pattern	RFLPs Differing from the DP-Specific Analogous Pattern
DPb	4			Msp I 6.71, 6.30 neg† 7.15 pos‡	IIIB	DPw5	Msp I 7.16 add§ 1.97 neg
DPc	3		DPw1	Msp I 6.71, 6.30 neg 7.15 pos Sst I 10.42 add			Msp I 7.16 add
9106*	1	IA		Msp I 6.71, 6.30 neg 7.15 pos Bgl II 7.24 neg, 5.29 pos	IVB	DPw2	Msp I 1.97, 2.17 neg, 3.11, 7.16 pos Bgl II 7.28 neg, 5.05 pos Bam HI 10.10 add
9066*	1		DPw5	Sst I 10.42 add Taq I 12.07 add			Msp I 3.11, 6.71 add
9060*	1			Msp I 7.15 neg Taq I 5.47 neg		0	Very specific pattern
9047*	1		DPw3	Bam HI 8.67 neg, 9.00 pos	IIIB	DPw3	Msp I 1.97 add
DPa	8	IIA	DPw4	Bgl II 5.29 neg, 7.24 pos	IVB	DPw4	Bgl II 5.05 neg, 7.28 pos
DPa Subtype	5			Bgl II 5.29 neg, 7.24 pos Bam HI 8.67 neg, 9.00 pos			Bgl II 5.05 neg, 7.28 pos

*Very rare DP blank alleles found with one single cell only; these blank alleles did not receive a new DP designation, but were called by their Workshop identification number. †Neg: absence of a fragment found in the analogous DP pattern. ‡Pos: presence of a fragment usually found in another pattern than the analogous DP pattern and supposed to replace the absence of fragment(s) mentioned in.† §Add: supplementary fragment. All fragments are given in kb.

Table 5. Informative DP-RFLP Clusters (Main Clusters not Included)

Probe	Taq I	Msp I	Bgl II	Bam HI	Others	Principal DP Associations	Missing Subspecificity(ies)	Supplementary Specificity(ies)
	4.66	4.87	3.41			1 + b + c		3.3 + 9106*
	5.47		*3.58*			5		1.2 + 9066
		6.30,6.71				1 + 5		1.2 + 9060 + 9066
	12.07					2		4.2 + 9066
DPA	13.21					3 + 4 + a + a sub	3.3	
				9.00		2 + a sub		4.2 + 9047
				8.67		3 + 4 + a	3.3 + 4.4	
			7.24			1 + 2 + 5 + a + a sub + b + c		3.2 + 3.3 + 9060 + 9066
			5.29			3 + 4	3.2 + 3.3	9047 + 9106
					Sst I 10.42	2 + 4 + c	4.4	9066
	7.66,3.99					a sub	9064 + 9016	
		4.98				1		
		6.71				1 + 5	5.2	9066
		3.11				3 + 5 + b		2.3 + 9066 + 9106
		2.17				2	2.3	9105 + 9066
		1.07				4 + a + a sub		
DPB		1.97				2 + 4 + 5 + a + a sub + c	2.3	9047 + 9066
		7.16				b + c		5.2 + 9106
			7.28			1 + 2 + 5 + a + a sub + b + c		9066
			5.05			3 + 4		9047 + 9066 + 9106
	2.64	*4.04*			*Hind III 1.81*	Only 9060		
	2.72				Hind III 1.86, Pst I 2.07	All except 9060		

Fragments printed in italic numerals are specific of one single DP allele (if one excludes the supplementary specificities).
*Workshop identification number of cells which did show a pattern referring to very rare blank alleles.

Table 6. DP Typing of 72 Core Cell Lines

DP Specificity and Subdivision	Cell ID No.	DNA RFLP Typing	PLT Typing			DP Specificity and Subdivision	Cell ID No.	DNA RFLP Typing	PLT Typing		
			Stand*	Eckels†	Hartzman†				Stand*	Eckels†	Hartzman†
DPw1.1	9006	1.1/1.1	1	1	1	DPw4.2	9057	4.2/4.2	4	4/+?	4L1
	9023	1.1/1.1	1	1	4	DPw4.3	9008	4.3/2.4	4/2	4/2	?
	9058	1.1/1.1	1	1	1≠		9017	4.3/2.1	4/2	4/2	4?
	9070	1.1/4.1	1/4	1/4	1/4		9026	4.3/4.3	4	4	4
DPw1.2	9021	1.2/ a'	1	1	1≠		9033	4.3/4.3	4	4	4
DPw1.3	9088	1.3/4.1	1/4	1/4	1/4≠		9061	4.3/4.3	4	4	4L1
							9075	4.3/4.3	4	4	4
DPw2.1	9011	2.1/4.1	2/4	2/4	2?/4	DPw4.4	9009	4.4/3.3	4/3	4/3	?
	9017	2.1/4.3	2/4	2/4	4?						
	9019	2.1/2.1	0	2B?	?L1	DPw5.1	9107	5.1/5.1	5	NT	NT
	9020	2.1/2.1	2	2W	2≠L1	DPw5.2	9055	5.2/5.2	5	5	5≠L1
	9029	2.1/2.1	2	2	2?						
	9036	2.1/2.1	2	2	?	DPa	9013	a / a	0	4B?	?
	9038	2.1/2.1	2	2	2≠L1		9042	a / a	0	4B?	?L1
	9039	2.1/2.1	2	2	?		9045	a /2.1	4/2	4B?/2	?
	9045	2.1/ a	2/4	2/4B?	?		9054	a / a	0	4B?/+?	?L1
	9050	2.1/2.1	2	2	2/6?		9071	a /3.3	3	4?/3	1?4?≠/3
	9056	2.1/b	2	2	2L1		9072	a / a	0	4B?	4?
	9068	2.1/2.1	2	2B?	2≠		9091	a / a	0	4B?	?
DPw2.2	9032	2.2/2.2	2	2	NT		9092	a / a	0	4B?	?
	9062	2.2/4.1	4	2w/4	2?/4	DPa sub	9010	a'/ a'§	0	?	?
DPw2.3	9063	2.3/2.3	2	2	2/6?L1	(a')	9016	a'/ a'	0	4B?	?
DPw2.4	9008	2.4/4.3	2/4	2/4	?		9021	a'/1.2	1	1	1≠
							9037	a'/ a'	0	4B?	4?≠L1
DPw3.1	9007	3.1/4.1	3/4	3/4	3/4		9064	a'/ a'	0	?	?≠L1
	9018	3.1/3.1	3	3	3,L1						
	9022	3.1/3.1	3	3	3≠	DPb	9003	b / b	0	?	?
	9030	3.1/3.1	3	3	?		9034	b / b	0	7	?
	9059	3.1/3.1	3	3	3/1?L1?		9043	b / b	0	28?	?L1
	9071	3.1/ a	3	3/4?	3/1?4?≠		9056	b /2.1	2	2	2
DPw3.2	9040	3.2/3.2	3	3	3≠L1						
DPw3.3	9009	3.3/4.4	3/4	3/4	?	DPc	9048	c / c	0	?	6
							9104	c /4.1	NT	NT	NT
DPw4.1	9002	4.1/4.1	4	4	?		9105	c / c	NT	NT	NT
	9004	4.1/4.1	4	4	4≠						
	9005	4.1/4.1	0	4	?	Other DP	9047	blank	0	4B?W	?
	9007	4.1/3.1	4/3	4/3	4/3	blank					
	9011	4.1/2.1	4/2	4/2	4/2?	allele	9060	blank‡	0	?	?≠L1
	9014	4.1/4.1	4	4	4						
	9025	4.1/4.1	4	4	4		9066	blank	0	7	?
	9028	4.1/4.1	4	4	4						
	9031	4.1/4.1	4	4	4?		9106	blank/4.1	NT	NT	NT
	9051	4.1/4.1	4	4	4≠						
	9052	4.1/4.1	4	4	4						
	9062	4.1/2.1	4	4/2W	4/2?						
	9065	4.1/4.1	4	4	4						
	9067	4.1/4.1	4	4	4						
	9069	4.1/4.1	4	4	4						
	9070	4.1/1.1	4/1	4/1	4/1						
	9082	4.1/4.1	4	4	4						
	9088	4.1/4.1	1/4	1/4	1/4#						
	9104	4.1/ c	NT	NT	NT						
	9106	4.1/blank	NT	NT	NT						

*Summary of the Workshop DP typing (obtained by PLT test) given to the participants by Hartzman at the time of the Workshop.
†DP typing (obtained by PLT test) as given in reference 5.
§a' = a subtype (see text, Determination of Different DP Blank Alleles).

DPw2 which belonged to the DPIIA-DPIIIB combination instead of the expected DPIIA-DPIVB (see Table 3, subdivision DPw2.3). The specific DP pattern was also different. The DPw3.3 subdivision had a DPw1 pattern with the DPA probe and the DPw2.3 subdivision a DPw3 pattern with the DPB probe. Minor modifications were more frequent and resulted from 1) a modification of the DP specific pattern inside the *same* DP main cluster: DPw1.2 subdivision had a specific DPw5 pattern instead of the DPw1 pattern and DPw4.2 and DPw4.4 subdivisions, a DPw2 and a DPw3 pattern respectively instead of a DPw4 pattern; and 2) smaller modifications inside a given DP-specific pattern where only a few fragments obtained with one or two enzymes differed from those found for the specific DP pattern (see Table 3). Taking all these modifications into consideration, three subdivisions were found for DPw1, four for DPw2, three for DPw3, four for DPw4, and two for DPw5. Data obtained recently in our laboratory lead us to think that at least some of these subdivisions have a cellular expression since some DP clones recognize only the cells that do not have a modification of their DP specific pattern.

Determination of Different DP Blank Alleles

About 30% of the DP alleles are typed as blank by using the cellular PLT technique. The DNA-RFLP analysis with only two enzymes (MspI and BstEII) allowed us in a previous work (2) to define three of them called DPa, DPb, DPc (total incidence higher than 80% of the blank alleles).

The Workshop data obtained with more enzymes and more DNA samples brought further information, as given in Table 4. DP blank alleles have been classified according to 1) the main DP cluster combination to which they belonged, 2) the DP-specific pattern with which they had analogies (analogous pattern), and 3) their specific DP pattern which allowed them to be differentiated from the other alleles belonging to the same DPA-DPB main cluster combination.

DPa, which could not be differentiated from DPw4 in our previous work, could be separated from it by fragments obtained with the BglII enzyme. Moreover, the DPa specificity could itself be subdivided into a DPa subtype by BamHI fragments.

DPb had, inside the DPIA cluster a pattern which was analogous to DPw1 and inside the DPIIIB cluster, a pattern analogous to DPw5 but differed from them by some MspI fragments.

DPc had, inside the DPIA cluster, a pattern identical to that described for the DPb allele except an additional SstI 10.42 kb fragment and, inside the DPIVB main cluster, a pattern analogous to that of DPw2 except an additional MspI 7.16 kb fragment found for DPc, but not for DPw2.

Finally 4 of the 72 cells studied (no. 9106, 9066, 9060, 9047) had RFLP patterns that did not fit in the previously defined patterns. Their characteristics are given in Table 4. A special mention has to be made for cell no. 9060, typed by the PLT test as Cp63 (3), which was particular since its behavior with the DPB probe did not fit any of the DPB main clusters described. The specific fragments found with the DPB probe (not mentioned in Table 4), were TaqI 2.64 kb, HindIII 1.81 kb, and MspI 4.04 kb. This cell was also the only one that did not show the "nonpolymorphic" TaqI 2.72 kb, PstI 2.07 kb and HindIII 1.85 kb fragments present in all other DNA samples analyzed.

Synthetic Analysis

Table 5 gives a summary of the informative DP RFLP clusters (main clusters excluded) that are necessary for DP DNA-RFLP typing. It can be seen that when a fragment reveals one or more DP specificities, it usually reveals all the different subdivision of this or these specificities. The previously published (4) extra and missing reactions do correspond in our opinion to, respectively, supplementary DP subdivision recognized by these fragments and DP subdivisions not revealed by these fragments.

Final DP DNA-RFLP typing as well as the results obtained by PLT typing (5) of each of the 72 core cell lines examined in this Workshop is given in Table 6.

References

1. Okada K, Prentice HL, Boss JM, et al. SB subregion of the human major histocompatibility complex: gene organization, allelic polymorphism and expression in transformed cells. EMBO J 1985;4:739.
2. Mitsuishi Y, Urlacher A, Tongio MM, Mayer S. Restriction fragment length polymorphism of DP genes defines three new DP alleles. Immunogenetics 1987;26:383.
3. Farrell C, Honeymann M. HLA-DP typing of a subset of the B-cell reference panel of the 10th International Histocompatibility Workshop. Tenth International Histocompatibility Workshop Newsletter 1987;1(5):22.
4. Hyldig-Nielsen JJ, Morling N, Odum N. Restriction fragment length polymorphism of the HLA-DP subregion and correlations to HLA-DP phenotypes. Proc Natl Acad Sci USA 1987;84:1644.
5. Yang SY, Milford EL, Hammerling U, Dupont B. Description of the reference panel for factors of the HLA system. In: Dupont B (ed): Immunobiology of HLA, Volume I: Histocompatibility Testing 1987. New York: Springer-Verlag, 1989.

Analysis of Class II RFLP Patterns and T-Cell Clone Reactions

Min Sik Park,[1] Aida Barbetti,[1] Toshinao Takenouchi,[1] and Paul I. Terasaki[1]

From the data distributed at the Princeton Workshop, we have been able to distinguish many alleles of the alpha and beta chains of the DR, DQ, and DP loci. In addition, alleles of the pseudogenes of the DO, DN, DQ 2nd locus, and DP 2nd locus have also been characterized. The alleles were established mainly by RFLP and were then compared to serologic and to T-cell clone reactions. The steps followed to arrive at the conclusions given in Figure 1 are detailed in this report. Pseudogenes are enclosed within the shaded areas of the figure and new alleles identified by this analysis are listed in lower case letters to separate them from the WHO recognized antigens.

The 67 core cell lines were analyzed beginning with serologic and cellular classifications (DW and DP antigens). Data from RFLP and T cells which corresponded with serology were segregated and the remaining reactions were subjected to further study. The data was first examined for allelism of each locus using locus-specific probes.

Participating Laboratory: US5TER[1]

Cross-hybridization bands, which had not been considered in the initial analysis, was postulated for any weak bands (4 or 6 in score), which appeared to fit as alleles of another locus. These were identified as "XH" in Table 2, and appear outside of the box or column in Table 3.

When very close kb bands, especially those in high kb band patterns, did not fit any of the alleles, the questionable bands were "compressed" into one band to determine if the new band fit. If so, then the new "compressed" band was used in the analysis.

In Tables 1A–H, the raw data with the initial cluster analysis are given for each locus or loci that show similar specificities with the RFLP bands and T-cell clones. The cell lines are listed vertically. The identification numbers given in the top row are for the probe-enzyme combinations or T-cell clones listed in Table 2. T-cell clones are designated with a "T" to separate them from the RFLP bands. A strong band is denoted as an 8, weaker bands as 6 or 4, with 1 being negative. A clear allelism of each locus can be seen. The groups of clusters represented in Tables 1A–H are: 1a and 1b for DRB1

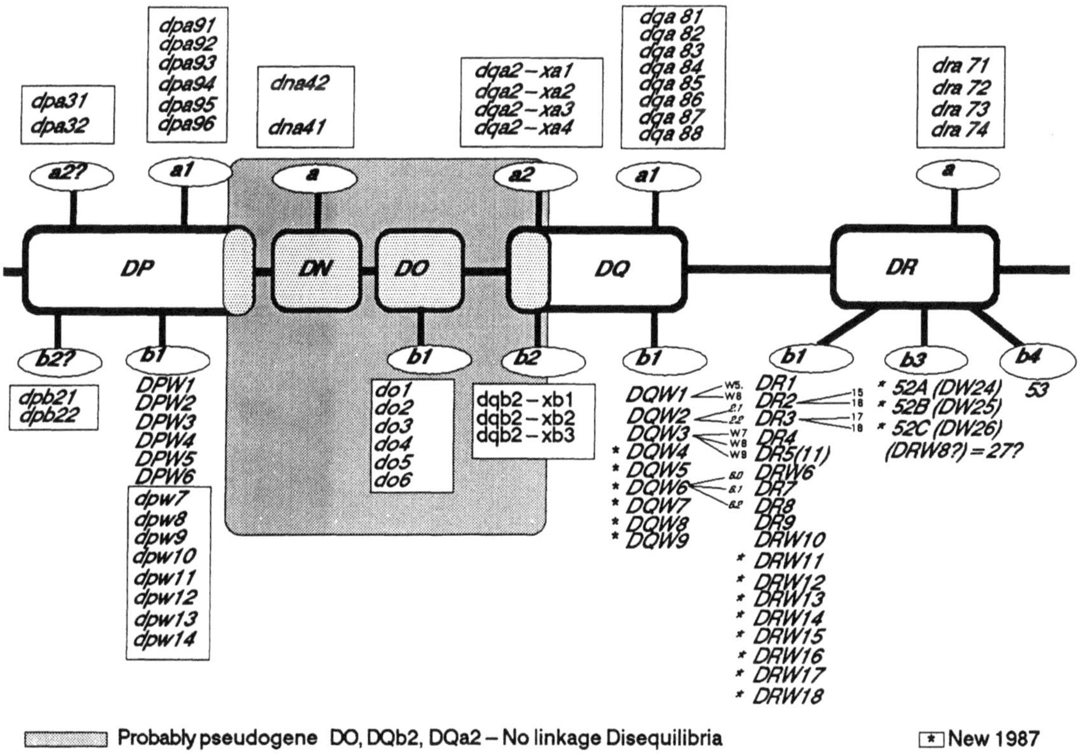

Figure 1. Simplified Class II polymorphism.

Table 1A. Allelism of DRB1 Locus

			TTTTTTTTTTT	T	T			T
			AAAAAAAAAAAAAAMLAAAAAAAAAL	MAAC	MCCCCCCCCM	CC	BBBBB	BLLMBMM
			52011111101113400033221001	3345	4111111120	11	11123	2002203
	DRB1	DRB1	22801234596792516343518544	4071	0123456907	87	68724	0987332
9002	DR 1		81886668866886888881848881	6688	8666816666	88	11111	1114111
9003	DR 1		16886648866686888886884888	8688	8668666666	88	11111	1111111
9004	DR 1		16886668866686888888888888	8188	8668866668	88	11111	1111111
9005	DR 1		16886668866886888806888888	8688	8648866068	88	11111	1111111
9006	DR 1		18886668866886888888880888	8688	8668866668	88	11111	1111111
9008	DR15		11111111111111111111111111	1111	8668666666	81	86816	6461666
9017	DR15		11111111111111111111111111	6111	8668866668	88	86886	8146866
9082	DR15		11111111111111111111111111	1111	8668661068	81	86886	6068866
9013	DR15		11111111111111111111111111	1111	8668861868	88	86886	8666866
9014	DR15		11111111111111111141111111	1111	8668868668	18	86886	8418886
9010	DR15		11111111111111111111111111	1111	8668866668	88	86886	8818866
9011	DR15		11111111111111181111111111	1111	8668866866	18	86881	6668846
9016	DR16		11111111111111111118111111	6611	8668866866	11	66881	8868866
9009	DR16		11111111111111111118111111	6114	8668664866	01	66886	4166686
9007	DR16	DR 4	11111111111111111111111111	6681	8668661666	11	66886	1116846
9018	DR17		11111111111111111111111111	1111	1111111111	11	11111	1111116
9019	DR17		11111111111111111111811111	1111	1111111111	11	11111	1111116
9020	DR17		11111111111111111111111111	1181	1111111111	11	11111	1111116
9022	DR17		11111111111111111111111111	1111	1111111161	11	11111	1111116
9023	DR17		11111111111111111111111111	6111	1111111161	11	11111	1111116
9088	DR17		11111111111111111111111111	4111	1111111161	11	11111	1111114
9021	DR18		11111111111111111111111111	1114	1111111661	11	11111	1111164
9031	DR 4		11111111111111111111111111	1111	1111111111	11	11111	1111111
9032	DR 4		11111111111111111111111111	1111	1111111111	11	11111	1111111
9033	DR 4		11111111111111111111111111	1111	1111111111	11	11111	1111111
9034	DR 4		11111111111111111111111111	1111	1111111161	11	11111	1111111
9029	DR 4		11111111111111111111111111	1111	1111111111	11	11111	1111111
9026	DR 4		11111111111111111111111111	1111	1111111111	11	11111	1111111
9025	DR 4		11111111111161111111111111	1111	1111111111	11	11111	1111111
9030	DR 4		11111111111111111111111111	1111	1111111111	11	11111	1111111
9091	DR 4		11111111111111111111111111	1111	1111111111	11	11111	1111111
9107	DR 4		11111111111111111111111110	1110	1111111111	11	11111	1111111
9092	DR 4		11111111111111111111111111	1111	1111111161	11	11111	1111111
9028	DR 4		11111111111111111111111111	1111	1111111811	11	11111	1111111
9038	DR12		11111111111111111111811111	1111	1111111611	11	11111	1111111
9045	DR12	DR11	11111111111111111111111111	1141	1111111611	81	11111	1111166
9037	DR11		11111111111111111111111111	1111	1611111166	81	11116	1111116
9039	DR11		11111111111111111111111111	1111	1111111111	11	11111	1111116
9040	DR11		11111111111111111111111111	1111	1111111111	11	11111	1111116
9042	DR11		11111111111111111111111111	1111	1111111411	11	11111	1111116
9043	DR11		11111111111111111111111111	1111	1111111111	11	11111	1111116
9036	DR11		11111111111111111111111111	1111	1111111116	11	11116	1116116
9105	DR11		11111111111111611111111110	1110	1111111116	11	11111	1111116
9064	DR11	DR14	11111111111111111118161111	1111	1111111861	11	11111	1111166
9058	DR13		11111111111111111111111111	6111	1111111168	11	11111	1114186
9060	DR13		11111111111111111111111111	1111	1111111114	11	11111	1116166
9062	DR13		11111111111116111111111111	1111	1111111166	11	11111	1116166
9065	DR13		11111111111116111111111111	1111	1111111166	11	11111	1116166
9055	DR13		11111111111111111111111111	1111	1111111116	11	11111	1116618
9059	DR13		11111111111111111111111111	1111	1111111118	11	11111	1116616
9063	DR13		11111111111111111111111111	1111	1111111116	11	11181	1116616
9054	DR14		11111111111111111111111111	1181	1111111116	11	11111	1111161
9056	DR14	DR13	11111111111111111111111111	1181	1111111116	11	11111	1111116
9057	DR14		11111111111111111111111111	1181	1111111116	11	11111	1111161
9061	DR14		11111111111111111111111111	1181	1111111118	11	11111	1111181
9050	DR 7		11111111111111111111111111	1111	1111111111	11	11111	1111116
9051	DR 7		11111111111111111111111111	1111	1111111111	11	11111	1111116
9047	DR 7		11111111111111111111111111	1111	1111111111	11	11111	1111116
9048	DR 7		11111111111111111111111111	1111	1111111111	11	11111	1111116
9068	DR 8		11111111111111111111111116	1118	1111111111	11	11111	1111111
9072	DR 8		11111111111111111111111111	1111	1111111111	11	11111	1111111
9067	DR 8		11111111111111111111111114	1118	1111111111	11	11111	1111111
9069	DR 8		11111111111111111111111111	1118	1111111111	11	11111	1111111
9071	DR 8		11111111111111111111111111	1111	1111111611	11	11111	1111111
9066	DR 8		11111111111111111184111111	1111	1111111116	11	11111	1111111
9070	DR 8		11111111111111111181111111	1111	1111111111	11	11111	1111111
9075	DR 9		11111111111111111111111111	1111	1111111161	11	11111	1111116

Table 1B. Allelism of DRB1 Locus

	DRB1	DRB1	TT TT T BBBBBBBBMLC 33222233410 01897653333	T ALB 540 811	TT AA 65 09	TT MFFFHFFMH 355526525 057960826	T BHIIM 54112 62253	T IHH 065 109	FF 11 78	J 1 5	M 1 3	TTT ELL 421 186	E 3 9	TTTT LLCL 2152 9526	DW	DW	HLA-B
9002	DR 1		11111111111	111	11	111111111	61111	111	11	1	1	111	1	1111	DW20		
9003	DR 1		11111111111	111	11	111111111	61111	111	11	1	1	110	1	1011	DW 1		
9004	DR 1		11111111111	111	11	111111111	61111	111	11	1	1	111	1	1111	DW 1		
9005	DR 1		11111111111	111	11	111111111	61111	111	11	1	1	110	1	1011	DW 1		
9006	DR 1		11111111111	111	11	111111111	61111	111	11	1	1	111	1	1411	DW 1		
9008	DR15		86688686648	111	11	111114111	11111	111	11	1	1	111	1	1111	DW 2		
9017	DR15		68688688668	116	11	111111111	11111	111	11	1	1	114	1	1411	DW 2		
9082	DR15		00688600608	111	11	111116111	11111	161	11	1	1	111	1	1111	DW 2		
9013	DR15		86688604648	111	11	111111111	11111	111	11	1	1	111	1	1101	DW 2		
9014	DR15		86688886668	111	11	111111111	11111	111	11	1	1	111	1	1111	DW 2		
9010	DR15		11688686648	111	11	111111111	11111	111	11	1	1	111	1	1111	DW 2		
9011	DR15		48688681111	111	11	111111111	11111	111	11	8	1	111	1	1111	DW12		
9016	DR16		11118111111	886	86	111111111	11111	111	11	1	8	111	1	1111	DW22		
9009	DR16		11111111111	846	84	111111111	61111	111	11	1	1	810	1	1011	DW21		
9007	DR16	DR 4	11111611111	886	11	111111111	61111	111	11	1	1	111	1	1111	DW14		
9018	DR17		11111111111	181	11	688818888	11111	111	11	1	1	110	1	1111	DW 3		B18
9019	DR17		11111111111	111	11	688868888	41111	111	11	1	1	114	1	1111	DW 3		B18
9020	DR17		11111161111	101	11	688816888	11111	111	11	1	1	110	1	1101	DW 3		B18
9022	DR17		11111111111	181	11	688848888	11111	111	11	1	1	110	1	1111	DW 3		B 8
9023	DR17		11111111111	116	11	680668888	11111	111	11	1	1	111	1	1111	DW 3		B 8
9088	DR17		11111111111	114	11	488068888	11111	111	11	1	1	111	1	1101	DW 3		B 8
9021	DR18		11111111111	161	11	611111111	11111	111	11	1	8	116	1	1111	DW N		B42
9031	DR 4		11111111111	111	11	111111111	11111	111	11	1	1	114	1	1111	DW 4		
9032	DR 4		11111111111	111	11	111111111	11111	111	11	1	1	110	1	1101	DW 4		
9033	DR 4		11111111111	111	11	111111111	11111	111	11	1	1	111	1	1111	DW 4		
9034	DR 4		11111111111	111	11	111111111	11111	111	11	1	1	111	1	1111	DW 4		
9029	DR 4		11111141111	111	11	111111111	11111	111	11	1	1	110	1	1101	DW 4		
9026	DR 4		11111111111	111	11	111111111	11111	111	11	1	1	111	1	1111	DW10		
9025	DR 4		11111161111	111	11	111111111	11111	111	11	1	1	111	1	1111	DW13		
9030	DR 4		11111181111	181	11	111111111	11111	111	11	1	1	164	1	1111	DW13		
9091	DR 4		11111111111	111	11	111111111	11111	111	11	1	1	111	1	1101	DW13		
9107	DR 4		11111111111	101	11	111111111	11111	111	11	1	1	111	1	1101	DW15		
9092	DR 4		11111101111	111	11	111111111	11111	111	11	1	1	110	1	1001	DW14		
9028	DR 4		41111161111	111	11	111111111	11111	111	11	1	1	111	1	1111	DW14		
9038	DR12		11111111111	101	11	111141111	11111	118	88	1	8	111	1	1118	DB 6		
9045	DR12	DR11	11111111111	141	11	111411181	11111	118	68	6	8	111	1	1114	DW 5		
9037	DR11		11181611111	111	11	111111111	11111	111	11	8	8	111	1	1118	DW 5		
9039	DR11		11111111611	111	11	111111161	18111	111	11	8	8	111	1	1111	DW N		
9040	DR11		11111111111	161	11	111161111	16111	111	11	1	8	111	1	1011	DW 5		
9042	DR11		11111111611	141	11	111111181	16111	111	11	6	8	114	1	1411	DW N		
9043	DR11		11111111611	111	11	111161181	11111	111	11	8	8	111	1	1111	DW 5		
9036	DR11		11111111801	111	11	111161181	61111	111	11	1	1	111	1	6111	DB 2		
9105	DR11		11111111604	111	11	111111161	60111	111	11	1	1	111	1	1101	DB 2		
9064	DR11	DR14	11111181111	101	11	111111818	11111	111	11	1	1	118	1	1111	DW16		
9058	DR13		11111111108	116	11	111166118	60111	111	11	1	1	110	1	1111	DW18		
9060	DR13		11111111668	111	11	111111118	68118	111	11	1	1	114	1	1111	DW18		
9062	DR13		11111111118	111	11	111111118	66111	111	11	1	1	111	1	1111	DW18		
9065	DR13		11111111168	111	11	111118118	68111	111	11	1	1	111	1	1111	DW18		
9055	DR13		11111141808	111	11	111111118	68668	111	11	1	1	111	1	1111	DW19		
9059	DR13		11111111818	181	11	111111118	66668	111	11	1	1	111	1	1111	DW19		
9063	DR13		11111111808	111	11	111111118	68668	111	11	1	1	111	1	1111	DW19		
9054	DR14		11111111101	111	11	111111118	61111	868	11	1	1	111	1	1101	DW 9		
9056	DR14	DR13	11111111608	111	11	111116118	68618	868	11	1	1	111	1	1111	DW 9	DW19	
9057	DR14		11111111101	111	11	111111118	61111	868	11	1	1	111	1	1111	DW 9		
9061	DR14		11111111111	111	11	111111118	61111	868	11	1	1	111	1	1111	DW 9		
9050	DR 7		11111111111	111	11	111111111	11111	111	11	1	1	111	1	1111	DW17		
9051	DR 7		11111111111	111	11	111111111	11111	111	11	1	1	111	1	1101	DW17		
9047	DR 7		11111111111	111	11	111111111	11111	111	11	1	1	111	1	1101	DB 1		
9048	DR 7		11111111111	111	11	111111111	11111	111	11	8	1	111	1	1111	DB 1		
9068	DR 8		11111611111	111	11	111111111	11111	111	11	1	8	666	8	8181	DW81		
9072	DR 8		11111601111	111	11	111111111	11111	111	11	1	8	866	8	8011	DW82		
9067	DR 8		11111611111	111	11	111116111	11111	111	11	1	8	848	8	1888	DW81		
9069	DR 8		11111641111	111	11	111111111	11111	111	11	1	8	848	8	6668	DW81		
9071	DR 8		11111641111	181	11	111111111	14111	111	11	1	8	864	8	8011	DW82		
9066	DR 8		11111111111	111	11	111111111	61111	111	11	8	1	111	8	1111	DB 7		
9070	DR 8		11111611111	111	11	111116111	11111	111	11	1	8	111	8	1001	DW83		
9075	DR 9		11111111111	141	11	111111111	11111	111	11	1	1	111	1	1111	DW23		

Table 1C. Allelism of DRB3 and DRB4 Loci

								T	T		T							
		HHHHHHHMHHH	HH	HHHHHHHHH	H	G	H	H	GGGGG		DDDDDDDDEDDDDDDEEMEEEEEEEEEME	ME	EE	EE	I	EEM		
		00123110111	44	445555522	5	1	1	3	11110		44444555155555500100011122O1	21	22	22	4	333		
	DRB3	34590015243	67	890143587	2	4	9	5	01239	DRB4	36789012645678912045634510 19	88	47	56	7	015	DRB1	DRB
--						B												
9022	DW24	84866666888	86	886666866	8	8	1	4	86688		11111111111111111111111111111	11	11	11	6	816	DR17	
9023	DW24	84866646888	86	886666866	8	8	1	1	86688		11111111111111111111111111111	11	11	11	6	816	DR17	
9088	DW24	86866666888	86	888666866	8	8	1	1	86688		11111111111111111111111111111	11	14	11	6	811	DR17	
9021	DW24	86866666818	86	888666866	8	8	1	1	66668		11111111111111111111111111811	11	14	11	1	816	DR18	
9064	DW24	86866666888	86	888666866	8	8	1	8	84688		11111111111111111111111111111	11	14	11	1	616	DR11	DR1·
9058	DW24	84866666688	86	886666866	8	8	1	1	86608		11111111111111111111111111111	11	81	11	1	116	DR13	
9062	DW24	86866666618	86	886616866	8	8	1	1	66688		11111111111111111111111111111	11	14	11	1	116	DR13	
9065	DW24	86866666818	86	888661866	8	8	1	1	66688		11111111111111111111111111111	11	16	11	1	116	DR13	
9018	DW25	88866866888	86	886666866	8	1	8	8	11111		11111111111111111111111111111	11	11	11	6	886	DR17	
9019	DW25	88866866888	86	888666866	8	1	8	8	11111		11111111111111111111111111111	11	11	11	6	886	DR17	
9020	DW25	88886866886	86	888666866	8	1	8	8	11111		11111111411111111111111111111	11	11	11	6	886	DR17	
9038	DW25	86166666816	86	688666866	8	1	8	8	11111		11111111111111111111111111111	11	11	11	1	886	DR12	
9045	DW25	86466866688	86	886666866	8	1	8	8	11111		11111111111111111111111111111	11	61	11	1	886	DR12	DR11
9037	DW25	88866666886	81	888616166	8	1	1	8	11111		11111111111111111111111116111	11	11	11	1	081	DR11	
9039	DW25	88866866886	86	888666866	8	1	8	8	11111		11111111111111111111111111111	11	11	14	1	881	DR11	
9040	DW25	88866866888	86	888666866	8	1	8	1	11111		11411111111111111111111111111	11	11	11	1	886	DR11	
9042	DW25	88866866688	86	886666866	8	1	8	8	11111		11111111111111111111111111111	11	11	11	1	886	DR11	
9043	DW25	88866866888	86	886666866	8	1	8	8	11481		11111111111111111111111111111	11	11	11	1	886	DR11	
9036	DW25	88866866888	86	888666866	8	1	8	8	11111		11111111111111111111111111161	81	11	11	1	881	DR11	
9105	DW25	86166866886	86	488666816	8	1	6	1	11111		11111111111111111111111111111	11	11	11	1	681	DR11	
9060	DW25	86866866886	86	688166866	8	1	8	8	11111		11111111111111111111111111111	11	11	11	1	881	DR13	
9054	DW25	86166866888	86	888666866	8	1	8	8	11111		11111111111111111111111111111	11	11	11	1	886	DR14	
9056	DW25	88866666816	86	888666866	8	1	8	8	11111		11111111111111111111111111416	11	11	11	1	886	DR14	DR13
9057	DW25	88866866888	86	888661866	8	1	8	8	11111		11111111111111111111111111111	11	11	11	1	886	DR14	
9061	DW25	86886866888	86	888661866	8	1	8	8	11111		11111111111111111111111111111	11	11	11	1	886	DR14	
9055	DW26	88866866888	86	888666866	1	8	1	1	11111		11111111111111111111111111118	11	11	11	1	811	DR13	
9059	DW26	88886866816	86	888661866	1	8	1	1	11111		11411111111111111111111111116	11	11	11	1	811	DR13	
9063	DW26	88866866818	86	886666866	1	8	1	1	11111		11111111111111111111111111118	11	11	11	1	811	DR13	
9068	D52D	86866666888	11	111111141	1	1	1	1	11111		11111111111111111111111111111	11	11	11	1	111	DR 8	
9072	D52D	86846116881	11	111111111	1	1	1	1	11111		11111111111111111111111114111	11	11	11	1	111	DR 8	
9067	D52D	86866666888	11	111111111	1	1	1	1	11111		11111111111111111111111111111	11	11	11	1	111	DR 8	
9069	D52D	84866116888	11	111111111	1	1	1	1	11111		11111111111111111111111111181	11	11	11	1	111	DR 8	
9071	D52D	86866661881	11	111116161	1	1	1	1	11111		11111111111111111111111111111	11	11	11	1	111	DR 8	
9066	D52D	88866661888	11	111111111	1	1	1	1	11111		11111111111111111111111111111	61	14	11	1	111	DR 8	
9070	D52D	88866666881	11	111111111	1	1	1	1	11111		11111111111111111111111111111	11	14	16	1	111	DR 8	
9007		11166116111	11	111111161	1	1	1	1	11111	DR53	16688666666688884686686 11866	68	86	66	1	666	DR16	DR 4
9031		11111111111	11	111111111	1	1	1	1	11111	DR53	86688666666888888888668686868	88	88	66	6	666	DR 4	
9032		11111111111	81	111111111	1	1	1	1	11111	DR53	86688666666688888664686 86866	68	16	64	6	666	DR 4	
9033		11111111111	11	111111111	1	1	1	1	11111	DR53	16688666666868888666686 46866	68	86	68	6	666	DR 4	
9034		11111111111	11	111111111	1	1	1	1	11111	DR53	86688666668688888686686 86866	68	86	66	6	666	DR 4	
9029		11111111111	11	111111111	1	1	1	1	11111	DR53	06688666668688888664686 86866	68	86	66	6	666	DR 4	
9026		11111111111	11	111111111	1	1	1	1	11111	DR53	81688666666888866864686 84866	88	86	66	6	666	DR 4	
9025		11111111111	11	111111111	1	1	1	1	11111	DR53	86688666666888888888668686868	88	86	66	6	666	DR 4	
9030		11111111111	11	111111111	1	1	1	1	11111	DR53	04688666666688888888668686668	88	88	68	1	666	DR 4	
9091		11811111111	11	111111111	1	1	1	1	11111	DR53	06688666666688888868668681066	68	86	66	6	666	DR 4	
9107		11111111111	11	111111111	1	1	1	1	11111	DR53	06688666868688888866686 86068	68	86	68	6	666	DR 4	
9092		11111111111	11	111111111	1	1	1	1	11111	DR53	06688666868688888888668686066	68	86	66	6	666	DR 4	
9028		11111111111	16	111111111	1	1	1	1	11111	DR53	06688866868888888888688686868	88	88	66	6	666	DR 4	
9050		11166111111	11	111111841	1	1	1	1	11111	DR53	86688666868688888888686868 68	88	86	86	6	666	DR 7	
9051		11166111111	11	111111811	1	1	1	1	11111	DR53	86688666868888888886668684868	88	86	86	6	666	DR 7	
9047		11166111111	11	111111811	1	1	1	1	11111	DR53	86688666868888888886668886868	88	86	86	6	666	DR 7	
9048		11166111111	11	111111811	1	1	1	1	11111	DR53	86688666666888888868868686068	88	86	88	6	666	DR 7	
9075		11111111111	11	111111861	1	1	1	1	11111	DR53	86688666668688888686686 86868	88	86	86	6	666	DR 9	
9002		11111111111	11	111111111	1	1	1	8	11111		11111111111111111111111111111	11	11	11	1	111	DR 1	
9003		11111111111	11	111111111	1	1	1	8	11111		11111111111111111111111111111	11	11	61	1	111	DR 1	
9004		11111111111	11	111111116	1	1	1	8	11111		11111111111111111111111111111	11	11	11	1	111	DR 1	
9005		11111111111	11	111111111	1	1	1	4	11111		11111111111111111111111111111	11	11	11	1	111	DR 1	
9006		11111111111	11	111111111	1	1	1	8	11111		11111111111111111111111111111	11	11	11	1	111	DR 1	
9008		11166666111	81	111111166	1	1	1	1	11111		11111111111111111111111111111	11	16	66	1	111	DR15	
9017		11166666411	11	111111186	1	1	1	1	11111		11111111111111111111111111111	11	88	46	1	111	DR15	
9082		11166666811	11	111111166	1	1	1	1	11111		11111111111111111111111111111	11	88	66	1	111	DR15	
9013		11166666411	11	111111186	1	1	1	1	11111		11111111111111111111111111111	11	88	46	1	111	DR15	
9014		11166666418	11	111111186	1	1	1	1	11111		11111111111111111111111111111	11	88	18	1	111	DR15	
9010		11166666416	11	111111186	1	1	1	1	11111		11111111111111111111111111111	11	88	11	1	111	DR15	
9011		11166666111	11	111161166	1	1	1	1	11111		11111111111111111111111114111	81	88	86	1	111	DR15	
9016		11866666118	11	111161161	1	1	1	1	11111		11111111111111111111111111111	18	88	11	1	111	DR16	
9009		11166616111	11	111111164	1	1	1	1	11111		11111111111111111111111111111	88	86	16	1	111	DR16	

Table 1D. Allelism of DQB1 Locus

	DQB1	DQB1	AAAAALAD 33343040 78906318	D 0 6	A 4 2	C 0 1	TT LA 34 48	BBD 440 787	BB 34 86	A 4 4	C 0 2	CCCCC 00000 46875	IMIF 1412 3642	JJJJJI 111114 423102	IIIIIE 555552 359678	JJJ 000 657	DDDDD 34333 90786	DD 33 45	FFFFFFFF 10000011 44568903	T D 3 3	DRB1	DRB1	DW	DW
9002	DQ51		88684668	1	8	1	81	688	66	6	1	11111	1111	111111	111111	111	11111	11	11111111	1	DR 1		DW20	
9004	DQ51		86686668	1	8	1	88	668	66	6	1	11111	1111	111111	111111	111	11111	11	11111111	1	DR 1		DW 1	
9005	DQ51		88686668	1	8	8	88	688	66	6	1	11111	1111	111111	111111	111	11111	11	11111111	1	DR 1		DW 1	
9006	DQ51		88686668	1	8	8	88	688	66	6	1	11111	1111	111111	111111	111	11111	11	11111111	1	DR 1		DW 1	
9009	DQ51		86686688	1	8	8	11	888	66	1	1	11111	1111	111111	111111	111	11111	11	11111161	1	DR16		DW21	
9057	DQ51		86686668	1	8	0	48	688	66	6	1	11111	1111	111111	111111	111	11111	11	11111111	1	DR14		DW 9	
9056	DQ51	DQ61	86681668	1	8	8	48	688	66	6	8	11111	6618	111111	111111	111	11111	11	11111111	1	DR14	DR13	DW 9	DW19
9007	DQ51	DQ8	86681668	1	8	8	11	688	66	1	1	11111	1111	111111	111116	111	48686	68	61111111	8	DR16	DR 4	DW14	
9003	DQ52		88686688	8	1	1	88	688	86	6	1	11111	1111	111111	111111	111	11111	11	11111111	1	DR 1		DW 1	
9036	DQ52		88686668	8	1	1	61	688	86	1	1	11111	1111	111111	111111	111	11111	11	11111111	1	DR11		DB 2	
9054	DQ52		88686668	8	1	1	68	688	66	6	1	11111	1111	111111	111111	111	11111	11	11111111	1	DR14		DW 9	
9061	DQ52		88686688	8	1	1	18	888	66	6	1	11111	1111	111111	111111	111	11111	11	11111111	1	DR14		DW 9	
9008	DQ60		11111111	8	1	8	41	688	66	1	8	88886	1111	111111	114118	111	11111	11	11111111	1	DR15		DW 2	
9017	DQ60		11111111	8	1	8	11	688	66	1	8	88886	1111	111111	111118	111	11111	11	11111111	1	DR15		DW 2	
9082	DQ60		11111111	8	1	8	11	688	66	1	8	88886	1111	111111	111118	111	11111	11	11111111	1	DR15		DW 2	
9013	DQ60		11111111	8	1	8	11	688	66	1	8	88886	1111	111111	111118	111	11111	11	11111111	1	DR15		DW 2	
9014	DQ60		11111111	8	1	8	01	688	66	1	8	88886	1111	111111	111118	111	11111	11	11111111	1	DR15		DW 2	
9010	DQ60		11111111	8	1	8	61	688	66	1	8	88886	1111	111111	111118	111	11111	11	11111111	1	DR15		DW 2	
9105	DQ60		11111111	8	1	0	11	668	66	1	8	88886	1111	111111	111111	111	11111	11	11111111	1	DR11		DB 2	
9058	DQ60		11111111	8	1	8	11	688	66	1	8	88886	1111	111111	111111	111	11111	11	11111111	1	DR13		DW18	
9060	DQ60		11111111	8	1	8	41	668	66	1	8	88886	1111	111111	111111	111	11111	11	11111111	1	DR13		DW18	
9062	DQ60		11111111	8	1	8	11	688	66	1	8	88886	1111	111111	111111	111	11111	11	11111111	1	DR13		DW18	
9065	DQ60		11111111	8	1	8	11	688	66	1	8	88886	1111	111111	111111	111	11111	11	11111111	1	DR13		DW18	
9055	DQ61		11111611	8	1	8	41	688	66	1	8	11111	6668	111111	111111	111	11111	11	11111111	4	DR13		DW19	
9059	DQ61		11111611	8	1	8	61	688	66	1	8	11111	6668	111111	111111	111	11111	11	11111111	1	DR13		DW19	
9063	DQ61		11111611	8	1	8	11	668	66	1	8	11111	6668	111111	111111	111	11111	11	11111111	1	DR13		DW19	
9011	DQ62		11111111	8	1	1	11	114	66	6	1	11111	1118	686868	111681	116	11111	11	11111111	1	DR15		DW12	
9066	DQ62		11111114	8	1	1	11	111	66	6	1	11111	1118	886868	111181	111	11111	11	11111111	1	DR 8		DB 7	
9022	DQ21		11111111	1	1	1	11	116	11	1	1	11111	1118	111111	668688	111	11111	11	11111118	1	DR17		DW 3	
9023	DQ21		11111111	1	1	1	11	116	11	1	1	11111	1118	111111	668888	111	11111	11	11111118	1	DR17		DW 3	
9088	DQ21		11111111	1	1	1	11	116	11	1	1	11111	1118	111111	868888	111	11111	11	61111118	1	DR17		DW 3	
9048	DQ21		11111161	1	1	1	41	111	11	1	1	11111	1118	111118	868888	111	11111	11	11111111	1	DR 7		DB 1	
9050	DQ22		11111161	1	1	1	41	111	11	1	1	11111	1118	111118	668688	868	11111	11	11111111	1	DR 7		DW17	
9051	DQ22		11111161	1	1	1	41	111	11	1	1	11111	1118	111118	668688	868	11111	11	11111111	1	DR 7		DW17	
9047	DQ22		11111161	1	1	1	11	111	11	1	1	11111	1118	111118	668688	861	11111	11	11111111	1	DR 7		DB 1	
9018	DQ22		11111111	1	1	1	11	118	11	1	1	11111	1118	111111	668888	668	11111	11	11111118	1	DR17		DW 3	
9019	DQ22		11111111	1	1	1	11	116	11	1	1	11111	1118	111111	868888	868	11111	11	11111118	1	DR17		DW 3	
9020	DQ22		11111111	1	1	1	11	116	11	1	1	11111	1118	111111	668888	866	11111	11	81111118	1	DR17		DW 3	
9021	DQ4	DQ22	11111118	8	1	1	11	118	11	1	1	11111	1111	111111	111118	868	68486	68	11111111	1	DR18		DW N	
9107	DQ4		11111118	8	1	1	11	118	11	6	1	11111	1111	111111	111118	111	68686	68	11111111	0	DR 4		DW15	
9067	DQ4		11111118	8	1	1	61	118	11	1	1	11111	1111	111111	116118	111	68686	68	11111111	1	DR 8		DW81	
9069	DQ4		11111118	8	1	1	11	118	11	1	1	11111	1111	111111	114118	111	61686	68	11111181	1	DR 8		DW81	
9071	DQ4		11111118	8	1	1	41	118	11	1	1	11111	1111	111111	111118	111	68686	68	11111181	0	DR 8		DW82	
9068	DQ4		11111118	8	1	1	41	118	11	1	1	11111	1111	111111	116118	111	68686	88	11111111	1	DR 8		DW81	
9072	DQ4		11111118	8	1	1	01	118	11	1	1	11111	1111	111111	114116	111	48686	48	61111111	0	DR 8		DW82	
9070	DQ7		11111111	1	1	1	11	111	11	1	1	11111	1118	111111	111118	111	11111	68	68886688	8	DR 8		DW83	
9016	DQ7		11111111	1	1	1	11	111	11	1	1	11111	1118	111111	111118	111	11111	48	68866668	8	DR16		DW22	
9025	DQ7		11111111	1	1	1	11	111	11	1	1	11111	1118	111111	111118	111	11111	88	68866668	1	DR 4		DW13	
9030	DQ7		11111111	1	1	1	11	111	11	1	1	11111	1118	111111	111118	111	11111	88	68866688	1	DR 4		DW13	
9091	DQ7		11111111	1	1	1	11	111	11	1	1	11111	1118	111111	111118	111	11111	68	68866668	1	DR 4		DW13	
9038	DQ7		11111111	1	1	1	61	111	11	1	1	11111	1118	111111	111118	111	11111	68	68866648	4	DR12		DB 6	
9045	DQ7		11111111	1	1	1	41	111	16	1	1	11111	1116	111111	111116	111	11111	68	18861188	6	DR12	DR11	DW 5	
9037	DQ7		11111111	1	1	1	81	161	11	1	1	18186	1118	111111	111118	111	11111	00	68816668	8	DR11		DW 5	
9039	DQ7		11111111	1	1	1	41	111	11	1	1	11111	1118	111111	111118	111	11111	48	68886668	4	DR11		DW N	
9040	DQ7		11111111	1	1	1	11	111	11	1	1	11111	1118	111111	111118	111	11111	68	68866688	8	DR11		DW 5	
9042	DQ7		11111111	1	1	1	01	111	11	1	1	11111	1118	111111	111118	111	11111	68	68866688	8	DR11		DW N	
9043	DQ7		11111111	1	1	1	61	111	11	1	1	11111	1118	111111	111118	111	11111	68	68886688	8	DR11		DW 5	
9064	DQ7		11111111	1	1	1	88	111	11	1	1	11111	1118	111111	111118	111	11111	68	18866668	8	DR11	DR14	DW16	
9031	DQ8		11111118	8	1	1	11	118	11	1	1	11111	1111	111111	111118	111	68686	88	11111111	8	DR 4		DW 4	
9032	DQ8		11111118	8	1	1	11	118	11	1	1	11111	1111	111111	111116	111	88686	88	61111111	8	DR 4		DW 4	
9033	DQ8		11111116	8	1	1	11	118	11	1	1	11111	1111	111111	111118	111	68686	68	11111111	8	DR 4		DW 4	
9034	DQ8		11111118	8	1	1	11	118	11	1	1	11111	1111	111111	111118	111	68486	88	11111111	8	DR 4		DW 4	
9029	DQ8		11111118	8	1	1	11	118	11	1	1	11111	1111	111111	111118	111	68686	88	11111111	8	DR 4		DW 4	
9026	DQ8		11111118	8	1	1	61	118	11	1	1	11111	1114	111111	111118	111	48486	88	11111111	8	DR 4		DW10	
9092	DQ8		11111118	8	1	1	18	118	11	1	1	11111	1111	111111	111118	111	68686	88	61111111	8	DR 4		DW14	
9028	DQ8		11111118	8	1	1	16	118	11	1	1	11111	1111	111111	111118	111	68686	88	11111111	8	DR 4		DW14	
9075	DQ9		11111118	8	1	1	11	118	11	1	1	11111	1111	111111	116118	111	68686	88	11111111	8	DR 9		DW23	

Table 1E. Allelism of DQA1, DQB2, and DQA2 Loci

	DQA1		EEEEEEE	HH	EBMEEEMEL	T EEEE	AAA	BBBBBBB	B	TTT BHHHHH	FFF
	|-------	--|	5444452	33	152012002	4444	444	4434444	5	533343	221
	|DQA	DQA|	0967819	34	046922873	3420	635	9493021	7	968907	019
9018	A82		8866866	68	111111111	1111	111	1111111	1	111111	111
9021	A82		8868668	11	111111116	1111	111	1111111	1	111111	111
9045	A82		8866866	68	111111111	1111	111	1111111	1	111111	111
9022	A82		8866866	68	111111111	1111	111	1111111	1	111111	111
9023	A82		8866866	68	111111111	1111	111	1111111	1	111111	111
9088	A82		8868668	68	111111111	1111	111	1111111	1	111111	111
9040	A82		8866861	68	111111111	1111	111	1111111	1	111111	111
9019	A82		8868668	68	111111111	1111	141	1111111	1	111111	111
9020	A82		8866668	66	111116111	1111	111	1111111	1	111111	111
9025	A83		1111118	11	888688866	1111	111	1111111	1	111111	111
9030	A83		1111111	11	888686666	1111	111	1111111	1	111111	111
9091	A83		1111118	11	886686666	1111	111	1111111	1	111111	111
9107	A83		1111116	11	886688866	6681	111	1111111	1	111111	111
9092	A83		1111118	11	686688860	1111	111	1111111	1	111111	111
9028	A83		1111118	11	888688866	6681	111	1111111	1	111111	111
9048	A83		1111118	11	886686861	1111	111	1111111	1	111111	111
9031	A83		1111118	11	888688866	1111	111	1111111	1	111111	111
9032	A83		1111118	11	686686860	1111	111	1111111	1	111111	111
9033	A83		1111116	11	686688866	1111	111	1111111	1	111111	111
9034	A83		1111116	11	686688866	1111	111	1111111	1	111111	111
9029	A83		1111118	11	686686860	1111	111	1111111	1	111111	111
9026	A83		1111116	11	686686860	1111	111	1111111	1	111111	111
9007	A83	A86	1111116	11	886686860	6681	161	6666666	1	611111	111
9050	A83		1111116	11	688688866	1111	111	1111111	1	111111	111
9051	A83		1111116	11	888688866	1111	111	1111111	1	111111	111
9047	A83		1111116	11	888688866	1111	111	1111111	1	111111	111
9075	A83		1111118	16	688688866	1111	111	1111111	1	111111	111
9064	A84		8868666	66	111111111	6681	111	1111111	1	111111	111
9068	A84		8866866	11	111111111	6688	111	1111111	1	111111	111
9072	A84		8866468	11	111111111	6664	111	1111111	1	111111	111
9067	A84		8866868	11	111111111	6686	111	1111111	1	111111	111
9069	A84		8868666	11	111111111	6688	111	1111111	1	111111	111
9071	A84		8868666	11	111111111	6686	111	1111111	1	111111	111
9070	A84		8866668	11	111111111	6684	111	1111111	1	111111	111
9004	A85		1111111	61	181111111	1111	866	6666688	6	111141	111
9005	A85		1111111	61	181111111	1111	866	6666686	6	111111	111
9006	A85		1111111	61	181111111	1111	866	6666686	6	111111	111
9061	A85		1111111	61	181111111	1111	868	6666688	6	111111	111
9056	A85		1111111	66	181111111	1111	666	6646686	6	611111	111
9057	A85		1111111	66	181111111	1111	616	6666668	6	111111	111
9054	A85		1111111	66	181111111	1111	616	6666688	6	111111	111
9002	A85		1111111	66	181111111	1111	668	6666688	6	111111	111
9003	A85		1111111	66	181111111	1111	666	6666686	6	111111	111
9082	A86		1111111	11	181111111	1111	111	6666888	1	811111	111
9013	A86		1111111	11	181111111	1111	111	6666686	1	811111	111
9014	A86		1111111	11	181111111	1111	111	6666888	1	811111	111
9009	A86		1111111	11	181111111	1118	111	6666686	1	611111	111
9036	A86		1111111	11	181111116	1111	061	6666888	1	611111	111
9008	A86		1111111	11	181111111	1111	111	6606686	1	811811	111
9011	A86		1111111	11	181116111	1111	111	6666666	6	611111	111
9059	A86		1111111	41	181111111	1111	111	6666886	1	611111	111
9063	A86		1111111	61	181111111	1111	111	6666686	1	811111	111
9010	A86		1111111	11	181111111	1111	111	6666686	1	811111	111
9017	A86		1111111	11	181111111	1111	111	6666686	1	611141	111
9055	A86		1111111	61	181111111	1111	111	6666688	1	811111	111
9105	A86		1111111	11	181111111	1111	111	1666688	6	616000	111
9066	A86		1111111	61	181111111	1114	141	6666888	6	611111	111
9062	A87		1111111	11	181111111	1111	111	1666688	6	666086	111
9058	A87		1111111	11	181116111	1111	111	1666688	6	866480	111
9060	A87		1111111	11	181111111	1111	111	1666688	6	666888	111
9065	A87		1111111	11	181111111	1111	111	1666688	6	866888	111
9016	A88		1168611	66	111111111	6688	111	1111111	1	111111	868
9037	A88		0168611	16	111111111	6661	111	6111111	1	111111	061
9042	A88		6866816	68	111111111	1111	111	1111111	1	111111	861
9043	A88		1166811	66	111118111	1111	111	1111111	1	111111	868
9039	A88		1866611	66	111111111	1111	111	1111111	1	111111	861
9038	A88		1866611	66	111111111	1111	111	1111111	1	111111	861

	DQB2		FMFFFF	K	K	G	DQA2		F	K	GGM	JB
	|-------	--|	333333	0	0	2	|-------	--|	4	1	232	00
	|DXB	DXB|	081234	1	4	8	|DXA	DXA|	8	0	915	88
9018	XB1		866668	1	1	1	XA1		8	8	111	11
9021	XB1		866668	1	1	1	XA1		8	8	111	11
9025	XB1		866668	1	1	1	XA1		8	8	111	11
9030	XB1		866666	1	1	1	XA1		8	8	118	11
9091	XB1		866866	1	1	1	XA1		8	8	111	11
9107	XB1		866668	1	1	1	XA1		8	8	111	11
9092	XB1		866668	1	1	1	XA1		8	8	111	11
9048	XB1		866866	1	1	1	XA1		8	8	111	11
9045	XB1		866668	1	1	1	XA1	XA3	6	6	111	61
9064	XB1		846666	1	1	1	XA1		8	8	111	11
9068	XB1		866668	1	1	1	XA1		8	8	111	11
9072	XB1		816666	1	1	1	XA1		8	8	111	11
9067	XB1		866868	1	1	1	XA1		8	8	111	11
9069	XB1		866668	1	1	1	XA1		8	8	111	11
9071	XB1		866666	1	1	1	XA1		8	8	111	11
9070	XB1		866666	1	1	1	XA1		8	8	111	11
9004	XB1		866618	1	6	1	XA1		8	8	111	11
9005	XB1		866648	1	6	1	XA1		8	8	111	11
9006	XB1		866618	1	6	1	XA1		8	8	111	11
9061	XB1		866668	1	1	1	XA1		8	8	111	11
9014	XB1		866868	1	1	1	XA1		8	8	111	11
9009	XB1		846668	1	1	1	XA1		8	8	111	11
9036	XB1		866886	1	1	1	XA1		8	8	111	11
9062	XB1		866668	1	1	1	XA1		8	8	111	11
9016	XB1		866466	1	1	1	XA1		8	8	111	11
9037	XB1		666666	1	1	1	XA1		8	8	111	11
9042	XB1		866668	1	1	1	XA1		8	8	111	11
9043	XB1		866668	1	1	1	XA1		8	8	111	11
9028	XB1	XB2	866668	6	4	1	XA1		8	8	111	11
9007	XB1	XB2	846666	6	6	1	XA1		8	8	111	11
9008	XB1	XB2	816660	6	1	1	XA1		8	8	111	11
9011	XB1	XB2	861166	8	1	1	XA1		8	8	111	11
9056	XB1		866666	1	1	1	XA1	XA3	8	8	111	61
9002	XB1		866666	1	1	1	XA2		8	1	688	11
9040	XB1		866668	1	1	1	XA2		8	1	688	11
9003	XB1		866666	1	1	1	XA3		1	8	111	81
9019	XB1		866668	1	1	1	XA3		1	8	111	81
9020	XB1		866666	1	1	1	XA3		1	8	111	81
9050	XB1		866668	1	1	1	XA3		1	8	111	81
9051	XB1		866668	1	1	1	XA3		1	8	111	81
9047	XB1		866668	1	1	1	XA3		1	8	111	81
9038	XB1		866666	1	1	1	XA3		1	8	111	81
9105	XB1		866666	1	1	1	XA3		1	8	111	81
9075	XB1		866868	1	1	1	XA4		1	8	111	18
9066	XB1		866866	1	1	1	XA4		1	8	111	18
9031	XB2		111111	8	6	1	XA1		8	8	111	11
9032	XB2		111111	8	6	1	XA1		8	8	111	11
9033	XB2		111111	6	6	1	XA1		8	8	111	11
9034	XB2		111111	8	6	1	XA1		8	8	111	11
9029	XB2		111111	8	6	1	XA1		8	8	111	11
9026	XB2		111111	8	6	1	XA1		8	8	111	11
9057	XB2		611111	8	6	1	XA1		8	8	111	11
9082	XB2		111111	6	6	1	XA1		8	8	111	11
9013	XB2		111111	6	6	1	XA1		8	8	111	11
9059	XB2		111116	8	6	1	XA1		8	8	111	11
9063	XB2		111111	6	6	1	XA1		8	8	111	11
9010	XB2		116116	6	6	1	XA1		8	8	111	11
9058	XB2		161111	6	6	1	XA1		8	8	111	11
9039	XB2		111111	8	6	1	XA1		8	8	111	11
9055	XB2		161111	6	6	1	XA2		8	1	688	11
9017	XB2	XB3	111111	6	6	6	XA2		8	6	666	11
9022	XB3		161111	1	6	8	XA2		8	1	668	11
9023	XB3		161111	1	6	8	XA2		8	1	668	11
9088	XB3		111111	1	6	8	XA2		8	1	688	11
9054	XB3		611111	1	6	8	XA2		8	1	688	11
9060	XB3		111111	1	6	8	XA2		8	1	688	11
9065	XB3		161111	1	6	8	XA2		8	1	688	11

Table 1F. Allelism of DRA and DOB Loci

	DRA	GG 11 98	A 2 3	E 3 8	EE 33 67	EEE 333 543	G 2 0	GGG 111 567	DRB1	DRB1	DW	DW
9004	A71	66	8	8	88	111	1	111	DR 1		DW 1	
9005	A71	66	8	8	88	111	1	111	DR 1		DW 1	
9006	A71	66	8	8	88	111	1	111	DR 1		DW 1	
9003	A71	66	8	8	86	111	1	111	DR 1		DW 1	
9056	A72	41	1	6	86	846	6	811	DR14	DR13	DW 9	DW19
9054	A72	11	1	1	86	866	8	111	DR14		DW 9	
9071	A72	61	1	1	86	866	8	181	DR 8		DW82	
9055	A72	11	1	1	88	866	8	111	DR13		DW19	
9061	A72	11	1	1	86	866	8	111	DR14		DW 9	
9017	A72	11	1	1	88	866	8	111	DR15		DW 2	
9082	A72	11	1	1	88	866	8	111	DR15		DW 2	
9014	A72	11	1	1	88	866	8	111	DR15		DW 2	
9105	A72	11	1	6	86	864	8	111	DR11		DB 2	
9011	A72	11	1	1	86	866	8	111	DR15		DW12	
9068	A72	11	1	1	86	866	8	111	DR 8		DW81	
9072	A72	11	1	1	81	846	8	116	DR 8		DW82	
9067	A72	11	1	1	86	866	8	111	DR 8		DW81	
9069	A72	11	1	1	81	866	8	116	DR 8		DW81	
9008	A72	11	1	1	86	846	8	111	DR15		DW 2	
9013	A72	11	1	1	88	866	8	111	DR15		DW 2	
9038	A72	11	1	1	86	866	8	111	DR12		DB 6	
9062	A73	66	1	1	11	111	8	866	DR13		DW18	
9065	A73	61	1	1	11	111	8	866	DR13		DW18	
9058	A73	66	1	1	11	111	8	866	DR13		DW18	
9063	A73	66	1	1	11	111	8	866	DR13		DW19	
9066	A73	66	1	1	11	111	8	866	DR 8		DB 7	
9022	A73	66	1	1	11	111	8	866	DR17		DW 3	
9023	A73	66	1	1	11	111	8	866	DR17		DW 3	
9088	A73	66	1	1	11	111	8	866	DR17		DW 3	
9021	A73	66	1	1	11	111	8	866	DR18		DW N	
9064	A73	66	1	1	11	111	8	866	DR11	DR14	DW16	
9059	A73	66	1	1	81	111	8	866	DR13		DW19	
9070	A73	66	1	1	81	111	8	866	DR 8		DW83	
9009	A74	11	1	8	86	861	1	111	DR16		DW21	
9007	A74	11	1	8	86	866	1	111	DR16	DR 4	DW14	
9036	A74	11	1	8	88	866	1	111	DR11		DB 2	
9019	A74	11	1	8	86	866	1	111	DR17		DW 3	
9045	A74	11	1	8	86	866	1	111	DR12	DR11	DW 5	
9042	A74	11	1	8	88	866	1	111	DR11		DW N	
9043	A74	11	1	8	88	866	1	111	DR11		DW 5	
9092	A74	11	1	8	86	866	1	111	DR 4		DW14	
9028	A74	11	1	8	88	886	1	111	DR 4		DW14	
9075	A74	11	1	8	86	866	1	111	DR 9		DW23	
9016	A74	11	1	8	86	866	1	111	DR16		DW22	
9030	A74	11	1	8	86	866	1	111	DR 4		DW13	
9091	A74	11	1	8	86	866	1	111	DR 4		DW13	
9037	A74	11	1	8	86	866	1	111	DR11		DW 5	
9039	A74	11	1	8	86	866	1	111	DR11		DW N	
9002	A74	41	1	8	86	860	1	111	DR 1		DW20	
9060	A74	11	1	8	86	866	1	111	DR13		DW18	
9048	A74	61	1	8	86	866	1	111	DR 7		DB 1	
9050	A74	11	1	8	88	866	1	111	DR 7		DW17	
9051	A74	11	1	8	88	866	1	111	DR 7		DW17	
9047	A74	66	1	8	88	811	1	111	DR 7		DB 1	
9020	A74	11	1	8	86	866	1	111	DR17		DW 3	
9107	A74	11	1	8	88	866	1	111	DR 4		DW15	
9025	A74	11	1	8	88	886	1	111	DR 4		DW13	
9032	A74	11	1	8	86	866	1	111	DR 4		DW 4	
9033	A74	11	1	8	88	861	1	111	DR 4		DW 4	
9034	A74	11	1	8	86	866	1	111	DR 4		DW 4	
9029	A74	11	1	8	86	866	1	111	DR 4		DW 4	
9057	A74	11	1	8	86	866	1	111	DR14		DW 9	
9010	A74	61	1	8	88	866	1	111	DR15		DW 2	
9018	A74	11	1	8	88	866	1	111	DR17		DW 3	
9040	A74	11	1	8	88	866	1	111	DR11		DW 5	
9031	A74	61	1	8	86	866	1	111	DR 4		DW 4	
9026	A74	11	1	8	86	866	1	111	DR 4		DW10	

	DOB	DOB	JJ 11 89	KKK 111 654	KK 11 32	K 1 7	F 4 3	F 4 2	F 3 9	FF 44 10	FF 44 45	F 4 6
9055	DO1		68	886	11	6	1	1	1	11	11	1
9058	DO2		11	886	88	6	1	1	1	11	11	1
9039	DO2		11	886	88	6	1	1	1	11	11	1
9019	DO2		11	886	88	6	1	1	1	11	11	1
9033	DO2		11	886	66	6	1	1	1	11	11	1
9091	DO2		11	886	88	6	1	1	1	11	11	1
9075	DO2		11	686	88	6	1	1	1	11	11	1
9008	DO2		11	486	88	6	1	1	1	11	11	1
9050	DO2		11	886	88	6	1	1	1	11	11	1
9026	DO2		11	886	88	6	1	1	1	11	11	1
9051	DO2		11	886	88	6	1	1	1	11	11	1
9082	DO2		11	886	88	6	1	1	1	11	11	1
9013	DO2		11	886	86	6	1	1	1	11	11	1
9029	DO2		11	886	88	6	1	1	1	11	11	1
9020	DO2		11	886	88	6	1	1	1	11	11	1
9057	DO2		11	886	86	6	1	1	1	11	11	1
9010	DO2		11	886	86	6	1	1	1	11	11	1
9105	DO2		11	886	66	6	1	1	1	11	11	1
9059	DO2		11	886	88	6	1	1	1	11	11	1
9063	DO2		11	886	86	6	1	1	1	11	11	1
9047	DO2		11	886	88	6	1	1	1	11	11	1
9011	DO2	DO3	11	866	86	6	4	6	6	66	11	1
9018	DO2	DO3	11	866	66	6	6	6	6	66	11	1
9017	DO2	DO4	11	866	66	6	1	1	6	86	66	1
9005	DO2	DO4	11	866	66	6	1	1	6	66	66	6
9037	DO2	DO6	11	116	88	6	1	1	1	68	18	6
9036	DO3		11	111	11	6	6	8	6	88	11	1
9040	DO3		11	111	11	6	6	6	8	88	11	1
9067	DO3		11	111	11	6	6	6	8	88	11	1
9014	DO3		11	111	11	6	6	8	8	88	11	1
9038	DO3		11	111	11	6	4	6	6	88	11	1
9061	DO3		11	111	11	6	6	6	8	88	11	1
9023	DO3		11	111	11	6	6	6	6	88	11	1
9088	DO3		11	111	11	6	6	6	6	88	11	1
9042	DO3		11	111	11	6	6	6	6	88	11	1
9003	DO3		11	111	11	6	4	6	8	88	11	1
9009	DO3		11	111	11	6	6	6	6	88	11	1
9045	DO3		11	111	11	6	6	6	6	88	11	1
9022	DO3		11	111	11	6	6	6	6	88	11	1
9025	DO3		11	111	11	6	6	6	8	88	11	1
9031	DO3		11	111	11	6	6	6	6	88	11	1
9034	DO3		11	111	11	6	6	6	6	88	11	1
9043	DO3		11	111	11	6	6	6	6	88	11	1
9069	DO4		11	111	11	6	1	1	8	88	68	1
9054	DO4		11	111	11	6	1	1	6	88	68	1
9068	DO4		11	111	11	6	1	1	8	88	66	1
9065	DO4		11	111	11	6	1	1	6	88	68	1
9066	DO4		11	111	11	6	1	1	6	88	68	1
9060	DO4		11	111	11	6	1	1	6	88	68	1
9028	DO4	DO5	11	111	11	6	1	6	8	88	68	1
9092	DO4	DO5	11	111	11	6	1	1	6	88	46	1
9007	DO4	DO6	11	111	11	6	1	6	8	88	68	6
9004	DO4	DO6	11	111	11	1	1	1	6	88	68	6
9048	DO4	DO6	11	111	11	1	1	1	6	88	68	8
9107	DO4	DO6	11	111	11	1	1	1	6	88	68	6
9006	DO4	DO6	11	811	11	1	1	1	6	88	66	6
9070	DO4	DO6	11	111	11	1	1	1	8	68	68	6
9072	DO4	DO6	11	111	11	1	1	1	6	88	68	6
9016	DO4	DO6	11	111	11	1	1	1	6	88	68	8
9032	DO5		11	111	11	6	1	6	8	88	18	1
9062	DO5		11	111	11	6	1	6	6	88	18	1
9064	DO5		11	111	11	6	1	6	6	88	18	1
9021	DO5		11	111	11	6	1	6	6	88	18	1
9030	DO5		11	111	11	6	1	8	6	88	18	1
9056	DO5	DO6	11	111	11	6	1	6	8	88	68	6
9071	DO5	DO6	11	111	11	6	1	6	6	88	48	6
9002	DO5	DO6	11	111	11	1	1	1	6	88	68	8

Table 1G. Allelism of DPB1 and DPB2 Loci

	DPB1 DPB	 DPB	T GG 22 45	G 2 1	III 000 324	T K 2 4	KK 22 32	CCCCCCC 2222222 4356879	C 3 0	C 4 5	M 3 7	HH 44 54	DPB2 DPB	 DPB	JJJJJ 44444 76589	KKK 000 789
9058	DP1	D--	88	6	888	1	11	1111111	1	1	1	11	DP21		88886	111
9006	DP1	D--	88	6	888	1	11	1111111	1	1	1	11	DP21		86886	111
9023	DP1	D--	88	6	888	1	11	1111111	1	1	1	11	DP21		86886	111
9070	DP1	DP4	68	6	668	1	11	6664666	6	6	1	11	DP21		66886	111
9088	DP1	DP4	80	1	688	1	11	6666666	6	6	1	11	DP21		66886	111
9021	DP1	dP4	81	6	688	1	11	6664666	6	6	1	11	DP21		68886	111
9020	DP2	D--	11	1	111	1	86	8886668	6	1	1	11	DP22		11111	666
9032	DP2	D--	11	1	111	4	86	8886668	6	1	1	11	DP22		11611	666
9029	DP2	D--	11	1	111	8	86	8886666	6	1	1	11	DP22		11111	666
9068	DP2	D--	11	1	111	8	86	8886668	6	1	1	11	DP22		11111	666
9036	DP2	D--	11	1	111	8	86	8886668	6	1	1	11	DP22		11111	666
9039	DP2	D--	11	1	111	8	86	8886668	6	1	1	11	DP22		11111	666
9105	DP2	d--	11	1	111	0	61	8886068	6	1	6	11	DP22		11111	666
9056	DP2	dP3	11	1	668	8	86	6664668	6	1	1	11	DP22		11811	666
9038	DP2	dP4	11	1	111	8	86	8886666	6	8	1	11	DP22		11111	666
9017	DP2	DP4	11	1	111	8	86	8886668	8	8	1	11	DP22		11111	666
9008	DP2	DP4	11	1	111	8	61	8886666	6	6	1	11	DP22		11111	666
9011	DP2	DP4	11	1	111	8	86	8886668	6	6	1	11	DP21	DP22	66881	666
9045	DP2	DP4	11	1	111	6	81	8886666	6	6	1	11	DP21	DP22	86681	666
9062	dP2	DP4	11	1	111	4	61	8886668	6	6	1	11	DP21	DP22	66614	666
9019	dP2	JET	11	1	111	1	86	8886668	6	1	1	11	DP22		11111	666
9050	DP2	JET	11	1	111	8	66	8886688	8	1	1	11	DP22		11111	666
9018	DP3	d--	11	1	888	1	11	1111111	1	1	1	11	DP22		11111	666
9022	DP3	d--	11	1	888	1	11	1111111	1	1	1	11	DP22		11111	666
9040	DP3	d--	11	1	888	1	11	1111111	1	1	1	11	DP21		88886	111
9030	DP3	d--	11	1	888	1	11	1111116	1	4	1	11	DP22		11111	666
9059	DP3	d--	11	1	888	1	11	1111111	1	1	1	11	DP22		11111	666
9007	DP3	DP4	11	1	668	1	11	6664668	6	6	1	11	DP21	DP22	68811	666
9071	DP3	dP4	11	1	688	1	11	8684668	6	6	1	11	DP21	DP22	66616	666
9009	DP3	DP4	11	1	668	1	11	6664168	6	6	1	11	DP21	DP22	64681	646
9067	DP4	D--	11	1	111	1	11	8886668	6	8	1	11	DP21		86886	111
9069	DP4	D--	11	1	111	1	11	8886168	6	8	1	11	DP21		66684	111
9026	DP4	D--	11	1	111	1	11	8886668	6	6	1	11	DP22		11111	666
9051	DP4	D--	11	1	111	1	11	8886688	8	8	1	11	DP21		88886	111
9033	DP4	D--	11	1	111	1	11	8886661	8	8	1	11	DP22		11181	666
9025	DP4	D--	11	1	111	1	11	8886668	6	8	1	11	DP21		68686	111
9031	DP4	D--	11	1	111	1	11	8886668	6	8	1	11	DP21		66886	111
9082	DP4	D--	11	1	111	1	11	8886668	6	6	1	11	DP21		66886	111
9004	DP4	D--	11	1	111	1	11	8886686	8	8	1	11	DP21		86881	111
9002	DP4	D--	11	1	111	1	11	8886668	6	8	1	11	DP21		66886	111
9014	DP4	D--	11	1	111	1	11	8886688	8	8	1	11	DP21		88886	111
9065	DP4	D--	11	1	111	1	11	8886668	6	8	1	11	DP21		66886	111
9061	DP4	D--	11	1	111	1	11	8886668	6	8	1	11	DP22		81111	666
9091	dp4	d--	11	1	111	1	11	8886668	6	6	1	11	DP21		66886	111
9092	dp4	d--	11	1	111	1	11	8886668	8	8	1	11	DP21		86886	111
9005	dP4	d--	11	1	111	1	11	8886688	8	8	1	11	DP21		86886	111
9057	DP4	d--	11	1	111	1	11	8886668	6	6	1	11	DP21		66886	111
9028	DP4	PET	11	1	111	1	11	8886668	8	8	1	11	DP21		68886	111
9075	DP4	PET	11	1	111	1	11	8886668	6	8	1	11	DP22		81111	666
9072	dP9	d--	11	1	111	1	11	8886666	6	6	1	11	DP21		66681	111
9054	dP9	d--	11	1	111	1	11	8886668	6	6	1	11	DP21		66886	111
9013	dP9	d--	11	1	111	1	11	8886061	8	8	1	11	DP21	DP22	86886	666
9010	dP9	d--	11	1	111	1	11	8886668	8	8	1	11	DP21		88886	111
9016	dP9	d--	11	1	111	1	11	8886668	6	6	1	11	DP21		66886	111
9037	dP9	d--	11	1	111	1	11	8886668	6	8	1	11	DP21		66886	111
9042	dP9	d--	11	1	111	1	11	8886681	8	8	1	11	DP21		88886	111
9064	dp9	d--	11	1	111	1	11	8886668	6	6	1	11	DP21		66886	111
9048	dP9	d--	11	1	111	1	11	8886668	6	8	6	11	DP22		11111	666
9107	DP5	d13	11	6	888	1	11	1111111	8	1	1	11	DP21		88886	111
9003	dP6	DHE	11	1	888	8	11	1111111	1	6	6	11	DP22		11111	666
9066	dP7	d--	11	6	111	1	86	8886668	6	1	1	11	DP22		11111	666
9063	dP8	d--	11	1	888	1	11	1111111	1	1	1	11	DP22		11111	666
9055	dP8	d13	11	1	888	1	11	1111111	8	1	6	11	DP21		86886	111
9060	d11	d--	11	4	888	1	11	1111161	1	1	1	88	DP22		11111	666
9047	d13	d--	11	1	888	1	11	1111111	8	1	1	11	DP22		11111	666
9034	d14	JET	11	1	888	1	11	1111111	1	1	6	11	DP22		11111	666
9043	d14	D--	11	1	888	1	11	1111111	1	1	6	11	DP22		11111	666

Table 1H. Allelism of DPA1, DNA, and DPA2 Loci

	DPA1 DPA	DPA1 DPA	JJJJ 3333 5786	JJJJJ 22223 56784	JJG 222 013	CC 44 98	CC 44 67	FFF 222 789	KK 22 78	KK 22 56	DNA DNA	DNA DNA	JJJJJ 33332 12309	CCCCCCCCCCCCC 4443343333333 1233509846721	DPA2 DPA	DPA2 DPA	K 2 1	C 5 0
9058	A91		8868	68668	111	11	11	111	11	14	N42		68688	1111111111111	A31		8	1
9006	A91		8868	68666	118	11	11	111	11	11	N42		68688	1111111111116	A31		8	1
9023	A91		8868	68666	118	11	11	111	11	11	N42		68688	1111111111116	A31		8	1
9105	A91		8868	06001	111	11	11	111	11	11	N41		66668	1100111111110	A31		8	1
9048	A91		8868	68666	111	11	11	111	11	11	N42		68688	1111111111111	A31		8	1
9003	A91		8868	68666	111	11	11	111	11	11	N42		68688	1111111111111	A31		8	1
9034	A91		8868	68666	111	11	11	111	11	11	N42		66688	1111111111111	A31		8	1
9043	A91		8868	68666	111	11	11	111	11	14	N42		68688	1111111111116	A31		8	1
9070	A91	A93	8868	68666	111	88	60	111	11	11	N41	N42	64106	6416646666686	A31	A32	8	8
9088	A91	A93	8868	68606	111	68	66	111	11	11	N41	N42	68666	6600868664686	A31	A32	8	8
9009	A91	A93	8848	68606	111	68	88	111	11	11	N41		11111	6606608664486	A31	A32	8	8
9056	A91	A95	8868	68660	111	11	11	111	66	86	N41	N42	11166	6466866666686	A31		8	1
																	*	
9107	A92		1111	68668	886	11	11	111	11	11	N42		68688	1111111111111	A31		8	1
9066	A92		1111	68666	886	41	11	111	11	41	N42		68688	1111111111111	A31		8	1
9055	A92		1161	68668	881	11	11	111	11	41	N42		68688	1111111111111	A31		8	1
9060	A92		1811	68666	886	11	11	111	11	11	N42		68688	1111111111111	A31		8	1
9021	A92	A94	1111	68661	886	11	11	888	66	11	N41	N42	66666	8466866666686	A31		8	1
																	*	
9018	A93		1111	11111	111	88	88	111	11	11	N41		11111	8666868666686	A32		1	8
9022	A93		1111	11111	111	88	88	111	11	11	N41		11111	8666868666686	A32		1	8
9040	A93		1111	11111	111	88	88	111	11	11	N41		11111	8666868666686	A31		8	1
9030	A93		1111	11111	111	88	88	111	11	11	N41		11111	8606868666686	A32		1	8
9059	A93		1111	11111	111	88	88	111	11	11	N41		11111	6666868660686	A32		1	8
9007	A93		1111	11111	111	88	88	111	11	11	N41		11111	6666868666686	A32		1	8
9071	A93		1111	11111	111	88	88	111	11	11	N41		11111	8666868666686	A31	A32	8	8
9067	A93		1111	11111	111	88	88	111	11	16	N41		11111	6666868666686	A32		1	8
9069	A93		1111	11111	111	88	88	111	11	11	N41		11111	8600868666681	A32		1	8
9026	A93		1111	11111	111	88	88	111	11	11	N41		11111	6666868666686	A32		1	8
9051	A93		1111	11110	111	88	88	111	11	11	N41		11111	8666868666686	A32		1	8
9033	A93		1111	11111	111	68	88	111	11	11	N41		11111	8666868666686	A32		1	8
9025	A93		1111	11111	111	88	88	111	11	11	N41		11111	8866886666686	A32		1	8
9031	A93		1111	11111	111	88	88	111	11	11	N41		11111	8666868666686	A32		1	8
9082	A93		1111	11111	111	88	88	111	11	11	N41		11111	6666868666686	A32		1	8
9004	A93		1111	11111	111	88	88	111	11	11	N41		11111	8666868666686	A32		1	8
9002	A93		1111	11111	111	88	88	111	11	11	N41		11111	8666868666686	A32		1	8
9014	A93		1111	11111	111	88	88	111	11	11	N41		11111	8660868666686	A32		1	8
9065	A93		1111	11111	111	88	88	111	11	11	N41		11111	8666868666686	A32		1	8
9061	A93		1111	11111	111	88	88	111	11	11	N41		11111	6466868666686	A32		1	8
9091	A93		1111	11116	111	88	88	111	11	11	N41		11111	8666868666686	A31		8	1
9092	A93		1111	11111	111	88	88	111	11	11	N41		11111	6666868666686	A31		8	1
9005	A93		1111	11111	111	88	88	111	11	11	N41		11111	8666868666686	A32		1	8
9028	A93		1111	11111	111	88	88	111	11	11	N41		11111	8866888666686	A32		1	8
9075	A93		1111	11111	111	88	88	111	11	11	N41		11111	6466868666686	A32		1	8
9072	A93		1111	11111	111	68	88	111	11	11	N41		11111	6606868666686	A31		8	1
9054	A93		1111	11111	111	88	88	111	11	11	N41		11111	8166868666686	A31		8	1
9013	A93		1111	11111	111	88	88	111	11	11	N41		11111	8600868666600	A31		8	1
9042	A93		1111	11111	111	88	88	111	11	11	N41		11111	8666868666686	A31		8	1
9017	A93	A95	1111	11111	111	68	66	111	66	86	N41		11111	8616868666686	A31	A32	8	8
9008	A93	A95	1111	11111	111	68	88	111	11	86	N41		11111	6466848666686	A31	A32	8	8
9011	A93	A95	1111	11111	111	68	66	111	66	86	N41		11111	6666868666686	A31	A32	8	8
9045	A93	A95	1111	11111	111	68	66	111	66	86	N41		11111	8400868666686	A31		8	1
9062	A93	A95	1111	11111	111	68	66	111	66	86	N41		11111	8666868666686	A31	A32	8	8
9016	A93	A96	1111	11116	111	88	11	111	88	11	N41		11111	6666868666686	A31		8	1
9064	A93	A96	1111	11111	111	88	11	111	88	11	N41		11111	8466868666686	A31		8	1
9047	A93	A96	1111	11110	111	88	11	111	88	11	N41		11111	8666868666686	A32		1	8
																	*	
9010	A94		1111	11111	111	11	11	888	88	11	N41		11111	8666868666686	A31		8	1
9037	A94	A96	1111	11111	111	88	10	886	80	11	N41		11111	8666868666686	A31		8	1
																	*	
9020	A95		1111	11111	111	11	11	111	88	88	N41		11111	6666868666686	A31		8	1
9032	A95		1111	11111	111	11	11	111	88	88	N41		11111	6666868666686	A31		8	1
9029	A95		1111	11111	111	11	11	111	88	88	N41		11111	6666868666686	A31		8	1
9068	A95		1111	11111	111	11	11	111	88	88	N41		11111	6666868666686	A31		8	1
9036	A95		1111	11111	111	11	11	111	88	88	N41		11111	8666868666686	A31		8	1
9039	A95		1111	11111	111	11	11	111	88	88	N41		11111	6666866666686	A31		8	1
9038	A95		1111	11111	111	11	11	111	88	88	N41		11111	6666868666686	A31		8	1
9019	A95		1111	11111	111	11	11	111	88	88	N41		11111	8666868666686	A31		8	1
9050	A95		1111	11110	111	11	11	111	88	88	N41		11111	8666868666686	A31		8	1
9057	A95		1111	11111	111	41	11	111	88	86	N41		11111	8666868666686	A32		1	8
9063	A95		1111	11111	111	11	11	111	88	88	N41		11111	8616868666686	A31		8	1

Table 2. Specificities of RFLP Band and T-Cell Clones

Data	Probe	Enzyme/kb	Specificity	False-Positive	False-Negative
J-18	DOB	BAM- 61	DO1	-	-
J-19	DOB	MSP- 24	DO1	-	-
K-14	DOB	BGL- 83	DO1,2	-	-
K-15	DOB	ECV- 37	DO1,2	-	-
K-16	DOB	ECI-128	DO1,2	-	-
K-17	DOB	MSP- 46	DO1,2,3,5,4	-	06,70,04,02
K-12	DOB	BAM- 73	DO2	-	-
K-13	DOB	MSP- 60	DO2	-	-
F-43	DOB	ECI-115	DO3	-	-
F-39	DOB	BAM- 51	DO3,4,5,6	-	-
F-40	DOB	ECV- 26	DO3,4,5,6	-	-
F-41	DOB	MSP- 14	DO3,4,5,6	-	-
F-42	DOB	BGL- 74	DO3,5	07	92
F-45	DOB	ECI-121	DO4,5,6	-	-
F-44	DOB	BGL- 77	DO4,6	-	37
F-46	DOB	MSP- 34	DO6	05	-
K-21	DPA	BGL- 72	DPA2/31	-	-
C-50	DPA	BGL- 53	DPA2/32	-	-
J-35	DPA	BGL- 34	DPA91	-	-
J-37	DPA	TAQ- 47	DPA91	60	-
J-38	DPA	MSP- 49	DPA91	-	55
J-36	DZA	BGL- 34	DPA91 XH	-	-
J-25	DPA	SST-188	DPA91,92/DNA42	-	-
J-26	DPA	TAQ- 64	DPA91,92/DNA42	-	-
J-27	DPA	PST- 26 WK	DPA91,92/DNA42	-	-
J-29	DPA	BAM- 69	DPA91,92/DNA42	-	09
J-31	DPA	HID- 49	DPA91,92/DNA42	-	09,56
J-28	DZA	SST-193 WK	DPA91,92/DNA42 XH	-	-
J-30	DZA	BAM- 71	DPA91,92/DNA42 XH	-	09
J-32	DZA	HID- 49	DPA91,92/DNA42 XH	-	09
J-33	DZA	ECI- 25 WK	DPA91,92/DNA42 XH	-	09
J-34	DZA	TAQ- 74	DPA91,92/DNA42 XH	91	21,105
G-23	DPA	MSP- 63	DPA92	06,23	55
J-20	DPA	BGL- 36	DPA92	-	-
J-21	DZA	BGL- 35	DPA92 XH	-	
C-46	DPA	BAM- 87	DPA93	-	16,47,64
C-47	DZA	BAM- 89	DPA93 XH	-	16,47,64
C-31	DPA	SST-161	DPA93,94,95/DNA41	-	06,23,43
C-32	DPA	SST- 44	DPA93,94,95/DNA41	-	-
C-34	DPA	HID- 32 WK	DPA93,94,95/DNA41	-	-
C-36	DPA	PST- 24 WK	DPA93,94,95/DNA41	-	-
C-37	DPA	MSP- 47 WK	DPA93,94,95/DNA41	-	-
C-38	DPA	MSP- 42 WK	DPA93,94,95/DNA41	-	-
C-39	DPA	MSP- 31	DPA93,94,95/DNA41	-	-
CM	DPA	BGL- 22	DPA93,94,95/DNA41	60	-
C-33	DZA	SST- 44 WK	DPA93,94,95/DNA41 XH	-	-
C-35	DZA	HID- 33	DPA93,94,95/DNA41 XH	-	-
C-40	DZA	MSP- 31 WK	DPA93,94,95/DNA41 XH	-	-
C-41	DZA	ECI- 42	DPA93,94,95/DNA41 XH	-	-
C-42	DZA	MSP- 46 WK	DPA93,94,95/DNA41 XH	-	-
C-43	DZA	SST-162 WK	DPA93.94,95/DNA41 XH	-	-
C-48	DPA	TAQ-132	DPA93,96	37	-
C-49	DZA	TAQ-134	DPA93,96 XH	37	-
F-27	DPA	TAQ- 77	DPA94	-	-
F-28	DPA	TAQ- 40	DPA94	-	-
F-29	DZA	TAQ- 83	DPA94 XH	-	-
K-25	DPA	TAQ-121	DPA95	-	-
K-26	DZA	TAQ-125	DPA95	67	-
K-27	DPA	BAM- 90	DPA95,94,96	-	08
K-28	DZA	BAM- 93	DPA95,94,96	-	08
G-24	DPB	MSP- 50	DPB1/DP 1	-	-
G-25	T6065		DPB1/DP 1	-	21
G-21	DPB	MSP- 67 WK	DPB1/DP 1,5,7,11	60	-
C-23	DPB	PST- 239	DPB1/DP 2,4,9,7	-	-
C-24	DPB	PST- 19	DPB1/DP 2,4,9,7	-	-
C-25	DPB	ECV- 69	DPB1/DP 2.4.9.7	-	-

Data	Probe	Enzyme/kb	Specificity	False-Positive	False-Negative
C-26	DPB	ECV- 34	DPB1/DP 2,4,9,7	-	-
C-29	DPB	HIC- 38	DPB1/DP 2,4,9,7	30	33,13',42
C-28	DPA	SST-104 WK	DPB1/DP 2,4,9,7 XH	-	-
C-27	DPB	BGL-150	DPB1/DP 2,4,9,7,11	-	-
C-30	DPB	MSP- 19	DPB1/DP 2,4,9,7,13	-	-
K-24	T6061		DPB1/DP 2,6	-	19,20
K-23	DPB	MSP- 217	DPB1/DP 2,7	-	-
K-22	DQB	MSP- 213WK	DPB1/DP 2,7 XH,WEAK	-	45,62,08,10
I-02	DPB	PST- 29	DPB1/DP 3,6,1,5,14,8	-	-
I-03	DPB	ECV-101	DPB1/DP 3,6,1,5,14,8	-	-
I-04	DPB	TAQ- 17	DPB1/DP 3,6,1,5,14,8	-	-
C-35	DPB	MSP- 11	DPB1/DP 4,9	03	-
N-37	DPB	MSP- 72 WK	DPB1/DP 6+14 WK	55,48,105	-
KM	DPB	KPN-244 WK	DPB1/DP 9,2,5,7	55,92,91,26,16	9105,08,45
H-44	DPB	TAQ- 26	DPB1/DP11/NEW	-	-
H-45	DPB	MSP- 40	DPB1/DP11/NEW	-	-
J-45	DPB	HID- 54	DPB2/21	56	-
J-46	DPB	ECI- 75	DPB2/21	-	-
J-47	DPB	MSP- 1	DPB2/21	61,75	-
J-48	DPB	HID- 53	DPB2/21	33	07,62,71
J-49	DPB	HIC- 71 WK	DPB2/21 WEAK	-	MANY FN
K-07	DPB	HID- 51 WK	DPB2/22	-	-
K-08	DPB	ECI- 72	DPB2/22	-	-
K-09	DPB	BAM- 90	DPB2/22	-	-
I-53	DQB	PVU- 63	DQ2	-	-
I-55	DQB	BAM- 38	DQ2	-	-
	DQB	SST- 26	DQ2 NOT CENTRAL DATA	NO DATA	NO DATA
	DQB	SST- 55	DQ2 NOT CENTRAL DATA	NO DATA	NO DATA
I-57	DQB	PVU- 16	DQ2+DQ6.2	-	-
I-59	DQB	PST- 52	DQ2+DR8.1	-	-
I-56	DRB	PST- 52 WK	DQ2+DW12 XH	-	-
J-05	DQB	MSP- 23	DQ2.2	-	-
J-06	DQA	BGL- 84	DQ2.2 XH	-	-
J-07	DQA	TAQ- 29	DQ2.2 XH	-	-
D-36	DQB	BAM-102	DQ4,DQ8,DQ9	-	-
D-37	DQB	KPN-191	DQ4,DQ8,DQ9	-	-
D-38	DQB	TAQ- 19	DQ4,DQ8,DQ9	-	-
D-39	DQB	BGL- 56	DQ4,DQ8,DQ9	-	-
D-40	DQB	TAQ- 25	DQ4,DQ8,DQ9	-	-
D-08	DQB	PVU- 34	DQ4,DQ8,DQ9,DQ5	-	-
D-06	DQB	HID- 330	DQ4,DQ8,DQ9,DQ6,DQ5.2	-	-
A-37	DQB	PST- 14	DQ5	-	-
A-38	DQB	BAM- 58	DQ5	-	-
A-39	DQB	ECV- 27	DQ5	-	-
A-40	DQB	ECV- 26	DQ5	-	-
A-36	DRB	PST- 13 WK	DQ5 XH	-	-
B-38	DQB	HID- 64	DQ5+6	-	-
B-46	DQB	KPN- 51	DQ5+6	-	-
B-47	DQB	ECV- 67	DQ5+6.0+6.1	-	-
D-07	DQB	PVU- 36	DQ5+6.0+6.1,4,8,2=DR3	-	-
	DQB	SST- 25	DQ5,6 NOT CENTRAL DATA	NO DATA	NO DATA
B-48	DQB	HIC- 65	DQ5,6.0+6.1	-	-
L-03	DQB	BAM- 25	DQ5,6.1	-	-
L-34	T6051		DQ5,DR 5,12,WK 8	10,59,	09,07,61,
A-41	DQB	BAM- 30	DQ5,DR 7WK	-	-
A-48	T6052		DQ5,DW14	64	02,07,09,
A-42	DQB	HID- 327	DQ5.1	-	-
C-04	DQB	PST-107	DQ6.0	-	-
C-05	DQB	BAM- 29	DQ6.0	-	-
C-06	DQB	TAQ- 29	DQ6.0	37	-
C-07	DQB	TAQ- 26	DQ6.0	37	-
C-08	DQB	PVU- 22	DQ6.0	37	-
C-02	DQB	ECV- 52	DQ6.0+6.1	-	-
C-01	DQB	TAQ- 19	DQ6.0+6.1+5	-	02,04
M-46	DQB	PST- 64 WK	DQ6.1	-	-
I-14	DQB	KPN-118 WK	DQ6.1	-	56

Table 2. *Continued*

Data Probe	Enzyme/kb	Specificity	False-Positive	False-Negative
I-13 DQA	BAM-239 WK	DQ6.1 XH	-	-
J-10 DQB	BAM- 87	DQ6.2	-	-
J-11 DQB	TAQ- 45	DQ6.2	-	-
J-12 DQB	ECV-118	DQ6.2	-	-
J-13 DQB	HIC-154	DQ6.2	-	-
J-14 DQB	PVU- 20	DQ6.2	-	-
DQB	SST-138	DQ6.2 EDIT 533	NO DATA	NO DATA
I-42 DQB	TAQ- 67	DQ6.2+DR7	-	-
F-04 DQB	HID- 34	DQ7	-	-
F-05 DQB	PVU- 37	DQ7	-	-
F-06 DQB	BAM- 35 WK	DQ7	-	-
F-08 DQB	KPN- 53 WK	DQ7	-	-
F-09 DQB	KPN- 12 WK	DQ7	-	-
F-10 DQB	BGL- 41	DQ7	-	-
F-14 DQA	PVU- 284WK	DQ7 XH	72,32,92,20,07	45,64
DQB	SST- 245	DQ7,8,9 NOT CENTRAL DATA	NO DATA	NO DATA
DQB	SST- 30	DQ7,8,9 NOT CENTRAL DATA	NO DATA	NO DATA
D-35 DQB	PST- 49	DQ7,8,9,4	-	-
D-34 DRB	PST- 493	DQ7,8,9,4	-	-
E-28 DQB	MSP- 36	DQ7,8,9,4,2,DR15=DW2	-	-
F-22 DQB	PVU- 24	DQ7,DQ2,DQ6.1+6.2(NEG TO 6.0)	-	-
F-13 DQB	TAQ- 46	DQ7,DR17 XH	-	-
D-33 T6019		DQ8,DQ9,DQ7	-	DW13
F-48 DQA	PST- 76	DQA2/XA1,XA2	-	-
K-10 DQA	KPN- 81	DQA2/XA1,XA3,XA4	17	-
M-25 DQA	KPN- 84	DQA2/XA2	30	-
G-31 DQA	BGL- 72	DQA2/XA2	-	-
G-29 DQB	MSP- 92 WK	DQA2/XA2 XH	-	-
J-08 DQA	PST- 68	DQA2/XA3	-	-
B-08 DQA	PST- 91	DQA2/XA4	-	-
B-40 DQA	PVU- 49 WK	DQA81(85,86,87)	-	-
B-41 DQA	ECI-134	DQA81(85,86,87)	-	-
B-43 DQA	PST-134 WK	DQA81(85,86,87)	-	-
B-44 DQA	MSP- 29 WK	DQA81(85,86,87)	-	-
B-42 DPA	ECI-144	DQA81(85,86,87)XH	-	-
B-39 DPB	KPN- 54 WK	DQA81(85,86,87)XH	-	-
B-54 DQA	BGL- 17	DQA81+83	-	-
E-50 DQA	PST- 42	DQA82+84	-	-
E-51 DQA	MSP- 474	DQA82+84	-	-
E-46 DQA	HID- 58 WK	DQA82+84+88	-	-
E-47 DQA	ECI-110 WK	DQA82+84+88	-	-
E-49 DQA	BGL- 37	DQA82+84+88	-	16,37,43
E-48 DPA	ECI-110	DQA82+84+88 XH	-	-
H-33 DQA	SST-165	DQA82+85+88	64	55,66,59,
H-34 DQA	TAQ- 37 WK	DQA82+88+85	64	DQA85 SHORT
E-29 DQA	PVU- 279	DQA82,83,84	42	40,30
E-07 DQA	BAM- 76	DQA83(DQA53)	-	-
M-26 DQA	KPN-117	DQA83(DQA53)	-	-
E-09 DQA	HID- 48	DQA83(DQA53)	-	-
E-10 DQA	ECI- 63	DQA83(DQA53)	-	-
E-12 DQA	PST- 23	DQA83(DQA53)	-	-
E-22 DQA	TAQ- 42	DQA83(DQA53)	-	-
L-23 DQA	SST-248	DQA83(DQA53)	21	48
M-08 DPA	ECI- 63	DQA83(DQA53)XH	-	-
E-42 DQA	BGL-131	DQA84	-	-
E-43 DQA	MSP- 325 ??	DQA84	-	-
E-44 DQA	MSP- 23 ??	DQA84	-	-
E-40 T6058		DQA84/DR 8	-	17,16,66
A-45 DQA	KPN-104	DQA85	-	-
A-46 DQA	TAQ- 25	DQA85	-	-
A-43 DPB	BGL- 69 WK	DQA85 XH	36	09,54,57
B-49 DQA	HID- 74 WK	DQA85+86	37	105
B-57 DQA	BAM-100 WK	DQA85+87		
B-59 DQA	KPN-142	DQA86+87	07,56	-
H-36 DQA	SST-155	DQA87(DQADW18)	-	-
H-38 DQA	HID- 77	DQA87(DQADW18)	-	-

Data Probe	Enzyme/kb	Specificity	False-Positive	False-Negative
H-40 T6028		DQA87(DQADW18)	-	-
H-37 T6038		DQA87(DQADW18)	-	-
H-39 T6075		DQA87(DQADW18)	-	-
F-20 DQA	PST- 43	DQA88	-	-
F-21 DQA	MSP- 473	DQA88	-	-
F-19 DQA	BGL- 38	DQA88 SHORT(DR5X2)	-	DQA88 SHORT
F-30 DQB	PST- 42	DQB2/XB1	54,57	-
F-31 DQB	MSP- 26 WK	DQB2/XB1	10	11
M-38 DQB	MSP- 82 WK	DQB2/XB1	-	08,72
F-32 DQA	PVU- 72 WK	DQB2/XB1 XH	-	11
F-33 DQA	ECI- 19 WK	DQB2/XB1 XH	-	04,06
F-34 DQA	TAQ- 15 WK	DQB2/XB1 XH	-	-
K-01 DQB	PST-112	DQB2/XB2	-	-
K-04 DQA	ECI- 13	DQB2/XB2,XB3	-	-
G-28 DQB	PST- 30	DQB2/XB3	-	-
A-08 DRB	HID- 61	DR 1	-	-
A-09 DRB	HID- 34 WK	DR 1	-	-
A-10 DRB	HID- 30	DR 1	-	-
A-14 DRB	TAQ- 58	DR 1	-	-
A-15 DRB	ECV- 63	DR 1	-	-
A-16 DRB	PVU- 66	DR 1	-	-
A-17 DRB	ECI-111	DR 1	-	-
A-19 DRB	BGL- 69	DR 1	-	-
L-14 T6045		DR 1	67,68	02
A-03 T6046		DR 1	-	-
A-04 T6047		DR 1	-	-
A-33 T6050		DR 1	64,66	02,07
A-05 T6053		DR 1	-	-
A-06 T6056		DR 1	-	-
A-21 T6059		DR 1	64	-
A-25 T6067		DR 1	19,38	-
A-18 T6068		DR 1	-	-
A-34 T6069		DR 1	66	-
L-01 T6070		DR 1	11	-
DRB	SST-112	DR 1 NOT CENTRAL	NO DATA	
DRB	SST-148	DR 1 NOT CENTRAL	NO DATA	
A-11 DPB	HID- 343WK	DR 1 XH	-	-
A-13 DPB	KPN-186 WK	DR 1 XH	-	-
A-12 DQB	KPN-177 WK	DR 1 XH	-	-
A-32 DQB	KPN-120 WK	DR 1 XH	62,65	-
M-45 DQB	PST- 63	DR 1 XH	105	-
A-44 DQB	PST- 51	DR		
		1,DQ6.2,DW15,DR12,DR14,DW18		
M-40 DRB	PST- 366	DR 1,DR 2	-	-
C-11 DRB	BAM- 48 WK	DR 1,DR 2	37	-
C-12 DRB	BGL-111 WK	DR 1,DR 2	-	-
C-13 DRB	PVU- 415	DR 1,DR 2	-	-
C-14 DRB	ECI- 51	DR 1,DR 2	-	-
C-15 DRB	ECI- 384WK	DR 1,DR 2	-	-
C-18 DRB	TAQ- 21	DR 1,DR 2	37,45	11,14
C-16 DPB	ECI- 39 WK	DR 1,DR 2 XH	-	07,13,82
C-19 T7009		DR 1,DR 2,DR12,DR18	28,64,71	-
C-20 DRB	BAM- 39 WK	DR 1,DR 2,DR52A	-	-
A-47 DRB	ECI-115	DR 1,DR14/DQA85	007,20	-
C-17 DRB	TAQ- 44	DR 1,DR15 ONLY	-	08,82
M-34 DRB	MSP- 49	DR 1,DR16(DW22 ONLY)	-	-
A-30 DRB	PST- 5 WK	DR 1,DR16(DW22 ONLY)	-	-
M-07 DRB	ECI- 383WK	DR 1,DR2,DR 6	-	-
C-51 T6057		DR 1,DR8.1 GOOD	-	-
A-22 DRB	PST- 14	DR 1.1(DW1 ONLY,NEG TO DW20)	-	-
A-52 DRB	PST- 373	DR 1.2 (ONLY 9002=DW20)	-	-
DRB	SST- 61	DR 2 NOT CENTRAL	NO DATA	
DRB	SST-129	DR 2 NOT CENTRAL	NO DATA	
DRB	SST- 99	DR 2 NOT CENTRAL	NO DATA	
B-20 DPB	PVU- 426	DR 2 XH	-	07
B-34 DQA	BAM-109 WK	DR 2 XH	36,37	11

Table 2. *Continued*

Data Probe	Enzyme/kb	Specificity	False-Positive	False-Negative	Data Probe	Enzyme/kb	Specificity	False-Positive	False-Negative
B-16 DRB	HID- 33	DR 2,	-	-	I-22 T7034		DR 7,DR 9	-	-
B-17 DRB	PVU- 45	DR 2,	-	-	I-21 DPB	PVU- 45	DR 7,DR 9 XH	-	-
B-18 DRB	ECI- 63 WK	DR 2,	-	-	I-31 DQB	ECV-200 WK	DR 7,DR 9 XH	-	-
B-22 DRB	PVU- 43	DR 2,	63	08	I-45 T7028		DR 7,DR8WK	03	70,66,68,72
L-08 DRB	BGL-145 WK	DR 2,	-	10,14	L-24 DRB	MSP- 15	DR 7-DB1,DR 4	-	32,34,26,30
L-09 T7013		DR 2,	-	09,17	L-29 T6012		DR 8 (DR8.1+8.2)	-	-
M-27 DRB	KPN-118 WK	DR 2,DR13	36	08,56.	E-41 T6055		DR 8 (DR8.1+8.2)	-	-
M-03 DRB	BGL- 31	DR 2,DR13(DW18),DR14	21	56	L-28 T7001		DR 8 (DR8.1+8.2)	-	-
B-23 DRB	ECI- 47	DR 2,DR13(DW19 ONLY)	-	-	L-16 T7002		DR 8 (DR8.1+8.2)	64	-
M-30 DQA	KPN-252 WK	DR 3 XH	-	-	DRB	SST-117	DR 8 NOT CENTRAL	NO DATA	
F-55 DQB	HID-114	DR 3 XH	-	-	E-39 DRB	TAQ- 87	DR 8,DR16	-	-
F-59 T6014		DR 3(DR17)	-	-	L-15 T7004		DR 8.1	CHECK	
F-57 T6073		DR 3(DR17)	-	-	C-52 T7007		DR 8.1	-	-
F-58 DQB	ECV-151	DR 3(DR17) XH	64	-	B-06 DRB	HID- 172	DR 9	-	-
F-60 DQA	TAQ- 69	DR 3(DR17)+ XH	56,65,58,	-	B-05 T7031		DR 9,DR7(DB1 ONLY)	68	-
H-26 DRB	MSP- 14 WK	DR 3(DR17)+WK	58,43,40,36	18,20	F-17 DRB	HID- 18	DR12	-	-
H-56 DRB	TAQ- 70	DR 3,DR 6	-	21	F-18 DRB	PVU- 59	DR12	-	-
DRB	SST-119	DR 3,DR 6 NOT CENTRAL	NO DATA		L-26 T6009		DR12+	67,69,37	-
M-22 DRB	HID- 255	DR 3=B18	45	-	H-59 DRB	PST- 273	DR12,DR14	-	-
D-13 DRB	PST- 41	DR 4	-	-	M-23 DRB	HID- 27	DR13 (DW19)	-	-
D-14 DRB	PST- 36 WK	DR 4	-	-	I-12 DRB	ECV-109 WK	DR13 (DW19)	-	-
D-15 DRB	PST- 205	DR 4	-	-	I-15 DRB	MSP- 16 WK	DR13 (DW19)	-	56
D-16 DRB	BAM- 53 WK	DR 4	-	-	C-03 DRB	ECI- 22	DR13+DR15 XH	-	-
D-19 DRB	TAQ- 52	DR 4	-	-	M-43 DRB	PST- 537	DR13+DR15/DQ6.0+6.1 XH	-	-
D-20 DRB	BGL-103 WK	DR 4	-	-	B-56 DRB	MSP- 41 WK	DR13,DQ5	66	-
D-21 DRB	ECV-229 WK	DR 4	-	-	H-42 T6013		DR13,DR 5S(DWNEW)	40	-
D-22 DRB	PVU- 57	DR 4	-	-	I-01 T6021		DR14 (DW9)	-	-
D-23 DRB	MSP- 37 WK	DR 4	-	-	H-60 DQB	KPN-134	DR14 (DW9) XH	-	-
D-24 DRB	HID- 158WK	DR 4	-	-	B-28 DRB	MSP- 38	DR15	-	-
D-29 DRB	ECI-151	DR 4	-	-	B-29 DRB	TAQ- 14	DR15	37	-
J-51 DRB	HIC- 41	DR 4 (DW13 ONLY)	31	-	B-30 T6011		DR15	-	-
L-19 T6023		DR 4 (DW4 ONLY)	-	-	B-31 T7012		DR15	-	10
L-20 T6081		DR 4 (DW4 ONLY)	-	-	B-33 T7019		DR15	-	11
D-18 DPB	BAM- 53 WK	DR 4 XH	-	-	B-27 DPB	HID- 337	DR15 XH	16	-
D-17 DQB	BAM- 53 WK	DR 4 XH	-	-	B-26 DQA	SST-194 WK	DR15,DR 8	07,37	66
D-44 DRB	HID- 84	DR 4,DR 7	-	-	B-35 T6082		DR15,DW16,DW13	20,28	91
D-45 T7026		DR 4,DR 7	-	-	L-13 T7016		DR15,DW2	65,60	11
D-41 DRB	ECI-182	DR 4,DR 7S,DR8	-	-	A-58 DRB	TAQ- 17	DR16	-	-
D-25 DPB	HID- 160	DR 4,DR 9 XH	-	07	B-01 DRB	MSP- 49 WK	DR16	17,58	-
D-27 DQA	MSP- 22	DR 4,DR 9 XH	-	-	A-59 T7006		DR16	-	07
D-26 DQB	ECV-145 WK	DR 4,DR 9 XH	-	-	A-60 T7017		DR16	-	07
D-28 DRB	TAQ- 27	DR 4,DR7(DB1 ONLY)	-	-	L-41 T6063		DR16+	59,40,71,18,22	-
J-15 DRB	ECI-165	DR 5	11,66,48	-	H-03 DRB	HID- 296	DR52	-	-
M-13 DQB	ECV-139	DR 5+DR 8+DR12	21,16	66,64,105,3	H-04 DRB	ECV-182	DR52	-	-
M-32 DRB	MSP- 24 WK	DR 5,DR13,DR 3,DR 7,DR 9,DR 2	-	-	H-13 DRB	BGL- 93	DR52	10,14,16	70,71,72
I-24 DRB	PST- 490	DR 7	-	-	H-14 DRB	PVU- 423	DR52	-	21,62,65
I-25 DRB	KPN- 94 WK	DR 7	-	-	H-15 DPB	PVU- 425	DR52 XH	16,91	38,54,105
I-34 DRB	PVU- 48	DR 7	-	-	H-10 DRB	BAM- 417	DR52,DR 2	-	07
L-43 T6016		DR 7	-	-	H-11 DPB	BGL- 94 WK	DR52,DR 2 XH	-	09,67
I-36 T7020		DR 7	-	-	M-05 DPB	ECI- 325WK	DR52,DR 2 XH	-	66,71
I-27 T7027		DR 7	-	-	H-12 DRB	ECI- 92	DR52,DR 2(DR15)	-	08,11
I-37 T7030		DR 7	-	-	H-30 DPB	BAM- 41 WK	DR52,DR 2,DR 7 XH	-	-
J-17 T7032		DR 7	70	-	H-29 DQB	BAM- 42 WK	DR52,DR 2,DR 7 XH	-	-
A-54 DQA	SST-235 WK	DR 7 (DB1 ONLY) XH	-	-	H-46 DRB	HID- 249	DR52-DR 8	32	-
I-38 DRB	TAQ- 72	DR 7 (NEG TO DB1)	-	DB1	H-47 DRB	BAM-113	DR52-DR 8	28	-
I-39 DRB	TAQ- 26	DR 7 (NEG TO DB1)	-	DB1	H-48 DRB	BGL- 92	DR52-DR 8	-	-
DRB	SST-170	DR 7 NOT CENTRAL	NO DATA		H-49 DRB	ECV- 75	DR52-DR 8	-	-
I-26 DPB	ECI-147 WK	DR 7 XH	-	-	H-50 DRB	ECI- 95	DR52-DR 8	-	-
I-35 DQA	MSP- 60 WK	DR 7 XH	-	-	H-51 DRB	ECV-153	DR52-DR 8	-	60
I-28 DQB	HID-128 WK	DR 7 XH	-	-	DRB	SST-107	DR52-DR 8 NOT CENTRAL	NO DATA	
I-29 DQB	KPN- 92	DR 7 XH	-	-	H-53 DPB	BGL- 89 WK	DR52-DR 8 XH	71	57,61,63
I-44 DRB	BAM- 405WK	DR 7+	69,72	-	H-54 DPB	ECI- 333WK	DR52-DR 8 XH	16	37,62
I-43 T7029		DR 7,DR 2(DW12 ONLY)	-	-	H-27 DRB	PST- 57 WK	DR52-DR 8+DR 2(DR15)	04	-
I-20 DRB	PST- 215	DR 7,DR 9	-	-	H-28 DRB	MSP- 35 WK	DR52-DR 8+DR 2,DR 9	71	105
I-30 DRB	ECV-204	DR 7,DR 9	-	-	H-55 DRB	TAQ- 41	DR52-DR 8+DR 7,DR 9	-	37
I-32 DRB	ECI-153 WK	DR 7,DR 9	-	-					

Table 2D. Specificities of RFLP Band and T-Cell Clones (*continued*)

Data	Probe	Enzyme/kb	Specificity	False Pos	False Neg
G-12	DPB	KPN- 77 WK	DR52A=DW24	-	-
G-09	DRB	PST- 11	DR52A=DW24	-	-
G-10	DRB	KPN- 77	DR52A=DW24	-	-
G-11	DRB	KPN- 27 WK	DR52A=DW24	-	-
G-13	T6072		DR52A=DW24	-	-
H-52	DRB	PVU- 98	DR52AB GOOD	60	-
G-14	DRB	HID- 285	DR52AC GOOD	-	-
H-19	DRB	TAQ-115	DR52B	-	-
H-35	T6060		DR52B,DR1	64	40,105
E-31	DRB	HID- 289	DR52B,DR53	-	-
D-43	DPB	TAQ- 40	DR53	-	
D-47	DRB	PST- 29 WK	DR53	-	-
D-48	DRB	HID- 72	DR53	-	-
D-49	DRB	HID- 31	DR53	-	-
D-50	DRB	HID- 14 WK	DR53	-	-
D-52	DRB	BAM- 66 WK	DR53	-	-
M-01	DRB	BAM- 46 WK	DR53	-	-
D-54	DRB	BAM- 43 WK	DR53	-	-
D-55	DRB	BGL- 56	DR53	-	-
D-56	DRB	ECV-165	DR53	-	-
D-57	DRB	PVU-135	DR53	-	-
D-58	DRB	PVU- 62	DR53	-	-
D-59	DRB	PVU- 52	DR53	-	-
E-01	DRB	PVU- 39	DR53	-	-
M-10	DRB	ECI-125	DR53	-	-
E-04	DRB	ECI- 41	DR53	-	-
E-05	DRB	ECI- 37 WK	DR53	-	-
E-06	DRB	MSP- 18 WK	DR53	-	-
E-13	DRB	PVU- 46	DR53	-	-
E-16	DRB	ECI-119	DR53	-	-
M-28	DRB	KPN-128	DR53	36	-
E-20	T7033		DR53	-	-
	DRB	SST-106	DR53 NOT CENTRAL	NO DATA	
	DRB	SST- 52	DR53 NOT CENTRAL	NO DATA	
D-46	DPB	ECI- 38 WK	DR53 XH	-	-
D-51	DPB	HID- 14 WK	DR53 XH	-	-
E-02	DPB	PVU- 52	DR53 XH	-	-
E-14	DPB	ECI- 40 WK	DR53 XH	-	-
E-15	DPB	PVU- 61	DR53 XH	-	-
E-21	DPB	BGL-116 WK	DR53 XH	-	31,91
E-24	DRB	HID- 249	DR53,DR 2	58	08,32
E-26	DRB	BGL- 39 WK	DR53,DR 2	-	10,14
E-27	DRB	MSP- 315	DR53,DR 2	65	-
E-25	DRB	BGL-117 WK	DR53,DR 2(DR15)	-	10,14
E-18	DRB	TAQ-138	DR53,DR 2(DR16)	-	-
I-47	DPB	BAM- 37 WK	DR53,DR 3(DR17) XH	-	07
E-19	DRB	KPN-105	DR53,DR13(DW19)	56	-
E-30	DRB	PST- 273	DR53,DR52-8	-	-
M-35	DRB	MSP- 52 WK	DR53,DR52-8	-	-
A-23	DRA	BGL- 39	DRA71	-	-
E-36	DRA	ECI- 32	DRA71,72,74	59,70	-
E-37	DRA	MSP- 31	DRA71,72,74	-	-
G-19	DRA	PVU-110 WK	DRA71,73	10,48	-
G-18	DRB	PVU-103 WK	DRA71,73 XH	-	-
E-38	DRA	TAQ- 194	DRA71,74	56,105	-
G-20	DRA	TAQ- 20	DRA72,73	-	-
E-34	DRA	PVU-119 WK	DRA72,74	-	-
E-35	DRA	BGL- 45	DRA72,74	-	-
E-33	DRB	PVU-114 WK	DRA72,74 XH	-	09,33,47
G-15	DRA	BGL- 42	DRA73	56	-
G-16	DRA	ECI- 31 WK	DRA73	-	-
G-17	DRA	MSP- 30 WK	DRA73	69,72	-

locus, 1c for DRB3 and DRB4, 1d for DQB1, 1e for DQA1,DQB2, and DQA2, 1f for DRA and DO, 1g for DPB1 and DPB2, 1h for DPA1, DPA2 and DNA.

In Table 1A, the first five lines are DR1. Fifteen different RFLP bands can be used to define DR1. There are, in addition, 11 T-cell clones that also clearly react with the DR1 cells. The DR 15 and 16 splits of DR2 are recognized by clusters of RFLP and T-cell clones. In Table 1B the DR17 split of DR3 is readily identified. Although DR11 is not well defined by RFLP or T-cell clones, DR12 is clearly seen with two probe-enzyme combinations, (i.e., F17 and F18). The DR14 split of DR6 is more readily defined than DR13, while one combination with E39 defines DR8.

On DRB3, DR52 is split into four parts, as shown in Table 1C. DW24 to DW26 are easily recognized by clusters of RFLP. Of interest is the apparent identity of "DR52D" with DR8. Whether the two antigens are closely linked or are identical will need to be established by examination of larger numbers of cells. As seen on the right side of Table 1C, DRB4 has only one specificity, DR53 as demonstrated by numerous RFLP bands.

Table 1D shows the separation of DQB1 into 11 alleles. DQW1 is split into five alleles, 5.1, 5.2 (given only in the raw data and not noted in the summary Fig. 1) and 6.0, 6.1, and 6.2. DQW2 is split into two alleles, 2.1 and 2.2. DQW3 is split into DQW7, 8 and 9, although the split of DQW9 is not very clear. DQW4 or DQWA is not distinctly defined by RFLP. Although there is no single band for DQW4, it may be defined by monoclonal antibodies.

DQA1 is shown to have seven alleles by RFLP (Table 1E). Four T-cell clones appear to react with two specificities.

In Table 1F, four alleles are found in the DRA locus by RFLP. Furthermore, six alleles of the DO locus are demonstrated by clusters of RFLP.

Table 1G shows DPB1 and DPB2. DPB1 has 12 alleles. Two large clusters, one with DP1,3,5, and 6 and the other with DP2 and 4, are defined. In addition, two T-cell clones (G25 and K24) are shown for DPW1 and DPW2, respectively. DPB2 contains two allelic series that appear to account for all the cells. Seven heterozygotes may be observed in the 67 lines.

In Table 1H, DPA1 has six clearly defined alleles by RFLP. DPA2, like DPB2 (Table 1G), has two alleles that account for all the cells.

Table 2 lists the probe-enzyme combinations that have been analyzed in detail in the previous tables. Information in this table includes the data identification, probe, enzyme, kb band, and specificities defined. The cell identifications of each false-positive and false-negative are listed instead of correlation values since the core lines were not randomly selected. The definition of false-positive/false-negative is based on the "best" phenotypes in this analysis as listed in Table 4.

Table 3 represents a simplified overview of informative probe-enzyme combinations, number of informative banding patterns and specificities of the different loci. Nonpolymorphic probe-enzymes, incomplete and missing data, and data obtained from outside the central data are indicated in Table 3. The asterisks denote unique bands for a given specificity, whereas the same letter in a row designates multispecificity in a given band. At the top of the table are specificities that can be detected by monoclonal antibodies, T-cell clones, or allogeneic antisera.

Summarizing the results given in the previous tables, Table 4 provides the allele assignments for all 67 cell lines of the Workshop listed first by the DRB1 locus assignments.

Table 5 represents some tentative linkage associations between the different loci. They are compared primarily with DRB1, DPB1 and DW antigens. Since the cell lines have not been randomly selected, these linkages may differ from those that may be revealed in future studies with different ethnic populations. A strong association between alpha and beta chain products, including pseudogenes, in a given locus is shown. The unexpected association between DXB2 and DR2 requires further study, as does the strong association between DN alpha and DP alpha. No association of DO alleles is observed. The linkage of DPw4 to two alleles of the DPB 2nd locus deserves further study in a random population. Furthermore, although it remains unclear, we postulate that DQW1 belongs to DQB instead of DQA.

Figures 2A–G presents the above associations between the loci. In the center of each circle are the new alleles. The next ring shows the old DR nomenclature, followed by the new DR splits and the DW specificities. Each pie-shaped segment shows the association of the new antigens to other loci. For example, in Figure 2C, DRA72 is associated with haplotypes of DR15-DW2 and DW12, DR8-DW8.1 and 8.2, DR14-DW9 and DR12-DB6. Another way to look at this figure is that DR14-DW9 is associated with DRA72, whereas DR14-DW16 is associated with DRA73.

In Figure 2D, alleles of DRB3 locus (MT2), DRB4 locus (MT3) and DR1 + DR2 locus (MT5), also referred to as "MT B locus," are given. These three loci, although mutually exclusive, can be viewed as a single locus system (see Terasaki PI (ed). Histocompatibility Testing 1980, Los Angeles: UCLA Tissue Typing Laboratory, 1980;573). This concept is based on the large number of DR1 + DR2 RFLP bands listed in Table 2B which are seen with eight different probe-enzymes as well as with serology. For practical purposes, these two loci and DR1 + DR2 may be typed as a single locus.

In summary, the RFLP data of the Workshop have been critical to the definition of new alleles on each of the class II loci. Many specificities which should qualify for official WHO designations have emerged. From the tables, the derivation of the summary Figure 1 should be evident, and it should now be possible to find antisera that react with any of the expressed gene products. It is the high linkage of alpha gene products with the beta 1 locus of DR, DQ, and DP specificities that has made earlier serologic work difficult.

Editor's Comments. This article represents an individual laboratory report. The Central Data Base for all Workshop components was subject to substantial updating, revisions and data verification during the period December 1987 to August 1988. The reader is referred to the two last papers in this volume: Simons et al., "Restriction Fragment Length Polymorphism of HLA Genes: Summary of the Tenth International Workshop Southern Blot Analysis" and Dupont et al., "A Display of Positive Correlations Obtained by Serology, Southern Blot Analysis, T-Lymphocyte Clones and Two-Dimensional Gel Analysis Tested on the Reference Panel of B-Lymphoblastoid Cell Lines for Factors of the HLA System," which is the definitive data base for the Workshop studies.

Table 3A. Simplified View of Polymorphic RFLP Bands with Different Probes and Enzymes, and Its Corresponding Alleles

	DRB1														DRB3/B4					DRA				DQB1										DQA1								DQB2			DQA2				DOB					
PROBE / ENZYME	DR1	DR15	DR16	DR17	DR18	DR4	DR5	DR12	DR13	DR14	DR7	DR8	DR9	DR10	DR52A	DR52B	DR52C	DR52D	DR53	DR71	DR72	DR73	DR74	DQ5	DQ6.0	DQ6.1	DQ6.2	DQ2.1	DQ2.2	DQ7	DQ8	DQ9	DQ4	DQ81	DQ82	DQ83	DQ84	DQ85	DQ86	DQ87	DQ88	DQXB1	DQXB2	DQXB3	DQXA1	DQXA2	DQXA3	DQXA4	DO1	DO2	DO3	DO4	DO5	DO6
MOAB	A	2	2	3	3		*	*	*		*		A	A	D	D	D		*			*	*	*	I	I	I	Q	Q	*	W	W	*	*	*	*		*	*	*		PROBABLE PSEUDOGENES												
T-CLONE	T	T	T	T		T		T	T	T	T	T	T		T	T	?		T	T	?			T										T			T	T	T	T														
SERA	S	S	S	S	S	S	S	S	S		S	S	S	S	?	S	S	S	S	S	?	S	S	S	I	I	I	Q	Q	S	S	S	S	S	S	S	S	S	S	S	S													
DRB.BAM	*	A	A			*					*				*	B	B		*																																			
DRB.BGL	*	A	A			*			B	B		*			D	D	D	*	*																																			
DRB.ECI	*	*	*			*	*		*	*	C	*	C		D	D	D	*	*																																			
DRB.ECV	*					*			*		C	*	C		D	D	D		*																																			
DRB.HIDIX	*	A	A	*		*		*	*		*	*	*		D	*	D	*	*																																			
DRB.KPN		A	A						A		*				*				*																																			
DRB.MSP	*	*	*			*	I		*						D	D	D		*						I	I	I																											
DRB.PST	*	*	*			*		*	*	*	*		*		*	B	B		*					*				Q	Q	W	W	W	W																					
DRB.PVU	*	A	A			*		*			*				D	D			*	E	B	E	B																															
drb.sst	*	A	A	C	C						*	*			D	D	D		*																																			
DRB.TAQ	*	*	*	C	C	*			C	C	*	*			D	*	D		*																																			
DRA.BGLIX																				*	A	*	A																															
DRA.ECI																				A	A	*	A																															
DRA.MSP																				A	A	*	A																															
DRA.PVU																				A	B	A	B																															
DRA.TAQ																				A	B	B	A																															
DQB.BAM		D	D			*					D				D	D	D							*	*	*	*	Q	Q	*	W	W	W																					
DQB.BGL																														*	W	W	W																					
DQB.ECV				*		*					*		*											*	*	*	*																											
DQB.HIC																								I	I	I	*																											
DQB.HID				C	C						*													*	I	I	I			*																								
DQB.KPN	*									*	*								*					I	I	*	I			*	W	W	W																					
DQB.MSP																													*	W	W	W	W									*				*								
DQB.PSTIX	*																							*	*	*		Q	Q	W	W	W	W									*	*	*										
DQB.PVU																									*	Q	*	Q	Q	*																								
dqb.sst																								I	I	I	*	Q	Q	W	W	W																						
DQB.TAQ				A							*													*	*	*	*			A	W	W	W																					
DQA.BAM		A	A																							*										*																		
DQA.BGL																													*					A	*	A	*				*					*								
DQA.ECI																																		*	C	*	C	I	I	I	C	*	A	A										
DQA.HID																																			C	*	C	D	D	*	C													
DQA.KPN				*																																*		*	A	A					C	*	C	C						
DQA.MSP						E					*		E																					*	*		*	I	I	I	*	*												
DQA.PSTIX																																		*	B	*	B	I	I	I	*				A	A	*	*						
DQA.PVU																														*				*	B	B	B	I	I	I		*												
DQA.SST		A										A																							B	*		B		*	B													
DQA.TAQ				*																									*						B	*		*			B	*												
DPB.BAM		D	D	C		*					D				A	D	D		C											*																								
DPB.BGL		A	A												D	D	D		*																			*																
DPB.ECI											*				D	D	D		*																																			
DPB.ECV																																																						
DPB.HIC																																																						
DPB.HID	*	*				E							E						*																																			
DPB.KPN	*														*																			*																				
DPB.MSPIX																																																						
DPB.PST																																																						
DPB.PVU		A	A								C		C		D	D	D		*																																			
DPB.TAQ																																																						
DPA.BAM																																																						
DPA.BGL																																																						
DPA.ECI																																		*	C	*	C				C													
DPA.HID																																																						
DPA.MSP																																																						
DPA.PST																																																						
DPA.SST																																																						
DPA.TAQIX																																																						
DZA.BAM																																																						
DZA.BGL																																																						
DZA.ECI																																																						
DZA.HID																																																						
DZA.MSP																																																						
DZA.SST																																																						
DZA.TAQIX																																																						
DOB.BAM																																																	*	*	B	B	B	B
DOB.BGL																																																	A	A	B	C	B	C
DOB.ECI																																																	A	A	*	B	B	B
DOB.ECV																																																	A	A	B	B	B	B
DOB.MSPIX																																																	*	*	B	B	B	*

Lower case probe-enzyme: Not part of the central data or incomplete.

No polymorphism: DRA(BAM, MIC, MID, KPW, PST), DQB(ECI), DQA(MIC), DPA(MIC,KPW,PVU), DOB(MIC,MID,KPW,PST,PVU,TAQ).

*Single specificity or can be defined by combinations of specificities, e.g. DR1+DR2,DR2+DR3,DRW, then * in DR1,DR2 and DR3.

Same letter in a row: multiple specificities in a given band.

MOAB: Already produced or published in the literature. T-Cell: evidence in this Workshop data (see Joint Report on T cell clones).

Table 3B. Simplified View of Polymorphic RFLP Bands with Different Probes and Enzymes, and Its Corresponding Alleles

BASE X100, SEE TABLE 3. FOR DETAIL
SOME KB MAY NOT BE NECESSARY IDENTICAL TO THE FINAL REPORT
DUE TO MODIFICATION AFTER THE PRINCETON MEETING
MISSING:SST-DRA,DPB, ECV-ALPHA

DPB1: DPB1	DPB1: DPB5	DPB1: DPB3	DPB1: DPB6	DPB1: DPB2	DPB1: DPB4	DPB1: DPB7	DPB1: DPB8	DPB1: DPB9	DPB1: DPB11	DPB1: DPB13	DPB1: DPB14	DPB2: DPB21	DPB2: DPB22	DPA2: DPA31	DPA2: DPA32	DNA: DNA42	DNA: DNA41	DPA1: DP91	DPA1: DP92	DPA1: DP93	DPA1: DP94	DPA1: DP95	DPA1: DP96	PROBE ENZYME	
		3	3	4	4								*												
T				T																					
4	4	3	3	4	4								$												
																								DRB.BAM	39,40,42,43,46,48,53,66,113
																								DRB.BGL	31,39,56,69,92,93,103,111,117,145
																								DRB.ECI	22,37,38,41,45,47,51,63,92,95,111,115,119,125,151,153,165,182
																								DRB.ECV	63,75,109,153,163,182,204,229
																								DRB.HID	14,16,17,18,25,27,28,29,30,31,33,34,61,72,84
																								DRB.KPN	27,77,94,105,118,128
																								DRB.MSP	16,18,25,31,35,37,38,41,49
																								DRB.PST	5,11,13,14,21,27,29,36,37,41,49,52,53
																								DRB.PVU	39,41,42,43,45,46,48,52,57,59,62,66,98,103,114,135
																								drb.sst	Not central data
																								DRB.TAQ	14,17,21,26,27,41,44,52,58,70,72,87,115,138
																								DRA.BGL	39,42,45
																								DRA.ECI	31,32
																								DRA.MSP	30,31
																								DRA.PVU	110,119
																								DRA.TAQ	19,20
																								DQB.BAM	25,29,30,35,38,42,53,58,87,102
																								DQB.BGL	41,56
																								DQB.ECV	26,27,52,67,118,145,151,200
																								DQB.HIC	65,154
																								DQB.HID	33,34,35,64,114,128
																								DQB.KPN	12,51,53,92,118,120,128,154,177,191
				B		B																		DQB.MSP	21,23,26,36,82,92
																								DQB.PST	14,30,36,42,49,52,63,64,107,112
																								DQB.PVU	16,20,22,24,34,37,63
																								dqb.sst	Not central data
																								DQB.TAQ	19,25,26,29,45,46,67
																								DQA.BAM	76,109,239
																								DQA.BGL	17,37,38,72,84,131
																								DQA.ECI	13,19,63,110,134
																								DQA.HID	48,58,74,77
																								DQA.KPN	81,84,104,117,142,252
																								DQA.MSP	22,23,29,33,47,57,60
																								DQA.PST	23,42,43,68,76,91,134
																								DQA.PVU	27,28,49,72
																								DQA.SST	155,165,194,248
																								DQA.TAQ	15,25,29,37,42,69
													*											DPB.BAM	35,37,41,53,90
				B	B	B		B	B															DPB.BGL	69,89,94,116,150
					*							*	*											DPB.ECI	33,38,39,40,72,75,147
A	A	A	A	B	B	B	A	B			A													DPB.ECV	34,69,101
				B	B	B		B				*												DPB.HIC	38,71
									*			*	*											DPB.HID	14,16,18,33,34,51,53,54
	C			C				C																DPB.KPN	54,77,186,244
*	*			*	B	*		B	*	*		*												DPB.MSP	11,19,22,40,50,67
A	A	A	A	B	B	B	A	B			A													DPB.PST	19,24,29
					*																			DPB.PVU	42,43,45,52,61
A	A	A	A				A		*		A													DPB.TAQ	17,26
																*		A	A	*		*	*	DPA.BAM	69,87,90
														*	*		*	*	*	C	C	C		DPA.BGL	22,34,36,53,72
																								DPA.ECI	63,110,144
																*	*	A	A	C	C	C		DPA.HID	32,49
																	*	*	*	C	C	C		DPA.MSP	31,42,47,49,63
																*	*	A	A	C	C	C		DPA.PST	24,26
				B	B	B		B								*	*	A	A	C	C	C		DPA.SST	44,104,161,188
				C		C	C											*	*	*	*		*	DPA.TAQ	40,47,64,77,121,132
																*		A	A	*		*	*	DZA.BAM	71,89,93
																		*	*					DZA.BGL	34,35
																*	*	A	A	C	C	C		DZA.ECI	25,42
																*	*	A	A	C	C	C		DZA.HID	33,49
																	*	*		C	C	C		DZA.MSP	31,46,49
																*	*	A	A	C	C	C		DZA.SST	44,162,193
																*		A	A	*	*	*	*	DZA.TAQ	74,83,125,134
																								DOB.BAM	51,61,73
																								DOB.BGL	74,77,83
																								DOB.ECI	115,121,128
																								DOB.ECV	26,37
																								DOB.MSP	14,24,34,60

Table 4A. The Phenotypes of the Tenth Workshop Core Lines Derived from Serology, T-Cell, and RFLP Band Patterns

ID	NAME	RACE	HLA-B	HLA-DW	DRB1													DRB3/B					DRA				DQB1									
					DR1	DR15	DR16	DR17	DR18	DR4	DR5	DR12	DR13	DR14	DR7	DR8	DR9	DR52A	DR52B	DR52C	DR52D	DR53	DR71	DR72	DR73	DR74	DQ5	DQ6.0	DQ6.1	DQ6.2	DQ2.2	DQ2.1	DQ7	DQ8	DQ9	DQ4
9002	MZO	CAU	B14	DW20	8																					8	8									
9003	KAS116	CAU	B51	DW1	8																		8				8									
9004	JES	CAU	B27	DW1	8																		8				8									
9005	HOM2	CAU	B27	DW1	8																		8				8									
9006	WT100	CAU	B35	DW1	8																		8				8									
9008	DO208	CAU	B18	DW2		8																		8				8								
9017	WT8	CAU	B 7	DW2		8																		8				8								
9082	HO104	CAU	B 7	DW2		8																		8				8								
7013	SCHU	CAU	B 7	DW2		8																		8				8								
9014	MGAR	HIS	B 8	DW2		8																		8				8								
9010	AMAL	IND	B53	DW2		8																				8		8								
9011	E41	CAU	B52	DW12		8																		8						8						
9016	RML	IND	B51	DW22			8																			8							8			
9009	KAS011	CAU	B37	DW21			8																			8	8									
9022	COX	CAU	B 8	DW3				8										8							8							8				
9023	VAVY	CAU	B 8	DW3				8										8							8							8				
9088	PFO	CAU	B 8	DW3				8										8							8							8				
9018	LOO	CAU	B18	DW3				8											8							8					8					
9019	DUCF	CAU	B18	DW3				8											8							8					8					
9020	QBL	CAU	B18	DW3				8											8							8					8					
9021	RSH	BLK	B42	DW_NEW					8									8							8						4					4
9031	BOLETH	CAU	B62	DW4						8												8				8								8		
9032	BSM	CAU	B62	DW4						8												8				8								8		
9033	BM14	CAU	B 7	DW4						8												8				8								8		
9034	SAVC	CAU	B 7	DW4						8												8				8								8		
9029	WT51	CAU	B65	DW4						8												8				8								8		
9007	DEM	CAU	B57	DW14,DW21?			4			4												8				8	4							4		
9026	YAR	JEW	B38	DW10						8												8				8								8		
9025	DEU	CAU	B35	DW13						8												8				8							8			
9030	JHAF	CAU	B51	DW13						8												8				8							8			
9091	MLF	CAU	B62	DW13						8												8				8							8			
9107	LKT3	JPN	B54	DW15						8												8				8										8
9092	BM92	CAU	B51,B62	DW14						8												8				8								8		
9028	PE117	IND	B60	DW14						8												8				8								8		
9045	TUBO	CAU	B51	DW5							4	4							8							8							8			
9037	SWE	CAU	B61	DW5							8								8							8							8			
9039	JVM	CAU	B18	DW_NEW							8								8							8							8			
9040	BM15	CAU	B49	DW5							8								8							8							8			
9042	TIS	CAU	B35	DW_NEW							8								8							8							8			
9043	BM21	CAU	B41	DW5							8								8							8							8			
9036	SPO	CAU	B44	DB2							8		?						8							8	4				4					
9105	FPAF	JEW	B35	DB2							8		?	?					8					8				8								
9038	BM16	CAU	B18	DB6								8							8					8									8			
9058	OMW	BLK	B45	DW18									8					8							8			8								
9062	WDV	CAU	B38	DW18									8					8							8			8								
9065	HHKB	CAU	B 7	DW18									8					8							8			8								
9060	CB68	CAU	B62	DW18									8						8							8		8								
9055	HO301	CAU	B14	DW19?									8							8				8					8							
9059	SLE	CAU	B60	DW19									8							8					8				8							
9063	WT47	CAU	B44	DW19									8							8					8				8							
9054	EK	CAU	B44	DW9										8					8					8			8									
9056	KOSE	CAU	B35	DW9,DW19									4	4					8					8			4		4							
9057	TEM	JEW	B38	DW9										8					8							8	8									
9061	ABO	CAU	B18	DW9										8					8					8			8									
9064	AMALA	IND	B62	DW16							?			8				8							8								8			
9050	MOU	CAU	B44	DW17											8							8				8					8					
9051	PIT	CAU	B44	DW17											8							8				8					8					
9047	PLH	CAU	B47	DB1											8							8				8					8					
9048	LBUF	CAU	B13	DB1											8							8				8						8				
9068	BM9	CAU	B35	DW8.1												8					8			8												8
9072	SPACH	IND	B??	DW8.2												8					8			8												8
9067	BTB	CAU	B23,B27	DW8.1												8					8			8												8
9069	MAD	CAU	B60	DW8.1												8					8			8												8
9071	OLGA	IND	B62	DW8.2												8					8			8												8
9070	LUY	CAU	B51	DW8.3												8					8				8								8			
9066	TAB	JPN	B46	DW8.3									?	?		8					8				8					8						
9075	DKB	CAU	B60	DW23													8					8				8									8	

8 = Homozygous; 4 = probably heterozygous; DPnew:P(PET),J(JET),D(DHE) defined by Hyldig-Nielson; 7 = doubtful specificity.

Table 4B. The Phenotypes of the Tenth Workshop Core Lines Derived from Serology, T-Cell, and RFLP Band Patterns

DQA1								DQB2			DQA2				DOB						DPB1													DPB2		DPA2		DNA		DPA1						
DQ81	DQ82	DQ83	DQ84	DQ85	DQ86	DQ87	DQ88	DQXB1	DQXB2	DQXB3	DQXA1	DQXA2	DQXA3	DQXA4	DO1	DO2	DO3	DO4	DO5	DO6	DP1	DP5	DP3	DP6	DP2	DP4	DP7	DP8	DP9	DP11	DP13	DP14	DPNEW	DP21	DP22	DP31	DP32	DN42	DN41	DP91	DP92	DP93	DP94	DP95	DP96	ID
				8				8				8							4	4						8								8			8		8			8				9002
				8				8					8				8							?									D		8	8		8		8						9003
				8				8			8							4		4						8								8			8		8			8				9004
				8				8			8					4		4								?								8			8		8			8				9005
				8				8			8							4		4	8													8		8		8		8						9006
					8			4	4		8					8									4	4									8	4	4		8			4		4		9008
					8				4	4		8				4		4							4	4									8	4	4		8			4		4		9017
					8				8		8					8										8								8			8		8			8				9082
					8				8		8					8													8					4	4	8			8			8				9013
					8			8			8						8									8								8			8		8			8				9014
					8				8		8					8													8					8		8			8				8			9010
					8			4	4		8					4	4								4	4								4	4	4	4		8			4		4		9011
							8	8			8							4		4									8					8		8			8			4			4	9016
					8			8			8						8						4			4								4	4	4	4		8	4		4				9009
	8									8		8					8						8												8		8		8			8				9022
	8									8		8					8				8													8		8		8		8						9023
	8									8		8					8				4					4								8		4	4	4	4	4		4				9088
	8							8			8					4	4						8												8		8		8			8				9018
	8							8					8			8									?								J		8	8			8					8		9019
	8							8					8			8									8										8	8			8					8		9020
	8							8			8								8		8					?								8		8		4	4		4		4			9021
		8							8		8						8									8								8			8		8			8				9031
		8							8		8								8						8										8	8			8					8		9032
		8							8		8					8										8									8		8		8			8				9033
		8							8		8						8															8	J		8	8		8		8						9034
		8							8		8					8									8										8	8			8					8		9029
		4			4			4	4		8							4		4			4			4								4	4		8		8			8				9007
		8							8		8					8										8									8		8		8			8				9026
		8						8			8						8									8								8			8		8			8				9025
		8						8			8								8				8												8		8		8			8				9030
		8						8			8					8										?								8		8			8			8				9091
		8						8			8							4		4		4									4			8		8		8			8					9107
		8						8			8							4	4							?								8		8			8			8				9092
		8						4	4		8							4	4							8							P	8			8		8			8				9028
	8							8			4		4				8								4	4								4	4	8			8			4		4		9045
							8	8			8					4				4									8					8		8			8			4			4	9037
							8		8		8					8									8										8	8			8					8		9039
	8							8				8					8						8											8		8			8			8				9040
							8	8			8						8												8					8		8			8			8				9042
							8	8			8						8															8			8	8		8		8						9043
					8			8			8						8								8										8	8			8					8		9036
					8			8					8			8									8										8	8			8	8						9105
							8	8					8				8								8	?									8	8			8					8		9038
						8			8		8					8					8													8		8		8		8						9058
						8		8			8								8						?	8								4	4	4	4		8			4		4		9062
						8				8		8						8								8								8			8		8			8				9065
						8				8		8						8												4					8	8		8			8					9060
					8				8			8			8													4			4			8		8		8			8					9055
					8				8		8					8							8												8		8		8			8				9059
					8				8		8					8												4							8	8			8					8		9063
				8						8		8						8											8					8		8			8			8				9054
				8				8			4		4						4	4			4		4										8	8		4	4	4				4		9056
				8					8		8					8										4								8			8		8					8		9057
				8				8			8						8									8									8		8		8			8				9061
			8					8			8								8										8					8		8			8			4			4	9064
		8						8					8			8									8								J		8	8			8					8		9050
		8						8					8			8										8								8			8		8			8				9051
		8						8					8			8															8				8		8		8			4			4	9047
		8						8			8							4		4									8						8	8		8		8						9048
			8					8			8							8							8										8	8			8					8		9068
			8					8			8							4		4									8					8		8			8			8				9072
			8					8			8						8									8								8			8		8			8				9067
			8					8			8							8								8								8			8		8			8				9069
			8					8			8								4	4			8			?								4	4	4	4		8			8				9071
			8					8			8							4		4	4					4								8		4	4	4	4	4		4				9070
					8			8						8				8									8								8	8		8			8					9066
		8						8						8		8										8							P		8		8		8			8				9075

Table 5. Associations to HLA-DR/DW/DQ/DP

DRW52A	DR17(B8),DR18,DRW13(DW18),DRW14(DW16)
DRW52B	DRW11,DRW12,DRW14(DW9),DR17(B18)
DRW52C	DRW13(DW19)
DRW52D	DRW8,NOT RECOGNIZED AS ALLELE
DRW53	DR4,DR7,DR9
DRA71	DR1(DW1)
DRA72	DRW15,DRW8(DW8.1,DW8.2),DRW14(DW9),DRW12
DRA73	DRW17(DW3-B8),DRW18,DRW13(DW18,DW19),DRW8(DW8.3)
DRA74	DR4,DR7,DR9,DRW11,DRW16,DR3(DW3-B18),DR1(DW20)
DQW1(DQW5)	DR1,DRW14(DW9),DRW16(DW21),DRW10?
DQW1(DQW6.0)	DRW15(DW2),DRW13(DW18),DRW11(DB2)
DQW1(DQW6.1)	DRW13(DW19)
DQW1(DQW6.2)	DRW8(DW8.3),DRW15(DW12)
DQW2.1	DR3(DW3-B8),DR7(DB1),DRW18
DQW2.2	DR3(DW3-B18),DR7(DW17)
DQW3(DQW7)	DRW11(DW5),DRW12,DR4(DW13),DRW14(DW16),DRW8(DW8.3),DRW16(DW22)
DQW3(DQW8)	DR4(DW4,W10,W14)
DQW3(DQW9)	DR9
DQW4	DRW8(DW8.1,DW8.2),DR4(DW15),DRW18
DQA81	DQW1/DQA85,DQA86,DQA87
DQA82	DR3,SOME DR5
DQA83	DRW53
DQA84	DRW8
DQA85	DQW5/DR1,DRW14(DW9),DRW10?
DQA86	DQW6/DRW15,DRW16(DW21),DRW13(DW19),DRW11(DB2),DRW8(DW8.3)
DQA87	DRW13(DW18)
DQA88	DRW11(DWNEW,SOME DW5),DRW12,DRW16(DW22)
DXB1	NO ASSOCIATION
DXB2	NO ASSOCIATION /DR2?,DW19?,DW4?
DXB3	NO ASSOCIATION
DXA1	DXB1,DXB2
DXA2	DXB2,DXB3
DXA3	DXB1/DQ2.2?
DXA4	DXB1
DO1	NO ASSOCIATION
DO2	NO ASSOCIATION
DO3	NO ASSOCIATION
DO4	NO ASSOCIATION/DRW8?
DO5	NO ASSOCIATION
DO6	NO ASSOCIATION
DNA41	DP2,DP4,DP3,DP6?/DPA93,94,95
DNA42	DP1,DP5/DPA91,92
DPA91	DPW1
DPA92	DPW5
DPA93	DPW3,DPW4
DPA94	DP NEW
DPA95	DPW2
DPA96	DP NEW
DPA31	DPW1,DPW5,DPW2
DPA32	DPW3,DPW4
DPB21	DPW1,DPW5,DPW4
DPB22	DPW2,DPW3,DPW4

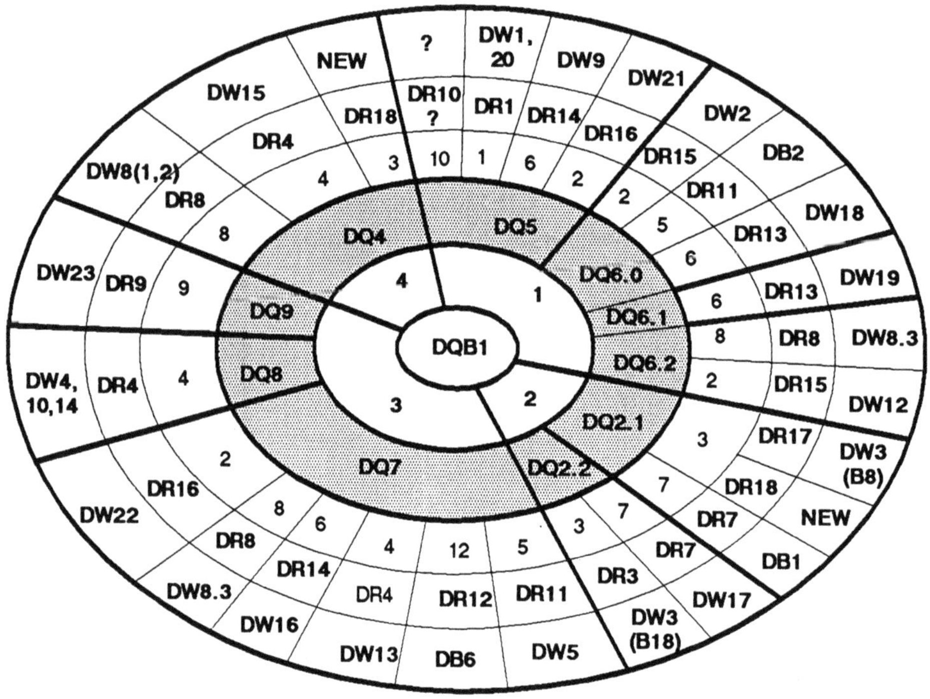

Figure 2A. DQB1 locus (DQW1–9).

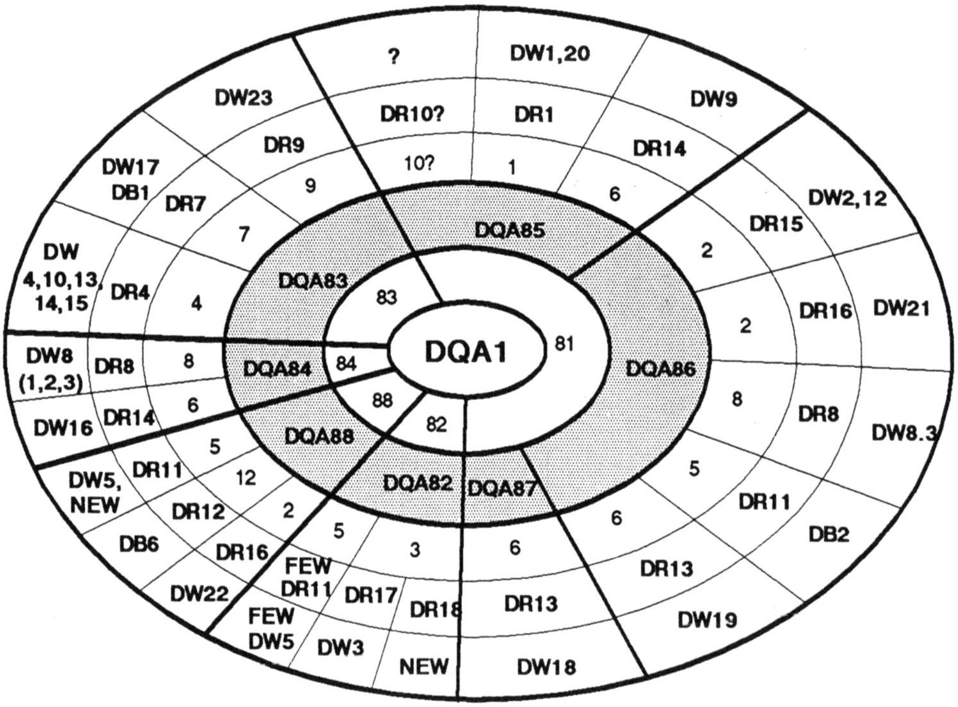

Figure 2B. DQA1 locus (81–88).

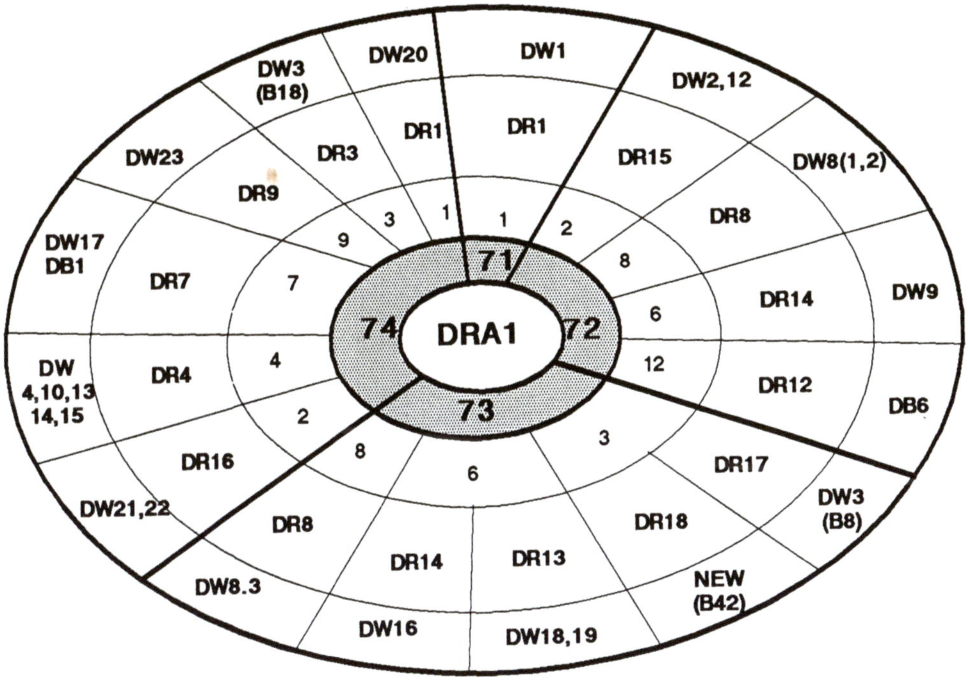

Figure 2C. DRA1 locus (71–74).

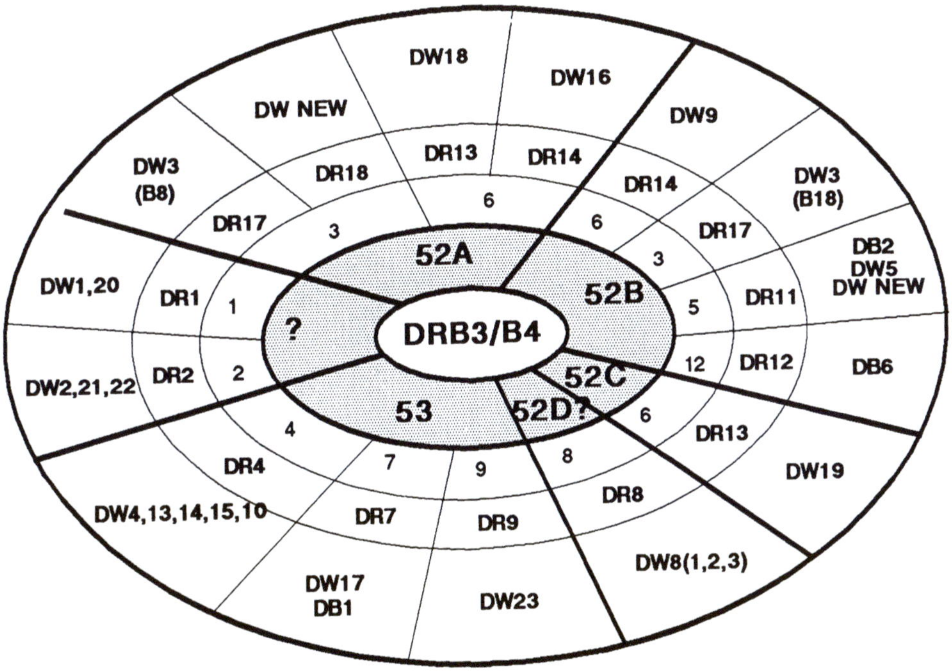

Figure 2D. DRB3/B4 locus.

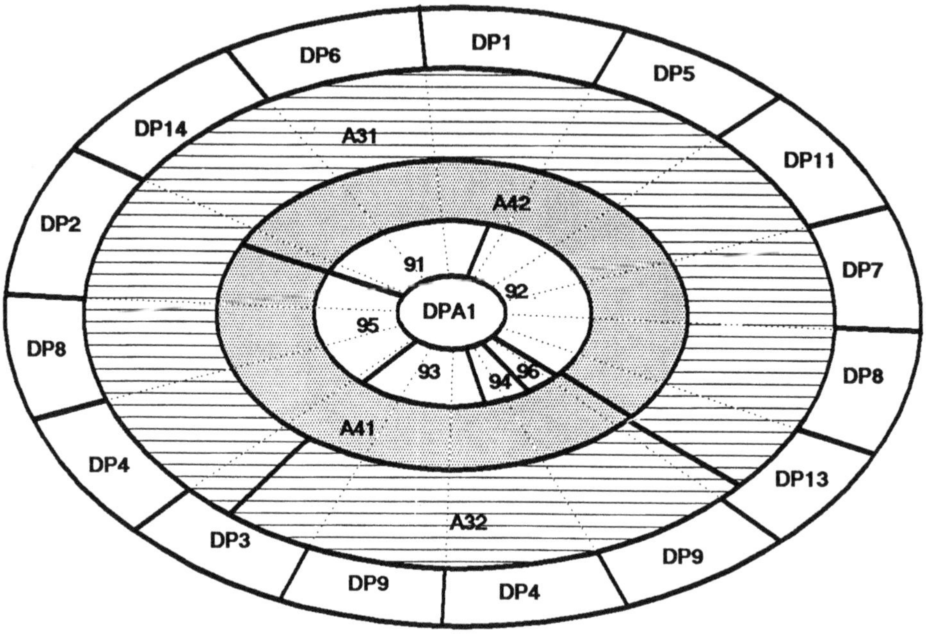

Figure 2E. DPA1 (91–96, DNA (41–42), DPA2 (31–32) loci.

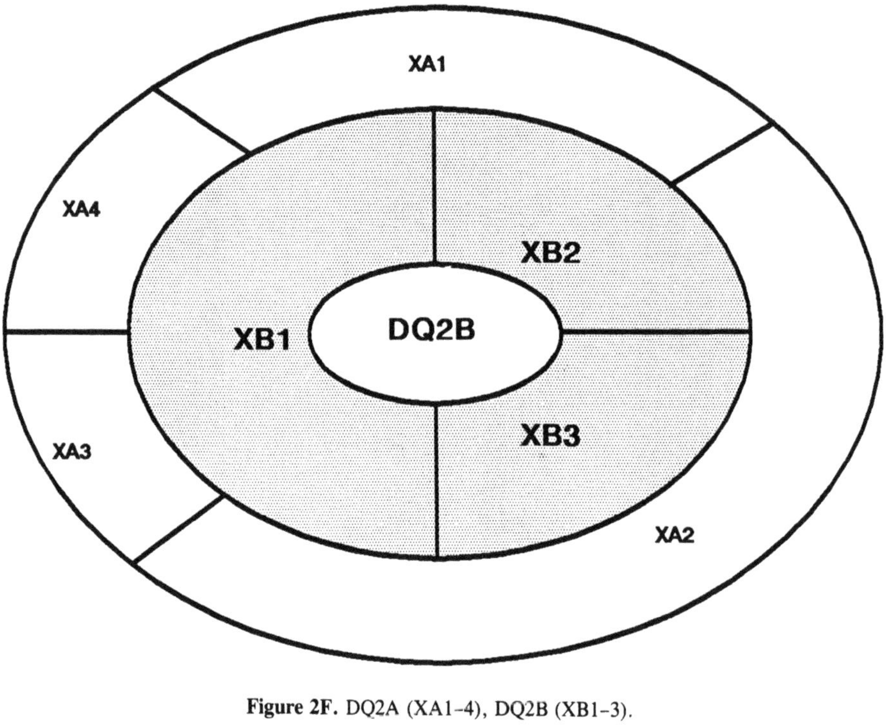

Figure 2F. DQ2A (XA1–4), DQ2B (XB1–3).

HLA Class II Typing by the Workshop Defined RFLP Specificities in Unrelated Individuals

M. Martell,[1] A. Marcadet,[1] and D. Cohen[1]

The Southern blot component of the Tenth International Workshop (WS) was based on the study of 70 cell lines, most of which were homozygous for the HLA complex. Thirty-four different, DR, DQ, Dw haplotypes emerged from serologic and Dw typings. The Tenth Workshop analysis demonstrates conclusively that almost all of these 34 haplotypes can be detected using RFLP techniques. This method, therefore, provides a convenient tool for accurate typing, which can now be based on a standard method. The aim of this chapter is to discuss the practical aspects of DR-DQ RFLP typing.

The Smallest Set of Systems Required for DR-DQ RFLP Typing: DRB/TaqI, DQB/TaqI, DQB/BamHI

In the course of the Tenth Workshop, 48 systems were tested, combining DRA, DQA, DRB, DQB probes and 12 restriction enzymes. Analyses and reports by J.D. Bignon et al. (1), M. Trucco et al. (2,3), and M. Segall et al. (4,5) show that by combining only the three systems, DRB/TaqI, DQB/TaqI, and DQB/BamHI, good resolution can be obtained. It must be emphasized that in most cases, the typing is not based only on specific fragments (i.e., those correlating with a single DR or DQ specificity), but rather on combinations of fragments that constitute patterns specific for single haplotypes.

In the 34 haplotypes tested, 24 serologically defined DR-DQ associations were present. Twenty-two of these are characterized by a specific pattern when the three systems, DRB/TaqI, DQB/TaqI, and DQB/BamHI, are combined (Table 1). Although the two remaining associations, DR4-DQw8 and DR4-DQw4, present a different pattern from all the others, they are indistinguishable from one another. Among the 22 DR-DQ associations well-defined with RFLPs, three can be split into two fragment patterns that coincide with Dw typing: DRw15, DQw6, Dw2 and Dw12, then DRw17, DQw2, Dw3, Dw24 and Dw25, and finally DRw13, DQw6, Dw18 and Dw19. Six other DR-DQ associations are split by Dw typing, but not by the above three systems. The first of these associations, DRw1, DQw5 can be divided into Dw20 and Dw1, using the system DRB/PstI (fragment 3.73 for Dw20 and fragment 1.42 for Dw1). Three other systems can be used: DRB/EcoRI, DRB/MspI, and DRB/PvuII (Table 1). However, DR4, DQw7, Dw4, and Dw13 cannot be separated, whatever the system, nor can DR4, DQw8, Dw4, Dw10 and Dw14, DRw8, DQw4 Dw8.1 and Dw8.2 or DR7 DQw2 Dw17 and DwDB1. All these data are shown in Figure 1. The first column lists DRB/TaqI fragments ordered according to their Tenth Workshop identification (WS ID) number, then the DQB/TaqI and DQB/BamHI fragments. The first column does not show all the fragments detected by the three systems. Fragments for DRB/TaqI are derived from the subset selected by Bignon et al. (1), with one modification in band 4 (9.65 kb). This band has been reported in some DR3, DRw6, DR7 and DR9 haplotypes (5). However, the band for the DR7 haplotype is extremely faint and, therefore, cannot be detected in any reproducible form. In the example given in Figure 2, band 4 is not detected in haplotype DR7 (lane 29), but is clearly present in haplotype DRw6 (lanes 31, 32, and 35). Thus, band DRB/TaqI 9.76 in haplotype DR7 has been omitted in Figure 1. Moreover, bands 3 and 27 have been added. Band 3 is reported to give a strong signal for DRw15 and some DRw11 haplotypes, and a faint signal for DR4 haplotypes (6) for which the band is not reported in Figure 1. Again, in Figure 2, band 3 is not seen for DR4 individuals (lanes 30, 31, 13, 36). Band 27, omitted in the DRB report (1), has been added in Figure 1. It should be remembered that band 13 in fact represents 3 or more bands which are difficult to differentiate (Fig. 2) (6). For DQB/TaqI and DQB/BamHI, the fragments listed in Figure 1 are identical to those selected by Trucco et al. (2). The distribution of each fragment is represented by symbols explained in the legend. It should be noted that only 66 cells instead of 70 are listed in Figure 1. Of the remaining four cells, three were heterozygous (9007, 9045, 9056). The fourth cell (9104) was probably wrongly identified. Figure 1 shows that with these three systems, 26 different RFLP patterns can be identified from the 34 haplotypes listed. Thus, these three systems do not discriminate among all haplotypes tested. Some fragments, usually from other systems (Table 2), can *sometimes* allow haplotypes, not detected by the three systems selected, to be distinguished. For example, the fragment TaqI DRB 11.11 is present in one of the three DRw11, DQw7, Dw5 haplotypes and absent in all three DwN. Therefore, the presence of this fragment in individuals RFLP typed as DRw11, DQw7 indicates that these individuals are probably Dw5. No information on the Dw type of DRw11, DQw7 individuals can be deduced when this fragment is absent. Table 2 lists all the *possible* discriminative systems.

Tenth Workshop RFLP Typing Nomenclature

Any workshop fragment will be reported by its system name and its Workshop molecular weight (the probe preceding the enzyme; the molecular weight in kb, to two decimal points, and without the letters "kb"): for example, DRB/TaqI 7.00. It is convenient to refer to

---

Participating Laboratory: FRACOH[1]

Immunobiology of HLA
Volume I

Table 1. List of Tenth Workshop D-DR, D-DQw, D-Dw RFLP Defined Haplotypes

D-DR,	D-DQw,	D-Dw	DRB/Taq I DQB/Taq I DQB/BamH I	Other Discriminative 10th WS Systems
D-DR1,	D-DQw5,		Y	
D-DR1,	D-DQw5,	D-Dw1	N	DRB PstI,EcoRI,MspI,PvuII
D-DR1,	D-DQw5,	D-Dw20	N	DRB PstI,EcoRI,MspI,PvuII
D-DRw15,	D-DQw5,	D-Dw2	Y	
D-DRw15,	D-DQw6,	D-Dw12	Y	
D-DRw16,	D-DQw5,	D-Dw21	Y	
D-DRw16,	D-DQw7,	D-Dw22	Y	
D-DRw17,	D-DQw2,	D-Dw3 (24)	Y	
D-DRw17,	D-DQw2,	D-Dw3 (25)	Y	
D-DRw18,	D-DQw4,	D-Dw"RSH"	Y	
D-DR4,	D-DQw7		Y	
D-DR4,	D-DQw7,	D-Dw4	N	N
D-DR4,	D-DQw7,	D-Dw13	N	N
D-DR4,	D-DQw8	or D-DQw4	Y	
D-DR4,	D-DQw8		N	N
D-DR4,	D-DQw4,	D-Dw15	N	N
D-DR4,	D-DQw8,	D-Dw4	N	N
D-DR4,	D-DQw8,	D-Dw10	N	N
D-DR4,	D-DQw8,	D-Dw14	N	N
D-DRw11,	D-DQw5,	D-DwDB2 (25)	Y	
D-DRw11,	D-DQw6,	D-Dw"FS"	Y	
D-DRw11,	D-DQw7		Y	DRB EcoRI,PvuII
D-DRw11,	D-DQw7,	D-Dw5 (25)	N	N
D-DRw11,	D-DQw7,	D-DwN (25)	N	N
D-DRw12,	D-DQw7,	D-DwDB6 (25)	Y	DRB EcoRI,PvuII
D-DRw13,	D-DQw5,	D-Dw18 (25)	Y	
D-DRw13,	D-DQw6,	D-Dw18 (24)	Y	
D-DRw13,	D-DQw6,	D-Dw19 (26)	Y	
D-DRw14,	D-DQw5,	D-Dw9 (25)	Y	
D-DRw14,	D-DQw7,	D-Dw16 (24)	Y	
D-DR7,	D-DQw2		Y	
D-DR7,	D-DQw2,	D-Dw17	N	N
D-DR7,	D-DQw2,	D-DwDB1	N	N
D-DR7,	D-DQw9,	D-Dw11	Y	DRB EcoRV,HindIII,MspI
D-DR7,	D-DQwX,	D-Dw"9106"	Y	
D-DR8,	D-DQw4		Y	
D-DR8,	D-DQw4,	D-Dw8.1	N	N
D-DR8,	D-DQw4,	D-Dw8.2	N	N
D-DR8,	D-DQw4,	D-Dw8.3"TAB"	Y	
D-DR8,	D-DQw7,	D-Dw8.3"LUY"	Y	
D-DR9,	D-DQw9,	D-Dw23	Y	DRB EcoRI,HindIII,MspI

This table lists all Tenth Workshop D-DR, D-DQw, D-Dw defined haplotypes, most of which are defined by the systems DRB/TaqI, DQB/TaqI and DQB/BamHI (Y=yes in the second column). As explained in the text, haplotypes DRw12 DQw7 and DR7 DQw9 are distinguishable form all other haplotypes, but homozygous haplotypes DRw11 DQw7 and DR9 DQw9 cannot be easily distinguished from heterozygous haplotypes DRw11 DQw7/DRw12 DQw7, and DR9 DQw9/DR7 DQw9, respectively, The third column indicates other Tenth Workshop discriminative systems which enable these particular cases to be solved (24, 25 and 26 refer to Dw24, Dw25 and Dw26).

each pattern of fragments by its probable immunologically defined corresponding type, preceded by D- (for *D*NA). For example, the pattern which associates the five fragments, DRB/TaqI 9.79, 7.00, 4.12 and DQB/TaqI 4.62 and 0.87, is specific for the haplotype DRw17, DQw2, Dw3 (24). Therefore, this pattern will be referred to as D-DRw17, D-DQw2, D-Dw3. The pattern that associates the fragments, DRB/TaqI 5.80, 4.40, 2.08, DQB/TaqI 5.34, 1.98, DQB/BamHI 5.76, 3.03 and DRB/PstI 3.73, will be referred to as D-DR1, D-DQw5, D-Dw20. Table 1 gives a list of the Tenth workshop D-DR, D-DQ, D-Dw RFLP-defined haplotypes.

Utilization of RFLP Patterns for HLA Typing: When?

This can be discussed in several different contexts.

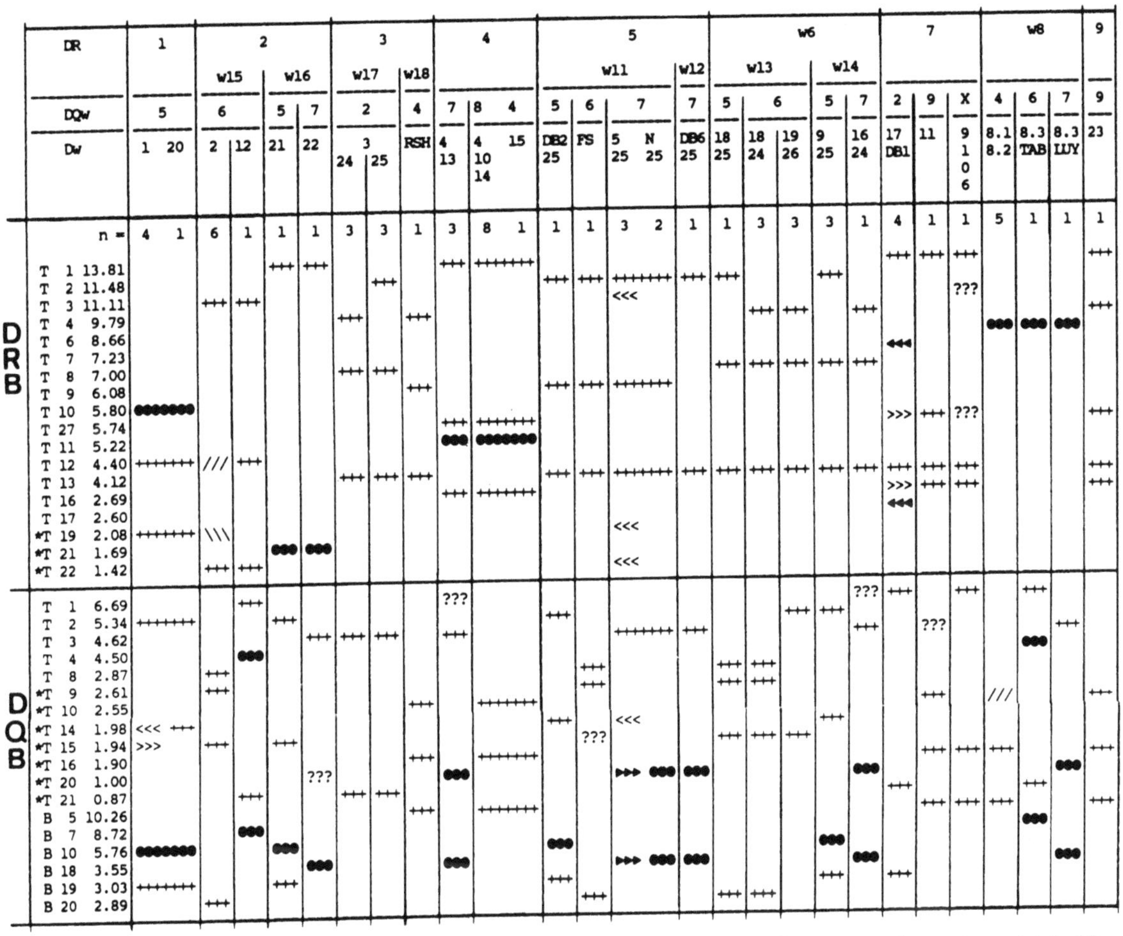

"n" = number of cells corresponding to a given type; * = fragment not used in the study (cf text); ●●● = fragment associated with a single DR or DQ allele, or with a subdivision of *one* DQ allele; +++ = fragment associated with several Class II specificities; ◀◀◀ or ▶▶▶ = when a fragment is found in only a subset of cells carrying all the same Class II specificity; <<< or >>> or /// or \\\ = when a fragment is found in a subset of cells carrying all the same Class II specificity and also in cells carrying other Class II specificities. For a given type, <<< (or ◀◀◀) *is mutually exclusive with* >>> (or ▶▶▶). For a given type, /// indicates a different cell distribution from \\\ but without mutual exclusion; ??? = not determined.

Figure 1. Diagram patterns of DRB and DQB hybridizing fragments according to the DR, DQw, and Dw cell typing. DRB TaqI, DQB TaqI, and DQB BamHI fragments are listed in the first column (T = TaqI; B = BamHI) with their Tenth Workshop ID numbers and their standard molecular weights (kb).

1. In Prenatal Diagnosis of 21-Hydroxylase Deficiency. Diagnosis is established by RFLP typing on DNA from chorionic villi (7). This technique, based on a family study of class I and II RFLPs, makes it possible to determine whether or not the fetus is HLA-identical to the affected sib. RFLP typing permits the diagnosis to be made at 10 to 12 weeks' gestation. In addition, it is a more sound method than HLA typing on amniotic fibroblasts. It is probable that current research on 21-hydroxylase gene mutation will soon permit direct detection of the abnormal gene.

2. In Disease Association and Population Studies. RFLP typing has already led to an improvement in our knowledge of the HLA and disease association. By standardizing the method and improving the polymorphism analysis, the Tenth Workshop offers to all scientists the possibility of systematic use of class II RFLP typing. The limitations experienced with serologic typing are essentially derived from the difficulty to accumulate sera for class II typing. The Workshop has demonstrated that with only two probes and two enzymes, it is possible to obtain a clear-cut analysis of class II polymorphism. The

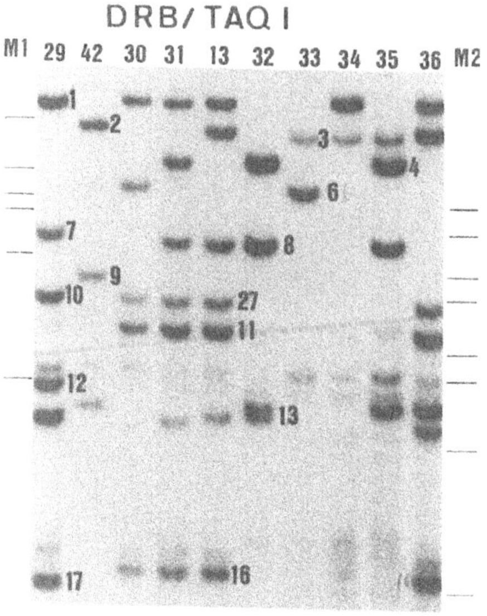

Figure 2. Example of an L-migration autoradiogram in the DRB/TaqI system. Molecular weight markers are positioned on both sides. Fragments observed are indicated by their Tenth Workshop ID numbers. The numbers at the top of the figure correspond to the different individuals listed in Table 3.

soundness of this method will allow a laboratory with no expertise in serology or cellular immunology to set up extensive typing programs. This conclusion is valid for class II typing but not for class I, which remains within the domain of serology, since RFLP typing, in the latter case, has not proved sufficiently informative. Moreover, the number of cells studied in the RFLP component of the Workshop was limited (the DRw10 specificity was not present in the panel tested) and the correlations found were, in some cases, based on only one cell. In most cases, however, these associations were already known from the literature. Therefore, in order to use RFLP typing in HLA and disease association studies, a certain procedure should be followed:

1. The Tenth workshop reference protocol (8) and reference probes should be used.
2. Both patients and controls should be RFLP-typed to compare the frequencies of markers defined by the same method. The use of controls typed only by immunologic methods must be avoided. In effect, this approach analyses the gene product, whereas the RFLP typing method is essentially based on linkage disequilibrium.
3. RFLP types should be reported in standard form (see above: Tenth Workshop RFLP typing nomenclature).

3. RFLP Typing for Matching in Transplantation.

1. *When serology is not possible*: In some cases, the serologic typing is difficult or impossible. RFLP typing has been used to select bone marrow donors in combined immunodeficiency syndrome (9) and could be used for patients suffering from chronic myeloid leukemia.
2. *Is it advantageous, in the current context, to replace class II serology by RFLP typing in transplantation?* RFLP typing is probably safer and less problematic than serology, and could prove valuable in transplan-

Table 2. Possible Tenth Workshop Discriminative Systems

D-DR	D-DQw	D-Dw	Enzyme	Probe	Fragments
1	5	1–20	TaqI	DQB	1.98, 1.94
			BglII	DRB	11.74
wll	7	5(25)-N(25)	TaqI	DRB	11.11, 2.08, 1.42
			TaqI	DQB	1.98, 1.00
			BamHI	DQB	3.55
			BamHI	DRB	11.33, 4.78, 3.92
			BglII	DRB	9.77
			EcoRI	DRB	3.31
			HindIII	DRB	2.60
			PstI	DRB	5.38, 5.11, 2.73
			MspI	DRB	8.93, 2.43
			SstI	DRB	19.53, 13.19, 12,38, 12.01, 11.25, 6.15
4	8-4		SstI	DQB	6.75
			BamHI	DRB	11.33, 3.92
			HincII	DRB	4.11
			HindIII	DRB	2.49
			MspI	DRB	1.52
			PvuII	DRB	11.43
4	8	4-10-14	/	/	/
4	7	4-13	/	/	/
7	2	17-DBl	EcoRI	DQB	16.51, 14.52
8	4	8.1-8.2	/	/	/

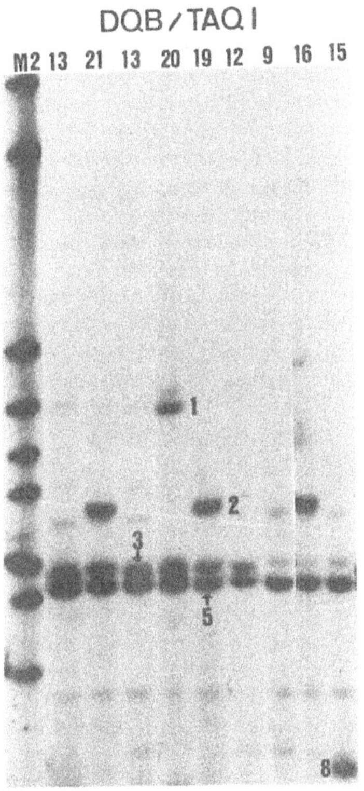

Figure 3. Example of an L-migration autoradiogram in the DQB/TaqI system. M2 molecular weight markers are shown on the left. Fragments observed are indicated by their Tenth Workshop ID numbers. The numbers at the top of the figure correspond to the different individuals listed in Table 3.

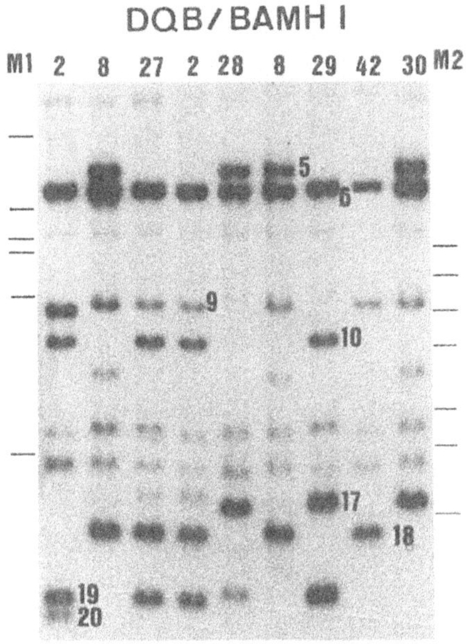

Figure 4. Example of an L-migration autoradiogram in the DQB/BamHI system. Molecular weight markers are positioned on both sides. Fragments observed are indicated by their Tenth Workshop ID numbers. The numbers at the top of the figure correspond to the different individuals listed in Table 3.

tation. Unfortunately, it is not yet possible to obtain the results of RFLP typing in a few hours, as is often required for donor typing. However, in bone marrow grafting, RFLP typing could prove extremely useful. Finally, retrospective studies will greatly benefit from DNA typing, which will allow identification of RFLP mismatches that influence graft or transplant survival.

Utilization of RFLP Patterns for HLA Typing: How?

1. The First Step Is to Establish the Tenth Workshop Standard RFLP Patterns. It is of the utmost importance to proceed in accordance with the instructions given in the Southern blot section of the reference protocol (8) and to use the reference probes of the Tenth Workshop. The use of other probes may lead to the detection of patterns not observed during the Tenth Workshop. It is probably more convenient to start with the three systems, DRB/TaqI, DQB/TaqI, DQB/BamHI. For each enzyme, both L and S migration conditions should be used; this will produce more reliable data. Once the bands are obtained, they have to be checked against the corresponding Workshop reference bands. To merely calculate the molecular weight has proved insufficient. The safest assignment is obtained by using the autoradiograms included in each system report (3,6,10). Any observed band can easily be *visually* assigned to a reference band by comparing by the relative mobility of both bands in relation to the marker mobility. It is, therefore, essential to utilize the molecular weight markers used for the Tenth Workshop. Figures 2, 3, and 4, respectively, show patterns with the three systems: DRB/TaqI, DQB/TaqI, and DQB/BamHI. DRB/TaqI is the only system that may present some difficulty in the assignment of observed bands to the Tenth WS reference bands. Bands 10 and 27 migrate so closely that it may not be easy to distinguish them. In fact, band 10 is specific for DR1 and, therefore, is always found to be associated with bands 12 and 19. On the other hand, band 27 is always associated with band 16 (Fig. 2: lane 29 with bands 10 and 12, lanes 30, 31, and 13 carrying bands 27 and 16). Figure 5, lane 4, on the DRB/TaqI system shows one individual, carrying bands 10 and 27, which actually comigrate. It must be remembered that very faint bands, even when polymorphic, cannot be taken into account.

2. How to proceed from fragment patterns to the D-type. Computer programs, not yet devised, could be used to deduce the D-type from the fragment patterns obtained. The D-type can also be easily obtained, manually, by following the steps described below:

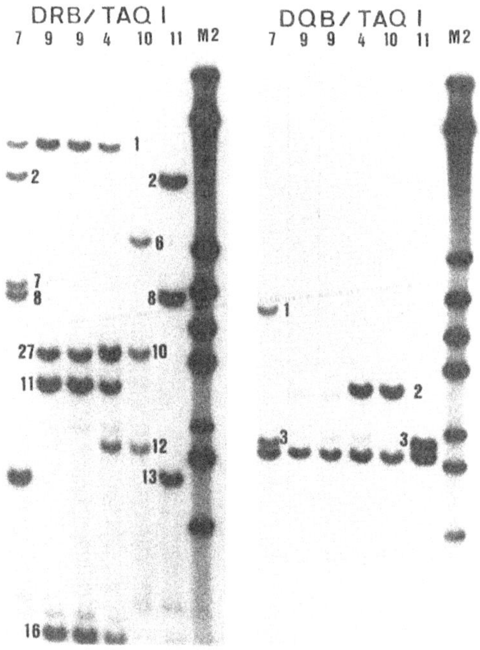

Figure 5. Example of L-migration autoradiogram in both DRB/TaqI and DQB/TaqI systems for the same six individuals. M2 molecular weight markers are shown on the left. Fragments observed are indicated by their Tenth Workshop ID numbers. The numbers at the top of the figure correspond to the different individuals listed in Table 3. This figure shows double intensity of DRB/TaqI fragments no. 2 and no. 8 for pattern 11, compared with the single intensity of the same fragments for pattern 7. DRB/TaqI no. 13 is present with double intensity for both patterns 11 and 7. Also shown is the double intensity of DQB/TaqI fragment no. 3 for pattern 11 compared with single intensity of the same band for pattern 7.

1. Use a photocopy of Figure 1 for each individual.
2. Hatch the distribution of each fragment found in the individual. Figure 6 gives an example of an individual carrying DRB/TaqI 1,27,13,16, DQB/TaqI 1,2,15,21, and DQB/BamHI 10,19.
3. Look for the columns carrying only hatched fragments. In the example given in Figure 6, these columns correspond to immunologically defined types: DRw15, DQw6, Dw2 and DR7, DQw2. In all other columns, there is at least one unhatched fragment.
4. The individual can, therefore, be typed. In the previous example, the individual is typed: D-DRw15, D-DQw6, D-Dw2/ D-DR7, D-DQw2.
5. Double intensity of fragments present at the homozygous state can be checked.

To test this procedure, we analyzed 80 unrelated IDDM patients for the three systems, DRB/TaqI, DQB/TaqI, and DQB/BamHI. Since only L conditions were used, some bands were not detected. These fragments are indicated by an asterisk in Figure 1. Forty-two different patterns were observed among these 80 individuals, all of whom had been D-typed (Table 3). Although most individuals were easily typed, a few cases required special attention (Table 4). For some individuals, only one column, in which all bands were hatched, was found (example: pattern 11 in Table 4). This corresponds to homozygosity. Double intensity of bands can be seen in Figure 5, lane 11, showing a homozygous individual.

In some cases, more than two columns were obtained. Pattern 17 (Table 4) occurs in three columns. Columns a and b correspond to the final type. Column c in fact comprises only bands found either in column a (DQB/TaqI no. 2) or column b (DRB/TaqI no. 1, 4, 8, 13). Pattern 26 is particularly intriguing: it shows six columns in which all bands are hatched. The final type is then given by columns a and e, i.e., the two columns sharing the smallest number of bands (only DRB/TaqI no. 13 is shared by columns a and e).

Thus, three simple rules can be applied:

1. One column means homozygosity.
2. Two columns means heterozygosity.
3. If there are more than two candidate columns, the two columns sharing the smallest number of bands will give the final type. There are two exceptions to Rule no. 2. The first is when the two columns are DRw11, DQw7 and DRw12, DQw7: in this case, the individual could be either D-DRw11 homozygous or D-DRw11/D-DRw12 heterozygous (e.g., individual 42 in Table 3). It is possible to check the band DRB/TaqI 6.08, the double intensity of which favors homozygosity. It is preferable to test such individuals with the system DRB/EcoRI to suppress any possible ambiguity. The second exception occurs when the two columns correspond to DR7, DQw9 and DR9, DQw9, in which case, the individual could be either D-DR9 homozygous or D-DR9/D-DR7,D-DQw9 heterozygous. Here, also, it is possible to check the intensity of fragment DRB/TaqI 9.79, the double intensity of which favors homozygosity or, preferably, to test these individuals with the system DRB/EcoRI or other systems (Table 1) to avoid ambiguity.

Finally, it should be borne in mind that a new RFLP pattern, not observed in this Workshop, could be found. This pattern could correspond to a DR,DQ haplotype not included in the analysis.

Conclusion

Thanks to the Tenth Workshop, RFLP class II typing will greatly improve the analysis of HLA polymorphism. It is possible to perform RFLP DP typing by using the five systems: DPA/TaqI, DPA/BamHI, DPA/BglII, DPB/BamHI, and DPB/BglII. Therefore, with three enzymes, TaqI, BamHI, and BglII and four probes, DRB, DQB, DPA, and DPB, complete RFLP D-typing can now be done following a standard procedure. To date, there are very few immunologically defined specificities that have not been revealed with this method. Therefore, if it proves worthwhile to undertake such a study, other methods, also standardized during the Tenth Workshop,

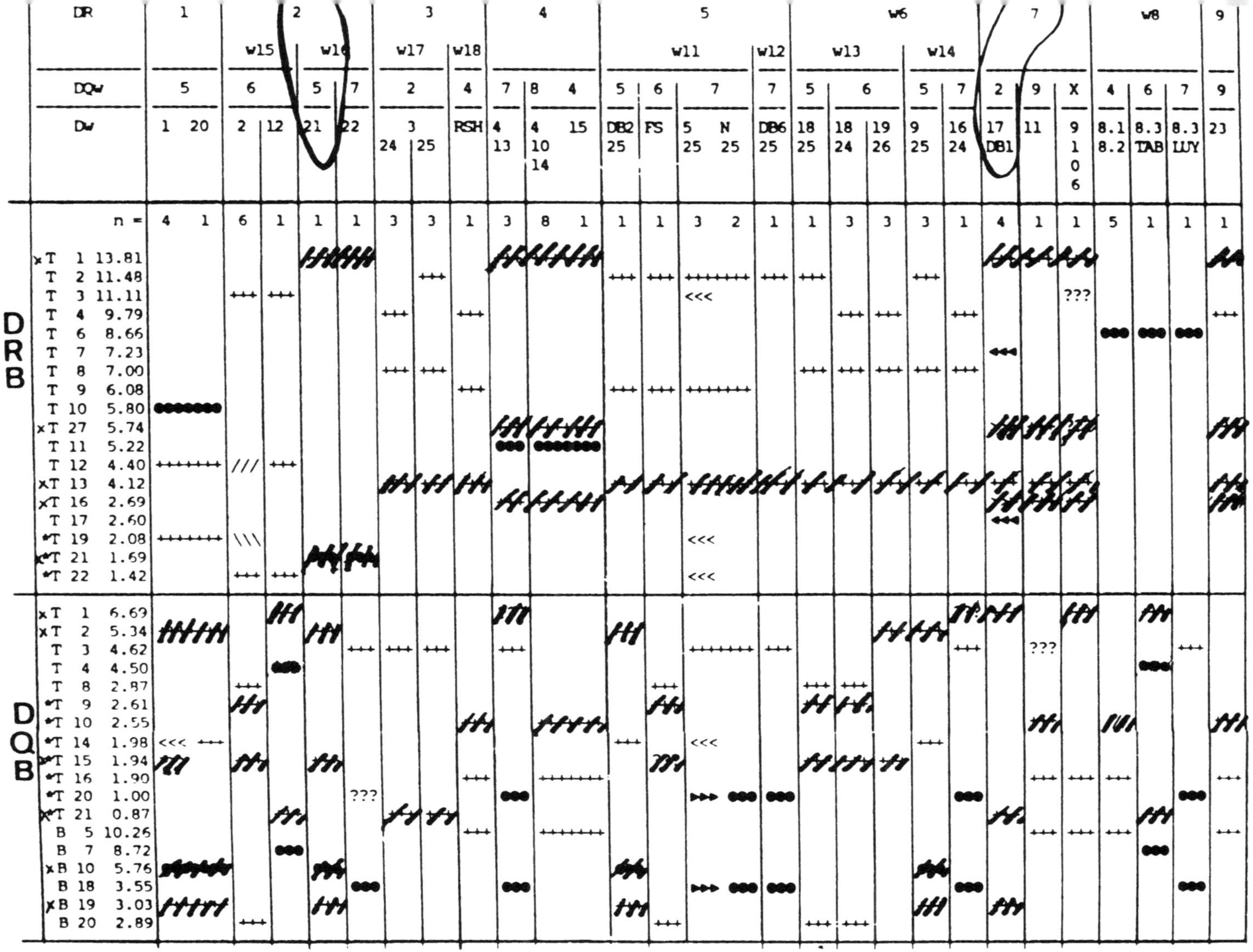

Figure 6. An example of RFLP typing using the three systems – DRB/TaqI, DQB/TaqI, DQB/BamHI – resulting, in this case, to a D-DRw16, D-DQw5, D-Dw21/D-DR7, D-DQw2 individual.

Table 3. List of the 42 Different RFLP Patterns

No.	DRB TaqI	DQB TaqI	DQB BamHI	D-DR	D-DQw	D-Dw
1	1,27,11		5	4	8/4	
	13,16			7	9	11
2	1,2,13	2,3	10,18,19	w12	7	DB6(25)
				w16	5	
3	2,4,8,13	3		w17	2	3 (24)
				w17	2	3 (25)
4	1,10,27	2	5,10,19	1	5	
	11,12,16			4	8/4	
5	1,7,27,11	1	5,19	7	2	
	13,16,17			4	8/4	
6	1,27,13,	1,2	10,19	7	2	
	16			w16	5	21
7	1,2,7,8,	1,3	19	7	2	
	13,17			w17	2	3 (25)
8	1,2,9,27,	3	5,18	4	8/4	
	11,13,16			w11	7	(25)
9	1,27,11,		5	4	8/4	
	16			4	8/4	
10	6,10,12	2	5,10,19	1	5	
				8	4	
11	2,8,13	3		w17	2	3 (25)
				w17	2	3 (25)
12	1,4,8,27,	3	5	4	8/4	
	11,13,16			w17	2	3 (24)
13	1,2,8,27,	3	5	4	8/4	
	11,13,16			w17	2	3 (25)
14	4,8,13	3		w17	2	3 (24)
				w17	2	3 (24)
15	1,3,27,	8	5,20	4	8/4	
	11,16			w15	6	2
16	1,4,8,27,	2	5	4	8/4	
	11,13,16			w13	6	19 (26)
17	1,4,8,13	2,3	10,19	w17	2	3 (24)
				w16	5	21
18	10,12	2	10,19	1	5	
				1	5	
19	1,10,27,	2,3	10,18,19	1	5	
	11,12,16			4	7	
20	1,4,7,8,	1,3	19	w17	2	3 (24)
	13,17			7	2	
21	1,4,8,27,	2,3	18	4	7	
	11,13,16			w13	6	19 (26)
22	2,6,13	3	5,18	8	4	
				w12	7	DB6(25)
23	4,8,10,	2,3	10,19	1	5	
	12,13			w17	2	3 (24)
24	1,4,8,27,	3	5	7	9	11
	13,16			w17	2	3 (24)
25	1,2,7,9,	1,3	18,19	7	2	
	13,17			w11	7	
26	2,4,8,9,	2,8	10,19,20	w11	5	DB2(25)
	13			w13	6	18 (24)
27	2,9,10,	2,3	10,18,19	1	5	
	12,13			w11	7	
28	1,6,7,13,	1	5,19	7	2	
	17			8	4	
29	1,7,10,12,	1,2	10,19	1	5	
	13,17			7	2	
30	1,6,27,		5	4	8/4	
	11,16			8	4	
31	1,4,8,27,	8	5,20	4	8/4	
	11,13,16			w13	6	18 (24)

Table 3. *Continued*

No.	DRB TaqI	DQB TaqI	DQB BamHI	D-DR	D-DQw	D-Dw
32	4,8,13	2,3		w17	2	3 (24)
				w13	6	19 (26)
33	3,6	8	5,20	8	4	
				w15	6	2
34	1,3	2,8	10,19,20	w15	6	2
				w16	5	21
35	3,4,8,13	2,8	20	w15	6	2
				w13	6	18 (24)
36	1,2,27,11,	3	18	4	7	
	13,16			w12	7	DB6(25)
37	1,7,27,	1	19	7	2	
	13,16,17			7	2	
38	1,2,9,27,	3	18	4	7	
	11,13,16			w11	7	
39	1,27,11	3	5,18	4	7	
	16			4	8/4	
40	1,3,7,13,	1,8	19,20	w15	6	2
	17			7	2	
41	2,8,10	2,3	10,19	1	5	
	12,13			w17	2	3 (25)
42	2,9,13	3	18	w11	7	
				w11/		
				w12	7	

Each individual is numbered from 1 to 42 in the left column. All fragments observed are indicated by their Tenth Workshop ID numbers for the three systems. The column on the right indicates the final class II RFLP type deduced by the method described in the text.

Table 4. Analysis of Four RFLP Patterns

	DRB TaqI	DQB TaqI	DQB BamHI	Possible type D-DR,D-DQw,D-Dw	Final type
Pattern 6	1,27,13,16	1,2	10,19		
Column a	1	2	10,19	w16,5,21	*
Column b	1,27,13,16	1	19	7, 2	*
Pattern 11	2,8,13	3			
Column a	2,8,13	3		w17,2,3(24)	**
Pattern 17	1,4,8,13	2,3	10,19		
Column a	1	2	10,19	w16,5,2	*
Column b	4,8,13	3		w17,2,3(24)	*
Column c	4,8,13	2		w13,6,9(26)	
Pattern 26	2,4,8,9,13	2,8	10,19,20		
Column a	2,9,13	2	10,19	w11,5,DB6	*
Column b	2,9,13	8	20	w11,6,"FS"	
Column c	2,8,13	8	20	w13,5,18(25)	
Column d	4,8,13	8	20	w13,6,18(24)	*
Column e	4,8,13	2		w13,6,19(26)	
Column f	2,8,13	2	10,19	w13,5,9(25)	

In the third case, the numbers underlined correspond to the fragments not shared by the three selected columns. * = Final haplotype; ** = final haplotype in homozygous state.

could be useful. Thanks to the efforts of the 80 participating laboratories, RFLP typing will prove to be of great importance in the next decade, before oligonucleotide typing becomes routinely available.

References

1. Bignon JD, Semana G, Tiercy JM, Simons M, Lalouel JM, Cohen D. DNA-RFLP analysis: HLA-DR beta workshop report. In: Dupont B (ed). Immunobiology of HLA. Volume I: Histocompatibility Testing, 1987. New York: Springer-Verlag, 1989.
2. Trucco M, Ball E. RFLP analysis of DQ-beta chain gene: workshop report. In: Dupont B (ed). Immunobiology of HLA. Volume I: Histocompatibility Testing 1987. New York: Springer-Verlag, 1989.
3. Marcadet A, Martell M, Cohen D, et al. RFLP standardization report for DQ beta/TaqI. In: Dupont B (ed). Immunobiology of HLA: Volume I: Histocompatibility Testing 1987. New York: Springer-Verlag, 1989.
4. Segall M. Specificity analysis of DQA fragments in core cell lines. In: Dupont B (ed). Immunobiology of HLA. Volume I: Histocompatibility Testing 1987. New York: Springer-Verlag, 1989.
5. Segall M. Specificity analysis of DRA fragments in core cell lines. In: Dupont B (ed). Immunobiology of HLA. Volume I: Histocompatibility Testing 1987. New York: Springer-Verlag, 1989.
6. Marcadet A, Le Gall I, Cohen D, et al. RFLP standardization report for DR beta/TaqI. In: Dupont B (ed). Immunobiology of HLA. Volume I: Histocompatibility Testing 1987. New York: Springer-Verlag, 1989.
7. Mornet E, Boué J, Raux-Demay M, et al. First trimester prenatal diagnosis of 21-hydroxylase deficiency by linkage analysis to HLA-DNA probes and by 17-hydroxyprogesterone determination. Hum Genet 1986;73:358.
8. Marcadet A, O'Connell P, Cohen D. Standardized Southern blot workshop technique. In: Dupont B (ed). Immunobiology of HLA. Volume I: Histocompatibility Testing 1987. New York: Springer-Verlag, 1989.
9. Marcadet A, Cohen D, Dausset J, Fischer A, Durandy A, Griscelli C. Genotyping with DNA probes in combined immunodeficiency syndrome with defective expression of HLA. N Engl J Med 1985;312:1287.
10. Hyldig-Nielsen JJ, Boehm BO, Wood KJ, Bushell AR, Singal DP, Svejgaard A. RFLP-standardization report for DQB/ BamHI. In: Dupont B (ed). Immunobiology of HLA. Volume I: Histocompatibility Testing 1987. New York: Springer-Verlag, 1989.

RFLP: Sequence Interrelations at the DPA and DPB Loci

Malcolm J. Simons[1] and Henry A. Erlich[2]

In the RFLP summary report (1), the RFLP reactivity patterns of 68 Workshop cells were reported. At the DPB1 locus, fragment patterns enabled current DPw1-5 specificities to be distinguished. None of the studied Workshop cells were assigned the DPw6 specificity. In addition, RFLP patterns indicative of subtypes of DPw2 and DPw4 specificities were identified, along with patterns associated with DPw "blank" cells representing new specificities.

The 68 cells are listed in Figure 1 according to the current assignment of DPw types based on cellular typing. It can be seen that 19 cells remain unassigned. Of the 49 cells in which there are DPw assignments, 41 bear only one DPw type.

Figure 1 also shows the reactivity patterns of 45 fragments detected by the DPA probe and 53 of the 176 fragments detected using the DPB probe. The total 98 fragments are arranged to show the patterns associated with the two mutually exclusive fragment clusters for each of the four DP region loci, namely DPB1, DPB2, DPA1, and DPA2. Recognition of locus-related fragments and fragment patterns associated with DPw specificities and DP beta chain types is aided by the use of column (fragment) and row (cell) lines. In the left half of Figure 1, the two clusters of the DPB1 locus are shown, with DPw1, DPw5, and DPw3 occurring in the group that is less frequent among the reference panel cells, and DPw2 and DPw4 occurring within the second, more frequent group. This division into two main categories is well known (2).

Fragments present in only one or a few cells or present in all but one or a few cells (HIC 341 to MSP 213) separate the fragments associated with the DPB1 locus (PST 285 to MSP 107) from the fragments associated with the DPB2 locus (HID 543 to SST 434). These are followed by fragments characterizing the DPA1 locus (MSP 715 to ECI 417). Finally, two fragments (BGL 529, ninth column from end; and BGL 724, last column) and associated fragments (BAM 867 to BGL 724) show a mutually exclusive pattern that is interpreted to represent the DPA2 locus.

In Figure 1, the eight cells at the bottom of the list (9056, 9021, 9070, 9088, 9071, 9007, 9009, 9106) are heterozygous for the two main clusters associated with the DPB1 locus. Cells 9071 and 9007 are also seen to be heterozygous at the DPB2 locus, while cells 9009 and 9106 have incomplete patterns for the same DPB2 cluster. Four additional cells (9013, 9011, 9045, 9062) are also heterozygous only at DPB2. Patterns of heterozygosity are more readily discernable in Figure 2, in which negative clusters of loci-defining fragments have

Participating Laboratories: US5SIM,[1] US3CTS[2]

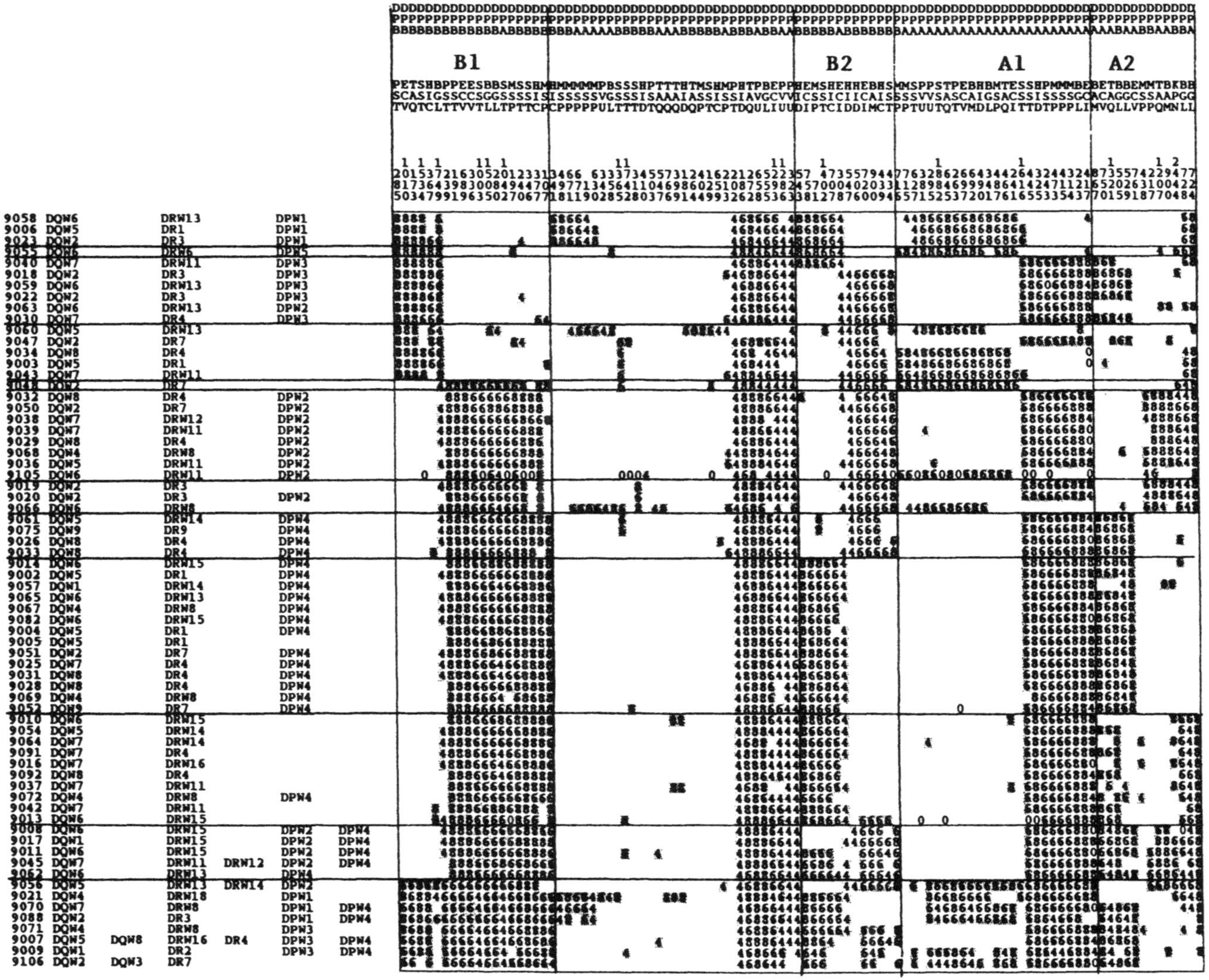

Figure 1. Tenth International Histocompatibility Workshop
Class II DPA,DPB loci RFLP selected fragments (n = 98); positive cluster.

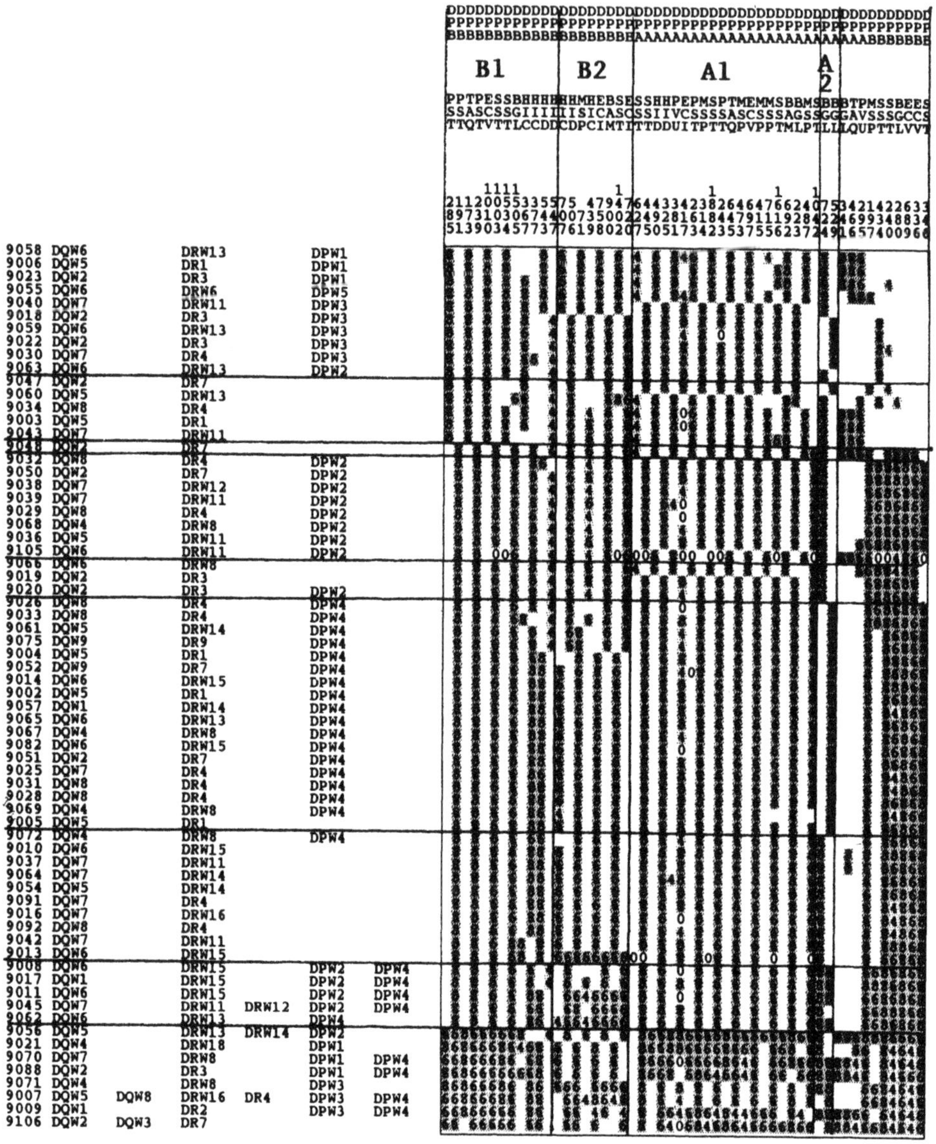

Figure 2. Tenth International Histocompatibility Workshop
Class II DPA,DPB loci-identifying RFLP fragments (n = 52); negative cluster.

been rearranged into DPB and DPA locus groups. The six cells heterozygous only at DPB1 and not at DPB2 (9056, 9021, 9070, 9088, 9009, 9106) are also heterozygous at the DPA1 locus. That the two DPA mutually exclusive BGL fragments (BGL 724, BGL 529) correspond to a DPA locus distinct from DPA1 is supported by the heterozygosity pattern involving cells 9008, 9017, 9011, 9062, 9070, 9088, 9071, and 9009. The six cells that are heterozygous either for DPB2 and/or DPA2, but not DPB1 and/or DPBA1 (9013, 9008, 9017, 9011, 9045, 9062) include four that are assigned as DPw2/w4 series heterozygotes (9008, 9017, 9011, 9045). These four are seen in the tree dendogram (Fig. 3) to have close relationships, along with 9013 and 9062. Cell 9062, and possibly also 9013, may be heterozygous for two alleles of the DPw2/w4 series.

Recent advances in sequence analysis of the DP beta second exon encoding the outer domain reveal 14 sequences at the DPB1 locus and 2 sequences at the DPA1 locus (3). These sequences are shown in Figure 4, along with one DPA2 locus sequence (called SXA in Fig. 4) and one DPB2 sequence (SXB). The two alleles of the DPA1 locus have been called DPA1 and DPA2.

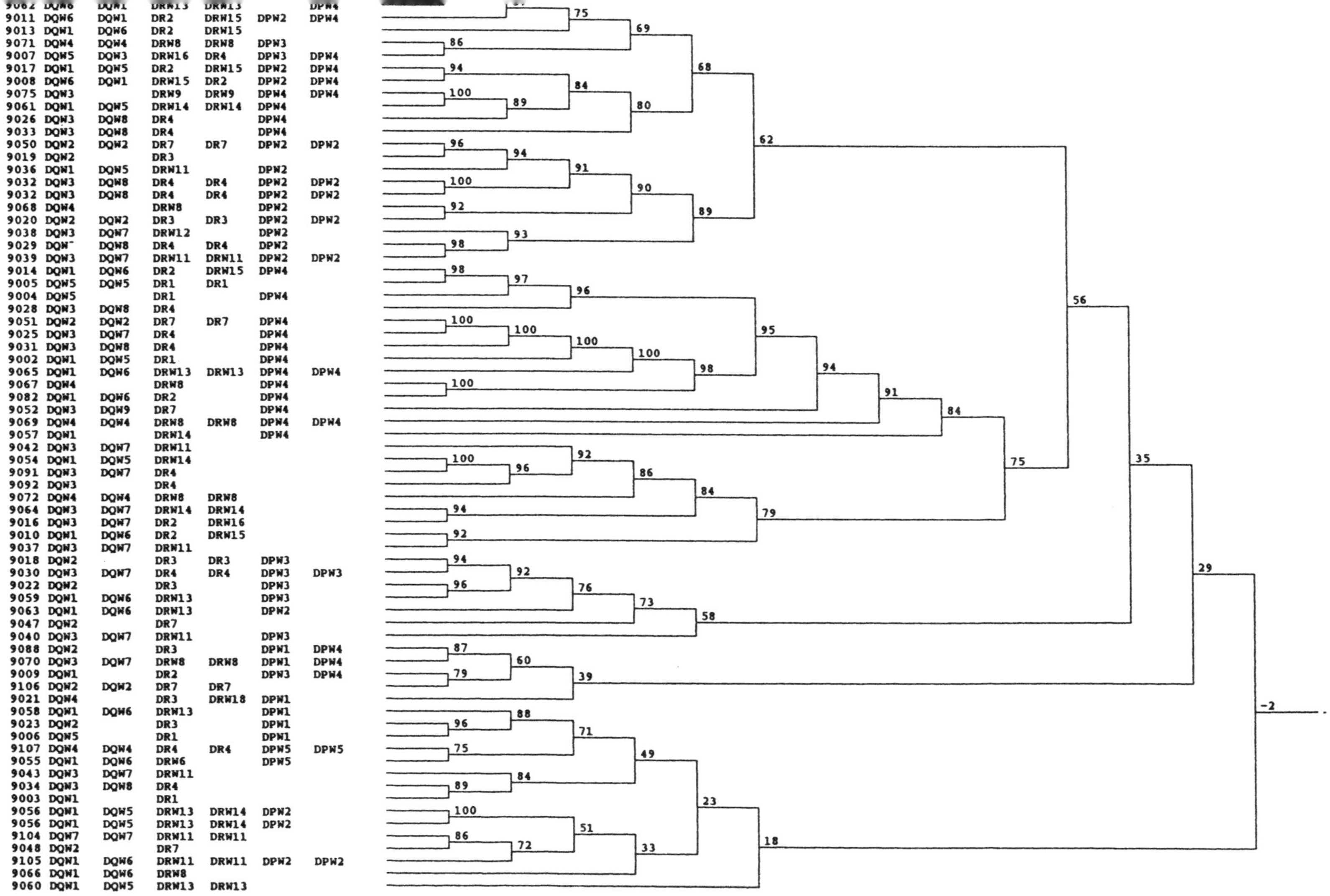

Figure 3. Tenth International Histocompatibility Workshop
Tree analysis of cell relationships: DPA,DPB loci RFLP selected fragments (45 + 53: total 98).

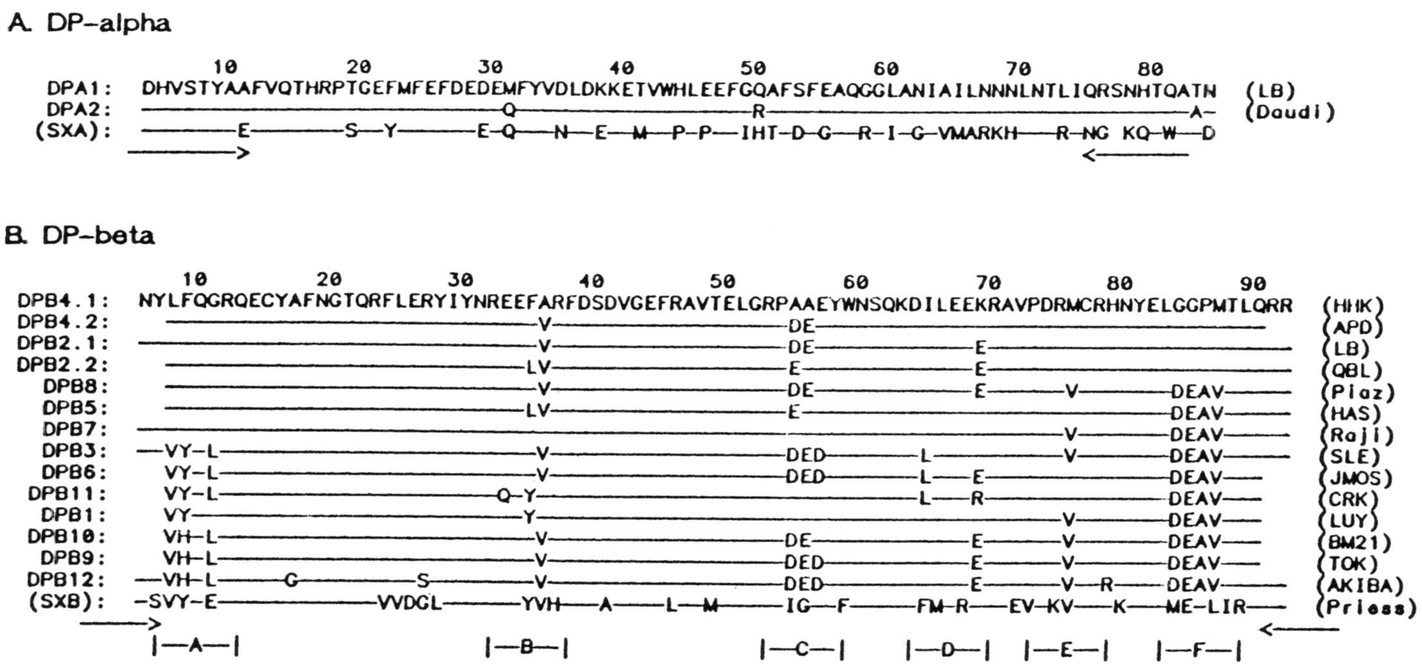

Figure 4. Alignment of the protein sequences of the HLA-DP genes. The nucleotide sequences were translated into the standard one-letter amino acid code and aligned to the DPB4.1 allele. A dash indicates identity with the prototype and a space indicates the sequence was not determined. The positions are for the mature peptide subunits, and the arrows indicate the positions of the PCR amplification primers. The designations of the alleles are shown at left, and representative sources indicated at the right. A = HLA DP-alpha alleles. The sequences of the two published DP-alpha alleles are shown. For comparison, the sequence of the DP-alpha pseudogene, SX-alpha, is also included. B = HLA DP-beta alleles. The sequence of the DP-beta pseudogene, designated SX-beta, is also included.

The first main contribution of sequence analysis has been to distinguish subtypes of DPw2 and DPw4. The listed DPw4 specificities are encoded by the DP beta 4.1 sequence for all but one cell (9072), which is DPB4.2. Cells cellularly typed as DPw2 are all of the DPB2.1 sequence subtype, again with one exception (9020, which is DPB2.2). Among the reference panel cells, DPw typing has predominantly identified the DPB2.1 and DPB4.1 sequence alleles of the DPB1 locus. Furthermore, sequence analysis of DPB1 alleles has revealed that 12 of the 19 cells lacking a DPw assignment are either DPB2.2 or DPB4.2 (9034, 9019, 9066, 9010, 9054, 9064, 9091, 9016, 9092, 9037, 9042, 9013). Apart from cell 9013, which is heterozygous at DPB2, all 12 cells appear to be homozygous at both DPB1 and DPB2 loci. Cell 9034 has been sequence typed as DPB4.2, DPB10. RFLP analysis reveals no evidence of heterozygosity for this cell, nor of a fragment pattern consistent with the existence of a DPw4 type. The only fragments present are those associated with the DPw1/w5/w3 cluster. This exception notwithstanding, it is apparent that much of the DPw blank includes cells with DPB2.2 and DPB4.2 sequence alleles. Since DPB2.2 and DPB4.2 sequence alleles were detected in cells of cellular typing specificities DPw2 and DPw4, respectively, they qualify as subtypes within the DPw2/w4 group.

The second main contribution of DPB1 second exon sequence analysis has been the identification of new alleles associated with the DPw1/w5/w3 group. Among the reference panel cells, the DPB1 sequence-defined alleles DPB9, DPB10, and DPB12 and four of the new RFLP patterns (9060, 9047, 9003, 9043) along with 9034 associated with the DPw1/w5/w3 group. Cells 9056 and 9106 are heterozygous for one specificity associated with the DPw1/w5/w3 group, and for one with the DPw2/w4 group. Based on beta chain sequencing, cell 9106 is DPB New, DPB4.1.

Two additional cells (9104, 9107) were not serologically typed and, therefore, were not included among the 68 reference panel cells tested both by serology and RFLP. Cell 9104 has been sequence typed as DPB9, DPB New heterozygous at the DPB1 locus, while cell 9107 has the DPB1 sequence allele DPB5 characteristic of DPw5 cells. The RFLP fragment pattern of cell 9107 is, in fact, most similar to that of the DPw5 cell 9055, as demonstrated in the tree dendogram (Fig. 3), but there is a significant pattern difference between the two. The similarity index between 9107 and 9055 is 75, an association of appreciably lower strength then exists for many other cell pairs. It is unlikely that both cells are homozygous for the same DPw5 allele. However, it must be recognized that differences at nonexpressed loci contribute to a low similarity index even where homozygosity exists for exon alleles.

In Figure 3, tree analysis utilizing 98 fragments selected for association with DPA1, DPA2, DPB1, and DPB2 loci has been applied to all 68 cells. In Figure 5, tree analysis has utilized only cells that were homozygous at the two DPA and two DPB loci. The eight cells having DPB1 locus New specificities associated with the DPw1/w5/w3 cluster (9104, 9060, 9047, 9034, 9003, 9043, 9056, 9106) have distinguishable RFLP fragment patterns. Their cell relationships, based on differences associated with pattern variation involving DPB fragments MSP 197, SST 240, MSP 107, SST 1365, SST 318, MSP 213, and BGL 728 among others, can be examined in Figures 3 and 5. Most obviously, the DPB New specificity represented by cell 9060 is associated with unique fragments and a unique pattern of fragment absence, distinguishing it from all other cells. It is positioned at the bottom of both tree dendograms. The next, least dissimilar, cell is that of 9066, but the similarity index between the two is very low. New DP specificity cells 9003 and 9034 pair together. Cell 9043 has the next highest similarity index to this pair. From the tree it can be seen that all the cells having DPB1 New specificities, except 9048, are associated with the DPw1/w5/w3 cluster. Tree analysis also indicates that cell 9063 appears to have been mistyped as DPw2, since it also is DPw1/w5/w3 cluster-associated and has no fragment pattern characteristics of DPw2.

The RFLP fragment patterns of the DPB12 cell 9048 and of cell 9104 are very similar (similarity index 86). Along with cells 9105 and 9056, they form a DPw2-associated cluster. This is consistent with the fragment pattern seen in Figure 1, which indicates an association with the DPw2/w4 cluster. Comparison of the fragment pattern similarity between cells 9048 and 9105 has to take account of the doubtful fragments associated with cell 9105 (scored as 0), At the DPA1 locus, both appear to be heterozygous for the mutually exclusive fragments SST 1042 and MSP 487. Cell 9105 is likely to be heterozygous for both of the two main DPA1 clusters. Cell 9056 is clearly heterozygous at both DPB1 and DPA1 loci. As mentioned previously, the non-DPw2-associated haplotype in cell 9056 is also a DPB New DPw1/w5/w3 group allele. Cell 9048 lacks the fragment SST 346, which is a characteristic of the DPB1 sequence subgroup, DPB2.2 (9019, 9020, 9066). There is, however, no close relation at a sequence level between DPB12 and either DPB4.2 or DPB2.2 (Fig. 4). The alignment is, in fact, closer to that of the DPw1/w5/w3/w6 series. This illustrates why caution should be exercised in comparing new specificities and currently known types at the level of sequence similarity patterns and RFLP fragment patterns until more information is available. The best association might be expected when restriction endonucleases that have intra-exonic restriction sites are employed. Although restriction sites for enzymes used in the Workshop exist in the second exon, size considerations indicate that they are unlikely to have contributed to the observed RFLP fragment patterns. For the present, it should be recognized that RFLP patterns only indirectly reveal allelic differences by linkage disequilibrial-based associations between sequences bearing restriction sites, on the one hand, and allele-distinguishing exon sequences, on the other.

The objective of HLA typing by RFLP is to distinguish between haplotypes bearing different allelic specificities. This requires the selection of enzymes that generate

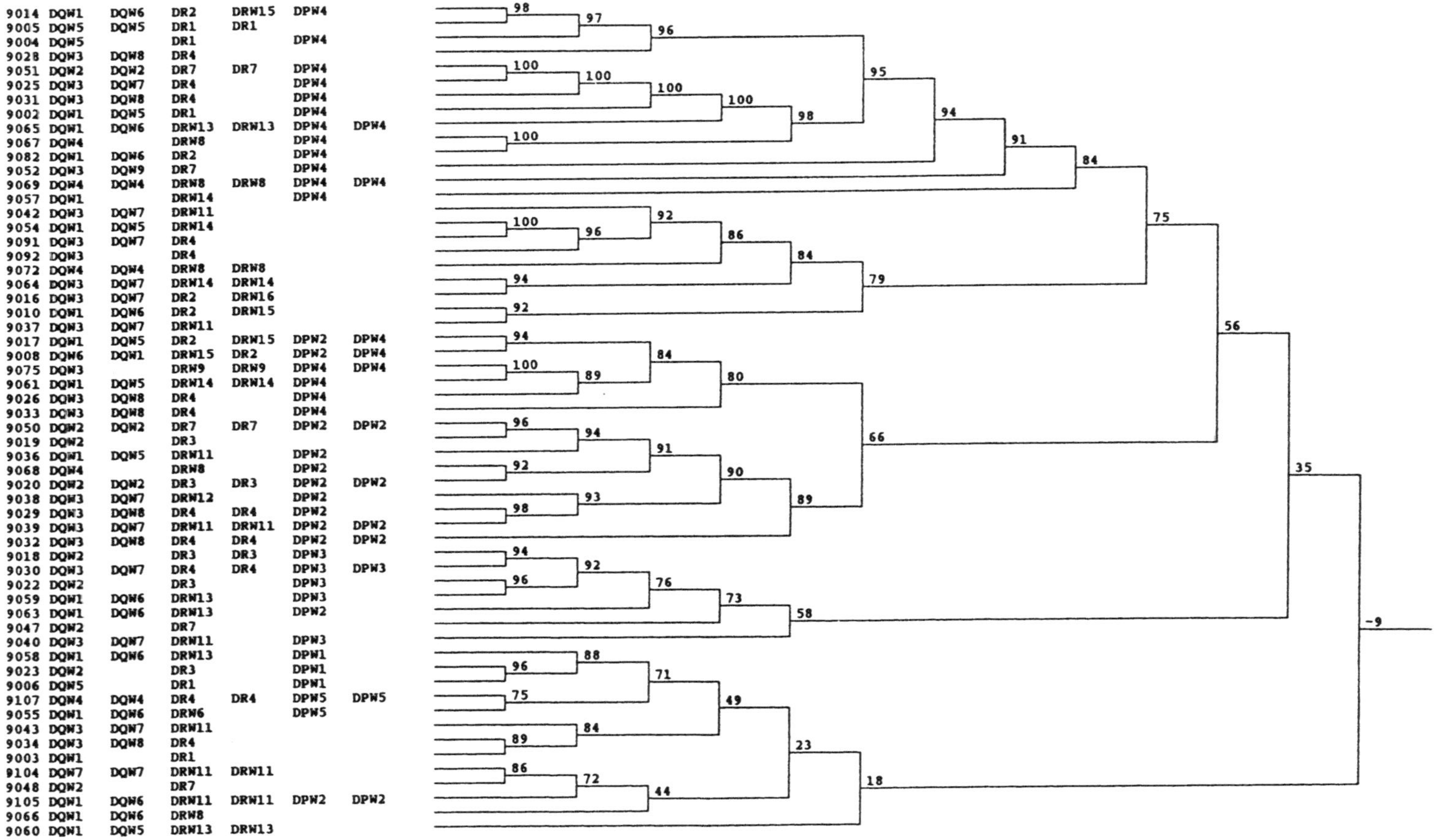

Figure 5. Tenth International Histocompatibility Workshop
Tree analysis of cell relationships: DPA,DPB loci RFLP selected fragments (n = 98), and DP homozygous cells (n = 58).

fragments associating with sequence variants due to linkage disequilibrium. In addition to the selection of informative enzymes, HLA typing is assisted by stratifying haplotypes for DPB type distinction by utilizing the contribution of the DPB2 and DPA2 loci. For example, the DPA2 fragment BGL 724 distinguishes between DPw4.1 and 4.2 cells. Reciprocally, the mutually exclusive DPA2 fragment BGL 529 is present only in DPw4.1 cells, and not in either DPw4.2 or DPw2 cells. The four DPW4.1 cells (9061, 9075, 9026, 9033) are distinguishable from the remaining DPw4.1 cells on the basis of DPB2 type. Similarly, for the DPw1/w5/w3 cluster, DPA1 typing enables DPw1 and DPw5 to be distinguished from DPw3. Detailed examination of Figure 1, in conjunction with cell relationships seen in Figures 3 and 5, reveal other examples where multilocus haplotype stratification enables allelic specificity assignment.

In conclusion, utilizing DPB and DPA locus probes, there are RFLP patterns for the five DPw specificities DPw1-w5 and for the corresponding DPB1 sequences represented among the reference panel cells. RFLP patterns have also been identified for the additional DPB1 sequence types (DPB2.2, DPB4.2, DPB9, DPB10, and DPB12) and for the DPB New specificities for which sequence information is not yet available. Tree analysis provides a useful indication of the similarity between haplotypes bearing these new specificities. In addition to the DPB types whose sequence is shown in Figure 4 and the DPB New specificities indicated by the RFLP patterns, unpublished results of Caucasian and Chinese cells reveal at least four additional DPB1 sequences. Thus, polymorphism at the DPB1 locus appears to involve at least 18 DPB1 sequence variants.

References

1. Simons MJ, Wheeler R, Cohen D, Lalouel J-M, Dupont B. Restriction fragment length polymorphism of HLA genes: Summary of the Tenth International Workshop Southern Blot Analysis. In: Dupont B (ed). Immunobiology of HLA, Volume I: Histocompatibility Testing 1987. New York: Springer-Verlag, 1989.
2. Hyldig-Nielsen J-J, Svejgaard A. Workshop report for the HLA-DP region. In: Dupont B (ed). Immunobiology of HLA, Volume I: Histocompatibility Testing 1987. New York: Springer-Verlag, 1989.
3. Bugawan TL, Horn GT, Long CM, et al. Analysis of HLA-DP allelic sequence polymorphism using the in vitro enzymatic DNA amplification of DP alpha and DP beta loci. J Immunol 1988;141:4024.

Restriction Fragment Length Polymorphism of HLA Genes: Summary of the Tenth International Workshop Southern Blot Analysis

Malcolm J. Simons,[1] Robert Wheeler,[2] Daniel Cohen,[3] Jean-Marc Lalouel,[2] and Bo Dupont[4]

Abstract. The Tenth Workshop represented the first international effort to evaluate the role of Southern blot technology in characterization of loci and haplotypes comprising the HLA system. The data generated from the use of 12 restriction endonucleases on DNA from 70 cells selected from the reference cell line panel were technically valid as revealed by fragment pattern consistency of hidden duplicates and by fragment pattern relations with known HLA specificities and haplotypes. This review details the total data, comprising more than 1100 fragments associated with 11 DNA probes.

From a practical standpoint, the Southern blot technique has been shown to be suitable for HLA typing of class II locus genes. One hundred eighty-seven "unique" fragments were identified using DRB, DQA, and DQB probes. Although RFLP "unique" fragments correlating with individual HLA specificities were not detected for all HLA class II antigens, the presence or absence of fragments correlating with two or more specificities provided the basis for comprehensive HLA class II DNA typing.

Participating Laboratories: US5SIM,[1] US7LAL,[2] FRACOH,[3] US7DUP[4]

DNA typing using the DPA and DPB locus probes has revealed RFLP patterns that correlate closely with known DPw1-DPw5 specificities. Two RFLP patterns corresponded to the nucleotide sequence subtypes DPB2.2 and DPB4.2 (Simons and Erlich, this volume), while seven additional patterns were associated with DPB "New" sequence variants in cells lacking assigned DPw specificities.

For HLA class I, 46 "unique" fragments were identified. The achievement of allele and haplotype discriminatory RFLP patterns might not have been expected, since DNA typing for HLA-A, -B, and -C alleles was performed using only a class-specific probe and not locus-specific probes. However, as for class II, the combination of the "unique" fragments, together with the presence or absence of broadly reactive fragments, did enable the different haplotypes to be distinguished. A new finding was the identification of 22 fragments correlating with HLA-C locus specificities, including RFLPs that are present only within the C locus "blank." It had been previously known that several HLA-A alleles correlated with "unique" fragments. Similarly, it was known that "unique" fragments only existed for a limited number of HLA-B specificities, using the commonly employed restriction endonucleases of the Tenth Workshop. The high degree of linkage disequilibrium between the alleles of the HLA-A, -B, and -C loci, and presumably also involving others of the class I loci, offset the paucity of B locus "unique" fragments, and enabled the identification of RFLPs that correlated strongly with most of the HLA-A, -B, and -C alleles as well as defining HLA-A, -B, and -C haplotypes.

More comprehensive information than previously existed was obtained on loci that are less relevant to the practice of HLA typing. However, RFLP patterns arising from the polymorphic loci (e.g., DQA2, DQB2, DOB, DNA,* DPA2, DPB2) contributed valuable information toward understanding the evolutionary relationships between haplotypes and their component loci. Tree dendogram analysis provided an indication of evolutionary relationships between the haplotypes of the cells studied in the Tenth Workshop.

This report is a comprehensive presentation of the total RFLP data on the reference panel cells. The figures and tables are presented in a manner that allows for ready inspection of the total database, and for further individual analysis. All data contributed by the Workshop standardization centers and all interpretations of the data as described in the individual HLA locus probe reports have been considered. This review should be read in conjunction with these component reports. It should be noted, however, that it has been necessary to revise the interpretation from some reports in view of central data analyses additional to those available to the rapporteurs.

Introduction

HLA gene restriction fragment length polymorphisms (RFLP) have been systematically examined using the standardized Southern blot Workshop protocol (1). This Workshop represents the first international multilaboratory effort towards the ultimate goal of single locus allele, and multilocus haplotype, characterization of the HLA system at the DNA level. The results reported here should be viewed as an attempt to evaluate the role of Southern blot technology in the characterization of the loci comprising the HLA system. This report provides an overview of the patterns associated with over 1100 fragments resulting from the use of up to 12 restriction endonucleases on DNA from 70 of the 107 reference panel cells of the Tenth International Workshop.

The design of the Southern blot component of the Workshop has been described elsewhere in this volume (2). In this report, the RFLP data are presented in a manner that is both comprehensible on inspection and amenable to further analysis. It takes into account information contributed by the working groups as reported in these proceedings, and should be read in conjunction with the individual reports.

The main criterion for inclusion of cells in the Tenth Workshop was homozygosity for DR and DQ specificities. Representation of Dw specificities was also a consideration. Additional cells that were suspected or known to be variants of the main types were included. As a result, some specificities were not included (e.g., DRw10), while some were present in only one cell (e.g., Dw20, Dw21, DB6, Dw13, Dw10).

In the final phase of data analysis, all cells were reassessed on the basis of serologic reactivity and the reassigned HLA specificities. Two cells (9104, 9107) were not serologically typed by the Workshop core serology serum set and, therefore, were dropped from analysis. Only one each of the duplicated cells (9032, 9056) was retained. Thus, RFLP analysis is based on 68 cells and 136 HLA haplotypes.

Although RFLP analysis by Southern blotting is a direct approach to gene detection, it is usually only an indirect method of identifying gene sequence variation. Restriction sites do occur within exons, but the large majority of RFLP sites for the restriction endonucleases employed in the Tenth Workshop occur in non-exon regions. Thus, the relationship between exon sequence variation and RFLP is a function of linkage disequilibrium between intron and flanking region fragments bearing polymorphic restriction sites, and exon segments with allelic variability. Careful selection of restriction enzyme/HLA locus probe combinations is critical to the identification of allelic differences.

Unlike serology, there are no "blank" cells in Southern blot analysis. All fragments generated by a particular endonuclease on DNA from a given cell can be expected to be present in the corresponding Southern blot lane. Furthermore, the intensity of detectable fragments provides information on gene dosage.

It is often stated that the disadvantage of RFLP for HLA typing is that polymorphic restriction sites are predominantly located outside of the exons encoding allelic variation. The argument contends that HLA typing should be directed at the exon sequences themselves or at expressed products of the genes. The former approach requires nucleotide sequence information on all relevant exons. While most allelic variation is encoded in the first domain exon, the possibility of a role for second domain sequences is suspected and cannot be eliminated in the interpretation of allele specific typing limited to first domain analysis. In principle, HLA typing technologies based on the detection of gene sequence-translated products will be the final arbiters of exon involvement. As reported elsewhere (3), this Workshop has advanced our understanding of HLA class II alpha and beta chain structural polymorphism. However, from a technological viewpoint, two-dimensional gel electrophoresis is not likely to be reduced to a routine procedure for HLA typing, and resolution of sequence-similar chains can be difficult. Epitope detection by T-cell cytotoxicity and proliferative in vitro procedures have proved to be highly

*The DNA probe for HLA DNA (previously DZA) did not function in the Workshop studies, and data obtained with this probe have not been included in the analysis.

Figure 1. Tenth International Histocompatibility Workshop class II DOB locus RFLP fragments (n=30). Fragment cluster.

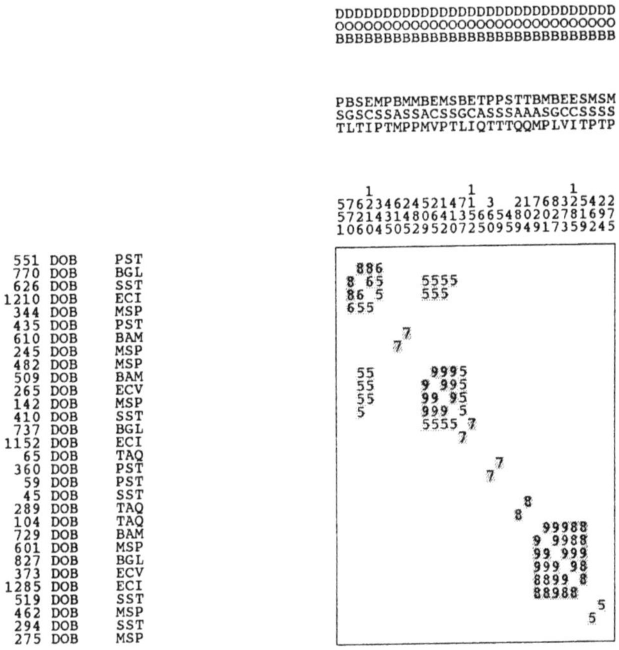

discriminatory (4), but are not yet realistic procedures for routine HLA typing.

The results reviewed here establish that serologically or cellularly detectable differences between HLA specificities are associated with discriminatory RFLP patterns based on the presence or absence of sites for restriction endonuclease cleavage. The results indicate that the fragments generated by combinations of 12 enzymes used in the Workshop, and even by single enzymes, detect differences between cells that have not been distinguished by serologic typing of the gene products. In addition to detection of differences within loci, cluster analysis of fragments generated at separate loci reveals patterns of linkage disequilibrium within HLA haplotypes. Thus, hybridization with full length locus probes in RFLP typing provides for distinction between cells whose exon products may not be easily distinguishable. The apparent disadvantage of non-exon polymorphism detection by RFLP in many instances becomes an asset. Some RFLPs provide very good approximations for identifying HLA alleles, and the RFLP patterns analysis is frequently discriminatory between haplotypes.

Fragment patterns associated with each cell have been investigated principally by cluster analysis involving assessment of successive pairs of cells for fragment pattern similarity. In this review, fragment pattern relationships between the panel cells have been presented as tree dendograms. By utilizing all fragment information, tree analysis reveals haplotype relationships which, in turn, can be presumed to reflect the evolutionary development of differences between the haplotypes.

One outcome of the locus and haplotype resolving power of RFLP is that differences have been identified between cells to which the same HLA specificities have been assigned. When the entire gamut of gene products, detectable as polypeptide chains or as epitopes detectable by serologic or cellular reagents, is established, and when exon nucleotide sequence analysis only infrequently detects new allelic varieties, it will again be necessary to reassess RFLP fragment patterns to determine the relationship between RFLP and "definitive" HLA haplotypes. The data have been presented in a manner that facilitates this future event. Unlike serology, in which many of the reagents, and the results obtained using those reagents, vary from workshop to workshop, the Southern blotting component has generated RFLP patterns that can be expected to be obtained each time the same enzymes and probes are applied to HLA RFLP analysis of the same cells. The sizes assigned to fragments may differ slightly, but the pattern relationships between fragments can be expected to be constant.

This report is divided into sections corresponding to the usual HLA classes and regions, namely class I, class III, class II alpha subregions and class II beta subregions. Virtually all fragment data are presented in the figures. The raw clustering outputs have been edited by rearranging the row sequence of listed cells and the column sequence of contiguous fragments both to simplify the complexity of pattern clusters and to produce an output in the form familiar to HLA serologists.

Results

Since the format for presenting the results for each individual locus, and for combinations of loci, is similar, it may be helpful to detail the approach taken using

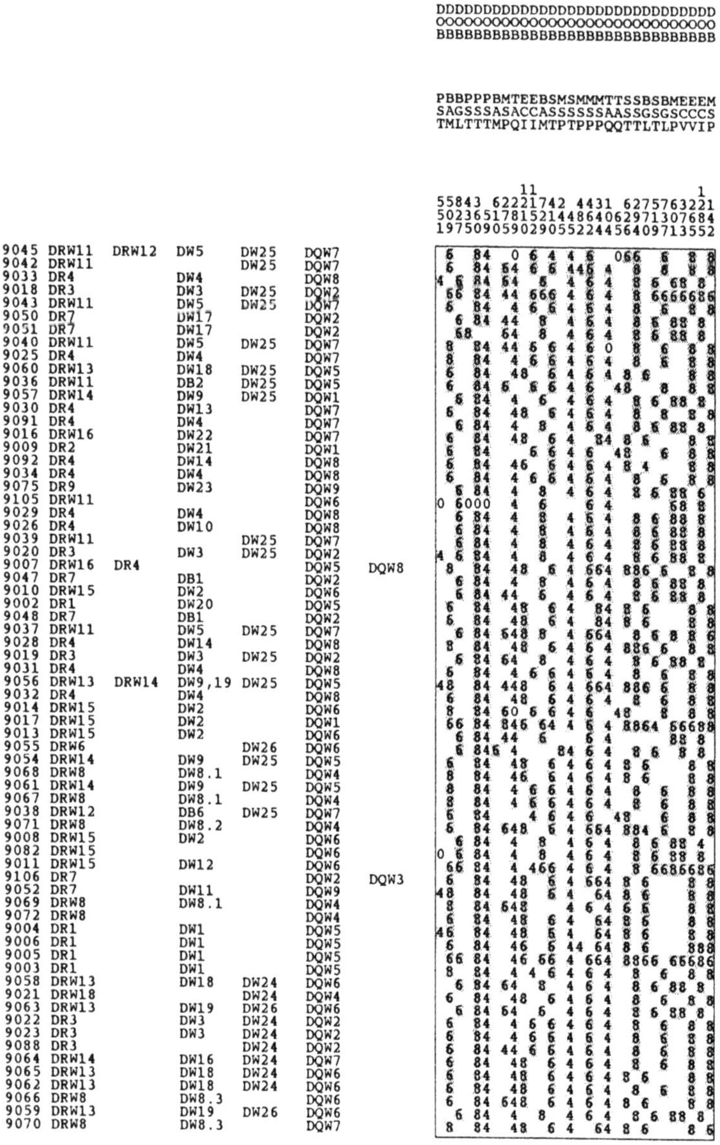

Figure 2. Tenth International Histocompatibility Workshop class II DOB locus RFLP fragments (n=30). Negative cluster.

a representative locus. The DOB locus is suitable for this purpose.

Class II DOB Locus. Figures 1 to 5 reveal the essential results. Figure 1 shows the pattern resulting from fragment by fragment cluster analysis. It can be seen that there are three clusters involving four or more fragments, and an additional five clusters, each comprising two fragments. Six of the total of 30 fragments generated by the 12 restriction endonucleases and detected using the DOB locus probe show reactivity patterns distinct from all the other fragments.

Two types of information can be obtained from the negative cluster analysis shown in Figure 2. First, the number of fragments associated with some cells is obviously different from that associated with other cells. Specifically, cells 9005, 9011, 9017, 9018, and 9037 are candidate heterozygotes. Similarly, but less obviously, cells 9007, 9056, 9106, 9028, 9069, and 9071 have features of heterozygosity. Second, mutually exclusive fragment pairs can be seen. Some such mutually exclusive fragments (e.g., BGL 827 and BAM 509; ECl 1285 and MSP 142; MSP 462 and MSP 364) are complete for all 68 cells. Others (BAM 729 and SST 410; SST 294 and BGL 770) are incomplete in that some cells lack both fragments.

Figure 3 shows the positive cluster analysis of the same 30 fragments. The fragments have been rearranged to illustrate the two clusters that characterize the DOB locus. The five cells recognized as candidate hetero-

Figure 3. Tenth International Histocompatibility Workshop class II DOB locus RFLP fragments (n=30). Cells listed by DQ specificity groups.

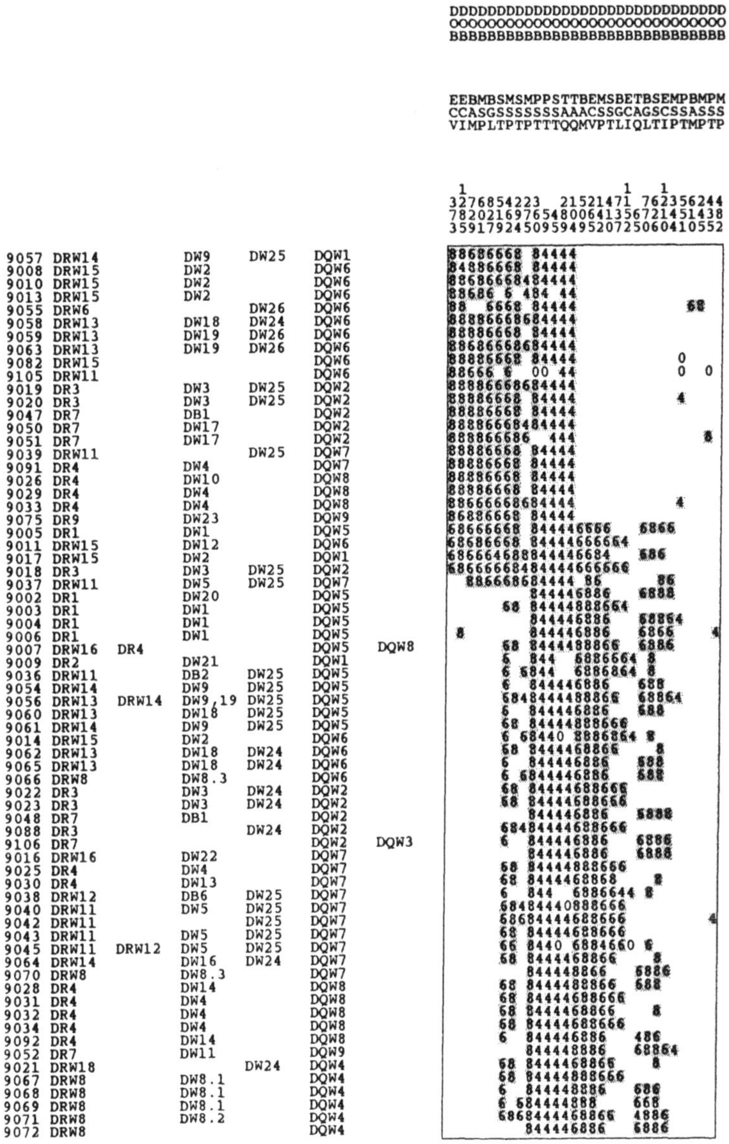

zygotes from negative clustering are seen to be true heterozygotes in that they have fragments associated with both of the two main clusters.

The cells in Figure 3 have been listed according to DQw type. Certain observations can be made about the distribution of DQw types in association with each of the two main clusters. For example, all DQw4 cells and all DQw5 cells apart from 9005 are associated with the second cluster in the lower half of the figure. Conversely, all but three of the DQw6 cells are associated with the cluster to the top left of the figure. Thus, it can be seen that DOB locus cluster type stratifies DQw types, in turn providing indications of evolutionary relationships.

Figure 4 depicts the tree analysis of relationships between the cells based on RFLP patterns associated with the presence or absence of the 30 fragments detected by the DOB locus fragments. Before looking in detail at the cell relationships, and at the degree of those relationships as expressed in the similarity index number, the quality of the RFLP data can be assessed by examining the similarity between the two "hidden" duplicate pairs, namely cells 9032 and 9056. At this locus and at all other loci, there was 100% concordance between the duplicates, providing confidence both in the quality of Southern blotting technical performance and in the validity of the computer-assisted transformation of the raw fragment data.

Figure 4 indicates that other cells have fragment patterns that are not distinguishable. These are also shown by the similarity index number "100." Subgroups of cells

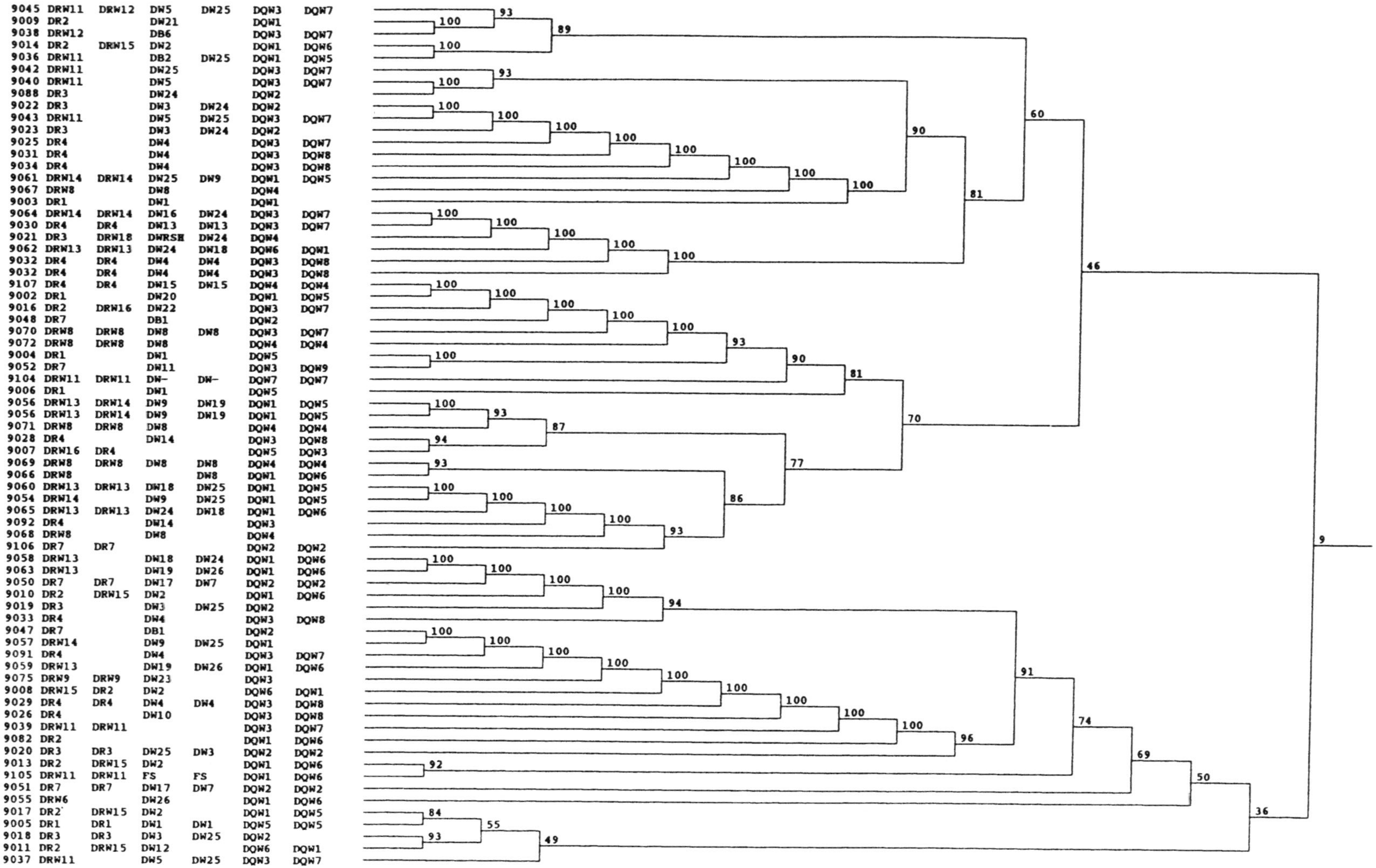

Figure 4. Tenth International Histocompatibility Workshop
Tree analysis of cell relationships: DOB locus.

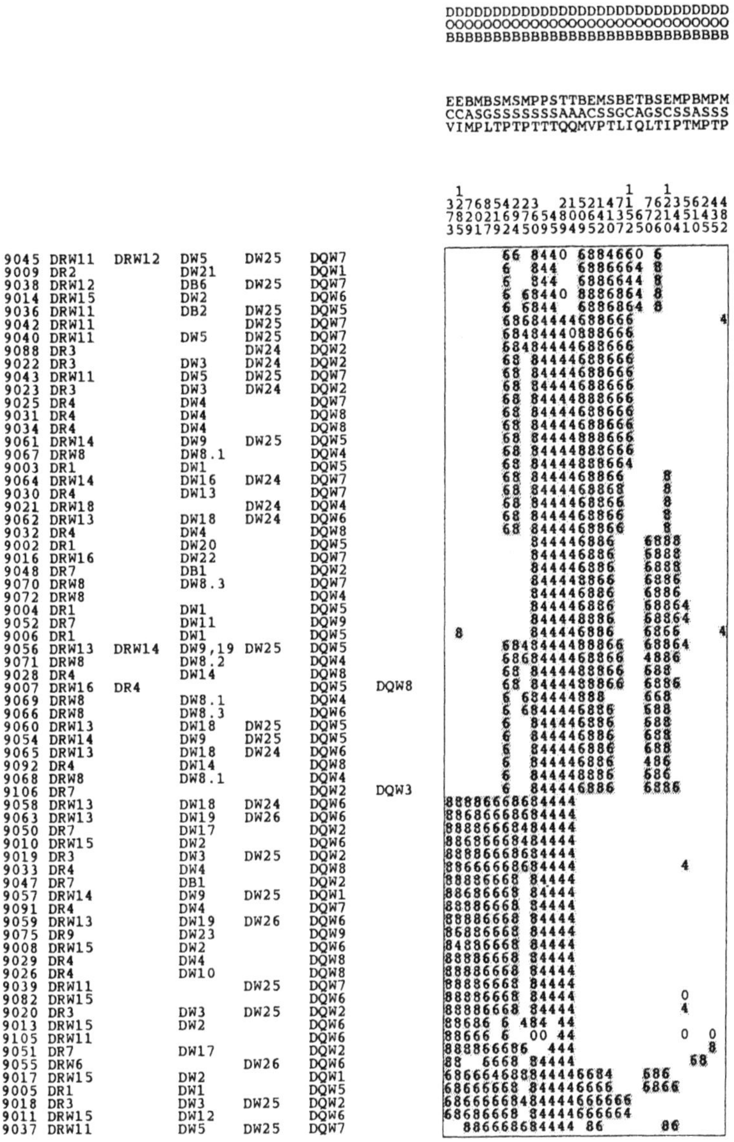

Figure 5. Tenth International Histocompatibility Workshop class II DOB locus RFLP fragments (n=30). Cells listed by tree relationships.

showing less than complete concordance form the basis of the tree. The number 9 at the extreme right of the figure indicates that there is virtually no association between the two major groups comprising the two main clusters.

Figure 5 depicts the positive cluster analysis in which the cells are listed based on the tree relationship. Comparison of Figure 5 with Figure 3 indicates that fragment reactivity in the former is more tightly blocked, confirming that the computer program has been more effective at grouping similar cells than the subjective attempt represented by Figure 3. The patterns of variation associated with both of the main clusters are more apparent from Figure 5.

Class I HLA-A, -B, -C Loci. In conventional HLA typing, it is the goal to seek reagents that are "unique" for individual HLA specificities. The practice of HLA serology evolves by identification of sera that react with one or a limited number of defined specificities. Similarly, RFLP fragments can be identified that are uniquely associated with discrete serologic specificities. Other fragments are generated by cells having one or more of the specificities within a given group and are comparable to the useful multispecific sera. (This comparison between HLA serology and RFLP analysis is qualified as described in the Introduction.) With time, reagents with specificities that progressively approach monoreactivity become identified. With nucleotide

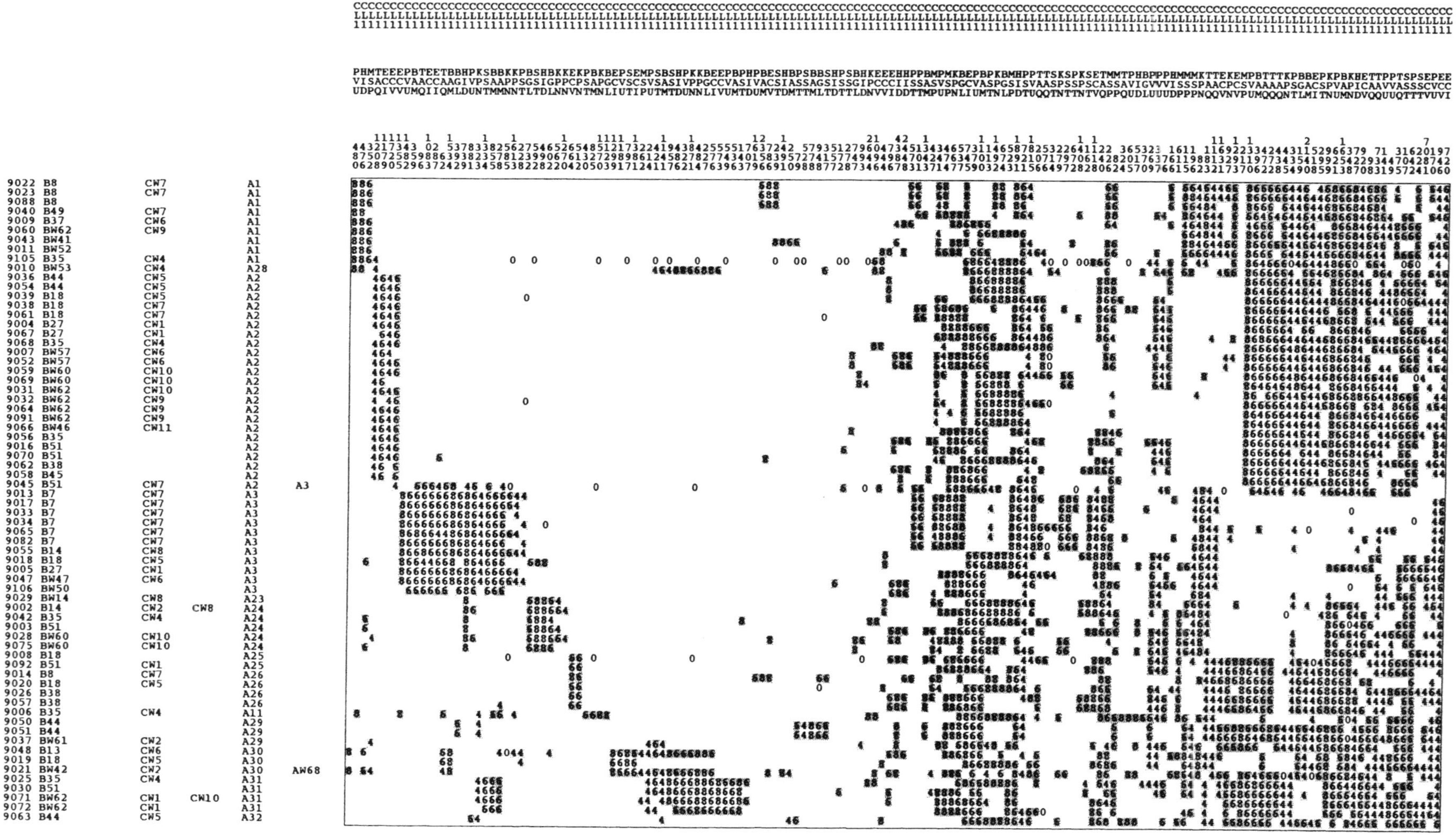

Figure 6A. Tenth International Histocompatibility Workshop
Class I loci RFLP fragments (n=204). Positive cluster. Cells listed by A locus type.

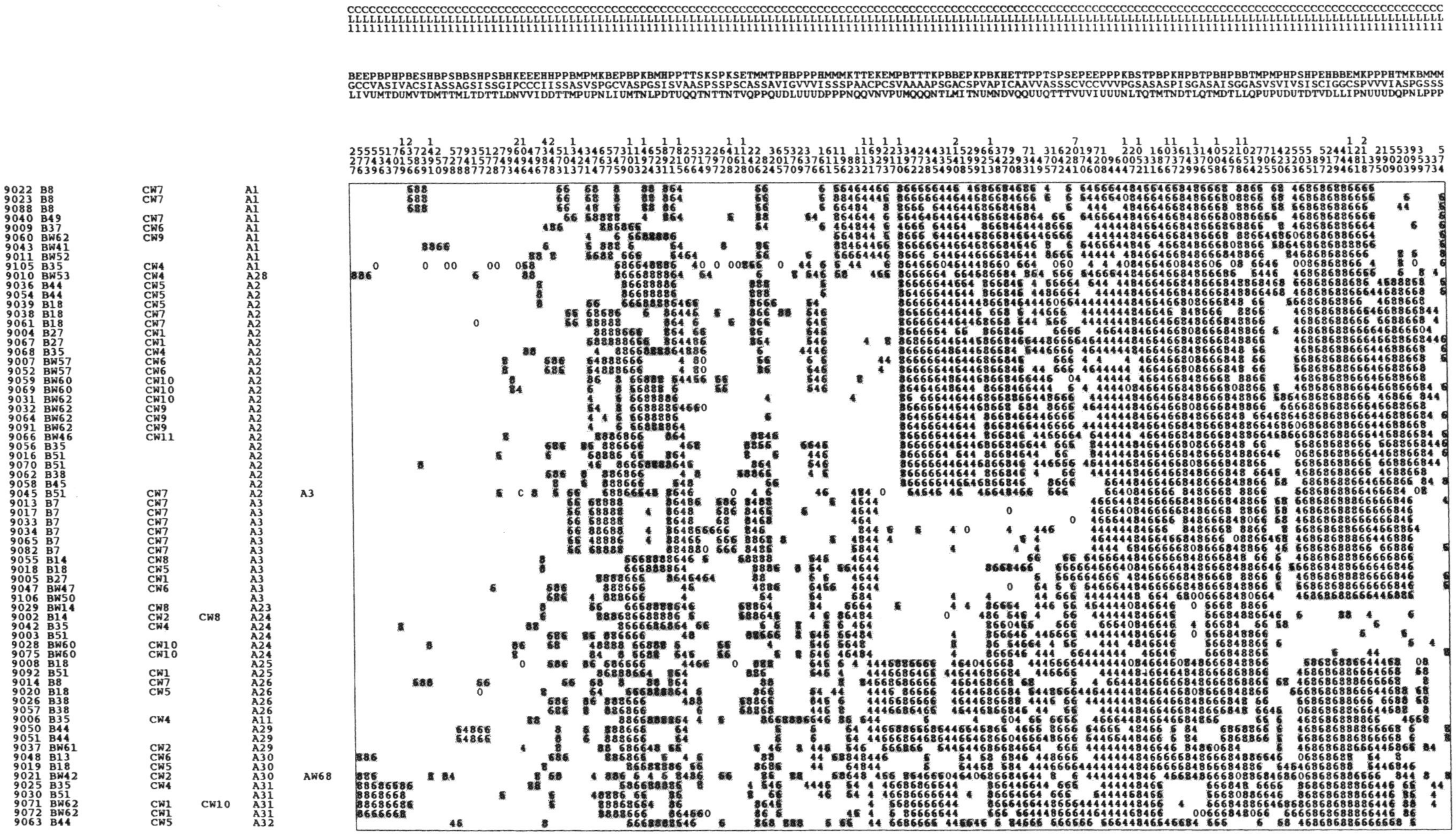

Figure 6B. Tenth International Histocompatibility Workshop
Class I loci RFLP fragments (n=204). Positive cluster. Cells listed by A locus type.

Table 1. "Unique" RFLP Fragments for Class I Locus Specificity Assignment

	BAM	BGL	ECI	ECV	HID	KPN	MSP	PST	PVU	SST	TAQ
RFLP Specificity A-											
A1					(476)						
A2			[1179]	[1355]							
A3			287 1083						(1482)		62 96
A23,A24		(602)			(498)	(562)				(592)	
A23,<A24						1270					
<A24				664							
A25,A26						512		430			
A11	825	1123	1299			1570					
A30			1381				262		1181	797	
A31	536				(707)			143	549		
<A31									(1619)		
RFLP Specificity B-											
B8	2356			(786)							
Bw41					1491						
Bw41,Bw42	250							(79)			
<B44	778	948								528	
Bw47								172			
<B51										278	
<Bw57		(747)									
Bw60					993						
Bw61,<Bw60					2644						
RFLP Specificity C-											
Cw4				1096							
Cw7								101			
<Cw7	1343										
C blank							472				

Unique RFLP fragments have been defined as those fragments that correlate completely with the latest HLA type assigned to each Workshop cell. Since the Workshop cells include those having variant/atypical fragment reactivity patterns, the definition has been extended to include fragments with either a single reactivity in excess of the stated specificity, or a single reactivity less than the stated specificity. Such fragments are identified by using () for additional reactivity and [] for deficient reactivity. The cells corresponding to both types of parentheses are listed below in this footnote.

Other fragments exist that have reactivity, or lack of reactivity, with two or more cells apart from that with the primary specificity. Some of these fragments, particularly those with reactivity to only a limited number of cell specificities, and those with higher frequency reactivity but in which the reactivity correlates closely with several specificities, are also useful for HLA specificity typing.

RFLP Specificity A-	Fragment	Cell Number
A1	HID 476	(6)
A2	ECI 1179	[32]
A2	ECV 1355	[7,69]
A3	PVU 1482	(6)
A23,A24	BGL 602	(18)
A23,A24	HID 498	(18)
A23,A24	SST 592	(18)
A23,A24	KPN 562	(48)
A31	HID 707	(42)
RFLP Specificity B-		
B8	ECV 786	(70)
Bw41,Bw42	PST 79	(63)
<Bw57	BGL 747	(66)

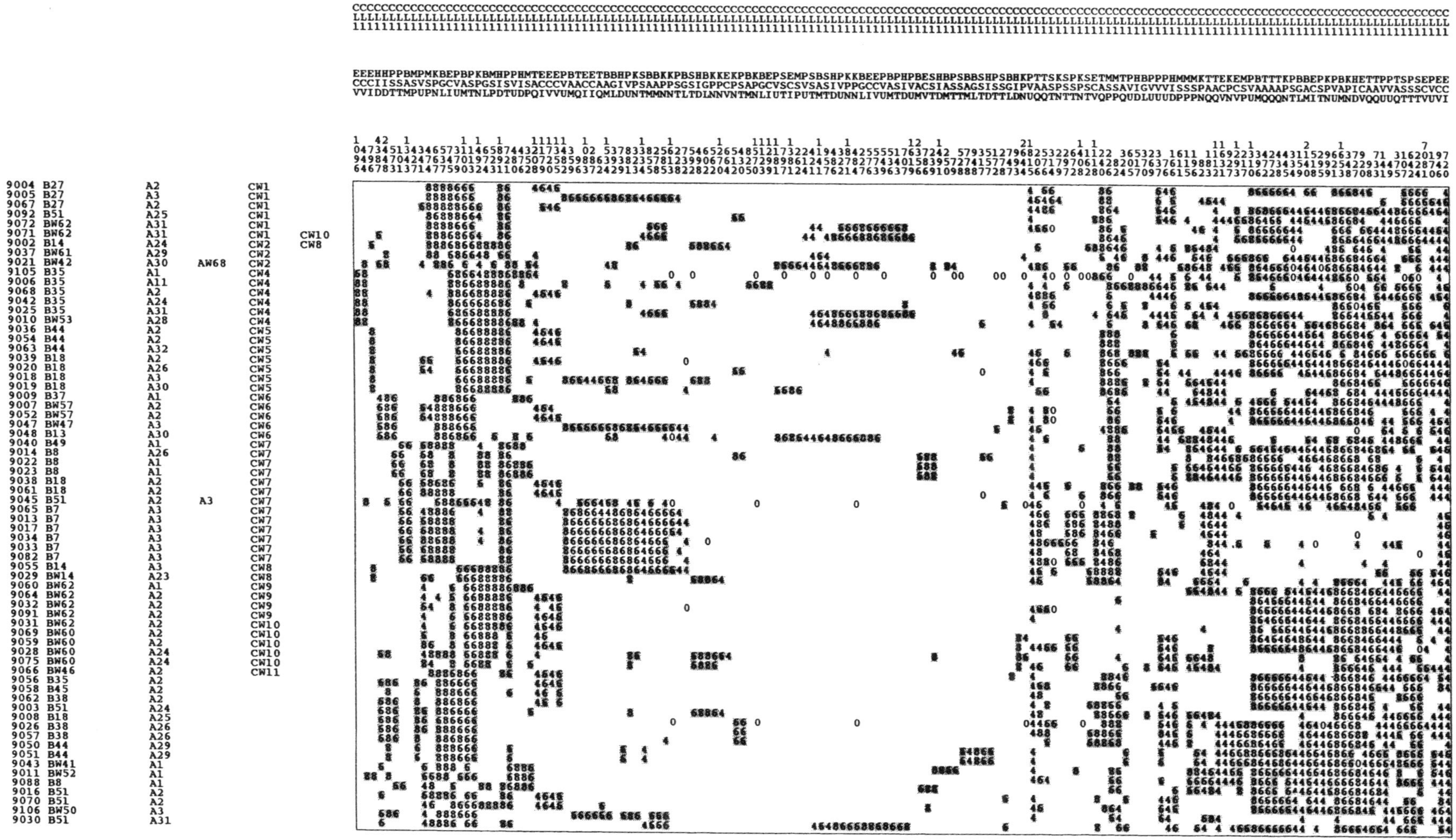

Figure 7A. Tenth International Histocompatibility Workshop
Class I loci RFLP fragments (n=204). Positive cluster. Cells listed by C locus type.

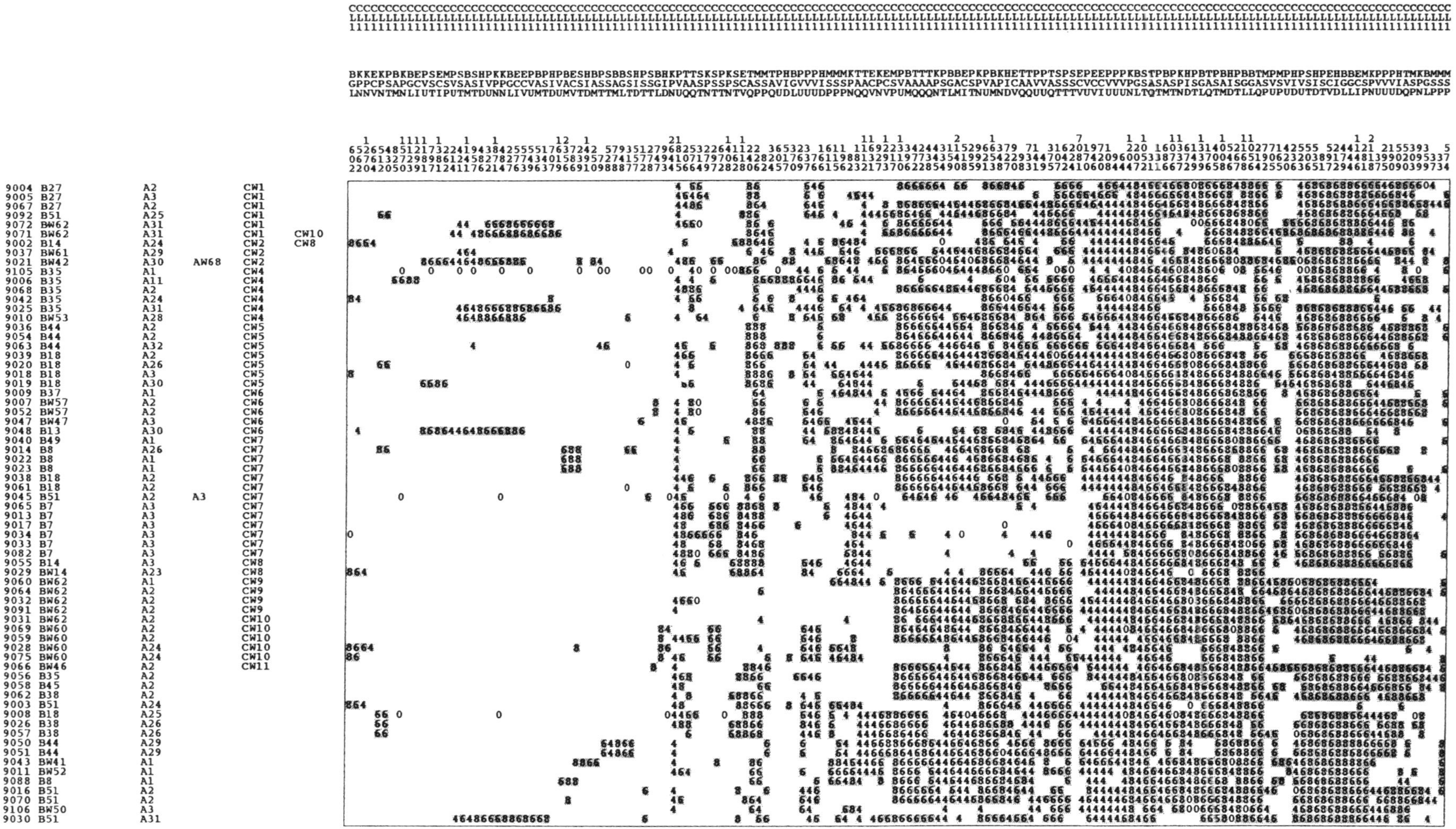

Figure 7B. Tenth International Histocompatibility Workshop
Class I loci RFLP fragments (n=204). Positive cluster. Cells listed by C locus type.

Figure 8. Tenth International Histocompatibility Workshop class I C locus-associated RFLP fragments (n=22). Negative cluster.

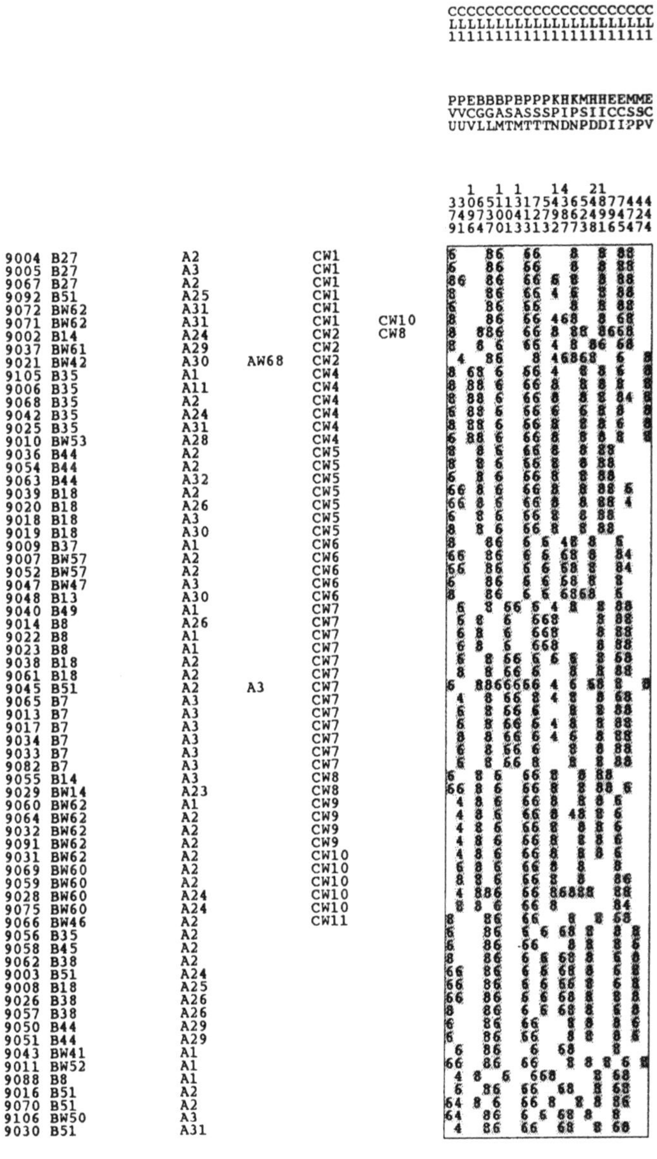

sequence information now becoming available for the HLA class I alleles at each locus, and with the advent of locus-specific probes, it will be possible to perform DNA typing for HLA class I alleles.

This situation is rapidly developing also for HLA class II typing. Nucleotide sequence information is available for many of the alleles of both class II alpha and class II beta genes. For the present, results of the HLA class I RFLP Workshop indicate that most of the fragment patterns corresponding to alleles at the HLA-A and -C loci can be detected despite using the HLA class I-specific DNA probe rather than locus-specific probes.

Two hundred and four (204) fragments were detected by the class I probe using 11 enzymes. Figure 6 shows the positive clustering of the total fragments against 68 cells listed by HLA-A specificities. Fragments associated with single or limited number of serologic specificities have been rearranged to illustrate the associations. To the left of the figure, from PVU 480 to PVU 1619, can be seen fragments that correspond well with the serologic HLA-A specificities. These are followed by fragments corresponding with the HLA-B specificities B8, Bw41, Bw42, Bw44, Bw47, B51, Bw57, Bw60, and Bw61 (BAM 2356 to KPN 2644). Then follow 22 fragments associated with HLA-C specificities (ECV 1096 to PST 721). Four of these "specificity-unique" C locus fragments along with 13 for HLA-B and 29 for HLA-A specificities are listed in Table 1. Fragments have been included that are present in one additional cell, or absent in one cell stated to bear

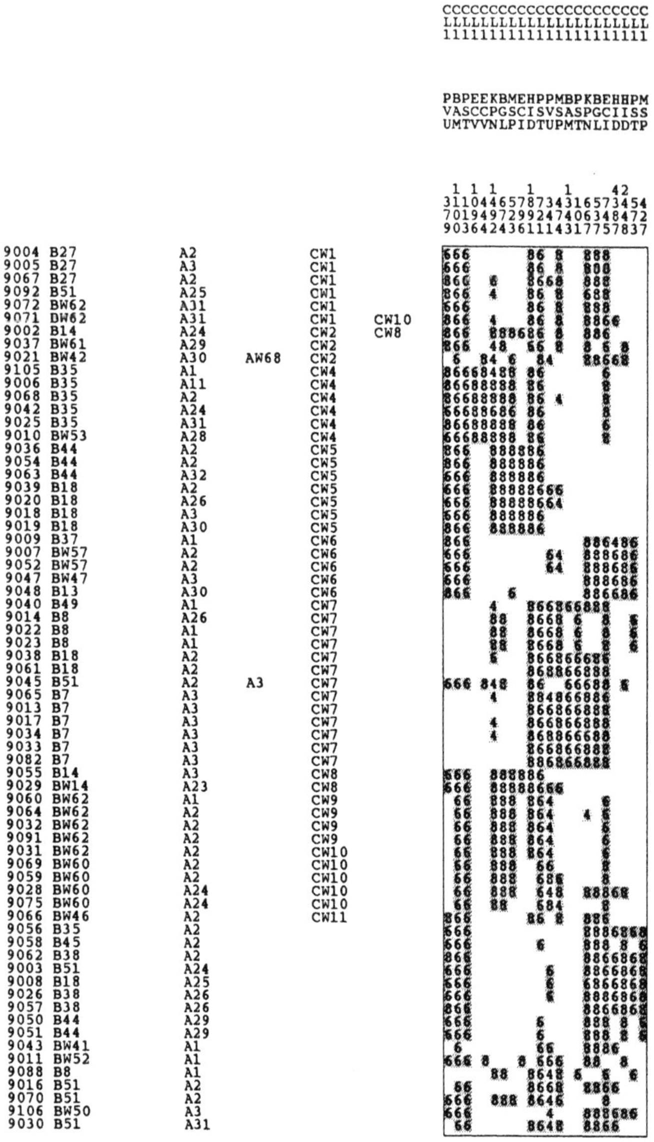

Figure 9. Tenth International Histocompatibility Workshop class I C locus-associated RFLP fragments (n=22). Positive cluster.

the relevant specificity. The 22 HLA-C-associated fragments are examined in more detail in Figures 7, 8, and 9. In Figure 6B, the remaining fragments are shown to the right of the HLA-C associated fragments as two major groups. The first group of 31 fragments (PVU 1815 to MSP 217), being present in many cells, showed patterns of co-clustering with individual specificity-associated fragments before the columns were rearranged to simplify recognition of the single specificity fragments. The second group (PVU 1310 to MSP 574) remain in their original position following cluster analysis of the total 204 fragments. Where haplotypes occur two or more times, the distinctive fragment patterns can be readily seen. It is not so obvious that all haplotypes are characterized by distinctive fragment patterns, since many of the haplotypes occur in the reference cell panel only once. In fact, all cells distinguishable by serology are also distinguishable by RFLP patterns.

In Figure 7, the HLA-C related fragments have been shifted to the first 22 columns (ECV 1096 to PST 721), and the cells rearranged according to HLA-C specificity. In Figure 8, negative clustering of the HLA-C-associated fragments reveals the basis for the association. As mentioned earlier, in RFLP, each enzyme generates a finite number of fragments. The mutually exclusive patterns of fragments produced by any one enzyme/probe combination are used for the characterization of HLA specificities being positively detectable by the presence of fragments, and confirmed by the absence of another group of fragment(s). Examination of the cell distribution of

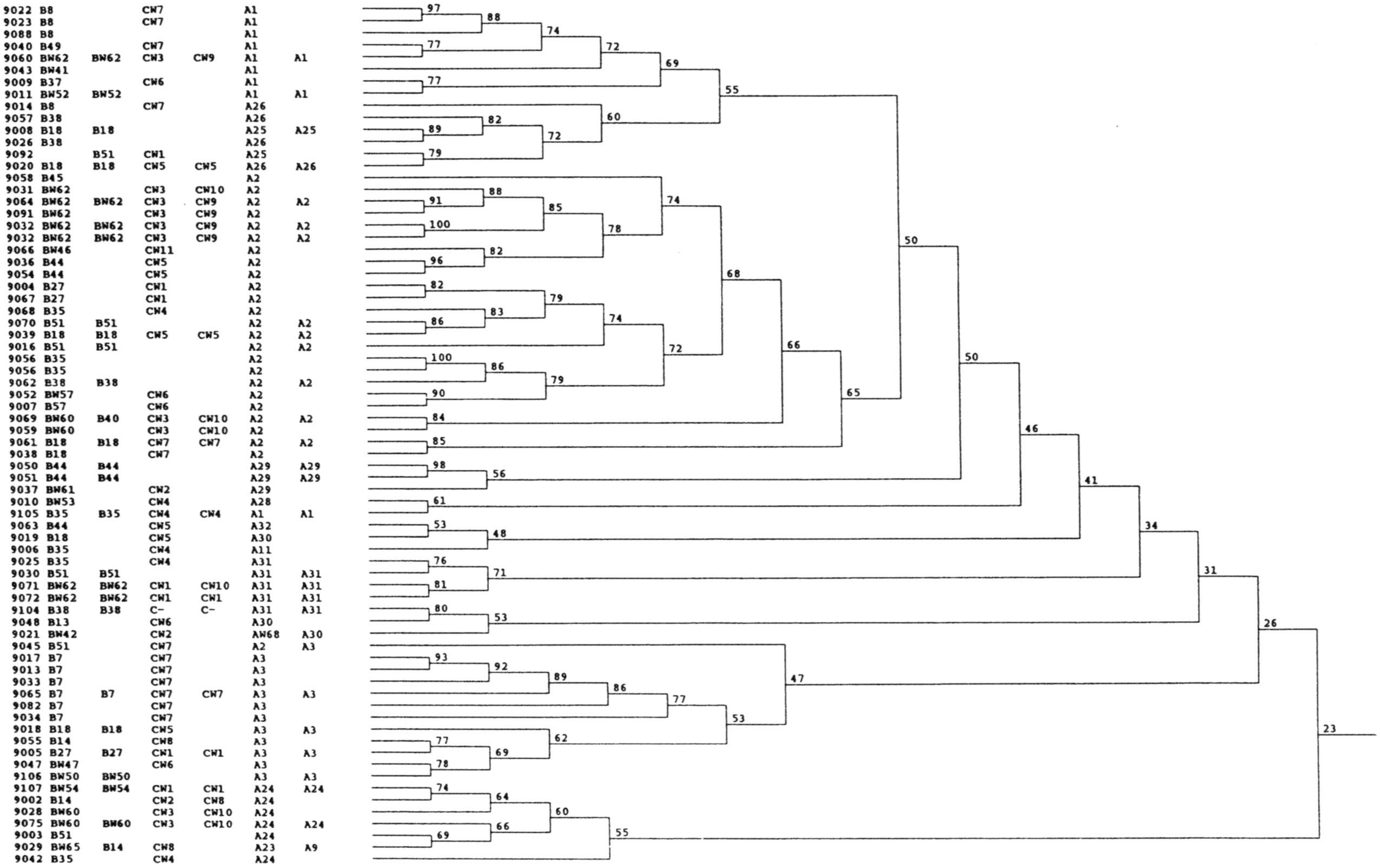

Figure 10. Tenth International Histocompatibility Workshop
Tree analysis of cell relationships: class I RFLP fragments (n=204).

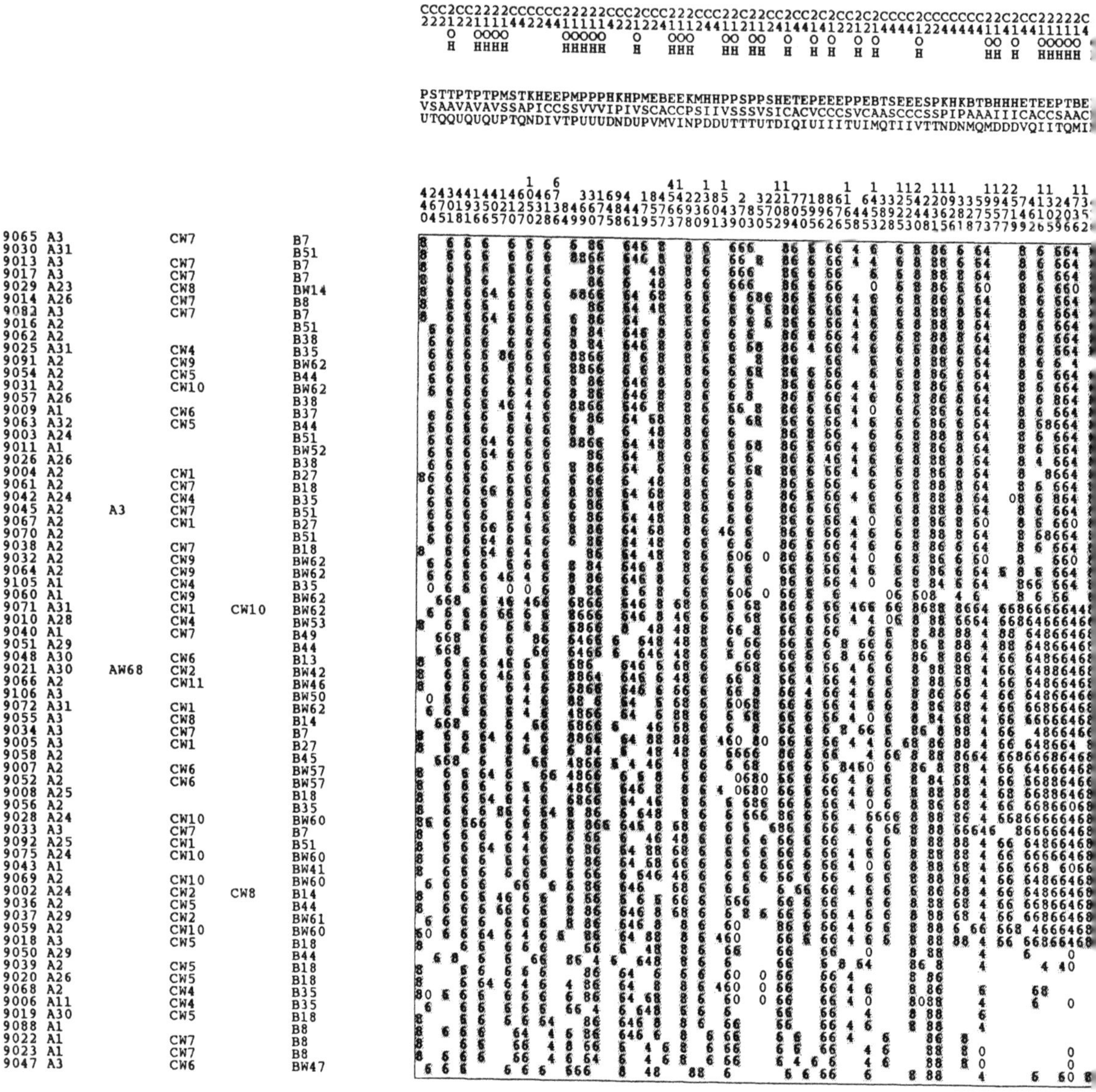

Figure 11. Tenth International Histocompatibility Workshop class III loci RFLP fragments (n=78). Negative cluster.

mutually exclusive fragments in Figure 9 reveals the relationship between RFLP patterns and serologic specificity. Fragments exist which, in serologic terms, would be regarded as "short" or "long" in that they are observed in less than all cells of a particular serologic specificity, or they arise in all cells having a particular specificity, plus additional cells. Since these varying degrees of correlation between fragments and exon product-detectable specificities reflect linkage disequilibrium between the restriction site-bearing sequence and the exon sequence, HLA typing by RFLP is based on the empiric identification of those enzymes that happen to generate fragments in linkage disequilibrium with allele-specific sequences.

From inspection of Figure 9, it is apparent that the main HLA-C specificities correspond to distinct RFLP patterns. Sixteen cells listed at the bottom of Figure 9 lack assigned HLA-C specificities. Some of the 16 cells have RFLP patterns that closely resemble those seen with known HLA-C specificities. It should be noted that fragment MSP 427 is present only in the 16 HLA-C "blank" cells, occurring nine times.

Figure 10 shows the tree dendogram analysis of cell relationships according to class I RFLPs. It can be seen that clusters within the tree reflect HLA haplotypes. Where individual haplotypes occur two or more times, the clustering is obvious and the clusters involve high

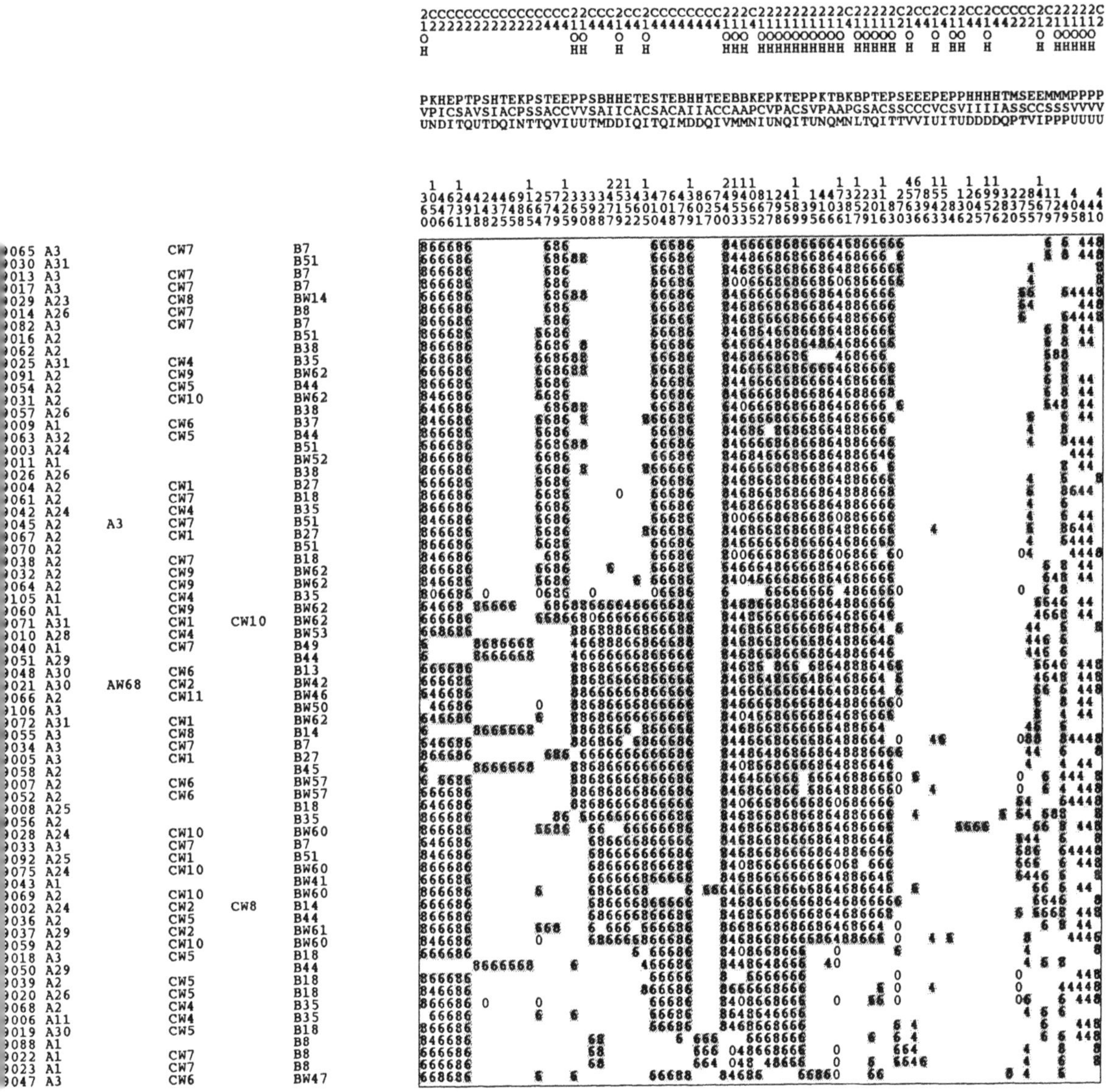

Figure 12. Tenth International Histocompatibility Workshop class III loci (C2,C4,210H) RFLP fragments (20,24,34). Positive cluster.

similarity indices. Where haplotypes occur only once, the similarity index between adjacent pairs is correspondingly lower. At face value, the clustering patterns suggest a predominant role of the HLA-A locus on the RFLP obtained with the HLA class I probe. It should, however, be realized that the reference panel cells represent a small number of HLA-A alleles relative to the number of HLA-B alleles.

Tree dendogram analysis (Fig. 10) confirms that the combined use of both fragment presence and specificity-associated fragment absence enables cells to be grouped according to established serospecificities, and to be distinguishable.

Class III Loci. A total of 78 fragments are available for analysis using class III locus probes (C2-20; C4-24; 210H-34). Negative cluster of these 78 fragments is shown in Figure 11. The column-ordering of fragments reveals mutually exclusive fragment pairs for single class III probes (e.g., C2: PVU 418, TAQ 491; C2: ECI 676, PST 1165; C2: KPN 988, HID 446. C4: SST 410, TAQ 627; C4: TAQ 707, ECI 750). For class III loci in partic-

9045	DQW3	DQW7	DW5	DW25	DRW11	DRW12
9061	DQW1	DQW5	DW25	DW9	DRW14	DRW14
9070	DQW3	DQW7	DW8	DW8	DRW8	DRW8
9067	DQW4		DW8		DRW8	
9003	DQW1		DW1		DR1	
9011	DQW6	DQW1	DW12		DR2	DRW15
9009	DQW1		DW21		DR2	
9026	DQW3	DQW8	DW10		DR4	
9031	DQW3	DQW8	DW4		DR4	
9054	DQW1	DQW5	DW9	DW25	DRW14	
9016	DQW3	DQW7	DW22		DR2	DRW16
9062	DQW6	DQW1	DW24	DW18	DRW13	DRW13
9064	DQW3	DQW7	DW16	DW24	DRW14	DRW14
9032	DQW3	DQW8	DW4	DW4	DR4	DR4
9032	DQW3	DQW8	DW4	DW4	DR4	DR4
9057	DQW1		DW9	DW25	DRW14	
9030	DQW3	DQW7	DW13	DW13	DR4	DR4
9065	DQW1	DQW6	DW24	DW18	DRW13	DRW13
9014	DQW1	DQW6	DW2		DR2	DRW15
9038	DQW3	DQW7	DB6		DRW12	
9029	DQW3	DQW8	DW4	DW4	DR4	DR4
9082	DQW1	DQW6			DR2	
9017	DQW1	DQW5	DW2		DR2	DRW15
9013	DQW1	DQW6	DW2		DR2	DRW15
9042	DQW3	DQW7	DW25		DRW11	
9004	DQW5		DW1		DR1	
9063	DQW1	DQW6	DW19	DW26	DRW13	
9025	DQW3	DQW7	DW4		DR4	
9091	DQW3	DQW7	DW4		DR4	
9105	DQW1	DQW6	FS	FS	DRW11	DRW11
9107	DQW4	DQW4	DW15	DW15	DR4	DR4
9028	DQW3	DQW8	DW14		DR4	
9033	DQW3	DQW8	DW4		DR4	
9043	DQW3	DQW7	DW5	DW25	DRW11	
9092	DQW3		DW14		DR4	
9075	DQW3		DW23		DRW9	DRW9
9036	DQW1	DQW5	DB2	DW25	DRW11	
9002	DQW1	DQW5	DW20		DR1	
9005	DQW5	DQW5	DW1	DW1	DR1	DR1
9104	DQW7	DQW7	DW-	DW-	DRW11	DRW11
9010	DQW1	DQW6	DW2		DR2	DRW15
9056	DQW1	DQW5	DW9	DW19	DRW13	DRW14
9056	DQW1	DQW5	DW9	DW19	DRW13	DRW14
9066	DQW1	DQW6		DW8	DRW8	
9021	DQW4		DWRSH	DW24	DR3	DRW18
9048	DQW2		DB1		DR7	
9052	DQW3	DQW9	DW11		DR7	
9008	DQW6	DQW1	DW2		DRW15	DR2
9007	DQW5	DQW3			DRW16	DR4
9059	DQW1	DQW6	DW19	DW26	DRW13	
9034	DQW3	DQW8	DW4		DR4	
9069	DQW4	DQW4	DW8	DW8	DRW8	DRW8
9037	DQW3	DQW7	DW5	DW25	DRW11	
9071	DQW4	DQW4	DW8		DRW8	DRW8
9106	DQW2	DQW2			DR7	DR7
9072	DQW4	DQW4	DW8		DRW8	DRW8
9018	DQW2		DW3	DW25	DR3	DR3
9019	DQW2		DW3	DW25	DR3	
9068	DQW4		DW8		DRW8	
9039	DQW3	DQW7			DRW11	DRW11
9020	DQW2	DQW2	DW25	DW3	DR3	DR3
9006	DQW5		DW1		DR1	
9047	DQW2		DB1		DR7	
9058	DQW1	DQW6	DW18	DW24	DRW13	
9055	DQW1	DQW6	DW26		DRW6	
9051	DQW2	DQW2	DW17	DW7	DR7	DR7
9040	DQW3	DQW7	DW5		DRW11	
9060	DQW1	DQW5	DW18	DW25	DRW13	DRW13
9050	DQW2	DQW2	DW17	DW7	DR7	DR7
9022	DQW2		DW3	DW24	DR3	
9023	DQW2		DW3	DW24	DR3	
9088	DQW2		DW24		DR3	

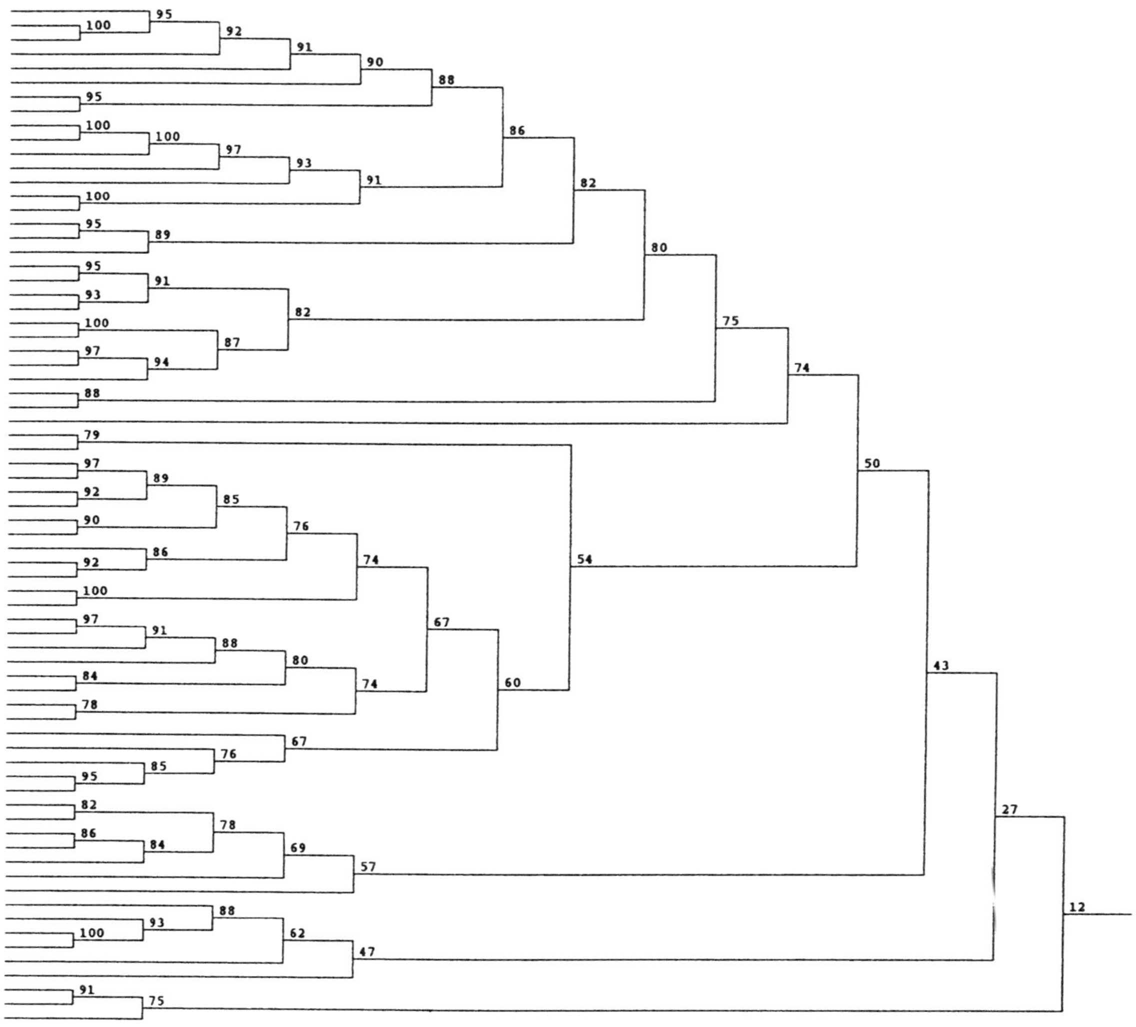

Figure 13. Tenth International Histocompatibility Workshop
Tree analysis of cell relationships: class III loci.

Figure 14. Tenth International Histocompatibility Workshop class III C4 RFLP fragments (n=26). First 25 cells listed by C4 phenotype.

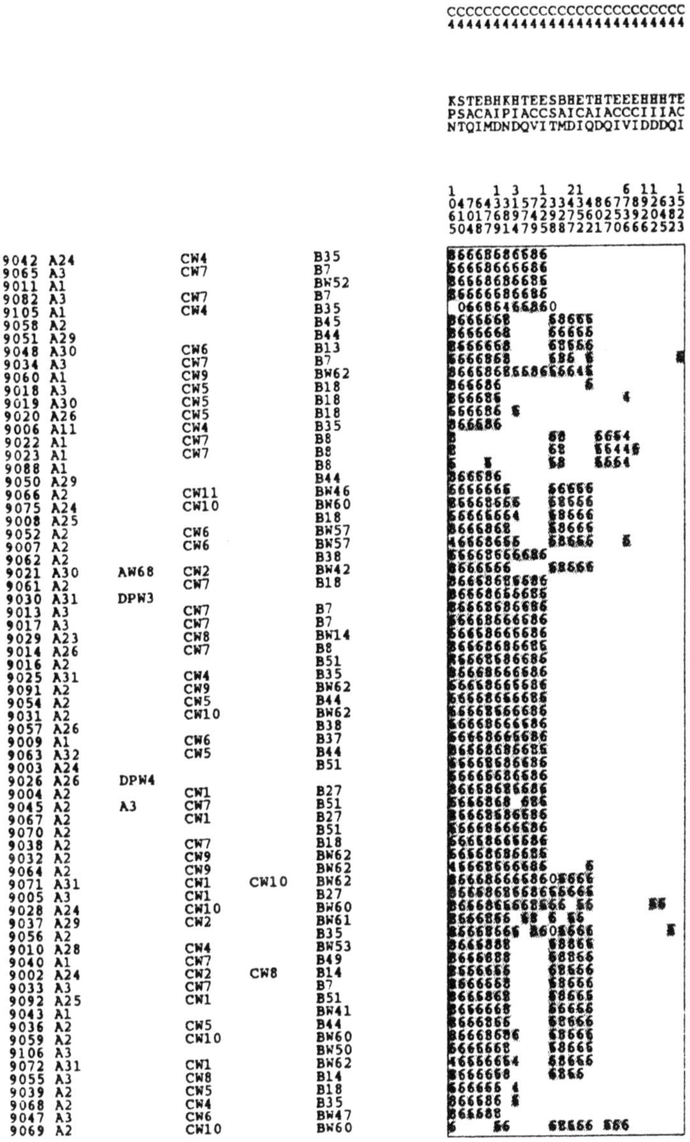

ular, caution needs to be exercised in the interpretation of pairs of fragments that are mutually exclusive since gene duplications, gene deletions, and the existence of pseudogenes are well established. In the case of the C4 and 210H loci, interpretation of the existence of mutually exclusive fragment pairs such as BAM 477 and ECV 4563 is uncertain. These issues are discussed in more detail in the RFLP report for the class III region (5) and in the overview of molecular biology of HLA class III presented in volume II of these proceedings (6).

The negative cluster (Fig. 11) indicates three broad groupings. The upper cells from 9065 to 9105 are relatively restricted in their heterogeneity and are the least complex. The second group of cells (9060 to 9059) have significantly increased number of fragments, particularly involving the 210H loci. The third group of cells (9018 to 9047) have an obvious block absence of fragments in the bottom right corner, while being similar to the cells of the first group for the remainder of the fragment pattern. These latter are known to include cells having deletions in the C4/210H region (5). This situation is revealed in a more obvious manner by Figure 12, in which the fragments are positively clustered against the same listing of cells.

Cell relationships involving these three loci are shown as a tree dendogram in Figure 13. In Figure 14, the cells have been regrouped so that those 25 for which C4A and C4B allotyping is available are shown at the top of the

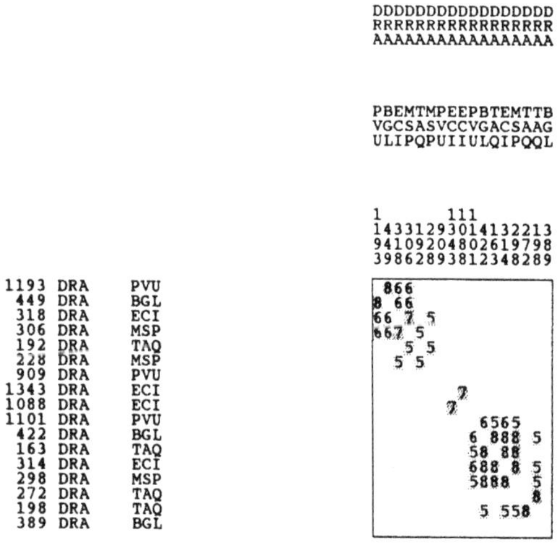

Figure 15. Tenth International Histocompatibility Workshop class II DRA locus RFLP fragments (n=17). Fragment cluster.

list. It is apparent that some RFLP patterns group according to C4 phenotype: C4A4, C4B2–9066, 9075, 9008; C4A3, C4BQ0–9018, 9019, 9020; C4A6, C4B1–9052, 9007). Cells typing for C4A3, C4B1 comprise two major RFLP pattern groups (9042, 9065, 9011, 9082; and 9058, 9051, 9048, 9034). Two cells (9105 and 9060) have unique patterns. For C4AQ0, C4B1, the prototype pattern is represented by three cells 9022, 9023, 9088 with the classical HLA haplotype A1, Cw7, B8, while the fourth cell of the group (9050) has a pattern indistinguishable from that of C4A3, C4BQ0 cells. The observed heterogeneity within C4 phenotypes probably reflects the frequent occurrence of gene duplication, gene deletion and other gene rearrangements (6), resulting in the phenomena of contraction and/or expansion within the HLA class III region (7).

Class II Alpha Loci. *DRA*: There were 17 informative fragments detected by the DRA probe. In Figure 15, the cluster pattern of these fragments is shown. Negative cluster of these fragments against cells listed by DR type is shown in Figure 16. Mutually exclusive fragments can be seen involving the three enzymes PVU, TAQ, and MSP. It is known from nucleotide sequence analysis of the DRA coding region that the expressed gene is nonpolymorphic (8). Thus, the DRA RFLP fragment patterns must reflect genetic linkage disequilibrium with polymorphic restriction sites in noncoding regions. Nonetheless, these DRA RFLPs do provide useful markers for DNA typing since they correlate strongly with certain DR specificities as shown in Figures 17 and 18. In Figure 17, the fragments have been positively clustered against cells listed by DR specificities. From cursory inspection, it is obvious that the block fragment patterns relate to DR specificities. The fragments are seen to comprise two main clusters. These patterns are more readily discerned when the cells are sorted according to their tree relationships (Fig. 18). The less frequent cluster is present in the last 12 cells in the list. These 12 cells comprise all 8 Dw24 cells, 2 of the 3 Dw26 cells, and both DRw8 cells assigned as Dw8.3. Although only six or seven DRA sequences have been determined, this limited number does include haplotypes of both the major RFLP clusters.

DQA: Of the 74 fragments associated with the DQA1 and DQA2 loci, two (KPN 716 and KPN 232) have been excluded. Figure 19 shows a cluster analysis of the remaining 72 fragments. Negative clustering of these fragments against cells listed by DR type is shown in Figure 20, in which the 10 cells that are heterozygous for either or both loci have been listed at the bottom of the figure. The last four cells (9008, 9007, 9028, 9106) are heterozygous at the DQA2 locus as revealed by the PVU fragments 819 and 722 (the first two columns) and by the TAQ fragments 209 and 155. Three of these four cells (9007, 9028, 9106) and the remaining six cells (9005, 9017, 9018, 9045, 9056, 9071) are heterozygous at DQA1.

Figure 21 presents the fragments as positive clusters against cells listed by DQ, Dw, and DR specificities. DQA-associated fragments occupy the left two-thirds of the figure (SST 1653 to PST 912). As with the DRA locus, the block patterns of fragment presence and absence are apparent, with greater resolution being achieved with the substantially larger number of DQA fragments. Six TAQ fragments (TAQ 817, 699, 619, 553, 285, 278) are of doubtful validity since they bear no apparent relation to DR or DQ specificities. Fragments reflecting the two major clusters associated with the DQA2 locus are shown at the extreme right of the figure (PVU 819 to MSP 568). These data are consistent with the results obtained by a detailed analysis of the HLA-DQ region demonstrating polymorphic restriction sites, both in the flanking regions of the DQA2 genes, as well as in the DQA2 exons (9).

In Figure 22, fragments have been positively clustered against cells listed by tree relationships. The reactivity pattern blocks are even more apparent, as is the lack of genetically meaningful association involving the six TAQ fragments.

Sequence analysis of DQA1 genes has identified eight distinct alleles (10). Six of the eight occur commonly, and have been designated DQA1.1, DQA1.2, DQA1.3, DQA2, DQA3, and DQA4. RFLP patterns associated with reference panel cells that have been DQA1 allele typed are shown in Figure 23. The first eight cells listed (9005 to 9061) have the DQA1.1 sequence type. The ninth cell, 9056, is heterozygous for DQA1.1 and DQA1.2. These nine cells have two fragments in common (KPN 1042; TAQ 256). The next eight cells (to 9008) are DQA1.2. The sequence type of cells 9008 and 9082 is not known. The next five cells (9011 to 9060) are DQA1.3. DQA2 cells are associated with DR7. MSP 605 is unique for DQA2. DQA3 is associated with DR4 and DR9 cells, and identified by MSP 215. DQA4 is present in cells expressing DR3, 5, and 8. Two fragments (HID 579; ECI 1107) characterize DQA4.

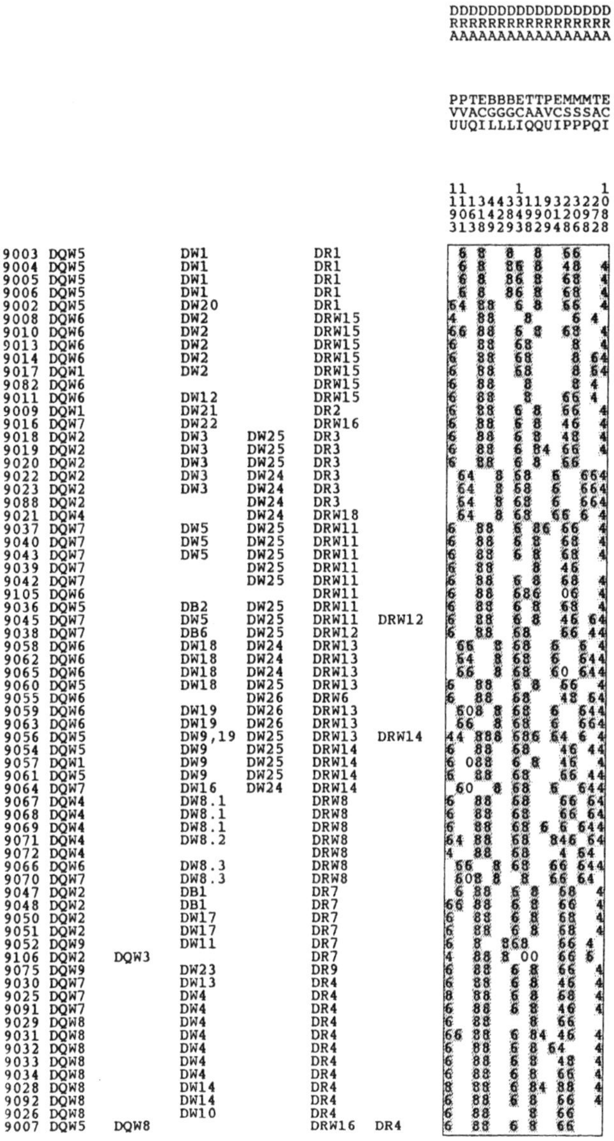

Figure 16. Tenth International Histocompatibility Workshop class II DRA locus RFLP fragments (n=17). Negative cluster.

```
                                             DDDDDDDDDDDDDDDDD
                                             RRRRRRRRRRRRRRRRR
                                             AAAAAAAAAAAAAAAAA

                                             EBPMTMETBEPEBTTMP
                                             CGVSASCAGCVCGAASV
                                             ILUPQPIQLIUILQQPU

                                                   1  11     1
                                             34921331301341221
                                             14029049880126799
                                             89982638981423283
9003 DQW5        DW1           DR1           8  686  8 6
9004 DQW5        DW1           DR1           8  4886 846
9005 DQW5        DW1           DR1           8  6886 846
9006 DQW5        DW1           DR1           8  6886 846
9002 DQW5        DW20          DR1           88 6866  44     6
9008 DQW6        DW2           DRW15         88   6 8      4 4
9010 DQW6        DW2           DRW15         88 6886  46     6
9013 DQW6        DW2           DRW15         88   868 4      6
9014 DQW6        DW2           DRW15         88   868 4    6 6
9017 DQW1        DW2           DRW15         88   868 4    6 6
9082 DQW6                      DRW15         88   8 8      4 6
9011 DQW6        DW12          DRW15         88 6 6 8      4 6
9009 DQW1        DW21          DR2           88 6866  4      6
9016 DQW7        DW22          DRW16         88 4866  4      6
9018 DQW2        DW3     DW25  DR3           88 4886  4      6
9019 DQW2        DW3     DW25  DR3           8846866  4      6
9020 DQW2        DW3     DW25  DR3           88 6866         6
9022 DQW2        DW3     DW24  DR3                 68 4668466
9023 DQW2        DW3     DW24  DR3                 68 4668466
9088 DQW2                DW24  DR3                 68 4668466
9021 DQW4                DW24  DRW18            6  68 46684 6
9037 DQW7        DW5     DW25  DRW11         8866866  4      6
9040 DQW7        DW5     DW25  DRW11         88 6886  4      6
9043 DQW7        DW5     DW25  DRW11         88 6886  4      6
9039 DQW7                DW25  DRW11         88 486          6
9042 DQW7                DW25  DRW11         88 6886  4      6
9105 DQW6                      DRW11         88 06668 4      6
9036 DQW5        DB2     DW25  DRW11         88 6886  4      6
9045 DQW7        DW5     DW25  DRW11  DRW12  88 4866  4    6 6
9038 DQW7        DB6     DW25  DRW12         88 6 668 4    4 6
9058 DQW6        DW18    DW24  DRW13               68 46686 6
9062 DQW6        DW18    DW24  DRW13               68 4668446
9065 DQW6        DW18    DW24  DRW13            0  68 4668646
9060 DQW5        DW18    DW25  DRW13         88 6866  4      6
9055 DQW6                DW26  DRW6          88 4 868 4    6 6
9059 DQW6        DW19    DW26  DRW13         8     68 4668046
9063 DQW6        DW19    DW26  DRW13               68 4668666
9056 DQW5        DW9,19  DW25  DRW13  DRW14  88 46 68 4468  64
9054 DQW5        DW9     DW25  DRW14         88 4 668 4    4 6
9057 DQW1        DW9     DW25  DRW14         88 4866  4   0  6
9061 DQW5        DW9     DW25  DRW14         88 6 668 4    4 6
9064 DQW7        DW16    DW24  DRW14               68 4668046
9067 DQW4        DW8.1         DRW8          88 6 668 4    6 6
9068 DQW4        DW8.1         DRW8          88 6 668 4    6 6
9069 DQW4        DW8.1         DRW8          8866  68 4    466
9071 DQW4        DW8.2         DRW8          88 4 668 448  6 6
9072 DQW4                      DRW8          88 4  68      464
9066 DQW6        DW8.3         DRW8             6  68 4668646
9070 DQW7        DW8.3         DRW8          8  6   8  668046
9047 DQW2        DB1           DR7           88 6886  46
9048 DQW2        DB1           DR7           88 6866  46     6
9050 DQW2        DW17          DR7           88 6888  4      6
9051 DQW2        DW17          DR7           88 6886  4      6
9052 DQW9        DW11          DR7           8  6 6688     4 6
9106 DQW2  DQW3                DR7           88 606 08     6 4
9075 DQW9        DW23          DR9           88 6866  4      6
9030 DQW7        DW13          DR4           88 4866  4      6
9025 DQW7        DW4           DR4           88 6886  4      6
9091 DQW7        DW4           DR4           88 4866  4      6
9029 DQW8        DW4           DR4           88 686          6
9031 DQW8        DW4           DR4           8844866  46     6
9032 DQW8        DW4           DR4           88 48 6  4 6    6
9033 DQW8        DW4           DR4           88 4886  4      6
9034 DQW8        DW4           DR4           88 6866  4      6
9028 DQW8        DW14          DR4           8848886  4      6
9092 DQW8        DW14          DR4           88 6866  4      6
9026 DQW8        DW10          DR4           88 686          6
9007 DQW5  DQW8                DRW16  DR4    88 6866         6
```

Figure 17. Tenth International Histocompatibility Workshop class II DRA locus RFLP fragments (n=17). Cells listed by DR specificity groups.

```
                                                           DDDDDDDDDDDDDDDDD
                                                           RRRRRRRRRRRRRRRRR
                                                           AAAAAAAAAAAAAAAAA

                                                           EBPMTMETBEPEBTTMP
                                                           CGVSASCAGCVCGAASV
                                                           ILUPQPIQLIUILQQPU

                                                                  1  11     1
                                                           34921331301341221
                                                           14029049880126799
                                                           89982638981423283
9045 DQW7         DW5     DW25    DRW11   DRW12    88 4866  4    6 6
9042 DQW7                 DW25    DRW11            88 6886  4      6
9033 DQW8         DW4             DR4              88 4886  4      6
9018 DQW2         DW3     DW25    DR3              88 4886  4      6
9043 DQW7         DW5     DW25    DRW11            88 6886  4      6
9050 DQW2         DW17            DR7              88 6886  4      6
9051 DQW2         DW17            DR7              88 6886  4      6
9040 DQW7         DW5     DW25    DRW11            88 6886  4      6
9025 DQW7         DW4             DR4              88 6886  4      8
9060 DQW5         DW18    DW25    DRW13            88 6866  4      6
9036 DQW5         DB2     DW25    DRW11            88 6886  4      6
9057 DQW1         DW9     DW25    DRW14            88 4866  4   0  6
9030 DQW7         DW13            DR4              88 4866  4      6
9091 DQW7         DW4             DR4              88 4866  4      6
9016 DQW7         DW22            DRW16            88 4866  4      6
9009 DQW1         DW21            DR2              88 6866  4      6
9092 DQW8         DW14            DR4              88 6866  4      6
9034 DQW8         DW4             DR4              88 6866  4      6
9075 DQW9         DW23            DR9              88 6866  4      6
9105 DQW6                         DRW11            88 06668 4      6
9029 DQW8         DW4             DR4              88 686          6
9026 DQW8         DW10            DR4              88 686          6
9039 DQW7                 DW25    DRW11            88 486          6
9020 DQW2         DW3     DW25    DR3              88 6866         6
9007 DQW5   DQW8                  DRW16   DR4      88 6866         6
9047 DQW2         DB1             DR7              88 6886  46
9010 DQW6         DW2             DRW15            88 6886  46     6
9002 DQW5         DW20            DR1              88 6866  44     6
9048 DQW2         DB1             DR7              88 6866  46     6
9037 DQW7         DW5     DW25    DRW11            8866866  4      6
9028 DQW8         DW14            DR4              8848886  4      8
9019 DQW2         DW3     DW25    DR3              8846866  4      6
9031 DQW8         DW4             DR4              8844866  46     6
9056 DQW5         DW9,19  DW25    DRW13   DRW14    88 46 68 4468  64
9032 DQW8         DW4             DR4              88 48 6  4 6    6
9014 DQW6         DW2             DRW15            88   868 4    6 6
9017 DQW1         DW2             DRW15            88   868 4    6 6
9013 DQW6         DW2             DRW15            88   868 4      6
9055 DQW6                 DW26    DRW6             88 4 868 4    6 6
9054 DQW5         DW9     DW25    DRW14            88 4 668 4    4 6
9068 DQW4         DW8.1           DRW8             88 6 668 4    6 6
9061 DQW5         DW9     DW25    DRW14            88 6 668 4    4 6
9067 DQW4         DW8.1           DRW8             88 6 668 4    6 6
9038 DQW7         DB6     DW25    DRW12            88 6 668 4    4 6
9071 DQW4         DW8.2           DRW8             88 4 668 448  6 6
9008 DQW6         DW2             DRW15            88   6 8      4 4
9082 DQW6                         DRW15            88   8 8      4 6
9011 DQW6         DW12            DRW15            88 6 6 8      4 6
9106 DQW2   DQW3                  DR7              88 606 08     6 4
9052 DQW9         DW11            DR7              8  6 6688     4 6
9069 DQW4         DW8.1           DRW8             8866  68 4    466
9072 DQW4                         DRW8             88 4  68      464
9004 DQW5         DW1             DR1              8  4886 846
9006 DQW5         DW1             DR1              8  6886 846
9005 DQW5         DW1             DR1              8  6886 846
9003 DQW5         DW1             DR1              8  686  8 6
9058 DQW6         DW18    DW24    DRW13                  68 46686 6
9021 DQW4                 DW24    DRW18               6  68 46684 6
9063 DQW6         DW19    DW26    DRW13                  68 4668666
9022 DQW2         DW3     DW24    DR3                    68 4668466
9023 DQW2         DW3     DW24    DR3                    68 4668466
9088 DQW2                 DW24    DR3                    68 4668466
9064 DQW7         DW16    DW24    DRW14                  68 4668046
9065 DQW6         DW18    DW24    DRW13               0  68 4668646
9062 DQW6         DW18    DW24    DRW13                  68 4668446
9066 DQW6         DW8.3           DRW8                6  68 4668646
9059 DQW6         DW19    DW26    DRW13            8     68 4668046
9070 DQW7         DW8.3           DRW8             8  6   8  668046
```

Figure 18. Tenth International Histocompatibility Workshop class II DRA locus RFLP fragments (n=17). Cells listed by tree relationships.

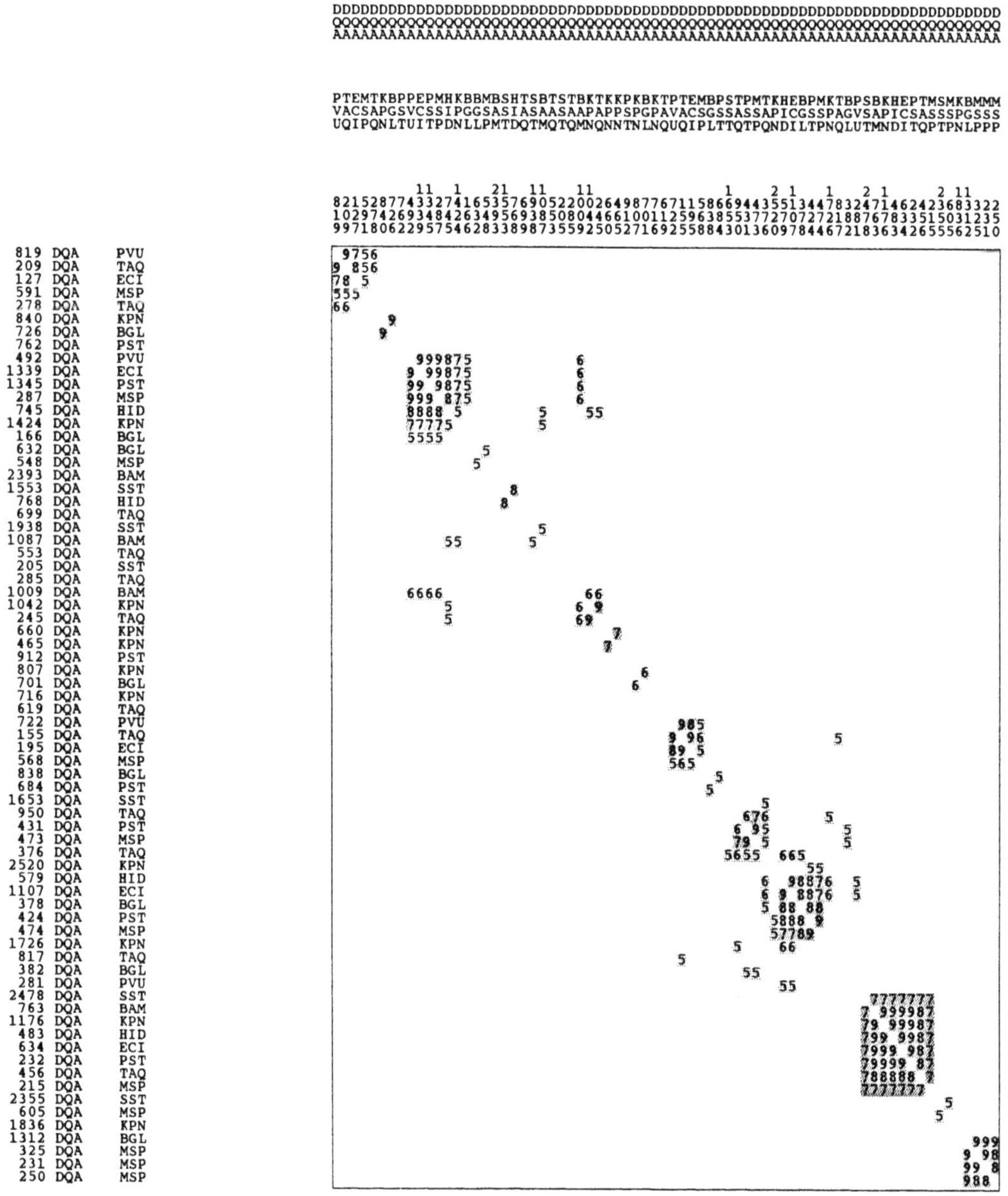

Figure 19. Tenth International Histocompatibility Workshop class II DQA RFLP fragments (n=72). Fragment cluster.

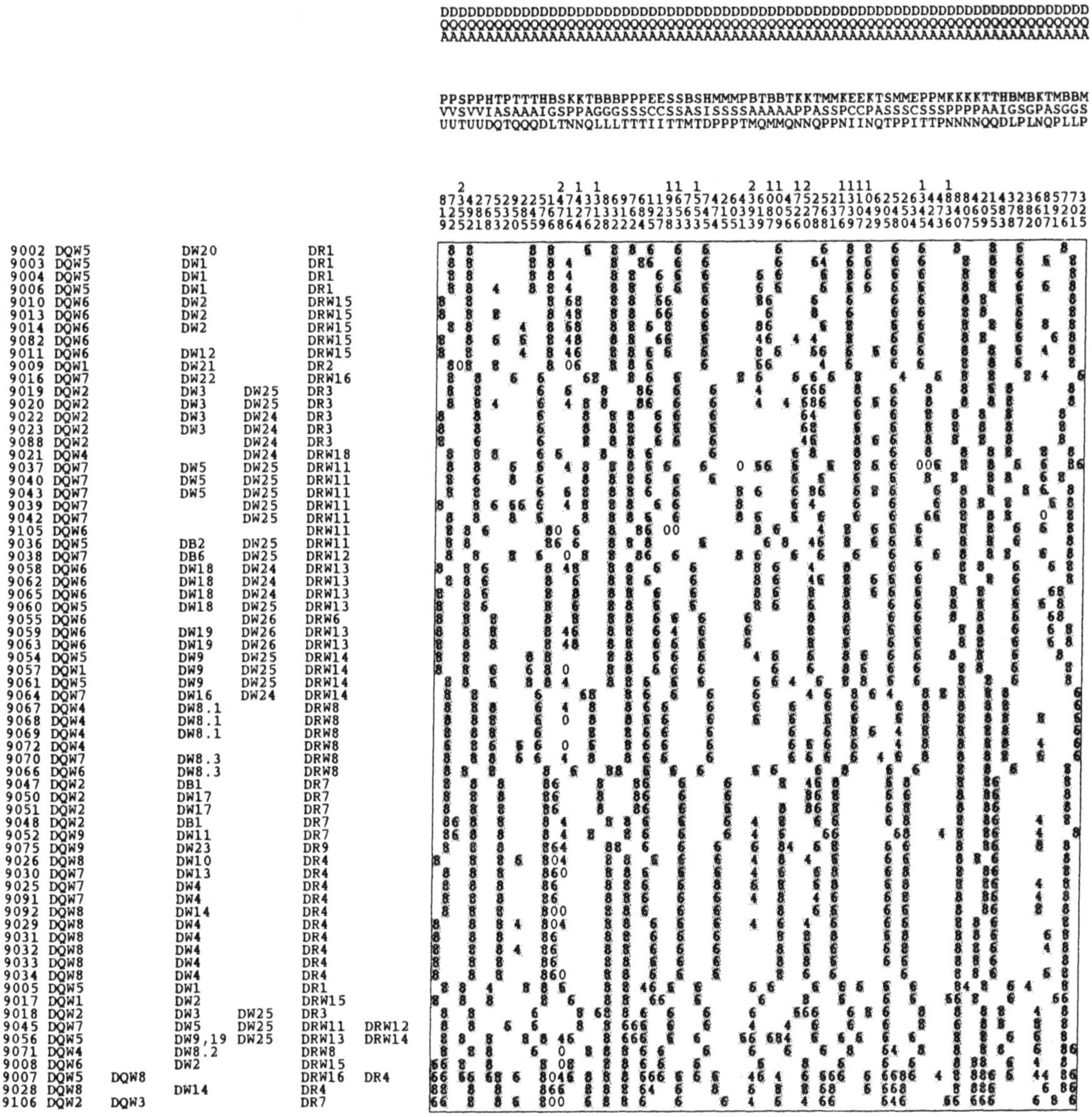

Figure 20. Tenth International Histocompatibility Workshop class II DQA locus RFLP fragments (n=72). Negative cluster.

9002	DQW5		DW20		DR1	
9003	DQW5		DW1		DR1	
9004	DQW5		DW1		DR1	
9006	DQW5		DW1		DR1	
9010	DQW6		DW2		DRW15	
9013	DQW6		DW2		DRW15	
9014	DQW6		DW2		DRW15	
9082	DQW6				DRW15	
9011	DQW6		DW12		DRW15	
9009	DQW1		DW21		DR2	
9016	DQW7		DW22		DRW16	
9019	DQW2		DW3	DW25	DR3	
9020	DQW2		DW3	DW25	DR3	
9022	DQW2		DW3	DW24	DR3	
9023	DQW2		DW3	DW24	DR3	
9088	DQW2			DW24	DR3	
9021	DQW4			DW24	DRW18	
9037	DQW7		DW5	DW25	DRW11	
9040	DQW7		DW5	DW25	DRW11	
9043	DQW7		DW5	DW25	DRW11	
9039	DQW7			DW25	DRW11	
9042	DQW7			DW25	DRW11	
9105	DQW6				DRW11	
9036	DQW5		DB2	DW25	DRW11	
9038	DQW7		DB6	DW25	DRW12	
9058	DQW6		DW18	DW24	DRW13	
9062	DQW6		DW18	DW24	DRW13	
9065	DQW6		DW18	DW24	DRW13	
9060	DQW5		DW18	DW25	DRW13	
9055	DQW6			DW26	DRW6	
9059	DQW6		DW19	DW26	DRW13	
9063	DQW6		DW19	DW26	DRW13	
9054	DQW5		DW9	DW25	DRW14	
9057	DQW1		DW9	DW25	DRW14	
9061	DQW5		DW9	DW25	DRW14	
9064	DQW7		DW16	DW24	DRW14	
9067	DQW4		DW8.1		DRW8	
9068	DQW4		DW8.1		DRW8	
9069	DQW4		DW8.1		DRW8	
9072	DQW4				DRW8	
9070	DQW7		DW8.3		DRW8	
9066	DQW6		DW8.3		DRW8	
9047	DQW2		DB1		DR7	
9050	DQW2		DW17		DR7	
9051	DQW2		DW17		DR7	
9048	DQW2		DB1		DR7	
9052	DQW9		DW11		DR7	
9075	DQW9		DW23		DR9	
9026	DQW8		DW10		DR4	
9030	DQW7		DW13		DR4	
9025	DQW7		DW4		DR4	
9091	DQW7		DW4		DR4	
9092	DQW8		DW14		DR4	
9029	DQW8		DW4		DR4	
9031	DQW8		DW4		DR4	
9032	DQW8		DW4		DR4	
9033	DQW8		DW4		DR4	
9034	DQW8		DW4		DR4	
9005	DQW5		DW1		DR1	
9017	DQW1		DW2		DRW15	
9018	DQW2		DW3	DW25	DR3	
9045	DQW7		DW5	DW25	DRW11	DRW12
9056	DQW5		DW9,19	DW25	DRW13	DRW14
9071	DQW4		DW8.2		DRW8	
9008	DQW6		DW2		DRW15	
9007	DQW5	DQW8			DRW16	DR4
9028	DQW8		DW14		DR4	
9106	DQW2	DQW3			DR7	

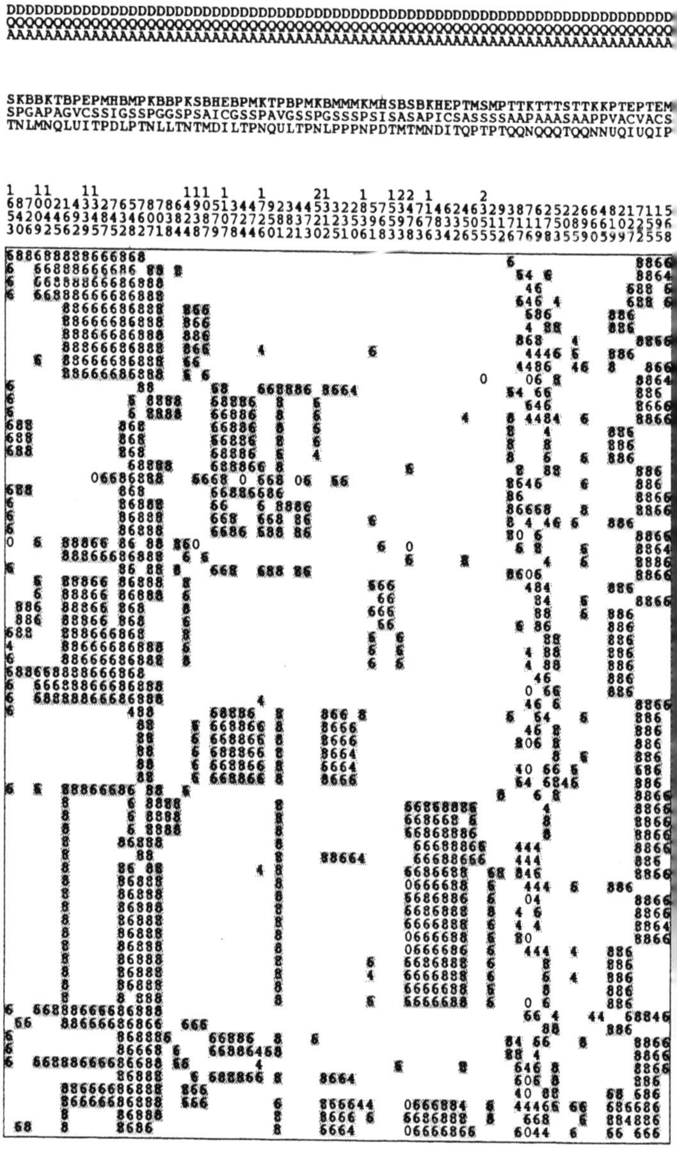

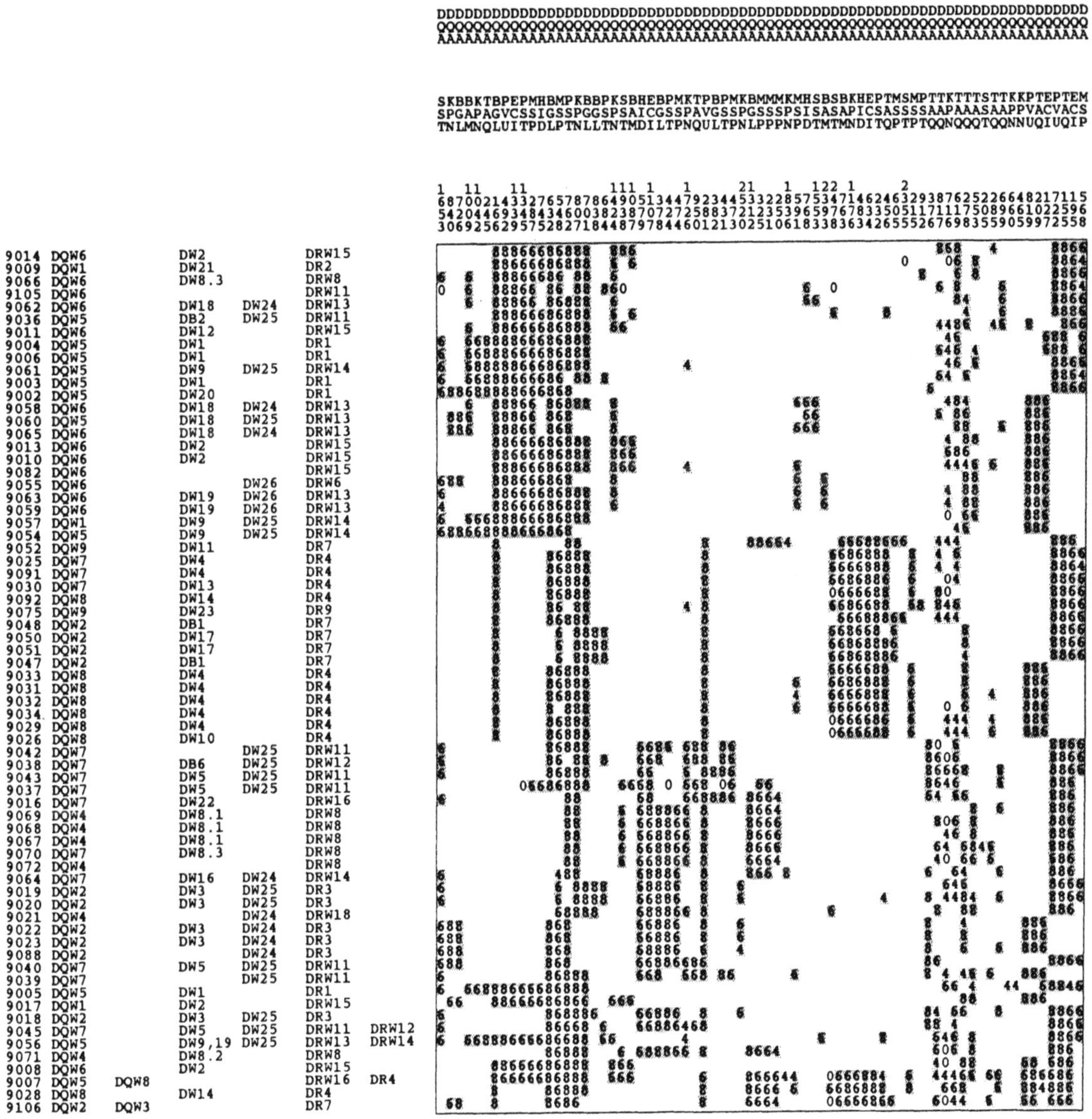

Figure 22. Tenth International Histocompatibility Workshop class II DQA RFLP fragments (n=72). Cells listed by tree relationships (60 homozygous cells).

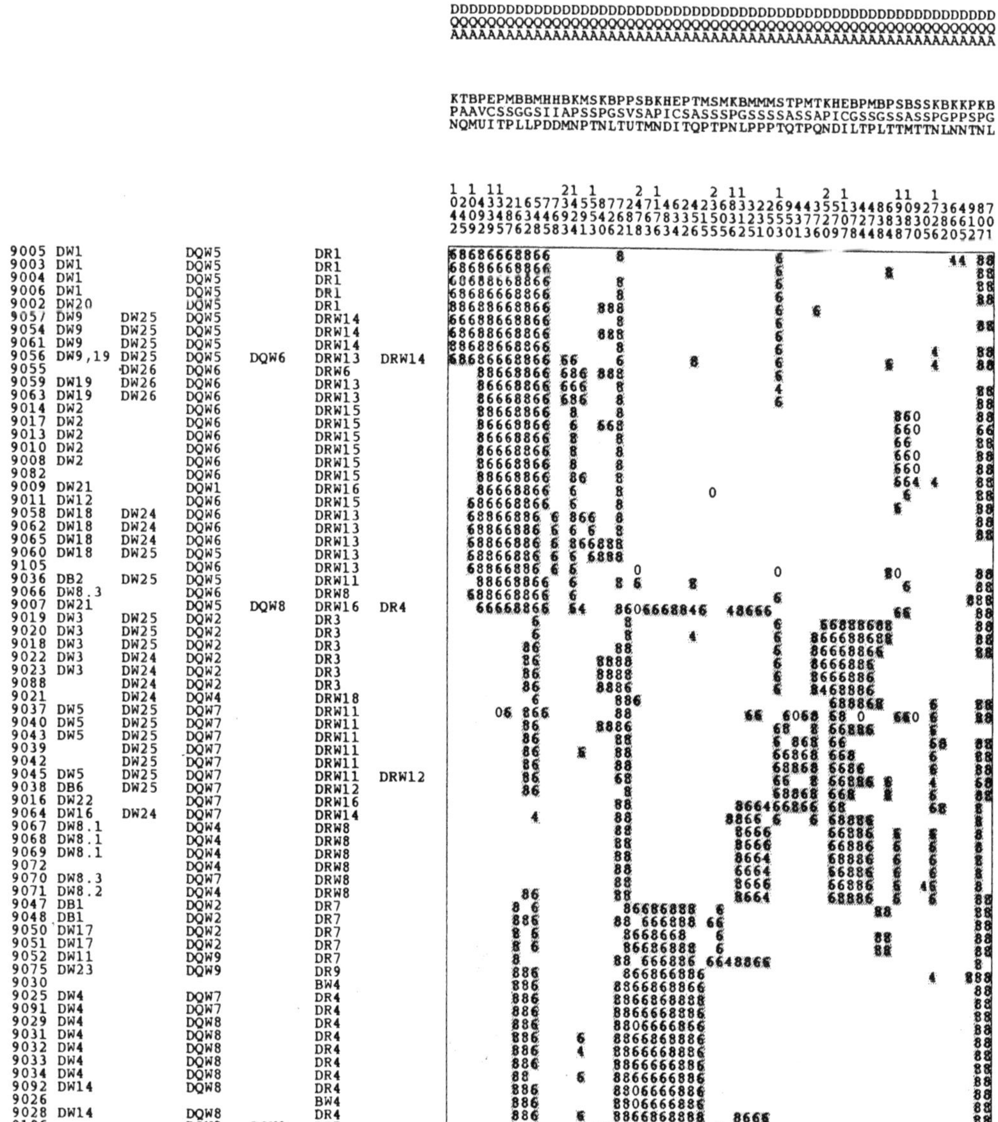

Figure 23. Tenth International Histocompatibility Workshop class II DQA1 locus RFLP fragments (n=59). Cells grouped according to available DQA1 sequence type. The following cell lines have been assigned DQA1 alleles by nucleotide sequence analysis.

1. DQA1.1 9005 - 9061
2. DQA1.2 9055 - 9008
3. DQA1.3 9011 - 9060
4. DQA2 9047 - 9052 (MSP 605)
5. DQA3 9075 - 9028 (MSP 215)
6. DQA4 9019 - 9071 (HID 579; ECI 1107)

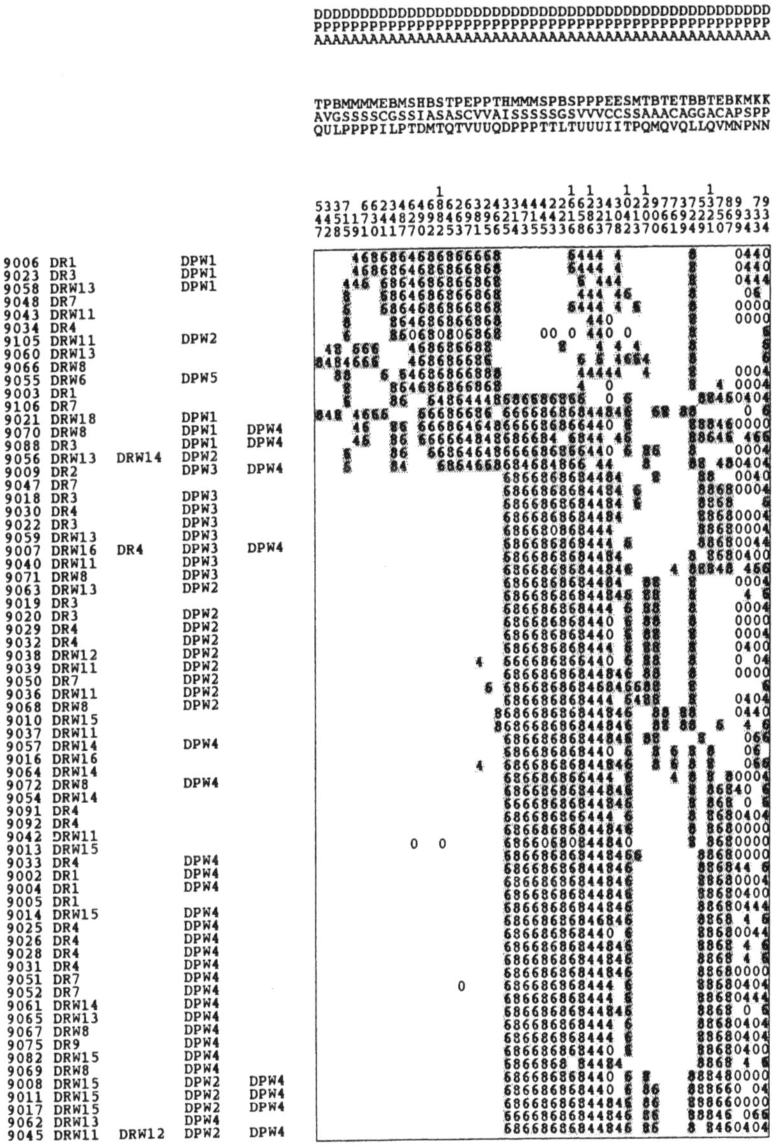

Figure 24. Tenth International Histocompatibility Workshop class II DPA RFLP fragments (n=49). W/S cells listed to show fragment patterns.

DPA: Forty-nine fragments were detected using the DPA locus probe. Positive cluster analysis of these 49 fragments is shown in Figure 24, in which the cells have been listed primarily by DPw type assigned on the basis of cellular typing. There are two major clusters at the DPA1 locus. Heterozygous cells are seen at the interface between the two clusters (9106, 9021, 9070, 9088, 9056, 9009). Equally apparent is the greater pattern fragment variation associated with the less frequent cluster (top left) relative to that of the more frequent cluster. The first cluster stratifies DPw types mainly into DPw1, DPw5, and DPw1/5-associated New alleles (11). The second cluster comprises DPw3, DPw2, and DPw4 types. Subgroups of this second cluster are delineated by fragment TAQ 1207, which correlates closely with DPw2-assigned cells, while fragment BAM 900 extends to include a subgroup of DPw4 cells that are independently known to have the DPB1 nucleotide sequence allele DPB4.2. (11). These DPw2 and DPw4 subgroup cells are seen to be associated with a "reciprocal" absence of fragments TAQ 1321, ECV 750, and BAM 867. The absence of fragment BGL 529 extends to include the remaining DPB4.2 cells.

The two BGL fragments 724 and 529 (Fig. 24, columns 8 and 9 from right side) comprise a mutually exclusive pattern. Since they do not correlate with the two main clusters of DPA1, they are interpreted to reflect polymorphism associated with the DPA2 locus.

Class II Beta Loci. It is necessary to describe the procedure that has been adopted in this summary report for selection of informative fragments for the three class II beta loci (DR, DQ, and DP). In the Tenth Workshop, the underlying concept for data analysis of class II beta probe patterns was to identify locus-specific fragments and the primary locus association of each cross-hybridizing fragment. This concept was incorporated into the data reporting, in that relative hybridization strengths were converted into a locus assignment pattern. These patterns are shown in each of the probe/enzyme system reports (this volume). Each class II beta probe summary report utilized the primary locus assignment in order to identify the locus-specific, serospecificity-associated fragment patterns. We have reexamined the entire fragment data generated by each probe to assess locus assignment of all 639 class II beta fragments independent of hybridization strengths and the resulting deduced locus assignment.

The three class II beta probe fragments were considered in turn. Two hundred thirty-eight (238) fragments were obtained with the DRB probe, 225 with the DQB probe, and 176 with the DPB probe. For DRB, then DQB, and finally DPB, all fragments were subjected to negative clustering with cells listed by DR, by DQ, and by DP specificities for each probe. Fragments were selected if their patterns correlated with cells grouped according to probe-related specificities. For example, based on the negative cluster of the 238 DRB fragments (Figs. 25A and B), 176 fragments were selected. The negative cluster of the selected 176 fragments is shown in Figures 26 A and B. Confirmation of the validity of fragment selection is depicted in the tree analysis (Fig. 27). The fragments selected for each probe were compared with those described in the probe summary reports. The selected fragments were also compared with the data presented in Tables 2 and 3, which represent a synthesis of presumed locus-specific "unique" fragments from system reports, probe reports, and central data analysis. There was good concordance between fragments selected by the Workshop standardization centers based on locus assignment and that obtained by our final analysis.

DRB. Figure 28 shows the positive cluster of 176 DRB locus-selected fragments with cells listed according to tree relationships revealed in Figure 27. The blocks of fragment patterns display a very close correlation with the serologically and cellularly defined DR and Dw specificities. Table 2 lists the "unique" fragments for DR specificity identification. Each of the 134 fragments, together with co-clustering fragments, can be identified in Figures 28 A and B. From Table 2, it can also be seen that there are fragments which correlate strongly with the DRw52 supertypic specificity and with the three cellularly defined DRw52 subtypes, Dw24, Dw25, and Dw26. DRw8-positive cells are also associated with the DRw52 supertypic specificity. It has been suggested that these cells may comprise an unique subtype within DRw52. However, there is no evidence from the negative cluster of DRB RFLP patterns (Fig. 26) of any unique RFLP cluster mutually exclusive with those fragments associating with Dw24, Dw25, and Dw26 specificities.

Concerning DRw53, there are some 30 fragments that react en bloc with the 19 DRw53-positive cells (Figs. 28A and B). A few fragments (e.g., BAM 780, ECI 1513) are not present in all DRw53-positive cells, raising the possibility of DRw53 polymorphism.

DQB. The 225 RFLP fragments detected by the DQB probe are shown as a negative cluster in Figures 29 A and B. Of the 225, 94 were selected as informative for the serologically detectable DQ specificities. Of the 101, 53 are "unique" fragments for particular DQ specificities (Table 3). In Figures 30 A and B, the co-clustering patterns among the 94 fragments can be seen. Figure 31 is a positive cluster with cells listed according to their tree relationships, in which the fragment columns have been repositioned to illustrate the block patterns associated with DQ serospecificities.

DRB, DQA and DQB. In this section, all fragments detectable by DRB, DQA, and DQB probes have been considered together. From these 536 fragments, 187 have been selected as "unique" RFLP fragments for DRB and DQB locus specificity assignment (see Tables 2 and 3). Figure 32 shows the tree relationship of the 70 cells using the total 536 fragments. Figure 33 shows all 536 fragments associated with the DRB, DQB, and DQA loci against the cells listed by tree relationships, showing DQ, Dw, and DR types for each cell. Comparison of Tables 2 and 3 enables the "unique" fragments to be identified. The well defined "block" fragment patterns are shown in Figures 33A, B, and the left half of Figure 33C.

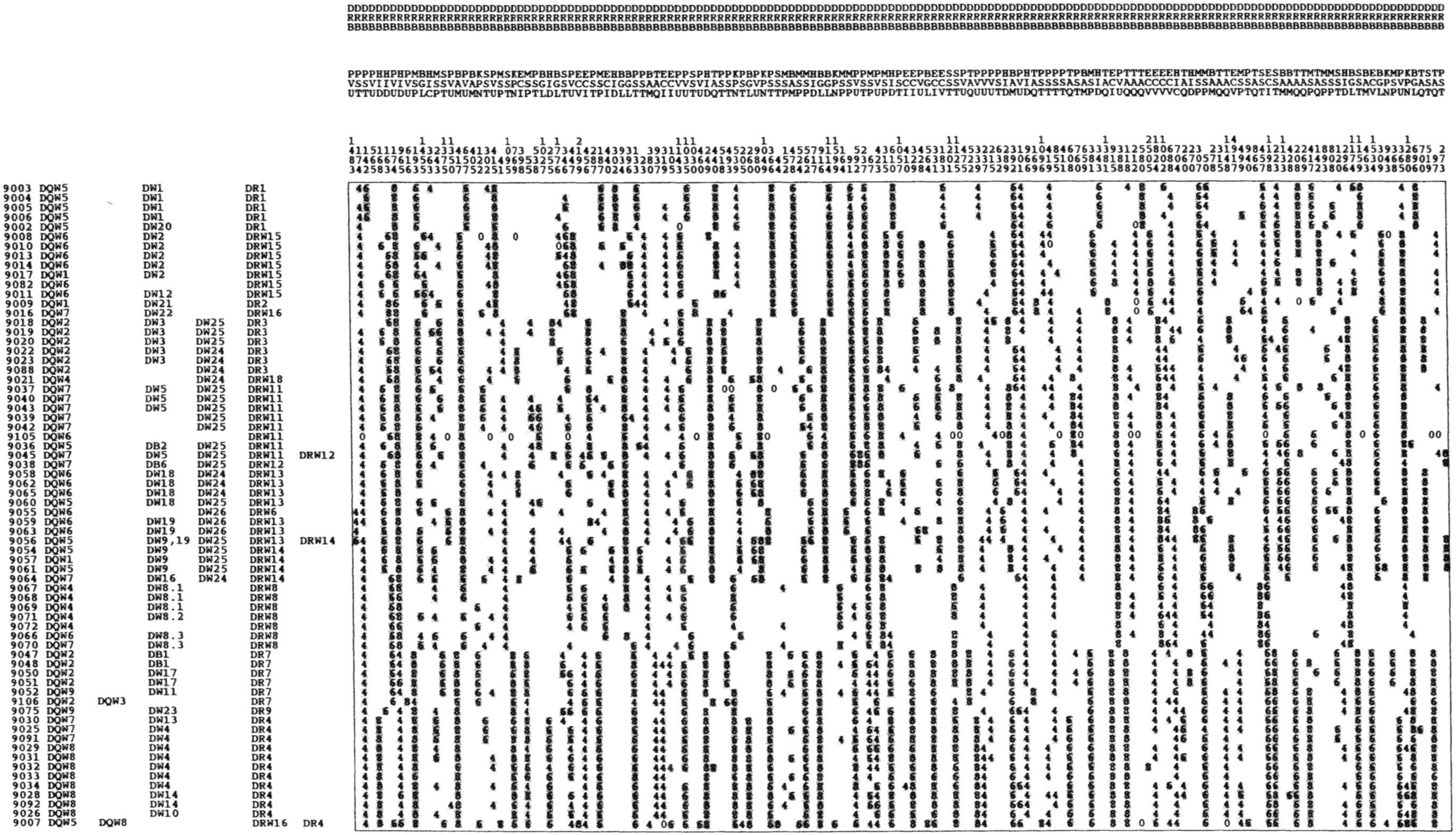

Figure 25A. Tenth International Histocompatibility Workshop Class II DRB loci RFLP fragments (n=238). Negative cluster.

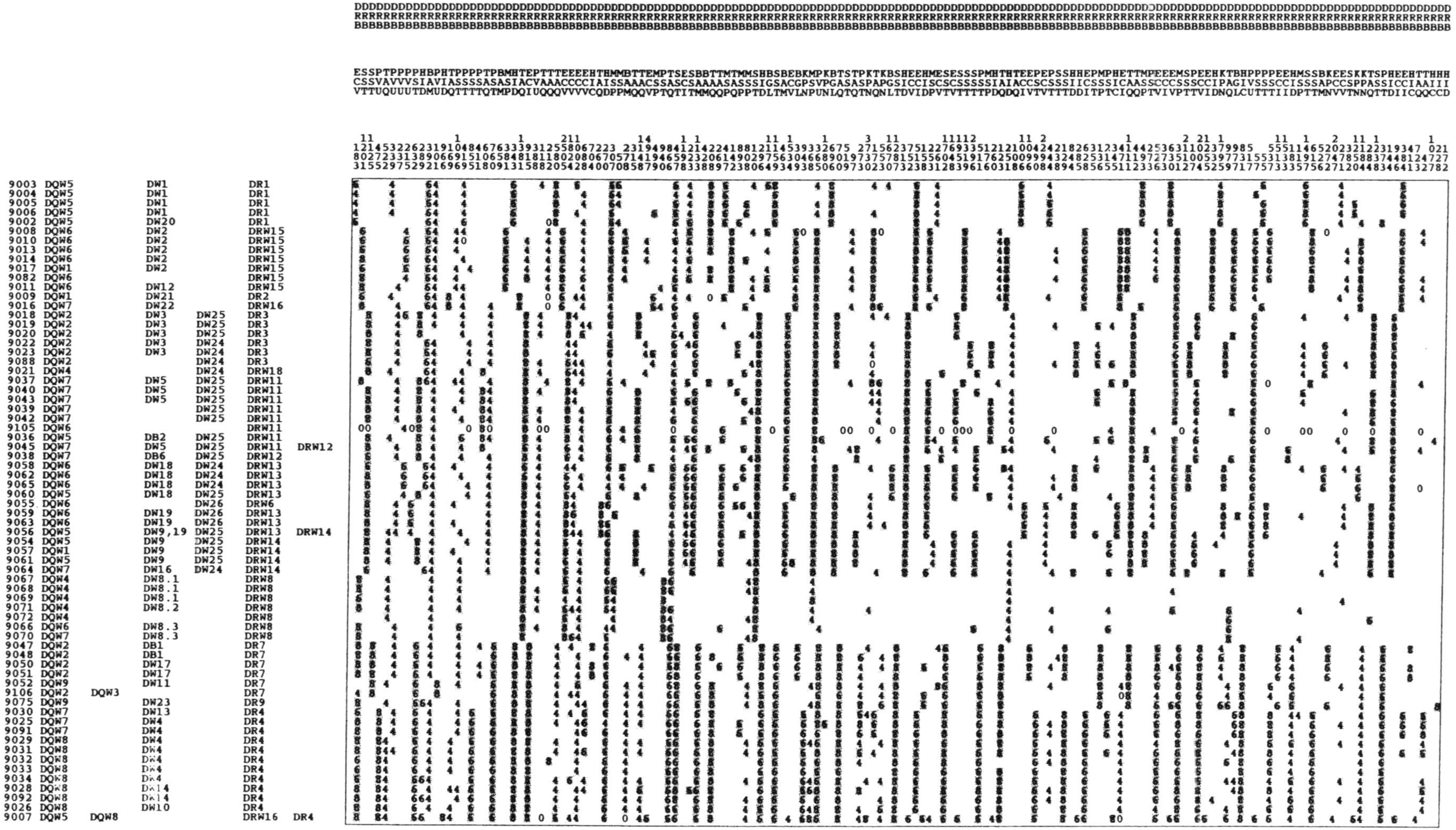

Figure 25B. Tenth International Histocompatibility Workshop Class II DRB loci RFLP fragments (n=238). Negative cluster.

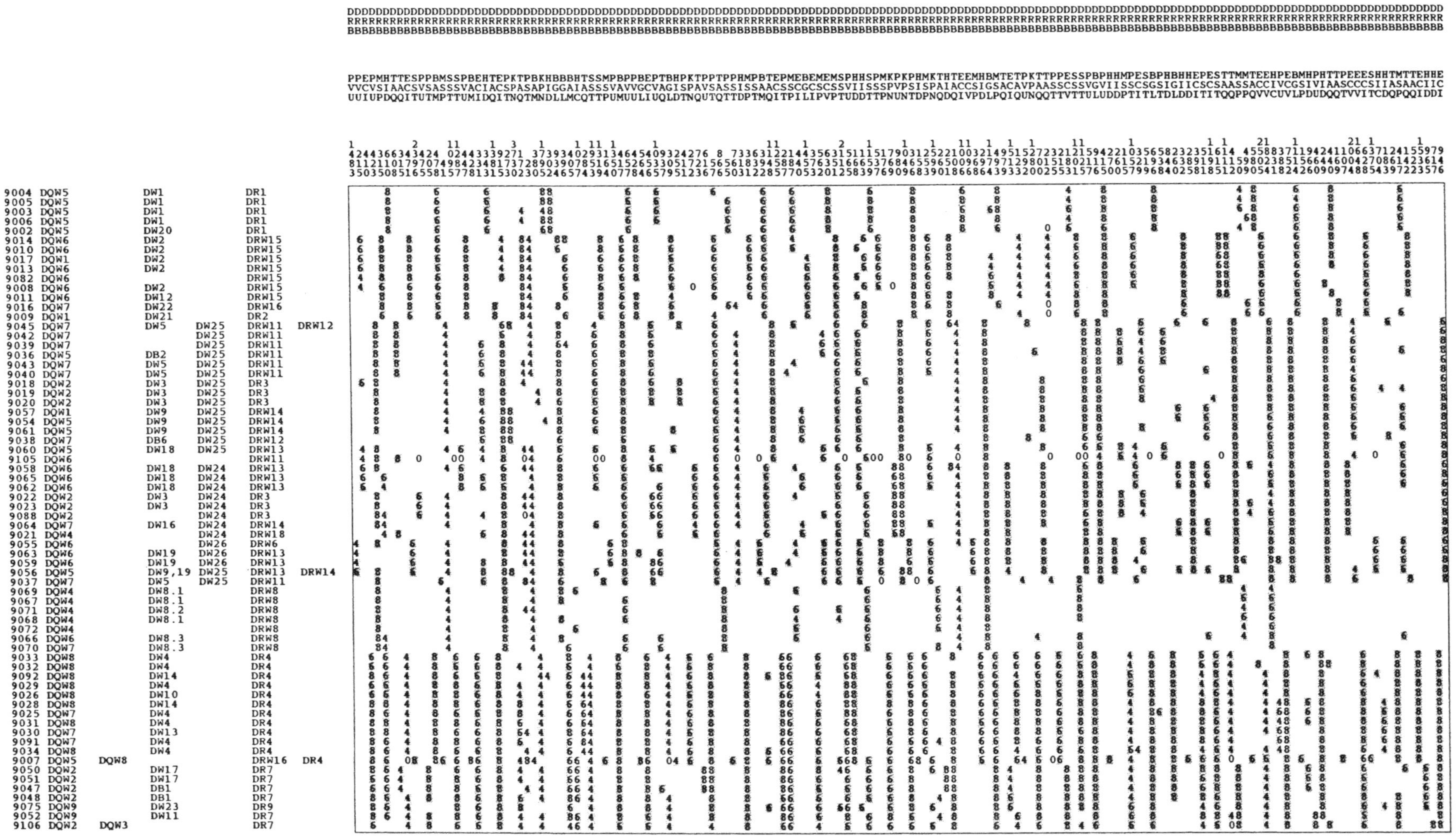

Figure 26A. Tenth International Histocompatibility Workshop Class II DRB loci RFLP fragments (n=176). Negative cluster.

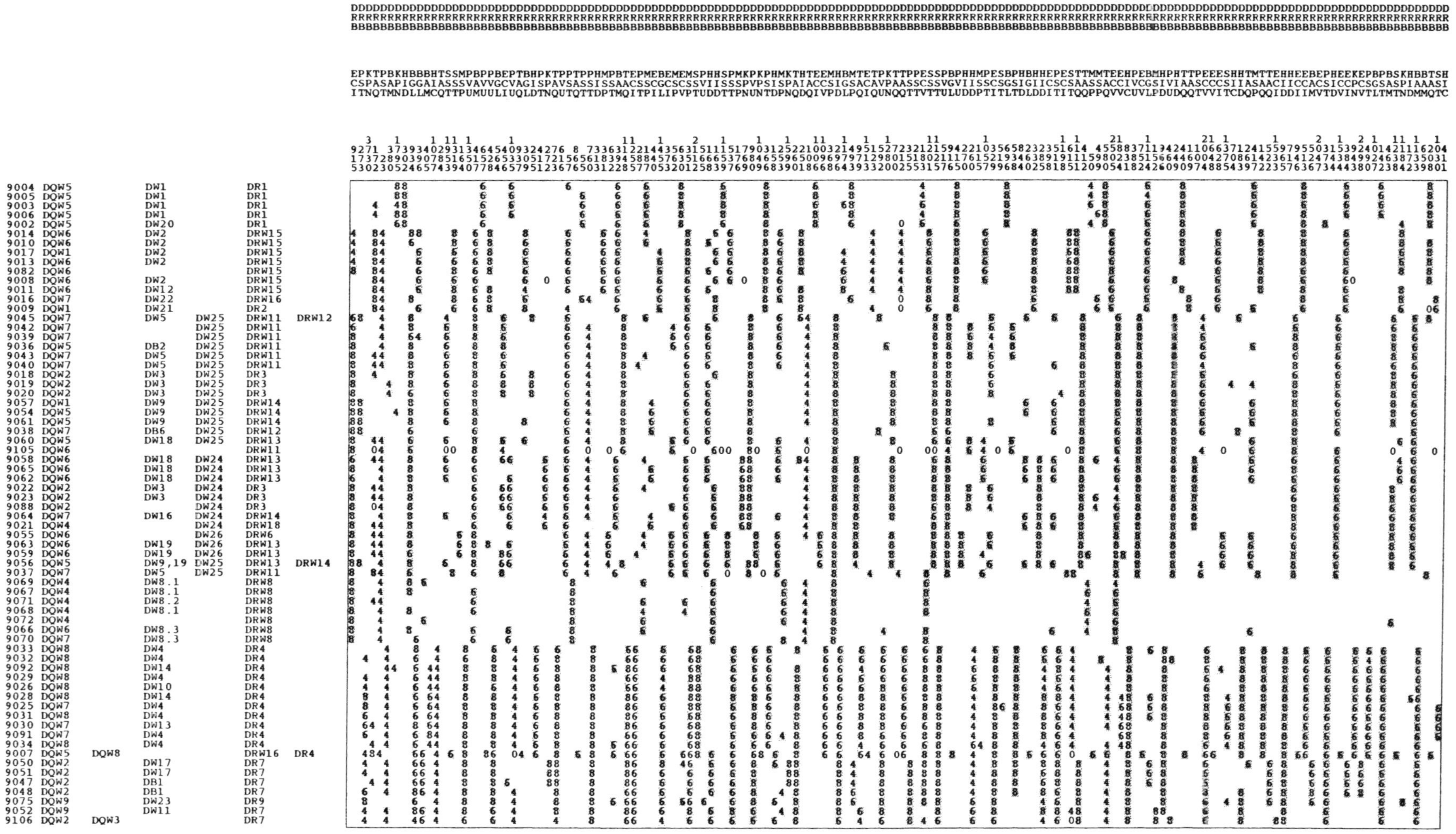

Figure 26B. Tenth International Histocompatibility Workshop Class II DRB loci RFLP fragments (n=176). Negative cluster.

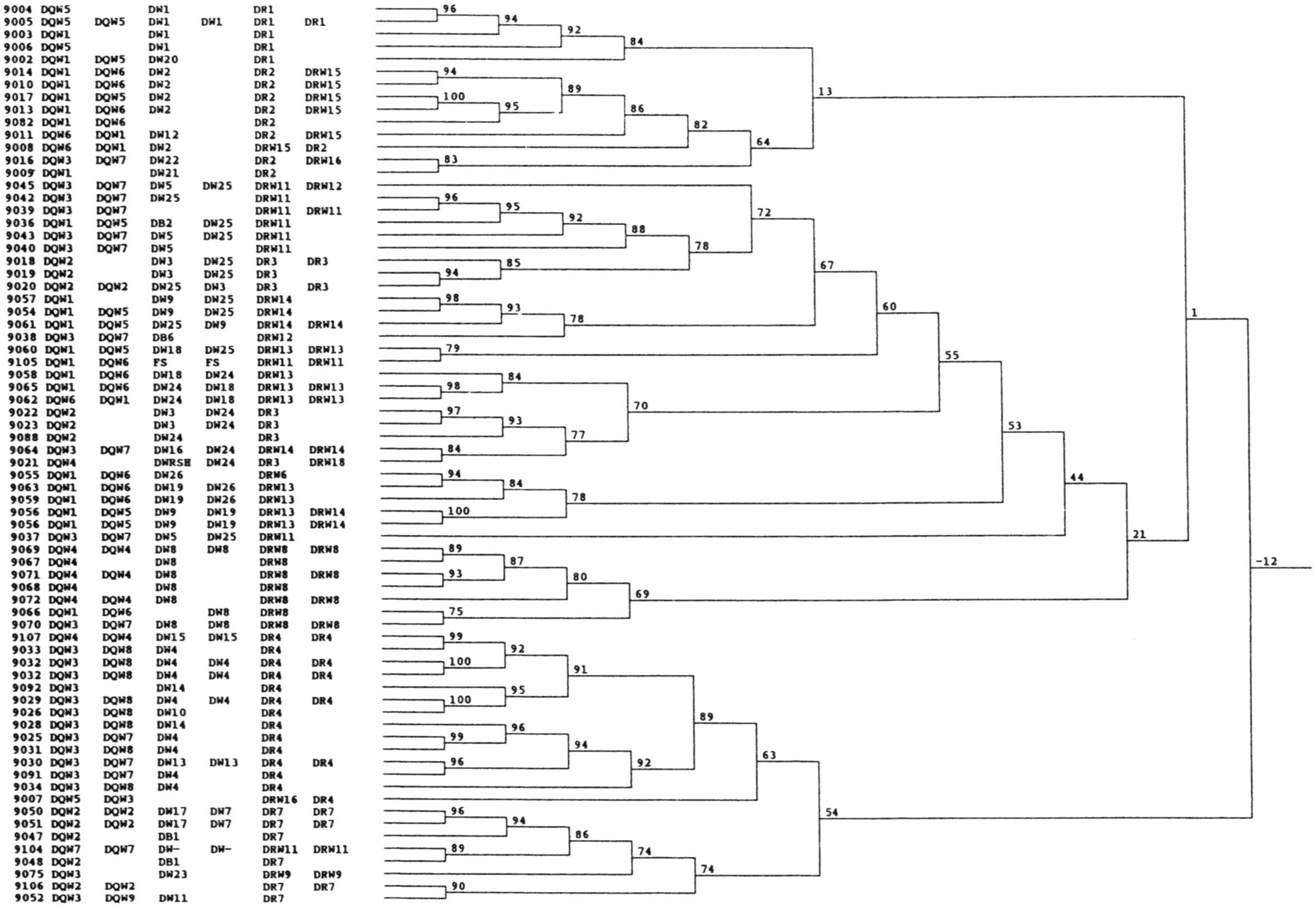

Figure 27. Tenth International Histocompatibility Workshop
Tree analysis of cell relationships: DRB loci RFLP fragments (n=176).

Table 2. "Unique" RFLP Fragments for DRB Locus Specificity Assignment

RFLP Specificity	D-	BAM	BGL	ECI	ECV	HIC	HID	KPN	MSP	PST	PVU	SST	TAQ
DR1, DR1,	Dw1 Dw20		(RB696)	**RB1006**	RB183 RB630		**RB302** **RB343** **RB610**	QB1775	(RB493)		**RB657**	**RB1159**	**(RB580)**
DR1,	Dw1									**RB142**			
DR1,	Dw20									**RB373**			
DR2				**RB628**			**RB332**				**RB447**	(RB1319)	
<DR2			RB1455										
DR15, DR15,	Dw2 Dw12								RB381			**(RB615)**	**(RB142)** (RB212)
DRw15, DRw15, DRw16,	Dw2 DW12 Dw21												(RB123)
DRw16, DRw16,	Dw21 Dw22											RB1030	**RB169**
DRw16,	Dw22		(QA382)										
DRw17,	Dw3				(QB1514)		QB1142	QA2520					
DRw17,	Dw3/24											**RB2396**	
DRw17,	Dw3/25						**RB255**						
<DRw11, DRw11,	Dw5 DB2									(RB538)			
DRw12							**RB183**				**RB592**		
DRw13,	Dw18						(QA768)					**RB1287** QA1553	
DRw13,	Dw19	(QA2393)			RB1096		RB270	[QB1175]		RB632 QB643	RB1483	(RB1375)	
DRw14,	Dw16							(QA1836) (QB1544)					
DRw8													**RB866**
DR7									QA605	(RB215)	**RB475**		
DR7, DR7,	DB1 Dw17			**(RB1533)**	(RB2047) (QB2001)		(QB1280)	RB943 (QB920)		**RB490**	(QB478)	QB540	
<DR7, DR7,	DB1 Dw17												RB260 RB723
DR7,	DW11				(QB1643)		**(RB966)**						
<DR7, DR7,	DB1 Dw11												QA2355
DRw9,	Dw23						RB172			(QA912)			
DR4		**RB526**	**RB1027**		**RB2298** (QB1452)	[RB1077]	**(RB158)**		**RB368** (QA215)	**RB205** **RB355** **RB411**	**RB565**		**RB522**
DR4,	DQw7					(RB411)							
DR4,	DQw8									[QB396] QB684			
DRw52					**RB1824**		**RB296**						
DRw52[a]	[DRw8]	[RB1133]	**RB916**	**RB946**	RB752 [RB1534]							[RB1125]	
DRw52,	Dw24							**RB272** RB769		**RB107**			
DRw52,	Dw25				[RB2004]								[RB1148]
DRw52,	Dw26	QA2393			RB1096		**RB270**	[QB1175]		RB632 [QB643]	RB1483	**(RB1375)**	

Table 2. *(Continued)*

	BAM	BGL	ECI	ECV	HIC	HID	KPN	MSP	PST	PVU	SST	TAQ
DRw53	**RB428** **RB463** **RB659** QA763	RB564	**RB373** **RB398** **RB410** **RB1255** (RB1198) QA634	RB1648		**RB144** **RB313** **RB717** QA483	QA1176	RB115 **RB176**	QA232	**RB388** **RB517** **RB616** **RB1350** [RB460]	**RB519** **RB1095** **RB2112**	RB284

[a]Signifies DR3, 5, 6, excluding DRw8.

Unique RFLP fragments have been defined as those fragments that correlate completely with the latest HLA type assigned to each Workshop cell. Since the Workshop cells include those having variant/atypical fragment reactivity patterns, the definition has been extended to include fragments with either a single reactivity in excess of the stated specificity, or a single reactivity less than the stated specificity. Such fragments are identified by using ( ) for additional reactivity and [ ] for deficient reactivity. The cells corresponding to both types of parentheses are listed below in this footnote.

Fragments typed in **bold** were identified in the DRB and DQB loci "antigen" reports. Fragments in regular type have been added during this review procedure. Fragments detected with the DQ beta probe that reflect cross-hybridization with DR beta specificity associated sequences have not been included.

Other fragments exist that have reactivity, or lack of reactivity, with two or more cells apart from that with the primary specificity. Some of these fragments, particularly those with reactivity to only a limited number of cell specificities, and those with higher frequency reactivity but in which the reactivity correlates closely with several specificities, are also useful for HLA specificity typing.

From the Workshop data, it is not possible to be certain of the primary locus of reactivity for three groups of fragments. They are:

1. DRw13, Dw19 – DRw52, Dw26
2. DR1, Dw1/20 – DQw5, DR1
3. DRw17, Dw3 – DQw2.3

The fragments for these three groups are listed in the corresponding two locations.

RFLP Specificity D-	Fragment		Cell Number
DR1	TAQ	RB580	(32)
	BGL	RB696	(25)
	MSP	RB493	(16)
DR2	SST	RB1319	(37)
DR15	SST	RB615	(37)
	TAQ	RB142	(37)
	TAQ	RB212	(37)
DRw15	TAQ	RB123	(37)
DRw16	BGL	QA382	(43)
DR17	ECI	QB1514	(64)
<DRw11	PST	RB538	(60)
DRw13	HID	QA768	(105)
	SST	RB1375	(92)
	KPN	QB1175	[56]
	PST	QB643	[56]
DRw14	KPN	QA1836	(52)
DRw8	PST	QA912	(75)
DR7	PST	RB215	(75)
	ECI	RB1533	(75)
	ECV	RB2047	(75)
	PVU	QB478	(52)
	HID	QB1280	(106)
	KPN	QB920	(106)
	ECV	QB1643	(106)
	HID	RB966	(106)
DRw9	PST	QA912	(66)
DR4	HIC	RB1077	[7]
	HID	RB158	(7)
	ECV	QB1452	(75)
	MSP	QA215	(75)
	HIC	RB411	(31)
	PST	QB396	[106]
DRw52	BAM	RB1133	[37],(28)
	SST	RB1125	[37]
	ECV	RB1534	[60]
	ECV	RB2004	[60]
	TAQ	RB1148	(37)
	KBN	QB1175	(56)
	SST	RB1375	(92)

RFLP Specificity D-	Fragment		Cell Number
DRw53	ECI	RB1198	(20)
	PVU	RB460	[106]
DQw1	KPN	QB515	(45)
	ECI	QB383	(37)
	HIC	QB850	(37)
	SST	QB228	(37)
DQw6.1	BAM	QB289	(37)
	TAQ	QB287	(37)
	TAQ	QB261	(37)
DQw2	PVU	QB63	(106)
	ECV	QB158	(106)
	PVU	QB166	(11,66)
	TAQ	QB87	(11,66)
DQw2.3	ECV	QB1514	(64)
DQw2.7	ECV	QB2001	[106],(75)
	HIC	QB766	[106]
	PVU	QB540	(106)
	PVU	QB478	(52)
	TAQ	QB669	(11,66)
DQw7	KPN	QB120	(45)
	KPN	QB533	(45)
	PVU	QB34	(45)
	SST	QB222	(37)
	SST	QB286	(37)
	TAQ	QB100	(37)
	BAM	QB355	(37)
DQw8	PST	QB396	(7)

Table 3. "Unique" RFLP Fragments for DQB Locus Specificity Assignment

RFLP Specificity D-	BAM	BGL	ECI	ECV	HIC	HID	KPN	MSP	PST	PVU	SST	TAQ
DQw1			(QB383)			**QB644**	**(QB515)**					
<DQw1[a]				**QB673**	**(QB850)**						**(QB228)**	**QB534**
DQw5	**QB576**			**QB26** **QB27**			[QA1042]		**QB136**			[QA245]
DQw6		(QB310)										
<6.Dw2/18/19[b]			**QB226** QB922	**QB524**								
<6.Dw2/18[b]	**(QB289)**								**QB1068**	**QB217**		(QB287) **(QB261)**
<6.Dw12/8.3[b]	**QB872**			QB1182	**QB185** **QB1536**					**QB202**		**QB450**
DQw2	QB383									[QB63] **(QB166)**		[QB158] **(QB87)**
DQw2.DR3[c]				**(QB1514)**		**QB1142**						
DQw2.DR7[c]				**(QB2001)**	**[QB766]**	**QB1280**	**QB920**			[QB478]	**[QB540]**	**(QB669)**
DQw7	**[QB355]**					**QB341**	**[QB120]** **[QB533]**			**QB372** **[QB34]**	**[QB222]** **[QB286]**	**[QB100]**
DQw8									[QB396] QB684			

[a]Four fragments identify a subgroup of DQw1. The four are present in all DQw5 and DQw6 cells, except for cells 9011 and 9066. These two DQw6 cells are identified as the DQw6.Dw12/8.3 subgroup (see below).
[b]There are unique RFLP fragments for three DQw6 subgroups. Since the three subgroups are associated with particular Dw types, they have been identifed as: DQw6.Dw2/18/19; DQw6.Dw2/18; and DQw6.Dw12/8.3.
[c]There are unique RFLP fragments for two DQw2 subtypes corresponding to the DR3 and the DR7 haplotypes.

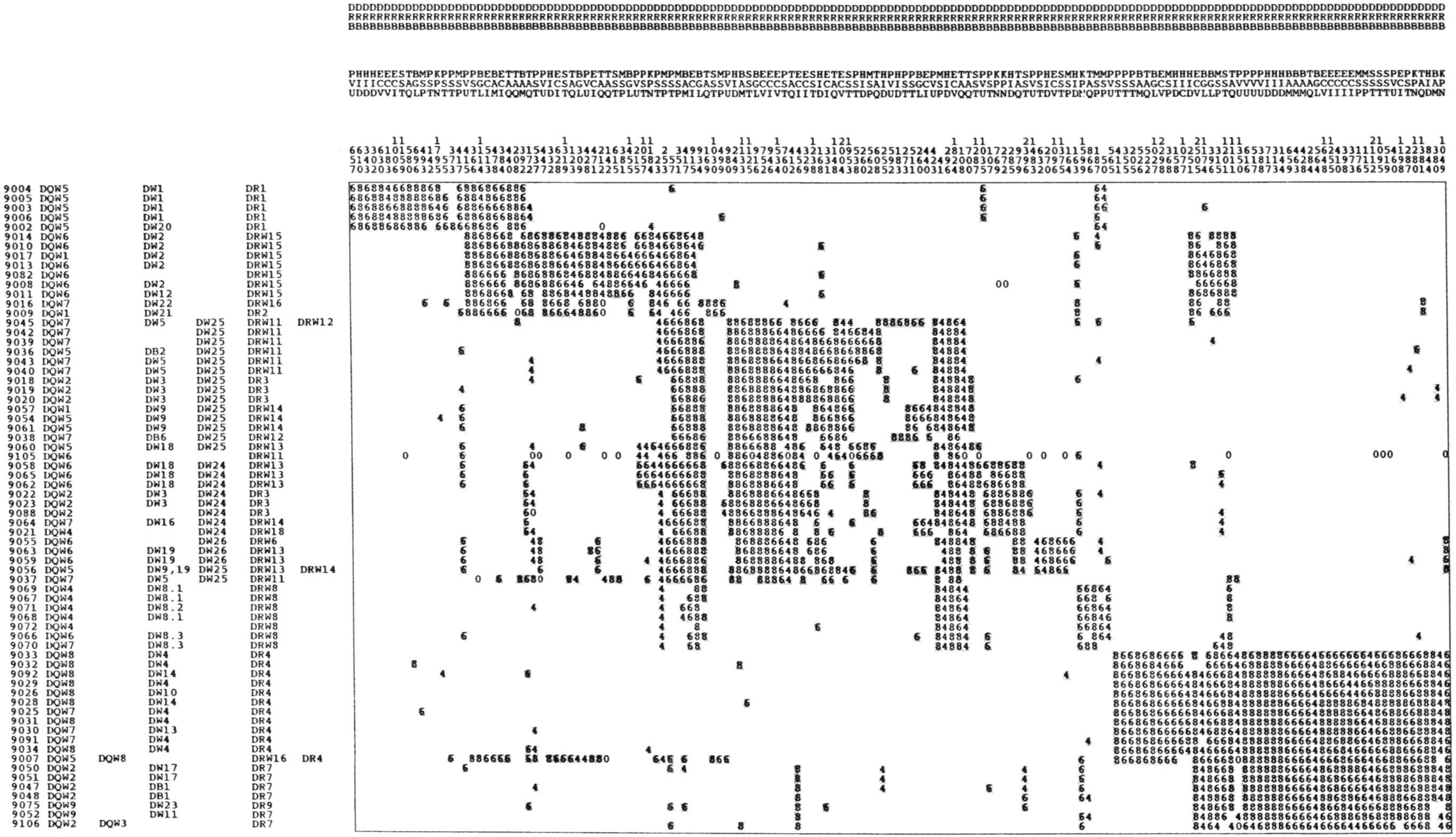

Figure 28A. Tenth International Histocompatibility Workshop
Class II DRB loci RFLP fragments (n=176). Cells listed by tree relationships.

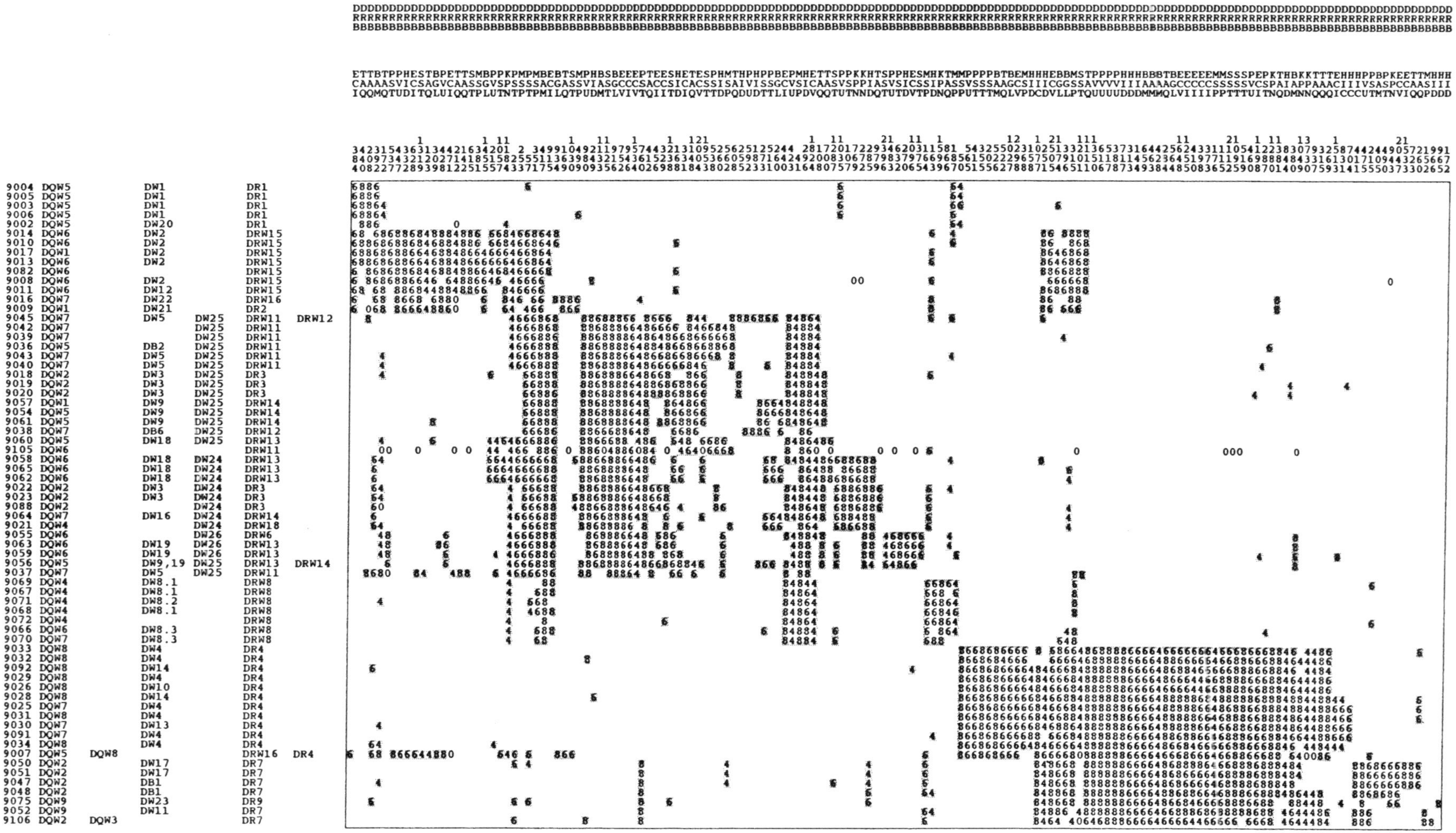

Figure 28B. Tenth International Histocompatibility Workshop
Class II DRB loci RFLP fragments (n=176). Cells listed by tree relationships.

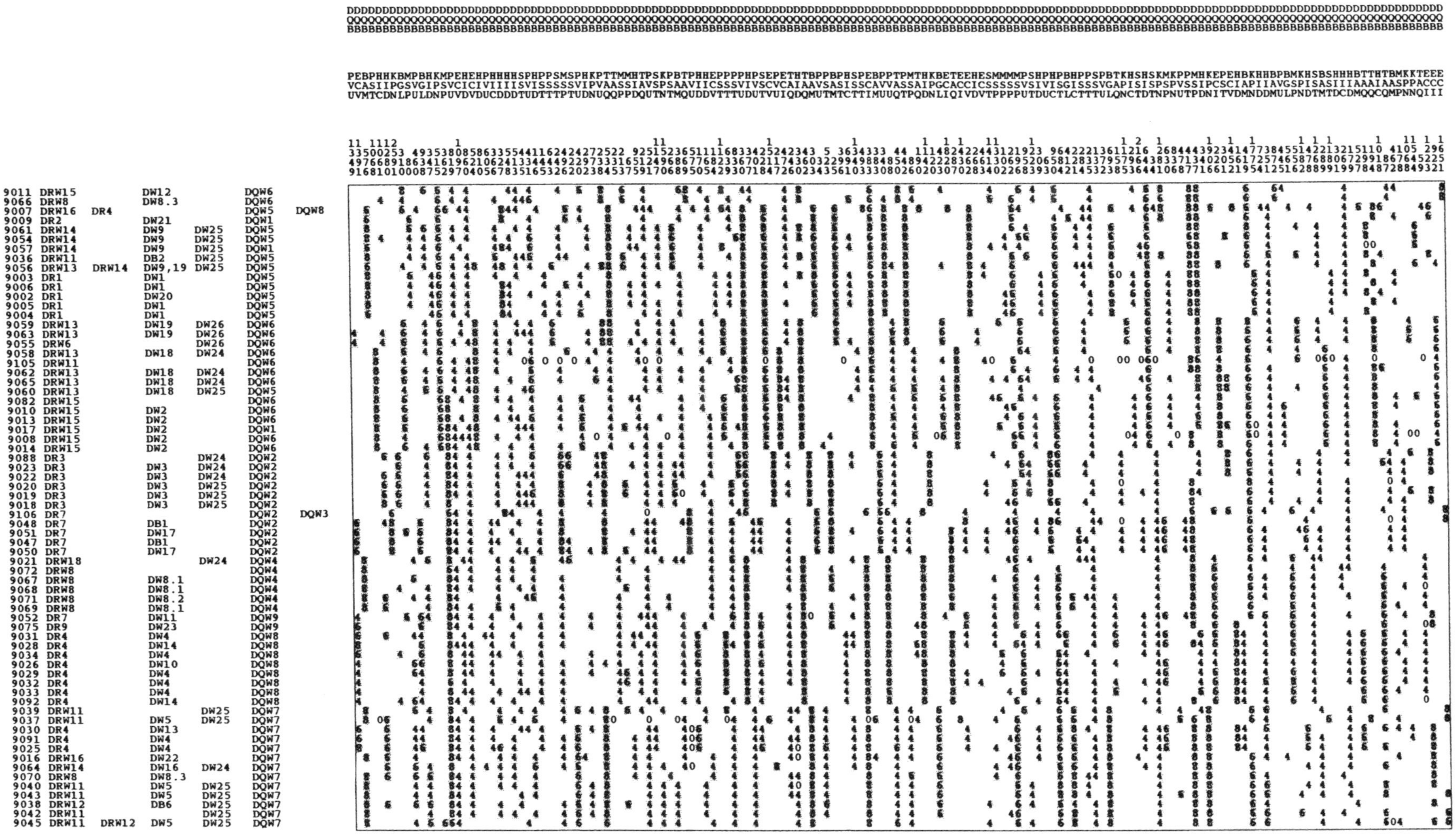

Figure 29A. Tenth International Histocompatibility Workshop
Class II DQB loci RFLP fragments (n=225). Negative cluster. Cells listed by tree relationships.

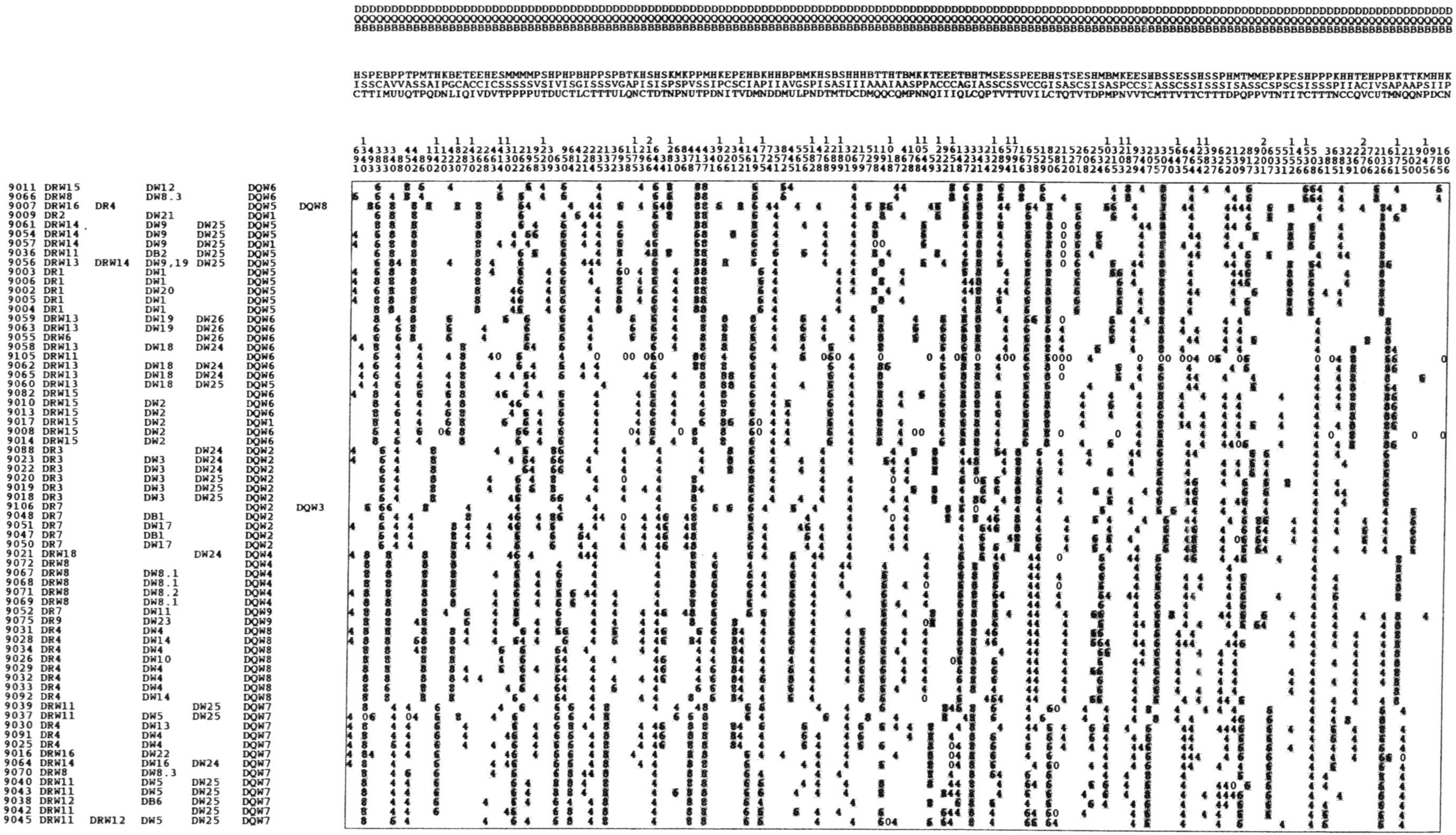

Figure 29B. Tenth International Histocompatibility Workshop
Class II DQB loci RFLP fragments (n=225). Negative cluster. Cells listed by tree relationships.

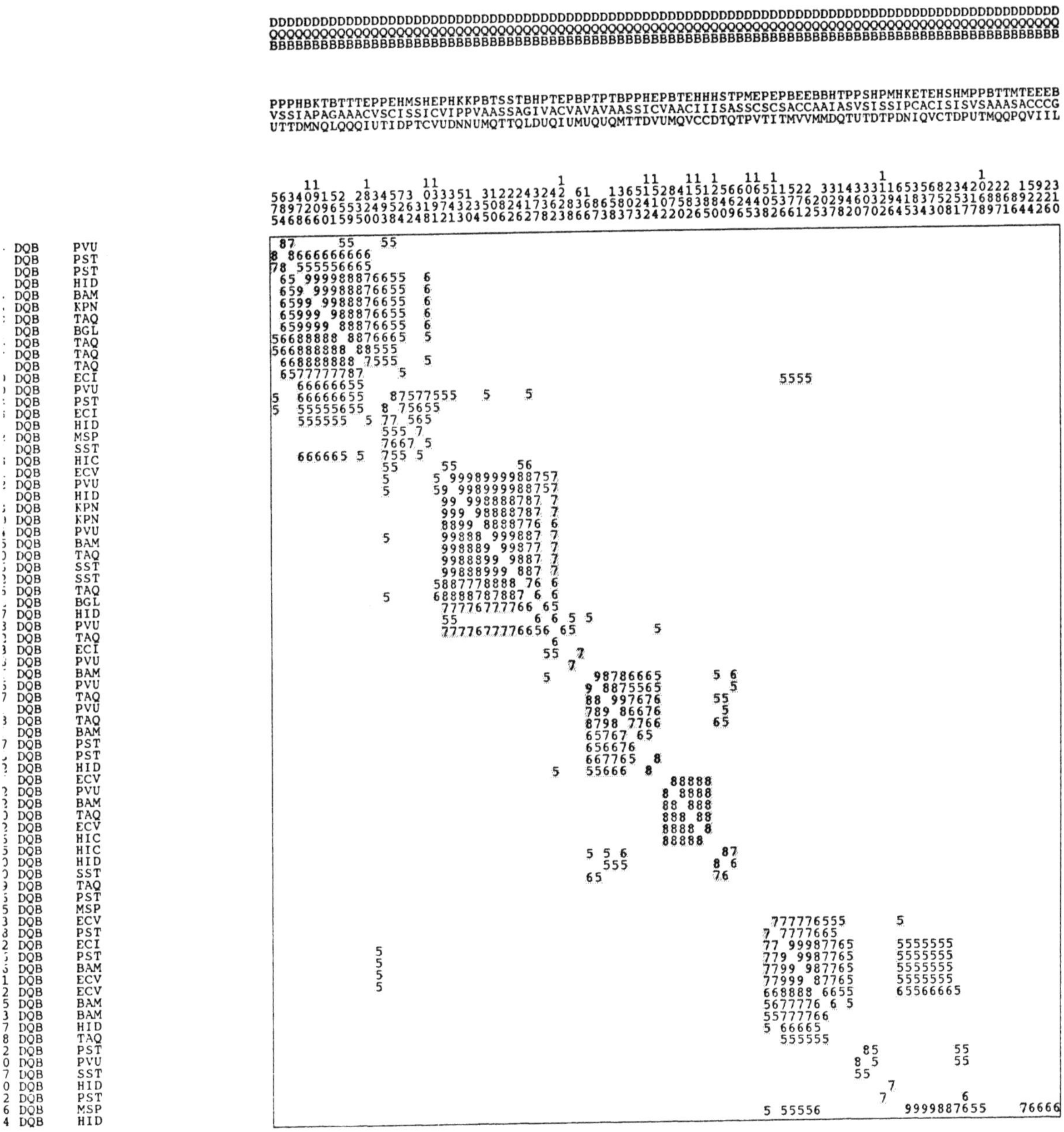

Figure 30A. Tenth International Histocompatibility Workshop class II DQB loci RFLP fragments (n=94). Fragment x fragment cluster.

Figure 30B. Tenth International Histocompatibility Workshop class II DQB loci RFLP fragments (n=94). Fragment x fragment clust

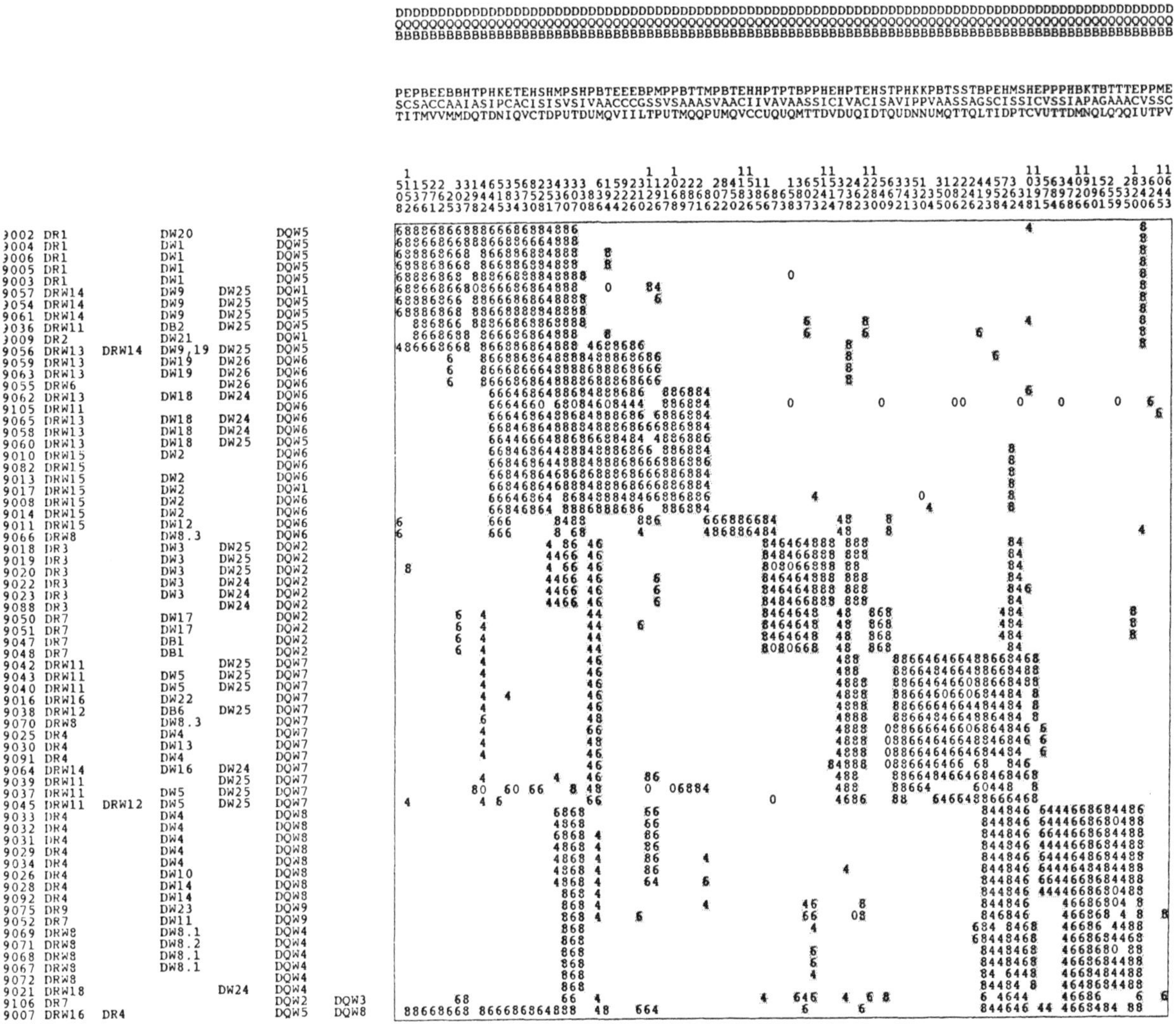

Figure 31. Tenth International Histocompatibility Workshop class II DQB loci RFLP fragments (n=94). Cell listing based on tree relationships.

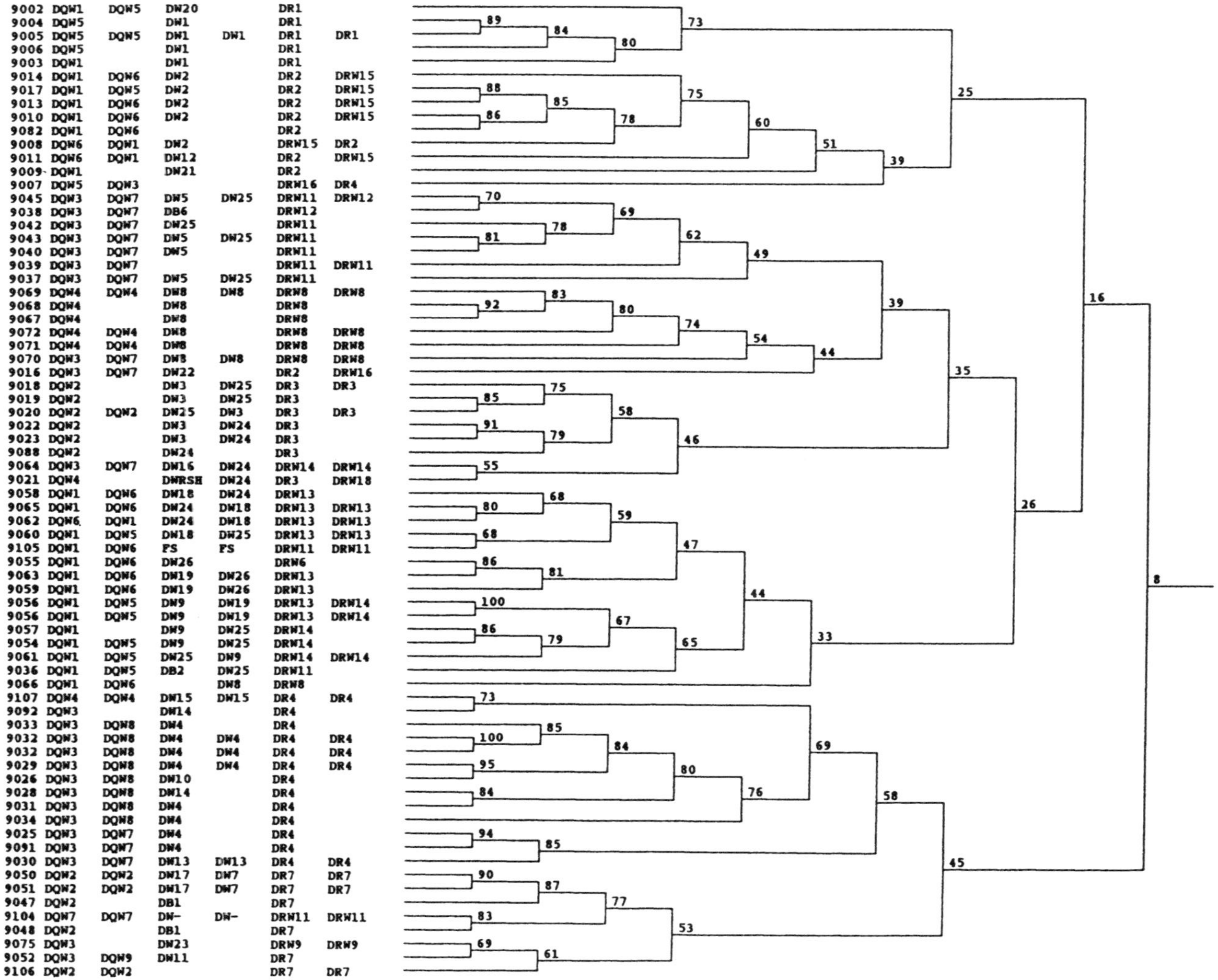

Figure 32. Tenth International Histocompatibility Workshop
Tree analysis of cell relationships: DRB,DQA,DQB loci RFLP fragments (n=536).

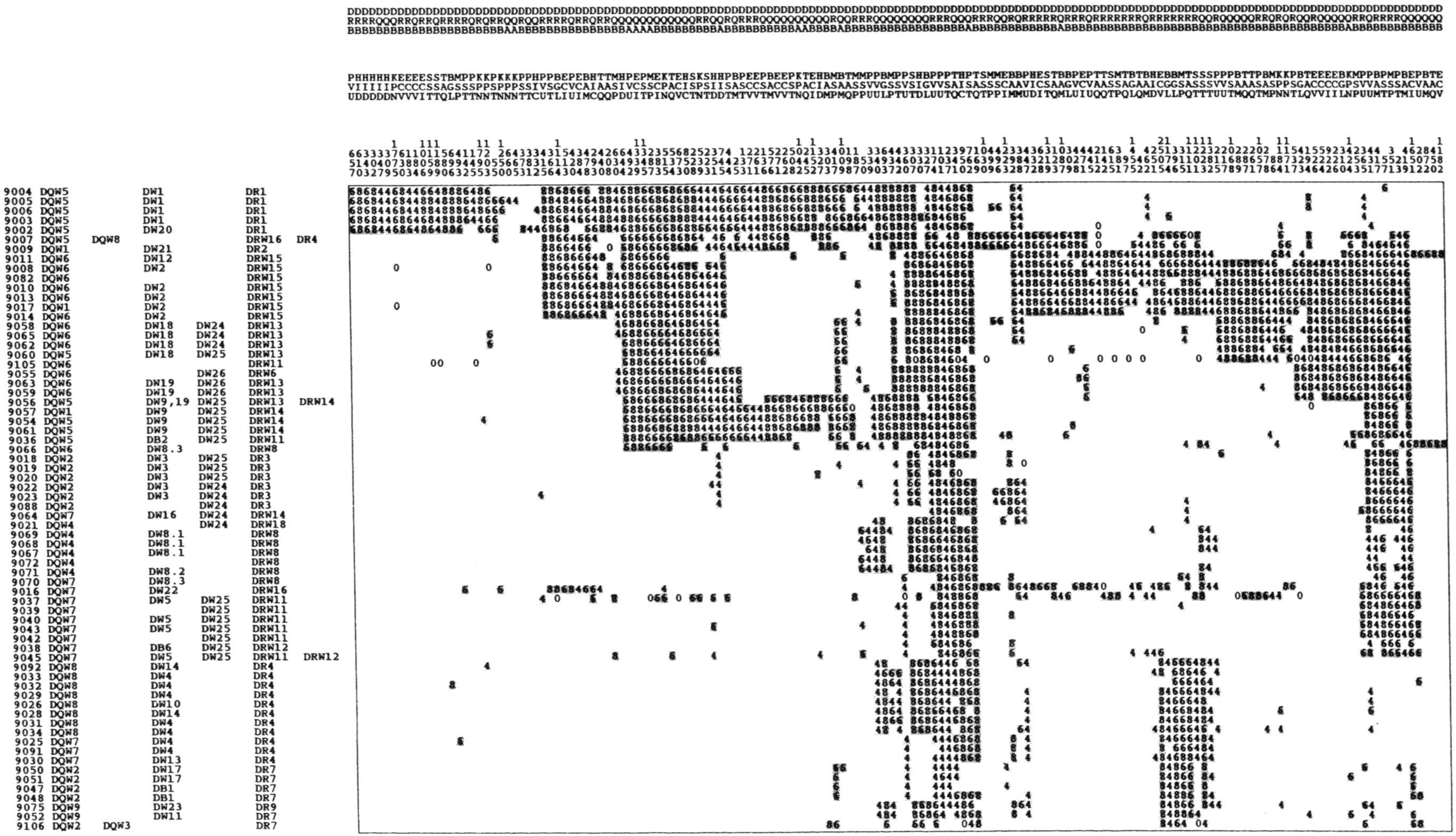

Figure 33A. Tenth International Histocompatibility Workshop
Class II DRB,DQA,DQB loci RFLP fragments (n=536). Cells listed by tree relationships.

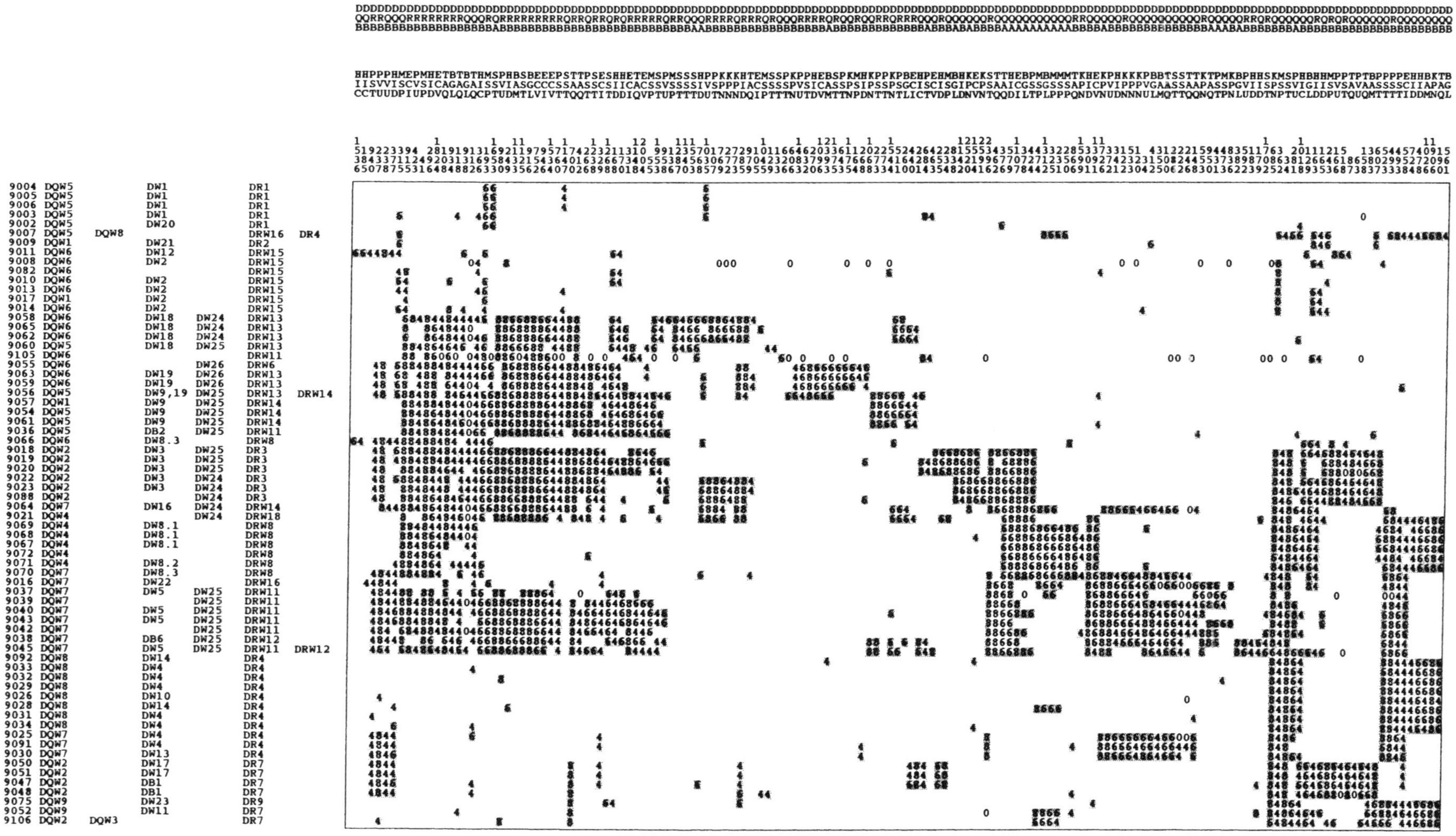

Figure 33B. Tenth International Histocompatibility Workshop
Class II DRB,DQA,DQB loci RFLP fragments (n=536). Cells listed by tree relationships.

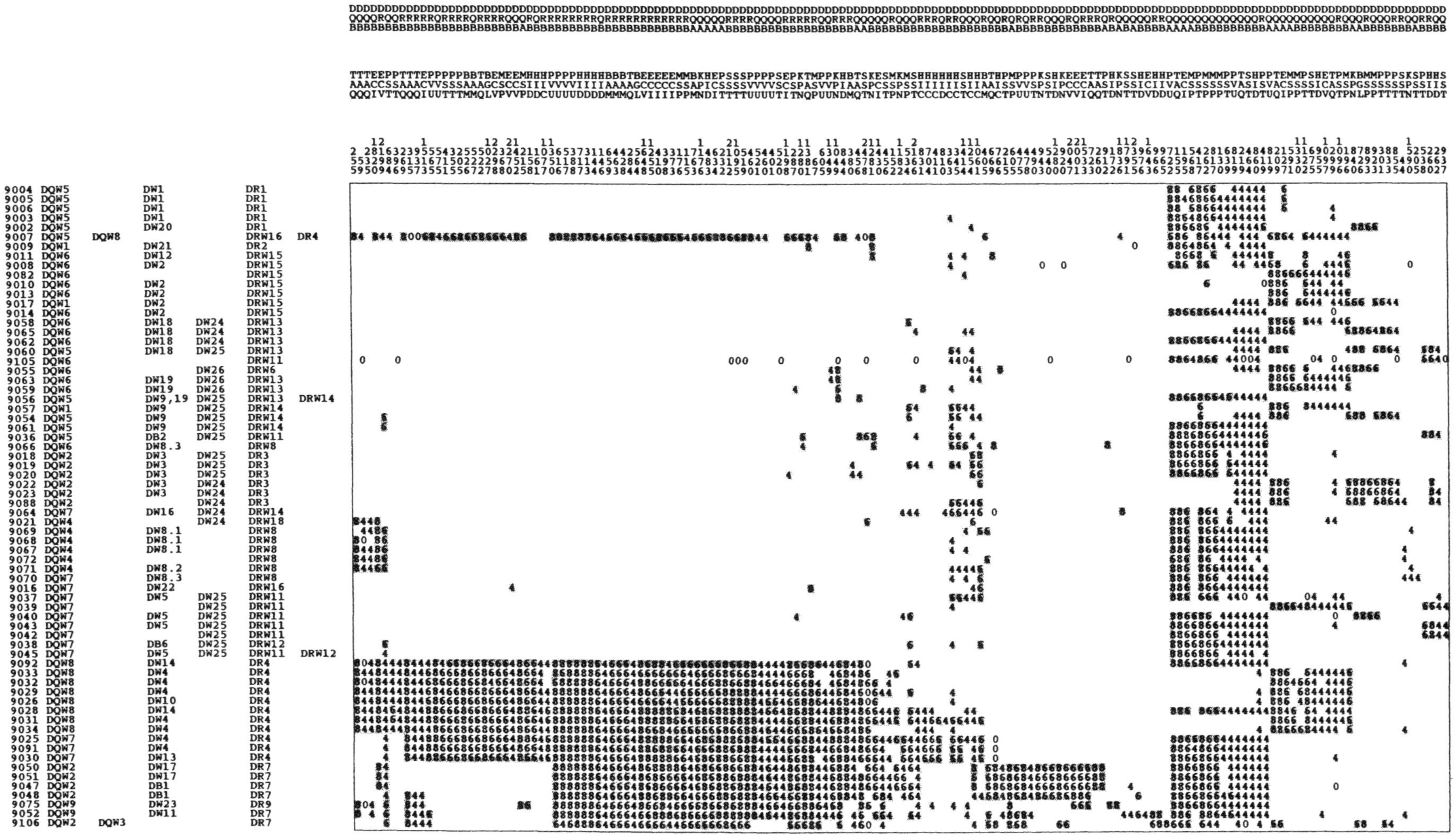

Figure 33C. Tenth International Histocompatibility Workshop
Class II DRB,DQA,DQB loci RFLP fragments (n=536). Cells listed by tree relationships.

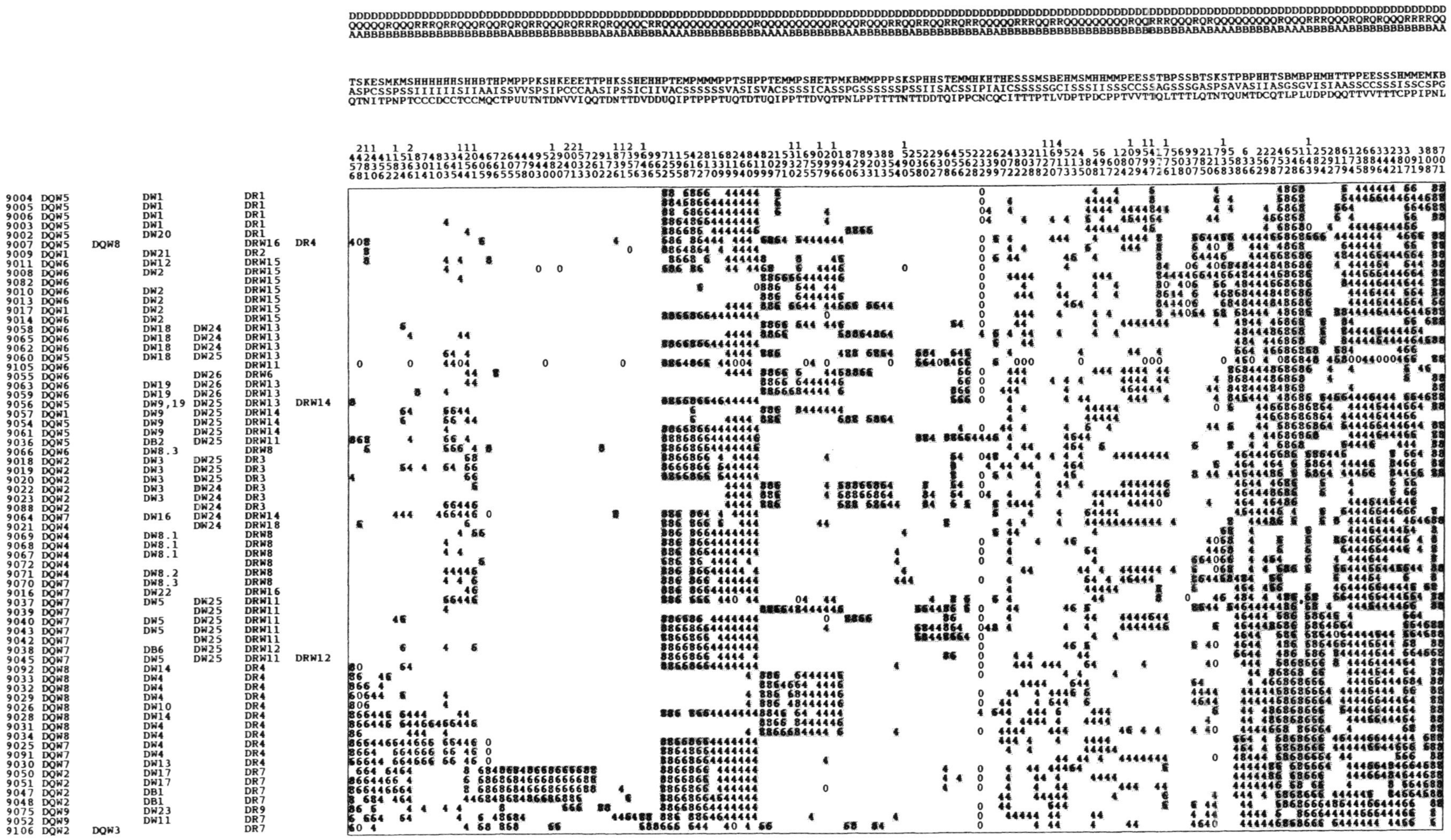

Figure 33D. Tenth International Histocompatibility Workshop
Class II DRB,DQA,DQB loci RFLP fragments (n=536). Cells listed by tree relationships.

Figure 34. Tenth International Histocompatibility Workshop class II DRB,DQB RFLP fragments (TAQ,BAM=72). Positive cluster.

The last columns of Figure 33D (HIC 228 to end) show fragments whose reactivity patterns are not well defined, but which may become so on further analysis. They are included for completion. Between the two groups of fragments are those associated with the DQB2 and DQA2 loci (PVU 722 to MSP 562). These patterns are considered in more detail in the section on class II DQA2 and DQB2 loci.

From the practical standpoint of DNA typing for DR and DQ specificities using RFLP, it is valuable to consider the cluster patterns created by combinations of the three most relevant probes (DRB, DQA, and DQB). The data presented in this section enable selection of a limited number of probe/enzyme combinations sufficient for definition of DR, Dw, and DQ haplotypes. The haplotype discriminatory power of any particular probe/enzyme combination can also be assessed. This enables a balance to be reached between minimizing probe/enzyme combinations, on the one hand, and achievement of adequate haplotype characterization, on the other. An example of such an approach can be found in Martell et al. (12), in which DRB TAQ, DQB TAQ, and DQB BAM were selected. Figure 34 is a positive cluster involving 72 fragments arising from these two probes and two enzymes. (Note that it has not been possible to delete the DRB BAM fragments in the computer-assisted analysis.)

DPB. The DPB probe recognized 176 fragments. Only some 53 fragments were specific for the DPB1 and DPB2 loci. The majority of the remaining 123 correlated with DR specificities. The most likely explanation for the DPB probe associations with DR specificities is cross-hybridization due to homology between the cDNA probe and DR region sequences. The possibility exists that any such sequence homology has arisen from donation of DR gene sequences to the DP gene by some gene rearrangement process. It is noteworthy that there is no comparable association with DQw specificities, except for DQw5. These patterns provide information for more detailed examination of evolutionary relationships between and within these three class II beta loci.

In Figure 35, 154 selected DPB probe-detectable fragments are shown against cells listed according to DR type. Figure 36 maintains the same order of the 154 fragments, but rearranges the cells according to DPw type. For the 54 DPw specificity-associated fragments, several groupings can be readily identified. The first two groups are those of the major clusters of the DPB1 locus. These are followed by fragments, the presence of which are informative in distinguishing between cells of different DPw and DPB nucleotide sequence type. Then follows fragments characterizing the two major clusters of the DPB2 locus.

The first 12 fragments are clearly associated with the DPw1/w5/w3/New DPB1 cluster. Three fragments specify DPw1 (HIC 341, MSP 498, MSP 671). Fragment 716 is present with DPw5 and with four of the six cells (9034, 9003, 9043, 9048) designated "New" in that no DPw specificity has been assigned. Fragment SST 1365 is present with the same four "New" cells plus a fifth (9047), and is also present in some additional cells. The remaining DPB "New" cell (9060) has a "unique" pattern of fragment presence and of fragment absence. From inspection, it can be seen that fragments PST 202, HID 186, TAQ 272, and ECI 1293 were absent only in this cell. Fragment SST 1742 is present only in cells 9047 and 9052. Both cells are DR7, and tree analysis of cell relationships (Figs. 37 and 38) would suggest them to have a close evolutionary association. It is, therefore, not possible at present to determine whether the fragment is subtyping DR7 or truly reflects a new DPw specificity.

Attention is drawn to two fragments (BGL 728 and KPN 2444) that are primarily observed with DPw2 and the DPB1 sequence of DPw4, DPB4.2 (11). These fragments are also seen in some cells in the DPw1/w5/w3 cluster. Fragments MSP 217 and MSP 438 appear to be specific for DPw2. Fragment SST 318 is associated with the DPB1 nucleotide sequence allele, DPB2.2 (11). Another pair of fragments (BGL 505 and MSP 107) have presence/absence patterns associated with DPw4. MSP 107 extends to include all 27 DPw4-associated homozygote cells, while BGL 505 is absent in all but two of the nine DPB4.2 cells.

Figure 37 depicts a tree dendogram based on the use of all cells, heterozygous as well as homozygous at the DPB loci. In Figure 38, DPB heterozygous cells have been removed, and the analysis repeated only on the 53 remaining homozygous cells. There is excellent stratification of cells into the two main clusters of DPw1/w5/w3-associated cells and of DPw2/w4 related cells. The three "blank" zones in which DPw specificities have not been assigned are clearly demarcated. The DPw4-related "blank" comprises cells of the DPB1 allele, DPB4.2 (11). Cells 9104 and 9048 are seen to be DPw2-associated.

The RFLP pattern relationships between these cells, again listed according to DPw type, is shown in Figure 39 which utilizes 45 fragments detectable by the DPA locus probe, as well as 53 detectable by the DPB probe. These relationships are considered in more detail elsewhere (11).

Class II Alpha and Beta Loci Fragment Patterns, with Comments on Linkage Disequilibrium

Class II DRA, DQA1, DQA2, DPA1, and DPA2 Loci. One hundred thirty-six RFLP fragments were associated with these five class II alpha loci. Figures 40 A and B show the fragment cluster pattern. It can be seen that fragments show intralocus co-clustering. In no instance are the clusters composed of fragments detectable by more than one locus probe. Figure 41 shows the tree analysis of cell relationships generated by these class II alpha loci RFLP fragments. Although the alpha chain products are not the primary determinant of serologically detectable specificities, it is apparent that the cells are stratified into recognizable groups corresponding to known DR and DQ relationships.

Class II DQA2 and DQB2 Loci. During analysis of the 536 fragments detectable by the DRB, DQA, and DQB probes, the similarity of mutually exclusive RFLP clusters for the two non-expressed DQ loci were apparent (see Figures 33C and D). In Figures 42 and 43 this association is presented in more detail. Figure 42 shows the negative cluster pattern with cells listed by tree relationships. Figure 43 is the corresponding positive cluster, in which it can be seen that the less frequent DQA2 cluster is associated with two DQB2 clusters. The more frequent cluster has a very similar pattern for both DQA2 and DQB2. There is very clear stratification of DQw type in relation to these clusters. For example, all DRw8, DQw4 cells, together with the two unusual DQw7 cells (9016 and 9064), are associated with a particular pattern. Again, the DQw8 cells are predominantly associated with one of the two DQB2 subgroups comprising the less frequent cluster. These and other patterns clearly suggest associations of evolutionary significance.

Class II Interlocus Co-Clustering Fragments. Of the total 781 fragments detectable by class II alpha and beta probes, 175 fragments participated in clusters involving

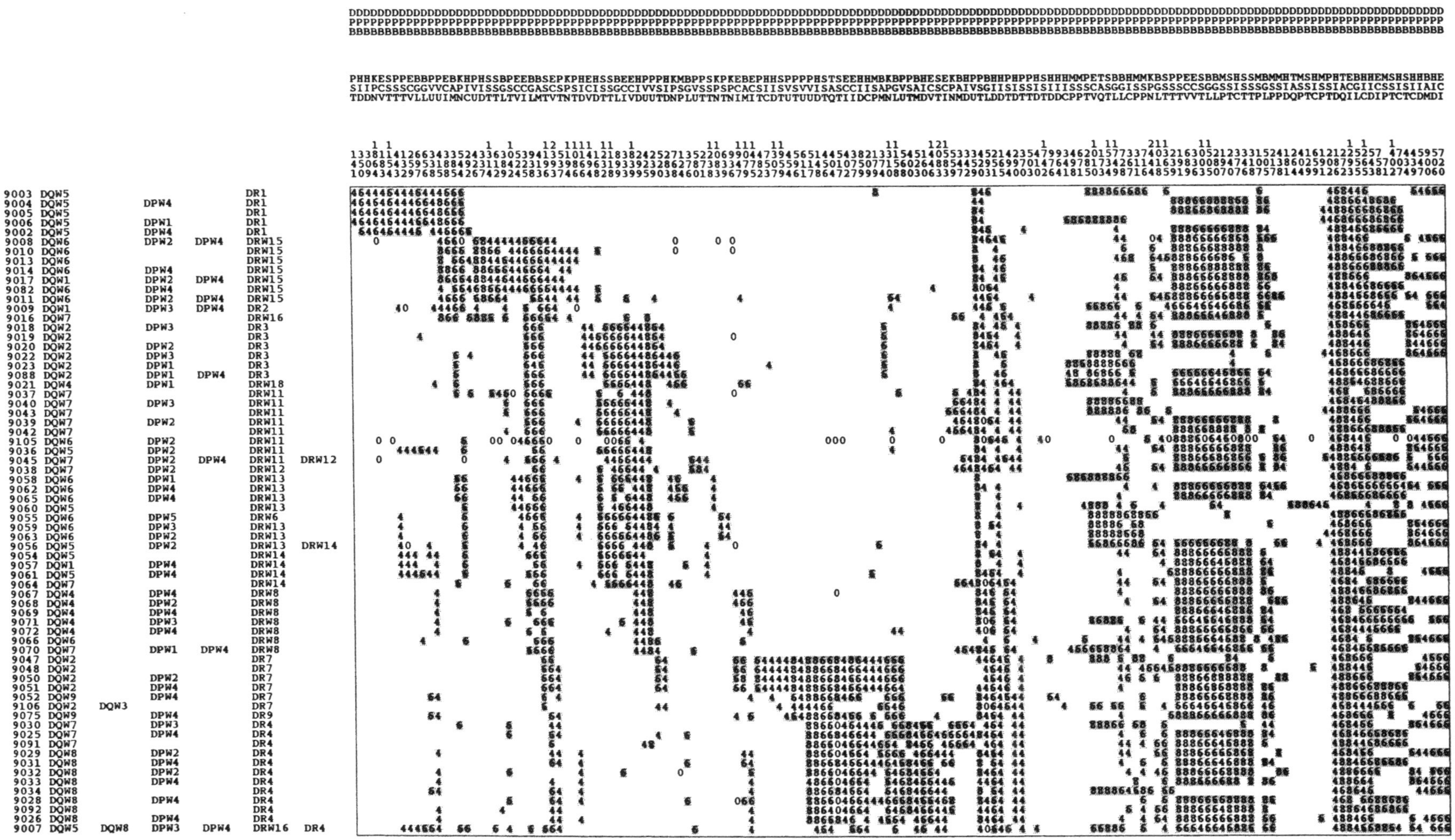

Figure 35. Tenth International Histocompatibility Workshop
Class II DPB locus RFLP fragments (n=154). Positive cluster. Cells listed by DRw type.

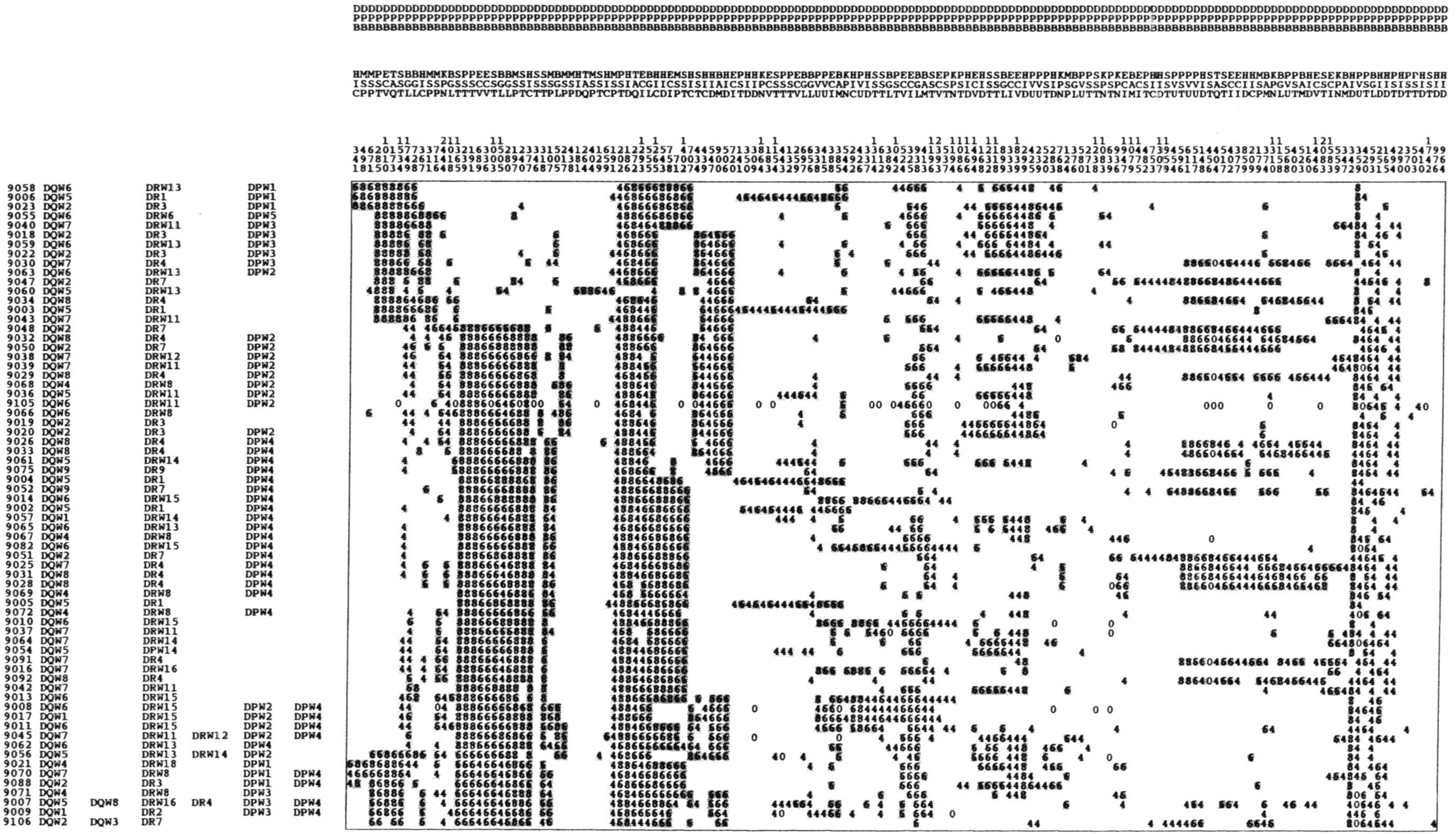

Figure 36. Tenth International Histocompatibility Workshop
Class II DPB locus RFLP fragments (n=154). Positive cluster. Cells listed by DPw type.

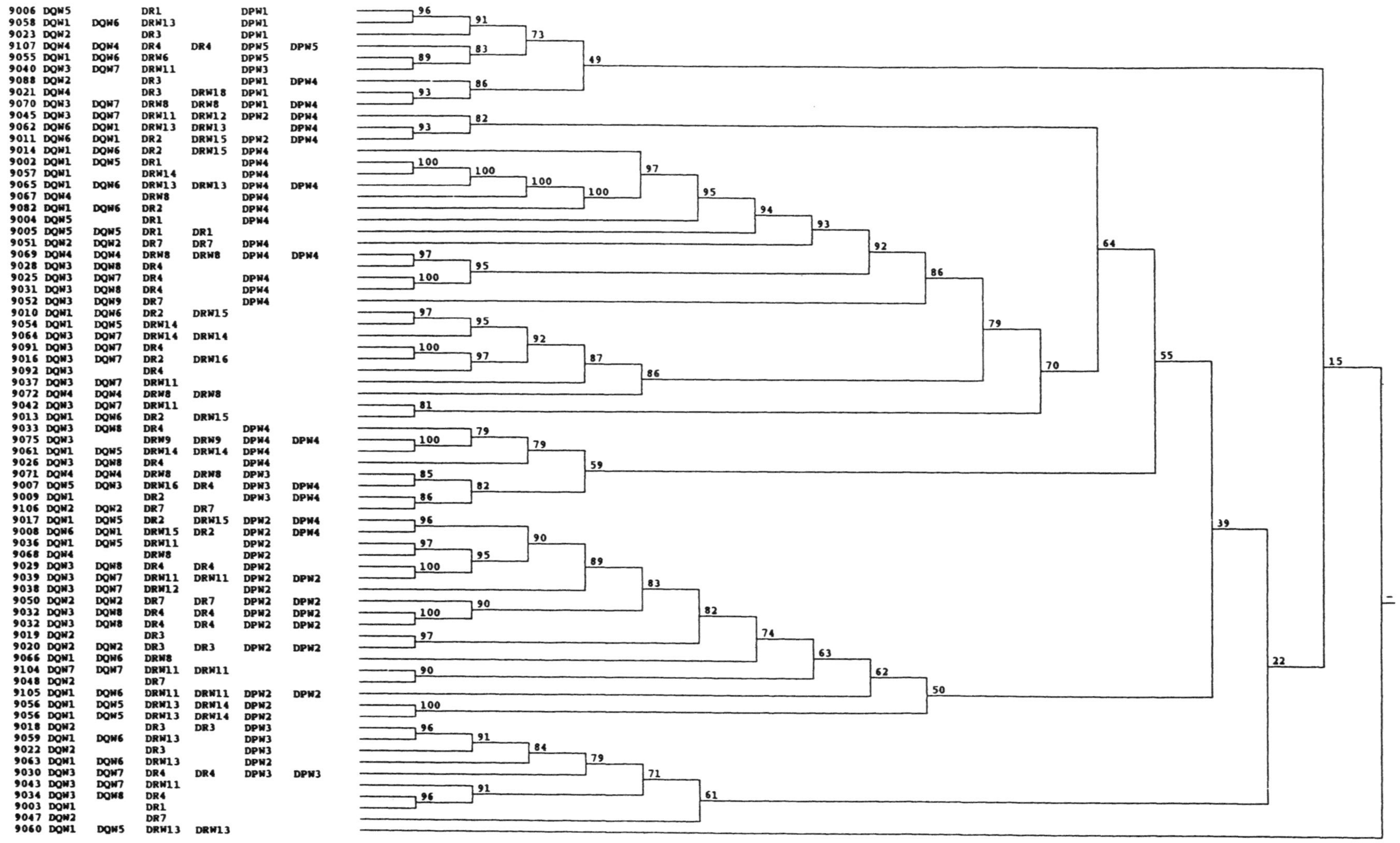

Figure 37. Tenth International Histocompatibility Workshop
Tree analysis of cell relationships: class II DPB loci RFLP fragments (n=57).

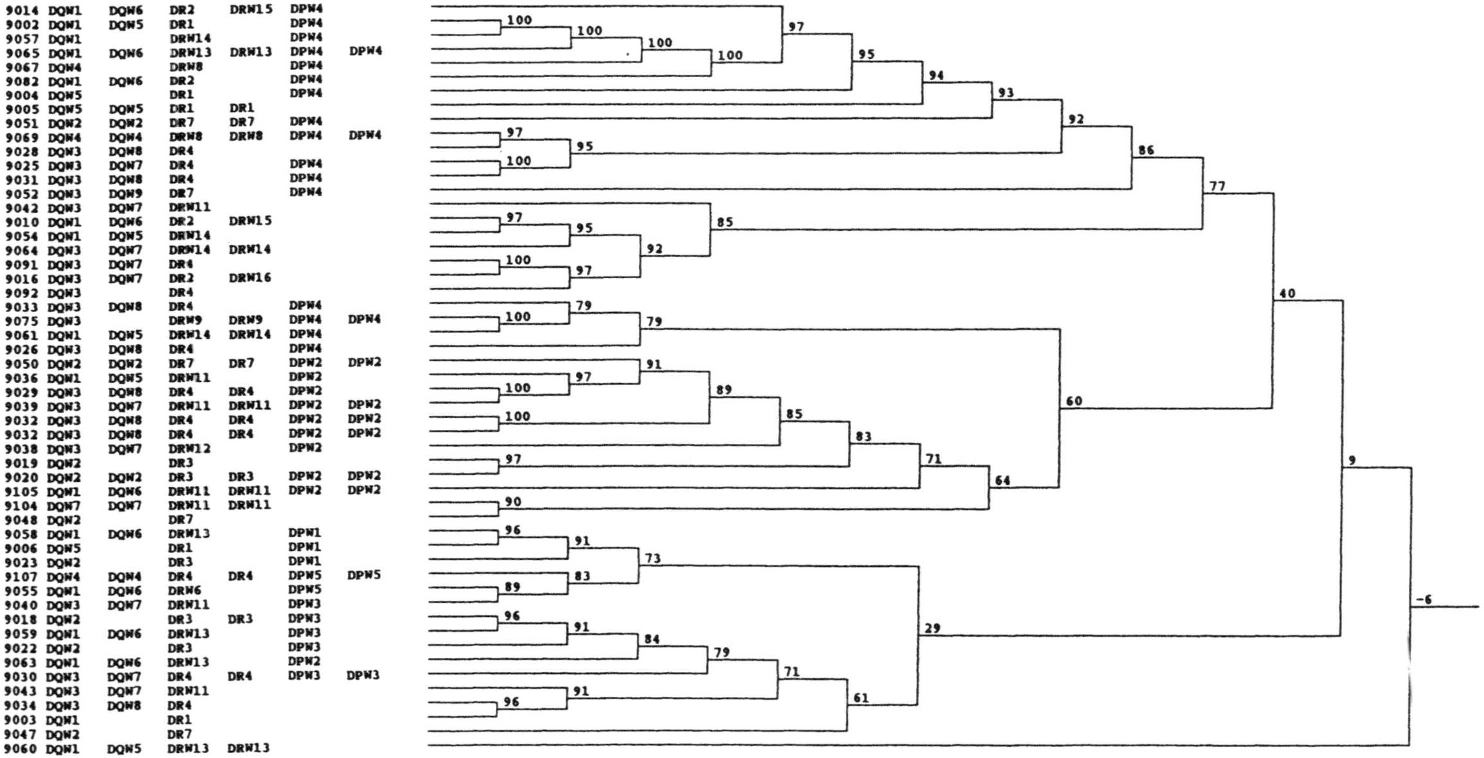

Figure 38. Tenth International Histocompatibility Workshop W/S cells (n=53) DPB1,DPB2 loci relationships using tree analysis.

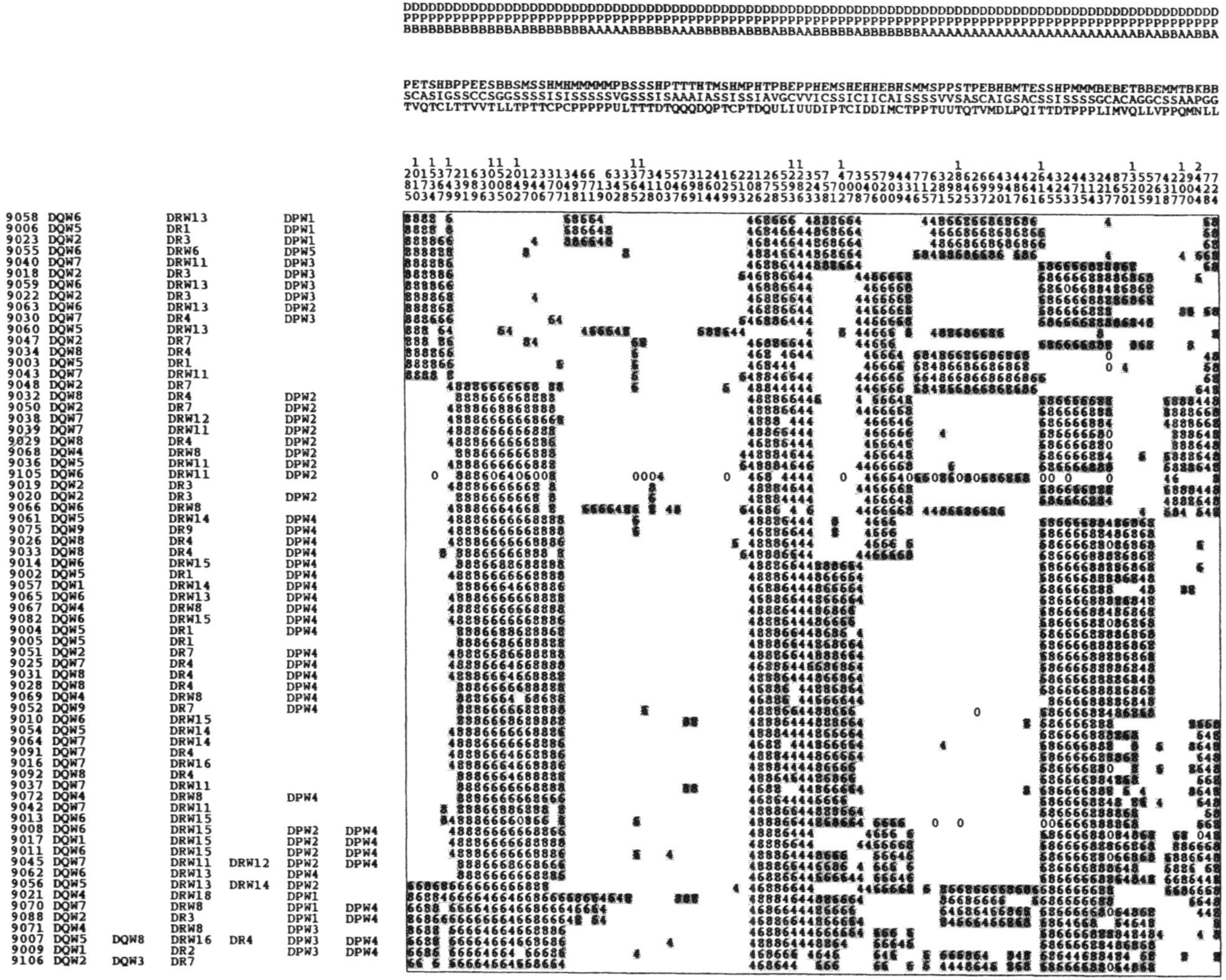

Figure 39. Tenth International Histocompatibility Workshop
Class II DPA,DPB loci RFLP selected fragments (n=98). Positive cluster.

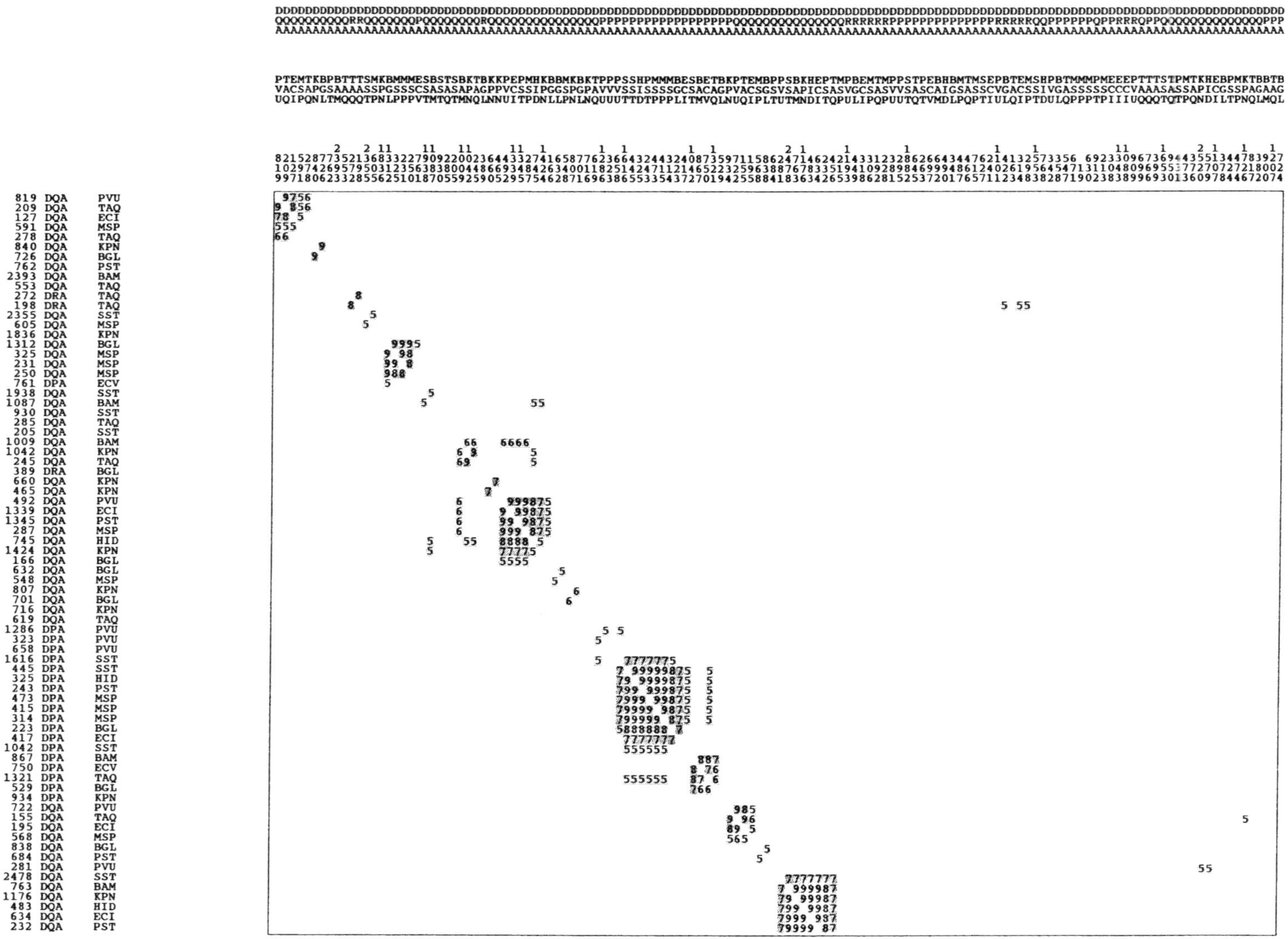

Figure 40A. Tenth International Histocompatibility Workshop
Class II DR, DQ, DP alpha loci RFLP fragments (n=136). Fragment cluster

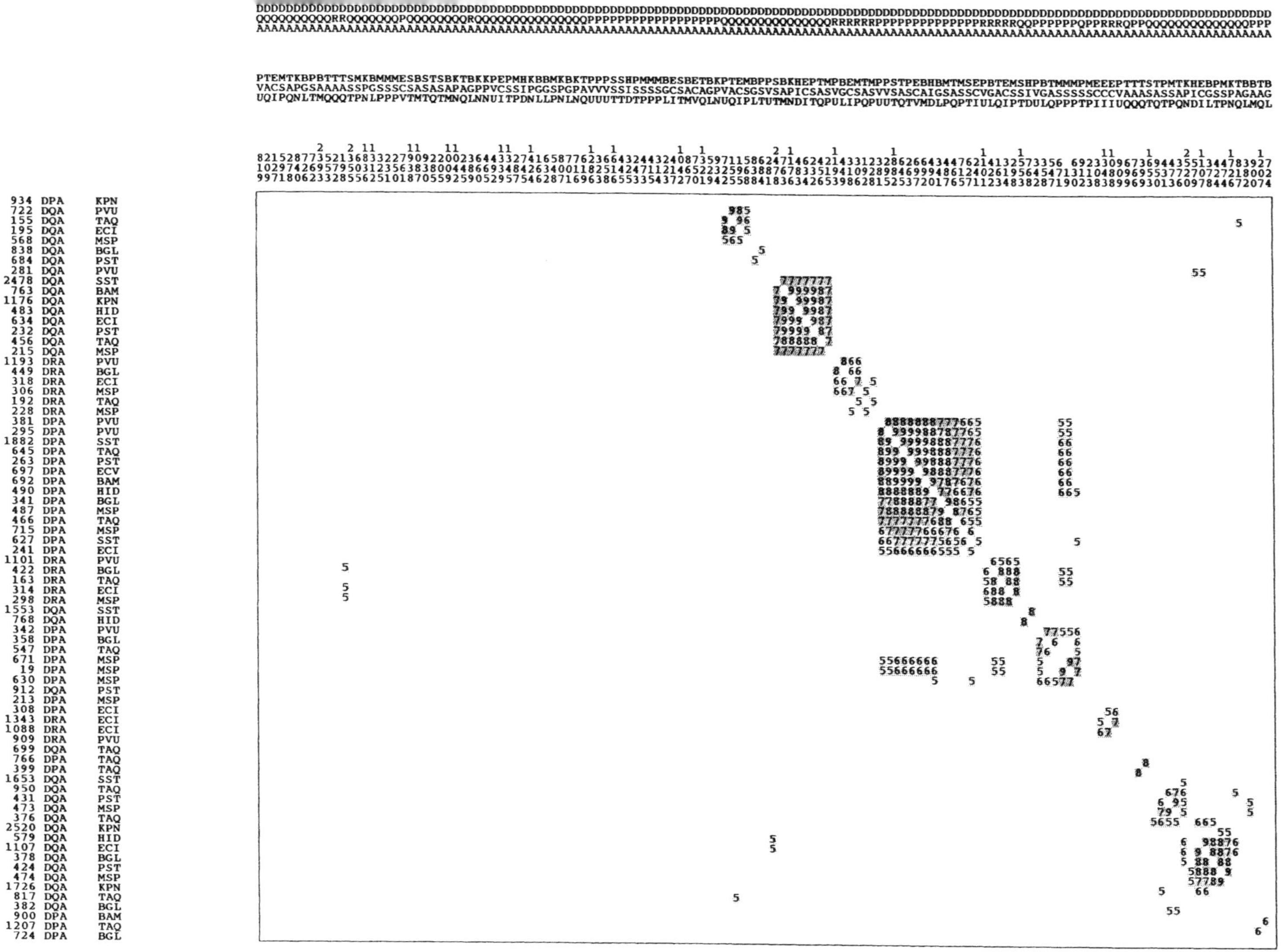

Figure 40B. Tenth International Histocompatibility Workshop
Class II DR,DQ,DP alpha loci RFLP fragments (n=136). Fragment cluster.

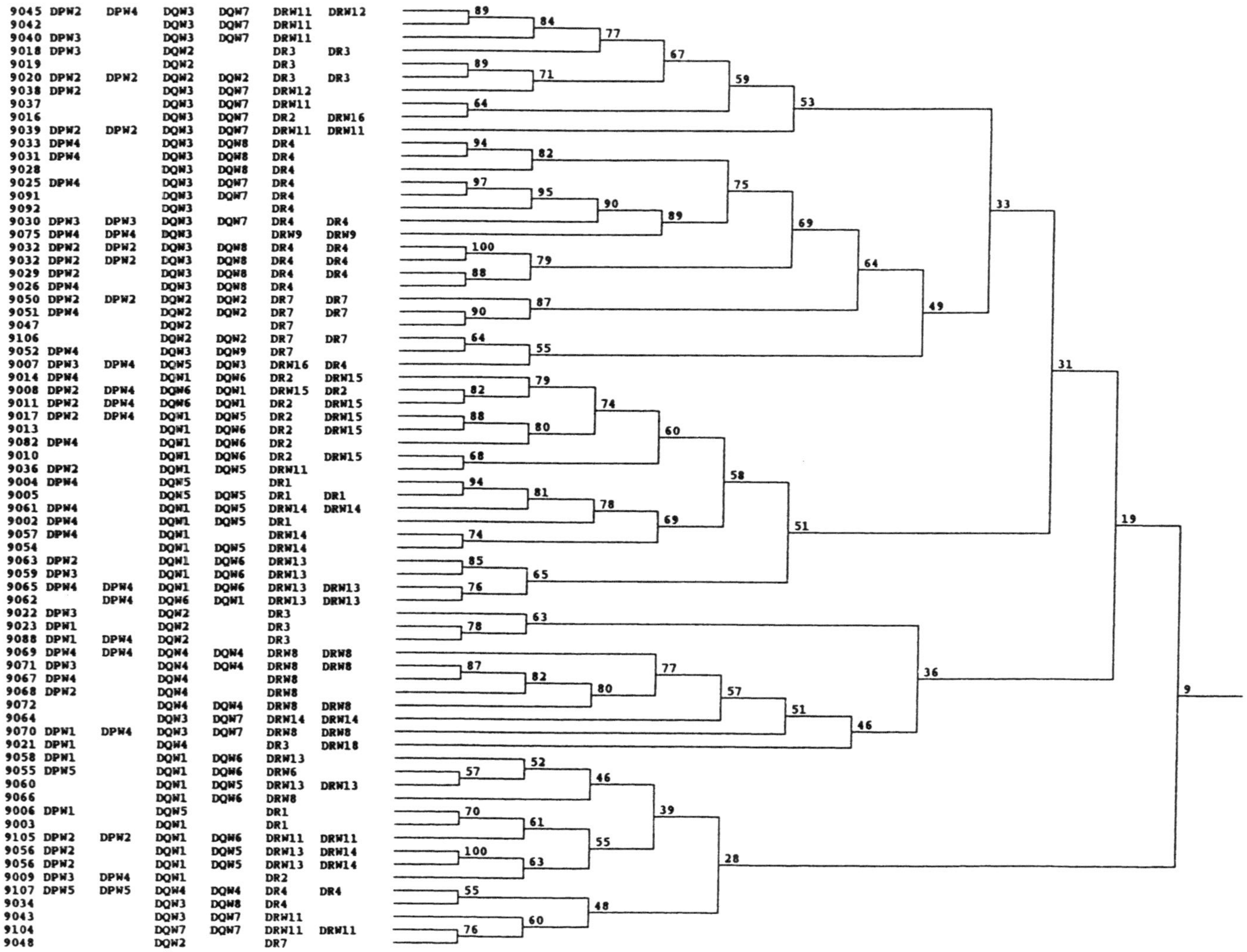

Figure 41. Tenth International Histocompatibility Workshop
Tree analysis of cell relationships: class II DR,DQ,DP alpha loci.

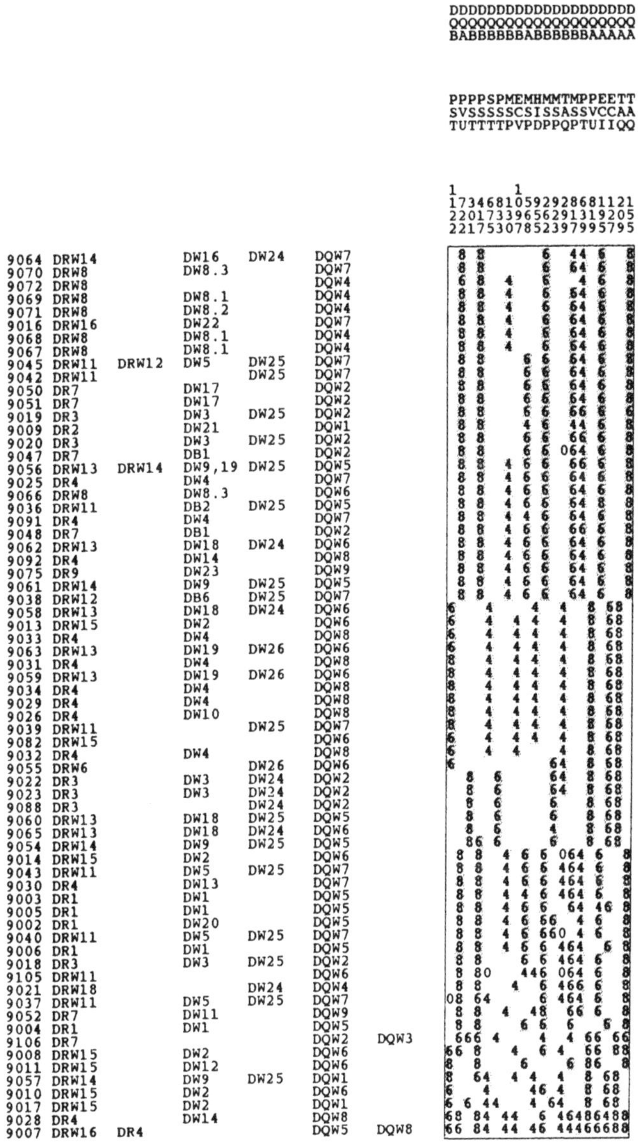

Figure 42. Tenth International Histocompatibility Workshop class II DQA2,DQB2 loci RFLP fragments (n=20). Negative cluster. Cells listed by tree relationships.

9045	DRW11	DRW12	DW5	DW25	DQW7	
9042	DRW11			DW25	DQW7	
9050	DR7		DW17		DQW2	
9051	DR7		DW17		DQW2	
9019	DR3		DW3	DW25	DQW2	
9009	DR2		DW21		DQW1	
9020	DR3		DW3	DW25	DQW2	
9047	DR7		DB1		DQW2	
9069	DRW8		DW8.1		DQW4	
9071	DRW8		DW8.2		DQW4	
9016	DRW16		DW22		DQW7	
9068	DRW8		DW8.1		DQW4	
9067	DRW8		DW8.1		DQW4	
9064	DRW14		DW16	DW24	DQW7	
9070	DRW8		DW8.3		DQW7	
9072	DRW8				DQW4	
9056	DRW13	DRW14	DW9,19	DW25	DQW5	
9025	DR4		DW4		DQW7	
9066	DRW8		DW8.3		DQW6	
9036	DRW11		DB2	DW25	DQW5	
9091	DR4		DW4		DQW7	
9048	DR7		DB1		DQW2	
9062	DRW13		DW18	DW24	DQW6	
9092	DR4		DW14		DQW8	
9075	DR9		DW23		DQW9	
9061	DRW14		DW9	DW25	DQW5	
9038	DRW12		DB6	DW25	DQW7	
9058	DRW13		DW18	DW24	DQW6	
9013	DRW15		DW2		DQW6	
9033	DR4		DW4		DQW8	
9063	DRW13		DW19	DW26	DQW6	
9031	DR4		DW4		DQW8	
9059	DRW13		DW19	DW26	DQW6	
9034	DR4		DW4		DQW8	
9029	DR4		DW4		DQW8	
9026	DR4		DW10		DQW8	
9039	DRW11			DW25	DQW7	
9082	DRW15				DQW6	
9032	DR4		DW4		DQW8	
9055	DRW6			DW26	DQW6	
9022	DR3		DW3	DW24	DQW2	
9023	DR3		DW3	DW24	DQW2	
9088	DR3			DW24	DQW2	
9060	DRW13		DW18	DW25	DQW5	
9065	DRW13		DW18	DW24	DQW6	
9054	DRW14		DW9	DW25	DQW5	
9002	DR1		DW20		DQW5	
9040	DRW11		DW5	DW25	DQW7	
9018	DR3		DW3	DW25	DQW2	
9014	DRW15		DW2		DQW6	
9043	DRW11		DW5	DW25	DQW7	
9030	DR4		DW13		DQW7	
9003	DR1		DW1		DQW5	
9005	DR1		DW1		DQW5	
9006	DR1		DW1		DQW5	
9021	DRW18			DW24	DQW4	
9105	DRW11				DQW6	
9037	DRW11		DW5	DW25	DQW7	
9004	DR1		DW1		DQW5	
9052	DR7		DW11		DQW9	
9106	DR7				DQW2	DQW3
9011	DRW15		DW12		DQW6	
9008	DRW15		DW2		DQW6	
9057	DRW14		DW9	DW25	DQW1	
9010	DRW15		DW2		DQW6	
9017	DRW15		DW2		DQW1	
9028	DR4		DW14		DQW8	
9007	DRW16	DR4			DQW5	DQW8

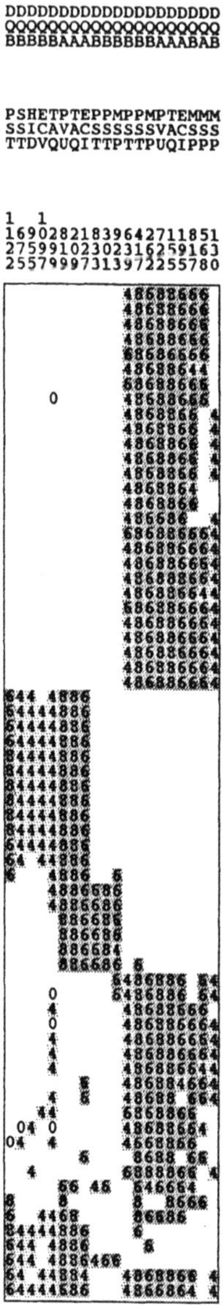

Figure 43. Tenth International Histocompatibility Workshop class II DQA2,DQB2 loci RFLP fragments (n=20). Positive cluster. Cells listed by tree relationships.

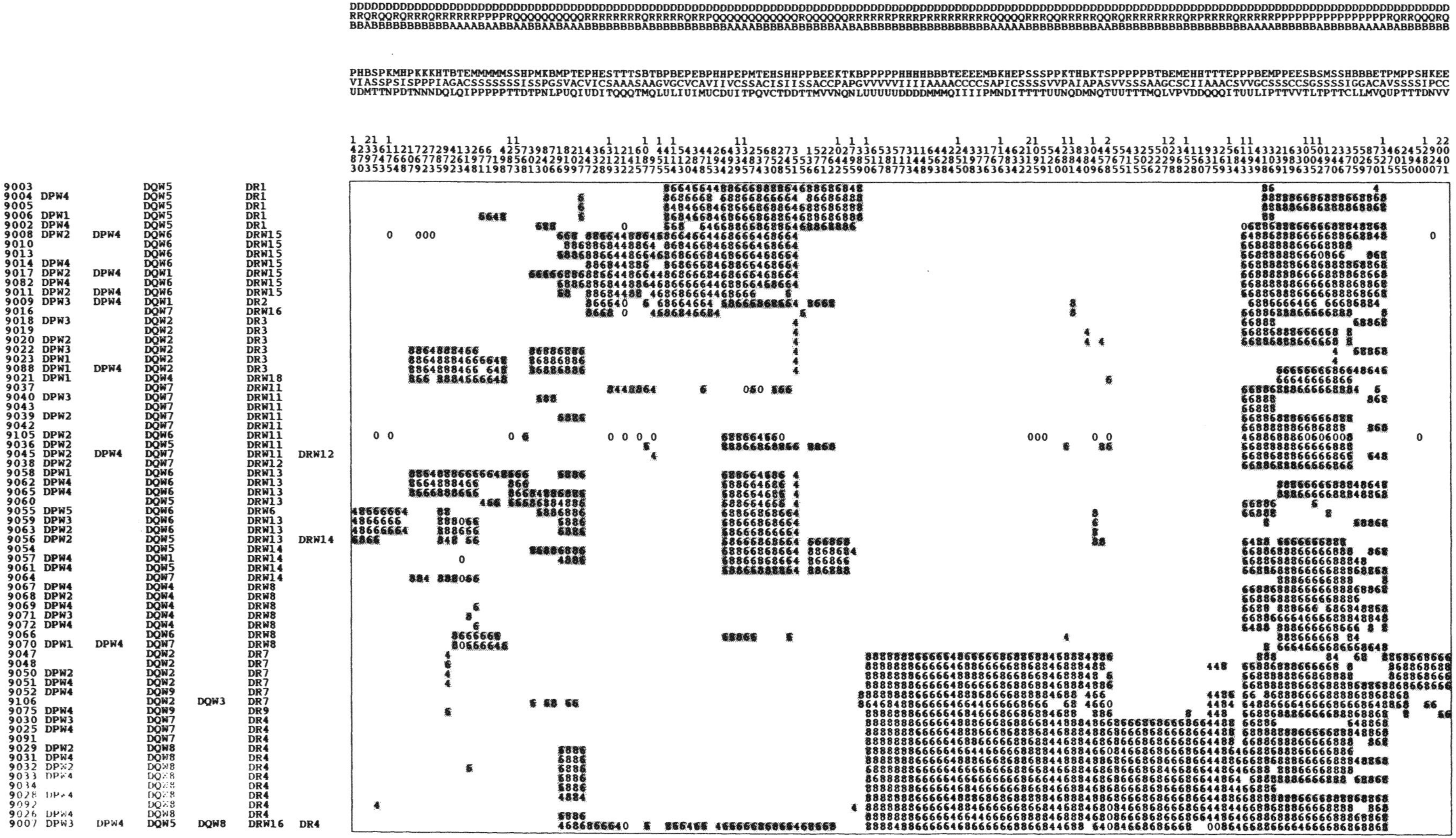

Figure 44A. Tenth International Histocompatibility Workshop Class II interlocus co-clustering fragments (n=175).

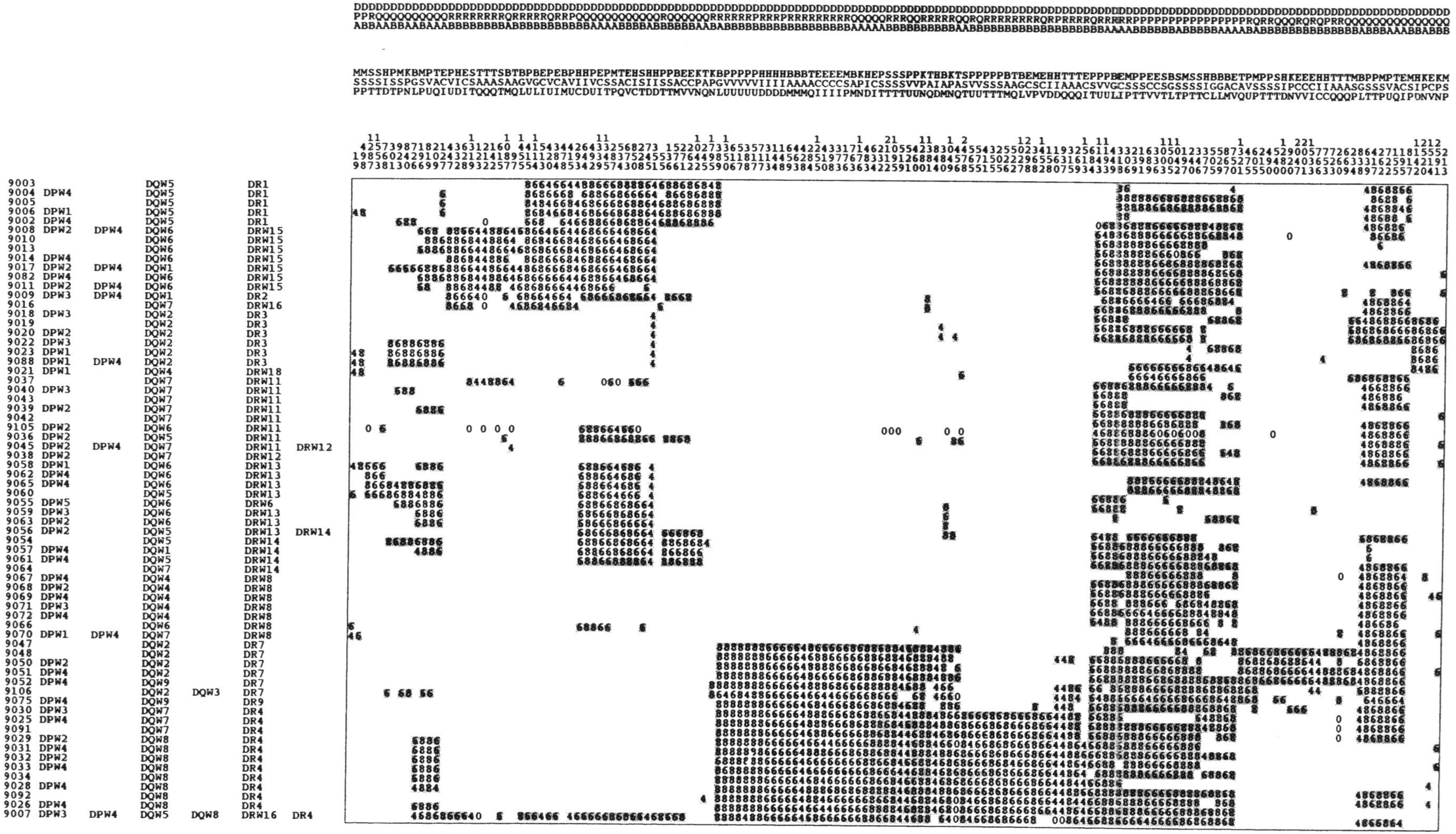

Figure 44B. Tenth International Histocompatibility Workshop Class II interlocus co-clustering fragments (n=175).

probes from two or more class II region loci. Figures 44 A and B show the positive cluster pattern of these fragments against cells listed by DR type. This figure is of particular use in distinguishing between likely interlocus hybridization, on the one hand, and genuine linkage disequilibrium, on the other.

Acknowledgments. The authors wish to acknowledge the invaluable contribution of Mr. Mark Ferguson of the Computing Center, Howard Hughes Medical Institute, University of Utah Medical Center, Salt Lake City, Utah. M.J.S. wants to record his appreciation for the enthusiastic secretarial support of Mrs. Percy Noble, which almost foundered on the "unique" fragment tables.

References

1. Marcadet A, O'Connell P, Cohen D. Standardized Southern blot workshop technique. In: Dupont B (ed). Immunobiology of HLA, Volume I: Histocompatibility Testing 1987. New York: Springer-Verlag, 1989.
2. Marcadet A, Dupont B, Cohen D. Organization and design of the Southern blot component of the tenth histocompatibility workshop. In: Dupont B (ed). Immunobiology of HLA, Volume I: Histocompatibility Testing 1987. New York: Springer-Verlag, 1989.
3. Knowles RW. Structural polymorphism of the HLA class II alpha and beta chains: summary of the tenth workshop 2-D gel analysis. In: Dupont B (ed). Immunobiology of HLA, Volume I: Histocompatibility Testing 1987. New York: Springer-Verlag, 1989.
4. Flomenberg N. Functional polymorphisms of HLA class II gene products detected by T-lymphocyte clones: summary of the Tenth International Histocompatibility Workshop cellular studies. In: Dupont B (ed). Immunobiology of HLA, Volume I: Histocompatibility Testing 1987. New York: Springer-Verlag, 1989.
5. Alper C, Andreas-Zietz A, Arnaiz-Villena A, et al. Workshop report of RFLP in C4, C2, BF and 21-OH genes. In: Dupont B (ed). Immunobiology of HLA, Volume I: Histocompatibility Testing 1987. New York: Springer-Verlag, 1989.
6. White P. Molecular genetics of the class III region of the HLA complex. In: Dupont B (ed). Immunobiology of HLA, Volume II: Immunogenetics and Histocompatibility. New York: Springer-Verlag, 1989.
7. Dawkins RL, Kay PH, Martin E, Christiansen FT. The genomic structure of ancestral haplotypes revealed by pulsed field gel electrophoresis. In: Dupont B (ed). Immunobiology of HLA, Volume I: Histocompatibility Testing 1987. New York: Springer-Verlag, 1989.
8. Trowsdale J, Young JAT, Kelly AP, et al. Structure, sequence and polymorphism in the HLA-D region. Immunol Rev 1985;85:5.
9. Jonsson A-K, Hyldig-Nielsen J-J, Servenius B, et al. Class II genes of the human major histocompatibility complex. J Biol Chem 1987;262:8767.
10. Horn GT, Bugawan TL, Long CM, Erlich HA. Allelic sequence variation of the HLA-DQ loci: relationship to serology and to insulin-dependent diabetes susceptibility. Proc Natl Acad Sci USA 1988;85:6012.
11. Simons MJ, Erlich HA. RFLP sequence interrelations at the DPA and DPB loci. In: Dupont B (ed). Immunobiology of HLA, Volume I: Histocompatibility Testing 1987. New York: Springer-Verlag, 1989.
12. Martell R, Marcadet A, Cohen D. HLA class II typing by the workshop-defined RFLP specificities in unrelated individuals. In: Dupont B (ed). Immunobiology of HLA, Volume I: Histocompatibility Testing 1987. New York: Springer-Verlag, 1989.

Joint Report: cDNA

First Domain Sequence Diversity of DR and DQ Subregion Alleles

Peter K. Gregersen,[1] John A. Todd,[2] Henry A. Erlich,[3] Eric Long,[4] Bo Servenius,[5] Ed Choi,[6] Hung Teh Kao,[7] and Janet S. Lee[8]

DR Beta First Domain Amino Acid Sequence

	10	20	30	40	50
DR1βI	GDTRPRFLWQ	LKFECHFFNG	TERVRLLERC	IYNQEESVRF	DSDVGEYRAV
DR1 NASC	----------	----------	----------	----------	----------
DR1 CETUS	----------	----------	----------	----------	----------
DR2 Dw2βI	--------Q-	D-Y-------	-----F-H-D	------DL--	----------
DR2 Dw12βI	--------Q-	D-Y-------	-----F-H-G	------N---	----------
DR2 AZHβI	-----C--Q-	D-Y-------	-----F-H-G	------N---	----------
DR3βI	--------EY	STS-------	-----Y-D-Y	FH----N---	------F---
DR3 DQwa	--------EY	STS-------	-----F---Y	FH----N---	----------
DR4 Dw4βI	--------E-	V-H-------	-----F-D-Y	F-H---Y---	----------
DR4 Dw10βI	--------E-	V-H-------	-----F-D-Y	F-H---Y---	----------
DR4 Dw13βI	--------E-	V-H-------	-----F-D-Y	F-H---Y---	----------
DR4 Dw14βI	--------E-	V-H-------	-----F-D-Y	F-H---Y---	----------
DR4 Dw15βI	--------E-	V-H-------	-----F-D-Y	F-H---Y---	----------
DR4 KT2	--------E-	V-H-------	-----F-D-Y	F-H-------	----------
DR4 CETUS	--------E-	V-H-------	-----F-D-Y	F-H---Y---	----------
DR5 Dw5βI	--------EY	STS-------	-----F-D-Y	F-----Y---	------F---
DR5 JVM	--------EY	STS-------	-----F-D-Y	F-----Y---	------F---
DRw6aβI	--------EY	STS-------	-----F-D-Y	FH----N---	------F---
DRw6bβI	--------EY	STS-------	-----F-D-Y	FH----F---	----------
DRw6 AMALA	--------EY	STS-------	-----F---Y	FH----N---	----------
DR7βI	---Q------	G-YK------	----QF---L	F-----F---	----------
DRw8βI	--------EY	STG--Y----	-----F-D-Y	F-----Y---	----------
DRw8-TAB	--------EY	STG--Y----	-----F-D-Y	F-----Y---	----------
DRw8-SPL	--------ET	STG--Y----	-----F-D-Y	F-----Y---	----------
DRw9βI	---Q----K-	D---------	-----Y-H-G	------N---	----------
DRw10βI	--------EE	V---------	---------R	VH----YA-Y	----------
Dw12βIII	----------	P-R-------	-----F-D-Y	F---------	------F---
Dw2βIII	----------	P-R-------	-----F-D-Y	F---------	------F---
AZHβIII	----------	P-R-------	-----F-D-Y	F---------	----------
DRw52a	--------EL	R-S-------	-----Y-D-Y	FH----FL--	----------
DRw52b	--------EL	--S-------	-----F---H	FH----YA--	----------
DRw52c	--------EL	--S-------	-----F---Y	FH----LA--	----------
DRw53	---Q----E-	A-C----L--	----WN-I-Y	------YA-Y	N--L---Q--

	60	70	80	90	
DR1βI	TELGRPDAEY	WNSQKDLLEQ	RRAAVDTYCR	HNYGVGESFT	VQRR
DR1 NASC	----------	----------	----------	----AV----	----
DR1 CETUS	----------	------I--D	E---------	----------	----
DR2 Dw2βI	----------	------F--D	----------	----------	----
DR2 Dw12βI	----------	------F--D	----------	----------	----
DR2 AZHβI	----------	------I---	A---------	----AV----	----
DR3βI	----------	----------	K-GR--N---	-----V----	----
DR3 DQwa	----------	----------	K-GR--N---	----------	----
DR4 Dw4βI	----------	----------	K---------	----------	----
DR4 Dw10βI	----------	------I--D	E---------	-----V----	----
DR4 Dw13βI	----------	----------	---E------	-----V----	----
DR4 Dw14βI	----------	----------	----------	-----V----	----
DR4 Dw15βI	------S---	----------	----------	----------	----
DR4 KT2	----------	----------	---E------	-----V----	----
DR4 CETUS	----------	----------	----------	----------	----
DR5 Dw5βI	-------E--	------F--D	----------	----------	----
DR5 JVM	-------E--	------I--D	E---------	-----V----	----
DRw6aβI	----------	------I--D	E---------	-----V----	----
DRw6bβI	------A--H	---------R	---E------	-----V----	----
DRw6 AMALA	----------	----------	----------	----------	----
DR7βI	------V--S	------I--D	--GQ---V--	----------	----
DRw8βI	------S---	------F--D	---L------	------	
DRw8-TAB	------S---	------I--D	---L------		
DRw8-SPL	----------	------F--D	--GL------		
DRw9βI	------V--S	------F--R	---E---V--	----------	----
DRw10βI	----------	---------R	----------	----------	----
Dw2βIII	----------	------I---	A---------	-----V----	----
Dw12βIII	----------	------I---	A---------	------	
AZHβIII	----------	------F--D	----------	----------	----
DRw52a	------V--S	----------	K-GR--N---	----------	----
DRw52b	R---------	----------	K-GQ--N---	-----V----	----
DRw52c	------V--S	----------	K-GQ--N---	----------	----
DR4 DRw53	----------	---------R	---E------	Y----V----	----

Participating Laboratories: US8RJW,[1] US3GRU,[2] US3CTS,[3] US1AID,[4] US3SCR,[5] US6BAL,[6] US8UBI,[7] US7JSL[8]

DQ Alpha First Domain Amino Acid Sequences

	10	20	30	40
DR1 DQw1.1	E D I V A D H V A S	C G V N L Y Q F Y G	P S G Q Y T H E F D	G D E E F Y V D L E
DR2 DQw1.2	- - - - - - - - - -	- - - - - - - - - -	- - - - - - - - - -	- - - Q - - - - - -
DR2 DQw1.AZH	- - - - - - - - - -	- - - - - - - - - -	- - - - - - - - - -	- - - Q - - - - - -
DRw6 DQw1.9		- - -	- - - - - - - - - -	- - - - - - - - - -
DRw6 DQw1.18		- - -	- - - - F - - - - -	- - - Q - - - - - -
DRw6 DQw1.19		- - -	- - - - - - - - - -	- - - Q - - - - - -
DRw8 DQw1		- - -	- - - - F - - - - -	- - - Q - - - - - -
DR5 DQw3.1	- - - - - - - - - -	Y - - - - - - S - -	- - - - - - - - - -	- - - Q - - - - - G
DRw8 ARC (WA)	- - - - - - - - - -	Y - - - - - - S - -	- - - - - - - - - -	- - - Q - - - - - G
DRw8 LUY (DQw3)		S - -	- - - - F - - - - -	- - - Q - - - - - G
DR4 DQw3.2	- - - - - - - - - -	Y - - - - - - S - -	- - - - - S - - - -	- - - - - - - - - -
DRw9 DQw3	- - - - - - - - - -	Y - - - - - - S - -	- - - - - S - - - -	- - - - - - - - - -
DR7 DQw2 and w3	- - - - - - - - - -	Y - - - - - - S - -	- - - - F - - - - -	- - - - - - - - - -

	50	60	70	80	
DR1 DQw1.1	R K E T A W R W P E	F S K F G G F D P Q	G A L R N M A V A K	H N L N I M I K R Y	N S T A A T N
DR2 DQw1.2	- - - - - - - - - -	- - - - - - - - - -	- - - - - - - - - -	- - - - - - - - - -	- - - - - - -
DR2 DQw1.AZH	- - - - - - - - - -	- - - - - - - - - -	- - - - - - - - - -	- - - - - - - - - -	- - - - - - -
DRw6 DQw1.9	- - - - - - - - - -	- - - - - - - - - -	- - - - - - - - - -	- - - - - - - - - -	- - - - - - -
DRw6 DQw1.18	K - - - - - - - - -	- - - - - - - - - -	- - - - - - - - - -	- - - - - - - - - -	
DRw6 DQw1.19	- - - - - - - - - -	- - - - - - - - - -	- - - - - - - - - -	- - - - - - - - - -	- - - - - - -
DRw8 DQw1	K - - - - - - - - -	- - - - - - - - - -	- - - - - - - - - -	- - - - - - - - - -	
DR5 DQw3.1	- - - - V - C L - V	L R Q - R[ ]- - - -	F - - T - I - - L -	- - - - S L - - - S	
DRw8 ARC (WA)	- - - - V - C L - V	L R Q - R[ ]- - - -	F - - T - I - - T -	- - - - - L - - - S	- - - - - - -
DRw8 LUY (DQw3)	- - - - V - C L - V	L R Q - R[ ]- - - -	F - - T - I - - T -	- - - - - L - - - S	
DR4 DQw3.2	- - - - V - Q L - L	- R R - R R - - - -	F - - T - I - - L -	- - - - - V - - - S	- - - - - - -
DRw9 DQw3	- - - - V - Q L - L	- R R - R R - - - -	F - - T - I - - L -	- - - - - V - - - S	- - - - - - -
DR7 DQw2 and w3	- - - - V - K L - L	- H R L R[ ]- - - -	F - - T - I - - L -	- - - - - L - - - S	

DQ Beta First Domain Amino Acid Sequences

	10	20	30	40	50
DR1 DQw1.1	R D S P E D F V Y Q	F K G L C Y F T N G	T E R V R G V T R H	I Y N R E E Y V R F	D S D V G V Y R A V
DR2 DQw1.2	- - - - - - - - F -	- - - M - - - - - -	- - - - - L - - - Y	- - - - - - - A - -	- - - - - - - - - -
DR2 DQw1.12	- - P - - - - - L -	- - A M - - - - - -	- - - - - Y - - - Y	- - - - - - D - - -	- - - - - - - - - -
DR2 DQw1.AZH	- - - - - - - - - -	- - - - - - - - - -	- - - - - - - - - -	- - - - - - - - - -	- - - - - - - - - -
DRw6 DQw1.9			- - - - - - - -	- - - - - - - - - -	- - - - - - - - - -
DRw6 DQw1.18			- - - L - - - -	- - - - - - - A - -	- - - - - - - - - -
DRw6 DQw1.19	- - - - - - - - - -	- - - M - - - - - -	- - - - - L - - - -	- - - - - - - A - -	- - - - - - - - - -
DRw10 DQw1.1	- - - - - - - - - -	- - - - - - - - - -	- - - - - - - - - -	- - - - - - - - - -	- - - - - - - - - -
DR3 DQw2	- - - - - - - - - -	- - - M - - - - - -	- - - - - L - S - S	- - - - - - I - - -	- - - - - E F - - -
DR4 DQw3.1	- - - - - - - - - -	- - A M - - - - - -	- - - - - Y - - - Y	- - - - - - - A - -	- - - - E - - - - -
DR4 DQw3.2	- - - - - - - - - -	- - - M - - - - - -	- - - - - L - - - Y	- - - - - - - A - -	- - - - - - - - - -
DRw9 DQw3.3	- - - - - - - - - -	- - - M - - - - - -	- - - - - L - - - Y	- - - - - - - A - -	- - - - - - - - - -
DR4 DQ WA	- - - - - - - - F -	- - - M - - - - - -	- - L - - - - - - Y	- - - - - - - A - -	- - - - - - - - - -
DRw8 DQ WA	- - - - - - - - F -	- - - M - - - - - -	- - - - - - - - - Y	- - - - - - - A - -	- - - - - - - - - -

	60	70	80	90	
DR1 DQw1.1	T P Q G R P V A E Y	W N S Q K E V L E G	A R A S V D R V C R	H N Y E V A Y R G I	L Q R R
DR2 DQw1.2	- - - - - - D - - -	- - - - - - - - - -	T - - E L - T - - -	- - - - - - F - - -	- - - -
DR2 DQw1.12	- - - - - - D - - -	- - - - - D I - - R	T - - E L - T - - -	- - - - - - F - - -	- - - -
DR2 DQw1.AZH	- - - - - - S - - -	- - - - - - - - - -	- - - - - - - - - -	- - - - - - - - - -	- - - -
DRw6 DQw1.9	- - - - - - D - - -	- - - - - - - - - -	- - - - - - - -		
DRw6 DQw1.18	- - - - - - D - - -	- - - - - - - - - -	T - - E L - T -		
DRw6 DQw1.19	- - - - - - - - - -	- - - - - - - - - R	T - - E L - T - - -	- - - - - G - - - -	- - - -
DRw10 DQw1.1	- - - - - - - - - -	- - - - - - - - - -	- - - - - - - - - -	- - - - - - - - - -	- - - -
DR3 DQw2	- L L - L - A - - -	- - - - - D I - - R	K - - A - - - - - -	- - - Q L E L - T T	- - - -
DR4 DQw3.1	- - L - P - D - - -	- - - - - - - - - R	T - - E L - T - - -	- - - Q L E L - T T	- - - -
DR4 DQw3.2	- - L - P - A - - -	- - - - - - - - - R	T - - E L - T - - -	- - - Q L E L - T T	- - - -
DRw9 DQw3.3	- - L - P - D - - -	- - - - - - - - - R	T - - E L - T - - -	- - - Q L E L - T T	- - - -
DR4 DQ WA	- - L - - L D - - -	- - - - - D I - - E	D - - - - - T - - -	- - - Q L E L - T T	- - - -
DRw8 DQ WA	- - L - - L D - - -	- - - - - D I - - E	D - - - - - T - - -	- - - Q L E L - T T	- - - -

Notes

Some DQ α and β sequences have been listed twice to avoid difficulty in the identification of the sequences from certain haplotypes and prevent confusion in nomenclature. The sequences are:

DQ α	DQ β
DR1 DQw1.1 = DRw6 DQw1.9	DR1 DQw1.1 = DRw10 DQw1.1
DR2 DQw1.2 = DR2 DQw1.AZH	
DR2 DQw1.2 = DRw6 DQw1.19	
DRw6 DQw1.18 = DRw8 DQw1	
DR4 DQw3.2 = DRw9 DQw3	

DR Beta First Domain Amino Acid Sequence

Allele Designation	Cell Line(s)	Tenth Workshop Cell Line
DR1βI	721.45.1, LG2	
DR1 NASC	NASC	
DR1 CETUS		
DR2 Dw2βI	PFG	
DR2 Dw12βI	DHO, BGE	
DR2 AZHβI	MN2, AZH	
DR3βI	AVL, WT49, DM28	
DR3 DQWa	RSH	9021
DR4 Dw4βI	GM3103, PRIESS, WT51, DM24	9029
DR4 Dw10βI	FS	
DR4 Dw13βI	SSTO	
DR4 Dw14βI	BIN40, LS40	
DR4 Dw15βI	KT3	9107
DR4 KT2	KT2	
DR4 CETUS		
DR5 Dw5βI	SWEI	9037
DR5 JVM	JVM	9039
DRw6aβI	HHK, APD	9065
DRw6bβI	DR4, 6	
DRw6 AMALA	AMALA	9064
DR7βI	BH, DBB, Burkhardt, MANN	9046, 9052
DRw8βI	Madura	9069
DRw8-TAB	TAB	9066
DRw8-SPL	SPL	9101
DRw9βI	DKB, ISK	9075
DRw10βI	NASC	
Dw12βIII	DHO, BGE	
Dw2βIII	PGF	
AZHβIII	MN2, AZH	
DRw52a	AVL, FS, KT3, HHK	9107, 9065
DRw52b	DR4, 6	
DRw52c		
DRw53	Burkhardt	

A discrepancy exists in the published sequences of the AZH/MNZ DRβI chain at position 6.

DQ Alpha First Domain Amino Acid Sequences

Allele Designation	Cell Line(s)	Tenth Workshop Cell Line
DR1 DQw1.1	LG2, BML	
DR2 DQw1.2	PGF	
DR2 DQw1.AZH	AZH	
DRw6 DQw1.9	LB	
DRw6 DQw1.18	WVB	
DRw6 DQw1.19	WT46	
DRw8 DQw1	TAB	9066
DR5 DQw3.1	MG3	
DRw8 ARC (WA)	ARC, Madura	9069
DRw8 LUY (DQw3)	LUY	9070
DR4 DQw3.2	JY, DM24, DM29	
DRw9, DQw3	DKB, ISK	9075
DR7 DQw2 and w3	LG10, Beiler, BH, DBB	9046, 9052

A second DRw9 DQw3 DQ α clone has been isolated and sequenced by Moriuchi et al. (1985). The gene from this haplotype is deleted for 15 base pairs, including the last codon in the signal sequence and the first four codons for the mature polypeptide. A silent mutation distinguishes the nucleotide sequences of DR4 DQw3.2 α and DRw9 DQw3 α.

DQ Beta First Domain Amino Acid Sequences

Allele Designation	Cell Line(s)	Tenth Workshop Cell Line
DR1 DQw1.1	721.45.1, BML, MVL	
DR2 DQw1.2	PGF	
DR2 DQw1.12	BGE	
(DRw8 DQw1.12)	TAB089	9066
DR2 DQw1.AZH	AZH	
DRw6 DQw1.9	WT52	
DRw6 DQw1.18	WVB, DMJB6	
DRw6 DQw1.19	Daudi, DM23	
DRw10 DQw1.1	NASC	
DR3 DQw2	WT49, DM24,28,29, VAVY	9023
(DR7 DQw2)	BH, PLH	9046, 9047
DR4 DQw3.1	BML, DM23	
DR5 DQw3.1	MG3	
DR4 DQw3.2	DR4i, 3102, FS, BIN40, DM24,29	
DRw9 DQw3.3	DKB	9075
(DR7 DQw3.3)	DBB	9052
DR4 DQ WA	KT3	9107
DRw8 DQ WA	Madura	9069

The DQ β (DQw2) chains found on DR3 and DR7 haplotypes are identical.

The DQ β (DQw3.1) chains found on DR4 and DR5 haplotypes are identical.

References

DR Beta Sequences

DR1β1
Tonnelle C, DeMars R, Long EO. DOβ: a new β chain gene in HLA-D with a distinct regulation of expression. EMBO J 1985;4:2839.
DR1 NASC
Merryman P, Gregersen PK, Lee S, Silver J, Nunez-Roldan A, Crapper R, Winchester RJ. Nucleotide sequence of a DRw10 β chain cDNA clone. J Immunol 1988;140:2447.
Hurley CK, Ziff BL, Silver J, Gregersen PK, Hartzman R, Johnson AH. Polymorphism of the HLA-DR1 haplotype in American Blacks. J Immunol 1988;140:4019.
DR1 CETUS
Erlich HA, manuscript in preparation.
DR2 Dw2β1, DR2 Dw12β1, DR2 AZHβ1
Wu S, Saunders TL, Bach FH. Polymorphism of human Ia antigens generated by reciprocal intergenic exchange between two DRβ loci. Nature 1986;324:676.
Lee BSM, Rust NA, McMichael AJ, McDevitt HO. HLA-DR2 subtypes for an additional supertypic family of DRβ alleles. Proc Natl Acad Sci USA 1987;84:4591.
Wu S, Yabe T, Madden M, Saunders TL, Bach F. cDNA cloning and sequencing reveals that the electrophoretically constant DRB2 molecules as well as the variable DRB1 molecules from the HLA DR2 subtypes have different amino acid sequences including a hypervariable region for a functionally important epitope. J Immunol 1987;138: 2953.
DR3β1
Gustafsson K, Wiman K, Emmoth E, et al. Mutations and selection in the generation of class II histocompatibility antigen polymorphism. EMBO J 1984;3:1655.
DR3 DQWa
Hurley CK, Gregersen PK, Gorski J, et al. The DR3 (w18), DQw4 haplotype differs from DR3 (w17), DQw2 haplotypes at multiple class II loci Human Immunol 1988; 25:37.
Knowles RW, Sartoris S, Briata P, et al. Recombination events suggested by the pattern of polymorphism in the DQβ, DQα, and DRβ chains from six distinct DR3, DR7, DRw8 and DR9 related haplotypes. Manuscript in preparation.
DR4 Dw4β1, DR4 Dw10β1, DW13β1, DR4 DRw14β1, DR4 DRw15β1
Gregersen PK, Shen M, Song Q, et al. Molecular diversity of HLA-DR4 haplotypes. Proc Natl Acad Sci USA 1986;83:2642.
Cairns JS, Curtsinger JM, Dahl CD, Freeman S, Alter BJ, Bach F. Sequence polymorphism of HLA-DRβ1 alleles relating to T-cell recognized determinants. Nature 1985; 317:166.
DR4 KT2
Gregersen PK, Goyert SM, Song Q, Silver J. Microheterogeneity of HLA-DR4 haplotypes: DNA sequence analysis of LD"KT2" and LD"TAS" haplotypes. Hum Immunol 1987;19;287.
DR5 Dw5β1
Tieber VL, Abruzzini LF, Didier DK, Schwartz BD, Rotwein P. Complete characterization and sequence of an HLA class II β chain from the DR5 haplotype. J Biol Chem 1986;261:2738.
DR5 JVM
Bell JI, Denney Jr D, Foster L, Belt T, Todd JA, McDevitt HO. Allelic variation in the DR subregion of the human major histocompatibility complex. Proc Natl Acad Sci USA 1987;84:6234.
DRw6aβ1, DRw6bβ1
Gorski J, Mach B. Polymorphism of human Ia antigens: Gene conversion between two DRβ loci results in a new HLA-D/DR specificity. Nature 1986;322:67.
DRw6 AMALA
Takahashi PC, Kao HT, Tang JC, Gregersen PK, Silver J, Wang CY. Molecular diversity of DRw6 haplotypes. In: Dupont B (ed): Immunobiology of HLA, Volume II: Immunogenetics and Histocompatibility. New York: Springer-Verlag, 1989.
DR7β1
Karr R, Gregersen PK, Obata F, et al. Analysis of DRβ and DQβ chain cDNA clones from a DR7 haplotype. J Immunol 1986;137:2886.
Gregersen PK, Moriuchi T, Karr RW, et al. Polymorphisms of HLA-DR β chains in DR4, −7, and −9 haplotypes: implications for the mechanisms of allelic variation. Proc Natl Acad Sci USA 1986;83:9149.
Knowles RW, et al., *op. cit.*
DRw8β1
Bell JI, et al., *op. cit.*
DRw8-TAB
Erlich HA, manuscript in preparation.
DRw8-SPL, DRw9β1
Bell JI, et al., *op. cit.*
DRw10β1
Merryman P, et al., *op. cit.*
Dw12βIII
Wu S, et al., *op. cit.*
Dw2βIII
Lee, et al., *op. cit.*
AZHβIII
DRw52a
Gosrki J, Mach B, *op. cit.*
DRw52b, DRw52c
Gorski J, manuscript in preparation.
DRw53
Karr R, et al., *op. cit.*

DQ Alpha Sequences

DR1 DQw1.1, DRw6 DQw1.9
Todd JA, Bell JI, McDevitt HO. HLA-DQβ gene contributes to susceptibility and resistance to insulin-dependent diabetes. Nature 1987;329:599.
Saiki RK, Bugawan TL, Horn GT, Mullis KB, Erlich HA. Analysis of enzymatically amplified β-globin and HLA-DQα DNA with allele-specific oligonucleotide probes. Nature 1986;324:163.
Horn GT, Bugawan TL, Long CM, Erlich HA. Allelic sequence variation of the HLA-DQ loci: relationship to serology and to insulin-dependent diabetes susceptibility. Proc Natl Acad Sci USA 1988;85:6012.
DR2 DQw1.2, DR2 DQw1.9
Auffray C, Lillie JW, Arnot D, Grossberger D, Kappes D, Strominger JL. Isotypic and allotypic variation of human class II histocompatibility antigen α-chain genes. Nature 1984;308:327.
DRw6 DQw1.19
Todd JA, et al.; Horn GT, et al., *op. cit.*
DRw6 DQw1.18
Horn GT, et al., *op. cit.*
DRw8 DQw1, DR5 DQw3.1
Horn GT, et al., *op. cit.*

DRw8 ARC (WA)
Horn GT, et al., *op. cit.*
DRw8 LUY (DQw3), DR4 DQw3.2
Moriuchi J, Moriuchi T, Silver J. Nucleotide sequence of an HLA-DQ α chain derived from a DRw9 cell line: genetic and evolutionary implications. Proc Natl Acad Sci USA 1985;82:3420.
DRw9 DQw3.3
Todd JA, et al., *op. cit.*
DR7 DQw2
Moriuchi J, et al., *op. cit.*
DR7 DQw3
Knowles RW, et al., *op. cit.*

DQ Beta Sequences

DR1 DQw1.1
Tonnelle C, et al., *op. cit.*
DRw10 DQw1.1, DR2 DQw1.2
Todd JA, et al., *op. cit.*
DR2 DQw1.12
Lee BSM, Bell JI, Rust NA, McDevitt HO. Structural and functional variability among DQβ alleles of DR2 subtypes. Immunogenetics 1987;26:85.
DR2 DQw1.AZH
Lee BMS, et al., *op. cit.*
DRw6 DQw1.9
Todd JA, et al., *op. cit.*
DRw6 DQw1.18
Todd JA, et al., *op. cit.*
DRw6 DQw1.19
Turco E, Care A, Comapgnone-Post P, Robinson C, Cascino I, Trucco M. Allelic forms of the alpha- and beta-chain genes encoding DQw1-positive heterodimers. Immunogenetics 1987;26:282.
DR3 DQw2
Boss JM, Strominger JL. Cloning and sequence analysis of the human major histocompatibility complex gene DC-3β. Proc Natl Acad Sci USA 1984;81:5199.
DR4 DQw3.1
Todd JA, et al., *op. cit.*
DR4 DQw3.2
Larhammar D, Hyldig-Nielsen JJ, Servenius B, Andersson G, Rask L, Peterson PA. Exon-intron organization and complete nucleotide sequence of a human major histocompatibility antigen DCβ gene. Proc Natl Acad Sci USA 1983;80:7313.
DRw9 DQw3.3
Todd JA, et al., *op. cit.*
DR4 DQ WA
Gregersen PK, Shen M, et al., *op. cit.*
DRw8 DQ WA
Todd JA, et al., *op. cit.*
Hurley CK, Gregersen PK, Steiner N, et al. Polymorphism of the HLA-D region in American blacks. J Immunol 1988;140:885.

Summary Cross-Correlation of the Tenth International Histocompatibility Workshop Data

A Display of Positive Correlations Obtained by Serology, Southern Blot Analysis, T-Lymphocyte Clones, and Two-Dimensional Gel Analysis Tested on the Reference Panel of B-Lymphoblastoid Cell Lines for Factors of the HLA System

Bo Dupont,[1] Malcolm Simons,[2] Robert Wheeler,[3] Mark Ferguson,[3] and Jean-Marc Lalouel[3]

The major scientific objective for the collaborative studies within the Tenth International Histocompatibility Workshop was to compare biochemical and molecular genetic techniques for characterization of polymorphic HLA determinants with HLA specificities defined by conventional serologic and cellular techniques.

A panel of 107 EBV-transformed B-lymphoblastoid cell lines was prepared for the purpose of serving as a common set of target cells for the different component projects within the Workshop. This Reference Panel of B-Lymphoblastoid Cell Lines for Factors of the HLA System has been described elsewhere in the proceedings (1). These B-cell lines have been extensively tested with serologic and cellular assays for HLA typing and reported in two reports (2,3). The cell lines were also used in the biochemical, cellular and molecular genetic studies as described in the summary reports (4–7).

The present report is not intended to provide a review of the results obtained in this Workshop. A detailed report and discussion of the results obtained in each Workshop component is presented in each of the summary reports, which also contain pertinent discussions of data in relation to other component studies (e.g., T-lymphocyte–defined epitopes detected on HLA class II molecules and their relationship to allelic sequence variations determined for the class II molecules) (7). Similarly, the HLA class I allelic variants detected in the one-dimensional isoelectric focusing studies (1D-IEF) and their relationship to serologically defined HLA specificities are described in details in the Workshop summary report (4), and will not be repeated in this report. The 1D-IEF study was designed as a population study, and the limited amount of 1D-IEF patterns present on the Reference Panel Cells do not allow for an informative data presentation in this summary paper.

This report contains a display of "positive clusters" obtained by the central data analysis when the data from the different Workshop components were compared and cross-correlated.

Data Analysis

The general concept for data collection, data processing, verification, and analysis is described in other studies in these proceedings (8–12). Each Workshop component generated a set of data on the Reference Panel of B-Lymphoblastoid Cell Lines. To compare data among the different components, it was necessary to convert all types of data to a common format. Unique data conversion systems were developed for each component study and the traditional HLA serologic scoring system was chosen as the final quantative measure (i.e., the scores 1 and 2 were negative [or not present], 4 were weak positive, and 6 and 8 were positive scores). The data displayed in the present report are a presentation of these scores obtained on the Reference Panel Cells. The score 1 is not shown and 0 indicates "not done."

The shaded area on the displays indicates clusters with statistically significant associations. The intensity of the shading reflects increasing significance levels. The limited number of target cells included in the cross-correlation analysis must, however, be recognized, and the statistical significance of the observations should be considered with caution.

The analytical program, called SERAN, was applied to the combined data set. The details for this software package is described in the HLA Serology section of these proceedings (10–12). The summary tables presented in the present report have been limited to display of "positive clusters." Each table lists the Reference Panel Cells included in the cross-correlation analysis and the positive/negative scores obtained with the particular reagent. For serologic data, this indicates positive or negative reactivity with the individual reagents; for RFLP, it indicates fragment presence or absence; for two-dimensional (2-D) gel analysis of HLA-DR or DQ antigens, it indicates "spot" presence or absence; and for T-cell clones, the scores indicate positive proliferative response or positive CTL response by the corresponding T-cell clone.

Results

The Reference Panel of B-lymphoblastoid Cell Lines consists of 107 cells (1), but all cell lines were not tested in all components. The number of target cells included in the individual display, therefore, varies according to the number of cells tested.

Each display lists the Reference Panel Cell ID#, followed by a display of selected HLA antigens present on the individual cell line and relevant to the particular comparison. Along the top of each table is given the code for each reagent included in the comparison:

HLA Serology: The number closest to the data display indicates the Workshop serum number.

Participating Laboratories: US7/DUP,[1] US5/SIM,[2] US7/LAL[3]

Immunobiology of HLA
Volume I

RFLP: The three lines at the top denote the DNA probe, e.g.: CL1 = HLA Class I probe, and DRB = HLA-DR beta probe.

The next three lines denote the restriction enzyme, e.g., BAM = BamHI.

The four lines closest to the data display indicate the kb for the fragment, e.g., 2356 = 23.56 kb.

T-Lymphocyte Clones: The four numbers closest to the data display indicate the T-cell clone number. All clones with 6000-numbers are proliferative clones; all clones with 7000-numbers are cytotoxic clones.

2-D Gel Data: All 2-D gel data are indicated by BII displayed in three lines located in the middle of the heading, i.e., the same position as the display of the restriction enzyme code.

A separate "Conversion Code" is provided for defining each "spot" detected in the 2-D gel analysis.

The following displays are included in this report:

1. RFLP-HLA class I with anti-HLA-A sera.
2. RFLP-HLA class I with anti-HLA-B sera.
3. RFLP-HLA class I with anti-HLA-C sera.
4. Anti-HLA class II sera with T-lymphocyte clones.
5. RFLP-HLA-DRA and HLA-DRB with T-lymphocyte clones; 2-D gel data and anti-HLA class II sera.
6. RFLP-HLA-DQA and HLA-DQB with T-lymphocyte clones, 2-D gel data and anti-HLA class II sera.
7. RFLP-HLA-DPA and HLA-DPB with T-lymphocyte clones, 2-D gel data, and anti-HLA class II sera.

Summary

This report contains a comprehensive display of "positive clusters" obtained in the Central Data Analysis by cross-correlation analysis between the individual Workshop component data. The data are provided as a permanent record of the results obtained in the collaborative Workshop studies of the Reference Panel of B-lymphoblastoid Cell Lines for Factors of the HLA System.

This report does not contain a summary of the 1-D IEF analysis of HLA class I alleles, which is presented elsewhere in these proceedings (4). The assignments of 1D-IEF alleles to the individual Reference Panel Cells are also provided elsewhere (13). The data presented in the individual displays have been reduced to a common scoring system, utilizing the conventional HLA serology scores. Therefore, the displays provide the assignment of positive reactivity versus lack of reactivity (for HLA serology and T-lymphocyte clone data) and fragment presence or absence for RFLP and 2-D gel data.

Conversion Code for HLA-DR and -DQ Two-Dimensional (2-D) Gel Data

DRβ1			
2-D Patterns			
DRβI-1.1	DR1	BII	1011
DRβ-a-2.2	DR2	BII	2
DRβ-a-2.12	DR2	BII	12
DRβ-a-2.16	DR2	BII	10
DRβI-3	DR1	BII	103
DRβI-4.1	DR1	BII	1041
DRβI-4.2	DR1	BII	1042
DRβI-4.3	DR1	BII	1043
DRβI-7	DR1	BII	107
DRβI-8.1	DR1	BII	1081
DRβI-8.2	DR1	BII	1082
DRβI-9	DR1	BII	109
DRβI-11.1	DR1	BII	1111
DRβI-12	DR1	BII	1112
DRβI-13	DR1	BII	1013
DRβI-14.1	DR1	BII	1141
DRβI-14.2	DR1	BII	1142
DRβ2			
2-D Patterns			
DRβ-b-2.1	DR1	BII	1021
DRβIII-1	DR3	BII	301
DRβIII-2	DR3	BII	302
DRβIII-3	DR3	BII	303
DRβIV-1	DR4	BII	401

DQβ			
2-D Patterns			
DQβ-1.1	DQB	BII	11
DQβ-1.2	DQB	BII	12
DQβ-1.12	DQB	BII	112
DQβ-1.18	DQB	BII	118
DQβ-2	DQB	BII	2
DQβ-3.1	DQB	BII	31
DQβ-3.2	DQB	BII	32
DQβ-3.3	DQB	BII	33
DQβ-Wa	DQB	BII	9
DQα			
2-D Patterns			
DQα-1.1	DQA	BII	11
DQα-1.2	DQA	BII	12
DQα-1.14	DQA	BII	14 α 114
DQα-1.18	DQA	BII	113
DQα-2	DQA	BII	2
DQα-3.1	DQA	BII	31
DQα-3.2	DQA	BII	32
DQα-3.7	DQA	BII	37
DQα-3.8	DQA	BII	38

```
                                                 AAAAAAAAAAACCAACCCCCCCCCCCCCCCCCCCCCCCCCCCCCCCCCCCCCCCCCCCCCCCCCAAACCCACAACCCCACCACCCCCCCCCAAACCCCACCCCCCCCAACCCCCCCCCCCAAAACCACCCCCCCCCCCCCCCCCCCCCCCCCCCCCCCCCCAAA
                                                 11223111111WLLW1LLLLLLLLLLLLLLLLLLLLLLLLLLLLLLLLLLLLLLLLLLLLLLLL333LLL3L33LLLL3LL2LLLLLLLL133LLLL1LLLLLLL33LLLLLLLLLL2222LL3LLLLLLLLLLLLLLLLLLLLLLLLLLLLLLLL113
                                                  460      3113111111111111111111111111111111111111111111111111111   111 1  1111 119111111111  11111111111  1111111111    11 1111111111111111111111111111111  1
                                                           6  6

                                                   AA  AAAAAPHA BESMPBTTTKPBBEPKPBKHETTPPTPBPSPSEEPEEPMPHKBMHP   TEE T  BPBB PK HSBBKKPB AATEKT SPBPKPM  TMMBPKHPBTPAAAATMAPPHPSHPEHBBEMKPPPPSTPKBBHPBBTM
                                                   2W  1WWW1VI1 ACSSVAAAAPSGACSPVAPICAAVVAVASSSSCCVCCVSVIPGSIS   ACC A  AVAG VP ISAAPPSG 11ACPA SSASPVS  ASSASPISGAS2333AS2VVIVSISCIGGCSPVVVVSASPGAISGGAS
                                                   36  1333 UD  MVTPUMQQQNTLMITNUMNDVQQUUQUMTTTTIVUVIUPUDNLPDT   QII Q  MUML UN DTMMNNTL 11QVNQ TTMTNUP  QPPMTNDTLQT8  1QP8UUDUTDTVDLLIPNUUUUTQTNLMDTLLQP
                                                    6   666

                                                                2  1        2   1          1    7          1  1  2 1 2  22 11          1   1   2 11  2  1 1         11  1  2222  2               1 2    1  1 1  11
                                                   11          44  372533424431152966379  71  331116270197714246587  0 020 003453 83173825627 0116 102631434   22 603613140000530215555 52441211   0 1220521027   1
                                                 603 1  1187155837197734354199254229344770104298742062397292441988060158395823335781235141388197400474542887374370344420499020389174481096353300046651 9  4
                                                 654067498250636669406228549085913870831959037246106055062431187363752909242113094585382547212377231814900626672996501260345093651729461884472114786786 42438

9022 B8    CW7        A1         86888880888806688 866666446 4686684686 4      6  646 68 88 86                                4 844468    6 68   6646684866               4686868866666464664868 8866884
9023 B8    CW7        A1         8088888 88882868886866666644644686684666 6    6  644 68 88 86                                8 844468    6 68   6646684866      6        4686868866666464660868088 6688
9088 B8               A1         8 888888888866688686 66 6446446866 84684       6   44 68 88 86                                6884 488    6 48   6646684866 46444         668686886664   4664868 886688
9009 B37   CW6        A1         2    888888886      64666 64464   866846444888666666    4  8        0    0 00                   0844480        62 6446684866 0000   0      46868688666444446648686886686
9011 BW52             A1         860  888 6888       88666 6446446666846 44 86666668 444   8    6                             0  4466       66  6646684866      8        4686868866644   46648684886688
9040 B49   CW7        A1              88888888       6664646446446866846864    66   64666  4  86                                6844468  66668  8846684866                  46868688666664646648680886688
9043 BW41             A1         64  28888888882  86866666644664686684646 8 6  6  646 686    64      2  2                       844468 8    6 2  8646684866     2         468686888866646 46 4868088 6688
9060 BW62  CW9        A1         4    88648888       68666 644644686684 66446 66666   44468688886                               4448      4        46684866                068686886664444466 4868 8866 8
9105 B35   CW4        A1         000000000008800  0686466660464448660 664  866060   4 46  4888 6000    0 00     0   0     0  000  4400    0  0066 4608486000008 0        0086868866 4 44466086 08 66000
9006 B35   CW4        A11                   8 8       6       4    604 66 6866666  46 6  88886            86      8 4 66 4 88844 686        88 8646684866  6 468       4686868888666464 6648     66
9031 BW62  CW10       A2          4                  86 6664464468668866 448 66666   44 68688886                4                4            4    4466848668888866846846868688666444 46648684886688
9032 BW62  CW9        A2             8               8666666446446866 8 684 8 66666  464  6688886                           0                 64     466808668088868668668686886664444466486848866888
9064 BW62  CW9        A2                            686466644644 8668466446 66666      4468688886                                             4    646684866 88886 46846868688666444444664868 48 66
9091 BW62  CW9        A2             2               86466644644 8668466446 66666  44446868888 6  0    0 00                      0    0       4     466848660000860468068686886664444466486848866888
9059 BW60  CW10       A2                             8666664864468668466446 66  0 4  4     888 6                                      8866    86     46684866888886866846868688666444444664 68 8866
9069 BW60  CW10       A2         6                  686464648644 8668466444 66  6  4 4 5  888 6                   4                   66    6     46684866888666 668468686886664444466086808866
9038 B18   CW7        A2               2             866666446446 668 6 446   66   444     6 86         8      6                     6 66 68  86646 848668888886866846868688666444444 66486  88668
9039 B18   CW5        A2                8            866666644644486684644466660668 4444   688886  0                      0               66  866466808668880868468668686 8866 444 466486848 668
9061 B18   CW7        A2                4            866666446446 8668 644 6   66   444       86  0    0 00                 0     0  6668844866466848660000860668468686886664 44 46648684886 6  8
9004 B27   CW1        A2                             8666664 66  866846    666666 6  4 6      86               2                          88686 46680866888 66 46846868688666664646648 6848866
9067 B27   CW1        A2          8          2  686866644644686684664488666666 6464      6  86                                   86 4     68688644668486688888686684686868866644444664868 48866 6
9056 B35              A2                44      68666666446446866846644 666666   84                                                        6 888664668086 68  8886 66866868688666444 466486848 662
9068 B35   CW4        A2          6             6866666486446866 84 6446866666   464     88886                                   46         4    6 46684866888886846846868688666444446648684 8 668
9016 B51              A2                        68666664 644 86684644    6666   8446      86                                             68      46684866688886 46806868688666444446648684 8866
9070 B51              A2          8           8  86666644644 6866846 446666666   464     88886  0   0600     2              80      0   46  8644668086600008606686686868866644 44 46668684 88668
9007 BW57  CW6        A2         2 64            686666644644686684 6     666666  4 4                                                     6486 66466808668888868468668686886 66   44664 6848 66488
9052 BW57  CW6        A2         8888             8666664464468668 46 44 666666  464                                          66          64    8646680866888886866866868688666444446 64 6848 668
9036 B44   CW5        A2          2     8       686666644664 866846 4 686666684 64468 88886         8                                      88888466848668888888846866868688 6864 4 466486848868
9029 B44   CW5        A2                6       68646664464 4 866846 4488666684  4468 88886  0    0 00                      0      0      88888466848660000860468468686886664444466486848866   4
9058 B45              A2                        8866666446646 86 6846   8666666     68       6                                0               6466848668888 66 668 6868688664646466486848866
9062 B38              A2           6  468 2      86666644644 866846 4  866 66    4446         0    0 00                     20      0      888664668486600008604684686868866644 4 466486848 662
9066 BW46  CW11       A2                        686666644644 866846 446866666 4 644686   86                                           8  8844668486 6688886 468668686886664444 4664 6848866
9045 B51   CW7        A2   A3     0     8          64646 46  46648466  666666        48 868886668486   68846   6 40    84 48   660  684 666 8486668884 8668668686886646646 4660868 88668
9013 B7    CW7        A3                        4                                  46 68    868886660688686888686466666448884446868666688848866684866 80046866686868688866664646648684 88666
9017 B7    CW7        A3               24   8                   0                  46 68 4  868886668688686888686466664 8884446068666688846666684866888846866846868688666646466 0868 88668
9033 B7    CW7        A3          8     0                               0          46 68    8688866686886868886864666 4 6884 4686866 688846866 84866888846866846868688666664646648684806688 8
9034 B7    CW7        A3                          6   4 0    4  446               44   8 4  86888666868868688868646 66  468844 886666 888846 6  8486688888686686686868866644444664868 886688 8
9065 B7    CW7        A3               286  6   2                   6 4            46468 4  88880866040068488868646666664 608444806666648888684 6684866000086046846868688666444444 66486 088668
9082 B7    CW7        A3                 2       6      4       4 4                44 46    888886668688686888866646 66  46884468866666688848666680866888886866866868688 6644 46668684 88668
9005 B27   CW1        A3                         6                6  666666 6646 6         868886666868868688866646 6664  8844462   88888 66684866 8824686684686868888666 64664868 8866
9018 B18   CW5        A3           888888                  8668466   66666688646 46 688886888644868886868886 8 4666   8844468      88888664848668088468468668684886666646460868488668 8
9047 BW47  CW6        A3          888 8          6             0   64 6666 6   646 6      888666868868688868646 66 44888 4446      8848866684866888 4686668686868866666464664868 48866 8
9055 B14   CW8        A3         2      6                      66   666 668 646 6  888868886868688686888686466 6 6446 88446 8  8888866684866888 866866846868688866666464664868 48 664
9106 BW50             A3                             4     4       44 666666   4444        8886668686 6 886 6 666         84 68      4 88 644 680066 886868468468686886664444 66486848066
9028 BW60  CW10       A24         8            8     8     86 64664 4 66 66     4 6  888 6              86       6 48866  48     446     66     6                        44  46648684886666
9075 BW60  CW10       A24         88   4        8    4     866646 444 66   6 6444    88 6               8        4 48866  84      46     66                   6    44444466   68488668 8
9002 B14   CW2   CW8  A24         8    0            0   486 646 4  866 666  4446    88886               86       4 48      88886446     660     6     6      88    4444 466486848866
9003 B51              A24         8                      4     866646 446666666  444         0    0 00    68     8024 480     6 8886646   0 6600006 0                 6 444 466486848866
9042 B35   CW4        A24         8    8                       8660466   666666     6 68 68686           8        4 46          6 46   4 66                          66464660 8684  664
9029 BW14  CW8        A23         86                     4 4   86664   446666 668 464    88886           8       84 66      66  86446  0 66                          444 4660868 8866688
9008 B18              A25              4      0      86666  464046668  4446666666 4444      6                 0  888444      06 8888846084866               8468 68686886644444466086848 8660
9092 B51   CW1        A25         8                  86466 44644686684  446866666   4    4  86                   88844         8  86 46484866              866846868686886664444446648686886 6888
9026 B38              A26         6                  86666  4644686688 4446666 66  444                  2  4     4 444         6  86646684866   4 8   8668668686886664 44 466486848 66688
9057 B38              A26               68           86466  4644686684 6 44 866 66   446         0    0 00        4   0844  0      888684668486600008  046806868688666444446648 6848 66
9014 B8    CW7        A26               22  2  4 688 86666  466468668 68         6   4  68 88 86                  86846 88    6 6868 8866684866 684    866846868688866646466486868866
9020 B18   CW5        A26        48888888            86666  4644686684 64486666666844 64   688886                  644         64  8664668086 6    8 8468666868686886664444466486848866
9010 BW53  CW4        A28                    88      8666664 664686684 864 666666  64646  88886                     4            6 6 46684866 88 6 6   46868688666664646648686   64
9050 B44              A29        88684             886668644664 6866 4666 8666666  646 6       6  0    0 00       864     08444 0               6 84    0000 6066846868688886 66 6466486868866
9051 B44              A29                          8864686446646866046664866666 66 644 6       6                  864      8444                 6 84     886 686668668686888686664646648686 8866
9037 BW61  CW2        A29                          866  6446446866 84664  866666  444 6  48 66                    8        48  46          8  6 446 84860 88666846846868688666444444664868 4
9019 B18   CW5        A30          88888             6     64468 684 444686666684444 64688886              68    0       4 884448           868666848668888468468468 68688  6444444664868 4886   4
9021 BW42  CW2        A30  AW68     2       8      888646660464068668 4644 8 6   6  44446864 6 8            48              8 448 66    4     8646684866 88 448   806868688 66644 4646 48680886 8882
9048 B13   CW6        A30        844 8      8          6     64 68 6846 448866666    4446     6             68         40446 84448           8866684866 88       806868688   6444446648 6848866 8
9071 BW62  CW1   CW10 A31         6                  86666644    8666466448866666 44444 6864   86                  4666    2 84   8      8  864   4     66888888846 868686888664444466486848866288
9072 BW62  CW1        A31                            86666644    666 664448666666 4464  66   86                   666    84 6        8  86       0066 88686846 6686868886644444666868480 66
9025 B35   CW4        A31                            86866644    866446644 866666    466  88886                  4666    84 4 8          64        66888866846 868686888664444466486848 66888
9030 B51              A31                            86666644 4  86664664 6 6666    6446    86                 6 4666  8 84 4 8     48     6        66888886846 8686868886644444666868088 66888
9063 B44   CW5        A32                   6        86666 446646 6 84666 6666666866 6 68 88886  0    0 00      6 4      0 44   06          8886846684 660000 606684686868866666464 66486    6    4
```

Tenth International Histocompatibility Workshop
class I RFLP fragments (n=204) with A locus sera (n=177). Positive cluster.

```
                                              AAAAAAACAAAAAAAAAAAAAAACCCAAAAAAAAAAAAAAAAAAAAACAACAAAACACAAACAAAAAAAAAAAAACCCAACACCCCACACACCCAAACCCACAAAAAAAAAAAAAAAAAAAAAAAAAAAACCAACCCCACCCCCCCCCACCCCCAAACCCC
                                              22212W2LW3222221222222LLL222222222222222222WL22L2222L2L922L221222212211LLL22L1LLLL1L2L2LLL229LLL2L9222932239W32322232W22WW2LL1WLLLL1LLLLLLLL2LLLLL333LLLL
                                              5661339130599 59    8   111                   61  1    1 1  5186066550860 111631 1111 131311134 11141  444 2532 32525552818866811041111111111111115111110001111
                                                  4  3                                      9                                                                   3           9  68    3

                                               A   A   AA      ABA A  EEEAAA AA AAAA AAAAAA TAABAA   KAM A M AA AAAAA A PPH   BAMMMBAT K EKH    SHB K  A      AAAAB          A AA ASKA PBPH BKBEPHPEE PPBPHAAAPKKB
                                               W   W   32      W82 W  CCCWWW WW 2222 22222W A2WGWW   P2S 2 S 2W W22WW W VVI   GWSSSG1A P CPI    SIG P  9      32211          1 WW WSPW SASI APGCVIVCC VVASI333VPPG
                                               4   4   16      3 8 6  IVV666 66 8888 888886 Q86L66   N8P 8 P 53 46633 3 UUD   L3PPPL1Q N VND    TDL N  .      24504          0 33 3TN3 TMTD MNLIUDUVV UUMTD111UNNL
                                               3   3           6   9     999 99              8   9 99           3 3  44 3        6                      3                    34 4  3

                                                    22           2       111  2       2  22222222122 2        2                   2                  2  1  2    2      2    2 22222       2     2                 1  111   1  1        1
                                              111   10111  1115 0        173  0       0  0000000020030     9  01        11       1 03 11  7 311303 2 669   0546 50      0 00000111111111 0   1   32  112 4  851265204165517111 3842
                                              1015604154922261924322725333123123222223223402353331923622338810998418037616415197026776497769907666676555556562512254434034178175795272901694214 3403342782
                                              9954967153278650042002905257891662730693174085699379957185392048147585376627712560448004431712826200523826852701964380583364493020912503957964099637 8942147

9022 B8     CW7         A1          0  8       88888        40                                  888 4   6888  6   8666                               8  88                   0      0                 6
9023 B8     CW7         A1        4 0  2       88868          24                             6  88862     88  6   8688                               8  2 20
9088 B8                 A1            468 22   88884         44       2                         488      4488  6  8666 8                             8  4          0
9009 B37    CW6         A1             00 88   80   4       0       0  00000000 00 0     6    0   6       20  6   8646 0          0     0    0 00000 2  0   0        0                8       2
9011 BW52               A1         00    0 2   8                                       6                   6      8666                                 8       0                          8
9040 B49    CW7         A1          8    8 8   88 888                             8      8  8     88488  48864    8 86 8                             888 8                                          8
9043 BW41               A1        682  0 6 824602886 86        0   4402               6    2    2 4 8868886088    8688 8                    2    2  2  2 2 0    02   2  668                  04
9060 BW62   CW9         A1                     6                                                        24        8666 8                             04          0
9105 B35    CW4         A1        0000000000000000000000     000000000000000000004 00 000000 000 0000000000004 600 066  0 0 0 0 0000   0 0000000000000000000000000000 000 0 0 0  0 4680 0   000  0
9006 B35    CW4         A11          8 8  888 8 2                                  8      6             88646    86886                            8           88                         8668888688
9031 BW62   CW10        A2        686488 8888 888888 8886460888888888888888888848888 8  8488846  828888888   8  2    0                             8
9032 BW62   CW9         A2         8 88888888888888888 46888888888888888888848888888868 88       888888888   8                                       6   8       0           0     8 0
9064 BW62   CW9         A2                8         8888 86468888888888888888808848888 88684 6              8                                          0              6 0
9091 BW62   CW9         A2        28   6800686 8 2608 88864686088888808800000000040080484 68 40   66 862280       2     0            0      022 0 00000       0     0
9059 BW60   CW10        A2               8  8       88 8 864688888888888888888848888888868 88    8       888646        8             8            8                0        6
9069 BW60   CW10        A2              26     4    8888886  88888888888888888 8848688 66664     8      6 6 6646                    48             4           6           0
9038 B18    CW7         A2          8   6  8 86888 888888864608888888888888888888848888888  6848884888888806888 64646                            8          6        4        0                       88
9039 B18    CW5         A2              88 8 88    8888888646088888888888888888884888888    60 68  88888888608864       06.          6             80          2
9061 B18    CW7         A2         2     000 6     40848888646860888880880000000004008026262 4004242          20646       0             0       0    0  00000       0         0        0
9004 B27    CW1         A2               6 8868 86 8886888864688888 888888888288488088886848884         6 88 8646                                     688888              0     686
9067 B27    CW1         A2            28688 4 2   488888888646068888888848888888888 8888 6 6448864   822866848864 68                        66   888884 6           0        6
9056 B35                A2              2  8 48   028888888646088888808888888888848888888866 48 448688      22088646      2                               2        2    0   022 0   2               6  2     0
9068 B35    CW4         A2             8488 6     2 888686864608888888686888888888848888 226  886  28      2628844686         4                      62        2            0        6               488
9016 B51                A2                 8        88 808646888888       88888880884 888688684              6446           6                                          80
9070 B51                A2        8  8800 8 884  808888886468808888808800000000040080886848044888888888806462              0             0    0 680  0000088  0    80       8
9007 BW57   CW6         A2          26 8888888 8   888488864  0688888888698888888848888 8 66 882     8666862886 6 28                                   4   88848   2     0        0                    8       2
9052 BW57   CW6         A2               886  8   8 08886888864688888888888888888848888 8868 6868 88088 8646888                                     0   888 8   0            40
9036 B44    CW5         A2          8 8888  88884888888688864608888888888888888888848888888888 880 8 8    64888 6                                    8       8 8        00      0
9054 B44    CW5         A2               00   8       808  888646880888880880000000004008088 26  80  6 8       0  68      0             0       0    0  00000       0         0       8
9058 B45                A2                8        22488  8886 688888888888888888888488888888 84   2          8           6                                             0       2
9062 B38                A2          2 4 00    8222  80848886 68808888808800000000040080888 68 802 888     2 26804 68      0          2 0     0       0  00000       0         20                     2
9066 BW46   CW11        A2                       4  48868864688888888888888888848888888888684           8              6                           8                  0       2
9045 B51    CW7         A2    A3               800   8888  406888888  88888808  0 06866   06          8  888  80 46      0         0                   0                  0                0       8       0     0
9013 B7     CW7         A3             6 6 8 82   00             0              0 00 0    6                2            0                              2    20            0                                  0
9017 B7     CW7         A3        8488488688 6       6 6      0       2          48 4        6    6     8688             6                             28  6   4          0                        6  8      0
9033 B7     CW7         A3        88888888888 8        8       8                             8  8   2 8686888 8       2    2                   8                 8     0           0
9034 B7     CW7         A3         8 88888888488       8       0     4                   48 84         88 8888      0                      0                             0          66
9065 B7     CW7         A3        6 8 600 68  66 08 8 2    0   4  0  00000000 00 0      0   068 2    6800    2      0            0       0    0  000004      0       60          02     2    8        6
9082 B7     CW7         A3         8 26 88 88   2   4 6       0    2                          0        6868                                      2  82                  0          0           8
9005 B27    CW1         A3           2 608 6   2  4 2                      842  2             2         626 6                         2          8682                   0          64
9018 B18    CW5         A3                   4    8888         0                          888 688 2   868864   6666886         6886888 888888 886888                                                     8
9047 BW47   CW6         A3          46   8           8                     4                8  8     8 8868646        6                                                0       6     6
9055 B14    CW8         A3            26  8 48  4   8868       0    2                      2    2 2 2882   64888646                             4         88            40    466
9106 BW50               A3                 4                                                                   64
9028 BW60   CW10        A24             6888888 8  446                          8    4      64886    2   8 6  8646      66 8 864468688688868  8488888886                0
9075 BW60   CW10        A24          0          888 8  8 8                                  8 8     0 8  888  8646      8646886   88868886888888888088                0
9002 B14    CW2   CW8   A24          0          888 8  88  4                                8       8        84 6       86 86 6 4   8868886808 88 888888                0
9003 B51                A24        2     00888 82 602 8           08     0  00000000 00 0    80      88   2 2   0646      666680684   806888860886080000088   0        0       6
9042 B35    CW4         A24                  08       2 88          2     2                  2 8     8    82  88  6     66 886      8868884888888888886 8            0        6                  88        8
9029 BW14   CW8         A23             868888  8   08 2                             8       8 8     8    482264884  8    6 8 848   888688 688     88888868
9008 B18                A25        886888 8888  8 88  8                            8    886  08      888 888888886 46   0 6      0                  88888888888880  6        6             0          8            0
9092 B51    CW1         A25        888888868888 88 88 6        0    2              8    8888 8 68 8  888888888888646  8 4  8                         68888888488864  0     88                        8
9026 B38                A26          8468888 8868 288  2                                88        68        688688888888 8 8468    6  8                                    8           222
9057 B38                A26          2 2 800   8    208 2       0          0  00000000 0080        0   888 8888488 04468   8  0          0       0    0  0000062   0         0                8
9014 B8     CW7         A26          82 6888888      88 28 2    0 4     2               46   888           88688886888  8    8  0                                28  8              0         4       2                   8
9020 B18    CW5         A26                80    8   8888                                   882   6        888888888 864 8   4
9010 BW53   CW4         A28                48 88 6     6868      668844 686888888848848888         4664868      86  646        688                                668            608868664                 886           8668
9050 B44                A29          6     00   88 8  0 8           0      0  00000000 00 0           0   2        2  806       6 068          0       0    0  000006   80       0
9051 B44                A29               8 44 8868                                  4                   8422     2 846  2    6  6                             0    8           0
9037 BW61   CW2         A29          6    84     88 8                              4  88      4       4          2 4464     68 6      4                                       0
9019 B18    CW5         A30               8    8      8868         0    6                      8 8       888          88844      6  6                  8      0            6                           6        888
9021 BW42   CW2         A30    AW68       88 8    4    888             8846888888888884888           84          8 6        68688                                       8068 86       84        8   8       8886668
9048 B13    CW6         A30               8828  2        6                      8      6     6 6     22 4  844       688                 4      8      64888                0  2                              8886668
9071 BW62   CW1   CW10  A31             6688 4     8 686                 22        8 2     4         8 4     2 88    2       86                           24 2       4         2    0      0                  86866888 86668
9072 BW62   CW1         A31               6888          0                  0                       4 66                                                                0         0                           66688886668
9025 B35    CW4         A31            8888888868    8  8                        8      84      4  686888886884 46    2  6 86                                    4             0                        48886668888 86668
9030 B51                A31       68   088888 88     8                   8                8      84         8888888888 46  0  6  6 86                     6       88888             0         8               8  86688886668
9063 B44    CW5         A32               00   8     08 8       0  8  0  00000000 00 0             0   4 8         880  6    6680          0       0    0800000 8 80   8 0                4       88
```

Tenth International Histocompatibility Workshop

class I RFLP fragments (n=204) with A locus sera (n=177). Positive cluster.

```
                                        CCCAACACCCCACACACCCAAACCCACAAAAAAAAAAAAAAAAAAAAAAAAAACCAACCCCACCCCCCCCCACCCCCAAACCCCCCCCCCAAAAAAACCCCACCCACCCCCCCCCCCCCAAAAAAACCAAAAAAAAAAAAAACCCAAACCCCCAAAA
                                        LLL22L1LLLL1L2L2LLL229LLL2L992229322 39W32322232W22WW2LL1WLLLL1LLLLLLLLL2LLLL333LLLLLLLLLL3333333LLLL3LLL9LLLLLLLLLLL2W11222LL2122122212W1WLLL122LLLLL22W2
                                        111631 1111 131311134 11141  444 2532 32525552818866811041111111111111151111100011111111110011000111111111 1111111111111530065511615506650660311109911111993 9
                                                                              3        9  68     3                                                    3                       6 4               3

                                        PPH  BAMMMBAT K EKH   SHB K  A     AAAAB          A  AA  ASKA PBPH BKBEPHPEE PPBPHAAAPKKBEESBSHAAA  A PSEM PMS PTTSEKBEHHPM   AAAAAKPAAAAAAAA  A  MKEA  BBSSHB
                                        VVI  GWSSSG1A P CPI   SIG P  9     32211          1  WW  WSPW SASI APGCVIVCC VVASI333VPPGCCSASI333  3 VSCS VSS VAASCPGCIISS   WWW22PS2122W2W2  W  SPCW  AGSSI4
                                        UUD  L3PPPL1Q N VND   TDL N  .     24504          0  33  3TN3 TMTD MNLIUDUVV UUMTD111UNNLIVTMTD110  1 UTIP UPT UQQTVNLIDDTP   33466NT50663536  3  PNV3  MLTTD5
                                               6                     3                       34  4  3                                                            443          3 4    4    4

                                                 2  1  2   2      2   2 22222   2    2            1  111   1  1         1      1  2       1 1         1    1  42                 2         2   1         2
                                        3 11  7 311303 2 669  0546 50  0 000001111111111 0  1   32 112 4  8512652041655171113842554194011111117321262 825116573454 11  1   541   01    110  292 11795351011
                                        37616415197026776497769907666676555556562512254434034178175795272901694214340334278277245853453438986511751076163484722588099130599609971069129823742153453
                                        76627712560448004431712826200523826852701964380583364493020912503957964099637894214763117607114251712041895662877578372368757205703749693268773992888773841
9022 B8    CW7          A1              6   8666             8  88           0   0        6       6        4   8 6 48 0     0 0   660     8
9023 B8    CW7          A1              6   8688             8  2 20                                       4   8 6 88 8      0    46      8
9088 B8                 A1              6   8666 8           8   4     0                                       6 6 8864      4    8       4
9009 B37   CW6          A1              6   8646 0     0   0  0 00000 2  0  0    0     8    2    0  4          886486  0    0  0  6    0
9011 BW52               A1             6    8666              8     0             8         0   464 88 8          0      46
9040 B49   CW7          A1             64   8 86 8            888 8                8           8 8   4  888       4   6       6
9043 BW41               A1                  8688 8     2  2  2  2 2 0  02 2 668     04   4  2  2  4  8886  22   6 2    2 66     2
9060 BW62  CW9          A1                  8666 8         04         0                 0         6                          6 0
9105 B35   CW4          A1             4 600 066  0 0 0 0 0000 0 000000000000000000000000000 000 0 0 0 0 4680 0 000 0 0 0 0000000 0 0 00 408 6 0000000 000000000000 6000 0 0000
9006 B35   CW4          A11            646    86886           8           88             8668888688          88     4 4  8     6    4
9031 BW62  CW10         A2               8  2  0              8                                           8            6
9032 BW62  CW9          A2               8                    6  8    0    0     8 0                          466   8          8   4
9064 BW62  CW9          A2                                         0      6 0                                     4 6
9091 BW62  CW9          A2                  2  0       0    022 0 00000   0  0                    0  2      4   6     2   0    0    0
9059 BW60  CW10         A2             646     8     8         8          0   6                  8          446  8
9069 BW60  CW10         A2             646          48     4    6         0                      6               8        4
9038 B18   CW7          A2             64646          8        6   4               88         24 0      446 686  2 02   8   0     4
9039 B18   CW5          A2             64      06       6    80   2                              2      466                2
9061 B18   CW7          A2             646      0       0   0 0 00000  0  0       0          0          24 6 888       0  2 0     0 0
9004 B27   CW1          A2             646              688888        0 686                  6488       4 6 888    2   2 8          4
9067 B27   CW1          A2             6468           66  888884 6    0  6            2      6 4       6448 888 2       842 6 82   6 4
9056 B35                A2             646  2          2  2  0 022 0 2         6 2   0        2       468 88866868     8           2
9068 B35   CW4          A2             44686    4      62     2       0                6    44  2     488  8      2   2 4
9016 B51                A2             446       6                 80           488                  6 4  88866
9070 B51                A2             6462     0      0   0 680 0000088 0 80  8                 08 2   46   8   8    0   0  8     0
9007 BW57  CW6          A2             6 6 28                4  88848  2  0   0    8 2     8        4 8  888686        22  6 44    2
9052 BW57  CW6          A2             646888              0  888 8  0    40                       48   4 8  888686 8 8 6  6 8   4
9036 B44   CW5          A2              6                 8    8 8       00  0                          88           88          8
9054 B44   CW5          A2              68      0      0   0 0 00000   0  0   8                  0          2        0  8 0       0
9058 B45                A2                       6                          0  2                          248  888 8 6             88
9062 B38                A2             4 68     0    2 0   0 0 00000   0  20      2             0 8 2   64 8688866868 8  0 20  0      0
9066 BW46  CW11         A2                8     6            8         0  2                                4  886                8
9045 B51   CW7          A2        A3    46      0   0            0     0              0  8  0 0         6 46  688 6   0     0
9013 B7    CW7          A3                      0            2   20    0                       0   2   6  486 8688        2  4
9017 B7    CW7          A3                      6  2        28  6  4   0              6 8   0  6 6     648 8888   46               4
9033 B7    CW7          A3                  2           8           8  0      0                  848     48  8888
9034 B7    CW7          A3                  0          0              0    66                     6        486 8688
9065 B7    CW7          A3                2     0      0   0 0 000004   0 60  02  2  8  6      0842     6 8466 8886  2  0  0 2 2 0 42   0
9082 B7    CW7          A3                                 2 82          0    0          8        2        6488 8888
9005 B27   CW1          A3             6 6           2       8682             64                   8        464  888        2   8
9018 B18   CW5          A3             64   6666886     6886888 888888 886888                8     8 8     4                 2
9047 BW47  CW6          A3             646       6                       0  6   6                  6     6 46  888686 888   6
9055 B14   CW8          A3             646            4  88         40  466                         4       46 68   26      2   2
9106 BW50               A3             64              8                                                    4  888686
9028 BW60  CW10         A24            646   66 8 8644686886888868 8488888886       0             8 6      88868         6
9075 BW60  CW10         A24            646   8646886    88888688868888888088                                 46  8     0
9002 B14   CW2   CW8    A24            4 6     86 86 6 4  8868886808 88 888888     0               88      6 68886         0
9003 B51                A24            646   666680684  8068886088608000088  0   0  6            0288     848  88866868    0  2 0        0
9042 B35   CW4          A24             6    66 886    8868884888888888886 8     0   6      88  8     8    84 6  8                       4
9029 BW14  CW8          A23            84 8   6 8 848  888688 688   88888868                               46 68                 6
9008 B18                A25            646   0 6      0          8888888888880  6   6     0   8   0    58  8446 686686888888886688888688888846      8
9092 B51   CW1          A25            646   8 4  8              68888888488864 0  88          8        8 8   4  8688  88888866688888 88 8 6460
9026 B38                A26            8468   6 8                  8            222        8           2 846  488688866868 888288668888828866 8 6462
9057 B38                A26            4468   8 0      0   0 0 0000062  0  0           8             0 6 4     6688866868 8888866688 08888208 6468
9014 B8    CW7          A26              8    8 8           28 8     0   4        2      8             4   64    8 6 8888888888888848886666688   66 68
9020 B18   CW5          A26            64 8   4                                                         4 4 6  888888886688888888888 8 468    0
9010 BW53  CW4          A28            646   688              668       608868664       886  86688646488 44    4  8    8    84    64 66       6
9050 B44                A29             6    6 068     0   0 0 000006  80    0                             0  4  888 8 62 8   20   608866 848666680 8
9051 B44                A29            6 2    6 6             0  8       8                          8   24  888 8 6       4  66628848666888 8
9037 BW61  CW2          A30            4464   68 8  4                    0                   464          8 8 8              666 88      4888
9019 B18   CW5          A30            44      6 6          8  0   6      0    6  888       888 8486686 66     68      0
9021 BW42  CW2          A30       AW68        58688               8                                                  
9048 B13   CW6          A30            44    688        4  8  64888     8068 86  84  8 8 888666886464888888888666446 486 88668 4  8  4 66
9071 BW62  CW1   CW10   A31              2      86         24 2     4  2 0 2       888666888664648888888888686846 4 6 886686 4466  2       6      8
9072 BW62  CW1          A31                    4 66                    0    0    86866888866688864 48888822   4    8866               6 6
9025 B35   CW4          A31            446   2  6 86                    0    0   66688888666886644  888868    6 466 888           68     6
9030 B51                A31             46 0 6  6 86         4             0    488868668888666886464888888 8    8 4 6                  646
9063 B44   CW5          A32              6     6680     0  0  6 0800000 8 80  8 0  8   4   88  8 866888866688864648888 6  6 8886  8  68 646
                                                                                                                4 0   46             0    0 6 6   6  0
```

Tenth International Histocompatibility Workshop

class I RFLP fragments (n=204) with A locus sera (n=177). Positive cluster.

```
BBBBBBBBBBBBBBBBBBBBBBBBBBBBCCCCBBBCCBBBBBBCC7BBBBBBBBBBBBBBBBBBBBBBBBBCCCBBBBBBBBBBBBCCBBBBCBBBCCCCCBBAAABBABBBBCCCBCCCCCCBBBBBCBBBBBBBBBBBCBBBBBBBBBBBBBBBBBBBBBC
1W24322222221338W2324WW34WLLLL11WLLWWW4W7LLC7777W777WWWWWW727W7W11LLL7WWWWWW4722LL73W5W3W5LLLLLWW322WW32W44LLLWLLLLLLL11WW1L1WW55555W55L5555554541145lW54311L
267477777777377 67774548051111775115645 5 11     6    754445 7 5 466111 5527549 7711 96 4751111111442444427444111511111113755717551111 5 111   11 919559172198661
0                5    7     7        7  736 0       7    352824   4 1     742351      4   2                        2           78    33      2          5  2

B BBB  BB   B   B       EEETB BKB B     PSBBBBBBB BB   BBBB BBBBS BPK B   BBB BB TE    C   PTTSK  ABB   BB BBPPH SHBKKE  BB B BB T    BSB  B BB    B    B    S
W 2WW  WW   W   W       CCCAW WPG W     SSW22247W W7   77W2 WWWWT ASP W   7WW W7 AC    W   VAASP  2WW   23 44VVI SIGPPC  WW G 44 E    WSW  W 55    1    W    S
4 744  44   5   4       IVVQ6 4NL 5     TT77770 7 4     57 4544. MTN 5    54 7  QV    6   UQQTN  344   77 55UUD TDLNNV  55 L 99 7    5T5  5 11    5    5    T
7  77  77   9   7           3 6   7       3     3 2     6  25221     8    62 3                                         87         6    2 2  2            2
                                                              6

 2 2 2                   1111                            2        1 1         22 1          1       222   22222           1
4304020222234444242444434173224293422442623333333323333333303333443142432333430021244154118253244000440010 03  115465263444474221111111122111111214224142244441
2572897888999901888289648725067295865347793448843723128204720385540077093084377417616498710717995669958099376899067697777472277776868716687826435487914655
1774118138272035097747190905863099577076279202375543021381157158431887372264406087935411656649387455640287766328220454892752339478275816461749276101803036
```

9031	A2		CW10		BW62	8 8 66486464888 88 868 2 84 8 86 8 444 8 68 6 88 88 8 8 8888 2
9060	A1		CW9		BW62	4 8 0 6 8
9064	A2		CW9		BW62	6 8 66464 868
9032	A2		CW9		BW62	8 8 4888 464888688886 0 2068 688 8 8 08 4660 6 8 8 8888 0 0 88888
9091	A2		CW9		BW62	2 0 0 0 6 22266464228686 0 2 00 4 2 4 0002600000 222 24284
9071	A31		CW1	CW10	BW62	8 8 88 2 2 2 6 2 6 8 6 42 2 6 48 4 22
9072	A31		CW1		BW62	6 8 688 4660 6
9039	A2		CW5		B18	0 864648886888 0 0 08 82 0 0 8 0 88 466 0 864 0 0 0
9038	A2		CW7		B18	2 8 8 2 286464888688 6 6 42 448886 8 66 68 8 8 68 8446 8 8646 62 4 6 08
9061	A2		CW7		B18	0 0 0 864646646844 0 2 666 008 62886 44 6 000 00000646 2
9008	A25				B18	0 08 866 868 08 0 6 8 0 2 8 8 88888 4466 88 8888888646 68 08 88 4 86 0 8
9020	A26		CW5		B18	8 4 68 8 8 8 8 0 4 6 864 0
9018	A3		CW5		B18	6 8 8 8 8 4 888888880 8864 6688 88
9019	A30		CW5		B18	8 2 8 2 8 0 8 84 66 844 0
9105	A1		CW4		B35	00000000000000000000000000 40000 000000 000000000000000000000000 00000000000006800000000 40 000000000004 600 00000 0000000000000000000000000000000
9006	A11		CW4		B35	8 68 6 6 88888 4 4 646 8
9056	A2				B35	2 2 2 2 2086464888866 2 4 2 66 0 8886886 468 22 2 8 646 2 2 2
9068	A2		CW4		B35	4 2 2 88646488868884 8 2 2 4 6 688 84886 66 68446 64 4 2 2
9042	A24		CW4		B35	822 2 2 4 88 664 66 88888888 6 6884 4 2
9025	A31		CW4		B35	8 2 8 24 4 4 8 2 8 6 0 6 6446 8 6
9010	A28		CW4		BW53	0 842 4 8 46868 4 6488888688 886646 4 66 6
9016	A2				B51	66464 68 8 88 64 8888888 886446 6868 8 88648666 8
9003	A24				B51	0 020 8 82 2 4 6 0 2 008886888 848 880008800000646868864 4888880888 88888488 688
9045	A2	A3	CW7		B51	4 08 6 6 660 4 8 046 88888 88 46 444646868880 488
9070	A2				B51	8 04080 6 8688 8 646488868 8 4 2 0 46 008 8 888 8846 880008800000006468 888 2888E888 88 88888888 84886486
9030	A31				B51	48 4 84 2 8 8 8 88 888888888888 468 4 48 8888888888688888 88868888 6
9092	A25		CW1		B51	688 88 8 68 8 8 4 6 88888888 886884 888888888888646 68 2886888888 888888888884888 8
9011	A1				BW52	6 0 0 88 464 8888886 8826 6 880 0 2 88 88888 86 2
9043	A1				BW41	208 2 22 2 20 2 228 2 2 2 6 488868 4 08 2 20 6 8 4 6 24 6 8 2 2 64 48 2 222222
9037	A29		CW2		BW61	6 8 4 8 6 8 8 6 446 0
9075	A24		CW10		BW60	888888 666888888 8 8 8884 8 8 46 88888888888 6486886 6
9069	A2		CW10		BW60	488 44 4 62686 488668 8 66666888 464 6 6 2 646 2
9028	A24		CW10		BW60	688 8 84 4 666 8888 66 8 888888888886468688664 4 46 0
9059	A2		CW10		BW60	888 8864648 86888 866 888 8 8 8 8 88 4466 6 0
9048	A30		CW6		B13	6864868 0 6 6 6 44 6 64 6 0 8 66 48 40 4 44 4 6 88888868688 44 4 8 6 6 44 6 6
9004	A2		CW1		B27	868886868688 8888 884864648886042 88888 8 28 888 888 888 8 66 4 66 8888888888 646 2
9067	A2		CW1		B27	2686868448886464868468868646 86868884 888886 40 84688 8468 464 6 888 8 8 04486 688888888886462 2 282 4 4 4
9005	A3		CW1		B27	868888888 2 8 88 888888 8 8 688 86 2 2 2 888 8 6464648888888 6 6
9047	A3		CW6		BW47	8888 8 88 884 8 4 8 4 8 446 8888888888 4646 8 4
9054	A2		CW5		B44	66080 0 8 8 864648286866 0 6 008 84888 880008800000 6 8
9050	A29				B44	8 080 0 8 8 8 0 00 8 4 88000880000006 8
9063	A32		CW5		B44	88080 0 88 8 2 6 0 008 8 888 46 880008800000 6 8 2 8
9051	A29				B44	82 84 8 8 8 2 2 4 6 48 4 8888888 8686 2
9036	A2		CW5		B44	88888 888 0 8 8 8864648886888 6 8 0 0 8 8888 4 0 8 88888 88 888888 8888 8 0 88
9058	A2				B45	8 8 6 64 2 8 8 0 8 248 6 8
9007	A2		CW6		BW57	24 6868844864 488868862 6 6282 8 8 62 8 888 64 80 888888888886 6 4884888 848 44 6 8
9052	A2		CW6		BW57	8 8 8888084864648886888 84 8 88 6 8 64 80 888888 8888646 8888888 88 6 8 0
9066	A2		CW11		BW46	6464 868 8 648 4 6 8 0
9106	A3				BW50	8 6 4 8 64
9040	A1		CW7		B49	888 0 4 84 88 6 2 8688 666 68 4 88888888888864 8 8 888866 8848
9009	A1		CW6		B37	0 0 0 88 8 4 8 6 0 00 8484 8 4 480008800000 6 42
9062	A2				B38	20 020 8 22 688 88686 648826828 4 222 0 4 4 08 688 22 8 0088888886464 8 880008800000004 6 4 4484 22 2 2 4886
9026	A26				B38	8 2 2 2 24 8 268 2 2 8 2688 2 2 2 8886 8 8 488 8888888 888 846 2 64 664 846
9057	A26				B38	0 0 0 8 8 8 8 0 8888 4 008888882 .2 6 880008800000446 286 4866
9065	A3		CW7		B7	6020 0 46 644646 8 8668888088888888 00888888666888888 8008884888468466 000 00000 2 8 26 6
9013	A3		CW7		B7	8 4 8868088888688488 488888 6662 488488 0486 2 24
9017	A3		CW7		B7	4286 4 2 228666 64888888888686868688888668666446666 6884888866 6248 6 8 4 2
9034	A3		CW7		B7	4 8668888888888 8888888888866 48888 4888488 66 448666 0 8
9033	A3		CW7		B7	0 84 8 0 8668888888888888888888888866 8864 68888488 86 48 8 0 68
9082	A3		CW7		B7	2 6 6 866888888868688888888 88666 2 86488 68 84880 8 4 2
9021	A30	AW68	CW2		BW42	4 2 86 86680888848888888 88888 2 4 84 8 480 8 2486 8 2 2
9023	A1		CW7		B8	2 2 2 6 68 0 80 4 8682 2 6 06 2 2 82 4 6 2 2 64 0
9088	A1				B8	4 2 2 4 8 2 6 66 6 2 42 8 6 4
9022	A1		CW7		B8	2 884 0 28 8488 4 6 06 8 4 6 0 4 8 26
9014	A26		CW7		B8	2 262 86 8 8 2 2 86 6 8 88886 4 48 6
9002	A24		CW2	CW8	B14	8 8 6 0 0 000 0 0 88888 88 6 488888 88884 68688664 8 0 6
9055	A3		CW8		B14	2844 2864 8 48 2 22 4 46 6 48 6 8 8884886 246 6 6 8646 6 2 6 0 8 2 6
9029	A23		CW8		BW14	8 828 6 2 8 8888 46 28888868888884 868864 0 2 4

Tenth International Histocompatibility Workshop
class I RFLP fragments (n=204) with B locus sera (n=304). Positive cluster.

```
                                                                  BBBBCCCCBBBCCCCCCCCBCCCCCBBBBBBBBCBCCCCCCCCCCBBCBBCBBCBBBBBBBBBBBBBBBBBBBBBBBBBBBBBBBBBBBBBBBCBBCBBBBBBBBBBBBBBBBBBBBBBBBABBBBCCCCCBBBBBBBBBBBBBBBBBBBCCCBBCCBBBB
                                                                  3131LLLLW1WLLLLLWLL2LLLLLWWWWWWWWWLWLLLLLLLLLL8WLW8L88L8W8331W81133W11W12121WW4WW255813WWWWLW4LWWWW4W444WWWW4W4W174442W4W7LLLLL85533W433555W353W3LLL3WLLWW13
                                                                  9696111115351111151171111116666666661611111111111 515 1  1 5 8964 3879488687877550461l1 4954461601666606000666606063 0009605 11111 1159595511  5515751115511555
                                                                          5 7                                    9 4        5    8      1  4       70 62        3110 0  1010 0    0001 1 0           7            3         3   1        3  06

                                                                      PHHMB   KPKEMCEE KBMHP      A     P  TPMMBPKHPBB B    E    S              B      B      B B     B B    H   KBBB BBBBBB BB     B BBBB     BMPPEE A     BBB BTBBBBBBBPEE BMMBBBB
                                                                      SIISW   PSPCSWCC PGSIS      N     V  ASSSASPISGW A    C    S              3      1      W W     8 3    I   P1WW 1W111W WW     4 WW11     2SVVCC 2     3W5 3E3WWWW35VCC WSSWW1W
                                                                      TDDP5   NTNVP8IV NLPDT            U  QTPPMTNDTL4 M    V    T              8      3      4 7       5    D   N366 363336 66     0 4422     7PUUVI 5     551 5755555651UVV 5PP4576
                                                                         4                                         6                                        7 0            10  1   1 10      17                   3   6 2332       2  15 2

                                                                          42          1   721  1   2    2        11   1 92 2   2               2                          2   99      2                  2 2222                                    1
                                                                  4444534433454922570046587415555155545142 6036132033074024334443032443224222244432241140410039336333333333333333333030001343431973512412222211211121120421  2322
                                                                  7565784209813291598797292900000002070288737436650880830196657897859771786878575237709690059654657666777555756557687940484747429670692101708991996940833313 2
                                                                  0255378716320737261224311967197020465626672993364961690292678743890024159266018980521835683304746098453593226160949693823410608425725382818155849646937271 8
9031 A2              CW10               BW62                      |             8     88886888888888866 4466848                6                       8888888888888          8                88 88 8688 4 44         8888                 8448888
9060 A1              CW9                BW62                      |           6       88888688888888866  466848                                           8                                      8    4 44                                  488
9064 A2              CW9                BW62                      |      6            88888688888888866 6466848                                           8                                      6    4  4           8           8         6 48 88
9032 A2              CW9                BW62                      |        0          88888868888888866  466808    8    8    0         0            880888888888888         80 0       0   0 808 888888864644644      88880        88         8  8888
9091 A2              CW9                BW62                      |                   088888808480 88866  466848   0  0   2    0           4   8886888866640464          2                   0 0000 88 4444  28  48 6         22         2 8488
9071 A31             CW1       CW10     BW62                      |   6    4    66  424  868888888888886664  4                            2 8  44 8  8 88 6                                     44488888648  4448 68 66                      6 44686
9072 A31             CW1                BW62                      |             6   4    868888888888869666       00                                   8                                       8868    8 464                         86          6668
9039 A2              CW5                B18                       |8        884888886888868888 866664668080            4  0 8 08   868    8 4868 866                                 88 8888 88664440088 88880   2
9038 A2              CW7                B18                       |2             86   8688868888 8666646 848            6    8 888888  888 22880808 8688        2   0 40           08 888  88864448888  88880  4           8644  2
9061 A2              CW7                B18                       |              0    8620628088 86666466848      0   0     0 0 8  6408 8    2   4  0                                0 0000 8488444    2                          4
9008 A25                                B18                       |848668866866466468  48      88868888 86688460848   4          688      68 6 8  08 8    8  88 8 88          0    4   08          4   888886  644468 6  6                                6
9020 A26             CW5                B18                       |8          86646 88488888888 8888 86666466808          0                 8   8     88888       888                                  88        6464648868
9018 A3              CW5                B18                       |8               868888886888888888 86688664848          4       6   2 8 8  88 8880          8888                                      868          6462888    2    0   6
9019 A30             CW5                B18                       |                8848888888888888848666668666848                   6  4868  86884 88      8  888                                      88 8         4448888         0
9105 A1              CW4                B35                       |0000        000    6 0   04888600000000 0606 46084800 00 000000000000000000000000000000000000000000  000000000000000000000000000000   4 00000000000000000046800  0000
9006 A11             CW4                B35                       |                         88888688828888 86686466848                  8886    68       8      88 8 88                          888        4688 8 8688808888888688          88
9056 A2                                 B35                       |   6688                    88888888 86666466808                   2        2    2        2028 0 42 0           0    0     62   0  0    82 82 6 84486868888808888888886  84446888
9068 A2              CW4                B35                       |4                        8888868888888 8666 466848                  62    4   444        66666 8 686842        4   2             6       468886884 4648468 668888 8 24 648864   888
9042 A24             CW4                B35                       |                         668686888 888888666 46  4                           8        868     8   8 8                          28     88888        68488888888 684 888 88     6 88
9025 A31             CW4                B35                       |                 466    888888688888 86664                       2    4    4    8888888868  888 6                6              88888 68      4 8 8   88880688     4488 8448888
9010 A28             CW4                BW53                      |                 66     888886 42 6    6666 466848            4       6    4                6  00                          2         8866      646 8  68886466   2   88 44
9016 A2                                 B51                       |  6                        86             66   466848                                         8  6                                       86 84 6     6 68888                    4
9003 A24                                B51                       | 426688                  0      00   808     666646   0       0    0   882  2 0 8          8      88 084                           2 0 0000        6444888      86 8 8888
9045 A2        A3    CW7                B51                       |  6          0       48 86           866 666 848                    8  8         888    8    8                 00                         80        6888 888848                  8
9070 A2                                 B51                       |262      8           088886  0   8088 86664466808      08  0   8   2 20 8          8   888     8888  0    6                            8         80 00002886446488888   8888888 8 88        8844   2
9030 A31                                B51                       |8   6  46   466      8    86       28 8   866 6                                8      04    888888888888888888                          8        88888    884 646888   888888888           4 4  880
9092 A25             CW1                B51                       | 2          6866466      84   86        8888 8666 464848             888        60          828888888882882286       8             8   8         8    88888        4088    888808    8                688
9011 A1                                 BW52                      |     8        46   8 8      6        8422 06666466848                                    68                                                      0  8                                 8
9043 A1                                 BW41                      |2    6        66  2   2    68888888888868666466848       2        8  4  8  00  880 8      86 88820   82  62       08       06240   66846680 0 6 88888 82 66468888888824400   2   2        22   6  22
9037 A29             CW2                BW61                      |     8        666      448 6688888888 860 446 848                      4                68               2  8448888888888888888888888888888   8 444     4                              4
9075 A24             CW10               BW60                      |                       6888    68888888 866   46                       8           8    888  8488  8 08       8888 888888888888888888880 888888888888848444   80        88         8
9069 A2              CW10               BW60                      |6                          888 6888886888    66  466848                   4             4   8              8888448484888888888884   86 888888    64 48 66 4                              4
9028 A24             CW10               BW60                      |   68                      8888 688886888 866 446                     8  4          88           8 8 84         4   88886 6686868888888884      888888888886884   4  6                           6 46
9059 A2              CW10               BW60                      |                         48888  68888888 866  466848                     88                     4           8888  888888888888888888888888888888888888868           4      8
9048 A30             CW6                B13                       |    6668  8     6        6      28  66 666886668488                     6        88684    8   8686 88       4048 88    0    264 66868888   8 868868              446  6   4                                    8
9004 A2              CW1                B27                       |   4               68    86        8    8666 466808                     8         6686        8 86  8   6   8                             6   88   8 8 866888     8                            4
9067 A2              CW1                B27                       |              8    686   862      684  86664466848                    442   6  8886   2      48 468       62         44  6      20      484864  2 28 6688888864648 8628     8      2                     644   22
9005 A3              CW1                B27                       |                   68    86      2668    668 666848                           2 28         6  8              2                          0   48           28 8 88    88 646   2
9047 A3              CW6                BW47                      |    6668                              4          86686666848                  6    8  2     88684       0  6  86 20      88  8             868    28886688          88    86       646
9054 A2              CW5                B44                       |                         884088886 0    0 8886688466848              0   0           0                           0                                                 0 0000 88       4                   6
9050 A29                                B44                       |     86        6688   0       6 0       088 8      6 84               0   0           082688   2     888   8          0  8     8   6            6      0 0000        646                        2
9063 A32             CW5                B44                       |               66886088886  0    0  686666686684                   0   0   2 2 2  0          4        6     0                            4      0 0000        6 666888
9051 A29                                B44                       |     86        6668   8       6    2  88      6 84                                                2        2             6           0   6   46    2 26886    2   644  26
9036 A2              CW5                B44                       |8         4      884888886              8886688466848                       0           88                              8    88             888    08 8688884      6488888  8    0   88
9058 A2                                 B45                       |      86            8   8        6882468888866 6466848                                                                          0   6                 802 4       2                                    4    8
9007 A2              CW6                BW57                      |   8668   8    44              8       86666466808              4        6       0   6  2       888 8  2 48   8                              884   88 488464 42 64   6688  6                           8
9052 A2              CW6                BW57                      |   6668        4                8       86686466808                             88 88  8       888 8 88     2            6             8       8   888  8846464   8   8888  8                           8
9066 A2              CW11               BW46                      |                    8  4     86   88  8888  66844668488                                         6                                                 8 6 8    64                                  4
9106 A3                                 BW50                      |     668                    4       88846888 466644  6800                                8  68                   6                                          4444                                   8 88
9040 A1              CW7                B49                       |                    6       84    86   8  8  8   86688466848                     8        84 0      828      8268688846                               8            88       8886646888888884                          8
9009 A1              CW6                B37                       |    8648              6       0          0     80  886664466848       0    0        8  0  0   8                                                    0 00004       4   2   8
9062 A2                                 B38                       |66686688              8       0          0   8088 86666466848 2      0    0     2 888  0  66862           2  2   2 06    2           2                  20 0000 82      4468888 4   6        2   2
9026 A26                                B38                       |48 86688  2866466       8        8 88  8 88 86666466848                  688       6   6       86846 6288 88           2     6            6 2  88 48 64442   8 6686         2             6
9057 A26                                B38                       |26286688    66466       8       0     8088 86668466848       0   0            888 0   8                2 06                            82                    0 0000     2       488488 2
9065 A3              CW7                B7                        |                 4          04  8880888088886666868466848     20    0  826 8460886862         2        8 68 068 42         4688668 88 2    808000044884 460888               8                           2
9013 A3              CW7                B7                        |4                           8   868868888884888688868848        0                          4                            4              0   0          2             8 8 8   886 46                                          4
9017 A3              CW7                B7                        |  6                         84  8688668888888866666668848                 6464688864 2       2    8 664684          2 0    246486       86668868886 464026                 6
9034 A3              CW7                B7                        |                6         84  868     8888888666  6  848                 44     888 8       88888888888 88          8      8 8 88        888888888888 448                2888
9033 A3              CW7                B7                        |              6   8       8   8688 888888866866 848                     88  6808 8      86888888888 88          8      0 8 884        888888888886 46                 8888
9082 A3              CW7                B7                        |    6                        8   88882 8888868866666808          4   242   2868              4           6   2            0     26286       284888826886 44     4
9021 A30       AW68  CW2                BW42                      | 2     68           66       84 6 888868888886864668482            2 8 4         8                 2           2 8 888       0                 688 8    8 4444    4                                 8
9023 A1              CW7                B8                        |   6            6  46       888 868840088880666646848886888888680086868 6428    66  88        28086          6   6  8   8  8 88888  6866448868 0822
9088 A1                                 B8                        | 2   6              8        688 8668886888866666466848886488888448866 8 8 44      66488       6488            2   6   8  28888  284  44884868
9022 A1              CW7                B8                        | 4   6      4    66       488 86 800888888866664668488888888888  88 04868 68      48 88      68888          4         6  8648888  88664688888888              4                           2
9014 A26             CW7                B8                        |    86       886686       888 86888 8888888666886666848886688886688 868 8   2   4    6  6 68886                  2   8     6888886286  4 8686  4 24   2
9002 A24             CW2       CW8      B14                       |                     06  888886 8  8888 8666446                     8 00 8   0    8  0  8                8488                         88  8 8   88    44   80        0
9055 A3              CW8                B14                       |                     88  888886884  8888 86688666848           2   66862648 848   48  2   6   2 68 8           0      4        464 6 8   6   6468668 2 42   2
9029 A23             CW8                BW14                      |                    688  88888688   8888 8666446   0                       8           84   86848  4  684844                      2    888846866464   8     866
```

Tenth International Histocompatibility Workshop
class I RFLP fragments (n=204) with B locus sera (n=304). Positive cluster.

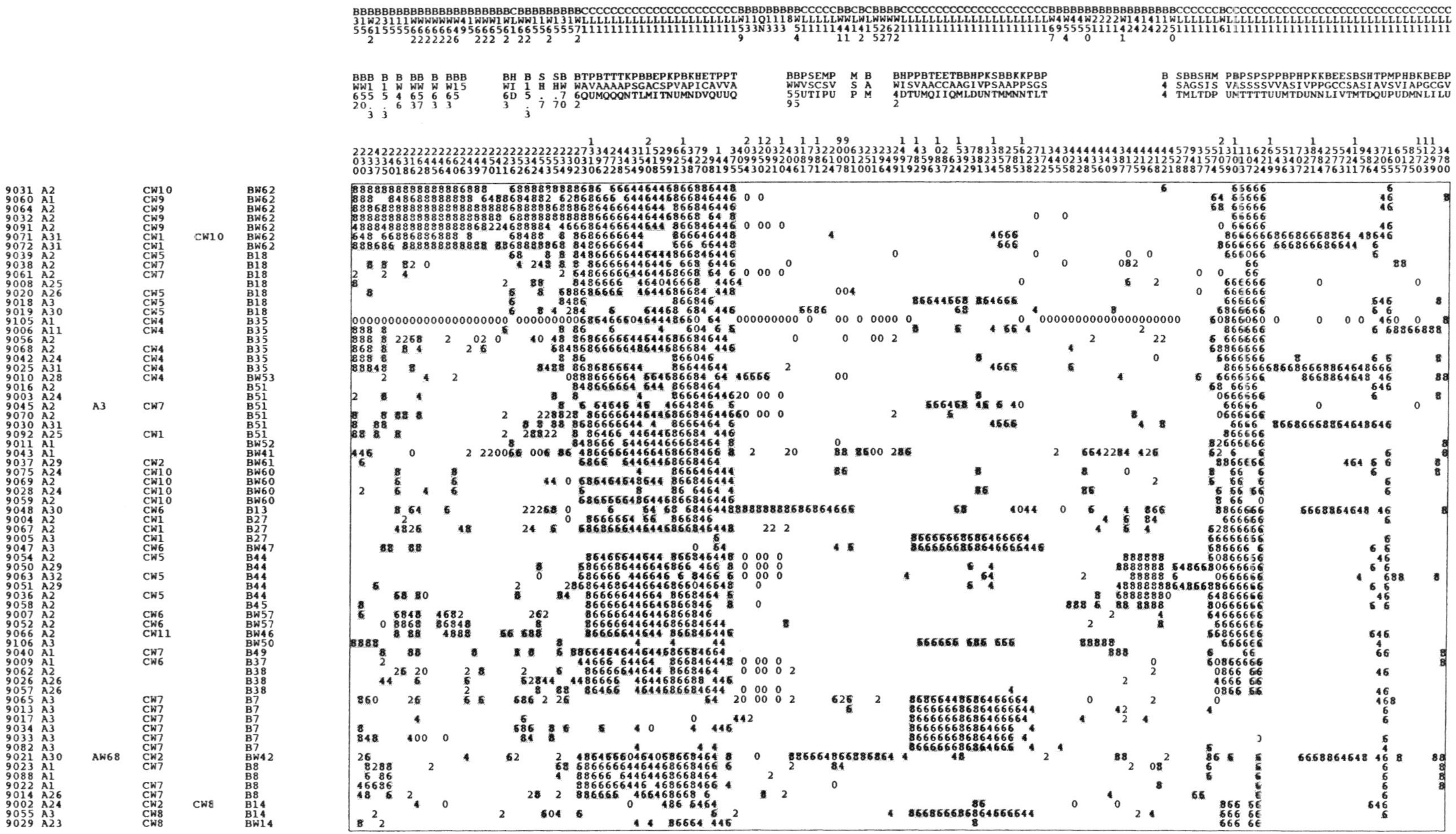

Tenth International Histocompatibility Workshop
class I RFLP fragments (n=204) with B locus sera (n=304). Positive cluster.

```
                               CCCCCCCCCCCBBBDBBBBBCCCCCBBCBCBBBBCCCCCCCCCCCCCCCCCCCCCBBBBBBBBBBBBBBBBBBCCCCCCBCCCCCCCCCCCCCCCCCCCCCCCCCCCCCCCCCCCCCCCCCCCCCCCCCCCCCCCCCCCCCCCCCCCCCC
                               LLLLLLLLLLLW11Q1118WLLLLLWWLWLWWWWLLLLLLLLLLLLLLLLLLLLLW4W44W2222W141411WLLLLLLWLLLLLLLLLLLLLLLLLLLLLLLLLLLLLLLLLLLLLLLLLLLLLLLLLLLLLLLLLLLLLLLLLLLLLL
                               11111111111533N333 51111144141526211111111111111111111116955551111424242251111116111111111111111111111111111111111111111111111111111111111111111111111111111
                                           9      4     11 2 5272                     7 4  0    1      0

                               PKPBKHETPPT       BBPSEMP  M B   BHPPBTEETBBHPKSBBKKPBP                B SBBSHM PBPSPSPPBPHPKKBEESBSHTPMPHBKBEBPHMMMKTTEPSHPEHBBEMKPPHTMPPPPSTPKBBHPBBTMKBE
                               SPVAPICAVVA       WWVSCSV  S A   WISVAACCAAGIVPSAAPPSGS                4 SAGSIS VASSSSVVASIVPPGCCSASIAVSVIAPGCGVISSSPAACVSISCIGGCSPVVIASVVVVSASPGAISGGASPGC
                               TNUMNDVQUUQ       55UTIPU  P M   4DTUMQIIQMLDUNTMMNNTLT                4 TMLTDP UMTTTTUUMTDUNNLIVTMTDQUPUDMNLILUDPPPNQQVUTDTVDLLIPNUUDQPUUUUTQTNLMDTLLQPNLI
                                                 95               2

                                  1        2 12 1  1 1  99       1 1  1  1    1   1                           2 1    1     1     1        111        11        1 2         1  1 1  11
                               2966379 1 3403203243173220063232324 43 02 53783382562713434444444344444445793551311162655173842554194371658512344311 116555 5244121215531   0 1220521027657
                               9925422944709959920089861001251949978598863938235781237440234334381212527415707010421434027827724582060127297875198813203891744819902009635330004665196 34
                               59138708195543021046171247810016491929637242913458538225558285609775968218887745903724996372147631176455575039006256232136517294618509038447211478678642775
9031 A2        CW10      BW62  46866886448                                                            6        66666                  6            4   4686868866646866444 466486848866  6
9060 A1        CW9       BW62  46866846446 0 0                                                              64 66666                46        88666484406868688666     44444664868 8866  6
9064 A2        CW9       BW62  4 866846446                                                                  68 66666                46                 468686886664688644444466486848 664 6
9032 A2        CW9       BW62  468668 64 8                                           0   0                     66666                                   6686868866666688644444466486848866 8
9091 A2        CW9       BW62  4 866846446 0 00 0                                                            0 66666                46                 068686886664688644444466486848866 6
9071 A31       CW1  CW10 BW62    866646448             4                      4666                            8666666866866688864 48646              4 8686868888646 8844444466486848866886
9072 A31       CW1       BW62    666 66448                                     666                           8666666 666866686644   6             46 4 6686868886646 8644444666868480668888
9039 A2        CW5       B18   44866846446                       0                  0            0  0         666066                6                  6686868866 46886444 466486848 66
9038 A2        CW7       B18   46 668 6446       0                                       0      082              66                    88              46868688666668864444466486  8866686
9061 A2        CW7       B18   468668 64 6 0 00 0                                                          0  0  66                                    4686868866666886444 466486848866888
9008 A25                 B18   046668  4464                                       0              6   2        666666        0             0      6 4 44 6868688664468  4444466086848866686
9020 A26       CW5       B18   4686684 448              004                                                0  666666                             4   4466868688666468 8 44444466486848866
9018 A3        CW5       B18     866846                             86644668 864666                            666666                646      8  666464466868488666468466646466086848866
9019 A30       CW5       B18   468 684 446         6686                  68         4          8             6866666                6 6           64844468686 88  6468464444446648684886
9105 A1        CW4       B35   448660 64  000000000 0   00 0 0000 0               0  000000000000000000000  60866060 0      0   0 0  460  0   8866 44  0086868866    8 4 44466086 08 66  8
9006 A11       CW4       B35   4   604 6 6                         8     6    4 66 4               2          866666                6 68866888 8 86 64446868688886   46664646648     66  8
9056 A2                  B35   4686684644         0      0   00 2                           2        22     6 666666                                   66868688666668886444 466486848 66886
9068 A2        CW4       B35   4686684 446                                               4                  68866666                                   468686886664688644444466486848 66  8
9042 A24       CW4       B35     866046                                     8                        0       6666666     8          6 6      8  66 464                664646608684  66  8
9025 A31       CW4       B35     86644644        2                            4666       6                    86666668666866688646486666          64 4 8686868888646 6644444466486848 66  6
9010 A28       CW4       BW53  4686684 64 46666         00                                      4          6 6666666     8668864648 46       888 68  4468686886666  6 664646648686  64  8
9016 A2                  B51   4 8668464                                                                    68 6666                 646                068686886664588644444466486848866886
9003 A24                 B51     86664644620 00 0                           8                                0666666                6        8  666484          6    6 444 4664868488668 86
9045 A2  A3    CW7       B51    4664846 6                            666468 46 6 40                            66666         0             0        484 668686886646684 66464660868 8866688
9070 A2                  B51   4686684644660 00 0               2       6                         8          0666666                                   6686868866666886444 466686848866  8
9030 A31                 B51     8666464 6                                    4666                 4  8        6666  86686668864648646            64 4 8686868886646 86444446668680886688 6
9092 A25       CW1       B51   4686684 446                                                                     866666                            4   4446868688666668  44444664868688666688
9011 A1                  BW52  466668464 8              0                                                   82666666                         88666664446868688666   8 44  4664868488668 8
9043 A1                  BW41  468668466 8  2    20     88 8600 286                    2   6642284 426      62 6  6                   6      88688464446868688886   664646 4868088 66888
9037 A29       CW2       BW61  468668464                                                                     8866666           464 6 6       8   646 468686886664686644444466486 84    8 6
9075 A24       CW10      BW60    866646444              86                  8              8     0          8  66  6                6        8  646484     6    4    4444466  6848866  8
9069 A2        CW10      BW60  4 866846444                                                           2      6  66  6                  6                4686868866666686644444466086808866  8
9028 A24       CW10      BW60    86 6464 4                                  86             86                6 66 66                  6          6648                 644  466486848866888
9059 A2        CW10      BW60  46866846446                                                                   8 66  0                                8  468686886666688644444664 68 8866  8
9048 A30       CW6       B13    68 68464488888888686864666               68      4044   0   6    4  866      8866666     6668864648 46       8 688484406868688  6  8  44444664868488668 86
9004 A2        CW1       B27     866846                                                        4  6  84        666666                  6                468686886664686666464664868488 66888
9067 A2        CW1       B27   46866846448    22 2                                            4   6  4      62866666                                 4 4686868866666886444444664868488 66888
9005 A3        CW1       B27           6                           8666666686864666664                      66666666                  6            46444686868888666846 66 64664868 8866888
9047 A3        CW6       BW47       0  64               4 6        86666666868646666446                     686666 6                6 6            46444668686886666684666464664868488 66888
9054 A2        CW5       B44   4 866846448 0 00 0                                                888888     60866666                 46                468686886664688644444466486848866
9050 A29                 B44   46866 466 8 0 00 0                         6  4                  8888888 6486680666666               6 6           64 44468686888886668 666 6466486868866888
9063 A32       CW5       B44   46 6 8466 6 0 00 0                 4          64          2      88888 6    0666666              4   688    8    66  4446868688666668 66646466486   6
9051 A29                 B44   46866046648    0                             6  4                 48888888864866886666666                6 6           64 44668686886866668 66646466486868866888
9036 A2        CW5       B44   4 8668464 6                                                   8  68888880    64866666                 46                66868688686468884 4 46648684 8868
9058 A2                  B45   46866846  8    0                                           888 6  88 8888     80666666                6 6                686868866466866664646648684 8866888
9007 A2        CW6       BW57  46866846                                                        2      4     64666666                                   6686868866646886   44664 6848 66888
9052 A2        CW6       BW57  4686684644        8                                          2         8      6666666                                   66868688666666886444444664 6848 66888
9066 A2        CW11      BW46  4 866846446                                                                  66866666                646                668686886664688644444664 6848866886
9106 A3                  BW50    4     44                           666666 686 666         88888              666666                 4             684 4686868866646886 4444 66486848066888
9040 A1        CW7       B49   4686684664                                                      888          6   66                   66       88 86464446868688666      664646648680886688 8
9009 A1        CW6       B37     866846448 0 00 0                                                    0      60866666                         886464844468686886666      44444664868688666 86
9062 A2                  B38   4 8668464   0 00 0 2                                                  2       0866 66                 46                4686868866646886444 466486848 66886
9026 A26                 B38   4686688 446                                                       2           4666 66                             6   44668686886666688 444 466486848 66888
9057 A26                 B38   4686684644  0 00 0                                4                           0866 66                 46          8   44068686886664688 4444466486848 66886
9065 A3        CW7       B7            64  20 00 0 2    626   2    86866448686466664        0        2        0                      468           484446868688866468864444466486 08866886
9013 A3        CW7       B7                               6        866666686864666644           42           4                        6            464466868688866668466646466486848866888
9017 A3        CW7       B7         0     442                      8666666686864666664     4     2  4                                 6            464446868688866668466646466 0868 8866888
9034 A3        CW7       B7         4  446                         8666666686864666 4                                                               8446686868886666688644444664868 8866688
9033 A3        CW7       B7                                        8666666686864666 4                               0                 6            464 468686886666684666464664868480668 88
9082 A3        CW7       B7         4 4                            8666666686864666  4  4                    6                        4            6844668686888666 688644  4666868488 66888
9021 A30 AW68  CW2       BW42  068668464 8   0    886664866886864 4      48           2         88     2    86 6   6     6668864648 46 8     88 68648 4068686886666  844444646 486808868886
9023 A1        CW7       B8    468668466 6        2     84                                        2  08      6      6                 6      88688464446868688666    66646466086808866  8
9088 A1                  B8    468668464       2                                                              6      6                 6      88666484 66868688666  444   4664868 8866  6
9022 A1        CW7       B8    468668466 4              0                                             8             6                 6      8866646444686868866 6    66464664868 8866  8
9014 A26       CW7       B8    468668 6       8  2                                             4           66       6                 6          8  84646868688866668  66464664868688 66  8
9002 A24       CW2  CW8  B14    486 6464                                    86            0     0              866 66                646          86484 6    88  4    6444 46648684886688 6
9055 A3        CW8       B14           6  2                      4  8668666868646666644             2 4        666 66                  6            46444686868886666866664646648684 8 66
9029 A23       CW8       BW14    86664 446                                  8                                  666 66                              6664                444 4660868 8866
```

Tenth International Histocompatibility Workshop
class I RFLP fragments (n=204) with B locus sera (n=304). Positive cluster.

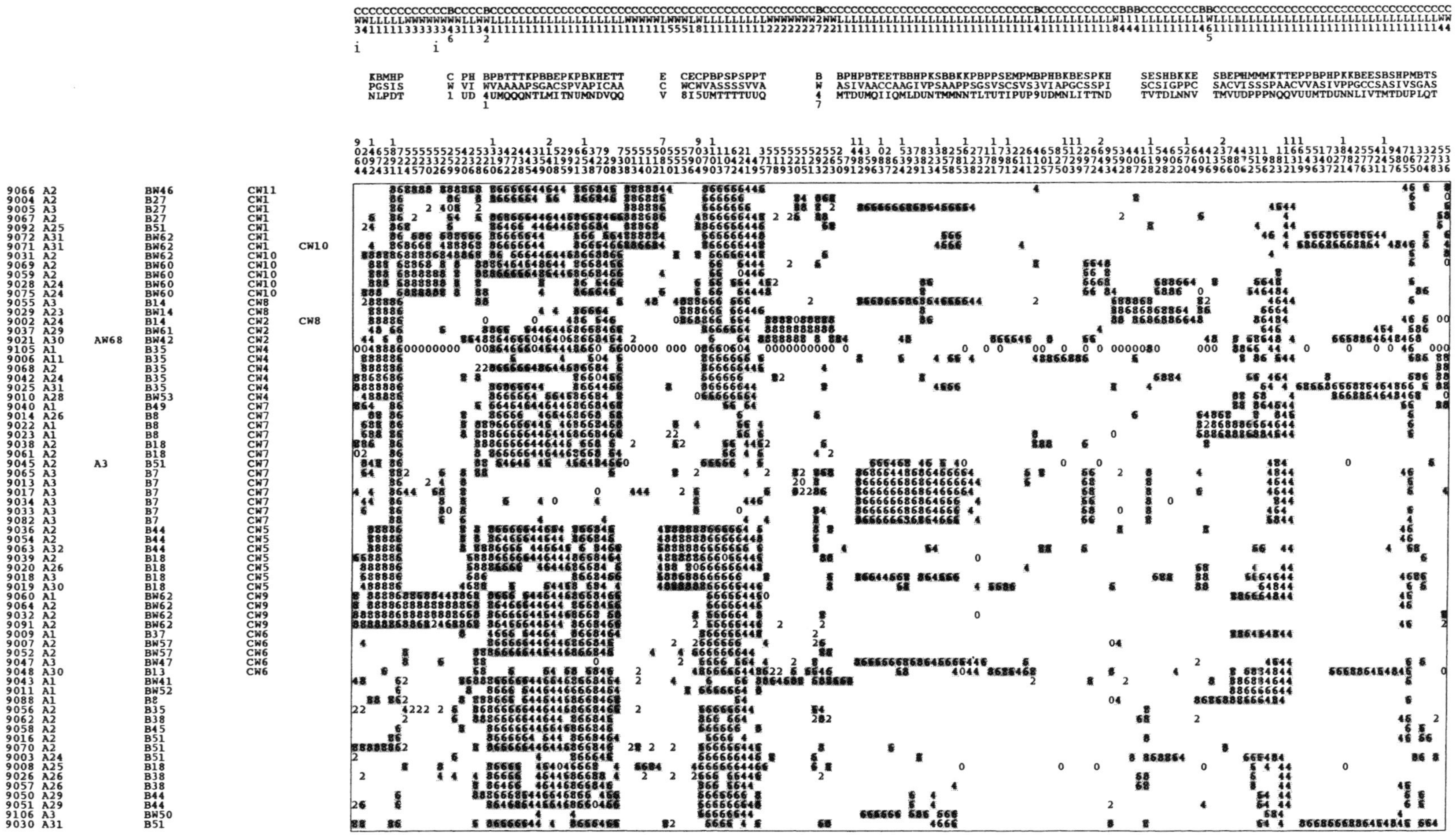

Tenth International Histocompatibility Workshop
class I RFLP fragments (n=204) with C locus sera (n=64). Positive cluster.

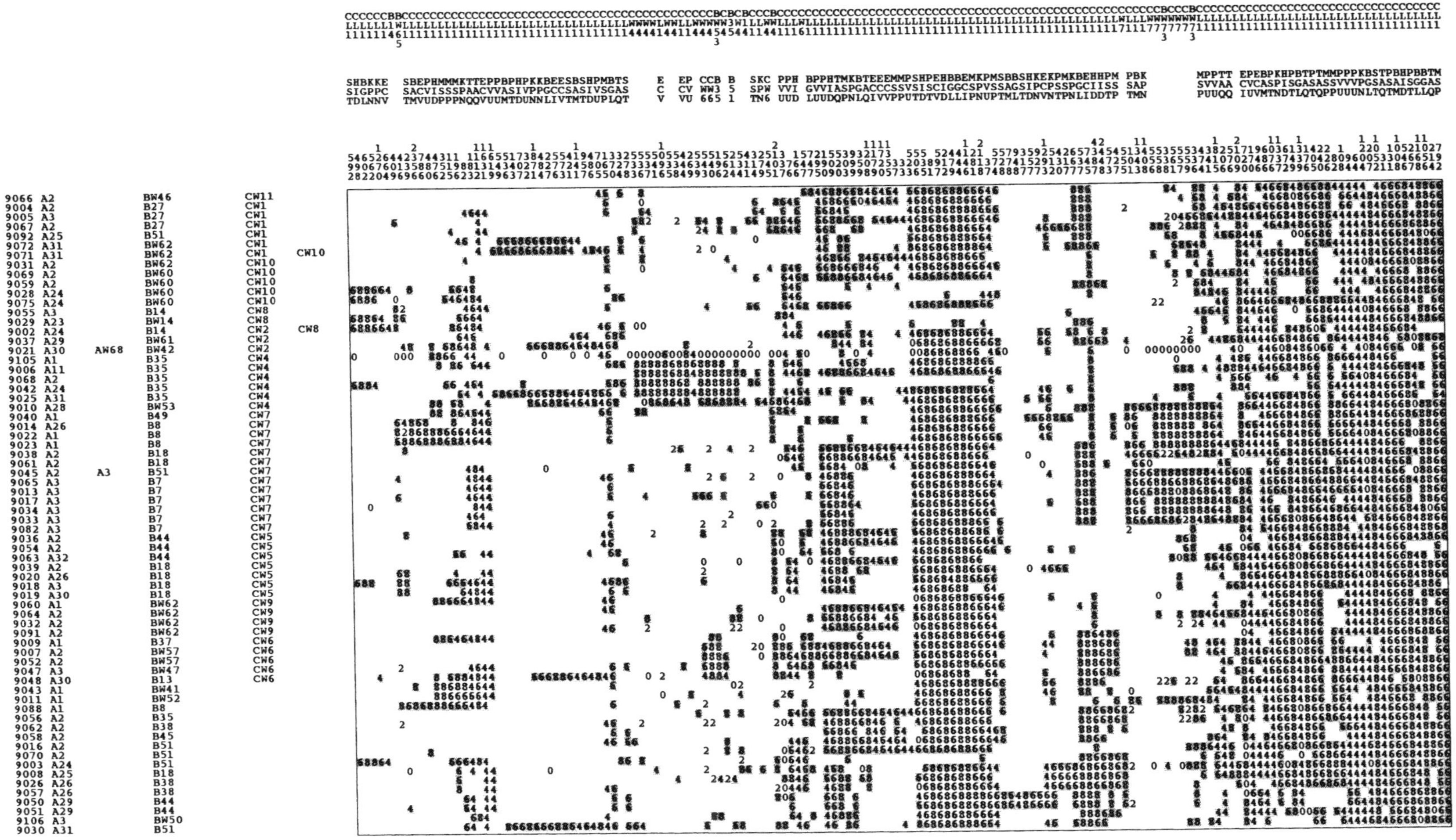

Tenth International Histocompatibility Workshop
class I RFLP fragments (n=204) with C locus sera (n=64). Positive cluster.

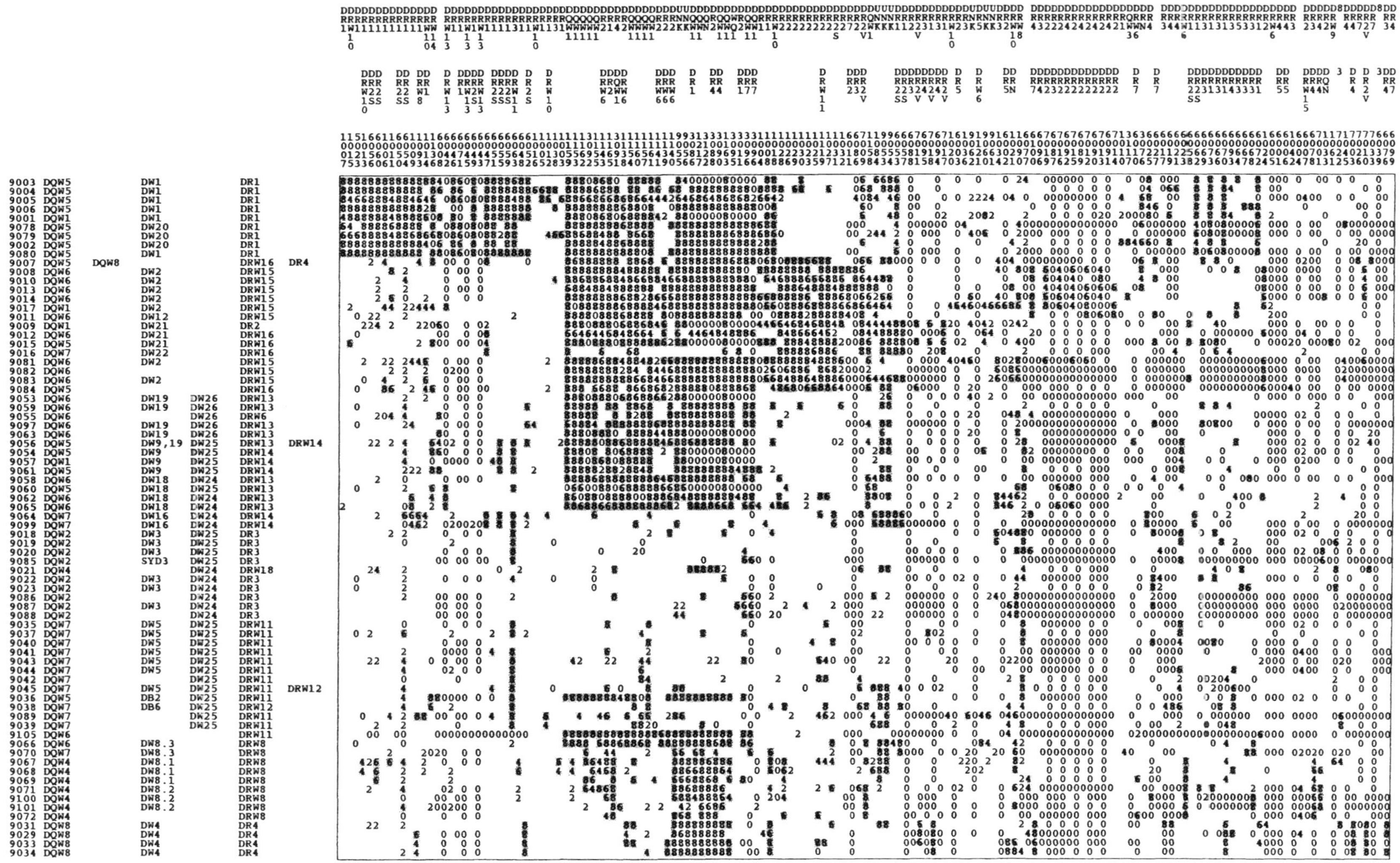

Tenth International Histocompatibility Workshop
class II sera (n=290) and T cells (n=134). Reordered positive cluster.

Cell	DQ		DW		DR		Reactions
9003	DQW5		DW1		DR1		[illegible]
9004	DQW5		DW1		DR1		[illegible]
9005	DQW5		DW1		DR1		[illegible]
9006	DQW5		DW1		DR1		[illegible]
9001	DQW1		DW1		DR1		[illegible]
9078	DQW5		DW20		DR1		[illegible]
9079	DQW5		DW20		DR1		[illegible]
9002	DQW5		DW20		DR1		[illegible]
9080	DQW5		DW1		DR1		[illegible]
9007	DQW5	DQW8			DRW16	DR4	[illegible]
9008	DQW6		DW2		DRW15		[illegible]
9010	DQW6		DW2		DRW15		[illegible]
9013	DQW6		DW2		DRW15		[illegible]
9014	DQW6		DW2		DRW15		[illegible]
9017	DQW1		DW2		DRW15		[illegible]
9011	DQW6		DW12		DRW15		[illegible]
9009	DQW1		DW21		DR2		[illegible]
9012	DQW6		DW21		DRW16		[illegible]
9015	DQW5		DW21		DRW16		[illegible]
9016	DQW7		DW22		DRW16		[illegible]
9081	DQW6		DW2		DRW15		[illegible]
9082	DQW6				DRW15		[illegible]
9083	DQW6		DW2		DRW15		[illegible]
9084	DQW5				DRW16		[illegible]
9053	DQW6		DW19	DW26	DRW13		[illegible]
9059	DQW6		DW19	DW26	DRW13		[illegible]
9055	DQW6			DW26	DRW6		[illegible]
9097	DQW6		DW19	DW26	DRW13		[illegible]
9063	DQW6		DW19	DW26	DRW13		[illegible]
9056	DQW5		DW9,19	DW25	DRW13	DRW14	[illegible]
9054	DQW5		DW9	DW25	DRW14		[illegible]
9057	DQW1		DW9	DW25	DRW14		[illegible]
9061	DQW5		DW9	DW25	DRW14		[illegible]
9058	DQW6		DW18	DW24	DRW13		[illegible]
9060	DQW5		DW18	DW25	DRW13		[illegible]
9062	DQW6		DW18	DW24	DRW13		[illegible]
9065	DQW6		DW18	DW24	DRW13		[illegible]
9064	DQW7		DW16	DW24	DRW14		[illegible]
9099	DQW7		DW16	DW24	DRW14		[illegible]
9018	DQW2		DW3	DW25	DR3		[illegible]
9019	DQW2		DW3	DW25	DR3		[illegible]
9020	DQW2		DW3	DW25	DR3		[illegible]
9085	DQW2		SYD3	DW25	DR3		[illegible]
9021	DQW4			DW24	DRW18		[illegible]
9022	DQW2		DW3	DW24	DR3		[illegible]
9023	DQW2		DW3	DW24	DR3		[illegible]
9086	DQW2			DW24	DR3		[illegible]
9087	DQW2		DW3	DW24	DR3		[illegible]
9088	DQW2			DW24	DR3		[illegible]
9035	DQW7		DW5	DW25	DRW11		[illegible]
9037	DQW7		DW5	DW25	DRW11		[illegible]
9040	DQW7		DW5	DW25	DRW11		[illegible]
9041	DQW7		DW5	DW25	DRW11		[illegible]
9043	DQW7		DW5	DW25	DRW11		[illegible]
9044	DQW7		DW5	DW25	DRW11		[illegible]
9042	DQW7			DW25	DRW11		[illegible]
9045	DQW7		DW5	DW25	DRW11	DRW12	[illegible]
9036	DQW5		DB2	DW25	DRW11		[illegible]
9038	DQW7		DB6	DW25	DRW12		[illegible]
9089	DQW7			DW25	DRW11		[illegible]
9039	DQW7			DW25	DRW11		[illegible]
9105	DQW6				DRW11		[illegible]
9066	DQW6		DW8.3		DRW8		[illegible]
9070	DQW7		DW8.3		DRW8		[illegible]
9067	DQW4		DW8.1		DRW8		[illegible]
9068	DQW4		DW8.1		DRW8		[illegible]
9069	DQW4		DW8.1		DRW8		[illegible]
9071	DQW4		DW8.2		DRW8		[illegible]
9100	DQW4		DW8.2		DRW8		[illegible]
9101	DQW4		DW8.2		DRW8		[illegible]
9072	DQW4				DRW8		[illegible]
9031	DQW8		DW4		DR4		[illegible]
9029	DQW8		DW4		DR4		[illegible]
9033	DQW8		DW4		DR4		[illegible]
9034	DQW8		DW4		DR4		[illegible]

Tenth International Histocompatibility Workshop
class II sera (n=290) with T cells (n=134). Reordered positive cluster.

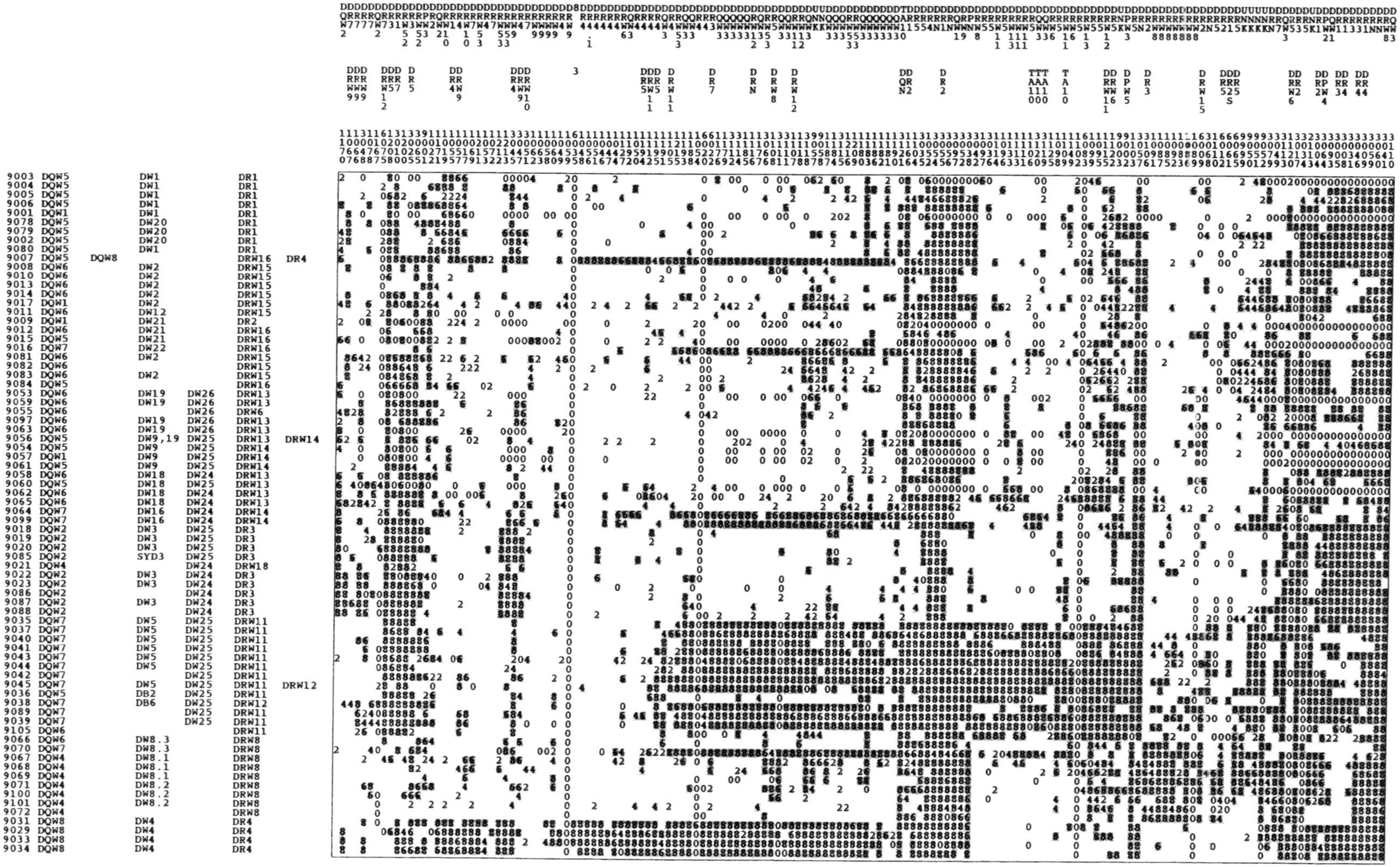

Tenth International Histocompatibility Workshop
class II sera (n=290) with T cells (n=134). Reordered positive cluster.

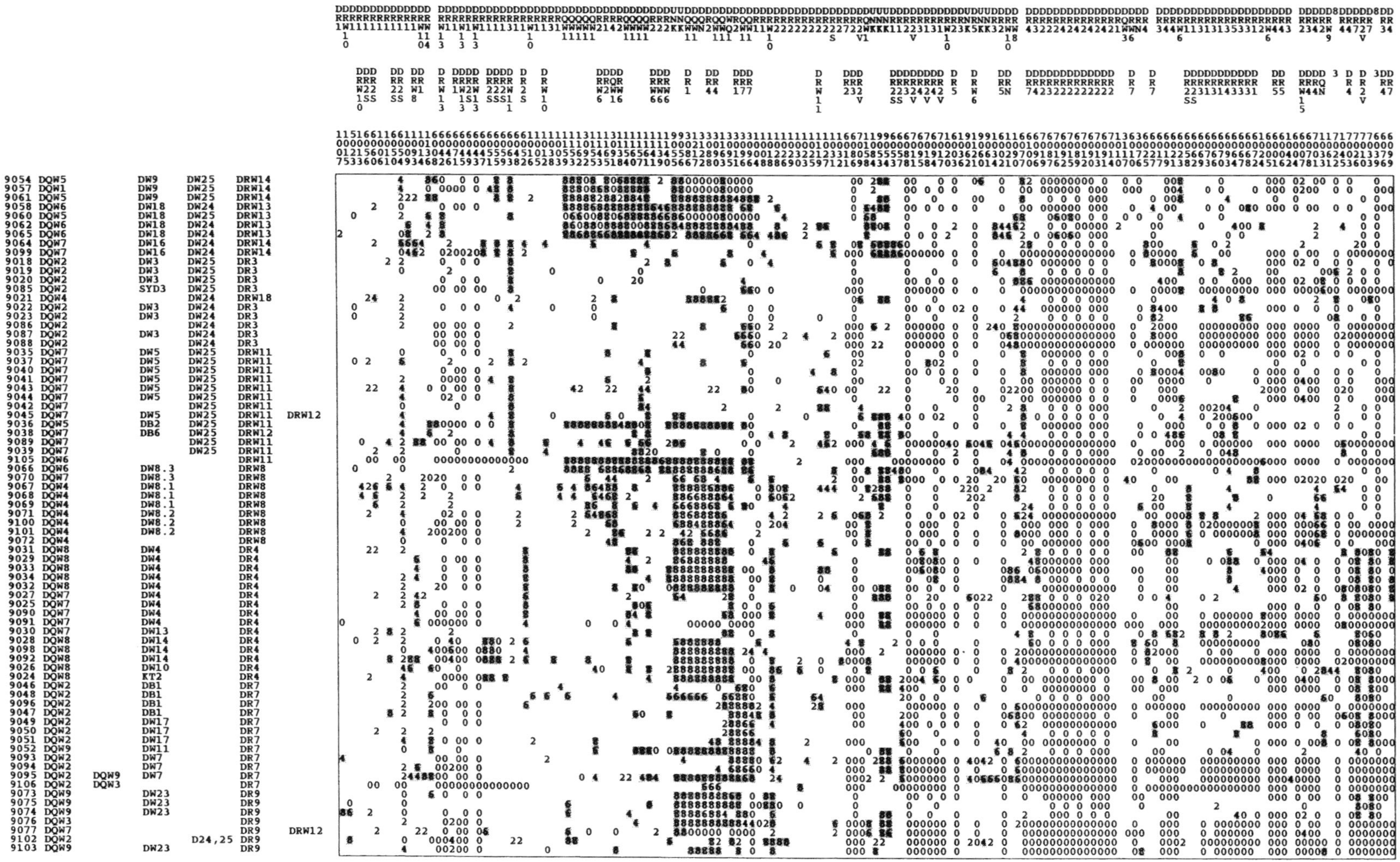

Tenth International Histocompatibility Workshop
class II sera (n=290) with T cells (n=134). Reordered positive cluster.

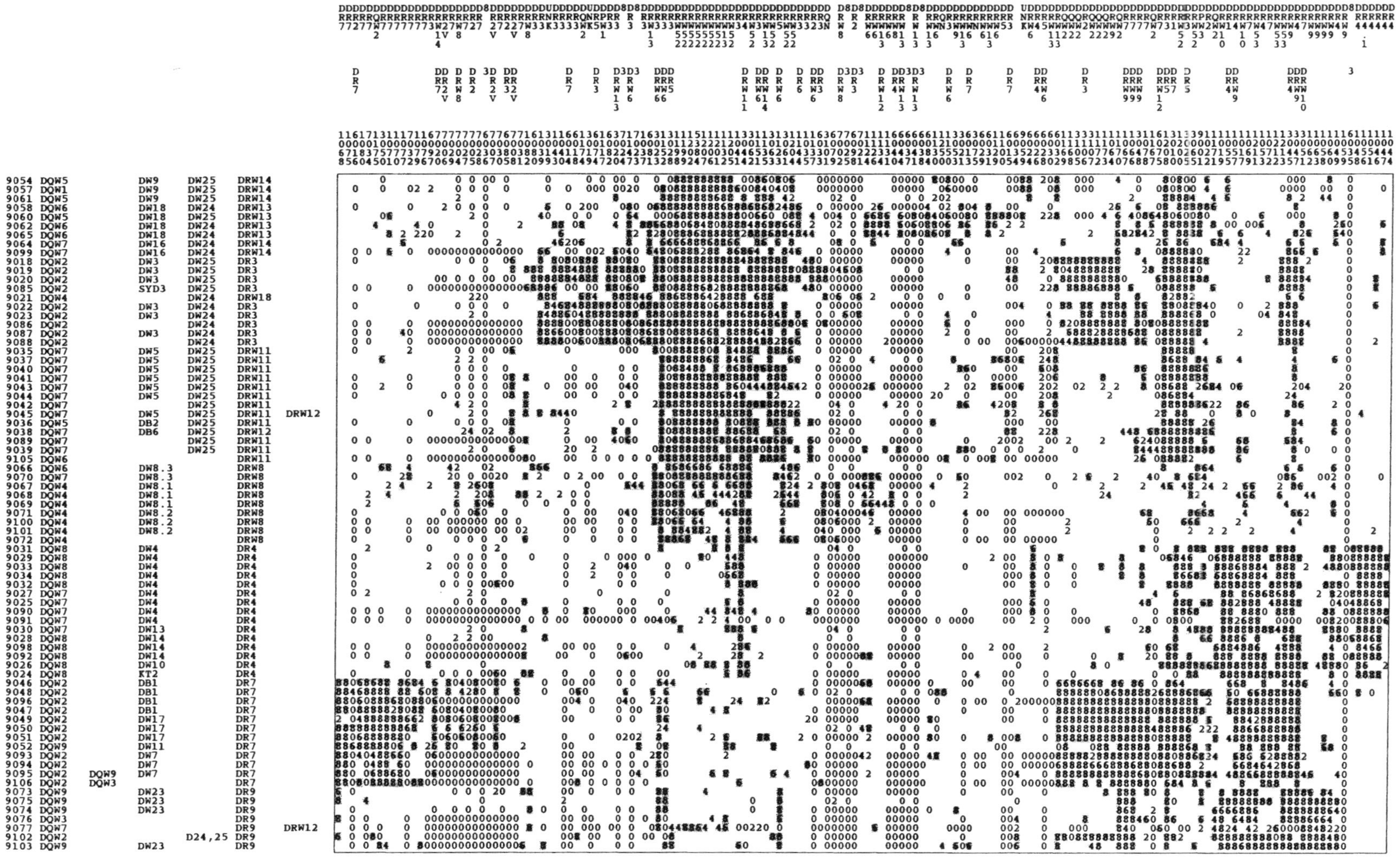

Tenth International Histocompatibility Workshop
class II sera (n=290) with T cells (n=134). Reordered positive cluster.

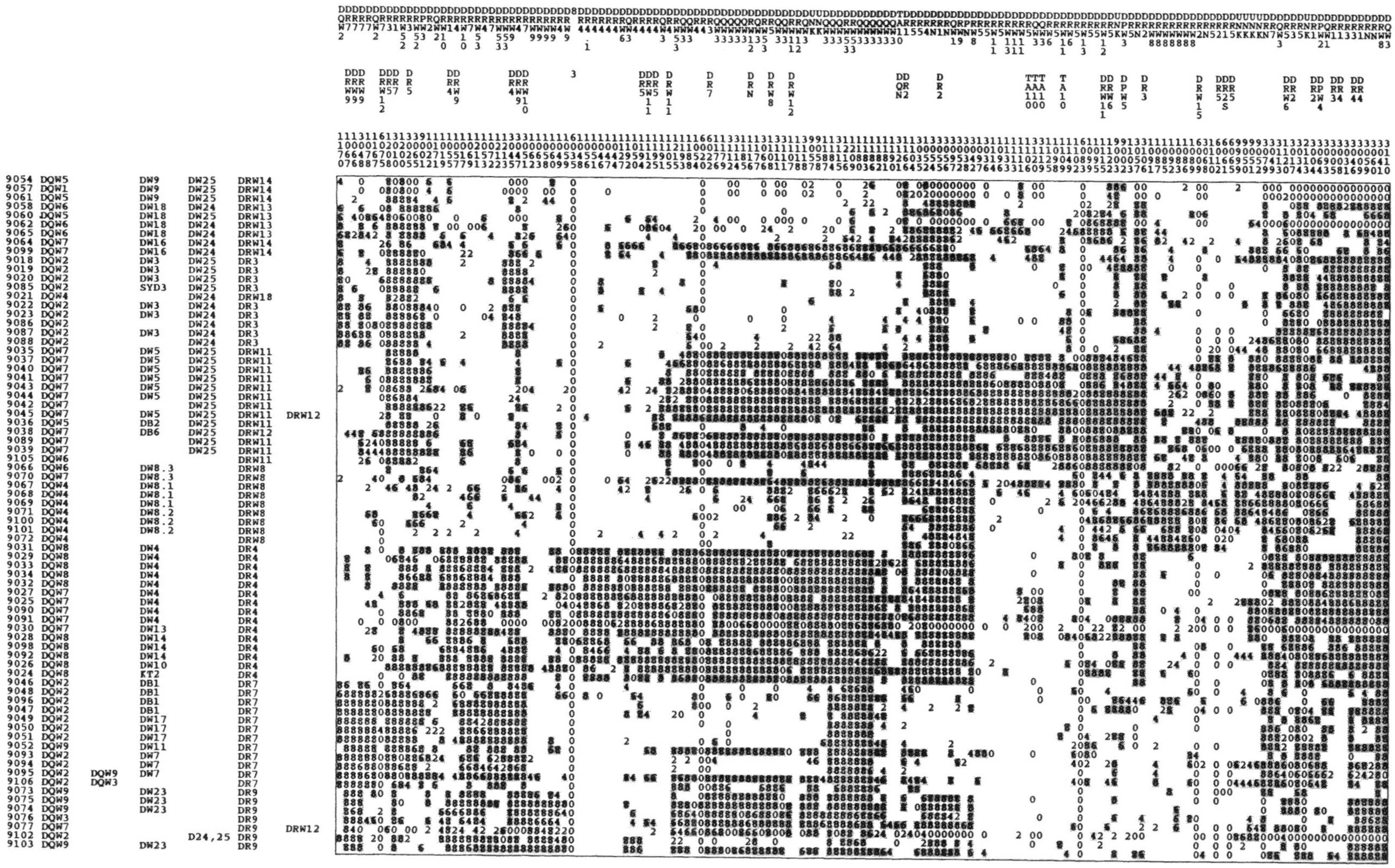

Tenth International Histocompatibility Workshop
class II sera (n=290) with T cells (n=134). Reordered positive cluster.

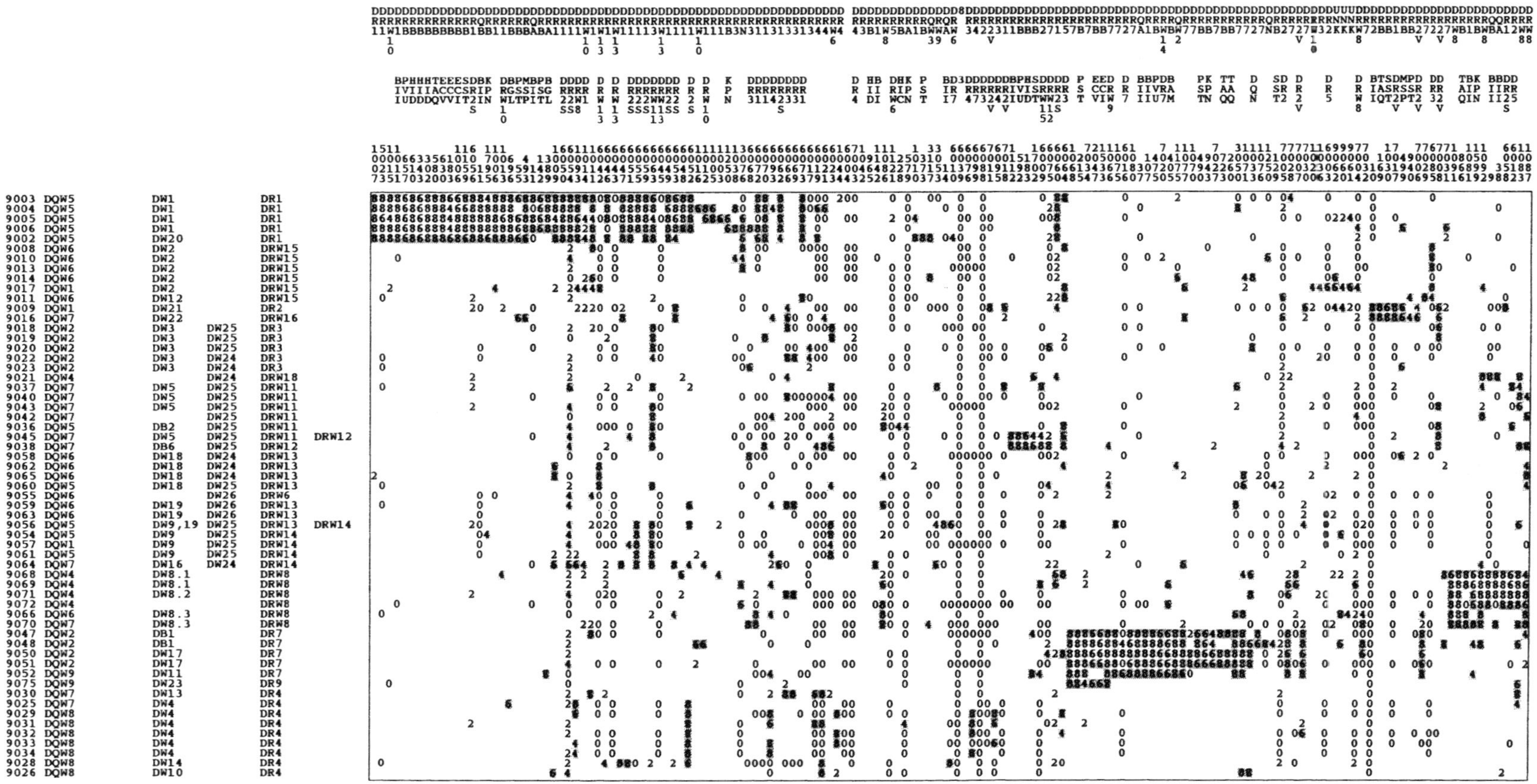

Tenth International Histocompatibility Workshop
class II DRA, DRB RFLP fragments (n=257) with T cells (n=134)/bio (n=42)/serol (n=290). DR listing.

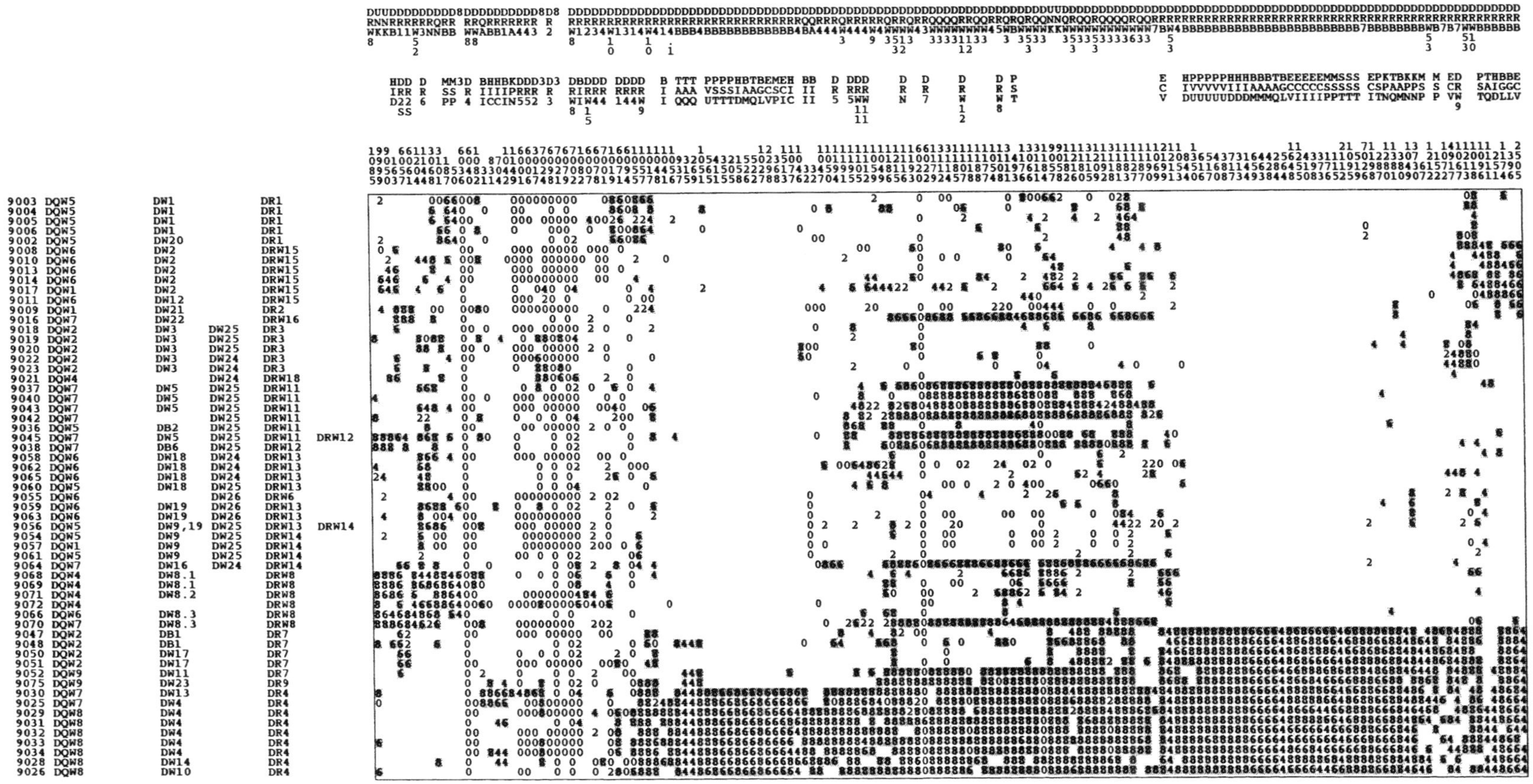

Tenth International Histocompatibility Workshop
class II DRA, DRB RFLP fragments (n=257) with T cells (n=134)/bio (n=42)/serol (n=290). DR listing.

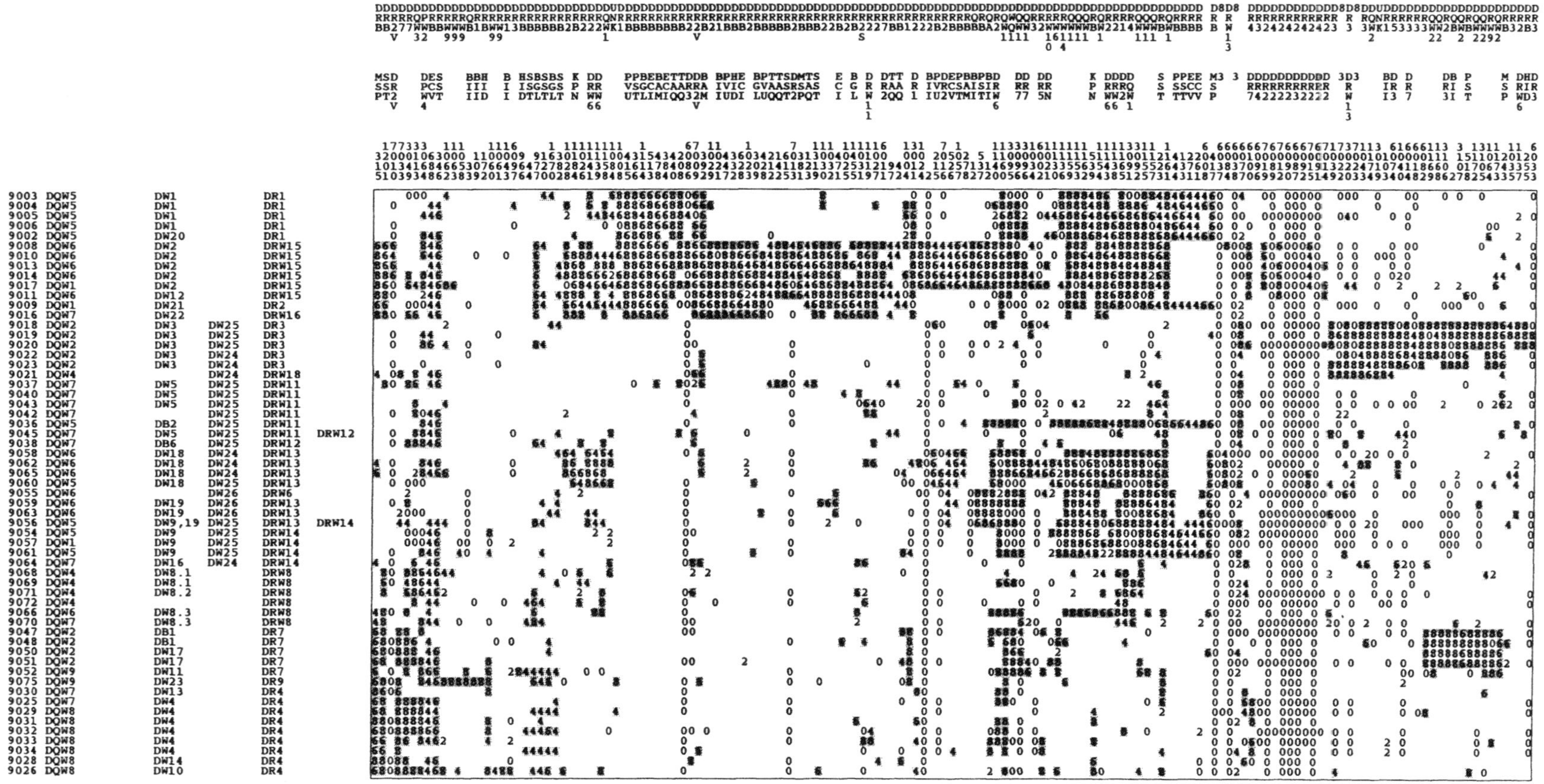

Tenth International Histocompatibility Workshop
class II DRA, DRB RFLP fragments (n=257) with T cells (n=134)/bio (n=42)/serol (n=290). DR listing.

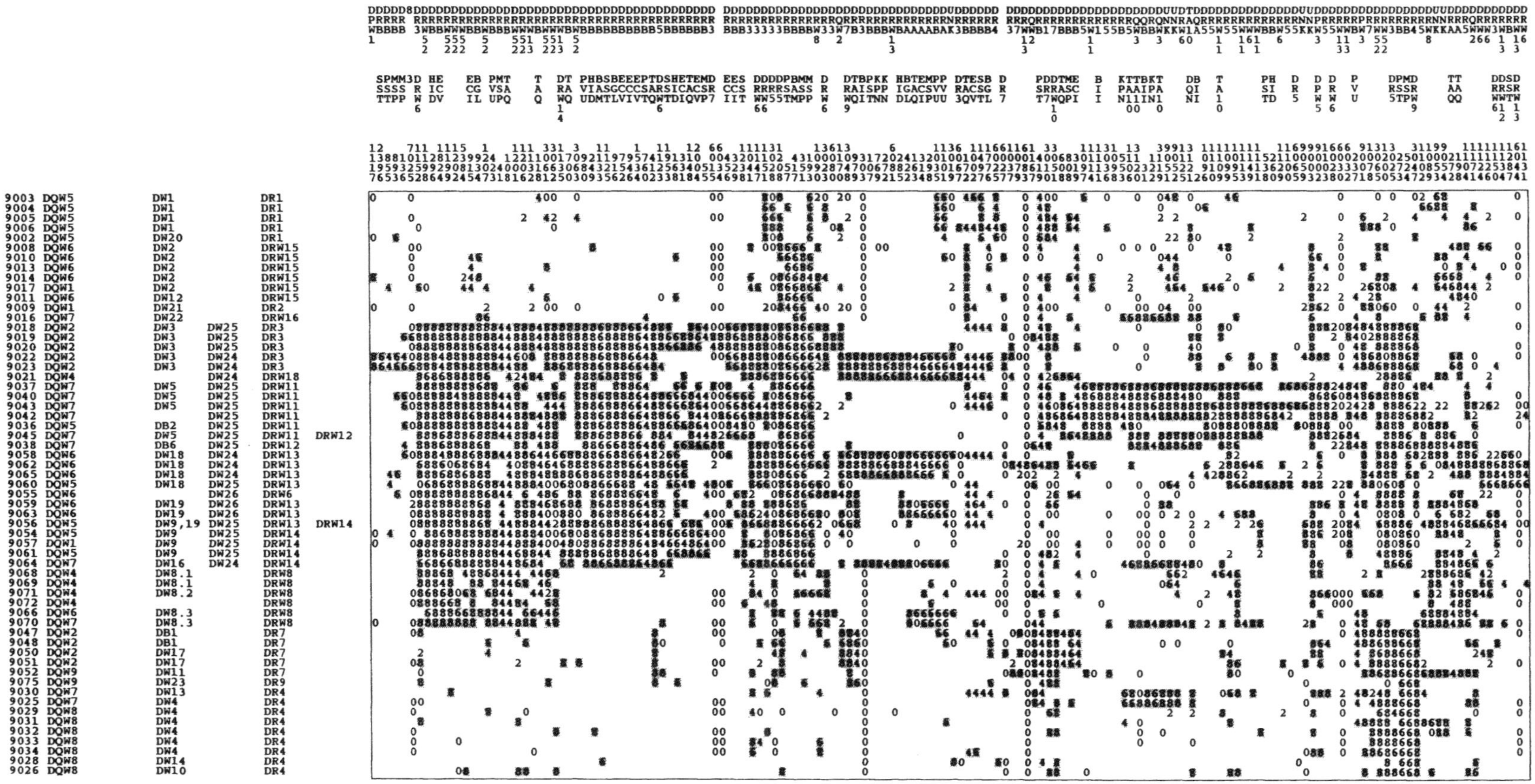

Tenth International Histocompatibility Workshop
class II DRA, DRB RFLP fragments (n=257) with T cells (n=134)/bio (n=42)/serol (n=290). DR listing.

```
DDDUUDTDDDDDDDDDDDDDDUUDDDDDDDDDDDDDDDDUUDDDDDDDDDDDDDDDDDDDDDDDDDDDDDDDDDUUDDDDDDDDDDDDDDDDDDDDDDDDDDDDDDDDDDDDDDDDDDDDDDDDD8DDDDDDDDDDDDDDDDDDDDDDDDDDDDDDDDDDDU
QRQNNRAQRRRRRRRRRRRRRNNPRRRRRPRRRRRRRRRNNRRRQRRRRRRRQRRRRRRRRRRRRRRRRRRRNNQQQRQQRRRRRPRRRRQRRRRRRRRRRRRRRRQRQRRQRRRRRRR RRRRRRRRRRRRRRRRRRRRRRRRRRRRRRRQN
BBWKKW1A55W55WWWBBW55KKW55WWBW7WW3BB45WKKAA5WWW3WBWWB1WWWWWB3BBBBBBBBWNBKKWWN2WWBW54WN1N1NW1W45W52B4BWABBB2ABBA1BBBBW4 253BBAAAAAANNBB5BBW4BWNNBBAA11313WWK
   3   60    1   161   6      3   11  3  55        8        266  1  16    16611                 1      11    11  1   9       1 5   8        6               6                          8  3              91
             1   1 1              33     22                       3  3     3   33                 3                  2                2

BKT     DB    T       PH    D     D  D     P       DPMD       TT       DDSD  BBDD     MBPPHESMPM   E    D    DD  PDD    D  D    D  D    DDP  H  PSSBBBHHBBPPEB  D3   BBHPPBEMTM   MM  SP  BP     PPEEDD  DDD
IPA     QI    A       SI    R     P  R     V       RSSR       AA       RRSR  IIRR     SIVSICSSSS   C    R    RR  VRR    R  R    R  R    RRV  I  VSSIIIIIISSCG   R    IIIVVGCSAS   SS  SS  IV     SVCCRR  RRR
IN1     NI    1       TD    5     W  W     U       5TPW       QQ       WWTW  IIW7     PIUTDVTPTP   I    1    44  UWW    2  2    2  W    W3U  C  UTTIIICCIITTIL   4    IIDUULIPQP   PP  TT  IU     TUII34  444
  0           0                   5  6               9                 61 1    1                                 16               1    6
                                                                        2 3    3                                  1               1

  39913  11111111   1169991666  91313   31199   111111161  166661   1   11      33  99313331  111331333331333131 6           1        6661     11        33   3   3   333   1133333332
 110011  011001115211000010002000202501000211111112011000001534620312001000210013110001000002000113040932113311254200003211433120030410411002130000000001
3215522  910991413620650002333076027240855790722538431114123360837976252705581289631255365559000434206410003111641471462223210849410925609357407157304800304491
1291251260995391809059323802718505347293428414604741831391322320654331555667280395222864567454520655159282233541310046520193398628907147411407971385819653
```

9003 DQW5	DW1		DR1		048 0 46 0 0 0 02 68 0 0 0 00 84000000 84 600 800000 000204 0 4 0 4 0 00 8686006606400800064 00000000
9004 DQW5	DW1		DR1		0 06 6688 8 4 0 8 888888466 88 888888 0 00 0 40600 88848866 4 8088884464868888 0
9005 DQW5	DW1		DR1		2 2 2 0 6 2 4 4 4 2 0 0 0 6 646864864 6482466884 286 24 0 0 00 8888688644646088864642828682
9006 DQW5	DW1		DR1		8 8 888 0 86 0 0 8888888846288888888 8884 0 20800 888680668668640888 6464888888
9002 DQW5	DW20		DR1		22 80 2 2 8 0 88888888 4288888888 8 00 0 0200 0686868686686480888 646488888888
9008 DQW6	DW2		DRW15		0 0 48 8 0 88 488 68 0 0 0 46 88888888 2486 886088 48888 0 4 0800 64886 886484480888064 8 88
9010 DQW6	DW2		DRW15		044 0 66 0 8 88 4 0 00 0 46 68888888 8 888888 668 0 4 0 0800 6688886 64064808888464 4 60
9013 DQW6	DW2		DRW15		4 8 4 8 4 0 8 0 8 4 0 0 00 0 46 88888888 888888 88 0 44 0800 66888 8866844808888464 88
9014 DQW6	DW2		DRW15		46 2 6 6 0 88 668 4 0 0 0 46 88888888 6468688888 86888 0 44 0800 66888 66648448088886464 8 80
9017 DQW1	DW2		DRW15		464 646 0 2 22 26808 4 646844 2 42 0 46 88888888 4488888888 86888 2 0600 66888 866844608888664 488 80
9011 DQW6	DW12		DRW15		2 8 2 4 28 4840 0 46 88888888 22 888842 88 8 2 8884 0800 66886 6 6668648068064 0 0
9009 DQW1	DW21		DR2		0 04 00 2862 0 88060 0 44 2 0 0 0 0 4 00 880000084644008000004000408 0 4 0 0 0 0880 6686860066044006000646400000000
9016 DQW7	DW22		DRW16		86688 88 6 6 4 8 6 88 4 0 46 6 88 68880 86668 0680 6686846646886480888464 0
9018 DQW2	DW3	DW25	DR3		0 80 8820848488888868 0 00 6 48 88 84 888 0 00 0 0808668388488668 8088806464888888888
9019 DQW2	DW3	DW25	DR3		88 8 8 2 8402888868 6 880 6 888 0444 0 64 0 088668868686684480888 846488888848
9020 DQW2	DW3	DW25	DR3		0 40 0 8 0 48 888868 0 0 0 6 888 88 888 0 000 0 0808668686848864 0888 6 6 88888888
9022 DQW2	DW3	DW24	DR3		0 88 6 8 8 4888 0 486808868 68 0 0 0 0 4 4 6 0 888 848688 0 00 040 88668648088064648888880
9023 DQW2	DW3	DW24	DR3		8 8 80 8 4 488688868 68 0 4 4 04 4880 8646888 0 0 0 886686648088064648888480
9021 DQW4		DW24	DRW18		2 28886 88 8 2 0 4 888824 88888 88 4 88 4 0 666 6884686488088864648686488648
9037 DQW7	DW5	DW25	DRW11		86888888886888668 8686888 24848 880 484 4 4 668 46 8 868886888 6888 888 6 0 66 0408866888686486644808888464 8
9040 DQW7	DW5	DW25	DRW11		86888480 888 868608486888868 0 06 4680 8 88 88888888888 0 00 0 6 0608866886 66864808880646886 60
9043 DQW7	DW5	DW25	DRW11		86888888886888888686868 8820 2428 888622 22 88282 00 00 8 8 46 2 22 868888888868068882 0 0 0608866888868686648088886464 0 8
9042 DQW7		DW25	DRW11		848888888828888886842 8888 848 88886882 82 24 6 84 468 2 8888888888888888 0 0608866888868686648088886464 0 8
9036 DQW5	DB2	DW25	DRW11		8088808888 8088 00 6888 80888 6 0 0 86 46 888888884 8088888888888 8888 0 80 66 0208866888686664808880846488844 88
9045 DQW7	DW5	DW25	DRW11	DRW12	8 8888808888888 8882684 8886 8 886 0 8 46 88 8888888888888 6488 4 00 0 86 8 0 088668868668448088864648 888
9038 DQW7	DB6	DW25	DRW12		84888688 886 8 22848 8888688884886 8 888888888888688884 0 4 866 080866886 6 8688448088064648888 80
9058 DQW6	DW18	DW24	DRW13		82 6 0 8 888 682888 886 22660 8800 88 4684488888888 8486 868084886 888 0 4 08 6 8 080 566686460888864648 60
9062 DQW6	DW18	DW24	DRW13		6 288646 6 2 6 8 888 6 6 0848888886868866 88 468 84888888 8882 28888886646884 44 08 666 0 0 8446864808888464 86 88
9065 DQW6	DW18	DW24	DRW13		4 428862 2 2 8888 68 2 488 8888848888 88 46 2 888866 8682 288866828 668 44 08 666 0 0 08 468648084888464
9060 DQW5	DW18	DW25	DRW13		064 0 86688888 888 228 880608 0 66686668888 866 4600 8600008 88 0080000008000608 0864 00088668688600660640080006464000000 0
9055 DQW6		DW26	DRW6		0 0 0 4 888 8 888 22 0 0808 66848666468 8 4 888888 6 0 8888868888288 0 4 00 060 66888 4 6 06480888864648886 80
9059 DQW6	DW19	DW26	DRW13		0 88 886 8 48 8888 8 8884868 0 08046 8684868666468884 8888888 8888888888888 00 4 080 8 8448088886464888 88 8
9063 DQW6	DW19	DW26	DRW13		0 0 2 0 4 688 8 0 4 0808 0 6 682 68 0 08088 86846866664600 48800000 8 8620008000008000 08 0 00 0 006 06 008000646400000000
9056 DQW5	DW9,19	DW25	DRW13	DRW14	0 2 2 2 26 688 2084 58886 488846868684 0008086 86844864 460 8868888 4886 8888 80 888 0 4 00 8888 8 06088868648646664808888 4064848
9054 DQW5	DW9	DW25	DRW14		0 0 0 886 208 080860 8848 8 0 0 0 00 880000084 880080000000000 08 0 00 668886668 0608866886 400660640080008464000000000
9057 DQW1	DW9	DW25	DRW14		0 0 0 88 08 080860 2 6 0 0 00 0 00 880000084 420080000008000208 0 4 00 668886468 04088668884006606400800064640000000
9061 DQW5	DW9	DW25	DRW14		2 2 8 8 48886 8848 4 2 0 8888888884 248888888 88 8 88 4 00 4 88866 8 0 08866886 68864648088886464828888
9064 DQW7	DW16	DW24	DRW14		86688480 8 86 8666 884866 6 466 8 66880 6 6 668 4 00660 646 0 0 8 6666480868886464 68 6 0
9068 DQW4	DW8.1		DRW8		662 4646 88 2 8 2888686 42 2 0 4 88868886428888 88864848884 0 4 0 00 66886 6866 8648068486464 8 60
9069 DQW4	DW8.1		DRW8		064 6 8 48 66 4 4 0 4 6668868 4 4688 88868 8 8668 6 0400 66888 68864 48088886464 8 460
9071 DQW4	DW8.2		DRW8		4 2 48 866000 668 8 82 686846 0 0400 0 4 8868888864688888888868684 0 44 0 44 000600 66888 488648648088864648 86 20
9072 DQW4			DRW8		0 8 8 000 8 488 0 00000 0 4 868 888 4 8688880 88666 4 0 00 00 000 00 6488 488 84480688864 0
9066 DQW6	DW8.3		DRW8		8 8 48 68888484 0 4 88888868 4466 88888 8884 4480866 8 0600 6 44864808888464 6 8
9070 DQW7	DW8.3		DRW8		8488848 2 8 8488 28 0 48 68 68888486 88 8 0 00 0 0 4 8 466 68 4 8668848488888 2 0 0 4 0400 8 8 66 86420888864 4 88
9047 DQW2	DB1		DR7		0 488888668 0 00 0 66 8 8 0 0 08006 8888686866648848 464 88 8
9048 DQW2	DB1		DR7		0 0 0 864 488888668 66 4 0 66666 8 8 8 0 0 0600668868686866 6488484884648 860
9050 DQW2	DW17		DR7		84 88 4 8688668 248 0 6 2 2 2 2 4 0 020066688886 6686488480848464 8880
9051 DQW2	DW17		DR7		86 8 8 6 0 4 88886682 2 0 00 0 48 4 286 2 0 0 0066688868864864864888084464 8880
9052 DQW9	DW11		DR7		80 0 886886668688884888 0 888888884 8 8 88888 848 4 44 0 0 00666 86 6 66864804 8 4 6 0 8
9075 DQW9	DW23		DR9		0 0 668 0 6 888888884 88 88888 088 4 0 08006668868686686864848 04464 88
9030 DQW7	DW13		DR4		86888 8 068 8 888 2 48248 6684 8 0 8 22888888888888888 60 0 66 0800666886868488668648848804464880 888 0
9025 DQW7	DW4		DR4		86888 8 8 0 0 4 4888668 0 0 0 886 888868688888 60 4 0 66 60066688888686448648888888886648888880
9029 DQW8	DW4		DR4		2 2 8 2 0 684668 0 0 0 868888888848888888848866884 0 0 4 80 00666888888686686488488064 888888 8
9031 DQW8	DW4		DR4		8 48888 6688888 8 0 0 888888888 8688888888888 888 864444 0 66 6080066688886868648648848888880
9032 DQW8	DW4		DR4		0 8 8 0 8888668 088 8 0 0 0 888888888 868888888888888848 0 44 0 80 0066688868680668648848804 6 88888880
9033 DQW8	DW4		DR4		0 0 8888668 0 00 0 8888888884 888888888888848 0 80066688868860686648488804 888888480
9034 DQW8	DW4		DR4		088 0 88686668 8 0 0 0 888888884 88 88888886888884 0 4 8080066 8888886668648864888804464888888880
9028 DQW8	DW14		DR4		0 8 668 8 0 6888888 4 88 888886 8 686 4 4 0 06006668886888686486486888864 66886 6 8
9026 DQW8	DW10		DR4		8 88 0 8 888668 8 86 8 0 888888884888 888888888884 0 0 4 0800666888688864864884880 4 6 6880

Tenth International Histocompatibility Workshop
class II DRA, DRB RFLP fragments (n=257) with T cells (n=134)/bio (n=42)/serol (n=290). DR listing.

Cell	DR	DR (2nd)	DW	DW (2nd)	DQ	RFLP fragment / T cell / serology pattern
9002	DR1		DW20		DQW5	[illegible]
9003	DR1		DW1		DQW5	[illegible]
9004	DR1		DW1		DQW5	[illegible]
9005	DR1		DW1		DQW5	[illegible]
9006	DR1		DW1		DQW5	[illegible]
9009	DR2		DW21		DQW1	[illegible]
9054	DRW14		DW9	DW25	DQW5	[illegible]
9057	DRW14		DW9	DW25	DQW1	[illegible]
9061	DRW14		DW9	DW25	DQW5	[illegible]
9036	DRW11		DB2	DW25	DQW5	[illegible]
9056	DRW13	DRW14	DW9,19	DW25	DQW5	[illegible]
9008	DRW15		DW2		DQW6	[illegible]
9010	DRW15		DW2		DQW6	[illegible]
9013	DRW15		DW2		DQW6	[illegible]
9014	DRW15		DW2		DQW6	[illegible]
9017	DRW15		DW2		DQW1	[illegible]
9011	DRW15		DW12		DQW6	[illegible]
9055	DRW6			DW26	DQW6	[illegible]
9059	DRW13		DW19	DW26	DQW6	[illegible]
9063	DRW13		DW19	DW26	DQW6	[illegible]
9058	DRW13		DW18	DW24	DQW6	[illegible]
9060	DRW13		DW18	DW25	DQW5	[illegible]
9062	DRW13		DW18	DW24	DQW6	[illegible]
9065	DRW13		DW18	DW24	DQW6	[illegible]
9066	DRW8		DW8.3		DQW6	[illegible]
9018	DR3		DW3	DW25	DQW2	[illegible]
9019	DR3		DW3	DW25	DQW2	[illegible]
9020	DR3		DW3	DW25	DQW2	[illegible]
9022	DR3		DW3	DW24	DQW2	[illegible]
9023	DR3		DW3	DW24	DQW2	[illegible]
9047	DR7		DB1		DQW2	[illegible]
9048	DR7		DB1		DQW2	[illegible]
9050	DR7		DW17		DQW2	[illegible]
9051	DR7		DW17		DQW2	[illegible]
9016	DRW16		DW22		DQW7	[illegible]
9038	DRW12		DB6	DW25	DQW7	[illegible]
9064	DRW14		DW16	DW24	DQW7	[illegible]
9037	DRW11		DW5	DW25	DQW7	[illegible]
9040	DRW11		DW5	DW25	DQW7	[illegible]
9042	DRW11			DW25	DQW7	[illegible]
9043	DRW11		DW5	DW25	DQW7	[illegible]
9045	DRW11	DRW12	DW5	DW25	DQW7	[illegible]
9070	DRW8		DW8.3		DQW7	[illegible]
9025	DR4		DW4		DQW7	[illegible]
9030	DR4		DW13		DQW7	[illegible]
9026	DR4		DW10		DQW8	[illegible]
9028	DR4		DW14		DQW8	[illegible]
9029	DR4		DW4		DQW8	[illegible]
9031	DR4		DW4		DQW8	[illegible]
9032	DR4		DW4		DQW8	[illegible]
9033	DR4		DW4		DQW8	[illegible]
9034	DR4		DW4		DQW8	[illegible]
9052	DR7		DW11		DQW9	[illegible]
9075	DR9		DW23		DQW9	[illegible]
9021	DRW18			DW24	DQW4	[illegible]
9068	DRW8		DW8.1		DQW4	[illegible]
9069	DRW8		DW8.1		DQW4	[illegible]
9071	DRW8		DW8.2		DQW4	[illegible]
9072	DRW8				DQW4	[illegible]

Tenth International Histocompatibility Workshop
class II DQA, DQB RFLP fragments (n=299) with T cells (n=134)/bio (n=42)/serol (n=290). DQ listing.

9002	DR1		DW20		DQW5
9003	DR1		DW1		DQW5
9004	DR1		DW1		DQW5
9005	DR1		DW1		DQW5
9006	DR1		DW1		DQW5
9009	DR2		DW21		DQW1
9054	DRW14		DW9	DW25	DQW5
9057	DRW14		DW9	DW25	DQW1
9061	DRW14		DW9	DW25	DQW5
9036	DRW11		DB2	DW25	DQW5
9056	DRW13	DRW14	DW9,19	DW25	DQW5
9008	DRW15		DW2		DQW6
9010	DRW15		DW2		DQW6
9013	DRW15		DW2		DQW6
9014	DRW15		DW2		DQW6
9017	DRW15		DW2		DQW1
9011	DRW15		DW12		DQW6
9055	DRW6			DW26	DQW6
9059	DRW13		DW19	DW26	DQW6
9063	DRW13		DW19	DW26	DQW6
9058	DRW13		DW18	DW24	DQW6
9060	DRW13		DW18	DW25	DQW5
9062	DRW13		DW18	DW24	DQW6
9065	DRW13		DW18	DW24	DQW6
9066	DRW8		DW8.3		DQW6
9018	DR3		DW3	DW25	DQW2
9019	DR3		DW3	DW25	DQW2
9020	DR3		DW3	DW25	DQW2
9022	DR3		DW3	DW24	DQW2
9023	DR3		DW3	DW24	DQW2
9047	DR7		DB1		DQW2
9048	DR7		DB1		DQW2
9050	DR7		DW17		DQW2
9051	DR7		DW17		DQW2
9016	DRW16		DW22		DQW7
9038	DRW12		DB6	DW25	DQW7
9064	DRW14		DW16	DW24	DQW7
9037	DRW11		DW5	DW25	DQW7
9040	DRW11		DW5	DW25	DQW7
9042	DRW11			DW25	DQW7
9043	DRW11		DW5	DW25	DQW7
9045	DRW11	DRW12	DW5	DW25	DQW7
9070	DRW8		DW8.3		DQW7
9025	DR4		DW4		DQW7
9030	DR4		DW13		DQW7
9026	DR4		DW10		DQW8
9028	DR4		DW14		DQW8
9029	DR4		DW4		DQW8
9031	DR4		DW4		DQW8
9032	DR4		DW4		DQW8
9033	DR4		DW4		DQW8
9034	DR4		DW4		DQW8
9052	DR7		DW11		DQW9
9075	DR9		DW23		DQW9
9021	DRW18			DW24	DQW4
9068	DRW8		DW8.1		DQW4
9069	DRW8		DW8.1		DQW4
9071	DRW8		DW8.2		DQW4
9072	DRW8				DQW4

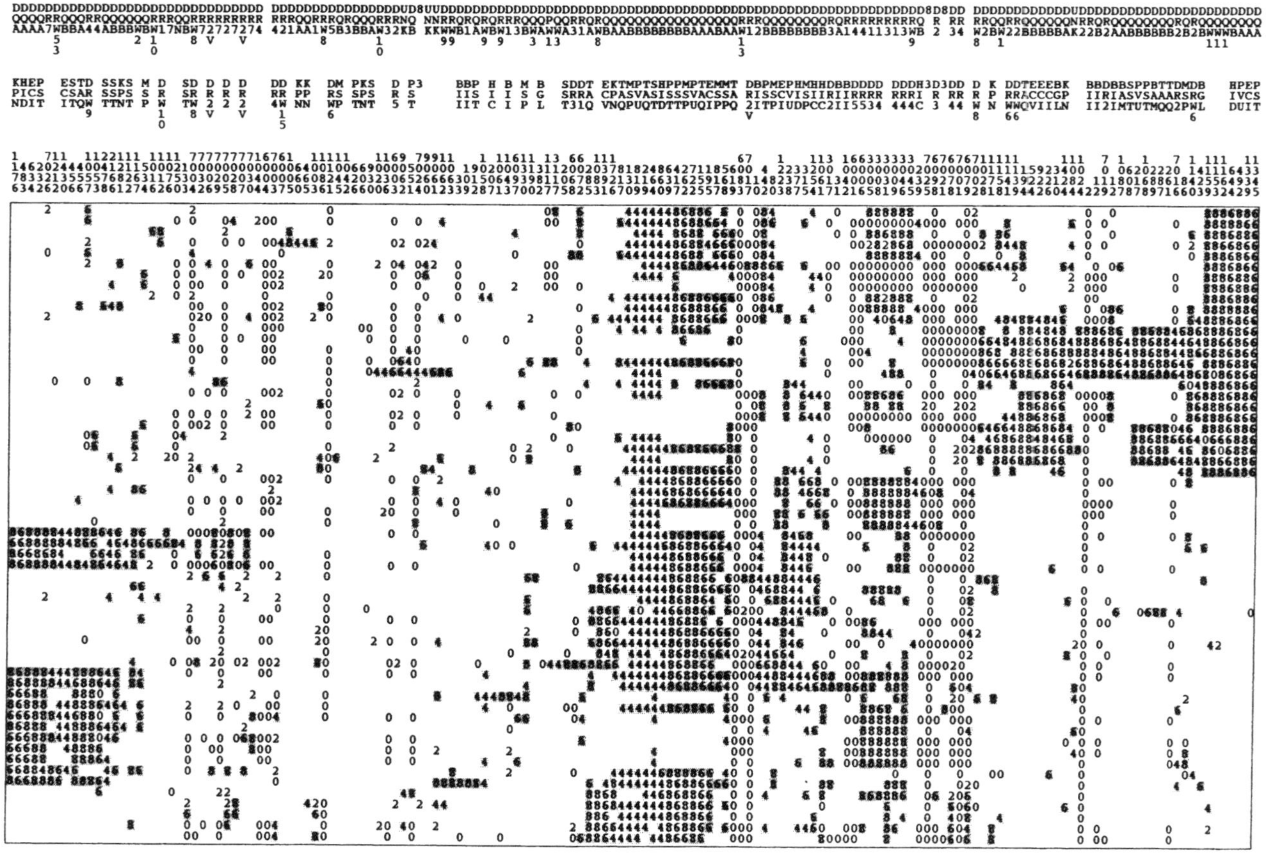

Tenth International Histocompatibility Workshop
class II DQA, DQB RFLP fragments (n=299) with T cells (n=134)/bio (n=42)/serol (n=290). DQ listing.

```
DDDDDDDDDDDDDDDDDDDDDDDDDDDDDDDDDDDDDDDDDDDDDDDDDDDDDDDDDDDDDD8D8 DDDDDDDDDD8DDDDDUDDDDDDDDDDDDDDDDDDDDDDDDD DDDDDDDDDDDDDDDDDDDDUUDDDDDD8DDD8DDDDDDDDDDDDDDD
QQQQQQQQQRRRRQQQQQQQQQQWQQRRRRRRRQRRRRRQQQRRQRRRQQQQQQQQQRQQQ R RRRRRRRRRR RQQRRNRRQQRRQQQQQQRQRRQRRRQQRQRRR QQRRRRQRRQQRRQQQQRQNNQRRRQQ RRR RRRRRRRRRRRRRRRQ
ABBBBBBAW2214WWBBWWWBWWQAWW5132W1AWWW3WBAAWWB1WWBBBBBBBA2BAA W 4424422423 3WB33K15BA33BAAAAA3B3WB323BB1B233 BA7W1WBWWBA12AAAA1BKKB11WNA W15 2231227413444B
 1     11   111 111 11      1   666 1    16   16          1            2                8                  8 8 88                  8      8     V   V     i
                    2       0       3    3    3           3

MEKTEHSH DDDD   HMDD M    B DDD DD K   DDSSHD BBDDBPBTEHHBBPBT3 3 DDDDDDDDD 3  E   BDHK DKHEBPMDT  S DDHMBHDBD EPD B H  BBDDBMMMBK  HDD  T3DDD3DDDDDD  BD   P
SCPACISI RRRQ   ISRR S    G RRR RR P   RRSSIR IIRRIVAACIIIISAA    RRRRRRRRR    C   IRIP RPICGSSRA  S RRISIIRIR CSR I I  IIRRGSSSIP  IRR  A RRR RRRRRR  IR   V
PINQVCTD WW2W   DP77 P    L WW2 5N N   WWTTDW IIW7IUMQVCCIITMQ    722222322    V   I3DN 7NDILTPWQ  T W3DPIDWI7 VTW I D  II25LPPPIN  D22  Q 423 324232  I4   U
         66 1                 16       61  1   1                                           6       6      1      1           S        SS     S  V V V
                              1          2   3   3                                                        5      2

         1111331   331 113 11131611 1111111 61 166     11     1 666666766766717111613 612662 1    3 61 116   1 6 66   6111 11   661    1199 6613 666666767677116111
23556827111100136001411011110000021111135701100012841511 5060000100000000000150001015005513440100401052110300460000300  003322059010001500000000000000000005
8813752454369955399536696612030203223534564311411075838110093837081991891324711410742819707272189333311170013687989088 3511235899565680534663819180344945
7354308573851251066473056152442142460040384183132220265328997487092014725923448093420241978443200557333299154345112175988225102090571213600498158581816661 5

9002 DR1              DW20          DQW5   8668688684688884488888888428  42    8                      0066 0 02  0 00  00                      2       0     2        0        2   8 0 8 00        2
9003 DR1              DW1           DQW5   6668888686 8008440088888088  0 0   0        0    0        066  0 04  0 00  000 0    0 0       0   0  2   04    00004€      2         2   80 0088000      0
9004 DR1              DW1           DQW5   68686666 8 888644808888888668    80   0                    0066 0 0   0 00  00        0 0          8          0    2        0              0  8 00             8
9005 DR1              DW1           DQW5   686868866666686844828866884 64  0460 24     0    0         066  0 00  0000  000 0           0    0  8     20     0000  0                       4008 000          2
9006 DR1              DW1           DQW5   6868688608688888448888888862€    00                        066  0 0   00 00  00                   0     0                                     8640 8 000      6   0 0
9009 DR2              DW21          DQW1   6668686668680084  008888088640  0240      0 0 0 0        0   066  0 02  0 00  00       0        0  0  4  0     000              008       4  488808004 008 8062
9054 DRW14            DW9     DW25  DQW5   666868668 680084  008888088 80 0        88  0 0 0 0        068 0 0820000  000 0       0          00   0     0800            0        2    0 00  000  0
9057 DRW14            DW9     DW25  DQW1   6668686686880084  008888088 40 0          8  0 0 000       066 0 0000 0 00  000 0      0  0       00   0     0800  0      0 0               0600 000 00
9061 DRW14            DW9     DW25  DQW5   686888864228888 4  88888888 2   80  22     0    0          066 0 08  0 00  00                         0   4  8              0                 60  00
9036 DRW11            DB2     DW25  DQW5   66686886084888 8  808888888808 0  4  8      0   0 8        0   60 08  0 00  000 0             0      0  4 8046  0800              8                 00 800    08
9056 DRW13  DRW14     DW9,19  DW25  DQW5   6668688668068888 4 808888888848  0   066 84   0000080     066 0008  00000000 0 2  0 0      00     002 4  046  080048            0                866002800  00
9008 DRW15            DW2           DQW6   6664686668488884 4 0 8 888824840 8066        0   0            08008600  006 0       0          0  0 8    04    000  0                  00 4   800 800     8
9010 DRW15            DW2           DQW6   58646866668488884808486888 68 0   0  0       0   00           0 0080 00 004 0      000         0           0  0   000  0 0               0  2   8  002 00     8     0
9013 DRW15            DW2           DQW6   6864686686884848448888888 8  0 0       0     0    00          0 0 0000600 004 0       0          0         404    000  0           0         44   8002 00     80
9014 DRW15            DW2           DQW6   6864686628688884488 88888864840 60 4          0   0            0 0 00 8600 004 0     20 0        0     4  8    0     000  0                     644   002 00     0
9017 DRW15            DW2           DQW1   686468 6888884488 868888844 8668800 2                          0 0 808000004          20                    64  046   0 0       4                 644   680  800     8       2
9011 DRW15            DW12          DQW6   6864688880     08888888 24  00               8668856888660 0 0 808008000                 0                    46    0    2        0              0 2800     4
9055 DRW6                     DW26  DQW6   866  6868668   8884488888828886  8  04 0222   0 08080       0    0 0 400000000 0     0 0      0      8488   0      0 00            0         2    8002 00     00
9059 DRW13            DW19    DW26  DQW6   686868666    88888448888888888    8       0 0 08040       0    0 02  0 00 00                          0      4                    0               880  00
9063 DRW13            DW19    DW26  DQW6   6668666688  80084400888808860 0   0 6 8   0 08080        0    0 0    0000000060       0          04 84 804     0 00              0         4    0800  00    00
9058 DRW13            DW18    DW24  DQW6   5864686 8888886446888888888844 0  0 2 264660 8800      80 6 040000 00  000 0 2   0 0     0      64    0       0 00   0                         002 000  00
9060 DRW13            DW18    DW25  DQW5   6464666 8868000 4 008888088880    0 6646846668888       80 6 0808 0 00 0084   0 0             0  64 4       0880   4      4                0 0   00
9062 DRW13            DW18    DW24  DQW6   6664686 00888884  8888808888888 44   88868466688866      80 660802 0000 00   88 08              60 6  2         2           44                 0  400        2
9065 DRW13            DW18    DW24  DQW6   6664686 868688844668688888868844644886844668848888       80 560802 0 00 006     6              20    44                    82              4     0  00  0      0
9066 DRW8             DW8.3         DQW6   666      6666888    86888888844            86888648066 0 02 0 00 00 6                       84844   46    0          888 88          64484 80   00
9018 DR3              DW3     DW25  DQW2              4   0   6    2 4  604 0     0     00           00  80 0800  00  0008088800888888866888684  844804   0800   0       0      0            4     00  00       08
9019 DR3              DW3     DW25  DQW2         0     4    4          6                              0   0 08  0 00  00 8688848888868866888684 8 488848   88848        8                  88 0  200       8
9020 DR3              DW3     DW25  DQW2        00     4 4      2      0  0     0 0 0 0          00   60 0860000000000080888488888868866888688 88  8846   0848486           00                8400  0000 000      8
9022 DR3              DW3     DW24  DQW2             04  0  46           0  0     0 0 0 0              0   0 04  0 00 000 04884688886886668884 44    04    0 00            0            4    00  000 00        8
9023 DR3              DW3     DW24  DQW2             04  0 4              0   4                        0   0  0  0 00  00 8848088888868866688884 2    0                                   4    0   00
9047 DR7              DB1           DQW2         0       84    688     6  0680      0    00            0   0 0000000 000 0      0     0             8    04    0000480       4           42    00  000   0
9048 DR7              DB1           DQW2              4  80    6 68              08                    0   0  0  0 00  00    6  0 0          8      6  04    0  0  2   8  8         642   0   00 0    8   008
9050 DR7              DW17          DQW2                 66  2 88     22      048                      0   0 04  0 00  00                              4  46    0   484                46    0 2800
9051 DR7              DW17          DQW2               8 88    88     840  40 28    0    00            0   0 0000000 000 0     0    0                  2  0    0000480      2      0       46    00  000   0
9016 DRW16            DW22          DQW7    4    6        06 68                                        0   0 02  0 00  00         0        68                                      8664      488   0  00  2 8
9038 DRW12            DB6     DW25  DQW7               6   0   8  8888   4  0            0             0   0 04  0 00  00               2  668              4684888488       88            88 8   0  800    8
9064 DRW14            DW16    DW24  DQW7               8   0      4           8          0             00  60 0280 0 00  00    46806     2  68888  0         2   4    0  866        46     0 2 00
9037 DRW11            DW5     DW25  DQW7  66 0  6664       6        886         4                      00  60 08  0 00  00                  68  0              8880      4    8 66                   00  8
9040 DRW11            DW5     DW25  DQW7              0            88 8  0  0              0     06    00  0 08  0 00 000  0    0           66886       0      0800          44  0             00 000 00
9042 DRW11                    DW25  DQW7               8      2    4888   2  0             24    6     0   0 08  0 00  00  2                6686 20            808          8488             8   0  00
9043 DRW11            DW5     DW25  DQW7               4   22   80    4886  02 028 2       00    00    0   80 0000 0 00  000 0    0  0  0     0 66          2 2 04   0800  0   2        0                002 00     08
9045 DRW11  DRW12     DW5     DW25  DQW7      6         06      8       8888                  0          00  0  08  0 00  00 808  04          4 66886 4  4         84488246 2        8 0            8844 8   0  800  0   8    0 4
9070 DRW8             DW8.3         DQW7              446       6       2   20   088  8  0   00          0   0 00  0000  00020 2        0     0  668868  6   0         0000   0088488  8   8666 4884848 800   0000 0
9025 DR4              DW4           DQW7           0        0     888  8  0 0           0   0            0   0 80  0 00  000 0                                  0      0000                                     8    00   000 00   4886888
9030 DR4              DW13          DQW7                      0   88888228                               0   0 80  0 00  00                                  44  4      0                                       8   0 2 00          8886
9026 DR4              DW10          DQW8            0    4  00848 888888  8  0                           0   0 02  0 00  00   4                                   4      00                  8 2                   0   00 0         866
9028 DR4              DW14          DQW8                      64    8    6  64                           0   60 0 60 00  00                                               00                  8666                   0 2000 0       6806888
9029 DR4              DW4           DQW8                      448  82888  0   0         0    0           0   000 40  00  000 0         00         0      046   0000                                           00 8000800       8888884
9031 DR4              DW4           DQW8                     0868888  8  8                               0   0 02  0 00  00                      0                 0000       0                                   0 2 00 6        8888886
9032 DR4              DW4           DQW8           0   8     0 48888  8  0  0           0    0           0   0 0  00000000 0          0                           046    0000                                           00 4000 000       8 8886
9033 DR4              DW4           DQW8                   00868888   8  08 0           0    00          0   0 0  00 00 000 0   2      0                     64  04    0000               8    0                  00   000600      8868888
9034 DR4              DW4           DQW8                     0 48 884  8  0800          0    0           0   0 04  0 00  00 0   2      04                    8   04    0000                                           00   00   00      888888
9052 DR7              DW11          DQW9           8    6  88  88888  86 8  0                           0   0 02  0 00  00     0        0                             400                     8866      4   2   0    00
9075 DR9              DW23          DQW9                  0    8  888        00                          0   0 0  0 00  00               2                       4      0                                        0    00
9021 DRW18                    DW24  DQW4             8 2        2              2            0          0   0 04  0 00  00 8886   884     68886  0                          8     88          84     880 4 00
9068 DRW8             DW8.1         DQW4          4  68 6    2 6 4  626     0042            2          0   0 0   0 00  00 2       020     46688680  8      46    0          68844888868666 884 8888088800        6
9069 DRW8             DW8.1         DQW4              886   80   66    48       660 4   4              0   60 0240 00  00        0 0      5888684                   80         888 6888868664 4884 8088086  00
9071 DRW8             DW8.2         DQW4           8 6864     6      688  0   46        0     04       0   0 0240000000 0      0  0  0    0 6888684  64  0        0000        8 8 888888664884684 6 8800  00    00
9072 DRW8                           DQW4                 48     0          0  0        0    000      00  0 0000 00 000 0          00      0 66886       0   0  0000        08084688088866640   4 468600   000 00     0
```

Tenth International Histocompatibility Workshop
class II DQA, DQB RFLP fragments (n=299) with T cells (n=134)/bio (n=42)/serol (n=290). DQ listing.

```
DDDDDDDDDDDDDDDDDDDDDDDDDDDDDDDDDDDDDDDDDDDDDDUDDDDDDDDDDDDDDDDDDDDDDDDDDDDDDDDDDDDDDDDDDDDUUDDDDDDDDUDDDDDDDDDDDDDDDDDDDDDUUDDDDDDD8DDDDDDDDDDDDDDDDDDDDDDDDDDDUUDD
QQQQQQQQQQRRRRRRRRRQQQQRRQRRQQRRQQQQQRQQNNQRQQRQQQQRQRQQRQQQQQQQQQQQQQQRQRQPQQQQQRQQNNQQQRQQQQNQQRRRRRRRRQQQQRRRRPRRRRNNRRRRQQQQ RRRRRQQRRRQQQQQQQQQQQQQQNNRQ
BBBBAWBBBA444144WWWWWWWWWW43WW45WBBBWWWWKKWWWWWWWWWWW4BW7BBBBBBBBBBBBA7W4BWBBBBBWBBKKWWN2WBBBKBB5NN21N1NWBBA5W4WNNN5WKKW3N5WAAA 5W155BB555WBBBBBBBBBBBWBKKWB
     3              9513333113   33   3    3533   3533533363   3                          3  2        9      11  1                    1     8 9      8  5    2      1         3              3   6
                     32     12                 3     3   3                                                                                               2             1
```

```
HBEEM BPPB  DDDD  D      D     D     D  PEH                  DH   HBKTBTTTEBMSP    SDPESSH PB    D    DDHPS PBB    DD D    THP           D     KBK3  B   SE   TTBPHBTSSPKKTB    K
IACCS ISSI  RRRR  R      R     R     R  SCI                  RI   IAPAGAAACGSSV    SPSCSSI VG    R    RRIVS VAI    RR R    AIS           R     PGP   I   SC   AAIVIAASSVPPAG    P
DMVVP ITTI  54W5  N      W     7     W  TID                  WC   DMNQLQQQILPTU    TWTVTTC UL    1    44DUT UMI    32 2    QCT           6     NLN   I   TI   11IUDMQTTUNN1L    N
             1           1           8                       1                      4                                                                   00          0
             1           2                                   1
```

```
  21 1      1111111111133111166111113    1331991113113111111111 11         1     33113      1   9931333    9     33313333    3313333119911311    611131 11113           3 991
152421 63 0010110211111111100110114571011001211211111110104091522 813 201040263230360002100333044 001100003970000000100021011877001100960111 333122 51140016
62751938934595195117118019228750195276185581810918828018672096535226384141638448649558128936003115706555945634553560855046070013911393591023745082332115528
17025124627045259569245783028748138466147826059281377586986601559092419394459421003566728000731907966456710252228902934448417165741687160601215062430921255
```

9002	DR1		DW20		DQW5
9003	DR1		DW1		DQW5
9004	DR1		DW1		DQW5
9005.	DR1		DW1		DQW5
9006	DR1		DW1		DQW5
9009	DR2		DW21		DQW1
9054	DRW14		DW9	DW25	DQW5
9057	DRW14		DW9	DW25	DQW1
9061	DRW14		DW9	DW25	DQW5
9036	DRW11		DB2	DW25	DQW5
9056	DRW13	DRW14	DW9,19	DW25	DQW5
9008	DRW15		DW2		DQW6
9010	DRW15		DW2		DQW6
9013	DRW15		DW2		DQW6
9014	DRW15		DW2		DQW6
9017	DRW15		DW2		DQW1
9011	DRW15		DW12		DQW6
9055	DRW6			DW26	DQW6
9059	DRW13		DW19	DW26	DQW6
9063	DRW13		DW19	DW26	DQW6
9058	DRW13		DW18	DW24	DQW6
9060	DRW13		DW18	DW25	DQW5
9062	DRW13		DW18	DW24	DQW6
9065	DRW13		DW18	DW24	DQW6
9066	DRW8		DW8.3		DQW6
9018	DR3		DW3	DW25	DQW2
9019	DR3		DW3	DW25	DQW2
9020	DR3		DW3	DW25	DQW2
9022	DR3		DW3	DW24	DQW2
9023	DR3		DW3	DW24	DQW2
9047	DR7		DB1		DQW2
9048	DR7		DB1		DQW2
9050	DR7		DW17		DQW2
9051	DR7		DW17		DQW2
9016	DRW16		DW22		DQW7
9038	DRW12		DB6	DW25	DQW7
9064	DRW14		DW16	DW24	DQW7
9037	DRW11		DW5	DW25	DQW7
9040	DRW11		DW5	DW25	DQW7
9042	DRW11			DW25	DQW7
9043	DRW11		DW5	DW25	DQW7
9045	DRW11	DRW12	DW5	DW25	DQW7
9070	DRW8		DW8.3		DQW7
9025	DR4		DW4		DQW7
9030	DR4		DW13		DQW7
9026	DR4		DW10		DQW8
9028	DR4		DW14		DQW8
9029	DR4		DW4		DQW8
9031	DR4		DW4		DQW8
9032	DR4		DW4		DQW8
9033	DR4		DW4		DQW8
9034	DR4		DW4		DQW8
9052	DR7		DW11		DQW9
9075	DR9		DW23		DQW9
9021	DRW18			DW24	DQW4
9068	DRW8		DW8.1		DQW4
9069	DRW8		DW8.1		DQW4
9071	DRW8		DW8.2		DQW4
9072	DRW8				DQW4

Tenth International Histocompatibility Workshop
class II DQA, DQB RFLP fragments (n=299) with T cells (n=134)/bio (n=42)/serol (n=290). DQ listing.

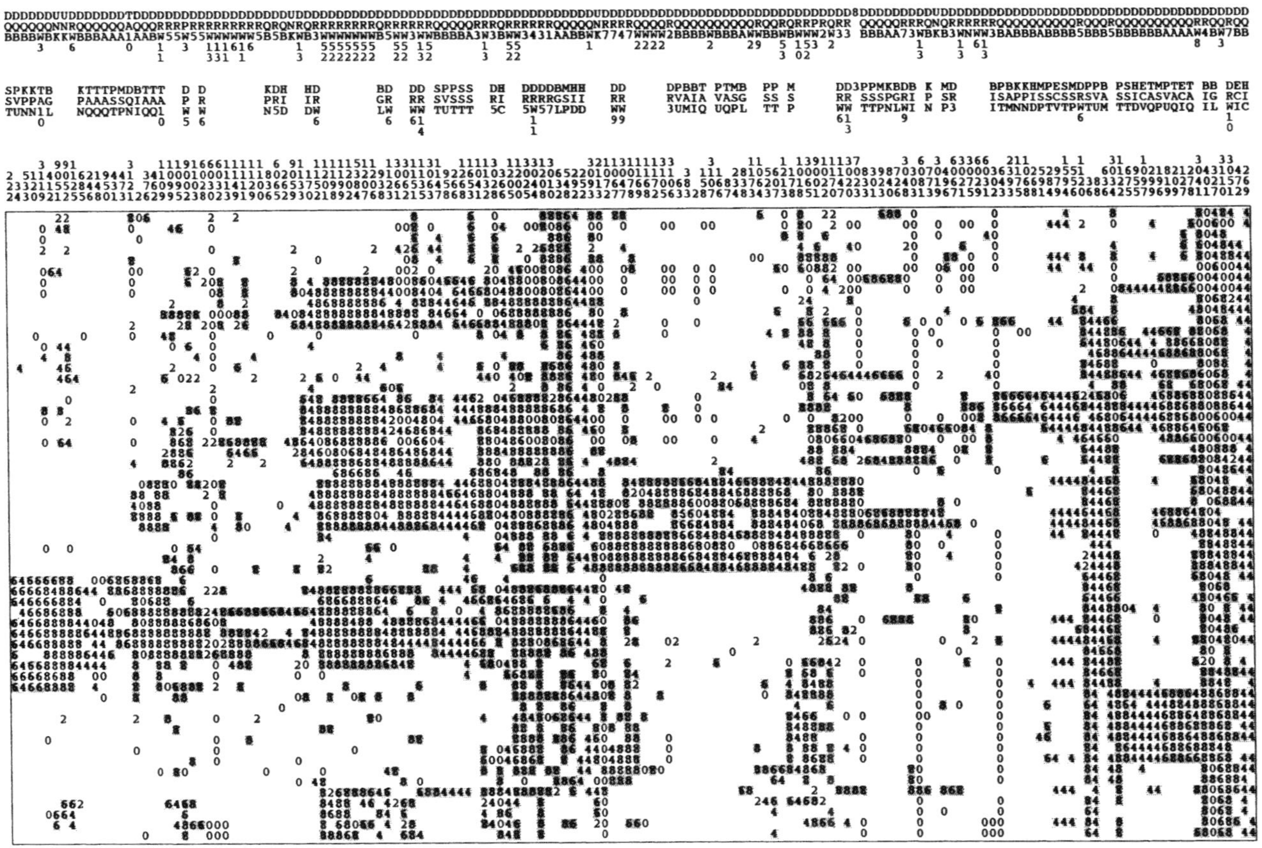

9002	DR1		DW20	DQW5
9003	DR1		DW1	DQW5
9004	DR1		DW1	DQW5
9005	DR1		DW1	DQW5
9006	DR1		DW1	DQW5
9009	DR2		DW21	DQW1
9054	DRW14		DW9 DW25	DQW5
9057	DRW14		DW9 DW25	DQW1
9061	DRW14		DW9 DW25	DQW5
9036	DRW11		DB2 DW25	DQW5
9056	DRW13	DRW14	DW9,19 DW25	DQW5
9008	DRW15		DW2	DQW6
9010	DRW15		DW2	DQW6
9013	DRW15		DW2	DQW6
9014	DRW15		DW2	DQW6
9017	DRW15		DW2	DQW1
9011	DRW15		DW12	DQW6
9055	DRW6		DW26	DQW6
9059	DRW13		DW19 DW26	DQW6
9063	DRW13		DW19 DW26	DQW6
9058	DRW13		DW18 DW24	DQW6
9060	DRW13		DW18 DW25	DQW5
9062	DRW13		DW18 DW24	DQW6
9065	DRW13		DW18 DW24	DQW6
9066	DRW8		DW8.3	DQW6
9018	DR3		DW3 DW25	DQW2
9019	DR3		DW3 DW25	DQW2
9020	DR3		DW3 DW25	DQW2
9022	DR3		DW3 DW24	DQW2
9023	DR3		DW3 DW24	DQW2
9047	DR7		DB1	DQW2
9048	DR7		DB1	DQW2
9050	DR7		DW17	DQW2
9051	DR7		DW17	DQW2
9016	DRW16		DW22	DQW7
9038	DRW12		DB6 DW25	DQW7
9064	DRW14		DW16 DW24	DQW7
9037	DRW11		DW5 DW25	DQW7
9040	DRW11		DW5 DW25	DQW7
9042	DRW11		DW25	DQW7
9043	DRW11		DW5 DW25	DQW7
9045	DRW11	DRW12	DW5 DW25	DQW7
9070	DRW8		DW8.3	DQW7
9025	DR4		DW4	DQW7
9030	DR4		DW13	DQW7
9026	DR4		DW10	DQW8
9028	DR4		DW14	DQW8
9029	DR4		DW4	DQW8
9031	DR4		DW4	DQW8
9032	DR4		DW4	DQW8
9033	DR4		DW4	DQW8
9034	DR4		DW4	DQW8
9052	DR7		DW11	DQW9
9075	DR9		DW23	DQW9
9021	DRW18		DW24	DQW4
9068	DRW8		DW8.1	DQW4
9069	DRW8		DW8.1	DQW4
9071	DRW8		DW8.2	DQW4
9072	DRW8			DQW4

Tenth International Histocompatibility Workshop
class II DQA, DQB RFLP fragments (n=299) with T cells (n=134)/bio (n=42)/serol (n=290). DQ listing.

DDDDDDDDDDDDDDDDDDDDDDDDDDDDDDDDDDDDDDDDDD DDDDDDDDDDDDDDDDDDDDDDDDDDDDDDDD DDDDDDDDDDDDDDDDDDDDDDDDDDDDDDDDDDDDDDDDDDDDDDDDDDDDDDDDDDDDDDDDDDDDDDDDDDDDDDUUD
RRRPPPPPPRQQPRRRRRRRRRRRRRRRRRRRRRRRPPPPP PRRRRRRRRRRRRRRRRRRRRRRRPPPRRRPR RRRPPPPPPRRRRRRRRPPRPPPPPPRPPPPRRPRPPRRRRPPRRRPPPQQQPRRRRPPRRRRRRRQRRQQQQRRQQRRQPQRQQNNQ
11W1BBBBB1BAB111111WW1W11113W1111BBBBB BWWW1111W133131133N11BBB222A4 3331BBBABW131W7BBWBBBBBB7BBBB44BWBB7B413BB44BBBWBAB4444BB444W4WWW43WWWWWWW45WBWWWWKW

Tenth International Histocompatibility Workshop
class II DPA, DPB RFLP fragments (n=226) with T cells (n=134)/bio (n=42)/serol (n=290). DP listing.

9058	DRW13		DW18	DW24	DPW1	[illegible]
9006	DR1		DW1		DPW1	[illegible]
9023	DR3		DW3	DW24	DPW1	[illegible]
9055	DRW6			DW26	DPW5	[illegible]
9040	DRW11		DW5	DW25	DPW3	[illegible]
9018	DR3		DW3	DW25	DPW3	[illegible]
9059	DRW13		DW19	DW26	DPW3	[illegible]
9022	DR3		DW3	DW24	DPW3	[illegible]
9063	DRW13		DW19	DW26	DPW2	[illegible]
9030	DR4		DW13		DPW3	[illegible]
9060	DRW13		DW18	DW25		[illegible]
9047	DR7		DB1			[illegible]
9034	DR4		DW4			[illegible]
9003	DR1		DW1			[illegible]
9043	DRW11		DW5	DW25		[illegible]
9048	DR7		DB1			[illegible]
9032	DR4		DW4		DPW2	[illegible]
9050	DR7		DW17		DPW2	[illegible]
9038	DRW12		DB6	DW25	DPW2	[illegible]
9029	DR4		DW4		DPW2	[illegible]
9068	DRW8		DW8.1		DPW2	[illegible]
9036	DRW11		DB2	DW25	DPW2	[illegible]
9019	DR3		DW3	DW25		[illegible]
9020	DR3		DW3	DW25	DPW2	[illegible]
9066	DRW8		DW8.3			[illegible]
9061	DRW14		DW9	DW25	DPW4	[illegible]
9075	DR9		DW23		DPW4	[illegible]
9026	DR4		DW10		DPW4	[illegible]
9033	DR4		DW4		DPW4	[illegible]
9014	DRW15		DW2		DPW4	[illegible]
9002	DR1		DW20		DPW4	[illegible]
9057	DRW14		DW9	DW25	DPW4	[illegible]
9065	DRW13		DW18	DW24	DPW4	[illegible]
9004	DR1		DW1		DPW4	[illegible]
9005	DR1		DW1			[illegible]
9051	DR7		DW17		DPW4	[illegible]
9025	DR4		DW4		DPW4	[illegible]
9031	DR4		DW4		DPW4	[illegible]
9028	DR4		DW14		DPW4	[illegible]
9069	DRW8		DW8.1		DPW4	[illegible]
9052	DR7		DW11		DPW4	[illegible]
9010	DRW15		DW2			[illegible]
9054	DRW14		DW9	DW25		[illegible]
9064	DRW14		DW16	DW24		[illegible]
9016	DRW16		DW22			[illegible]
9037	DRW11		DW5	DW25		[illegible]
9072	DRW8				DPW4	[illegible]
9042	DRW11			DW25		[illegible]
9013	DRW15		DW2			[illegible]
9008	DRW15		DW2		DPW2 DPW4	[illegible]
9017	DRW15		DW2		DPW2 DPW4	[illegible]
9011	DRW15		DW12		DPW2 DPW4	[illegible]
9045	DRW11	DRW12	DW5	DW25	DPW2 DPW4	[illegible]
9062	DRW13		DW18	DW24	DPW4	[illegible]
9056	DRW13	DRW14	DW9,19	DW25	DPW2	[illegible]
9021	DRW18			DW24	DPW1	[illegible]
9070	DRW8		DW8.3		DPW1 DPW4	[illegible]
9071	DRW8		DW8.2		DPW3	[illegible]
9009	DR2		DW21		DPW3 DPW4	[illegible]

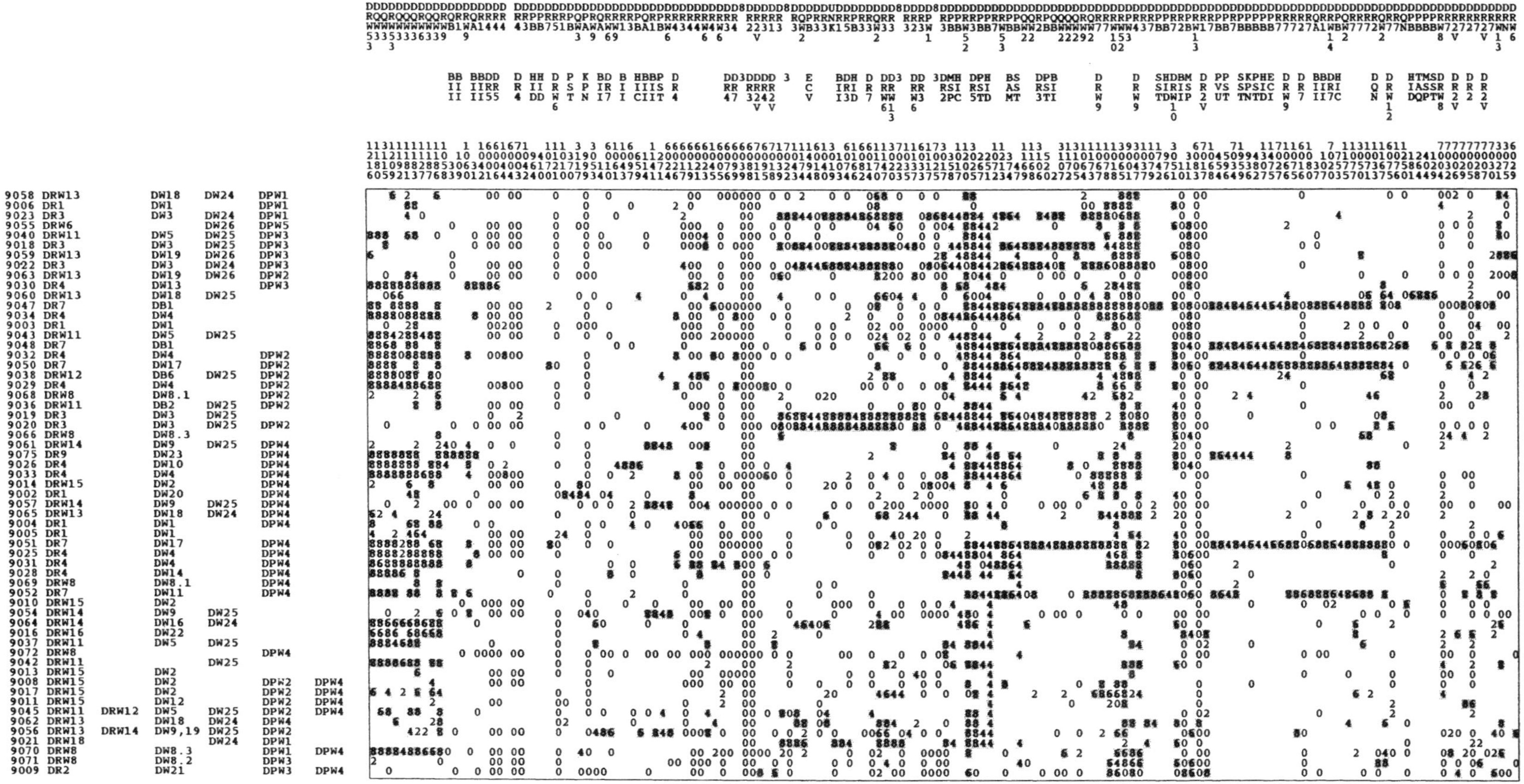

Tenth International Histocompatibility Workshop

class II DPA, DPB RFLP fragments (n=226) with T cells (n=134)/bio (n=42)/serol (n=290). DP listing.

Cell	DR	DR (2nd)	DW	DW (supertype)	DPW	DPW (2nd)	RFLP / T cell / bio / serol reactivity pattern
9058	DRW13		DW18	DW24	DPW1		[illegible]
9006	DR1		DW1		DPW1		[illegible]
9023	DR3		DW3	DW24	DPW1		[illegible]
9055	DRW6			DW26	DPW5		[illegible]
9040	DRW11		DW5	DW25	DPW3		[illegible]
9018	DR3		DW3	DW25	DPW3		[illegible]
9059	DRW13		DW19	DW26	DPW3		[illegible]
9022	DR3		DW3	DW24	DPW3		[illegible]
9063	DRW13		DW19	DW26	DPW2		[illegible]
9030	DR4		DW13		DPW3		[illegible]
9060	DRW13		DW18	DW25			[illegible]
9047	DR7		DB1				[illegible]
9034	DR4		DW4				[illegible]
9003	DR1		DW1				[illegible]
9043	DRW11		DW5	DW25			[illegible]
9048	DR7		DB1				[illegible]
9032	DR4		DW4		DPW2		[illegible]
9050	DR7		DW17		DPW2		[illegible]
9038	DRW12		DB6	DW25	DPW2		[illegible]
9029	DR4		DW4		DPW2		[illegible]
9068	DRW8		DW8.1		DPW2		[illegible]
9036	DRW11		DB2	DW25	DPW2		[illegible]
9019	DR3		DW3	DW25			[illegible]
9020	DR3		DW3	DW25	DPW2		[illegible]
9066	DRW8		DW8.3				[illegible]
9061	DRW14		DW9	DW25	DPW4		[illegible]
9075	DR9		DW23		DPW4		[illegible]
9026	DR4		DW10		DPW4		[illegible]
9033	DR4		DW4		DPW4		[illegible]
9014	DRW15		DW2		DPW4		[illegible]
9002	DR1		DW20		DPW4		[illegible]
9057	DRW14		DW9	DW25	DPW4		[illegible]
9065	DRW13		DW18	DW24	DPW4		[illegible]
9004	DR1		DW1		DPW4		[illegible]
9005	DR1		DW1				[illegible]
9051	DR7		DW17		DPW4		[illegible]
9025	DR4		DW4		DPW4		[illegible]
9031	DR4		DW4		DPW4		[illegible]
9028	DR4		DW14		DPW4		[illegible]
9069	DRW8		DW8.1		DPW4		[illegible]
9052	DR7		DW11		DPW4		[illegible]
9010	DRW15		DW2				[illegible]
9054	DRW14		DW9	DW25			[illegible]
9064	DRW14		DW16	DW24			[illegible]
9016	DRW16		DW22				[illegible]
9037	DRW11		DW5	DW25			[illegible]
9072	DRW8				DPW4		[illegible]
9042	DRW11			DW25			[illegible]
9013	DRW15		DW2				[illegible]
9008	DRW15		DW2		DPW2	DPW4	[illegible]
9017	DRW15		DW2		DPW2	DPW4	[illegible]
9011	DRW15		DW12		DPW2	DPW4	[illegible]
9045	DRW11	DRW12	DW5	DW25	DPW2	DPW4	[illegible]
9062	DRW13		DW18	DW24	DPW4		[illegible]
9056	DRW13	DRW14	DW9,19	DW25	DPW2		[illegible]
9021	DRW18			DW24	DPW1		[illegible]
9070	DRW8		DW8.3		DPW1	DPW4	[illegible]
9071	DRW8		DW8.2		DPW3		[illegible]
9009	DR2		DW21		DPW3	DPW4	[illegible]

Tenth International Histocompatibility Workshop

class II DPA, DPB RFLP fragments (n=226) with T cells (n=134)/bio (n=42)/serol (n=290). DP listing.

ID	DR	DR (2nd)	DW	DW (2nd)	DPW	DPW (2nd)
9058	DRW13		DW18	DW24	DPW1	
9006	DR1		DW1		DPW1	
9023	DR3		DW3	DW24	DPW1	
9055	DRW6			DW26	DPW5	
9040	DRW11		DW5	DW25	DPW3	
9018	DR3		DW3	DW25	DPW3	
9059	DRW13		DW19	DW26	DPW3	
9022	DR3		DW3	DW24	DPW3	
9063	DRW13		DW19	DW26	DPW2	
9030	DR4		DW13		DPW3	
9060	DRW13		DW18	DW25		
9047	DR7		DB1			
9034	DR4		DW4			
9003	DR1		DW1			
9043	DRW11		DW5	DW25		
9048	DR7		DB1			
9032	DR4		DW4		DPW2	
9050	DR7		DW17		DPW2	
9038	DRW12		DB6	DW25	DPW2	
9029	DR4		DW4		DPW2	
9068	DRW8		DW8.1		DPW2	
9036	DRW11		DB2	DW25	DPW2	
9019	DR3		DW3	DW25		
9020	DR3		DW3	DW25	DPW2	
9066	DRW8		DW8.3			
9061	DRW14		DW9	DW25	DPW4	
9075	DR9		DW23		DPW4	
9026	DR4		DW10		DPW4	
9033	DR4		DW4		DPW4	
9014	DRW15		DW2		DPW4	
9002	DR1		DW20		DPW4	
9057	DRW14		DW9	DW25	DPW4	
9065	DRW13		DW18	DW24	DPW4	
9004	DR1		DW1		DPW4	
9005	DR1		DW1			
9051	DR7		DW17		DPW4	
9025	DR4		DW4		DPW4	
9031	DR4		DW4		DPW4	
9028	DR4		DW14		DPW4	
9069	DRW8		DW8.1		DPW4	
9052	DR7		DW11		DPW4	
9010	DRW15		DW2			
9054	DRW14		DW9	DW25		
9064	DRW14		DW16	DW24		
9016	DRW16		DW22			
9037	DRW11		DW5	DW25		
9072	DRW8				DPW4	
9042	DRW11			DW25		
9013	DRW15		DW2			
9008	DRW15		DW2		DPW2	DPW4
9017	DRW15		DW2		DPW2	DPW4
9011	DRW15		DW12		DPW2	DPW4
9045	DRW11	DRW12	DW5	DW25	DPW2	DPW4
9062	DRW13		DW18	DW24	DPW4	
9056	DRW13	DRW14	DW9,19	DW25	DPW2	
9021	DRW18			DW24	DPW1	
9070	DRW8		DW8.3		DPW1	DPW4
9071	DRW8		DW8.2		DPW3	
9009	DR2		DW21		DPW3	DPW4

Tenth International Histocompatibility Workshop
class II DPA, DPB RFLP fragments (n=226) with T cells (n=134)/bio (n=42)/serol (n=290). DP listing.

9058	DRW13		DW18	DW24	DPW1		[illegible]
9006	DR1		DW1		DPW1		[illegible]
9023	DR3		DW3	DW24	DPW1		[illegible]
9055	DRW6			DW26	DPW5		[illegible]
9040	DRW11		DW5	DW25	DPW3		[illegible]
9018	DR3		DW3	DW25	DPW3		[illegible]
9059	DRW13		DW19	DW26	DPW3		[illegible]
9022	DR3		DW3	DW24	DPW3		[illegible]
9063	DRW13		DW19	DW26	DPW2		[illegible]
9030	DR4		DW13		DPW3		[illegible]
9060	DRW13		DW18	DW25			[illegible]
9047	DR7		DB1				[illegible]
9034	DR4		DW4				[illegible]
9003	DR1		DW1				[illegible]
9043	DRW11		DW5	DW25			[illegible]
9048	DR7		DB1				[illegible]
9032	DR4		DW4		DPW2		[illegible]
9050	DR7		DW17		DPW2		[illegible]
9038	DRW12		DB6	DW25	DPW2		[illegible]
9029	DR4		DW4		DPW2		[illegible]
9068	DRW8		DW8.1		DPW2		[illegible]
9036	DRW11		DB2	DW25	DPW2		[illegible]
9019	DR3		DW3	DW25			[illegible]
9020	DR3		DW3	DW25	DPW2		[illegible]
9066	DRW8		DW8.3				[illegible]
9061	DRW14		DW9	DW25	DPW4		[illegible]
9075	DR9		DW23		DPW4		[illegible]
9026	DR4		DW10		DPW4		[illegible]
9033	DR4		DW4		DPW4		[illegible]
9014	DRW15		DW2		DPW4		[illegible]
9002	DR1		DW20		DPW4		[illegible]
9057	DRW14		DW9	DW25	DPW4		[illegible]
9065	DRW13		DW18	DW24	DPW4		[illegible]
9004	DR1		DW1		DPW4		[illegible]
9005	DR1		DW1				[illegible]
9051	DR7		DW17		DPW4		[illegible]
9025	DR4		DW4		DPW4		[illegible]
9031	DR4		DW4		DPW4		[illegible]
9028	DR4		DW14		DPW4		[illegible]
9069	DRW8		DW8.1		DPW4		[illegible]
9052	DR7		DW11		DPW4		[illegible]
9010	DRW15		DW2				[illegible]
9054	DRW14		DW9	DW25			[illegible]
9064	DRW14		DW16	DW24			[illegible]
9016	DRW16		DW22				[illegible]
9037	DRW11		DW5	DW25			[illegible]
9072	DRW8				DPW4		[illegible]
9042	DRW11			DW25			[illegible]
9013	DRW15		DW2				[illegible]
9008	DRW15		DW2		DPW2	DPW4	[illegible]
9017	DRW15		DW2		DPW2	DPW4	[illegible]
9011	DRW15		DW12		DPW2	DPW4	[illegible]
9045	DRW11	DRW12	DW5	DW25	DPW2	DPW4	[illegible]
9062	DRW13		DW18	DW24	DPW4		[illegible]
9056	DRW13	DRW14	DW9,19	DW25	DPW2		[illegible]
9021	DRW18			DW24	DPW1		[illegible]
9070	DRW8		DW8.3		DPW1	DPW4	[illegible]
9071	DRW8		DW8.2		DPW3		[illegible]
9009	DR2		DW21		DPW3	DPW4	[illegible]

Tenth International Histocompatibility Workshop
class II DPA, DPB RFLP fragments (n=226) with T cells (n=134)/bio (n=42)/serol (n=290). DP listing.

References

1. Yang SY, Milford EL, Hammerling U, Dupont B. Description of the reference panel of B-lymphoblastoid cell lines for factors of the HLA system. In: Dupont B (ed): Immunobiology of HLA, Volume I: Histocompatibility Testing 1987. New York: Springer-Verlag, 1989.
2. Milford EL, Kennedy L, Yang SY, Dupont B, Yunis EJ. Serological characterization of the reference panel of B-lymphoblastoid cell lines for factors of the HLA system. In: Dupont B (ed): Immunobiology of HLA, Volume I: Histocompatibility Testing 1987. New York: Springer-Verlag, 1989.
3. Mickelson E, Reinsmoen N, Robbins FM, et al. HLA-Dw and HLA-DP typing of the reference panel. In: Dupont B (ed): Immunobiology of HLA, Volume I: Histocompatibility Testing 1987. New York: Springer-Verlag, 1989.
4. Yang SY. Population analysis of class I HLA antigens by one-dimensional isoelectric focusing gel electrophoresis. In: Dupont B (ed): Immunobiology of HLA, Volume I: Histocompatibility Testing 1987. New York: Springer-Verlag, 1989.
5. Knowles RW. Structural polymorphism of the HLA class II alpha and beta chains: summary of the 10th Workshop 2-D gel analysis. In: Dupont B (ed): Immunobiology of HLA, Volume I: Histocompatibility Testing 1987. New York: Springer-Verlag, 1989.
6. Simons MJ, Wheeler R, Cohen D, Lalouel J-M, Dupont B. Restriction fragment length polymorphism of HLA genes: summary of the Tenth international workshop Southern blot analysis. In: Dupont B (ed): Immunobiology of HLA, Volume I: Histocompatibility Testing 1987. New York: Springer-Verlag, 1989.
7. Flomenberg N. Functional polymorphisms of HLA class II gene products detected by T-lymphocyte clones: summary of the Tenth International Histocompatibility Workshop cellular studies. In: Dupont B (ed): Immunobiology of HLA, Volume I: Histocompatibility Testing 1987. New York: Springer-Verlag, 1989.
8. Lalouel J-M, Ferguson M, Wheeler R, King N. Overview of data processing and analysis. In: Dupont B (ed): Immunobiology of HLA, Volume I: Histocompatibility Testing 1987. New York: Springer-Verlag, 1989.
9. King N, Wheeler R, Ferguson M, Lalouel J-M. Display of correlation structures. In: Dupont B (ed): Immunobiology of HLA, Volume I: Histocompatibility Testing 1987. New York: Springer-Verlag, 1989.
10. Ferguson M, Milford EL, Wheeler R, Dupont B, Lalouel J-M. Data processing and serum analysis. In: Dupont B (ed): Immunobiology of HLA, Volume I: Histocompatibility Testing 1987. New York: Springer-Verlag, 1989.
11. Wheeler R, Milford EL, Lalouel J-M. Pattern analysis. In: Dupont B (ed): Immunobiology of HLA, Volume I: Histocompatibility Testing 1987. New York: Springer-Verlag, 1989.
12. Wheeler R, Milford EL, Kennedy L, Lalouel J-M. Cell typing. In: Dupont B (ed): Immunobiology of HLA, Volume I: Histocompatibility Testing 1987. New York: Springer-Verlag, 1989.
13. Yang SY. Assignment of HLA-A and HLA-B antigens for the reference panel of B-lymphoblastoid cell lines determined by 1D-IEF. In: Dupont B (ed): Immunobiology of HLA, Volume I: Histocompatibility Testing 1987. New York: Springer-Verlag, 1989.

Workshop Participants

Acton, R.T., US5UAB
Adluri, V., US7KNO
Aizawa, M., JAPAIZ
Akaza, T., JAPJUJ
Albert, E.D., GERALB
Alekseyev, L., EAESER
Alosco, S., NCYCBR
Alper, C.A., NCYCBR
Altshuler, R., GERGAZ
Alvarez, C., NCYCAR
Amemiya, H., JAPAMY
Amesland, F., FRADDC
Amoroso, A., IT1CEP
Amos, D.B., US1AMO
Andersen, R., US2WOL
Ando, A., JAPTSU
Andreas-Zietz, A., GERALB
Andrien, M., BENDPT
Angelini, G., IT2FER
Aparico, J., JAPAIZ
Arce-Gomez, B., LATBAG
Arnaiz-Villena, A., FRAARN
Atoh, M., JAPSEK
Attatippaholkun, W., ANZDCH
Avakian, H., UKIJAM
Awad, J., UKIFES
Bach, F.H., US2BAC
Badenhoop, K., UKISTB
Balakrishnan, K., US5BAL
Balboni, S., IT2FER
Ball, E.J., US4STA
Ballas, M., GERGOL
Barbalho, T.P., LATTBA
Barbanti, M., IT1MRA
Barbetti, A., US5TER
Barboni, F., IT2BAR
Barger, B.O., US5UAB
Bashir, S., ANZBSH
Batchelor, J.R., UKIBAT
Battat, S., GERBRA
Baur, F.H., GERBAU
Beach, M., NCYCAR
Beatty, P.G., US7HAN
Belcher, R., US6CBC
Bellane, C., FRACOH
Belvedere, M., IT1MRA
Benson, J.K., NCYRAG
Berg-Loonen, E.van den, BENBER
Betuel, H., FRABET
Bidwell, J.L., UKIBID
Biegel, A.A., US1IND
Bignon, J.-D., FRABIG
Biro, A., UKIFES
Bodmer, J.B., UKIBOD
Bodmer, W., UKIBOD
Boehm, B.O., GERBOE
Bontrop, R., BENROO
Bontrop-Elferink, D., BENROO
Boral, L.I., US5KWC
Bouhallier, O., FRAFAU
Bow, L., US5SWE
Bradley, B., UKIBRS
Brahmi, Z., US5IND
Bratlie, A.S., SCATSB
Braun, W.E., US6BRN
Brautbar, C., GERBRA
Breur-Vriesendorp, B.S., BENIVA
Bruyere, M. de, BENBRU
Buc, M., EAEBUC
Bunce, M., UKIMOR
Bushell, A.R., UKIMOR
Call, E., SAFDUT
Callaway, C., US7POL
Calot, M., FRAOHA
Cambon-Thomsen, A., FRAOHA
Cappelacci, S., IT2GAN
Capper, J., US3GVY
Caraballo, L., LATCOL
Carcassi, C., IT1CON
Carpenter, C.B., NCYCAR
Carpenter, C., US8RUB
Carter, C., UKIMOR
Caruso, C., IT1MRA
Cascino, I., IT1CNG
Casilio-Lonardo, E., US6BRN
Cassell, P., UKIFES
Castagnoli, C., IT1CEP
Ceppellini, R., IT1CEP
Chan, S.H., ANZCSH
Chandanayingyong, D., ANZDCH
Charron, D.J. FRACHA
Chen, S.N., US5FTW
Chertkoff, L.P., LATFAI
Cheung, K.W.-K., US5KWC
Chiewsilp. P., ANZCHI
Choi, E., US5BAL
Choo, Y., US7HAN
Christiansen, F.T., ANZDAW
Clarholm, B., SCALOW
Coates, D., UKIAST
Coffing, M.J.NMN, US5KWC
Cohen, D., FRACOH
Cohen, N,. US3GRU
Colombe, B., US3GVY
Conighi, C., IT1CNG
Contu, L., IT1CON
Coppin, H., FRAOHA
Crafford, S., SAFDUT
Crepaldi, T., IT1CEP
Crichton. B., UKIGLA
Cross, D.E., US5CRO
Cross, R., ANZCRS
Cuccia, M., IT1MRA
Cullen, C. UKIMID
Curtoni, S., IT1CEP
Cutbush, S., UKIFES
D'Amaro, J., BENROO
Darke, C., UKIDRK
Davey, N.J., UKIBAT
Dawkins, R.L., ANZDAW
Decary, F., US8DEC
Degos, L., FRADDC
Dejelo, C., US6BRN
Dejour, G., FRAFAU
De Lange, P., BENROO
Demopulos, J., US5UAB
DeTotero, D., IT1FER
DeVries, R.P., BENROO
De Waal, L.P., BENENG
DiLonardo, M., LATDIL
Dohi, K., JAPDOH
Dombrausky, L., US4STA
Doran, T., ANZBSH
Dos Santos, A., UKIFES
Doxiadis, G., GERGSW
Doxiadis, I., GERGSW
Drummond, V., UKISTB
Duffy, B., NCYSTL
Duffy, C.C., US2PEL
Dunston, G., US8DUN
Dupont, B., US7DUP
Dupont, E., BENDPT
Duquesnoy, R.J., US1DUQ
Du Toit, E., SAFDUT
Dwyer, E., US8RJW
Dyer, P.A., UKIDYR
Eckels, D., US2MBC
Eiermann, T., GERGOL
Ejlersen, E., SCAKIS
Elbaek, A., SCAJER
Ellis, S., UKIMCM
Elsen, van der. P., BENROO
Englefriet, C.P., BENENG
Enomoto, Z., US5PAK
Erlich, H.A., US3CTS
Fainboim, L., LATFAI
Falk, J.A., NCYFAL
Fan, L., US7HAN
Farrell, C., ANZBSH
Fauchet, R., FRAFAU
Ferguson, M., US7LAL
Fernandez, N., UKIFES

Ferrara, G.B., IT2FER
Festenstein, H., UKIFES
Flomenberg, N., US7NFL
Flores, J., LATGOR
Fong, R., ANZFON
Fong, Y.M., US7DWS
Fotino, M., US8FOT
Francisco, E., US3GVY
Frazer, I., ANZFRA
Frenz, G., GERGSW
Fuggle, S., UKIMOR
Fukunishi, T., JAPTKB
Fuller, A., NCYFUL
Fuller, T., NCYFUL
Gagne, J., NCYCAR
Gandini, E., IT2GAN
Gao, X., US4STA
Garcia, L.F., LATGAR
Garovoy, M.R., US3GVY
Garrido, F., FRAGRA
Gazit, E., GERGAZ
Gebel, H., US2RSH
Gebuhrer, L., FRABET
Geczy, A., ANZBSH
Gelsthorpe, K. UKIGEL
Gerbase-De-Lima, M., LATGER
Gill, D., US5MIT
Gin, W., ANZDAW
Giphart, M.J., BENROO
Giraudo, C., LATGIR
Goeken, N., NCYGOE
Goldmann, S., GERGOL
Good, R.A., NCYRAG
Gorodezky, C., LATGOR
Grappa, M.T., IT2FER
Graugaard, B., SCAKIS
Gregersen, P.K., US8RJW
Grosse-Wilde, H., GERGSW
Groves, V., UKIBOD
Grumet, F.C., US3GRU
Guttridge, M.G., UKIBRS
Gyodi, E., EAEPTR
Haas, E., LATHAA
Hajek-Rosenmayr, A., EAEMYR
Hall, P. UKIKNT
Hämmerling, U., US7UHA
Hammond, M.G., SAFHAM
Hansen, J.A., US7HAN
Hartzman, R., US1HJH
Hasekura, H., JAPHAS
Hawkins, B.R. ANZHAW
Heard, R. UKIBAT
Heise, E.R., US1HEI
Henry, M.-A., FRAPRR
Herbut, B., UKIFES
Hermans, P., FRACHA
Herrero, M., LATFAI
Hirata, R., JAPJUJ
Hogan, V., US8DEC
Holland, P., NCYSAC
Holtkamp, B. GERBIO
Honeyman, M., ANZBSH
Hoover, M., US1HOO
Hors, J., FRADDC
Horst, A. van der, BENENG
Horton, B., ANZDAW
Hsia, S., US4SND
Hunter, J., US2MBC
Hurley, C., US1HJH
Hyldig-Nielsen, J.J., SCASVE
Ikeda, H., JAPAIZ
Inoko, H., JAPTSU
Irle, C., FRAJEA
Ivanyi, P., BENIVA
Ivaskova, E., EAEIVS
Jakobsen, B., SCASVE
Jalais, E., FRAFAU
James, D.C.O. UKIJAM
Jeannet, M., FRAJEA
Jenkin, J., US7HAN
Jersild, C., SCAJER
Jingjuan, X., CHICHE
Johnson, A.H., US1HJH
Johnson, U., SCALOW
Jones, E.H., UKILAW
Jones, M., UKILAW
Joysey, V.C., UKIJOY
Juji, T. JAPJUJ
Kakaiya, R., US2WOL
Kalil, J., LATKAL
Kamoun, M., US1ZMI
Kanatani, M., US5TER
Kaneko, T., JAPKSH
Kao, H.T., US8UBI
Kao, Y.S., US1NWO
Kaplan, C., FRAMUL
Kashiwagi, N., JAPKSH
Kastelan, A., EAEKAS
Kato, K., US7DUP
Kay, P.H., ANZDAW
Kazumi, O., JAPSEK
Kazutuka, M., JAPTSU
Keller, E., GERALB
Kennedy, L.J. UKIBOD
Keyserlingk, H. von, GERKEY
Khan, R., US7DUP
Kienzle, B., US7KNO
Kilaru, S., US7KNO
King, L.S., US5KWC
King, N., US7LAL
Kirnbauer, M., EAEMYR
Klein, J.S., US7NFL
Klouda, P.T., UKIBRS
Knight, S., UKIKNT
Knitter-Jack, N., US7VMC
Knowles, R.W., US7KNO
Ko, W.J., US7DWS
Kobayashi, K., JAPSEK
Konoeda, Y., JAPSEK
Koskimies, S., SCAKOS
Kostyu, D.D., US1AMO
Kreisler, J.M., FRAKRE
Kristof, S., NCYSTL
Krisunas, L., US5SWE
Krom, F.E.J.M., BENIVA
Kulova, L., US7SYY
Kurnick, J., NCYKUR
Kwan, K.H., US7DWS
Kwan, L., US5MLS
Kwong, P.C., NCYFAL
Labonne, M.P., FRABET
Lalouel, J.-M., US7LAL
Lamm, L.U., SCAKIS
Lanchbury, J.S.S., UKIWEL
Lane, J.T.L., US5ABS
Langley, J.F., ANZCRS
Lardy, N.M., BENIVA
Larsen, B., NCYMAR
Latinne, D., BENBRU
Laundy, G.J., UKIBRS
Layrisse, Z., LATLAY
Lazda, V., US1JON
Le Gall, I., FRACOH
Lecchi, M., IT1CEP
Lechler, R., UKIBAT
Lee, J.S., US7JSL
Lee, T.D.(J.), US8JLE
Lems, S.P.M., BENREE
Lenhard, V., GERLEN
Leone, A.L., IT1CON
Lepage, V., FRADDC
Lepetit, J.-C., FRALEP
Leverenz, G., EAERIC
Leverenz, S., EAERIC
Lima, M.G., LATGER
Lin, J., US5LIN
Lin, N.H., US7DWS
Lindsay, J., UKIBOD
Little, A.-M., UKIDYR
Long, E., US1AID
Longo, A., IT2FER
Lopez-Larrea, C., FRAARN
Lovisone, E., IT1CEP
Low, B., SCALOW
Lui, E., NCYFAL
Lulli, P., IT2GAN
Lundin, K., SCATSB
Mackintosh, P., UKIMAC
Madrigal, J.A., UKIBOD
Maeda, H., JAPJUJ
Maeda, M., JAPTSU
Magli, C., LATLAY
Mahoney, R., NCYMAH
Marcadet, A., FRACOH
Markussen, G., SCATSB
Marrari, M., US1DUQ
Marsh, S.G.E., UKIBOD
Marshall, W.H., NCYMAR
Martell, M., FRACOH
Martell, R.W., SAFDUT
Martin, A., UKIBRS
Martin, E., ANZDAW
Martinetti, M., IT1MRA
Marwick, J.R., UKICXH
Masaewicz, S., US7HAN
Masaru, K., JAPSEK
Masniuk, J., US5TIL
Matsuki, K., JAPJUJ
Matthews, E., US8RJW

Mattiuz, P.L., IT1MTT
Mayer, S., FRAMYE
Mayr, W.R., EAEMYR
Mazzilli, M.C., IT2GAN
McAuley, J.E., US5KWC
McCloskey, D.J., UKIFES
McDonald, W.I., UKIBAT
Mehra, N.K., ANZMEH
Menicucci, A., IT1MTT
Mervart, H., NCYMRV
Mickelson, E., US7HAN
Middleton, D., UKIMID
Milford, E.L., NCYCAR
Minev, M., EAEMIN
Mitsuishi, Y., FRAMYE
Mittal, K.K., US5MIT
Miyako, A., JAPSEK
Mizuno, S., US7DUP
Monos, D., US1ZMI
Moraes, J.R., LATJRM
Moraes, M.E., LATJRM
Morel, M.-C., FRAMUL
Morellini, M., IT2GAN
Moreno, M.E., FRAKRE
Morishima, Y., JAPMRI
Morling, N., SCASVE
Morris, P., UKIMOR
Moses, J.H., UKIBAT
Mota, A.H., JAPNAI
Mueller-Eckhardt, G., GERMUE
Müller, C., GERMUC
Nagao, N., JAPAMY
Naito, S., JAPNAI
Nakata, S., JAPMAT
Nakatsuji, T., JAPTSU
Navarette, C., UKIFES
Nelson, K., US7GSC
Nepom, B., US7GSC
Nepom, G.T., US7VMC
Neumeyer, H., GERNEU
Nikaein, A., NCYNIK
Nisbet-Brown, E., NCYFAL
Nisperos, B., US7HAN
Noreen, H., US2BAC
Nuñez-Roldan, A., FRANRD
Obata, F. JAPKSH
O'Connell, P., US7LAL
Ødum, N., SCASVE
Ogahara, S., JAPNAI
Ohayon, E., FRAOHA
Ohkubo, M., JAPSEK
Oishi, H., JAPTSU
Oldfather, J.W., US1KYL
Olivetti, E., IT1CEP
Olivo, A., US8RUB
Opelz, G., GERALB
Ota, M., JAPHAS
Otani, F., JAPKSH
Otting, N., BENROO
Oudshoorn, M., SAFDUT
Paik, Y.K., US5PAK
Palavecino, E.A.L., LATFAI
Pan, N., US7DWS
Park, M.S., US5TER
Parodi, B., IT2FER
Partanen, J., SCAKOS
Pask, S.L., US3GRU
Patel, H., UKIJAM
Paulsen, G., SCATSB
Payne, R., US3PAY
Perchotte, M., US5MLS
Perkins, H.A., US3PER
Perrier, P., FRAPRR
Peruccio, D., IT1CEP
Petranyi, G.G., EAEPTR
Phelan, D., NCYSTL
Piccioli, P., IT2FER
Piscone, A., FRACOH
Plätke, R., US7LAL
Pollack, M.S., US7POL
Prigoshin-Godelman, N., LATPRI
Purpura, M., IT1PUR
Quillivic, F., SAFHAM
Rabiasz, G.J., UKIBOD
Radka, S.F., US7GSC
Radvany, R., US4RAD
Raffoux, G., FRACOH
Ratner, L.J., NCYCAR
Reekers, P., BENREE
Regueiro, J.R., FRAARN
Reid, B., US5SIN
Reinsmoen, N., US2BAC
Richiardi, P., IT1CEP
Riley, M. US1CHR
Rinke de Wit, T.F., BENROO
Robbins, F.M., US1HJH
Rodey, G., NCYROD
Rodriguez de Cordoba, S., US8RUB
Rubinstein, P., US8RUB
Ruggiero, G., IT2ZAP
Rush, D,. US5TIL
Rust, N.A., UKIMOR
Sacks, S., UKIMOR
Sada, M., JAPAMY
Saji, H., JAPJUJ
Sakahara, N., US3PER
Salerno, A., IT1SAL
Salvaneschi, L., IT1MRA
Samaniego, M.C., LATSAM
Sammons, D.W., US7DWS
Sanchez, B., FRANRD
Sanchez, F., FRANRD
Sano, N., JAPTSU
Saracco, P., IT1CEP
Sasazuki, T., JAPSAS
Sasportes, M., FRADDC
Satz, L., LATFAI
Saueracher, G., ANZDAW
Saunders, P., US4SND
Savage, D., UKIMID
Savi, M., IT1SAV
Schalke, B., GERWUE
Schluender, L., US2BAC
Scholz, S.G., GERALB
Schreuder, G.M.Th., BENROO
Schroeder, M.L., US5MLS
Schunter, F., GERSCN
Schwarz, G., UKISTB
Schweizer, R., US5SWE
Segall, M., US2BAC
Sekiguchi, S., JAPSEK
Sell, T., US2MBC
Semana, G., FRAFAU
Semeijn, J., BENIVA
Serova, L., EAESER
Servenius, B., US3SCR
Shabalin, V.N., EAESHA
Sharrock, C., UKIBAT
Sierp, G.M., GERALB
Siminovitch, K., NCYFAL
Simmonney, N., FRAMUL
Simons, M.J., US5SIM
Singal, D.P., US5SIN
Sirotina, A., US7NFL
Small, T., US7NFL
Smeysters, L., BENREE
Smith, A., US7HAN
Somani, A.K., NCYFAL
Sommer, E., FRAOHA
Stastny, P., US4STA
Sterkers, G., FRALEV
Stewart, D., US1STE
Straachan, N., ANZCRS
Stratton, A., UKIAST
Strong, D.M., US7GSC
Strothman, R., US5FTW
Suarez, A., LATSUA
Sudmeyer, G., GERSUD
Sujirachato, K., ANZCHI
Sullivan, K., US1TUL
Sun, Y.P., CHIBCH
Svejgaard, A., SCASVE
Tait, B., ANZTAI
Tait, S.M., UKIJOY
Tajima, Y., JAPAIZ
Takada, A., JAPAIZ
Takata, H., JAPTAK
Takenouchi, T., US5TER
Taljaard, D.G., SAFDUT
Tanaka, A., JAPAIZ
Tanigaki, N., NCYRSP
Tate, D.G., UKIDYR
Taylor, C.J., UKIMOR
Teran, L., LATTRN
Terasaki, P., US5TER
Termijtelen, A., BENROO
Thompson, J.S., US1THP
Thomsen, M., FRAOHA
Thorsby, E., SCATSB
Tiercy, J.-M., FRAJEA
Tijssen, H., BENREE
Tingl A., UKIMOR
Tkaczuk, J., FRAOHA
Todd, J.A., US3GRU
Tokashige, N., US5PAK
.Tokunaga, K., JAPTOK
Tonai, R., US5TER
Tongio, M.-M., FRAMYE
Tonks, S., UKIBOD

Laboratory Addresses

Region ANZ

ANZBSH
Tissue Typing Laboratory
NSW Red Cross Blood Transfusion Service
153 Clarence Street
Sydney, NSW 2000, Australia

ANZCHI
Department of Pathology
Fac. Medicine/Ramathibodi Hospital
Rama Vi Road
Bangkok, 10400, Thailand

ANZCRS
Tissue Typing Laboratory
Red Cross Blood Bank
Kavanagh & Balston Streets
So. Melbourne, Vic 3205, Australia

ANZCSH
WHO Immunology Research & Training Center
National University of Singapore
Lower Kent Ridge Road
Singapore, 0511

ANZDAW
Dept. of Clinical Immunology
Royal Perth Hospital
GPO Box X2213
Perth, WA 6001, Australia

ANZDCH
Blood Bank
Siriraj Hospital and Medical School
2 Prannok Road
Bangkok, 10700, Thailand

ANZFON
Immunohaematology Dept/Lab Service
Wellington Public Hospital
Wellington, New Zealand

ANZFRA
Clinical Immunology
Princess Alexandra Hospital
Ipswich Road
Woolloongabba, Qnl., 4102, Australia

ANZHAW
Department of Pathology
Queen Mary Hospital
Hong Kong

ANZMEH
Cellular Immunol. Lab/Dept Anatomy
All-India Institute of Medical Sciences
Ansari Nagar
New Delhi, 110029, India

ANZTAI
Tissue Typing Labs, Rm N923
Royal Melbourne Hospital
Grattan Street
Parkville, Vic 3052, Australia

ANZWOO
Tissue Typing Laboratory
NZ Blood Transfusion Service
Park Avenue
Auckland, 1, New Zealand

Region BEN

BENBER
Tissue Typing Laboratory
Academic Hospital, Maastricht
P.O. Box 1918
Maastricht, 6201 CX, The Netherlands

BENBRU
Tissue Typing Laboratory
Universite de Louvain
Clos Chapelle aux Champs 30, BP 3052
Brussels, B-1200, Belgium

BENDPT
Department of Immunology
Hopital Erasme
Route de Lennik 808
Brussels, B-1070, Belgium

BENENG
Central Laboratory
Netherlands Red Cross Blood Trans. Svc.
Plesmanlaan 125
Amsterdam, 1066 CX, The Netherlands

BENIVA
Central Laboratory
Netherlands Red Cross Blood Trans. Svc.
Plesmanlaan 125
Amsterdam, 1066 CX, The Netherlands

BENREE
Lab Transplantation Serology
Radboud Hospital
Geert Grooteplein Z 12
Nijmegen, 6500 BH, The Netherlands

BENROO
Dept. Hematol. & Blood Bank
University Hospital, Bldg 1, E3-Q
Rijnsburgerweg 10
Leiden, 2333 AA, The Netherlands

BENVER
Tissue Typing Laboratory
Bloedtransfusiecentrum
O.L. Vrouwstraat 42
Leuven, 3000, Belgium

Region CHI

CHIBCH
HLA Laboratory
Beijing Children's Hospital
Nan Li Shi Road
Beijing, Peoples Rep. of China

CHICHE
Immunogenetics Laboratory
Shanghai Institute of Immunology
280 Chongqing Road
Shanghai, 200025, Peoples Rep. of China

CHIYEG
Immunology Laboratory
Institute of Basic Medical Sciences
P.O. Box 130
Beijing, Peoples Rep. of China

CHIZHA
HLA Typing Laboratory
Shanghai Blood Center
Nanking Rd. (w) 479-10
Shanghai, 200003, Peoples Rep. of China

Region EAE

EAEBUC
Transplantation Immunology Laboratory
Medical Faculty, Comenius University
Sasinkova 4
Bratislava, Czechoslovakia

EAEIVS
Dept. of Immunogenetics
Inst. for Clin. and Exp. Medicine
Vedenska 800
Prague 4, 14000, Czechoslovakia

EAEKAS
Tissue Typing Center
Fac of Med/Univ. of Zagreb
Kispaticeva 12 Rebro
Zagreb, 41000, Yugoslavia

EAEMIN
Lab for Transplant Immunity
Medical Inst. Pirogov
21 Boul. Totleben
Sofia, 1606, Bulgaria

EAEMYR
Abteilung Transfusionsmedizin
RWTH Aachen
Pauwelsstrasse
Aachen, D-5100, West Germany

EAEMYR
(Hajek-Rosenmayr) HLA Laboratory
Inst. for Blood Group Serology
Spitalgasse 4
Vienna, A-1090, Austria

EAEPTR
HLA Typing Laboratory
Nat. Inst. of Hematol. & Blood Trans.
Daroczi Str. 24
Budapest, H1113, Hungary

EAERIC
Blood Transf.
Bezirksinstitut
Atzpodinstr. 9–11
Berlin, DDR 113, East Germany

EAEWEG
HLA Laboratory
Bezirks Inst. fur Blutspendewesen
Robert Koch Str. 10
Rostock, DDR 2500, East Germany

Region FRA

FRAARN
Immunologia
Hospital Primero de Octubre
Carretera Andalucia
Madrid, 28041, Spain

FRABET
Histocompatibilite Laboratoire
Blood Transfusion Centre
1–3, Rue du Vercors
Lyon, 69007, France

FRABIG
Laboratoire HLA
Centre Transfusion Sanguine
Allee Ch. Mirallie,BP 349
Nantes, 44011 CX, France

FRACHA
Lab Immunogenet. Molec.
Institute de Cordeliers
15 Rue de l'Ecole de Medecine
Paris, 75006, France

FRACOH
C.E.P.H.
College du France
3 rue d'Ulm
Paris, 75005, France

FRADDC
Lab d'Immunol. et Transp.
Hopital St. Louis, U93 INSERM
1, rue Claude Vellefaux
Paris, 75475, France

FRAFAU
Histocompatibilite
C.T.S.
Rue Pierre Jean Gineste
Rennes, 35000, France

FRAGRA
Analisis Clinicos Servicio
C.S. "Virgen de las Nieves"
Avda. de Calvo Solelo s/n
Granada, 18014, Spain

FRAJEA
Unit Transplantation Immunology
Hopital Cantonal Universitaire
24, rue Micheli-du-Crest
Geneva 4, 1211, Switzerland

FRAKRE
Department of Immunology
Clinica Puerta de Hierro
4, San Martin de Porres
Madrid, 20035, Spain

FRALEP
Lab Immunol. du C.H.R.
C.T.S.
25, Boul. Pasteur
St. Etienne, 42022, France

FRALEV
Lab d'Immunol. & Virol. des Tumeurs
INSERM U152
27, Rue du Faubourg St. Jacques
Paris, 75014, France

FRAMUL
Serv d'Immunologie Leucoplaquettare
CNTS Institut
6 rue Alexandre Cabanel
Paris, 75015, France

FRAMYE
Lab Histocompatibilite
Centre de Transfusion Sanguine
10, rue Spielmann
Strasbourg, 67085, France

FRANRD
Lab Immunologia
Hospital Univ. Virgen del Rocio
Avda Manuel Siurot s/n
Seville, 41013, Spain

FRAOHA
Immunol. & Immunogenet Humaine
U100 INSERM
Chu Purpan
Toulouse, 31052 CX, France

FRAPRR
Laboratoire d'Histocompatibilite
Centre Regional de Transf. Sanguine
9–11, rue Lionnois
Nancy, 54000, France

Region GER

GERALB
Labor Immunogenetics
Kinderpoleklinik der LMU
Pettenkofer Str. 8
Munich 2, 8000, West Germany

GERALB
(Opelz) Inst. of Immunology
Univ. of Heidelberg
Neuenheimer Feld 305
Heidelberg, D-6900, West Germany

GERBAU
Inst. f. Med. Statistics
Univ. of Bonn
Sigmund Freud Str. 25
Bonn, 5300, West Germany

GERBIO
Biotest AG
Geleisstrasse 103
Offenbach, 6050, West Germany

GERBOE
Dept. Int. Med./Div. Endocrinology
University Hospital/Frankfurt Med School
Theodor Stern Kai 7
Frankfurt/Main, 6000, West Germany

GERBRA
Clin. Microbiol/Tissue Typing Unit
Hadassah Univ. Hospital
Hadassah, Ein Carem
Jerusalem, 91120, Israel

GERGAZ
Tissue Typing Lab/Dept. Pediatrics
Chaim Sheba Medical Center
Tel-Hashomer
Ramat-Gan, 52621, Israel

GERGOL
Dept. Transpl. Immunologie
Red Cross Blood Bank/Univ. Ulm
Oberer Eselsberg 10
Ulm, D-7900, West Germany

GERGSW
Inst. für Immunogenetics
Univ. Essen
Virchowstr. 171
Essen, 4300, West Germany

GERGSW (Doxiadis)
BIOTEST AG
Landsteinerstrasse 5
Dreieich 4, 6072
West Germany

GERKEY
Labor für Gewebetypisierung
Klinikum Steglitz (FBZ)
Hindenburgdamm 30
Berlin 45, 1000, West Germany

GERLEN
HLA Laboratory
Biotest AG
Landsteiner Str. 5d 305
Dreieich 4, 6072, West Germany

GERMUC
Immunology Lab/HLA Serology
Med. Univ. Clinic, Dept II
Olfried Muller Str. 10
Tubingen, D-7400, West Germany

GERMUE
Institute for Clinical Immunology
Justus-Liebig University
Langhansstr. 7
Giessen, D-6300, West Germany

GERNEU
Abt. für Immunologie
Gewebetypisierungs Labor (HLA)
Robert Koch Str. 40, TL 137
Gottingen, D-3400, West Germany

GERSCN
Red Cross Blood Transfusion Niedersachse
Soltauer Strasse 160
Rotenburg (Wumme), D-2720, West Germany

GERSUD
Gewebetypisierungs Labor.
Zentral Krankenhaus
St. Juergen Str.
Bremen 1, D-2800, West Germany

GERWAN
Immunogenetisches Labor.
Inst. fur Immunol/Univ. Munich
Goethestr. 31
Munich, D-8000, West Germany

GERWUE
Department of Neurology
Univ. of Wuerzburg
Wuerzburg, West Germany

Region IT1

IT1CEP
Inst. di Genetica Medica
Univ. di. Torino, CNR
Via Santena 19
Torino, 10126, Italy

IT1CNG
Inst. Di. Genetica Medica
Univ. Di Ferrara
Via L. Borsari 46
Ferrara, 44100, Italy

IT1CON
Tissue Typing Laboratory
Inst. Clin. Med., Univ. Cagliari
Via S. Giorgio 12
Cagliari, 09100, Italy

IT1MRA
Laboratorio HLA
Centro Transfusionale A.V.I.S.
Largo Volontari del Sangue 2
Pavia, 21700, Italy

IT1MTT
Cen. Reg. di Tipizzazione Tissutale
Genetica Umana
Largo Palagi 1
Firenze, 50134, Italy

IT1PUR
Tissue Typing/Immunol., CNTS
Croce Rossa Italiana
Via Ramazzini 15
Rome, 00151, Italy

IT1SAL
Generale Patologia Inst.
Univ. di Palermo
Corso Tukory 211
Palermo, 90134, Italy

IT1SAV
Tissue Typing Lab, Clin. Med. Nefrol.
Ospedale Maggiore
Via A. Gramsci 14
Parma, 43100, Italy

Region IT2

IT2FER
Centro Richerche Immunoematologiche
Inst. Naz. Ric. sul Cancro
Via Benedetto XV, N, 10
Genova, 16132, Italy

IT2GAN
Cattedra di Genetica Medica
Ospedale Spallanzani
Viale Portuense 292
Rome, 00149, Italy

IT2GAN
Instituto di Genetica Medica
Univ. of Ferrara
Via Borsari 46
Ferrara, 44100, Italy

IT2TOS
Instituto di Biologia Cellulare
CNR 00106
Roma, Italy

Region JAP

JAPAIZ
Department of Pathology
Hokkaido Univ. School of Medicine
N15 W7, Kita-ku
Sapporo, 060, Japan

JAPAMY
Department of Surgical Research
National Cardiovascular Center
5-7-1 Fujishirodai, Suite A
Osaka, 565, Japan

JAPDOH
2nd Department of Surgery
Hiroshima Univ. School of Medicine
1-2-3 Kasumi, Minami-ku
Hiroshima, 734, Japan

JAPHAS
Department of Legal Medicine
Shinshu Univ. School of Medicine
3-1-1 Asahi, Matsumoto
Nagano, 390, Japan

JAPJUJ
Blood Transfusion Service
Tokyo University Hospital
Hongo 7-3-1, Bunkyo-ku
Tokyo, 113, Japan

JAPJUJ
(Akaza) Central Blood Center
Japan Red Cross
4-1-31, Hiroo, Shibuya-ku
Tokyo, 150, Japan

JAPJUJ
(Saiji)
Kyoto Red Cross Blood Center
Sanjusangendo-mawari, Higashiyama-ku
Kyoto, 605, Japan

JAPKSH
Dept. Transplantation Immunology
Kitasato Univ. School of Medicine
1-15-1 Kitasato, Sagamihara Shi
Kanagawa, 228, Japan

JAPMAT
1st Dept. of Surgery
Osaka Univ. School of Medicine
1-1-50 Fukushima, Fukushima-ku
Osaka, 553, Japan

JAPMRI
HLA Typing Laboratory
Aichi Red Cross Blood Center
Minamiyamaguchi-cho 593-3, Seto,
Aichi, 489, Japan

JAPNAI
Department of Internal Medicine
Fukuoka Univ. School of Medicine
7-45-1 Nanakuma, Jonan-ku
Fukuoka, 814-01, Japan

JAPSAS
Department of Genetics
Med Inst. of Bioregulation/Kyushu Univ.
3-1-1 Maidashi, Higashi-ku
Fukuoka, 812, Japan

JAPSEK
Tissue Typing Lab/Dept Lab Medicine
National Defense Medical College
Namiki 3-2, Tokorozawa
Saitama, 359, Japan

JAPTAK
Lab Histocompatibility/Clin. Immunol.
Keio Univ. School of Medicine
35 Shinano-machi, Shinjuki-ku
Tokyo, 160, Japan

JAPTKB
Renal Transplantion Center
Hyogo Prefectural Nishinomiya Hospital
13-9 Rokutanji-Cho, Hyogo-ken
Nishinomiya-Shi, 662, Japan

JAPTOK
2nd Res. Laboratory
Fukuoka Red Cross Blood Center
232-11 Ooaza Kamikoga
Chikushino-Shi, 818, Japan

JAPTSU
Dept. of Transplantation
Tokai Univ. School of Medicine
Bohseidai, Isehara
Kanagawa, 259-11, Japan

JAPYOS
Dept. Microbiology & Immunology
Hamamatsu Univ. School of Medicine
3600 Handa-Cho
Hamamatsu, 431-31, Japan

Region LAT

LATBAG
Sector Ciencias Biologicas
Curitiva, Brazil

LATCOL
Lab de Immunologia
Universidad de Cartagena
Cartagena, Colombia

LATDIL
Hospital Durand
Buenos Aires, Argentina

LATFAI
Immunogenetica
Cimae
Luis Viale 2831
Buenos Aires, 1416, Argentina

LATGAR
Universidad de Antioquia
Medellin, Colombia

LATGER
Immunologia
Escola Paulista De Medicine
Rua Botacatu 862, 4 Andar
Sao Paulo, 04023, Brazil

LATGIR
Histocompatibilidad
Hospital Privado
Naciones Unidas 346
Cordoba, 5016, Argentina

LATGOR
Dept. De Immunogenetica
ISET, SSA
Carpio 470-1
Mexico, D.F. 11340, Mexico

LATHAA
Laboratorio de Histocompatibilidad
Sanatorio Guemes
Cordoba 3933
Capital Federal, 1189, Argentina

LATJRM
Lab De Immunologia
Hosp. Dos Servidores Do Estado
Rua Sacadura Cabral 178, 4 Andar
Rio De Janeiro, 20226, Brazil

LATKAL
Lab Immunologia de Transplantes
Instituto do Coracao da Fac Med, USP
Av. Dr. Arnaldo 455, 3 andar/sala 33
São Paulo, 01246, Brazil

LATLAY
Fisiopatologia/Dept. Med. Exp.
I.V.I.C.
Km 11 Carretera Panamerica, Apt 1827
Caracas DF, 1020A, Venezuela

LATPRI
Lab De Histocompatibilidad
Inst. Medicos Antartida
Rosario 437, Piso 6
Buenos Aires, 1424, Argentina

LATSAM
Universidad de Buenos Aires
Buenos Aires, Argentina

LATSUA
Lab De Histocompat.Y Hematol.
Centro De Educacion
Larrea 871, Piso 8, Dep. B
Buenos Aires, 1425, Argentina

LATTBA
Hemope
Cx Postal 1715
Recife, PE 52011, Brazil

LATTRN
Inst. Nac. de Enfermedade Respiratorias
Mexico City, Mexico

Region NCY

NCYCAR
Immunogenetics Laboratory
Brigham & Women's Hospital
75 Francis St.
Boston, MA 02115, USA

NCYCBR
Center for Blood Research
800 Huntington Avenue
Boston, MA 02115, USA

NCYFAL
Histocompatibility Laboratory
Toronto Western Hospital
399 Bathurst Street
Toronto, Ont, M5T 2S8, Canada

NCYFUL
Blood Transfusion Service
Massachusetts General Hospital
Boston, MA 02115, USA

NCYGOE
Iowa Reg. Tissue Typing Laboratory
VA Medical Center
Hiway 6 West
Iowa City, IA 52240, USA

NCYKAP
Histocompatibility Lab/Pathology
Case Western Reserve University
2085 Adelbert Road
Cleveland, OH 44106, USA

NCYKUR
Cox 5
Massachusetts General Hospital
Fruit Street
Boston, MA 02114, USA

NCYMAH
Immunogenetics Laboratory
Foundation for Blood Research
69 US Route 1
Scarborough, ME 04074, USA

NCYMAR
Clin. Immunology Diagnostic Laboratory
Memorial Univ. of Newfoundland
St. John's, Newfoundland, A1B 3V6, Canada

NCYMRV
HLA Sec./National Reference Laboratory
The Canadian Red Cross Society
1800 Alta Vista Drive
Ottawa, Ont, K1G 4J5, Canada

NCYNIK
Transplant Immunology
Baylor University Med Center
3500 Gaston Avenue
Dallas, TX 75246, USA

NCYRAG
Immunoparameter Laboratory
All Children's Hospital
801 6th Street South/P.O. Box 31020
St. Petersburg, FL 33731, USA

NCYROD
Clinical Laboratories
Emory Univ. School of Med. Hospital
1364 Clifton Road, NE
Atlanta, GA 30322, USA

NCYRSP
Department of Molecular Immunology
Roswell Park Memorial Institute
Buffalo NY, 14263, USA

NCYSAC
Blood Center
Sacramento Medical Foundation
1625 Stockton Blvd.
Sacramento, CA 95816, USA

NCYSTL
American Red Cross
4050 Lindell Blvd.
St. Louis, MO 63108, USA

NCYTOM
HLA Laboratory
So. Florida Blood Service
1675 NW 9th Avenue
Miami, FL 33136, USA

NCYYUN
Dept. of Pediatric Oncology
Dana Farber Cancer Institute
44 Binney Street
Boston, MA 02115, USA

NCYZIE
Dept of Pediatrics, Rm 9097A
Children's Hospital
34th St. & Civic Center Blvd.
Philadelphia, PA 19104, USA

Region SAF

SAFDUT
Provincial Lab for Tissue Immunology
Private Bag 4, Observatory
Cape Province, 7935, South Africa

SAFHAM
Transplantation Unit
Natal Inst. of Immunology
10 Robin Road, P.O. Box 2356
Durban, Natal, 4000, South Africa

Region SCA

SCAJER
Regional Centre for Blood Transfusion
Aalborg Hospital
P.O. Box 561
Aalborg, DK 9100, Denmark

SCAKIS
Tissue Typing Lab/Clin. Immunology
Aarhus Univ. Hospital
Aarhus, DK 8000, Denmark

SCAKOS
Nat. Ref. Lab for Tissue Typing
Finnish Red Cross Blood Transf. Service
Kivihaantie 7
Helsinki, 00310, Finland

SCALOW
Tissue Typing Laboratory
Univ. Hospital Blood Center
Lund, S-22185, Sweden

SCASVE
Tissue Typing Laboratory
State Univ. Hospital
Tagensvej 18-20
Copenhagen, DK 2200, Denmark

SCASVE
(Hyldig-Nielsen)
Harboe's Laboratorium ApS
Guldborgvej 22
Copenhagen F-2000, Denmark

SCATSB
Inst. of Transplantation Immunology
The National Hospital
Oslo, 0027, Norway

Regin UKI

UKIAST
Regional Transfusion Center
General Hospital
Holland Dr., Barrack Road
Newcastle-Tyne, NE2 4NQ, England

UKIBAT
Dept. Immunol/Royal Postgraduate Med Sch
Hammersmith Hospital
DuCane Road
London, W12 OHS, England

UKIBAT
(McDonald) Institute of Neurology
Nat. Hospital for Nervous Diseases
Queen Square
London, WC1, England

UKIBID
UK Transplant Service
Southmead Hospital
Southmead Road
Bristol, BS10 5ND, England

UKIBOD
Tissue Antigen Laboratory, Rm 307
Imperial Cancer Research Fund Labs
Lincoln's Inn Fields, PO Box 123
London, WC2A 3PX, England

UKIBRS
UK Transplant Service
Southmead Hospital
Southmead Road
Bristol, BS10 5ND, England

UKICXH
Tissue Typing Laboratory
Charing Cross Hospital
Fulham Palace Road, Hammersmith
London, W6 8RF, England

UKIDRK
Tissue Typing Laboratory
Cardiff Blood Transfusion Center
Rhydlafar, St. Fagans
Cardiff, Wales, CF5 6XF, England

UKIDYR
Medical Genetics Dept.
St. Mary's Hospital
Hathersage Road
Manchester, M13 OJH, England

UKIFES
Department of Immunology
London Hospital Medical College
56-76 Ashfield Street
London, E1 2AD, England

UKIGEL
Tissue Typing Laboratory
Regional Transfusion Center
Longley Lane
Sheffield, S5 7JN, England

UKIGLA
Immunology/Tissue Typing
Royal Infirmary
10 Alexandra Parade
Glasgow, G31 2ER, Scotland, UK

UKIJAM
Anthony Nolan Laboratories
St. Mary Abbot's Hospital
Marloes Road, Kensington
London, W8 5LQ, England

UKIJOY
Tissue Typing Laboratory
Addenbrookes Hospital
Hills Road
Cambridge, CB2 2QQ, England

UKIKNT
Div. of Immunological Medicine
MRC Clinical Research Centre
Watford Road, Harrow
Middlesex, HA1 3JU, England

UKILAW
Institute of Cancer Research
The Royal Marsden Hospital
Fulham Road
London, SW3 6JJ, England

UKILAW
(Wakeling) Sec. of Human Cancer Genetics
Haddow Labs, Inst. of Cancer Research
15 Clifton Avenue, Belmont
Surrey, SM2 5NG, England

UKIMAC
Tissue Typing Laboratory
Regional Transfusion Service
Vincent Drive
Birmingham, B15 2SG, England

UKIMCM
Institute for Molecular Medicine
John Radcliffe Hospital
Headington
Oxford, OX3 9DU, England

UKIMID
N. Ireland Tissue Typing Laboratory
Belfast City Hospital
Lisburn Road
Belfast, BT9 7AD, Northern Ireland

UKIMOR
Nuffield Department of Surgery
John Radcliffe Hospital
University of Oxford
Oxford, OX3 9DU, England

UKISTB
Dept. of Diabetes & Immunogenetics
St. Bartholomew's Hospital
59 Bartholomew Close, Dominion House
London, EC1A 7ED, England

UKIWEL
Tissue Typing Laboratory
Guy's Hospital
St. Thomas Street
London, SE1 9RT, England

UKIWOR
Department of Medicine
Univ. of Liverpool
PO Box 147
Liverpool, L69 3BX, England

Region US1

US1AID
Laboratory of Immunogenetics
NIAID
National Institutes of Health
Bethesda, MD 20892, USA

US1AMO
Div. of Immunology
Duke Univ. Medical Center
Box 3010
Durham, NC 27710, USA

US1CHR
Blood Bank
Charlotte Memorial Hospital
Charlotte, NC 28211, USA

US1DUQ
HLA Laboratory
Univ. of Pittsburgh School of Medicine
Children's Tower, 812 Fifth Avenue
Pittsburgh, PA 15213, USA

US1DUQ
(Trucco) Pittsburgh Cancer Institute
3343 Forbes Avenue
Pittsburgh, PA 15213, USA

US1HEI
Dept. of Microbiol/Immunology
Bowman Gray School of Medicine
300 S. Hawthorne Road
Winston-Salem, NC 17103, USA

US1HJH
Clinical Histocompatibility Laboratory
Georgetown Hospital, Gorman 224
3800 Reservoir Road, NW
Washington, DC 20007, USA

US1HJH
(Hartzman) Navy Medical Res. Inst.
Naval Med Command, Mailstop 36
8901 Wisconsin Ave., Bldg 18, Rm 128
Bethesda, MD 20814, USA

US1HOO
Dept. of Microbiology & Immunology
University of Louisville
Louisville, KY 40292, USA

US1IND
Histocompatibility Laboratory
Indiana Univ. Med Cen/Riley Hosp SO9
702 Barnhill Drive
Indianapolis, IN 46223, USA

US1JON
Dept. of Surgery, Tissue Typing Lab
Univ. of Illinois Medical Center
840 S. Wood St., Rm 518H, CSB
Chicago, IL 60612, USA

US1KYL
Immunogenetics Laboratory
Jewish Hospital
217 E. Chestnut Street
Louisville, KY 40202, USA

US1NWO
Histocompatibility Lab/Blood Bank
Charity Hospital of Louisiana
1532 Tulane Avenue
New Orleans, LA 70140, USA

US1STE
Histocompatibility Laboratory
Alton Ochsner Medical Foundation
1516 Jefferson Highway
New Orleans, LA 70121, USA

US1THP
Department of Medicine
Univ. Kentucky, Chandler Medical Center
800 Rose Street
Lexington, KY 40502, USA

US1TUL
Histocompatibility & Immunogen/Medicine
Tulane Univ. Medical School
1430 Tulane Avenue
New Orleans, LA 70112, USA

US1ZMI
Dept. Pathol. & Lab Medicine
Hospital of Univ. of Pennsylvania
3400 Spruce Street
Philadelphia, PA 19104, USA

US1ZMI
(Monos) Dept. Biochemistry & Molec. Bio.
Harvard University
77 Divinity St.
Cambridge, MA 02138, USA

Region US2

US2BAC
Immunobiology Research Center
Univ. of Minnesota, Box 724 Mayo Bldg.
420 Delaware Street. NE
Minneapolis, MN 55455, USA

US2MBC
Immunogenetics Research Section
Blood Center of SE Wisconsin
1701 W. Wisconsin Avenue
Milwaukee, WI 53233, USA

US2PEL
Pel-Freez
9099 N. Deerbrook Trail
Brown Deer, WI 53223, USA

US2RSH
Dept. of Immunology & Microbiology
Rush University
Chicago, IL 60612, USA

US2WOL
Tissue Typing Lab/Dept. of Surgery
Northwestern Univ. Medical School
320 E. Superior, Searle 13-489
Chicago, IL 60611, USA

US2WOL
(Kakaya) American Red Cross
209 Farmington Avenue
Farmington, CT 06032, USA

Region US3

US3CTS
Dept. of Human Genetics
Cetus Corporation
1400 53rd Street
Emeryville, CA 94608, USA

US3GRU
Dept. of Pathology/Blood Center
Stanford Univ. School of Medicine
800 Welch Road
Palo Alto, CA 94306, USA

US3GVY
Immunogenetics & Transplantation Lab
Univ. of California, HSE 520
3rd and Parnassus Avenues
San Francisco, CA 94143, USA

US3PAY
Tissue Typing Lab
Stanford Univ. Blood Bank
800 Welch Road
Palto Alto, CA 94305, USA

US3PER
Scientific Services/HLA Laboratory
Irwin Memorial Blood Bank
270 Masonic Avenue
San Francisco, CA 94118, USA

US3SCR
Department of Immunology
Scripps Clinic and Research Foundation
10660 N. Torry Pines Road
La Jolla, CA 92307, USA

Region US4

US4RAD
The Genetic Center
Southwest Biomedical Research Institute
6401 E. Thomas Road
Scottsdale, AZ 85251, USA

US4SND
Southwest Immunodiagnostics, Inc.
7614 Louis Pasteur Dr., Suite 210A
San Antonio, TX 78229, USA

US4SND
(Hsia) Histocompat. Lab, Dept. Lab Med
Allegheny General Hospital
320 E N Avenue
Pittsburgh, PA 15212, USA

US4STA
Tissue Typing Laboratory
Univ. of Texas Health Sciences Center
5323 Harry Hines Blvd.
Dallas, TX 75235, USA

Region US5

US5ABS
Histocompatibility Laboratory
Blood Systems Inc.
6220 E. Oak Street
Scottsdale, AZ 85257, USA

US5ABS
(Lane) Gulf Stream Blood System, Inc.
1700 N. State Road 7
Ft. Lauderdale, FL 85257, USA

US5BAL
Tissue Typing Laboratory
Univ. of Cincinnati/Hoxworth Blood Cen.
3231 Burnet Ave., M.L. #055
Cincinnati, OH 45267, USA

US5CRO
Histocompatibility Laboratory
Midwest Organ Bank
305 W. 43rd Street
Kansas City, MO 64111, USA

US5FTW
HLA Lab
Fort Wayne Regional Blood Center
1212 E. California Road
Fort Wayne, IN 46825, USA

US5KWC
Clinical Histocompatibility Testing Lab
Central Indiana Regional Blood Center
3450 N. Meridian Street
Indianapolis, IN 46208, USA

US5LIN
American Red Cross Blood Services
2225 Third Avenue North/P.O. Box 11003
Birmingham, AL 35202, USA

US5MIT
Transpl. Biol. Laboratory
FDA/CDB/Office of Biologic Res. & Review
8800 Rockville Pike, Bldg 29, Rm 232
Bethesda, MD 20817, USA

US5MLS
HLA Lab/Paediatrics CS116
Univ. of Manitoba
840 Sherbrook St./CS104
Winnipeg, Man., R3A 1S1, Canada

US5PAK
Dept. of Pathology
Univ. of Hawaii/St. Francis Hospital
2230 Liliha Street
Honolulu, HA 96817, USA

US5SIM
Simons GeneType Corporation
851 Burlway Road, Suite 704
Burlingame, CA 94010, USA

US5SIN
Histocompatibility Lab/Dept. Pathology
McMaster University
1200 Main St., W./HSC-4H20
Hamilton, Ont. L8N 3Z5, Canada

US5SWE
Histocompatibility/Surgical Research
Hartford Hospital
80 Seymour Street
Hartford, CT 06115, USA

US5TER
Tissue Typing Laboratory
UCLA, 15-22 Rehabilitation Center
1000 Veteran Avenue
Los Angeles, CA 90024, USA

US5TIL
GF320 Transplant Immunology Laboratory
Health Sciences Center
700 William Street
Winnipeg, Man., R3E 0Z3, Canada

US5UAB
Immunogenetics Program
Univ. of Alabama
1808 7th Ave., S., Rm 802
Birmingham, AL 35294, USA

US5WAL
Dept. of Pathology
Univ. of California School of Medicine
Center for the Health Sciences
Los Angeles, CA 90024, USA

Region US6

US6BRN
Histocompat. & Immunogenetics Labs
Cleveland Clinic Foundation
9500 Euclid Avenue
Cleveland, OH 44106, USA

US6CBC
HLA Laboratory
Community Blood Center
349 South Main Street
Dayton, OH 45402, USA

US6TRP
Dept. of Pathology
Univ. of New Mexico School of Medicine
Albuquerque, NM 87131, USA

Region US7

US7DUP
Immunogenetics Laboratory
Memorial Sloan-Kettering Cancer Center
1275 York Avenue
New York, NY 10021, USA

US7DWS
Center for Separation Sciences
University of Arizona
Tucson, AZ 85621, USA

US7GSC
HLA Laboratories
Genetic Systems Corporation
3005 First Avenue
Seattle, WA 98121, USA

US7HAN
Human Immunogenetics Program
Fred Hutchinson Cancer Research Center
1124 Columbia Street
Seattle, WA 98104, USA

US7JSL
Molecular Immunobiology Laboratory
Memorial Sloan-Kettering Cancer Center
1275 York Avenue
New York, NY 10021, USA

US7KNO
Cell Surface Immunochem. Laboratory
Memorial Sloan-Kettering Cancer Center
1275 York Avenue
New York, NY 10021, USA

US7LAL
Howard Hughes Medical Institute
Univ. of Utah Medical Center
Wintrobe Bldg, 7th Floor
Salt Lake City, UT 84132, USA

US7NFL
Effector Lymphocyte Biology Laboratory
Memorial Sloan-Kettering Cancer Center
1275 York Avenue
New York, NY 10021, USA

US7NTC
Northwest Tissue Center
Seattle, WA 98121, USA

US7POL
Histocompatibility & Clin. Immunol.
Baylor Col. of Med/Methodist Hospital
6565 Fannin
Houston, TX 77030, USA

US7SYY
Immunogenetics Laboratory
Memorial Sloan-Kettering Cancer Center
1275 York Avenue
New York, NY 10021, USA

US7VMC
Immunology Program
Virginia Mason Research Center
1000 Seneca Street, R1-RC
Seattle, WA 98101, USA

US7UHA
Laboratory of Immunochemistry
Memorial Sloan-Kettering Cancer Center
1275 York Avenue
New York, NY 10021, USA

Region US8

US8DEC
Blood Transfusion Service
The Canadian Red Cross Society
3131 Sherbrooke Street E.
Montreal, Que., H1W 1B2, Canada

US8FOT
Immunogenetics Transplant. Center
Rogosin Institute
430 E. 71st Street
New York, NY 10021, USA

US8DUN
Human Immunogenetics, Surgery/4B39
Howard University College of Medicine
520 W Street, NW
Washington, DC 20059, USA

US8JLE
Holland Laboratories
American Red Cross National Headquarters
16501 Crabbs Branch Way
Rockville, MD 20855, USA

US8RJW
Dept. of Rheumatic Diseases
Hospital for Joint Diseases
301 E. 17th Street
New York, NY 10003, USA

US8RUB
Immunogenetics
The New York Blood Center
310 E. 67th Street
New York, NY 10021, USA

US8UBI
Molecular Biology Division
United Biomedical Inc.
2 Nevada Drive
Lake Success, NY 11042, USA

www.ingramcontent.com/pod-product-compliance
Ingram Content Group UK Ltd.
Pitfield, Milton Keynes, MK11 3LW, UK
UKHW050934280826
474908UK00005B/193

* 9 7 8 1 4 6 1 2 8 1 5 4 2 *